BURGER'S MEDICINAL CHEMISTRY, DRUG DISCOVERY AND DEVELOPMENT

Seventh Edition

Volume 8: CNS Disorders

Edited by

Donald J. Abraham
Virginia Commonwealth University

David P. Rotella
Wyeth Research

Burger's Medicinal Chemistry, Drug Discovery and Development
is available Online in full color at
http://mrw.interscience.wiley.com/emrw/9780471266945/home/

A JOHN WILEY & SONS, INC., PUBLICATION

Copyright © 2010 by John Wiley & Sons, Inc. All rights reserved

Published by John Wiley & Sons, Inc., Hoboken, New Jersey
Published simultaneously in Canada

No part of this publication may be reproduced, stored in a retrieval system, or transmitted in any form or by any means, electronic, mechanical, photocopying, recording, scanning, or otherwise, except as permitted under Section 107 or 108 of the 1976 United States Copyright Act, without either the prior written permission of the Publisher, or authorization through payment of the appropriate per-copy fee to the Copyright Clearance Center, Inc., 222 Rosewood Drive, Danvers, MA 01923, (978) 750-8400, fax (978) 750-4470, or on the web at www.copyright.com. Requests to the Publisher for permission should be addressed to the Permissions Department, John Wiley & Sons, Inc., 111 River Street, Hoboken, NJ 07030, (201) 748-6011, fax (201) 748-6008, or online at http://www.wiley.com/go/permission.

Limit of Liability/Disclaimer of Warranty; While the publisher and author have used their best efforts in preparing this book, they make no representations or warranties with respect to the accuracy or completeness of the contents of this book and specifically disclaim any implied warranties of merchantability or fitness for a particular purpose. No warranty may be created or extended by sales representatives or written sales materials. The advice and strategies contained herein may not be suitable for your situation. You should consult with a professional where appropriate. Neither the publisher nor author shall be liable for any loss of profit or any other commercial damages, including but not limited to special, incidental, consequential, or other damages.

For general information on our other products and services or for technical support, please contact our Customer Care Department within the United States at (800) 762-2974, outside the United States at (317) 572-3993 or fax (317) 572-4002.

Wiley also publishes its books in a variety of electronic formats. Some content that appears in print may not be available in electronic formats. For more information about Wiley products, visit our web site at www.wiley.com.

Library of Congress Cataloging-in-Publication Data:

Abraham, Donald J., 1936-
 Burger's medicinal chemistry, drug discovery, and development/Donald J. Abraham, David P. Rotella. – 7th ed.
   p. ; cm.
 Other title: Medicinal chemistry, drug discovery, and development
 Rev. ed. of: Burger's medicinal chemistry and drug discovery. 6th ed. / edited by Donald J. Abraham. c2003.
   Includes bibliographical references and index.
   ISBN 978-0-470-27815-4 (cloth)
1. Pharmaceutical chemistry. 2. Drug development. I. Rotella, David P. II. Burger, Alfred, 1905-2000. III. Burger's medicinal chemistry and drug discovery. IV. Title. V. Title: Medicinal chemistry, drug discovery, and development.
 [DNLM: 1. Chemistry, Pharmaceutical–methods. 2. Biopharmaceutics–methods. 3. Drug Compounding–methods.   QV 744 A105b 2010]
 RS403.B8 2010
 615'.19–dc22   2010010779

Printed in Singapore

10 9 8 7 6 5 4 3 2 1

BURGER'S MEDICINAL CHEMISTRY, DRUG DISCOVERY AND DEVELOPMENT

BURGER'S MEDICINAL CHEMISTRY, DRUG DISCOVERY AND DEVELOPMENT

Editors-in-Chief
Donald J. Abraham
Virginia Commonwealth University

David P. Rotella
Wyeth Research

Consulting Editor
Al Leo
BioByte Corp

Editorial Board
John H. Block
Oregon State University

Robert H. Bradbury
AstraZeneca

Robert W. Brueggemeier
Ohio State University

John W. Ellingboe
Wyeth Research

William R. Ewing
Bristol-Myers Squibb Pharmaceutical Research Institute

Richard A. Gibbs
Purdue University

Richard A. Glennon
Virginia Commonwealth University

Barry Gold
University of Pittsburgh

William K. Hagmann
Merck Research Laboratories

Glen E. Kellogg
Virginia Commonwealth University

Christopher A. Lipinski
Melior Discovery

John A. Lowe III
JL3Pharma LLC

Jonathan S. Mason
Lundbeck Research

Andrea Mozzarelli
University of Parma

Bryan H. Norman
Eli Lilly and Company

John L. Primeau
AstraZeneca

Paul J. Reider
Princeton University

Albert J. Robichaud
Lundbeck Research

Alexander Tropsha
University of North Carolina

Patrick M. Woster
Wayne State University

Jeff Zablocki
CV Therapeutics

Editorial Staff
VP & Director, STMS Book Publishing: **Janet Bailey**
Editor: **Jonathan Rose**
Production Manager: **Shirley Thomas**
Production Editor: **Kris Parrish**
Illustration Manager: **Dean Gonzalez**
Editorial Program Coordinator: **Surlan Alexander**

CONTENTS

PREFACE		vii
CONTRIBUTORS		ix
1	Chemical Perspective on Neurotransmission and Drug Discovery for the Central Nervous System	1
2	Cognition	15
3	Cholinergics/Anticholinergics/Muscarinics/Nicotinics	61
4	CNS Stimulants	89
5	Anticonvulsants	121
6	Antipsychotic Agents	161
7	Antidepressants	219
8	Therapeutic Agents for the Treatment of Migraine	265
9	Approaches to Amyloid Therapies for the Treatment of Alzheimer's Disease	329
10	Non-amyloid Approaches to Alzheimer's Disease	405
11	Stroke Therapy	447
12	Therapeutic and Diagnostic Agents for Parkinson's Disease	529
13	Opioid Receptor Ligands	569
14	PET and SPECT in Drug Development	737
INDEX		749

PREFACE

The seventh edition of Burger's Medicinal Chemistry resulted from a collaboration established between John Wiley & Sons, the editorial board, authors, and coeditors over the last 3 years. The editorial board for the seventh edition provided important advice to the editors on topics and contributors. Wiley staff effectively handled the complex tasks of manuscript production and editing and effectively tracked the process from beginning to end. Authors provided well-written, comprehensive summaries of their topics and responded to editorial requests in a timely manner. This edition, with 8 volumes and 116 chapters, like the previous editions, is a reflection of the expanding complexity of medicinal chemistry and associated disciplines. Separate volumes have been added on anti-infectives, cancer, and the process of drug development. In addition, the coeditors elected to expand coverage of cardiovascular and metabolic disorders, aspects of CNS-related medicinal chemistry, and computational drug discovery. This provided the opportunity to delve into many subjects in greater detail and resulted in specific chapters on important subjects such as biologics and protein drug discovery, HIV, new diabetes drug targets, amyloid-based targets for treatment of Alzheimer's disease, high-throughput and other screening methods, and the key role played by metabolism and other pharmacokinetic properties in drug development.

The following individuals merit special thanks for their contributions to this complex endeavor: Surlan Alexander of John Wiley & Sons for her organizational skills and attention to detail, Sanchari Sil of Thomson Digital for processing the galley proofs, Jonathan Mason of Lundbeck, Andrea Mozzarelli of the University of Parma, Alex Tropsha of the University of North Carolina, John Block of Oregon State University, Paul Reider of Princeton University, William (Rick) Ewing of Bristol-Myers Squibb, William Hagmann of Merck, John Primeau and Rob Bradbury of AstraZeneca, Bryan Norman of Eli Lilly, Al Robichaud of Wyeth, and John Lowe for their input on topics and potential authors. The many reviewers for these chapters deserve special thanks for the constructive comments they provided to authors. Finally, we must express gratitude to our lovely, devoted wives, Nancy and Mary Beth, for their tolerance as we spent time with this task, rather than with them.

As coeditors, we sincerely hope that this edition meets the high expectations of the scientific community. We assembled this edition with the guiding vision of its namesake in mind and would like to dedicate it to Professor H.C. Brown and Professor Donald T. Witiak. Don collaborated with Dr. Witiak in the early days of his research in sickle cell drug discovery. Professor Witiak was Dave's doctoral advisor at Ohio State University and provided essential guidance to a young scientist. Professor Brown, whose love for chemistry infected all organic graduate students at Purdue University, arranged for Don to become a medicinal chemist by securing a postdoctoral position for him with Professor Alfred Burger.

It has been a real pleasure to work with all concerned to assemble an outstanding and up-to-date edition in this series.

DONALD J. ABRAHAM
DAVID P. ROTELLA

March 2010

CONTRIBUTORS

Ross J. Baldessarini, Harvard Medical School, Boston, MA
James C. Barrow, Merck & Co., Inc., West Point, PA
Raymond G. Booth, University of Florida, Gainesville, FL
Jeffrey M. Brown, Bristol-Myers Squibb Company, Wallingford, CT
Gene M. Dubowchik, Bristol-Myers Squibb Research and Development, Wallingford, CT
Robert L. Hudkins, Cephalon, Inc., West Chester, PA
Rosaria Gitto, Università di Messina, Messina, Italy
Richard A. Glennon, Virginia Commonwealth University, Richmond, VA
Robert K. Griffith, West Virginia University, Morgantown, WV
Leslie Iversen, University of Oxford, Oxford, UK
J. Steven Jacobsen, Wyeth Research, Princeton, NJ
Ji-In Kim, Wyeth Research, Princeton, NJ
Jie Jack Li, Bristol-Myers Squibb Company, Wallingford, CT
Laura De Luca, Università di Messina, Messina, Italy
John E. Macor, Bristol-Myres Squibb Research and Development, Wallingford, CT
Richard B. Mailman, Pennsylvania State University, Hershey, PA
Michael J. Marino, Cephalon, Inc., West Chester, PA
Robert J. Mark, Merck Research Laboratories, West Point, PA
Christopher R. Mccurdy, The University of Mississippi, University, MS
Philippe G. Nantermet, Merck Research Laboratories, West Point, PA
John L. Neumeyer, Harvard Medical School, Boston, MA
David E. Nichols, Purdue University, West Lafayette, IN
Sally Pimlott, NHS Greater Glasgow and Clyde, Glasgow, UK
Thomas E. Prisinzano, University of Kansas, Lawrence, KS
Albert J. Robichaud, Wyeth Research, Princeton, NJ
Giovanbattista De Sarro, Università Magna Græcia, Catanzaro, Italy
Mary J. Savage, Merck & Co., Inc., West Point, PA
Evan F. Shalen, Merck Research Laboratories, West Point, PA
Michael W. Sinz, Bristol-Myers Squibb Company, Wallingford, CT
Lone Veng, Merck & Co., Inc., West Point, PA
Shankar Venkatraman, Merck Research Laboratories, West Point, PA
Kent E. Vrana, Pennsylvania State University, Hershey, PA
Andy Welch, University of Aberdeen, Aberdeen, UK
Michael Williams, Cephalon, Inc., West Chester, PA
Celina V. Zerbinatti, Merck & Co., Inc., West Point, PA
Hong Zhu, Merck Research Laboratories, West Point, PA

CHEMICAL PERSPECTIVE ON NEUROTRANSMISSION AND DRUG DISCOVERY FOR THE CENTRAL NERVOUS SYSTEM

RICHARD B. MAILMAN*
KENT E. VRANA
Pennsylvania State University College of Medicine, Hershey Medical Center, Departments of Pharmacology and Neurology and Drug Discovery Core, Hershey, PA

1. PAST, CURRENT, AND FUTURE CENTRAL NERVOUS SYSTEM "DRUG DISCOVERY"

The past three decades have been notable for major thematic changes in the search for novel central nervous system (CNS) drugs. One might argue that Ehrlich's concept of the magic bullet became a reality with the development [1] and widespread application of radioreceptor assays in the mid-1970s. This, for the first time, permitted rapid screening for drugs aimed at specific CNS targets, as well as providing a rational way to avoid or predict side effects. Indeed, this led to a sentiment that the field was on the cusp of being able to bypass the physiologically and behaviorally based approaches that had been dominant in prior decades. In the decade following, the impact of molecular biology and genetics became palpable, and there was a heady sense that a new modern era of CNS drug discovery had arrived. With the cloning of the whole-genome only a matter of time, and 100,000 or more genes to provide targets for magic bullets, surely there were to be impending cures for all of the difficult to treat CNS disorders. The reality, however, is that while these advances have markedly impacted the understanding of the presumed mechanisms of action of both existing and new drugs, they have not yet produced the expected (predicted?) explosion of novel (and improved) CNS drugs.

The past 10 years have reminded us that these magnificent advances have only provided the tools needed for truly innovative drug discovery. What has been clear is that disciplines like physiology and behavior that had been felt by many to be passé have again become central parts of the science needed for truly innovative drug discovery. What we would like to do in this chapter is to take a historical perspective of certain neglected aspects of nervous system function and relate it to new concepts that may be relevant to advances of the next decade. One of the points that will pervade this discourse is the critical importance that drugs and medicinal chemistry have played in the science of drug discovery as well as the discovery of drugs themselves. We shall discuss the current and near-future pharmacological concepts that we think will help to guide CNS medicinal chemistry. These include magic bullets, multi-targeting drugs (magic shotguns), allosteric modulation, and functional selectivity.

2. EARLY CNS "DRUG DISCOVERY": RECEPTOR-SELECTIVE DRUGS AS MAGIC BULLETS

The science of CNS drug discovery is centuries old. The utility of natural products has been known for millennia, for example, as illustrated by reference to the use of opium in the Ebers Papyrus, an important medical reference of ancient Egypt (about 1500 BC). The study of opium, its chemical constituents, and their mechanisms of action actually provides examples of the development of major concepts in drug discovery. One of the key first developments in modern pharmacology began with Friedrich Serturner's isolation of morphine in a purified state in the early 1800s. Although this permitted the precise administration of the active ingredient, opium extract actually had been used for centuries as an anesthetic as well as for other indications. These initial studies on morphine were followed by the identification of codeine and

---

*Richard Mailman has a potential conflict of interest as regards a financial interest in Biovalve Technologies and Effipharma, Inc. who have licenses for some cited compounds or their offshoots. This conflict of interest is monitored by Pennsylvania State University. All opinions in this manuscript are those of the authors alone, and not of Pennsylvania State University, Biovalve Technologies, or Effipharma, Inc.

other molecules as additional active ingredients from the opium plant, and subsequent development of heroin as the first semisynthetic opiate. The manifold actions of a single pure compound markedly influenced the conceptualization of receptors.

Paul Ehrlich and John Langley independently developed the concept of the "receptive substance" near the turn of the twentieth century (reviewed in Refs [2,3]). Over the next decades, the appreciation for the concept of a receptor led to models that conceptualized the drug-receptor interaction, and provided theoretical frameworks for study. Clark, adapting models used in enzyme kinetics, proposed the occupancy model that linked the actions of a drug to the proportion of receptors occupied at equilibrium (for review, see Ref. [4]). Clark and Gaddum also introduced the concept of the log concentration–effect curve, and described the now-familiar "parallel shift" of the log concentration–effect curve (e.g., a Schild analysis) produced by a competitive antagonist. This was followed by Ariens and Stephenson addressing the notion of intrinsic activity (efficacy) as a way of explaining the degree of effect that resulted from drug binding. The final piece of the puzzle for early modern pharmacology was a demonstration of the complexity of receptor selectivity when Ahlquist showed that adrenaline had differential effects on two distinct receptor populations [5].

It was decades after Ehrlich and Langley's concept of the receptor that crude preparations of opioid receptors were first isolated and characterized. The very stringent structural and stereospecific requirements for opioid ligands made it highly likely that there would be different receptors, differentially affected by drug that were involved in the transduction of opioid signals [6–8]. *In vivo* pharmacology suggested four main classes of opioid receptors designated as μ, δ, κ, and σ, although it is now accepted that σ receptors are not members of the opioid family [9,10]. It is noteworthy that the concept of μ-, δ-, and κ-opioid receptors was established long before their genes were actually cloned, although the genome elucidation led to the identification of a fourth opioid-like receptor, ORL1. Each of the opioid receptor subtypes enjoys a unique anatomical distribution and subserves distinct pharmacological functions [11]. As importantly, the recent work has led to the elucidation of the endogenous ligands (enkephalins, β-endorphin, dynorphins, etc.) that, in turn, has helped form conceptualizations that have better explained the actions of morphine and synthetic opioid ligands (reviewed in Ref. [9]). As an aside, it was functional effects differentiating "full" μ-opioid agonists [12] that provided an early demonstration of the concept of functional selectivity (see below).

The opioids were not the only natural products that influenced modern CNS drug discovery and medicinal chemistry. The amphetamines and related stimulants were another arena of early CNS research. Amphetamine (phenylisopropylamine) itself first was synthesized in 1887 as a compound related to ephedrine, the active species isolated from the Chinese traditional medicine Ma Huang. It was decades later that amphetamine was resynthesized and eventually marketed by SmithKline as a nasal decongestant, with nonmedical uses quickly becoming common. The effects of amphetamine and other derivatives on feelings of fatigue and mood were soon noted, and there was widespread use of amphetamine by military of all camps during World War II, of particular note was use by Adolf Hitler. Amphetamines were still in use in the US Air Force as late as 2003 to combat pilot fatigue, a practice that may have played a role in the deaths of four Canadian soldiers in Afghanistan involved in a friendly fire incident. The study of the actions of these stimulants and their derivatives has been a major tool for understanding roles of catecholamine and serotonin receptors.

A third interesting CNS area is that of the hallucinogen, a class of agent in which early medicinal chemists also did impressive research and biological trials [13]. Albert Hofmann first synthesized LSD in 1938 while working at Sandoz on medically useful ergot derivatives. His work led to a lysergic acid derivative, lysergic acid diethylamide, as a potential improvement over nikethamide as an analeptic agent (i.e., a nonspecific stimulant of respiration and circulation). LSD did not seem useful in this regard and was shelved. Several years later, while working with the compound again, Hofmann became dizzy and was forced to stop working. He wrote that he was affected by a "remarkable restlessness, combined with a slight dizziness" and an "intoxicated-like condition" characterized by an extremely stimulated imagination. Hoffman correctly deduced that these psychoactive effects were a result of accidentally absorbing a tiny amount of LSD into his skin, a hypothesis he tested a few days later with deliberate ingestion of a larger dose. This 1943 clinical study clearly demonstrated that drugs could profoundly alter states of consciousness and perception. Nearly 70 years later, there remain unanswered questions about the unique effects of LSD versus other drugs targeting the same (5-HT_{2A} serotonin) receptor [13].

These examples of centrally active agents underscore one of the foundations of modern pharmacology and medicinal chemistry—that drugs acting as magic bullets can have highly specific effects, and that their targets may have therapeutic utility. When tools became available (first binding and later functional assays) to perform high-throughput screening, the search for new magic bullets for both older and newly discovered targets accelerated markedly. Indeed, there has been a sentiment, engendered a decade or more ago, that available high-throughput tools would meet all of the needs in pharmaceutical research. Although the search for magic bullets remains a cogent part of the drug discovery armamentarium, as will be discussed below, there is an impressive new complexity that has impacted CNS drug discovery. One major lesson from this early history should not be lost. Many major advances were a result of the hand-in-hand cooperation of medicinal chemists and pharmacologists (indeed, in some cases the medicinal chemist was also the biologist!). The current emphasis in biomedical research on "translation" has long been a hallmark of how medicinal chemists work. Indeed, we should not succumb to the hubris that a single independent field or technology will provide all of our discovery and development needs.

3. FROM MAGIC BULLET TO MAGIC SHOTGUN

The serendipitous observations about the beneficial actions of chlorpromazine in schizophrenia [14] ultimately led to the finding that these antipsychotic effects were due to antidopaminergic actions [15]. Numerous antipsychotics subsequently were developed that, like chlorpromazine, worked primarily as dopamine D_2 receptor antagonists. Drugs of this type (first called typical and now named first-generation antipsychotics) include a variety of different chemical classes. With these typical drugs, the clinical potency highly correlated with their affinity for the dopamine D_2 receptor [16]. Thus, actions at dopamine systems have been a critical mechanism in all currently approved antipsychotic drugs and many of the compounds in the discovery and development pipeline.

Central nervous system disorders present a particular problem in drug discovery, and the value of the concept of multitargeting when considering CNS drug discovery can be gleaned by considering the example of antipsychotic drugs. There have been many attempts to develop magic bullets for a receptor that would give salutatory effects of the best antischizophrenic drugs (see below), without the toxicity or side effects. Many such directions have been enthusiastically explored, yet with results that have often been dispiriting. One direction was based on what was thought to be the essential pharmacology of clozapine (the best current drug) [17], and focused on the dopamine D_4 receptor, with either a selective antagonist [18] or a dual-acting $5\text{-HT}_{2A}/D_4$ antagonist [19]. Both were ineffective. Another rational direction that unfortunately has not yielded an effective drug was a 5-HT_{2A} selective antagonist [20,21]. Although there is hope that there may be success in developing a single magic bullet for schizophrenia or other CNS disorders, to date the results are disappointing.

The study of the genomics and genetic determinants of most of the common central nervous system disorders (e.g., schizophrenia, depression, bipolar disorder, Parkinson's disease, epilepsy, etc.) has provided a clue. The etiology of these conditions is generally polygenic, and may involve interaction with environmental, developmental, and/or epigenetic factors [22–24]. Other conditions (e.g., stroke) may involve insult to different and diffuse areas of the brain. For these reasons, it is not surprising that drugs selective for single molecular targets (magic bullets) may not be successful. It has been argued that discovery of "selectively nonselective" drugs ("magic shotguns" by analogy to magic bullets) can be a useful research direction. Roth et al. [25] called this idea "rich pharmacology," a euphemism for the "dirty drug" nomenclature that described the fact that many CNS drugs (e.g., antipsychotic drugs) engage multiple receptor targets [26]. The notion is that a magic shotgun could provide a composite action from effects on several targets that would result in a more effective medication than would a magic bullet.

3.1. Why "Rich Pharmacology" ("Dirty Drugs") Might be Useful

The medicinal chemistry and pharmacology of the antipsychotic drugs [26,27] provide an excellent, but not the only, example of why multitargeting drugs may have unique utility. The original antipsychotic drugs (e.g., chlorpromazine and haloperidol) are known to be potent antagonists of dopamine D_2 receptors [16]. These compounds, called typical or first-generation antipsychotics, were notable also for causing neurological side effects, both acute (parkinsonism, extrapyramidal side effects) and chronic (tardive dyskinesia), and they have been succeeded by drugs (atypical or second generation) that are largely devoid of these undesired actions. One of the most useful drugs today is clozapine that, a half-century after its discovery, remains the "gold standard" antipsychotic drug (see review in Ref. [17]). Clozapine has demonstrated clinical superiority in terms of both positive and negative/cognitive aspects of the disorder, but fails to cause debilitating neurological and endocrine side effects. On the other hand, clozapine has its own severe side effects ranging from significant risk of potentially fatal agranulocytosis to seizures and metabolic syndrome. Clozapine has a complex pharmacological profile, with significant affinity for several members of the serotonin, dopamine, muscarinic cholinergic, and adrenergic families, among others. For decades, scientists have attempted to ascertain which combination of the many targets of clozapine provides the desirable clinical effects. While still unclear, the prevailing view is that the pleiotropic actions of clozapine underlie its exceptional clinical profile. This has resulted in several drugs (e.g., olanzapine, quetiapine, ziprasidone, asenapine, risperidone, iloperidone, paliperidone) that, like clozapine, do not cause neurological side effects, and like clozapine have "rich pharmacology." These compounds (atypical or second-generation antipsychotics) clearly represent an advance over first-generation drugs in terms of neurological side effects, and possibly some minimal improvement in overall efficacy. This has also led to specific hypotheses about what makes these multitargeting drugs effective, such as the need for concomitant antagonism of 5-HT_{2A} as well as D_2 receptors [28,29]. It is fair to say, however, that other mechanisms are what make clozapine still the gold standard of treatment. This example underscores the future opportunities that yet exist in this and other areas.

3.2. Complications and Opportunities with "Rich Pharmacology"

Although drugs with action at several receptors have yielded improved clinical effects (either increased efficacy, decreased side effects, or both), there is an inherent limitation of this strategy of multitargeting. Although designing compounds with very high selectivity is often challenging, even when the target is of clear CNS importance (e.g., see Ref. [30]), success in designing a magic bullet leads to the expectation that side effects will be predictable, and largely due to unwanted actions at the target receptor. The situation changes with a magic shotgun. The chemical features that allow a drug to be nonselective markedly increase the chances that the compound will have off-site actions at undesired receptors, as well as desired receptors [31]. An example of this can be seen in the clinical pharmacology of the antipsychotic drugs noted earlier. Although the newer atypical (second-generation drugs) are largely devoid of neurological side effects, their clinical pharmacology is very complex, and can include side effects such as weight gain and metabolic syndrome, endocrine effects, arrhythmias, sedation, and so on.

In many cases, the source of the undesired effects is known, and appears to be at undesired off-site targets that are involved because of the promiscuous consequences of multitargeting [25,32].

4. AFFECTING COTRANSMISSION: ANOTHER NOVEL DIRECTION FOR CNS DRUG DISCOVERY

One of the major precepts in CNS-focused medicinal chemistry has been to target the machinery of synaptic transmission. Thus, targeting the proteins that regulate the synthesis, degradation, and synaptic availability of neurotransmitters, as well as the synaptic and extrasynaptic receptors has been a classic drug discovery strategy. More recently, these efforts extended into the proteins that constitute the signaling cascades affected by changes in neurotransmission. In our view, there is another aspect of neurophysiology that may be a future productive drug discovery direction—modulation of cotransmission. A brief history may make clear why this could be potentially important. For many years, a major dogma in neuroscience was termed "Dale's principle," an idea drawn from Sir Henry Dale's work and developed and promulgated by Sir John Eccles. The common formulation of Dale's principle was that a neuron will release one and only one transmitter at all of its synapses (although Dale himself actually only proposed that a neuron released the same chemical transmitter at all of its terminals, not that it released only a single transmitter) [33].

The problem with this conceptualization of Dale's principle was shown most clearly by studies of parasympathetic salivary neurons [34,35]. These neurons were shown to release both acetylcholine and the neuropeptide VIP (vasoactive intestinal polypeptide) in proportion to their firing rate, and each transmitter caused physiological effects that interacted synergistically. It is not generally appreciated that rather than being an exception, cotransmission is probably the rule with excitable tissues [36–38]. Moreover, in systems where cotransmission has been studied in detail, there seems to be a marked interaction between the actions of cotransmitters and the physiological effects being studied (see above). These observations may be another reason why it is difficult to design agonists that completely mirror the actions of a neurotransmitter—even if the targeted receptor subtype is "right," the actions of a selective drug may require the synergistic effects of a cotransmitter. In some cases, it may be useful to exploit this physiology with the design of dual-targeting (bioptic) ligands.

5. TARGETING ALLOSTERIC SITES OF RECEPTORS

G-protein-coupled receptors (GPCRs) or 7-transmembrane receptors (7TM) are the largest class of cell-surface receptor, and play important roles in virtually every cell and organ system. GPCRs are encoded by more than 1000 genes [39], yet synthetic ligands for only a small fraction of these are available, and, for many receptors, intense efforts have failed to yield highly selective ligands that could ultimately be used as drug leads. Not surprisingly, they are the target of many current therapeutic agents, and are the largest single target group mediating the actions of current drugs. The endogenous ligands for GPCRs range from neurotransmitters/modulators and hormones to odorants and photons of light. The intracellular signaling of GPCRs is diverse and includes a broad range of effectors. Despite the many GPCR drug targets, useful drugs do not exist for the majority of GPCRs, and there are also limitations for existing GPCR-targeted drugs.

Decades ago, the study of benzodiazepine action led to the notion that drugs acting at sites distant from the binding site of the endogenous ligand (in this case, GABA) had important modulatory roles that resulted in effective drug action. Concomitant with the discovery of the benzodiazepines, the idea of allosteric protein effects became an intrinsic aspect of biochemical thought [40,41]. Although the $GABA_A$ receptors are from a different superfamily than the GPCR receptors, in theory it might be possible to find ligands (allosteric ligands) that act at sites distinct from the endogenous ligand or current drugs (orthosteric ligands).

One may conceptualize several situations where allosteric ligands might have specific utility. First, the conservation in ligand binding domains of members of a given family in some cases has made it difficult to discover ligands with adequate pharmacological selectivity. This problem could be overcome if less conserved aspects of the receptor could be targeted. Second, the ligand binding domains of some classes of 7TM receptors (e.g., those with peptide or protein ligands) may make the search for small molecule ligands difficult, because the orthosteric ligands might need to recapitulate the complex chemical aspects of the endogenous ligands that would, for example, engender pharmacokinetic problems. Third, allosteric ligands can in theory have actions far more diverse than orthosteric ligands. It is possible that allosteric ligands could cause conformation changes that directly activate the receptor (allosteric agonist), or conformational changes that block the actions of orthosteric ligands (allosteric antagonist). Moreover, in addition to these direct actions on receptor function, allosteric ligands could have modulatory functions in interacting with orthosteric ligands, either synergizing (positive allosteric modulation) or attenuating (negative allosteric modulation) the actions of the endogenous ligand. Indeed, in theory allosteric ligands could affect the potency and/or the maximal activity elicited by an orthosteric ligand. Since the first demonstration of this principle with GPCRs [42], this area has expanded markedly and has been the subject of several excellent reviews [43].

Thus, the discovery of allosteric ligands (as agonists, antagonists, and modulators) is a new venue that may offer exciting opportunities, particularly in CNS-related areas. In theory, allosteric ligands offer a great deal of potential for fine-tuning cellular signaling, especially in situations where discovery of orthosteric ligands has been unsuccessful (see Fig. 1). There are now examples that this mechanism can lead to an approved drug for a GPCR (e.g., cinacalcet [44]). Future medicinal chemistry research may very well lead to exciting new allosteric ligands for receptor targets for which orthosteric ligand discovery has not yet fulfilled current needs. Of particular relevance is the fact that the concept of allosteric action must integrate with the concept of functional selectivity that is discussed in the next section. Together, this provides an unprecedented level of theoretical issues, yet also provides unprecedented drug discovery opportunities.

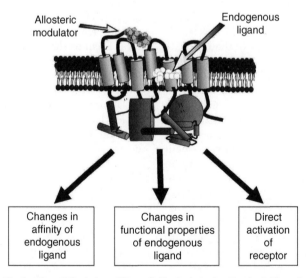

Figure 1. Schematic illustration of the interactions of allosteric and orthosteric ligands with a Class A GPCR (unfolded for the sake of clarity). The manifold possible consequences of allosteric ligands are shown in the boxes.

6. FUNCTIONAL SELECTIVITY AND RECEPTOR TARGETING

Despite a huge literature that has emerged in the past decade, many medicinal chemists and pharmacologists are unaware that the concept of functional selectivity has changed our understanding of how both orthosteric and allosteric ligands may affect function. As was recently reviewed [45], the field seems to have coalesced around the term functional selectivity that we first used few years ago [46]. The reader should be aware, however, that earlier reviews (and even a few current ones) still use equivalent terminology that includes "biased agonism," "agonist-directed trafficking (of signaling)," and so on. These and other issues have been reviewed recently [45].

As of the writing of this chapter, functional selectivity is a clearly established pharmacological *phenomenon*, with the focus now on study of the involved mechanisms, and ways in which it can be harnessed in drug discovery. Moreover, although it was elucidated and has been most intensively studied as regards orthosteric ligands, it is of equal relevance to the actions of allosteric drugs. Although it is now widely recognized that functional selectivity must be considered in attempting to understand differential actions of drugs targeting the same receptor(s), it is often unclear how to harness this phenomenon to design better screens for novel compounds. The following sections will attempt to elaborate on these points.

6.1. The Notion of Intrinsic Efficacy

The dogma inherent in our teaching and practice of pharmacology for most of the past century is that a ligand will always cause a particular type of response when interacting with a single-receptor target [47]. These effects are governed by two important properties: affinity, the property of attraction between a ligand and its receptor, and efficacy, the property that allows ligands, once bound, to produce a response [48]. Thus, based on these preconceived concepts, a full agonist will always activate the targeted receptor to the same degree as the endogenous ligand, a partial agonist will always cause incomplete activation even when saturating the receptor, and an antagonist will always cause no response and block the actions of full and partial agonists. The notion was that each drug had innate "intrinsic efficacy," a measure of the stimulus per receptor molecule produced by a ligand at its target receptor [49]. A corollary of this idea was that the ability of a ligand to impart (or reduce) stimulus once that ligand is bound to the receptor is an inherent property of the ligand–receptor complex. Intrinsic efficacy is thus differentiated from the more operational term "intrinsic activity" (Ariens, 1954) that refers to the maximal effect (E_{max}) of a ligand relative to a reference agonist in any single given experimental system. When inverse agonism was reported in some systems, it was accommodated because it depended on a high level of constitutive activity. Indeed, receptor classification prior to the molecular era relied upon agonist potency ratios, in which rank orders of potency were immutable scales through which drugs could be used to identify/classify the target receptors.

Even today, this concept is inherent in the design of high-throughput screens (HTS) that search for compounds that have the desired effect in such a functional assay. Although it has been known for decades that the signaling of receptors involves accessory proteins, in general the receptor alone itself was considered as the unit of signal transduction even when receptor models were updated to include states that had other proteins (e.g., G-proteins) associated with the receptor [50–52]. What was intrinsic to these models was the notion of "intrinsic efficacy" [49] in which it was possible for any drug (agonist or antagonist) to produce responses that were only different by their intensity.

6.2. History of Functional Selectivity

There were many anomalies in the literature that provided clues that the classical theory could not always explain data. For example, the accepted dogma was that long-term administration of receptor antagonists would induce a compensatory upregulation of the target receptor [53], whereas chronic administration of agonists would tend to cause receptor downregulation. Lesions to presynaptic neurons also could cause upregulation [54]. These changes seemed to have a teleological basis as an apparent cellular attempt to compensate for increases or decreases in signaling of the targeted receptor. Yet, there were findings that were paradoxical to this concept. For example, it was known that some 5-HT$_{2A}$ antagonists/inverse agonists would cause paradoxical receptor internalization and downregulation (for review, see Ref. [55]). Such findings, and the burgeoning understanding of the complexity of 7TM receptor signaling, even led to the suggestion "that selective agonists and antagonists might be developed, which have specific effects on a particular receptor-linked effector system" [56]. This was, to our knowledge, the first explicit speculation that the phenomenon now called functional selectivity existed and might impact both receptor pharmacology and drug discovery.

6.3. Explicit Demonstration of Functional Selectivity

When our group discovered the first full D$_1$ dopamine agonist dihydrexidine [57], its off-site action at dopamine D$_2$ receptors was quickly noted [58,59]. Because of its full D$_1$ agonist properties, it was hypothesized that dihydrexidine (see Fig. 2) would also be a D$_2$ full agonist. Indeed, both binding and canonical functional assays suggested this was true. Dihydrexidine competed for D$_2$ receptors in heterologous systems or in brain tissue with shallow, guanine-nucleotide-sensitive curves similar to typical agonists [59–61], and it was also a full agonist at D$_2$ receptor mediated inhibition of adenylate cyclase [60,61], and inhibited prolactin secretion *in vivo*. Yet, the behavioral effects of dihydrexidine were not typical for a drug (e.g., apomorphine) that supposedly had high intrinsic activity at both D$_1$ and D$_2$ receptors [62]. As further functional studies were conducted, it was found that although dihydrexidine bound to D$_2$-like presynaptic/autoreceptors that mediate inhibition of dopamine neuron firing, dopamine release, or dopamine synthesis [60], it had antagonist effects at these functions. These findings were replicated in single-cell systems [60,61]. With the exclusion of off-site actions to explain these findings [59], this raised two questions: how could a single drug have both agonist and antagonist actions at a single receptor, and if true, did it predict unique physiological effects of such a drug?

The answer to the former question was clear: one needed to consider complexes of signaling proteins ("signalsomes") as opposed

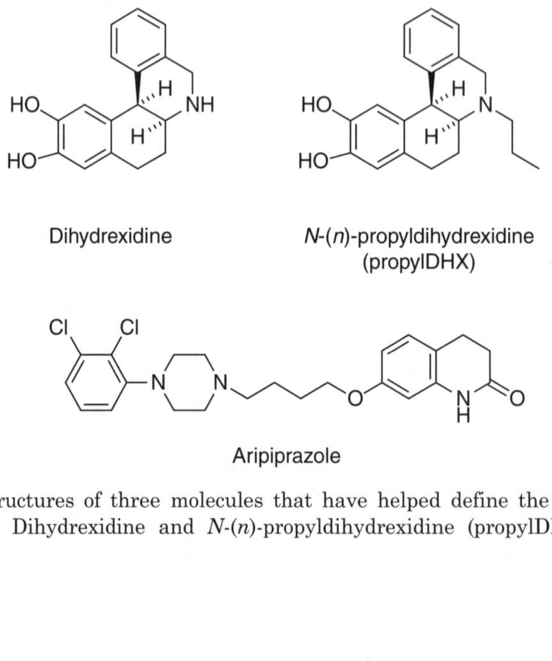

Figure 2. Chemical structures of three molecules that have helped define the concept and impact of "functional selectivity." Dihydrexidine and *N*-(*n*)-propyldihydrexidine (propylDHX) (top); aripiprazole (bottom).

to the receptor alone [63]. To address the second issue, we studied N-(n)-propyldihydrexidine (propylDHX, Fig. 2), an analog of dihydrexidine with slightly higher D_2 affinity, and markedly lower D_1 affinity. This made it a D_2:D_1 selective ligand with only weak partial agonist properties at the D_1 receptor [64]. Like dihydrexidine, propylDHX looked like a typical D_2 agonist in binding assays, was a full agonist at D_2-mediated inhibition of adenylate cyclase [59–61,64], and inhibited prolactin secretion *in vivo*. Yet also like dihydrexidine, it was an antagonist at other D_2-mediated functions in both brain tissue and heterologous systems [60,61]. Studies of propylDHX, like those with dihydrexidine [62], showed atypical behavioral properties. Selective D_2-like agonists are known to have biphasic behavioral effects, causing locomotor inhibition at low doses and locomotor stimulation and stereotypies at higher doses [65]. Yet, propylDHX only caused modest locomotor inhibition across a wide range of doses, with no competing behaviors that might have interfered with locomotion [66]. These data were consistent with the hypothesis that atypical signaling properties will predict unusual behavioral (or physiological) actions.

These data may have been the earliest that took the speculative hypothesis of Roth and Chuang [56] and demonstrated that such effects could occur, and most importantly, that they had pharmacological relevance. There were soon other reports of similar functional "mismatches" (see Ref. [67] for another early example with the serotonin system), and it is historically gratifying that another of these early examples was with the very opioid system that has played such a major role in our understanding of drug mechanisms [12]. For these reasons, some years ago, we proposed that functional selectivity "... allows drug effects to be refined to a degree not possible just by targeting specific receptor isoforms. This could yield important therapeutic advances, although it introduces a new level of complexity that will require significantly greater understanding of receptor dynamics and the interaction with transduction mechanisms" [63]. What remained was to determine if these interesting laboratory phenomena had clinical relevance.

6.4. Functional Selectivity is Clinically Significant: The Case of Aripiprazole

Aripiprazole (Fig. 2) has been accepted as the first "third-generation" antipsychotic based on the fact that it has an unusual clinical profile [68]. It does not have a favorable 5-HT$_{2A}$:D_2 affinity ratio, and would be predicted to cause significant extrapyramidal side effects, yet does not. Moreover, it is also devoid of other side effects of the second-generation drugs including endocrine and metabolic issues [69]. Although the compound has effects on some other target receptors, the consensus view of many leading figures in schizophrenia biology [68,70,71] is that aripiprazole acts as a "dopamine stabilizer" due to its D_2 dopamine receptor partial agonist properties in some assay systems [72,73]. We have recently reviewed how partial agonism was thought to explain the unique clinical effects of aripiprazole [26].

What was ignored in this conceptualization, however, was the body of data showing that the actions of aripiprazole were more likely a consequence of functionally selective D_2 signaling, not simple D_2 partial agonism. For example, unlike typical partial agonists, the intrinsic activity of aripiprazole is markedly dependent on the functional endpoint studied [73–75]. Moreover, the intrinsic activity of aripiprazole can be affected when the same functional endpoint is studied in different cellular milieus of the D_2 receptor [73–75]. Lastly, there are behavioral data such as in the unilaterally lesioned 6-hydroxydopamine (6-OHDA) treated rat [76], which are wholly inconsistent with the partial agonist mechanism [77]. Thus, the current data with aripiprazole appears to be evidence that functionally selective signaling properties can lead to novel behavioral effects in humans, as well as in laboratory animals. The way that differential signaling might lead to unexpected clinical profiles is illustrated in Fig. 3.

7. CONCLUSION: IMPACT OF THESE CONCEPTS ON FUTURE CNS DRUG DISCOVERY

With many predicting that we will soon have personalized medicine based on the unique

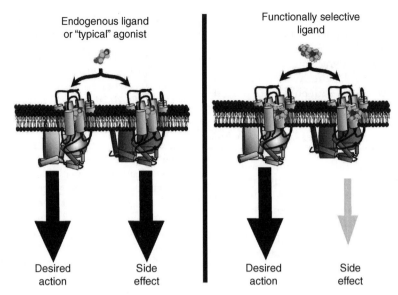

Figure 3. Schematic illustration of functional selectivity with a Class A GPCR as the example receptor. The functional effects of a typical ligand (in this case, the endogenous ligand) are shown on the left, and that of a functionally selective ligand on the right.

chemistry within each of us, recent experience may suggest that this view be tempered, at least as regard CNS disorders and other illness that do not have clear and simple molecular etiologies. Yet, our pessimism can be obviated somewhat by the fact that basic pharmacological advances have opened doors to better or even truly novel drugs. We have attempted to provide an overview of several conceptual approaches that may impact on the future drug discovery. Each of these arenas has many unanswered questions and is a major basic science research front. For the medicinal chemist, the question is how to harness these ideas in useful ways. One obvious approach relevant to all of these arenas is how to use in vitro approaches to assess differential actions of compounds, and to guide medicinal chemists in the process of optimizing selective pathway activation. In vitro assays remain the major tool pharmacologists and chemists use to define the intrinsic activities of ligands, and in one sense, the answer is operational: if one does not ask a question, one will never get an answer. For example, propranolol, a β-blocker first described nearly a half-century ago [78], is one of the most extensively studied drugs in the world, yet availability of new assays indicated that this antagonist/inverse agonist was actually an agonist at the β-arrestin-dependent ERK signaling [79].

The task that remains is how to do this effectively in a way that can positively impact medicinal chemistry. In recent years, the biotechnology industry has been providing many turnkey assay systems to permit more in-depth characterization of existing drugs and novel ligands across multiple signaling pathways, allowing study of molecules using multiple HTS assays. Another area of great interest has been high-content screening that takes advantage of spatially and/or temporally resolved signals to provide parallel information. The availability of automated high-resolution microscopy, robotics, and sophisticated software has already shown itself to be very useful in studying potential drug effect aspects of cell function involved in or that describe a disease [80–83]. One of the major issues often not addressed, however, is that of receptor milieu. There is sometimes the assumption that a convenient HTS or even high-content assay system will reflect accurately on the actions of a drug when interacting with the same receptor in situ. As shown by the example discussed above with aripiprazole [73,74,84], this assumption may not always be well grounded.

These approaches do not offer a quick and uncomplicated route to discovery of novel drugs. These concepts do, however, indicate that there is a rich, if complicated, future for medicinal chemistry. Allosteric modulation, functional selectivity, and biotic ligands have the potential to address medical problems that previously were thought to be solvable only by genetic methods. The horizons have opened for discovering novel drugs that engage "old," or what were once thought to be exhaustively mined, targets. Moreover, these same concepts can be applied to novel targets in which desired and undesired actions were thought to be obligatorily linked. Finally, the impact of interdisciplinary collaboration is even more essential if the benefits of research are to be translated into novel drugs.

REFERENCES

1. Devynck MA, Pernollet MG, Meyer P, Fermandjian S, Fromageot P. Angiotensin receptors in smooth muscle cell membranes. Nat N Biol 1973;245:55–58.
2. Maehle AH, Prull CR, Halliwell RF. The emergence of the drug receptor theory. Nat Rev Drug Discov 2002;1:637–641.
3. Langley JN. On the reaction of cells and of nerve-endings to certain poisons, chiefly as regards the reaction of striated muscle to nicotine and to curari. J Physiol 1905;33:374–413.
4. Ross EM, Kenakin TP. Pharmacodynamics: mechanisms of drug action and the relationship between drug concentration and effect. In: Hardman GJG, Limbird, LE, editors. Goodman & Gilman's The Pharmacological Basis of Therapeutics. New York: McGraw-Hill; 2001.
5. Ahlquist RP. A study of the adrenotropic receptors. Am J Physiol 1948;153:586–600.
6. Beckett AH, Casy AF. Synthetic analgesics: stereochemical considerations. J Pharm Pharmacol 1954;6:986–1001.
7. Beckett AH, Casy AF. Stereochemistry of certain analgesics. Nature 1954;173:1231–1232.
8. Portoghese PS. Stereochemical factors and receptor interactions associated with narcotic analgesics. J Pharm Sci 1966;55:865–887.
9. Corbett AD, Henderson G, McKnight AT, Paterson SJ. 75 years of opioid research: the exciting but vain quest for the Holy Grail. Br J Pharmacol 2006;147(Suppl 1): S153–S162.
10. Snyder SH, Pasternak GW. Historical review: opioid receptors. Trends Pharmacol Sci 2003;24:198–205.
11. Trescot AM, Datta S, Lee M, Hansen H. Opioid pharmacology. Pain Physician 2008;11: S133–S153.
12. Whistler JL, Chuang HH, Chu P, Jan LY, von Zastrow M. Functional dissociation of mu opioid receptor signaling and endocytosis: implications for the biology of opiate tolerance and addiction. Neuron 1999;23:737–746.
13. Nichols DE. Hallucinogens. Pharmacol Ther 2004;101:131–181.
14. Delay J, Deniker P, Harl JM. Therapeutic method derived from hiberno-therapy in excitation and agitation states. Annu Med Psychol 1952;110:267–273.
15. Carlsson A, Lindqvist M. Effect of chlorpromazine and haloperidol on formation of 3-methoxytyramine and normetanephrine in mouse brain. Acta Pharmacol Toxicol 1963;20:140–144.
16. Creese I, Burt DR, Snyder SH. Dopamine receptor binding predicts clinical and pharmacological potencies of antischizophrenic drugs. Science 1976;192:481–483.
17. Miyamoto S, Duncan GE, Marx CE, Lieberman JA. Treatments for schizophrenia: a critical review of pharmacology and mechanisms of action of antipsychotic drugs. Mol Psychiatry 2005;10:79–104.
18. Bristow LJ, Kramer MS, Kulagowski J, Patel S, Ragan CI, Seabrook GR. Schizophrenia and L-745,870, a novel dopamine D4 receptor antagonist. Trends Pharmacol Sci 1997;18:186–188.
19. Truffinet P, Tamminga CA, Fabre LF, Meltzer HY, Riviere ME, Papillon-Downey C. Placebo-controlled study of the D4/5-HT2A antagonist fananserin in the treatment of schizophrenia. Am J Psychiatry 1999;156:419–425.
20. de Paulis T. M-100907 (Aventis). Curr Opin Investig Drugs 2001;2:123–132.
21. Meltzer HY, Arvanitis L, Bauer D, Rein W. Placebo-controlled evaluation of four novel compounds for the treatment of schizophrenia and schizoaffective disorder. Am J Psychiatry 2004;161:975–984.
22. Gill M, Donohoe G, Corvin A. What have the genomics ever done for the psychoses? Psychol Med 2010;40:529–540.
23. Roth TL, Lubin FD, Sodhi M, Kleinman JE. Epigenetic mechanisms in schizophrenia. Biochim Biophys Acta 2009;1790:869–877.

24. Schwab SG, Wildenauer DB. Update on key previously proposed candidate genes for schizophrenia. Curr Opin Psychiatry 2009;22:147–153.
25. Roth BL, Sheffler DJ, Kroeze WK. Magic shotguns versus magic bullets: selectively non-selective drugs for mood disorders and schizophrenia. Nat Rev Drug Discov 2004;3:353–359.
26. Mailman RB, Murthy V. Third generation antipsychotic drugs: partial agonism or receptor functional selectivity? Curr Pharm Des 2010;16:488–501.
27. Gray JA, Roth BL. The pipeline and future of drug development in schizophrenia. Mol Psychiatry 2997;12:904–922.
28. Altar CA, Wasley AM, Neale RF, Stone GA. Typical and atypical antipsychotic occupancy of D2 and S2 receptors: an autoradiographic analysis in rat brain. Brain Res Bull 1986;16: 517–525.
29. Colpaert FC. Discovering risperidone: the LSD model of psychopathology. Nat Rev Drug Discov 2003;2:315–320.
30. Mailman RB, Huang X. Dopamine receptor pharmacology. In: Koller WC, Melamed E editors. Parkinson's disease and related disorders, Part 1. Handbook of Clinical Neurology. Elsevier. 2007; 83:77–105.
31. Kroeze WK, Sheffler DJ, Roth BL. G-protein-coupled receptors at a glance. J Cell Sci 2003;116:4867–4869.
32. Setola V, Roth BL. Screening the receptorome reveals molecular targets responsible for drug-induced side effects: focus on 'fen-phen'. Expert Opin Drug Metab Toxicol 2005;1:377–387.
33. Eccles JC. Chemical transmission and Dale's principle. Prog. Brain Res 1986;68:3–13.
34. Lundberg JM, Anggard A, Fahrenkrug J, Hokfelt T, Mutt V. Vasoactive intestinal polypeptide in cholinergic neurons of exocrine glands: functional significance of coexisting transmitters for vasodilation and secretion. Proc Natl Acad Sci USA 1980;77:1651–1655.
35. Hokfelt T, Lundberg JM, Schultzberg M, Johansson O, Skirboll L, Anggard A, Fredholm B, Hamberger B, Pernow B, Rehfeld J, Goldstein M. Cellular localization of peptides in neural structures. Proc R Soc Lond B Biol Sci 1980;210:63–77.
36. Trudeau LE, Gutierrez R. On cotransmission and neurotransmitter phenotype plasticity. Mol Interv 2007;7:138–146.
37. Teschemacher AG, Johnson CD. Cotransmission in the autonomic nervous system. Exp Physiol 2009;94:18–19.
38. Burnstock G. Cotransmission. Curr Opin Pharmacol 2004;4:47–52.
39. May LT, Leach K, Sexton PM, Christopoulos A. Allosteric modulation of G protein-coupled receptors. Annu Rev Pharmacol Toxicol 2007;47:1–51.
40. Levitzki A, Koshland DE Jr. Negative cooperativity in regulatory enzymes. Proc Natl Acad Sci USA 1969;62:1121–1128.
41. Monod J, Changeux JP, Jacob F. Allosteric proteins and cellular control systems. J Mol Biol 1963;6:306–329.
42. Jakubik J, Bacakova L, Lisa V, el-Fakahany EE, Tucek S. Activation of muscarinic acetylcholine receptors via their allosteric binding sites. Proc Natl Acad Sci USA 1996;93: 8705–8709.
43. Conn PJ, Christopoulos A, Lindsley CW. Allosteric modulators of GPCRs: a novel approach for the treatment of CNS disorders. Nat Rev Drug Discov 2009;8:41–54.
44. Block GA, Martin KJ, de Francisco AL, Turner SA, Avram MM, Suranyi MG, Hercz G, Cunningham J, Abu-Alfa AK, Messa P, Coyne DW, Locatelli F, Cohen RM, Evenepoel P, Moe SM, Fournier A, Braun J, McCary LC, Zani VJ, Olson KA, Drueke TB, Goodman WG. Cinacalcet for secondary hyperparathyroidism in patients receiving hemodialysis. N Engl J Med 2004;350:1516–1525.
45. Urban JD, Clarke WP, von Zastrow M, Nichols DE, Kobilka B, Weinstein H, Javitch JA, Roth BL, Christopoulos A, Sexton PM, Miller KJ, Spedding M, Mailman RB. Functional selectivity and classical concepts of quantitative pharmacology. J Pharmacol Exp Ther 2007;320:1–13.
46. Lawler CP, Watts VJ, Booth RG, Southerland SB, Mailman RB. Discrete functional selectivity of drugs: OPC-14597, a selective antagonist for post-synaptic dopamine D2 receptors. Soc Neurosci Abstr 1994;20:525.
47. Stephenson RP. A modification of receptor theory. Br J Pharmacol 1956;11:379–393.
48. Kenakin T. Pharmacologic Analysis of Drug-Receptor Interaction. Philadelphia: Lippincott-Raven; 1997.
49. Furchgott RF. The use of -haloalkyamines in the differentiation of receptors and in the determination of dissociation constants of receptor–agonist complexes. In: Harper NJ, Simmonds AB, editors. Advances in Drug Research. New York: Academic Press; 1966. p 21–55.

50. Leff P, Scaramellini C, Law C, McKechnie K. A three-state receptor model of agonist action. Trends Pharmacol Sci 1997;18:355–362.
51. Leff P. The 2-state model of receptor activation. Trends Pharmacol Sci 1995;16:89–97.
52. Kenakin T. Collateral efficacy in drug discovery: taking advantage of the good (allosteric) nature of 7TM receptors. Trends Pharmacol Sci 2007;28:407–415.
53. Burt DR, Creese I, Snyder SH. Antischizophrenic drugs: chronic treatment elevates dopamine receptor binding in brain. Science 1977;196:326–328.
54. Creese I, Burt DR, Snyder SH. Dopamine receptor binding enhancement accompanies lesion-induced behavioral supersensitivity. Science 1977;197:596–598.
55. Gray JA, Roth BL. Paradoxical trafficking and regulation of 5-HT(2A) receptors by agonists and antagonists. Brain Res Bull. 2001;56:441–451.
56. Roth BL, Chuang DM. Multiple mechanisms of serotonergic signal transduction. Life Sci 1987;41:1051–1064.
57. Lovenberg TW, Brewster WK, Mottola DM, Lee RC, Riggs RM, Nichols DE, Lewis MH, Mailman RB. Dihydrexidine, a novel selective high potency full dopamine D-1 receptor agonist. Eur J Pharmacol 1989;166:111–113.
58. Brewster WK, Nichols DE, Riggs RM, Mottola DM, Lovenberg TW, Lewis MH, Mailman RB. trans-10,11-Dihydroxy-5,6,6a,7,8,12b-hexahydrobenzo[a]phenanthridine: a highly potent selective dopamine D1 full agonist. J Med Chem 1990;33:1756–1764.
59. Mottola DM, Brewster WK, Cook LL, Nichols DE, Mailman RB. Dihydrexidine, a novel full efficacy D_1 dopamine receptor agonist. J Pharmacol Exp Ther 1992;262:383–393.
60. Mottola DM, Kilts JD, Lewis MM, Connery HS, Walker QD, Jones SR, Booth RG, Hyslop DK, Piercey M, Wightman RM, Lawler CP, Nichols DE, Mailman RB. Functional selectivity of dopamine receptor agonists. I. Selective activation of postsynaptic dopamine D2 receptors linked to adenylate cyclase. J Pharmacol Exp Ther 2002;301:1166–1178.
61. Kilts JD, Connery HS, Arrington EG, Lewis MM, Lawler CP, Oxford GS, O'Malley KL, Todd RD, Blake BL, Nichols DE, Mailman RB. Functional selectivity of dopamine receptor agonists. II. Actions of dihydrexidine in D2L receptor-transfected MN9D cells and pituitary lactotrophs. J Pharmacol Exp Ther 2002;301:1179–1189.
62. Darney KJ Jr, Lewis MH, Brewster WK, Nichols DE, Mailman RB. Behavioral effects in the rat of dihydrexidine, a high-potency, full-efficacy D1 dopamine receptor agonist. Neuropsychopharmacology 1991;5:187–195.
63. Mailman RB, Nichols DE, Lewis MM, Blake BL, Lawler CP. Functional effects of novel dopamine ligands: dihydrexidine and Parkinson's disease as a first step. In: Jenner P, Demirdemar R, editors. Dopamine Receptor Subtypes: From Basic Science to Clinical Application. IOS Stockton Press; 1998. p 64–83.
64. Knoerzer TA, Watts VJ, Nichols DE, Mailman RB. Synthesis and biological evaluation of a series of substituted benzo[a]phenanthridines as agonists at D1 and D2 dopamine receptors. J Med Chem 1995;38:3062–3070.
65. Eden RJ, Costall B, Domeney AM, Gerrard PA, Harvey CA, Kelly ME, Naylor RJ, Owen DA, Wright A. Preclinical pharmacology of ropinirole (SK&F 101468-A) a novel dopamine D2 agonist. Pharmacol Biochem Behav 1991;38:147–154.
66. Smith HP, Nichols DE, Mailman RB, Lawler CP. Locomotor inhibition, yawning and vacuous chewing induced by a novel dopamine D2 postsynaptic receptor agonist. Eur J Pharmacol 1997;323:27–36.
67. Berg KA, Maayani S, Goldfarb J, Scaramellini C, Leff P, Clarke WP. Effector pathway-dependent relative efficacy at serotonin type 2A and 2C receptors: evidence for agonist-directed trafficking of receptor stimulus. Mol Pharmacol 1998;54:94–104.
68. Stahl SM. Dopamine system stabilizers, aripiprazole, and the next generation of antipsychotics. Part 1. "Goldilocks" actions at dopamine receptors. J Clin Psychiatry 2001;62:841–842.
69. Goodnick PJ, Jerry JM. Aripiprazole: profile on efficacy and safety. Expert Opin Pharmacother 2002;3:1773–1781.
70. Tamminga CA. Partial dopamine agonists in the treatment of psychosis. J Neural Transm 2002;109:411–420.
71. Lieberman JA. Dopamine partial agonists: a new class of antipsychotic. CNS Drugs 2004;18:251–267.
72. Burris KD, Molski TF, Xu C, Ryan E, Tottori K, Kikuchi T, Yocca FD, Molinoff PB. Aripiprazole, a novel antipsychotic, is a high-affinity partial agonist at human dopamine D2 receptors. J Pharmacol Exp Ther 2002;302:381–389.
73. Lawler CP, Prioleau C, Lewis MM, Mak C, Jiang D, Schetz JA, Gonzalez AM, Sibley DR, Mailman RB. Interactions of the novel antipsy-

chotic aripiprazole (OPC-14597) with dopamine and serotonin receptor subtypes. Neuropsychopharmacology 1999;20:612–627.
74. Shapiro DA, Renock S, Arrington E, Sibley DR, Chiodo LA, Roth BL, Mailman RB. Aripiprazole, a novel atypical antipsychotic drug with a unique and robust pharmacology. Neuropsychopharmacology 2003;28:1400–1411.
75. Urban JD, Vargas GA, von Zastrow M, Mailman RB. Aripiprazole has functionally selective actions at dopamine D(2) receptor-mediated signaling pathways. Neuropsychopharmacology 2007;32:67–77.
76. Ungerstedt U, Arbuthnott GW. Quantitative recording of rotational behavior in rats after 6-hydroxy-dopamine lesions of the nigrostriatal dopamine system. Brain Res 1970;24:485–493.
77. Kikuchi T, Tottori K, Uwahodo Y, Hirose T, Miwa T, Oshiro Y, Morita S. 7-(4-[4-(2,3-Dichlorophenyl)-1-piperazinyl]butyloxy)-3,4-dihydro-2(1H)-qui nolinone (OPC-14597), a new putative antipsychotic drug with both presynaptic dopamine autoreceptor agonistic activity and postsynaptic D2 receptor antagonistic activity. J Pharmacol Exp Ther 1995;274:329–336.
78. Black JW, Duncan WA, Shanks RG. Comparison of some properties of pronethalol and propranolol. Br J Pharmacol Chemother 1965;25:577–591.
79. Azzi M, Charest PG, Angers S, Rousseau G, Kohout T, Bouvier M, Pineyro G. Beta-arrestin-mediated activation of MAPK by inverse agonists reveals distinct active conformations for G protein-coupled receptors. Proc Natl Acad Sci USA 2003;100:11406–11411.
80. Milligan G. High-content assays for ligand regulation of G-protein-coupled receptors. Drug Discov Today 2003;8:579–585.
81. Bleicher KH, Bohm HJ, Muller K, Alanine AI. Hit and lead generation: beyond high-throughput screening. Nat Rev Drug Discov 2003;2:369–378.
82. Eggert US, Mitchison TJ. Small molecule screening by imaging. Curr Opin Chem Biol 2006;10:232–237.
83. Giuliano KA, Haskins JR, Taylor DL. Advances in high content screening for drug discovery. Assay Drug Dev Technol 2003;1:565–577.
84. Mailman RB. GPCR functional selectivity has therapeutic impact. Trends Pharmacol Sci 2007;28:390–396.

COGNITION

Robert L. Hudkins,
Michael J. Marino,
Michael Williams
Cephalon, Inc., Worldwide
Discovery Research, West
Chester, PA

1. INTRODUCTION

The aging process and traumatic insults to the brain frequently result in a decline in cognitive performance with an associated decrease in the individual quality of life. At the tissue level, this decline involves neuronal/glial dysfunction and/or overt neurodegeneration with marked cell loss, disruption of discrete synaptic pathways, and altered neural network function [1].

Drugs to treat cognitive dysfunction are a high priority in biomedical research as the elderly population grows, showing an increased incidence of what has been termed "benign senescent forgetfulness" that progresses to mild cognitive impairment (MCI) and Alzheimer's disease (AD). Aspects of cognitive dysfunction also occur in association with Parkinson's disease, AIDS-associated dementia, stroke, schizophrenia, stress, sleep deprivation, depression, anxiety, recreational, prescription drug usage, and surgical procedures.

The degree of cognitive deficit and its underlying causality differs both anatomically and, to the extent known, mechanistically in this broad spectrum of CNS disorders, making it somewhat naïve to expect that new chemical entities acting via a single mechanism will have efficacy in all CNS disorders that involve some aspect of cognitive dysfunction. The molecular and anatomical substrates of cognition and associated behaviors that include attention and memory are complex and diffuse and suggest that drugs addressing this area will necessarily be polypharmic [2]. As an example, given the relative ease of diagnosis of AIDS and its relatively short progression, AIDS dementia was for a period of time viewed as a surrogate disease state for testing compounds that might have potential utility in the treatment of AD. However, equating the cognitive dysfunction associated with the slow progressive timeline of AD neurodegeneration with the cerebral atrophy occurring due to opportunistic viral, fungal, protozoal, and bacterial infections in AIDS to a common mechanistic pathway is, based on present knowledge, an optimistic stretch.

1.1. Cognition

Cognition is defined by the Merriam-Webster Online Dictionary as "to become acquainted with, know." Cognition can also be defined as the ability to process contemporary information in the context of existing knowledge to appropriately respond—in terms of decision making—to a given situation. In either definition, cognition involves several distinct behavioral domains. These include attention, perception, emotion, memory, action, and problem solving [2].

Therapeutic approaches to ameliorating cognitive dysfunction can be simplistically viewed as involving two ends of a continuum—at one end the prophylatic enhancement of cognitive function and at the other end the restoration of function and/or arrest of the decline occurring to the aging or traumatized brain. Cognition enhancers include psychostimulants such as caffeine, the most widely ingested drug in the world in the form of coffee and soft drinks [3], amphetamines [4], nootropics such as piracetam **1**, and the antioxidant, idebenone **2** [5]. Drugs currently approved for use in the treatment of neurodegenerative disorders such as AD include the cholinesterase inhibitors, tacrine **3**, donepezil **4**, rivastigmine **5**, and galanthamine **6** [6] and memantine **7**, a partial agonist at glutamate receptors [7]. None of these drugs works especially well in treating the symptoms of AD [8] probably reflecting the advanced stage of the disease when it is diagnosed than the intrinsic efficacy of these drugs. To restore function to a dead cell, a "Lazarus-like" effect, is a phenomenon that has yet to be achieved outside the realm of fiction.

A variety of new chemical entities (NCEs) are currently being explored to identify new cognitive enhancers. Some have discrete mechanistic targets against that they are/were optimized. Others involve a heuristic

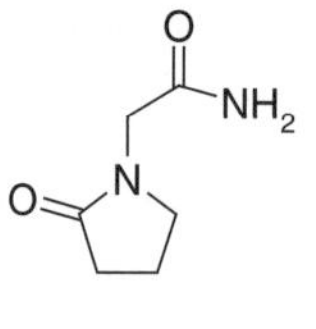

1 Piracetam

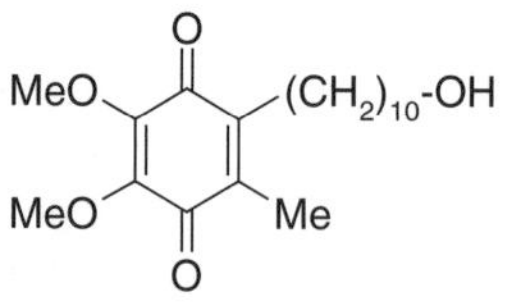

2 Idebenone

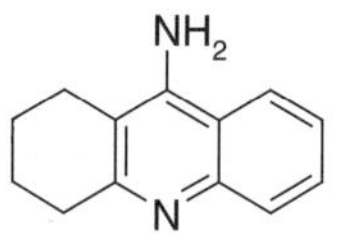

3 Tacrine

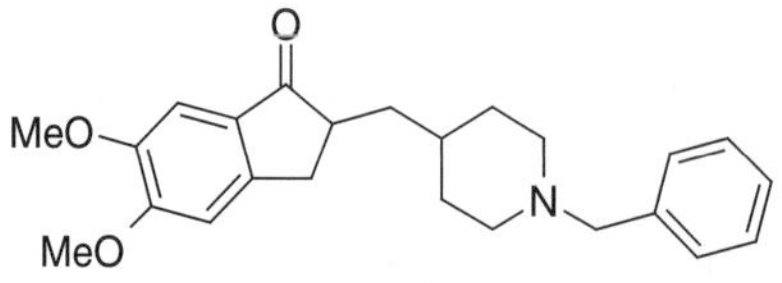

4 Donepezil

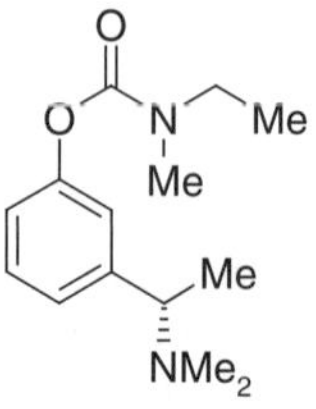

5 Rivastigmine

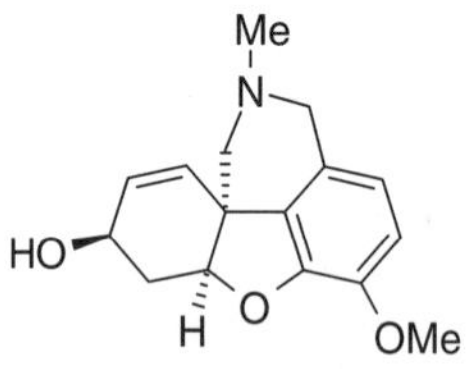

6 Galantamine

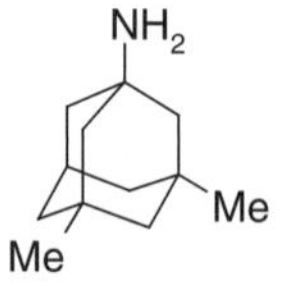

7 Memantine

postrationalization of a target based on a phenotypic behavioral response while yet others have their origins in ayurvedic medicine or folklore and currently lack a robust mechanism of action (MoA), for example, *Ginko biloba* and the nootropics [5].

In neurodegeneration-associated cognitive deficit, significant efforts have been focused on the β-amyloid [9] and tau hyperphosophorylation [10] hypotheses of AD that reflect the involvement of amyloid plaques and neurofibrillary tangles (NFTs), respectively, in disease pathophysiology. Both are found in postmortem AD brain although whether they are causative, or a result, of the disease remains to be determined [11]. Prevention of β-amyloid (Aβ) formation by altering cleavage products via inhibition/alteration of secretase enzyme processing [12], prevention of Aβ deposition or enhancing its removal by the use of metal chelators or vaccination will eventually provide information on the causative role of this peptide fragment in AD. However, the path forward is far from certain as initial positive results have, more often than not, been confounded by results from subsequent studies, the more important of which have been Phases II and III clinical trials where approaches to reducing brain amyloid have failed to show efficacy in altering disease progression [11]. Approaches to altering the hyperphosphorylation state of tau as a target for AD treatment are similarly complex. There are some 80 serine and threonine residues on tau that are potential substrates for kinase activity [10]. With only 30 of these functioning as phosphorylation sites under normal physiological conditions and 25 of these being identified as sites of "abnormal phosphorylation" there is significant redundancy in the ability to alter tau phosphorylation. Candidate kinases for tau phosphorylation are the proline-directed kinases, GSK3β, CDK5, and ERK2 that can phosphorylate 13 residues in tau associated with proline residues. Microtubule-affinity-regulating kinase (MARK) and cAMP-dependent protein kinase A (PKA) are nonproline-directed kinases that may also affect tau hyperphosphorylation. Inhibitors of "tau kinase," dual GSK3/CDK5 inhibitors such as the indirubicins—while effective in altering phosphorylation of key serine and threonine sites on tau in cellular and transgenic animal systems—have yet to demonstrate robust biochemical or phenotypic effects in native systems.

One inevitable outcome from research into cognition enhancers for the potential treatment of disease-associated cognitive dysfunction is that of nootropics or "smart drugs." [14] In the same way that caffeine and amphetamines are viewed as improving normal function via their stimulant actions, newer generations of cognition enhancers are anticipated to improve performance [14].

1.1.1. Domains of Cognition and Memory

Historically, the understanding of cognitive processes is based on the empirical school of psychology known as behaviorism [15]. Behaviorist theories, championed most notably by James Watson and B.F. Skinner, evolved from the early learning and memory studies of Pavlov and Thorndike. These behaviorists were primarily concerned with defining stimulus–response relationships and focused on precise observation and mathematical formalism that viewed the intervening processing of stimuli as irrelevant as only directly observable behaviors were believed to be amenable to scientific investigation. While this approach was successful in rigorously characterizing certain behaviors, it was soon apparent that an understanding of mental processing, the realm of cognitive psychology [16], was essential for interpreting the full spectrum of behavior, and of human behavior in particular.

A useful framework for understanding cognitive processes is provided by the information processing models of cognitive psychology. These models range in complexity from the black-box model of the behaviorists (Fig. 1a) to extremely complex connectionist models [17]. However, common themes are apparent that allow the derivation of a simplified model (Fig. 1b). Importantly, this framework allows for the deconstruction of the cognitive process into experimentally addressable domains that can be studied both preclinically and clinically.

Perception A detailed description of the processes underlying perception is outside of the scope of cognition. However, deficits in preattentive processing and attention play significant roles in defining the scope and consequences of the cognitive deficits observed in a number of CNS disorders including schizophrenia [18]. Since these early processing events are critical for subsequent cognitive processing, a brief treatment is necessary.

Preattentive processing refers to the nonconscious events that occur at the earliest stage of information processing. Information

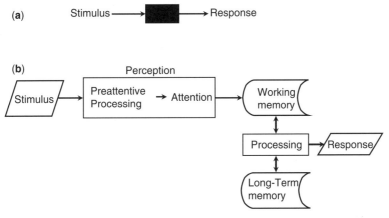

Figure 1. Schematic representation of information processing models of cognitive function. (a) The black-box model employed in behaviorist theory. Stimulus and response are quantifiable while knowledge of the inner workings of the cognitive process is considered intractable to experimental methods. (b) The more complex model if information processing based on probing cognitive function in both animal and human using the tools of psychology and neurobiology. See text for explanation of labels.

enters the CNS as a sensory stream (e.g., visual or auditory input from the eyes or ears, respectively) representing a massive amount of data that would quickly saturate downstream systems. A large number of distractors exist in the environment that must be quickly filtered out of the sensory stream to pass along information that is behaviorally and contextually relevant. This is accomplished by a system of filters that determine the saliency of stimuli based on temporal–spatial frequency and importance [19]. The preattentive processing and filtering of the sensory stream occurs rapidly, within 100 ms of the appearance of a stimulus, and has been demonstrated by simple visual search experiments in which a subject is asked to identify a unique target in a field of distractors [20]. For simple targets and dissimilar distractors, the time required for target identification is independent of the number of distractors, indicating that a conscious search through the field is unnecessary and therefore the process is preattentive.

Once preattentive processing filters the sensory stream to a manageable number of stimuli that then enter into consciousness, it is necessary that the system attends and responds to the relevant stimulus. Attention is the cognitive process of selectively focusing on a relevant stimulus while ignoring other stimuli in the environment. This focusing of consciousness may be an overt process in which attention is directed at a particular stimulus by focusing the sensory apparatus on that stimulus, or covert, where attention is mentally focused independently of sensation. In addition, attention may be focused on cognitive processing independent of sensory input. This executive attention or executive function is discussed in greater detail below.

Attention is far from a simple linear selection process as it is subject to a number of complex modulatory interactions [21]. Notably, significant top–down processing occurs during when information in working memory, conscious determination of the importance of stimuli, and motor control of the sensory stream, all play a role in directing and maintaining attention.

Memory The processes of information acquisition (learning) and recall (memory) form the core of cognitive processing. These processes are significantly impaired in the aged and/or traumatized brain. Memories from years past appear facile to recall, while events that are more recent cannot be recalled. Whether the problem is in the acquisition of recent memories or the ability to recall these is unknown and has considerable impact on studying the problem of cognitive dysfunction from a drug discovery perspective. If a memory is not acquired, searching for drugs to recall that memory at a point distal to its acquisition is an exercise in futility.

Memory is divided into two main categories: working, or short-term memory, and long-term memory.

Working Memory Working memory is classically described as consisting of three subcomponents: a central executive, a verbal or phonological rehearsal loop, and a visuospatial sketchpad [22]. The central executive function, which maintains control over voluntary activities, is described below. Both verbal rehearsal and visuospatial sketchpad are methods of temporarily holding information arising from the perceptive process or which is recalled from long-term memory. For example, keeping a phone number in mind by verbally rehearsing until the number is dialed, or imagining the landmarks on a street corner when relaying directions. These working storage elements provide a dynamic representation of both the inner and the outer world that constitutes the substrates for further cognitive processing. The key characteristics of working memory are a limited duration of only a few seconds, and a relatively small capacity known to range between 4 and 10 items [23,24]. Because of its dynamic nature, working memory is subject to rapid change and can be easily disrupted. Therefore, information can only be maintained through consolidation into long-term memory.

The process of memory consolidation in which an item in working memory is transferred into long-term memory is essential to all forms of learning. Memories are consolidated over time, and with repetition or rehearsal, in a process that is facilitated by sleep [25]. Much is being uncovered regarding the molecular substrates of memory consolidation discussed further below.

Long-Term Memory In contrast to working memory, long-term memory is static, long lasting, and resistant to disruption and can be divided into declarative and nondeclarative memory [26]. Nondeclarative or nonconscious memory includes items learned by nonassociative means such as habituation or sensitization, innate motor and cognitive skills, and dispositions. While it is debatable as to whether these items are learned without conscious input, they all clearly involve unconscious processing. Declarative or conscious memory is the type of memory more obviously associated with cognitive processes and can be further divided into semantic and episodic memories.

Semantic memories are factual recollections. They involve facts about the world that are independent of a particular place or time. Semantic memories include facts about the individual, others, or shared knowledge. For example, one's height, a friend's birthday, or the capital of a particular country all are items of semantic memory. Semantic memories are distinguished from episodic memories in that the former are things that are "known" rather than reexperienced or "remembered."

Episodic memories hold specific relevance in time and space and are often autobiographical and involve life experience. As such, these memories are self referential and frequently context dependent involving relationships to times, places, other individuals and other events. Episodic memories are typically organized around particular periods of time, and are recalled in a manner in which the individual mentally recreates the events.

It should be obvious that there is a close relationship between semantic and episodic memories. Both types of memories may contribute to a given recall event [27] and, over time episodic memories are believed to be transformed into semantic memories. For example, a memory of an event that occurred several years ago may not be recalled as a mental re-creation of the event but rather a knowledge of attendance at the event and a list of related occurrences.

Processing The box labeled processing in Fig. 1 is the cornerstone of cognitive function. It represents a generalization of a number of complex, and, to date, poorly understood processes that take the simple inputs and storage devices described above to create conscious cognition. These are the processes that behaviorists considered intractable in terms of the scientific method, and while they have begun to yield to well-designed experimentation, there is still much to be learnt.

As mentioned above, the concept of a central executive is a necessary component of working memory and consolidation required to oversee verbal rehearsal and the visuospatial sketchpad. Furthermore, there is implicit in the concept of attention the need for a process that directs the focusing of consciousness. This observation led Broadbent [28] to distinguish between automatic and controlled processes ultimately leading to the concept of the central executive as a type of orchestra conductor or CEO that oversees cognitive processes (for review, see Ref. [29]).

The central executive oversees a variety of functions often termed "higher order" processes. These include process such as planning, goal setting, and the initiation of actions. Furthermore, the central executive is responsible for planning, motivation, problem solving, language processing, and a variety of other higher order functions.

From this brief overview of the processes involved in cognition, it is obvious that drugs to potentially improve cognitive function will target particular domains of cognition. As the disruption of cognitive domains in disease states will be heterogeneous with respect to the underlying pathologies, it is important to understand the different indications for which cognitive enhancing drugs are being developed.

1.1.2. Indications and Diagnostic Criteria The *Diagnostic and Statistical Manual of Mental Disorders*, Fourth Edition, Text Revised (DSM-IV-TR, 2004) includes cognitive dysfunction under the general heading "Delirium, Dementia, and Amnestic Disorders." [30] This is a broad spectrum of CNS disorders that includes delirium (Diagnostic code 293), AD (294.1), vascular dementia (290), (Delirium) and amnestic disorders/Other Cognitive Disorders (294). Delirium, also known as acute confusional state or reversible madness is a behavioral response to widespread disturbances in cerebral metabolism that can be caused among other events by: substance intoxication or withdrawal, head trauma, dehydration, congestive heart failure, sleep deprivation, endocrine dysfunction, etc. Acetylcholine (ACh) is the primary neurotransmitter involved in delirium with the primary neuroanatomical site being the reticular formation. The clinical abnormalities associated with delirium occur in the domains of arousal, language and cognition, perception, orientation;, mood, sleep and wakefulness, and neurological functioning. Dementia involves multiple cognitive deficits that are differentiated on the basis of etiology [30] and include dementia due to AD, Parkinson's disease (PD), Pick's disease, Huntington's disease, HIV disease, head trauma, vascular disease, substance-induced persisting, other medical conditions, and medications. Quite clearly, the ability to diagnose the precise form of dementia, for example, cognitive dysfunction, will dictate both defining the potential molecular lesion and treatment which has a major impact on identifying viable drug discovery approaches and effective patient treatments. Evaluation of medical history, physical examination, routine and specialized laboratory tests and brain scans [30] are all pertinent to diagnosis underlying a major need for definitive biomarkers for cognitive dysfunction [31–33].

Cognition in Alzheimer's Disease AD and its variants are neurodegenerative diseases where cell death plays a major role in the cognitive dysfunction phenotype. Memory impairment, difficulty in solving problems and decreases in spontaneity, reaction speed and accuracy are early signs of AD. The inability to remember names, misplacing items, forgetting what one was about to be done and difficulty in word finding (anomia) are early signs of AD that are usually dismissed as signs of "getting older" or "needing more sleep" that can lead to depression, aggression, confusion, and wandering that can further exacerbate cognitive function. The memory loss in early stage AD is most obvious with newly acquired memories that also lead the individual to avoiding new, unfamiliar situations. AD can progress to overt memory loss, for example, dementia with accompanying physical traits that lead to the AD patient being bedridden and ultimately to death. AD is a major healthcare challenge with current estimates of 25–34 MM individuals being currently affected worldwide [11] with projections of triple this number by 2050. Since diagnosis of AD is exclusionary, a major challenge in finding effective treatments for AD is the ability to diagnose the disease at a stage early enough when drugs may act to arrest and potentially reverse the disease process, for example, before the cells are dead and the cellular substrates for drug action are no longer present [34]. Current biomarkers, including plasma and CSF amyloid levels and brain imaging [31] are insufficiently robust to be useful in disease diagnosis [11]. In October 2008, the FDA's Peripheral and Central Nervous System Drugs Advisory Committee in reviewing the use of radionucleotide imaging for the detection of cerebral amyloid to assist in AD diagnosis "felt that too many questions remain about the relationship between cerebral amyloid and AD to diagnose AD or predict risk of the disease based on a positive test for the marker." [35] Given that the currently approved treatments for AD have questionable efficacy [8] and the major research focus on the amyloid hypothesis is also in question [11], there is considerable effort ongoing to identify drug candidates that restore brain function via effects in enhancing neurotransmitter release, reducing brain inflammation and oxidative stress and in enhancing mitochondrial function [11]. These approaches are discussed in detail below.

Schizophrenia Cognitive impairment in schizophrenia occurs before the manifestation of overt psychotic symptoms, remaining severe through the course of the disease. Schizophrenia-associated cognitive dysfunction involves multiple domains including executive function, attention, processing, vigilance, verbal learning and memory, verbal and spatial working memory, semantic memory, and social cognition [30,36]. There is a major unmet medical need in finding effective treatments to treat the cognitive domain of schizophrenia (CDS) since it is considered to be of equal or greater importance than positive or negative symptoms in predicting the functional consequences of schizophrenia, especially in regard to work status and quality of life [37]. The US federal initiative, Measurement and Treatment Research to Improve Cognition in Schizophrenia (MATRICS) is discussed further below.

Parkinson's Disease While the major clinical features of PD reflect tremor, rigidity and bradykinesia that result from the degeneration of dopaminergic neurons in the locus coeruleus and substantia nigra, dementia occurs in approximately 40% of PD patients older than 70 [38]. Like AD, the cerebral cortex of PD patients contains amyloid plaques, NFTs, and also eosinophilic Lewy bodies. The latter also occur in Lewy body dementia (LBD), a degenerative brain disease that is a leading cause of dementia in the elderly population and accounts for up to 20% of all dementia cases. PD is a form of LBD. Treatment approaches to PD-associated dementia and LBD, such as those for AD are currently focused on plaque removal with additional initiatives in enhancing dopamine formation and release, reducing brain inflammation and oxidative stress and enhancing mitochondrial function.

HIV-Associated Dementia Dementia associated with the HIV infection in AIDS results from the decline in immune competence associated with viral load. In approximately 10% of HIV-positive individuals, the neurological symptoms that lead to dementia reflect the first sign of AIDS [30]. Cognitive disorders in AIDS dementia include memory impairment, lack of concentration and confusion. These are accompanied by apathy, depression, anhedonia, delusions, and hallucinations as well as motor dysfunction that can lead to PD-like symptoms. Neuropathological abnormalities occur in 90% of brains of individuals that reflect opportunistic viral, fungal, protozoal, and bacterial infections that lead to cerebral atrophy and nonspecific white matter loss [39]. Additionally, several of the antiviral agents used to treat AIDS have their own effect on CNS function leading to anxiety, depression, and confusion. In addition to antiviral therapy to reduce the cause of the neuropathological abnormalities in AIDS, treatment approaches to PD-associated dementia and LBD, such as those for AD are currently focused on plaque removal with additional initiatives in enhan-cing dopamine formation and release, reducing brain inflammation and oxidative stress and enhancing mitochondrial function.

Stroke/Vascular Dementia Vascular dementia is the second most common form of dementia after AD and is caused by a major or multiple cerebrovascular accidents (CVAs), now being more commonly known as stroke or "brain attack." A stroke is defined by the WHO as a "neurological deficit of cerebrovascular cause that persists beyond 24 hours or is interrupted by death within 24 hours" and occurs when blood flow to the brain is attenuated by the ischemia resulting from thrombosis or an embolism or from hemorrhage. Some four million cases of stoke occur each year highlighting the need for effective therapies [40]. Memory impairment following a stroke is accompanied by aphasia, apraxia, agnosis, and deficits in executive function, spasticity, ataxia, hemiparesis, and white matter lesions [30]. While t-PA is effective as antithrombolytic therapy, the search for drugs to treat the consequences of the excitotoxic insult associated with stoke has been frustrating due to a lack of predictive animal models and compounds advanced to clinical trials that lacked efficacy or had side effects that limited their use or made the consequences of the stroke worse [41,42]. As with AIDS and PD, there are expectations that medications effective in treating the cognitive dysfunction associated with AD will have utility in the treatment of the cognitive decline following stroke.

Life Style-Associated Cognitive Dysfunction In addition to stress, sleep deprivation, depression, and anxiety, life style choices associated with impaired cognitive function include recreational and prescription drug usage and severe alcohol dependence which is the third leading cause of dementia. The latter occurs late in life following 15–20 years of heavy drinking [30]. Additionally, the use of sedatives, hypnotics, and/or anxiolytics can lead to dementia.

1.1.3. Disease Diagnosis

ADAS-COG The most commonly used rating instrument in trials of cognition enhancers is the cognitive subscale of the AD Assessment Scale (ADAS-cog) [43,44]. This scale was designed to reliably identify and rate the major characteristics of

AD across a range of dysfunction from mild to severe dementia. The ADAS-cog is a relatively brief test taking approximately 30 min to complete. The test consists of 11 parts that primarily rate memory, language, and praxis (performance of an action). Notably, nearly half of the items in the ADAS-cog relate to memory making this scale a very sensitive measure of memory dysfunction ideally suited for the assessment of AD patients [44].

CANTAB While the ADAS-cog is a useful instrument for the study of AD, it is limited in the domains of cognitive function assessed, and is subject to rater errors that can be compounded during long disease progression trials where the raters may turnover several times during the course of the study [45]. The Cambridge Neurophysiological Test Automated Battery (CANTAB) was developed as a computer-based tool that has high precision, speed, and reliability and can produce objective feedback [46]. The battery is administered at a computer terminal and relies entirely on nonverbal stimuli and responses. CANTAB measures different aspects of cognitive function including visual memory, attention, and planning. As such, CANTAB can measure certain aspects of executive function that are not addressed in detail in the ADAS-cog. The CANTAB instrument was designed as a battery of tests that mimic behavioral paradigms used in preclinical studies that have helped to establish the neuronal substrates of cognitive function. Therefore, the battery has the unique potential to not only identify cognitive dysfunction but also indicate possible sites of disease action in the brain. The CANTAB has proven to be a sensitive measure effectively detecting cognitive dysfunction in a number of disorders including AD, PD, schizophrenia, and mood disorders, as well as the cognitive decline associated with normal aging.

MATRICS Schizophrenia is often thought of in terms of the positive symptoms of the disorder including hallucinations and paranoia. However, it is now understood that the cognitive dysfunction associated with schizophrenia is both severe and poorly treated by current therapies. Cognitive impairment in schizophrenia begins before the onset of the psychosis and can worsen throughout the course of the illness. Schizophrenia is associated with widespread, multifaceted impairments in cognitive function, including executive function, attention, processing, vigilance, verbal learning and memory, verbal and spatial working memory, semantic memory, and social cognition. Cognitive impairment in schizophrenia may be of equal or greater importance than positive or negative symptoms in predicting functional outcomes and quality of life [37].

The recognition of the prevalence of cognitive dysfunction in schizophrenia together with the realization that there was a lack of a consensus as to how cognition in schizophrenia should be measured, led the National Institute of Mental Health to sponsor the Measurement and Treatment Research to Improve Cognition in Schizophrenia (MATRICS) initiative that has produced the MATRICS Consensus Cognitive Battery (MCCB) [47]. The MCCB employs 10 tests that assess performance in 7 cognitive domains: speed of processing, attention, working memory, verbal learning, visual learning, reasoning and problem solving, and social cognition. The battery is intended to be a comprehensive assessment of cognitive dysfunction in schizophrenia, and therefore requires more than 1 h for administration. The MCCB relies on neuropsychological tests that are standardized and widely used making it somewhat user friendly for the clinician. However, it lacks the translational power of CANTAB in that many of the tests that have no preclinical analog. In an attempt to remedy this, efforts are ongoing to match domains of clinical efficacy with appropriately predictive animal models [48].

1.2. Substrates of Cognition

1.2.1. Neuroanatomical Our understanding of the neuroanatomical substrates of cognition began in the 1920s with relatively crude lesion studies pioneered by Karl Lashley during his search for the elusive memory trace termed the engram. While Lashely's work was influential and set the stage for all future research on the neural basis of cognition, his focus on the reductionist approach led him to famously conclude after 30 years of research that he discovered "nothing directly of the real nature of the engram" and that he sometimes felt in reviewing the evidence that "the necessary conclusion is that learning just is not possible." [49]. The development of functional imaging methods along with the ability to produce discrete anatomical, pharmacological, and molecular lesions has greatly advanced the field now known as cognitive neuroscience.

Neuroanatomical Substrates of Memory As first identified in the work of Penfield [50], focal electrical stimulation delivered during brain surgery to the cortex, the temporal lobes in particular, can evoke strong conscious memories. These studies suggested that memories are somehow stored in the cortex, possibly in a distributed network formed by altering synaptic connections and weights that can be selectively activated by electrical stimulation. The medial temporal lobe that includes the hippocampus, entorhinal, and perirhinal cortex is the key brain complex responsible for processing information from the neocortical and limbic regions and integrating the information into a memory that encodes for various aspects of an event [51]. The medial temporal lobe interconnects with a number of brain regions including sensory cortical regions, for example, the superior temporal gyrus and insular cortex, regions that encode for fear and emotion responses, including the amygdale and cingulated cortex, and executive regions of the neocortex. The medial temporal lobe is therefore ideally situated to receive information regarding the multiple components of an experience (sensory, emotional, and cognitive) and integrating these. The "bound" memory is then consolidated into long-term memory in the neocortex [52].

Perhaps the most dramatic proof of the role of the medial temporal lobe in the formation of new memories comes from the classic case of the patient known as HM [53] who died in 2008. HM suffered from severe intractable epilepsy that was treated by bilateral removal of the medial temporal lobes [54]. This procedure was performed prior to an understanding of the role of this brain structure, and produced a surprising and remarkable change in HM's ability to encode memory in that he developed complete anterograde amnesia. While appearing otherwise normal, in terms of problem solving ability, IQ, and language comprehension, HM was unable to incorporate any new memories. Interestingly, HM and other patients with damage to the medial temporal lobes also experienced a temporally graded retrograde amnesia with more recently encoded memories of events prior to the damage being lost, but older memories being maintained. This has led to the suggestion that the process of memory consolidation is a slow multicomponent process with the initial encoding of long-term memories residing in the medial temporal complex followed by a slow transfer to the neocortex over time [52].

Neuroanatomical Substrates of Executive Function As described above, the central executive function is a necessary yet poorly defined function that oversees the bulk of what is considered conscious processing. While the complexity of the overall roll of the central executive makes it difficult to accurately define, it is clear that it resides in prefrontal cortex [55]. The prefrontal cortex is connected to all functional units within the central nervous system [56] implying that it exerts a broad controlling influence on behavior. Of particular interest when considering central executive function is that the prefrontal cortex has extensive interconnections with the dorsomedial thalamic nucleus, a key integrator of information from the thalamus a region encoding emotional states that includes the amygdale and cingulate cortex and the key memory areas in the medial temporal lobe.

The prefrontal cortex, and the dorsolateral prefrontal cortex in particular, is involved in the processing of goals, actions, and planning, executive functions that operate with working memory [57]. These functions include both the ability to guide behavior by internal representations and the ability to adapt to changes that require subsequent alterations in behavior. This cognitive flexibility, or the ability to shift cognitive set, is a key aspect of central executive function that is disrupted in number of neuropsychiatric disorders involving impairments in prefrontal cortical function [58].

Damage to the prefrontal cortex induced by stroke or traumatic brain injury can produce a variety of syndromes that underscore the importance of this structure in central executive function. Resultant symptoms include perseverative behaviors, flat affect or dramatic changes in personality, a loss of ability to follow internal plans or maintain attention, and a loss of cognitive flexibility. These same behaviors are observed in neurological and neuropsychiatric disorders associated with prefrontal dysfunction as discussed below.

1.2.2. Molecular Substrates of Memory As discussed above, memory is believed to exist in a distributed network created by altering synaptic connections and weights. The concept that neuronal activity can lead to changes in synaptic coupling efficiency was first proposed in Hebb's postulate [59]. Hebb proposed that when a given neuron excites a second neuron repeatedly and persistently, some change occurs such that the efficiency of coupling between the two cells is increased. This activity-dependent synaptic plasticity has been the focus of considerable research in an attempt to identify the molecular substrates of memory.

Long-Term Potentiation The finding that repetitive activation of glutamatergic synapses in the hippocampus led to a long-lasting increase in synaptic strength provided evidence for the first potential physiological substrate for memory in a mammals [60]. This phenomenon, long-term potentiation (LTP) has since been observed at a number synapses and has been the subject of intensive research efforts aimed at understanding the molecular events underlying synaptic plasticity. The best studied of these synapses remains the Schaffer collateral-CA1 synapse of the hippocampal formation.

There are two major types of ionotropic glutamate receptors that contribute to fast postsynaptic response at glutamatergic synapses [61] named for their selective agonists, AMPA (α-amino-3-hydroxy-5-methyl-4-isoxazole propionic acid) and NMDA (N-methyl-D-aspartate). AMPA receptors families are monovalent cation channels that are gated by glutamate and produce rapid excitatory responses. NMDA receptors are permeant to monovalent cations and to calcium. A critical feature of the NMDA receptor channel is that it is blocked by extracellular magnesium in a voltage-dependent fashion. This magnesium block essentially renders the NMDA receptor inactive at normal membrane potential, but allows it to open at more depolarized potentials. Induction of LTP at the Schaffer-collateral-CA1 synapse requires activation of NMDA receptors [62]. Strong synaptic stimulation, such as those produced by a high-frequency (100 Hz) electrical stimulation of the Schaffer collateral fibers, leads to sufficient AMPA receptor-mediated depolarization of the CA1 pyramidal neurons to relieve the magnesium block of the NMDA receptor. This allows an influx of calcium through the NMDA receptor that triggers the biochemical cascades that lead to a persistent change in synaptic efficacy. While a large number of putative signaling molecules have been implicated in this process, the exact biochemical pathway underlying LTP remains unknown [63]. However, there are several key events that have been identified, which provide an understanding of molecular substrates. The calcium/calmodulin-dependent kinase, CAMKII appears central to LTP induction. CAMKII is activated by calcium influx and autophosphorylates with LTP induction [64]. Knockout of CAMKII or replacement with an autophosphorylation-deficient variant prevents LTP induction [65,66]. Other kinases have also been implicated in LTP induction including cAMP-dependent protein kinase [67], extracellular signal-related kinase [68], Src kinase [69], and protein kinase C [70]. The biochemical steps linking CAMKII activation and autophosphorylation to the expression of LTP are not fully understood. Enhanced synaptic efficacy appears to involve an increase in postsynaptic AMPA receptors produced by a modulation of receptor trafficking [71]. Intracellular AMPA receptors are trafficked from recycling endosomes to the plasma membrane by a process that requires Rab11a, a small GTP binding protein [72]. The AMPA receptors interact with a family of transmembrane AMPA receptor regulatory proteins (TARPs) that provide an interaction with postsynaptic density proteins (PSDs) [73] directing the AMPA receptors to their appropriate location within the synaptic membrane. CAMKII phosphorylation of TARPs appears to be important for LTP expression [74,75] providing a plausible link between CAMKII activation, an increase in postsynaptic AMPA receptor density, and resultant enhanced synaptic efficacy.

Long-Term Depression The finding that the Schaffer collateral-CA1 synapse and other synapses can also express long-term depression (LTD) suggests that bidirectional control of synaptic strength is possible [76]. LTD can be induced by a prolonged low-frequency (1 Hz) stimulation. Interestingly, this form of LTD, like LTP is NMDA receptor-dependent and relies on calcium influx suggesting that the temporal and spatial dynamics of spike timing and calcium influx into the postsynaptic dendritic spine are critical in determining the form of plasticity expressed [77]. NMDA receptor-dependent LTD appears to involve the activation of the protein phosphatases, PP1 and PP2B [78]. Subsequent calcium-dependent dephosphorylation of AMPA receptors and associated proteins lead to a dissociation of the receptors from the postsynaptic complex and a triggering of clathrin and dynamin-dependent endocytosis [79].

1.2.3. Other Forms of Synaptic Plasticity

This brief overview of the molecular basis of plasticity has focused on NMDA-dependent plasticity in the hippocampal CA1 region. While this represents the best-studied form of synaptic modulation, plasticity occurs at many other synapses within and outside of the hippocampus. The events may be NMDA-independent [80], have a presynaptic locus of action [81], be mediated and expressed by GPCRs [82], or involve retrograde synaptic transmission [83]. Furthermore, plasticity is not solely a property of excitatory synapses also occurring at inhibitory GABAergic synapses [84]. Therefore, memory is very likely encoded in a distributed fashion across both excitatory and inhibitory synapses expressing multiple forms of plasticity. This suggests that the identification of a single molecular substrate of memory, or even of a particular type of memory may be as elusive as Lashley's engram.

1.3. Preclinical Behavioral Assessment of Cognition

The facts that most, if not all, of the disorders discussed above are uniquely human represent a significant challenge for drug discovery efforts to identify therapeutically relevant cognitive enhancing agents. While it is relatively simple to ask a human subject if they recall a particular word or number, rodents and nonhuman primates are typically less cooperative. In addition, preclinical species do not have the capability for certain aspects of cognitive function, especially those involving language. The importance of assay and model selection and careful interpretation of results is underscored by a large increase in preclinical studies reporting cognitive enhancement in animal models that is not reflected in the paucity of new therapeutics emerging from clinical trials, a relationship that may be in part due to an underappreciation of the complexity of measuring cognitive function in preclinical species [32]. Despite these hurdles, a number of assays have been developed that allow a reliable and informative assessment of cognition in both rodents and nonhuman primates.

1.3.1. Disease Models

The validity of an animal model for any disorder can be rated on three scales; predictive, construct, and face validity. *Predictive validity* focuses on how well results produced in the animal model are borne out in the clinic. More often, animal models are back-validated using clinical benchmarks to provide a basis for arguing for future predictive validity. This is particularly problematic in the field of cognition because there are few approved drugs, and those that are approved have modest efficacy [85]. *Construct validity* concerns the theoretical rationale underlying the model. A model with a high degree of construct validity would disrupt the same neurotransmitter systems and engage the same neuronal circuitry as the human disorder. Unfortunately, understanding of the underlying pathophysiology of cognitive dysfunction is far from complete, suggesting that construct validity is difficult to ascertain. *Face validity* is a measure of how accurately the model reproduces the symptoms of the human disorder. As noted, face validity can be a challenge because there are certain aspects of cognitive function that are not expressed in preclinical species.

Because of the problems with model validity, and construct validity in particular, animal models of cognitive dysfunction typically model some aspect of a disorder rather than recapitulate the human syndrome. For example, attempts at generating an AD model by genetically recreating alterations in the amyloid system have recapitulated certain aspects of the disease including cognitive dysfunction and altered synaptic plasticity [86], the approach has failed to produce a mouse that faithfully produces plaque deposition, tangle formation, neurodegeneration, and cognitive dysfunction and recent data have questioned the amyloid hypothesis [11].

A large number of genetic and pharmacological models of cognitive dysfunction exist. Some commonly used models include scopolamine-induced impairments to mimic cholinergic dysfunction in AD [87] and phencyclidine (PCP) 8-induced impairments thought to replicate NMDA-hypofunction associated with schizophrenia [88]. A major caveat to the use of these models in drug discovery is the particular pharmacological or genetic insult used to disrupt function may not be involved in the clinical pathology, and even if it is, may not be the sole cause of cognitive dysfunction. Therefore, a risk exists that an NCE will be developed with utility restricted to treatment of the animal model. Perhaps the best path for developing novel cognitive enhancers with broad therapeutic potential is lies in testing NCEs for cognitive enhancing effects in normal animals, or in aged animals exhibiting age-associated cognitive impairment. The use of such models requires careful attention to verifying both the presence and the function of the target in the disease state.

1.3.2. Preclinical Cognitive Assays As mentioned above, there are aspects of human cognition such as verbal memory and fluency that cannot be modeled in preclinical species. Furthermore, due to the complexity and variability inherent in animal studies, the assessment of long-term memory is typically limited to a simple increase in intertrial time that occurs on a time scale on the order of hours to days rather than weeks or years. Below some of the assays more commonly used in drug discovery research are described. While the focus is on a basic overview of the tasks, it is import to keep in mind that these assays exist in multiple forms and often are conducted with significant procedural differences between laboratories. Most of the assays described can be used with short or long intertrial intervals to assess both short- and long-term memories, and can be employed to assess function in genetically or pharmacologically impaired models. Furthermore, by varying the timing of treatment, it is often possible to tease out the aspect of memory formation impacted by the treatment such as acquisition or consolidation.

The 5-Choice Serial Reaction Time Task as a Measure of Attention and Impulsivity In the clinic, attention, and impulsivity are typically measured using the continuous performance task (CPT) in which the subject is asked to attend to a stream of stimuli and respond to an infrequent target (e.g., the number 9) while withholding response to irrelevant distractors (e.g., the numbers 0–8). The 5-choice serial reaction time task (5-CSRTT) has been developed as a preclinical analog of the CPT that allows the assessment of attention and impulsivity in rodents [89]. In its simplest form, the task requires the animal to attend to five small openings in the wall of a test chamber and wait for one of the openings to become illuminated. The animal responds to the illumination with a nose poke into the appropriate opening and receives a food reward. The openings are illuminated in a random fashion, and the animal is required to respond within approximately 1 s. Errors may be measured as incorrect responses in which the animal pokes the wrong opening, misses in which the animal fails to respond, or premature responses in which the animal responds before the illumination. An error produces a short period of darkness (time-out) that serves as a negative reinforcement. Treatments that enhance attention would be expected to decrease the number of errors, with premature response rate serving as an indication of impulsivity.

The 5-CSRTT has been used to asses the proattentive properties of a number of clinical and preclinical compounds including the psychostimulant methylphenidate **9**, the selective norepinephrine uptake (NET) inhibitor atomoxetine **10** [90], the H_3 histamine receptor inverse agonist ciproxifan **11** [91], and the $\alpha_4\beta_2$ nicotinic agonist ABT-418 **12** [92].

Social Recognition and Novel Object Recognition as Measures of Working Memory Recognition memory has become an increasingly popular way of assessing the effect of cognitive enhancing agents in rodents. These methods rely on the animal's natural curiosity regarding a novel stimulus. The social recognition assay is based on the finding that social recognition memory of adult rats for juvenile rats decreases as the time interval between presentations of the same juvenile rats to adult rats is increased [93]. The assay is typically performed by exposing an adult rat to a juvenile and measuring the time the adult spends investigating the juvenile during a short (3–5 min) trial. Reexposure to the same juvenile after a short intertrial interval (5–15 min) will result in significantly decreased investigation time suggesting that the adult rat remembers and recognizes the juvenile. Longer (>2 h) intertrial intervals produce investigation times similar to those observed on the initial presentation suggesting a lack of memory for the juvenile. Administration of a test compound that improves working memory would be expected to decrease the investigation time observed after a long intertrial interval.

In the novel object recognition paradigm, the animal is exposed to two objects and exploration is assessed as in the social recognition assay. After an appropriate intertrial interval, the animal is reexposed to one of the original objects plus a novel object. The animal's memory of the familiar object is reflected in an increased duration of exploration of the novel object [94]. As with social recognition memory, longer intertrial intervals lead to a loss of memory that can be rescued with the administration of cognitive enhancing agents.

A large number of preclinical and clinical compounds produce promnesic effects in recognition memory models including the histamine H_3 receptor inverse agonist GSK189254 **13** [95], the cholinesterase inhibitor tacrine **3** [96], and the ampakine CX-546 **14** [97].

The Morris Water Maze as a Measure of Spatial Memory Working Memory One of the most commonly used memory tasks in preclinical drug discovery research is the Morris water maze (MWM) [98]. The assay employs a large circular pool filled with opaque water in which a small escape platform is hidden. A rat or mouse placed in the pool for the first time will swim and randomly encounter the escape platform. In the simplest form of the assay,

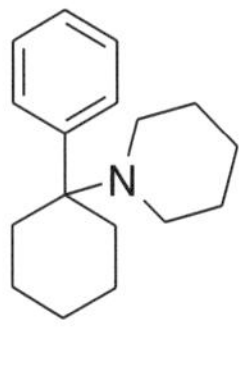

8 Phencyclidine

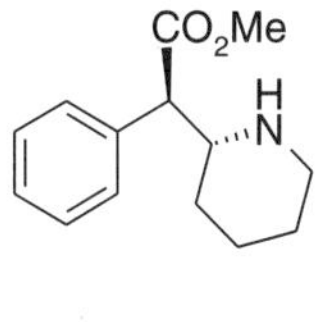

9 Methylphenidate

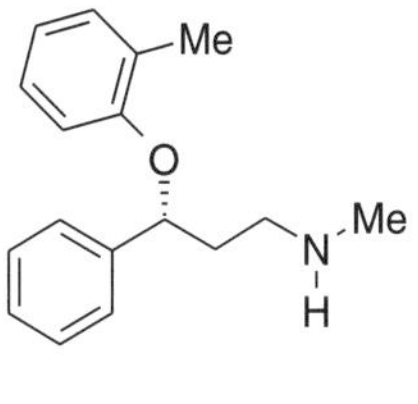

10 Atomoxetine

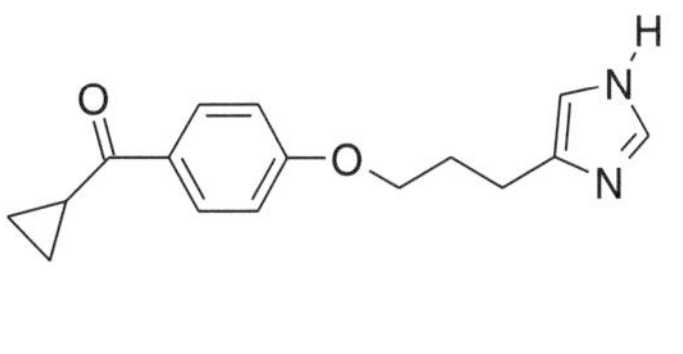

11 Ciproxifan

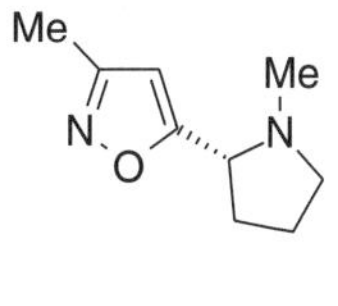

12 ABT-418

the animal learns over successive trials to employ spatial cues to find the platform resulting in shorter escape latencies. Treatments that enhance spatial memory are expected to accelerate the learning process such that the escape latency decreases significantly with fewer trials. Importantly, swim speed should be measured and should not vary with treatment to avoid assigning cognitive enhancing properties to an agent that simply produces hyperlocomotion. Similar information on spatial learning and memory can be obtained using other maze-based assays, for example, the radial arm maze [99], however, these methods are more complex involving the use of food reward while the MWM relies on the animals native escape response. A large literature exists describing the effects of pharmacological treatment on learning in the MWM. A number of clinical and preclinical compounds are effective in this model including the cholinesterase inhibitors donepezil **4** and rivastigmine **5** [100], the 5-HT$_6$ receptor antagonist SB-271046 **15** [101], and the α_7 nicotinic receptor positive allosteric modulator, NS1738 **16** [102].

Conditioned Fear as a Measure of Contextual Memory A number of conditioned fear assays have been employed in drug discovery research as a test for cognitive enhancement. The most common variants of these methods are the passive avoidance task [103] and contextual freezing [104]. In the passive avoidance task, the animal is placed in a chamber that is divided between an open illuminated side and an enclosed dark side. Rodents have a natural preference for dark spaces, so the will quickly move into the dark side of the chamber. Entry into the dark side of the chamber is accompanied by a foot shock. On subsequent trials, the latency to enter the dark chamber increases as the animal learns to associate the shock with the context of the dark compartment. Contextual freezing is measured by placing the animal in a box with obvious contextual cues, and administering a foot shock. The animal is later placed in the same context and the amount of time the animal remains immobile (freezes) is recorded. This freezing behavior is a natural response in expectation of receiving a foot shock. Therefore, the time spent in the frozen posture is related to the memory for the context in which the shock occurred. Conditioned fear assays are relatively facile, and therefore in common use in drug discovery research. However, it must be noted that these assays are primarily measuring fear and emotional memory processed through the amygdala that may not represent other forms of memory. Pharmacological manipulations improve conditioned fear memory including the cholinesterase inhibitor physostigmine **17** [105], the 5-HT$_{1A}$ antagonist, WAY-100635 **18** [106], and the PDE4 inhibitor, rolipram **19** [107].

Set shifting as a Measure of Executive Function A commonly used clinical test of executive function is the Wisconsin card sorting test

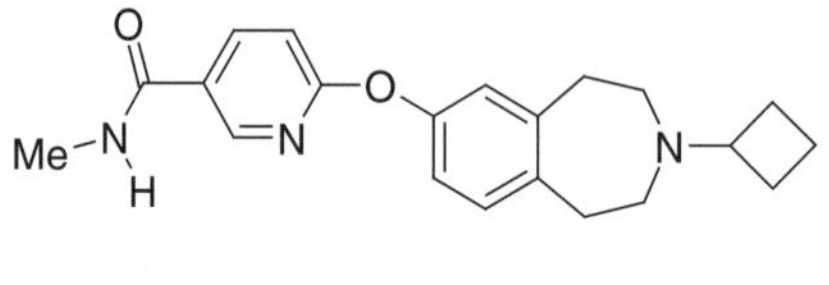

13 GSK189254

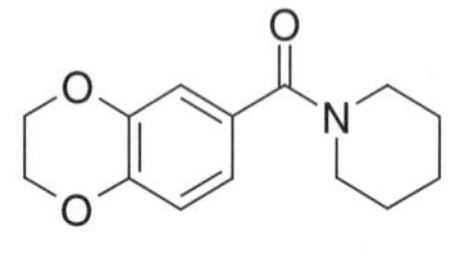

14 CX-546

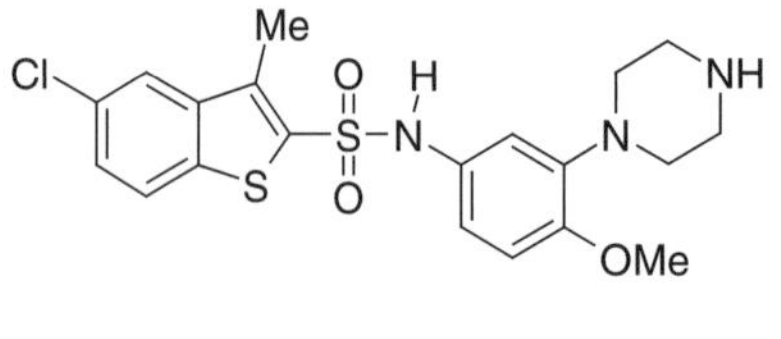

15 SB-271046

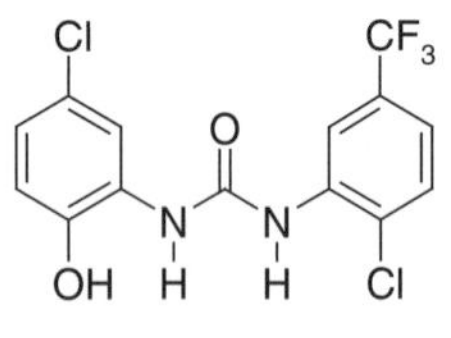

16 NS1738

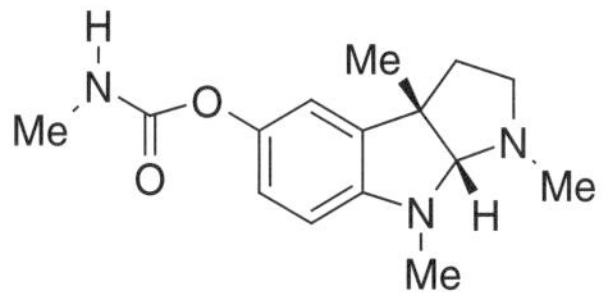

17 Physostigmine

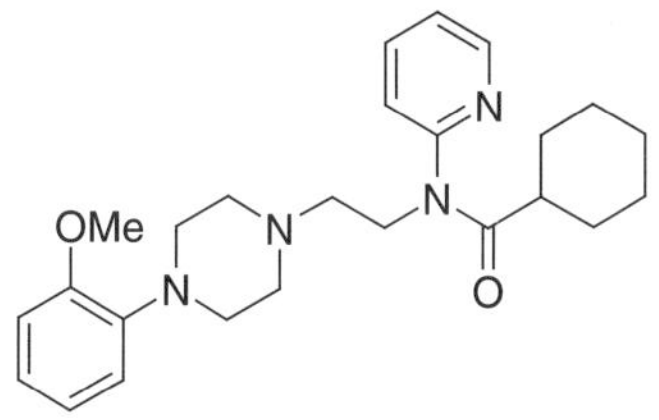

18 WAY-100635

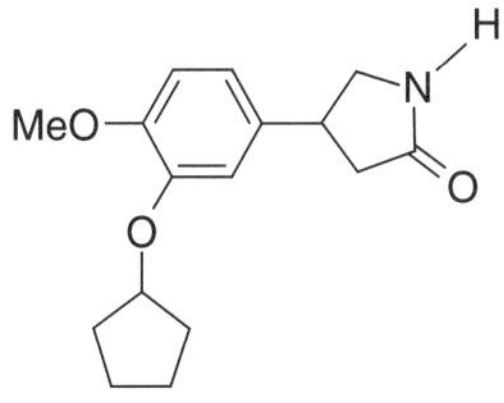

19 Rolipram

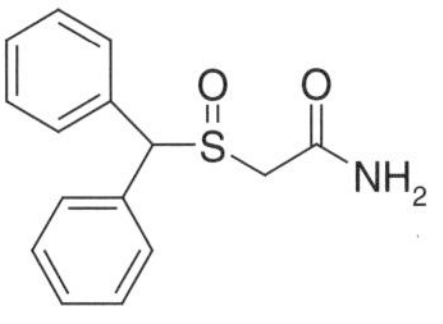

20 Modafinil

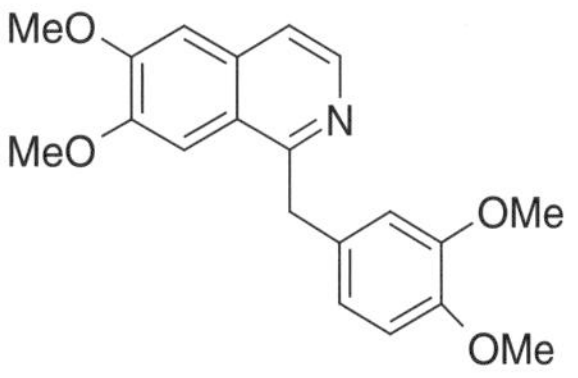

21 Papaverine

(WCST) [108]. The test employs a series of cards containing shapes (stars, circles, triangles, etc.) present in different number and in different colors. The subject is asked to sort the cards and is rewarded for discovering the appropriate sorting rule (e.g., sort by shape). The experimenter then changes the rule without telling the subject. The ability to shift set to the new rule is considered a measure of cognitive flexibility. Patients with deficits in prefrontal executive function, for example, those with schizophrenia, will exhibit difficulty shifting set, and will make numerous perseverative errors [109]. The rodent analog of the WCST is termed attentional set shifting [110]. In this assay, rodents are trained to dig in one of the two containers to obtain a hidden food reward. The reward may be paired with a particular odor (e.g., lavender), or a particular digging medium (e.g., sawdust). Once an animal learns to correctly identify the rewarded container, the rules are changed, and the animal's performance is measured. The change may be an intradimensional shift (e.g., the correct container was associated with lavender and is now associated with a clove odor), or extradimensional (e.g., the correct container was associated with lavender and is now associated with shredded paper). The experimenter can then measure the number of trials needed for the animal to learn the new rule. Compounds active in improving attentional set shifting include modafinil **20** [111], the 5-HT$_6$ receptor antagonist, SB-271046 **15** [112], and the nonselective PDE10 inhibitor, papaverine **21** [113].

2. SMALL-MOLECULE APPROACHES TO COGNITIVE ENHANCEMENT

2.1. Historical

Amyloid plaques, neurofibrillary tangles, and neuronal loss characterize AD. This loss of neurons produces insufficiencies in several neurotransmitter systems; one of the more prominent is the loss of cholinergic neurons in the basal forebrain, which project into the hippocampus and cortex, brain regions that play an important role in memory and cognitive function. Loss of cholinergic neurons results in up to 90% reduction in the activity of choline acetyltransferase (ChAT) needed for synthesis of ACh. These findings led to the cholinergic hypothesis of AD, some 50 years old, that the dementia resulted from cholinergic dysfunction and/or loss [114,115]. This hypothesis was supported with animal studies of cholinergic dysfunction that impaired learning and memory tasks. Although the exact relationship between the cholinergic neuron loss and dementia is not well understood, it is accepted that the cholinergic system is involved in cognitive (attention and memory) and noncognitive (apathy, depression, psychosis, sleep disturbances, aggression) processes of AD. Patients with MCI have increased ChAT in frontal cortex and hippocampus at autopsy, suggesting it may be a compensatory mechanism to slow disease progression [116]. Early treatment strategies have therefore focused on cholinergic replacement therapy with the development of cholinesterase inhibitors (ChEIs) and cholinomimetic agents.

2.1.1. Cholinesterase Inhibitors Cholinesterase (ChE) enzymes function to hydrolyze and terminate the action of ACh at postsynaptic sites. There are two forms of the enzyme, acetylcholinesterase (AChE) and butyrylcholinesterase (BuChE, also known as pseudocholinesterase), with different CNS and PNS localization. Both forms are found in the brain, although AChE represents approximately 95% of the ChE activity in the normal human brain and colocalizes with cholinergic synapses. BuChE is in lower abundance in brain, primarily in glial or satellite cells, and is virtually absent in neurons. BuChE is synthesized in the liver and highly distributed in the plasma, hematopoetic cells, intestine, liver, heart, and lung. Although AChE selectively hydrolyses ACh, BuChE hydrolyses ACh, in addition to multiple xenobiotic esters. Two forms of AChE exist in the brain, a tetrameric extracellular membrane-anchored G4 isoform, and a monomeric, intracellular cytoplasmic G1 isoform [117]. During AD progression, levels of AChE decrease to approximately 10–15% with preferential loss of the G4 isoform, while BuChE increases to approximately 120%. With the preferential loss of the G4 isoform of AChE, an ideal AChEI for treating AD may require high specificity for the G1 form. Inhibition of AChE results in accumulation of ACh in the synapse, thus producing effects equivalent to enhanced stimulation of cholinergic receptors. The AChEIs currently approved for treatment of mild to moderate AD include donepezil **4**, rivastigmine **5**, and galantamine **6** (Table 1). The initially approved AChEI, tacrine **3**, has been removed from the market and replaced by the newer compounds due to modest efficacy, significant side effects and hepatotoxicity.

Tacrine Tacrine **3** was the first generation AChEI being approved in 1993 for treatment of mild to moderate AD. It is a reversible, noncompetitive inhibitor with similar potency for AChE and BuChE and the G1 and G4 isoforms. Tacrine also displays multiple biochemical activities, including interaction with potassium channels, inhibition of histamine N-methyltransferase, and weak blockade of muscarinic receptors. The use of tacrine was limited by poor oral bioavailability, poor pharmacokinetics (qid dosing), and limiting adverse drug reactions (nausea, diarrhea, urinary incontinence, hepatotoxicity) such that few patients could tolerate therapeutic doses, which led its withdrawal from the market [118].

Table 1. Summary of Marketed Cholinesterase Inhibitors

	Donepezil	Rivastigmine	Galantamine
Chemotype	Piperidine	Carbamate	Alkaloid
Inhibition	Reversible, noncompetitive; AChE > BuChe	Pseudo-irreversible, noncompetitive; AChE ≈ BuChe	Reversible competitive; AChE > BuChE; nAChR
Human dose (mg)	5–10	6–12	16–32
Dosage interval	qd	bid	bid
Plasma half-life (h)	50-70	1-1.5	7
Plasma protein binding (%)	>90	40	10–20
Metabolism	CYP2D6, CYP3A4	Nonhepatic esterases	CYP2D6, CYP3A4
Route of elimination	Parent and metabolites in urine	Metabolites in urine	Unchanged parent in urine
Improvement in ADAS-cog scale	2.9	4.9	3.9

Donepezil Donezepil **4** is a benzyl piperidine-based reversible, noncompetitive inhibitor approved in 1996 for treatment of mild to moderate AD and dementia. It had greater than 570-fold selectivity for human AChE over BuChE and greater potency for brain over peripheral AChE with no inhibition in cardiac and smooth muscle. It had a long elimination half-life (70 h), excellent brain permeability, and oral bioavailability (%F = 100) with no hepatotoxicity. Donezepil is metabolized in the liver by CYP2D6 and CYP3A4 and thus has potential for drug–drug interactions with drugs such as fluoxetine and paroxetine. In clinical efficacy and safety studies evaluating both short- and long-term effects, donepezil showed significant improvement in ADAS-cog measures (2.49 at 5 mg; 2.88 at 10 mg) and the CIBIC-plus rating compared to placebo [119]. The double blind study was followed by a blind washout phase that found, on all measures, a decline to values not different from placebo, indicating the treatment was symptomatic in nature. The most serious issues noted were insomnia, agitation, nausea, and leg cramps with GI disturbances. In addition to increasing ACh levels, donepezil may also affect cellular and molecular processes involved in early stage AD pathogenesis [120].

Rivastigmine Rivastigmine **5** is a pseudo-irreversible carbamate type AChEI approved in 2000 for the treatment of mildly to moderately severe AD. It is also approved for the treatment of Parkinson's dementia. Rivastigmine is a centrally and brain region selective inhibitor that facilitates cholinergic activity in the cortex and hippocampus [121]. Increased levels of ACh were seen with rivastigmine without changes in levels of NE, 5-HT, DA, and their metabolites. Rivastigmine showed equal activity for AChE and BuChE, and displayed greater potency for the G1 verse the G4 isoform of the enzyme. Since rivastigmine is metabolized by esterases rather than liver cytochrome P450 enzymes, there is minimal potential for drug–drug interactions. It had a short plasma half-life (1–1.5 h), although the pharmacodynamic half-life (10 h) was much longer due to the pseudo-irreversible, slow dissociation from the enzyme. Pharmacokinetic/pharmacodynamic results showed a 6 mg dose reduced AChE activity by 50% for up to 7 h postdosing [122]. Preclinical studies also showed that rivastigmine had selectivity for central versus peripheral AChE. In the clinic, rivastigmine (6–12 mg/day) was efficacious as assessed by the ADAS-cog, CIBIC-plus and activities of daily living scale compared to placebo. No blood pressure changes were observed, and the most serious adverse events reported were cholinergic side effects, agitation, nausea, anorexia, and GI disturbances.

Galantamine Galantamine **6** is a tertiary amine alkaloid from the bulbs of the Caucasian snowdrop, *Galanthus woronowi*, approved in 2001 for treatment of mild to moderate AD. Galantamine has a unique dual mechanism of action combining reversible, competitive inhibition of AChE, with positive, albeit weak, allosteric modulation of the nicotinic acetylcholine receptor (nAChR) [123]. It is the only actively marketed drug approved for AD that has demonstrated both mechanisms of action. Galantamine has been proposed to increase ACh levels and to facilitate glutamate, 5-HT and norepinephrine release in key brain regions involved in cognition [124]. In healthy volunteers, galantamine was rapidly absorbed after oral dosing and showed high oral bioavailability (%F = 100) and a plasma half-life of 7 h. The plasma protein binding was low (10–20%) with no accumulation of drug observed after 2–6 months of dosing. Galantamine is metabolized by CYP2D6 and CYP3A4, and thus has the potential for drug–drug interactions. In double-blinded placebo-controlled clinical studies, galantamine showed improvements in ADAS-cog, the disability assessment for dementia (DAD), and the AD cooperative study/activities of daily living (ADCS/ADL) [124].

Miscellaneous AChEIs Only huperzine A **22**, an alkaloid isolated from the Chinese moss *Huperzia serrata* that has been used for centuries in Chinese folk medicine, is currently in clinical trials [125]. It is a reversible AChEI with additional weak NMDA antagonist activity. Phenserine **23**, a dual AChE/β-amyloid (Aβ) protein inhibitor, had been in phase III trials to evaluate its potential to lower levels of β-amyloid precursor protein (β-APP) and Aβ levels in patients with mild to moderate

Alzheimer's disease [126]. Clinical development of this compound was halted when results from a phase III trial showed no benefit over placebo in the primary efficacy endpoints.

As noted retrospective meta-analyses of clinical trials has indicated that none of the ACHEIs works especially well in treating the symptoms of AD [8].

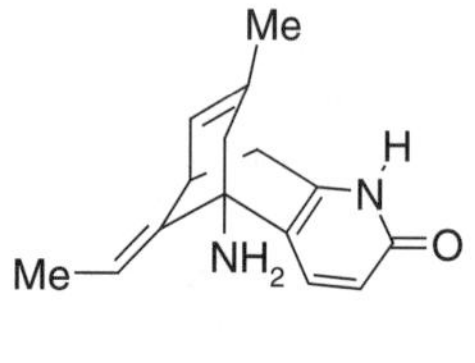

M_1 receptors are postsynaptic and abundant in cortex and hippocampus and M_2 receptors are presynaptic inhibiting ACh release. At least 12 putative M_1 receptor selective agonists have advanced to the clinic. However, due to the broad tissue distribution of M_1 receptors and dubious selectivity and efficacy, a plethora of undesirable cholinergic side ef-

22 Huperzine A

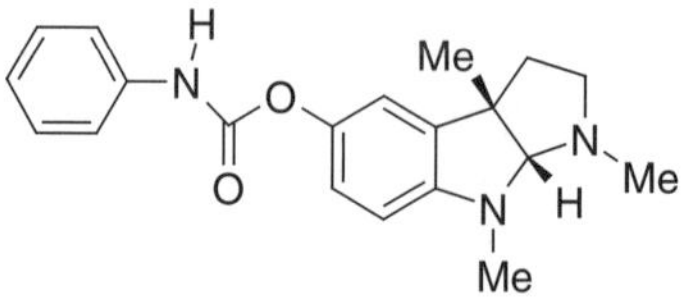

23 Phenserine

AChEIs in Schizophrenia Current marketed AChEIs have modest benefit on cognition and global functioning but do not alter the course of the disease, raising concerns about their cost-effectiveness [127,128]. AChEIs have been evaluated for cognitive enhancing activities in schizophrenia patients in trials to broaden clinical indications [129,130]. In double blind placebo-controlled clinical trials, donezepil **4** and rivastigmine **5**, failed to demonstrate a benefit in treating cognitive deficits, while galantamine **6** did exert a benefit for processing speed and verbal memory, but failed to show a significant difference in global composite score [131]. Galantamine **6** differs from donezepil **4** and rivastigmine **5** in its dual mechanism; it acts also as a positive allosteric modulator of both α_7 and α_4/β_2 nAChRs [123]. The allosteric nicotinic properties of galantamine **6** could lead to increased release of ACh or indirectly affect cognition through effects on glutamate and DA. Galantamine enhanced DA release in prefrontal cortex and hippocampus via allosteric modulation of nAChRs [132].

2.1.2. Muscarinics The muscarinic receptor family is composed of five GPCRs, M_1–M_5. M_1, M_3, and M_5 receptors are coupled with Gq linked to phopholipase C and are associated with an increase in intracellular Ca^{2+}. M_2 and M_4 subtypes are associated with Gi subunits coupled to adenylyl cyclase. The M_1 and M_2 receptor subtypes are involved in cognition.

fects have limited the utility of these NCEs. Xanomeline **24**, the most widely studied M_1 agonist had good M_1/M_4 potency but modest functional selectivity over M_2 and M_3 receptors [133]. The M_1 receptor was associated with the procognitive effects of xanomeline, while the M_4 subtype may be responsible for its antipsychotic-like actions [134]. Clinical data with xanomeline reported cognitive improvements, but with a high incidence of cholinergic side effects that led to considerable patient dropout and discontinuation of the clinical program [135]. Cevimeline (AF-102B) **25** is an M_1/M_3 agonist originally studied for the treatment of AD. It later gained approval for symptoms of dry mouth in patients with Sjögren's syndrome, an autoimmune rheumatic disease [136]. Alvameline (LU-25-109T) **26** had higher affinity and a unique M_1 agonist/M_2 antagonist profile, but was discontinued due to adverse events including salivation, dizziness, GI disturbances, and cardiovascular side effects [137]. AF267B (NGX267) **27** improved cognitive symptoms, cholinergic markers, and tau hyperphosphorylation *in vivo* and reportedly advanced into phase I for cognitive impairment in schizophrenia and for the treatment of AD dementia. It is currently in Phase II for treatment of xerostomia. *In vitro*, M_1 agonists including AF267B increase secreted APP and decrease $A\beta$ levels and tau hyperphosphorylation [138]. Norclozapine (ACP-104) **28**, the desmethyl metabolite of clozapine **29**, a partial agonist at M_1 and M_5 receptors failed in Phase II clinical trials for cognitive deficits in schizophrenia [139]. Due to the high sequence conservation in the M_1–M_5 orthosteric binding site, recent interest has focused on allosteric M_1 agonist to achieve improved selectivity and side effect profiles [140,141]. The positive allosteric modulators, VU0090157 **30** and VU0029767 **31** were identified using a functional screen and function by potentiation of agonist activation at the M_1 orthosteric site [142].

2.1.3. NMDA Modulators *Memantine* Glutamate is the main excitatory neurotransmitter in the human brain and exerts its effects via a number of receptors, including the NMDA subtype. Under normal conditions, NMDA receptor activation results in long-term potentiation of neuronal activity, a process believed to be the basis of learning and memory [60]. In neurodegeneration of AD, an increase in extracellular glutamate leads to excessive activation of NMDA receptors with consequent deficits in cognitive function and ultimately, neuronal death. Memantine **7**, an orally active, weak uncompetitive NMDA receptor antagonist, was approved in 2002 for treatment of moderate to severe AD [7]. Memantine binds to the open channel state of the NMDA receptor and blocks tonic pathological activation induced by excessive glutamate concentrations. Preclinically, it was shown to permit physiological activation of the NMDA receptor while protecting against cytotoxicity under conditions of chronic glutaminergic stimulation [143]. The clinical efficacy of memantine has been questioned [8] and it is increasingly being used as an adjunct to ChEI therapy rather than replacement.

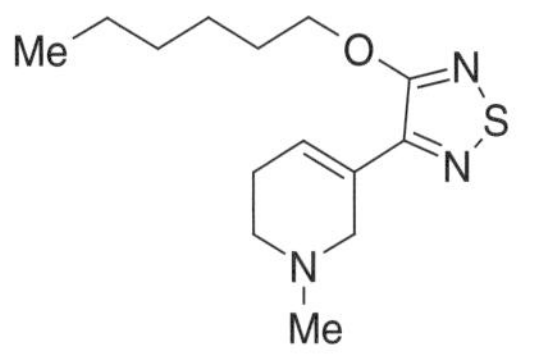

24 Xanomeline

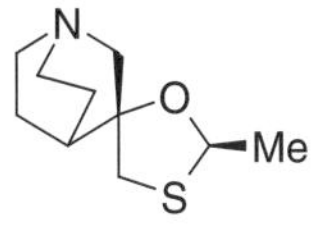

25 Cevimeline

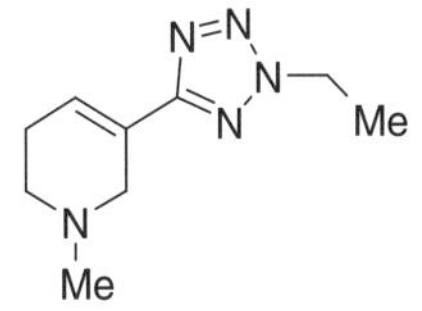

26 Alvameline

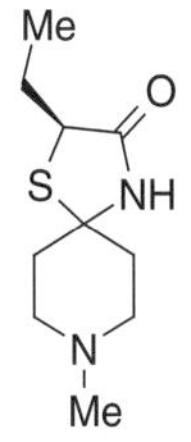

27 AF267B

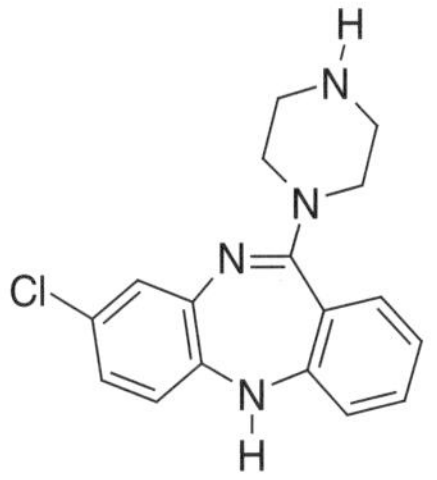

28 Norclozapine

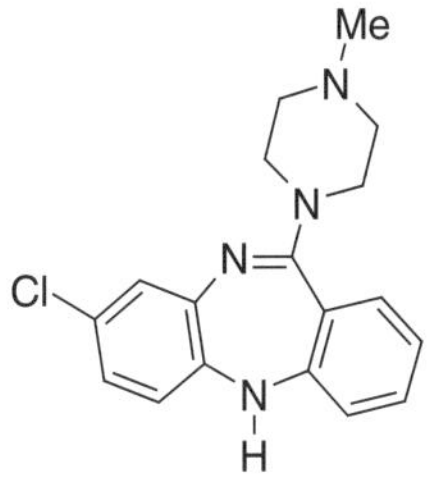

29 Clozapine

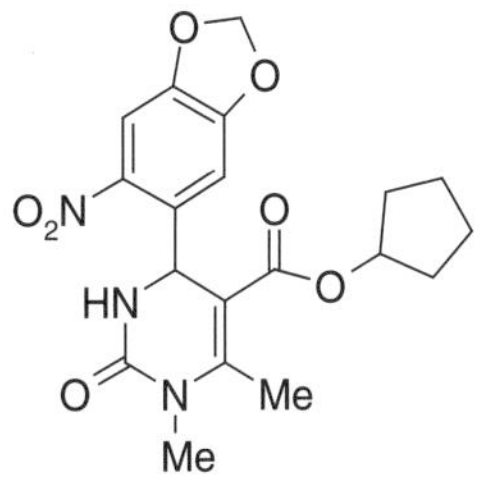

30 VU0090157

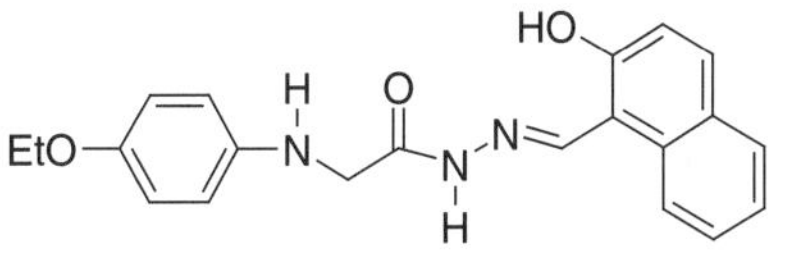

31 VU0029767

2.1.4. Nootropics

The term nootropic was coined by Giurgea in 1972 from the Greek *noos* (mind) and *tropos* (turn) to describe the properties of the first substance, piracetam 1, which had positive effects in the treatment of memory loss, age related memory decline and lack of concentration. Piracetam, which has no known mechanism of action, stimulated the design and synthesis of a large number of structural "acetam" analogs that had a similar pharmacological profile with modest clinical benefit. Nootropics have been proposed to work through modulation of AMPA, NMDA, and cholinergic signaling and the preclinical *in vivo* effects are normally associated with conditions of impaired brain function such as aging, hypoxia, glucose deprivation, injury, or neurodegeneration, for example, normal healthy animals show little benefit from treatment with nootropics such as piracetam [5]. Studies have suggested effects on membrane fluidity and mitochondrial dysfunction could explain the effects of piracetam [144]. Aniracetam 32 was reported as more potent than piracetam, with essentially no side effects. In a 10-patient clinical trial, it demonstrated enhanced vigilance based on EEG analysis similar to piracetam (2 g p.o.). Oxiracetam 33, an analog of piracetam reversed scopolamine-induced deficits in the radial arm maze in rats, but failed to show benefit in AD patients and produced insomnia, agitation, headaches, and occasional GI upsets in the clinic [5]. Nebracetam (WEB-1881 FU) 34 blocked scopolamine-induced disruption of spatial cognition at 10 mg/kg p.o. and enhanced oxotremorine-induced tremors, indicating a cholinergic enhancing mechanism. It also decreased the Δ^9-tetrahydrocannabinol-induced disruption of spatial cognition, suggesting additional limbic and hippocampal noradrenergic mechanisms in its cognition enhancing profile [145]. It was abandoned in Phase III. Levetiracetam 35, the S-enantiomer of etiracetam, an antiepileptic approved in the European Union, significantly improved cognitive function and QoL in patients with refractory partial seizures based on performance time on the WCST and scores on cognitive and social function [146,147].

32 Aniracetam   **33** Oxiracetam   **34** Nebracetam   **35** Levetiracetam

2.2. Emerging Targets

2.2.1. Neuronal Nicotinic Agonists

Neuronal nicotinic acetylcholine receptors (nAChRs) are pentameric ligand-gated cationic channels, highly expressed in the CNS. Twelve subunit genes, designated α_2–α_{10} and β_2–β_4 have been identified, with five additional subunits expressed in the peripheral nervous system (α_1, β_1, γ, δ, and ϵ). Various subunits can combine to provide a diversity of receptor subtypes with unique brain and neuron-specific distribution [148,149]. The α_7 and α_4/β_2 subtypes are involved in cognition and attention, and are highly expressed in brain regions involved in learning and memory, including the hippocampus, thalamus, and cortex [150,151]. nAChRs are involved directly and indirectly in release of neurotransmitters including ACh, dopamine (DA), glutamate, and norepinephrine (NE), and in AD and schizophrenia, levels of cortical nAChRs are decreased [152]. In schizophrenia, a link between loss of nAChRs and sensory gating deficits has been proposed as a self-medicating phenomenon for the higher incidence of cigarette smoking in schizophrenia patients. One neurophysiological abnormality in schizophrenia patients, P50 auditory gating deficit, indicates an impaired information processing and a diminished ability to filter unimportant information or repetitive sensory information. These deficits can be normalized by nicotine **36** and studies indicate that the α_7 subtype is involved in the cognition [153]. Nicotine **36** is not selective and its clinical utility was limited by side effects including seizures, irregular heartbeat, hypertension, and GI effects including nausea. DMXB-A (GTS-21) **37**, a weak α_7 partial agonist, had preclinical efficacy in rodent models of auditory gating [154]. An initial proof-of-concept trial in schizophrenia involving single-day administration showed positive cognitive effects on attention [155]. A second clinical trial with 4-week administration (75 and 150 mg bid) showed significant improvement on the negative symptoms but not in the cognitive battery assessment [156]. Adverse events included mild tremor and nausea. DMXB-A generates several active metabolites in humans that may contribute to a mixed pharmacological profile. A-582941 **38** is a pyridazine α_7 partial agonist with good physiochemical and pharmacokinetic properties, including CNS penetration. It had good efficacy in several rodent and primate cognition models thought to reflect memory and learning [157]. SSR-180711 **39**, a partial α_7 agonist that advanced to Phase II, displayed high affinity for rat and human α_7 (22 and 14 nM, respectively) with 330-fold selectivity over other nAChRs. *In vivo* it was active in object recognition memory with no tolerance, in the MWM, and reversed MK-801 deficits on short-term memory and novelty learning [158]. TC-5619 **40**, a moderately potent α_7 agonist, is one of the four additional NCEs where the structure is known reported to be in Phase I for treatment of CDS. The selective α_7 partial agonist JN403 **41**, showed rapid CNS penetration after oral administration in mice and rats and was active in the social recognition test over a broad dose range [159]. The urea analog, NS-1738 **16**, a positive allosteric modulator of the α_7 nAChR, has shown cognition-enhancing activity [160]. NS-1738 was unable to activate α_7 nAChR alone. It effectively enhanced ACh-evoked currents in cells transfected with human α_7 nAChR as well as in rat hippocampal neurons. NS-1738 showed little effect on desensitization kinetics of α_7 nAChRs. In rats, NS-1738 (10 and 30 mg/kg i.p.) improved cognitive performance in the social recognition test and reversed the scopolamine-induced performance impairment in the MWM (MED = 30 mg/kg i.p.).

Varenicline **42**, a $\alpha_4\beta_2$ partial agonist/α_7 full agonist launched for treatment of smoking cessation. It reversed nicotine withdrawal induced deficits in learning and memory preclinically, and improved mood and cognition during smoking abstinence in the clinic [161,162]. Varenicline is being evaluated in Phase III for CDS. The full α_4/β_2 agonist ABT-418 **12**, showed efficacy in acute studies in AD patients, but failed to show efficacy in a double blind, placebo-controlled AD trial. Ispronicline (TC-1734; AZD-3480) **43** [163], showed preclinical activity in several models of cognition (step through passive avoidance, object recognition, radial arm maze) and was well tolerated up to 320 mg in Phase I, but similarly failed to show benefit in Phase IIb [164]. The pyridyl ether ABT-089 **44** and ABT-894 **45** are $\alpha_4\beta_2$ agonists that advanced to Phase II for cognitive impairment in AD and ADHD, respectively. ABT-089 was effective in preclinical models of impaired cognitive function, including aging, septal lesion, and scopolamine-induced deficits in the MWM. Clinically, it showed positive signs of cognitive activity in a reaction time test [165]. ABT-894 was efficacious in an adult ADHD Phase II trial, comparable with atomoxetine **10**. The primary endpoint was the total score of the Connors Adult ADHD Rating Scale (CAARS). Overall, progress in this field continues to be slow despite years of research, due mainly with the lack of efficacy and selectivity-related side effects including emesis, motor dysfunction, and hallucinations. Current drug discovery directions are focused on positive allosteric modulators with the goal of improved selectivity and avoiding agonist-induced receptor desensitization [166].

2.2.2. Histamine The histamine GPCR family consists of four members: H_1–H_4. H_3Rs are expressed predominately in the brain, localized to the cerebral cortex, amygdala, hippocampus, striatum, thalamus and hypothalamus, where they are expressed on presynaptic terminals and function as inhibitory auto- and heteroreceptors [167]. H_3 antagonists increase the release of various neurotransmitters, including histamine, ACh, NE, 5-HT, and DA, and have potential utility in treating cognitive deficits associated with various dementias and schizophrenia [168]. H_3R knockout mice and H_3R inhibitors demonstrate enhanced learning and memory in various animal models of cognitive function. A nuance of the H_3R and its ligands is the high degree of constitutive activity *in vitro* and *in vivo* [169].

The search for H_3 antagonists with drug-like properties has focused exclusively on amine-based compounds as NCEs with reduced side effect liabilities have been identified and advanced into clinical evaluation.

ABT-239 **46** is currently the most widely studied H_3R inhibitor [168]. It is a potent H_3R inverse agonist and effective at low doses (0.1 mg/kg sc) in a repeat trial inhibitory avoidance task, the rat social recognition model of short-term memory and in a water maze model [170]. Although ABT-239 had an impressive *in vivo* profile for cognition enhancement, its development was halted due to cardiovascular liabilities [168]. GSK189254 **13** had high affinity for recombinant human H_3R ($K_i = 0.9$ nM) and for rat H_3R blockade *in vivo* ($ID_{50} = 0.05$ mg/kg p.o.), with greater than 10,000-fold selectivity for H_3 versus other receptors [95]. GSK189254, currently in

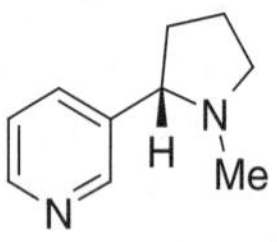

36 Nicotine

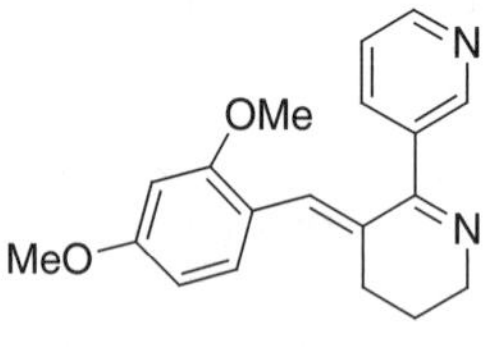

37 DMXB-A

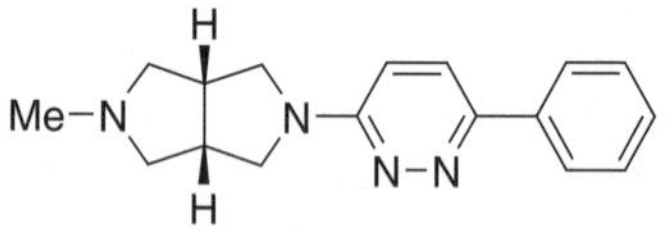

38 A-582941

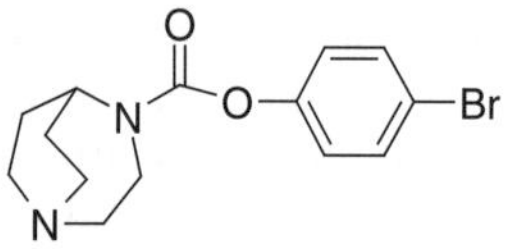

39 SSR-80711

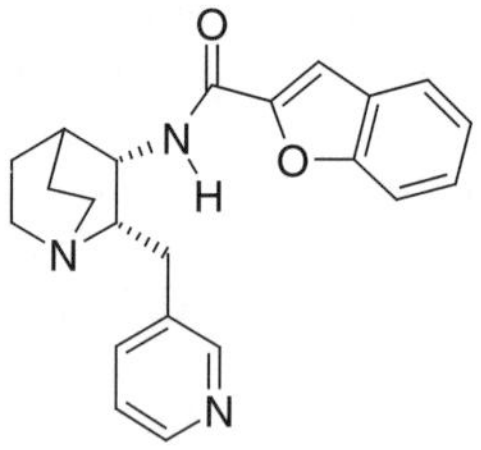

40 TC-5619

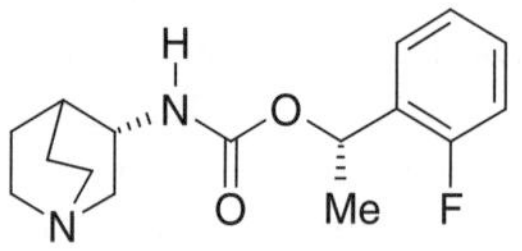

41 JN403

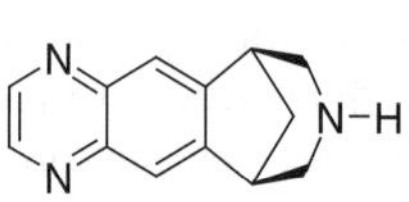

42 Varenicline

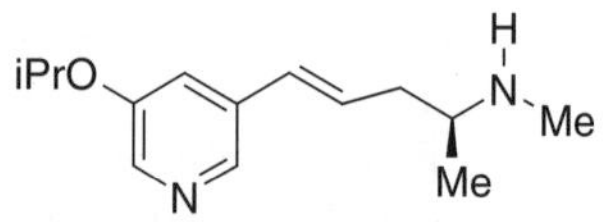

43 TC-1734 (AZD-3480)

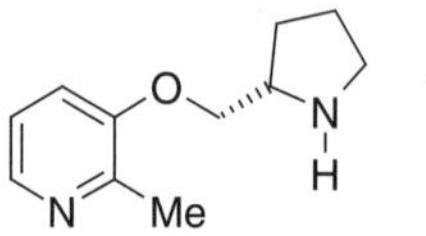

44 ABT-089 (Pozanicline)

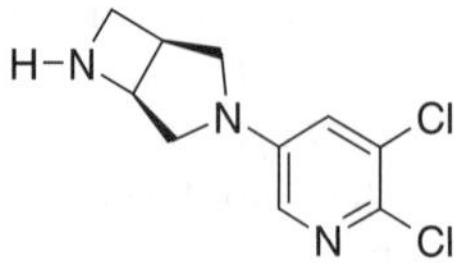

45 ABT-894 (Sofinicline)

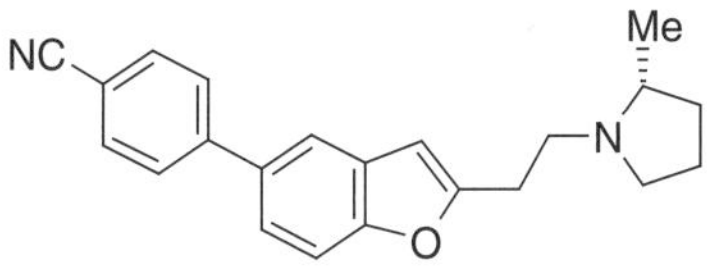

46 ABT-239

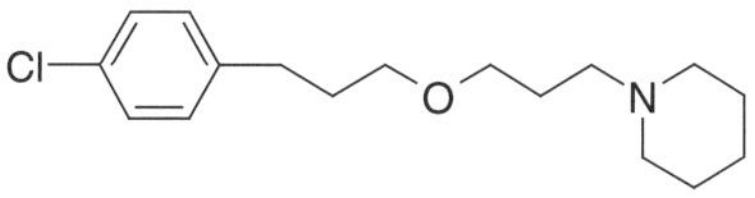

47 BF2.649

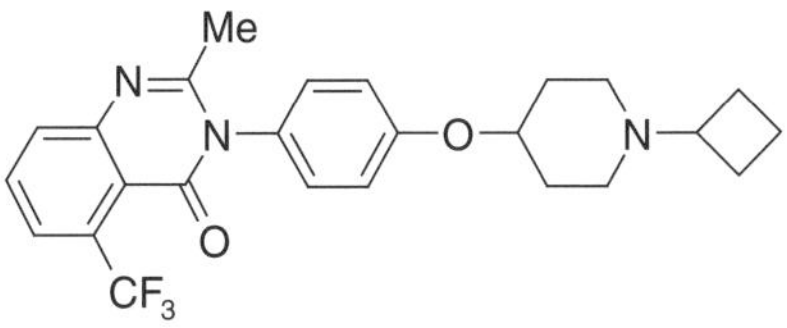

48 Merck

Phase II, was efficacious preclinically across a panel of models designed to test different cognitive domains in rodent at 0.3–3 mg/kg p.o., reversing scopolamine-induced deficits in passive avoidance tasks, improving performance of aged rats in a water maze model and improving memory in an object recognition task. GSK239512 (undisclosed structure) is in Phase II for the treatment of AD. BF2.649 **47**, despite its unusual pharmacokinetics and safety profile is in late stage clinical trials for various indications, including cognitive enhancement [168]. Quinazolone **48** is a compound of further interest from Merck [171]. H_3 antagonists continue to be an area of intense interest for drug development as the field awaits clinical data [168].

2.2.3. Serotonin

The serotonin (5-hydroxytryptamine; 5-HT) **49** system originates from the raphe nucleus in the mid- and hindbrain regions and projects to virtually all brain regions, including the cortex, hippocampus, amygdala, hypothalamus, and thalamic nuclei. There are presently 14 5-HT receptor subtypes, some of which exist as multiple splice variants that are classified into 7 families according to their primary structures, signal transduction coupling, and pharmacology [172,173]. Except for the 5-HT$_3$ subtype, which is a ligand-gated cation channel, 5-HT receptors are GPCRs. While all subtypes have been linked to learning and memory, the 5-HT$_{1A}$, 5-HT$_{2A}$, 5-HT$_4$, 5-HT$_6$, and 5-HT$_7$ subtypes are current targets of interest for drug discovery and have resulted in several clinical candidates for cognition [173,174].

5-HT$_{1A}$ 5-HT$_1$ receptors are grouped into five major subtypes and are negatively coupled to adenylyl cyclase via Go/i [172,176]. Evidence suggests that 5-HT$_{1A}$ antagonists may reverse the cognitive deficits seen in AD [176]. 5-HT$_{1A}$ receptors are most highly concentrated in cortical and hippocampal pyramidal neurons and provides inhibitory tone to cholinergic and glutamatergic neurons. 5-HT$_{1A}$ antagonists facilitate glutamate and cholinergic transmission [172–175]. Serotonergic neurons also provide inhibitory tone to the cortical pyramidal pathway via 5-HT$_{1A}$ receptors. Thus, 5-HT$_{1A}$ antagonists may reverse AD-associated cognitive deficits both by enhancing excitatory cholinergic and glutamate neurotransmission and by blocking direct inhibitory 5-HT input [176]. 5-HT$_{1A}$ antagonists facilitate glutamatergic activation and signal transduction by blocking the hyperpolarization and Ca^{2+} flux induced by inhibitory 5-HT tone. Except for clozapine **29**, antipsychotics have mixed results on CDS [177,178]. The atypical antipsychotics aripiprazole, clozapine, olanzapine, and quetiapine are partial 5-HT$_{1A}$ agonists while risperidone **50** and sertindole **51** are full antagonists [174]. WAY-100635 **18** was the first well-characterized, selective 5-HT$_{1A}$ antagonist [179]. The first clinical compound from this series, lecozotan **52**, was a potent full antagonist with greater than

100-fold selectivity against 50 other receptors except the D4 DA receptor, and was active in several models of learning and memory [180,181]. It completed a Phase II/III study in combination with donepezil in patients with mild to moderate AD. The ability to distinguish full 5-HT_{1A} antagonists from partial in **54** showed improvement in the ANAM but not the WCST test of cognition in patients stabilized with antipsychotic therapy. M100907 (volinanserin) **55**, active in preclinical models of schizophrenia and cognition, and showed fewer errors in the WCST, was discontinued in Phase III.

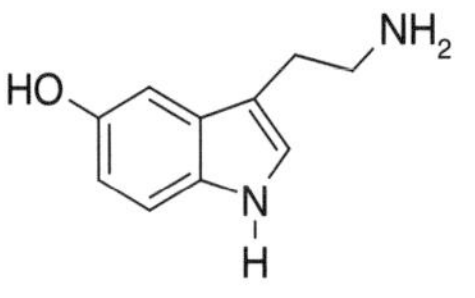

49 Serotonin

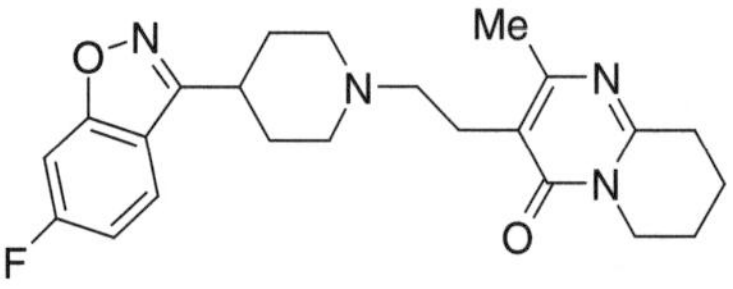

50 Risperidone

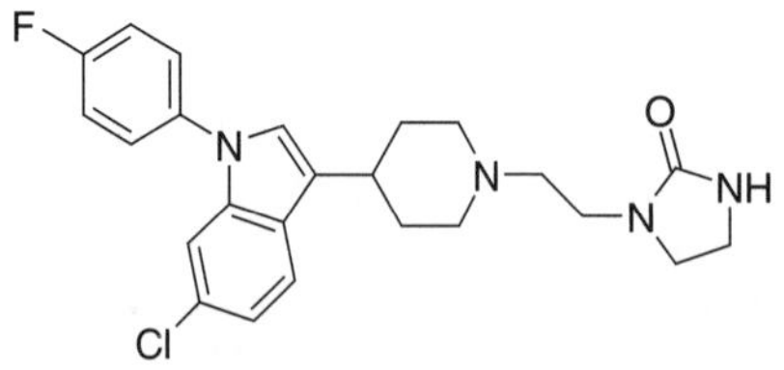

51 Sertindole

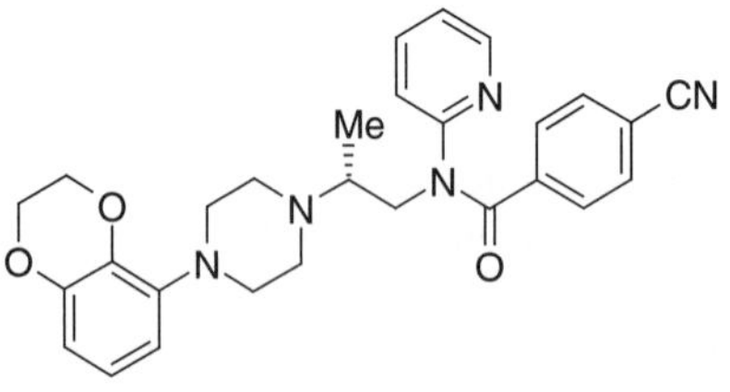

52 Lecozotan

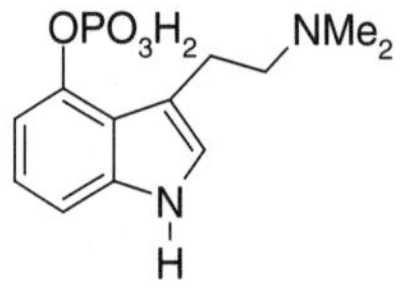

53 Psilocybin

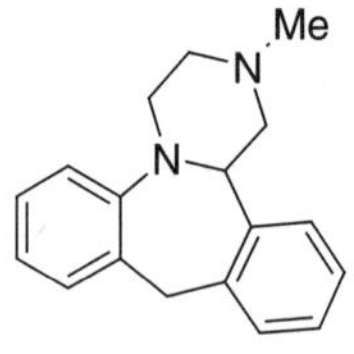

54 Mianserin

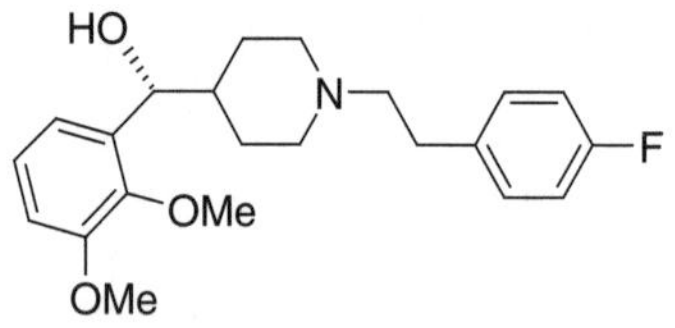

55 Volinanserin

agonists in the assay systems has been problematic and remains a key issue in developing antagonists.

5-HT_{2A} Postsynaptic 5-HT_{2A} receptors are highly localized in cortical pyramidal neurons and may play a role in integrating cognitive and perceptual information [174]. The 5-HT_{2A} receptor is colocalized with the NR1 subunit of the NMDA receptor, suggesting that an antagonist may potentially be beneficial in treating cognition in schizophrenia by normalizing NMDA-receptor function. Limited clinical data on 5-HT_{2A} antagonists are available. The 5-HT_{2A} agonist, psilocybin **53** produced a deficit in a continuous performance test in healthy volunteers. The antagonist, mianser-

5-HT_4 5-HT_4 receptors are enriched in the nigrostriatal and mesolimbic regions. They are positively coupled to adenylyl cyclase and modulate the release of ACh, DA, GABA, and 5-HT. The partial agonist, RS17017 **56**, improved performance in rodent tests of social, olfactory-associated learning, and spatial memory, and also improved delayed matching-to-sample responses in young and old primates [175]. SL650155 (capeserod) **57**, a partial agonist with high affinity for 5-HT_4 ($K_i = 0.4$ nM) had greater than 100-fold selectivity over other 5-HT receptor subtypes. *In vivo*, SL650155 improved performance in rodent models of learning and memory including novel object recognition (0.001–0.1 mg/kg i.p. or p.o.), the linear maze task in aged rats (0.01 and 0.1 mg/kg i.p.), and the MWM in mice, where it reversed scopolamine-induced deficits at 0.1 and 0.3 mg/kg i.p. SL650155 had a greater than additive effect on cognitive performance in the Y-maze in combination with rivastigmine **5** [182]. SL650155 (0.1 mg/kg s.c.) improved performance in the 5-CSRT task [183] and was advanced to Phase II before being terminated. Nonetheless, 5-HT_4 receptors remain an active area in drug discovery, largely due to actions on amyloid deposition, although GI side effects of agonists may ultimately limit their use.

5-HT_6 The 5-HT_6 receptor is positively coupled to adenylyl cyclase via Gs with expression almost exclusively in the CNS in the olfactory tubercles, cerebral cortex, nucleus accumbens, and hippocampus. Blockade of 5-HT_6 receptor function increases cholinergic and glutaminergic transmission and *in vivo* cognitive efficacy in rodent behavior models [174,175]. Atypical antipsychotics, such as clozapine **29** and olanzapine **58**, bind with high affinity as inverse agonists at 5-HT_6 receptors, which coupled with its distribution in key brain areas involved in learning and memory has enhanced interest in identifying clinical candidates for this target [174,175]. Early 5-HT_6 receptor antagonists had high affinity and good selectivity, but were very hydrophilic and had poor brain penetration that limited their utility. The preclinical *in vivo* data on SB-271046 **15** led to its being the first NCE in clinical trials, but it was discontinued after Phase I due to poor brain partitioning [184]. Further efforts led PRX-07034 **59**, SB-742457 **60**, and SAM-315 (undisclosed structure) advancing into the clinic. SB-742457 showed clinical proof-of-concept with results from a Phase II trial demonstrating that treatment of patients with 35 mg of SB-742457 for 24 weeks resulted in a significant improvement in global function compared to placebo. A second Phase II trial comparing SB-742457 and donepezil **4** to placebo demonstrated that patients receiving SB-742457 had similar improvements in global function and cognitive function compared to donepezil-treated patients [185]. Dimebon **61**, an orally active NCE approved in Russia as an antihistaminic that has emerged as a novel treatment for AD binds with nanomolar potency to the 5-HT_6 receptor [186]. Additionally, it also interacts with 5-HT_7 receptors, butyryl- and acetylcholinesterase, L-type calcium channels, the mitochondrial permeability transition pore, AMPA and NMDA receptors [187]. In a cohort of 183 patients with mild-to-moderate AD, dimebon demonstrated a significant improvement in ADAS-cog and CIBIC-plus scores [188]. If these initial findings are substantiated in the second pivotal trial, this compound will represent a major milestone in AD treatment.

5-HT_7 The 5-HT_7 receptor is the newest member of the 5-HT GPCR family. In addition to depression, schizophrenia, and migraine, antagonists of the 5-HT_7 receptor may find utility in sleep disorders and cognitive dysfunction, although the role of the 5-HT_7 receptor in learning and memory processes is still under investigation [189]. Interest in 5-HT_7 receptors for cognition is based on localization in the brain (thalamus, hypothalamus, and hippocampus, with lower levels in cortex and amygdale) and behavioral pharmacology [189,190]. A lack of selective 5-HT_7 ligands has slowed progress with this target. The sulfonamide, SB-258719 **62** was one of the first 5-HT_7 antagonists reported. Lead optimization produced SB-269970A **63** ($K_i = 1$ nM), a compound with improved affinity and selectivity. SB-269970A enhanced working and reference memory in a radial arm maze task but had poor pharmacokinetic properties with low oral bioavailability [191]. SB-656104A **64,** a potent and selective 5-HT_7 receptor antagonist had a K_i value of 2 nM with low affinity for α_{1B} adrenoceptors ($K_i = 220$ nM) and greater than 100-fold selectivity over other 5-HT receptor subtypes [192]. The 5-HT_7 receptor remains a target of interest in drug discovery for cognition and also as an antipsychotic agent to address positive symptoms of schizophrenia. However, better NCEs are needed for clinical proof of concept.

2.2.4. Dopamine Interest in the role of dopamine in cognitive function has focused on its ability to modulate executive function, including working memory, planning, and attention [193]. In schizophrenia, positive symptoms are hypothesized to be due to

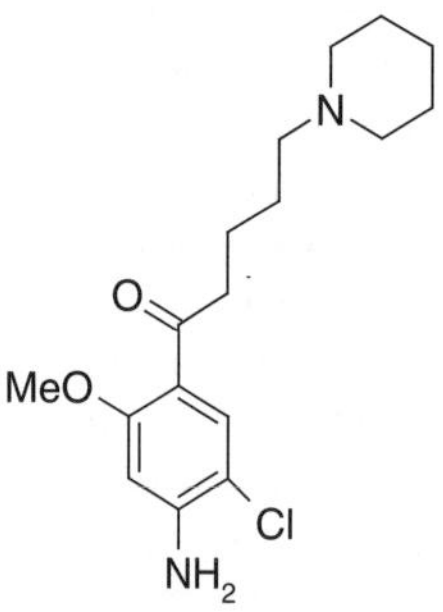

56 RS17017

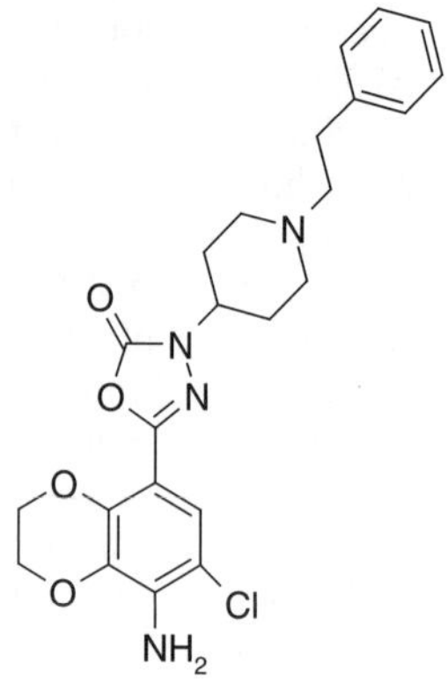

57 SL650155

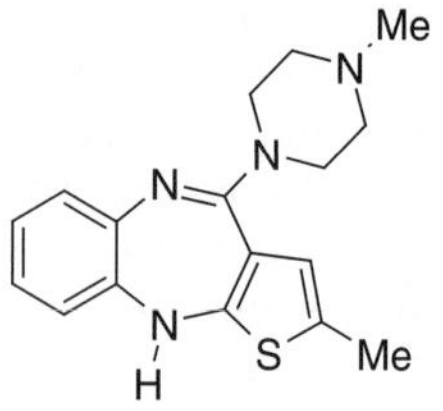

58 Olanzapine

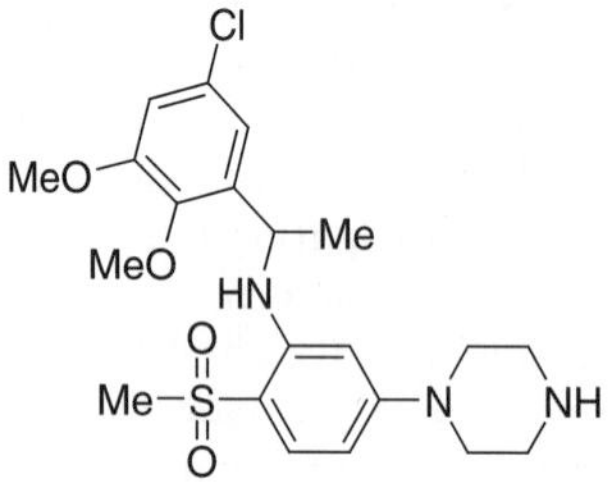

59 PRX-07034

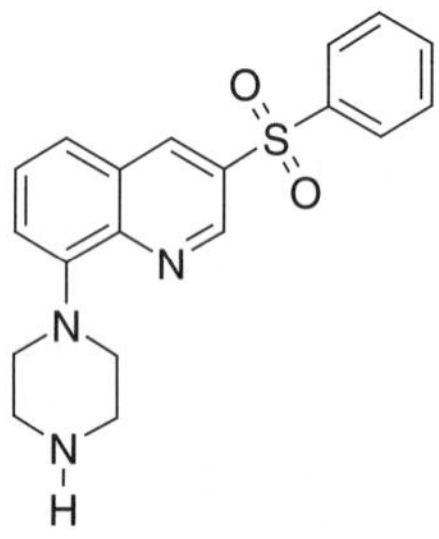

60 SB-742457

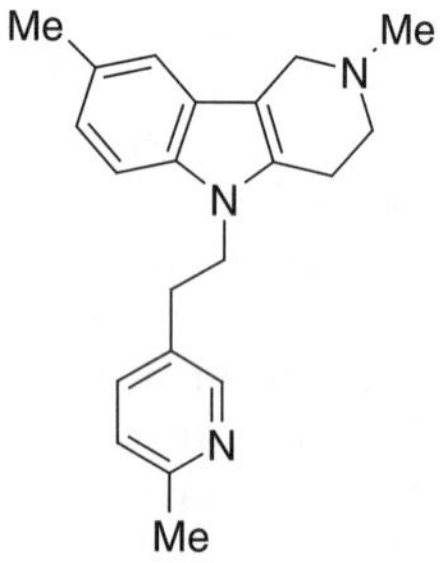

61 Dimebon

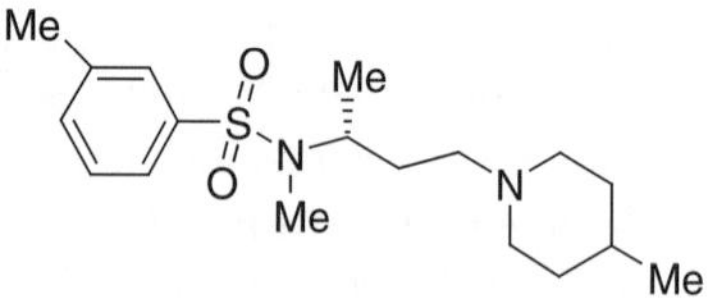

62 SB-258719

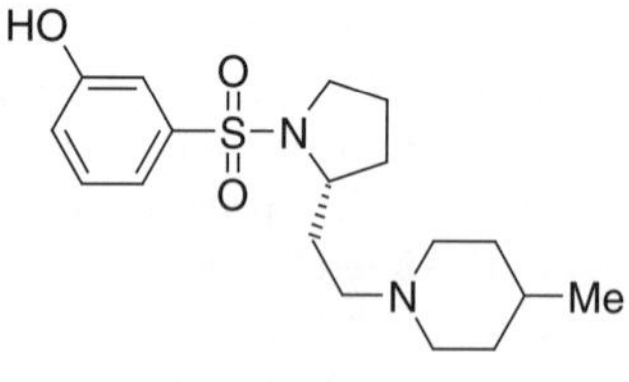

63 SB-269970A

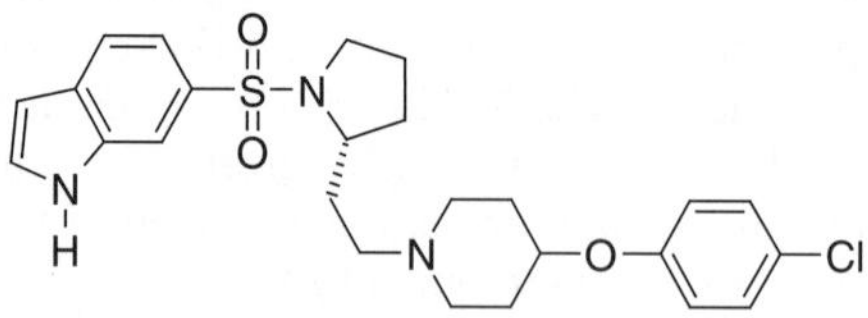

64 SB-656104A

subcortical hyperdopamineric function, while cognitive deficits result from hypodopaminergic activity in the prefrontal cortex [194]. Consistent with this view are clinical reports that D1 receptors in prefrontal cortex are upregulated in schizophrenia due to a localized decrease in DA activity, and that D1 antagonists worsen psychotic symptoms

[195]. In monkeys, DA levels in prefrontal cortex increase during working memory in a delayed alternation task and inhibition of prefrontal DA decreases working memory performance [196]. Five distinct DA receptor subtypes are known: D1-like (D1 and D5) and D2-like (D2, D3, and D4). The D1 receptors interact with the Gs complex to activate adenylyl cyclase, while D2 receptors interact via Gi to inhibit cAMP production [197]. D1 agonists are of interest as targets for cognition, while D4 agonists have cognitive efficacy in animal models [195,196]. The isochroman, A-68930 **65**, and the benzazepine, SKF-81297 **66**, the first full D1 agonists reported had procognitive effects in animal models [196]. Dihydrexidine **67**, under evaluation for cocaine dependency, improved cognitive performance in rodents and primates and was under development for treatment of CDS. It is a potent full D1 agonist ($K_i = 5.5$ nM) with modest 11-fold selectivity over D2 receptors and 23-fold over α_2 receptors [198]. Although safe and well tolerated in man, poor oral bioavailability and short half-life hindered the advancement of this NCE. Adrogolide (ABT-431/DAS-431) **68**, a prodrug of the di-phenol A-86929, reversed working memory deficits associated with chronic antipsychotic drug therapy in primates but was inactive in the MWM. Adrogolide failed in clinical trials for PD and is still being studies for cocaine dependence. The DA D4 agonist, A-412997 **69** was efficacious in a social recognition test of short-term memory and in a 5-trial repeated acquisition inhibitory avoidance model, while the nonselective agonist, PD168077 **70** was active only in short term memory [199]. A-412997 increased ACh and DA levels in the rat medial prefrontal cortex but not in the dorsal hippocampus. A major issue that has plagued advancement of full D1 agonists is receptor desensitization.

2.2.5. Norepinephrine Noradrenergic neurotransmission in the prefrontal cortex plays a key role in attention and cognitive processing [200–202]. Moderate increases in NE levels can enhance cognitive function through activation of postsynaptic α_{2A} receptors [203]. The α_{2A} agonists clonidine **71** and guanfacine **72** improved cognitive function in humans [202]. Selective NE reuptake inhibitors (SNRIs) are an additional approach to elevate extracellular levels of NE in the brain. These include reboxetine **73**, approved for treatment of depression in Europe and purportedly under evaluation for treatment of CDS in the United States [204], and atomoxetine **10**, the first nonstimulant approved for use in ADHD. In a phase II study, adjunctive atomoxetine treatment to second-generation antipsychotics showed no improvements on prefrontal cognitive ability and function in schizophrenics [205]. Nicergoline **74**, an ergot alkaloid, had been used to treat symptoms of cognitive decline in elderly patients with cerebrovascular insufficiencies. It has a broad spectrum of activity, including α_1-adrenoceptor antagonism, vasodilation and increased arterial blood flow, and enhancement of cholinergic and catecholaminergic neurotransmitter function [206].

2.2.6. Glutamate The excitatory neurotransmitter glutamate mediates its effects via both ionotropic and metabotropic receptors. ionotropic glutamate receptors (iGluRs) include NMDA, AMPA, and kainate subtypes. The NMDA receptor is a ligand-gated ion channel composed of a combination of two NR1 and two NR2 subunits, and requires concomitant binding of glutamate at the NR2 subunit and glycine or a glycine site coagonist for activation. AMPA receptors mediate fast excitatory transmission in the CNS and exist as hetero- and homotetrameric receptors composed of $GluA_1$–$GluA_4$ subunits, with each subunit comprised of one of the two splice variants. NMDA and AMPA receptors operate in an independent, complementary fashion in controlling excitatory neurotransmission. The AMPA receptor conducts primarily Na^+ ions, while NMDA receptors are high conductance, slow activating nonselective cationic channels that are permeable to calcium [61,207,208]. At normal membrane potentials, the NMDA receptor channel is subject to voltage-dependant Mg^{2+} blockade and its opening requires membrane depolarization by AMPA receptors. Thus, increasing AMPA receptor activity can increase NMDA receptor function. As discussed above, NMDA receptor activation is involved in membrane trafficking of AMPA receptors, a

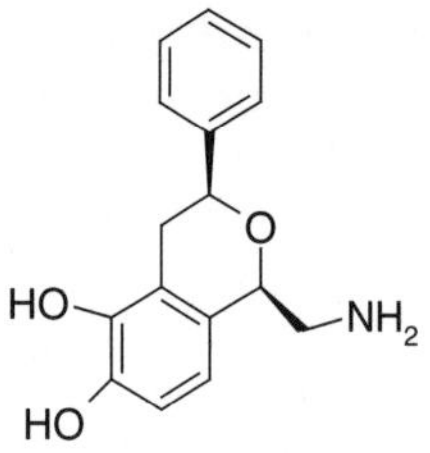

65 A-68930

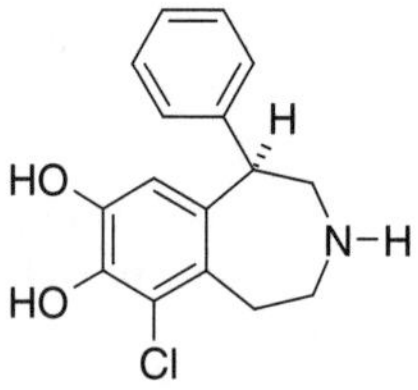

66 SKF-81297

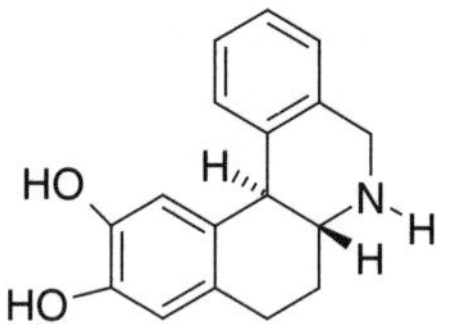

67 Dihydrexidine

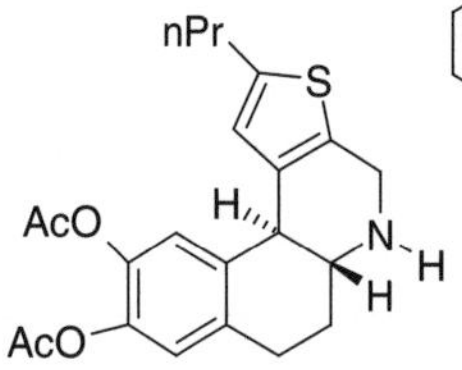

68 Adrogolide (ABT-431)

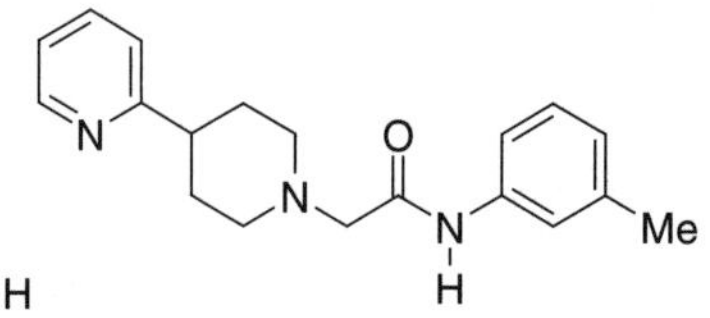

69 A-412997

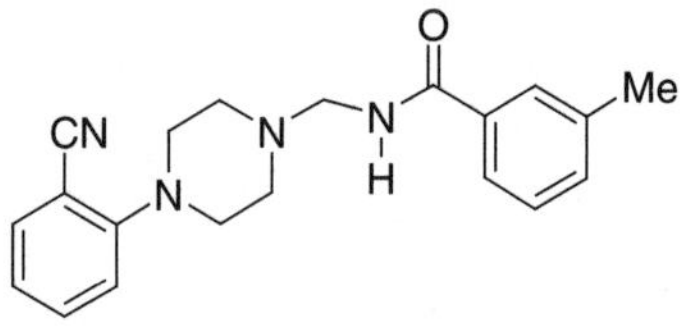

70 PD-168077

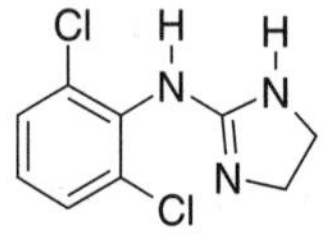

71 Clonidine

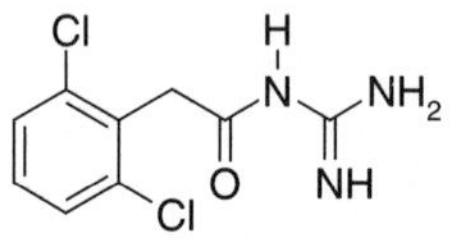

72 Guanfacine

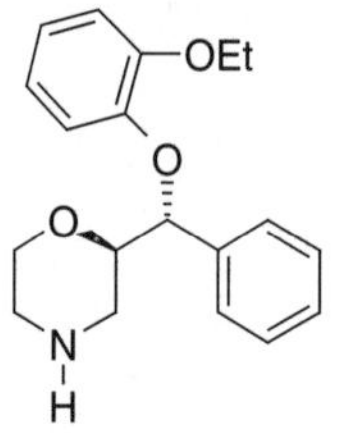

73 Reboxetine

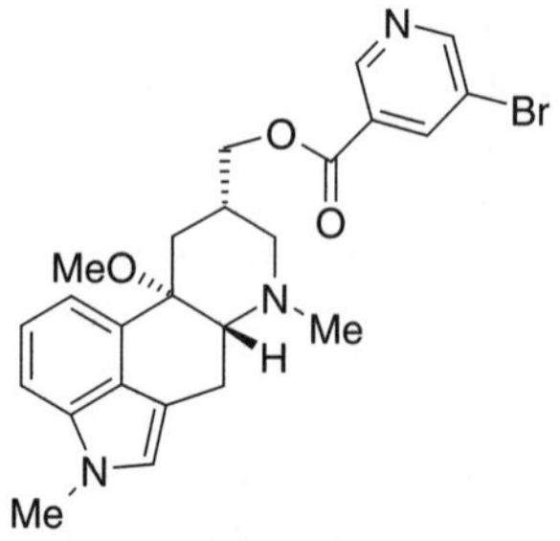

74 Nicergoline

process believed to underlie the basis of neuroplasticity (LTP and LTD) [61].
Metabotropic Glutamate Receptors (mGluRs)
Metabotropic glutamate receptors are GPCRs comprised of eight receptor subtypes grouped into three families: Group I (mGluR1, mGluR5), Group II (mGluR2, mGluR3), and Group III (mGluR4, mGluR6–mGluR8). mGluRs have important roles in synaptic activity in the CNS and are targets of current interest in treating schizophrenia and cognitive dysfunction. Group I receptors are linked to Gq and increase phosphatidylinositol turnover via phospholipase C activation to elevate intracellular Ca^{2+}. Group II and Group III receptors are located presynaptic, inhibit adenylyl cyclase activity via Gi and modulate glutamate release. The potential for group II

mGluR2/3 receptor agonists to treat positive and negative symptoms of schizophrenia has been established with LY2140023 **75**, the orally active prodrug of LY404039 **76** [209]. LY404039 also increased cortical DA turnover in rats, an event predictive of procognitive activity. Conversely, mGluR2/3 antagonists, such as LY341495 **77** reduce memory performance. mGluR5 receptors, in addition to being a potential antipsychotic target, potentiate NMDA receptor currents in a number of brain regions, indicating that activation of this target would result in cognitive enhancement. mGluRs have subtype-specific allosteric sites. Positive allosteric modulation offers several advantages over classically orthosteric competitive agonists, including subtype selectivity and lower risk of toxicity by avoiding agonist overstimulation [210]. The pyrazole, CDPPB **78** was the first sufficiently selective mGluR5 positive allosteric modulator for *in vivo* testing. CDPPB shifted the glutamate-induced Ca^{2+} increase fourfold with an EC_{50} of 20 nM, and reduced amphetamine-induced locomotor activity and normalized amphetamine-induced disruption of prepulse inhibition. The oxadiazole, ADX-47273 **79** increased novel object recognition and reduced impulsivity in the 5-CSRT test at 1 and 10 mg/kg i.p., respectively [211]. The selective mGluR5 positive allosteric modulator ADX-63365 reportedly advanced to preclinical development for treatment of mild cognitive impairment.

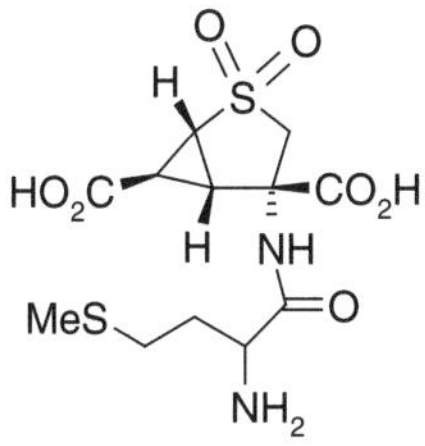

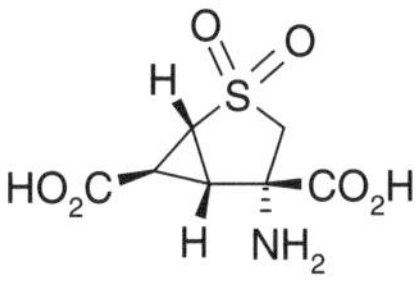

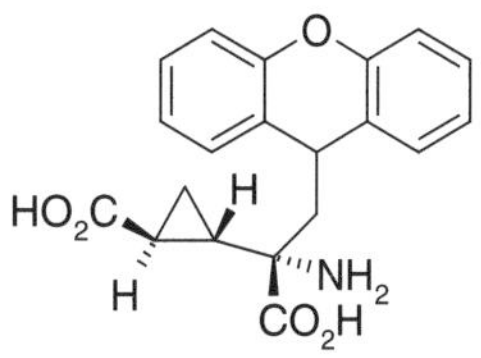

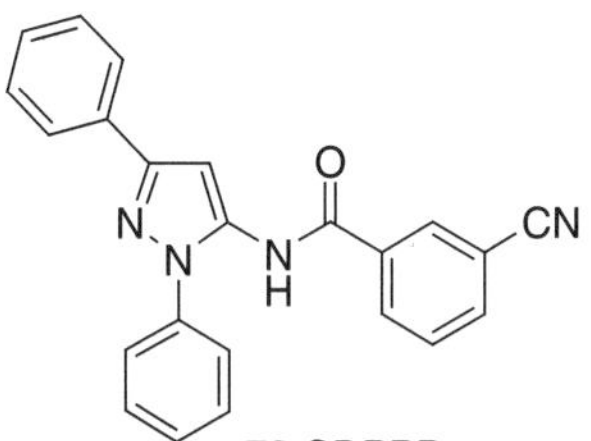

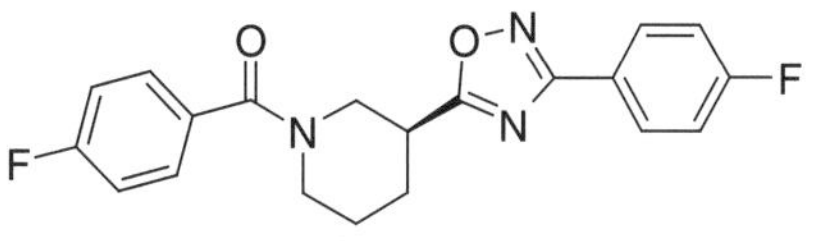

AMPA Potentiators Positive modulation of AMPA receptors may have therapeutic potential in the treatment of cognitive deficits and potentially avoid many of the issues of direct AMPA receptor activation, for example, seizures, excitotoxicity, and loss of efficacy due to desensitization [207,208]. Ampakines are a drug class that enhance attention, alertness, and facilitate learning and memory by allosteric activation of the AMPA receptor [212]. AMPA receptor potentiators include the pyrrolidone nootropics, for example, piracetam **1** and aniracetam **32**, benzothiazides (cyclothiazide **80**), benzylpiperidines (CX-516 **81**, CX-546 **14** and CX-691 **82**), and biarylpropylsulfonamides (LY404187 **83** and LY503430 **84**). These compounds enhance cognitive function in rodents, which appears to correlate with increased hippocampal activity. In addition to directly enhancing glutamatergic synaptic transmission, AMPA receptor activation can increase neurotrophin expression *in vitro* and *in vivo*, which may contribute to the functional and neuroprotective effects of LY404187 and LY503430 [207,208,212]. CX-516 had been in numerous Phase II trials for the treatment of autism, schizophrenia and AD dementia but was discontinued for the treatment of MCI due to lack of efficacy. The toxicity of CX-516 limited the ability to achieve dose levels comparable to the efficacious doses in animal studies. CX-691/ORG-24448 (faramptator) **82** is being evaluated as an adjunct therapy for CDS [213]. It improved short-term memory, but appeared to impair episodic memory in a group of 16 healthy elderly volunteers. Side effects included headache, somnolence, and nausea.

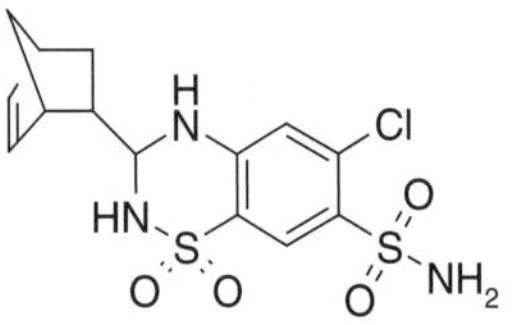

toms, possibly due to the presence of the antipsychotic agent.

80 Cyclothiazide

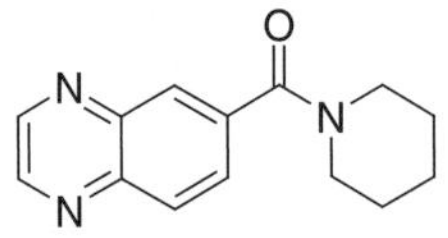

81 CX-516

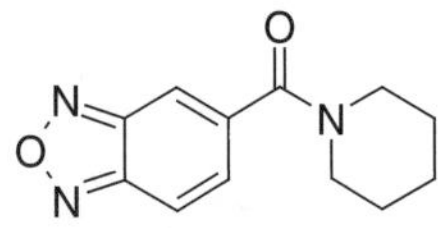

82 CX-691/ORG-24448

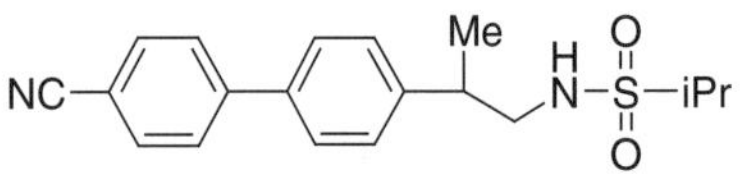

83 LY404187

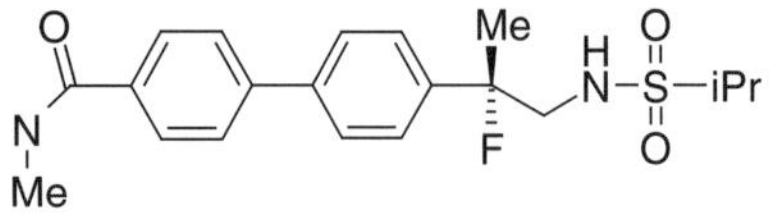

84 LY503430

GlyT1 Glycine is a major inhibitory neurotransmitter in the cerebellum, brainstem, and spinal cord, acting via ligand-gated strychnine-sensitive glycine-A receptors. It also acts as a required positive allosteric modulator of glutamate by binding to the glycine-B site on the NMDA receptor, which facilitates glutamate binding to the NR2 subunit of the NMDA complex enhancing excitatory glutamateric transmission in cortex and hippocampus [214]. The NMDA receptor antagonist PCP **8** mimics the positive, negative, and cognitive symptoms of schizophrenia in man [215]. Glycine **85** was efficacious in improving negative symptoms and some aspects of cognitive dysfunction as an add-on therapy in schizophrenia [216,217]. Furthermore, NMDA and glycine agonists such as D-serine **86** and D-cycloserine **87** improved negative symptoms in schizophrenics undergoing conventional antipsychotic therapy, with apparent decreased EPS/tardive dyskinesia [215]. These studies demonstrated improvements in negative or cognitive symptoms, but not in positive symp-

Blockade of the Type 1 glycine transporter (GlyT1), a member of the sodium/chloride-dependent transporter family is responsible for regulation of synaptic glycine levels. Its distribution mirrors that of NMDA receptor expression, suggesting colocalization with the NMDA receptor. The GlyT2 transporter colocalizes with inhibitory strychnine-sensitive glycine-A receptors. Inhibitors of GlyT1 are either substrate-based (sarcosine series) or nonsubstrate-based compounds [218]. Sarcosine **88** was efficacious as an add-on therapy against positive, negative and cognitive systems of schizophrenia in two trials, but had weak GlyT1 inhibitory activity ($IC_{50} = 38\,\mu M$) with poor pharmacokinetic properties and brain penetration, thus limiting its clinical utility as a drug candidate. Newer sarcosine analogs include NFPS (ALX-5407) **89**, ORG24461 **90**, and ORG-25935 **91**. Preclinically, ORG-25935 elevated glycine levels and reversed PCP-induced deficits in novel object recognition. It is reportedly in Phase II for the treatment of psychosis. While these potent and selective inhibitors have been instrumental in studying the role of GlyT1 in schizophrenia and cognition, a variety of serious side effects such as ataxia, hypoactivity, and decreased respiration have been observed with sarcosine-based inhibitors in rodents [219]. SSR504734 **92**, a selective and reversible nonsarcosine GlyT1 inhibitor had

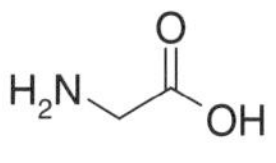

85 Glycine
86 D-Serine 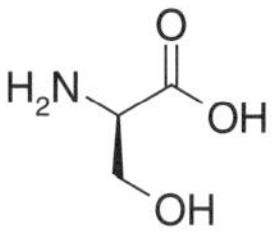
87 D-cycloserine 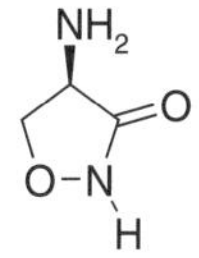
88 Sarcosine 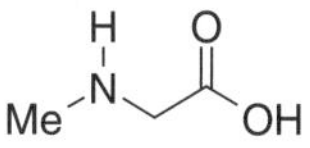
89 NFPS 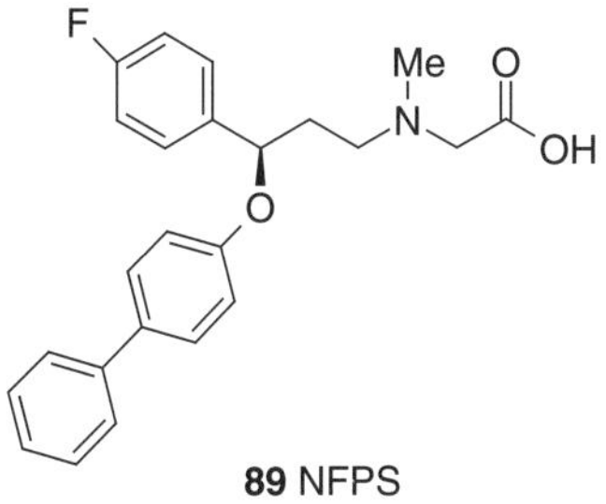
90 ORG-24461 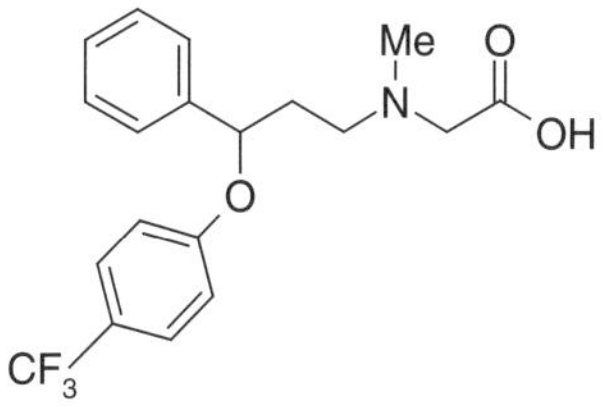
91 ORG-25935 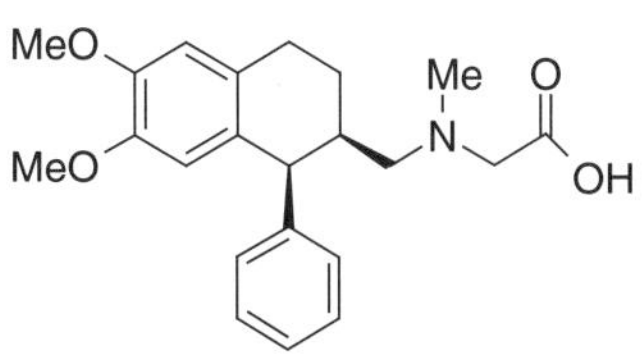
92 SSR504734 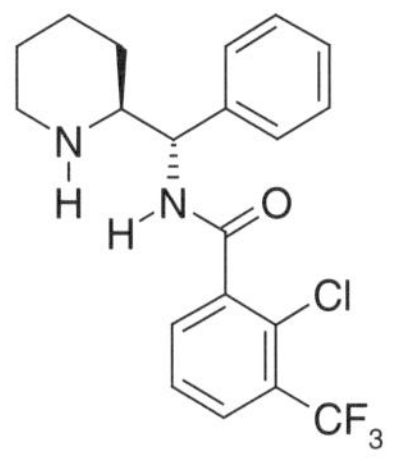
93 DCCCyB 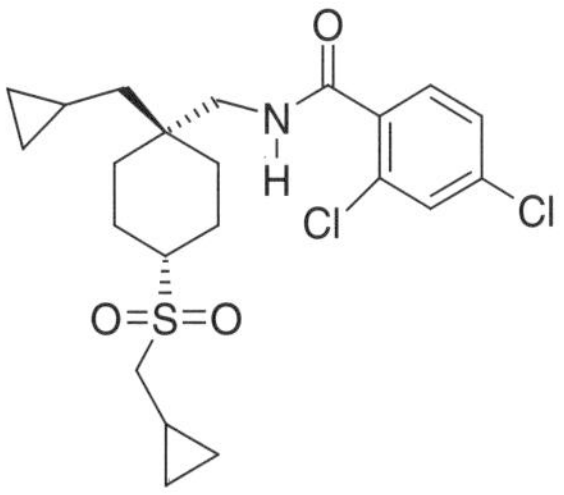

a human GlyT1 IC$_{50}$ value of 18 nM [220]. SSR103800, an analog of SSR504734 (undisclosed structure) is more potent (human GlyT1 IC$_{50}$ = 1.9 nM) and reportedly has advanced into the clinic. SSR103800 and SSR504734 were active in the social recognition model and potentiated MK-801- and amphetamine-induced disruption of latent inhibition, models believed to be predictive of positive, negative, and cognitive aspects of schizophrenia, respectively [221]. Two newer nonsarcosine-based NCEs are DCCCyB **93** and GSK1018921 (undisclosed structure) that have advanced into early clinical development.

2.2.7. GABA$_A$ Receptor γ-Aminobutyric acid (GABA) is the major inhibitory neurotransmitter in the CNS. GABA$_A$ receptors are pentameric GABA-gated chloride channels, composed of four transmembrane subunits ($α1–6$, $β1–3$, $γ1–3$, $δ$, $ε$, $θ$, and $π$) [222]. Nineteen GABA$_A$ receptor subunits have been identified, with the majority of receptors in the brain comprising α-, β-, and γ- subunits in a 2:2:1 stoichiometry. Binding of GABA to the GABA$_A$ receptor can modulate simultaneous binding of various modulators to allosteric sites on the ion channel complex; the most studied is the benzodiazepine (BZ) site. Positive allosteric modulators, for example, diazepam **94**, of the BZ site enhance the action of GABA on GABA$_A$ receptors. Negative allosteric modulators or inverse agonists reduce GABA effects on GABA$_A$ receptors, whereas agents that block the actions of both positive and negative allosteric modulators are categorized as neutralizing allosteric modulators, for example, the BZ antagonist flumazenil (Ro-151788) **95**. BZ site agonists produce their anxiolytic, sedative, anticonvulsant, and cognitive-impairing effects via GABA$_A$ receptors containing β- and γ- and α1–3 or α5 subunits. Diazepam **94**, which has been used as an anxiolytic, hypnotic, and muscle relaxant for nearly 50 years enhanced the inhibitory effects of GABA and impair learning and memory in man [223]. DMCM **96,** a full nonselective inverse agonist

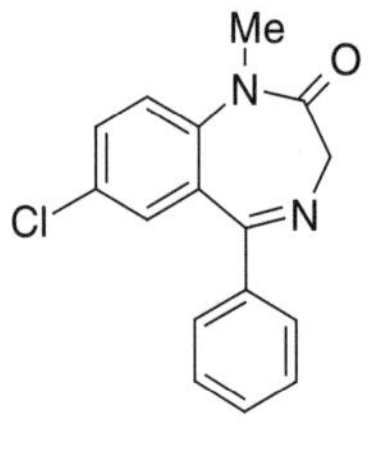

94 Diazepam
95 Flumazenil 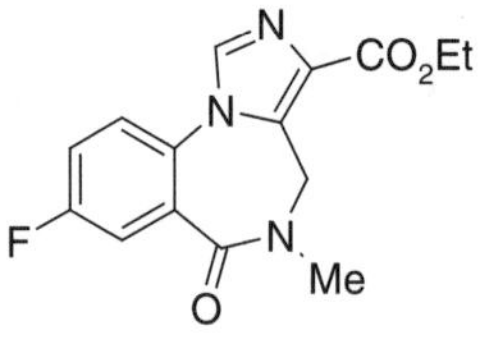
96 DMCM 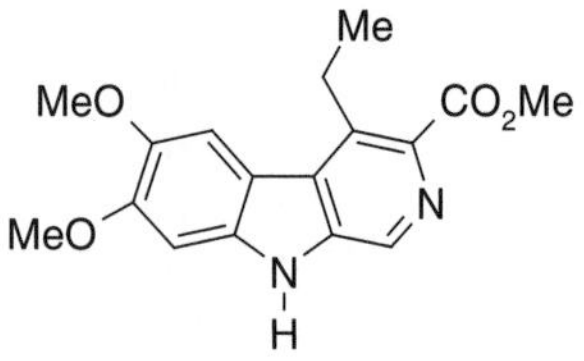
97 L-655708 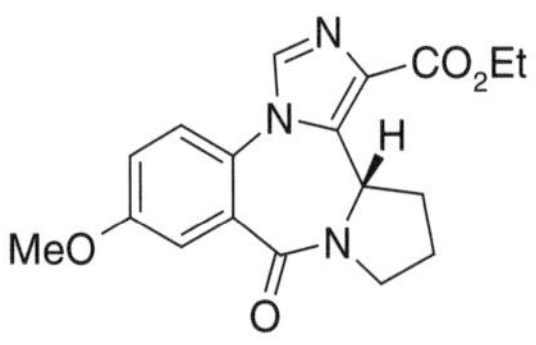
98 α-5IA 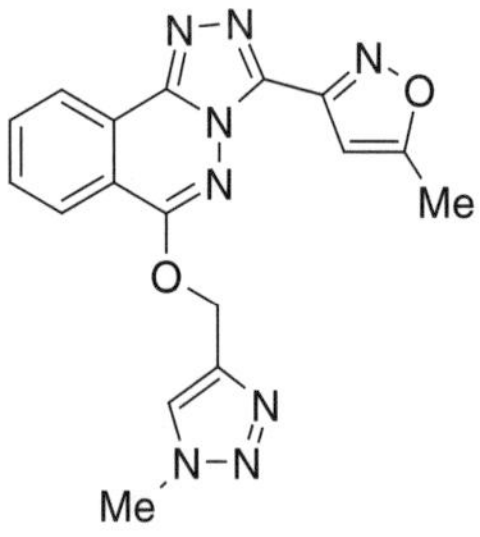
99 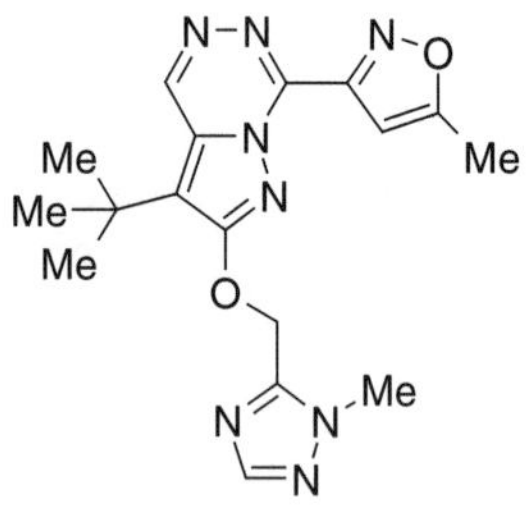

not only enhanced cognitive performance in rats but also produced anxiogenic and proconvulsant activity and altered attentional processing [224]. Pharmacological and genetic research suggests that the α5-subunit-selective inverse agonists may enhance cognition. Mice lacking the α5 gene show improved performance in the MWM, whereas performance in nonhippocampal-dependent learning and anxiety tasks was unaltered compared to wild type [225]. The α5 inverse agonist, L-655708 **97** enhanced spatial learning in the MWM [226]. The clinical candidate α-5IA **98** robustly enhanced LTP in mouse hippocampal slices and performed in a rat hippocampal-dependent test of learning and memory. In humans, α-5IA was toxic due to the formation of the hydroxymethyl isoxazole metabolite that precluded its use in long-term studies. In healthy volunteers, α-5IA reversed the memory-impairing effects of alcohol [227]. Pyrazolotriazine **99** had a better preclinical efficacy and safety profile compared to α-5IA, but questions remain regarding its overall BZ subtype selectivity profile. While the field awaits clinical efficacy data to validate the GABA$_A$ subtype selective inverse agonist approach to cognitive impairment, concerns remain regarding sedation, and the potential for proconvulsant activity of α5-inverse agonists.

2.2.8. Other Approaches

Adenosine The neuromodulator adenosine plays a major role in the regulation of synaptic transmission and neuronal excitability in the CNS. Four adenosine GPCRs (A$_1$, A$_{2A}$, A$_{2B}$, and A$_3$) are known [228]. A$_1$ and A$_3$ receptors are coupled to the inhibitory G-proteins Gi and Go, and A$_{2A}$ and A$_{2B}$ receptors are coupled to stimulatory Gs proteins. A$_1$ receptors are highly expressed throughout the CNS, including the cortex, hippocampus, and cerebellum—important areas for cognitive function. A$_{2A}$ receptors are localized in the striatum where they are coexpressed with dopamine D2 receptors in GABAergic striatopallidal neurons and play important roles in DA neuromodulation. In contrast, A$_{2B}$ and A$_3$ receptors have low abundance in the brain.

Preclinical pharmacological and genetic studies support the involvement of adenosine receptors in learning and memory [3,229]. Adenosine modulates cognition primarily through A$_1$ receptors, but there is now emerging evidence for a role of A$_{2A}$ receptors. Administration of selective A$_1$ receptor agonists disrupt learning and memory in rodents, while nonselective antagonists such as caffeine **100** or theophylline **101**, or selective A_1 (DPCPX **102**) or A_{2A} antagonists (ZM241835 **103**) facilitate rodent learning and memory in diverse behavioral tasks [230]. Clinical results on the cognitive effects of caffeine in nondemented humans were inconclusive [229]. Apaxifylline **104**, a selective and potent A_1 antagonist ($K_i = 5$ nM) that had 100-fold selectivity over A_{2A}, reversed scopolamine-induced behavioral deficits in rats, increased vigilance and enhanced ACh release in cats. Apaxifylline advanced to Phase II for treatment of AD but was discontinued due to its short half-life and the extensive formation of CNS active metabolites.

A_{2A} receptor antagonists such as istradefylline (KW-6002) **105** have been assessed for the treatment of PD acting as indirect DA agonists [230]. They have also shown neuroprotective and cognitive enhancing activity [230]. A_{2A} receptor knockout mice were resistant to motor impairment and MPTP-induced neurotoxicity and had improved spatial recognition memory, while overexpression of A_{2A} receptors resulted in working memory deficits in rats [231]. While istradefylline was not approved for the treatment of PD due to inconclusive Phase III trials, the selective A_{2A} antagonists SCH-420814 **106**, BIIB014 **107**, and SYN-115 (undisclosed structure) are in clinical trials for PD and have shown beneficial effects on cognitive-related functions including motivation, attention and reward-related behavior [230]. The dual A_1/A_{2A} antagonist ASP5854 **108** is under investigation to treat both motor disabilities and cognitive deficits in PD and AD [232]. As a class, the main issues with development of A_1 antagonists for CNS diseases have been poor water solubility, pharmaceutical properties, poor brain penetration, and cardiovascular side effects.

Neurotrophic Agents Neurotrophic factors are polypeptides that support the growth, differentiation, and survival of neurons in development and sustain neurons in the mature adult nervous system. Nerve growth factor (NGF) has selective, survival promoting properties for cholinergic neurons in the CNS as well as neurite outgrowth promoting properties on sympathetic and sensory neurons of the dorsal root ganglia [233,234]. A large body of evidence indicates that NGF promotes survival of basal forebrain cholinergic neurons [235]. NGF reversed reductions in ChAT and AChE activity in nucleus basalis magnocellularis (NBM) lesioned rats, promoted survival of septal cholinergic neurons and improved learning after fimbria–fornix transection, supporting its rationale for evaluation in the clinic [235]. In the clinic NGF was infused intraventricularly in one patient over 3 months resulting in an increase in uptake and binding of [^{11}C]nicotine in the frontal and temporal cortex and improved verbal episodic memory [236]. In an alternate strategy to enhance delivery of NGF to the brain by *ex vivo* gene transfer, a Phase 1 trial in six patients with mild AD revealed the rate of cognitive decline slowed by 36–51% based on ADAS-cog and MMSE assessments, with no reported adverse effects; however, the study did not include a placebo group [237]. Although showing encouraging results, the therapeutic potential of polypeptides remains limited due to their size and pharmacokinetic characteristics, which prevent their systemic administration for treatment of CNS diseases.

Brain-derived neurotrophic factor (BDNF) and its receptor tyrosine kinase TrkB are highly expressed in the hippocampus, cortex, and basal forebrain and are targets for the treatment of AD [238]. BDNF supports survival and differentiated function of ACh and DA neurons, and improved learning and memory in animals via activation of TrkB and the low affinity NGF receptor p75 [239]. Low plasma levels of BDNF mRNA have been suggested as a marker for therapeutic monitoring in AD [240]. Insulin-like growth factor I (IGF-I) deficiency has also been implicated in cognitive deficits seen in AD. IGF-I levels were investigated for associated cognitive performance and decline, and were related to information processing speed, memory, and MMSE score [241].

Cannabinoids The endocannabinoid system consists of two GPCRs, CB-1 and CB-2. CB-1 receptors are abundant in brain regions associated with memory and learning, while CB-2 receptors are confined to cells of the immune system. CB-1 receptor antagonists may have therapeutic utility for the treatment

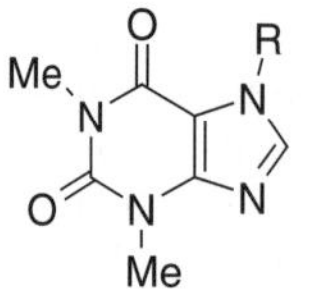

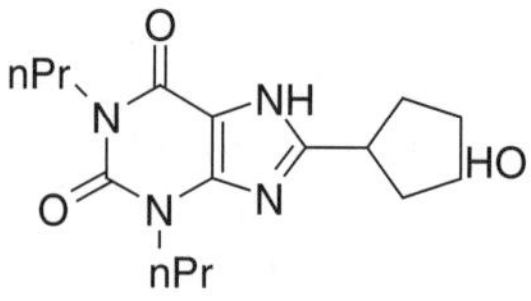

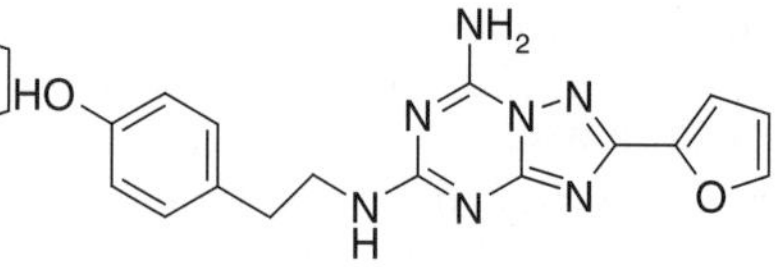

100 Caffeine R = methyl
101 Theophylline R = H
102 DPCPX
103 ZM241835
104 Apaxifylline 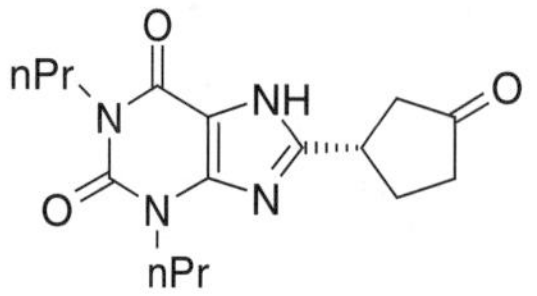
105 KW-6002 (Istradefylline) 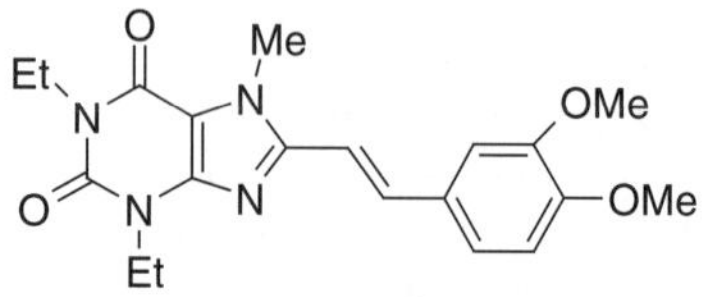
106 SCH-420814 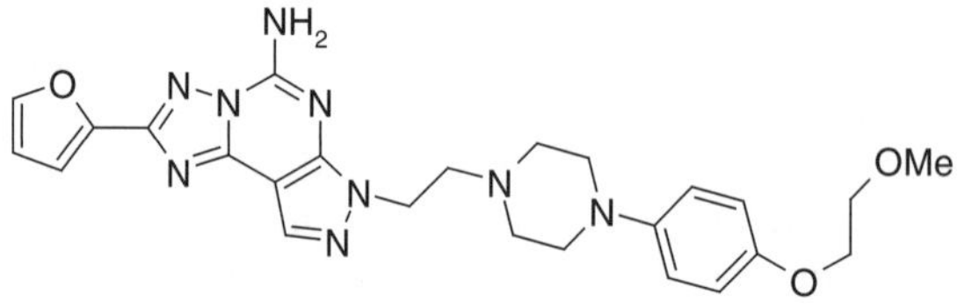
107 BIIB-014 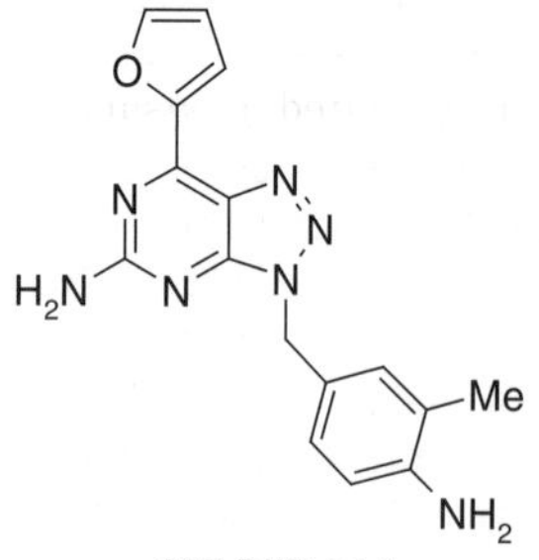
108 ASP5854 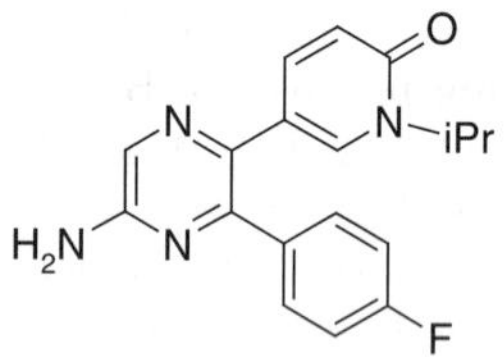

of cognitive deficits associated with AD or schizophrenia [242]. The CB-1 antagonist rimonabant (SR141716) **109** improved olfactory short-term memory assessed by the social recognition test and enhanced spatial memory in the radial-arm maze task in rodents. In addition, rimonabant reversed amnesia induced by i.c.v injections of β-amyloid fragments in mice. *In vivo* rimonabant selectively increased NE, DA, and ACh efflux in the prefrontal cortex, suggesting a potential role for CB-1 antagonists in treatment of attention and ADHD [242,243]. The future of CB-1 receptor antagonists as drugs is confounded by the psychiatric side effects associated with Rimonabant use that has led to its non-approval as an antiobesity agent in the United States and its withdrawal from the market in the European Union for the same indication [244].

Neuropeptides Neuropeptides and their receptors have represented novel targets for CNS disorders for more than half a century, with minimal success as evidenced by the inability to find improvements on morphine as analgesic acting via the opioid receptor

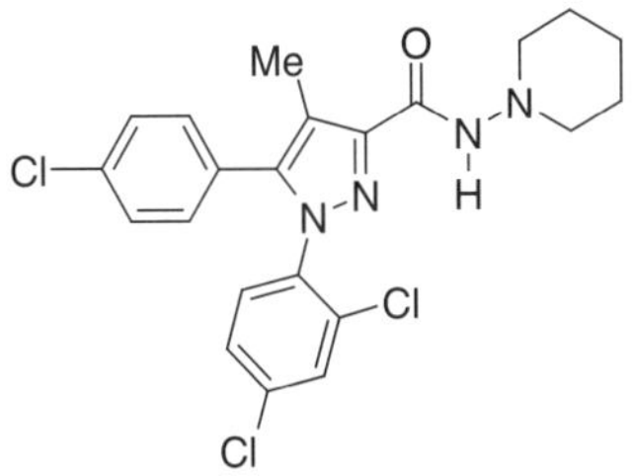

109 Rimonabant

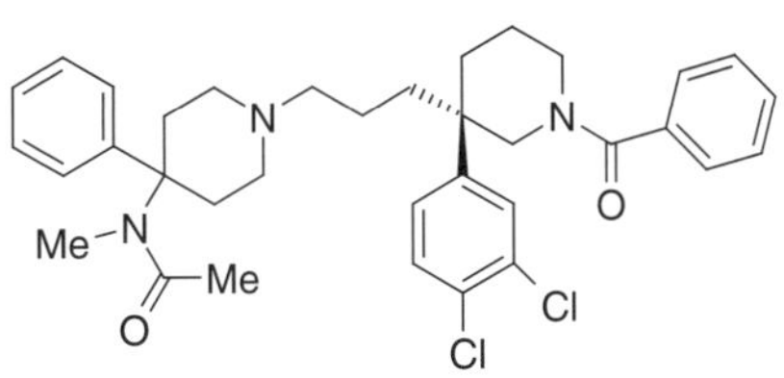

110 Osanetant

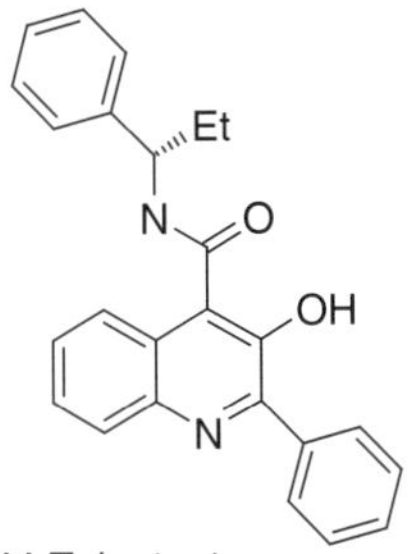

111 Talnetant

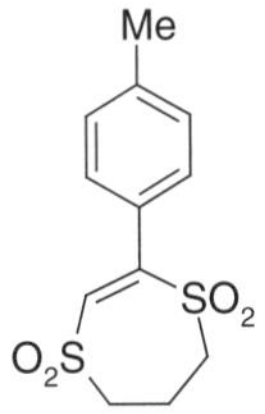

112 RWJ-57408

family and the failure of neurokinin-1 (NK-1) antagonists as analgesics and antidepressants. Nonetheless, several neuropeptides or their inhibitors including neurokinin B, angiotensin IV, galanin (GAL), adrenocorticotrophic hormone (ACTH), oxytocin (OT), arginine vasopressin (AVP), and thyrotrophin-releasing hormone (TRH) have shown efficacy in cognition models.

The neurokinins (NKs) are a family of three neuropeptides, substance P (SP), neurokinin A (NKA), and neurokinin B (NKB), that mediate their biological effects via activation of NK1, NK2, and NK3 GPCRs, respectively [245]. NK3 receptors are expressed in brain regions involved in emotion and cognition, and stimulation of NK3 receptors can enhance DA, NE, 5-HT, GABA, and ACh release. Antagonists may have therapeutic value in treating psychosis and CDS. NK3 knockout mice displayed deficits compared to wild-type mice in several cognition tests, including passive avoidance, acquisition of conditioned avoidance responding and MWM [246]. Two NK3 antagonists, osanetant **110** and talnetant **111** displayed antipsychotic activity the clinic; however, both compounds suffer from poor pharmacokinetics and were abandoned [247].

GAL and its receptors (GALR1–GALR3) are distributed in basal forebrain, cortex, hippocampus, and amygdala, where it modulates ACh, NE, and 5-HT pathways [248]. GAL inhibits ACh release *in vitro* and *in vivo*, and impairs cognitive performance in models of spatial learning and memory [249]. Mice overexpressing GAL have selective search and spatial navigation deficits with impaired learning and memory. The GALR1 antagonist, RWJ-57408 **112**, reversed the GAL-inhibited release of ACh *in vitro* [250] and the selective peptide antagonists, M35 and M40, reversed GAL-induced deficits in various models of learning and memory in rats [248]. Taken together, the data suggest GALR1 as a promising target for cognitive deficits in AD. However, to date, high-throughput screening and drug discovery efforts have failed to identify potent, drug-like NCEs for this target.

ACTH modulates cognition and attention in humans. $ACTH_{4-10}$ is a potent modulator of attention in humans. It may also have neurotrophic or neuroprotective properties, and was studied for the treatment of memory disturbances in AD [251]. ORG2766 (*H*-Met (O$_2$)-Glu-His-Phe-D-Ly-Phe-OH) enhanced recovery in behavioral models of forebrain lesioned animals and ebiratide, an analog of ACTH$_{4-9}$, increased ChAT and AChE activities in aged rats, suggesting potential therapy in AD [252].

Angiotensin IV (Ang IV), initially thought to be an inactive product of Ang II degradation, enhanced learning and memory in normal rodents and reverse memory deficits in models of amnesia. The CNS effects are mediated by the AT4 receptor, found in high levels in brain regions involved in cognition. The AT4 receptor was identified as the transmembrane enzyme, insulin-regulated membrane aminopeptidase (IRAP). Inhibition of IRAP via the AT4 receptor may inhibit degradation of neuropeptides involved in cognitive enhancement [253].

There is growing body of evidence that the neuropeptides OT and AVP modulate complex social behavior and social cognition [254,255]. OT knockouts display social amnesia in the social recognition test, despite a normal ability to recognize familiar nonsocial scents. The deficit in social recognition can be completely reversed with OT infusion [256]. AVT increased glutamate release and intracellular Ca^{2+} concentration in hippocampal and cortical astrocytes and blocked Aβ-induced impairment of LTP in rat hippocampus *in vivo* [257].

In addition to its role in regulating thyroid function, TRH also modulates ACh activity in the CNS. TRH administration (0.3 mg/kg i.v.) in 10 AD patients showed statistically significant increases in arousal and improvement in affect, with modest improvement in semantic memory [258].

The successful transition of peptide hormones to small-molecule mimetics or antagonists remains a major challenge in CNS drug discovery perhaps due to the fact that as labile peptides, these neuromodulators have very short half-lives whereas small molecules may be too long lasting in their effects to be efficacious and side effect free. Similarly, antagonists may provide prolonged blockade of autocrine signaling that is deleterious to tissue function. As examples, NK1 receptor antagonists MK-869 (Aprepitant) **113** and CP-122721 **114** despite good drug-like properties and copious preclinical efficacy, data have singularly failed in the clinic. Likewise the considerable efforts to improve on morphine as an analgesic have yet to yield viable NCEs that provide convincing evidence that opiod receptor subtype selectivity will avoid the side effects of morphine and its congeners including addiction, respiratory depression, constipation, and euphoria.

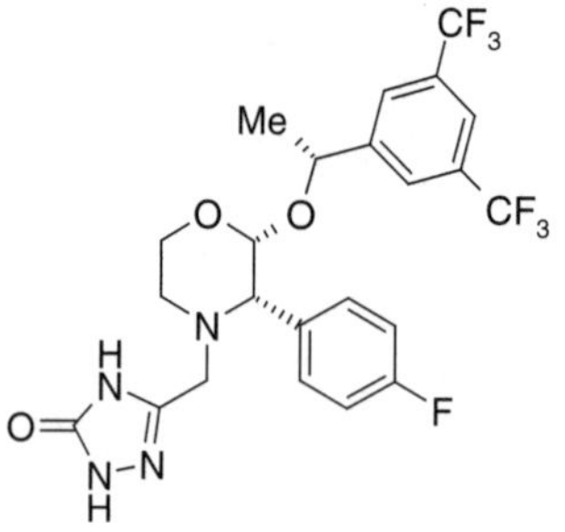

113 MK-869

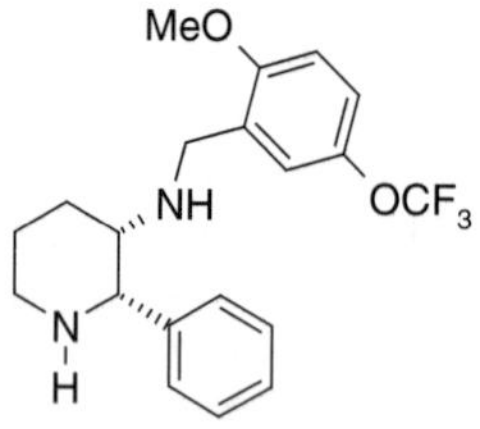

114 CP-122721

Phosphodiesterase Inhibitors Phosphodiesterases (PDEs) function to hydrolyze the phosphodiester bond and degrade the key second messengers, cyclic adenosine monophosphate (cAMP), and cyclic guanosine monophosphate (cGMP) to control their intracellular levels [259]. At least 11 distinct PDE families (PDE1–PDE11) are known that are classified by their substrate specificity. PDE3, PDE4, PDE7, and PDE8 are specific for cAMP; PDE5, PDE6, and PDE9 are cGMP selective enzymes; and PDE1, PDE2, PDE10, and PDE11 function as dual-substrate PDEs [259].

Essentially all PDEs are expressed in the CNS with evidence suggesting that PDE2, PDE4, PDE5, PDE9, and PDE10 may have therapeutic potential for cognitive disorders. A substantial body of genetic and pharmacological evidence demonstrates that the cAMP response element binding (CREB) protein is a required process in formation of long-term memory [260,261]; lower levels of CREB result in memory impairment, while higher levels facilitate memory formation [262,263]. cAMP signaling through PDE4 inhibition has been associated with consolidation and retention of LTP. The prototype PDE4 inhibitor rolipram **19** facilitated LTP in the hippocampus and increased phosphorylation of CREB and expression of the cAMP-dependent, memory-related protein, Arc [264]; changes linked to retention of long-term memory. *In vivo*, rolipram reversed scopolamine deficits in object recognition and radial arm maze tasks [265]. PDE4 inhibition and activation of cAMP/CREB signaling cascade appear to specifically facilitate long-term, but not short-term memory formation [264]. HT-0712 **115**, MK-0952 **116**, and MEM1414 (undisclosed structure) are PDE4 inhibitors that reportedly advanced to the clinic. The primary obstacle with the development of PDE4 inhibitors has been dose-limiting side effects of emesis and nausea. The four PDE4 subtypes (PDE4A–PDE4D) are differentially expressed in the CNS [266]. PDE4 knockout and behavioral studies hypothesized that the PDE4D isozyme was specifically involved in emesis [267] and PDE4B implicated in the regulation of LTP in the hippocampus [268].

The PDE2A expression pattern in CNS stimulated interest in the study of PDE2 inhibitors for treatment of cognitive disorders. BAY 60-7550 **117** (PDE2A $IC_{50} = 4.7$ nM) had selectivity versus PDE3B, PDE7B, PDE8A, PDE9A, PDE11A, and PDE1, increased cGMP levels in culture and enhanced LTP in hippocampal slices [269]. BAY 60-7550 improved performance in social and object recognition memory tasks, and reversed MK-801-induced deficits in T-maze spatial alteration. Although PDE5 expression in the brain is low the PDE5 inhibitor, sildenafil **118** facilitated memory consolidation in rodent novel object recognition tasks and reversed scopolamine-induced deficits in performance in a T-maze task [270]. PDE9A is a cGMP-specific PDE widely distributed in the CNS. The selective PDE9 inhibitors, BAY 73-6691 **119** enhanced long-term memory in social recognition and object recognition tasks and attenuated scopolamine-induced deficit in passive avoidance and MK-801-induced deficit in a T-maze alteration task [271]. BAY 73-6691 is reportedly under development for AD [265]. PDE10A is widely expressed in brain regions associated with DA and glutamate function. Increasing interest in the discovery of PDE10A inhibitors stems from reports of the antipsychotic activity of papaverine **21** and the potential for PDE10A inhibitors to treat CDS [265,272]. Papaverine enhanced CREB phosphorylation, reversed PCP-induced deficits in the EDID-set shift assay, a test of executive function, and was efficacious in the novel object recognition assay [273,274]. PF-02545920 **120**, a selective, picomolar PDE10A inhibitor, reportedly entered Phase II clinical trials for schizophrenia [275].

3. SMART DRUGS

As noted in the introduction [14], a natural extension of the use of drugs intended to treat cognitive impairment in disease states such as AD, PD, schizophrenia, and ADHD and in situations following brain trauma and stroke, is the use of such agents to improve cognitive performance and quality of life in healthy individuals. Stimulant compounds such as caffeine **100** (in coffee and soft drinks), amphetamine **121**, methylphenidate **9**, and modafinil **20** are frequently used by students to improve examination performance, by shift workers (including those in the medical professions), airline crews, and by active military personnel [14]. Their potential use in airport-security screeners has also been suggested. This has led to an ethical debate on the use of cognition enhancers to improve "brain energy" and overcome the effects of sleep deprivation in "an overworked 24/7 society pushed to the limits of human endurance." [14] In the United States, the nonprescription use of the

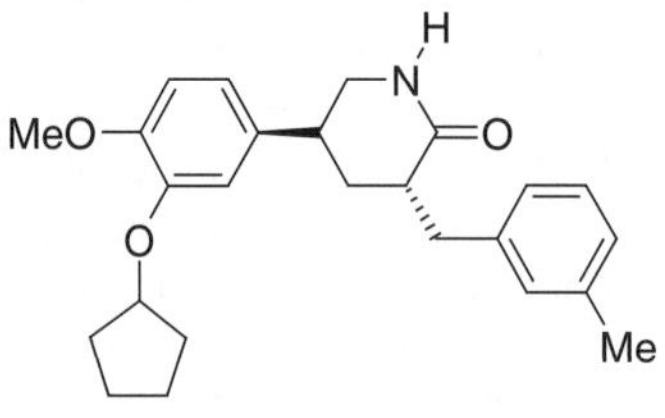

115 HT-0712

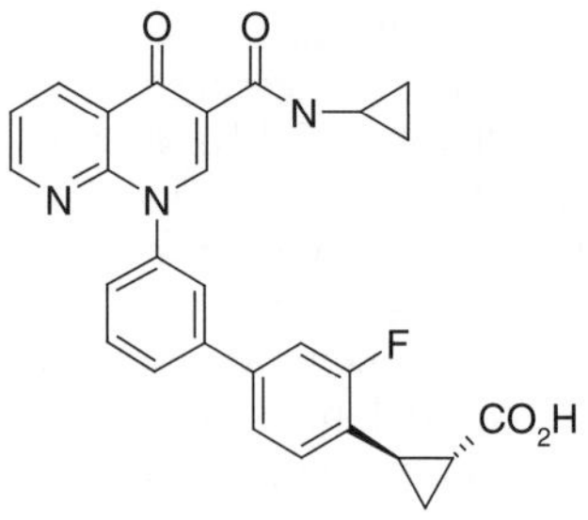

116 MK-0952

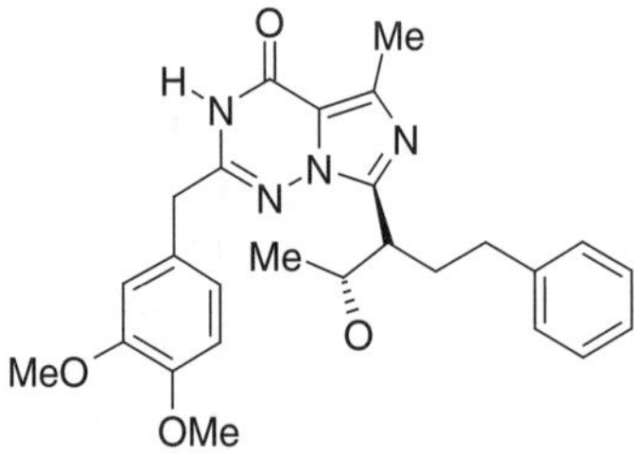

117 BAY 60-7550

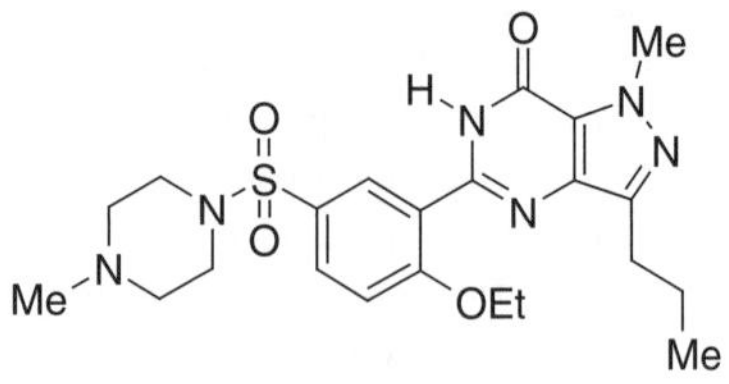

118 Sildenafil

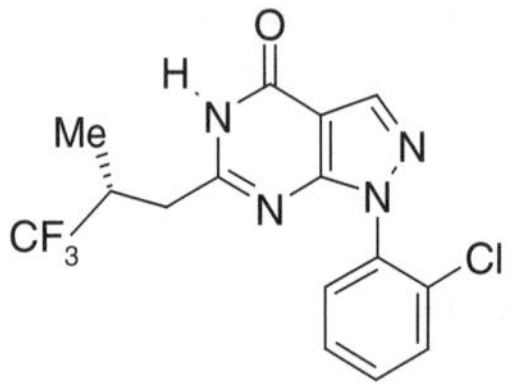

119 BAY 73-6691

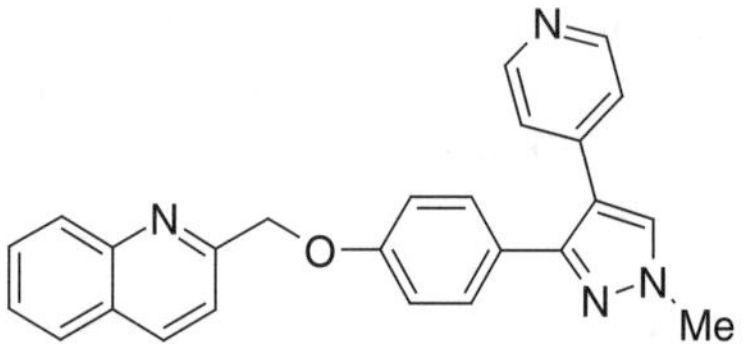

120 PF-02545920

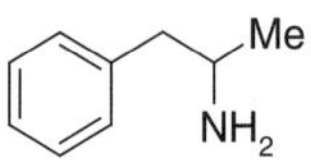

121 Amphetamine

stimulants amphetamine and methylphenidate is a crime [276] creating a black market in their sale. Yet questions have been raised as to whether the responsible use of cognition enhancers in the healthy is actually a "good health habit" and how, in these circumstances, taking a prescription psychostimulant differs from consuming a double espresso. Indeed, the increased consumption of coffee may be related to the fact that an 8 oz cup of Starbucks coffee contains nearly three times the caffeine as a regular coffee [277]. Modern-day society is excessively sleep deprived, stressed, and overloaded with information via the Internet such that quality of life is at a premium. Accordingly, there have been calls [276,278] for further research to generate an evidence base on the risk and benefits associated with the use of cognition enhancers in the healthy with the argument that these modalities may in time be viewed in the same prophylatic manner as vitamins or vaccines.

4. SUMMARY AND FUTURE CHALLENGES

The need for drugs to treat cognitive dysfunction in all its forms has been clearly established and represents a major challenge in drug discovery given the many failures of compounds robustly active in animal models of attention and memory in the clinic. It may be argued that nearly every class of NCE active on CNS targets will, at one point or another, be examined for effects on cognitive function. Similarly, the lack of clinical success may be argued as reflecting either the inadequacy of the animal models used to triage CNS compounds for advancement to the clinic [32,279] or the inability to diagnose patients at a sufficiently early stage in their disease progression to show benefit [11].

With an understanding of these limitations, a better temporal and informational interface between preclinical and clinical research activities [280] and the use of appropriate biomarkers, it may be anticipated that newer compounds advanced to the clinic may show improved efficacy and patient benefit and add significantly to the quality of life in patients with currently untreatable neurodegenerative and traumatic disorders of the brain.

REFERENCES

1. Uhlhass PJ, Singer W. Neuron 2006;52: 155–168.
2. Squires LR, Kandel ER. Memory: From Mind to Molecules. W.H. Freeman; 2000.
3. Fredholm BB, Bättig K, Holmén J, Nehlig A, Zvartau EE. Pharmacol Rev 1999;51:83–133.
4. Barch DM, Carter CS. Schizophr Res 2005;77: 43–58.
5. Gouliaev AH, Senning A. Brain Res Rev 1994; 19:180–222.
6. Birks J. Cochrane Database Syst Rev 2006;1: CD005593.
7. Rogawski MA, Wenk GL. CNS Drug Rev 2003; 9:275–308.
8. Raina P, Santaguida P, Ismaila A, Patterson C, Cowan D, Levine M, Booker L, Oremus M. Ann Intern Med 2008;148:379–397.
9. Castellani RJ, Lee H-G, Zhu X, Nunomura A, Perry G, Smith MA. Acta Neuropathol 2006; 111:503–509.
10. Mazanetz MP, Fischer PM. Nat Rev Drug Discov 2007;6:464–479.
11. Williams M. Curr Opin Invest Drugs 2009;10: 19–30.
12. Olson RE, Marcin LR. Ann Rep Med Chem 2007;42:27–47.
13. Schneider A, Mandelkow E. Neurotherapeutics 2008;5:443–457.
14. Sahakian B, Morein-Zamir S. Nature 2007; 450:1157–1159.
15. Thompson RF. Psychology Rev 1994;101: 259–265.
16. Albright TD, Kandel ER, Posner MI. Curr Opin Neurobiol 2000;10:612–624.
17. Mai TV, Cleeremans A. Trends Cog Sci 2005;9: 397–404.
18. Geyer MA. Neurotox Res 2006;10:211–220.
19. Koch C, Ullman S. Hum Neurobiol 1985;4: 219–227.
20. Smilek D, Frischen A, Reynolds MG, Gerritsen C, Eastwood JD. Percept Psychophys 2007;69: 1105–1116.
21. Knudsen EI. Ann Rev Neurosci 2007;30:57–78.
22. Repovs G, Baddeley A. Neuroscience 2006;139: 5–21.
23. Miller GA. Psychol Rev 1956;63:81–97.
24. Cowan N. Behav Brain Sci 2001;24:86–14.
25. Stickgold R. Nature 2005;437:1272–1278.
26. Tulving E.In: Tulving E, Donaldson W, Ower GH, editors. Organization of Memory. New York: Academic Press; 1972. p 381–403.
27. Westmacott R, Black SE, Freedman M, Moscovitch M. Neuropsychologia 2004;42:25–48.
28. Broadbent DE. Psychol Rev 1957;64:205–215.
29. Goldberg E, The Executive Brain. New York: Oxford University Press; 2001.
30. First MB, Tasman AC. DSM-IV-TR Mental Disorders: Diagnosis, Etiology and Treatment. Chichester, UK: Wiley; 2004. p 263–309.
31. Shaw LM, Korecka M, Clar CM, Lee WM-Y, Trojanowski JQ. Nat Rev Drug Discov 2007; 6:295–303.
32. Day M, Balci F, Wan HI, Fox GB, Rutkowski JL, Feuerstein G. Curr Opin Invest Drugs 2008;9:696–707.
33. Chin A-V, Robinson DJ, O'Connell H, Hamilton F, Bruce I, Coen R, Walsh B, Coakley D, Molloy A, Scott J, Lawlor BA, Cunningham CJ. Age Ageing 2008;37:559–564.
34. Vellas B, Andrieu S, Sampaio C, Coley N, Wilcock G. Lancet Neurol 2008;7:436–450.
35. News in brief. Nature Rev Drug Disc 2008;7: 964–965.
36. Green MF. Ann Rev Clin Psychol 2007;3: 159–180.
37. McGurk SR. J Psychiatr Pract 2006;6: 190–196.
38. Kaplan H, Saddock B, Grebb J. Kaplan and Saddock's Synopsis of Psychiatry. 7th ed. Baltimore MD: Williams & Wilkins; 2003.
39. Gray F, Adle-Biassette H, Chrétien F, Lorin de la Grandmaison G, Force G, Keohane C. Clin Neuropathol 2001;20:146–155.
40. Zaleska MM, Mercado ML, Chavez J, Feuerstein GZ, Pangalos MN, Wood A. Neuropharmacology 2009;56:329–341.

41. Hall ED. Comp Med Chem II 2007;6:253–277.
42. O'Collins VE, Macleod MR, Donnan GA, Horky LL, van der Worp BH, Howells DW. Ann Neurol 2005;59:467–477.
43. Rosen WG, Mohs RC, Davis KL. Am J Psychiatry 1984;141:1356–1364.
44. Mohs RC, Rosen WG, Davis KL. Psychopharmacol Bull 1983;19:448–450.
45. Connor DJ, Sabbagh MN. J Alzheimer Dis 2008;15:461–464.
46. Sahakian BJ, Owen AM. J Royal Soc Med 1992;85:399–402.
47. Nuechterlein KH, Green MF, Kern RS, Baade LE, Barch DM, Cohen JD, Essock S, Fenton WS, Frese FJ, Gold JM, Goldberg T, Heaton RK, Keefe RS, Kraemer H, Mesholam-Gately R, Seidman LJ, Stover E, Weinberger DR, Young AS, Zalcman S, Marder SR. Am J Psychiatry 2008;165:203–213.
48. Geyer MA. Neurotox Res 2008;14:71–78.
49. Lashley KS. Symp Soc Exp Biol 1950;4:454–482.
50. Penfield W. AMA Arch Neurol Psychiatry 1952;67:178–198.
51. Lipton PA, Eichenbaum H. Neural Plasticity 2008;2008: 258467.
52. Squire LR, Alvarez P. Curr Opin Neurobiol 1995;5:169–177.
53. Corkin S. Nature Rev Neurosci 2002;3:153–160.
54. Scoville WB, Milner B. J Neurol Neurosurg Psychiatr 1957;20:11–21.
55. Goldman-Rakic PS. Philos Trans R Soc Lond 1996;351:1445–1453.
56. Nauta WJ. Acta Neurobiol Exp 1972;32:125–140.
57. Petrides M. Exp Brain Res 2000;133:44–54.
58. Hanes KR, Andrewes DG, Pantelis C. J Int Neuropsychol Soc 1995;1:545–553.
59. Hebb DO. The Organization of Behavior: A Neuropsychological Theory. New York: John Wiley & Sons; 1949.
60. Bliss TV, Lomo T. J Physiol 1973;232:331–356.
61. Dingledine R, Borges K, Bowie D, Traynelis SF. Pharmacol Rev 1999;51:7–61.
62. Harris EW, Ganong AH, Cotman CW. Brain Res 1984;323:132–137.
63. Sanes JR, Lichtman JW. Nat Neurosci 1999;2:597–604.
64. Fukunaga K, Muller D, Miyamoto E. J Biol Chem 1995;270:6119–6124.
65. Silva J, Paylor R, Wehner JM, Tonegawa S. Science 1992;257:206–211.
66. Giese KP, Fedorov NB, Filipkowski RK, Silva AJ. Science 1998;279:870–873.
67. Greengard P, Jen J, Nairn AC, Stevens CF. Science 1991;253:1135–1138.
68. English JD, Sweatt JD. J Biol Chem 1996;271:24329–24332.
69. Lu YM, Roder JC, Davidow J, Salter MW. Science 1998;279:1363–1367.
70. Malenka RC, Madison DV, Nicoll RA. Nature 1986;321:175–177.
71. Malenka RC. Ann NY Acad Sci 2003;1003:1–11.
72. Park M, Penick EC, Edwards JG, Kauer JA, Ehlers MD. Science 2004;305:1972–1975.
73. Nicoll RA, Tomita S, Bredt DS. Science 2006;311:1253–1256.
74. Tsui J, Malenka RC. J Biol Chem 2006;281:13794–13804.
75. Tomita S, Stein V, Stocker TJ, Nicoll RA, Bredt DS. Neuron 2005;45:269–277.
76. Dudek SM, Bear MF. Proc Natl Acad Sci USA 1992;89:4363–4367.
77. Dan Y, Poo MM. Physiol Rev 2006;86:1033–1048.
78. Mulkey RM, Herron CE, Malenka RC. Science 1993;261:1051–1055.
79. Ashby MC, De La Rue SA, Ralph GS, Uney J, Collingridge GL, Henley JM. J Neurosci 2004;24:5172–5176.
80. Harris EW, Cotman CW. Neurosci Lett 1986;70:132–137.
81. Nicoll RA, Malenka RC. Nature 1995;377:115–118.
82. Jin Y, Kim SJ, Kim J, Worley PF, Linden DJ. Neuron 2007;55:277–287.
83. Gerdeman GL, Ronesi J, Lovinger DM. Nature Neurosci 2002;5:446–451.
84. Nugent FS, Kauer JA. J Physiol 2008;586:1487–1493.
85. Raina P, Santaguida P, Ismaila A, Patterson C, Cowan D, Levine M, Booker L, Oremus M. Ann Intern Med 2008;148:379–397.
86. Jacobsen JS, Wu CC, Redwine JM, Comery TA, Arias R, Bowlby M, Martone R, Morrison JH, Pangalos MN, Reinhart PH, Bloom FE. Proc Natl Acad Sci USA 2006;103:5161–5166.
87. Patel S, Tariot PN. Psychiatr Clin North Am 1991;14:287–308.
88. Amitai N, Semenova S, Markou A. Psychopharmacology 2007;193:521–537.

89. Robbins TW. Psychopharmacology 2002;163: 362–380.
90. Navarra R, Graf R, Huang Y, Logue S, Comery T, Hughes Z, Day M. Prog Neuropsychopharm Biol Psych 2008;32:34–41.
91. Day M, Pan JB, Buckley MJ, Cronin E, Hollingsworth PR, Hirst WD, Navarra R, Sullivan JP, Decker MW, Fox GB. Biochem Pharmacol 2007;73:1123–1134.
92. Hahn B, Sharples CG, Wonnacott S, Shoaib M, Stolerman IP. Neuropharmacology 2003;44: 1054–1067.
93. Thor DH, Wainwright KL, Holloway WR. Dev Psychobiol 1982;15:1–8.
94. Ennaceur A, Cavoy A, Costa JC, Delacour J. Behav Brain Res 1989;33:197–207.
95. Medhurst D, Atkins AR, Beresford IJ, Brackenborough K, Briggs MA, Calver AR, Cilia J, Cluderay JE, Crook B, Davis JB, Davis RK, Davis RP, Dawson LA, Foley AG, Gartlon J, Gonzalez MI, Heslop T, Hirst WD, Jennings C, Jones DN, Lacroix LP, Martyn A, Ociepka S, Ray A, Regan CM, Roberts JC, Schogger J, Southam E, Stean TO, Trail BK, Upton N, Wadsworth G, Wald JA, White T, Witherington J, Woolley ML, Worby A, Wilson DM. J Pharmacol Exp Ther 2007;321: 1032–1045.
96. Scali C, Giovannini MG, Prosperi C, Bartolini L, Pepeu G. Pharmacol Res 1997;36:463–469.
97. Shimazaki T, Kaku A, Chaki S. Eur J Phamacol 2007;575:94–97.
98. Morris R. J Neurosci Methods 1984;11:47–60.
99. Olton DS. Physiol and Behav 1987;40:793–797.
100. Van Dam D, Abramowski D, Staufenbiel M, De Deyn PP. Psychopharmacology 2005;180: 177–190.
101. Marcos B, Chuang TT, Gil-Bea FJ, Ramirez MJ. Br J Pharmacol 2008;155:434–440.
102. Timmermann DB, Grønlien JH, Kohlhaas KL, Nielsen EØ, Dam E, Jørgensen TD, Ahring PK, Peters D, Holst D, Chrsitensen JK, Malysz J, Briggs CA, Gopalakrishnan M, Olsen GM. J Pharmacol Exp Ther 2007;323: 294–307.
103. Banfi S, Cornelli U, Fonio W, Dorigotti L. J Pharmacol Methods 1982;8:255–263.
104. Fanselow MS. Anim Learn Behav 1990;18: 264–270.
105. Sienkiewicz-Jarosz H, Maciejak P, Krzaścik P, Członkowska AI, Szyndler J, Bidziński A, Kostowski W, Płaźnik A. Phamacol Biochem Behav 2003;75:491–496.
106. Misane I, Ogren SO. Neuropsychophamacology 2003;28:253–264.
107. Egawa T, Mishima K, Matsumoto Y, Iwasaki K, Iwasaki K, Fujiwara M. Jpn J Pharmacol 1997;75:275–281.
108. Eling P, Derckx K, Maes R. Brain Cogn 2008; 67:247–253.
109. Everett J, Lavoie K, Gagnon JF, Gosselin J Psychiatry Neurosci 2001;26:123–130.
110. Birrell JM, Brown VJ. J Neurosci 2000;20: 4320–4324.
111. Goetghebeur P, Dias R. Psychopharmacology 2008; PMID:18392753.
112. Rodefer JS, Nguyen TN, Karlsson JJ, Arnt J. Neuropsychopharmacology 2008;33:2657–2666.
113. Rodefer JS, Murphy ER, Baxter MG. Eur J Neurosci 2005;21:1070–1076.
114. Bartus RT, Dean RL, Beer B, Lippa AS. Science 1982;217:408–414.
115. Bartus RT. Exp Neurol 2000;163:495–529.
116. DeKosky ST, Ikonomovic MD, Styren SD, Beckett L, Wisniewski S, Bennett DA, Cochran EJ, Kordower JH, Mufson EJ. Ann Neurol 2002;51:145–155.
117. Davis KL, Mohs RC, Marin D, Purohit DP, Perl DP, Lantz M, Austin G, Haroutunian V. JAMA 1999;281:1401–1406.
118. Watkins PB, Zimmerman HJ, Knapp MJ, -Gracon SI, Lewis KW. JAMA 1994;271: 992–998.
119. Shigeta M, Homma A. CNS Drug Rev 2001;7: 353–368.
120. Jacobson SA, Sabbagh MN. Exp Opin Drug Metab Toxicol 2008;4:1363–1369.
121. Gottwald MD, Rozanski RI. Exp Opin Investig Drugs 1999;8:1673–1682.
122. Cutler NR, Polinsky RJ, Sramek JJ, Enz A, Jhee SS, Mancione L, Hourani J, Zolnouni P. Acta Neurol Scand 1998;97:244–250.
123. Samochocki M, Höffle A, Fehrenbacher A, Jostock R, Ludwig J, Christner C, Radina M, Zerlin M, Ullmer C, Pereira EF, Lübbert H, Albuquerque EX, Maelicke A. J Pharmacol Exp Ther 2003;305:1024–1036.
124. Lilienfeld S. CNS Drug Rev 2002;8:159–176.
125. Bai DL, Tang XC, He XC. Curr Med Chem 2000;7:355–374.
126. Greig NH, Sambamurti K, Yu QS, Brossi A, Bruinsma GB, Lahiri DK. Curr Alzheimer Res 2005;2:281–290.
127. Vellas B, Andrieu S, Sampaio C, Wilcock G. Lancet Neurol 2007;6:56–62.

128. Benzi G, Moretti A. Eur J Pharmacol 1988; 346:1–16.
129. Ferreri F, Agbokou nad C, Gauthier S. J Psychiatry Neurosci 2006;31:369–376.
130. Friedman JL. Psychopharmacology 2004; 174:45–53.
131. Buchanan RW, Freedman R, Javitt DC, Abi-Dargham A, Lieberman JA. Schizophr Bull 2007;33:1120–1130.
132. Schilström B, Ivanov VB, Wiker C, Svensson TH. Neuropsychopharmacology 2007;32:43–53.
133. Jeppesen L, Olesen PH, Hansen L, Sheardown MJ, Thomsen C, Rasmussen T, Jensen AF, Christensen MS, Rimvall K, Ward JS, Whitesitt C, Calligaro DO, Bymaster FP, Delapp NW, Felder CC, Shannon HE, Sauerberg P. J Med Chem 1999;42:1999–2006.
134. Stanhope KJ, Mirza NR, Bickerdike MJ, Bright JL, Harrington NR, Hesselink MB, Kennett GA, Lightowler S, Sheardown MJ, Syed R, Upton RL, Wadsworth G, Weiss SM, Wyatt A. J Pharmacol Exp Ther 2001; 299:782–792.
135. Graul A, Castaner J. Drugs Future 1996;21:911–916.
136. Mizobe F, Nakahara N, Ogane N, Saito Y, Ise M, Iga Y, Fukui K, Fujise N, Haga T, Kawanishi G.In: Nagatsu T, Fisher A, Yoshida M, editors. Basic, Clinical, and Therapeutic Aspects of Alzheimer's and Parkinson's Disease. New York: Plenum Press; 1990. p. 351–358.
137. Mucke HAM, Castaner J. Drugs Future 1998; 3:843
138. Caccamo A, Oddo S, Billings LM, Green KN, Martinez-Coria H, Fisher A, LaFerla FM. Neuron 2006;49:671–682.
139. DailyDrugNewscom (Daily Essentials) June 17, 2008.
140. Conn PJ, Tamminga C, Schoepp DD, Lindsley C. Mol Interv 2008;8:99–107.
141. Thomas RL, Mistry R, Langmead CJ, Wood MD, Challiss RA. J Pharmacol Exp Ther 2008; 327:365–374.
142. Marlo JE, Niswender CM, Days EL, Bridges TM, Xiang Y, Rodriguez AL, Shirey JK, Brady AE, Nalywajko T, Luo Q, Austin CA, Williams MB, Kim K, Williams R, Orton D, Brown HA, Lindsley CW, Weaver CD, Conn PJ. Mol Pharmacol 2009;75:577–588.
143. Danysz W, Parsons CG. Int J Geriatr Psychiatry 2003;18(Suppl 1): S23–S32.
144. Müller WE, Eckert GP, Eckert A. Pharmacopsychiatry 1999;32(Suppl 1): 2–9.
145. Iwasaki K, Matsumoto Y, Fujiwara M. Jpn J Pharmacol 1992;58:117–126.
146. Abou-Khalil B. Neuropsychiatr Dis Treat 2008;4:507–523.
147. Zhou B, Zhang Q, Tian L, Xiao J, Stefan H, Zhou D. Epilepsy Behav 2008;12:305–310.
148. Gotti C, Moretti M, Gaimarri A, Zanardi A, Clementi F, Zoli M. Biochem Pharmacol 2007;74:1102–1111.
149. Lippiello PM, Bencherif M, Hauser TA, Jordan KG, Letchworth SR, Mazurov AA. Exp Opin Drug Discov 2007;2:1185–1203.
150. Levin ED. J Neurobiol 2002;53:633–640.
151. Lloyd GK, Williams M. J Pharmacol Exp Ther 2000;292:461–467.
152. Cincotta SL, Yorek MS, Moschak TM, Lewis SR, Rodefer JS. Curr Opin Investig Drugs 2008;9:47–56.
153. Freedman R, Coon H, Myles-Worsley M, Orr-Urtreger A, Olincy A, Davis A, Polymeropoulos M, Holik J, Hopkins J, Hoff M, Rosenthal J, Waldo MC, Reimherr F, Wender P, Yaw J, Young DA, Breese CR, Adams C, Patterson D, Adler LE, Kruglyak L, Leonard S, Byerley W. Proc Natl Acad Sci USA 1997;94:587–592.
154. Broad LM, Sher E, Astles PC, Zwart R, O'Neill MJ. Drugs Future 2007;32:161–170.
155. Olincy A, Harris JG, Johnson LL, Pender V, Kongs S, Allensworth D, Ellis J, Zerbe GO, Leonard S, Stevens KE, Stevens JO, Martin L, Adler LE, Soti F, Kem WR, Freedman R. Arch Gen Psychiatry 2006;63:630–638.
156. Freedman R, Olincy A, Buchanan RW, Harris JG, Gold JM, Johnson L, Allensworth D, Guzman-Bonilla A, Clement B, Ball MP, Kutnick J, Pender V, Martin LF, Stevens KE, Wagner BD, Zerbe GO, Soti F, Kem WR. Am J Psychiatry 2008;165:1040–1047.
157. Tietje KR, Anderson DJ, Bitner RS, Blomme EA, Brackemeyer PJ, Briggs CA, Browman KE, Bury D, Curzon P, Drescher KU, Frost JM, Fryer RM, Fox GB, Gronlien JH, Håkerud M, Gubbins EJ, Halm S, Harris R, Helfrich RJ, Kohlhaas KL, Law D, Malysz J, Marsh KC, Martin RL, Meyer MD, Molesky AL, Nikkel AL, Otte S, Pan L, Puttfarcken PS, Radek RJ, Robb HM, Spies E, Thorin-Hagene K, Waring JF, Ween H, Xu H, Gopalakrishnan M, Bunnelle WH. CNS Neurosci and Ther 2008; 14:65–82.
158. Pichat P, Bergis OE, Terranova JP, Urani A, Duarte C, Santucci V, Gueudet C, Voltz C, Steinberg R, Stemmelin J, Oury-Donat F, Avenet P, Griebel G, Scatton B. Neuropsychopharmacology 2007;32:17–34.

159. Feuerbach D, Lingenhoehl K, Olpe HR, Vassout A, Gentsch C, Chaperon F, Nozulak J, Enz A, Bilbe G, McAllister K, Hoyer D. Neuropharmacology 2009;56:254–263.
160. Timmermann DB, Groenlien JH, Kohlhaas KL, Nielsen EO, Dam E, Joergensen TD, Ahring PK, Peters D, Holst D, Chrsitensen JK, Malysz J, Briggs CA, Gopalakrishnan M, Olsen GM. J Pharmacol Exp Ther 2007; 323:294–307.
161. Patterson F, Jepson C, Strasser AA, Loughead J, Perkins KA, Gur RC, Frey JM, Siegel S, Lerman C. Biol Psychiatry 2009;65:144–149.
162. Raybuck JD, Portugal GS, Lerman C, Gould TJ. Behav Neurosci 2008;122:1166–1171.
163. Gatto GJ, Bohme GA, Caldwell WS, Letchworth SR, Traina VM, Obinu MC, Laville M, Reibaud M, Pradier L, Dunbar G, Bencherif M. CNS Drug Rev 2004;10:147–166.
164. Morrison T. *BioWorld* September 17 2008.
165. Rueter LE, Anderson DJ, Briggs CA, Donnelly-Roberts DL, Gintant GA, Gopalakrishnan M, Lin NH, Osinski MA, Reinhart GA, Buckley MJ, Martin RL, McDermott JS, Preusser LC, Seifert TR, Su Z, Cox BF, Decker MW, Sullivan JP. CNS Drug Rev 2004;10:167–168.
166. Arneric SP, Holladay M, Williams M. Biochem Pharmacol 2007;74:1092–1101.
167. Hill SJ, Ganellin C, Timmermans H, Schwartz JC, Shankley N, Young JM, Schunack W, Levi R, Haas HL. Pharmacol Rev 1997;49:253–278.
168. Hudkins RL, Raddatz R. Ann Rep Med Chem 2007;42:49–63.
169. Wieland K, Bongers G, Yamamoto Y, Hashimoto T, Yamatodani A, Menge WM, Timmerman H, Lovenberg TW, Leurs R. J Pharmacol Exp Ther 2001;299:908–914.
170. Fox GB, Esbenshade TA, Pan JB, Radek RJ, Krueger KM, Yao BB, Browman KE, Buckley MJ, Ballard ME, Komater VA, Miner H, Zhang M, Faghih R, Rueter LE, Bitner RS, Drescher KU, Wetter J, Marsh K, Lemaire M, Porsolt RD, Bennani YL, Sullivan JP, Cowart MD, Decker MW, Hancock AA. J Pharmacol Exp Ther 2005;313:176–190.
171. Nagase T, Mizutani T, Sekino E, Ishikawa S, Ito S, Mitobe Y, Miyamoto Y, Yoshimoto R, Tanaka T, Ishihara A, Takenaga N, Tokita S, Sato N. J Med Chem 2008;51:6889–6901.
172. Barnes NM, Sharp T. Neuropharmacology 1999;38:1083–1152.
173. Hannon J, Hoyer D. Behav Brain Res 2008; 195:198–213.
174. Roth BL, Hanizavareh SM, Blum AE. Psychopharmacology 2004;174:17–24.
175. King MV, Marsden CA, Fone KC. Trends Pharmacol Sci 2008;29:482–492.
176. Ogren SO, Eriksson TM, Elvander-Tottie E, D'Addario C, Ekström JC, Svenningsson P, Meister B, Kehr J, Stiedl O. Behav Brain Res 2008;195:54–77.
177. Galletly CA, Clark CR, McFarlane AC. Eur Neuropsychopharmacol 2005;15:601–608.
178. Meltzer HY, McGurk SR. Schizophr Bull 1999;2:233–255.
179. Forster EA, Cliffe IA, Bill DJ, Dover GM, Jones D, Reilly Y, Fletcher A. Eur J Pharmacol 1995; 281:81–88.
180. Childers WE, Jr, Abou-Gharbia MA, Kelly MG, Andree TH, Harrison BL, Ho DM, Hornby G, Huryn DM, Potestio L, Rosenzweig-Lipson SJ, Schmid J, Smith DL, Sukoff SJ, Zhang G, Schechter LE. J Med Chem 2005;48: 3467–3470.
181. Childers WE, Jr, Harrison BL, Abou-Gharbia MA, Raje S, Parks V, Pangalos MN, Schechter LE. Drugs Future 2007;32:399–407.
182. Moser PC, Bergis OE, Jegham S, Lochead A, Duconseille E, Terranova JP, Caille D, Berque-Bestel I, Lezoualc'h F, Fischmeister R, Dumuis A, Bockaert J, George P, Soubrié P, Scatton B. J Pharmacol Exp Ther 2002;302: 731–741.
183. Hille C, Bate S, Davis J, Gonzalez MI. Behav Brain Res 2008;195:180–186.
184. Robichaud AJ. Curr Top Med Chem 2006;6: 553–568.
185. Johnson NC. Drugs Future 2008;33 (Suppl A): Abst L05.
186. Wu J, Li Q, Bezprozvanny I. Mol Neurodegener 2008;3:15.
187. Bachurin S, Bukatina E, Lermontova N, Tkachenko S, Afanasiev A, Grigoriev V, Grigorieva I, Ivanov Y, Sablin S, Zefirov N. Ann NY Acad Sci 2001;939:425–435.
188. Doody RS, Gavrilova SI, Sano M, Thomas RG, Aisen PS, Bachurin SO, Seely L, Hung D. Lancet 2008;372:207–215.
189. Pérez-García G, Gonzalez-Espinosa C, Meneses A. Behav Brain Res 2006;169:83–92.
190. Cifariello A, Pompili A, Gasbarri A. Behav Brain Res 2008;195:171–179.
191. Gasbarri A, Cifariello A, Pompili A, Meneses A. Behav Brain Res 2008;195:164–170.
192. Forbes IT, Douglas S, Gribble AD, Ife RJ, Lightfoot AP, Garner AE, Riley GJ, Jeffrey P, Stevens AJ, Stean TO, Thomas DR. Bioorg Med Chem Lett 2002;12:3341–3344.

193. El-Ghundi M, O'Dowd BF, George SR. Rev Neurosci 2007;18:37–66.
194. Guillin O, Abi-Dargham A, Laruelle M. Int Rev Neurobiol 2007;78:1–39.
195. Goldman-Rakic PS, Castner SA, Svensson TH, Siever LJ, Williams GV. Psychopharmacology 2004;174:3–16.
196. Rogers BN, Schmidt CJ. Ann Rep Med Chem 2006;41:3–21.
197. Alexander SPH, Mathie MA, Peters JA. Br J Pharmacol. 2008;153(Suppl 2): S34–S35.
198. Salmi P, Isacson R, Kull B. CNS Drug Rev 2004;10:230–242.
199. Moreland RB, Patel M, Hsieh GC, Wetter JM, Marsh K, Brioni JD. Pharmacol Biochem Behav 2005;82:140–147.
200. Briand LA, Gritton H, Howe WM, Young DA, Sarter M. Prog Neurobiol 2007;83:69–91.
201. Ramos BP, Arnsten AF. Pharmacol Ther 2007;113:523–536.
202. Coull JT. Drugs Aging 1994;5:116–126.
203. Arnsten AF. Psychopharmacology 2004;174:25–31.
204. ClinicalTrials.gov Web site (NCT00409201).
205. Friedman JL, Carpenter D, Lu J, Fan J, Tang CY, White L, Parrella M, Bowler S, Elbaz Z, Flanagan L, Harvey PD. J Clin Psychopharmacol 2008;28:59–63.
206. Winblad B, Fioravanti M, Dolezal T, Logina I, Milanov IG, Popescu DC, Solomon A. Clin Drug Investig 2008;28:533–552.
207. O'Neill MJ, Bleakman D, Zimmerman DM, Nisenbaum ES. Curr Drug Targets CNS Neurol Disord 2004;3:181–194.
208. Black MD. Psychopharmacology 2005;179:154–163.
209. Patil ST, Zhang L, Martenyi F, Lowe SL, Jackson KA, Andreev BV, Avedisova AS, Bardenstein LM, Gurovich IY, Morozova MA, Mosolov SN, Neznanov NG, Reznik AM, Smulevich AB, Tochilov VA, Johnson BG, Monn JA, Schoepp DD. Nat Med 2007;13:1102–1107.
210. Williams DL, Jr, Lindsley CW. Curr Top Med Chem 2005;5:825–846.
211. Liu F, Grauer S, Kelley C, Navarra R, Graf R, Zhang G, Atkinson PJ, Wantuch C, Popiolek M, Day M, Khawaja X, Smith D, Olsen M, Kouranova E, Gilbert A, Lai M, Pausch MH, Pruthi F, Pulicicchio C, Brandon NJ, Comery TA, Beyer CE, Logue S, Rosenzweig-Lipson S, Marquis KL. J Pharmacol Exp Ther 2008; 327:827–839.
212. Arai AC, Kessler M. Curr Drug Targets 2007;8:583–602.
213. Wezenberg E, Verkes RJ, Ruigt GS, Hulstijn W, Sabbe BG. Neuropsychopharmacology 2007;32:1272–1283.
214. Bergeron R, Meyer TM, Coyle JT, Greene RW. Proc Natl Acad Sci USA 1998;95:15730–15734.
215. Marino MJ, Knutsen LJ, Williams M. J Med Chem 2008;51:1077–1107.
216. Heresco-Levy U, Ermilov M, Lichtenberg P, Bar G, Javitt DC. Biol Psychiatry 2004;55: 165–171.
217. Javitt DC, Silipo G, Cienfuegos A, Shelley AM, Bark N, Park M, Lindenmayer JP, Suckow R, Zukin SR. Int J Neuropsychopharmacol 2001; 4:385–391.
218. Sur C, Kinney GG. Expert Opin Investig Drugs 2004;13:515–521.
219. Perry KW, Falcone JF, Fell MJ, Ryder JW, Yu H, Love PL, Katner J, Gordon KD, Wade MR, Man T, Nomikos GG, Phebus LA, Cauvin AJ, Johnson KW, Jones CK, Hoffmann BJ, Sandusky GE, Walter MW, Porter WJ, Yang L, Merchant KM, Shannon HE, Svensson KA. Neuropharmacology 2008;55:743–754.
220. Depoortère R, Dargazanli G, Estenne-Bouhtou G, Coste A, Lanneau C, Desvignes C, Poncelet M, Heaulme M, Santucci V, Decobert M, Cudennec A, Voltz C, Boulay D, Terranova JP, Stemmelin J, Roger P, Marabout B, Sevrin M, Vigé X, Biton B, Steinberg R, Françon D, Alonso R, Avenet P, Oury-Donat F, Perrault G, Griebel G, George P, Soubrié P, Scatton B. Neuropsychopharmacology 2005;30:1963–1985.
221. Black MD, Varty GB, Arad M, Barak S, De Levie A, Boulay D, Pichat P, Griebel G, Weiner I. Psychopharmacology 2009;202:385–396.
222. Barnard EA, Skolnick P, Olsen RW, Mohler H, Sieghart W, Biggio G, Braestrup C, Bateson AN, Langer SZ. Pharmacol Rev 1998;50: 291–313.
223. Ghoneim MM, Mewaldt SP. Anesthesiology 1990;72:926–938.
224. Maubach K. Curr Drug Targets CNS Neurol Disord 2003;2:233–239.
225. Collinson N, Kuenzi FM, Jarolimek W, Maubach KA, Cothliff R, Sur C, Smith A, Otu FM, Howell O, Atack JR, McKernan RM, Seabrook GR, Dawson GR, Whiting PJ, Rosahl TW. J Neurosci 2002;22:5572–5580.
226. Atack JR, Bayley PJ, Seabrook GR, Wafford KA, McKernan RM, Dawson GR. Neuropharmacology 2006;51:1023–1029.
227. Atack JR. CNS Neurosci Ther 2008;14:25–35.
228. Fredholm BB, Jzerman IAP, Jacobson KA, Klotz KN, Linden J. Pharmacol Rev 2001; 53:527–552.

229. Takahashi RN, Pamplona FA, Prediger RD. Front Biosci 2008;13:2614–2632.
230. Simola N, Morelli M, Pinna A. Curr Pharm Des 2008;14:1475–1489.
231. Giménez-Llort L, Schiffmann SN, Shmidt T, Canela L, Camón L, Wassholm M, Canals M, Terasmaa A, Fernández-Teruel A, Tobeña A, Popova E, Ferré S, Agnati L, Ciruela F, Martínez E, Scheel-Kruger J, Lluis C, Franco R, Fuxe K, Bader M. Neurobiol Learn Mem 2007;87:42–56.
232. Mihara T, Mihara K, Yarimizu J, Mitani Y, Matsuda R, Yamamoto H, Aoki S, Akahane A, Iwashita A, Matsuoka N. J Pharmacol Exp Ther 2007;323:708–719.
233. Thoenen H, Barde YA. Physiol Rev 1980;60:1284–1335.
234. Cuello AC, Bruno MA, Bell KF. Curr Alzheimer Res 2007;4:351–835.
235. Hefti F, Schneider LS. Clin Neuropharmacol 1991;14(Suppl 1):S62–S66.
236. Seiger A, Nordberg A, vonHolst H, Backman L, Ebendal T, Alafuzoff I, Amberla K, Hartvig P, Herlitz A, Lilja A, Lundqvist H, Langstrom B, Meyerson B, Persson A, Viitanen M, Winblad B, Olson L. Behav Brain Res 1993;57:255–261.
237. Tuszynski MH, Thal L, Pay M, Salmon DP, Hoi SU, Bakay R, Patel P, Blesch A, Vahlsing HL, Ho G, Tong G, Potkin SG, Fallon J, Hansen L, Mufson EJ, Kordower JH, Gall C, Conner J. Nat Med 2005;11:551–555.
238. Pezet S, Malcangio M. Expert Opin Ther Targets 2004;8:391–399.
239. Tapia-Arancibia L, Aliaga E, Silhol M, Arancibia S. Brain Res Rev 2008;59:201–220.
240. Gunstad J, Benitez A, Smith J, Glickman E, Spitznagel MB, Alexander T, Juvancic-Heltzel J, Murray L. J Geriatr Psychiatry Neurol 2008;21:166–170.
241. Dik MG, Pluijm SM, Jonker C, Deeg DJ, Lomecky MZ, Lips P. Neurobiol Aging 2003;24:573–581.
242. Grotenhermen F. Curr Drug Targets CNS Neurol Disord 2005;4:507–530.
243. Wotjak CT. Mini Rev Med Chem 2005;5:659–670.
244. Jones D. Nat Rev Drug Discov 2008;7:961–962.
245. Holmes A, Heilig M, Rupniak NM, Steckler T, Griebel G. Trends Pharmacol Sci 2003;24:580–588.
246. Siuciak JA, McCarthy SA, Martin AN, Chapin DS, Stock J, Nadeau DM, Kantesaria S, Bryce-Pritt D, McLean S. Psychopharmacology 2007;194:185–195.
247. Spooren W, Riemer C, Meltzer H. Nat Rev Drug Discov 2005;4:967–975.
248. Counts SE, Perez SE, Kahl U, Bartfai T, Bowser RP, Deecher DC, Mash DC, Crawley JN, Mufson EJ. CNS Drug Rev 2001;7:445–470.
249. Vrontakis ME. Curr Drug Targets CNS Neurol Disord 2002;1:531–541.
250. Scott MK, Ross TM, Lee DH, Wang HY, Shank RP, Wild KD, Davis CB, Crooke JJ, Potocki AC, Reitz AB. Bioorg Med Chem 2000;8:1383–1391.
251. Hol EM, Gispen WH, Bär PR. Peptides 1995;16:979–993.
252. Matsumoto T, Tsuda S, Nakamura S. J Neural Transm Gen Sect 1995;100:1–15.
253. Albiston AL, Morton CJ, Ng HL, Pham V, Yeatman HR, Ye S, Fernando RN, De Bundel D, Ascher DB, Mendelsohn FA, Parker MW, Chai SY. FASEB J 2008;22:4209–4217.
254. Donaldson ZR, Young LJ. Science 2008;322:900–904.
255. Marazziti D, Dell'osso C. Curr Med Chem 2008;15:698–704.
256. Ferguson JN, Young LJ, Hearn EF, Insel TR, Winslow JT. Nature Genetics 2000;25:284–288.
257. Jing W, Guo F, Cheng L, Zhang JF, Qi JS. Neurosci Lett 2009;450:306–310.
258. Mellow AM, Sunderland T, Cohen RM, Lawlor BA, Hill JL, Newhouse PA, Cohen MR, Murphy DL. Psychopharmacology 1989;98:403–437.
259. Rotella DP. In: Moos WH, editor. Comprehensive Medicinal Chemistry II. Amesterdam: Elsevier Press; 2007. p 919–957.
260. Silva AJ, Kogan JH, Frankland PW, Kida S. Annu Rev Neurosci 1998;21:127–148.
261. Dash PK, Hochner B, Kandel ER. Nature 1990;345:718–721.
262. Barco A, Alarcon JM, Kandel ER. Cell 2002;108:689–703.
263. Josselyn SA, Shi C, Carlezon WA, Jr, Neve RL, Nestler EJ, Davis M. J Neurosci 2001;21:2404–2412.
264. Monti B, Berteotti C, Contestabile A. Neuropsychopharmacology 2006;31:278–286.
265. Halene TB, Siegel SJ. Drug Discov Today 2007;12:870–878.
266. Menniti FS, Faraci WS, Schmidt CJ. Nat Rev Drug Discov 2006;5:660–670.
267. Robichaud A, Stamatiou PB, Jin SL, Lachance N, MacDonald D, Laliberté F, Liu S, Huang Z, Conti M, Chan CC. J Clin Invest 2002;110:1045–1052.

268. Ahmed T, Frey JU. Neuroscience 2003;117: 627–638.
269. Boess FG, Hendrix M, van der Staay FJ, Erb C, Schreiber R, van Staveren W, de Vente J, Prickaerts J, Blokland A, Koenig G. Neuropharmacology 2004;47:1081–1092.
270. Devan BD, Sierra-Mercado D, Jr, Jimenez M, Bowker JL, Duffy KB, Spangler EL, Ingram DK. Pharmacol Biochem Behav 2004;79:691–699.
271. van der Staay FJ, Rutten K, Bärfacker L, Devry J, Erb C, Heckroth H, Karthaus D, Tersteegen A, van Kampen M, Blokland A, Prickaerts J, Reymann KG, Schröder UH, Hendrix M. Neuropharmacology 2008;55: 908–918.
272. Schmidt CJ, Chapin DS, Cianfrogna J, Corman ML, Hajos M, Harms JF, Hoffman WE, Lebel LA, McCarthy SA, Nelson FR, Proulx-LaFrance C, Majchrzak MJ, Ramirez AD, Schmidt K, Seymour PA, Siuciak JA, Tingley FD 3rd, Williams RD, Verhoest PR, Menniti FS. J Pharmacol Exp Ther 2008;325:681–690.
273. Roderfer JS, Murphy ER, Baxter MG. Eur J Neurosci 2005;21:1070–1076.
274. Siuciak JA, Chapin DS, Harms JF, Lebel LA, McCarthy SA, Chambers L, Shrikhande A, Wong S, Menniti FS, Schmidt CJ. Neuropharmacology 2006;51:386–396.
275. Verhoest RP. 236th ACS National Meeting, Philadelphia, PA, August 17–21, 2008. MEDI 219.
276. Greely H, Sahakian B, Harris J, Kessler RC, Gazzaniga M, Campbell P, Farah MJ. Nature 2008;456:702–705.
277. Mitler MM, O'Malley MB. In: Kryger MH, Roth T, Diment WC, editors. Principles and Practice of Sleep Medicine. 4th ed. Philadelphia, PA: Elsevier Saunders; 2005. p 485.
278. Lanni C, Lenzken SC, Pascale A, Del Vecchio I, Racchi M, Pistoia F, Govoni S. Pharmacol Res 2008;57:196–213.
279. Markou A, Chiamulera C, Geyer MA, Tricklebank M, Steckler T. Neuropsychopharmacology 2009;34:74–89.
280. Enna S, Williams M. J Pharmacol Exp Ther 2009;329:404–411.

CHOLINERGICS/ ANTICHOLINERGICS/ MUSCARINICS/NICOTINICS

Robert K. Griffith
West Virginia University School of Pharmacy, Morgantown, WV

1. INTRODUCTION

The transmission of impulses from neuron to effector cell throughout the cholinergic nervous system is mediated by the neurotransmitter acetylcholine 1. Cholinergic neurons are integral components of the peripheral autonomic and somatic nervous systems [1] and play important roles in many parts of the central nervous system (CNS) [2]. For excellent reviews of the basics of cholinergic pharmacology, see Refs [1–4]. Chemicals that produce pharmacological effects by mimicking acetylcholine are called cholinergics or parasympathomimetics.

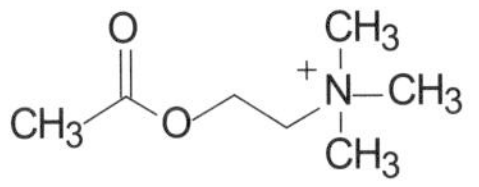

1

Choline is taken into a cholinergic nerve terminal by a sodium-dependent high-affinity transporter where the enzyme choline acetyltransferase acetylates the choline with acetyl coenzyme A to form **1**. The acetylcholine is then taken up into storage vesicles by an ATP-dependent transporter where it is kept until released. Depolarization of the neuron leads to an influx of calcium ions that promotes fusion of the storage vesicle with the cell membrane releasing the neurotransmitter into the synaptic junction with the effector cell. The acetylcholine then binds reversibly to cholinergic receptors on the surface of the effector cell triggering a response by that cell. The released acetylcholine may also bind to presynaptic cholinergic autoreceptors that modulate neurotransmitter release. Removal of acetylcholine from the junction is carried out by the enzyme acetylcholinesterase, anchored in the junction, which very rapidly hydrolyzes the neurotransmitter back to choline and acetate. Inhibition of acetylcholinesterase results in continued stimulation of the cholinergic receptors by acetylcholine and such inhibitors constitute important classes of drugs discussed elsewhere in these volumes. A comprehensive review of the cholinergic synapse is available [5].

Nearly 100 years ago Dale divided acetylcholine actions on different tissues into muscarine-like and nicotine-like based on their selective responses to two alkaloids: muscarine **2**, isolated from the mushroom *Amanita muscaria* and nicotine **3**, isolated from tobacco [6]. He noted that the effects of both muscarine and acetylcholine were blocked by atropine **4**, isolated from *Atropa belladonna*, but the effects of nicotine were not blocked by atropine. At that time Dale postulated, but did not prove, that acetylcholine might be a neurotransmitter. A brief history of discoveries concerning mechanisms of cholinergic transmission and acetylcholine receptors was recently published [7].

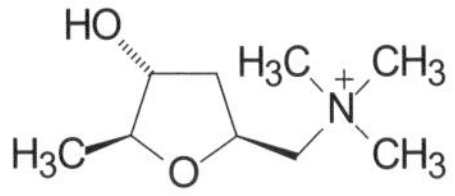

2

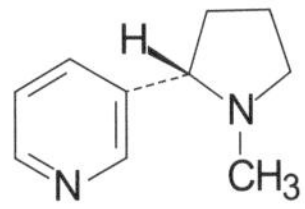

3

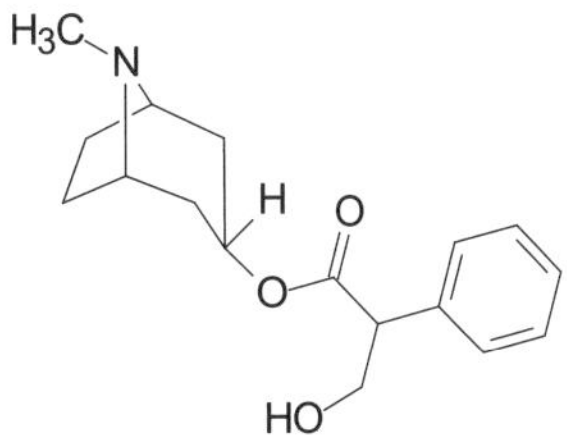

4

In general peripheral muscarinic receptors mediate the actions of parasympathetic target organs such as increasing salivation, gastric secretion, intestinal peristalsis, bladder constriction, pupillary constriction, bronchoconstriction, and pupillary constriction, and decreasing heart rate. Peripheral nicotinic receptors are found postsynaptically on skeletal (striated) muscle fibers and in all autonomic ganglia, both sympathetic and parasympathetic. It was recognized early on that the nicotinic receptor found on skeletal muscle in the neuromuscular junction differed from that located on neurons in the ganglia. There are also numerous muscarinic and nicotinic pathways in the central nervous system of interest in current research efforts as will be discussed below. CNS muscarinic pathways are of particular importance in Alzheimer's disease.

Through the mid-twentieth century and later, differential tissue responses to specific chemicals provided evidence that muscarinic receptors were not all alike, and that there were at least two types of nicotinic receptors, one at the neuromuscular junction and the other on ganglionic neurons. The history of these discoveries is beyond the scope of this chapter.

1.1. Muscarinic Receptors

Both pharmacological techniques and advances in molecular biology and gene cloning led to the discovery of subclasses of muscarinic receptors and determinations of their amino acid sequences. Currently, there are five recognized muscarinic receptors, designated M_1–M_5, reviewed in Refs [8–11]. All muscarinic receptors are members of the superfamily of G-protein-coupled receptors consisting of a single polypeptide chain that threads back and forth through the cell membrane seven times in a roughly circular pattern to form the binding pocket for the neurotransmitter. Muscarinic receptors M_1, M_3, and M_5 are typically coupled to α-subunits of the $G_{q/11}$ family while M_2 and M_4 couple to α-subunits of the $G_{i/o}$ family. The "odd-numbered" receptors generally activate phospholipase C_β mobilizing phosphoinositides and increasing intracellular calcium, while the "even-numbered" receptors inhibit adenylyl cyclase.

M_1 receptors are primarily located in the CNS, particularly in the cerebral cortex, hippocampus, and striatum, and are implicated in the cholinergic hypothesis of Alzheimer's disease [12–14]. Therefore, development of selective M_1 agonists as potential Alzheimer's therapies has been an area of intense research. M_1 receptors may also play a role in schizophrenia [15]. However, M_1 receptors are not exclusively located the CNS and have been found in exocrine glands. They also have a role in modulating ganglionic neurotransmission in the periphery [16].

M_2 receptors are very widespread, found both in the CNS and the heart, smooth muscle, and autonomic nerve terminals. M_2 receptors in the CNS and on autonomic nerve terminals appear to be primarily autoreceptors, but they also can act as inhibitory receptors on dopaminergic neurons and may provide a target for new treatments of schizophrenia [17,18].

M_3 receptors are also widespread in the CNS, although less so than M_1 and M_2. The M_3 receptor is also abundant in smooth muscle, the heart, and salivary glands. Many smooth muscle tissues express both M_2 and M_3 receptors but the M_3 receptors appear to play a predominant role, particularly in the bladder. For this reason selective M_3 antagonists are an attractive target for designing drugs to treat urinary disorders such as overactive bladder (OAB) [19].

M_4 receptors in the CNS are largely associated with dopamine receptors and modulate the effects of dopamine [20]. These receptors may have a role in psychoses [17] and selective M_4 antagonists could play a role in treating Parkinson's disease [21] and perhaps as novel analgesics [22].

M_5 receptors are expressed in low levels in both the CNS and periphery, but are expressed in dopaminergic neurons [23] and appear to play an important role in enhancing dopamine release in the reward areas of the brain. M_5 antagonists may provide novel approaches to treating drug addiction [24]. M_5 receptors also play a role in controlling CNS microvasculature [25] and it has been suggested [10] that selective M_5 agonists might be useful in Alzheimer's therapy.

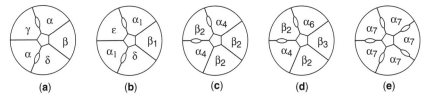

Figure 1. Examples of subunit arrangements of pentameric nicotinic acetylcholine receptors (nAChRs). Arrangement of subunits as seen looking down central ion channel from outside a cell. Small ovals indicate the acetylcholine binding sites. (a) From electric organ of *Torpedo*, first nAChR structure determined, also structure of mammalian fetal skeletal muscular receptor. (b) Adult mammalian skeletal muscle receptor. (c) $\alpha_4\beta_2$ Neuronal receptor, thought responsible for nicotine addiction, most common in CNS. (d) $\alpha_4\alpha_6\beta_2\beta_3$ Neuronal receptor found in the superior colliculus of mammalian brain, one of the highly localized nAChRs. (e) Homomeric α_7 neuronal receptor, also widespread in primate brain.

1.2. Nicotinic Receptors

Nicotinic acetylcholine receptors (nAChRs) are the prototypical ligand-gated ion channel and consist of five protein subunits arranged around a central pore (Fig. 1). nAChRs are members of a family of ligand-gated ion channels known as the Cys-loop family that includes receptors for GABA, glycine, and 5-HT [26]. The Cys-loop receptors all consist of a complex of five subunits arranged around a central ion channel. The subunits are classified into α-subunits, which contain a highly conserved Cys–Cys disulfide bridge between cysteines located 13 amino acids apart, and several additional subunits, β, γ, δ, and ε, that do not possess a Cys-loop. Acetylcholine always binds at the interface between an α unit and an adjoining unit. There are excellent recent reviews that provide details on the nAchR structure and function and leads into the primary literature [27,28]. Unless otherwise noted the following discussion is taken from these reviews.

The initial determination of nicotinic receptor structure and nomenclature, consisting of two α-subunits and β-, γ-, and δ-subunits (Fig. 1a) came from studies of electric fish such as the marine ray *Torpedo* [29]. In mammals, the fetal nAChR has the same α-, β-, γ-, δ-subunits as the electric organ nAChR, but in adult mammals the γ unit is replaced by subunit designated ε, Fig. 1b [30]. To date, the α_1, β_1, δ, and ε units have only been found in muscle cells, all other nAChRs consist of various combinations of α_{2-10}- and β_{2-4}-subunits. Ten different types of α-subunits have been discovered, all having the characteristic Cys-loop, and designated α_1–α_{10}. Only four additional β-subunits have been discovered, designated β_1–β_4, but numerous combinations of α and β pentamers have been discovered throughout the central nervous system [31–34] and have generated enormous interest as therapeutic targets for a variety of CNS disorders, including Alzheimer's, schizophrenia, and drug addiction [35,36]. Figure 1 illustrates just a few of the many ways in which the α- and β-subunits can be combined in addition to the initially determined muscular nAChRs (Fig. 1a and b). For example, Fig. 1c illustrates the $(\alpha_4)_2(\beta_2)_3$ receptor known to be involved in nicotine addiction [37,38]. Figure 1d is a more complex arrangement of $\alpha_4\alpha_6(\beta_2)_2\beta_3$-subunits found in part of the brain (superior colliculus) involved in vision [39], and finally, Fig. 1e illustrates the homopentamer of α_7-subunits found in many regions of the primate brain [40].

Although the vast majority of recent literature on nAChRs is concerned with neuronal receptors, nicotinic receptors have been found in a variety of other cell types including cancer, immune cells, airway epithelial cells, and vascular endothelial cells, raising the possibility of drug design targeting these nonneuronal receptors [41–43].

2. CHOLINERGIC AGONISTS

Over the years, so many thousands of chemicals have been investigated for cholinergic activity that a comprehensive discussion would require a large book and is well beyond the scope of this chapter. Instead, a summary of salient discoveries will be presented with references to reviews of the older literature. Caution must be employed in interpretation of much of the older literature since the pharmacology was usually done on whole tissues or animals. Few conclusions can be drawn beyond muscarinic or nicotinic activity. More detailed discussion will be limited to the more recent literature.

2.1. Acetylcholine Analogs

Ing's paper [44] nicely summarized much of the work from previous years and established his five-atom rule for parasympathetic activity of acetylcholine analogs. The five-atom rule states that for a parasympathetic drug of the type R-$^+$N(CH$_3$)$_3$, in any homologous series, maximal cholinergic activity requires five non-hydrogen atoms attached in a chain to the quaternary ammonium head. The carbonyl is not strictly necessary, nor is the ether oxygen. Maximal activity is obtained with a trimethylammonium. Larger groups on the nitrogen diminish activity and as does replacement of N with different positively charged atoms such as S, P, or As.

Replacement of the acetate ester with a carbamate ester (Table 1) generates carbamoylcholine 5, carbachol, which fits the five-atom rule and but is highly resistant to hydrolysis by acetyllcholinesterase, and is active at both muscarinic and nicotinic receptors [3]. Carbachol is an ophthalmic drug used to treat glaucoma through causing pupillary constriction that opens the canal of Schlemm, relieving intraocular pressure.

Methyl groups beta to the quaternary nitrogen, compounds 6, methacholine and 8, bethanechol provide selectivity for muscarinic sites over nicotinic. Methacholine has been used as a diagnostic agent for asthma and bethanechol to relieve atony of the gut and bladder following surgery. Neither drug is widely used today, although bethanechol remains a useful pharmacological tool because of its muscarinic selectivity and resistance to hydrolysis. Addition of the methyl group introduces stereochemistry not present in 1. The S-(+)-isomer of methacholine is 240 times more active on guinea pig ileum than the R-(−)-isomer. For bethanechol, the S-isomer is 915 times more active than the R-isomer [45]. The α-methyl analogs 7 and 9 are more potent at nicotinic sites than muscarinic, although much less active than 1 [46]. However, stereoselectivity at nicotinic sites was minimal and not consistent between preparations [47]. Similarly, the weak muscarinic activity showed only an eightfold difference in favor of the S-stereoisomer.

2.2. Muscarinic Agonists

In recent years most of the interest in muscarinic agonists has centered on finding potent, but highly selective M$_1$ agonists for use in treating Alzheimer's disease. Muscarine itself, 2, stimulates all five mAChRs.

The natural stereoisomer of muscarine (2S,3R,5S), as drawn above, is the only one of the eight possible isomers that has significant cholinergic activity [48]. As expected, this has the same stereochemistry as the more active S-isomer of methacholine 6. Muscarone 10, even more closely resembles 1 but has nicotinic activity not seen in muscarine.

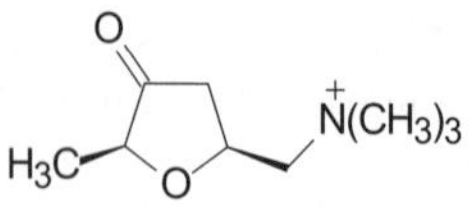

Table 1. Acetylcholine 1 and Functional Analogs

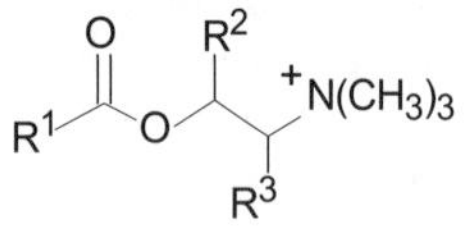

	R^1	R^2	R^3
1	CH$_3$	H	H
5	NH$_2$	H	H
6	CH$_3$	CH$_3$	H
7	CH$_3$	H	CH$_3$
8	NH$_2$	CH$_3$	H
9	NH$_2$	H	CH$_3$

10

Muscarinic activity of dioxolane 11 resides in the *cis*-isomers with the C4-R enantiomer (drawn) more than 100 times as potent as the C4-*S* [49].

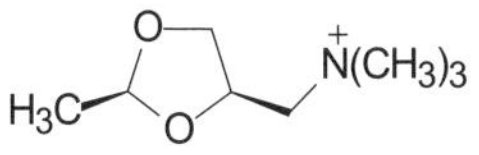

11
H₃C—[dioxolane]—CH₂N⁺(CH₃)₃

Oxathiolane analog **12** was synthesized and both the *cis*- and *trans*-isomers evaluated for muscarinic activity [50]. Again, the (+)-*cis*-isomer was the most potent in the series and has the same absolute configuration as muscarine and the most potent isomer of dioxolane **11**.

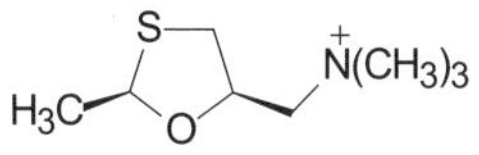

12
H₃C—[oxathiolane]—CH₂N⁺(CH₃)₃

The same group also made the sufoxide derivatives of **12**, compounds **13**, and **14** [51]. Since the sulfoxides are chiral, an additional stereocenter is introduced. Compound **14**, the (2R,3R,5R) isomer has the same absolute configuration as muscarine, **2**, and is a potent muscarinic agonist but without subtype selectivity.

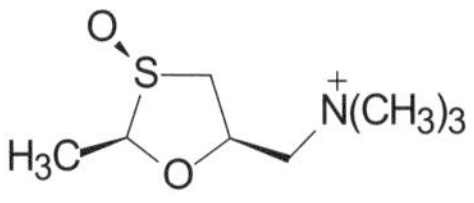

13

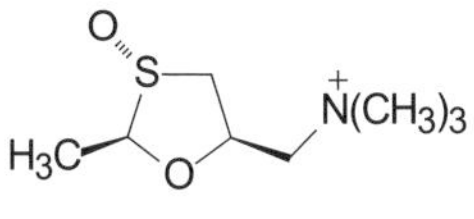

14

This Italian group continued synthesizing and evaluating oxathiolane and dioxolane and wrote extensive reviews of their work in 1995 and 1998 [52]. More recently, in an attempt to improve receptor selectivity they synthesized a series of *N*-methyl pyrrolidine substituted dioxolanes and oxathiolanes, **15** and **16**, and quaternized analogs **17** and **18** [53].

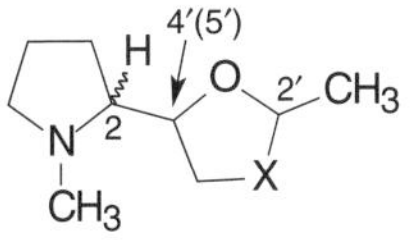

15 X = O
16 X = S

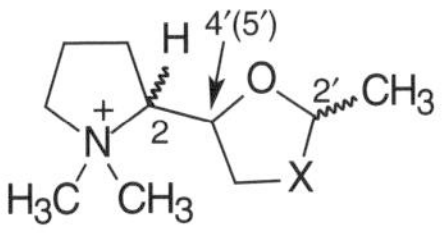

17 X = O
18 X = S

Two members of the series, **17** with (2R,4′S,2′S) stereochemistry and **18**, stereochemistry (2R, 5′S,2′S) showed some promise with M_2 selectivity as a partial agonist. They continued these studies with the corresponding sulfoxides, which introduced another chiral center and obtained another partial agonist selective at M_2 receptors [54]. In a more recent study, the group substituted furan rings for the dioxolane and oxathiolane rings of the previous study and obtained compounds that had no nicotinic activity, but no selectivity for a particular muscarinic receptor [55].

An oxathiolane attached to a quinuclidine bicyclic system, **20**, first reported as AF102B [56], was identified initially as an agonist of muscarinic heteroreceptors facilitating the release of dopamine. Compound **20** was subsequently identified as an M_1 agonist that attenuated chemically induced cognitive dysfunctions in animal models of Alzheimer's disease [57,58]. Eventually, following observations of the effect of AF102B on salivation [59,60], which is likely an M_3 agonist effect, the compound became marketed as cevimeline for the treatment of xerostomia (dry mouth) and Sjogren's syndrome. Cevimeline has entered clinical trials as a treatment for Alzheimer's disease.

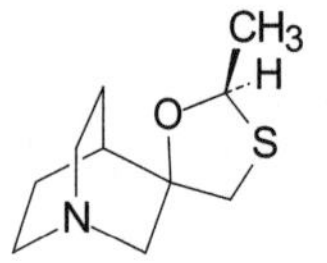

20

The quinuclidine dioxolane **21** was synthesized [61] and found to be a partial muscarinic agonist.

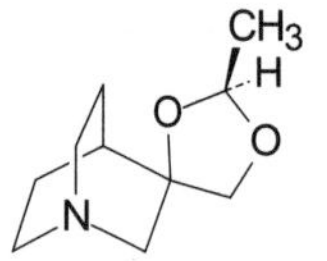

21

Arecoline, **22**, is isolated from the betel nut, seeds of *Areca catechu* [62]. The plant is cultivated in Southeast Asia and people often chew the seeds which causes profuse salivation due to arecoline's muscarinic agonist activity. Arecoline has been demonstrated to improve cognitive functions when given i.v. to persons with presenile dementia [63–65]. However, the rapid ester hydrolysis of arecoline leaves much room for improvement and has led to an extensive program of synthesis of non-ester arecoline bioisosteres in attempt to overcome this problem.

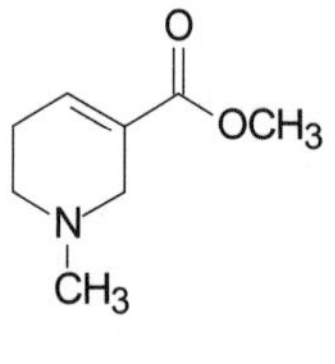

22

An series of compounds employing substituted isoxazole bioisosteres of arecoline, was prepared and evaluated as muscarinic agonists [66]. Compound **23** was the most active in the series.

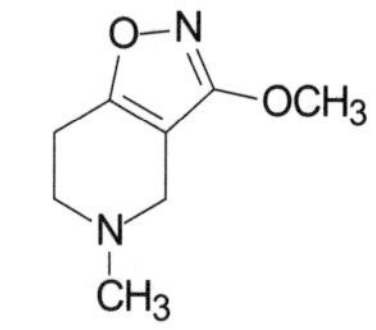

23

Spirodioxolane analogs of arecoline were prepared and in the series compound **24** was found to be an excellent muscarinic agonist comparable in efficacy to arecoline [67].

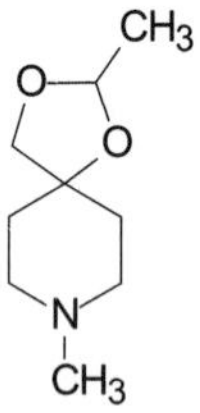

24

An extensive series of quinuclidinyl heterocycles were prepared and oxadiazole **25** was the most efficacious muscarinic agonist [68].

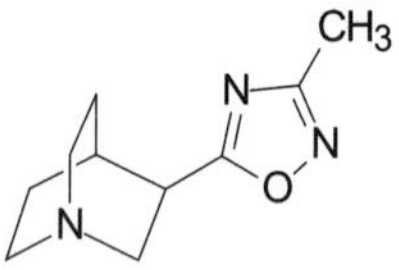

25

Two additional analogs of arecoline that have demonstrated great potential in Alzheimer's therapy are compounds **26**, AF150(S) and **27**, AF267B (sometimes referred to as NGX267), revealed together in 2000 [69,70].

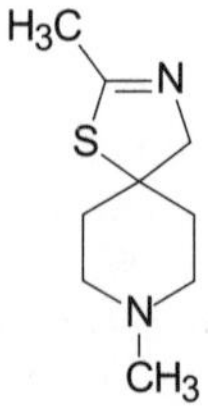

26

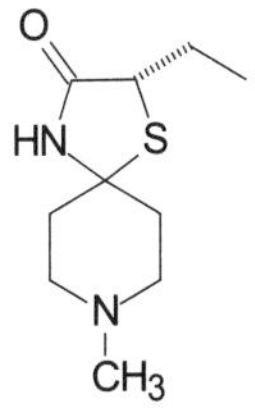

27

Both **26** and **27** are exceptionally effective in animal models of Alzheimer's disease [71] and **27** has recently entered clinical trials in humans [72].

In examining a series of oxime analogs of arecoline, compound **28** was the most potent in the series, being two or three orders of magnitude more potent than arecoline [73]. After promising tests in animals including primates, **28** was given the generic name milameline and entered clinical trials as potential therapy for Alzheimer's disease [74,75].

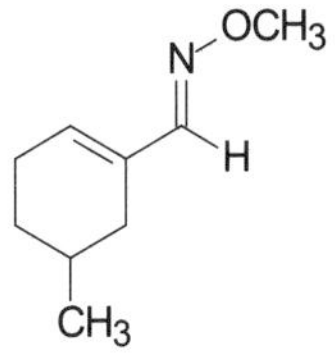

28

Another series of arecoline analogs, thiadiazoles, afforded the potent muscarinic agonist **29** [76]. This compound was also sufficiently potent as a selective M_1 agonist [77] to warrant a generic name xanomeline and enter clinical trial for Alzheimer's disease [78]. In addition to promising activity in Alzheimer's, xanomeline has recently shown promise in treating schizophrenia [79].

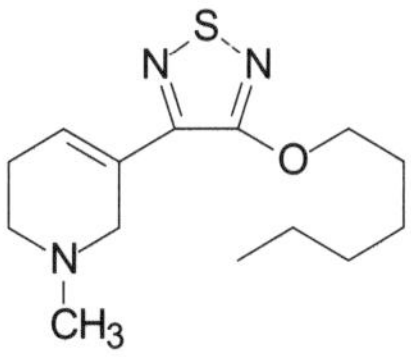

29

Modification of the aldoxime ether of **28** with an electron withdrawing nitrile group, and attaching the oxime to a quinuclidine afforded a series of compounds in which **30** was a selective partial agonist of M_1 receptors [80]. Although **30** was promising enough in animal experiments to warrant the generic name sabcomeline [81], it never entered clinical trials in humans.

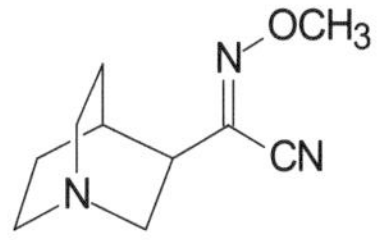

30

Compound **31** also demonstrated excellent selectivity for M_1 receptors [82] and quickly entered clinical trials in humans [83]. As talsaclidine, compound **31** continued to show promise in Alzheimer's disease [84], but was never brought to market.

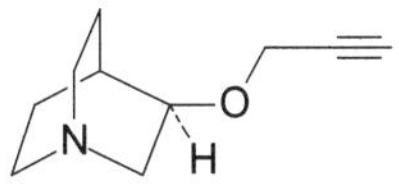

31

In a departure from the recent emphasis on muscarinic agonists for Alzheimer's therapy, Sauerberg et al. investigated a large series of thioether analogs of xanomeline as analgesics [85]. The most potent of the compounds synthesized was butylthioether **32**. The analgesic effect was clearly mediated by central muscarinic receptors, but the type of mAChR was not clear.

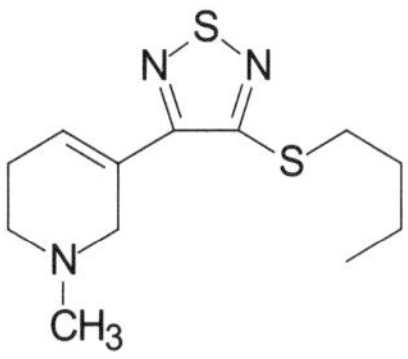

32

Pilocarpine, **33**, isolated from the South American plant *Pilocarpus jaborandi* and other *Pilocarpus* species, has been used as an ophthalmic for treating glaucoma since the late nineteenth century [3]. Pilocarpine is still in use, although far from a first-line therapy for glaucoma. Like carbachol, **5**, it acts on the cilary muscle of the eye to cause pupillary constriction that opens the canal of Schlemm and relieves intraocular pressure. Pilocarpine is primarily muscarinic in its actions but has not led to an extensive analog synthesis program.

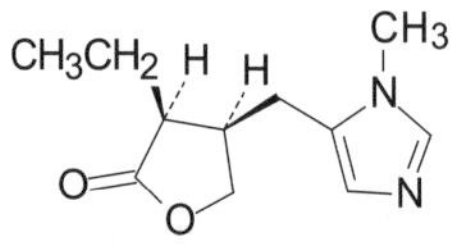

33

Oxotremorine **34** is another classic general muscarinic agonist that is equipotent with acetylcholine but lacking any nicotinic effects [86]. Oxotremorine has no therapeutic utility, but has long been a useful pharmacological tool. The potential for selective muscarinic M_1 agonists to have therapeutic potential in treating Alzheimer's disease led to a renewed interest in oxotremorine analogs. The previous edition of this work summarized the preparation of hundreds of analogs, none of which had therapeutic potential [87].

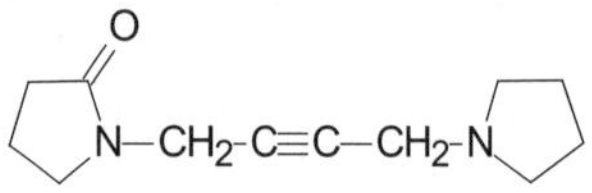

34

None of the muscarinic agonists derived from known natural agonists have proven as successful in the clinic as hoped, a newer approach through allosteric activators shows promise. The chemical called TBPB **35** by its discoverers shows exceptionally specific agonism at M_1 receptors [88]. The authors suggest that since the orthosteric binding site of all mAChRs are similar that enhanced selectivity can best be obtained through allosteric activation.

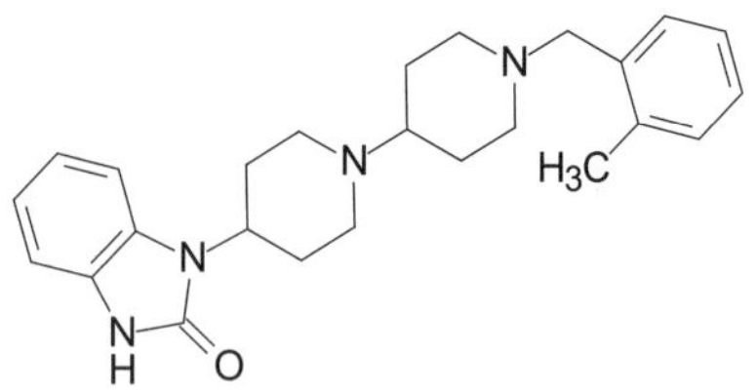

35

Two subsequent papers described the synthesis of numerous additional analogs of **35** [89,90]. Although none of the analogs were markedly superior to TBPB, many of them were comparable and provide leads to additional compounds.

2.3. Nicotinic Agonists

Natural nicotine, **3**, is the *S*-stereoisomer and the *R*-isomer is notably less potent [91]. Enlarging the pyrrolidine ring to a piperidine ring **36** affords anabasine, a minor component of tobacco that has approximately 1/10 the activity of nicotine [92]. Contracting the ring to an azetidine **37** affords a compound with the same affinity as nicotine in a CNS binding assay, but shows greater behavioral effects in the rat [93].

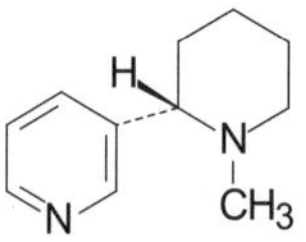

36

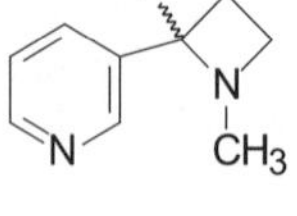

37

The pyrido[3,4-*d*]azepines **38a** and **38b** showed high affinity for nicotinic receptors in rat brain [94], but the pyridyl azepines **39a** and **39b** had much less affinity for the receptors. The same authors reported additional nicotine analogs including aminoalkyl, aminopropenyl, aminopropynyl, and aminoethoxy.

Some bound CNS nAChRs but no structure–activity conclusions could be reached.

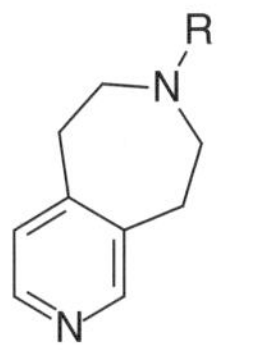

38a. R = H
38b. R = CH$_3$

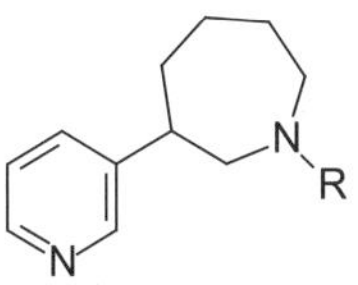

39a. R = H
39b. R = CH$_3$

Epibatidine, **40**, was isolated from the skin of the Equadoran poison frog *Epibatobades tricolor* by Daly's group and reported to have analgesic activity 200–500 times morphine, an effect not blocked by the narcotic antagonist naloxe [95]. Subsequently, **40** was found to be a powerful nicotinic agonist binding about 30 times more strongly to neuronal nicotinic receptors than does nicotine [96]. The extraordinary potency and selectivity for neuronal nAChRs gave a new impetus to nicotinic agonist research. Epibatidine binds to the $\alpha_4\beta_2$ nAChR with a 1–2 nM affinity, but also binds other nAChRs such as the ganglionic $\alpha_3\beta_4$ and homomeric α_7 [97]. A detailed review of the neuronal nicotinic agonists measured for their activity at $\alpha_4\beta_2$ through about the year 2000 is available [98].

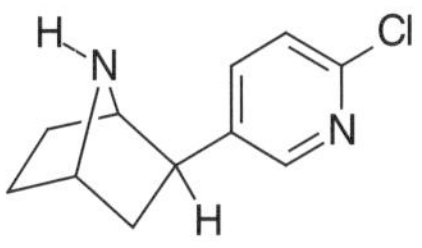

40

The novel activity of epibatidine inspired the synthesis of numerous analogs including the azetidine **41** ABT594 [99,100]. ABT594 with a K_i at $\alpha_4\beta_2$ nAChRs of 0.05 nM is much more specific for those receptors and so lacking many of the toxicities of epibatidine yet remains some 70 times as potent as morphine. Compound **41** entered clinical trials as tebanicline.

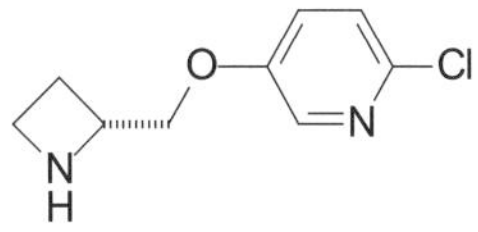

41

The chlorines on epibatidine and ABT594 are not necessary for binding activity as compounds **42** and **43** both have subnanomolar binding to brain receptors but the chlorine does contribute to analgesic activity as these two compounds lack analgesic activity [101].

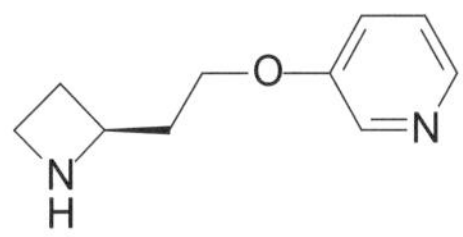

42

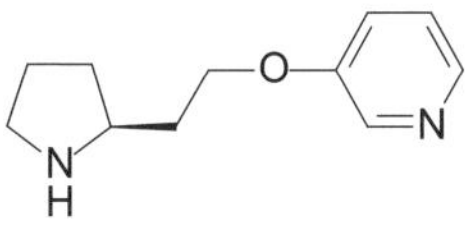

43

Another analog of epibatidine, **44**, DBO-83 with a $K_i = 4$ nM at $\alpha_4\beta_2$ nAChRs [102] showed strong antinociceptive activity.

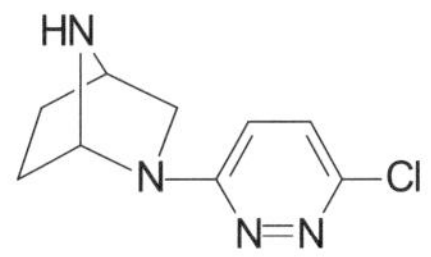

44

An isoxazole analog of epibatidine, **45**, epiboxidine, retained many of the properties of epibatidine binding [103].

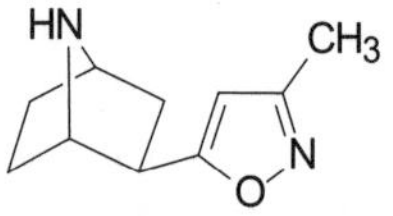

45

The toxin produced by certain freshwater cyanobacterium, anatoxin-a **46** is a potent agonist at nicotinic receptors [104]. Anatoxin-a binds most tightly to the $\alpha_4\beta_2$ form with a K_i of 0.61 nM [98].

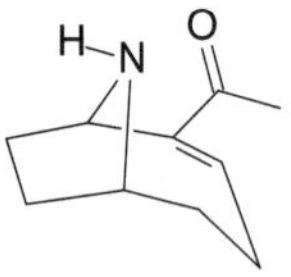

46

An interesting compound that combines the chloropyridine feature of epibatidine with the bicyclic ring system of anatoxin-a is UB-165, **47** [105,106]. UB-165 binds the $\alpha_4\beta_2$ receptor at K_i 0.17 nM [98]. Several analogs have been made and evaluated against other cloned neuronal nAChRs (see Refs 107 and 108).

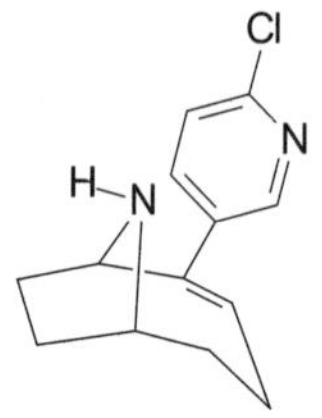

47

Another toxic plant alkaloid is cytisine, **48**, isolated from *Laburnum anagyroides* is a selective $\alpha_4\beta_2$ partial agonist, $K_i = 0.12$ nM, but like epibatidine does bind $\alpha_3\beta_4$ receptors, $K_i = 54$ nM, and α_7 K_i 250 nM [109]. A number of analogs have been prepared [110–112]. Cytisine has been used for years in Eastern Europe in smoking cessation programs that made it the lead compound in the development varenicline **49**, recently approved in the United States for smoking cessation. The rationale for using a partial agonist and the reasoning behind varenicline has been thoroughly reviewed [113–115].

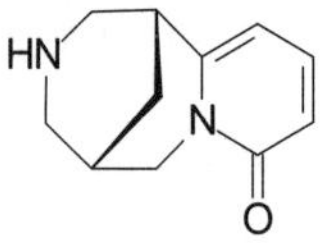

48

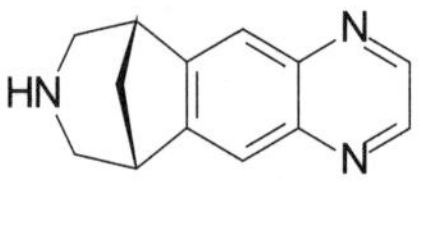

49

Another series of modified cytisine analogs was prepared and led to racemic methyl cytisine **50a** and hydroxymethyl analog **50b** [116]. These simple modifications of cytisine afforded the most selective $\alpha_4\beta_2$ compounds known. The affinity ratio of K_i values for $\alpha_3\beta_2/\alpha_4\beta_2$ is over 3500.

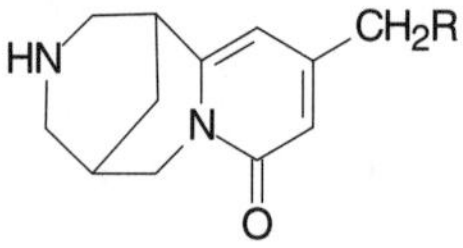

50a R = H

50b R = OH

More recently still another series of *N*-substituted cytisine analogs have been reported [117]. The most potent in this series were derivatives **51a–d**, all of which had K_i values less than 10 nM, and several thousand-fold selectivity for $\alpha_4\beta_2$ receptors over α_7 homopentamers.

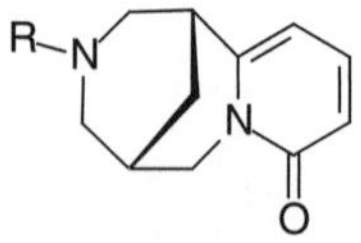

51a R = (CH$_2$)$_2$–4-pyridyl
51b R = (CH$_2$)$_2$–CO-C$_6$H$_5$
51c R = (CH$_2$)$_2$–CO-C$_6$H$_4$-4F
51d R = (CH$_2$)$_2$–CO-CH$_3$

Another simpler series of selective $\alpha_4\beta_2$ agonists was found that afforded **52**, the most selective $\alpha_4\beta_2$ agonist discovered at that time with a $K_i = 0.95$ nM. This compound and a few others in the series were reported to be under consideration for brain PET studies [118].

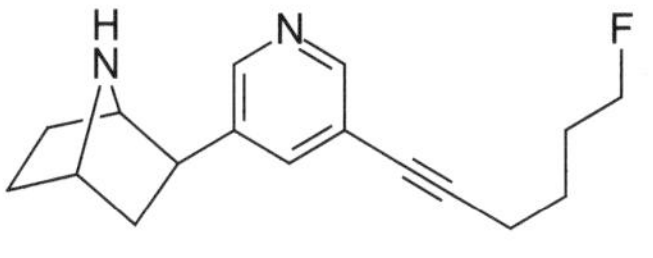

52

There is substantial evidence that α_7 nAChR agonists may have utility in treating schizophrenia [119]. Therefore considerable effort has been expended developing such agonists. Quinuclidine **53** was the first compound discovered that was highly selective α_7 agonist [120]. In rat behavioral models **53** improved sociability, a marker for potential use in schizophrenia [121].

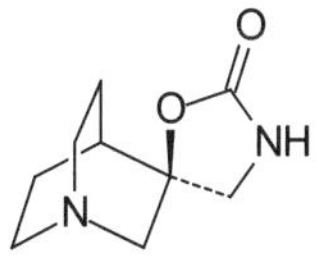

53

Further substitutions on the oxazolidine ring of **53** afforded a new series of which thiophene derivative **54** was the most potent, $K_i = 9$ nM and very selective for α_7 receptors over other nAChRs but inhibited CYP2D6 which made it an unlikely candidate for clinical trial [122]. Further modification in another series afforded **55** that is also a potent α_7 agonist, $K_i = 3$ nM, but does not exhibit problematic inhibitions of metabolic enzymes [123].

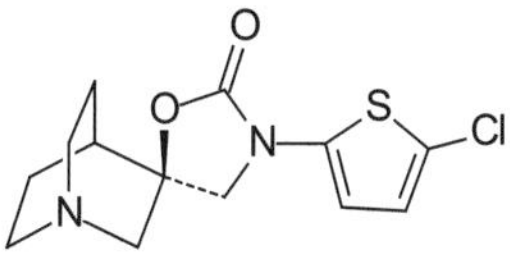

54

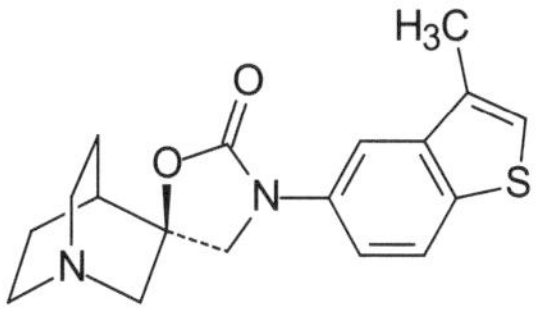

55

Another group prepared a series of quinuclidine amides that resulted in additional α_7 agonists such as **56** (PNU-282987), which was the best in the series measured as an EC_{50} of 154 nM [124]. It was highly selective versus ganglionic nAChR $\alpha_3\beta_4$, and neuromuscular receptor form $\alpha_1\beta_1\gamma\delta$.

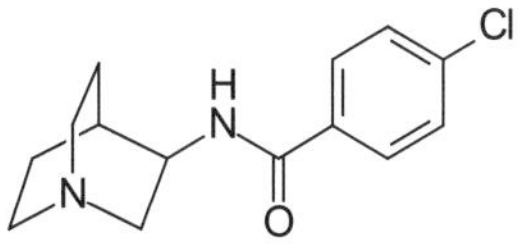

56

Continuing the same series, the same group discovered compound **57** that had an EC_{50} of 65 nM and also showed excellent selectivity for α_7 receptors versus ganglionic and neuromuscular nAChRs [125].

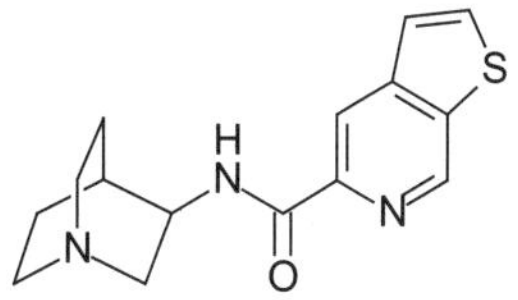

57

A *Journal of Medicinal Chemistry* Perspective article was written in 2008 and provides a good overview of the field of allosteric α_7 selective agonists [126].

More recently diazabicyclo[3.0.3] compound **58**, R5 = R6 = H was used as the starting point for a study that found some very interesting structure–activity relationships for selective binding of either $\alpha_4\beta_2$ or α_7 binding [126]. Incorporation of substituents such as a halogen or small alkyl ether at R5 provides highly selective $\alpha_4\beta_2$ ligands whereas substituting an aryl group provides α_7 selectivity.

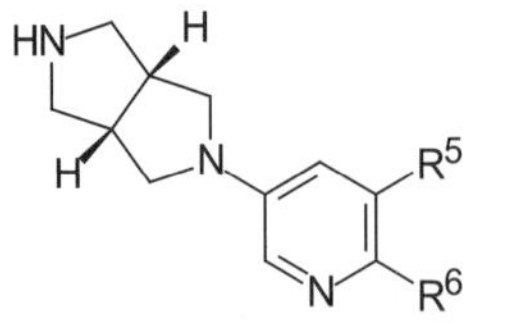

58

3. CHOLINERGIC ANTAGONISTS

From the early twentieth century muscarinic and nicotinic receptors were differentiated by their responses to antagonists. Although both broad classes respond to acetylcholine as the natural agonists, there does not seem to be any compound that is an antagonist at both classes.

3.1. Muscarinic Antagonists

The prototypical muscarinic antagonist, or parasympatholytic, is the racemic compound atropine 4, an alkaloid found in *A. belladonna*, deadly nightshade, as the chiral compound *S* (−)-hyoscyamine, **59**. *Atropa belladonna* is a member of the family of Solanaceae that includes *Hyocsyamus niger* (black henbane) and *Datura stramonium* (jimson weed, jamestown weed, or thornapple). Hyoscyamine becomes racemized to atropine during the typical extraction process. Atropine, hyoscyamine, and related natural products and synthetic analogs that are in current clinical use are shown in Fig. 2. Scopolamine **60** is also found in the solanaceous plants. It has more CNS effects than atropine, and is commonly used to prevent motion sickness. Hyoscyamine, atropine, and scopolamine are esters of tropic acid and tropinol. Homatropine **61** is a semisynthetic derivative made from tropinol and mandelic acid. Homatropine is employed primarily as a mydriatic for eye exams. The methylated quaternary salts **62** and **63** of atropine and scopolamine retain the antimuscarinic properties of the parent, but do not cross the blood–brain barrier.

Compound **64** is ipratropium, a bronchodilator administered by inhalation, used to treat asthma and chronic obstructive pulomonary disease (COPD). Tiotropium **65** is used for the same indications as **64**, but is both more potent and has a much longer duration of action. The latter is attributed to a much slower dissociation from lung M_3 receptors than ipratropium [127].

In a very recent report atropine analog **66** was reported as part of a series of novel mAChR antagonists intended for use in COPD [128]. The authors reported promising preclinical studies in mice favorable for an inhaled drug.

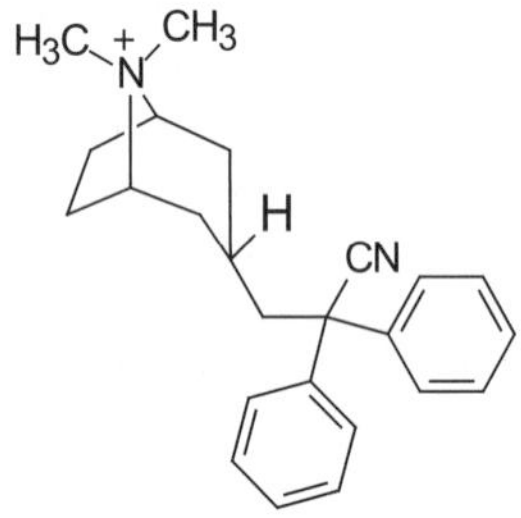

66

The tropane ring is not necessary for antimuscarinic activity and most synthetics do not contain it. The SAR for "classic" antimuscarinics follows the general pattern of **67** in Fig. 3 and best illustrated with examples.

In classical antimuscarinics R^1 is a hydrogen bonding group, usually OH and X is an ester, both of which are present in the prototypical antimuscarinic, atropine. Oxyphenonium **68**, penthienate **69**, and glycopyrrolate **70** fit this pattern. However, both are not necessary as can be seen in procyclidine **71** and biperiden **72** that have only the alcohol. Methantheline **73** and piperidolate **75** have only the ester. The X group may lay outside the direct line between the ends as in procyclidine **74**, R^2 and R^3 must be lipophilic rings, and one is often aromatic and the other aliphatic, but both may be aromatic as in **73**, **74**, and **75** piperidolate. The chain to the nitrogen is often just two methylenes, but can be part of a ring as in **70** and **75**. Finally, R^4, R^5, and R^6 can

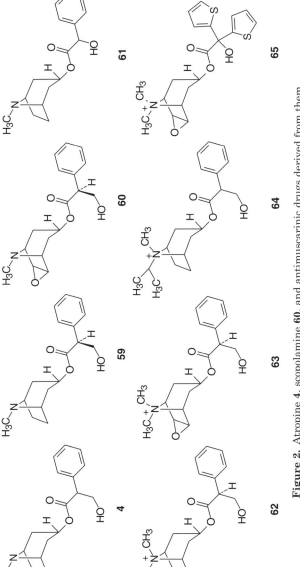

Figure 2. Atropine 4, scopolamine 60, and antimuscarinic drugs derived from them.

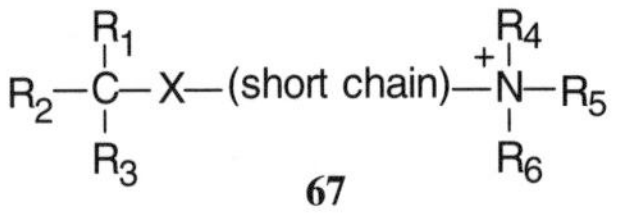

Figure 3. Structure–activity relationships for synthetic antimuscaringics (parasympatholytics). See text for discussion.

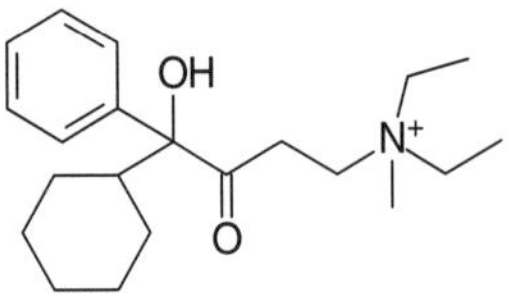

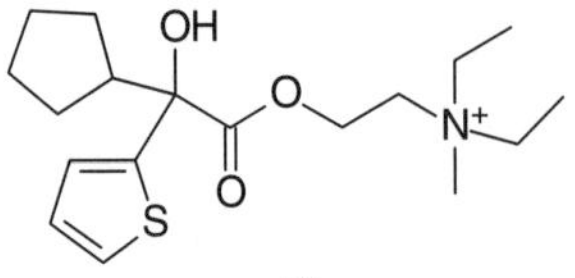

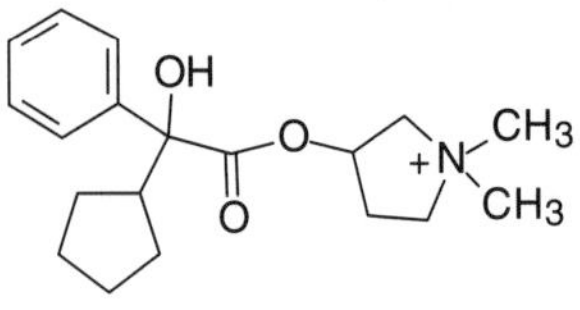

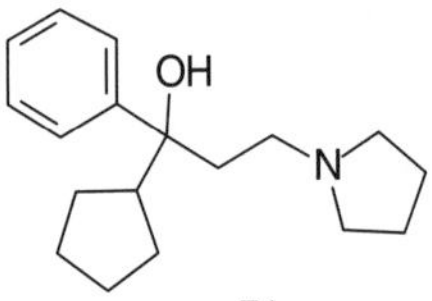

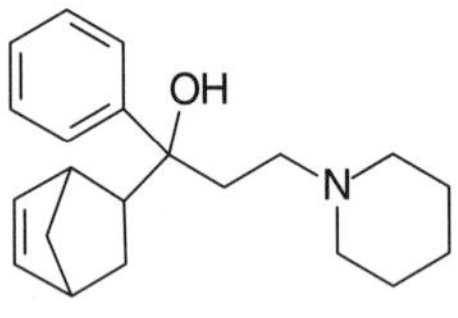

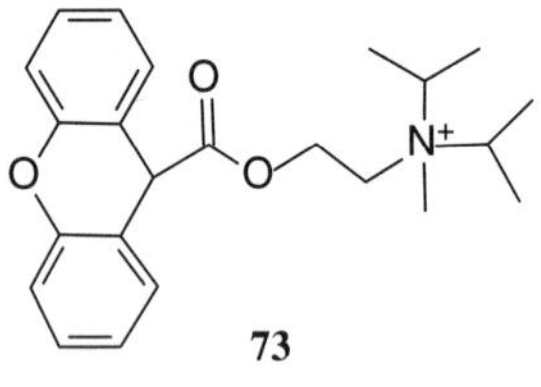

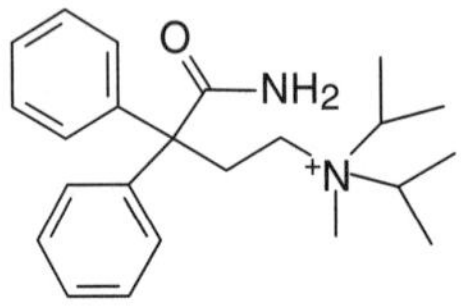

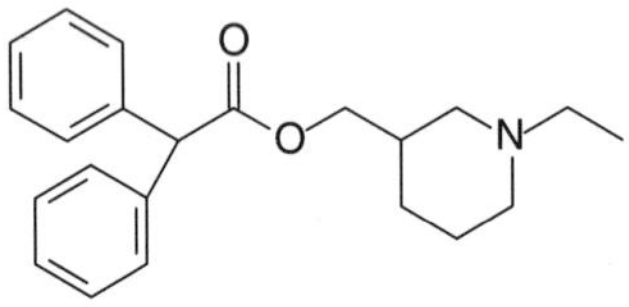

be three alkyl groups forming a quaternary salt, or one of them can be the hydrogen of the ionized form of a tertiary aliphatic amine. The nitrogen must be positively charged for binding to the cholinergic receptor. The nitrogen may be part of an aliphatic ring. The compounds drawn here are merely illustrative of the dozens of classical antimuscarinics that came on the market in the middle of the twentieth century. As better drugs have been invented, many of the classical antimuscarinics have fallen by the wayside. They are not selective for specific mAChRs, and have many side effects including dry mouth, constipation, and blurred vision, all as a result of muscarinic antagonism at nontarget tissues.

Among the first selective mAChR antagonists to be developed are the selective M_3 antagonists employed to treat overactive bladder, a serious quality of life issue for many in an aging population. The drugs currently employed to treat OAB are oxybutynin **76**, tolterodine **77**, darifenacin **78**, solifenacin **79**, and trospium **80**. In a recent review, Hegde reported that **77** and **80** do not show selectivity for M_3 receptors in lab screenings compared to the others, and only **78** is truly selective [129]. Nevertheless **77** does seem to be efficacious in the clinic. Dry mouth due to inhibition of salivary secretion is a problem with all of them. Fesoterodine **81** has just been approved for treatment of OAB, but it also does not appear to be a selective agent having antagonist action at all five mAChRs [130]. It is a prodrug and the ester will be hydrolyzed to the active compound.

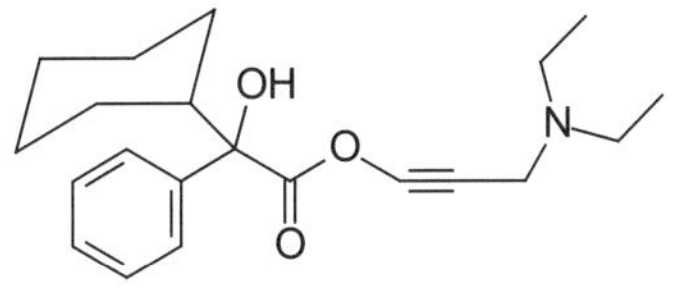

76

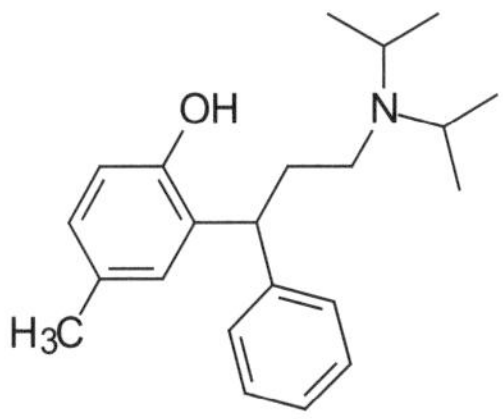

77

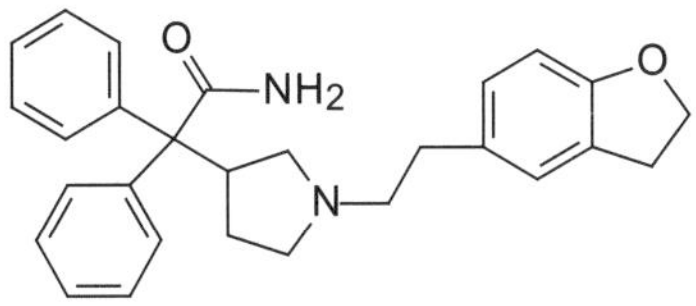

78

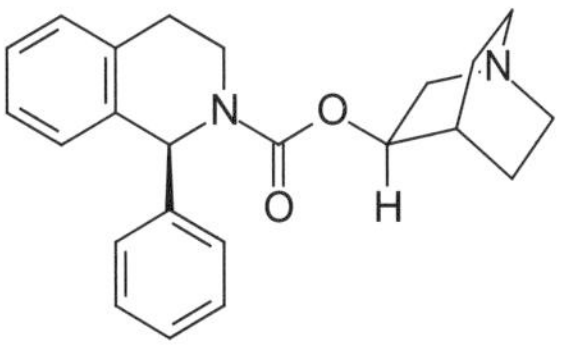

79

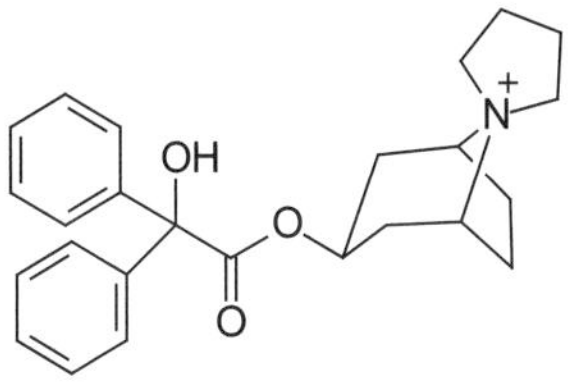

80

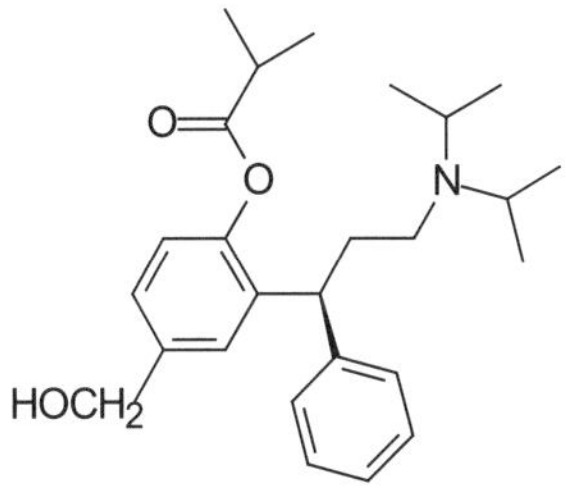

81

There are additional M₃ antagonists being developed for overactive bladder. A series of alkyne-quinuclidines has been prepared with the most active being compound **82** [131]. In tests on isolated guinea pig bladder, **82** compared favorably with the drugs currently on the market for treating OAB.

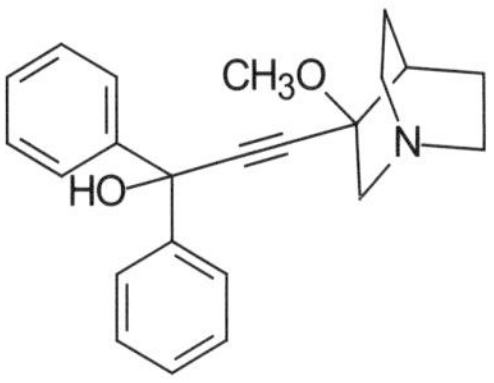

82

Since M₃ receptors also mediate bronchoconstriction, M₃ antagonists have potential application in asthma or COPD. The same group that made **82** synthesized another series of alkyne-quinuclidines targeted toward the lung [132]. One of the most active in this series was **83**. Replacement of the aromatic rings of **82** with large cycloalkanes as in **83** provided longer duration of action through slower diffusion off the receptor.

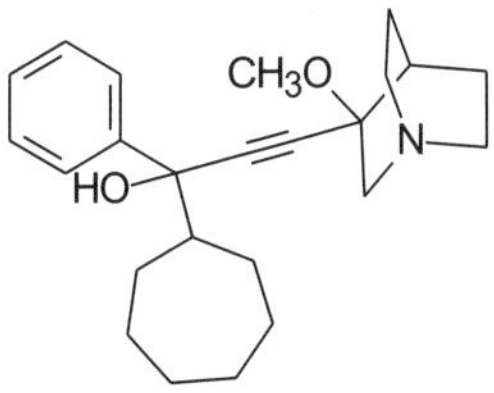

83

Another group searching for selective M_3 antagonists developed a series of biphenyl piperazines, reported in two papers [133,134]. Their lead compound **84** was discovered through high-throughput screening of their in-house compound collection. A solid phase synthesis procedure was used to synthesize analogs leading eventually to **85** that was reported to have 500-fold selectivity for M_3 over M_2. The compounds are being targeted for treatment of COPD.

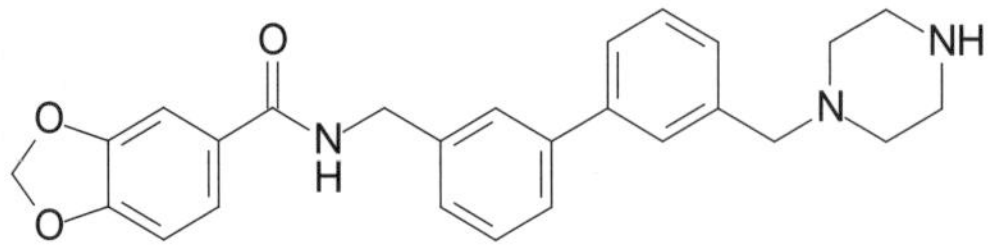

84

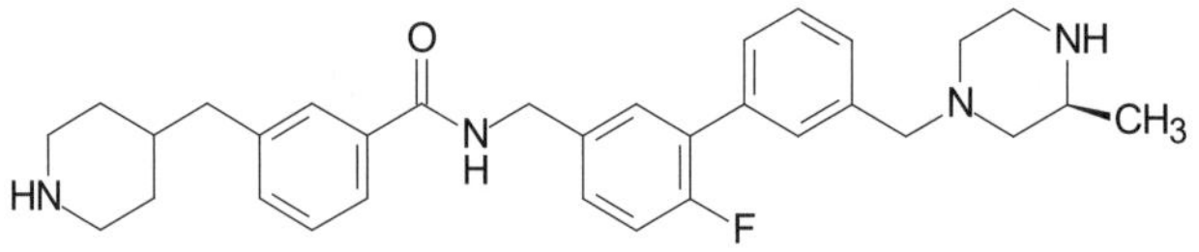

85

Still another approach to selective M_3 antagonists was through a series of tyrosine ureas [135]. Once again a high-throughput screening of an in-house compound library afforded lead **86**. Subsequent optimization led to substituted thiophene **87**, with reported 100-fold selectivity for M_3 over M_2.

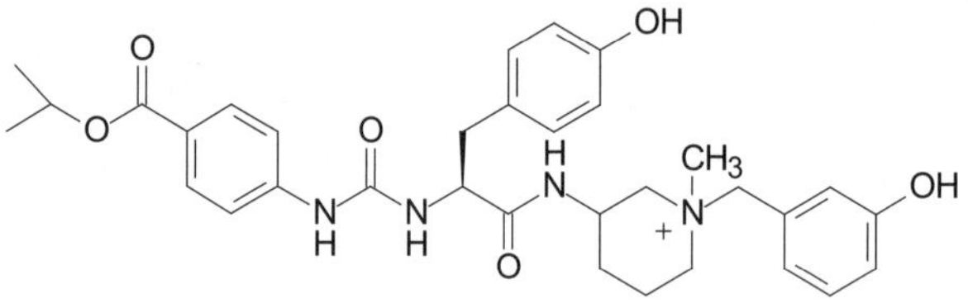

86

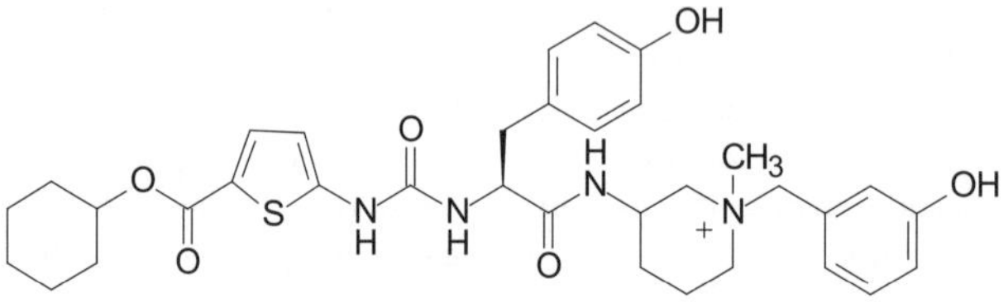

87

In 1987, a group developed a set of very large substituted polyamines as muscarinic antagonists, one of which, **88** became known as methoctramine [136]. In subsequent years, the group worked on developing more selective antagonists and created FC-20-94 **89**, spirotramine **90**, and tripitramine **91** [137].

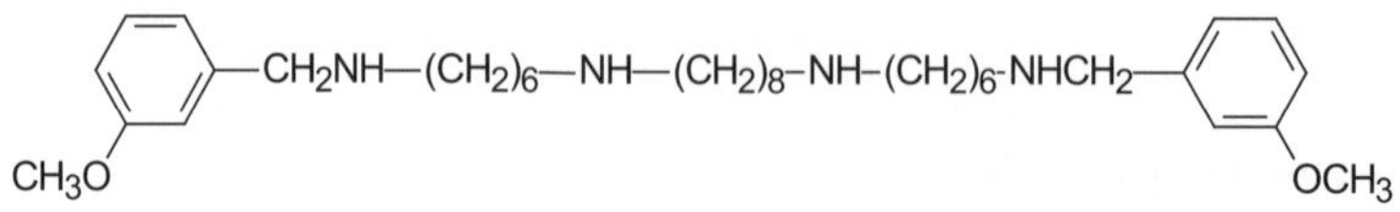

88

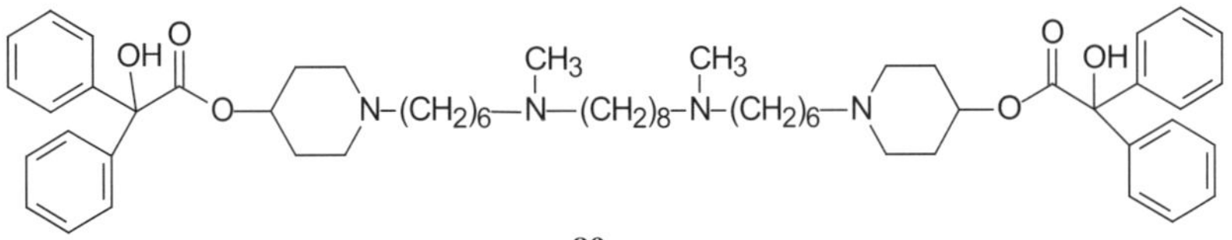

89

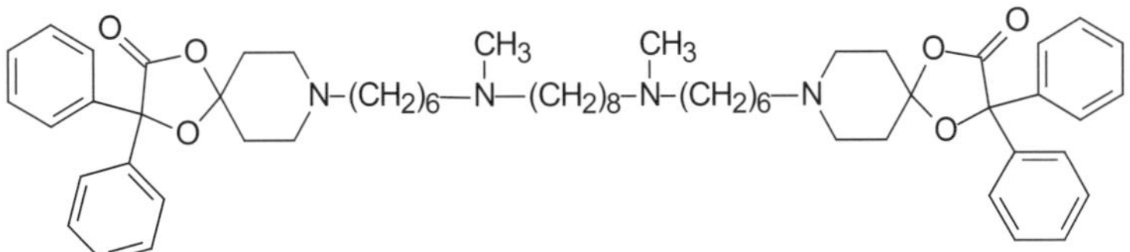

90

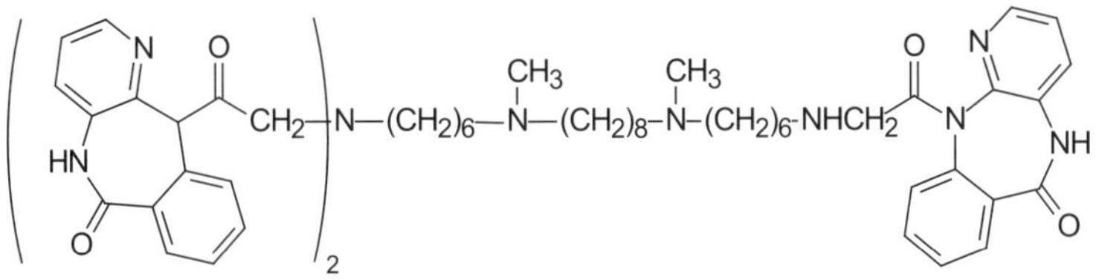

91

Methoctramine **88** is a selective M$_2$ antagonist. But while selective for M$_2$ over other muscarinic receptors, it also bound to a variety of nonmuscarinic receptors. Compound **89** is selective for M$_2$ and has a better profile than **88**. The best compounds in their series were **90**, selective for M$_1$ receptors while **91** is a selective M$_2$ antagonist. The polyamine compounds have proven useful tools in identifying receptors in specific tissues [138].

3.2. Nicotinic Antagonists

The first definitive differentiation of subclasses of nicotinic receptors was between the neuromuscular nicotinic receptor stimulated and then blocked by decamethonium **92** and the ganglionic nicotinic receptor blocked by hexamethonium **93** [7,139]. Of course the prototypical neuromuscular blocker is D-tubocurarine **94a**, isolated from the South American plant *Chondrodenron tomentosum*. Crude extracts of the plant, called curare, were being used by South American Indians to make poison darts and arrows at the time Europeans first encountered them in the early sixteenth century [140]. The distance between the positively charged nitrogens was the principle factor in separating a ganglionic blocker from a neuromuscular blocker. However, the smaller decamethonium produces an initial depolarization followed by block, whereas tubocurarine is nondepolarizing.

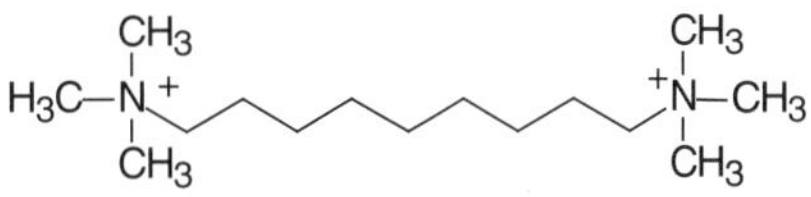

92

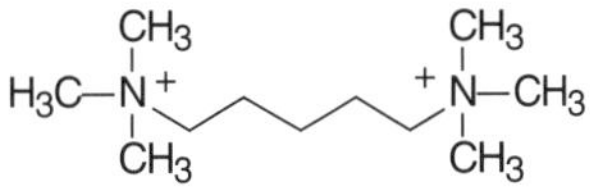

93

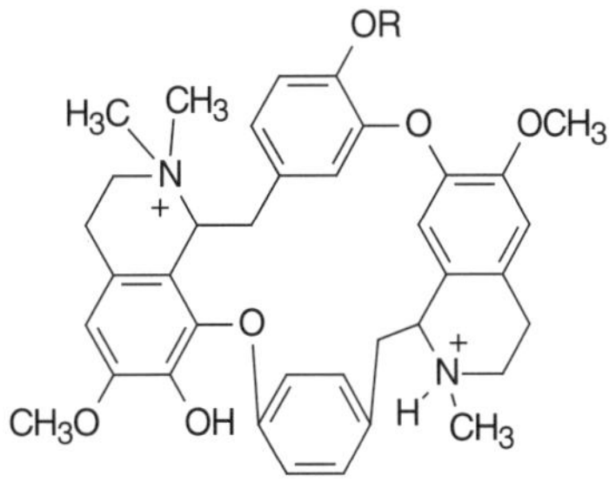

94a R = H, D-tubocurarine
94b R = Ch3, Metocurine

Table 2. Aminosteroid-Based Neuromuscular Blockers Currently in Clinical Use

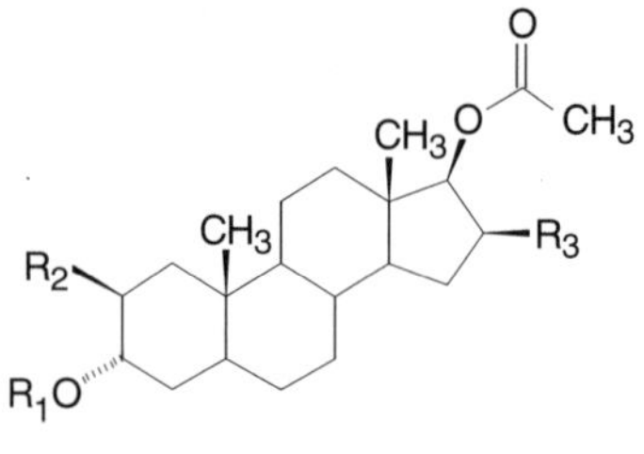

	Compound #	R_1	R_2	R_3
Pancuronium	98	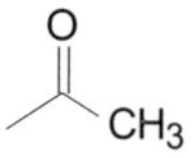 -OC(O)CH₃	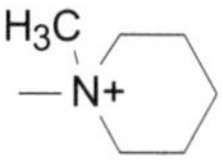 H₃C-N+(piperidine)	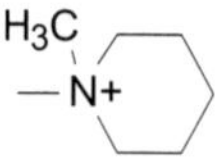 H₃C-N+(piperidine)
Pipercuronium	99	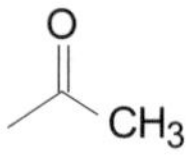 -OC(O)CH₃	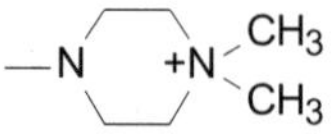 -N(piperazine)-+N(CH₃)₂	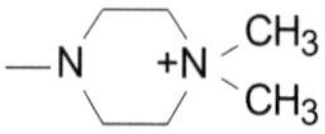 -N(piperazine)-+N(CH₃)₂
Vercuronium	100	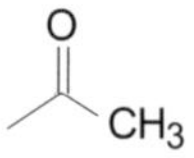 -OC(O)CH₃	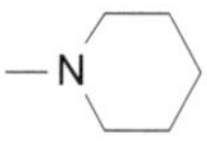 -N(piperidine)	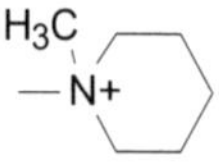 H₃C-N+(piperidine)
Rocuronium	101	H	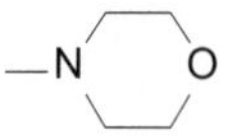 -N(morpholine)	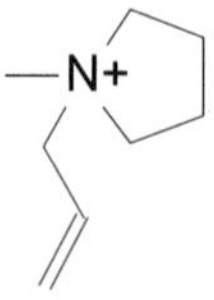 -N+(pyrrolidine with allyl)

3.2.1. Ganglionic Nicotinic Antagonists

Blockade of autonomic ganglia decreases sympathetic output in the periphery, and can thus lower blood pressure. In the 1950s, several ganglionic blockers in addition to 93 were developed and marketed as antihypertensives. These included pentolinium 95 and mecamylamine 96. Unfortunately blockade of autonomic ganglia also decreases output of the parasympathetic nervous system and the ganglionic blockers were quickly found to have severe side effects primarily due to this parasympathetic inhibition [141]. As better antihypertensives with a different mechanism of action were developed, ganglionic blockers fell out of use.

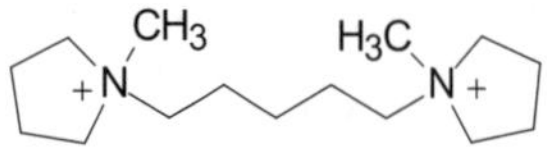

95 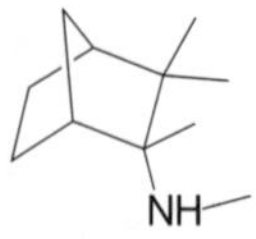 **96**

However, after disappearing from use as an antihypertensive, interest in mecamylamine has been renewed because of its properties as an antagonist of neuronal nicotine receptors [142,143]. Mecamylamine has shown promise in smoking cessation programs and as an antidepressant.

There is also evidence that there is more than one ganglionic nicotinic receptor. A recent paper indicates that the nicotinic receptor at the sympathetic ganglia is different from the receptors at parasympathetic ganglia [144]. It seems likely that α_7 receptors mediate at sympathetic ganglia and $\alpha_4\beta_2$ receptors at parasympathetic ganglia.

3.2.2. Neuromuscular Nicotinic Antagonists

The paralyzing effects of neuromuscular blockers has given them wide use in surgery and they are a mainstay of modern surgical procedures. Tubocurarine has a long duration of action, and a shorter acting derivative was needed. Succinylcholine 97, essentially two acetylcholine molecules linked back to back, was also introduced in the 1950s [140]. The esters of succinylcholine are rapidly hydrolyzed, providing a short duration of action, but like decamethonium, succinylcholine is a depolarizing neuromuscular blocker that causes undesirable effects on the patient.

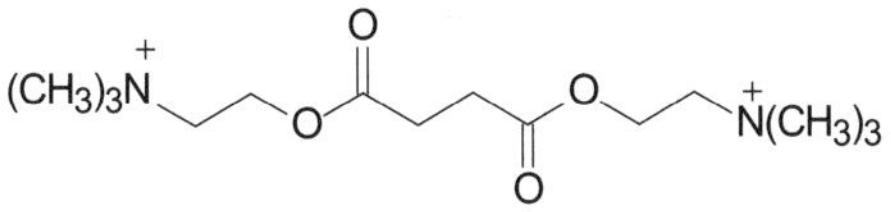

Over the years, a number of additional ganglionic blockers have been developed [145]. Based on the activity of a natural aminosteroid, a series of steroidal neuromuscular blockers was developed, compounds **98–101** (Table 2). Another series, compounds **102–105** based on quaternary isoquinolines has also been developed (Fig. 4). Atracurium **102** is a mixture of 10 geometrical and optical isomers with differing

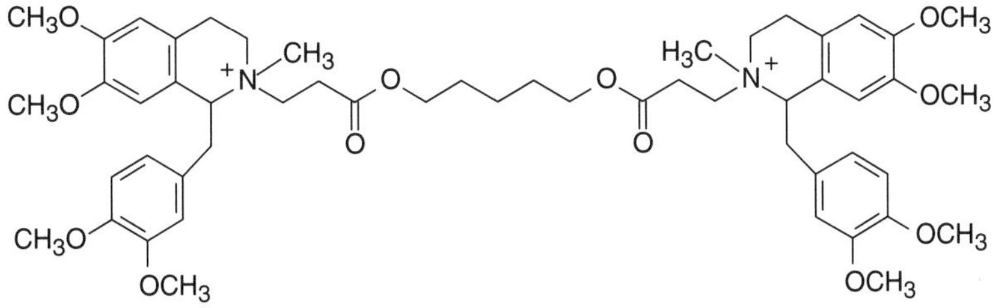

97

Atracurium **102**

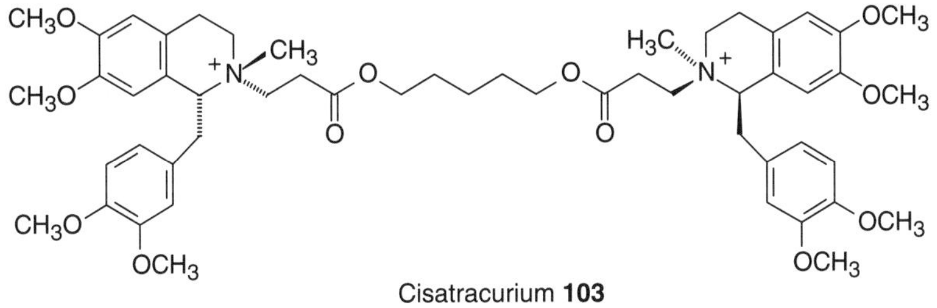

Cisatracurium **103**

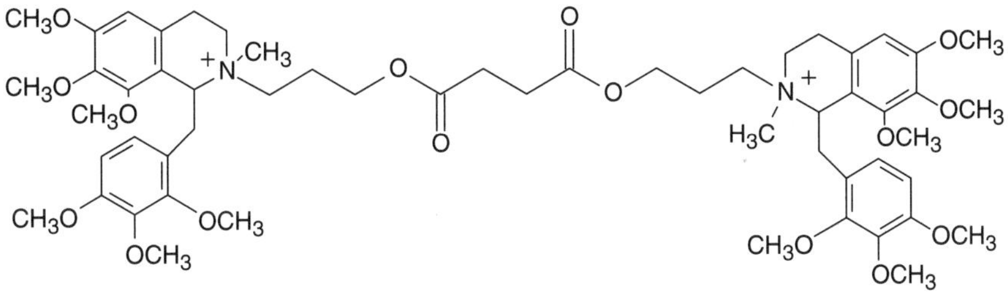

Doxacurium **104**

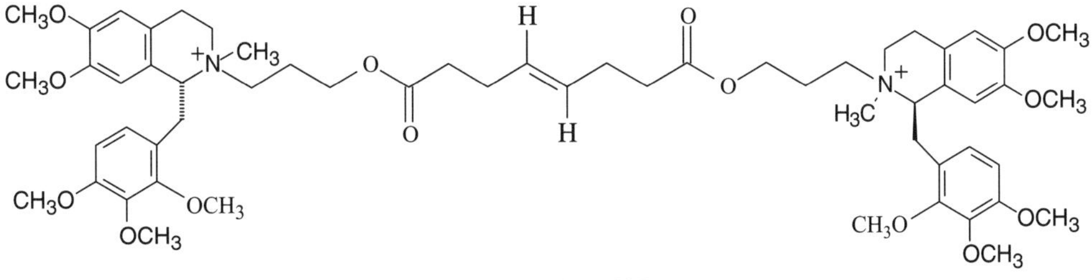

Mivacurium **105**

Figure 4. Isoquinoline-based neuromuscular blockers currently in clinical use.

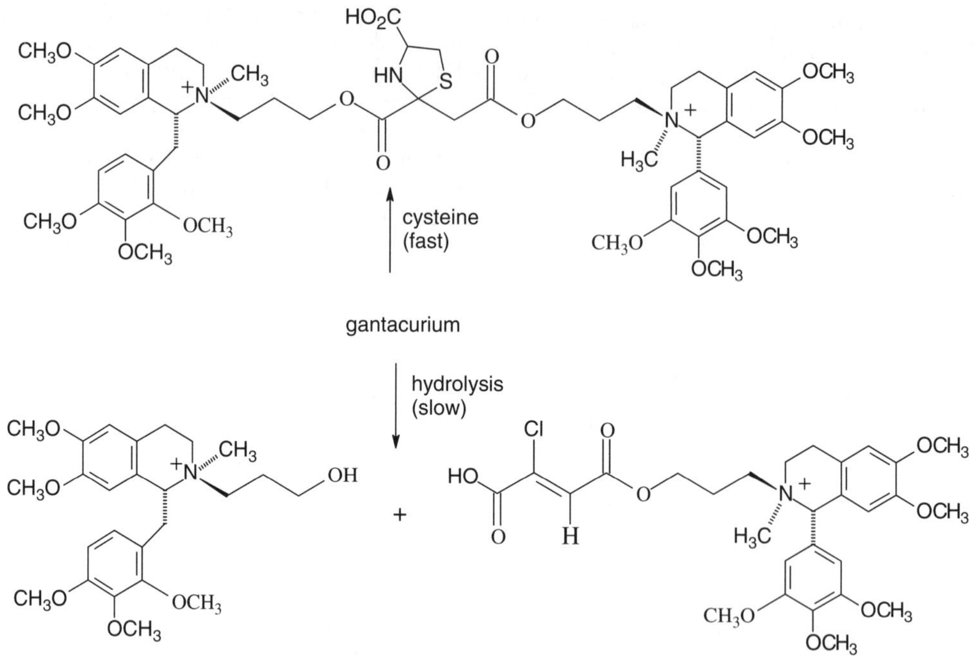

Figure 5. Gantacurium **106** inactivation. In aqueous solution at physiological pH gantacurium undergoes a slow spontaneous hydrolysis, but reacts rapidly with the amino acid cysteine to generate an inactive adduct.

properties that makes controlling response difficult [146]. Cisatracurium **103** is a single isomer, the *R-cis*, with much more predictable responses and does not stimulate histamine release as does atracurium.

Much of the research into new neuromuscular blocker has been attempts to develop a nondepolarizing ultrashort acting agent to replace the troublesome succinylcholine. A very recent effort that shows promise is the synthesis and evaluation of gantacurium **106** [147,148]. A unique property of gantacurium that provides short and controllable duration is shown in Fig. 5. The amino acid cysteine adds to the chlorofumarate that makes up the linker between the onium ends to form a heterocyclic adduct that is inactive. The ester hydrolysis is slow as it appears not to be an enzymatic reaction. In animal trials, the duration of action can be shortened by intravenous infusion of additional cysteine. Gantacurium is currently in clinical trials.

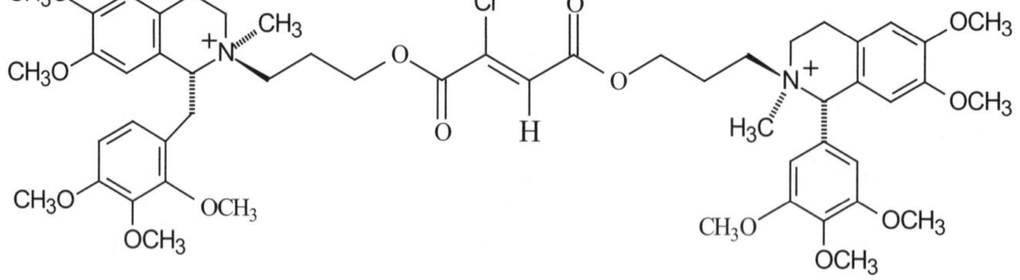

106

There is also evidence that the neuromuscular blockers are not only selective for the αβδε receptor of muscle cells but also can affect neuronal nicotinic receptors [149].

REFERENCES

1. Westfall TC, Westfall DP. Neurotransmission: the autonomic and somatic motor nervous systems. In: Brunton LL, Lazo JS, Parker KL, editors. Goodman & Gilman's Pharmacological Basis of Therapeutics. 11th ed. New York: McGraw-Hill; 2006. p 137–181.
2. Bloom FE. Neurotransmission and the central nervous system. In: Brunton LL, Lazo JS, Parker KL, editors. Goodman & Gilman's Pharmacological Basis of Therapeutics. New York: McGraw-Hill; 2006. p 317–339.
3. Brown JH, Palmer T. Muscarinic receptor agonists and antagonists. In: Brunton LL, Lazo JS, Parker KL, editors. Goodman & Gilman's Pharmacological Basis of Therapeutics. 11th ed. New York: McGraw-Hill; 2006. p 183–200.
4. Taylor P. Agents acting at the neuromuscular junction and autonomic ganglia. In: Brunton LL, Lazo JS, Parker KL, editors. Goodman & Gilman's Pharmacological Basis of Therapeutics. 11th ed. New York: McGraw-Hill; 2006. p 217–236.
5. Whittaker VP. The Cholinergic Synapse. New York: Springer-Verlag; 1988.
6. Dale HH. The action of certain esters and ethers of choline and their relation to muscarine. J Pharmacol Exp Ther 1914;6:147–190.
7. Brown DA. Acetylcholine. Br J Pharmacol 2006;147(Suppl 1): S120–S126.
8. Caulfield MP. Muscarinic receptors—characterization, coupling and function. Pharmacol Ther 1993;58:319–379.
9. Caulfield MP, Birdsall NJ. International Union of Pharmacology. XVII. Classification of muscarinic acetylcholine receptors. Pharmacol Rev 1998;50:279–290.
10. Eglen RM. Muscarinic receptor subtype pharmacology and physiology. Prog Med Chem 2005;43:105–136.
11. Ferreira VF, da Rocha DR, Lima Araujo KG, Santos WC. Advances in drug discovery to assess cholinergic neurotransmission: a systematic review. Curr Drug Discov Technol 2008;5:236–249.
12. Bartus RT, Dean RL 3rd Beer B, Lippa AS. The cholinergic hypothesis of geriatric memory dysfunction. Science 1982;217:408–414.
13. Levey AI. Muscarinic acetylcholine receptor expression in memory circuits: implications for treatment of Alzheimer disease. Proc Natl Acad Sci USA 1996;93:13541–13546.
14. Fisher A. M1 muscarinic agonists target major hallmarks of Alzheimer's disease—the pivotal role of brain M1 receptors. Neurodegener Dis 2008;5:237–240.
15. Liao DL, Hong CJ, Chen HM, Chen YE, Lee SM, Chang CY, Chen H, Tsai SJ. Association of muscarinic m1 receptor genetic polymorphisms with psychiatric symptoms and cognitive function in schizophrenic patients. Neuropsychobiology 2003;48:72–76.
16. Lechner SG, Mayer M, Boehm S. Activation of M1 muscarinic receptors triggers transmitter release from rat sympathetic neurons through an inhibition of M-type K^+ channels. J Physiol 2003;553:789–802.
17. Felder CC, Porter AC, Skillman TL, Zhang L, Bymaster FP, Nathanson NM, Hamilton SE, Gomeza J, Wess J, McKinzie DL. Elucidating the role of muscarinic receptors in psychosis. Life Sci 2001;68:2605–2613.
18. Rasmussen T, Fink-Jensen A, Sauerberg P, Swedberg MD, Thomsen C, Sheardown MJ, Jeppesen L, Calligaro DO, DeLapp NW, Whitesitt C, Ward JS, Shannon HE, Bymaster FP. The muscarinic receptor agonist BuTAC, a novel potential antipsychotic, does not impair learning and memory in mouse passive avoidance. Schizophr Res 2001;49:193–201.
19. Abrams P, Andersson KE, Buccafusco JJ, Chapple C, de Groat WC, Fryer AD, Kay G, Laties A, Nathanson NM, Pasricha PJ, Wein AJ. Muscarinic receptors: their distribution and function in body systems, and the implications for treating overactive bladder. Br J Pharmacol 2006;148:565–578.
20. Wolfe BB, Yasuda RP. Development of selective antisera for muscarinic cholinergic receptor subtypes. Ann NY Acad Sci 1995;757:186–193.
21. Bohme TM, Augelli-Szafran CE, Hallak H, Pugsley T, Serpa K, Schwarz RD. Synthesis and pharmacology of benzoxazines as highly selective antagonists at M(4) muscarinic receptors. J Med Chem 2002;45:3094–3102.
22. Wess J. Muscarinic acetylcholine receptor knockout mice: novel phenotypes and clinical implications. Annu Rev Pharmacol Toxicol 2004;44:423–450.
23. Zhang W, Yamada M, Gomeza J, Basile AS, Wess J. Multiple muscarinic acetylcholine receptor subtypes modulate striatal dopamine release, as studied with M1-M5 muscarinic receptor knock-out mice. J Neurosci 2002;22:6347–6352.

24. Yang G. Muscarinic receptors: a novel therapeutic target for drug addiction. Trends Pharmacol Sci 2002;23:551.
25. Elhusseiny A, Cohen Z, Olivier A, Stanimirovic DB, Hamel E. Functional acetylcholine muscarinic receptor subtypes in human brain microcirculation: identification and cellular localization. J Cereb Blood Flow Metab 1999; 19:794–802.
26. Lester HA, Dibas MI, Dahan DS, Leite JF, Dougherty DA. Cys-loop receptors: new twists and turns. Trends Neurosci 2004;27: 329–336.
27. Albuquerque EX, Pereira EF, Alkondon M, Rogers SW. Mammalian nicotinic acetylcholine receptors: from structure to function. Physiol Rev 2009;89:73–120.
28. Millar NS, Harkness PC. Assembly and trafficking of nicotinic acetylcholine receptors (review). Mol Membr Biol 2008;25:279–292.
29. Unwin N. The nicotinic acetylcholine receptor of the Torpedo electric ray. J Struct Biol 1998;121:181–190.
30. Mishina M, Takai T, Imoto K, Noda M, Takahashi T, Numa S, Methfessel C, Sakmann B. Molecular distinction between fetal and adult forms of muscle acetylcholine receptor. Nature 1986;321:406–411.
31. Lindstrom J, Anand R, Peng X, Gerzanich V, Wang F, Li Y. Neuronal nicotinic receptor subtypes. Ann NY Acad Sci 1995;757: 100–116.
32. Picciotto MR, Caldarone BJ, Brunzell DH, Zachariou V, Stevens TR, King SL. Neuronal nicotinic acetylcholine receptor subunit knockout mice: physiological and behavioral phenotypes and possible clinical implications. Pharmacol Ther 2001;92:89–108.
33. Gotti C, Moretti M, Gaimarri A, Zanardi A, Clementi F, Zoli M. Heterogeneity and complexity of native brain nicotinic receptors. Biochem Pharmacol 2007;74:1102–1111.
34. Gotti C, Clementi F, Fornari A, Gaimarri A, Guiducci S, Manfredi I, Moretti M, Pedrazzi P, Pucci L, Zoli M. Structural and functional diversity of native brain neuronal nicotinic receptors. Biochem Pharmacol 2009;78:703–711.
35. Schmitt JD, Bencherif M. Targeting nicotinic acetylcholine receptors: advances in molecular design and therapeutics. Ann Rep Med Chem 2000;35:41–61.
36. Jensen AA, Frolund B, Liljefors T, Krogsgaard-Larsen P. Neuronal nicotinic acetylcholine receptors: structural revelations, target identifications, and therapeutic inspirations. J Med Chem 2005;48:4705–4745.
37. Picciotto MR, Zoli M, Rimondini R, Lena C, Marubio LM, Pich EM, Fuxe K, Changeux JP. Acetylcholine receptors containing the beta2 subunit are involved in the reinforcing properties of nicotine. Nature 1998;391:173–177.
38. Watkins SS, Epping-Jordan MP, Koob GF, Markou A. Blockade of nicotine self-administration with nicotinic antagonists in rats. Pharmacol Biochem Behav 1999;62:743–751.
39. Champtiaux N, Gotti C, Cordero-Erausquin M, David DJ, Przybylski C, Lena C, Clementi F, Moretti M, Rossi FM, Le Novere N, McIntosh JM, Gardier AM, Changeux JP. Subunit composition of functional nicotinic receptors in dopaminergic neurons investigated with knock-out mice. J Neurosci 2003;23:7820–7829.
40. Han ZY, Le Novere N, Zoli M, Hill JA Jr, Champtiaux N, Changeux JP. Localization of nAChR subunit mRNAs in the brain of Macaca mulatta. Eur J Neurosci 2000;12: 3664–3674.
41. Kurzen H, Wessler I, Kirkpatrick CJ, Kawashima K, Grando SA. The non-neuronal cholinergic system of human skin. Horm Metab Res 2007;39:125–135.
42. Wessler I, Kirkpatrick CJ. Acetylcholine beyond neurons: the non-neuronal cholinergic system in humans. Br J Pharmacol 2008;154:1558–1571.
43. Kawashima I, Fujii T. Basic and clinical aspects of non-neuronal acetylcholine: overview of non-neuronal cholinergic systems and their biological significance. J Pharmacol Sci 2008;106:167–173.
44. Ing HR. The structure–action relationships of the choline group. Science 1949;109:264–266.
45. Schworer H, Lambrecht G, Mutschler E, Kilbinger H. The effects of racemic bethanechol and its (R)- and (S)-enantiomers on pre- and postjunctional muscarine receptors in the guinea-pig ileum. Naunyn Schmiedebergs Arch Pharmacol 1985;331:307–310.
46. Simonart A. On the action of certain derivatives of choline. J. Pharmacol. 1932;46:157–193.
47. Lesser E. The stereospecificity of acetyl-alpha-methylcholine. Br J Pharmacol 1965;25: 213–216.
48. Waser PG. Chemistry and pharmacology of muscarine, muscarone, and some related compounds. Pharmacol Rev 1961;13:485–515.
49. Belleau B, Puranen J. Stereochemistry of the interaction of enantiomeric 1,3-doxolane ana-

logs of muscarone with cholinergic receptors. J Med Chem 1963;6:325–328.

50. Teodori E, Gualtieri F, Angeli P, Brasili L, Giannella M, Pigini M. Resolution, absolute configuration, and cholinergic enantioselectivity of (+)- and (−)-cis-2-methyl-5-[(dimethylamino)methyl]-1,3-oxathiolane methiodide. J Med Chem 1986;29:1610–1615.

51. Teodori E, Gualtieri F, Angeli P, Brasili L, Giannella M. Resolution, absolute configuration, and cholinergic enantioselectivity of (−)- and (+)-c-2-methyl-r-5-[(dimethylamino)methyl]-1,3-oxathiolane t-3-oxide methiodide and related sulfones. J Med Chem 1987; 30:1934–1938.

52. Angeli P. Pentatomic cyclic agonists and muscarinic receptors: a 20 years review. Farmaco 1995;50:565–577.

53. Dei S, Angeli P, Bellucci C, Buccioni M, Gualtieri F, Marucci G, Manetti D, Matucci R, Romanelli MN, Scapecchi S, Teodori E. Muscarinic subtype affinity and functional activity profile of 1-methyl-2-(2-methyl-1,-3-dioxolan-4-yl)pyrrolidine and 1-methyl-2-(2-methyl-1,3-oxathiolan-5-yl)pyrrolidine derivatives. Biochem Pharmacol 2005;69:1637–1645.

54. Scapecchi S, Matucci R, Bellucci C, Buccioni M, Dei S, Guandalini L, Martelli C, Manetti D, Martini E, Marucci G, Nesi M, Romanelli MN, Teodori E, Gualtieri F. Highly chiral muscarinic ligands: the discovery of $(2S,2'R,3'S,5'R)$-1-methyl-2-(2-methyl-1,3-oxathiolan-5-yl)pyrrolidine 3-sulfoxide methyl iodide, a potent, functionally selective, M2 partial agonist. J Med Chem 2006;49:1925–1931.

55. Scapecchi S, Nesi M, Matucci R, Bellucci C, Buccioni M, Dei S, Guandalini L, Manetti D, Martini E, Marucci G, Romanelli MN, Teodori E, Cirilli R. Synthesis and pharmacological characterization of chiral pyrrolidinylfuran derivatives: the discovery of new functionally selective muscarinic agonists. J Med Chem 2008;51:3905–3912.

56. Ono S, Saito Y, Ohgane N, Kawanishi G, Mizobe F. Heterogeneity of muscarinic autoreceptors and heteroreceptors in the rat brain: effects of a novel M1 agonist, AF102B. Eur J Pharmacol 1988;155:77–84.

57. Fisher A, Brandeis R, Pittel Z, Karton I, Sapir M, Dachir S, Levy A, Heldman E. (+/−)-cis-2-Methyl-spiro(1,3-oxathiolane-5,3′) quinuclidine (AF102B): a new M1 agonist attenuates cognitive dysfunctions in AF64A-treated rats. Neurosci Lett 1989;102:325–331.

58. Fisher A, Brandeis R, Karton I, Pittel Z, Gurwitz D, Haring R, Sapir M, Levy A, Heldman E. (+/−)-cis-2-Methyl-spiro(1,3-oxathiolane-5,3′) quinuclidine, an M1 selective cholinergic agonist, attenuates cognitive dysfunctions in an animal model of Alzheimer's disease. J Pharmacol Exp Ther 1991;257:392–403.

59. Iwabuchi Y, Masuhara T. Sialogogic activities of SNI-2011 compared with those of pilocarpine and McN-A-343 in rat salivary glands: identification of a potential therapeutic agent for treatment of Sjorgen's syndrome. Gen Pharmacol 1994;25:123–129.

60. Iga Y, Arisawa H, Ogane N, Saito Y, Tomizuka T, Nakagawa-Yagi Y, Masunaga H, Yasuda H, Miyata N. (+/−)-cis-2-Methylspiro[1,3-oxathiolane-5,3′-quinuclidine] hydrochloride, hemihydrate (SNI-2011, cevimeline hydrochloride) induces saliva and tear secretions in rats and mice: the role of muscarinic acetylcholine receptors. Jpn J Pharmacol 1998;78: 373–380.

61. Saunders J, Showell GA, Baker R, Freedman SB, Hill D, McKnight A, Newberry N, Salamone JD, Hirshfield J, Springer JP Synthesis and characterization of all four isomers of the muscarinic agonist 2′-methylspiro[1-azabicyclo [2.2.2]octane-3,4′-[1,3]dioxolane]. J Med Chem 1987;30:969–975.

62. Arjungi KN. Areca nut: a review. Arzneimittelforshung 1976;26:951–956.

63. Tariot PN, Cohen RM, Welkowitz JA, Sunderland T, Newhouse PA, Murphy DL, Weingartner H. Multiple-dose arecoline infusions in Alzheimer's disease. Arch Gen Psychiatry 1988;45:901–905.

64. Raffaele KC, Berardi A, Morris PP, Asthana S, Haxby JV, Schapiro MB, Rapoport SI, Soncrant TT. Effects of acute infusion of the muscarinic cholinergic agonist arecoline on verbal memory and visuo-spatial function in dementia of the Alzheimer type. Prog Neuropsychopharmacol Biol Psychiatry 1991;15:643–648.

65. Asthana S, Greig NH, Holloway HW, Raffaele KC, Berardi A, Schapiro MB, Rapoport SI, Soncrant TT. Clinical pharmacokinetics of arecoline in subjects with Alzheimer's disease. Clin Pharmacol Ther 1996;60: 276–282.

66. Krogsgaard-Larsen P, Falch E, Sauerberg P, Freedman SB, Lembol HL, Meier E. Bioisosteres of arecoline as novel CNS-active muscarinic agonists. Trends Pharmacol Sci 1988; 69–74.

67. Saunders J, Showell GA, Snow RJ, Baker R, Harley EA, Freedman SB. 2-Methyl-1,3-dioxaazaspiro[4.5]decanes as novel muscarinic

cholinergic agonists. J Med Chem 1988;31: 486–491.
68. Saunders J, Cassidy M, Freedman SB, Harley EA, Iversen LL, Kneen C, MacLeod AM, Merchant KJ, Snow RJ, Baker R. Novel quinuclidine-based ligands for the muscarinic cholinergic receptor. J Med Chem 1990; 33:1128–1138.
69. Fisher A. Therapeutic strategies in Alzheimer's disease: M1 muscarinic agonists. Jpn J Pharmacol 2000;84:101–112.
70. Fisher A, Brandeis R, Bar-Ner RH, Kliger-Spatz M, Natan N, Sonego H, Marcovitch I, Pittel Z. AF150(S) and AF267B: M1 muscarinic agonists as innovative therapies for Alzheimer's disease. J Mol Neurosci 2002; 19:145–153.
71. Fisher A. M1 muscarinic agonists target major hallmarks of Alzheimer's disease—an update. Curr Alzheimer Res 2007;4:577–580.
72. Ivanova A, Murphy M. An adaptive first in man dose-escalation study of NGX267: statistical, clinical, and operational considerations. J Biopharm Stat 2009;19:247–255.
73. Toja E, Bonetti C, Butti A, Hunt P, Fortin M, Barzaghi F, Formento ML, Maggioni A, Nencioni A, Galliani G. 1-Alkyl-1,2,5,6-tetrahydropyridine-3-carboxaldehyde-O-alkyl-oximes: a new class of potent orally active muscarinic agonists related to arecoline. Eur J Med Chem 1991;26:853–868.
74. Trollor JN, Sachdev PS, Haindl W, Brodaty H, Wen W, Walker BM. Combined cerebral blood flow effects of a cholinergic agonist (milameline) and a verbal recognition task in early Alzheimer's disease. Psychiatry Clin Neurosci 2006;60:616–625.
75. Schwarz RD, Callahan MJ, Coughenour LL, Dickerson MR, Kinsora JJ, Lipinski WJ, Raby CA, Spencer CJ, Tecle H. Milameline (CI-979/RU35926): a muscarinic receptor agonist with cognition-activating properties: biochemical and in vivo characterization. J Pharmacol Exp Ther 1999;291:812–822.
76. Sauerberg P, Olesen PH, Nielsen S, Treppendahl S, Sheardown MJ, Honore T, Mitch CH, Ward JS, Pike AJ, Bymaster FP, et al. Novel functional M1 selective muscarinic agonists. Synthesis and structure–activity relationships of 3-(1,2,5-thiadiazolyl)-1,2,5,6-tetrahydro-1-methylpyridines. J Med Chem 1992;35:2274–2283.
77. Shannon HE, Bymaster FP, Calligaro DO, Greenwood B, Mitch CH, Sawyer BD, Ward JS, Wong DT, Olesen PH, Sheardown MJ, et al. Xanomeline: a novel muscarinic receptor agonist with functional selectivity for M1 receptors. J Pharmacol Exp Ther 1994;269:271–281.
78. Veroff AE, Bodick NC, Offen WW, Sramek JJ, Cutler NR. Efficacy of xanomeline in Alzheimer disease: cognitive improvement measured using the computerized neuropsychological test battery (CNTB). Alzheimer Dis Assoc Disord 1998;12:304–312.
79. Shekhar A, Potter WZ, Lightfoot J, Lienemann J, Dube S, Mallinckrodt C, Bymaster FP, McKinzie DL, Felder CC. Selective muscarinic receptor agonist xanomeline as a novel treatment approach for schizophrenia. Am J Psychiatry 2008;165:1033–1039.
80. Bromidge SM, Brown F, Cassidy F, Clark MS, Dabbs S, Hadley MS, Hawkins J, Loudon JM, Naylor CB, Orlek BS, Riley GJ. Design of [R-(Z)]-(+)-alpha-(methoxyimino)-1-azabicyclo[2.2.2]octane-3-acetonitri le (SB 202026), a functionally selective azabicyclic muscarinic M1 agonist incorporating the N-methoxy imidoyl nitrile group as a novel ester bioisostere. J Med Chem 1997;40:4265–4280.
81. Hodges H, Peters S, Gray JA, Hunter AJ. Counteractive effects of a partial (sabcomeline) and a full (RS86) muscarinic receptor agonist on deficits in radial maze performance induced by S-AMPA lesions of the basal forebrain and medial septal area. Behav Brain Res 1999; 99:81–92.
82. Ensinger HA, Doods HN, Immel-Sehr AR, Kuhn FJ, Lambrecht G, Mendla KD, Muller RE, Mutschler E, Sagrada A, Walther G, et al. WAL 2014—a muscarinic agonist with preferential neuron-stimulating properties. Life Sci 1993;52:473–480.
83. Adamus WS, Leonard JP, Troger W, Phase I clinical trials with WAL 2014, a new muscarinic agonist for the treatment of Alzheimer's disease Life Sci 2014;56:883–890.
84. Hock C, Maddalena A, Raschig A, Muller-Spahn F, Eschweiler G, Hager K, Heuser I, Hampel H, Muller-Thomsen T, Oertel W, Wienrich M, Signorell A, Gonzalez-Agosti C, Nitsch RM. Treatment with the selective muscarinic m1 agonist talsaclidine decreases cerebrospinal fluid levels of A beta 42 in patients with Alzheimer's disease. Amyloid 2003; 10:1–6.
85. Sauerberg P, Olesen PH, Sheardown MJ, Suzdak PD, Shannon HE, Bymaster FP, Calligaro DO, Mitch CH, Ward JS, Swedberg MD. Muscarinic agonists as analgesics. Antinociceptive activity versus M1 activity: SAR of alkylthio-

TZTP's and related 1,2,5-thiadiazole analogs. Life Sci 1995;56:807–814.
86. Cho AK, Haslett WL, Jenden DJ. The peripheral actions of oxotremorine, a metabolite of tremorine. J Pharmacol Exp Ther 1962;138:249–257.
87. Cannon JG. Cholinergics. In: Abraham DJ, editor. Burger's Medicinal Chemistry and Drug Discovery. Vol. 6. New York: John Wiley & Sons; 2003. p 71–78.
88. Jones CK, Brady AE, Davis AA, Xiang Z, Bubser M, Tantawy MN, Kane AS, Bridges TM, Kennedy JP, Bradley SR, Peterson TE, Ansari MS, Baldwin RM, Kessler RM, Deutch AY, Lah JJ, Levey AI, Lindsley CW, Conn PJ. Novel selective allosteric activator of the M1 muscarinic acetylcholine receptor regulates amyloid processing and produces antipsychotic-like activity in rats. J Neurosci 2008;28: 10422–10433.
89. Bridges TM, Brady AE, Kennedy JP, Daniels RN, Miller NR, Kim K, Breininger ML, Gentry PR, Brogan JT, Jones CK, Conn PJ, Lindsley CW. Synthesis and SAR of analogues of the M1 allosteric agonist TBPB. Part I: Exploration of alternative benzyl and privileged structure moieties. Bioorg Med Chem Lett 2008; 18:5439–5442.
90. Miller NR, Daniels RN, Bridges TM, Brady AE, Conn PJ, Lindsley CW. Synthesis and SAR of analogs of the M1 allosteric agonist TBPB. Part II: Amides, sulfonamides and ureas—the effect of capping the distal basic piperidine nitrogen. Bioorg Med Chem Lett 2008; 18:5443–5447.
91. Aceto MD, Martin BR, Uwaydah IM, May EL, Harris LS, Izazola-Conde C, Dewey WL, Bradshaw TJ, Vincek WC. Optically pure (+)-nicotine from (.+−.)-nicotine and biological comparisons with (−)-nicotine. J Med Chem 1979;22:174–177.
92. Glennon RA, Maarouf A, Fahmy S, Martin B, Fan E. Structure–affinity relationships of simple nicotine analogs. Med Chem Res 1993;2:546–551.
93. Abood LG, Lu X, Banerjee S. Receptor binding characteristics of a 3H-labeled azetidine analogue of nicotine. Biochem Pharmacol 1987;36:2337–2341.
94. Cheng Y-X, Dukata M, Dowda M, Fiedlera W, Martin B, Damaj MI, Glennon RA. Synthesis and binding of 6,7,8,9-tetrahydro-5H-pyrido[3,4-d]azepine and related ring-opened analogs at central nicotinic receptors. Eur J Med Chem 1999;34:177–190.
95. Spande TF, Garraffo HM, Edwards MW, Yeh HJ, Pannell L, Daly JW. Epibatidine: a novel (chloropyridyl)azabicycloheptane with potent analgesic activity from an Equadoran poison frog. J Am Chem Soc 1992;114: 3475–3478.
96. Dukat M, Damaj MI, Glassco W, Dumas D, May EL, Martin BR, Glennon RA. Epibatidine: a very high affinity nicotine-receptor ligand. Med Chem Res 1994;4:131–139.
97. Badio B, Garraffo HM, Spande TF, Daly JW. Epibatidine: discovery and definition as a potent analgesic and nicotinic agonist. Med Chem Res 1994;4:440–448.
98. Tonder JE, Olesen PH. Agonists at the alpha4-beta2 nicotinic acetylcholine receptors: structure–activity relationships and molecular modelling. Curr Med Chem 2001;8:651–674.
99. Bannon AW, Decker MW, Curzon P, Buckley MJ, Kim DJ, Radek RJ, Lynch JK, Wasicak JT, Lin NH, Arnold WH, Holladay MW, Williams M, Arneric SP. ABT-594 [(R)-5-(2-azetidinylmethoxy)-2-chloropyridine]: a novel, orally effective antinociceptive agent acting via neuronal nicotinic acetylcholine receptors: II. *In vivo* characterization. J Pharmacol Exp Ther 1998;285:787–794.
100. Bannon AW, Decker MW, Holladay MW, Curzon P, Donnelly-Roberts D, Puttfarcken PS, Bitner RS, Diaz A, Dickenson AH, Porsolt RD, Williams M, Arneric SP. Broad-spectrum, non-opioid analgesic activity by selective modulation of neuronal nicotinic acetylcholine receptors. Science 1998;279:77–81.
101. Abreo MA, Lin NH, Garvey DS, Gunn DE, Hettinger AM, Wasicak JT, Pavlik PA, Martin YC, Donnelly-roberts DL, Anderson DJ, Sullivan JP, Williams M, Arneric SP, Holladay MW. Novel 3-pyridyl ethers with subnanomolar affinity for central neuronal nicotinic acetylcholine receptors. J Med Chem 1996; 39:817–825.
102. Ghelardini C, Galeotti N, Barlocco D, Bartolini A. Antinociceptive profile of the new nicotinic agonist DBO-83. Drug Dev Res 1997; 40:251–258.
103. Badio B, Garraffo HM, Plummer CV, Padgett WL, Daly JW. Synthesis and nicotinic activity of epiboxidine: an isoxazole analogue of epibatidine. Eur J Pharmacol 1997;321:189–194.
104. Spivak CE, Witcop B, Albuquerque EX. Anatoxin-a: a novel potent agonist at the nicotinic receptor. Mol Pharmacol 1980;18: 384–394.
105. Wright E, Gallagher T, Sharpless CG, Wonnacott S. Synthesis of UB-165: a novel nicotinic ligand and anatoxin-a/epibatidine hybrid. Bioorg Med Chem Lett 1997;7:2867–2870.
106. Sharples CG, Kaiser S, Soliakov L, Marks MJ, Collins AC, Washburn M, Wright E, Spencer

JA, Gallagher T, Whiteaker P, Wonnacott S. UB-165: a novel nicotinic agonist with subtype selectivity implicates the alpha4beta2* subtype in the modulation of dopamine release from rat striatal synaptosomes. J Neurosci 2000;20:2783–2791.
107. Sharples CG, Karig G, Simpson GL, Spencer JA, Wright E, Millar NS, Wonnacott S, Gallagher T. Synthesis and pharmacological characterization of novel analogues of the nicotinic acetylcholine receptor agonist (+/−)-UB-165. J Med Chem 2002;45:3235–3245.
108. Karig G, Large JM, Sharples CG, Sutherland A, Gallagher T, Wonnacott S. Synthesis and nicotinic binding of novel phenyl derivatives of UB-165. Identifying factors associated with alpha7 selectivity. Bioorg Med Chem Lett 2003;13:2825–2828.
109. Daly JW. Nicotinic agonists, antagonists, and modulators from natural sources. Cell Mol Neurobiol 2005;25:513–552.
110. Imming P, Klaperski P, Stubbs MT, Seitz G, Gundisch D. Syntheses and evaluation of halogenated cytisine derivatives and of bioisosteric thiocytisine as potent and selective nAChR ligands. Eur J Med Chem 2001; 36:375–388.
111. Carbonnelle E, Sparatore F, Canu-Boido C, Salvagno C, Baldani-Guerra B, Terstappen G, Zwart R, Vijverberg H, Clementi F, Gotti C. Nitrogen substitution modifies the activity of cytisine on neuronal nicotinic receptor subtypes. Eur J Pharmacol 2003;471:85–96.
112. Slater YE, Houlihan LM, Maskell PD, Exley R, Bermudez I, Lukas RJ, Valdivia AC, Cassels BK. Halogenated cytisine derivatives as agonists at human neuronal nicotinic acetylcholine receptor subtypes. Neuropharmacology 2003;44:503–515.
113. Coe JW, Brooks PR, Vetelino MG, Wirtz MC, Arnold EP, Huang J, Sands SB, Davis TI, Lebel LA, Fox CB, Shrikhande A, Heym JH, Schaeffer E, Rollema H, Lu Y, Mansbach RS, Chambers LK, Rovetti CC, Schulz DW, Tingley FD 3rd O'Neill BT. Varenicline: an alpha4beta2 nicotinic receptor partial agonist for smoking cessation. J Med Chem 2005; 48:3474–3477.
114. Rollema H, Coe JW, Chambers LK, Hurst RS, Stahl SM, Williams KE. Rationale, pharmacology and clinical efficacy of partial agonists of alpha4beta2 nACh receptors for smoking cessation. Trends Pharmacol Sci 2007;28: 316–325.
115. Coe JW, Rollema H, O'Neill BT. Case history: Chantix/Champix (varenicline tartarate), a nicotinic acetylcholine receptor partial agonist as a smoking cessation aid. In: Macor J, editor. Annual Reports in Medicinal Chemistry. Vol. 44. New York: Academic Press; 2009. p 71–101.
116. Chellappan SK, Xiao Y, Tueckmantel W, Kellar KJ, Kozikowski AP. Synthesis and pharmacological evaluation of novel 9- and 10-substituted cytisine derivatives. Nicotinic ligands of enhanced subtype selectivity. J Med Chem 2006;49:2673–2676.
117. Tasso B, Canu Boido C, Terranova E, Gotti C, Riganti L, Clementi F, Artali R, Bombieri G, Meneghetti F, Sparatore F. Synthesis, binding, and modeling studies of new cytisine derivatives, as ligands for neuronal nicotinic acetylcholine receptor subtypes. J Med Chem 2009;52:4345–4357.
118. Wei ZL, Xiao Y, Yuan H, Baydyuk M, Petukhov PA, Musachio JL, Kellar KJ, Kozikowski AP. Novel pyridyl ring C5 substituted analogues of epibatidine and 3-(1-methyl-2(S)-pyrrolidinylmethoxy)pyridine (A-84543) as highly selective agents for neuronal nicotinic acetylcholine receptors containing beta2 subunits. J Med Chem 2005;48:1721–1724.
119. Gray JA, Roth BL. Molecular targets for treating cognitive dysfunction in schizophrenia. Schizophr Bull 2007;33:1100–1119.
120. Mullen G, Napier J, Balestra M, DeCory T, Hale G, Macor J, Mack R, Loch J 3rd Wu E, Kover A, Verhoest P, Sampognaro A, Phillips E, Zhu Y, Murray R, Griffith R, Blosser J, Gurley D, Machulskis A, Zongrone J, Rosen A, Gordon J. (−)-Spiro[1-azabicyclo[2.2.2]octane-3,5′-oxazolidin-2′-one], a conformationally restricted analogue of acetylcholine, is a highly selective full agonist at the alpha 7 nicotinic acetylcholine receptor. J Med Chem 2000;43:4045–4050.
121. Van Kampen M, Selbach K, Schneider R, Schiegel E, Boess F, Schreiber R. AR-R 17779 improves social recognition in rats by activation of nicotinic alpha7 receptors. Psychopharmacology (Berl) 2004;172:375–383.
122. Tatsumi R, Fujio M, Satoh H, Katayama J, Takanashi S, Hashimoto K, Tanaka H. Discovery of the alpha7 nicotinic acetylcholine receptor agonists. (R)-3′-(5-Chlorothiophen-2-yl)spiro-1-azabicyclo[2.2.2]octane-3,5′-[1′,3′]oxazolidin-2′-one as a novel, potent, selective, and orally bioavailable ligand. J Med Chem 2005;48:2678–2686.
123. Tatsumi R, Fujio M, Takanashi S, Numata A, Katayama J, Satoh H, Shiigi Y, Maeda J, Kuriyama M, Horikawa T, Murozono T, Hashimoto K, Tanaka H. (R)-3′-(3-Methylbenzo[b]thiophen-5-yl)spiro[1-azabicyclo[2,2,2]oc-

tane-3,5′-oxazolidin]-2′-one, a novel and potent alpha7 nicotinic acetylcholine receptor partial agonist displays cognitive enhancing properties. J Med Chem 2006;49:4374–4383.

124. Bodnar AL, Cortes-Burgos LA, Cook KK, Dinh DM, Groppi VE, Hajos M, Higdon NR, Hoffmann WE, Hurst RS, Myers JK, Rogers BN, Wall TM, Wolfe ML, Wong E. Discovery and structure–activity relationship of quinuclidine benzamides as agonists of alpha7 nicotinic acetylcholine receptors. J Med Chem 2005; 48:905–908.

125. Wishka DG, Walker DP, Yates KM, Reitz SC, Jia S, Myers JK, Olson KL, Jacobsen EJ, Wolfe ML, Groppi VE, Hanchar AJ, Thornburgh BA, Cortes-Burgos LA, Wong EH, Staton BA, Raub TJ, Higdon NR, Wall TM, Hurst RS, Walters RR, Hoffmann WE, Hajos M, Franklin S, Carey G, Gold LH, Cook KK, Sands SB, Zhao SX, Soglia JR, Kalgutkar AS, Arneric SP, Rogers BN. Discovery of N-[(3R)-1-azabicyclo[2.2.2] oct-3-yl]furo[2,3-c]pyridine-5-carboxamide, an agonist of the alpha7 nicotinic acetylcholine receptor, for the potential treatment of cognitive deficits in schizophrenia: synthesis and structure–activity relationship. J Med Chem 2006;49:4425–4436.

126. Faghih R, Gopalakrishnan M, Briggs CA. Allosteric modulators of the alpha7 nicotinic acetylcholine receptor. J Med Chem 2008;51:701–712.

127. Barnes PJ, Belvisi MG, Mak JC, Haddad EB, O'Connor B. Tiotropium bromide (Ba 679 BR), a novel long-acting muscarinic antagonist for the treatment of obstructive airways disease. Life Sci 1995;56:853–859.

128. Wan Z, Laine DI, Yan H, Zhu C, Widdowson KL, Buckley PT, Burman M, Foley JJ, Sarau HM, Schmidt DB, Webb EF, Belmonte KE, Palovich M. Discovery of (3-endo)-3-(2-cyano-2,2-diphenylethyl)-8,8-dimethyl-8-azoniabicyclo[3.2.1]octane bromide as an efficacious inhaled muscarinic acetylcholine receptor antagonist for the treatment of COPD. Bioorg Med Chem Lett 2009;19:4560–4562.

129. Hegde SS. Muscarinic receptors in the bladder: from basic research to therapeutics. Br J Pharmacol 2006;147(Suppl 2): S80–S87.

130. Ney P, Pandita RK, Newgreen DT, Breidenbach A, Stohr T, Andersson KE. Pharmacological characterization of a novel investigational antimuscarinic drug, fesoterodine, in vitro and in vivo. BJU Int 2008;101:1036–1042.

131. Starck JP, Talaga P, Quere L, Collart P, Christophe B, Lo Brutto P, Jadot S, Chimmanamada D, Zanda M, Wagner A, Mioskowski C, Massingham R, Guyaux M. Potent anti-muscarinic activity in a novel series of quinuclidine derivatives. Bioorg Med Chem Lett 2006;16: 373–377.

132. Starck JP, Provins L, Christophe B, Gillard M, Jadot S, Lo Brutto P, Quere L, Talaga P, Guyaux M. Alkyne-quinuclidine derivatives as potent and selective muscarinic antagonists for the treatment of COPD. Bioorg Med Chem Lett 2008;18:2675–2678.

133. Jin J, Budzik B, Wang Y, Shi D, Wang F, Xie H, Wan Z, Zhu C, Foley JJ, Webb EF, Berlanga M, Burman M, Sarau HM, Morrow DM, Moore ML, Rivero RA, Palovich M, Salmon M, Belmonte KE, Laine DI. Discovery of biphenyl piperazines as novel and long acting muscarinic acetylcholine receptor antagonists. J Med Chem 2008;51:5915–5918.

134. Budzik B, Wang Y, Shi D, Wang F, Xie H, Wan Z, Zhu C, Foley JJ, Nuthulaganti P, Kallal LA, Sarau HM, Morrow DM, Moore ML, Rivero RA, Palovich M, Salmon M, Belmonte KE, Laine DI, Jin J. M3 muscarinic acetylcholine receptor antagonists: SAR and optimization of biaryl amines. Bioorg Med Chem Lett 2009;19: 1686–1690.

135. Jin J, Wang Y, Shi D, Wang F, Fu W, Davis RS, Jin Q, Foley JJ, Sarau HM, Morrow DM, Moore ML, Rivero RA, Palovich M, Salmon M, Belmonte KE, Busch-Petersen J. Muscarinic acetylcholine receptor antagonists: SAR and optimization of tyrosine ureas. Bioorg Med Chem Lett 2008;18:5481–5486.

136. Melchiorre C, Cassinelli A, Quaglia W. Differential blockade of muscarinic receptor subtypes by polymethylene tetraamines. Novel class of selective antagonists of cardiac M-2 muscarinic receptors. J Med Chem 1987; 30:201–204.

137. Melchiorre C, Minarini A, Budriesi R, Chiarini A, Spampinato S, Tumiatti V. The design of novel methoctramine-related tetraamines as muscarinic receptor subtype selective antagonists. Life Sci 1995;56:837–844.

138. Budriesi R, Cacciaguerra S, Toro RD, Bolognesi ML, Chiarini A, Minarini A, Rosini M, Spampinato S, Tumiatti V, Melchiorre C. Analysis of the muscarinic receptor subtype mediating inhibition of the neurogenic contractions in rabbit isolated vas deferens by a series of polymethylene tetra-amines. Br J Pharmacol 2001;132:1009–1016.

139. Paton WDM, Zaimes EJ. The pharmacological actions of polymethylene bistrimethyl-ammoniium salts. Br J Pharmacol 1949; 4:381–400.

140. Raghavendra T. Neuromuscular blocking drugs: discovery and development. J Royal Soc Med 2002;95:363–367.

141. Sears HTN, Snow PJD, Houston IB. Treatment of hypertension with pentolinium and mecamylamine. Br Med J 1959;1:462–465.
142. Shytle RD, Penny E, Silver AA, Goldman J, Sanberg PR. Mecamylamine (Inversine): an old antihypertensive with new research directions. J Hum Hypertens 2002;16: 453–457.
143. Andreasen JT, Redrobe JP. Antidepressant-like effects of nicotine and mecamylamine in the mouse forced swim and tail suspension tests: role of strain, test and sex. Behav Pharmacol 2009;20:286–295.
144. Li Y. Specific subtypes of nicotinic cholinergic receptors involved in sympathetic and parasympathetic cardiovascular responses. Neurosci Lett 2009;462:20–23.
145. Brown WC. Neuromuscular Block. Br J Pharmacol 2006;147:S277–S286.
146. Welch RM, Brown A, Ravitch J, Dahl R. The *in vitro* degradation of cisatracurium, the *R,cis-R'*-isomer of atracurium, in human and rat plasma. Clin Pharmacol Ther 1995;58: 132–142.
147. Boros EE, Bigham EC, Boswell GE, Mook RA Jr, Patel SS, Savarese JJ, Ray JA, Thompson JB, Hashim MA, Wisowaty JC, Feldman PL, Samano V. Bis- and mixed-tetrahydroisoquinolinium chlorofumarates: new ultra-short-acting nondepolarizing neuromuscular blockers. J Med Chem 1999;42: 206–209.
148. Lien CA, Savard P, Belmont M, Sunaga H, Savarese JJ. Fumarates: unique nondepolarizing neuromuscular blocking agents that are antagonized by cysteine. J Crit Care 2009; 24:50–57.
149. Jonsson M, Gurley D, Dabrowski M, Larsson O, Johnson EC, Eriksson LI. Distinct pharmacologic properties of neuromuscular blocking agents on human neuronal nicotinic acetylcholine receptors: a possible explanation for the train-of-four fade. Anesthesiology 2006; 105:521–533.

CNS STIMULANTS

David E. Nichols
Department of Medicinal Chemistry and Molecular Pharmacology, School of Pharmacy and Pharmaceutical Sciences, Purdue University, West Lafayette, IN

1. INTRODUCTION

Natural products that have stimulant properties have been known for millennia, and their active species (including ephedrine and cocaine) are now well known. Central nervous system (CNS) Stimulants, also called psychostimulants, are drugs that lead to increased arousal, improved performance on tasks of vigilance and alertness, and a sense of self-confidence and well being. High doses can produce feelings of elation or euphoria and because of these effects stimulants have reinforcing properties, and can produce dependence. That is, because they make users feel "good," they are sometimes taken for extended periods of time in an attempt to maintain an elevated mood. Tolerance develops to the mood elevating properties of psychostimulants, however, and more and more of the drug must be taken to maintain the effect. Increased doses also prevent sleep, and continued use can result in symptoms of psychosis. Cessation of the drug after one of these binges (abstinence) may lead to an emotional and physical "crash" (the result of poor nutrition, lack of sleep, and increased physical activity), and severely depressed mood. In dependent individuals, intense craving for the drug occurs, resulting in another period of drug seeking, extensive drug use, and subsequent crash. This cycle is repeated in chronic psychostimulant dependence.

The reader should be aware that in this chapter the use of the term "CNS stimulation" encompasses several physiological mechanisms of action, and many different types of biologically active substances. A number of different agents, such as caffeine, affect these pharmacological mechanisms and cause CNS stimulation. Other diverse examples include strychnine (causing CNS stimulants by blockade of inhibitory glycine receptors), and benzodiazepine inverse agonists (causing CNS stimulation by attenuating the inhibitory effects of GABA on chloride channels). It is not the intent of this chapter to provide an encyclopedic treatment of all the possible substances that can cause "CNS stimulation," but rather to focus primarily on the psychostimulants (i.e., drugs that affect brain monoaminergic systems).

Many CNS stimulants also have appetite suppressant effects that led to their use in treating obesity. In short-term studies, amphetamine-like drugs have been shown to be more effective than placebo in promoting weight loss. Long-term (>20 weeks) weight loss has not been shown, however, unless the drug is taken continuously [1]. At one time, stimulants were widely prescribed for appetite control, but they lose efficacy rather quickly through the development of tolerance. Thus, it was not uncommon for patients to become dependent on them, with symptoms of withdrawal upon abrupt cessation. The increased awareness of the addictive potential of stimulants, coupled with their widespread abuse, has led to much more extensive restrictions over their availability. These drugs are much more carefully controlled today, and are rarely used for weight control except in a few special instances.

There remain some important medical uses for this class of drug, yet as noted later, the therapeutic actions of psychostimulants must be balanced against their undesirable actions. Issues of dependence, tolerance, and potential abuse must be considered when deciding whether treatment with a psychostimulant is an appropriate therapy. Nonetheless, new generations of drugs that have sprung from an understanding of the classic stimulants may open important therapeutic horizons for the future.

1.1. Ephedra and Khat

The Chinese drug ma huang (*Ephedra sinica* Stapf) has been used in China for more than 5000 years. The alkaloid that is responsible for the CNS stimulant effects is ephedrine. The levorotatory erythro isomer (**1**) is the most active of the four possible stereoisomers with that structural formula. Khat (kat, or qat) or Abyssinian tea (*Catha edulis* Forskal) is the product from a small tree or shrub indigenous to tropical East Africa. Khat leaves are chewed habitually by peoples in East Africa and certain other Arabian countries, and produce a mild CNS stimulant effect [2]. The principle active component in Khat is a substituted phenethylamine derivative known as (−)-cathinone **2** [3].

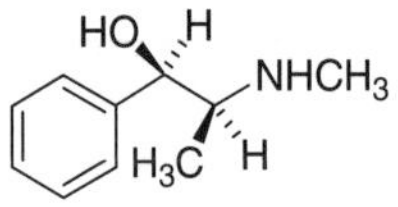

Both of these compounds possess a beta-phenethylamine framework, a common structural theme that occurs in many related CNS stimulants. In general, these compounds have similar mechanisms of action.

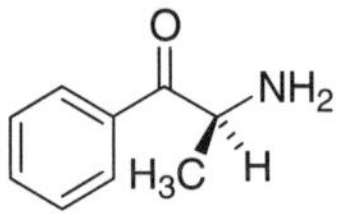

(−)-Cocaine **3** has a completely different structure, and as we shall see later, its mechanism of action is also somewhat different from the structurally simpler **1** and **2**. Nevertheless, all of these natural prototype CNS stimulants have the common action of exerting powerful effects on brain pathways that utilize dopamine as the neurotransmitter.

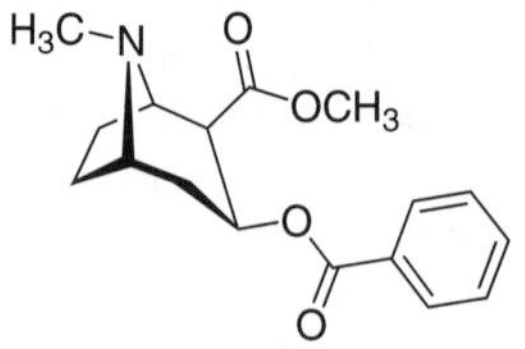

1.2. Caffeine

From an economic standpoint, the most important central nervous system stimulant is caffeine, 1,3,7-trimethylxanthine **4**. It occurs naturally and is a product of kola (cola) nuts (*Cola nitida*, where it occurs to the extent of approximately 3.5%, by weight), of coffee beans (*Coffea arabica* where it comprises approximately 1–2% by weight), and tea (*Camellia sinesis* where it makes up 1–4% of the mass of dried leaves). The annual consumption of caffeine has been estimated at 120 million kilograms, the approximate equivalent of one caffeine-containing beverage per day for each of the world's five billion plus inhabitants. As a beverage, the worldwide consumption of tea is surpassed only by water. The structurally related dimethylxanthines theophylline **5** and theobromine **6** have less of a CNS stimulant effect, and are principally important for their ability to relax smooth muscle. Cocoa and chocolate have little caffeine, but do contain theobromine.

A regular cup of coffee contains between 40 and 176 mg of caffeine, with a mean content of approximately 85 mg. Tea contains less caffeine, with an average of approximately 27 mg per cup, and an ounce of sweet chocolate typically contains between 75 and 150 mg of combined methylxanthines [4]. Caffeinated "energy drinks" are increasing dramatically in popularity, and these can contain anywhere from 75 to 505 mg of caffeine, depending on the brand [5].

2. HISTORY

The historical development of amphetamine and methamphetamine is described in interesting detail by Angrist and Sudilovsky [6]. The discovery of psychostimulants differs somewhat from the usual drug discovery process because there was a long folkloric history of use of khat, coca leaves, and *ma huang*

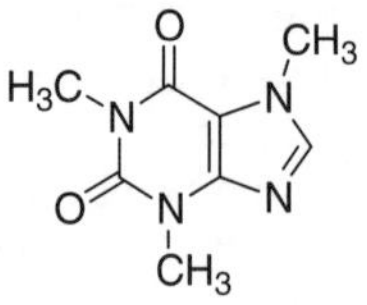

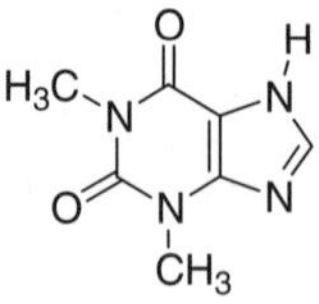

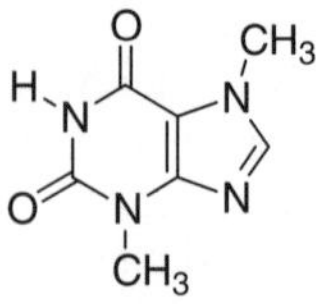

(ephedra). Although there may not have been a formal pharmacological classification of CNS stimulants at that time, the ability of these agents to alleviate fatigue was certainly well recognized.

Amphetamine itself was first synthesized in 1887 and studied as early as 1910, but its stimulant effects were not discovered for another approximately 20 years. Amphetamine was independently resynthesized in 1927 by the noted psychopharmacologist Gordon Alles in a program to develop synthetic substitutes for ephedrine, a drug then being used as a bronchodilator for the treatment of asthma [7]. The central stimulant effects of amphetamine were probably noted about 1930, when it appeared in nasal inhalers in Germany. The first medical use for amphetamine was in the treatment of narcolepsy [8] and by 1936 orally active Benzedrine® tablets were available without prescription [9]. By 1937, it was being used recreationally by the general population, with particular popularity among American college students [10].

It is not clear when or by whom methamphetamine was first synthesized. Various accounts indicate its first preparation somewhere between 1888 and 1934 [6]. In any case, Hauschild [11] published the first studies of the pharmacology of methamphetamine in 1938, characterized its stimulant effects in animals, and also carried out a self-experiment.

3. CLINICAL USE OF AGENTS

3.1. Therapeutic Applications

Psychostimulants generally increase the level of activity, alleviate fatigue, increase alertness, and elevate mood (or cause euphoria in high doses). Unfortunately, the ability to produce euphoria leads these compounds to have a high potential for abuse and dependence. The principal clinical indications for psychostimulants are in the treatment of attention deficit hyperactivity disorder (ADHD) and the sleep disorder known as narcolepsy. A less commonly recognized use, but one that is gaining importance, is in the treatment of depression in terminal patients or the chronically ill [12–14]. There also is need for psychostimulants in certain occupations, for example, in the military, as a countermeasure to fatigue from irregular or prolonged work hours, where a high level of vigilance and alertness must be maintained [15,16]. Some specific clinical applications will now be discussed.

Attention deficit hyperactivity disorder is a diagnosis applied mostly to children, but one that persists into adulthood for many people. It is reflected in a persistent pattern of inattention and/or hyperactivity–impulsivity that is more frequent and severe than typically observed in individuals at a comparable level of development [17]. Inattention prevents ADHD patients from keeping their mind on one thing and focusing their attention; they are easily bored with a task after only a short while. They have no difficulty devoting attention to activities they enjoy, but find it hard to focus conscious attention to organizing or completing a task, or learning something new. They may forget to plan ahead and tasks are rarely completed, or are filled with errors.

Children with ADHD (particularly of school age) have great difficulty being still, they may be in and out of their seats, and talk incessantly. The inability to focus makes learning tasks boring, and exacerbates the desire to move around and become involved in distractions. ADHD children may squirm, shake their legs, touch everything, or make distracting noises. Hyperactive teens and adults may feel intensely restless, and may try to do several things at once, going from one activity to the next. Impulsivity is another characteristic of ADHD, with patients often acting without thinking about the consequences. They may have difficulty curbing their immediate reactions to situations, making inappropriate remarks without thinking what they are saying. They find it hard to wait for things they want or to wait to take their turn.

In normal subjects, psychostimulants can increase activity and talkativeness, especially at higher doses. Paradoxically, in ADHD sufferers, stimulants appear to have a calming effect, and allow an increased focus and attention to tasks. While appearing paradoxical, it is now believed that the decreases in activity in ADHD are secondary to improvements in attention. This beneficial effect of low doses of the stimulants has led to a large number of children being prescribed methylphenidate (Ritalin®) or various amphetamine preparations for the treatment of ADHD. That, in turn, led to great concern about these drugs being overprescribed for ADHD, and that children who are merely highly energetic were routinely being given them for behavior management. The reader should be aware of this social issue, but it requires no further comment in the context of this chapter.

Narcolepsy is a condition that includes as its predominant symptoms excessive daytime sleepiness (EDS), persistent drowsiness, and daytime sleep attacks that may occur without warning and are often irresistible. Another hallmark symptom of narcolepsy is cataplexy, which is a sudden loss of voluntary muscle control, often triggered by emotions such as laughter or surprise. Cataplexy occurs more frequently during stress or fatigue. The attack may involve only a feeling of weakness and limp muscles or it may result in total muscular collapse, during which the person can appear unconscious, but actually remains awake and alert. Attacks may be very brief or may last for tens of minutes. Another characteristic symptom of narcolepsy is hypnagogic hallucinations. These are vivid, realistic, and often frightening, reminiscent of nightmares, and usually accompanied by sleep paralysis, a temporary inability to move. Whereas the psychostimulants can have a beneficial effect, they are likely to be supplanted by newer drugs that are more specific and have fewer side effects.

Use for depression in terminal illness. Although this indication for psychostimulants is not as widely recognized, agents such as amphetamine and methylphenidate are preferred because they do not suffer from the weeks-long delay in onset of action that is characteristic of traditional antidepressant medications. Thus, a rapid antidepressant response can be achieved in severely ill patients, who in some cases may not survive the several weeks necessary for a traditional antidepressant medication to begin to have an effect [12,18–20].

Use in obesity. As noted earlier, many of the psychostimulants also have been used as anorectics (anorexics; anorexigenics) (Table 1), that is, as appetite suppressants. A few of them are still useful in this regard, but the high abuse potential of psychostimulants, coupled with the development of tolerance to their anorectic effects, has meant that prescribing psychostimulants for weight control has generally fallen into disfavor.

Apnea in premature infants. Apnea of prematurity (AOP) occurs in approximately 90% of premature neonates weighing less than 1 kg at birth, and in 25% of infants with a weight of less than 2.5 kg [21]. The first line pharmacological therapies for the management of AOP, to stimulate respiration, are the methylxanthines, with theophylline (**5**) presently being most extensively used. Recent studies suggest, however, that caffeine (**4**) should be considered the drug of choice because of similar efficacy, longer half-life, fewer adverse effects, and better brain penetration than theophylline [22].

3.2. Side Effects, Adverse Effects, and Drug Interactions

Generally, psychostimulants such as amphetamine **7** and methylphenidate **8** can be used safely with most classes of medications and with few contraindications [23]. The acute adverse reactions to stimulants can generally be understood from the perspective of their pharmacology. Psychostimulants act as indirect sympathomimetic agents; they either directly release stored catecholamines, including those in peripheral adrenergic neurons responsible for vascular tone, or block their reuptake. These actions affect the cardiovascular system in fairly predictable ways. In addition, cocaine produces a local anesthetic effect by the blockade of sodium channels [24]. Although that would normally be the pharmacological basis for a class I antiarrhythmic drug, it paradoxically induces proarrhythmia [25].

7

Table 1. Psychostimulant and Anorexigenic Preparations

Generic Name (Structure)	Trade Name	Originator	Chemical Class	Dose (mg/day)[b]
Psychostimulants				
Cocaine HCl (3)	Cocaine HCl powder	Mallinckrodt	Ecgonine methyl ester benzoate	NA
Amphetamine (7)	Adderall	Shire Richwood	Phenethylamine	5–30
Amphetamine sulfate (7)	Amphetamine sulfate	Lannett		5
Dextroamphetamine sulfate (10)	Dextroamphetamine sulfate	Various		5–10
	Dexedrine	SmithKline Beecham		
	Dextrostat	Richwood		
Methamphetamine HCl (11)	Desoxyn	Abbott		5
	Desoxyn Gradumet	Abbott		5–10
Methylphenidate HCl (8)	Methylphenidate HCl	Various	α-Phenyl-2-piperidineacetic acid methyl ester	5–20
	Ritalin	Ciba-Geigy		
	Methylin	Mallinckrodt		
	Metadate ER	Medeva		10 mg ER
	Concerta	Alza		18 mg ER
Modafinil (44)	Provigil	Cephalon	Diphenylmethyl-sulfinyl-2-acetamide	100–200
Pemoline (9)	Pemoline	Apothecan	2-Amino-5-phenyl-4(5H)oxazolone	18.75–75
	Cylert	Abbott		
	PemADD	Mallinckrodt		
Caffeine (4)	Quick Pep; Caffedrine; NoDoz; Stay Awake; Vivarin; Stay Alert; Enerjets; Starbucks	Thompson; Thompson; Bristol-Myers; Major; SK-Beecham; Apothecary; Chilton	Trimethylxanthine	75–200
Anorexiants				
Benzphetamine (22)	Didrex	Upjohn	Phenethylamine	25–50
Diethylpropion (16)	Diethylpropion	Various	Phenethylamine	25–75
	Tenuate	Aventis	Phenethylamine	
Phendimetrazine (21)	Phendimetrazine	Various	Phenethylamine	35
	Bontril PDM	Carnrick		
	Plegine	Wyeth-Ayerst		
	Phendimetrazine	Various		
	Adipost	Jones		105 mg SR

Table 1 (*Continued*)

Generic Name (Structure)	Trade Name	Originator	Chemical Class	Dose (mg/day)[b]
	Bontril Slow-Release	Carnrick		
	Dital	UAD		
	Dyrexan-OD	Trimen		
	Melfiat-105 Unicelles	Numark		
	Prelu-2	Boehringer-Ingelheim		
	Rexigen Forte	ION Labs		
Phentermine (**19**)	Phentermine	Various	Phenethylamine	8–37.5
	Ionamin	Medeva		
	Fastin	SmithKlineBeecham		
	Zantryl	Ion		
	Adipex-P	Lemmon		
	Obe-Nix 30-P	Holloway		
Decongestants and Bronchodilators				
Ephedrine sulfate (**1**)	Pretz-D 0.25% spray	Parnell	Phenethylamine	NA
	Ephedrine sulfate	West-Ward	Phenethylamine	25
	Ephedrine sulfate	Various	Phenethylamine	50 mg/mL

[a]Administered orally unless otherwise noted.

In addition to acute effects, however, prolonged usage of amphetamines (and other psychostimulants) can produce an "amphetamine psychosis." This syndrome was first clearly documented by Connell [26], and is regarded as very similar to paranoid schizophrenia, comprising "paranoid psychosis with ideas of reference, delusions of persecution, auditory and visual hallucinations in a setting of clear consciousness" [26]. The psychosis clears quickly after the drug is withdrawn. Psychosis has been induced experimentally in normal subjects by continuous amphetamine administration [27]. Amphetamine psychosis has been discussed extensively by Angrist [28]. Interestingly, psychostimulants can induce a psychotogenic response in schizophrenics, in doses that are subpsychotogenic in normal subjects, and methylphenidate was found to have greater potency in that regard [29]. Activation of psychotic symptoms by methylphenidate was found to be a predictor of relapse risk [30]. These, and other similar studies, are all consistent with the dopamine hypothesis of schizophrenia.

Methylphenidate. Methylphenidate **8**, (Ritalin) is widely prescribed for the treatment of ADHD. Indeed, methylphenidate remains the most common drug therapy for treatment of ADHD [31], and approximately 90% of children treated for ADHD are given this drug [32], representing approximately 2.8% of all US children aged 5 to 18 [33]. It is both well tolerated and efficacious in the treatment of attention deficit hyperactivity disorder, and is associated with few serious adverse effects [34]. Although there are rare reports of drug interactions between methylphenidate and certain other drugs, they are so infrequent that there is no consistent pattern that can be identified. Toxic concerns with methylphenidate would principally revolve around the abuse of this drug to obtain a stimulant high, and the consequent possibility of developing dependence. A further concern with the long-term use of methylphenidate is the possibility that patients may be at increased risk for psychostimulant abuse. Although when taken orally methylphenidate has a low euphorigenic potential [35], when used intravenously it has an abuse pattern and symptoms of toxicity similar to cocaine and amphetamine [36]. Recent studies in rats also have shown that animals treated with methylphenidate develop behavioral sensitization, suggesting that human users may have increased susceptibility to psychostimulant abuse [37].

Methylphenidate has chiral centers, and exists as enantiomers, although the racemate is used clinically. Studies have shown that the (+)-isomer is responsible for the effect **8** of the racemate [38,39].

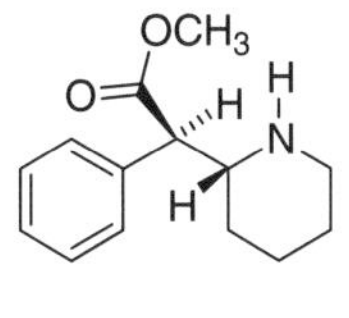

8

Pemoline. Pemoline (**9**), an agent used in treatment of ADHD, has been associated with hepatotoxicity, with the majority of cases occurring in pediatric patients. From its marketing in 1975 up to 1989, 12 cases of acute hepatic failure and 6 deaths associated with pemoline hepatotoxicity had been reported to the FDA [40]. Death generally occurred within 4 weeks of the onset of signs and symptoms of liver failure. In two recent cases, pemoline-induced liver failure required liver transplantation [41].

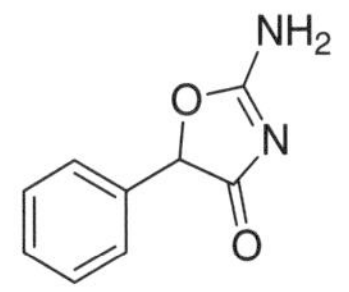

9

Although the absolute number of reported cases is not large, the rate of reporting is 4–17 times higher than that expected in the general population. This estimate may be conservative because of underreporting and because the long latency between initiation of pemoline treatment and the occurrence of hepatic failure may limit recognition of the association. If only a portion of actual cases was recognized and reported, the risk could be substantially higher. By contrast, a meta-analysis of the literature by Shevell and Schreiber [42] suggests that the risk of acute hepatic failure may be an overestimate. Nevertheless, because of its association with life-threatening hepatic failure, pemoline should not be considered as a first-line therapy for ADHD. In fact, pemoline has been withdrawn from the Canadian market as a result of this toxicity [43].

Cocaine. The coca plant is a small shrub or tree that is indigenous to South America, where for centuries the leaves have been chewed by the local native populations. The dried leaves of *Erythroxylum coca* Lamarck, or *E. truxillense* Rusby, commercially known as Huanuco coca, or Truxillo coca [4], respectively, serve as the raw material for the production of (−)-cocaine (**3**). Cocaine was first isolated in 1860, and became medically important as an excellent local anesthetic agent, but which is a potent and highly addictive CNS stimulant. The acute toxicity of cocaine derives primarily from its intense sympathomimetic actions. In 1991, an attempt was made to assess the intrinsic toxicity of cocaine by computing the incidence of adverse health outcomes per population of drug abusers. The rates of emergency room visits and deaths were estimated at 15.1 and 0.5, respectively, per 1000 persons using drugs [44].

Acutely, cocaine can cause anxiety or panic reactions. Used chronically, cocaine can induce a psychosis that closely resembles that produced by amphetamine. It is generally considered that amphetamine psychosis predominantly mimics the positive symptoms of schizophrenia, but in fact stimulant-induced psychosis can mimic a broad range of symptoms, including negative and bizarre symptoms [45]. Paranoid behavior has been produced in experienced cocaine users by continuous (4 h) cocaine infusion [46].

Cocaine can have marked effects on the heart and cardiovascular system. Adverse actions may include myocardial ischemia, cardiac arrhythmias, cardiotoxicity, hypertensive effects, cerebrovascular events, and a hypercoagulable state [25,44]. By 1997, more than 250 cases of myocardial infarction related to the recreational use of cocaine had been documented in the literature [47]. Although less common, aortic dissection related to the use of cocaine-free base ("crack cocaine") has been reported [48]. Seizures also can be associated with cocaine use [49].

In addition to these physiological toxicities, cocaine addicts suffer from a variety of social and economic problems that result in tremendous costs to society. Many of the estimated 1.5 million cocaine addicts in the United States (see www.nida.nih.gov), are underemployed, if they are employed at all, are likely to be involved in drug distribution activities, and typically perform only marginal roles in the legal economic system [50,51]. Adults in such drug-using households rarely engage in conventional behaviors, and often parent children using conduct norms that are structured to produce individuals who have reduced chances to become conventional adults [52].

Caffeine. The psychostimulant action of caffeine generally is accepted as well established. Caffeine quickens reaction time and enhances vigilance, increases self-rated alertness, and improves mood. There is, however, little unequivocal evidence to show that regular caffeine use is likely to benefit substantially either mood or performance. Indeed, one of the significant factors motivating caffeine consumption appears to be "withdrawal relief." [53] The most common symptom of caffeine withdrawal is headache, which typically begins 12–24 h after the last dose of caffeine [54]. Other symptoms of caffeine withdrawal include fatigue, drowsiness, dysphoria, difficulty in concentrating, decreased cognitive performance, depression, irritability, nausea or vomiting, and muscle aches or stiffness [54].

Caffeine can produce adverse and unpleasant effects if doses are increased. Caffeine also has weak reinforcing properties, but with little or no evidence for upward dose adjustment, possibly because of the adverse effects of higher doses. Withdrawal symptoms, although relatively limited with respect to severity, do occur, and may contribute to continued caffeine consumption [55]. Health hazards are small, if any, and caffeine use is not associated with any type of incapacitation [56]. Acute intake of caffeine does increase blood pressure, with the strongest pressor response in hypertensive subjects. Thus, regular caffeine consumption may be harmful to some hypertension-prone subjects [57]. Some studies with repeated administration of caffeine have shown a persistent pressor effect, whereas in others chronic caffeine ingestion did not increase blood pressure [57]. Epidemiologic studies have produced contradictory findings regarding the association between blood pressure and coffee consumption. During regular use, tolerance to the cardiovascular responses develops in some people, and therefore no systematic elevation of blood pressure can be shown either in long-term or in population studies. The hemodynamic effects of chronic coffee and caffeine consumption have not been sufficiently studied. Finally, caffeine may provoke a panic attack in individuals who suffer from panic or anxiety disorders [58–61].

The psychostimulant properties of caffeine are now generally attributed to antagonism of adenosine A1 and A2 receptors, although the contribution of the respective receptors is still under debate. Recent studies suggest that heterodimers of A1 and A2 receptors may provide a concentration-dependent "switch" by which low and high concentrations of synaptic adenosine produce opposite effects on glutamate release [62]. Recently, an adenosine A_{2A} receptor knockout mouse has been developed that has behavioral symptoms that correspond to functional antagonism of this receptor, similar to the effects of caffeine [63].

3.3. Absorption, Distribution, Metabolism, and Elimination

All substituted amphetamines are strong organic bases, with pK_a values ranging from 9.5 to 10 [64]. The pK_a of both cocaine and phenmetrazine is somewhat lower, at 8.5, and methylphenidate has a pK_a of 8.8 [64]. Thus, these bases are all significantly protonated at physiological pH, and binding to their biological targets probably occurs with the protonated species (for example, see Ref. [65]). These drugs are all administered as their water-soluble salts, usually hydrochlorides or sulfates. Of course, at physiological pH these bases exist in an equilibrium between the protonated ionized form, and the unprotonated unionized species. The latter free bases are lipid soluble and readily penetrate the brain, where they exert their CNS stimulant effects. Many of these drugs are eliminated in the urine unchanged because acidic urine leads to a higher fraction of protonated species, thus decreasing reabsorption of the unchanged drug in the renal tubules. Decreasing urinary pH by, for example, administering ammonium chloride leads to the anticipated increased urinary excretion and reduced duration of action [66]. A comparison has been reported of the urinary excretion pattern of methamphetamine in humans, guinea pig, and rat [67]. In humans, 23% of the dose was excreted unchanged. Ring-hydroxylated and N-demethylated metabolites were excreted as 18% and 14% of the dose, respectively.

Amphetamine metabolism. The metabolism of (+)-amphetamine **10** is variable, depending on the species studied. Possible metabolic transformations involve hydroxylation at the alpha- or beta-side-chain carbon atoms, on the nitrogen atom, and at the *para* position of the aromatic ring. These metabolites can then be further oxidized, or conjugated and excreted. One or more of these pathways predominates, depending on which animal species is being studied. In man, the half-life of (+)-amphetamine has been reported as 7h [68]. Approximately 30% of the dose of racemic amphetamine is excreted unchanged, and acidification of the urine can decrease the half-life significantly [69,70]. In man, the principal metabolite is benzoic acid [71]. The details of the sequences of metabolic reactions of amphetamine that lead to benzoic acid have not been elucidated [68], but the beta-hydroxylated metabolite, norephedrine, also has been identified as a metabolite in man [72].

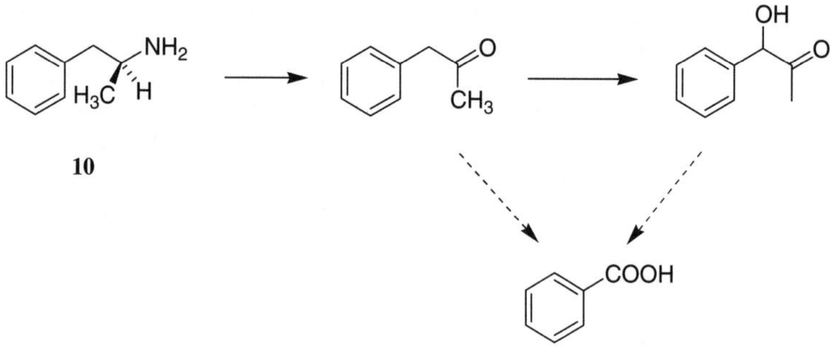

The metabolism of methamphetamine **11** involves both *N*-demethylation and ring hydroxylation. Caldwell et al. [67] reported that in humans 22% of the administered dose was excreted as unchanged drug, and 15% as the 4-hydroxylated compound.

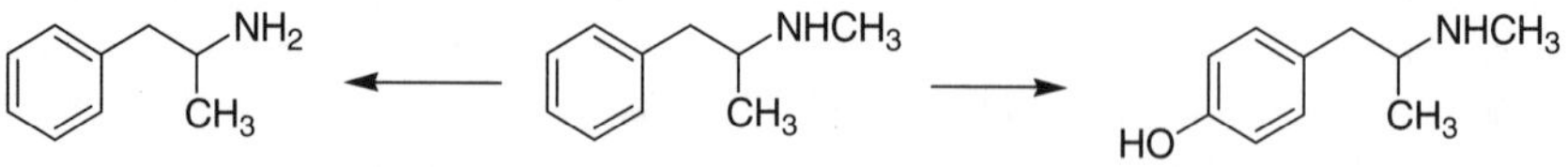

Methylphenidate metabolism. The metabolism and pharmacokinetics of methylphenidate have been studied extensively. Methylphenidate **8** is administered as the racemic threo isomer, but the (−)-threo enantiomer is more rapidly metabolized.

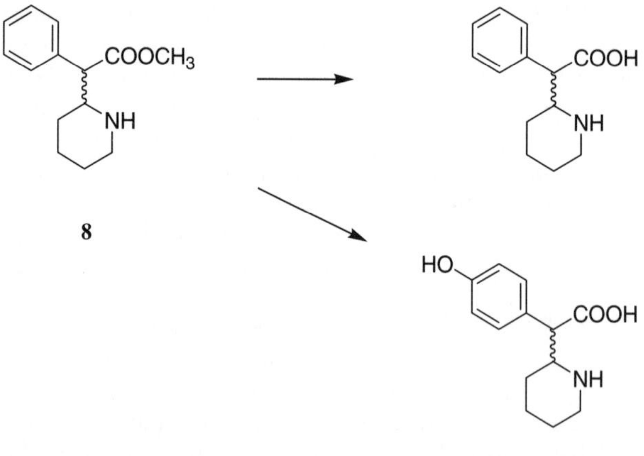

metabolites after an orally administered dose [73]. Ritalinic acid is not pharmacologically active. The lability of the ester function is probably the major factor limiting the oral bioavailability of methylphenidate to between 10% and 50% [74]. A ring-hydroxylated ritalinic acid metabolite (2%) also has been identified. Other minor pathways involving oxidation of the piperidine ring (oxo-ritalin) and conjugation reactions represent less than approximately 1% of the administered dose [32].

Methylphenidate is an ester, and the methyl ester is rapidly cleaved. The ester hydrolysis product, called ritalinic acid, comprises approximately 80% of the urinary

Cocaine metabolism. Cocaine **3** has two ester functions, and both can be hydrolyzed *in vivo* to generate metabolites. Hydrolysis of the methyl ester leads to benzoylecgonine **12**, and hydrolysis of the benzoyl ester leads to ecgonine methyl ester **13**. Tropan-3β-ol-2β-carboxylic acid is known as ecgonine **14**. In cocaine users who also consume significant amounts of ethanol, a transesterification product (cocaethylene **15**) is also detected. Cocaethylene is also a potent psychostimulant, with approximately four times higher potency as a local anesthetic than cocaine itself [75], and can enhance the cardiotoxicity associated with cocaine use.

and thus, the principal metabolite is the N-deethylated compound **17**, comprising approximately 35% of the administered dose. Reduction of the carbonyl is less important, with approximately 20% of the dose going that route to afford N,N-diethylnorephedrine. Approximately 30% of the dose cannot be accounted for as an amine product in the urine and is probably a deaminated metabolite [76]. Studies by Yu et al. [77] found that the N-deethylated metabolite **17** was probably

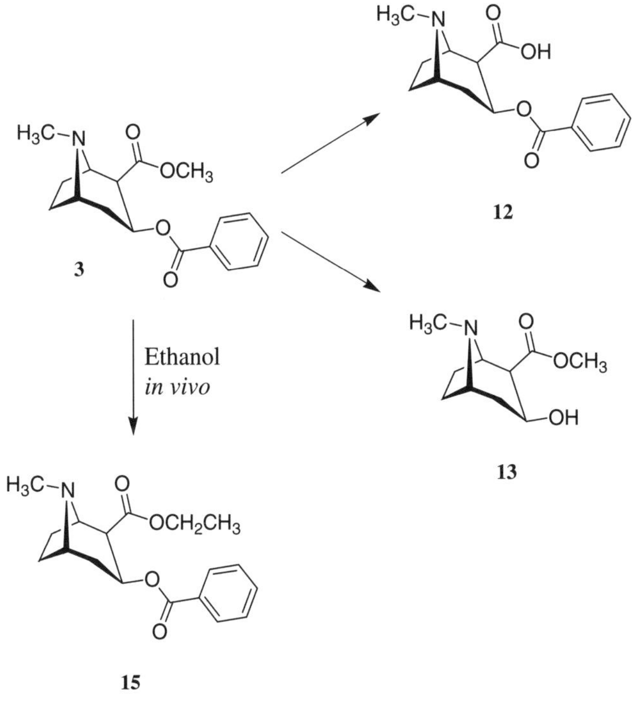

Diethylpropion metabolism. Diethylpropion **16**, is used most extensively as an appetite suppressant. It possesses the core phenethylamine structure characteristic of many psychostimulants, but is a tertiary aminoketone. It is extensively metabolized in man, with only approximately 3–4% of the drug excreted unchanged. Alpha-alkylaminoketones are apparently N-dealkylated readily

responsible for the pharmacological effects of diethylpropion **16**. These workers reported that the N-monoethyl metabolite **17** was a substrate at the norepinephrine and serotonin transporters and an inhibitor at the dopamine uptake transporter, whereas (1R,2S) and (1S,2R)-N,N-diethylnorephedrine as well as diethylpropion itself were inactive in those assays.

4. PHYSIOLOGY AND PHARMACOLOGY

4.1. Where and How These Drugs Work

Neurons in the central nervous system communicate by chemical transmission. Of relevance to the present discussion are monoamine neurons that release dopamine, norepinephrine, or serotonin as one of their transmitters in response to an action potential. Reuptake transporter proteins embedded in the neuronal plasma membrane then clear the synapse of monoamines, typically taking up 70–80% of the released transmitter. This reuptake is thought to be the major termination mechanism for the monoamine chemical signaling process.

All psychostimulants appear to elevate synaptic levels of dopamine and norepinephrine. In addition, cocaine, and to a lesser extent some of the other agents also raise synaptic levels of serotonin. It is the current consensus that elevated dopamine levels lead to CNS stimulation and are responsible for the reinforcing properties of stimulants [78–84]. It is now widely accepted that mesocortical dopamine plays a critical role in drug reward [85].

There are two principal mechanisms for increasing synaptic monoamine levels. One is to block the reuptake of neurotransmitter after its excitation-coupled release from the neuronal terminal. Thus, blocking the action of the uptake carrier protein prevents clearance of the neurotransmitter from the synapse, leaving high concentrations in the synaptic cleft that can continue to exert a signaling effect. This mechanism is the one invoked to explain the action of cocaine, a potent inhibitor of monoamine reuptake at the dopamine, serotonin, and norepinephrine transporters, and of methylphenidate, which is a reuptake inhibitor at the dopamine and norepinephrine transporters [86]. It should be noted, however, that methylphenidate also has the ability to induce the release of catecholamines stored in neuronal vesicles [87,88].

The second mechanism is the one more relevant to the action of amphetamine and related agents. This mechanism is illustrated in Fig. 1. Biogenic amine transporters, which include the dopamine transporter (DAT), norepinephrine transporter (NET), and serotonin transporter (SERT) are the targets for psychostimulant drugs. These proteins are members of the neurotransmitter/sodium symporter (NSS) family [89]. Amphetamine, and other small molecular weight compounds with similar structures, are substrates at the monoamine uptake carriers and are transported into the neuron (Fig. 2). The uptake carrier has an extracellular and intracellular face, and after transporting a substrate (amphetamine, etc.) into the neuron, the intracellular carrier face can bind to dopamine and transport it back to the extracellular face.

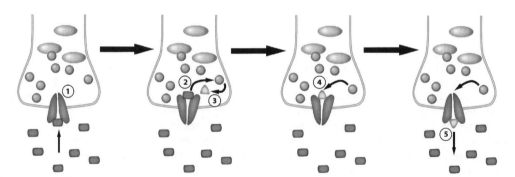

Figure 1. Amphetamine interacts with the dopamine transporter protein, 1, and is transported inside. After being transported inside the terminal, high concentrations of amphetamine can displace dopamine from vesicular storage sites 2, leading to elevated cytoplasmic levels of dopamine 3. After amphetamine dissociates on the intraneuronal surface, dopamine binds to the carrier 4. The carrier then transports dopamine to the extracellular face 5, driven by the favorable concentration gradient, where the dopamine dissociates and leaves the carrier available for another cycle. This overall process is referred to as the alternating access model of molecular transport [90,91].

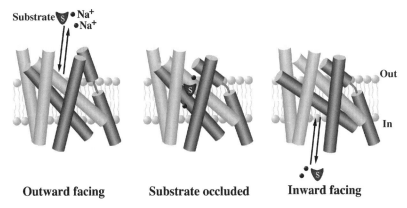

Figure 2. This illustration shows the normal functioning of monoamine uptake transporters. The monoamine is taken up from the synaptic space and transported into the neuron terminal. Na^+ and Cl^- are cotransported, and K^+ is countertransported in the process.

This exchange diffusion mechanism is calcium independent, and is capable of robustly increasing synaptic transmitter levels. This process is often described as a "reversal" of the normal uptake carrier process.

The substrate binding site is located near the center of the protein in the membrane, and has "alternating access" to either side of the membrane, resulting from reciprocal opening and closing of the cavities that connect the binding site to either side of the membrane. The energy for transport is primarily derived from the Na^+ gradient. Intracellular levels of Na^+ are kept low by active pumping out of the cell. The resulting inward Na^+ gradient, coupled with the negative membrane potential, is the source of the energy used to drive substrates into the cell. A more detailed look into the mechanics of this process has been developed from recent studies of crystal structures for the NCS1 benzyl-hydantoin tansporter from *Microbacterium liquefaciens* [92].

Whereas the CNS stimulant effects of these molecules depend on an action in the brain, uptake inhibitors and substrates at peripheral monoamine carrier sites obviously can exert other physiological effects. Cocaine is an excellent local anesthetic agent. Furthermore, its potent inhibition of norepinephrine reuptake leads to stimulation of alpha-adrenergic receptors, causing local vasoconstriction that delays the diffusion of the anesthetic agent out of the tissue. Similarly, users who chronically insufflate cocaine into their nasal passages often develop necrotic lesions due to the local vasoconstricting effect of cocaine, again due to the blockade of norepinephrine reuptake. Not surprisingly, cocaine and amphetamines have effects on the cardiovascular system, by virtue of their ability to enhance indirect adrenergic transmission at peripheral sites. Knowledge of the physiology of the sympathetic nervous system and the functions of peripheral adrenergic nerve terminals allows a relatively straightforward prediction of the types of drug effects produced by monoamine uptake inhibitors or releasing agents.

Nevertheless, recent studies have begun to focus attention on glutamate systems as potential key components of the actions of psychostimulants. For example, Swanson et al. [93] have shown that repeated cocaine administration leads to long-term attenuation of group I metabotropic glutamate receptor function in the nucleus accumbens. In particular, this functional reduction was related to significantly reduced mGluR5 immunoreactivity in the medial nucleus accumbens. Even more exciting is the report that mGluR5 knockout mice do not display the reinforcing and locomotor effects of cocaine, in spite of the fact that cocaine administration increases extracellular dopamine in the nucleus accumbens of these mice to levels that do not differ from wild type animals [94].

Group III mGluR agonists have also been shown to have profound effects when administered with psychostimulants. For example, a nonselective group III mGluR agonist attenuated the amphetamine- or cocaine-induced increases in locomotion and striatal dopamine [95–97]. Among the group III mGluRs, the metabotropic glutamate receptor 7 (mGluR7) has the highest expression in brain regions involved in reward phenomena [98–100]. AMN082, a selective mGluR7 agonist, dose-dependently inhibited the rewarding effects of cocaine, as assessed by electrical brain stimulation reward and cocaine self-administration in rats [101]. The effect was blocked by coadministration of a selective mGluR7 antagonist. In the near future, the role of glutamate systems in the actions of psychostimulants likely will be more fully elucidated, resulting in new approaches to the treatment of conditions that now respond to classical stimulants.

4.2. Biochemical Pharmacology: Receptor Types and Actions

The monoamine reuptake carrier proteins (targets of the psychostimulants) are members of a larger Na^+/Cl^- symporter family that includes a number of other proteins, including the GABA transporters, amino acid transporters, and orphan transporters [102]. The primary amino acid sequence of the monoamine transporters is highly conserved, with several regions of these proteins having high homology [103]. It is presently believed that all of the members of this family possess a membrane-spanning 12 alpha-helix motif, with a single large loop containing glycosylation sites on the external face of the membrane (Fig. 3). Members of this family of proteins have been identified not only in mammalian species but also in eubacteria and archeobacteria, indicating their very early emergence in the evolution of life.

The human norepinephrine uptake transporter was first sequenced and then expressed in HeLa cells in 1991 [104] and found to have properties identical to those of the native transporter. The cloning and sequencing of the dopamine transporter [105–107] and the serotonin transporter [108,109] were reported in the same year. There are a number of excellent review articles written about monoamine transporters [102,103,110–112]. The most exciting recent development is the construction of homology models of the dopamine [113,114] and serotonin [115,116] transporters, derived from the 1.65 Å X-ray crystal structure of the bacterial Leucine transporter (LeuT), from *Aquifex aeolicus* [117].

Pharmacological studies of the mechanism of action for psychostimulants in animals have

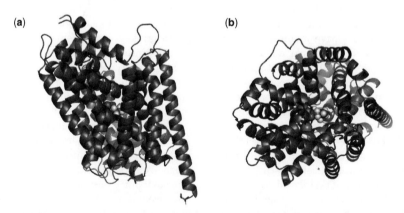

Figure 3. Representation of the 12 helix transmembrane transporter protein family from the X-ray crystal structure of the *Aquifex aeolicus* leucine transporter [117]. (a) Side view, looking in the plane of the cell membrane, with the extracellular side toward the top. (b) The top view, looking from the extracellular face of the transporter. Both the amino terminus and carboxyl terminus are intracellular, with the second extracellular loop being larger, and possessing glycosylation sites. For clarity, all of the loops are not displayed. The extracellular side is toward the top of the figure, and the intracellular surface toward the bottom. Leucine and two sodium ions are visible in the center of the bundle, represented as space-filling models.

almost uniformly pointed to the importance of dopamine pathways for the increases of locomotor activity and reinforcing properties [118,119]. The conclusions of those studies have generally been extrapolated to humans, with little clinical evidence until recently to support these ideas. In the past several years, however, clinical studies of several stimulants, using *in vivo* brain imaging either with SPECT techniques or with PET techniques, has provided evidence for elevated extracellular dopamine in response to psychostimulant administration. In essence, these studies employ either a single photon- or a positron-emitting dopamine receptor antagonist. The labeled antagonist is administered both in the absence and in the presence of the stimulant drug of interest. The imaging technique then is used to determine how much of the labeled ligand has been displaced from its receptors by competition from increased extracellular endogenous dopamine. Based on the known affinity of the labeled ligand for its dopamine receptor, calculations can determine the increased concentration of dopamine that must have been available at the receptors. These definitive studies have clearly established a role for dopamine in the effects of stimulants in humans [78,80].

This type of approach has recently been applied to the study of methylphenidate. For example, Booij et al. [120] used SPECT imaging and an [^{123}I]benzamide dopamine D_2 receptor ligand antagonist ([^{123}I]IBZM) to measure significant displacement of the ligand by endogenous dopamine that had been released in response to administration of methylphenidate. In related work, Volkow et al. [80] used [^{11}C]-(+)-threo-methylphenidate to show that greater than 80% occupancy of the dopamine transporter was required to produce the stimulant "high." With the dopamine D_2 receptor antagonist [^{11}C]raclopride, Volkow et al. [121,122] showed that the intensity of the methylphenidate "high" was quantitatively correlated with the levels of released dopamine and dopamine D_2 receptor occupancy. Subjects who perceived the most intense high had the highest increases in extracellular dopamine. Conversely, no high was experienced by subjects when methylphenidate did not increase dopamine levels. In a second study, using the same methodology, this same group [123] found that subjects who "liked" the effects of methylphenidate had significantly lower dopamine D_2 receptor levels than subjects who disliked its effects. The authors speculated that lower D_2 receptor density might be a factor contributing to psychostimulant abuse by providing a more pleasurable response. These imaging studies illustrate how data from animal research can now be validated in humans.

Because the stimulants cause increased synaptic levels of dopamine, and other monoamine neurotransmitters, they indirectly lead to stimulation of various postsynaptic receptors through the increased concentrations of neurotransmitter. A large number of animal studies have been reported using various agonists and antagonists to elucidate the role of different dopamine receptor isoforms. Until recently, however, only nonspecific ligands (i.e., with effects on both the D_1-like and D_2-like families) were available. In drug discrimination studies, rats have been trained to recognize and discriminate the interoceptive cues produced by injection of amphetamine or cocaine [124]. Administration of the partial but selective D_1-receptor agonist SKF 38393 was partially recognized by the cocaine-trained rats, but not by amphetamine-trained animals. Yet, both amphetamine- and cocaine-trained rats discriminated the cue produced by the dopamine D_2 agonist bromocriptine as being similar to their training drugs. A dopamine D_3-selective agonist produced cocaine responses, but was only partially recognized by amphetamine-trained rats. Following additional experiments with dopamine receptor subtype selective antagonists, the authors concluded that the dopamine D_2 receptor played an essential role, but that both the D_1 and D_3 receptors might have some less important function. There is an extensive present research effort underway in many laboratories that is attempting to elucidate both the anatomical substrates and the specific postsynaptic receptor isoforms that are important in the various actions of psychostimulants.

A role for serotonin? Although the conventional wisdom is that stimulants elevate synaptic dopamine, it is not at all clear that this sole mechanism is responsible for the spectrum of effects produced by the psychostimulants. In addition, it is becoming evident that animal models used to understand the stimulants must consider mechanisms underlying effects on locomotor activity somewhat differently from those that govern either reward or drug discrimination phenomena [125,126]. The use of mice genetically deficient for the serotonin or dopamine transporter ("knockout mice") has produced some particularly interesting findings. For example, knockout mice lacking the DA transporter have high levels of extracellular dopamine, a condition that would presumably mimic the pharmacological action of cocaine. Consistent with this prediction, they display spontaneous hyperlocomotion [127]. Surprisingly, however, these mice still self-administered cocaine [128]. Further experiments in these mice indicated the probable involvement of the serotonin transporter. In addition, conditioned place preference, another animal model of the reinforcing quality of a drug, could be established for cocaine in mice lacking either the dopamine transporter or the serotonin transporter [129]. Similarly, place preference also could be established for methylphenidate, another stimulant that is thought to work through dopamine mechanisms, in mice lacking the dopamine transporter.

Experiments with knockout mice often produce unexpected results. It must be kept in mind, however, that when a key protein is missing during neural development, the offspring may have some type of adaptation that does not occur in the wild type organism. Some caution, therefore, must be exercised in interpreting the results. For example, Belzung et al. [130] found that mice lacking the serotonin 5-HT_{1B} receptor failed to display conditioned place preference. When, however, these knockout mice were compared in studies using classical pharmacological antagonists of the 5-HT_{1B} receptor, divergent results were obtained. The 5-HT_{1B} receptor knockout mice had an increased locomotor response and increased propensity to self-administer cocaine [131]. By contrast, a 5-HT_{1B} receptor antagonist attenuated cocaine-induced locomotor effects but had no effect on cocaine self-administration [132]. The authors point out that compensatory changes during development of the knockout mice may have rendered them more vulnerable to the effects of cocaine.

Nonetheless, there is a vast body of literature documenting interactions between dopamine and serotonin pathways in the brain [133–137]. Clearly, however, if a drug (e.g., cocaine) releases multiple transmitters, then a behavioral interaction is not surprising. Inhibition of presynaptic reuptake of serotonin, for example, no doubt leads to postsynaptic activation of a variety of other receptors, some of which could modulate dopamine function. In addition to potential effects on 5-HT_{1B} receptors, other studies have implicated serotonin 5-HT_4 receptors [138], 5-HT_{2A} receptors [139], and 5-HT_{2C} receptors [140]. 5-HT_{1A} receptors also can modulate the locomotor effects of cocaine [141].

A role for norepinephrine? Although the vast majority of studies of psychostimulants have focused on the role of dopamine and/or serotonin, the importance of norepinephrine (thought to be paramount 35 years ago) generally has been overlooked. Details of the mechanism of action of psychostimulants have been developed primarily using animal models, in which dopamine seems to be the key player, and these results then have been extrapolated to man. Yet cocaine also is a potent NE uptake inhibitor, and the potency of amphetamine for norepinephrine release is similar to that for dopamine release. Indeed, in the rat prefrontal cortex, amphetamine and cocaine increased extracellular norepinephrine to an extent that was quantitatively similar to that of dopamine [142]. Further, it appeared that the increase in prefrontal cortical norepinephrine was actually due to the blockade of the norepinephrine transporter by both drugs. Recently, Rothman et al. [143] reported that the oral doses of several stimulants required to produce amphetamine-like subjective effects in humans were most closely correlated with their ability to release NE, and not DA. Further, their ranking in subjective effects did not correlate with decreased plasma prolactin, a response that is mediated by dopamine receptor stimulation. These authors suggested that NE might contribute to the amphetamine-like psychopharmacology of stimulants, at least in humans.

Until clinical studies are carried out using receptor blockers and specific norepinephrine transporter inhibitors, this area will remain muddy, at best. In virtually every example, from amphetamine to cocaine, the compounds have significant effects at the norepinephrine transporter, in some cases equal to or even greater, than at the dopamine transporter. When behavioral or mood changes are correlated with levels of extracellular dopamine, and dopamine is highly correlated with changes in extracellular norepinephrine, one cannot be certain which underlying pharmacology is ultimately more important without experiments using specific blockers of both dopamine and norepinephrine transporters and receptors. It may be that effects on dopamine are necessary, but not sufficient, and that both norepinephrine and serotonin play modulatory roles. Because the stimulants have such diverse effects, including increasing activity, mood, appetite suppression, etc., it seems likely that serotonin and norepinephrine play more or less important modulatory roles, depending on which aspect of the specific drug's effects are being studied.

In a related vein, the subjective psychostimulant effects of amphetamine were attenuated following a 2 h pretreatment with a tyrosine- and phenylalanine-free amino acid mixture [144]. These amino acids are biosynthetic precursors of the catecholamines, and deprivation would be expected to produce transient reductions in endogenous dopamine and norepinephrine. The authors concluded that tyrosine depletion attenuates the release of dopamine required for the psychostimulant effect. Interestingly, the pretreatment did not reduce the subjective appetite suppressant (anorectic) effect of amphetamine. The study authors attributed this latter finding to a continued release of norepinephrine by amphetamine. Tyrosine depletion, however, would also attenuate norepinephrine biosynthesis and it may be more reasonable to conclude that the anorectic effect might be related to the often-overlooked ability of amphetamine to release neuronal serotonin.

This chapter will make no attempt to review all the literature that focuses on the role of norepinephrine and serotonin in the actions of psychostimulants. At the time of this writing, the general consensus seems to be that effects on dopamine systems are necessary, but perhaps not sufficient conditions to explain all the different actions of stimulants. There appears to be increasing awareness, spurred initially by studies of cocaine, that serotonin may be a much more important player than was heretofore recognized. In the next few years, this role likely will be studied and elucidated in much greater detail.

Psychostimulant sensitization. Repeated administration of amphetamine over a short period of time to rats, monkeys, or humans leads to a persistent behavioral sensitization to a subsequent low dose amphetamine challenge [145–148]. Long-lasting changes in brain neurochemistry and morphology also are observed following amphetamine sensitization [149]. In rodents, repeated exposure to amphetamine or cocaine leads to persistent changes in dendrite structure and dendritic spines in brain areas involved in incentive behavior, motivation, and judgment and inhibitory control of behavior [150]. This reorganization of synaptic connectivity is speculated to contribute to persistent behaviors associated with psychostimulant abuse, including addiction.

Caffeine and the other methylxanthines inhibit phosphodiesterases, the enzymes that degrade cAMP. For many years, it was believed that the stimulant effect of caffeine was due to this enzyme inhibition. At the plasma concentrations obtained after two to three cups of coffee (ca. 10 µM), however, antagonism of adenosine A_{2A} (and A_1) receptors in brain is believed to be the most relevant action to explain the stimulant effects of caffeine [151,152]. Perhaps not surprisingly, in view of earlier discussion in this chapter, caffeine administration has been shown to lead to elevated levels of brain dopamine [153,154]. It is thought that adenosine receptor stimulation facilitates GABA- or acetylcholine-mediated inhibition of dopamine receptors in striatopallidal and striatonigral neurons [155], with the end result of decreased dopaminergic function; adenosine antagonists would thus have a reverse action. Many studies have examined the interaction between adenosine A_{2A} receptors and dopamine receptors, both of which are highly concentrated and colocalized in the striatum and have reciprocal antagonistic interactions [156–159]. There is abundant evidence for pre- and postsynaptic interactions between adenosine and dopamine receptors, by which adenosine inhibits dopaminergic activity (for example, see Ref. [160]).

With respect to stimulation of locomotor activity in animal models, studies have implicated the dopamine D_2 receptor [161,162]. It has not been clear, however, whether effects mediated by striatal adenosine A_{2A} receptors absolutely depend on the presence of dopamine D_2 receptors. To study this problem, Chen et al. [163] employed either genetic knockout mice deficient in dopamine D_2 receptors, adenosine A_{2A} receptors, or a double knockout mouse deficient in both types of receptors. These studies found that A_{2A} receptors may affect neuronal activity in a manner that is partially independent of the presence of dopamine D_2 receptors, such that endogenous adenosine may be most accurately viewed as a facilitative modulator of striatal neuronal activity rather than simply as an inhibitory modulator of D_2 receptor neurotransmission.

These studies, and many others, conclude that the acute locomotor stimulant effects of caffeine in animal models are mediated in part by dopaminergic systems and dopamine receptors. Recent studies suggest that tolerance to the locomotor stimulant effects of chronic caffeine also may be related to specific changes in dopaminergic function [160]. Thus, in spite of the fact that methylxanthines are structurally different from other psychostimulants, and do not directly affect dopamine transporters or receptors, in fact their stimulant action appears to be derived from effects on central dopamine pathways.

5. STRUCTURE–ACTIVITY RELATIONSHIPS

Examination of the structure–activity relationships of several of the classic stimulants provides not only an understanding of the development of other drugs but also important clues as to the underlying mechanisms involved in interaction with the target protein(s). The following sections will hopefully illustrate both of these points.

5.1. Amphetamine

There are a number of related structures that are often referred to as "amphetamines" although the name amphetamine refers to one specific molecular entity. Grouped in this class would be (+)-amphetamine **10**, *N*-methylamphetamine (*S*-(+)-methamphetamine **18**), phentermine **19**, phenmetrazine (Preludin, **20**), and phendimetrazine **21**. Diethylpropion (Tenuate®; **16**) is used as an appetite suppressant and, although it has the amphetamine skeleton, its effects are much weaker as a stimulant than the other structures listed here. The stereochemistry at the alpha-side-chain methyl group is the same for the most potent enantiomer of each structure, although the pure enantiomer has not generally been marketed except for the cases of (+)-amphetamine **10** and (+)-methamphetamine **18**.

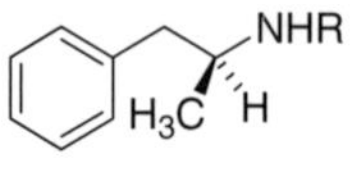

10 R = H
18 R = CH₃

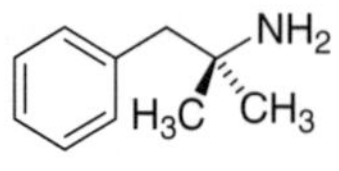

19

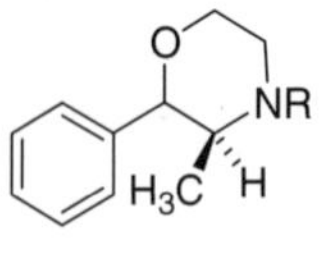

20 R = H
21 R = CH₃

The structural requirements of the dopamine (and norepinephrine) transporter for substrates appear to be fairly rigid. There is very little molecular variation that is tolerated without significant loss of activity. The relatively limited information that is available, mostly from animal studies, can be summarized by considering the various areas of substitution for a general phenethylamine structure. These structure–activity relationships recently have been surveyed [164], and an extensive and comprehensive review by Biel and Bopp [165] covered the older literature.

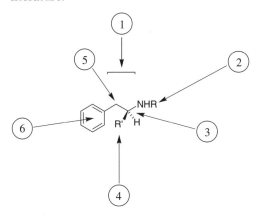

5.1.1. Length of the Side Chain The length of the side chain is limited to two carbon atoms [166,167]. That is, for transporter substrates, the optimum pharmacophoric template appears to be a basic nitrogen two carbon atoms removed from an aromatic ring system. This observation of course is not too surprising because the transporter substrates dopamine, norepinephrine, and serotonin all bear this essential core.

5.1.2. Nitrogen Substituents Tolerated nitrogen substituents are very limited. The primary amine (amphetamine) and the N-methylamine (methamphetamine) are the most potent compounds [167]. An N-methyl increases the potency of both amphetamine and cathinone **2** [168]. Larger alkyl groups [167,169] or N,N-dialkylation, either dramatically attenuate or completely abolish stimulant activity [170]. Nevertheless, N,N-dimethylamphetamine has appeared on the illicit market [171], and does appear to have behavioral effects in rats and monkeys similar to amphetamine [170,172]. The rapid onset of action suggested that the N,N-dimethyl compound itself had pharmacological effects, rather than the N-demethylated metabolite, methamphetamine, although the latter is one of the known metabolites of N,N-dimethylamphetamine [173].

Active metabolites may be much more important in N,N-dialkylated compounds that possess a beta-keto function, as in cathinone **2**. In that case, the N,N-dimethyl compound is nearly as active as the N-monomethyl compound [174]. It is known, however, that the alkyl groups of beta-aminoketones are readily cleaved metabolically. Thus, the N,N-dimethyl cathinone analog is likely rapidly converted *in vivo* to the N-monomethyl compound methcathinone. This argument is consistent with the finding that the N-monoethyl metabolite is the active species following administration of diethylpropion, the N,N-diethyl congener of cathinone [76,77].

Although longer N-alkyl groups lead to less active compounds, one exception to this generalization is benzphetamine (Didrex), N-benzyl-N-methylamphetamine **22**. Despite the N,N-dialkyl groups in benzphetamine, in humans it produces subjective effects characteristic of amphetamine-like drugs such as phenmetrazine **20** [175]. Although *para*-hydroxy-N-benzylamphetamine is a major metabolite of benzphetamine, methamphetamine and amphetamine also are detectable in urine and hair following administration of benzphetamine [176–178]. It is not clear from the literature whether the reinforcing effects of benzphetamine are due to metabolic formation of amphetamine or methamphetamine. Based on the studies with N,N-dimethylamphetamine by Witkin [170], however, one would predict that the parent molecule has some pharmacological activity.

22

5.1.3. Stereochemistry at the Alpha-Carbon

The stereochemistry at the alpha-carbon atom, when enantiomers exist, is homochiral to that of S-($+$)-amphetamine **10**, shown earlier. Both the releasing actions at dopamine and norepinephrine transporters in isolated rat brain slices [179], and the locomotor and stereotypic effects in rodents [180] are more potently affected by the S-($+$) isomer of amphetamine than by the R-($-$) isomer. In this latter study, the ($+$) enantiomer was approximately five times more potent than the ($-$) isomer, paralleling the potency difference found with the enantiomers *in vitro*, using rat brain striatal synaptosomes [181]. The two isomers were of nearly equal potency in their effects on norepinephrine accumulation by rat hippocampal synaptosomes [181]. This stereochemical requirement applies to beta-keto derivatives as well; the corresponding active isomer has the S-($-$) configuration [168].

5.1.4. The Alpha-Alkyl Substituent

The alpha-alkyl group cannot be much larger than a methyl. Phenethylamine itself, lacking the side-chain alpha-methyl group, is inactive *in vivo*, as a result of rapid inactivation by monoamine oxidase. Addition of the alpha-methyl group retards metabolism by this route, leading to the orally bioavailable drug amphetamine. The uptake transporter, however, cannot tolerate large groups in this region and the alpha-ethyl analogs of both amphetamine and methamphetamine had markedly attenuated activity in a drug discrimination assay with rats trained to discriminate ($+$)-amphetamine [182]. Alpha, alpha-dimethyl groups, as in phentermine **19**, result in an active compound, but one with reduced activity.

Attempts to incorporate the side chain into ring structures also lead to compounds with attenuated activity. For example, in drug discrimination assays using rats trained to recognize the effect of ($+$)-amphetamine **10**, compounds **23** and **24** either failed to produce amphetamine-like effects or had much lower potency [182,183]. When $n=3$, the compound lacked any amphetamine-like action.

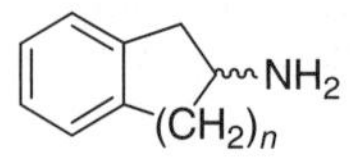

23 $n = 1$
24 $n = 2$

5.1.5. Other Side-Chain Substitutions

Limited substitution of the side chain is tolerated. A beta-hydroxy group on methamphetamine gives ephedrine **1**, shown earlier. Although ephedrine is a CNS stimulant, its effects are much weaker than those of methamphetamine. Similarly, addition of a beta-hydroxy to amphetamine gives phenylpropanolamine, a compound that is nearly devoid of CNS stimulant effects. One may speculate that the polar hydroxy group reduces the hydrophobicity of these compounds such that CNS penetration is dramatically reduced. The N-methyl of ephedrine increases lipid solubility, so ephedrine has greater CNS action than phenylpropanolamine. Addition of a keto function to the structure of amphetamine or methamphetamine gives cathinone **2** or its corresponding N-methyl derivative, methcathinone, the latter which also has greater potency than the primary amine [174]. It should be noted that an oxygen at the beta-position can be incorporated into a heterocyclic ring as in phenmetrazine **20** and phendimetrazine **21**. Methyl aminorex, **25**, is also a potent stimulant that incorporates the essential features of the amphetamine template into an oxazoline ring. The $4S,5S$-trans isomer shown (**25**) is the more potent of the four possible stereoisomers [184,185].

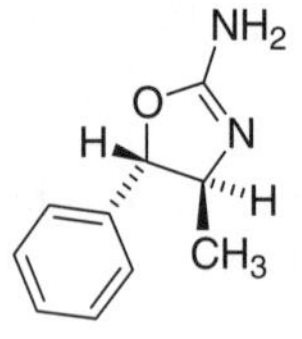

25

5.1.6. Aromatic Ring Substitution

Simple ring substituents can change the targets of the amphetamines from one monoamine uptake carrier to another. The dopamine and norepinephrine uptake carrier proteins have the most stringent structural demands, and any ring substitution decreases their potency at these sites. The serotonin carrier is relatively promiscuous, and tolerates a variety of ring substituents, many of which dramatically increase the potency at the serotonin carrier relative to that of amphetamine itself. No ring modifications are known that give rise to a substituted amphetamine that completely retains amphetamine-like psychostimulant activity. *Para*-fluoroamphetamine (**26**; X = F) has been reported to have effects in rats resembling those of amphetamine, but substitution with larger halogens (e.g., chloro or iodo) leads to compounds that have significant serotonin releasing potency, and which produce behavioral effects different from those of amphetamine itself [186].

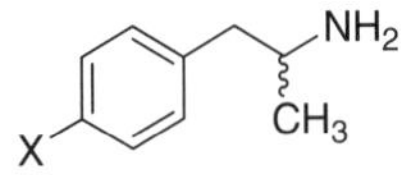

26 X = F, Cl, I

5.2. Methylphenidate

The R,R-(+)-stereoisomer of methylphenidate **8** is known to be the more active [187], and is often referred to as the active "threo" isomer. The (−)-enantiomer and the erythro stereoisomers are much less potent. One study has reported a series of aromatic ring-substituted analogs. The most potent compounds in that report were halogen substituted in the 3-, or 3,4- positions of the ring. For example, the dichloro compound **27** was 32-fold more potent than methylphenidate itself in inhibiting dopamine reuptake [188]. That finding parallels a recent report by Deutsch et al. [189], that replacing the phenyl ring with a beta-naphthyl moiety [190] gave a compound with approximately eightfold higher affinity for the dopamine transporter. Those workers also reported that the corresponding alpha-naphthyl analog had only approximately one-tenth the potency of methylphenidate at the DAT. Taken together, these latter observations indicate that the DAT must have a hydrophobic region that generally extends from the 3,4-positions of the aromatic phenyl ring of methylphenidate.

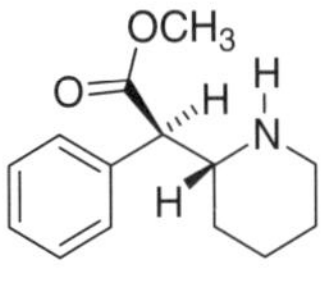

(+)-**8** 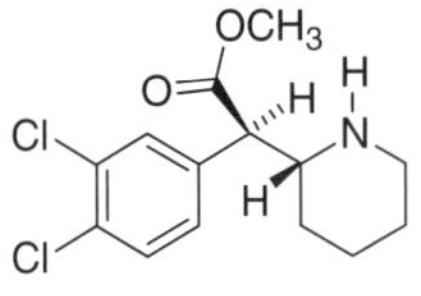   **27**

Deutsch et al. [189] also examined the effect of heterocyclic ring size. The pyrrolidyl and azepino, as well as the azacyclooctane congeners were significantly less potent than methylphenidate itself. That report also contained data for the morpholine analog of methylphenidate [190], which had approximately 15-fold lower affinity at the DAT. Beyond the studies cited here, very little additional SAR work has been done with methylphenidate.

5.3. Cocaine

Of all the psychostimulants, cocaine probably has been most studied, particularly within the last decade, as a result of its widespread abuse. Structure–activity studies have been carried out with numerous analogs, not only to elucidate the molecular requirements for interaction with the various monoamine transporters but also in attempts to develop treatments that might be useful for cocaine addiction. Ideally, understanding the structure–activity relationships will be useful to understanding the functional topography of the binding site of the transporters. Nevertheless, because the topic of this chapter is stimulants, and not the structure–activity relationships of monoamine transporters, an exhaustive summary of the more numerous studies that have appeared on the SAR of cocaine and its analogs will not be presented. A useful perspective on the SAR of cocaine analogs as it was understood in 1992 has been presented by Carroll et al. [191], with more a recent update in 1997 [192].

An attempt will be made here to distill down the essence of the SAR of cocaine as it relates to its stimulant properties. In many cases, compounds have been reported that have not been tested *in vivo*, but have been compared only for affinity at the monoamine transporters, or in an *in vivo* assay. Some of these data will be summarized if they are reported in the context of the stimulant effects of cocaine. Similarly, there have been numerous attempts to develop cocaine analogs that may bind to the dopamine transporter and actually block the stimulant or reinforcing effects of cocaine itself, in efforts to develop treatments for cocaine addiction. This chapter largely ignores many of those studies unless they contain *in vivo* data suggesting they are relevant to a discussion of stimulant effects. Nevertheless, because stimulant properties have been associated with binding to the DAT, a good deal of the SAR discussion here must be discussed in the context of *in vitro* DAT affinity.

A consideration of the structure–activity relationships of cocaine can focus on a number of key elements in the structure, as indicated below. Each of the following sections will include a discussion of the particular numbered structural element.

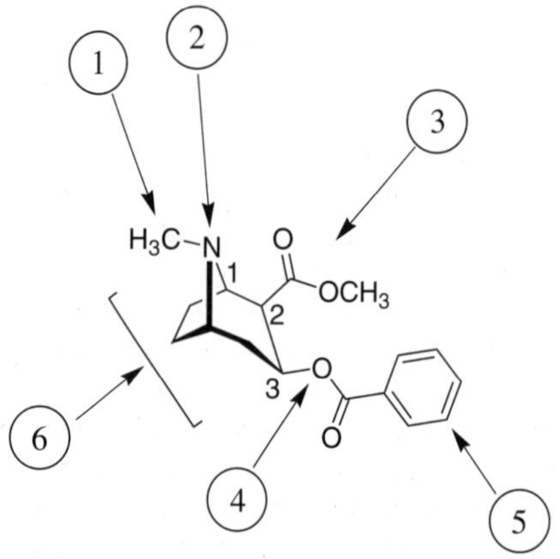

28 R = CH_3, n-C_3H_8, n-C_4H_9

5.3.1. N-Substituents
N-demethylation of cocaine has only a minor effect on affinity at monoamine transporters [193]. In phenyltropane analogs where the ester linkage has been removed (e.g., **28**), extensions of the *N*-alkyl out to *n*-butyl have no effect on dopamine transporter affinity [194]. Effects at the serotonin transporter are variable, but affinity only decreases modestly. At the norepinephrine transporter, affinity drops approximately threefold with the longer *N*-alkyl group.

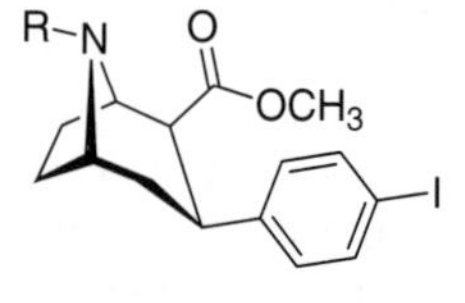

5.3.2. Basic Nitrogen Atom
For many years it was assumed that the basic nitrogen of cocaine was required for activity. It seemed logical to believe that the nitrogen, protonated at physiological pH, would interact with an anionic site such as an aspartate residue in the transporter [195]. It was surprising, therefore, when nonbasic *N*-sulfonyl cocaine analogs such as **29** were found to possess high affinity for the dopamine transporter [196]. These compounds are not protonated at physiological pH, and if hydrogen bonding were required for activity, these analogs could only serve as hydrogen bond acceptors. Even then, the low electron density remaining on a nitrogen with the powerfully electron-withdrawing trifluoromethylsulfonyl group attached, would suggest that this interaction should be very weak.

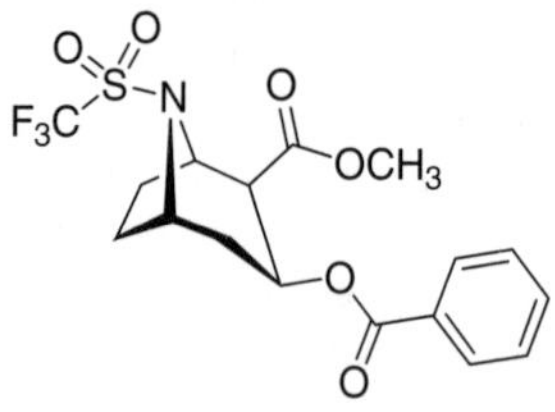

29

Replacement of the nitrogen atom with oxygen as in **30** gives compounds that retain high affinity for the dopamine transporter [197]. This finding was accommodated by proposing that the oxygen atom could act as a hydrogen bond acceptor at the transporter [197], a conclusion that would at least be consistent with the activity of the *N*-sulfonated derivatives **29**.

It was even more surprising; therefore, when the report appeared that even a polar oxygen was not required for good uptake inhibitors. Carbocyclic compounds such as **31** proved to have transporter affinities nearly equal to their amine-containing counterparts [198]!

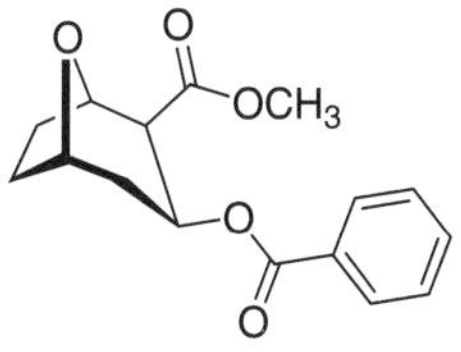

30

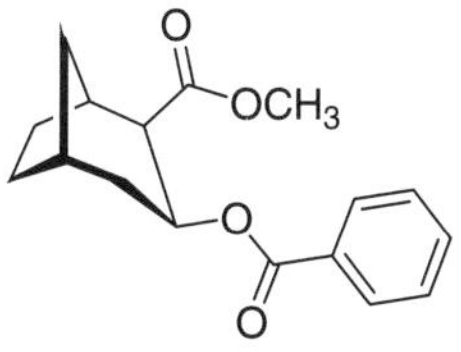

31

These authors postulated that there are various acceptor sites in the dopamine transporter, where an inhibitor may bind and cause dopamine uptake inhibition. The topography of these sites is probably different in the three monoamine transporters.

5.3.3. Substituent at C(2) Epimerization of the ester function to give pseudo-cocaine **32** results in approximately a 150-fold loss in affinity for the dopamine transporter [193]. In compounds lacking the ester linkage (see Section 5.3.4) the effect is more dramatic, resulting in a more than 1000-fold lower potency.

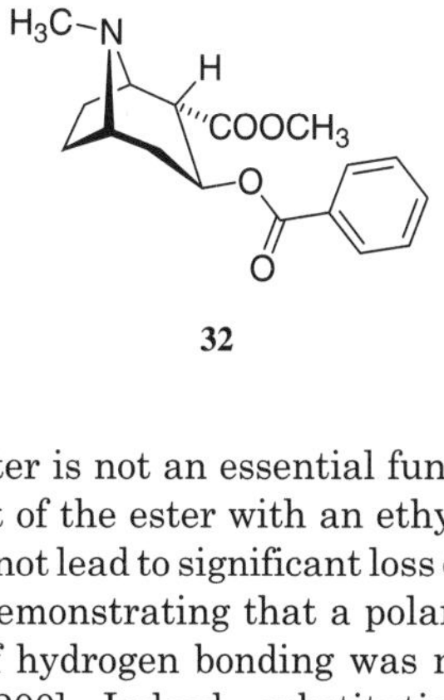

32

The ester is not an essential function. Replacement of the ester with an ethyl or vinyl group did not lead to significant loss of binding affinity, demonstrating that a polar function capable of hydrogen bonding was not essential [199,200]. Indeed, substitution at the 2β-position with alkyl groups as long as n-butyl, 2-phenethyl, or 2-stryl gave compounds (e.g., **33**) with exceptionally high affinities at the dopamine transporter [200].

Kelkar et al. [201] have extended the 2β-alkyl group to include a polar hydroxy or methyl ester function at the distal end of a three carbon chain, with no significant loss of affinity compared to a simple carbomethoxy function. They concluded that this region of the cocaine binding site must be either a large cleft in the transporter protein or exterior to the binding site. They also noted that this region is relatively insensitive to electrostatic interactions. Chang et al., [202], found that the 2β-phenyl analog **34** was equipotent to the 2-carbomethoxy compound, but had enhanced selectivity for the dopamine transporter over the serotonin transporter. These authors also concluded that a hydrophobic group at this region of the molecule might be a contributing factor for binding at the dopamine transporter.

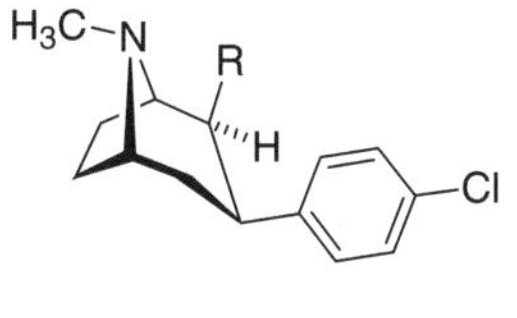

33 R = alkyl 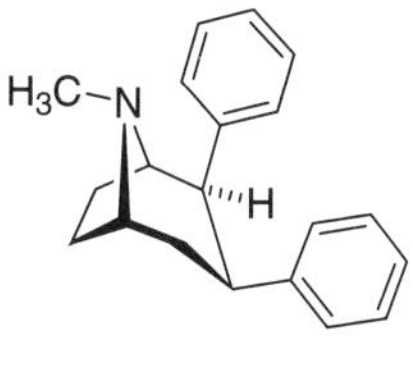   **34**

Esters larger than a methyl are quite potent. In the 3-benzoyl series of tropane esters, both the isopropyl and phenyl esters had high affinity and selectivity for the dopamine transporter [203]. The phenyl ester (**35**; RTI-15) dose-dependently substituted in the drug discrimination paradigm in rats trained to discriminate the effects of cocaine. In contrast, whereas cocaine increased locomotor activity in mice, RTI-15 had no effect on activity and at high doses even decreased this measure [204]. This compound was a potent inhibitor of the dopamine transporter, suggesting that high selectivity for the dopamine transporter may lead to differential retention of cocaine-like effects.

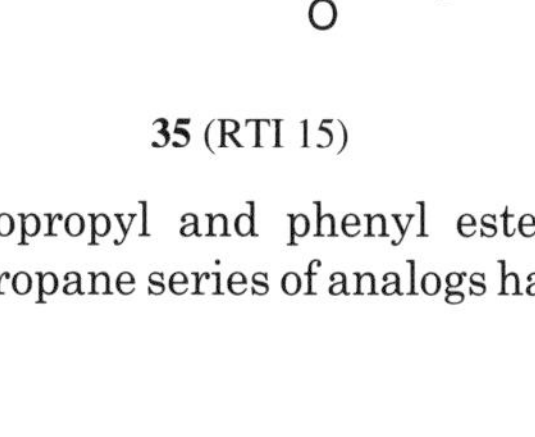

35 (RTI 15)

The isopropyl and phenyl esters in the 3-phenyltropane series of analogs have higher affinity for the dopamine transporter than the methyl ester [205]. Similarly, tertiary amide analogs of cocaine and phenyltropane analogs are more potent than secondary or primary amides, and also have enhanced selectivity for binding at the dopamine transporter over the norepinephrine or serotonin transporters [205]. Replacement of the ester or amide function with a substituted carboethoxy isoxazole substituent gave **36**, a highly potent inhibitor with selectivity for the dopamine transporter [206,207]. This compound had approximately twice the affinity of cocaine at the DAT.

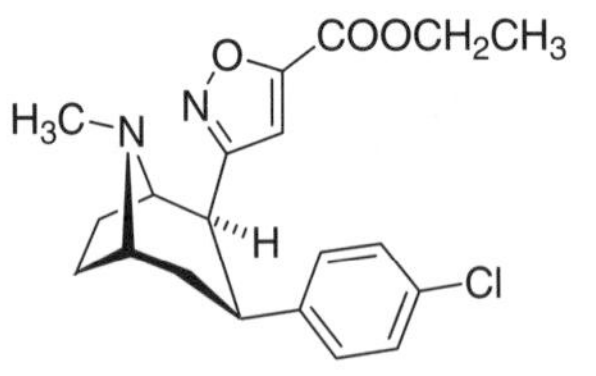

Similarly, Carroll et al. [208] reported that the 1,2,4-oxadiazoles (e.g., **37**) that are bioisosteres of ester groups, are potent cocaine analogs. Compound **37** had low nanomolar affinity for the dopamine uptake transporter with greater than 100-fold selectivity for the dopamine transporter over the norepinephrine and serotonin transporters.

5.3.4. The Ester Linkage at C(3) In cocaine, the 3α-epimer "allococaine" **38** has considerably reduced activity when compared to cocaine itself [209]. This structural change, however, causes the tropane ring to favor the pseudo-chair, rather than the boat conformation that occurs in natural (−)-3β-cocaine.

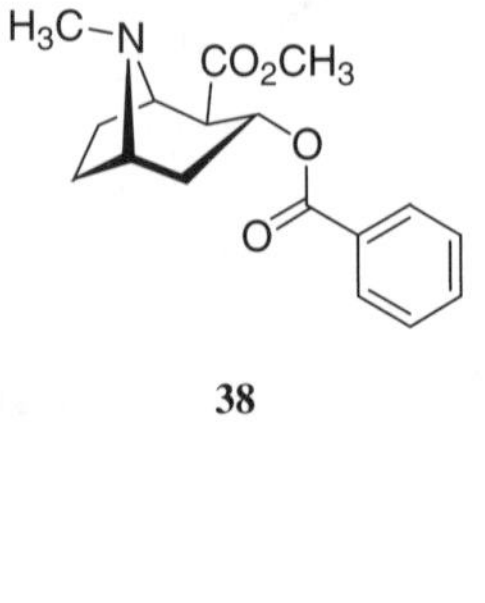

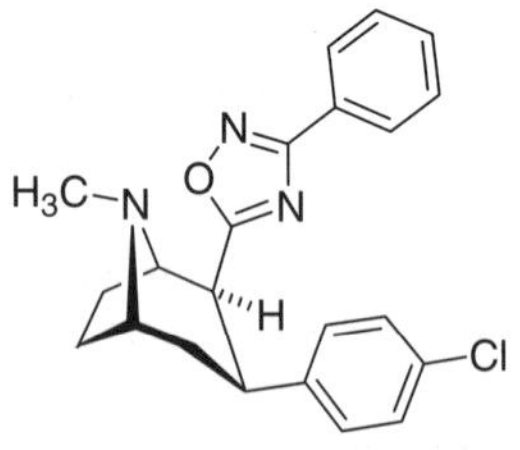

It was first reported by Clarke et al. [210] that removal of the ester linkage from cocaine, to give a compound with the phenyl ring directly attached to the tropane ring (WIN 35,065-2; **39**), possessed higher affinity for the dopamine transporter than did cocaine itself. By contrast to benzoyl esters, however, the configuration at the 3-position is not so critical in phenyltropane compounds. That is, in the WIN series where the ester has been removed, the 3β-phenyl orientation **39** was only approximately twofold more potent than the 3α-phenyl **40** at the dopamine transporter. At the serotonin transporter, however, the 3β-compound was significantly more potent [197]. A similar trend was observed in the 8-oxa analogs, leading to the conclusion that the dopamine transporter is able to accommodate the 3-phenyl ring when the bicyclic ring is in either the boat or chair conformation, whereas the serotonin transporter is less accommodating.

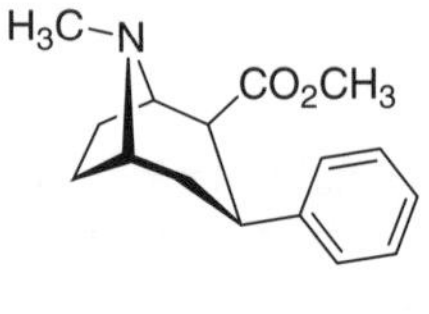

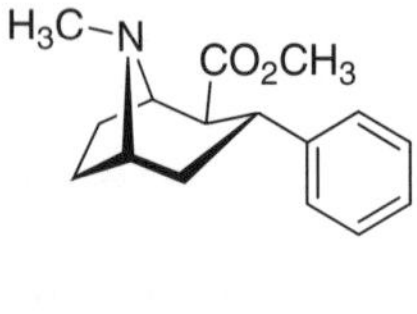

5.3.5. Substitutions on the Aromatic Ring at Position 3 In the phenyltropane analogs of cocaine, where the ester linkage has been removed and the phenyl ring is attached directly to the tropane ring (WIN and RTI compounds), substitution at the *para* ring position with halogens or a methyl group gave compounds (**41**) with increased affinities at the dopamine transporter compared with the unsubstituted compound, and with much increased affinity compared to cocaine itself [211]. Behavioral potency paralleled the affinity increases, with all of the phenyltropanes being considerably more potent in elevating locomotor activity in mice [212] and in substituting for a cocaine stimulus in the drug discrimination paradigm in rats [213].

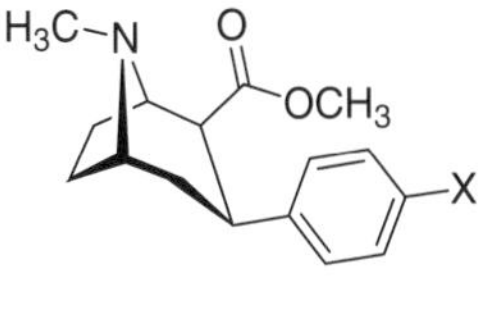

41; X = F, Cl, Br, I, CH$_3$

The rank order of affinity for aromatic ring substituents in the WIN series was 3,4-Cl$_2$ > I > Cl > F > H, whereas in the 8-oxa (3β) analogs it was 3,4-Cl$_2$ > Br > Cl > I > F > H [197]. Replacing the 3β-phenyl with either a 1- or 2-naphthyl substituent gave significantly enhanced affinity at all three monoamine transporters, with the 2-naphthyl **43** being approximately five- to sixfold more potent than the 1-naphthyl **42** [214]. This result parallels similar findings reported by Deutsch et al. [189], where replacing the phenyl ring of methylphenidate with a 2-naphthyl moiety gave an analog with approximately 70-fold higher potency than when the phenyl ring was replaced with a 1-naphthyl.

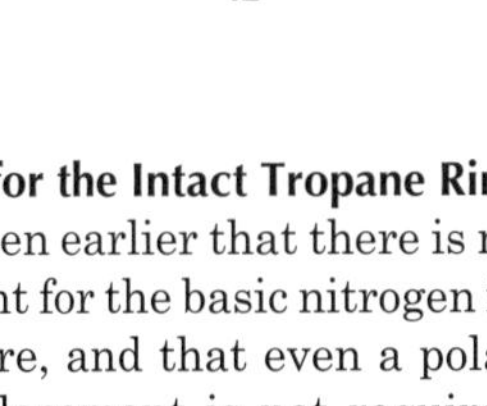

42

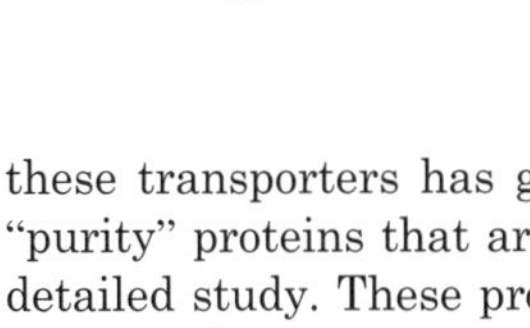

5.3.6. Requirement for the Intact Tropane Ring System
We have seen earlier that there is no absolute requirement for the basic nitrogen in the tropane structure, and that even a polar oxygen isostere replacement is not required for cocaine congeners to possess potent mono-

amine reuptake inhibition. It is perhaps not too surprising, therefore, that the bridged bicyclic tropane ring is not an essential structural feature. In a series of 4-arylpiperidine carboxylic acid methyl esters, several of the compounds were significantly better uptake inhibitors than cocaine [215]. The most potent compound in the series, **44**, was approximately 20 times more potent than cocaine at the dopamine uptake transporter.

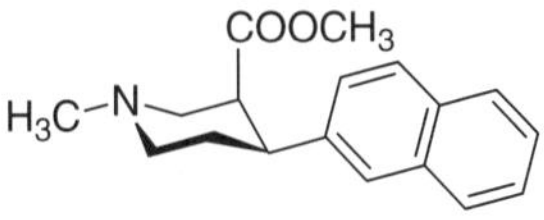

44

6. RECENT AND FUTURE DEVELOPMENTS

As reviewed above, the drugs that have been used for their stimulant properties were largely the result of compounds that were discovered empirically over many centuries. Understanding the active principles of these drugs has led to major advances in medicinal chemistry. For example, one of the most exciting recent findings has to do with understanding the monoamine transporters at the molecular level. Whereas it has been known in pharmacology for many years that cocaine targeted an energy-dependent reuptake pump, the cloning and expression of

43

these transporters has given access to high "purity" proteins that are amenable to more detailed study. These proteins have all been sequenced, and shown to be membrane bound with twelve membrane-spanning helical segments. A large number of site-specific mutations have been used to correlate specific residues in the protein with specific functions. A major breakthrough occurred with the solution of the crystal structure of the *Aquifex aeolicus* bacterial leucine transporter in 2005. Since that time, a number of related transporter proteins have been crystallized and their structures determined by X-ray crystallography. Combined with site-directed mutagenesis, cysteine-scanning accessibility methods, high field NMR, homology modeling, and continued development of structure–activity relationships, these advances are now leading to continually improved models of the monoamine transporters. At the moment, these approaches are having the greatest impact on studies of uptake inhibitors, such as antidepressants, rather than studies of substrates.

Particularly interesting recent advances indicate that ligands from different chemical classes may bind in novel ways to the dopamine transporter. Using site-directed mutagenesis and photoaffinity labeling probes, investigators have produced results suggesting that the substrate uptake and cocaine binding sites are probably not identical [216]. Indeed, it also now seems likely that different chemical classes of uptake inhibitors may even bind to distinct regions of the transporter [217,218], leading to different overall conformations in the transporter protein and perhaps subtly altered mechanisms of inhibition. These different transporter conformations would explain the observed differences in the pharmacology of different chemical classes of DAT inhibitors. New derivatives that have selective affinity at these alternate binding sites may block the actions of cocaine without markedly affecting the normal transport function of the protein. Hence, there is presently intense interest in such compounds because they may provide new avenues for the treatment of cocaine and psychostimulant addiction.

There also is a need for improved drugs to replace existing CNS stimulants, as treatments for medical conditions such as narcolepsy, ADHD, obesity, and for general attentional purposes. Yet virtually all of the existing stimulants have the capacity to produce enhanced mood, or euphoria. This side effect means that they all possess abuse potential to a greater or lesser degree. Advances in medicinal chemistry and molecular neurobiology have provided hope, however, for new generations of drugs. For example, the nonamphetamine, nonstimulant drug modafinil (Provigil **45**) was recently approved for use in narcolepsy (for example, see Ref. [219]). This drug has been shown to be more effective and have fewer side effects than amphetamine, although its mechanism of action has not yet been fully elucidated. A number of double-blind placebo controlled studies have shown efficacy of modafinil in narcolepsy [220]. Modafinil also has shown efficacy in shift work sleep disorder [221], and there is increasing evidence that modafinil can improve working memory, episodic memory, and processes requiring cognitive control.

Studies with mutant dopamine D_1 and D_2 knockout mice have recently shown that both the D_1 and D_2 receptors are necessary for the wakefulness induced by modafinil [222]. A recent PET study showed significant binding of the DAT and NET, and DAT occupancy comparable to methylphenidate at clinically relevant doses [223]. Modafinil seems to have multiple effects on brain catecholaminergic systems, including NET and DAT inhibition, elevating catecholamines, glutamate, and serotonin, activating the orexinergic system, and decreasing GABA. Alpha-adrenergic, D_1, and D_2 receptors appear to mediate its effects [220].

45

The discovery that mutations of either the gene for the novel neuromodulator orexin or the orexin receptor can cause narcolepsy, leads to the hope for even better and more specific drugs to treat that disorder, as well as to the possibility of better treatments for obesity [224–227].

Even those actions of the stimulants that are absolutely dependent on activation of monoamine systems also may be amenable to breakthroughs in medicinal chemistry. For example, the beneficial effects of stimulants on ADHD may be due primarily to activation of only certain receptors. Recently, the anatomical and functional substrates of attention, learning, and memory have begun to yield their secrets. This work has suggested that certain drugs (e.g., selective D_1 dopamine agonists) [228,229] may provide all or much of the beneficial effects of the stimulants without the abuse potential. Finally, the cracking of the human genome, and the prediction of more than 100,000 human proteins, offers the hope for novel targets for future efforts in medicinal chemistry. Whereas these may be known neurotransmitter pathways (e.g., GABA or glutamate receptors), new targets may be novel proteins whose function is not understood today, but will be tomorrow, and current uses of psychostimulants may be no more than historical artifacts in a decade.

7. WEB SITE ADDRESSES AND RECOMMENDED READING

http://www.nida.nih.gov/DrugAbuse.html
http://www.mentalhealth.com/
http://www.psyweb.com/indexhtml.html
http://www.fda.gov/medwatch/

REFERENCES

1. Bray GA. Ann Intern Med 1993;119:707–713.
2. Luqman W, Danowski TS. Ann Intern Med 1976;85:246–249.
3. Halbach H. NIDA Res Monogr 1979;27: 318–319.
4. Robbers JE, Speedie MK, Tyler VE. Pharmacognosy and Pharmacobiotechnology. Baltimore: Williams & Wilkins; 1996.
5. Reissig CJ, Strain EC, Griffiths RR. Drug Alcohol Depend 2009;99:1–10.
6. Angrist B, Sudilovsky A. Central nervous system stimulants: historical aspects and clinical effects. In: Iversen LL, Iversen SD, Snyder SH, editors. Handbook of Psychopharmacology. New York: Plenum Press; 1978. p 99–165.
7. Leake CD, The Amphetamines, Their Actions and Uses. Springfield, IL: Charles C. Thomas; 1958.
8. Prinzmetal M, Bloomberg W. J Am Med Assn 1935;105:2051–2054.
9. Myerson A. Arch Neurol Psychiat 1936;36: 916–922.
10. Editorial. J Amer Med Assn 1937;108: 1973–1974.
11. Hauschild F. Arch Exp Path U Pharmacol 1938;191:465–481.
12. Macleod AD. J Pain Symptom Manage 1998;16:193–198.
13. Olin J, Masand P. Psychosomatics 1996;37: 57–62.
14. Rothenhausler HB, Ehrentraut S, von Degenfeld G, Weis M, Tichy M, Kilger E, Stoll C, Schelling G, Kapfhammer HP. J Clin Psychiatry 2000;61:750–755.
15. Akerstedt T, Ficca G. Chronobiol Int 1997;14: 145–158.
16. Caldwell JA, Jr, Caldwell JL, Smythe NK, III, Hall KK. Psychopharmacology 2000;150: 272–282.
17. American Psychiatric Association. Diagnostic and Statistical Manual of Mental Disorders. DSM-IV. 4th ed. Washington, DC: American Psychiatric Association; 1994.
18. Block SD. Ann Intern Med 2000;132:209–218.
19. Lloyd-Williams M, Friedman T, Rudd N. Palliat Med 1999;13:243–248.
20. Sullivan MD. Semin Clin Neuropsychiatry 1998;3:151–156.
21. Comer AM, Perry CM, Figgitt DP. Paediatr Drugs 2001;3:61–79.
22. Bhatia J. Clin Pediatr 2000;39:327–336.
23. Markowitz JS, Morrison SD, DeVane CL. Int Clin Psychopharmacol 1999;14:1–18.
24. Rump AF, Theisohn M, Klaus W. Forensic Sci Int 1995;71:103–115.
25. Williams RG, Kavanagh KM, Teo KK. Can J Cardiol 1996;12:1295–1301.
26. Connell PH. Amphetamine Psychosis, Maudsley Monographs Number Five. London: Oxford University Press; 1958.
27. Angrist BM, Gershon S. Biol Psychiatry 1970;2:95–107.
28. Angrist B. Amphetamine psychosis: clinical variations of the syndrome. In: Cho AK, Segal DS, editors. Amphetamine and its Analogs. San Diego: Academic Press; 1994. p 387–414.
29. Lieberman JA, Kane JM, Alvir J. Psychopharmacology 1987;91:415–433.
30. Lieberman JA, Alvir J, Geisler S, Ramos-Lorenzi J, Woerner M, Novacenko H, Cooper

T, Kane JM. Neuropsychopharmacology 1994;11: 107–118.
31. Markowitz JS, Straughn AB, Patrick KS. Pharmacotherapy 2003;23:1281–1299.
32. Kimko HC, Cross JT, Abernethy DR. Clin Pharmacokinet 1999;37:457–470.
33. Foley R, Mrvos R, Krenzelok EP. J Toxicol Clin Toxicol 2000;38:625–630.
34. Gadow KD, Sverd J, Sprafkin J, Nolan EE, Grossman S. Arch Gen Psychiatry 1999;56: 330–336.
35. Aanonsen NO. Tidsskr Nor Laegeforen 1999;119:4040–4042.
36. Parran TV, Jr, Jasinski DR. Arch Intern Med 1991;151:781–783.
37. Meririnne E, Kankaanpaa A, Seppala T. J Pharmacol Exp Ther 2001;298:539–550.
38. Markowitz JS, Patrick KS. J Clin Psychopharmacol 2008;28:S54–S61.
39. Quinn D. J Clin Psychopharmacol 2008;28: S62–S66.
40. Safer DJ, Zito JM, Gardner JE. J Am Acad Child Adolesc Psychiatry 2001;40:622–629.
41. Adcock KG, MacElroy DE, Wolford ET, Farrington EA. Ann Pharmacother 1998;32: 422–425.
42. Shevell M, Schreiber R. Pediatr Neurol 1997;16:14–16.
43. Hogan V. CMAJ 2000; 162 106–110.
44. Benowitz NL. Ciba Found Symp 1992;166: 125–143.
45. Harris D, Batki SL. Am J Addict 2000;9:28–37.
46. Sherer MA, Kumor KM, Cone EJ, Jaffe JH. Arch Gen Psychiatry 1988;45:673–677.
47. Galasko GI. J Cardiovasc Risk 1997;4: 185–190.
48. Madu EC, Shala B, Baugh D. Angiology 1999;50:163–168.
49. Winbery S, Blaho K, Logan B, Geraci S. Am J Emerg Med 1998;16:529–533.
50. Swartz JA, Lurigio AJ, Goldstein P. Arch Gen Psychiatry 2000;57:701–707.
51. Cross JC, Johnson BD, Davis WR, Liberty HJ. Drug Alcohol Depend 2001;64:191–201.
52. Johnson BD, Dunlap E, Maher L. Subst Use Misuse 1998;33:1511–1546.
53. Rogers PJ, Dernoncourt C. Pharmacol Biochem Behav 1998;59:1039–1045.
54. Juliano LM, Griffiths RR. Psychopharmacology 2004;176:1–29.
55. Griffiths RR, Chausmer AL. Nihon Shinkei Seishin Yakurigaku Zasshi 2000;20:223–231.
56. Daly JW, Fredholm BB. Drug Alcohol Depend 1998;51:199–206.
57. Nurminen ML, Niittynen L, Korpela R, Vapaatalo H. Eur J Clin Nutr 1999;53: 831–839.
58. Nutt DJ, Bell CJ, Malizia AL. J Clin Psychiatry 1998;59(Suppl 17): 4–11.
59. Bourin M, Baker GB, Bradwejn J. J Psychosom Res 1998;44:163–180.
60. Lin AS, Uhde TW, Slate SO, McCann UD. Depress Anxiety 1997;5:21–28.
61. Bourin M, Malinge M, Guitton B. Therapie 1995;50:301–306.
62. Ferre S, Ciruela F, Borycz J, Solinas M, Quarta D, Antoniou K, Quiroz C, Justinova Z, Lluis C, Franco R, Goldberg SR. Front Biosci 2008;13: 2391–2399.
63. Deckert J. Int J Neuropsychopharmcol 1998;1:187–190.
64. Newton DW, Kluza RB. Drug Intell Clin Pharm 1978;12:546–554.
65. Nettleton J, Wang GK. Biophys J 1990;58: 95–106.
66. Beckett AH, Rowland M. J Pharm Pharmacol 1965;17:(Suppl): 109S–114S.
67. Caldwell J, Dring LG, Williams RT. Biochem J 1972;129:11–22.
68. Cho AK, Kumagai Y. Metabolism of amphetamine and other arylisopropylamines. In: Cho AK, Segal DS, editors. Amphetamine and its analogs San Diego: Academic Press; 1994. p. 43–77.
69. Rowland M. J Pharm Sci 1969;58:508–509.
70. Davis JM, Kopin IJ, Lemberger L, Axelrod J. Ann NY Acad Sci 1971;179:493–501.
71. Dring LG, Smith RL, Williams RT. Biochem J 1970;116:425–435.
72. Caldwell J, Dring LG, Williams RT. Biochem J 1972;129:23–24.
73. Faraj BA, Israili ZH, Perel JM, Jenkins ML, Holtzman SG, Cucinell SA, Dayton PG. J Pharmacol Exp Ther 1974;191:535–547.
74. Chan YM, Soldin SJ, Swanson JM, Deber CM, Thiessen JJ, Macleod S. Clin Biochem 1980;13: 266–272.
75. Schuelke GS, Terry LC, Powers RH, Rice J, Madden JA. Pharmacol Biochem Behav 1996;53:133–140.
76. Beckett AH, Stanojcic M. J Pharm Pharmacol 1987;39:409–415.
77. HanYu H, Rothman RB, Dersch CM, Partilla JS, Rice KC. Bioorg Med Chem 2000;8:2689–2692.

78. Volkow ND, Wang GJ, Fischman MW, Foltin RW, Fowler JS, Abumrad NN, Vitkun S, Logan J, Gatley SJ, Pappas N, Hitzemann R, Shea CE. Nature 1997;386:827–830.
79. Leshner AI, Koob GF. Proc Assoc Am Physicians 1999;111:99–108.
80. Volkow ND, Wang GJ, Fowler JS, Gatley SJ, Ding YS, Logan J, Dewey SL, Hitzemann R, Lieberman J. Proc Natl Acad Sci USA 1996;93:10388–10392.
81. Ritz MC, Lamb RJ, Goldberg SR, Kuhar MJ. Science 1987;237:1219–1223.
82. Self DW, Nestler EJ. Annu Rev Neurosci 1995;18:463–495.
83. Wise RA, Rompre PP. Annu Rev Psychol 1989;40:191–225.
84. Kuhar MJ, Ritz MC, Boja JW. Trends Neurosci 1991;14:299–302.
85. Di CG, Bassareo V. Curr Opin Pharmacol 2007;7:69–76.
86. Patrick KS, Caldwell RW, Ferris RM, Breese GR. J Pharmacol Exp Ther 1987;241:152–158.
87. Russell V, de Villiers A, Sagvolden T, Lamm M, Taljaard J. Behav Brain Res 1998;94:163–171.
88. Seiden LS, Sabol KE, Ricaurte GA. Annu Rev Pharmacol Toxicol 1993;33:639–677.
89. Saier MH, Jr. J Cell Biochem 1999; (Suppl 32–33): 84–94.
90. Jardetzky O. Nature 1966;211:969–970.
91. Tanford C. Proc Natl Acad Sci USA 1983;80:3701–3705.
92. Weyand S, Shimamura T, Yajima S, Suzuki S, Mirza O, Krusong K, Carpenter EP, Rutherford NG, Hadden JM, O'Reilly J, Ma P, Saidijam M, Patching SG, Hope RJ, Norbertczak HT, Roach PC, Iwata S, Henderson PJ, Cameron AD. Science 2008; 322:709–713.
93. Swanson CJ, Baker DA, Carson D, Worley PF, Kalivas PW. J Neurosci 2001;21:9043–9052.
94. Chiamulera C, Epping-Jordan MP, Zocchi A, Marcon C, Cottiny C, Tacconi S, Corsi M, Orzi F, Conquet F. Nat Neurosci 2001;4:873–874.
95. David HN, Abraini JH. Neuropharmacology 2003;44:717–727.
96. Mao L, Lau YS, Wang JQ. Eur J Pharmacol 2000;404:289–297.
97. Mao L, Wang JQ. Pharmacol Biochem Behav 2000;67:93–101.
98. Bradley SR, Rees HD, Yi H, Levey AI, Conn PJ. J Neurochem 1998;71:636–645.
99. Corti C, Restituito S, Rimland JM, Brabet I, Corsi M, Pin JP, Ferraguti F. Eur J Neurosci 1998;10:3629–3641.
100. Kinoshita A, Shigemoto R, Ohishi H, van der PH, Mizuno N. J Comp Neurol 1998;393:332–352.
101. Li X, Li J, Peng XQ, Spiller K, Gardner EL, Xi ZX. Neuropsychopharmacology 2009.
102. Nelson N. J Neurochem 1998;71:1785–1803.
103. Beuming T, Shi L, Javitch JA, Weinstein H. Mol Pharmacol 2006;70:1630–1642.
104. Pacholczyk T, Blakely RD, Amara SG. Nature 1991;350:350–354.
105. Kilty JE, Lorang D, Amara SG. Science 1991;254:578–579.
106. Shimada S, Kitayama S, Lin CL, Patel A, Nanthakumar E, Gregor P, Kuhar M, Uhl G. Science 1991;254:576–578.
107. Usdin TB, Mezey E, Chen C, Brownstein MJ, Hoffman BJ. Proc Natl Acad Sci USA 1991;88:11168–11171.
108. Blakely RD, Berson HE, Fremeau RT, Jr, Caron MG, Peek MM, Prince HK, Bradley CC. Nature 1991;354:66–70.
109. Hoffman BJ, Mezey E, Brownstein MJ. Science 1991;254:579–580.
110. Rudnick G, Clark J. Biochim Biophys Acta 1993;1144:249–263.
111. Blakely RD, Ramamoorthy S, Qian Y, Schroeter S, Bradley CC. Regulation of antidepressant-sensitive serotonin transporter. In: Reith EA, editor. Neurotransmitter Transporters: Structure, Function, and Regulation Totowa, NJ: Humana Press; 1997. p 29–72.
112. Amara SG, Sonders MS. Drug Alcohol Depend 1998;51:87–96.
113. Kniazeff J, Shi L, Loland CJ, Javitch JA, Weinstein H, Gether U. J Biol Chem 2008;283:17691–17701.
114. Indarte M, Madura JD, Surratt CK. Proteins 2008;70:1033–1046.
115. Jorgensen AM, Tagmose L, Jorgensen AM, Topiol S, Sabio M, Gundertofte K, Bogeso KP, Peters GH. ChemMedChem 2007;2:815–826.
116. Celik L, Sinning S, Severinsen K, Hansen CG, Moller MS, Bols M, Wiborg O, Schiott B. J Am Chem Soc 2008;130:3853–3865.
117. Yamashita A, Singh SK, Kawate T, Jin Y, Gouaux E. Nature 2005;437:215–223.
118. Tzschentke TM. Prog Neurobiol 2001;63:241–320.
119. Ikemoto S, Panksepp J. Brain Res Brain Res Rev 1999;31:6–41.

120. Booij J, Korn P, Linszen DH, van Royen EA. Eur J Nucl Med 1997;24:674–677.
121. Volkow ND, Wang GJ, Fowler JS, Logan J, Gatley SJ, Wong C, Hitzemann R, Pappas NR. J Pharmacol Exp Ther 1999;291:409–415.
122. Volkow ND, Wang G, Fowler JS, Logan J, Gerasimov M, Maynard L, Ding Y, Gatley SJ, Gifford A, Franceschi D. J Neurosci 2001;21: RC121.
123. Volkow ND, Wang GJ, Fowler JS, Logan J, Gatley SJ, Gifford A, Hitzemann R, Ding YS, Pappas N. Am J Psychiatry 1999;156: 1440–1443.
124. Filip M, Przegalinski E. Pol J Pharmacol 1997;49:21–30.
125. Tzschentke TM, Schmidt WJ. Crit Rev Neurobiol 2000;14:131–142.
126. Gainetdinov RR, Mohn AR, Bohn LM, Caron MG. Proc Natl Acad Sci USA 2001;98:11047–11054.
127. Giros B, Jaber M, Jones SR, Wightman RM, Caron MG. Nature 1996;379:606–612.
128. Rocha BA, Fumagalli F, Gainetdinov RR, Jones SR, Ator R, Giros B, Miller GW, Caron MG. Nat Neurosci 1998;1:132–137.
129. Sora I, Wichems C, Takahashi N, Li XF, Zeng Z, Revay R, Lesch KP, Murphy DL, Uhl GR. Proc Natl Acad Sci USA 1998;95: 7699–7704.
130. Belzung C, Scearce-Levie K, Barreau S, Hen R. Pharmacol Biochem Behav 2000;66: 221–225.
131. Rocha BA, Ator R, Emmett-Oglesby MW, Hen R. Pharmacol Biochem Behav 1997;57: 407–412.
132. Castanon N, Scearce-Levie K, Lucas JJ, Rocha B, Hen R. Pharmacol Biochem Behav 2000;67:559–566.
133. Broderick PA, Phelix CF. Neurosci Biobehav Rev 1997;21:227–260.
134. Cunningham KA, Bradberry CW, Chang AS, Reith ME. Behav Brain Res 1996;73:93–102.
135. Lieberman JA, Mailman RB, Duncan G, Sikich L, Chakos M, Nichols DE, Kraus JE. Biol Psychiatry 1998;44:1099–1117.
136. Bubar MJ, Cunningham KA. Curr Top Med Chem 2006;6:1971–1985.
137. Nic Dhonnchadha BA, Cunningham KA. Behav Brain Res 2008;195:39–53.
138. McMahon LR, Cunningham KA. J Pharmacol Exp Ther 1999;291:300–307.
139. McMahon LR, Cunningham KA. J Pharmacol Exp Ther 2001;297:357–363.
140. McMahon LR, Cunningham KA. Neuropsychopharmacology 2001;24:319–329.
141. Przegalinski E, Filip M. Behav Pharmacol 1997;8:699–706.
142. Tanda G, Pontieri FE, Frau R, Di Chiara G. Eur J Neurosci 1997;9:2077–2085.
143. Rothman RB, Baumann MH, Dersch CM, Romero DV, Rice KC, Carroll FI, Partilla JS. Synapse 2001;39:32–41.
144. McTavish SF, McPherson MH, Sharp T, Cowen PJ. J Psychopharmacol 1999;13: 144–147.
145. Robinson TE, Becker JB. Brain Res 1986;396: 157–198.
146. Castner SA, Goldman-Rakic PS. Neuropsychopharmacology 1999;20:10–28.
147. Castner SA, al-Tikriti MS, Baldwin RM, Seibyl JP, Innis RB, Goldman-Rakic PS. Neuropsychopharmacology 2000;22:4–13.
148. Strakowski SM, Sax KW. Biol Psychiatry 1998;44:1171–1177.
149. Selemon LD, Begovic A, Goldman-Rakic PS, Castner SA. Neuropsychopharmacology 2007;32:919–931.
150. Robinson TE, Kolb B. Neuropharmacology 2004;47(Suppl 1): 33–46.
151. Fredholm BB, Battig K, Holmen J, Nehlig A, Zvartau EE. Pharmacol Rev 1999;51:83–133.
152. El Yacoubi M, Ledent C, Menard JF, Parmentier M, Costentin J, Vaugeois JM. Br J Pharmacol 2000;129:1465–1473.
153. Kirch DG, Taylor TR, Gerhardt GA, Benowitz NL, Stephen C, Wyatt RJ. Neuropharmacology 1990;29:599–602.
154. Morgan ME, Vestal RE. Life Sci 1989;45: 2025–2039.
155. Ferre S. Psychopharmacology 1997;133: 107–120.
156. Ferre S, Fuxe K, von Euler G, Johansson B, Fredholm BB. Neuroscience 1992;51:501–512.
157. Ongini E, Fredholm BB. Trends Pharmacol Sci 1996;17:364–372.
158. Ferre S, Fredholm BB, Morelli M, Popoli P, Fuxe K. Trends Neurosci 1997;20:482–487.
159. Diaz-Cabiale Z, Hurd Y, Guidolin D, Finnman UB, Zoli M, Agnati LF, Vanderhaeghen JJ, Fuxe K, Ferre S. Neuroreport 2001;12: 1831–1834;.
160. Powell KR, Iuvone PM, Holtzman SG. Pharmacol Biochem Behav 2001;69:59–70.
161. Fink JS, Weaver DR, Rivkees SA, Peterfreund RA, Pollack AE, Adler EM, Reppert SM. Brain Res Mol Brain Res 1992;14:186–195.

162. Schiffmann SN, Vanderhaeghen JJ. J Neurosci 1993;13:1080–1087.
163. Chen JF, Moratalla R, Impagnatiello F, Grandy DK, Cuellar B, Rubinstein M, Beilstein MA, Hackett E, Fink JS, Low MJ, Ongini E, Schwarzschild MA. Proc Natl Acad Sci USA 2001;98:1970–1975.
164. Nichols DE. Medicinal chemistry and structure–activity relationships. In: Cho AK, Segal DS, editors. Amphetamine and Its Analogues. San Diego: Academic Press; 1994. p 3–41.
165. Biel JH, Bopp BA. Amphetamines: structure–activity relationships. In: Iversen LL, Iversen SD, Snyder SH, editors. Handbook of Psychopharmacology. New York: Plenum Press; 1978. p 1–39.
166. Daly JW, Creveling CR, Witkop B. J Med Chem 1966;9:273–280.
167. Van der Schoot JB, Ariens EJ, Van Rossum JM, Hurkmans JA. Arzneim Forsch 1961;9:902–907.
168. Glennon RA, Young R, Martin BR, Dal Cason TA. Pharmacol Biochem Behav 1995;50:601–606.
169. Woolverton WL, Shybut G, Johanson CE. Pharmacol Biochem Behav 1980;13:869–876.
170. Witkin JM, Ricaurte GA, Katz JL. J Pharmacol Exp Ther 1990;253:466–474.
171. Abercrombie TJ. Tieline 1987;12:39–53.
172. Katz JL, Ricaurte GA, Witkin JM. Psychopharmacology 1992;107:315–318.
173. Inoue T, Suzuki S. Xenobiotica 1987;17:965–971.
174. Dal Cason TA, Young R, Glennon RA. Pharmacol Biochem Behav 1997;58:1109–1116.
175. Chait LD, Uhlenhuth EH, Johanson CE. J Pharmacol Exp Ther 1987;242:777–783.
176. Cody JT, Valtier S. J Anal Toxicol 1998;22:299–309.
177. Fujinami A, Miyazawa T, Tagawa N, Kobayashi Y. Biol Pharm Bull 1998;21:1207–1210.
178. Kikura R, Nakahara Y. Biol Pharm Bull 1995;18:1694–1699.
179. Heikkila RE, Orlansky H, Mytilineou C, Cohen G. J Pharmacol Exp Ther 1975;194:47–56.
180. Segal DS. Science 1975;190:475–477.
181. Steele TD, Nichols DE, Yim GK. Biochem Pharmacol 1987;36:2297–2303.
182. Oberlender R, Nichols DE. Pharmacol Biochem Behav 1991;38:581–586.
183. Glennon RA, Young R, Hauck AE, McKenney JD. Pharmacol Biochem Behav 1984;21:895–901.
184. Glennon, RA. NIDA Research Monograph. In: Ashgar K, De Souza E, editors. NIDA Research Monograph Series. Washington, DC: Government Printing Office; 1989. p 43–67.
185. Kankaanpaa A, Ellermaa S, Meririnne E, Hirsjarvi P, Seppala T. J Pharmacol Exp Ther 2002;300:450–459.
186. Marona-Lewicka D, Rhee GS, Sprague JE, Nichols DE. Eur J Pharmacol 1995;287:105–113.
187. Shafi'ee A, Hite G. J Med Chem 1969;12:266–270.
188. Deutsch HM, Shi Q, Gruszecka-Kowalik E, Schweri MM. J Med Chem 1996;39:1201–1209.
189. Deutsch HM, Ye X, Shi Q, Liu Z, Schweri MM. Eur J Med Chem 2001;36:303–311.
190. Axten JM, Krim L, Kung HF, Winkler JD. J Org Chem 1998;63:9628–9629.
191. Carroll FI, Lewin AH, Boja JW, Kuhar MJ. J Med Chem 1992;35:969–981.
192. Carroll FI, Lewin AH, Kuhar MJ. Dopamine transporter uptake blockers: structure–activity relationships. Reith MEA, editor. Neurotransmitter Transporters: Structure, Function, and Regulation Towata: Humana; 1997. p 263–295.
193. Ritz MC, Cone EJ, Kuhar MJ. Life Sci 1990;46:635–645.
194. Scheffel U, Lever JR, Abraham P, Parham KR, Mathews WB, Kopajtic T, Carroll FI, Kuhar MJ. Synapse 1997;25:345–349.
195. Kitayama S, Shimada S, Xu H, Markham L, Donovan DM, Uhl GR. Proc Natl Acad Sci USA 1992;89:7782–7785.
196. Kozikowski AP, Saiah MK, Bergmann JS, Johnson KM. J Med Chem 1994;37:3440–3442.
197. Meltzer PC, Liang AY, Blundell P, Gonzalez MD, Chen Z, George C, Madras BK. J Med Chem 1997;40:2661–2673.
198. Meltzer PC, Blundell P, Yong YF, Chen Z, George C, Gonzalez MD, Madras BK. J Med Chem 2000;43:2982–2991.
199. Kozikowski AP, Roberti M, Xiang L, Bergmann JS, Callahan PM, Cunningham KA, Johnson KM. J Med Chem 1992;35:4764–4766.
200. Kozikowski AP, Eddine Saiah MK, Johnson KM, Bergmann JS. J Med Chem 1995;38:3086–3093.
201. Kelkar SV, Izenwasser S, Katz JL, Klein CL, Zhu N, Trudell ML. J Med Chem 1994;37:3875–3877.
202. Chang AC, Burgess JP, Mascarella SW, Abraham P, Kuhar MJ, Carroll FI. J Med Chem 1997;40:1247–1251.

203. Carroll FI, Abraham P, Lewin AH, Parham KA, Boja JW, Kuhar M. J Med Chem 1992;35:2497–2500.
204. Cook CD, Carroll FI, Beardsley PM. Drug Alcohol Depend 1998;50:123–128.
205. Carroll FI, Kotian P, Dehghani A, Gray JL, Kuzemko MA, Parham KA, Abraham P, Lewin AH, Boja JW, Kuhar MJ. J Med Chem 1995;38:379–388.
206. Kotian P, Abraham P, Lewin AH, Mascarella SW, Boja JW, Kuhar MJ, Carroll FI. J Med Chem 1995;38:3451–3453.
207. Kotian P, Mascarella SW, Abraham P, Lewin AH, Boja JW, Kuhar MJ, Carroll FI. J Med Chem 1996;39:2753–2763.
208. Carroll FI, Gray JL, Abraham P, Kuzemko MA, Lewin AH, Boja JW, Kuhar MJ. J Med Chem 1993;36:2886–2890.
209. Carroll FI, Lewin AH, Abraham P, Parham K, Boja JW, Kuhar MJ. J Med Chem 1991;34:883–886.
210. Clarke RL, Daum SJ, Gambino AJ, Aceto MD, Pearl J, Levitt M, Cumiskey WR, Bogado EF. J Med Chem 1973;16:1260–1267.
211. Boja JW, Cline EJ, Carroll FI, Lewin AH, Philip A, Dannals R, Wong D, Scheffel U, Kuhar MJ. Ann N Y Acad Sci 1992;654:282–291.
212. Cline EJ, Scheffel U, Boja JW, Carroll FI, Katz JL, Kuhar MJ. J Pharmacol Exp Ther 1992;260:1174–1179.
213. Cline EE, Terry PP, Carroll FF, Kuhar MM, Katz JJ. Behav Pharmacol 1992;3:113–116.
214. Bennett BA, Wichems CH, Hollingsworth CK, Davies HM, Thornley C, Sexton T, Childers SR. J Pharmacol Exp Ther 1995;272:1176–1186.
215. Tamiz AP, Zhang J, Flippen-Anderson JL, Zhang M, Johnson KM, Deschaux O, Tella S, Kozikowski AP. J Med Chem 2000;43:1215–1222.
216. Giros B, Wang YM, Suter S, McLeskey SB, Pifl C, Caron MG. J Biol Chem 1994;269:15985–15988.
217. Vaughan RA, Gaffaney JD, Lever JR, Reith ME, Dutta AK. Mol Pharmacol 2001;59:1157–1164.
218. Vaughan RA, Agoston GE, Lever JR, Newman AH. J Neurosci 1999;19:630–636.
219. Ferraro L, Antonelli T, O'Connor WT, Tanganelli S, Rambert FA, Fuxe K. Biol Psychiatry 1997;42:1181–1183.
220. Minzenberg MJ, Carter CS. Neuropsychopharmacology 2008;33:1477–1502.
221. Czeisler CA, Walsh JK, Roth T, Hughes RJ, Wright KP, Kingsbury L, Arora S, Schwartz JR, Niebler GE, Dinges DF. N Engl J Med 2005;353:476–486.
222. Qu WM, Huang ZL, Xu XH, Matsumoto N, Urade Y. J Neurosci 2008;28:8462–8469.
223. Madras BK, Xie Z, Lin Z, Jassen A, Panas H, Lynch L, Johnson R, Livni E, Spencer TJ, Bonab AA, Miller GM, Fischman AJ. J Pharmacol Exp Ther 2006;319:561–569.
224. Date Y, Nakazato M, Matsukura S. Nippon Rinsho 2001;59:427–430.
225. Krahn LE, Black JL, Silber MH Mayo Clin Proc 2001;76:185–194.
226. de Lecea L, Sutcliffe JG. Cell Mol Life Sci 1999;56:473–480.
227. Sutcliffe JG, de Lecea L. J Neurosci Res 2000;62:161–168.
228. Granon S, Passetti F, Thomas KL, Dalley JW, Everitt BJ, Robbins TW. J Neurosci 2000;20:1208–1215.
229. Robbins TW. Exp Brain Res 2000;133:130–138.

ANTICONVULSANTS

Rosaria Gitto[1]
Laura De Luca[1]
Giovanbattista De Sarro[2]
[1] Dipartimento Farmaco-Chimico, Università di Messina, Italy
[2] Dipartimento di Medicina Sperimentalee Clinica, Università Magna Græcia, Catanzaro, Italy

1. INTRODUCTION

Epilepsy is commonly considered the result of an imbalance between excitatory and inhibitory "tone" leading to periodic and unpredictable seizures related to an abnormal discharge of cerebral neurons. Epilepsy is one of the most common neurological disorders, affecting approximately 1% of the population worldwide, and the incidence increases to 3% by the age of 75 years. Although most people become seizure free with drug therapy, there are still a significant number of patients (30%) who are resistant to the currently available antiepileptic drugs (AEDs) whether used alone or in combination.

In addition, continuous medication is necessary for most patients with epilepsy to provide a lasting treatment because the drugs currently available generally suppress only the occurrence of epileptic seizures, but do not cure the pathology itself; AEDs should thus be considered anticonvulsants but not antiepileptogenic drugs. Because the discontinuation of therapies after a long seizure-free period may induce seizures, the ideal drug should control epilepsy by preventing the formation and development of epileptic foci and contrast epileptogenesis without interfering with physiological neuronal activity and without producing side effects.

Epilepsy is not a single disorder, but a series of conditions, known as epilepsy syndromes, classified according to characteristic seizure manifestations, etiology, age relationships, response to therapeutic treatment, and prognosis. The classification is subject to frequent revision by the International League Against Epilepsy (ILAE, www.ilae-epilepsy.org). It is a different classification from that of seizures, the latter concerns the external and principal manifestation of epilepsy, with many syndromes sharing the most common seizure patterns, known as tonic–clonic, absence, myoclonic, atonic, simple partial, or complex partial with or without secondary generalization (Table 1). A distinction can also be drawn between generalized and localized-related epileptic syndromes (Table 2). Generalized epileptic seizures originate at some point within bilaterally and simultaneously distributed structures, including cortical and subcortical. Focal epileptic seizures originate within one or more epileptogenic network limited to one hemisphere with a discrete localization or a wider distribution in a focal area of the cerebral cortex.

A proposed new classification is currently under consideration that takes into account a number of additional factors related to age, manifestations, and causative factors of the syndrome. Different mechanisms have to be considered as regards the pathogenesis within the spectrum of genetic or acquired causes (e.g., trauma, inflammation, tumor, vascular disorders, inborn errors of metabolism, or cortical malformation). The pathophysiology of epilepsies is complex and several mechanisms play a role in epileptogenesis and epileptogenicity. Neurotransmitters are important mediators for the generation of epilepsy, as they are needed to transfer signals from one neuron to another. A plethora of phenomena can contribute to development of diverse phenotypes of epilepsy, but the main ones are perturbations to the GABA-ergic system as well as alterations in the glutamate system, which play a significant role.

2. MOLECULAR TARGETS FOR ANTICONVULSANTS

The efficacy of anticonvulsant drugs may be connected with different molecular targets to reduce the excitability of neurons involved in seizure onset [1,2]. Anticonvulsant activity can generally be achieved by modifying the firing properties and reducing synchronization in neurons [3]. Historically, ion-channel activity is considered important for neuronal

Table 1. Classification of Epileptic Seizures

Partial onset seizures:
- Simple partial seizures
- Complex partial seizures
- Secondarily generalized tonic–clonic seizures

Generalized onset seizures:
- Absence seizures
- Tonic seizures
- Clonic seizures
- Myoclonic seizures
- Primary generalized tonic–clonic seizures
- Atonic seizures

---

signaling, through the movement of ions producing voltage changes across the membrane and in turn generating the synaptic potential; the ion channel opening and closing processes being influenced by binding of excitatory and inhibitory neurotransmitters as well as by voltage modifications mediated by voltage-gated ion channels [3]. The ion flux is controlled by the permeability and gating of the ion channels; we could summarize this by saying sodium and calcium channels mediate excitation, whereas potassium and chloride promote inhibition. Moreover, nonspecific cation channels may decrease the sensitivity of neurons to firing. The ion channels are divided into two broad categories, depending on their mechanism of activation. Voltage-gated ion channels are controlled by changes in membrane potential, whereas ligand-gated ion channels are activated by binding of neurotransmitters as ligands. Until recently, there has been a consensus that the major targets of anticonvulsants are voltage-gated sodium channels, voltage-gated calcium channels, and the inhibitory neurotransmitter γ-aminobutyric acid (GABA) system.

The mechanism of action of many currently prescribed agents falls into the following classification:

1. Modulation of voltage-dependent ion-channel activity,
2. Enhancement of synaptic inhibition,
3. Reduction of synaptic excitation.

However, new targets have been identified using electrophysiological and molecular approaches resulting in the development of newer anticonvulsants that act via nontraditional mechanisms and through different targets (see Table 3) [1]. Moreover, genetic studies of familial idiopathic epilepsies have identified numerous genes encoding both ion channels and other targets associated with diverse epilepsy syndromes; thus, confirming the mechanism of action of anticonvulsants on the market as well as suggesting new and unexpected targets [2,4,5].

2.1. Voltage-Gated Ion Channels

Voltage-gated ion channels control the excitability in the central nervous system. They are a superfamily, also called "voltage-gated-like ion channel," in which the pore-forming subunits consist of six-transmembrane segments (S1–S6), the fourth of which (S4) contains a highly positively charged voltage-sensing element, for this reason these kinds of ion

Table 2. Classification of Epilepsy Syndromes

(1) Generalized
- Primary—cause unknown (idiopathic or genetic epilepsy)
- Symptomatic due to structural/metabolic causes (e.g., head trauma, stroke, infection, malformation of cortical development, and inborn errors of metabolism)

(2) Localization-related (i.e., partial, local, and focal) seizures, simple or complex—anatomically defined, in terms of the origin of events
- Frontal
- Temporal
- Parietal
- Occipital

(3) Undetermined whether focal or generalized

(4) Special syndromes
- Isolated events
- Febrile seizures

Table 3. Molecular Targets for Anticonvulsants

- Voltage-gated ion channels (sodium, calcium, potassium, etc.)
- Ligand-gated ion channels
- Neurotransmitter transporters and synaptic proteins
- Enzymes
- Other targets

channels are called the S4 family [6]. There are some differences in the membrane topology of the different voltage-gated ion channels. In voltage-gated Na^+ and Ca^{2+} channels, four S1–S6 domains are formed from a single polypeptide chain arranged around a central pore that controls the ion current. In voltage-gated K^+ channel, however, the channel is a tetrameric combination of four individual subunits, each of which contains a single S1–S6 domain. Furthermore, four reentrant pore loops between the S5 and the S6 segments form the external ring surrounding the ion channel in all voltage-gated channels. A highly conserved sequence within these loops confers specificity for Na^+, Ca^{2+}, or K^+. The tetrameric organization of voltage-gated K^+ channels also characterizes calcium-activated K^+ channels and hyperpolarization-activated cyclic nucleotide-modulated cation channels (HCN).

2.1.1. Voltage-Gated Sodium Channels Voltage-gated sodium channels are essential for the propagation of action potentials. Indeed, they control membrane depolarization by allowing the influx of sodium ions into the cell. Mammalian voltage-gated Na^+ channels consist of α-subunits (four S1–S6 domains) (Fig. 1) responsible for selectivity and voltage gating [7]. However, some sodium channels also have one or two smaller auxiliary β1- and β2-subunits with modulatory functions. Nine genes encode the pore-forming α subunit leading different isoforms, some of which play a significant role in epilepsy and in the action of anticonvulsants ($Na_v1.1$, $Na_v1.2$, and $Na_v1.6$). For example, various forms of idiopathic generalized epilepsy are related to mutations of genes encoding α-subunits [8] and auxiliary β-subunits [9].

By binding inactivated state of the voltage-dependent Na^+ channels, the anticonvulsants (phenytoin, carbamazepine, lamotrigine, zonisamide, and topiramate, Table 4) produce a voltage- and use-dependent channel blockade; thus, suppressing high-frequency, repetitive action potential firing. These drugs demonstrate different clinical profiles related to (a) differences in terms of binding kinetics and selectivity for the different gating states of the channel, (b) different selectivity for Na^+ channel isoforms and their expression in brain

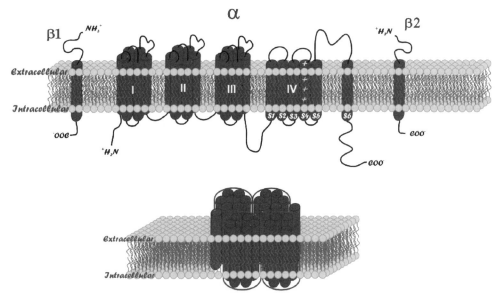

Figure 1. Schematic representation of the voltage-gated sodium channel architecture. There are four homologous domains for α-subunits (I–IV), formed from a single polypeptide chain, and the two auxiliary transmembrane subunits β1 and β2. The four α-subunits are assembled to form the ion selective channel pore with auxiliary subunits. (This figure is available in full color at http://mrw.interscience.wiley.com/emrw/9780471266945/home.)

Table 4. Voltage-Gated Ion Channels: Subunits, Drugs, and Anticonvulsant Mechanism

Channel	Channel Subunit	Drug	Anticonvulsant Mechanism
Na$^+$ channel	Na$_v$1.1, Na$_v$1.2, Na$_v$1.6	Phenytoin, carbamazepine, lamotrigine, zonisamide, topiramate	Stabilize inactivated state
Ca^{2+} channel	Cav3.1 (T-type)	Ethosuximide	Decrease Ca^{2+} current
K$^+$ channel	K$_v$7.2	Retigabine	Channel activator
	K$_{Ca}$	—	—
HCN channels (I_h currents)	HCN1–4	Lamotrigine	—
Cl$^-$ channel	ClC-2	—	—

regions, and (c) different secondary effects on other voltage-gated or ligand-gated ion channels. Site-directed mutagenesis studies in recombinant α-subunits made it possible to identify the critical aminoacid residues on specific domains involved in the binding of some sodium channels blockers such as phenytoin and lamotrigine [10].

2.1.2. Voltage-Gated Calcium Channels

Voltage-gated calcium channels control the membrane potential as well as membrane electrical events related to other important cellular functions including exocytosis of some neurotransmitters. The channel pore-forming α1-subunits of the voltage-gated Ca^{2+} channels are homologous to the voltage-gated Na$^+$ channel α-subunit (Fig. 2). Moreover, the α1-subunits might be associated with some auxiliary subunits, including the intracellular β-subunits, the intramembranal γ-subunits and the intramembranal/extracellular α2-δ-subunits (Fig. 2). Voltage-gated calcium channels are classified [7] into the following three subfamilies based on their different biological and pharmacological properties: (a) L-type (high-voltage activated), (b) N-, P/Q-, and R-type (responsible for the Ca^{2+} entry that controls neurotransmitter release), and (c) T-type (with low-voltage threshold). Like sodium channels, some mutations in voltage-activated Ca^{2+} channel subunits are linked with human epilepsy syndromes. Many anticonvulsant agents have been reported as Ca^{2+} current blockers, but the T-type channels are the primary target only in the case of ethosuximide and succinimide analogs (Table 4). The efficacy of other anticonvulsants (phenobarbital and lamotrigine) could be partially due to the blockade of calcium current through the N-type channel [11]. More recently, it has been demonstrated that the auxiliary subunits of voltage-gated calcium channels α2-δ could be the targets of gabapentinoids such as gabapentin and pregabalin [12].

2.1.3. Voltage-Gated Potassium Channels

Voltage-gated potassium channels reduce excitability in neuronal cells. These voltage-gated ion channels are formed from α-subunits that comprise the ion-conduction pore, selectivity filter, gating apparatus, as well as some auxiliary subunits that have modulatory effects (Fig. 3). The membrane topology of the K$^+$ channel contains a tetrameric combination of α-subunits, each one of which is homologous to a S1–S6 domain of the pore-forming α-subunits of the other voltage-gated channels (sodium and calcium voltage-gated channels). The architecture of the voltage-gated potassium channels is characterized by a P-loop that controls ion selectivity. Different families of voltage-gated potassium channels are known [13,14], but only the six-transmembrane-helix voltage-gated (K$_v$) channels and the Ca^{2+}-activated K$^+$ channels (K$_{Ca}$) are of particular interest for epilepsy (Table 4). The different K$_v$ channels include two channel subfamilies: (a) A-current (K$_v$1, K$_v$3, and K$_v$4) and (b) M-current (KNCQ, also called K$_v$7). M-current plays a particular role in regulating the dynamics of the neuronal firing and this channel has long been known to control excitability of neuronal cells. Because reducing

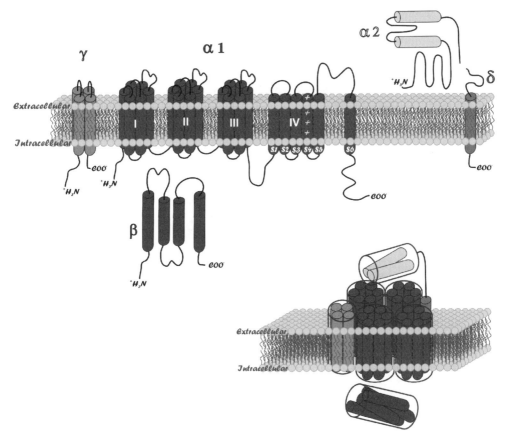

Figure 2. Schematic representation of the voltage-gated calcium channel architecture. Likely voltage-gated sodium channel type, a single polypeptide chain forms the four homologous domains for α1-subunits; they could be associated with auxiliary modulatory subunits (α2-δ, β, and γ). The combination of four α1-subunits with auxiliary subunits forms the ion selective channel pore. (This figure is available in full color at http://mrw.interscience.wiley.com/emrw/9780471266945/home.)

M-current enhances neuronal excitability and predisposes to seizures, the enhancement of M-current might be expected to protect from seizures. Furthermore, a mutation of $K_v7.2$ and $K_v7.3$ channel subunits has been identified in a hereditary epilepsy syndrome. The anticonvulsant retigabine [12,15] activates the KCNQ subfamily ($K_v7.2$–$K_v7.5$) of voltage-gated potassium channels via specific binding with S5, S6, and pore loop region.

2.1.4. Hyperpolarization-Activated Cyclic Nucleotide Cation (HCN) Channels HCN channels are Na^+/K^+ permeable channels present in neurons and cardiac cells. The channel opens by hyperpolarization to negative membrane potential and by cAMP binding to a cyclic nucleotide binding domain in the carboxy terminal domain (CTD). Binding of cAMP shifts the voltage dependence of activation to more positive potentials; thus, opening the channel. The current produced by the opening of the HCN channels is known as hyperpolarization-activated current I_h. There are four known subunits (HCN1–4) expressed in different brain regions, each subunit is formed by six-transmembrane segments. A combination of four subunits generates homo- or hetero-tetrameric channels. A role for HCN channels in epilepsy has been widely proposed [16,17] and it has been suggested that lamotrigine acts also by modifying the I_h current. New evidence suggests that different

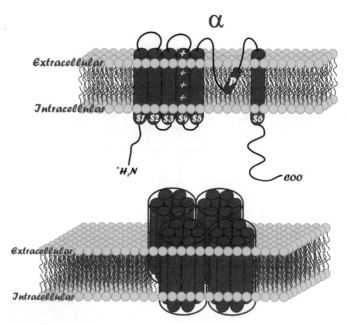

Figure 3. Schematic representation of the voltage-gated potassium channel architecture. A tetrameric combination of four different α-subunits forms the ion channel. For each subunit the P-loop intramembranal domain controls ion selectivity. (This figure is available in full color at http://mrw.interscience.wiley.com/emrw/9780471266945/home.)

neurotransmitters could control I_h current through a cAMP modulator site.

2.1.5. Voltage-Gated Chloride Channels Voltage-gated chloride channels are considered to play a role in the GABA-mediated inhibitory neurotransmission [18]. There are nine different voltage-gated chloride channels and one of these, that is, the ClC-2 found in neurons and glia, has been implicated in animal models of epilepsy [19]. The ClC-2 channel is not a target of currently available anticonvulsants, but it does represent an attractive target for future investigations.

2.2. Ligand-Gated Ion Channels

Ligand-gated ion-channel activation is controlled by the binding of neurotransmitters or exogenous ligands that open the channel through conformational changes; thus, eliciting inhibitory or excitatory phenomena in neuronal cells [20–22]. In terms of the receptor membrane topology, ligand-gated ion channels fall into two main subfamilies, namely, the cys-loop pentameric receptors (GABA$_A$, glycine, and nicotinic acetylcholine receptors) and the tetrameric glutamate ionotropic receptors (AMPA, NMDA, and kainate receptors). These different classes of ligand-gated ion channels are also named on the basis of specific ligands (Fig. 4).

The cys-loop receptors are characterized by different ion selectivity [22]; GABA$_A$ and glycine receptors are permeable to Cl$^-$, whereas nicotinic acetylcholine receptors (nAChRs) are permeable to Na$^+$, K$^+$, and Ca^{2+}. The ionotropic glutamate receptors (iGluRs) could be Na$^+$ and Ca^{2+} permeable.

2.2.1. GABA$_A$ Receptors The γ-aminobutyric acid (GABA) is the most abundant inhibitory neurotransmitter in the mammalian central nervous system where about 20% of all receptors are GABAergic. It interacts with different receptor subtypes distinguished by two ligand-gated ion channels [23], ionotropic GABA$_A$ and GABA$_C$, and G-protein-coupled metabotropic receptor subtypes (GABA$_B$). The GABA$_A$ receptor is a hetero pentameric complex with five subunits arranged around a central Cl$^-$ selective pore (Fig. 5) [24]. There are different genes encoding nineteen

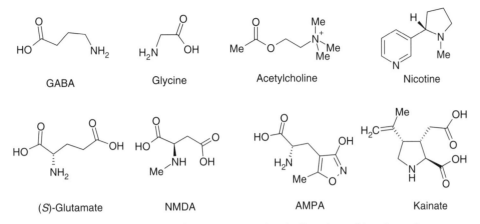

Figure 4. Endogenous and exogenous agonists for ligand-gated ion channels.

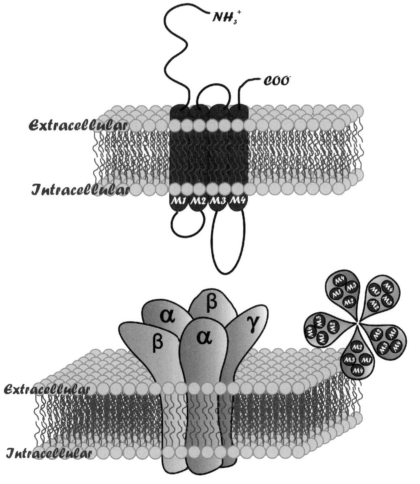

Figure 5. Schematic representation of the architecture of GABA$_A$ receptor complex. A pentameric combination of different subunits forms the ion channel. (This figure is available in full color at http://mrw.interscience.wiley.com/emrw/9780471266945/home.)

different subunits (α1–6, β1–3, γ1–3, δ, ε, θ, π, ρ1–3), but the majority of GABA$_A$ receptors are composed of α-, β-, and γ- or δ-subunits. Each subunit contains four α-helical transmembrane segments (M1–4), with a large cytoplasmatic loop between M3 and M4; both the N-terminal domain (NTD) and C-terminal domain (CTD) are located extracellulary. These extracellular segments form the recognition sites for the endogenous agonist GABA (two per channel) and for allosteric modulators (one per channel). GABA$_A$ receptors mediate synaptic inhibition through the generation of fast, transient, rapidly desensitizing current in postsynaptic neurons in response to synaptically released GABA. GABA$_A$-channel opening results in Cl$^-$ influx and hyperpolarization of the neuronal membrane.

The different subunit compositions correspond to different biological and pharmacological properties of the receptor-channel complexes. In fact, the composition plays an important role in the anticonvulsant actions of drugs (Table 5) interacting with GABA$_A$ receptors, especially in the case of benzodiazepine-like agonists [25].

A large variety of molecules (benzodiazepines, barbiturates, anesthetics, and neurosteroids) are able to interact with GABA$_A$ receptors, influencing their Cl$^-$ conductance or modulating the GABA binding. Those that mimic or improve the effects of GABA are generally sedative or anesthetic, and also show anticonvulsant action. Several GABA analogs acting as GABA agonists are known but they are not useful as systemic anticonvulsants.

The majority of anticonvulsants interact on the allosteric modulatory sites of the GABA$_A$ receptor complex. Benzodiazepine-like anticonvulsants act as positive allosteric modulators of GABA$_A$ receptors containing two α- (α1-, α2-, α3-, or α5-), two β- (β2- and/or β3-), and γ2-subunits; the binding of benzodiazepines increases the frequency of channel opening thus requiring the presence of the neurotransmitter GABA. In contrast, barbiturates affect GABA$_A$ activity by increasing the duration of channel opening. Benzodiazepine sedative actions are mediated by receptors containing α1-subunits, anxiolytic actions by receptors containing α2-subunit, and myorelaxant actions by receptors comprising α2-, α3-, and α5-subunits. Therefore, benzodiazepine-like anticonvulsants that selectively target GABA$_A$ receptors containing α2 or α3 can be expected to avoid sedative side effects. Substantial efforts have been devoted

Table 5. Ligand-Gated Ion Channels: Subunits, Drugs, and Anticonvulsant Mechanism

Receptor Complex	Ligand	Channel Subunits	Drug	Anticonvulsant Mechanism
GABA$_A$ receptor	GABA	α1, α2, α3, α5, β2, and/or β3, γ2	Benzodiazepine	Positive allosteric modulators
			barbiturates, neurosteroids felbamate, topiramate	
iGluRs	Glutamate			
	AMPA	GluR1–4	Talampanel	Negative allosteric modulator
	Coagonist glycine	NR1	Felbamate	Antagonist
	NMDA	NR2A–D		(glycine site, noncompetitive)
		NR3A–B		
	Kainate	GluR5, GluR6,	Topiramate	GluR5 selective antagonist
		GluR7, KA1, KA2		
nAChRs	ACh	α4 or α7	Carbamazepine	Noncompetitive inhibitor

to obtain $GABA_A$ receptor positive allosteric modulators that have reduced activity on $GABA_A$ receptors containing α1-subunits, to avoid the sedation that is believed to be mediated by this specific receptor subunit. The compounds developed are generally nonbenzodiazepines that bind to the benzodiazepine site on all benzodiazepine-sensitive isoforms, but are only partial agonists with reduced efficacy on certain isoforms.

2.2.2. Ionotropic Glutamate Receptors

By binding the neurotransmitter glutamate (Glu), the ionotropic glutamate receptors (iGluRs) mediate most of the fast excitatory transmission that control all brain functions especially learning and memory processes [26]. Glu can also activate G-protein-coupled metabotropic receptors (mGluRs, see below) leading to inhibitory and excitatory effects [27,28]. iGluRs are a superfamily of ligand-gated cation channels comprising three receptor families defined by the agonists that selectively activate them: α-amino-3-hydroxyl-5-methyl-4-isoxazole-propionate (AMPA), N-methyl-D-aspartate (NMDA), and kainate (KA) (Fig. 4) [29,30]. Their excessive activation plays an important role in epileptic phenomena. Glu activates all ionotropic glutamate receptors at synapses, but NMDA receptors also require the presence of the coagonist glycine. The crystal structures of some iGluR subunits have been determined, thereby describing the binding modes of agonist/antagonist ligands. The iGluRs are a heteromeric tetrameric combination of more than one type of subunit (Fig. 6). Each family of iGluRs comprises a specific composition of subunits: four for the AMPA receptors (GluR1–4), five for the kainate receptors (GluR5–7 and KA1–2), and seven for the NMDA receptors (NR1, NR2A–D, and NR3A–B) [29]. Alternative splicing and RNA editing of the subunits further contributes to diversity in ion channel properties. Each subunit has an extracellular N-terminal domain (NTD), three transmembrane segments (M1–M3), a reentrant M2 segment and an intracellular CTD. Each iGluR subunit contains an agonist recognition site formed by two lobes (S1 and S2) linked to transmembrane domains, as well as allosteric modulator binding sites with a different localization for the different families. The ion-selective filter is the reentrant pore loop (M2) between the M1 and the M3 domains.

All ionotropic glutamate receptors are permeable to Na^+ and K^+ ions, but differ in Ca^{2+} permeability based on the RNA-editing process that controls the aminoacid composition in the M2 domain.

NMDA receptor types contain NR1 and one or more NR2A–D subunits depending on the regional localization in the brain. Each NR1 subunit binds the coagonist glycine whereas the NR2 subunit constitutes the binding site for the agonist glutamate. The NMDA receptor complex is a tetramer in which the binding of two glycine and two glutamate molecules is required for channel opening. It is thought that the tetrameric functional channel comprises two dimers (dimer of dimers), one composed of NR1 subunits and the other with NR2 subunits. NMDA receptors are inactivated in a voltage-dependent manner by the presence of Mg^{2+}. AMPA receptors are composed of four types of subunits (GluR1–4) which combine to form homotetramers or symmetric dimer of dimers. The binding of two or more molecules of glutamate opens the channel. In the case of the KA receptor subtype, five subunits (GluR5–8, KA1–2) form the heterodimer receptor complex.

There is evidence that some anticonvulsants (Table 5) currently used in therapy (felbamate and topiramate) as well as newer agents under development (talampanel) exert the blockade of iGluR-mediated excitatory neurotransmission [31–33]. Their mechanism of action encompasses both antagonism and allosteric modulation of receptor activity [34,35].

2.2.3. Nicotinic Acetylcholine Receptors

nAChRs are pentameric structures constituted by a combination of 16 subunits (α1–7, α9, α10, β1–4, δ, ε, and γ) in which each subunit contains four transmembrane domains [36]. Neuronal receptors in the brain are composed of α1- or α4-subunits; they are presynaptic and modulate neurotransmitter release by increasing Na^+ and Ca^{2+} entry. They are thought to play an important role in cognitive functions and it seems that the noncompetitive blockade of nicotinic receptors is related to the

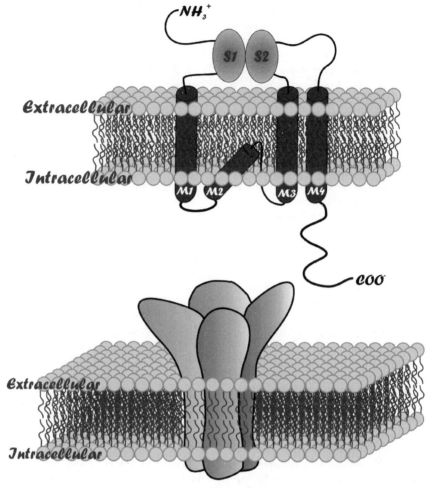

Figure 6. Schematic representation of the architecture of ionotropic glutamate receptors (iGluRs). For each subunit the S1/S2 extracellular lobes constitute the agonist ligand binding site. The reentrant M2 segment controls the ion selectivity. A tetrameric combination of different subunits (e.g., dimer of dimers for NMDA receptor subtype) forms the ion channel. (This figure is available in full color at http://mrw.interscience.wiley.com/emrw/9780471266945/home.)

anticonvulsant effect of carbamazepine (Table 5) in partial seizures [37].

2.2.4. Glycine Receptors Glycine receptors are ligand-gated Cl⁻ channels that are similar to $GABA_A$ receptors. They are pentameric and are composed of α- and β-subunits that have been encoded by different genes [38]. Several endogenous ligands are able to open glycine receptors [39]. Moreover, glycine receptors are positively modulated by volatile anesthetics. In contrast, strychnine is a powerful selective antagonist of glycine receptors that binds selectively to glycinergic synapses, this being the reason for this glycine receptor subtype to be defined strychnine sensitive. Although, strychnine induces convulsions, no human epilepsy syndromes have been associated with these receptors that lack a functional role.

2.3. Neurotransmitter Transporters and Synaptic Proteins

There are different neurotransmitter transporter families that can be distinguished as

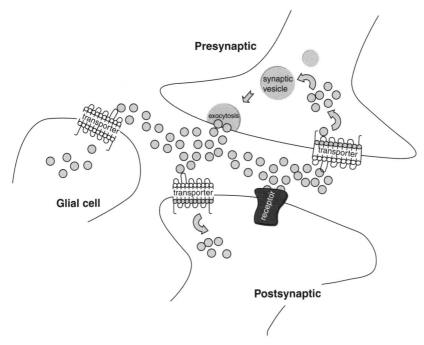

Figure 7. Synaptic membrane neurotransmitters. (This figure is available in full color at http://mrw.interscience.wiley.com/emrw/9780471266945/home.)

vesicular and plasma membrane transporters on the basis of their different role and localization (Fig. 7) [40]. Vesicular transporters move neurotransmitters into synaptic vesicles, whereas plasma membrane transporters sequester the neurotransmitters into cells after they are released from nerve terminals. In view of these roles, it has been suggested that changes in function of the inhibitory or excitatory neurotransmitter transporters could control synaptic potential, thereby producing pharmacological effects.

The four GABA transporters (GAT1–4), located on presynaptic nerve endings and glial cells, remove synaptically released GABA, to limit or terminate its inhibitory action [41]. Reuptake into terminals permits immediate recycling by vesicular uptake, whereas reuptake into astrocytes leads to neurotransmitter metabolism carried out by GABA-transaminase and successively succinic semialdehyde dehydrogenase. The anticonvulsant tiagabine is considered a GABA reuptake inhibitor that increases the synaptic availability of a neurotransmitter via a highly selective GAT-1 inhibition (Table 6); whereby GABA concentration in the synaptic cleft is prolonged; thus, enhancing its inhibitory effects [42].

In contrast, the activators of excitatory neurotransmitter glutamate transporters could be potential anticonvulsant agents. Furthermore, recent evidence has demonstrated that several anticonvulsants cause upregulation of glutamate transporters in addition to their effects on the GABA-ergic system (zonisamide, clobazam, and topiramate).

The five excitatory amino acid transporters (EAAT1–5), expressed in neuronal and astrocytic plasma membranes, influence neurotransmitter uptake by controlling its extracellular concentration [43]. It is well known that various alterations in glutamate transporters have been associated with seizures; in humans, EAAT1 mutation has been found in a patient with epileptic syndrome. Interestingly, the potent EAATs inhibitor TFB-TBOA (chemically, (3S)-3-[[3-[[4-(trifluoromethyl)benzoyl]amino]phenyl]methoxy]-L-aspartic acid) and analogs (Table 6), have produced epileptiform discharges in hippocampal slices and severe convulsions in mice by leading to an excessive extracellular glutamate tone [44].

Table 6. Promising Targets for Anticonvulsants

Target		Drug	Mechanism	Pharmacology
GABA transporters	GAT1	Tiagabine	GAT-1 inhibition	Anticonvulsant
Excitatory aminoacid transporters	EAAT1–5	TFB-TBOA	Glutamate transporter inhibition	Convulsant
Synaptic vesicle proteins SV2	SV2A	Levetiracetam Brivaracetam	Unknown	Anticonvulsant
GABA-transaminase	GABA-T	Vigabatrin	Enzyme inhibition	Anticonvulsant
Carbonic anhydrase	CAII	Acetazolamide	Enzyme inhibition	Anticonvulsant
	CAV	Zonisamide		
	CAVII	Topiramate		
Protein kinases	Ion channels and membrane transporter molecules	Topiramate	Enzyme modulation	Anticonvulsant
Gap junctions	Connexin	Carabersat	Connexon blockade	Anticonvulsant
G-protein-coupled receptors	GABA$_B$ receptors	Baclofen	GABA$_B$ agonist	
	mGluRs	AIDA	Group I antagonist	Anticonvulsants
		ACPT-1	Group III agonist	

There are also three vesicular glutamate transporters (VGLUT1–3) that have higher specificity for this neurotransmitter, but lower affinity than EAAT1–5. There is evidence that the expression of VGLUT1 may be related to epilepsy and for this reason vesicular glutamate transporters represent potential research targets in the anticonvulsant field [45].

The synaptic vesicle proteins, SV2s, also represent the target of some anticonvulsants. SV2s control vesicle formation/function and enhance low-frequency neurotransmission by priming docked vesicles in quiescent neurons and by coordinating synaptic vesicle exocytosis. SV2A, an ubiquitous synaptic vesicle glycoprotein, is the target of the anticonvulsant levetiracetam (Table 6) and its analog brivaracetam. This interaction seems to mediate their anticonvulsant activity. Despite the fact that levetiracetam and brivaracetam bind SV2 synaptic vesicles, the way in which they influence seizure susceptibility is not understood. However, the identification of SV2A as an anticonvulsant target confirms the potential of using proteins as targets as well as targeting ion channels [12].

2.4. Enzymes

Some enzymes that control certain metabolic processes within neuronal cell constitute molecular targets for anticonvulsant agents.

The mitochondrial enzyme GABA-transaminase (4-aminobutyrate-2-oxoglutarate aminotransferase, GABA-T) metabolizes synaptically released GABA, reducing the available concentration in the synaptic cleft (Fig. 8). The GABA analog, γ-vinyl GABA (vigabatrin), irreversibly inhibits GABA-T, increasing brain GABA content [46–48] and consequently it produces anticonvulsant effects (Table 6). The initial hypothesis that the anticonvulsant action of vigabatrin is a consequence of enhanced neurotransmitter concentration has recently been reconsidered, but both the desensitization of synaptic GABA$_A$ receptors and the activation of nonsynaptic GABA$_A$ receptors also contribute to vigabatrin activity. Moreover, it seems that the increase in extracellular GABA is not dependent on vesicular release, but rather on reverse transport of GABA by neurotransmitter transporters GAT-1 and GAT-3 [48].

2.4.1. Carbonic Anhydrase Carbonic anhydrase (CA) is a metalloenzyme that is present

γ-aminobutyric acid → Succinic semialdehyde

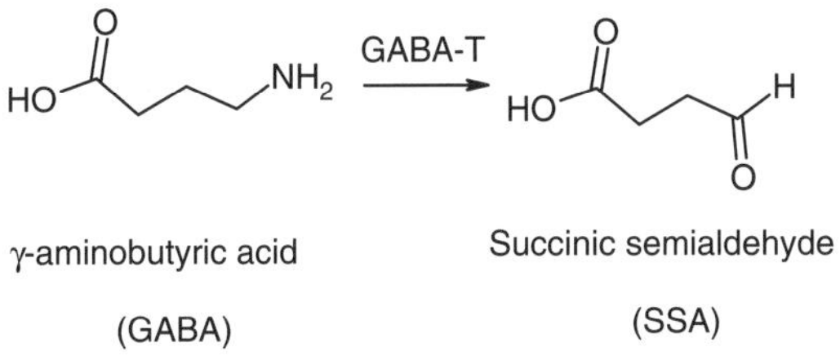

(GABA) (SSA)

Figure 8. GABA metabolism: GABA-T is the GABA transaminase.

in most tissues in different isoforms. These enzymes control the rate of hydration/dehydration of CO_2 (Fig. 9) by modulating intracellular and extracellular pH. It is well known that neuronal excitability is influenced by pH and that, in particular, $GABA_A$ receptor activation can be modulated by the intraneuronal generation of HCO_3^- by a specific carbonic anhydrase isozyme (CAVII). In astrocytes, carbonic anhydrase controls neuronal excitability by regulating extracellular pH. Decreasing pH depresses cell firing through a variety of indirect actions via neurotransmitter receptors such as glutamate NMDA receptors.

The diuretic drug acetazolamide and two recently introduced anticonvulsants, zonisamide, and topiramate [49] are CA inhibitors and their anticonvulsant action (Table 6) [50] can also be related to cytosolic carbonic anydrase isozyme CAII and mitochondrial CAV.

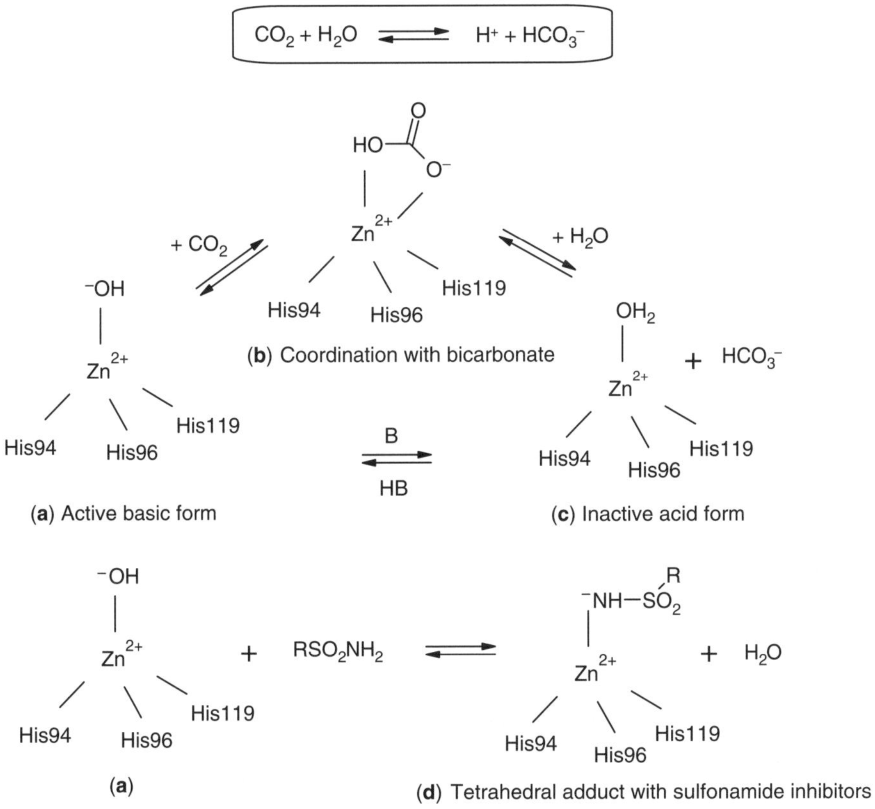

$$CO_2 + H_2O \rightleftharpoons H^+ + HCO_3^-$$

(a) Active basic form

(b) Coordination with bicarbonate

(c) Inactive acid form

(d) Tetrahedral adduct with sulfonamide inhibitors

Figure 9. Carbonic anhydrase catalysis and inhibition mechanism: the Zn^{2+} ion is essential for catalytic process; the metal is coordinated by three histidine residues (His94, His96, and His119) (a) and hydroxide ions or water in the active/inactive form, respectively. When the enzyme is in the active state a strong attack on CO_2 occurs, leading to the formation of bicarbonate ions coordinated to Zn^{2+} (b) The bicarbonate is then displaced by a water molecule and liberated. The inactive water/Zn^{2+} coordinated form (c) is successively converted into the corresponding active hydroxide form (a). The tetrahedral adduct (d) with sulfonamide derivatives determines enzymatic inhibition.

Carbonic anhydrase is a zinc-containing enzyme that is inhibited by interaction of sulfonamide or sulfamate compounds [51] with the Zn(II) ion and specific residues of the protein (see Fig. 9). However, the problem of tolerance and certain side effects have compromised the development of some CA inhibitors as anticonvulsants.

Moreover, the phosphorylation state of specific sites of some ion channels and membrane transporter neurotransmitter may modulate their activity. There is a great deal of evidence to suggest that functional changes in voltage-gated ion channels, GABA and glutamate receptors, and transporters in brain tissue result from altered phosphorylation. Mutations involving kinase or phosphatase systems would therefore be associated with epilepsy. Pharmacological agents that influence these enzymes may not be expected to have specific enough actions to be useful anticonvulsants. However, it seems that the anticonvulsant topiramate may act indirectly on ion channels by affecting the phosphorylation state.

2.5. Other Targets and New Mechanisms for Anticonvulsants

Connexins (Cx) are proteins that build the gap junctions forming two hemichannels (connexons) with a central pore that controls the ion flux, second messengers, and other small molecules. These hemichannels also modulate the release of glutamate and ATP. Gap junctions can be between neurons or between astrocytes (homocellular) or between astrocytes and neurons (heterocellular) [52]. It has been demonstrated that mutations in the encoding of connexins might play a role in epileptogenesis and it seems that the anticonvulsant carabersat (Table 6) is a gap junction blocker [53].

2.5.1. G-Protein-Coupled Receptors (GPCRs)

GPCRs are the largest family of receptors having an extracellular N-terminal domain (NTD), an intracellular C-terminal domain (CTD), seven α-helical transmembrane segments and an intracellular loop that binds G-proteins to produce second messenger signals. GPCRs are characterized by a "venus flytrap" extracellular ligand binding domain that shows a characteristic mechanism of activation. Metabotropic glutamate receptors (mGluRs) and $GABA_B$ receptors are GPCRs that by controlling excitatory synapses could be useful in developing new anticonvulsants (Table 6) [54]. mGluRs are classified into three groups differentiated by their sequence homology, second messenger effects, and pharmacology. The postsynaptic Group I (mGluR1 and mGluR5) receptors control excitatory neurotransmission, instead Group II (mGluR2 and mGluR3) and III receptors (mGluR4, mGluR7, and mGluR8) are located presynaptically, and their activation reduces glutamate release. There is some experimental evidence that Group I antagonists and Group III agonists show anticonvulsant properties. $GABA_B$ receptors, expressed presynaptically and postsynaptically at GABAergic synapses, mediate the actions of GABA, but they also decrease glutamate release at glutamatergic synapses. Baclofen, the classic $GABA_B$ receptor agonist, is selective for $GABA_B$ receptors and shows activity in epilepsy animal models; thus, suggesting that $GABA_B$ receptors represent a promising target for the development of new agents.

3. CLASSIFICATION OF ANTICONVULSANT AGENTS

Anticonvulsants can be classified into different classes, each comprising drugs that act via a specific mechanism of action. However, this very simple approach is difficult to apply in view of the following: (a) that there are remarkable advances in the understanding of epileptic phenomena, (b) that there is an increasing number of newly identified molecular targets, and (c) that in some cases, different mechanisms of action can contribute to pharmacological efficacy for the same drug.

The older antiepileptic drugs were detected by chance, but from 1990 onward new agents have been introduced, the so-called second-generation anticonvulsants, some of which are advantageous in terms of pharmacokinetic properties, tolerability, and advantage for drug interaction. In recent years, a new era has begun of anticonvulsants acting via nontraditional molecular targets identified using innovative approaches. In fact, studies on the mechanism of seizure onset and propagation have identified new targets for potential anticonvulsants [4,55,56]. The strategy of searching for mechanisms that are not affected by existing drugs is an attractive one because a novel model of action might be useful in identifying agents for refractory epilepsies. Even if nowadays drugs with new mechanism are discovered by chance, several compounds have been identified through high-throughput screenings that use animal models of epilepsy; while, other compounds have emerged from molecular modeling approaches.

Anticonvulsants on the market predominantly target voltage-gated cation channels (the α-subunits of voltage-gated Na^+ channels and also T-type voltage-gated Ca^{2+} channels) or influence GABA-mediated inhibition. Recently, α2-δ voltage-gated Ca^{2+} channel subunits and the SV2A synaptic vesicle protein have been recognized as useful targets. Moreover, recent advances have suggested other potential molecular targets, including various voltage-gated Ca^{2+} channel subunits and auxiliary proteins, A- or M-type voltage-gated K^+ channels, and ionotropic glutamate receptors (AMPA and NMDA) as well as enzymes, and important other targets are still emerging for the development of future drugs. Figures 10–12 report some selected antiepileptic drugs currently in use and under development.

1. "First-generation" or older agents are phenobarbital (**1**), primidone (**2**), phenytoin (**3**), ethosuximide (**4**), trimethadione (**5**), carbamazepine (**6**), benzodiazepine derivatives (**7,8**), and valproic acid (**9**) (see Fig. 10).
2. The class of "second generation" (Fig. 11) consists of oxcarbazepine (**10**), lamotrigine (**11**), vigabatrin (**12**), gabapentin (**13**), pregabalin (**14**), levetiracetam (**15**), zonisamide (**16**), topiramate (**17**), tiagabine (**18**), and felbamate (**19**). These agents were generally

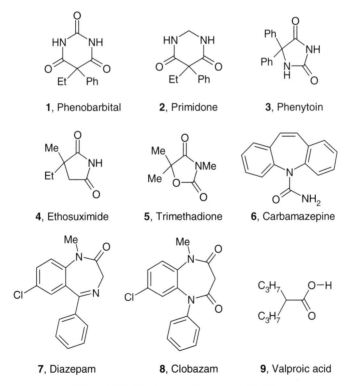

Figure 10. First-generation agents (**1–9**).

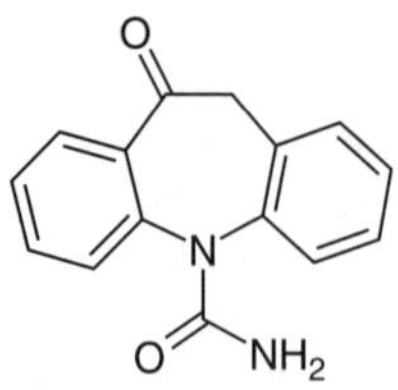

10, Oxcarbazepine

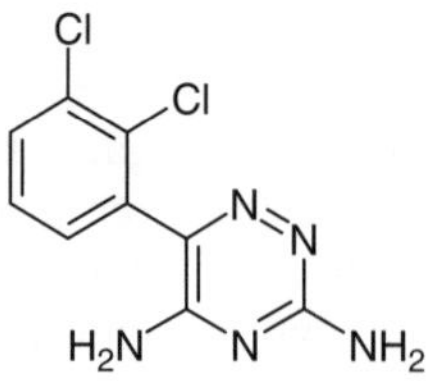

11, Lamotrigine

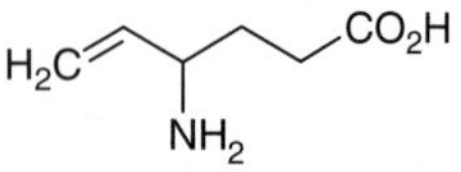

12, Vigabatrin

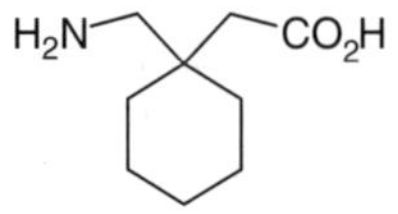

13, Gabapentin

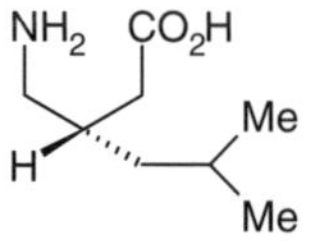

14, Pregabalin

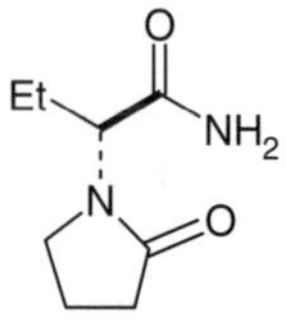

15, Levetiracetam

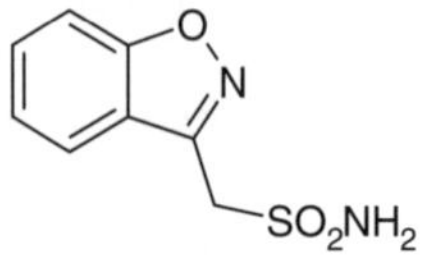

16, Zonisamide

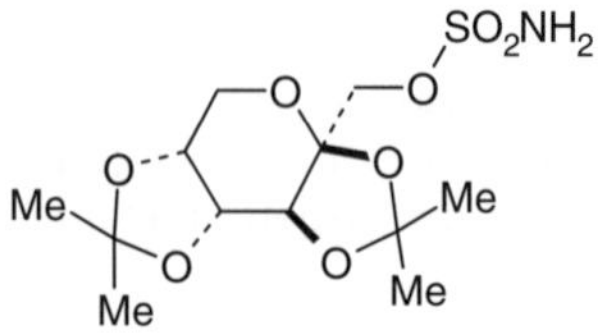

17, Topiramate

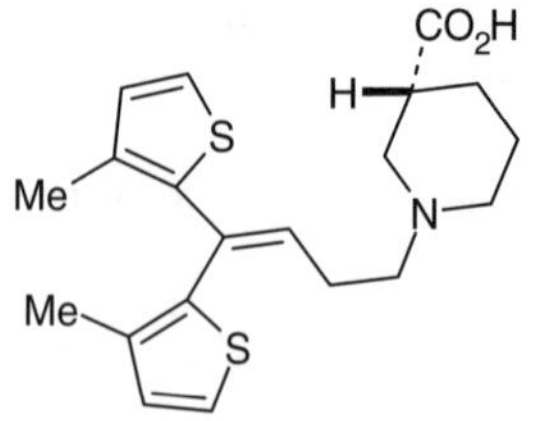

18, Tiagabine

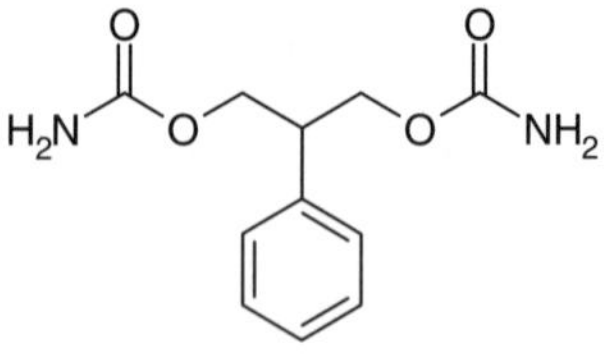

19, Felbamate

Figure 11. Second-generation agents (**10–19**).

the result of rational design, rather than serendipity. They are based on GABA-related molecules, or enzymes involved in GABA metabolism or blocking GABA reuptake or are designed as modifications of other agents currently being used for different diseases.

3. "Third-generation" agents (**20–30**) (Fig. 12) are currently in preclinical or clinical development: these include drugs that are derived from effective compounds on the market, but with improvements made to their properties in terms of safety and efficacy; other compounds contain chemical structures that do not remind existing drugs.

In the past decades, different chemical classes of anticonvulsants have been identified (Tables 7–9). In some cases, certain chemical classes have been widely explored, providing large series of derivatives containing various

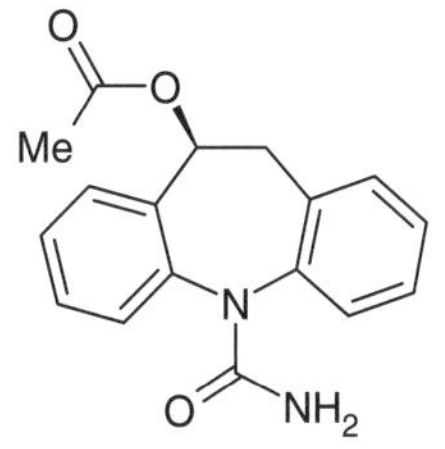

20, Eslicarbazepine acetate
21, Brivaracetam 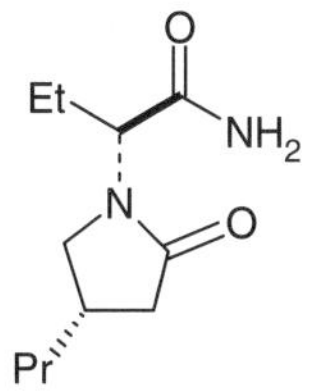
22, Valrocemide 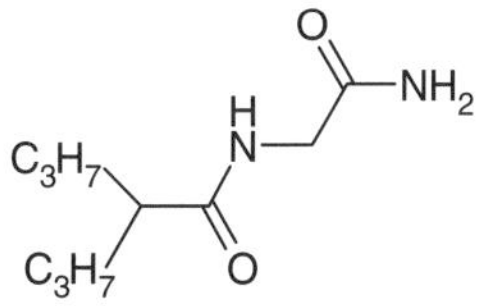
23, Talampanel 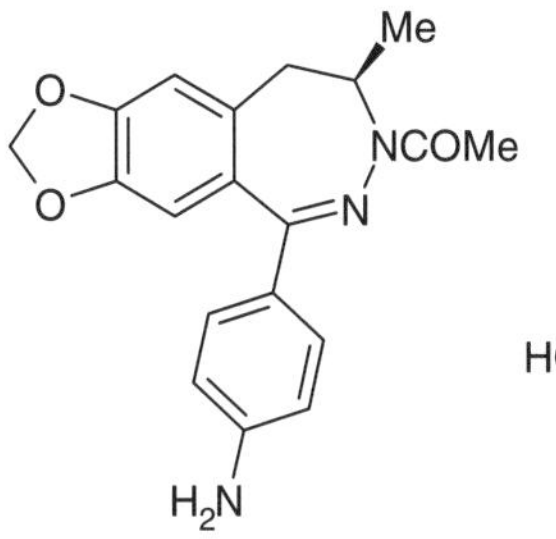
24, Ganaxolone 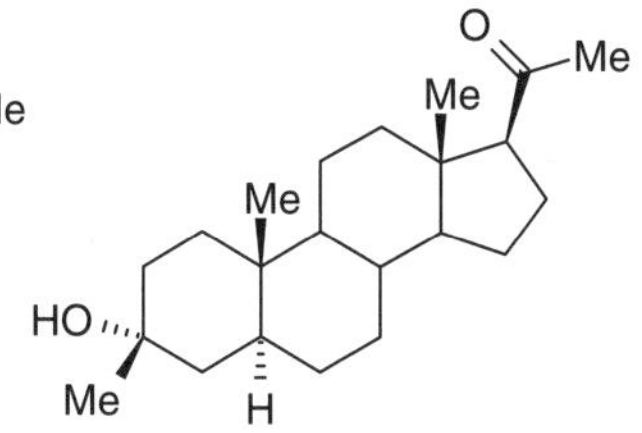
25, Rufinamide 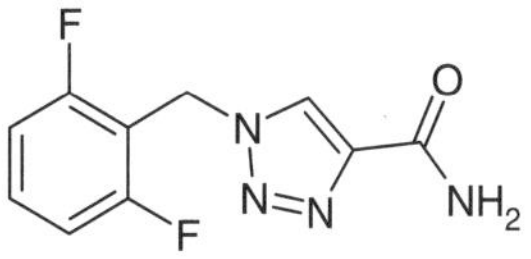
26, Retigabine 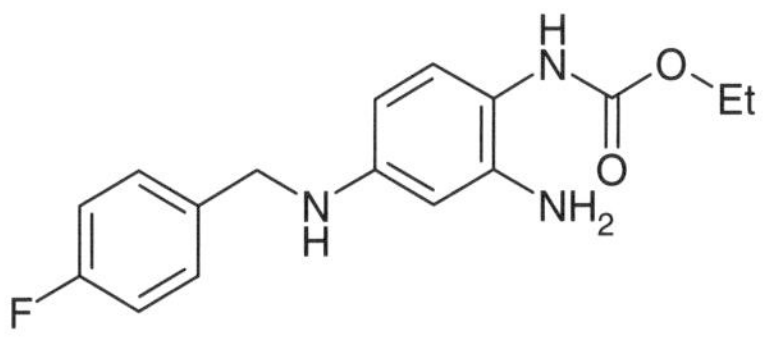
27, Carabersat 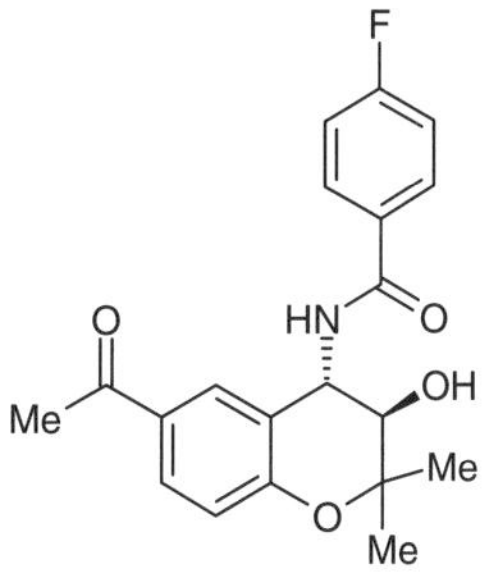
28, Lacosamide 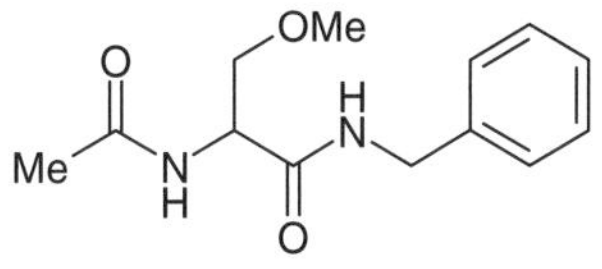
29, Stiripentol 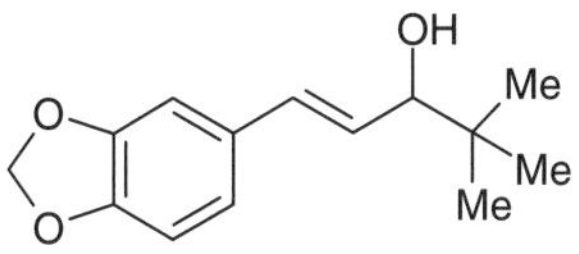
30, Carisbamate 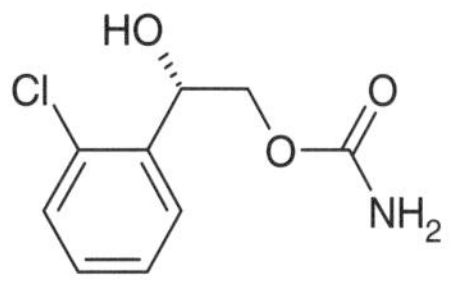

Figure 12. Third-generation agents (**20–30**).

substituents on a common skeleton, providing structure–activity relationship considerations (barbiturates and benzodiazepines). Nevertheless, anticonvulsants are developed from different chemical classes; thus, making it very difficult to list the anticonvulsants following an accurate chemical classification. Furthermore, considering that the anticonvulsants can act via different mechanism of action related to diverse molecular targets, what is lacking is a general two or three-dimensional pharmacophore that is useful to describe pharmacodynamic properties. Finally, the observation of serious side effects has led to the withdrawal of several different drugs over the years. The above reported

Table 7. First Generation: Proprietary and Nonproprietary Names, Chemical Classes, and Mechanism of Action of Anticonvulsants in Current Use

Generic Name (Proprietary Name)	Chemical Classes	Mechanism of Action
Phenytoin (Dilantin)	Hydantoin	Inhibition of voltage-gated sodium channels
Mephenytoin (Mesontoin)	Hydantoin	Inhibition of voltage-gated sodium channels
Ethotoin (Peganone)	Hydantoin	Inhibition of voltage-gated sodium channels
Fosphenytoin sodium (Cerebyx)	Hydantoin	Inhibition of voltage-gated sodium channels
Carbamazepine (Tegretol)	Iminostilbene	Inhibition of voltage-gated sodium channels
Valproic acid (Depakene)	Aliphatic acid	Positive allosteric $GABA_A$ modulation
Clorazepate dipotassium (Tranxene)	Benzodiazepine	Positive allosteric $GABA_A$ modulation
Clonazepam (Klonopin)	Benzodiazepine	Positive allosteric $GABA_A$ modulation
Diazepam (Valium)	Benzodiazepine	Positive allosteric $GABA_A$ modulation
Ethosuximide (Zarontin)	Succinimide	T-type calcium ion channel blockade
Phensuximide (Milontin)	Succinimide	T-type calcium ion channel blockade
Phenobarbital (Luminal)	Barbiturate	Positive allosteric $GABA_A$ modulation
Primidone (Mysoline)	Dihydrobarbiturate	Positive allosteric $GABA_A$ modulation
Trimethadione (Tridione)	Oxazolidinedione	T-type calcium ion channel blockade

Table 8. Second Generation: Proprietary and Nonproprietary names, Chemical Classes, and Mechanism of Action of Anticonvulsants in Current Use

Generic Name (Proprietary Name)	Chemical Classes	Mechanism of Action
Vigabatrin (Sabril)	GABA analog	Inhibition of GABA-T
Gabapentin (Neurontin)	Cyclohexaneacetic acid	Inhibition of $\alpha2$-δ-subunits of P/Q type calcium channels
Felbamate (Felbatol)	Carbamate	Inhibition of voltage-gated sodium channels. Indirect antagonism of NMDA receptors. Inhibition of voltage-activated calcium currents. Increase of GABA transmission through $GABA_A$
Lamotrigine (Lamictal)	Triazine	Inhibition of voltage-gated sodium channels. Inhibition of voltage-activated calcium channels
Oxcarbazepine (Trileptal)	Iminostilbene	Inhibition of voltage-gated sodium channels
Zonisamide (Zonegran)	Sulfonamide	Inhibition of voltage-gated sodium channels. Inhibition of T-type calcium channels. Inhibition of carbonic anhydrase
Tiagabine hydrochloride (Gabitril)	Nipecotic acid analog	Inhibition of GABA transporters
Topiramate (Topamax)	Sulfamate	Inhibition of voltage-gated sodium channels. Inhibition of voltage-activated calcium channels. Positive modulation of $GABA_A$ receptors. Inhibition of kainite and AMPA receptors. Inhibition of carbonic anhydrase
Levetiracetam (Keppra)	Pyrrolidin-2-one	Binding to synaptic vesicle protein 2A (SV2A).

Table 9. Third Generation: Generic Names, Chemical Classes, and Proposed Mechanism of Action of Emerging Anticonvulsants in Phase II or Phase III

Generic Name	Chemical Classes	Mechanism of Action
Brivaracetam	Pyrrolidin-2-one	Binding to synaptic vesicle protein 2A (SV2A) with additional sodium channel-blocking properties
Fluorofelbamate	Carbamate	NMDA/NR2B receptor selective blockade. Slight decrease in voltage-dependent sodium currents
Ganaxolone	Steroid	Positive allosteric modulation of $GABA_A$ receptors
Lacosamide	Amide	Binding to strychnine-insensitive glycine recognition site of the NMDA receptor
Retigabine	Carbamate	Selective M-current K^+ channel opening at KCNQ2.3 and KCNQ3.5
Rufinamide	Triazole	Interaction with the inactive state of sodium channels
Carisbamate	Carbamate	Unknown
Carabersat	Benzopyrane	Selective binding to gap junctions
Stiripentol	Allylic alcohol	Inhibition of GABA uptake
Talampanel	Benzodiazepine	Selective noncompetitive antagonism of AMPA receptors
Valrocemide	N-Valproyl glycinamide	Unknown (possibly related to valproic acid)

anticonvulsants were selected on the basis of both their historical significance and attractive new molecular targets.

4. ANTICONVULSANT AGENTS IN THERAPY AND IN DEVELOPMENT

4.1. Barbiturates

Some 5,5-disubstituted barbituric acid derivatives (chemically 2,4,6(1H,3H,5H)-pyrimidinetriones or malonylureas), also called barbiturates, display anticonvulsant effects mainly by facilitating GABA mediated inhibition via allosteric modulation of postsynaptic $GABA_A$ receptor subtypes. They enhance the activation of $GABA_A$ receptors by increasing the mean duration of channel opening without affecting open frequency or conductance. The resulting Cl⁻ influx then produces the hyperpolarization process for the postsynaptic neuronal cell membrane, thus reducing the transmission of epileptic firing. It has also been demonstrated that barbiturates activate $GABA_A$ receptor directly in the absence of neurotransmitter, this effect is related to the sedatives effects that characterize some barbiturates.

An extensive series of barbiturates have been studied and some derivatives are currently employed in therapy as anticonvulsants (Fig. 13) as well as hypnotics and sedatives. The most active anticonvulsant barbitutates contain some main chemical features that control both the allosteric modulation of $GABA_A$ receptor and pharmacokinetic properties: (a) two alkyl or aryl substituents on C-5 position and (b) suitable substituent on one of the two nitrogen atoms.

As anticonvulsant agents, barbiturates are currently relegated to third-line status because of enzyme-inducing properties, their potential for habituation, somnolence, physical slowing, and potential use in suicide.

Nevertheless, the phenobarbital, 5-ethyl-5-phenylbarbituric acid (1) [57], is generally recommended by the World Health Organization as first-line treatment for generalized tonic–clonic seizures in developing countries (even if it also remains a popular choice in many parts of industrialized world). Phenobarbital is a 5,5-disubstituted barbituric acid with a pK_a approximating physiological pH, it is relatively insoluble in water, but is readily soluble as the sodium salt. Phenobarbital has good bioavailability and a long half-life. Like other barbiturates, phenobarbital induces sedation and other cognitive and behavioral side effects. It is mainly metabolized in the liver forming two inactive metabolites, p-hydroxyphenobarbital by aromatic hydroxylation

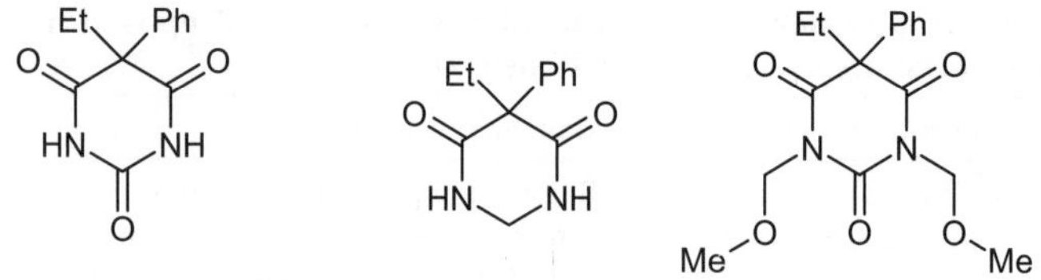

1, Phenobarbital

2, Primidone

31, Eterobarb

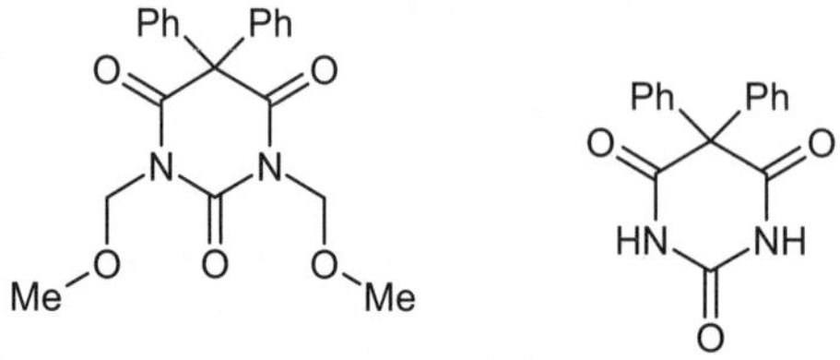

32, T2000

33, 5,5-Diphenyl-barbituric acid

Figure 13. Phenobarbital and analogs.

(cytochrome P450, CYP) and 9-D-glucopyranosylphenobarbital by glucosidation; the remaining is excreted, unchanged, in the urine.

Generally, barbiturates are the target of hepatic enzymes but it is well known that their administration also produces the induction of CYP enzymes, increasing the clearance of many commonly used lipid-soluble drugs that undergo hepatic biotransformation. In fact, phenobarbital increases the expression of a range of CYP and phase II metabolizing enzymes, thus producing significant drug interactions.

The analog primidone (2-desoxyphenobarbital, chemically 5-ethyldihydro-5-phenyl-4,6 (1H,5H)pyrimidinedione **2**) undergoes biotransformation in phenobarbital by oxidation. Nevertheless, it has been demonstrated that primidone is active as an anticonvulsant before undergoing biotransformation.

The eterobarb (N,N'-dimethoxymethyl phenobarbital, chemically 5-ethyl-1,3-bis(methoxymethyl)-5-phenyl-2,4,6(1H,3H,5H) pyrimidinetrione, **31**) (Fig. 13) may be considered a prodrug of phenobarbital (**1**), its metabolism rapidly gives the active intermediate N-monomethoxy phenobarbital, then the complete biotransformation occurs (Fig. 14). Eterobarb possesses reduced sedative and hypnotic properties when compared to the parent compound.

Over the last few years, new barbiturates (Fig. 13) [58] have been under development for the treatment of epilepsy: T2000 (**32**), chemically 1,3-bis(methoxymethyl)-5,5-diphenyl-2,4,6(1H,3H,5H)pyrimidinetrione, and the corresponding active metabolite 5,5-diphenylbarbituric acid (**33**). The pharmacological effects of 5,5-diphenylbarbituric acid and its prodrug T2000 are similar to phenytoin and barbiturates, and it has been demonstrated

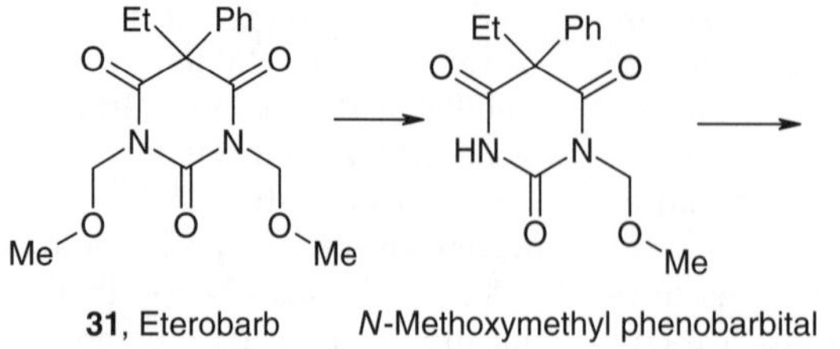

31, Eterobarb → N-Methoxymethyl phenobarbital → **1**, Phenobarbital

Figure 14. Metabolic pathway of eterobarb to give phenobarbital.

that they interact with voltage-gated sodium channels, $GABA_A$ receptors, and/or other molecular targets; but their exact mechanism of anticonvulsant activity remains to be clarified. The interest in these newer anticonvulsant agents derives from the consideration that T2000 and its metabolite exhibit useful pharmacological properties at dosages not limited by sedative side effects.

4.2. Hydantoins (2,4-Imidazolidinediones)

Phenytoin (5,5-diphenyl-2,4-imidazolidindione, **3**) is the most widely studied hydantoin derivative [59]. The mechanism of action of phenytoin (**3**) and the related anticonvulsant hydantoins [60] (**34**–**36**) (mephenytoin, ethotoin, and fosphenytoin) (Fig. 15) involves the inhibition of voltage-gated sodium channels, thus resulting in increased inhibition of action–potential firing activity.

Phenytoin (**3**) is metabolized in the liver through cytochrome P450 into inactive hydroxylated metabolites (Fig. 16) via enantioselective metabolic pathway. Mephenytoin (3-methyl-5-ethyl-5-phenylhydantoin, **34**) undergoes biotransformation in the corresponding *N*-desmethyl active metabolite (*R*-enantiomer) or inactive *p*-hydroxyderivative (*S*-enantiomer) (Fig. 16). The derivative ethotoin (**35**) furnishes two main inactive metabolites by *p*-hydroxylation and *N*-dealkylation pathways. On the contrary, the main product of the metabolization of the water-soluble prodrug fosphenytoin (**36**) is the corresponding active metabolite phenytoin (**3**).

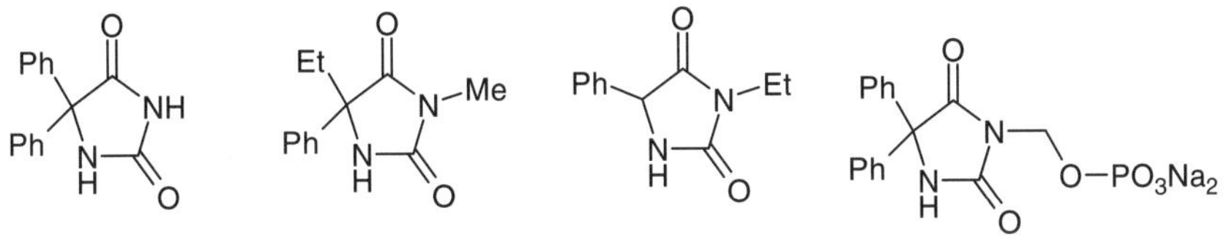

3, Phenytoin **34**, Mephenytoin **35**, Ethotoin **36**, Fosphenytoin

Figure 15. Hydantoins (2,4-imidazolidinedione derivatives).

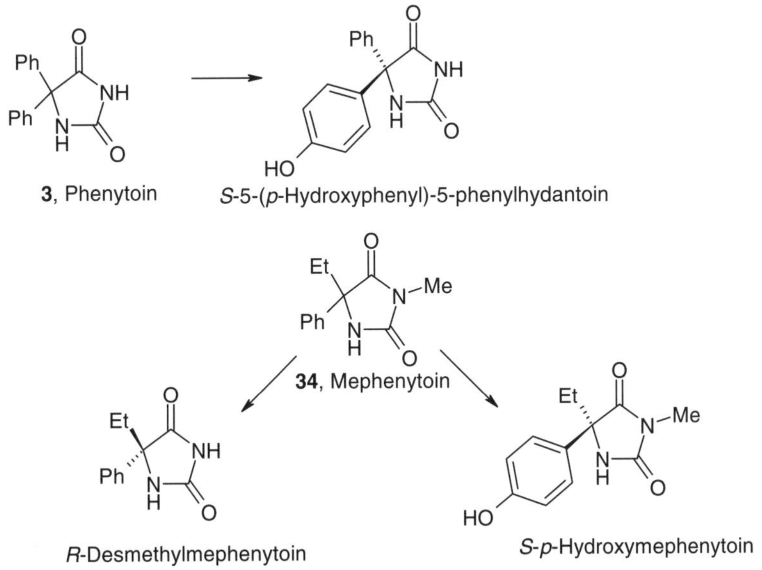

Figure 16. Biotransformation of hydantoins.

4.3. Succinimides (Pyrrolidine-2,5-Diones)

Ethosuximide (3-ethyl-3-methyl-pyrrolidine-2,5-dione, **4**) and methsuximide (1,3-dimethyl-3-phenyl-pyrrolidine-2,5-dione, **37**) are succinimide derivatives that exert anticonvulsant effects via the blockading of T-type Ca^{2+} channels [61]. Ethosuximide shows clinical efficacy in generalized seizures (absence and myoclonic seizures), it is relatively lacking in side effects and drug interactions (it increases phenytoin levels, and decreases those of primidone and phenobarbital). Ethosuximide undergoes biotransformation into the inactive hydroxyethyl derivatives, while the N-desmethyl-metabolite of methsuximide is an active metabolite (Fig. 17).

4.4. Trimethadione (3,5,5-Trimethyloxazolidine-2,4-Dione)

The 3,5,5-trimethyloxazolidine-2,4-dione (**5**), named trimethadione (Fig. 18), is a first-generation anticonvulsant that is useful in controlling absence seizures by acting as T-type Ca^{2+} channel blocker. However, it shows the potential to produce fetal malformations and serious adverse effects; for these reasons, it should be used only in cases where other drugs are ineffective in controlling seizures [62]. The main metabolite of trimethadione is the active N-desmethyl derivative (Fig. 18).

4.5. Iminostilbenes

The carbamazepine (5H-dibenz[b,f]azepine-5-carboxamide, **6**) and its analogs are iminostilbene derivatives acting as anticonvulsants via the blockade of voltage-gated sodium channels and consequent inhibition of action potentials. The iminostilbenes share the same mechanism of action as phenytoin, and the prototype, carbamazepine, is generally used to control generalized tonic–clonic and partial seizures. Carbamazepine is biotransformed in the liver to form inactive *trans*-10,11-diol through the 10,11-epoxide reactive intermediate (Fig. 19) [63]. The 10,11-epoxidecarbamazepine is considered the principal metabolite that plays a key role in enzyme autoinduction, inducing its own metabolism, as well as in enzyme induction thereby interfering with other xenobiotics (drug–drug interactions) metabolized by the cytochrome-P450 system.

The main side effects of carbamazepine are the potential to produce aplastic anemia and agranulocytosis. Furthermore, carbamazepine can lead to allergic skin reactions occurring.

A large series of iminostilbenes structurally related to carbamazepine have been developed with the aim of identifying more stable and less toxic derivatives; thus second-generation and newer iminostilbenes are being studied [64].

Oxcarbazepine (chemically 10,11-dihydro-10-oxo-5H-dibenz[b,f]azepine-5-carboxamide or 10,11-dihydro-10-oxocarbamazepine, **10**) is the keto analog of carbamazepine. Oxcarbazepine differs from its parent compound in terms of its metabolism; it being extensively metabolized (Fig. 19) in the corresponding

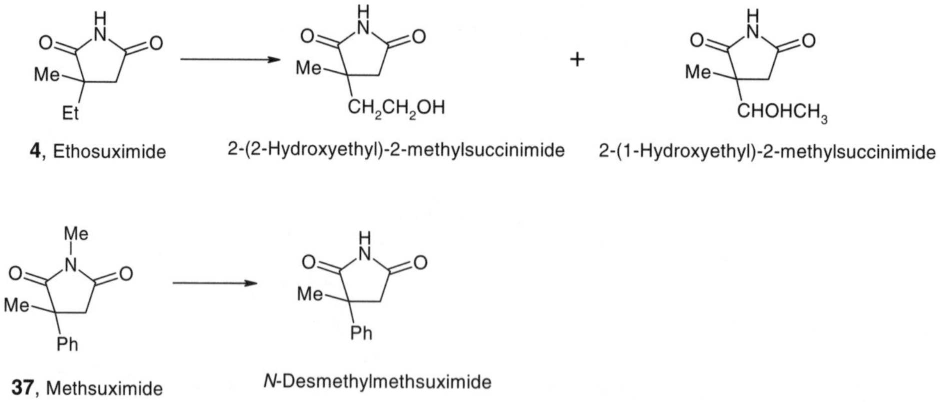

Figure 17. Succinimide derivatives and their main metabolic pathway.

5, Trimethadione → **N-Desmethyltrimethadione**

Figure 18. Trimethadione and its N-desmethyl metabolite.

racemic mixture of alcohols S-(+)-10,11-dihydro-10-hydroxycarbamazepine (**38**) and R-(-)-10,11-dihydro-10-hydroxycarbamazepine (**39**) in the ratio 4:1. Subsequently the active metabolite S-monohydroxy derivative (S-licarbazepine **38**) undergoes the biotransformation in the inactive *trans*-diol, the same metabolite of carbamazepine. Oxcarbazepine was designed to have similar efficacy, but fewer adverse effects than carbamazepine. For these reasons, it is a relatively safe anticonvulsant and shows a reduced tendency to induce oxidative metabolism compared with carbamazepine. Furthermore, after oral administration oxcarbazepine is rapidly and almost completely absorbed by the gastrointestinal tract with peak concentrations being obtained after approximately 1 h. While carbamazepine shows a plasma half-life of 3–5 h, the pharmacokinetics of the main active metabolite 10,11-dihydro-10-hydroxycarbamazepine ((S)-(+)-licarbazepine, **38**) presents a considerably slower elimination half-life (8–10 h). Moreover, oxcarbazepine (**10**) inhibits several types of voltage-gated calcium channels, potentially resulting in a broader therapeutic spectrum. In addition to its anticonvulsant action, oxcarbazepine has a positive effect on psychiatric diseases.

Eslicarbazepine acetate (**20**), chemically (S)-(−)-10-acetoxy-10,11-dihydro-5H-dibenz[b,f]azepine-5-carboxamide, may be regarded as the third-generation derivatives of carbamazepine. It is currently completing phase III

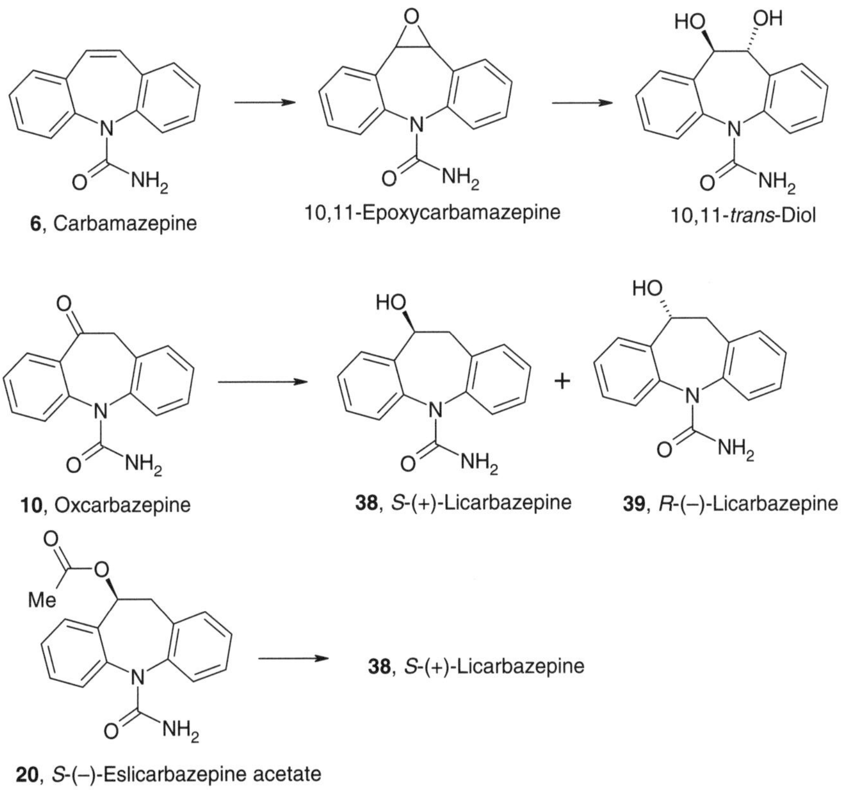

6, Carbamazepine → **10,11-Epoxycarbamazepine** → **10,11-*trans*-Diol**

10, Oxcarbazepine → **38, S-(+)-Licarbazepine** + **39, R-(−)-Licarbazepine**

20, S-(−)-Eslicarbazepine acetate → **38, S-(+)-Licarbazepine**

Figure 19. Metabolism of carbamazepine and related iminostilbenes.

144 ANTICONVULSANTS

clinical trials as adjunctive therapy in partial epilepsy.

The structural optimization of eslicarbazepine acetate gave a different metabolism; it is not metabolized into carbamazepine-10,11-epoxide and is not susceptible to induction of its own metabolism. Unlike oxcarbazepine (**10**), which is metabolized into both (*S*)-(+)-licarbazepine (**38**) and (***R***)-(−)-licarbazepine (**39**), then eslicarbazepine acetate (**20**) is a prodrug of eslicarbazepine (**38**), which is the molecule responsible for anticonvulsant effects [65,66]. Currently, there are ongoing studies to evaluate its efficacy for the treatment of neuropathic pain.

4.6. Benzodiazepines

A large series of benzodiazepines (Fig. 20) are currently among the most frequently prescribed drugs all over the world. They act as anxiolytics, sedatives, hypnotics, antiepileptics and muscle relaxants [67,68]. Despite their common chemical scaffold, these molecules differ in relation to their pharmacokinetic and metabolic properties [69,70]. Benzodiazepines are considered to be medications of first choice to block seizures. Even if they are effective in about 60% of all episodes, a sizeable proportion of patients remain in status epilepticus (SE). Benzodiazepines gained popularity among medical professionals for

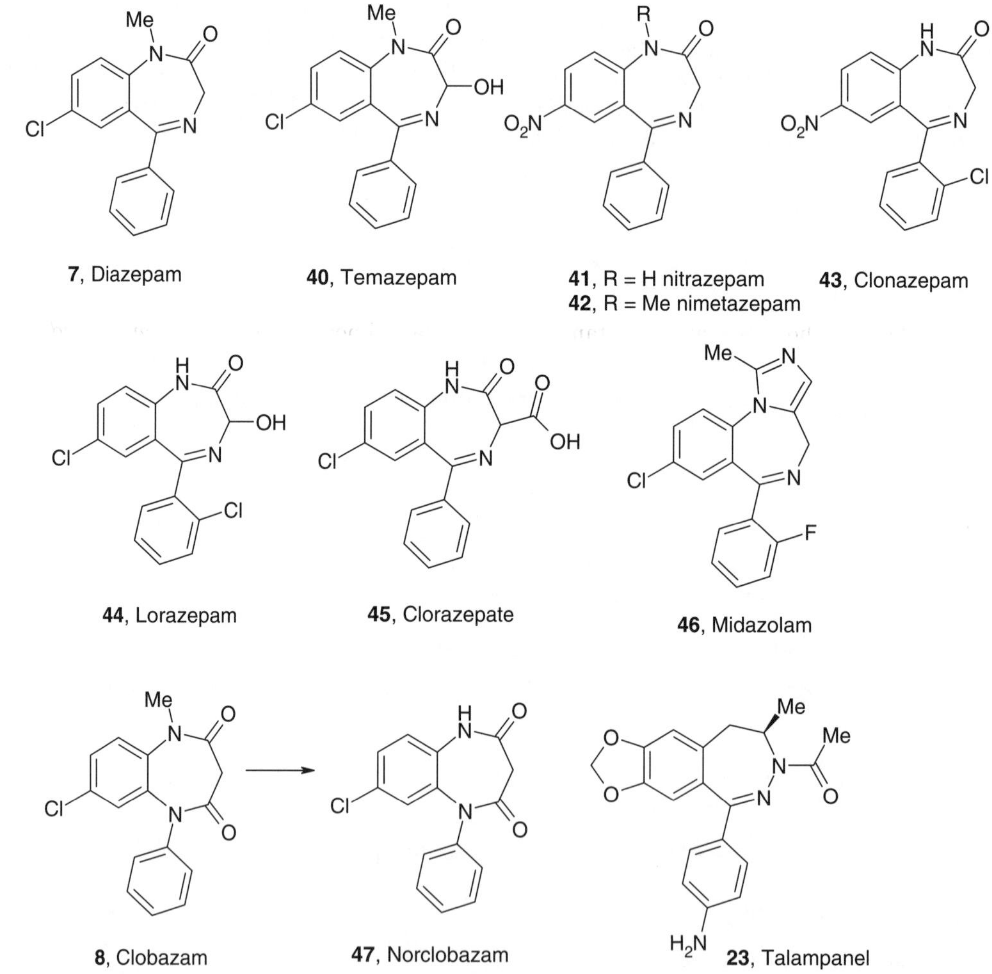

7, Diazepam   **40**, Temazepam   **41**, R = H nitrazepam   **43**, Clonazepam
42, R = Me nimetazepam

44, Lorazepam   **45**, Clorazepate   **46**, Midazolam

8, Clobazam   **47**, Norclobazam   **23**, Talampanel

Figure 20. Benzodiazepine derivatives.

being an improvement upon barbiturates, which have a comparatively narrow therapeutic index, and are far more sedating at therapeutic doses. Benzodiazepines are also far less dangerous; death rarely results from a benzodiazepine overdose, except in cases where it is consumed with large amounts of other depressants [71,72].

Their actions are due to the potentiation of the neural inhibition that is mediated by GABA neurotransmitter through the ionotropic $GABA_A$ receptors in the central nervous system [73]. Benzodiazepines do not activate $GABA_A$ receptors directly requiring GABA to do so. They have affinity for the allosteric binding site, called "the benzodiazepine site" (BZR), which is found on the α_1-subtype of the $GABA_A$ receptor. It is a specific subunit on the $GABA_A$ receptor at a site that is distinct from the binding site of endogenous GABA.

Benzodiazepines exert their inhibitory activity as positive allosteric modulators causing an enhancement of the electric effect of GABA binding on neurons; thus, resulting in an increased influx of chloride ions into the neurons. Benzodiazepines, do not have any effect on the levels of GABA in the brain and no effect on GABA-transaminase (GABA-T). Some of them do, however, affect glutamate decarboxylase (GAD) activity.

Benzodiazepine receptors (BZRs) are found in the central nervous system, but are also found in a wide range of peripheral tissues such as lungs, liver, and kidney as well as mast cells, platelets, lymphocytes, heart, and numerous neuronal and nonneuronal cell lines. Long-term use of benzodiazepine can lead to tolerance and dependency [74].

Of the many drugs in this class, only clobazam (**8**), clonazepam (**43**), and clorazepate (**45**) are used to treat epilepsy. The following benzodiazepines are mainly used to treat status epilepticus: diazepam (**7**), midazolam (**46**) (as an alternative to diazepam by the buccal mucosa), and lorazepam (**44**). Lastly, temazepam (**40**), nitrazepam (**41**), and nimetazepam (**42**) are powerful anticonvulsant agents. However, their use is rare due to an increased incidence of side effects and their strong sedative and motor-impairing properties.

Diazepam (**7**), chemically is 7-chloro-1,3-dihydro-1-methyl-5-phenyl-2H-1,4-benzodiazepin-2-one commonly used for treating anxiety, insomnia, seizures, and muscle spasms. It may also be used before certain medical procedures (such as endoscopies), and surgical procedures to reduce tension and anxiety. Diazepam can be administered orally, intravenously, intramuscularly, or as a suppository. Diazepam is highly lipid soluble, and is widely distributed throughout the body after administration. It easily crosses both the blood–brain barrier and the placenta, and is excreted into breast milk. It undergoes oxidative metabolism (Fig. 21) in the liver and its main active metabolite is N-desmethyldiazepam (also known as *nordiazepam* or *nordazepam*). Its other active metabolites include temazepam (**40**) and oxazepam (**48**) (Fig. 21). These metabolites are conjugated with glucuronide, and are excreted primarily in the urine.

Diazepam (**7**) is rarely used for the long-term treatment of epilepsy because if it is taken for 6 weeks or longer it, like other benzodiazepine drugs, can cause tolerance, physical dependence, addiction, and what is

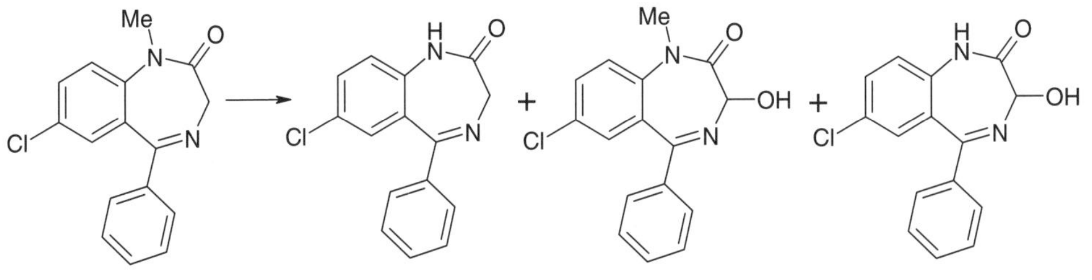

7, Diazepam   Nordiazepam   **40**, Temazepam   **48**, Oxazepam

Figure 21. Metabolic pathway of diazepam: demethylation and oxidation processes.

known as the benzodiazepine withdrawal syndrome.

The 3-hydroxy metabolite of diazepam is the commercially sold agent temazepam (**40,** 7-chloro-1,3-dihydro-3-hydroxy-1-methyl-5-phenyl-2H-1,4-benzodiazepin-2-one). It is generally prescribed for the short-term treatment of severe or debilitating sleeplessness in patients who have difficulty falling asleep or staying asleep. In addition, temazepam possesses anticonvulsant, anxiolytic, and skeletal muscle relaxant properties. It is only very slightly soluble in water and oral administration of temazepam results in rapid absorption. This drug has very good bioavailability and is metabolized through demethylation (cf. diazepam) prior to excretion as O-conjugate. The metabolite is the active derivative oxazepam (**48**). There is clinical evidence that chronic or excessive use of temazepam (**40**) may cause physical dependence and drug tolerance, which can develop rapidly; this drug is therefore not recommended for long-term use. Unlike many benzodiazepines, pharmacokinetic interactions involving the P450 system have not been observed with temazepam. Nitrazepam (**41**), chemically 1,3-dihydro-7-nitro-5-phenyl-2H-1,4-benzodiazepin-2-one, is a nitrobenzodiazepine showing long lasting activity; it is lipophilic and has a good cerebral uptake; it is also bound to plasma proteins. It is an agonist for both central benzodiazepine receptors and to peripheral type benzodiazepine receptors. Nitrazepam is most often used to treat short-term sleeping problems but is also useful in the management of myoclonic seizures and seizure disorders in children as well as for treating refractory epilepsies. However, the loss of antiepileptic efficacy due to tolerance renders nitrazepam ineffective in prolonged therapy. Nitrazepam is mainly metabolized hepatically via oxidative pathways.

Clonazepam, the 5-(2-chlorophenyl)-1,3-dihydro-7-nitro-2H-1,4-benzodiazepin-2-one (**43**), is a chlorinated derivative of nitrazepam with highly potent anticonvulsant efficacy. It is also characterized by muscle relaxant and anxiolytic properties. Clonazepam is sometimes used for refractory epilepsies, in the acute control of nonconvulsive status epilepticus, and in the management of seizure disorders in children. However, the usefulness of clonazepam is limited due to its dose limiting side effects, especially its negative effect on cognition.

Clonazepam passes through the blood–brain barrier easily in the blood and rapidly into the central nervous system with levels in the brain corresponding with levels of unbound clonazepam in blood serum. However, clonazepam is mainly bound to plasma proteins.

The inactive metabolites of clonazepam are 7-aminoclonazepam, 7-acetamidoclonazepam, and 3-hydroxy clonazepam. Tolerance to the anticonvulsant effects of clonazepam occurs frequently in humans. Discontinuation of, or reduction in, dosage after regular use may result in the clonazepam withdrawal syndrome with dysphoric manifestations, irritability, aggressiveness, anxiety, and hallucinations. For these reasons, clonazepam should be reduced slowly and gradually when discontinuing the drug to reduce the effects of withdrawal.

The 3-hydroxy 1,4-benzodiazepine derivative lorazepam (**44**), chemically 7-chloro-5-(2-chlorophenyl)-1,3-dihydro-3-hydroxy-2H-1,4-benzodiazepin-2-one, has a role in long-term prophylactic treatment of resistant forms of petit mal epilepsy but not as first-line therapy, mainly because of the development of resistance to its effects. Lorazepam is a highly potent benzodiazepine showing advantageous and disadvantageous characteristics related to its pharmacokinetic properties. Because of its poor lipid solubility lorazepam is absorbed relatively slowly by mouth and is unsuitable for rectal administration, but its poor lipid solubility and high degree of protein binding mean that its volume of distribution is mainly the vascular compartment, causing relatively prolonged peak effects replacing diazepam as the *intravenous* agent of choice in status epilepticus. Lorazepam serum levels are proportional to the dose administered. Approximately 2 h later, half of which is lorazepam, half its inactive metabolite, lorazepam-glucuronide. This metabolism does not involve hepatic oxidation and is therefore relatively unaffected by reduced liver function.

The 3-carboxy derivative clorazepate (**45**) (7-chloro-2,3-dihydro-2-oxo-5-phenyl-1H-1,4-benzodiazepine-3-carboxylic acid) was principally prescribed in the treatment of alcohol withdrawal, insomnia, and epilepsy, although a useful anxiolytic because of its long half-life; however, clorazepate was discontinued in February 2006 in the United Kingdom. Clorazepate produces the metabolite desmethyldiazepam that has a half-life of 36–200 h. Regular use of clorazepate by humans causes the development of dependence characterized by tolerance to the therapeutic effects of benzodiazepines.

Midazolam (**46**) is a cyclofunctionalized 1,4-benzodiazepine derivative, chemically 8-chloro-6-(2-fluorophenyl)-1-methyl-4H-imidazo[1,5-a][1,4]benzodiazepine indicated for the acute management of seizures such as status epilepticus, whereas its long term use for the management of epilepsy is not recommended due to tolerance. Midazolam is metabolized almost completely by the cytochrome P450-3A4, more specifically, it is hydroxylated to the active metabolite 1-hydroxy-midazolam and then glucuronidated before being renally excreted.

Clobazam (**8**), the 7-chloro-1-methyl-5-phenyl-1H-1,5-benzodiazepine-2,4(3H,5H)-dione, is used in tonic–clonic, complex partial, and myoclonic seizures as well as in tonic varieties and nonstatus absence seizures and certain types of status epilepticus. The biotransformation of clobazam gives N-desmethyl-clobazam (norclobazam, **47**) and 4'-hydroxyclobazam, the former being active and this also acts by enhancing GABA-activated chloride currents at GABA$_A$-receptor-coupled Cl$^-$ channels. In humans, tolerance to the anticonvulsant effects of clobazam frequently occurs and withdrawal seizures can occur during abrupt or overly rapid withdrawal. Clobazam, like other benzodiazepine drugs, can lead to physical dependence, addiction, and what is known as the benzodiazepine withdrawal syndrome. Withdrawal symptoms can, however, occur from standard dosages and even after short-term use.

More recently, other molecules containing benzodiazepine scaffold have been developed as anticonvulsants. Some 2,3-benzodiazepine derivatives have demonstrated pharmacological efficacy through a new and attractive mechanism of action [75–80]. In fact, these compounds were able to produce neuroprotective effects via noncompetitive antagonism at glutamate AMPA receptor and without affecting the GABA$_A$-receptor benzodiazepine site. The most interesting molecule is the (8R)-7-acetyl-5-(4-aminophenyl)-8,9-dihydro-8-methyl-7H-1,3-dioxolo[4,5-h][2,3]benzodiazepine, named talampanel (**23**) [32], which is currently under clinical trials in patients with severe epilepsy not responsive to other drugs.

Continuing efforts are being addressed to noncompetitive AMPA receptor antagonists [34], which could have the advantage of remaining effective independently of the level of neurotransmitter glutamate or the polarization state of the synaptic membrane during neurological diseases.

4.7. Valproic Acid

The simple aliphatic derivative valproic acid and the corresponding sodium salt are anticonvulsants probably acting as voltage-gated sodium channel blockers. Valproic acid (2-propylpentanoic acid **9**) (Fig. 22) is also an effective drug in migraine prophylaxis and in the treatment of bipolar disorder [81].

Valproic acid is the drug of first choice for patients with idiopathic and symptomatic generalized epilepsy despite its disadvantages, particularly weight gain, hepatotoxicity, and teratogenicity. It is extensively metabolized and supplies the active metabolite 2-ene-derivative and metabolites (Fig. 22) with a terminal double-bond that are considered responsible for hepatotoxicity.

Recently, second-generation valproic acid analogs (**22, 49–51**) have been synthesized aimed at enhancing brain penetration, improving central nervous system activity, and preventing the formation of toxic metabolites.

It was found that the corresponding CNS-active amide of valproic acid, named valpromide (2-propylpentanamide, **49**), is more potent than its parent compound and it was not teratogenic in animal models. Nevertheless, in humans valpromide represented a prodrug of valproic acid; thus, it always showed teratogenic side effects. In contrast, valproyl glycinamide (2-propylpentyl glycinamide, **22**), also called valrocemide, displays promising activity in various animal models for epilepsy and is currently in phase III clinical trials. It is

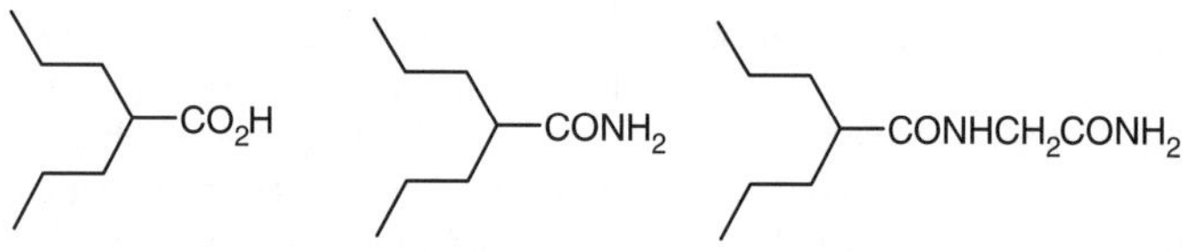

9, Valproic acid **49**, Valpromide **22**, Valrocemide

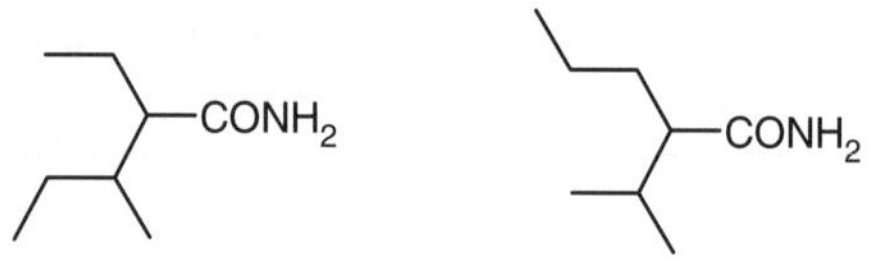

50, Valnoctamide **51**, Propylisopropylactamide

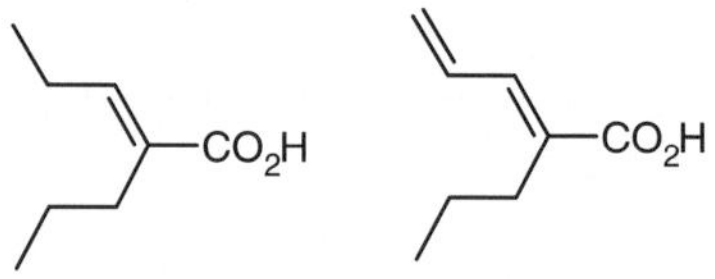

2-Ene derivative 2,4-Diene derivative

Figure 22. Valproic acid and related compounds.

more stable than valpromide and the fraction of valrocemide metabolized to valproic acid is approximately 4–6%, but the implications of this have not yet been established. Also the racemate, phase I (2-ethyl-3-methylpentanamide, **50**), currently sold for bipolar disorder, is a potent second-generation valproic acid nonteratogenic derivative with a broad spectrum of anticonvulsant activity in animal models. More recently, propylisopropyl acetamide (**51**) has been under development as a new and potentially nonteratogenic and nonhepatotoxic valproic acid analog with a mechanism that is still not clear. Unlike valpromide, the propylisopropylacetamide is not biotransformed in animals into its corresponding acid; therefore, it can be considered as a metabolically stable valpromide isomer (**49**).

4.8. Lamotrigine

Lamotrigine (3,5-diamino-6-[2′,3′-dichlorophenyl]-1,2,4-triazine, **11**) (Fig. 23) is a phenyltriazine derivative approved for clinical use in 1993 for the treatment of focal and generalized epilepsy. It has been demonstrated that lamotrigine acts primarily postsynaptically reducing the excitability of neurons through sodium channel blockade; moreover presynaptically, it also reduces the release of excitatory neurotransmitter glutamate [82]. There is also evidence that lamotrigine may also reduce high-voltage-activated calcium currents.

Lamotrigine (**11**) is rapidly absorbed showing bioavailability of approximately 98% and a minimal first-pass metabolism. It induces its own metabolism predominantly by glucuronidation into inactive 2-*N*-glucuronide and the 5-*N*-glucuronide.

The side effects of lamotrigine administration are exanthema, vertigo, gastrointestinal symptoms, somnolence, nausea, diplopia, ideo-syncratic thrombocytopenia, leukopenia, and increased transaminase levels. Lamotrigine is also a useful therapeutic agent for the treatment of Parkinson's disease and bipolar disorder. The lamotrigine analog JZP-4 (3-(2,3,5-trichlorophenyl)-pyrazine-2,6-diamine, **52**) (Fig. 23) seems to block voltage-activated

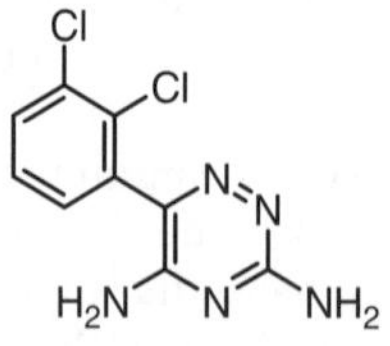

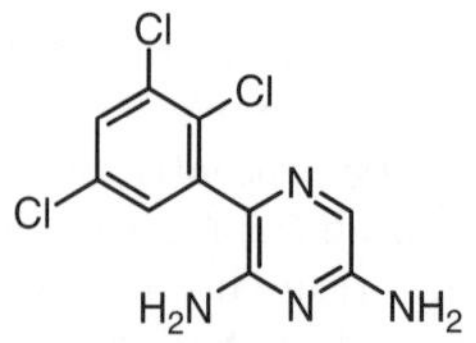

11, Lamotrigine **52**, JZP-4

Figure 23. Lamotrigine and JZP-4.

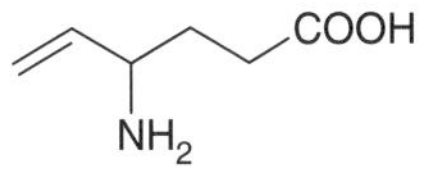

12, Vigabatrin 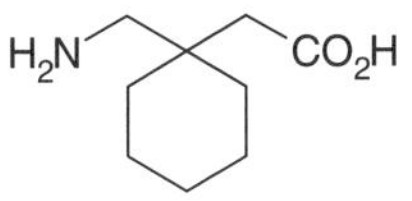 **13,** Gabapentin 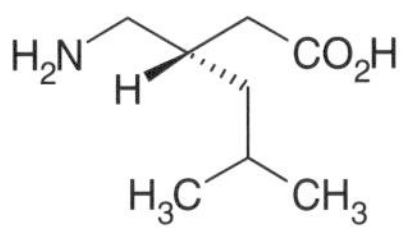 **14,** Pregabalin

Figure 24. GABA analogs.

sodium and calcium channels. This newer compound [64] lacks enzyme induction/inhibition and the formation of reactive metabolites and thus carries a low risk of cutaneous side effects.

4.9. GABA Analogs

Vigabatrin, the 4-amino-5-hexenoic acid (**12**) (Fig. 24) [83,84], exists as a racemic mixture of R-(–)- and S-(+)-isomers, where the S-(+)-isomer is eutomer. The inactive isomer is considered to be an impurity that does not influence the kinetics and the action of the S-(+)-isomer. Vigabatrin (**12**) mainly acts by blocking GABA-transaminase (GABA-T), the enzyme responsible for the breakdown of GABA, as an enzyme-activated irreversible inhibitor thus raising neurotransmitter levels. Experimental evidence suggests that vigabatrin exerts several other actions that are not related to GABA; for instance, it controls brain levels of the excitatory neurotransmitters such as glutamate.

Vigabatrin was initially marketed for treating complex partial epilepsy, but was virtually abandoned due to certain side effects causing retinal pathologies as a consequence of GABA-transaminase and glutamic acid decarboxylase inhibition.

The pharmacokinetic profile of vigabatrin (**12**) is favorable, given its low protein binding, lack of hepatic metabolism, renal excretion, and extended functional half-life [85].

Gabapentin, chemically 1-(aminomethyl)-cyclohexaneacetic acid (**13**) (Fig. 24), was originally developed as a lipid-soluble GABA analog that would cross the blood–brain barrier while retaining many of the chemical and physical properties of GABA. Despite the fact that gabapentin is a GABA analog, its mechanism of action is generally unrelated to this inhibitory neurotransmitter. The most important mechanism of action of gabapentin is probably due to the binding with high affinity to the auxiliary α2-δ-subunits of voltage-gated calcium channels. These represent the main site of action of gabapentin that thereby reduces the synaptic release of some neurotransmitters (glutamate, serotonine, dopamine, etc.) and decreases postsynaptic Ca^{2+} influx resulting in diminished excitation. However, several other possible mechanisms via which gabapentin affects neurotransmission have also been considered, but their impact on antiepileptic activity is not completely clear. Gabapentin is an anticonvulsant agent that is useful in the treatment of partial seizures and also in the elderly. Its use is contraindicated in primary generalized epilepsy. Its short half-life requires multiple daily doses. It is neither metabolized nor protein bound. Gabapentin exhibits dose-dependent bioavailability, in that the plasma concentration of the drug is not directly proportional to the dose throughout the therapeutic range of dosage. However, owing to its linear relation to glomerular filtration rate and creatinine clearance, it must be administered with great care and at the appropriate dosage in patients with impaired renal function. Although gabapentin does not interact with other drugs, the concomitant administration of antacids decreases its bioavailability by 20%.

Like gabapentin, pregabalin (**14**), S-(+)-isobutyl GABA (Fig. 24), is another specific ligand of the α2-δ-subunits of voltage-gated calcium ion channels. Preliminary evidence indicates that pregabalin is also useful to treat generalized anxiety disorders and neuropathic pain. Pregabalin is not significantly metabolized in humans and shows 90% oral bioavailability.

4.10. Pyrrolidin-2-Ones

Piracetam (2-oxo-1-pyrrolidine acetamide, **53**) (Fig. 25) was the precursor of this class of

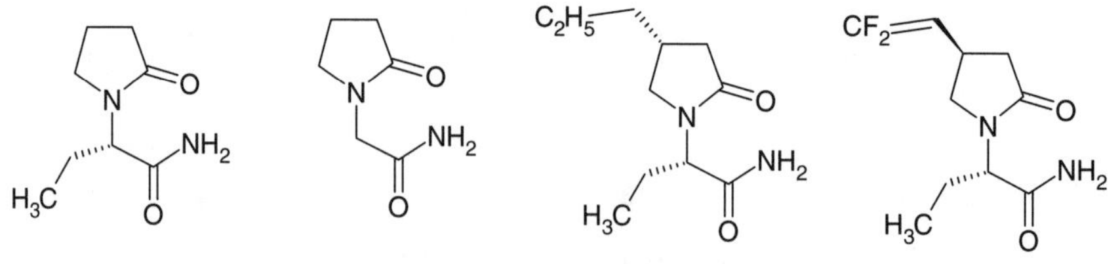

15, Levetiracetam **53, Piracetam** **21, Brivaracetam** **54, Seletracetam**

Figure 25. Pyrrolidin-2-ones.

neuroactive compounds. Other pyrrolidinone derivatives (**15, 21,** and **54**) have been developed [86,87] starting from this "lead compound." Levetiracetam (**15**) is the (*S*)-(−)-ethyl-2-oxo-1-pyrrolidine acetamide, which was approved for clinical use as an anticonvulsant in 2002 [88,89]. It mainly acts with a new and innovative mechanism of action by modulating the exocytotic function of synaptic vesicle protein SV2A, thereby enhancing the release of inhibitory neurotransmitters. Levetiracetam also significantly reduces N- and partially P/Q-type high-voltage-activated calcium currents, whereas sodium currents are unaffected.

Levetiracetam (**15**) was initially approved as an add-on therapy in partial seizures with and without secondary generalization. However, subsequent experimental data demonstrated its efficacy even when used alone and in epilepsies with pharmacoresistance to other anticonvulsants.

As regards its pharmacokinetic properties, levetiracetam (**15**) is rapidly absorbed and oral bioavailability is 100% [90]. It produces several inactive metabolites (Fig. 26): levetiracetam carboxylic acid (24%), diastereomeric 3-hydroxy levetiracetam (2%), and ring-opened butanamide (1%). Studies have indicated that levetiracetam to be well tolerated and no drug interaction. However, the FDA has recently revised its safety labeling to warn of psychiatric symptoms (depression, agitation, or hostility) when starting with medication. Dizziness and somnolence were found to occur to a statistically significant extent in patients receiving levetiracetam in combination with other anticonvulsants. In a patient with impaired renal function, a dose reduction may be needed. Furthermore, it must not be administered during pregnancy and breast-feeding.

A series of levetiracetam analogs have subsequently been developed and are in clinical studies. Like levetiracetam, brivaracetam

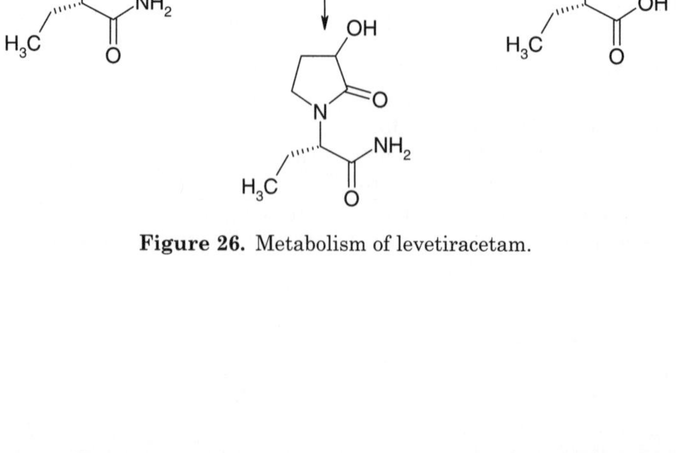

Figure 26. Metabolism of levetiracetam.

((2S)-2-[(4R)-2-oxo-4-propylpyrrolidin-1-yl] butanamide **21**) and seletracetam ((2S)-2-[(4S)-4-(2,2-difluoroethenyl)-2-oxopyrrolidin-1-yl]butanamide,**54**) bind to the synaptic vesicle protein 2A with a higher affinity (10-fold) and greater potency than parent compounds in animal models of epilepsy [91–93]. However, there is no clear mechanism to explain how pyrrolidinones binding to SV2A produce anticonvulsant effects; it is probably related to synaptic vesicle exocytosis and neurotransmitter release. Brivaracetam (**21**) also displays inhibitory activity at voltage-dependent Na^+ channels.

4.11. Sulfonamides

Acetazolamide, chemically the N-(5-aminosulfonyl)-1,3,4-thiadiazol-2-ylacetamide (**55**), and methazolamide (**56**) (Fig. 27) are old members of the sulfonamide class that show anticonvulsant activity but they are considered to belong to a minor class of antiepileptic agents. Several newer derivatives, such as zonisamide (**16**) and topiramate (**17**), are effective antiepileptics [50]. The anticonvulsant effects of these sulfonamides/sulfamates are due to CO_2 retention secondary to the enzyme carbonic anhydrase inhibition in the brain, but other mechanisms of action, such as blockade of sodium channels and kainate/AMPA receptors, as well as the enhancement of GABA-ergic transmission, have also been hypothesized to explain the efficacy of these drugs.

Acetazolamide, topiramate and zonisamide contain a sulfamate/sulfamide moiety that is essential for their enzymatic inhibition properties.

Topiramate, chemically 2,3:4,5-bis-O-(1-methylethylidene)-β-D-fructopyranose sulfamate (**17**), is an anticonvulsant approved since 1998. It is characterized by a very complex mechanism of action [49,94–99]; it blocks sodium channels, inhibits high-voltage-activated calcium channels, attenuates the effects of excitatory neurotransmitters, enhances GABAergic neurotransmission, and inhibits carbonic anhydrase isoenzyme CAII. Topiramate is generally employed for partial-onset primary generalized tonic–clonic and myoclonic seizures [100,101]. The administration of this drug produces various side effects including word-finding disturbances, cognitive slowing, ataxia, poor concentration, dizziness, fatigue, paresthesia, somnolence, reduced appetite, and weight loss. There is also evidence of a small increased risk of kidney stone formation, hypohydrosis, and acute angle closure glaucoma. Some of these side effects could be related to carbonic anhydrase inhibition activity related to the presence of a sulfamate moiety, which is considered a chelating functionality for the zinc metal of this metalloenzyme. The pharmacokinetic properties of

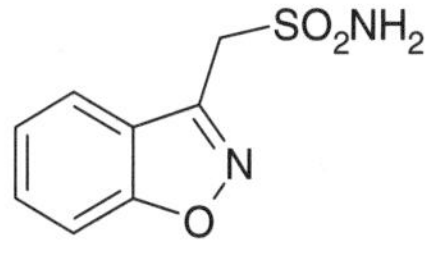

16, Zonisamide

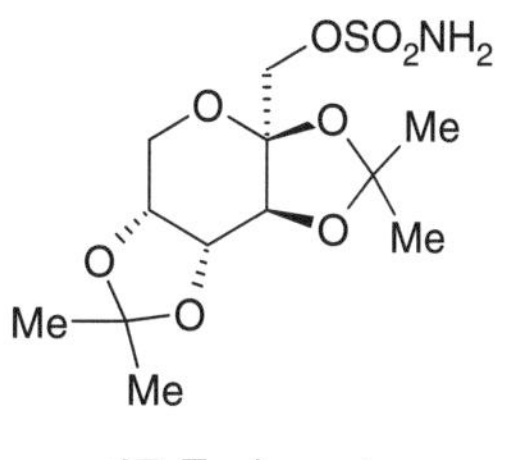

17, Topiramate

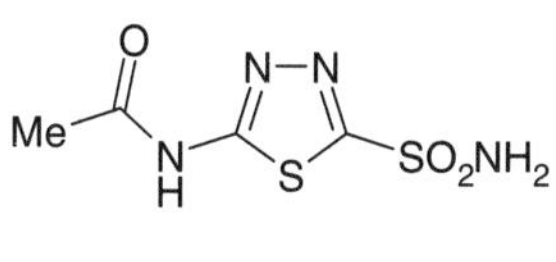

55, Acetazolamide

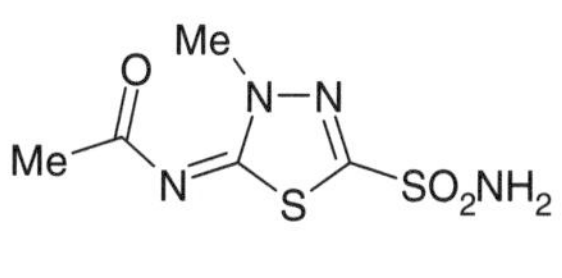

56, Methazolamide

Figure 27. Sulfonamides.

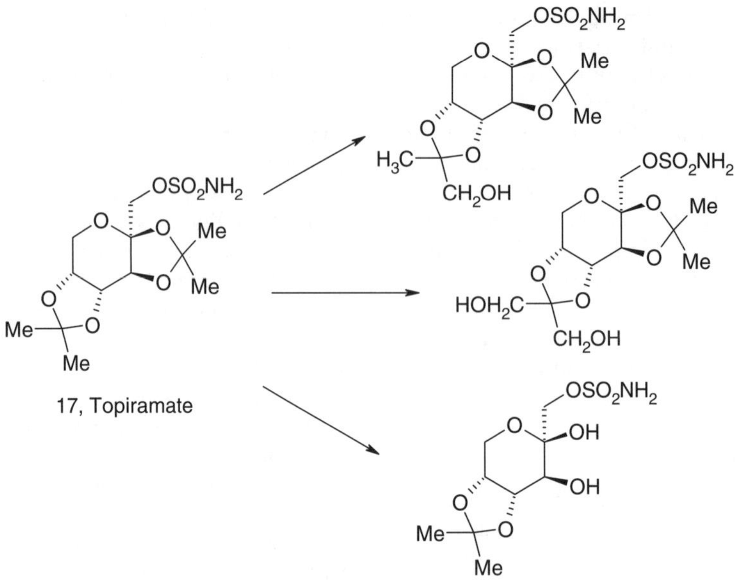

Figure 28. Metabolism of topiramate.

topiramate are good as its absorption, producing 85% of bioavailability without metabolization. The identified metabolites di- and monohydroxymethyl derivatives and ring cleavage as well as glucuronidation metabolites are of no clinical relevance (Fig. 28).

The anticonvulsant zonisamide (**16**) (Fig. 27) is a sulfonamide derivative that blocks voltage-dependent sodium channels as well as T-type calcium channels, and inhibits carbonic anhydrase. Zonisamide is effective against partial-onset seizures and generalized seizure subtypes, tonic–clonic, tonic, atonic, atypical absence, and myoclonic seizures. It has no effect on hepatic metabolism, but its half-life is reduced by enzyme-inducing drugs. Side effects include anorexia, dizziness, ataxia, fatigue, somnolence, confusion, and poor concentration.

4.12. Tiagabine

Tiagabine, (R)-(–)-1-[4,4-bis(3-methyl-thienyl)-3-butenyl]-3-piperidinecarboxylic acid hydrochloride (**18**) (Fig. 29), is a selective GABA reuptake inhibitor [102], it increases neurotransmitter synaptic availability through inhibition of the GABA transporter GAT-1 on presynaptic neurons and glial cells. Tiagabine prevents the reuptake of GABA by interacting specifically with the uptake carrier. The presence of elevated GABA concentrations in the synaptic cleft is therefore prolonged enhancing its inhibitory effects [103]. Tiagabine (**18**) is effective as an add-on therapy for adults and children aged 12 years and older in the treatment of partial seizures. It possesses clinically significant dose-dependent side effects related to: impaired concentration, speech or language problems, confusion, somnolence, and fatigue.

4.13. Felbamate

Felbamate, chemically 2-phenyl-1,3-propanediol dicarbamate (**19**) (Fig. 30), is a second-generation anticonvulsant [104] chemically related to the older drug meprobamate. It exerts multiple effects as an anticonvulsant, but the blocking effects on voltage-dependent Na^+ channels is the main mechanism of

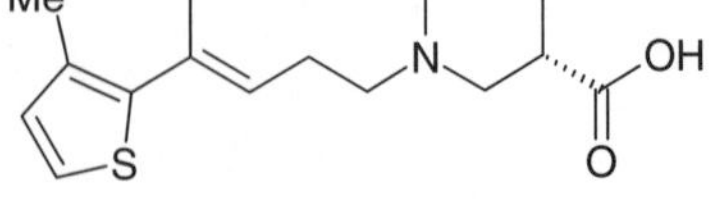

Figure 29. Tiagabine.

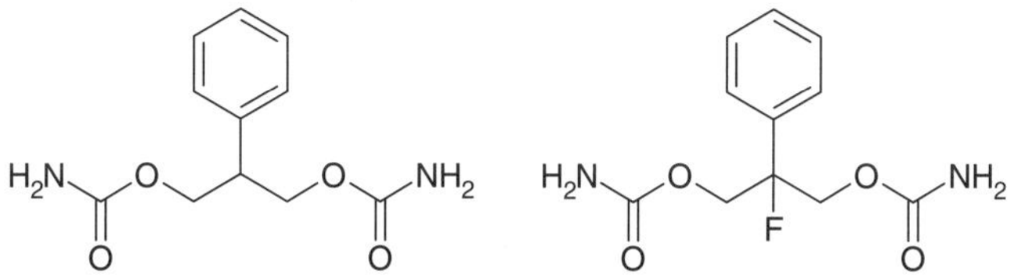

19, Felbamate

57, Fluorofelbamate

Figure 30. Felbamate and fluorofelbamate.

action. However, felbamate has been found to interact with NMDA-associated glycine binding sites in concentrations within the therapeutic range; thus, resulting in a reduction in excitatory amino acid transmission. Additionally, felbamate blocks NMDA responses in a low affinity noncompetitive site [105], or open channel manner. It can also enhance $GABA_A$ receptor-mediated responses [106].

Felbamate (**19**) (Fig. 30) [107] is well absorbed after oral administration, distributed to all tissues and undergoes biotransformation via several pathways (Fig. 31): (a) aromatic and aliphatic hydroxylation to produce *p*-hydroxyfelbamate and 2-hydroxyfelbamate, and (b) hydrolysis to form the primary alcohol, 2-phenyl-1,3-diol monocarbamate. These metabolites are devoid of anticonvulsant efficacy.

Felbamate (**19**) is an adjuvant anticonvulsant, containing the warning that its use is associated with severe side effects, namely, a marked increase in the incidence of aplastic anemia and the warning that patients being started on the drug should have liver function tests performed before therapy is initiated. Felbamate is recommended as second or third-line therapy and is indicated for those patients who respond inadequately to alternative treatments.

Fluorofelbamate (2-phenyl-2-fluoro-1,3-propanediol dicarbamate, **57**) (Fig. 30) is a fluorine analog of felbamate designed to produce the same broad-spectrum anticonvulsant activity without the serious adverse effects because the presence of a fluorine atom could prevent the production reactive

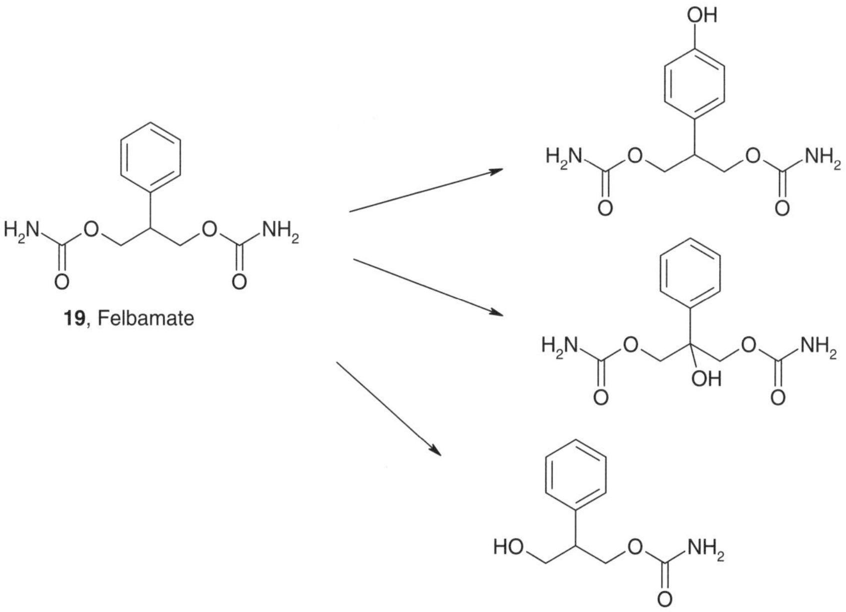

19, Felbamate

Figure 31. Metabolic route of felbamate.

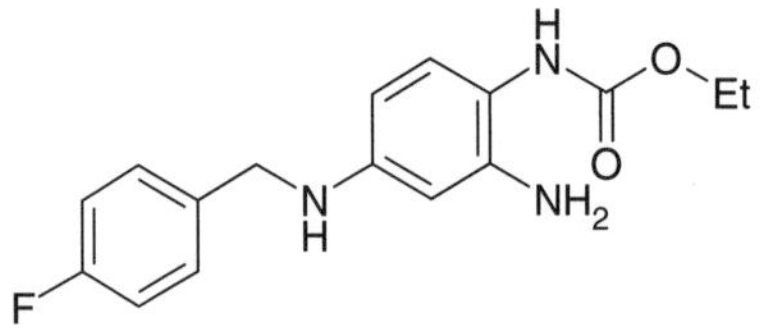

26, Retigabine

Figure 32. Retigabine.

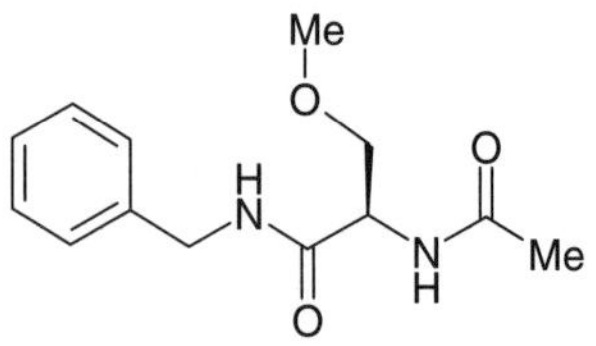

28, Lacosamide

Figure 33. Lacosamide.

toxic metabolites responsible for the aplastic anemia and liver toxicity. Preliminary evidence suggests that fluorofelbamate acts, at least in part, by decreasing responses to receptor ligands, GABA, kainate and NMDA, and by reducing voltage-dependent sodium currents.

4.14. Retigabine

Retigabine is the N-(2-amino-4-(4′-fluorobenzylamino)phenyl)carbamic acid ethyl ester (**26**) (Fig. 32) under development as an anticonvulsant agent [108,109]. The mechanism of action of retigabine is related to the activation of M-current trough opening of KCNQ2 and KCNQ3 (also named $K_v7.2$ and $K_v7.3$) potassium channels.

By interaction with K_v7 channel types, in addition to its anticonvulsant activity, retigabine may be useful in treating neuropathic pain, strokes, and other neurodegenerative disorders. Secondary mechanisms of action include potentiation of GABA-evoked currents in cortical neurons via the activation of $GABA_A$ receptors. Retigabine was generally well tolerated and the side effects were sedation accompanied by hyperexcitability and decreased body temperature. It is extensively metabolized by hydrolysis/acetylation and glucuronidation resulting in the formation of two inactive N-glucuronides and an N-acetyl metabolite, renally excreted.

4.15. Lacosamide

Lacosamide (formerly harkoseride, (2R)-2-(acetylamino)-3-methoxy-N-(phenylmethyl) propanamide, **28**) (Fig. 33) is a member of a class of functionalized amino acids screened for their anticonvulsant properties [110,111].

Lacosamide has demonstrated broad anticonvulsant effects in seizure models for generalized seizures, complex partial-onset seizures, and status epilepticus. The mechanism of action of lacosamide has not yet been clarified. Some studies suggest that lacosamide selectively enhances the slow inactivation component of voltage-gated sodium channels, exerting no effects on fast inactivation, which represents the target of some antiepileptics such as carbamazepine, phenytoin, and lamotrigine. The enhancement of slow inactivation induced by lacosamide normalizes activation thresholds and decreases pathophysiological neuronal activity, thereby controlling neuronal hyperexcitability. It has also been suggested that lacosamide may act as an antagonist of NMDA receptors through interaction with the strychnine-insensitive glycine site. Experimental data show its promising clinical potential. Lacosamide is well absorbed, shows low protein binding, and does not produce significant drug–drug interactions. The most common side effects are dizziness, nausea, and vomiting, which seem to be dose related.

4.16. Rufinamide

Rufinamide [1-(2,6-difluorophenyl)methyl-1H-1,2,3-triazole-4-carboxamide] (**25**) (Fig. 34) is a triazole derivative approved in 2005 for two epilepsy treatments [112–115]: (1) as adjunctive treatment of partial-onset seizures with and without secondary generalization in adults and adolescents aged 12 years or older

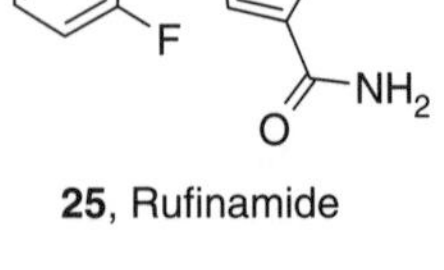

25, Rufinamide

Figure 34. Rufinamide.

and (2) as adjunctive treatment of seizures associated with Lennox–Gastaut syndrome in children aged 4 years and older and in adults.

The antiepileptic effects of rufinamide have been assessed in several animal models of generalized and partial seizures. While the precise mechanism by which rufinamide exerts its antiepileptic effect is unknown, *in vitro* studies have suggested that the main mechanism of action of rufinamide is the modulation of sodium-channel activity and, in particular, prolongation of the inactivated state, thus limiting high-frequency firing of action potentials in neurons. The interaction of rufinamide with neurotransmitter systems has also been investigated, but no effects have been, or only weak activity has been observed for the following: GABA, adenosine, monoaminergic, cholinergic binding sites, as well as excitatory amino acid binding sites.

4.17. Stiripentol

Stiripentol is the allylic alcohol 4,4-dimethyl-1-[(3,4-methylenedioxy)phenyl]-1-penten-3-ol (**29**) (Fig. 35) that has been investigated and employed in France and in Canada for over a decade. Its clinical development was delayed because of drug interactions resulting from the inhibitory effect of stiripentol on hepatic cytochrome P450 (CYP450) enzymes [116].

The anticonvulsant properties of stiripentol are related to different mechanisms of potentiation of GABAergic neurotransmission [117]: it inhibits the synaptosomal uptake of GABA and enhances both central GABAergic transmission and the release of GABA as well as the duration of the activation of GABA$_A$ receptors in a concentration-dependent manner.

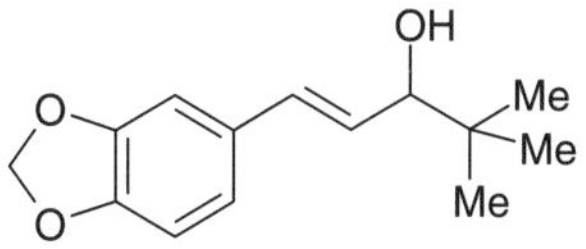

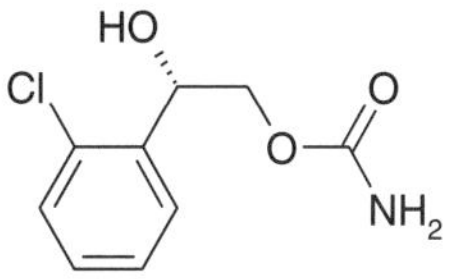

30, Carisbamate

Figure 36. Carisbamate.

4.18. Carisbamate

Carisbamate (RWJ-333369), chemically (2*S*)-2-(2-chlorophenyl)-2-hydroxyethyl carbamate (**30**) (Fig. 36), is a monocarbamate derivative that shows a broad spectrum of activity in preclinical models of epilepsy and has demonstrated efficacy and tolerability profile in clinical trials. The mechanism of action that contributes to its broad spectrum of anticonvulsant activity has not been deciphered and remains under investigation [64].

4.19. Carabersat

Carabersat (SB-204269, *N*-[(3*R*,4*S*)-6-acetyl-3-hydroxy-2,2-dimethyl-4-chromanyl]-*p*-fluorobenzamide **27**) (Fig. 37) and the chloroanalog tonabersat (SB-220453, **58**) are benzoylaminobenzopyran derivatives that display anticonvulsant effects, selectively and specifically binding a stereoselective site on neuronal gap junctions [64]. Carabersat and tonabersat (Fig. 37) [53,118] have shown potent oral anticonvulsant properties in animal seizure models. It has been demonstrated that these benzoylaminobenzopyran derivatives show no significant interaction with

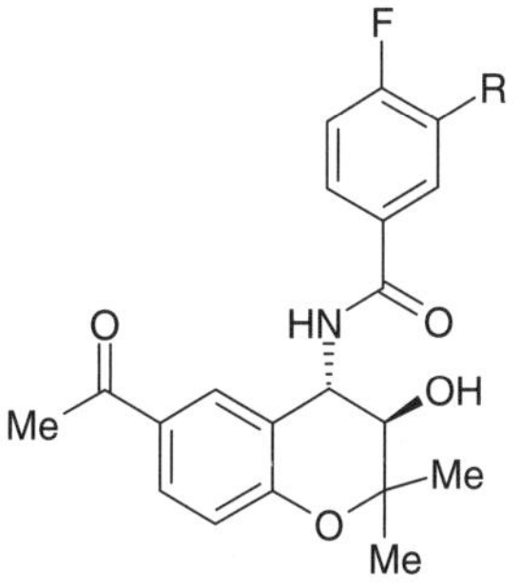

29, Stiripentol

Figure 35. Stiripentol.

27, R = H carabersat
58, R = Cl tonabersat

Figure 37. Carabersat and tonabersat.

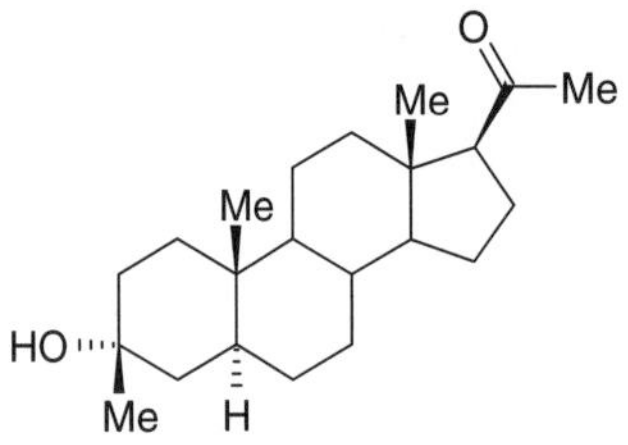

24, Ganaxolone

Figure 38. Ganaxolone.

well-known anticonvulsant molecular targets (e.g., sodium channels, glutamatergic, and GABAergic neurotransmission).

4.20. Ganaxolone

Ganaxolone (3a-hydroxy-3b-methyl-5a-pregnan-20-one, **24**) (Fig. 38) is the 3-methylated synthetic analog of the neurosteroid allopregnanolone, a metabolite of progesterone. It is a neuroactive steroid that allosterically modulates the GABA$_A$ receptor complex through a recognition site, that differs from benzodiazepine or barbiturate binding sites [119,120]. This compound was developed after observations that endogenous metabolites of progesterone showed anticonvulsant effects in different experimental models of epilepsy. Ganaxolone does not possess nuclear hormone activity and phase I and phase II studies in refractor infantile spasms are currently being undertaken. Experimental evidence suggests good safety and tolerability [64,121].

REFERENCES

1. Meldrum BS, Rogawski MA. Molecular targets for antiepileptic drug development. Neurotherapeutics 2007;4(1):18–61.
2. Rogawski MA. Molecular targets versus models for new antiepileptic drug discovery. Epilepsy Res 2006;68(1):22–28.
3. Soderpalm B. Anticonvulsants: aspects of their mechanisms of action. Eur J Pain 2002; 6(Suppl A):3–9.
4. Stefan H, Steinhoff BJ. Emerging drugs for epilepsy and other treatment options. Eur J Neurol 2007;14(10):1154–1161.
5. Rogawski MA. Diverse mechanisms of antiepileptic drugs in the development pipeline. Epilepsy Res 2006;69(3):273–294.
6. Yu FH, Yarov-Yarovoy V, Gutman GA, Catterall WA. Overview of molecular relationships in the voltage-gated ion channel superfamily. Pharmacol Rev 2005;57(4):387–395.
7. Catterall WA, Perez-Reyes E, Snutch TP, Striessnig J. International Union of Pharmacology. XLVIII. Nomenclature and structure–function relationships of voltage-gated calcium channels. Pharmacol Rev. 2005;57(4):411–425.
8. Fujiwara T, Sugawara T, Mazaki-Miyazaki E, Takahashi Y, Fukushima K, Watanabe M, Hara K, Morikawa T, Yagi K, Yamakawa K, Inoue Y. Mutations of sodium channel alpha subunit type 1 (SCN1A) in intractable childhood epilepsies with frequent generalized tonic–clonic seizures. Brain 2003;126(Pt 3):531–546.
9. Wallace RH, Scheffer IE, Parasivam G, Barnett S, Wallace GB, Sutherland GR, Berkovic SF, Mulley JC. Generalized epilepsy with febrile seizures plus: mutation of the sodium channel subunit SCN1B. Neurology 2002;58 (9):1426–1429.
10. Yarov-Yarovoy V, Brown J, Sharp EM, Clare JJ, Scheuer T, Catterall WA. Molecular determinants of voltage-dependent gating and binding of pore-blocking drugs in transmembrane segment IIIS6 of the Na$^+$ channel alpha subunit. J Biol Chem 2001;276(1):20–27.
11. Wang SJ, Huang CC, Hsu KS, Tsai JJ, Gean PW. Inhibition of N-type calcium currents by lamotrigine in rat amygdalar neurones. Neuroreport 1996;7(18):3037–3040.
12. Rogawski MA, Bazil CW. New molecular targets for antiepileptic drugs: alpha(2)delta, SV2A, and K$_v$7/KCNQ/M potassium channels. Curr Neurol Neurosci Rep 2008;8(4):345–352.
13. Wei AD, Gutman GA, Aldrich R, Chandy KG, Grissmer S, Wulff H. International Union of Pharmacology. LII. Nomenclature and molecular relationships of calcium-activated potassium channels. Pharmacol Rev 2005;57 (4):463–472.
14. Gutman GA, Chandy KG, Grissmer S, Lazdunski M, McKinnon D, Pardo LA, Robertson GA, Rudy B, Sanguinetti MC, Stuhmer W, Wang X. International Union of Pharmacology. LIII. Nomenclature and molecular relationships of voltage-gated potassium channels. Pharmacol Rev 2005;57(4):473–508.
15. Schenzer A, Friedrich T, Pusch M, Saftig P, Jentsc TJ, Grotzinger J, Schwake M. Molecu-

15. lar determinants of KCNQ (K_v7) K^+ channel sensitivity to the anticonvulsant retigabine. J Neurosci 2005;25(20):5051–5060.
16. Poolos NP. h-Channels and seizures: less is more. Epilepsy Curr 2005;5(3):89–90.
17. Poolos NP. The h-channel: a potential channelopathy in epilepsy?. Epilepsy Behav 2005;7(1):51–56.
18. Jentsch TJ, Neagoe I, Scheel O. CLC chloride channels and transporters. Curr Opin Neurobiol 2005;15(3):319–325.
19. Heils A. CLCN2 and idiopathic generalized epilepsy. Adv Neurol 2005;95:265–271.
20. Collingridge GL, Olsen RW, Peters J, Spedding M. A nomenclature for ligand-gated ion channels. Neuropharmacology 2009;56(1):2–5.
21. Collingridge GL, Olsen R, Peters JA, Spedding M. Ligand gated ion channels. Neuropharmacology 2009;56(1):1
22. Absalom NL, Schofield PR, Lewis TM. Pore structure of the cys-loop ligand-gated ion channels Neurochem Res 2009;34(10):1805–1815.
23. Chebib M, Johnston GA. GABA-activated ligand gated ion channels: medicinal chemistry and molecular biology. J Med Chem 2000;43(8):1427–1447.
24. Johnston GA. $GABA_A$ receptor channel pharmacology. Curr Pharm Des 2005;11(15):1867–1885.
25. Henschel O, Gipson KE, Bordey A. GABAA receptors, anesthetics and anticonvulsants in brain development. CNS Neurol Disord Drug Targets 2008;7(2):211–224.
26. Mayer ML. Glutamate receptor ion channels. Curr Opin Neurobiol 2005;15(3):282–288.
27. Neugebauer V. Glutamate receptor ligands. Handb Exp Pharmacol 2007;177217–249.
28. Genoux D, Montgomery JM. Glutamate receptor plasticity at excitatory synapses in the brain. Clin Exp Pharmacol Physiol 2007;34(10):1058–1063.
29. Mayer ML, Armstrong N. Structure and function of glutamate receptor ion channels. Annu Rev Physiol 2004;66161–181.
30. Madden DR. The structure and function of glutamate receptor ion channels. Nat Rev Neurosci 2002;3(2):91–101.
31. Kleckner NW, Glazewski JC, Chen CC, Moscrip TD. Subtype-selective antagonism of N-methyl-D-aspartate receptors by felbamate: insights into the mechanism of action. J Pharmacol Exp Ther 1999;289(2):886–894.
32. Howes JF, Bell C. Talampanel. Neurotherapeutics 2007;4(1):126–129.
33. Aujla PK, Fetell MR, Jensen FE. Talampanel suppresses the acute and chronic effects of seizures in a rodent neonatal seizure model. Epilepsia. 2009;50(4):694–701.
34. De Sarro G, Gitto R, Russo E, Ibbadu GF, Barreca ML, De Luca L, Chimirri A. AMPA receptor antagonists as potential anticonvulsant drugs. Curr Top Med Chem 2005;5(1):31–42.
35. Catarzi D, Colotta V, Varano F. Competitive AMPA receptor antagonists. Med Res Rev 2007;27(2):239–278.
36. Dajas-Bailador F, Wonnacott S. Nicotinic acetylcholine receptors and the regulation of neuronal signalling. Trends Pharmacol Sci 2004;25(6):317–324.
37. Picard F, Bertrand S, Steinlein OK, Bertrand D. Mutated nicotinic receptors responsible for autosomal dominant nocturnal frontal lobe epilepsy are more sensitive to carbamazepine. Epilepsia 1999;40(9):1198–1209.
38. Webb TI, Lynch JW. Molecular pharmacology of the glycine receptor chloride channel. Curr Pharm Des 2007;13(23):2350–2367.
39. Lynch JW. Molecular structure and function of the glycine receptor chloride channel. Physiol Rev 2004;84(4):1051–1095.
40. Gether U, Andersen PH, Larsson OM, Schousboe A. Neurotransmitter transporters: molecular function of important drug targets. Trends Pharmacol Sci 2006;27(7):375–383.
41. Bragina L, Marchionni I, Omrani A, Cozzi A, Pellegrini-Giampietro DE, Cherubini E, Conti F. GAT-1 regulates both tonic and phasic GABA(A) receptor-mediated inhibition in the cerebral cortex. J Neurochem 2008;105(5):1781–1793.
42. Dalby NO, Thomsen C, Fink-Jensen A, Lundbeck J, Sokilde B, Man CM, Sorensen PO, Meldrum B. Anticonvulsant properties of two GABA uptake inhibitors NNC 05-2045 and NNC 05-2090, not acting preferentially on GAT-1. Epilepsy Res 1997;28(1):51–61.
43. Shigeri Y, Seal RP, Shimamoto K. Molecular pharmacology of glutamate transporters, EAATs and VGLUTs. Brain Res Brain Res Rev 2004;45(3):250–265.
44. Bridges RJ, Esslinger CS. The excitatory amino acid transporters: pharmacological insights on substrate and inhibitor specificity of the EAAT subtypes. Pharmacol Ther 2005;107(3):271–285.

45. Dunlop J. Glutamate-based therapeutic approaches: targeting the glutamate transport system. Curr Opin Pharmacol 2006;6(1):103–107.
46. Petroff OA, Rothman DL. Measuring human brain GABA in vivo: effects of GABA-transaminase inhibition with vigabatrin. Mol Neurobiol 1998;16(1):97–121.
47. Wang QP, Jammoul F, Duboc A, Gong J, Simonutti M, Dubus E, Craft CM, Ye W, Sahel JA, Picaud S. Treatment of epilepsy: the GABA-transaminase inhibitor, vigabatrin, induces neuronal plasticity in the mouse retina. Eur J Neurosci 2008;27(8):2177–2187.
48. Sarup A, Larsson OM, Schousboe A. GABA transporters and GABA-transaminase as drug targets. Curr Drug Targets CNS Neurol Disord 2003;2(4):269–277.
49. Leniger T, Thone J, Wiemann M. Topiramate modulates pH of hippocampal CA3 neurons by combined effects on carbonic anhydrase and Cl^-/HCO_3^- exchange. Br J Pharmacol 2004;142(5):831–842.
50. Thiry A, Dogne JM, Supuran CT, Masereel B. Carbonic anhydrase inhibitors as anticonvulsant agents. Curr Top Med Chem 2007;7(9):855–864.
51. Thiry A, Dogne JM, Supuran CT, Masereel B. Anticonvulsant sulfonamides/sulfamates/sulfamides with carbonic anhydrase inhibitory activity: drug design and mechanism of action. Curr Pharm Des 2008;14(7):661–671.
52. Nakase T, Naus CC. Gap junctions and neurological disorders of the central nervous system. Biochim Biophys Acta 2004;1662(1–2):149–158.
53. Nicolson A, Leach JP. Future prospects for the drug treatment of epilepsy. CNS Drugs 2001;15(12):955–968.
54. Foster AC, Kemp JA. Glutamate- and GABA-based CNS therapeutics. Curr Opin Pharmacol 2006;6(1):7–17.
55. Stefan H, Feuerstein TJ. Novel anticonvulsant drugs. Pharmacol Ther 2007;113(1):165–183.
56. Bialer M, Johannessen SI, Kupferberg HJ, Levy RH, Perucca E, Tomson T. Progress report on new antiepileptic drugs: a summary of the Seventh Eilat Conference (EILAT VII). Epilepsy Res 2004;61(1–3):1–48.
57. Brodie MJ, Kwan P. Phenobarbital: a drug for the 21st century?. Epilepsy Behav 2004;5(6):802–803.
58. Bialer M, Johannessen SI, Kupferberg HJ, Levy RH, Perucca E, Tomson T. Progress report on new antiepileptic drugs: a summary of the Eighth Eilat Conference (EILAT VIII). Epilepsy Res 2007;73(1):1–52.
59. Yaari Y, Selzer ME, Pincus JH. Phenytoin: mechanisms of its anticonvulsant action. Ann Neurol 1986;20(2):171–184.
60. Wolf HH, Swinyard EA, Goodman LS. Anticonvulsant properties of some N-substituted hydantoins. J Pharm Sci 1962;5174–76.
61. Chen G, Weston JK, Bratton AC Jr. Anticonvulsant activity and toxicity of phensuximide, methsuximide and ethosuximide. Epilepsia 1963;466–76.
62. Loewe S. Anticonvulsant actions of trimethadione-phenobarbital and trimethadione-diphenylhydantoin combinations. Fed Proc 1948;7(1 Pt 1):240
63. Ramsay RE, McManus DQ, Guterman A, Briggle TV, Vazquez D, Perchalski R, Yost RA, Wong P. Carbamazepine metabolism in humans: effect of concurrent anticonvulsant therapy. Ther Drug Monit 1990;12(3): 235–241.
64. Bialer M, Johannessen SI, Levy RH, Perucca E, Tomson T, White HS. Progress report on new antiepileptic drugs: a summary of the Ninth Eilat Conference (EILAT IX). Epilepsy Res 2009;83(1):1–43.
65. Bonifacio MJ, Sheridan RD, Parada A, Cunha RA, Patmore L, Soares-da-Silva P. Interaction of the novel anticonvulsant, BIA 2-093, with voltage-gated sodium channels: comparison with carbamazepine. Epilepsia 2001;42(5):600–608.
66. Almeida L, Soares-da-Silva P. Eslicarbazepine acetate (BIA 2-093). Neurotherapeutics 2007;4(1):88–96.
67. Pages KP, Ries RK. Use of anticonvulsants in benzodiazepine withdrawal. Am J Addict 1998;7(3):198–204.
68. Shader RI, Greenblatt DJ. Benzodiazepines: some aspects of their clinical pharmacology. Ciba Found Symp 1979;74141–155.
69. Breimer DD, Jochemsen R, von Albert HH. Pharmacokinetics of benzodiazepines. Short-acting versus long-acting. Arzneimittelforschung 1980;30(5a):875–881.
70. Curry SH, Whelpton R. Pharmacokinetics of closely related benzodiazepines. Br J Clin Pharmacol 1979;8(1):15S–21S
71. Oreland L. The benzodiazepines: a pharmacological overview. Acta Anaesthesiol Scand Suppl 1988;8813–16.
72. Iversen LL. The present status of benzodiazepines in psychopharmacology. Arzneimittelforschung 1980;30(5a):907–910.

73. Costa E. Benzodiazepines and neurotransmitters. Arzneimittelforschung 1980;30(5a):858–861.
74. Gent JP, Feely MP, Haigh JR. Differences between the tolerance characteristics of two anticonvulsant benzodiazepines. Life Sci 1985;37(9):849–856.
75. Tarnawa I, Vize ES. 2,3-Benzodiazepine AMPA antagonists. Restor Neurol Neurosci 1998;13(1–2):41–57.
76. Andrasi F, Horvath K, Sineger E, Berzsenyi P, Borsy J, Kenessey A, Tarr M, Lang T, Korosi J, Hamori T. Neuropharmacology of a new psychotropic 2, 3-benzodiazepine. Arzneimittelforschung 1987;37(10):1119–1124.
77. Solyom S, Tarnawa I, Non-competitive AMPA antagonists of 2, 3-benzodiazepine type. Curr Pharm Des 2002;8(10):913–939.
78. Micale N, Ritz M, Grasso S, Niu L. Mechanism of inhibition of the GluR2 AMPA receptor channel opening by 2, 3-benzodiazepine derivatives. Biochemistry 2008;47(3):1061–1069.
79. Weiser T, Herrmann A, Wienrich M. Interactions of the dye Evans Blue and GYKI 52466 a 2,3-benzodiazepine, with (S)-alpha-amino-3-hydroxy-5-methyl-4-isoxazolepropionic acid receptors in cultured rat cortical neurons: electrophysiological evidence for at least two different binding sites for non-competitive antagonists. Neurosci Lett 1996;216(1):29–32.
80. Gitto R, Orlando V, Quartarone S, De Sarro G, De Sarro A, Russo E, Ferreri G, Chimirri A. Synthesis and evaluation of pharmacological properties of novel annelated 2, 3-benzodiazepine derivatives. J Med Chem 2003;46(17):3758–3761.
81. Bialer M, Yagen B. Valproic Acid: second generation. Neurotherapeutics 2007;4(1): 130–137.
82. Lees G, Leach MJ. Studies on the mechanism of action of the novel anticonvulsant lamotrigine (Lamictal) using primary neurological cultures from rat cortex. Brain Res 1993;612(1–2):190–199.
83. Wild JM, Ahn HS, Baulac M, Bursztyn J, Chiron C, Gandolfo E, Safran AB, Schiefer U, Perucca E. Vigabatrin and epilepsy: lessons learned. Epilepsia 2007;48(7):1318–1327.
84. Wheless JW, Ramsay RE, Collins SD. Vigabatrin. Neurotherapeutics 2007;4(1):163–172.
85. Tong X, Ratnaraj N, Patsalos PN. The pharmacokinetics of vigabatrin in rat blood and cerebrospinal fluid. Seizure 2007;16(1):43–49.
86. Klitgaard H. Levetiracetam: the preclinical profile of a new class of antiepileptic drugs?. Epilepsia 2001;42(Suppl 4):13–18.
87. Loscher W, Honack D, Rundfeldt C. Antiepileptogenic effects of the novel anticonvulsant levetiracetam (ucb L059) in the kindling model of temporal lobe epilepsy. J Pharmacol Exp Ther 1998;284(2):474–479.
88. De Smedt T, Raedt R, Vonck K, Boon P. Levetiracetam. Part II. The clinical profile of a novel anticonvulsant drug. CNS Drug Rev 2007;13(1):57–78.
89. De Smedt T, Raedt R, Vonck K, Boon P. Levetiracetam: the profile of a novel anticonvulsant drug. Part I. Preclinical data. CNS Drug Rev 2007;13(1):43–56.
90. Luszczki JJ, Andres MM, Czuczwar P, Cioczek-Czuczwar A, Ratnaraj N, Patsalos PN, Czuczwar SJ. Pharmacodynamic and pharmacokinetic characterization of interactions between levetiracetam and numerous antiepileptic drugs in the mouse maximal electroshock seizure model: an isobolographic analysis. Epilepsia 2006;47(1):10–20.
91. Sargentini-Maier ML, Espie P, Coquette A, Stockis A. Pharmacokinetics and metabolism of 14C-brivaracetam, a novel SV2A ligand, in healthy subjects. Drug Metab Dispos 2008;36(1):36–45.
92. Rogawski MA. Brivaracetam: a rational drug discovery success story. Br J Pharmacol 2008;154(8):1555–1557.
93. von Rosenstiel P. Brivaracetam (UCB 34714). Neurotherapeutics 2007;4(1):84–87.
94. Shank RP, Gardocki JF, Vaught JL, Davis CB, Schupsky JJ, Raffa RB, Dodgson SJ, Nortey SO, Maryanoff BE. Topiramate: preclinical evaluation of structurally novel anticonvulsant. Epilepsia 1994;35(2):450–460.
95. Ben-Menachem E. Topiramate: current status and therapeutic potential. Expert Opin Investig Drugs 1997;6(8):1085–1094.
96. Bittermann HJ, Steinhoff BJ. Topiramate: an effective new anticonvulsant. An open prospective study. Nervenarzt 1997;68(10):836–838.
97. Perucca E. A pharmacological and clinical review on topiramate, a new antiepileptic drug. Pharmacol Res 1997;35(4):241–256.
98. White HS, Brown SD, Woodhead JH, Skeen GA, Wolf HH. Topiramate enhances GABA-mediated chloride flux and GABA-evoked chloride currents in murine brain neurons and increases seizure threshold. Epilepsy Res 1997;28(3):167–179.

99. Zona C, Ciotti MT, Avoli M. Topiramate attenuates voltage-gated sodium currents in rat cerebellar granule cells. Neurosci Lett 1997;231(3):123–126.
100. De Sarro G, Gratteri S, Bonacci F, Musumeci SA, Elia M, De Sarro A. Topiramate potentiates the antiseizure activity of some anticonvulsants in DBA/2 mice. Eur J Pharmacol 2000;388(2):163–170.
101. van Passel L, Arif H, Hirsch LJ. Topiramate for the treatment of epilepsy and other nervous system disorders. Expert Rev Neurother 2006;6(1):19–31.
102. Nielsen EB, Suzdak PD, Andersen KE, Knutsen LJ, Sonnewald U, Braestrup C. Characterization of tiagabine (NO-328), a new potent and selective GABA uptake inhibitor. Eur J Pharmacol 1991;196(3):257–266.
103. Suzdak PD, Jansen JA. A review of the preclinical pharmacology of tiagabine: a potent and selective anticonvulsant GABA uptake inhibitor. Epilepsia 1995;36(6):612–626.
104. White HS, Wolf HH, Swinyard EA, Skeen GA, Sofia RD. A neuropharmacological evaluation of felbamate as a novel anticonvulsant. Epilepsia 1992;33(3):564–572.
105. Subramaniam S, Rho JM, Penix L, Donevan SD, Fielding RP, Rogawski MA. Felbamate block of the N-methyl-D-aspartate receptor. J Pharmacol Exp Ther 1995;273(2):878–886.
106. Ticku MK, Kamatchi GL, Sofia RD. Effect of anticonvulsant felbamate on GABAA receptor system. Epilepsia 1991;32(3):389–391.
107. Roecklein BA, Sacks HJ, Mortko H, Stables J. Fluorofelbamate. Neurotherapeutics 2007;4(1):97–101.
108. Hirano K, Kuratani K, Fujiyoshi M, Tashiro N, Hayashi E, Kinoshita M. K_v7.2-7.5 voltage-gated potassium channel (KCNQ2-5) opener, retigabine, reduces capsaicin-induced visceral pain in mice. Neurosci Lett 2007;413(2):159–162.
109. Porter RJ, Nohria V, Rundfeldt C. Retigabine. Neurotherapeutics 2007;4(1):149–154.
110. Beyreuther BK, Freitag J, Heers C, Krebsfanger N, Scharfenecker U, Stohr T. Lacosamide: a review of preclinical properties. CNS Drug Rev 2007;13(1):21–42.
111. Doty P, Rudd GD, Stoehr T, Thomas D. Lacosamide. Neurotherapeutics 2007;4(1):145–148.
112. Arroyo S. Rufinamide. Neurotherapeutics 2007;4(1):155–162.
113. Hakimian S, Cheng-Hakimian A, Anderson GD, Miller JW. Rufinamide: a new anti-epileptic medication. Expert Opin Pharmacother 2007;8(12):1931–1940.
114. Heaney D, Walker MC. Rufinamide. Drugs Today 2007;43(7):455–460.
115. Kluger G, Bauer B. Role of rufinamide in the management of Lennox-Gastaut syndrome (childhood epileptic encephalopathy). Neuropsychiatr Dis Treat 2007;3(1):3–11.
116. Lin HS, Levy RH. Pharmacokinetic profile of a new anticonvulsant, stiripentol, in the rhesus monkey. Epilepsia 1983;24(6):692–703.
117. Wojtal K, Trojnar MK, Trojnar MP, Czuczwar SJ. Stiripentol. A novel antiepileptic drug. Pharmacol Rep 2005;57(2):154–160.
118. Parsons AA, Bingham S, Raval P, Read S, Thompson M, Upton N. Tonabersat (SB-220453) a novel benzopyran with anticonvulsant properties attenuates trigeminal nerve-induced neurovascular reflexes. Br J Pharmacol 2001;132(7):1549–1557.
119. Monaghan EP, McAuley JW, Data JL. Ganaxolone: a novel positive allosteric modulator of the GABA$_A$ receptor complex for the treatment of epilepsy. Expert Opin Investig Drugs 1999;8(10):1663–1671.
120. Snead OC 3rd. Ganaxolone, a selective, high-affinity steroid modulator of the gamma-aminobutyric acid-A receptor, exacerbates seizures in animal models of absence. Ann Neurol 1998;44(4):688–691.
121. Laxer K, Blum D, Abou-Khalil BW, Morrell MJ, Lee DA, Data JL, Monaghan EP. Assessment of ganaxolone's anticonvulsant activity using a randomized, double-blind, presurgical trial design. Ganaxolone Presurgical Study Group Epilepsia 2000;41(9):1187–1194.

ANTIPSYCHOTIC AGENTS

Jeffrey M. Brown
Jie Jack Li
Michael W. Sinz
Drug Discovery, Bristol-Myers Squibb Company, Wallingford, CT

1. INTRODUCTION

Schizophrenia is a devastating and complex psychiatric illness that remains a major medical problem in our society. Although the manifestations of this disorder have been known and described for over a century, our understanding of the biology and pathology is still in its infancy. Pharmacological treatments for schizophrenia remain limited in mechanism and largely address only one aspect or symptom domain of this disorder. This fact stems largely from three challenges. First, it is a general lack of understanding of the complex pathological mechanisms that contribute to the disease. Second, there is a need for updated diagnostic criteria that includes evaluation of additional symptom domains. Third, it is a lack of preclinical models that capture multiple domains of the disease. Efforts over the last decade have begun to address these issues and have opened the door for new hypotheses regarding the integration of neurotransmitter systems and neuronal circuits. These efforts have lead to the development and testing of antipsychotic medications with novel mechanisms of action, which may address multiple domains of this complex disorder. As these new pharmacological tools become available they will also help in the development and validation of new preclinical models.

2. DEFINING SCHIZOPHRENIA

In the early twentieth century, it was believed that all forms of psychiatric illness represented a single common disease or pathology. This idea was challenged by Emil Kraepelin who distinguished a manic-depressive state, "dementia praecox," from other chronic psychiatric illnesses. The term schizophrenia (meaning split brain) was coined by Swiss psychiatrist Paul Eugen Bleuler to describe a psychiatric illness associated with an impaired perception of reality manifested by hallucinations as well as disorganized thought and speech. Schizophrenia is seen worldwide with prevalence rates reported in the range of 0.5–1.5%. [1] Initial onset usually occurs in the mid to late twenties and is often noted by family members, friends and/or coworkers as bizarre or inappropriate behavior [1]. Although genetic links have been seen in schizophrenia, no biological test has been validated for diagnosis. Instead diagnosis depends on the presence of specific behavioral signs that present for at least 6 months and are not associated with schizoaffective disorder or other mood disorders [1]. These behavioral signs can be categorized into three symptom domains: positive symptoms, negative symptoms, and cognitive deficits. In order to understand the rationale for past, current, and future antipsychotic drug design, it is important to have an understanding of these symptom domains as well as the underlying pathologies associated with schizophrenia.

2.1. Symptom Domains [1]

2.1.1. Positive Positive symptoms in schizophrenia are by far the most obvious and most associated with the schizophrenic condition. These symptoms include delusions, hallucinations, and/or disorganized speech. Clear rational thought and perceptions of reality become distorted. The manifestation of these hallucinations may be unrealistic paranoia, usually auditory (paranoid schizophrenia subtype) or disorganized speech and inappropriate behavior (disorganized schizophrenia subtype). Catatonic schizophrenia refers to abnormal motor behaviors and can include a lack of responsiveness to danger or environmental situations, rigid or bizarre postures, or excessive purposeless movements. The term psychotic is often used to describe this cluster of positive symptoms. Antipsychotic medications used in the treatment of schizophrenia are most effective in treating this symptom domain.

2.1.2. Negative Negative symptoms of schizophrenia are characterized by a loss or decrease in normal function. Specific conditions related to negative symptoms include affective flattening, alogia, and avolition. Affective flattening refers to abnormal facial immobility, lack of eye contact or a general lack of or reduced body language. Alogia is a general poverty of speech or lack of normal conversational dialog. Individuals often show a paucity of responses to questions or an unwillingness to speak. Avolition describes an inability to initiate or maintain goal-directed activities or a general lack of motivation or desire. Antipsychotic medications are less effective at treating negative symptoms of schizophrenia.

2.1.3. Cognitive Cognitive deficits represent on overall decrease in normal executive function. Executive function is a general term used to describe the integration of multiple sensory inputs, which contributes to a specific behavioral response and includes aspects of working memory, inhibitory control, and information processing. A number of studies have demonstrated that cognitive deficits are a pervasive and central component of schizophrenia. These deficits precede the development of the actual schizophrenic illness and are not associated with antipsychotic treatment. Initial estimates of the prevalence of cognitive deficits in schizophrenia range from 40% to 80% of the patient population. Although cognitive function is believed to be the largest contributor of morbidity associated with schizophrenia it is not part of the overall diagnostic requirements in the DSM-IV [1]. In addition, antipsychotic medications, both typical and atypical (see below), are less effective in treatment of cognitive deficits in schizophrenia. Development of new medications specifically targeting cognitive deficits has been hindered by the lack of a clear standardized test or tests for evaluation of cognitive function and the effect of drugs in clinical populations. In an attempt to standardize measures of cognitive functions, members of the MATRICS (Measurement and Treatment Research to Improve Cognition in Schizophrenia) initiative recommended evaluating six specific measures: working memory, attention/vigilance, verbal learning and memory, visual learning and memory, reasoning and problem solving, and speed of processing and social cognition [2]. These efforts are critical as they lay the foundation for establishment of standards for evaluating cognitive deficits in schizophrenia.

2.2. Pathology of Schizophrenia

The pathology of schizophrenia is complex and likely involves changes in multiple neurochemical substrates as well as structural reorganization within the brain. Although a detailed description of this pathology is beyond the scope of this chapter, understanding the key neurochemical substrates, particularly dopaminergic and glutaminergic/GABAergic systems, will help to define the rational for past, present, and future drug development efforts.

2.2.1. Dopamine Design and development of antipsychotic agents are often tailored to address a specific pathology or set of pathologies seen in schizophrenia patients. Typical antipsychotics that were developed based on the serendipitous finding that chlorpromazine, initially developed as an antihistamine, was effective in an animal model of conditioned avoidance [3]. Later, it was demonstrated that chlorpromazine and another antipsychotic agent haloperidol increased the turnover of monoamines in the brain [4]. In 1966, Van Rossum hypothesized that dopamine receptor blockade is an important factor in the mode of action of antipsychotic agents [5]. This hypothesis was later validated by two groups that demonstrated the ability of antipsychotic agents to block stimulated dopamine release and haloperidol binding (D2 receptors) was correlated with clinical potencies and dosages [6,7]. Although initial studies suggested elevation in D2 receptors in schizophrenic patients, this finding remains controversial and may be related to previous antipsychotic treatment or a subset of schizophrenia patients [8–12]. However, evidence to support a hyperdopaminergic state has been demonstrated. Amphetamine is a psychostimulant that causes a rapid and pronounced dopamine release within the brain and can lead to a psychosis mimicking the behavioral manifestations seen in schizophrenic patients. Amphetamine-induced dopamine release is increased in schizophrenic patients and increases in D2 receptor displacement correlates with a worsening of positive, but not negative symptoms [13]. One conclusion from this study is that this increased dopamine release induced by amphetamine likely results from alterations in dopamine synthesis and/or degradation and not due to differential affinity or expression of the D2 receptor [14]. Collectively, these and other findings have lent support to a primary hypothesis regarding the pathology of schizophrenia, the dopamine hypothesis. This hypothesis states that schizophrenia results from a hyperdopaminergic tone within the CNS. This hypothesis has been a primary driver for the development of dopaminergic antagonists, particularly D2 antagonists, termed "typical psychotics" over the past half century. However, a major drawback to typical antipsychotic is that potent inhibition of D2 receptors can cause unwanted adverse events (see below). To lessen the D2 associated adverse events, atypical antipsychotic medications that show overall lower D2 affinity and blockade of nondopaminergic (e.g., serotonin) receptors have been developed.

As our understanding of the pathology of schizophrenia has developed so too has the dopamine hypothesis. Originally, though to be a disorder of hyperdopaminergic tone, it is now believed that both a hypodopminergic and a hyperdopaminergic tone may exist in schizophrenia. Specifically, Bannon and Roth hypothesized that a hypodopaminergic tone in prefrontal cortical areas may result in a corresponding increase or hyperdopaminergic tone in subcortical areas [14]. This hypothesis was echoed by Davis et al. (1991) who stated that schizophrenia results from a hypodopaminergic tone in the mesocortical pathway and a hyperdopaminergic tone in the mesolimbic pathway [15]. As the positive symptoms of schizophrenia are believed to be related to the hyperdopaminergic state, the negative and cognitive aspects may result from the hypodopaminergic state in the prefrontal cortex. This has led some to evaluate both dopamine receptor agonists and positive modulators for the treatment of schizophrenia. Most notable is the new antipsychotic aripiprazole that acts as a partial agonist of the D2 receptor and may help to balance a disrupted dopamine tone.

2.2.2. Glutamate/GABA As our understanding of schizophrenia develops, so to does the list of possible new pharmacological targets. As early as the 1959, Luby et al. reported that administration of phencyclidine (PCP; Sernyl) to nonschizophrenic patients caused a psychopathology similar to that seen in schizophrenic individuals [16]. When administered to schizophrenic patients, PCP intensified the primary behavioral pathology seen in this cohort [16]. PCP is an antagonist at the ionotropic glutamate receptor (NMDA receptor) leading to the speculation that hypofunction of the NMDA receptor contributes to the core symptoms of schizophrenia [17,18]. This NMDA hypofunction may result from decreased glutamate levels in the brain (i.e., decrease receptor activation). However, glutamate levels have been shown to be unaltered in schizophrenic patients [19,20]. In addition to glutamate, NMDA receptor activation requires the cofactor glycine and blockade of this glycine site inhibits NMDA function. Kynurenic acid, a breakdown product of tryptophan, is elevated in the CSF and cortex of schizophrenic patients [21–23]. In a study by Schwarcz et al. (2001) it was demonstrated that, in rats, chronic treatment with haloperidol decreased kynurenic acid levels suggesting increases seen in schizophrenic patients was not a result of drug treatment [21]. This is of relevance to NMDA function as kynurenic acid may be an antagonist at the coagonist glycine site [24]. Moreover, administration of the kynurenic acid precursor kynurenine decreases the PPI response in rats, an animal model of schizophrenia. This effect was reversed by both a typical and an atypical antipsychotic [25]. These findings would support the hypotheses that decreased NMDA function, due to a kynurenic acid-mediated blockade of the glycine site of the NMDA receptor, contributes to the pathology of schizophrenia.

In addition to kynurenic acid, NMDA function can be inhibited by other endogenous NMDA antagonists. For example, activity and binding of glutamate carboxypeptidase, the enzyme that breaks down N-acetyl-aspartyl glutamate (NAAG) to NAA and glutamate, is reduced in schizophrenic patients [26,27] an effect unlikely mediated by changes in mRNA [28]. This is of relevance since NAAG antagonizes NMDA receptor function [29,30]. These data support a hypothesis that hypofunction of the NMDA receptor may result from endogenous, orthosteric, or allosteric inhibitors of NMDA receptor function. As treatment with orthosteric NMDA agonists may result in neurotoxicity, targeting the allosteric site of the NMDA receptor would enhance NMDA responses to glutamate without direct activation [31].

Although modulation of NMDA receptor function remains an interesting pharmacological target, it is not the only glutamate receptor that has links to schizophrenia. Recent genetic data suggests genetic associations between schizophrenia and the metabotropic glutamate receptors (mGluRs), most notable the mGLuR3 [32,33]. Early clinical studies with an mGluR 2/3 agonist have shown promise. Specifically, administration of a prodrug of the mGluR2/3 agonist, LY404039, was shown to have positive clinical effects over placebo and a comparable effect to olanzapine [34]. This finding is of significance as it suggests that nondopaminergic treatments can be developed for the treatment of schizophrenia.

Although the dopamine and glutamate hypotheses arose from different findings, the two may be interrelated. Within the prefrontal cortex GABAergic parvalbumin positive (PV+) interneurons provide inhibitory input to glutaminergic pyramidal neurons. These same PV+ interneurons receive input from dopaminergic afferents [35] and are innervated by local glutaminergic axon collaterals [36]. Thus, under the control of dopaminergic and glutaminergic input, these GABAergic interneuorns are in a position to regulate the activity of prefrontal cortical glutaminergic neurons. One of the most reproducible findings in schizophrenia is a decrease in GABAergic markers in the prefrontal cortex and hippocampus. Specifically, it has been demonstrated that GAD67 (the enzyme responsible for GABA synthesis) and the GABA transporter (GAT), both of which are localized to GABAergic interneurons, are decreased in schizophrenic patients [37–40]. Since these GABAergic interneurons are poised to regulate glutaminergic pyramidal neuron activity, it would suggest a general dysfunction of a prefrontal neuronal circuit within schizophrenic patients. This has lead some to speculate that modulation of this circuit by enhancing GABAergic transmission via GABA-A receptor modulators may represent a viable alternative for the treatment of schizophrenia [41]. Since the glutaminergic pyramidal neurons in the prefrontal cortex are important integrators of sensory input and overall executive function, targeting these deficits may represent a pharmacologic mechanism to address an area of unmet medical need in schizophrenia, cognitive deficits.

2.2.3. Other Neurotransmitters As dopamine, glutamate, and GABA may be important in the pathophysiology of schizophrenia; other neurotransmitters systems likely contribute either directly or indirectly to the pathology and/or efficacy of antipsychotic medications. In a review by Meltzer et al. (2003), it was stated that the superior effects of atypical antipsychotics are due to there effect on serotonin receptors, particularly 5-HT_{1a} and 5-HT_{2a} [42]. Preclinical as well as clinical data supports this hypothesis. For example, administration of atypical (clozapine, olazapine, and ziprasidone) but not a typical (haloperidol) antipsychotic caused greater increases in dopamine release in the prefrontal cortex of wild type but not 5-HT_{1a} knockout mice [43]. Clinical studies using a 5-HT1a partial agonist to augment current therapy have shown some positive results, particularly with cognitive deficits, in schizophrenia patients [44]. Finally, polymorphisms within the promoter region of the 5-HT_{1a} gene are reported to mediate the response to antipsychotic treatment, particularly in negative and depressive symptoms of schizophrenia [45].

5-HT_{2a} receptors have been localized to glutaminergic pyramidal and PV+ GABAergic interneurons in the prefrontal cortex and thus are in a position to regulate the same prefrontal cortical circuit outline above [46,47]. Similar to that seen with the 5-HT_{1a} receptor, levels of 5-HT_{2a} receptors may also predict response to atypical antipsychotics, particularly with negative symptoms [48]. Thus, serotonin receptor function contributes to the effectiveness of antipsychotic medications and may underlie the pathology of negative and/or cognitive aspects of schizophrenia. The mechanism of these effects may be related to the regulation of dopaminergic and glutaminergic circuits in the brain.

Nicotinic receptors may also play an important role particularly in relation to cognitive deficits [49]. Schizophrenia patients show high rates of smoking and nicotine consumption, a phenomenon that may represent a form of self-medication. In fact, administration of a nicotine patch with haloperidol attenuates haloperidol-induced cognitive deficits in schizophrenics [50]. Nicotinic receptors are composed of both alpha and beta subunits. In schizophrenia patient's, reports of decreases in alpha4 beta2 and decrease binding of alpha bungarotoxin, presumably alpha7 subunits, in multiple brain regions has been reported [51–53]. Preliminary studies in schizophrenia patients with an alpha7 partial agonist have shown promising results in terms of a cognitive benefit [54]. This has led many to speculate that agonists, particularly alpha7 agonists may be useful in the treatment of cognitive deficits in schizophrenia.

To date multiple neurotransmittrer and neuropeptide systems (e.g., norepinephrine and neurotensin [55,56]) have been implicated in schizophrenia. However, the contribution of these systems to the basic pathology, symptomology, and/or drug responsiveness remains to be determined. Given the fact that schizophrenia is a polygenic (see below) and multisystem disorder, determining how these individual systems contribute to the overall regulation of neuronal circuits involved in schizophrenia is vital to understanding the basic pathology and the mechanism of current and future antipsychotic medication.

2.2.4. Genetics The association of genetic alterations and chromosomal aberrations is a vital component to understanding any disease. Schizophrenia clearly has a strong genetic component as first-degree relatives of schizophrenics show up to a 10-fold increased risk for schizophrenia. To date no specific mutation or dysfunction of a specific "schizophrenia gene" has been identified. Results evaluating specific chromosomal regions have shown some promising results. For example, there is a 20- to 80-fold higher prevalence of the 22q11.2 deletion in patients with schizophrenia relative to that of the general population [57]. This is of interest as this chromosomal region contains some genes showing relevance to schizophrenia [57].

Analysis of specific genes, mutations and single nucleotide polymorphisms (SNPs) has failed to identify a specific schizophrenic trait. This leads one to speculate that the genetic nature of schizophrenia is likely polygenic in nature with specific groups or individuals showing certain genetic abnormalities that are absent in others. Thus, identifying a genetic "silver bullet" underlying schizophrenic pathology is highly unlikely. Although genetic association studies have been variable, efforts to examine these studies in total to identify a common set of traits specific to schizophrenia has been published. In 2008, Allen et al. published results of a meta-analysis of 118 polymorphic variants for 52 genes. The results identified four genes with a "strong degree of epidemiological credibility," the D1 dopamine receptor, dysbindin, 5,10-methylenetetrahydrofolate reductase and tryptophan hydroxylase [58].

These are only two examples of numerous published reports of genetic association studies with schizophrenia. As this list of genes continues to rapidly grow, so too do questions of validity and reproducibility of results. This is not surprising given the numerous confounding factors such as genetic variability, differential diagnostic, and statistical methodologies and lack of diagnostic criteria for certain symptoms domains (e.g., cognitive deficits). Schizophrenia clearly is a polygenic disorder and as with evaluation of neurotransmitter system, understanding how these and other genetic alterations contribute to an overall "systems deficit" in schizophrenia will greatly advance our understanding of the pathology and subsequent drug development.

3. ANTIPSYCHOTIC AGENTS

3.1. Side Effects

As with all drugs, administration of antipsychotic agents is often associated with adverse events. These effects can range from the unpleasant, which can contribute to compliance issues, to serious life threatening events. Some of the adverse events may be directly linked to the mechanism of action (e.g., inhibition of D2 receptors) and as such the frequency and severity is related to dose and potency. These adverse events include extrapyramidal symptoms and anticholinergic effects. With other antipsychotic-associated adverse events the mechanism is less clear and may result from off target effects (e.g., potassium channel inhibition) or a complex array of pathological, biological and/or genetic factors. For example, the elderly population tend to be more susceptible to adverse events, women are more prone to clozapine-induced agranulocytosis and individuals with a history of epilepsy may be more prone to seizures associated with antipsychotic treatment. Of particular relevance is a black box warning placed on both typical and atypical antipsychotic medications stating that antipsychotic drugs of both types (typical and atypical) are associated with an increased risk of death when used in elderly patients for dementia-related psychosis. Therefore, patient characteristics must be considered when deciding on the best antipsychotic regimen.

3.1.1. Extrapyramidal Symptoms [59]

The most common and challenging side effect associated with use of antipsychotic medication is the emergence of extrapyramidal symptoms (EPSs). EPS represent abnormal and uncontrollable movements and are classified into two groups, acute and tardive syndromes. Acute EPS include akathisia (motor restlessness due to feelings of distress or discomfort), dystonias (sustained muscle contractions cause twisting and repetitive movements or abnormal postures) and parkinsonian symptoms (bradykinesia; slowness of movement, rigidity; tremor; and postural instability). Akathisia are most often acute, developing within a week of treatment initiation, but are a reversible phenomenon. Parkinsonian symptoms, however, can persist for months after cessation of treatment. This is of particular relevance to older patients as the presence of parkinsonianism can reflect a purely drug-induced effect, an exacerbation of previously undiagnosed parkinsonian pathology or an actual pathology unrelated to drug treatment. The acute EPS effects result from blockade of D2 receptors in the basal ganglia, a phenomenon linked to the mechanism of most typical antipsychotics [59]. In fact, it has been estimated that occupancy of 65% of the D2 receptors is needed for clinical efficacy whereas 78% occupancy leads to EPS [60].

To prevent or minimize acute EPS, particularly akathisia, typical antipsychotics with lower affinity for the D2 receptor such as chlorpromazine can be used. As low affinity typical antipsychotics tend to have higher affinity for cholinergic receptors, using an anticholinergic agent in combination with a low affinity typical antipsychotic is often employed. Alternatively, atypical antipsychotic can be used. In a direct comparison of risperidone (atypical), olanzapine (atypical), quetiapine (atypical), ziprasidone (atypical), and perphenazine (typical), it was demonstrated that more patients discontinued treatment with perphenazine due to EPS [61]. Drug-induced parkinsonianism is often treated by reduced dosage or switching to clozapine or quetiapine, both of which are less associated with parkinsonianism or left untreated if effects are tolerable.

Tardive dyskinesia (TD) is a late onset phenomena characterized by stereotypic involuntary movements most often seen in the face, lips and tongue but can also include the extremities. Prevalence rates of TD range from 10% to 25% of schizophrenic patients although older patients may show a higher frequency. Although still speculative, the underlying pathology is believed to result from a hypersensitivity of D2 receptor and possible structural alterations within the brain following prolonged antipsychotic treatment. While there remains no effective way to treat TD, atypical agents such as clozapine and quetiapine are believed to have a lower incidence of TD. While use of anticholinergics may aggravate the symptoms, use of GABA receptor agonist (i.e., benzodiazepines) may be helpful.

In general, it is believed that EPS are more often associated with typical antipsychotics, particularly high potency agents. However, the magnitude of benefit between typical and atypical antipsychotics in terms of EPS remains difficult to ascertain. This ambiguity is due largely to multiple factors such as; EPS are more often associated with younger patients, dosage can affect the presence or absence of EPS with both typical and atypical agents and not all studies discriminate between acute and tardive types of EPS.

3.1.2. Neuroleptic Malignant Syndrome Neuroleptic malignant syndrome (NMS) resembles a severe form of parkinsonianism and is characterized by catatonia, tremors, hyperpyrexia, autonomic system instability, and/or altered mental status. The reported incidence rate of NMS is 1% of all patients treated with antipsychotics and can be life threatening. Like EPS, NMS is associated with antagonism of dopamine receptors, possibly those in the striatum and hypothalamus although the precise pathological mechanism remains to be determined. Of the available agents, haloperidol is most often associated with NMS reports although cases with other antipsychotic agents have been seen. Treatment of NMS may involve cessation of the current treatment, supportive therapy and possibly, benzodiazepines, muscle relaxants or dopamine agonists although questions remain as to the viability of pharmacotherapy options [62].

3.1.3. Cardiovascular Effects [63] Antipsychotic agents can negatively affect the cardiovascular system. Orthostatic hypotension is one of the most common adverse cardiac effects and is present in approximately three quarters of patients. In elderly patients and those with previous cardiovascular or autonomic system dysfunction, this risk of hypotension is greater. Complications resulting from antipsychotic-induced hypotension include dizziness, visual disturbances, cognitive impairment and syncope. Syncope is of particular concern in the elderly population due to the risk of falls and associated bone fractures. The risk of antipsychotic-induced hypotension decreases with continued use as patients will develop tolerance to these effects. Hypotension is often managed by titration of dose, particularly during initial treatment and minimizing use of agents associated with higher potential for orthostatic hypotension.

As a general rule, typical antipsychotics that show lower potency are more likely to cause hypotension. All atypical antipsychotics are associated with increased risk of hypotension, however, the risk varies. Clozapine is associated with the highest risk followed by quitiapine, risperidone, olazapine, and finally aripiprazole [64]. With atypical antipsychotics the propensity to cause hypotension may be related to the affinity for the alpha-1 adrenergic receptors.

Prolongation of the Q–T interval, the time from ventricular depolarization to ventricular repolarization, can lead to cardiac arrhythmias, torsades de pointes, and possibly death. Drug-induced prolongation of Q–T interval can result from blockade of a delayed rectifier potassium channel that is important for depolarization of the ventricles. Typical and atypical antipsychotics are associated with blockade of these potassium currents. For typical antipsychotics, the most notable is thioridazine that has been associated with reversible prolongation of Q–T interval and an increased rate of sudden death. For atypicals ziprasidone and quetiapine appear to have the largest effect on Q–T interval whereas aripiprazole has not been associated with an increase of the QT interval [64]. Management of adverse cardiac effects involves an initial determination of patient risk, selecting agents with lower propensity to cause adverse cardiac effects, using lower doses when possible and periodic monitoring of cardiac function.

Other adverse cardiovascular effects include cardiomyopathy and myocarditis which have been associated with use of the atypical clozapine. Cardiovascular disease has a higher incidence in schizophrenics than that seen in the general population. This fact may be due to the higher rates of smoking and a general lack of physical activity or exercise. Antipsychotics do have negative metabolic effects including weight gain, hyperlipidemia, and altered glucose homeostasis (see below). However, the role of antipsychotic medications in this increased risk is still controversial.

3.1.4. Anticholinergic Effects In addition to affinities at D2 and alpha1 receptors, most antipsychotic medications also antagonize muscarinic receptors leading to atropine-like effects. These include dry mouth, blurred vision, constipation, and urinary retention and can be more pronounced in elderly populations. Of the typical antipsychotics thioridazine, which has a higher affinity for M1 receptors, is associated with more pronounced anticholinergic effects. For the atypical agents clozapine and olanzapine have higher antimuscarinic effects whereas aripiprazole, risperidone and ziprasidone may have less anticholinergic effects. In a 2004 article, Watanabe et al. compared 50 schizophrenia patients treated with either an atypical (risperidone, olanzapine, quetiapine) or a typical (haloperidol, chlorpromazine) antipsychotic. Following treatment each patient completed a survey relating the side effects including anticholinergic effects. Results from the survey distinguished no overall differences between typical and atypical antipsychotic medication in terms of anticholinergic effects [65].

3.1.5. Metabolic Effects

Psychiatric illnesses, particularly schizophrenia, are associated with overall poor physical health. Weight gain is common in schizophrenia and may result from a combination of factors including disease pathology, lifestyle (smoking), and use of antipsychotic medications. Weight gain is common with antipsychotic treatment, particularly the newer atypical antipsychotics. In a review by Haddad and Sharma, they stated that of the newer atypical antipsychotics olanzapine and clozapine cause the greatest weight gain whereas aripiprazole causes less weight gain, although increase weight was seen with all atypical agents when compared to placebo [66]. Significant weight gain with olanzapine compared with risperidone and ziprasidone was also reported by Stroup et al. [67]. In addition, olanzapine was associated with increases in serum levels of cholesterol and triglycerides [67], an effect that may be directly or indirectly linked to weight gain.

Both typical and atypical drug can worsen glycemic control. In general, olazapine and clozapine appear to have a higher risk hyperglycemia when compared to other antipsychotics [66]. The effects of this altered glycemic control in relation to development of diabetes is unclear. Collectively, studies appear to suggest a link between newer antipsychotics and diabetes, however, the effects are variable and may be confounded by such factors as age, increased incidence of smoking and weight gain in schizophrenic populations. Lambert et al. (2006) suggested that the newer antipsychotics olanzapine, risperidone and quetiapine increase the risk of developing diabetes by 60–70% in comparison with haloperidol [68]. Overall, it is suggested that monitoring blood glucose levels, particularly during early treatment, as well as being vigilant for signs suggesting development of diabetes is recommended when using antipsychotic medication [66].

3.1.6. Seizures

Although increased risk of seizures has been reported particularly with older antipsychotics, the actual incidence of seizures is estimated to be less than 1%. One notable exception is clozapine that may have a slightly higher incidence of seizures up to 5%. For the typical antipsychotics, chlorpromazine is regarded as the most epileptogenic [69]. Caution should be taken in administration of antipsychotic medication in patients with epilepsy.

3.1.7. Sedation

Sedation associated with antipsychotic medications likely results from antagonism of histamine receptors. Agents with lower D2 affinity tend to have higher affinity for the histamine receptor and are therefore more sedating. Clozapine and olanzapine have the highest affinity for histamine receptors and tend to be the most sedating while aripiprazole may be the least sedating. These sedating effects may be beneficial in patients who are agitated but may also worsen or confound evaluation of cognitive and negative symptoms. Sedation associated with antipsychotic medications is not permanent as tolerance to these effects does develop.

3.1.8. Hematologic Effects

Neutropenia is a condition resulting from decreased neutrophil counts. Agranulocytosis, a severe form of neutropenia, has been associated with clozapine and is more often seen in Asian populations, women and older patients [66]. As such it is recommended that clozapine be reserved for treatment resistant patients or for reducing the risk of suicidal behavior in patients with schizophrenia or schizoaffective disorder. Patients treated with clozapine must have a baseline and regular blood cell counts as well as follow-up blood counts 4 weeks after cessation of treatment [70].

3.1.9. Sexual Side Effects
Sexual dysfunction is a psychologically distressing event that can seriously hamper compliance with antipsychotic agents. It is estimated that up to 43% of individuals using antipsychotic medications report sexual dysfunction [71]. However, evaluation and classification of what represents sexual dysfunction is unclear. Based on the different methodologies of data collection, differential interpretation of sexual dysfunction and potential underlying pathologies it is unclear as to the rates of sexual dysfunction among the two classes of antipsychotic medications [66].

Sexual dysfunction may result from various factors associated with antipsychotic medications including hyperprolactemia. Dopamine inhibits prolactin release and therefore agents that block dopamine receptors would be expected to increase prolactin levels. Typical antipsychotic drugs appear to increase prolactin levels in a dose-dependent manner consistent with the idea of D2 receptor blockade. Atypical agents including clozapine, quetiapine, and olanzapine show little to no effects on prolactin while risperidone can increase serum prolactin levels [72].

3.1.10. Hepatic Effects
Elevations in liver enzymes have been reported with many antipsychotics. Although these increases are common they are largely asymptomatic and rarely warrant discontinuation of treatment. Direct studies of olanzapine, risperidone, and quetiapine demonstrated some increases in liver enzymes associated with these antipsychotics but significant liver enzyme elevations are rare during atypical antipsychotic treatment [73]. Similar results were found with risperidone treatment in a pediatric population [74].

3.1.11. Antipsychotic Discontinuation
Development of adverse events or lack of efficacy may require discontinuation of antipsychotic treatment. As with most CNS related drugs, cessation of the medication may result in adverse effects possibly related to withdrawal. Both motor (restlessness) and nonmotor (vomiting) effects have been reported upon termination of antipsychotic medications. For example, within days of stopping clozapine treatment dystonias, motor restlessness, anxiety, nausea, and altered consciousness have been reported [66].

3.2. Clinical Applications

Currently, there are a number of agents approved for the treatment of schizophrenia, see Section 5. Determining which antipsychotic is best depends on multiple factors including currently prescribed medications, cost, side effects, and patient populations. Much has been written and debated regarding the methodology for selection of the most appropriate antipsychotic agent. In 2003, Kane et al. presented results from a survey of experts regarding the selection of antipsychotic agents. It was concluded that atypical antipsychotics, particularly risperidone, were endorsed as the primary medications for first-line treatment of schizophrenia [75]. In a review by Tandon et al. (2008), it was suggested that as the number of prescriptions for the more costly antipsychotics increased, based on earlier recommendations, concerns were raised by governments regarding the usefulness of these new more expensive antipsychotic medications compared to older typical antipsychotics [76]. These concerns were the catalysts for two government sponsored studies, CUtLASS (Cost Utility of the Latest Antipsychotic Drugs in Schizophrenia Study) and CATIE (Clinical Antipsychotic Trial of Intervention Effectiveness). A primary conclusion from the CUtLASS study was that there was no disadvantage to using typical antipsychotic medications when compared to atypical antipsychotic medications in the outcome measures evaluated [77]. In a second study termed CATIE, Lieberman et al. (2005) reported that there were no significant differences in effectiveness between the typical and the newer atypical antipsychotic medications, although olanzapine may have demonstrated some early benefit [61]. Results from these studies have been questioned based on a number of factors including methodological differences [76]. To date, the question regarding the cost benefit of typical versus atypical antipsychotics for the treatment of schizophrenia remains unresolved

3.3. Other Indications

In addition to schizophrenia, antipsychotic medications are effective for the treatment of various psychiatric and nonpsychiatric indications. D2 dopamine receptors in the chemoreceptor trigger zone and in the stomach are believed to participate in the emetic response. Therefore, antipsychotic agents that antagonize D2 receptors (e.g., chlorpromazine) are effective antiemetic agents. Promethazine, an antipsychotic with affinity for the H1 receptor has been used as a preoperative sedative. Other indications include chlorpromazine and haloperidol for the treatment of intractable hiccups. Haloperidol has also been approved for the treatment of tics associated with Tourette's syndrome and agitation associated with dementia. Chlorpromazine has been used for the treatment of psychosis resulting from PCP and amphetamine use. In addition to FDA approved indications, antipsychotic medications are also used "off label" at the discretion of the physician. Off label uses have included the use of antipsychotics for the treatment of dementia related psychosis, particularly in the elderly population. However, this practice has come under scrutiny largely due to black box warnings placed on all antipsychotics indicating an increased risk of death in elderly patients.

4. ANIMAL MODELS

Development and testing of antipsychotic medications is dependent on the evaluation of both effectiveness and side effects at the preclinical level. Because schizophrenia represents a complex disorder involving multiple neurochemical and structural alterations in the brain modeling this disease in an animal is challenging. In developing or validating any animal model of psychiatric disease it is important to utilize models that; recapitulate a behavioral symptom or symptoms associated with the disease, possesses pathology or pathologies which mimics the disease and responds to clinically relevant drugs. Efforts to develop animal models of psychiatric disease include pharmacological and genetic manipulations to model specific aspects of the disease relating to efficacy, side effects or in some cases both. For schizophrenia, animal models are based largely on pathological findings in schizophrenia patients, behavioral correlates between human and animal models and/or knowledge of specific neurochemical circuits invoiced in the pathology of schizophrenia. For the purpose of this chapter, models of efficacy have been divided into pharmacological, behavioral, and genetic. However, it should be noted that this division does not imply these models are mutually exclusive or even independent. For example, the behavioral measure of prepulse inhibition (PPI) can be used in combination with NMDA antagonists, amphetamine or in mice genetically predisposed to PPI deficits. As no one animal model truly recapitulates the schizophrenia pathology, using multiple models to evaluate individual aspects remains the most utilized approach for screening and evaluating antipsychotic medications.

4.1. Animal models of Efficacy

4.1.1. Pharmacology Based Models

Dopamine Model As stated above, disruption of normal dopamine tone is believed to underlie the pathophysiology of schizophrenia, particularly the positive symptoms (see dopamine hypothesis in Section 2.2.1). Reproducing this altered dopaminergic state in animals represents a plausible method for evaluation of antipsychotic efficacy. Amphetamine and apomorphine increase locomotor activity in rats whereas apomorphine can induce a climbing behavior. Both amphetamine- and high-dose apomorphine-induced locomotor effects are mediated by the mesolimbic dopamine system [78–80]. Both typical and atypical antipsychotics including haloperidol, clozapine, and aripiprazole have shown efficacy at blocking dopamine agonist-induced increases in locomotor activity and apomorphine-induced climbing [81–83]. At higher doses, both apomorphine and amphetamine can induce stereotypic behavior. Unlike the locomotor effects, stereotypic behavior may be mediated by mesostriatal dopamine pathways. For example, direct injection of apomorphine into the caudate induced stereotypic behavior whereas infusion into the nucleus accumbens increased locomotor activity [78]. Interestingly, typical and atypical antipsychotics show differential effects on mesostriatal and mesolimbic dopamine pathways. For example, Chiodo and Bunney (1983) demonstrated that chronic treatment with antipsychotics can differentially affect mesolimbic and mesostriatal dopamine neurons such that typical neuroleptics which induce EPS inactivate mesolimbic and mesostriatal neurons whereas clozapine, which lacks EPS, increases mesostriatal and decreases mesolimbic cell firing [84]. This finding is in line with the idea that mesolimbic dopamine systems may be hyperactive in schizophrenia whereas inhibition of mesostriatal dopamine, indicative of typical antipsychotics, may cause extrapyramidal side effects. Thus when evaluating agents with antipsychotic potential, one method employed has been to couple locomotor activity with stereotypic behavior in an attempt to differentiate a mesostriatal versus a mesolimbic mechanism of action.

Dopamine-based models have long been used for the evaluation of antipsychotic medications. These models capture some of the behavioral aspects of schizophrenia, particularly positive symptoms. In addition, the use of amphetamine, which causes release of dopamine, may mimic a hyperdopaminergic state hypothesized to be important in the pathology. Finally, the use of antipsychotic medications, particularly dopamine antagonists are effective in these models. However, these models do posses some drawbacks. For example, they may not capture negative or cognitive aspects of schizophrenia. This model may show bias toward antipsychotics with a direct dopamine action. Finally, both amphetamine and apomorphine-induced behavioral effects can be altered by nondopaminergic agents including serotonin agonists, picrotoxin, H3 antagonist and sigma1 receptor modulators [83,85–87]. Since the effectiveness of these compounds in schizophrenia has not been proven it is unclear as to the true predicative validity of these models.

NMDA Model [88] The finding that PCP administration caused a psychopathology similar to that seen in schizophrenic individuals [16] paved the way for a new hypothesis of schizophrenia, NMDA hypofunction. Accordingly, use of the NMDA antagonist PCP and later ketamine have been utilized in rodents as a model of schizophrenia. Although it is impossible to completely recapitulate the behavioral effects of schizophrenia in any single animal model, the PCP/ketamine model may have some interesting advantages. Unlike amphetamine-induced psychosis, inhibition of NMDA receptors may address some aspects of both negative and cognitive deficits in schizophrenia. For example, inhibition of NMDA receptor function in humans can disrupt cognitive function as measured by performance on the Wisconsin card sort and deficits in delayed recall and verbal fluency. Similarly, both MK-801 [89] and PCP [90] have been shown to disrupt attentional set-shifting, a rodent analog to the Wisconsin card sorting task. Although both acute and chronic administration of NMDA antagonists result in cognitive deficits, chronic administration may induce a hypofunction of the prefrontal cortex and therefore be more representative of cognitive deficits seen in schizophrenic patients [88]. NMDA antagonists also disrupt behaviors in rodents that can be linked to negative symptoms in schizophrenia such as social withdrawal and depression. Chronic PCP administration enhances immobility in the forced swim test [91,92], a model of behavioral despair [93]. This effect was reversed by administration of atypical but not typical antipsychotics [92]. In addition, it was shown that PCP administration, but not amphetamine, caused social withdrawal in rats. It has been suggested that this model may detect the ability of an antipsychotic to treat negative symptoms in schizophrenia [94].

One way to validate animal models of NMDA hypofunction is to block the behavioral effect of acute PCP administration in humans or alleviate behavioral symptoms in schizophrenics. However, data obtained show limited and conflicting efficacy in human studies. For example, data from Lahti et al. demonstrated that haloperidol failed to block ketamine-induced psychosis [95]. Malhotra et al. showed that clozapine blunted ketamine-induced positive symptoms but had no effect on negative symptoms in schizophrenic individuals [96]. Interestingly, clozapine may be more effective in reversing PCP-induced deficits in rodents when compared to haloperidol [88] consistent with animal studies.

4.1.2. Behavioral Models

Prepulse Inhibition Abnormalities in information processing and attention are characteristics of schizophrenia. Startle reflex, a measure of information processing, refers to the ability of a prepulse signal to inhibit a subsequent startle evoked by a secondary signal. This process involves an inhibitory sensory/motor mechanism mediated by the prefrontal cortex. PPI, an experimental measure of this phenomenon, is commonly used to evaluate sensory motor gating in animals and humans. Schizophrenia patients show a deficit in the PPI response [97]. In animal models, PPI can be measured by delivering a prepulse (tone) followed by a louder tone to induce a startle. Decreasing the PPI can be induced by pharmacological agents (amphetamine and NMDA inhibition), developmental manipulations (isolation rearing) and genetic differences among rodent strains (C57Bl6 mice show PPI deficits).

Antipsychotic medications can reverse PPI deficits in schizophrenia patients. Most notable are clozapine and risperidone that may be superior to typical antipsychotics medications at reversing this PPI deficit. This finding may extend to animal models. For example, PPI deficits caused by social isolation and amphetamine are reversed by most typical and atypical antipsychotics with equal efficacy. However, atypical antipsychotics appear superior at reversing PPI deficits induced by serotonin and NMDA antagonists [98]. This is consistent with the hypothesis that schizophrenia results from a disruption of multiple neurotransmitter systems and atypical antipsychotic medications antagonize a number of receptors to these transmitters. Overall, PPI remains a primary model to evaluate the efficacy of novel antipsychotic medications, particularly those with differing mechanisms of action.

Latent Inhibition [99] Attention is an aspect of executive function that allows an individual to filter out or inhibit unnecessary or irrelevant information. Cognitive deficits in schizophrenia involve multiple aspects of executive function including attention. Latent inhibition is a term used to describe a phenomena whereby an individual or animal, when repeatedly presented with a stimulus in the absence of a reinforcer (positive or negative) will decrease interest in that stimulus. A primary mechanism driving latent inhibition involves attention and the ability to filter out sensory information from the continually presented stimuli. Elevations in the basal dopamine state decrease measures of latent inhibition, consistent with the dopamine hypothesis of schizophrenia. For example, in healthy volunteers, dopamine agonists disrupt latent inhibition [100,101]. Acutely psychotic schizophrenia patients show a disruption in this measure [102,103]. As in humans, an alteration in the latent inhibition following treatment with dopamine agonists and antipsychotic medications is seen in animals [104]. Animal models of latent inhibition also have predictive ability as both typical and atypical antipsychotics enhance latent inhibition [99]. Unlike PPI, blockade of NMDA receptors does not decrease measures of latent inhibition. For example, Weiner and Feldon (1992) showed no effects of PCP in an animal model of latent inhibition [105]. Similar results have been seen with other NMDA antagonists including ketamine and MK-801 [106,107]. Based on these findings it can be suggested that NMDA function does not contribute to the biological mechanism underlying this measure of attention. However, it has been proposed that NMDA inhibition does affect the latent inhibition by inducing a preservative response caused by an inability to "switch" from the initial stimulus to the conditioned stimulus [108]. Thus, based on this hypothesis and the observation that PPI does not require a condition stimulus it has been suggested that these models are not mutually exclusive but may utilize discretely different neurotransmitter systems and neuronal circuits.

Conditioned Avoidance [109] As stated previously, the first suggestion that chlorpromazine had antipsychotic properties was captured in an animal model of condition avoidance [3]. In a conditioned avoidance model, animals are trained to perform a specific response to avoid an adverse event (e.g., shock). These responses can involve active (e.g., level pressing) or passive (remaining in a specific chamber) behaviors. This behavioral paradigm is believed to be mediated largely by the mesocorticolimbic dopaminergic system. Antipsychotic medications reduce the avoidance response at doses that do not impair normal motor function. Both typical antipsychotics (e.g., haloperidol) and atypical antipsychotics (e.g., olanzapine, risperidone, ziprasidone and aripiprazole) have all shown positive results in an animal model of conditioned avoidance [110–112]. Response in the conditioned avoidance and catalepsy (both D2-mediated responses) can be differentiated based on the extent of occupancy of the D2 receptor (i.e., >80% causing catalepsy [113,114]). Although the conditioned avoidance response has been used as a primary screening tool for development of antipsychotic medication it does present some issues. First, false positives (e.g., morphine) have been identified in this model. Second, this model does not clearly differentiate between typical and atypical antipsychotics. Third, antipsychotic medications are effective following acute dosing whereas chronic use is needed for efficacy in schizophrenic patients. Finally, dopaminergic mechanism are likely the primary driver of this behavior and therefore this model may be limited in predicting efficacy for negative and cognitive domains of the disease.

Drug Discrimination Drug discrimination is a commonly used model for evaluating drug receptor interactions. This model uses a reference compound to train an animal to give a specific response when the reference drug is administered (e.g., lever press). Following this training a new drug is substituted for the reference drug. If the same behavioral response is produced it can be said the test compound pharmacologically acts at the same receptor(s) or in a similar manner. Although this model does not have direct pathophysiological or biological links to schizophrenia it is an effective way to determine if a test compound has a similar pharmacology to a reference compound.

4.1.3. Genetic Based Models
Pharmacological manipulation of neurotransmitter systems is one approach to modeling the pathology of schizophrenia. Alternatively, candidate genes associated with schizophrenia can be deleted in mice to determine the effect on behavior. This type of genetic knockout approach has been used in an attempt to model schizophrenia. Below is a list of selected genes that have been associated with schizophrenia and have been deleted in mice. For further review, see Ref. [115].

Dysbindin Dysbindin (DTNBP1, dystrobrevin binding protein1), a protein that interacts with dystrobrevins, was identified in 2001 by Benson et al. This protein was shown to be expressed in neurons within the brain particularly the hippocampus [116]. An association between schizophrenia and single nucleotide polymorphisms within the dysbindin gene has been reported [117]. In postmortem studies of schizophrenic brain samples significant reductions in dysbindin immunoreactivity within glutaminergic fields in the hippocampus has been seen [118]. In this latter study, it was hypothesized that disruption in dysbindin may alter normal neurotransmission within the hippocampus an effect that may contribute to cognitive deficits in schizophrenia [118]. A spontaneous mutation in the dysbindin gene has been identified in mice (*sdy* mouse). These mice display decreased locomotor activity and decreased social interaction and may represent an animal model of the negative symptoms associated with schizophrenia [119]. However, additional links to schizophrenia or the effect of antipsychotic medications in these mice has yet to be reported.

DISC1 Previous studies have identified specific chromosomal regions associated schizophrenia. One of these regions is on chromosome 1 (1q42) where a balanced (1;11) (q42.1; q14.3) translocation event has been associated with a psychopathology seen in schizophrenia as well as depression [120]. Two genes that fall within this disrupted region have been termed disrupted-in-schizophrenia 1 and 2 (DISC1 and DISC2 [121]). Multiple studies have linked alterations in the DISC1 gene with schizophrenia. For example, a frame shift mutation in the DISC1 gene resulting in a nonfunctional truncated product was associated with individuals having schizophrenia or schizoaffective disorder [122].

In an attempt to determine the effects of DISC1 disruption on behavior, multiple groups have generated mice with an altered DISC1 gene. For example, Li et al. (2007) generated transgeneic mice with an inducible c-terminal fragment of the DISC1 gene that, when expressed, can inhibit normal function of DISC1 [123]. Behavioral analysis of these mice demonstrated that when the DISC1 mutant was expressed during early postnatal development, deficits in spatial working memory, depressive traits, and social deficits were present [123]. Clapcote et al. generated two DISC1 mutants using N-nitroso-N-ethylurea mutagenesis. Of these mutants one showed profound deficits in PPI and LI latent inhibition both of which were reversed by haloperidol and clozapine [124].

Other Genes A number of other susceptibility genes have also been identified in schizophrenic populations. These include, but are not limited to, catechol-O-methyltransferase (COMT(), neuregulin 1 (NGR1), proline dehydrogenase (ProDH), regulator of G-protein signaling 4 (RGS-4) and trance amine receptor 4 (TAR4), [115,125]. As with the pharmacological models outlined above, no one gene likely represents the sole causative factor for schizophrenia. Moreover, the findings related to specific genes are often difficult to reproduce in genetic population studies adding additional complexity to the issue. For animal models, the mechanism of gene alteration (i.e., condition knockout, embryonic knockout, or regional knockout) may results in confounding artifacts such as compensatory up regulation of alternate genes or effects in multiple cells types. In addition, expression and subsequent protein function can be altered in a manner independent of any DNA anomalies (i.e., mutations or truncations). For example, epigenetic mechanisms (e.g., DNA methylation or histone acetylation) can affect overall protein expression without any detectable changes in the DNA sequence. For example, Costa et al. hypothesize that decreased expression of specific genes (e.g., GAD67 and reelin) seen in schizophrenia results not from alteration in the DNA but from hypermethylation of the promoter regions of these genes [126].

In summary, the genetic models represent a powerful tool for understanding pathological mechanisms and antipsychotic action in relation to schizophrenia. As with any experimental tool, interpretation of findings and prediction of antipsychotic efficacy data must be done with an awareness of limitations and caveats of these models.

4.1.4. Electroencephalogram (EEG) [127]

An emerging area of study, relating to cognitive deficits, is EEG changes reported in schizophrenic patients. EEG recordings reflect synchronization of neuronal firing within specific cortical areas. Resting EEG recordings can be divided into multiple frequencies ranging from low frequency theta (<5 Hz) to high frequency gamma (~ 40 Hz) oscillations. It has been proposed that changes in gamma oscillations may be a measure of attention and maintenance of working memory in humans [128]. In schizophrenic patients, evaluation of evoked gamma band oscillations shows an overall deficit in EEG power at 40 Hz when compared to nonschizophrenic patients, an effect hypothesized to result from deficits in GABAergic interneurons that help to synchronize neuronal firing [129,130].

Changes in positive and negative voltage deflections in response to presentation of a stimulus (visual or auditory) can also been measured by EEG. These evoked related potentials (ERPs) can be further divided into early events (deflections that occur within the first 50 ms which may reflect sensory-evoked responses) and later events (deflections that occur within 300–400 ms and may reflect cognitive-related components) [127]. Like gamma oscillations alterations in ERP both early (positive deflection at 50 ms; P50) and late (positive deflections at 300 ms; P300) have been reported in schizophrenic patients. As these measures represent a nonevasive tool for evaluating cognitive function in patients, the potential to develop a translatable animal model remains intriguing; however, issues with regard to animal models remain. For example, no deflection is seen in animals at 50 ms although deflections can be recorded at other time points (e.g., 40 milliseconds). However, the relationship between these changes and those seen at 50 ms in human remains to be determined [131].

4.2. Model of Side Effects

4.2.1. Catalepsy
Catalepsy refers to a drug-induced behavioral state whereby an animal placed in an awkward state will remain in that position. Typical protocols for evaluation of catalepsy in rodents involve placing the front paws on an elevated platform and measuring the length of time the animal remains in this fixed position. This effect can be induced in animals by administration of a D2 receptor antagonist, with haloperidol being the most commonly used agent. All typical antipsychotics induce a cataleptic state, likely reflecting their potent effects at the D2 receptor. Newer atypical antipsychotics have been shown to induce catalepsy only at high doses. Thus, catalepsy may represent a method for linking D2 receptor antagonism with a behavioral effect. As potent inhibition of D2 receptors may represent the mechanism for EPS (see above) this model is often used as a predictor of EPS potential with novel antipsychotics. Newer atypical agents appear to have minimal effects on catalepsy measures possibly due to their decreased affinity for the D2 receptor.

4.2.2. Paw Test
The paw test relies on the natural tendency of an animal to retract an extended limb. The test is conducted by placing the two forelimbs and two hind limbs into four separate holes and measuring the forelimb and hind limb retraction time (FRT and HRT, respectively). Typical antipsychotics such as haloperidol increased FRT and HRT with equal potencies whereas clozapine increased HRT at lower doses than are needed to increase FRT. Agents with no antipsychotic activity including diazepam, morphine and desipramine show no effect in the paw test [132]. With this model it is believed that the ability of drugs to prolong FRT is associated with a more potent inhibition of dopamine receptors in the dorsal striatum. In contrast, increased HRT may be associated with inhibition of dopamine receptors within the dorsal striatum, accumbens and olfactory tubercle [133]. Although most antipsychotics show activity in this assay it is not widely used for evaluation and screening of novel antipsychotic medications.

4.2.3. Vacuous Chewing Movements
In patients treated with antipsychotic medications, development of a late onset phenomenon, tardive dyskensia (TD; see above) characterized by repetitive, choreic, involuntary, and stereotypic behavior can develop. Development of TD is hypothesized to be related to both the percent and manner in which the D2 receptors are inhibited [134]. In rodents vacuous chewing movements (VCMs), purposeless movements of the mouth in the vertical plane with or without protrusion of the tongue, are believed to represent a model of TD. As with patients, chronic treatment with typical antipsychotics can result in the development of these abnormal facial movements. Typical antipsychotics show more development of VCM than classical atypical antipsychotics. For example, Gao et al. showed that chronic treatment with haloperidol to rats results in development of VCM whereas olanzapine and sertindole, administered for the same time period, did not result in a high incidence of oral dyskinesias, an effect consistent with human data [135]. Based on the relative low cost of this assay it remains a viable option for evaluation of development of TDS for antipsychotic medications.

4.2.4. Prolactin Response [72]
Under basal conditions dopamine inhibits prolactin release. This effect is mediated by D2 receptors in the pituitary. As such, administration of agents that inhibit D2 receptor activity will result in increased prolactin release. Typical antipsychotics tend to increase prolactin levels in a dose-dependent manner reflecting their potent inhibition of D2 receptors. Most, but not all, atypical antipsychotics, show little effect of prolactin levels in schizophrenic populations. Of note are risperidone and amisulpride that cause significant increases in serum prolactin (for review, see Ref. [72]). In animal models the effect on prolactin is less

clear. For example, following acute administration olazapine and risperidone both increase prolactin to a level comparable to haloperidol, although risperidone was more potent. Following chronic administration both risperidone and olazapine increased prolactin although the magnitude of effect may have been greater with risperidone [136]. In an alternate study it was demonstrated that olazapine but not aripiprazole increased prolactin secretion in rats, an effect consistent with the lack of effect of aripiprazole on prolactin in humans [137]. Thus, it would appear that animal models mimic, but do not replicate findings in schizophrenic populations. This could be due to numerous factors including acute versus chronic administration, stress or differences in drug exposure.

5. FDA APPROVED ANTIPSYCHOTIC DRUGS

5.1. Introduction

Older antipsychotic drugs included opiates, belladonna derivatives, bromides, barbiturates, antihistamines, and chloral hydrates. The first conventional antipsychotic drug chlorpromazine became available in 1952. Prochlorperazine, an analog of chlorpromazine, is now available as a generic drug. Another conventional antipsychotic drug haloperidol was discovered in 1958. Distinctive from chlorpromazine, haloperidol is a butyrophenone derivative which is more potent and has fewer side effects. Haloperidol remained one of the most prescribed neuroleptics 40 years after its discovery until the emergence of atypical antipsychotics. Unfortunately, conventional antipsychotic drugs are all liable to cause severe extrapyramidal symptoms (EPSs) including parkinsonian symptoms, akathisia, dyskinesia, and dystonia.

Atypical antipsychotics, also known as serotonin–dopamine antagonists, effectively reduce EPS. They are also believed to reduce the negative, cognitive, and affective symptoms of schizophrenia more effectively. All atypical antipsychotics are potent antagonists of serotonin 5-HT_{2A} and dopamine D_2 receptors, however, they also act on many other receptors including multiple serotonin receptors (5-HT_{1A}, 5-$HT_{1B/1D}$, 5-HT_{2C}, 5-HT_3, 5-HT_6, and 5-HT_7), the noradrenergic system (α_1 and α_2), the cholinergic system (M_1), and the histamine receptors (H_1). It has been postulated that the additional 5-HT_{1A} agonist activity shown by several atypical antipsychotic agents could reduce EPS and alleviate the anxiety that often precipitates psychotic episodes in schizophrenia patients. The challenge remains to determine which of these secondary pharmacological properties may lead to improved efficacy and which are undesired and account for the side effects. Clozapine, the first atypical antipsychotic, became available in 1959 and was followed by risperidone in 1993 and olanzapine in 1996. Additional atypical antipsychotics include quetiapine, ziprasidone, aripiprazole, zotepine, sertindole, and paliperidone. Among the FDA approved antipsychotics drugs, there are two types, namely, conventional antipsychotic agents and atypical antipsychotics [138–142].

5.2. Conventional Antipsychotic Drugs

There are currently three conventional antipsychotic agents on the market, chlorpromazine (**1**), prochlorperazine (**2**), and haloperidol (**3**).

1, Chloropromazine

The genesis of chlorpromazine (**1**) can be traced back to antihistamines: diphenhydramines in general and Benadryl in particular. The first was promethazine, an antihistamine. Systematic structure–activity relationship(SAR) investigations by enhancing promethazine's "side effects" in the >central nervous system (CNS) led to the synthesis of chlorpromazine in 1950. The structure of chlorpromazine differed only slightly from that of promethazine. Chlorpromazine (**1**) has an extra chlorine atom and a slight difference in the diamine side chain. It was introduced in 1952, in France and in 1954 in the United States under the trademark "Thorazine®." In the first 8 months, more than 2 million patients were administered the drug. It contributed to an 80% reduction of the resident population in mental hospitals. Thorazine added a great impetus to the beginning of the psychopharmacological revolution. Subsequently, chlorpromazine (1) was shown to be a potent dopamine 2 receptor (D_2), antagonist ($K_i = 3$ nM) with other pharmacological properties that were thought to cause unwanted side effects. Thus, the D_2-receptor antagonism of the conventional antipsychotic mediates not only their therapeutic effects but also some of their side effects.

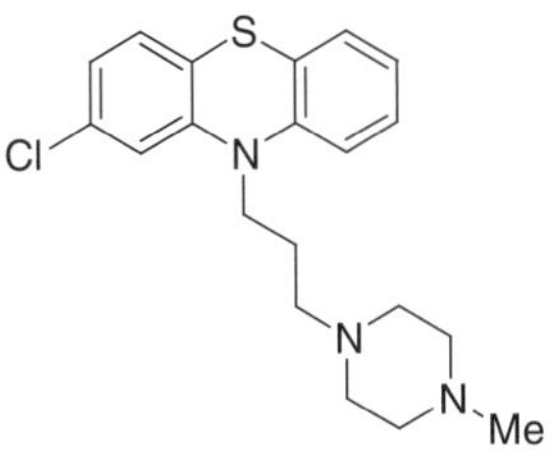

2, Prochlorperazine

Prochlorperazine (2) has been approved for the control of severe nausea and vomiting and for management of the manifestations of psychotic disorders. It is effective for the short-term treatment of generalized nonpsychotic anxiety.

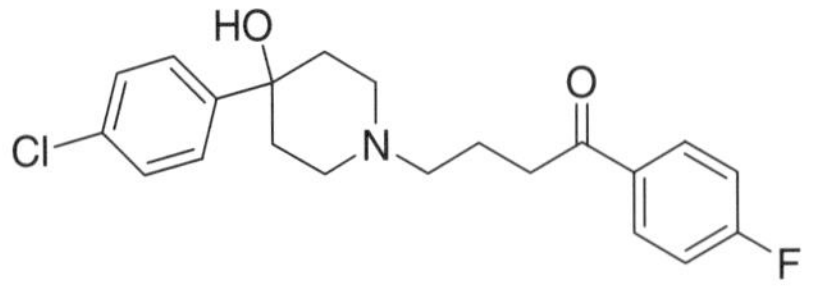

3, Haloperidol

Haloperidol (3) is 50–100 times more potent than chlorpromazine (1) and a more selective D_2 antagonist. The D_2-receptor blockade in the mesolimbic pathway is believed to reduce the positive symptoms of schizophrenia. More importantly, it was almost devoid of the antiadrenergic and other autonomic effects of chlorpromazine (1). However, haloperidol (3) is ineffective in treating the negative symptoms and neurocognitive deficits of schizophrenia. In addition, administration of the drug typically causes EPS. Thus, the D_2-receptor antagonism of the conventional antipsychotics mediates not only their therapeutic effects but also some of their side effects.

With the discovery of newer atypical antipsychotics, the older conventional antipsychotics are no longer used for first-line therapy, but can still be effective as second-line or add-on treatments.

5.3. Atypical Antipsychotic Drugs

Atypical antipsychotics, sometimes called serotonin–dopamine antagonists (SDAs) reduce EPS compared with conventional antipsychotics and are also believed to more effectively reduce negative, cognitive, and affective symptoms of schizophrenia. All atypical antipsychotics are potent antagonists of serotonin 5-HT_{2A} and dopamine D_2 receptors, however, they also act on many other receptors including multiple serotonin receptors (5-HT_{1A}, 5-$HT_{1B/1D}$, 5-HT_{2C}, 5-HT_3, 5-HT_6, and 5-HT_7), the noradrenergic system (α_1 and α_2), the cholinergic system (M_1), and the histamine receptors (H_1). The challenge remains to determine which of these secondary pharmacologic properties may be synergistic leading to improved efficacy, and which are undesired and account for the side effects. It is generally accepted that an atypical antipsychotic should combine a minimum of 5-HT_{2A} antagonism with D_2 antagonism in order to provide increased efficacy with fewer side effects. Serotonin–dopamine antagonists, but not conventional antipsychotics (dopamine antagonist without 5-HT_{2A} antagonism), increase dopamine release in the mesocortical pathway. This provides a possible explanation for the improved efficacy of atypical antipsychotics in the treatment of negative symptoms of schizophrenia. Furthermore, 5-HT_{2A} antagonism in the nigrostriatal pathway is believed to reduce EPS and tardive dyskinesia because dopamine release from this pathway is regulated by serotonin. If serotonin is not present at its 5-HT_{2A}-receptor on the nigrostriatal dopaminergic neuron, then dopamine is released. Compounds **4–11** represent atypical antipsychotics currently available on the market.

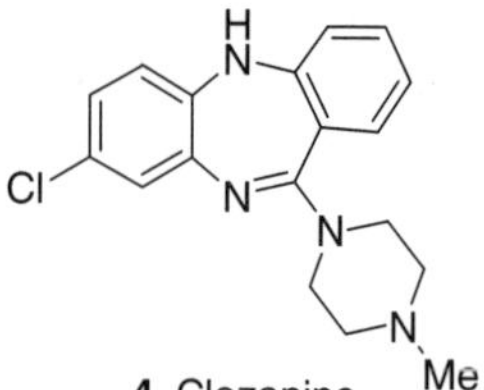

4, Clozapine

Clozapine (**4**), the first atypical antipsychotic was synthesized in 1967 by Sandoz–Wander chemists and marketed in 1972. It was removed from the market in 1975 due to drug-associated agranulocytosis, a potentially fatal blood disorder that results in lowered white-cell counts, which occurred in approximately 2–3% of patients. Additional side effects of clozapine therapy include sedation (H_1), weight gain (5-HT_{2C}) and orthostatic hypotension (α_1). Clozapine (**4**) was reintroduced in 1990 and is now relegated as a second-line treatment with extensive monitoring of the patient's blood cell count. However, over the years it has demonstrated efficacy against treatment-resistant schizophrenia and some still consider it to be the gold standard for treatment-refractory patients.

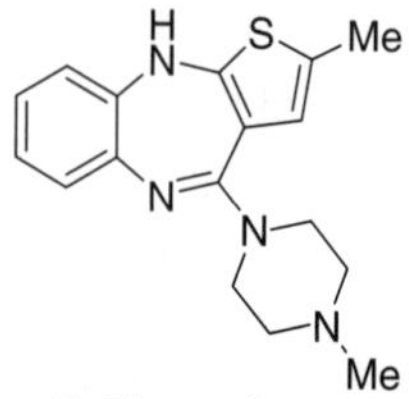

5, Olanzapine

Olanzapine (**5**) is a close analog of clozapine (**4**) where one of the benzene rings of the tricyclic nucleus is replaced with a methylthiophene ring. It has high affinity for the 5-HT_{2A}, 5-HT_{2C}, H_1 and M_1 receptors and moderate affinity for the D_2 and α_1 receptors. Olanzapine (**5**) is associated with high levels of weight gain (second only to clozapine) and it also causes some EPS at higher doses.

Similar to clonazapine (**4**) and olanzapine (**5**), risperidone (**6**), paliperidone (**7**), ziprasidone (**8**), and quetiapine (**9**) are currently considered four additional first-line drugs for psychosis and they will be highlighted in detail in this chapter. The newest antipsychotic to make its way to the market is aripiprazole (**10**). It has a slightly different mechanism of action than the atypical antipsychotic drugs in that it is a D_2 partial agonist rather than a full antagonist. Each of these drugs has a unique pharmacological and clinical profile therefore the clinician must balance the benefit-risk factors for each patient in determining which drug to prescribe.

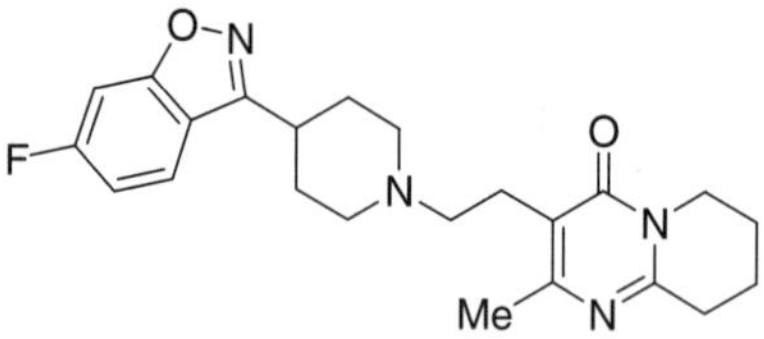

6, Risperidone

Risperidone (**6**) has high affinity for D_2, 5-HT_{2C} and α_1 receptors and a very high affinity for the 5-HT_{2A} receptor. It is the most likely of the atypical antipsychotics to cause prolactin increases, but has a lower weight gain liability than olanzapine (**5**) or quetiapine (**9**). Risperidone (**6**) has a relatively narrow therapeutic window since doses above 6 mg/day cause EPS in a dose-dependent manner.

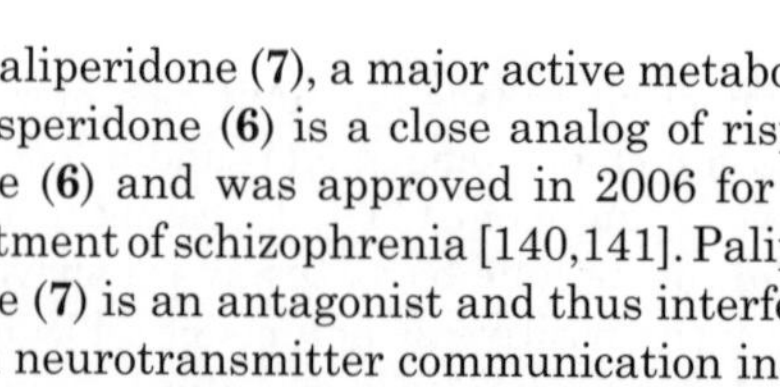

7, Paliperidone

Paliperidone (**7**), a major active metabolite of risperidone (**6**) is a close analog of risperidone (**6**) and was approved in 2006 for the treatment of schizophrenia [140,141]. Paliperidone (**7**) is an antagonist and thus interferes with neurotransmitter communication in the brain. It blocks D_2, 5-HT, and α_2 adrenergic receptors, all of which have been implicated in schizophrenia.

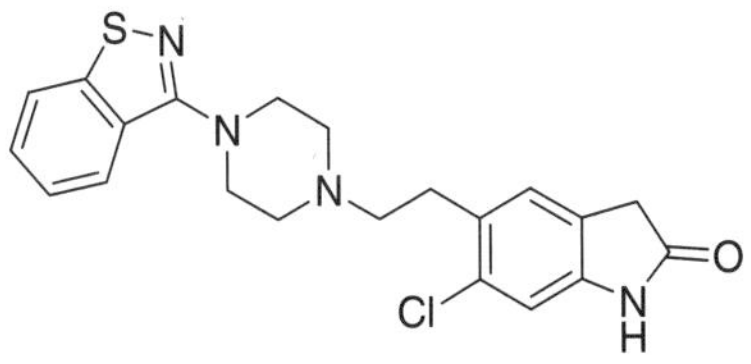

FDA APPROVED ANTIPSYCHOTIC DRUGS

8, Ziprasidone

Ziprasidone (**8**) has high affinity for the D_2 receptor, but even higher affinity for $5\text{-}HT_{2A}$ and $5\text{-}HT_{2C}$ receptors. Unlike other atypical antipsychotics, ziprasidone (**8**) also has potent $5\text{-}HT_{1B/1D}$ antagonist and $5\text{-}HT_{1A}$ partial agonist activity, as well as moderate SRI/NRI activity. This receptor profile suggests that ziprasidone (**8**) may be useful in relieving some of the depressive/anxious symptoms of schizophrenia. It has moderate affinity for the H_1 and α_1 receptors and negligible affinity for the M_1 receptor. Ziprasidone (**8**) is more likely to increase the QTc interval than other atypical antipsychotics, but it appears to have the lowest liability for body weight gain.

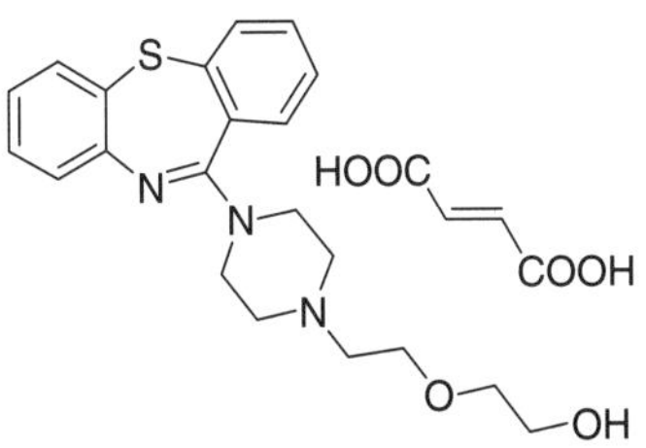

Aripiprazole (**10**) is a D_2 partial agonist with an intrinsic activity of approximately 30%. Therefore, it acts as an agonist on presynaptic autoreceptors, which have a high receptor reserve, and as an antagonist on D_2 postsynaptic receptors, where significant levels of endogenous dopamine exist and there is no receptor reserve. The intrinsic activity of 30% for aripiprazole (**10**) prevents D_2 blockade from rising more than 70%, which is more than the 65% D_2 occupancy needed for a clinical response but lower than the 80% D_2 occupancy where EPS is observed. Consistent with this partial agonist mechanism, EPS was not observed with aripiprazole even when striatal D_2 receptor occupancy values where more than 90%. Aripiprazole (**10**) can be considered atypical since it is also an antagonist at $5\text{-}HT_{2A}$ receptors. It is also a partial agonist at $5\text{-}HT_{1A}$ receptors which may provide some benefit against some of the negative symptoms of schizophrenia. Preliminary clinical studies have demonstrated that aripiprazole (**10**) is well tolerated and does not significantly induce EPS, weight gain, QT prolongation, or increase plasma prolactin levels.

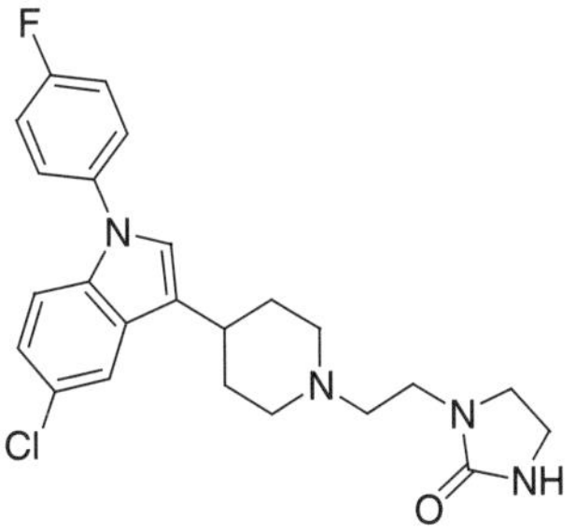

9, Quetiapine fumarate

Quetiapine (**9**) has the lowest affinity for the D_2 and $5\text{-}HT_{2A}$ receptors among the atypicals, and therefore relatively high doses are required for maximal efficacy. It causes significant weight gain, but less than that of olanzapine (**5**). Other side effects include sedation, dizziness, and hypotension.

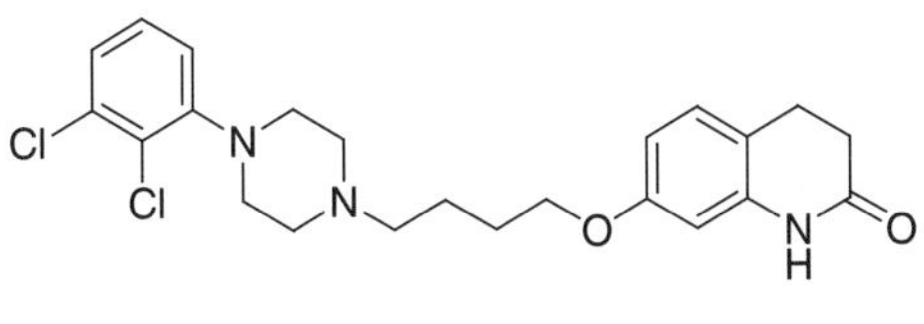

10, Aripiprazole

11, Sertindole

Sertindole (**11**), introduced in 1996, also belongs to this class of atypical antipsychotics, however, it is used less frequently. It has been shown to be efficacious for the treatment of positive and negative symptoms of schizophrenia. However, sertindole (**11**) has recently been withdrawn from the market because it causes significant prolongation of the QTc interval that may lead to a ventricular arrhythmia known as *torsades des pointes*.

Much remains to be discovered about the underlying pathophysiology of schizophrenia and there is still a great need for medicinal

180 ANTIPSYCHOTIC AGENTS

chemists to develop more selective drugs that are devoid of clinically limiting side effects and also address the cognitive impairment symptoms.

6. STRUCTURE–ACTIVITY RELATIONSHIPS

6.1. Drug Classes

In this section, we only summarize the structure–activity relationship of atypical antipsychotics because little effort is on-going in both academia and industry on conventional antipsychotic drugs. Old drug classes are not covered to give room for more coverage of atypical antipsychotics. The reader is referred to earlier editions for the structure–activity relationship of phenothiazines, thioxanthene, and butyrophenones. The SAR discussion is focused on atypical tricyclic neuroleptics and benzisoxazole, benzithiazole and related atypical neuroleptics. In comparison to conventional antipsychotic drugs known as potent D_2 antagonists, atypical neuroleptics have at least two distinctive features: an affinity for both the serotonin 2 receptor (5-HT_2) and D_2 receptor, thus the name serotonin–dopamine antagonists.

6.2. Atypical Tricyclic Neuroleptics

Further scrutiny of the full spectrum of pharmacology revealed that both conventional antipsychotic drugs and atypical neuroleptics modulate many G-protein coupled receptors (GPCRs) with different degrees of potency. Conventional antipsychotic drugs have at least four actions: blockade of D_2, blockade of muscarinic cholinergic receptors (M_1), blockade of α adrenergic receptors (α1), and blockade of histamine receptors (H_1), which explains the antihistaminic actions of the conventional antipsychotic drugs. On the other hand, atypical neuroleptics have even more complicated pharmacology. In addition to 5-HT_2 and D_2, they are known to have at least four other pharmacological actions: blockade of D (D_4 and D_1), α ($α_1$, $α_2$), M_1, and H_1.

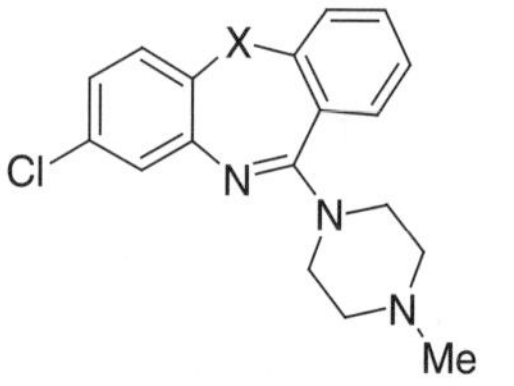

4, X = NH, clozapine
12, X = O, loxapine
13, X = CH_2, perlapine

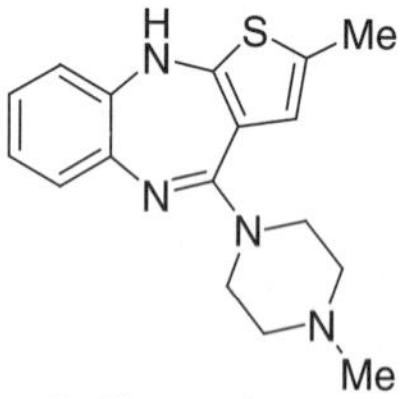

5, Olanzapine

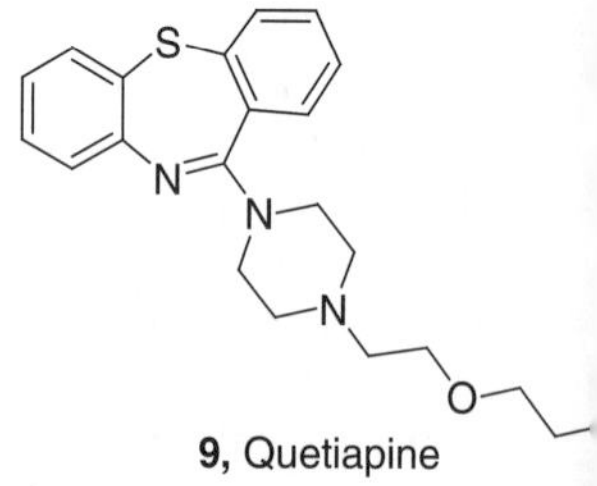

9, Quetiapine

Shown in Table 1 are the dissociation constants (pK_i) of atypical tricyclic neuroleptics. Because the values of IC_{50} and dissociation constants are dependent upon the radio-labeled ligands used and the methods employed; therefore, intrinsic dissociation constants are a better indication of the structure–activity relationship. Shown in Table 1 is the structure–activity relationship of many popular atypical tricyclic neuroleptics including clozapine (**4**), olanzapine (**5**), quetiapine (**9**), loxapine (**12**), and perlapine (**13**) [143–146].

Clozapine (**4**) is the prototype of atypical antipsychotics. According to Meltzer's "$5TH_{2A}/D_2$ hypothesis [147]," compounds having a pK_i ratio for $5TH_{2A}/D_2$ higher than 1.2 fall into the category of atypical antipsycho-

Table 1. Dissociation Constants (K_i) of Atypical Tricyclic Neuroleptics

	D_2 (nM)	D_4 (nM)	5-HT_{2A} (nM)	K_{D2}/K_{5-HT2A}	K_{D2}/K_{D4}
Clozapine (**4**)	44	1.6	11	4.00	28.00
Loxapine (**12**)	5.2	7.8	10.2	0.51	0.67
Perlapine (**13**)	60	30	30	2.00	2.00
Olanzapine (**5**)	3.7	2	5.8	0.64	1.85
Quetiapine (**9**)	310	1600	120	2.58	0.19

tics; therefore, clozapine (4) is an atypical antipsychotic. Because the affinity of clozapine (4) for a 5-HT$_2$ receptor is twice as high as for the dopamine D$_2$ receptors suggests a favorable property of atypical antipsychotics in that development of EPS caused by blockade of dopamine D$_2$ receptors is countered by blockade of central 5-HT$_2$ receptors. Blockade of 5-HT$_2$ receptors has also been suggested to be beneficial for treating the negative symptoms of schizophrenia. Clozapine (4) has an exceedingly complex pharmacology, interacting with high to moderate potency at the D$_4$ receptor, 5-HT$_{2A}$, 5-HT$_{2C}$ receptors, acetylcholine (muscarinic) receptors, adrenergic (α_1, α_2, and α_3) receptors, and histamine (H$_1$) receptors, as well as other receptors. Its analogs 5–19 possess an array of pharmacological activities. For example, loxapine (12) and clothiapine (17) are typical antipsychotics; and perlapine (13) is a hypnotic [148].

Loxapine (12) is the direct analog of clozapine (4) whose NH fragment is replaced by oxygen. Loxapine (12) is more potent than clozapine (4) for the D$_2$ receptor (5.2 nM versus 44 nM), but is less potent at the D$_4$ receptor (7.8 nM versus 1.6 nM). As a consequence, the ratios of K$_{D2}$/K$_{D4}$ are drastically different (0.67 versus 28.00), which also reflect their respective pharmacological and side effect profiles. Interestingly, chronic administration of loxapine (12) and clozapine (4) to rats for 4 weeks or 10 weeks did not produce enhancement of striatal dopamine receptor density [149]. However, there was a marked reduction (50–60%) of cortical serotonin receptor density associated with loxapine (12) or clozapine (4) administration. Acute doses of loxapine (12) and clozapine (4) produced the same potent effect. The possibility that these two antipsychotic drugs act via the serotonin system in the brain has been proposed. Perlapine (13), an old drug discovered in the 1970s as a sleep-promoting and sedative agent, is the carbon-analog of clozapine (4) that has weaker binding to the D$_2$, D$_4$, and 5-HT$_{2A}$ receptors. Although it was initially reported to lack antipsychotic efficacy, comparison of perlapine (13) with chlorpromazine (1), haloperidol (3) and clozapine (4) was carried out with regard to dopamine metabolism [150]. All four drugs produced a dose-dependent increase in levels of 3,4-dihydroxyphenylacetic acid (DOPAC) in two dopamine-rich structures, striatum and tuberculum olfactorium of rat. The potency of perlapine (13) was similar to that of chlorpromazine (1).

Olanzapine (5) is a close analog of clozapine (4) where one of the benzene rings of the tricyclic nucleus is replaced with a thiophene ring. It has high affinity for the 5-HT$_{2A}$, 5-HT$_{2C}$, H$_1$ ($K_i = 0.65$ nM) and M$_1$ receptors and moderate affinity for the D$_2$ ($K_i = 3.7$ nM) and α_1 receptors which fits well with the "5TH$_{2A}$/D$_2$ hypothesis." Olanzapine (5) is less potent as a D$_4$ antagonist relative to its D$_2$-antagonist properties, compared to clozapine (4). It is also much weaker than clozapine (4) as an α_1 (K_i 19 nM versus 7 nM), and α_2 antagonist relative to D$_2$, D$_4$, or 5TH$_{2A}$ antagonism [151,152]. Like clozapine (4), olanzapine (5) has high affinity for all five muscarinic (M) receptor subtypes. The anti-M$_1$ ($K_i = 26$ nM) appears to play an important role in the suppression of EPS although it does cause some EPS at higher doses. In addition, it is associated with high levels of weight gain [second only to clozapine (4)]. Recently, olanzapine (5) has been expanded to treat other psychiatric disorders such as bipolar disorder, anorexia nervosa and mood disorder.

Quetiapine (9) has the lowest affinity for the D$_2$ and 5-HT$_{2A}$ receptors among the atypicals, and therefore relatively high doses are required for maximal efficacy. However, it is still a *bona fide* atypical antipsychotic because it has a higher affinity for serotonin (5-HT$_{2A}$) receptors relative to dopamine (D$_2$) receptors in the brain [153–155]. Quetiapine's (9) pharmacological effects appear selective for mesolimbic and mesocortical dopamine systems, which are believed to be the area of the brain responsible for the therapeutic effects of antipsychotics. In contrast to most conventional antipsychotics and some atypical antipsychotics, quetiapine's (9) effects on the nigrostriatal dopamine system, which is responsible for the extrapyramidal (or motor) side effects, are minimal. Furthermore, quetiapine (9) also has minimal activity on dopamine receptors in the tuberinfundibular system, thereby avoiding the problem of hyperprolactinemia, which is common with the standard antipsychotics and some atypical antipsychotics. Because of these unique pharmacological properties, quetiapine (**9**) is an effective atypical antipsychotic agent with a relatively benign side effect profile.

EPS and less prolactin elevation than risperidone (**6**). Since amoxapine (**14**) is off-patent, it may be a valuable alternative to new atypical antipsychotics, especially in low-income coun-

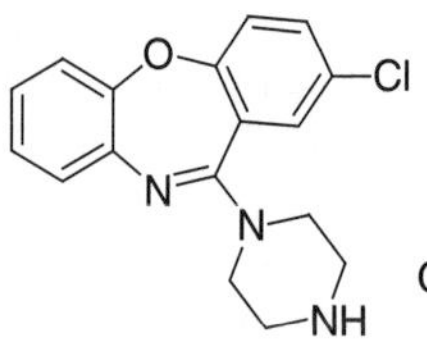

14, Amoxapine

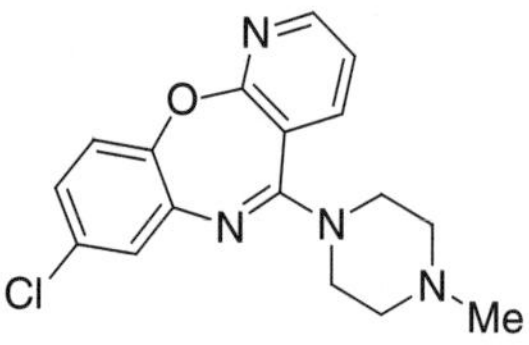

15, JL-13

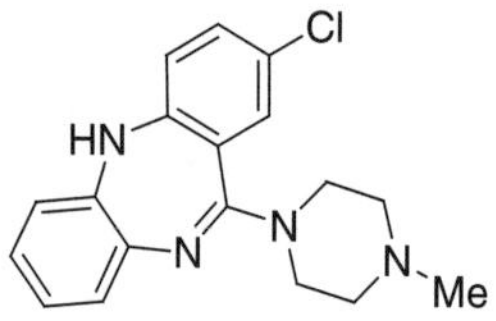

16, Isoclozapine

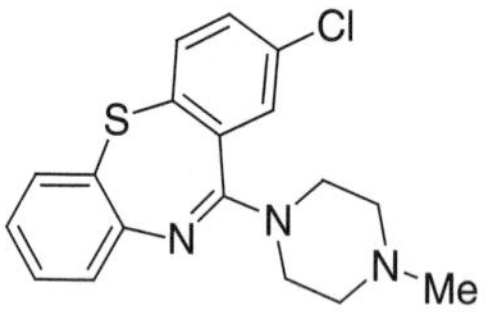

17, Clothiapine

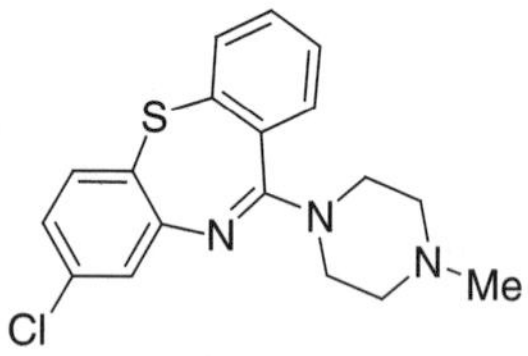

18, Isoclothiapine

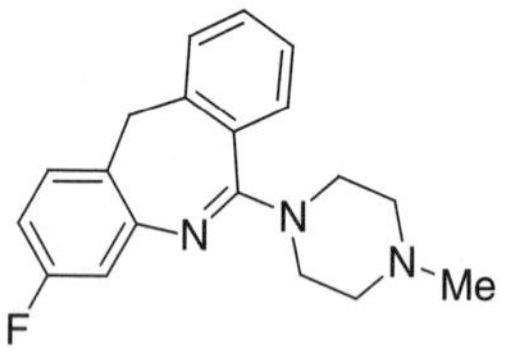

19, Fluperlapine

Shown in Table 2 are the pharmacological profiles of additional atypical tricyclic neuroleptics. Amoxapine (**14**) was initially marketed as an antidepressant, however, its *in vitro* profile, receptor occupancy and preclinical effects were found to be very similar to atypical antipsychotics. Meltzer et al. suggested that a combination of high affinity for serotonin (5-HT$_2$) antagonism along with modest dopamine (D$_2$) antagonism may provide one basis for atypical antipsychosis activity [151]. A comparative clinical study of amoxapine with risperidone (**6**) [156,157] demonstrated that amoxapine (**14**) showed efficacy as an atypical antipsychotic, improving positive, negative, and depressive symptoms. Amoxapine (**14**) was also associated with less

tries where the majority of patients are still treated with typical antipsychotics.

Pyridobenzoxazepine JL-13 (**15**) is the pyridyl analog of amoxapine (**14**) with an additional methyl group on the piperazine ring. It is slightly more potent than amoxapine (**14**) although chloro-substitution offered greater potency [158]. JL-13 (**15**) possessed an interesting preclinical profile that demonstrated a clozapine-like profile with less side effects; therefore, it could be a potential successor to clozapine (**4**).

Clothiapine (**17**), an older drug discovered in the late 1960s, has a 5TH$_{2A}$/D$_2$ ratio of 1.10, hence it is a borderline atypical antipsychotic and is considered as an atypical antipsychotic in some literature [159]. A comparative study

Table 2. K_i Values of D$_1$, D$_2$, and 5-HT Receptor Binding Site and Ratios for Atypical Tricyclic Neuroleptics

	D$_1$ (nM)	D$_2$ (nM)	5-HT$_2$ (nM)	D$_2$/D$_1$	5-HT$_2$/D$_2$
Amoxapine (**14**)	7.2	7.7	8.9	1.08	1.02
JL-13 (**15**)	5.04	4.96	4.65	0.98	0.94
Isoclozapine (**16**)	7.55	7.90	8.75	1.05	1.11
Clothiapine (**17**)	7.85	8.35	9.23	1.06	1.10
Isoclothiapine (**18**)	6.62	7.52	8.38	1.14	1.11
Fluperlapine (**19**)	6.85	6.88	8.41	1.00	1.22

of the effect of clozapine (4) and clothiapine (17) using different preparations of guinea pig and rat isolated organs was carried out [160]. It was found that clothiapine (17) was a competitive antagonist at the 5-HT receptor, a noncompetitive antagonist at the dopamine and histamine receptors while clozapine (4) was a noncompetitive antagonist at 5-HT, dopamine, and histamine receptors. Interestingly, isoclothiapine (18) has a similar pharmacological profile as clothiapine (17) although the position of the chlorine atom has been switched between the two benzene rings. In addition, isoclothiapine (18) is found to be a potent antagonist of the histamine H_4 receptor and isoclothiapine (18, $pK_i = 5.69$) is significantly less potent as an antagonist for the histamine H_4 receptor in comparison to clozapine (4, $pK_i = 6.75$) [161].

Fluperlapine (19) is the carbon-analog of clozapine (4). In the striatum of rats, it binds less to dopamine D_2 receptor sites, but it enhances dopamine-turnover more than clozapine [162,163]. Like clozapine (4) and unlike haloperidol (3), it is equally active in the striatum, the nucleus accumbens and the cortex. Unlike clozapine (4), it does not significantly enhance noradrenaline (NA) or 5-HT turnover, and it does not increase prolactin blood levels significantly.

PCP and analogs such as ketamine have long been known to induce schizophrenia-like symptoms via antagonism of the N-methyl-D-aspartate (NMDA) subtype of glutamate receptors in normal adult subjects. Fluperlapine (19), along with olanzapine (5), loxapine (12), and amoxapine (14), can prevent NMDA antagonist neurotoxicity in the rat to mimic the antipsychotic properties of clozapine (4) [164].

6.3. Benzisoxazole, Benzithiazole, and Related Atypical Neuroleptics

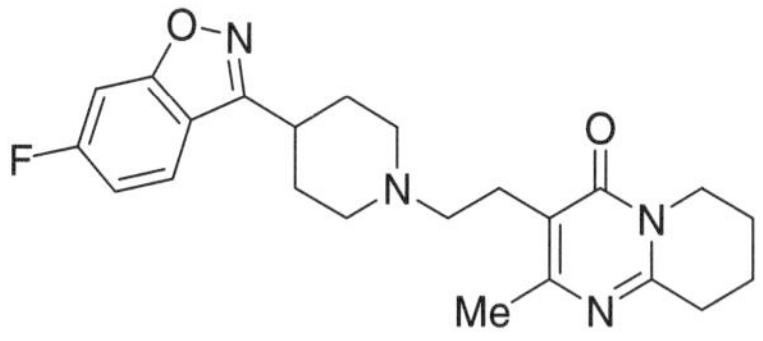

6, Risperidone

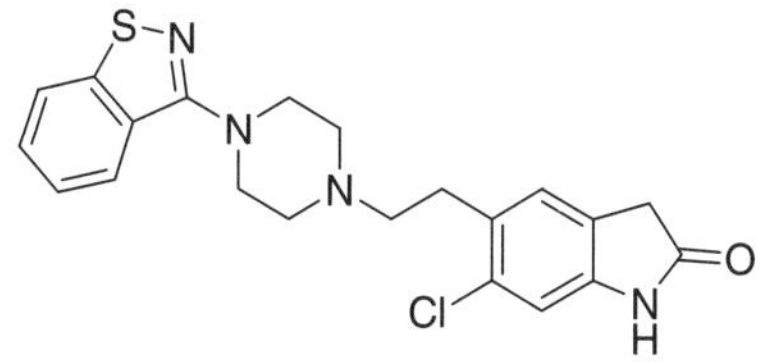

8, Ziprasidone

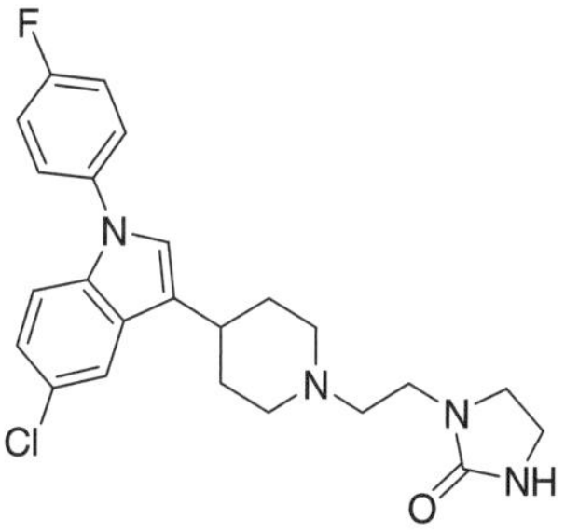

11, Sertindole

In addition to common pharmacological features, most novel atypical neuroleptics have high affinities for additional receptors [165]. While clozapine (4) and olanzapine (5) are most potent for the dopamine D_1 receptors with K_i values of 53 and 10 nM, respectively, ziprasidone (8) and sertindole (11) have high relative affinities for 5-HT$_{2C}$ receptor with K_i values of 0.55 and 0.51 nM, respectively (Table 3). On the other hand, risperidone (6) has a high affinity for α_2-adrenoceptors with a K_i value of 1.8 nM.

Risperidone (6), synthesized in 1984 by Janssen Pharmaceuticals, was one of the first benzisoxazole atypical neuroleptics and is a mixed serotonin–dopamine antagonist. In comparison to clozapine (4), risperidone (6) has much higher affinity for dopamine D_2 and

Table 3. K_i Values of D_1, D_2, and 5-HT Receptor Binding Site and Ratios for Benzisoxazole, Benzithiazole, and Related Atypical Neuroleptics

	D_2 (nM)	5-HT_{2A} (nM)	D_2/5-HT_{2A}	D_1 (nM)	5-HT_{2C} (nM)
Risperidone (6)	0.44	0.39	1.13	21	6.4
Ziprasidone (8)	2.8	0.25	11.2	9.5	0.55
Sertindole (11)	0.45	0.20	2.25	12	0.51

5-HT_2 receptors while clozapine (4) only has moderate affinity for dopamine D_2 and 5-HT_2 receptors. Therefore, risperidone (6) is used in lower dose than clozapine (4) because the doses that are recommended for the treatment of schizophrenia are related to the dopamine D_2 receptor affinity. Resperidone studies in dogs reveal potent dopamine-D_2 antagonistic activity with excellent oral bioavailability and a relatively long duration of action [166]. Risperidone (6) possesses the complementary clinical effects of a ritanserin (20)-like serotonin-5-HT_2 and a haloperidol-like dopamine-D_2 antagonist. 5-HT_2 antagonism may improve the quality of sleep, reduce negative and affective symptoms in schizophrenic patients, and decrease extrapyramidal symptoms induced by classical neuroleptics. Since risperidone (6) is a dopamine-D_2 antagonist, antidelusional, antihallucinatory, and antimanic actions are expected. Clinical studies indicate that two additional therapeutic targets, which are not achieved with classical neuroleptics, may be obtained with risperidone (6) in the monotherapy of schizophrenia and related disorders, that being very important contact and mood-elevating properties and EPS-free maintenance therapy.

20, Ritanserin

Ziprasidone (8) [167,168] appears to discriminate from other atypical antipsychotics by its low propensity for weight gain and by the availability of a short-acting intramuscular formulation. In concert with most atypical antipsychotics, it has a receptor binding profile characterized by high affinity to serotonin (5-HT_{2A}) and dopamine (D_2) receptors. However, in contrast to other atypical antipsychotics, its binding is much more substantial (~11 times) to 5-HT_{2A} than D_2. Ziprasidone's (8) binding profile for serotonin receptors is complex and it also has potent affinity for 5-HT_{1A}, 5-HT_{1D}, and 5-HT_{2c}. It functions as an agonist at the 5-HT_{1A} receptor and as an antagonist at the remaining 5-HT receptors. As mentioned before, the additional 5-HT_{1A} agonist activity shown by several atypical antipsychotic agents could reduce EPS and alleviate the anxiety that often precipitates psychotic episodes in schizophrenia patients. Furthermore, ziprasidone (8) is also an antagonist at α_1-adrenoreceptors. In a manner that is to some extent analogous to the newer antidepressant medications, it also inhibits reuptake at serotonin and norepinephrine receptor sites. It has relatively little affinity for histaminergic receptors.

As a consequence of the aforementioned pharmacological profile, ziprasidone (8) is effective in decreasing the positive and negative symptoms of schizophrenia, as well as treating symptoms of anxiety and depression that are often associated with schizophrenia. The receptor binding profile (e.g., agonist for 5-HT_{1A}) also predicts a low propensity for extrapyramidal side effects. However, it is not simple to pinpoint which pharmacological attributes are responsible for ziprasidone's (8) weight neutral profile.

Sertindole (11) [169] is a phenylindole-derived atypical antipsychotic that has marked affinity for dopamine D_2 receptors, serotonin 5-HT_2 receptors and α_1-adrenoceptors. Indeed, the fundamental mechanism of sertindole action is considered to be selective inhibition of dopamine D_2 receptors in the mesolimbic system (ventral tegmental area) versus the nigrostriatum, together with inhibition of CNS serotonin 5-HT$_2$ receptors and α_1-adrenoceptors. *In vitro* studies revealed sertindole (**11**) to have high affinity for dopamine D$_2$ receptors, 5-HT$_{2A}$ receptors, 5-HT$_{2C}$, and α_1-adrenoceptors, with binding affinities of 0.45, 0.20, 0.51, and 1.4 nM, respectively [165]. Sertindole is an inverse agonist at the 5-HT$_{2C}$ receptor and has reduced agonist binding to these receptors [170,171]. It has been suggested that this inverse agonism may be a mechanism by which atypical antipsychotics such as sertindole (**11**) improve negative symptoms in patients with schizophrenia. Dopaminergic neurons in the ventral tegmentum mediate the antipsychotic effects of antipsychotic agents, whereas in the nigrostriatum, such neurons can mediate EPS associated with antipsychotic compounds. The neuropharmacological profile of sertindole (**11**) therefore suggests antipsychotic activity, but with fewer EPS than conventional antipsychotics.

In patients with schizophrenia, sertindole (**11**) occupancy of striatal D$_2$ receptors were significantly ($p < 0.05$) lower (61% occupancy) than haloperidol (**3**, 87%) or high-dose risperidone (**6**, 75%) and higher than clozapine (**4**, 33%) in one single-photon emission computerized tomography (SPECT) study [172]. In another SPECT study, sertindole occupancy was significantly ($p < 0.05$) higher than clozapine (**4**) and olanzapine (**5**) and similar to risperidone (**6**) [36,173]

In Europe, sertindole (**11**) marketing was voluntarily suspended by the manufacturer in 1998. Approval has not been achieved in the United States because of concerns over risk of QTc prolongation that could lead to cardiac arrhythmia and sudden death.

In summary, these atypical neuroleptics may be divided into four groups according to their pharmacological profiles [143].

(a) *"Loose" Neuroleptics Displaceable by Endogenous Dopamine*: This group includes those atypical neuroleptics that have low affinity at D$_2$ and thus may be readily displaced by high endogenous concentration of dopamine in the caudate/putmen. This group includes clozapine (**4**) and perlapine (**13**).

(b) *Combined Block of D$_2$- and Muscarinic Receptors*: This small group includes clozapine (**4**) and thioridazine, which strongly block both D$_2$- and muscarinic receptors. Clozapine (**4**), for example, is in the order of 20–50-fold more potent in blocking muscarinic acetylcholine receptors than blocking dopamine D$_2$ receptors.

(c) *Combined Block of D$_2$- and Serotonin 5-HT$_{2A}$-Receptors*: The blockade of serotonin increases the release of dopamine. Clozapine (**4**) and olanzapine (**5**) could be viewed as selective inhibitors of the serotonin 5-HT$_{2A}$ receptor.

(d) *Selective Block of Dopamine D$_4$ Receptor*: A fourth possible mechanism for atypical neuroleptic action may be the selective blockade of dopamine D$_4$ receptors. Again clozapine (**4**) belongs to this group, as do perlapine (**13**) and olanzapine (**5**). Not surprisingly, different binding affinities towards these GPCRs are also responsible for their different pharmacological and side-effect profiles.

7. PHARMACOKINETICS, BIODISTRIBUTION, AND DRUG–DRUG INTERACTIONS

7.1. General Considerations

The collection of antipsychotic agents available today is chemically varied; however, their absorption, distribution, metabolism, and elimination properties are similar. In general, these drugs are well absorbed and highly distributed. The extent of plasma protein binding typically ranges from 74% to 99% with many of the drugs having binding in excess of 90%. The drugs are extensively metabolized typically by cytochrome P450 (CYP) oxidation and the drugs are predominately eliminated as metabolites. Many of the drugs have been shown to be substrates of CYP2D6 and CYP3A4, as well as inhibitors of CYP2D6 and aldehyde oxidase. In several cases, circulating metabolites are pharmacologically active and contribute to the overall antipsychotic activity. Two notable exceptions to these generic characteristics are the atypical antipsychotics amisulpride and paliperidone that are both eliminated predominately intact in urine; moreover, amisulpride is not significantly bound to plasma proteins (protein binding 17%). In addition, due to extensive first pass metabolism and/or in order to prolong exposure, many of the antipsychotics are available in oral and intramuscular formulations. In many cases, the pharmacokinetic parameters are highly variable between subjects due to first pass metabolism and extensive metabolism. Therefore, the pharmacokinetic parameters provided are representative and wherever possible, multiple references have been provided which span the range of values. In addition to the information provided for each antipsychotic agent, there are several excellent review articles describing the pharmacokinetics [174–177], pharmacogenetics [178], metabolism [179–182], and drug–drug interaction potential [183–188] of past and present antipsychotic agents.

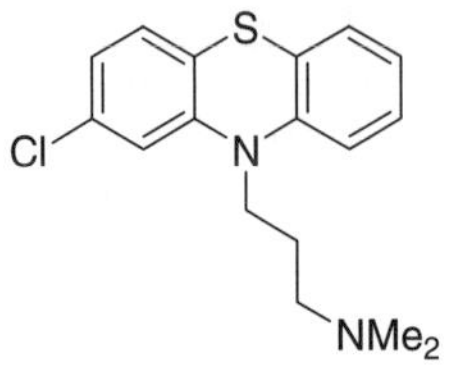

7.2. Conventional (Traditional) Antipsychotics

7.2.1. Phenothiazines The first class of antipsychotic agents was the phenothiazines that consisted of several chemical analogs. Chlorpromazine (**1**) (Thorazine) was the original drug in the phenothiazine class. The average dosages of this drug range from 400 to 600 mg/day (in one to four divided doses). It has been observed that there is a wide interpatient variability of chlorpromazine exposure with dose; hence, the following pharmacokinetic parameters are highly variable [166]. The time to peak plasma concentration is 1–4 h with a C_{max} of 25–150 ng/mL (100 mg bid, 33 days of dosing). The oral clearance of chlorpromazine is 0.52 L/h/kg with a half-life of 30 h. Chlorpromazine has an oral bioavailability of 32%, a volume of distribution of 21 L/kg, and its plasma protein binding is 95–98% [174,189–191].

1, Chlorpromazine

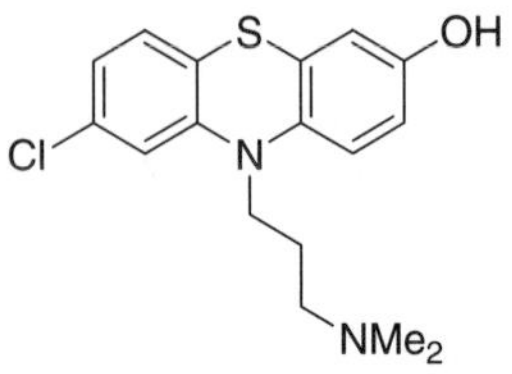

21, 7-Hydroxychloropromazine

Chlorpromazine (**1**) undergoes extensive first pass hepatic metabolism and <1% of the parent drug is eliminated in urine. Metabolic pathways involve N-demethylation, N-oxide, and S-oxide formation, hydroxylation to multiple phenols (that are subsequently glucuronidated), reductive dechlorination, and quaternary N-glucuronide formation on the aliphatic amine. The major metabolites are the N-desmethylchlorpromazine, chlorpromazine N-oxide, chlorpromazine S-oxide, N,N-didesmethylchlorpromazine, and 7-hydroxy chlorpromazine (**21**) (along with its glucuronide conjugate) [178,180,181]. Although many metabolites are present in the systemic circulation, the 7-hydroxy chlorpromazine metabolite is the only metabolite that likely contributes significantly to the overall activity of the drug. *In vitro* studies indicate that CYP2D6 is responsible for formation of the 7-hydroxy chlorpromazine metabolite with a minor contribution from CYP1A2. Flavin monooxygenase (FMO) and CYP3A4 are predominately involved in the formation of the N- and S-oxide metabolites, respectively [192–194]. Only minor increases in peak plasma concentration and exposure were observed when chlorpromazine was coadministered with quinidine (CYP2D6 inhibitor) indicating that this enzyme is not predominately involved in the elimination of chlorpromazine [172,195]. *In vitro* studies with human liver microsomes indicate that chlorpromazine is a weak inhibitor of CYP2D6 ($IC_{50} = 1.7\,\mu M$) and that chlorpromazine is an inhibitor of cytosolic aldehyde oxidase ($IC_{50} = 0.6\,\mu M$) [185,187,196].

There are several other notable examples of the phenothiazine chemotype, such as thioridazine (**22**), fluphenazine (**24**), trifluoperazine (**25**), and perphenazine (**26**). The average thioridazine (**22**) (Mellaril®) dosages range from 200 to 800 mg/day (in divided doses). Similar to chlorpromazine there is a wide interpatient variability of thioridazine exposure with dose. The time to peak plasma concentration is 1.9 hours with a C_{max} of 48 ng/mL (25 mg qd). The oral half-life of thioridazine is variable with an average approximately 24 h [174,197,198].

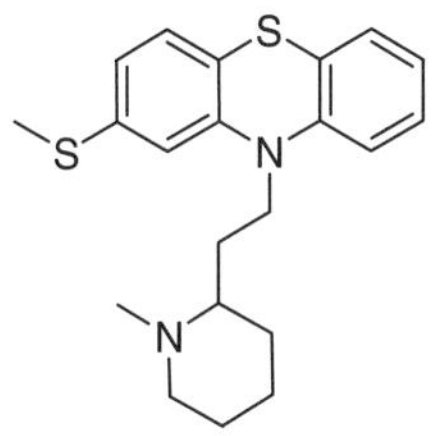

Thioridazine is metabolized to the 2-sulfoxide (**23**) (mesoridazine) and the 2-sulfone (sulforidazine), as well as the 5-sulfoxide (ring sulfoxide) and N-desmethylthioridazine. Both the 2-sulfoxide (mesoridazine, also marketed as an antipsychotic) and the 2-sulfone (sulforidazine) contribute significantly to the overall activity of the drug and the 5-sulfoxide has been shown to cause cardiotoxicity [178,180,181,199,200]. In vitro studies indicate that CYP2D6 is responsible for formation of the 2-sulfoxide and the 2-sulfone metabolites with a minor contribution from CYP3A4 in the formation of the 2-sulfone. CYP1A2 and CYP3A4 are predominately involved in the formation of the N-desmethyl and 5-sulfoxide metabolites [201]. The pharmacokinetics of thioridazine was studied in slow and rapid metabolizers of the CYP2D6 phenotype. Slow metabolizers demonstrated 2.4- and 4.5-fold higher C_{max} and AUC levels of thioridazine compared to rapid metabolizers, respectively. In addition, lower exposures of the two active metabolites (2-sulfoxide/sulfone) were observed in slow metabolizers indicating the importance of CYP2D6 in this metabolic pathway [175]. In a second study, fluvoxamine (CYP1A2 inhibitor) was found to increase thioridazine (and 2-sulfoxide/sulfone) plasma levels approximately threefold, most likely due to reduced metabolism of thioridazine by CYP1A2 and increased metabolism through the alternate CYP2D6 pathway [179]. In vitro studies with human liver microsomes indicate that thioridazine is an inhibitor of CYP2D6 ($IC_{50} = 0.36\,\mu M$) and cytosolic aldehyde oxidase ($IC_{50} = 0.16\,\mu M$) [185,187,196].

Another phenothiazine, similar to the aforementioned compounds is fluphenazine (Prolixen®, Permitil®) that also exhibits a great deal of interindividual variability in pharmacokinetics. The dose range for fluphenazine is 0.5–30 mg/day. The immediate release formulation gives a C_{max} of 2.3 ng/mL and a T_{max} of 2.8 h after a 12 mg dose. The oral half-life is ~18 h and the compound has a clearance of 0.6 L/h/kg. The volume of distribution is 11 L/kg and the bioavailability is low, 2.7% [174,202–206]. Fluphenazine is extensively metabolized and negligible parent drug is found in urine. The circulating metabolites identified to date include the S- and N-oxides, as well as a 7-hydroxy metabolite [205]. Although little is known about the enzymes involved in the biotransformation of fluphenazine, a clinical study did show increased clearance of fluphenazine in smokers compared to nonsmokers [207]. These results indicate that CYP1A2 most likely has some involvement in the elimination of fluphenazine. In addition, coadministration of fluoxetine (CYP2D6 inhibitor) increased the plasma exposure of fluphenazine 65% indicating that CYP2D6 may be involved in the biotransfor-

22, Thiordiazine

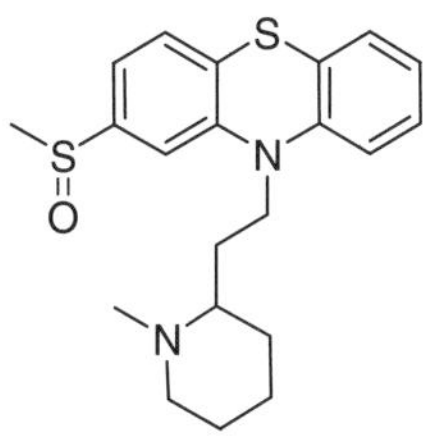

23, Thioridiazine S-Oxide

mation of this compound [208]. *In vitro* studies with human liver microsomes indicate that fluphenazine is a moderate inhibitor of CYP2D6 ($K_i = 9.4\,\mu M$) [210].

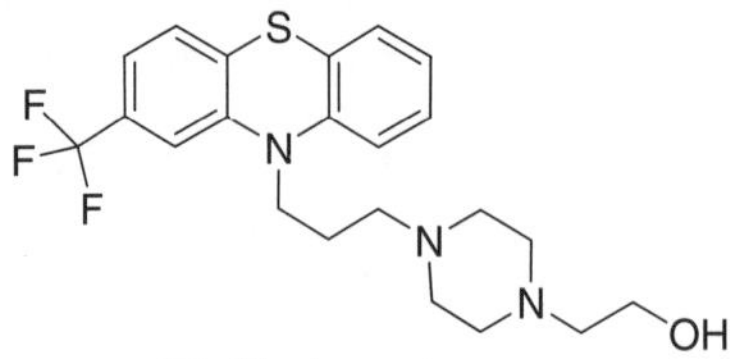

24, Fluphenazine

The dose range for perphenazine (**25**) (Trilafon®) is 8–32 mg/day. The drug gives a C_{max} of 0.28 ng/mL and a T_{max} of 2–4 h after a 6 mg dose [174,209–211]. Perphenazine metabolism is similar to the other phenothiazines (*N*-desmethyl, *S*- and *N*-oxides, and a 7-hydroxy metabolite which is eliminated as a glucuronide conjugate) [203,212]. The 7-hydroxy metabolite is active; however, it is unknown how much it contributes to the antipsychotic activity. The *N*-demethylation reaction has been shown to be catalyzed by several CYP enzymes: CYP1A2, 2C19, 2D6, and 3A4 [212]. Coadministration of paroxetine (CYP2D6 inhibitor) increased the C_{max} and AUC exposure of perphenazine by 6- and 7-fold, respectively [211]. In addition, the C_{max} and AUC of perphenazine were increased in CYP2D6 poor metabolizers by 3.3- and 4-fold, respectively [209]. Combined, these *in vivo* studies indicate that CYP2D6 is involved in the biotransformation of this compound. *In vitro* studies with human liver microsomes indicate that perphenazine is an inhibitor of CYP2D6 ($IC_{50} = 0.33\,\mu M$) and a very potent inhibitor of aldehyde oxidase ($IC_{50} = 0.033\,\mu M$) [185,187,196].

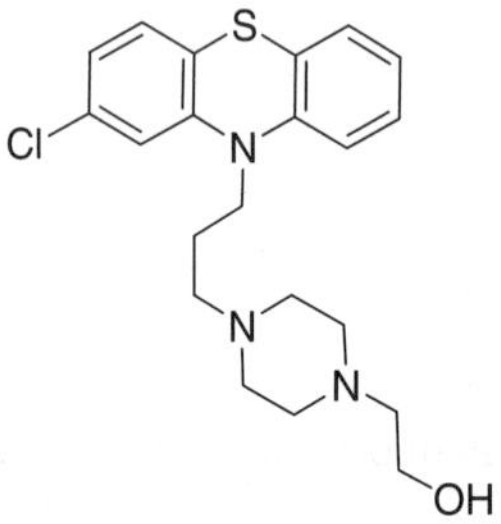

The final compound in this chemical class is trifluoperazine (**26**) (Stelazine®) that has a dose range 15–40 mg/day (qd or bid). The drug gives a C_{max} of 1.1 ng/mL and a T_{max} of 2.5 h after a 5 mg dose. The drug has an oral half-life of 12.5 h, a clearance of 8.6 L/h/kg, and a volume of distribution of 122 L/kg [174,213]. Trifluoperazine is metabolized by *N*-demethylation, *S*- and *N*-oxidation, 7-hydroxylation, and direct glucuronidation to form a quaternary *N*-glucuronide [214]. Little is known about the enzymes involved in the oxidative metabolism of trifluoperazine, however, formation of the *N*-glucuronide has been shown to be catalyzed by UGT1A4 [215]. *In vitro* studies indicate that trifluoperazine is a potent inhibitor of aldehyde oxidase ($IC_{50} = 0.24\,\mu M$) [215].

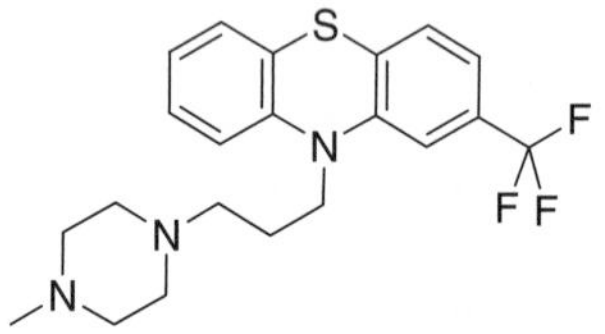

25, Perphenazine

26, Trifluoperazine

7.2.2. Thioxanthenes The thioxanthene class is similar in structure to the phenothiazines with the addition of a double bond in the *cis* configuration. Thiothixene (**27**) (Navane®) is given in dosages of 20–60 mg/day. Very little information is available about the pharmacokinetics or metabolism of thiothixene. After oral administration of thiothixene the time to peak plasma concentration was 2.2 h with a C_{max} of 27 ng/mL (20 mg dose). The clearance of thiothixene is 7.23 L/h/kg, the half-life is 13.7 h, and both values were found to be highly variable [174,216,217]. Thiothixene undergoes metabolism to two identified metabolites: thiothixene sulfoxide and *N*-desmethyl thiothixene [218,219]. The clearance of thiothixene is higher in patients who smoke, higher in males versus females, and higher in the young population versus the elderly population [220]. The clearance is also higher in patients who are coadministered carbamazepine possibly indicating involvement of CYP3A4 in the elimination of the compound [220]. *In vitro* studies with human liver microsomes indicate little potential for thiothixene to inhibit the major CYP enzymes [187].

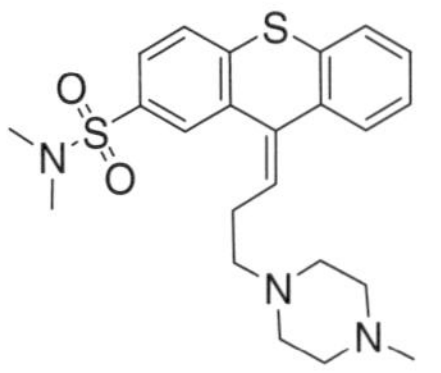

27, Thiothixene

7.2.3. Dibenzoxazepines Loxapine (**12**) (Loxitane®) is a conventional antipsychotic that is structurally very similar to the atypical antipsychotic clozapine. The standard dose of loxapine ranges 20–100 mg/day, taken in divided doses. Loxapine is rapidly absorbed with a time to peak plasma concentration of 2 h and a C_{max} of 9.8 ng/mL (25 mg solution dose). The half-life of the drug is reported to be biphasic with values of 5 and 12–19 h [174,221]. Additional information regarding volume of distribution, protein binding, and drug–drug interactions are not well documented for loxapine. The drug is extensively metabolized with no parent drug found in urine or feces. Routes of metabolism include N-demethylation to form amoxepine (an antidepressant), hydroxylation to the 7- and 8-hydroxy metabolites which undergo further glucuronidation and sulfation, N-oxidation at the 4-piperidyl nitrogen, and formation of a quarternary N-glucuronide [221–223]. The drug is predominately eliminated in urine as a variety of conjugates derived from the hydroxylated metabolites. Several of the metabolites (amoxepine and the hydroxylated metabolites) have been shown to have antipsychotic properties and most likely contribute to the pharmacology of the drug. Loxapine has been shown to be an inhibitor of aldehyde oxidase ($IC_{50} = 2.3\,\mu M$), however, the efficacious plasma concentrations of loxapine are generally low and drug interactions are not anticipated [185].

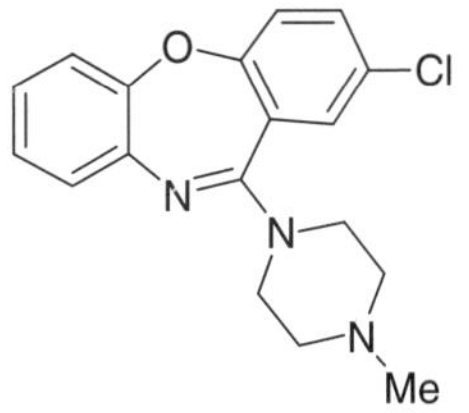

12, Loxapine

7.2.4. Butyrophenones Haloperidol (**3**) (Haldol®) dosages range from 2–20 mg/day. The time to peak plasma concentration is variable and ranges 1.7–6.1 h with a C_{max} of 9.2 ng/mL (20 mg dose). The oral clearance of haloperidol is 0.71 L/h/kg with a half-life of 18.5 h. Haloperidol has an oral bioavailability of 60%, a volume of distribution of 18 L/kg, and its plasma protein binding is 92% [174,189,224,225].

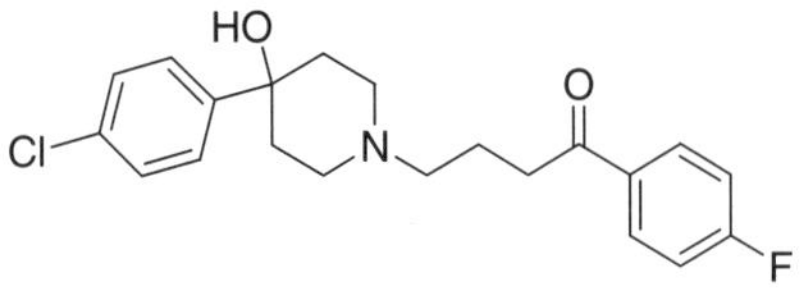

3, Haloperidol

Haloperidol undergoes extensive first pass hepatic metabolism to a variety of metabolites via several metabolic pathways. These metabolic pathways include: N-dealkylation (with formation of *para*-fluorobenzoylpropionic acid and 4-(4-chlorophenyl)-4-hydroxypiperidine), formation of a pyridinium metabolite, stereospecific reduction of the ketone to the alcohol (S-(−) isomer), and glucuronidation of the hydroxyl. All of these metabolites are inactive; however, the reduced haloperidol can be reoxidized to form the parent drug [178,180,181,189,224,225]. The major circulating metabolites are the glucuronide conjugate and reduced haloperidol. Approximately 30% of a dose of haloperidol is excreted in urine, predominately as the glucuronide conjugate and < 1% of the dose is excreted as parent drug. *In vitro* studies indicate that CYP3A4 is responsible for formation of the *para*-fluorobenzoylpropionic acid and 4-(4-chlorophenyl)-4-hydroxypiperidine metabolites, the pyridinium metabolite, and the reoxidation of reduced haloperidol; moreover, the formation of reduced haloperidol is by cytosolic carbonyl reductase [225–227]. Although CYP2D6 has been implicated in formation of the *para*-fluorobenzoylpropionic acid and 4-(4-chlorophenyl)-4-hydroxypiperidine metabolites, as well as the pyridinium metabolite, the literature is not well defined in this regard [225–227]. However, increases in peak plasma concentration and exposure were observed when haloperidol was coadministered with fluoxetine or quinidine (CYP2D6 inhibitors) and when administrered to poor metabolizers of the CYP2D6 phenotype [225]. Therefore, although the *in vitro* data are not clear in regards to CYP2D6-mediated metabolism of haloperidol, the *in vivo* studies clearly indicate that CYP2D6 is involved in the elimination of haloperidol. *In vitro* studies with human liver microsomes indicate that haloperidol and reduced haloperidol are reasonably potent inhibitors of CYP2D6 (K_i 0.89 and 0.11 μM, respectively) [187,228].

7.2.5. Diarylbutylamines Pimozide (**28**) (Orap® [229]) is not approved in the United States for the treatment of psychosis but is approved for the treatment of Tourette's syndrome. Pimozide is approved in Europe as an antipsychotic where the dosage of pimozide ranges 2–10 mg/day (qd). Pimozide is slowly absorbed after oral administration with a time to peak plasma concentration (T_{max}) of 6–8 h and a C_{max} of ~4 ng/mL (6 mg dose). Although the drug undergoes extensive first pass metabolism, it has a low oral clearance of 0.25 L/h/kg and a long half-life that ranges from 55 to 111 h (accumulation of the drug is observed upon multiple dosing). The half-life appears to segregate into two populations (~50 h or ~100 h) depending on the study and patient population. Pimozide has an oral bioavailability of <50%, an apparent volume of distribution of 28.2 L/kg, and its plamsa protein binding is 99% [174,230–233].

28, Pimozide

Pimozide (**28**) undergoes hepatic metabolism to two metabolites via *N*-dealkylation at the center of the molecule to form 1,3-dihydro-1-(4-piperidinyl)-2*H*-benzimadazole-2-one and 4,4-bis(4-fluorophenyl)butanoic acid. Both metabolites appear to be pharmacologically inactive and the drug and its metabolites are predominately eliminated in the urine. *In vitro* studies indicate that CYP3A4 is the predominate enzyme responsible for the *N*-dealkylation of pimozide, however, CYP1A2 also contributes to the metabolism of pimozide [234,235]. Whereas pimozide is predominately eliminated by metabolism mediated by CYP3A4, potent inhibitors of CYP3A4, such as azole antifungals, macrolide antibiotics, and protease inhibitors are all contraindicated when taking pimozide. For example, clarithromycin (a macrolide antibiotic and CYP3A4 inhibitor) has been shown to significantly increase pimozide C_{max}, half-life, and AUC during coadministration [230]. Pimozide is associated with prolongation of the QT interval and fatal ventricular arrhythmia, therefore elevations in pimozide plasma exposures (especially when coadminstered with CYP3A4 inhibitors) should be avoided [229,235].

7.2.6. Dihydroindolone Molindone (**29**) (Moban®, [236]) is structurally unrelated to the other conventional antipsychotics and its pharmacokinetic and metabolic properties are somewhat unique [174,237–239]. Maintenance doses of molindone for mild–moderate symptoms are 5–25 mg (three to four times a day). Molindone is rapidly absorbed after oral administration with a time to peak plasma concentration of 1.1 h and a C_{max} of 347 ng/mL (100 mg dose). The drug undergoes extensive metabolism to more than 30 metabolites and less than 3% of the drug is eliminated as parent drug in urine and feces. Plasma concentrations of molindone are negligible at 12 h after dosing and the drug is almost completely (90%) eliminated within 24 h. Molindone has an extremely short half-life of 2 h; however, the pharmacological effects of molindone continue 24–36 h after a single dose. It has been suggested that the majority of antipsychotic activity of molindone is not from the parent drug itself but one or more active metabolites. Molindone is not very lipophilic and its plasma protein binding is only 76%. Unfortunately, the identification of molindone metabolites and any potential drug–drug interactions are not well documented.

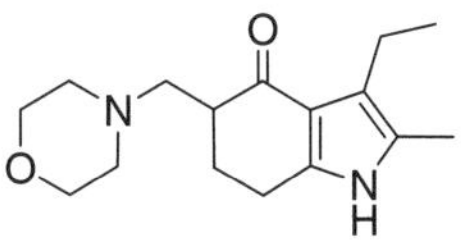

29, Molindone

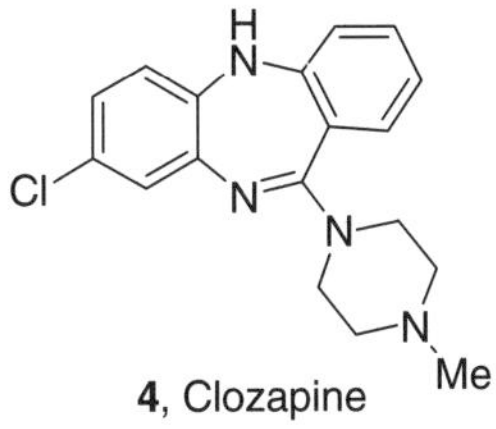

4, Clozapine

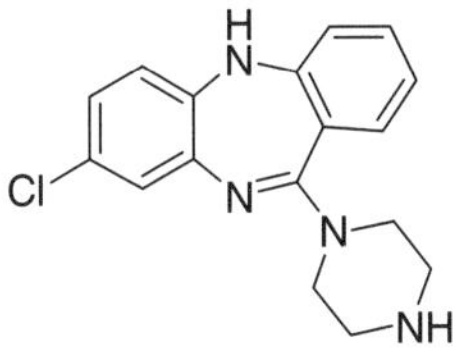

30, *N*-Desmethylclozapine

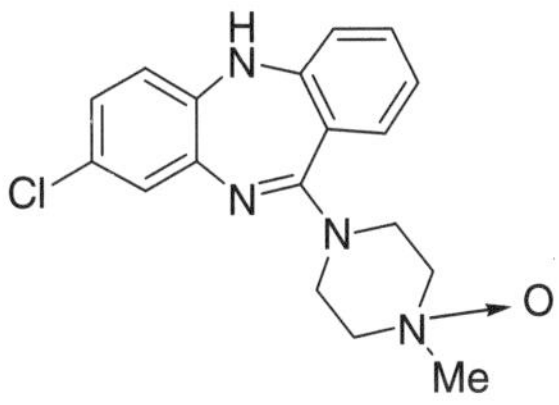

31, Clozapine *N*-oxide

7.3. Atypical Antipsychotics

7.3.1. Dibenzodiazepines Clozapine (**4**) (Clozaril® [240]) is similar in chemical structure to the conventional antipsychotic, loxapine. Dosages of clozapine range 250–400 mg/day (bid), although doses as high as 900 mg/day have been safely administered. It has been observed that there is a wide interpatient variability of clozapine exposure with dose; hence, the following pharmacokinetic parameters are averages, typically over a significant range [174,177,241–244]. The time to peak plasma concentration is ~2.5 h with a C_{max} of 319 ng/mL (100 mg bid). The oral clearance of clozapine is 0.36 L/h/kg and appears to exhibit a biphasic half-life (consistent with a 2-compartment model) with values of 3 and 16 h. There is a linear dose proportionality relationship between clozapine plasma concentrations and clinical doses. Clozapine has an oral bioavailability of 55% and the bioavailability is unaffected when given with food. Clozapine has an apparent volume of distribution of 5.4 L/kg and its plasma protein biding is 97%.

Clozapine (**4**) undergoes extensive hepatic metabolism to the following metabolites: *N*-desmethyl clozapine (**30**), clozapine *N*-oxide (**31**), hydroxylated clozapine, and dehalogenated clozapine [180,182,189]. The major circulating metabolites are the *N*-desmethyl and *N*-oxide that have been reported to be 60% and 15% of circulating clozapine concentrations. Although present in the systemic circulation, it is believed that neither metabolite contributes significantly to the overall activity of the drug; however, the *N*-desmethyl metabolite does have modest antipsychotic activity. The drug and its metabolites are eliminated in urine (50% of dose) and feces (30% of dose) with only trace amounts of parent drug in urine or feces. *In vitro* studies indicate that CYP1A2 is responsible for formation of the major metabolite, the *N*-desmethyl while CYPs 2D6, 2C9, 2C19, and 3A4 have also been implicated in the metabolism of clozapine [245]. There are a multitude of drug interactions involving clozapine and other coadministered drugs [178,183,184,186,188]. Fluvoxamine, a potent inhibitor of CYP1A2, has been shown to significantly increase the exposure of clozapine and two of its metabolites threefold (N-desmethyl and N-oxide). There are also several case reports of significant lowering of clozapine plasma concentrations when coadministered with carbamazepine (50% decrease) or phenytoin (65–85% decrease). Each of these studies implicates CYP1A2 and CYP3A4 in the metabolism of clozapine. In addition, a modest lowering of clozapine plasma concentrations (18%) was observed in smokers compared to nonsmokers and the clearance of clozapine was found to be lower (~30%) in woman compared to men. Minor increases in exposure were observed when clozapine was coadministered with fluoxetine and paroxetine

7.3.2. Thienobenzodiazepine
Olanzapine (5) (Zyprexa®, Symbyax® [246]), similar in structure to clozapine, has a dose range from 10–30 mg/day (qd). Olanzapine is well absorbed after oral administration with a time to peak plasma concentration of 6 h and a C_{max} of 18.3 ng/mL (7.5 mg qd for 8 days). The oral clearance of olanzapine is 0.36 L/h/kg with a fairly long half-life of 33 h. Accumulation of drug is observed and steady state is achieved after one week where day 7 plasma exposures are approximately twofold higher compared to day 1. The kinetics of olanzapine are linear throughout the clinical dose range and plasma exposures are unaffected by food. Olanzapine undergoes extensive first pass metabolism and the oral bioavailability of olanzapine is 60%. Olanzapine has an apparent volume of distribution of 14.3 L/kg and its plasma protein binding is 93% [174,177,247–250].

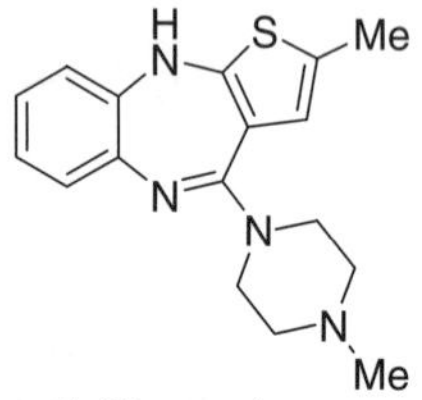

5, Olanzapine

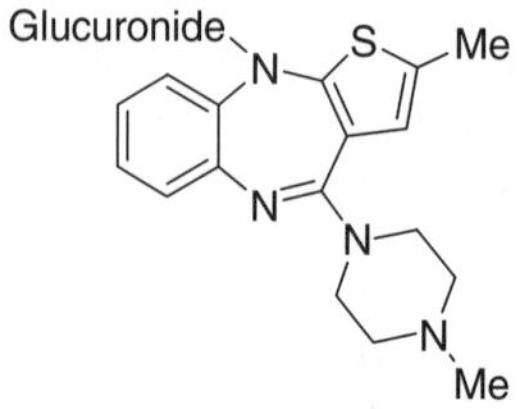

32, Olanzapine N-glucuronide

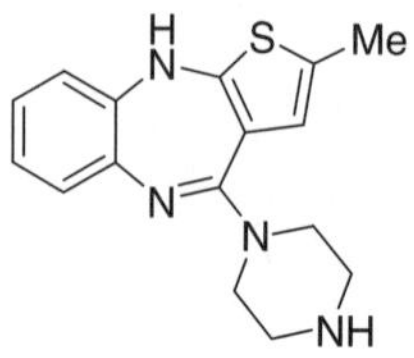

33, N-Desmethyl olanzapine

(CYP2D6 inhibitors). In vitro studies with human liver microsomes indicate that clozapine is a weak inhibitor of CYPs 2C9 and 2D6 (IC_{50} values > 19 μM) and another in vitro study indicated clozapine to be a weak inhibitor of aldehyde oxidase ($IC_{50} = 4.4$ μM) [185,187]. Albeit, these inhibitory parameters are well above the efficacious plasma concentrations of clozapine (350 ng/mL), hence drug–drug interactions with clozapine are not anticipated.

Olanzapine (5) undergoes extensive hepatic metabolism to the following metabolites: a tertiary N-glucuronide (32, 10-N-glucuronide), 4′-N-desmethyl olanzapine (33), 4′-N-oxide, 2-hydroxymethyl, 2-carboxylic acid, 7-hydroxy, and a minor 4′-quaternary N-glucruonide metabolite [180,182,248]. The major circulating metabolites are the 10-N-glucuronide and the 4′-N-desmethyl that have been reported to be 44% and 31% of circulating olanzapine concentrations. Although present in the systemic circulation, it is believed that neither metabolite contributes significantly to the overall activity of the drug, however, each metabolite does have a long half-life similar to the parent drug (glucuronide ~40 h/N-desmethyl ~93 h). The drug and its metabolites are eliminated in urine (57% of dose) and feces (30% of dose) with 7% of the parent drug eliminated in urine. *In vitro* studies indicate that CYP1A2 is responsible for formation of the 4'-N-desmethyl and 7-hydroxy metabolites; CYP2D6 is responsible for the 2-hydroxy metabolite; FMO3 is responsible for formation of the 4'-N-oxide; and UGT1A4 is responsible for formation of the 10-N-glucuronide conjugate [251,252]. Similar to clozapine, olanzapine has the potential for multiple drug-drug interactions [178,183,184,186,188]. Potent inhibitors of CYP1A2, such as fluvoxamine and ciprofloxacin have been shown to significantly increase the exposure of olanzapine (84% increase in C_{max} and 119% increase in AUC, respectively). Also, coadministration of carbamazepine increased the clearance of olanzapine by ~50% and the clearance was also increased (23%) in smokers compared to nonsmokers. Minor increases in exposure were observed when olanzapine was coadministered with fluoxetine (CYP2D6 inhibitor), but there was no change in exposure when olanzapine was given to poor and extensive metabolizers of CYP2D6 substrates. When olanzapine was coadministered with probenecid (UGT inhibitor) there was a modest increase in olanzapine AUC and C_{max} [226]. Combined, these studies imply that UGT, CYP1A2, and CYP3A4 are involved in the metabolism of olanzapine. Interestingly, the clearance of olanzapine is lower (~30%) in woman, elderly, and Japanese patients. Finally, *in vitro* studies with human liver microsomes indicate little potential for olanzapine to inhibit the major CYP enzymes.

7.3.3. Dibenzthiazepines Quetiapine (9) (Seroquel® (254)) is similar in structure to the earlier phenothiazines, fluphenazine, and perphenazine. The typical dose range of quetiapine is from 300–500 mg/day (bid or tid). Quetiapine is rapidly absorbed after oral administration with a time to peak plasma concentration of 1–2 h and a C_{max} of 625 ng/mL (150 mg tid). The oral clearance of quetiapine is 1.2 L/h/kg with a half-life 3–6 h. The pharmacokinetic parameters of quetiapine are linear throughout the therapeutic dose range and a high fat meal marginally increases the AUC and C_{max} of quetiapine (15% and 25%, respectively). Quetiapine has an apparent volume of distribution of 10 L/kg, low oral bioavailability (9%), and its binding to plasma proteins is moderate, 83% [174,253–257].

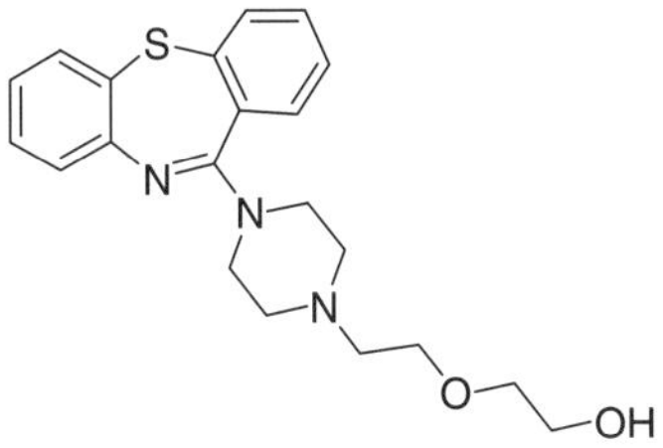

9, Quetiapine

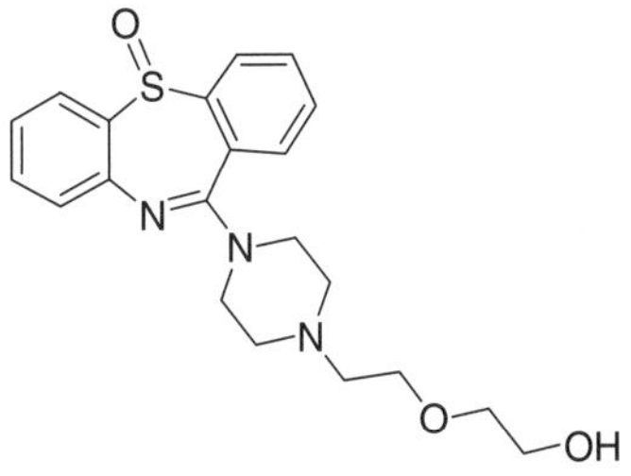

34, Quetiapine *S*-Oxide

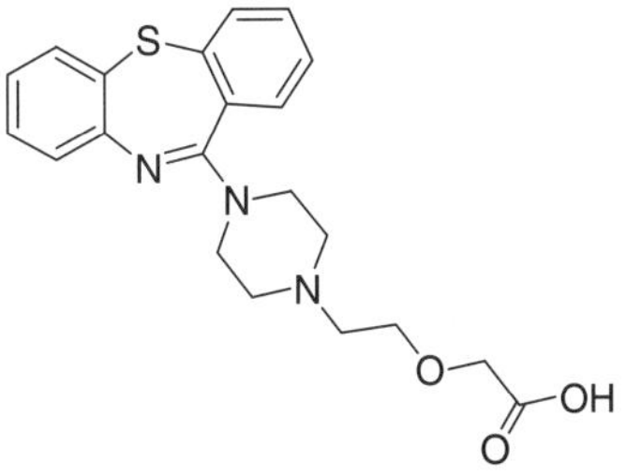

35, Quetiapine carboxylic acid derivative

Quetiapine (**9**) undergoes extensive hepatic metabolism to two major circulating metabolites, a sulfoxide (**34**) and a carboxylic acid derivative (**35**). Multiple other metabolites (11 in total) have been identified, such as the 7-hydroxy, the *N*- and *O*-dealkylated metabolites, and several combinations of these pathways [177,180,182,258,259]. The drug and its metabolites are predominately eliminated in urine (73% of dose) and partially in feces (21% of dose) with less than 1% of the dose eliminated as parent drug. Most of the metabolites are pharmacologically inactive; however, the 7-hydroxy and the 7-hydroxy-*N*-dealkylated metabolites do have some antipsychotic activity, albeit their exposures are low and most likely do not contribute to the pharmacology of the drug [177]. *In vitro* studies indicate that CYP3A4 is the predominate enzyme responsible for the *S*-oxidation and the *N/O*-dealkylated metabolites of quetiapine, however, CYP2D6 also contributes to the formation of the 7-hydroxy metabolite [182,253]. Whereas quetiapine is predominately eliminated by metabolism mediated by CYP3A4, potent inhibitors and inducers of CYP3A4 have been shown to significantly alter the exposure of quetiapine when coadministered. For example, ketoconazole (CYP3A4 inhibitor) has been shown to significantly decrease quetiapine oral clearance by 84% (335% increase in C_{max}) and both phenytoin and carbamazepine (CYP3A4 inducers) increased the oral clearance of quetiapine by five- and sevenfold, respectively [257]. *In vitro* studies have shown that quetiapine is not expected to significantly inhibit the cytochromes P450, however, it is a moderate inhibitor of aldehyde oxidase ($IC_{50} = .4\,\mu M$). The oral exposure of quetiapine was 30–50% lower in the elderly volunteers compared to the young volunteers, but smoking status had no affect on quetiapine exposure [254].

7.3.4. Dibenzothiepine Zotepine (**36**) (Lodopin® [260]) is not approved in the United States; however, it is approved in Europe and Japan where the standard dose of zotepine is 75–150 mg/day (tid). Zotepine is rapidly absorbed after oral administration with a time to peak plasma concentration (T_{max}) of 3–4 h and a C_{max} of 30–249 ng/mL (100 mg dose). The drug has an oral clearance of 4.1–5.2 L/h/kg and an oral terminal half-life of 14.8–24 h. The oral bioavailability of zotepine is unknown but anticipated to be low based on its oral pharmacokinetics. Zotepine has an apparent volume of distribution of 80–131 L/kg and its binding to plasma proteins is 97% [175,260–264].

Zotepine (**36**) undergoes extensive hepatic metabolism to metabolites such as: *N*-desmethyl zotepine (norzotepine), 2- and 3-hydroxyzotepine, zotepine *S*-oxide, and zotepine *N*-oxide [179,261]. The norzotepine metabolite has been shown to be pharmacologically active and possibly contributes to the overall pharmacological effects [263]. *In vitro* studies indicate that CYP3A4 is predominately responsible for the formation of norzotepine and the *S*-oxide metabolites while CYP1A2 and CYP2D6 are involved in formation of the 2- and 3-hydroxyzotepine metabolites [265]. The urinary elimination of parent drug and zotepine metabolites is minimal.

36, Zotepine

Very few clinical studies have evaluated the drug–drug interaction potential of zotepine. There have been several reports which indicate concurrent use of benzodiazepines with zotepine significantly elevate zotepine exposure. For example, one study observed a modest increase in steady state zotepine plasma concentration (13.8–17.5 ng/mL) when coadministered with diazepam [266]. The authors concluded that the interaction may have been caused by diazepam inhibition of CYP3A4 that would have affected the metabolic elimination of zotepine. Finally, zotepine half-life and exposure have been shown to increase significantly in the elderly populations as compared to the younger populations [263].

7.3.5. Benzisoxazoles Risperidone (**6**) (Risperdal® [267]) dosages range 2–8 mg/day (bid). Risperidone is completely and rapidly absorbed after oral administration with a time to peak plasma concentration of 1 h and a C_{max} of 10 ng/mL (8 mg bid). The oral clearance of risperidone is 0.32 L/h/kg with a half-life of ~3 h. Multiple dose pharmacokinetics of risperidone is linear from 0.5–25 mg/day and plasma exposures are not changed when taken with food. Risperidone undergoes first pass metabolism and the oral bioavailability of risperidone is 66%. Risperidone has an apparent volume of distribution of 1–2 L/kg and its plasma protein binding is 90% [174,175,177,189,267–270].

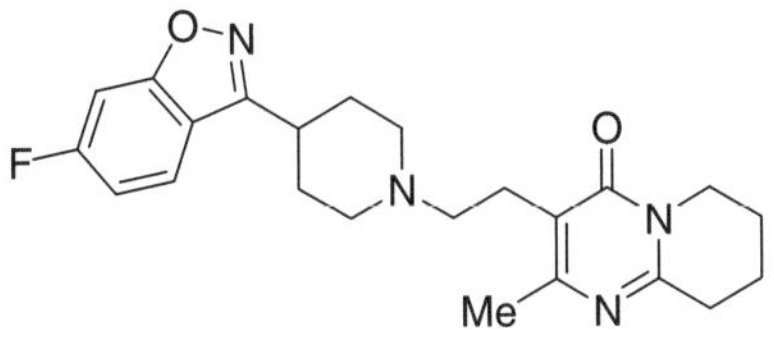

or paroxetine have been shown to increase the plasma concentration of risperidone [178,183,184,186,188]. For example, fluoxetine increases the mean plasma concentration of risperidone by 4.7-fold with no change in the concentration of 9-hydroxyrisperidone [272]. Risperidone and 9-hydroxyrisperidone exposures were evaluated in groups of poor and extensive metabolizer of the CYP2D6 phenotype. Although plasma concentrations of risperidone increase in poor metabolizers, the ratio of risperidone to 9-hydroxyrisperidone did not change significantly [273]. Therefore, CYP2D6 genotype status will cause a pharmacokinetic change in risperidone exposures, but the genotype

6, Risperidone

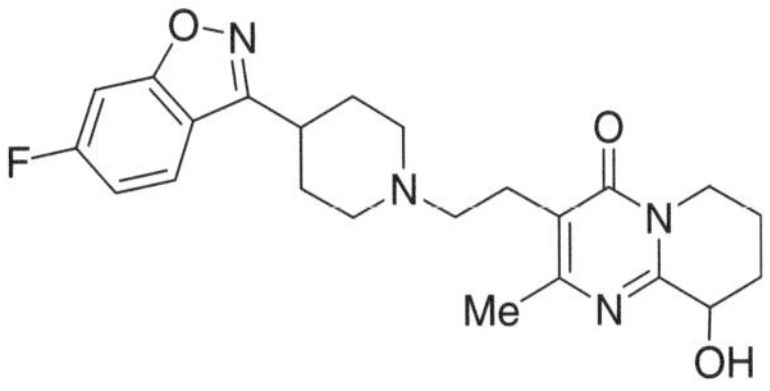

37, 9-Hydroxy Risperidone

Risperidone (**6**) undergoes extensive hepatic metabolism by hydroxylation to 7- and 9-hydroxyrisperidone and *N*-dealkylation. The predominate metabolite is 9-hydroxyrisperidone (**37**) that is also active (known as paliperidone (**37**)) [179,181,182,271]. The major circulating metabolite, 9-hydroxyrisperidone, is predominately eliminated in urine (31% of a resperidone dose) and contributes to the overall antipsychotic effects of resperidone. The drug and its metabolites are predominately eliminated in urine (70% of dose) with a much smaller proportion in feces (14% of dose) and ~10% of the parent drug is eliminated in urine while no parent drug is found in feces. *In vitro* studies indicate that CYP2D6 is responsible for the formation of 9-hydroxyrisperidone while CYP3A4 plays a smaller role. Hydroxylation at the 9-position of risperidone forms a chiral center and CYP2D6 has been shown to form the (+)-9-hydroxyrisperidone while CYP3A4 forms the (−)-9-hydroxyrisperidone (both isomers are active). As risperidone is metabolized predominately by CYP2D6, potent inhibitors of CYP2D6 such as fluoxetine

does not result in a pharmacodynamic effect and dose adjustments may or may not be necessary. The CYP3A4 inducer carbamazepine decreases the plasma concentrations of both risperidone and 9-hydroxyrisperidone which results in a significant lowering of antipsychotic activity [274]. The disposition of risperidone appears to be lower in the elderly and in the hepatic or the renal impaired patients. *In vitro* studies with human liver microsomes indicate little potential for risperidone to inhibit the major CYP enzymes, albeit risperidone is a weak inhibitor of CYP2D6 that is probably not clinically relevant at therapeutic doses [187,196]. Risperidone is also a substrate and weak inhibitor of P-glycoprotein [252].

Paliperidone (**37**) (9-hydroxyrisperidone, Invega® [276]) is a racemic mixture of (+) and (−) 9-hydroxyrisperidone with dosages of 3–12 mg/day (qd) delivered in an extended release formulation. Following administration of paliperidone, the (+) and (−) enantiomers interconvert reaching a steady state AUC ratio (+/−) of ~1.6. Paliperidone is well absorbed after oral administration with a time to peak plasma concentration of 24 h and a C_{max} of 9 ng/mL (1 mg dose). The clearance of paliperidone is 0.08 L/hr/kg and the half-life of paliperidone is 25 h which is much longer than risperidone. Multiple dose pharmacokinetics of paliperidone is linear with steady state concentrations achieved after 4–5 days and the C_{max} and AUC of paliperidone are increased when taken with food (60% and 54%, respectively). Paliperidone has an oral bioavailability of 28%, a volume of distribution of 7 L/kg, and its binding to plasma proteins is 74% [175,276,277]. Paliperidone does not undergo extensive hepatic metabolism, however, several metabolites have been identified, none of which are active or detected in the systemic circulation [179,182,278]. The identified metabolic pathways of paliperdone include: N-dealkylation, hydroxylation, alcohol dehydrogenation, and benzisoazole ring scission. The drug and its metabolites are predominately eliminated in urine (80% of dose) with a much smaller proportion in feces (11% of dose) and 59% of the parent drug is eliminated in urine [276]. Therefore, unlike many of the other antipsychotic agents, the predominate elimination pathway of paliperidone is renal elimination with a small proportion of the dose eliminated as metabolites. *In vitro* studies indicate that CYP2D6 and CYP3A4 are responsible for the metabolism of paliperidone [278]. The disposition of paliperidone is much lower in patients with renal impairment; however, no changes were noted due to age, race, smoking status, or hepatic impairment [276]. *In vitro* studies with human liver microsomes indicate little potential for paliperidone to inhibit the major CYP enzymes, albeit the drug is a substrate and weak inhibitor of P-glycoprotein [183,184,276].

7.3.6. Benzoisothiazoles Ziprasidone (8) (Geodon® [279]) dosages range 40–160 mg/day (bid). Ziprasidone is well absorbed after oral administration with a time to peak plasma concentration of 6–8 h and a C_{max} of 68 ng/mL (40 mg/day bid for 8 days). The oral clearance of ziprasidone is 0.45 L/h/kg with a half-life of 6.6 h. Multiple-dose pharmacokinetics of ziprasidone is linear throughout the clinical dose range and plasma exposures are increased by twofold when taken with food. Ziprasidone undergoes extensive first pass metabolism and the oral bioavailability of ziprasidone is 60%. Ziprasidone has an apparent volume of distribution of 1.5 L/kg and its binding to plasma proteins is extensive at ~99.9% [175,177,279–281].

Ziprasidone (8) undergoes extensive hepatic metabolism to several metabolites: there is a sulfoxide (38) that is further oxidized to a sulfone, the parent drug undergoes N-demethylation at the piperazine nitrogen to form two metabolites (benzisothiazole piperazine (39) and an oxindole aldehyde that is further oxidized to the chloro-oxindole acetic acid) (40), the parent drug is also reduced at the N–S bond of the benzisothizole ring to form the dihydroziprasidone which undergoes S-methylation to form the S-methyldihydroziprasidone (41) [179,180,182,280]. The predominant metabolic pathway is reduction of ziprasidone followed by the various oxidation reactions. The major circulating metabolite is the S-methyldihydroziprasidone which is mostly eliminated in the feces. Other minor circulating metabolites include the benzisothiazole piperazine sulfoxide and sulfone metabolites, as well as ziprasidone sulfoxide. Although present in the systemic circulation, it is believed that none of the metabolites contribute significantly to the activity of the drug. The drug and its metabolites are eliminated in feces (66% of dose) and less in urine (20% of dose) with <4% of the parent drug eliminated in urine and feces. *In vitro* studies indicate that aldehyde oxidase is involved in the reductive cleavage of the benzisothiazole ring and the oxidation of the oxindole aldehyde while CYP3A4 is predominately involved in the oxidation of ziprasidone with a minor contribution from CYP1A2 [185,282]. Because ziprasidone is metabolized predominantly by aldehyde oxidase, potent inhibitors and inducers of CYP3A4 have only modest affects on the plasma concentration of ziprasidone [178,183,186,188]. For example, ketoconazole (CYP3A4 inhibitor) increases both the AUC and C_{max} of ziprasidone by about 33% and carbamazepine (CYP3A4 inducer) decreases the AUC and C_{max} of ziprasidone by 36% and 27%, respectively [183,283]. The disposi-

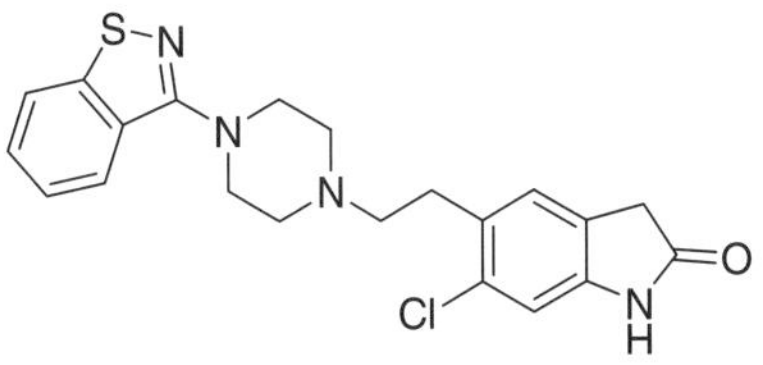

8, Ziprasidone

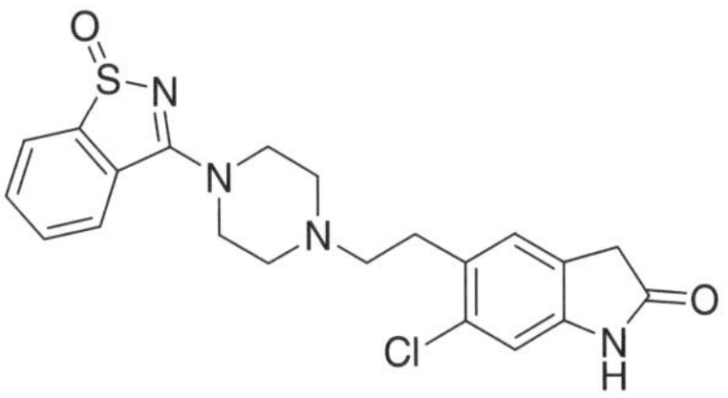

38, Ziprasidone S-oxide

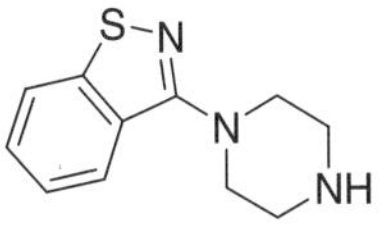

39, Benzisothiazolepiperazine

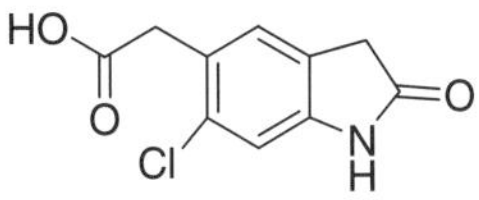

40, Chloro-oxindole acetic acid

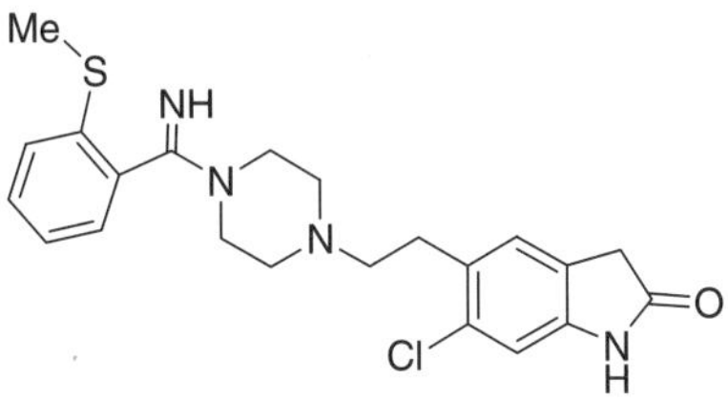

41, S-Methyldihydroziprasidone

tion of ziprasidone does not appear to be affected by age, gender, hepatic or renal impairment [279]. In addition, *in vitro* studies with human liver microsomes indicate little potential for ziprasidone to inhibit the major CYP enzymes, albeit ziprasidone is a weak inhibitor of CYP2D6 ($IC_{50} = 11\ \mu M$), however, clinical studies have not demonstrated an interaction [284].

Perospirone (**42**) (Lullan® [285]) is structurally similar to ziprasidone and currently approved in Japan but not in the United States. The dosages of perospirone range 12–48 mg/day (bid). Perospirone is rapidly absorbed after oral administration with a time to peak plasma concentration of ~2 h and a C_{max} of 9 ng/mL (16 mg/day bid). The oral half-life appears to be bi-phasic with values of 1–3 h followed by 5–8 h. The C_{max} and AUC of perospirone are increased by 1.6- and 2.4-fold, respectively, when taken with food. Perospirone undergoes extensive first pass metabolism, has an apparent volume of distribution of 24.8 L/kg, and its binding to plasma proteins is 92% [175,285–289].

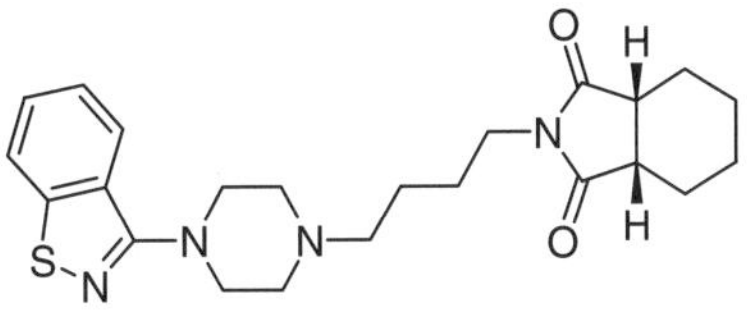

42, Perospirone

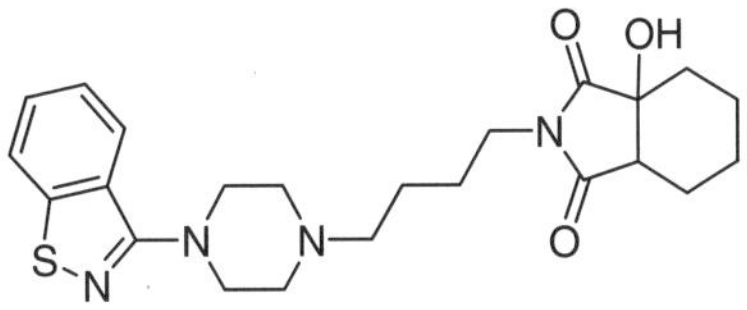

43, 1-Hydroxyperospirone

Perospirone (**42**) undergoes extensive hepatic metabolism to several metabolites: hydroxylation on the cyclohexane dicarboxyimide (1-hydroxy perospirone, (**43**)), N-dealkylation at the piperazine nitrogen, and S-oxidation of the isothiazole ring [179,285,290]. Hydroxylated perospirone (**43**) is the major metabolite present in the systemic circulation at approximately threefold the parent drug and is believed to contribute to the activity of the drug. *In vitro* studies indicate that CYP3A4 is predominately involved in the metabolism of perospirone [290]. Because perospirone is metabolized predominately by CYP3A4, inducers and inhibitors of this enzyme significantly affect the plasma concentration of perospirone. For example, itraconazole (CYP3A4 inhibitor) increases the AUC and C_{max} of perospirone by seven- and sixfold, respectively, and coadministration of carbamazepine (CYP3A4 inducer) decreases the exposure of perospirone from 4 ng/mL to below the limit of detection (~0.1 ng/mL) [286,287].

7.3.7. Quinolinone Aripiprazole (**10**) (Abilify® [291]) dosages range from 10–30 mg/day (qd). Aripiprazole is rapidly and well absorbed after oral administration with a time to peak plasma concentration of 3–5 h and a C_{max} of 242 ng/mL (15 mg qd for 14 days). The oral clearance of aripiprazole is low at 0.05 L/h/kg with a long half-life of 75 h. Accumulation of drug is observed and steady state is achieved in 14 days. The pharmacokinetic parameters of aripiprazole are linear throughout the therapeutic dose range and a high fat meal does not affect AUC or C_{max}, but T_{max} is increased by 3 h. Aripiprazole has an oral bioavailability of 87%, an apparent volume of distribution of 4.9 L/kg, and its binding to plasma proteins is greater than 99% [175,177,291–294].

10, Aripiprazole

44, Dehydroaripiprazole

Aripiprazole (**10**) undergoes extensive hepatic metabolism to the dehydroaripiprazole (**44**) while other minor metabolic reactions include *N*-dealkylation and hydroxylation [182,292,294]. The drug and its metabolites are predominately eliminated in feces (55% of dose) and somewhat in urine (25% of dose) with 1% of the parent drug eliminated in urine and 18% in feces. The dehydroaripiprazole (**44**) is pharmacologically active and its exposure is 40% that of the parent drug. The half-life of dehydroaripiprazole is also long (94 h) and it is 99% bound to plasma proteins; therefore, the dehydroaripiprazole likely contributes to the overall effects of the drug. *In vitro* studies indicate that CYP3A4 and CYP2D6 are predominately responsible for formation of the dehydroaripiprazole and the hydroxylated metabolite while CYP3A4 is responsible for the *N*-dealkylated metabolite. Potent inhibitors of CYP3A4 and CYP2D6 have been shown to significantly increase the exposure of aripiprazole [178,183,185,188]. For example, ketoconazole (CYP3A4 inhibitor) significantly increased the AUC of aripiprazole and dehydroaripiprazole by ~70% each and quinidine (CYP2D6 inhibitor) increased the AUC of aripiprazole by 112% and decreased the AUC of dehydroaripiprazole by 35%. When coadministered with the CYP3A4 inducer carbamazepine, the AUC and C_{max} of both aripiprazole and dehydroaripiprazole decreased by 70%. In all of these instances, CYP3A4 inhibitors/inducers or CYP2D6 inhibitors, dosage adjustments of aripiprazole are recommended. As would be expected, poor metabolizers of CYP2D6 substrates have altered aripiprazole pharmacokinetics. The half-life of aripiprazole increased from 75 h in extensive metabolizers to 196 hours in poor metabolizers. In addition, there was an 80% increase in aripiprazole exposure along with a 30% decrease in dehydroaripiprazole exposure. Hence in patients with the poor metabolizer CYP2D6 genotype, the overall exposure of pharmacologically active components increases by ~60%. Finally, aripiprazole does not inhibit the elimination of several drugs metabolized by CYP's 2C9, 2C19, 2D6, or 3A4; therefore, aripiprazole does not appear to interact with most coadministered drugs metabolized by these enzymes [291].

7.3.8. Phenylindole

Sertindole (**11**) (Serdolect® [295]) is not approved in the United States; however, clinical trials are active in the United States and the drug is approved in Europe where the standard dose of sertindole is 12–20 mg/day (qd). Sertindole is slowly absorbed after oral administration with a time to peak plasma concentration (T_{max}) of 8–h and a C_{max} of 53–64 ng/mL (12 mg dose). The drug has an oral clearance of 0.2–0.6 L/h/kg and an oral half-life of 50–111 h. Accumulation of the drug is observed and 1–2 weeks are necessary to achieve steady state plasma concentrations. The oral bioavailability of sertindole is 75% and although there is a small food effect, no dosage adjustments are necessary. Sertindole has an apparent volume of distribution of 25 L/kg and its binding to plasma proteins is extensive at 99.5% [177,295–298].

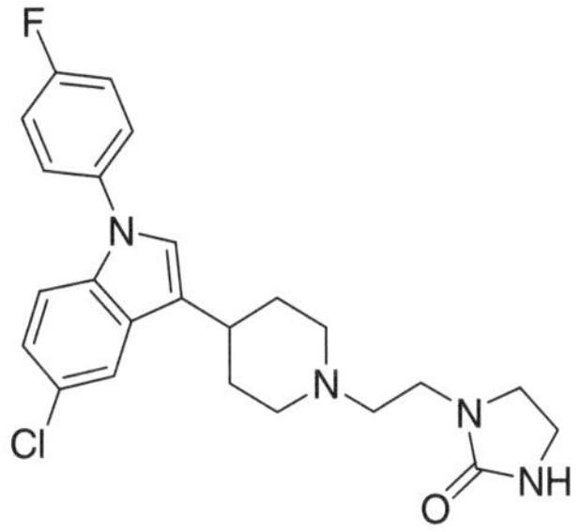

11, Sertindole

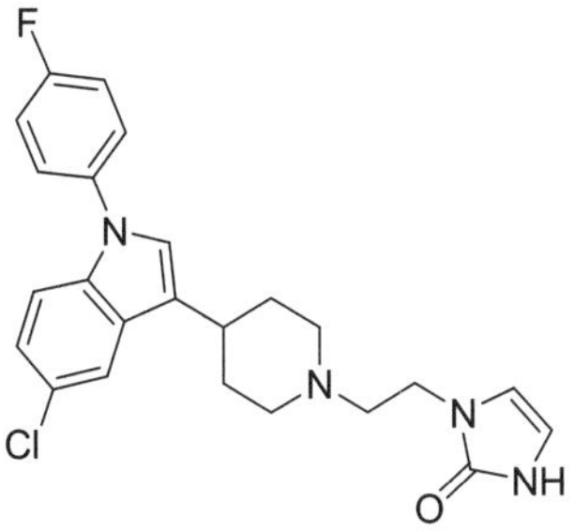

45, Dehydrosertindole

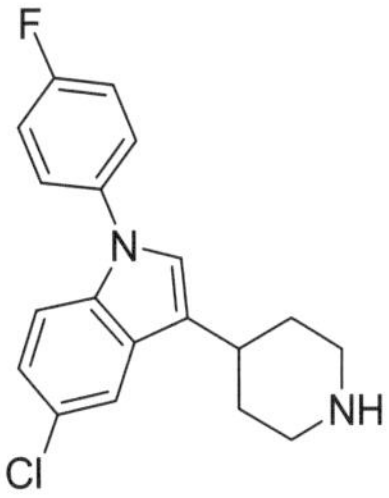

Sertindole (**11**) undergoes hepatic metabolism to dehydrosertindole (**45**) and norsertindol (**46**) and less than 4% of the drug (parent and metabolites) is eliminated in the urine [177,188,295,297]. The dehydrosertindole (**45**) metabolite has been shown to be pharmacologically active and at steady state its plasma concentration is 80% that of the parent drug suggesting that the metabolite may contribute to the pharmacological effects. *In vitro* studies indicate that CYP3A4 and CYP2D6 are the predominate enzymes responsible for the metabolism of sertindole and potent inhibitors of these enzymes are contraindicated when administering sertindole. Drugs such as ketoconazole (CYP3A4 inhibitor) and paroxetine or fluoxetine (CYP2D6 inhibitors) have all been shown to significantly increase the plasma exposure of sertindole. In addition, inducers of CYP3A4, such as carbamazepine and phenyotin have been shown to reduce the plasma exposure of sertindole during coadministration, sometimes requiring an increase in dosage. Moreover, sertindole exposures are significantly increased in CYP2D6 poor metabolizers where plasma concentrations increase two- to three-fold compared to extensive metabolizers [295].

7.3.9. Substituted Benzamide

Racemic amisulpride (**47**) (Solian® [299]) is not approved in the United States; however, it is approved in Europe where the standard dose of amisulpride ranges from 50–800 mg/day. Amisulpride is rapidly absorbed after oral administration with two plasma concentration peaks at 1 and 4 h (T_{max}) of 42 and 56 ng/mL, respectively (50 mg dose). The oral half-life of amisulpride is ~12 h. The oral bioavailability of

46, Norsertindole

amisulpride is 43–48% and there is a significant food effect that reduces the bioavailability by 50%. Amisulpride has an apparent volume of distribution of 5.8 L/kg and its binding to plasma proteins is very low at 17% [175,177,300]. Amisulpride (**47**) undergoes very little metabolism and is predominately excreted in the urine. Minor pathways of amisulpride metabolism that have been identified include N-demethylation, oxidation of the pyrolidine ring, and hydroxylation. The renal clearance of amisulpride is 17–20 L/h that is greater than glomerular filtration perhaps indicating active drug secretion [127]. After IV and oral dosing, 50% and 23% of the dose is excreted as unchanged parent in the urine, respectively. Given the low protein binding and lack of metabolism, there are no reported drug-drug interactions with amisulpride, however, amisulpride exposure has been shown to increase significantly in the elderly and in the renally impaired patients [179,182,188].

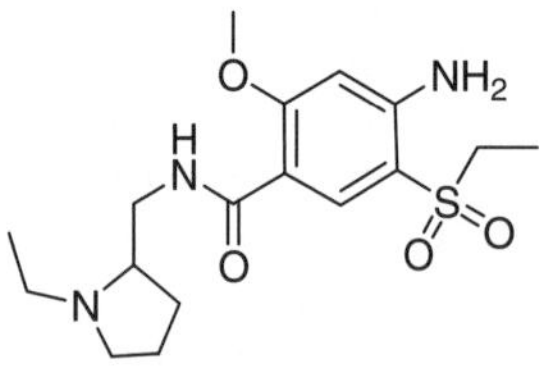

47, Amisulpride

8. STRATEGIES FOR DRUG DISCOVERY

In terms of the mechanism of action (MOA), the "dopamine theories" of antipsychotic and AMPA receptors); histamine H_3 receptor antagonists; muscarinic cholinergic agonists; neurokinin$_3$ (NK$_3$) receptor antagonists; PDE inhibitors and cannabinoids. Newer strategies include nicotine acetylcholine receptor agonists, positive allosteric modulators of D_1 or D_5 agonists, and 5-HT$_6$ antagonists.

8.1. Dopamine Receptor Subtype Approaches

There are five subtypes of dopamine receptors, D_1–D_5. While historically the D_2 dopamine receptor has attracted the most attention, more and more evidence suggest that D_1, D_3, and D_5 dopamine receptors are important for the therapeutic actions of antipsychotics as well. For instance, many of the cognitive abnormalities in schizophrenia are similar to those resulting from damage to the prefrontal cortex (PFC) including attention abnormalities, problems in reasoning and judgment and working memory deficits. Clinical reports suggest that PFC D_1 receptors are upregulated in schizophrenia due to a localized decrease in dopaminergic activity and that D_1 receptor antagonists aggravate psychotic symptoms.

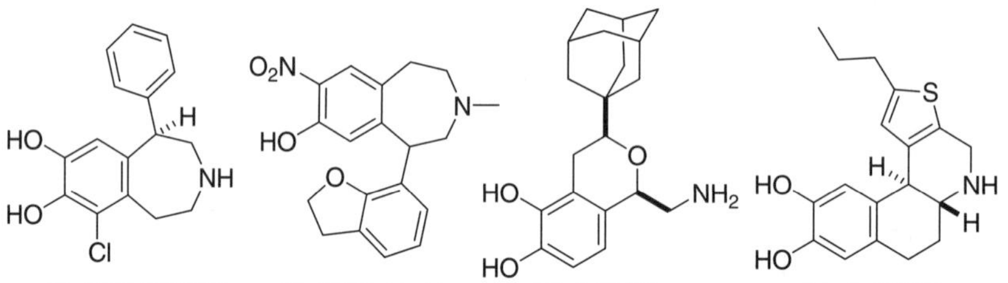

48, SKF-812197   **49**, NNC-01-687   **50**, A-7763   **51**, A-86929

agents have dominated the field for decades. Conventional antipsychotic drugs are dopamine antagonists, whereas atypical antipsychotic drugs are dopamine antagonists with additional actions on serotonin receptors such as 5-HT$_{1A}$ and 5-HT$_{2A}$. Thanks to the diverse pharmacological profile of clozapine (**4**), it provides a useful roadmap to the discovery of next generation antipsychotics. Multiple approaches being pursued include: dopamine receptor subtypes; serotonin receptor subtypes; glutamatergic agents (mGlu, NMDA

48 (SKF-812197) is a full agonist of the D_1 receptor [301,302]. Unfortunately, in three open clinical trials of **48** in schizophrenia, no appreciable antipsychotic activity was observed although it had precognitive effects in animal models. A clinical study of the D_1 receptor antagonist **49** (NNC-01-0687) reported efficacy on the negative symptoms of schizophrenia. Compound **50**, with a half-life of 37 h, was reported to produce significant receptor desensitization following a single dose, while tid dosing with **51** maintained activity for up to

30 days of treatment. For some antipsychotic agents with more complex pharmacological profile, D_1 receptor modulation plays an important role. For example, zuclopenthixol (**52**) is a D_2/D_1 antagonist while asenapine (**53**) is a $D_1/D_2/5\text{-}HT_{2A}$ antagonist [303].

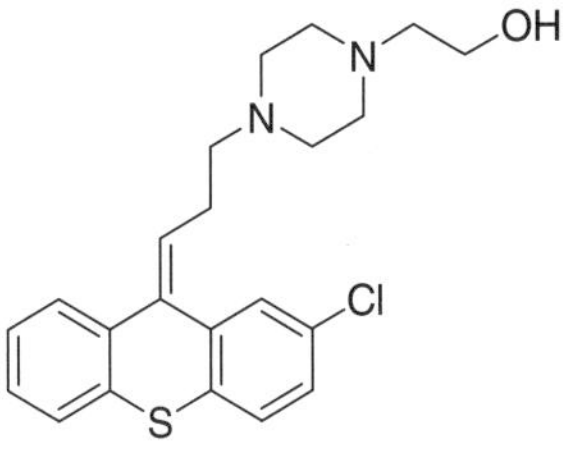

52, Zuclopenthixol

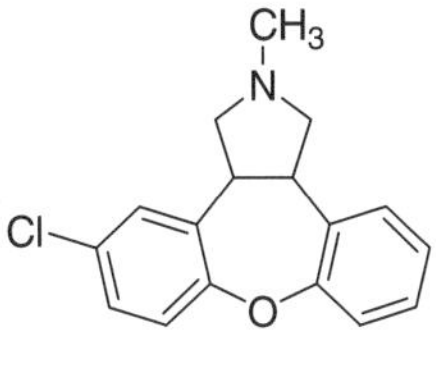

53, Asenapine

Whereas nearly all antipsychotic agents are D_2 receptor antagonists to some degree, selective dopamine D_3 receptor antagonists have attracted much attention in the area of schizophrenia during the last decade [304–307]. Selective dopamine D_3 receptor antagonists are anticipated to have reduced EPS liability compared to D_2 receptor antagonists. Recent examples include tetrahydrobenzazepine **54**, arylalkylpiperazine **55** and benzazepinone **56**.

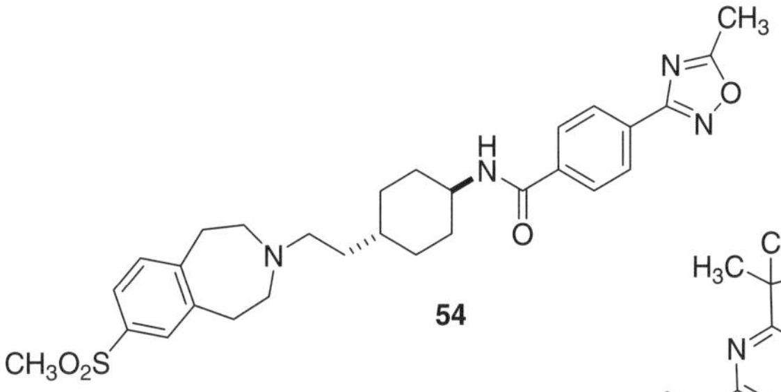

and at least 15 subpopulations have been cloned to date [308]. Among the known subtypes, both $5\text{-}HT_{2C}$ and $5\text{-}HT_6$ are implicated to have a relationship to schizophrenia in addition to $5\text{-}HT_{1A}$ and $5\text{-}HT_{2A}$ as we discussed before.

Unlike $5\text{-}HT_{2A}$ and $5\text{-}HT_{2B}$ receptor subtypes, the $5\text{-}HT_{2C}$ receptor is distributed almost exclusively in the brain. $5\text{-}HT_{2C}$ has now been established as a potential target for treating schizophrenia [302,309,310] in additional to anxiety, depression, and obesity. The teracyclic indoline **57** (WAY-163909) is a potent and selective $5\text{-}HT_{2C}$ agonist with a K_i of 10.5 nM ($EC_{50} = 8$ nM) while for $5\text{-}HT_{2A}$ its K_i is 485 nM and for $5\text{-}HT_{2B}$ its K_i is 212 nM [311]. **57** also worked in several antipsychotic animal models including rat conditioned avoidance ($ID_{50} = 1.3$ mg/kg), mouse PCP-induced hyperactivity (0.1 mg/kg), and MK-801-induced disruption of prepulse inhibition (MED 5.4 mg/kg). Encouragingly, it did not show cataleptic potential in mouse. A full $5\text{-}HT_{2C}$ agonist, VER-2692 (**58**), was active *in vivo* at 3 mg/kg dose in a feeding model [312]. These positive results suggest that $5\text{-}HT_{2C}$ receptor agonists could be atypical neuroleptics.

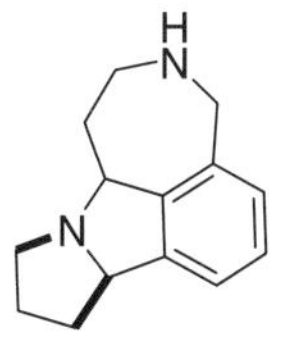

54

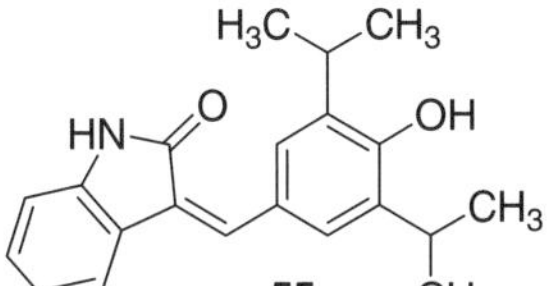

55

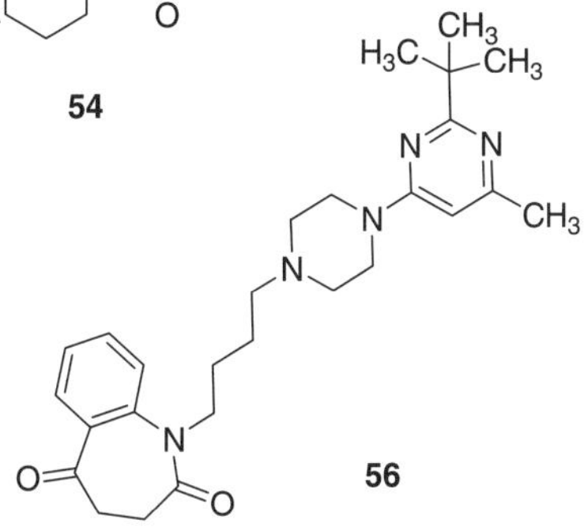

56

8.2. Serotonin Receptor Subtype Approaches

Seven distinct families of serotonin receptors (5-HT) have been identified ($5\text{-}HT_1$–$5\text{-}HT_7$),

57, WAT-163909

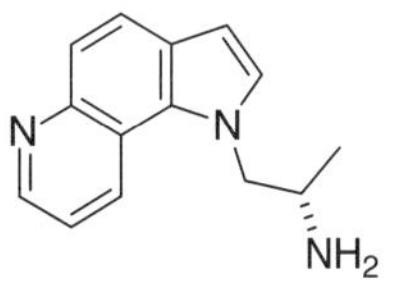

58, VER-2692

Meanwhile, the 5-HT$_6$ receptor, identified in the 1990s, has recently emerged as an attractive target for the treatment of schizophrenia. The 5-HT$_6$ receptor exhibits only 36–41% transmembrane homology with 5-HT$_{1A}$, 5-HT$_{1B}$, 5-HT$_{1D}$, 5-HT$_{1E}$, 5-HT$_{2A}$, and 5-HT$_{2C}$ receptors. The 5-HT$_6$ receptor was the first cloned 5-HT receptor shown to be coupled positively to activation of adenylate cyclase, similar to 5-HT$_4$ and 5-HT$_7$ [313]. A number of antipsychotic agents, typical and atypical, as well as antidepressants, have been shown to bind at rat 5-HT$_6$ receptors with low nanomolar affinity. Since the 5-HT$_6$ receptor has been known for a decade, several 5-HT$_6$ receptor anatagonists have been advanced to the clinic for the treatment of CNS diseases. Indolylsulfonate LY-483518 (**59**) is a potent ($K_i = 1.3$ nM) and selective 5-HT$_6$ receptor antagonist [314]. It is now in phase II clinical trials for the treatment of cognitive impairment associated with schizphrenia. Thiephene-sulfonamide **60** (SB-271046), also a potent ($K_i = 1.3$ nM) and selective 5-HT$_6$ receptor antagonist, is currently in phase I clinical trials for the treatment of schizophrenia and Alzheimer's disease [315].

8.3. Glutamatergic Agents

L-Glutamate (**61**) is a major excitatory neurotransmitter in the mammalian central nervous system and plays a fundamental role in the control of motor function, cognition, and mood. The physiological and pathophysiological effects of glutamate are mediated through two receptor families: metabolic and ionotropic. The metabotropic (G-protein-coupled) glutamate receptors are responsible for mediating slow metabolic response whereas the ionotropic (ligand-gated ion channel) glutamate receptors mediate the fast synaptic response to extracellular L-glutamate. The ionotropic glutamate receptors are further divided into three subclasses according to their molecular and pharmacological differences and are named after the agonists that were originally identified to selectively activate them: NMDA (**62**, *N*-methyl-D-aspartate), AMPA (**63**, alpha-amino-3-hydroxy-5-methyl-4-isoxazole-propionic acid), and kainate (**64**, 2-carboxy-3-carboxymethyl-4-isopenylpyrrolidine). mGluRs, NMDA, and AMPA receptors have emerged as promising therapeutic targets for a variety of clinical indications including schizophrenia, depression, and Alzeimer's disease.

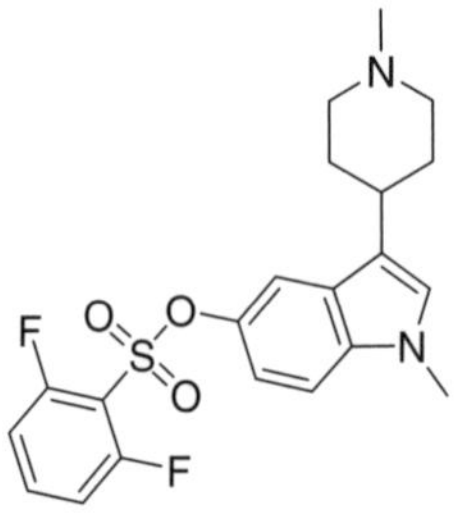

59, LY-483518

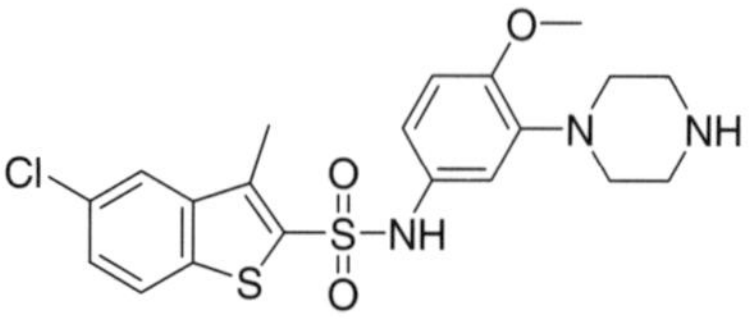

60, SB-271046

61, (*S*)-Glutamic acid

62, NMDA

63, AMPA

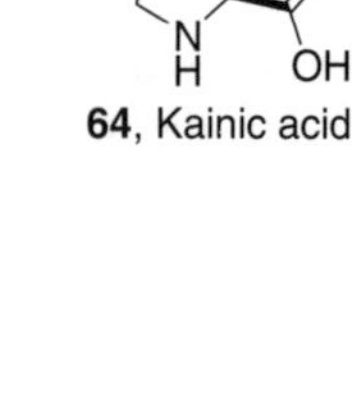

64, Kainic acid

8.3.1. mGluR

The mGluRs belong to the family of GPCRs and eight subtypes of mGluRs have been identified. Based on sequence homology, pharmacological selectivity and primary G-protein coupling, mGluRs have been divided into three groups: group I mGluRs include $mGluR_1$ and $mGluR_5$, both of which are coupled to $G_{q/11}$ to activate phospholipase C (PLC); group II mGluRs ($mGluR_2$ and $mGluR_3$) and group III mGluRs ($mGluR_4$, $mGluR_6$, $mGluR_7$ and $mGluR_8$) are coupled to $G_{i/o}$ and associated effectors such as ion channels or adenylyl cyclase. The mGluRs provide a mechanism by which glutamate can modulate activity at the same synapses at which it elicits fast excitatory synaptic responses. Because of the ubiquitous distribution of glutamatergic synapses, mGluRs participate in a wide variety of functions in the CNS [316,317]. Recent genetic data suggests genetic associations between the schizophrenia and the metabotropic glutamate receptors, most notable with the $mGLuR_3$ receptor [32,33]. In contrast to ligand-gated ion-channels (NMDA, AMPA, and kainite receptors), which are responsible for fast excitory transmission, mGluRs have a more modulatory role, contributing to the fine-tuning of synaptic efficacy and control of the accuracy and sharpness of transmission.

In early clinical studies mGluR 2/3 agonists have shown promise. Administration of a prodrug of the mGluR2/3 agonist that mimics glutamate, **65** (LY404039) was shown to have positive clinical effects over placebo and comparable effect to olanzapine (**4**) [34]. The corresponding methionine prodrug **66** (LY2140023), had increased oral bioavailability. In phase II clinical trials, **66** relieved schizophrenia patients' positive and negative symptoms to a greater extent than a placebo. In addition, the compound did not cause weight gain, a common side effect of several established schizophrenia drugs. The phase II trial results validated the idea that normalizing glutamate neurotransmission might alleviate some of the negative symptoms.

65, LY-404039

66, LY-2140023

On the other hand, $mGluR_5$ is expressed ubiquitously in the mammalian CNS and is primarily localized postsynaptically, although it also displays some presynaptic localization. Activation of $mGluR_5$ elicits slow synaptic responses and modulates neuronal excitability through downstream signaling pathways. Previous studies have led to the hypothesis that $mGluR_5$-selective ligands may have potential utility as novel therapeutic agents for multiple psychiatric or neurological disorders including schizophrenia [319,320]. Therefore, $mGluR_5$ positive allosteric modulators represent an exciting approach for the treatment of schizophrenia. Three distinct series of $mGluR_5$ positive allosteric modulators exist. The first is difluorobenzaldazine (DFB, **67**), the second series is N-[4-chloro-2-[(1,3-dioxo-1,3-dihydro-2H-isoindol-2-yl)methyl]phenyl]-2-hydroxybenzamide (CPPHA, **68**) and the third is 3-cyano-N-(1,3-diphenyl-1H-pyrazol-5-yl)benzamide (CDPPB, **69**). Compound **67** was the first $mGluR_5$ positive allosteric modulators identified. It does not activate $mGluR_5$ when added alone, but increases the sensitivity of $mGluR_5$ to orthosteric agonists, and thereby shifts the concentration–response curve of orthosteric agonists to the left. It is highly selective for $mGluR_5$ and has little activity at other mGluR subtypes. The discovery of **67** provided a major breakthrough in demonstrating the possibility of developing $mGluR_5$-selective positive allosteric modulators. However, the poor potency, efficacy and solubility of this compound prevented its further study in native tissue preparations [316,317]. The validity of the $mGluR_5$-selective positive allosteric modulators needs to be further demonstrated by molecules with superior potency, efficacy and pharmacokinetic profiles.

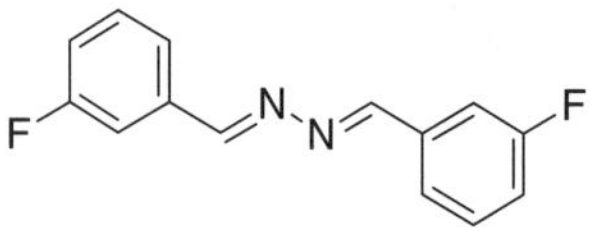

67, DFB 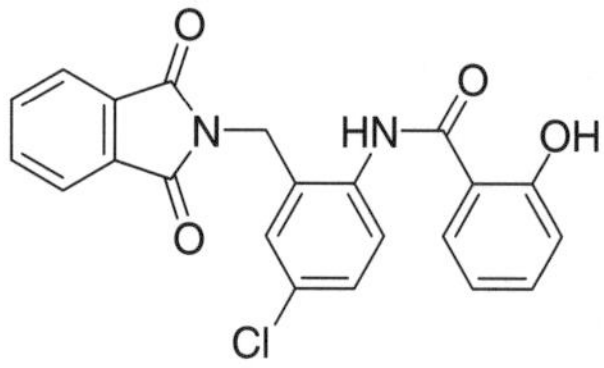   **68**, CPPHA 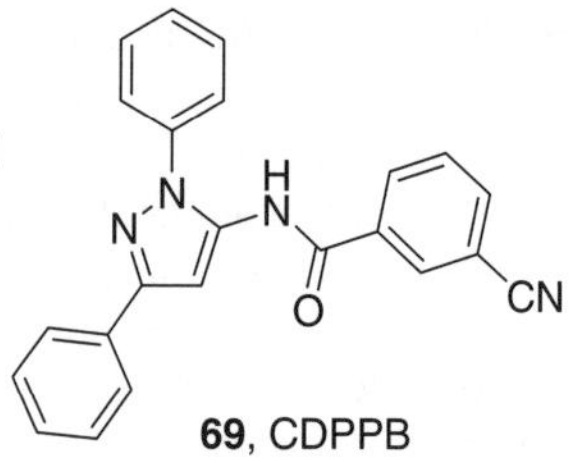   **69**, CDPPB

8.3.2. NMDA Receptor

In Section 2.2.2, we described experimental data linking the ionotropic glutamate receptor (NMDA receptor) with the core symptoms of schizophrenia [17,18]. PCP can induce all three types of schizophrenia symptoms and works by blocking the NMDA receptor. The NMDA hypofunction may result from decreased glutamate levels in the brain. However, glutamate levels have been shown to be unaltered in schizophrenic patients [19,20]. In addition to glutamate, NMDA receptor activation requires the cofactor glycine and blockade of this glycine site inhibits NMDA function. Other findings would support the hypotheses that decreased NMDA function, due to a kynurenic acid (**70**)-mediated blockade of the glycine site of the NMDA receptor contributes to the pathology of schizophrenia. In addition to kynurenic acid, NMDA function can be inhibited by other endogenous NMDA antagonists. For example, activity and binding of glutamate carboxypeptidease, the enzyme that breaks down NAAG (**71**) to NAA and glutamate, is reduced in schizophrenic patients [26,27], an effect unlikely mediated by changes in mRNA [28]. This is of relevance since NAAG antagonizes NMDA receptor function [29,30]. These data support a hypothesis that hypofunction of the NMDA receptor may result from endogenous, orthosteric, or allosteric inhibitors of NMDA receptor function.

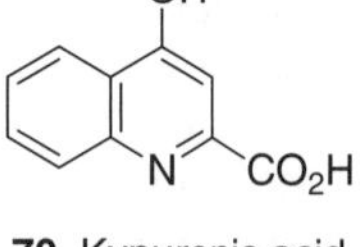

70, Kynurenic acid 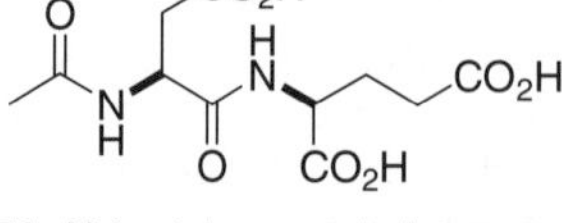   **71**, N-Acetyl-aspartyl glutamate

When given in combination with antipsychotic drugs, positive modulators of glycine/D-serine site of the NMDA receptor such as D-serine, glycine, and D-alanine significantly improve symptoms in patients with schizophrenia [318]. As orthosteric NMDA agonists result in neurotoxicity, targeting the allosteric site of the NMDA receptor would therefore enhance NMDA responses to glutamate without direct activation [31]. This is a more indirect way to target glutamate neurotransmission—blocking a glycine transporter known as GlyT1. Normally, GlyT1 decreases the concentration of glycine in synapses, the interneuron structures through which neurochemical messages propagate from neuron to neuron. Glycine, together with glutamate, is required to activate the NMDA receptor, which mediates glutamate signaling in the brain. A drug that blocks GlyT1 would allow higher concentrations of glycine to remain in the synapse and could aid in normalizing glutamate signaling in patients. Sarcosine (**72**), a GlyT1 antagonist, was more effective at reducing both positive and negative symptoms than was direct activation of the glycine/D-serine site of the NMDA receptor by D-serine [319].

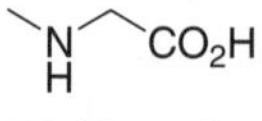

72, Sarcosine

Among the newer GlyT1 inhibitors, there are glycine and sarcosine (**72**) derivatives such as **73** (ALX-5407) and **74** [320]. There are also potent and selective nonamino-acid-containing GlyT1 inhibitors such as **75** (SSR504734), **76**, and **77** (DCCCyB). Sulfone **77** was derived from a HTS hit through methodical SAR to increase the GlyT1 selectivity as well as to decrease the liability of the P-glycoprotein transporter. Its phase I clinical trial has been recently completed.

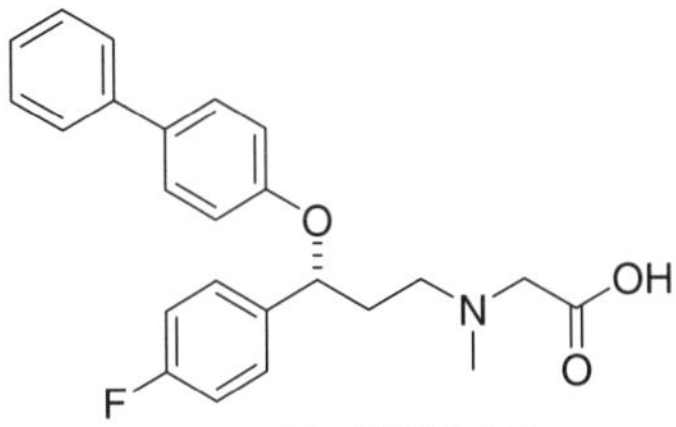

73, ALX-5407

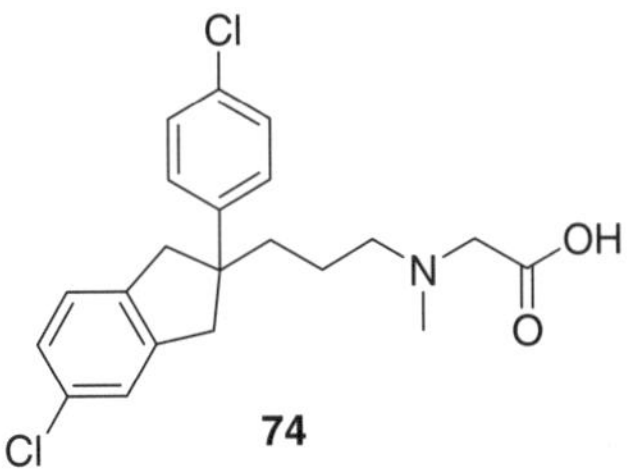

74

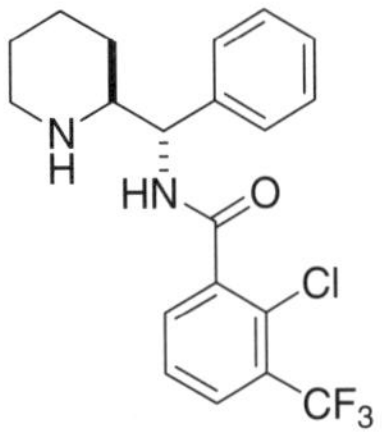

75, SSR504734

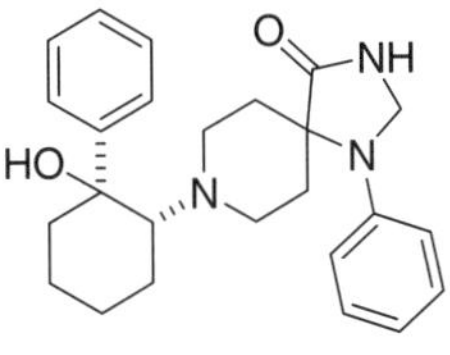

76

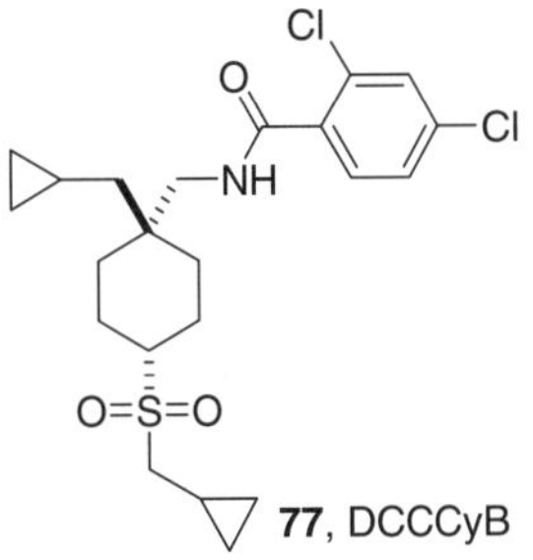

77, DCCCyB

8.3.3. AMPA Receptor

Ligands showing competitive antagonistic action at the AMPA type of Glu receptors were first reported in 1988, and the systemically active 2,3-dihydroxy-6-nitro-7-sulphamoyl-benzo[f]quinoxaline (**78**, NBQX) was first shown to have useful therapeutic effects in animal models of neurological disease in 1990. Since then, the quinoxaline template has represented the backbone of various competitive AMPA receptor antagonists belonging to different classes which had been developed in order to increase potency, selectivity and water solubility, but also to prolong the *in vivo* action [321].

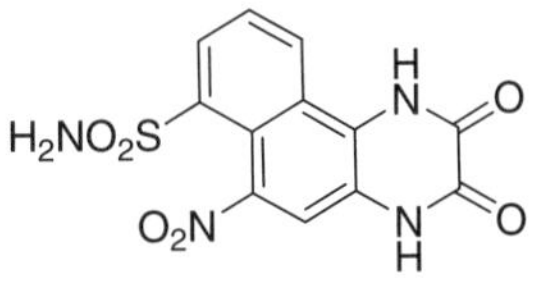

Unfortunately, direct activation of AMPA receptors carries the risk of producing seizures, excitotoxicity and a loss of efficacy due to desensitization. However, the discovery of positive allosteric modulators (PAMs) offers a mechanism for enhancing receptor activity while avoiding these issues [322]. There are two major structural classes of AMPA positive allosteric modulators [302,323].

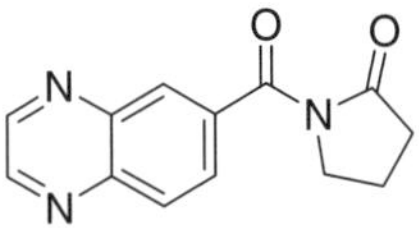

The benzamide series were the first class of AMPA positive allosteric modulators to be investigated when aniracetam (**79**) was found to potentiate AMPA receptor-mediated transmission in the hippocampus.

78, NBQX

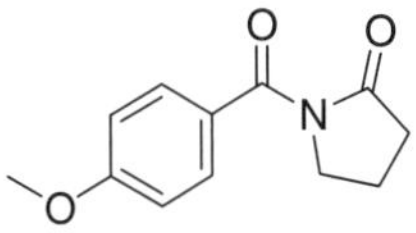

79, aniracetam

80, CX-516

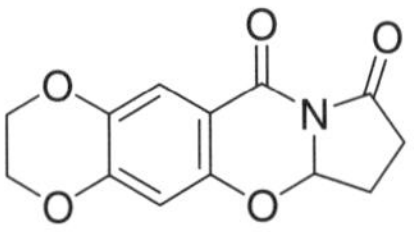

81, CX-614

The second major class of AMPA positive allosteric modulators are the benzothiadiazines derived from cyclothiazide (**82**). **82** is highly selective for flipoforms, giving a twofold (EC2×) potentiation value of the AMPA current at a concentration of 1.6 μM. S-18986 (**83**, EC2× = 60 μM) is in phase I clinical trials for the treatment of cognitive deficits. Benzothiadiazine **84** (EC2× = 8.8 μM) with even higher potency than **83** has also been reported.

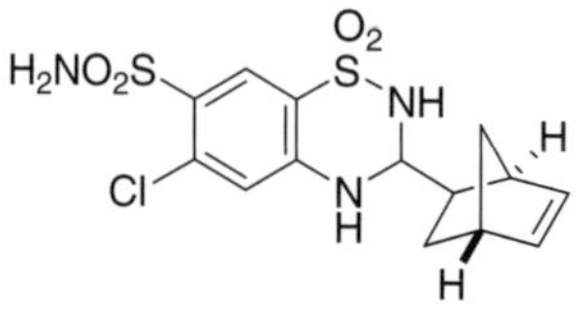

modulate the release of other neurotransmitters such as norepinephrine, dopamine, acetylcholine, serotonin, and GABA. Because of the effects of H_3 signaling on multiple neuronal transmitters, it has been suggested that H_3 antagonists/inverse agonists could be effective therapeutics for several CNS-related disorders. For instance, the potent and selective

82, cyclothiazide 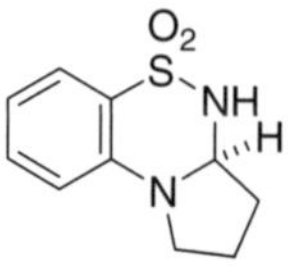   **83**, S-18986 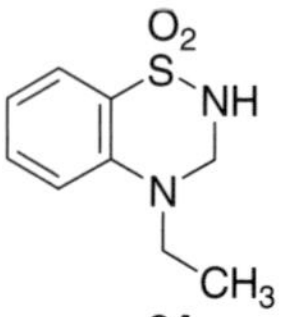   **84**

Additional AMPA positive allosteric modulators may be represented by the sulfonamides **85–87**.

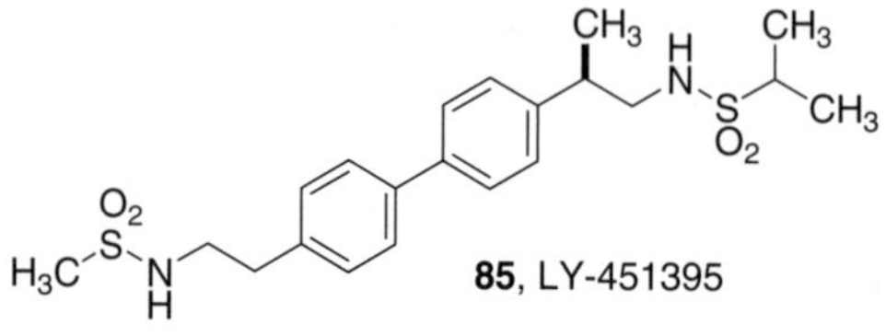

histamine H_3 receptor inverse agonist **88** (BF2.649) suppressed the excessive daytime sleep of narcoleptic patients [324].

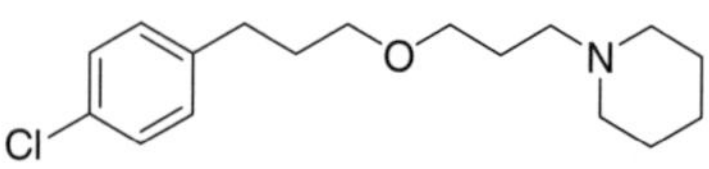

85, LY-451395

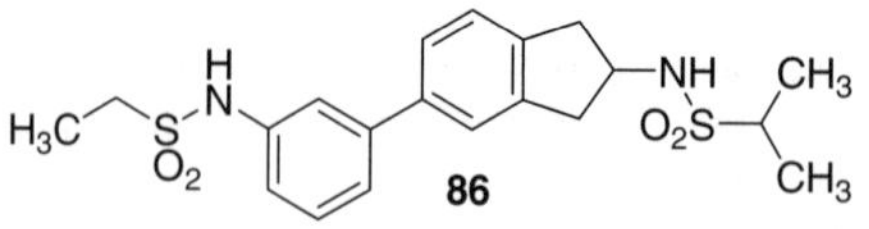

86    **87**

8.4. Histamine H_3 Receptor

Antagonists of the histamine H_1 and H_2 receptors have been successful as blockbuster drugs for treating allergic conditions and gastric ulcers, respectively. The histamine H_3 receptor was discovered pharmacologically in 1983 and genetically identified in 1999. Signaling through the H_3 receptor activates G proteins that inhibit adenylate cyclase activity and reduce intracellular cAMP levels. In the CNS, the H_3 receptor is localized on the presynaptic membrane as an autoreceptor and negatively regulates the release and synthesis of histamine. In addition, the H_3 receptor is known to

88, BF2.649, hH_3 (IC$_{50}$) 12 ± 3 nM

Several histamine H_3 receptor antagonists have been advanced into clinical trials [325–327]. **89** (ABT-239), a potent histamine H_3 receptor antagonist, was effective at a low dose (0.1 mg/kg s.c.) in a repeat trial inhibitory avoidance task in SHR pups, a model involving aspects of attention, impulsivity and learning that is thought to be related to characteristics of ADHD. Compound **89** was active in a social recognition model of short-term memory in aged and adult rats and in the water maze model, demonstrating effects on different aspects of cognitive impairment. Furthemore, **89** was also active in the prepulse inhibition startle model, a model of sensory gating proposed to be related to schizophrenia [325]. Unfortunately, despite its impressive *in vivo* profile for cognition enhancement, the development of **89** was halted due to cardiovascular liabilities. It inhibits [^{3}H]-dofetilide binding to the hERG potassium channel with a K_i of 195 nM. Its liabilities also include a high $c\log P$ of 5.2, which is believed to contribute to the high brain partitioning (B/P > 34) and high plasma protein binding. Another potent histamine H$_3$ receptor antagonist **90** (GSK189254) was in phase I clinical trials for the treatment of dementia. The potential of H$_3$ antagonists/inverse agonists as antipsychotics still needs to be validated in clinical trials.

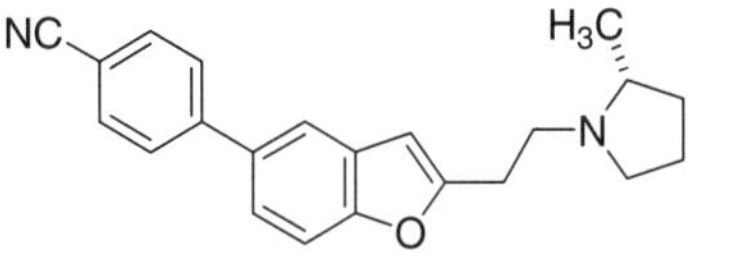

macokinetics [328]. Another selective $\alpha_4\beta_2$ agonist **92** (ABT-089) underwent phase I clinical trials and was reported to have an excellent pharmacokinetic profile in humans, good cardiovascular and gastrointestinal tolerability, and positive signs of cognitive effects as measured by decrease in reaction time.

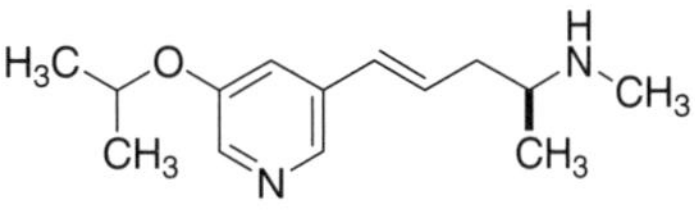

91, Ispronicline

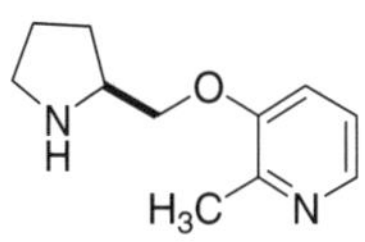

92, ABT-089

Positive allosteric modulators of subtype α_7 neuronal nicotinic acetylcholine receptor represent an important therapeutic potential. Positive allosteric modulators increase the probability of channel opening, while decrease the inherent agonist potential for receptor desensitization [302]. The selective $\alpha 7$ nAChR positive allosteric modulator **93** is in phase II

89, ABT-239, hH$_3$ (IC$_{50}$) 0.63 ± 0.24 nM

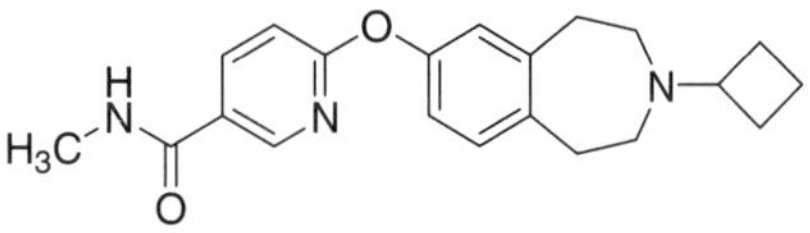

90, GSK189254, hH$_3$ (IC$_{50}$) 0.55 ± 0.07 nM

8.5. Neuronal Nicotinic Acetylcholine Receptor

Smoking is three times higher for schizophrenics than that in the general population and nicotine can produce modest transient improvements in cognitive and sensory deficits. Meanwhile, neuronal nicotinic acetylcholine receptors (nAChRs) are involved in a variety of attention and cognitive processes. nAChRs are Ca^{2+}-permeable, ligand-gated ion channels that modulate synaptic transmission of key regions of the central nervous systems involved in learning and memory [302].

The highly selective $\alpha_4\beta_2$ nAChR agonist **91** (isproniciline, TC-1734) is active in several animal models indictive of cogniotive enhancement. In phase I clinical trials, **91** was well tolerated and demonstrated linear phar-

clinical trials and several other positive allosteric modulators **94**, **95**, and **96** are in phase I clinical trials [306].

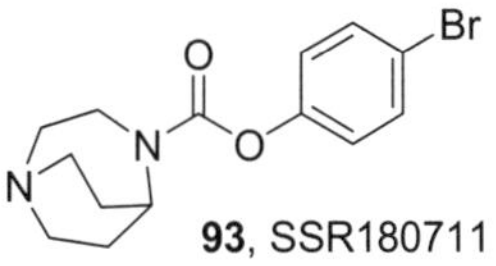

93, SSR180711

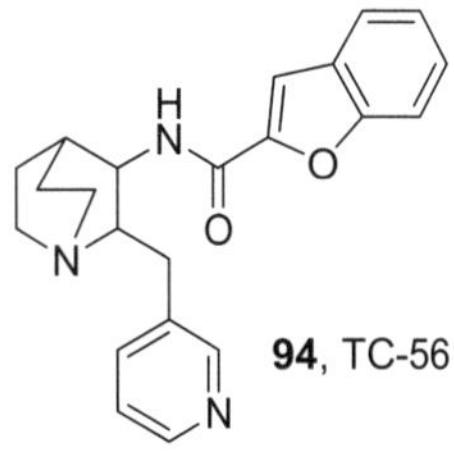

94, TC-5619

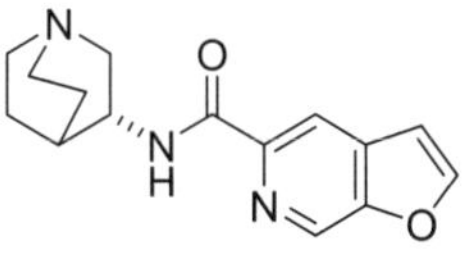

95, PHA-543613

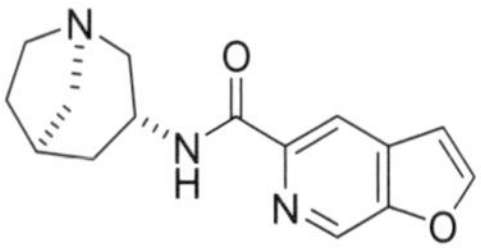

96, PHA-709829

8.6. Neurokinin₃ (NK₃) Receptor Antagonists

Antagonists of the neurokinin-1 receptor (NK1r) and NK2r have been extensively explored for both peripheral indications (e.g., urinary incontinence, asthma) and central indications (e.g., pain, depression). Recently, neurokinin receptor expression was systematically compared in rodent, primate, and human brain. These studies verified the presence of NK3r in primate brain although at exceptionally low levels compared with NK1r. NK₃ receptors are present on dopamine neurons in the A9 and A10 groups and modulate dopamine release and cholinergic tone. Overall, support for NK3r as a critical mediator in the human CNS remains equivocal based on expression and localization studies. However, two clinical studies disclosed two structurally unrelated NK3r antagonists that are reported to be well tolerated and show antipsychotic efficacy. These reports have accelerated further exploration of NK3r antagonists and heightened speculation about the potential for such antagonists to be a new generation of antipsychotics [329].

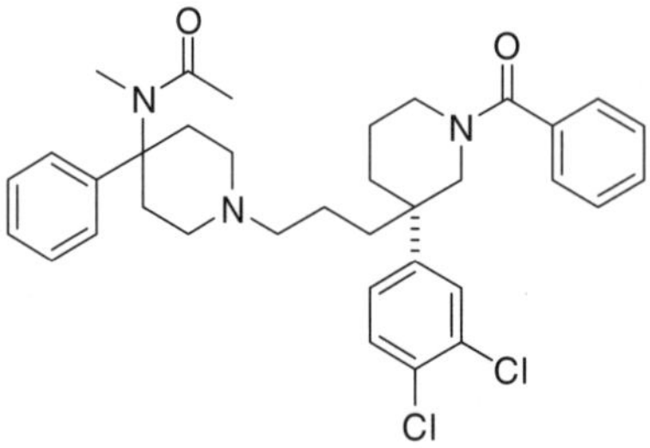

In animal models, the NK3r antagonist **97** (SSR146977) prevented NK3r agonist-induced release of Ach, 5-HT, and DA. At 200 mg/day, **97** (osanetant), an NK3r antagonist, showed antipsychotic efficacy similar to haloperidol (**3**). But unlike haloperidol (**3**), Compound **97** is devoid of EPS and weight gain. It was discontinued in August 2005 possibly because it failed to show dependent related efficacy [306]. Another selective NK3r antagonist **98** (talnetant) was active in phase II trials for schizophrenia. In a study involving 236 patients over 6 weeks, **98** showed significant antipsychotic effects in 40% of patients at 200 mg. Furthermore, it was better tolerated than resperidone (**6**) with reduced prolactin elevation, decreased tetesterone reduction, no weight gain, no QT_C effects and no EPS [329].

97, Osanetant
NK3r K_i = 1.2 nM
NK2r K_i = 40 nM
NK1r K_i = 774 nM

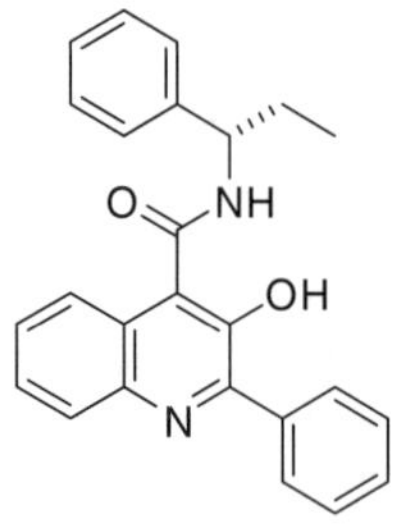

98, Talnetant
NK3r K_i = 1.0 nM
NK2r K_i = 144 nM
NK1r K_i = > 10 μM

8.7. PDE Inhibitors

Phosphodiesterase (PDE) inhibitors such as sildenafil, tadalafil and vardenafil as PDE5 inhibitors have had distinctive clinical benefit and notable commercial success for the treatment of erectile dysfunction.

PDEs are a class of key enzymes within the intracellular signal transduction cascade that follow activation of many types of membrane-bound receptors. PDEs degrade cyclic adenosine mono phosphate (cAMP) and/or cyclic guanosine mono phosphate (cGMP) by hydrolysis of phosphodiester bonds. Twenty-one genes are currently known to encode at least 11 different PDE families (PDE1–PDE11). Modifying the rate of cyclic nucleotide formation or degradation via PDEs will change the activation state of related pathways. Therefore, PDE inhibitors can prolong or enhance the effects of physiological processes mediated by cAMP or cGMP. It is noteworthy that PDEs are the most important means of inactivating intracellular cAMP in the brain, suggesting that PDE inhibitors present a potentially powerful means to manipulate second messengers involved in learning, memory, and mood.

PDE4 is a cGMP-specific phosphodiesterase that is widely distributed in humans. In animal models (acoustic startle and PPI), the PDE4 inhibitor **99** (rolipram) appears to have a pharmacological profile similar to that of some newer antipsychotics that claim to reduce motor side effect liability at therapeutic doses. However, **99** produces nausea and emesis at doses that overlap the therapeutic range, suggesting that more selective agents may be needed [330]. Another PDE4 inhibitor MEM1414 is in clinical trials for potential use in treating Alzheimer's disease [331]. In mice and rats, cognitive improvements were observed at doses as low as 1 mg/kg i.p. Phase II clinical trials for MEM1414 were initiated in 2006.

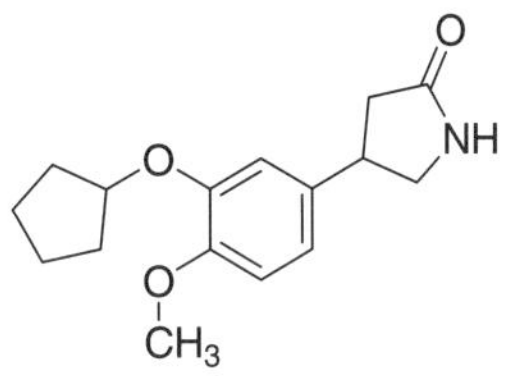

completed and it now has entered into phase II trials [333]. This compound will help determine whether PDE10A inhibitors are effective against schizophrenia.

8.8. Cannabinoids

The association between cannabinoids and psychosis has been long recognized. Cannabinoids can induce acute transient psychotic symptoms or an acute psychosis in a small portion of individuals. Similar to smoking and lung cancer, it is more likely that cannabis exposure is a component cause that interacts with other factors, for example, genetic risk, to "cause" schizophrenia. There is also tantalizing evidence from postmortem, neurochemical and genetic studies suggesting CB_1 receptor dysfunction (endogenous hypothesis) in schizophrenia that warrants further investiga-

99, rolipram

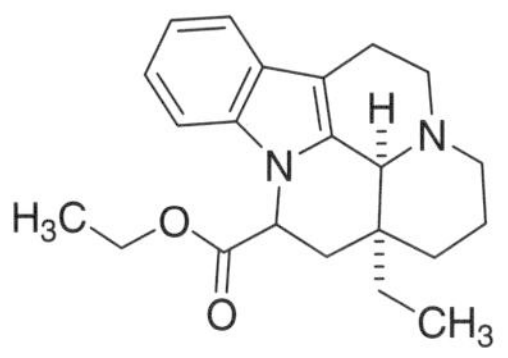

100, papaverine

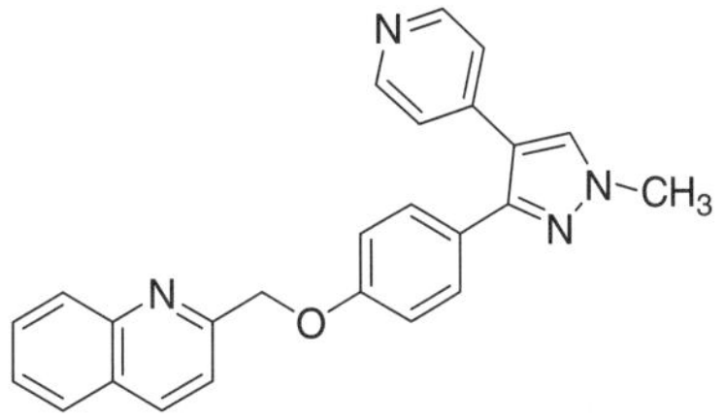

101, PF-2545920

PDE10 is a dual substrate cyclic nucleotide phosphodiesterase (cAMP and cGMP) that has preferential expression in the spiny neurons of the striatum, a portion of the brain that is believed to have impaired regulation in schizophrenia. Inhibiting PDE10 in rodents leads to increases in cyclic nucleotides in the striatum and a behavioral antipsychotic phenotype. In mice, **100** (papaverine), a PDE10A inhibitor, is associated with increased cGMP levels in the striatum and increased phosphorylation of CREB, which are both crucial for striatal function [332]. Interestingly, **100** was also found to reduce deficits caused by chronic phencyclidine treatment, a recognized animal model for schizophrenia. The findings led to intense interest in exploring this approach for schizophrenia treatment. Another PDE10A selective inhibitor **101** (PF-2545920) was effective at treating schizophrenia symptoms in animals. Phase I clinical trials with **101** were

tion [334]. The central cannabinoid (CB_1) receptor is a G protein-coupled receptor that is widely distributed in the central nervous system and several peripheral tissues.

A potent and selective CB_1 receptor antagonist **102** (rimonabant) was approved in Europe for obesity and is currently suspended for neurological and psychiatric side effects. Compound **102** was shown to reduce stimulant-induced hyperactivity [335]. Compound **103** (AVE-1625) is a highly potent, selective antagonist for the CB_1 receptor with K_i values of 0.16–0.44 nM. At 1–3 mg/kg, AVE-1625 significantly improves the performance of rodents in working memory tasks [336]. **104** (SLV-319) 319 is a potent and selective CB_1 receptor antagonist with K_i values of 7.8 and 7943 nM for CB_1 and peripheral cannabinoid (CB_2), respectively [337]. **104** is less lipophilic (log P 5.1) and therefore more water soluble than other known CB_1 receptor ligands.

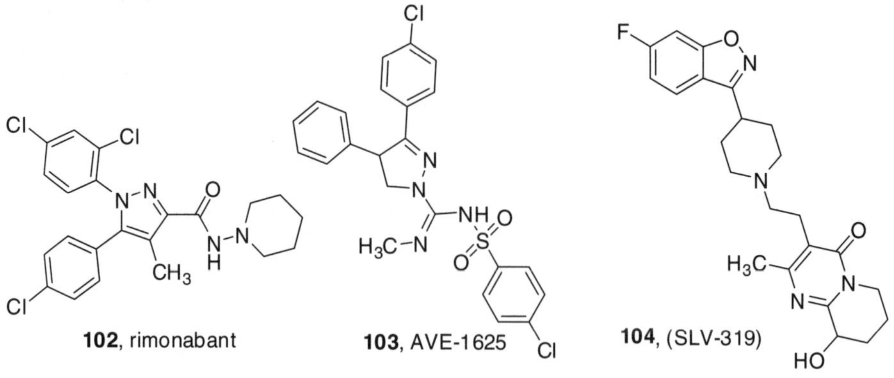

102, rimonabant

103, AVE-1625

104, (SLV-319)

In summary, many targets exist for the treatment of schizophrenia beyond the traditional dopamine/serotonin receptor mechanism. Hopefully, some of these mechanisms will provide novel antipsychotics that are as efficacious as the existing antipsychotics but yet devoid of the side effects associated with them such as EPS and weight gain.

REFERENCES

1. *Diagnostic and Statistical Manual of Mental Disorders*. 4th ed. Text Rev. Washington, DC: American Psychiatric Association; 2000.
2. Green MF, Nuechterlein KH, Gold JM, Barch DM, Cohen J, Essock S, Fenton WS, Frese F, Goldberg TE, Heaton RK, Keefe RS, Kern RS, Kraemer H, Stover E, Weinberger DR, Zalcman S, Marder SR Biol Psychiatry 2004;56:301–307.
3. Courvoisier S. J Clin Exp Psychopathol 1956;17:25–37.
4. O'Keeffe R, Sharman DF, Vogt M. Br J Pharmacol 1970;38:287–304.
5. Van Rossum JM. Arch Int Pharmacodyn Ther 1966;160:492–494.
6. Seeman P, Lee T. Science 1975;188:1217–1219.
7. Creese I, Burt DR, Snyder SH. Science 1976;30:481–483.
8. Cross AJ, Crow TJ, Owen F. Psychopharmacology 1981;74:122–124.
9. Wong DF, Wagner HN, Jr, Tune LE, Dannals RF, Pearlson GD, Links JM, Tamminga CA, Broussolle EP, Ravert HT, Wilson AA, Toung JK, Malat J, Williams JA, O'Tuama LA, Snyder SH, Kuhar MJ, Gjedde A. Science 1986;234:1558–1563.
10. Farde L, Wiesel FA, Stone-Elander S, Halldin C, Nordström AL, Hall H, Sedvall G. Arch Gen Psychiatry 1990;47:213–219.
11. Hietala J, Syvälahti E, Vuorio K, Någren K, Lehikoinen P, Ruotsalainen U, Räkköläinen V, Lehtinen V, Wegelius U. Arch Gen Psychiatry 1994;51:116–123.
12. Nordström AL, Farde L, Eriksson L, Halldin C. Psychiatry Res 1995;61:67–83.
13. Laruelle M, Abi-Dargham A, Van Dyck CH, Gil R, D'Souza CD, Erdos J, McCance E, Rosenblatt W, Fingado C, Zoghbi SS, Baldwin RM, Seibyl JP, Krystal JH, Charney DS, Innis RB. PNAS 1996;93:9235–9240.
14. Bannon JM, Roth RH. Pharmacol Rev 1983;35:53–68.
15. Davis KL, Kanh RS, Ko G, Davidson M. Am J Psychiatry 1991;148:1474–1486.
16. Luby ED, Cohen BD, Rosenbaum G, Gottlieb JS, Kelley AMA Arch Neurol Psychiatry 1959;81:363–369.
17. Javitt DC, Zukin SR. Am J Psychiatry 1991;148:1301–1308.
18. Olney JW, Farber NB. Arch Gen Psychiatry 1995;52:998–1007.
19. Perry TL. Neurosci Lett. 1982;28:81–85.
20. Korpi ER, Kaufmann CA, Marnela KM, Weinberger DR. Psychiatry Res 1987;20:337–45.
21. Schwarcz R, Rassoulpour A, Wu HQ, Medoff D, Tamminga CA, Roberts RC. Biol Psychiatry 2001;50:521–530.
22. Erhardt S, Schwieler L, Engberg G. Adv Exp Med Biol 2003;527:155–165.
23. Nilsson LK, Linderholm KR, Engberg G, Paulson L, Blennow K, Lindström LH, Nordin C, Karanti A, Persson P, Erhardt S. Schizophr Res 2005;80:315–322.

24. Parsons CG, Danysz W, Quack G, Hartmann S, Lorenz B, Wollenburg C, Baran L, Przegalinski E, Kostowski W, Krzascik P, Chizh B, Headley PM. J Pharmacol Exp Ther 1997;283:1264–1275.
25. Erhardt S, Schwieler L, Emanuelsson C, Geyer M. Biol Psychiatry 2004;56:255–260.
26. Tasi G, Passani LA, Slusher BS, Carter R, Baer L, Kleinman JE, Coyle JT. Arch Gen Psychiatry 1995;52:829–836.
27. Guilarte TR, Hammoud DA, McGlothan JL, Caffo BS, Foss CA, Kozikowski AP, Pomper MG. Schizophr Res 2008;99:324–332.
28. Ghose S, Weickert CS, Colvin SM, Coyle JT, Herman MM, Hyde TM, Kleinman JE. Neuropsychopharmacology 2004;29:117–125.
29. Grunze HC, Rainnie DG, Hasselmo ME, Barkai E, Hearn EF, McCarley RW, Greene RW. J Neurosci. 1996 16:2034–2043.
30. Bergeron R, Imamura Y, Frangioni JV, Greene RW, Coyle JT. J Neurochem. 2007;100:346–357.
31. Coyle JT. Cell Mol Neurobiol 2006;26:365–384.
32. Fujii Y, Shibata H, Kikuta R, Makino C, Tani A, Hirata N, Shibata A, Ninomiya H, Tashiro N, Fukumaki Y. Psychiatr Genet 2003;13:71–76.
33. Egan MF, Straub RE, Goldberg TE, Yakub I, Callicott JH, Hariri AR, Mattay VS, Bertolino A, Hyde TM, Shannon-Weickert C, Akil M, Crook J, Vakkalanka RK, Balkissoon R, Gibbs RA, Kleinman JE, Weinberger DR. Proc Natl Acad Sci 2004;101:12604–12609.
34. Patil ST, Zhang L, Martenyi F, Lowe SL, Jackson KA, Andreev BV, Avedisova AS, Bardenstein LM, Gurovich IY, Morozova MA, Mosolov SN, Neznanov NG, Reznik AM, Smulevich AB, Tochilov VA, Johnson BG, Monn JA, Schoepp DD. Nat Med 2007;13:1102–1107.
35. Sesack SR, Hawrylak VA, Melchitzky DS, Lewis DA. Cereb Cortex 1998;8:614–622.
36. Melchitzky DS, Lewis DA. Cereb Cortex 2003;13:452–460.
37. Volk DW, Austin MC, Pierri JN, Sampson AR, Lewis DA. Arch Gen Psychiatry 2000;57:237–245.
38. Volk D, Austin M, Pierri J, Sampson A, Lewis D. Am J Psychiatry 2001;158:256–265.
39. Guidotti A, Auta J, Davis JM, Di-Giorgi-Gerevini V, Dwivedi Y, Grayson DR, Impagnatiello F, Pandey G, Pesold C, Sharma R, Uzunov D, Costa E. Arch Gen Psychiatry 2000;57:1061–1069.
40. Schleimer SB, Hinton T, Dixon G, Johnston GA. Neuropsychobiology 2004;50:226–230.
41. Guidotti A, Auta J, Davis JM, Dong E, Grayson DR, Veldic M, Zhang X, Costa E. Psychopharmacology 2005;180:191–205.
42. Meltzer HY, Li Z, Kaneda Y, Ichikawa J. Prog Neuropsychopharmacol Biol Psychiatry 2003;27:1159–1172.
43. Díaz-Mataix L, Scorza MC, Bortolozzi A, Toth M, Celada P, Artigas F. J Neurosci 2005;25:10831–10843.
44. Sumiyoshi T, Park S, Jayathilake K, Roy A, Ertugrul A, Meltzer HY. Schizophr Res 2007;95:158–168.
45. Reynolds GP, Arranz B, Templeman LA, Fertuzinhos S, San L. Am J Psychiatry 2006;163:1826–1829.
46. Jakab RL, Goldman-Rakic PS. Proc Natl Acad Sci 1998;20:735–740.
47. Willine DL, Deutch AY, Roth BL. Synapse 1997;27:79–82.
48. Arranz B, Rosel P, San L, Ramírez N, Dueñas RM, Salavert J, Centeno M, Del Moral E. Psychiatry Res 2007;153:103–109.
49. Levin ED, McClernon FJ, Rezvani AH. Psychopharmacology 2006;184:523–539.
50. Levin ED, Wilson W, Rose JE, McEvoy J. Neuropsychopharmacology 1996;15:429–36.
51. Freedman R, Hall M, Adler LE, Leonard S. Biol Psychiatry 1995;38:22–33.
52. Durany N, Zöchling R, Boissl KW, Paulus W, Ransmayr G, Tatschner T, Danielczyk W, Jellinger K, Deckert J, Riederer P. Neurosci Lett 2000;287:109–112.
53. Guan ZZ, Zhang X, Blennow K, Nordberg A. Neuroreport 1999;10:1779–1782.
54. Olincy A, Harris JG, Johnson LL, Pender V, Kongs S, Allensworth D, Ellis J, Zerbe GO, Leonard S, Stevens KE, Stevens JO, Martin L, Adler LE, Soti F, Kem WR, Freedman R. Arch Gen Psychiatry 2006;63:630–638.
55. Yamamoto K, Hornykiewicz O. Prog Neuropsychopharmacol Biol Psychiatry 2004;28:913–922.
56. Binder EB, Kinkead B, Owens MJ, Nemeroff CB. Biol Psychiatry 2001;50:856–72.
57. Arinami T. J Hum Genet 2006;51:1037–1045.
58. Allen NC, Bagade S, McQueen MB, Ioannidis JP, Kavvoura FK, Khoury MJ, Tanzi RE, Bertram L. Nat Genet 2008;40:827–834.
59. Dayalu P, Chou KL. Expert Opin Pharmacother 2008;9:1451–1462.

60. Kapur S, Zipursky R, Jones C, Remington G, Houle S. Am J Psychiatry 2000;157:514–520.
61. Lieberman JA, Stroup TS, McEvoy JP, Swartz MS, Rosenheck RA, Perkins DO, Keefe RS, Davis SM, Davis CE, Lebowitz BD, Severe J, Hsiao JK. Clinical Antipsychotic Trials of Intervention Effectiveness (CATIE) Investigators. N Engl J Med 2005;353:1209–1223.
62. Strawn JR, Keck PE, Jr, Caroff SN. Am J Psychiatry 2007;164:870–876.
63. Mackin P. Cardiac Hum Psychopharmacol 2008; (Suppl) 1: 3–14.
64. Drici MD, Priori S. Pharmacoepidemiol Drug Saf 2007;16:882–890.
65. Watanabe A, Shibata I, Kato T. Psychiatry Clin Neurosci 2004;58:268–273.
66. Haddad PM, Sharma SG. CNS Drugs 2007;21:911–936.
67. Stroup TS, Lieberman JA, McEvoy JP, Swartz MS, Davis SM, Rosenheck RA, Perkins DO, Keefe RS, Davis CE, Severe J, Hsiao JK. Am J Psychiatry 2006;163:611–622.
68. Lambert BL, Cunningham FE, Miller DR, Dalack GW, Hur K. Am J Epidemiol 2006;164:672–681.
69. Haddad PM, Dursun SM. Hum Psychopharmacol 2008;Supp 1: 15–26.
70. *Clozapine package inset.* Teva Pharmaceudicals Ind. Ltd; Jerusalem, Israel.
71. Wallace M. Int Clin Psychopharmacol 2001;16 (Suppl) 1: S21–24.
72. Haddad PM, Wieck A. Drugs 2004;64:2291–2314.
73. Atasoy N, Erdogan A, Yalug I, Ozturk U, Konuk N, Atik L, Ustundag Y. Prog Neuropsychopharmacol Biol Psychiatr 2007;31:1255–1260.
74. Erdogan A, Atasoy N, Akkurt H, Ozturk D, Karaahmet E, Yalug I, Yalug K, Ankarali H, Balcioglu I. Prog Neuropsychopharmacol Biol Psychiatry 2008;32:849–857.
75. Kane JM, Leucht S, Carpenter D, Docherty JP. J Clin Psychiatry 2003;64:5–19.
76. Tandon R, Belmaker RH, Gattaz WF, Lopez-Ibor JJ, Jr, Okasha A, Singh B, Stein DJ, Olie JP, Fleischhacker WW, Moeller HJ. Section of Pharmacopsychiatry, World Psychiatric Association. Schizophr Res 2008;100:20–38.
77. Jones PB, Barnes TR, Davies L, Dunn G, Lloyd H, Hayhurst KP, Murray RM, Markwick A, Lewis SW. Arch Gen Psychiatry 2006;63:1079–1087.
78. Van Ree JM, Elands J, Király I, Wolterink G. Eur J Pharmacol 1989;166:441–452.
79. Gold LH, Swerdlow NR, Koob GF. Behav Neurosci 1988;102:544–552.
80. Jones GH, Robbins TW. Pharmacol Biochem Behav 1992;43:887–895.
81. Costall B, Naylor RJ, Nohria V. Eur J Pharmacol 1978;50:39–50.
82. Arnt J. Eur J Pharmacol 1983;90:47–55.
83. Nordquist RE, Risterucci C, Moreau JL, Von Kienlin M, Künnecke B, Maco M, Freichel C, Riemer C, Spooren W. Neuropharmacology 2008;54:405–446.
84. Chiodo LA, Bunney BS. J Neurosci 1983;3:1607–1619.
85. Akhtar M, Uma Devi P, Ali A, Pillai KK, Vohora D. Fundam Clin Pharmacol 2006;20:373–378.
86. Skuza G, Rogóz Z. Pharmacol Rep 2006;58:626–635.
87. Siuciak JA, Chapin DS, McCarthy SA, Guanowsky V, Brown J, Chiang P, Marala R, Patterson T, Seymour PA, Swick A, Iredale PA. Neuropharmacology 2007;52:279–590.
88. Jentsch JD, Roth RH. Neuropsychopharmacology 1999;20:201–225.
89. Stefani MR, Moghaddam B. Ann NY Acad Sci 2003;1003;464–467.
90. Rodefer JS, Nguyen TN, Karlsson JJ, Arnt J. Neuropsychopharmacology. 2008;33:2657–2666.
91. Noda Y, Yamada K, Furukawa H, Nabeshima T. Br J Pharmacol 1995;116:2531–2537.
92. Noda Y, Kamei H, Mamiya T, Furukawa H, Nabeshima T. Neuropsychopharmacology 2000;23:375–387.
93. Porsolt RD, Le Pichon M, Jalfre M. Nature 1977;266:730–732.
94. Sams-Doss F. Behav Pharmacol 1998;7:3–23.
95. Lahti AC, Koffel B, LaPorte D, Tamminga CA. Neuropsychopharmacology 1995;13:9–19.
96. Malhotra AK, Adler CM, Kennison SD, Elman I, Pickar D, Breier A. Biol Psychiatry 1997;42:664–668.
97. Braff D, Stone C, Callaway E, Geyer M, Glick I, Bali L. Psychophysiology 1978;15:339–343.
98. Kumari V, Sharma T. Psychopharmacology 2002;162:97–101.
99. Lubow RE. Schizophr Bull 2005;31:139–153.
100. Gray NS, Pickering AD, Hemsley DR, Dawling S, Gray JA. Psychopharmacology 1992;107:425–430.

101. Swerdlow NR, Stephany N, Wasserman LC, Talledo J, Sharp R, Auerbach PP. Psychopharmacology 2003;169:314–320.
102. Baruch I, Hemsley DR, Gray JA. J Nerv Ment Discov 1988;176:598–606.
103. Gray NS, Pilowsky LS, Gray JA, Kerwin RW. Schizophr Res 1995;17:95–107.
104. Weiner I, Lubow RE, Feldon J. Psychopharmacology 1984;83:194–199.
105. Aguado L, San Antonio A, Pérez L, del Valle R, Gómez J. Behav Neural Biol 1994;61:271–281.
106. Turgeon SM, Auerbach EA, Duncan-Smith MK, George JR, Graves WW. Pharmacol Biochem Behav 2000;66:533–539.
107. Weiner I, Feldon J. J Pharmacol Biochem Behav 1992;42:625–631.
108. Weiner I. Psychopharmacology 2003;169:257–297.
109. Wadenberg ML, Hicks PB. Neurosci Biobehav Rev 1999;23:851–862.
110. Moore NA, Tye NC, Axton MS, Risius FC. J Pharmacol Exp Ther 1992;262:545–551.
111. Seeger TF, Seymour PA, Schmidt AW, Zorn SH, Schulz DW, Lebel LA, McLean S, Guanowsky V, Howard HR, Lowe JA, 3rd, Heym J. J Pharmacol Exp Ther 1995;275:101–113.
112. Natesan S, Reckless GE, Nobrega JN, Fletcher PJ, Kapur S. Neuropsychopharmacology 2006:31:1854–1863.
113. Takano A, Suhara T, Maeda J, Ando K, Okauchi T, Obayashi S, Nakayama T, Kapur S. Psychiatry Clin Neurosci 2004;58:330–332.
114. Wadenberg ML, Kapur S, Soliman A, Jones C, Vaccarino F. Psychopharmacology 2000;150:422–429.
115. O'Tuathaigh CM, Babovic D, O'Meara G, Clifford JJ, Croke DT, Waddington JL. Neurosci Biobehav Rev 2007;31:60–78.
116. Benson MA, Newey SE, Martin-Rendon E, Hawkes R, Blake DJ. J Biol Chem 2001;276:24232–24241.
117. Straub RE, Jiang Y, MacLean CJ, Ma Y, Webb BT, Myakishev MV, Harris-Kerr C, Wormley B, Sadek H, Kadambi B, Cesare AJ, Gibberman A, Wang X, O'Neill FA, Walsh D, Kendler KS. Am J Hum Genet 2002;71:337–348.
118. Talbot K, Eidem WL, Tinsley CL, Benson MA, Thompson EW, Smith RJ, Hahn CG, Siegel SJ, Trojanowski JQ, Gur RE, Blake DJ, Arnold AE. J Clin Invest 2004;113:1353–1363.
119. Hattori S, Murotani T, Matsuzaki S, Ishizuka T, Kumamoto N, Takeda M, Tohyama M, Yamatodani A, Kunugi H, Hashimoto R. Biochem Biophys Res Commun. 2008;22: 298–302.
120. St Clair D, Blackwood D, Muir W, Carothers A, Walker M, Spowart G, Gosden C, Evans HJ. Lancet 1990;336:13–16.
121. Millar JK, Wilson-Annan JC, Anderson S, Christie S, Taylor MS, Semple CA, Devon RS, Clair DM, Muir WJ, Blackwood DH, Porteous DJ. Hum Mol Genet 2000;22:1415–1423.
122. Sachs NA, Sawa A, Holmes SE, Ross CA, DeLisi LE, Margolis RL. Mol Psychiatry 2005;10:758–764.
123. Li W, Zhou Y, Jentsch JD, Brown RA, Tian X, Ehninger D, Hennah W, Peltonen L, Lönnqvist J, Huttunen MO, Kaprio J, Trachtenberg JT, Silva AJ, Cannon TD. Proc Natl Acad Sci USA 2007;13:18280–18285.
124. Clapcote SJ, Lipina TV, Millar JK, Mackie S, Christie S, Ogawa F, Lerch JP, Trimble K, Uchiyama M, Sakuraba Y, Kaneda H, Shiroishi T, Houslay MD, Henkelman RM, Sled JG, Gondo Y, Porteous DJ, Roder JC. Neuron 2007;54:387–402.
125. Gogos JA, Gerber DJ. Trends Pharmacol Sci 2006;27:226–233.
126. Costa E, Dong E, Grayson DR, Guidotti A, Ruzicka W, Veldic M. Epigenetics 2007;2:29–36.
127. van der Stelt O, Belger A. Schizophr Bull 2007;33:955–970.
128. Jensen O, Kaiser J, Lachaux JP. Trends Neurosci 2007;30:317–324.
129. Kwon JS, O'Donnell BF, Wallenstein GV, Greene RW, Hirayasu Y, Nestor PG, Hasselmo ME, Potts GF, Shenton ME, McCarley RW. Arch Gen Psychiatry 1999;56:1001–1005.
130. Gallinat J, Winterer G, Herrmann CS, Senkowski D. Clin Neurophysiol 2004;115:1863–1874.
131. Hajós M, Hoffmann WE, Kocsis B. Biol Psychiatry 2008;63:1075–1083.
132. Ellenbroek B, Cools AR. Life Sci 1988;42:1205–1213.
133. Prinssen EP, Ellenbroek BA, Stamatovic B, Cools AR. Brain Res 1995;673:283–289.
134. Turrone P, Remington G, Nobrega JN. Neurosci Biobehav Rev 2002;26:361–380.
135. Gao XM, Sakai K, Tamminga CA. Neuropsychopharmacology 1998;19:428–433.
136. Rourke C, Starr KR, Reavill C, Fenwick S, Deadman K, Jones DN. Psychopharmacology 2006;184:107–114.

137. Kalinichev M, Rourke C, Daniels AJ, Grizzle MK, Britt CS, Ignar DM, Jones DN. Psychopharmacology 2005;182:220–231.
138. The FDA website: http://www.centerwatch.com/patient/drugs/DRUGLIST.html.
139. Li JJ, Johnson DS, Sliskovic DR, Roth BD. Contemporary Drug Synthesis. Hoboken, NJ: Wiley & Sons; 2004. p 89–111.
140. Li JJ. *Laughing Gas, Viagra, and Lipitor, The Human Stories behind the Drugs We Use*. New York, NY: Oxford University Press; 2006. p 150–158.
141. De Oliveira IR, Juruena MF. J Clin Pharm Therap 2006;31:523–534.
142. Kramer M, Simpson G, Maciulis V, Kushner S, Vijapurkar U, Lim P, Eerdekens M. J Clin. Psychopharm 2007;27:6–14.
143. Seeman P, Corbett R, Nam D, Van Tol HHM. Jpn J Pharmacol 1996;71:187–204.
144. Seeman P, Corbett R, Nam D, Van Tol HHM. Neuropsychopharmacol 1997;16:93–110.
145. Seeman P. Can J Psychiatry 2002;47:27–38.
146. Miyamoto S, Duncan GE, Marx CE, Lieberman JA. Mol Psychiatry 2005;10:79–104.
147. Meltzer HY, Matsubara S, Lee JC. J Pharmacol Exp Therap 1989;251:238–246.
148. Lee T, Tang SW. Psychiatry Res 1984;12:277–285.
149. Kiss B, Bitter I. Structural Analogues of Clozapine. In: Fischer J, Ganellin CR, editors. Analogue-Based Drug Discovery. Weinheim, Germany: Wiley-VCH; 2006. p 297–313.
150. Wilk S, Stanley M. Eu J Pharmacol 1977;41:65–726.
151. Meltzer HY, Fibiger HC. Neuropsychopharmacol 1976;14:83–85.
152. Bhana N, Foster RH, Olney R, Plosker GL. Drugs 2001;61:111–161.
153. Gunasekara NS, Spencer CM. CNS Drugs 1998;9:325–340.
154. Goldstein JM. Drugs Today 1999;35:193–210.
155. Nemeroff CB, Kinkead B, Goldstein J. J Clin Psychiatry 2002;63 (Suppl 13): 5–11.
156. Apiquian R, Fresan A, Ulloa R-E, de la Fuente-Sandoval C, Herrera-Estrella M, Vazquez A, Nicolini H, Kapur S. Neuropsychopharmacol 2005;30:72236–72244.
157. Lydiard RB, Gelenberg AJ. Pharmacotherapy 1981;1:163–178.
158. Liegeois JFF, Rogister FA, Bruhwyler J, Damas J, Nguyen TP, Inarejos MO, Chleide EMG, Mercier MGA, Delarge JE. J Med Chem 1984;37:519–525.
159. Seminara G, Trassari V, Prestifilippo N, Chiavetta R. Calandra Minerva Psichiatrica 1993;34:95–99.
160. Velasco A, Lerida MT, Mayo R, Perez-Accino C, Alamo C. Gen Pharmacol 1998;30:521–524.
161. Smits RA, Lim HD, Stegink B, Bakker RA, De Esch IJP, Leurs R. J Med Chem 2006;49:4512–4516.
162. Eichenberger E. Arzneimittel-Forschung 1984;34:5110–113.
163. Andersen PH, Braestrup C. J Neurochem 1986;47:1822–1831.
164. Farber NB, Foster J, Duhan NL, Olney JW. Schizophrenia Res 1996;21:33–37.
165. Arnt J, Skarsfeldt T. Neuropsychopharmacol 1998;18:63–101.
166. Janssen PAJ, Niemegeers CJE, Awouters F, Schellekens KHL, Megens AAHP, Meert TF. J Pharmacol Exp Ther 1988;244:685–693.
167. Buckley PF. Drugs Today 2000;36:583–589.
168. Schmidt AW, Lebel LA, Howard HR, Zorn SH. Eur J Pharmacol 2001;425:197–201.
169. Murdoch DK, Gillian M. CNS Drugs 2006;20:233–255.
170. Herrick-Davis K, Grinde E, Teitler M. J Pharmacol Exp Therapeutics 2000;295(1): 226–232.
171. Hietala J, Kuoppamäki M, Majasuo H, Palvimaki E-P, Laakso A, Syvalahti E. Psychopharmacol 2001;157(2): 180–187.
172. Kasper S, Tauscher J, Küfferle B, Barnas C, Hesselmann B, Asenbaum S, Podreka I, Brucke T. Psychopharmacol 1998;136:367–373.
173. Pilowsky LS, O'Connell PN, et al. Psychopharmacol 1997;130:152–158.
174. Baldessarini R, Tarazi F. Goodman and Gilman's the Pharacological Basis of Therapeutics. McGraw Hill: New York; 2006.
175. Caccia S. Curr Opin Investig Drugs 2002;3:1073.
176. Mathews M, Muzina DJ. Cleve Clin J Med 2007;74:597.
177. Mauri MC, Volonteri LS, Colasanti A, Fiorentini A, De Gaspari IF, Bareggi SR. Clin Pharmacokinet 2007;46:359.
178. Murray M. J Pharm Pharmacol 2006;58:871.
179. Caccia S. Clin Pharmacokinet 2000;38:393.
180. DeVane C, Markowitz J. Metabolic Drug Interactions. Philadelphia: Lippincott Williams & Wilkins; 2000. p 245.
181. Hubbard JW, Midha KK, Hawes EM, McKay G, Marder SR, Aravagiri M, Korchinski ED. Br J Psychiatry (Suppl) 1993;19.

182. Urichuk L, Prior TI, Dursun S, Baker G. Curr Drug Metab 2008;9:410.
183. Conley RR, Kelly DL. Psychopharmacol Bull 2007;40:77.
184. Meyer J. CNS Spectr 2007;12:6.
185. Obach RS, Walsky RL. J Clin Psychopharmacol 2005;25:605.
186. Prior TI, Baker GB. J Psychiatry Neurosci 2003;28:99.
187. Shin JG, Soukhova N, Flockhart DA. Drug Metab Dispos 1999;27:1078.
188. Spina E, de Leon J. Basic Clin Pharmacol Toxicol 2007;100:4.
189. DeVane CL. Am J Health Syst Pharm 1995;52:S15.
190. Dahl SG, Strandjord RE. Clin Pharmacol Ther 1977;21:437.
191. Yeung PK, Hubbard JW, Korchinski ED, Midha KK. Eur J Clin Pharmacol 1993;45:563.
192. Cashman JR, Yang Z, Yang L, Wrighton SA. Drug Metab Dispos 1993;21:492.
193. Green DE, Forrest IS. Can Psychiatr Assoc J 1966;11:299.
194. Yoshii K, Kobayashi K, Tsumuji M, Tani M, Shimada N, Chiba K. Life Sci 2000;67:175.
195. Muralidharan G, Cooper JK, Hawes EM, Korchinski ED, Midha KK. Eur J Clin Pharmacol 1996;50:121.
196. Yamamoto T, Suzuki A, Kohno Y. Xenobiotica 2003;33:823.
197. Ravic M, Warrington S, Boyce M, Dunn K, Johnston A. Br J Clin Pharmacol 2004;58(Suppl1): 34.
198. von Bahr C, Movin G, Nordin C, Liden A, Hammarlund-Udenaes M, Hedberg A, Ring H, Sjoqvist F. Clin Pharmacol Ther 1991;49:234.
199. Cohen BM, Lipinski JF, Waternaux C. Psychopharmacology 1989;97:481.
200. Lin G, Hawes EM, McKay G, Korchinski ED, Midha KK. Xenobiotica 1993;23:1059.
201. Wojcikowski J, Maurel P, Daniel WA. Drug Metab Dispos 2006;34:471.
202. Carrillo JA, Ramos SI, Herraiz AG, Llerena A, Agundez JA, Berecz R, Duran M, Benitez J. J Clin Psychopharmacol 1999;19:494.
203. Jann MW, Ereshefsky L, Saklad SR. Clin Pharmacokinet 1985;10:315.
204. Koytchev R, Alken RG, McKay G, Katzarov T. Eur J Clin Pharmacol 1996;51:183.
205. Marder SR, Midha KK, Van Putten T, Aravagiri M, Hawes EM, Hubbard JW, McKay G, Mintz J. Br J Psychiatry 1991;158:658.
206. Midha KK, Hubbard JW, Marder SR, Marshall BD, Van Putten T. J Psychiatry Neurosci 1994;19:254.
207. Ereshefsky L, Jann MW, Saklad SR, Davis CM, Richards AL, Burch NR. Biol Psychiatry 1985;20:329.
208. Goff DC, Midha KK, Sarid-Segal O, Hubbard JW, Amico E. Psychopharmacology 1995;117:417.
209. Dahl-Puustinen ML, Liden A, Alm C, Nordin C, Bertilsson L. Clin Pharmacol Ther 1989;46:78.
210. Hansen LB, Elley J, Christensen TR, Larsen NE, Naestoft J, Hvidberg EF. Br J Clin Pharmacol 1979;7:75.
211. Ozdemir V, Naranjo CA, Herrmann N, Reed K, Sellers EM, Kalow W. Clin Pharmacol Ther 1997;62:334.
212. Olesen OV, Linnet K. Br J Clin Pharmacol 2000;50:563.
213. Midha KK, Hawes EM, Hubbard JW, Korchinski ED, McKay G. Psychopharmacology 1988;95:333.
214. Aravagiri M, Hawes EM, Midha KK. J Pharmacol Exp Ther 1986;237:615.
215. Uchaipichat V, Mackenzie PI, Elliot DJ, Miners JO. Drug Metab Dispos 2006;34:449.
216. Guthrie SK, Hariharan M, Kumar AA, Bader G, Tandon R. J Clin Pharm Ther 1997;22:221.
217. Hobbs DC, Welch WM, Short MJ, Moody WA, Van der Velde CD. Clin Pharmacol Ther 1974;16:473.
218. Hobbs DC. J Pharm Sci 1968;57:105.
219. Weissman A. Adv Biochem Psychopharmacol 1974;9:471.
220. Ereshefsky L, Saklad SR, Watanabe MD, Davis CM, Jann MW. J Clin Psychopharmacol 1991;11:296.
221. Heel RC, Brogden RN, Speight TM, Avery GS. Drugs 1978;15:198.
222. Cheung SW, Tang SW, Remington G. J Chromatogr 1991;564:213.
223. Luo H, Hawes EM, McKay G, Korchinski ED, Midha KK. Xenobiotica 1995;25:291.
224. Froemming JS, Lam YW, Jann MW, Davis CM. Clin Pharmacokinet 1989;17:396.
225. Kudo S, Ishizaki T. Clin Pharmacokinet 1999;37:435.
226. Avent KM, DeVoss JJ, Gillam EM. Chem Res Toxicol 2006;19:914.
227. Kalgutkar AS, Taylor TJ, Venkatakrishnan K, Isin EM. Drug Metab Dispos 2003;31:243.

228. Shin JG, Kane K, Flockhart DA. Br J Clin Pharmacol 2001;51:45.
229. Gate. http://www.Gatepharma.Com/orap/script_info.html; 2008.
230. Desta Z, Kerbusch T, Flockhart DA. Clin Pharmacol Ther 1999;65:10.
231. McCreadie RG, Heykants JJ, Chalmers A, Anderson AM. Br J Clin Pharmacol 1979;7:533.
232. Pinder RM, Brogden RN, Swayer R, Speight TM, Spencer R, Avery GS. Drugs 1976;12:1.
233. Sallee FR, Pollock BG, Stiller RL, Stull S, Everett G, Perel JM. J Clin Pharmacol 1987;27:776.
234. Desta Z, Kerbusch T, Soukhova N, Richard E, Ko JW, Flockhart DA. J Pharmacol Exp Ther 1998;285:428.
235. Desta Z, Soukhova N, Flockhart DA. J Clin Psychopharmacol 2002;22:162.
236. Endo. http://www.Endo.Com/pdf/moban_pack_insert.Pdf.
237. Owen RR, Jr, Cole JO. J Clin Psychopharmacol 1989;9:268.
238. Peper M. J Clin Psychiatry 1985;46:26.
239. Zetin M, Cramer M, Garber D, Plon L, Paulshock M, Hoffman HE, Schary WL. Clin Ther 1985;7:169.
240. Novartis. http://www.Pharma.Us.Novartis.Com/product/pi/pdf/clozaril.Pdf; 2008.
241. Choc MG, Lehr RG, Hsuan F, Honigfeld G, Smith HT, Borison R, Volavka J. Pharm Res 1987;4:402.
242. Fitton A, Heel RC. Drugs 1990;40:722.
243. Jann MW, Grimsley SR, Gray EC, Chang WH. Clin Pharmacokinet 1993;24:161.
244. Marder S, Wirshing D. Textbook of psychopharmacology. Washington, DC: American Psychiatric Publishing; 2004. p 443.
245. Chetty M, Murray M. Curr Drug Metab 2007;8:307.
246. Lilly E. http://pi.Lilly.Com/us/zyprexa-pi.Pdf; 2008.
247. Callaghan JT, Bergstrom RF, Ptak LR, Beasley CM. Clin Pharmacokinet 1999;37:177.
248. Kassahun K, Mattiuz E, Nyhart E, Jr, Obermeyer B, Gillespie T, Murphy A, Goodwin RM, Tupper D, Callaghan JT, Lemberger L. Drug Metab Dispos 1997;25:81.
249. Markowitz JS, Devane CL, Liston HL, Boulton DW, Risch SC. Clin Pharmacol Ther 2002;71:30.
250. Schulz S, Olson S, Kotlyar M. Textbook of Psychopharmacology. Washington, DC: American Psychiatric Publising; 2004. p 457.
251. Linnet K. Hum Psychopharmacol 2002;17:233.
252. Ring BJ, Catlow J, Lindsay TJ, Gillespie T, Roskos LK, Cerimele BJ, Swanson SP, Hamman MA, Wrighton SA. J Pharmacol Exp Ther 1996;276:658.
253. Zeneca A. http://www1.Astrazeneca-us.Com/pi/seroquel.Pdf; 2008.
254. DeVane CL, Nemeroff CB. Clin Pharmacokinet 2001;40:509.
255. Goren JL, Levin GM. Pharmacotherapy 1998;18:1183.
256. Grimm SW, Richtand NM, Winter HR, Stams KR, Reele SB. Br J Clin Pharmacol 2006;61:58.
257. Keating GM, Robinson DM. Drugs 2007;67:1077.
258. Gunasekara N, Spencer C. CNS Drugs 1998;9:325.
259. Lieberman J. Textbook of Psychopharmacology. Washington, DC: American Psychiatric Publishing; 2004. p 473.
260. Astella. http://www.E-search.Ne.Jp/~jpr/pdf/astellas52.Pdf; 2008.
261. Noda K, Suzuki A, Okui M, Noguchi H, Nishiura M, Nishiura N. Arzneimittelforschung 1979;29:1595.
262. Otani K, Kondo T, Kaneko S, Hirano T, Mihara K, Fukushima Y. Human Psychopharmacology 1992;7:331.
263. Prakash A, Lamb H. CNS Drugs 1998;9:153.
264. Tanaka O. Nihon Shinkei Seishin Yakurigaku Zasshi 1996;16:49.
265. Shiraga T, Kaneko H, Iwasaki K, Tozuka Z, Suzuki A, Hata T. Xenobiotica 1999;29:217.
266. Kondo T, Tanaka O, Otani K, Mihara K, Tokinaga N, Kaneko S, Chiba K, Ishizaki T. Psychopharmacology 1996;127:311.
267. Janssen. 2008.
268. Goff D. Textbook of Psychopharmacology. Washington, DC: American Psychiatric Publishing; 2004. p 495.
269. Heykants J, Huang ML, Mannens G, Meuldermans W, Snoeck E, Van Beijsterveldt L, Van Peer A, Woestenborghs R. J Clin Psychiatry 1994;55 (Suppl): 13.
270. Huang ML, Van Peer A, Woestenborghs R, De Coster R, Heykants J, Jansen AA, Zylicz Z, Visscher HW, Jonkman JH. Clin Pharmacol Ther 1993;54:257.
271. Mannens G, Huang ML, Meuldermans W, Hendrickx J, Woestenborghs R, Heykants J. Drug Metab Dispos 1993;21:1134.

272. Spina E, Avenoso A, Scordo MG, Ancione M, Madia A, Gatti G, Perucca E. J Clin Psychopharmacol 2002;22:419.
273. Scordo MG, Spina E, Facciola G, Avenoso A, Johansson I, Dahl ML. Psychopharmacology 1999;147:300.
274. Spina E, Avenoso A, Facciola G, Salemi M, Scordo MG, Giacobello T, Madia AG, Perucca E. Ther Drug Monit 2000;22:481.
275. Zhu HJ, Wang JS, Markowitz JS, Donovan JL, Gibson BB, DeVane CL. Neuropsychopharmacology 2007;32:757.
276. Janssen. http://www.Invega.Com/invega/shared/pi/invega.Pdf#zoom=100; 2008.
277. Dolder C, Nelson M, Deyo Z. Am J Health Syst Pharm 2008;65:403.
278. Vermeir M, Naessens I, Remmerie B, Mannens G, Hendrickx J, Sterkens P, Talluri K, Boom S, Eerdekens M, van Osselaer N, Cleton A. Drug Metab Dispos 2008;36:769.
279. Pfizer. http://www.Pfizer.Com/files/products/uspi_geodon.Pdf; 2008.
280. Daniel D, Copeland L, Tamminga C. Textbook of psychopharmacology. Washington, DC: American Psychiatric Publishing; 2004. p 507.
281. Stimmel GL, Gutierrez MA, Lee V. Clin Ther 2002;24:21.
282. Prakash C, Kamel A, Cui D, Whalen RD, Miceli JJ, Tweedie D. Br J Clin Pharmacol 2000; 49 (Suppl 1): 35S.
283. Miceli JJ, Anziano RJ, Robarge L, Hansen RA, Laurent A. Br J Clin Pharmacol 2000; 49 (Suppl 1): 65S.
284. Wilner KD, Demattos SB, Anziano RJ, Apseloff G, Gerber N. Br J Clin Pharmacol 2000; 49 (Suppl 1): 43S.
285. Sumitomo. http://www.E-search.Ne.Jp/~jpr/pdf/sumito12.Pdf; 2008.
286. Masui T, Kusumi I, Takahashi Y, Koyama T. Prog Neuropsychopharmacol Biol Psychiatry 2006;30:1330.
287. Masui T, Kusumi I, Takahashi Y, Koyama T. Ther Drug Monit 2006;28:73.
288. Onrust SV, McClellan K. CNS Drugs 2001;15:329.
289. Yasui-Furukori N, Furukori H, Nakagami T, Saito M, Inoue Y, Kaneko S, Tateishi T. Ther Drug Monit 2004;26:361.
290. Kitamura A, Mizuno Y, Natsui K, Yabuki M, Komuro S, Kanamaru H. Biopharm Drug Dispos 2005;26:59.
291. Squibb B-M. http://packageinserts.Bms.Com/pi/pi_abilify.Pdf; 2008.
292. Lieberman J. Textbook of psychopharmacology. Washington, DC: American Psychiatric Publishing; 2004. p 487.
293. Mallikaarjun S, Salazar DE, Bramer SL. J Clin Pharmacol 2004;44:179.
294. Swainston Harrison T, Perry CM. Drugs 2004;64:1715.
295. EMEA. http://www.Emea.Europa.Eu/pdfs/human/referral/sertindole/285202en.Pdf; 2008.
296. Ereshefsky L. J Clin Psychiatry 1996;57 (Suppl11): 12.
297. Murdoch D, Keating GM. CNS Drugs 2006;20:233.
298. Wong SL, Cao G, Mack RJ, Granneman GR. Clin Pharmacol Ther 1997;62:157.
299. Aventis S. http://www.Medicines.Ie/emc/industry/; 2008.
300. Rosenzweig P, Canal M, Patat A, Bergougnan L, Zieleniuk I, Bianchetti G. Hum Psychopharmacol 2002;17:1.
301. Schaus JM, Bymaster FP. Ann Rep Med Chem 1998;33:1–10.
302. Rogers BN, Schmidt CJ. Ann Rep Med Chem 2006;41:3–21.
303. Nikam SS, Awasthi AK. Curr Opin Investig Drugs 2008;9:37–46.
304. Micheli F, Heidbreder C. Exp Opin Therap Pat 2008;18:821–840.
305. Strange PG. Trends Pharm Sci 2008;29: 314–321.
306. Marino MJ, Knutsen LJS, Williams M. J Med Chem 2008;51:1077–1107
307. Rowley M, Bristow LJ, Hutson PH. J Med Chem 2001;44:477–501.
308. Childers WE, Jr, Robichaud AJ. Ann Rep Med Chem 2005;40:17–33.
309. Roth BL, Shapiro DA. Exp Opin Therap Targets 2001;5:685–695.
310. Fitzgerald LW, Ennis MD. Ann Rep Med Chem 2006;37:21–30.
311. Dunlop J, Sabb AL, Mazandarani H, Zhang J, Kalgaonker S, Shukhina E, Sukoff S, Vogel RL, Stack G, Schechter L, Harrison BL, Rosenzweig-Lipson S. J Pharmacol Exp Ther 2005;313:862–869.
312. Adams DR, Bentley JM, Benwell KR, Bickerdike MJ, Bodkin CD, Cliffe IA, Dourish CT, George AR, Kennett GA, Knight AR, Malcolm CS, Mansell HL, Misra A, Quirk K, Roffey JRA, Vickers SP. Bioorg Med Chem Lett 2006;16:677–680.
313. Holenz J, Pauwels PJ, Diaz JL, Merce R, Codony X, Buschmann H. Drug Discov Today 2006;11:283–299.

314. Filla SA, Flaugh ME, Gillig JR, Heinz LJ, Krushinski JH, Jr, Liu B, Pineiro-Nunez MM, Schaus JM, Ward JS.WO patent 2002060871.
315. Bromidge SM, Brown AM, Clarke SE, Dodgson K, Gager T, Grassam HL, Jeffrey PM, Joiner GF, King FD, Middlemiss DN, Moss SF, Newman H, Riley G, Routledge C, Wyman P. J Med Chem 1999;42:202–205.
316. Chen Y, Conn PJ. Drugs Fut. 2008;33: 355–360.
317. Gasparini F, Bilbe G, Gomez-Mancilla B, Spooren W. Curr Opin Drug Discov Dev. 2008;11:655–665.
318. Kristiansen LV, Huerta I, Beneyto M, Meador-Woodruff JH. Cur Opin Pharm 2007;7:48–55.
319. Lane HY, Chang YC, Liu YC, Chiu CC, Tsai GE. Arch Gen Psychiatry 2005;62:1196–1204.
320. Lechner SM, Sandra M. Curr Opin Pharm 2006;6:75–81.
321. Catarzi D, Colotta V, Varano F. Med Res Rev 2007;27:239–278.
322. Staubi U, Rogers G, Lynch G. Proc Natl Acad Sci USA 1994;91:777–781.
323. Morrow JA, Maclean JKF, Jamieson C. Curr Opin Drug Discov Dev 2006;9:571–579.
324. Nagase T, Mizutani T, Sekino E, Ishikawa S, Ito S, Mitobe Y, Miyamoto Y, Yoshimoto R, Tanaka T, Ishihara A, Takenaga N, Tokita S, Sato N. J Med Chem 2008;51: nnnn–nnn.
325. Hudkins RL, Raddatz R. Ann Rep Med Chem 2007;42:49–62.
326. Berlin M, Boyce CW. Exp Opin Therap Pat 2007;17:675–687.
327. Celanire S, Wijtmans M, Talaga P, Leurs R, de Esch IJP. Drug Discov Today 2005;10: 1613–1627.
328. Breining SR, Mazurov AA, Miller CH. Ann Rep Med Chem 2005;40:3–16.
329. Albert JS, Potts W. Exp Opin Therap Pat 2006;16:925–937.
330. Halene TB, Siegel SJ. Drug Discov Today 2007;12:870–878.
331. Siuciak JA, et al. Neuropharmacol. 2006;51: 386–396.
332. Brandon NJ, Rottela DP. Ann Rep Med Chem 2008;42:3–12.
333. Verhoest PR.Abstracts of Papers, 236th ACS National Meeting, Philadelphia, PA; 2008 August 17–21; 2008; MEDI-219.
334. D'Souza DC. Int Rev Neurobiol 2007;78: 289–326.
335. Poncelet M, Barnouin MC, Breliere JC, Le Fur G, Soubrie P. Neuropharmacol 1999;144: 144–150.
336. Borowsky B, Stevens R, Mark B, et al. Neuropharmacol 2005;30:S77–S142.
337. Lange JHM, Coolen HKAC, van Stuivenberg HH, Dijksman JAR, Herrmans AHJ, Ronken E, Keizer HG, McCreary AC, Veerman W, Wals HC, Stork B, Vereer PC, den Hartog AP, de Jong NMJ, Adolfs TJP, Hoogendoorn J, Kruse CG. J Med Chem 2004;47: 627–643.

ANTIDEPRESSANTS

RICHARD A. GLENNON[1]
LESLIE IVERSEN[2]
[1] Virginia Commonwealth University, Richmond, VA
[2] University of Oxford, Oxford, UK

1. INTRODUCTION

The term *depression* refers to a depression of mood than can be a transient but, more commonly, is a chronic condition. The symptoms are distressing; patients report an ineffable and all-pervading depth of despair and hopelessness far beyond normal experience [1,2]. Depressed mood is typically accompanied by low energy or fatigue, and sleep disturbances. Physical, social, and personal functioning are greatly impaired, leading to at least as much handicap as such physical illnesses as diabetes or arthritis. The risk of suicide is greatly increased. Major depression is an important and common disorder affecting as much as 5% of the population: each year approximately 100 million people worldwide develop depression [2,3], with the prevalence in women being approximately twice that in men. Depression can also be associated with other psychiatric disorders, such as bipolar illness and anxiety disorders, or with physical illnesses that can worsen the prognosis. Although a partial remission of symptoms rather than full recovery is the usual outcome of drug treatment, the availability of safe and effective antidepressants is a major achievement of twentieth century psychopharmacology. Worldwide sales of antidepressants exceed US$10 billion annually, making them the single most important group of psychopharmaceuticals, and include some of the most widely used prescription drugs (e.g., venlafaxine, duloxetine, and fluoxetine).

The new wave of safer antidepressants introduced in the 1990s, led by fluoxetine (Prozac), was dominated by drugs that are serotonin-selective reuptake inhibitors (SSRIs). The SSRIs exhibit some adverse side effects, notably in impaired sexual function in both men and women, but because they are relatively safe, physicians were less inhibited about using them. This has extended the clinical recognition and treatment of depression to a much greater number of people than before. At the same time, new indications have been demonstrated and officially approved for the SSRIs, including syndromes associated with anxiety and panic, extending still further the wide use of these agents in psychopharmacology. The development of SSRIs was followed by a new generation of mixed norepinephrine/serotonin reuptake inhibitors that lacked the cardiac and other toxicity associated with the earlier tricyclic antidepressants, and some of these drugs and SSRIs have been approved also for the treatment of generalized anxiety disorder.

2. CLINICAL USE

2.1. Available Agents and Classification

Table 1 lists the antidepressant drugs currently registered for use in the United States and/or Europe. All are administered orally, usually on a once-a-day dosing regime. A typical course of treatment lasts 3 months. Why do we have so many antidepressants, and why does there still exist a need for additional agents? There are four primary reasons.

1. The initial selection of an antidepressant for the treatment of depressed patients depends, to a large extent, on diagnostic criteria. Often, however, the first agent selected is not the most therapeutically beneficial or best tolerated by the patient. Subsequent selection is empirical and involves trials with different agents from different classes (and sometimes even from within the same class), to identify the most suitable agent (i.e., the most effective agent with the fewest undesirable side effects) for a given patient.

2. Although effective, many of the presently available agents, particularly the tricyclic antidepressants (TCAs), produce undesirable side effects. Patients who initiate therapy on an SSRI, for example, are more likely to complete a course of adequate dose and duration of antidepressant therapy than patients

Table 1. Antidepressants: Drugs Currently on the Market [6,7]

Generic Name	Proprietary Name United States	Proprietary Name Europe	Manufacturer	Daily Dose Range (mg)
Amitriptyline (1)	Elavil	Lentizol	Astra Zeneca	50–200
Amoxepine (2)	Asendin	Asendis	Wyeth Ayerst	100–250
Bupropion (3)	Wellbutrin	Wellbutrin	GlaxoSmithKline	300–450
Citalopram (4)	Celexa	Cipramil	Lundbeck	20–60
Clomipramine (5)	Anafranil	Anafranil	Novartis	30–150
Desipramine (6)	Norpamin	—	HMR	100–300
Desvenlafaxine (7)	Pristiq	—	Wyeth	50-100
Dothiepin (8)	—	Prothiaden	Knoll	75–150
Doxepine (9)	Sinequan	Sinequan	Pfizer	10–100
Duloxetine (10)	Cymbalta	Ariclaim	Eli Lilly	60–80
Escitalopram (11)	Lexapro	Cipralex	Lundbeck	10–20
Fluoxetine (12)	Prozac	Prozac	Eli Lilly	20–60
Fluvoxamine (13)	Luvox	Faverin	Solvay	100–200
Imipramine (14)	Tofranil	Tofranil	Novartis	50–200
Lofepramine (15)	—	Gamanil	Merck	140–210
Maprotiline (16)	Ludiomil	Ludiomil	Novartis	25–75
Mirtazapine (17)	Remeron	Zispin	Organon	15–45
Moclobemide (18)	—	Manerix	Roche	300–600
Nefazodone (19)	Serzone	Dutonin	BMS	100–400
Nortriptyline (20)	Aventyl	Allegron	King	75–150
Paroxetine (21)	Paxil	Seroxat	GlaxoSmithKline	20–60
Phenelzine (22)	Nardil	Nardil	Parke Davis	30–45
Protriptyline (23)	Vivactil	—	Merck	15–60
Reboxetine (24)	Edronax	Edronax	Pharmacia/Upjohn	8–12
Sertraline (25)	Zoloft	Lustral	Pfizer	50–200
Tranylcypromine (26)	Parnate	Parnate	GlaxoSmithKline	20–30
Trazodone (27)	Desyrel	Molipaxan	HMR	150–300
Trimipramine (28)	Surmontil	Surmontil	Wyeth	150–300
Venlafaxine (29)	Effexor	Effexor	Wyeth	75–150

who initiate therapy with a TCA [4]. In a review of more than 100 cases, patients receiving SSRIs tended to discontinue medication less often than those on TCAs or heterocyclic antidepressants. The difference was statistically significant when SSRI use was compared with the use of TCAs, but differences were not as significant when the SSRIs were compared to the heterocyclic antidepressants. The findings were attributed more to the side effects associated with the TCAs than with the therapeutic efficacy of the agents [5].

3. Most antidepressants have a delayed time of onset; currently available antidepressants require administration for at least 2–4 weeks (or more) before effects are evident. Newer agents with shorter onset times, or newer strategies for enhancing the onset time of existing agents, are required to overcome this problem.

4. Finally, there is a certain population (ca. 30%) of patients who are resistant to current therapies.

Most antidepressants in clinical use today act by enhancing the neurotransmission of serotonin (5-hydroxytryptamine (5-HT)), norepinephrine (NE; noradrenaline (NA)), or both. They do so either by blocking the reuptake (transport) of neurotransmitter, blocking the metabolism of neurotransmitter (i.e., monoamine oxidase (MAO) inhibitors), or by direct action on a neurotransmitter receptor.

Hence, the antidepressants can be classified on the basis of their putative mechanisms of action (Table 2 and Figs 1–4). Agents that block neurotransmitter reuptake can be further divided into those that are nonselective norepinephrine/serotonin reuptake inhibitors (e.g., tricyclic antidepressants or newer drugs with mixed action), SSRIs, and norepinephrine-selective reuptake inhibitors (NSRIs; referred to in Europe as noradrenergic-selective reuptake inhibitors or NaRIs). Nomenclature has recently become a bit unwieldy, in that the newer nonselective agents are referred to as serotonin- and norepinephrine-selective reuptake inhibitors (SNRIs) to distinguish them from the earlier tricyclic antidepressants. Furthermore, some investigators now classify the norepinephrine-favoring tricyclic antidepressants as NSRIs; however, this latter nomenclature will not be used here to refer to the tricyclic antidepressants. A newer category of antidepressants is the "heterocyclic antidepressants," to differentiate them from other known antidepressants (most of which also happen to be heterocyclic!). For the most part, their mechanism(s) of action is not known with certainty; there is evidence, however, that many of these agents interact directly with serotonin receptors in the brain. The term "heterocyclic antidepressants" will not be used herein. Finally, some newer agents, with potentially unique mechanisms of action, are currently being explored; some of these agents are discussed.

2.2. Clinical Efficacy of Antidepressants

Antidepressants are given orally, usually for extended periods of treatment, ranging from a few weeks to many months in duration. Clinical trials invariably employ self-reporting of symptoms, using standardized questionnaires, the tool most often used being the 17- or 21-item Hamilton rating scale for depression (HAM-D). A positive response to drug treatment is usually defined as a decrease of at least 50% in the baseline HAM-D score.

There are several puzzling features of antidepressant drug action. The first is that regardless of which drug is used, one-third or more of those treated fail to show any significant response. All drugs seem effective in approximately 60–70% of those treated, but placebo response rates range from 30% to 50%. The reasons for the high placebo response are partly because some patients show a spontaneous remission from their depression during the 6–8 weeks of the drug trial, but partly also

Table 2. Mechanistic Classification of Antidepressants and Other Types of Agents/Mechanisms Currently Being Explored for the Treatment of Depression

Transport blockers (reuptake inhibitors)	
Mixed 5-HT/NE	Amitriptyline (**1**)
	Amoxepine (**2**)
	Clomipramine (**5**)
	Desvenlafaxine (**7**)
	Dothiepin (**8**)
	Doxepine (**9**)
	Duloxetine (**10**)
	Imipramine (**14**)
	Trimipramine (**28**)
NE-favoring	Desipramine (**6**)
	Lofepramine (**15**)
	Maprotiline (**16**)
	Nortriptyline (**20**)
	Protriptyine (**23**)
SSRIs	
	Citalopram (**4**)
	Escitalopram (**11**)
	Fluoxetine (**12**)
	Fluvoxamine (**13**)
	Paroxetine (**21**)
	Sertraline (**25**)
	Venlafaxine (**29**)
NSRIs; NaRIs	
	Reboxetine (**24**)
Monoamine oxidase (MAO) inhibitors	
Irreversible MAO inhibitors	
	Phenelzine (**22**)
	Tranylcypromine (**26**)
Reversible (RIMAs)	
	Moclobemide (**18**)
Serotonergic agents	
	Nefazodone (**19**)
	Trazodone (**27**)
Other agents	
	Bupropion (**3**)
	Mirtazepine (**17**)

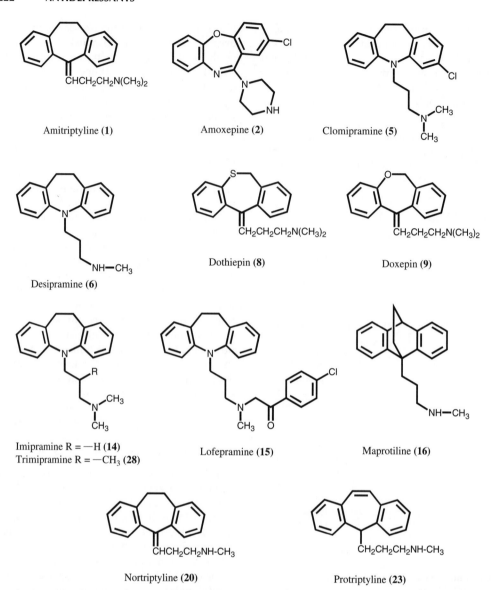

Figure 1. Structures of tricyclic antidepressants. (This figure is available in full color at http://mrw.interscience.wiley.com/emrw/9780471266945/home.)

because of the genuine power of the placebo effect, which is particularly noticeable in the treatment of psychiatric illnesses. The magnitude of the placebo effect is exceptionally large in trials of antidepressants and is at least as great as that attributable to the antidepressant drug [8]. The placebo effect is variable; it is larger, for example, in patients with milder depression than in those with more severe forms of the illness. It is not uncommon for clinical trials of antidepressant drugs to fail to show a statistically significant difference between the drug-treated and the placebo groups. However, this is not obvious from the published literature since negative trial results are often not published. A review of 74 FDA registered clinical trials, involving 12 antidepressant drugs and 12,564 patients in the period 1987–2004, makes this point. Only 51% of the trials had a statistically significant

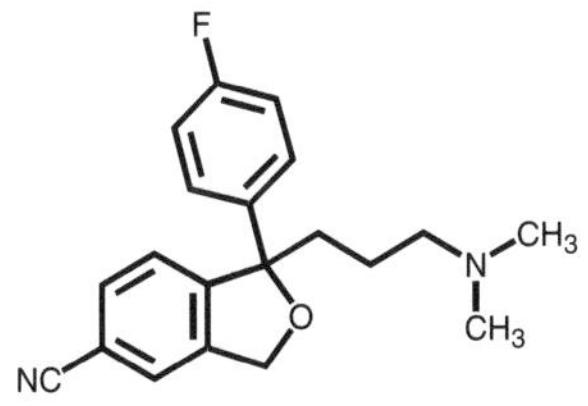

Citalopram (**4**)

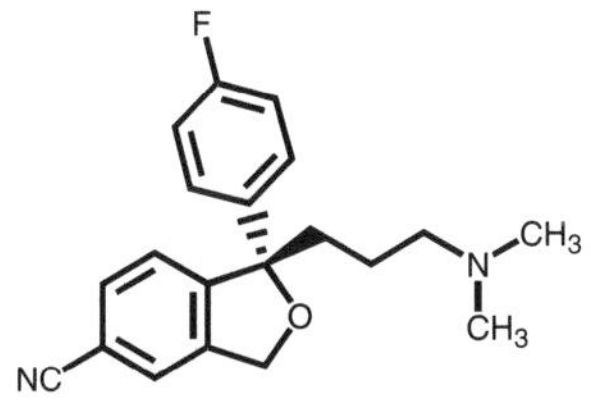

Escitalopram (**11**)

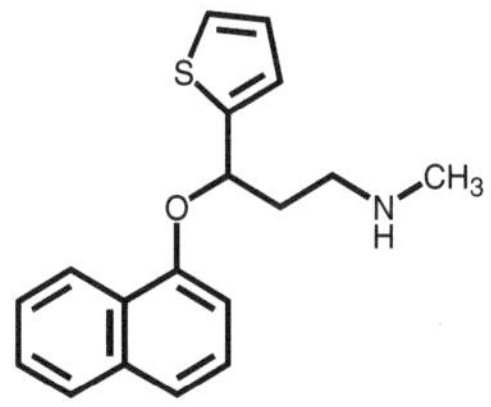

Duloxetine (**10**)

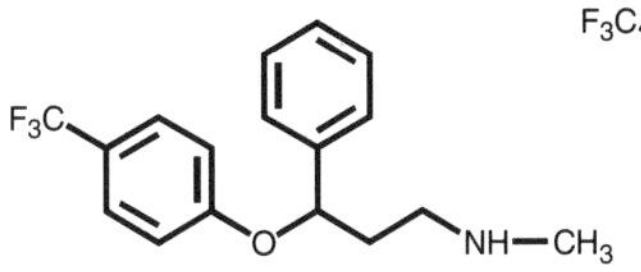

Fluoxetine (**12**)

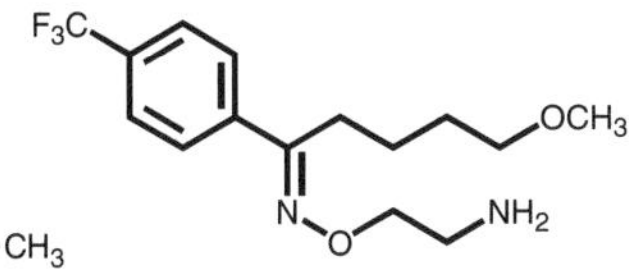

Fluvoxamine (**13**)

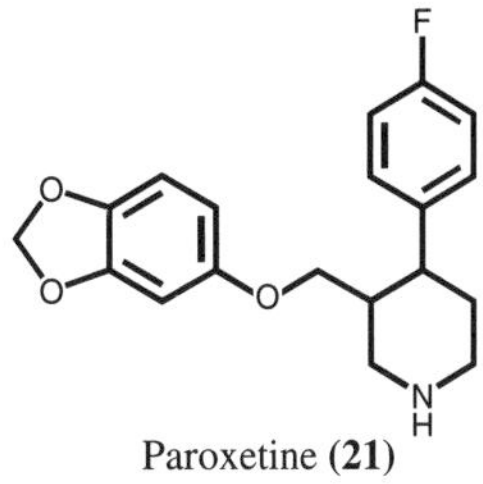

Paroxetine (**21**)

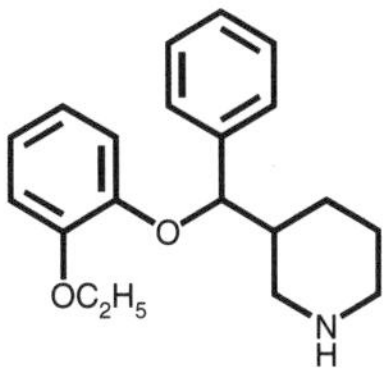

Reboxetine (**24**)

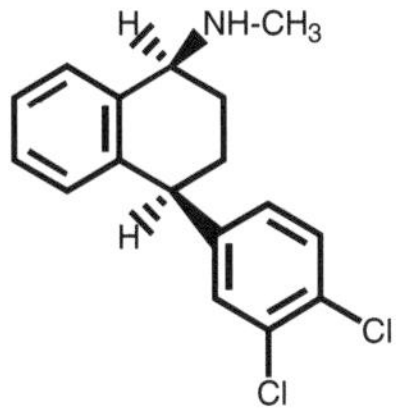

Sertraline (**25**)

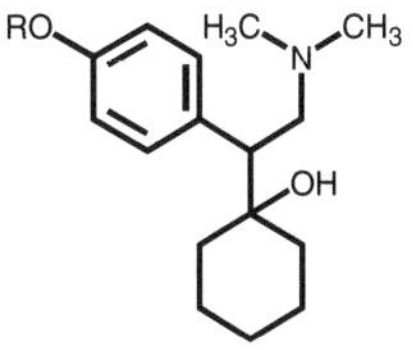

Venlafaxine R = CH$_3$ (**29**)
Desvenlafaxine R = H (**7**)

Figure 2. Structures of representative SSRIs, NSRIs, and SNRIs.

positive outcome, and many of the trials with a negative or questionable outcome were not published [9]. The various classes of antidepressant drugs exhibit few differences in their clinical efficacy; the advantages of the newer compounds are related to their improved side-effect profiles rather than to a more powerful antidepressant action [8,10]. The results of a meta-analysis of antidepressant drug trials illustrate this point (Table 3).

A meta-analysis of 186 randomized controlled trials with amitriptyline (**1**) indicated that although this drug is less well tolerated than other tricyclic or SSRI antidepressants there was a small but significant 2.5% higher proportion of responders compared to that of the other drugs [12]. A systematic search of 108 other meta-analyses suggested that combined serotonin/norepinephrine reuptake inhibitors have slightly superior efficacy to that of the SSRIs [13]. Recent studies of the norepinephrine-selective reuptake inhibitor reboxetine (**24**) showed it to have comparable efficacy to that of other antidepressants, but it

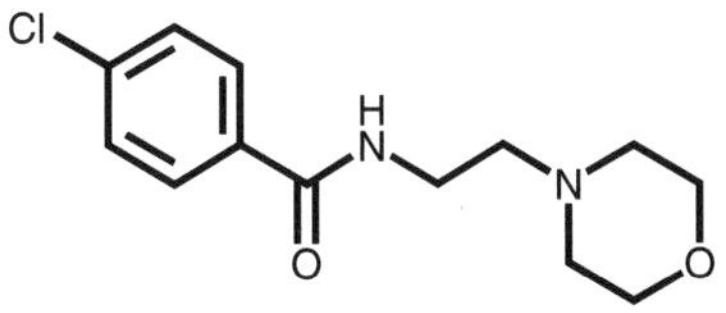

Moclobemide (**18**)

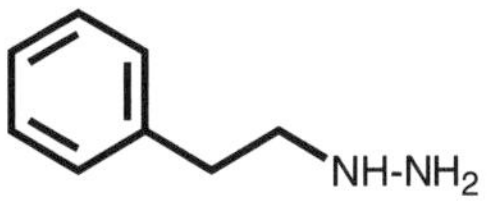

Phenelzine (**22**)

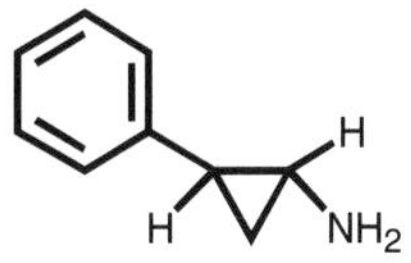

Tranylcypromine (**26**)

Figure 3. Structures of common monoamine oxidase inhibitors.

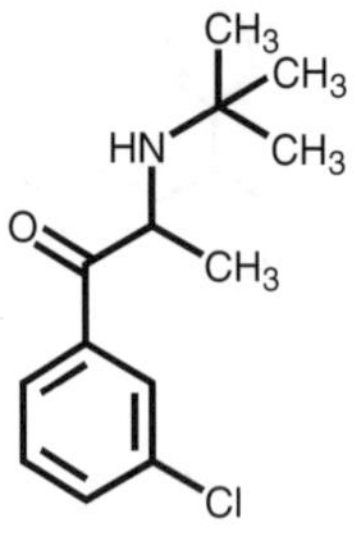

Bupropion (**3**)

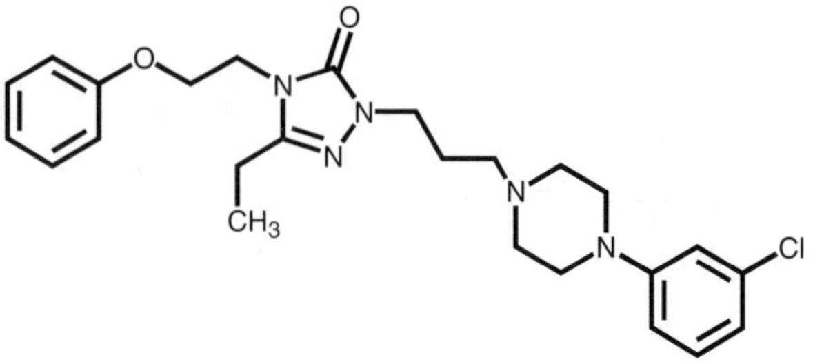

Nefazodone (**19**)

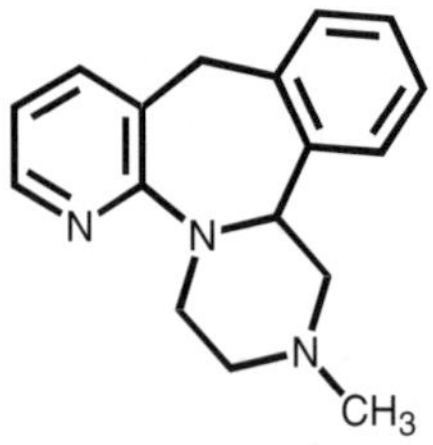

Mirtazepine (**17**)

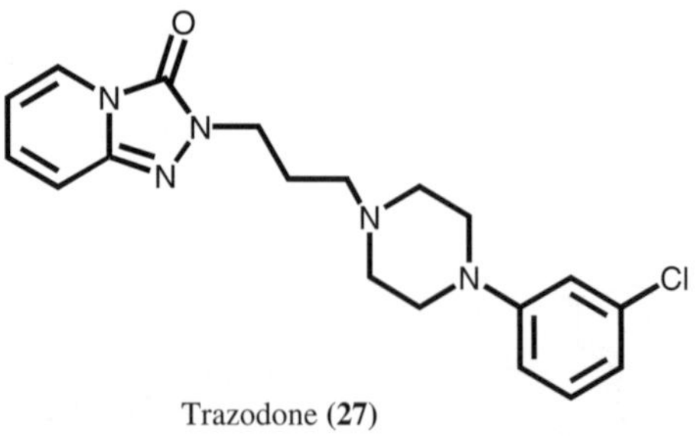

Trazodone (**27**)

Figure 4. Structures of some of the newer atypical antidepressants: bupropion, mirtazepine, nefazodone, and trazodone.

improved social functioning more than did the SSRIs [14,15].

A second unexplained feature of antidepressant drug action is that the maximum clinical benefit is not seen until treatment has been continued for several weeks [10]. Although a significant improvement in HAM-D scores can sometimes be detected after 2 weeks of treatment, it takes 4–6 weeks to obtain the maximum response. Boyer and Feighner [16] per-

Table 3. Efficacy of Antidepressants Compared with Placebo in Controlled Clinical Trials [11]

Antidepressant	Drug-Treated Percent Responders	Placebo Percent Responders	Drug–Placebo Percent Treated
Tricyclics ($n = 3327$)			
Amitriptyline (**1**)	60	25	35
Amoxepine (**2**)	67	49	18
Imipramine (**14**)	68	40	28
SSRIs ($n = 2463$)			
Paroxetine (**21**)	45	23	22
Fluoxetine (**12**)	60	33	27
Fluvoxamine (**13**)	67	39	28
Sertraline (**25**)	79	48	31
MAO inhibitors ($n = 1944$)			
Phenelzine (**22**)	64	30	34
Moclobemide (**18**)	64	24	40
Other ($n = 277$)			
Mirtazapine (**17**)	48	20	28

formed a meta-analysis of six trials to determine the predictive value of nonresponse to medication early in a clinical trial. They found that patients who failed to achieve at least a 20% reduction in HAM-D scores at any point during the first 4 weeks of a study had less than a 5% chance of becoming a "responder," as defined by a 50% or more reduction in HAM-D score by the end of the 6-week trial. The authors concluded that a full 6 weeks' trial of antidepressant medication is usually not justified if patients fail to respond during the first 4 weeks. A number of explanations have been proposed to explain the delayed clinical response to antidepressant drugs.

It has been hypothesized that the delay might be related, in part, to the initial elevation in synaptic 5-HT levels, which reduces the firing of serotonergic neurons by activating autoreceptors, mainly of the 5-HT$_{1A}$ subtype [17]. During treatment with antidepressants these autoreceptors are desensitized and proper firing of 5-HT neurons is restored; many believe this to be one of the key changes elicited by antidepressant drugs of many different categories. 5-HT$_{1A}$ serotonin receptors are found both presynaptically and postsynaptically. Agents that behave as antagonists at presynaptic 5-HT$_{1A}$ receptors (i.e., somatodendritic autoreceptors) could, in theory, shorten the time of onset of those antidepressants that act by increasing synaptic levels of serotonin. A series of potent and selective 5-HT$_{1A}$ antagonists is now available, including NAN-190 (**30**), WAY-100635 (**31**), robalzotan (NAD-299; **32**), and LY-426965 (**33**). In animal studies, the combination of acute treatment with an SSRI together with 5-HT$_{1A}$ antagonists potentiated postsynaptic serotonergic function, as predicted [18–20]. Clinical trials of these compounds in combination with antidepressants have not yet been reported. However, clinical trial data are available for the beta-blocker pindolol, which has appreciable affinity as an antagonist at 5-HT$_{1A}$ receptors (K_d value of approximately 10 nM). To date, the results of 15 placebo-controlled clinical trials, involving some 800 patients, using pindolol in treating depression have been published [21,22]. Pindolol significantly accelerated the onset of action of SSRIs in five out of seven trials designed to test this concept. The combination of pindolol with fluoxetine (**12**), for example, reduced the median period required to obtain a clinical response (50% reduction in baseline score) from 29 to 19 days [20]. A similar acceleration of rate of onset has been shown with a combination of pindolol and paroxetine (**21**), and citalopram (**4**) (see Refs. [23–26] and references therein). A role for β-adrenergic involvement in these actions of pindolol has been ruled out [27].

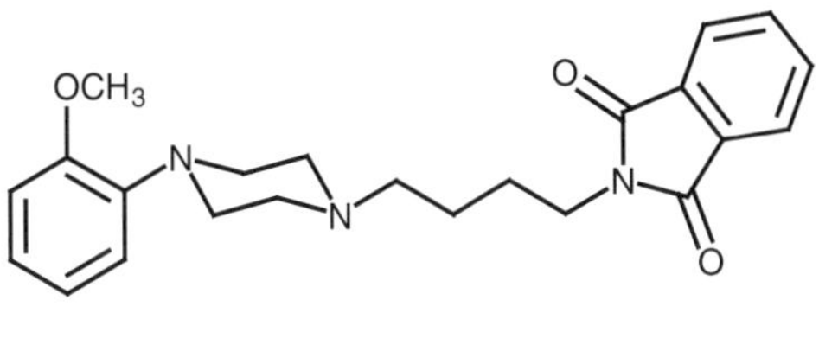

NAN-190 (**30**)

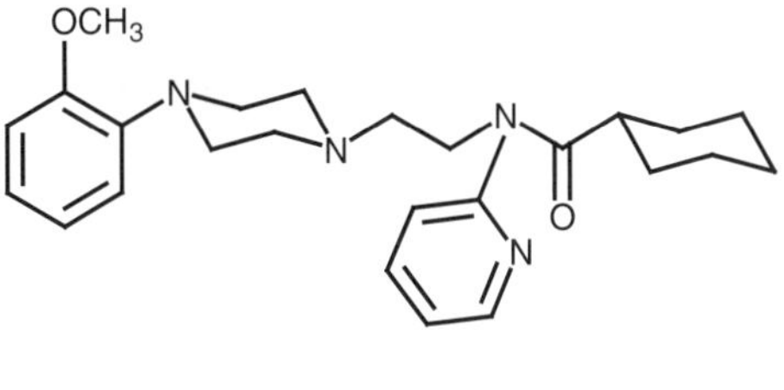

WAY-100,635 (**31**)

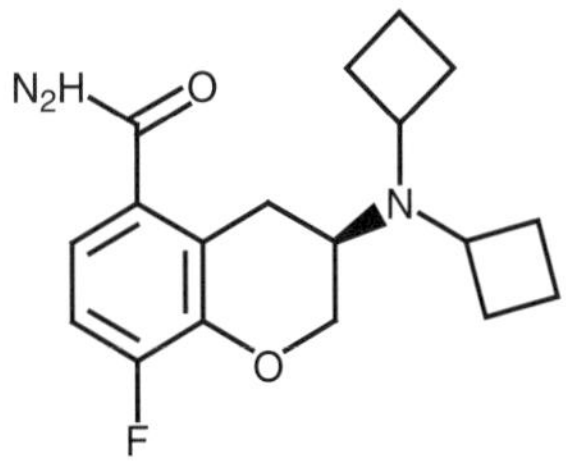

Robalzotan (**32**)

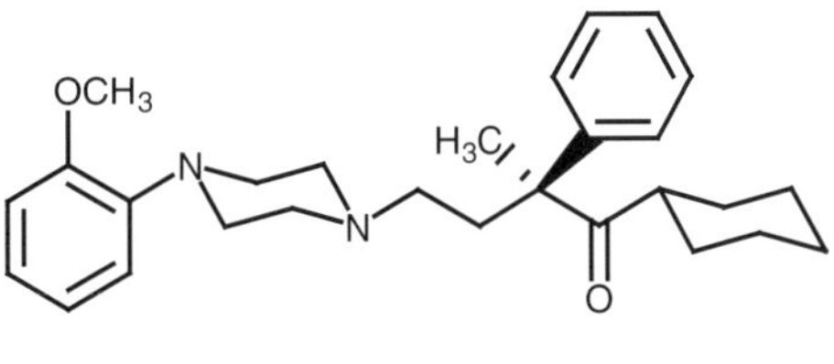

LY-426965 (**33**)

However, there are other alternative explanations for the delayed clinical response. There have been many studies of the neurochemical changes caused in animal brain by chronic treatment with antidepressants. One of the changes elicited by many antidepressants is a downregulation in the expression of β-adrenergic receptors (β-adenoceptors) [28–30]. This is of interest because one of the most consistent findings in depressed patients has been that β-adrenergic receptors are upregulated in peripheral lymphocytes and in the brains of suicide victims [31]. It was proposed that the downregulation of β-receptors represents a marker of antidepressant efficacy [28,29]. The validity of this concept was soon challenged, however, because it was found that the newer SSRIs did not consistently downregulate β-adrenergic receptors, and citalopram (4) actually caused an increase [26]. β-Adrenergic receptors are coupled to cyclic AMP formation but, although the receptors may be downregulated by antidepressant drugs, other components of cellular signaling that are regulated by cyclic AMP are upregulated, notably the "cyclic AMP response element protein" (CREB), a prominent transcription factor in the brain [32]. Another means of increasing levels of cyclic AMP is to inhibit its degradation by phosphodiesterases (see Fig. 5). The phosphodiesterase inhibitor rolipram has been found to have clinical antidepressant activity [33]. Although rolipram is not well tolerated because it causes severe nausea, it is possible that inhibitors with more selectivity for phosphodiesterase subtypes could be of future interest. In animals chronic treatment with SSRIs leads to increased expression of the PDE4A and PDE4B subtypes [34].

Among the many genes that may be regulated by CREB are those encoding neurotrophic factors [35]. An important new finding is that chronic treatment with antidepressants leads to increased expression of brain-derived neurotrophic factor (BDNF) in rat hippocampus, and this may underlie the recent finding that chronic antidepressant treatment causes a proliferation of progenitor cells in the rat hippocampus [32,36,37]. There is general agreement that neuronal plasticity or neuronal remodeling, particularly in the adult hippocampus, plays a seminal role in depression, stress, cognition, and the actions of many antidepressants [38]. That is, there is atrophy or loss of hippocampal neurons associated with stress and depression, while treatment with antidepressants produces the opposite effect. More specifically, *gene transcription factors* and *neurotrophic factors* appear to

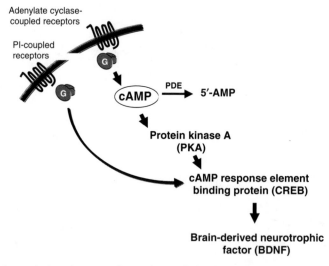

Figure 5. General associations between G-protein-coupled neurotransmitter receptors, cAMP, PKA, the transcription factor CREB, and the neurotrophic factor BDNF. See text for additional discussion. (This figure is available in full color at http://mrw.interscience.wiley.com/emrw/9780471266945/home.)

be major determinants of neuronal plasticity. This has led to concepts variously referred to as the *gene expression theory of depression*, the *stress-induced neurodegeneration theory of depression*, and the *neurotrophic (or neurogenic) theory of depression*. In the most general sense, the transcription factor CREB (cyclic adenosine monophosphate response element binding protein) and BDNF, and possibly other neurotrophic factors, seem to mechanistically underlie this process [38].

Serotonin, norepinephrine, and several other neurotransmitters, bind at G-protein-coupled receptors (see Fig. 5) to initiate and promote a complex cascade of events that is still under investigation. These interactions, and their consequent receptor association with G-proteins, results in activation of, for example, 3′,5′-cyclic adenosine monophosphate (cAMP) and the subsequent phosphorylation of protein kinase A (PKA). Levels of cAMP are regulated by the phosphodiesterase (<genSeq>PDE</genSeq>) family of enzymes that converts cAMP to adenosine 5′-monophosphate. Activated PKA elicits a number of effects (e.g., see Figs 5 and 6) including the phosphorylation of CREB to increases its transcriptional activity. CREB is also regulated via neurotransmitter receptors coupled to phosphoinositide (PI) metabolism (e.g. phospholipase C, via activation of 5-HT_2 receptors), via receptor tyrosine kinase (Trk), and via other pathways. CREB regulates gene expression and the production/activation of several neurotrophic factors including BDNF. BDNF is a nerve growth factor that promotes growth of new neurons and neuronal connections.

Because depression is associated with a variety of symptoms that include altered mood and cognitive function, and because the hippocampus has been shown to be directly or indirectly involved in these functions, it is thought that dysregulation of the hippocampus and its associated structures or anatomical targets modulates the core symptoms of major depressive disorders, most notably alterations in cognitive function and mood [39]. Numerous preclinical and clinical studies support this idea [39]. For example, brain imaging studies with depressed patients reveal structural changes (i.e., a decrease in volume) in

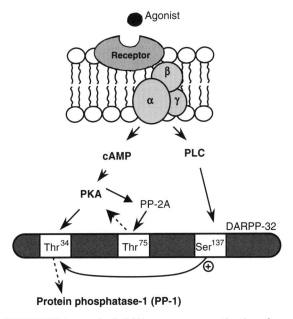

Figure 6. The PKA/DARPP-32/PP-1 cascade. Solid lines represent activation whereas broken lines indicate inhibition. Phosphorylation of Ser-137 inhibits dephosphorylation of Thr-34. (Adapted from Svenningsson et al. [179].)

key limbic nuclei, including the hippocampus and regions with which it shares connections such as the amygdala and prefrontal cortex, as being involved in the underlying pathophysiology of depression; antidepressant treatment reverses or reduces hippocampal atrophy.

Protein kinase A (Fig. 5) levels are also reduced in postmortem hippocampal tissue of depressed patients; as a target of PKA activation, expression of CREB binding protein is also decreased [40]. The transcription factor CREB, which is found throughout the brain, is important for receptor-stimulus/transcription coupling. CREB-mediated gene transcription involves dimerization, binding with cAMP response elements (CREs) on DNA, and phosphorylation; CRE sites are found within the promoter region of numerous genes and, if a promoter contains CREs, the gene could be subject to regulation by CREB [40]. Induction of CREB in the hippocampus by antidepressants could contribute to their therapeutic efficacy; however, modulation of CREB in the nucleus accumbens and certain other brain regions seems to be related to increased episodes of depression and anxiety, some adverse actions of drugs of abuse (e.g. stimulants such as cocaine and methylphenidate, and opioids), and disruption of cognitive performance [40]. This has led to the suggestion that strategies to exploit CREB should be directed to regional differences in upstream factors, or to specific CREB-related genes, rather than CREB itself [40].

BDNF, a member of the neurotrophin family of growth factors, is widely expressed throughout mammalian brain including the hippocampus, and binds at several different receptors, most notably a particular receptor TrkB. Chronic antidepressant regimens regulate expression of BDNF in the hippocampus; altered expression of BDNF can influence hippocampal neurogenesis [39]. Evidence suggests that activation of TrkB signaling might be a common effect of all antidepressants regardless of pharmacological classification, and is both sufficient and necessary for antidepressant-like behaviors in rodents [41]. Administration of BDNF directly into specific brain areas can mimic the effects of antidepressants in animal models of depression. Decreased mitogen-activated protein (MAP) kinase signaling could also be involved in the pathophysiology of depressive disorders [42]. Neurotrophic factors such as BDNF, are directly coupled to MAP kinase signaling and evidence suggests that decreased expression of BDNF contributes to depression, and that upregulation of BDNF is involved in the actions of antidepressants [43].

The kinase inhibitors U0126 (**34**) and PD184161 (**35**) have been shown to attenuate the antidepressant-like effects of BDNF and the behavioral effects of antidepressants [39].

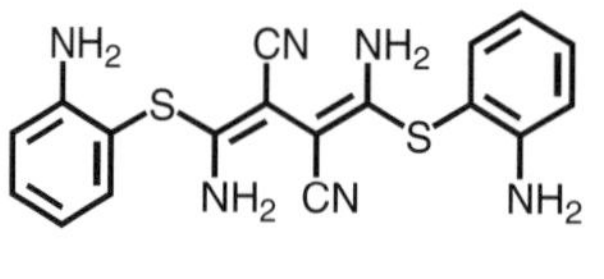

U0126 (**34**) 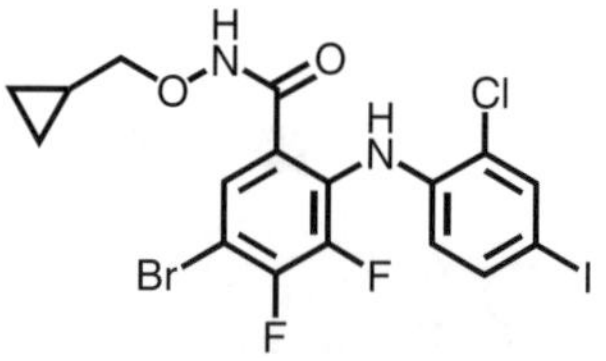   PD184161 (**35**) 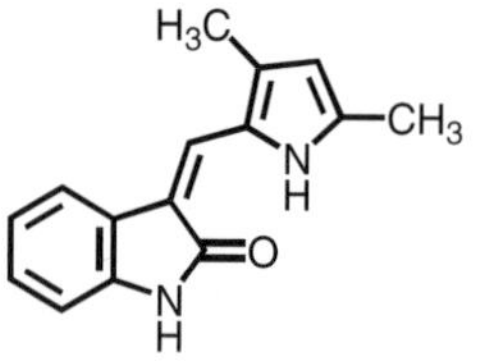   SU5416 (**36**)

Levels of other neurotrophic and growth factors (e.g., vascular endothelial growth factor or VEGF—of which at least six types have been identified) have been recently implicated as playing a role in depression and stress. VEGF binds to high-affinity receptor tyrosine kinases such as fetal liver kinase (Flk-1) in the hippocampus. Intraventricular infusion of VEGF increases basal rates of neurogenesis in the hippocampus. Administration of antidepressants also increases hippocampal neurogenesis. Chronic stress decreases neurogenesis and the expression of VEGF in the hippocampus, whereas VEGF levels are increased following chronic administration of antidepressants [44]. Intraventricular injection of the Flk-1 receptor antagonist SU5416 (semaxanib; **36**) blocked the effect of antidepressant treatment.

Associations between depression and stress cannot be ignored. Environmental factors, such as stress, decrease neurogenesis in the hippocampus of rodent and nonhuman primate brain, and decrease BDNF levels. The hypothalamic–pituitary–adrenal (HPA) axis might also be involved in both events and certain depressed patients (ca. 50%) display increased cortisol levels [39]. Hippocampal neurogenesis is stimulated by environmental enrichment and exercise, and inhibited by stress [45].

Overall, the neurotrophic hypothesis is supported by evidence that (1) antidepressants promote hippocampal neurogenesis, (2) antidepressants increase the expression of neurotrophic factors (e.g., BDNF and VEGF), and (3) central infusions of these trophic factors exert an antidepressant-like effect in animal models of depression [46]. These theories do not supplant the original 40-year-old "biogenic amine hypothesis" but, rather, strengthen it. Further, they account for the long delay in onset of action seen with antidepressants, account for the relationship between stress, depression, and increased cortisol levels associated with depression, explain problems that depressed patients have with cognition, and account for the antidepressant actions of certain PDE inhibitors. Although studies are still ongoing, the mechanistic underpinnings of depression are becoming clearer, and new targets for the treatment of depression are being identified.

2.2.1. Additional Indications for Antidepressants Depression is often associated with anxiety or other forms of psychiatric disorder, and the SSRIs and the SNRIs have come to be used increasingly in a variety of conditions other than major depression. Table 4 summarizes the uses for which SSRIs have been approved, based on the finding of significant beneficial effects in controlled clinical trials. Apart from the use of these drugs in various phobic conditions, a major change in recent years has been the recognition of their effectiveness in the treatment of generalized anxiety disorder, and the first FDA approval of antidepressant compounds for this indication

Table 4. Additional FDA-Approved Indications for SSRI and mixed NE/5-HT Antidepressants

Bulimia nervosa
Panic disorder
Obsessive compulsive disorder (OCD)
Generalized anxiety disorder
Social anxiety disorder
Posttraumatic stress syndrome

(venlafaxine (**29**), duloxetine (**10**), and escitalopram (**11**)). In addition, agents in this class in controlled trials have shown usefulness in treating premenstrual dysphoria, borderline personality disorder, obesity, smoking cessation, and alcoholism [46].

The remarkable success of the SSRIs has prompted the question of whether genetic defects in the serotonin transporter (SERT) gene or in the regulation of its expression might explain the etiology of mood disorders. The gene coding for human SERT is localized on chromosome 17q11.2. It spans more than 35 kb and is organized in 14 introns. No genetic variations have been found in the coding region of the *SERT* gene in depressed patients, but a number of studies have found that certain variants in the promoter region are associated with depressive illness, anxiety-related personality traits, or suicidal alcoholism [47].

2.3. Adverse Side Effects

2.3.1. Introduction The antidepressants include a wide range of compounds with differing modes of action (Table 2). It is not surprising to find that they display a plethora of differing side effects [48,49]. These range from adverse effects that can be unpleasant but relatively harmless to rarer and often unpredictable serious adverse reactions. Particularly in the older, so-called first-generation antidepressants, these can be life threatening. Because depressed patients are often suicidal, it is not surprising that these drugs were often implicated in deaths resulting from intentional overdose [50,51]. Some antidepressants caused rare, idiosyncratic adverse effects that were, nevertheless, so severe as to lead to withdrawal of the drugs from the market (e.g., nomifensine, zimelidine).

2.3.2. Monoamine Oxidase Inhibitors The first-generation monoamine oxidase inhibitors (MAOIs) inhibitors tended to cause hypotension (often causing dizziness), headache, and mild anticholinergic effects such as dry mouth, constipation, blurred vision, and difficulty in micturition. Phenelzine (**22**) can cause mild sedation, but tranylcypromine (**26**) is more likely to act as a psychostimulant with mild amphetamine-like properties, thereby leading to agitation and insomnia [48]. These effects are dose dependent and tend to lessen in severity with time. In fact, although antidepressants are not typically associated with abuse potential, 16 of 21 case reports of antidepressant abuse involved tranylcypromine [52]. More serious adverse effects are related to the fact that the first-generation MAO inhibitors are compounds that irreversibly inhibit the enzyme in the brain and other organs. With chronic drug treatment, inhibition of enzyme activity is cumulative and may become almost complete. The enzyme is abundant in the liver, where it serves the function of detoxifying a variety of pharmacologically active organic amines that are absorbed from the diet.

Inhibition of liver MAO leaves the patient vulnerable to the so-called wine-and-cheese syndrome, with adverse cardiovascular effects caused by absorption of such vasoactive amines as tyramine into the general circulation (for review, see Blackwell et al. [53]). The syndrome can include severe headache and hypertension and may lead to cerebral hemorrhage and death. Although this is a real risk, it seems likely that fears of MAO-food interactions may have been grossly exaggerated. Pare [54] reviewed the evidence in 1985 and noted that despite the widespread use of MAO inhibitors in the previous decade there had only been 17 reports of food interactions with phenelzine (**22**) and none of these proved fatal. With tranylcypromine (**26**) seven deaths had been reported, but in only two of these could a definite relationship with diet be established.

MAO inhibitors when taken alone in overdose can be fatal, with death usually resulting from pulmonary and cerebral edema. Despite the relatively low risk of food interactions, a wide range of foodstuffs is prohibited to patients taking MAO inhibitors, including mature cheeses, meat or yeast extracts, mature fish, pickled fish, smoked foods, and broad bean pods. Because of the long-lasting inhibition of MAO caused by these irreversible enzyme inhibitors, the dietary precautions have to be maintained for at least 14 days after cessation of drug treatment. These restrictions make these drugs unpopular both with patients and with physicians. At least as important as the food-interaction risk is that of drug interactions. MAO inhibitors are dangerous to use in conjunction with a number of other clinically important drugs [48,54]. Serious interactions (usually hypertensive crises) may occur with pethidine, levodopa, sympathomimetic amines such as amphetamine and ephedrine, and other antidepressants including the tricyclics and SSRIs.

Moclobemide (**18**), a second-generation MAOI, was designed to lessen the risk of food and drug interactions seen with earlier MAO inhibitors. Moclobemide selectively targets one form of the enzyme, monoamine oxidase A (MAO-A), leaving monoamine oxidase B (MAO-B) in the liver active and capable of detoxifying tyramine and other dietary vasoactive amines [55]. Moclobemide is also a reversible inhibitor of the enzyme, so its effects are not cumulative and are more rapidly reversible on termination of drug treatment. Adverse effects are few and infrequent, with dropouts because of side effects in clinical trials being uncommon. Nausea, insomnia, headache, and dizziness occurred in some patients taking the drug, and others experienced agitation and restlessness [55,56]. Despite the improved selectivity of moclobemide, there have been a small number of reports of hypertension when moclobemide was combined with tyramine-rich foods [57]. Moclobemide exhibits fewer adverse drug interactions than first-generation compounds, although combination with SSRIs is not recommended because of the danger that a "serotonin syndrome" could result [58]. Moclobemide does, however, appear to be safe when given in conjunction with sympathomimetic amines found in many common cough and cold remedies. The drug appears to be safe if taken in overdose. Moclobemide represents a real advance in terms of its safety profile over the earlier MAO inhibitors, but the class as a whole remains largely out of favor.

2.3.3. Tricyclic Antidepressants

The older antidepressant drugs in this class exhibit a variety of adverse side effects, most of which are related to their secondary pharmacological actions on targets other than the monoamine transporters [59,60]. Although some of these side effects are uncomfortable but not serious, others are life threatening, particularly the effects on cardiac function. The tricyclic antidepressants have only a narrow therapeutic index, and serious toxicity or death can occur at doses that are only 2–6 times therapeutic. Thus, a single bottle of tablets can prove lethal. Given the increased risk of suicide in depressed patients, it is not surprising that tricyclic antidepressant overdose was among the commonest causes of drug-related death in the United States in the early 1980s [50]. Leonard [51] analyzed the number of deaths attributable to antidepressant overdose in England and Wales for the period 1977–1983, a total of more than 1500 deaths. It was possible to calculate an estimated death rate per million patients treated for each antidepressant. The rates were amitriptyline = 166, dothiepin = 147, maprotiline = 115, imipramine = 105, doxepine 99, trimipramine = 93, and clomipramine 34. To put these data in perspective, however, the deaths associated with tricyclic antidepressants represent only approximately 10% of the risk of death through overdose with barbiturates, which were then widely used as hypnotics [61]. The most serious effects of the older drugs are attributed to direct quinidine-like actions on the heart, interfering with normal conduction and causing prolongation of the QRS or QT interval. Death is most commonly the result of cardiac arrhythmia and arrest [62]. Other toxic effects include respiratory depression, delirium, seizures, shock, and coma. It is worth noting that the newer tricyclic antidepressant lofepramine (15) has very little cardiotoxicity. In a review of fatal poisonings associated with antidepressants in England and Wales for the period 1993–1995, it was associated with fatal overdose in only 4.3 per million treatment episodes (of 3 months' duration), comparable to the fatalities associated with SSRIs (2.4 per million treatment episodes) [63,64].

In therapeutic doses, the tricyclic antidepressants also cause a variety of less serious unwanted side effects. Many of these are associated with the ability of many of these drugs to act as antagonists at various monoamine receptors (see Table 6 below). Antidepressants interact with a large number of neurotransmitter receptors. Some of the targets listed in Table 6 represent multiple subtypes; for example, "5-HT2" is in fact a composite of effects of 5-HT_{2A}, 5-HT_{2B}, and 5-HT_{2C}. (For a more complete and up-to-date summary of drug interactions with neurotransmitter receptors visit the online NIMH Psychoactive drug database http://pdsp.med.unc.edu.) Several of the tricyclic agents bind with high affinity at muscarinic cholinergic receptors, and the blockade of these receptors causes a range of side effects; again, there are five different subtypes of muscarinic receptors. Anticholinergic side effects include dry mouth, blurred vision, constipation, urinary retention, and sinus tachycardia. The most potent effects of all are seen at histamine H1 receptors, where several tricyclic drugs bind with subnanomolar affinities. Blockade of these receptors probably contributes to the sedative effects of some of these agents. Sedation is also a side effect caused by blockade of α-adrenergic receptors, another common feature in this class of drugs. Orthostatic hypotension is another side effect related to α_1-adrenergic blockade, leading to dizziness on rising in young adults, but possibly causing syncope and falls in the elderly. Another unwanted feature of the older drugs is their tendency to cause significant weight gain, which may reduce patient compliance with the drug treatment regime. Tricyclic antidepressants also reduce seizure threshold, and this can lead to drug-induced seizures in 0.1–0.5% of patients, usually early in treatment [48,49]. CNS side effects include a propensity to cause mania or hypomania in patients suffering from bipolar depression. Sexual dysfunction may be associated with tricyclic use. Decreased libido and delayed orgasm can be seen in men and women, and men may experience erectile dysfunction. These effects, however, are much more common in patients treated with the newer SSRIs (see Table 5).

Table 5. Summary of Side Effect Profiles of Antidepressant Drugs [65]

Generic Name	Sedation	Anticholinergic Effects	Hypotension	Cardiac Effects	Weight Gain
Amitriptyline (1)	+++	++++	+++	+++	+++
Amoxepine (2)	+	++	+	++	+
Bupropion (3)	0	0	0	+	0
Citalopram (4)	+	+	0/+	0/+	0
Clomipramine (5)	++	+++	++	+++	+
Desipramine (6)	+	+	+	++	+
Doxepine (9)	+++	++	+++	++	++
Fluoxetine (12)	0/+	0	0	0	0
Fluvoxamine (13)	0	0	0	0	0
Imipramine (14)	+++	++	++	+++	++
Maprotiline (16)	++	++	++	++	+
Mirtazapine (17)	+++	+	0/+	+	+
Moclobemide (18)	0	0	0	0	0
Nefazodone (19)	+	0	+	0/+	0/+
Nortriptyline (20)	+	+	+	++	+
Paroxetine (21)	+	+	0	0	0
Phenelzine (22)	+	0	+++	0	++
Protriptyline (23)	0/+	++	+	+++	+
Sertraline (25)	0	0	0	0	0
Tranylcypromine (26)	+	0	++	0	0/+
Trazodone (27)	+++	0	++	0/+	+
Trimipramine (28)	+++	+++	++	+++	++
Venlafaxine (29)	0/+	0/+	0	+	0

The tricyclic drugs are extensively metabolized in the liver by the cytochrome p450 enzymes. Consequently, other drugs that induce such enzymes or compete with the tricyclic antidepressants for metabolism may alter their actions. Calcium channel blockers, cimetidine, phenothiazines, haloperidol, methylphenidate, glucocorticoids, oral contraceptives, and most SSRIs may inhibit the metabolism of the tricyclics. This will tend to exacerbate the adverse side effects of the tricyclic antidepressants. Conversely, carbamazepine, phenytoin, barbiturates, primidone, and alcohol may induce liver cytochrome p450 enzymes and accelerate metabolism, making tricyclic antidepressants less effective [8,48,49].

2.3.4. Serotonin-Selective Reuptake Inhibitors

Since their introduction in the mid-1980s, SSRIs became the most widely used of all antidepressants. This is largely because of their improved safety and tolerability in clinical use. Although the SSRIs are no more efficacious or rapid in onset of action than the tricyclic antidepressants, they lack most of the serious toxicity and adverse side effects associated with the first-generation drugs. The relative absence of cardiac toxicity makes the SSRIs relatively safe in overdose [66]. Fatal overdose has, however, been reported in six patients taking citalopram (4) [67], although the cause of death was disputed [68]. The symptoms of SSRI overdose include nausea, agitation, seizures, and sometimes loss of consciousness [48,49]. The relative safety of the SSRIs has led to their being prescribed more freely than the earlier antidepressants, and their use in a number of indications in addition to the treatment of major depression (Table 4).

There are, however, some hazards associated with the use of these drugs. SSRIs decrease serotonin uptake from the blood by platelets. Because platelets cannot synthesize serotonin, which is involved in platelet aggregation, SSRIs may impair platelet function. A case-controlled study found that the risk of gastrointestinal bleeding was three times greater in SSRI users than in controls [69]. This conclusion was confirmed by a retrospective cohort study of 317,824 SSRI users, which emphasized that the risk of gastrointestinal

Table 6. Affinities of Antidepressant Drugs for Human Monoamine Receptors[a]

Generic Name	Histamine H1	Muscarinic	α_1-Adrenergic	α_2-Adrenergic	Dopamine D2	5-HT$_{1A}$	5-HT$_2$
Amitriptyline	1.1	18	27	940	1,000	450	18
Amoxepine	25	1,000	50	2,600	160	—	—
Bupropion	6,600	48,000	4,600	81,000	—	>35,000	>35,000
Citalopram	470	2,200	1,900	15,300	—	—	—
Clomipramin	31	37	38	3,200	190	—	—
Desipramine	110	198	130	7,200	3,300	6,400	350
Dothiepin	3.6	25	470	2,400	—	—	—
Doxepine	0.24	80	24	1,100	2,400	276	27
Duloxetine	2,300	3,000	8,300	8,600	14,000	>5,000	>500
Escitalopram	1,973	1,242	3,870	—	—	—	2,531
Fluoxetine	6,200	2,000	5,900	13,000	—	32,400	280
Fluvoxamine	>100	24,000	7,500	15,000	—	—	—
Imipramine	11	90	90	3,200	2,000	5,800	150
Lofepramine	360	67	100	2,700	2,000	4,600	200
Maprotiline	2	570	90	9,400	350	—	—
Nefazodone	24,000	11,000	48	640	910	80	26
Nortriptyline	10	150	60	2,500	1,200	294	41
Paroxetine	22,000	108	4,600	17,000	32,000	>35,000	19,000
Protriptyline	25	25	130	6,600	2,300	—	—
Reboxetine	1,400	3,900	10,000	4,300	9,000	—	—
Sertraline	24,000	630	380	4,100	10,700	>35,000	9,900
Trazodone	350	>100	36	490	3,800	96	25
Trimipramine	0.2	58	24	680	180	—	—
Venlafaxine	>35,000	>35,000	>35,000	>35,000	>35,000	>35,000	>35,000

[a] Equilibrium dissociation constants are nanomolar. Data were obtained from binding studies using the following: histamine H1: ^{3}H-doxepin or ^{3}H-pyrilamine; muscarinic: ^{3}H-quinuclidinyl benzilate; α_1-adrenergic: ^{3}H-prazosin; α_2-adrenergic: ^{3}H-rauwolscine; dopamine D2: ^{3}H-spiperone; 5HT$_{1A}$: ^{3}H-8-OH-DPAT; 5-HT$_2$: ^{3}H-ketanserin (data from refs 59,60,86). For a more complete and up-to-date summary of drug interactions with neurotransmitter receptors, see the NIMH Psychoactive Drug Data Web site (http://pdsp.med.unc.edu).

bleeding was particularly important for elderly patients [69].

The SSRIs exhibit only low affinities for muscarinic and most other monoamine receptors (Table 6). Consequently at therapeutic doses they are relatively free from the cholinergic and anti-histamine side effects associated with the TCAs, and are less likely to cause sedation and drowsiness or hypotension. Instead of promoting weight gain, the SSRIs tend to suppress appetite and this can lead to weight loss. The most common side effect in the acute use of SSRIs is nausea, which clinical trial data indicate affects approximately 20% of patients taking fluoxetine (**12**), fluvoxamine (**13**), paroxetine (**21**), and citalopram (**4**) [48,49]. The risk of nausea is reduced if the SSRIs are taken with a meal. Most SSRIs tend to cause CNS "activation," leading to insomnia, agitation, or anxiety. Paradoxically, paroxetine (**21**) tends to cause sedation and drowsiness to such an extent that in several countries the drug is prescribed with a warning not to drive [48]. All SSRIs tend to cause sexual dysfunction, including loss of libido, erectile dysfunction, and delayed or absent orgasm. Both men and women are affected, and the incidence of these side effects is quite high. Although initial clinical trial data suggested that only a small proportion of patients suffered sexual dysfunction, more recent reports suggest that these side effects may occur in as many as 50–70% of patients taking SSRIs, and 30–40% of all those on antidepressant medication of any kind [70,71]. Patients are reluctant to volunteer information on sexual dysfunction, but when asked specifically a truer picture emerges.

A debate has raged for several years over the alleged association between the use of SSRIs and the occurrence of suicidal thoughts and suicide [72,73]. Some studies, have shown significant increases in self-harm or suicidal thoughts in young people treated with these drugs [74], and this led to regulatory agencies on both sides of the Atlantic insisting on the inclusion of a warning in patient information leaflets about the possibility of suicidal thoughts. However, this appears to have had the effect of inhibiting physicians from using antidepressants in young people, and this may have contributed perversely to an increase in suicide rates in this group. There is a well-established inverse relation between antidepressant use and suicide rates [74].

SSRIs also are metabolized by cytochrome p450 enzymes in the liver. Most SSRIs inhibit CYP2D6, fluvoxamine (10) inhibits CYP1A2, and fluoxetine (9) inhibits CYP3A4. Consequently, these drugs may interfere with the metabolism of a number of other agents. Given concurrently with TCAs they may cause serious adverse effects. Fluvoxamine may raise levels of caffeine and theophylline, and fluoxetine can interfere with the metabolism of clozapine, cyclosporin, and tefenadine [48,75]. SSRIs should never be given with MAO inhibitors because a fatal "serotonin syndrome" has been reported with fluoxetine in this combination [48].

2.3.5. Serotonin/Norepinephrine Reuptake Inhibitors The new generation of mixed monoamine uptake inhibitors, venlafaxine (29) and the recently introduced single isomer form desvenlafaxine (7), and duloxetine (10) are largely devoid of the cardiac toxicity associated with the TCAs. Consequently these drugs, such as the SSRIs, are much safer to use. Unlike the TCAs they have little affinity for muscarinic, histamine, or alpha-adrenergic receptors [76] (Table 6). Consequently, they do not exhibit the cholinergic, sedative, or hypotensive side effects seen with the earlier compounds [48,49]. Nevertheless, they tend to be associated with a number of less serious adverse side effects, of which nausea may be the most common, although dry mouth, sweating, dizziness, fatigue, constipation, anorexia, and sexual dysfunction may also occur.

2.3.6. Other Agents

Bupropion Bupropion (3) is a weak inhibitor of dopamine reuptake that has not been widely used outside the United States as an antidepressant. It has received a new worldwide lease on life as an effective means of treatment for tobacco smoking cessation [77]. It has little sedative, cholinergic, hypotensive, or cardiotoxic properties [78]. There are some adverse side effects, however, related to the ability of this drug to enhance dopaminergic function. These include insomnia, agitation, nausea, weight loss, and sometimes psychosis [79]. The drug decreases seizure threshold and so should not be given to those at risk of seizures [77]. In overdose acute toxicity is less serious than that seen with tricyclics [80]. Bupropion (3) should not be given with MAO inhibitors, levodopa, or dopaminergic receptor agonists [79].

Trazodone Trazodone (24) is a weak inhibitor of serotonin uptake and is an antagonist at 5-HT and alpha$_1$-adrenergic receptors (Table 6). These properties appear to be related to the side effects of sedation and hypotension (leading to dizziness) seen with the drug, and common at high doses [81]. Trazodone lacks cardiac toxicity, although dysrhythmias have been reported [48]; the drug appears safe in overdose. Trazodone sometimes causes an unusual type of sexual dysfunction, which can include increased libido, priapism, and spontaneous orgasm [82,83]. These symptoms, although rare, are dramatic and have received considerable attention. Combination with SSRIs or MAO inhibitors should be avoided because of the risk of "serotonin syndrome" [48].

Nefazodone Nefazodone (16) is chemically related to trazodone but acts in a different manner, largely through inhibition of serotonin uptake and antagonism at 5-HT$_2$ receptors. Adverse effects are mild and infrequent. They include sedation, dry mouth, and dizziness in around 10% of patients [84]. The drug causes less hypotension than does trazodone, and is unlikely to cause sexual dysfunction [48]. It is considered safe to use in epilepsy and there appears to be no overdose risk [48]. Nefazodone is a potent inhibitor of cytochrome p450 CYP3A4 and so should not be given with alprazolam, astenizole, terfenadine, cisapride, or cyclosporin [85].

Reboxetine Reboxetine (21) is a norepinephrine-selective reuptake inhibitor that lacks affinity for most of the monoamine receptors [86]. It thus does not exhibit the typical side-effect profile of the tricyclics. Nevertheless, side effects include increased sweating, postural hypotension (leading to dizziness), dry mouth, constipation, blurred vision, impotence, and dysuria. Tachycardia and urinary retention have also been reported [87]. There is no evidence of cardiotoxicity and sexual dysfunction seems to be rare. In contrast to some of the earlier tricyclics that are sedative, reboxetine is nonsedating and can cause insomnia [88,89].

2.4. Pharmacokinetics

2.4.1. Tricyclic Antidepressants Although time to peak plasma concentration can vary from 1 to 12 h according to both the drug and the individual, the tricyclic antidepressants are, by and large, well absorbed following oral administration, Most of these agents have long half-lives, and many are metabolized by demethylation in the liver, to yield biologically active desmethyl metabolites which further extends their duration of action [90,91] (Table 7). For example, imipramine (14) is metabolized to desipramine (6). In the case of lofepramine (15), its metabolite, desipramine (6), plays an important role in the overall actions of the drug in that desipramine has a considerably longer elimination half-life

Table 7. Pharmacokinetic Parameters for Antidepressant Drugs[a]

Generic Name	Time to Peak Plasma Concentration (h)	Elimination Half-Life (h)	Percentage Plasma Protein Binding	Important Metabolite
Amitriptyline (1)	1–5	10–26	94	Nortriptyline (20)
Amoxepine (2)	1–2	8–30	90	8-Hydroxyamoxapine
Bupropion (3)	3	10–21	85	BP-threoamino-alcohol
Citalopram (4)	1–6	33	80	Desmethylcitalopram
Clomipramine (5)	2–6	21–31	97	Desmethylclomipramine
Desipramine (6)	3–6	11–31	90	2-OH-desipramine
Dothiepin (8)	n.a.	14–24	n.a.	Desmethyldothiepin
Doxepine (9)	1–4	11–23	80	Desmethyldoxepine
Duloxetine (10)	4–6	10–12	—	Various
Fluoxetine (12)	4–8	24–120	94	Norfluoxetine
Fluvoxamine (13)	2–8	15–26	77	None
Imipramine (14)	1–3	11–25	92	Desipramine (6)
Lofepramine[b] (15)	n.a.	4–6	n.a.	Desipramine (6)
Maprotiline (16)	4–12	28–58	88	Desmethylmaprotiline
Mirtazapine (17)	2–3	20–40	85	None
Moclobemide (18)	1–1.5	1.4	n.a.	Numerous
Nefazodone (19)	1	2–4	99	mCPP (37)
Nortriptyline (20)	3–12	18–44	92	10-OH-nortriptyline
Paroxetine (21)	5–7	24–31	95	None
Phenelzine (22)	2–4	n.a.	n.a.	n.a.
Protriptyline (23)	6–12	67–89	93	None
Reboxetine[c] (24)	2–4	12	97	Various
Sertraline (25)	6–8	27	99	N-Desmethylsertraline
Tranylcypromine (26)	1.5–3	1.5–3.5	n.a.	n.a.
Trazodone (27)	1–2	6–11	92	mCPP (37)
Trimipramine (28)	3	9–11	95	None
Venlafaxine (29)	2	5	30	O-Desmethylvenlafaxine

[a] Data from Refs [59,94].
[b] Data from Ref. [57].
[c] Data from Ref. [93]. n.a.: data not available.

than that of the parent compound [57,65] (Table 7). The drugs are extensively metabolized in the liver by demethylation, hydroxylation, and glucuronide conjugation of the hydroxy metabolites. The lipid-soluble drugs are thus converted to water-soluble conjugates that are readily excreted by the kidney [92]. There is increased renal clearance in children and decreased clearance in older people, factors that need to be taken into account in determining optimum dosage levels. Clearance is also reduced in patients with compromised liver or kidney function.

There is considerable individual variation in the metabolism of the tricyclic antidepressants, and this is largely attributed to genetically determined differences in liver enzymes. Some 7 to 9% of Caucasians are classified as "slow metabolizers," measured by the rate of hydroxylation of the drug debrisoquin. This is caused by genetic polymorphism in the cytochrome p450 enzyme CYP2D6 [95]. This enzyme plays an important role in the aromatic hydroxylation of tricyclic antidepressants. The tricyclic drugs have a narrow therapeutic window, so individual variations in drug metabolism can be important in determining the correct therapeutic dose and avoiding toxic overdose [96,97].

2.4.2. SSRIs All the SSRIs are well absorbed and most have long half-lives, compatible with their use as once-a-day drugs [98–100]. The formation of biologically active desmethyl metabolites is again a factor in prolonging the duration of action of some of these drugs. This is particularly important for fluoxetine (**12**), which is metabolized in part to form norfluoxetine, an active metabolite that has a half-life of 4–16 days [101,102]. The desmethyl metabolite of sertraline (**25**), although less potent than the parent drug [98], also has an extended half-life [103,104]. The desmethyl metabolite of citalopram (**4**) is formed in relatively small amounts and appears to contribute less importantly [65,105]. The long duration of action of fluoxetine (**12**) and sertraline (**25**) make long drug-free periods necessary when switching patients to other drugs. All of the SSRIs are metabolized by cytochrome p450 CYP2D6 in the liver, so individual genetically determined differences exist in rates of drug clearance and there is the potential for interaction with other drugs that are metabolized by this enzyme [99,100]. Citalopram (**4**) and fluvoxamine (**13**) are also substrates of p450 CYP 2C19, which exhibits a particularly high rate of genetic polymorphism in Asians [106]. The pharmacokinetics of fluoxetine and citalopram are complicated by the fact that they are racemic compounds and the individual enantiomers may be metabolized and eliminated differently [100,107].

2.4.3. MAO Inhibitors The MAO inhibitors are rapidly absorbed and are extensively degraded by first-pass metabolism in the liver. For the irreversible enzyme inhibitors phenelzine (**22**) and tranylcypromine (**26**) the elimination half-lives are relatively unimportant, given that enzyme inhibition persists for many days after the drug has been eliminated. A minimum of 7–14 days is needed after stopping treatment with these drugs before it is safe to switch to other agents. The reversible MAO inhibitor moclobemide (**18**) is rapidly absorbed and extensively metabolized in the liver. It has an elimination half-life of approximately 12 h [108]. Because enzyme inhibition is reversible, the time to recover after stopping moclobemide treatment is much shorter, 16–24 h [109].

2.4.4. Other Antidepressants The other antidepressant drugs in Table 7 are rapidly absorbed and vary in their half-lives, with some requiring multiple daily dosing. Trazodone (**27**) and nefazodone (**19**) are metabolized in part to m-chlorophenylpiperazine (m-CPP; (**37**)), a compound that acts as an agonist at some serotonin receptors [110]. The metabolite may, thus, contribute to the biological action of these drugs. m-CPP (**37**) also has a longer half-life than the parent drugs and readily penetrates the CNS.

*m*CPP (**37**)

The norepinephrine selective uptake inhibitor reboxetine (24) is rapidly and completely absorbed, and is metabolized mainly by the cytochrome p450 3A4; because it does not interact with CYP 2D6 there is less risk of interactions with other drugs [111–113]. Mirtazepine (17) also shows little interaction with p450 cytochrome isozymes and there is only a low risk of drug interactions [114]. Mirtazepine is a racemate, and the two enantiomers are eliminated at different rates, with a two-fold higher rate of elimination of the (S)-enantiomer than of the (R)-enantiomer [114].

3. PHYSIOLOGY AND PHARMACOLOGY

3.1. Monoamine Transporters

3.1.1. Discovery The majority of both old and new antidepressants act by virtue of their ability to inhibit monoamine transporter mechanisms in the brain. The concept that neurotransmitters are inactivated by uptake of the released chemical into the nerve terminal from which it had been released or into adjacent cells is less than 40 years old. Before this it was generally assumed that the inactivation of norepinephrine and the other monoamine neurotransmitters after their release from nerves was likely to involve rapid enzymatic breakdown, akin to that seen with acetylcholinesterase. The degradation of monoamines by the enzyme monoamine oxidase was known early on, and in the 1950s a second enzyme catechol-O-methyl transferase (COMT) was discovered and was thought to play a key role in inactivating norepinephrine and other catecholamines.

It was not until high specific activity tritium-labeled radioactive catecholamines became available in the late 1950s, however, that experiments could be performed using quantities of monoamine small enough to mimic the very low concentrations of epinephrine or norepinephrine normally encountered in body fluids. When the first experiments were performed in the Axelrod laboratory at the National Institutes of Health with ^{3}H-epinephrine [113] and later with ^{3}H-norepinephrine [114], they yielded an unexpected result. Although in laboratory animals most of the injected dose of labeled catecholamine was rapidly metabolized (mainly by COMT), a substantial proportion of the injected monoamine (30–40%) was removed from the circulation by a rapid uptake into tissues, where it remained for some time unchanged. A key observation was that the uptake of ^{3}H-norepinephrine into the heart was virtually eliminated in animals in which the sympathetic innervation had been destroyed by surgical removal of the superior cervical ganglion [115]. This led Hertting and Axelrod [116] to propose that the reuptake of norepinephrine by the same nerves from which it had been released might represent a novel mechanism for inactivating this neurotransmitter.

The discovery of norepinephrine uptake was followed by the finding that similar but distinct transporters were involved in the inactivation of 5-HT and dopamine, and that similar mechanisms existed for the inactivation of the amino acid neurotransmitters GABA, glycine, and L-glutamate [117,118]. Research interest has focused on these mechanisms, including in recent years the identification and cloning of the genes encoding the transporter proteins involved and the development of knockout strains of genetically engineered mice lacking one or other of these gene products. The family of neurotransmitter transporters has turned out to be far more extensive than previously imagined, with more than 20 different members (for review, see Refs [119,120]).

3.1.2. Monoamine Transporters The norepinephrine transporter (NET) was cloned by Pacholczyk et al. in 1991 [121] and this soon led to the discovery of other related members of the transporter gene family. Separate transporters exist for SERT and DAT. The monoamine transporters are dependent on sodium and chloride ions for their function. They use the electrochemical gradient of sodium between the outside and inside surfaces of the cell membrane to provide the thermodynamic energy required to pump neurotransmitters from low concentrations outside the cell to the much higher concentrations inside the cell. Chloride ions accompany the entry of neurotransmitter and sodium, and there is a net movement of positively charged ions into the cell, although not in sufficient amounts to appreciably alter the resting membrane potential of the cell.

The vesicular neurotransmitter transporters represent another family [119] whose function is to maintain the very high concentrations of monoamine and amino acid neurotransmitters in storage vesicles. They use the proton gradient that exists across the vesicular membrane as the motive force. The vesicular monoamine transporters (VMAT) recognize serotonin, dopamine, norepinephrine, epinephrine, and histamine. VMAT-1 is present chiefly in amine-containing endocrine and paracrine cells in peripheral organs, whereas VMAT-2 is the predominant form found in monaminergic neurons in the CNS. It is also expressed in the histamine-containing cells of the stomach, and in the adrenal medulla and in blood cells. The Na^+/Cl^+-dependent transporters and the vesicular transporters are membrane proteins consisting of a single polypeptide chain of 5–600 amino acid residues, with a 12 a-helical membrane-spanning domain [119]. New insight into the mechanism by which antidepressants block transporter function was provided by studies of the crystal structure at 2.9 Å of the bacterial leucine transporter, a homolog of SERT, NET and DAT, in complex with leucine and desipramine. Desipramine binds to a site on the extracellular cavity of the transporter, near the leucine binding site, effectively locking the gate and preventing the conformational changes needed to effect leucine transport. Mutagenesis studies on human SERT and DAT confirmed that both the desipramine binding site and its inhibition mechanism are probably conserved in the human transporters [122]. Immunocytochemical and *in situ* hybridization techniques have been used to study the cellular distribution of the transporters [119]. Whereas NET and DAT are expressed exclusively in monoaminergic neurons SERT is expressed in both neurons and glia.

3.1.3. Drugs as Inhibitors of Monoamine Transporters By far the most important group of CNS drugs that target the NE and serotonin neurotransmitter transporters (i.e., NET and SERT, respectively) is the tricyclic antidepressants and their modern counterparts. The discovery by the Axelrod group in 1961 [123] that imipramine (**14**) potently inhibited the uptake of norepinephrine led to the first understanding of the mechanism of action of the first-generation tricyclic antidepressants. After the discovery of the serotonin uptake system in the brain it soon became apparent that the classical tricyclic drugs imipramine (**14**) and amitriptyline (**1**) were potent as inhibitors of both NE and 5-HT uptake (Table 2). This reinforced the monoamine hypothesis of depression as a monoamine-deficiency state, and stimulated much further research in the pharmaceutical industry to discover new inhibitors of monoamine uptake. The debate as to whether inhibition of either NE or 5-HT was the more important in conferring antidepressant efficacy has swung one way and the other over the past 40 years and there is no definitive answer to this question. An early effort to improve the selectivity of antidepressants was made in the 1970s by scientists at the Ciba-Geigy Company in Switzerland (now Novartis), who developed the selective NE uptake inhibitor maprotiline (**13**) [124]. This proved to be clinically effective as an antidepressant but it was not a great success commercially and had few clear advantages over the classical TCAs. This idea was also swept away by the wave of enthusiasm for SSRIs that started with the success of fluoxetine [125,126]. Table 8 summarizes the affinities of currently used antidepressants on cloned human monoamine transporters expressed in tissue culture cell lines [127]. The availability of the human transporter proteins for screening represents a considerable advance. Although there are many published accounts of the effects of antidepressants on monoamine transporter mechanisms, most of these employed animal tissues and there are few reported studies in which a large number of drugs were tested under the same experimental protocols.

Some of the most recently introduced antidepressants, the SNRIs, hark back to the nonselective compounds of the earlier era. Thus, duloxetine (**10**) and venlafaxine (**29**) combine both NE and serotonin reuptake inhibition [76,128]. However, estimates of the degree of transporter occupancy at clinical doses suggested that whereas venlafaxine

Table 8. Antidepressants: Inhibition of Human SERT, NET, and DAT[a]

Generic Name	Human SERT, K_d (nM)	Human NET, K_d (nM)	Human DAT, K_d (nM)	Selectivity: SERT versus NET
Amitriptyline (1)	4.3	35	3250	8
Amoxepine (2)	58	16	4310	0.3
Bupropion (3)	9100	52,000	520	5.7
Citalopram (4)	1.2	4070	28,100	3500
Clomipramine (5)	0.3	38	2190	130
Desipramine (6)	17.6	0.8	3190	0.05
Dothiepin (8)	8.6	46	5310	5.3
Doxepine (9)	68	29.5	12,100	0.4
Duloxetine (10)	0.8	7.5	240	9.4
Escitalopram (11)	2.5	2514	>100,000	>1000
Fluoxetine (12)	0.8	240	3600	300
Fluvoxamine (13)	2.2	1300	9200	580
Imipramine (14)	1.4	37	8500	27
Lofepramine (15)	70	5.4	18,000	0.08
Maprotiline (16)	5800	11.1	1000	0.002
Mirtazapine (17)	>100,000	4600	>100,000	—
Nefazodone (19)	200	360	360	1.8
Nortriptyline (20)	18	4.4	1140	0.24
Paroxetine (21)	0.13	40	490	300
Protriptyline (23)	19.6	1.4	2100	0.07
Reboxetine[b] (24)	129	1.1	—	0.008
Sertraline (25)	0.29	420	25	1400
Trazodone (27)	160	8500	7400	53
Trimipramine (28)	149	2450	780	16
Venlafaxine (29)	8.9	1060	9300	120

[a] Data from Refs [76,127].
[b] Data from Ref. [102].

achieved >90% occupancy of SERT, the occupancy of NET was only 60%. The authors questioned whether this was sufficient to affect noradrenergic function [129]. The compound sibutramine (38) is also an inhibitor of both NE and serotonin uptake, but it has been approved for use as an antiobesity agent rather than an antidepressant [130]. At the same time, reboxetine (24) is the first antidepressant drug in a new class of NET-selective inhibitors [130]. Reboxetine is reported to be as effective as the SSRIs or older tricylics, but is not associated with sexual dysfunction [88,89]. It is claimed to be more effective than fluoxetine in improving the social adjustment of depressed patients [131]. The older antidepressant bupropion (3) acts as a weak inhibitor of NE and dopamine uptake, with little effect on serotonin uptake, but it and some of its metabolites may indirectly activate noradrenergic mechanisms (see Section 5.4). The compound has had little success as an antidepressant, but has been approved in the United States and Europe as an aid to smoking cessation [77].

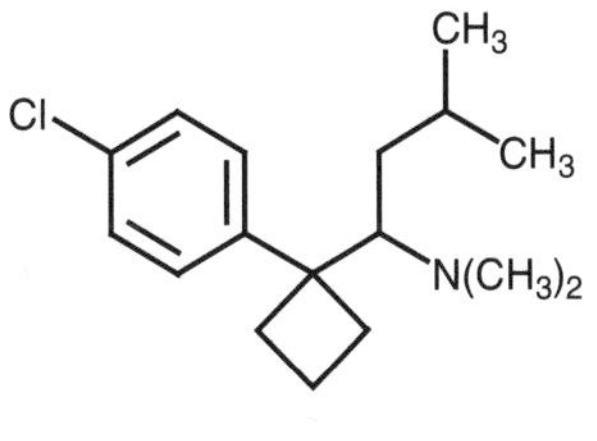

Sibutramine (38)

What are we to make of these twists and turns? How can drugs that are selective NE reuptake inhibitors be equally effective as those that selectively target SERT? In practice, it is difficult to know how selective the monoamine uptake inhibitors are *in vivo*. None of the antidepressants is completely selective for either NET or SERT. The

SSRIs have some affinity for NET, and some (e.g., paroxetine) are quite potent inhibitors of NET [130]. In some cases the formation of active metabolites alters the drug selectivity profile. Thus the nonselective compound imipramine (14) and the partially NET-selective compound lofepramine (15) are extensively metabolized to desipramine (6), a highly potent and selective NE reuptake inhibitor. Similarly, whereas amitriptyline (1) has little selectivity for either NET or SERT, the metabolite nortriptyline (20) is a selective NET inhibitor. It seems likely that both NET-selective agents and SSRIs exert their effects through some common final pathway in the brain. Perhaps the SSRIs act indirectly to modulate noradrenergic function [132,133]. Experimental data from animal experiments using microdialysis probes showed increased levels of extracellular norepinephrine in rat hippocampus after chronic treatment with paroxetine (18) [132]. The original monoamine hypothesis of depression as formulated by Schildkraut in 1965 [134] stated

Some, if not all, depressions are associated with an absolute or relative deficiency of catecholamines, particularly norepinephrine, at functionally important adrenergic receptor sites in the brain. Elation conversely may be associated with an excess of such amines.

Opinion currently seems to be swinging back in support of the view that an upregulation of noradrenergic function may be a key element underlying the efficacy of antidepressant drugs [132,133]. There is some evidence that NET-selective drugs are subtly different in their clinical profiles from those of SSRIs. Healy and McMonagle [135] have suggested that these drugs affect overlapping clinical domains. They suggest that NET-selective agents tend to promote levels of energy and interest, whereas SSRIs affect impulse control and both categories of drug treat mood, anxiety, and irritability. Some antidepressants, notably mazindol and bupropion (3), inhibit the dopamine transporter (DAT) as well as NET or SERT. The DAT is best known, however, as one of the principal sites of action of the psychostimulant drug cocaine. Mice that are genetically engineered to knock out the expression of the DAT gene are profoundly hyperactive and fail to show any further stimulation of activity in response to cocaine or (+)-amphetamine [136]. Such animals, nevertheless, will continue to self-administer cocaine [137], suggesting that the rewarding properties of the drug cannot be explained entirely by its ability to inhibit DAT. Cocaine (32) is also a potent inhibitor of both serotonin and NE reuptake. It retains some rewarding properties even in combined SERT and DAT knockout mice [138], suggesting that inhibition NE reuptake may also contribute importantly to its pharmacology. A corollary of the understanding that cocaine owes important parts of its overall CNS profile to mechanisms other than inhibition of DAT is that more selective inhibitors of dopamine reuptake might be useful and free of dependency liability. One such compound, brasofensine, is in clinical development for the treatment of Parkinson's disease [139]. Other selective DAT inhibitors have been proposed for the treatment of the withdrawal phase of CNS drug abuse. On the other hand, the structure of cocaine (39) has been modified in such a manner (e.g., (40) and (41)) that the resulting agents behave primarily as selective 5-HT reuptake inhibitors [140,141].

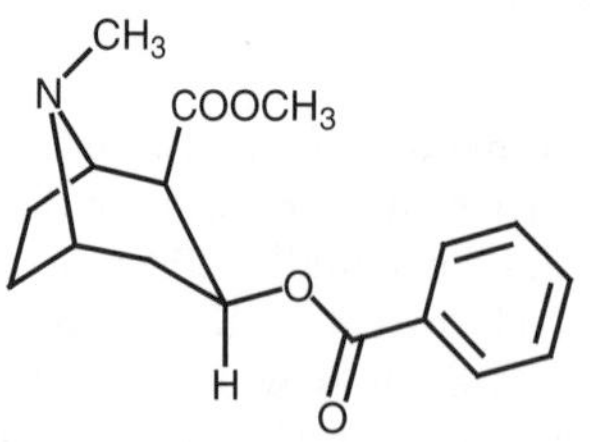

Cocaine (39) 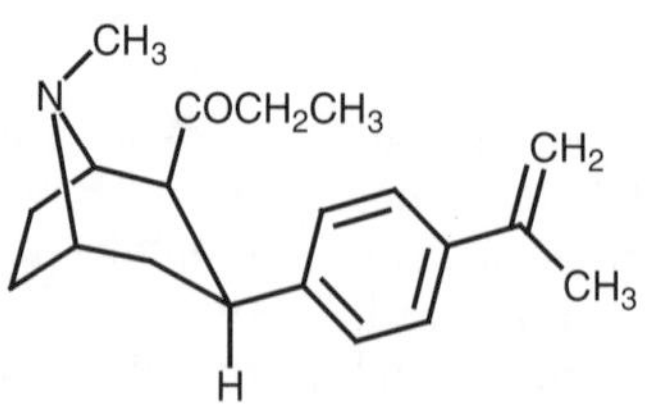   40 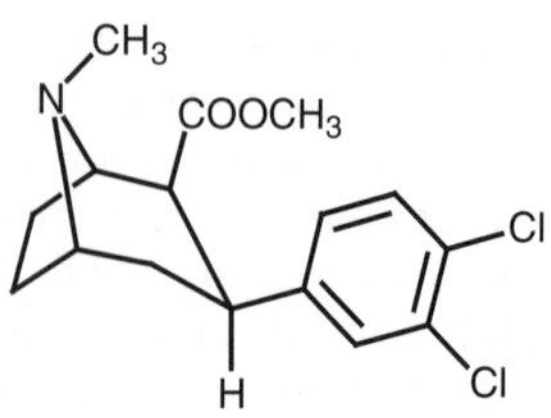   41

Some have suggested that a supersensitivity of central dopaminergic mechanisms may play an important role in the actions of antidepressant drugs [142]. Animals treated chronically with antidepressants become sensitized to the behavioral stimulant effects of dopaminergic drugs, and there is evidence for increased dopamine release in the brain [143].

The neurotransmitter transporter family has provided many valuable targets for psychopharmacology. There is every prospect that this will continue. It might seem that the monoamine transporters had already been fully exploited, but the reemergence of NET-specific antidepressants and SNRIs, and the possible applications of selective inhibitors of DAT suggest that there may still be room for innovation even in such a crowded field.

3.2. Serotonergic Agents

3.2.1. Receptor Populations Seven major families or populations of serotonin receptors have been identified: 5-HT_1– 5-HT_7 receptors. Several of these populations are divided into subpopulations [144–146]. For example, 5-HT_1 (5-HT_{1A}, 5-HT_{1B}, 5-HT_{1D}, 5-HT_{1E}, and 5-HT_{1F}) and 5-HT_2 (5-HT_{2A}, 5-HT_{2B}, and 5-HT_{2C}) receptors represent some serotonin receptor populations for which subpopulations exist. With the exception of the 5-HT_3 receptors, which are directly linked to an ion channel, all the other 5-HT receptors belong to the G-protein-coupled superfamily of receptors. 5-HT_1 receptors are negatively coupled to an adenylate cyclase second-messenger system, whereas the 5-HT_4, 5-HT_6, and 5-HT_7 receptors are positively coupled. 5-HT_2 receptors are coupled to a phosphatidylinositol second-messenger system. There is evidence that 5-HT_2 receptors and 5-HT_{1A} receptors are involved in depression. Certain of the other 5-HT receptors may also play a role in depression and there is evidence for functional interactions between 5-HT receptors.

3.2.2. 5-HT$_2$ Receptors The exact mechanism of action of some antidepressants is currently unknown. For example, the atypical antidepressant trazodone (**24**) is a weak 5-HT reuptake inhibitor, whereas nefazodone (**19**) is a weak inhibitor of both 5-HT and NE reuptake (Table 7). Inhibition of neurotransmitter reuptake does not seem to account for the antidepressant actions of these two agents. Both bind at 5-HT_{2A} receptors with high ($K_i < 25\,\text{nM}$) affinity [60] and are 5-HT_2 antagonists [147] (Table 6). A plausible mechanism of action for these drugs is that they enhance noradrenergic and serotonergic function by their ability to block presynaptic 5-HT receptors that normally exert an inhibitory effect on monoamine release in the brain, or by blockade of postsynaptic 5-HT_2 receptors (see below). Mianserin ("Tolvan") [42] is a nonselective 5-HT_1 and 5-HT_2 antagonist with antidepressant activity [148]; it binds with high affinity ($K_i < 10\,\text{nM}$) at 5-HT_{2A} receptors [149]. Hence, trazodone (**27**), nefazodone (**19**), and mianserin (**42**) represent atypical antidepressants that have in common a high affinity for 5-HT_{2A} receptors and 5-HT_2 antagonist action. It might be noted that certain tricyclic antidepressants also bind at 5-HT_{2A} receptors; imipramine (**14**), desipramine (**6**), nortriptyline (**20**), and maprotiline (**16**), for example, bind with submicromolar K_i values [149] (Table 6).

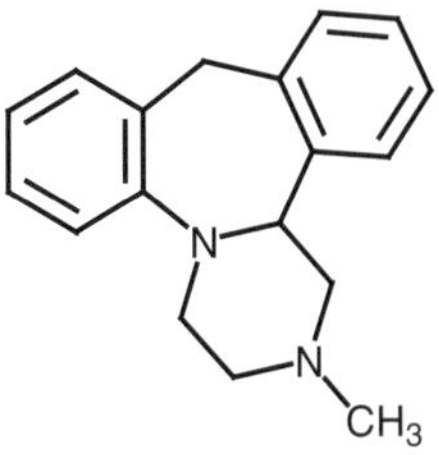

Mianserin (**42**)

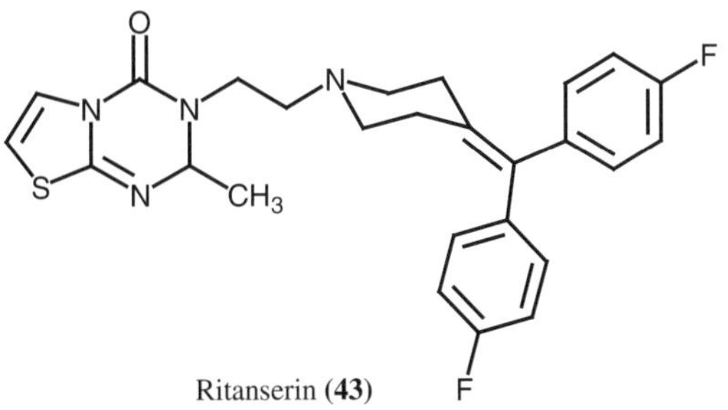

Ritanserin (**43**)

5-HT receptors are upregulated in depression. Hence, agents that downregulate 5-HT receptors might be of benefit in the treatment of this disorder. According to receptor adaptation theory, neurotransmitter antagonists should upregulate neurotransmitter receptors. However, paradoxically, 5-HT$_2$ antagonists generally downregulate 5-HT$_2$ receptors (e.g., see Ref. [150]). Mechanistically, then, it is reasonable that 5-HT$_2$ antagonists display antidepressant activity. Ritanserin (**43**) is an example of a 5-HT$_{2A}$ antagonist. In humans, ritanserin has been found to be as effective as amitriptyline (**1**), and superior to trazodone (**27**), as an antidepressant [151]. The SSRI fluoxetine (**12**) is metabolized to norfluoxetine (**44**).

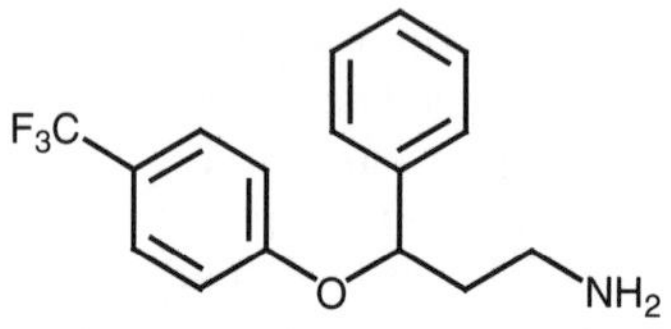

and anxiolytic actions in several animal models.

3.2.3. 5-HT$_{1A}$ Receptors 5-HT$_{1A}$ receptors have been implicated as playing roles both in depression and in anxiety [155]. Postsynaptic 5-HT$_{1A}$ (partial) agonist effects may be more important for antidepressant action, whereas agonist effects at presynaptic 5-HT$_{1A}$ receptors may be more important for antianxiety activity [156–158]. Furthermore, postsynaptic 5-HT$_{1A}$ receptors have been shown to be hypersensitive in depressed patients, whereas presynaptic receptors are hyposensitive [159]. Also, electroconvulsive therapy has been demonstrated to upregulate cortical 5-HT$_{1A}$ receptors [160].

Norfluoxetine (**44**)

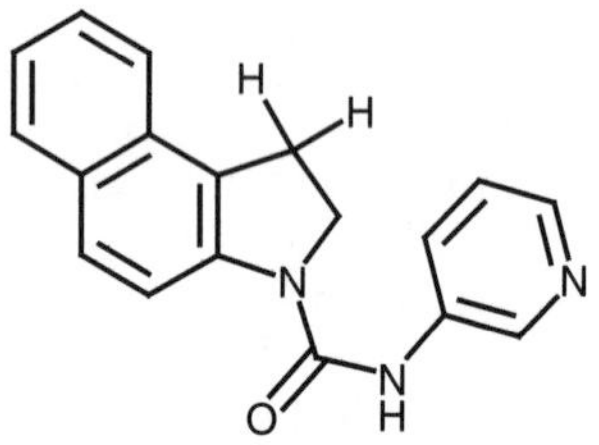

S32006 (**45**)

(−)-Norfluoxetine retains antidepressant activity and displays only slightly lower affinity for 5-HT$_{2C}$ receptors than it displays for the 5-HT transporter [152]. Results from an animal behavioral model predictive of antidepressant activity (the forced swim test) suggest that compounds that activate 5-HT$_{2C}$ receptors have antidepressant-like profiles [153]. *m*-Chlorophenylpiperazine (**37**), which has significant affinity as an agonist at 5-HT$_{2C}$ receptors, was among the compounds that were positive in this test. Because *m*CPP is an important and long-lasting metabolite of both trazodone (**27**) and nefazodone (**19**), it might contribute to their antidepressant profiles. In contrast to the above findings with agonists, the 5-HT$_{2C}$ receptor antagonist S32006 (**45**) [154] displayed antidepressant

Certain long-chain arylpiperazines (LCAPs) [161] have been demonstrated to possess both anxiolytic and antidepressant actions. For example, the anxiolytic agents buspirone (**46**), gepirone (**47**) [162], and ipsapirone (**48**) [163] showed antidepressant activity in clinical studies.

These agents are full agonists at presynaptic 5-HT$_{1A}$ receptors, but partial agonists at postsynaptic 5-HT$_{1A}$ receptors [151]. An alternative approach is to develop postsynaptic 5-HT$_{1A}$ agonists with greater efficacy than that of those currently available. Flesinoxan (**49**), an example of such an agent, is currently in clinical trials. A newer agent of this type is exemplified by (**50**) [164], which binds at 5-HT$_{1A}$ receptors with high affinity ($K_i = 1$ nM).

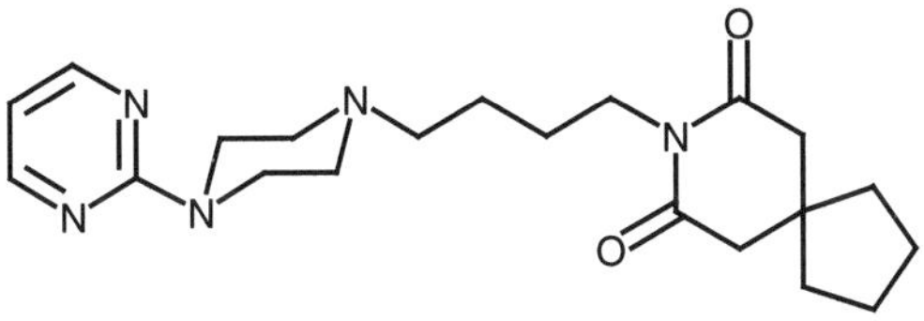

Buspirone (**46**)

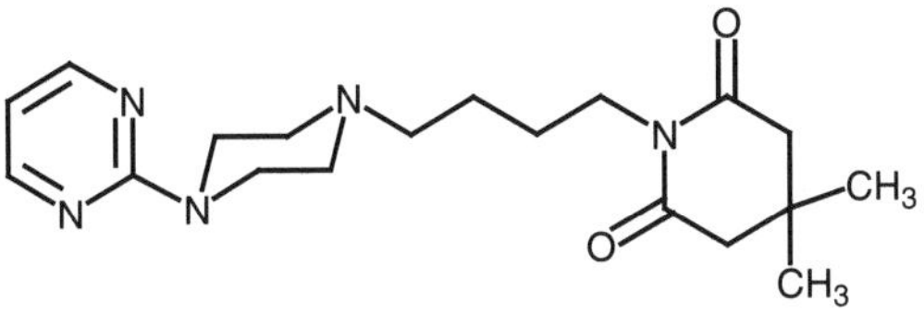

Gepirone (**47**)

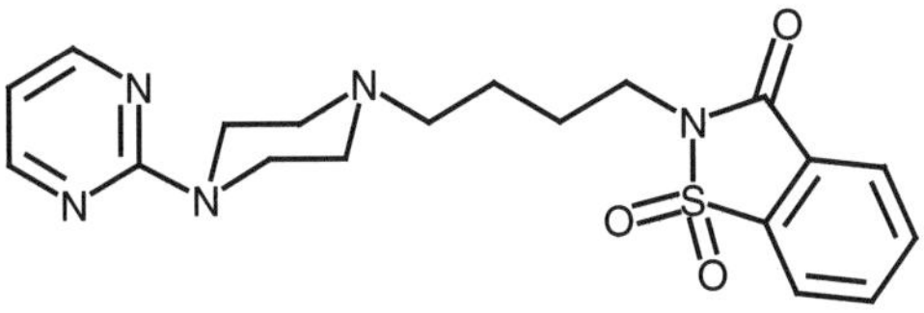

Ipsapirone (**48**)

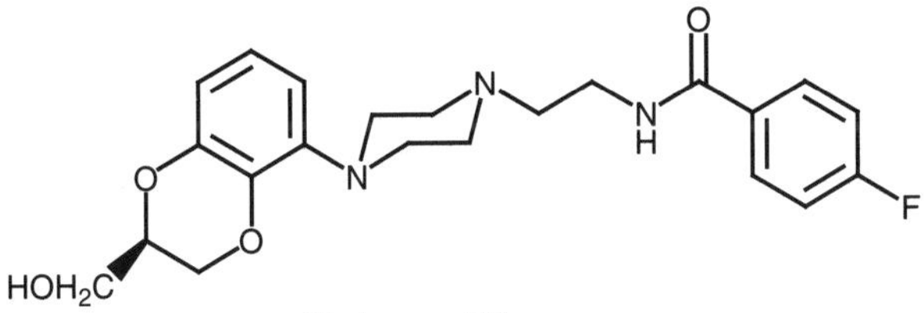

Flesinoxan (**49**)

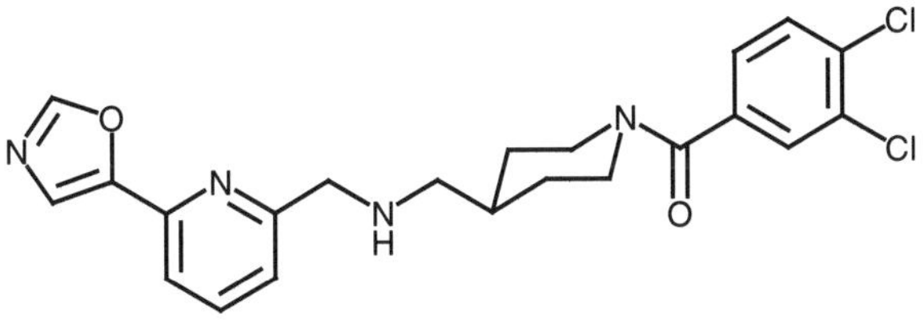

50

The actions of 5-HT on presynaptic 5-HT$_{1A}$ autoreceptors has been proposed as a possible explanation of the delayed onset of the clinical effects of antidepressants, and the potential application of 5-HT$_{1A}$ antagonists to accelerate the onset of clinical benefits has been discussed in Section 2.2 [17–22].

3.2.4. Mixed-Function Ligands

Recently, there have been attempts to incorporate multiple actions into the same molecule. For example, compound **51** (YM-35992) is a 5-HT$_2$ antagonist (IC$_{50}$ = 86 nM) and a 5-HT reuptake inhibitor (SERT IC$_{50}$ = 20 nM) [165]. Compound **52** is a 5-HT$_{1A}$ (K_i = 2.3 nM) antagonist with high affinity for the 5-HT transporter (SERT $K_i = 12$ nM) [166], whereas compound **53** displays somewhat higher affinity for both sites (5-HT$_{1A}$ IC$_{50}$ = 4.6 nM; SERT IC$_{50}$ = 1.7 nM) [167].

Compounds combining high affinity both for 5-HT$_{1A}$ and for 5-HT$_{2A}$ receptors would be potentially useful for the treatment of depression. RK-153 (**54**) [168,169] and adatanserin (**55**) [153] are examples of such agents. RK-153 (**54**) binds at 5-HT$_{1A}$ and 5-HT$_{2A}$ receptors with high affinity (K_i values of 0.4 and 34 nM, respectively) and displays reduced affinity for α_1-adrenergic receptors ($K_i > 1000$ nM) with high affinity for dopamine D2 receptors ($K_i = 2.7$ nM), whereas adatanserin (**55**) binds at 5-HT$_{1A}$ and 5-HT$_{2A}$ receptors with comparable affinity (K_i values of 1 and 73 nM, respectively), but displays substantially reduced affinity for dopamine D2 receptors ($K_i = 166$ nM). Both agents are 5-HT$_{1A}$ partial agonists, and adatanserin (**55**) has been demonstrated to be a 5-HT$_{2A}$ antagonist and is being developed as an antidepressant [170].

Strategies for the development of other mixed-function agents as potential antidepressants have been recently reviewed [171].

3.2.5. Other 5-HT Receptors Although presynaptic somatodendritic serotonin autoreceptors are of the 5-HT$_{1A}$ type, terminal autoreceptors are of the 5-HT$_{1D/1B}$ type. These latter receptors are involved in the modulation of serotonin release, and blockade of these receptors has been proposed as a possible means of developing novel antidepressants or of enhancing the effects of SSRIs (reviewed in Refs. [172,173]). SSRIs require several weeks before their antidepressant actions are evident; this delay corresponds to the time required to desensitize 5-HT$_{1A}$ and 5-HT$_{1D/1B}$ receptors [174]. Hence, a 5-HT$_{1D/1B}$ autoreceptor antagonist might hasten the onset of effects of an SSRI by mimicking its desensitizing action. Compound (**56**), a 5-HT$_{1D/1B}$ antagonist, has been demonstrated to augment citalopram-induced effects in rat ventral hippocampus and provides some support for this concept [175].

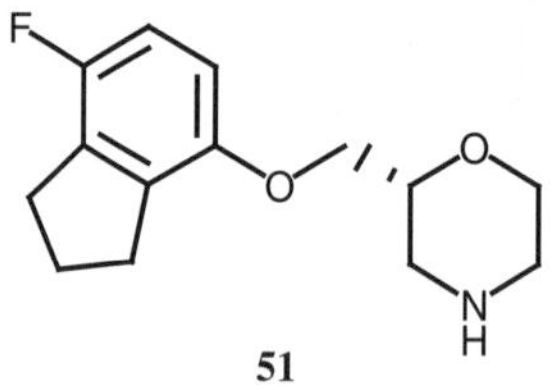

51

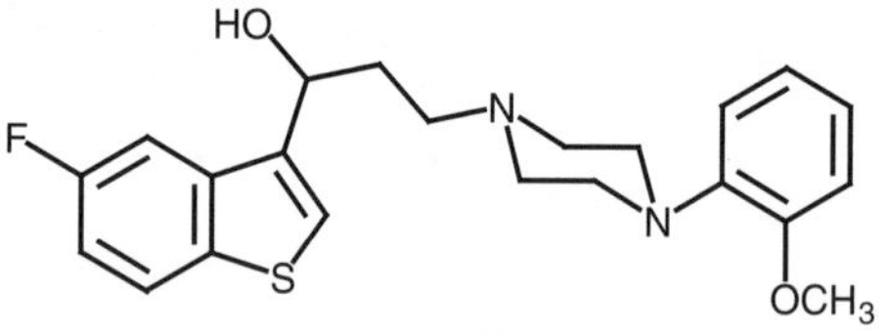

52

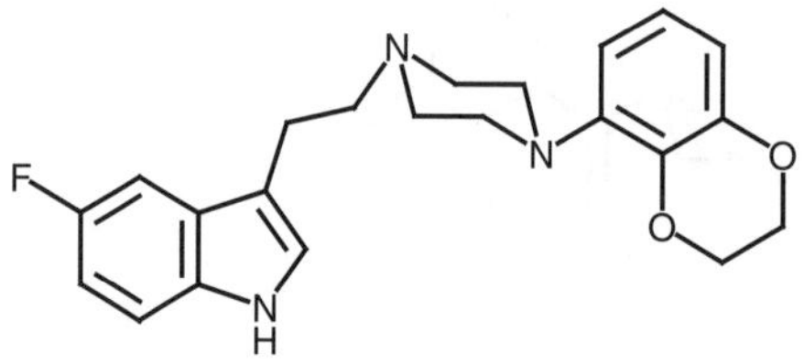

53

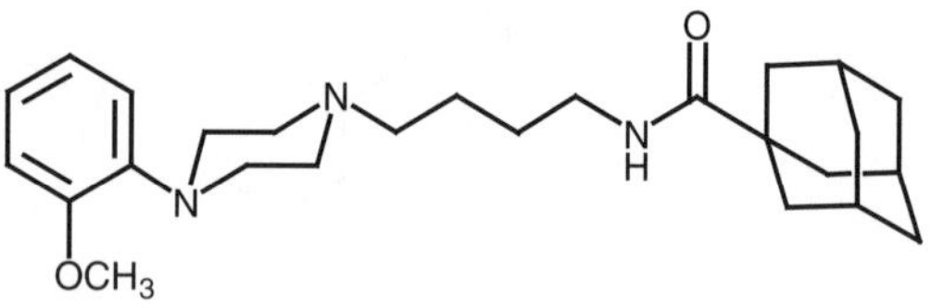

RK-153 (**54**)

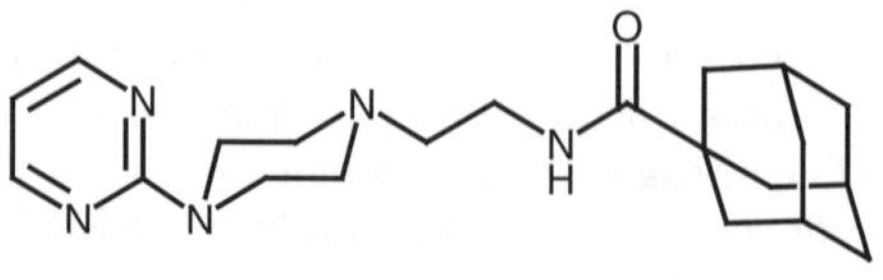

Adatanserin (**55**)

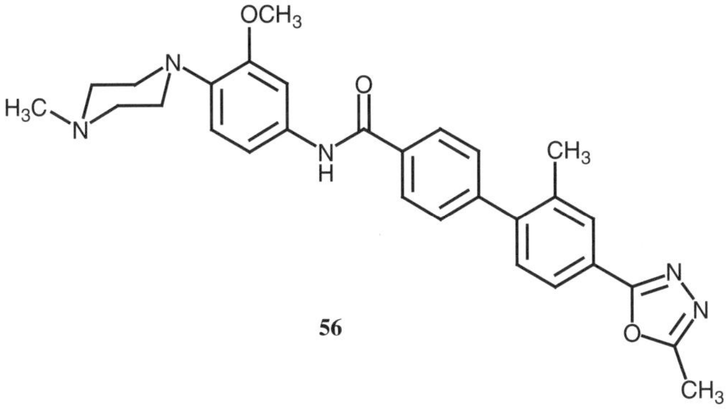

56

Anpirtoline (**57**)

Various antidepressants increase synaptic concentrations of serotonin, and the available serotonin can interact with one or more of a number of serotonin receptor subtypes. Modulation of 5-HT$_{1B}$ receptors might play a heretofore underappreciated role in the action of antidepressants, and this action seems to be substantially augmented by the protein p11 (also known as S100A10, calpactin I light chain, or annexin II light chain) in those regions of the brain where the two are colocalized [176]. 5-HT$_{1B}$ receptors represent presynaptic autoreceptors on serotonergic neurons and serve as heteroceptors on some nonserotonergic neurons; protein p11 regulates translocation of certain other proteins, including 5-HT$_{1B}$ receptors, to the cell surface [177]. Several lines of evidence support possible involvement of p11 in depression, and a modulatory role for p11 on 5-HT$_{1B}$ receptor function: long-term imipramine administration increased the amount of p11 in the forebrain, imipramine and electroconvulsive therapy increase p11 levels in cortex, p11 levels are reduced in patients who had suffered from depression, transgenic mice in which p11 is overexpressed behaved in animal models as if they had been treated with antidepressants, p11 knockout mice showed a reduced number of 5-HT$_{1B}$ receptor binding sites and the reduced number was reflected by the reduced ability of anpirtoline (**57**) (a 5-HT$_{1B}$ receptor agonist) to increase GTP-γ-S in functional studies, and p11 knockout mice displayed increased levels of serotonin turnover and/or metabolism [177].

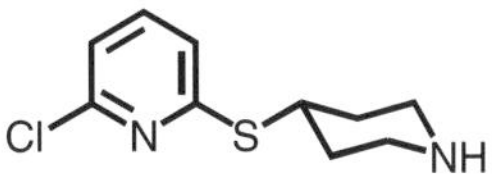

5-HT$_{1B}$ receptors are only one population of serotonin receptors implicated in the actions of antidepressants. Serotonin, as do some other neurotransmitters, interacts with receptors that are associated with a 3′,5′- cyclic-adenosine monophosphate/protein kinase A second messenger pathway (see Section 3.5). One of the major targets of this pathway is dopamine- and cAMP-regulated phosphoprotein of molecular weight 32,000 (i.e., DARPP-32, hereafter referred to simply as DARPP). The function of DARPP is regulated by its phosphorylation state [178]. Activation of PKA by stimulation of adenylate cyclase results in phosphorylation of DARPP (Fig. 6). Phosphorylation of a particular threonine residue (Thr-34) converts DARPP to phospho-Thr^{34}-DARPP, which is a potent inhibitor of protein phosphatase-1 (PP-1). Hence, the interaction of a neurotransmitter or agonist with a receptor positively coupled to adenylate cyclase initiates a cascade of events that results in inhibition of PP-1 and thereby produces an enhanced signaling effect. Phosphorylation of DARPP by a cyclin-dependent kinase (i.e., Cdk5) at a different residue, Thr-75, produces phospho-Thr^{75}-DARPP which converts DARPP into an inhibitor of PKA and antagonizes the PKA/Thr^{34}-DARPP cascade [178]. Phosphorylation of DARPP at serine-137 (Ser-137) by casein kinase (i.e., CK1) potentiates the PKA/Thr^{34}-DARPP cascade by preventing dephosphorylation of phospho-Thr^{34}-DARPP. Fluoxetine (12), for example, increases inhibition of PP-1 in brain (prefrontal cortex, hippocampus, striatum) by increasing phosphorylation of DARPP at Thr-34 and Ser-137, and decreasing phosphorylation at Thr-75 [178]. Although early studies focused on the dopaminergic system (hence, derivation of the name DARPP), it has been found that stimulation of 5-HT$_4$ and 5-HT$_6$ serotonin receptors—receptors that are positively coupled to adenylate cyclase—causes increased phosphorylation at Thr-34 and decreased phosphorylation at Thr-75 [179]. The 5-HT$_6$ receptor agonist EMDT (58) [180] has been shown to produce fluoxetine-like effects in a rodent model of depression (tail-suspension test) and the effect was blocked by pretreatment of the animals with a 5-HT$_6$ receptor antagonist (i.e., SB271046) [181]. Interestingly, the behavioral effects of fluoxetine in this same animal model were only partially antagonized by SB271046 leading to the suggestion that other serotonin receptors, in addition to 5-HT$_6$ receptors, contribute to the actions of fluoxetine [181]. Indeed, it has been demonstrated that stimulation of 5-HT$_2$ receptors increases phosphorylation of DARPP at Ser-137 through a phospholipase C/CK1 mechanism [179]. Furthermore, the 5-HT$_4$ receptor agonists RS 67333 (59) and prucalopride (60), display antidepressant-like actions in cellular and behavioral models of depression with an onset to action of several times less than that of classical antidepressants [182]. It has been suggested that 5-HT$_4$ receptor agonists might be a "silver bullet" for the treatment of depression [183]. These findings indicate that activation of 5-HT$_6$, and perhaps 5-HT$_4$, receptors offer new approaches to developing novel antidepressants.

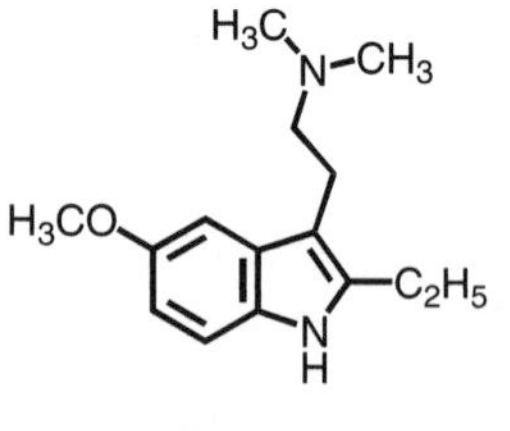

EMDT (58)

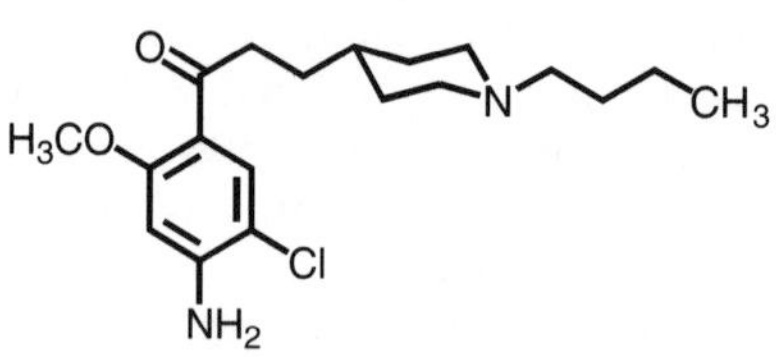

RS 67333 (59)

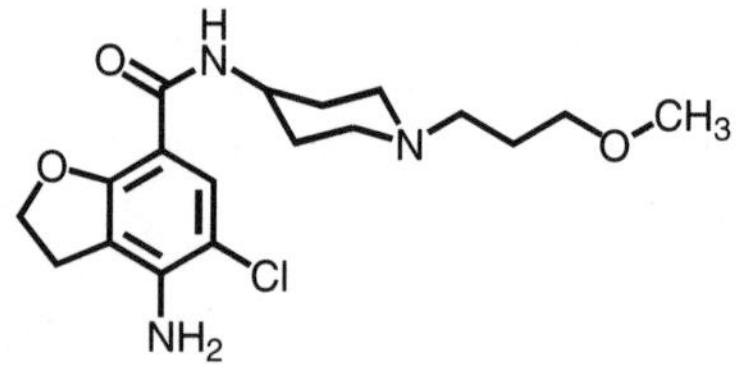

Prucalopride (60)

A population of receptors yet to receive extensive investigation from an antidepressant perspective are the 5-HT$_3$ receptors. 5-HT$_3$ receptors are known to be involved in the release of several neurotransmitters, including norepinephrine, serotonin, and dopamine, and there are preclinical data indicating 5-HT$_3$ antagonists might play a role in anxiety and, to a lesser extent, in depression (reviewed in Ref. [184]). The clinically available 5-HT$_3$ antagonist ondansetron, for example, has been found active in several rodent models of depression, and augmented the actions of fluoxetine and venlafaxine, but not those of desipramine [185]. It might also be noted that mirtazepine (14), although its antidepressant properties have been attributed primarily to its antagonist action at α-adrenergic receptors and 5-HT_2 serotonin receptors, is also a 5-HT_{1A} agonist and a 5-HT_3 receptor antagonist [186].

One of the most recently discovered populations of 5-HT receptors are the 5-HT_7 receptors. Of interest is that certain typical and atypical antidepressants, as well as typical and atypical antipsychotic agents, bind at 5-HT_7 receptors [187,188]. A role for these receptors in the actions of these agents has yet to be firmly established, but the nanomolar affinity of the agents for these receptors has heightened interest in them. Recently, several novel 5-HT_7 receptor antagonists have been shown to display antidepressant-like and/or anxiolytic-like actions in animal behavioral models [189–191].

3.3. α-Adrenergic Receptors (α-Adrenoceptors)

Although several antidepressant drugs have weak affinity for alpha-adrenergic receptors (Table 6) only one compound is thought to owe its antidepressant activity to such an interaction. This is mirtazepine (**14**), which has nanomolar affinity as an $α_2$-adrenergic receptor antagonist [192,193]. The $α_2$-adrenergic receptors are located presynaptically on both noradrenergic and serotonergic nerve endings and cell bodies, and exert an inhibitory effect on monoamine neuronal firing and monoamine release. Administration of mirtazepine to animals has been reported to increase the spontaneous rate of firing of noradrenergic neurons in rat locus coeruleus, and serotonin neurons in the raphe nucleus [194], and to increase levels of 5-HT release in hippocampus [195], although this was not confirmed in another publication [196]. Clinical trials have shown that mirtazepine is equivalent to amitriptyline and other tricyclics in antidepressant efficacy [197]. Some of these studies showed a clinical improvement within the first week of treatment [197]. Mianserin (**34**) is another agent whose effects, at least in part, might involve $α_2$-adrenergic antagonism [198]. Many adrenergic agents possess an imidazoline ring. Imidazolines have been found to bind at nonadrenergic imidazoline binding sites and this has led to speculation that such sites might play a role in depression (see Section 5.4).

Only recently are subpopulation-selective adrenergic agents available for evaluation in animal models of depression. The $β_3$-adrenoceptor agonist SR58611A (amibegron; **61**) [199–201] and the $α_{2C}$-adrenoceptor antagonist JP-1302 (**62**) [202] displayed antidepressant-like actions in preclinical studies.

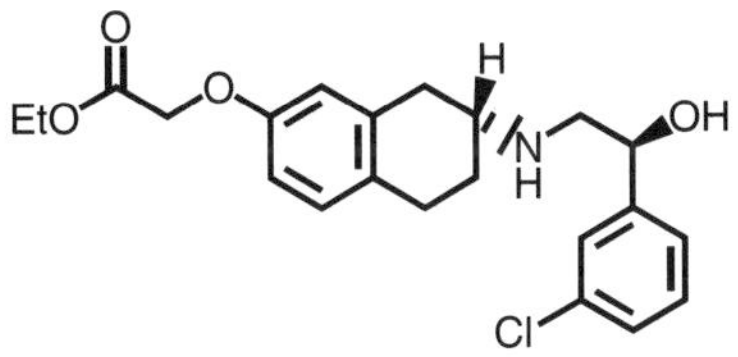

SR58611A (**61**)

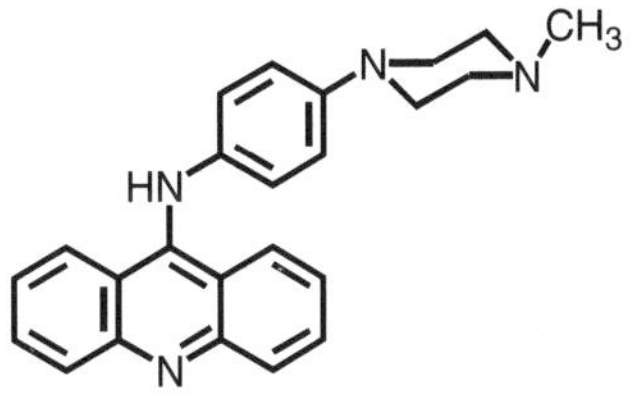

JP-1302 (**62**)

3.4. Monoamine Oxidase

Monoamine oxidase is an enzyme located in the outer mitochondrial membrane. As its name implies, it catalyzes the oxidative deamination of a variety of monoamines. The enzyme is particularly abundant in liver, where it serves to detoxify a variety of amines that are absorbed from the diet; these include the vasoactive monoamines tyramine, octopamine, phenylethanolamine, and phenylethylamine. The enzyme is also present in monoamine-containing neurons, where it serves to regulate levels of cytoplasmic monoamine neurotransmitters [203]. There are two forms of MAO: MAO-A and MAO-B [204]. Both are present in the liver, and in most monoaminergic neurons, although MAO-A is predominant in norepinephrine and dopamine-containing neurons, and MAO-B in serotonergic cells. Both forms of the enzyme have a wide and overlapping range of substrates, but MAO-A shows some preference for the catecholamines norepinephrine and epinephrine, and serotonin, and MAO-B for tyramine, phenylethylamine, phenylethanolamine, and benzylamine. Both enzymes metabolize dopamine and tryptamine [203–205].

The first-generation MAO inhibitors phenelzine (**22**) and tranylcypromine (**26**) act as substrates for MAO-A and MAO-B but are converted by the enzyme to highly reactive intermediates that then react irreversibly with the enzyme to cause an irreversible inhibition of activity. Recovery of MAO activity after exposure to these MAO inhibitors requires the synthesis of new enzyme protein, a process that takes some weeks to completely restore activity [206,207]. A clinical antidepressant response is associated with an inhibition of platelet MAO activity of approximately 80%, and measurement of platelet MAO activity can be used to monitor treatment dose regimes [208].

A new series of MAO inhibitors are selective for MAO-A and cause reversible inhibition of the enzyme, thus leaving MAO-B in the liver intact to detoxify dietary amines, and showing rapid recovery of enzyme activity after discontinuation of drug treatment. The only drug of this type so far available for human use is moclobemide (**18**).

4. HISTORY

4.1. Discovery of the First Antidepressants

Before 1954, except for the use of electroconvulsive therapy, there were no effective treatments for depression. The two major classes of antidepressants, the monoamine oxidase inhibitors and inhibitors of monoamine transport, were discovered by accident in the 1950s. The drug iproniazid used for the treatment of tuberculosis was found to have a mood-elevating property [209] and clinical studies by George Crane and Nathan Kline in the United States showed it to be effective in treating major depression [210,211]. Its actions were traced to its ability to inhibit the monoamine-degrading enzyme MAO [212]. Although this and other subsequently developed MAO inhibitors proved highly effective in the treatment of depression, the possible dangers associated with their use led to their being largely replaced by the safer inhibitors of monoamine transport. The first examples of the latter drugs to be widely used were imipramine (**14**) in Europe and amitriptyline (**1**) in the United States. Imipramine was synthesized originally by the Swiss company Geigy as a chlorpromazine-like molecule with potential as an antipsychotic drug. The Swiss psychiatrist Roland Kuhn, however, found it to be an effective antidepressant [213,214]. On the other side of the Atlantic, Merck first made amitriptyline also as a chlorpromazine-like molecule. It was shown subsequently to be an antidepressant by Frank Ayd [215]. For a detailed and entertaining account of the history of the discovery of antidepressant drugs, see Healy's *The Antidepressant Era* (1997) [216].

4.2. Case History: Fluoxetine (Prozac) (12)

One of the earliest theories of how antidepressant drugs work was that they caused an increased availability of serotonin in the brain. This was supported by data from the British psychiatrist Alec Coppen, that combining an MAO inhibitor with the serotonin precursor tryptophan was a more effective antidepressant treatment than the MAO inhibitor alone [217], a result repeated by the Dutch psychiatrist Herman van Praag. The "serotonin hypothesis" was largely lost sight of, however, after the discovery that imipramine (**14**) and related tricyclic antidepressants were potent inhibitors of norepinephrine uptake [123] and the idea that norepinephrine was the key player in antidepressant drug actions dominated thinking, particularly in the United States [218]. The norepinephrine-selective uptake inhibitor desipramine (**6**) from Merck proved highly successful, and in Europe the Swiss company Ciba launched its norepinephrine-selective uptake inhibitor maprotiline (**16**) [124].

Nevertheless, the "mixed" norepinephrine/ serotonin uptake inhibitors imipramine (**14**) and amitriptyline (**1**) continued to be very popular, particularly in Europe. In Sweden, the neuropharmacologist Arvid Carlsson, originally a champion of the norepinephrine hypothesis, interested in the idea of developing selective inhibitors of serotonin uptake. Having failed to interest any major pharmaceutical company in this idea, he collaborated with the chemist Hans Corrodi at the Swedish company Astra. They produced the first SSRI, zimelidine (**63**), which was more potent than clomipramine (**5**), the best SSRI then available, and unlike clomipramine zimelidine did not break down in the body to form a norepinephrine-selective active metabolite. Astra reported the first positive clinical trials with zimelidine as an antidepressant in 1980 [219]. Zimelidine was initially successful in Europe and was to be marketed in the United States by Merck, who had completed extensive clinical trials and submitted the results to the FDA in 1983, although the drug was withdrawn shortly thereafter following reports of drug-induced Guillain-Barré syndrome (peripheral nerve damage) in Europe. Carlsson also took the SSRI idea to the Danish company Lundbeck, which subsequently launched the highly selective and potent compound citalopram (**4**) in 1986. Meanwhile at Ciba-Geigy in Switzerland, Peter Waldmeier and colleagues had been working on the serotonin idea since the early 1970s and discovered several potent SSRIs, but were unable to persuade the company to develop any of them further [124]. David Wong and colleagues at Eli-Lilly were more fortunate, but only after a long delay [125,126]. They had also been stimulated by Carlsson's ideas and started screening for SSRIs in 1972. They discovered fluoxetine (**12**) and reported its biochemical profile in 1974 [125], but Lilly was not at all clear what the drug would be used for. At one of the meetings of clinical experts that the company convened, Alec Coppen suggested that it might be tested in depression, only to be told that this was definitely not the target the company had in mind! [216]. It was only in the 1980s, when the antidepressant profile of zimelidine was reported, that Lilly began to speed up the development of fluoxetine (**12**), and it was 1985 until the first positive clinical trial results in depression were reported [126], followed by registration in the United States in 1987.

Zimelidine (**63**)

Several other SSRIs were registered: fluvoxamine (**13**) in 1983, sertraline (**25**) in 1990, and paroxetine (**21**) in 1991. It was fluoxetine (**12**), however, that captured the public imagination and became the single most important psychopharmaceutical product of the late twentieth century. Physicians liked it because it was safe in overdose, the dosage regime was simple (not requiring any gradual titration) and the side-effect profile was an improvement over that of the earlier tricyclics. Prozac was on the cover of *Time* and *Newsweek* and it gained a number of additional medical uses (Table 4). Some people also used it for non-medical purposes, just to make them feel better, or to enjoy the opera more! A whole literature was spawned around the drug, most famously in Peter Kramer's book, *Listening to Prozac* [220]. A new market was even found for the drug in the treatment of depression in companion animals. By the turn of the century, sales of Prozac were earning Lilly in excess of US$2 billion annually.

5. STRUCTURE–ACTIVITY RELATIONSHIP AND METABOLISM

5.1. Reuptake Inhibitors

The SAR of the tricyclic antidepressants has been studied for over 40 years and numerous reviews are readily available on these and related agents [221–224]. Only the highlights of some of the pertinent SAR studies are presented here. Tricyclic antidepressants have in common a tricyclic ring structure consisting of a six-membered or, more commonly, a seven-membered ring flanked, typically, by two benzene rings. The exact composition and orientation of the central ring is relatively inconsequential (see Fig. 1). Tricyclic antipsychotics and tricyclic antidepressants appear to exist on a structural continuum. For example, introduction of an electron-withdrawing group to the aromatic ring typically enhances the antipsychotic profile of the tricyclic agent. Replacement of a ring nitrogen atom (as in the phenothiazines) or an sp^2-hybridized carbon atom (as with the thioxanthenes) to which the aminopropyl substituent is attached, with an sp^3-hybridized carbon atom detracts from the antipsychotic profile but can enhance the antidepressant profile of an agent. That is, a ring nitrogen atom or an sp^2-hybridized carbon atom is not required at this position for antidepressant activity; for example, compare the structures of nortriptyline (20) and protriptyline (23).

The tricyclic system is generally attached to a basic terminal amine by a three-atom chain. The chain can be shortened to two atoms with retention of antidepressant activity and a decrease in antipsychotic character. The terminal amine is typically a secondary or tertiary amine, and the amine can be part of an alicyclic ring. In general, tertiary amines are more nonselective with respect to inhibition of 5-HT and NE reuptake, whereas the secondary amines are typically more selective at inhibiting NE reuptake. *In vivo*, however, one of the major routes of metabolism of the tertiary amines is by demethylation to a secondary amine.

Doxepine (9) is related in structure to the antipsychotic agent pinoxepin (64). However, because doxepine lacks the *N*-hydroxyethyl group of pinoxepin (a functionality known to enhance antipsychotic activity) and the electron-withdrawing chloro group, it behaves more as an antidepressant than as an antipsychotic agent.

Because the presence of a ring nitrogen atom or an sp^2-hybridized carbon atom in the central ring of these tricyclic structures is not a requirement for antidepressant activity, protriptyline (23) retains antidepressant action. Certain tricyclic antidepressants with a six-membered central ring have been shown to possess an antidepressant profile. If the central ring is capable of aromatization, by oxidation for example, to afford a completely planar structure, the resulting compound is inactive. If aromatization can be prevented, however, antidepressant activity is retained. Maprotiline (16) is an example of a tricyclic agent where aromatization is prevented by introduction of an alkyl bridge. Aromatization is prevented according to Bredt's rule. Because it is a secondary amine, maprotiline is fairly selective for inhibition of NE reuptake. Note that maprotiline does not possess either a nitrogen atom or an sp^2-hybridized carbon atom in the central ring. Although originally considered as a member of a new class of antidepressants—the "tetracyclic antidepressants"— maprotiline is now classified as a TCA.

Amoxepine is a structurally interesting agent because of the introduction of an electron-withdrawing group. As already mentioned, the orientation of the tricyclic system does not seem crucial to antidepressant (or antipsychotic) activity. Loxapine (65), for example, is an antipsychotic agent. Amoxepine is the *N*-desmethyl analog of loxapine. Because secondary amines show a greater antidepressant profile than that of tertiary amines, amoxepine possesses greater antidepressant character than that of loxapine. However, amoxepine also possesses some antipsychotic character attributed to the pre-

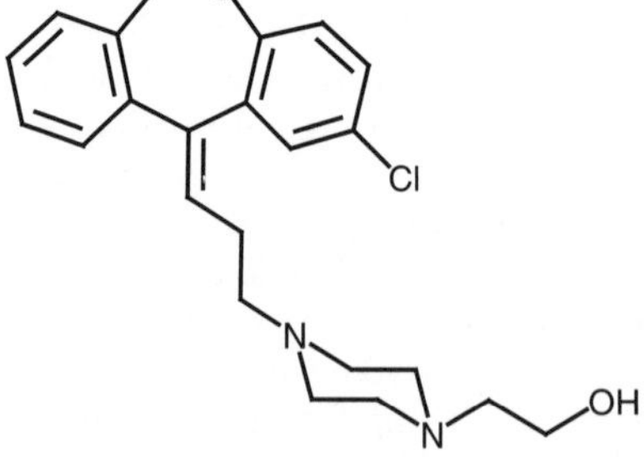

Pinoxepin (64)

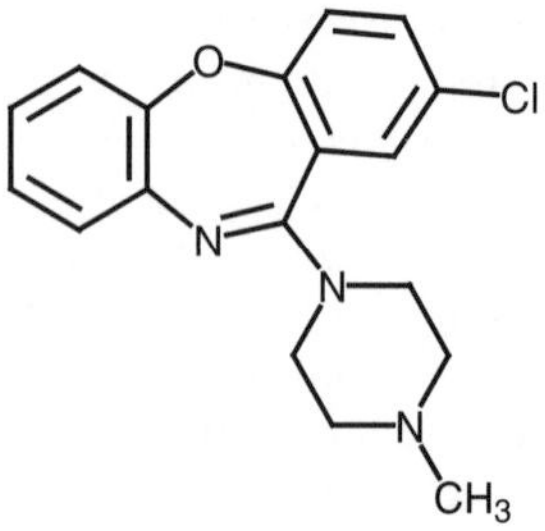

Loxapine (65)

sence of the electron-withdrawing chloro group.

The tricyclic antidepressants typically undergo multiple routes of metabolism. The most common, depending on the particular ring system, are N-demethylation of the terminal amine, aromatic hydroxylation, and benzylic or "bridge" hydroxylation (see Table 7). In general, with the exception of the secondary amine metabolites of N,N-dimethylamino TCAs, the metabolites are usually less active or inactive as antidepressants.

Remarkably, little has been published on the structure–activity relationships of reuptake inhibitors. Close inspection of these inhibitors reveals that many of them are related in structure to ring-opened tricyclic antidepressants. Indeed, small structural changes can result in shifts in selectivity. For example, fluoxetine (**12**) is a serotonin-selective reuptake inhibitor, whereas nisoxetine (**66**) is a norepinephrine-selective reuptake inhibitor.

of most other SSRIs, the desmethyl metabolite of fluoxetine, norfluoxetine (**44**), retains the ability to inhibit serotonin reuptake [226]. Hence, fluoxetine takes longer to achieve steady-state levels, and it retains activity after metabolism. The half-life of fluoxetine is approximately 1 day, but elimination of norfluoxetine is prolonged (7–15 days) [226]. In addition to its actions as an SSRI, norfluoxetine also binds at 5-HT$_2$ receptors [152].

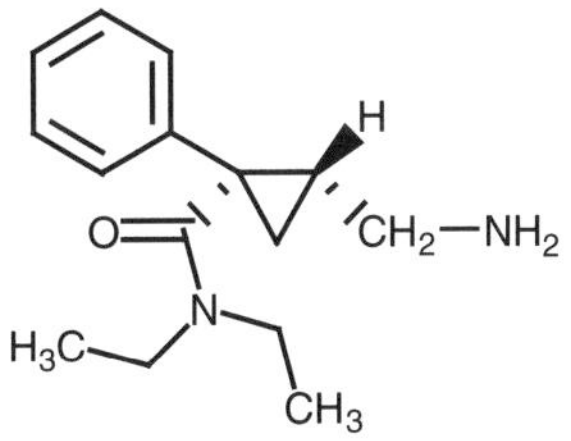

5.2. Monoamine Oxidase Inhibitors

Because of undesirable side effects associated with monoamine oxidase inhibitor therapy (see Section 2.3.2), pharmaceutical companies nearly abandoned research on new analogs in the 1960s. The early MAO inhibitors were nonselective and irreversible. Today, efforts toward the development of monoamine oxidase inhibitors are focused on selective MAO-A or MAO-B inhibitors. Selective MAO-B inhibitors

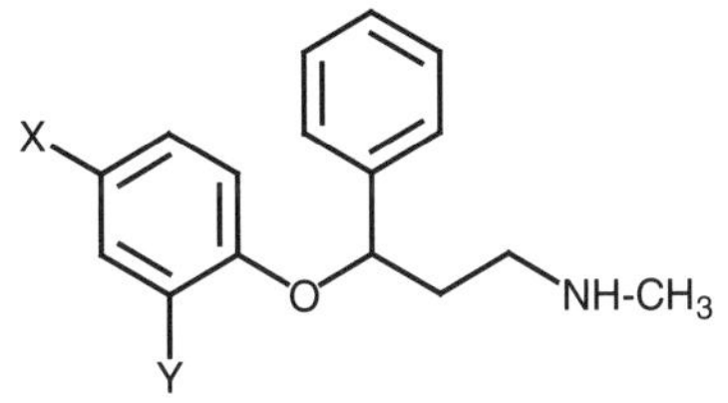

Fluoxetine (**12**) X = —CF$_3$, Y = —H
Nisoxetine (**66**) X = —H, Y = —OCH$_3$

Milnacipran (**67**)

A newer norepinephrine-selective reuptake inhibitor currently in clinical trials is milnacipran (**67**). An analog of a bupropion metabolite, (**68**), has been shown to act as a norepinephrine-selective reuptake inhibitor [225].

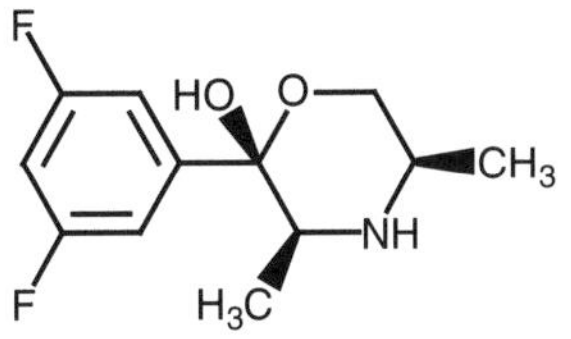

are being examined in the treatment of, for example, schizophrenia, Alzheimer's disease, and Parkinson's disease. MAO-B inhibitors might be effective in the treatment of depression, but relatively little work has been done in this area. Selegiline (**69**) or (−)-deprenyl, a selective irreversible MAO-B inhibitor, is one of the few agents that has been examined in this regard and the clinical results are mixed (reviewed in Ref. [227]).

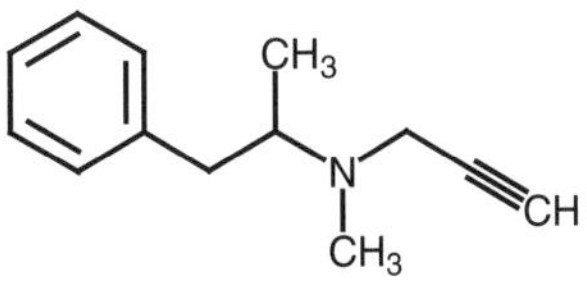

68

Sertraline and fluoxetine undergo metabolic demethylation. Unlike the metabolites

Selegiline (**69**)

Of greater application to the treatment of depression are the reversible inhibitors of monoamine oxidase-A or RIMAs. Brofaromine (**70**) is a "tight binding" but reversible inhibitor of MAO-A with approximately 100-fold selectivity versus MAO-B. Moclobemide (**18**) is a RIMA with approximately 5- to 10-fold selectivity for MAO-A. Although both agents display micromolar (or lower) affinity for adrenergic, serotonin, and most other receptors, brofaromine, interestingly, binds at the 5-HT and NE transporters with modest affinity (IC_{50} values of 150 and 500 nM) [228]. For comparison, the half-life for disappearance of MAO-A inhibition in brain is phenelzine (**22**), 11 days; tranylcypromine (**26**), 2.5 days; brofaromine (**70**), 12 h; and moclobemide (**18**), 6 h (reviewed in Ref. [228]). Another advantage of RIMAs over the older MAO inhibitors is that a shorter washout time is required; whereas the older MAO inhibitors typically require approximately 2 weeks for washout, the washout period for moclobemide is approximately 48 h [229]. This is an important consideration when switching antidepressant therapies from a MAO inhibitor to, for example, a TCA or SSRI.

Moclobemide (**18**) undergoes multiple routes of metabolism with formation of a lactam (**71**) being a major metabolite. Other metabolites include 3-hydroxymoclobemide, meclobemide N-oxide, and compounds (**72**) and (**73**) [209]. Lactam (**71**) is devoid of activity as a MAO inhibitor, whereas the primary amine (Ro 16-6491, **73**) and several other minor metabolites are inhibitors of MAO-B [228]. Analogs of the latter compound have been developed, including the selective MAO-B inhibitor Ro 19-6327 (**74**) and the selective MAO-A inhibitor Ro 41-1049 (**75**) [224].

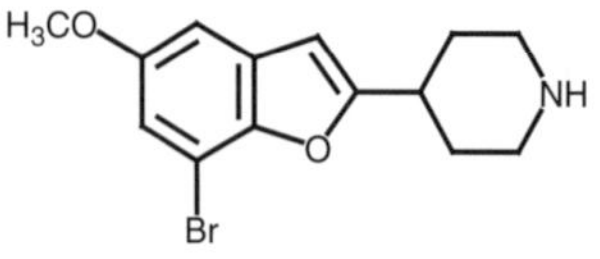

Brofaromine (**70**)

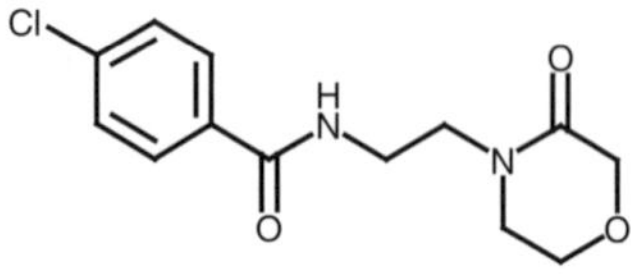

Moclobemide lactam (**71**)

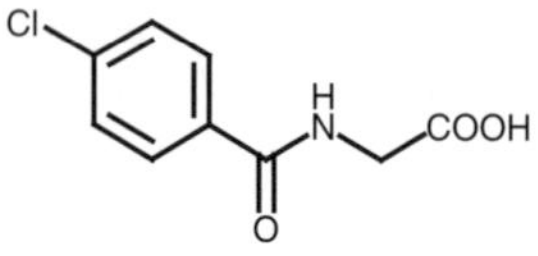

72

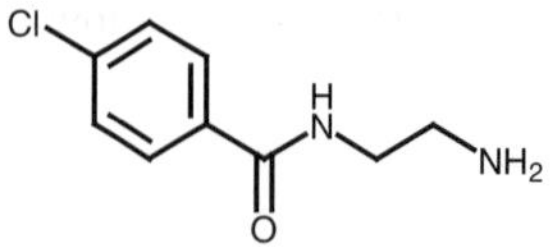

Ro 16-6491 (**73**)

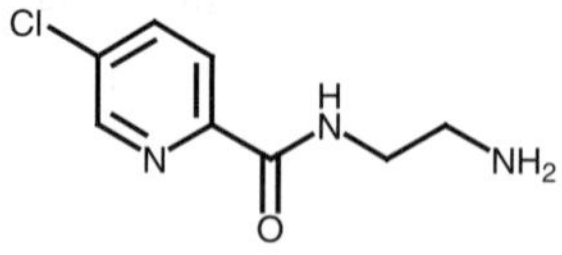

Ro 19-6327 (**74**)

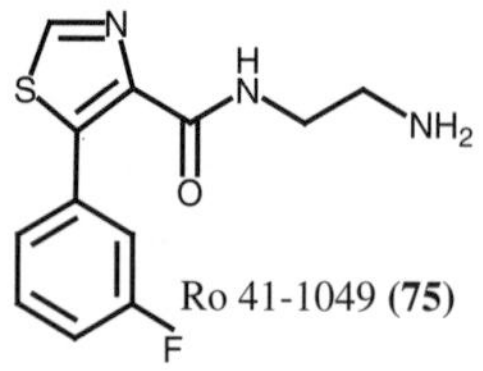

Ro 41-1049 (**75**)

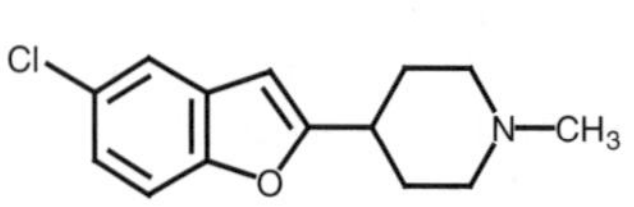

Sercloremine (**76**)

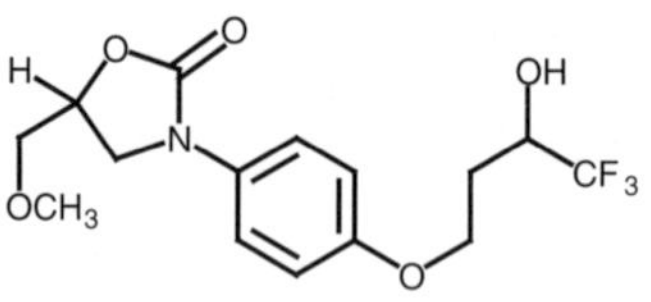

Befloxatone (**77**)

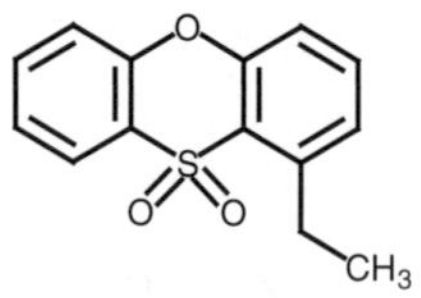

BW 1370U87 (**78**)

Newer RIMAs include the brofaromine analog sercloremine (**76**) (which, incidentally, also behaves as a 5-HT reuptake inhibitor), befloxatone (**77**), and BW 1370U87 (**78**) [221]. The latter two agents are somewhat more MOA-A selective than moclobemide and the relative order of potency is befloxatone > brofaromine > moclobemide > BW 1370U87, whereas duration of action decreases in the order brofaromine > BW 1370U87 > befloxatone > moclobemide [228].

5.3. Serotonergic Agents

The two populations of 5-HT receptors that have been definitely linked to depression are the 5-HT$_2$ and the 5-HT$_{1A}$ receptors. Numerous 5-HT$_2$ antagonists are available and belong to a multitude of chemical classes; structure–activity relationships are well beyond the scope of this chapter and several reviews are available (e.g., see Refs [145,168,230,231]). 5-HT$_2$ receptors consist of three families: 5-HT$_{2A}$, 5-HT$_{2B}$, and 5-HT$_{2C}$. An important issue at this time is which one (or more) of these subpopulations is most involved in depression. This question remains to be answered. Another issue is that 5-HT$_{2B}$ receptors have been linked to cardiac valvulopathy [232]. Hence, future development of 5-HT$_2$ ligands might wish to avoid agents with high affinity for 5-HT$_{2B}$ receptors.

5-HT$_{1A}$ receptor involvement is complicated by the issue of functional activity; and, more importantly, by agonist versus partial agonist versus antagonist actions. Evidence suggests that presynaptic 5-HT$_{1A}$ antagonism and postsynaptic 5-HT$_{1A}$ agonism might be the most desirable features for agents to target the treatment of depression. The structure–activity relationships of 5-HT$_{1A}$ agents have been reviewed [145,168,233]. Of particular interest are the arylpiperazines and, more specifically, the long-chain arylpiperazines (LCAPs) (Fig. 7). Simple arylpiperazines, those bearing only a small or no N$_4$-substituent, bind with modest affinity at 5-HT$_{1A}$ receptors and display little to no selectivity, whereas the LCAPs, arylpiperazines with elaborated N$_4$ substituents, bind with higher affinity at 5-HT$_{1A}$ receptors and can display considerable selectivity [161,168].

The *aryl* portion of the LCAPs (Fig. 7) can be widely varied, and the unbranched or branched *spacer* can be of two to five (or more, depending on the nature of the terminus) atoms in length. The *terminus* is usually an aromatic, heteroaromatic, imide, or amide function, and is associated with a region of considerable bulk tolerance [161]. Given this pharmacophore model, hundreds, if not thousands, of possible analogs can be envisioned. Many have been synthesized and evaluated (e.g., reviewed in Ref. [168]). Advantage has been taken of the region of bulk tolerance to develop some very bulky analogs [234]. Inspection of the structures of buspirone (**46**), gepirone (**47**), and ipsapirone (**48**) shows that they meet these criteria. Alteration of the substituents modifies functional activity, but this has not yet been investigated in a systematic manner. For example, buspirone, gepirone, and ipsapirone are partial agonists; BMY-7378 (**79**) and NAN-190 (**30**) were the first examples of LCAPs with 5-HT$_{1A}$ receptor antagonist actions (i.e., they are very low efficacy partial agonists); and WAY-100,635 (**31**) was the first example of a 5-HT$_{1A}$ "silent antagonist" (reviewed in Ref. [234]).

BMY-7378 (**79**)

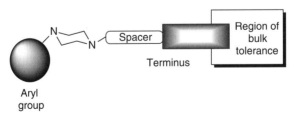

Figure 7. General structural features of long-chain arylpiperazines (LCAPs). Adapted from Refs. [161,234].

Structural modification, particularly with respect to the spacer and terminus groups, also influences selectivity; LCAPs show varying degrees of affinity toward several receptor populations, notably, 5-HT$_{1A}$, 5-HT$_2$, dopamine D2, and α-adrenergic receptors [161,168,234]. This has led to attempts to develop agents selective for each of these receptor populations by the appropriate structural modification. RK-153 (**54**) and adatanserin (**55**) are just two examples of such agents. Continued exploration of LCAPs should prove bountiful for the development of agents with the desired mix of selectivity and functional activity.

5.4. Other Agents

Several antidepressants work via mechanisms that are not yet fully understood. Perhaps the best example of this is bupropion (**3**). Although the mechanism of action of bupropion is most commonly attributed to inhibition of dopamine reuptake, its actions are diverse [235]. In addition, one or more bupropion metabolites also seem to be active. In humans, bupropion is metabolized to two amino alcohols, the racemic threo amino alcohol (*R,R*-threo; **80**) and the racemic erythro amino alcohol (*R,S*-erythro; **81**), threohydroxybupropion and erythrohydroxybupropion, respectively; and a morpholinol, hydroxybupropion (**82**; BW 306U). The levels and half-lives of these metabolites have been quantitated in healthy volunteers [236–238]). Hydroxybupropion (**82**) is the major metabolite in depressed patients [239]. Interestingly, all three metabolites predominated over the parent compound in plasma and cereobrospinal fluid at steady state, and high levels of these metabolites were suggested to be associated with poor clinical outcome resulting from toxic effects possibly involving the dopaminergic system [239]. In contrast, bupropion metabolites showed various degrees of activity in several animal models of antidepressant activity [240]. However, the routes of metabolism in rodents and dogs are known to be somewhat different from those in humans [241,242]; in fact, routes of metabolism in mice differ from those in rats [243]. Nevertheless, at least hydroxybupropion is considered to be an active metabolite of bupropion [243]. Both threohydroxybupropion (**80**) and hydroxybupropion (**82**) inhibit norepinephrine reuptake with potencies comparable to that of bupropion but are 10 to 25 times less potent than bupropion with respect to inhibition of dopamine reuptake [245].

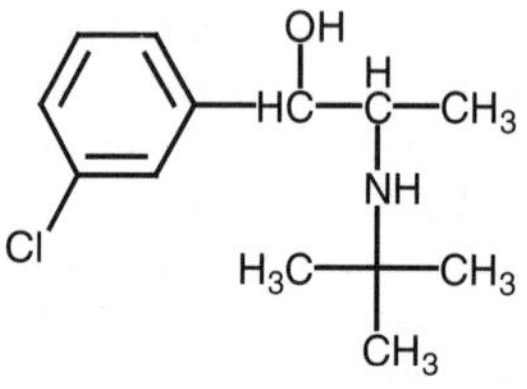

80 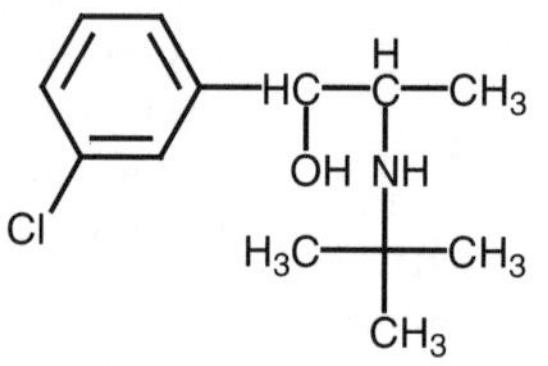 **81** 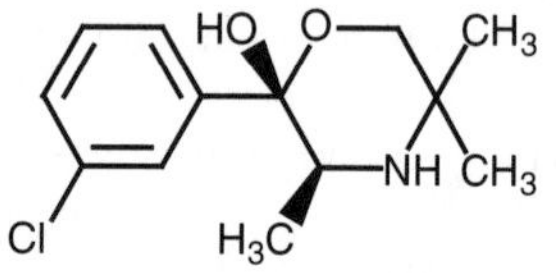 Hydroxybupropion (**82**)

Although there is evidence for dopaminergic involvement in the antidepressant actions of bupropion as determined by measuring, for example, extracellular levels of dopamine in the striatum and nucleus accumbens in rats [244], or by examining the increase in dopamine release in striatal synaptosomes upon treatment with bupropion [245], alternative mechanisms have been proposed. Bupropion and its metabolites produced noradrenergic-like effects on the firing rates of noradrenergic neurons located in the locus coeruleus of rat [246] and it has been suggested that bupropion might work through some yet unidentified adrenergic mechanism (reviewed in Ref. [235]). More recently, it has been shown that sustained administration of bupropion by the use of minipumps decreases the firing rate of norepinephrine neurons arising from increased activation of somatodendritic a_2-adrenoceptors, suggesting that the actions of bupropion are mainly attributable to enhancement of norepinephrine release, not to reuptake inhibition [247]. This latter study also showed that bupropion can act on the serotonergic system.

One of the latest theories to account for the action of bupropion is its influence on imidazoline I_1 receptors. Agmatine (decarboxylated arginine) may be an endogenous ligand for I_1 sites. Evidence indicates that plasma agmatine levels are significantly elevated in depressed patients relative to those in healthy controls and that treatment with bupropion normalized these levels [248,249] and references therein). Peripheral norepinephrine neurons possess nonadrenergic imidazoline binding sites that mediate inhibitory effects on the release of norepinephrine [250]. Platelet I_1 receptors are also downregulated after antidepressant treatment; this has led to conjecture that elevation in brain imidazoline receptors might lead to greater inhibition of norepinephrine release [249].

Another curious agent is tianeptine (**83**). *In vitro*, neither tianeptine nor any of its major metabolites has any effect on monoamine reuptake, release, or neurotransmitter binding; the biochemical effects (acute or chronic) of this agent *in vivo* indicate enhanced serotonin uptake in the cortex and hippocampus [250,251].

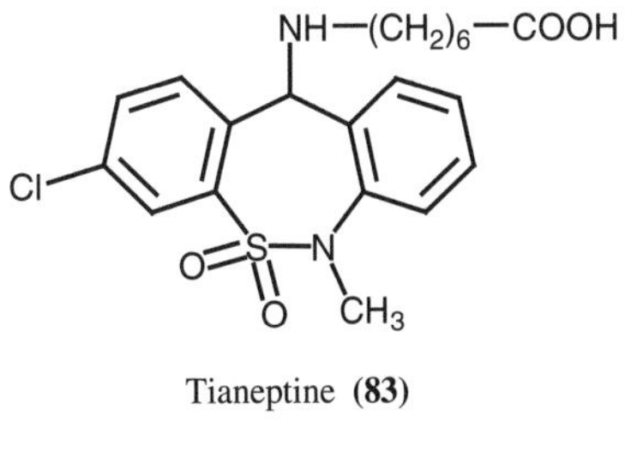

Tianeptine (**83**)

Tianeptine has been evaluated in more than 3000 patients and has been found to be at least as effective as the TCAs (e.g., amitriptyline, imipramine, maprotiline), SSRIs (e.g., fluoxetine, sertraline), and mianserin [251,252].

6. RECENT DEVELOPMENTS AND THINGS TO COME

For an up to date list of compounds in development as antidepressants, see http://neurotransmitter.net/newdrugs.html.

6.1. Monoaminergic Drugs

All of the existing antidepressant drugs act through monoaminergic mechanisms, and further refinements are still possible in this arena. The concept of developing dual norepinephrine/serotonin uptake inhibitors with improved safety profiles has proved very successful, with duloxetine and venlafaxine taking major shares of the antidepressant market. Triple inhibitors of norepinephrine/serotonin and dopamine uptake are being developed, such as NS2359 and DOV 21,947 (**84**). There is also a development compound (Serdaxin) that combines SERT and DAT inhibition, and another NET-selective inhibitor (LY 2216684). Another commercially successful idea has been to separate the individual enantiomers of existing racemic antidepressant drugs, in the hope that the individual isomers may show some improvement over the parent drug. A number of existing antidepressants are racemates, which include fluoxetine (**12**), citalopram (**4**), mirtazepine (**17**), and reboxetine (**24**). The active enantiomer of citalopram (**4**), escitalopram (**11**), which is reported to be slightly more efficacious and to have fewer side effects than citalopram [253,254], has been successfully marketed. This approach will not always succeed, however. Lilly's attempt to develop (*R*)-fluoxetine (***R*-12**) failed because the compound caused small increases in cardiac QT intervals [255]. Another approach is to develop an active metabolite of a parent drug, such as the imipramine metabolite desipramine (**6**), and the venlafaxine metabolite desvenlafaxine (**7**).

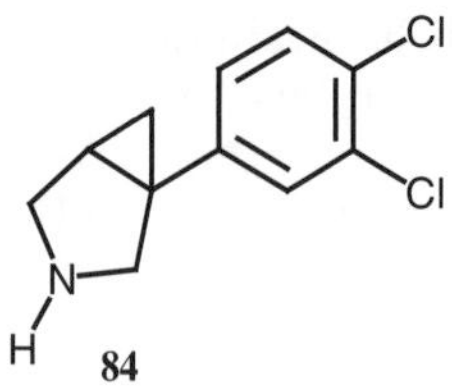

84

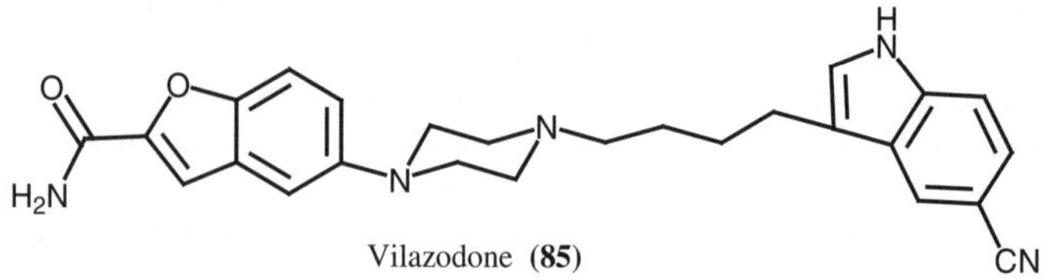

Vilazodone (**85**)

The search for more selective antagonists or partial agonists acting at the 5-HT$_{1A}$ receptor continues. The compound vilazodone (SB649915; **85**), combines activity as an SSRI with a partial agonist profile at the 5-HT$_{1A}$ receptor, and is in development as an antidepressant [256]. Other 5-HT$_{1A}$ partial agonists or antagonists in development include LU AA21004, GSK 588045 and GSK163090. Other 5-HT-related targets include the 5-HT$_7$ antagonist JNJ 18038683, the 5-HT$_{2C}$ agonist vabicaserin (SCA-136) and the 5-HT$_{2C}$, 5-HT$_{2B}$ antagonist agomelatine which combines these properties with agonist action at melatonin M$_1$/M$_2$ receptors [257].

6.2. NMDA Receptor Antagonists

Since the early 1990s evidence has accumulated to show that a variety of drugs that block the glutamate receptor of the NMDA subtype have antidepressant-like profiles in behavioral tests thought to predict antidepressant activity [258,259]. Chronic treatment of animals with NMDA antagonists causes a downregulation of β-adrenergic receptors, as seen with some antidepressant drugs [260]. Skolnick and colleagues further showed that treatment of animals with clinically active antidepressants almost invariably downregulated NMDA receptor function, by reducing the affinity of glycine for an important modulatory site on the receptor [261,262]. Interest in the NMDA receptor as a target for novel antidepressants has been greatly stimulated by the finding that the NMDA antagonist ketamine acts very rapidly as an antidepressant in the clinic [263,264]. In a double blind placebo-controlled trial a single intravenous dose of ketamine caused a rapid and relatively sustained antidepressant effect [264]. Studies of sub-anesthetic doses of ketamine in animals have suggested that blockade of the NMDA receptor might exert rapid antidepressant effects by enhancing AMPA receptor-mediated glutamate function relative to NMDA in critical neuronal circuits [265]. Unfortunately, the NMDA antagonist drugs that are currently available are not well tolerated in humans; they cause a range of serious adverse effects, including psychosis [266,267]. It is possible, however, that novel drugs that targeted particular NMDA subunit combinations could prove more benign and potentially useful as antidepressants [258].

6.3. Drugs Acting at Neuropeptide Receptors

6.3.1. Substance P-NK1 Receptor Antagonists

Substance P is a neuropeptide that is widely distributed in small sensory nerve fibers and in a variety of neural pathways in the brain. During the 1990s a range of potent nonpeptide drugs were developed that act as potent and selective antagonists at the NK1 receptor, which recognizes substance P. Although these drugs were targeted initially at the treatment of pain, clinical trials in this indication proved disappointing [268]. A breakthrough came when a team at Merck discovered that NK1 antagonists were active in animal behavioral models predictive of antidepressant action, and undertook a successful clinical trial [269]. The results showed the NK1 antagonist aprepitant (MK-869) (**86**) to be as effective as the SSRI paroxetine (**18**), and to be associated with fewer sexual dysfunction side effects. This result triggered a high level of interest in what is potentially the first series of antidepressant drugs that are not directly related to monoaminergic function [270], but larger Phase III clinical trials were unable to confirm the antidepressant effects of MK-869, and no other successful NK1 antagonist trials have been reported, although the NK1 antagonist orvipetant (GW823396) continues to be listed as in development. Antagonists selective for the NK2 neurokinin receptor are in development as antidepressants; these include saredutant (SR-48968) and SAR102279.

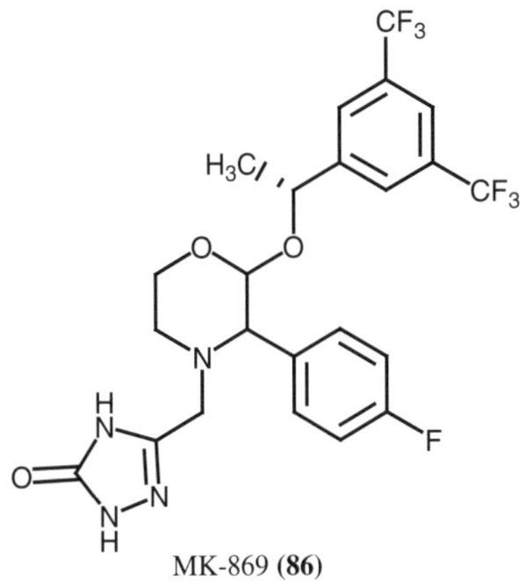

MK-869 (**86**)

6.3.2. CRF Antagonists

Corticotrophin releasing factor (CRF, or CRH) is a peptide secreted by hypothalamic neurons to control the release of ACTH from the anterior pituitary, but in addition, is present in a variety of neural pathways within the brain [271]. A number of animal experiments suggest that increased release of CRF within the brain and from the hypothalamus may represent a final common pathway in mediating the effects of stress on the brain and body [271–273]. CRF elicits behavioral signs of fear and anxiety when administered into the brain, and CRF antagonists conversely have anxiolytic/antidepressant profiles in behavioral tests [271]. The chronic administration of antidepressant drugs leads to a downregulation of CRF expression in rat brain [274]. These findings have led to an increased interest in the potential use of CRF antagonist drugs as agents to combat stress, anxiety, and depression. A variety of small-molecule nonpeptide antagonists have been discovered and shown to possess antidepressant activity in animal behavioral tests [272–274]. Several CRF_1 antagonists are in development as antidepressants; these include pexacerfont (BMS-562086), ONO-2333, JNJ 19567470, and SSR 125543 (**87**) [275,276].

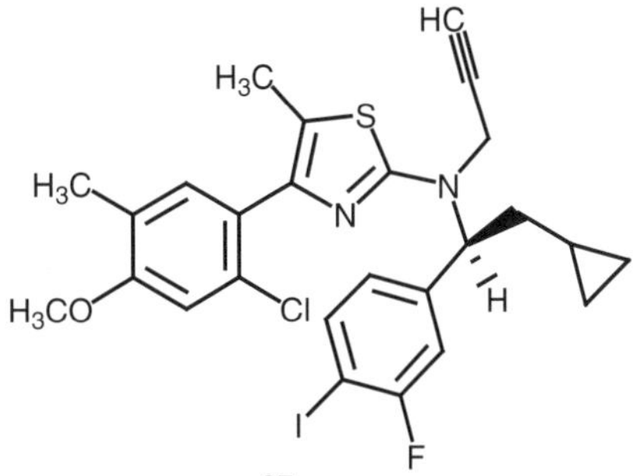

87

6.3.3. Galanin Receptor Ligands

The neuropeptide galanin coexists with norepinephrine and with 5-HT in CNS and a number of animal studies have linked it to a role in the control of mood [277]. Galanin analogs which act as agonists at galanin receptor subtype 2 (GalR2) increase levels of forebrain 5-HT and are active in animal models predictive of antidepressant action [277,278]. Conversely antagonists acting at the GalR3 receptor subtype also

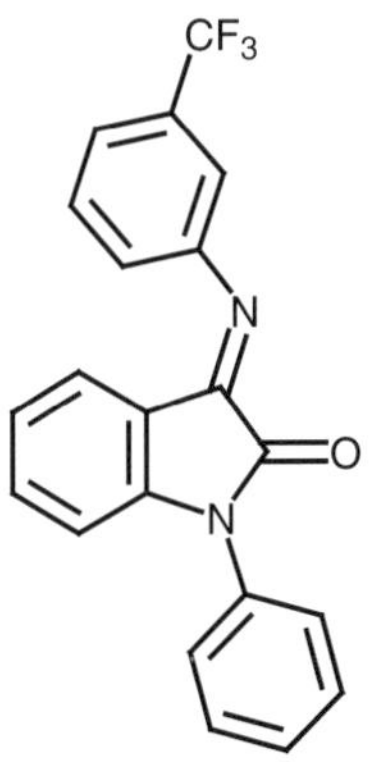

88

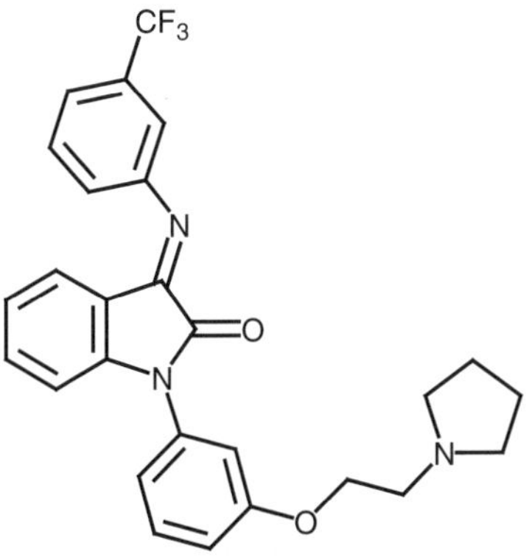

89

possess antidepressant activity in animal models [279,280] and two small-molecule GalR3 selective antagonists have been described, SNAP 37889 (**88**) and SNAP 398299 (**89**) [280]. This field is still in the early stages of development, but offers another novel approach to modulating CNS noradrenergic and seritonergic function.

REFERENCES

1. American Psychiatric Association. *Diagnostic and Statistical Manual of Mental Disorders.* 4th ed. Washington, DC: American Psychiatric Press; 1994. p 317–391.
2. Lader MH. In: Checkley S, editor. *The Management of Depression.* Oxford, UK: Blackwell Press; 1998. p 18–211.
3. Kessler RC, McGonagle KA, Zhao S, Nelseon CB, Hughes M, Eshelman S, Wittchen HU, Kendler KS. Arch Gen Psychiatry 1994;51:8–19.
4. Donoghue J. Acta Psychiatr Scand 2000;403 (Suppl.):57–61.
5. Barbui C, Hotopf M, Freemantle N, Boynton J, Churchill R, Eccles MP, Geddes JR, Hardy R, Lewis G, Mason JM. Cochrane Database Syst Rev 2000;4:CD002791.
6. Parfitt K, editor. *Martindale. The Complete Drug Reference.* 32nd ed. London: Pharmaceutical Press; 2000. p 271–312.
7. Walsh P. *Physician's Desk Reference.* 55th ed. Montvale, NJ: Medical Economics Co., Inc.; 2001.
8. Andrews G. Br J Psychiatry 2001;178: 192–194.
9. Turner EH, Matthews AM, Linardatos E, Tell RA, Rosenthal R. New Engl J Med 2008;358: 252–260.
10. Fawcett J, Barkin RL. J Clin Psychiatry 1997;58 (Suppl 6): 32–39.
11. Davis JM, Wang Z, Janicak PG. Psychopharmacol Bull 1993;29:175–181.
12. Barbui C, Hotopf M. Br J Psychiatry 2001;178:129–144.
13. Anderson IM. Br Med Bulletin 2001;57: 161–178.
14. Bosc M. Compr Psychiatry 2000;41:63–69.
15. Montgomery SA. J Psychopharmacol 1997;11 (Suppl 4): S9–S15.
16. Boyer WF, Feighner JP. Depression 1994;2: 32–35.
17. Rutter JJ, Gundlah C, Auerbach SB. Synapse 1995;20:225–233.
18. Hervas I, Quieroz CM, Adell A, et al. Br J Pharmacol 2000;130:160–166.
19. Arborelius L, Wallsten C, Ahlenius S, Svensson TH. Eur J Pharmacol 1999;382:1323–138.
20. Tordera R, Pei Q, Newson M, Gray K, Sprakes M, Sharp T. Neuropharmacology 2003;44: 893–902.
21. Artigas F, Celada P, Laruelle M, Adell A. Trends Pharmacol Sci 2001;22:224–228.
22. Perez V, Puigdemont D, Gilaberte I, et al. J Clin Psychopharmacol 2001;21:36–45.
23. Artigas F, Perez V, Alvarez E. Arch Gen Psychiatry 1994;51:248–251.
24. Tome MB, Cloniger CR, Watson JP, McIsaac MT. J Affect Disord 1997;44:101–109.
25. Puzantian T, Kawase K. Pharmacotherapy 1999;19:205–212.
26. Perez V, Soler J, Puigdemont D, Alvarez E, Artigas F. Arch Gen Psychiatry 1999;56: 375–379.
27. Vetulani J, Nalepa I. Eur J Pharmacol 2000;405:351–363.
28. Editorial. J Clin Psychiatry 2001;62:380–382.
29. Vetulani J, Stawarz RJ, Dingell JV, Sulser F. Naunyn Schmiedebergs Arch Pharmacol 1976;293:109–114.
30. Bruinello N, Racagni G. Hum Psychopharmacol 1998;13:S13–S19.
31. Leonard BE. Neuropsychopharmacology 1997;7 (Suppl 1): S11–S16.
32. Vaidya V, Duman R. Br Med Bull 2001;57: 61–79.
33. Fleischhacker WW, Hinterhuber H, Bauer H, et al. Neuropsychobiology 1992;26:59–64.
34. Takahashi M, Terwilliger R, Lane C, Mezes PS, Conti M, Duman RS. J Neurosci 1999;19: 610–618.
35. Tardito D, Perez J, Tiraboschi E, Musazzi L, Racagni G, Popoli M. Pharmacol Rev 2006;58: 115–134.
36. Nivuya M, Nestler EJ, Duman RS. J Neurosci 1996;16:2365–2372.
37. Ongur D, Drevets WC, Price JL. Proc Natl Acad Sci USA 1998;95:13290–13295.
38. Duman RS. Eur Psychiat 2002;17 (Suppl 3): 306–310.
39. Schmidt HD, Duman RR. Behav Pharmacol 2007;18:391–418.
40. Dwivedi Y, Rizavi HS, Shukla PK, Lyons J, Faludi G, Palkovits M, Sarosi A, Conley RR, Roberts RC, Tamminga CA, Pandey GN. Biol Psychiatry 2004;55:234–243.

41. Duman, Nestler EJ. Trends Neurosci 2005;28: 436–445.
42. Rantamaki T, Hendolin P, Kankaanpaa A, Mijatovic J, Piepponen P, Domenici E, Chao MV, Mannisto PT, Castren E. Neuropsychopharmacology 2007;32:2152–2162.
43. Dwivedi Y, Rizavi HS, Teppen T, Sasaki N, Chen H, Zhang H, Roberts RC, Conley RR, Pandey GN. Neuropsychopharmacology 2007;32: 2338–2350.
44. Duman RS, Monteggia LM. Biol Psychiatry 2006;59:1116–1127.
45. (a) Warner-Schmidt JL, Duman RS. Curr Opin Pharmacol 2008;8:14–18. (b) Schatzberg A. J Clin Psychiatry 2000;61 (Suppl 11): 9–17.
46. Pittenger C, Duam RS. Neuropsychopharmacology 2008;33:88–109.
47. Young KA, Bonkale W, Holcomb LA, Hicks PB, German DC. Brit J Psychiatry 2008;192: 285–289.
48. Taylor D. In: Checkley S, editor. *The Management of Depression*. Oxford, UK: Blackwell Press; 1998. p 212–236.
49. Papakostas GI. J Clin Pyschiatry 2008;69 (Suppl E): 8–13.
50. Callahan M, Kassel D. Ann Emerg Med 1985;14:1–9.
51. Leonard BE. Lancet 1986;ii:1105.
52. Haddad P. J Psychopharmacol 1999;13: 300–307.
53. Blackwell B, Marley E, Price J, Taylor D. Br J Psychiatry 1967;113:349–365.
54. Pare CMB. Br J Psychiatry 1985;146:576–584.
55. Fitton A, Faulds D, Goa KL. Drugs 1992;43: 561–596.
56. Moll E, Neumann N, Schmid-Burgk M, Stabl M, Amrein R. Clin Neuropharmacol 1994;17 (Suppl 1):S74–S87.
57. Coulter DM, Pillaus PI. Lancet 1995;346:1032.
58. Neuvonen PJ, Pojhola-Sintonen S, Tacke U, Vuori E. Lancet 1993;342:1419.
59. Richelson E, Nelson A. J Pharmacol Exp Ther 1994;230:94–102.
60. Cusack B, Nelson A, Richelson E. Psychopharmacology 1994;114:559–565.
61. Lapierre YD, Anderson K. Am J Psychiatry 1983;140:493–494.
62. Zermak WR, Kenna GA. Am J Health Syst Pharmacol 2008;65:1029–1038.
63. Mason J, Freemantle N, Eccles M. Br J Gen Pract 2000;50:366–370.
64. Buckley NA, McManus PR. Drug Safety 1998;18:369–381.
65. Heninger GR. In: Gelder N, editor. *Oxford Textbook of Psychiatry*. London: Oxford University Press; 2000. p 1293–1305.
66. Parfitt K, editor. *Martindale. The Complete Drug Reference*. 32nd ed. London: Pharmaceutical Press; 2000. p 285.
67. Oström M, et al. Lancet 1996;348:339–340.
68. Brion F, et al. Lancet 1996;348:1380.
69. De Abajo FJ, Rodriguez LA, Montero D. Br Med J 1999;319:1106–1109.
70. van Walraven C, Mamdani MM, Wells PS, Williams JI. Br Med J 2001;323:655–658.
71. Ferguson JM. J Clin Psychiatry 2001;62 (Suppl 3): 22–34.
72. Beasley CM, Jr, Dornseif BE, Bosomworth JC. Br Med J 1991;303:658–662.
73. Montgomery SA. Drugs 1992;43 (Suppl 2): 24–31.
74. Dudley M, Hadzi-Pavlovic D, Andrews D, Perich T. Aust NZ J Psychiatr 2008;42: 456–466.
75. Taylor D, Lader M. Br J Psychiatry 1996;167: 529–532.
76. Bymaster FP, Dreshfield-Ahmad LJ, Threlkeld PG, Shaw JL, Thompson L, Nelson DL, Hemrick-Lueke SK, Wong DT. Neuropsychopharmacol. 2001;25:871–880.
77. Taylor D, Lader M. Br J Psychiatry 1996;167: 529–532.
78. Rudorfer MV, Potter WZ. Drugs 1989;37: 713–738.
79. Parfitt K, editor. *Martindale. The Complete Drug Reference*. 32nd ed. London: Pharmaceutical Press; 2000. p 280.
80. Hayes PE, Kristoff CA. Clin Pharm 1986;5: 471–480.
81. Brogden RN, Heel RC, Speight TM, Avery GS. Drugs 1981;21:401–429.
82. Thompson JW, Jr, Ware MR, Blashfield RK. J Clin Psychiatry 1990;51:430–433.
83. Purcell P, Ghurye R. J Clin Psychopharm 1995;15:293–295.
84. Preskorn SH. J Clin Psychiatry 1995;56 (Suppl 6): 12–21.
85. Robertson DS, Roberts DL, Smith JM, et al. J Clin Psychiatry 1996;57(Suppl 2): 31–38.
86. Wong EHF, Sonders MS, Amara SG, Tinholt PM, Piercey MFP, Hoffmann WP, Hyslop DK, Franklin S, Porsolt RD, Bonsignori A, Carfag-

na N, McArthur RA. Biol Psychiat 2000;47: 818–829.
87. Parfitt K, editor. *Martindale. The Complete Drug Reference.* 32nd ed. London: Pharmaceutical Press; 2000. p 307.
88. Versiani M, Amin M, Chouinard G. J Clin Psychopharmacol 2000;20:28–34.
89. Schatzberg AF. J Clin Psychiatry 2000;61 (Suppl):1031–1038.
90. Bertilsson L, Mellstrom B, Sjoqvist F. Life Sci 1964;25:1285–1291.
91. Potter WZ, Calil HM, Manian AA, Zavadil AP, Goodwin FK. Biol Psychiatry 1979;14: 601–613.
92. Potter WZ, Bertilsson L, Sjoqvist F. In: van Praag HM, Lader MH, Rafaelson OJ, Sachar EJ, editors. *The Handbook of Biological Psychiatry. Part IV. Practical Applications of Psychotropic Drugs and Other Biological Treatments.* New York: Marcel Dekker; 1981. p 71–134.
93. Garattini S. In: Ban TA, Silvestrini T, editors. *Trazadone: Modern Problems in Pharmacopsychiatry.* Vol. 9. Basel, Switzerland: Karger; 1974. p 29–46.
94. Dostert P, Benedetti MS, Pogessi I. Eur Neuropsychopharmacol 1997.7 (Suppl 1): S23–S35.
95. Lin KM, Polland RE, Silver B.In: Lin KM, Polland RE, Nakasaki G, editors. *Psychopharmacology and Psychobiology of Ethnicity: Progress in Psychiatry.* Vol. 39.Washington, DC: American Psychiatric Press; 1993, p 16–17.
96. Preskorn SH, Irwin HA. J Clin Psychiatry 1982;43:151–156.
97. Rudorfer MV, Young RC. Am J Psychiatry 1980;137:984–986.
98. Goodnick PJ, Goldstein BJ. J Psychopharmacol 1998;12 (Suppl B): S5–S20.
99. Hiemke C, Hartter S. Pharmacol Ther 2000;85:11–28.
100. Baumann P, Rochat B. Int Clin Psychopharmacol 1995;10 (Suppl 1): 15–21.
101. Lemberger L, Bergstrom RF, Wolen RL, Farid NA, Enas GG, Arnoff GR. J Clin Psychiatry 1985;46:14–19.
102. Wong DT, Bymaster FT, Horng JS, Molly BB. J Pharmacol Exp Ther 1975;193:804–811.
103. Doogan DP, Caillard V. J Clin Psychiatry 1988;49 (Suppl): 46–51.
104. Heym J, Koe BK. J Clin Psychiatry 1988;49 (Suppl): 40–45.
105. Linnet K, Olesen OV. Ugeskr Laeger 1996;158: 4920–4923.
106. Poolsup N, Li-Wan-Po A, Knight TL. J Clin Pharm Ther 2000;25:197–200.
107. Sidhu J, Priskorn M, Poulsen M, Segonzac A, Gollier G, Larsen F. Chirality 1997;9:686–692.
108. Haefely W, Burkhard WP, Cesura A, et al. Clin Neuropharmacol 1993;16 (Suppl 2): S8–S18.
109. Cesura AM, Pletscher A. Prog Drug Res 1992;38:171–297.
110. Fleishaker JC. Clin Pharmacokinet 2000;39: 413–427.
111. Avenoso A, Facciola G, Scordo MG, Spina E. Ther Drug Monit 1999;21:577–579.
112. Timmer CJ, Sitsen JM, Delbressine LP. Clin Pharmacokinet 2000;38:461–474.
113. Axelrod J, Weil-Malherbe H, Tomchick R. J Pharmacol Exp Ther 1959;127:251–256.
114. Whitby LG, Axelrod J, Weil-Malherbe H. J Pharmacol Exp Ther 1961;132:193–201.
115. Hertting G, Axelrod J, Kopin IJ, Whitby LG. Nature 1961;189:66.
116. Hertting G, Axelrod J. Nature 1961;192: 172–173.
117. Iversen LL. Br J Pharmacol 1971;41:571–591.
118. Iversen LL. In: Iversen L, Iversen S, Snyder S, editors. *Handbook of Psychopharmacology.* Vol. 3. New York: Plenum; 1975. p 381–432.
119. Masson J, Sagńe C, Hamon M, El Mestikawy S. Pharmacol Rev 1999;514:439–464.
120. Iversen LL. Br J Pharmacol 2006;147: S82–S88.
121. Pacholczyk T, Blakely RD, Amara SG. Nature 1991;350:350–353.
122. Zhou Z, Zhen J, Karpowich N, Goetz RM, Law CJ, Reith MEA, Wang D-N. Science 2007;317:1390–1393.
123. Axelrod J, Whitby L, Hertting G. Science 1961;133:383.
124. Waldmeier P. In: Healy D, editor. *The Psychopharmacologists.* London: Chapman & Hall; 1996. p 565–586.
125. Wong DT, Horng JS, Bymaster FP, Hauser KL, Malloy BB. Life Sci 1974;15:471–479.
126. Wong DT, Bymaster FP, Engelman E. Life Sci 1995;57:411–441.
127. Tatsumi M, Groshan K, Blakely RD, Richelson E. Eur J Pharmacol 1997;340:249–258.
128. Mendlewicz J. Int J Clin Psychopharmacol 1995;10 (Suppl 2): 5–13.
129. Owens MJ, Krulewicz S, Simon JS, Sheehan DV, Thase ME, Carpenter DJ, Plott SJ, Nemeroff CB. Neuropsychopharmacol 2008; 33:3201–3212.

130. Owens MJ, Morgan WN, Plott SJ, Nemeroff CB. J Pharmacol Exp Ther 1997;283: 1305–1322.
131. Schatzberg AF. J Clin Psychiatry 2000;61 (Suppl 10): 31–38.
132. Svensson TH. Acta Psychiatr Scand 2000;101 (Suppl 402): 18–27.
133. Gorman JM, Sullivan G. J Clin Psychiatry 2000;61 (Suppl 1): 13–16.
134. Schildkraut J. Am J Psychiatry 1965;122: 509–522.
135. Healy D, McMonagle T. J Psychopharmacol 1997;11 (Suppl 4): S25–S31.
136. Giros B, Jaber M, Jones SR, Wightman RM, Caron MG. Nature 1996;379:606–612.
137. Rocha BA, Fumagalli F, Gainetdinov RR, Jones SR, Ator R, Giros B, Miller GW, Caron MG. Nat Neurosci 1998;1:132–137.
138. Gainetdinov RR, Wetsel WC, Jones SR, Levin ED, Jaber M, Caron MG. Science 1999;283: 397–401.
139. Graul A, Castaner J. Drugs Future 1999;24: 128–132.
140. Davies HML, Kuhn LA, Thomley C, Matasi JJ, Sexton T, Childers SR. J Med Chem 1996;39: 2554–2558.
141. Blough BE, Abraham P, Mills AC, Lewin AH, Boja JW, Scheffel U, Kuhar MJ, Carroll FI. J Med Chem 1997;40:3861–3864.
142. D'Aquila PS, Collu M, Gessa GL, Serra G. Eur J Pharmacol 2000;405:365–373.
143. Baldessarini RJ. In: Hardman JG, Limbird LE, editors. *Goodman and Gilman's The Pharmacological Basis of Therapeutics*. 10th ed. New York: McGraw-Hill; 2001. p 447–483.
144. Hoyer D, Clarke DE, Fozard JR, Hartig PR, Martin GR, Mylecharane EJ, Saxena PR, Humphrey PPA. Pharmacol Rev 1994;46: 157–203.
145. Glennon RA, Dukat M. In: Williams DA, Lemke T, editors. *Foye's Textbook of Medicinal Chemistry*. Baltimore, MD: Williams & Wilkins; 2008; p 417–443.
146. Olivier B, van Wijngaarden I, Soudijn W, editors. *Serotonin Receptors and Their Ligands*. Amsterdam: Elsevier; 1997.
147. Eison AS, Eison MS, Torrente JR, et al. Psychopharmacol Bull 1990;26:311–315.
148. Mitchell PB, Mitchell MS. Aust Fam Physician 1994;23:1771–1773.
149. Roth BL, Kroeze WK, Patel S, Lopez E. Neuroscientist 2000;6:252–262.
150. Barker EL, Sanders-Bush E. Mol Pharmacol 1993;44:725–730.
151. Murphy DL, Mitchell PB, Potter WZ. In: Bloom FE, Kupfer DJ, editors. *Psychopharmacology: The Fourth Generation of Progress*. New York: Raven; 1995. p 1143–1153.
152. Wong DT, Threlkeld PG, Robertson DW. Neuropsychopharmacology 1991;5:43–47.
153. Cryan JF, Lucki I. J Pharmacol Exp Ther 2000;295:1120–1126.
154. Dekeyne A, la Cour CM, Gobert A, Brocco M, Lejune F, Serres F, Sharp T, Daszuta A, Soumier A, Papp M, Rivet J-M, Flik G, Cremers TI, Muller O, Lavielle C, Milan MJ. Psychopharmacology 2008;199:549–568.
155. Blier P, Abbott FV. J Psychiatr Neurosci 2001;26:37–43.
156. Blier P, deMontigny C. J Clin Psychopharmacol 1990;10:13S–20S.
157. Luscombe GP, Martin KF, Hutchins LJ, Grosen J, Heal DJ. Br J Pharmacol 1993; 108:669–677.
158. Schreiber R, de Vry J. Prog Neuropsychopharmacol Biol Psychiatry 1993;17:87–104.
159. Blier P, deMontigny C. Trends Pharmacol Sci 1994;15:220–226.
160. Nowak G, Dulinski J. Pharmacol Biochem Behav 1991;38:691–698.
161. Glennon RA. Drug Dev Res 1992;26:251–262.
162. Jenkins SW, Robinson DS, Fabre LF, Andary JJ, Messina MD, Reich LN. J Clin Psychopharmacol 1990;10:77S–85S.
163. Vacher B, Bonnaud B, Funes P, Jubault N, Koek W, Assie M-B, Cosi C. J Med Chem 1998;41:5070–5083.
164. Rutter JJ, Gundlah C, Auerbach SB. Synapse 1995;20:225–233.
165. Takeuchi H, Yatsugi S, Hatanaka K, Nakato K, Hattori H, Sonoda S, Koshiya K, Fujii M, Yamaguchi T. Eur J Pharmacol 1997;329: 27–35.
166. Martinez J, Perez S, Oficialdegui AM, Heras B, Orus L, Villanueva H, Palop JA, Roca J, Mourelle M, Bosch A, Del Castillo J-C, Lasheras B, Yordera R, del Rio J, Monge A. Eur J Med Chem 2001;36:55–61.
167. Evrard DA, Harrison BL. Ann Rep Med Chem 1999;34:1–10.
168. Glennon RA, Dukat M. Curr Drugs Serotonin 1992;1:1–45.
169. Raghupathi RK, Rydelek-Fitzgerald L, Teitler M, Glennon RA. J Med Chem 1991;34: 2633–2638.

170. Abou-Gharbia MA, Childers WE, Fletcher H, McGauhey G, Patel U, Webb MB, Yardley J, Andree T, Boast C, Kucharik RJ, Marquis K, Morris H, Scerni R, Moyer JA. J Med Chem 1999;42:5077–5094.
171. Schechter LF, Ring RH, Beyer CE, Hughes ZA, Khawaja X, Malberg JE, Rosenzweig-Lipson S. NeuroRx 2005;2:590–611.
172. Briley M, Moret C. Clin Neuropharmacol 1993;16:387–400.
173. Glennon RA, Westkaemper RB. Drug News Perspect 1993;6:390–405.
174. Blier P. J Clin Psychiatry 2001;62 (Suppl 4): 7–11.
175. Liao Y, Bottcher H, Harting J, Greiner H, van Amsterdam C, Cremers T, Sundell S, Marz J, Rautenberg W, Wikstrom H. J Med Chem 2000;43:517–525.
176. Svenningsson P, Greengard P. Curr Opin Pharmacol 2007;7:27–32.
177. Svenningsson P, Chergui K, Rachleff I, Flajolet M, Zhang X, El Yacoubi M, Vaugeois J-M, Nomikos G, Greengard P. Science 2006; 311:77–80.
178. Svenningsson P, Tzavara ET, Witkin JM, Fienberg AA, Nomikos GG, Greengard P. Proc Nat Acad Sci USA 2002;99:3182–3187.
179. Svenningsson P, Tzavara ET, Liu F, Fienberg AA, Nomikos GG, Greengard P. Proc Nat Acad Sci USA 2002;99:3188–3193.
180. Glennon RA, Lee M, Rangisetty JB, Dukat M, Roth BL, Savage JF, McBride A, Ufeisen L, Lee DK. J Med Chem 2000;43:1011–1018.
181. Svenningsson P, Tzavara ET, Qi H, Carruthers R, Witkin JM, Nomikos GG, Greengard P. J Neurosci 2007;27:4201–4209.
182. Lucas G, Rymar VV, Du J, Mnie-Filali O, Bisgaard C, Manta S, Lambas-Senas L, Wiborg O, Haddjeri N, Pineyro G, Sadikot AF, Debonnel G. Neuron 2007;55:712–725.
183. Duman RS. Neuron 2007;55:679–681.
184. Gozlan H.In: Olivier B, van Wijngaarden I, Soudijn W, editors. Serotonin Receptors and Their Ligands. Amsterdam: Elsevier; 1997. p 221–258.
185. Ramamoorthy R, Radhakrishnan M, Borah M. Behav Pharmacol 2008;19:29–40.
186. Naughton M, Mulrooney JB, Leonard BE. Hum Psychopharmacol Clin Exp 2000;15: 397–415.
187. Monsma FJ, Shey Y, Ward RP, et al. Mol Pharmacol 1993;43:320–327.
188. Roth BL, Craigo SC, Choudhary MS, et al. J Pharmacol Exp Ther 1994;268:1403–1410.
189. Hedlund PB, Huitron-Resendiz S, Henriksen SJ, Sutcliffe JG. Biol Psychiat 2005;58: 831–837.
190. Wesolowska A, Nikiforuk A, Stachowicz K, Tatarczynska E. Neuropharmacology 2006; 51:578–596.
191. Bonaventure P, Kelly L, Aluisio L, Shelton J, Lord B, Galici R, Miller K, Atack J, Lovenberg TW, Duzovic C. J Pharmacol Exp Ther 2008;321:690–698.
192. Kent JM. Lancet 2000;355:911–918.
193. Wheatley DP, van Moffaert M, Timmerman L, et al. J Clin Psychiatry 1998;59:306–312.
194. Haddjeri N, Blier P, De Montigny C. J Pharmacol Exp Ther 1996;277:861–871.
195. De Boer TH, Maura G, Raiteri M, De Vos CJ, Wieringa J, Pinder RM. Neuropharmacology 1998;27:399–408.
196. Bengtsson HJ, Kele J, Johansson J, Hjorth S. Naunyn Schmiedebergs Arch Pharmacol 2000;362:406–412.
197. Fawcett J, Barkin RL. J Affect Disord 1998;51: 267–285.
198. Broekkamp CLE, Leysen D, Peeters BWMM, Pinder RM. J Med Chem 1995;38:4615–4633.
199. Consoli D, Leggio GM, Mazzola C, Micale V, Drago F. Eur J Pharmacol 2007;573:139–147.
200. Overstreet DH, Stemmelin J, Griebel G. Pharmacol Biochem Behav 2008;89:623–626.
201. Stemmelin J, Cohen C, Terranova JP, Lopez-Grancha M, Pichat P, Bergis O, Decobert M, Santucci V, Francon D, Alonso R, Stahl SM, Keane P, Avenet P, Scatton B, le Fur G, Griebel G. Neuropsychopharmacology 2008; 33:574–587.
202. Sallinen J, Hoglund I, Engstrom M, Lehtimaki J, Virtanen R, Sirvio J, Wurster S, Savola JM, Haapalinna A. Br J Pharmacol 2007;150: 391–402.
203. Youdim MB, Sourkes TL. Can J Biochem 1965;43:1305–1318.
204. Cesura AM, Pletscher A. Prog Drug Res 1992; 38:171–297.
205. Fowler CJ, Ross SB. Med Res Rev 1984; 4:323–358.
206. Larsen JK. Acta Psychiatr Scand 1988; 78 (Suppl 345): 74–80.
207. McDaniel KD. Clin Neuropharmacol 1986;9: 207–234.
208. Liebowitz MR, Quitkin FM, Stewart JW. Pharmacol Bull 1981;17:159–161.
209. Bloch RG, Doonief AS, Buchberg AS, Spellman S. Ann Intern Med 1954;40:881–900.

210. Crane GE. Psychiatr Res Rep Am Psychiatr Assoc 1957;8:142–152.
211. Kline NS. J Clin Exp Psychopathol 1958;19 (Suppl 1): 72–78.
212. Zeller EA, Barsky J, Fouts JP, Kircheimer WF, Van Orden LS. Experientia 1952;8:349.
213. Kuhn R. Schweiz, Med Wochenschr 1957;87: 1135.
214. Kuhn R. Am J Psychiatry 1958;115:459–464.
215. Ayd FJ, Psychosomatics 1960;1:320–325.
216. Healy D, *The Antidepressant Era*. Cambridge, MA: Harvard University Press; 1997.
217. Coppen A, Shaw DA, Farrell JP. Lancet 1963; i:79–80.
218. Bunney WE, Davis JM. Arch Gen Psychiatry 1965;13:483–494.
219. Montgomery SA, McAuley R, Rani SJ, Roy D, Montgomery DB. Acta Psychiatr Scand 1981;63Suppl290: 314–327.
220. Kramer P. *Listening to Prozac*. New York: Viking Press; 1993.
221. Broekkamp CLE, Leysen D, Peeters BWMM, Pinder RM. J Med Chem 1995;38:4615–4633.
222. Biel JH, Bopp B.In: Gordon M, editor. *Psychopharmacological Agents*. New York: Academic Press; 1974. p 1143–1153.
223. Fielding S, Lal H, editors. *Antidepressants*. Mt. Kisco, NY: Futura; 1975.
224. Pinder RM, Wieringa JH. Med Res Rev 1993;13:259–325.
225. Kelley JL, Musso DL, Boswell GE, Soroko FE, Cooper BR. J Med Chem 1996;39:347–349.
226. Stanford SC. Trends Pharmacol Sci 1996;17: 150–154.
227. Yu PH, Boulton AA. In: Kennedy SH, editor. *Clinical Advances in Monoamine Oxidase Inhibitor Therapies*. Washington, DC: American Psychiatric Press; 1994. p 61–82.
228. Waldmeier PC, Amrein R, Schmid-Burgk W. In: Kennedy SH, editor. *Clinical Advances in Monoamine Oxidase Inhibitor Therapies*. Washington, DC: American Psychiatric Press; 1994. p 33–59.
229. Kennedy SH, Glue P. In: Kennedy SH, editor. *Clinical Advances in Monoamine Oxidase Inhibitor Therapies*. Washington, DC: American Psychiatric Press; 1994. p 279–290.
230. van Wijngaarden I, Soudin W. In: Olivier B, van Wijngaarden I, Soudin W, editors. *Serotonin Receptors and Their Ligands*. Amsterdam: Elsevier; 1997. p 161–197.
231. Roth BL, Willins DL, Kristiansen K, Kroeze WK. Pharmacol Ther 1998;79:231–257.
232. Rothman RB, Baumann MH, Savage JE, Rauser L, McBride A, Hufeisen SJ, Roth BL. Circulation 2000;102:2836–2841.
233. van Wijngaarden I, Soudin W, Tulp MTM. In: Olivier B, van Wijngaarden I, Soudin W, editors. *Serotonin Receptors and Their Ligands*. Amsterdam: Elsevier; 1997. p 17–43.
234. Glennon RA, Dukat M. Serotonin ID Res Alerts 1997;2:351–372.
235. Cooper BR, Hester TJ, Maxwell RA. J Pharmacol Exp Ther 1980;215:127–134.
236. Devane CL, Laizure SC, Stewart JT, Kolts BE, Ryerson EG, Miller RL, Lai AA. J Clin Psychopharmacol 1990;10:328–332.
237. Laizure SC, De Vane CL, Stewart JT, Dommisse CS, Lai AA. Clin Pharmacol Ther 1985;38:586–589.
238. Posner J, Bye A, Dean K, Peck AW, Whiteman PD. Eur J Clin Pharmacol 1985;29: 97–103.
239. Golden RN, DeVane CL, Laizure SC, Rudorfer MV, Sherer MA, Potter WZ. Arch Gen Psychiatry 1988;45:145–149.
240. Martin P, Massol J, Colin JN, Lacomblez L, Puech AJ. Pharmacopsychiatry 1990;23: 187–194.
241. Schroeder DH. J Clin Psychiatry 1983; 44:79–81.
242. Suckow RF, Smith TM, Perumal AS, Cooper TB. Drug Metab Dispos 1986;14:692–697.
243. Welch RM, Lai AA, Schroeder DH. Xenobiotica 1987;17:287–298.
244. Nomikos GG, Damsma G, Wenkstern D, Fibiger HC. Neuropsychopharmacology 1989;2: 273–279.
245. Sitges M, Reyes A, Chiu LM. J Neurosci Res 1994;39:11–22.
246. Cooper BR, Wang CM, Cox RF, Norton R. Shea V, Ferris RM. Neuropsychopharmacology. 1994;11:133–141.
247. Dong J, Blier P. J Psychopharmacol 2001;155: 52–57.
248. Halaris A, Zhu H, Feng Y, Piletz JE. Ann NY Acad Sci 1999;881:445–451.
249. Zhu H, Halaris A, Madakasira S, Pazzaglia P, Goldman N, De Vane CL, Andrew M, Reis D, Piletz JE. J Psychiatr Res 1999;33: 323–333.
250. Molderings GH, Gothert M. Naunyn Schmiedebergs Arch Pharmacol 1995;351:507–516.
251. Mennini T, Mocaer E, Garattini S. Naunyn Schmiedebergs Arch Pharmacol 1987;336: 478–482.

252. Loo H, Deniker P. Clin Neuropharmacol 1988;11 (Suppl 2): S97–S102.
253. Hogg S, Sanchez C. Eur Neuropsychopharmacol 1999;9:S213.
254. Montgomery SA, Loft H, Sanchez C, Reines EH, Papp M. Pharmacol Toxicol 2001;88: 282–286.
255. SCRIP. No. 2586. 24: (2000).
256. de Paulis T. Drugs 2007;10:193–201.
257. Eser D, Baghai TC, Möller HJ. Int Clin Psychopharmacol 2007;22 (Suppl 2): S15–9.
258. Skolnick P. Eur J Pharmacol 1999;375:31–40.
259. Papp M, Moryl E. Eur J Pharmacol 1994;263: 1–7.
260. Paul IA, Trullas R, Skolnick P, Nowak G. Psychopharmacology 1992;106:285–287.
261. Paul IA, Nowak G, Layer RT, Popik P, Skolnick P. J Pharmacol Exp Ther 1994;269:95–102.
262. Skolnick P, Layer RT, Popik P, Nowak G, Paul IA, Trullas R. Pharmacopsychiatry 1996;29: 23–26.
263. Berman RM, Cappiello A, Anand A, Oren DA, Heninger GR, Charney DS, Krystal JH. Biol Psychiatry 2000;47:351–354.
264. Zarate CA, Jr, Singh JB, Carlson PJ, Brutsche NE, Ameli R, Luckenbaugh DA,et al. Arch Gen Psychiatry 2006;63:856–864.
265. Maeng S, Zarate CA, Jr, Jing D, Schloesser RJ, McCammon J, chen G, Manji HK. Biol Psychiatry 2008;63:349–352.
266. Hickenbottom SL, Grotta J. Semin Neurol 1998;18:485–492.
267. Rogawski MA. Amino Acids 2000;19:133–149.
268. Rupniak NMJ, Kramer MS. Trends Pharmacol Sci 1999;20:485–490.
269. Kramer MS, Cutler N, Feighner J, et al. Science 1998;281:1640–1645.
270. Stout SC, Owens MJ, Nemeroff CB. Ann Rev Pharmacol 2001;41:877–906.
271. Koob GF. Biol Psychiatry 1999;46:1167–1180.
272. Holsboer F. J Psychiatr Res 1999;33:181–214.
273. Gilligan PJ, Robertson DW, Zaczek R. J Med Chem 2000;43:1641–1660.
274. Brady LS, Whitfield HJ, Jr, Fox RJ, Gold PW, Herkenham M. J Clin Invest 1991;87:831–837.
275. Kehne JH. CNS Neurol Disord Drug Targets 2007;6:163–82.
276. Valdez GR. CNS Drugs 2006;20:887–896.
277. Lu X, Sharkey L, Bartfai T. CNS Neurol Disord Drug Targets 2007;6:183–192.
278. Bartfai T, Lu X, Badie-Mahdavi H, Barr AM, Mazarati A, Hua X-Y, Yaksh T, Haberhauer G, Conde Ceide S, Trembleau L, Somogyi L, Krock L, Rebek J, Jr. Proc Natl Acad Sci USA 2004;101:1047–10475.
279. Barr AM, Kinney JW, Hill MN, Lu X, Biros S, Rebek J, Jr, Bartfai T. Nerosci Lett 2006;405: 111–115.
280. Swanson CJ, Blackburn TP, Zhang X, Xu ZQ, Hökfelt T, Wolinsky TD, Konkel MJ, Chen H, Zhong H, Walker MW, Craig DA, Gerald CP, Branchek TA. Proc Natl Acad Sci USA 2005;102:17489–17494.

THERAPEUTIC AGENTS FOR THE TREATMENT OF MIGRAINE

GENE M. DUBOWCHIK
JOHN E. MACOR

Neuroscience Discovery Chemistry, Bristol-Myers Squibb Research and Development, Wallingford, CT

1. INTRODUCTION

Migraine is one of the oldest diseases to have been described in detail, possibly as early as about 3000 BC in Babylon. Certainly, the Ebers Papyrus (Egypt, about 1830 BC) describes in detail symptoms that most migraineurs would recognize along with some simple, and possibly effective, remedies [1]. The medieval abbess Hildegaard of Bingen, who was a composer, painter and writer, incorporated her migraine auras, which she described in detail, into her artwork [2]. Nevertheless, for the vast majority of primary headache sufferers, migraine is a debilitating condition, often with nearly unbearable throbbing pain (often unilateral), nausea, and greatly reduced sensitivity thresholds to sound (phonophobia), light (photophobia), smell, and tactile sensations. It is estimated that migraine affects 12–15% of the Western population, with more then twice as many women sufferers. In contrast to most other neurological diseases, migraine mostly afflicts those in middle age or youth. In women, the prevalence of migraine (as well as the divergence from prevalence in males) increases until about age 40 [3]. A recent study found a robust association between migraine and a reduced risk of postmenopausal estrogen and progesterone receptor-positive breast cancer [4], emphasizing the strong hormonal component in migraine attacks. Although frequency of attacks varies widely, almost 3% of Americans are considered to have chronic migraine, defined as more than 15 days with headache per month [5]. Duration of attack varies widely too, from several hours to 3 days or more [6]. However, in a recent large study (>1200 patients), 34.4% of patients had a headache that lasted 24–72 h [7]. Since everyday tasks usually aggravate the symptoms, sufferers often require immobility and isolation during attacks. As a result, severe migraine is considered by the World Health Organization to be comparable to quadriplegia in its debilitating impact on the daily life of the individual.

There are considered to be four or five phases in a typical migraine attack [8]. In the prodrome phase, which can last hours or even days before headache onset, the patient is aware that an attack is imminent. This occurs in about 80% of patients. A large proportion (20–30%) of migraineurs visualize auras—bright rings or spots of light—before and/or during the initial phase of headache pain. Migraine with aura has been shown to be a risk factor for early onset stroke. However, the mechanistic connections have yet to be uncovered [9]. The overall headache phase usually lasts from 4–72 h and is often accompanied by photo- and phonophobia. It is sometimes divided into early and late (resolution) periods. The postdromal phase is marked by fatigue, both physical and mental, and residual signs of neuronal hyperexitability such as allodynia and photophobia. Overuse of several types of acute migraine medications has been shown to result in progression to increasingly chronic migraine conditions [10].

There is a strong hereditary component in migraine with many reports of positive family histories. This is especially true of familial hemiplegic migraine (FHM) where approximately 50% of sufferers have been shown to possess a mutation in a voltage-gated Ca^{2+} channel gene, strong evidence for the involvement of ion channels in the auras that are prevalent in this condition [11].

Migraine therapy can be divided into two classes: abortive (or acute) and preventative (or prophylactic). The most widely used medications are shown in Table 1. The former has been dominated by over-the-counter analgesics (primarily acetaminophen and NSAIDs) and, since the early 1990s, by the triptans (5-$HT_{1B/1D}$ agonists) that, until very recently, were only available by prescription. However, despite efforts to motivate patients to seek physician care and treatment, less than half of migraine sufferers (48%) are diagnosed by physicians [12]. The majority of patients remain undiagnosed or misdiagnosed, commonly with either sinus- or tension-type head-

Table 1. Currently Available Therapies for Migraine by Class [13]

Indication	Class	Examples
Abortive	Analgesics	Acetominophen
		NSAIDs
	Antimigraine agents	Ergot alkaloids
		5-HT$_{1B/1D}$ agonists
Preventative	Anticonvulsants	Topiramate
		Valproate
		Gabapentin
	Antidepressants	Tricyclics
		Duloxetine
	ACE inhibitors	Lisinopril
	Calcium channel blockers	Verapamil
		Flunarizine
	β-Adrenergic blockers	Atenolol
		Propranolol
	Melatonin agonists (sleep)	Ramelteon
	Miscellaneous	Vitamin B$_{12}$
		Botulinum toxin
		Coenzyme Q10
		Magnesium

ache. Among the undiagnosed population, there is a high degree of self-treatment with over-the-counter treatments. Preventative treatments are used especially by frequent migraine sufferers along with acute treatments, and include a wide range of pharmacologic mechanisms. As might be expected, most of these prophylactic agents are older drugs whose mechanisms of action are often incompletely understood and whose mechanism of action and/or side effect profile determines who can use them (see Section 11).

2. PATHOPHYSIOLOGY

The stimuli or triggers that bring on a migraine attack vary greatly between individuals and include certain foods (chocolate, cheeses and red wine, in particular), weather, hormone levels, disrupted sleep patterns, and stress. Migraine trigger mechanisms have been studied but are still largely a mystery. In addition, many sufferers apparently need no obvious trigger to bring on an attack. Migraineurs may have an increased tendency toward intermittent neuronal excitability, possibly due to inherited mutations in, for example, ion channels [14,15], which lowers their threshold for neuronal firing in reaction to modest physiological changes such as menstrual hormone fluctuations, emotional stress and even awakening from sleep.

There presently exist several theories concerning the basic causal mechanisms that lead to migraine pain and associated symptoms. An early "vascular-only" hypothesis suggested that migraine symptoms could be explained by the change in diameter of dural blood vessels. Briefly, the migraine trigger results in vasoconstriction leading to localized ischemia. This is followed by an excessive compensatory response by the trigeminovascular system (TGVS) in which the vessels are dilated by secretion of vasoactive substances such as calcitonin gene-related peptide (CGRP) by neuronal afferents, leading to activation of the perivascular sensory nerve projecting to the brain [16].

It is generally accepted that activation of the TGVS, which innervates the face and the meningeal blood vessels surrounding the brain, is responsible for migraine pain. In addition, cortical spreading depression (CSD), a self-propagating wave of neuronal depolarization in the cortex of the brain, is thought to be directly responsible for the visual aura that precedes headache for approximately one-third of migraineurs. Many researchers now believe that a combination of environmental factors (triggers) and a lowered threshold for cortical hyperexcitability leads to CSD, possibly in all migraineirs since aura may only manifest itself when CSD occurs in visual areas of the brain [15]. CSD, which may be aided by low magnesium concentrations leading to enhanced NMDA receptor activation, results in extracellular release of a number of neuronally active substances such as potassium ions, nitric oxide (NO), glutamate (Glu), prostaglandins, and arachidonic acid. These may activate or sensitize trigeminal afferents in the cortex [17]. The activated TGVS secretes neuropeptides such as CGRP, causing vasodilation, and substance P that may increase blood vessel permeability and cause plasma extravasation and mast cell degranulation, leading to dural inflammation [18]. Other vasoactive substances have been identified by immunohistochemical studies in perivascular nerves supplying intracranial blood vessels including serotonin (5-HT), NO, substance P, neurokinin A, and vasoactive intestinal peptide (VIP) [19]. Most of these substances, such as NO, 5-HT, and CGRP also mediate neurotransmission in the CNS, and thus may have multiple functions in migraine pathophysiology, complicating the interpretation of mechanism-based therapies. Neurogenic inflammation may further activate and sensitize TGVS afferents in the dura. The resulting overstimulation of the trigeminal system, which synapses with second-order neurons in the trigeminal nucleus caudalis (TNC) in the brainstem, is postulated to cause central sensitization [20].

Recent findings, especially from brain imaging studies, have stressed the potential importance of brainstem dysfunction in migraine, in particular, the periaqueductal gray (PAG), which contains various neurons involved in the pain desensitization pathway. A number of researchers now believe that altered brainstem activity may be critically important both in triggering and in lowering the threshold for migraine attacks [21].

3. ERGOT ALKALOIDS

Ergotamine had been used as an acute antimigraine agent since 1925, soon after its isolation in 1918 from the ergot fungus [22]. However, unpleasant side effects such as exacerbated nausea and chest tightness related to vasoconstriction were common until the development of dihydroergotamine (DHE) in 1945 [23]. Ergotamines (Fig. 1) display complex pharmacology through modulation of a variety of neurotransmitter receptors, including the 5-HT_{1B} and 5-HT_{1D} receptors (IC_{50} values ≤ 1 nM) known to be involved in the antimigraine activity of triptans, but also 5-HT_{2B} and 5-HT_{2C} receptors ($IC_{50} < 10$ nM) that may also be of relevance [24]. They also bind with high affinity ($IC_{50} < 10$ nM) to serotonin 5-HT_{1A}, and 5-HT_{2A} receptors, noradrenaline a_{1A}, a_{2A}, and a_{2B} receptors, and dopamine D_{2L}, D_3, and D_4 receptors. It is believed that 5-HT_{1A} and dopamine D_2 activities are responsible for many of the side effects seen with ergotamines. Clinical antimigraine activity was long thought to be solely due to constriction of meningeal arteries, but newer evidence suggests that, like the triptans, ergotamines exert additional effects on neurotransmission and neurogenic inflammation [25]. DHE is used both as an abortive and as a preventative migraine therapy. The compound has very low oral bioavailability due to extensive oxidative metabolism in the liver and poor gut penetration (Table 2). However, upon repeated oral administration, a prominent active metabolite, 8′-hydroxydihyrdoergotamine (8′-OH-DHE), is formed which has a relatively long half-life (10–13 h), and has been postulated to account for the low rate of headache recurrence seen with DHE. An intranasal formulation (1–4 mg) avoids first pass hepatic metabolism and provides potentially effective plasma levels within the timeframe of intramuscular delivery [26]. In clinical trials, adverse events reported for the nasal spray were mild to moderate, and were generally related to nasal delivery (i.e., congestion, irritation, unusual taste) rather than the drug itself.

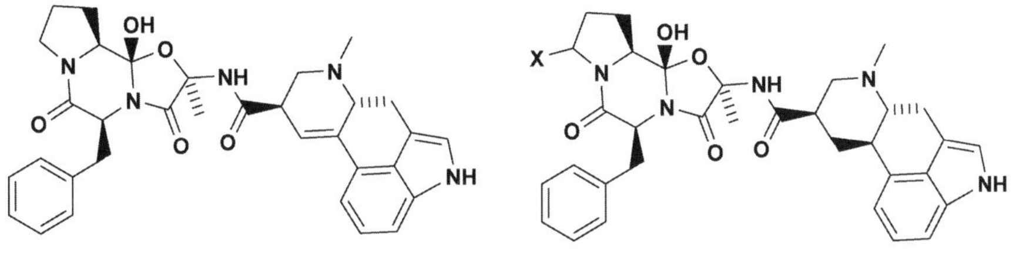

Figure 1. Ergotamine, DHE, and the prominent metabolite of DHE, 8′-hydroxydihydroergotamine.

Table 2. Pharmacokinetics of DHE in Healthy Volunteers

Mode of Administration	Bioavailability (%)	T_{max} (min)
Oral	<1	—
Intramuscular	100	24
Intranasal	40	30–60
Intravenous	—	1–2

4. NONSTEROIDAL ANTI-INFLAMMATORY DRUGS (NSAIDS) AND PARACETAMOL (ACETAMINOPHEN)

Nonsteroidal anti-inflammatory drugs are nonselective inhibitors of cyclooxygenase enzymes (COX-1 and COX-2 isoforms) that catalyze the cyclization of arachidonic acid to form prostagalandins. Paracetamol (acetominiphen), although not formally a member of this drug class, is often grouped with the NSAIDs. It was long thought to work through different mechanisms but there is recent evidence that it is a fairly selective COX-2 inhibitor in humans [27]. These agents are popular over-the-counter, nonaddicting pain killers and anti-inflammatory medications. Aspirin, which has been available since the late 19th century, has been shown to prevent neurogenic inflammation attributable to TGVS activation [28], a potential mechanism for migraine prophylaxis. The NSAIDs shown in Fig. 2 are the most frequently used therapies for treating migraine, especially among those whose headaches are classified as "mild to moderate" [29]. However, a few, particularly ketorolac, are also dosed intravenously to patients in an emergency room setting alone or in combination "cocktails" with caffeine and/or antihistamines [30]. NSAIDs are also used for migraine prophylaxis, either intermittently or daily, especially if migraine triggers are predictable [31]. However, their adverse effects on the GI tract, a result of COX-1 inhibition, can make chronic use problematic.

5. 5-HT$_{1B/1D}$ AGONISTS (TRIPTANS)

5.1. Introduction: Sumatriptan

Serotonin (5-hydroxytryptamine, 5-HT) (Fig. 3) was discovered in 1948 as an extract from blood serum, and its name was derived from the fact the this extract was capable of inducing contractions on isolated blood vessels and certain tissues (serotonin) [32].

As described earlier in this chapter, dilation of cranial blood vessels is one of the early hallmarks of the onset of a migraine headache. Thus, drug discovery efforts at a number of pharmaceutical companies in the late 1980s

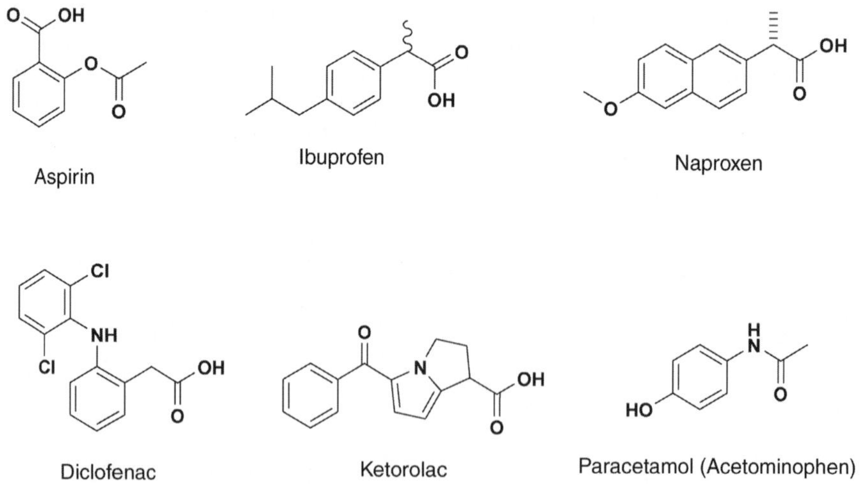

Figure 2. Nonsteroidal anti-inflammatory drugs and paracetamol (acetominophen), commonly used to treat migraine.

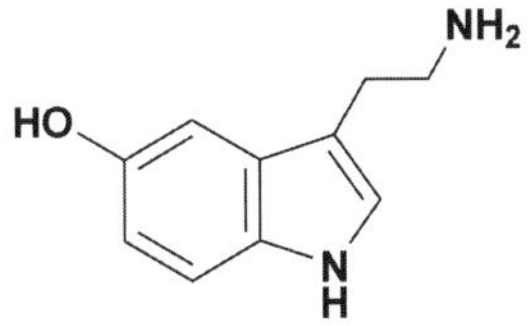

5-hydroxytryptamine, 5-HT

Figure 3. Serotonin (5-hydroxytryptamine, 5-HT).

focused on the role of serotonin in migraine, and 5-HT receptors, specifically the subtypes involved in vasoconstriction—5-HT$_{1B/1D}$ receptors [33], as a new target for treating migraine headaches [34]. The introduction of sumatriptan in 1996 (Table 3) literally revolutionized the treatment of migraine headaches [35,36]. Sumatriptan is a direct analog of serotonin with N,N-dimethylamine repla-

Table 3. Marketed Triptans with Launch Dates

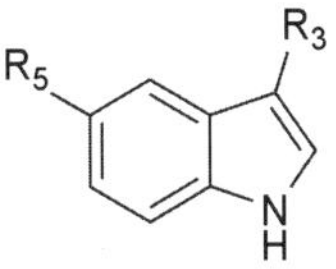

Triptan	R$_5$	R$_3$	Launched
Sumatriptan	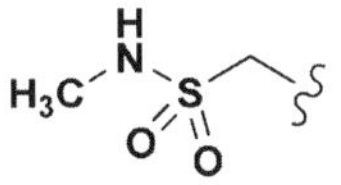 H$_3$C-NH-SO$_2$-CH$_2$-	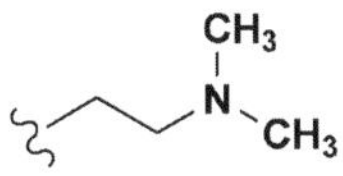 -CH$_2$CH$_2$-N(CH$_3$)$_2$	1991
Naratriptan	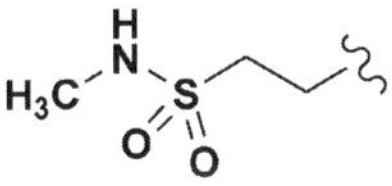 H$_3$C-NH-SO$_2$-CH$_2$CH$_2$-	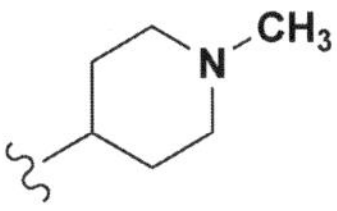 N-methylpiperidine	1997
Rizatriptan	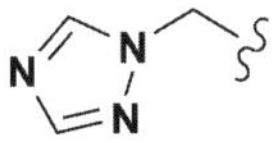 1,2,4-triazol-1-ylmethyl	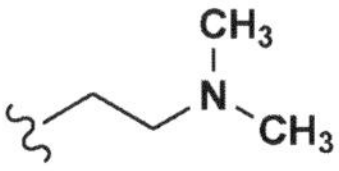 -CH$_2$CH$_2$-N(CH$_3$)$_2$	1998
Zolmitriptan	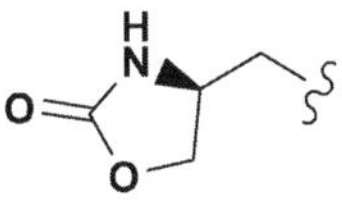 oxazolidinone-CH$_2$-	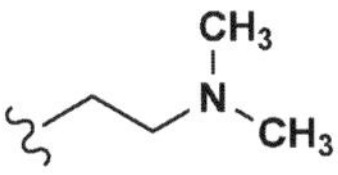 -CH$_2$CH$_2$-N(CH$_3$)$_2$	1997
Almotriptan	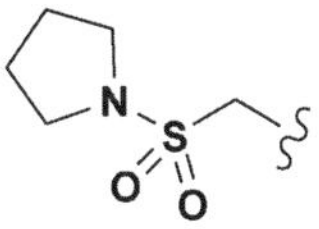 pyrrolidinyl-SO$_2$-CH$_2$-	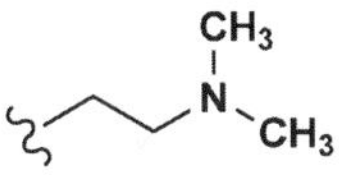 -CH$_2$CH$_2$-N(CH$_3$)$_2$	2000
Frovatriptan	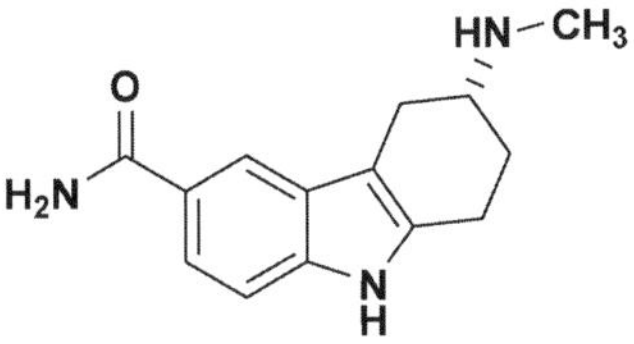 (carboxamide/methylamino tetrahydrocarbazole structure)		2002
Eletriptan	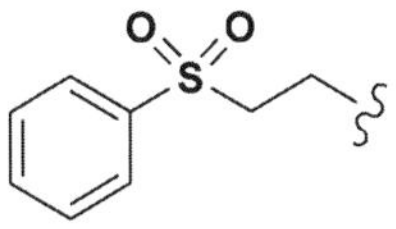 PhSO$_2$-CH$_2$CH$_2$-	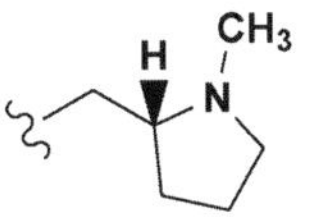 (S)-N-methylpyrrolidin-2-ylmethyl	2001

cing the primary amine of 5-HT, and with the hydroxyl group of 5-HT being replaced by an N-methylaminosulfonylmethyl group at the C5-position of the indole in sumatriptan. These changes made sumatriptan selective for $5\text{-HT}_{1B/1D}$ receptors, some of the receptors involved with the vasoconstrictive effects of serotonin. It is noteworthy that these small modifications to serotonin which led to sumatriptan made the new molecule selective for only a small subset of the over dozen different receptors in the 5-HT family (5-HT_{1A}, 5-HT_{1B}, 5-HT_{1D}, 5-HT_{1E}, 5-HT_{1F}, 5-HT_{2A}, 5-HT_{2B}, 5-HT_{2C}, 5-HT_{3A}, 5-HT_{3B}, 5-HT_{4}, 5-HT_{5A}, 5-HT_{5B}, 5-HT_{6}, and 5-HT_{7}). With its introduction in 1996, sumatriptan is the most prescribed triptan and numerous reviews on its discovery, development and use have been written [37–39].

5.2. Second-Generation Triptans

With the initial reports of the success of sumatriptan in early preclinical and clinical studies, the pharmaceutical industry as a whole focused its attention on "fast follow-on" analogs of sumatriptan with improved pharmacological and/or pharmaceutical properties. This effort led to six additional approved triptans (Table 1), often referred to as the "second-generation" triptans. Table 4 [40,41] summarizes the pharmacokinetic and clinical properties of the marketed triptans. The "second-generation" triptans with the exception of naratriptan were all tryptamine derivatives with rizatriptan, zolmitriptan, and almotriptan containing the indole C3-dimethylaminoethyl group found in sumatriptan. They only differed in their indole C5-substituent. Frovatriptan and eletriptan contained conformationally restricted C3-aminoethyl groups—frovatriptan with its N-methylaminocyclohexeno group fused at C2 and C3 of the indole scaffold, and eletriptan with its indole C3-N-methylpyrrolidin-2-ylmethyl sidechain. Both also had indole C5-substituents differing from the other triptans. Naratriptan was unique among the "second-generation" triptans in that it contained an indole C3-piperidin-4-yl substituent, a conformationally restricted aminopropyl group. All "second-generation" triptans were potent 5-HT_1 recepor agonists, but they differed in their physiochemical and AMDE properties (Table 4). At the time of their discovery and development, there was significant speculation that these differences would lead to differences in their clinical effectiveness and a reduction in adverse events when compared to sumatriptan. However, in retrospect, there is still insufficient data to show a clear difference between the triptans as a class [42]. As shown it Table 4, the triptan class is generally effective in approximately 60% of migraine attacks with significant headache recurrence for all members of the class. Differences experienced in patients are likely due to individual differences in metabolism or other genetic differences, and often patients will try a number of different triptans until they find the most effective one for them.

5.3. Triptans' Mechanism of Action: Anti-Inflammatory or Vasoconstriction

With the discovery of sumatriptan and its therapeutic efficacy in treating migraine headaches, a number of researchers tried to more deeply understand the mechanism of action of this new entity. Two theories emerged in the early 1990s that, while not mutually exclusive, were nonetheless very different in what they hypothesized as to the mechanism of action of the triptans. Moskowitz put forth a compelling theory that sumatriptan blocked the neurogenic inflammation of the trigeminal sensory nerve terminals and subsequent protein plasma extravasation (release of nociceptive and other inflammatory proteins) that characterized a migraine headache [43]. The alternate hypothesis centered on the direct, dose-dependent vasoconstrictive effect of sumatriptan on intracranial blood vessels that correlated with its antimigraine activity [44]. There was significant preclinical research effort to support both hypotheses. **CP-122,288** (Fig. 4), a conformationally restricted analog of sumatriptan [45] was found to be an exceptionally potent inhibitor of plasma protein extravasation response within dura mater following trigeminal ganglion stimulation in guinea pigs (7000-fold more potent than sumatriptan in their minimally effective doses) [46]. In addition, the

Table 4. Comparison of Pharmacokinetic and Clinical Properties of Triptans

Generic Name	Trade Name	Marketing Company	Available Doses (mg)	Oral Bioavailability (%)	Half-Life ($t_{1/2}$) (h)	Onset of Action (h)	T_{max} (h)	Pain Free at 2 h (%)	Headache Recurrence in 24 h (%)
Sumatriptan	Imitrex	Glaxo-SmithKline	25, 50, 100	15	2	0.5–1	2.5	21–23	40
Rizatriptan	Maxalt	Merck	5, 10	70	3	1–3	3–4	25–44	41
Naratriptan	Amerge	Glaxo-SmithKline	1, 2.5	40–48	6	0.75	2–3	Not done	27
Zolmitriptan	Zomig	AstraZeneca	2.5, 5	45	2	0.5–2	1–1.5	9–47	30
Almotriptan	Axert	Johnson & Johnson	6.25, 12	70	3–4	1–3	1–3	30–39	30
Frovatriptan	Frova	Elan	2.5	20–30	26	1–2	1–2	Not done	Not done
Eletriptan	Relpax	Pfizer	20, 40, 80	50	4	1–2	2	57	11–25

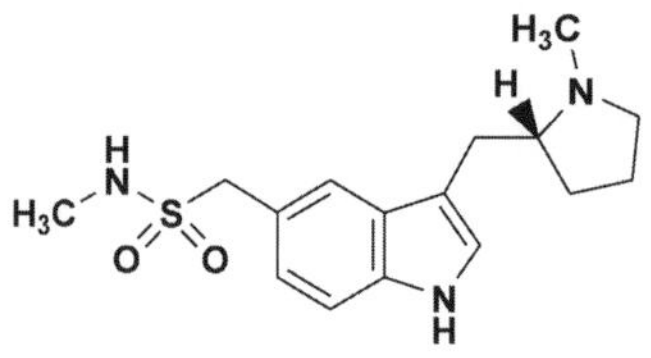

CP-122,288

Figure 4. **CP-122,288**, a conformationally restricted analog of sumatriptan.

potent effects of **CP-122,288** on inhibiting protein plasma extravasation was found not be mediated through 5-HT$_{1B}$ receptors, as the potent activity of the compound was demonstrated in 5-HT$_{1B}$ receptor "knock out" mice, whereas sumatriptan was not active in the 5-HT$_{1B}$ "knock out" mice [47] in inhibiting protein plasma extravasation. **CP-122,288** provided a clinical tool to test the competing hypotheses, and a potential opportunity of antimigraine activity in a triptan-like molecule at doses devoid of vasoconstriction.

However, in human clinical trials, **CP-122,288** was not effective at doses and plasma concentrations in excess of those required to inhibit neurogenic plasma extravasation in animals. It was concluded that neurogenic plasma extravasation was unlikely to play a crucial role in the pathophysiology of migraine headaches [48], and that the vasoconstrictive action of sumatriptan (and correspondingly the subsequent triptans) on the cranial blood vessels was the primary mode of action in its antimigraine behavior. All triptans share this vasoconstrictive action and accordingly, the majority of adverse events associated with triptan use is a result of vasoconstriction of tissues other than cranial arteries, including cardiac vessels and other vasoactive tissues. In rare cases, ischemic cardiac events have been reported with triptan use [49]. Secondary vessel and tissue vasoconstriction is the major adverse event in triptan use, limiting the number of doses allow per migraine attack and during the course of a month. Newer mechanisms that are not active vasoconstrictors (i.e., CGRP antagonists) are devoid of these side effects and potentially represent safer antimigraine agents.

5.4. Summary

The triptans represent an improved and alternate treatment of migraine headache in comparison to the drugs that were used before their introduction. While some of the triptans have been used in alternate formulations (i.e., subcutaneous injection and intranasal), the primary delivery form of these agents is oral. The triptans are active vasoconstricting agents working primary via 5-HT$_{1B/1D}$ receptors on cranial arteries. Because of their vasoconstriction, the major side effects of triptans are vasoconstriction of other vessels and tissues that have led to a limitation of their usage both on an event basis and a monthly basis.

Numerous reviews of the triptans are available in the primary literature [50–52]. In short, there have been no significant separation between the individual triptans from patient and clinical point of view, and often patients end up trying different members of the class until one is found to be consistent in efficacy. Regardless, the triptan class of antimigraine agents has tremendously advanced the treatment of migraine headaches and offer significant relief to a large number of patients to whom treatment previously was not available.

6. CALCITONIN GENE-RELATED PEPTIDE ANTAGONISTS

CGRP is a naturally occurring 37-amino-acid peptide first identified in 1982 [53]. Two forms of the peptide (αCGRP and βCGRP) are expressed that differ by one and three amino acids in rats and humans, respectively. The peptide is widely distributed in both the peripheral and the central nervous system, is principally localized in sensory afferent and central neurons, and displays a number of biological effects, including vasodilation.

CGRP is thought to play a causal role in migraine [54]. Serum levels of CGRP are elevated during migraine in humans, and treatment with antimigraine drugs returns CGRP levels to normal coincident with the alleviation of headache [55]. Intravenous CGRP infusion produces lasting headache in normal individuals [56] and migraineurs [57]. In addition, nitric oxide, which has been shown to be released as a result of CSD-activation of the TGVS [17], mediates release of CGRP [58]. Treatment with a CGRP receptor antagonist would relieve migraine by blocking CGRP-induced neurogenic inflammation and neurogenic vasodilation, returning dilated intracranial arteries to normal (an approach that avoids the cardiovascular liabilities of active vasoconstriction associated with triptans). Preclinical studies in dog and rat report that systemic CGRP blockade with the peptide antagonist CGRP(8-37) does not alter resting systemic hemodynamics nor regional blood flow [59]. Indeed, clinical studies by Boehringer Ingelheim demonstrate proof of concept for CGRP antagonist treatment of migraine with an injectable formulation (Olcegepant, **BIBN4096BS**) with an absence of cardiovascular liabilities [60].

6.1. The CGRP Receptor

When released from the cell, CGRP binds to specific cell surface G-protein-coupled receptors and exerts its biological action predominantly by activation of intracellular adenylate cyclase [61,62]. The CGRP receptor has three components: (i) a seven transmembrane calcitonin receptor-like receptor (CRLR), (ii) the single transmembrane receptor-activity modifying protein-type 1 (RAMP1), and (iii) the intracellular receptor component protein (RCP) [63]. RAMP1 is required for transport of CRLR to the plasma membrane and for ligand binding to the CGRP receptor [64]. The RCP is required for signal transduction [63]. There are known species-specific differences in binding of small-molecule antagonists to the CGRP receptor with greater affinity seen for antagonism of the primate receptor than for other species [65]. The amino acid sequence of RAMP1 determines the species selectivity; in particular, the amino acid residue Trp74 is responsible for the phenotype of the human receptor [66]. Only primates, not lower species, possess CGRP receptors that are comparable to the human receptor.

For this reason, typical neurovascular rodent models of migraine such as the intravital model (for a review of animal models of migraine, see Ref. [67]), in which differences in cranial blood vessel diameter can be observed directly, have generally not been used for efficacy evaluation of small-molecule CGRP antagonists. Instead, primate models have been employed in which differences in blood flow, as a surrogate measure of changes in vasodilation induced by CGRP and antagonism of the CGRP receptor, are measured by laser Doppler flowmetry [68].

6.2. BIBN4096 BS

The discovery of the first clinically efficacious CGRP antagonist, **BIBN4096 BS** (olcegepant) (Fig. 5), began with a high-throughput screening effort at Boehringer-Ingelheim against the human CGRP receptor expressed in SK-N-MC cells (a human neuroblastoma cell line) that resulted in the identification of a dipeptide motif, 3,5-dibromo-(R)-Tyr-(S)-Lys, in compound **1** ($IC_{50} = 17\,\mu M$) (Table 5) as a critical pharmacophore [69]. The corresponding dibromoanilines (compounds **2–4**) were comparably active. For this series, the unnatural D-configuration of the aromatic

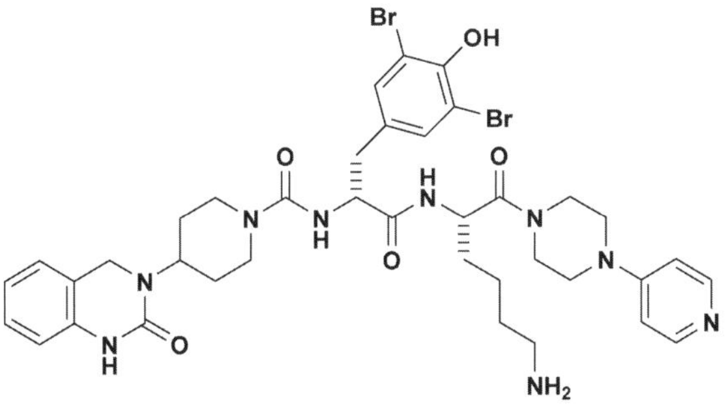

BIBN4096 BS

Figure 5. **BIBN4096 BS**, the first CGRP antagonist to be tested in the clinic.

Table 5. Human CGRP Receptor Binding Affinities of Dipeptide CGRP Antagonists

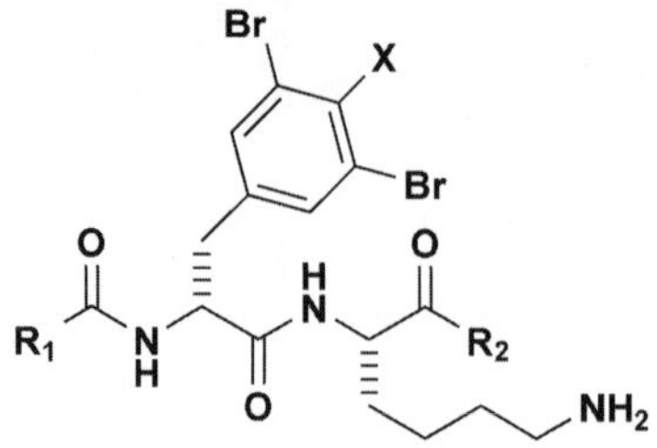

Compound	R$_1$	R$_2$	X	CGRP IC$_{50}$ (nM)
1	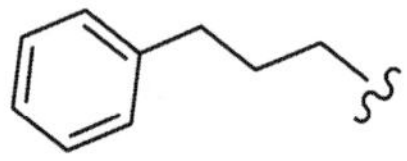 3-phenylpropyl	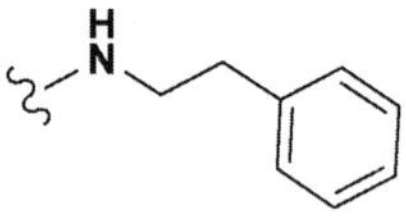 NH-CH$_2$CH$_2$-phenyl	OH	17,000
2	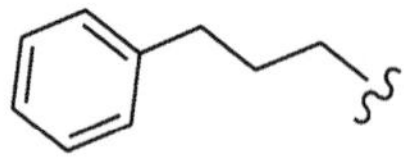 3-phenylpropyl	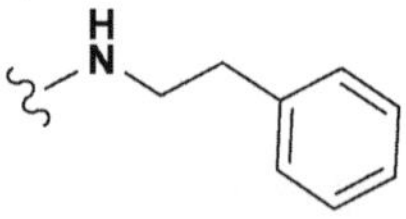 NH-CH$_2$CH$_2$-phenyl	NH$_2$	40,000
3	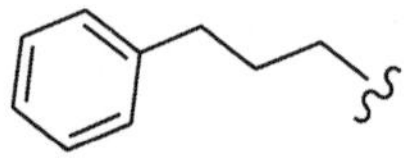 3-phenylpropyl	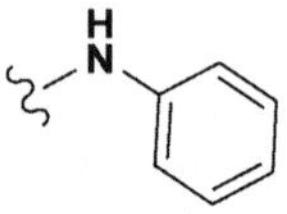 NH-phenyl	NH$_2$	50,000
4	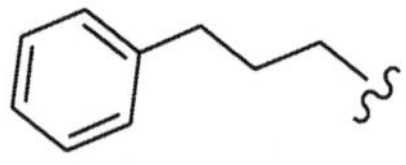 3-phenylpropyl	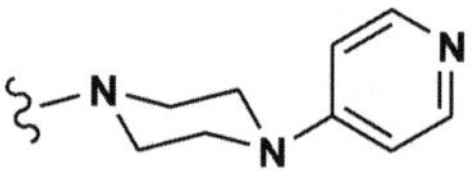 4-(pyridin-4-yl)piperazin-1-yl	NH$_2$	10,300
5	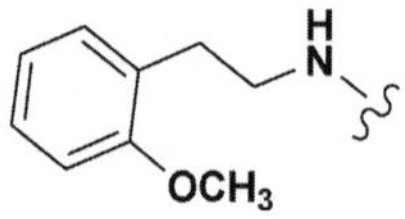 2-methoxyphenethylamino	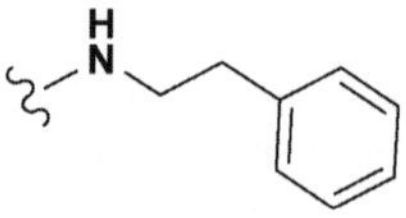 NH-CH$_2$CH$_2$-phenyl	OH	13,000
6	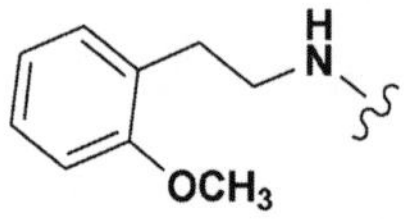 2-methoxyphenethylamino	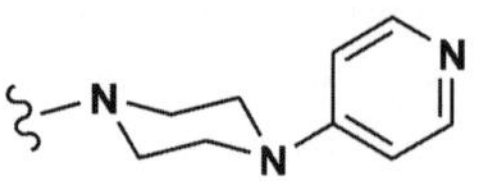 4-(pyridin-4-yl)piperazin-1-yl	OH	1,000
7	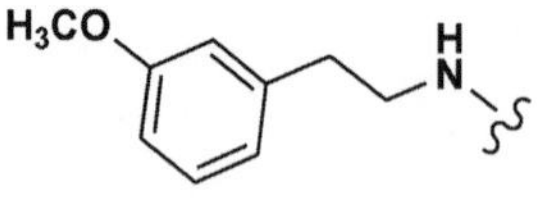 3-methoxyphenethylamino	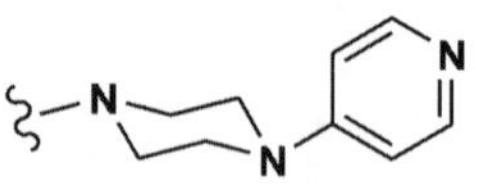 4-(pyridin-4-yl)piperazin-1-yl	OH	200
8	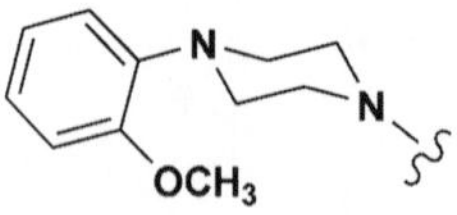 4-(2-methoxyphenyl)piperazin-1-yl-methyl	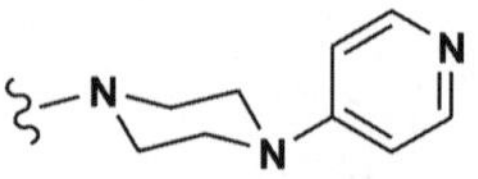 4-(pyridin-4-yl)piperazin-1-yl	OH	226
9	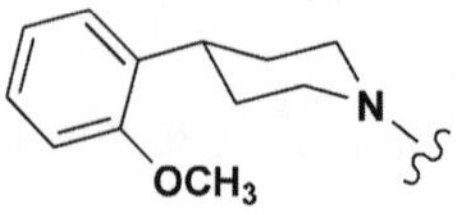 4-(2-methoxyphenyl)piperidin-1-yl-methyl	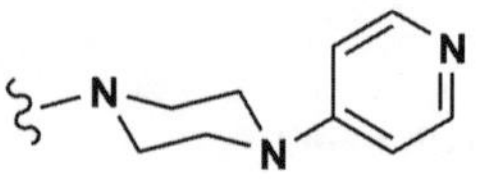 4-(pyridin-4-yl)piperazin-1-yl	OH	250

Table 5. (Continued)

Compound	R₁	R₂	X	CGRP IC$_{50}$ (nM)
10	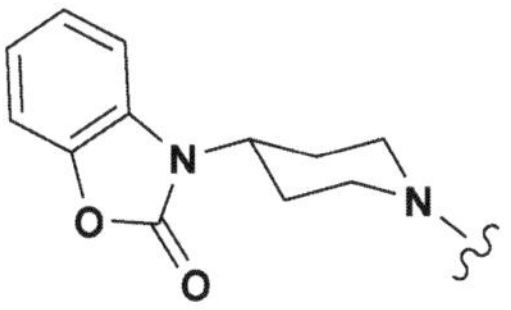 (benzoxazolone-piperidine)	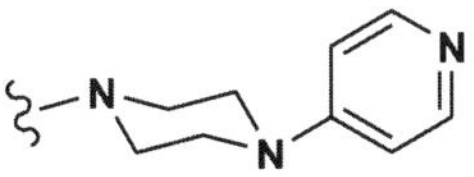 (piperazine-pyridine)	OH	44
11	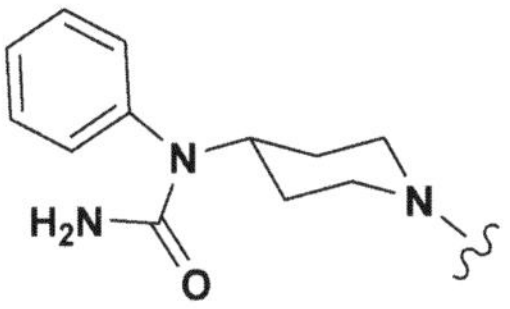 (phenylurea-piperidine)	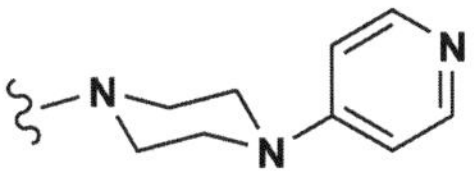 (piperazine-pyridine)	OH	4.7
12	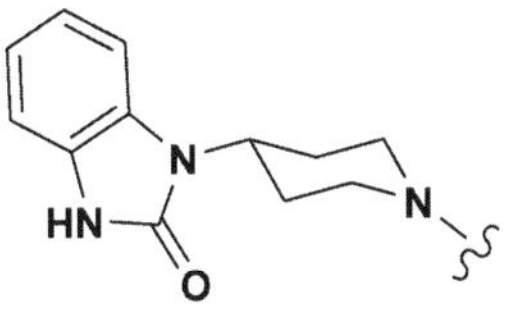 (benzimidazolone-piperidine)	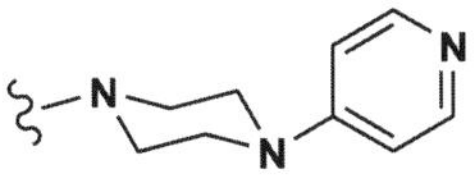 (piperazine-pyridine)	OH	0.2
13	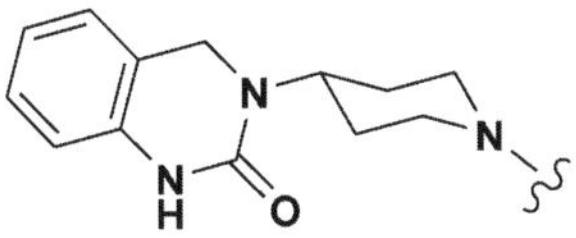 (tetrahydroquinazolinone-piperidine)	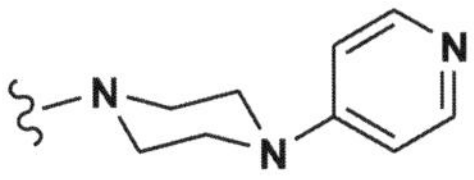 (piperazine-pyridine)	OH	0.03
14	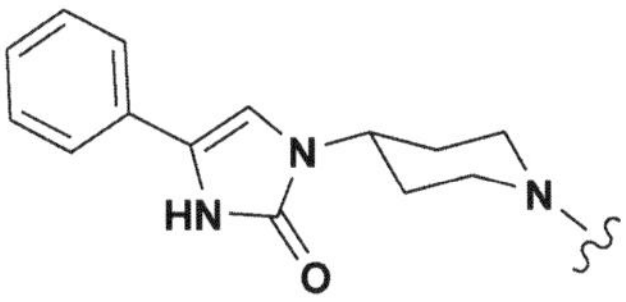 (phenylimidazolone-piperidine)	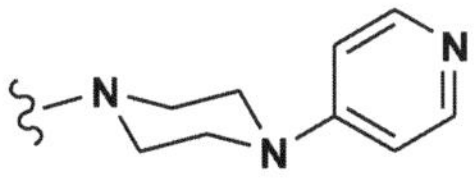 (piperazine-pyridine)	OH	0.05

amino acid and the natural L-configuration of the lysine were absolutely necessary for activity (IC$_{50}$ > 300 µM for either amino acid antipode).

Constraining the R$_2$ aromatic ring with a piperazine resulted in a large gain in potency with compound **6** (IC$_{50}$ = 1000 nM). Further addition of a properly placed hydrogen-bond acceptor anisole attached to a piperazine (**8**, IC$_{50}$ = 226 nM) or piperidine (**9**, IC$_{50}$ = 250 nM) ring on R$_1$ was also advantageous. With a cyclic urethane carbonyl as an enhanced hydrogen-bond acceptor at R$_1$, CGRP receptor binding was increased fivefold (**10**, IC$_{50}$ = 44 nM), while the further addition of a hydrogen-bond donating urea NH resulted in single-digit nanomolar affinity (**11**, IC$_{50}$ 4.7 nM). When this urea was incorporated into an R$_1$ imidazolone (**12**, IC$_{50}$ = 0.2 nM) and then into a tetrahydroisoquinolone (**13**, **BIBN-4096 BS**, IC$_{50}$ = 0.03 nM), a >150-fold further increase in affinity was observed. **BIBN 4096 BS** possesses higher affinity for the human CGRP receptor than the endogenous ligand and 150-fold better than the peptide antagonist CGRP(8-37). Compound **14**, is comparably potent. However, the phenylbenzimidazolone has been shown to be oxidatively unstable.

BIBN4096 BS (Fig. 5) was shown to be a highly potent, full antagonist (pA2 = 11.1) in a cyclic AMP assay in SK-N-MC cells. Interestingly, the compound is three log units less potent against rat CGRP receptors

($IC_{50} = 6.4$ nM) [68]. This is not seen with the peptide antagonist CGRP(8-37), but has been found to be a common feature among "small-molecule" CGRP antagonists. The divergence has been ascribed to the critical difference in residue 74 of the RAMP1 component protein of the CGRP receptor mentioned above. Interestingly, **BIBN4096 BS** appears to be a competitive antagonist at the human receptor, expressed in subcutaneous arteries, at low concentrations (about 0.01 nM), but noncompetitive at higher concentrations (0.1–1 nM) [70].

BIBN4096 BS has been used as a biochemical tool compound in a large number of studies elucidating the mechanisms and consequences of CGRP antagonism (reviewed in Ref. [71]). In order to differentiate it from the triptans as a "nonvasoconstricting" antimigraine agent, **BIBN4096 BS** was tested in *ex vivo* preparations of human and dog coronary arteries and, in contrast to sumatriptan, was found to have no inherent vasoconstrictive properties [72].

In a marmoset model of neurogenic vasodilation in which TGVS-mediated secretion of CGRP was stimulated by direct electrical stimulation of the trigeminal ganglion, **BIBN4096 BS** was found to potently inhibit the resulting increases in facial blood flow (a surrogate marker for vasodilation) in a dose-dependent manner with $ED_{50} = 0.003$ mg/kg (intravenous dosing) [68]. The compound was similarly tested in rats in both a facial blood flow and intravital models. Electrical stimulation resulted in dilation of both the middle meningeal artery and the cortical pial arterioles. However, **BIBN4096 BS** only caused a significant reduction in dilation of the former, which is not protected by the blood–brain barrier, suggesting a purely peripheral action of the compound [73].

As might be expected for a fairly large (MW 870) and highly polar molecule containing more than 10 hydrogen-bonding groups and several ionizable functionalities, **BIBN4096 BS** showed very low oral bioavailability in animals (<1% in dog) and was almost unbound by plasma proteins [69]. As a result, clinical proof-of-concept and safety studies were carried out using an intravenous formulation.

A Phase I study in healthy volunteers specifically addressed the potential cardiovascular safety issues around CGRP antagonism. No significant changes in electrocardiogram, blood pressure, pulse rate, or forearm blood flow were seen up to 10 mg (i.v.) [74]. Other Phase I studies examined cerebrovascular effects of **BIBN4096 BS** (2.5 or 10 mg, i.v.) using brain imaging [75] as well as the ability of the compound (2.5 mg) to prevent CGRP-induced headache in healthy volunteers [76]. Again, no undesired effects were observed. The compound effectively inhibited agonist-induced headache in the 6 of 10 patients in which it manifested itself, and imaging appeared to confirm that pain relief, at least in this model, occurred without significant effects on CNS blood vessels (e.g., middle cerebral artery).

A multicenter, double-blind, randomized Phase II clinical trial of **BIBN4096 BS** was carried out in 126 patients with migraine receiving placebo or total doses of 0.25, 0.5, 1.0, 2.5, 5.0 or 10.0 mg (i.v., total dose) infused over a period of 10 min [60]. No serious adverse events were observed up to the highest dose. Mild paresthesia (8%) was the only significant adverse event reported. Mean maximal C_{max} levels were dose-proportional. Human V_d (20 L) generally reflects a highly polar molecule with limited bilayer membrane permeability. Total plasma clearance (C_L) was 12 L/h and terminal $t_{1/2} = 2.5$ h.

Two hours following a 2.5 mg dose, 66% of patients reported significant migraine pain relief. The therapeutic response was similar to that observed for oral triptans in previous clinical trials, although less than that reported for subcutaneous (SC) sumatriptan [77]. No further improvement in the primary or secondary endpoints was seen at higher doses and no significant efficacy was seen at the next lower dose (1 mg). **BIBN4096 BS**-treated patients also exhibited significant superiority over the placebo group with regard to most secondary endpoints, including pain-free rate at 2 h, rate of sustained response over 24 h, rate of headache recurrence, improvement in nausea, photophobia, phonophobia, and functional capacity, and time to meaningful relief. It is worth noting that a total plasma C_{max} 200 nM was necessary for **BIBN4096 BS** ($K_i = 0.014$ nM) to demonstrate robust clinical efficacy. If indeed it is peripheral CGRP receptors that are the target for migraine, this potentially represents at least a 1000-fold required coverage over binding K_i. On the other hand, if the target receptors are on the abluminal side of the dilated blood vessels then this may partly reflect a permeability barrier for the highly polar **BIBN4096 BS**. A further possibility involves the hypothesis that the blood–brain barrier (BBB) is somewhat compromised during migraine attack. **BIBN4096 BS** on it own would not be expected to be significantly brain penetrant. However, a leaky BBB might allow a fraction of the compound in plasma into the CNS to act on central CGRP receptors. This is a controversial issue and it has been argued that it is not necessary to invoke a leaky BBB in this regard and that if even a small fraction of the 200 nM total plasma **BIBN4096 BS** entered the CNS, it might be sufficient to act on central CGRP receptors [78]. In May 2006, Boehringer-Ingelheim announced that it has discontinued development of **BIBN4096 BS** [79].

From a marketing point of view, an injectable CGRP antagonist may be unlikely to be more than a niche product. There is evidence in the patent literature that Boehringer has pursued noninjectable formulations of **BIBN4096 BS**, most prominently a powder inhalation form (see Ref. [80] and others).

6.3. MK-0974

Merck scientists developed the first oral CGRP antagonist, **MK-0974** (telcagepant), that has reached clinical trials. Their effort originated with the discovery of a benzodiazepine screening lead **15** (Fig. 6) having a K_i value of 4.3 µM in a high-throughput human CGRP receptor binding assay [81]. Previously, they had made use of a moderately potent D-dibromotyrosine analog, "Compound 1" (**16**), in a number of studies concerned mainly with elucidating aspects of CGRP receptor pharmacology [82].

Recognizing that there was good three-dimensional overlap between the amide donor–acceptor pair in the hydantoin contained in **15** with that of the GPCR-priviledged benzimidazolone in **16**, they prepared a series of piperidine-linked heterocycles (Table 6). Although the benzimidazolone (**17**) itself was inactive, ring expansion to the dihydroquinazolinone (**19**, $K_i = 2.2$ µM) resulted in activity comparable to the screening hit. The opposite enantiomer (**20**) was four to fivefold less active ($K_i = 8.4$ µM). Comparing the peptide based antagonists [A34, F35]CGRP(28–37)-OH ($K_i >$ 30,000 nM), [A34, F35]CGRP(28-37)-NH$_2$ ($K_i = 495$ nM), **BIBN4096 BS** and the SAR of benzodiazepine-based CGRP antagonists, they hypothesized that the dihydroquinazolinone substructure may be a conformationally constrained surrogate of the Ala-Phe-NH$_2$ dipeptide amide (Fig. 7).

Increasing the size of the *N*-1-substituent from methyl to ethyl or isopropyl resulted in modest increases in binding potency (**21** and **22**, $K_i = 897$ nM and 1060 nM, respectively). Note that covering the hydrogen-bond donating NH with methyl (**25**) caused a complete

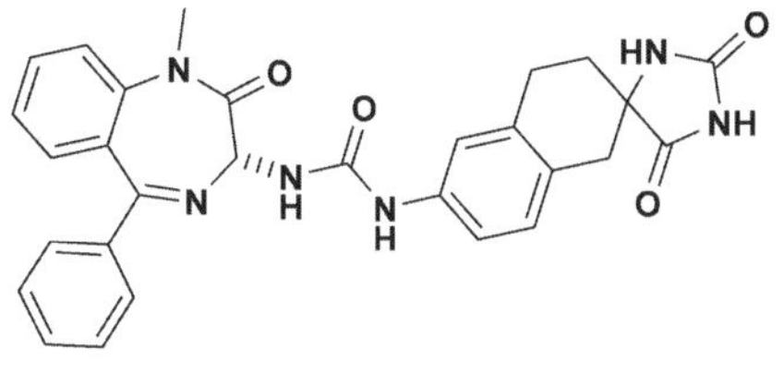

15
$K_i = 4.8$ µM

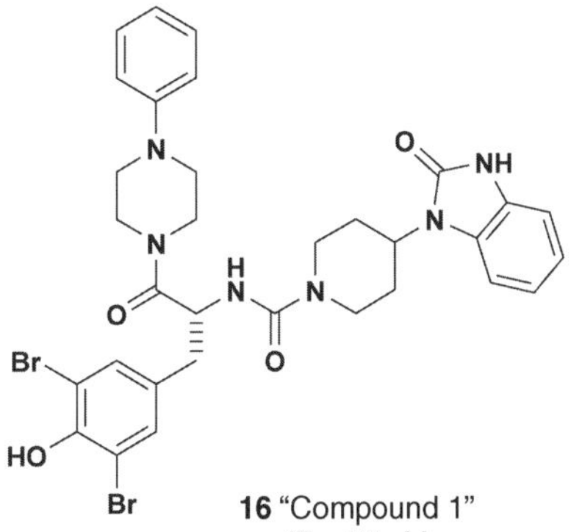

16 "Compound 1"
$K_i = 83$ µM

Figure 6. Merck benzodiazepine oral CGRP antagonist screening lead (**15**), and tool compound (**16**).

Table 6. Human CGRP Binding Affinities of Benzodiazepine CGRP Antagonists

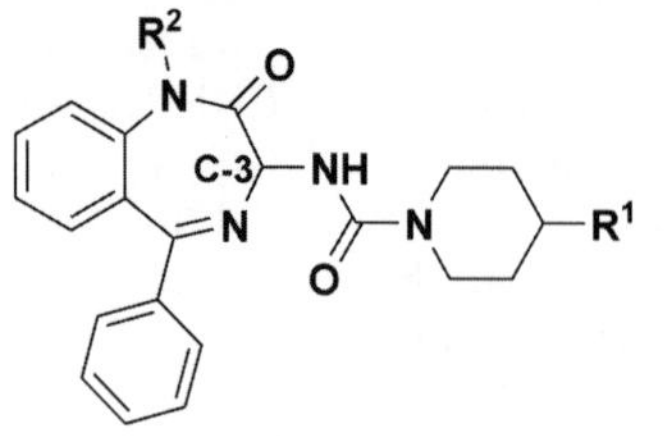

Compound	R^1	R^2	C-3 Stereochemistry	K_i (nM)
17	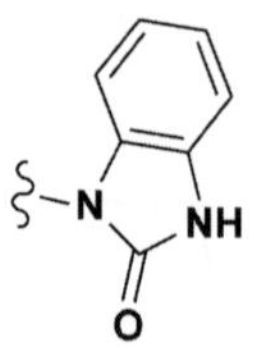 benzimidazolone-N-	CH_3	R,S	>13,000
18	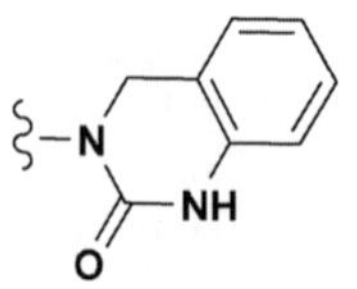 dihydroquinazolinone-N-	CH_2CF_3	R	2,370
19	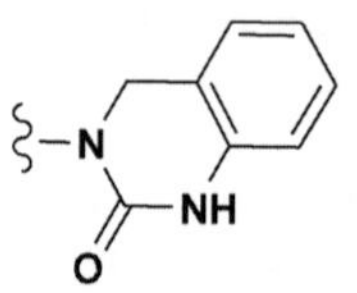 dihydroquinazolinone-N-	CH_3	R	2,200
20	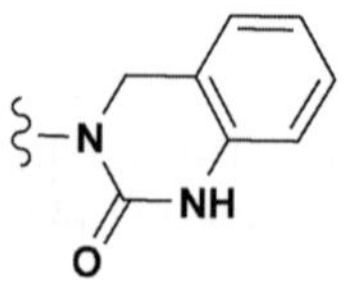 dihydroquinazolinone-N-	CH_3	S	8,400
21	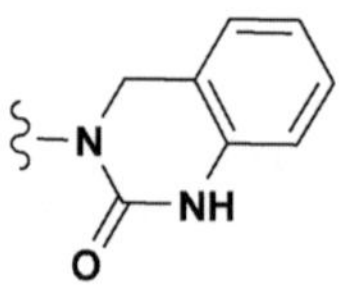 dihydroquinazolinone-N-	CH_2CH_3	R	897
22	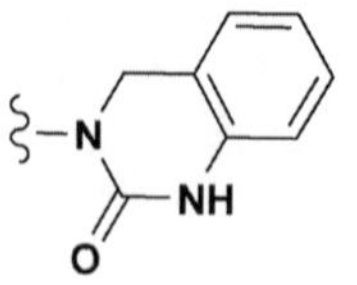 dihydroquinazolinone-N-	i-Pr	R,S	1,060
23	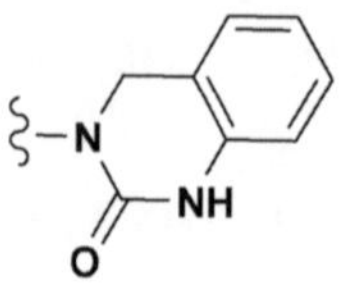 dihydroquinazolinone-N-	CH_2CF_3	R	44

Table 6. (Continued)

Compound	R¹	R²	C-3 Stereochemistry	K_i (nM)
24	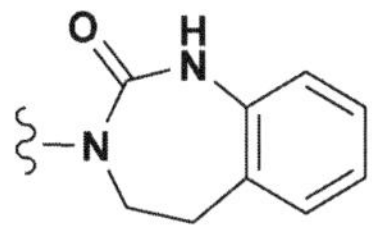 (benzazepinone-N)	CH_2CF_3	R	61
25	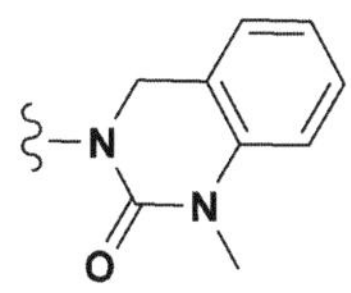 (N-methyl dihydroquinazolinone)	i-Pr	R,S	>100,000

loss of activity. A 50-fold improvement was seen with the trifluoromethyl group (23, $K_i = 44$ nM), with activity equivalent to "Compound 1." Functional antagonist activity ($IC_{50} = 38$ nM) was also comparable to the unconstrained 16 ($IC_{50} = 75$ nM). The dihydroquinazolinone ring was susceptible to air oxidation at the benzylic methylene while the further ring-expanded benzodiazepinone (24) was stable under ambient conditions and equipotent to 23 ($K_i = 61$ nM). Despite this, compound 23 was chosen for in vivo PK studies. Compound 23 demonstrated low oral exposure in rats ($F_{po} = 10\%$, 10 mg/kg) and was undetected in plasma when dosed at 2 mg/kg in beagle dogs.

Further optimization of the N-1 trifluoroethyl-benzodiazepinone core was carried out with the intention of improving oral exposure and stability [83]. Introduction of nitrogen into all four positions of the dihydroquinazolinone phenyl ring in an effort to increase electron-deficiency led to >sevenfold reductions in binding potency, suggesting a very limited tolerance for polarity in that portion of the binding pocket. In the substituted benzimidazolone series (Table 7) this point was emphasized along with apparently narrow requirements for relative spatial geometries, electronic polarizabilities, and/or substituent sizes. Notably, an electron-withdrawing substituent at the 4-position, either fluorine (26), chlorine (32), or 4-aza-substitution (37, 41, and 42), resulted in significantly improved binding affinities in comparison with 18.

A series of substituted piperidyl imidazolinones was also synthesized (Table 8). Aryl derivatives (i.e., 43–46) were potent but, as mentioned above, also subject to air oxidation under ambient conditions. Nonaryl derivatives 48–52 were more stable but showed significantly reduced binding potency. A set of imidazolidinones, in which the double bond that is the target of oxidation was reduced, yielded stable but less active compounds (not shown). On the other hand, compound 53, the phenyltriazolinone derivative of 43, lost little activity ($K_i = 26$ nM) and was air stable.

In fact, despite the omnipresence of aromatic rings in the GPCR-privileged structures used in these CGRP antagonists, it was found that for the most potent molecules they probably serve primarily as scaffolds for hydrogen-bond accepting functionality (as in compounds

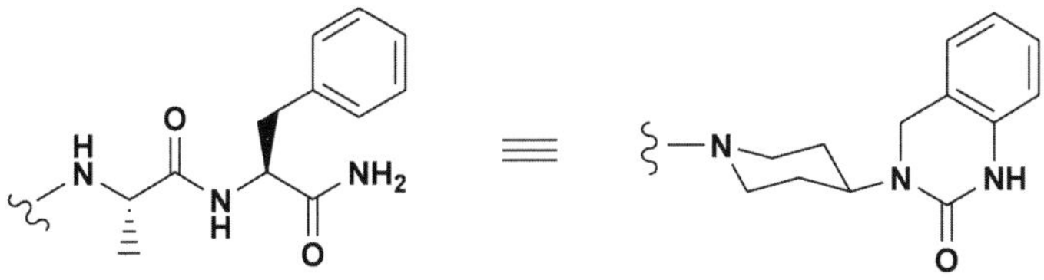

Figure 7. Dihydroquinazolinone as a structural mimetic of Ala-Phe.

Table 7. Human CGRP Binding Affinities of Benzodiazepine CGRP Antagonists

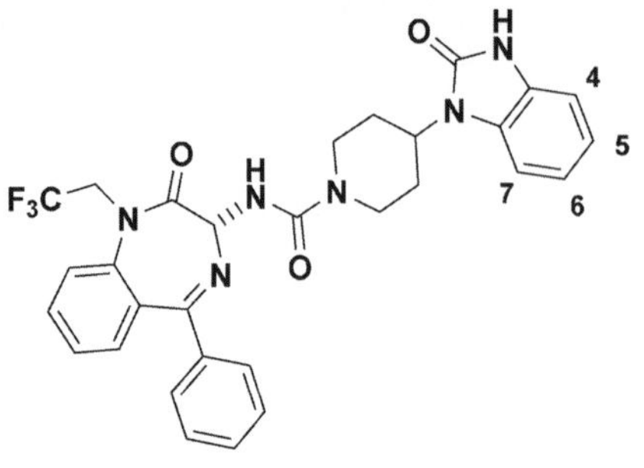

Compound	R	K_i (nM)
26	4-F	375
27	5-F	2,800
28	6-F	983
29	4-Me	6,900
30	5-Me	6,100
31	7-Me	20,000
32	4-Cl	143
33	5-Cl	>10,000
34	7-Cl	6,400
35	5-COOH	810
36	5-SO$_2$Me	220
37	4-Aza	23
38	5-Aza	685
39	6-Aza	853
40	7-Aza	427
41	4,6-Diaza	51
42	4,7-Diaza	156

Table 8. Human CGRP Binding Affinities of Benzodiazepine CGRP Antagonists

Compound	R	K_i (nM)
43	phenyl	11
44	3-methoxyphenyl	1.1
45	4-methoxyphenyl	205
46	2-pyridyl	27
47	benzyl	1125
48	–C(O)OCH$_3$	110
49	–C(O)OH	2900
50	–CH$_2$OH	20000
51	CN	575
52	H	583

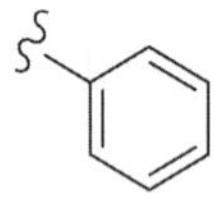

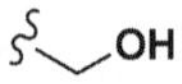

37, 44, and 46). This was shown most explicitly by compound 54 ($K_i = 29$ nM) (Fig. 8) which contains no fused or pendant aryl group but a potential hydrogen bond accepting carbonyl roughly in the position occupied by the pyridine nitrogen in 37 ($K_i = 23$ nM).

Although benzodiazepine derivatives displayed good activity in the CGRP binding assay, in general they displayed poor physicochemical properties and had low oral bioavailabilities. The Merck group explored removal of the two aryl rings in the core structure along with simplification to a substituted azepinone skeleton (Fig. 9) [84]. The related five- and six-membered lactam analogs were found to be considerably less active. Deletion of both phenyl groups resulted in a large loss of potency but the chemotype retained the stereochemical preference of the

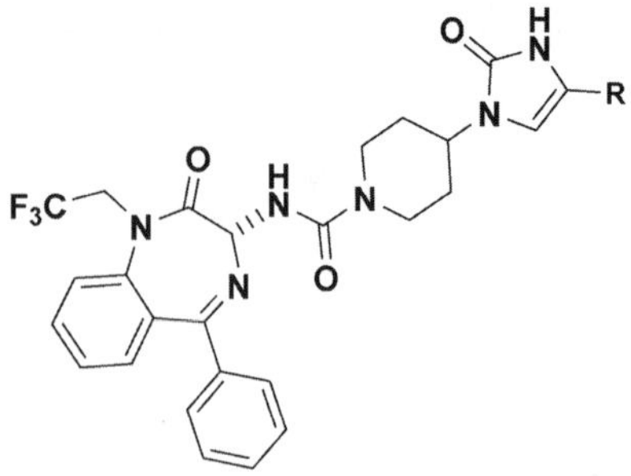

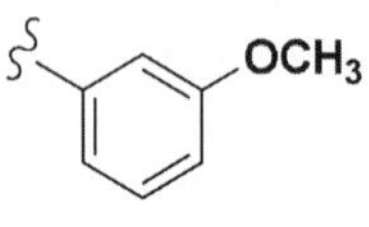

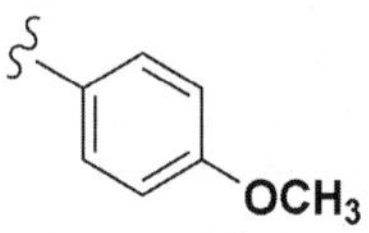

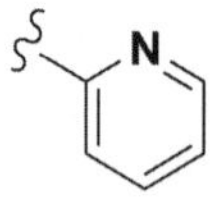

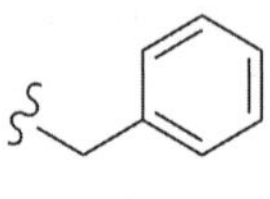

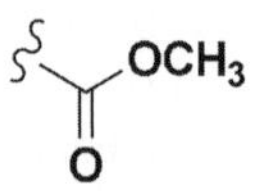

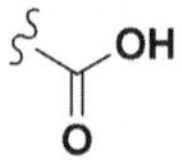

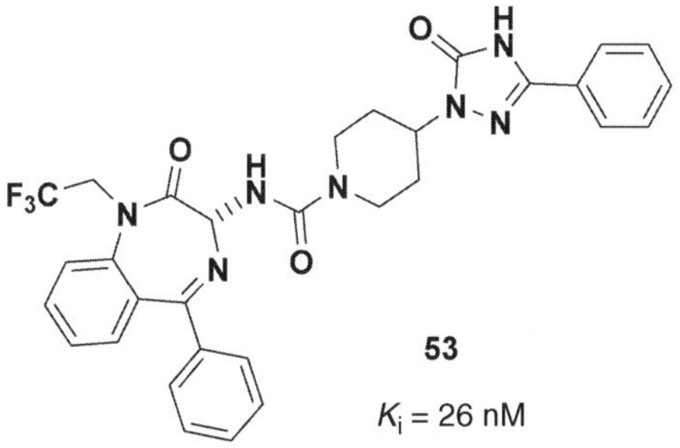

53
$K_i = 26$ nM

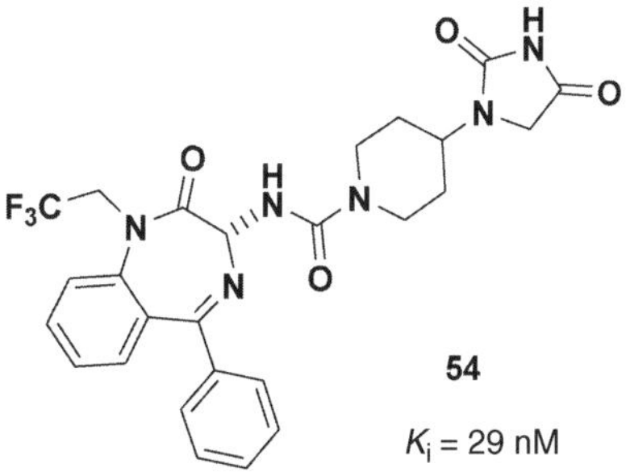

54
$K_i = 29$ nM

Figure 8. Benzodiazepinone CGRP antagonists containing air-stable GPCR-privileged structures.

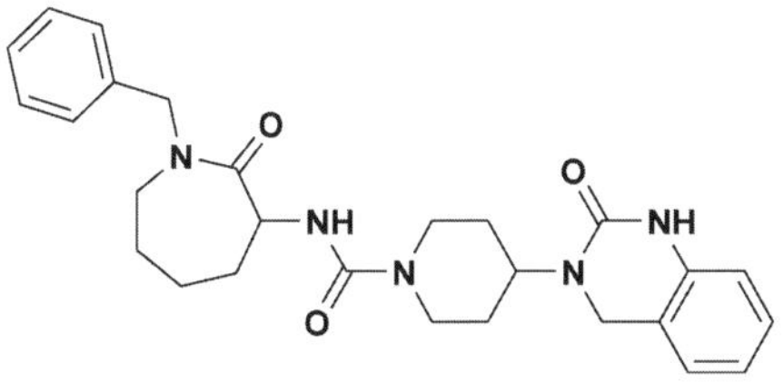

55 (R)-Enantiomer, $K_i = 1.8$ µM
56 (S)-Enantiomer, $K_i = 11$ µM

Figure 9. Simplified azepinone CGRP antagonist core.

benzodiazepinone (i.e., compound **55**, $K_i = 1.8$ µM, versus compound **56**, $K_i = 11$ µM).

Table 9 illustrates the effects of phenyl substitution and stereochemistry using a handful of GPCR-privileged structures with a further optimized N-cyclopropylmethyl azepinone. The effect of relative cis–trans stereochemistry appeared to be larger (in favor of trans) with phenyl at the 6-position (compound **58**, $K_i = 31$ nM, versus compound **57**, $K_i = 695$ nM) in comparison with 5-phenylazepinones (compound **60**, $K_i = 266$ nM, versus compound **59**, $K_i = 544$ nM). The trans (3R)-amino-(6S)-phenyl caprolactam scaffold consistently provided the most active compounds with compound **63** showing single-digit nanomolar binding ($K_i = 2$ nM) and functional antagonist ($IC_{50} = 4$ nM) activities.

Compound **63**, although highly potent, contained the oxidatively unstable phenylimidazolone GPCR-privileged structure. Perhaps, as a result, it demonstrated low oral bioavailability in dogs ($F_{po} = 6$%) and moderate oral exposure in rats ($F_{po} = 27$%). In addition, plasma free fractions for this compound as well as **61** and **62**, all containing relatively lipophilic GPCR-privileged structures, were apparently low, as demonstrated by >25-fold serum shifts in the cell-based CGRP functional assay. However, use of the more polar azabenzimidazolone component in compound **64** ($K_i = 11$ nM) resulted in a functional serum-adjusted CGRP activity of only ninefold, largely mitigating the loss of receptor potency. Importantly, the more chemically stable compound **64** showed good oral bioavailability in dogs ($F_{po} = 41$%).

Retaining the piperidyl-azabenzimidazolone along with 6-substitution on the azepinone core, the group set out to further optimize the series for oral exposure and serum-adjusted CGRP activity [85] (Table 10). In the absence of a nitrogen substituent (R_2) on the caprolactam, the group on C6 was varied. Although highly speculative, it is tempting to think that this group binds in a part of the pocket occupied by the D-dibromotyrosine residue in **BIBN4096 BS**, although the spatial relationship with the amide backbone is clearly different (three atoms versus one atom). SAR of the Boehringer series suggested the presence of a hydrogen-bond acceptor for the tyrosine hydroxyl in that otherwise lipophilic pocket. Clearly, the hydroxyl group of phenol **66** ($K_i = 770$ nM) does not interact with that acceptor. Pyridyl compounds **67** ($K_i = 4400$ nM) and **68** ($K_i = 9900$ nM) emphasize the hydrophobicity of the pocket; as do thiophene **69** ($K_i = 470$ nM), isopropyl **70** ($K_i = 470$ nM), and benzyl **71** ($K_i = 610$ nM) that show increased affinity in comparison with compounds containing polar groups. Significantly improved binding affinity was seen with 2-fluorophenyl- (**72**, $K_i = 22$ nM) and, to a lesser extent, with 3-fluorophenyl substitu-

Table 9. Human CGRP Receptor Binding Affinities and Serum Shift Values of Azepinone CGRP Antagonists

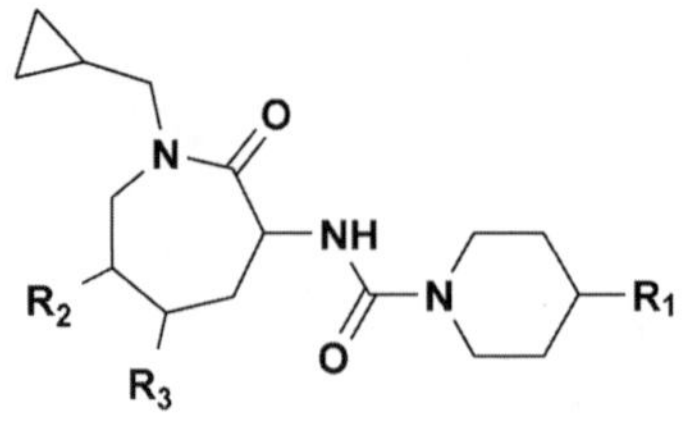

Compound	R$_1$	R$_2$	R$_3$	Stereochemistry	K$_i$ (nM)	Serum Shift[a]
57	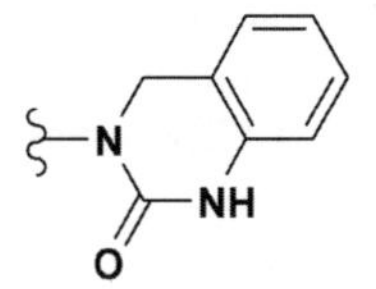 (dihydroisoquinolinone-N)	Ph	H	cis-racemic	695	
58	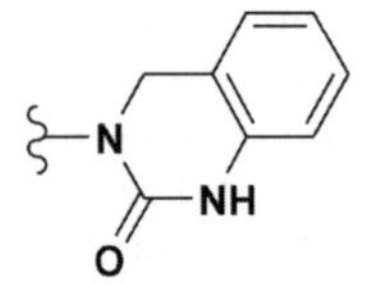 (dihydroisoquinolinone-N)	Ph	H	trans-racemic	31	
59	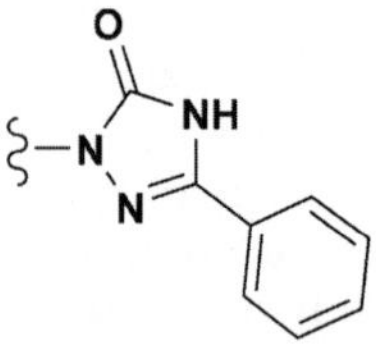 (phenyl-triazolone-N)	H	Ph	cis-racemic	544	
60	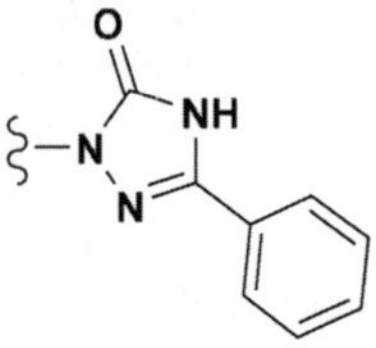 (phenyl-triazolone-N)	H	Ph	trans-racemic	266	
61	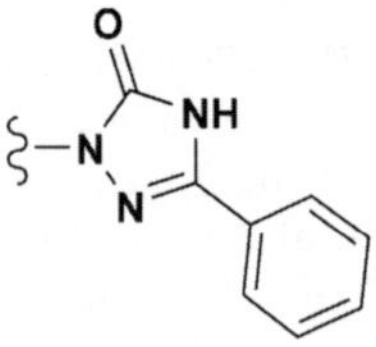 (phenyl-triazolone-N)	Ph	H	trans-(3R,6S)	29	>42
62	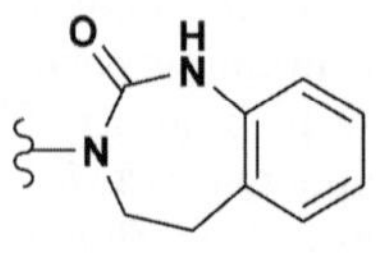 (benzodiazepinone-N)	Ph	H	trans-(3R,6S)	29	39

Table 9. (Continued)

Compound	R₁	R₂	R₃	Stereochemistry	K_i (nM)	Serum Shift[a]
63	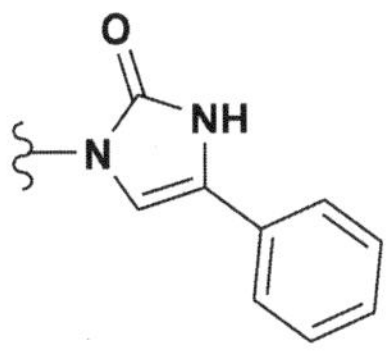 (imidazolinone with phenyl substituent)	Ph	H	trans-(3R,6S)	2	28
64	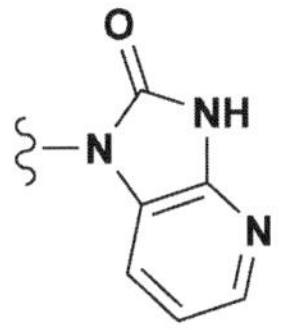 (pyrido-imidazolinone)	Ph	H	trans-(3R,6S)	11	9

[a] Fold shift upon addition of 50% human serum.

tion (**73**, $K_i = 51$ nM). However, there was a greater than additive effect with 2,3-difluorophenyl substitution (**74**, $K_i = 3.6$ nM), which is >20-fold more potent than unsubstituted phenyl **65** ($K_i = 83$ nM). This compound displayed moderate oral bioavailability in dogs ($F_{po} = 30\%$) but poor exposure in rats ($F_{po} = 5\%$). Again, it is interesting to think that a portion of the electron-deficient 2,3-difluorophenyl ring is acting as at least a weak hydrogen-bond donating group.

Further enhancement of binding potency and oral exposure were obtained by modifying the R₂-substituent. Ethyl (**76**, $K_i = 2.4$ nM) and methyl (**75**, $K_i = 2.7$ nM) gave comparable receptor activities but modest boosts in exposures. Additions of an oxygen atom in the form of methoxyethyl (**78**, $K_i = 0.3$ nM) or trifluoromethoxyethyl (**80**, $K_i = 0.19$ nM) resulted in subnanomolar activities. However, oral bioavailabilities were low, possibly due to O-dealkylation. If so, the hydroxyethyl derivative, **79** ($K_i = 4.2$ nM), was much less active, suggesting that it might be a rather poor active metabolite. The presence of a basic amino group (**81**, $K_i = 4.9$ nM) did not improve activity over the unsubstituted analog, **74**, but when basicity was modulated by incorporation into a morpholine (**82**, $K_i = 0.5$ nM), better potency was obtained. Morpholine **82** also demonstrated very low oral exposure, an example of the difficulty of introducing polar atoms into a chemotype that already possesses two ureas and an amide as hydrogen-bonding groups. Fluorine substitution at R₂ was much more successful in this respect, with the trifluoroethyl analog (**85**, **MK-0974**, $K_i = 0.77$ nM) showing the best balance of CGRP receptor activity, plasma free fraction, and oral exposures in this series.

A select group of compounds were tested in a pharmacodynamic model of neuropeptide-induced blood flow increase [86] for further differentiation. The in vivo assay involves topical application of capsaicin to the forearm of rhesus monkeys resulting in release of CGRP and increased local blood flow that can be measured by laser Doppler imaging. Upon i.v. dosing, compounds **75**, **76**, and **85** demonstrated significant activity in the model (Table 11). However, **MK-0974** (**85**) was clearly superior, in line with its improved serum-adjusted CGRP activity. **MK-0974** also showed low clearance in an independent PK study ($C_L = 7.0$ mL/min/kg). Against a panel of 160 receptors, transporters and enzymes, **MK-0974** was highly selective (>10,000-fold). The compound appeared to be a competitive antagonist at the human receptor with very rapid association and dissociation rates, especially in comparison with **BIBN4096 BS**. In addition, although **MK-0974** showed less evidence of noncompetitiveness than the Boehringer compound, peptide agonists did not

Table 10. Human CGRP Receptor Binding Affinities, Functional Activities, Serum Shift Values, and Oral Bioavailabilities of Azepinone CGRP Antagonists

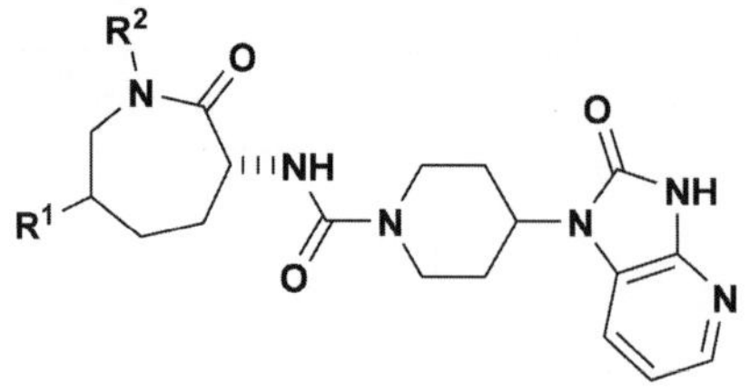

Compound	R_1	Stereochemistry	R_2	K_i (nM)	cAMP IC_{50} (nM)	Shift (Fold)	Rat F_{po} (%)	Dog F_{po} (%)
65	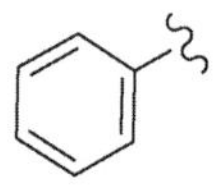 phenylmethyl	S	H	83	520	1.3		
66	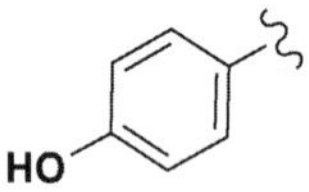 4-hydroxyphenylmethyl	R,S	H	770	1000			
67	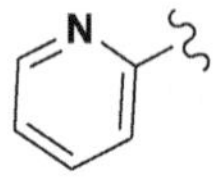 2-pyridylmethyl	R,S	H	4400				
68	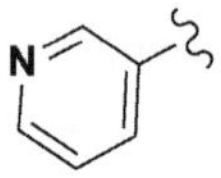 3-pyridylmethyl	R,S	H	9900				
69	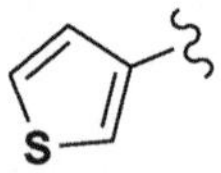 3-thienylmethyl	R,S	H	470	750	2.4		
70	 isopropyl	R,S	H	470	380	1.6		
71	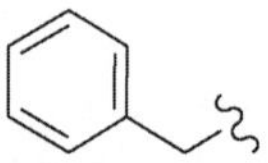 phenylethyl	R,S	H	610	920			
72	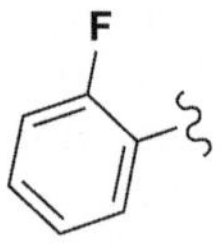 2-fluorophenylmethyl	S	H	22	65	1.8		
73	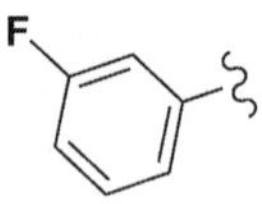 3-fluorophenylmethyl	S	H	51	220	1.1		
74	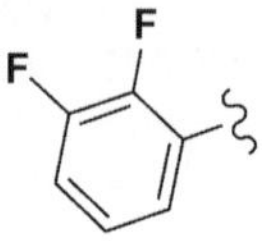 2,3-difluorophenylmethyl	S	H	3.6	14	1.6	5	30

Table 10. (Continued)

Compound	R₁	Stereochemistry	R₂	K_i (nM)	cAMP IC$_{50}$ (nM)	Shift (Fold)	Rat F$_{po}$ (%)	Dog F$_{po}$ (%)
75	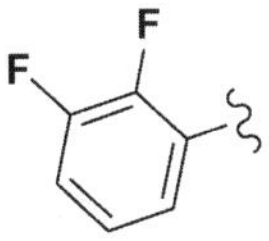 2,3-diF-phenyl	S	Me	2.7	8	2.8	12	60
76	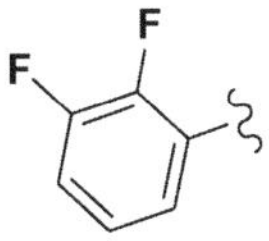 2,3-diF-phenyl	S	Et	2.4	6	5	30	61
77	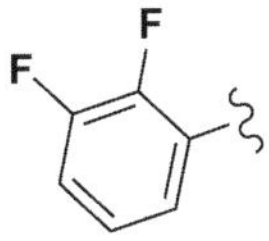 2,3-diF-phenyl	S	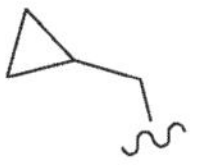 CH₂-cyclopropyl	1.4	2	11	8	17
78	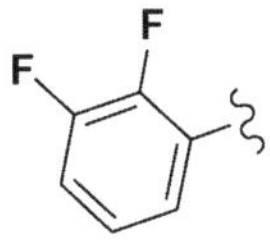 2,3-diF-phenyl	S	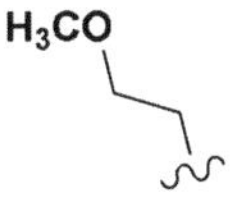 H₃CO(CH₂)₂–	0.3	2	1.5	6	6
79	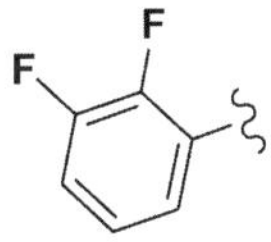 2,3-diF-phenyl	S	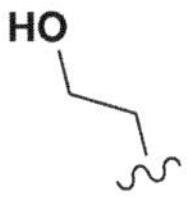 HO(CH₂)₂–	4.2	15	1.5	<1	
80	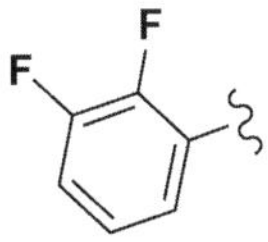 2,3-diF-phenyl	S	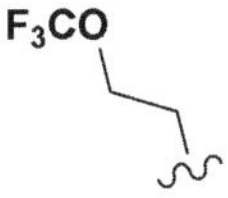 F₃CO(CH₂)₂–	0.19	1	6	3	7
81	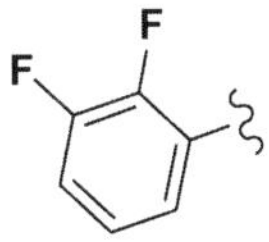 2,3-diF-phenyl	S	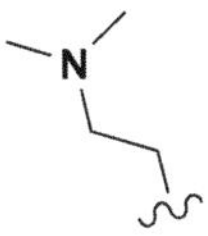 Me₂N(CH₂)₂–	4.9	16	1	2	
82	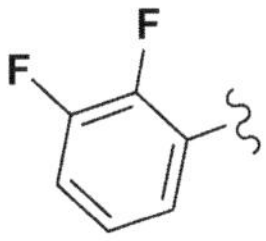 2,3-diF-phenyl	S	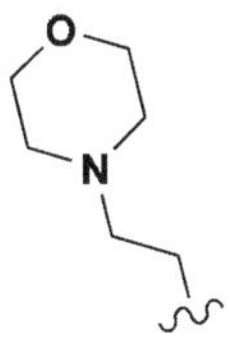 morpholinoethyl	0.5	4	1	2	1
83	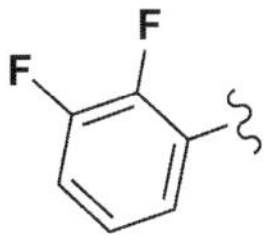 2,3-diF-phenyl	S	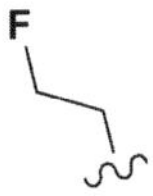 F(CH₂)₃–	1.4	5	2	17	14

Table 10. (Continued)

Compound	R₁	Stereochemistry	R₂	K_i (nM)	cAMP IC_{50} (nM)	Shift (Fold)	Rat F_{po} (%)	Dog F_{po} (%)
84	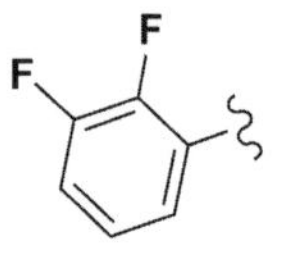 (difluorophenyl)	S	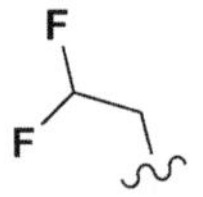 (difluoroethyl)	0.9	5	3	25	11
85 MK-0974	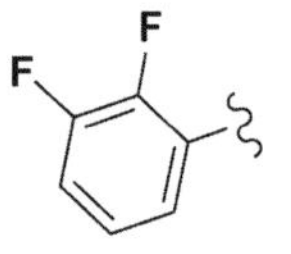 (difluorophenyl)	S	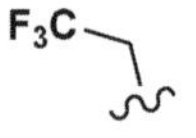 F_3C–	0.77	2.2	5	20	35

fully displace it in competition binding studies, suggesting the possibility of some allosteric behavior at the receptor [87].

The safety and efficacy of **MK-0974** in relieving migraine pain were assessed in a randomized, double-blind, placebo-, and active-controlled dose ranging Phase II trial in two stages [88]. In stage 1, either placebo, rizatriptan (10 mg) or **MK-0974** at oral doses of 25, 50, 100, 200, 300, 400, and 600 mg was used to treat moderate-to-severe migraineurs, with or without aura. The four lowest doses of **MK-0974** (25, 50, 100, and 200 mg) were discontinued due to insufficient efficacy. For the remaining treatment groups (300, 400, and 600 mg) in stage 2, 68% of patients taking 300 mg of **MK-0974** reported pain relief at the end of 2 h ($n = 38$), 48% taking 400 mg of **MK-0974** ($n = 45$), 68% taking 600 mg of **MK-0974**, 70% with rizatriptan and 46% of patients reported pain relief on placebo. A similar pattern of sustained pain relief was seen at 24 h. **MK-0974** was generally well tolerated. Adverse events included nausea; dizziness and somnolence. Occurrences were similar to the placebo group and did not increase with dose.

In a larger Phase III trial, **MK-0974** was tested at 150 mg ($n = 333$) and 300 mg ($n = 354$) doses (liquid-filled gelcaps) in comparison with zolmitriptan (5 mg, $n = 345$) and placebo ($n = 348$) [89]. The primary endpoints were relief of pain, and freedom from pain, nausea, photophobia, and phonophobia at 2 h. A secondary endpoint was 24 h pain freedom without resort to further medication. A majority of enrolled patients were women (85%) with a median age of 43. Efficacy for patients taking the 300 mg dose of **MK-0974** was comparable to zolmitriptan and superior to placebo for all primary endpoints (Table 12). The 150 mg dose of **MK-0974** was somewhat less effective than both the 300 mg dose and the zolmitriptan. Secondary 24 h efficacy endpoints also demonstrated superiority of 300 mg **MK-0974** over placebo (again, comparable to zolmitriptan). Perhaps importantly, an exploratory 48 h pain-free endpoint showed superiority of **MK-0974** over both zolmitriptan and placebo. Adverse events in this trial were similar to those seen in Phase II with no serious events reported for either agent.

Table 11. Activity of Selected CGRP Antagonists in a Pharmacodynamic model of Neuropeptide-Induced Blood Flow Increase

Compound	Serum-Shifted cAMP CGRP Functional Activity (nM)	Rhesus PD EC_{50} (nM)	Rhesus C_L (mL/min/kg)	Rhesus $t_{1/2}$ (h)
75	22	4000	7.1	1.9
76	30	3600	8.7	2.3
85	11	1000	7.0	2.8

Table 12. Phase III Efficacy and Adverse Event Reports for MK-0974 (300 mg) Versus Active Comparator, Zolmitriptan, and Placebo

Endpoint at 2 h	MK-0974 (300 mg) (%)	Zolmitriptan (5 mg) (%)	Placebo (%)
Pain relief	55	56	28
Pain freedom	27	31	10
Absence of nausea	65	71	55
Absence of photophobia	51	50	29
Absence of phonophobia	58	55	37
Overall adverse events	37	51	32

At the time of writing, several other Phase III trials with **MK-0974** are ongoing including a study in combination with sumatriptan (two 300 mg capsules [90]), and one in which the compound is coadministered with ibuprofen or acetaminophen (280 mg tablets [91]). In addition, the compound was being evaluated in a Phase II study for migraine prophylaxis at two doses (140 and 280 mg, twice daily for 3 months) [92], suggesting that chronic CGRP receptor inhibition had been deemed safe in preclinical studies. However, on April 21 2009, Merck announced the termination of this trial because of elevated (more than three times normal) liver transaminases in a small subset of drug-treated patients.

Several other chemotypes were explored as further alternatives to benzodiazepines to address PK-related issues. A simple achiral pyridinone core (Fig. 10) was investigated as a structural mimic to capture the spatial orientation of the substituents [93]. Other features of the chemotype included ease of synthesis and resistance to oxidative metabolism. SAR studies of the pyridinone chemotype showed good binding affinity in the human CGRP binding and functional cAMP assays. In general, compounds with lipophilic aromatic groups such as **86** ($K_i = 17$ nM, cAMP IC$_{50}$ 18 nM) displayed good bioavailability in both rat ($F_{po} = 34\%$) and dog ($F_{po} = 65\%$) with long half-lives and low clearance values.

A further series of spirohydantoin-containing benzimidazolones were elaborated into the indanylspiroazaoxindole **87** ($K_i = 0.23$ nM) (Fig. 11). This highly potent compound showed poor rat oral bioavailability (F_{po} 1%), but good exposure in dog ($F_{po} = 58\%$) and rhesus monkey ($F_{po} = 35\%$) [94].

In a recent ACS National meeting [95], Merck scientists reported preclinical data for CGRP antagonist **MK-2918** (Fig. 12), which is a potent follow-up compound to **MK-0974**. **MK-2918** exhibited an IC$_{50}$ value of 0.2 nM against cAMP and displayed a K_i value of 0.04 nM against the human CGRP receptor. Pharmacokinetics studies following oral dosing with **MK-2918** in rats, dogs, and monkeys revealed bioavailability levels of 16, 10, and 5%, respectively. Interestingly, an active metabolite of **MK-2918** resulting from O-demethylation of the ether was generated in substantial levels in these studies.

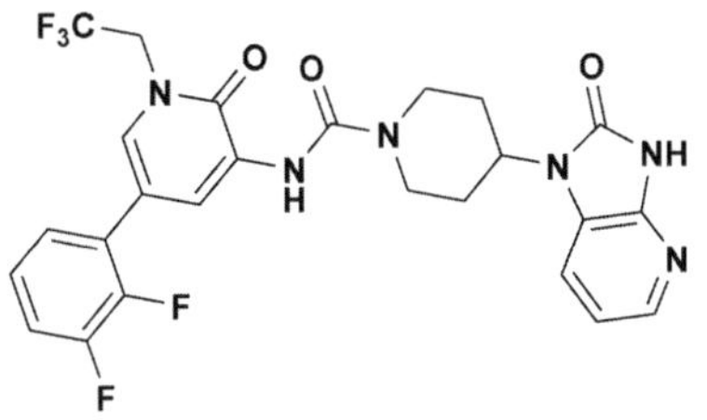

86
$K_i = 17$ nm

Figure 10. Lead pyridone-based CGRP antagonist.

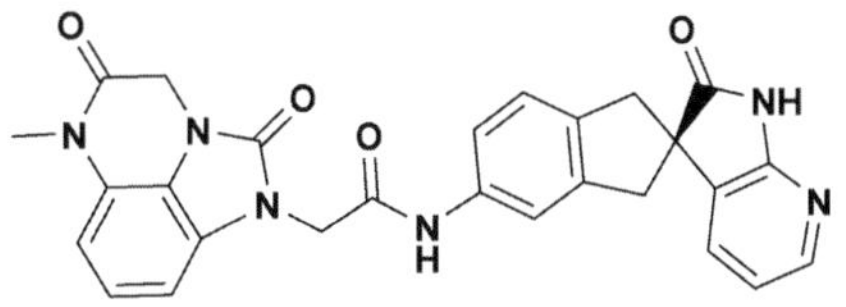

87
$K_i = 0.23$ nm

Figure 11. Indanylspiroazaoxindole CGRP antagonist.

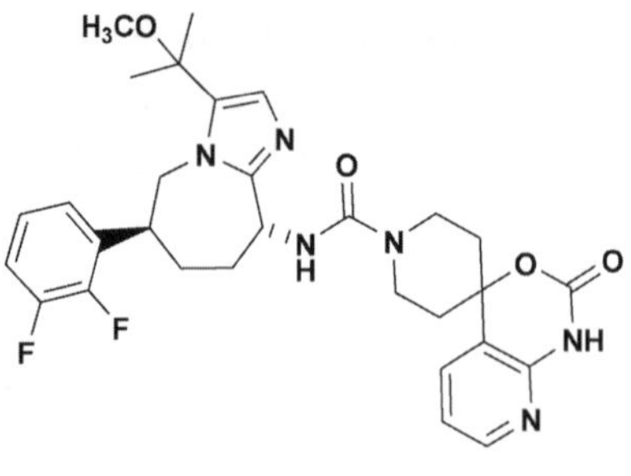

Figure 12. Merck oral CGRP antagonist backup compound, **MK-2918**.

Finally, in a 2008 business presentation, Merck announced a second CGRP antagonist backup, **MK-3207** (Fig. 13), being tested in Phase IIb clinical studies for acute migraine at five doses (2.5–100 mg) [96]. In 2008, the compound had completed a Phase I study in which it was dosed between and during migraine attacks to study the effects of gastric stasis, a condition that might delay efficacy of oral antimigraine agents in some patients (results not yet disclosed) [97]. The chemical structure and some preclinical data for **MK-3207** were presented at a 2009 ACS National Meeting [98]. The compound shows excellent CGRP receptor binding ($K_i = 0.020$ nM), low plasma protein binding (human serum-adjusted $IC_{50} = 0.17$ nM), and very favorable activity in their rhesus efficacy model ($ED_{50} < 20$ nM). In addition, by increasing $c \log p$, but also adding a polar amino group, the Merck scientists was able to overcome extensive first-pass intestinal metabolism in rhesus monkeys encountered with **MK-0974** [99], resulting in a compound possessing very good oral exposure (rhesus $F_{po} = 93\%$, rat $F_{po} = 74\%$, dog $F_{po} = 67\%$).

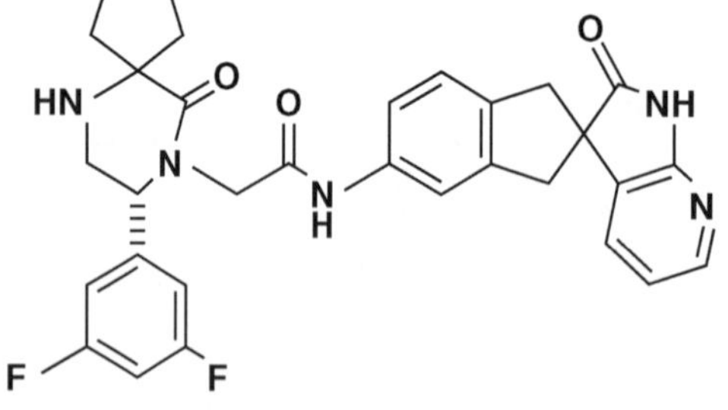

6.4. BMS-694153

Traditionally formulated oral medications generally require some time (often 1–2 h) before they reach effective concentrations in plasma or target tissues due to passage through the gut. For migraine, it would be greatly advantageous to have a convenient, faster acting formulation. This would not only shorten the duration of headache pain but also by short-circuiting the attack early, might also reduce its severity [100]. In addition, migraine-associated nausea can limit the effectiveness of oral delivery [101]. Injectable agents are, by their very nature, fast acting but are not convenient for individual patient use. Intranasal delivery would seem ideal for migraine, given an agent with appropriate target potency and ADME properties. Two triptans, sumatriptan and zolmitriptan, are available in intranasal formulations. Sumatriptan nasal spray has been reported to have inconsistent results and an unpleasant taste, which limits its use [102]. In contrast, zolmitriptan nasal spray has shown a rapid onset of action in comparison with oral triptans [103], high response rates, and an improved tolerability profile, although "unusual taste" remains the most prominent adverse event [104].

Researchers at Bristol-Myers Squibb developed the first intranasal CGRP antagonist, **BMS-694153** (Table 13) [105]. Compound requirements for intranasal delivery are somewhat more stringent than for oral. The molecule must be very potent at the target receptor and demonstrate low plasma protein binding because of the need to deliver a low dose to the nasal cavity [106]. Also, the compound must have very good aqueous solubility, as the maximum volume of liquid that can be delivered in each human nostril is only about 0.1 mL [107], but sufficient permeability at the nasal membrane.

Benzothiophene **88** (Table 13) was prepared in an attempt to balance polarity and lipophilicity. Although the compound demon-

Figure 13. Merck oral CGRP antagonist backup compound, **MK-3207**.

Table 13. Human CGRP Receptor Binding Affinities, Recombinant CYP3A4 Inhibitory Activities, and Aqueous Solubility Values for Unnatural Amino Acid-Based CGRP Antagonists

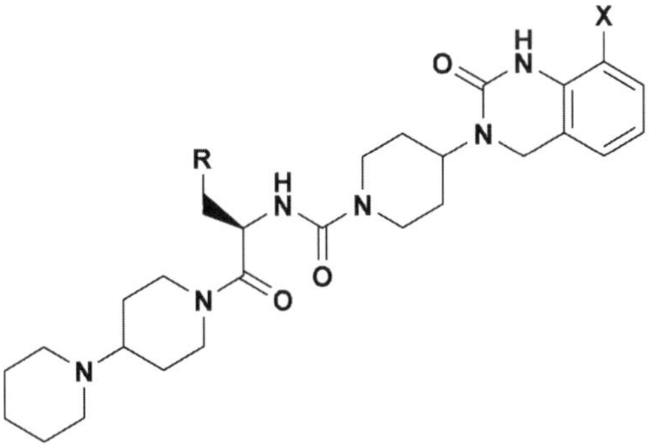

Compound	R	X	K_i (nM)	CYP3A4 BFC, IC_{50} (µM)	CYP3A4 BZR, IC_{50} (µM)	Aqueous Solubility (mg/mL)
88	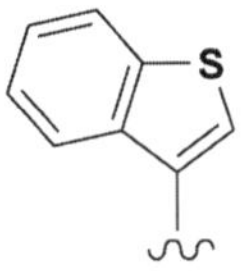 benzothiophene	H	0.55	0.87	0.084	ND[a]
89	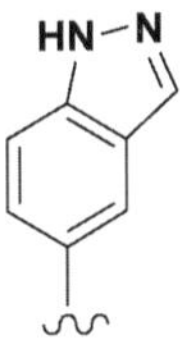 5-indazolyl	H	0.26	>40	4.0	ND
90	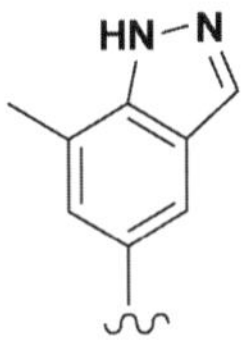 7-methyl-5-indazolyl	H	0.010	36	6.0	15
91 BMS-694153	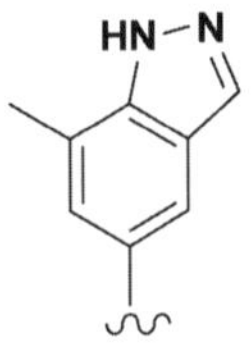 7-methyl-5-indazolyl	F	0.013	26	7.8	>500

[a] ND = Not determined.

strated subnanomolar CGRP binding affinity ($K_i = 0.55$ nM), it presented an obvious liability with extremely potent CYP3A4 inhibition of 7-benzyloxyresorufin (BZR) ($IC_{50} = 84$ nM). Subsequent SAR studies identified the benzothiophene as the residue primarily responsible for this strong interaction. A number of analogs containing various D-amino acid side chains were prepared and the indazole substituted at the 5-position (compound 89) was found to confer very good CGRP binding potency ($K_i = 0.26$ nM) with much improved

CYP3A4 inhibition ($IC_{50} = 4.0\,\mu M$). Further work led to 7-methyl substitution on the indazole (compound **90**), which was >20-fold more potent at the CGRP receptor ($K_i = 0.010\,nM$) with no increase in CYP3A4 inhibitory activity ($IC_{50} = 6.0\,\mu M$, $IC_{50} > 30\,\mu M$ for other representative CYP isoforms). However, the aqueous solubility of compound **90** (15 mg/mL at pH 5.0 from amorphous solid) was not sufficient to support intransasal dosing. Interestingly, simple addition of a fluorine atom at C8 of the quinazolinone (compound **91**, **BMS-694153**) had the dramatic effect of increasing aqueous solubility to >500 mg/mL from crystalline free base (pH 1–6.8) while preserving CGRP binding potency ($K_i = 0.013\,nM$) and preventing stronger CYP3A4 inhibition ($IC_{50} = 6.7\,\mu M$). The compound was shown to be a full antagonist at the human CGRP receptor in SK-N-MC cells (cAMP $EC_{50} = 0.034\,nM$) but, like other small-molecule CGRP antagonists, it showed no activity against the rat receptor.

In two experiments using *ex vivo* preparations of human intracranial arteries, **BMS-694153** both reversed ($EC_{50} = 0.06\,nM$) and prevented ($K_b = 0.028\,nM$) CGRP-induced blood vessel dilation. In contrast, **BMS-694153** did not cause active vasoconstriction of *ex vivo* human coronary artery at concentrations up to $10\,\mu M$, while sumatriptan produced concentration-dependent vessel contraction up to 59% of maximum (induced by $10\,\mu M$ serotonin) with $EC_{50} = 311\,nM$. As a surrogate for the cardiovascular side effects seen with triptans, these results suggest that **BMS-694153** and other CGRP antagonists may be free of the mechanism-based CV liabilities displayed by the triptans.

As a measure of *in vivo* efficacy, the BMS group employed a marmoset facial blood flow model similar to that used by Boehringer-Ingelheim. However, instead of electrical stimulation-induced CGRP release, alpha-CGRP was given IV ($10\,\mu g/kg$) once before, and at three time points following subcutaneous dosing of **BMS-694153**. This allowed assessment of individual baseline response to CGRP as well as duration of *in vivo* inhibition by the antagonist. At 0.03 mg/kg, **BMS-694153** demonstrated robust inhibition (>60%) of CGRP-induced increases in facial blood flow at 15, 60, and 105 min. At 0.01 mg/kg, 50% inhibition was observed at 60 and 105 min, while modest activity (approximately 30%) was seen at 15 min. No inhibition was noted at 0.003 mg/kg. Comparing exposure with efficacy, total plasma levels of about 10 nM were associated with robust activity in this model.

Intranasal pharmacokinetics was assessed in the rabbit. When sprayed as an aqueous solution, **BMS-694153** was efficiently absorbed from the nasal cavity. The intranasal bioavailabilities at 1.0 and 0.3 mg/kg were 59% ($C_{max} = 1200\,nM$) and 55% ($C_{max} = 360\,nM$), respectively, with T_{max} occurring within 10 min for both doses, suggesting the potential for rapid onset of efficacy. When dosed i.v. (0.5 mg/kg), mean clearance was 13 mL/min/kg, volume of distribution (steady state) was 2.7 L/kg, and $t_{1/2}$ was 11 h. **BMS-694153** also showed good bioavailability when dosed intratracheally ($F_{IT} = 73\%$) and subcutaneously (43–100%, depending on vehicle) in rats. Oral bioavailability in cynomolgus monkey and rat was very low ($F_{po} < 0.3\%$), probably due primarily to poor bilayer permeability (Caco-2 $P_c < 15\,nm/s$). At $10\,\mu M$, the compound demonstrated very low binding to plasma proteins from several species including human (fu = 35–77%).

Along with the reduced risk of "triptan-like" cardiovascular liabilities predicted by the *ex vivo* blood vessel studies mentioned above, **BMS-694153** also showed a low likelihood for ion channel-mediated CV adverse events. At $30\,\mu M$, the compound produced about 19% inhibition of hERG channel activity, and showed minimal effect on sodium channel activity. In addition, **BMS-694153** showed no significant potential for other off-target liabilities in a broad panel of receptors, ion channels and enzymes.

7. GLUTAMATE MODULATORS

Glutamate (Glu) is the most important excitatory neurotransmitter in the CNS. Glu and, to a lesser extent, aspartate have been shown to be significantly elevated in migraine patients, especially in those with aura, and most prominently during attacks. These increased levels may underlie cortical hyperexcitability in migraine patients and suggests a defective cellular reuptake mechanism [108]. As mentioned above, Glu is involved in the propagation of CSD, and most likely plays a role in TGVS activation, the neuroinflammatory response and central sensitization [109]. A close connection has been noted between 5-HT$_{1B/1D}$ receptors (the target of triptans) and Glu receptors, and one mechanism of triptan therapy may be to block release of Glu from trigeminal afferents [110].

7.1. Glutamate Receptors

There are two major classes of Glu receptors, ionotropic, transmitter ligand-gated ion channels (iGluR), and metabotropic, G-protein coupled, with multiple subtypes of each. Ionotropic receptors are further classified by their binding preference for nonendogenous, subtype-selective agonists, N-methyl-D-aspartate (NMDA), alpha-amino-3-hydroxy-5-methyl-4-isoxazolepropionic acid (AMPA) and kainate (KA) (Fig. 14). They are composed of four homologous subunits, each with three transmembrane segments [111]. Like all ligand-gated ion channels, they contain a ligand binding domain with an associated ion channel. In addition, they have several additional modulatory regions. The KA receptor has been implicated in pain transmission in the CNS, and is found in the trigeminal ganglion, although preclinical data can be used to support functional antagonism of any of the ionotropic receptors to treat migraine [112].

7.2. Ionotropic Glutamate Receptor Antagonists

Scientists at Eli Lilly have pursued ionotropic glutamate receptor (iGluR) antagonists based on the decahydroisoquinoline-3-carboxylic acid skeleton as pain treatments for several years [113,114]. In the process, they have discovered nonselective AMPA/KA and then later selective iGluR5 KA antagonists and evaluated them as potential antimigraine agents.

7.2.1. LY293558
This highly constrained amino acid core structure was evolved from compounds such as the arylethylphosphonic acid (Fig. 15), both of which were originally prepared as competitive NMDA antagonists [115]. A great deal of SAR was generated in the investigation of stereochemistry, length and makeup of the connecting chain to the acidic functionality, as well as its identity.

There was little tolerance for stereochemical variation in the core structure of **92** (Fig. 16). Amino acid **93**, the C6 epimer of **92**, was about 12-fold less potent against the AMPA receptor and roughly equipotent at the KA receptor. In a cortical slice preparation, **93** was only a very weak antagonist. Compound **94**, the C3 epimer of **92**, was inactive in both binding and functional assays as was **95**, essentially the "unnatural" amino acid version of **92**, **96**, the tetrahydroquinoline analog of **92**. Compound **97**, in which the carboxyl group was transposed from C-3 to C-1, regained

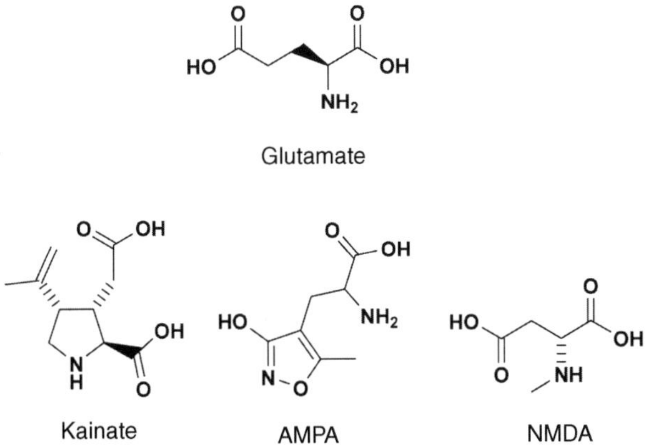

Figure 14. Glutamate and nonendogenous, subtype-selective GluR agonists, NMDA, AMPA, and KA.

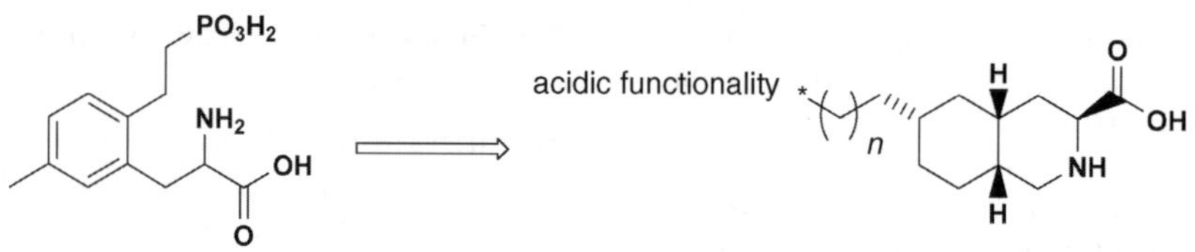

Figure 15. Arylethylphosphonic acid origin of iGluR antagonist core structure.

NMDA receptor binding activity (with only mild functional activity) but was inactive at AMPA and KA receptors.

Table 14 shows the effects of linker variations on binding and function. With no linker (compound **98**) or with one methylene group (compound **99**), very good selectivity for the NMDA receptor is observed, while a simple two-carbon spacer (compound **100**) significantly improves both AMPA binding and func-

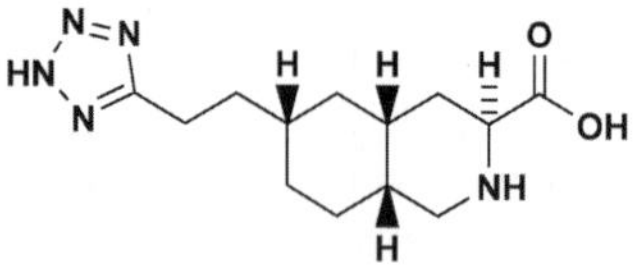

92

Binding:
NMDA IC$_{50}$ = 26.4 µM
AMPA IC$_{50}$ = 4.8 µM
KA IC$_{50}$ = 247 µM

Function:
NMDA IC$_{50}$ = 61.3 µM
AMPA IC$_{50}$ = 6.0 µM
KA IC$_{50}$ = 31.7 µM

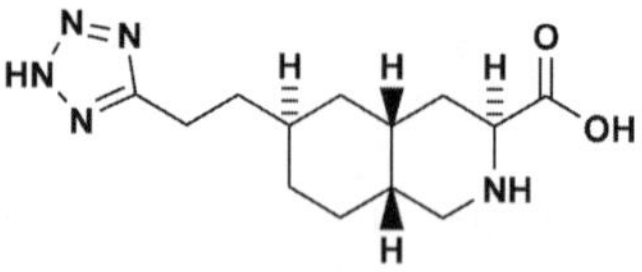

93

Binding:
NMDA IC$_{50}$ = 60.6 µM
AMPA IC$_{50}$ = 59.6 µM
KA IC$_{50}$ = 180 µM

Function:
NMDA IC$_{50}$ >100 µM
AMPA IC$_{50}$ <100 µM
KA IC$_{50}$ >100 µM

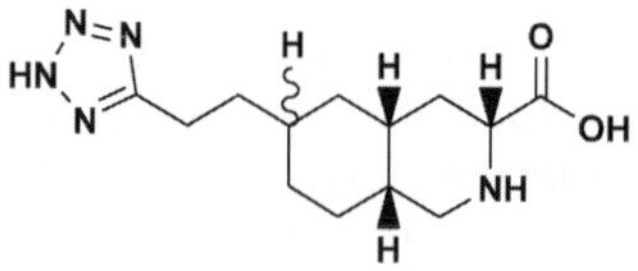

94

Binding:
NMDA IC$_{50}$ >10 µM
AMPA IC$_{50}$ >100 µM
KA IC$_{50}$ >100 µM

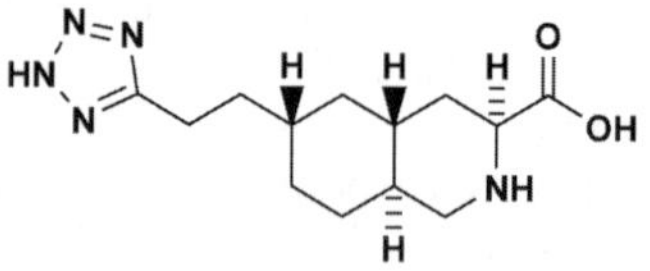

Function:
NMDA IC$_{50}$ >100 µM
AMPA IC$_{50}$ >100 µM
KA IC$_{50}$ >100 µM

95

Binding:
NMDA IC$_{50}$ >100 µM
AMPA IC$_{50}$ >100 µM
KA IC$_{50}$ >100 µM

Function:
NMDA IC$_{50}$ >100 µM
AMPA IC$_{50}$ >100 µM
KA IC$_{50}$ >100 µM

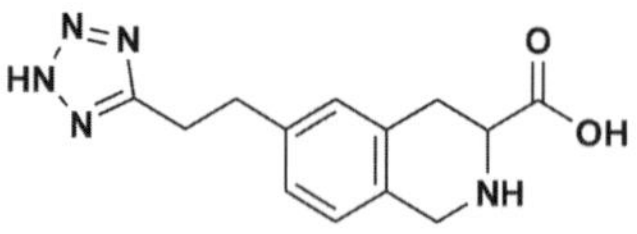

96

Binding:
NMDA IC$_{50}$ >10 µM
AMPA IC$_{50}$ >100 µM
KA IC$_{50}$ >10 µM

Function:
NMDA IC$_{50}$ >100 µM
AMPA IC$_{50}$ >100 µM
KA IC$_{50}$ >100 µM

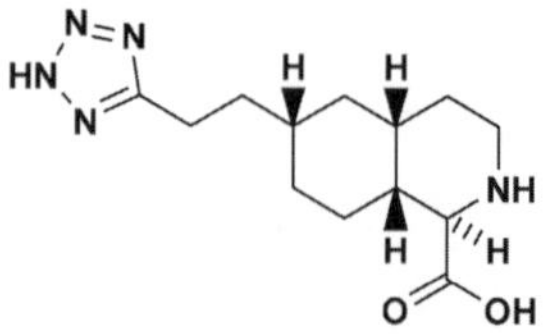

97

Binding:
NMDA IC$_{50}$ = 12.8 µM
AMPA IC$_{50}$ >100 µM
KA IC$_{50}$ >100 µM

Function:
NMDA IC$_{50}$ = 100 µM
AMPA IC$_{50}$ >100 µM
KA IC$_{50}$ >100 µM

Figure 16. iGluR binding and functional activities of decahydroisoquinoline-3-carboxylic acids.

Table 14. iGluR Binding and Functional Activities of Decahydroisoquinoline-3-Carboxylic Acids

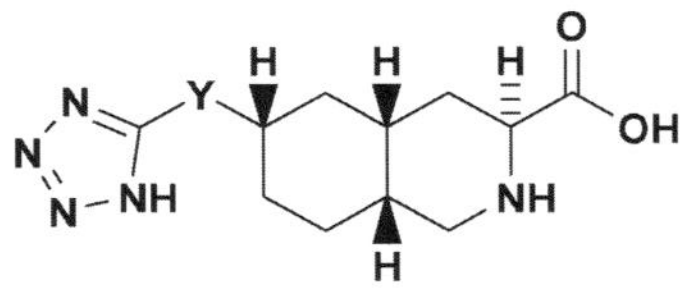

Compound	Linker Y	NMDA Binding IC$_{50}$ (μM)[a]	AMPA Binding IC$_{50}$ (μM)[b]	Kainic Acid Binding IC$_{50}$ (μM)[c]	NMDA Function IC$_{50}$ (μM)	AMPA Function IC$_{50}$ (μM)	Kainic Acid Function IC$_{50}$ (μM)
Effects of chain length							
98	None	1.6	12.8	31.8	7.5	40.9	>100
99	CH$_2$	0.94	84.5	>100	1.4	>100	>100
100	(CH$_2$)$_2$	26.4	4.8	247	61.3	6.0	31.7
101	(CH$_2$)$_3$	27.0	16.8	18.7	>100	22.0	>100
102	(CH$_2$)$_4$	>100	57.0	>100	>100	27.6	<100
Effects of heteroatom substitution adjacent to hydroisoquinoline core							
103	CH$_2$O	5.9	24.0	55	28.3	29.5	>100
104	CH$_2$NH	11.5	>100	>100	33.4	69.3	>100
105	CH$_2$NMe	65.8	>100	>100	>100	>100	>100
106	CH$_2$NCHO	>100	23.1	>100	NT[d]	NT	NT
Effects of substitution on the two-carbon chain							
107	CH$_2$CHMe	53.2	3.0	35.0	>100	6.0	58
108	CHMeCH$_2$	>100	4.21	27.3	NT	NT	NT
109	CH$_2$CHPh	15.1	9.8	>100	100	9.0	23.6

[a] Affinity at NMDA receptors determined using tritiated CGS19755 [116].
[b] Affinity at AMPA receptors determined using tritiated AMPA.
[c] Affinity at kainic acid receptors determined using tritiated kainic acid.
[d] NT = Not tested.

tion while also resulting in moderate KA functional antagonism. Homologation to the propylene- and butylene-spaced compounds (**101** and **102**) diminished activity.

Holding the two-atom spacer constant, the effect of heteroatom substitution adjacent to the bicyclic amino acid was investigated. Oxygen (compound **103**) reduced AMPA binding and functional activity by ca. fivefold and NH substitution (compound **104**) abolished activity. Similarly, the *N*-methyl analog, **105**, was inactive while the *N*-formyl derivative, **106**, had weak affinity for the AMPA receptor. Compounds with a methyl group on either of the carbons in the spacer (**107** and **108**) were comparable to compound **92** in both AMPA receptor affinity and antagonist potency (and moderate KA activity), while the phenyl-substituted analog, **109**, regained NMDA receptor affinity.

When the methylene group adjacent to the tetrazole of compound **92** and its C-6 epimer was replaced with sulfur the resulting derivatives (compounds **110** and **111**) (Fig. 17) showed somewhat improved AMPA activity. Interestingly, the single-carbon-linked thiotetrazole **112** displayed weak agonism in the cortical slice functional assay.

A series of analogs in which the tetrazole of **92** was replaced with other acidic functionalities were also prepared. These included carboxyl, phosphonate, sulfonate, methanesulfonamide, benzenesulfonamide, amidotetrazole, and 3-isoxazolone. Of these, only the

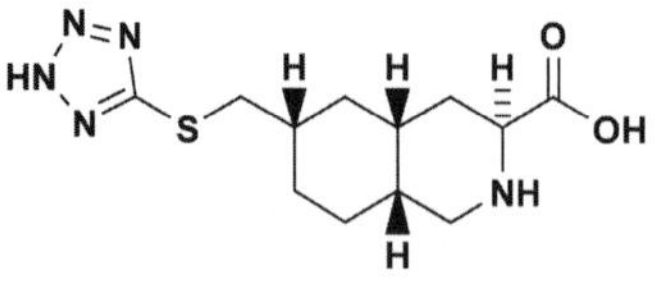

110

Binding:
NMDA IC$_{50}$ = 29.4 μM
AMPA IC$_{50}$ = 0.9 μM
KA IC50 = 30.1 μM

Function:
NMDA IC$_{50}$ >100 μM
AMPA IC$_{50}$ = 19.8 μM
KA IC$_{50}$ >31.6 μM

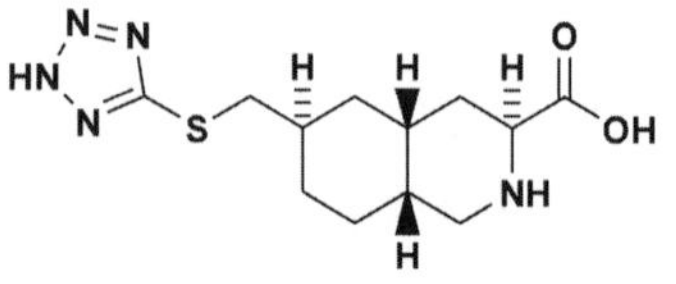

111

Binding:
NMDA IC$_{50}$ = 47.9 μM
AMPA IC$_{50}$ = 4.9 μM
KA IC$_{50}$ = 37.3 μM

Function:
NMDA IC$_{50}$ >100 μM
AMPA IC$_{50}$ = 6.1 μM
KA IC$_{50}$ >100 μM

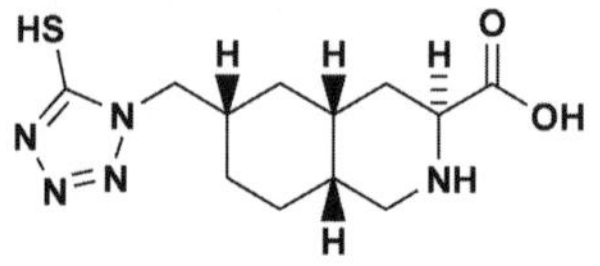

112
Weak agonist in cortical slice assay

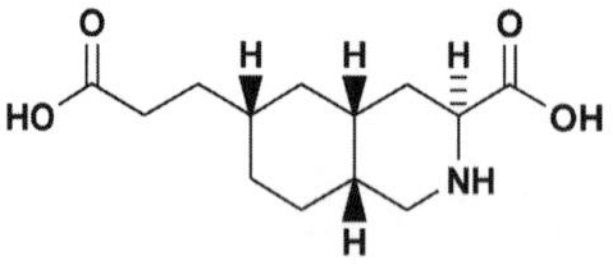

113

Binding:
NMDA IC$_{50}$ = 43.9 μM
AMPA IC$_{50}$ = 27.8 μM
KA IC$_{50}$ >100 μM

Function:
NMDA IC$_{50}$ = 57 μM
AMPA IC$_{50}$ = 42 μM
KA IC$_{50}$ >100 μM

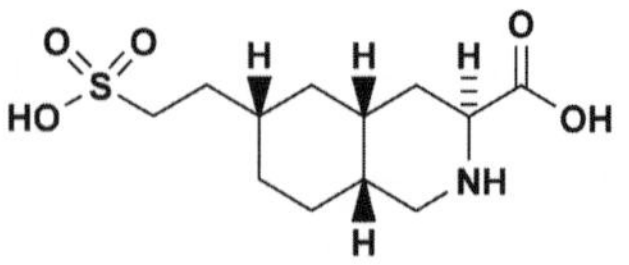

114

Binding:
NMDA IC$_{50}$ = 61.8 μM
AMPA IC$_{50}$ = 11.6 μM
KA IC$_{50}$ >100 μM

Function:
NMDA IC$_{50}$ >100 μM
AMPA IC$_{50}$ = 31.4 μM
KA IC$_{50}$ = 100 μM

Figure 17. iGluR binding and functional activities of decahydroisoquinoline-3-carboxylic acids.

carboxyl and sulfonate derivatives (**113** and **114**) (Fig. 17) showed significant activity.

Compound **92** (**LY293558,** Tezampanel) was chosen to move forward into two preclinical migraine models. When dosed i.v., **LY293558** demonstrated efficacy in two accepted models of acute migraine: a rat plasma protein extravasation (PPE) model ($ID_{100}=30$ ng·kg) and a nucleus caudalis c-fos expression model (maximal inhibition at 3 mg/kg). The former depends on inhibition of neurotransmitter release from the peripheral branches of trigeminal sensory neurons following activation of the trigeminal nerve. The c-fos model assesses inhibition of central neurotransmitter release. Importantly, for potential cardiovascular safety, **LY293558** did not show evidence of inherent vasoconstrictive activity in rabbit saphenous veins [117].

LY293558 was tested in a small placebo-controlled, triple-blinded migraine proof-of-concept trial [118]. The compound was dosed i.v. at 1.2 mg/kg, the maximal-tolerated dose based on an earlier study in healthy volunteers [119] and compared with 6 mg sumatriptan (subcutaneous). Headache improvement response rates 2 h after infusion, the primary outcome measure, were 69% for **LY293558**, 87% for sumatriptan and 25% for placebo. Pain-free rates at 2 h were 54% for **LY293558**, 60% for sumatriptan and 6% for placebo. No patients receiving **LY293558** reported headache recurrence (2–24 h), compared with one in the sumatriptan group and two in the placebo group. Adverse events were lowest (2/13) for the **LY293558** group (dizziness and sedation) compared with 8/15 in the sumatriptan group and 5/31 in the placebo group. There was no indication of unwanted cardiovascular effects with **LY293558** while 2/13 patients receiving sumatriptan reported chest symptoms. Mild and reversible visual distortion had been the most consistent side effect of intravenous **LY293558** reported in healthy volunteers. However in this study it was much less prominent (8%) than among those taking sumatriptan (27%). The sustained pain relief (69%) and pain-free (60%) responses and especially the lack of headache recurrence over 24 h were especially impressive given the compound's short plasma half-life (about 1 h). Besides the possibility of especially favorable receptor kinetics, it was speculated that **LY293558** may suppress release of proinflammatory neuropeptides that prolong migraine attack. **LY293558** (Tezampanel) has been licensed to Torrey Pines who have been testing it as a subcutaneous antimigraine agent [120].

7.2.2. LY382884 Further work in this area focused on selective iGluR5 kainate receptor antagonists based on the same decahydroisoquinoline-3-carboxyl core structure and incorporating cyclic spacers (aromatic or saturated) into the linker. **LY382884** (compound **116**) (Fig. 18) [121] demonstrated very good selectivity for the iGluR5 receptor. When tested in a rat, formalin-induced model of persistent pain [122], the nonselective AMPA/kainate antagonist **LY293558** (0.1–5 mg/kg, i.p.) caused both antinociception and ataxia. In contrast, the iGluR2-preferring antagonist **LY302679** (compound **117**, 5 mg/kg, i.p.) caused ataxia but had no significant effect on pain. However, the iGluR5-selective antagonist, **LY382884** (5–100 mg/kg, i.p.), exhibited antinociceptive activity without causing ataxia.

Further optimization of this series led to compound **115**, an antagonist with iGluR5 binding $K_i = 156$ nM and >300-fold selectivity over other Glu receptors [123]. Because **115**, an amino dicarboxylic acid, had low oral bioavailability in rat, the corresponding diethyl ester prodrug (**118**, $K_i > 40\,\mu$M against all iGluR receptors) was used for oral evaluation. When dosed in rats up to 30 mg/kg, **118** was not detected in plasma at 1 h. The plasma half-life of **115**, following oral dosing of **118**, was 2–3 h, and its oral bioavailability was 50%. Oral dosing of **118** and i.v. dosing of **115** in the neurogenic dural PPE model, both resulted in dose-related activity with ID_{50} values of 100 and 0.03 pg/kg, respectively. In the c-fos expression model, a 10 mg/kg dose of **118** produced an approximate 30% decrease in central fos expression. Since the two models measure peripheral and central effects, respectively, the difference in potencies may reflect a low degree of CNS penetration of **115**.

7.2.3. LY466195 Even better receptor subtype selectivity, due to a threefold increase in binding affinity at iGluR5, was achieved with

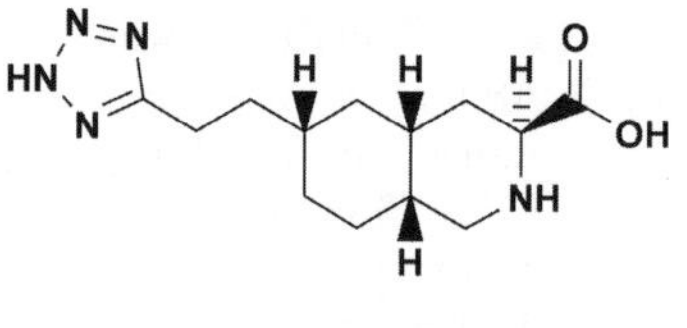

92
Tezampanel (Ly-293558)

AMPA
- iGluR1: K_i = 9.2 µM
- iGluR2: K_i = 3.2 µM
- iGluR3: K_i = 50 µM

KA
- iGluR4: K_i = 4.2 µM
- iGluR5: K_i = 100 µM

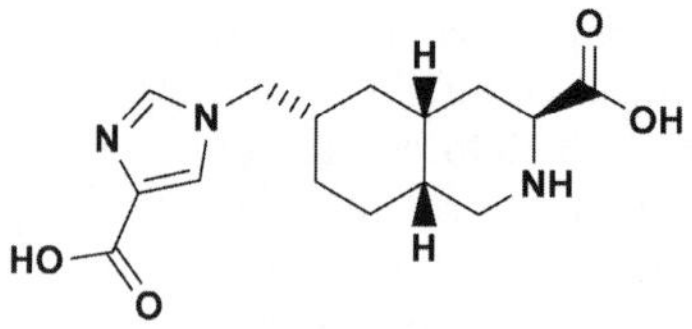

115

AMPA
- iGluR1: K_i = 134µM
- iGluR2: K_i = 117 µM
- iGluR3: K_i = 247 µM
- iGluR4: K_i = 262 µM

KA
- iGluR5: K_i = 0.156 µM
- iGluR6: K_i = 1000 µM
- iGluR7: K_i = 48.5 µM

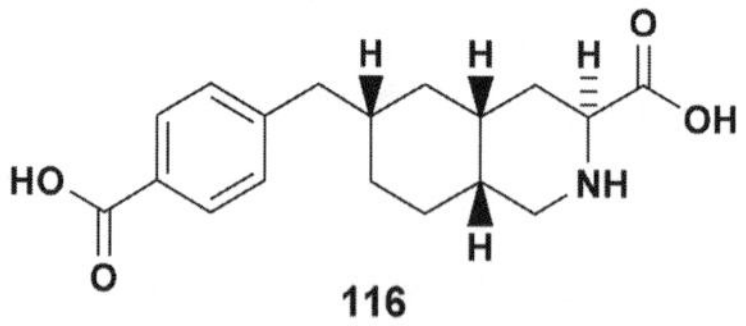

116
LY382884

AMPA
- iGluR1: K_i = 100 µM
- iGluR2: K_i = 100 µM
- iGluR3: K_i = 100 µM

KA
- iGluR4: K_i = 6.8 µM
- iGluR5: K_i = 100 µM

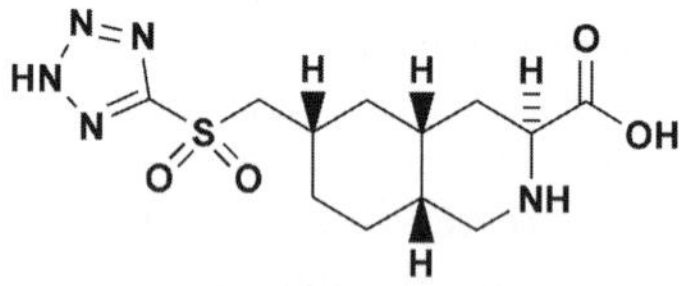

117
LY-302679

AMPA
- iGluR1: K_i = 7.9 µM
- iGluR2: K_i = 0.6 µM
- iGluR3: K_i = 14.8 µM

KA
- iGluR4: K_i = 4.7 µM
- iGluR5: K_i = 100 µM

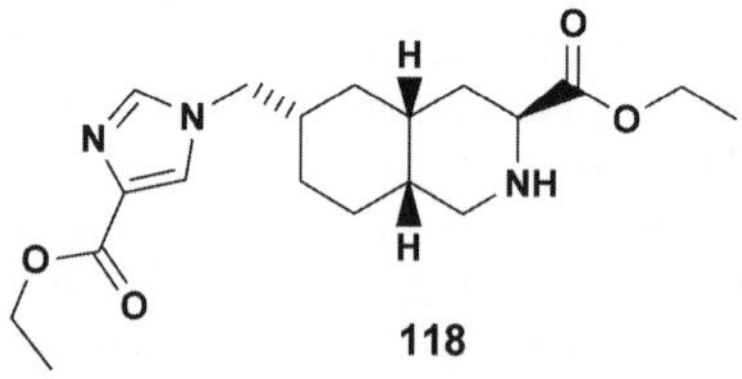

118

Figure 18. iGluR Binding Activities of decahydroisoquinoline-3-carboxylic Acids.

LY466195, (compound **119**) (Fig. 19) [117]. In the rat, PPE and c-fos models, **LY466195** and its corresponding ethyl ester prodrug (structure not shown) demonstrated good efficacy when dosed i.v. and orally, respectively. In addition, the compound showed no inherent contractile activity in a rabbit saphenous vein preparation.

Recently, a crystal structure of the iGluR5 ligand binding domain in complex with **LY466195** has appeared [124]. In this structure (Fig. 20), the decahydroisoquinoline-3-carboxyl group forms a salt bridge with Arg95 and the most prominent interaction that the amino NH appears to make is a hydrogen bond to the hydroxyl group of Thr90 and the backbone carbonyl oxygen of Pro88. At the other end of the molecule, the difluoroproline carboxyl group appears to interact with two backbone amide NH groups (Glu190 and Thr142) as well as the hydroxyl group of Thr142.

7.2.4. Perampanel Perampanel (**E2007**, Eisai) (Fig. 21) is an AMPA/kainate glutamate receptor antagonist from a very different chemical class that was evaluated as a symptomatic treatment for Parkinson's disease and terminated in Phase III [125]. A recent Phase

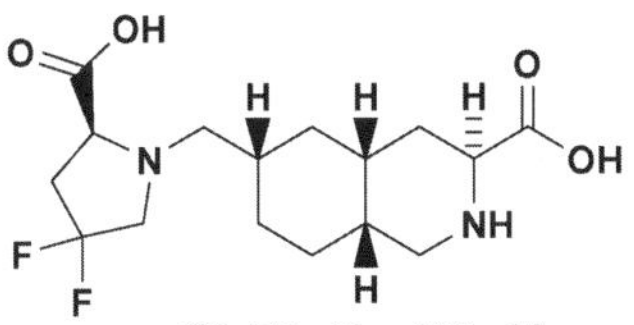

119
LY466195

AMPA
- iGluR1: K_i = 7.5 μM
- iGluR2: K_i = 269 μM
- iGluR3: K_i = 312 μM
- iGluR4: K_i = 432 μM

KA
- iGluR4: K_i = 0.05 μM
- iGluR4: K_i = 89 μM

iGluR5: K_i = 432 μM

Figure 19. iGluR binding activities of **LY466195**.

II clinical study for migraine prophylaxis has been completed but results have not been publicly disclosed [126].

8. ANTIEMETICS/ANTIPSYCHOTICS

Older antiemetic drugs with dopamine antagonist activity such as prochlorperazine (Fig. 23) (also used as an antipsychotic) and metaclopramide (Fig. 22) have been employed primarily in an emergency room setting, given intravenously, for acute migraine therapy [127]. Thus, rapid onset is achieved by bypassing relatively slow oral absorption and, for prochlorperazine, extensive first-pass metabolism [128]. However, injectable delivery is generally not considered convenient for unsupervised patient use.

8.1. Inhaled Prochlorperazine (AZ-001) and Loxapine (AZ-104)

A new inhalation system developed by Alexza Pharmaceuticals has been successfully used to deliver two compounds, prochlorperazine and loxapine (Fig. 23), a typical antipsychotic dopamine antagonist that also has antiemetic activity [129], to migraineurs in clinical trials. The Staccato® system consists of a hand-held device in which a thin film of active drug coated on stainless steel is heated (about 350°C) to form an aerosol (with no excipients) that is inhaled by the patient [130]. The aerosol condenses into 1–3 μm particles and is cooled prior to inhalation that delivers the particles in a single breath to alveoli in the lungs. Because the drug coating is very thin (<3 μm), exposure time to the high temperature is short, resulting in minimal degradation. Clearly, many polar, and/or large molecular weight compounds would be incompatible with this system. However, additional drugs successfully aerosolized and recovered (≥ 94%) using this technology include midazolam, zolpidem, rizatriptan, atropine, and sildenafil.

In a study of healthy volunteers using the prochlorperazine formulation with the Stac-

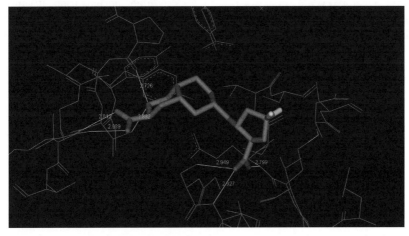

Figure 20. Detail of PDB entry 2qs4. **LY466195** bound in the iGluR5 ligand binding domain showing prominent hydrogen-bonding interactions. (This figure is available in full color at http://mrw.interscience.wiley.com/emrw/9780471266945/home.)

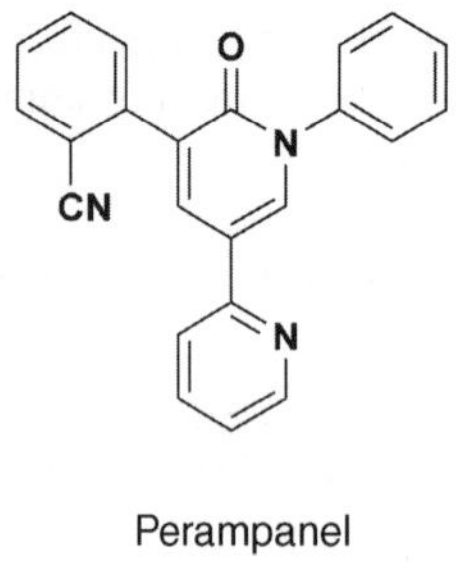

Figure 21. Perampanel.

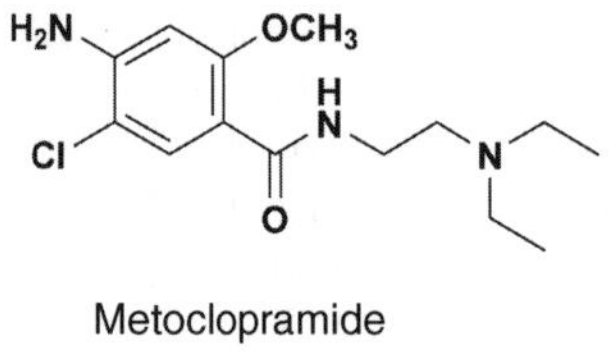

Figure 22. Metoclopramide.

cato system (**AZ-001**) in direct comparison with i.v. dosing, aerosol delivery resulted in comparable peak plasma drug concentrations and T_{max} values with essentially 100% bioavailability (Table 15) [131].

AZ-001 was tested in three doses (5, 7.5, and 10 mg) in a placebo-controlled (Staccato device containing no drug) Phase IIb clinical study in patients ($n = 400$) with migraine headache [132]. All doses met the primary endpoint of pain relief at 2 h ($p < 0.01$) (Table 16). For the secondary endpoints, pain-free rate at 2 and 24 h, only the 5 mg dose failed to reach significance in comparison with placebo. In addition, patients receiving the 7.5 mg dose showed statistically significant pain relief at 15 min, while at 30 min, all subjects receiving active drug achieved a pain response. There were no serious adverse events. Side effects were dose related and included taste (25–33%), throat irritation (18–30%), cough (16–30%), and somnolence (6–10%).

The Staccato loxapine formulation (**AZ-104**) was similarly successful at 2.5 and 5 mg doses in 2 h migraine pain relief [133]. **AZ-104** in higher doses is also in development for treatment of acute agitation in schizophrenia.

9. NITRIC OXIDE SYNTHASE (NOS) INHIBITORS

NO is a remarkable neurotransmitter since it is a reactive gas. However, in the absence of a catalyst it can last for seconds in the presence of oxygen and water and its small size and "lipophilicity" allow it to diffuse rapidly among synapses and cross bilayer membranes.

9.1. Nitric Oxide Synthases

NO is generated by nitric oxide synthases, a family of cytochrome P450-like hemoproteins that convert L-arginine to NO and L-citrulline (Fig. 24) and are highly regulated [133].

There are three known isoforms of NOS in mammals:

(1) Inducible NOS (iNOS), found in the immune as well as the cardiovascular systems; secreted by macrophages in high concentration. It can react with superoxide, also secreted by macrophages, forming highly toxic peroxynitrite as a defense against pathogens.
(2) Endothelial NOS (eNOS, constitutive) is involved in regulating vascular function within blood vessels, primar-

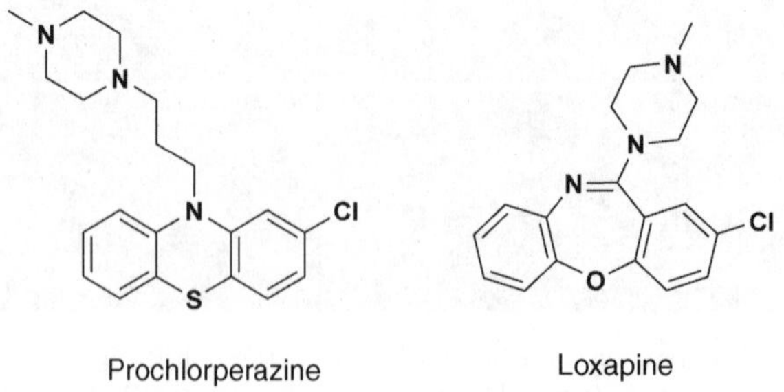

Figure 23. Antiemetic/antipsychotic compounds delivered as pure aerosols by the Alexza Staccato® system.

NITRIC OXIDE SYNTHASE (NOS) INHIBITORS

Table 15. Initial Peak Plasma Drug Characteristics of 0.5 mg of Prochlorperazine Delivered as a Thermally Generated Aerosol or 5 s Intravenous Infusion

Route of Administration	Initial T_{max} (min)	Initial C_{max} (ng/mL)
Intravenous over 5 s	3.50 ± 2.88	1.06 ± 0.84
Aerososol (single breath)	2.00 ± 0.76	1.38 ± 0.56

ily by inducing vasodilation. Activated by increases in intracellular calcium levels.

(3) Neuronal NOS (nNOS, constitutive) is found in both the CNS and the periphery and produces NO as a neurotransmitter. It is also activated by calcium ions resulting from synaptic activity.

NO is known to induce release of the potent vasodilator CGRP and CGRP antagonism has been shown to inhibit NO-induced vasodilation [135]. While it is clear that NO is intimately involved in migraine pathophysiology, there is uncertainty around the identity of the isoform(s) involved. From the descriptions above, a rationale could be made for any or all of them and it may be possible that more than one is implicated. However, eNOS inhibition has generally been avoided in drug design since it has been shown to result in blood pressure changes.

Nitroglycerin (Fig. 25), from which NO is liberated *in vivo*, has been known to cause migraine-like headaches for almost a century. Interestingly, in several studies in which nitroglycerin was given i.v. to migraine patients along with matched normal controls, the resulting headache in migraineurs was longer lasting, nearly always without aura, and did not return to baseline as it did for control subjects [136,137]. In addition, 80% of migraineurs suffered a second headache (maximum about 7 h following infusion) that had all characteristics of migraine without aura. These and other studies suggest that aura, which is probably caused by CSD, may produce headache at least partly through the resulting production of NO [138].

9.2. Nitric Oxide Synthase Inhibitor Tool Compounds

Inhibition of NOS was first shown to be a potential antimigraine therapy in a placebo-controlled, double-blind study of L-N^G-methylarginine (L-NMMA, **546C88**) (Fig. 26). Two hours after i.v. dosing of L-NMMA, 10 of 15 migraineurs reported significant headache relief as compared with 2 of 14 receiving placebo [139].

Table 16. Phase IIb efficacy results for AZ-001

	Pain Relief at 2 h (%)	Pain Free at 2 h (%)	Pain Free at 24 h (%)
Prochlorperazine 10 mg	66.0	35.0	30.1
Prochlorperazine 7.5 mg	63.7	29.7	23.1
Prochlorperazine 5 mg	60.2	21.4	Not reported
Placebo	40.8	15.3	10.2

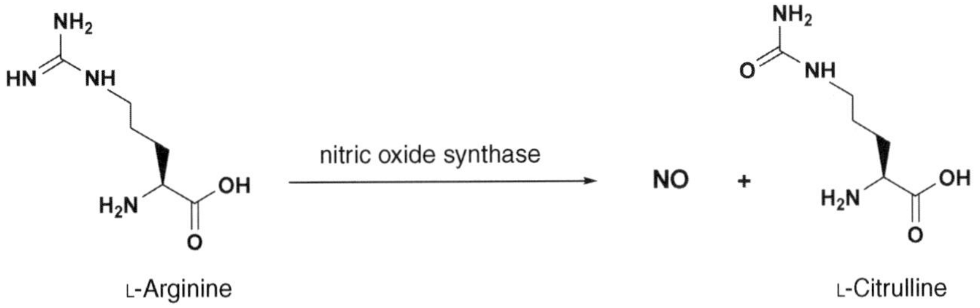

Figure 24. Nitric oxide synthases convert L-Arg to L-citrulline, releasing NO.

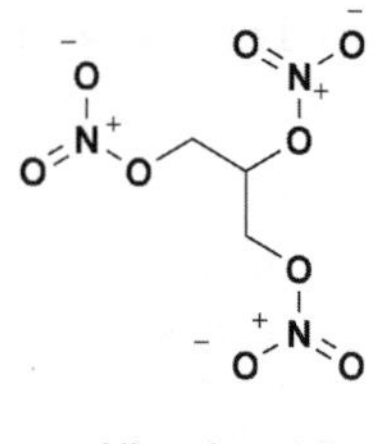

Figure 25. Nitroglycerin.

Other amino acid-based tool compounds (Fig. 26) that have been used to elucidate the biochemical and pharmacological consequences of NOS inhibition include N^G-nitro-L-arginine methyl ester (L-NAME) [140], and N^ε-L-(1-iminoethyl)lysine (L-NIL) [141]. A number of dipeptide inhibitors of NOS have been prepared, suggesting that there may be endogenous peptide inhibitors of NOS [142]. These and most other existing inhibitors are competitive with the L-arginine substrate and are assumed to bind in the arginine binding site [143].

Several groups have attempted to develop nonpeptide iNOS or nNOS inhibitors for migraine and other conditions in which excess NO is implicated. The benzyl acetamidine, **1400W**, is noteworthy in this respect since it is a potent ($IC_{50} = 0.23\,\mu M$), highly selective iNOS inhibitor (>4000-fold versus eNOS) [144]. However, **1400W** has shown unexplained toxic effects upon bolus dosing >10 mg/kg.

9.3. iNOS Inhibitors: GW273629 and GW274150

Scientists from Glaxo Wellcome investigated N-(1-iminoethyl)lysine-based systems in their search for iNOS-selective inhibitors with the goal of improving on the approximately 30-fold eNOS/iNOS selectivity found with NIL [145]. Simple homologation to the homolysine derivative, **121** (Fig. 27), reduced iNOS affinity at least fivefold ($IC_{50} = 20\,\mu M$). Replacement of a central methylene with oxygen (**122**) resulted in comparable affinity at iNOS ($IC_{50} = 1.2\,\mu M$)

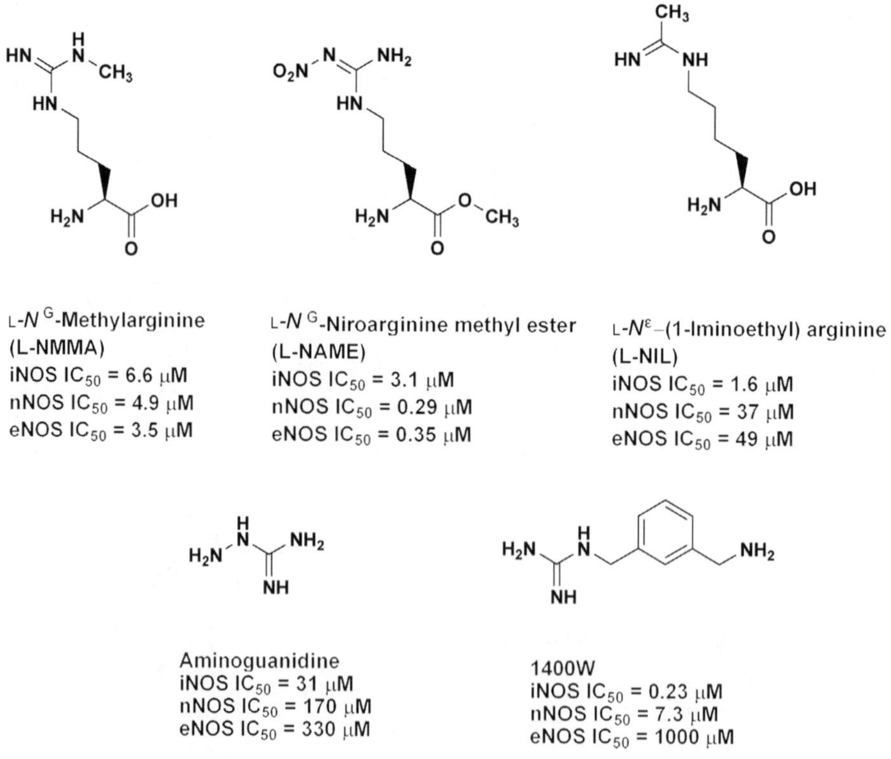

L-N^G-Methylarginine (L-NMMA)
iNOS IC_{50} = 6.6 μM
nNOS IC_{50} = 4.9 μM
eNOS IC_{50} = 3.5 μM

L-N^G-Niroarginine methyl ester (L-NAME)
iNOS IC_{50} = 3.1 μM
nNOS IC_{50} = 0.29 μM
eNOS IC_{50} = 0.35 μM

L-N^ε-(1-Iminoethyl) arginine (L-NIL)
iNOS IC_{50} = 1.6 μM
nNOS IC_{50} = 37 μM
eNOS IC_{50} = 49 μM

Aminoguanidine
iNOS IC_{50} = 31 μM
nNOS IC_{50} = 170 μM
eNOS IC_{50} = 330 μM

1400W
iNOS IC_{50} = 0.23 μM
nNOS IC_{50} = 7.3 μM
eNOS IC_{50} = 1000 μM

Figure 26. NOS inhibitor tool compounds.

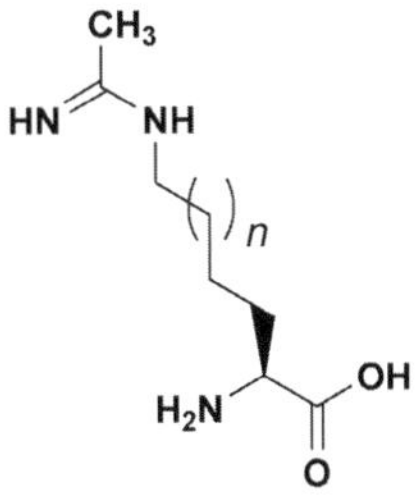

120: n = 2 (L)
iNOS IC$_{50}$ = 1.6 μM
eNOS IC$_{50}$ = 78 μM
nNOS IC$_{50}$ = 37 μM

121: n = 3 (DL)
iNOS IC$_{50}$ = 20 μM
eNOS IC$_{50}$ >100 μM
nNOS IC$_{50}$ >100 μM

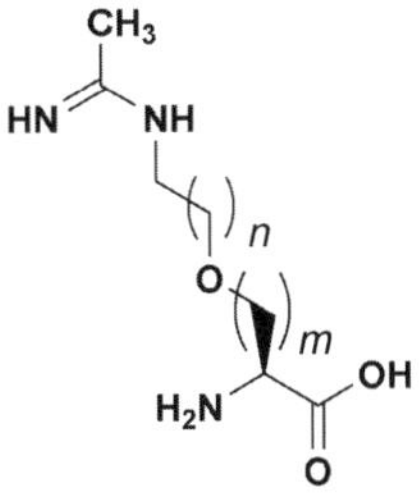

122: n = 1, m = 1
iNOS IC$_{50}$ = 1.2 μM (L)
eNOS IC$_{50}$ = 13 μM
nNOS IC$_{50}$ = 11.6 μM

123: n = 1, m = 2
iNOS IC$_{50}$ = 1.4 μM (DL)
eNOS IC$_{50}$ = 72 μM
nNOS IC$_{50}$ = 6.0 μM

124: n = 2, m = 1
iNOS IC$_{50}$ >100 μM (L)

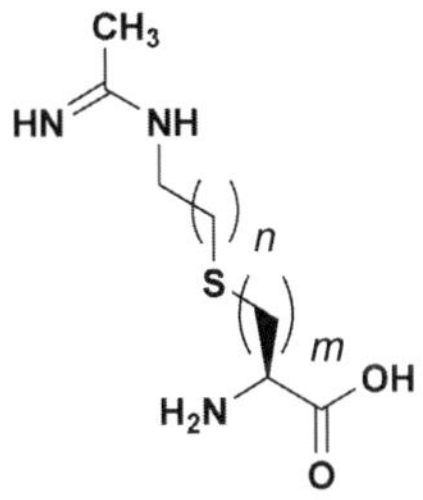

125: n = 1, m = 1
iNOS IC$_{50}$ = 0.7 μM (L), 5.4 μM (D)
eNOS IC$_{50}$ = 40 μM (L), >100 μM (D)
nNOS IC$_{50}$ = 13.2 μM (L), 53 μM (D)

126: n = 1, m = 2 (L = GW274150)
iNOS IC$_{50}$ = 1.4 μM (L), 5.9 μM (D)
eNOS IC$_{50}$ = 466 μM (L), >100 μM (D)
nNOS IC$_{50}$ = 145 μM (L), >100 μM (D)

127: n = 2, m = 1
iNOS IC$_{50}$ = 62 μM (L)

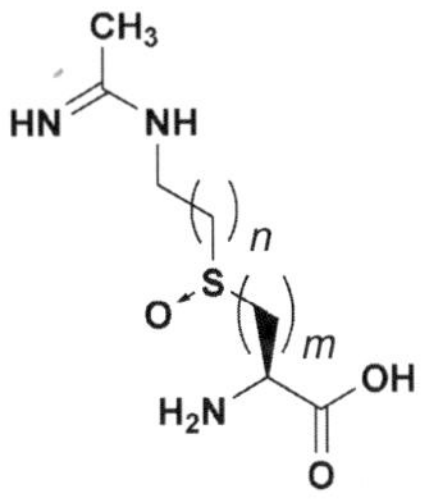

128: n = 1, m = 1
iNOS IC$_{50}$ = 8.0 μM (L)
eNOS IC$_{50}$ = 155 μM
nNOS IC$_{50}$ = 68 μM (L)

129: n = 1, m = 2
iNOS IC$_{50}$ >100 μM (DL)

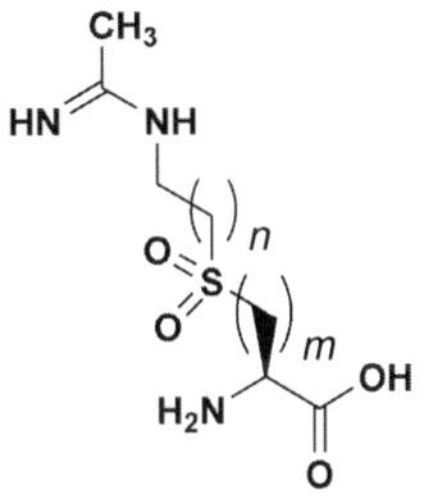

130: n = 1, m = 1 (GW273629)
iNOS IC$_{50}$ = 8.0 μM (L)
eNOS IC$_{50}$ >1000 μM (L)
nNOS IC$_{50}$ = 630 μM (L)

131: n = 1, m = 2
iNOS IC$_{50}$ >100 μM (DL)

132: n = 2, m = 1
iNOS IC$_{50}$ >100 μM (DL)

Figure 27. N-(1-iminoethyl)lysine-based NOS inhibitors.

and unwanted increases in eNOS (IC$_{50}$ = 13 μM) and nNOS (IC$_{50}$ = 11.6 μM) binding. The addition of one methylene group on the amino acid side (**123**) resulted in no real change in iNOS (IC$_{50}$ = 1.4 μM) and nNOS (IC$_{50}$ = 6.0 μM) affinity, but a diminution in eNOS activity (IC$_{50}$ = 72 μM). Homologation toward the amidine abolished iNOS binding (**124**, IC$_{50}$ > 100 μM).

A series of thioether and corresponding sulfoxides and sulfones had the most interesting and useful enzyme subtype selectivities. The thioether linkage in the lysine series (**125**) showed a modest increase in NOS potency against all isoforms (iNOS IC$_{50}$ = 0.7 μM). Homologation toward the aminoacid portion (**126, GW274150**) gave iNOS activity (IC$_{50}$ 1.4 μM) identical to L-NIL (IC$_{50}$ = 1.6 μM) but greatly improved selectivity over eNOS (333-fold versus 49-fold). Alternative homologation (**127**) resulted in a >40-fold loss in iNOS potency (IC$_{50}$ = 62 μM). Oxidation of **125**

Table 17. Pharmacokinetcs of GW274150 and GW273629 in Rats and Mice Following i.v. Dosing

Compound	Species	Dose (mg/kg)	AUC (h μg/mL)	C_{max} (μg/mL)	T_{max} (h)	$t_{1/2}$ (h)	C_L (L/h/kg)	V_{ss} (kg^{-1})
GW274150	Mouse	1	6.0	1.6	0.08	5.7	0.17	1.32
	Rat	10	74.2	13.4	0.08	6.5	0.14	1.04
GW273629	Mouse	1	15.5	49.1	0.08	0.14	0.65	0.11
	Rat	10	25.9	36.7	0.08	3.0	0.39	0.65

to sulfoxide **128** reduced iNOS activity by approximately 10-fold (IC$_{50}$ = 8.0 μM) and eNOS/iNOS selectivity by 3-fold. The sulfoxide analog of **GW274150** was inactive, as was the corresponding sulfone **131**. However, the sulfone in the lysine series, **130** (**GW273629**), retained single-digit micromolar binding potency (IC$_{50}$ = 8.0 μM) while exhibiting no detectable eNOS activity (IC$_{50}$ > 1000 μM).

The two compounds in this series possessing both single-digit micromolar binding potency against human iNOS and >100-fold selectivity over eNOS, **GW274150** (**126**) and **GW273269** (**130**), were chosen for further investigation *in vivo* in pharmacokinetic studies in mice and rats [146]. In rats, both compounds given i.v. displayed biphasic kinetics with **GW274150** eliminated somewhat more slowly ($t_{1/2}$ = 6.5 h) than **GW273629** ($t_{1/2}$ = 3.0 h) (Table 17). In mice, elimination of **GW273629** appeared monophasic and was much more rapid ($t_{1/2}$ approximately 10 min). Perhaps, because it is more lipophilic, **GW274150** had a higher volume of distribution in both species. Both compounds were reported to have high oral bioavailabilities in rats and mice (approximately 90%).

GW274150, and to a lesser extent **GW273629**, have been tested in several animal models of pain and inflammation [146–148]. Although there are no published results in specific migraine models, **GW274150** has been tested in clinical trials both as an acute (5–180 mg) [149] and a preventative migraine therapies (up to 120 mg daily for 12 weeks) [150]. No results have yet been publicly disclosed.

9.4. nNOS Inhibitors: NeurAxon

Researchers at NeurAxon have targeted nNOS for migraine treatment with a series of aminobenzothiazoles [151]. The group sought to use fairly lipophilic guanidine isosteric groups, thiophene amidine and S-ethyl isothiourea, perhaps to increase chances of CNS penetration, according to the general NOS inhibitor pharmacophore model (Fig. 28) elucidated by previous workers (i.e., **GW274150** and **1400W**).

As shown in Table 18, the thiophene amidine appeared to be superior interms of both potency and isoform selectivity in comparison with S-ethyl isothiourea on the basic aminobenzothiazole scaffold. In addition, little difference was seen between 5- and 6-substitution.

Keeping the thiophenoamidine constant, the group then investigated substitution on the thiazole amino group (Table 19). Although nNOS inhibitory potency did not depend to a great extent on 5- (type I) or 6-substitution (type II), selectivity over both eNOS and iNOS was somewhat improved by the former. A basic nitrogen was found to be beneficial, optimally linked through 2–3 carbon atoms. Best potency was achieved by the compound with probably the most basic nitrogen, the racemic pyrrolidine **141** (IC$_{50}$ = 0.03 nM). This compound also showed good selectivity over eNOS (53-fold) and iNOS (43-fold). Compounds without basic nitrogens such as **147** and **148** were somewhat less active.

Further optimization was carried out on a small set of *N*-benzyl-4-aminopiperidines, keeping the amidine at the 6-position of the benzothiazole (Table 20). Placement of a methoxy group on either the 3- or the 4-posi-

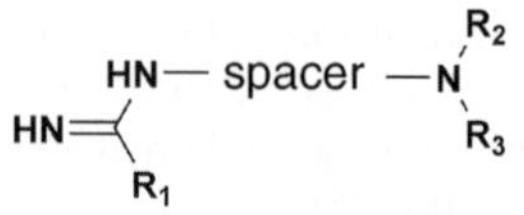

Figure 28. General pharmacophore model for NOS inhibitors.

Table 18. Inhibition of Human NOS Isoforms by Substituted Benzothiazoles

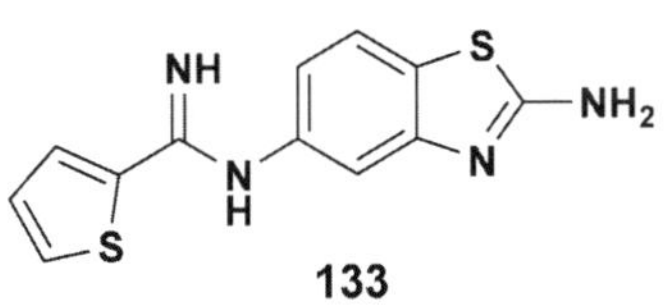

Compound	nNOS IC$_{50}$ (μM)	eNOS IC$_{50}$ (μM)	iNOS IC$_{50}$ (μM)
133	0.2	2.3	79
134	0.2	2.9	25
135	4.6	12	97
136	1.4	9.2	46

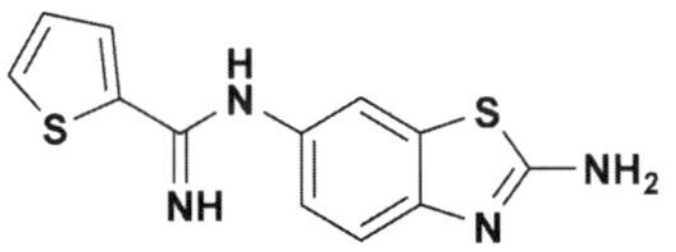

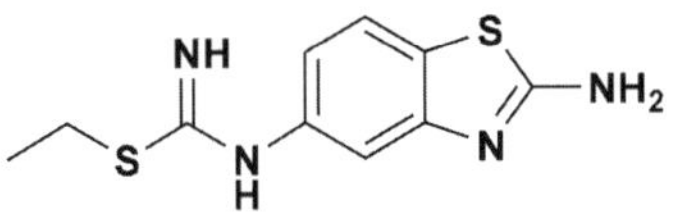

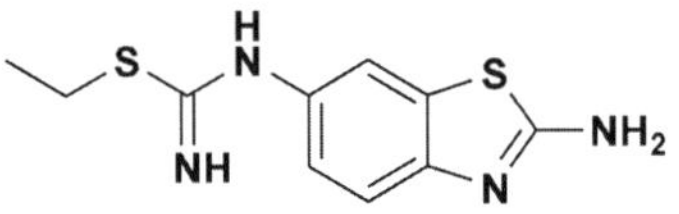

tion of the benzyl group (compounds **149** and **150**, respectively) resulted in almost identical nNOS and eNOS inhibition values (IC$_{50}$ = 0.1 and 0.2 μM, respectively) but very different levels of iNOS selectivity (>1000-fold versus 45-fold, respectively). The methyl ester (**151**) is also tolerated by iNOS (IC$_{50}$ = 0.3 μM), but not the corresponding carboxylic acid (**152**), suggesting electrostatic repulsion.

At a recent chemistry conference, NeurAxon reported that an unsubstituted derivative, **153** (Fig. 29), was active in animal models of chronic pain (unspecified in the abstract) without eNOS-related blood pressure effects [152].

At the 2008 European Headache and Migraine Trust Symposium in London, NeurAxon announced that **NXN-188**, a compound that combines nNOS inhibition with 5-HT agonism (unspecified), demonstrated significant sustained pain relief in patients suffering from migraine with aura in a small (60 patients) randomized, multicenter, double-blind, placebo-controlled parallel-group Phase II trial [153]. Perhaps significantly, since NO may be more prominent in migraine with aura as a result of CSD, the overall effect for the primary endpoint did not reach statistical significance (15% greater than placebo) for the entire study, only for patients with aura (47% of the study subjects) at the 100 mg dose of **NXN-188**. The company claimed that, if validated in larger trials, these data would represent a 30% increase in sustained pain relief compared with sumatriptan. No serious adverse events were reported.

Although the structure of **NXN-188** has not been publicly disclosed, it is tempting to assume that it comes from a chemical series that has so far only appeared in a published patent application [154]. By replacing the aminobenzothiazole in their previous series with substituted 5-aminoindoles, the NeurAxon researchers were able to integrate "triptan-like" activity in their selective nNOS inhibitors (Table 21). Compound **154** contains the dimethylaminoethyl side chain found on sumatriptan, zolmitriptan, rizatriptan and almotriptan. Against human nNOS, it is about 10-fold less potent (IC$_{50}$ = 2.6 μM) than the best benzothiazoles reported with only approximately 10-fold selectivity against eNOS. Binding activity against bovine 5-HT$_{1D}$ (IC$_{50}$ = 0.36 μM) is also somewhat inferior to single- and

Table 19. Inhibition of Human NOS Isoforms by Substituted Thiophenoamidine-Benzothiazoles

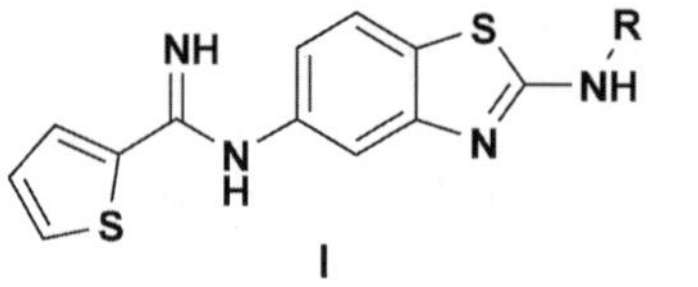

Compound	Substitution	R	nNOS IC$_{50}$ (μM)	eNOS IC$_{50}$ (μM)	iNOS IC$_{50}$ (μM)
137	I	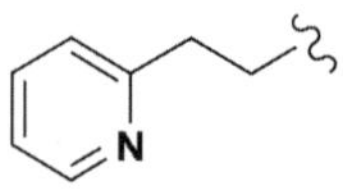 2-pyridylethyl	1.1	6	139
138	I	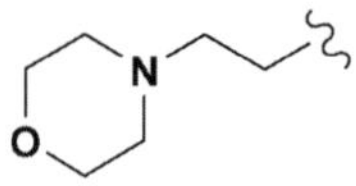 morpholinoethyl	1.3	12	141
139	I	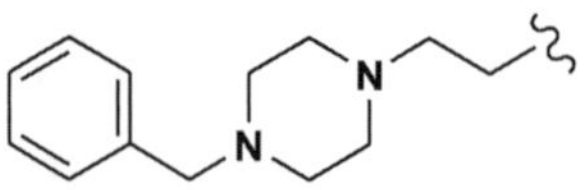 4-benzylpiperazinylethyl	0.4	16	48
140	I	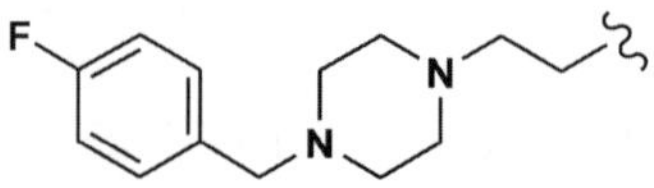 4-(4-fluorobenzyl)piperazinylethyl	0.8	13	91
141	I	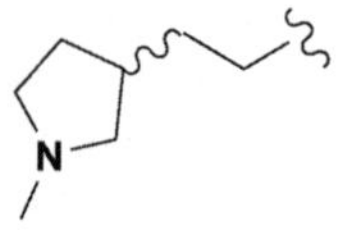 (1-methylpyrrolidin-3-yl)methyl	0.03	1.6	15
142	II	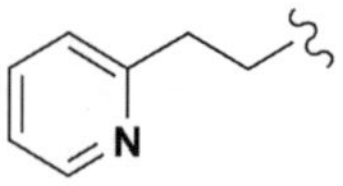 2-pyridylethyl	1.3	0.9	13
143	II	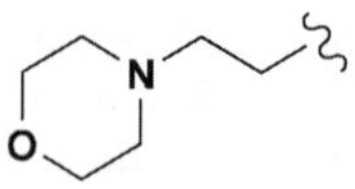 morpholinoethyl	2.0	1.5	37
144	II	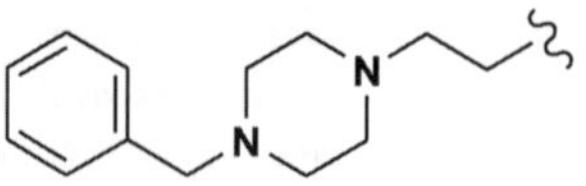 4-benzylpiperazinylethyl	0.5	6	14
145	II	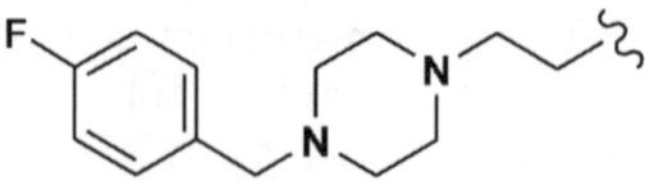 4-(4-fluorobenzyl)piperazinylethyl	0.2	8	14
146	II	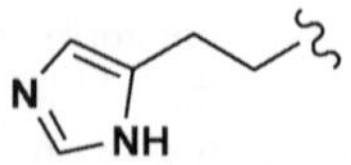 imidazolylethyl	0.3	0.7	6

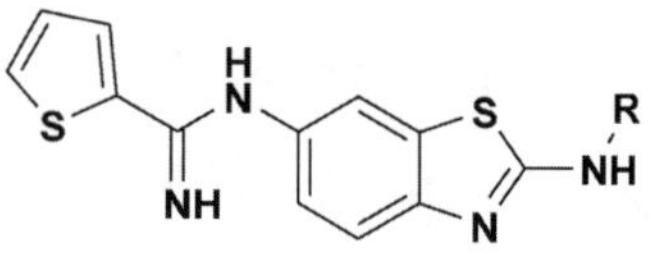

Table 19. (*Continued*)

Compound	Substitution	R	nNOS IC$_{50}$ (µM)	eNOS IC$_{50}$ (µM)	iNOS IC$_{50}$ (µM)
147	II	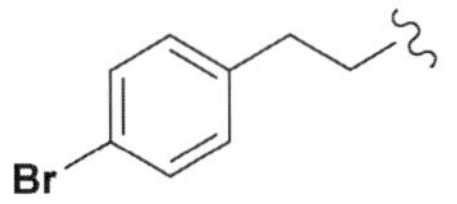 4-Br-phenethyl	4.4	2.1	75
148	II	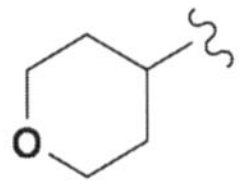 tetrahydropyran-4-yl	2	16	64

Table 20. Inhibition of Human NOS Isoforms by Substituted *N*-Benzyl-4-Aminopiperidines

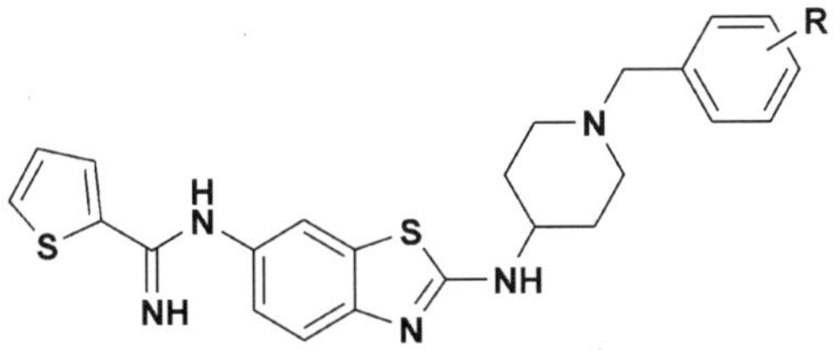

Compound	R	nNOS IC$_{50}$ (µM)	eNOS IC$_{50}$ (µM)	iNOS IC$_{50}$ (µM)
149	3-OMe	0.1	3.8	>100
150	4-OMe	0.2	5.7	9
151	4-CO$_2$Me	0.3	5.2	>100
152	4-CO$_2$H	5.1	59	>100

double-digit nanomolar values normally seen for the marketed triptans. However, it is possible with dual-acting compounds where synergy may be desired that 5-HT potency may be intentionally reduced to avoid triptan side effects. Using the *N*-methyl-4-piperidyl side chain (**155**) found in naratriptan, selectivity versus iNOS increased from 5- to 30-fold. How-ever, the corresponding 3,4-unsaturated analog (**156**) exhibits much improved eNOS selectivity (55-fold), submicromolar nNOS affinity (IC$_{50}$ = 0.92 µM), and potent 5-HT$_{1D}$ (IC$_{50}$ = 51 nM) and 5-HT$_{1B}$ (IC$_{50}$ = 160 nM) activity. Utilization of the chiral *N*-methylpyrrolidin-2-ylmethyl side chain (**158**) found in eletriptan resulted in activities and selectivity comparable to **156**. The enantiomer (**160**) was similar except for an eightfold reduction in 5-HT$_{1D}$ potency. The best nNOS inhibition (IC$_{50}$ = 0.43 µM) and eNOS/nNOS selectivity (90-fold) was achieved by the racemic *N*-methylpyrrolidin-3-ylindole, **161**. Activity against 5-HT$_{1D}$ for this compound, however, was sevenfold lower than for **156**. It is noteworthy that the replacement of thiophene with furan appears to have only a marginal effect on binding potencies across the board (**156** versus **157**, **158** versus **159**, and **161** versus **162**).

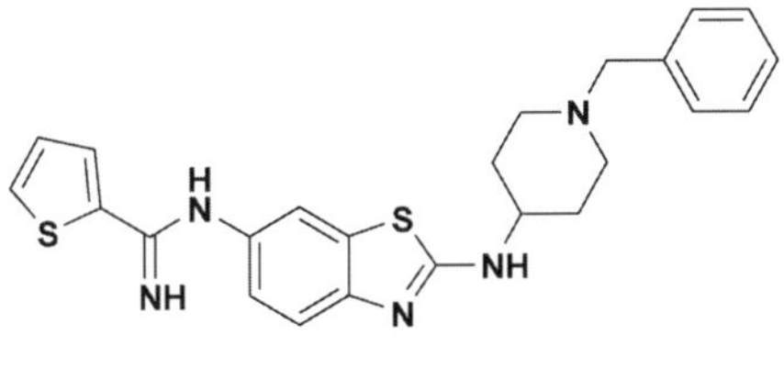

153

Figure 29. NeurAxon clinical nNOS inhibitor.

Table 21. Inhibition of Human NOS Isoforms and 5-HT$_{1D}$ and 5-HT$_{1B}$ Binding Activity of Substituted Thiophenoamidine-Indoles

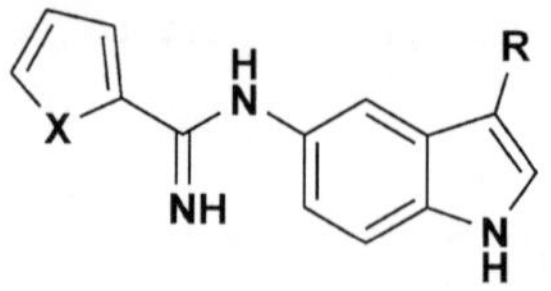

Compound	X	R	nNOS IC$_{50}$ (μM)	eNOS IC$_{50}$ (μM)	iNOS IC$_{50}$ (μM)	5-HT$_{1D}$[a] IC$_{50}$ (μM)	5-HT$_{1B}$[b] IC$_{50}$ (μM)
154	S	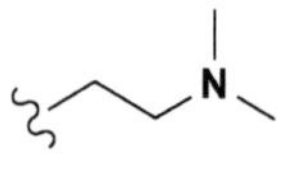 ethyl-N-methyl	2.6	26	12	0.36	—
155	S	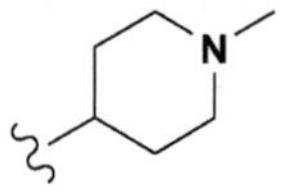 4-(N-methyl)piperidine	1.9	33	58	0.57	—
156	S	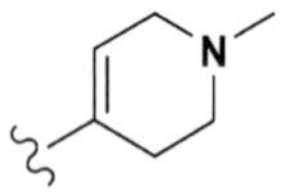 4-(N-methyl)tetrahydropyridine	0.92	51.1	20	0.051	0.16
157	O	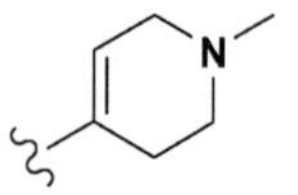 4-(N-methyl)tetrahydropyridine	1.8	54	58	0.050	—
158	S	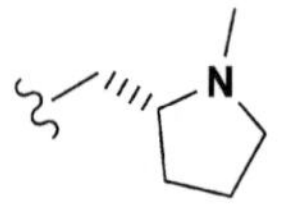 (S)-2-(N-methyl)pyrrolidine	0.84	34.5	—	0.13	0.31
159	O	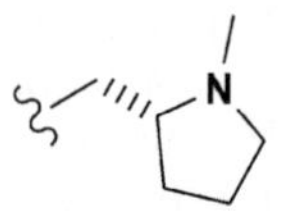 (S)-2-(N-methyl)pyrrolidine	2.08	27.1	—	—	—
160	S	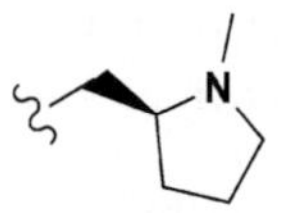 (R)-2-(N-methyl)pyrrolidine	0.82	23	—	1.1	0.29
161	S	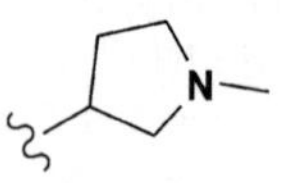 3-(N-methyl)pyrrolidine	0.43	39	—	0.35	0.49
162	O	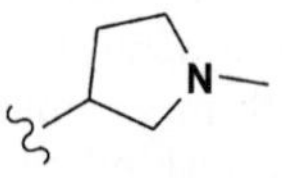 3-(N-methyl)pyrrolidine	1.2	34	—	0.56	1.1

[a] Bovine caudate.
[b] Rat cerebral cortex.

10. VANILLOID RECEPTOR ANTAGONISTS

The transient receptor potential vanilloid 1 (TRPV1) receptor is a six-transmembrane-spanning ligand-gated cation channel that is a member of a superfamily of TRP channels. TRPV1 can be activated by a wide variety of endogenous and exogenous ligands as well as by heat (>43°C), acidic pH, and physical abrasion [155]. Most prominently, TRPV1 is activated by capsaicin (Fig. 30), a vanilloid metabolite of chili peppers that causes a burning sensation in tissues with which it comes into contact. However, the list of agents that activate or sensitize the receptor includes such diverse compounds as nerve growth factor, anandamide (endogenous cannabinoid), adenosine, prostaglandins, polyamines, bradykinin, and insect and jellyfish venoms.

TRPV1-knockout mice lack thermal hypersensitivity that wild-type mice display as a result of hind-paw injection of proinflammatory agents. Inflammatory conditions serve also to lower the activation threshold of the receptor leading to increased peripheral sensitization. TRPV1 is found in neurons in both the periphery and the CNS, and is involved in the transmission of pain signals and the initiation and integration of neurogenic inflammation, making it an important target for pain relief in numerous disease states including migraine.

Consistent with its role in neuroinflammation, TRPV1 is found in peripheral neuronal afferents such as the trigeminal and dorsal root ganglia where proinflammatory peptides such as CGRP and substance P are released [156]. In rats, i.v. capsaicin causes plasma-protein extravasation [157]. Topical capsaicin also causes acute dermal vasodilation [158]. Interestingly, the CGRP peptide antagonist CGRP8-37, but not a NOS inhibitor (L-NMMA) (Fig. 27), a prostaglandin inhibitor (indomethacin), or a substance P antagonist (arepitant) blocked capsaicin-induced vasodilation in humans [159], suggesting strong interaction between CGRP and TRPV1, and supporting the premise that TRPV1 antagonism may have an antimigraine effect.

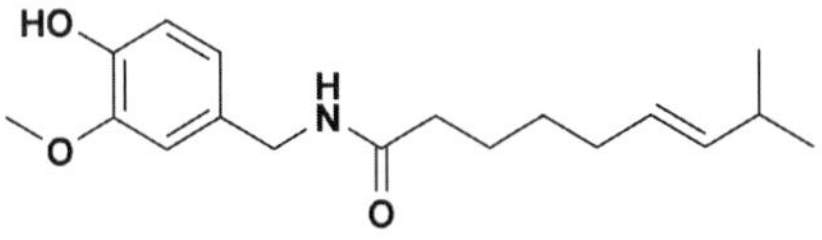

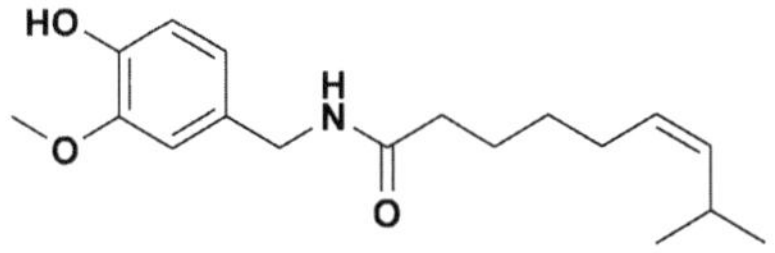

10.1. Capsazepine, Civamide, and AMG-517

TRPV1 has been found to be involved in the regulation of body temperature, and antagonism of the receptor has resulted in hyperthermia in several species, including humans [160]. This has resulted in the termination of a promising highly selective TRPV1 antagonist, **AMG-517** (Fig. 31), when it was found to cause undesirable levels of hyperthermia [161], representing a potential side effect of this mechanism.

Long-term TRPV1 activation (e.g., prolonged exposure to capsaicin) results in receptor depletion leading to reduction in pain sensation and blockade of neuroinflammation (i.e., release of CGRP and substance P). Indeed, intranasal and topical capsaicin has been marketed as an "all-natural" migraine remedy to reduce the frequency and severity of attacks. In addition, civamide (Fig. 30), the cis-isomer of capsaicin, has been tested in a small ($n = 34$), double-blinded, nonplacebo controlled clinical trial as an acute therapy [162]. When dosed intranasally at either 20 or 150 μg, 55.6 % of subjects experienced pain relief at 2 h (22% pain free). At 4 h postdose, 72.7% of patients reported pain relief (33% pain free). The most prominent adverse effects reported by civamide-treated subjects (91.2%) were a localized burning sensation in the nasal passages, and lacrimation (44.1%).

In an effort to avoid the initial pain sensations and delayed onset of action that are a prelude to the analgesia induced by overactivation of TRPV1, several groups have sought selective TRPV1 antagonists that can be

Figure 30. Capsaicin and civamide.

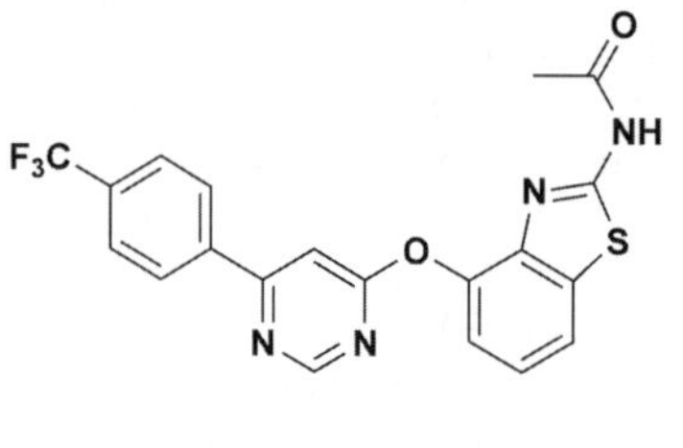

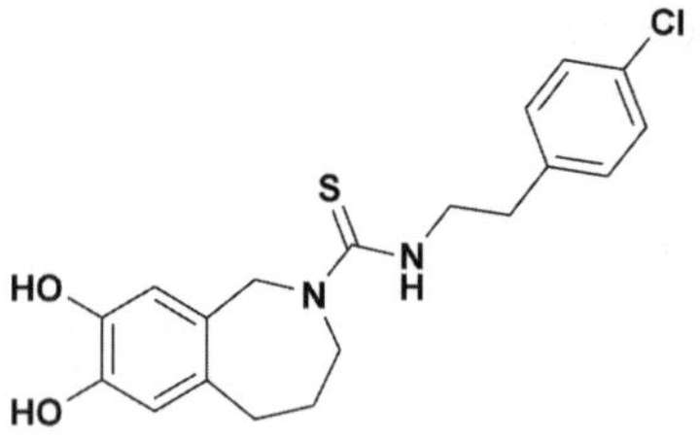

Figure 31. AMG-517 and capsazepine.

dosed orally. The first competitive TRPV1 antagonist, capsazepine, was reported by Sandoz in 1992 [163]. The compound can be seen as a constrained analog of capsaicin possessing antagonist activity. It has served as a useful tool compound but its nonspecific actions against acetylcholine [164] and voltage-gated calcium channels [165] have precluded its use as a therapeutic agent.

10.2. SB-705498

Scientists at GSK developed a series of selective TRPV1 antagonists that culminated in SB-705498, the first, and so far only, in this class of agents to reach clinical trials for migraine. In 2004, GSK reported the discovery of SB-366791 (Fig. 32), a highly potent antagonist ($pK_b = 7.7$, binding $IC_{50} = 5.7$ nM against the human receptor) that was identified directly through high-throughput screening using a FLIPR (fluorescence imaging plate reader)-based calcium ion assay [166]. The compound was reported to be highly selective (<25% inhibition at 1 µM) in a large commercial panel of binding assays (Cerep, Poitiers, France) that included numerous ion channels and GPCRs. As a result, unlike capsazepin, SB-366791 had no measurable ion channel-related effects in rat sensory neurons. However, it remained a tool compound due to poor pharmacokinetic properties (no details reported) [167].

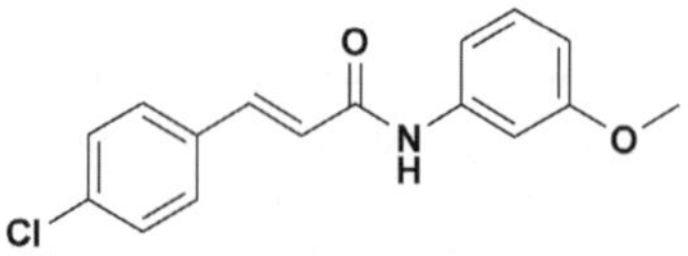

Further screening efforts at GSK revealed compound **167** (**SB-452533**), a highly potent ($pK_b = 7.8$) N-anilinoethyl-N-arylurea [168]. A follow-up investigation of SAR generated a series of TRPV1 antagonist ureas (Table 22). Replacement of the 2-bromo group on **SB-452533** with significantly smaller substituents such as H (**163**, $pK_b = 7.2$), methyl (**164**, $pK_b = 6.9$) or fluorine (**165**, $pK_b = 6.8$) resulted in an obvious loss of potency. The chloro analog (**166**) was nearly equipotent ($pK_b = 7.7$). Enlarging the aryl moiety to 1-naphthyl (**170**) also reduced activity ($pK_b = 7.1$), as did moving the bromo-substituent around the ring (**168** and **169**, $pK_b = 7.4$ and 7.2, respectively). Adding a second chlorine to **166** at either the 3- or the 5-position essentially preserved the potency of that compound ($pK_b = 7.7$ and 7.6, respectively). Removal of the anilino N-ethyl group on the screening hit, **SB-452533**, abolished activity ($pK_b < 6.8$), while the smaller methyl group had diminished it modestly ($pK_b = 7.4$). Neither the larger benzyl group (**175**) nor the more polar acetyl group (**176**) was tolerated by the receptor ($pK_b < 6.7$ and 6.8, respectively), suggesting a small, lipophilic pocket. Moving the anilino 3-methyl group to the 2-position (**178**) reduced activity significantly ($pK_b = 6.9$), while at the 4-position (**179**) potency was increased ($pK_b = 8.1$). Replacement of the 3-methyl group on **SB-452533** by fluorine (**180**) or addition of fluorine at the 4-position (**181**) maintained activity ($pK_b = 7.6$ and 7.8, respectively), while the 3-fluoro-4-methyl analog (**182**) exhibited the best potency in the series ($pK_b = 8.2$).

Using a quaternary salt (**183**) (Fig. 33) with moderate activity ($pK_b = 7.0$) that was pre-

Figure 32. SB-366791.

Table 22. Antagonist Activity (FLIPR) Versus Capsaicin Against Human TRPV1 Receptors

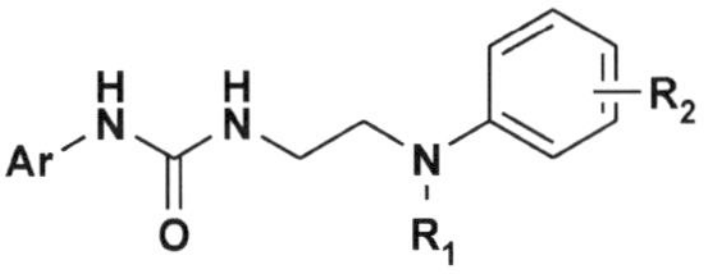

Compound	Ar	R_1	R_2	pK_b
163	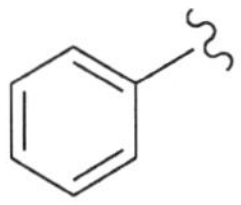 phenyl	Et	3-Me	7.2
164	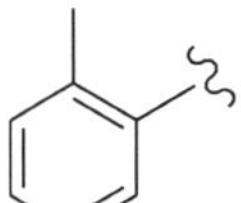 2-methylphenyl	Et	3-Me	6.9
165	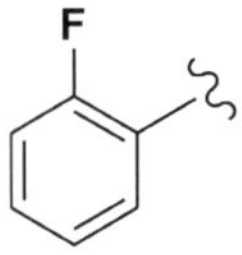 2-fluorophenyl	Et	3-Me	6.8
166	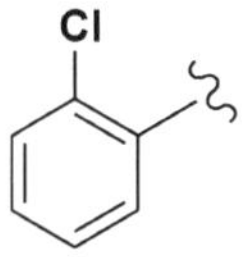 2-chlorophenyl	Et	3-Me	7.7
167 SB-452533	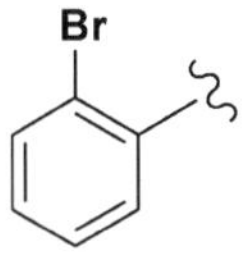 2-bromophenyl	Et	3-Me	7.8
168	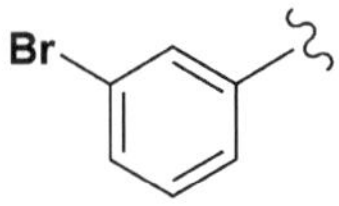 3-bromophenyl	Et	3-Me	7.4
169	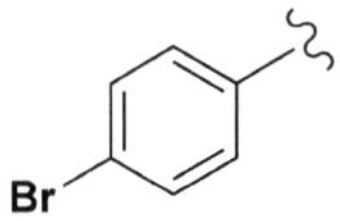 4-bromophenyl	Et	3-Me	7.2
170	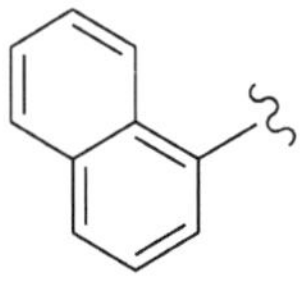 1-naphthyl	Et	3-Me	7.1

Table 22. (*Continued*)

Compound	Ar	R_1	R_2	pK_b
171	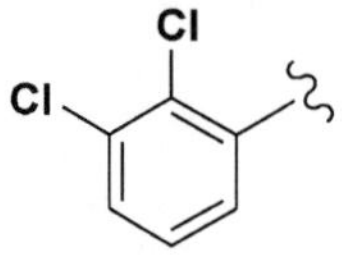 2,3-diCl-phenyl	Et	3-Me	7.7
172	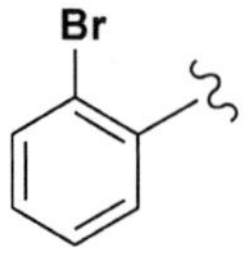 2-Br-phenyl	Et	3-Me	7.6
173	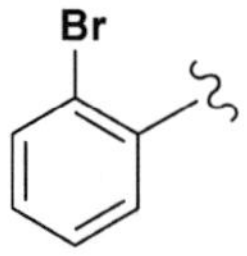 2-Br-phenyl	H	3-Me	<6.8
174	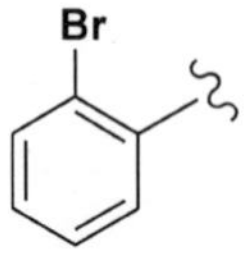 2-Br-phenyl	Me	3-Me	7.4
175	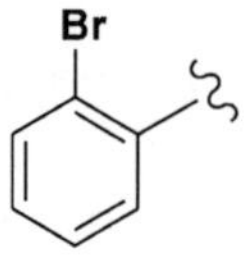 2-Br-phenyl	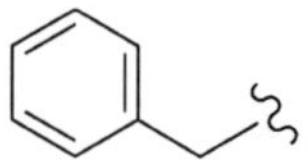 CH$_2$Ph	3-Me	<6.7
176	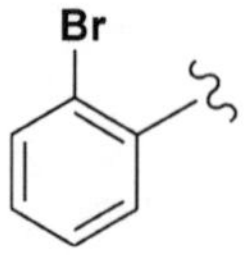 2-Br-phenyl	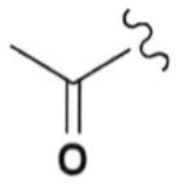 C(O)Me	3-Me	<6.6
177	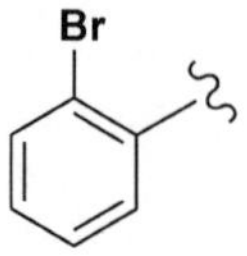 2-Br-phenyl	Et	H	7.2
178	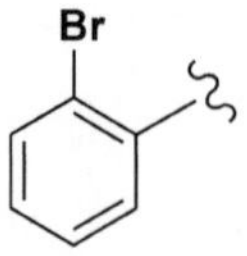 2-Br-phenyl	Et	2-Me	6.9
179	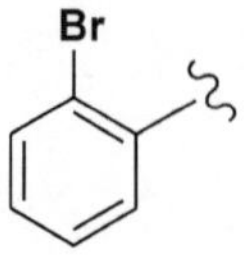 2-Br-phenyl	Et	4-Me	8.1

Table 22. (Continued)

Compound	Ar	R_1	R_2	pK_b
180	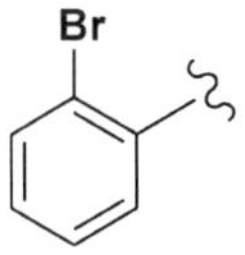 2-Br-phenyl	Et	3-F	7.6
181	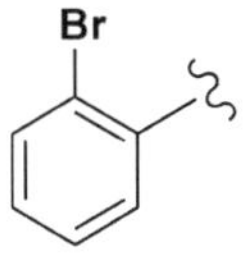 2-Br-phenyl	Et	3,4-di-F	7.8
182	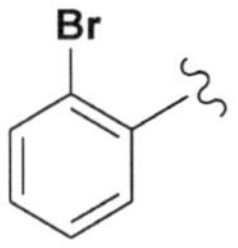 2-Br-phenyl	Et	3-F, 4-Me	8.2
Capsazepine				7.5
SB-366791				7.7

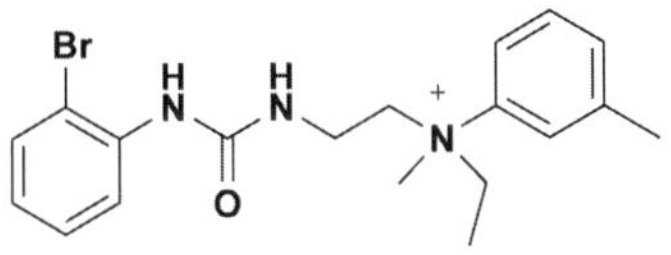

Figure 33. N-Methyl quaternary salt of SB-452533.

pared from **SB-452533** by treatment with methyl iodide, the group investigated the capsaicin binding site on TRPV1. Early reports suggested that the site was accessible from the cytosol [169]. However, when **183** was applied intracellularly it showed no activity up to 100 µM, whereas extracellular application provided full blockade of receptor activity at 1 µM, suggesting that the binding site for this class of compounds is only accessible from the cell surface.

The original screening hit, **SB-452533**, was chosen for further investigation in electrophysiology experiments where it was found to be a competitive, reversible inhibitor at the capsaicin binding site against capsaicin itself as well as heat and low pH. However, **SB-452533** showed high predicted intrinsic clearance in rat and human liver microsomes (>50 and 41 mL/min/g liver, respectively), effectively ruling it out for *in vivo* studies.

Further optimization of this chemotype was undertaken with the goal of improving pharmacokinetics, specifically high liver clearance which was assumed to be due to dealkylation of the N-ethyl group of **SB-452533** [167,170]. Initial work on a series of dihydroindoles, in which the N-ethyl group is cyclized onto the anilino-aryl ring resulted in a series of modestly potent antagonists (Table 23). The parent dihydroindole, **184**, was 10-fold less potent ($pK_b = 6.8$) than **SB-452533**. However, introduction of methyl or fluoro at either the 4- (**185**: $pK_b = 7.4$ and **187**: $pK_b = 7.4$, respectively) or the 5-position (**186**: $pK_b = 7.3$ and **188**: $pK_b = 7.1$, respectively) regained a good deal of activity. The ring-expanded 6-methyltetrahydroquinoline, **189**, was equipotent with **185** ($pK_b = 7.4$). However, the predicted liver clearance of this series was no better than the linear derivatives. For **188** and **189**, *in vitro* clearance values were >50 mL/min/g liver and >45 mL/min/g liver in rat and human liver microsomes, respectively.

The alternative cyclization of the ethyl group onto the ethylene spacer was next ex-

Table 23. Antagonist Activity (FLIPR) Versus Capsaicin Against Human TRPV1 Receptors

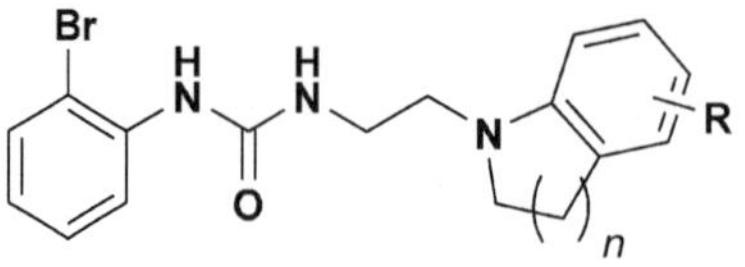

Compound	n	R	pK_b
184	1	H	6.8
185	1	4-Me	7.4
186	1	5-Me	7.3
187	1	4-F	7.4
188	1	5-F	7.1
189	2	6-Me	7.4

plored in a series of aminopyrrolidines. The stereochemical preference of the receptor was first identified: the S-isomer of the N-3-methylphenyl-aminopyrrolidine was at least 10-fold less active ($pK_b < 6.7$) than the R-isomer **190** ($pK_b = 7.7$). The SAR of further substitution on the phenyl ring (compounds **190–193**) closely tracked with that seen in the linear series (Table 24) in terms of both substitution preferences and level of potency. A series of N-(2-pyridyl)pyrrolidines was prepared in which the parent unsubstituted

Table 24. Antagonist Activity (FLIPR) Versus Capsaicin Against Human TRPV1 Receptors

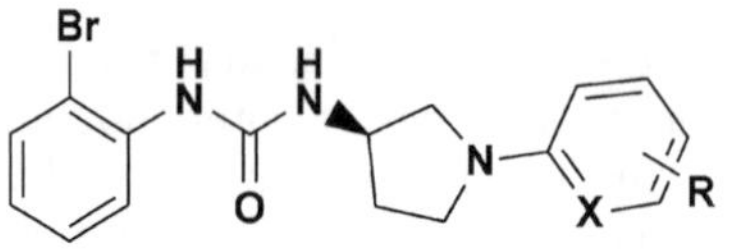

Compound	X	R	pK_b
190	C	3-Me	7.7
191	C	3-F	7.8
192	C	3,4-di-F	8.1
193	C	3-F, 4-Me	8.3
194	N	H	<6.6
195	N	3-CF_3	7.1
196	N	4-CF_3	6.9
197 SB-705498	N	5-CF_3	7.5
198	N	6-CF_3	7.2
199	N	5-Br	7.2
200	N	5-Cl	7.3
201	N	5-CN	6.7
202	N	5-Me	<6.7

pyridine (**194**) as well as the 5-methyl analog (**202**) were inactive ($pK_b < 6.7$), suggesting a lack of tolerance in the active site for the additional basic nitrogen. However, upon addition of electron-withdrawing groups to pyridine receptor activity approached that seen for phenyl derivatives. The exception is nitrile (**201**, $pK_b = 6.7$), which serves to emphasize the hydrophobic nature of the aryl binding pocket. When trifluoromethyl was moved around the ring, the 5-CF_3 analog (**197**) was found to be optimal ($pK_b = 7.5$). The corresponding chloro- (**200**) and bromo-pyridines were less active.

Although phenyl-pyrrolidines **190–193** exhibited the best antagonist activities against TRPV1 and showed much improved rat and human liver microsomal stabilities (≤ 11 and ≤ 4 mL/min/g liver) over previous series, their overall lipophilicity resulted in poor aqueous solubilities (<0.1 mg/mL). With significantly better solubility (0.4 mg/mL in simulated gastric juice), the 5-CF_3-pyridyl analog, **197** (**SB-705498**), was therefore chosen for further investigation [171]. The compound showed good selectivity against a broad panel of ion channels, receptors and enzymes (Cerep, Poitiers, France) and displayed rapid, potent, and reversible inhibition of capsaicin-, heat- and, low pH-mediated human TRPV1 activation in electrophysiology studies. With an attractive pharmacokinetic profile in three species (Table 25), including good oral bioavailabilities (39–86%), SB-705498 was evaluated in two in vivo pain models, capsaicin-induced hyperalgesia in the rat, and mechanical hyperalgesia in the guinea pig. In the rat model, SB-705498 showed dose-related activity from 3–30 mg/kg with good reversal of allodynia. Similarly, in the guinea pig the compound showed 80% reversal of allodynia.

A randomized, placebo-controlled Phase II study in acute migraine was begun in 2006 for SB-705498 [172]. The study has been terminated and no results have been publicly disclosed.

11. PREVENTATIVE MIGRAINE THERAPIES

Prophylactic therapy, in addition or as an alternative to acute therapy, is recommended

Table 25. Antagonist Activity Versus Capsaicin and Pharmacokinetic Profile for SB-705498 in Rat, Guinea Pig, and Dog

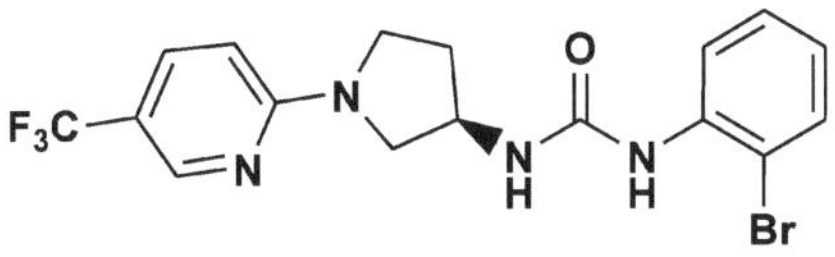

	pK_b	C_{Li} (mL/min/g liver)	$t_{1/2}$ (h)	V_{ss} (L/kg)	C_L (mL/min/kg)	F_{po} (%)
Rat	7.5	<0.5	3.1	0.9	4	86[a]
Guinea pig	7.3	4.2	2.6	3.6	34	39[b]
Dog	—	0.9	1.9	1.9	11	42[a]

[a] 3 mg/kg dose.
[b] 1 mg/kg dose.

for patients who have two or more acute migraine attacks per month, especially if these attacks are severe, prolonged and resistant to abortive medications [173–175]. Patients with less common migraine conditions such as familial hemiplegic migraine and those with prolonged aura are also candidates for preventative therapy. As discussed already, many or most migraineurs may be considered to have a low threshold for neuronal hyperexcitability, possibly due to ion channel abnormalities, that can be triggered by what could be considered minor day-to-day stresses. Repeated migraine attacks could result in a chronic state of dysregulation of pain pathways that might explain the refractoriness to acute therapy that some migraineurs experience [176].

All of the agents used for preventative migraine therapy were developed for other indications and most would be considered older drugs. All have potential limiting side effects and, since there are no helpful comparative clinical studies that show superiority of one drug, or class of drugs, over another, the choice of preventative therapy is usually based on potential side effects and patient comorbidities [177]. For this reason also, therapy is usually begun with a potential minimally effective dose which is slowly titrated up. There are also no studies of combination therapies for prophylaxis, so none can be recommended even though there may be reasons to expect they may be beneficial [178]. Mechanisms of action for these drugs in migraine are incompletely understood, if at all. However, since most older drugs can cause a substantial amount of polypharmacology, a connection with known migraine pathophysiology can usually be inferred. Using current preventative therapies, roughly two-thirds of patients can expect to experience a 50% reduction in headache frequency [179]. It is recommended that patients remain on a given drug for at least 2–3 months to adequately assess level of efficacy. Following 6 months or so of successful treatment, patients are often advised to break off therapy to determine whether a degree of "remission" has been achieved.

11.1. β-Blockers

β-Adrenergic receptor (adrenoreceptor) blockers are the most widely used class of drugs for migraine prophylaxis. Developed in the 1950s, for treatment of hypertension and cardiac arrhythmia, they modulate the sympathetic nervous system by blocking the action of endogenous amines such as epinephrine and norepinephrine on β-adrenoreceptors [180]. There are three β-adrenergic receptor subtypes: β1, β2, and β3. Subtypes β1 and, to a lesser extent, β2 are primarily involved in the sympatholytic actions of β-blockers. Upon activation, these GPCRs have an ionotropic effect by increasing intracellular calcium levels. Some β-blockers, such as pindolol, alprenolol, oxprenolol, and acebutolol, possess some partial agonist activity on β-adrenoreceptors (sympathomimetic activity) while still blocking epinephrine binding. These have not shown efficacy in migraine prophylaxis [179].

The mechanism of action of β-blockers in migraine is unclear. Some (but not all) have 5-HT$_{2B}$ and 5-HT$_{2C}$ receptor antagonist activity at pharmacologically relevant concentrations that may contribute to efficacy (IC$_{50}$ 200 nM for (-)-propranolol (Fig. 34) against the rat 5-HT$_2$ receptor [181]. It may be more likely that these compounds are acting on central β-adrenoreceptors [182]. Perhaps significantly, migaineurs show enhanced levels of centrally mediated secretion of epinephrine following light exposure that returns to normal after propranolol treatment [183]. Arguing against a "central-only" hypothesis is the fact that atenolol, a hydrophilic, brain-impenetrant β-blocker [184] has shown efficacy in several trials of migraine prophylaxis [185].

Adverse effects associated with β-blockers include fatigue, sleep disturbances, dizziness, and nausea [173]. In addition, a 2007 study suggested that β-blockers used to control hypertension significantly increase an individual's risk of developing type 2 diabetes [186]. β-Adrenoreceptors also stimulate cellular uptake of glucose. All β-blockers can cause an increase in serum levels of plasma glucose and triglycerides and a decrease in HDL cholesterol levels [187].

Propranolol, was the first β-blocker to be developed and shows no selectivity between the β1- and the β2-receptor subtypes. It has shown the most consistent efficacy in migraine prevention at a daily dose of 120–240 mg [188]. An analysis of 53 clinical trials showed that propranolol demonstrated a 44% reduction in attacks, based on daily headache readings, and a 65% reduction based on clinical ratings of improvement [189]. In addition, propranolol is the only agent that has been shown to be effective in preventing migraine in children (66% reduction in headache frequency at 60 mg for children weighing <77 lb or 120 mg for those >77 lb) [190].

Propranolol is a lipophilic β-blocker with a half-life of 4–6 h. It is often taken in divided doses twice a day, although a longer acting form is available [179]. The compound suffers extensive first-pass metabolism with a promi-

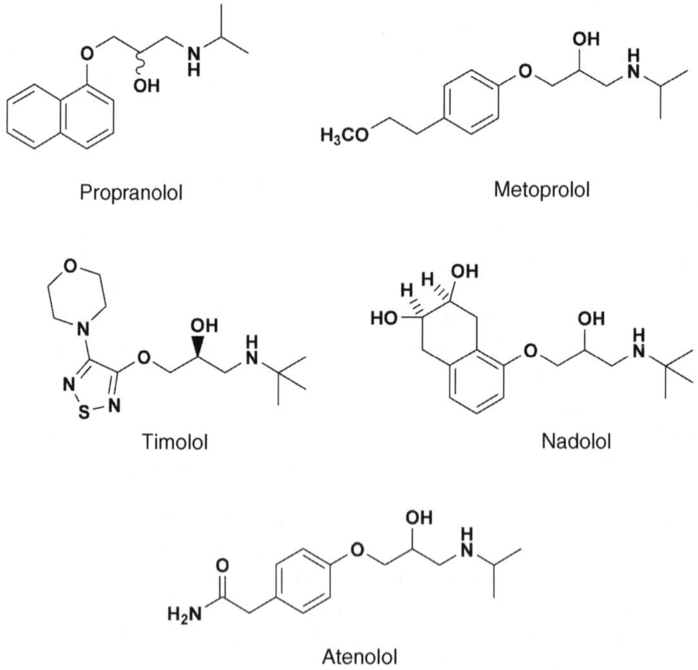

Figure 34. β-Blockers used for migraine prophylaxis.

nent, active metabolite, 4-hydroxypropranolol, most likely contributing to efficacy.

Other agents in this class have demonstrated efficacy in migraine prophylaxis but in fewer clinical trials. Timolol, another nonselective β-blocker, is also considered a first-line agent [173]. Nadolol, also nonselective for β1- and β2-receptor subtypes, is more hydrophilic than propanolol. Presumably because of low brain penetrance, nadolol has fewer CNS side effects than propranolol [191]. At 160 mg/day in one study, nadolol demonstrated superiority over propranolol in headache frequency, intensity, and use of acute relief medications. Metoprolol, a β1-selective agent, has been shown to be effective at 100–200 mg/day in reducing headache frequency, total number of migraine days and use of other analgesics [192]. Like other lipophilic β-blockers, metoprolol has a relatively short half-life (3–7 h), although it too is available in a long-acting formulation.

11.2. Anticonvulsants

The use of anticonvulsants (antiepileptic drugs) in migraine may have a more intuitive link to known pathophysiology since the primary goal of antiepileptic treatment is to suppress uncontrolled and excessive neuronal activation associated with the initial stages of seizures [193]. Indeed, a close pathophysiological connection between migraine and epilepsy has been noted, especially in the neuronal hyperexcitability that makes individuals susceptible to migraine attacks [194]. It is therefore not surprising that antiepileptic drugs are commonly used in migraine prophylaxis. There exist at least two dozen drugs marketed for epilepsy encompassing a wide range of chemical classes from very simple compounds such as potassium bromide, the earliest effective treatment, and small fatty amides and acids such as valproic acid, to more complex entities such as the benzodiazepines and topiramate (Fig. 35). Biological targets-of-action include voltage-gated calcium and sodium channels, and various receptors associated with GABA signaling (reviewed in Ref. [193]). There is also evidence that blockade of AMPA and KA ionotropic glutamate receptors are potential antiepileptic targets, a further connection with migraine (see Section 7) [195].

At the time of writing, anticonvulsants represent the largest class of drugs being tested in the clinic for migraine prophylaxis, according to ClinicalTrials.gov. The most-studied so far has been topiramate (Ortho-McNeil Pharmaceuticals), a monosaccharide derivative that was approved for this indication in 2004. It has multiple pharmacologic activities, most prominently suppression of neuronal hyperexcitability through AMPA and KA glutamate receptors and enhancement of GABA-activated chloride channel activity [196]. Abnormal GABA metabolism has been reported in migraineurs [106]. Adverse events in patients receiving 200–400 mg/day were CNS-related and included somnolence and fatigue, as well as psychomotor slowing, difficulty with concentration, speech hesitation, and difficulty finding words. These were generally mild to moderate in intensity. In addition, an increased frequency of kidney stones among patients taking topiramate has been reported, and may be at least in part due to modest inhibition of carbonic anhydrase [197].

In a number of 6-month trials for migraine prevention ($n > 1000$), topiramate at 100 mg/

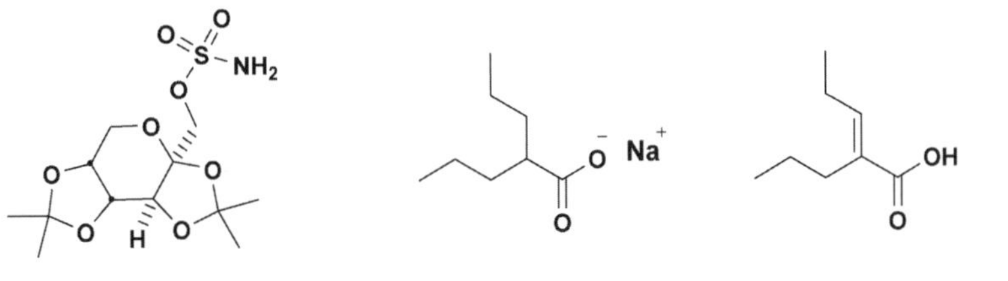

Figure 35. Anticonvulsants used for migraine prophylaxis.

day (50 mg/day, bid) demonstrated at least a 50% reduction in headache frequency in 52% of patients compared with 23% of those taking placebo [198]. Headache frequency was reduced by 75% for 27% of topiramate users. Paresthesia was a prominent side effect, occurring in 51% of patients. Although most (71%) were rated as mild, 20% of patients withdrew from studies due to adverse events, most because of paresthesias. Interestingly, patients taking topiramate in these trials experienced a dose-related weight loss, losing an average of 2.5 kg over 6 months. A clinical trial is planned to use MRI to study the occipital cerebral cortex of migraineurs taking topiramate (25–50 mg, bid) to determine whether the drug has lowered their excitability threshold [199].

Sodium valproate is another anticonvulsant with multiple mechanisms of action that has been in use for over 40 years. Like topiramate, it is thought to potentiate GABAergic functions (at least partly through inhibition of GABA transaminase) and modulate glutamate receptor activity (through NMDA receptors). However, it has a very broad spectrum of activity as an antiepileptic that can probably only be explained by extended polypharmacology (reviewed in Ref. [200]).

Valproic acid is a multifunctional molecule that has been used as an anticonvulsant and mood stabilizer since the 1960s [201]. In addition, it has demonstrated anticancer activity most likely through its inhibitory activity against histone deacetylases [202]. Divalproex sodium is a 1:1 complex of sodium valproate and valproic acid that was approved for migraine prophylaxis in the United States in 1996. A number of placebo-controlled clinical studies provide strong support for the efficacy of this formulation in migraine prophylaxis [203]. Valproate has high oral bioavailability (approximately 80%) and readily enters the brain. The concentration of valproate required to effectively inhibit GABA transaminase is higher than that likely to be achieved by patients, even at the average dose of almost 1 g/day. However, a major active metabolite, E-2-en-valproic acid, has shown comparable CNS penetration along with anticonvulsant and GABA potentiation *in vivo* [204]. Other possible mechanisms of action include direct effects on neuronal membranes, inhibition of aspartate release and kindling, and modulation of the central 5-HT system (reviewed in Ref. [179]). The most common side effects reported for valproate are nausea (almost 50%), drowsiness and weight gain. It is contraindicated in pregnant women because of teratogenicity.

Of particular interest is tonabersat (**SB-220453**) (Table 26), a newer anticonvulsant that, perhaps uniquely, was designed to be a migraine prophylactic agent by scientists at SKB [205]. The compound inhibits abnormal levels of neuronal excitability ("gap junction blocker") and is much more selective with a better defined mechanism of action in comparison with other anticonvulsants [206]. Tonabersat showed good activity in animal models of CSD and also inhibits neurogenic plasma extravasation [207] and showed no inherent *ex vivo* vasoconstrictive activity in comparison with sumatriptan [208]. Licensed by Minster Pharmaceuticals, the compound has shown early promise in a small Phase II clinical study ($n = 31$, 12 weeks, 20–40 mg/day), according to a company news item [209]. Results of a larger ($n = 500$) ongoing study are expected to appear in 2009 [210].

Other anticonvulsants have recently been tested in clinical trials for migraine prophylaxis (Table 26). Zonisamide has shown significant efficacy and has been further studied in patients refractory to topiramate where it has shown some benefit [211]. However, in a separate study in patients who had failed on an average of six other preventative agents, zonisamide did not show a significant effect [212]. In a small pediatric study ($n = 19$), levetiracetam (125–750 mg bid) eliminated headache in 10 patients and reduced severity and frequency in 7 patients. Isovaleramide (**NPS-1776**) was studied in a moderate-sized trial ($n = 189$) for acute migraine therapy but failed to meet the primary endpoint, pain relief at 2 h [213].

11.3. Calcium Channel Blockers

The initial rationale for use of voltage-gated calcium channel blockers (Fig. 36) for migraine prevention was that they might prevent vasoconstrictive hypoxia that was

PREVENTATIVE MIGRAINE THERAPIES 317

Table 26. Anticonvulsant Drugs in Recent Clinical Trials for Migraine Prevention

Compound	Structure	Company	Phase, Year (Clinical trials.gov Identifier)
SB-220453 Tonabersat	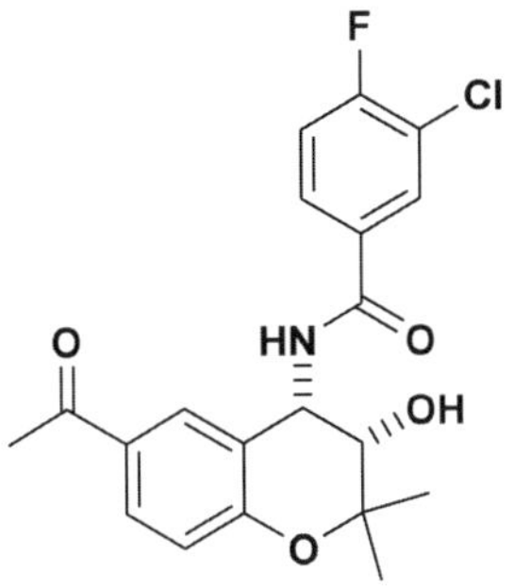	Minster Pharma	All Phase II, 2006 (NCT00311662); 2006 (NCT00332007); 2007-10 (NCT00534560)
Zonisamide	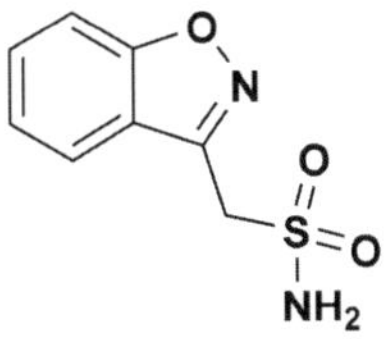	Elan Eisai (after 2003)	Phase II, 2002 (NCT00055484) Phase IV, 2004 (NCT00259636)
Levetiracetam	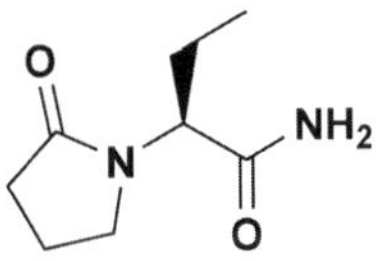	UCB Pharma	Phase II, 2002 (NCT00203216)
NPS-1776 Isovaleramide	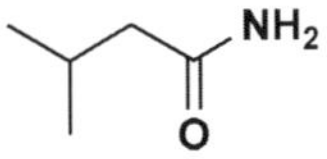	NPS Pharma	Phase II, 2003 (NCT00172094)
RWJ-333369 Carisbamate	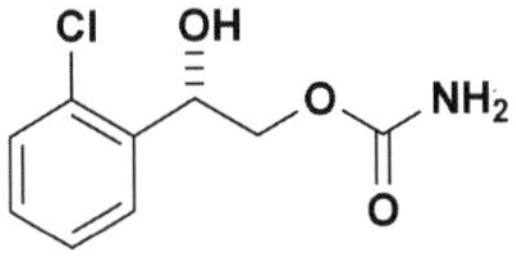	Johnson & Johnson	Phase II, 2005 (NCT00109083)
Lacosamide	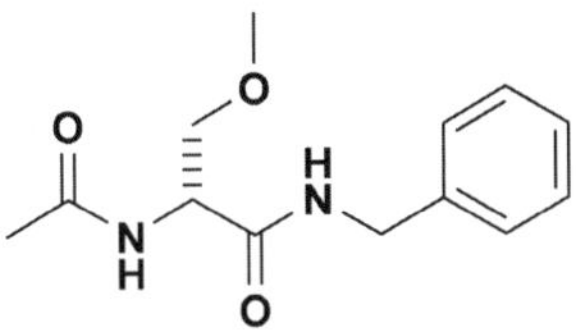	UCB Pharma	Phase II, 2008 (NCT00440518)

thought to be an initiating factor in the pathologic cascade. Subsequently, it has become clear that there are multiple potential mechanisms for their activity, including inhibition of CGRP release from neurons on which L-type calcium channels are colocalized [214]. The anticonvulsant, gabapentin is an N- and P/Q-Type calcium channel blocker that is often prescribed for neuropathic pain [215], an indication that was approved in the United States in 2002. Although it was originally synthesized as a GABA mimic, and does indeed modulate GABA function, it apparently does so without high affinity binding to GABA$_A$ or GABA$_B$ receptors [216]. The basis for its efficacy in prevention of migraine is

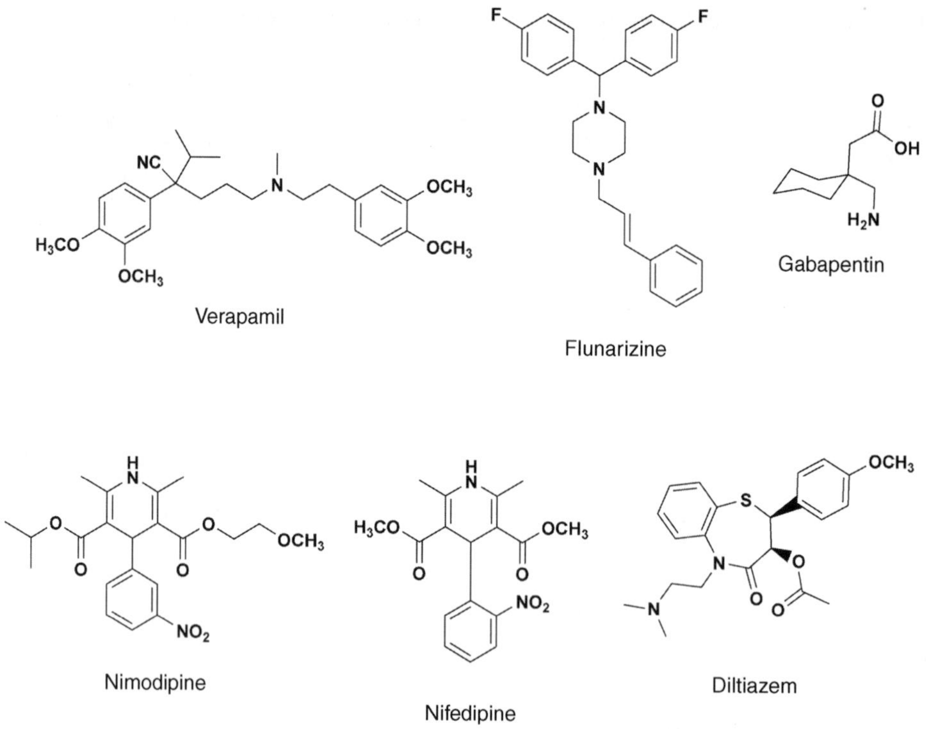

Figure 36. Calcium channel blockers used for migraine prophylaxis.

mainly based on one moderate-sized ($n = 143$) clinical trial that found that 2.4 g/day of gabapentin may be an effective preventive drug for patients with migraine headaches. However, there were frequent withdrawals due to adverse events such as dizziness and fatigue [217].

Flunarizine (outside of the United States) and verapamil are established antimigraine agents despite narrow therapeutic indices. There have been numerous trials comparing calcium channel blockers with drugs from other classes and most appear to show little or no differences in efficacy between various agents (summarized in Ref. [179]). Diltiazem, nifedipine, and nimodipine have also seen extensive use despite similarly mixed clinical results. Onset of migraine improvement often requires weeks of treatment with calcium channel blockers and headache frequently intensifies during the initial phase. In addition, adverse events such as dizziness, depression, weight gain, and flushing are relatively frequent with this class of drugs.

11.4. Antidepressants

Older tricyclic antidepressants, most prominently amitriptyline, nortriptyline, imipramine, and clomipramine (Fig. 37), have demonstrated benefits in migraine prevention that is apparently distinct from their efficacy in depression [218]. This idea is supported partly by the fact that onset of pain relief generally occurs more quickly than onset of antidepressant effects. In fact, tricyclic antidepressants are effective in the treatment of many types of chronic pain, including diabetic neuropathy, arthritic pain, postherpetic neuralgia and cancer pain [219]. The tricyclics are particularly beneficial for patients who suffer from both depression and migraine. However, they usually have significant side effects, including sedation, tachycardia, weight gain and cognitive disturbances. Their mechanism of action in migraine, and pain in general, is not well understood but may involve a combination of effects on inhibition of serotonin and norepinephrine uptake as well as sodium

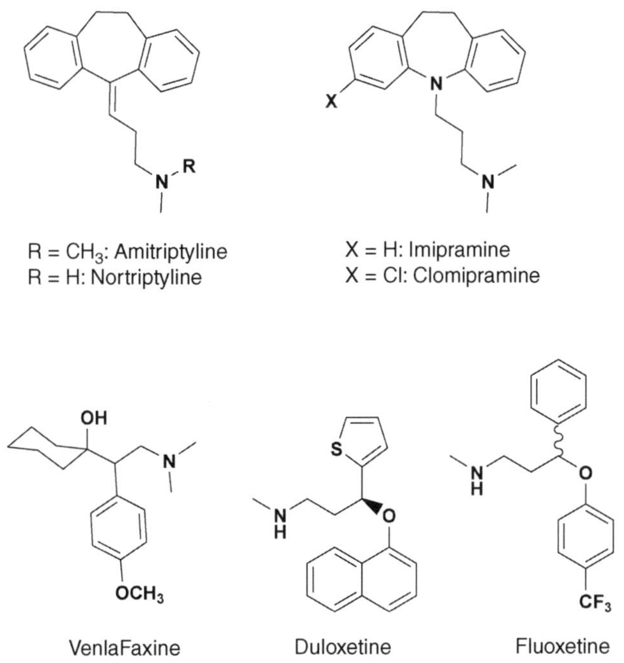

R = CH$_3$: Amitriptyline
R = H: Nortriptyline

X = H: Imipramine
X = Cl: Clomipramine

VenlaFaxine Duloxetine Fluoxetine

Figure 37. Antidepressants used for migraine prophylaxis.

channel blockade [220]. The tricyclics have been shown to be more effective in migraine prevention than the monoamine reuptake inhibitors, and this is generally true for other pain indications [221].

Serotonin-reuptake inhibitors (SSRIs) such as fluoxetine are sometimes helpful in reducing migraines, although in general their effects have been considered disappointing. Newer antidepressants targeting additional neurotransmitters, such as norepinephrine, alone or in addition to serotonin such as venlafaxine are showing some promise in preventing migraines. A retrospective analysis of patients taking duloxetine, a relatively new selective serotonin and norepinephrine reuptake inhibitor (SNRI), suggested a minimal effect in migraine prevention [222]. However, clinical trials continue to be carried out for this class of drugs [223].

11.5. Angiotensin Converting Enzyme Inhibitors

Angiotensin converting enzyme (ACE) catalyses the conversion of angiotensin I to angiotensin II, a potent vasoconstrictor, and is involved in the inactivation of bradykinin, a potent vasodilator [224]. ACE inhibitors (Fig. 38) are primarily used to control cardiovascular risk factors such as hypertension and treat congestive heart failure. It may seem counterintuitive that inhibition of this system would be useful in treatment of migraine. However, an associative genetic study has linked increased ACE activity to migraine frequency and prevalence [225], and other evidence has appeared to suggest it is a pathogenic mechanism common to arterial hypertension and migraine [226]. Lisinopril, a comparatively hydrophilic ACE inhibitor, was tested in a small ($n = 60$) placebo-controlled 12-week trial for migraine prevention. At 10 mg twice daily, it showed modest efficacy but was well tolerated [227].

A related mechanism that has been investigated is angiotensin II receptor inhibition. Both candesartan [228] and olmesartan [229] have shown efficacy in small migraine prophylaxis trials. ACE and angiotensin II inhibitors may be especially useful for migraineurs with comorbid hypertension.

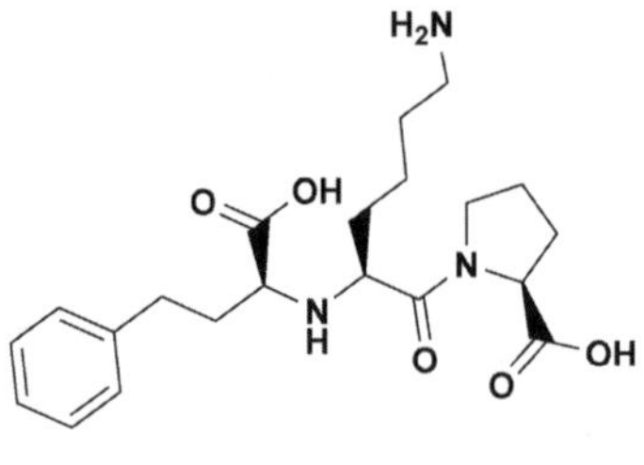

Figure 38. ACE inhibitors used for migraine prophylaxis.

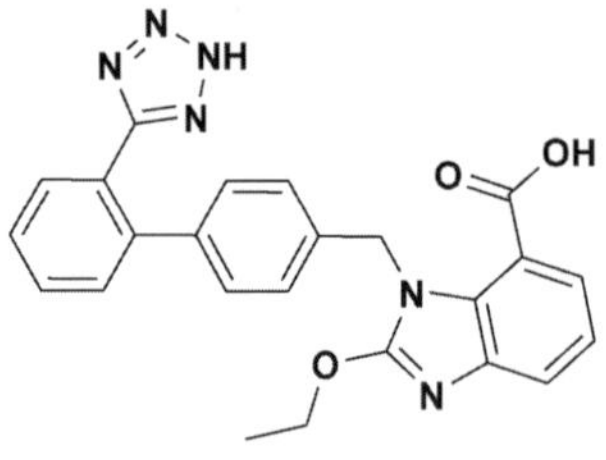

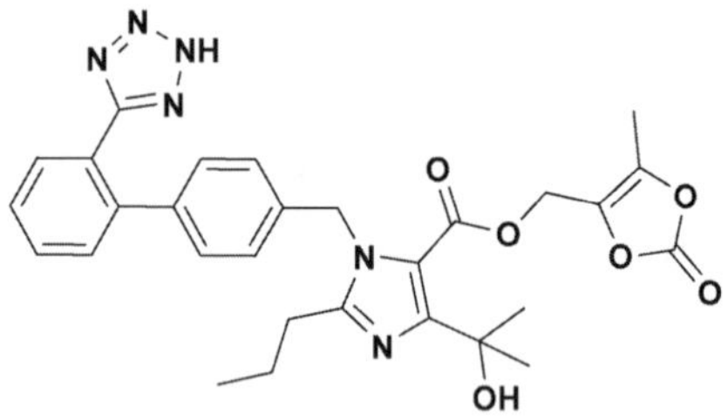

12. SUMMARY AND PERSPECTIVES

Migraine remains a severely undertreated condition, despite the availability of specific acute medications, the triptans, for over 15 years. Although representing a revolution in primary headache therapy at the time, the triptans have a number of shortcomings. These include cardiovascular side effects resulting from their inherent vasoconstrictive activities, and a significant proportion of refractory patients. The former, along with cost, are probably the most important factors that have limited the number of doses that can be taken per month, leading to many patients using them as a last resort at the height of headache pain, when they are likely to be least helpful.

The CGRP antagonists appear to represent the next revolution in migraine therapy, with clinical safety issues shown so far to be comparable to placebo. However, 2 h response rates for the two compounds that have been tested so far, **BIBN4096 BS** and **MK-0974**, are essentially identical to the triptans, although better results in 24 h pain measures and headache recurrence are more promising for the mechanism. For abortive therapy, there may be room for improvement, either with further optimized CGRP antagonists or with other mechanisms. However, it is also promising that **MK-0974** was being tested for migraine prophylaxis, presumably testifying to preclinical results showing that chronic CGRP inhibition is safe. Preventative migraine therapy has, until now, been dominated by older drugs developed for other indications. These agents often have significant side effects or are unsuitable for large classes of patients, such as blood pressure lowering medications for normotensive migraineurs.

Drugs that act on other relevant mechanisms such as ion channels and nitric oxide generation have been developed and tested in the clinic. These have yet to emerge as rivals to the triptans. However, the pathophysiology of migraine is complex and combination therapies that target multiple mechanisms, including one or more of these, may turn out to be more efficacious than monotherapy. An additional migraine target that has attracted recent interest is the orexinergic system as a regulator of trigeminovascular signaling [230].

An important aspect of migraine therapy that has been addressed with only partial success is speed of onset of pain relief. This may be improved to some extent by patients' ability to take safer agents such as CGRP antagonists orally at the first signs of a headache, and by the advent of rapidly disintegrating oral tablets. Intranasal delivery offers the potential for more rapid systemic exposure. In the case of zolmitriptan nasal spray, this has led to reports of significant reductions in treatment latency; and the highly potent CGRP antagonist **BMS-694153** has demonstrated C_{max} levels within 10 min in rabbits when dosed intranasally. Pulmonary delivery, as with the aerosolized vapor systems, **AZ-001** and **AZ-104**, promise even faster onset of efficacy, comparable to intravenous injection, although reports of clinical trials so far have not addressed this. Migraine patients are in particular need of improved nonoral drug delivery systems that can provide rapid relief of headache pain along with associated symptoms. It will be interesting to see how the medical marketplace, which has reflexively seen traditional oral delivery as the gold standard, accepts these potentially fast-acting formulation technologies.

REFERENCES

1. Borchardt JK. Drug News Perspect 1999;12(2): 123–127.
2. Sachs O. The Man Who Mistook His Wife for a Hat and Other Clinical Tales. New York: Perennial Library; 1987. Chapter 20.
3. Stewart WF, Shechter A., Rasmussen BK. Neurology 1994;44(6 Suppl 4): S17–S23.
4. Mathes RW, Malone KE, Daling JR, Davis S, Lucas SM, Porter PL, Li CI. Cancer Epidemiol Biomarkers Prev 2008;17(11): 3116–3122.
5. Scher AI, Stewart WF, Liberman J, Lipton RB. Headache 1998;38(7): 497–506.
6. The International Classification of Headache Disorders. 2nd ed. Available at http://ihs-classification.org/en/.
7. Kelman L. Headache 2006;46(6): 942–953.
8. Linde M. Acta Neurol Scand 2006;114(2): 71–83.
9. Scher AI, Terwindt GM, Picavet HS, Verschuren WM, Ferrari MD, Launer LJ. Neurology 2005;64(4): 614–620.
10. Bigal ME, Lipton RB. Neurology 2008;71(22): 1821–1828.
11. Catterall WA, Dib-Hajj S, Meisler MH, Pietrobon DJ. Neuroscience 2008;28(46): 11768–11777.
12. Tepper SJ, Dahlöf CG, Dowson A, Newman L, Mansbach H, Jones M, Pham B, Webster C, Salonen R. Headache 2004;44(9): 856–864.
13. Goadsby PJ. Drugs Fut 2006;31(11): 969–977.
14. Welch KM. Neurology 2003; 61 (8 Suppl 4): S2–S8.
15. Pietrobon D, Striessnig J. Nat Rev Neurosci 2003;4(5): 386–398.
16. Parsons AA, Strijbos PJ. Curr Opin Pharmacol 2003;3(1): 73–77.
17. Pietrobon D. Neuroscientist 2005;11(4): 373–386.
18. Longoni M, Ferrarese C. Neurol Sci 2006;27 (Suppl 2): S107–S110.
19. Edvinsson L, Jansen I, Cunha e Sá M, Gulbenkian S. Cephalalgia 1994;14(2): 88–96.
20. Dodick D, Silberstein S. Headache 2006;46 (Suppl 4): S182–S191.
21. Goadsby PJ, Lipton RB, Ferrari MD. N Engl J Med 2002;346(4): 257–270.
22. Rothlin E. Int Arch Allergy Appl Immunol 1955;7(4–6): 205–209.
23. Colman I, Brown MD, Innes GD, Grafstein E, Roberts TE, Rowe BH. Ann Emerg Med 2005;45(4): 393–401.
24. Schaerlinger B, Hickel P, Etienne N, Guesnier L, Maroteaux L. Br J Pharmacol 2003;140(2): 277–284.
25. Goadsby PJ. Prog Neurobiol 2000;62(5): 509–525.
26. Silberstein SD, McCrory DC. Headache 2003;43(2): 144–166.
27. Hinz B, Cheremina O, Brune K. FASEB J 2008;22(2): 383–390.
28. Buzzi MG, Sakas DE, Moskowitz MA. Eur J Pharmacol 1989;165(2–3): 251–258.
29. Nebe J, Heier M, Diener HC. Cephalalgia 1995;15(6): 531–535.
30. Blumenthal HJ, Weisz MA, Kelly KM, Mayer RL, Blonsky J. Headache 2003;43(10): 1026–1031.
31. MacGregor EA. Neurol Clin 1997;15(1): 125–141.
32. Rapport MM, Green AA, Page IH. Science 1948;108(2804): 329–330.
33. Muller-Schweinitzer E, Fanchamps A. Adv Neurol 1982;33:343–356.

34. Johnson KW, Phebus LA, Cohen ML. Prog Drug Res 1998;51:219–244.
35. Doenicke A, Brand J, Perrin VL. Lancet 1988;331(8598): 1309–1311.
36. Perrin VL, Färkhilä M, Goasguen J, Doenicke A, Brand J, Tfelt-Hansen P. Cephalalgia 1989;9(Suppl 9): 63–72.
37. Cady R, Schreiber C. Exp Opin Pharmacother 2006;7(11): 1503–1514.
38. Dahlöf C. Therapy 2005;2(3): 349–356.
39. Humphrey PPA., The discovery of sumatriptan and a new class of drug for the acute treatment of migraine. In: Humphrey P, Ferrari M, Olesen J,editors. Frontiers in Headache Research. Vol. 10:Oxford University Press, Inc.; 2001. p 3–10.
40. Tortorice K, Good CB.Drug Class Review: Oral 5-HT$_1$ Receptor Agonists. Available at http://www.pbm.va.gov/reviews/5ht1review.pdf. 2001. Accessed 2008 Dec 31.
41. Elkind AH, Ishkanian G, Mereddy SR. Drug Dev Res 2007;68:441–448.
42. Mett A, Tfelt-Hansen P. Curr Opin Neurol 2008;21(3): 331–337.
43. Buzzi MG, Moskowitz MA. Br J Pharmacol 1990;99(1): 202–206.
44. Humphrey PPA, Goadsby PJ. Cephalalgia 1994;14(6): 401–410.
45. Macor JE, Blank DH, Post RJ, Ryan K. Tetrahedron Lett 1992;33(52): 8011–8014.
46. Lee WS, Moskowitz MA. Brain Res 1993;626 (1–2):303–305.
47. Yu XJ, Waeber C, Castanon N, Scearce K, Hen R, Macor JE, Chauveau J, Moskowitz MA, Chaveau J. Mol Pharmacol 1996;49(5): 761–765.
48. Roon KI, Olesen J, Diener HC, Ellis P, Hettiarachchi J, Poole PH, Christianssen I, Kleinermans D, Kok JG, Ferrari MD. Ann Neurol 2000;47(2): 238–241.
49. Mueller L, Gallagher RM, Ciervo CA. Headache 1996;36(5): 329–331.
50. Tfelt-Hansen P, De Vries P, Saxena PR. Drugs 2000;60(6): 1259–1287.
51. Connor HE, Salonen R, Humphrey PPA. Sumatriptan and related 5-HT$_{1B/1D}$ receptor agonists: novel treatments for migraine. In: Bountra C, Munglani R, Schmidt WK,editors. Pain: Current Understanding, Emerging Therapies and Novel Approaches to Drug Discovery. Marcel Dekker, Inc.; 2003. p 743–748.
52. Mondell BE. Clin Therapeut 2003;25(2): 331–341.
53. Amara SG, Jonas V, Rosenfeld MG, Ong ES, Evans RM. Nature 1982;298(5871): 240–244.
54. Edvinsson L. Cephalalgia 2004;24(8): 611–622.
55. Goadsby PJ, Edvinsson L. Ann Neurol 1993;33 (1): 48–56.
56. Petersen KA, Lassen LH, Birk S, Lesko L, Olesen J. Clin Pharmacol Ther 2005;77(3): 202–213.
57. Lassen LH, Haderslev PA, Jacobsen VB, Iversen HK, Sperling B, Olesen J. Cephalalgia 2002;22:54–61.
58. Strecker T, Dux M, Messlinger K. J Vasc Res 2002;39(6): 489–496.
59. Shen YT, Pittman TJ, Buie PS, Bolduc DL, Kane SA, Koblan KS, Gould RJ, Lynch JJ Jr. J Pharmacol Exp Ther 2001;298(2): 551–558.
60. Olesen J, Diener HC, Husstedt IW, Goadsby PJ, Hall D, Meier U, Pollentier S, Lesko LM. BIBN 4096 BS Clinical Proof of Concept Study Group. N Engl J Med 2004;350:1104–1110.
61. Poyner DR, Andrew DP, Brown D, Bose C, Hanley MR. Br J Pharmacol 1992;105(2): 441–447.
62. Van Valen F, Piechot G, Jurgens H. Neurosci Lett 1990;119(2): 195–198.
63. Evans BN, Rosenblatt MI, Mnayer LO, Oliver KR, Dickerson IM. J Biol Chem 2000;275(40): 31438–31443.
64. McLatchie LM, Fraser NJ, Main MJ, Wise A, Brown J, Thompson N, Solari R, Lee MG, Foord SM. Nature 1998;393(6683): 333–339.
65. Brain SD, Poyner DR, Hill RG. Trends Pharmacol Sci 2002;23(2): 51–53.
66. Mallee JJ, Salvatore CA, LeBourdelles B, Oliver KR, Longmore J, Koblan KS, Kane SA. J Biol Chem 2002;277(16): 14294–14298.
67. Arulmani U, Gupta S, VanDenBrink AM, Centurión D, Villalón CM, Saxena PR. Cephalalgia 2006;26(6): 642–659.
68. Doods H, Hallermayer G, Wu D, Entzeroth M, Rudolf K, Engel W, Eberlein W. Br J Pharmacol 2000;129(3): 420–423.
69. Rudolf K, Eberlein W, Engel W, Pieper H, Entzeroth M, Hallermayer G, Doods H. J Med Chem 2005;48(19): 5921–5931.
70. Sheykhzade M, Lind H, Edvinsson L. Br J Pharmacol 2004;143(8): 1066–1073.
71. Doods H, Arndt K, Rudolf K, Just S. Trends Pharmacol Sci 2007;28(11): 580–587.
72. Doods H. Curr Opin Investig Drugs 2001;2(9): 1261–1268.

73. Petersen KA, Birk S, Doods H, Edvinsson L, Olesen J. Br J Pharmacol 2004;143(6): 697–704.
74. Iovino M, Feifel U, Yong CL, Wolters JM, Wallenstein G. Cephalalgia 2004;24(8): 645–656.
75. Petersen KA, Birk S, Lassen LH, Kruuse C, Jonassen O, Lesko L, Olesen J. Cephalalgia 2005;25(2): 139–147.
76. Petersen KA, Lassen LH, Birk S, Lesko L, Olesen J. Clin Pharmacol Ther 2005;77(3): 202–213.
77. The Subcutaneous Sumatriptan International Study Group. N Engl J Med 1991;325(5): 316–321.
78. Edvinsson L, Tfelt-Hansen P. Cephalalgia 2008;28(12): 1245–1258.
79. Boehringer-Ingelheim Announcement (IDdb3), May 15, 2006.
80. US20030191068A1, Published October 9, 2003.
81. Williams TM, Stump CA, Nguyen DN, Quigley AG, Bell IM, Gallicchio SN, Zartman CB, Wan BL, Penna KD, Kunapuli P, Kane SA, Koblan KS, Mosser SD, Rutledge RZ, Salvatore C, Fay JF, Vacca JP, Graham SL. Bioorg Med Chem Lett 2006;16(10): 2595–2598.
82. Edvinsson L, Sams A, Jansen-Olesen I, Tajti J, Kane SA, Rutledge RZ, Koblan KS, Hill RG, Longmore J. Eur J Pharmacol 2001;415(1): 39–44.
83. Burgey CS, Stump CA, Nguyen DN, Deng JZ, Quigley AG, Norton BR, Bell IM, Mosser SD, Salvatore CA, Rutledge RZ, Kane SA, Koblan KS, Vacca JP, Graham SL, Williams TM. Bioorg Med Chem Lett 2006;16(19): 5052–5056.
84. Shaw AW, Paone DV, Nguyen DN, Stump CA, Burgey CS, Mosser SD, Salvatore CA, Rutledge RZ, Kane SA, Koblan KS, Graham SL, Vacca JP, Williams TM. Bioorg Med Chem Lett 2007;17(17): 4795–4798.
85. Paone DV, Shaw AW, Nguyen DN, Burgey CS, Deng JZ, Kane SA, Koblan KS, Salvatore CA, Mosser SD, Johnston VK, Wong BK, Miller-Stein CM, Hershey JC, Graham SL, Vacca JP, Williams TM. J Med Chem 2007;50(23): 5564–5567.
86. Salvatore CA, Hershey JC, Corcoran HA, Fay JF, Johnston VK, Moore EL, Mosser SD, Burgey CS, Paone DV, Shaw AW, Graham SL, Vacca JP, Williams TM, Koblan KS, Kane SA. J Pharmacol Exp Ther 2008;324(2): 416–421.
87. Moore EL, Burgey CS, Paone DV, Shaw AW, Tang YS, Kane SA, Salvatore CA. Eur J Pharmacol 2009;602(2–3): 250–254.
88. Ho F TW, Mannix F LK, Fan F X, Assaid F C, Furtek F C, Jones F CJ, Lines F CR, Rapoport F AM. MK-0974 Protocol 004 study group. Neurology 2008;70(16): 1304–1312.
89. Ho TW, Ferrari MD, Dodick DW, Galet V, Kost J, Fan X, Leibensperger H, Froman S, Assaid C, Lines C, Koppen H, Winner PK. Lancet 2008;2472(9656): 2115–2123.
90. ClinicalTrials.gov Identifier: NCT00701389.
91. ClinicalTrials.gov Identifier: NCT00758836.
92. ClinicalTrials.gov Identifier: NCT00797667.
93. Nguyen DN, Paone DV, Shaw AW, Burgey CS, Mosser SD, Johnston V, Salvatore CA, Leonard YM, Miller-Stein CM, Kane SA, Koblan KS, Vacca JP, Graham SL, Williams TM. Bioorg Med Chem Lett 2008;18(2): 755–758.
94. Stump CA, Bell IM, Bednar RA, Bruno JG, Fay JF, Gallicchio SN, Johnston VK, Moore EL, Mosser SD, Quigley AG, Salvatore CA, Theberge CR, Zartman CB, Zhang XF, Kane SA, Graham SL, Vacca JP, Williams TM. Bioorg Med Chem Lett 2009;19(1): 214–217.
95. Paone DV. 235th ACS National Meeting, April 6–10, New Orleans LA, 2008. Abstract MEDI 11.
96. ClinicalTrials.gov Identifier: NCT00712725.
97. ClinicalTrials.gov Identifier: NCT00548353.
98. Bell IM 237th ACS National Meeting, March 22–26, Salt Lake City, UT, 2009. Abstract MEDI 28.
99. Roller S, Cui D, Laspina C, Miller-Stein C, Rowe J, Wong B, Prueksaritanont T. Xenobiotica 39:(1): 2009; 33–45.
100. Goadsby PJ, Zanchin G, Geraud G, de Klippel N, Diaz-Insa S, Gobel H, Cunha L, Ivanoff N, Falques M, Fortea J. Cephalalgia 2008;28(4): 383–391.
101. Rapoport AM, Bigal ME, Tepper SJ, Sheftell FD. CNS Drugs 2004;18(10): 671–685.
102. Dahlöf C. Neurology 2003;61(8 Suppl 4): S31–S34.
103. Dodick D, Brandes J, Elkind A, Mathew N, Rodichok L. CNS Drugs 2005;19(2): 125–136.
104. Dowson AJ, Charlesworth BR, Green J, Färkkilä M, Diener HC, Hansen SB, Gawel M., INDEX Study Group. Headache 2005;45 (1): 17–24.
105. Degnan AP, Chaturvedula PV, Conway CM, Cook DA, Davis CD, Denton R, Han X, Macci R, Mathias NR, Moench P, Pin SS, Ren SX, Schartman R, Signor LJ, Thalody G, Widmann KA, Xu C, Macor JE, Dubowchik GM. J Med Chem 2008;51(16): 4858–4861.

106. Wermeling DP, Miller JL, Rudi AC. Drug Del Technol 2002;2(1): 56–61.
107. Merkus P, Guchelaar HJ, Bosch DA, Merkus FW. Neurology 2003;60(10): 1669–1671.
108. Ferrari MD, Odink J, Bos KD, Malessy MJ, Bruyn GW. Neurology 1990;40(10): 1582–1586.
109. Ramadan NM. CNS Spectr 2003;8(6): 446–449.
110. Ma QP. Neuroreport 2001;12(8): 1589–1591.
111. Wollmuth LP, Sobolevsky AI. Trends Neurosci 2004;27(6): 321–328.
112. Sahara Y, Noro N, Iida Y, Soma K, Nakamura Y. J Neurosci 1997;17(17): 6611–6620.
113. Ornstein PL, Arnold MB, Allen NK, Bleisch T, Borromeo PS, Lugar CW, Leander JD, Lodge D, Schoepp DD. J Med Chem 1996; 39(11): 2219–2231.
114. Ornstein PL, Arnold MB, Allen NK, Bleisch T, Borromeo PS, Lugar CW, Leander JD, Lodge D, Schoepp DD. J Med Chem 1996;39 (11): 2232–2244.
115. Ornstein PL, Arnold MB, Augenstein NK, Lodge D, Leander JD, Schoepp DD. J Med Chem 1993;36(14): 2046–2048.
116. Murphy DE, Hutchison AJ, Hurt SD, Williams M, Sills MA. Br J Pharmacol 1988;95(3): 932–938.
117. Weiss B, Alt A, Ogden AM, Gates M, Dieckman DK, Clemens-Smith A, Ho KH, Jarvie K, Rizkalla G, Wright RA, Calligaro DO, Schoepp D, Mattiuz EL, Stratford RE, Johnson B, Salhoff C, Katofiasc M, Phebus LA, Schenck K, Cohen M, Filla SA, Ornstein PL, Johnson KW, Bleakman D. J Pharmacol Exp Ther 2006;318(2): 772–781.
118. Sang CN, Ramadan NM, Wallihan RG, Chappell AS, Freitag FG, Smith TR, Silberstein SD, Johnson KW, Phebus LA, Bleakman D, Ornstein PL, Arnold B, Tepper SJ, Vandenhende F. Cephalalgia 2004;24(7): 596–602.
119. Sang CN, Hostetter MP, Gracely RH, Chappell AS, Schoepp DD, Lee G, Whitcup S, Caruso R, Max MB. Anesthesiology 1998;89(5): 1060–1067.
120. ClinicalTrials.gov Identifier: NCT00567086.
121. Bleisch TJ, Ornstein PL, Allen NK, Wright RA, Lodge D, Schoepp DD. Bioorg Med Chem Lett 1997;7(9): 1161–1166.
122. Simmons RM, Li DL, Hoo KH, Deverill M, Ornstein PL, Iyengar S. Neuropharmacology 1998;37(1): 25–36.
123. Filla SA, Winter MA, Johnson KW, Bleakman D, Bell MG, Bleisch TJ, Castaño AM, Clemens-Smith A, del Prado M, Dieckman DK, Dominguez E, Escribano A, Ho KH, Hudziak KJ, Katofiasc MA, Martinez-Perez JA, Mateo A, Mathes BM, Mattiuz EL, Ogden AM, Phebus LA, Stack DR, Stratford RE, Ornstein PL. J Med Chem 2002;45(20): 4383–4386.
124. PDB entry: 2qs4. X-ray crystal structure available at: www.rcsb.org/.
125. Gottwald MD, Aminoff MJ. Drugs Today 2008;44(7): 531–545.
126. ClinicalTrials.gov Identifier: NCT00154063.
127. Friedman BW, Esses D, Solorzano C, Dua N, Greenwald P, Radulescu R, Chang E, Hochberg M, Campbell C, Aghera A, Valentin T, Paternoster J, Bijur P, Lipton RB, Gallagher EJ. Ann Emerg Med 2008;52(4): 399–406.
128. Isah AO, Rawlins MD, Bateman DN. Br J Clin Pharmacol 1991;32(6): 677–684.
129. Shen WW. Psychosomatics 1989;30(1): 118–119.
130. Rabinowitz JD, Wensley M, Lloyd P, Myers D, Shen W, Lu A, Hodges C, Hale R, Mufson D, Zaffaroni A. J Pharmacol Exp Ther 2004;309 (2): 769–775.
131. Avram MJ, Spyker DA, Henthorn TK, Cassella JV. Clin Pharmacol Ther 2009;85(1): 71–77.
132. Alexza Pharmaceuticals Company Announcement, March 28, 2007.
133. Alexza Pharmaceuticals Company Announcement, March 7, 2008.
134. Schulz R, Rassaf T, Massion PB, Kelm M, Balligand JL. Pharmacol Ther 2005;108(3): 225–256.
135. Wei EP, Moskowitz MA, Boccalini P, Kontos HA. Circ Res 1992;70(6): 1313–1319.
136. Olesen J, Iversen HK, Thomsen LL. Neuroreport 1993;4(8): 1027–1030.
137. Thomsen LL, Kruuse C, Iversen HK, Olesen J. Eur J Neurol 1994;1:73–80.
138. Olesen J. Pharmacol Ther 2008;120(2): 157–171.
139. Lassen LH, Ashina M, Christiansen I, Ulrich V, Grover R, Donaldson J, Olesen J. Cephalalgia 1998;18(1): 27–32.
140. Furfine ES, Harmon MF, Paith JE, Garvey EP. Biochemistry 1993;32(33): 8512–8517.
141. Moore WM, Webber RK, Jerome GM, Tjoeng FS, Misko TP, Currie MG. J Med Chem 1994;37(23): 3886–3888.
142. Park JM, Higuchi T, Kikuchi K, Urano Y, Hori H, Nishino T, Aoki J, Inoue K, Nagano T. Br J Pharmacol 2001;132(8): 1876–1882.

143. Alderton WK, Cooper CE, Knowles RG. Biochem J 2001;357(Pt 3): 593–615.
144. Garvey EP, Oplinger JA, Furfine ES, Kiff RJ, Laszlo F, Whittle BJ, Knowles RG. J Biol Chem 1997;272(8): 4959–4963.
145. Young RJ, Beams RM, Carter K, Clark HA, Coe DM, Chambers CL, Davies PI, Dawson J, Drysdale MJ, Franzman KW, French C, Hodgson ST, Hodson HF, Kleanthous S, Rider P, Sanders D, Sawyer DA, Scott KJ, Shearer BG, Stocker R, Smith S, Tackley MC, Knowles RG. Bioorg Med Chem Lett 2000;10(6): 597–600.
146. Alderton WK, Angell AD, Craig C, Dawson J, Garvey E, Moncada S, Monkhouse J, Rees D, Russell LJ, Russell RJ, Schwartz S, Waslidge N, Knowles RG. Br J Pharmacol 2005;145(3): 301–312.
147. De Alba J, Clayton NM, Collins SD, Colthup P, Chessell I, Knowles RG. Pain 2006;120(1–2): 170–181.
148. Dugo L, Marzocco S, Mazzon E, Di Paola R, Genovese T, Caputi AP, Cuzzocrea S. Br J Pharmacol 2004;141(6): 979–987.
149. ClinicalTrials.gov Identifier: NCT00319137.
150. ClinicalTrials.gov Identifier: NCT00242866.
151. Patman J, Bhardwaj N, Ramnauth J, Annedi SC, Renton P, Maddaford SP, Rakhit S, Andrews JS. Bioorg Med Chem Lett 2007;17(9): 2540–2544.
152. Patman J, Renton P, Rakhit S, Maddaford S, Andrews JS. 89th Canadian Chemistry Conference, CSC2006, Halifax, NS. Abstract 1004.
153. NeurAxon company news item: September 5, 2008.
154. US 2006/0258721 A1, Published November 16, 2006.
155. Szallasi A, Cortright DN, Blum CA, Eid SR. Nat Rev Drug Discov 2007;6(5): 357–372.
156. Ichikawa H, Sugimoto T. Brain Res 2001;890(1): 184–188.
157. Markowitz S, Saito K, Moskowitz MA. J Neurosci 1987;7(12): 4129–4136.
158. Van der Schueren BJ, de Hoon JN, Vanmolkot FH, Van Hecken A, Depre M, Kane SA, De Lepeleire I, Sinclair SR. Br J Clin Pharmacol 2007;64(5): 580–590.
159. Van der Schueren BJ, Rogiers A, Vanmolkot FH, Van Hecken A, Depré M, Kane SA, De Lepeleire I, Sinclair SR, de Hoon JN. J Pharmacol Exp Ther 2008; (1): 248–255.
160. Gavva NR, Bannon AW, Surapaneni S, Hovland DN Jr, Lehto SG, Gore A, Juan T, Deng H, Han B, Klionsky L, Kuang R, Le A, Tamir R, Wang J, Youngblood B, Zhu D, Norman MH, Magal E, Treanor JJ, Louis JC. J Neurosci 2007;27(13): 3366–3374.
161. Gavva NR, Treanor JJ, Garami A, Fang L, Surapaneni S, Akrami A, Alvarez F, Bak A, Darling M, Gore A, Jang GR, Kesslak JP, Ni L, Norman MH, Palluconi G, Rose MJ, Salfi M, Tan E, Romanovsky AA, Banfield C, Davar G. Pain 2008;136(1–2): 202–210.
162. Diamond S, Freitag F, Phillips SB, Bernstein JE, Saper JR. Cephalalgia 2000;20(6): 597–602.
163. Bevan S, Hothi S, Hughes G, James IF, Rang HP, Shah K, Walpole CS, Yeats JC. Br J Pharmacol 1992;107(2): 544–552.
164. Liu L, Simon SA. Neurosci Lett 1997;228(1): 29–32.
165. Docherty RJ, Yeats JC, Piper AS. Br J Pharmacol 1997;121(7): 1461–1467.
166. Gunthorpe F MJ, Rami F HK, Jerman F JC, Smart F D, Gill F CH, Soffin F EM, Luis Hannan F S, Lappin F SC, Egerton F J, Smith F GD, Worby F A, Howett F L, Owen F D, Nasir F S, Davies F CH, Thompson F M, Wyman F PA, Randall AD, Davis F JB. Neuropharmacology 2004;46(1): 133–149.
167. Rami HK, Thompson M, Stemp G, Fell S, Jerman JC, Stevens AJ, Smart D, Sargent B, Sanderson D, Randall AD, Gunthorpe MJ, Davis JB. Bioorg Med Chem Lett 2006;16(12): 3287–3291.
168. Rami HK, Thompson M, Wyman P, Jerman JC, Egerton J, Brough S, Stevens AJ, Randall AD, Smart D, Gunthorpe MJ, Davis JB. Bioorg Med Chem Lett 2004;14(14): 3631–3634.
169. Jung J, Hwang SW, Kwak J, Lee SY, Kang CJ, Kim WB, Kim D, Oh U. J Neurosci 1999;19(2): 529–538.
170. Westaway SM, Chung YK, Davis JB, Holland V, Jerman JC, Medhurst SJ, Rami HK, Stemp G, Stevens AJ, Thompson M, Winborn KY, Wright J. Bioorg Med Chem Lett 2006;16(17): 4533–4536.
171. Gunthorpe MJ, Hannan SL, Smart D, Jerman JC, Arpino S, Smith GD, Brough S, Wright J, Egerton J, Lappin SC, Holland VA, Winborn K, Thompson M, Rami HK, Randall A, Davis JB. J Pharmacol Exp Ther 2007;321(3): 1183–1192.
172. ClinicalTrials.gov indentifier: NCT00269022.
173. Modi S, Lowder DM. Am Fam Physician 2006;73(1): 72–78.

174. Silberstein SD, Lipton RB. Neurology 1994;44 (10 Suppl 7): S6–S16.
175. Schürks M, Diener HC, Goadsby PJ. Curr Treat Options Neurol 2008;10(1): 20–29.
176. Ramadan NM. Headache 2007;47(Suppl 1): S52–S57.
177. Evers S. Exp Opin Pharmacother 2008;9(15): 2565–2573.
178. Peterlin BL, Calhoun AH, Siegel S, Mathew NT. Headache 2008;48(6): 805–819.
179. Silberstein SD, Goadsby PJ. Cephalalgia 2002;22(7): 491–512.
180. Taira CA, Carranza A, Mayer M, Di Verniero C, Opezzo JA, Höcht C. Curr Clin Pharmacol 2008;3(3): 174–184.
181. Tinajero JC, Fabbri A, Dufau ML. Endocrinology 1993;133(1): 257–264.
182. Shields KG, Goadsby PJ. Brain 2005;128(Pt 1): 86–97.
183. Stoica E, Enulescu O. Eur Neurol 1990;30(1): 19–22.
184. Mahar Doan KM, Humphreys JE, Webster LO, Wring SA, Shampine LJ, Serabjit-Singh CJ, Adkison KK, Polli JW. J Pharmacol Exp Ther 2002;303(3): 1029–1037.
185. Forssman B, Lindblad CJ, Zbornikova V. Headache 1983;23(4): 188–190.
186. Elliott WJ, Meyer PM. Lancet 2007;369(9557): 201–207.
187. Pollare T, Lithell H, Mörlin C, Präntare H, Hvarfner A, Ljunghall S. J Hypertens 1989; 7(7): 551–559.
188. Andersson KE, Vinge E. Drugs 1990;39(3): 355–373.
189. Holroyd KA, Penzien DB, Cordingley GE. Headache 1991;31(5): 333–340.
190. Wasiewski WW. J Child Neurol 2001;16(2): 71–78.
191. Sudilovsky A, Elkind AH, Ryan RE Sr, Saper JR, Stern MA, Meyer JH. Headache 1987; 27(8): 421–426.
192. Olsson JE, Behring HC, Forssman B, Hedman C, Hedman G, Johansson F, Kinnman J, Pålhagen SE, Samuelsson M, Strandman E. Acta Neurol Scand 1984;70(3): 160–168.
193. Rogawski MA, Löscher W. Nat Rev Neurosci 2004;5(7): 553–564.
194. Rogawski MA. Arch Neurol 2008;65(6): 709–714.
195. Rogawski MA, Gryder D, Castaneda D, Yonekawa W, Banks MK, Lia H. Ann NY Acad Sci 2003;985:150–162.
196. Rosenfeld WE. Clin Ther 1997;19(6): 1294–1308.
197. Kossoff EH, Pyzik PL, Furth SL, Hladky HD, Freeman JM, Vining EP. Epilepsia 2002;43 (10): 1168–1171.
198. Silberstein SD. Headache 2005;45(Suppl 1): S57–S65.
199. ClinicalTrials.gov Identifier: NCT00286923.
200. Löscher W. CNS Drugs 2002;16(10): 669–694.
201. Henry TR. Psychopharmacol Bull 2003;37 (Suppl 2): 5–16.
202. Neri P, Tagliaferri P, Di Martino MT, Calimeri T, Amodio N, Bulotta A, Ventura M, Eramo PO, Viscomi C, Arbitrio M, Rossi M, Caraglia M, Munshi NC, Anderson KC, Tassone P. Br J Haematol 2008;143(4): 520–531.
203. Silberstein SD. Headache 1996;36(9): 547–555.
204. Nau H, Löscher W. J Pharmacol Exp Ther 1982;220(3): 654–659.
205. Chan WN, Evans JM, Hadley MS, Herdon HJ, Jerman JC, Parsons AA, Read SJ, Stean TO, Thompson M, Upton N. Bioorg Med Chem Lett 1999;9(2): 285–290.
206. Upton N, Thompson M. Prog Med Chem 2000;37:177–200.
207. Smith MI, Read SJ, Chan WN, Thompson M, Hunter AJ, Upton N, Parsons AA. Cephalalgia 2000;20(6): 546–553.
208. MaassenVanDenBrink A, van den Broek RW, de Vries R, Upton N, Parsons AA, Saxena PR. Cephalalgia 2000;20(6): 538–545.
209. Minster Pharmaceuticals press release, PRNewswire, October 24, 2008.
210. ClinicalTrials.gov Identifier: NCT00534560.
211. Pascual-Gómez J, Alañá-García M, Oterino A, Leira R, Láinez-Andrés JM. Rev Neurol 2008;47(9): 449–451.
212. Ashkenazi A, Benlifer A, Korenblit J, Silberstein SD. Cephalalgia 2006;26(10): 1199–1202.
213. NPS Pharmaceuticals press release, October 12, 2004.
214. Akerman S, Williamson DJ, Goadsby PJ. Br J Pharmacol 2003;140(3): 558–566.
215. Backonja MM, Serra J. Pain Med 2004;5 (Suppl 1): S28–S47.
216. Eckstein-Ludwig U, Fei J, Schwarz W. Br J Pharmacol 1999;128(1): 92–102.
217. Mathew NT, Rapoport A, Saper J, Magnus L, Klapper J, Ramadan N, Stacey B, Tepper S. Headache 2001;41(2): 119–128.
218. Galer BS. Neurology 1995;45(12 Suppl 9): S17–S25.

219. Guay DR. Pharmacotherapy 2001;21(9): 1070–1081.
220. Silberstein SD. Cephalalgia 1998;18(Suppl 21): 50–55.
221. Perrot S, Javier RM, Marty M, Le Jeunne C, Laroche F. Rheum Dis Clin North Am 2008;34 (2): 433–453.
222. Taylor AP, Adelman JU, Freeman MC. Headache 2007;47(8): 1200–1203.
223. ClinicalTrials.gov Identifier: NCT00443352.
224. Niu T, Chen X, Xu X. Drugs 2002;62(7): 977–993.
225. Paterna S, Di Pasquale P, D'Angelo A, Seidita G, Tuttolomondo A, Cardinale A, Maniscalchi T, Follone G, Giubilato A, Tarantello M, Licata G. Eur Neurol 2000;43 (3): 133–136.
226. Fanciullacci M, Alessandri M, De Cesaris F, Pietrini U. J Headache Pain 2004;5(Suppl 2): S85–S87.
227. Schrader H, Stovner LJ, Helde G, S and T, Bovim G. Br Med J 2001;322(7277): 19–22.
228. Tronvik E, Stovner LJ, Helde G, Sand T, Bovim G. J Am Med Assoc 2003;289(1): 65–69.
229. Charles JA, Jotkowitz S, Byrd LH. Headache 2006;46(3): 503–507.
230. Holland P, Goadsby PJ. Headache 2007;47(6): 951–962.

APPROACHES TO AMYLOID THERAPIES FOR THE TREATMENT OF ALZHEIMER'S DISEASE

ALBERT J. ROBICHAUD[1]
JI-IN KIM[1]
J. STEVEN JACOBSEN[2]
[1]Chemical Sciences, Wyeth Research, Princeton, NJ
[2]Neuroscience, Wyeth Research, Princeton, NJ

1. INTRODUCTION

Recognized as the most common of the neurodegenerative disorders, Alzheimer's disease (AD) affects greater than 18 million people worldwide and is predicted to become the largest socioeconomic burden of the 21st century [1–3]. Conservatively, it is estimated to affect the elderly, at approximately 60 years of age, at a rate of 1–1.5%, with the risk doubling every 5 years to greater than 40% incidence among those at 80–85 years of age. Characterized by a gradual and increasing loss of memory and cognitive function, AD is accompanied, in the later stages, by severe behavioral abnormalities, including psychosis, depression, physical and verbal aggression, agitation, sleep disturbances, and hyperphagia [4,5].

The underlying mechanism of Alzheimer's disease is as yet unknown, but the growing body of pathophysiologic evidence is beginning to point to two main causes. The hallmark neuropathological changes of the disease are the formation of β-amyloid plaques, neurofibrillary tangles (NFT) and ultimately neuronal death (Fig. 1) [5–9]. This is commonly referred to as the amyloid hypothesis of Alzheimer's disease.

The prevention of the accumulation, aggregation, and deposition of β-amyloid peptide (Aβ) monomers, leading first to multiple oligomeric species and then to fibrils and ultimately senile plaques, is the center of focus for an approach to disease modification. The extracellular deposits, comprised of a proteinacious core of insoluble aggregated Aβ peptides [10], is typically surrounded by neurites, activated microglia, and reactive astrocytes. The Aβ peptide monomers are hydrophobic 39–42 amino acid peptides, derived from the sequential cleavage of the type 1 membrane protein, amyloid precursor protein (APP), by two key enzymes (discussed in Section 2). There is a wealth of evidence to implicate this peptide and its subsequent aggregation in the etiology of the disease [11]. The abundant presence of these deposits in brains of elderly Alzheimer's disease patients as well as the well-documented neurotoxic effects of the aggregated and soluble oligomeric species strongly suggests that excessive production or the inability to clear this peptide from the brain is detrimental to the viability of neurons [12,13]. Irrespective of the cause of the toxic species or the exact nature of the peptide involved in this deposit, it is likely that there exists a delicate balance between Aβ production and catabolism, either through peptidases or direct clearance, in normal aging individuals. It is apparent that through the aging process this equilibrium is changed causing an imbalance and overproduction, or decreased catabolism and/or clearance, of the polymeric Aβ peptide. The result is an accumulation of the synaptotoxic Aβ species, subsequent amyloid deposits, and ultimately neuronal degeneration. It would follow that approaches to shift this equilibrium back to normal, through the multitude of possibilities presented in the amyloid cascade, would represent a viable approach to disease modification.

The vast majority of research over the last decade has been, and continues to be, focused on these disease-modifying, antiamyloidogenic approaches to AD. It is this subject, and the various approaches, past and present to amelioration of AD through the various agents being explored, that will be the focus of this chapter [14–24]. Thus, any approach that addresses the amyloid hypothesis, that is, Aβ formation, Aβ clearance, Aβ protein aggregation, neurofibrillary tangle formation, amyloid plaque deposition, or neuronal degradation or neuronal cell death due to Aβ will be considered disease-modifying and will be discussed in this review.

Nonetheless, one must not neglect the importance and potential value of palliative approaches to the treatment of the disease, those targeted at the symptoms of the disease and not the presumed underlying causes. There

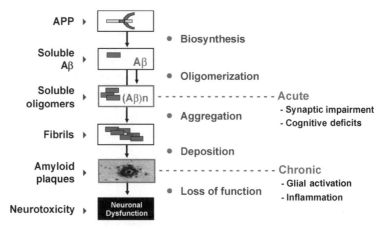

Figure 1. Amyloid cascade hypothesis of Alzheimer's disease. (This figure is available in full color at http://mrw.interscience.wiley.com/emrw/9780471266945/home.)

are a plethora of approaches to cognitive dysfunction and synaptic deficiencies that have been the focus of vast research efforts for the past 2 decades [25]. In many instances the targeted approach may have some ancillary data to support a disease modifying role, but at present the validation has not been made and for the purposes of this review they will be considered as symptomatic treatments of the disease. Presently, the only approved treatments for AD are from this class, and the modest efficacy realized with these entities serves as the main cause for the search for disease modifying medicaments. A detailed discussion of this approach to AD can be found in Chapter xx of this book series and will not be discussed in any detail here.

2. THE ROLE OF THE SECRETASES IN APP PROCESSING

Although the native function of APP remains unknown, the protein is highly conserved throughout evolution and is expressed widely in many different cell types [26]. It is a membrane bound protein of which the predominant brain derived species is 695 amino acids in length. Figure 2 shows the partial sequence of APP and the respective cleavage sites of the three known secretases within the peptide backbone. As can be seen, BACE (β-site cleaving enzyme) and α-secretase cleave in the luminal region on the N-terminal end, while γ-secretase is a membrane-embedded protein complex that cleaves within the membrane spanning region. The sequence of cleavage and possible toxic and nontoxic routes in the amyloid cascade are represented in Fig. 3. From this, one can see there are two possible fates of the catabolism of APP, toxic and nontoxic. In the latter pathway, APP is first cleaved by α-secretase to afford the membrane bound C83 fragment and the sAPPα fragment released into the lumen and is ultimately secreted. Subsequent cleavage of C83 by γ-secretase affords the nonamyloidogenic p3

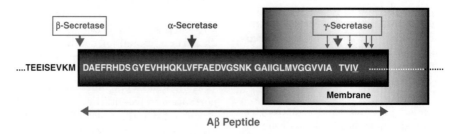

Figure 2. Sequence of APP and respective cleavage sites of secretases. (This figure is available in full color at http://mrw.interscience.wiley.com/emrw/9780471266945/home.)

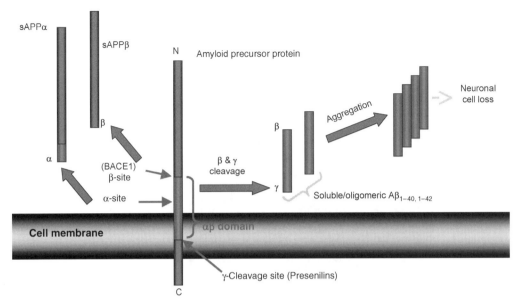

Figure 3. Formation of Aβ peptide by secretase processing. (This figure is available in full color at http://mrw.interscience.wiley.com/emrw/9780471266945/home.)

fragment. This pathway is preferred and leads to a harmless catabolism of the APP peptide. Alternatively, the toxic pathway is initiated by the rate determining cleavage of APP by β-secretase, yielding the C99 membrane bound fragment and the luminal sAPPβ species. Further processing of C99 by γ-secretase, at two predominant sites, results in the formation of the undesirable $A\beta_{40-42}$ monomers (40 and 42 amino acid peptide fragments). The relevance of the toxicological differences of the closely related $A\beta_{40}$ and $A\beta_{42}$ species is complicated and will not be detailed here, other than to say both species have been implicated in the progression of the disease and the details of their respective roles is the subject of much effort and debate [27–30]. As was described earlier, agents that work to prevent the formation of the toxic $A\beta_{40-42}$ species have the potential to effect disease modification. Therefore, inhibition of either β- or γ-secretase, or increase in the activity and efficiency of α-secretase, presents viable targets for the approach to the treatment of AD.

2.1. β-Secretase

BACE is a Type-1 aspartyl protease discovered in 1999 by multiple groups [31–33]. The enzyme contains two key aspartic acid residues as the central warhead and the vast majority of inhibitors have been shown to bind here extending to both the prime and unprimed side of the catalytic region (Fig. 4). In humans, there are two known β-secretase enzymes, BACE1 (beta-site APP cleaving enzyme 1, ASP2, memapsin 2) and BACE2 (beta-site APP cleaving enzyme 2, ASP1, memapsin1), colocalized with APP in the endosomal compartment [34]. Whereas both process the APP protein at the same site, only BACE1 is significantly expressed in brain, particularly in neurons indicating that these cells are the major source of Aβ peptides in brain. The two subtypes are highly homologous with subtle differences around the active site and a disulfide bond unique to BACE1. While it is known that BACE2 is expressed in heart, kidney, and placenta, ligands targeting BACE1 will likely need to demonstrate some level of selectivity against the BACE2 subtype to prevent potential unwanted side effects in the periphery [35,36]. The interest in BACE as a target was bolstered in 2000 when the BACE1 knockout (KO) mouse was cloned and shown to develop normally [37]. Furthermore, it was shown that these mice lacked the ability to form $A\beta_{40-42}$ and subsequently were devoid

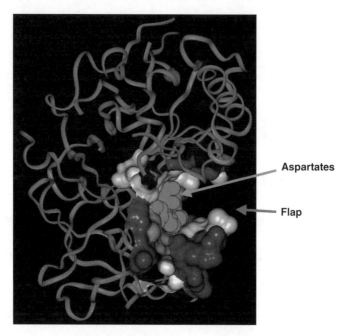

Figure 4. BACE1 enzyme. (See color insert.)

of plaque formation resulting from peptide accumulation. Later on it was reported that BACE1 knockout mice have developmental issues associated with decreased myelin formation, but as this is a necessary component of maturation from birth, the relevance of BACE inhibition for elderly subjects may be without this adverse consequence [38,39].

From the outset, the availability of three-dimensional structural information has enabled the rapid development of potent and selective ligands for this site. The first reported X-ray structure of BACE1 complexed with an inhibitor, by Ghosh and Tang in 2000 (Fig. 5), allowed for a greater understanding of the enzyme binding properties of BACE1, and served as the launching point for a multitude of research groups [40]. The inhibitor, OM99-2, an 8 residue substrate-based peptidomimetic, spans the P4 to P4′ binding

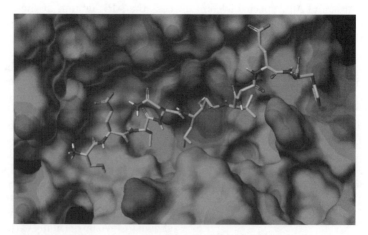

Figure 5. BACE1 enzyme crystallized with OM00-3. (See color insert.)

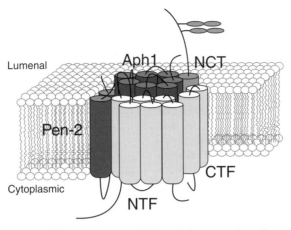

Figure 6. γ-Secretase complex. (This figure is available in full color at http://mrw.interscience.wiley.com/emrw/9780471266945/home.)

pockets of the enzyme site. From this structure it can be seen that the BACE enzyme possesses strong similarities to pepsin-like aspartyl proteases, with an extensive β-sheet organization, and two key aspartyl residues (Asp32 and Asp228) comprising the catalytic machinery for the peptide bond cleavage of APP. As a drug target, BACE1 possesses some unique issues and challenges. Like other aspartyl proteases, BACE1 has a large and shallow active site, and peptide ligands of high affinity usually require 8–10 residues (i.e., OM00-3). The need for an extensive hydrogen bonding network of these large ligands does not necessarily lend itself to the design of small-molecule hydrophobic entities with adequate drug-like properties. Furthermore, the obvious similarities to other aspartyl proteases, such as renin, pepsin, and Cathepsin D, necessitates the requirement for selectivity over these essential enzymes, as inhibition presents the likelihood of unwanted side effects and potential toxicity. Since the seminal work of Ghosh and Tang nearly a decade ago, the plethora of crystallographic data and structure-based design advances in ligands for the BACE1 site has been critical in allowing for the understanding of this important target and paved the way for entities that will explore the feasibility of BACE inhibitors for the treatment of AD. Section 3 describes in more detail the efforts and advances made in the various drug design approaches to BACE1 inhibitors.

2.2. γ-Secretase

γ-Secretase is a multiprotein heterooligomer comprised of at least four protein components: presenilin (PS1/PS2), nicastrin (NCT), APH-1, and PEN-2, in a high molecular weight complex of unknown stoichiometry (Fig. 6) [41]. This complex is reportedly assembled in the endoplasmic reticulum (ER) and rapidly locates to the plasma membrane. The contributions made by each of the subunits are only now being realized, and it is possible that each of the subunits may be a target for therapeutic intervention. The inability to crystallize the complex and to readily access stable preparations has added to the difficulty in exploring this target as a potential treatment for AD. However, the inhibition of γ-secretase complex must be considered as a viable target as the processing of the membrane bound C83 and C99 fragments, products of α- and β-secretase cleavage, respectively, is completed by this catabolism and represents a direct mechanism to reduce the generation of Aβ. There is, however, a potential liability of this target as it is known that a number of other proteins are also substrates of this enzyme complex. In particular, the processing of the Notch receptor has been reported to be inhibited by γ-secretase blockers [17]. Inhibition of Notch processing has been implicated in various gastrointestinal toxicities and clearly would diminish the value of such an agent [42]. Therefore, any such inhibitors will most likely need to demonstrate some level of selectivity against Notch and any other γ-secretase targets.

2.3. α-Secretase

α-Secretase is an ADAM family metalloproteinase and cleaves at the Lys16-Leu17 site within the Aβ domain of APP [43]. As was mentioned, this is the nontoxic catabolic pathway of APP processing, resulting in fragments that have not been shown to be associated with the generation of AD. Enhanced cleavage at this site may represent a disease modifying strategy [44], however, approaches targeting elevation of aspartyl protease activity are less mature and would represent an area worthy of additional focus.

3. APPROACHES TO DISEASE-MODIFYING TREATMENT IN AD

The following are areas with significant research activity as potential approaches to AD centered on the aforementioned amyloid cascade. This chapter will examine both the preclinical and clinical approaches in each respective targeted area. In many of the approaches, clinical entities have yet to be identified, although there may be significant efforts underway to identify candidates for the particular approach. A multitude of ligands are currently in clinical trials, based on amyloid-driven therapy and the data from these compounds should help test this mechanism as a disease-modifying therapy. Each of these potential therapeutics is discussed in their respective section.

As much of the progress toward AD is still preclinical, selected key examples in each of these areas from the last several years will be detailed. Several of the sections are focused on the extensive preclinical efforts and separately highlight the clinical progress in that particular area, while others are combined discussions. It should be noted that this overview is by no means exhaustive in the reference to specific research efforts, particularly in the preclinical sections, but is meant to highlight areas where a more concentrated effort has been evolving. The interested reader is referred to the relevant references and various reviews cited throughout this chapter for additional detail in each of the target areas.

3.1. γ-Secretase Inhibitors—Preclinical

There has been an extensive effort directed toward the discovery and subsequent development of γ-secretase inhibitors (GSIs). Two very recent reviews have appeared giving detailed accounts of this research [45,46]. Several of these campaigns have resulted in clinical candidates that will be detailed in this section.

3.1.1. Early Peptide Isostere γ-Secretase Inhibitors

Transition-State Analogs The earliest investigations to identify GSIs included peptide isosteres that were reported to bind to the active site in the γ-secretase (GS) complex. Binding of peptide isostere ligands provided important insight into GS-mediated cleavage of the substrate, APP. The identification that presenilin is the catalytic subunit of γ-secretase and that APP cleavage mechanism is similar to aspartyl proteases, ultimately enhanced the design of GSIs [47–49].

Merck's L-685458 (**1**) and related derivative **2** (Scheme 1) are some of the early examples of peptide isostere transition-state analogs (containing an APP-CTF scissile bond mimetic). After numerous modifications involving cross-linking, photolabeling and altering of the helical nature of the ligands by variations in the backbone structures and stereochemistry, it was determined that the overall understanding of the substrate requirement was still minimal [50–52]. This poorly defined understanding was attributed to many possibilities; that the model may be inadequate, that the complexity of sequential cleavage reactions made direct structure–activity relationship (SAR)-based approaches difficult, or that the transition-state isosteres may not interact at the γ-secretase active site as predicted [46].

Additional derivatives identified through modifications included hydroxyethyl ureas (HEUs), such as compound **3**, that have an altered substrate "shape" for probing the γ-secretase active site. Other examples such as peptide aldehyde MDL-21870 (**4**) and urea

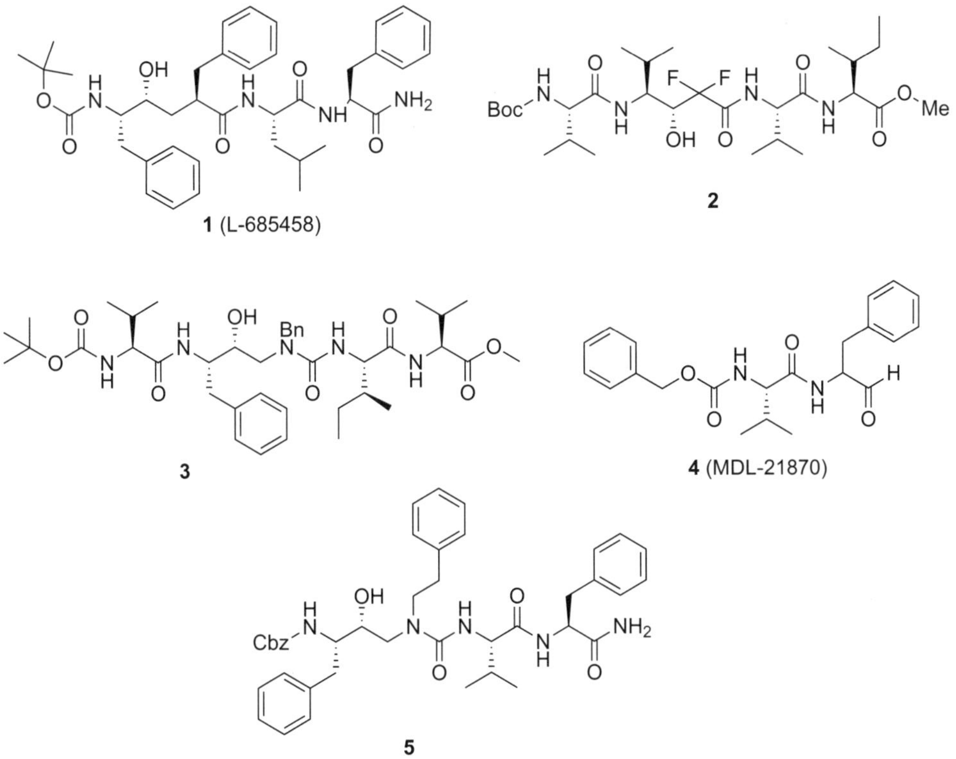

Scheme 1.

analog (5) from Merck along with L-685458 were aimed at understanding differential inhibition of APP versus Notch cleavage [53,54].

There are two key caveats for development of peptide isostere-based GSIs for therapy. First, due to the presence of many peptidic bonds, their pharmacokinetic (PK) properties are predicted to be poor and their utility via oral dosing may be severely limited. [55–57] Second, and most importantly for this target, these peptide-derived GSIs are highly potent but do not provide selectivity between APP processing and Notch processing. To date, not a single peptide isostere GSI has advanced into clinical development.

In addition to the peptide isosteres, the identification of difluorophenacylalanine (DAPT) (6) and DAPT-related structures as potent GSIs allowed for these to be used as powerful tools in understanding the presenilin binding mode (Scheme 2). Cross-linking studies and competition experiments revealed that photoactivable derivatives of DAPT were shown to bind presenilin at a different site than peptide isosteres [58–60]. Researchers at Bristol-Myers Squibb (BMS) identified a related analog 7, resulting from optimization of a high-throughput screen (HTS) hit, which has features that hybridized the peptide isostere backbone and a benzyl moiety of DAPT. This derivative, with reduced peptide-like features, possesses a substituted benzyl group as a replacement of the P2 Val or P2 NHBoc group present in earlier peptide isosteres, such as 1, 2, and 4, affording a highly potent small-molecule inhibitor (IC$_{50}$ = 1 nM). This work has added to the understanding of the scaffold requirement for potent GSIs, however derivative 7 was not further pursued because it reached only submicromolar brain exposure level when Tg2576 mice (mouse line expressing human APP transgene) [61] were dosed orally with no significant lowering of brain Aβ levels [62].

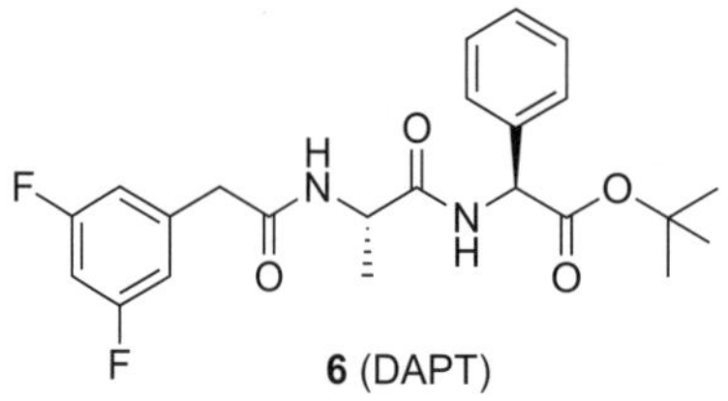

336 APPROACHES TO AMYLOID THERAPIES FOR THE TREATMENT OF ALZHEIMER'S DISEASE

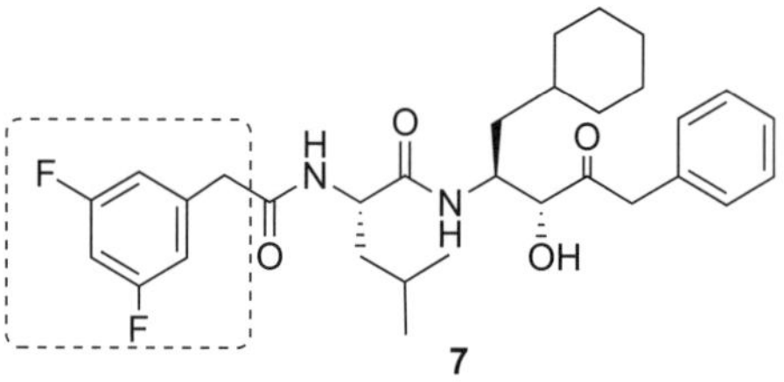

Scheme 2.

Azepine and Diazepine-Derived γ-Secretase Inhibitors Research insights gained through the development of peptide isosteres also helped to shape the SAR around a different class of GSIs, the benzazepines (Scheme 3). This novel class of GSIs, initially reported by Lilly and Elan possesses a dipeptide-like moiety linked to a lactam. Researchers at Lilly successfully identified a lead azepine inhibitor **8** (LY-411,575) from the series that was the basis for their first clinical candidate LY-450,139, semagacestat (discussed later). Azepine **8** is a 30 pM inhibitor, which also demonstrated *in vivo* efficacy in PDAPP mice ((ED$_{50}$ < 1 mg/kg (mpk) oral (p.o.)) by lowering brain Aβ [63]. Unfortunately, at high doses,

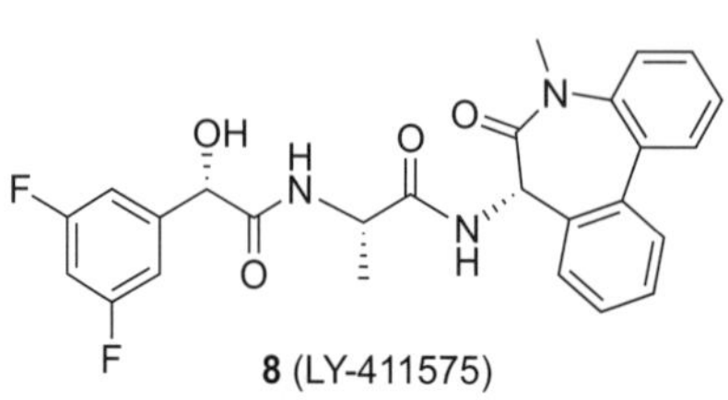

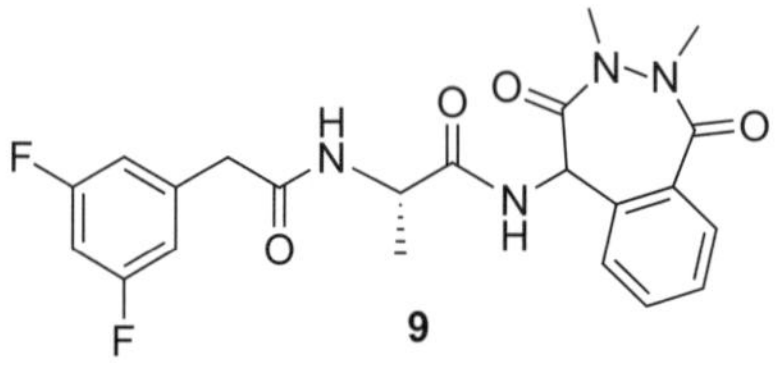

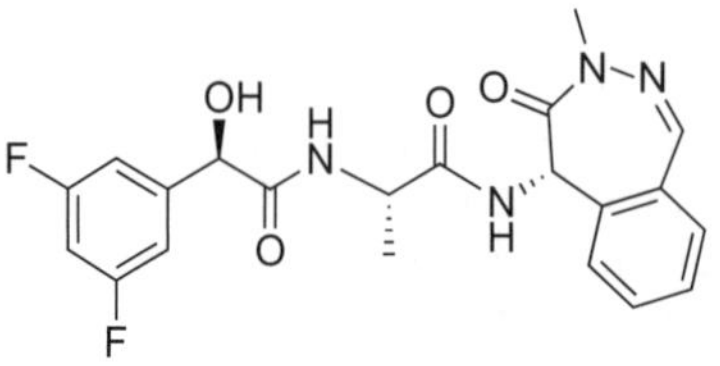

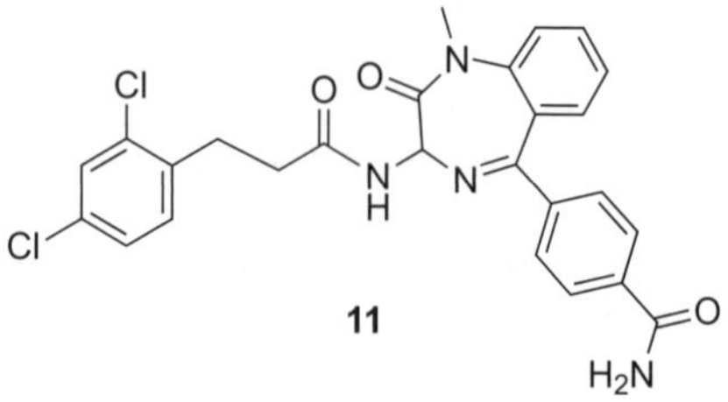

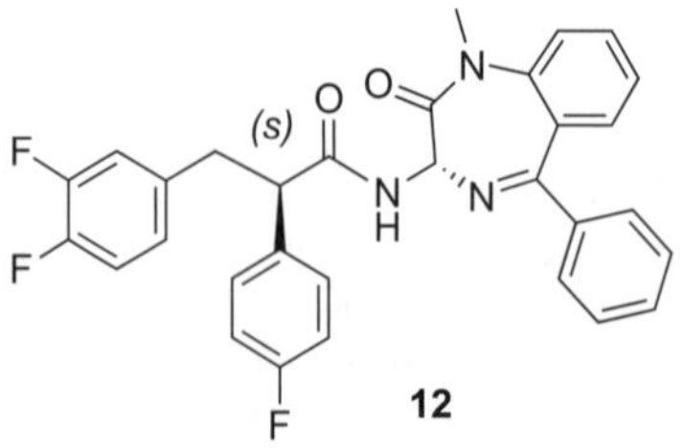

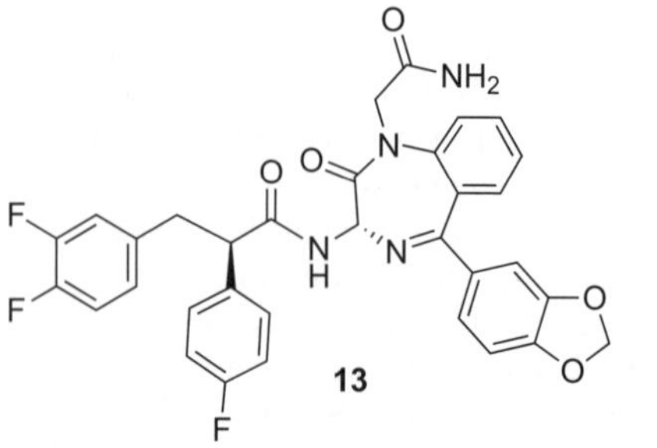

Scheme 3.

severe gastrointestinal toxicity and lymphocyte maturation in mice, which was believed to be due to the lack of appreciable Notch selectivity, were observed [64–66]. However, the impact of this "lack of selectivity" will only be answered through advanced *in vivo* pharmacology studies directed at understanding side effects relative to the inhibition of Notch processing.

Researchers at Bristol-Myers Squibb identified an HTS lead that was optimized to afford benzodiazepinedione **9**, which is related to their peptide isostere **6** where the C-terminus has been replaced by a benzodiazepinone-dione. This modification afforded improved potency ($IC_{50} = 5$ nM) and *in vivo* efficacy showing 43% reduction in brain Aβ in Tg2576 mice at a 200 µmol/kg dose [62]. Extensive work on this scaffold led to the constrained diazepine **10** (BMS-433796, $IC_{50} = 0.3$ nM), with greatly improved *in vivo* efficacy but in a chronic dosing study in mice, intestinal goblet cell hyperplasia toxicity was observed [67].

A direct comparison of the Lilly analog LY-450,139 and the BMS analog BMS-433796 shows the resemblance of these derivatives. One can easily see that the N-terminus contains isopropyl in semagacestat versus difluorophenyl in **10**, while the C-terminus contains a modified *N*-methylated azepine versus diazepine moiety. Both of these early GSIs lack Notch-sparing selectivity and the observed hyperplasia toxicity from BMS-433796 (**10**) is believed to be Notch-related. However, subsequent *in vivo* data from Merck and BMS on rodent models suggested that Aβ inhibition and Notch "toxicity" may be separable with some GSIs [64]. Future additional data from clinical trials on compounds with different Notch-sparing selectivity should hopefully be able to confirm if selectivity will translate to appreciable margins of safety in human subjects.

Interesting SAR analysis around Merck's benzodiazepines led to the discovery of a potent series of GSIs. Starting with an HTS hit **11** ($IC_{50} = 33$ nM in SH-SY5Y cell), the researchers were able to make several improvements. The identification of substitution at the alpha position of the side-chain amide by introduction of an (S)-Me group ($IC_{50} = 1.9$ nM) led to optimization at this site. SAR studies showed that the removal of the undesirable *p*-carboxamide group could be offset by addition of an alpha methyl substituent. Further optimization at this position led to replacement of the (S)-Me with a substituted (S)-aryl group and the introduction of fluorophenyl group, affording compound **12** ($IC_{50} = 3.8$ nM). Ultimately, efforts to improve the *in vivo* exposure led to the introduction of a carbox-amido methyl on the benzazepine N-1 to afford **13** ($IC_{50} = 1.2$ nM) whose brain C_{max} level reached 0.17 µM after an oral dosing at 5 mpk [68–70].

Additional work on the benzodiazepine class was reported by Merck on an alcohol analog (**14**) and an (*R*)-alkyl substituted alcohol by BMS (**15**) while benzazepine modifications afforded carbamate-derived GSI (**16**) from Roche and sulfur containing benzazepine (**17**) from AstraZeneca [68,71–73]. Efforts directed toward replacement of benzazepine ring by a five-membered heteroaryl ring system have been reported by Elan (**18**) and Pfizer (**19**), respectively (Scheme 4) [74,75].

3.1.2. Nonpeptidomimetic γ-Secretase Inhibitors

Due to the obvious, and often mentioned, deficiencies of peptide derived entities as potential drug candidates, particularly those aimed at oral delivery, the focus on nonpeptide derivatives in this area was significant and well funded. There have been a plethora of reports of nonpeptide GSIs over the last decade and a half [45,46]. The state of this area is quite promising as several efforts have reportedly advanced compounds to clinical development. The discussion of the preclinical efforts leading up to these hopeful drug candidates will describe some of the more mature work in this arena.

Sulfonamide-Based Inhibitors Sulfonamide-based inhibitors yielded multiple candidates and had the interest of many groups as shown in Scheme 5. Identified as allosteric binders of γ-secretase (those that bind at an alternate site than the native ligand, in this case APP), the sulfonamides were a product of high-throughput screening efforts by various companies such as Merck, BMS, Wyeth, and others. One common motif present in many of the earlier analogs from this series is the *p*-chlorophenyl sulfonamide group. In general,

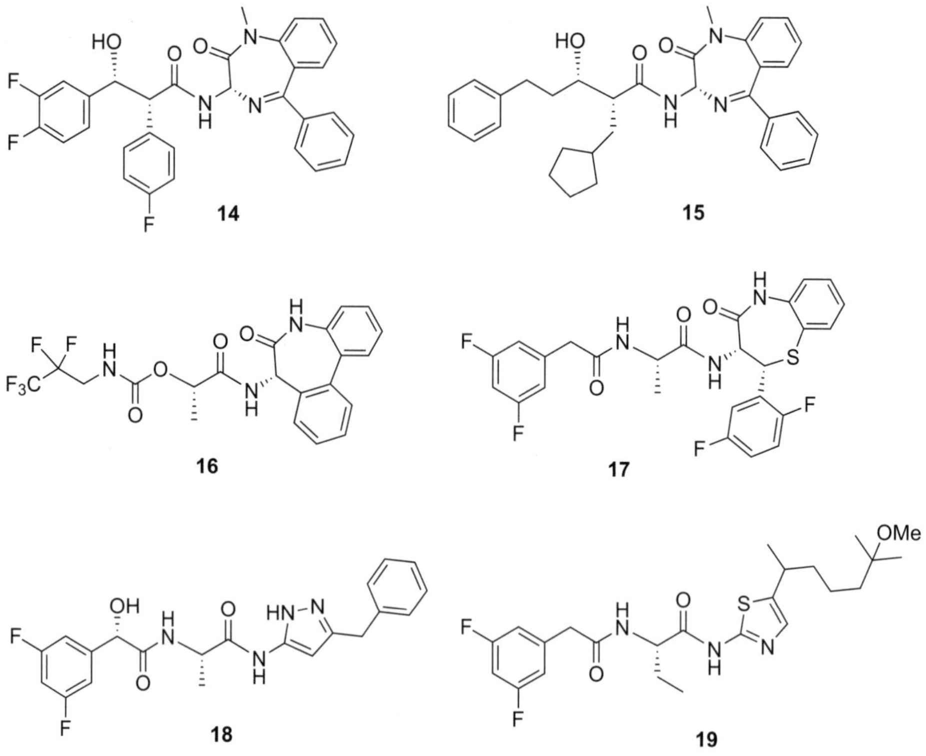

Scheme 4.

earlier hits and many of the optimized analogs possess an aryl or a benzyl group on the sulfonamide nitrogen along with a branched group, containing either a substituted aryl or a modified carboxyl group (BMS **20**, Pfizer **21**, Schering Plough **22**, Elan **23**, and Wyeth **24**). [76–82]

Researchers at Bristol-Myers Squibb were able to optimize their lead **25** by addition of a branching group on the alkyl chain to afford **26**, which greatly improved the potency (from 850 to 5 nM) and resulted in a modest reduction of brain Aβ (25% reduction at 500 μmol/kg) despite good brain exposure level (14 μM) [80,83,84]. Subsequent introduction of a terminal sulfonamide group afforded subnanomolar inhibitors, although their brain to plasma (B/P) ratio was poor. Replacement of the second sulfonamide group with a tetrazole, a method often employed by medicinal chemists to improve the physicochemical properties of a molecule containing a polar group such as a carboxyl or sulfoxyl group, resulted in derivative **27** that was quite potent ($IC_{50} = 510$ pM). As expected compound **27** not only had improved PK and central nervous system (CNS) exposure but also markedly improved *in vivo* efficacy (41% reduction of brain Aβ at 200 μmol/kg) [80]. From this work one of the early sulfonamide clinical candidates (BMS-299897) was identified by BMS in collaboration with SIBIA Neurosciences (discussed in Section 3.2) [85].

Merck's exploration of the sulfonamide class that began from an HTS hit led to the identification of a bicyclic aryl sulfonamide containing a 2-chloro thiophene group as a replacement for the *p*-chlorophenyl group (**28** versus **29**), shown in Scheme 6. Compound **29** provided an order of magnitude boost in potency and served as the launching point for additional SAR [86].

In further modifications, researchers at Merck discovered that it was possible to retain the potency of the sulfonamides when replacing the *p*-chlorophenyl or chlorothiophene

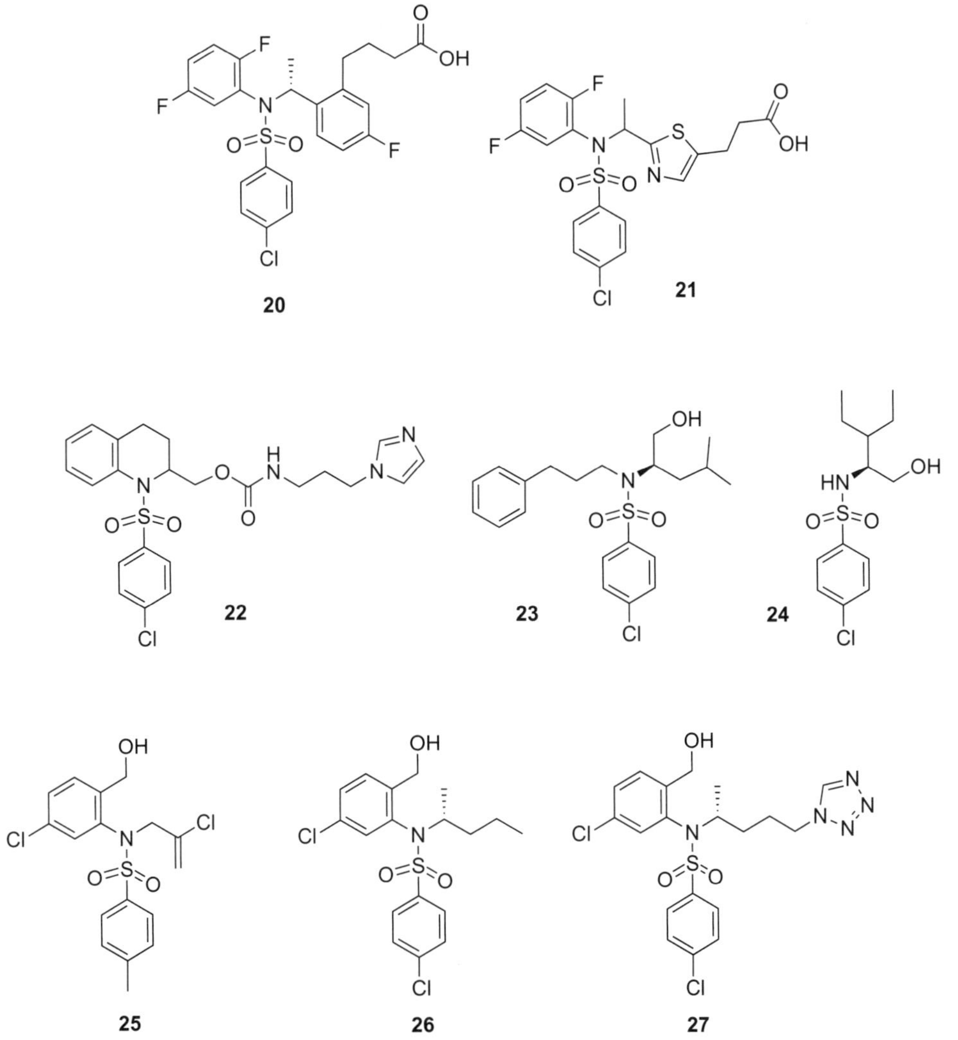

Scheme 5.

group with substituted alkyl amine. Subsequently, it was possible to design conformationally constrained spirosulfonamides that are very potent (down to 60 pM) or showed greatly improved PK (i.e., compound **30**) while some showed *in vivo* efficacy in mouse model to lower brain Aβ_{40} at 17 mpk p.o. However, none of these analogs were able to demonstrate appreciable Notch-sparing selectivity [87].

Similar to BMS, the idea of replacing the nitrogen in the sulfonamide group with a carbon atom was utilized to afford potent GSIs especially when the benzylic carbon was substituted with a branched group such as isopropyl to boost potency (from IC$_{50}$ of 4.4 µM to 70 nM in Compound **31**). Conformationally constraining the alkyl group into a cyclohexyl ring provided additional potency culminating in the identification of Merck's first GSI clinical candidate **32** (MK-560) that is a picomolar GSI reported in 2006 (discussed in Section 3.2) [88–90]. Subsequently, the metabolic stability and PK properties of **32** were greatly improved when the sulfone group was fused onto a sulfonamide ring as in **33**. This compound was shown to demonstrate reduction of

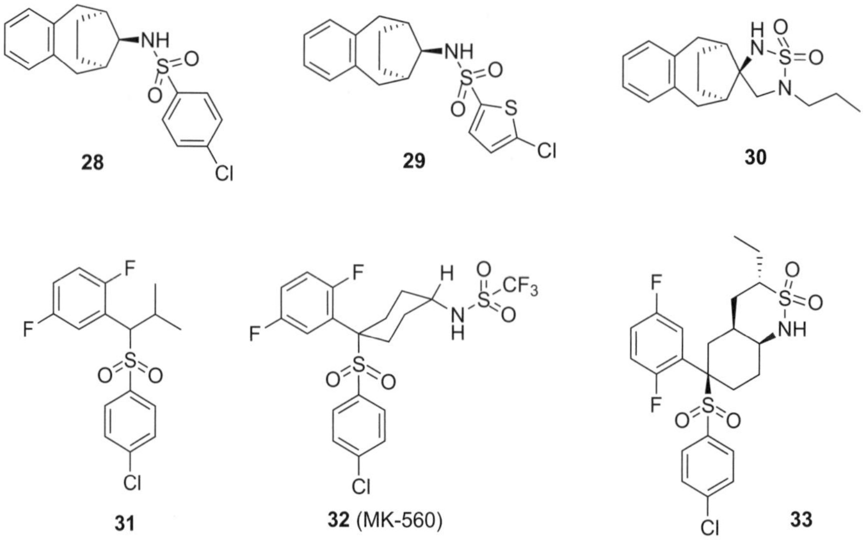

Scheme 6.

Aβ in mice (with brain ED_{50} of 2.8 mpk) although similar to earlier analogs no Notch selectivity was demonstrated with this molecule [91].

In preclinical studies, MK-560 (**32**) (IC_{50} 650 pM) demonstrated efficacy in a variety of *in vivo* models; reduced Aβ$_{40}$ (1 mpk p.o.) in APP-YAC (yeast artificial chromosome) mouse, chronically reduced brain Aβ levels by 43% in Tg2576 mice, reduced amyloid plaque (3 mpk p.o.), and reduced brain and cerebral spinal fluid (CSF) Aβ levels in rat (at 6 and 10 mpk, respectively). Although it lacked Notch selectivity, it was carried forward into the clinic (discussed in Section 3.2).

An extensive program at Wyeth has afforded a number of advanced compounds targeting γ-secretase inhibition. Internal HTS hit modification afforded compound **34** that had promising GSI activity ($IC_{50} = 3.5\,\mu M$ in hAPP$_{swe}$CHO, Chinese hamster ovary, cells). However, the team identified a better starting point with hydroxyethyl sulfonamide analogs **35** and **36** in collaboration with ArQule (IC_{50} 5.5 μM). As with Merck, Wyeth also observed an order of magnitude boost in potency with the introduction of the 5-chlorothiophene moiety (i.e., **37**) with an excellent Notch-sparing selectivity (13.9-fold) and focused on the optimization of the alkyl appendage [86,92,93]. In an effort to improve the PK properties, the bis-CF3 analog **38** was designed that demonstrated improved metabolic stability while retaining potency ($IC_{50} = 16\,nM$) and Notch selectivity (15-fold) [94]. This analog reportedly was shown in preclinical studies to lower brain Aβ$_{40}$ by 27% in Tg2576 mice at 5 mpk p.o. In addition, advanced analogs from this series, reportedly have moderate selectivity (~20-fold) over Notch processing, a potential advantage over the compounds by Merck and Lilly. As a culmination of these efforts, in 2007 Wyeth reported the advancement of their first compound into human clinical trials (Section 3.2) (**47**, Scheme 10).

Miscellaneous Preclinical GSIs Pfizer reported development of a potent series of dipeptide GSIs containing a thiazole C-terminus, similar to BMS's dipeptide isosteres. Introduction of a C5 substituted thiazole ring was shown to improve *in vitro* potency dramatically as in compound **39** ($IC_{50} = 80\,pM$ in hAPP$_{swe}$H4 cells) shown in Scheme 8 [75]. In addition, a few key additional modifications such as replacing the second amide group and changing the C5 lipophilic group may have attributed to analogs with improved stability. Ultimately, heterocyclic replacement of the thiazole to a pyrazole, led to their development compound (discussed in Section 3.2). Other

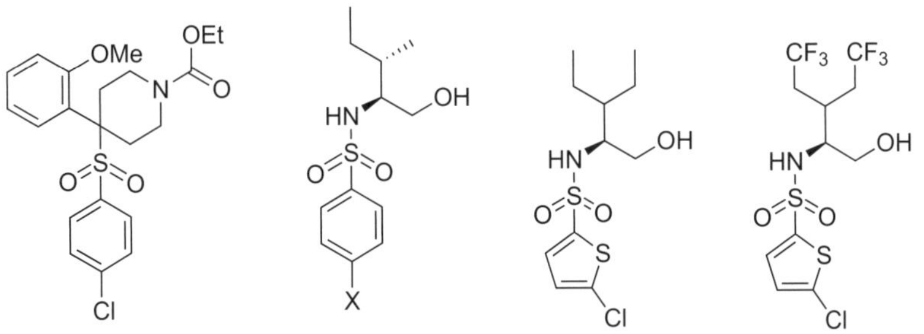

34 **35** (X = Cl) **36** (X = Br) **37** **38**

Scheme 7.

companies such as Schering-Plough **40** [95], Hoffman-LaRoche **41** [96], and Elan **42** (ELN-475516) [97] have reported on alternative binding site sulfonamides [95,98].

3.2. γ-Secretase Inhibitors—Clinical

A number of GSIs have entered into clinical trials as potential disease-modifying agents for the treatment of AD. Most notably from Lilly, Merck, Bristol-Myers Squibb, Pfizer, and Wyeth, these clinical candidates fall into two different categories of chemical structure: benzazepines or sulfonamides. The hypothesis that by blocking Aβ formation and decreasing the levels of Aβ aggregates by this mechanism potentially slows the progression of Alzheimer disease can now be tested in the clinic with these molecules.

The most advanced GSI in human clinical trials at present is semagacestat (**43**) (LY-450,139; $IC_{50} = 15\,nM$), shown in Scheme 9 [99,100]. Semagacestat reportedly entered Phase III in March 2008 with 1500 enrolled patients (aged 55 years and above) that exhibited mild-to-moderate symptoms of AD. This study was designed (i) to monitor the ability of semagacestat (100 and 140 mg

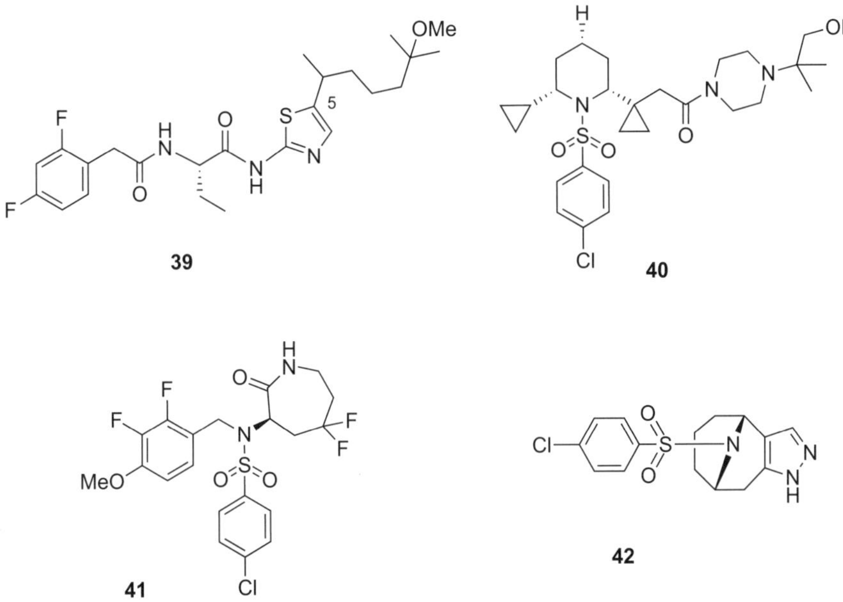

39 **40** **41** **42**

Scheme 8.

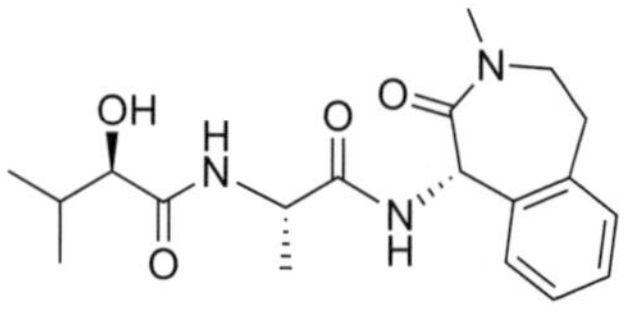

43 (LY-450,139 or semagacestat)

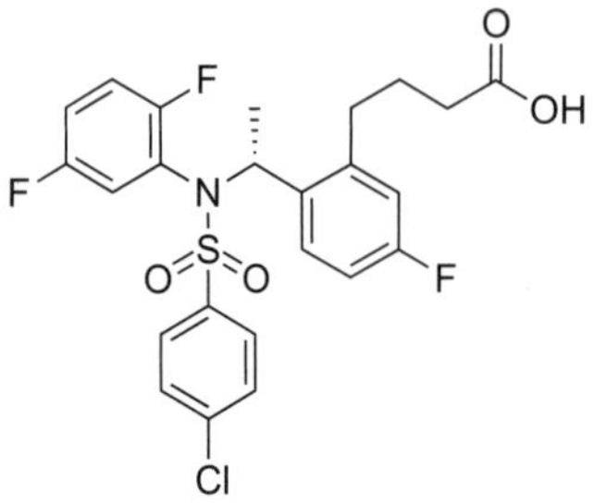

44 (BMS-299897) 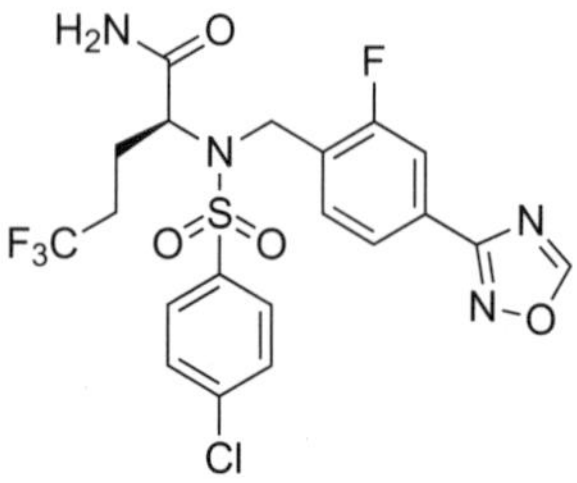   **45** (BMS-708163)

Scheme 9.

p.o. daily) to slow the rate of cognitive and functional decline and (ii) to determine secondary end points such as tau levels and brain volume of amyloid plaque. Previous clinical data from a number of Phase I and II studies (doses ranging from 5 to 280 mg) revealed a demonstrable lowering of plasma Aβ (up to 65%). However, various side effects have also been reported from these studies including one fatality wherein the patient developed endocarditis, an inflammation of the inner layer of the heart.

Consistent with preclinical reports, a rebound effect of plasma Aβ levels (up to 300%) was reported when patients were dosed with semagacestat while the concentration of CSF Aβ was consistently found to be unchanged. With GSIs the rebound effect is seen as a rapid rise in plasma Aβ$_{42}$ level after the initial reduction of Aβ. This rebound effect is a phenomenon that has commonly been seen preclinically with many GSIs, but not with BACE inhibitors or other known mechanisms targeting Aβ [101–104].

Positive evidence for the capacity of semagacestat to inhibit γ-secretase was reported by Randall Bateman's group in 2008 [105]. This study quantified the synthesis and clearance rates of Aβ in healthy volunteers via the administration of ^{13}C-labeled leucine. More specifically, monitoring of CSF and blood samples containing ^{13}C-leucine labeled Aβ confirmed a lowering of newly synthesized Aβ when semagacestat was administered at the highest dose (280 mg) The researchers found that Aβ formation in CSF occurs with one of the fastest measured production rates known for *in vivo* protein biosynthesis (fractional synthesis rate of 8% per hour (h)). This enabled the measurement of Aβ change with semagacestat within the "newly" synthesized pool of protein. However, there is some continuing discussion as to whether lowering of newly synthesized Aβ is an adequate measure of Aβ lowering versus when affecting the lowering in the total Aβ level [103]. The further understanding of this awaits clinical trial results, and will no doubt guide the direction of all future studies based on the amyloid hypothesis.

As mentioned earlier, BMS identified the sulfonamide **44** (BMS-299897) as their first clinical candidate. This compound possessed moderate selectivity over Notch processing (15-fold) and was ultimately terminated in clinical development due to poor PK and

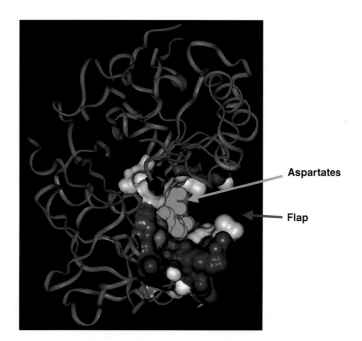

Figure 4 (Chapter 9). BACE1 enzyme.

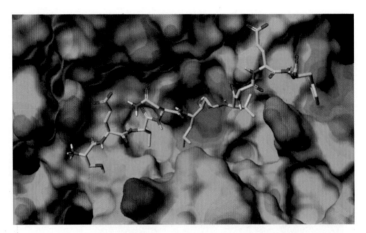

Figure 5 (Chapter 9). BACE1 enzyme crystallized with OM00-3.

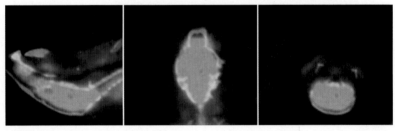

Figure 4 (Chapter 14). Hybrid PET/CT images of FDG uptake in a mouse brain, acquired on a Suinsa Argus preclinical PET/CT scanner.

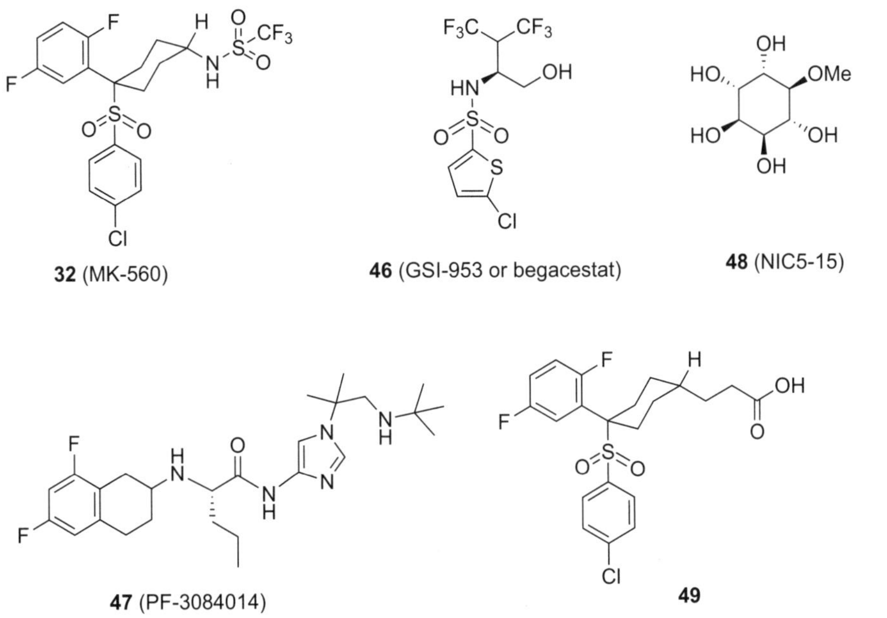

32 (MK-560)

46 (GSI-953 or begacestat)

48 (NIC5-15)

47 (PF-3084014)

49

Scheme 10.

insufficient exposure levels [85]. Subsequently, BMS identified a highly Notch-selective GSI clinical candidate **45** (BMS-708163; $IC_{50} = 300$ pM in hAPPsweH4 cells, >190-fold Notch selective) whose structure and clinical data were first disclosed in 2009 [106]. BMS-708163 appears to have an excellent PK profile with an extended half-life in human subjects when measured at the 400 mg dose. More importantly, it was reported that the administration of BMS-708163 led to a 40% decrease in CSF Aβ levels. Although no additional detail regarding plaque lowering or cognitive improvements have been detailed, BMS reported that it was shown to chronically lower brain Aβ levels in rat and dog without significant side effects. Notably, a lack of gastrointestinal and lymphoid toxicity observed in preclinical studies with rats and dogs may be related to the capacity for BMS-708163 to improve Notch selectivity.

As discussed earlier, Merck's first advanced MK-560 (**32**), shown in Scheme 10, into clinical trials in 2006. In reports of preclinical studies, MK-560 showed 49% decrease in Aβ$_{40}$ levels in Tg2576 mice and lower amyloid deposits in a chronic study (3 mpk p.o. for 12 months), while remaining plaque size was unchanged. This data seemed to suggest that MK-560 does not alter amyloid deposit growth after the initiation has begun. Based on this observation, an early intervention with a GSI was recommended as a more effective therapy for AD. In addition to MK-560, it has been reported that the company has advanced other GSIs into human clinical trials, such as Compound **49** and MK-0752 whose structure has yet to be reported [107]. Early reports state that MK-0752 was effective in reducing both plasma and CSF Aβ levels greater than 35% for doses greater than 500 mg. [108,109] Similar concentrations of estimated "free" MK-0752 are reported both in CSF and plasma in these studies, although a 2 h-delayed shift in T_{max} was observed in CSF, relative to plasma T_{max}. Similar to Lilly's semagacestat, MK-0752 is reported to be nonselective for Notch processing and exhibits a plasma rebound effect. However, the company reports that MK-0752 appears to have superior tolerability than semagacestat in safety studies up to 1000 mg. While originally developed as an AD treatment, MK-0752 is also being explored as a potential breast cancer treatment as it is highly potent against Notch signaling (IC_{50} of 55 nM). In Phase I oncology trials in patients with T-cell acute lymphoblastic leukemia, only a modest clinical efficacy was observed at 450 mg dose with 46% mean decrease in plasma Aβ. Due to the observation of GI toxicity and fatigue, an intermittent dosing schedule was being explored in future studies. Based on more recent data available on clinical study Web sites, no additional studies are planned for AD currently beyond a biomarker study that recently completed this year. However, patients are actively being recruited for multiple Phase I/II studies with MK-0752 for breast cancer and other solid tumors (clinicalTrials.gov; NCT00756717 for Breast Cancer in Combination with Tamoxifen or Letrozole; NCT00106145 for Notch Signaling Pathway Inhibitor).

Wyeth investigated optimization of sulfonamide analogs related to begacestat (**46**) (GSI-953) and has afforded another ligand to test the viability of GSIs as treatments for AD [110]. In safety and tolerability studies of begacestat in healthy subjects, Wyeth reported it was well tolerated and showed lowering of plasma Aβ by 40% following an oral dose of 600 mg. However, CSF concentration of begacestat was determined to be 10-fold lower than plasma level and no CSF Aβ lowering was observed. PK/PD (pharmacodynamic) biomarker studies in Tg2576 mice and humans suggested that a 450 mg dose of GSI-953 in AD patients should lead to optimal exposure to produce brain Aβ lowering, similar to what was observed in plasma (∼28%) [111]. A second compound, GSI-136, is also reported to have entered its Phase I study in 2008 although its structure is yet to be disclosed.

The structure of Pfizer's GSI clinical compound **47** (PF-3084014) and some *in vivo* data were recently disclosed [112]. Based on preclinical data, PF-3084014 showed high levels of Aβ lowering in plaque-free Tg2576 mice (92%, 78%, and 72% in plasma, brain, and CSF, respectively) with no rebound effect. PF-3084014 reportedly entered Phase I trials in 2006, however, at the same meeting it was reported that PF-3084014 was no longer in development as it did not reach "targeted" compound exposure levels required.

Additional GSIs belonging to the structural subtypes mentioned previously have reportedly entered clinical development. Elan has reportedly advanced a sulfonamide, ELND006, whose structure is still undisclosed but is reported to be related to **42** (ELN-475516; Section "Miscellaneous Preclinical GSIs") into Phase I clinical development in 2008 [113]. As of this report, no additional clinical data have been reported on ELND006 (no studies are reported in ClinicalTrials.gov under ELND006). Instead, Elan is currently enrolling AD patients for a long-term study with an alternate entity, ELND005, which is not a sulfonamide GSI but a derivative of scyllo-inositol. This compound is reportedly related to **48** (NIC5-15), also known as D-pinitol. This natural product antidiabetic agent was taken to Phase II clinical trial by Humanetics and Mt. Sinai School of Medicine in early 2007 as an "alternative" medicine for AD and the trial is ongoing (Clinical Trials.gov NCT00470418). In preclinical studies, NIC5-15 was shown to prevent the formation of plaques through selective inhibition of certain insulin mechanisms and by inhibiting γ-secretase activity. Interestingly, both NIC5-15 and ELND005 are hexanol analogs that seem to be eliciting favorable effects through a combination of multiple mechanisms and they may not be acting as selective GSIs (also see Section "Natural Product Related and Other Small-Molecule Inhibitors") [114].

As was described in several cases above, quantification of the degree of Notch selectivity as a guide to limit *in vivo* toxicity has not been well defined. Since the lack of Notch-sparing selectivity has been associated with untoward side effects, much of the research in identifying the best GSIs has been directed toward shifting the inhibition of APP and Notch processing toward levels favoring APP inhibition. So far, the SAR-driven design for Notch selectivity is not well understood and only limited rodent *in vivo* data exists to directly correlate observed toxicity with lack of Notch selectivity. However, it is noteworthy that researchers at Merck and Pfizer have been able to take advantage of, or opportunistically utilize, this potential issue by targeting these Notch active entities toward cancer therapy.

In addition to the Notch selectivity question, other issues still remain unresolved both

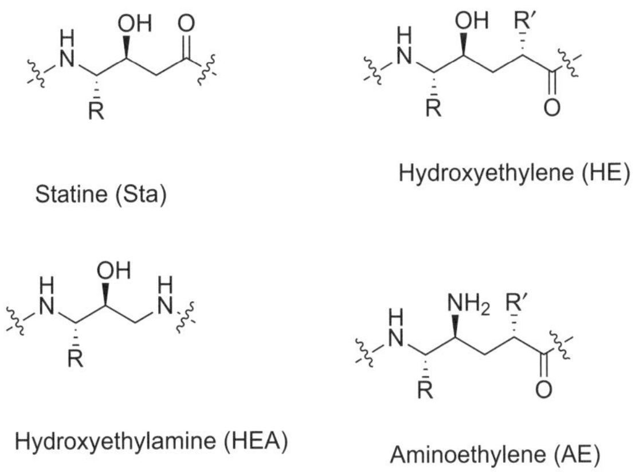

Statine (Sta)

Hydroxyethylene (HE)

Hydroxyethylamine (HEA)

Aminoethylene (AE)

Scheme 11.

at the discovery level and in the clinic. Some of these questions are related to the principal causes of AD relating to the potential toxic species of oligomers, the role of plaque, and the observation of the rebound effect, seen with many of these GSIs [45,115,116].

3.3. β-Secretase Inhibitors—Preclinical

One of the most intensely researched areas for the last decade has been the identification and potential utility of BACE1 inhibitors [117–123]. The progress made to date has been astounding, owing to the dedication and tenacity of the research community, although clinical validation has yet to be realized. Notwithstanding these focused and abundant efforts, there are a paucity of clinical candidates for this target. However, the advancement of preclinical candidates and contribution to the understanding of the amyloid cascade justifies a detailed examination of the progress in this area.

3.3.1. Early Peptidomimetic BACE1 Inhibitors

Due to the knowledge gained from decades of work on development of aspartyl protease inhibitors for enzymes such as HIV protease, renin, and cathepsin, the initial efforts at BACE1 inhibitor design logically followed this paradigm [124–128]. Namely, transition-state inhibitor design based on replacement of the substrate scissile bond with an isostere such as a secondary alcohol or other tetrahedral intermediate surrogate. There are countless reports of a variety of these peptidomimetics shown to be potent inhibitors of BACE enzyme activity. The use of statine (Sta), hydroxyethylene (HE), hydroxyethylamine (HEA), aminoethylene (AE), and other related isosteres was the focus at the outset by many groups, and significant potency was realized with a variety of structure types (Scheme 11) [120,123]. The mode of hydrogen bonding engagement of the catalytic aspartic acid residues of the enzyme backbone varies with each of these arrays. However, X-ray studies have shown that the hydroxyl groups of the various isostere types form the key hydrogen bond interactions with one or both of these residues [129].

The initial reports on substrate-based inhibitors were from the Elan and University of Oklahoma groups [130]. The strategy was to identify truncated polypeptides, based on the native sequence, with a noncleavable transition-state mimic as described above. The Elan compound, **50**, was the P10-P4′ APP substrate containing a statine residue replacing the S1 of the sequence (Scheme 12). This tool compound, bearing IC_{50} of approximately 300 nM, paved the way for more drug-like entities as it allowed for purified enzyme isolation by affinity chromatography. Shortly thereafter, Ghosh and Tang reported their crystal structure of a recombinant BACE1 enzyme with **51** (OM99-2), possessing a K_i of 1.6 nM. Slight modification of the side-chain residues to this polypeptide analog afforded the picomolar ligand, **52** (OM00-3) [130,131].

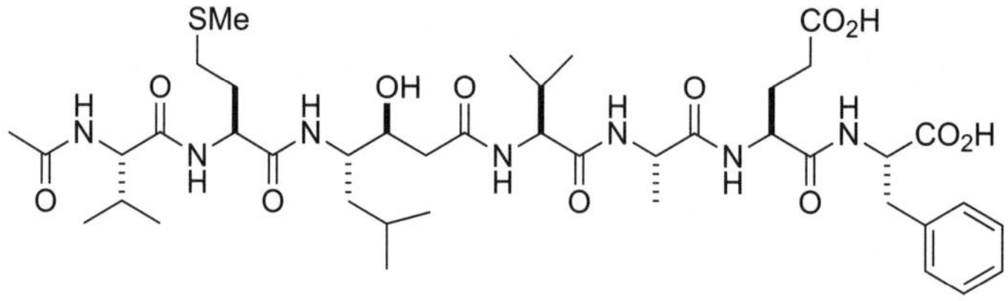

50

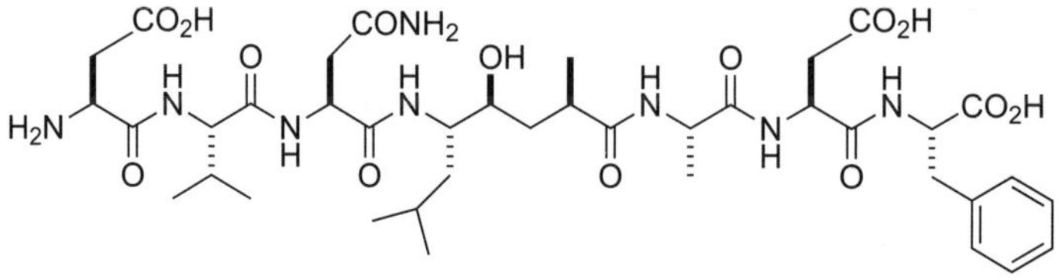

51 (OM99-2)

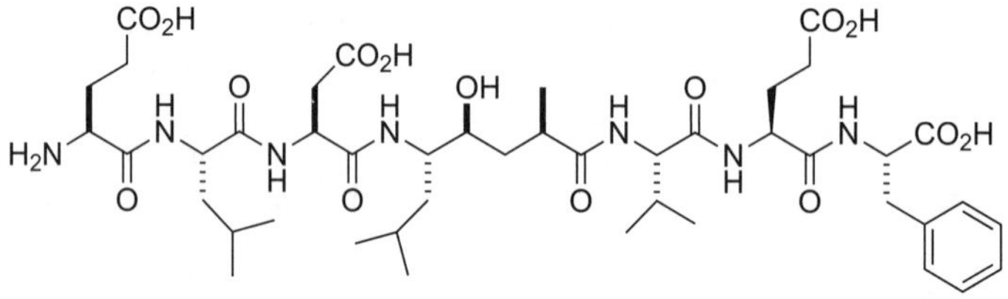

52(OM00-3)

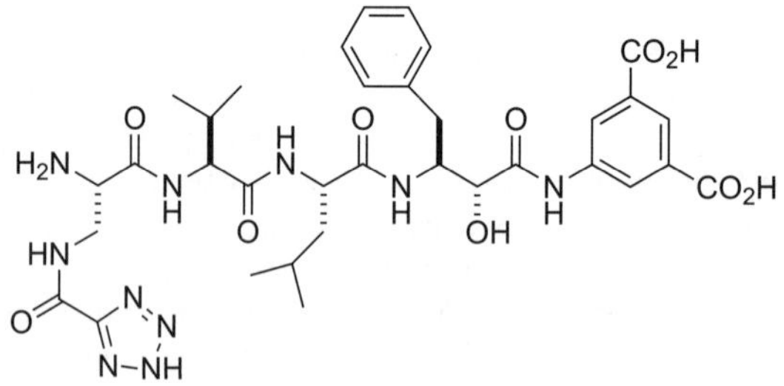

53 (KMI-429)

Scheme 12.

With these potent peptidomimetics reported, the search was heightened for smaller, lower molecular weight entities, with more drug-like properties. The work of Kimura et al. showed that this sequence could be reduced affording optimized analogs, such as **53** (KMI-429), with good potency ($IC_{50} = 3.9$ nM) and efficacy in Aβ lowering assays [132–135]. KMI-429 reduced the levels of soluble and insoluble $A\beta_{40}$ and $A\beta_{42}$ in wild-type and Tg2576 mice when injected directly into the hippocampus [136]. Although these ligands helped to validate the target and advance the understanding of BACE1 inhibition as a potential treatment for Aβ reduction, entities based on the peptidomimetic approach suffered from the well-documented issues associated with peptides [55–57]. Low oral bioavailability, poor metabolic stability, insufficient capability to cross the blood–brain barrier, minimal solubility, and susceptibility to efflux due to p-glycoprotein (pgp) affinity and transport are all properties that highlight the need for small-molecule nonpeptidomimetic-based ligands. In addition, many of these peptidomimetic ligands suffered from

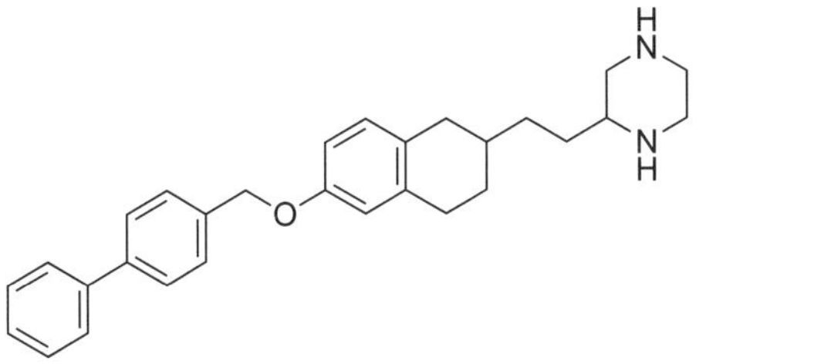

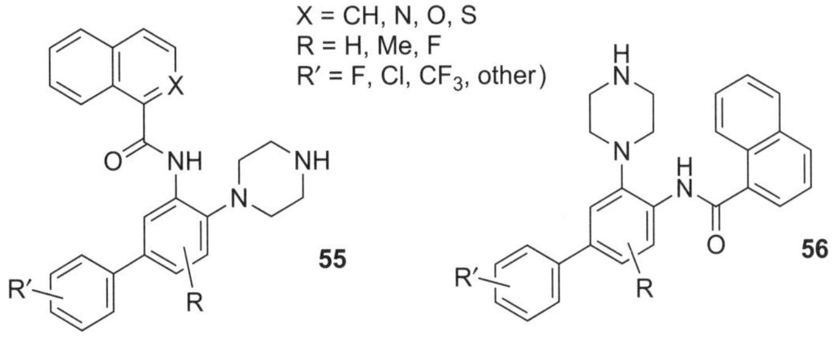

Scheme 13.

an inability to cross cell membranes, a necessary property for this target as BACE1 is an intracellular enzyme operating on the golgi and endoplasmic reticulum [137,138].

3.3.2. Nonpeptidomimetic β-Secretase Inhibitors

First Reported Leads The first report of nonpeptidomimetic BACE1 inhibitors was in 2001 by Miyamoto et al. at Takeda Chemical Industries (Scheme 13) [139]. Among a series of tetraline derivatives, compound **54** was shown to have an IC_{50} of 350 nM at the BACE1 enzyme. There was not much done to extend this work, but it was now evident that the identification of nonpeptide-based ligands with drug-like properties for this aspartyl protease enzyme was possible. Whether by coincidence or happenstance, this marked the launch of countless research groups toward the design of small-molecule inhibitors of BACE1. Researchers at Vertex Pharmaceuticals subsequently disclosed an array of heterocyclic inhibitors, the more potent of which is represented by classes such as **55** and **56**, with K_i values of 3 μM or less [140]. In this patent application they further proposed a three-dimensional pharmacophore mapping approach to the design of potent BACE1 inhibitors. Key interactions with the aspartates in the catalytic site and hydrophobic interactions of the aromatic portions of the ligands with the enzyme pocket was the core of this hypothesis.

Transition-State Isosteres: Statines and Related Derivatives Based on the early reports of transition-state inhibitors utilizing statines and statine-based isosteres, many groups continued in this direction with smaller, rationally designed entities utilizing a traditional SAR approach. The now ready-supply of enzyme and demonstrated ability to cocrystallize ligands in the active site brought the design of potential BACE1 inhibitors to the fore and gave medicinal chemists the tools needed to attack this objective.

In an effort to improve the pharmacokinetic and cell penetrant properties of their early leads, the Elan group developed a series of isophthalamide-based ligands that incorporated the hydroxyethylene core in place of the statine (Scheme 14) [141–143]. These compounds (**57**) possessed improved enzyme potency (30–300 nM) and metabolic stability, relative to their statine counterparts, as well as a reduced shift (C/E ~100-fold) in the cell to enzyme-based (C/E) assay potency (C/E

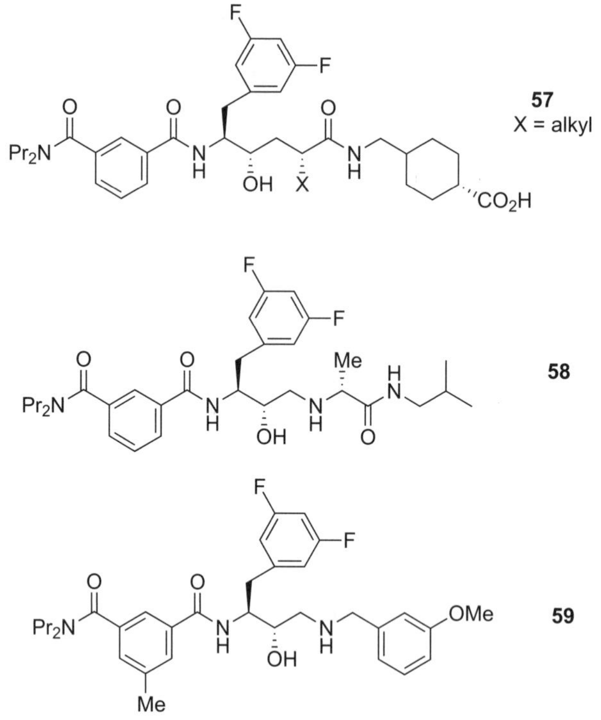

Scheme 14.

~1000-fold with the statines). This improvement in cell penetration properties affirmed that progress was being made, however there was much work to be done before potent cell-based ligands were realized. Toward that end, the Elan group again switched the catalytic isostere to the basic hydroxyethylamine core [144]. This change had additional benefits for potency as the S1' site of the enzyme pocket had been shown to prefer basic residues on the ligand. Optimization of analogs with the HEA core led to the identification of **58**, an isophthalamide derivative with IC_{50} of 130 nM against the enzyme and an EC_{50} of 230 nM in the cell-based assay (C/E 1.8) [145]. Additional optimization afforded **59**, $EC_{50} = 15$ nM, an analog with one less amide bond and an excellent ratio (C/E < 1). Elan, alone or in conjunction with Pharmacia, subsequently reported a diverse array of potent ligands modeled around the HE or HEA isostere in a number of patents from 2002 through 2007 [121].

There was a significant effort on the HE- and the HEA-type derivatives in the ensuing several years by a number of others as well. Coburn et al., at Merck Research labs, reported in 2004 the identification of a small-molecule nonpeptide BACE inhibitor as a result of a multimillion compound HTS [146,147]. The potency of the 25 µM lead compound, **60**, was improved by replacing the tertiary amide group with a sulfonate ester and the addition of a fluorine to the pendant aryl ring, affording **61** ($IC_{50} = 1.4$ µM) (Scheme 15). X-ray crystallographic data showed that this ligand occupied the S1–S4 pocket and possessed a somewhat nontraditional interaction with the catalytic warhead of the enzyme. Apparently, the NH of the amide on the ligand interacts with the catalytic center of the enzyme through a hydrogen-bonded water molecule. Efforts to optimize the potency of this novel lead centered on effecting a more efficient bond to the aspartates, namely through direct interaction of an HEA isostere core with these residues of the enzyme backbone. Toward that end, preparation of isophthalamide HEA derivatives with a

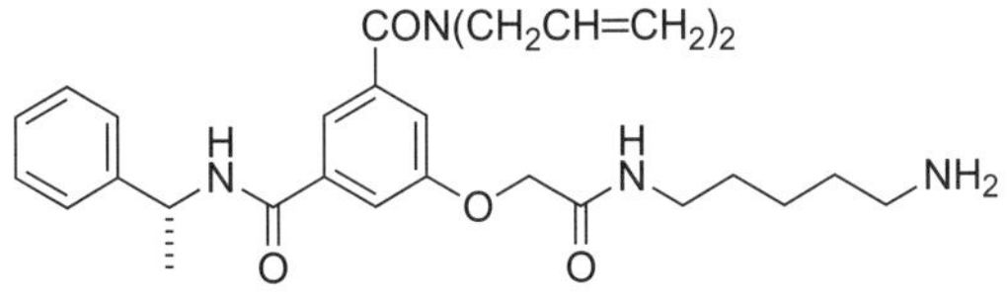

APPROACHES TO DISEASE-MODIFYING TREATMENT IN AD 349

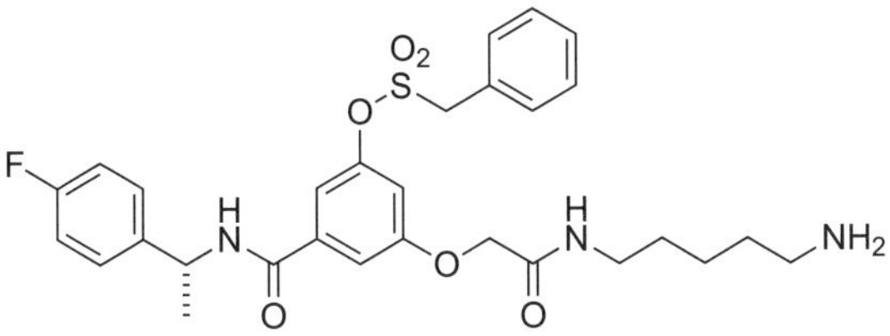

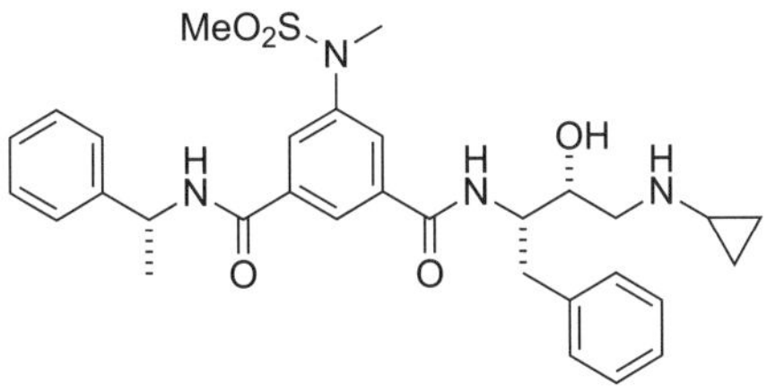

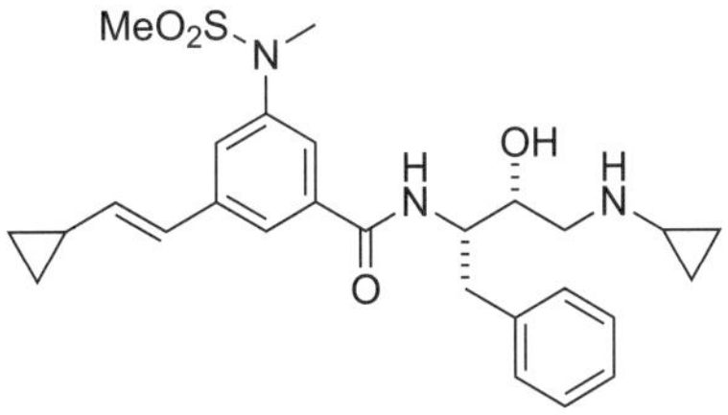

Scheme 15.

sulfonamide in place of the sulfonate ester, that is **62**, demonstrated good potency (IC$_{50}$ 15 nM) toward BACE1 with selectivity versus BACE2, renin and cathepsin D of 15-fold, >1000-fold, and >500-fold, respectively [148,149]. The resultant >100-fold improvement in affinity was attributed to a number of key interactions in the S1–S3, and S1' sites. Additional optimization through the identification of truncated analogs, such as **63** (IC$_{50}$ = 35 nM), led to compounds with improved pharmacokinetic properties [150]. These monoamide derivatives showed improved cell permeability but were still quite susceptible to p-glycoprotein transporter-mediated efflux.

Others have attempted to utilize the isophthalamide core to optimize potency and efficacy in their respective leads (Scheme 16). Ghosh and Tang at OMRF identified **64** (GRL-8234) with an IC$_{50}$ of 1.8 nM and efficacy in plasma Aβ$_{40}$ reduction (60%) in transgenic Tg2576 mice, when dosed at 8 mpk intraperitoneally (i.p.) [151]. Researchers at Takeda disclosed compounds such as **65** with affinity of ~50 nM at the BACE1 enzyme [152], and the group at BMS have filed applications on related HEA derivatives such as **66** with reportedly less than 10 nM potency [153]. The former SmithKline-Beecham group prepared hydroxyethylene derivatives of the type exemplified by **67** with BACE1 enzyme affinity of

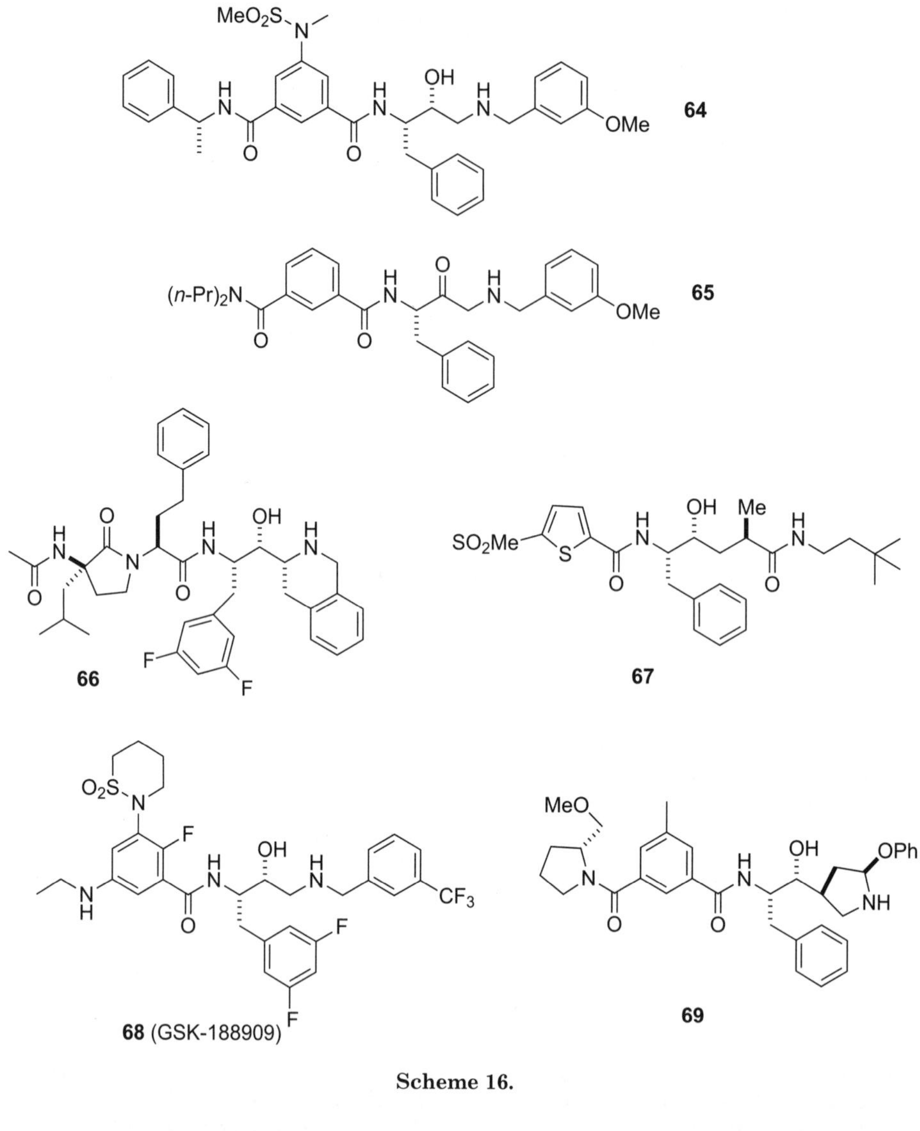

Scheme 16.

180 nM [154–156]. A series of publications outlining the progress of Demont and coworkers at GlaxoSmithKline (GSK) has identified **68** (GSK-188909) as a potent and efficacious orally active BACE1 inhibitor [157–160]. This phthalyl hyroxyethylamine derivative, with IC$_{50}$ of 4 nM against BACE1, was dosed orally at 250 mpk twice daily (bid) in TASTPM mice (double transgenic mice overexpressing the amyloid precursor protein and presenilin-1 genes) for 5 days. Examination of brain Aβ_{40} and Aβ_{42} levels showed a significant reduction of 18% and 23%, respectively. When coadministered with a pgp inhibitor, the brain exposure was markedly increased and the levels of the Aβ_{40} and Aβ_{42} reduction were improved to 68% and 55%, respectively [161]. This is but another example of the propensity of this class of compounds to be high affinity pgp substrates and underscores their limitations and liabilities. Finally, researchers at Schering-Plough reported an orally efficacious BACE1 inhibitor **69**, with IC$_{50}$ of 0.7 nM at the BACE1 enzyme and moderate to excellent selectivity versus BACE2 and several other aspartyl proteases [162,163]. When dosed to preplaque

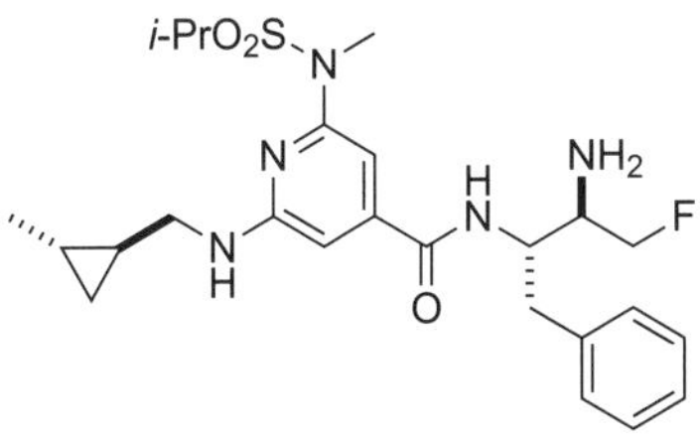

Scheme 17.

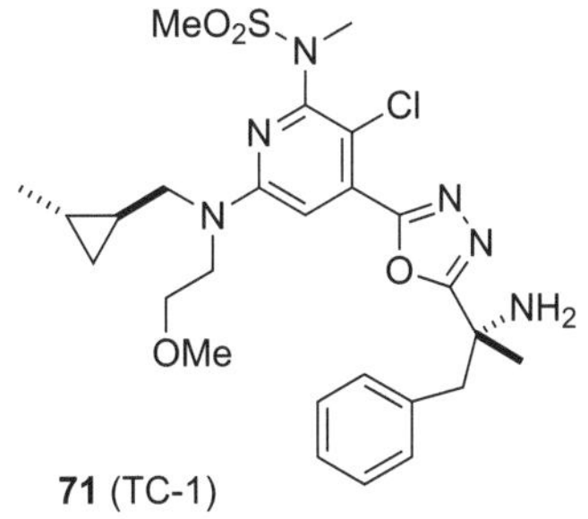

bearing CRND8 mice at 10–100 mpk orally, **69** exhibited plasma Aβ$_{40}$ reduction of 4–70% in a dose response fashion. Not surprisingly, when dosed subcutaneously the reduction of plasma Aβ$_{40}$ was significantly better in comparison and although brain exposures were 50-fold greater than the IC$_{50}$, there was no reduction of cortical Aβ$_{40}$ levels.

Further attempts to improve the CNS and pharmacokinetic properties of their equity (i.e., compounds **62** and **63**), led the Merck group to investigate a related series of isonicotinamides (Scheme 17) [164,165]. From their extensive SAR of the various related systems, and the use of crystallography and modeling data, the group identified **70** as a key representative from this class. When dosed at 50 mpk i.v., **70** showed a maximal reduction of Aβ$_{40}$ of 34% in the brain at the 3 h-time point. However, the poor overall PK properties of these derivatives, as seen with other prior series, limited their potential utility. The culmination of this SAR appears to be evident in the recently identified oxadiazole analog **71** (TC-1), with an IC$_{50}$ of 0.4 nM at BACE1 and ∼40-fold selectivity versus BACE2 and minimal affinity for the pgp efflux transporter [166]. Sankaranarayanan and coworkers have reported *in vivo* efficacy in non-human primates with this advanced analog **71**, albeit with one caveat. When rhesus monkeys were coadministered orally bid for 3.5 days with a CYP3A4 (cytochrome P450) inhibitor, ritonavir, TC-1 effected a significant and sustained reduction of plasma and CSF Aβ$_{40}$ levels of 61% and 43%, respectively. The CYP3A4 inhibitor was necessary due to the rapid metabolism and clearance of TC-1 and allowed for the needed exposures to demonstrate oral efficacy.

Macrocyclization of the isophthalamide derivatives (i.e., **62**), in an attempt to improve potency and permeability, afforded the simple variant **72** (IC$_{50}$ = 3 µM), with moderate activity against the BACE1 enzyme (Scheme 18) [167,168]. Further refinement led to optimized analog **73**, with an IC$_{50}$ of 4 nM and noticeably improved cell permeability. When lactam **73** was dosed intravenously (i.v.) to APP-YAC mice, the compound showed moderate central exposure (B/P ∼ 0.25) and a robust 25% decrease in Aβ$_{40}$ levels in brain at the 1 h-time point. This reduction in Aβ40 level was diminished significantly (∼10%) at the 3 h-time point highlighting the apparent rapid clearance of the compound. Other examples of macrocyclic BACE1 inhibitors by researchers at BMS (**74**) (IC$_{50}$ < 100 nM) and Novartis (**75**) (IC$_{50}$ = 2 nM) are also shown in Scheme 18 [169,170].

Although much progress was made with this class of inhibitors in general, the lack of oral efficacy and central exposure attributed to high pgp substrate affinity, a common denominator among these derivatives, limited their viability as potential drug candidates. Clearly, the goal of obtaining potent, brain penetrant ligands for the BACE1 enzyme site would only be accomplished through the use of smaller, more drug-like entities.

Acylguanidines and Related Compounds Increased interest in BACE as a target and the multitude of issues associated with the transition-state isosteres of the statine genre led to the search for true "small-molecule" drug-like leads through both high-throughput

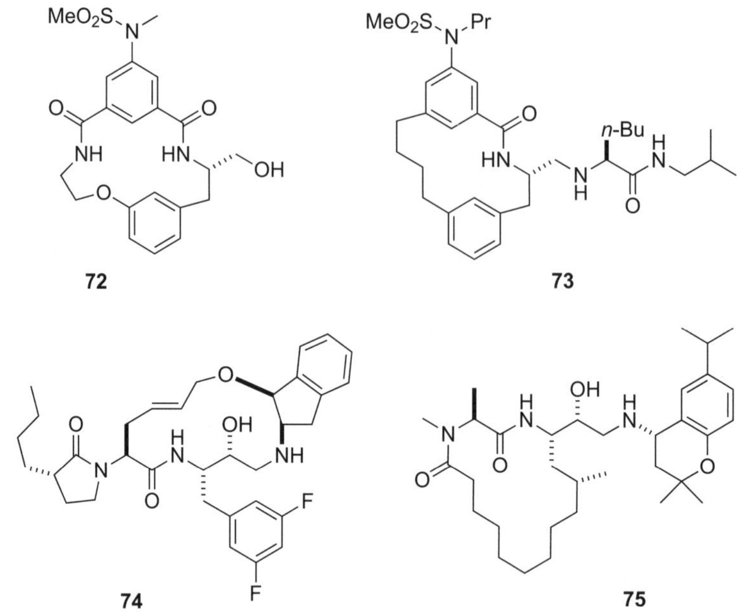

Scheme 18.

screening and fragment-based design. A variety of research groups have reported on novel structure types centered on a guanidine core backbone and extension into both the prime and nonprime side of the catalytic dyad.

Through a series of applications and later disclosures, the Wyeth group reported the identification of a novel, fairly simple acylguanidines, typified by **76** (WY-25105) with IC$_{50}$ of 3.7 μM from a screen of their corporate compound collection (Scheme 19) [171]. This prototype BACE inhibitor served as the start for a structure-based design approach to potent ligands from this and related classes [172–174]. Through the use of X-ray data and molecular modeling studies, the Wyeth team was able to improve the potency to 110 nM with derivative **77**, a ligand that reportedly interacts with multiple sites of the enzyme pocket [175,176]. This SAR was extended later to other heterocyclic core derivatives [173,177]. The thiophene and phenyl analogs, exemplified by **78** (WAY-211816) and **79**, respectively, were of comparable molecular potency to the pyrrole lead, and highlighted the utility of the core heterocycle to position the guanidine at the catalytic site of the enzyme. Unfortunately, as with other related derivatives, these compounds suffered from poor cell penetration and a corresponding lower affinity (>2 μM) in cellular assays. The Wyeth group further expanded on the potential of the acylguanidine moiety as a key binding motif by preparing aminopyridine derivatives **80** [178]. This core has contained within it an aromatic variant of this bifunctional guanidine group and has afforded potent derivatives of ~100 nM.

Similar to these structures, Astex, through *in silico* library generation and high-throughput X-ray crystallography, has identified an aminopyridine motif **81** [179,180]. Through a structure-based design approach, they quickly improved the potency to submicromolar levels (IC$_{50}$ = 690 nM) with the more complex derivative **82**. As with the vast majority of small-molecule BACE1 inhibitors, key interactions with the S1, S3, and S2′ region, as well as the catalytic dyad, of the enzyme were noted.

There have been several other reports of acylguanidine-type derivatives in the patent and journal literature with claims to inhibitory activity at the BACE1 enzyme from 0.1 to 30 μM (Scheme 20). Worthy of note is the work done by Bristol-Myers Squibb, where Gerritz

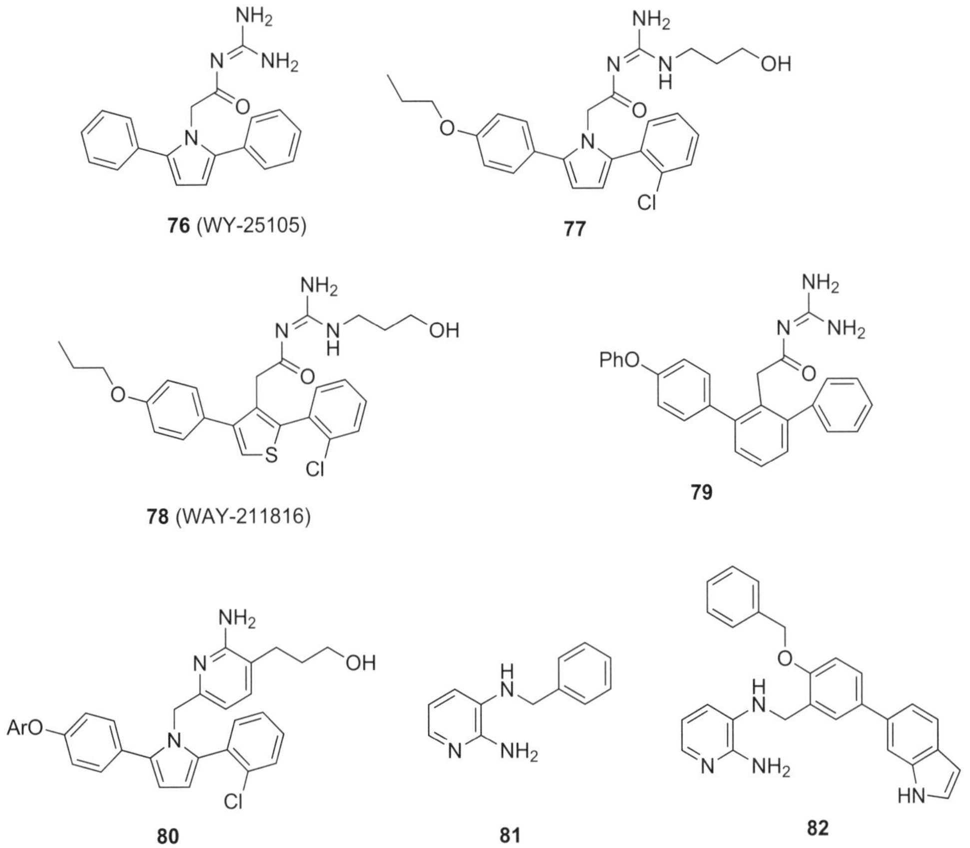

APPROACHES TO DISEASE-MODIFYING TREATMENT IN AD   353

Scheme 19.

and coworkers highlighted the optimization of a series of moderately potent *N*-benzylic substituted acylguanidines of types **83** and **84** to potencies <10 nM [181,182]. Additional diversity was obtained by preparing structurally constrained analogs, that is **85**, $IC_{50} = 5$ nM, via macrolide formation [183–186]. As with many compounds from this class, the high total polar surface area (TPSA) and affinity for the pgp efflux transporter system obviated their utility as potential drug candidates.

Guanidines, Aminohydantoins, and the Like
Concomitant with their report of the acyclic acylguanidine **76** (WY-25105), the group at Wyeth also identified **86** (WY-24544) as a viable small-molecule lead in their initial HTS (Scheme 21) [187–189]. This cyclic variant of the guanidine moiety, with a reported IC_{50} of 38 μM, served as an alternate starting point for what has become a large effort by the group with several significant compounds being reported in the last few years [188,190–193]. The initial revelation of removing the pendant ring of WY-24454, to afford the aminohydantoin **87**, served several purposes. Along with the simplified synthetic preparation of the monocyclic amine analogs, in general these derivatives were 10-fold more potent, had improved CNS exposure, and afforded another series to define the SAR.

Although there was a significant effort on the original bicyclic series, which covered much chemical space, it became apparent that the smaller monocyclic aminohydantoins possessed superior drug-like properties. Through an extensive SAR and structure-based design approach, Malamas and colleagues have identified **88** (WAY-258131) as a potent ($EC_{50} = 24$ nM), orally bioavailable BACE1 inhibitor that showed efficacy in a number of biochemical

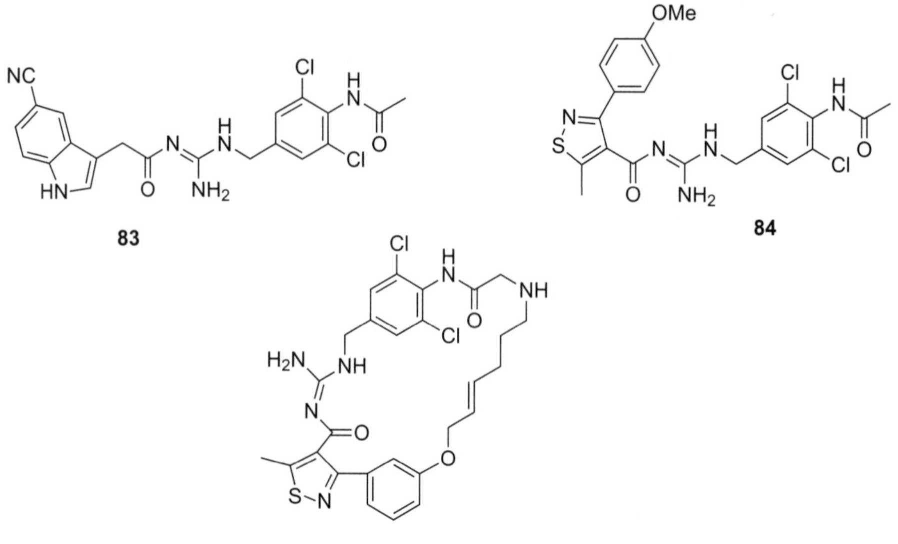

83

84

85

Scheme 20.

and behavioral end points [194]. When dosed at 10 and 30 mpk orally, subchronic administration (7–14 days bid) of WAY-258131 reduced $A\beta_{40}$ and $A\beta_{42}$ levels by 30–60% in plasma and 13–22% in brain, respectively. Although the overall B/P was low (0/05–0.10), the oral bioavailability was sufficient (25%) to afford exposure levels high enough to elicit a biochemical efficacy end point. In addition, WAY-258131 showed a significant improvement in reversing the cognitive deficit of Tg2576 mice, as measured in a rodent behavioral contextual fear conditioning (CFC) assay, at doses of 30–100 mpk p.o. acutely, and 10–30 mpk subchronically (5 days p.o.). This important tool compound demonstrated the potential of this class of BACE1 inhibitors, but also showed the PK limitations

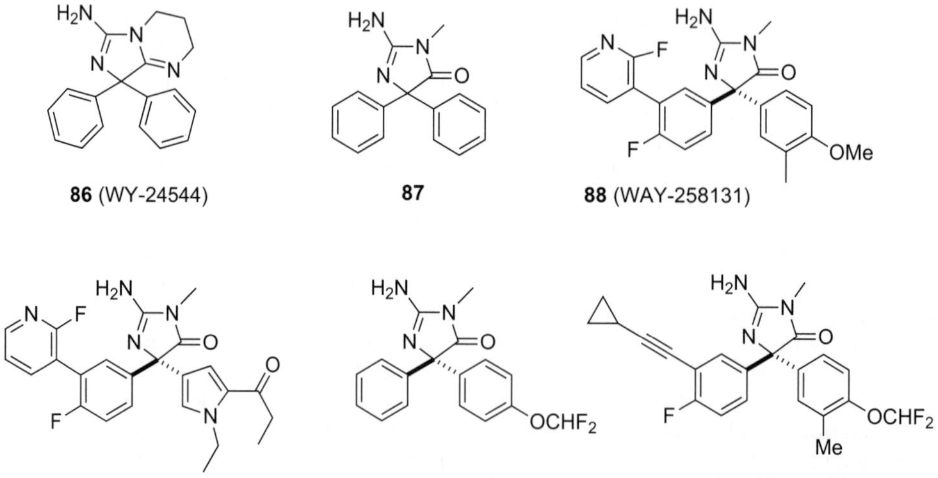

86 (WY-24544) **87** **88** (WAY-258131)

89 **90** (WAY-264116) **91** (WYE-106531)

Scheme 21.

in efficient delivery to the target organ. The pyrrole analog **89** was reported later and was the first advanced compound with good PK properties and potency at the enzyme [195]. Pyrrole **89** was reported to be 63% bioavailable in rats, with significantly improved microsomal stability, and had cellular potency of 9 nM at the BACE1 enzyme. This compound possesses excellent selectivity versus BACE2 as well as the related aspartyl proteases, pepsin, renin, and Cathepsin D. However, as with this and other key derivatives from this series, the ability to balance the optimized potency with sufficient brain exposure was difficult if not impossible. It became apparent that the increasing TPSA of optimized compounds afforded an increased affinity for pgp transporters, resulting in rapid efflux from the brain.

In an effort to improve the B/P and overall exposure to the brain of analogs of this class, the Wyeth group identified the truncated and simplified analog **90** (WAY-264116) [196–198]. Although this derivative had reduced affinity for the enzyme (cellular $EC_{50} = 189$ nM), the significantly lower TPSA and excellent PK properties afforded a candidate with excellent B/P (~1.0), and bioavailability (89% in rats) and was advanced to *in vivo* characterization. WAY-264116 reversed the cognitive deficits of Tg2576 mice at doses of 30 and 100 mpk p.o., acutely, as measured in the CFC behavioral assay. In wild-type guinea pigs, the reduction of $A\beta_{40}$ in both the plasma and CSF was >75% at a dose of 200 mpk p.o. Perhaps most compelling is the report of reduction of overall plaque load in the brain of PSAPP mice (a cross between Tg2576 and mice overexpressing human Presenilin-1) dosed twice daily with WAY-264116 at 45 mpk p.o. After 3 months of dosing, the mice showed an approximate 25% reduction of average plaque area, as measured by thioflavin-S staining in the cortical region. A second compound with good potency and central exposure was **91** (WYE-106531), which showed lower levels of CYP450 inhibition and hERG channel affinity than the parent compound WAY-264116.

Other groups have identified derivatives with this embedded guanidine moiety as well. There have been multiple claims among various patents and patent applications, with limited details of *in vivo* efficacy validation [199–201]. In a collaboration between AstraZeneca and Astex, closely related aminohydantoins **92** and **93** have been claimed in several patent applications, with reported potencies <100 nM (Scheme 22) [202–204]. Scientists at Schering-Plough have also claimed several aminohydantoin examples **94**, as well as the unusual cyclic derivative **95**, with potencies <1 µM [205–207]. Finally, patents by Roche have claimed derivatives with the related aminooxazole heterocycle **96** as potent BACE inhibitors [208].

Baxter and coworkers at Johnson & Johnson reported a 900 nM novel cyclic guanidine lead **97** identified in an HTS for the BACE1 enzyme (Scheme 23) [209–211]. Through a structure-based design approach the group optimized the interactions with the enzyme pocket and identified a collection of potent derivatives, exemplified by **98** ($IC_{50} = 11$ nM). Although selectivity versus cathepsin (IC_{50} 110 nM) and the hERG channel ($IC_{50} = 140$ nM) was moderate, **98** demonstrated efficacy in lowering plasma $A\beta_{40}$ in rats when dosed orally at 30 mpk. The poor CNS exposure, reportedly due to the high pgp transporter affinity, prevented further development of this class of compounds.

In a further extension of the collaboration between AstraZeneca and Astex, Edwards and coworkers reported the identification of a related cyclic guanidine derivative through a fragment-based search [179]. The initial hit **99**, identified through NMR binding was minimally potent with ~1 mM affinity as measured by surface plasmon resonance (SPR). Utilizing their high-throughput X-ray crystallographic capabilities, the group was quickly able to optimize this modest lead to derivative **100**, a compound with IC_{50} of 80 nM at the BACE1 enzyme.

There are a multitude of variations on the previous examples reported in the patent and journal literature over the last few years. The challenges highlighted in this section underscore the difficulty in identifying compounds with the requisite properties to make the significant advancement to clinical trials. The search for potent, bioavailable, CNS penetrant BACE1 inhibitors has led to many excellent advances that will surely light the way

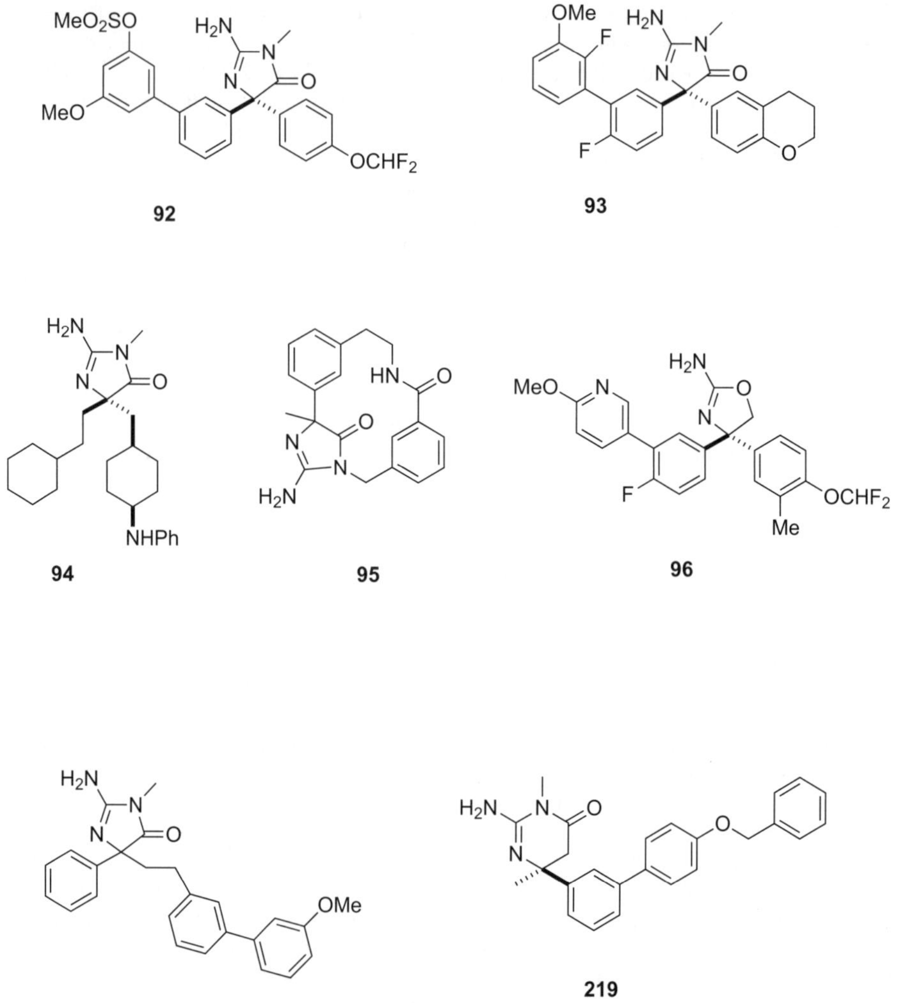

Scheme 22.

for future clinical candidates utilizing this approach. While proof of concept experiments in human clinical trials remains unrealized, research in many groups continues in earnest for this approach to the treatment of AD.

3.4. β-Secretase Inhibitors—Clinical

While there has been an intense effort directed toward the identification of β-secretase inhibitors in the last decade, there is but a single confirmed report of a clinical development program as of this publication. Astellas Pharma in conjunction with CoMentis have reported the advancement of CTS-21166 (ASP1702), from a series of peptidomimetic ligands, into Phase I human clinical trials. Although the structure of this ligand has not been revealed, the pharmacologic profile and early clinical results have been disclosed in a series of conferences in 2008–2009 [212–217]. CTS-21166 is a potent ($IC_{50} = 2.5\,nM$) and selective BACE1 inhibitor with reported oral bioavailability of 12–60% in rodents and dogs. When dosed at 100 mpk (i.p.), CTS-21166 reduced Aβ levels by 20–30% in plasma, CSF, and brain of wild-type rats. Treatment of Tg2576 mice, containing the Swedish and

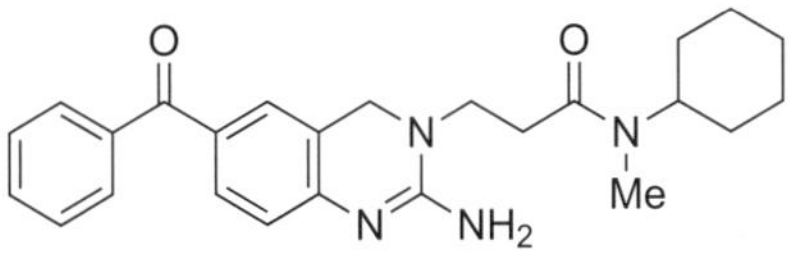

Scheme 23.

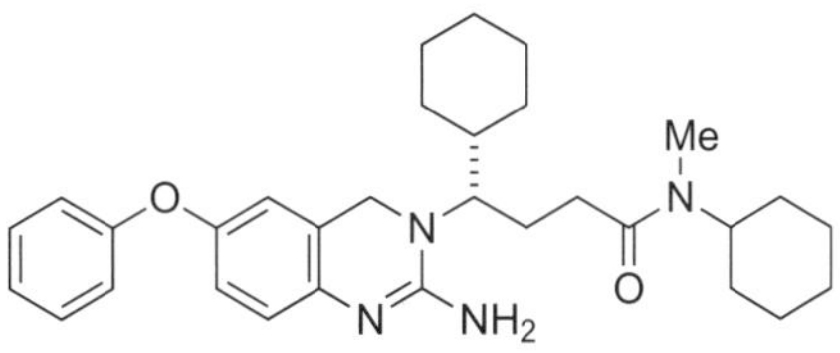

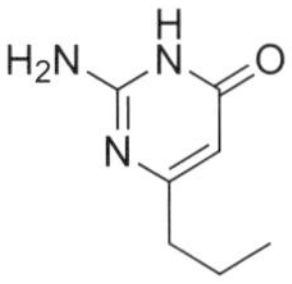

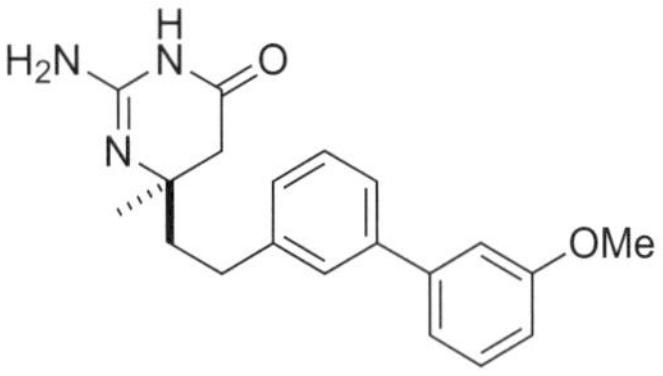

London mutations, for 6 weeks with 4 mpk i.p. resulted in a 35–38% reduction in brain Aβ$_{40-42}$. Brain and plasma exposures of 9 and 20 nM, respectively, were reached with this chronic dosing regimen (measured 3 h post dose), and a 38% reduction of brain plaque load in cortex and hippocampus, as measured by immunohistochemical staining with a 6E10 antibody, was reported. Single-dose administration at 100 mpk p.o. or subchronic (5 days bid) dosing at 10 and 30 mpk p.o. reversed memory deficits in Tg2576 mice with concomitant reductions of plasma and brain Aβ$_{40}$. Soluble cortical Aβ$_{40}$ was reduced 23% either acutely at a dose of 300 mpk p.o., or 15–20% subchronically (5 days bid) at doses from 10 to 100 mpk p.o. with CTS-21166 in these same mice.

On the basis of the preclinical package, CTS-21166 was advanced into human clinical trials. The data from the Phase I trial, initiated in June 2007, showed that a single i.v. administration of CTS-21166 in healthy males induced a 60% reduction of Aβ$_{40}$ in plasma exposure over a 24 h period. The average EC$_{50}$ required for plasma reduction was approximately 24 nM and was consistent at all four i.v. doses of 45–225 mg. Results also showed that the high dose regimen of CTS-21166 produced a sustained 40% reduction of plasma Aβ$_{40}$ over a 72 h-time frame. Pharmacokinetics showed good dose proportionality of i.v. CTS-21166, and it was reportedly well tolerated at doses up to 225 mg i.v. and 200 mg p.o. Although these Phase I trials were run with i.v. dosing, Astellas reported that they were considering initiating clinical studies with an oral formulation based on the favorable pharmacokinetics observed.

The AD research community anxiously awaits further results from this very important clinical investigation. The potential report of a proof of concept for this mechanism will no doubt accelerate the already frenetic search for BACE1 inhibitors and other amyloid hypothesis approaches for the treatment of AD.

3.5. α-Secretase Modulators—Preclinical and Clinical

Very few cases of α-secretase enhancers or modulators are reported in the literature. One such α-secretase stimulator is **101** (EHT 0202) developed by ExonHit Therapeutics, shown in Scheme 24 (Clinical Trials in Alzheimer's Disease Meeting Las Vegas, NV, USA, October, 2009). EHT 0202 is a pyrazolopyridine that has been shown to induce stimulation of sAPPα in rat cortical neurons *in vitro* and *in vivo* (10 mpk for 15 days in guinea pig). Furthermore, it was reported in preclinical studies to improve memory and cognition as demonstrated in Morris water maze and

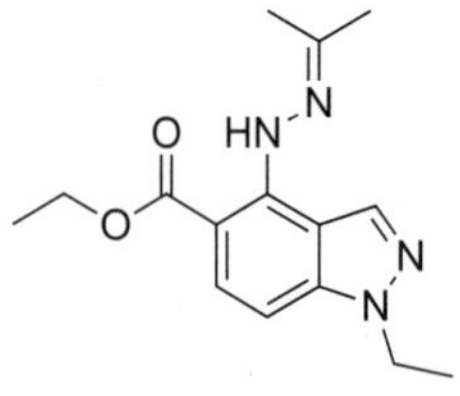

101 (EHT 0202)

Scheme 24.

Barnes behavioral assays. EHT 0202 entered clinical trials and Phase IIa is ongoing, as an add-on therapy to AChE inhibitor in 135 mild to moderate AD patients (3-month trial with 40 and 80 mg bid dosing). Based on the preclinical data that shows that EHT 0202 stimulates long-term potentiation (LTP) but is neuroprotective in a GABA-dependent manner, it is not clear that the observed positive effects are derived strictly from its α-secretase activity. In fact, EHT 0202 is reported to modulate GABA-A receptors and weakly inhibit PDE4. Therefore, it is fair to postulate that the observed cognitive enhancement may be derived from affecting a combination of targets especially if cognitive effects cannot be directly correlated to the levels of sAPPα [218].

3.6. γ-Secretase Modulators—Preclinical and Clinical

Although most of the focus of the last fifteen years on inhibiting γ-secretase has been directed at isosteric ligands (ligands interacting at the native site of cleavage for APP), a significant more recent effort has begun to develop in the area of γ-secretase modulators. γ-Secretase modulators are defined as compounds that effect Aβ levels without affecting the rate of APP processing. Although the preclinical landscape of this area has yet to evolve into the depth that inhibitors have, there has been some results on investigations of potential modulators in the clinic.

To that end, there was a report in *Nature* that some nonsteroidal anti-inflammatory (NSAID) compounds, such as fenofibrate **102** and tarenflurbil **103**, shown in Scheme 25, lower the secretion of Aβ$_{42}$ independently of cyclooxygenase (COX) activity by altering APP processing and may provide a different therapeutic approach [219,220]. These compounds sometimes called SALAs, (selective Aβ$_{42}$ lowering agents) may be acting as γ-secretase modulators without inhibiting γ-secretase-mediated APP cleavage nor Notch processing [221]. They were shown to bind to the substrate APP, rather than binding to the secretase enzyme and this binding alters Aβ production and inhibits Aβ aggregation to provide a combined overall benefit [222].

Myriad pharmaceutical examined tarenflurbil **103** (Flurizan), the inactive (*R*)-enantiomer of the NSAID in a Phase III clinical trial as a therapy for AD. The clinical trial was reportedly terminated in 2008 when no efficacy (800 mg bid) was demonstrated in improving cognitive behavior in patients with *moderate* AD. However, a slight benefit to the daily living activity index was observed among these *mild* AD patients [223]. Due to its weak GS inhibitory activity (IC$_{50}$ > 10 μM) and poor brain penetration, the validity of using tarenflurbil in this key proof of concept study is deemed somewhat dubious [224].

Eisai was also developing a γ-secretase modulator, E-2012B, whose structure has yet to be disclosed, as a treatment for AD. In 2007, a Phase I trial was suspended temporarily following the observation of lenticular opacity, a type of cataract in the eye, in a preclinical safety study in rats. Additional data submitted to the FDA by Eisai from their

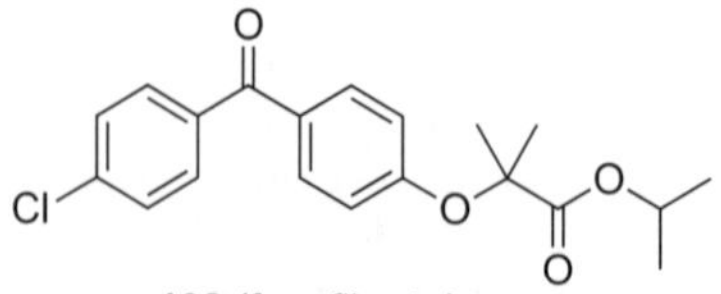

102 (fenofibrate) 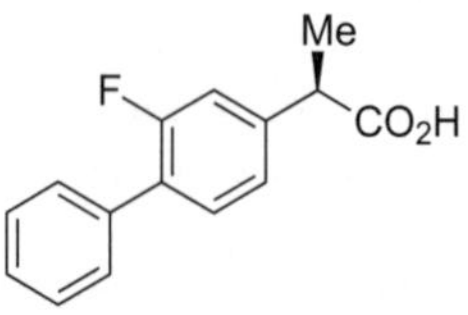   **103 (tarenflurbil)**

Scheme 25.

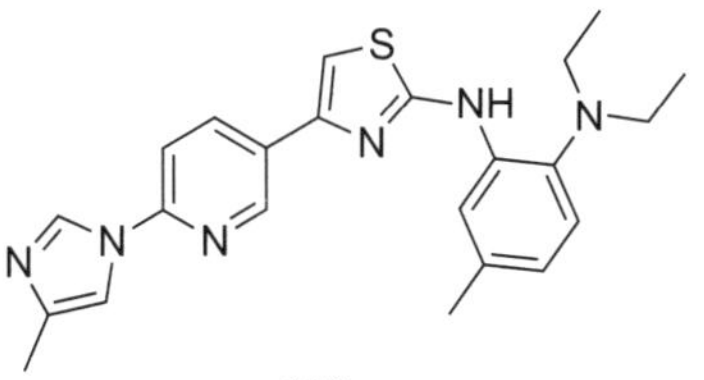

Scheme 26.

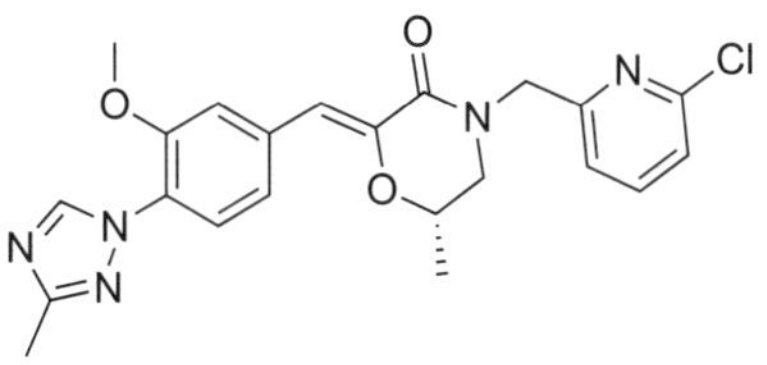

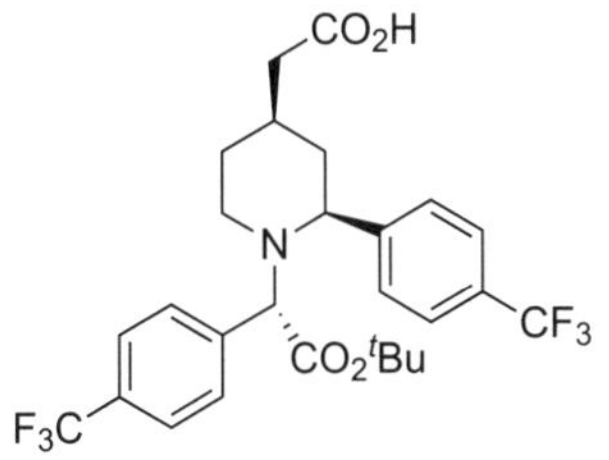

preclinical and Phase I studies provided Eisai an opportunity in 2008 to continue with its trial of E-2012. Additional results on tolerance and efficacy of E-2012 could not be found on Eisai's company Web site and no studies are listed under US clinical trials Web site.

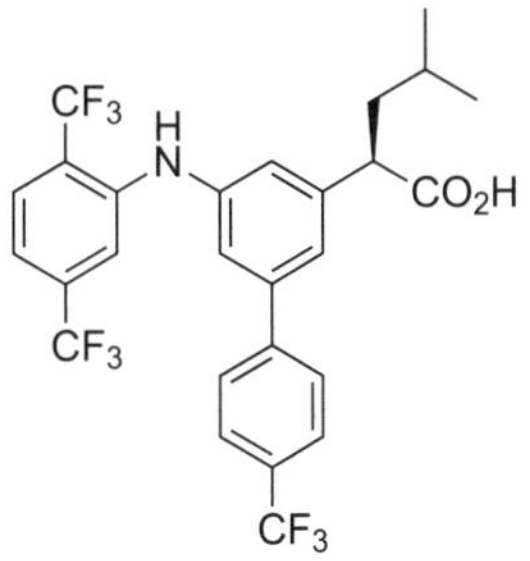

Other companies such as Neurogenetics and Eisai have filed patent applications for a series of imidazoles (i.e., **104**) and triazoles (i.e., **105**) as modulators (Scheme 26) while Merck reported on compounds such as **106** that are claimed to inhibit Aβ_{42} over Aβ_{40} by 50-fold [225–227]. One notable claim of *in vivo* efficacy by a modulator is reported by Cellzome in collaboration with Ortho-McNeil Pharma on biphenyl carboxylic acid derivative **107** that is closely related to the ibuprofen analogs [228]. The compound **107** was shown to be efficacious in lowering Aβ_{42} by 30% in rat model (30 mpk p.o.) [229].

3.7. HMG-CoA Reductase Inhibitors—Preclinical and Clinical

There are many questions around cholesterol's role and its association to neurodegeneration and AD. The clearest genetic risk factor for late-onset AD is the presence of the e4 allele of the apolipoprotein E (ApoE) gene that has a role in both cholesterol metabolism and Aβ clearance. With Aβ deposition being a hallmark of AD, another rational approach to amyloid-driven therapy could be the identification of processes that *indirectly* affect the level of amyloid deposition [230].

To that end, many different possible targets have been under consideration. Brain lipid metabolism targets such as acyl-CoA: cholersteroyl acyltransferase (ACAT) and liver-X receptor-β (LXR-β) as well as protein–protein interaction of lipoprotein receptor protein–apolipoprotein E (LRP-ApoE) may be promising. Another viable option may be to alter cholesterol homeostasis where ApoE plays a fundamental role as a carrier of cholesterol in the brain. However, the role of plasma cholesterol level and the brain lipid homeostasis in AD is unclear since it has been reported that the lipoproteins do not cross the blood–brain barrier [231,232].

There is some evidence to suggest that increased levels of cholesterol in middle aged population is associated with an increased risk of developing AD. 3-Hydroxy-3-methyl-glutaryl (HMG)-CoA reductase is a rate-limiting enzyme in cholesterol biosynthesis and statins are a class of HMG-CoA reductase inhibitors that are widely used as cholesterol

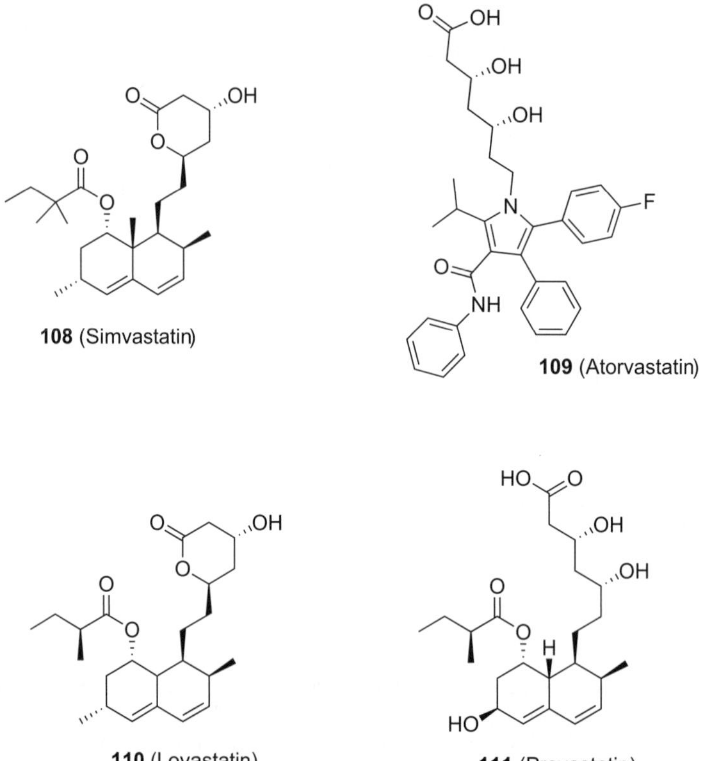

108 (Simvastatin)

109 (Atorvastatin)

110 (Lovastatin)

111 (Pravastatin)

Scheme 27.

lowering medications. As such, the ability of statins to reduce cholesterol synthesis and influence the antiamyloidogenic path has been investigated in the field by clinical investigation with several available HMG-CoA reductase inhibitors [233–235].

Initial data suggested that a statin, such as Merck's simvastatin **108** (Zocor or MK-733), shown in Scheme 27, may be associated with reduced risk of AD dementia. [236–238] Other promising data from guinea pig and Tg2576 mice studies demonstrated that chronic treatment with statins reduced cerebral amyloid production. Therefore, the potential benefits of three different statins (atorvastatin **109**, simvastatin **108**, and lovastatin **110**) to alter APP proteolysis and produce less amyloidogenic peptides are being explored in Phase III trials with the idea of utilizing them as a disease-modifying AD therapy [239,240].

Initial clinical results were reported by Pfizer utilizing **109** (atorvastatin) in a Phase III trial LEAD (Lipitor's Effect on Alzheimer's Dementia) with 80 mg of Lipitor plus 10 mg of donepezil versus donepezil alone. A 72-week trial with more than 600 patients from 50 to 90 years old was concluded in 2007 [241]. This study revealed no significant difference in cognitive or behavioral end points between the treatment group and patients treated with donepezil only, despite the fact that magnetic resonance imaging (MRI) showed that a subset of treated patients showed slower decline in hippocampal volume. To further complicate the data, there seems to be a lower rate of decline in cognition among men in the treated population whereas no evidence of such effect was observed in women [231].

Two other marketed statins, Merck's Zocor **108** (simvastatin) and Mevacor **110** (also known as MK-803, lovastatin) being marketed by Andrx Corp., have also been tested in Phase III trials. Unfortunately, both studies have been discontinued and no explanation was provided for the termination.

To further complicate the promise of statins as AD therapy, in some preclinical comparative studies, it was concluded that only some statins lowered total level of cholesterol (only lovastatin, but neither simvastatin nor atorvastatin) while all statins affected translocation of CNS cholesterol, which may be considered more important to Aβ production in the CNS than the total cholesterol level [242]. Riekse et al. compared biomarker levels in CSF in 14-week study following administration of 80 mg/day of **108** (simvastatin) and **111** (pravastatin) [235]. They observed no difference in the total levels of Aβ and sAPPβ in CSF of the subjects. However, only simvastatin, which has the best CNS exposure of the two, was able to lower the level of phosphorylated tau181 in the CSF. Their conclusion is that the reduced risk of AD and lowering Aβ deposition in animal models of AD may be associated with the ability of the statins to penetrate into CNS [235].

In follow-up studies, the CNS permeability of various statins using *in situ* rat brain perfusion technique and other permeability assays both *in vitro* and *in vivo* were performed. Pravastatin, marketed by Bristol-Myers Squibb, was shown to have a much lower transport into the brain than lovastatin or simvastatin (greater than 1 to 2 orders of magnitude difference) [243]. Based on this data, pravastatin is predicted to have a lower potential CNS side effect than more lipophilic and blood–brain barrier permeable statins such as lovastatin and simvastatin that are inactive lactone prodrugs that convert *in vivo* into the active acids. Unfortunately, corresponding toxicity data is not available in the literature to directly correlate and test these observations.

Overall, results from various studies on these statins seem to contradict each other in several ways. First, not all the statins affect the total level of cholesterol while all statins affect translocation of CNS cholesterol while some statins easily penetrate the brain while others do not, depending on the structure. Secondly, total levels of Aβ and sAPPβ do not change regardless of the statins while tau levels are affected by simvastatin, but not pravastatin. To complicate the matter even further, there is a report of a clinical trial of a nonstatin-based cholesterol lowering drug's *inability* to prevent AD that suggests a cholesterol-independent, but statin-mediated anti-AD effect [244].

Based on these mixed results, many of the following putative mechanisms can be considered for the statin's ability to reduce risk of AD and lower Aβ deposition: (i) a pathway related to nitric oxide bioavailability, (ii) a mechanism involving heat shock protein and survival-related phosphoinositide 3-kinase serine-threonine kinase (PI-3K/Akt) pathway, as well as, (iii) an endocytosis dependent pathway reducing APP level and β-secretase activities [244–246].

Finally, another possibility is that the effective neuroprotection of statins may result from an anti-inflammatory mechanism, independent of the cholesterol lowering characteristics. If the positive effect of statin is mostly due to inhibition of the HMG-CoA reductase by-product, mevalonic acid, an anti-inflammatory agent, this also cannot be ruled out. As such, the use of statins as a disease-modifying therapy for AD is still unproven while their use as a possible "palliative therapy" for AD could still be a promising option.

3.8. Fibrillization Inhibitors—Preclinical and Clinical

Identification of antiamyloid agents or fibrillization inhibitors as an AD therapy has also been an actively explored mechanism over the last decade. Nearly 10 different entities are reported to be in clinical trials with half of them currently in Phase I. The rationale behind inhibition of fibrillization is to prevent AD by interrupting or inhibiting the self-assembly of soluble nontoxic Aβ polypeptides into β-pleated sheets containing toxic fibrils, one of the recognized hallmarks of the disease.

A combination of NMR, CD, FT-IR, and microscopic data were used to shed light on the 3D structure of β-strand motif and its packing strain of Aβ self-assembly, which is required for fibril growth [247,248]. Spectroscopic studies of Aβ have revealed the presence of a remarkably high frequency of aromatic residues (i.e., phenylalanine, tyrosine, and tryptophan) in short, functional Aβ fragments involved in molecular recognition and self-assembly. These three aromatic amino acid residues usually have the lowest frequency of occurrence in proteins. As such, it has been postulated by some research groups that these residues may be present to facilitate planar aromatic ring stacking interaction in a "parallel" manner similar to the π-stacking interaction [249,250]. This information provided some structural basis for the design of novel fibrillation inhibitors.

3.8.1. Peptide-Derived Fibrillization Inhibitors

Similar to γ-secretase and BACE1 inhibitor research efforts, peptide isosteres of Aβ also provided an important starting point for the design of fibrillization inhibitors. Peptidomimetic analogs derived from the central residues of the Aβ sequence (see Fig. 2), congeners of KLVFF (Aβ$_{16-20}$), LVFFA (Aβ$_{17-21}$), were identified as potential inhibitors. Other peptide conjugates that alter the aggregate formation such as β-sheet breaker peptides have also been reported [248,251–254].

Praecis Pharmaceuticals, currently part of GlaxoSmithKline, identified peptide analogs based on chemically stable, substituted D-amino acid analogs of LVFFA (Aβ$_{17-21}$). [255,256] These peptides were designed to interfere with the Aβ protein–protein interaction by binding to Aβ, which alters its conformation. Although these small penta- or hexapeptides were shown to be somewhat stable in CSF and had moderate CNS exposure capabilities, due to rapid hepatic elimination, these analogs were considered unsuitable for further development.

Other peptide analogs include β-sheet epitopes that target Aβ and α,α-disubstituted amino acid mimetics of Aβ that interact with the β-sheet assemblies by altering the rate of Aβ aggregation and its morphology [251]. Another novel approach is the identification of positively charged analogs of CLAC (collagenous Alzheimer amyloid plaque component) that prevent further Aβ peptide incorporation into β-sheet rich plaques by binding to Aβ-deposits [257]. Based on the perusal of recent literature, there are no reported cases of peptide-based fibrillization inhibitors in the clinic.

3.8.2. Small-Molecule Inhibitors of Fibrillization or Aβ Inhibitors

The current state of structure-based design of fibrillization inhibitors seems to be limited primarily to in vitro demonstration of the inhibition of Aβ protein with structures derived from natural products, metal chelators, and small peptides related to Aβ.

Glycosaminoglycan Mimetics One of the most advanced fibrillization inhibitors in development was NC-531 that was reported to have entered clinical trials in 1999 by Neurochem Pharma in collaboration with Lundbeck [258]. NC-531 is a sulfated glycosaminoglycan (GAG) mimetic that inhibits the binding of the GAG protein to Aβ by affecting its conformation during aggregation into β-sheet fibrils. In a 3-month Phase II study in mild AD patients (58 patients), an impressive 70% reduction of Aβ in CSF was observed in the 200 or 300 mpk dose groups. However, no improvement in the ADAS-Cog score (Alzheimer's Disease Assessment Scales-Cognition) was detected after 12 weeks while dose-dependent side effects (i.e., nausea and vomiting) were observed [259]. As there are no current clinical trials reported on NC-531, development of this compound may have been discontinued.

Tramiprosate **112** (Alzhemed), shown in Scheme 28, is another small-molecule inhibitor that has reportedly demonstrated efficacy in reducing brain and plasma Aβ levels in preclinical studies. However, in a Phase III clinical study, tramiprosate, another GAG mimetic, did not demonstrate a statistically significant difference in the primary end points over the placebo group in 18-month trial in AD patients. And yet, similar to the results observed from the study with EGB-761, a naturally derived fibrillization inhibitor, a significant difference in hippocampal volume was observed in the tramiprosate treated group suggesting some protection in the AD brain, as measured by MRI [259,260].

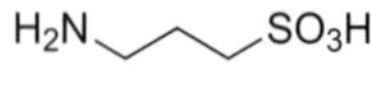

$H_2N\diagdown\diagup SO_3H$

112 (Tramiprosate)

Scheme 28.

Metal–Protein Attenuating Compounds Another class of compounds actively being pursued as a possible therapeutic intervention to AD is metal–protein attenuating compounds (MPACs). Use of these MPACs is based on the "metal theory" of AD in which the accumulation of metals (such as Cu^{2+}, Fe^{3+}, and Zn^{2+}) in the brain is the principal cause of the advancement of AD and not plaque formation. Researchers at Prana Biotech believe that the accumulation of these metals causes Aβ protein to convert into a "rogue" enzyme. This enzyme is reportedly associated with the production of hydrogen peroxide, which causes damage in brain cells responsible for amyloid plaques (related to zinc) and neurodegeneration (related to damage of COX-2, one of the key inflammation enzymes). Through some focused studies they demonstrated that MPAC-derived compounds compete with Aβ by directly binding to excess metal ions in the brain and by attenuating their damaging actions. [261–263]

Prana Biotech successfully advanced two compounds **113** (PBT-1, clioquinol), shown in Scheme 29, and a follow-up compound, PBT-2, which is an 8-hydroxyquinoline derivative (although unspecified) into Phase II/III studies. PBT-2 surpassed PBT-1 in the development process due to its improved safety and efficacy profile, having demonstrated significantly greater effectiveness in lowering plaque in preclinical studies [264]. PBT-2 that progressed to Phase II is reported to target abnormal protein–metal interaction. Positive Phase II results were reported in 2008 whereby patients treated with 250 mg daily dose of PBT-2 showed 13% lowering of CSF $Aβ_{42}$ while showing cognitive improvement in both neuropsychological test battery (NTB) and in executive function tests versus placebo [264].

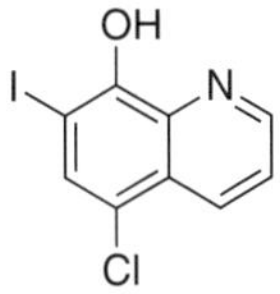

113 (PBT-1, clioquinol)

Scheme 29.

Also significant is the fact that PBT-2 did not seem to alter serum concentration of metals such as Zn^{2+} and Cu^{2+} concentrations, suggesting a central effect of the drug on Aβ metabolism. The current status of the compound is unknown.

This positive data is in contrast to much of the reported clinical data on compounds derived from HMG-CoA reductase inhibitors and other Aβ aggregation inhibitors. The key questions of mechanism-specific selectivity of the MPACs, and the safe use of these metal chelators in a chronic dosing regimen for AD, will need to be addressed in future studies.

Natural Product Related and Other Small-Molecule Inhibitors As referred to earlier, many of the small-molecule-based approaches seem to be centered on "naturally occurring compounds" such as flavones and curcumins [265]. One of the most extensively studied compounds is EGb-761, an extract of ginkgo biloba. In a preclinical study, EGb-761 was shown to "protect" aging brain, especially the hippocampus, but its precise mechanism of action (MOA) is unclear since it is also reported to have a broad spectrum of pharmacological effects, including free-radical scavenging properties. EGb-761 entered Phase III clinical trials to treat dementia and results from various studies with different dosing regimen have been reported (120 mg tid or 240 mg bid) [266]. When EGb-761, enriched in flavonoids and terpenes, was given to AD patients with mild cognitive impairment, some modest improvements were reported on various cognition readouts (i.e., Skilled Knowledge Test (SKT), ADAS-cog). However, no pharmacological biomarker data is available from these results to correlate the extent of improvement to the compound's ability to inhibit Aβ aggregation. [267,268] Ipsen is currently recruiting mild AD patients in a Phase II trial to monitor the effect of EGb-761 on brain glucose metabolism (ClinicalTrials.gov NCT0081436).

Another small-molecule indole-3-propionic acid **114** (oxigon) is reported to be an antioxidant that was shown to prevent Aβ aggregation (Scheme 30) [269]. It is being developed by Intellect Neurosciences in collaboration with New York University. Phase I trials were completed in 2007 and because the compound

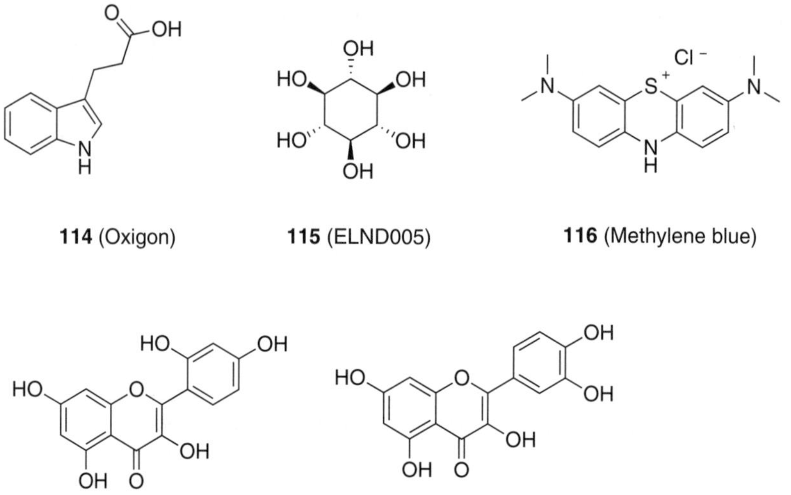

114 (Oxigon)   **115** (ELND005)   **116** (Methylene blue)

117 (morin)   **118** (quercetin)

Scheme 30.

was well tolerated, it was expected to continue in Phase II studies during 2008. However, at this time, there are no scheduled or active clinical trials using the compound. In preclinical transgenic mouse models, it reportedly showed some potential to reduce brain amyloid burden and improve cognition in AD model. However, as oxigon was also found to have potent antioxidant properties *in vitro* and *in vivo*, and protected neurons from β-amyloid-induced toxicity, its mechanism of action and specificity remain unclear.

Elan began a Phase II trial of polyalcohol **115** (ELND005; AZD-103) in 2007 based on positive results from a Canadian Phase I study carried out by the originator, Transition Therapeutics. The clinical trial data showed that it was well tolerated by all subjects. ELND005, scyllo-inositol, is reportedly a brain-penetrant small-molecule inhibitor of beta-amyloid peptide. In preclinical studies, ELND005 has demonstrated lowering of Aβ and slowing down of plaque growth in Tg2576 mice while showing improvements in a spatial memory function in rodent behavioral assay (i.e., Morris water maze) [270]. Currently, the company is enrolling patients to evaluate the safety and efficacy of ELND005 in mild-to-moderate AD patients (ClinicalTrials.gov, NCT00568776 Phase II ongoing; NCT00934050 is enrolling patients for a Phase II study in collaboration with Transition Therapeutics).

Other oligomerization inhibitors include a class of dyes that reportedly have the ability to result in π-stacking of aromatic residues [247]. Methylene blue **116** (MB), a compound of the phenothiazine class, seems to promote fibril formation by both depleting the monomer pool required for oligomerization and by promoting filament elongation. Because MB has some drug-like properties, that is, good physiochemical properties, low toxicity, and brain penetration, it was suggested that it would act as a tool to shed light on the pathological significance of amyloid oligomers and fibrils in AD animal models. However, some caution needs to be exercised in the utility of MB since it is reported to affect multiple pathways thus making it difficult to tease out the specific activity for this ligand. A few pathways that MB seems to affect include blocking production of reactive oxygen species, affecting oxygen homeostasis, and its role around tau-filaments formation [271].

Other small molecules also include examples of the flavone class of compounds, such as morin **117** and quercetin **118**, shown in

Scheme 30. These small molecules are reported not only to inhibit the formation and extension of β-amyloid fibrils (fAβ) but also destabilize fAβ dose-dependently *in vitro* [272,273]. Although many antiamyloidogenic compounds have been reported and some reported results include *in vivo* activities in HT22 cells or protection in other cytotoxicity assays how they inhibit β-amyloid fibril formation remains poorly understood [274].

ProteoTech has reported on the identification of a small molecule for AD, PTI-703 (Exebryl-1) [275]. Although its structure is unknown, it is a plant derivative extracted from the Amazonian plant *Uncaria tomentosa* (Cats Claw) and is in development as a potential therapy for Alzheimer's disease. PTI-703 entered into Phase I development after having demonstrated 27–45% lowering of brain Aβ and amyloid plaque load after a 50 mpk i.p. dose for 90 days in APP transgenic mice in addition to demonstrating improvement in memory deficit in the rodent Morris water maze model [276]. The compound is reported to be orally bioavailable and to have excellent brain penetration, reaching a ratio greater than 1 for cortex to plasma concentration 4 h after oral dosing (100 mpk) and maintaining that ratio over a 24 h period. Based on efficacy data showing that PTI-703 was able to inhibit and reduce tau protein from forming paired helical filaments, in addition to affecting amyloid plaque, the two major characteristic and pathology of Alzheimer's disease, a US Phase II trial was launched. There are a few patents from Castillo et al. and ProteoTech that describe the extraction of this class of compounds and their use in treatment for AD [277,278].

3.9. Receptor for Advanced Glycosylation End Products Inhibitors

In a collaboration between Pfizer and TransTech Pharmaceuticals, PF-04494700 (TTP 488) whose structure has yet to be disclosed, is reported as a small-molecule antagonist of receptor for advanced glycosylation end products (RAGE) for the treatment of AD and diabetic nephropathies [279–281]. RAGE has been implicated in the pathophysiology of AD since cells expressing RAGE are activated in the vicinity of the senile plaque. As such, antagonism of the RAGE/ligand interaction has been shown to reduce amyloid plaque formation in mouse preclinical models. [282,283] PF-04494700, identified by TransTech using their genomic and proteomic data, was shown to be well tolerated in a Phase I clinical trial. The secondary outcome from the trial did not show high significance on plasma biomarkers or on cognitive function improvements. However, in preclinical studies it has been shown to lower amyloid load in the brain after a 90-day treatment in addition to a reduction in latency and traveled distance in an *in vivo* mouse model, the Morris water maze. To date, PF-04494700 is the only small-molecule RAGE antagonist reported to be in the clinical development, and is currently recruiting mild-to-moderate AD patients for a Phase IIa study. There have been sporadic reports of other small-molecule RAGE antagonists, but none have advanced beyond preclinical investigation [284–286].

3.10. Miscellaneous Mechanisms for Aβ Reduction

In addition to the various methods mentioned so far, other pathways that modulate microglial activation and influence Aβ accumulation or toxicity have also been targeted. Although some relationship between Aβ and cerebral vasculature pathology seems to exist, very little data seems to be available to firmly establish the key correlates. [287,288] In addition to the profuse numbers of Aβ plaques and neurofibrillary tangles, the pathology of AD identified the presence of a type of chronic microglial inflammation. Microglia are the first to respond to any perturbation or injury within the CNS, and may be related to the formation of insoluble plaques and tangles. There is some data to suggest that inflammation seems to occur early in the pathogenesis of the disease. In addition, association between cerebral atherosclerosis and AD pathology development has been established. [289–291]

Based on these data, an inflammatory hypothesis for AD was formulated, which states that the microglial activation by Aβ could be partially responsible for loss of healthy neurons or damage to the synapses.

To this end, many different chemical series seem to be associated to this inflammatory response, such as NSAIDs, curcumines, and polyalcohols that were already mentioned in the previous sections [287,292]. A data set in the literature to support the inflammatory hypothesis as a disease-modifying therapy seems to be somewhat limited. As such, a few questions still remain to be answered; to what extent do inflammatory factors in the brain *directly* lead to synaptic damage and neurodegeneration and could future anti-inflammatory agents *alone* be useful as a therapy.

4. ALZHEIMER'S DISEASE IMMUNOTHERAPY—ACTIVE AND PASSIVE APPROACHES

The pioneering discovery by Elan Pharmaceuticals was the first to suggest that approaches using anti-Aβ antibodies may result in slowing or halting the progression of Aβ-mediated pathology in APP transgenic mouse models of AD [293]. This seminal work has led to the suggestion that both "active" immunization against injected Aβ-derived antigen (to *mount* a response in host), and "passive" immunization with anti-Aβ monoclonals (to *provide* a response in host), may be useful treatment strategies for AD. The aim of this section of the review is to highlight the core concepts having evolved from the early preclinical observations, the lessons learned from first-in-man studies, the rationale of different clinical strategies, and the status of current clinical candidates for Aβ immunotherapy.

4.1. Immunization Against Aβ

In the first of several studies, young 6-week-old PDAPP transgenic mice were actively immunized monthly with i.p. injections until 13 months of age with aggregated Aβ$_{1-42}$ to stimulate a polyclonal antibody response directed toward Aβ. In a preplaque model, the brains of immunized animals had developed significantly reduced amyloid deposits and neuritic pathology in comparison to vehicle-treated animals. Even more exciting was the observation that similar immunization of older 11-month-old plaque-bearing PDAPP mice for a period of 6 months also resulted in the significant reduction of brain Aβ levels, amyloid deposits, and neuritic pathology. These studies demonstrated that active immunization with Aβ peptide permitted the animals to mount an immune response and the generation of anti-Aβ antibodies that were efficacious. [293,294]

Subsequent studies demonstrated that the direct passive transfer of monoclonal antibodies engineered to bind Aβ into transgenic PDAPP mice by chronic i.p. administration was equally efficacious in preventing or reversing the progression of amyloid plaque formation [294]. Further *in vitro* characterization of efficacious antibodies demonstrated common properties of binding aggregated and soluble Aβ species, in contrast to antibodies recognizing only soluble forms of Aβ [295]. The clear preference for the binding to epitopes located at or near the amino-terminus of Aβ and specific isotype specificities contributed toward understanding the mechanism(s) of plaque clearance [295]. That treatment to actively induce the generation of polyclonal antibodies, or passive treatment with engineered monoclonal antibodies, not only slowed the progression of pathology, but even reversed it, suggesting the relevance of testing Aβ immunotherapy in Alzheimer's patients [293].

Numerous groups have subsequently replicated these initial reports in other transgenic APP mouse [296–298] and aged non-human primate amyloid plaque-bearing models [299,300]. There is concurrence that antibodies cross the blood–brain barrier in preclinical animal models, facilitate the removal of preexisting amyloid plaque therapeutically, reduce the formation of amyloid plaque, neuritic dystrophy [301,302] and synaptic degeneration [303,304] prophylactically, and possibly even reduce early hyperphosphorylated tau aggregates [305].

4.2. Functional Improvement of Behavioral and Learning Deficits in TgAPP Mice

APP transgenic mouse models display biochemical, behavioral and pathological markers consistent with many aspects of AD. In general, a reduction of synaptophysin occurs in young preplaque bearing mice [306] that appears to correlate with deficits in basal synaptic transmission and long-term potentiation in hippocampal CA1 and dentate gyrus prior to increases in soluble Aβ levels [307–309]. Impaired synaptic plasticity precedes an elevation of Aβ aggregates and oligomers [310] and deficits of spatial memory in a modified water maze [311], contextual memory in young mice [312,313], prior to the development of amyloid plaque and neuritic pathology in older mice [314].

In addition to the clearance of brain pathologies, active and passive Aβ immunization have resulted in functional improvements [315,316] including the reduction of cognitive deficits, altered cognitive function, and inhibition of hippocampal LTP of APP transgenic mice [317,318]. A single passive immunization with a monoclonal antibody, m266, that recognizes epitopes within the central domain of Aβ and binds to soluble Aβ, but not amyloid plaques or fibrils, has been shown to acutely ameliorate memory deficits without the reduction of amyloid plaques [319–321]. The apparent lack of correlation with plaque deposition is consistent with the notion that different behavioral deficits are related to plaque while others are more acute and thought to be caused by soluble Aβ oligomers [318]. Neutralizing antibodies that bind to synaptic oligomeric species of Aβ have been shown to acutely improve or reverse deficits in synaptic plasticity, LTP in vitro and in vivo [322,323]. These observations suggest such antibodies may provide a rapid effect on learning and memory in addition to the more chronic and prolonged disease slowing or reversing effects regarding the clearance or prevention of amyloid plaque formation.

4.3. Potential Mechanisms of Action for Anti-Aβ Antibodies

Although there is a general agreement as to the preclinical effectiveness of both active and passive immunization approaches, several hypotheses exist for how these approaches elicit efficacy. It is quite possible that several of these hypothetical mechanisms are important for mediating the benefits observed in preclinical animal models.

The microglial-mediated phagocytosis model is based on antibodies that bind the N-terminal Aβ epitope, cross the blood–brain barrier, target Aβ contained in fibrils and amyloid plaque, and the subsequent recruitment of phagocytosing microglia [293,294]. Murine monoclonal antibodies of the IgG2a isotype with highest affinity for the FcγR receptors expressed on microglia have been correlated with greater potency in an ex vivo model for plaque removal from tissue [294], and in vivo efficacy for plaque clearance from the brains of TgAPP mice [294,295,324,325].

The direct dissolution model, based on in vitro studies, suggests that anti-Aβ-antibodies inhibit or disrupt the formation of toxic Aβ aggregates or fibrils, as well as dissolve pre-existing fibrils [326,327]. Furthermore, the model suggests that antibodies binding the N-terminal $A\beta_{3-6}$ epitope (amino acids EFRH) are most potent [328,329]. Additional in vivo studies demonstrate that antibodies binding the N-terminal $A\beta_{4-10}$ epitope inhibit fibrillogenesis and cell death in transgenic TgCRND8 mice [330] and are capable of attenuating amyloid deposition and neuritic dystrophy [295,301]. This data is also consistent with observations using F(ab′)$_2$ fragments [331] or Fcγ-receptor KO mice [332] to demonstrate that efficacy is independent of Fc effector function. This suggests the possibility of two phases for amyloid clearance including an initial microglial-dependent phase and second a microglial-independent process in which diffuse Aβ deposits are cleared [333].

The third proposed mechanism is based on the peripheral amyloid sink hypothesis [319] and was developed using m266, a murine monoclonal antibody, that is directed to epitopes within the central domain of Aβ and binds only to soluble forms of Aβ [334]. Peripheral administration of m266 resulted in the rapid increase in plasma Aβ and a corresponding reduction in total brain Aβ. This observation led to the hypothesis whereby sequestration of peripheral plasma Aβ shifts the equilibrium between CNS and plasma Aβ pools, resulting in a net efflux of Aβ from the CNS into the peripheral circulation, where Aβ is degraded [319,335–337]. Consistent with the observations described above, administration of m266 has also been shown to rapidly improve cognitive behavior in APP transgenic mice, likely by the reduction of brain levels of soluble synaptotoxic species of Aβ [320]. Furthermore, the receptor-transport efflux of antibody:Aβ immune complexes across the blood–brain barrier to the periphery by the neonatal Fc receptor (FcRn) has been proposed [338].

4.4. First in Human Aβ Immunization

Successful preclincial studies demonstrating efficacy following immunotherapy in PDAPP mice with aggregated $A\beta_{1-42}$ [293] led to the development of the first human active immunization trial with AN1792, a synthetic Aβ peptide in combination with a QS-21 adjuvant [339–341]. Despite extensive preclinical safety and tolerability studies in animals, reports of an acute meningoencephalitis in 6% of the patients treated with AN1792 halted an early Phase IIa trial [341]. Of the nearly 300 AN1792-treated subjects, analysis showed that approximately 20% developed good antibody responses to Aβ. However, there was no correlation between severity of meningoencephalitis and the level of antibody titer produced. From the interrupted Phase II AN1792 trial, a small subset of AN1792-treated patients were evaluated by ADAS-cog and Mini-Mental State Examination (MMSE) and the initial data was suggestive of a slowing of cognitive decline among those patients generating the highest antibody titers [342]. However, a more comprehensive analysis of all treated patients demonstrated no significant effects on exploratory measures of cognition or disability [340]. In contrast, significant effects were reported in several cognitive tests associated with a nine component neuropsychological test battery, particularly immediate and delayed memory, and these data were suggestive of a dose response when correlated with antibody titer [340]. CSF analyses were performed in a small subset of patients, and while no differences in Aβ levels were observed, a decrease in total tau levels was measured potentially indicating the reduction of degenerating neurons [340]. A more recent report, following up on an even smaller subset of previous AN1792 responders, has suggested that, although the clearance of amyloid plaque might be long lasting, the cognitive benefit associated with plaque clearance might be more limited in duration [343]. However, another report has suggested more long-term functional benefits [344]. These conclusions and implications for the larger AD patient population will need to be confirmed by the outcome of future and ongoing studies since both of these reports are based on the follow up of very few patients several years after the AN1792 trial had been terminated.

Other clinical investigators used serial MRI to examine cerebral volume changes in patients treated with AN1792. Comparison of patient scans at baseline and 12 months after dosing of AN1792 demonstrated greater ventricular enlargement and greater hippocampal volume loss in those patients classified as antibody responders. Interestingly, increased brain volume loss was correlated with cognitive improvement using the NTB measures, rather than with cognitive decline [345]. The exact mechanism accounting for these opposing changes in loss of brain volume and improved cognition remain unclear, but may be related to clearance of Aβ deposits or amyloid plaque, and therefore diminished inflammation and water content [346]. The examination of brains from several post mortem cases of AN1792 responders have demonstrated clear evidence of Aβ phagocytosing microglia and the lack of amyloid plaques in those brain regions, supporting a positive effect of AN1792 via amyloid clearance [343,347–349]. Although preclinical data suggesting that passive immunization can attenuate early tau pathology in transgenic animals [305], no significant effects were observed on either vascular amyloid deposits or neurofibrillary tangles in humans [347,349].

Post mortem analysis of the two early cases of meningoencephalitis revealed a marked CD4-positive T-cell infiltration suggestive of a T-cell response to Aβ [347–349]. The mapping of T-cell epitopes to the C-terminus of Aβ supports the hypothesis that meningoencephalitis might have been the result of an undesired T-cell-based inflammatory response to those epitopes in an unfolded fraction of the aggregated $A\beta_{1-42}$ (AN1792) [350,351]. Since these events were not observed in the Phase I trial, it is thought that the addition of polysorbate-80 detergent in the formulation of AN1792 materials used in Phase II may have contributed to enhanced T-cell response, particularly in the presence of a potent T-cell adjuvant QS-21 [340,352]. The T-cell epitope mapping data and the observed robust antibody response to the Aβ N-terminus in preclinical PDAPP studies [295] and in AN1792 responders [353], has provided a path forward and the basis for next generation active and passive immunotherapy strategies, such as those currently being development for the treatment of AD by Wyeth and Elan/Janssen as described below.

4.5. Additional Preclinical and Clinical Studies

Aside from the approaches described above, additional reports have been published demonstrating the utility of other active and passive immunization strategies, including shorter Aβ fragments as immunogen, alternate routes of administration and the selection of different adjuvants. For example, the use of shorter Aβ fragments lacking T-cell epitopes has been described thereby avoiding unwanted Th1 cellular responses [354–357]. Intranasal administration of Aβ peptide has successfully lowered central Aβ levels and pathology in a mouse model of AD [298,358,359]. Other groups have treated APP transgenic mice with Aβ peptide sequences expressed on recombinant adeno-associated virus (AAV) vectors, or using phage display and demonstrated significant reductions in plaque burden and neuroinflammation, as well as improved cognitive performance [360,361]. The intranasal administration of an adenovirus vectors encoding either 11 tandem repeats of Aβ_{1-6} [362–364], or 4 tandem repeats of Aβ_{1-15} [365], can induce anti-inflammatory Th2 immune responses in mice resulting in high titers of anti-Aβ antibodies that are predominantly of the IgG1 isotype. Similarly, herpes simplex virus (HSV)-derived amplicons expressing either Aβ_{1-42} alone, or Aβ_{1-42} fused with the molecular adjuvant tetanus toxin Fragment C [365,366], or virus-like particles (VLPs) to induce weak or negligible T-cell response against Aβ [367,368], have been employed. In other approaches alternative adjuvants were used to prevent the unwanted inflammatory Th1 T-cell response [352,369]. Others have utilized Aβ single-chain antibody (scFv) that were injected into the corticohippocampal [370], or ventricles [371] of AD mouse models. Vaccines using short peptides that mimick epitopes of the native Aβ sequence as the antigenic component has been reported [372]. This approach provides a "non-self" epitope that circumvents breaking tolerance against self proteins, incorporates only six amino acids precluding the activation of Aβ-specific autoreactive T-cells, and does not cross-react with APP [372]. Oral immunization of Aβ-derived peptides loaded into microparticles have been shown to elicit a stronger immune response in mice by inducing anti-Aβ antibodies for prolonged time [373]. The observation that a small percentage of antibodies in human immunoglobulin preparations are directed against Aβ peptide sequences has led to an in infusion of natural antibody approaches [374]. Intravenous infusion of immunoglobulins in five AD patients over a 6-month period prevented further cognitive decline, suggesting this approach could potentially act like a passive Aβ-directed immunotherapy approach [374–376]. These results, irrespective of the approach taken, have consistently demonstrated that active or passive immunization strategies, targeting sequences within the Aβ peptide, are able to slow disease pathology and reverse memory deficits in preclinical models of AD [295,315,316,319,325,330–332,335].

4.6. General Advantages and Disadvantages of Active and Passive Immunotherapies

Although active immunization is likely to be less invasive and easier to administer to patients (by i.m. injection), passive immunotherapy (by i.v. administration) using humanized monoclonal anti-Aβ antibodies does confer some potential advantages. For example, the passive immunization approach eliminates the risk of potential T-cell-mediated responses to Aβ as observed with active immunization with AN1792. In addition, the passive strategy enables the rapid delivery of antibody, as well as the ability to quickly stop dosing should adverse events be observed during the course of a clinical trial. In contrast, active immunization is slower to respond to immunogen and more difficult to stop since individuals may continue to respond to the drug months after the last dose and patients may require plasmaphoresis or dialysis.

Concerns have been raised about the potential for both active and passive immunotherapy approaches to cause microhemorrhage. It has been reported that passive immunization with antibodies recognizing a variety of Aβ epitopes in transgenic APP mice with preexisting evidence of cerebral amyloid angiopathy (CAA) resulted in an increased severity and/or incidence of CAA-associated microhemorrhages [377–379]. The physiological implications of these findings remain unclear given the doses of antibody used were in some instances extremely high and the animals used had severe and preexisting cerebral amyloid angiopathy. More recent studies have demonstrated that certain antibodies prevented or cleared vascular Aβ (VAβ) and were associated with microhemorrhage in a dose-dependent manner [380]. Interestingly, the administration of antibody by intracerebroventricular (ICV) injection was efficacious in clearing parenchymal plaque in APP transgenic mice without altering CAA levels or increasing the incidence of microhemorrhage as compared to peripheral antibody administration [381]. However, microhemorrhage has not been observed by others working in the field or reported in clinical trials to date [382].

5. PASSIVE AND ACTIVE IMMUNOTHERAPY STRATEGIES IN CLINICAL EVALUATION

Following the initial reports from Elan/Wyeth regarding AN1792, a number of reports have emerged suggesting that alternative strategies are being aggressively pursued to find a safer and more tolerated Aβ immunotherapy approach for the treatment of AD. Several of these new strategies are well into clinical evaluation and a brief overview of approach and status has been included to illustrate ongoing efforts. Additional summaries of pre- clinical and clinical progress are highlighted in Table 1 (passive immunization) and Table 2 (active immunization).

5.1. Bapineuzamab

Bapineuzumab (AAB-001; Janssen/Pfizer, formerly Elan and Pfizer, formerly Wyeth) is a humanized IgG1 monoclonal antibody that binds to the N-terminal amino acids residues 1–5 of Aβ and was derived from the murine monoclonal antibody 3D6. In Tg2576 mice, 3D6 was shown to cross the blood–brain barrier, directly bind amyloid plaques and reduce amyloid burden. Antibody 3D6 binds soluble and insoluble aggregates of Aβ *in vitro*, and participates in the clearance of full length and N-terminally truncated species of aggregated Aβ within amyloid plaques through Fc receptor-mediated phagocytosis by cultured microglial cells in *ex vivo* assays [294,295].

Phase III 18-month studies have been initiated for the evaluation of efficacy and safety of bapineuzumab administered i.v. in patients with mild to moderate AD. Parallel-group studies are multicenter, randomized, double-blind, placebo-controlled, and will be conducted in over 5000 patients who are either apolipoprotein E4 carriers or E4 noncarriers. Studies sponsored by Janssen will include cognitive and functional assessments as primary outcome measures, and cognitive and global assessments as secondary outcome measures (ClinicalTrials.gov; NCT00574132, NCT00575055). Those sponsored by Pfizer will include Alzheimer's Disease Assessment Dementia (DAD) and Alzheimer's Disease Assessment Scale-Cognitive Subscale (ADAS-cog) as primary outcome measures, and neuropsychological test battery and Clinical Dementia Rating Scale (CDRS) as secondary outcome measures (ClinicalTrials.gov; NCT00667810, NCT00676143, NCT00909623, NCT00909675).

A Phase II 6-month study to evaluate the safety and effectiveness of bapineuzumab administered by weekly subcutaneous injections for the treatment of mild to moderate Alzheimer disease was initiated recently. This is a multicenter, randomized, double-blind, placebo-controlled, multiple ascending dose safety, tolerability, reactogenicity, and pharmacokinetic study in 160 subjects. The pri-

Table 1. Clinical Candidates for Passive Immunization

Drug (Sponsor)	Immunization Strategy	Study Design and Dosage	Program Status and Outcome	Comments and References [ClinicalTrials.gov]
Bapineuzumab AAB-001 (i.v.) (Janssen/Pfizer (Elan/Wyeth))	N-terminal, end-specific $A\beta_{1-5}$, humanized IgG1 monoclonal derived from murine 3D6	*Phase I*, randomized, double-blind, multicenter, single ascending dose (0.5–5 mpk, i.v., $n = 30$) safety and tolerability study in patients with mild-moderate AD	Increase of plasma $A\beta$; improved MMSE at 1.5 mpk, $p < 0.047$; VE	[NCT00397891] [383]
		Phase II, randomized, double-blind, multicenter, single ascending dose (0.15, 0.5, 1.0, or 2.0 mpk, i.v., each 13 weeks for 65 weeks, $n = 234$) in patients with mild-moderate AD; final assessment at 78 weeks	Prespecified primary outcome measures were not statistically achieved, however, post hoc analysis revealed positive trends (ADAS-cog, NTB)	[385,400]
		Phase III (see text)	Currently recruiting participants.	[NCT00676143]; [NCT00909675]; [NCT00575055]; [NCT00937352]
Solaneuzumab LY2062430 (Lilly)	Central-domain $A\beta_{16-23}$, humanized IgG1 monoclonal derived from murine 266	*Phase I*, single i.v. doses of 0.5, 1.5, 4.0, or 10 mg/kg	Increase of plasma and CSF $A\beta$ levels	[401]
		Phase II, randomized, double-blind, multicenter, placebo-control, multiple-dose safety in subjects with mild-moderate AD (100 or 400 mg, i.v., weekly or each 4 weeks for 12 weeks, $n = 52$) and single-dose safety in healthy volunteers (100 mg, i.v., $n = 13$)	Increase in plasma $A\beta$ and pyroE3-$A\beta$ levels; increase in CSF $A\beta_{1-40}$ and $A\beta_{1-42}$ levels; no change in ADAS-cog at 3 months; PET (IMPY) signal partially correlated with $A\beta_{1-42}$	[NCT00329082] [402–405]
		Phase III, randomized, double-blind, single dose (400 mg, i.v., each 4 weeks for 80 weeks), and placebo-control, parallel assignment, safety and efficacy study in patients with AD; final assessment at 80 weeks	Currently recruiting participants	[NCT00905372, Expedition]; [NCT00904683, Expedition2]

(Continued)

Table 1. (*Continued*)

Drug (Sponsor)	Immunization Strategy	Study Design and Dosage	Program Status and Outcome	Comments and References [ClinicalTrials.gov]
IVIg Gammagard (Baxter/NIH/Cornell)	Human polyclonal immunoglobulin	*Phase I*, open-label in patients with mild to moderate AD ($n = 8$)	Increased plasma $A\beta_{1-40}$ and $A\beta_{1-42}$ levels; decreased CSF $A\beta_{1-40}$ and $A\beta_{1-42}$ levels (after 6 months); improved MMSE scores at 3 (1.4 pts) and 6 (2.5 pts) months	[NCT00299988]; [NCT00812565]
		Phase II, randomized, double-blind, multicenter, placebo-control, parallel assignment study for safety and efficacy for the treatment of mild-moderate AD. Gammagard (100 or 400 mg, i.v., weekly or each 4 weeks for 12 weeks, $n = 52$) and single-dose safety in healthy volunteers (100 mg, i.v., $n = 13$)		[406] [NCT00299988]; [NCT00812565]
		Phase III, randomized, double-blind, placebo-control, two dose-arm parallel study for safety and effectiveness human globulin. Gammagard, one of two doses (200 or 400 mg/kg i.v.) every 2 weeks for 9 and 18 months	Study is currently recruiting participants	[NCT00818662] [394,395]
Bapineuzumab AAB-001 (s.c.) (Janssen/Pfizer (Elan/Wyeth))	N-terminal, end-specific $A\beta_{1-5}$, humanized IgG1 monoclonal derived from murine 3D6 (s.c.)	*Phase II*, multicenter, randomized, double-blind placebo-control, multiple ascending dose safety, tolerability and PK study for subcutaneous administration of bapineuzumab	Study is currently recruiting participants	[NCT00663026]

PF-04360365 (RN1219) (Pfizer (Rinat))	C-terminal Aβ$_{33-40}$, humanized monoclonal (i.v.)	*Phase I*, randomized, placebo-control, double-blind dose-escalation study for safety, tolerability, PK, PD, and immunogenicity of a single i.v. dose in subjects with mild-to-moderate AD	[NCT00733642]; [NCT00455000]
		Phase II, randomized, multicenter, double-blind, placebo-control, safety, tolerability, and PK/PD study of multiple doses (0.1, 0.5, 1, 3, and 8.5 mg/kg, or 7.5 and 10 mg/kg in a second trial) in patients with mild-to-moderate AD	Well tolerated over the 0.1–10 mg/kg dose range; plasma Aβ increased dose proportionally [NCT00722046]; [NCT00945672] [407,408]
Gantenerumab R1450 (Roche/Morphosys)	N-terminal and central domain Aβ, humanized monoclonal (i.v.)	*Phase IB*, multicenter, multiple ascending dose, randomized, double-blind placebo-control, parallel group study for safety, tolerability, PK, and PD following i.v. administration in patients	Study is ongoing, but not recruiting participants [NCT00531804]
GSK933776A (GSK)	Aβ, humanized monoclonal (i.v.)	*Phase I*, 52-week randomized, single-blind, placebo-control, single-group assessment study to investigate the safety, immunogenicity, PK, and PD following i.v. infusion in patients with AD (total enrollment = 50)	Assessments included safety and tolerability, MRI, cognitive status (primary), and pharmacokinetics parameters and PD effects on plasma and CSF biomarkers, and titers and neutralizing activity of antibodies (secondary) Currently recruiting participants [NCT00459550]

(Continued)

373

Table 1. (*Continued*)

Drug (Sponsor)	Immunization Strategy	Study Design and Dosage	Program Status and Outcome	Comments and References [ClinicalTrials.gov]
MABT5102A (ACI-01-Ab7) (Genentech/AC Immune)	Aβ, humanized monoclonal	*Phase I*, randomized, double-blind, placebo-control, multicenter, study to assess the safety, PK, PD, and immunogenicity following i.v. administered in a single-dose, dose-escalation stage followed by a multidose, parallel-treatment stage in patients with mild-moderate AD (total enrollment = 50)	Assessments included safety and tolerability of singe and multiple doses (primary) and PK and PD after single and multiple doses (secondary). Currently recruiting participants	[NCT00736775]
MRKxxxx/ACU-5A5 (Merck (Acumen))	Targets oligomeric species of Aβ		Preclinical	
IN-N01 (Intellect Neuroscience)	Targets N- and C-terminal epitopes of Aβ, thereby preventing the binding of antibody to the APP protein		Preclinical, undergoing humanization	
BAN-2401 (Eisai/BioArtic)	Targets oligomeric species of Aβ		Preclinical	
(Ablynx and Boehringer Ingelheim)	Biologic "nanobody" platform to target Aβ and improve blood–brain barrier penetration in comparison to antibodies		Preclinical	
AAB-002 (Janssen/Pfizer (Elan/Wyeth))	Conformational Aβ epitope, humanized monoclonal (i.v.)		Preclinical	
(BiogenIdec/ Neuroimmune)	Conformational Aβ epitope, humanized monoclonal (i.v.)		Preclinical	

Table 2. Clinical Candidates for Active Immunization

Drug (Sponsor)	Immunization Strategy	Program Status	Outcome Measures	Comments and Reference [ClinicalTrials.gov ID]
		Phase I		
ACC-001 (Janssen/Pfizer (Elan/Wyeth))	N-terminal Aβ$_{1-7}$ conjugated to CRM197	Phase II, randomized, double-blind, multicenter, placebo-control, parallel assignment multiple-dose safety, tolerability, and immunogenicity study in subjects with mild-moderate AD. Two studies conducted each with four arms comprised of ACC-001 (3 cohorts of 3, 10 or 30 μg) with or without QS-21 adjuvant, QS-21 alone, or PBS placebo comparator, each administered i.m. at day 1, month 1, 3, 6, and 12 with a total enrollment of either 80 (study ID 3134K1-200) or 240 (study ID 3134K1-2201) patients	Study 3134K1-200 completion estimated 05/2012 with assessments for safety and tolerability of multiple doses (primary) and immunogenicity of each dose level (secondary); study 134K1-2201 completion estimated 05/2012 with assessments for adverse events and tolerability of multiple doses (primary), and cognitive and functional measures (secondary)	[NCT00479557]; [NCT00498602]
CAD-106 (Novartis/Cytos)	N-terminal Aβ$_{1-6}$ coupled to Qb virus-like particles	Phase I, 52-week multicenter, randomized, double-blind, placebo-control time-lagged, parallel group study in patients with mild to moderate Alzheimer's disease to investigate the safety, tolerability and Aβ-specific response following three injections (s.c.)	The study has completed with assessments for safety and tolerability (MRI, AE/SAE monitoring, and IgG and IgM antibody titers against Aβ and carrier protein; primary), and immune response, cognitive and functional assessments (secondary)	[409,410] [NCT00411580]

(Continued)

Table 2. (Continued)

Drug (Sponsor)	Immunization Strategy	Program Status	Outcome Measures	Comments and Reference [ClinicalTrials.gov ID]
		Phase II, randomized, double-blind, multicenter, placebo-control, parallel group study in patients with mild-moderate AD to investigate the safety and tolerability of repeated subcutaneous injections. Two studies comparing CAD106 against placebo were reported with total enrollment of 27 and 30 (ongoing) patients	Assessments were similar in both studies included safety and tolerability at multiple time points (primary) and immune response, cognitive and function assessments at multiple time points (secondary). Completion of both studies is anticipated 1Q2010. Preliminary reports suggest that titers were lower and short lived and that exploratory outcome measures did not differ significantly	[411,412] [NCT00733863]; [NCT00795418]
V950 (Merck)	Bispecific binding to both N-terminal and central Aβ fragments	*Phase I*, randomized, double-blind, dose-escalating study to evaluate the safety, tolerability and immunogenicity of V950 formulated in aluminum-containing adjuvant with or without Iscomatric™ in patients with Alzheimer's disease; single center, parallel group, single dosage ($n=12$, administered four times in 4-week intervals, with or without aluminum hydroxide adjuvant) in patients with mild-moderate AD for a tolerability and safety of repeated subcutaneous administration	Study is currently recruiting participants; assessments include general safety and tolerability dose (primary) and immunogenicity (secondary) after each dose	[NCT00464334]

Affitope AD01/AD02 Affiris/GSK	Mimotopes of N-terminal Aβ	*Phase I*, randomized, patient-blinded, single center, parallel group, single dosage ($n = 12$, administered four times in 4-week intervals, with or without aluminum hydroxide adjuvant) in patients with mild-moderate AD for a tolerability and safety of repeated subcutaneous administration	Safe and well tolerated based on initial reports	[372,413,414] [NCT00495417]; [NCT00711139]; [NCT00633841]; [NCT00711321]
Mimovax MV01 and MV02 (Affiris/GSK)	Mimotopes of truncated and PyroGlu3 near N-terminal Aβ (Alum)		Preclinical	[414]
PADRE (Lundbeck/Pharmexa)	Aβ fragment and helper cell PADRE epitope		Preclinical	
Boehringer/Ablynx	Nanobody to Aβ epitope		Preclinical	
ACI-24 (ACImmune)	Aβ$_{1-15}$ (liposome-based delivery)		Preclinical	
RECALL-VAX™ (Intellect Neuroscience)	Short chimeric peptide with free-end fragments of Aβ linked to T-helper cell epitope (tetanus toxoid)		Preclinical	

mary outcome measures include the incidence of treatment emergent adverse events; secondary outcome measures include pharmacokinetic parameters including serum concentration of bapineuzumab, terminal half-life of elimination, observable area under the concentration–time curve (AUC), and steady state serum concentration (ClinicalTrials.gov; NCT00663026).

In addition, a Phase II 1-year study was initiated in patients who participated in the preceding double-blind study, for long-term evaluation of safety and tolerability of bapineuzumab when administered by subcutaneous injection at 5, 10, or 20 mpk/week. The primary outcome measures include safety end points and brain volumetric MRI, and secondary outcome measures include pharmacokinetic parameters and maximal serum drug concentration, time to maximal serum drug concentration, and terminal half-life of elimination (ClinicalTrials.gov; NCT00916617).

Initial data presented for a Phase I single ascending study of 0.1, 1.5, and 5.0 mpk (i.v.) generated interest in that a significant improvement in MMSE scores were observed at 1.5 mpk ($p<0.047$) at 4 months following treatment, but not in the lower and higher dose cohorts. It was noted that three patients in the higher dose cohort developed MRI abnormalities consistent with vasogenic edema (VE) that spontaneously resolved [383]. In a Phase IIa study, subjects were randomized (8 bapineuzumab:7 placebo) and received a total of six treatments at 0.15, 0.5, 1.0, or 2.0 mpk (i.v.). MRI scans were performed 6 weeks following each infusion and it was determined that 9.7% (12/124) of those subjects receiving bapineuzumab had developed signal abnormalities. The study concluded that VE was reversible and appeared to be both dose-dependent and associated with the ApoE4 genotype. The mechanism underlying VE and pathophysiology was not elucidated, but may have been associated with vascular amyloid deposition and congophilic amyloid angiopathy (CAA) lesions [384–387].

5.2. Solaneuzumab

Solaneuzumab (Lilly; LY2062430) is a humanized monoclonal antibody that targets the central domain of $A\beta_{16-23}$ and is derived from the murine monoclonal antibody 266. Preclinical studies have demonstrated that antibody 266 does not bind to brain $A\beta$ fibrils of amyloid plaque, but binds to $A\beta$ monomer resulting in the efflux of peptide to the periphery for elimination of immune complex. The "amyloid sink" hypothesis explains the putative depletion of brain levels of $A\beta$ and a corresponding reversal of behavioral deficits, and the prevention of accumulating plaque [319,335,336].

A Phase II trial for safety and tolerability involving only 52 patients in a 12-week study has been completed. While improvement in cognition performance was not expected, a change in efficacy surrogates with a rise in $A\beta$ levels in blood and CSF was observed. However, the demonstration of plaque clearance by PET imaging (IMPY probe) was negative [388].

Based on the Phase II $A\beta$ biomarker data, two Phase III trials were initiated. They are randomized, double-blind for safety and efficacy studies in 2000 patients evaluating the effect of passive immunization on the progression of AD. Patients will receive 400 mg of solaneuzumab or placebo by i.v. administration every 4 weeks for 80 weeks. Primary outcome measures at 80 weeks will include change from baseline to end point using ADAS-Cog and Alzheimer's Disease Cooperative Study-Activities of Daily Living Inventory (ADCS-ADL); secondary outcome measures will include, among others, change from baseline to end point in Clinical Dementia Rating-Sum of Boxes (CDR-SB), NPI, volumetric magnetic resonance imaging (vMRI), MMSE, Resource Utilization in Dementia-Lite (RUD-Lite), and plasma $A\beta$ levels (ClinicalTrials.gov; NCT00904683, NCT00905372).

5.3. PF-04360365

PF-04360365, formerly RN-1219 from Rinat Neuroscience licensed by Pfizer, is a humanized IgG2 monoclonal antibody that targets the C-terminal amino acid residues 33–40 of $A\beta_{1-40}$, and requires the free carboxy terminus of $A\beta$ for antibody binding. The lowering of brain $A\beta$ levels and improvement of performance in various APP transgenic mouse models of learning and memory has been demonstrated with murine 2H6 antibody. The same preclinical models also demonstrated increased levels of vascular amyloid and microhemorrhage [378,389], however, both were prevented without the loss of efficacy by treating with an antibody, de-2H6, that was enzymatically deglycosylated [390]. Removing or preventing antibody glycosylation decreases the ability of the antibody to bind complement and may reduce the risk of inflammation. A humanized monoclonal antibody, possibly derived by modification of 2H6, has been characterized for PK and PD in young and aged cynomolgus monkeys and has advanced into clinical development [391,392].

Two multiple dose Phase II double-blinded, randomized, placebo-controlled, multicenter studies evaluating safety and tolerability, and PK and PD, in patients with mild to moderate AD for PF-04360365 is in progress. Primary outcome measures include brain MRI, cognitive assessments, brain amyloid burden, and CSF Aβ; secondary outcome measures include Alzheimer's Disease Assessment Scale-Cognitive Subscale (ADAS-cog); Disability Assessment for Dementia (DAD); plasma Aβ, CSF tau and phospho-tau (ClinicalTrials.gov; NCT00722046, NCT00945672).

5.4. Gammaguard Liquid

Gammaguard Liquid (Baxter Healthcare Corporation) is based on the observation that levels of both Aβ and anti-Aβ antibody increase in human CSF suggesting the attempted natural clearance of Aβ from the brain. Studies using pools of natural antibodies isolated from human plasma have shown the prevention of Aβ aggregation *in vitro* and the improvement of cognitive performance when injected in the brains of APP transgenic mice [376,393–397].

The first of two pivotal Phase III trials has been initiated as an 18-month, randomized, double-blind, placebo-controlled study with two dose-arms. A parallel study in 360 subjects with mild-to-moderate AD will evaluate Gammaguard Liquid for decline in dementia. Subjects will receive either 200 or 400 mpk every 2 weeks for 70 weeks. Primary outcome measures for efficacy will be evaluated at 9 and 18 months and will include assessment by ADAS-Cog for cognition, and the Alzheimer's Disease Cooperative Study-Clinical Global Impression of Change rating (ADCS-CGIC) for global clinical outcome; secondary outcome measures will be assessed at 18 months and include behavioral, functional, and quality-of-life outcome measures, as well as several plasma, cerebrospinal fluid, and imaging biomarkers for the assessment of disease progression and response to therapy (ClinicalTrials.gov; NCT00818662).

5.5. R1450

R1450 (Hoffmann-La Roche/Morphosis) is a bifunctional humanized IgG1 monoclonal antibody that targets both N-terminal and internal amino acid residues of Aβ, and binds to both soluble and insoluble species of Aβ. R1450 has been shown to cross the blood–brain barrier of TgAPP mice and lower brain Aβ levels and depolymerize aggregated Aβ (presented by Bohrmann et al. [415]).

R1450 is currently undergoing a multidose Phase IB investigation for safety, tolerability, pharmacokinetics and pharmacodynamics in patients with mild to moderate AD. This multiple ascending dose-escalation study will enroll less than 100 patients who will receive a total of seven doses using an adaptive design. Primary outcome measures will include safety and pharmacokinetic parameters; secondary outcome measures will include CSF biomarkers and clinical efficacy parameters (ClinicalTrials.gov; NCT00531804).

5.6. MABT5102A

MABT5102A has been reported by a collaboration between Genentech and AC Immune. Using AC Immune's technology designed to identify antibodies that recognize conformational epitopes, ACI-01-Ab7 was found to block Aβ aggregation *in vitro* and deposition *in vivo*. It was also shown to shift Aβ from an insoluble to soluble form, thereby promoting the clearance of soluble Aβ and improvement of memory in APP transgenic mice. Two i.p. injections of ACI-01-Ab7 (300 µg) in APP transgenic mice was reported to significantly improve memory in a novel object recognition

(NOR) model. However, the actual clearance of soluble Aβ and the prevention of reaggregation into synaptotoxic Aβ species was not demonstrated [398]. PK parameters were as expected in normal mice and cynomolgus monkeys and a PK/PD model utilizing plasma Aβ as a PD marker was developed for dose-ranging [399]. MABT5102A is currently in a Phase 1 study to assess the safety PK, PD, and immunogenicity with i.v. administration (ClinicalTrials.gov; NCT00736775).

The number of preclinical and clinical reports describing the immunization against Aβ has rapidly expanded demonstrating the utility of various active and passive strategies, targeted Aβ epitopes, modes of administration, and selection of adjuvants for the treatment of Alzheimer's disease. With each improvement or refinement of strategy, it is hopeful that our understanding will enable the successful implementation of a safe, well tolerated and clinically efficacious solution for the treatment of this devastating neurodegenerative disease.

6. SUMMARY

For the past two decades the search for treatments for Alzheimer's disease has led to a number of important discoveries. The advances made by the varied academic and industrial research teams have made possible the identification of a growing number of preclinical and clinical candidates that give hope to the search for a treatment to this horrible and debilitating disease. There now exists a multitude of diverse approaches, both palliative and potentially disease modifying, to potentially improving the cognitive dysfunction and neurodegeneration that are the hallmark deficits associated with the disease. Many of these have evolved to clinical development and the future holds great promise for alternate and improved treatments, either as monotherapies or in conjunction with other targeted approaches.

Given that the ultimate goal in the field is a disease-modifying therapy, the interest in palliative therapies will most probably wane as the understanding of the pathogenesis and etiology of the disease further evolves. That being said, research in disease modifying approaches has yet to provide a clinically validated mechanism and may not do so in the next few years. In addition, the millions of patients currently suffering from the disease already demonstrate neurological impairment and the currently available and potential future palliative treatments will continue to be necessary. Furthermore, the gradual advancement of the disease, complicated by the paucity of diagnostic tools, necessitates an early and effective treatment for the initial cognitive dysfunction that may precede the neurodegeneration that underscores disease progression. Finally, effective cessation of the disease, when realized, may not result in a reversal of any preexisting neurobehavioral deficits and beneficial improvements resulting from these potential palliative treatments may be possible. However, the need to stop, and even possibly reverse the disease, is paramount for the future of the aging population, if there is to be a future with an improved prognosis for aging. Thus, the search for AD medicines continues and will no doubt be the focal point for neuroscience groups all over the world until this disease can be assaulted by multiple mechanisms and drug entities.

REFERENCES

1. Plosker GL, Keating GM. Management of mild to moderate Alzheimer disease: defining the role of rivastigmine. Disease Manage Health Outcomes 2004;12(1): 55–72.
2. Selkoe DJ. Alzheimer disease: mechanistic understanding predicts novel therapies. Ann. Intern. Med. 2004;140(8): 627–638.
3. Hebert LE, Scherr PA, Bienias JL, Bennett DA, Evans DA. Alzheimer disease in the US population: prevalence estimates using the 2000 census. Arch Neurol 2003;60(8): 1119–1122.
4. Cummings JL. Cognitive and behavioral heterogeneity in Alzheimer's disease: seeking the neurobiological basis. Neurobiol Aging 2000; 21(6): 845–861.
5. Hardy J, Selkoe DJ. The amyloid hypothesis of Alzheimer's disease: progress and problems on the road to therapeutics. Science (Washington, DC, United States) 2002;297(5580): 353–356.
6. Hope T, Keene J, Fairburn CG, Jacoby R, McShane R. Natural history of behavioural

changes and psychiatric symptoms in Alzheimer's disease. A longitudinal study. Br J Psychiatry: J Mental Sci 1999;174:39–44.

7. Klein WL, Krafft GA, Finch CE. Targeting small Abeta oligomers: the solution to an Alzheimer's disease conundrum? Trends Neurosci 2001;24(4): 219–224.

8. Lacor PN, Buniel MC, Chang L, Fernandez SJ, Gong Y, Viola KL, Lambert MP, Velasco PT, Bigio EH, Finch CE, Krafft GA, Klein WL. Synaptic targeting by Alzheimer's-related amyloid beta oligomers. J Neurosci 2004; 24(45): 10191–10200.

9. Selkoe DJ. Alzheimer's disease: genes, proteins, and therapy. Physiol Rev 2001;81(2): 741–766.

10. Glenner GG, Wong CW. Alzheimer's disease: initial report of the purification and characterization of a novel cerebrovascular amyloid protein. Biochem Biophys Res Commun 1984;120 (3): 885–890.

11. Robichaud A, Tattersall FD, Choudhury I, Rodger IW. Emesis induced by inhibitors of type IV cyclic nucleotide phosphodiesterase (PDE IV) in the ferret. Neuropharmacol 1999;38(2): 289–297.

12. Walsh DM, Klyubin I, Fadeeva JV, Rowan MJ, Selkoe DJ. Amyloid-beta oligomers: their production, toxicity and therapeutic inhibition. Biochem Soc Trans 2002;30(4): 552–557.

13. Marlatt MW, Webber KM, Moreira PI, Lee H-G, Casadesus G, Honda K, Zhu X, Perry G, Smith M.A. Therapeutic opportunities in Alzheimer disease: one for all or all for one? Curr Med Chem 2005;12(10): 1137–1147.

14. Marlatt MW, Webber KM, Moreira PI, Lee H-G, Casadesus G, Honda K, Zhu X, Perry G, Smith MA. Therapeutic opportunities in Alzheimer disease: one for all or all for one? Curr Med Chem 2005;12(10): 1137–1147.

15. Pangalos MN, Jacobsen SJ, Reinhart PH. Disease modifying strategies for the treatment of Alzheimer's disease targeted at modulating levels of the beta-amyloid peptide. Biochem Soc Trans 2005;33(4): 553–558.

16. Roggo S. Inhibition of BACE, a promising approach to Alzheimer's disease therapy. Curr Top Med Chem (Hilversum, Netherlands) 2002;2(4): 359–370.

17. Selkoe D, Kopan R. Notch and presenilin: regulated intramembrane proteolysis links development and degeneration. Annu Rev Neurosci 2003;26:565–597.

18. Thompson LA, Bronson JJ, Zusi FC. Progress in the discovery of BACE inhibitors. Curr Pharmaceut Design 2005;11(26): 3383–3404.

19. Bullock R. Future directions in the treatment of Alzheimer's disease. Expert Opin Investig Drugs 2004;13(4): 303–314.

20. Churcher I, Beher D. gamma-Secretase as a therapeutic target for the treatment of Alzheimer's disease. Curr Pharmaceut Design 2005;11(26): 3363–3382.

21. Citron M. Strategies for disease modification in Alzheimer's disease. Nat Rev Neurosci 2004;5(9): 677–685.

22. Czech C, Adessi C. Disease modifying therapeutic strategies in Alzheimer's disease targeting the amyloid cascade. Curr Neuropharmacol 2004;2(3): 295–307.

23. Jacobsen JS. Alzheimer's disease: an overview of current and emerging therapeutic strategies. Curr Top Med Chem 2002;2(4): 343–352.

24. Frolich L, Fox J, Padberg F, Maurer K, Moller H.-J, Hampel H. Targets of antidementive therapy: drugs with a specific pharmacological mechanism of action. Curr Pharmaceut Design 2004;10(3): 223–229.

25. Robichaud AJ. Approaches to palliative therapies for Alzheimer's disease. Curr Top Med Chem (Sharjah, United Arab Emirates) 2006;6(6): 553–568.

26. Haass C, Schlossmacher MG, Hung AY, Vigo-Pelfrey C, Mellon A, Ostaszewski BL, Lieberburg I, Koo EH, Schenk D, et al. Amyloid beta-peptide is produced by cultured cells during normal metabolism. Nature (London, United Kingdom) 1992;359(6393): 322–325.

27. Bernstein SL, Dupuis NF, Lazo ND, Wyttenbach T, Condron MM, Bitan G, Teplow DB, Shea J-E, Ruotolo BT, Robinson CV, Bowers MT. Amyloid-beta protein oligomerization and the importance of tetramers and dodecamers in the etiology of Alzheimer's disease. Nat Chem 2009;1 (4): 326–331, S326/1–S326/14.

28. Jan A, Gokce O, Luthi-Carter R, Lashuel HA. The ratio of monomeric to aggregated forms of A.beta 40 and A.beta 42 is an important determinant of amyloid-beta aggregation, fibrillogenesis, and toxicity. J Biol Chem 2008;283(42): 28176–28189.

29. Sami SK, Good TA. Structure and stability of beta-amyloid oligomers and fibrils: implications in the toxicity to neurons in Alzheimer's disease. Abstracts of Papers, 238th ACS National Meeting; 2009. August 16–20; Washington, DC, United States; 2009, BIOT-178.

30. Welander H, Fraanberg J, Graff C, Sundstroem E, Winblad B, Tjernberg LO. A.beta 43 is more frequent than A.beta 40 in amyloid plaque cores from Alzheimer disease brains. J Neurochem 2009;110(2): 697–706.

31. Sinha S, Anderson JP, Barbour R, Basi GS, Caccavello R, Davis D, Doan M, Dovey HF, Frigon N, Hong J, Jacobson-Croak K, Jewett N, Keim P, Knops J, Lieberburg I, Power M, Tan H, Tatsuno G, Tung J, Schenk D, Seubert P, Suomensaari SM, Wang S, Walker D, Zhao J, McConlogue L, John V. Purification and cloning of amyloid precursor protein beta-secretase from human brain. Nature (London) 1999;402(6761): 537–540.

32. Cai H, Wang Y, McCarthy D, Wen H, Borchelt DR, Price DL, Wong PC. BACE1 is the major beta-secretase for generation of Abeta peptides by neurons. Nat Neurosci 2001;4(3): 233–234.

33. Nilsberth C, Westlind-Danielsson A, Eckman CB, Condron MM, Axelman K, Forsell C, Stenh C, Luthman J, Teplow DB, Younkin SG, Naeslund J, Lannfelt L. The 'Arctic' APP mutation (E693G) causes Alzheimer's disease by enhanced Abeta protofibril formation. Nat Neurosci 2001;4(9): 887–893.

34. Vassar R, Bennett BD, Babu-Khan S, Kahn S, Mendiaz EA, Denis P, Teplow DB, Ross S, Amarante P, Loeloff R, Luo Y, Fisher S, Fuller J, Edenson S, Lile J, Jarosinski MA, Biere AL, Curran E, Burgess T, Louis J-C, Collins F, Treanor J, Rogers G, Citron M. beta-Secretase cleavage of Alzheimer's amyloid precursor protein by the transmembrane aspartic protease BACE. Science (Washington, DC) 1999;286(5440): 735–741.

35. Bennett BD, Babu-Khan S, Loeloff R, Louis J-C, Curran E, Citron M, Vassar R. Expression analysis of BACE2 in brain and peripheral tissues. J Biol Chem 2000;275(27): 20647–20651.

36. Vattemi G, Engel WK, McFerrin J, Pastorino L, Buxbaum JD, Askanas V. BACE1 and BACE2 in pathologic and normal human muscle. Exp Neurol 2003;179(2): 150–158.

37. Roberds SL, Anderson J, Basi G, Bienkowski MJ, Branstetter DG, Chen KS, Freedman SB, Frigon NL, Games D, Hu K, Johnson-Wood K, Kappenman KE, Kawabe TT, Kola I, Kuehn R, Lee M, Liu W, Motter R, Nichols NF, Power M, Robertson DW, Schenk D, Schoor M, Shopp GM, Shuck ME, Sinha S, Svensson KA, Tatsuno G, Tintrup H, Wijsman J, Wright S, McConlogue L. BACE knockout mice are healthy despite lacking the primary beta-secretase activity in brain: implications for Alzheimer's disease therapeutics. Hum Mol Genet 2001; 10(12): 1317–1324.

38. Hu X, Hicks CW, He W, Wong P, Macklin WB, Trapp BD, Yan R. Bace1 modulates myelination in the central and peripheral nervous system. Nat Neurosci 2006;9(12): 1520–1525.

39. Willem M, Garratt AN, Novak B, Citron M, Kaufmann S, Rittger A, DeStrooper B, Saftig P, Birchmeier C, Haass C. Control of Peripheral Nerve Myelination by the beta-Secretase BACE1. Science (Washington, DC, United States) 2006;314(5799): 664–666.

40. Hong L, Koelsch G, Lin X, Wu S, Terzyan S, Ghosh AK, Zhang XC, Tang J. Structure of the protease domain of memapsin 2 (beta-secretase) complexed with inhibitor. Science (Washington, D. C.) 2000;290(5489): 150–153.

41. Shirotani K, Edbauer D, Prokop S, Haass C, Steiner H. Identification of Distinct gamma-Secretase Complexes with Different APH-1 Variants. J Biol Chem 2004;279(40): 41340–41345.

42. Searfoss GH, Jordan WH, Calligaro DO, Galbreath EJ, Schirtzinger LM, Berridge BR, Gao H, Higgins MA, May PC, Ryan TP. Adipsin, a biomarker of gastrointestinal toxicity mediated by a functional gamma-secretase inhibitor. J Biol Chem 2003;278(46): 46107–46116.

43. Zhong Z, Higaki J, Murakami K, Wang Y, Catalano R, Quon D, Cordell B. Secretion of beta-amyloid precursor protein involves multiple cleavage sites. J Biol Chem 1994;269(1): 627–632.

44. Nitsch RM, Slack BE, Wurtman RJ, Growdon JH. Release of Alzheimer amyloid precursor derivatives stimulated by activation of muscarinic acetylcholine receptors. Science (Washington, DC, United States) 1992;258(5080): 304–307.

45. Kreft AF, Martone R, Porte A. Recent Advances in the identification of gamma-secretase inhibitors to clinically test the abeta oligomer hypothesis of Alzheimer's disease. J. Med. Chem. 2009;52(20): 6169–6188.

46. Olson RE, Albright CF. Recent progress in the medicinal chemistry of gamma-secretase inhibitors. Curr Top Med Chem (Sharjah, United Arab Emirates) 2008;8(1): 17–33.

47. Wolfe MS, De Los Angeles J, Miller DD, Xia W, Selkoe DJ. Are presenilins intramembrane-cleaving proteases? Implications for the molecular mechanism of Alzheimer's disease. Biochemistry 1999;38(35): 11223–11230.

48. Wrigley JDJ, Nunn EJ, Nyabi O, Clarke EE, Hunt P, Nadin A, de Strooper B, Shearman MS, Beher D. Conserved residues within the putative active site of gamma-secretase differentially influence enzyme activity and inhibitor binding. J Neurochem 2004;90(6): 1312–1320.

49. Doan A, Thinakaran G, Borchelt DR, Slunt HH, Ratovitsky T, Podlisny M, Selkoe DJ, Seeger M, Gandy SE, Price DL, Sisodia SS. Protein topology of presenilin 1. Neuron 1996;17(5): 1023–1030.

50. Wolfe MS, Citron M, Diehl TS, Xia W, Donkor IO, Selkoe DJ. A substrate-based difluoro ketone selectively inhibits Alzheimer's γ-secretase activity. J Med Chem 1998;41(1): 6–9.

51. Shearman MS, Beher D, Clarke EE, Lewis HD, Harrison T, Hunt P, Nadin A, Smith AL, Stevenson G, Castro JL. L-685,458, an aspartyl protease transition state mimic, is a potent inhibitor of amyloid beta-protein precursor gamma-secretase activity. Biochemistry 2000;39 (30): 8698–8704.

52. Nadin A, Owens AP, Castro JL, Harrison T, Shearman MS. Synthesis and gamma-secretase activity of APP substrate-based hydroxyethylene dipeptide isosteres. Bioorg Med Chem Lett 2003;13(1): 37–41.

53. Lewis HD, Perez Revuelta BIP, Nadin A, Neduvelil JG, Harrison T, Pollack SJ, Shearman MS. Catalytic site-directed gamma-secretase complex inhibitors do not discriminate pharmacologically between Notch S3 and beta-APP cleavages. Biochemistry 2003;42(24): 7580–7586.

54. Barten DM, Meredith JE Jr, Zaczek R, Houston JG, Albright CF. gamma-secretase inhibitors for Alzheimer's disease: balancing efficacy and toxicity. Drugs R&D 2006;7(2): 87–97.

55. Hamman JH, Enslin GM, Kotze AF. Oral delivery of peptide drugs: barriers and developments. BioDrugs 2005;19(3): 165–177.

56. Nestor JJ. Peptide and protein drugs: issues and solutions. Comprehensive Med Chem II 2006;2:573–601.

57. Pauletti GM, Gangwar S, Siahaan TJ, Aube J, Borchardt RT. Improvement of oral peptide bioavailability: peptidomimetics and prodrug strategies. Adv Drug Deliv Rev 1997;27(2–3): 235–256.

58. Seiffert D, Bradley JD, Rominger CM, Rominger DH, Yang F, Meredith JE Jr, Wang Q, Roach AH, Thompson LA, Spitz SM, Higaki JN, Prakash SR, Combs AP, Copeland RA, Arneric SP, Hartig PR, Robertson DW, Cordell B, Stern AM, Olson RE, Zaczek R. Presenilin-1 and -2 are molecular targets for γ-secretase inhibitors. J Biol Chem 2000;275(44): 34086–34091.

59. Dovey HF, John V, Anderson JP, Chen LZ, De Saint Andrieu P, Fang LY, Freedman SB, Folmer B, Goldbach E, Holsztynska EJ, Hu KL, Johnson-Wood KL, Kennedy SL, Kholodenko D, Knops JE, Latimer LH, Lee M, Liao Z, Lieberburg IM, Motter RN, Mutter LC, Nietz J, Quinn KP, Sacchi KL, Seubert PA, Shopp GM, Thorsett ED, Tung JS, Wu J, Yang S, Yin CT, Schenk DB, May PC, Altstiel LD, Bender MH, Boggs LN, Britton TC, Clemens JC, Czilli DL, Dieckman-McGinty DK, Droste JJ, Fuson KS, Gitter BD, Hyslop PA, Johnstone EM, Li WY, Little SP, Mabry TE, Miller FD, Ni B, Nissen JS, Porter WJ, Potts BD, Reel JK, Stephenson D, Su Y, Shipley LA, Whitesitt CA, Yin T, Audia JE. Functional gamma-secretase inhibitors reduce beta-amyloid peptide levels in brain. J Neurochem 2001;76(1): 173–181.

60. Morohashi Y, Kan T, Tominari Y, Fuwa H, Okamura Y, Watanabe N, Sato C, Natsugari H, Fukuyama T, Iwatsubo T, Tomita T. C-terminal fragment of presenilin is the molecular target of a dipeptidic γ-secretase-specific inhibitor DAPT (N-[N-(3,5-difluorophenacetyl)-L-alanyl]-S-phenylglycine t-butyl ester). J Biol Chem 2006;281(21): 14670–14676.

61. Hsiao KK. Understanding the biology of beta-amyloid precursor proteins in transgenic mice. Neurobiol Aging 1995;16(4): 705–706.

62. Prasad CVC, Vig S, Smith DW, Gao Q, Polson CT, Corsa JA, Guss VL, Loo A, Barten DM, Zheng M, Felsenstein KM, Roberts SB. 2,3-Benzodiazepin-1, 4-diones as peptidomimetic inhibitors of gamma-secretase. Bioorg Med Chemistry Lett 2004;14(13): 3535–3538.

63. May P. Chronic treatment with a functional gamma-secretase inhibitor reduces ABeta burden and plaque pathology in PDAPP mice. Neurobiol Aging 2003;23(1): S133.

64. Wolfe MS. Inhibition and modulation of gamma-secretase for Alzheimer's disease. Neurotherapeutics 2008;5(3): 391–398.

65. Wolfe MS. gamma-Secretase: structure, function, and modulation for Alzheimer's disease. Curr Top Med Chem (Sharjah, United Arab Emirates) 2008;8(1): 2–8.

66. Wong GT, Manfra D, Poulet FM, Zhang Q, Josien H, Bara T, Engstrom L, Pinzon-Ortiz M, Fine JS, Lee H-JJ, Zhang L, Higgins GA,

Parker EM. Chronic treatment with the gamma-secretase inhibitor LY-411575 inhibits beta-amyloid peptide production and alters lymphopoiesis and intestinal cell differentiation. J Biol Chem 2004;279(13): 12876–12882.

67. Prasad CVC, Zheng M, Vig S, Bergstrom C, Smith DW, Gao Q, Yeola S, Polson CT, Corsa JA, Guss VL, Loo A, Wang J, Sleczka BG, Dangler C, Robertson BJ, Hendrick JP, Roberts SB, Barten DM. Discovery of (S)-2-(S)-2-(3,5-difluorophenyl)-2-hydroxyacetamido)-N-(S,Z)-3-methyl-4-oxo-4,5-dihydro-3H-benzo[d][1,2]diazepin-5-yl)propanamide (BMS-433796): a gamma-secretase inhibitor with Abeta lowering activity in a transgenic mouse model of Alzheimer's disease. Bioorg Med Chem Lett 2007;17(14): 4006–4011.

68. Churcher I, Ashton K, Butcher JW, Clarke EE, Harrison T, Lewis HD, Owens AP, Teall MR, Williams S, Wrigley JDJ. A new series of potent benzodiazepine gamma-secretase inhibitors. Bioorg Med Chem Lett 2003;13(2): 179–183.

69. Owens AP, Nadin A, Talbot AC, Clarke EE, Harrison T, Lewis HD, Reilly M, Wrigley JDJ, Castro JL. High affinity, bioavailable 3-amino-1,4-benzodiazepine-based gamma-secretase inhibitors. Bioorg Med Chem Lett 2003;13(22): 4143–4145.

70. Keerti AR, Kumar BA, Parthasarathy T, Uma V. QSAR studies-potent benzodiazepine γ-secretase inhibitors. Bioorg Med Chem 2005;13(5): 1873–1878.

71. Becker C, Dembofsky B, Jacobs R, Kang J, Ohnmacht C, Rosamond J, Shenvi AB, Simpson T, Woods J.Preparation of peptidyl lactams for treatment of neurological disorders. 2003-SE1534 2004031154, 20031002. 2004.

72. Olson RE, Liu H, Thompson LA III.Preparation and use of hydroxyalkanoyl aminolactams and related structures as inhibitors of beta-amyloid protein production. 2001-805645 2002052360, 20010314. 2002.

73. Flohr A, Galley G, Jakob-Roetne R, Kitas EA, Peters J-U, Wostl W.Preparation of carbamic acid alkyl ester derivatives. 2004-951229 2005075327, 20040927. 2005.

74. Tung JS, Garofalo A, Pleiss MA.Preparation of acylated amino acid amidyl pyrazoles and related compounds. 2004-US18202 2005009344, 20040604. 2005.

75. Chen YL, Cherry K, Corman ML, Ebbinghaus CF, Gamlath CB, Liston D, Martin B-A, Oborski CE, Sahagan BG. Thiazole-diamides as potent gamma-secretase inhibitors. Bioorg Med Chem Lett 2007;17(20): 5518–5522.

76. Anderson JJ, Holtz G, Baskin PP, Turner M, Rowe B, Wang B, Kounnas MZ, Lamb BT, Barten D, Felsenstein K, McDonald I, Srinivasan K, Munoz B, Wagner SL. Reductions in β-amyloid concentrations in vivo by the γ-secretase inhibitors BMS-289948 and BMS-299897. Biochem Pharmacol 2005;69(4): 689–698.

77. Zhang D, Hanson R, Roongta V, Dischino DD, Gao Q, Sloan CP, Polson C, Keavy D, Zheng M, Mitroka J, Yeola S. In vitro and in vivo metabolism of a gamma-secretase inhibitor BMS-299897 and generation of active metabolites in milligram quantities with a microbial bioreactor. Curr Drug Metab 2006;7(8): 883–896.

78. Zhang L.Preparation of thiazole sulfonamides as inhibitors of gamma-secretase for the treatment of neurodegenerative disorders. 2005-78741 2005222223 20050311. 2005.

79. Asberom T, Bara TA, Clader JW, Greenlee WJ, Guzik HS, Josien HB, Li W, Parker EM, Pissarnitski DA, Song L, Zhang L, Zhao Z. Tetrahydroquinoline sulfonamides as gamma-secretase inhibitors. Bioorg Med Chem Lett 2007;17(1): 205–207.

80. Bergstrom Carl P, Sloan Charles P, Lau W-Y, Smith David W, Zheng M, Hansel Steven B, Polson Craig T, Corsa Jason A, Barten Donna M, Felsenstein Kevin M, Roberts Susan B. Carbamate-appended N-alkylsulfonamides as inhibitors of gamma-secretase. Bioorg Med Chem Lett 2008;18(2): 464–468.

81. Neitzel M, Dappen MS, Marugg J.Preparation of N-(oxoazepanyl) benzenesulfonamides and related derivatives as γ-secretase inhibitors for treating Alzheimer's disease. 2004-US35951 2005042489, 20041029. 2005.

82. Kreft A, Harrison B, Aschmies S, Atchison K, Casebier D, Cole DC, Diamantidis G, Ellingboe J, Hauze D, Hu Y, Huryn D, Jin M, Kubrak D, Lu P, Lundquist J, Mann C, Martone R, Moore W, Oganesian A, Porte A, Riddell DR, Sonnenberg-Reines J, Stock JR, Sun S-C, Wagner E, Woller K, Xu Z, Zhou H, Steven Jacobsen J. Discovery of a novel series of Notch-sparing gamma-secretase inhibitors Bioorg Med Chem Lett 18(14): 4232–4236.

83. Bergstrom CP, Sloan CP, Wang HH, Parker MF, Smith DW, Zheng M, Hansel SB, Polson CT, Barber LE, Bursuker I, Guss VL, Corsa JA, Barten DM, Felsenstein KM, Roberts SB. Nitrogen-appended N-alkylsulfonamides as inhibitors of gamma-secretase. Bioorg Med Chem Lett 2008;18(1): 175–178.

84. Parker MF, Barten DM, Bergstrom CP, Bronson JJ, Corsa JA, Deshpande MS, Felsenstein

KM, Guss VL, Hansel SB, Johnson G, Keavy DJ, Lau WY, Mock J, Prasad CVC, Polson CT, Sloan CP, Smith DW, Wallace OB, Wang HH, Williams A, Zheng M. N-(5-Chloro-2-(hydroxymethyl)-N-alkyl-arylsulfonamides as gamma-secretase inhibitors. Bioorg Med Chem Lett 2007;17(16): 4432–4436.

85. Barten DM, Guss VL, Corsa JA, Loo A, Hansel SB, Zheng M, Munoz B, Srinivasan K, Wang B, Robertson BJ, Polson CT, Wang J, Roberts SB, Hendrick JP, Anderson JJ, Loy JK, Denton R, Verdoorn TA, Smith DW, Felsenstein KM. Dynamics of beta-amyloid reductions in brain, cerebrospinal fluid, and plasma of beta-amyloid precursor protein transgenic mice treated with a gamma-secretase inhibitor. J Pharmacol Exp Ther 2005;312(2): 635–643.

86. Lewis SJ, Smith AL, Neduvelil JG, Stevenson GI, Lindon MJ, Jones AB, Shearman MS, Beher D, Clarke E, Best JD, Peachey JE, Harrison T, Castro JL. A novel series of potent gamma-secretase inhibitors based on a benzobicyclo[4.2.1]nonane core. Bioorg Med Chem Lett 2005;15(2): 373–378.

87. Sparey T, Beher D, Best J, Biba M, Castro JL, Clarke E, Hannam J, Harrison T, Lewis H, Madin A, Shearman M, Sohal B, Tsou N, Welch C, Wrigley J. Cyclic sulfamide gamma-secretase inhibitors. Bioorg Med Chem Lett 2005;15 (19): 4212–4216.

88. Teall M, Oakley P, Harrison T, Shaw D, Kay E, Elliott J, Gerhard U, Castro JL, Shearman M, Ball RG, Tsou NN. Aryl sulfones: a new class of gamma-secretase inhibitors. Bioorg Med Chem Lett 2005;15(10): 2685–2688.

89. Churcher I, Beher D, Best JD, Castro JL, Clarke EE, Gentry A, Harrison T, Hitzel L, Kay E, Kerrad S, Lewis HD, Morentin-Gutierrez P, Mortishire-Smith R, Oakley PJ, Reilly M, Shaw DE, Shearman MS, Teall MR, Williams S, Wrigley JDJ. 4-Substituted cyclohexyl sulfones as potent, orally active gamma-secretase inhibitors. Bioorg Med Chem Lett 2006;16(2): 280–284.

90. Best JD, Jay MT, Otu F, Churcher I, Reilly M, Morentin-Gutierrez P, Pattison C, Harrison T, Shearman MS, Atack JR. In vivo characterization of Abeta (40) changes in brain and cerebrospinal fluid using the novel gamma-secretase inhibitor N-[cis-4-[(4-chlorophenyl)sulfonyl]-4-(2,5-difluorophenyl)cyclohexyl]-1,1,1-trifluoromethanesulfonamide (MK-560) in the rat. J Pharmacol Exp Ther 2006;317 (2): 786–790.

91. Shaw D, Best J, Dinnell K, Nadin A, Shearman M, Pattison C, Peachey J, Reilly M, Williams B, Wrigley J, Harrison T. 3,4-Fused cyclohexyl sulfones as gamma-secretase inhibitors. Bioorg Med Chem Lett 2006;16(11): 3073–3077.

92. Cole DC, Stock JR, Kreft AF, Antane M, Aschmies SH, Atchison KP, Casebier DS, Comery TA, Diamantidis G, Ellingboe JW, Harrison BL, Hu Y, Jin M, Kubrak DM, Lu P, Mann CW, Martone RL, Moore WJ, Oganesian A, Riddell DR, Sonnenberg-Reines J, Sun S-C, Wagner E, Wang Z, Woller KR, Xu Z, Zhou H, Jacobsen J. S.(S)-N-(5-Chlorothiophene-2-sulfonyl)-beta,beta-diethylalaninol a Notch-1-sparing gamma-secretase inhibitor. Bioorg Med Chem Lett 2009;19(3): 926–929.

93. Mayer SC, Kreft AF, Harrison B, Abou-Gharbia M, Antane M, Aschmies S, Atchison K, Chlenov M, Cole DC, Comery T, Diamantidis G, Ellingboe J, Fan K, Galante R, Gonzales C, Ho DM, Hoke ME, Hu Y, Huryn D, Jain U, Jin M, Kremer K, Kubrak D, Lin M, Lu P, Magolda R, Martone R, Moore W, Oganesian A, Pangalos MN, Porte A, Reinhart P, Resnick L, Riddell DR, Sonnenberg-Reines J, Stock JR, Sun S-C, Wagner E, Wang T, Woller K, Xu Z, Zaleska MM, Zeldis J, Zhang M, Zhou H, Jacobsen JS. Discovery of begacestat, a Notch-1-sparing gamma-secretase inhibitor for the treatment of Alzheimer's disease. J Med Chem 2008; 51(23): 7348–7351.

94. Zhang M, Porte A, Diamantidis G, Sogi K, Kubrak D, Resnick L, Mayer SC, Wang Z, Kreft AF, Harrison BL. Asymmetric synthesis of novel alpha-amino acids with beta-branched side chains. Bioorg Med Chem Lett 2007;17(9): 2401–2403.

95. McBriar MD, Clader JW, Chu I, Del Vecchio RA, Favreau L, Greenlee WJ, Hyde LA, Nomeir AA, Parker EM, Pissarnitski DA, Song L, Zhang L, Zhao Z. Discovery of amide and heteroaryl isosteres as carbamate replacements in a series of orally active gamma-secretase inhibitors. Bioorg Med Chem Lett 2008;18(1): 215–219.

96. Kitas EA, Galley G, Jakob-Roetne R, Flohr A, Wostl W, Mauser H, Alker AM, Czech C, Ozmen L, David-Pierson P, Reinhardt D, Jacobsen H. Substituted 2-oxo-azepane derivatives are potent, orally active gamma-secretase inhibitors. Bioorg Med Chem Lett 2008; 18(1): 304–308.

97. Brigham B, Liao A, Bova M, Chen KS, Cabrera C, Chen X-H, Cole T, Eichenbaum T, Goldbach E, Hu K, Keim P, Kondrei A, Lee M, Ni H, Nguyen L, Mattson MN, Mutter L, Quincy D, Santiago P, Sauer J-M, Shopp G, Soriano F,

Webb S, Wehner N, Basi G. Treatment with a novel sulfonamide γ-secretase inhibitor reduces β-amyloid production without effects on Notch related toxicity in wildtype mice. In: Presented at the 37th Annual Meeting of the Society of Neuroscience; 2007; San Diego, CA. 2007, Abstract 486.

98. Li H, Asberom T, Bara TA, Clader JW, Greenlee WJ, Josien HB, McBriar MD, Nomeir A, Pissarnitski DA, Rajagopalan M, Xu R, Zhao Z, Song L, Zhang L. Discovery of 2,4,6-trisubstituted N-arylsulfonyl piperidines as gamma-secretase inhibitors. Bioorg Med Chem Lett 2007;17(22): 6290–6294.

99. Siemers ER, Dean RA, Friedrich S, Ferguson-Sells L, Gonzales C, Farlow MR, May PC. Safety, tolerability, and effects on plasma and cerebrospinal fluid amyloid-beta after inhibition of gamma-secretase. Clin Neuropharmacol 2007;30(6): 317–325.

100. Fleisher AS, Raman R, Siemers ER, Becerra L, Clark CM, Dean RA, Farlow MR, Galvin JE, Peskind ER, Quinn JF, Sherzai A, Sowell BB, Aisen PS, Thal LJ. Phase 2 safety trial targeting amyloid beta production with a gamma-secretase inhibitor in Alzheimer disease. Arch Neurol 2008;65(8): 1031–1038.

101. Siemers ER. Commentary on "Optimal design of clinical trials for drugs designed to slow the course of Alzheimer's disease". Alzheimer's Dementia 2006;2(3): 140–142.

102. Siemers ER, Quinn JF, Kaye J, Farlow MR, Porsteinsson A, Tariot P, Zoulnouni P, Galvin JE, Holtzman DM, Knopman DS, Satterwhite J, Gonzales C, Dean RA, May PC. Effects of a gamma-secretase inhibitor in a randomized study of patients with Alzheimer disease. Neurology 2006;66(4): 602–604.

103. Bateman RJ, Munsell LY, Chen X, Holtzman DM, Yarasheski KE. Stable isotope labeling tandem mass spectrometry (SILT) to quantify protein production and clearance rates. J Am Soc Mass Spectrom 2007;18(6): 997–1006.

104. Bateman RJ, Siemers ER, Mawuenyega KG, Wen G, Browning KR, Sigurdson WC, Yarasheski KE, Friedrich SW, De Mattos RB, May PC, Paul SM, Holtzman DM. A gamma-secretase inhibitor decreases amyloid-beta production in the central nervous system. Ann Neurol 2009;66(1): 48–54.

105. Bateman R. Validation and optimization of stable isotope labeling kinetic analysis of amyloid beta from human cerebrospinal fluid. In: 9th International Conference on AD Drug Discovery; 2008; New York, NY.

106. Macor JE, Albright CF, Meredith JE, Zaczek RC, Barten DM, Toyn JH, Slemmon R, Lentz K, Wang J-S, Denton R, Pilcher G, Wang O, Gu H, Dockens R, Berman R, Tong G, Bronson JJ, Parker MF, Mate RA, McElhone K, Starrett JE Jr, Gillman KW, Olson RE. Discovery of BMS-708163: a potent and selective gamma-secretase inhibitor which lowers CSF beta-amyloid in humans. In: Abstracts of Papers. 237th ACS National Meeting; 2009, March 22–26; Salt Lake City, UT, United States. 2009; MEDI-027.

107. Matthews CZ, Woolf EJ. Determination of a novel gamma-secretase inhibitor in human plasma and cerebrospinal fluid using automated 96 well solid phase extraction and liquid chromatography/tandem mass spectrometry. J Chromatogr, B: Anal Technol Biomed Life Sci 2008;863(1): 36–45.

108. Deangelo DJ, Stone RM, Silverman LB, Stock W, Attar EC, Fearen I, Dallob A, Matthews C, Stone J, Freedman SJ, Aster J. A phase I clinical trial of the notch inhibitor MK-0752 in patients with T-cell acute lymphoblastic leukemia/lymphoma (T-ALL) and other leukemias. J Clin Oncol 2006;24(18s): 6585.

109. Rosen LB, Stone JA, Plump A, Yuan J, Harrison T, Flynn M, Dallob A, Matthews C, Stevenson D, Schmidt D, Palmieri T, Leibowitz M, Jhee S, Ereshefsky L, Salomon R, Winchell G, Shearman MS, Murphy MG, Gottesdiener KM. The gamma secretase inhibitor MK-0752 acutely and significantly reduces CSF Abeta40 concentrations in humans. Alzheimers Dement 2006;2(3S1): S79.

110. Martone RL, Zhou H, Atchison K, Comery T, Xu JZ, Huang X, Gong X, Jin M, Kreft A, Harrison B, Mayer SC, Aschmies S, Gonzales C, Zaleska MM, Riddell DR, Wagner E, Lu P, Sun S-C, Sonnenberg-Reines J, Oganesian A, Adkins K, Leach MW, Clarke DW, Huryn D, Abou-Gharbia M, Magolda R, Bard J, Frick G, Raje S, Forlow SB, Balliet C, Burczynski ME, Reinhart PH, Wan HI, Pangalos MN, Jacobsen JS. Begacestat (GSI-953): a novel, selective thiophene sulfonamide inhibitor of amyloid precursor protein gamma-secretase for the treatment of Alzheimer's disease. J Pharmacol Exp Ther 2009;331(2): 598–608.

111. Wan HI, Day M, Hurko O, Rutkowski JL. Biomarkers for Alzheimer's disease: translational medicine approaches in development of disease modifying therapeutics. Am Pharm Rev 2008;11 (7): 83–84, 88–92.

112. Brodney M. Discovery of PF-3084014: a novel gamma-secretase inhibitor for the treatment

of Alzheimer's disease. In: 238th ACS National Meeting; 2009; Washington, DC. 2009, Abstract and a lecture.

113. Zhao B, Yu M, Neitzel M, Marugg J, Jagodzinski J, Lee M, Hu K, Schenk D, Yednock T, Basi G. Identification of gamma-secretase inhibitor potency determinants on presenilin. J Biol Chem 2008;283(5): 2927–2938.

114. Nitz M, Fenili D, Darabie AA, Wu L, Cousins JE, McLaurin J. Modulation of amyloid-beta aggregation and toxicity by inosose stereoisomers. FEBS 2008;275(8): 1663–1674.

115. Olson RE, Marcin LR. Secretase inhibitors and modulators for the treatment of Alzheimer's disease. Annu Rep Med Chem 2007;42:27–47.

116. Luistro L, He W, Smith M, Packman K, Vilenchik M, Carvajal D, Roberts J, Cai J, Berkofsky-Fessler W, Hilton H, Linn M, Flohr A, Jakob-Rotne R, Jacobsen H, Glenn K, Heimbrook D, Boylan JF. Preclinical profile of a potent gamma-secretase inhibitor targeting notch signaling with *in vivo* efficacy and pharmacodynamic properties. Cancer Res 2009;69 (19): 7672–7680.

117. Evin G, Kenche VB. BACE inhibitors as potential therapeutics for Alzheimer's disease. Recent Patents CNS Drug Discov 2007;2(3): 188–199.

118. Hamada Y, Kiso Y. Recent progress in the drug discovery of non-peptidic BACE1 inhibitors. Expert Opin Drug Discov 2009;4(4): 391–416.

119. Huang W-H, Sheng R, Hu Y-Z. Progress in the development of nonpeptidomimetic BACE 1 inhibitors for Alzheimer's disease. Curr Med Chem 2009;16(14): 1806–1820.

120. John V. Human beta-secretase (BACE) and BACE inhibitors: progress report. Curr Top Med Chem (Sharjah, United Arab Emirates) 2006;6(6): 569–578.

121. Silvestri R. Boom in the development of non-peptidic beta-secretase (BACE1) inhibitors for the treatment of Alzheimer's disease. Med Res Rev 2009;29(2): 295–338.

122. Stachel SJ. Progress toward the development of a viable BACE-1 inhibitor. Drug Dev Res 2009;70(2): 101–110.

123. Ghosh AK, Kumaragurubaran N, Hong L, Koelsh G, Tang J. Memapsin 2 (beta-secretase) inhibitors: drug development. Curr Alzheimer Res 2008;5(2): 121–131.

124. Cooper JB. Aspartic proteinases in disease: a structural perspective. Curr Drug Targets 2002;3(2): 155–173.

125. Ghosh AK. Harnessing nature's insight: design of aspartyl protease inhibitors from treatment of drug-resistant HIV to Alzheimer's disease. Journal of Medicinal Chemistry 2009;52(8): 2163–2176.

126. Guruprasad K, Dhanaraj V, Groves M, Blundell TL. Aspartic proteinases: the structures and functions of a versatile superfamily of enzymes. Perspect Drug Discov Design 1995;2(3): 329–341.

127. Nguyen J-T, Hamada Y, Kimura T, Kiso Y. Design of potent aspartic protease inhibitors to treat various diseases. Arch Pharm (Weinheim, Germany) 2008;341(9): 523–535.

128. Thompson LA, Tebben AJ. Pharmacokinetics and design of aspartyl protease inhibitors. Annu Rep Med Chem 2001;36:247–256.

129. Holloway MK, Hunt P, McGaughey GB. Structure and modeling in the design of beta - and gamma-secretase inhibitors. Drug Dev Res 2009;70(2): 70–93.

130. Ghosh AK, Bilcer G, Harwood C, Kawahama R, Shin D, Hussain KA, Hong L, Loy JA, Nguyen C, Koelsch G, Ermolieff J, Tang J. Structure-based design: potent inhibitors of human brain memapsin 2 (beta-secretase). J Med Chem 2001;44(18): 2865–2868.

131. Ghosh AK, Devasamudram T, Hong L, DeZutter C, Xu X, Weerasena V, Koelsch G, Bilcer G, Tang J. Structure-based design of cycloamide-urethane-derived novel inhibitors of human brain memapsin 2 (beta-secretase). Bioorg Med Chem Lett 2005;15(1): 15–20.

132. Kimura T, Shuto D, Hamada Y, Igawa N, Kasai S, Liu P, Hidaka K, Hamada T, Hayashi Y, Kiso Y. Design and synthesis of highly active Alzheimer's beta-secretase (BACE1) inhibitors, KMI-420 and KMI-429, with enhanced chemical stability. Bioorg Med Chem Lett 2005;15(1): 211–215.

133. Kimura T, Shuto D, Kasai S, Liu P, Hidaka K, Hamada T, Hayashi Y, Hattori C, Asai M, Kitazume S, Saido TC, Ishiura S, Kiso Y. KMI-358 and KMI-370, highly potent and small-sized BACE1 inhibitors containing phenylnorstatine. Bioorg Med Chem Lett 2004;14 (6): 1527–1531.

134. Shuto D, Kasai S, Kimura T, Liu P, Hidaka K, Hamada T, Shibakawa S, Hayashi Y, Hattori C, Szabo B, Ishiura S, Kiso Y. KMI-008, a novel beta-secretase inhibitor containing a hydroxymethylcarbonyl isostere as a transition-state mimic: design and synthesis of substrate-based octapeptides. Bioorg Med Chem Lett 2003;13(24): 4273–4276.

135. Kimura T, Hamada Y, Stochaj M, Ikari H, Nagamine A, Abdel-Rahman H, Igawa N, Hidaka K, Nguyen J-T, Saito K, Hayashi Y, Kiso Y. Design and synthesis of potent beta-secretase (BACE1) inhibitors with P1' carboxylic acid bioisosteres. Bioorg Med Chem Lett 2006;16(9): 2380–2386.

136. Asai M, Hattori C, Iwata N, Saido TC, Sasagawa N, Szabo B, Hashimoto Y, Maruyama K, Tanuma S-I, Kiso Y, Ishiura S. The novel beta-secretase inhibitor KMI-429 reduces amyloid beta peptide production in amyloid precursor protein transgenic and wild-type mice. J Neurochem 2006;96(2): 533–540.

137. Huse JT, Liu K, Pijak DS, Carlin D, Lee VMY, Doms RW. beta-Secretase processing in the trans-Golgi network preferentially generates truncated amyloid species that accumulate in Alzheimer's disease brain. J Biol Chem 2002;277(18): 16278–16284.

138. Yan R, Han P, Miao H, Greengard P, Xu H. The transmembrane domain of the Alzheimer's beta-secretase (BACE1) determines its late Golgi localization and access to beta-amyloid precursor protein (APP) substrate. J Biol Chem 2001;276(39): 36788–36796.

139. Miyamoto M, Matsui J, Fukumoto H, Tarui N. Preparation of 2-[2-amino- or 2-(N-heterocyclyl)ethyl]-6-(4-biphenylylmethoxy)tetralin derivatives as beta-secretase inhibitors. 2001-JP4144 2001087293, 20010518. 2001.

140. Bhisetti GR, Saunders JO, Murcko MA, Lepre CA, Britt SD, Come JH, Deninger DD, Wang T. Preparation of beta-carbolines and other inhibitors of BACE-1 aspartic proteinase useful against Alzheimer's and other BACE-mediated diseases. 2002-US13741 2002088101, 20020429. 2002.

141. John V, Beck JP, Bienkowski MJ, Sinha S, Heinrikson RL. Human beta-Secretase (BACE) and BACE Inhibitors. J Med Chem 2003;46(22): 4625–4630.

142. Hom RK, Gailunas AF, Mamo S, Fang LY, Tung JS, Walker DE, Davis D, Thorsett ED, Jewett NE, Moon JB, John V. Design and Synthesis of hydroxyethylene-based peptidomimetic inhibitors of human beta-secretase. J Med Chem 2004;47(1): 158–164.

143. Hom R, Mamo S, Tung J, Gailunas A, John V, Fang L.Preparation of hydroxyethylenes with peptide subunits for pharmaceutical use in the treatment of Alzheimer's disease. 2001-US9501 2001070672, 20010323. 2001.

144. Tung JS, Davis DL, Anderson JP, Walker DE, Mamo S, Jewett N, Hom RK, Sinha S, Thorsett ED, John V. Design of substrate-based inhibitors of human beta-secretase. J Med Chem 2002;45(2): 259–262.

145. Maillard MC, Hom RK, Benson TE, Moon JB, Mamo S, Bienkowski M, Tomasselli AG, Woods DD, Prince DB, Paddock DJ, Emmons TL, Tucker JA, Dappen MS, Brogley L, Thorsett ED, Jewett N, Sinha S, John V. Design, synthesis, and crystal structure of hydroxyethyl secondary amine-based peptidomimetic inhibitors of human beta-secretase. J Med Chem 2007;50(4): 776–781.

146. Coburn CA, Stachel SJ, Li Y-M, Rush DM, Steele TG, Chen-Dodson E, Holloway MK, Xu M, Huang Q, Lai M-T, DiMuzio J, Crouthamel M-C, Shi X-P, Sardana V, Chen Z, Munshi S, Kuo L, Makara GM, Annis DA, Tadikonda PK, Nash HM, Vacca JP. Identification of a small molecule nonpeptide active site beta-secretase inhibitor that displays a nontraditional binding mode for aspartyl proteases. J Med Chem 2004;47(25): 6117–6119.

147. Coburn CA, Steele TG, Vacca JP, Annis DA Jr, Makara GM, Nash HM, Tadikonda PK, Wang T.Preparation of aminopentylaminooxoethyl-phenylcarboxamides as beta-secretase inhibitors for the treatment of Alzheimer's disease. 2005-US15949 2005113484, 20050509. 2005.

148. Stachel SJ, Coburn CA, Steele TG, Jones KG, Loutzenhiser EF, Gregro AR, Rajapakse HA, Lai M-T, Crouthamel M-C, Xu M, Tugusheva K, Lineberger JE, Pietrak BL, Espeseth AS, Shi X-P, Chen-Dodson E, Holloway MK, Munshi S, Simon AJ, Kuo L, Vacca JP. Structure-based design of potent and selective cell-permeable inhibitors of human beta-secretase (BACE-1). J Med Chem 2004;47(26): 6447–6450.

149. Coburn CA, Stachel SJ, Vacca JP.Preparation of phenylcarboxamide derivatives as beta-secretase inhibitors for the treatment of Alzheimer's disease. 2003-US35316 2004043916, 20031106. 2004.

150. Stachel SJ, Coburn CA, Steele TG, Crouthamel M-C, Pietrak BL, Lai M-T, Holloway MK, Munshi SK, Graham SL, Vacca JP. Conformationally biased P3 amide replacements of beta-secretase inhibitors. Bioorg Med Chem Lett 2006;16(3): 641–644.

151. Ghosh AK, Kumaragurubaran N, Hong L, Kulkarni S, Xu X, Miller HB, Reddy DS, Weerasena V, Turner R, Chang W, Koelsch G, Tang J. Potent memapsin 2 (beta-secretase) inhibitors: Design, synthesis, protein–ligand X-ray structure, and in vivo evaluation. Bioorg Med Chem Lett 2008;18(3): 1031–1036.

152. Uchikawa O, Aso K, Koike T, Tarui N, Hirai K. Preparation of benzamide derivatives as beta-secretase inhibitors. 2003-JP10045 2004014843, 20030807. 2004.

153. Decicco CP, Tebben AJ, Thompson LA, Combs AP. Preparation of novel alpha-amino-gamma-lactams as beta-secretase inhibitors. 2003-US24407 2004013098, 20030805. 2004.

154. Demont EH, Redshaw S, Walter DS. Preparation of hydroxyethylamine derivatives for the treatment of Alzheimer's disease. 2004-EP2644 2004080376, 20040311. 2004.

155. Faller A, MacPherson DT, Milner PH, Stanway SJ, Trouw LS. Preparation of benzamide derivatives as inhibitors of Asp-2. 2002-EP13515 2003045913, 20021129. 2003.

156. Faller A, Milner PH, Ward JG. N-Carbamoylalkylcarboxamides and -sulfonamides with Asp-2 inhibitory activity. 2002-EP13517 2003045903, 20021129. 2003.

157. Beswick P, Charrier N, Clarke B, Demont E, Dingwall C, Dunsdon R, Faller A, Gleave R, Hawkins J, Hussain I, Johnson CN, MacPherson D, Maile G, Matico R, Milner P, Mosley J, Naylor A, O'Brien A, Redshaw S, Riddell D, Rowland P, Skidmore J, Soleil V, Smith KJ, Stanway S, Stemp G, Stuart A, Sweitzer S, Theobald P, Vesey D, Walter DS, Ward J, Wayne G. BACE-1 inhibitors part 3: identification of hydroxy ethylamines (HEAs) with nanomolar potency in cells. Bioorg Med Chem Lett 2008;18(3): 1022–1026.

158. Clarke B, Demont E, Dingwall C, Dunsdon R, Faller A, Hawkins J, Hussain I, MacPherson D, Maile G, Matico R, Milner P, Mosley J, Naylor A, O'Brien A, Redshaw S, Riddell D, Rowland P, Soleil V, Smith KJ, Stanway S, Stemp G, Sweitzer S, Theobald P, Vesey D, Walter DS, Ward J, Wayne G. BACE-1 inhibitors part 2: identification of hydroxy ethylamines (HEAs) with reduced peptidic character. Bioorg Med Chem Lett 2008;18(3): 1017–1021.

159. Clarke B, Demont E, Dingwall C, Dunsdon R, Faller A, Hawkins J, Hussain I, MacPherson D, Maile G, Matico R, Milner P, Mosley J, Naylor A, O'Brien A, Redshaw S, Riddell D, Rowland P, Soleil V, Smith KJ, Stanway S, Stemp G, Sweitzer S, Theobald P, Vesey D, Walter DS, Ward J, Wayne G. BACE-1 inhibitors Part 1: identification of novel hydroxy ethylamines (HEAs). Bioorg Med Chem Lett 2008;18(3): 1011–1016.

160. Demont EH, Redshaw S, Walter DS. Preparation of 3-(1,1-dioxotetrahydro-1,2-thiazin-2-yl) or 3-(1,1-dioxo-isothiazolidin-2-yl) substituted benzamide compounds for treatment of Alzheimer's disease. 2004-EP6594 2004111022, 20040617. 2004.

161. Hussain I, Hawkins J, Harrison D, Hille C, Wayne G, Cutler L, Buck T, Walter D, Demont E, Howes C, Naylor A, Jeffrey P, Gonzalez MI, Dingwall C, Michel A, Redshaw S, Davis JB. Oral administration of a potent and selective non-peptidic BACE-1 inhibitor decreases beta-cleavage of amyloid precursor protein and amyloid-beta production *in vivo*. J Neurochem 2007;100(3): 802–809.

162. Iserloh U, Pan J, Stamford AW, Kennedy ME, Zhang Q, Zhang L, Parker EM, McHugh NA, Favreau L, Strickland C, Voigt J. Discovery of an orally efficacious 4-phenoxypyrrolidine-based BACE-1 inhibitor. Bioorg Med Chem Lett 2008;18(1): 418–422.

163. Iserloh U, Wu Y, Cumming JN, Pan J, Wang LY, Stamford AW, Kennedy ME, Kuvelkar R, Chen X, Parker EM, Strickland C, Voigt J. Potent pyrrolidine- and piperidine-based BACE-1 inhibitors. Bioorg Med Chem Lett 2008;18(1): 414–417.

164. Stanton MG, Stauffer SR, Gregro AR, Steinbeiser M, Nantermet P, Sankaranarayanan S, Price EA, Wu G, Crouthamel M-C, Ellis J, Lai M-T, Espeseth AS, Shi X-P, Jin L, Colussi D, Pietrak B, Huang Q, Xu M, Simon AJ, Graham SL, Vacca JP, Selnick H. Discovery of isonicotinamide derived beta-secretase inhibitors: *in vivo* reduction of beta-amyloid. J Med Chem 2007;50(15): 3431–3433.

165. Stauffer SR, Stanton MG, Gregro AR, Steinbeiser MA, Shaffer JR, Nantermet PG, Barrow JC, Rittle KE, Colussi D, Espeseth AS, Lai M-T, Pietrak BL, Holloway MK, McGaughey GB, Munshi SK, Hochman JH, Simon AJ, Selnick HG, Graham SL, Vacca JP. Discovery and SAR of isonicotinamide BACE-1 inhibitors that bind beta-secretase in a N-terminal 10s-loop down conformation. Bioorg Med Chem Lett 2007;17(6): 1788–1792.

166. Sankaranarayanan S, Holahan MA, Colussi D, Crouthamel M-C, Devanarayan V, Ellis J, Espeseth A, Gates AT, Graham SL, Gregro AR, Hazuda D, Hochman JH, Holloway K, Jin L, Kahana J, Lai M-T, Lineberger J, McGaughey G, Moore KP, Nantermet P, Pietrak B, Price EA, Rajapakse H, Stauffer S, Steinbeiser MA, Seabrook G, Selnick HG, Shi X-P, Stanton MG, Swestock J, Tugusheva K, Tyler KX, Vacca JP, Wong J, Wu G, Xu M, Cook JJ, Simon AJ. First demonstration of cerebrospinal fluid and plasma Abeta lowering with oral administration of a beta-site amyloid precursor protein-cleaving

enzyme 1 inhibitor in nonhuman primates. J Pharmacol Exp Ther 2009;328(1): 131–140.

167. Stachel SJ, Coburn CA, Sankaranarayanan S, Price EA, Pietrak BL, Huang Q, Lineberger J, Espeseth AS, Jin L, Ellis J, Holloway MK, Munshi S, Allison T, Hazuda D, Simon AJ, Graham SL, Vacca JP. Macrocyclic inhibitors of beta-secretase: functional activity in an animal model. J Med Chem 2006;49(21): 6147–6150.

168. Coburn C, Stachel SJ, Vacca JP. Preparation of macrocyclic beta-secretase inhibitors for the treatment of Alzheimer's disease. 2004-US25791 2005018545, 20040810. 2005.

169. Thompson LA, Boy KM, Shi J, Macor JE, Good AC, Marcin LR. Tetrahydroisoquinoline derivatives as beta-secretase inhibitors and their preparation, pharmaceutical compositions and use in the treatment of neurological diseases. 2007-951516 2008153868, 20071206. 2008.

170. Machauer R. Preparation of macrocyclic compounds useful as BACE inhibitors. 2007-EP57492 2008009734, 20070719. 2008.

171. Cole DC, Manas ES, Stock JR, Condon JS, Jennings LD, Aulabaugh A, Chopra R, Cowling R, Ellingboe JW, Fan KY, Harrison BL, Hu Y, Jacobsen S, Jin G, Lin L, Lovering FE, Malamas MS, Stahl ML, Strand J, Sukhdeo MN, Svenson K, Turner MJ, Wagner E, Wu J, Zhou P, Bard J. Acylguanidines as small-molecule beta-secretase inhibitors. J Med Chem 2006;49(21): 6158–6161.

172. Cole DC, Manas ES, Jennings LD, Lovering FE, Stock JR, Moore WJ, Ellingboe JW, Condon JS, Sukhdeo MN, Zhou P, Wu J, Morris KM. Preparation of azolylacylguanidines as beta-secretase inhibitors. 2006-352820 2006183790, 20060213. 2006.

173. Fobare WF, Solvibile WR. Preparation of thienyl and furyl acylguanidines as beta-secretase (BACE) inhibitors. 2006-352646 2006183792, 20060213. 2006.

174. Hu B. Preparation of terphenylguanidines as beta-secretase inhibitors for inhibition of the formation of beta-amyloid deposits and neurofibrillary tangles present in neurodegenerative diseases. 2006-352887 2006183943, 20060213. 2006.

175. Cole DC, Stock JR, Chopra R, Cowling R, Ellingboe JW, Fan KY, Harrison BL, Hu Y, Jacobsen S, Jennings LD, Jin G, Lohse PA, Malamas MS, Manas ES, Moore WJ, O'Donnell M-M, Olland AM, Robichaud AJ, Svenson K, Wu J, Wagner E, Bard J. Acylguanidine inhibitors of beta-secretase: optimization of the pyrrole ring substituents extending into the S1 and S3 substrate binding pockets. Bioorg Med Chem Lett 2008;18(3): 1063–1066.

176. Jennings LD, Cole DC, Stock JR, Sukhdeo MN, Ellingboe JW, Cowling R, Jin G, Manas ES, Fan KY, Malamas MS, Harrison BL, Jacobsen S, Chopra R, Lohse PA, Moore WJ, O'Donnell M-M, Hu Y, Robichaud AJ, Turner MJ, Wagner E, Bard J. Acylguanidine inhibitors of beta-secretase: optimization of the pyrrole ring substituents extending into the S'1 substrate binding pocket. Bioorg Med Chem Lett 2008;18(2): 767–771.

177. Fobare WF, Solvibile WR, Robichaud AJ, Malamas MS, Manas E, Turner J, Hu Y, Wagner E, Chopra R, Cowling R, Jin G, Bard J. Thiophene substituted acylguanidines as BACE1 inhibitors. Bioorg Med Chem Lett 2007;17(19): 5353–5356.

178. Malamas MS, Fobare WF, Solvibile WR, Lovering FE, Condon JS, Fan KY, Turner J, Bard J, Barnes K, Hui Y, Johnson M, Hu Y, Chopra R, Manas E, Robichaud AJ. Pyrrolyl 2-aminopyridines as potent BACE1 inhibitors. In: Abstracts of Papers, 238th ACS National Meeting; 2009, August 16–20; Washington, DC, United States. 2009, MEDI-039.

179. Congreve M, Aharony D, Albert J, Callaghan O, Campbell J, Carr RAE, Chessari G, Cowan S, Edwards PD, Frederickson M, McMenamin R, Murray CW, Patel S, Wallis N. Application of fragment screening by x-ray crystallography to the discovery of aminopyridines as inhibitors of beta-secretase. J Med Chem 2007;50(6): 1124–1132.

180. Murray CW, Callaghan O, Chessari G, Cleasby A, Congreve M, Frederickson M, Hartshorn MJ, McMenamin R, Patel S, Wallis N. Application of fragment screening by x-ray crystallography to beta-secretase. J Med Chem 2007;50(6): 1116–1123.

181. Thompson LA, Shi J, Zusi FC, Dee MF, Macor JE. Preparation of indoleacetic acid acyl guanidines as beta-secretase (BACE) inhibitors. 2006-508481 2007049589, 20060823. 2007.

182. Gerritz S, Zhai W, Shi S, Zhu S, Good AC, Thompson LA III. Preparation of isoxazolylcarbonyl- and isothiazolylcarbonylguanidines as beta-secretase (BACE) inhibitors. 2006-US24172 2007002214, 20060620. 2007.

183. Gerritz SW, Shi S, Zhai W, Zhu S, Thompson LA, Toyn JH, Meredith JE, Good AC, Muckelbauer JK, Camac DM, Dodd DS, Cook LS, Padmanabha R, Albright CF, Sofia MJ, Poss

MA, Macor JE. Discovery and optimization of acyl guanidines as novel BACE-1 inhibitors. In: Abstracts of Papers. 235th ACS National Meeting; 2008, April 6–10; New Orleans, LA, United States. 2008, MEDI-186.

184. Shi S, Gerritz S, Zhu S.Preparation of macrocyclic acylguanidines as beta-secretase (BACE) inhibitors. 2008-99334 2008262055, 20080408. 2008.

185. Wu Y-J, Gerritz S, Shi S, Zhu S.Preparation of oxime-containing acyl guanidines as beta-secretase inhibitors. 2007-693026 2007232581, 20070329. 2007.

186. Wu Y-J, Gerritz S, Shi S, Zhu S.Oxime-containing macrocyclic acyl guanidines as beta-secretase inhibitors and their preparation. 2007-940597 2008139523, 20071115. 2008.

187. Malamas MS, Erdei J, Gunawan I, Pawel N, Chlenov M, Robichaud AJ, Turner J, Hu Y, Wagner E, Aschmies S, Comery T, Di L, Fan K, Chopra R, Oganesian A, Huselton C, Bard J. Aminohydantoins as highly potent, selective and orally active BACE1 inhibitors. In: Abstracts of Papers. 233rd ACS National Meeting; 2007, March 25–29; Chicago, IL, United States. 2007, MEDI-234.

188. Malamas MS, Erdei JJ, Gunawan IS, Zhou P, Yan Y, Quagliato DA.Preparation of amino-5,5-diphenylimidazolone derivatives for inhibition of beta-secretase and treatment of beta-amyloid-related diseases. 2005-153633 2005282825, 20050615. 2005.

189. Malamas MS, Fobare WF, Solvibile WR, Lovering FE, Condon JS, Robichaud AJ.Amino-pyridines as inhibitors of beta-secretase and their preparation, and pharmaceutical compositions. 2006-344432 2006173049, 20060131. 2006.

190. Malamas MS, Erdei JJ, Gunawan IS, Barnes KD, Johnson MR, Hui Y.Preparation of diphenylimidazopyrimidine and -imidazole amines as selective inhibitors of beta-secretase for use against Alzheimer's disease and other disorders. 2005-152925 2005282826, 20050615. 2005.

191. Malamas MS, Erdei JJ, Gunawan IS, Nowak P, Harrison BL.Preparation of 8,8-diphenyl-2,3,4,8-tetrahydroimidazo[1,5-a]pyrimidin-6-amines as beta-secretase inhibitors for the treatment of Alzheimer's disease and related disorders. 2006-US656 2006076284, 20060109. 2006.

192. Zhou P, Bard J, Chopra R, Fan KY, Hu Y, Li Y, Magolda RL, Malamas MS, Pangalos M, Reinhart P, Turner J, Wang Z, Robichaud AJ. Pyridinylaminohydantoins as small molecule BACE1 inhibitors: Exploration of the S3 pocket. In: Abstracts of Papers. 234th ACS National Meeting; 2007, August 19–23; Boston, MA, United States. 2007, MEDI-294.

193. Zhou P, Yan Y, Fobare WF, Malamas M, Solvibile WR, Chopra R, Fan KY, Hu Y, Turner J, Wagner E, Magolda RL, Abougharbia MA, Reinhart P, Pangalos M, Bard J, Robichaud A. Substituted-pyrrole 2-amino-3,5-dihydro-4H-imidazol-4-ones as highly potent BACE1 inhibitors: optimization of the S3 pocket. In: Abstracts of Papers. 235th ACS National Meeting; 2008, April 6–10; New Orleans, LA, United States. 2008, MEDI-259.

194. Malamas MS, Erdei J, Gunawan I, Barnes K, Johnson M, Hui Y, Turner J, Hu Y, Wagner E, Fan K, Olland A, Bard J, Robichaud AJ. Aminoimidazoles as potent and selective human beta-secretase (BACE1) inhibitors. J Med Chem 2009;52(20): 6314–6323.

195. Fobare WF, Malamas M, Robichaud A, Solvibile WR, Zhou P, Quagliato DA, Erdei J, Gunawan I, Yan Y, Andrae PM, Turner J, Wagner E, Hu Y, Fan KY, Aschmies S, Chopra R, Bard J. Substituted-pyrrole 2-amino-3,5-dihydro-4H-imidazol-4-ones as highly potent and selective BACE1 inhibitors. In: Abstracts of Papers, 235th ACS National Meeting; 2008, April 6–10; New Orleans, LA, United States. 2008, MEDI-182.

196. Malamas MS, Robichaud AJ, Porte AM, Morris KM, Solvibile WR, Kim J-I.Preparation of amino-5-[4-(difluoromethoxy)phenyl]-5-phenyl-imidazolone derivatives as inhibitors of beta-secretase. 2008-US3783 2008118379, 20080320. 2008.

197. Malamas MS, Robichaud AJ, Porte AM, Solvibile WR, Morris KM, Antane SA, Kim J-L, McDevitt RE.Amino-5-[substituted-4-(difluoromethoxy)phenyl]-5-phenylimidazolone compounds as beta-secretase inhibitors and their preparation, and use in the treatment of beta-amyloid deposits and neurofibrillary tangles. 2008-US3681 2008115552, 20080320. 2008.

198. Robichaud AJ. The structure based design of potent, selective BACE1 inhibitors as potential disease modifying treatments for Alzheimer's disease. In: International Symposium on Advances in Synthetic and Medicinal Chemistry; 2009, August 26; Kiev, Ukraine. 2009.

199. Barrow JC, Coburn CA, Egbertson MS, McGaughey GB, McWherter MA, Neilson LA, Selnick HG, Stauffer SR, Yang Z-Q, Yang W, Lu W, Fahr B, Rittle KE.Preparation of spiropi-

peridine compounds as beta-secretase inhibitors for the treatment of Alzheimer's disease. 2005-US36752 2006044497, 20051012. 2006.

200. Coburn CA, Egbertson MS, Graham SL, McGaughey GB, Stauffer SR, Rajapakse HA, Nantermet PG, Stachel SR, Yang W, Lu W, Fahr B.Preparation of triazaspirodecenones as beta-secretase inhibitors for the treatment of Alzheimer's disease. 2006-US27594 2007011833, 20060714. 2007.

201. Coburn CA, Egbertson MS, Graham SL, McGaughey GB, Stauffer SR, Yang W, Lu W, Fahr B.Preparation of 1,3,8-triazaspiro[4.5]decane derivatives as beta-secretase inhibitors for treatment of Alzheimer's disease. 2006-US27544 2007011810, 20060714. 2007.

202. Burrows JN, Hellberg S, Hogdin K, Karlstrom S, Kolmodin K, Lindstrom J, Slivo C.Preparation of substituted 2-amino-3,4-dihydro-4*H*-imidazol-4-ones for treating and preventing cognitive impairment, Alzheimer disease, neurodegeneration and dementia. 2008-120608 2008287460, 20080514. 2008.

203. Berg S, Burrows J, Hellberg S, Hoegdin K, Kolmodin K.Substituted 2-aminopyrimidine-4-ones, their preparation, pharmaceutical compositions and their use in the treatment and/or prevention of Abeta-related pathologies. 2006-SE1434 2007073284, 20061218. 2007.

204. Berg S, Burrows J, Chessari G, Congreve MS, Hedstroem J, Hellberg S, Hoegdin K, Kihlstroem J, Kolmodin K, Lindstroem J, Murray C, Patel S.New 2-amino-3,5-dihydro-4*H*-imidazol-4-one derivatives and their use as BACE inhibitors for treatment of cognitive impairment, Alzheimer disease, neurodegeneration and dementia. 2006-SE1317 2007058602, 20061120. 2007.

205. Zhu Z, McKittrick B, Sun Z-Y, Ye YC, Voigt JH, Strickland C, Smith EM, Stamford A, Greenlee WJ, Mazzola RD Jr, Caldwell J, Cumming JN, Wang L, Wu Y, Iserloh U, Liu X, Huang Y, Li G, Pan J, Misiaszek JA, Guo T, Le TXH, Saionz KW, Babu SD, Hunter RC, Morris ML, Gu H, Qian G, Tadesse D, Lai G, Duo J, Qu C, Shao Y. Preparation of imidazolidin-2-imines and their analogs as aspartyl protease inhibitors for treating various diseases. 2008-XA2182 2008103351, 20080220. 2008.

206. Zhu Z, McKittrick B, Sun Z-Y, Ye YC, Voigt JH, Strickland CO, Smith EM, Stamford A, Greenlee WJ, Mazzola RD, Caldwell JP, Cumming JN, Wang L, Wu Y, Iserloh U, Liu X, Guo T, Le TXE, Saionz KW, Babu SD, Hunter RC, Morris ML, Gu H, Qian G, Tadesse D, Huang Y, Li G, Pan J, Misiaszek JA, Lai G, Duo J, Qu C, Shao Y.Preparation of imidazolidin-2-imines and their analogs as aspartyl protease inhibitors for treating various diseases. 2007-710582 2008200445, 20070223. 2008.

207. Iserloh U, Zhu Z, Stamford A, Voigt JH.Macrocyclic heterocyclic aspartyl protease inhibitors. 2006-451064 2006281729, 20060612. 2006.

208. Andreini M, Gabellieri E, Guba W, Marconi G, Narquizian R, Power E, Travagli M, Woltering T, Wostl W.Preparation of 4,4-diphenyl-4,5-dihydro-oxazol-2-ylamine derivatives as beta-secretase inhibitors. 2009-369782 2009209529, 20090212. 2009.

209. Baxter E, Bischoff FP, Boyd R, Braeken M, Coats S, Huang Y, Jordan A, Luo C, Mercken MH, Reynolds CH, Ross TM, Tounge BA, Schulz M, De Winte HLJ, Pieters SMA, Reitz AB.Novel 2-aminoquinazoline derivatives, their preparation and use as inhibitors of beta-secretase for treating Alzheimer's disease and related disorders. 2005-US28191 2006017836, 20050808. 2006.

210. Baxter EW, Conway KA, Kennis L, Bischoff F, Mercken MH, De Winter HL, Reynolds CH, Tounge BA, Luo C, Scott MK, Huang Y, Braeken M, Pieters SMA, Berthelot DJC, Masure S, Bruinzeel WD, Jordan AD, Parker MH, Boyd RE, Qu J, Alexander RS, Brenneman DE, Reitz AB. 2-Amino-3,4-dihydroquinazolines as inhibitors of BACE-1 (beta-site APP cleaving enzyme): use of structure based design to convert a micromolar hit into a nanomolar lead. J Med Chem 2007;50(18): 4261–4264.

211. Reitz AB, Luo C, Huang Y, Ross TM, Baxter EE, Tounge BA, Parker MH, Strobel ED, Reynolds CH.Preparation of 2-amino-3,4-dihydropyrido[3,4-*d*]pyrimidines as inhibitors of beta-secretase (BACE). 2006-US41487 2007050612, 20061024. 2007.

212. Bilcer GM. Pharmacological profile of BACE1 inhibitor CTS-21166 (ASP1702). In: Abstract, International Congress on Alzheimer's Disease; 2008; Chicago IL, USA. 2008.

213. Hey JA. Single dose administration of the B-secretase inhibitor CTS-21166 (ASP1702) reduces plasma AB40 in human subjects. In: Abstract, International Congress on Alzheimer's Disease; 2008; Chicago IL, USA. 2008.

214. Hsu HH. Phase I safety and pharmacokinetic profile of single doses of CTS21166 (ASP1702) in healthy males. In: Abstract, International Congress on Alzheimer's Disease; 2008; Chicago IL, USA. 2008.

215. Koelsch G. BACE1 inhibitor CTS21166 (ASP1702) penetrates brain and reduces AB pathology in a transgenic mouse model of advanced AD. In: Abstract, International Congress on Alzheimer's Disease; 2008; Chicago IL, USA. 2008.

216. Shitaka Y. Oral administration of the BACE1 inhibitor CTS21166 (ASP1702) improves cognition and reduces brain AB in Tg2576 transgenic mice. In: Abstract, International Congress on Alzheimer's Disease; 2008; Chicago IL, USA. 2008.

217. Yu J. Pharmacokinetic/pharmacodynamic analysis of plasma AB40 reduction in Human subjects produced by the B-secretase inhibitor CTS21166 (ASP1702). In: Abstract, International Congress on Alzheimer's Disease; 2008; Chicago IL, USA. 2008.

218. Pando MMM, Rayer, Peillon A, Drouin D, Desire L. An alpha-secretase stimulator drug for cognitive disorders associated with neurodegeneration. In: Alzheimer's and Parkinson's Diseases; Prague; Czech Republic. 2009, ExonHit Therapeutics

219. Stewart WF, Kawas C, Corrada M, Metter EJ. Risk of Alzheimer's disease and duration of NSAID use. Neurology 1997;48(3): 626–632.

220. Anthony JC, Breitner JCS, Zandi PP, Meyer MR, Jurasova I, Norton MC, Stone SV, Burke J, Calvert T, Gau B, Helms M, Khachaturian A, Leslie C, Newman T, Plassman B, Steffens DC, Steinberg M, Tschanz JT, Welsh-Bohmer KA, West N, Wyse B. Reduced prevalence of AD in users of NSAIDs and H2 receptor antagonists: the Cache County Study. Neurology 2000;54(11): 2066–2071.

221. Weggen S, Eriksen JL, Das P, Sagl SA, Wang R, Pletrzik CU, Findlay KA, Smith TE, Murphy MP, Bulter T, Kang DE, Marquez-Sterling N, Golde TE, Koo EH. A subset of NSAIDs lower amyloidogenic Abeta 42 independently of cyclooxygenase activity. Nature (London, UK) 2001;414(6860): 212–216.

222. Kukar TL, Ladd TB, Bann MA, Fraering PC, Narlawar R, Maharvi GM, Healy B, Chapman R, Welzel AT, Price RW, Moore B, Rangachari V, Cusack B, Eriksen J, Jansen-West K, Verbeeck C, Yager D, Eckman C, Ye W, Sagi S, Cottrell BA, Torpey J, Rosenberry TL, Fauq A, Wolfe MS, Schmidt B, Walsh DM, Koo EH, Golde TE. Substrate-targeting gamma-secretase modulators. Nature (London, UK) 2008;453(7197): 925–929.

223. Green RC. Safety and efficacy of tarenflurbil in subjects with mild Alzheimer's disease: results from an 18-month multi-center phase 3 trials. In: 8th International Conference for Alzheimer's and Parkinson's Disease; 2008, March; Chicago, IL. Myriad Pharmaceuticals, Inc.; 2008.

224. Williams M. Progress in Alzheimer's disease drug discovery: an update. Curr Opin Investig Drugs 2009;10(1): 23–34.

225. Cheng S, Comer DD, Mao L, Balow GP, Pleynet D.Aryl compounds and uses in modulating amyloid beta. 2004-US15239 2004110350, 20040514. 2004.

226. Kimura T, Kawano K, Doi E, Kitazawa N, Miyagawa T, Sato N, Kaneko T, Shin K, Ito K, Takaishi M, Sasaki T, Hagiwara H.Preparation of urea compounds for inhibiting amyloid-beta production. 2007-JP60187 2007135969, 20070518. 2007.

227. Madin A, Ridgill MP, Kulagowski JJ.Preparation of arylpiperidinylacetates for the treatment of Alzheimer's disease. 2007-GB50176 2007116228, 20070402. 2007.

228. Wilson F, Reid A, Reader V, Harrison RJ, Sunose M, Hernandez-Perni R, Major J, Boussard C, Smelt K, Taylor J, Le Formal A, Cansfield A, Burckhardt S.Preparation of substituted biphenyl carboxylic acids as gamma-secretase modulators for treatment of Alzheimer's disease. 2006-112938 1849762, 20060421. 2007.

229. Garofalo AW. Patents targeting gamma-secretase inhibition and modulation for the treatment of Alzheimer's disease: 2004–2008. Expert Opin Ther Patents 2008;18(7): 693–703.

230. Vega GL, Weiner MF. Plasma 24S hydroxycholesterol response to statins in Alzheimer's disease patients: effects of gender, CYP46, and ApoE polymorphisms. J. Mol. Neurosci. 2007;33(1): 51–55.

231. Guimaraes HC, Caramelli P. Statins and Alzheimer disease. Int J Atheroscler 2007;2(2): 151–155.

232. Martins IJ, Hone E, Foster JK, Suenram-Lea SI, Gnjec A, Fuller SJ, Nolan D, Gandy SE, Martins RN. Apolipoprotein E, cholesterol metabolism, diabetes, and the convergence of risk factors for Alzheimer's disease and cardiovascular disease. Mol Psychiatry 2006;11(8): 721–736.

233. Kivipelto M, Helkala EL, Laakso MP, Hanninen T, Hallikainen M, Alhainen K, Soininen H, Tuomilehto J, Nissinen A. Midlife vascular risk factors and Alzheimer's disease in later life: longitudinal, population based study. BMJ 2001;322(7300): 1447–1451.

234. Wang H, Lynch JR, Song P, Yang H-J, Yates RB, Mace B, Warner DS, Guyton JR, Laskowitz DT. Simvastatin and atorvastatin improve behavioral outcome, reduce hippocampal degeneration, and improve cerebral blood flow after experimental traumatic brain injury. Exp Neurol 2007;206(1): 59–69.

235. Riekse RG, Li G, Petrie EC, Leverenz JB, Vavrek D, Vuletic S, Albers JJ, Montine TJ, Lee VMY, Lee M, Seubert P, Galasko D, Schellenberg GD, Hazzard WR, Peskind ER. Effect of statins on Alzheimer's disease biomarkers in cerebrospinal fluid. J Alzheimer's Dis 2006;10(4): 399–406.

236. Kandiah N, Feldman HH. Therapeutic potential of statins in Alzheimer's disease. J Neurol Sci 2009;283(1–2): 230–234.

237. Rebollo A, Pou J, Alegret M. Cholesterol lowering and beyond: role of statins in Alzheimer's disease. Aging Health 2008;4(2): 171–180.

238. Wolozin B. Cholesterol, statins and Alzheimer's disease: past, present and future. Res Prog Alzheimer's Dis Dementia 2007;2:123–135.

239. Fassbender K, Simons M, Bergmann C, Stroick M, Lutjohann D, Keller P, Runz H, Kuhl S, Bertsch T, Von Bergmann K, Hennerici M, Beyreuther K, Hartmann T. Simvastatin strongly reduces levels of Alzheimer's disease beta-amyloid peptides Abeta 42 and Abeta 40 in vitro and in vivo. Proc Natl Acad Sci USA 2001;98(10): 5856–5861.

240. Petanceska SS, De Rosa S, Olm V, Diaz N, Sharma A, Thomas-Bryant T, Duff K, Pappolla M, Refolo LM. Statin therapy for Alzheimer's disease. Will it work? J Mol Neurosci 2002;19 (1/2): 155–161.

241. Jones RW, Kivipelto M, Feldman H, Sparks L, Doody R, Waters DD, Hey-Hadavi J, Breazna A, Schindler RJ, Ramos H. The atorvastatin/donepezil in Alzheimer's disease study (LEADe): design and baseline characteristics. Alzheimer's Dementia 2008;4(2): 145–153.

242. Botti RE, Triscari J, Pan HY, Zayat J. Concentrations of pravastatin and lovastatin in cerebrospinal fluid in healthy subjects. Clin Neuropharmacol 1991;14(3): 256–261.

243. Saheki A, Terasaki T, Tamai I, Tsuji A. In vivo and in vitro blood–brain barrier transport of 3-hydroxy-3-methylglutaryl coenzyme A (HMG-CoA) reductase inhibitors. Pharm. Res. 1994;11(2): 305–311.

244. Green RC, McNagny SE, Jayakumar P, Cupples LA, Benke K, Farrer LA. Statin use and the risk of Alzheimer's disease: the MIRAGE study. Alzheimer's Dementia 2006;2(2): 96–103.

245. Christensen Daniel D. Changing the course of Alzheimer's disease: anti-amyloid disease-modifying treatments on the horizon. Prim Care Companion J Clin Psychiatry 2007;9(1): 32–41.

246. Christensen Daniel D. Alzheimer's disease: progress in the development of anti-amyloid disease-modifying therapies. CNS Spectr 2007;12 (2): 113–116, 119–123.

247. Necula M, Breydo L, Milton S, Kayed R, van der Veer WE, Tone P, Glabe CG. Methylene Blue inhibits amyloid Ab oligomerization by promoting fibrillization. Biochemistry 2007;46 (30): 8850–8860.

248. Luhrs T, Ritter C, Adrian M, Riek-Loher D, Bohrmann B, Dobeli H, Schubert D, Riek R. 3D structure of Alzheimer's amyloid-beta (1-42) fibrils. Proc Natl Acad Sci USA 2005;102(48): 17342–17347.

249. Gazit E. Global analysis of tandem aromatic octapeptide repeats: the significance of the aromatic-glycine motif. Bioinformatics 2002;18(6): 880–883.

250. Gazit E. A possible role for pi-stacking in the self-assembly of amyloid fibrils. FASEB J 2002;16(1): 77–83.

251. Etienne MA, Aucoin JP, Fu Y, McCarley RL, Hammer RP. Stoichiometric inhibition of amyloid beta-protein aggregation with peptides containing alternating alpha, alpha-disubstituted amino acids. J Am Chem Soc 2006;128 (11): 3522–3523.

252. Smith TJ, Stains CI, Meyer SC, Ghosh I. Inhibition of β-amyloid fibrillization by directed evolution of a β-sheet presenting miniature protein. J Am Chem Soc 2006;128(45): 14456–14457.

253. Balbach JJ, Ishii Y, Antzutkin ON, Leapman RD, Rizzo NW, Dyda F, Reed J, Tycko R. Amyloid fibril formation by Abeta 16-22, a seven-residue fragment of the Alzheimer's beta-amyloid peptide, and structural characterization by solid state NMR. Biochemistry 2000;39(45): 13748–13759.

254. Tenidis K, Waldner M, Bernhagen J, Fischle W, Bergmann M, Weber M, Merkle M-L, Voelter W, Brunner H, Kapurniotu A. Identification of a penta- and hexapeptide of islet amyloid polypeptide (IAPP) with amyloidogenic and cytotoxic properties. J Mol Biol 2000;295(4): 1055–1071.

255. Findeis MA. Peptide inhibitors of beta amyloid aggregation. Curr Top Med Chem (Hilversum, Netherlands) 2002;2(4): 417–423.
256. Findeis MA, Gefter ML, Musso G, Signer ER, Wakefield J, Molineaux S, Chin J, Lee J-J, Kelley M, Komar-Panicucci S, Arico-Muendel CC, Phillips K, Hayward NJ.D-Amino acid-containing peptide modulators of beta-amyloid peptide aggregation. 96-703675 6303567, 19960827. 2001.
257. Osada Y, Hashimoto T, Nishimura A, Matsuo Y, Wakabayashi T, Iwatsubo T. CLAC binds to amyloid beta peptides through the positively charged amino acid cluster within the collagenous domain 1 and inhibits formation of amyloid fibrils. J Biol Chem 2005;280(9): 8596–8605.
258. Geerts H. NC-531 (Neurochem). Curr Opin Invest Drugs (Thomson Sci) 2004;5(1): 95–100.
259. Bayes M, Rabasseda X, Prous JR. Gateways to clinical trials. Methods Find Exp Clin Pharmacol 2006;28(8): 533–591.
260. Greenberg SM, Rosand J, Schneider AT, Pettigrew LC, Gandy SE, Rovner B, Fitzsimmons B-F, Smith EE, Edip Gurol M, Schwab K, Laurin J, Garceau D. A Phase 2 study of tramiprosate for cerebral amyloid angiopathy. Alzheimer Dis Assoc Disord 2006;20(4): 269–274.
261. Doraiswamy PM. Non-cholinergic strategies for treating and preventing Alzheimer's disease. CNS Drugs 2002;16(12): 811–824.
262. Ritchie CW, Bush AI, Mackinnon A, Macfarlane S, Mastwyk M, MacGregor L, Kiers L, Cherny R, Li Q-X, Tammer A, Carrington D, Mavros C, Volitakis I, Xilinas M, Ames D, Davis S, Beyreuther K, Tanzi Rudolph E, Masters Colin L. Metal–protein attenuation with iodochlorhydroxyquin (clioquinol) targeting Abeta amyloid deposition and toxicity in Alzheimer disease: a pilot phase 2 clinical trial. Arch Neurol 2003;60(12): 1685–1691.
263. Kharkar PS, Dutta AK. Metal–protein attenuating compounds (MPACs): an emerging approach for the treatment of neurodegenerative disorders. Curr Bioact Compd 2008;4(2): 57–67.
264. Lannfelt L, Blennow K, Zetterberg H, Batsman S, Ames D, Harrison J, Maters CL, Targum S, Bush AI, Murdoch R, Wilson J, Ritchie CW. Safety, efficacy, and biomarker findings of PBT2 in targeting Abeta as a modifying therapy for Alzheimer's disease: a phase IIa, double-blind, randomized, placebo-controlled trial. Lancet Neurol 2008;7 (9): 779–786.
265. Kim H, Park B-S, Lee K-G, Choi CY, Jang SS, Kim Y-H, Lee S-E. Effects of naturally occurring compounds on fibril formation and oxidative stress of beta-amyloid. J Agric Food Chem 2005;53(22): 8537–8541.
266. Napryeyenko O, Sonnik G, Tartakovsky I. Efficacy and tolerability of Ginkgo biloba extract EGb 761 by type of dementia: Analyses of a randomised controlled trial. J Neurol Sci 2009;283(1–2): 224–229.
267. Wu Y, Wu Z, Butko P, Christen Y, Lambert MP, Klein WL, Link CD, Luo Y. Amyloid-beta-induced pathological behaviors are suppressed by Ginkgo biloba extract EGb 761 and ginkgolides in transgenic *Caenorhabditis elegans*. J Neurosci 2006;26(50): 13102–13113.
268. Schneider LS, DeKosky ST, Farlow MR, Tariot PN, Hoerr R, Kieser M. A randomized, double-blind, placebo-controlled trial of two doses of ginkgo biloba extract in dementia of the Alzheimer's type. Curr Alzheimer Res 2005;2(5): 541–551.
269. Bendheim PE, Poeggeler B, Neria E, Ziv V, Pappolla MA, Chain DG. Development of indole-3-propionic acid (OXIGON) for Alzheimer's disease. J Mol Neurosci 2002;19 (1/2): 213–217.
270. Townsend M, Cleary JP, Mehta T, Hofmeister J, Lesne S, O'Hare E, Walsh DM, Selkoe DJ. Orally available compound prevents deficits in memory caused by the Alzheimer amyloid-beta oligomers. Ann Neurol 2006;60(6): 668–676.
271. Oz M, Lorke DE, Petroianu GA. Methylene blue and Alzheimer's disease. Biochem Pharmacol 2009;78(8): 927–932.
272. Ono K, Hasegawa K, Naiki H, Yamada M. Curcumin has potent anti-amyloidogenic effects for Alzheimer's beta-amyloid fibrils *in vitro*. J Neurosci Res 2004;75(6): 742–750.
273. Ono K, Yoshiike Y, Takashima A, Hasegawa K, Naiki H, Yamada M. Potent anti-amyloidogenic and fibril-destabilizing effects of polyphenols *in vitro*: implications for the prevention and therapeutics of Alzheimer's disease. J Neurochem 2003;87(1): 172–181.
274. Hirohata M, Hasegawa K, Tsutsumi-Yasuhara S, Ohhashi Y, Ookoshi T, Ono K, Yamada M, Naiki H. The anti-amyloidogenic effect is exerted against Alzheimer's beta-amyloid fibrils *in vitro* by preferential and reversible binding of flavonoids to the amyloid fibril structure. Biochemistry 2007;46(7): 1888–1899.

275. Snow A, Exebryl-1: A novel small-molecule drug that markedly reduces amyloid plaque load and improves memory, enters human clinical trials. Alzheimer's Dementia 2008;4(4S): T463.

276. Snow A. Phase 1A human clinical trial results with the small molecule exebryl-1 for the treatment of mild-to-moderate Alzheimer's disease. In: 9th International Conference on Alzheimer's and Parkinson's Disease; 2009, Prague, Czech Republic. 2009.

277. Castillo G, Snow AD.Blended compositions for treatment of Alzheimer's disease and other amyloidoses. 99-US19721 2000012102, 19990830. 2000.

278. Castillo G, Choi PY, Nguyen B, Snow AD. Methods of isolating amyloid-inhibiting compounds and use of compounds isolated from *Uncaria tomentosa* and related plants. 2001-US51131 2002042429, 20011102. 2002.

279. Mjalli AMM, Andrews R, Wysong C.Preparation of aromatic carboxamides as modulators of receptor for advanced glycated end products (RAGE). 2002-US6707 2002070473, 20020305. 2002.

280. Mjalli AMM, Andrews RC, Gopalaswamy R, Hari A, Avor K, Qabaja G, Guo X-C, Gupta S, Jones DR, Chen X.Preparation of imidazole and benzimidazole derivatives that inhibit the interaction of ligands with RAGE. 2003-US6749 2003075921, 20030305. 2003.

281. Mjalli AMM, Andrews RC, Shen JM, Rothlein R.RAGE antagonists as agents to reverse amyloidosis and associated diseases. 2004-US16104 2005000295, 20040520. 2005.

282. Origlia N, Capsoni S, Cattaneo A, Fang F, Arancio O, Yan SD, Domenici L. Abeta-dependent inhibition of LTP in different intracortical circuits of the visual cortex: the role of RAGE. J Alzheimer's Dis 2009;17(1): 59–68.

283. Origlia N, Arancio O, Domenici L, Yan SSD. MAPK, beta-amyloid and synaptic dysfunction: the role of RAGE. Expert Rev Neurother 2009;9(11): 1635–1645.

284. Bierhaus A, Yan SD, Bierhaus A, Nawroth PP, Stern DM. RAGE and Alzheimer's disease: a progression factor for amyloid-beta-induced cellular perturbation? J. Alzheimer's Dis 2009;16(4): 833–843.

285. Deane R, Bell RD, Sagare A, Zlokovic BV. Clearance of amyloid-beta peptide across the blood–brain barrier: implication for therapies in Alzheimer's disease. CNS Neurol Disorders Drug Targets 2009;8(1): 16–30.

286. Zlokovic BV, Deane R, Miller BL.Tertiary amide inhibition of amyloid-beta peptide/RAGE receptor interaction at the blood–brain barrier, and therapeutic uses. 2007-US2220 2007089616, 20070126. 2007.

287. Imbimbo BP. An update on the efficacy of non-steroidal anti-inflammatory drugs in Alzheimer's disease. Expert Opin Investig Drugs 2009;18(8): 1147–1168.

288. Walker D, Lue L-F. Anti-inflammatory and immune therapy for Alzheimer's disease: current status and future directions. Curr Neuropharmacol 2007;5(4): 232–243.

289. Beach TG, Wilson JR, Sue LI, Newell A, Poston M, Cisneros R, Pandya Y, Esh C, Connor DJ, Sabbagh M, Walker DG, Roher AE. Circle of Willis atherosclerosis: association with Alzheimer's disease, neuritic plaques and neurofibrillary tangles. Acta Neuropathol 2007;113(1): 13–21.

290. Choi SH, Lee DY, Chung ES, Hong YB, Kim SU, Jin BK. Inhibition of thrombin-induced microglial activation and NADPH oxidase by minocycline protects dopaminergic neurons in the substantia nigra *in vivo*. J Neurochem. 2005;95(6): 1755–1765.

291. Roher AE, Esh C, Kokjohn TA, Kalback W, Luehrs DC, Seward JD, Sue LI, Beach TG. Circle of Willis atherosclerosis is a risk factor for sporadic Alzheimer's disease. Arterioscler Thromb Vasc Biol 2003;23(11): 2055–2062.

292. Liu H, Qiu D, Lei Q, Li Z. Mechanism of curcumin in Alzheimer's disease treatment. Zhonghua Shenjingke Zazhi 2009;42(2): 135–138.

293. Schenk D, Barbour R, Dunn W, Gordon G, Grajeda H, Guido T, Hu K, Huang J, Johnson-Wood K, Khan K, Kholodenko D, Lee M, Liao Z, Lieberburg I, Motter R, Mutter L, Soriano F, Shopp G, Vasquez N, Vandevert C, Walker S, Wogulis M, Yednock T, Games D, Seubert P. Immunization with amyloid-beta attenuates Alzheimer-disease-like pathology in the PDAPP mouse. Nature 1999;400(6740): 173–177.

294. Bard F, Cannon C, Barbour R, Burke RL, Games D, Grajeda H, Guido T, Hu K, Huang J, Johnson-Wood K, Khan K, Kholodenko D, Lee M, Lieberburg I, Motter R, Nguyen M, Soriano F, Vasquez N, Weiss K, Welch B, Seubert P, Schenk D, Yednock T. Peripherally administered antibodies against amyloid beta-peptide enter the central nervous system and reduce pathology in a mouse model of Alzheimer disease. Nat Med 2000;6(8): 916–919.

295. Bard F, Barbour R, Cannon C, Carretto R, Fox M, Games D, Guido T, Hoenow K, Hu K, Johnson-Wood K, Khan K, Kholodenko D, Lee C, Lee M, Motter R, Nguyen M, Reed A, Schenk D, Tang P, Vasquez N, Seubert P, Yednock T. Epitope and isotype specificities of antibodies to beta-amyloid peptide for protection against Alzheimer's disease-like neuropathology [erratum appears in Proc Natl Acad Sci USA 2004 Aug 3;101(3):11526]. Proc Natl Acad Sci USA 2003;100(4): 2023–2028.

296. Lemere CA, Maron R, Selkoe DJ, Weiner HL. Nasal vaccination with beta-amyloid peptide for the treatment of Alzheimer's disease. DNA Cell Biol 2001;20(11): 705–711.

297. Lemere CA, Spooner ET, Leverone JF, Mori C, Clements JD. Intranasal immunotherapy for the treatment of Alzheimer's disease: *Escherichia coli* LT and LT(R192G) as mucosal adjuvants. Neurobiol Aging 2002;23(6): 991–1000.

298. Leverone JF, Spooner ET, Lehman HK, Clements JD, Lemere CA. Abeta1-15 is less immunogenic than Abeta1-40/42 for intranasal immunization of wild-type mice but may be effective for "boosting". Vaccine 2003;21 (17–18): 2197–2206.

299. Lemere CA, Beierschmitt A, Iglesias M, Spooner ET, Bloom JK, Leverone JF, Zheng JB, Seabrook TJ, Louard D, Li D, Selkoe DJ, Palmour RM, Ervin FR. Alzheimer's disease abeta vaccine reduces central nervous system abeta levels in a non-human primate, the Caribbean vervet. Am J Pathol 2004;165(1): 283–297.

300. Lemere CA, Maier M, Jiang L, Peng Y, Seabrook TJ. Amyloid-beta immunotherapy for the prevention and treatment of Alzheimer disease: lessons from mice, monkeys, and humans. Rejuvenation Res 2006;9(1): 77–84.

301. Lombardo JA, Stern EA, McLellan ME, Kajdasz ST, Hickey GA, Bacskai BJ, Hyman BT. Amyloid-beta antibody treatment leads to rapid normalization of plaque-induced neuritic alterations. J Neurosci 2003;23(34): 10879–10883.

302. Brendza RP, Bacskai BJ, Cirrito JR, Simmons KA, Skoch JM, Klunk WE, Mathis CA, Bales KR, Paul SM, Hyman BT, Holtzman DM. Anti-Abeta antibody treatment promotes the rapid recovery of amyloid-associated neuritic dystrophy in PDAPP transgenic mice. J Clin Investig 2005;115(2): 428–433.

303. Oddo S, Caccamo A, Shepherd JD, Murphy MP, Golde TE, Kayed R, Metherate R, Mattson MP, Akbari Y, LaFerla FM. Triple-transgenic model of Alzheimer's disease with plaques and tangles: intracellular Abeta and synaptic dysfunction. Neuron 2003;39(3): 409–421.

304. Buttini M, Masliah E, Barbour R, Grajeda H, Motter R, Johnson-Wood K, Khan K, Seubert P, Freedman S, Schenk D, Games D. Beta-amyloid immunotherapy prevents synaptic degeneration in a mouse model of Alzheimer's disease. J Neurosci 2005;25(40): 9096–9101.

305. Oddo S, Billings L, Kesslak JP, Cribbs DH, LaFerla FM. Abeta immunotherapy leads to clearance of early, but not late, hyperphosphorylated tau aggregates via the proteasome [see comment]. Neuron 2004;43(3): 321–332.

306. Hsia AY, Masliah E, McConlogue L, Yu GQ, Tatsuno G, Hu K, Kholodenko D, Malenka RC, Nicoll RA, Mucke L. Plaque-independent disruption of neural circuits in Alzheimer's disease mouse models. Proc Natl Acad Sci USA 1999;96(6): 3228–3233.

307. Larson J, Lynch G, Games D, Seubert P. Alterations in synaptic transmission and long-term potentiation in hippocampal slices from young and aged PDAPP mice. Brain Res 1999;840(1–2): 23–35.

308. Moechars D, Dewachter I, Lorent K, Reverse D, Baekelandt V, Naidu A, Tesseur I, Spittaels K, Haute CV, Checler F, Godaux E, Cordell B, Van Leuven F. Early phenotypic changes in transgenic mice that overexpress different mutants of amyloid precursor protein in brain. J Biol Chem 1999;274(10): 6483–6492.

309. Mucke L, Masliah E, Yu GQ, Mallory M, Rockenstein EM, Tatsuno G, Hu K, Kholodenko D, Johnson-Wood K, McConlogue L. High-level neuronal expression of abeta 1-42 in wild-type human amyloid protein precursor transgenic mice: synaptotoxicity without plaque formation. J Neurosci 2000;20(11): 4050–4058.

310. Westerman MA, Cooper-Blacketer D, Mariash A, Kotilinek L, Kawarabayashi T, Younkin LH, Carlson GA, Younkin SG, Ashe KH. The relationship between Abeta and memory in the Tg2576 mouse model of Alzheimer's disease. J Neurosci 2002;22(5): 1858–1867.

311. Chapman PF, White GL, Jones MW, Cooper-Blacketer D, Marshall VJ, Irizarry M, Younkin L, Good MA, Bliss TV, Hyman BT, Younkin SG, Hsiao KK. Impaired synaptic plasticity and learning in aged amyloid precursor protein transgenic mice. Nat Neurosci 1999;2(3): 271–276.

312. Corcoran KA, Lu Y, Turner RS, Maren S. Overexpression of hAPPswe impairs rewarded alternation and contextual fear conditioning in a transgenic mouse model of Alzheimer's disease. Learning Memory 2002;9(5): 243–252.
313. Dineley KT. Beta-amyloid peptide–nicotinic acetylcholine receptor interaction: the two faces of health and disease. Front Biosci 2007;12:5030–5038.
314. Hsiao K, Chapman P, Nilsen S, Eckman C, Harigaya Y, Younkin S, Yang F, Cole G. Correlative memory deficits, Abeta elevation, and amyloid plaques in transgenic mice [see comment]. Science 1996;274(5284): 99–102.
315. Janus C, Pearson J, McLaurin J, Mathews PM, Jiang Y, Schmidt SD, Chishti MA, Horne P, Heslin D, French J, Mount HT, Nixon RA, Mercken M, Bergeron C, Fraser PE, St George-Hyslop P, Westaway D. A beta peptide immunization reduces behavioural impairment and plaques in a model of Alzheimer's disease [see comment]. Nature 2000;408 (6815): 979–982.
316. Morgan D, Diamond DM, Gottschall PE, Ugen KE, Dickey C, Hardy J, Duff K, Jantzen P, DiCarlo G, Wilcock D, Connor K, Hatcher J, Hope C, Gordon M, Arendash GW. A beta peptide vaccination prevents memory loss in an animal model of Alzheimer's disease [see comment] [erratum appears in Nature 2001 Aug 9;412(6847):660]. Nature 2000;408(6815): 982–985.
317. Chen G, Chen KS, Knox J, Inglis J, Bernard A, Martin SJ, Justice A, McConlogue L, Games D, Freedman SB, Morris RG. A learning deficit related to age and beta-amyloid plaques in a mouse model of Alzheimer's disease. Nature 2000;408(6815): 975–979.
318. Walsh DM, Klyubin I, Fadeeva JV, Cullen WK, Anwyl R, Wolfe MS, Rowan MJ, Selkoe DJ. Naturally secreted oligomers of amyloid beta protein potently inhibit hippocampal long-term potentiation in vivo [see comment]. Nature 2002;416(6880): 535–539.
319. DeMattos RB, Bales KR, Cummins DJ, Dodart JC, Paul SM, Holtzman DM. Peripheral anti-A beta antibody alters CNS and plasma A beta clearance and decreases brain A beta burden in a mouse model of Alzheimer's disease [see comment]. Proc Natl Acad Sci USA 2001;98 (15): 8850–8855.
320. Dodart JC, Bales KR, Gannon KS, Greene SJ, DeMattos RB, Mathis C, DeLong CA, Wu S, Wu X, Holtzman DM, Paul SM. Immunization reverses memory deficits without reducing brain Abeta burden in Alzheimer's disease model. Nat Neurosci 2002;5(5): 452–457.
321. Kotilinek LA, Bacskai B, Westerman M, Kawarabayashi T, Younkin L, Hyman BT, Younkin S, Ashe KH. Reversible memory loss in a mouse transgenic model of Alzheimer's disease. J Neurosci 2002;22(15): 6331–6335.
322. Klyubin I, Walsh DM, Lemere CA, Cullen WK, Shankar GM, Betts V, Spooner ET, Jiang L, Anwyl R, Selkoe DJ, Rowan MJ. Amyloid beta protein immunotherapy neutralizes Abeta oligomers that disrupt synaptic plasticity in vivo. Nat Med 2005;11(5): 556–561.
323. Hartman RE, Izumi Y, Bales KR, Paul SM, Wozniak DF, Holtzman DM. Treatment with an amyloid-beta antibody ameliorates plaque load, learning deficits, and hippocampal long-term potentiation in a mouse model of Alzheimer's disease [see comment]. J Neurosci 2005;25(26): 6213–6220.
324. Wilcock DM, Munireddy SK, Rosenthal A, Ugen KE, Gordon MN, Morgan D. Microglial activation facilitates Abeta plaque removal following intracranial anti-Abeta antibody administration. Neurobiol Dis 2004;15(1): 11–20.
325. Bacskai BJ, Kajdasz ST, Christie RH, Carter C, Games D, Seubert P, Schenk D, Hyman BT. Imaging of amyloid-beta deposits in brains of living mice permits direct observation of clearance of plaques with immunotherapy. Nat Med 2001;7(3): 369–372.
326. Solomon B, Koppel R, Frankel D, Hanan-Aharon E. Disaggregation of Alzheimer beta-amyloid by site-directed mAb. Proc Natl Acad Sci USA 1997;94(8): 4109–4112.
327. Solomon B, Koppel R, Hanan E, Katzav T. Monoclonal antibodies inhibit in vitro fibrillar aggregation of the Alzheimer beta-amyloid peptide. Proc Natl Acad Sci USA 1996;93(1): 452–455.
328. Frenkel D, Balass M, Solomon B. N-terminal EFRH sequence of Alzheimer's beta-amyloid peptide represents the epitope of its anti-aggregating antibodies. J Neuroimmunol 1998;88(1–2): 85–90.
329. Frenkel D, Katz O, Solomon B. Immunization against Alzheimer's beta-amyloid plaques via EFRH phage administration. Proc Natl Acad Sci USA 2000;97(21): 11455–11459.
330. McLaurin J, Cecal R, Kierstead ME, Tian X, Phinney AL, Manea M, French JE, Lambermon MH, Darabie AA, Brown ME, Janus C, Chishti MA, Horne P, Westaway D, Fraser PE, Mount HT, Przybylski M, St George-Hy-

slop P. Therapeutically effective antibodies against amyloid-beta peptide target amyloid-beta residues 4-10 and inhibit cytotoxicity and fibrillogenesis. Nat Med 2002;8(11): 1263–1269.

331. Bacskai BJ, Kajdasz ST, McLellan ME, Games D, Seubert P, Schenk D, Hyman BT. Non-Fc-mediated mechanisms are involved in clearance of amyloid-beta *in vivo* by immunotherapy. J Neurosci 2002;22(18): 7873–7878.

332. Das P, Howard V, Loosbrock N, Dickson D, Murphy MP, Golde TE. Amyloid-beta immunization effectively reduces amyloid deposition in FcRgamma−/− knock-out mice. J Neurosci 2003;23(24): 8532–8538.

333. Wilcock DM, DiCarlo G, Henderson D, Jackson J, Clarke K, Ugen KE, Gordon MN, Morgan D. Intracranially administered anti-Abeta antibodies reduce beta-amyloid deposition by mechanisms both independent of and associated with microglial activation. J Neurosci 2003;23(9): 3745–3751.

334. Seubert P, Vigo-Pelfrey C, Esch F, Lee M, Dovey H, Davis D, Sinha S, Schlossmacher M, Whaley J, Swindlehurst C, et al. Isolation and quantification of soluble Alzheimer's beta-peptide from biological fluids. Nature 1992;359(6393): 325–327.

335. DeMattos RB, Bales KR, Cummins DJ, Paul SM, Holtzman DM. Brain to plasma amyloid-beta efflux: a measure of brain amyloid burden in a mouse model of Alzheimer's disease. Science 2002;295(5563): 2264–2267.

336. DeMattos RB, Bales KR, Parsadanian M, O'Dell MA, Foss EM, Paul SM, Holtzman DM. Plaque-associated disruption of CSF and plasma amyloid-beta (Abeta) equilibrium in a mouse model of Alzheimer's disease. J Neurochem 2002;81(2): 229–236.

337. Lemere CA, Spooner ET, LaFrancois J, Malester B, Mori C, Leverone JF, Matsuoka Y, Taylor JW, DeMattos RB, Holtzman DM, Clements JD, Selkoe DJ, Duff KE. Evidence for peripheral clearance of cerebral Abeta protein following chronic, active Abeta immunization in PSAPP mice. Neurobiol Dis 2003;14(1): 10–18.

338. Deane R, Sagare A, Hamm K, Parisi M, LaRue B, Guo H, Wu Z, Holtzman DM, Zlokovic BV. IgG-assisted age-dependent clearance of Alzheimer's amyloid beta peptide by the blood–brain barrier neonatal Fc receptor. J Neurosci 2005;25(50): 11495–11503.

339. Bayer AJ, Bullock R, Jones RW, Wilkinson D, Paterson KR, Jenkins L, Millais SB, Donoghue S. Evaluation of the safety and immunogenicity of synthetic Abeta42 (AN1792) in patients with AD [see comment]. Neurology 2005;64(1): 94–101.

340. Gilman S, Koller M, Black RS, Jenkins L, Griffith SG, Fox NC, Eisner L, Kirby L, Rovira MB, Forette F, Orgogozo JM, Team ANS. Clinical effects of Abeta immunization (AN1792) in patients with AD in an interrupted trial. Neurology 2005;64(9): 1553–1562.

341. Orgogozo JM, Gilman S, Dartigues JF, Laurent B, Puel M, Kirby LC, Jouanny P, Dubois B, Eisner L, Flitman S, Michel BF, Boada M, Frank A, Hock C. Subacute meningoencephalitis in a subset of patients with AD after Abeta42 immunization. Neurology 2003;61 (1): 46–54.

342. Hock C, Konietzko U, Streffer JR, Tracy J, Signorell A, Muller-Tillmanns B, Lemke U, Henke K, Moritz E, Garcia E, Wollmer MA, Umbricht D, de Quervain DJ, Hofmann M, Maddalena A, Papassotiropoulos A, Nitsch RM. Antibodies against beta-amyloid slow cognitive decline in Alzheimer's disease. Neuron 2003;38(4): 547–554.

343. Holmes C, Boche D, Wilkinson D, Yadegarfar G, Hopkins V, Bayer A, Jones RW, Bullock R, Love S, Neal JW, Zotova E, Nicoll JA. Long-term effects of Abeta42 immunisation in Alzheimer's disease: follow-up of a randomised, placebo-controlled phase I trial [see comment]. Lancet 2008;372(9634): 216–223.

344. Vellas B, Black R, Thal LJ, Fox NC, Daniels M, McLennan G, Tompkins C, Leibman C, Pomfret M, Grundman M, Team ANS. Long-term follow-up of patients immunized with AN1792: reduced functional decline in antibody responders. Curr Alzheimer Res 2009;6(2): 144–151.

345. Fox NC, Black RS, Gilman S, Rossor MN, Griffith SG, Jenkins L, Koller M, Study AN. Effects of Abeta immunization (AN1792) on MRI measures of cerebral volume in Alzheimer disease. Neurology 2005;64(9): 1563–1572.

346. Bussiere T, Bard F, Barbour R, Grajeda H, Guido T, Khan K, Schenk D, Games D, Seubert P, Buttini M. Morphological characterization of Thioflavin-S-positive amyloid plaques in transgenic Alzheimer mice and effect of passive Abeta immunotherapy on their clearance. Am J Pathol 2004;165(3): 987–995.

347. Ferrer I, Boada Rovira M, Sanchez Guerra ML, Rey MJ, Costa-Jussa F. Neuropathology and pathogenesis of encephalitis following amyloid-beta immunization in Alzheimer's disease. Brain Pathol 2004;14(1): 11–20.

348. Masliah E, Hansen L, Adame A, Crews L, Bard F, Lee C, Seubert P, Games D, Kirby L, Schenk D. Abeta vaccination effects on plaque pathology in the absence of encephalitis in Alzheimer disease. Neurology 2005;64(1): 129–131.

349. Nicoll JA, Wilkinson D, Holmes C, Steart P, Markham H, Weller RO. Neuropathology of human Alzheimer disease after immunization with amyloid-beta peptide: a case report [see comment]. Nat Med 2003;9(4): 448–452.

350. Glaser R, Kiecolt-Glaser JK, Malarkey WB, Sheridan JF. The influence of psychological stress on the immune response to vaccines. Ann NY Acad Sci 1998;840:649–655.

351. Monsonego A, Zota V, Karni A, Krieger JI, Bar-Or A, Bitan G, Budson AE, Sperling R, Selkoe DJ, Weiner HL. Increased T cell reactivity to amyloid beta protein in older humans and patients with Alzheimer disease [see comment]. J Clin Investig 2003;112(3): 415–422.

352. Cribbs DH, Ghochikyan A, Vasilevko V, Tran M, Petrushina I, Sadzikava N, Babikyan D, Kesslak P, Kieber-Emmons T, Cotman CW, Agadjanyan MG. Adjuvant-dependent modulation of Th1 and Th2 responses to immunization with beta-amyloid. Int Immunol 2003;15:505–514.

353. Lee M, Bard F, Johnson-Wood K, Lee C, Hu K, Griffith SG, Black RS, Schenk D, Seubert P. Abeta42 immunization in Alzheimer's disease generates Abeta N-terminal antibodies. Ann Neurol 2005;58(3): 430–435.

354. Li Q, Cao C, Chackerian B, Schiller J, Gordon M, Ugen KE, Morgan D. Overcoming antigen masking of anti-amyloidbeta antibodies reveals breaking of B cell tolerance by virus-like particles in amyloidbeta immunized amyloid precursor protein transgenic mice. BMC Neurosci 2004;5:21.

355. Agadjanyan MG, Ghochikyan A, Petrushina I, Vasilevko V, Movsesyan N, Mkrtichyan M, Saing T, Cribbs DH. Prototype Alzheimer's disease vaccine using the immunodominant B cell epitope from beta-amyloid and promiscuous T cell epitope pan HLA DR-binding peptide. J Immunol 2005;174(3): 1580–1586.

356. Solomon B. Generation of anti-beta-amyloid antibodies via phage display technology towards Alzheimer's disease vaccination. Vaccine 2005;23:2327–2330.

357. Zurbriggen R, Amacker M, Kammer AR, Westerfeld N, Borghgraef P, Van Leuven F, Van der Auwera I, Wera S. Virosome-based active immunization targets soluble amyloid species rather than plaques in a transgenic mouse model of Alzheimer's disease. J Mol Neurosci 2005;27:157–166.

358. Weiner HL, Lemere CA, Maron R, Spooner ET, Grenfell TJ, Mori C, Issazadeh S, Hancock WW, Selkoe DJ. Nasal administration of amyloid-beta peptide decreases cerebral amyloid burden in a mouse model of Alzheimer's disease. Ann Neurol 2000;48(4): 567–579.

359. Lemere CA, Maron R, Spooner ET, Grenfell TJ, Mori C, Desai R, Hancock WW, Weiner HL, Selkoe DJ. Nasal A beta treatment induces anti-A beta antibody production and decreases cerebral amyloid burden in PD-APP mice. Ann NY Acad Sci 2000;920:328–331.

360. Lavie V, Becker M, Cohen-Kupiec R, Yacoby I, Koppel R, Wedenig M, Hutter-Paier B, Solomon B. EFRH-phage immunization of Alzheimer's disease animal model improves behavioral performance in Morris water maze trials. J Mol Neurosci 2004;24(1): 105–113.

361. Zhang J, Wu X, Qin C, Qi J, Ma S, Zhang H, Kong Q, Chen D, Ba D, He W. A novel recombinant adeno-associated virus vaccine reduces behavioral impairment and beta-amyloid plaques in a mouse model of Alzheimer's disease. Neurobiol Dis 2003;14(3): 365–379.

362. Kim HD, Maxwell JA, Kong FK, Tang DC, Fukuchi K. Induction of anti-inflammatory immune response by an adenovirus vector encoding 11 tandem repeats of Abeta1-6: toward safer and effective vaccines against Alzheimer's disease. Biochem Biophys Res Commun 2005;336(1): 84–92.

363. Kim HD, Jin JJ, Maxwell JA, Fukuchi K. Enhancing Th2 immune responses against amyloid protein by a DNA prime-adenovirus boost regimen for Alzheimer's disease. Immunol Lett 2007;112(1): 30–38.

364. Kim HD, Tahara K, Maxwell JA, Lalonde R, Fukuiwa T, Fujihashi K, Van Kampen KR, Kong FK, Tang DC, Fukuchi K. Nasal inoculation of an adenovirus vector encoding 11 tandem repeats of Abeta1-6 upregulates IL-10 expression and reduces amyloid load in a Mo/Hu APPswe PS1dE9 mouse model of Alzheimer's disease. J Gene Med 2007;9(2): 88–98.

365. Frazer ME, Hughes JE, Mastrangelo MA, Tibbens JL, Federoff HJ, Bowers WJ. Reduced pathology and improved behavioral performance in Alzheimer's disease mice vaccinated with HSV amplicons expressing amyloid-beta and interleukin-4. Mol Ther: J Am Soc Gene Ther 2008;16(5): 845–853.

366. Bowers WJ, Mastrangelo MA, Stanley HA, Casey AE, Milo LJ Jr, Federoff HJ. HSV amplicon-mediated Abeta vaccination in Tg2576 mice: differential antigen-specific immune responses [erratum appears in Neurobiol Aging 2005. Apr;26(4):391]. Neurobiol Aging 2005; 26(4): 393–407.

367. Chackerian B, Rangel M, Hunter Z, Peabody DS. Virus and virus-like particle-based immunogens for Alzheimer's disease induce antibody responses against amyloid-beta without concomitant T cell responses. Vaccine 2006;24 (37–39): 6321–6331.

368. Zamora E, Handisurya A, Shafti-Keramat S, Borchelt D, Rudow G, Conant K, Cox C, Troncoso JC, Kirnbauer R. Papillomavirus-like particles are an effective platform for amyloid-beta immunization in rabbits and transgenic mice. J Immunol 2006;177(4): 2662–2670.

369. Maier M, Seabrook TJ, Lemere CA. Modulation of the humoral and cellular immune response in Abeta immunotherapy by the adjuvants monophosphoryl lipid A (MPL):cholera toxin B subunit (CTB) and E. coli enterotoxin LT(R192G). Vaccine 2005;23(44): 5149–5159.

370. Fukuchi K, Tahara K, Kim HD, Maxwell JA, Lewis TL, Accavitti-Loper MA, Kim H, Ponnazhagan S, Lalonde R. Anti-Abeta single-chain antibody delivery via adeno-associated virus for treatment of Alzheimer's disease. Neurobiol Dis 2006;23(3): 502–511.

371. Levites Y, Jansen K, Smithson LA, Dakin R, Holloway VM, Das P, Golde TE. Intracranial adeno-associated virus-mediated delivery of anti-pan amyloid beta, amyloid beta40, and amyloid beta42 single-chain variable fragments attenuates plaque pathology in amyloid precursor protein mice [erratum appears in J Neurosci. 2006 Dec 6;26(49):preceding 12847]. J Neurosci 2006;26(46): 11923–11928.

372. Schneeberger A, Mandler M, Otawa O, Zauner W, Mattner F, Schmidt W. Development of AFFITOPE vaccines for Alzheimer's disease (AD)—from concept to clinical testing. J Nutr Health Aging 2009;13(3): 264–267.

373. Rajkannan R, Arul V, Malar EJ, Jayakumar R. Preparation, physiochemical characterization, and oral immunogenicity of Abeta(1-12):Abeta (29-40):and Abeta(1-42) loaded PLG microparticles formulations. J Pharm Sci 2009;98(6): 2027–2039.

374. Dodel R, Hampel H, Depboylu C, Lin S, Gao F, Schock S, Jackel S, Wei X, Buerger K, Hoft C, Hemmer B, Moller HJ, Farlow M, Oertel WH, Sommer N, Du Y. Human antibodies against amyloid beta peptide: a potential treatment for Alzheimer's disease. Ann Neurol 2002;52(2): 253–256.

375. Dodel RC, Du Y, Depboylu C, Hampel H, Frolich L, Haag A, Hemmeter U, Paulsen S, Teipel SJ, Brettschneider S, Spottke A, Nolker C, Moller HJ, Wei X, Farlow M, Sommer N, Oertel WH. Intravenous immunoglobulins containing antibodies against beta-amyloid for the treatment of Alzheimer's disease. J Neurol Neurosurg Psychiatry 2004;75(10): 1472–1474.

376. Szabo P, Relkin N, Weksler ME. Natural human antibodies to amyloid beta peptide. Autoimmun Rev 2008;7(6): 415–420.

377. Pfeifer M, Boncristiano S, Bondolfi L, Stalder A, Deller T, Staufenbiel M, Mathews PM, Jucker M. Cerebral hemorrhage after passive anti-Abeta immunotherapy. Science 2002;298 (5597): 1379.

378. Wilcock DM, Rojiani A, Rosenthal A, Subbarao S, Freeman MJ, Gordon MN, Morgan D. Passive immunotherapy against Abeta in aged APP-transgenic mice reverses cognitive deficits and depletes parenchymal amyloid deposits in spite of increased vascular amyloid and microhemorrhage. J Neuroinflammation 2004;1(1): 24.

379. Racke MM, Boone LI, Hepburn DL, Parsadainian M, Bryan MT, Ness DK, Piroozi KS, Jordan WH, Brown DD, Hoffman WP, Holtzman DM, Bales KR, Gitter BD, May PC, Paul SM, DeMattos RB. Exacerbation of cerebral amyloid angiopathy-associated microhemorrhage in amyloid precursor protein transgenic mice by immunotherapy is dependent on antibody recognition of deposited forms of amyloid beta. J Neurosci 2005;25(3): 629–636.

380. Schroeter S, Khan K, Barbour R, Doan M, Chen M, Guido T, Gill D, Basi G, Schenk D, Seubert P, Games D. Immunotherapy reduces vascular amyloid-beta in PDAPP mice. J Neurosci 2008;28(27): 6787–6793.

381. Chauhan NB, Siegel GJ. Intracerebroventricular passive immunization with anti-Abeta antibody in Tg2576. J Neurosci Res 2003;74(1): 142–147.

382. Goni F, Sigurdsson EM. New directions towards safer and effective vaccines for Alzheimer's disease. Curr Opin Mol Ther 2005;7(1): 17–23.

383. Black RS, Sperling R, Kirby L, Safirstein B, Matter R, Pallay A, Nichols A, Grundman M. A single ascending dose study of bapineuzumab,

a humanized monoclonal antibody to Abeta, in AD. In: International Geneva/Springfield Symposium on Advances in Alzheimer Therapy; 2006; Geneva, Switzerland. 2006.

384. Grundman M, Gilman S, Black RS, Fox NC, Koller M. An alternative method for estimating efficacy of the AN1792 vaccine for Alzheimer disease. Neurology 2008;71(9): 697–698.

385. Grundman M, Black R. O3-04-05 Clinical trials of bapineuzumab, a beta-amyloid-targeted immunotherapy in patients with mild to moderate Alzheimer's disease. Alzheimer's Dementia 2008;4(4, S1): T166.

386. Sperling RA, Salloway S, Fox NC, Barackos J, Morris K, Francis G, Black RS, Grundman M. S32.001 Risk factors and clinical course associated with vasogenic edema in a phase II trial of bapineuzumab. In: American Academy of Neurology; 2009; Seattle, Washington, USA. 2009, Abstract.

387. Honig LS, Gilman S, Morris K, Black R, Grundman M, Francis G. S32.005 Safety profile of bapineuzumab in a phase II trial of mild-to-moderate Alzheimer's disease (AD). In: American Academy of Neurology; 2009 Seattle, Washington, USA. 2009, Abstract.

388. Siemers ER, Demattos RB, Stuart F, Dean RA, Sethuraman G, Margaret R, Jennings D, Tamagnan G, Marek K, Seibyl J, Paul SM. IN3-2.009 Use of a monoclonal anti-A antibody with biochemical and imaging biomarkers to determine amyloid plaque load in patients with Alzheimer's disease (AD) and control subjects. In: American Academy of Neurology; 2009; Seattle, Washington, USA. 2009, Abstract.

389. Wilcock DM, Rojiani A, Rosenthal A, Levkowitz G, Subbarao S, Alamed J, Wilson D, Wilson N, Freeman MJ, Gordon MN, Morgan D. Passive amyloid immunotherapy clears amyloid and transiently activates microglia in a transgenic mouse model of amyloid deposition. J Neurosci 2004;24(27): 6144–6151.

390. Wilcock DM, Alamed J, Gottschall PE, Grimm J, Rosenthal A, Pons J, Ronan V, Symmonds K, Gordon MN, Morgan D. Deglycosylated anti-amyloid-beta antibodies eliminate cognitive deficits and reduce parenchymal amyloid with minimal vascular consequences in aged amyloid precursor protein transgenic mice. J Neurosci 2006;26(20): 5340–5346.

391. Freeman GB, Lin JC, Pons J, Raha N. P3-272 39-week toxicity and toxicokinetic study of PF-04360365 in cynomolgus monkeys with 12-week recovery period. Alzheimer's Dementia 2009; 5 (4, S1): P423.

392. Wang EQ, Wentland J-A, Rajadhyaksha M, Durham RA, Freeman GB. P3-273 Pharmacokinetic and pharmacodynamic evaluation of PF-04360365, a humanized monoclonal anti-amyloid antibody, in young and aged cynomolgus monkeys. Alzheimer's Dementia 2009;5(4, S1): P423.

393. Relkin NR, Szabo P, Rotondi M, Mujalli D. P3-288 Antibodies in the dimer fraction of IVIG have the capacity to bind beta amyloid. Alzheimer's Dementia 2009;5(4, S1): P427–P428.

394. Shankle WR, Hara J. P06.067 Longitudinal measure of IVIG treatment effect in patients with Alzheimer's and Lewy Body disease. In: American Academy of Neurology; 2009; Seattle, Washington, USA. 2009, Abstract.

395. Shankle WR, Hara J. P3-297 Longitudinal measure of IVIG treatment effect in patients with Alzheimer's and Lewy Body disease. Alzheimer's Dementia 2009;5(4, S1): P430.

396. Szabo P, Mujalli DM, Rotondi ML, Relkin NR. P1-365 A method for measuring anti-beta amyloid (Aβ) oligomer antibodies in human plasma and intravenous immunoglobulin. Alzheimer's Dementia 2008;4(4, S1): T326.

397. Szabo P, Rotondi M, Mujalli D, Rospigliosi C, Eliezer D, Relkin N. P2-312 Human antibodies in intravenous immunoglobulin alter the assembly of amyloidogenic peptides into soluble oligomers. Alzheimer's Dementia 2008;4(4, S1): T464.

398. Watts RJ, Chen M, Atwal J, Greve JM, Wu Y, Mortensen DL, Varisco Y, Aldolfsson O, Pihlgren M, Pfeifer A, Muhs A. P3-283 Selection of an anti-Abeta antibody that binds various forms of Abeta and blocks toxicity both *in vitro* and *in vivo*. Alzheimer's Dementia 2009;5(4, S1): P426.

399. Mortensen D, Deng R, Adolfsson O, Tam S, Peng K, Lu Y, Erickson R, Schofield C, DuPree K, Pastuskovas CV, Boswell CA, Goldman H, Bormans G, Iyer S, Khawli LA, Theil F-P, Watts R, Prabhu S. P3-259 Characterization of the pharmacokinetics, pharmacodynamics, and distribution of anti-abeta antibody MAB-T5102A. Alzheimer's Dementia 2009;5(4, S1): P419.

400. Salloway S, Sperling R, Gregg K, Black R, Grundman M. S32.002 Cognitive and functional outcomes from a phase II trial of bapineuzumab in mild to moderate Alzheimer's disease. In: American Academy of Neurology; 2009; Seattle, Washington, USA. 2009, Abstract.

401. Siemers ER, Benson C, Gonzales C, Hansen R, Dean RA, Farlow M, DeMattos R. 04-03-04 Safety assessments and biomarker changes following a monoclonal Ap antibody given to subjects with Alzheimer's disease. Alzheimer's Dementia 2006;2 (3, S1).

402. Siemers ER, Ferguson-Sells L, Waters DG, DeMattos RB, Hansen RJ. P4-312 A model-based comparison of the nonclinical and clinical effects of anti-Aβ antibodies on plasma Aβ1-40 levels. Alzheimer's Dementia 2006;2 (3, S1): S608.

403. Siemers ER, Friedrich S, Dean RA, Sethuraman G, Demattos R, Jennings D, Tamagnan G, Marek K, Seibyl J. P4-346 Safety, tolerability and biomarker effects of an Abeta monoclonal antibody administered to patients with Alzheimer's disease. Alzheimer's Dementia 2008;4(4, S1): T774.

404. Siemers E. S2-02-05 Biochemical biomarkers as endpoints in clinical trials: applications in Phase 1, 2 and 3 studies. Alzheimer's Dementia 2009;5(4, S1): P95.

405. Siemers ER, Dean RA, Lachno DR, Zajac JJ, Sethuraman G, DeMattos RB, May PC. P1-275 Measurement of cerebrospinal fluid total tau and phospho-tau in phase 2 trials of therapies targeting Aβ. Alzheimer's Dementia 2009; 5(4, S1): P258.

406. Tsakanikas D, Shah K, Flores C, Assuras S, Relkin NR. P4-351 Effects of uninterrrupted intravenous immunoglobulin treatment of Alzheimer's disease for nine months. Alzheimer's Dementia 2008;4:(4, S1): T776.

407. Nicholas T, Knebel W, Gastonguay MR, Bednar MM, Billing CB, Landen JW, Kupiec JW, Corrigan B, Laurencot R, Zhao Q. P1-262 Preliminary population pharmacokinetic modeling of PF-04360365, a humanized anti-amyloid monoclonal antibody, in patients with mild-to-moderate Alzheimer's disease. Alzheimer's Dementia 2009;5(4, S1): P253.

408. Bednar Zhao Q, Landen JW, Billing CB, Rohrbacher K, Kupiec JW. O4-04-03 Safety and pharmacokinetics of the anti-amyloid monoclonal antibody PF-04360365 following a single infusion in patients with mild-to-moderate Alzheimer's disease: preliminary results Alzheimer's Dementia 2009;5(4, S1): P157.

409. Imbert G, Marrony S, Ulrich P, Goldsmith P. O2-05-05 Antibody immune response in cynomolgus monkeys following treatment with the active Aβ immunotherapy CAD106. Alzheimer's Dementia 2009;5(4, S1): P427.

410. Staufenbiel M, Beibel M, Danner S, Reichwald J, Wiederhold K-H. O2-05-06 Strong parenchymal amyloid reduction following CAD106 immunotherapy is associated with an increase in vascular Aβ but not microhemorrhages. Alzheimer's Dementia 2009;5(4, S1): P114.

411. Winblad BG. S2-04-06 Safety, tolerability and immunogenicity of the Aβ immunotherapeutic vaccine CAD106 in a first-in-man study in Alzheimer patient. Alzheimer's Dementia 2008;4(4, S1): T128.

412. Winblad BG, Minthon L, Floesser A, Imbert G, Dumortier T, He Y, Maguire P, Karlsson M, Östlund H, Lundmark J, Orgogozo JM, Graf A, Andreasen N. O2-05-05 Results of the first-in-man study with the active Aβ immunotherapy CAD106 in Alzheimer patients. Alzheimer's Dementia 2009;5(4, S1): P113–P114.

413. Schneeberger A, Mandler M, Zauner W, Mattner F, Schmidt W. P1-273 Development of Alzheimer AFFITOPE vaccines—from concept to clinical testing. Alzheimer's Dementia 2009;5(4, S1): P257.

414. Mandler M, Santic R, Weninger H, Kopinits E, Schmidt W, Mattner F. O2-05-08 The MimoVax vaccine: a novel Alzheimer treatment strategy targeting truncated Aβ40/42 by active immunization. Alzheimer's Dementia 2009;5 (4, S1): P114.

415. Bohrmann B, et al. 10th International Hong Kong/Springfield Pan-Asian Symposium on Advances in Alzheimer Therapy; Hong Kong; 2008.

NON-AMYLOID APPROACHES TO ALZHEIMER'S DISEASE

Lone Veng
Mary J. Savage
James C. Barrow
Celina V. Zerbinatti

Merck Research Laboratories, Merck & Co., Inc., Sumneytown Pike, West Point, PA

1. INTRODUCTION

The amyloid cascade is a prevalent hypothesis to explain the underlying pathology of Alzheimer's disease (AD), characterized by accumulation of amyloid-β peptides (Aβ) within extracellular plaques in the hippocampal and cortical brain regions. Aβ is released from the transmembrane amyloid precursor protein (APP) by two cellular proteases, β- and γ-secretases [1–3]. Aβ-mediated toxicity to both neurons and glia cells has been well established *in vitro* and derives from both soluble dimeric and oligomeric Aβ species, as well as insoluble intracellular and extracellular Aβ deposits. As part of its toxicity, it is also believed that amyloid triggers hyperphosphorylation of the microtubule associated protein Tau, causing it to aggregate and form intracellular neurofibrillar tangles (NFTs), another hallmark of AD pathology [4–6].

Together, amyloid plaques and neurofibrillary tangles define the AD brain. AD therapies have focused primarily on intervening at brain amyloid levels, including β- and γ-secretase inhibitors, and active and passive immunization directed to Aβ. However, since Aβ immunization has so far failed to show improvement of cognitive function in AD patients [7], interest in new venues for intervention that can provide additional benefits is on the rise. This review will focus on therapeutic approaches addressing metabolic risk factors and development of targets that reduce Tau pathology.

2. TARGETING METABOLIC SYNDROME

Epidemiological evidence suggests that individuals with metabolic syndrome are at greatly increased risk for dementia [8–12]. Metabolic syndrome is defined by the coexistence of at least three cardiovascular risk factors including large waist circumference, hypertriglyceridemia, low high-density lipoprotein cholesterol level, elevated blood pressure, and fasting hyperglycemia [13]. Several interventional approaches for dysfunctions in cholesterol and glucose metabolism validated in preclinical AD models and under evaluation or proposed for AD patients are discussed below.

2.1. Cholesterol-Related Approaches

The brain corresponds to only 2% of body weight, but it contains about 23% of the total body cholesterol [14]. In the normal adult brain cholesterol is relatively inert and vastly deposited in myelin that surrounds axons; only a small pool of cholesterol is active as part of neuronal and glial cellular membranes [14]. After myelination is completed during development, cholesterol is primarily made by astrocytes and secreted in the form of plasma high-density lipoprotein (HDL)-like particles containing apolipoprotein E (apoE). These particles can be internalized by several apoE receptors expressed in neurons to provide cholesterol needed for synaptic plasticity and membrane maintenance. Daily cholesterol synthesis in the brain exceeds requirements; therefore, excess cholesterol is routinely eliminated from the brain in the form of 24-S-hydroxycholesterol [15]. This brain-specific oxysterol produced by the catalytic activity of cholesterol 24-hydroxylase has much improved aqueous solubility over its cholesterol precursor and diffuses freely in the circulation to undergo further hepatic metabolization and excretion. Interestingly, cholesterol 24-hydroxylase deletion in mice causes severe impairment in learning and memory in the Morris Water Maze test [16], suggesting that inability to eliminate excess brain cholesterol via this pathway may affect brain function. In addition, excess cholesterol is partially eliminated from the brain by unclear mechanisms involving apoE [14]. Deletion of apoE in mice also leads to behavioral deficits [17], further suggesting an important role for cholesterol transport within and out of the brain for proper neuronal function.

Even though all brain cholesterol is made locally and is not affected by circulating plasma lipoproteins that do not cross the blood–brain barrier (BBB) [14], elevated plasma low-density lipoprotein (LDL) cholesterol has been associated with increased age-related cognitive decline [18]. There is also a striking reduction—up to 70%—of AD incidence in hypercholesterolemic patients treated with cholesterol lowering drugs named statins [19]. Follow-up *in vitro* and *in vivo* studies showed that reducing cellular membrane cholesterol with statins or cholesterol sequesters such as cyclodextrin decreases processing of APP to toxic Aβ peptides [20,21]. In contrast, increasing cholesterol levels in cellular membranes leads to augmented processing of APP to Aβ [21]. Therefore, the AD-cholesterol link was first proposed based on a role for membrane cholesterol in modulating Aβ production and the subsequent amyloid-mediated cascade of pathological events causing AD. However, other potential underlying mechanisms by which cholesterol is connected to AD are independent of amyloid and related to the crucial roles of membrane cholesterol in vesicle fusion and neurotransmitter release, signal transduction initiated in specialized membrane regions named lipid rafts, and receptor recycling [22]. In addition, because the ε4 allele of apoE is a well-established genetic risk factor for AD [23], the roles of apoE-mediated cholesterol transport in neurite outgrowth, synaptogenesis, remodeling, and inflammatory response are all potential mechanisms that can explain the AD-cholesterol connection [24]. Four main interventions in cholesterol metabolism that have been proposed and/or tested to treat AD patients include both inhibition of cholesterol synthesis and cholesterol esterification, as well as enhancing cholesterol turnover and HDL function.

2.1.1. Cholesterol Synthesis Statins are inhibitors of 3-hydroxy-3-methylglutaryl-coenzyme A (HMGCoA) reductase clinically approved to reduce plasma LDL cholesterol and coronary heart disease (CHD) incidence. HMGCoA reductase is the rate-limiting enzyme in the cholesterol biosynthetic pathway, and its inhibitors have been proven effective and relatively safe to reduce LDL-cholesterol levels and CHD in hypercholesterolemic patients [25]. Even though it is unclear that brain cholesterol is increased in hypercholesterolemic patients or that statins have the ability to effectively reduce brain cholesterol levels [26,27], several clinically approved statins (Fig. 1) have been tested in trials with AD patients based on the evidence discussed above. Several of these trials have been completed with atorvastatin **1** (Lipitor) and with simvastatin **2** (Zocor). Other statins such as lovastatin **3** (Mevacor), pravastatin **4** (Pravachol), and pitavastatin **5** were also evaluated in AD trials. Outcomes from these trials include evaluation of cognitive function and/or changes in brain amyloid based on the premise that reducing brain cholesterol leads to diminished processing of APP to toxic Aβ species (Table 1).

Overall, results from these clinical trials have been unable to demonstrate that statin treatment at approved doses reduced Aβ in human patients [28]. An alternative and perhaps more likely hypothesis proposed to explain the protective effect of statins in AD in both longitudinal studies and AD clinical trials are mechanisms associated to their pleiotropic actions [29]. These cholesterol-independent actions of statins are mostly related to decreased synthesis of isoprenoids and resulting anti-inflammatory and antioxidant activities, with improved endothelial function and better circulation in the brain [29]. The Cholesterol Lowering Agent to Slow Progression (CLASP) of Alzheimer's Disease Phase III Study sponsored by the National Institutes of Health was designed to help elucidate the mechanisms by which statins are protective against AD incidence by excluding patients with previous history of CHD or high LDL cholesterol. CLASP is a multicenter randomized, double-blind, placebo controlled 18-month study to evaluate both efficacy and safety of simvastatin treatment in slowing disease progression in 400 mild to moderate AD patients. Trial has been completed but results have not yet been disclosed.

2.1.2. Cholesterol Esterification Acyl-coenzyme A:cholesterol acyltrasferase (ACAT) converts excess free cholesterol to less toxic cholesteryl esters in the endoplasmic reticu-

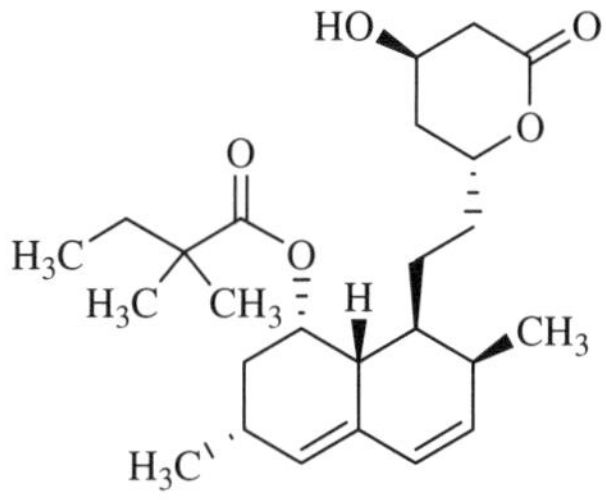

1 Simvastatin

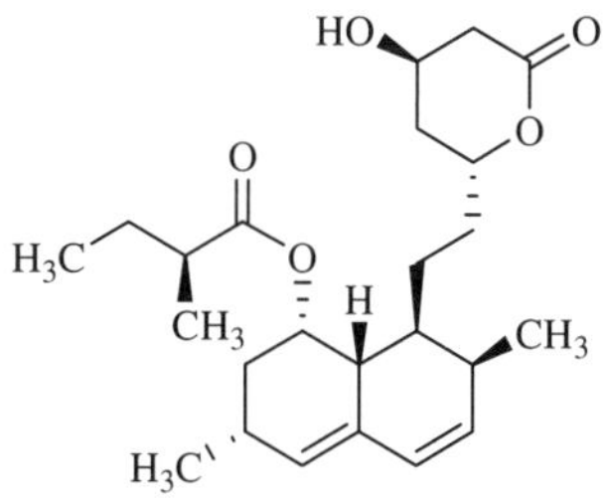

2 Lovastatin

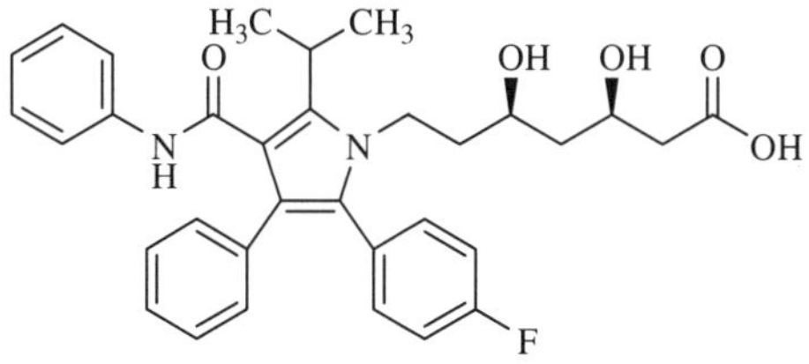

3 Atorvastatin

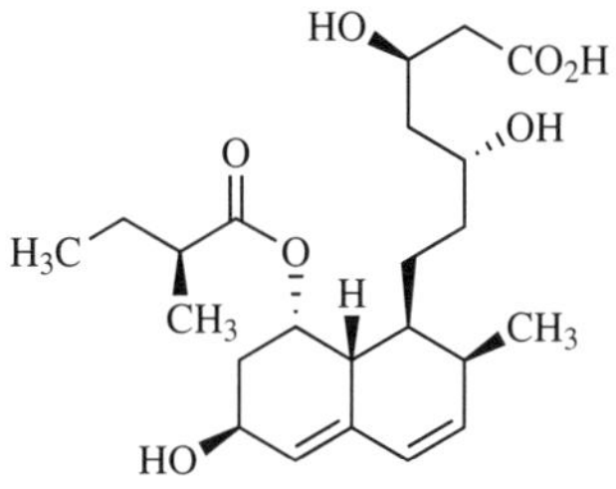

4 Pravastatin

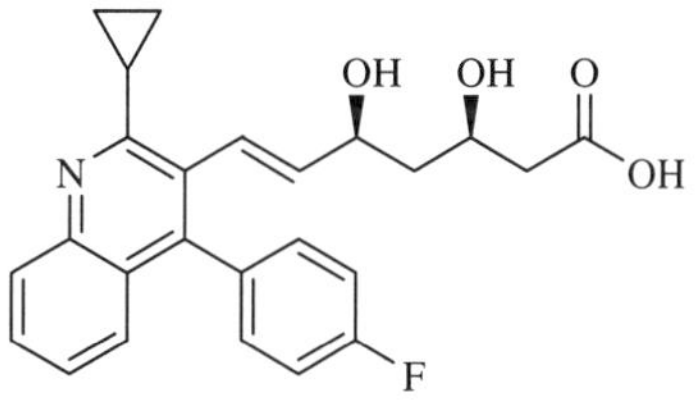

5 Pitavastatin

Figure 1. Statins evaluated in Alzheimer's disease clinical trials.

lum (ER) [30]. Cholesteryl esters can be safely stored in droplets within the cytoplasm and hydrolyzed back to cholesterol as needed. ACAT is a regulatory enzyme essential for the tight control of free cholesterol levels in the ER membranes. ER cholesterol is directly involved in two important mechanisms that maintain proper cholesterol levels in cells: (1) regulation of cell surface LDL receptor (LDLR) and cholesterol uptake and (2) regulation of *de novo* synthesis of cholesterol [30].

Under certain stress conditions, this classic regulatory mechanism is bypassed. For example, when LDL undergoes modification such as aggregation and/or oxidation, which prevents recognition by LDLR, cholesterol uptake by unregulated scavenger mechanisms leads to excessive accumulation of cholesteryl esters and the formation of foam-like cells that initiate atherosclerotic lesions in the vessel walls [31]. Therefore, ACAT inhibition was initially proposed as a clinical strategy to reduce CHD incidence by preventing uncontrolled accumulation of cholesteryl esters in macrophages within the arterial wall [32]. Alternatively, it was also hypothesized that inhibition of cholesterol esterification in tissues such as the intestine and liver could reduce the amount of cholesteryl esters available for packing into very-low-density lipoprotein (VLDL), the precursor for the highly atherogenic LDL [32].

Table 1. Summary of Nonamyloid Approaches Evaluated in Clinical Trials for Alzheimer's Disease

Company	Phase	Compound	Dates of Initiation/Completion	Size, Disease Stage	Study Design and Strategy	Major Outcomes
National Institute on Aging/Paul Beeson Faculty Scholars Program/The John A. Hartford Foundation/The Atlantic Philanthropies/Starr Foundation/American Federation for Aging Research/Merck NIA-ESPRIT	Ph2	Simvastatin	Completed—June 2009	100 middle-aged adults (35–69 years) who have a parent with documented Alzheimer's disease	Randomized, double-blind, placebo control, parallel assignment, efficacy study with 40 mg tablet each night for 1 month, then 80 mg for 8 months; placebo for 9 months	CSF levels of Aβ, inflammatory markers and cholesterol; changes in cognitive function; MRI substudy for effects on blood flow
National Institute of Aging (NIA)/Alzheimer's Disease Cooperative Study (ADCS) NIH-CLASP	Ph3	Simvastatin	Completed—October 2007	400 patients with mild to moderate AD	Rrandomized, double-blind, placebo-controlled, parallel group design trial with 20 mg of simvastatin or matching placebo for 6 weeks, followed by 40 mg of simvastatin or matching placebo for the remainder of the 18-month study period	Disease progression measured by ADCS-CGIC, mental status, functional ability, behavioral disturbances, and quality of life
Pfizer LEADe	Ph3	Atorvastatin	Completed—July 2007	641 patients with mild to moderate AD	Randomized, multicenter, parallel-group, double-blind study to evaluate efficacy and safety of atorvastatin 80 mg plus an acetylcholinesterase inhibitor versus acetylcholinesterase inhibitor alone for 72 weeks	Preliminary results showed no significant benefits

Seattle Institute for Biomedical and Clinical Research	Ph4	Simvastatin/ Pravastatin	Completed— April 2005	60 cognitively normal middle-aged and older persons with hypercholesterolemia	Randomized, double-blind active control, parallel assignment, prevention trial with simvastatin (40 mg) and pravastatin (80 mg) for 12 week with 30 subjects in each group	Simvastatin, but not pravastatin, reduced CSF levels of phospho-tau-181 in all subjects. CSF Aβ40, Aβ42, soluble APP, total tau, 24S-hydroxycholesterol, apoE, total cholesterol and F2-isoprostanes were unchanged
Institute for the Study of Aging (ISOA)/Pfizer NIA-PIT-ROAD	Ph2	Atorvastatin	Completed— August 2004	63 individuals with mild to moderate AD	Randomized, double-blind, placebo control, efficacy study with atorvastatin for 12 months	GDS and ADCS-cog scores improved significantly with drug at 6 but not at 12 months
ResVerlogix	Ph1a	RVX-208	Completed— November 2008	24 subjects	Double-blind, dose-escalation (2, 3, and 8 mg/Kg per day), placebo controlled trial for 1 week	12–14% increase in plasma Aβ40 levels with highest dose
GlaxoSmithKline/ University of Washington	Ph2	Rosiglitazone	Completed	30 subjects with mild AD or amnestic mild cognitive impairment	Placebo-controlled, double-blind, parallel-group study with rosiglitazone (4 mg QD) or placebo for 6 months	Stabilization of plasma Aβ levels and improvement on some cognitive subdomains in patients receiving rosiglitazone
GlaxoSmithKline	Ph3	Rosiglitazone	Completed	511 patients with mild to moderate AD	Randomized, double-blind parallel treatment with placebo or rosiglitazone (2, 4, and 8 mg) for 24 weeks	No significant effects on ADAS-cog or CIBIC+ in general study group, but significant improvement on ADAS-cog in non-APOE4 carriers
GlaxoSmithKline	Ph3	Rosiglitazone	Completed	862 patients with mild to moderate AD	Randomized, double-blind parallel treatment with placebo, donepezil or rosiglitazone for 24 weeks	Preliminary analysis found no significant effects of rosiglitazone on primary outcomes ADAS-cog, CIBIC+; analysis of secondary outcome measures has not yet been reported

(*continued*)

Table 1. (Continued)

Company	Phase	Compound	Dates of Initiation/Completion	Size, Disease Stage	Study Design and Strategy	Major Outcomes
Takeda Pharmaceuticals	Ph2	Pioglitazone	Completed	25 patients with mild AD	Randomized, double-blind parallel treatment with placebo or pioglitazone (15–45 mg qd) for 18 months	Improvement in some cognitive functions at 18 months in patients treated with pioglitazone
Takeda Pharmaceuticals	Ph2	Pioglitazone	November 2008–August 2011	300 patients with mild cognitive impairment	Randomized, double-blind parallel treatment with placebo, exercise or pioglitazone (30–45 mg qd) for 6 months	Cognitive function and insulin resistance will be measured
National Institute on Aging/Institute for the Study of Aging/Colombia University	Ph2	Metformin	February 2008–January 2010	40 participants with amnestic mild cognitive impairment	Randomized double-blind placebo or Metformin 1000 mg BID via oral route for 12 months	Study will asses effects of metformin on ADAS-cog, plasma Aβ, FGD PET and MRI
Allon Therapeutics, Inc.	Ph2a	NAPVSIPQ AL-108	January 2007–January 2008	120 patients with amnestic mild cognitive impairment AD	5 mg qd or 15 mg bid for 12 weeks intranasal; randomized, double-blind, placebo to control safety/efficacy study	Safe and well-tolerated; improved recognition, short-term and working memory
TauRx Therapeutics, Ltd.	Ph2	Methylene Blue, Rember	Completed	321 patients with mild to moderate AD	30, 60, and 100 mg/kg tid for 50 weeks; study may not have been blinded as compound turns urine blue	Dose-dependent improvement in ADAS-cog except at 100 mg with AEs found; prevented decline in cerebral blood flow; improved glucose uptake; however, no placebo group after 24 weeks

In relationship to AD, ACAT inhibitors have been tested preclinically as a mean to change the distribution of free and esterified intracellular cholesterol pools in neurons and therefore influence the trafficking and processing of APP to toxic Aβ peptides. Initial *in vitro* studies validated the prediction that ACAT inhibition reduces intracellular levels of esterified cholesterol and decreases Aβ secretion [33,34]. *In vivo*, the ACAT inhibitor CP-113818 was able to reduce brain cholesteryl esters, decrease brain amyloid burden and improve behavior deficits in human APP transgenice (APP-Tg) mice [35].

Although promising, the results above have not been widely reproduced. Furthermore, ACAT inhibitors Pactimice **6** (CS-505) and Avasimibe **7** (CI-1011) that were in clinical trials for CHD (Fig. 2), ended with disappointing results since they were not efficacious in reducing atherosclerotic plaque volume in hypercholesterolemic patients. Most importantly, preliminary results from the ACTIVATE (ACAT intravascular atherosclerosis treatment evaluation) trial revealed that pactimibe was also associated with detrimental effects at secondary plaque-related endpoints that led to the premature termination of clinical Phase II trials [36]. There is still speculation in the literature regarding the validity of ACAT inhibitors as efficacious drugs for the treatment of atherosclerosis [37]. Points are raised around the potential toxicity related to ACAT1 inhibition in macrophages leading to elevated intracellular levels of toxic free cholesterol, and that selective inhibition of ACAT2, primarily expressed in the liver and small intestine [32], could still prove beneficial to reduce LDL cholesterol and CHD incidence and potentially be evaluated in AD patients.

2.1.3. Cholesterol Turnover Liver X receptors (LXRs) are nuclear receptors involved in regulating cholesterol homeostasis and inflammatory responses throughout the body [38]. The endogenous ligands of LXR are hydroxylated metabolites of cholesterol—oxysterols—such as the brain-specific 24-S-hydroxycholesterol and 27-hydroxycholesterol made peripherally [39]. Two isoforms of LXR are found in higher species: LXRα, expressed primarily in the liver and adipose tissue; and LXRβ, which is ubiquitously expressed. In the liver, activation of LXRα predominantly upregulates the expression of genes involved in fatty acid synthesis. Activation of LXRs also increases expression of genes involved in cholesterol efflux, including the ATP binding cassette transporter proteins A1 and G1 (ABCA1 and ABCG1), and the cholesterol acceptor apoE. In addition, LXRs suppress the expression of inflammatory genes [40].

The ability of LXRs to stimulate cholesterol efflux from cells and its excretion from the body, a process named reverse cholesterol transport, as well as to decrease inflammation has led to the active pursuit of LXR as a drug target for CHD [41,42]. Synthetic LXR agonists are effective in raising plasma HDL cholesterol and decreasing the progression of atherosclerosis in animal models [43–45]. Plasma HDL cholesterol has also been positively correlated with cognitive function in aging humans and low HDL cholesterol is

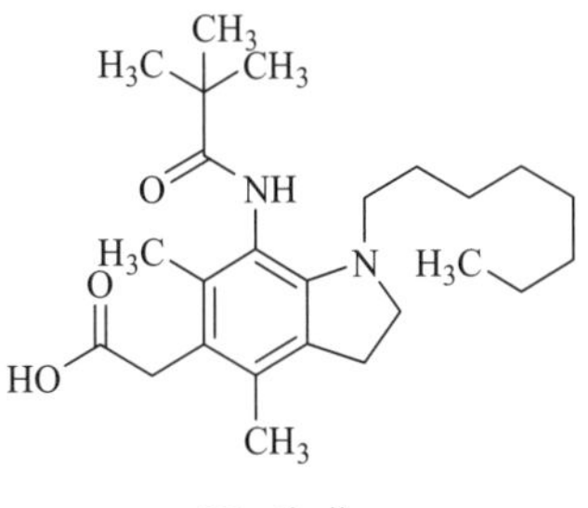

6 Pactimibe

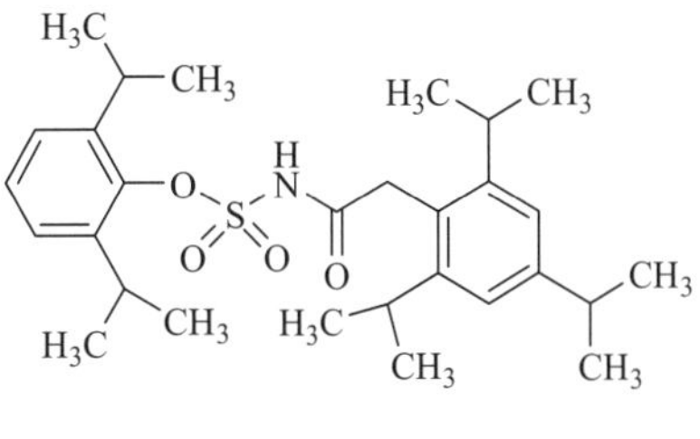

7 Avasimibe

Figure 2. ACAT inhibitors evaluated in clinical trials for reducing the progression of CHD.

associated with increased dementia and AD incidence [46]. Genetic manipulation of LXR and LXR target genes demonstrated that this pathway also plays relevant role in regulating brain cholesterol homeostasis [47–49]. Deletion of the major LXR-induced cholesterol-efflux pump ABCA1 decreased astrocyte-secreted apoE and apoE lipidation [50,51], while overexpression of ABCA1 increased the lipidation status of astrocyte-secreted apoE [52]. ApoE can form a stable complex with Aβ *in vitro* and also appears to participate in Aβ metabolism *in vivo*, since it is associated with amyloid extracted from AD brains [53,54]. It has been proposed that apoE lipidation influences the fate of Aβ, that is, more lipidated apoE favoring Aβ degradation and/or clearance versus lipid poor apoE promoting Aβ aggregation and deposition into amyloid plaques. In support of this hypothesis, increased amyloid burden in APP-Tg mice was observed when ABCA1 was deleted [55–57]. Conversely, ABCA1 overexpression led to a marked decrease in amyloid deposition in the brain of APP-Tg mice [52].

More recently, deletion of LXRs themselves also showed to increase amyloid pathology in the brain of APP-Tg mice [58] and, conversely, pharmacological activation of LXRs with potent synthetic agonists T0901317 **8** and GW3965 **9** (Fig. 3) significantly reduced brain amyloid levels in these mice [59–62]. Besides favoring the clearance/degradation of amyloid, LXR agonists could also prove beneficial as a mean to improve synaptic plasticity and tissue remodeling. LXRs are primarily expressed in glia cells and LXR agonists increase apoE-cholesterol secretion by astrocytes [63]. ApoE-cholesterol produced by astrocytes is likely the major cholesterol source to neurons, which have limited cholesterol biosynthetic capacity, but express many apoE receptors. LXR agonists have also shown to decrease lipopolysaccharide (LPS)-induced release of inflammatory cytokines by microglia in culture [58,60]. Inflammation is a major component of AD, where it has been implicated in neuronal cell death. Therefore, reduction of inflammatory responses in the brain to prevent neurodegeneration can be added as another potentially positive effect of LXR agonists in AD.

In summary, LXRs are a promising target to treat and/or prevent AD and other neuro-

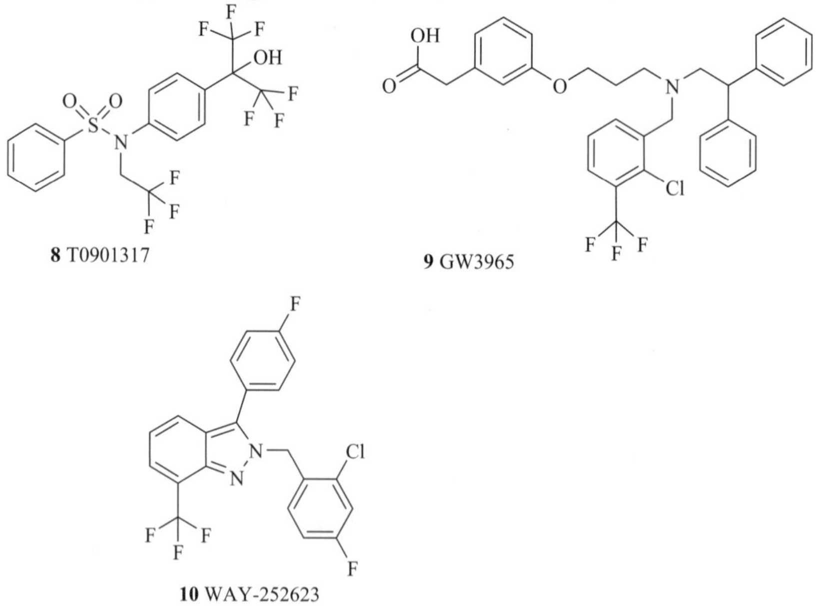

Figure 3. Potent nonselective LXR agonists (**8, 9**). WAY-252623 (**10**) entered Phase I clinical trials for CHD but was discontinued due to undesirable side effects.

degenerative disorders. However, the clinical development of LXR synthetic agonists has been greatly hindered because activation of LXRα in the liver leads to triglyceride accumulation and hepatic steatosis [64]. To avoid this undesirable side effect, development of LXRβ-selective agonists is essential; however, identification of such molecules has been proven difficult since the binding pockets of LXRα and LXRβ are virtually identical. Nonetheless, recent advances toward selectivity appear promising as β-selective LXR agonists **40** are beginning to advance to Phase I clinical trials for CHD [65] (Fig. 3).

2.1.4. Apolipoproteins E and AI ApoE is a 34 KDa protein highly expressed in the central nervous system (CNS) where it plays a major role in lipid transport [66]. Three human apoE isoforms arise from polymorphisms within the APOE gene, named, apoE2, E3, and E4. The APOE4 gene is a well-established risk factor for AD [23]. While only 15% of the normal population has APOE4, up to 70% of AD patients carry one or two copies of APOE4. The mechanism by which APOE4 increases the risk for AD is still unknown, but it has been reported that both normal individuals and AD patients expressing apoE4 have increased brain amyloid burden [67,68]. Likewise, APP-Tg mice expressing human apoE4 have increased brain amyloid deposits when compared to human apoE3-expressing mice [69,70].

ApoE forms a stable complex with Aβ [53,54], impacting both clearance and deposition of amyloid within the CNS [71]. Because apoE4 mice have increased brain amyloid load and reduced apoE levels when compared to apoE3 mice [70,72,73], therapies to increase apoE levels have been proposed to ameliorate amyloid pathology in AD patients [70]. As discussed above, activation of LXRs represents a major therapeutic approach to increase CNS apoE levels, leading to concomitant brain amyloid reduction [59–62]. Alternative approaches to increase apoE levels in the CNS include apoE mimetic peptides containing only the receptor binding region of apoE [74]. However, despite providing benefits associated with decreased inflammatory response in brain injury models [75,76], these peptides do not appear to display other apoE properties known to alter amyloid deposition in APP-Tg mice.

Apolipoprotein AI (apoAI) is a 28 kDa protein made in the liver and intestine that plays a major role in the formation, metabolism and catabolism of plasma HDL [77]. Because circulating HDLs have many antiatherogenic properties including reverse cholesterol transport, antioxidant and anti-inflammatory activities, apoAI elevating drugs and apoAI mimetic peptides are current strategies pursued as treatment for CHD [78]. In mouse models of atherosclerosis, small peptides containing the apoAI ligand-binding domain are effective in reducing atherosclerotic lesions by as much as 79% [79]. Since plasma HDL cholesterol levels have been correlated with cognitive function in aging adults and low HDL/ApoAI levels were associated with reduced cognitive function and higher AD incidence [80], there is current interest in considering apoAI-elevating strategies to treat AD as well.

ResVerlogix has completed a Phase Ia clinical trial for AD with its first-in-class **11** ApoAI/prebeta-HDL elevating drug RVX-208 (Fig. 4). In this double-blind, dose escalation, placebo-controlled trial with 24 subjects, the group receiving the highest dose of RVX-208 (8 mg/Kg per day) showed 12–14% increase in plasma Aβ40 after 7 days of dosing (Table 1). ApoAI is not made by neuronal or glial cells, but it is present in the cerebral spinal fluid (CSF) lipoproteins and it is likely produced by endothelial cells in the brain [81]. Similarly to apoE, apoAI has also been shown to bind Aβ and to localize to amyloid plaques deposited

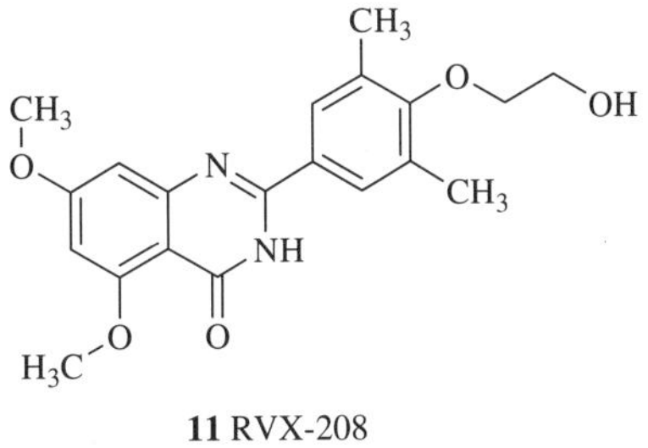

11 RVX-208

Figure 4. Potential Structure of RVX-208 extracted from a published ResVerlogix patent WO-2008092231.

within the brain [82–84]. Even though apoAI deletion had no significant effect on brain amyloid burden in APP-Tg mice [85], results from the ResVerlogix trial support the hypothesis that plasma apoAI works by drawing Aβ from the brain to the circulation for subsequent elimination from the body. Future larger clinical trials should provide further insights into the proposed mechanism of action, and additional efficacy endpoints will help to evaluate this strategy for prevention and treatment of AD.

2.2. Insulin-Related Approaches

Recent epidemiological studies have found that individuals with insulin resistance or type II diabetes mellitus (T2DM) have a two- to threefold greater risk than age-matched nondiabetics of developing dementia, most frequently AD, followed by vascular dementia [86–89]. While AD patients have increased peripheral levels of insulin, levels of the hormone in the CSF are decreased, suggesting that transport of insulin into the brain may be impaired in this disease [90]. Indeed, these findings are consistent with observations that peripheral insulin resistance, accompanied by hyperinsulinemia, leads to decreased binding to and signaling by the insulin receptor (IR). While the peripheral and central IR are not fully homologous, both the insulin binding site and the major signaling moieties are conserved and thus peripheral actions of insulin are apparently conserved in the brain [91,92]. Importantly, IR is expressed in all regions of the brain in both neurons and glia, rendering them susceptible to the effects of circulating insulin, which readily crosses the BBB via active transport at the choroid plexus, where IR is highly expressed [93].

The hypothalamus was originally found to be the main site of insulin action in the brain, where insulin signaling regulates a number of neuropeptides that contribute to the regulation of food intake and energy metabolism [94]. The role of insulin signaling in other brain areas is less well understood, although several interesting findings regarding the role of insulin in memory have recently emerged. In both humans and animals, acute insulin administration enhances a variety of cognitive processes [93,95]. Furthermore, numerous animal studies addressing the potential role of insulin in synaptic plasticity suggest that insulin signaling may modulate neuronal excitability and structural plasticity, at least in the hippocampus, where the IR is highly expressed and which therefore has been frequently studied due to its role in memory [93,96–98].

Clinical evidence suggests that the memory enhancing effects of insulin are highly dependent on its optimal level in the CNS, and this may be disturbed in conditions wherein the insulin balance is disrupted. Indeed several studies have described cognitive deficits in elderly diabetic patients relative to age-matched controls [99–101] and it has been shown that cognitive symptoms improve with diabetes treatment, presumably due to normalized levels of circulating insulin or improved insulin signaling [99]. Furthermore, excessive insulin levels, achieved by experimentally induced hyperinsulinemia in normal volunteers, resulted in an increase in soluble brain Aβ and inflammatory cytokines in the CSF, both of which are found to be elevated in AD patients and potentially linked to disease pathogenesis [102,103]. Interestingly, insulin administration to AD patients either intravenously concomitant with a euglycemic clamp or via intranasal delivery, which does not affect peripheral insulin or glucose levels, has been shown to enhance performance on several cognitive tasks, suggestive of improvements in functioning following normalization of central insulin levels [104,105]. In AD patients, higher doses of insulin are required to improve cognitive performance than in normal elderly volunteers [106] and it has been shown that IR activity is decreased in AD brain, despite increases in IR binding, leading to the hypothesis that AD is a disorder of central insulin resistance [107].

Importantly, potential links between insulin action and the clinical manifestations and molecular mechanisms of AD pathology have been uncovered pointing toward several druggable targets of interest for the treatment of AD. For instance, there is a great deal of interest in exploring the efficacy of insulin sensitizing treatment in alleviating both symptoms and disease progression in AD patients. Several classes of insulin sensitizing agents are being pursued such as agonists of the nuclear receptor peroxisome proliferator-activated receptor (PPAR)-γ and activators of the AMP-activated protein kinase (AMPK). Additionally, signaling molecules that act downstream of the IR, such as glycogen synthase kinase 3 (GSK3) have been found to play a prominent role in AD pathology and inhibitors of this enzyme are therefore also of interest for the treatment of AD. In addition to GSK3 effects downstream of the IR, many interactions between insulin signaling and protein kinase C (PKC) have also been described, that may contribute to the action of insulin on synaptic plasticity and memory [98]. However, there are currently no treatments under investigation that are aimed at directly modifying PKC signaling in AD as this kinase is an upstream regulator of many important cellular functions. Other molecules further downstream in the insulin signaling cascade may be more suitable drug targets as their inhibition would confer greater specificity.

2.2.1. GSK3 Insulin effects on cells are mediated by insulin binding to the IR, which leads to autophosphorylation of IR cytoplasmic tyrosine residues that provide docking sites for several adaptor proteins, such as insulin receptor substrate (IRS) 1 and 2 [94]. On docking to the phosphorylated IR, IRS 1 and 2 can recruit and activate multiple other effector proteins, thereby initiating the cascade of activation of IR responsive intracellular second messengers, including PI3 kinase, PKA, PKC, MAPK, JAK/STAT, and mTOR. Downstream effects mediated by PI3 kinase are manifold and generally effectuate the metabolic consequences of insulin release. For instance, PI3 kinase mediates the recruitment of Akt (also called PKB) and PKC to the plasma membrane where they can be activated by other upstream activators [94]. Akt has many important cellular targets, including GSK3, and phosphorylation of GSK3 by Akt leads to inactivation of the kinase. Inactivated GSK3 allows for an increased rate of glycogen synthesis by glycogen synthase, which is otherwise inhibited by the activity of GSK3 [108]. This cascade links increased IR signaling to increased glycogen synthesis and storage within

the target cells as well as other downstream pathways affected by GSK3 activity. In conditions of both peripheral and central hyperinsulinemia, or other conditions where IR signaling may be blunted, GSK3 activity would be increased leading to excess phosphorylation of the target substrates of this kinase.

GSK3 exists as two analogous isoforms, GSK3α and GSK3β, both expressed in the brain. GSK3β is most closely implicated in the pathogenesis of AD, as this kinase is one of the most prominent kinases responsible for the pathological phosphorylation of Tau found in AD brain [109]. The hyperphosphorylation of Tau is a seminal event in precipitating the neurofibrilary tangles characteristic of AD brain pathology [110]. Furthermore, GSK3 was also shown to regulate the production of Aβ in cellular systems [111,112] and in APP-Tg mice [113]. Interestingly, it was recently reported that peripheral insulin deficiency in an animal model of type 1 diabetes resulted in reduced phosphorylation of brain IR, AKT, and GSK3. The changes in insulin signaling were associated with increases in Tau phosphorylation and Aβ levels as well as reduced insulin degrading enzyme (IDE) protein expression [114]. If confirmed, these results directly implicate GSK3 as a downstream effector of diabetes-induced AD pathology. The discovery and development of GKS3 inhibitors for AD is described in relation to Tau phosphorylation later in this chapter.

2.2.2. PPAR-γ PPAR-γ belongs to a class of nuclear receptors activated by various lipoproteins and lipids that are either obtained through the diet or are products of normal metabolism. Once activated, PPARs act as transcriptional activators for a number of genes involved in lipid metabolism and storage [115]. Based on PPARs pivotal role in regulating the body's metabolic response to dietary intake of macronutrients and their resulting ability to sensitize the body to an insulin response, PPAR-γ agonists have been developed for the treatment of T2DM (Fig. 5). In addition to their insulin-sensitizing properties, PPAR-γ agonists also inhibit expression of proinflammatory genes, another potential therapeutic benefit in AD patients as this disorder is characterized by a high degree of

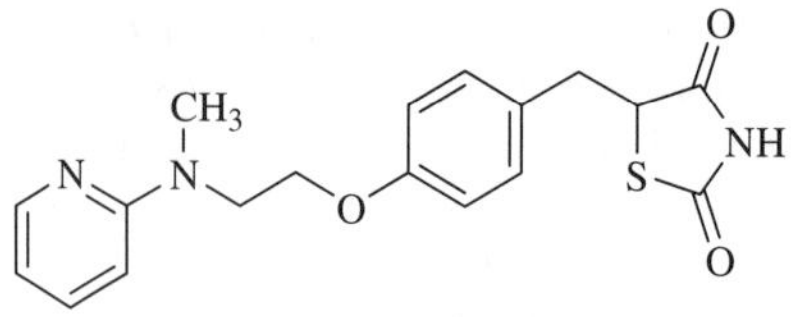

Figure 5. Drugs approved for the treatment of TT2DM, which are currently being investigated for potential efficacy against the symptoms and progression of AD.

12 Rosiglitazone

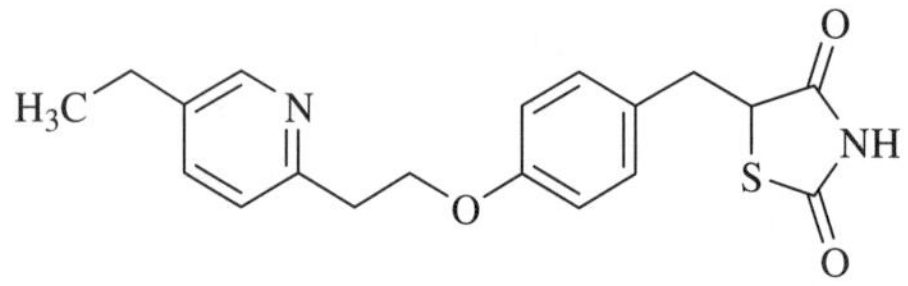

13 Pioglitazone

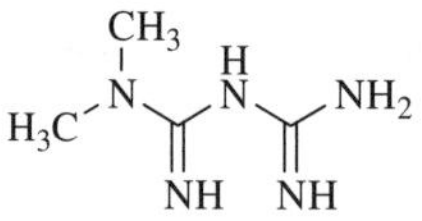

14 Metformin

brain inflammation. Two thiazolidinedione agonists of the PPAR-γ receptor have been approved for the treatment of T2DM: rosiglitazone **12** and pioglitazone **13** (Fig. 5). The discovery and development of PPAR-γ agonists has recently been reviewed [115,116]. Although both rosiglitazone and pioglitazone may somewhat poorly penetrate the BBB [117], there is now great interest in exploring their efficacy for the treatment of AD based on the epidemiological and experimental evidence implicating insulin resistance as a risk factor for AD [118].

Encouragingly, both rosiglitazone and pioglitazone have shown some promise toward alleviating certain aspects of disease pathology in APP-Tg mice. For instance, chronic pioglitazone treatment in APP-Tg mice resulted in a small reduction in the brain levels of soluble Aβ, although plaque burden and microglial activation were unaffected by the treatment [119]. In another study, a higher dose of pioglitazone was administered to APP-Tg mice and in this study the pioglitazone treatment resulted in a 20–25% reduction in plaque burden and soluble Aβ in the brain, accompanied by a significant reduction in the number of activated microglia [120]. Others have reported that pioglitazone treatment attenuated age-related decreases in cerebrovascular function, oxidative stress and microglial activation but failed to improve spatial memory in APP-Tg mice [121]. The efficacy of chronic rosiglitazone treatment has also been studied in APP-Tg mouse models and has been found to normalize hyperinsulinemia [122], reduce circulating levels of corticosterone and hippocampal glucocorticoid receptor upregulation, and alleviate learning and memory deficits that develop in APP-Tg mice as they age [123,124]. Additionally, rosiglitazone treatment was found to normalize both mRNA levels and activity of IDE as well as soluble Aβ in the brain in these mice [123].

The link between PPAR-γ and Aβ has been further strengthened by evidence implicating PPAR-γ activity in the expression or activity of two major enzymes involved in the production and degradation of Aβ: the β-secretase β-site APP cleavage enzyme (BACE) and the IDE. It has been shown that BACE transcription, and thereby Aβ production, is upregulated in response to inflammatory cytokines and that this effect is suppressed by PPAR-γ activity [125], as BACE transcription is under the regulation of a PPAR-γ response element [126]. Another study showed that PPAR-γ activity suppressed the expression of APP and resulted in higher levels of APP degradation via ubiquitination [127]. Thus, PPAR-γ activation may influence Aβ levels directly by transcriptional or posttranslational regulation of BACE or APP. PPAR-γ activation may also influence Aβ levels indirectly by normalizing insulin levels and thereby restoring the balance between insulin and Aβ clearance by IDE as this enzyme is known to effectively degrade Aβ both in cellular neuronal systems and in the brain [128–131]. Furthermore, mutations in IDE that cause diabetes also decrease Aβ degradation [132], which could accelerate accumulation and deposition of these peptides in the brain. Thus, it has been postulated that the excess insulin levels in patients with hyperinsulinemia overwhelms the enzymatic capacity of IDE in the brain and that Aβ clearance by IDE therefore becomes shunted [133]. However, an alternative hypothesis is that central insulin resistance caused by peripheral hyperinsulinemia may cause downregulation of IDE in the brain as decreased insulin signaling has been associated with decreased IDE activity and increased Aβ levels in the AD brain [134] and in the brains of APP-Tg mice fed a high fat diet [135]. Nevertheless, restoring insulin sensitivity and thereby also peripheral insulin levels by PPAR-γ activation may normalize central IDE activity and thereby Aβ clearance regardless of the exact mechanism.

The evidence described above has ignited interest in PPAR-γ agonists as disease modifying therapies for AD and, combined with clinical evidence for insulin's role in memory that was suggestive of insulin-lowering therapies providing symptomatic benefit in AD, has spurred a number of clinical trials in AD (Table 1). Several recent studies in AD patients have shown some benefit of rosiglitazone or pioglitazone treatment in improving cognitive function. Recently, it was reported that rosiglitazone treatment in a small cohort of mild AD patients improved several domains of cognitive function [136]. Additionally, in this study plasma Aβ levels deceased in placebo treated subjects at the 6-month endpoint but remained unchanged in subjects on active treatment, consistent with the idea that active treatment prevented the decrease in plasma Aβ sometimes observed with AD progression. A larger follow-on study found no effect of rosiglitazone treatment on cognitive or functional measures in a general AD population but when stratified for APOE genotype, rosiglitazone significantly improved both measures of cognitive and daily functioning in APOE2 and APOE3 but not in APOE4 carriers [137]. These results are consistent with an emerging body of literature suggesting that insulin resistance is not a risk factor for AD in APOE4 carriers and that AD patients who are also APOE4 carriers are less susceptible to insulin-mediated memory enhancement and Aβ modulation than non-APOE4 carriers [90,106,138,139]. However promising, these initial findings need confirmation by further clinical investigation and rosiglitazone is currently undergoing a Phase II open label extension study in AD patients, due to conclude in 2009.

While only preliminary results have been disclosed for pioglitazone treatment in AD, a pilot study demonstrated that pioglitazone improved cognition in a small cohort of diabetic AD patients and patients with mild cognitive impairment [140]. A larger trial of pioglitazone in patients with mild cognitive impairment is ongoing (Table 1). Results from further studies could provide a potential path forward for PPAR-γ agonist treatment for AD or for populations at high risk for developing AD, such as APOE4 negative T2DM patients.

Another class of insulin sensitizer is exemplified by the biguanide compound metformin 14 (Fig. 5), an activator of AMPK [141]. Metformin was discovered and developed in the 1970s [142] and is now available as a generic drug. While this compound is well tolerated and effective for the treatment of T2DM, its potential for treating AD is less clear. The AMPK pathway has not been specifically implicated in any aspect of AD pathogenesis, either directly or indirectly, and only one preclinical study has been published that examined the effect of metformin on Aβ peptide production [143]. This study reported that BACE transcription was upregulated in a cellular system following metformin treatment, suggesting the metformin could potentially accelerate AD pathology if the cellular effects replicate *in vivo*. However, these results clearly warrant further investigation. Nevertheless, a Phase II trial is under way that will study the effects of metformin on cognitive function and plasma Aβ levels in overweight patients with mild cognitive impairment (Table 1). This trial is due to conclude in 2010.

In summary, mounting evidence points to a possible role for insulin in AD pathology and associated cognitive dysfunction and several strategies are being pursued to investigate both novel and existing treatments that may modify the impact of insulin imbalance on the symptoms and progression of AD.

3. TARGETING TAU PATHOLOGY

The neurofibrillary tangle (NFT) is composed primarily of intraneuronal deposits of hyperphosphorylated Tau protein thought to result from an imbalance of kinase and phosphatase activities. Efforts to reduce Tau-related toxicity in AD have included kinase inhibition [144–146], increased glycosylation [147], inhibition of Tau fibrillization [140–150], microtubule (MT) stabilization [151,152], and reduction of Tau protein level [153,154]. Tau kinases have been a primary focus of pharmaceutical and academic efforts for the past 10 years; however, other post-translational modifications such as improving glycosylation and inhibiting higher order Tau fibril formation are receiving increased attention. Tau protein reduction has also been suggested as a therapeutic approach, as certain Tau haplotypes that may increase AD risk are associated with increased Tau protein levels [155]. Specific targets that reduce Tau protein levels include Hsp90; however, few pharmaceutical efforts have been reported for Hsp90 specifically and Tau protein reduction in general. MT stabilization, while not a direct Tau target, has been pursued to bolster an axonal transport system compromised by Tau loss-of-function.

3.1. Tau Phosphorylation

One of the primary challenges in targeting Tau phosphorylation has been the uncertainty regarding which kinase activity and phosphorylation site(s) drives Tau pathology. Within the AD brain, Tau is phosphorylated at approximately 40 sites, where most are Ser/Thr-Pro motifs [156]. Tau normally functions as a MT binding protein, both stabilizing the MT network and providing tracks for the transport of organelles and vesicles [157–159]. When Ser/Thr residues in or around the MT binding "repeat domains" of Tau are hyperphosphorylated, the binding of Tau to the MT becomes unstable [160–163] and neuronal death is thought to ensue via a breakdown in axonal transport. It is unclear whether Tau phosphorylation is a prerequisite for Tau aggregation, though it appears that abnormal Tau phosphorylation may precede full NFT formation [164]. Importantly, Tau hyperphosphorylation in the absence of NFTs can also lead to neuronal loss [165–167]. Cruz et al. [168] also reported neuronal loss that correlated better with Tau hyperphosphorylation, prior to NFT formation. Whether it is a loss of Tau function within the axon that leads to neuronal loss or a toxic response that is gained due to hyperphosphorylation is unclear. Since Tau knockout (KO) animals have no observable phenotype, the notion of loss-of-function toxicity is less attractive [169]. Axonal transport is also unaffected in Tau KO mice [170], also supporting no loss-of-function toxicity. Conditional Tau KO animals would confirm this as several MT proteins may functionally compensate for the absence of Tau [169].

Tau phospho-epitopes that appear earliest during AD progression may initiate the pathology. An early role for pThr231 and pSer262 was supported in a study of AD brains, staged according to the degree of Tau pathology [171], with pSer396 linked to later stages. pThr231 is a key phosphorylation site that reduced Tau/MT interaction following phosphorylation by GSK3 [172]. In fact, when pThr231 is mutated, GSK3 phosphorylation failed to reduce Tau/MT binding. Both pThr231 and pSer396 flank the MT binding domain and disruption of this region by phosphorylation destabilizes the MT-Tau interaction [172]. Both CDK5 [172] and GSK3 [144,173] contribute to phosphorylation at pThr231 and pSer396, with phosphorylation at an upstream priming site demonstrated for CDK5 [174]. In addition, evidence supports reduced phosphatase activity in AD [175,176], which could drive hyperphosphorylation in the absence of increased kinase activity. Phosphatase inhibition in cell models allows the formation of specific AD-type pTau epitopes [177].

3.1.1. CDK5 and GSK3

CDK5 and GSK3 have been actively targeted as Tau kinases over the last decade for the development of AD therapeutics due to their ability to phosphorylate these key Tau residues. CDK5 and GSK3 are brain enriched, proline-directed Serine/Threonine kinases [178,179], which phosphorylate Tau in cell-free [180] and transfected cell systems [174,181–183] as well as animal models. Mice that have been genetically engineered to have an increased GSK3 activity [184–186] or increased CDK5 activity via overexpression of p25 [168,187,188] develop hyperphosphorylated Tau, neuronal loss and behavioral abnormalities. There is also evidence for a role of both CDK5 and GSK3 in the toxic effects of Aβ acting directly on the neuron. Aβ treatment in culture stimulates the cleavage of the CDK5 accessory protein p35 to p25 and the inhibition of CDK5 activity attenuates Aβ42-induced neuronal loss [189–191]. GSK3β is also implicated in Aβ-induced Tau phosphorylation [190,192,193].

Few CDK5 and GSK3 inhibitors have been tested clinically; however, KO models can give clues as to their tolerability. CDK5 KO mice die at birth [194,195], resulting in concerns related to chronic CDK5 inhibition in humans. Several Phase I and II clinical trials have been completed with flavopiridol, a pan-CDK antagonist [196]. Dose-limiting toxicities were reversible. There are no reports of clinical trials with GSK3 inhibitors; however, GSK3β homozygous KO mice die during development likely from liver degeneration due to enhanced TNF-α toxicity [197]. GSK3β heterozygous mice develop normally. Inhibition of GSK3β may also negatively impact cardiovascular function [198].

Few pharmaceutical companies have developed CDK5 inhibitors with AD as the targeted disease. These companies (including AstraZeneca and Pfizer) and Inje University have programs in preclinical development [199]. The crystal structure of CDK5 complexed with p25 was reported [200] demonstrating an unprecedented mechanism for activation of a CDK family member in that the typical two step process for activation, cyclin binding and activation loop phosphorylation, was replaced by binding of the p25 coactivator, without a phosphorylation step. A higher resolution crystal structure of CDK5/p25 was also reported, complexed with three ATP binding site inhibitors: (*R*)-roscovitine, aloisine-A, and indirubin-3'-oxime [201]. Olomoucine **15**, roscovitine **16** (Seliciclib, CYC-202) and flavopiridol **17** (Alvocidib, L-86-8275) are the three most widely studied CDK inhibitors to date [201,202] (Fig. 6). CP-681301 **18** (Pfizer) is the only selective CDK5 inhibitor reported in preclinical development. Both CP-681301 and the structurally related CP-668863 **19** are potent and selective CDK5 inhibitors with K_i values of 2.9 and 13.7 nM, respectively [203] (Fig. 6). CP-681301 was evaluated in a forebrain specific inducible p25 Tg mouse model [204], where it reduced gliosis in the hippocampus relative to control and rescued performance in a fear conditioning assay. CP-681301 also reversed the effects of increased Aβ levels following overexpression of p25 [205].

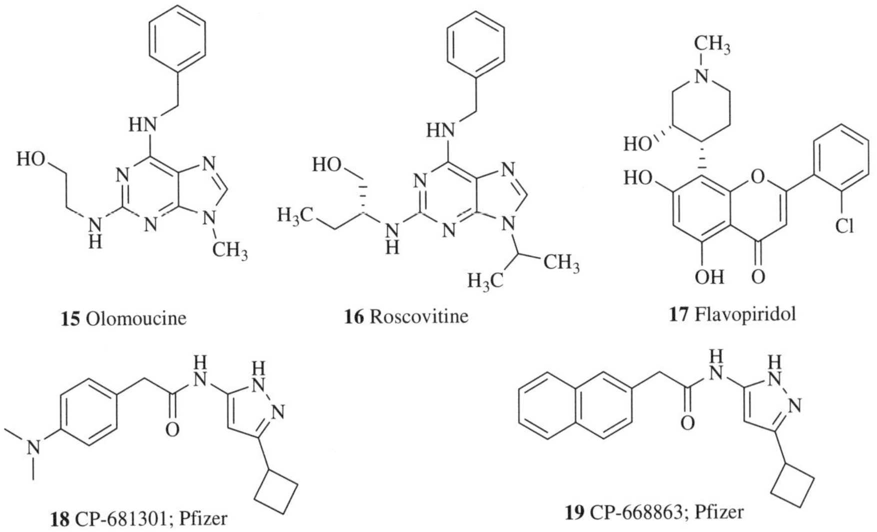
15 Olomoucine **16** Roscovitine **17** Flavopiridol **18** CP-681301; Pfizer **19** CP-668863; Pfizer

Figure 6. CDK5 Inhibitors.

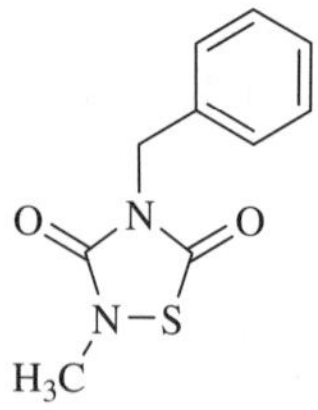

20 NP-12

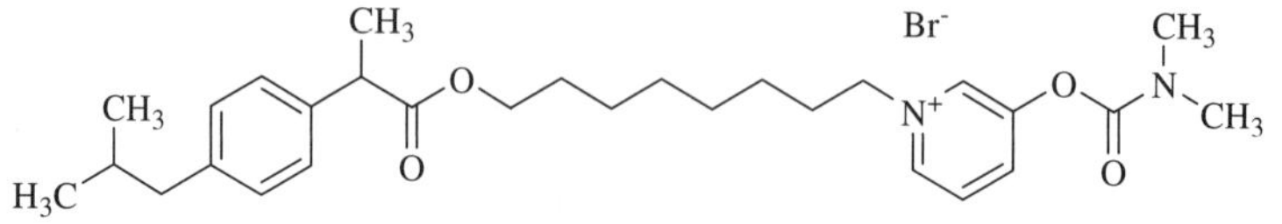

21 IBU-PO

Figure 7. GSK3β inhibitors in the clinic.

While several companies reported GSK3 inhibitors in preclinical drug development for AD, only NeuroPharma (NP-12/TDZD-8) **20** and the Israel Institute for Biological Research (IBU-PO) **21** have GSK3β inhibitors in Phase I for AD (Fig. 7). The crystal structure for GSK-3β was obtained in 2001 [206], supporting rational inhibitor design. The general GSK3 for inhibitor motif possesses a hinge binding region comprised of a hydrogen bond donor and acceptor, a potential interaction with a salt bridge located in hydrophilic portion of the active site, and a solvent-exposed hydrophobic region. Medicinal chemists have used this pharmacophore in the design of novel inhibitors. Small molecule inhibitors such as lithium [207,208], AR-A014418 **22** [209], the paullones **23** [210], indirubins **24** [211], maleimide analogs **25** [146,212,213], hymenialdisine **26** [214,215], 2,4-disubstituted thiadiazolidine-dione **27** [216], aloisines **28** [217], and amino pyrazoles **29** [218] have all been reported as GSK3 inhibitors (Fig. 8), some also inhibit CDK5.

3.1.2. MAPK Inhibitors
Activation of three MAPK pathways has been demonstrated in AD brain: p38 [219,220], JNK [221,222], and ERK [223,224]. Similarly, p38 and JNK were also activated in the brain of APP-Tg animals, coincidently with amyloid pathology [225–227]. While both p38 and JNK kinases co-purify with NFTs from AD brain [220], recent studies suggest that JNK activation alone may not be sufficient for Tau hyperphosphorylation [228]. Several reports support the ability of p38 and JNK to phosphorylate Tau in cell-free conditions [229], with phosphorylation by p38-γ and -δ resulting in the greatest reduction of Tau binding to MTs [230] when compared to JNK or p38-α or -β. Consistent with this observation, overexpression of p38-γ (SAPK3), but not JNK or other p38 kinases in neuroblastoma, reduced the association of Tau with the cytoskeleton [231]. Therefore, the γ- or δ-isoforms of p38 might be an appropriate target for intervention in Tau phosphorylation in AD. While specific JNK inhibitors have not progressed into development, many p38 inhibitors have been tested in clinical trials.

Inhibitors of p38-α and p38-β SB203580 **30**, SB202190 **31** are not selective between these isoforms, but are selective against p38-γ and p38-δ, while BIRB796 **32** is a novel allosteric diaryl urea inhibitor of all four p38 isoforms [232] (Fig. 9). Despite the fact that more than 20 p38 inhibitors are in clinical trials for rheumatoid arthritis and asthma and it has been over 10 years since p38 inhibitors have been studied clinically, there are still no approved drugs. However, since all p38 inhibitors to date have targeted p38-α and -β and knocking out p38-α but not the other isoforms leads to lethality in mice, targeting the other p38 isoforms may increase the probability of success for drug development. The dual p38 and JNK inhibitor Semapimod **33** (CNI-1493, Fig. 10) has been reported to improve pathol-

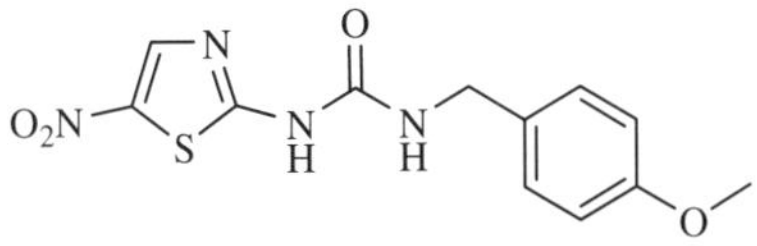

22 AR-A014418 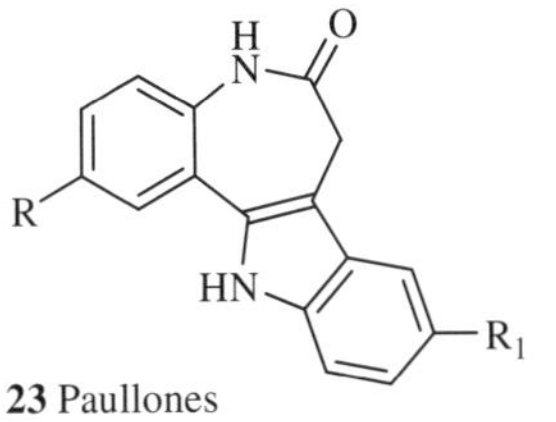   **23** Paullones 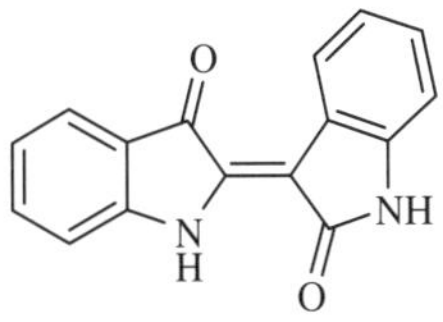   **24** Indirubin

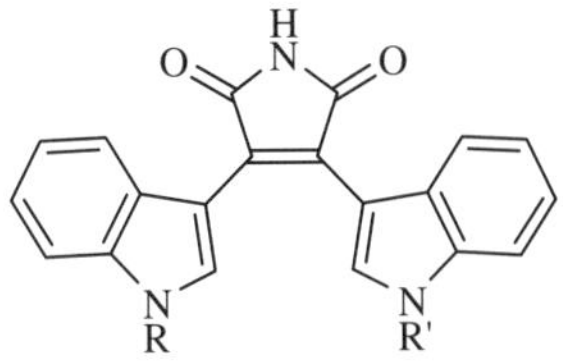

25 Maleimide analogs 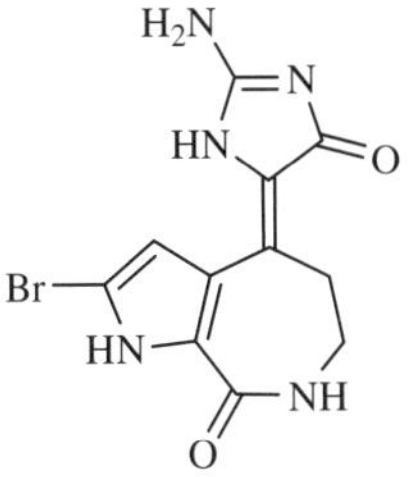   **26** Hymenialdisine 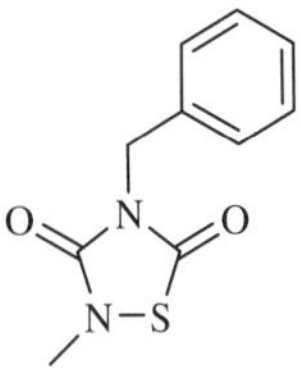   **27** Thiadiazolidine-dione

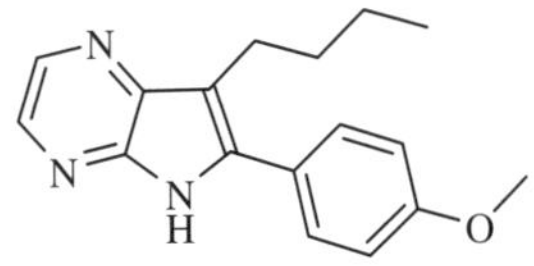

28 Aloisine, RP106 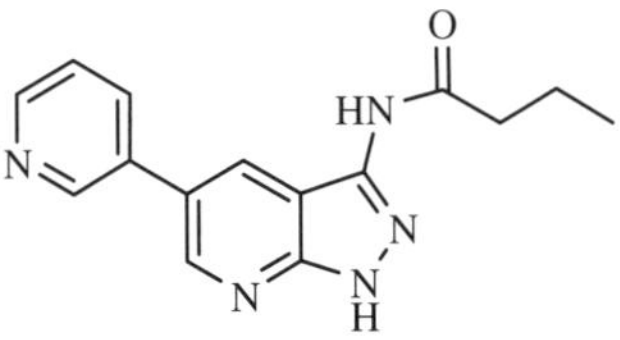   **29** 3-Aminopyrazole

Figure 8. GSK3β inhibitor.

ogy and cognitive function in APP-Tg mice [233]. Semapimod is a synthetic guanylhydrazone with wide-ranging kinase inhibitory activities including Raf kinase, which is involved in the MAPK/ERK signaling cascade. Semapimod decreases MEK phosphorylation and ultimately inhibits phosphorylation of p38 MAP kinase.

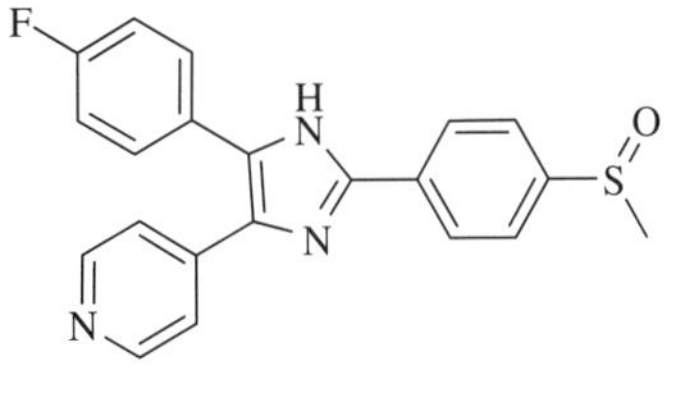

30 SB-203580 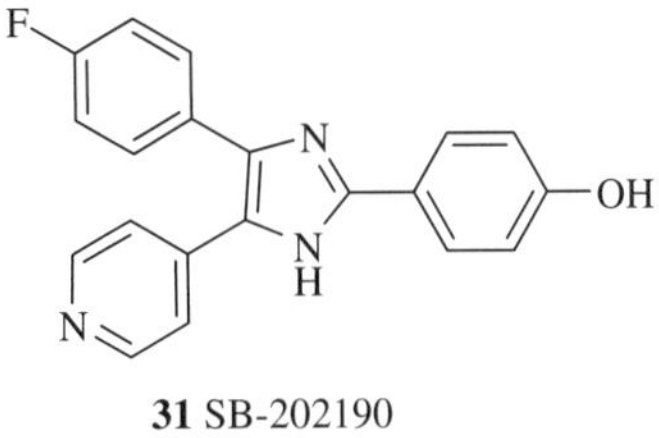   **31** SB-202190

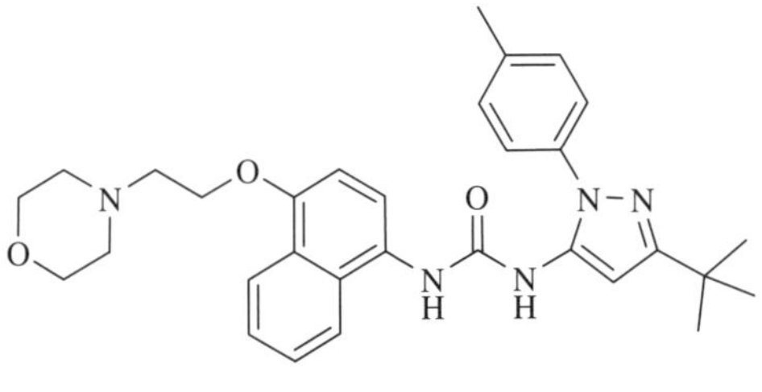

32 BIRB-796, Doramapimod

Figure 9. Inhibitors of p38.

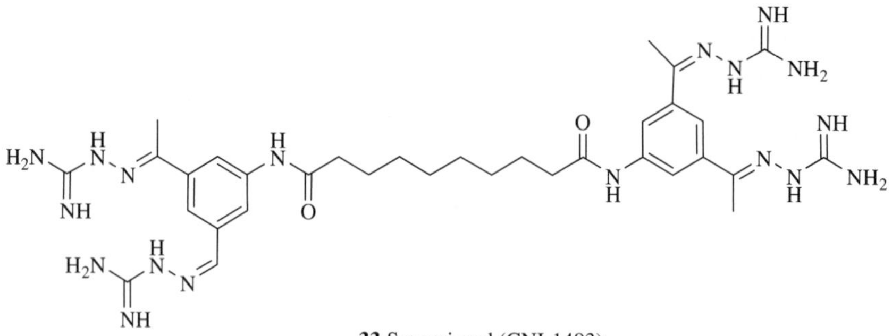

33 Semapimod (CNI-1493)

Figure 10. Dual JNK and p38 inhibitor.

One of the major challenges facing kinase inhibitor development is the extent of selectivity versus off-target kinase inhibition. Since most targeted kinase inhibitors involve competitive inhibition with the ATP site, multiple targets may be inhibited as these regions exhibit high homology. The specificity of a select number of inhibitors was evaluated and it was found that many are not as selective as reported [234]. Kinase selectivity in the context of AD drugs is even more challenging since patients will require chronic treatment. Because the initial stages of the disease are not life threatening, the disease modifying kinase inhibitors need minimal liability due to off-target inhibition. An additional challenge for kinases inhibitors for AD is the requirement for brain penetration. Greater than 98% of small molecule inhibitors are unable to penetrate the BBB [235].

3.1.3. Neurotrophic Peptides The neurotrophic peptide NAP (NAPVSIPQ, AL-108) reduced Tau phosphorylation in an animal model of Tau pathology [236,237] and also provided neuroprotection following intranasal delivery [238]. NAP enters the central nervous system via intranasal administration and can be detected concurrently in the CSF and plasma [238]. In addition to a few GSK3 inhibitors, NAP is the only other inhibitor targeted to Tau phosphorylation in clinical trials for AD (Table 1). NAP interacts with neuronal and glial tubulin, promotes MT assembly, influences MT dynamics in postmitotic cells, and reduces phosphorylated Tau in vitro at fentomolar concentrations. Despite these observations, the precise mechanism of action for NAP in affecting Tau phosphorylation is unknown. Preclinical in vivo studies also suggest this peptide has very potent effects, with efficacy between 0.5–2 μg/day/mouse. Phase I clinical trials indicate that intranasal NAP is well tolerated and produces blood levels of compound that are consistent with preclinical efficacy [237].

3.2. Tau Glycosylation

In addition to phosphorylation, Tau is post-translationally modified by addition of carbohydrate side chains. In the AD brain, the level of the side chain O-linked β-N-acetylglucosamine (O-GlcNAc) is lower than that in control brains, indicating compromised glycosylation [239]. O-GlcNAc moieties are found on the hydroxyl side chains of serine and threonine residues. As phosphorylation and glycosylation can occur on the same residues of Tau, and these two events are mutually exclusive, glycosylation and phosphorylation exist in equilibrium. In a study suggesting a link between glucose metabolism and AD pathology, reduced blood glucose levels in starved mice led to reduction in O-GlcNAc on Tau, and a parallel increase in Tau hyperphosphorylation in the brains of mice [240]. This study suggests that glycosylation abnormalities might trigger Tau pathology and that preserving O-GlcNAc residues on Tau, thereby preventing excess phosphorylation, is another approach to maintain Tau function.

The glycoside hydrolase O-GlcNAcase removes GlcNAc units from Tau, while uridine diphosphate GlcNAc polypeptidyl transferase (OGT) transfers GlcNAc from the donor sugar

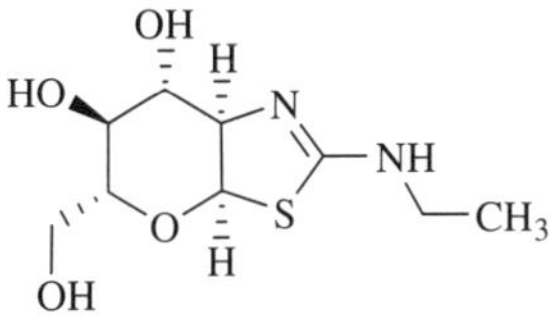

34 Thiamet G

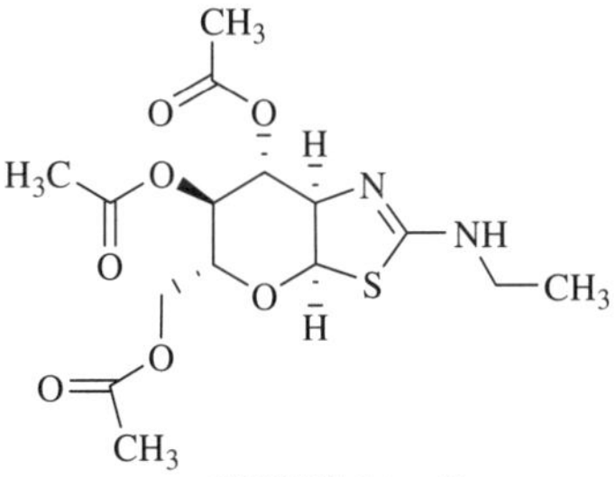

35 NAG-thiazoline

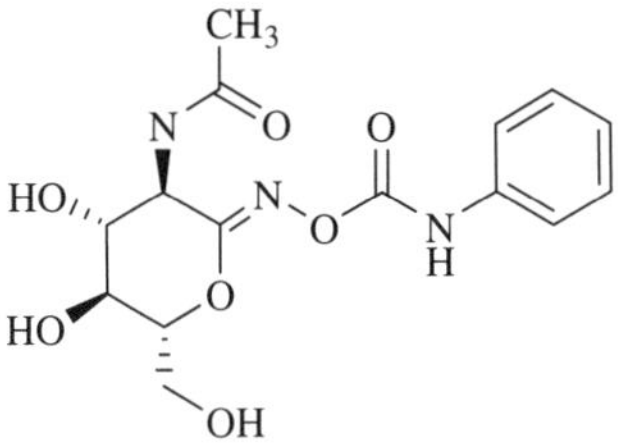

36 PUGNAc

Figure 11. O-GlcNAcase inhibitors.

uridine 5'-diphospho-GlcNAc to the target hydroxyl groups of acceptor proteins. Reduced activity of OGT, as demonstrated with shRNA KO of OGT [241], or increased activity of O-GlcNAcase shift the equilibrium towards phosphorylation. Therefore, modulating the activity of these enzymes could complement efforts to balance Tau kinase and phosphatase activities.

O-GlcNAcase inhibitors have been identified that effectively maintain serine and threonine glycosylation while reducing Tau phosphorylation, such as Thiamet G **34** [147], (Fig. 11). Other O-GlcNAcase inhibitors have been identified, however, these compounds and most derivatives are less selective (also inhibit β-hexosaminidase), less stable, and less brain-penetrant compared to Thiamet G, including NAG-thiazoline **35** and PUGNAc **36** (Fig. 11) [242]. To date, published results support the ability of Thiamet G to reduce Tau phosphorylation in healthy systems without disease-relevant pathology [147]. Long-term studies in disease models will be critical to further assess the potential of this approach to reverse pathological Tau toxicity.

3.3. Tau Aggregation

Tau filament accumulation correlates well with the degree of cognitive impairment in AD patients [243,244]. In addition, cell and animal models engineered to produce Tau fibrils demonstrate coincident neurodegeneration consistent with the toxic nature of these structures [168,188,245]. Tau protein in the AD brain assembles into highly ordered polymers as neuropil threads and NFTs within dystrophic neurites. While neuropil threads are composed of straight filaments, NFTs are composed of paired helical filaments (PHFs) that assemble in the somatodendritic compartment of neurons. Filament formation is thought to begin within the short hydrophobic residues of exposed MT binding repeats; however, the process leading to mature filaments containing antiparallel β-pleated sheets is poorly understood [246]. Hyperphosphorylation may initiate the process, driving Tau off the MT and increasing cytoplasmic concentrations. Tau fibril formation is facilitated *in vitro* following protease truncation at the C-terminus of Tau and the presence of anionic cofactors such as heparin or fatty acids. Tau truncation and fatty acid association with NFTs are also found in AD brain. The mechanisms by which Tau aggregates are toxic may include a reduction in proteosome activity, sequestration of normal Tau resulting in loss of normal function, or some physical obstruction in axonal transport due to the presence of the insoluble aggregates [247].

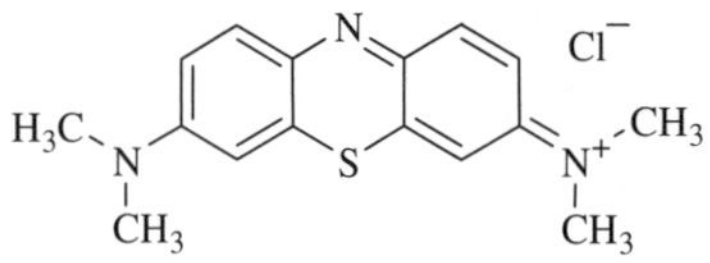

37 Methylene blue

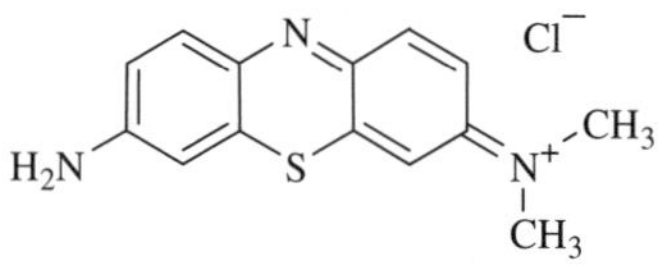

38 Azure A

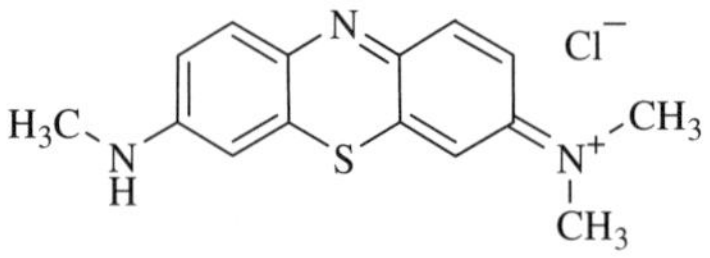

39 Azure B

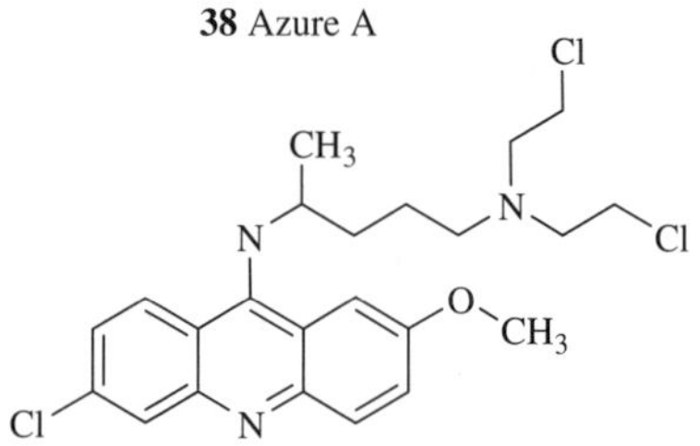

40 Quinicrine mustard

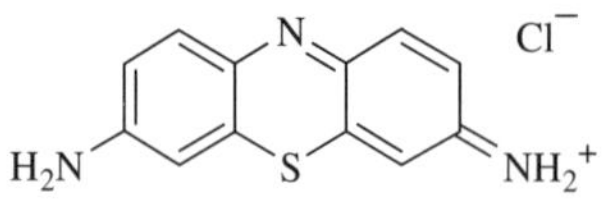

41 Thionine

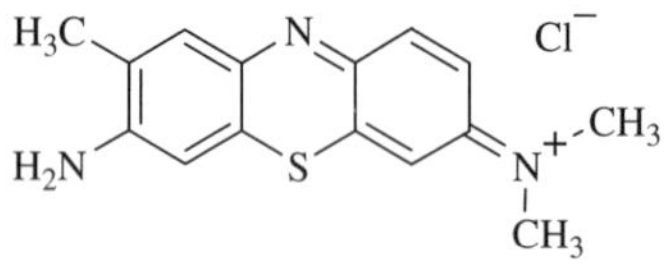

42 Toluidine blue

Figure 12. Phenothiazines.

Compounds that inhibit Tau polymerization have been identified using *in vitro* Tau fibrillization screening assays [148,248–250]; including phenothiazines [148,250], polyphenols and porphyrins [250], thiocarbocyanines [248], N-phenylamines, anthraquinones [249], phenylthiazolyl-hydrazides [251], thioxothiazolidinones (rhodanines) [252], aminothienopyridizines [253], and an extract of olive oil, oleocanthal [254] (Figs 12–19). These screening assays have used a range of wild-type Tau constructs with either 3- or 4-MT binding repeats, or Tau constructs containing mutations present in individuals with fronto-temporal dementia with Parkinsonism linked to chromosome 17 (FTDP-17) [255]. Relative fluorescence of Tau aggregates is measured typically using thioflavin S or T, which is sensitive to the interaction between the dye and the assembled β-structured fibrils. A review of the structure-activity relationship (SAR) and modeling of several families of Tau aggregation inhibitors has been recently published [256].

3.3.1. Phenothiazines The phenothiazine dye methylene blue (MB) **37** is orally active and has been used clinically for vasoplegia, [257], urinary tract infections, malaria, and in met-hemoglobulinemia [258] (Fig. 12). The active ingredient is methylthioninium chloride (MTC), the chloride salt of the methylthioninium moiety. As reported in 2008, MB was evaluated in a Phase II clinical trial for AD by Tau Rx Therapeutics [259]. In Phase II clinical trials, 30 and 60 mg tid doses demonstrated efficacy, while the highest dose (100 mg tid) was not efficacious and produced side effects, particularly on red blood cells and gut (Table 1). Whether these side effects are due to the ability of MTC to interfere with MT assembly in addition to Tau aggregation is unknown. At high dose, MTC cross-linked with the gelatin capsule material resulting in differential dissolution and absorption, which was not observed at the two lower doses. Subsequent studies demonstrated that the most effective form of MTC was the uncharged and colorless leuco-MT, or LMT. LMT is in preclinical development by Tau Rx.

The Phase II efficacy data gathered from 321 AD subjects over 6 months treatment with MB demonstrated 81% improvement in cognitive function compared to the placebo group [260]. Phase III studies are planned

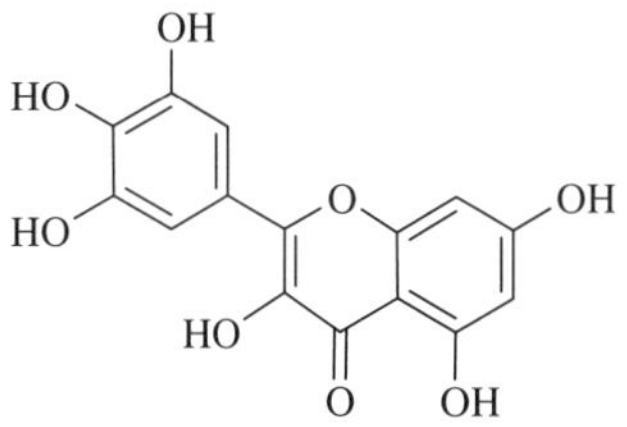

Figure 13. Polyphenols and porphyrin.

43 Myricetin
44 Exifone 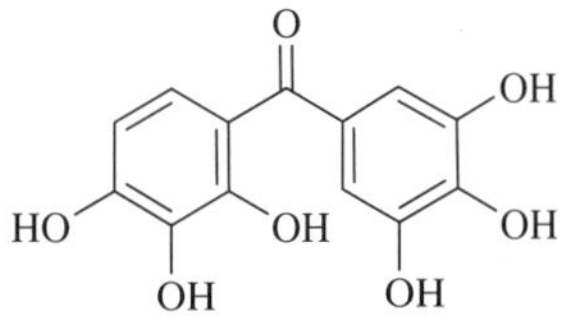
45 Gossypetin 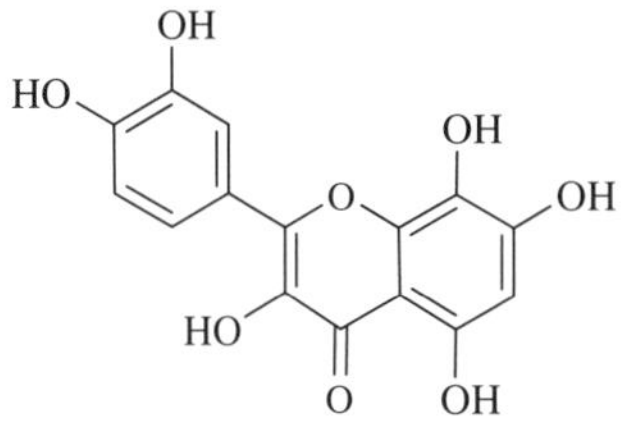
46 Hypericin 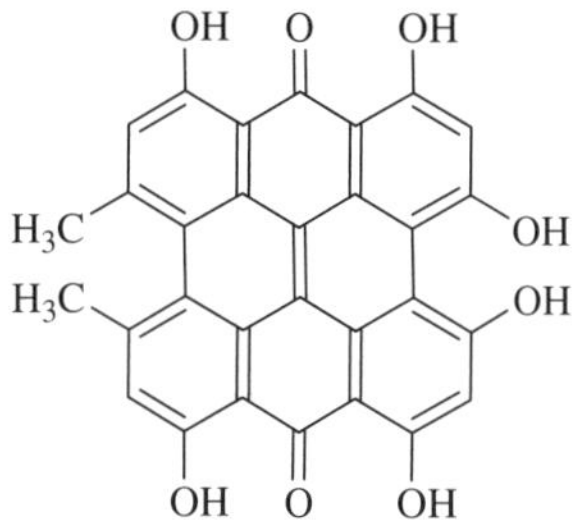
47 Phthalocyanine 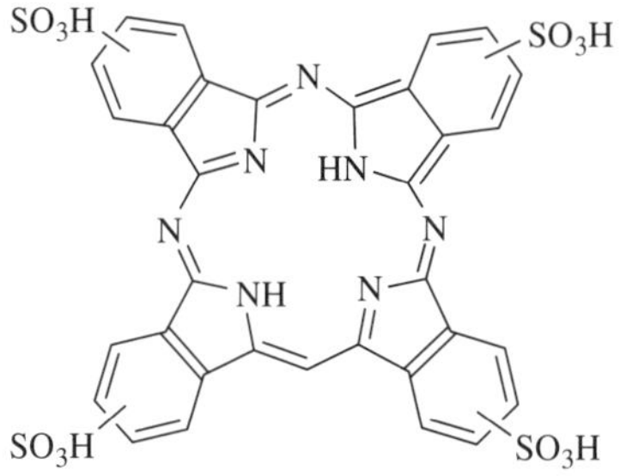

with the MTC form or the LMT form. Tau Rx is also planning studies in other closely related neurodegenerative disorders that have similar Tau pathology to AD, including progressive supranuclear palsy, cortico basal degeneration, and FTDP-17. Despite encouraging clinical data, few *in vivo* preclinical studies have been reported with MB. A zebrafish model of Tau-induced neurodegeneration demonstrated no reduction in Tau hyperphosphorylation or neuronal rescue following treatment with MB [261]. Additional active phenothiazines were confirmed by two groups, including MB, azure A **38**, azure B **39**, quinacrine mustard **40**, thionine **41**, and toluidine blue or tolonium chloride **42** [148,250] (Fig. 12). MB, azure B, azure A were studied by both groups and while the potency of MB was comparable, azure B and azure A were approximately 10-fold more potent as previously reported [260], likely due to assay differences. Of this class, only MB has advanced to the clinic as an inhibitor targeting Tau aggregation.

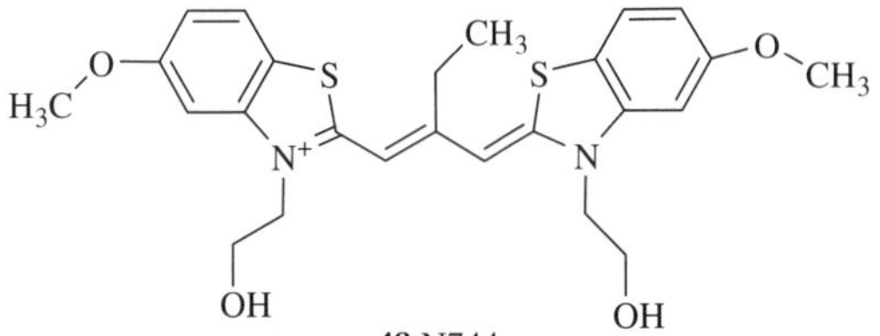

48 N744

Figure 14. Thiocarbocyanine N744.

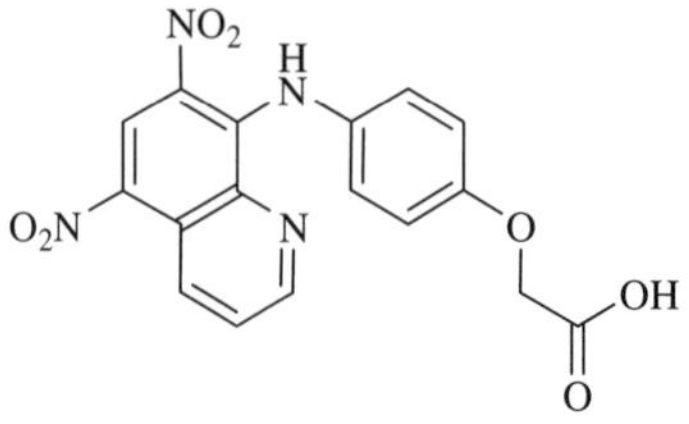

49 B4D5

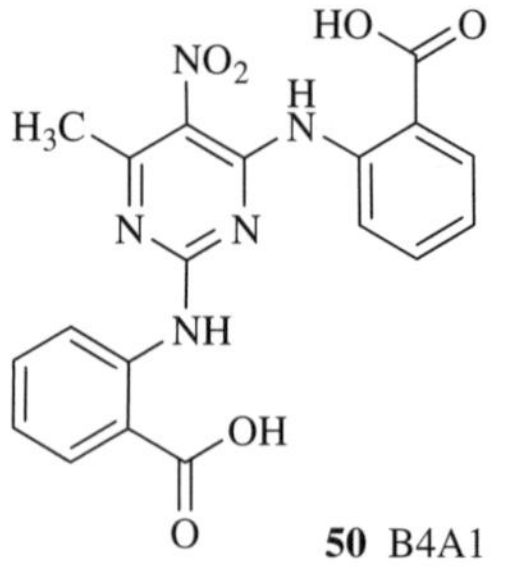

50 B4A1

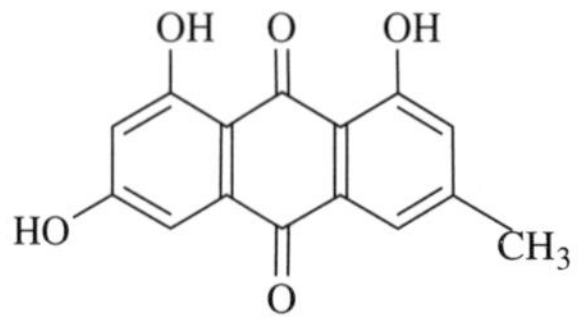

51 Emodin

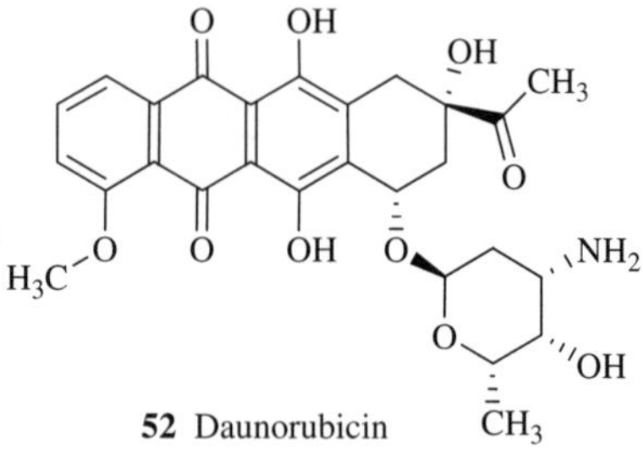

52 Daunorubicin

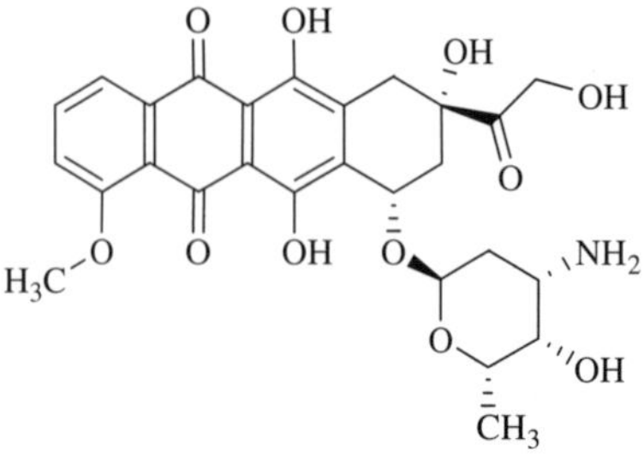

53 Adriamycin (doxorubicin)

Figure 15. *N*-Phenyl hydrazine derivatives and anthraquinones.

3.3.2. Polyphenols and Porphyrins Polyphenols identified as Tau aggregation inhibitors also inhibited Aβ peptide aggregation [250]. These compounds included myricetin **43**, exifone **44**, and gossypetin **45** as most potent for Tau aggregation inhibition with IC50 of 1.2, 1.8, and 2.0 µM, respectively (Fig. 13). Of the nine additional polyphenols described, the potency observed for Tau was not always similar to that for amyloid, highlighting differences between Tau and Aβ aggregates despite the common β-pleated sheet structure. For example, both myricetin and hypericin **46** (Fig. 13) demonstrated an IC_{50} of 0.9 µM against amyloid aggregation, while their potencies against Tau aggregation were 1.2 and 26.8 µM IC_{50}, respectively. Interestingly, MB had different effects on Aβ fibrils and oligomers, promoting Aβ fibril formation while reducing Aβ oligomer formation [262]. While exifone inhibited Tau:Tau interactions, it did not inhibit Tau:MT interactions, or the process of MT assembly at concentrations between 0.2–20 µM. Regarding the mechanism by which polyphenols inhibit Tau aggregation, exifone did not block Tau binding to heparin. This suggests that the inhibition does not stem from interference in heparin's polyanion function in the Tau aggregation process. In addition, myricetin bound more strongly to aged and filamentous soluble Tau than to the monomeric protein, suggesting an effect at the level of higher order aggregation species. The porphyrin phthalocyanine **47** bound all three Tau species and demonstrated particular efficacy at depolymerizing Tau aggregates [250] (Fig. 13). The

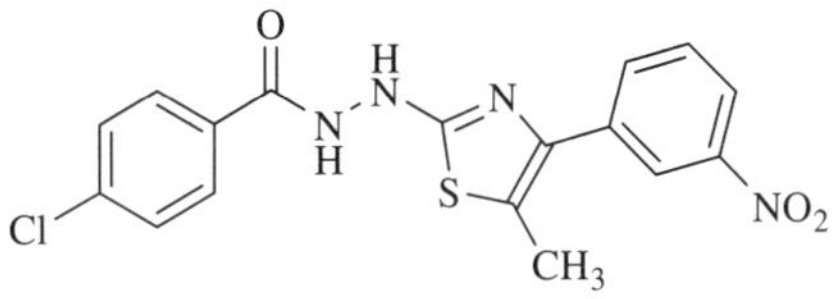

54 BSc2463

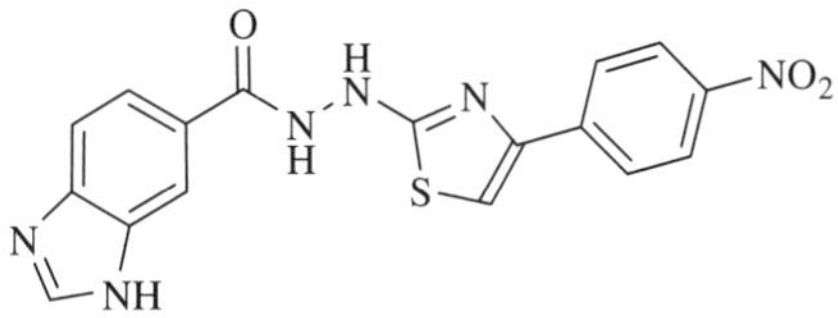

55 BSc3094

Figure 16. Phenylthiazolyl-hydrazines.

polyphenols and porphyrins do not cross the BBB, precluding their clinical development.

3.3.3. Thiocarbocyanines
Substochiometric levels of the cyanine dye N744 **48** (Fig. 14) inhibited the elongation phase of Tau fibrillization in the presence of 100-fold excess fatty acid inducer. This compound also drove end depolymerization of preformed fibrils in

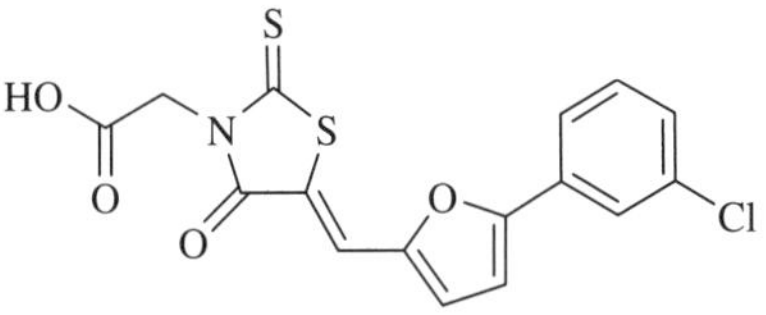

56

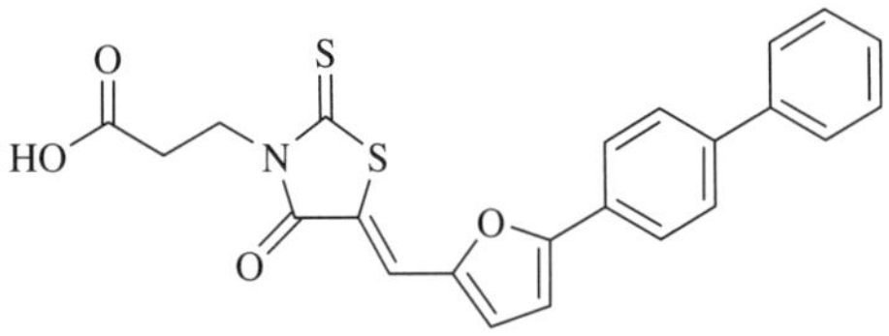

57

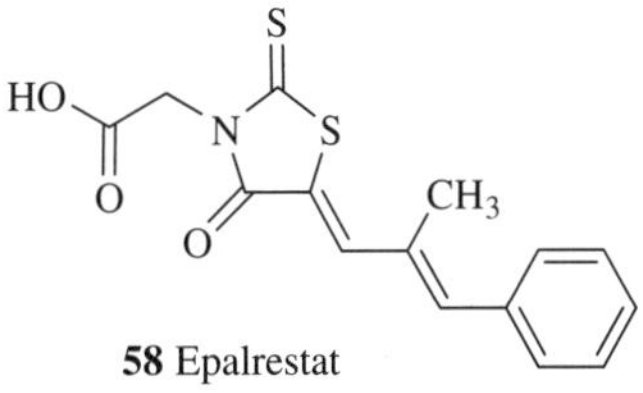

58 Epalrestat

Figure 17. Rhodanines.

a substochiometric manner, with an IC_{50} of approximately 0.3 µM [263]. While a thiocarbocyanine dye is approved for use in human angiography (Cardio-green, FDA application 011525) [264], N744 demonstrated complex kinetics, highlighted by a narrow efficacious range in the presence of posttranslational Tau modifications such as phosphorylation and C-terminal truncation that may preclude this family of Tau aggregation inhibitors from use *in vivo* [265].

3.3.4. N-Phenylamines and Anthraquinones
The N-phenylamines B4D5 **49** and B4A1 **50** (Fig. 15) inhibited Tau aggregate formation and fostered aggregate disassembly with an IC_{50} of 10 µM *in vitro*. These effects also correlated with a reduction in cellular toxicity associated with aggregation of Tau repeat regions in N2a cells engineered to overexpress these hydrophobic fragments of Tau in an inducible system [245,249]. The anthraquinones were identified from a 200,000 compound library screened for inhibitors of Tau aggregation [149] and were represented by emodin **51**, daunorubicin **52**, and adriamycin **53** (Fig. 15). These compounds inhibited filament formation with IC_{50} values ranging 1–5 µM and also disassembled preformed filaments at EC_{50} values ranging 2–4 µM. These compounds did not interfere with Tau/MT stabilization yet inhibited formation of Tau aggregates and cytotoxicity in neuroblastoma cells overexpressing Tau. Compounds from the N-phenylamine series demonstrated significantly lower *in vitro* toxicity compared to the anthraquinones [149]; however, the toxicity associated with both of these series prevented *in vivo* testing.

3.3.5. Phenylthiazoyly-Hydrazides
Phenylthiazolyl-hydrazide BSc2463 **54** was identified from an *in silico* screen of an initial library hits [249,251] (Fig. 16). This compound was used as a template for synthesis of 49 additional compounds leading to BSc3094 **55**, with a 6-fold improved IC_{50} of 1.6 µM for inhibition of Tau polymerization, and a 15-fold improved potency of 0.7 µM for Tau aggregate depolymerization. While these compounds were noted as interesting leads, further optimization to improve oral bioavailability, metabolic stability,

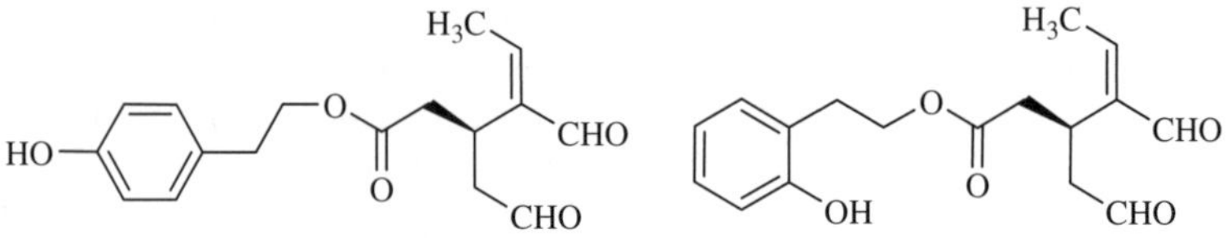

59 (−)Oleocanthal

60

Figure 18. Dialdehydes.

aqueous solubility, and BBB permeability would be required for further development. The phenylthiazolyl-hydrazides displayed comparable activity in cellular assays but reduced biochemical potency compared to the rhodanines described below.

3.3.6. Rhodanines The rhodanine scaffold identified by Bulic et al. [256] resulted in the most potent Tau aggregation inhibitors reported, leading to structures **56** and **57** (Fig. 17) with potencies below 200 nM, both to inhibit aggregation and to dissolve preformed fibrils. The SAR analysis for this group of potent Tau aggregation inhibitors revealed that the central rhodanine scaffold preserved compound efficacy better than alternate scaffolds including thiohydantoin, thioxooxazolidine, oxazolidinedione, or hydantoin [256]. The carboxylic acid substituent was also important for preserving efficacy, although increasing the length of the carbon chain improved potency as evident from compound **56** to compound **57**. The presence of an aromatic side chain was also necessary for activity. The rhodanine-based compound epalrestat **58** (Fig. 17), an aldose reductase inhibitor, was safe in a long-term clinical study to treat diabetic peripheral neuropathy [266]. While these inhibitors demonstrated minimal *in vitro* toxicity at 10 µM, compound **57** inhibited Tau aggregation and enhanced Tau disassembly in the cell-free system at EC_{50} values of 0.17 and 0.13 µM, respectively. These compounds also had minimal effect at 60 µM on Tau-MT binding, as measured by a cell-free tubulin polymerization assay [252]. The $c \log p$ of most compounds (between 0.59 and 5.3) suggests low membrane permeability, which is reflected in the large shift between cell-free and cell-based potency. Compound **56** had submicromolar IC_{50} against

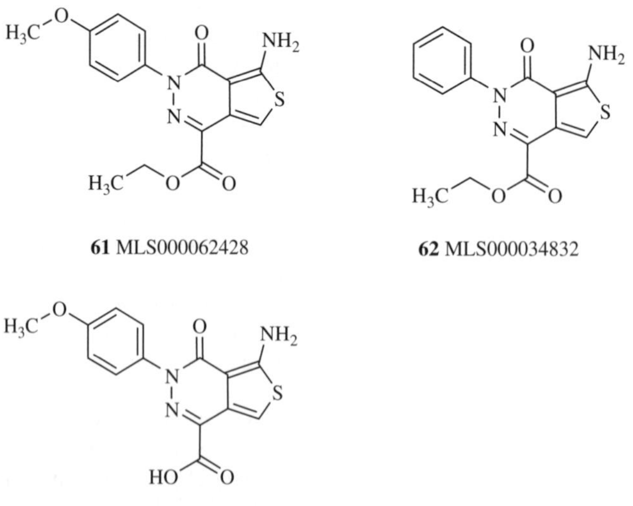

61 MLS000062428

62 MLS000034832

63 NCGC00182500

Figure 19. Aminothienopyridizines.

cell-free Tau aggregation and inhibition of Tau fibril formation (820 and 100 nM, respectively), while cellular Tau aggregation was inhibited by only 20% at 15 µM. Improved ADME properties would be needed to allow development of these compounds as AD therapeutics.

3.3.7. Recently Reported Series The most recently described Tau aggregation inhibitors include the polyphenol (-)-oleocanthal **59** (OC) and the aminothienopyridizine family (ATPZ) [253,254]. OC is the dialdehyde of (-)-deacetoxy-ligstroside aglycone found in extra virgin olive oil (Fig. 18). While this compound demonstrated properties of a nonsteroidal anti-inflammatory drug (NSAID) [267], it also blocked aggregation of both Aβ and Tau protein while other NSAIDs did not [254]. In the case of Tau, OC interacted with a random coil configuration and prevented conversion to β-pleated sheets as visualized using Fourier transform IR. The mechanism by which OC interfered with Tau fibril formation was via dialdehyde interaction and adduct formation with lysine moieties, likely within the third MT binding domain. Interestingly, when one of these lysines was removed or modified, Tau lost the ability to form filaments [268]. OC potency against aggregation of mutant truncated Tau was approximately 3 µM IC$_{50}$, while 10 µM OC achieved 50% inhibition of aggregation against full length Tau. However, it should be considered that compound activity was assessed for 15 days in full length Tau assay versus 24 h in the assay utilizing mutant truncated Tau. Ten additional OC analogs, including the dialdehyde preserved in **60** (Fig. 18), were also potent inhibitors of *in vitro* Tau aggregation. However, as OC apparently induces Tau cross-linking as its inhibitory mechanism, this series would likely lead to toxicity associated with covalent modification of other proteins in the body. The mechanism of targeting these lysine residues in Tau is one approach to future therapeutic development.

The ATPZ series was identified from a screen of 292,000 compounds using a truncated four MT binding repeat Tau carrying a FTDP-17 mutation coupled with both thioflavin T (ThT) and fluorescence polarization (FP) readouts [253] (Fig. 19). Special attention was paid to compounds that reduced both ThT and FP signal with comparable IC$_{50}$ values of between 1 to 10 µM. In this assay, FP signal represented the smaller Tau oligomers, while ThT signal represented Tau filaments forming β-pleated sheets. The maximal inhibition observed in the ThT assay was typically greater than that observed with FP, suggesting the ATPZ compounds caused greater inhibition of cross-β-fibril structures with less inhibition of Tau oligomer formation. Size exclusion chromatography in the presence of Tau aggregates treated with MLS000062428 **61** (Fig. 19) also suggested an increase in the Tau oligomer-to-fibril ratio in the presence of compound. Since 15 µM Tau was present in the screen, a stoichiometry of approximately 1:1 compound-to-Tau molar ratio was required for maximal inhibition. In addition to the ATPZ family, catecholamines, thiourea, hydrazinyl-dihydrothiozole, and benzimidazole compounds were identified as hits in the screen. Neither of the ATPZ compound hits MLS000062428 or MLS000034832 **62** (both with IC$_{50}$ of 10 µM) interfered with the ability of Tau to support MT assembly nor were they potent inhibitors of Aβ fibril formation. Twenty-one additional ATPZ analogs with drug-like properties were made with nine demonstrating improved potency compared to **61** or **62**. The most potent analog with 2.5 µM IC$_{50}$ in the cell-free assay was NCGC00182500 **63** (Fig. 19). As several reports have speculated to whether the toxic Tau species is the oligomer or fibril and whether hyperphosphorylated Tau is sufficient, it will be important to assess if the increase in oligomer load with ATPZ compounds leads to any evidence of *in vitro* or *in vivo* toxicity.

All approaches to Tau aggregation target Tau protein-to-protein interactions (PPIs). Although there are approved antibody therapeutics targeting PPI such as Herceptin [269], small molecules targeting PPI are still being evaluated in clinical trials [270]. Also, because many of the above mentioned agents are effective only at concentrations approaching that of the aggregated Tau, these compounds would have to be extremely selective to maintain adequate safety margins for therapeutic use.

3.4. Microtubule Stabilization

Though it is still unclear whether the toxicity attributed to Tau in AD stems from a loss-of-function regarding support of MT structure or a gain of toxic function as discussed above, evidence from the literature supports the idea that MT stabilizers would improve neuronal function in AD [271,272]. Despite the well-documented synaptic deficiency in AD [273,274], there are contradictory results regarding whether axonal transport deficits exist in AD-relevant models [275–277]. Adalbert et al. [278] demonstrated that an overwhelming majority of dystrophic axons remain continuous 4 months after dystrophy first arises in APP-Tg mouse model. Because neuronal cell bodies appeared normal despite evidence for a partial block in axonal transport and synaptic loss, the authors concluded that the therapeutic window for rescuing individual neurons could be relatively long.

Paclitaxel **64**, a tubulin binding agent and marketed anticancer drug (Paxceed™) (Fig. 20), restored fast axonal transport and improved motor dysfunction in a Tau transgenic mouse [272]. Paclitaxel, a complex diterpene from the bark of the Pacific yew (*Taxus brevifolia*) [279], stabilizes MTs in the mitotic spindle in a substochiometric manner (1 molecule per 100 molecules of tubulin) leading to apoptosis of rapidly replicating tumor cells [280]. Paclitaxel is frequently prescribed for a number of malignancies including ovarian, breast, and lung cancers. Additional agents with similar mechanisms of action have been approved for oncology indications, including the epothilone ixabepilone **65**, with several newer classes in preclinical development including derivatives of discodermolide **66**, elutherobin and laulimalide [281]. Epothilone A, discodermolide, as well as the paclitaxel analog UK-100 showed similar neuroprotection to paclitaxel following Aβ challenge [271]. Despite these promising results, this class of compounds inhibits normal MT dynamics, demonstrates limited brain bioavailability due to P-glycoprotein transport [282] and, since many are natural products with only partial synthetic routes, supply is limiting. In addition, neuropathy is a major adverse effect of MT stabilizing agents, with severe peripheral neuropathy occurring in as many as 30% of treated patients. Advance of these compounds into the clinic for AD will be dependent on whether these challenges can be overcome.

3.5. Tau Protein Levels

Both genetic evidence and studies from Tau KO animals suggest that reducing Tau protein levels may be a viable therapeutic approach for AD. Since Tau haplotypes driving slightly higher Tau expression increase AD risk [155], reducing Tau levels might be protective. Roberson et al. [154] determined the effect of

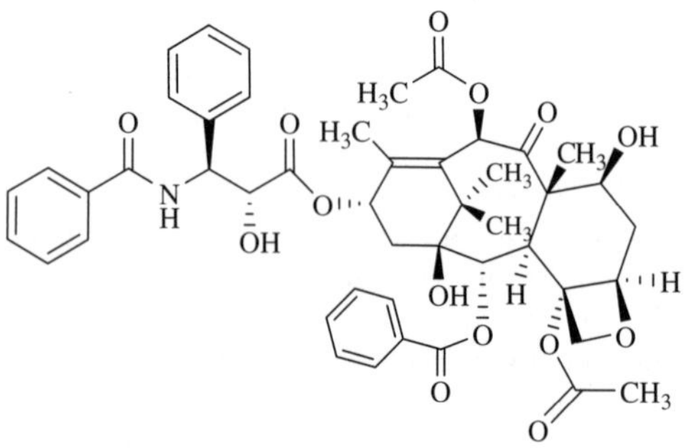

64 Paclitaxel

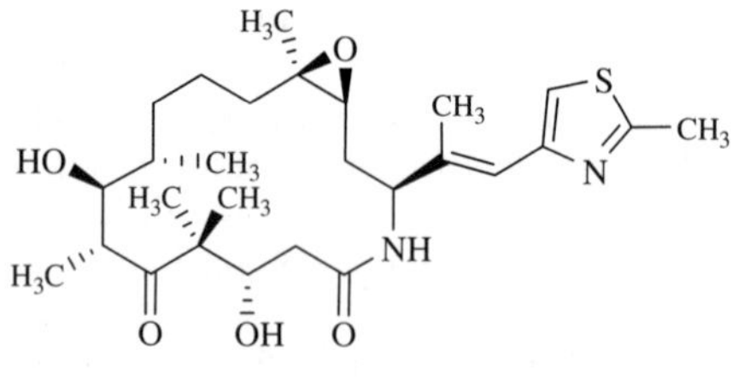

65 Ixabepilone

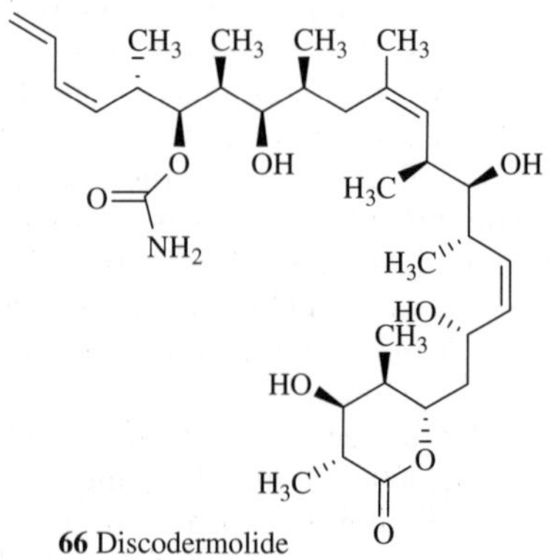

66 Discodermolide

Figure 20. Microtubule stabilizers.

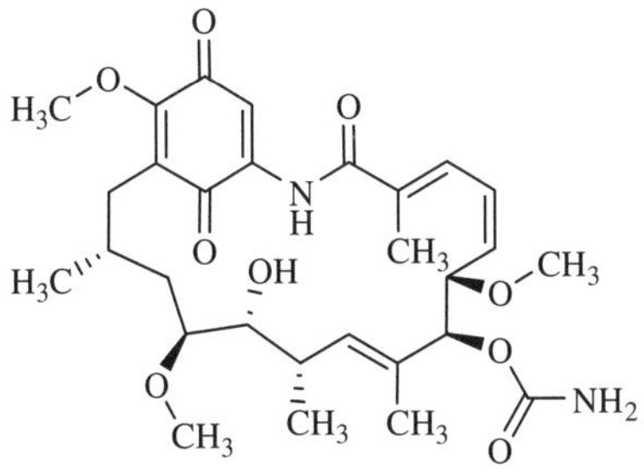

67 Geldanamycin 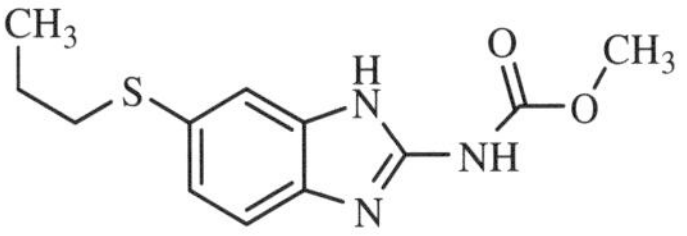   **68** Albendazole

Figure 21. Tau lowering agents.

reducing endogenous Tau expression on cognitive deficits in APP-Tg mice. In this case, Tau elimination was achieved via breeding strategies between Tau KO and APP-Tg mice. Interestingly, the behaviors and death phenotype were rescued without changes in levels of either Aβ protein or amyloid pathology in the Tau KO × APP-Tg line, suggesting that rescue occurred downstream of Aβ through changes associated with reduced Tau level. Additional support for this idea came from *in vitro* studies of primary neurons from Tau KO mice [283]. These neurons were resistant to Aβ toxicity compared to wild-type neurons expressing mouse Tau or neurons overexpressing human Tau, again suggesting that Tau protein is necessary for neurotoxicity downstream of Aβ.

One approach to reduce Tau levels is by increasing its clearance. Tau is normally present as a soluble and natively unfolded protein with a random coil structure [284]. Although the mechanism by which Tau changes from a soluble state to an insoluble aggregate is unclear, molecular chaperones such as heat shock proteins (Hsps) have been implicated [285,286]. Hsps function primarily to protect proteins, but they also present misfolded proteins for turnover to either the ubiquitin-proteosome pathway or lysosome-mediated autophagy. While *in vitro* evidence supported a role for Hsp-90 and Hsc-70-interacting chaperone protein (CHIP) in Tau degradation [287–289], deletion of CHIP in mice did not induce Tau aggregation [290,291] confirming additional pathways for Tau turnover.

A compound screen to identify Tau lowering agents employed human H4 glioma cells with robust Tau expression and both three- and four-repeat microtubule binding domains [153,292]. This screen was validated using Hsp-90 inhibitors including geldanamycin (GA) **67** (Fig. 21), which reduces Hsp-90 function by interacting with the ATP binding pocket and inhibiting ATP hydrolysis. Induction of Hsp response is also triggered by compounds like GA via activation of the heat shock factor-1 (HSF-1) transcription [293]. A small library of predominantly Hsp-90 inhibitors as well as marketed compounds was screened for Tau inhibition [153,292], but the individual structures of these Hsp-90 inhibitors are not available. Tau protein reduction caused by these inhibitors was accompanied by an increase in the chaperones Hsp-70, Hsp-40, and Hsp-27 levels, consistent with a role for these proteins in mediating Tau protein turnover. Nine marketed compounds (diazaquone, MB, alexidine HCl, colchicine, albendazole, chelidonine, rotenone, lasalocid salt, and norethindrone) were able to reduce Tau protein levels by >25%, while GAPDH control protein levels are reduced by <10% [292]. Only two, MB and albendazole **68**, were sufficiently potent, nontoxic (10-fold separation between the IC_{50} for Tau reduction and cell toxicity) and selective (reduced Tau and not microtubule-associated protein MAP2) (Fig. 20). Albendazole, a benzimidazole used to treat worm infestation, acts to reduce glucose uptake in the worm via binding to the colchicine-sensitive site of tubulin inhibiting MT assembly. Not surprisingly, most of the identified compounds act either via effects on MT dynamics/cell proliferation or protein filament formation/

aggregation. Albendazole and paclitaxel also reduced Tau transcription, however, the other compounds either elevated Tau transcription, such as MB, or had no effect. One caveat to these experiments is that the Tau was not present in an aggregated or toxic state so that efficacy would have to be confirmed in a more disease-like system.

A second mechanism for Tau clearance is via immunotherapy. Immunogens against pTau 396-404 have demonstrated efficacy when injected into animals with the FTDP-17 mutation P301L [294]. It is intriguing that this approach was effective since Tau is an intracellular protein. Since there are reports of antibody uptake into cells [295,296], intracellular tangles and pretangles may be cleared in addition to extracellular NFTs that are no longer contained within a neuronal membrane. Because previous Aβ immunotherapy approach for AD evaluated in clinical trials showed meningoencephalitis in 6% of the active treatment group [297], special care must be taken to ensure safety of future immunotherapy approaches in this elderly population.

REFERENCES

1. Selkoe DJ. The molecular pathology of Alzheimer's disease. Neuron 1991;6(4): 487–498.
2. Sisodia SS, St George-Hyslop PH. Gamma-Secretase, Notch, Abeta and Alzheimer's disease: where do the presenilins fit in? Nat Rev Neurosci 2002;3(4):281–290.
3. Vetrivel KS, Thinakaran G. Amyloidogenic processing of beta-amyloid precursor protein in intracellular compartments. Neurology 2006;66(2 Suppl 1):S69–73.
4. Lee VM, Trojanowski JQ. Progress from Alzheimer's tangles to pathological tau points towards more effective therapies now. J Alzheimers Dis 2006;9(3 Suppl):257–262.
5. Iqbal K, Grundke-Iqbal I. Discoveries of tau, abnormally hyperphosphorylated tau and others of neurofibrillary degeneration: a personal historical perspective. J Alzheimers Dis 2006;9(3 Suppl):219–242.
6. Hyman BT, Augustinack JC, Ingelsson M. Transcriptional and conformational changes of the tau molecule in Alzheimer's disease. Biochim Biophys Acta 2005;1739(2–3): 150–157.
7. Holmes C, Boche D, Wilkinson D, Yadegarfar G, Hopkins V, Bayer A, Jones RW, Bullock R, Love S, Neal JW, Zotova E, Nicoll JA. Long-term effects of Abeta42 immunisation in Alzheimer's disease: follow-up of a randomised, placebo-controlled phase I trial. Lancet 2008;372(9634):216–223.
8. Yaffe K. Metabolic syndrome and cognitive decline. Curr Alzheimer Res 2007;4(2): 123–126.
9. Craf S. The role of metabolic disorders in Alzheimer disease and vascular dementia: two roads converged. Arch Neurol 2009;66(3): 300–305.
10. Sparks DL, Martin TA, Gross DR, Hunsaker JC 3rd. Link between heart disease, cholesterol, and Alzheimer's disease: a review. Microsc Res Tech 2000;50(4):287–290.
11. Austen B, Christodoulou G, Terry JE. Relation between cholesterol levels, statins and Alzheimer's disease in the human population. J Nutr Health Aging 2002;6(6):377–382.
12. Luchsinger JA, Gustafson DR. Adiposity and Alzheimer's disease. Curr Opin Clin Nutr Metab Care 2009;12(1):15–21.
13. Cornier MA, Dabelea D, Hernandez TL, et al The metabolic syndrome. Endocr Rev 2008;29 (7):777–822.
14. Dietschy JM. Central nervous system: cholesterol turnover, brain development and neurodegeneration. Biol Chem 2009;390(4): 287–293.
15. Lund EG, Xie C, Kotti T, Turley SD, Dietschy JM, Russell DW. Knockout of the cholesterol 24-hydroxylase gene in mice reveals a brain-specific mechanism of cholesterol turnover. J Biol Chem 2003;278(25):22980–22988.
16. Kotti TJ, Ramirez DM, Pfeiffer BE, Huber KM, Russell DW. Brain cholesterol turnover required for geranylgeraniol production and learning in mice. Proc Natl Acad Sci USA 2006;103(10):3869–3874.
17. Veinbergs I, Mallory M, Mante M, Rockenstein E, Gilbert JR, Masliah E. Altered long-term potentiation in the hippocampus of apolipoprotein E-deficient mice. Neurosci Lett 1999;265 (2–3):218–222.
18. Yaffe K, Barrett-Connor E, Lin F, Grady D. Serum lipoprotein levels, statin use, and cognitive function in older women. Arch Neurol 2002;59(3):378–384.
19. Wolozin B. Cholesterol, statins and dementia. Curr Opin Lipidol 2004;15(6):667–672.
20. Fassbender K, Simons M, Bergmann C, Stroick M, Lutjohann D, Keller P, Runz H, Kuhl S, Bertsch T, von Bergmann K, Hennerici

M, Beyreuther K, Hartmann T. Simvastatin strongly reduces levels of Alzheimer's disease beta-amyloid peptides Abeta 42 and Abeta 40 in vitro and in vivo. Proc Natl Acad Sci USA 2001;98(10):5856–5861.

21. Simons M, Keller P, De Strooper B, Beyreuther K, Dotti CG, Simons K. Cholesterol depletion inhibits the generation of beta-amyloid in hippocampal neurons. Proc Natl Acad Sci USA 1998;95(11):6460–6464.

22. Liu JP, Tang Y, Zhou S, Toh BH, McLean C, Li H. Cholesterol involvement in the pathogenesis of neurodegenerative diseases. Mol Cell Neurosci 2010; 43(1):33–42.

23. Strittmatter WJ, Saunders AM, Schmechel D, Pericak-Vance M, Enghild J, Salvesen GS, Roses AD. Apolipoprotein E: high-avidity binding to β-amyloid and increased frequency of type 4 allele in late-onset Alzheimer disease. Proc Natl Acad Sci USA 1993;90(5): 1977–1981.

24. Bu G. Apolipoprotein E and its receptors in Alzheimer's disease: pathways, pathogenesis and therapy. Nat Rev Neurosci 2009;10(5): 333–344.

25. Walker JF, Tobert JA. The clinical efficacy and safety of lovastatin and MK-733: an overview. Eur Heart J 1987;8(Suppl E):93–96.

26. Evans BA, Evans JE, Baker SP, Kane K, Swearer J, Hinerfeld D, Caselli R, Rogaeva E, St George-Hyslop P, Moonis M, Pollen DA. Long-term statin therapy and CSF cholesterol levels: implications for Alzheimer's disease. Dement. Geriatr Cogn Disord 2009;27(6): 519–524.

27. Hoglund K, Thelen KM, Syversen S, Sjogren M, von Bergmann K, Wallin A, Vanmechelen E, Vanderstichele H, Lutjohann D, Blennow K. The effect of simvastatin treatment on the amyloid precursor protein and brain cholesterol metabolism in patients with Alzheimer's disease. Dement Geriatr Cogn Disord 2005;19 (5–6):256–265.

28. Sparks DL, Sabbagh M, Connor D, Soares H, Lopez J, Stankovic G, Johnson-Traver S, Ziolkowski C, Browne P. Statin therapy in Alzheimer's disease. Acta Neurol Scand Suppl 2006;185:78–86.

29. van der Most PJ, Dolga AM, Nijholt IM, Luiten PG, Eisel UL. Statins: mechanisms of neuroprotection. Prog Neurobiol 2009;88(1):64–75.

30. Buhman KF, Accad M, Farese RV. Mammalian acyl-CoA:cholesterol acyltransferases. Biochim Biophys Acta 2000;15;1529(1–3): 142–154.

31. Tabas I. Nonoxidative modifications of lipoproteins in atherogenesis. Annu Rev Nutr 1999;19:123–139.

32. Rudel LL, Lee RG, Cockman TL. Acyl coenzyme A: cholesterol acyltransferase types 1 and 2: structure and function in atherosclerosis. Curr Opin Lipidol 2001;12(2):121–127.

33. Puglielli L, Konopka G, Pack-Chung E, Ingano LA, Berezovska O, Hyman BT, Chang TY, Tanzi RE, Kovacs DM. Acyl-coenzyme A: cholesterol acyltransferase modulates the generation of the amyloid beta-peptide. Nat Cell Biol 2001;3(10):905–912.

34. Huttunen HJ, Greco C, Kovacs DM. Knockdown of ACAT-1 reduces amyloidogenic processing of APP. FEBS Lett 2007;581(8):1688–1692.

35. Hutter-Paier B, Huttunen HJ, Puglielli L, Eckman CB, Kim DY, Hofmeister A, Moir RD, Domnitz SB, Frosch MP, Windisch M, Kovacs DM. The ACAT inhibitor CP-113,818 markedly reduces amyloid pathology in a mouse model of Alzheimer's disease. Neuron 2004;44 (2):227–238.

36. Meuwese MC, de Groot E, Duivenvoorden R, Trip MD, Ose L, Maritz FJ, Basart DC, Kastelein JJ, Habib R, Davidson MH, Zwinderman AH, Schwocho LR, Stein EA, CAPTIVATE Investigators ACAT inhibition and progression of carotid atherosclerosis in patients with familial hypercholesterolemia: the CAPTIVATE randomized trial. JAMA 2009;301(11): 1131–1139.

37. Farese RV Jr. The nine lives of ACAT inhibitors. Arterioscler Thromb Vasc Biol 2006;26 (8):1684–1686.

38. Zelcer N, Tontonoz P. Liver X receptors as integrators of metabolic and inflammatory signaling. J Clin Invest 2006;116(3):607–614.

39. Chen W, Chen G, Head DL, Mangelsdorf DJ, Russell DW. Enzymatic reduction of oxysterols impairs LXR signaling in cultured cells and the livers of mice. Cell Metab 2007;5(1):73–79.

40. Beaven SW, Tontonoz P. Nuclear receptors in lipid metabolism: targeting the heart of dyslipidemia. Annu Rev Med 2006;57:313–329.

41. Tangirala RK, Bischoff ED, Joseph SB, Wagner BL, Walczak R, Laffitte BA, Daige CL, Thomas D, Heyman RA, Mangelsdorf DJ, Wang X, Lusis AJ, Tontonoz P, Schulman IG. Identification of macrophage liver X receptors as inhibitors of atherosclerosis. Proc Natl Acad Sci USA 2002;99(18):11896–11901.

42. Lund EG, Menke JG, Sparrow CP. Liver X receptor agonists as potential therapeutic agents for dyslipidemia and atherosclerosis. Arterioscler Thromb Vasc Biol 2003;23(7): 1169–1177.

43. Joseph SB, McKilligin E, Pei L, Watson MA, Collins AR, Laffitte BA, Chen M, Noh G, Good-

man J, Hagger GN, Tran J, Tippin TK, Wang X, Lusis AJ, Hsueh WA, Law RE, Collins JL, Willson TM, Tontonoz P. Synthetic LXR ligand inhibits the development of atherosclerosis in mice. Proc Natl Acad Sci USA 2002;99(11): 7604–7609.

44. Lund EG, Peterson LB, Adams AD, Lam MH, Burton CA, Chin J, Guo Q, Huang S, Latham M, Lopez JC, Menke JG, Milot DP, Mitnaul LJ, Rex-Rabe SE, Rosa RL, Tian JY, Wright SD, Sparrow CP. Different roles of liver X receptor alpha and beta in lipid metabolism: effects of an alpha-selective and a dual agonist in mice deficient in each subtype. Biochem Pharmacol 2006;71(4):453–463.

45. Bradley MN, Hong C, Chen M, Joseph SB, Wilpitz DC, Wang X, Lusis AJ, Collins A, Hseuh WA, Collins JL, Tangirala RK, Tontonoz P. Ligand activation of LXR beta reverses atherosclerosis and cellular cholesterol overload in mice lacking LXR alpha and apoE. J Clin Invest 2007;117(8):2337–2346.

46. Singh-Manoux A, Gimeno D, Kivimaki M, Brunner E, Marmot MG. Low HDL cholesterol is a risk factor for deficit and decline in memory in midlife: the Whitehall II study. Arterioscler. Thromb Vasc Biol 2008;28(8):1556–1562.

47. Wang L, Schuster GU, Hultenby K, Zhang Q, Andersson S, Gustafsson JA. Liver X receptors in the central nervous system: from lipid homeostasis to neuronal degeneration. Proc Natl Acad Sci USA 2002;99(21):13878–13883.

48. Whitney KD, Watson MA, Collins JL, Benson WG, Stone TM, Numerick MJ, Tippin TK, Wilson JG, Winegar DA, Kliewer SA. Regulation of cholesterol homeostasis by the liver X receptors in the central nervous system. Mol Endocrinol 2002;16(6):1378–1385.

49. Repa JJ, Li H, Frank-Cannon TC, Valasek MA, Turley SD, Tansey MG, Dietschy JM, Liver X receptor activation enhances cholesterol loss from the brain, decreases neuroinflammation, and increases survival of the NPC1 mouse. J Neurosci 2007;27 (52):14470–14480.

50. Wahrle SE, Jiang H, Parsadanian M, Legleiter J, Han X, Fryer JD, Kowalewski T, Holtzman DM. ABCA1 is required for normal central nervous system ApoE levels and for lipidation of astrocyte-secreted apoE. J Biol Chem 2004;279(39):40987–40993.

51. Hirsch-Reinshagen V, Zhou S, Burgess BL, Bernier L, McIsaac SA, Chan JY, Tansley GH, Cohn JS, Hayden MR, Wellington CL. Deficiency of ABCA1 impairs apolipoprotein E metabolism in brain. J Biol Chem 2004;279 (39):41197–41207.

52. Wahrle SE, Jiang H, Parsadanian M, Kim J, Li A, Knoten A, Jain S, Hirsch-Reinshagen V, Wellington CL, Bales KR, Paul SM, Holtzman DM. Overexpression of ABCA1 reduces amyloid deposition in the PDAPP mouse model of Alzheimer disease. J Clin Invest 2008;118(2): 671–682.

53. Manelli AM, Stine WB, Van Eldik LJ, LaDu MJ. ApoE and Abeta1-42 interactions: effects of isoform and conformation on structure and function. J Mol Neurosci 2004;23(3):235–246.

54. Naslund J, Thyberg J, Tjernberg LO, Wernstedt C, Karlstrom AR, Bogdanovic N, Gandy SE, Lannfelt L, Terenius L, Nordstedt C. Characterization of stable complexes involving apolipoprotein E and the amyloid beta peptide in Alzheimer's disease brain. Neuron 1995;15:219–228.

55. Koldamova R, Staufenbiel M, Lefterov I. Lack of ABCA1 considerably decreases brain ApoE level and increases amyloid deposition in APP23 mice. J Biol Chem 2005;280(52): 43224–43235.

56. Wahrle SE, Jiang H, Parsadanian M, Hartman RE, Bales KR, Paul SM, Holtzman DM. Deletion of Abca1 increases Abeta deposition in the PDAPP transgenic mouse model of Alzheimer disease. J Biol Chem 2005;280 (52): 43236–43242.

57. Hirsch-Reinshagen V, Maia LF, Burgess BL, Blain JF, Naus KE, McIsaac SA, Parkinson PF, Chan JY, Tansley GH, Hayden MR, Poirier J, Van Nostrand W, Wellington CL. The absence of ABCA1 decreases soluble ApoE levels but does not diminish amyloid deposition in two murine models of Alzheimer's disease. J Biol Chem 2005;280(52):43243–43256.

58. Zelcer N, Khanlou N, Clare R, Jiang Q, Reed-Geaghan EG, Landreth GE, Vinters HV, Tontonoz P. Attenuation of neuroinflammation and Alzheimer's disease pathology by liver X receptors. Proc Natl Acad Sci USA 2007;104 (25):10601–10606.

59. Koldamova RP, Lefterov IM, Staufenbiel M, Wolfe D, Huang S, Glorioso JC, Walter M, Roth MG, Lazo JS. The liver X receptor ligand T0901317 decreases amyloid beta production *in vitro* and in a mouse model of Alzheimer's disease. J Biol Chem 2005;280 (6):4079–4088.

60. Lefterov I, Bookout A, Wang Z, Staufenbiel M, Mangelsdorf D, Koldamova R. Expression profiling in APP23 mouse brain: inhibition of Abeta amyloidosis and inflammation in response

to LXR agonist treatment. Mol Neurodegener 2007;2:20–35.
61. Riddell DR, Zhou H, Comery TA, Kouranova E, Lo CF, Warwick HK, Ring RH, Kirksey Y, Aschmies S, Xu J, Kubek K, Hirst WD, Gonzales C, Chen Y, Murphy E, Leonard S, Vasylyev D, Oganesian A, Martone RL, Pangalos MN, Reinhart PH, Jacobsen JS. The LXR agonist TO901317 selectively lowers hippocampal Abeta42 and improves memory in the Tg2576 mouse model of Alzheimer's disease. Mol Cell Neurosci 2007;34(4):621–628.
62. Jiang Q, Lee CY, Mandrekar S, Wilkinson B, Cramer P, Zelcer N, Mann K, Lamb B, Willson TM, Collins JL, Richardson JC, Smith JD, Comery TA, Riddell D, Holtzman DM, Tontonoz P, Landreth GE. ApoE promotes the proteolytic degradation of Abeta. Neuron 2008;58(5):681–693.
63. Liang Y, Lin S, Beyer TP, Zhang Y, Wu X, Bales KR, DeMattos RB, May PC, Li SD, Jiang XC, Eacho PI, Cao G, Paul SM. A liver X receptor and retinoid X receptor heterodimer mediates apolipoprotein E expression, secretion and cholesterol homeostasis in astrocytes. J Neurochem 2004;88(3):623–634.
64. Bryan J, Goodwin William J, Zuercher Jon L. Collins. Recent advances in liver X receptor biology and chemistry. Curr Top Med Chem 2008;8:781–791.
65. Katz A, Udata C, Ott E, Hickey L, Burczynski ME, Burghart P, Vesterqvist O, Meng X. Safety, pharmacokinetics, and pharmacodynamics of single doses of LXR-623, a novel liver X-receptor agonist, in healthy participants. J Clin Pharmacol 2009;49(6):643–649.
66. Bu G. Apolipoprotein E and its receptors in Alzheimer's disease: pathways, pathogenesis and therapy. Nat Rev Neurosci 2009;10(5):333–344.
67. Walker LC, Pahnke J, Madauss M, Vogelgesang S, Pahnke A, Herbst EW, Stausske D, Walther R, Kessler C, Warzok RW. Apolipoprotein E4 promotes the early deposition of Abeta42 and then Abeta40 in the elderly. Acta Neuropathol 2000;100(1):36–42.
68. Reiman EM, Chen K, Liu X, Bandy D, Yu M, Lee W, Ayutyanont N, Keppler J, Reeder SA, Langbaum JB, Alexander GE, Klunk WE, Mathis CA, Price JC, Aizenstein HJ, DeKosky ST, Caselli RJ. Fibrillar amyloid-beta burden in cognitively normal people at 3 levels of genetic risk for Alzheimer's disease. Proc Natl Acad Sci USA 2009;106(16):6820–6825.
69. Holtzman DM, Bales KR, Tenkova T, Fagan AM, Parsadanian M, Sartorius LJ, Mackey B, Olney J, McKeel D, Wozniak D, Paul SM. Apolipoprotein E isoform-dependent amyloid deposition and neuritic degeneration in a mouse model of Alzheimer's disease. Proc Natl Acad Sci USA 2000;97(6):2892–2897.
70. Bales KR, Liu F, Wu S, Lin S, Koger D, DeLong C, Hansen JC, Sullivan PM, Paul SM. Human APOE isoform-dependent effects on brain beta-amyloid levels in PDAPP transgenic mice. J Neurosci 2009;29(21):6771–6779.
71. Kim J, Basak JM, Holtzman DM. The role of apolipoprotein E in Alzheimer's disease. Neuron 2009;63(3):287–303.
72. Ramaswamy G, Xu Q, Huang Y, Weisgraber KH. Effect of domain interaction on apolipoprotein E levels in mouse brain. J Neurosci 2005;25(46):10658–10663.
73. Riddell DR, Zhou H, Atchison K, Warwick HK, Atkinson PJ, Jefferson J, Xu L, Aschmies S, Kirksey Y, Hu Y, Wagner E, Parratt A, Xu J, Li Z, Zaleska MM, Jacobsen JS, Pangalos MN, Reinhart PH. Impact of apolipoprotein E (ApoE) polymorphism on brain ApoE levels. J Neurosci 2008;28(45):11445–11453.
74. Laskowitz DT, Fillit H, Yeung N, Toku K, Vitek MP. Apolipoprotein E-derived peptides reduce CNS inflammation: implications for therapy of neurological disease. Acta Neurol Scand Suppl 2006;185:15–20.
75. Lynch JR, Tang W, Wang H, Vitek MP, Bennett ER, Sullivan PM, Warner DS, Laskowitz DT. APOE genotype and an ApoE-mimetic peptide modify the systemic and central nervous system inflammatory response. J Biol Chem 2003;278(49):48529–48533.
76. Wang H, Durham L, Dawson H, Song P, Warner DS, Sullivan PM, Vitek MP, Laskowitz DT. An apolipoprotein E-based therapeutic improves outcome and reduces Alzheimer's disease pathology following closed head injury: evidence of pharmacogenomic interaction. Neuroscience 2007;144(4):1324–1333.
77. Mahley RW, Innerarity TL, Rall SC Jr, Weisgraber KH. Plasma lipoproteins: apolipoprotein structure and function. J Lipid Res 1984;25(12):1277–1294.
78. Wong NC. Novel therapies to increase apolipoprotein AI and HDL for the treatment of atherosclerosis. Curr Opin Investig Drugs 2007;8(9):718–728.
79. Navab M, Anantharamaiah GM, Reddy ST, Van Lenten BJ, Fogelman AM. Apo A-1 mimetic peptides as atheroprotective agents in

79. murine models. Curr Drug Targets 2008;9(3):204–209.
80. Merched A, Xia Y, Visvikis S, Serot JM, Siest G. Decreased high-density lipoprotein cholesterol and serum apolipoprotein AI concentrations are highly correlated with the severity of Alzheimer's disease. Neurobiol Aging 2000;21(1):27–30.
81. Möckel B, Zinke H, Flach R, Weiss B, Weiler-Güttler H, Gassen HG. Expression of apolipoprotein AI in porcine brain endothelium in vitro. J Neurochem 1994;62(2):788–798.
82. Koudinov A, Matsubara E, Frangione B, Ghiso J. The soluble form of Alzheimer's amyloid β protein is complexed to high density lipoprotein 3 and very high density lipoprotein in normal human plasma. Biochem Biophys Res Commun 1994;205(2):1164–1171.
83. Golabek A, Marques MA, Lalowski M, Wisniewski T. Amyloid (binding proteins in vitro and in normal human cerebrospinal fluid. Neurosci Lett 1995;191(1–2):79–82.
84. Wisniewski T, Golabek AA, Kida E, Wisniewski KE, Frangione B. Conformational mimicry in Alzheimer's Disease. Role of apolipoproteins in amyloidogenesis. Am J Pathol 1995;147(2):238–244.
85. Fagan AM, Christopher E, Taylor JW, Parsadanian M, Spinner M, Watson M, Fryer JD, Wahrle S, Bales KR, Paul SM, Holtzman DM. ApoAI deficiency results in marked reductions in plasma cholesterol but no alterations in amyloid-(pathology in a mouse model of Alzheimer's disease-like cerebral amyloidosis. Am J Pathol 2004;165(4):1413–1422.
86. Fontbonne A, Berr C, Ducimetière P, Alpérovitch A. Changes in cognitive abilities over a 4-year period are unfavorably affected in elderly diabetic subjects: results of the Epidemiology of Vascular Aging Study. Diabetes Care 2001;24(2):366–370.
87. Yaffe K, Blackwell T, Kanaya AM, Davidowitz N, Barrett-Connor E, Krueger K. Diabetes, impaired fasting glucose, and development of cognitive impairment in older women. Neurology 2004;63(4):658–663.
88. Rönnemaa E, Zethelius B, Sundelöf J, Sundström J, Degerman-Gunnarsson M, Berne C, Lannfelt L, Kilander L. Impaired insulin secretion increases the risk of Alzheimer disease. Neurology 2008;71(14): 1065–1071.
89. Xu W, Qiu C, Gatz M, Pedersen NL, Johansson B, Fratiglioni L Mid- and late-life diabetes in relation to the risk of dementia: a population-based twin study. Diabetes 2009;58(1):71–77.
90. Craft S, Peskind E, Schwartz MW, Schellenberg GD, Raskind M, Porte D Jr. Cerebrospinal fluid and plasma insulin levels in Alzheimer's disease: relationship to severity of dementia and apolipoprotein E genotype. Neurology 1998;50(1):164–168.
91. Heidenreich KA, Zahniser NR, Berhanu P, Brandenburg D, Olefsky JM. Structural differences between insulin receptors in the brain and peripheral target tissues. J Biol Chem 1983;258(14):8527–8530.
92. Zahniser NR, Goens MB, Hanaway PJ, Vinych JV. Characterization and regulation of insulin receptors in rat brain. J Neurochem 1984;42(5):1354–1362.
93. Zhao WQ, Alkon DL. Role of insulin and insulin receptor in learning and memory. Mol Cell Endocrinol 2001;177(1–2):125–134.
94. Schwartz MW, Woods SC, Porte D Jr, Seeley RJ, Baskin DG. Central nervous system control of food intake. Nature 2000;404(6778): 661–671.
95. Watson GS, Craft S. Modulation of memory by insulin and glucose: neuropsychological observations in Alzheimer's disease. Eur J Pharmacol 2004;490(1–3):97–113.
96. Zhao WQ, Chen H, Quon MJ, Alkon DL. Insulin and the insulin receptor in experimental models of learning and memory. Eur J Pharmacol 2004;490(1–3):71–81.
97. Zhao WQ, Townsend M. Insulin resistance and amyloidogenesis as common molecular foundation for type 2 diabetes and Alzheimer's disease. Biochim Biophys Acta 2009;1792(5): 482–496.
98. Nelson TJ, Sun MK, Hongpaisan J, Alkon DL, Insulin PKC signaling pathways and synaptic remodeling during memory storage and neuronal repair. Eur J Pharmacol 2008;585(1): 76–87.
99. Meneilly GS, Cheung E, Tessier D, Yakura C, Tuokko H. The effect of improved glycemic control on cognitive functions in the elderly patient with diabetes. J Gerontol 1993;48(4): M117–121.
100. Gregg EW, Yaffe K, Cauley JA, Rolka DB, Blackwell TL, Narayan KM, Cummings SR. Is diabetes associated with cognitive impairment and cognitive decline among older women? Study of Osteoporotic Fractures Research Group. Arch Intern Med 2000;160(2): 174–180.
101. Ryan CM, Geckle MO. Circumscribed cognitive dysfunction in middle-aged adults with type 2 diabetes. Diabetes Care 2000;23(10): 1486–1493.

102. Watson GS, Peskind ER, Asthana S, Purganan K, Wait C, Chapman D, Schwartz MW, Plymate S, Craft S. Insulin increases CSF Abeta42 levels in normal older adults. Neurology 2003;60(12):1899–1903.

103. Fishel MA, Watson GS, Montine TJ, Wang Q, Green PS, Kulstad JJ, Cook DG, Peskind ER, Baker LD, Goldgaber D, Nie W, Asthana S, Plymate SR, Schwartz MW, Craft S. Hyperinsulinemia provokes synchronous increases in central inflammation and Aβ in normal adults. Arch Neurol 2005;62(10):1539–1544.

104. Craft S, Asthana S, Newcomer JW, Wilkinson CW, Matos IT, Baker LD, Cherrier M, Lofgreen C, Latendresse S, Petrova A, Plymate S, Raskind M, Grimwood K, Veith RC. Enhancement of memory in Alzheimer disease with insulin and somatostatin, but not glucose. Arch Gen Psychiatry 1999;56(12):1135–1140.

105. Reger MA, Watson GS, Green PS, Wilkinson CW, Baker LD, Cholerton B, Fishel MA, Plymate SR, Breitner JC, DeGroodt W, Mehta P, Craft S. Intranasal insulin improves cognition and modulates Aβ in early AD. Neurology 2008;70(6):440–448.

106. Craft S, Asthana S, Cook DG, Baker LD, Cherrier M, Purganan K, Wait C, Petrova A, Latendresse S, Watson GS, Newcomer JW, Schellenberg GD, Krohn AJ. Insulin dose-response effects on memory and plasma amyloid precursor protein in Alzheimer's disease: interactions with apolipoprotein E genotype. Psychoneuroendocrinology 2003;28(6):809–822.

107. Frölich L, Blum-Degen D, Riederer P, Hoyer S. A disturbance in the neuronal insulin receptor signal transduction in sporadic Alzheimer's disease. Ann NY Acad Sci 1999;893:290–293.

108. Cross DA, Alessi DR, Cohen P, Andjelkovich M, Hemmings BA. Inhibition of glycogen synthase kinase-3 by insulin mediated by protein kinase B. Nature 1995;378(6559):785–789.

109. Balaraman Y, Limaye AR, Levey AI, Srinivasan S. Glycogen synthase kinase 3beta and Alzheimer's disease: pathophysiological and therapeutic significance. Cell Mol Life Sci 2006;63(11):1226–1235.

110. Roder HM, Hutton ML. Microtubule-associated protein Tau as a therapeutic target in neurodegenerative disease. Expert Opin Ther Targets 2007;11(4):435–442.

111. Phiel CJ, Wilson CA, Lee VM, Klein PS. GSK-3 alpha regulates production of Alzheimer's disease amyloid-beta peptides. Nature 2003;423 (6938):435–439.

112. Ryder J, Su Y, Liu F, Li B, Zhou Y, Ni B. Divergent roles of GSK3 and CDK5 in APP processing. Biochem Biophys Res Commun 2003;312(4):922–929.

113. Su Y, Ryder J, Li B, Wu X, Fox N, Solenberg P, Brune K, Paul S, Zhou Y, Liu F, Ni B. Lithium, a common drug for bipolar disorder treatment, regulates amyloid-beta precursor protein processing. Biochemistry 2004;43(22):6899–6908.

114. Jolivalt CG, Lee CA, Beiswenger KK, Smith JL, Orlov M, Torrance MA, Masliah E. Defective insulin signaling pathway and increased glycogen synthase kinase-3 activity in the brain of diabetic mice: parallels with Alzheimer's disease and correction by insulin. J Neurosci Res 2008;86(15):3265–3274.

115. Feldman PL, Lambert MH, Henke BR. PPAR modulators and PPAR pan agonists for metabolic diseases: the next generation of drugs targeting peroxisome proliferator-activated receptors? Curr Top Med Chem 2008;8(9):728–749.

116. Cho N, Momose Y. Peroxisome proliferator-activated receptor gamma agonists as insulin sensitizers: from the discovery to recent progress. Curr Top Med Chem 2008;8(17):1483–1507.

117. Strum JC, Shehee R, Virley D, Richardson J, Mattie M, Selley P, Ghosh S, Nock C, Saunders A, Roses A. Rosiglitazone induces mitochondrial biogenesis in mouse brain. J Alzheimers Dis 2007;11(1):45–51.

118. Landreth G, Jiang Q, Mandrekar S, Heneka M. PPARgamma agonists as therapeutics for the treatment of Alzheimer's disease. Neurotherapeutics 2008;Jul 5(3):481–489.

119. Yan Q, Zhang J, Liu H, Babu-Khan S, Vassar R, Biere AL, Citron M, Landreth G. Anti-inflammatory drug therapy alters beta-amyloid processing and deposition in an animal model of Alzheimer's disease. J Neurosci 2003;23(20):7504–7509.

120. Heneka MT, Sastre M, Dumitrescu-Ozimek L, Hanke A, Dewachter I, Kuiperi C, O'Banion K, Klockgether T, Van Leuven F, Landreth GE. Acute treatment with the PPARgamma agonist pioglitazone and ibuprofen reduces glial inflammation and Abeta1-42 levels in APP-V717I transgenic mice. Brain 2005;128(Pt 6):1442–1453.

121. Nicolakakis N, Aboulkassim T, Ongali B, Lecrux C, Fernandes P, Rosa Neto P, Tong XK, Hamel E. Complete rescue of cerebrovascular function in aged Alzheimer's disease transgenic mice by antioxidants and pioglitazone,

a peroxisome proliferator-activated receptor gamma agonist. J Neurosci 2008;28(37): 9287–9296.

122. Pedersen WA, Flynn ER. Insulin resistance contributes to aberrant stress responses in the Tg2576 mouse model of Alzheimer's disease. Neurobiol Dis 2004;17(3):500–506.

123. Pedersen WA, McMillan PJ, Kulstad JJ, Leverenz JB, Craft S, Haynatzki GR. Rosiglitazone attenuates learning and memory deficits in Tg2576 Alzheimer mice. Exp Neurol 2006;199(2):265–273.

124. Escribano L, Simón AM, Pérez-Mediavilla A, Salazar-Colocho P, Del Río J, Frechilla D. Rosiglitazone reverses memory decline and hippocampal glucocorticoid receptor down-regulation in an Alzheimer's disease mouse model. Biochem Biophys Res Commun 2009;379(2): 406–410.

125. Sastre M, Dewachter I, Landreth GE, Willson TM, Klockgether T, van Leuven F, Heneka MT. Nonsteroidal anti-inflammatory drugs and peroxisome proliferator-activated receptor-gamma agonists modulate immunostimulated processing of amyloid precursor protein through regulation of beta-secretase. J Neurosci 2003;23(30):9796–9804.

126. Sastre M, Dewachter I, Rossner S, Bogdanovic N, Rosen E, Borghgraef P, Evert BO, Dumitrescu-Ozimek L, Thal DR, Landreth G, Walter J, Klockgether T, van Leuven F, Heneka MT. Nonsteroidal anti-inflammatory drugs repress beta-secretase gene promoter activity by the activation of PPARgamma. Proc Natl Acad Sci USA 2006;103(2):443–448.

127. d'Abramo C, Massone S, Zingg JM, Pizzuti A, Marambaud P, Dalla Piccola B, Azzi A, Marinari UM, Pronzato MA, Ricciarelli R. Role of peroxisome proliferator-activated receptor gamma in amyloid precursor protein processing and amyloid beta-mediated cell death. Biochem J 2005;391(Pt 3):693–698.

128. Kurochkin IV, Goto S. Alzheimer's beta-amyloid peptide specifically interacts with and is degraded by insulin degrading enzyme. FEBS Lett 1994;345(1):33–37.

129. Qiu WQ, Walsh DM, Ye Z, Vekrellis K, Zhang J, Podlisny MB, Rosner MR, Safavi A, Hersh LB, Selkoe DJ. Insulin-degrading enzyme regulates extracellular levels of amyloid beta-protein by degradation. J Biol Chem 1998;273 (49):32730–32738.

130. Vekrellis K, Ye Z, Qiu WQ, Walsh D, Hartley D, Chesneau V, Rosner MR, Selkoe DJ. Neurons regulate extracellular levels of amyloid beta-protein via proteolysis by insulin-degrading enzyme. J Neurosci 2000;20(5):1657–1665.

131. Farris W, Mansourian S, Chang Y, Lindsley L, Eckman EA, Frosch MP, Eckman CB, Tanzi RE, Selkoe DJ, Guenette S. Insulin-degrading enzyme regulates the levels of insulin, amyloid beta-protein, and the beta-amyloid precursor protein intracellular domain in vivo. Proc Natl Acad Sci USA 2003;100(7):4162–4167.

132. Farris W, Mansourian S, Leissring MA, Eckman EA, Bertram L, Eckman CB, Tanzi RE, Selkoe DJ. Partial loss-of-function mutations in insulin-degrading enzyme that induce diabetes also impair degradation of amyloid beta-protein. Am J Pathol 2004;164(4):1425–1434.

133. Craft S. Insulin resistance and Alzheimer's disease pathogenesis: potential mechanisms and implications for treatment. Curr Alzheimer Res 2007;4(2):147–152.

134. Cook DG, Leverenz JB, McMillan PJ, Kulstad JJ, Ericksen S, Roth RA, Schellenberg GD, Jin LW, Kovacina KS, Craft S. Reduced hippocampal insulin-degrading enzyme in late-onset Alzheimer's disease is associated with the apolipoprotein E-epsilon4 allele. Am J Pathol 2003;162(1):313–319.

135. Ho L, Qin W, Pompl PN, Xiang Z, Wang J, Zhao Z, Peng Y, Cambareri G, Rocher A, Mobbs CV, Hof PR, Pasinetti GM. Diet-induced insulin resistance promotes amyloidosis in a transgenic mouse model of Alzheimer's disease. FASEB J 2004;18(7):902–904.

136. Watson GS, Cholerton BA, Reger MA, Baker LD, Plymate SR, Asthana S, Fishel MA, Kulstad JJ, Green PS, Cook DG, Kahn SE, Keeling ML, Craft S. Preserved cognition in patients with early Alzheimer disease and amnestic mild cognitive impairment during treatment with rosiglitazone: a preliminary study. Am J Geriatr Psychiatry 2005;13(11):950–958.

137. Risner ME, Saunders AM, Altman JF, Ormandy GC, Craft S, Foley IM, Zvartau-Hind ME, Hosford DA, Roses AD, Rosiglitazone in Alzheimer's Disease Study Group Efficacy of rosiglitazone in a genetically defined population with mild-to-moderate Alzheimer's disease. Pharmacogenomics J 2006;6(4): 246–254.

138. Craft S, Asthana S, Schellenberg G, Cherrier M, Baker LD, Newcomer J, Plymate S, Latendresse S, Petrova A, Raskind M, Peskind E, Lofgreen C, Grimwood K. Insulin metabolism in Alzheimer's disease differs according to apolipoprotein E genotype and gender. Neuroendocrinology 1999;70(2):146–152.

139. Craft S, Asthana S, Schellenberg G, Baker L, Cherrier M, Boyt AA, Martins RN, Raskind M, Peskind E, Plymate S. Insulin effects on glucose metabolism, memory, and plasma amyloid precursor protein in Alzheimer's disease differ according to apolipoprotein-E genotype. Ann NY Acad Sci 2000;903:222–228.

140. Hanyu H, Sato T, Kiuchi A, Sakurai H, Iwamoto T. Pioglitazone improved cognition in a pilot study on patients with Alzheimer's disease and mild cognitive impairment with diabetes mellitus. J Am Geriatr Soc 2009;57(1): 177–179.

141. Zhou G, Myers R, Li Y, Chen Y, Shen X, Fenyk-Melody J, Wu M, Ventre J, Doebber T, Fujii N, Musi N, Hirshman MF, Goodyear LJ, Moller DE. Role of AMP-activated protein kinase in mechanism of metformin action. J Clin Invest 2001;108(8):1167–1174.

142. Schäfer G. Biguanides. A review of history, pharmacodynamics and therapy. Diabetes Metab 1983;9(2):148–163.

143. Chen Y, Zhou K, Wang R, Liu Y, Kwak YD, Ma T, Thompson RC, Zhao Y, Smith L, Gasparini L, Luo Z, Xu H, Liao FF. Antidiabetic drug metformin (Glucophage®) increases biogenesis of Alzheimer's amyloid peptides via up-regulating BACE1 transcription. Proc Natl Acad Sci USA 2009;106(10):3907–3912.

144. Bhat RV, Budd SL. GSK3beta signalling: casting a wide net in Alzheimer's disease. Neurosignals 2002;11(5):251–261.

145. Meijer L, Flajolet M, Greengard P. Pharmacological inhibitors of glycogen synthase kinase 3. Trends Pharmacol Sci 2004;25(9):471–480.

146. Engler TA, Henry JR, Malhotra S, Cunningham B, Furness K, Brozinick J, Burkholder TP, Clay MP, Clayton J, Diefenbacher C, Hawkins E, Iversen PW, Li Y, Lindstrom TD, Marquart AL, McLean J, Mendel D, Misener E, Briere D, O'Toole JC, Porter WJ, Queener S, Reel JK, Owens RA, Brier RA, Eessalu TE, Wagner JR, Campbell RM, Vaughn R. Substituted 3-imidazo[1,2-a]pyridin-3-yl- 4-(1,2,3,4-tetrahydro-[1,4]diazepino-[6,7,1-hi]indol-7-yl)pyrrole-2,5-dion es as highly selective and potent inhibitors of glycogen synthase kinase-3. J Med Chem 2004;47(16):3934–3937.

147. Yuzwa SA, Macauley MS, Heinonen JE, Shan X, Dennis RJ, He Y, Whitworth GE, Stubbs KA, McEachern EJ, Davies GJ, Vocadlo DJ. A potent mechanism-inspired O-GlcNAcase inhibitor that blocks phosphorylation of tau in vivo. Nat Chem Biol 2008;4(8):483–490.

148. Wischik CM, Edwards PC, Lai RY, Roth M, Harrington CR. Selective inhibition of Alzheimer disease-like tau aggregation by phenothiazines. Proc Natl Acad Sci USA 1996;93 (20):11213–11218.

149. Pickhardt M, Gazova Z, von Bergen M, Khlistunova I, Wang Y, Hascher A, Mandelkow EM, Biernat J, Mandelkow E. Anthraquinones inhibit tau aggregation and dissolve Alzheimer's paired helical filaments in vitro and in cells. J Biol Chem 2005;280(5):3628–3635.

150. Necula M, Chirita CN, Kuret J. Cyanine dye N744 inhibits tau fibrillization by blocking filament extension: implications for the treatment of tauopathic neurodegenerative diseases. Biochemistry 2005;44(30): 10227–10237.

151. Ross JL, Santangelo CD, Makrides V, Fygenson DK. Tau induces cooperative Taxol binding to microtubules. Proc Natl Acad Sci USA 2004;101(35):12910–12915.

152. Trojanowski JQ, Smith AB, Huryn D, Lee VM. Microtubule-stabilising drugs for therapy of Alzheimer's disease and other neurodegenerative disorders with axonal transport impairments. Expert Opin Pharmacother 2005;6(5): 683–686.

153. Dickey CA, Eriksen J, Kamal A, Burrows F, Kasibhatla S, Eckman CB, Hutton M, Petrucelli L. Development of a high throughput drug screening assay for the detection of changes in tau levels: proof of concept with HSP90 inhibitors. Curr Alzheimer Res 2005;2(2):231–238.

154. Roberson ED, Scearce-Levie K, Palop JJ, Yan F, Cheng IH, Wu T, Gerstein H, Yu GQ, Mucke L. Reducing endogenous tau ameliorates amyloid beta-induced deficits in an Alzheimer's disease mouse model. Science 2007;316(5825):750–754.

155. Myers AJ, Pittman AM, Zhao AS, Rohrer K, Kaleem M, Marlowe L, Lees A, Leung D, McKeith IG, Perry RH, Morris CM, Trojanowski JQ, Clark C, Karlawish J, Arnold S, Forman MS, Van Deerlin V, de Silva R, Hardy J. The MAPT H1c risk haplotype is associated with increased expression of tau and especially of 4 repeat containing transcripts. Neurobiol Dis 2007;25(3):561–570.

156. Goedert M, Spillantini MG. A century of Alzheimer's disease. Science 2006;314 (5800):777–781.

157. Lee VM, Goedert M, Trojanowski JQ. Neurodegenerative tauopathies. Annu Rev Neurosci 2001;24:1121–1159.

158. Forman MS. Genotype-phenotype correlations in FTDP-17: does form follow function? Exp Neurol 2004;187(2):229–234.

159. Skovronsky DM, Lee VM, Trojanowski JQ. Neurodegenerative diseases: new concepts of pathogenesis and their therapeutic implications. Annu Rev Pathol 2006;1:151–170.
160. Terry RD. The pathogenesis of Alzheimer disease: an alternative to the amyloid hypothesis. J Neuropathol Exp Neurol 1996;55(10):1023–1025.
161. Iqbal K, Grundke-Iqbal I. Molecular mechanism of Alzheimer's neurofibrillary degeneration and therapeutic intervention. Ann NY Acad Sci 1996;777:132–138.
162. Cho JH, Johnson GV. Glycogen synthase kinase 3 beta induces caspase-cleaved tau aggregation in situ. J Biol Chem 2004;279(52):54716–54723.
163. Lovestone S, Hartley CL, Pearce J, Anderton BH. Phosphorylation of tau by glycogen synthase kinase-3 beta in intact mammalian cells: the effects on the organization and stability of microtubules. Neuroscience 1996;73(4):1145–1157.
164. Iqbal K, Alonso Adel C, Grundke-Iqbal I. Cytosolic abnormally hyperphosphorylated tau but not paired helical filaments sequester normal MAPs and inhibit microtubule assembly. J Alzheimers Dis 2008;14(4):365–370.
165. Wittmann CW, Wszolek MF, Shulman JM, Salvaterra PM, Lewis J, Hutton M, Feany MB. Tauopathy in Drosophila: neurodegeneration without neurofibrillary tangles. Science 2001;293(5530):711–714.
166. Bu B, Li J, Davies P, Vincent I. Deregulation of cdk5, hyperphosphorylation, and cytoskeletal pathology in the Niemann-Pick type C murine model. J Neurosci 2002;22(15):6515–6525.
167. Mudher A, Shepherd D, Newman TA, Mildren P, Jukes JP, Squire A, Mears A, Drummond JA, Berg S, MacKay D, Asuni AA, Bhat R, Lovestone S. GSK-3beta inhibition reverses axonal transport defects and behavioural phenotypes in Drosophila. Mol Psychiatry 2004;9(5):522–530.
168. Cruz JC, Tseng HC, Goldman JA, Shih H, Tsai LH. Aberrant Cdk5 activation by p25 triggers pathological events leading to neurodegeneration and neurofibrillary tangles. Neuron 2003;40(3):471–483.
169. Harada A, Oguchi K, Okabe S, Kuno J, Terada S, Ohshima T, Sato-Yoshitake R, Takei Y, Noda T, Hirokawa N. Altered microtubule organization in small-calibre axons of mice lacking tau protein. Nature 1994;369(6480):488–491.
170. Yuan A, Kumar A, Peterhoff C, Duff K, Nixon RA. Axonal transport rates in vivo are unaffected by tau deletion or overexpression in mice. J Neurosci 2008;28(7):1682–1687.
171. Augustinack JC, Schneider A, Mandelkow EM, Hyman BT. Specific tau phosphorylation sites correlate with severity of neuronal cytopathology in Alzheimer's disease. Acta Neuropathol 2002;103(1):26–35.
172. Cho JH, Johnson GV. Glycogen synthase kinase 3beta phosphorylates tau at both primed and unprimed sites. Differential impact on microtubule binding. J Biol Chem 2003;278(1):187–193.
173. Li T, Paudel HK. Glycogen synthase kinase 3beta phosphorylates Alzheimer's disease-specific Ser396 of microtubule-associated protein tau by a sequential mechanism. Biochemistry 2006;45(10):3125–3133.
174. Li T, Hawkes C, Qureshi HY, Kar S, Paudel HK. Cyclin-dependent protein kinase 5 primes microtubule-associated protein tau site-specifically for glycogen synthase kinase 3beta. Biochemistry 2006;45(10):3134–3145.
175. Tian Q, Wang J. Role of serine/threonine protein phosphatase in Alzheimer's disease. Neurosignals 2002;11(5):262–269.
176. Vogelsberg-Ragaglia V, Schuck T, Trojanowski JQ, Lee VM. PP2A mRNA expression is quantitatively decreased in Alzheimer's disease hippocampus. Exp Neurol 2001;168(2): 402–412.
177. Mudher AK, Perry VH. Using okadaic acid as a tool for the in vivo induction of hyperphosphorylated tau. Neuroscience 1998;85(4): 1329–1332.
178. Woodgett JR. Molecular cloning and expression of glycogen synthase kinase-3/factor A. EMBO J 1990;9(8):2431–2438.
179. Leroy K, Brion JP. Developmental expression and localization of glycogen synthase kinase-3beta in rat brain. J Chem Neuroanat 1999;16(4):279–293.
180. Sengupta A, Wu Q, Grundke-Iqbal I, Iqbal K, Singh TJ. Potentiation of GSK-3-catalyzed Alzheimer-like phosphorylation of human tau by cdk5. Mol Cell Biochem 1997;167(1–2): 99–105.
181. Paudel HK. Phosphorylation by neuronal cdc2-like protein kinase promotes dimerization of Tau protein in vitro. J Biol Chem 1997;272(45):28328–28334.
182. Patrick GN, Zukerberg L, Nikolic M, de la Monte S, Dikkes P, Tsai LH. Conversion of p35 to p25 deregulates Cdk5 activity and promotes neurodegeneration. Nature 1999;402(6762):615–622.

183. Hong M, Chen DC, Klein PS, Lee VM. Lithium reduces tau phosphorylation by inhibition of glycogen synthase kinase-3. J Biol Chem 1997;272(40):25326–25332.
184. Lucas JJ, Hernandez F, Gomez-Ramos P, Moran MA, Hen R, Avila J. Decreased nuclear beta-catenin, tau hyperphosphorylation and neurodegeneration in GSK-3beta conditional transgenic mice. EMBO J 2001;20(1–2):27–39.
185. Hernandez F, Borrell J, Guaza C, Avila J, Lucas JJ. Spatial learning deficit in transgenic mice that conditionally over-express GSK-3beta in the brain but do not form tau filaments. J. Neurochem 2002;83(6):1529–1533.
186. Jackson GR, Wiedau-Pazos M, Sang TK, Wagle N, Brown CA, Massachi S, Geschwind DH. Human wild-type tau interacts with wingless pathway components and produces neurofibrillary pathology in *Drosophila*. Neuron 2002;34(4):509–519.
187. Ahlijanian MK, Barrezueta NX, Williams RD, Jakowski A, Kowsz KP, McCarthy S, Coskran T, Carlo A, Seymour PA, Burkhardt JE, Nelson RB, McNeish JD. Hyperphosphorylated tau and neurofilament and cytoskeletal disruptions in mice overexpressing human p25, an activator of cdk5. Proc Natl Acad Sci USA 2000;97(6):2910–2915.
188. Santacruz K, Lewis J, Spires T, Paulson J, Kotilinek L, Ingelsson M, Guimaraes A, DeTure M, Ramsden M, McGowan E, Forster C, Yue M, Orne J, Janus C, Mariash A, Kuskowski M, Hyman B, Hutton M, Ashe KH. Tau suppression in a neurodegenerative mouse model improves memory function. Science 2005;309(5733):476–481.
189. Lee MS, Kwon YT, Li M, Peng J, Friedlander RM, Tsai LH. Neurotoxicity induces cleavage of p35 to p25 by calpain. Nature 2000;405 (6784):360–364.
190. Alvarez A, Toro R, Caceres A, Maccioni RB. Inhibition of tau phosphorylating protein kinase cdk5 prevents beta-amyloid-induced neuronal death. FEBS Lett 1999;459(3):421–426.
191. Milton NG. Phosphorylation of amyloid-beta at the serine 26 residue by human cdc2 kinase. Neuroreport 2001;12(17):3839–3844.
192. Ferreira A, Lu Q, Orecchio L, Kosik KS. Selective phosphorylation of adult tau isoforms in mature hippocampal neurons exposed to fibrillar A beta. Mol Cell Neurosci 1997;9(3):220–234.
193. Takashima A, Honda T, Yasutake K, Michel G, Murayama O, Murayama M, Ishiguro K, Yamaguchi H. Activation of tau protein kinase I/glycogen synthase kinase-3beta by amyloid beta peptide (25–35) enhances phosphorylation of tau in hippocampal neurons. Neurosci Res 1998;31(4):317–323.
194. Ohshima T, Ward JM, Huh CG, Longenecker G, Veeranna Pant HC, Brady RO, Martin LJ, Kulkarni AB. Targeted disruption of the cyclin-dependent kinase 5 gene results in abnormal corticogenesis, neuronal pathology and perinatal death. Proc Natl Acad Sci USA 1996;93(20):11173–11178.
195. Gilmore EC, Ohshima T, Goffinet AM, Kulkarni AB, Herrup K. Cyclin-dependent kinase 5-deficient mice demonstrate novel developmental arrest in cerebral cortex. J Neurosci 1998;18(16):6370–6377.
196. Zhai S, Senderowicz AM, Sausville EA, Figg WD. Flavopiridol, a novel cyclin-dependent kinase inhibitor, in clinical development. Ann Pharmacother 2002;36(5):905–911.
197. Hoeflich KP, Luo J, Rubie EA, Tsao MS, Jin O, Woodgett JR. Requirement for glycogen synthase kinase-3beta in cell survival and NF-kappaB activation. Nature 2000;406 (6791):86–90.
198. Sugden PH, Fuller SJ, Weiss SC, Clerk A. Glycogen synthase kinase 3 (GSK3) in the heart: a point of integration in hypertrophic signalling and a therapeutic target? A critical analysis. Br J Pharmacol 2008;153(Suppl 1): S137–S153.
199. Sorbera LA, Bozzo J, Serradell N. Alzheimer's disease one century later: the search for effective therapeutic targets continues. Drugs Fut 2007;32(7):625–641.
200. Tarricone C, Dhavan R, Peng J, Areces LB, Tsai LH, Musacchio A. Structure and regulation of the CDK5-p25(nck5a) complex. Mol Cell 2001;8(3):657–669.
201. Mapelli M, Massimiliano L, Crovace C, Seeliger MA, Tsai LH, Meijer L, Musacchio A. Mechanism of CDK5/p25 binding by CDK inhibitors. J Med Chem 2005;48 (3):671–679.
202. Fischer PM, Gianella-Borradori A. Recent progress in the discovery and development of cyclin-dependent kinase inhibitors. Expert Opin Investig Drugs 2005;14(4):457–477.
203. Karran E, Palmer AM. Neurodegenerative disorders and their treatment. Drug News Perspect 2007;20(6):407–412.
204. Tsai LH. The effect of CDK inhibitors in a mouse model for Alzheimer's disease. http://www.pasteur.fr/applications/euroconf/proteinkinase/10Tsaiabstract.pdf. 2006.

205. Wen Y, Planel E, Herman M, Figueroa HY, Wang L, Liu L, Lau LF, Yu WH, Duff KE. Interplay between cyclin-dependent kinase 5 and glycogen synthase kinase 3 beta mediated by neuregulin signaling leads to differential effects on tau phosphorylation and amyloid precursor protein processing. J. Neurosci 2008;28(10):2624–2632.

206. Dajani R, Fraser E, Roe SM, Young N, Good V, Dale TC, Pearl LH. Crystal structure of glycogen synthase kinase 3 beta: structural basis for phosphate-primed substrate specificity and autoinhibition. Cell 2001;105(6):721–732.

207. Gould TD, Manji HK. Glycogen synthase kinase-3: a putative molecular target for lithium mimetic drugs. Neuropsychopharmacology 2005;30(7):1223–1237.

208. Aghdam SY, Barger SW. Glycogen synthase kinase-3 in neurodegeneration and neuroprotection: lessons from lithium. Curr. Alzheimer Res. 2007;4(1):21–31.

209. Bhat R, Xue Y, Berg S, Hellberg S, Ormö M, Nilsson Y, Radesäter AC, Jerning E, Markgren PO, Borgegård T, Nylöf M, Giménez-Cassina A, Hernández F, Lucas JJ, Díaz-Nido J, Avila J. Structural insights and biological effects of glycogen synthase kinase 3-specific inhibitor AR-A014418. J Biol Chem 2003;278(46): 45937–45945.

210. Kunick C, Lauenroth K, Leost M, Meijer L, Lemcke T. 1-Azakenpaullone is a selective inhibitor of glycogen synthase kinase-3 beta. Bioorg Med Chem Lett 2004;14(2):413–416.

211. Leclerc S, Garnier M, Hoessel R, Marko D, Bibb JA, Snyder GL, Greengard P, Biernat J, Wu YZ, Mandelkow EM, Eisenbrand G, Meijer L. Indirubins inhibit glycogen synthase kinase-3 beta and CDK5/p25, two protein kinases involved in abnormal tau phosphorylation in Alzheimer's disease. A property common to most cyclin-dependent kinase inhibitors? J Biol Chem 2001;276(1):251–260.

212. Engler TA, Malhotra S, Burkholder TP, Henry JR, Mendel D, Porter WJ, Furness K, Diefenbacher C, Marquart A, Reel JK, Li Y, Clayton J, Cunningham B, McLean J, O'toole JC, Brozinick J, Hawkins E, Misener E, Briere D, Brier RA, Wagner JR, Campbell RM, Anderson BD, Vaughn R, Bennett DB, Meier TI, Cook JA. The development of potent and selective bisarylmaleimide GSK3 inhibitors. Bioorg Med Chem Lett 2005;15(4):899–903.

213. Shen L, Prouty C, Conway BR, Westover L, Xu JZ, Look RA, Chen X, Beavers MP, Roberts J, Murray WV, Demarest KT, Kuo GH. Synthesis and biological evaluation of novel macrocyclic bis-7-azaindolylmaleimides as potent and highly selective glycogen synthase kinase-3 beta (GSK-3 beta) inhibitors. Bioorg Med Chem 2004;12(5):1239–1255.

214. Meijer L, Thunnissen AM, White AW, Garnier M, Nikolic M, Tsai LH, Walter J, Cleverley KE, Salinas PC, Wu YZ, Biernat J, Mandelkow EM, Kim SH, Pettit GR. Inhibition of cyclin-dependent kinases, GSK-3beta and CK1 by hymenialdisine, a marine sponge constituent. Chem Biol 2000;7(1):51–63.

215. Wan Y, Hur W, Cho CY, Liu Y, Adrian FJ, Lozach O, Bach S, Mayer T, Fabbro D, Meijer L, Gray NS. Synthesis and target identification of hymenialdisine analogs. Chem Biol 2004;11(2):247–259.

216. Castro A, Encinas A, Gil C, Bräse S, Porcal W, Pérez C, Moreno FJ, Martínez A. Non-ATP competitive glycogen synthase kinase 3beta (GSK-3beta) inhibitors: study of structural requirements for thiadiazolidinone derivatives. Bioorg Med Chem 2008;16(1):495–510.

217. Mettey Y, Gompel M, Thomas V, Garnier M, Leost M, Ceballos-Picot I, Noble M, Endicott J, Vierfond JM, Meijer L. Aloisines, a new family of CDK/GSK-3 inhibitors. SAR study, crystal structure in complex with CDK2, enzyme selectivity, and cellular effects. J Med Chem 2003;46(2):222–236.

218. Witherington J, Bordas V, Garland SL, Hickey DMB, Ife RJ, Liddle J, Saunders M, Smith DG, Ward RW. 5-Aryl-pyrozolo [3,4-b] pyridines: potent inhibitors of glycogen synthase kinase-3 (GSK-3). Bioorg Med Chem Lett 2003; 13(9): 1577–1580.

219. Hensley K, Floyd RA, Zheng NY, Nael R, Robinson KA, Nguyen X, Pye QN, Stewart CA, Geddes J, Markesbery WR, Patel E, Johnson GV, Bing G. p38 kinase is activated in the Alzheimer's disease brain. J Neurochem 1999;72(5):2053–2058.

220. Zhu X, Rottkamp CA, Boux H, Takeda A, Perry G, Smith MA. Activation of p38 kinase links tau phosphorylation, oxidative stress, and cell cycle-related events in Alzheimer disease. J Neuropathol Exp Neurol 2000;59(10): 880–888.

221. Shoji M, Iwakami N, Takeuchi S, Waragai M, Suzuki M, Kanazawa I, Lippa CF, Ono S, Okazawa H. JNK activation is associated with intracellular beta-amyloid accumulation. Brain Res Mol Brain Res 2000;85(1–2): 221–233.

222. Zhu X, Raina AK, Rottkamp CA, Aliev G, Perry G, Boux H, Smith MA,2001; Activation and redistribution of c-jun N-terminal kinase/

222. ...stress activated protein kinase in degenerating neurons in Alzheimer's disease. J Neurochem 2001;76(2):435–441.
223. Perry G, Roder H, Nunomura A, Takeda A, Friedlich AL, Zhu X, Raina AK, Holbrook N, Siedlak SL, Harris PL, Smith MA. Activation of neuronal extracellular receptor kinase (ERK) in Alzheimer disease links oxidative stress to abnormal phosphorylation. Neuroreport 1999;10(11):2411–2415.
224. Ferrer I, Blanco R, Carmona M, Ribera R, Goutan E, Puig B, Rey MJ, Cardozo A, Vinals F, Ribalta T. Phosphorylated map kinase (ERK1, ERK2) expression is associated with early tau deposition in neurones and glial cells, but not with increased nuclear DNA vulnerability and cell death, in Alzheimer disease, Pick's disease, progressive supranuclear palsy and corticobasal degeneration. Brain Pathol 2001;11(2):144–58.
225. Savage MJ, Lin YG, Ciallella JR, Flood DG, Scott RW. Activation of c-Jun N-terminal kinase and p38 in an Alzheimer's disease model is associated with amyloid deposition. J Neurosci 2002;22(9):3376–3385.
226. Hwang DY, Cho JS, Lee SH, Chae KR, Lim HJ, Min SH, Seo SJ, Song YS, Song CW, Paik SG, Sheen YY, Kim YK. Aberrant expressions of pathogenic phenotype in Alzheimer's diseased transgenic mice carrying NSE-controlled APPsw. Exp Neurol 2004;186(1):20–32.
227. Bellucci A, Rosi MC, Grossi C, Fiorentini A, Luccarini I, Casamenti F. Abnormal processing of tau in the brain of aged TgCRND8 mice. Neurobiol Dis 2007;27(3):328–338.
228. Sahara N, Murayama M, Lee B, Park JM, Lagalwar S, Binder LI, Takashima A. Active c-jun N-terminal kinase induces caspase cleavage of tau and additional phosphorylation by GSK-3beta is required for tau aggregation. Eur J Neurosci 2008;27(11):2897–2906.
229. Feijoo C, Campbell DG, Jakes R, Goedert M, Cuenda A. Evidence that phosphorylation of the microtubule-associated protein Tau by SAPK4/p38delta at Thr50 promotes microtubule assembly. J Cell Sci 2005;118(Pt 2): 397–408.
230. Goedert M, Hasegawa M, Jakes R, Lawler S, Cuenda A, Cohen P. Phosphorylation of microtubule-associated protein tau by stress-activated protein kinases. FEBS Lett 1997;409 (1):57–62.
231. Jenkins SM, Zinnerman M, Garner C, Johnson GV. Modulation of tau phosphorylation and intracellular localization by cellular stress. Biochem J 2000;345(Pt 2):263–270.
232. Pargellis C, Tong L, Churchill L, Cirillo PF, Gilmore T, Graham AG, Grob PM, Hickey ER, Moss N, Pav S, Regan J. Inhibition of p38 MAP kinase by utilizing a novel allosteric binding site. Nat Struct Biol 2002;9(4):268–272.
233. Bacher M, Dodel R, Aljabari B, Keyvani K, Marambaud P, Kayed R, Glabe C, Goertz N, Hoppmann A, Sachser N, Klotsche J, Schnell S, Lewejohann L, Al-Abed Y. CNI-1493 inhibits Abeta production, plaque formation, and cognitive deterioration in an animal model of Alzheimer's disease. J Exp Med 2008;205(7): 1593–1599.
234. Bain J, Plater L, Elliott M, Shpiro N, Hastie CJ, McLauchlan H, Klevernic I, Arthur JS, Alessi DR, Cohen P. The selectivity of protein kinase inhibitors: a further update. Biochem J 2007;408(3):297–315.
235. Pardridge WM. Drug targeting to the brain. Pharm Res 2007;24(9):1733–1744.
236. Matsuoka Y, Gray AJ, Hirata-Fukae C, Minami SS, Waterhouse EG, Mattson MP, LaFerla FM, Gozes I, Aisen PS. Intranasal NAP administration reduces accumulation of amyloid peptide and tau hyperphosphorylation in a transgenic mouse model of Alzheimer's disease at early pathological stage. J Mol Neurosci 2007;31(2):165–170.
237. Matsuoka Y, Jouroukhin Y, Gray AJ, Ma L, Hirata-Fukae C, Li HF, Feng L, Lecanu L, Walker BR, Planel E, Arancio O, Gozes I, Aisen PS. A neuronal microtubule-interacting agent, NAPVSIPQ, reduces tau pathology and enhances cognitive function in a mouse model of Alzheimer's disease. J. Pharmacol Exp Ther 2008;325(1):146–153.
238. Gozes I, Morimoto BH, Tiong J, Fox A, Sutherland K, Dangoor D, Holser-Cochav M, Vered K, Newton P, Aisen PS, Matsuoka Y, van Dyck CH, Thal L. NAP: research and development of a peptide derived from activity-dependent neuroprotective protein (ADNP). CNS Drug Rev. 2005;11(4):353–368.
239. Gong CX, Liu F, Grundke-Iqbal I, Iqbal K. Impaired brain glucose metabolism leads to Alzheimer neurofibrillary degeneration through a decrease in tau O-GlcNAcylation. J Alzheimers Dis 2006;9(1):1–12.
240. Liu F, Iqbal K, Grundke-Iqbal I, Hart GW, Gong CX. O-GlcNAcylation regulates phosphorylation of tau: a mechanism involved in Alzheimer's disease. Proc Natl Acad Sci USA 2004;101(29):10804–10809.
241. Liu F, Shi J, Tanimukai H, Gu J, Gu J, Grundke-Iqbal I, Iqbal K, Gong CX. Reduced

O-GlcNAcylation links lower brain glucose metabolism and tau pathology in Alzheimer's disease. Brain 2009;132(Pt 7):1820–1832.
242. Whitworth GE, Macauley MS, Stubbs KA, Dennis RJ, Taylor EJ, Davies GJ, Greig IR, Vocadlo DJ. Analysis of PUGNAc and NAG-thiazoline as transition state analogues for human O-GlcNAcase: mechanistic and structural insights into inhibitor selectivity and transition state poise. J Am Chem Soc 2007;129(3):635–644.
243. Arriagada PV, Growdon JH, Hedley-Whyte ET, Hyman BT. Neurofibrillary tangles but not senile plaques parallel duration and severity of Alzheimer's disease. Neurology 1992;42(3 Pt 1):631–639.
244. Bierer LM, Hof PR, Purohit DP, Carlin L, Schmeidler J, Davis KL, Perl DP. Neocortical neurofibrillary tangles correlate with dementia severity in Alzheimer's disease. Arch Neurol 1995;52(1):81–88.
245. Khlistunova I, Biernat J, Wang Y, Pickhardt M, von Bergen M, Gazova Z, Mandelkow E, Mandelkow EM. Inducible expression of Tau repeat domain in cell models of tauopathy: aggregation is toxic to cells but can be reversed by inhibitor drugs. J Biol Chem 2006;281(2):1205–1214.
246. Gamblin TC, Berry RW, Binder LI. Modeling tau polymerization *in vitro*: a review and synthesis. Biochemistry 2003;42(51): 15009–15017.
247. Gendron TF, Petrucelli L. The role of tau in neurodegeneration. Mol Neurodegener 2009;4:13–31.
248. Chirita C, Necula M, Kuret J. Ligand-dependent inhibition and reversal of tau filament formation. Biochemistry 2004;43(10): 2879–2887.
249. Pickhardt M, von Bergen M, Gazova Z, Hascher A, Biernat J, Mandelkow EM, Mandelkow E. Screening for inhibitors of tau polymerization. Curr Alzheimer Res 2005;2(2): 219–226.
250. Taniguchi S, Suzuki N, Masuda M, Hisanaga S, Iwatsubo T, Goedert M, Hasegawa M. Inhibition of heparin-induced tau filament formation by phenothiazines, polyphenols, and porphyrins. J Biol Chem 2005;280(9): 7614–7623.
251. Pickhardt M, Larbig G, Khlistunova I, Coksezen A, Meyer B, Mandelkow EM, Schmidt B, Mandelkow E. Phenylthiazolyl-hydrazide and its derivatives are potent inhibitors of tau aggregation and toxicity *in vitro* and in cells. Biochemistry 2007;46(35):10016–10023.
252. Bulic B, Pickhardt M, Khlistunova I, Biernat J, Mandelkow EM, Mandelkow E, Waldmann H. Rhodanine-based tau aggregation inhibitors in cell models of tauopathy. Angew Chem Int Ed Engl 2007;46(48):9215–9219.
253. Crowe A, Huang W, Ballatore C, Johnson RL, Hogan AM, Huang R, Wichterman J, McCoy J, Huryn D, Auld DS, Smith AB, Inglese J, Trojanowski JQ, Austin CP, Brunden KR, Lee VM. Identification of aminothienopyridazine inhibitors of tau assembly by quantitative high-throughput screening. Biochemistry 2009;48 (32):7732–7745.
254. Li W, Sperry JB, Crowe A, Trojanowski JQ, Smith AB, 3rd Lee VM. Inhibition of tau fibrillization by oleocanthal via reaction with the amino groups of tau. J Neurochem 2009;110(4):1339–1351.
255. Hutton M, Lendon CL, Rizzu P, Baker M, Froelich S, Houlden H, Pickering-Brown S, Chakraverty S, Isaacs A, Grover A, Hackett J, Adamson J, Lincoln S, Dickson D, Davies P, Petersen RC, Stevens M, de Graaff E, Wauters E, van Baren J, Hillebrand M, Joosse M, Kwon JM, Nowotny P, Che LK, Norton J, Morris JC, Reed LA, Trojanowski J, Basun H, Lannfelt L, Neystat M, Fahn S, Dark F, Tannenberg T, Dodd PR, Hayward N, Kwok JB, Schofield PR, Andreadis A, Snowden J, Craufurd D, Neary D, Owen F, Oostra BA, Hardy J, Goate A, van Swieten J, Mann D, Lynch T, Heutink P. Association of missense and 5′-splice-site mutations in tau with the inherited dementia FTDP-17. Nature 1998;393(6686):702–705.
256. Bulic B, Pickhardt M, Schmidt B, Mandelkow EM, Waldmann H, Mandelkow E. Development of tau aggregation inhibitors for Alzheimer's disease. Angew Chem Int Ed Engl 2009;48(10):1740–1752.
257. Shanmugam G. Vasoplegic syndrome: the role of methylene blue. Eur J Cardiothorac Surg 2005;28(5):705–710.
258. Wainwright M, Byrne MN, Gattrell MA. Phenothiazinium-based photobactericidal materials. J Photochem Photobiol B 2006;84(3): 227–230.
259. Gura T. Hope in Alzheimer's fight emerges from unexpected places. Nat Med 2008;14(9): 894.
260. Wischik CM, Bentham P, Wischik DJ, Seng KM.Tau aggregation inhibitors (TAI) therapy with Rember™ arrests disease progression in mild and moderate Alzheimer's disease over 50 weeks. 2008; ICAD. Abstract 03-04-07.

261. Paquet D, Bhat R, Sydow A, Mandelkow EM, Berg S, Hellberg S, Fälting J, Distel M, Köster RW, Schmid B, Haass C. A zebrafish model of tauopathy allows *in vivo* imaging of neuronal cell death and drug evaluation. J Clin Invest 2009;119(5):1382–1395.

262. Necula M, Breydo L, Milton S, Kayed R, van der Veer WE, Tone P, Glabe CG. Methylene blue inhibits amyloid Abeta oligomerization by promoting fibrillization. Biochemistry 2007;46 (30):8850–8860.

263. Necula M, Chirita CN, Kuret J. Cyanine dye N744 inhibits tau fibrillization by blocking filament extension: implications for the treatment of tauopathic neurodegenerative diseases. Biochemistry 2005;44(30): 10227–10237.

264. Brancato R, Trabucchi G. Fluorescein and indocyanine green angiography in vascular chorioretinal diseases. Semin Ophthalmol 1998;13 (4):189–198.

265. Congdon EE, Necula M, Blackstone RD, Kuret J. Potency of a tau fibrillization inhibitor is influenced by its aggregation state. Arch Biochem Biophys 2007;465(1):127–135.

266. Hotta N, Kawamori R, Atsumi Y, Baba M, Kishikawa H, Nakamura J, Oikawa S, Yamada N, Yasuda H, Shigeta Y, ADCT Study Group Stratified analyses for selecting appropriate target patients with diabetic peripheral neuropathy for long-term treatment with an aldose reductase inhibitor, epalrestat. Diabet Med 2008;25(7):818–825.

267. Beauchamp GK, Keast RS, Morel D, Lin J, Pika J, Han Q, Lee CH, Smith AB, Breslin PA. Phytochemistry: ibuprofen-like activity in extra-virgin olive oil. Nature 2005;437(7055): 45–46.

268. Li W, Lee VM. Characterization of two VQIXXK motifs for tau fibrillization *in vitro*. Biochemistry 2006;45(51):15692–701.

269. Hudis CA. Trastuzumab: mechanism of action and use in clinical practice. N Engl J Med 2007;357(1):39–51.

270. Busschots K, De Rijck J, Christ F, Debyser Z. In search of small molecules blocking interactions between HIV proteins and intracellular cofactors. Mol Biosyst 2009;5(1):21–31.

271. Michaelis ML, Ansar S, Chen Y, Reiff ER, Seyb KI, Himes RH, Audus KL, Georg GI. β-Amyloid-induced neurodegeneration and protection by structurally diverse microtubule-stabilizing agents. J Pharmacol Exp Ther 2005;312(2):659–668.

272. Zhang B, Maiti A, Shively S, Lakhani F, McDonald-Jones G, Bruce J, Lee EB, Xie SX, Joyce S, Li C, Toleikis PM, Lee VM, Trojanowski JQ. Microtubule-binding drugs offset tau sequestration by stabilizing microtubules and reversing fast axonal transport deficits in a tauopathy model. Proc Natl Acad Sci USA 2005;102(1):227–231.

273. Terry RD, Masliah E, Salmon DP, Butters N, DeTeresa R, Hill R, Hansen LA, Katzman R. Physical basis of cognitive alterations in Alzheimer's disease: synapse loss is the major correlate of cognitive impairment. Ann Neurol 1991;30(4):572–580.

274. Scheff SW, Price DA, Schmitt FA, DeKosky ST, Mufson EJ. Synaptic alterations in CA1 in mild Alzheimer disease and mild cognitive impairment. Neurology 2007;68(18): 1501–1508.

275. Roy S, Zhang B, Lee VM, Trojanowski JQ. Axonal transport defects: a common theme in neurodegenerative diseases. Acta Neuropathol 2005;109(1):5–13.

276. Stokin GB, Goldstein LS. Axonal transport and Alzheimer's disease. Annu. Rev. Biochem. 2006;75:607–627.

277. Pigino G, Morfini G, Atagi Y, Deshpande A, Yu C, Jungbauer L, LaDu M, Busciglio J, Brady S. Disruption of fast axonal transport is a pathogenic mechanism for intraneuronal amyloid beta. Proc Natl Acad Sci USA 2009;106(14): 5907–5912.

278. Adalbert R, Nogradi A, Babetto E, Janeckova L, Walker SA, Kerschensteiner M, Misgeld T, Coleman MP. Severely dystrophic axons at amyloid plaques remain continuous and connected to viable cell bodies. Brain 2009;132(Pt 2):402–416.

279. Miles D, von Minckwitz G, Seidman AD. Combination versus sequential single-agent therapy in metastatic breast cancer. Oncologist 2002;7(Suppl 6):13–19.

280. Wilson L, Panda D, Jordan MA. Modulation of microtubule dynamics by drugs: a paradigm for the actions of cellular regulators. Cell Struct Funct 1999;24 (5):329–335.

281. Kingston DG. Tubulin-interactive natural products as anticancer agents (1). J Nat Prod 2009;72(3):507–515.

282. Lee JJ, Swain SM. Peripheral neuropathy induced by microtubule-stabilizing agents. J Clin Oncol 2006;24(10):1633–1642.

283. Rapoport M, Dawson HN, Binder LI, Vitek MP, Ferreira A. Tau is essential to beta-amyloid-induced neurotoxicity. Proc Natl Acad Sci USA 2002;99(9):6364–6369.

284. Schweers O, Mandelkow EM, Biernat J, Mandelkow E. Oxidation of cysteine-322 in the repeat domain of microtubule-associated protein tau controls the *in vitro* assembly of paired helical filaments. Proc Natl Acad Sci USA 1995;92(18):8463–8467.

285. Dabir DV, Trojanowski JQ, Richter-Landsberg C, Lee VM, Forman MS. Expression of the small heat-shock protein alphaB-crystallin in tauopathies with glial pathology. Am J Pathol 2004;164(1):155–166.

286. Nemes Z, Devreese B, Steinert PM, Van Beeumen J, Fésüs L. Cross-linking of ubiquitin, HSP27, parkin, and alpha-synuclein by gamma-glutamyl-epsilon-lysine bonds in Alzheimer's neurofibrillary tangles. FASEB J 2004;18(10):1135–1137.

287. Hatakeyama S, Matsumoto M, Kamura T, Murayama M, Chui DH, Planel E, Takahashi R, Nakayama KI, Takashima A. U-box protein carboxyl terminus of Hsc70-interacting protein (CHIP) mediates poly-ubiquitylation preferentially on four-repeat Tau and is involved in neurodegeneration of tauopathy. J Neurochem 2004;91(2):299–307.

288. Petrucelli L, Dickson D, Kehoe K, Taylor J, Snyder H, Grover A, De Lucia M, McGowan E, Lewis J, Prihar G, Kim J, Dillmann WH, Browne SE, Hall A, Voellmy R, Tsuboi Y, Dawson TM, Wolozin B, Hardy J, Hutton M. CHIP and Hsp70 regulate tau ubiquitination, degradation and aggregation. Hum Mol Genet 2004;13(7):703–714.

289. Shimura H, Schwartz D, Gygi SP, Kosik KS. CHIP-Hsc70 complex ubiquitinates phosphorylated tau and enhances cell survival. J Biol Chem 2004;279(6):4869–4876.

290. Sahara N, Murayama M, Mizoroki T, Urushitani M, Imai Y, Takahashi R, Murata S, Tanaka K, Takashima A. *In vivo* evidence of CHIP up-regulation attenuating tau aggregation. J Neurochem 2005;94(5): 1254–1263.

291. Dickey CA, Yue M, Lin WL, Dickson DW, Dunmore JH, Lee WC, Zehr C, West G, Cao S, Clark AM, Caldwell GA, Caldwell KA, Eckman C, Patterson C, Hutton M, Petrucelli L. Deletion of the ubiquitin ligase CHIP leads to the accumulation, but not the aggregation, of both endogenous phospho- and caspase-3-cleaved tau species. J Neurosci 2006;26 (26):6985–6996.

292. Dickey CA, Ash P, Klosak N, Lee WC, Petrucelli L, Hutton M, Eckman CB. Pharmacologic reductions of total tau levels; implications for the role of microtubule dynamics in regulating tau expression. Mol Neurodegener 2006;1: 6–14.

293. Dou F, Netzer WJ, Tanemura K, Li F, Hartl FU, Takashima A, Gouras GK, Greengard P, Xu H. Chaperones increase association of tau protein with microtubules. Proc Natl Acad Sci USA 2003;100(2):721–726.

294. Asuni AA, Boutajangout A, Quartermain D, Sigurdsson EM. Immunotherapy targeting pathological tau conformers in a tangle mouse model reduces brain pathology with associated functional improvements. J Neurosci 2007;27 (34):9115–9129.

295. Aihara N, Tanno H, Hall JJ, Pitts LH, Noble LJ. Immunocytochemical localization of immunoglobulins in the rat brain: relationship to the blood–brain barrier. J Comp Neurol 1994;342(4):481–496.

296. Mohamed HA, Mosier DR, Zou LL, Siklós L, Alexianu ME, Engelhardt JI, Beers DR, Le WD, Appel SH. Immunoglobulin Fc gamma receptor promotes immunoglobulin uptake, immunoglobulin-mediated calcium increase, and neurotransmitter release in motor neurons. J Neurosci Res 2002;Jul 1 69(1): 110–116.

297. Orgogozo JM, Gilman S, Dartigues JF, Laurent B, Puel M, Kirby LC, Jouanny P, Dubois B, Eisner L, Flitman S, Michel BF, Boada M, Frank A, Hock C. Subacute meningoencephalitis in a subset of patients with AD after Abeta42 immunization. Neurology 2003;61(1):46–54.

STROKE THERAPY

Philippe G. Nantermet
Evan F. Shalen
Shankar Venkatraman
Hong Zhu
Robert J. Mark
Merck Research Laboratories, West Point, PA

1. INTRODUCTION

Although many clinical trials in stroke have been run, and many more reviews have been written on the subject, there remains only one FDA-approved medicine for the acute treatment of stroke—tissue plasminogen activator [1].

This chapter will not attempt to give comprehensive coverage to the myriad of targets that have been the focus of the above-mentioned trials and reviews. We have rather chosen to focus on a few of the acute neuroprotection targets that have had sufficient medicinal chemistry activity to warrant a commentary.

Ischemic damage due to stroke is clearly a process and not an instantaneous event [2] (Fig. 1). There are a number of points at which intervention is possible, yet the therapeutic efficacy of most of these intervention points remains to be proven.

Patients suffering cerebral ischemia are a heterogeneous population in terms of age and comorbid conditions. Likewise the severity and localization of the brain injury varies among patients. In contrast, most clinical trials in acute stroke have ignored this heterogeneity when enrolling patients for experimental drug treatment. Perhaps, the investigators were hoping to discover a treatment that would be useful in all cerebral ischemia patients, but ignoring the heterogeneous characteristics of the trial participants may be an important contributor to the failure of so many stroke trials.

Certainly, it can be said that the best way to treat a stroke is to not have one. Unfortunately, despite the many advances in stroke prevention treatments, stroke remains the third most common cause of death and a major cause of long-term disability (American Stroke Association Web site). Despite intensive activity in the study of stroke prevention, a real challenge persists in the unmet medical need for an efficacious, therapeutically meaningful acute stroke medicine. In the United States, there are approximately 750,000 stroke cases reported annually.

Pharmacological therapy for acute stroke promises the opportunity to reduce the cerebral injury and hence disabilities resulting from stroke in patients. Therapeutic approaches can be split into categories based on the physiological event they aim to target. While in targeting acute stroke therapy it is easiest to think of postischemic pathophysiology as distinct steps or stages, as in most complex biological systems, these events overlap both spatially and temporally. The first broad category targets the events in the arterial lumen. Because most ischemic strokes are secondary to arterial thromboembolism, thrombolysis offers the most direct treatment for thrombotic-type ischemic strokes. According to the pivotal National Institutes of Neurological Disorders and Stroke study, intravenous tissue plasminogen activator (tPA) improved the clinical outcomes of all types of ischemic stroke if treatment began within 3 h of symptom onset [1].

Other intravenous antithrombotic agents have been tested in clinical trials, but none of them has received FDA approval [3].

Since the approval of tPA, no less than six Phase III clinical trials have explored alternative methods for selecting patients and extending time windows. Attempts with new recombinant protein, thrombolytics have included streptokinase, prourokinase (rpro-UK), and desmoteplase. The goal was an effective treatment beyond the strict, and perhaps arbitrary, time limit of 3 h.

Streptokinase (SK) was tested in two prominent trials, MAST I and E studies [4] MAST I enrolled 622 patients with acute ischemic stroke within 6 h of symptom onset. Patients were randomized with a 2×2 factorial design to (i) a 1 h intravenous infusion of streptokinase,

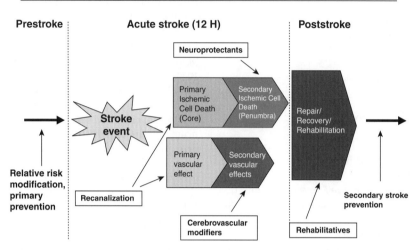

Figure 1. Schematic of potential intervention points in stroke therapy. (This figure is available in full color at http://mrw.interscience.wiley.com/emrw/9780471266945/home.)

(ii) 300 mg/day buffered aspirin for 10 days, (iii) both of these active treatments, or (iv) neither of these treatments. Analyses of the data collected indicated that streptokinase may have increased the risk of early death in acute stroke patients significantly if combined with aspirin and not significantly if given alone. Moreover, although thrombolytic and antithrombotic treatments both seem to reduce 6-month case fatality and disability, no statistically significant evidence for effectiveness of either streptokinase or aspirin could be obtained.

Recombinant prourokinase (rpro-UK), was tested in two prospective, randomized, double-blind, placebo controlled trials: PROACT and PROACT 2 [5].

A total of 180 patients with acute ischemic stroke of less than 6 h postsymptom onset, caused by angiographically proven occlusion of the MCA and without hemorrhage or major early infarction signs on CT scan were enrolled. Patients were randomized to receive intra-arterial r-proUK plus heparin ($n = 121$) or heparin only ($n = 59$). Treatment with intra-arterial r-proUK within 6 h of the onset of acute ischemic stroke caused by MCA occlusion significantly improved clinical outcome at 90 days, however, it was also associated with an increased frequency of symptomatic intracranial hemorrhage.

Desmodus rotundus salivary plasminogen activator alpha1 (desmoteplase) was a promising candidate for the treatment of acute ischemic stroke. In the DIAS-2 (Desmoteplase in Acute Ischemic Stroke) study, time was a key focus [6]. The investigators preferred to extend the window from 3 to 9 h after symptom onset to test this novel fibrinolytic drug for late reperfusion. DIAS-2 incorporated sophisticated trial design with the latest CT and MRI techniques to confirm benefit of desmoteplase to treat acute ischemic stroke up to 9 h after symptom onset. In this randomized, placebo-controlled, double-blind, dose-ranging study, 186 patients with acute ischemic stroke and tissue at risk seen on either MRI or CT imaging were randomly assigned (1:1:1) to one of the two doses of desmoteplase or placebo within 3–9 h after the onset of symptoms of stroke. DIAS-2 did not show a beneficial effect for either dose of desmoteplase given 3–9 h after the onset of stroke symptoms. The rate of symptomatic intracranial hemorrhage was low and consistent with rates seen in the earlier studies. Despite these results, other Phase III clinical trials (DIAS-3 and DIAS-4) are currently under way with a planned sample size of 320 patients having vessel occlusion or high-grade stenosis on MRI or CT-angiography in the proximal cerebral arteries.

The available data did not provide sufficient evidence to determine the magnitude of treatment effect, nor did it yield answers to many of the key questions that exist and limit the use of thrombolytic therapies. It is hard to see how thrombolytic therapies will gain wider acceptance until several questions are answered including: What is the latest time window in which the treatment is still beneficial? Which types of strokes and which grades of stroke severity will respond favorably? Which types of patients are most likely to be harmed?

Until reproducible data is generated, recent efforts to extent the thrombolytic treatment window beyond 3 h, may lead to the disappointing conclusion that vessel recanalization after 4.5–5 h poststroke onset may generally be inefficacious for tissue salvage.

The second category for therapeutic intervention targets the consequences of ischemia/reperfusion on the neuronal environment. Often referred to as neuroprotection, as our understanding of postischemic pathophysiological cascade has improved, this approach has grown to include targets that are nonneuronal, including glia, inflammatory cells, and cells in the blood vessel walls.

Because cell loss following cerebral ischemia is a process and not an instantaneous event, there is the potential to modify the process by intervening at key points in the cascade and thus favorably affect the outcome/functionality of the patient (Fig. 2).

As imperfect as they may be argued to be, a major accomplishment for the animal models of cerebral ischemia is the elucidation of details of the pathophysiological events that constitute the brain's response to this injury [7].

Due to the breadth of approaches and large volume of published research on stroke therapy, in this chapter, we have summarized chosen medicinal chemistry efforts representative of many of the events depicted in Fig. 1 (e.g., Ca^{2+}-activated proteases, VDCC, and antioxidants).

We neither cover the many stroke prevention strategies that are clinically employed nor we cover in any detail the many theories that are being pursued in the areas of enhanced functional recovery and physical therapy protocols [8]. Without a doubt, poststroke physical therapy will find a vital place in any stroke recovery plan, but to date, controlled, blinded, clinical studies are in their infancy.

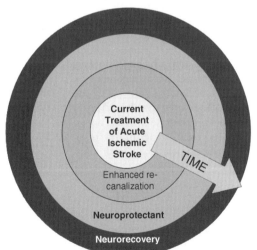

Figure 2. Schematic of infarct progression with time, postischemia. (This figure is available in full color at http://mrw.interscience.wiley.com/emrw/9780471266945/home.)

2. EXCITATORY AMINO ACID RECEPTORS

2.1. NMDA Antagonists

Calcium entry through the NMDA receptors became a target of intense interest for stroke protection following the report by Simon et al. [9] that blockade of the NMDA receptors attenuated infarct volume in rats following transient vessel occlusion.

However, the early excitement around NMDA blockers was soon clouded by conflicting results due to effects on body temperature, blood flow, and the time window of treatment necessary for protection. Eventually, the failure of clinical efficacy in stroke trials and drug and dose-related adverse experiences, have dampened enthusiasm for the likelihood that attenuation of excitotoxicity is a clinically viable approach. The most common adverse experiences in the NMDA antagonist trials were agitation, hallucinations, confusion, somnolence, and nausea. For further reading on some of the key NMDA antagonist trials, the reader can review Refs. [10] (selfotel), [11] (cerestat), and [12] (dextrorphan).

A large number of NMDA antagonists have been prepared and evaluated as neuroprotective agents and then against various CNS disorders. The reader is directed to read the recent reviews by Monaghan and Jane [13], Muir [14], Albensi [15], Gerber and Vallano [16], and Childers and Baudy [17]. Most of the findings reported in those reviews have been summarized here, in addition to the findings from the most recent reports.

2.1.1. Noncompetitive Ion Channel Blockers

Ion channel blockers act as antagonists by binding the NMDA receptor channels in the open state, at the same sites that bind Mg^{2+}. Magnesium has been shown to prevent specific NMDA-mediated excitotoxic injury in rodents [18,19]. Stroke clinical trials with magnesium have been unsuccessful [20–22]. Ion channel blockers typically comprise lipophilic groups and basic amine moieties.

High-Affinity Blockers Ketamine (**1**) and dizocilpine (MK-801, **2**) are considered high affinity, slow off-rate, blockers, also known as "trapping blockers." They have been extensively studied as NMDA antagonists but their therapeutic potential is limited by the prevalence of serious CNS related side effects including hallucinations and loss of motor coordination, even below therapeutic dose [23].

1: Ketamine **2**: Dizocilpine MK-801

Memantine and Derivatives Spider toxins of the arylalkylamine class have displayed open channel blockade that is relieved more readily than the MK-801 block [24]. The recent search for low affinity, uncompetitive antagonist with rapid off-rates has been geared toward the identification of channel blockers with a greater therapeutic index [25,26]. Amantadine (**3**) and memantine (**4**) are two prototypical examples of "partial trappers." Memantine has received FDA approval for the treatment of moderately severe to severe AD. Several reports of animal model studies suggest the potential of memantine to treat ischemic stroke [27–32]. A second generation of memantine derivatives, the nitromemantines have been displayed better efficacy in animal models of neuroprotection [33]. Additional low-affinity channel blockers (**6–9**) have been disclosed and studied for various CNS disorders, including stroke [34–36].

3: Amantadine **4**: Memantine **5**: Bicifadine **6**: Neramexane

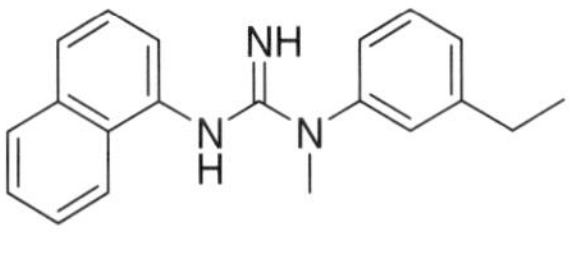

7: Aptiganel 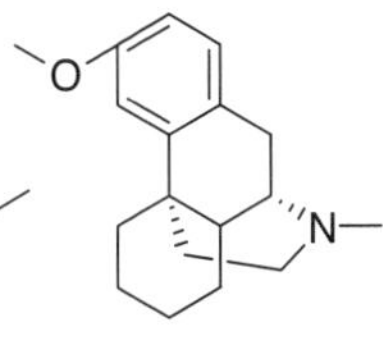   **8**: Dextromethorphan 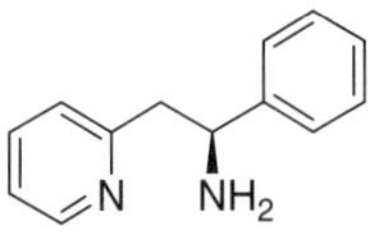   **9**: AR-R15896AR

2.1.2. Competitive Receptor Antagonists

Glutamate Site Antagonists Prototypical NMDA receptor antagonists that bind to the glutamate recognition site display alpha-amino acid moieties and a phosphonic acid derivative. As such, they often suffer from poor bioavailability and poor brain penetration. When used in clinical trials, compounds such as selfotel (**12**) were found ineffective due to unacceptable psychotomimetic side effects [37,38]. Replacement of the amino acid moiety with a squaric acid moiety provided perzinfotel and its prodrug WAY-129 (**13** and **14**), which have been described to be 10-fold selective for rodent NMDA receptors having the NR2A subunit over NR2B and have a superior therapeutic index [39–42].

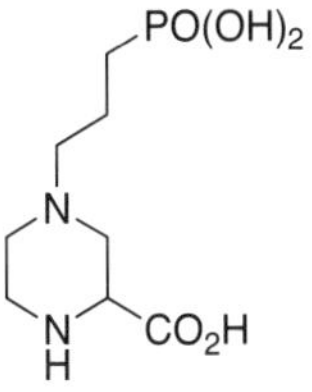

10: CPP 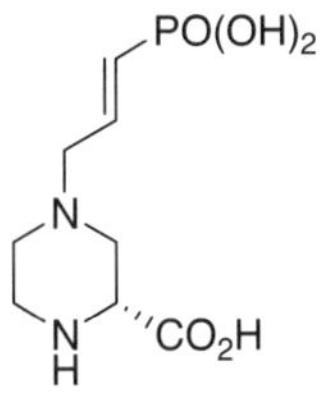   **11**: EAA-494 D-CP Pene   **12**: Selfotel

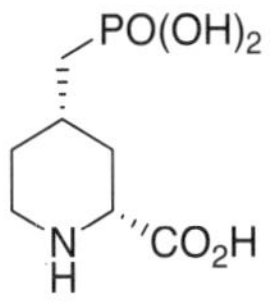

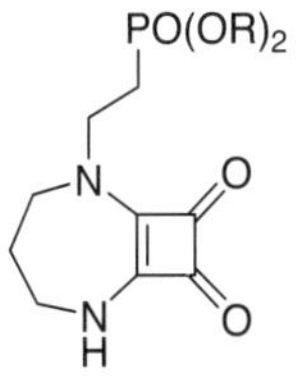

R = H: **13** EAA-090, Perzinfotel
R = CH2OBz: **14** WAY-129

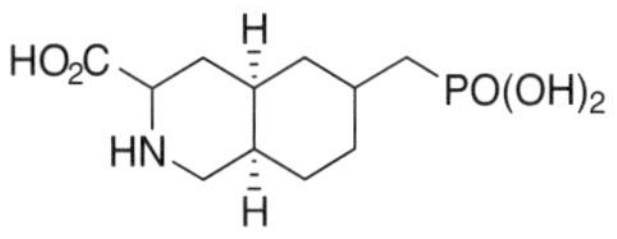

15: LY-235959

Alternative mimics of glutamic acid or NMDA such as **16** and **17** have been suggested to interact at the glutamate site [43,44].

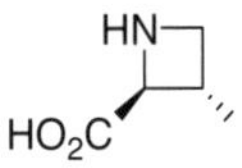

16 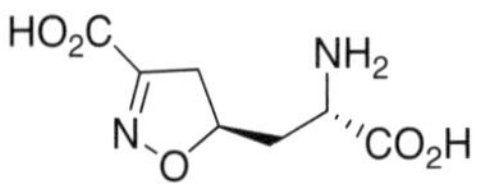   **17**

A family of small peptides with neuroprotective properties, the conantokins, has been isolated from cone snail venom. They have been suggested as glutamate binding site NMDA receptor antagonist [45].

Glycine Site Antagonists Binding of glycine to the NMDA glycine binding site on the NR1 subunit reduces the rate of NMDA receptor desensitization; thus, potentiating receptor response. Glycine site antagonists have therefore attracted considerable attention and have been reviewed quite extensively [46–49]. Most of the findings reported in those reviews have been summarized here, with special focus on antagonists that have been used in stroke applications. Although glycine site antagonists do not cause psychotomimetic side effects they are plagued with sedative and other adverse side effects. A number of compounds have demonstrated efficacy in animal ischemia models and some have been advanced to stroke clinical trials but without success.

Kynurenic acid (**18**) represents the first example of Gly/NMDA receptor antagonist. The 7-chloro derivative **19** has displayed efficacy in various models of stroke. Further elaboration resulted in the discovery of carboxytetrahydroquinoline antagonist as exemplified by L-689,560 (**20**) that remains one of the most potent Gly/NMDA receptor antagonists.

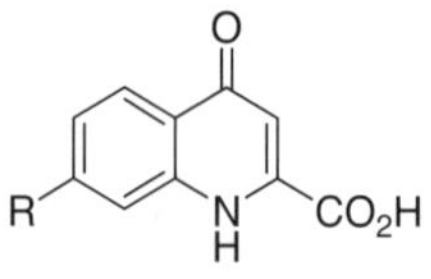

R = H: **18**: KYN
R = Cl: **19**

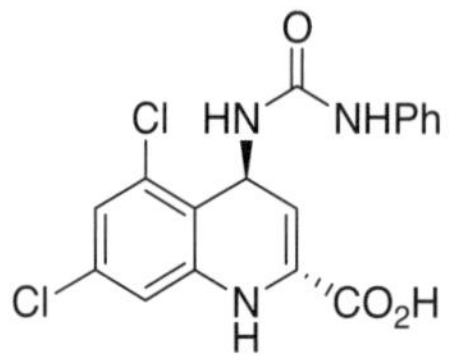

20: L-689,560

Indole-2-carboxylic derivatives have been extensively investigated. Gavestinel (**21**, GV-150526) was advanced to Phase III clinical trials for stroke [50–52]. The use of CoMFA has led scientists at Aventis to the discovery of derivatives such as **22** that is active in neuroprotection animal models [53]. Tricyclic derivatives have been designed, the most representative example being SM-31900 (**23**) that also displayed efficacy in stroke animal models [54].

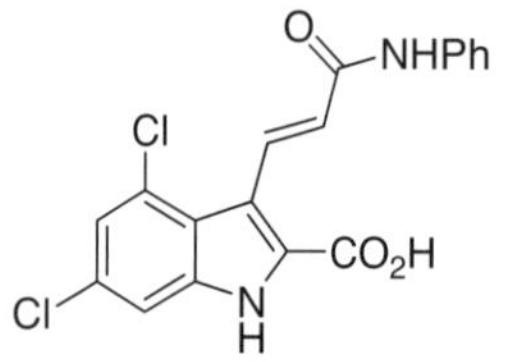

21: GV-150526
Gavestinel

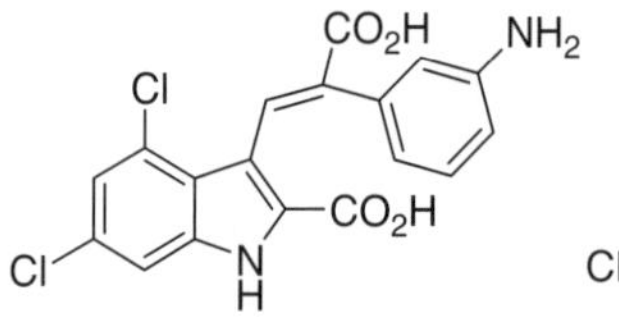

22

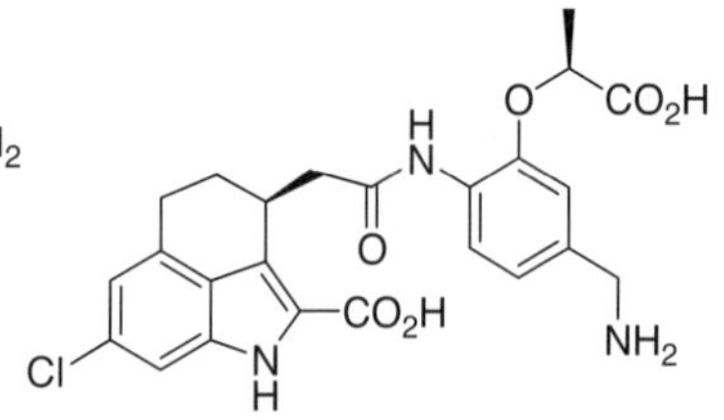

23: SM-31900

Quinaxoline-2,3-dione represent yet another important class of NMDA glycine site antagonists. Licostinel (**24**, ACEA-1021), the prototype in this series, was shown to be neuroprotective in animal models and was studied in human clinical trials although development was halted [55]. Tricyclic derivatives with neuroprotective properties in animals, such as SM-18400 (**25**), have also been prepared [56].

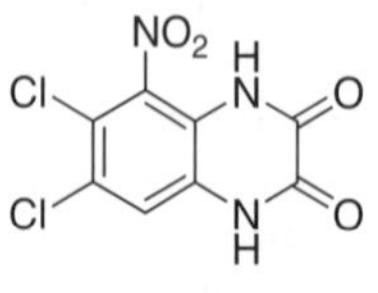

24: ACEA-1021
Licostinel

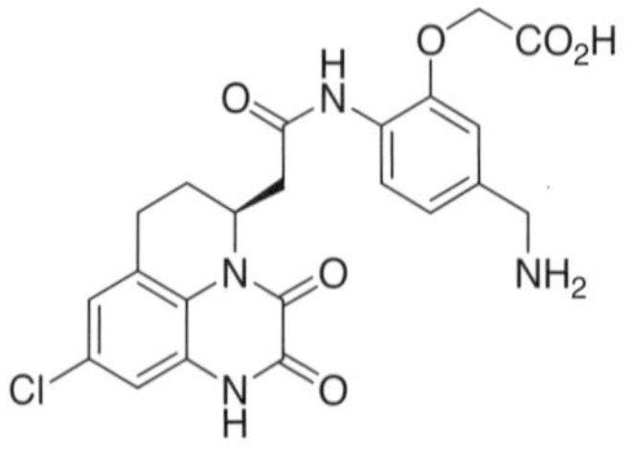

25: SM-18400

The fourth class of NMDA glycine site antagonists, built around a 1,2-dihydropyridazinoquinoline scaffold, has displayed improved pharmacokinetic properties. ZD-9379 (**26**) and MRZ 2/576 (**27**) have both demonstrated efficacy in ischemia models. Additional, recent developments have been reported by AstraZeneca scientists [57].

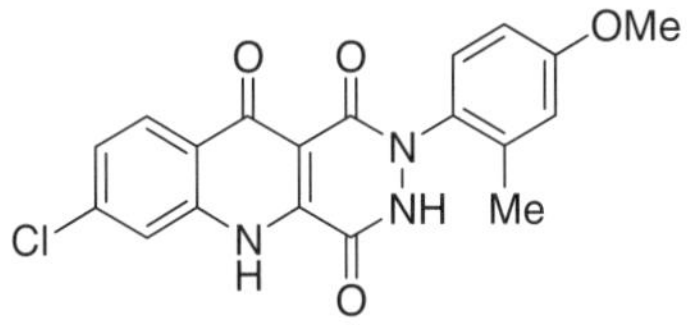

26: ZD-9379

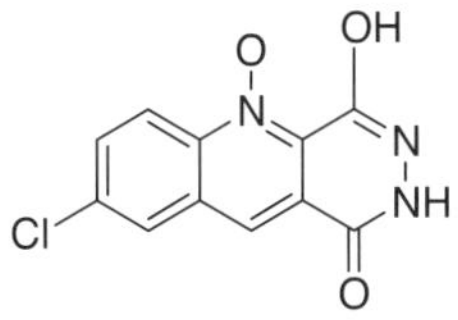

27: MRZ 2/576

Docking of a random selection of molecules from the GSB-11 chemical universe database into the NMDA glycine site receptor resulted in the identification of weak but novel and small antagonists such as piperazine dione **28** [58].

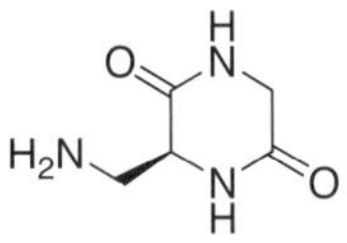

28

2.1.3. NR2B-Specific Noncompetitive Receptor Antagonists

Polyamines such as spermine and spermidine and others are known to positively modulate the NMDA receptor at the "polyamine site" on the NR2B subunit [59]. The search for compounds that would antagonize this effect and hopefully yield neuroprotective NMDA antagonists with reduce side effects yielded the phenylethanolamine class of antagonists. Although they were originally thought to compete directly at the polyamine site, they were recently shown to be use-dependent, selective blockers of receptors displaying the NR2B subunit. They seem to exert their effect via a noncompetitive allosteric interaction that results in modulation of the polyamines effect. Because of the restriction of the NR2B subunit containing receptors to specific regions of the brain it was hoped that NR2B subtype selective NMDA receptor antagonists would be devoid of cognitive and locomotor side effects that have plagued previous antagonists. While early compounds were typically targeted toward stroke indications, the more recent antagonists have been mostly studied in pain and movement disorders. Several specific NR2B reviews have appeared recently [60–62]. Most of the findings reported in those reviews have been summarized here, with special focus on recent disclosures or antagonists that have been used in stroke applications.

The first generation of NR2B antagonists was derived from phenylethanolamine and the prototypes were by ifenprodil (**29**) and eliprodil (**30**) followed by the potent and selective analogs Ro 25-6981 (**31**) and traxoprodil (**32**, CP-101,606). A number of these compounds have shown efficacy in animal models of cerebral ischemia with attenuated locomotor and CNS side effects [63–75]. Clinical information from stroke human trials has been limited. However, traxoprodil has been shown to be well tolerated in head trauma patients and efficacious for pain in patients with pathological pain [76–78].

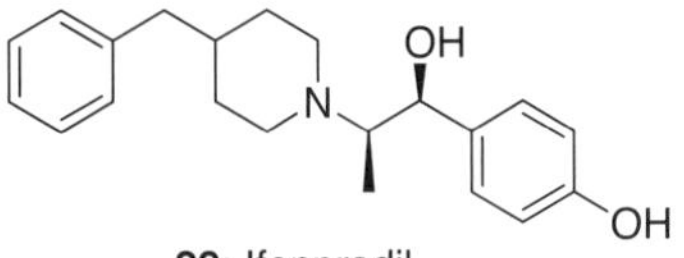

29: Ifenprodil

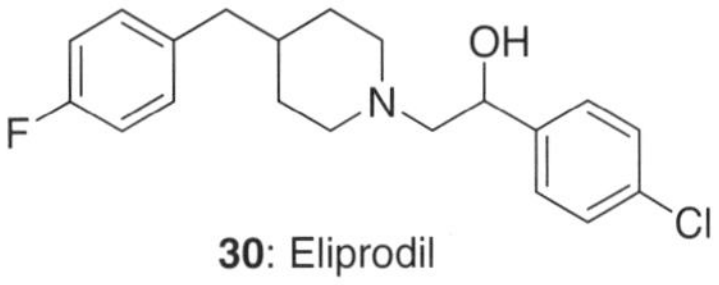

30: Eliprodil

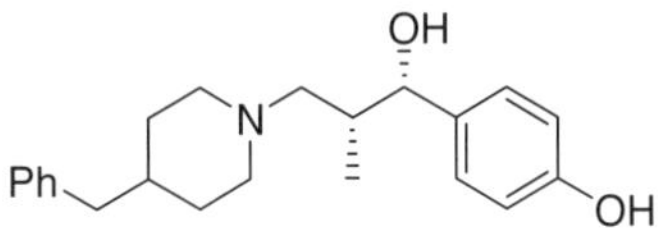

31: Ro 25-6981

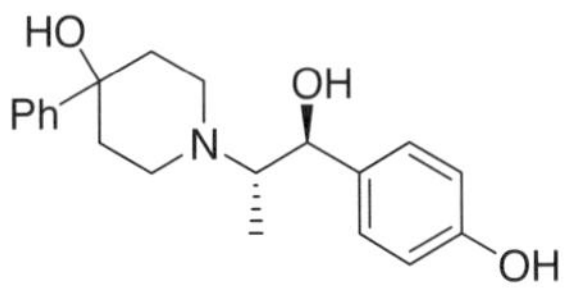

31: Traxoprodil CP-101,606

Based on some of the early NR2B antagonists, the pharmacophore was defined as incorporating an aryl group connected to a phenolic group via a basic amine containing linker. Many of the second-generation antagonists were designed to find phenolic group mimics and replace the basic amine linker that often induced hERG cross-reactivity. Besonprodil (**32**) represents one of the first attempts at phenol isostere identification. Compounds **33–35** illustrate antagonists with similar phenol isostere or even devoid of phenolic hydroxyl group [79–81]. Antagonist **36** exemplifies an early example of nonbasic central linker with improved hERG profile [82].

32: Besonprodil

33

34

35

36

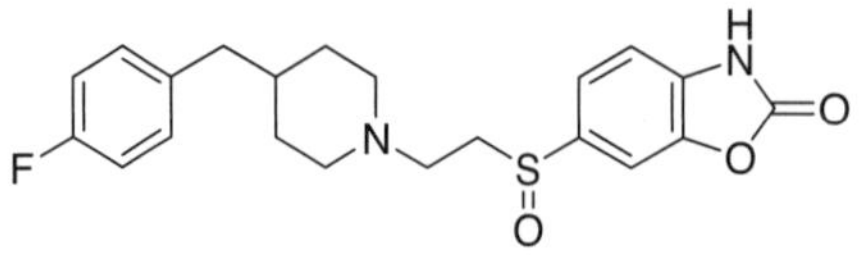

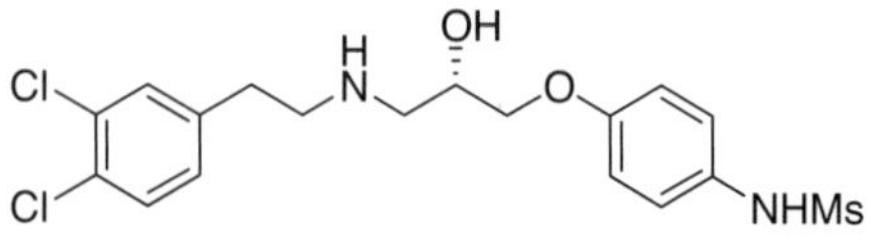

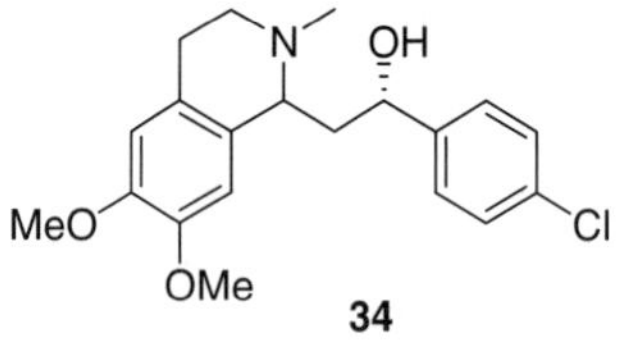

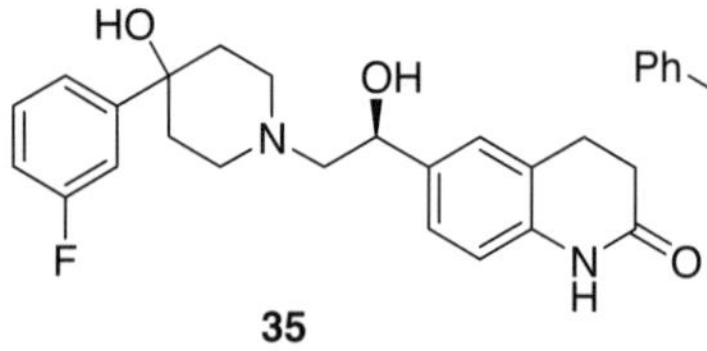

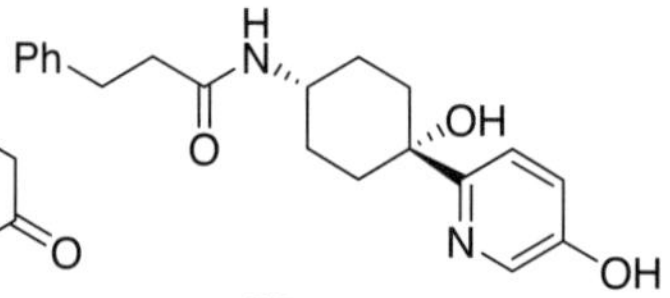

Additional examples of either phenol or basic amine linker replacement, such as **37** and **38** have been discovered, starting from a benzimidazole HTS hit [83]. They were mostly studied in pain animal models.

37

38

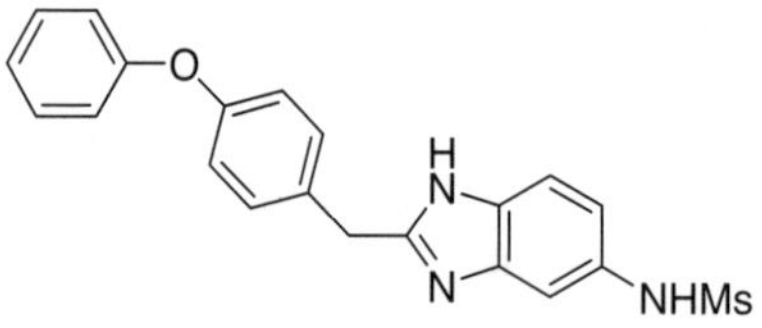

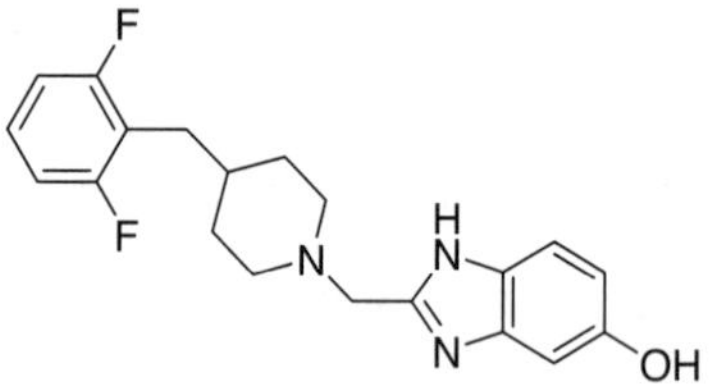

A number of acylated piperidine have been prepared by various groups as one of the simplest means to eliminate the basic amine linker. Cinamide piperidine or oxamide piperidine derivatives such as **39–40** have been reported to have a better selectivity profile [84,85]. A subsequent series of indole- and benzimidazole-derived NR2B antagonists was reported [86,87]. A benzyl carbamate series illustrated by **43** has been reported to display efficacy in pain and Parkinson's disease model [88].

39

40

X = CH_2 : **41**
X = N: **42**

43

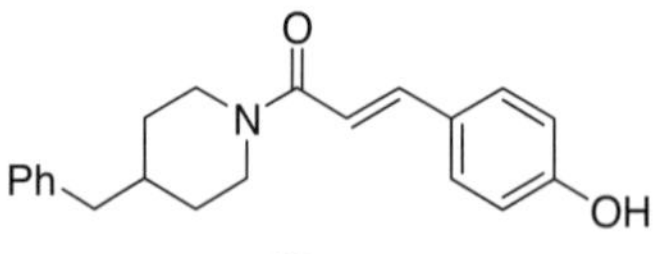

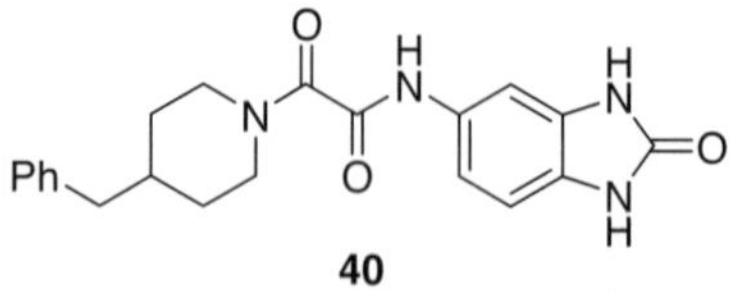

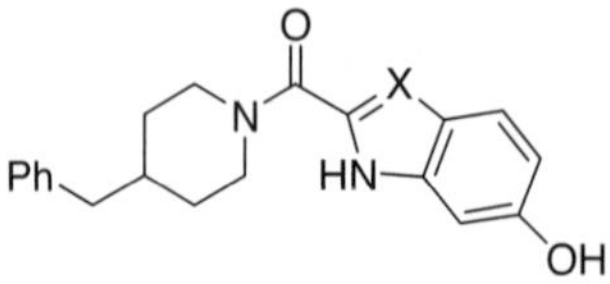

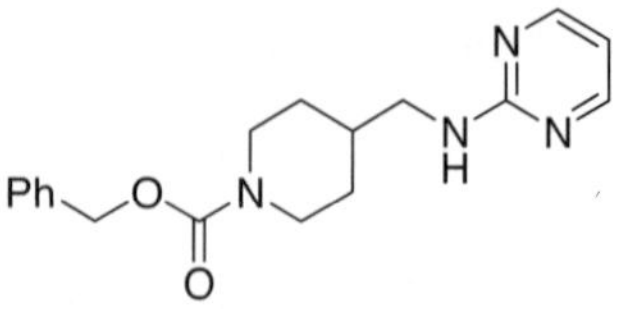

A number of amidine- and cinamidine-based NR2B antagonist such as compounds **44–47** have been reported by various groups, mostly for pain indications [89–92]. Antagonist **48**, based on a dihydroimidazoline scaffold has been studied in a sound-induced audiogenic seizure assay [93].

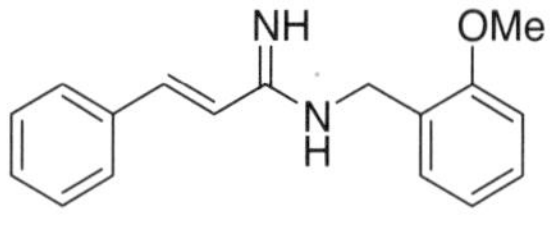

NR2B antagonists designed around a 4-aminoquinoline (**49**) or simplified 4-aminopyridine (**50**) scaffold have been disclosed [94–96].

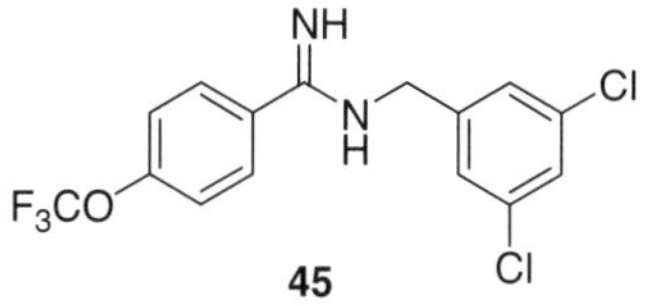

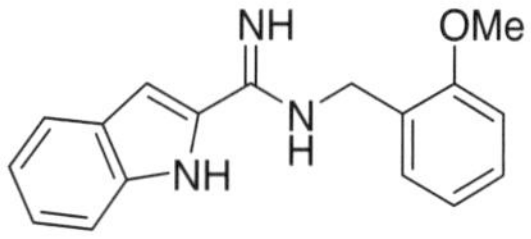

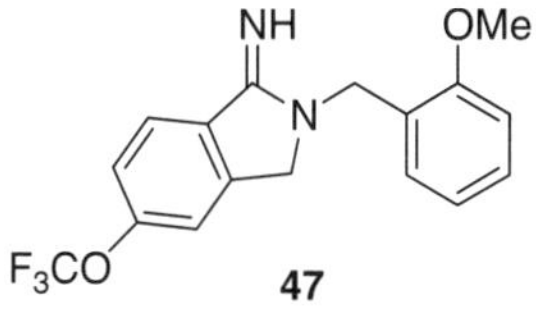

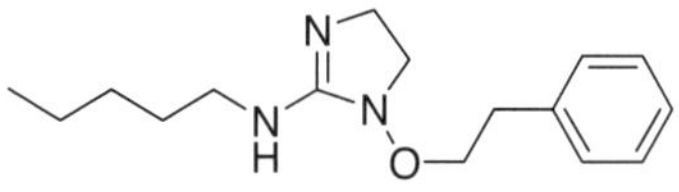

Kynurenic acid amide derivatives such as **51** have also been reported as NR2B selective antagonists [97].

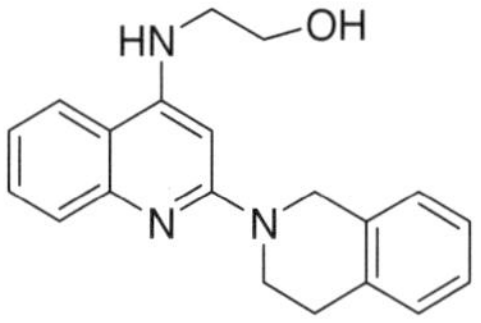

A few reports of computational studies aimed at the discovery of NR2B receptor antagonists have recently been disclosed [98,99].

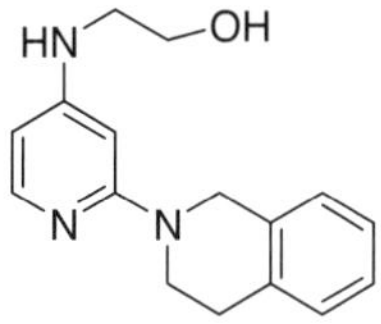

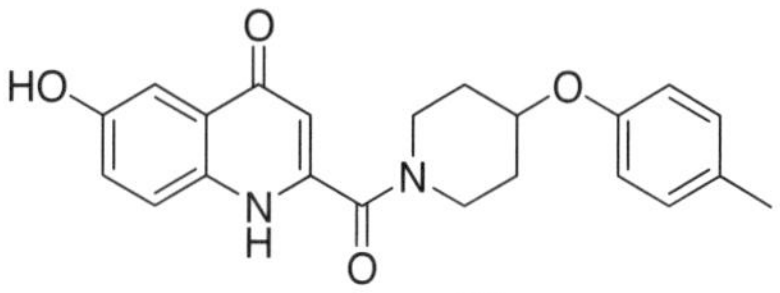

Recently, compounds that inhibit NMDA receptor and acetylcholinesterase activity have been disclosed [100]. Indeed, they display antioxidant properties and inhibit AChE-induced Aβ aggregation.

2.2. VDCC-Voltage-Dependent Calcium Channels

Hypotheses generally agreed that the neurotoxic ischemic cascade is initiated by a pathological release of excitatory amino acids and subsequent disregulation of intraneuronal calcium concentrations. The belief in the importance of restoring intracellular ion homeostasis led voltage-gated cation channels to be frequent targets in stroke drug development efforts. Dihydropyridine blockers of L-type Ca^{2+} VDCC (e.g., nimodipine) have been tested in several preclinical neuroptrotection studies. In in vitro neuronal cultures and hippocampal slice cultures, the dihydropyridine-like blockers attenuated cell death induced by excitotoxic challenge or oxygen–glucose deprivation.

In rodent models of ischemia, the dihydropyridines had a mixed success rate. Some studies reported efficacy, while others reported no effect or even a worsening of ischemic damage [101]. The outcome of human clinical trials with L-type VDCC blockers has been less ambiguous. The results of 28 completed clinical trials (of 7521 patients) showed no significant improvement in either neurological or functional endpoints [102].

Many reviews have recently appeared in the literature on antagonists of VDCC—Winquist [103], Takahara [104], Niimi [105], and Nishizawa [106]. VDCCs are generally grouped into three major families (Ca_v1, Ca_v2, and Ca$_v$3) based on the gene of α$_1$-pore-forming subunits. These are further divided into 10 different subtypes in the α$_1$-subunit in mammalian cells (L-type Ca$_v$1.1–1.4, P/Q-type Ca$_v$2.1, N-type Ca$_v$2.2, R-type Ca$_v$2.3, and T-type Ca$_v$3.1–3.3). Many peptides (20–30 C-terminal amino acid amide) have been isolated from different cone snails and shown to be highly specific and potent antagonists of various VDCC subtypes [107]. A major challenge in this area has been finding small molecules that specifically inhibit a particular channel with excellent selectivity against voltage-gated sodium and potassium channels. Classified on the basis of types of calcium channel, this chapter will highlight various small molecules as an antagonist of voltage-gated calcium channel with emphasis on structural diversity. Presently, there are no significant reports in literature on the role of P/Q-, R-, and T-type calcium channels in *in vitro* or *in vivo* models of stroke. Hence, these channels will not be reviewed in this chapter.

2.2.1. N-Type VDCC The blockade of N-type VDCC has been established as a promising therapeutic target for neuronal ischemia following stroke, due to inhibition of the excitatory neurotransmitter release as well as blockade of entering calcium reaching toxic levels. High concentrations of glutamate induce excitotoxicity in neuronal cells. The ω-conotoxins have shown to be specific and potent antagonists of the N-type calcium channel. Synthetic analog of ω-conotoxin, Ziconotide, is currently used as a drug to treat severe pain [108]. These suffer from numerous drawbacks including being administered as intrathetical injections with significant adverse effects. There are reports of N-type VDCC antagonist small molecules that show efficacy in stroke models. However, these compounds are nonselective VDCC antagonists with activity in sodium and potassium channels.

Peptide-Based Ligands A series of peptide-based small molecules have been identified as N-type calcium channel blockers.

PD151307 (**1**), possessing a leucine-tyrosine dipeptide, was shown to be an antagonist of N-type channel in IMR-32 a human neuroblastoma cell (IC$_{50}$ = 0.32 uM). It had moderate *in vivo* potency in the audiogenic seizure model. This compound also had very high C log*p* value (6.50) preventing its further development. Additional structure activity studies resulted in identification of **2** with improved potency, and was active at lower doses in the audiogenic seizure model of DBA/2 mice [109–111]. The same group also identified 4-arylalkylmino piperidine derivatives (**3**) as N-type antagonists with sixfold selectivity over voltage-dependent sodium channel. Although compound **3** demonstrated excellent *in vivo* activity, it suffered from high clearance (>90 mL/min/kg) and large volumes of distribution (20 L/kg). Recently, Merck and Neuromed filed number of patents with varying peptide motif (**4**) as antagonists for N-type VDCC. Ono has reported thiazolidine derivatives **5** with excellent antagonist activity (IC$_{50}$ = 0.14 uM) and 15–20-fold selectivity over L-type channel [112].

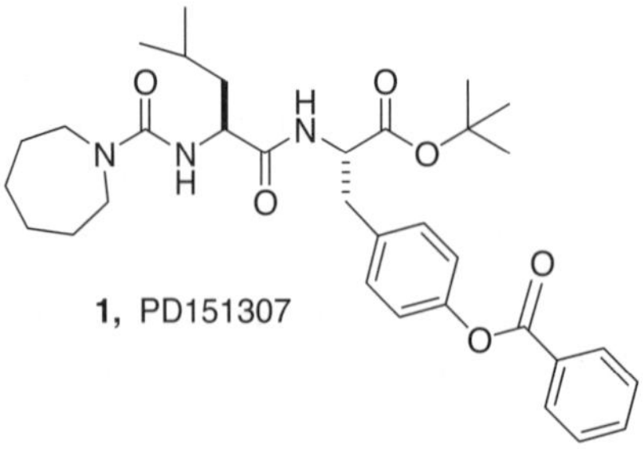

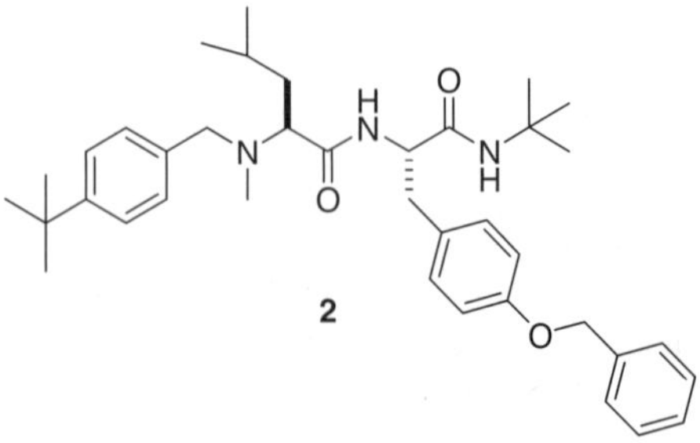

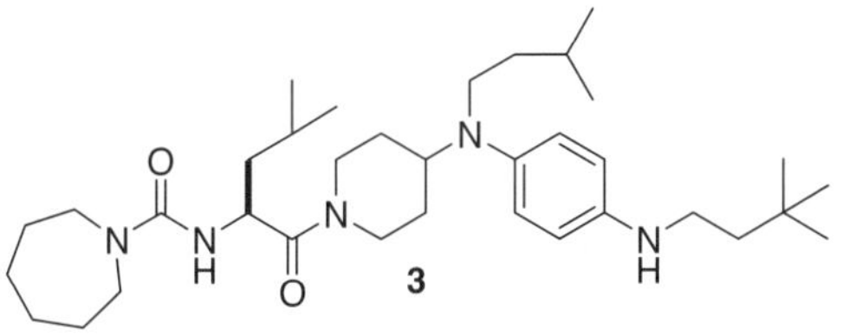

Pyrimidine-Based Scaffolds

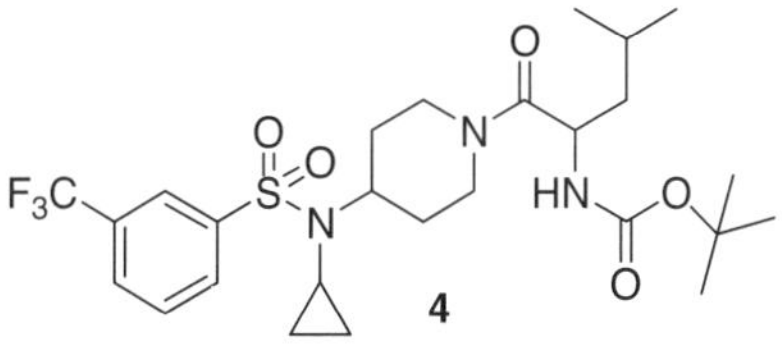

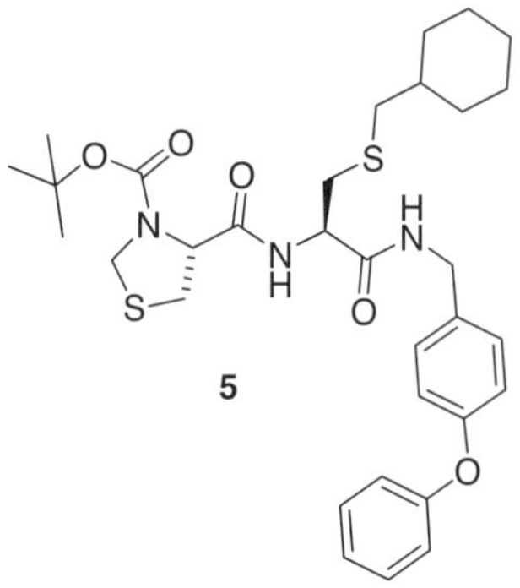

NS-7, a dual blocker for VDCC (**7**, IC$_{50}$ = 4.5, 7.3, 171 uM for L-, N-, P/Q-, and T-type calcium channels) and voltage-dependent sodium channel (IC$_{50}$ = 7.8 uM), with good selectivity for voltage-dependent potassium channel (IC$_{50}$ = 160 uM) showed inhibitory activity against veratridine-induced glutamate release from cortical slices and KCl-induced NOS activity. Intravenous administration of NS-7 significantly decreased infarct volumes in both transient and permanent MCAO in rat models. Development of NS-7 for stroke was terminated in Phase II. Although the exact reason is unknown, there was one case of QT prolongation in clinical trials [113–116]. N-type antagonist, AR-R18565 (**8**), was efficacious in the rat transient model with excellent pharmacokinetics in rats, large volume of distribution and β-half-life of 4 h [117].

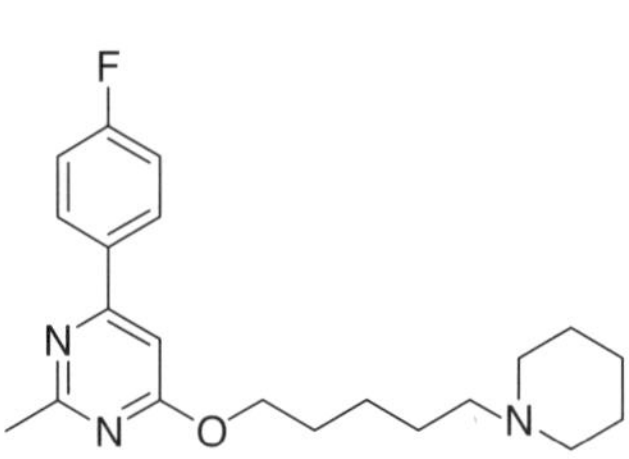

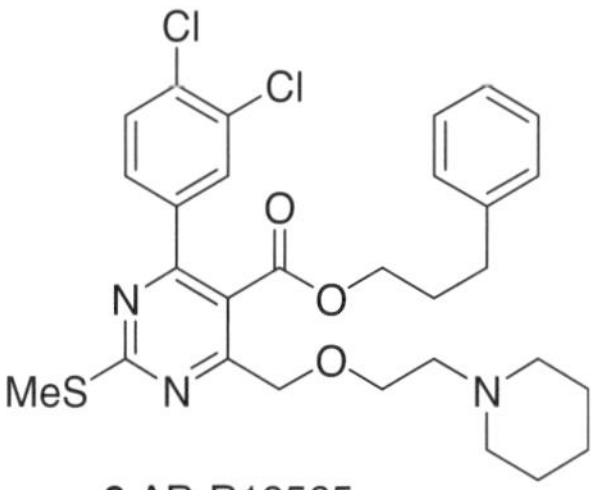

Piperidine/Piperazine-Based Scaffolds

E-2050 (**9**) from Esai was shown to be a potent inhibitor of N-type calcium channel (IC$_{50}$ = 0.8 uM) with moderate selectivity against sodium and potassium channel. This compound inhibited glutamate release in rat cortical slices (IC$_{50}$ = 3.7 uM) and inhibited KCl-induced free calcium concentration in rat cortical synaptosomes (IC$_{50}$ = 5.9 uM). Intravenous administration of E-2050 reduced infarct volumes in both transient and permanent MCAO in rats [118]. Currently, in Phase I clinical trials, no progress has since been reported on this compound. Recently, Merck and Neuromed took MK-6721 (**10**, NMED-160) to clinical trails for chronic pain. This compound was a potent inhibitor of the N-type channel but had little selectivity over L- and P/Q-type channels [119,120]. Neuromed also announced that the clinical development of NMED-160 was discontinued due to unacceptable pharmacological characteristics, although no adverse effects were observed in clinical trials at the maximum dose of 1600 mg/day. Merck and Neuromed have filed number of patents with varying structural motif (**12**) as antagonists for N-type VDCC. A number of these compounds were demonstratively active in *in vivo* models of pain [121]. The piperidine oxime **16** reported by Euro-Celtique (Purdue Pharmaceuticals) had excellent antagonist activity ($IC_{50} = 80$ nM) with greater than 250-fold selectivity versus L-type VDCC [122]. However, there are no reports on its *in vivo* efficacy in stroke models. Structure–activity-based studies on Cyproheptadine, an orally active antiallergic drug in the market, led to identification of **14** by Ajinomoto Pharmaceuticals. This compound is effective in IMR-32 neuronal blastoma cells (IC_{50} 3.2 uM) with improved selectivity over histamine 1, $5HT_{2a}$, and L-type VDCC channels [123].

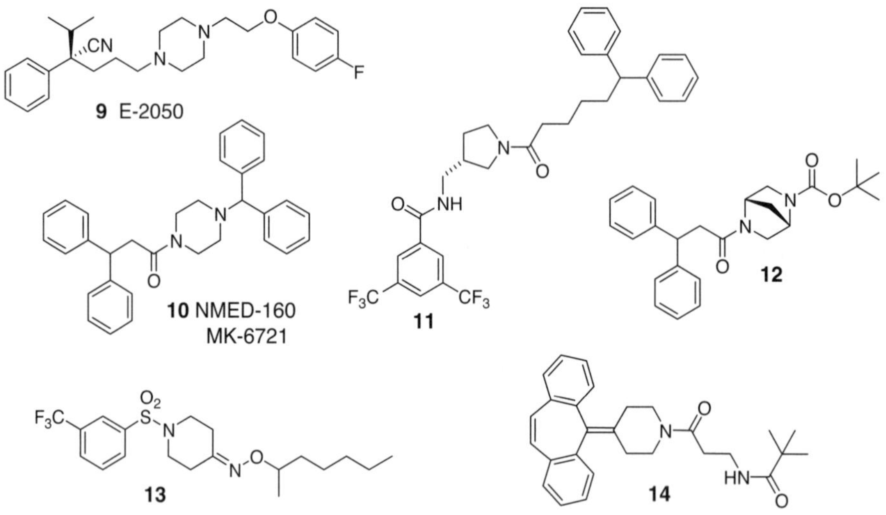

Novel Heterocycles SB-221420-A (**15**, $IC_{50} = $ 2.2 uM) a nonselective N-type calcium channel antagonist, demonstrated neuroprotective effects in focal and global models of neuronal ischemia [124]. Compound LY393615 (**16**) from Eli Lilly was identified as a nonselective VDCC antagonist ($IC_{50} = $ 1.9, 4.0, and 5.2 for N, P/Q, and R, respectively) and showed neuroprotective effects in many *in vitro* and *in vivo* models [125]. Pfizer has identified a series of structural motif as antagonists of N-type VDCC, most notable being **17** with excellent potency (IC_{50} 40 nM in human balstoma cells, [126]. Astellas has identified tetrahydroquinoline (**18**) as N-type VDCC blockers that have shown to be active in formalin-induced pain and Chung models when administered orally [127]. Thiazolidinone derivative **19** from Ionix Pharmaceuticals effectively blocked activity against N-type VDCC ($IC_{50} = $ 0.73 uM) with more than 30-fold selectivity over the L-type channel [128]. Cilindipine (**20**), a potent dual blocker for N/L-type VDCC, is currently used in Japan and South Korea for treating hypertension. A structure–activity relationship study of cilindipine led to identification of **21** with improved selectivity over the L-type channel. These compounds have terminal carboxylic acid rendering them with high solubility and low liphophilicity. No clinical reports have been reported for this compound [129].

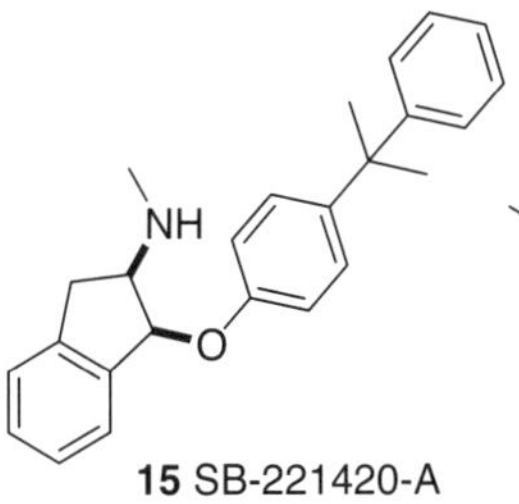

15 SB-221420-A 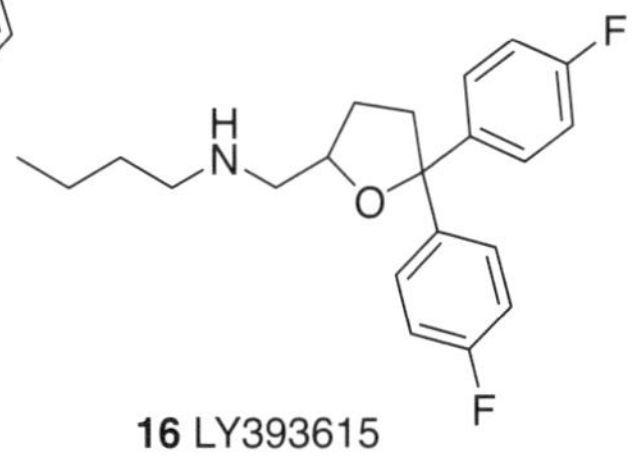   **16** LY393615 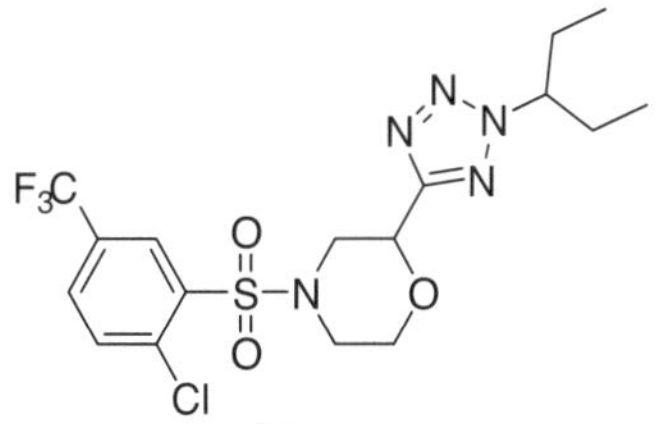   **17**

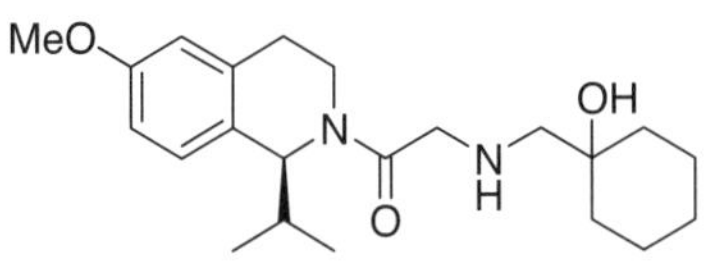

18 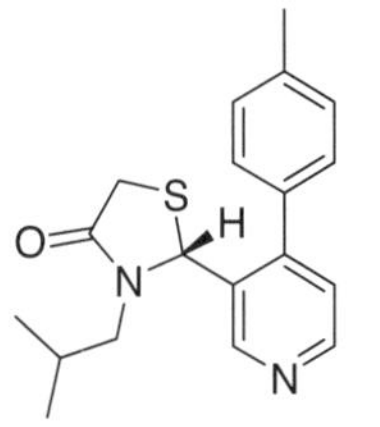   **19**

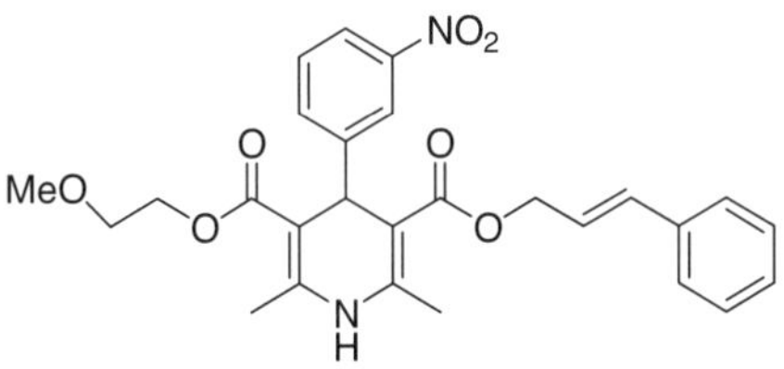

20 Cilnidipine 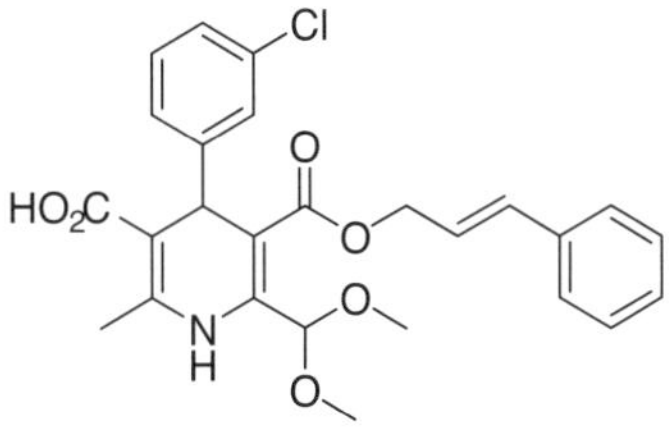   **21**

2.2.2. L-Type Calcium Channel A series of nonselective L-type voltage-gated calcium channel antagonists have been approved as drugs for hypertension. Verapamil (**22**) is an L-type calcium channel blocker of the phenylalkylamine class approved by the U.S. Food and Drug Administration in 1981. A series of dihydropyridine derivatives such as clevidipine (**23**), amlodipine (**24**, Norvasac, Istin), and Lacidipine (**25**) are used as an antihypertensive to treat angina. A number of these L-type calcium channel blockers such as isradipine (**26**, 2.5 mg/kg), when administered subcutaneously 30 min following permanent MCA occlusion, significantly reduced the volume of ischemic brain damage in the cerebral hemisphere (25%; $p = 0.0001$), cerebral cortex (18%; $p = 0.0034$), and caudate nucleus (33%; $p = 0.0002$) when assessed at 24 h post-MCA occlusion. In SHR, undergoing transient (2 h) MCA occlusion, isradipine administered 30 min post-MCA occlusion produced a significant reduction (47%; $p = 0.001$) in hemispheric infarct volume, whereas isradipine administered at the onset of reperfusion did not confer any significant neuroprotection. No change in functional deficit was seen with isradipine with either dosing paradigm at 24 h post-MCA occlusion. This result substantiates the neuroprotective efficacy of isradipine and other L-type calcium channel antagonists in permanent and transient focal cerebral ischemia [130].

If VGCC antagonists were to produce vasodilation, it should increase blood flow into the regions of penumbra after the onset of stroke. Many clinical trials were carried out with nimodipine (**27**) in patients with stroke symptoms. The results of these studies demonstrated no significant improvement in neurological or functional aspects of life [131–134]. It should be noted that, in most of these studies, patients were allowed to enroll at 24–48 h after the onset of stroke. Current studies suggest that neuroprotectant needs to be dosed within 6 h after the onset of stroke, which may explain the lack of efficacy in clinical trials of stroke associated with nimodipine [135].

Nimodipine functions as a neuroprotective through its action as an L-type calcium channel blocker, the mechanism of which as been previously. While a large number of animal studies in several species were performed to determine efficacy, a subsequent meta-analysis of this set of data suggests that the studies collectively did not demonstrate efficacy, and if they did were of poor experimental design [136]. Furthermore, Horn et al. make the observation that many of the animal studies were performed concurrently to human clinical studies.

While one early clinical trial showed a decrease in mortality among the group receiving nimodipine [137], a number of subsequent trials were performed from the 1980s through 2001 [138–140], all of which failed to show efficacy, even while increasing the dose and decreasing the time between the onset of symptoms and enrollment in the study to less than 6 h. As a result, development was discontinued.

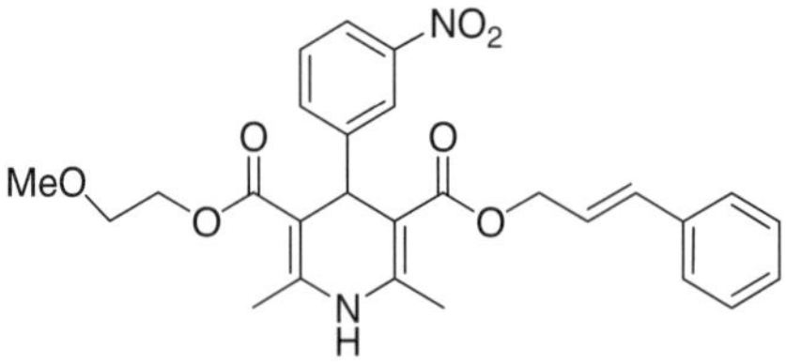

20 Cilnidipine

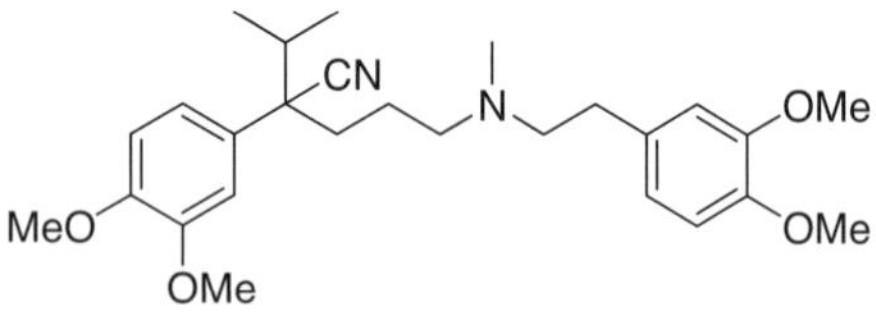

22 Verapamil

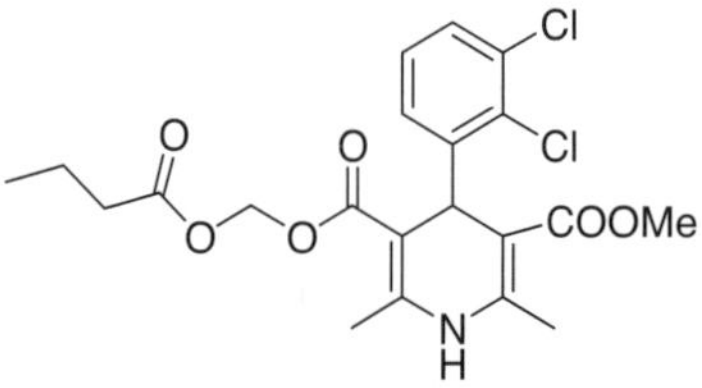

23 Clevidipine

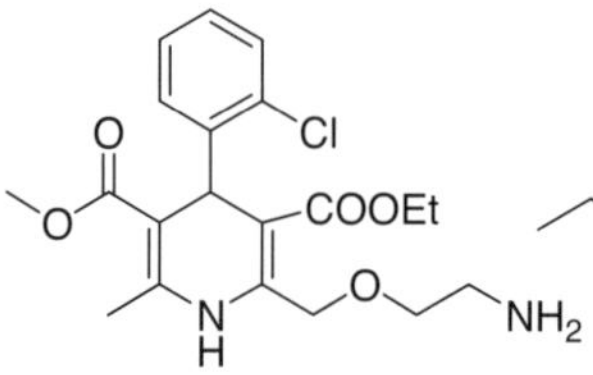

24 Amlodipine

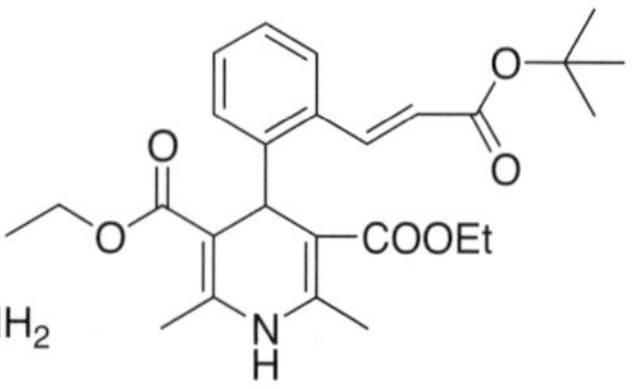

25 Lacidipine

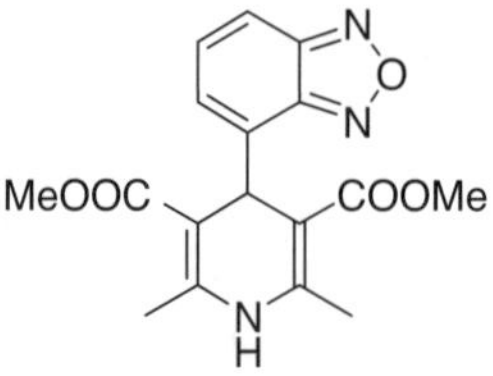

26 isradipine

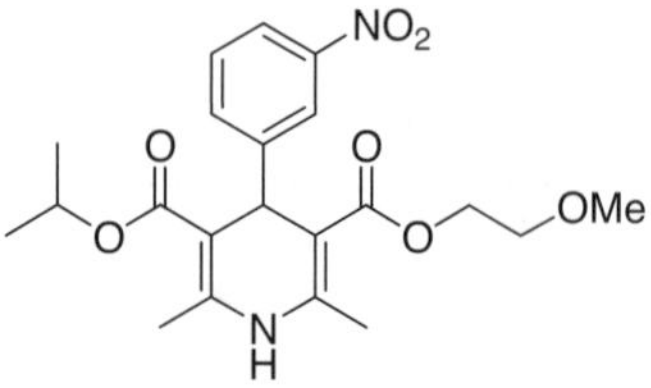

27 nimodipine

2.3. Calpain Inhibitors

Two ubiquitous calpains, μ-calpain (I) and m-calpain (II) are highly expressed in the central nervous system. Calpain has been reported to be activated during apoptosis in many *in vitro* systems, such as in thymocytes, cerebellar granular neurons, in NGF-deprived rat PC12 cells, and neuroblastoma SH-SY5Y cells [141]. Animal models predict that calpain activation is an obligatory downstream event in the ischemic cell death cascade. It has therefore been perceived as a good target in stroke therapy. A significant focus of research has been on calpain-mediated proteolysis and its contribution to neuronal death in ischemic neuronal injury. In the brain, potential substrates for calpains include a number of cytoskeletal proteins such as actin binding proteins (e.g., spectrin, actinin, gelsolin, gephyrin, ankyrin, talin, tubulin, microtubules-associated proteins, and neurofilaments) (reviewed in Ref. [142]). Potential substrates for calpains are also membrane proteins such as growth factors receptors and calcium channels, adhesion molecules, ion transporters, enzymes, such as kinases, phosphatases (calcineurin), and phospholipases, cytokines, and transcription factors [143]. Growth factors may help ameliorate some of the penumbral cell death as well as to promote the initiation of a recovery process. Therefore, calcium-activated proteases directly damage a cell by degrading cytoskeleton-associated functions and indirectly through destruction of the intracellular signaling components of growth factor receptors.

The strongest evidence for the direct participation of calpains in neurodegeneration has been found in rodent models of acute brain ischemia and was highlighted by the cytoprotective effect of calpain inhibitors [144]. Generally, inhibitors of calpain have been examined in many models of ischemia, where they exhibited a high neuroprotective potency [145]. The hope has been that targeting a "downstream" event in the ischemic cascade, such as interfering with activated μ-calpain, can prolong the time for the initiation of therapy. The therapeutic time window for calpain inhibition stretches at least the first 2 h after an insult as delineated in the model of global ischemia [146]. This therapeutic window was even as long as 6 h after the insult in a reversible focal cerebral ischemia model [147].

As μ-calpain activation seems to be an obligatory downstream event in the ischemic cell death cascade, it is potentially a good target in stroke therapy.

A large number of peptidic, peptidomimetic, and nonpeptidic inhibitors have been described in the literature. Their potential role of neuroprotection has been reviewed extensively by Wells and Bihovsky in 1998 [148] and by Ray in 2006 [149]. Thiol proteases as therapeutic targets were also reviewed recently by Leung-Toung and coworkers [150]. Most of the findings reported in those reviews have been summarized here. Additional and/or more recent work in the field of calpain inhibition has been reported here regardless of its implication in stroke. These molecules, while not presently tested in CNS injury models, may be useful jump-off structures for future research in stroke.

2.3.1. Reversible Inhibitors Reversible inhibitors aim to mimic the tetrahedral intermediate derived from the enzymatic hydrolysis of a substrate amide bond. As such, they often display a carbonyl functionality to interact with the active site cysteine sulfhydryl.

Aldehydes Early peptidyl aldehydes such as leupeptin 1, isolated from *Strepomyces* strains [151] display moderate potency against calpain and various levels of selectivity versus other cysteine proteases. Leupeptin provided significant protection to hippocampal neurons in rat brain ischemia model [152] and prevented neuronal degeneration in rat occlusion models [153]. Despite those results, the presence of a charged side chain such as arginine is perceived as a barrier to cell membrane permeation, leading the field to the examination of neutral analogs such as calpeptin 2 [154] and MDL-28170 3 [155]. Calpeptin has been shown to be neuroprotective in cell cultures [156–158] but has not been evaluated in brain ischemia models. MDL-28170, on the other hand, has demonstrated positive effects in several animal models [159–162].

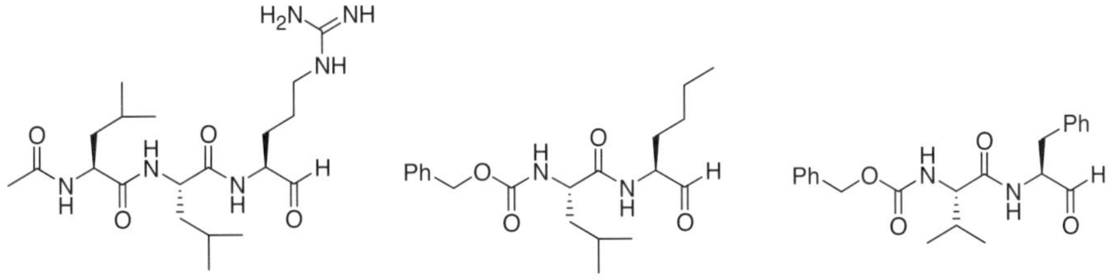

1: Leupeptin

2: Calpeptin

3: MDL-28170
IC_{50} = 10 nM

Replacement of the traditional Boc and Cbz N-terminus caps with sulfonamides caps has led to the discovery of SJA-6017 **4** that has demonstrated functional neuroprotection in a mouse model of brain injury [163–166]. Based on a similar concept, Cephalon scientists have generated potent clapain inhibitors derived from D-serine (**5**) [167]. Superimposition of the P2 and P3 groups inspired the preparation of bezothiazine dioxides such as **6** [168]. Effects of such inhibitors in animal models of neuroprotection have not been reported. The most recent development in the area of sulfonamides consists of the addition of water solubilizing photoswitchable diazo- and triazene arenes such as illustrated with inhibitors **7** and **8** [169]. Inhibitor **10** retards calpain-induced cataract formation in lens culture.

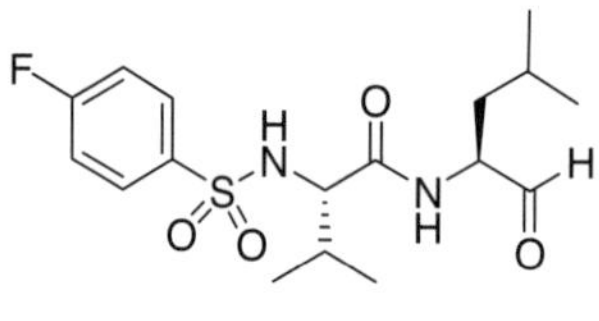

4: SJA-6017
IC_{50} = 7 nM

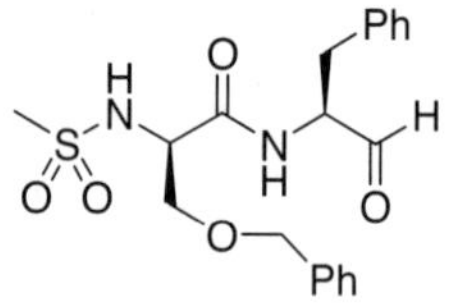

5: Cephalon
IC_{50} = 11 nM

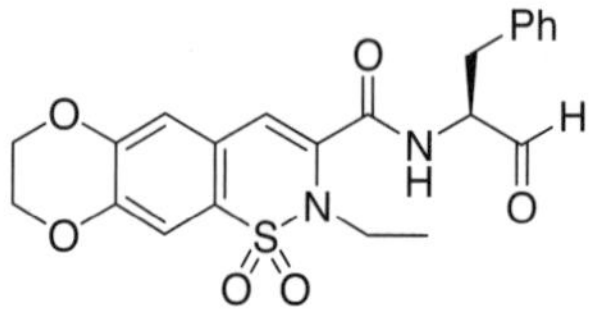

6: Cephalon
IC_{50} = 8 nM

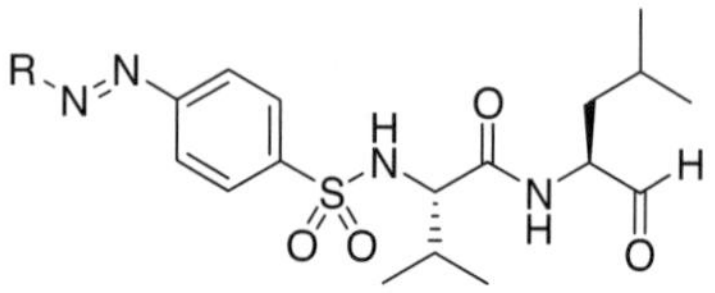

R = Ph: **7** Diazo
IC_{50} = 45 nM
R = Piperidine: **8** Triazene
IC_{50} = 90 nM

Recently, urea-based peptidomimetic inhibitors with improved selectivity over other cysteine proteases have been prepared, as exemplified by **9** [170]. In a similar quest for more selective inhibitors, the same group reported on the synthesis of inhibitors with constrained amino acids (**10**) [171]. Evaluation in stroke related models was not reported.

Yet another constraint based on D-proline was evaluated by the same research group. Inhibitor **11** displayed modest cytotoxicity in leukemia and solid tumor cell lines but no correlation was observed between calpain inhibition and cytotoxicity [172].

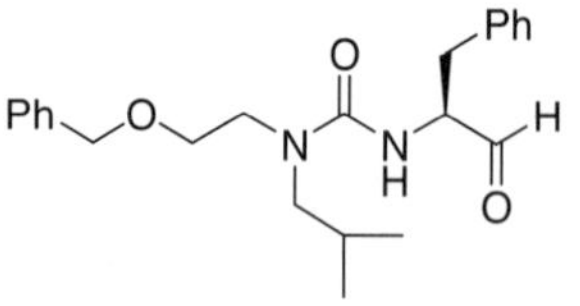

9
IC_{50} = 150 nM

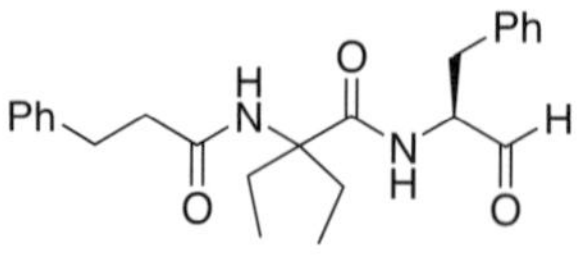

10
IC_{50} = 80 nM

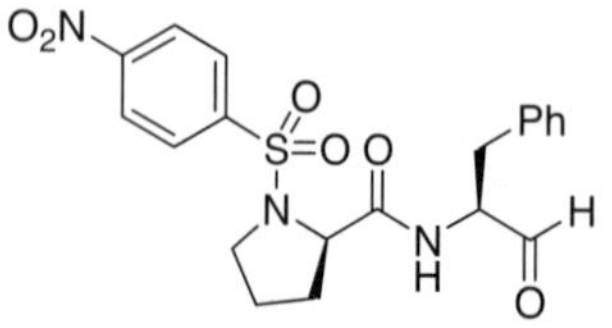

11
IC_{50} = 20 nM

In general, peptidyl aldehydes are potent inhibitors of various cysteine and serine proteases. This lack of selectivity makes it difficult to correlate results in animal models with the inhibition of calpain itself or any specific protease.

Masked Aldehydes In an effort to address inherent liabilities of aldehydes regarding metabolism and cell/tissue permeation, a number of research groups have studied masked aldehydes in the form of hemiacetal or the like. Cyclocondensation of MDL-28170 (**3**) with formaldehyde provided MDL-104903 (**12**) [173]. More recently, work from the Ipsen laboratories has described the combination of a similar masked aldehyde concept with an antioxidant pharmacophore to provide dual inhibitors of calpain and lipid peroxidation. BN-8270 (**13**) provided synergistic protection against maitotoxin-induced glial cell death, twofold more efficiently than either a calpain inhibitor or an antioxidant alone, including full protection at 100 µM [174,175]. Further *in vivo* evaluation in animal models of neuroprotection has not been reported. Researchers at Senju Pharmaceuticals recently reported the application of a similar aldehyde masking strategy to their inhibitor SJA-6017 (**4**). Following additional sulfonamide replacement with a thiourea group, they identified SNJ-1715 (**14**) that displayed 66% oral bioavailability and 2 h half-life in rat, and neuroprotective efficacy in a rat retinal ischemia model [176].

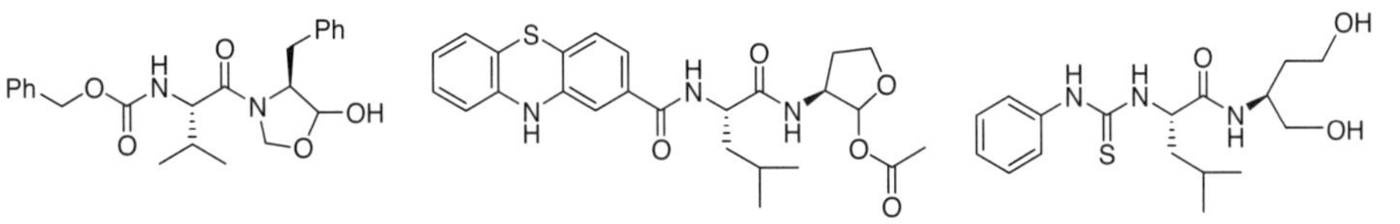

12: MDL-104903
IC_{50} = 33 nM

13: BN-82270
IC_{50} > 1000 nM (prodrug)

14: SNJ-1715
IC_{50} = 86 nM

Ketones and Ketoamides Unactivated ketones such as methyl ketones are generally weak calpain inhibitors. Numerous efforts to increase the reactivity of the carbonyl group have been reported. Trifluoromethylketones, nitriles, α-ketoheteroaryl, and semicarbazones are generally weak cysteine protease inhibitors, especially in comparison to their potency as serine protease inhibitors. α-Ketocarboxylic acids are potent cysteine protease inhibitors (**15**) but suffer from inadequate cell permeation properties while their ester counterparts are less potent (**16**). Corresponding phosphate esters, such as **17**, display comparable activity [177]. Interesting cyclopropenone alcohols such as **18** have been disclosed but their mechanism of inhibition has not been elucidated [178].

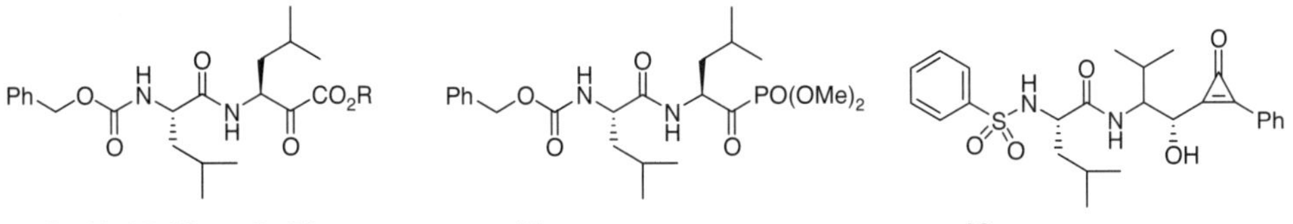

R =H: **15**: IC_{50} = 7 nM
R =Et: **16**: IC_{50} = 600 nM

17
IC_{50} = 430 nM

18
IC_{50} = 350 nM

The most studied and most successful activated ketones have been α-ketoamides. We will again focus here on inhibitors that have displayed functional results or have been disclosed recently. Conversion of the early neutral dipeptidyl aldehydes to the corresponding α-ketoamides yielded AK-275 (**19**) and the more water soluble AK-295 (**20**). Following supracortical perfusion, AK-275 displayed robust neuroprotection to in a rat focal brain ischemia model [179,180]. AK-295 also offered neuroprotection in a similar model [181,182].

Researchers at Senju pharmaceuticals have also evaluated water soluble, cell membrane permeable α-ketoamides such as SNJ-1945 (**21**) [183] and SNJ-2008 (**22**) [184]. SNJ-1945 has been selected as a development candidate for the treatment of retinal diseases [185,186]. In addition, SNJ-1945 was reported to offer neuroprotection at up to 72 h when administered i.p. at 100 mpk up to 6 h post-MCAO in mice [187].

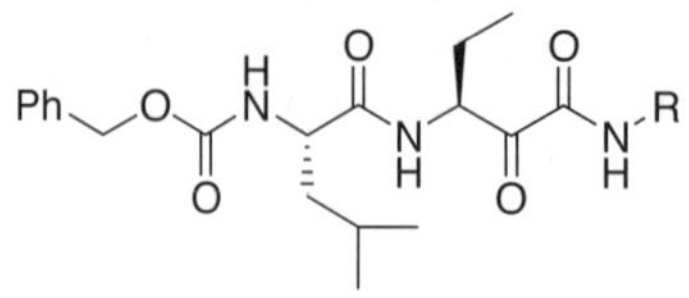

R = Et:
19: AK-275
IC_{50} = 250 nM

R = (morpholinoethyl)
20: AK-295
IC_{50} = 140 nM

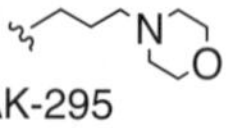

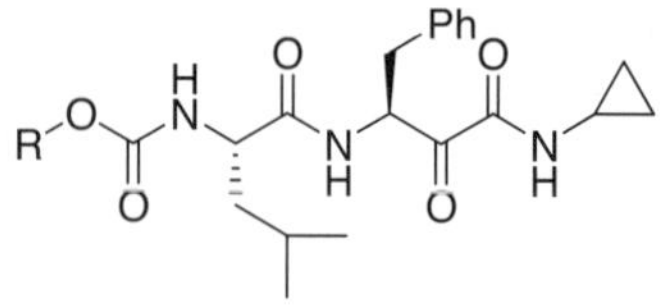

R = ((CH$_2$)$_2$O)$_2$Me
21: SNJ-1945
IC_{50} = 170 nM

R = ((CH$_2$)$_2$-2-Pyr
22: SNJ-2008
IC_{50} = 38 nM

Recent work from Lee and coworkers describes the evaluation of chromone and quinolones carboxamides as selective calpain inhibitors (**23a** and **23b**) [188,189]. Acyclic analogs such as **24** were also evaluated [190]. Lascop and coworkers have studied α-ketoamides calpain inhibitors for the treatment of muscular dystrophy [191]. The specific introduction of a nonpolar lipophilic residue, as illustrated with lipoyl **25**, significantly improves muscle cell uptake and such inhibitors have displayed positive effects on histological parameters in an animal model of Duchenne muscular dystrophy.

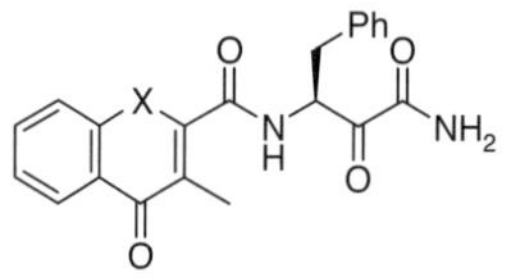

X = O: **23a**   X = NH: **23b**
IC_{50} = 250 nM IC_{50} = 710 nM

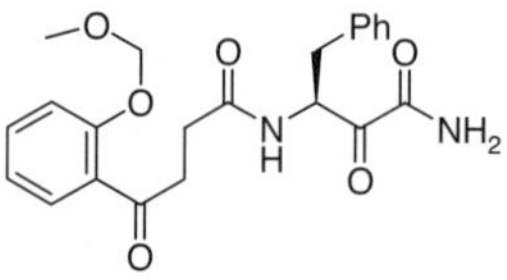

24
IC_{50} = 340 nM

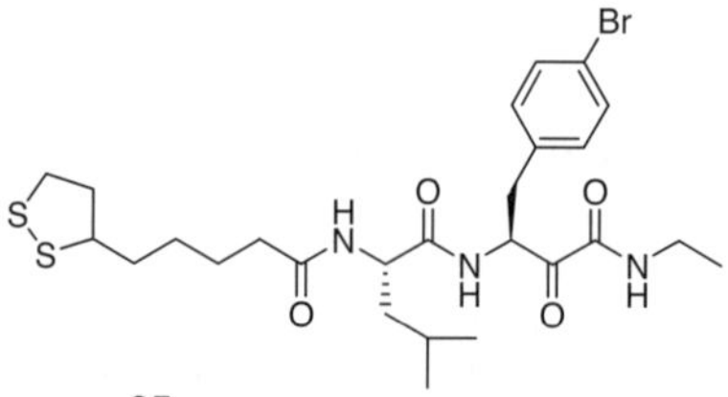

25
IC_{50} = 20 nM

Michael Acceptors Isoquinoline derivatives such as **26a** are potent calpain inhibitors, presumably via reaction with the active site thiol [192]. Peptidic hybrids such as **26b** do not offer significant potency advantages.

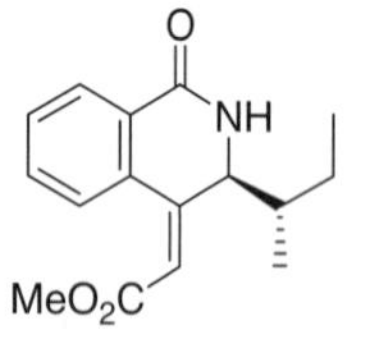

X = O: **26a**
IC_{50} = 25 nM

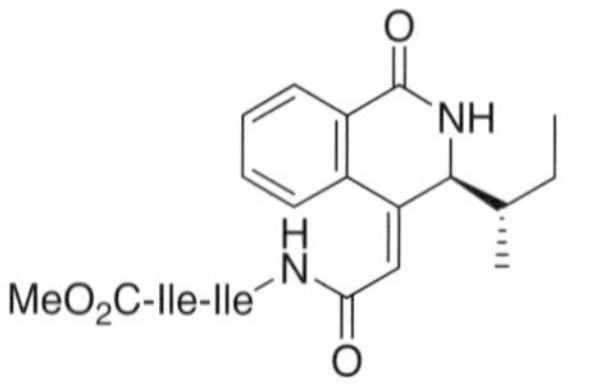

X = O: **26b**
IC_{50} = 159 nM

2.3.2. Irreversible Inhibitors

Epoxysuccinates Isolated from natural sources, E-64 (**27**) represents the prototypical epoxysuccinate cysteine protease inhibitor [193]. Replacement of the guanidino moiety with nonpolar substituent provided irreversible inhibitors such as E-64c (**28**) that inhibits the degradation of the cytoskeletal protein MAP2 in the rat MCAO model [194].

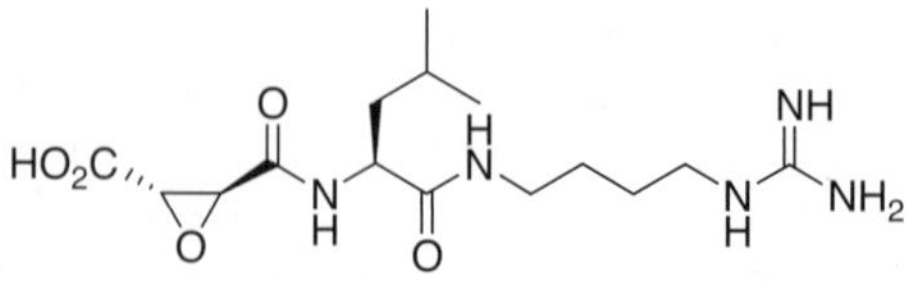

27: E-64

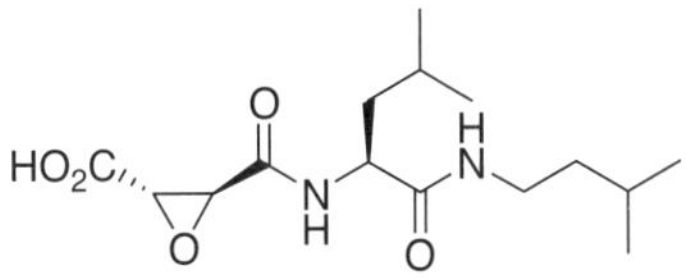

28: E-64c

Halomethyl Ketones and Similar Electrophiles

As observed with epoxysuccinates, halomethyl ketones, and diazomethyl ketones are irreversible inhibitors of various cysteine proteases. The concern of undesired inhibition of numerous endogenous nucleophiles has directed the field toward the evaluation of less elecrophilic fluoromethyl ketones such as **29** [195]. Alternative leaving groups such as benzotriazole (**30**) [196] and phosphinate (**31**) [177,197] have also been prepared. Fluoromethyl ketones and the like have received more attention as caspase or cathepsin inhibitors and will be discussed later in this chapter. β-Lactam irreversible inhibitor **32** represents an interesting example lacking the standard peptidic recognition elements [198].

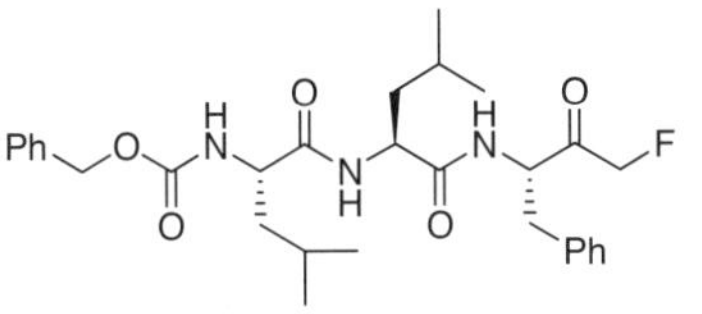

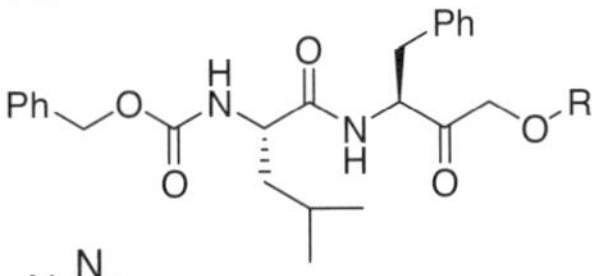

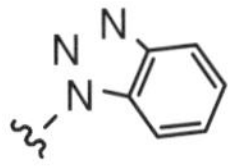

2.3.3. Unconventional Inhibitors

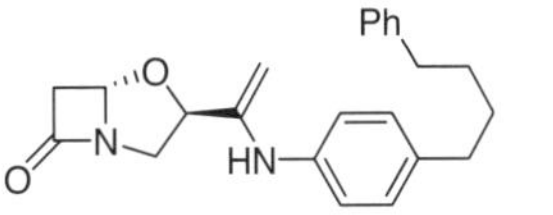

PD-150606 (**33**) is a potent, selective, reversible and membrane permeable calpain inhibitor that was found to bind to the noncatalytic calcium binding domain VI remote from any calcium binding loop [199]. Upon intracerebroventicular administration, PD-150606 has demonstrated positive effects in various brain ischemia models [200]. A class of quinolones exemplified by **34** has been described and is likely to bind outside the active site. Diketopiperazine **35** is also likely in the same situation.

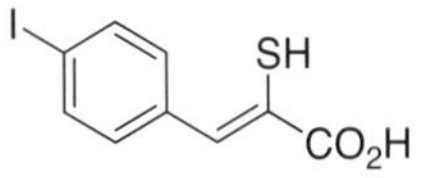

33: PD-150606
IC_{50} = 210 nM

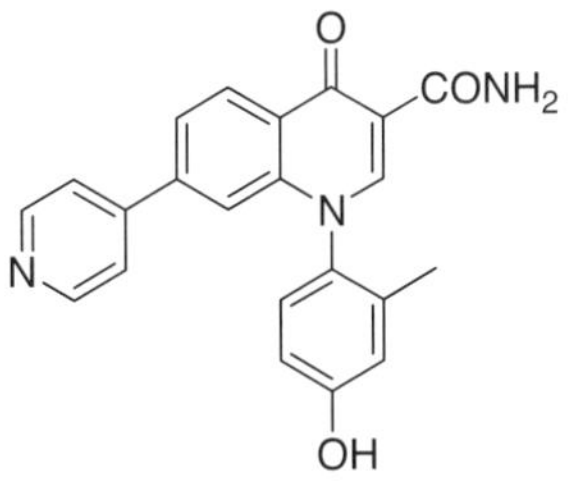

34
IC_{50} = 510 nM

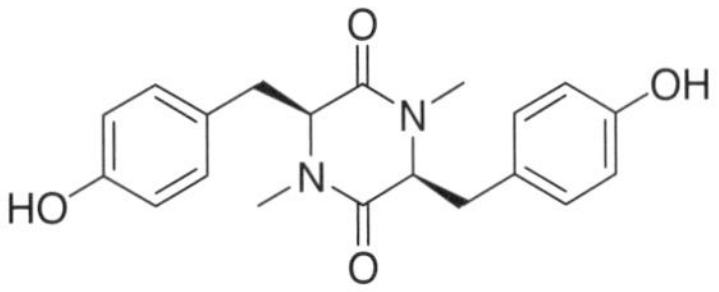

35
IC_{50} = 800 nM

2.4. Caspase 3/9 Inhibitors

Caspases belong to a family of highly conserved cysteine-dependent aspartate-specific acid proteases that use a cysteine residue as the catalytic nucleophile and share a stringent specificity for cleaving their substrates after aspartic acid residues in target proteins [201].

The caspase gene family consists of 15 mammalian members that are grouped into two major subfamilies, namely, inflammatory caspases and apoptotic caspases. The apoptotic caspases are further subdivided into two subgroups, initiator caspases and executioner caspases. These apoptotic caspases have been the subject of over 10,000 research papers and numerous transgenic and knockout animals. Despite the intense scientific activity around them, small-molecule inhibitors of caspases are yet to make an impact on neurodegenerative disorders.

A large number of peptidic, peptidomimetic, and nonpeptidic inhibitors have been described in the literature. Their potential role in neuroprotection has been reviewed extensively by Ray in 2006 [149]. Thiol proteases as therapeutic targets were also reviewed recently by Linton [202], Leung-Toung and coworkers [203], and Weber and coworkers [204]. Most of the findings reported in those reviews have been summarized here. Limited applications to stroke therapy have been reported; recent work regarding caspase inhibition outside the stroke has been summarized here as well, with special emphasis on caspase-3 and -9 inhibition.

A tetrapeptidic motif is sufficient for specific substrate binding and a lot of peptidic inhibitors have been developed on this premise and a fixed aspartic residue as the P1 substituent.

2.4.1. Reversible Inhibitors

Aldehydes Numerous peptidic aldehyde inhibitors have been studied and served as early leads to fluoromethylketones and the like. Ac-LEHD-CHO (**1**), a caspase-9 inhibitor was shown to reduce neuronal loss after hypoxic ischemia in rat cortex, upon i.c.v. administration [205]. The search for caspase-1 (ICE) inhibitors has led to the discovery of nonpeptidic aldehydes such as **2a** and **2b** displaying modest potency [206,207]. Hydantoin-based peptidomimetic caspase-3 inhibitors have recently been reported [208]. Researchers at Sunesis have reported on the discovery of potent and selective caspase-3 and -7 inhibitors such as **4** that were derived from their extended tethering technology enabled hits [209–211].

1

R = 2-Tol: **2a**
Casp1 IC_{50} = 20 μM

2b
Casp1 IC_{50} = 1 μM

3
Casp3 IC_{50} = 3.9 μM

4
Casp3 IC_{50} = 50 nM

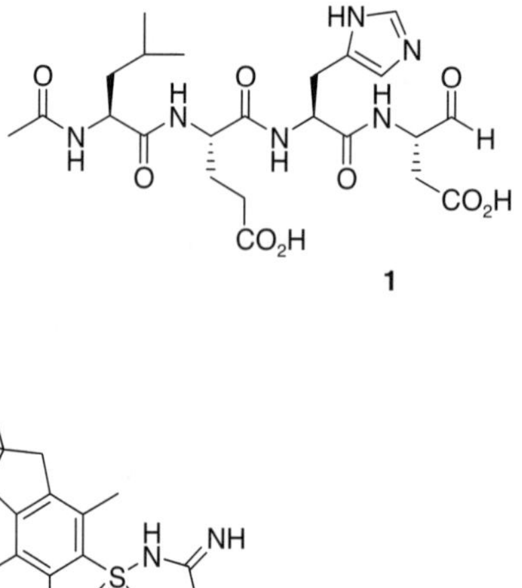

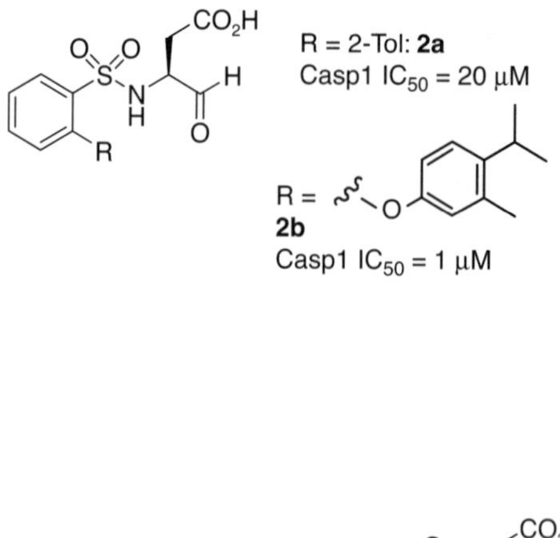

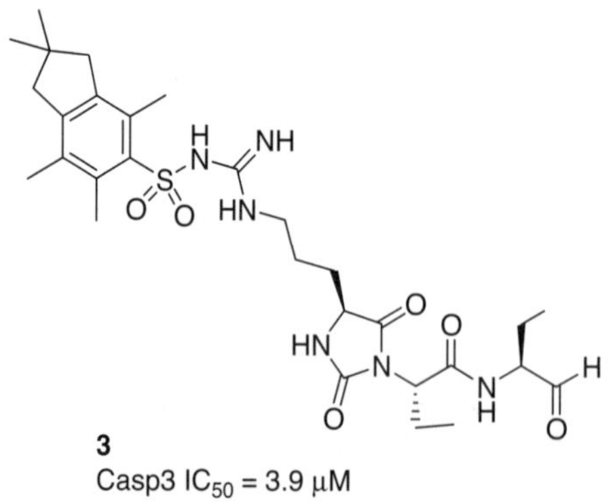

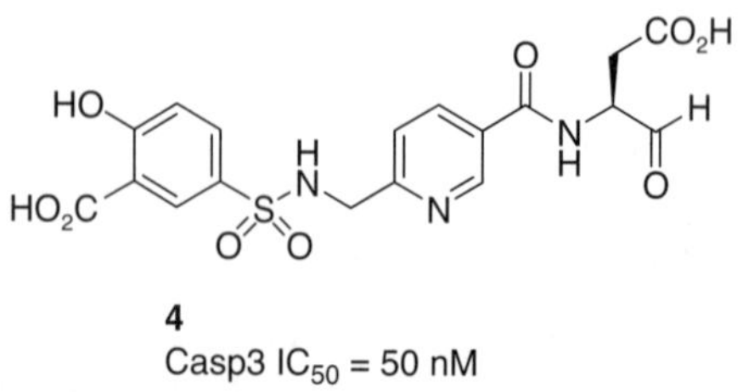

EXCITATORY AMINO ACID RECEPTORS

Potent pyrazinone-based ketone inhibitors of caspase-3 such as 5 have been disclosed by Merck scientist and have served as starting leads for the discovery of more potent inhibitors to be discussed later.

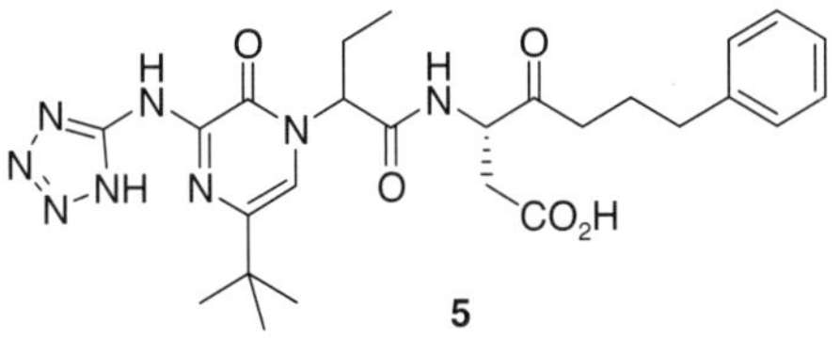

5
Casp3 IC_{50} = 5 nM

Masked Aldehydes In an effort to address inherent liabilities of aldehydes regarding metabolism and cell/tissue permeation masked aldehydes resulting from the cyclization of the P1 carboxylic acid onto the warhead aldehyde have been studied. Acyl dipeptide pan-caspase inhibitors such as 6 are potent caspase-3 inhibitors [212]. Pralnacasan (7, VX-740) is a potent caspase-1 (ICE) inhibitor that entered clinical development for arthritis (RA and OA) and psoriasis [213]. Liver fibrosis in dogs resulted in clinical trials discontinuation in November 2003. The acyl-hemiketal prodrug allowed for a 10-fold enhancement in bioavailability, from 4%F to 40–60%F in rodents. VX-765 (8), a related analog, was also advanced into clinical trials for psoriasis.

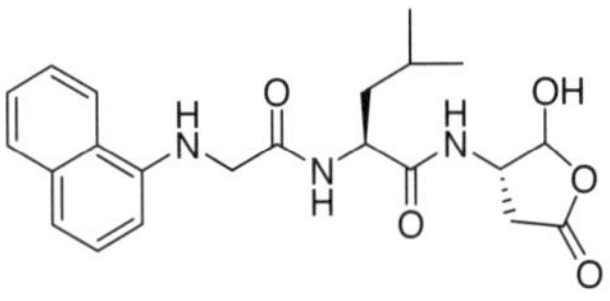

6
Casp3 IC_{50} = 13 nM

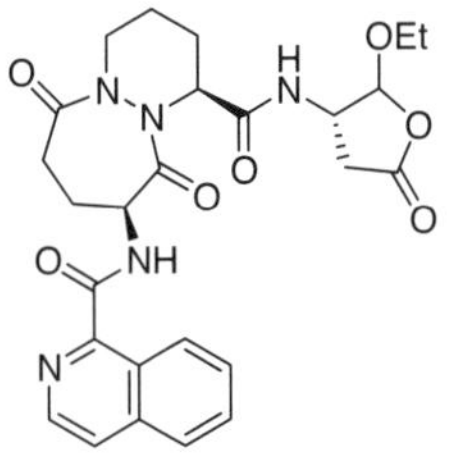

7: VX-740
Casp1 IC_{50} = 13 nM
Casp3 IC_{50} = 2300 nM

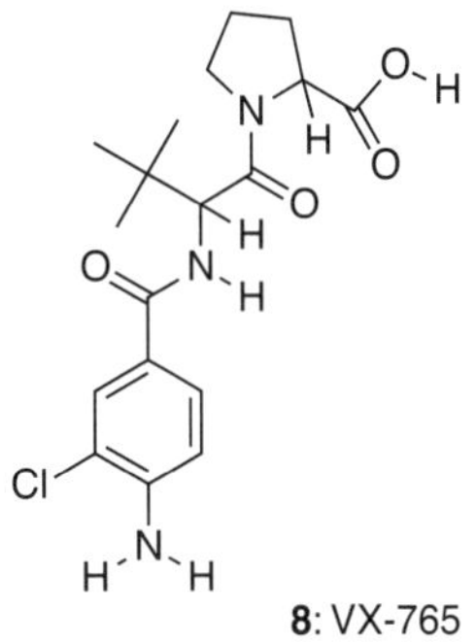

8: VX-765

Michael Acceptors Selective inhibitors of caspase-3 have been derived from an azapeptide template incorporating a Michael acceptor as cysteine ligand [214].

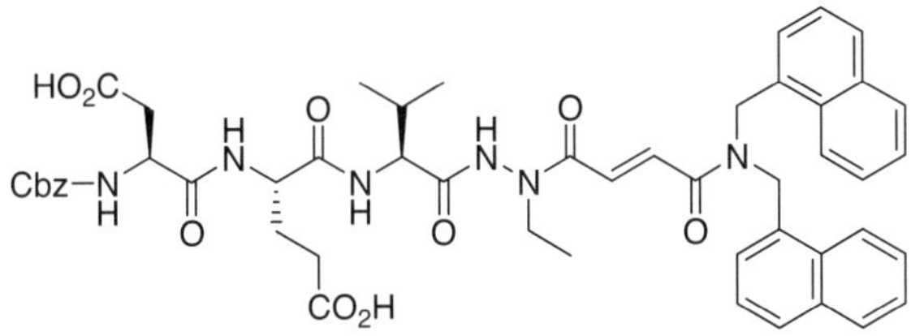

9

2.4.2. Irreversible Inhibitors

Halomethyl Ketones Fluoromethyl ketone as a moderately electophilic cysteine trap has received the most attention. Inhibitors 10–12 have been evaluated in some models of stroke. Tripetptide 10, a pan-csapase inhibitor relatively specific for caspase-1 was shown to reduce neuronal injury after focal but not global cerebral ischemia [215]. Evaluation of another caspase-1 inhibitor (11, i.c.v. administration) in the rat MCAO demonstrated prolonged neuroprotection [216]. Pre- or postischemic treatment with 12, up to 9 h after injury, yielded 21-day-long neuroprotection in a model of ischemia [217]. Independent studies with **12** have also shown that functional plasticity in neurons protection cannot be the result of caspase inhibition solely [218]. More recently, it was suggested that neuroprotection with **12** reflects inhibition of calpain-related necrotic cell death [219].

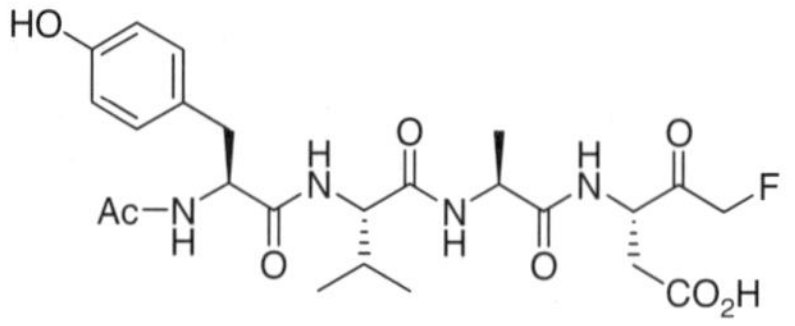

11
Casp1 K_i = 0.76 nM

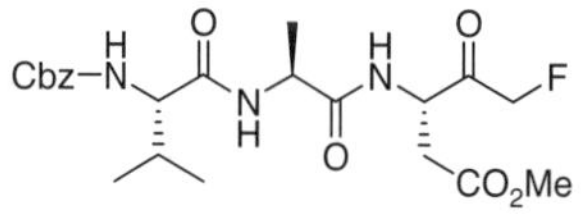

10
Casp1 K_i = 0.76 nM

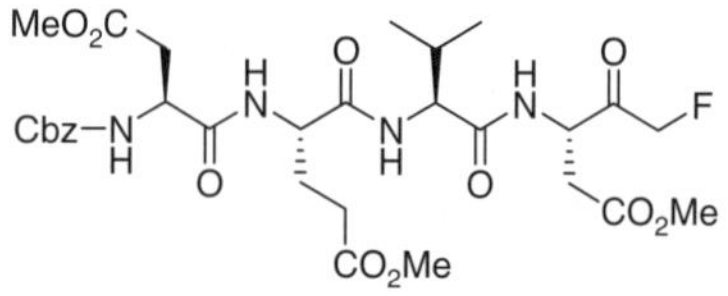

12

A series of dipeptide fluoromethyl ketone inhibitors has been described by researcher at Cytovia/Maxim Pharmaceuticals. MX-1013 (**13**) has displayed some efficacy in various models of stroke as well as liver injury and myocardial infarction. In a rat MCAO model, upon i.v. bolus/infusion administration, MX-1013 yielded 45–55% reduction in infarct size [220]. Minor structural modifications led to the discovery of MX-1122 (**14**) that displayed a twofold improvement in cellular activity [221]. Additional modifications included MX-1153 (**15**), a reversed urethane derivative with a >5000-fold selectivity for caspase-3 over other cysteine proteases [222]. Further optimization provided the peptidomimetic caspase-3 inhibitor EP-1113 (**16**) that displays good activity in cell apoptosis protection assays [223].

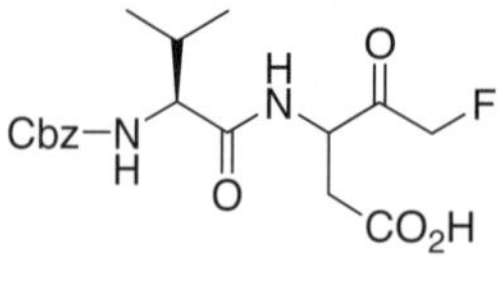

13: MX-1013
Casp3 IC$_{50}$ = 30 nM

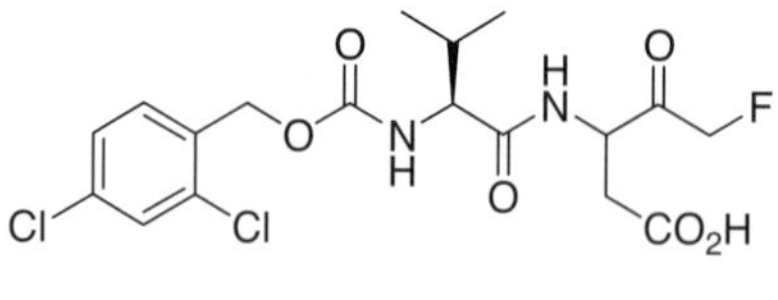

14: MX-1122
Casp3 IC$_{50}$ = 25 nM

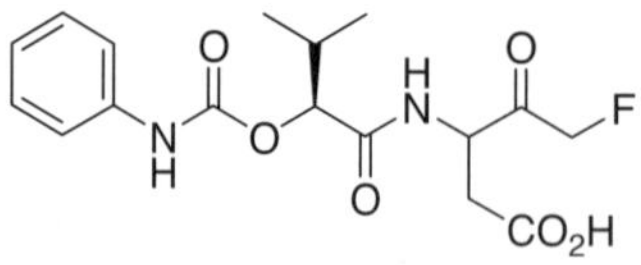

15 MX-1153
Casp3 IC$_{50}$ = 17 nM

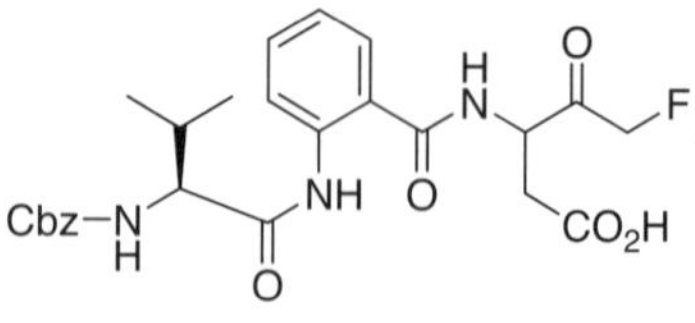

16: EP-1113
Casp3 IC$_{50}$ = 33 nM

The IDUN/Novartis collaboration has yielded a number of interesting fluoromethyl ketones that have also been optimized as other activated methylketone inhibitors. The oxamyl dipeptide series exemplified by **17** was originally derived from acyl dipeptide reversible aldehyde inhibitors. IDN-1965 (**18**) and IDN-5370 (**19**) are, respectively, based on indole and oxoazepino scaffolds. IDN-1965 has been shown to be effective in liver injury models believed to operate via caspase-3 activation [224]. IDN-5370 has displayed brain tissue protection in a rat permanent MCAO model, up to 28 days after injury [225].

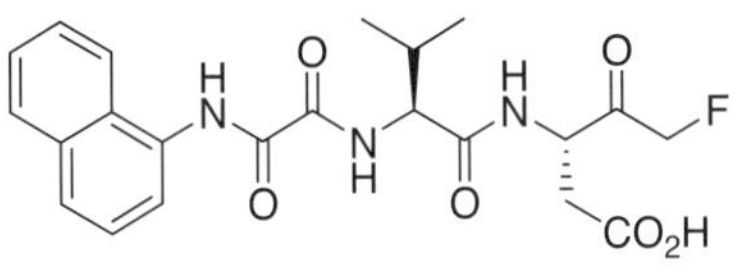

17

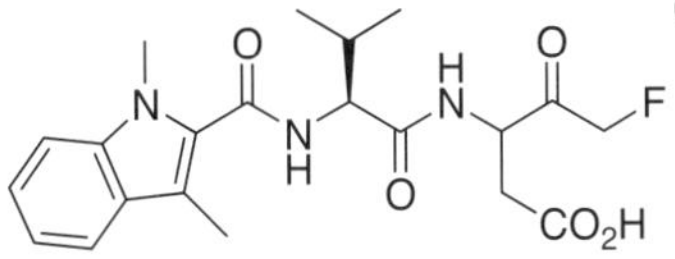

18: IDN-1965

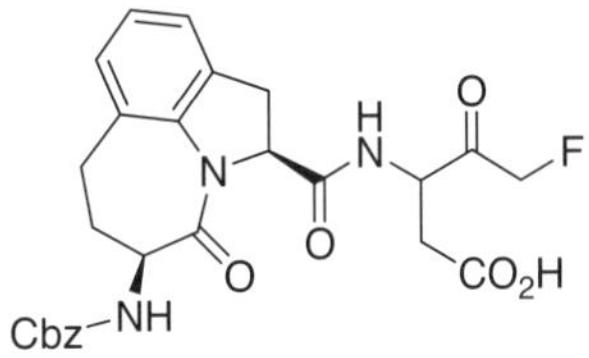

19: IDN-5370

Fluoromethylketones are known to metabolically degrade to fluoroacetates that can interfere with ATP production; thus, limiting the clinical development potential of fluoromethyl ketone-based inhibitors [226–228].

Alpha Oxo/Thio/Amino-Methyl Ketones This class of inhibitors in which the fluorine from previously described fluoromethyl ketones has been replaced by a heteroatom seemed like the logical evolution regarding modulation of the carbonyl electrophilicity, and indeed it has been studied by many in the field of caspase inhibition.

Sunesis researchers converted their aldehyde-based inhibitors such as **4** to the corresponding chlorobenzyl thiomethyl ketone derivatives **20** and **21**, and tetrafluorophenoxy methyl ketone **22** and **23** [229–231]. These inhibitors were tested in Fas-induced apoptosis assays, shedding some light on the importance of caspase-3 and -8 inhibition in this context.

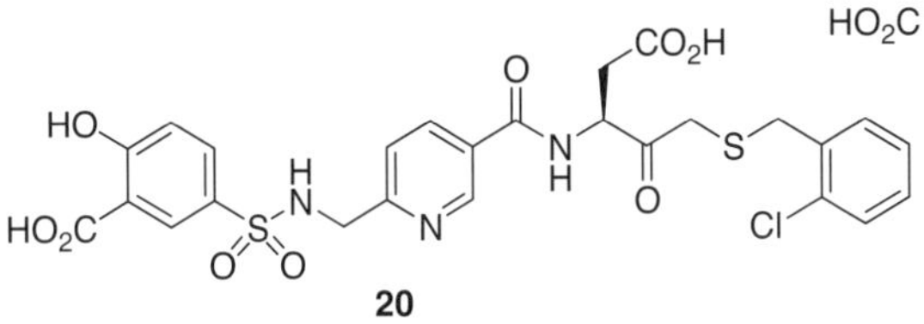

20
Casp3 IC_{50} = 30 nM

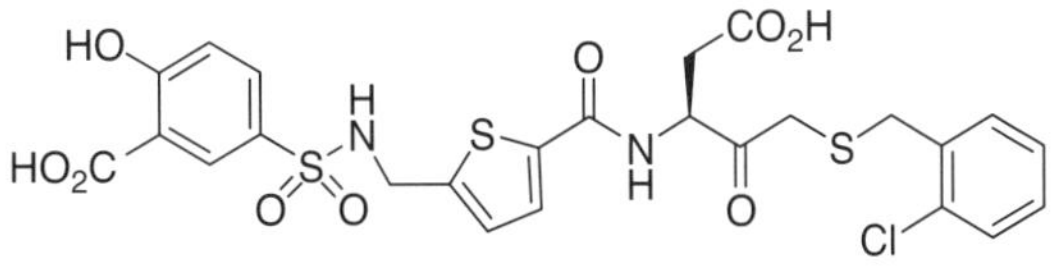

21
Casp3 IC_{50} = 20 nM

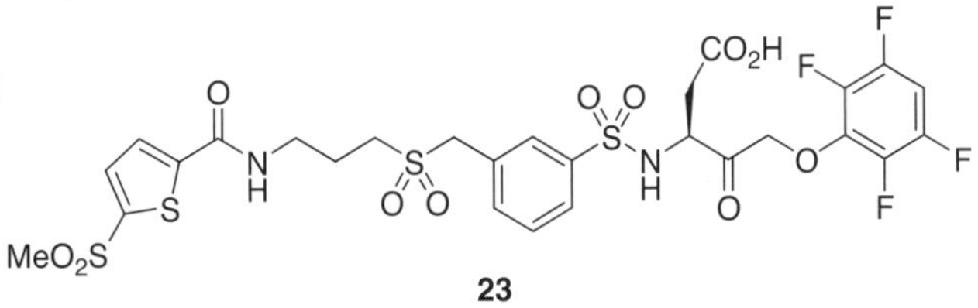

23

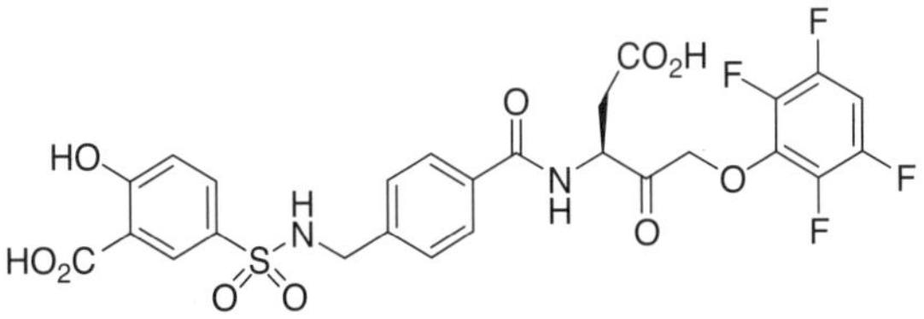

22

The IDUN group also converted their fluoromethyl ketone inhibitors to the corresponding tetrafluorophenoxy methyl ketones. In the oxamyl didpeptide series, inhibitor **24** yielded an 18% infarct size reduction in a rat permanent MACO model, upon i.v. administration [232]. In the same series, IDN-6556 (**25**) has been extensively studied [233–237]. It was advanced into Phase IIb clinical trials to treat HCV patients and patients undergoing liver or other solid organ transplantation, showing local therapeutic protection against cold ischemia/warm reperfusion injury, and received orphan drug label from the FDA [238–242].

Additional related studies around the oxamyl dipeptide template have been published recently [243,244].

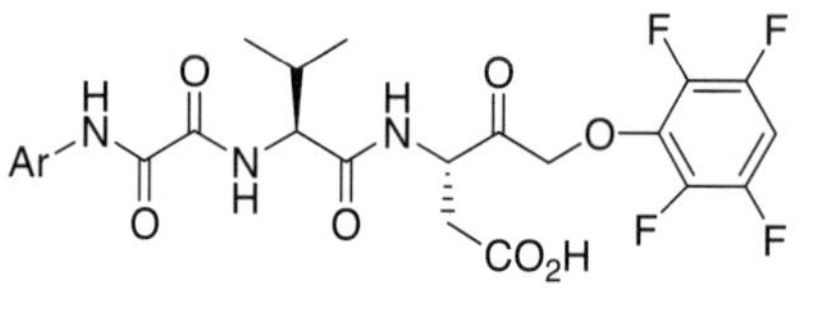

Ar = 2,3,5,6-Cl$_4$-Ph: **24**
Ar = 2-*t* Bu-Ph: **25**: IDN-6556

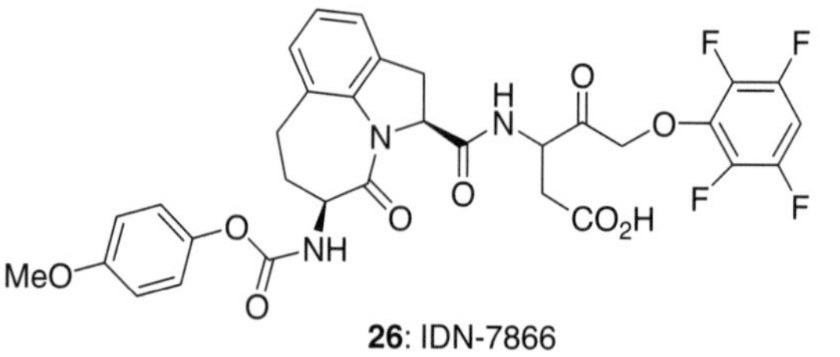

26: IDN-7866

Merck scientists expanded their work on peptidomimetic ketone **5** to thiomethyl ketones and aminomethyl ketones (**27–29**) [245–250]. M-826, more specifically a caspase-3 and -7 inhibitor, was evaluated in a rat neonatal brain injury model. M-826 provided significant protection against hypoxia-ischemia-induced brain injury by reducing apoptosis related DNA fragmentation and brain tissue loss, but it did not stop calpain activation in the cortex that is associated with excitotoxic/necrotic cell death [251]. This result would therefore indicate that caspase-3 inhibition alone might not be sufficient to treat the consequences of cerebral ischemia. Recent studies support the idea that calpain and caspase-3 are operating in synergy to acute and chronic neuronal cell death following ischemia, with early activation of calpain and subsequent activation of caspase-3 [252–256].

Studies of M-826 in a Huntington disease model have also demonstrated positive effects on cell death [257–260].

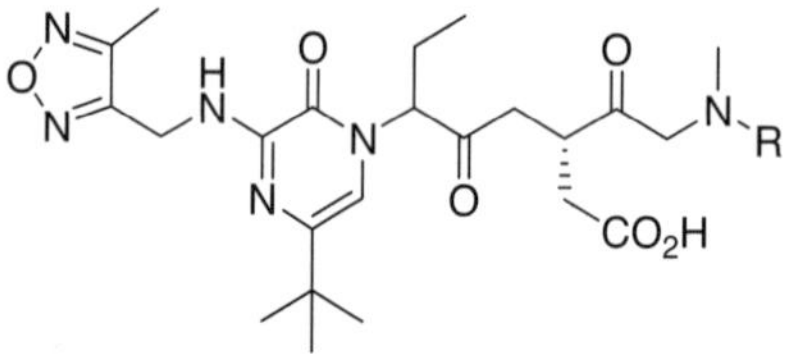

R = hex: **27**: M-826
Casp3 IC$_{50}$ = 6 nM

R = pent: **28**: M-867
Casp3 IC$_{50}$ = 1 nM

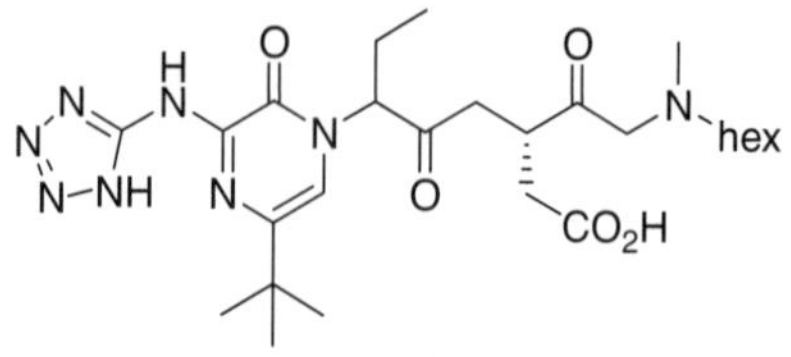

29
Casp3 IC$_{50}$ = 7 nM

2.4.3. Isatin-Derived Caspase Inhibitors 5-Nitroisatin was uncovered during an HTS campaign geared toward caspase inhibitors [261]. Structural optimization led to the discovery of reversible, selective caspase-3 and -7 inhibitors such as **30–31** [262]. While there is no P1 Asp group, the pyrrolidine ring fills S3, an area of structural heterogeneity for caspasse, thus explaining the inhibitory specificity toward caspase-3 and -7. Isatin derivative **30**

showed positive effects in a rabbit model of myocardial infarction [263]. In order to alleviate the inherent high reactivity of such ketone containing inhibitors toward any cysteine residue, isatin Michael acceptors such as **32** were also developed [264].

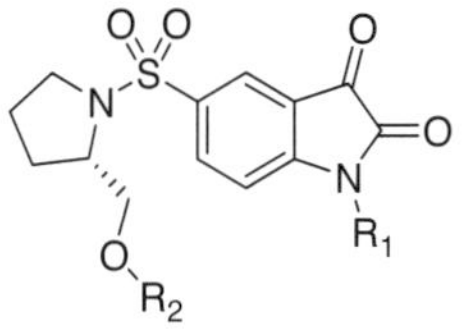

R$_1$ = H, R$_2$ = Me: **30**
Casp3 IC$_{50}$ = 60 nM

R$_1$ = Bn, R$_2$ = Ph: **31**
Casp3 IC$_{50}$ = 1 nM

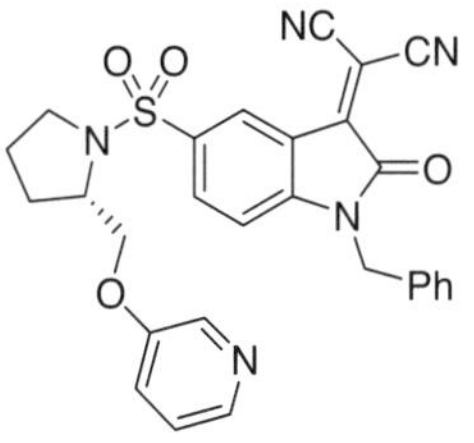

32
Casp3 IC$_{50}$ = 8 nM

2.4.4. Quinoline-Derived Caspase Inhibitors

Novel, nonpeptidic caspase-3 inhibitors such as **33** have been prepared from a quinoline scaffold. It is not clear at the moment if they react with the active site cysteine via the phthalimide carbonyls or if they bind outside the active site [265–268].

Another set of isoquinoline-derived caspase-3 and -7 is illustrated by **34**. These compounds were shown to be irreversible, noncompetitive inhibitors, and to modulate apoptosis induced by amyloid-beta, via caspase-3 activation, in neuronal cells [269–271].

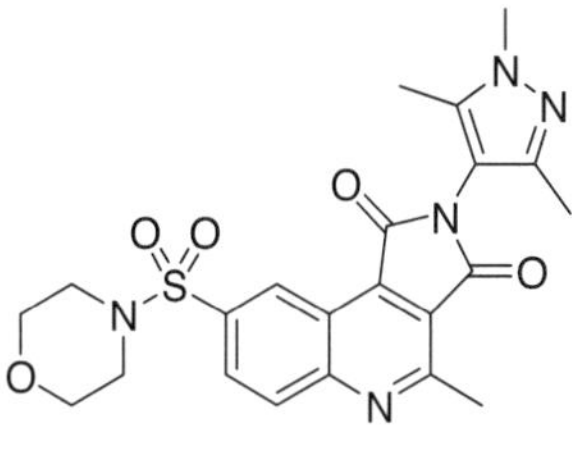

33
Casp3 IC$_{50}$ = 4 nM

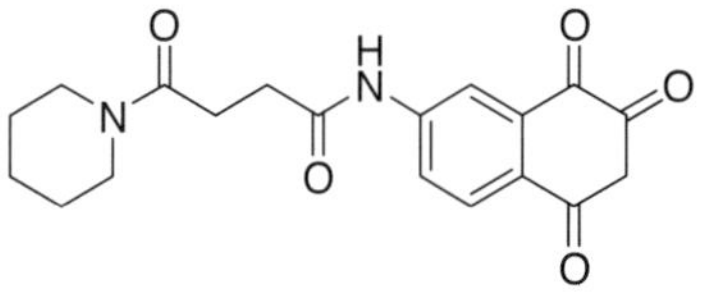

34

2.4.5. Other Caspase Inhibitors

Scientists at the Burnam Institute recently reported on a class of nonpeptidic, reversible, caspase inhibitors, as illustrated by compound **35** [272]. Researchers at the University of Queensland have identified competitive and reversible azide-based inhibitors such as **36** [273].

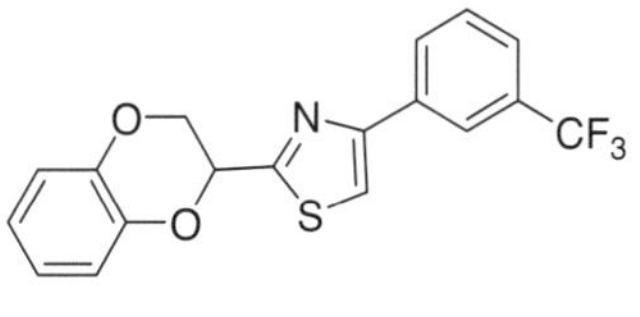

35
Casp3 IC$_{50}$ = 4.3 µM

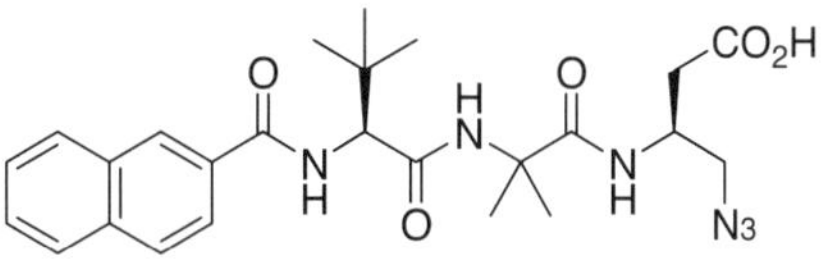

36
Casp3 IC$_{50}$ = 8 nM

2.5. Cathepsin B Inhibitors

Lysosomal cysteine proteases, generally known as the cathepsins, are localized in lysosomes and endosomes, and degrade intracellular or endocytosed proteins. There are 11 known human family members and they are optimally active in the slightly acidic, reducing environment typically found in the lysosomes. The lysosomal membrane is a physical barrier preventing the lysosomes hydrolytic enzymes from digesting the cellular proteins, but its disruption can cause cell death in the pathologic states. Brunk et al. [274] suggested a quantitative relationship between the amount of lysosomal rupture and the mode of cell death: low intensity stresses would trigger a limited release of lysosomal enzymes to the cytoplasm followed by apoptosis, whereas high intensity stresses would provoke a generalized lysosomal rupture followed by necrosis. The spreading of hydrolytic enzymes into the cytoplasm through the lysosomal membrane injury or rupture was confirmed in the brain [275] following ischemic injuries.

Immunohistochemistry studies in the monkey experimental cerebral ischemia paradigm, reported translocation of cathepsins B and L, in postischemic CA1 neurons [276]. The author suggested that the sustained calpain activation in these postischemic neurons may cause long-standing lysosomal membrane disruption with the resultant leakage of lysosomal enzymes including cathepsins B and L for as long as 5 days postischemia. The strongest evidence for a promising role for cathepsin B inhibition in stroke therapy comes from Yoshida et al. [277]. These authors demonstrated neuroprotection in nonhuman primate brain, 5 days postischemia using the peptide-based, cathepsin inhibitors CA-074 and E-64c.

Inhibitors of cathepsin B and their roles against therapeutic targets have recently been reviewed by Frlan [278] and Leung-Toung and coworkers [279]. Most of the findings reported in those reviews have been summarized here. Limited applications to stroke therapy have been reported; recent work regarding cathepsin B inhibition specifically, outside the stroke field has also been summarized here.

2.5.1. Reversible Inhibitors

Aldehydes Numerous peptidic aldehyde inhibitors of cathepsin B have been prepared, some displaying selectivity for cathepsin B are illustrated here. Modifications of leupeptin has led to the discovery of tripeptide inhibitor **2** [280]. The original search for cruzain inhibitors afforded dipeptides **3** and **4** as modest cathepsin B inhibitors [281].

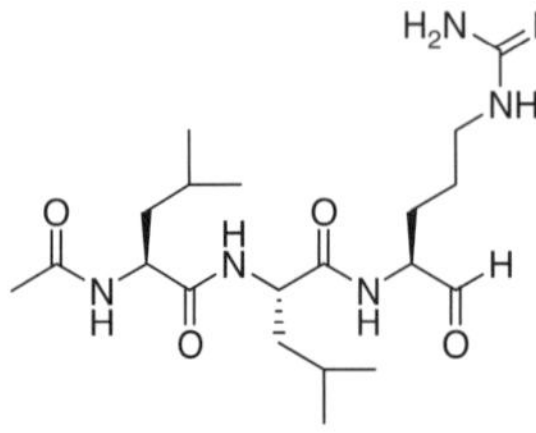

1: Leupeptin
IC$_{50}$ = 310 nM

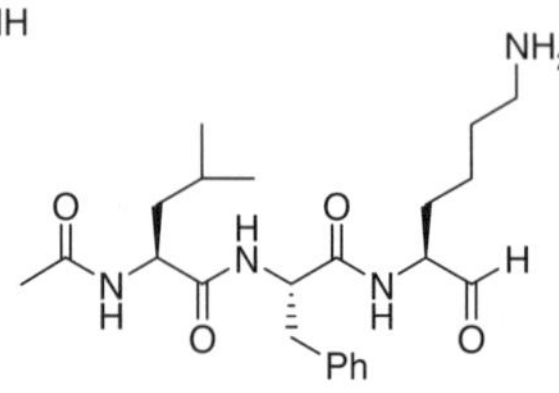

2
IC$_{50}$ = 4 nM

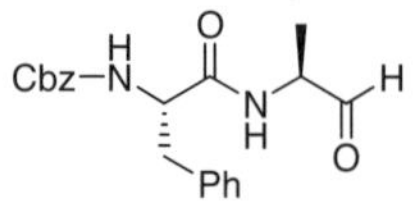

3
IC$_{50}$ = 50 nM

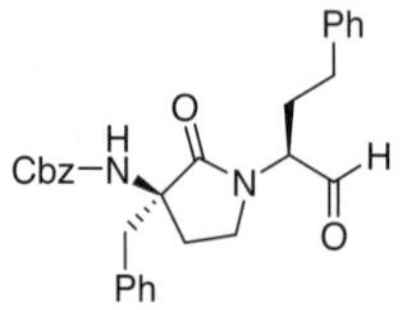

4
IC$_{50}$ = 1000 nM

Yamashima and coworkers have reported the neuroprotective effects of pyridoxal (**5**, PL) and pyridoxal phosphate (**6**, PLP), active coenzyme forms of vitamin B$_6$ [282]. Following ischemic insult in monkeys, PL and PLP (daily injections, 15 mpk/day) provided 54% and 17% neuronal death protection, respectively. Cathepsin B inhibition has been linked to the aldehyde functionality [283].

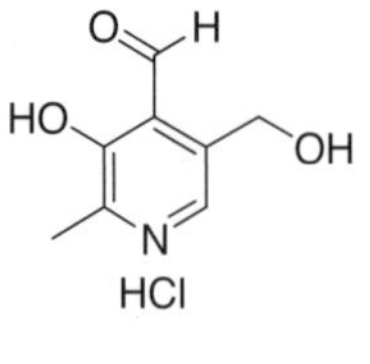

5: PL
30% inhibition
@ 1 mM

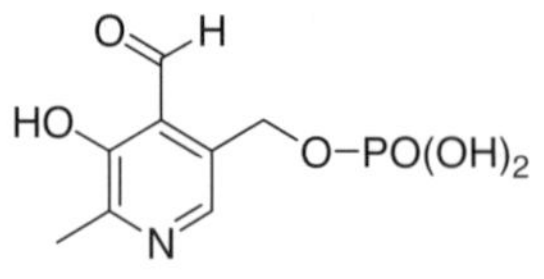

6: PLP
IC$_{50}$ = 100 μM

Ketones and Cyclopropenones Weak cathepsin B inhibitors such as **7** have been identified from combinatorial work based on a cyclohexanone scaffold [284]. Cyclopropenone derivatives such as **8** have been disclosed [285]. They are closely related to previously discussed calpain inhibitors (see **18**, Section 2.3.1, ketones and ketonamides, p.7) and may act as electrophiles or precursor of cyclopropenium cations [286].

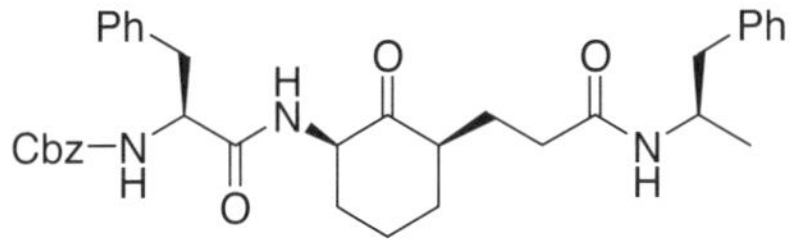

7
K_i = 1.1 mM

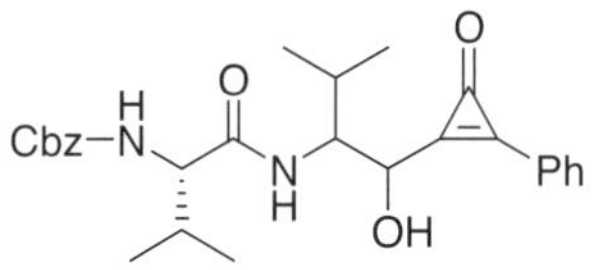

8
IC_{50} = 44 nM

Cyclometallated Complexes N,N-Dimethyl-1-phenethylamine/dppf complexes have been demonstrated to reversibly inhibit cathepsin B and to provide 90% tumor growth progression in Walker tumor-bearing rats [287].

Nitriles Nitriles function as reversible cysteine protease inhibitors via the formation of a thioimidate intermediate. Potent didpeptide nitrile cathepsin B inhibitors such as **9** have been disclosed; this particular compound displayed 81-fold and 138-fold selectivity against cathepsin L and S, respectively [288]. Further structural modifications aimed at improving the pharmacokinetic profile of such inhibitors led to compounds such as **10** which displayed excellent oral bioavailability in rats [289]. Weak inhibitors but displaying significant reduction in peptidic character were constructed around a cyanopyrrolidine scaffold, as illustrated by **11** [290].

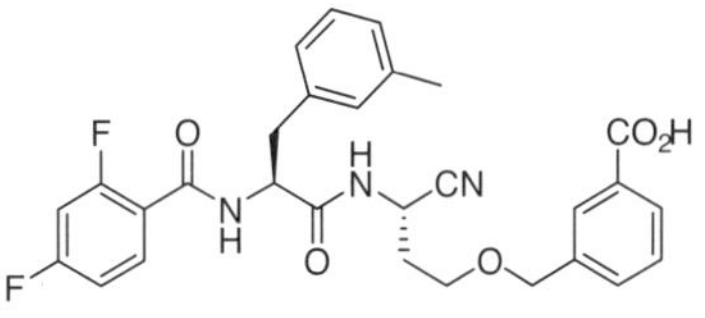

9
IC_{50} = 7 nM

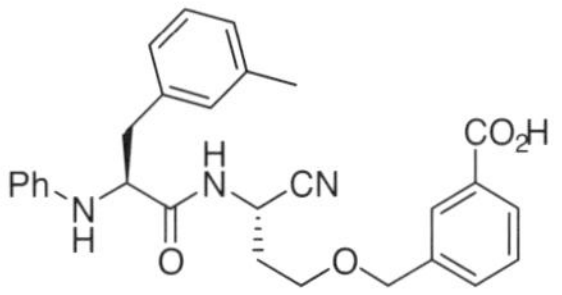

10
IC_{50} = 12 nM

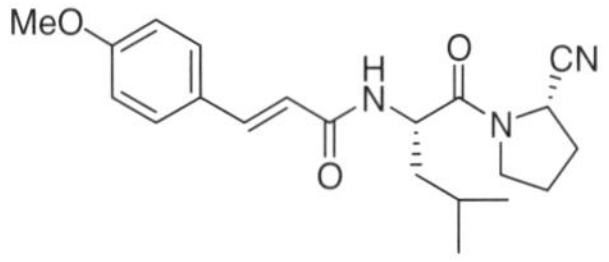

11
K_i = 92 μM

Oxorhenium Complexes The synthesis of novel oxorhenium complexes and their inhibitory activity against cathepsin B and L has been reported. Compounds such as **12** and **13** are active site directed tight binding, reversible inhibitors. Compound **13** is 260-fold more active against cathepsin B than cathepsin L [291,292].

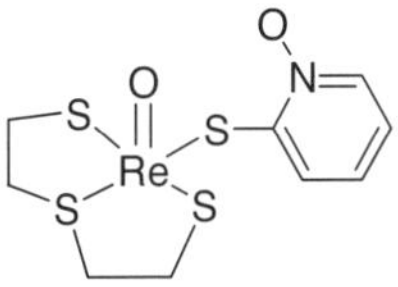

12
IC_{50} = 30nM

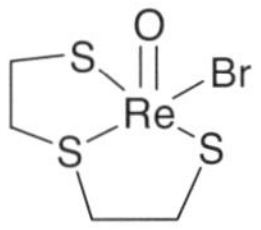

13
IC_{50} = 1nM

2.5.2. Irreversible Inhibitors

Epoxysuccinate Inhibitors Epoxysuccinyl peptide-based inhibitors have received the most attention, especially regarding their potential as neuroprotectives. Applications as calpain inhibitors have been discussed earlier (Section 2.3). E-64 (**14**) was isolated from natural source and displays marginal selectivity among various cysteine proteases [293]. Synthetic analogs such as E-64c (**15**) and its prodrug E-64d (**16**) [294], and CA-074 (**17**) [295] display better membrane permeability and, in the case of CA-074, superior selectivity toward cathepsin B. Numerous studies with these synthetic inhibitors have been reported. Studies with E-64 compounds may lead to inconclusive target-therapeutic effect relationships due to the activity of these inhibitors against other cysteine proteases such as calpain. Studies with the selective cathepsin B inhibitor CA-074, however, are more conclusive regarding the role of cathepsin B *in vivo*. E-64c, administered i.v. to monkeys immediately after 20 min whole brain ischemia, protected 84% of CA1 neurons from delayed neuronal death on day 5, while CA-074 provided 67% protection in the same model [296,297]. The results with CA-074 suggest that calpain-induced cathepsin B release is crucial for the development of ischemic neuronal death, highlighting the potential importance of cathepsin B inhibitors. Similar results were reported later [298]. More recently, in a rat model of stroke, E-64d provided significant brain protection against ischemic/reperfusion injury by attenuating neuronal and endothelial apoptosis [299]. The authors suggest that E64-d treatment provides neuroprotection by inhibiting the upregulation of MMP-9 [300]. Continuous administration of E-64d at 1 mpk to a rat model of spinal cord injury resulted in inhibition of calpain activity and other factors contributing to apoptosis in the lesion and surrounding areas [301]. E-64d entered clinical trials in Japan for muscular dystrophy but development was discontinued in Phase III [302]. E-64d is hepatotoxic in rats [303] while E-64 is teratogenic in rats [304].

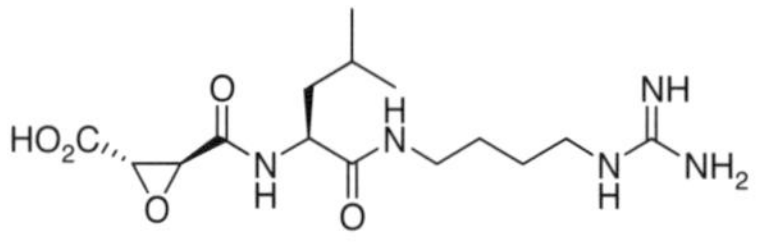

14: E-64
IC_{50} = 55 nM

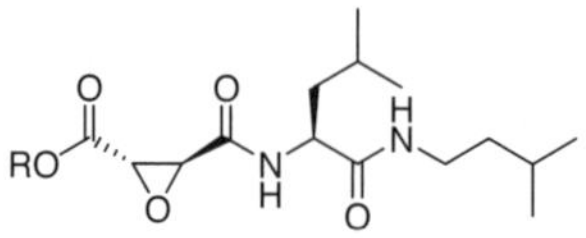

R = H: 15: E-64c
IC_{50} = 9 nM

R = Et: 16:
E-64d, loxistatin

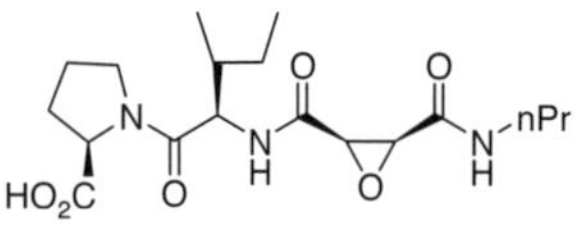

17: CA-074
IC_{50} = 2 nM

Piperazine derivatives **18** and **19**, in the epoxysuccinate class, have been reported as calpain and cathepsin B inhibitors [305]. They were originally studied for myocardial infarction but then were shown to have in vitro and in vivo anticancer activities leading to the advancement of NCO-700 to Phase II clinical trial as an oncolytic [306].

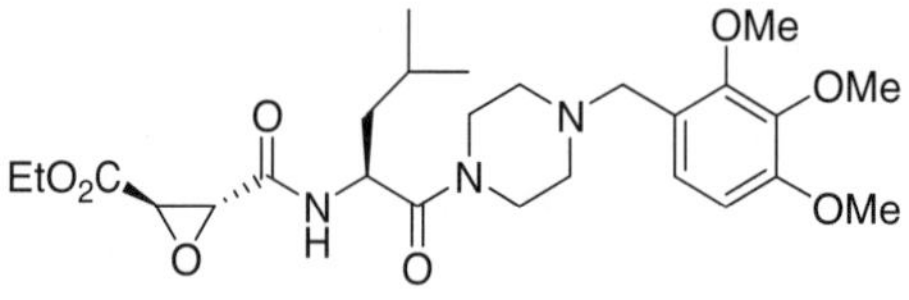

18 NCO-700

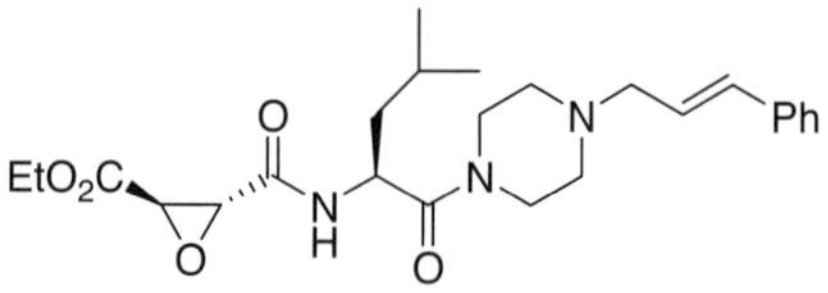

19 TOP-008

Aziridine Inhibitors Derived from the epoxysuccinate series, the aziridine inhibitors present the advantage of an additional point of diversification via substitution at the aziridine nitrogen. Cathepsin B or L selectivity can be obtained, depending on substitution pattern, as illustrated by **20** and **21** [307–310].

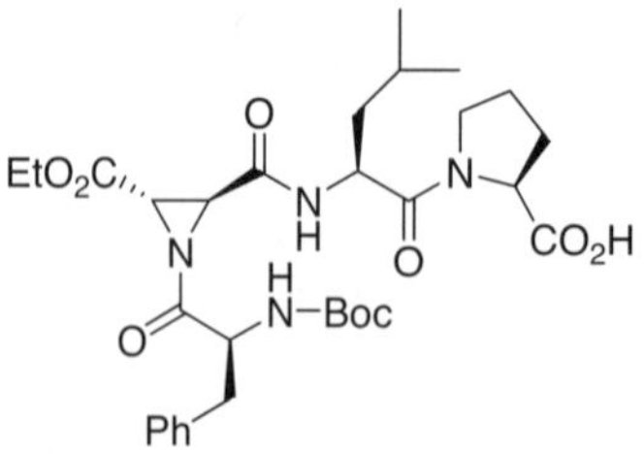

20
cathB k_2 = 6860 $M^{-1}min^{-1}$
cathL k_2 = 210 $M^{-1}min^{-1}$

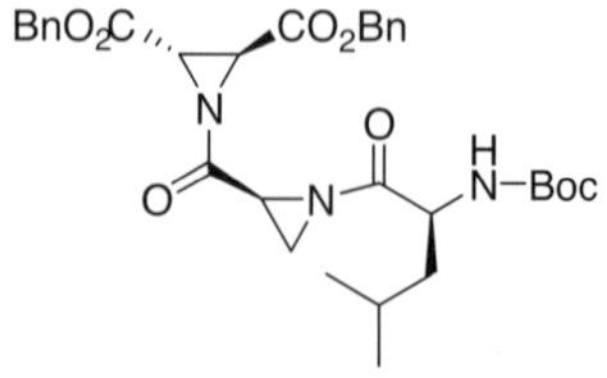

21
cathB K_i = 9400 nM
cathL K_i = 13 nM

Thiadiazole Inhibitors 1,2,4-Thiadiazole has been introduced by Leung-Toung and coworkers, as a novel electrophilic cysteine trap [311]. Attack of the cysteine thiol group at the 1-position, results in disulfide formation and opening of the heterocycle, leading finally to irreversible inactivation of the enzyme. This ring-opening event is specific to thiols, as opposed to alcohols and amines. This mechanism can only operate in the extracellular space due to the presence of competing gluthatione intracellularly; this allows for selective extracellular cathepsin B inhibition, with no disruption of cathepsin housekeeping functions.

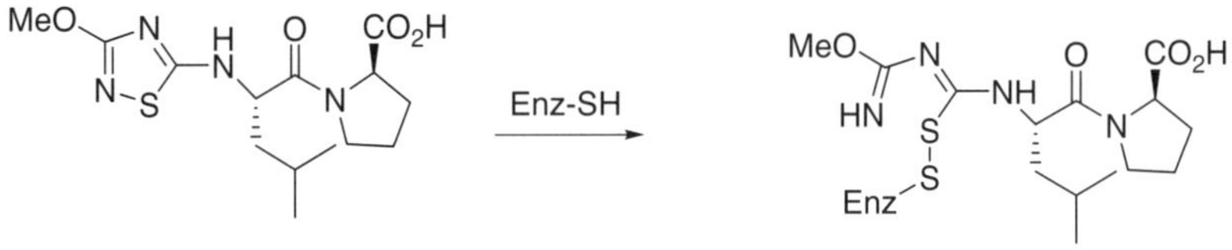

22
cathB K_i = 2600 nM

Acyloxymethyl, Chloromethyl, and Diazomethyl Ketones As described earlier with calpain and caspases ihibitors, acyloxymethyl ketones serve as irreversible cathepsin B inhibitors. Representative examples are shown below [312–314].

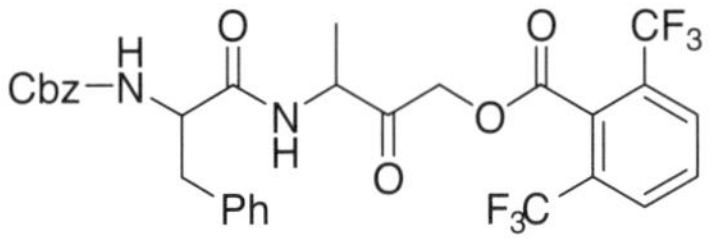

23

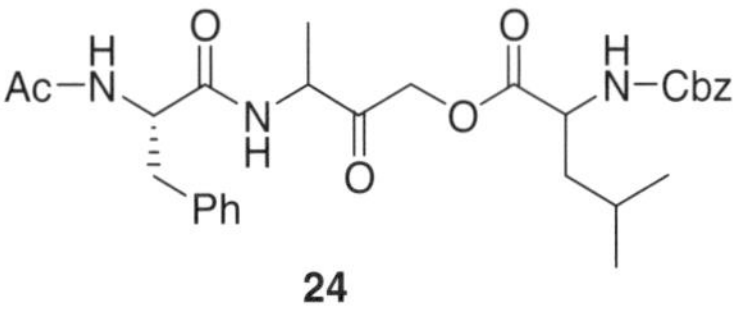

24

Tetrapeptide chloromethyl ketone **25** was demonstrated to rescue neuronal cells from cell death in response to oxidative stress and oxygen/glucose deprivation. Subsequent affinity purification and sequencing work identified the biological target to be cathepsin B [315].

Diazomethyl ketone CP-1 (**26**), when tested in a rat MCAO model, was found to reduce the infarct volume, neurological deficits, cathepsin B activity, and the amount of heat shock proteins and albumin in the brain, suggesting lessened secondary ischemic damage, which is related to lessened cerebral tissue damage and improved neurological recovery [316,317].

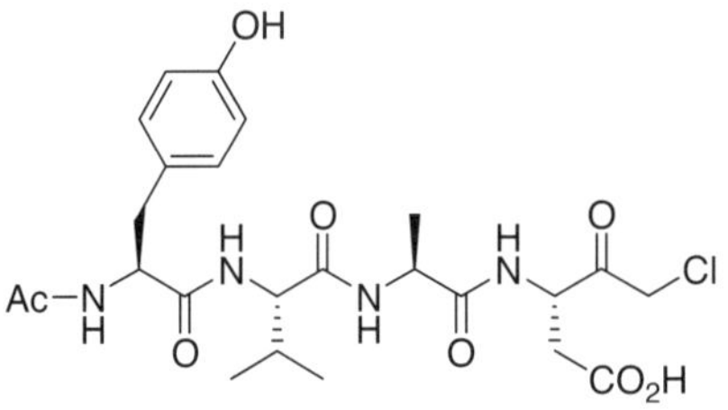

25

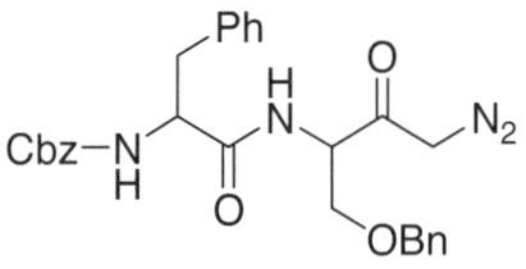

26: CP-1

β-Lactams and Cyclic Carbamates Inhibitors β-Lactam-based cathepsin inhibitors such as **27** have been disclosed [318]. They typically are less selective for cathepsin B than other cathepsins. Applications to a penam scaffold (**28a–28d**) yielded cathepsin inhibitors with a similar profile [319,320].

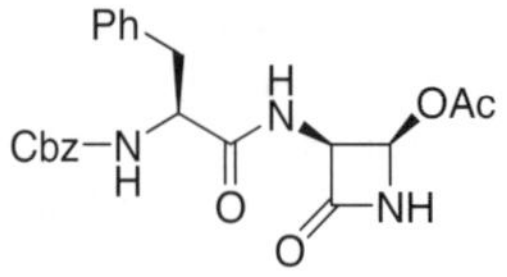

27
catB IC$_{50}$ = 470 nM
catL IC$_{50}$ = 42 nM
catK IC$_{50}$ = 100 nM
catS IC$_{50}$ = 207 nM

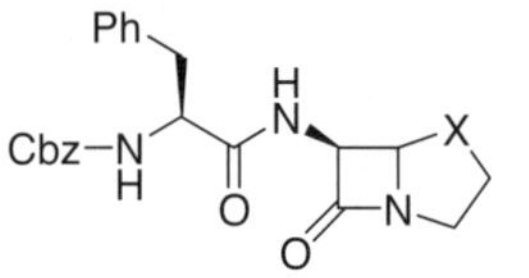

28a–d

IC$_{50}$ (μM)

X	O	S	SO	SO$_2$
catB	1.7	>50	16	0.35
catK	1.6	>2.5	0.04	0.0005
catL	0.09	>6.7	0.07	0.001

Cyclic carbamates such as **29** can afford potent and selective cathepsin B inhibitors [321]. The inherent ring strain present in this system results in minimal Π-overlap between the bridgehead nitrogen and the carbonyl that renders the later more electrophilic and susceptible to attack by the active site cysteine thiol.

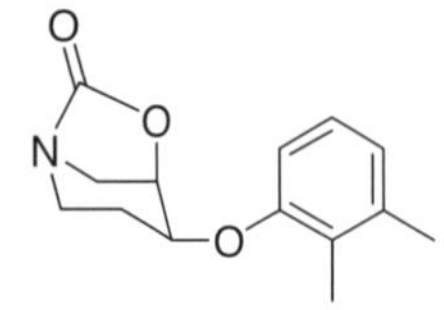

29
catB IC$_{50}$ = 1 nM
catL IC$_{50}$ = 22 nM
catK IC$_{50}$ = 38 nM
catS IC$_{50}$ = 16 nM

Organotellurium and Organoruthenium Inhibitors A series of organotellurium and organoruthenium complexes have been described as cathepsin B inhibitors, as illustrated by **30** [322]. They are believed to be irreversible inhibitors resulting from the displacement of one of the chloride ligands by the active site cysteine thiol. They offer the potential advantage of being specifically extracellular inhibitors due to the reductive nature of the cytoplasm that would allow for intracellular inactivation of such inhibitors. Ruthenium complexes such as **31** have been prepared, providing cathepsin B inhibitors that function according to a similar displacement mechanism [323].

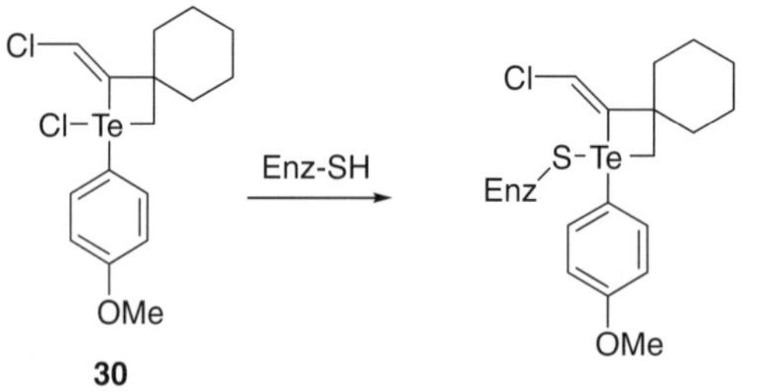

30

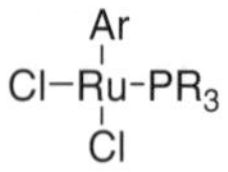

31

Other Irreversible Inhibitors Lim and coworkers have designed a series of irreversible inhibitors such as **32** constructed around a central acylated hydroxamate [324]. The acylation of the enzyme is assisted by the presence of an acetate group that in turn is rendered more labile by the interaction with His199 from the enzyme.

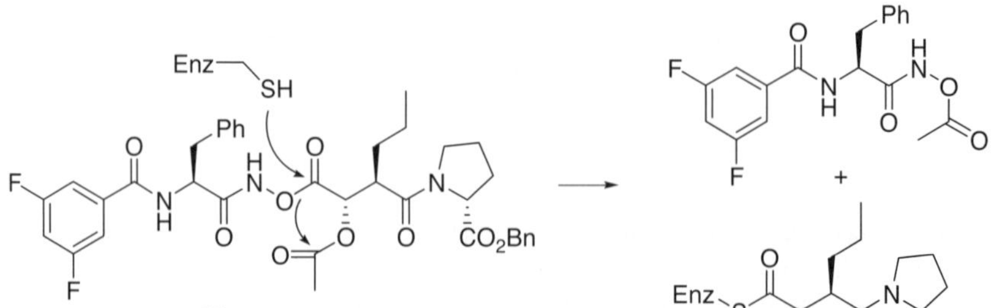

32
K_i = 4.4 μM

Inhibitors based on Michael acceptor as cysteine trap have been designed and prepared, as illustrated by vinyl ester **33** [325] and vinyl sulfone **34** [326].

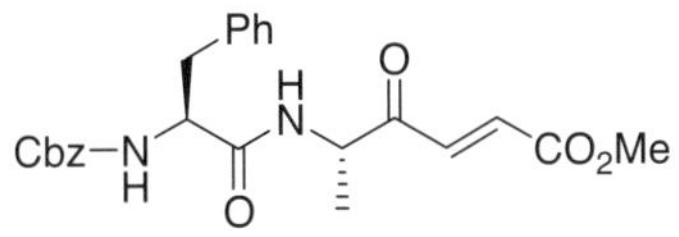

33
K_i = 900 nM

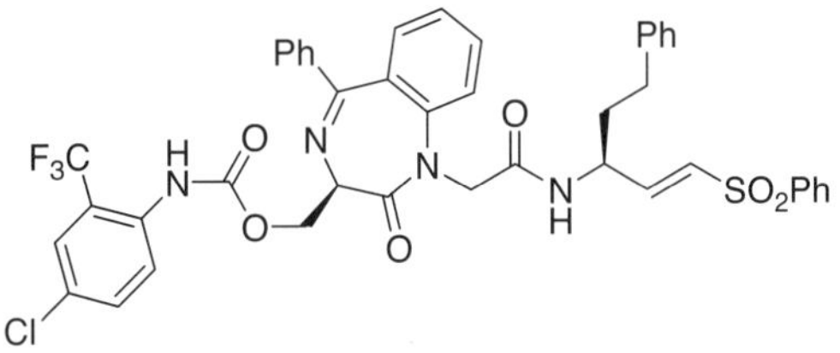

34
K_i = 3100 nM

Carbamate and ester derivatives of the 1,1-dioxobenzo[b]thiophene-2ylmethyloxycarbonyl (Bsmoc) can serve as Michael acceptors and be derivatized as irreversible inhibitors of cysteine proteases, as illustrated by compound **35** [327]. Michael addition, followed by rearomatization, leads to covalent attachment to the cysteine protease.

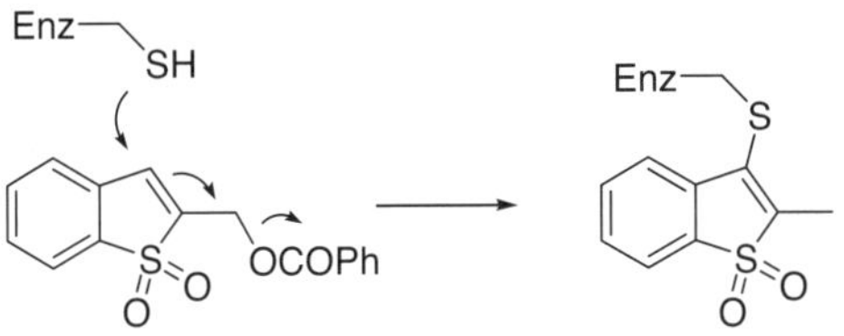

35

2.5.3. Reversible Inhibitors Lacking a Cysteine Trap

Using the structural information obtained with well-characterized inhibitors such as E-64c and CA-074, new inhibitors that lack a cysteine trap and span the prime and non-prime subsites have been designed. CAA0445 spans P3 to P2′ and inhibits cathepsin B with an IC$_{50}$ below 100 nM [328]. Azapeptides derived from an azaglycine scaffold have yielded some of the most potent cathepsin B inhibitors [328,330]. Z-Arg-Leu-Arg-Agly-Ile-Val-OMe (**37**), a 480 pM inhibitor is also 2310-fold more active against cathepsin B than cathepsin L.

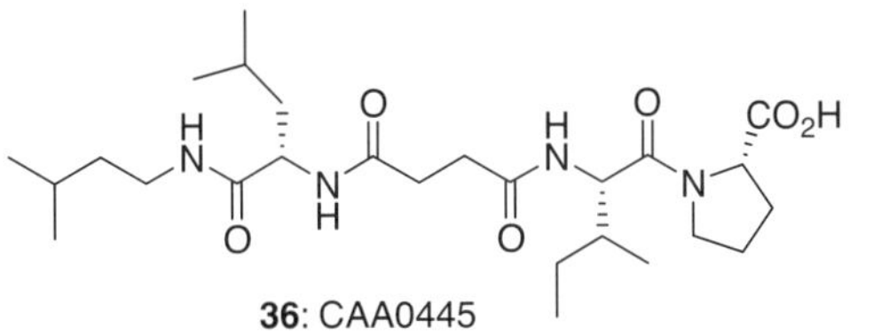

36: CAA0445

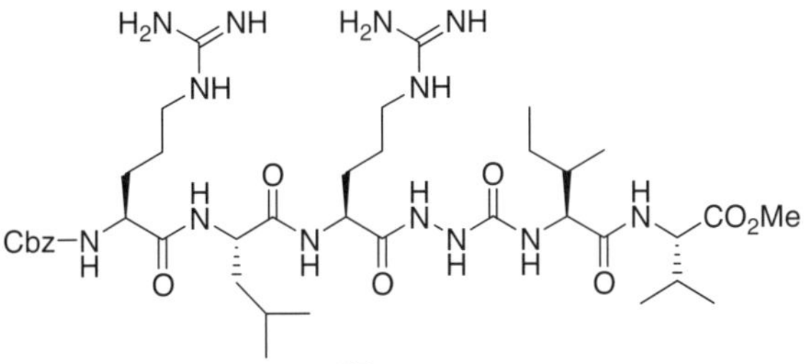

37
K_i = 0.48 nM

Thiocarbazate-based cathepsin L inhibitors such as **38** and **39** have recently been described. Although there was no evidence of reactivity with the protease, the close proximity of the activated carbonyl to the active site cysteine calls for caution as to the mechanism by which these inhibitors operate [331].

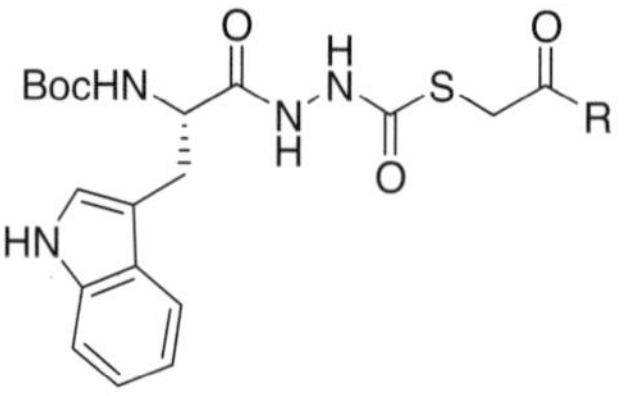

R = NH-2Et-Ph: **38**
catL IC$_{50}$ = 56 nM

R = (1,2,3,4-tetrahydroisoquinolin-2-yl)
39
catL IC$_{50}$ = 7 nM

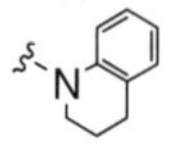

A double-headed cathepsin B inhibitor devoid of a cysteine trap (**40**) has recently been reported on [332]. This inhibitor competitively and reversibly inhibits cathepsin B with its occluding loop stabilized in its closed conformation. Asteropterin, a natural product isolated from marine sponge source, was identified in a cathepsin-B screen [333]. Partial structures have been proposed (**41** and **42**). Phthalate derivatives **43** and **44** have also been isolated from marine sources and found to be modest cathepsin B inhibitors [334,335].

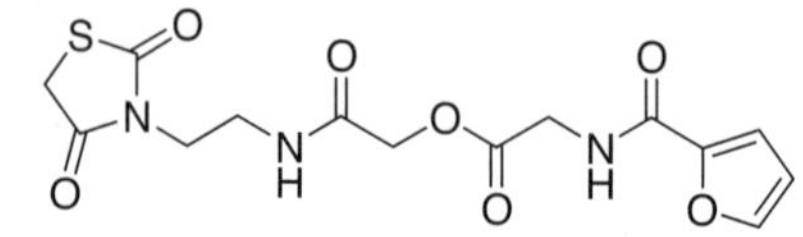

40
cat L K_i = 4800 nM

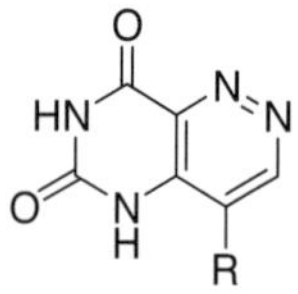

41 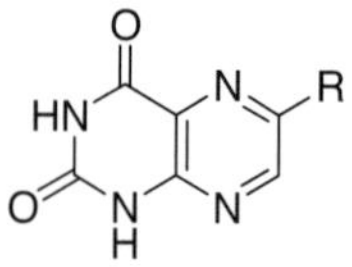   **42** 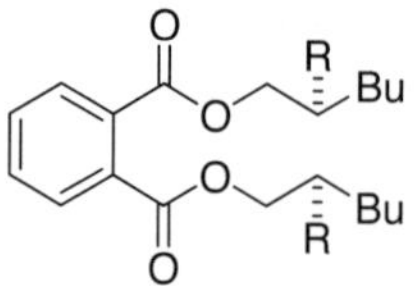   R = H: **43**
                         R = Et: **44**

Starting from previously reported cathepsin K inhibitors and while looking for cathepsin L inhibitors researchers at GSK identified azepanone-derived cathepsin B inhibitors such as **45** [336]. Pyrazole derivative **46** was identified during an NIH HTS screen as an alternated substrate for cathepsin B, rather than as an inhibitor [337].

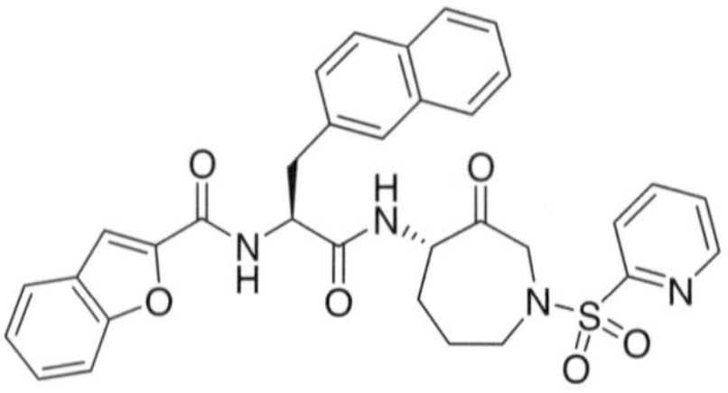

45
catB K_i approximately 24 nM

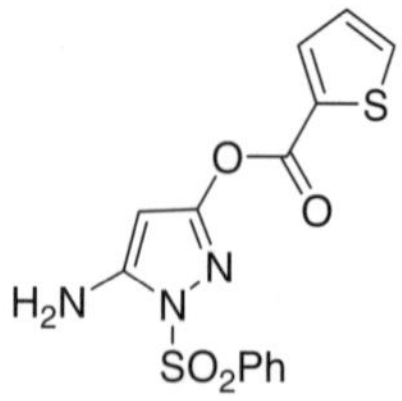

46

3. ANTIOXIDANTS

There is significant evidence that reactive oxygen species (ROS) are important mediators of tissue injury following ischemia and reperfusion. ROS are capable of causing oxidative damage to macromolecules; thus, leading to lipid peroxidation, the oxidation of amino acid side chains, and DNA damage [338]. The brain is especially vulnerable to ROS damage because of its high oxygen consumption rate, abundant lipid content, and relative paucity of antioxidant enzymes in comparison to other organs [339]. Under such pathological conditions as cerebral ischemia, ROS may be produced by a variety of mechanisms, including mitochondrial electron transport dysfunction, glutamate excitotoxicity, intracellular calcium overload, nitric oxide synthase, and xanthine oxidase [340].

ROS including hydrogen peroxide (H_2O_2) and superoxide radical ($O_2^{\bullet-}$) are produced by a number of cellular oxidative metabolic processes involving xanthine oxidase, NAD(P)H oxidases, metabolism of arachidonic acid by cyclooxygenases and lipoxygenases, monoamine oxidases, and the mitochondrial respiratory chain. Endogenous defenses that detoxify ROS include enzymatic systems such as superoxide dismutase (SOD), catalase, and glutathione peroxidase (GSH-Px), and nonenzymatic antioxidants such as ascorbic acid, vitamin E, β-carotene, and glutathione (GSH). While ROS are proposed to play important roles in coordinating and regulating a number of cellular signaling pathways (redox signaling), oxidative stress results when the formation of ROS exceeds the capacity of antioxidant defense systems [341]. Evidence that ROS contribute to ischemic brain injury has been obtained from studies with transgenic mice and gene knockout animal models. Mice overexpressing either SOD1 or GSH-Px-1 showed significantly smaller infarcts compared to wild-type counterparts following focal cerebral ischemia. Conversely, ischemic injury was increased in mice deficient in SOD1 [342]. Taken together, these studies show that both SOD and GSH-Px have essential roles in detoxifying ROS following stroke.

The highly reactive hydroxyl radical (•OH) is not produced as a by-product of any known enzymatic reaction, but is formed from H_2O_2 in the presence of divalent metal ions, especially Fe^{2+} and Cu^{2+}, via the Fenton reaction [343]. Once formed, •OH reacts quickly with many cellular components, including polyunsaturated fatty acids of membrane lipids. The initial reaction of •OH with polyunsaturated fatty acids produces an alkyl radical, which in turn reacts with molecular oxygen to form a peroxyl radical (ROO•). The ROO• can abstract hydrogen from an adjacent fatty acid to produce a lipid hydroperoxide (ROOH) and a second alkyl radical, propagating a chain reaction of lipid oxidation [343].

This body of scientific literature on a role for ROS in ischemia/reperfusion injury has led to a wide ranging, and chemically varied, number of antioxidant molecules that have been tested in preclinical and clinical studies.

3.1. NXY-059

Based on the evidence that radical oxygen participates in the ischemic cascade [344], drugs such as the spin trap, NXY-059, were developed with the hope of quenching radical species formed during the ischemic period. A study on MCAO rats by Kuroda et al. confirmed this hypothesis showing significant dose-dependent decreases in infarct area among rats treated with NXY-059 [345]. They were also able to determine an optimal therapeutic window of 3–6 h postocclusion. In accordance with STAIR guidelines, additional studies confirmed these results in rat and primate models [346]. Marshall et al. demonstrated that treatment 4 h after pMCAO not only decreased infarct size by 28% in marmosets but also showed positive effects versus a control group in tasks designed to test motor skills.

Based on promising *in vivo* data, NXY-059 was brought into the clinic for the SAINT I trial. In that study, a large randomized population of stroke patients was treated with NXY-059 within 6 h from the onset of symptoms, while maintaining a mean of less than 4 h postonset. NXY-059 was found to be efficacious against its primary endpoint of reducing disability at 90 days postinfarct as determined by the modified Rankin scale, with an overall decrease amounting to 0.13 points per patient [347]. Nevertheless, the trial failed to demonstrate efficacy against secondary endpoints, including disability based on the NIHSS, and decreased mortality as compared to a placebo group. The second, and significantly larger, SAINT II trial, however, failed to reproduce the positive results of the first, and showed no improvement against the placebo toward any of its endpoints. As a result of these findings, development was discontinued.

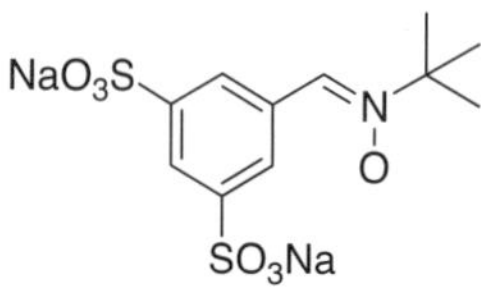

NXY-059

3.2. Tirilazad

Tirilazad is a 21-aminosteroid designed to inhibit lipid peroxidation without corticosteroid activity that has previously been the subject of an in-depth review [348]. A number of animal studies have shown efficacy [349–351] in MCAO models, reducing overall infarct sizes, while one study in MCAO rats measured the sodium and potassium ion concentration shifts and water entry as an indicator of infarct size [352], and arrived at a similar conclusion. These results were expected as glucocorticosteroids have long been known to act as neuroprotectives against lipid peroxidation in cases of CNS trauma, a subject that has been reviewed previously [353]. A thorough reexamination of all animal studies of tirilazad has been conducted by Sena et al. [354] who determined that, based on the original data of 18 studies involving a total of 544 animal subjects, tirilazad did, in fact, significantly reduce infarct volumes.

Promising animal data across a number of species suggested that tirilazad should reduce infarct volumes in human cases of AIS. The RANTTAS trials were designed to test efficacy against the two primary endpoints of disability based on the Glasgow Outcome Scale, and daily living as determined by the Barthel Index 3 months after the ischemic event [355] by administering an initial dose of 150 mg within 6 h of the onset of symptoms, followed by 1.5 mg/kg every 6 h over three days for a total of 12 doses. Secondary endpoints included NIHSS score and cerebral infarct volume as measured by CT scan at 7–10 days postischemia. Between May 1993 and December 1994, 556 eligible patients were enrolled in the trial, with an ultimate goal of 1130 participants. However, interim analysis performed on 500 patients enrolled before August 23, 1994 showed no benefit from tirilazad mesylate as compared to placebo and the trial was discontinued. The RANTTAS II trial was of similar design, but doubled doses to a 300 mg initial dose followed by 10 mg/kg per day for men and 12 mg/kg per day for women, due to varying pharmacokinetics. This study, however, was discontinued after only 126 patients were enrolled. The TESS I and II trials were run concurrently to the RANTTAS I and II, respectively, in Europe and Australasia with similar experimental design, dosing, and outcome. Interestingly, a secondary analysis of the TESS CT scan data showed a significant decrease in infarct size among men with cortical lesions, though as a whole, the patient population showed no benefit from tirilazad as compared to placebo [356]. Furthermore, a combined analysis of clinical data from all 1757 enrolled patients showed a significant increase in death or disability among those receiving tirilazad.

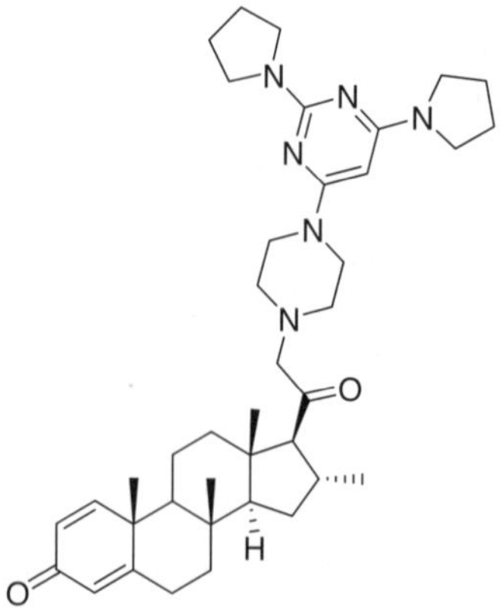

Tirilazad

3.3. Ebselen

Ebselen is a seleno-organic compound that represents a novel mechanism of neuroprotection against lipid peroxidation. It has been previously reviewed extensively [357]. Unlike radical scavengers, ebselen's neuroprotective properties result from glutathione peroxidase-like activity, which allows it to catalytically reduce hydroperoxides [358] thereby preventing lipid peroxidation. There has also been some evidence that ebselen is a redox modulator NMDA and may exhibit some neuroprotective effect through that pathway [359]. A number of studies have shown that ebselen is efficacious as a neuroprotective in rat *in vivo* and *in vitro* reperfusion [360], permanent ischemia [361], and spinal cord injury models [362]. Similarly, rabbit MCAO models showed a significant effect, but only when administered shortly after occlusion [363]. An *in vitro* study of mouse hippocampal neuroblastoma HT22 cells showed that ebselen protected against cell death induced by glutamate and reactive oxygen species, but not TNFα [364]. Interestingly, one study in MCAO rats receiving a single 10 or 100 mg/kg oral dose of ebselen 1 h before the inducement of severe ischemia did not show a decrease in infarct volume at 1 week [365].

The first Phase III clinical trial of ebselen enrolled 302 patients with AIS, and measured efficacy based primarily on the Glasgow Outcome Scale score at 1 and 3 months [366]. Secondary endpoints were neurological status as determined by the modified Mathew Scale, and functional status as determined by the modified Barthel Index. Patients were administered ebselen 150 mg bid or placebo as a suspension in water, orally, for 2 weeks. The trial stipulated that dosing was to occur within 48 h of the onset of ischemia, though the mean approximately 29 h for both groups. The trial found a significant improvement in outcome among patients who received ebselen within 24 h of the onset of symptoms, but not within the larger population of participants. Similarly, a second trial enrolled 99 patients and administered ebselen or placebo within 12 h of the onset of ischemia, but failed to demonstrate significant efficacy [367].

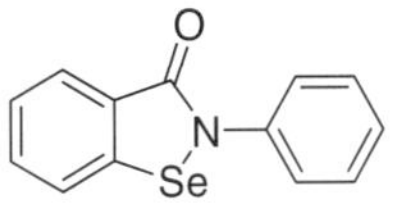

Ebselen

3.4. Edaravone

Edaravone is a radical scavenger that has shown neuroprotective effects in a number of species, including rats [368,369] and humans when administered within 72 h of the onset of ischemia [370]. It was approved in 2001 for the treatment of acute brain infarction and has been marketed as Radicut in Japan, after a relatively small clinical trial that enrolled only 250 patients, showed a significant improvement in functional outcome at 3 months. A second, smaller clinical trial enrolled 61 patients with carotid territorial stroke and concluded that patients receiving edaravone exhibited decreased infarct after 2 days but not after 14 days [371]. The findings of the second trial were not available until after approval. No other Phase III clinical trials have been completed, and edaravone has not been considered for approval or further development elsewhere [372]. Curiously, however, Feng et al. [373] cite what they call a "systematic review of eight trials of edaravone for acute ischemic stroke" that could not be identified by this author. Furthermore, no reference to any additional trials has been found elsewhere. While the Phase III trial did not observe any increase in adverse effects as compared to the placebo group, subsequent reports suggest a link to severe renal disorders [374], and liver failure [375]. In a comprehensive review of the development of edaravone, however, Watanabe et al. suggest that reports of acute renal failure have been overblown, and have not considered comorbidities [376].

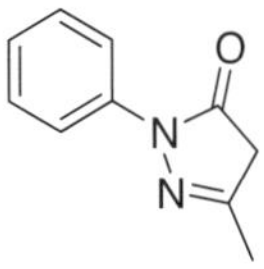

Edaravone

3.5. Deferoxamine

Deferoxamine has been marketed since 1962 as an iron chelator for transfusional iron overload [377]. Since then, deferoxamine and other iron chelators have been found to be effective as anticancer and antifungal agents, presumably by inactivating iron-dependent enzymatic processes [378] and there is evidence that they may be effective against a number of other diseases [379].

At least one excellent comprehensive review of the role of iron in ischemic stroke has been written [380]. It has been shown that iron catalyzes conversion of hydrogen peroxide and superoxide to the hydroxyl radical [381], resulting in neuronal injury through an oxidative pathway. Castellanos et al. demonstrated that rats fed high-iron diets displayed significantly increased infarct volume in MCAO models [382], while there is clinical evidence that patients with higher iron levels had worse outcomes after treatment with tPA [383]. Deferoxamine therapy has been proposed as a method of limiting oxidative damage and has shown efficacy in MCAO rat models when administered after the ischemic event [384,385]. A Phase II clinical trial designed to test the tolerability and efficacy of deferoxamine in patients treated with t-PA is ongoing (http://clinicaltrials.gov/ct2/show/record/NCT00777140).

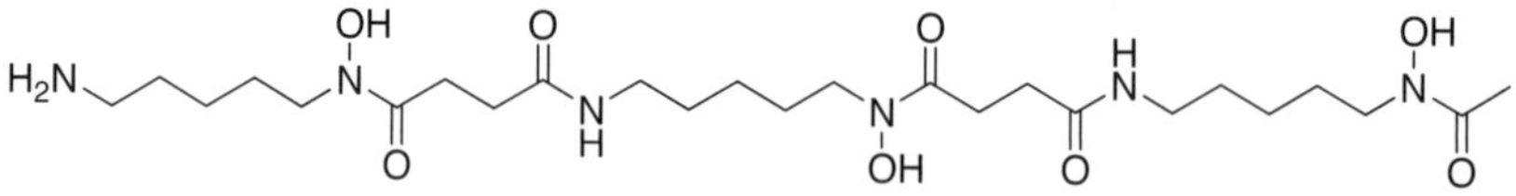

Deferoxamine

4. OTHER CLINICALLY TESTED MECHANISMS

4.1. Maxi K-Channel

Another mechanism of neuronal damage during the ischemic cascade results from high intracellular Ca^{2+}, leading to hyper excitability and eventual cell death. The role of calcium in ischemic cell death has previously been comprehensively reviewed [386]. It has long been known that hypoxia leads to the entry of Ca^{2+} into intracellular space [387], however, mitigating this problem through the use of calcium channel antagonists has the potential for significant adverse effects as known compounds are not sufficiently specific for ischemic neurons [388]. The maxi-K channel is one of the numbers of voltage-gated potassium channels (VGKCs), which is more structurally diverse than other voltage-gated ion channels, and thus can be more specifically targeted. By activating the maxi-K channel, hyperpolarization of the ischemic neuron by K^+ influx decreases Ca^{2+} influx, thereby preventing Ca^{2+}-mediated apoptosis [389].

4.1.1. BMS 204352/MaxiPost
BMS-204352 is a fluorooxindole voltage and Ca^{2+}-dependent modulator of the VGKC maxi-K, such that it is most potent as an activator at high Ca^{2+} concentrations or during depolarization. Preclinical results showed a high brain:plasma ratio of 9.6 and no effect on blood pressure or cerebral blood flow [388]. In rat pMCAO models, BMS-204352, 0.1 and 0.3 mg/kg administered intravenously 2 h after occlusion, significantly decreased infarct volume at 5.5 and 24 h by MRI. Unfortunately, BMS-204352 showed no effect in the clinic, and its development was discontinued in 2001 [390].

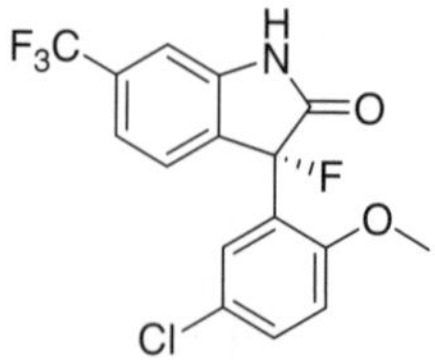

BMS-204352

4.1.2. BMS-191011
BMS-191011 is the follow-up effort to MaxiPost and is thought to

function through the same mechanism of neuroprotection. Despite poor aqueous solubility of the parent compound, *in vivo* efficacy was observed in animal models [391]. As a result, a deoxycarnitine ester prodrug was prepared, which significantly improved physical properties [392].

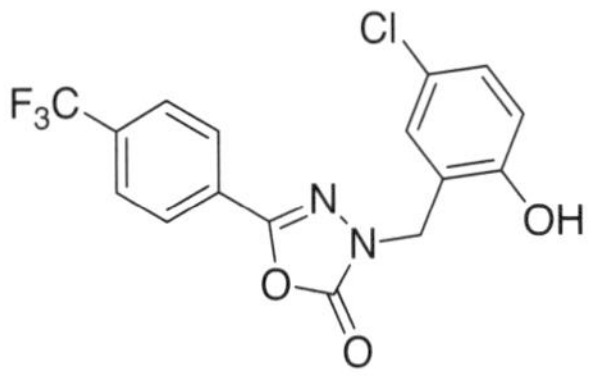

BMS-191011

4.2. GABA$_A$ Agonist

Gamma-aminobutyiric acid (GABA) is a neurotransmitter that has been shown to decrease excitotoxicity through a cascade originating in the activation of the GABA$_A$ receptor complex, and ultimately leading to the inhibition of glutamate-inducted neuronal activity that leads to exitotoxicity and apoptosis. The GABA mechanism has previously been reviewed extensively [393,395]. There is also evidence that GABA inhibits NMDA-mediated responses to stroke [395].

Clomethiazole is a GABA-mimetic modulator of the GABA$_A$ receptor complex. It is currently marketed as a sedative under the name Himenevrin. A large number of animal studies have shown clomethiazole to have neuroprotective properties; the methods and outcomes of greater than 25 of those studies have been summarized in detail by Green et al. [393]. Notably, however, Cross et al. showed that a single administration of 600 μmol/kg 1 h after 5 min transient ischemia reduced hippocampal damage by greater that 50% as compared to saline at 21 days post-ischemia with no sedative effect [396].

The Clomethiazole Acute Stroke Study (CLASS) sought to determine the efficacy and safety of clomethiazole in patients with AIS [397]. Patients were administered clomethiazole (75 mg/kg) or placebo over 24 h as an intravenous infusion, although in some cases the full dose could not be administered due to sedation of the patient. The primary outcome measure was functional independence at 90 days as determined by the Barthel Index. Overall, 1360 patients were included in the analysis, though the clomethiazole group was not significantly different from placebo at 90 days. Interestingly, subgroup analysis showed a significant improvement among patients diagnosed with total anterior circulation syndrome (TACS) with 40.8% reaching functional independence as compared to 29.8% in the placebo group. As a result, a second clinical trial with similar design, with the exception that the dose of clomethiazole was lowered to 68 mg/kg, know as CLASS-I was designed to test efficacy among TACS patients. Unfortunately, no significant differences between the clomethiazole and placebo group were observed for any of the outcomes, including Barthel Index, NIHSS, Scandinavian Stroke Scale, or CT infarct volume [398].

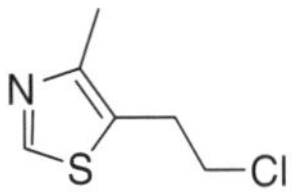

Clomethiazole

4.3. ONO2506

ONO-2506, (R)-(-)-2-propyloctanioic acid, has been found to inhibit the formation of activated astrocytes after an ischemic event; thus, maintaining GABA$_A$ activity and suppressing glial cell formation [399]. Furthermore, ONO-2506 has been shown to inhibit expression of S-100β [400] in pMCAO rats, the overproduction of which has been implicated in neuronal death [401]. Preclinical studies showed efficacy in improving functional outcomes and decreasing infarct volume in pMCAO rat models when ONO-2506 (10 mg/kg i.v.) was administered immediately after the onset of ischemia and repeated daily for 3 or 7 days, though it is noted that administration beginning at 24 and 48 h after the onset of ischemia also yielded significant decreases [400].

A phase I dose-escalating trial in patients with AIS showed that ONO-2506 was well tolerated when administered intravenously over 1 h daily for 7 days at doses ranging from 2–12 mg/kg/h [402]. Patients receiving 8 mg/kg/h showed significant decreases in NIHSS as compared to placebo at 3, 7, 10, and 40 days after the ischemic event. It was also demonstrated that patients receiving ONO-2506 had significantly decreased serum S-100β levels, which correlated to an improvement in NIHSS [403]. Clinical trials are ongoing.

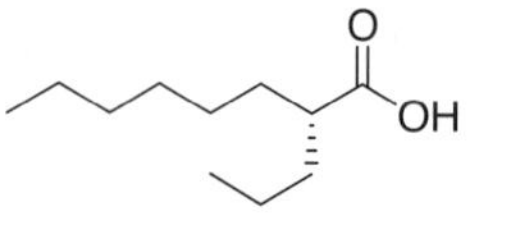

ONO-2506

5. PARP INHIBITORS

Poly-ADP-ribose polymerase is a nuclear enzyme that is activated by single-stranded DNA breaks. In the context of ischemia-reperfusion, these single-stranded breaks are likely caused by reactive oxygen species, and the breaks appear to require reperfusion. In a study by Chen et al. [404], there are no breaks after 60 min focal ischemia, but numerous breaks were detected within 1–5 min of reperfusion.

In preclinical studies, activation of PARP appears to play a critical role in damage following transient ischemia. Cortical cell cultures from PARP −/− mice were nearly completely resistant to 60 min of oxygen/glucose deprivation [405]. Studies in knockout mice [406] or that used specific PARP inhibitors [407], reported reduction in infarct volumes of 50–80%. Despite extensive study and development of numerous inhibitors, the mechanism of PARP-mediated cell death remains poorly defined.

To date, large numbers of inhibitors for poly (ADP-ribose) polymerase (PARP) have been published and extensively reviewed by Woon [408,409]. Among seven isoforms of PARP that have been identified, PARP-1 has been well studied and extensively characterized. This review will highlight select PARP inhibitors and efforts to address brain penetration, improved physical properties and compounds in clinic.

The earliest known inhibitor of PARP is 3-aminobenzamide (1) with an IC$_{50}$ of 22 uM, designed from substrate NAD +. In the case of benzamide, the amide NH is *cis* to the carbonyl group and picks up a key hydrogen bond interaction with glycine 863 at the active site [410]. Based on this observation, a number of restricted conformation analogs were evaluated (Fig. 3), leading to a number of heterocycles such as isoindolines (2), isoquinolines (3 and 5), quinzolinones (4), and pthalizinones (6) and variation thereof in these structures. Additionally, the benzene in benzamide has been replaced by thiophenes leading to theino(3,4-*c*)pyridones (7), theino(3,4-*d*)pyrimidonones (8), and so on. Among these heterocycles, unsatu-

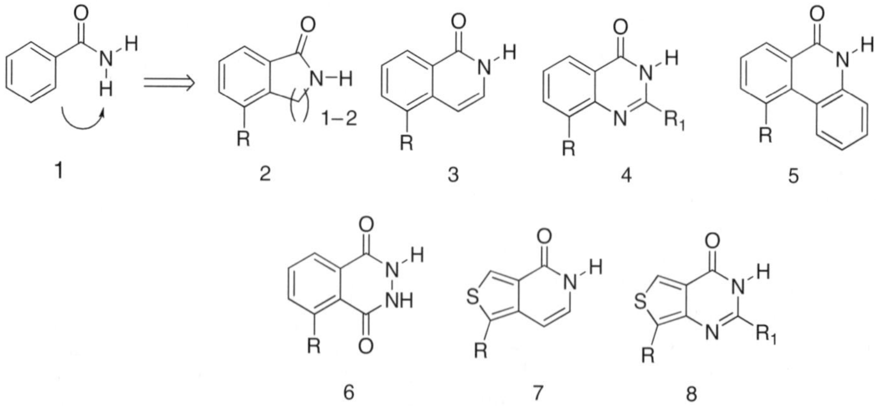

Figure 3. Restricted conformation analogs of benzamides.

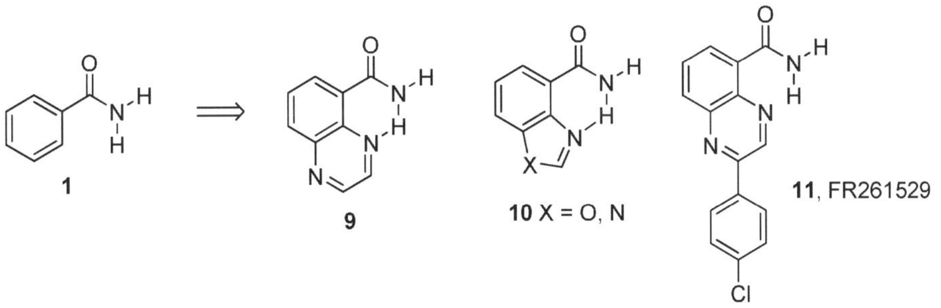

Figure 4. Restricted conformation through hydrogen bonding.

rated ring systems such as 3–5 were found to be more potent than other heterocycles [411–416].

Crystal structure of the catalytic domain of human PARP is known in literature and a number of these inhibitors have been cocrystallized or modeled in the active site [417]. The aromatic rings in these inhibitors stack between phenylalanine 907 and phenylalanine 896 at the active site of PARP. These inhibitor interaction studies also partially support the notion that planar electron-rich aromatic ring is needed to enhance the ability of the carbonyl group to participate in hydrogen bonding within the active site. Alternatively, a heterocyclic system such as 9 and 10 can be incorporated *ortho* to the amide, thereby restricting the conformation through hydrogen bonding (Fig. 4) [413,418,419].

A combination of the above two features have been utilized in the following tricyclic structures as shown in Fig. 5 [413,418,419].

High degree of homology, with greater than 90% identity in the proximity of the NAD binding site, has led to few selective PARP-1 and PARP-2 inhibitors. Iwashita et al. have reported that the quinazolinone derivative as shown in FR247304 (15) exhibited high selectivity for PARP-1 (IC_{50} 13 nM) over PARP-2 (IC_{50} 500 nM). This compound was cocrystallized with human PARP-1 catalytic domain and docked into the homology model of PARP-2. The preferential activity was ascribed to leu 769 in PARP-1 and gly 314 in PARP-2. This Leu/Gly substitution results in loss of hydrophobicity preventing the side chain of quinazolinone to bind PARP-2 [420].

A series of quinoxaline, as exemplified by compound 16 demonstrated appreciable selectivity for PARP-2 (IC_{50} 8 nM) over PARP-1 (IC_{50} 101 nM) [421]. This was ascribed to the single glu 763/gln 308 mutation resulting in increased side chain mobility in PARP-2, and further leading to optimal interaction with the distal *para*-chlorophenyl moiety of quinoxaline derivatives. Similarly, introduction of benzoyl group in isoquinoline derivatives 17 and 18 (PARP-2: IC_{50} 150 nM, PARP-1: IC_{50} 9 uM), led to PARP-2 selective inhibitors [422]. At present, it is not clear whether selective inhibition of PARP-1 or PARP-2 is a valuable therapeutic strategy for any disease.

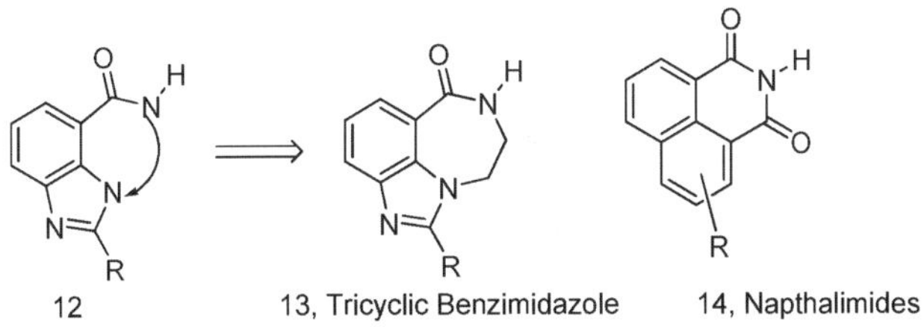

Figure 5. Tricyclic pharmacophores.

Many of these above-mentioned potent compounds suffer from poor solubility, greatly impacting bioavailability, formulations, and preventing them from being dosed intravenously. An improvement in solubility has been achieved in a number of scaffolds by incorporating amine containing side chain, or hydroxyl group containing side chain as shown in Fig. 6.

In many cases, the compounds retained potency while improving solubility in water. For example, FR247304 (**15**) exhibited strong potency (IC_{50} 35 nm), good water solubility (5.6 mg/mL) and a brain to plasma ratio of 5. In addition, this compound exhibited high bioavailability (70%) in rats, making it suitable for stroke studies.

Numerous PARP inhibitors have been evaluated in animal models of stroke [423,424]. This is highlighted with FR247304 in *in vitro* and *in vivo* models of stroke. In cell death model, treatment with FR247304 (**15**) significantly reduced NAD depletion by PARP inhibition and attenuated cell death after hydrogen peroxide (100 uM) exposure. After 90 min

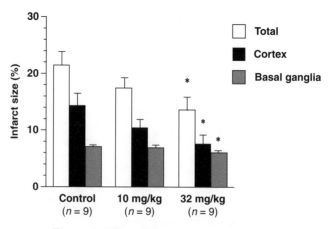

Figure 6. Effect of FR247304 on stroke model.

of middle cerebral artery occlusion in rats, PARP activity and NAD depletion were markedly increased in the cortex and stratium from 1 h after reperfusion. The increased PARP immunoreactivity and NAD depletion were attenuated by 32 mpk i.p. treatment and significantly decreased ischemic brain damage measured at 24 h after reperfusion (Fig. 6).

While there is sufficient evidence in literature that PARP inhibition leads to reduction in infarct size in animal models of stroke, there are few reports on whether a decrease in the infarct size leads to improvement in cognition, memory, and so on. While a number of PARP inhibitors (KU-0059436 (**23**), ABT-888 (**24**), BSI-201, and INO-1001) have entered clinic as an add-on therapy for treatment of breast, pancreatic, melanoma, and ovarian cancer, there are no reports of any PARP inhibitors in stroke related clinical trials [425–428].

23, KU-0059436   **24**, ABT-888

PARP-1 knockout mice, showed an increased mutation rates after challenge with a variety of alkylating agents [429]. Long-term consequence of PARP inhibitors on chronic treatments such as stroke and other neurodegenerative diseases is unknown. Alternatively, PARP inhibitor can be envisioned as a short-term therapy in stroke that may ameliorate toxicity issues associated with chronic PARP inhibitors.

6. HDAC INHIBITORS

The eukaryotic genome is packed into chromatin. The term chromatin refers to the cell's DNA that is wound around core histone proteins to form the basic unit of a nucleosome. Changes in the tertiary packing structures of nucleosomes can regulate gene expression of the underlying DNA. The core histone proteins are enriched in lysine residues that are the target of a variety of posttranslational modifications. These posttranslational modifications result in shifts from more "open" to "compacted" local structures of the chromatin. Histone deacetylases (HDACs) catalyze the removal of acetyl groups from histones resulting in compaction of the local chromatin environment; thus, preventing transcription factor access to the DNA, and effectively, attenuating gene expression from that region of DNA.

Following ischemia/reperfusion, neurons become oxidatively stressed due to overproduction of reactive oxygen species. Oxidative damage is believed to play a role in numerous neurodegenerative disorders. A therapeutically beneficial effect of HDAC inhibitors has been reported in *in vitro* and animal model of neuronal injury.

In cultured primary cortical neurons, HDAC inhibitors protected against oxidation-induced cell death [430]. This protection was shown to be dependent upon transcription factor Sp-1 activity.

Several groups have reported that HDAC inhibition reduces infarct volume in middle cerebral artery occlusion (MCAO) studies in rats [431,432]. Transcriptional profiling studies report an upregulation of a broad array of potentially neuroprotective genes in response to HDAC inhibition. However, as the gene target responsible for the protection seen with HDAC inhibitors is still unknown, it has made it difficult to precisely tune the dosing and timing of HDAC inhibitor that will be most beneficial to ischemia protection.

A large number of reviews have recently appeared in literature on HDAC inhibitors. Fattorri [433], Miyata [434], Ottow [435], Blumcke [436], Monneret [437], and Glaser [438]. Most inhibitors have been designed and evaluated as anticancer agents and the findings reported in those reviews are summarized here in an effort to represent key structural classes and clinical findings. Few cases using rodent MCAO model show that HDAC inhibition suppressed ischemia induced p53 expression as well as neuronal caspase-3 activation and induced expression of HSP70 and bcl-2. They both are known to protect from neuronal death in an ischemic brain. HDAC

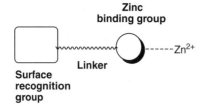

Figure 7. Structural aspect of HDAC inhibitors.

inhibitors were also shown to exhibit anti-inflammatory effects that might contribute to beneficial effects in MCAO model [439].

In general, these classes of compounds have a structure comprising zinc chelating moiety, a linker and an external motif—the so-called "surface recognition motif" (Fig. 7). They are classified based on the zinc chelating moiety. Many of these inhibitors have been primarily evaluated as anticancer agents.

6.1. Hydroxamate as HDAC Inhibitors

Hydroxamates represent the major class of HDAC inhibitors. The only approved drug for HDAC and large numbers of inhibitors in clinical trails belongs to this class of molecules. Depending on the nature of the linker, they can be further subdivided into linear side chain, cinnamoyl and aromatic, heteroaromatic, or based on the zinc binding motif.

6.1.1. Hydroxamate Inhibitors with Linear Side-Chain Linker A well-known inhibitor in this class of compounds is suberoylanilide hydroxamic acid (1, SAHA or vorinostat). SAHA from Merck (Aton) is approved by FDA as a treatment for cutaneous T-cell lymphoma. In stroke studies, treatment of SAHA (50 mpk, i.p.) increased histone-H3 acetylation within the normal brain by approximately eightfold after 6 h and prevented histone deacetylation in ischemic brain [440]. When mice were injected with SAHA at 25 and 50 mpk, there was a 30% reduction in infarct volumes compared to placebo. At higher doses it did not provide significant ischemic neuroprotection. In an effort to improve intrinsic potency, numerous analogs of SAHA with α-substitution on the terminal amide 2–4 have been reported by Merck (Fig. 8) [441].

In addition to amide, a number of other connection units such as oxazole **5** [442], thiazole **7** [443], urea **8** [444], and thiols **9–11** [445] have been evaluated and shown to be potent inhibitors of HDAC (Fig. 9).

6.1.2. Hydroxamate Inhibitors with Cinnamoyl Linker Replacement of the linear linker by cinnamoyl was shown to be beneficial to the inhibitory activity of HDAC. Saturation of the C=C bond of cinnamic moiety significantly reduced HDAC inhibition. Many compounds in this series have entered clinical trials for various cancer treatments.

1: SAHA
IC$_{50}$ = 48 nM

2
IC$_{50}$ = 12.1 nM

3
IC$_{50}$ = 13 nM

4
IC$_{50}$ = 36 nM

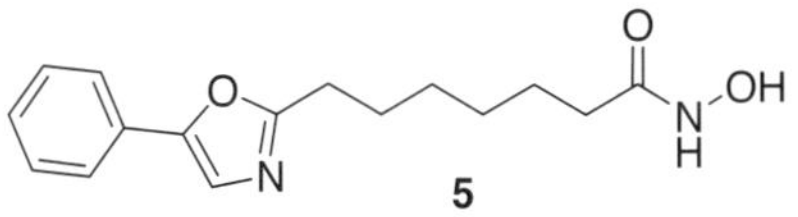

5
IC$_{50}$ = 10 nM

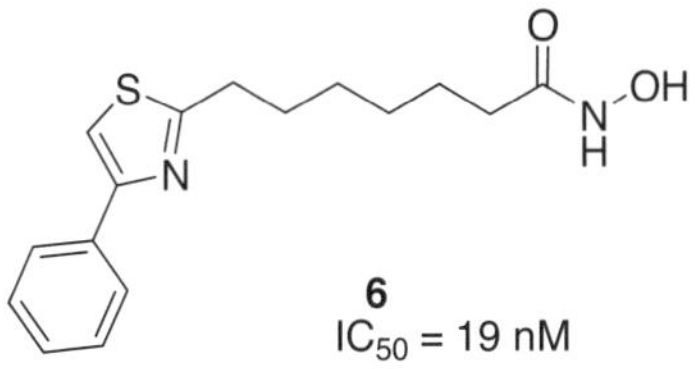

6
IC$_{50}$ = 19 nM

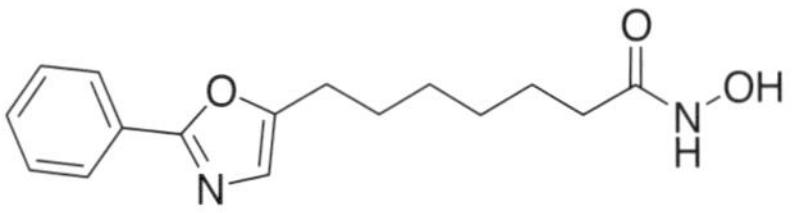

7
IC$_{50}$ = 5.4 nM

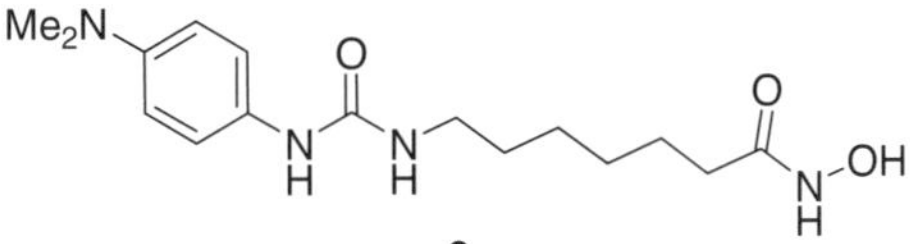

8
IC$_{50}$ = 800 nM

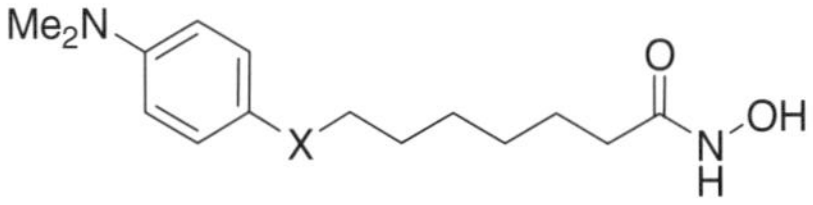

9: S, IC$_{50}$ = 120 nM
10: SO, IC$_{50}$ = 60 nM
11: SO$_2$, IC$_{50}$ = 40 nM

Novartis claimed aminomethyl-substituted cinnamoyl hydroxamates as inhibitors of HDAC. LBH-589 (**12**) and LAQ-824 (**13**) from this series have entered clinical trials for solid tumors. LAQ-824 was discontinued in Phase II, while LBH-589 has entered Phase III clinical trial for malignant diseases [446]. Similarly, TopoTarget (Prolifix) reported the design and synthesis of a new class of HDAC inhibitors wherein reverse and forward sulfonamides were explored as a connecting unit [447]. Based on this work, Curagen and TopoTarget have initiated Phase II clinical trials for Belinostat (**14**, PXD101, belinostat) in patients with advanced multiple myeloma [448]. SBio has utilized benzimidazole derivatives as cinnamoyl hydroxamic acid analogs. These compounds showed excellent HDAC potency and anticancer activity. Based on these data, SB-939 (**15**) has entered Phase I clinical trial for solid tumors [449].

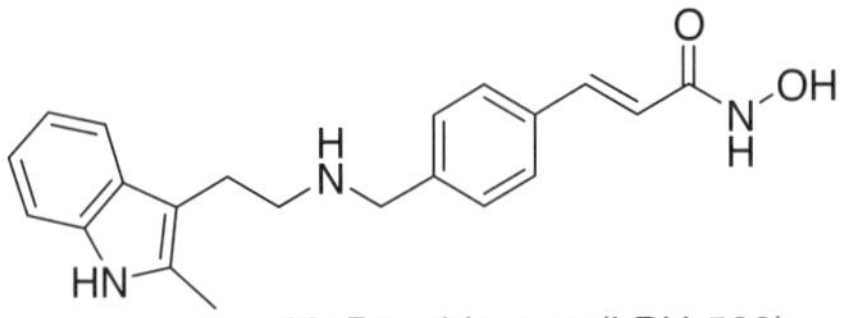

12: Panobinostat (LBH-589)
Phase III (Novartis)

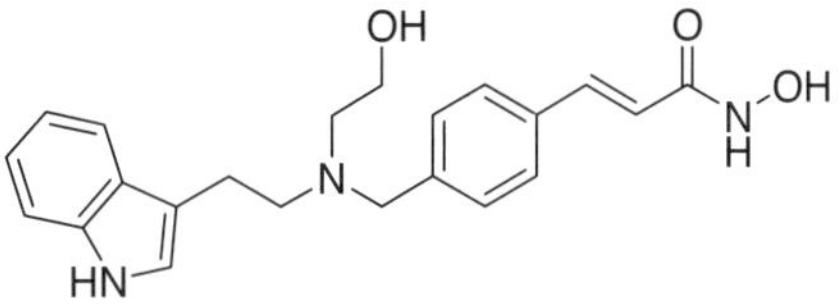

13: LAQ 824
Phase II discontinued (Novartis)

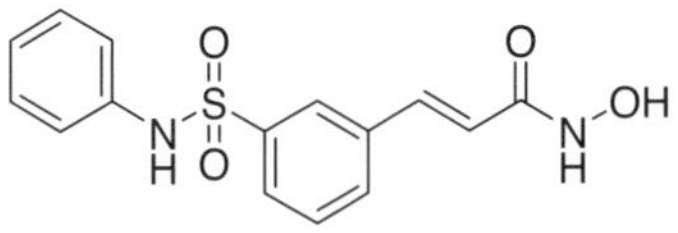

14: Belinostat (PXD 101)
Phase III (Topotarget)

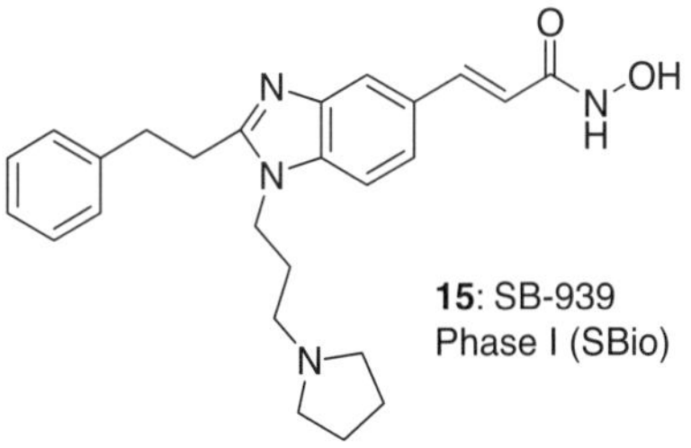

15: SB-939
Phase I (SBio)

Fujisawa has claimed IP on a large set of cinnamic-type hydroxamates (**16**) with benzimidazole as hydrophobic and methyl linker as a connection unit [450]. Takeda (Syrrix) has identified closely related benzimidazole derivatives (**17**) as inhibitors of HDACs [451]. SK chemicals and In2Gen Co., Ltd. have reported a series of cinnamoyl derivatives where the connecting unit is a benzylamide (**18**) [452]. Many compounds in this series were potent against various cancer cell lines and efficacious in *in vivo* assay. Due to these encouraging results, a number of compounds were selected as preclinical candidates. Celera Genomics has identified hydroxyethylamide as a linker with various heterocycles as hydrophobic units [453]. These compounds demonstrated excellent *in vitro* potency and were efficacious in animal models. From these studies, CRA024781 was selected as a clinical candidate and is currently in Phase I. Scientists from Myachi & Co., Japan have identified a series of cyclic amide/imide bearing hydroxamic acids (**20** and **21**) as HDAC inhibitors [454,455].

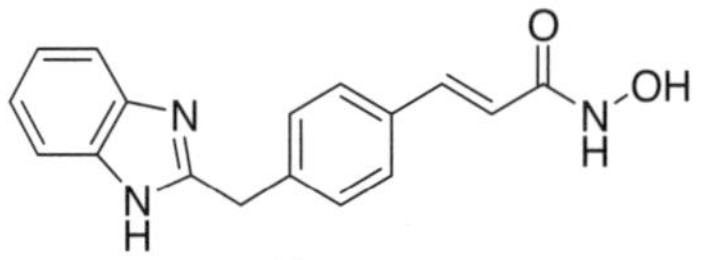

16
IC_{50} = 280 nM

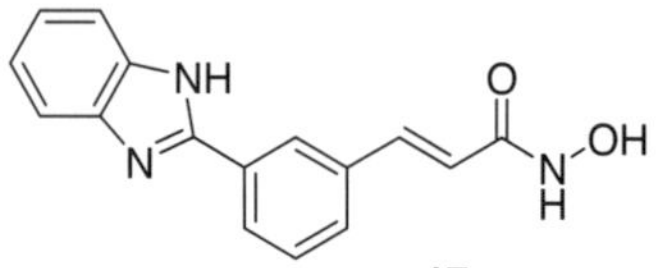

17
IC_{50} = 80 nM

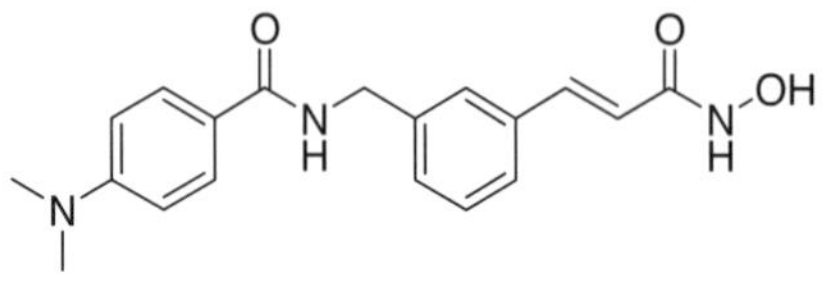

18
IC_{50} = 53 nM

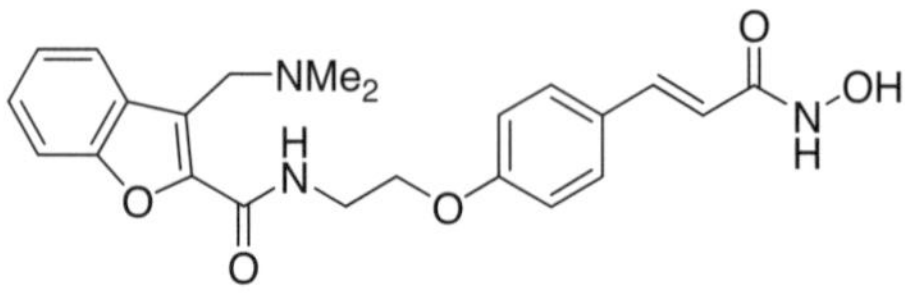

19: CRA024781, IC_{50} = 7 nM
Phase 1 (Celera Genomics)

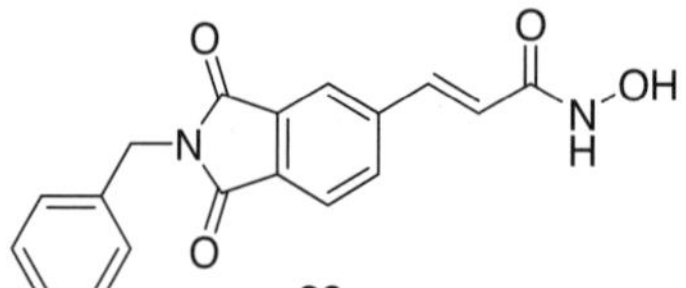

20
IC_{50} = 150 nM

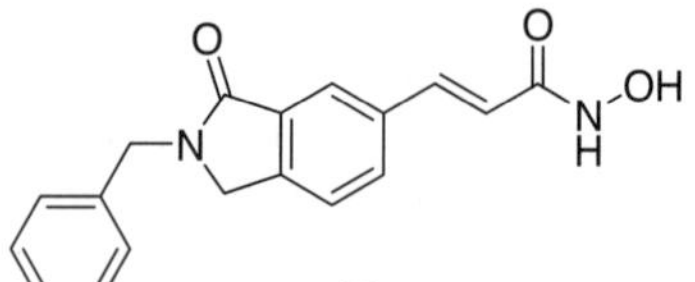

21
IC_{50} = 38 nM

6.1.3. Hydroxamate Inhibitors with Aryl and Heteroaryl Linker

A number of HDAC inhibitors with aryl and heteroaryl moiety as linker units have appeared in literature and patents. Italfarmaco has identified phenyl hydroxamates with carbamate derivatives as linkers and napthyl as a hydrophobic capping unit. ITF-2357 (**22**) from Italfarmaco is currently in Phase II and being evaluated for solid tumors [456]. Merck has shown a simple phenyl ring (**23**) as a linker along with amide and sulfonamide as connecting units. The capping hydrophobic units were derived from a large number of aromatic, heteroaromatic, biaryl, and napthyl systems. This series led to a number of compounds with *in vitro* potency in cellular assays [457].

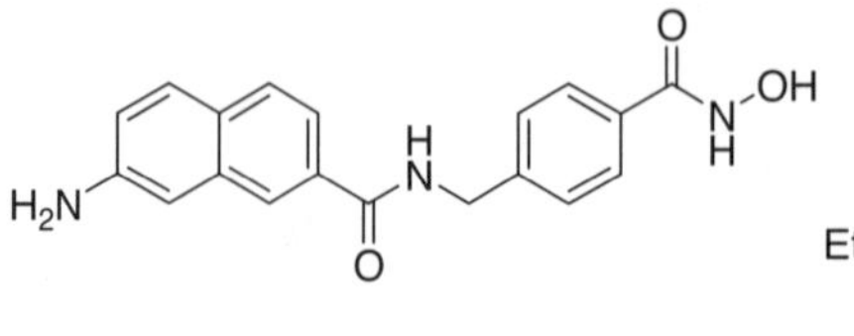

22

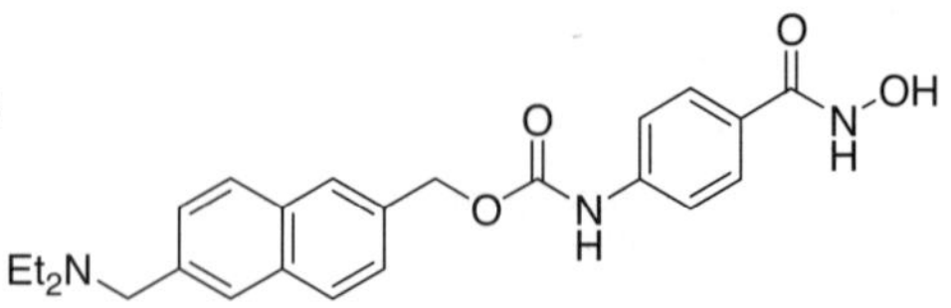

23: ITF-2357
Phase II (Italfarmaco)

Janssen Pharmaceuticals has many patents wherein phenyl is replaced by pyrimidine as a linker. Currently, Janssen is evaluating R306465 (**24**, JNJ-16241199) in Phase I for safety and human pharmacokinetic properties. The thiophene scaffold represented by **25** was initially reported by Roche and represented a novel class of compounds as HDAC inhibitors [458]. Argenta Discovery published similar structures, where thiophene was linked to various other heterocycles such as pyrazole, pyrimidine, and so on (**26**) [459]. SBio found that incorporating a methylene unit between thiophene and hydroxamic acid yielded selective HDAC-8 inhibitors as shown in **27** [460].

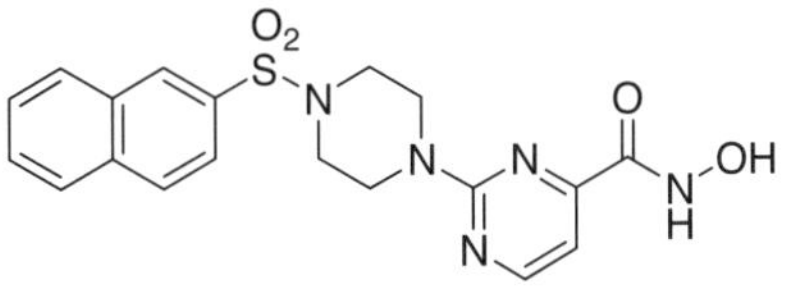

24: (R306465, JNJ-16241199) Phase I (J&J)

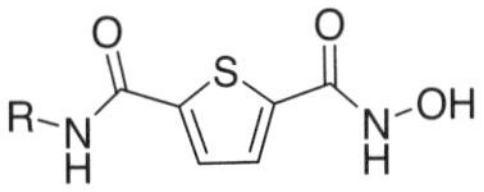

25

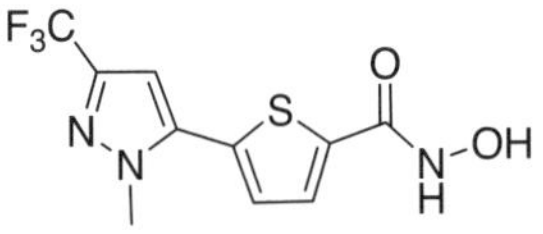

26 IC_{50} = 750 nM

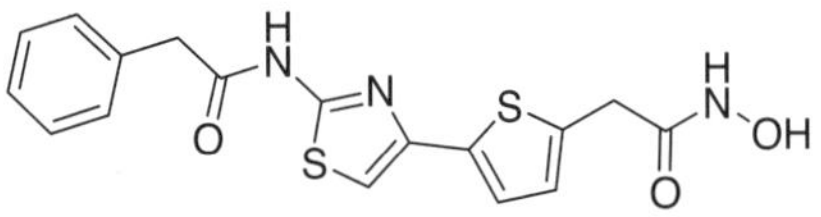

27

6.2. Thiols as HDAC Inhibitors

Thiols are well-known inhibitors of zinc-dependent enzymes such as ACE and MMP. First reports of thiols **28** as inhibitors of HDACs were reported by Mitsubishi Pharmaceuticals [461]. Replacement of hydroxamic acid by thiol in SAHA led to equipotent HDAC inhibitors. However, these compounds were weak inhibitors on human lung cancer cell line NCI-H460. The capping of sulfur as its isobutyryl derivative (**29**), and replacing phenyl by phenylthiazole yielded compound NC-51 (**30**) with similar potency as SAHA. Gu & Co. scientists identified α-mercaptoketones and showed that they were more potent than the corresponding free thiols. The α-isobutyryl derivatives of these mercaptoketones (**31**) demonstrated excellent cellular activity [462].

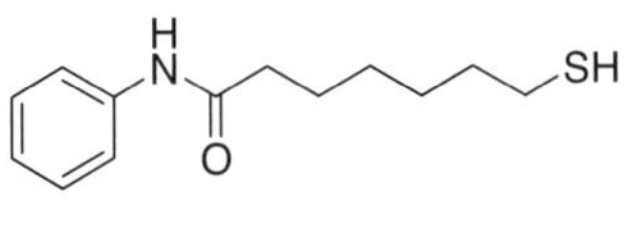

28: HDAC IC_{50} = 0.21
EC_{50} > 50 uM

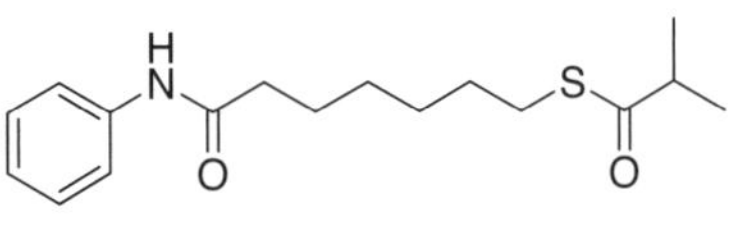

29: EC_{50} = 20 uM

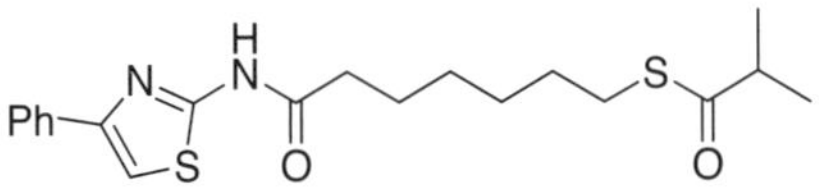

30: EC_{50} = 2.1 uM

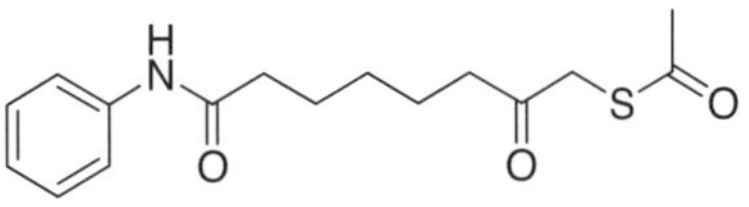

31: HDAC IC_{50} = 0.081
EC_{50} = 0.19

6.3. 2-Aminophenylamides as HDAC Inhibitors

2-Phenylaminoamides is formally considered as the zinc chelating group although there are no crystal structures in any of the zinc binding enzymes. The simplest version of this scaffold has been evaluated by Pfizer and CI-994 identified (**32**) as a clinical candidate [463]. This has demonstrated significant antitumor activity against a broad spectrum of murine, rat, and human tumor models. Currently in Phase II, this compound is being evaluated in combination with other anticancer agents such as carboplatin, gemcitabine, and so on.

Compound CI-994 induces adverse events such as thrombocytophenia, anemia, and neutropenia in patients. Glucuronide prodrugs have been evaluated to improve solubility and increase its therapeutic window [464].

Methylgene has identified MGCD-0103 (**33**) as a clinical candidate in this class of compounds. It selectively inhibits HDAC 1, 2, 3, and 11 with an IC_{50} value of 0.1, 0.2, 2.0, and 20 uM, respectively. Tumor growth was dose-dependently inhibited by MGCD-0103 (15 and 30 mpk, i.p.) in A549 xenografts models by 53% and 83%, respectively. Introducing cinnamoyl linker with *ortho*-phenyldiamine led to potent 2-phenylaminoamides as HDAC inhibitors. Furthermore, introduction of *para*-aromatic or *para*-heteroaromatic substitutions on the aniline ring improved HDAC inhibitory potency significantly as shown for **34**. Low solubility of this compound might explain its low cellular activity in various cellular assays. However, compound **34** had a tumor volume inhibition of 78% when administered to mice bearing xenografts of the HCT-116 tumor cell line [465].

Mitsui Pharmaceuticals identified a series of synthetic benazmides as shown in MS-275 (**35**), currently in Phase II clinical trials [466]. This compound displayed HDAC potency of 4.8 uM, albeit with much lower potency than the earlier class of compounds. However, this compound demonstrated to be antiproliferative in several cancer cell lines and had significant *in vivo* activity against solid tumors and lymphomas. In addition, MS-275, at lower doses (15 umol/kg), increases the content of acetyl-histone (Ac-H3) in the frontal cortex. At 60 umol/kg, it inhibits HDAC at the frontal cortex and hippocampus. However, no significant inhibition is seen at striatum even at the highest doses (120 umol/kg). These results suggest that MS-275 is a potent brain-selective HDAC inhibitor and may be efficacious in the treatment of epigenetically induced psychiatric disorders [467].

Merck has recently published a systemic study on SARs, isoform selectivity and PK data on a series of 2-aminophenylamides. In this series, **36** was found to be inactive on HDAC-6 and -8 ($IC_{50} > 10$ uM), almost 10-fold selective for HDAC-1 (36 nM) over HDAC-2 ($IC_{50} = 300$ nM) and 20-fold selective over HDAC-3 ($IC_{50} = 697$ nM). This compound had excellent bioavailability in rats and dogs, but was a little low on rhesus [468]. Plasma clearance was low on all species. There are a large number of reports by various pharmaceutical companies in this class as shown in **37–42**.

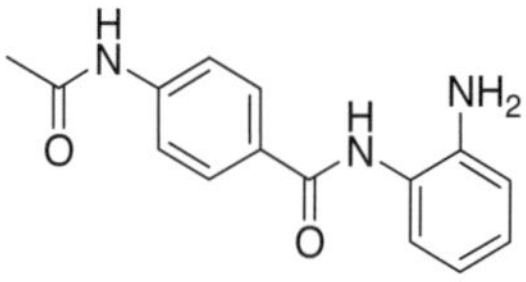

32: CI-994
Phase II (Pfizer)

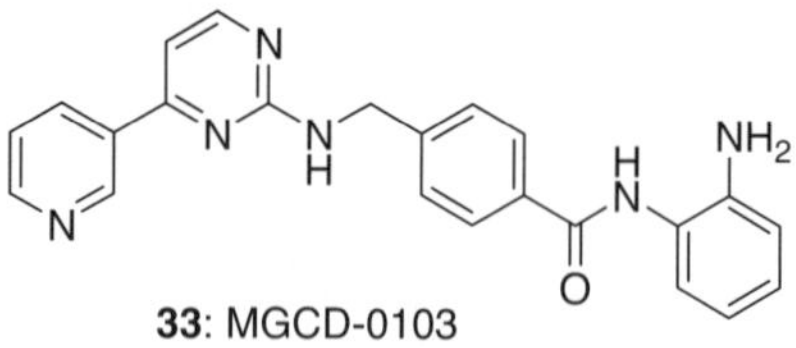

33: MGCD-0103
Phase II (Methylgene)

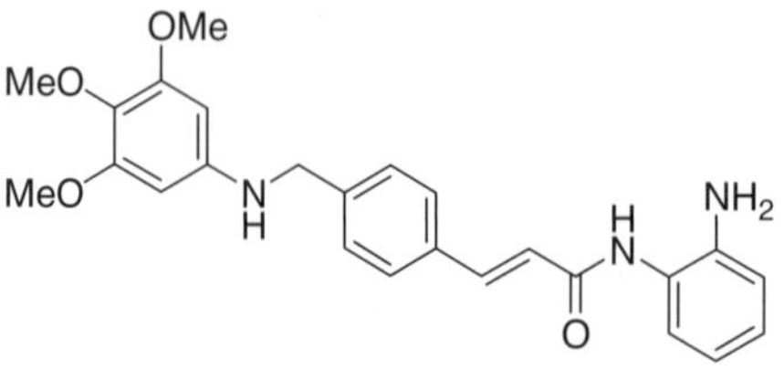

34

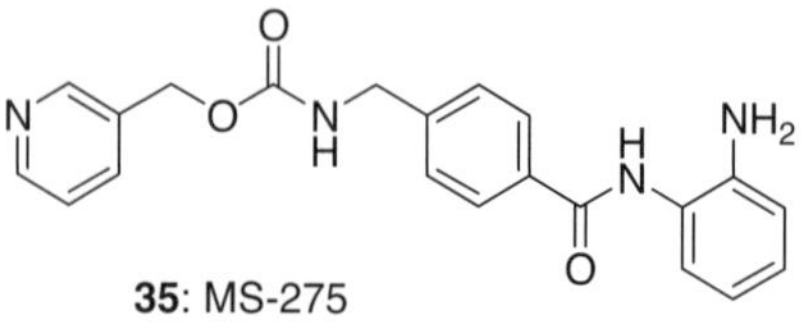

35: MS-275
Phase II (Bayer/Syndex)

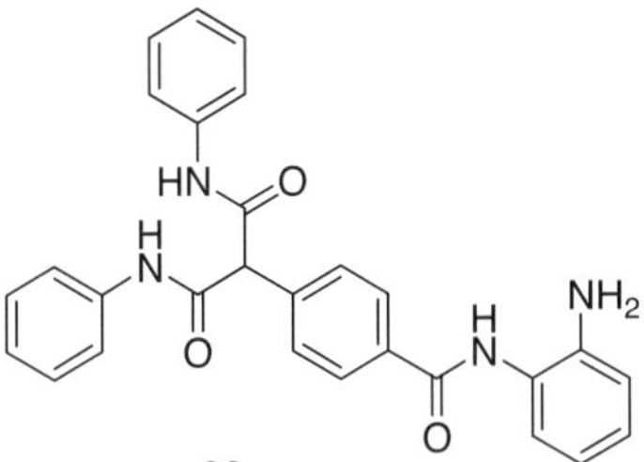

36

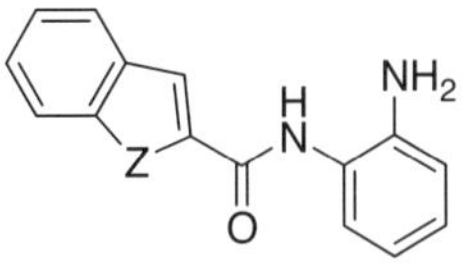

37: (Z= O, NR, S etc)

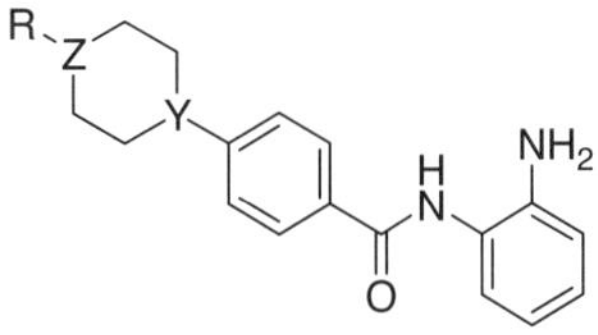

38: (Y= Z= CH, N)

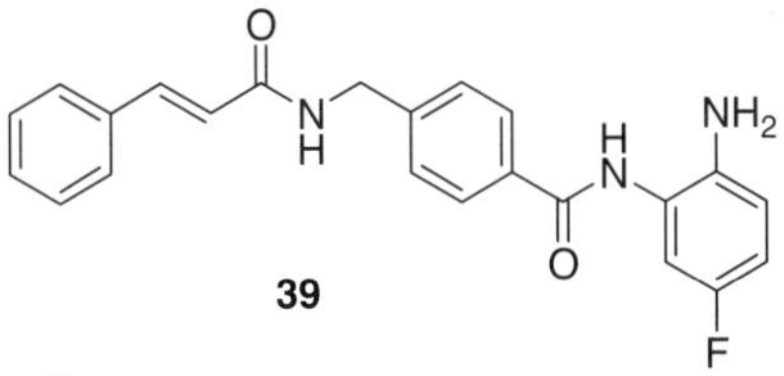

39

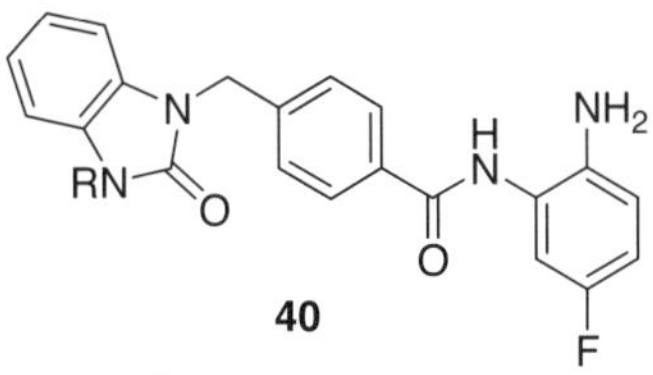

40

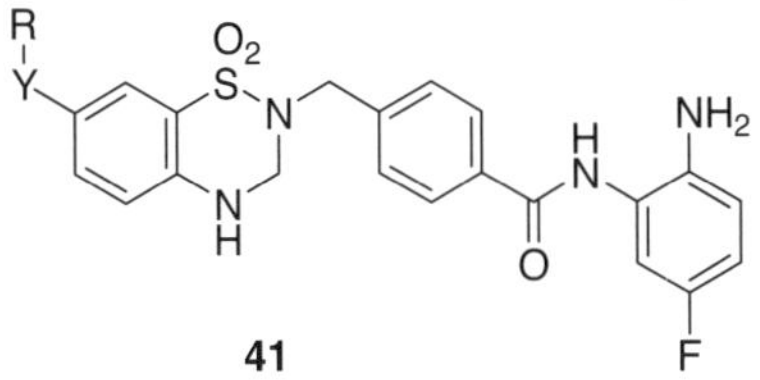

41

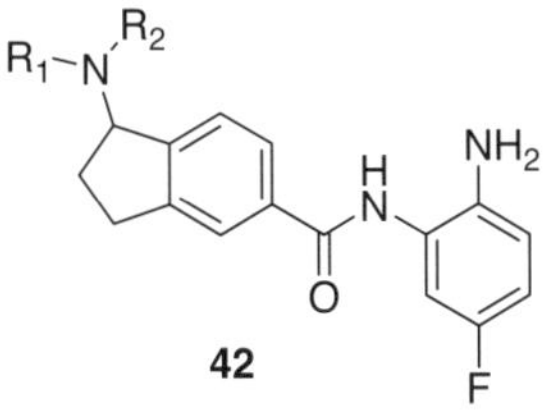

42

6.4. Ketones as HDAC Inhibitors

Ketones including trifluoromethyl ketones, [470] α-ketoamides [470], α-ketoesters, and α-keto heterocycles [471] have been reported as HDAC inhibitors. In all these cases, SARs correlated well with hydroxamates and indicated a similar binding mode. However, it is well known that a number of ketone moieties are readily reduced to the corresponding alcohol *in vivo*. To date, there are no clinical candidates from this class of molecules.

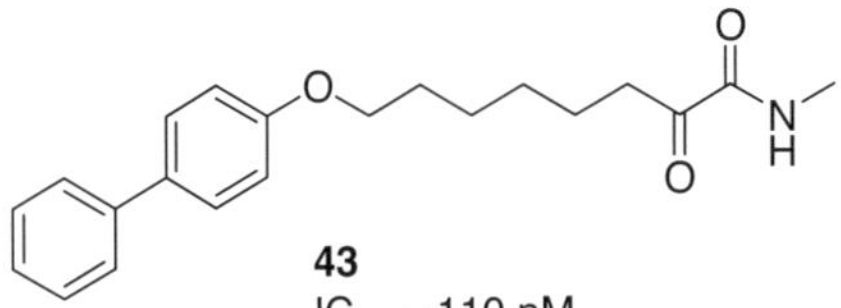

43
IC_{50} = 110 nM

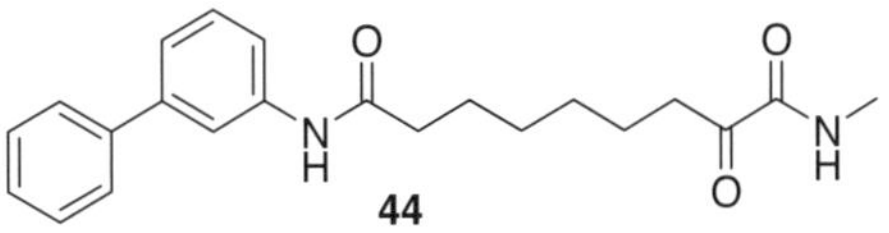

44
IC_{50} = 3.7 nM

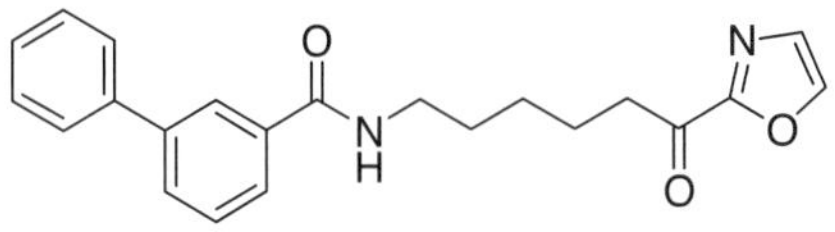

44
IC_{50} = 60 nM

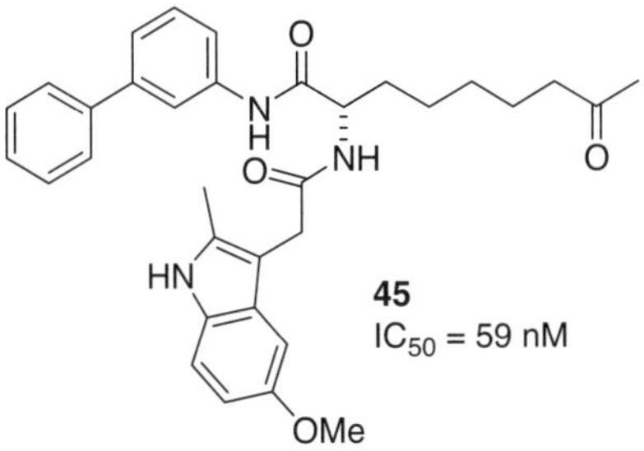

45
IC_{50} = 59 nM

6.5. Boronates and Silanediols as HDAC Inhibitors

Recently, the catalytic mechanism of deacetylation of the acetylated lysine residues by HDAC was proposed. Boronates and silanediols have been designed as transition state mimics of HDAC inhibitors. In the case of boronic acid, **46** is threefold less potent than SAHA in enzyme assay but is equipotent in number of cellular assays [472]. This is attributed to the difference in log D between SAHA (-1.6) and boronate (1.0). Log D calculations suggest that boronic acid is more lipophilic than hydroxamate and may permeate through cell membrane more efficiently. Furthermore, utilizing western blot analysis on HCT116 cell lines, it was shown that these compounds inhibit HDAC activity by elevating acetylated histone in a dose-dependent fashion. Similarly, silanediols have been evaluated as transition state inhibitors of HDACs. However, no biological data has been reported for silanediols by Axys Pharmaceuticals [473].

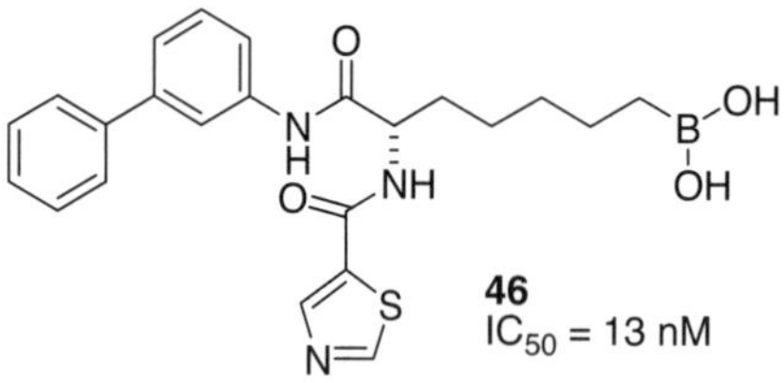

46
IC$_{50}$ = 13 nM

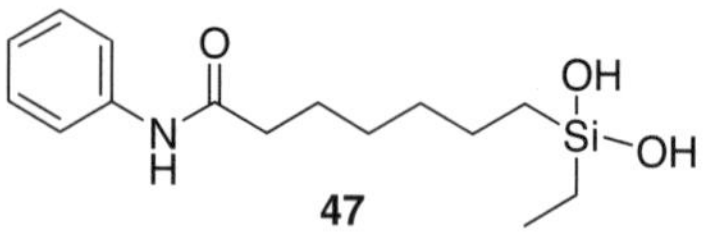

47

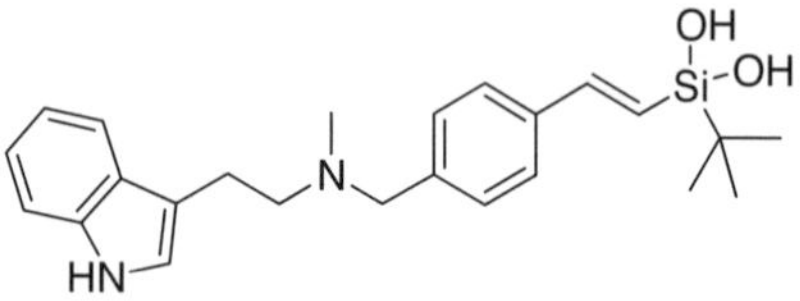

48

6.6. Carboxylic Acids as HDAC Inhibitors

Valproic acid, a drug commonly used to treat seizure and bipolar mood disorders, has been shown to have neuroprotective properties at therapeutic levels in cellular and animal models. Rats were treated with 300 mpk VPA by s.c. injection every 12 h after the onset of insult in rat permanent transient middle occlusion artery (pMCAO) model [474]. VPA treatment decreased infarct size by approximately 37% under these conditions. It was also shown that VPA at this dose inhibits HDAC activity by a threefold increase in H3 acetylation in the intact contra lateral brain hemisphere. Strikingly, delayed administration of VPA after 3 or 6 h onset of stroke still reduced infarct volumes and improved neurologic scores suggesting a comparatively long therapeutic window for potential stroke treatment. At 24 h after ischemic onset, plasma concentration of valproic acid was in the lower range of therapeutic plasma levels used to treat bipolar disorder. Similar results have been demonstrated for butyric acid or sodium butyrate [474].

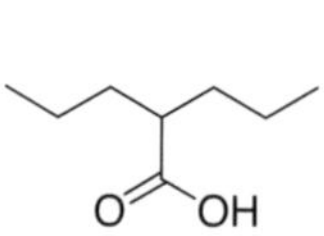

49: Valproic acid
Phase II

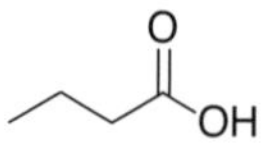

50: Butyric acid

6.7. Miscellaneous HDAC Inhibitors

A number of novel HDAC inhibitors have been identified as such from high-throughput screens as shown in **51** and **52** [475]. Natural product cyclostellettamines and its closely related analogs were identified through screening of natural product extracts from marine sponge [476]. Further, SAR study led to identification of bispyridinium diene **53** as HDAC-1 selective inhibitors. Similarly, trapoxin **54** and its analog were identified through screening of natural product extracts for antiproliferative activity [477]. Trapoxin are believed to be irreversible inhibitors of HDAC [478].

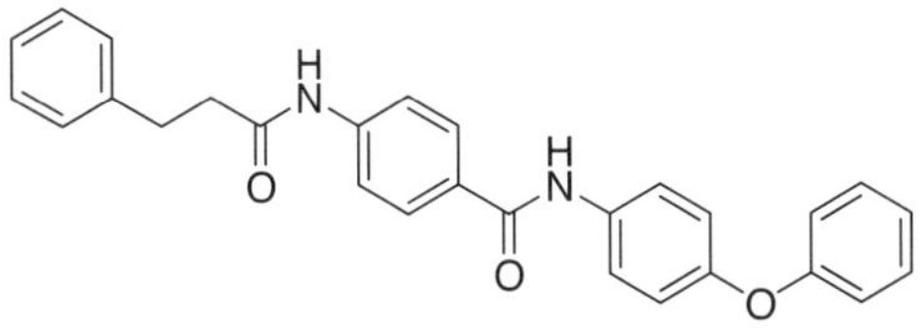

51: SB-429201

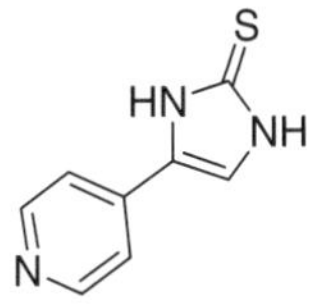

52: SB-379278A

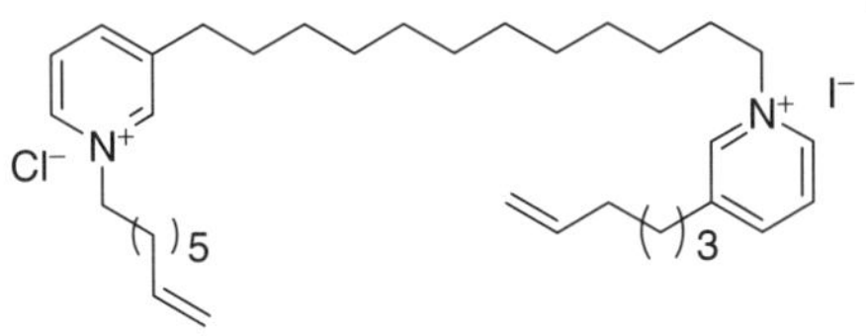

53

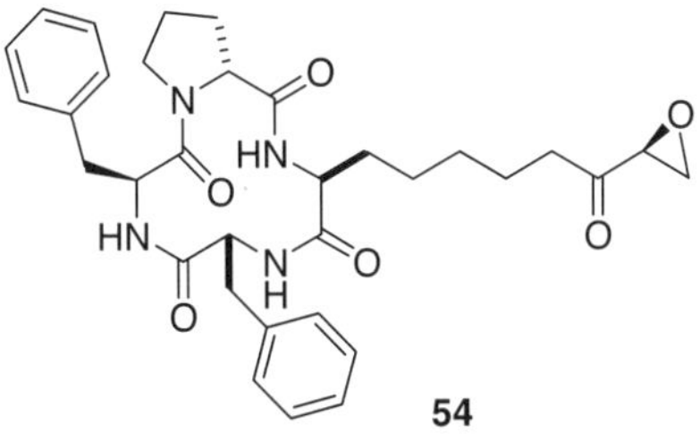

54

7. INOS/NNOS INHIBITORS

NO is a physiological messenger in the central nervous system [479] and is synthesized by the NO synthase (NOS)-catalyzed reaction. Activation of this enzyme forms NO and L-citrulline from L-arginine; thus, participating in the transduction pathway leading to elevations in intracellular cyclic GMP levels. While the exact role of NO production in the brain remains to be fully elucidated, it seems that this free radical is involved in important functions, such as the regulation of cerebral blood flow. Three isoforms of NOS have been characterized in brain cells. Neurons produce NO mainly by Ca^{2+}-dependent activation of neuronal NOS (nNOS), which is constitutively expressed in these cells. Glial cells (astrocytes, microglia, and oligodendrocytes) synthesize NO mainly after calcium-independent inducible NOS expression (iNOS). Finally, endothelial cells produce NO by the constitutive, Ca^{2+}-dependent activity of endothelial NOS (eNOS). The mechanism through which excess NO formation within the brain leads to cell death has been reported to involve energy depletion, lipid and protein peroxidation, protein nitrosylation, and DNA damage [480]. However, while NO has been shown to cause some cell damage *in vitro*, increasing evidence suggests that the endogenous formation of peroxynitrite anion (ONOO3), from the reaction of NO with superoxide ion (O3-) [481], may be a possible mechanism through which neurotoxicity is induced. The relevance of these mechanisms for NO-mediated brain damage in hypoxia-ischemia has been tested, preclinically via the creation of small-molecule NOS inhibitors.

Inhibitors of NOS and their use in the field of stroke and inflammation have been reviewed recently [482–495]. Some of the findings related to NOS inhibition (more specifically nNOS and iNOS) described in these reviews have been summarized here, along with recently published work. Due to the availability of large number of publications regarding NOS inhibition, this section will focus on structural diversity and applications to the stroke indication.

7.1. L-Arginine Analogs

L-Arginine being the substrate of NOS oxidation at the guanidine moiety, early inhibitors were substituted guanidine derivatives of L-arginine. Inhibitors such as l-NNA (**1**), L-NMA (**2**), and L-NAME (**31**) prevent the binding of L-arginine to the binding site on NOS and as such are potent but show little selectivity for any NOS isoform. Preclinical evaluation of these early inhibitors in experimental models of stroke has been contradictory. L-NNA was reported to have beneficial effects on infarct volume, neuroprotection, cerebral eodema, and neuronal death in some studies and deleterious effects on infarct volume, blood pressure, and cerebral blood flow in other studies. Similar contradictory results were reported with L-NAME [482]. Human clinical studies with L-NMA suggested systemic vasoconstriction [486,487] and dose-dependent reduction of cerebral blood flow [488]. Amidine analogs derived from ornithine (**4**, L-NIO) or lysine (**5**, L-NIL) are also well-known inhibitors that display some selectivity toward iNOS inhibition [489,490]. It was therefore concluded that these nonspecific inhibitors of NOS have limited utility in stroke neuroprotection.

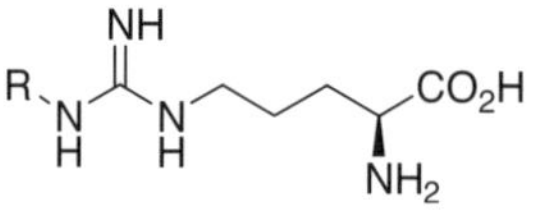

R = NO$_2$: **1**: L-NNA
R = Me: **2**: L-NMA

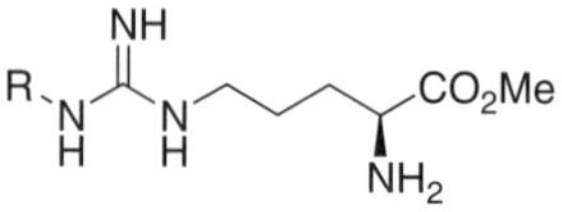

R = NO$_2$: **3**: L-NAME

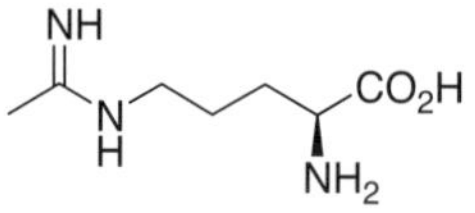

4: L-NIO

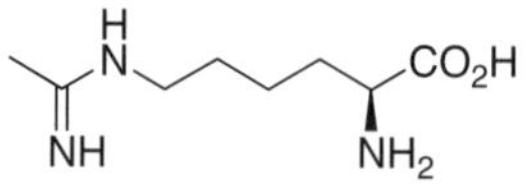

5: L-NIL

Numerous groups have explored structural modifications of these early inhibitors with the common goal to improve isoform selectivity. Conformational restriction of the aminoacid side chain with an aryl group (**6** and **7**) only marginally improves selectivity [491,492]. However, such conformational restriction, when utilized with an amidine tail end is more successful: W1400 (**8**) is 50-fold selective for iNOS versus eNOS [493]. Although W1400 displayed promising results in experimental stroke models, its toxicity at higher dose precluded further development [494]. N-Benzylacetamidine, a simplified analog of W1400, has also been described recently as a potent and selective iNOS inhibitor [495]. L-NIL acetamidine derivatives that incorporate a thiol (**9**, GW274150) or sulfone (**10**, GW273629) in the side chain have been demonstrated to be iNOS selective as well but their time-dependent mode of inhibition also impeded their potential for development [496,497].

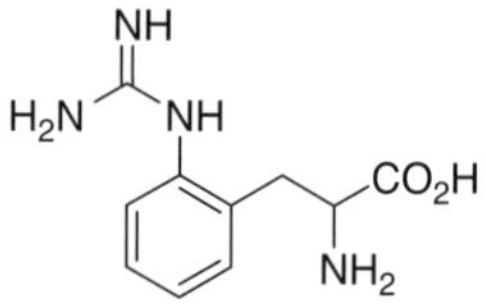

6

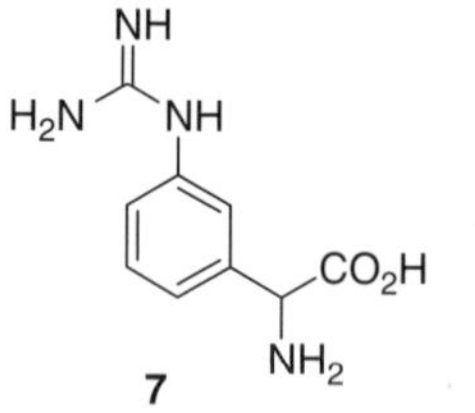

7

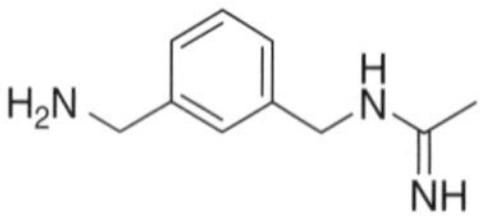

8: W1400

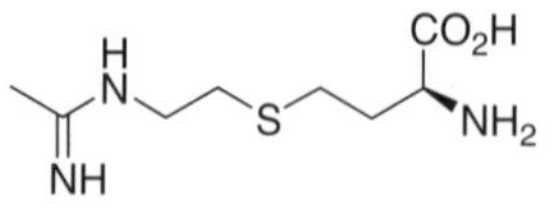

9: GW274150

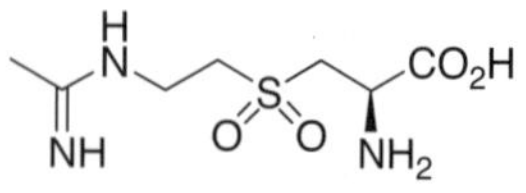

10: GW273629

Based on the nNOS over iNOS selectivity of L-nitroarginine, a large number of dipeptide analogs or peptidomimetic analogs have been evaluated as selective nNOS inhibitors. A few representative examples are illustrated below. The aromatic reduced peptidomimetic derivative **15** demonstrated some efficacy in a rabbit model of fetal neurodegeneration [498].

12

11

13

14

15

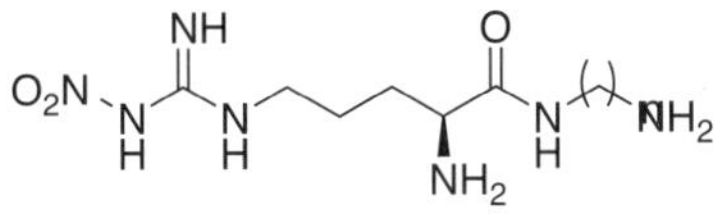

7.2. Guanidine and Isothiourea Analogs

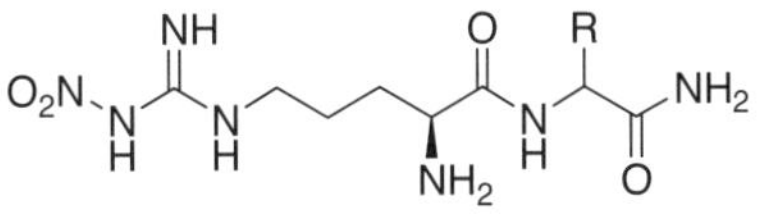

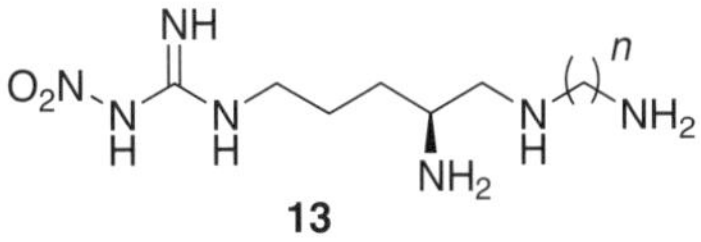

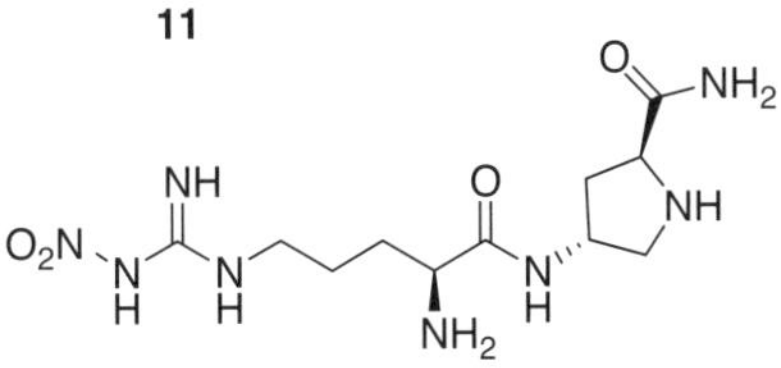

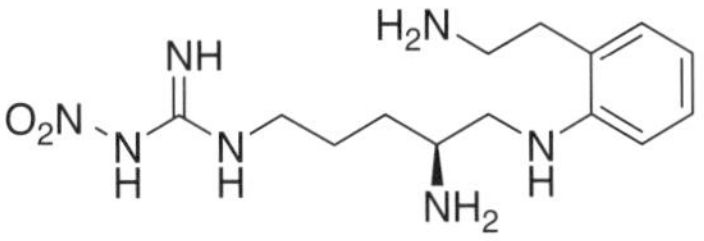

Aminoguanidine (**16**, pimagedine) is a selective inhibitor of iNOS over eNOS and slightly over nNOS [599]. It has shown some efficacy in animal models of stroke and is well tolerated in human [500–503]. The aminoguanidine moiety has recently been incorporated into an arginine template and the resulting inhibitor (**17**) has been described as iNOS selective without irreversible inactivation of the enzyme [504]. This finding reflects a general trend among NOS inhibitors. The smallest amidine, guanidine, and isothiourea derivatives are often the most potent and least selective versus various isoforms while the larger, more complex analogs display increased selectivity and attenuated potency. The same is observed with isothioureas analogs such as **18** and **19** that have proved too toxic for *in vivo* evaluation [505]. Cyclic isothiourea analogs (**20**), along with oxo and selenium derivatives (**21**) have also been reported on and shown to provide marginal selectivity between isoforms [506–508].

16

17

18

19

20

21
X = S, O, Se

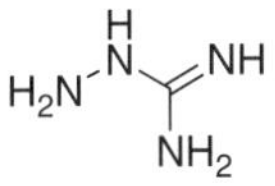

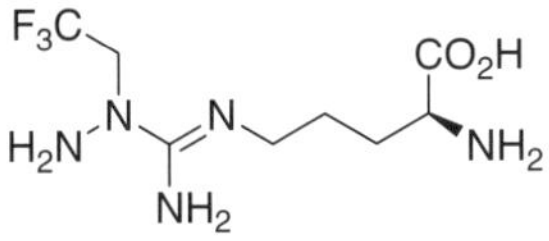

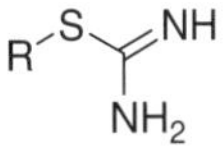

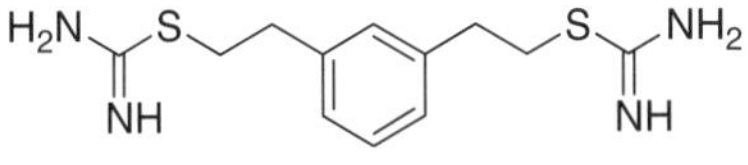

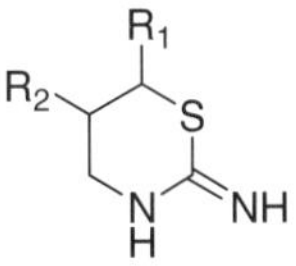

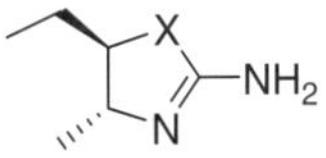

7.3. Amidines

Heterocyclic amidines such as **22** are potent NOS inhibitors although typically displaying poor selectivity among isoforms [499,509]. W1400 (**8**) represents a slightly more evolved amidine and was described earlier as a selective iNOS inhibitor. Cyclic amidines such as **22** and **23** have received ample attention resulting in potent inhibitors but with marginal selectivity.

22

23
$n = 1–3$

24
$n = 1,2$

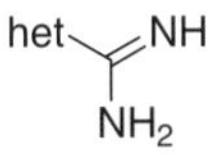

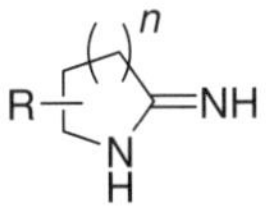

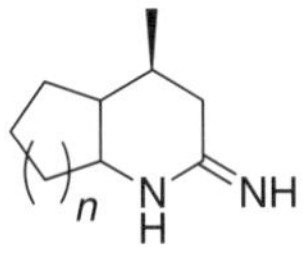

Incorporation of cyclic amidine moiety into fused bicyclic systems led to the discovery of dihydroquinolines such as **25** as potent and somewhat selective iNOS inhibitors [510]. Optimization of this series resulted in the identification of AR-C102222 (**26**), an orally bioavailable, selective, iNOS inhibitor that displays efficacy in animal models of inflammation and arthritis [511,512].

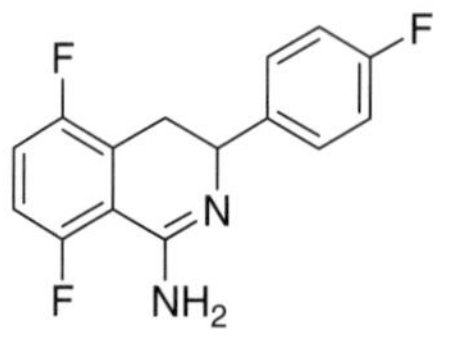

Further extension from the amino group of heterocyclic amidines **22** or elaboration of the early methyl acetamidine derivatives has led different research groups to the discovery of selective nNOS inhibitors based on a 2-thienylcarbamidine scaffold. AR-R17477 (**27**) is a potent and selective nNOS inhibitor demonstrating significant reduction of infarct volume in animal model of stroke at 1 mg/kg, but not at higher or lower doses making the preclinical data difficult to interpret [513,514]. Recent modifications in this series are illustrated by inhibitor **28** [515]. Further development of the previous 2-thienylcarbamidine scaffold led another group of researcher to design a dual acting inhibitor. BN 80933 (**29**) is the result of appending an antioxidant fragment to the nNOS selective pharmacophore displayed by R-R17477. This nNOS selective inhibitor displayed beneficial synergistic efficacy on infarct volume and neurological outcome, at up to 8 h postischemic insult [516,517].

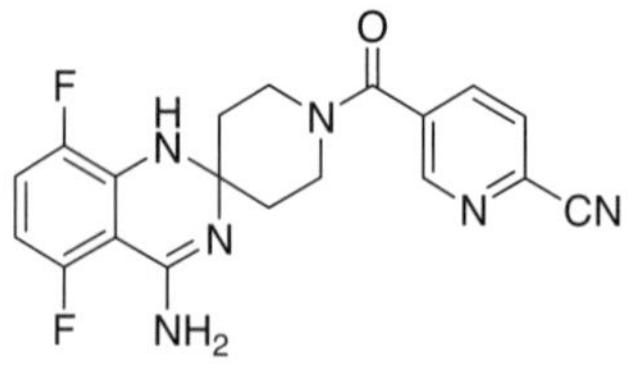

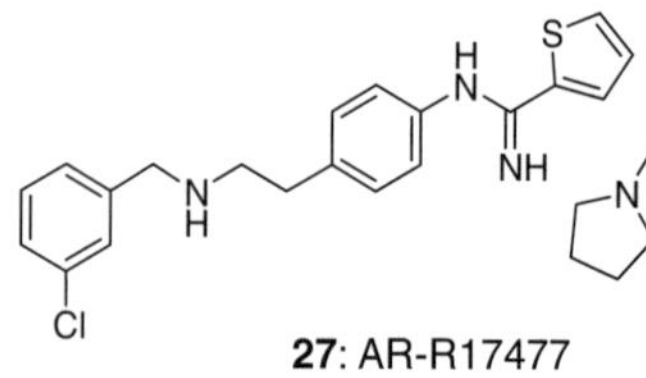

25

26: AR-C102222

27: AR-R17477

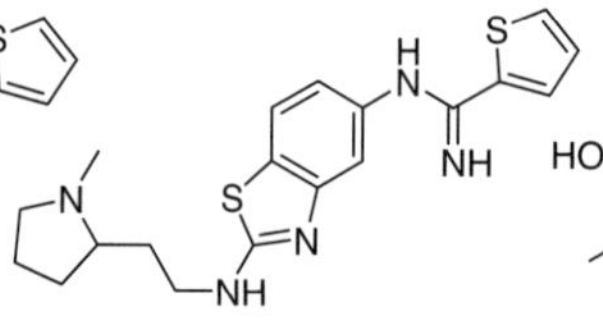

28

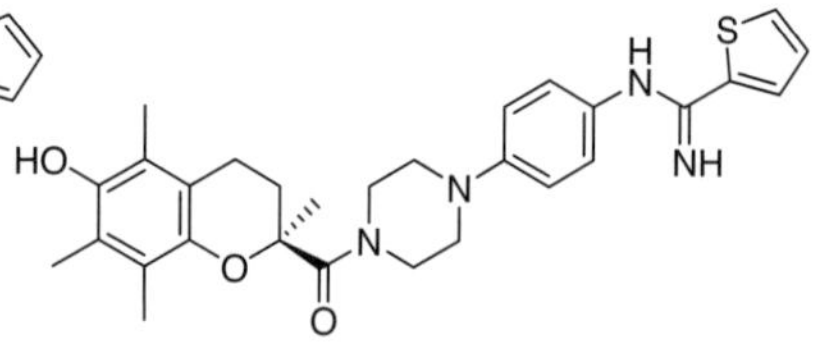

29: BN 80933

7.4. Aminopyridines

2-Aminopyridine derivatives represent the logical extension from cyclic amidines to their aromatized analogs, and as such, have received ample attention. 2-Amino-4-methylpyridine (**30**) is one of the simplest representatives and displays good potency but modest potency [518,519]. Further studies suggest that larger substituents are only tolerated at the 6-position but still do not confer selectivity [520]. Despite the marginal selectivity obtained with inhibitor **31**, its ^{18}F isotope has recently been prepared as a potential PET ligand to image iNOS activation [521]. Application of the aminopyridine pharmacophore to the L-NIL template has led to the preparation of potent, but not selective NOS inhibitors such as **32** [522]. More elaborated aminopyridines such as **33** and **34** were found to be potent and display some selectivity toward nNOS [523–525]. Although 2-amino group substitution is usually not tolerated, derivatives such as AR-C133057 (**35**) were shown to be potent and selective toward iNOS [526].

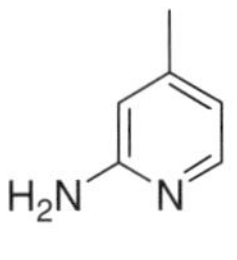

Recent exploration of the 6-substituent using a "fragment hopping" strategy yielded potent nNOS selective aminopyridine inhibitors such as **36** and **37**, which demonstrate activity in a rabbit model for the prevention of neurobehavioral symptoms of cerebral palsy. The approach is centered around the derivation of the minimal pharmacophoric element for each pharmacophore [498,527–529].

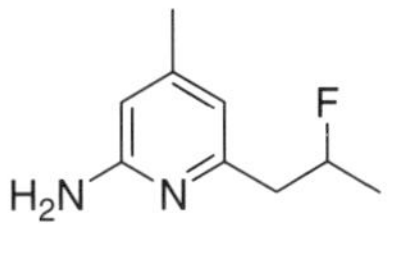

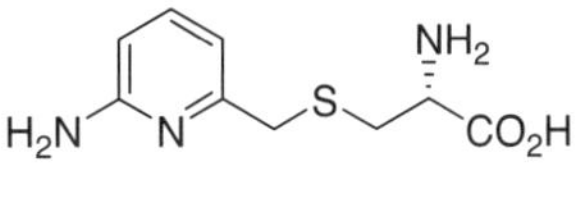

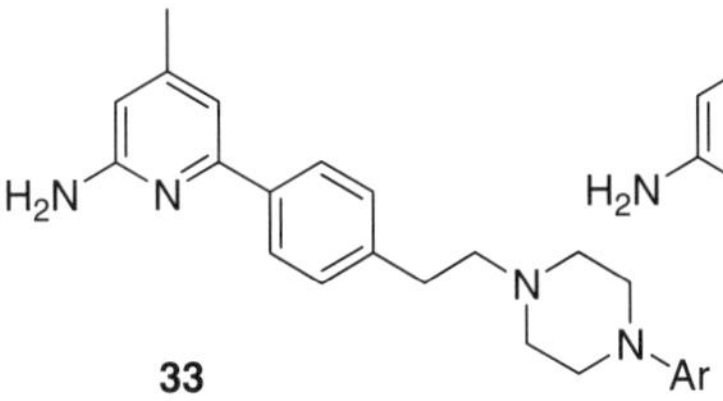

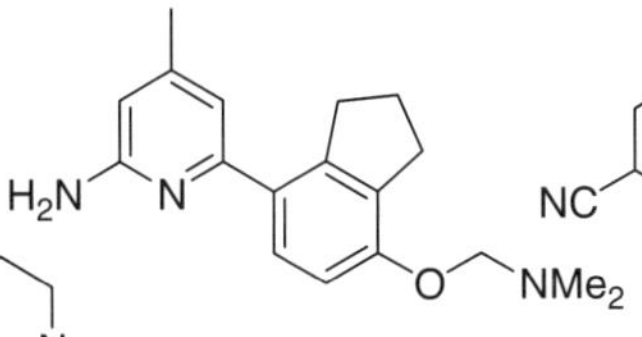

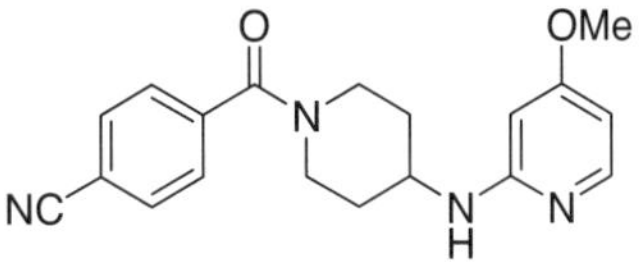

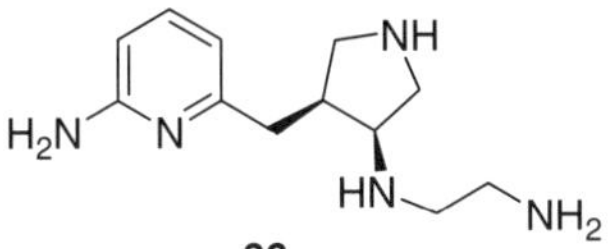

7.5. Coumarin Derivatives

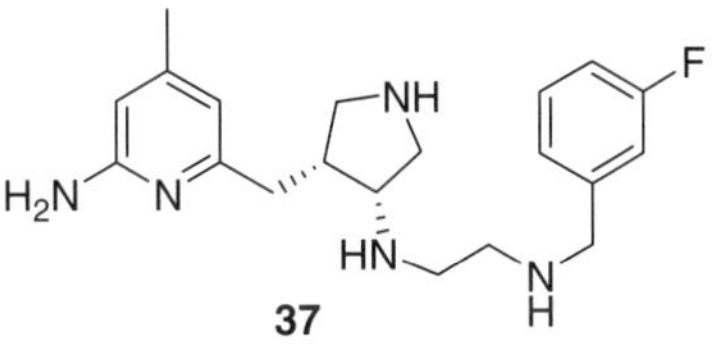

Screening efforts geared toward the discovery of small-molecule inhibitors of iNOS identified thiocoumarin **38**, as an interesting iNOS inhibitor lacking amidine or guanidine functionality. SAR studies led to the discovery of the potent and selective iNOS inhibitor **39**. This coumarin derivative was found to be competitive with arginine at the substrate binding site and without time-dependency inhibition [530].

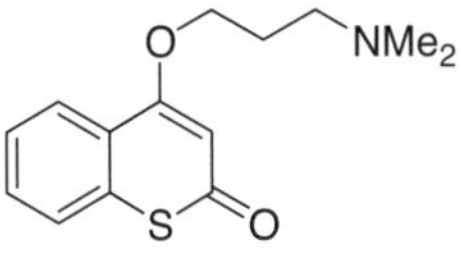

7.6. Inhibitors not Acting at the Arginine Site

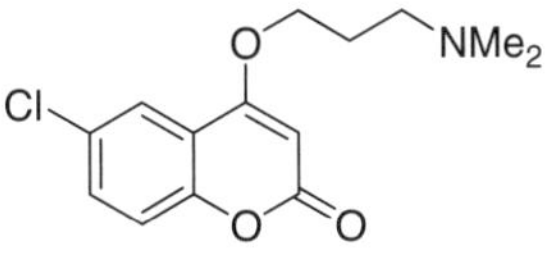

Because NOS functions in the presence of multiple cofactors, inhibitors that interact with the enzyme outside the arginine binding site have been studied as well. The three isoforms share the various cofactors and therefore selectivity issues are likely to be encountered with such inhibitors.

7.6.1. Compounds Acting at the Calmodulin Site

It has been suggested that melatonin (**40**) modulates glutamate-mediated responses in rat striatum via nNOS inhibition in a dose- and calmodulin-dependent manner. It was also shown that the kynurenamine derivative AMK (**41**) is the nNOS active metabolite of melatonin, and is responsible for NOS activity. A number of conformationally rigid analog of kynurenamine (**42–44**) have been prepared and found to be moderate inhibitors of both iNOS and nNOS, with some iNOS selectivity within the **44** series. Selected compounds from the general structure **44** reduce *in vivo* NOS activity in cytosol and mitochondria in the MTP model of Parkinson's disease [531].

DY-9760e (**45**) is a calmodulin antagonist displaying nNOS inhibition and demonstrated efficacy in stroke animal models, up to 30 min postischemic challenge. The *in vivo* neuroprotective properties have been attributed to NOS inhibition despite the *in vitro* activity of DY-9760e on CaMKII and calcineurin [532–537].

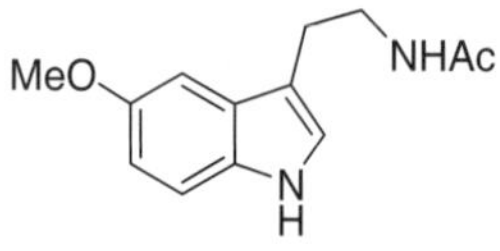

40
Melatonin

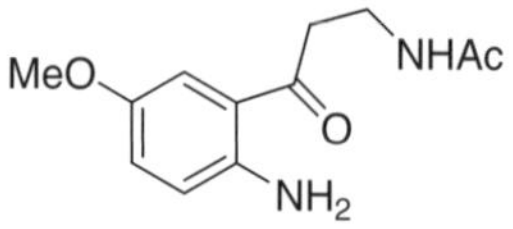

41:
AMK

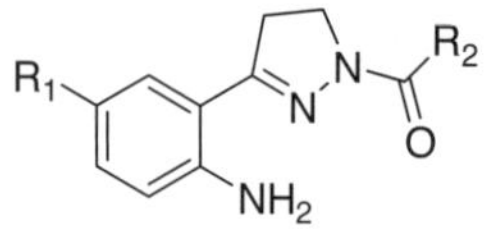

42

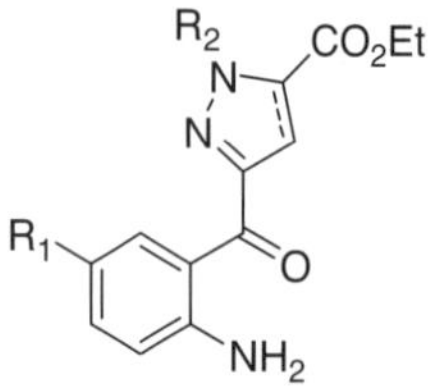

43

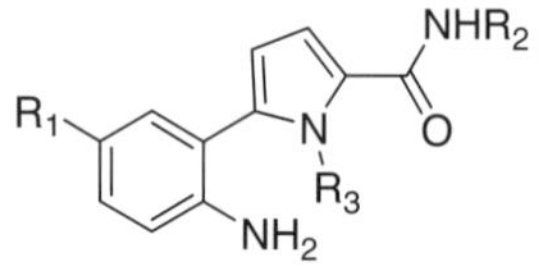

44

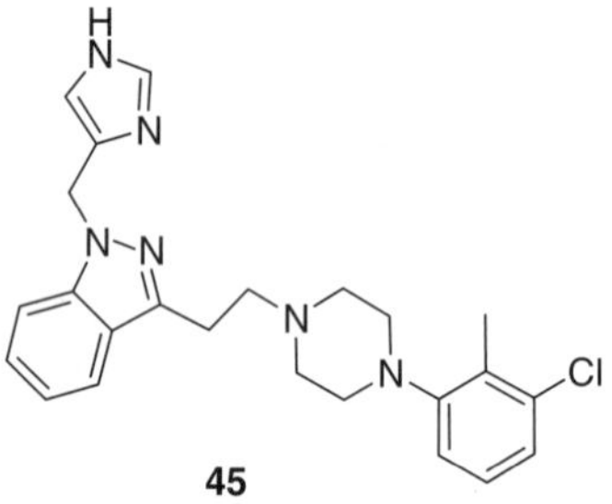

45
DY-9760e

7.6.2. Compounds Acting at the Tetrahydrobiopterin Site

The NOS family of enzymes requires the binging of the cofactor tetrahydrobiopterin for maximal activation, a unique phenomenon in heme-containing enzymes. Pteridine derivatives NOS inhibitors binding at the tetrahydrobiopterin site have been reviewed recently [538]. 4-Aminotetrahydrobiopterin (**46**) and PHS-32 (**47**) are examples of synthetic pterin derivatives that are active as NOS inhibitors. More surprisingly, nitroindazole 7-NI (**48**) has also been claimed to compete at the tetrahydrobiopterin site although the actual site of binding seems to differ between NOS isoforms. Additionally, 7-NI shows no isoform selectivity *in vitro* although *in vivo* it appears more selective toward nNOS [497,539]. 7-NI has demonstrated efficacy in animal models of stroke, although sometime with deleterious effects on cerebral blood flow [540–543]. SAR studies on 4-substituted indazoles and their effect in antinociceptive studies have recently been reported [544].

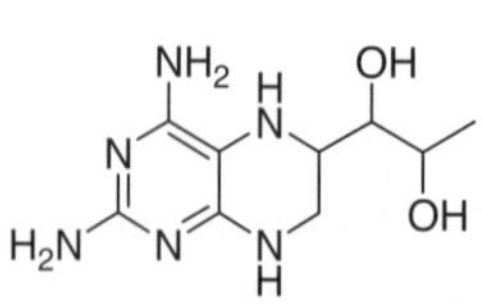

46

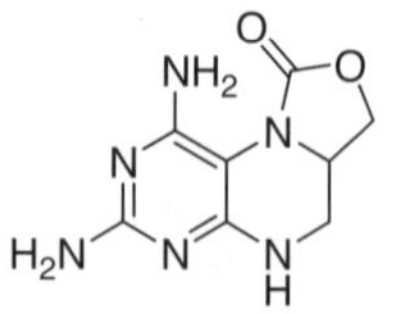

47: PHS-32

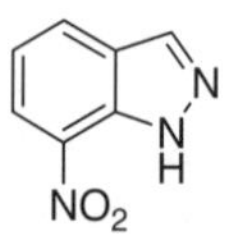

48: 7-NI

7.6.3. Compounds Acting at the Heme Site

Due to mechanistic similarities between CYP450 and NOS enzymes, imidazoles have been studied as NOS inhibitors via their ability to bind the heme iron present in those mono-oxygenase enzymes. As such, TRIM (**49**) is a well-known nNOS selective inhibitor with neuroprotective properties. Concomitant competition at the tetrahydrobiopeterin and arginine binding sites, however, makes its mode of inhibition unclear [545–547].

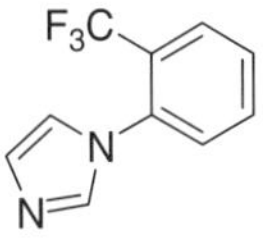

49: TRIM

7.6.4. NOS Dimerization Inhibitors

Dimerization inhibitors have been studied for iNOS exclusively at the moment. Although they do carry an imidazole moiety that can interact with the heme site, they are much more complex in structure, implying a more elaborate binding mode and the potential for selectivity. Typically, they do not inhibit the activity or production of NOS enzymes but rather prevent the dimerization of the monomeric form into the functionally active homodimers. Because iNOS protein turnover is higher than for the other isozymes and because the iNOS dimer is the least stable among the NOS isozymes, inhibiting its dimerization could provide a selectivity advantage, however, likely not in a chronic therapy. The early compounds discovered to operate as dimerization inhibitors were the antifungal clotromazole (**50**) and miconazole (**51**), and displayed weak potency [548]. Recent studies have uncovered nanomolar iNOS inhibitors such as FR-260330 (**52**) that displays activity in *in vivo* models of organ transplant rejection [549,550]. Scientists at Berlex uncovered iNOS dimerization inhibitors by whole cell screening of a library of imidazole containing compounds originally designed to operate as direct inhibitors. BBS-1 (**53**), an early prototype, displayed efficacy in *in vivo* models of organ transplant rejection, sepsis, and lung injury [551–555]. Structure optimization led to the discovery of BBS-4 (**54**) that displays activity in a rat model of arthritis [556]. X-ray structural analysis demonstrates that the imidazole moiety does indeed interact with the heme but also suggest that the binding of the rest of the molecule to remote parts of the enzyme might be responsible for the dimerization inhibiting conformational changes. The available structural data allowed for the rational design of additional dimerization inhibitors [557]. Cell-based uHTS led Kalypsys scientist to the discovery of quinolone based iNOS dimerization inhibitors such as **55**. This inhibitor was found to be selective over eNOS and to display efficacy in neuropathic pain models upon oral dosing [558]. An undisclosed development candidate (KD7332) was claimed to be derived from this series.

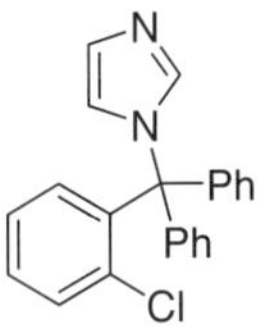

50: Clotrimazole

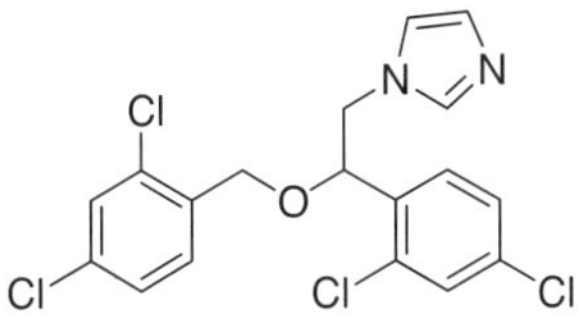

51: Miconazole

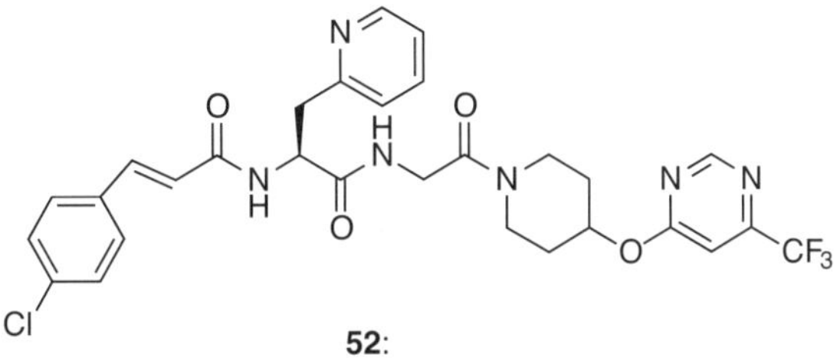

52: FR-260330

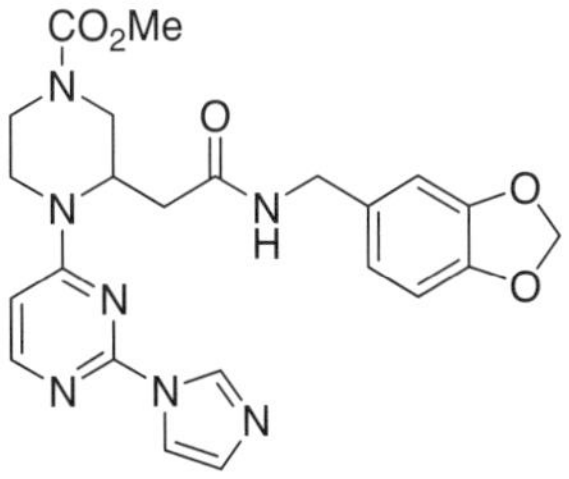

53: BBS-1

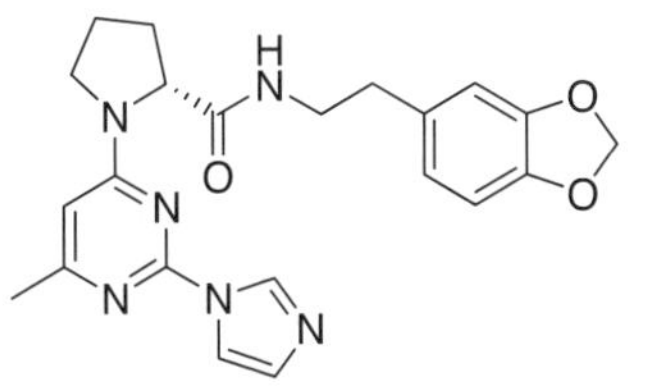

54: BBS-4

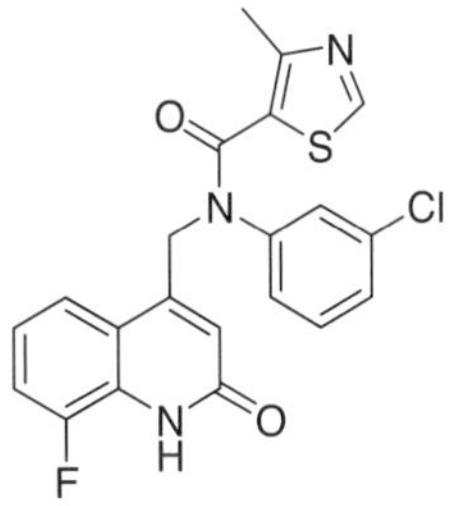

55

7.7. NOS Inhibitor/NO Donor Compounds

Based on the concept that NOS inhibition prevents high NO output and can therefore be beneficial in inflamed tissues, provided that low level of NO are maintained, a new class of compounds that combine an iNOS inhibitor scaffold with an NO-donor moiety were designed and prepared. NI-NOD1 (**56**) is such an example that inhibits IL-1β production, modulates PGE$_2$ production, and protects against apoptosis. NI-NOD1 also demonstrated some activity in an animal model of colitis [559].

56
NI-NOD1

8. MMP-9 INHIBITORS

Matrix metalloproteinases (MMPs) are proteolytic enzymes that function to remodel the pericellular environment by degrading and remodeling the extracellular matrix (ECM) proteins. The MMP family constitutes more than 20 members, which are all Zn^{2+}-dependent endopeptidases. Based on their primary structures, domain organization, substrate specificity and cellular localization, MMPs have been classified into six major subfamilies; these include collagenases (MMP-1, MMP-8, MMP-13, and MMP-18), gelatinases (MMP-2 and MMP-9), stromelysins (MMP-3, MMP-10, and MMP-11), membrane-type MMPs (MT-MMPs, MMP-14, and MMP-17), and other MMPs such as matrilysin (MMP-7), and metalloelastase (MMP-12) [560]. Together, these MMPs can catalyze the degradation of all the protein constituents of the ECM. The activities of MMPs are kept under tight control. All of the MMPs are synthesized as proenzymes and the removal of a propeptide is a prerequisite for activation. Once activated, the MMPs are subject to inhibition by the tissue inhibitors of metalloproteinases (TIMPs). The TIMPs are small (approximately 20 kDa) proteins that can bind MMPs noncovalently and block their activities [561].

Physiologically, MMPs play an important role in wound healing, ovulation, blastocyst implantation, bone growth, and angiogenesis. Pathologically, they are involved in a variety of disease processes such as tumor invasion and metastasis, rheumatoid arthritis, periodontal disease, and atherosclerosis [562]. MMPs are expressed in the CNS; the expression pattern of MMPs and the presence of their substrates in the brain suggest that MMPs are involved in neuronal differentiation and plasticity, such as neurite extension and synaptic reorganization. MMPs have also been shown to play a role in brain injury, such as multiple sclerosis, amyotrophic lateral sclerosis, Alzheimer's disease, and cerebral hemorrhage [563].

MMPs are rapidly upregulated after cerebral ischemia in animal models and clinical stroke patients [564]. By degrading neurovascular matrix, MMPs may mediate blood–brain barrier leakage, contributing to edema and hemorrhage [565]. Although there are many MMPs, a particularly important role for MMP-9 has been suggested. MMP-9 knockout mice showed reduced infarct size after focal cerebral ischemia and transient global cerebral ischemia [566]. Additionally, MMP-9 knockout mice develop less brain edema compared with wild-type mice. Gu et al. employed a relatively specific inhibitor of MMP-9, SB-3CT, and reported reduced infarct volume after transient focal ischemia in mice [567].

Emerging clinical data also supports the relevance of MMP-9. It is increased in plasma of acute stroke patients and demonstrates a correlation with clinical outcomes [568]. Also increased MMP-9 activity has detected in brain tissue sections following ischemic and hemorrhagic stroke [569].

MMP inhibitors and their utility against various therapeutic targets, mostly oncology and arthritis, have recently been reviewed by Fingleton [570], Rowan [571], Skiles [572] Whittaker [573], Matter [574], and Tu [575]. Most of the findings related to MMP-9 inhibition specifically have been summarized here, along with recently published work, focusing mostly on structural diversity.

8.1. Hydroxamic Acid-Derived Inhibitors

A large number of hydroxamates have been described and only a small representative selection based on structural diversity id represented here.

8.1.1. Prototypical Hydroxamate Inhibitors

Batimastat (**1**) was the first succinic acid based, peptidic hydroxamate MMP inhibitor to be evaluated in cancer clinical trials [576]. Marimastat (**2**), an orally active derivative, was advanced to Phase III trials for oncology as well, without success [577]. Trocade (**3**) was one of the first examples of nonpeptidic succinic acid-based hydroxamate; it was evaluated in clinical trials for RA, without success [578–580].

CGS-27023 (**4**) is the first example of glycine-derived sulfonamide and as such was the first nonpeptidic hydroxamate to be advanced to oncology clinical trials [581]. Prinomastat (**5**) represents one of the first cyclic sulfonamides and was also advanced to human clinical trials without success [582].

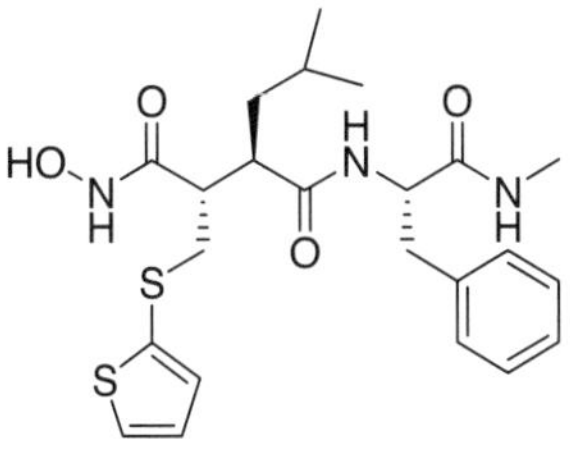

1: BB-94/Batimastat
IC_{50} = 1 nM

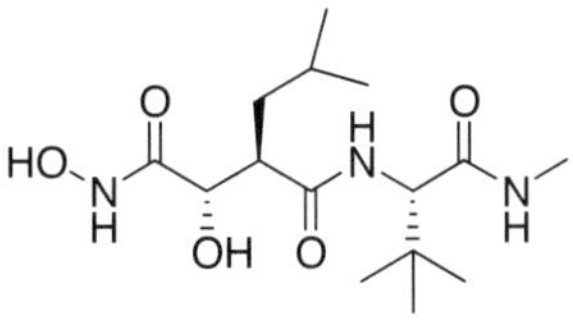

2: BB-2516/Marimastat
IC_{50} = 3 nM

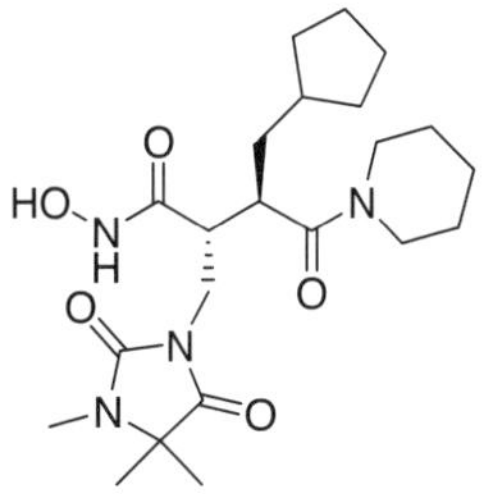

3: Ro 32-3555/Trocade
IC_{50} = 59 nM

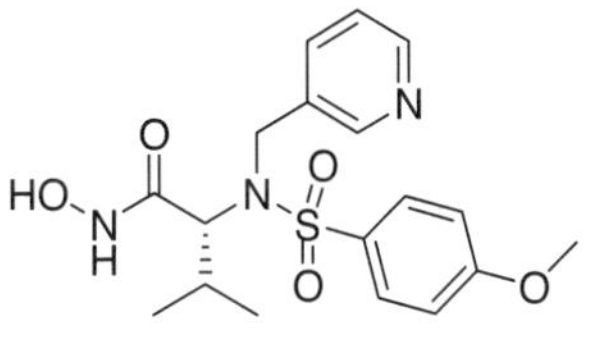

4: CGS-27023
IC_{50} = 8 nM

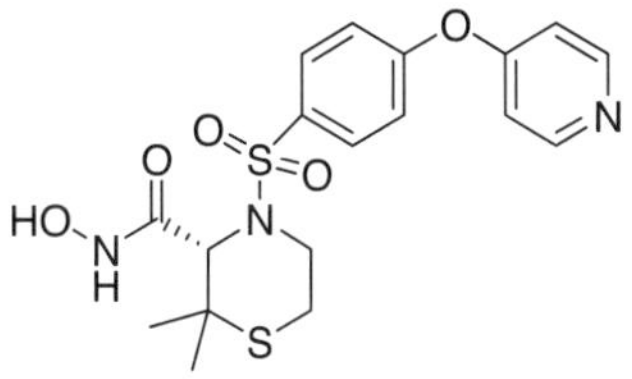

5: AG-3340/Prinomastat
IC_{50} = 0.26 nM

8.1.2. Evolution of Acyclic Hydroxamates

Deviating from the succinic acid template, a series of N-benzoyl 4-aminobutyric acid hydroxamate analogs (**6** and **7**) were evaluated and found to be potent MMP-9 inhibitors [583]. Further deviation and using combinatorial techniques resulted in the discovery of potent 2-substituted-3-bisarylthiopropionic acid hydroxamates such as **8** that displayed efficacy in a tumor reduction animal model [584].

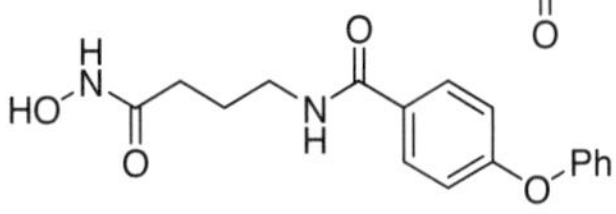

6
IC_{50} = 12 nM

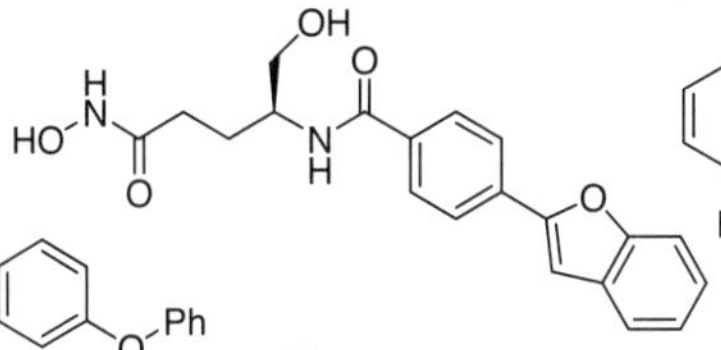

7
IC_{50} = 12 nM

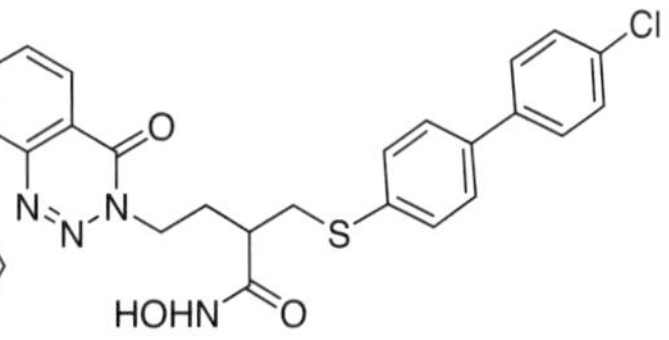

8
IC_{50} = 0.5 nM

Following the original discovery of sulfonamide-based MMP inhibitors, a large number of analogs have been prepared [585]. Compounds **9–12** represent examples of inhibitors based on the aminoacid sulfonamides [586–589]. The N-*i*-propoxy derivative **10** displayed antiangiogenic properties in HUVEC cells. Inhibitor **13** illustrates how the sulfonamide moiety can be replaced with a phosphamide functional group, without any significant loss in potency.

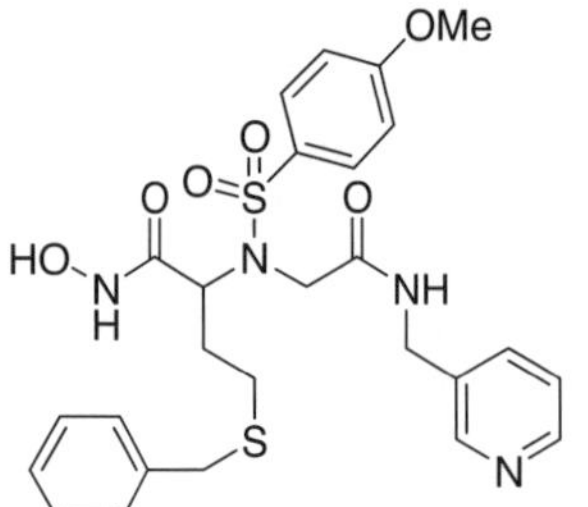

9
IC$_{50}$ = 0.2 nM

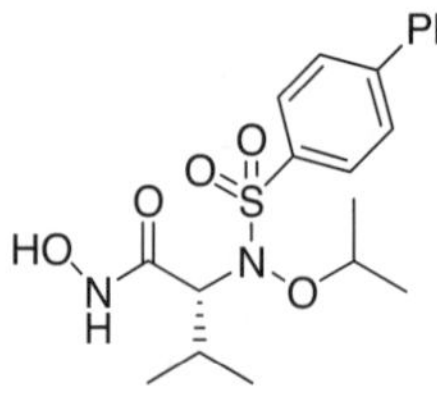

10
IC$_{50}$ = 6.7 nM

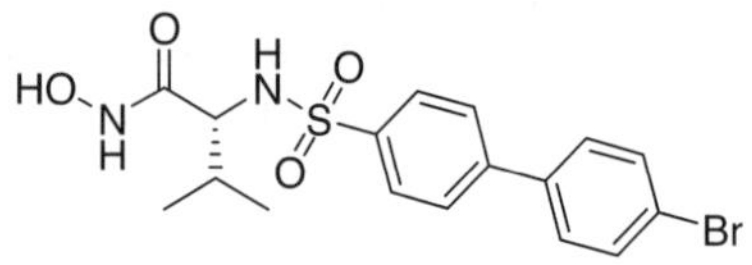

11
IC$_{50}$ = 18 nM

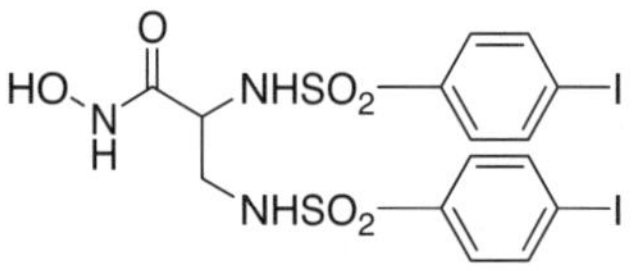

12
IC$_{50}$ = 2.6 nM

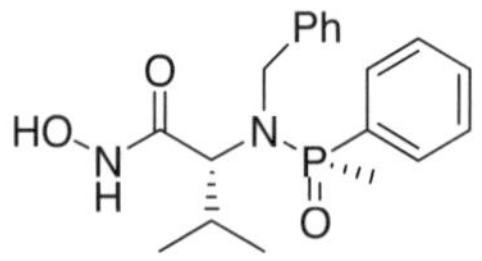

13
IC$_{50}$ = 21 nM

The utilization of a beta-aminoacid template and the relocation of the sulfonamide moiety led to the discovery of MMP-9 inhibitors such as **14** [590]. The introduction of the morpholino group resulted in improved PK profile and the compound displayed efficacy in a rat stroke model.

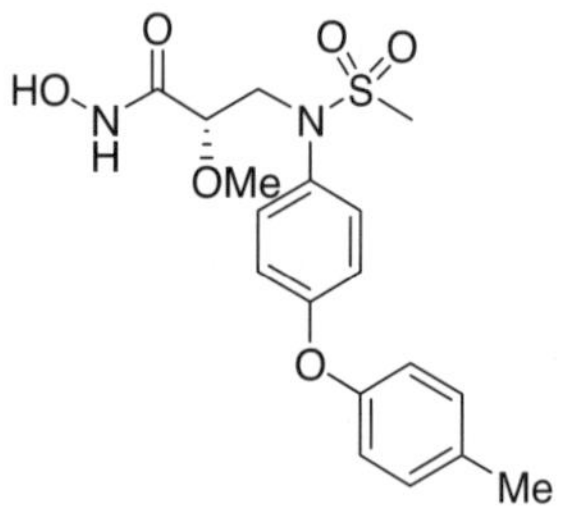

14
IC$_{50}$ = 4 nM

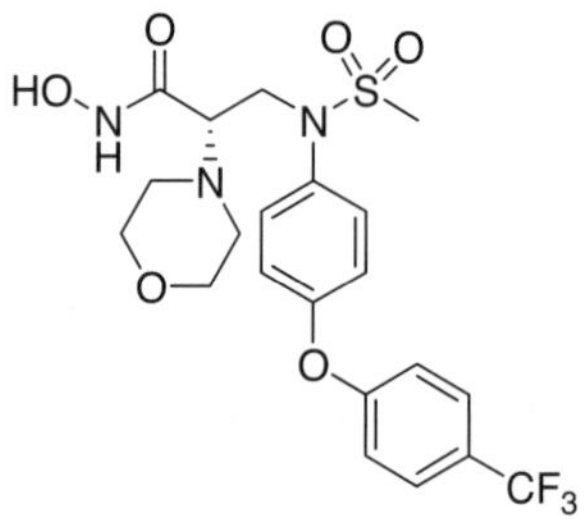

15
IC$_{50}$ = 3 nM

8.1.3. Evolution of Cyclic Hydroxamates

Numerous proline-derived sulfonamides such as examples **16** and **17** have been evaluated as MMP inhibitors [591]. Several derivatives have demonstrated efficacy in animal models of arthritis or cancer.

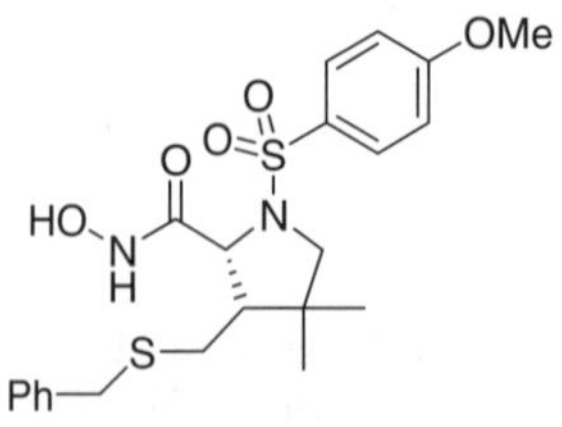

16
IC$_{50}$ = 0.9 nM

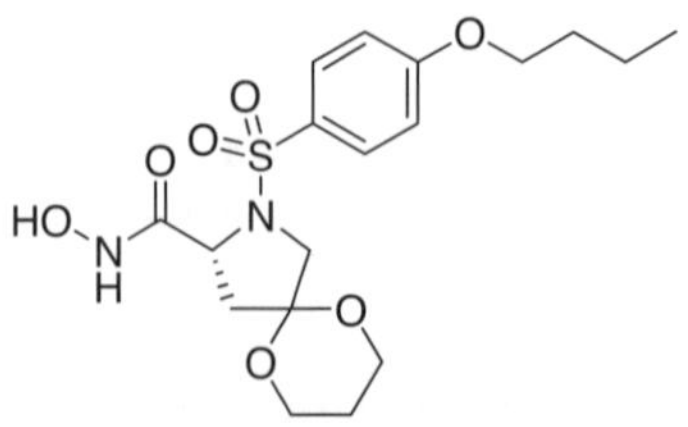

17: PGE-3321996
IC$_{50}$ = 0.8 nM

The concept has also been applied to piperidine acid derivatives (Prinomastat) and a variety of nitrogen containing six-membered rings, as exemplified by the following inhibitors. Structural diversity includes ketopiperazine 18 [592], azasugar derivative 19 [593], bis-sulfonamide 20 [594], piperazine 21 [595], and tetrahydroisoquinoline phosphamide 22 [596].

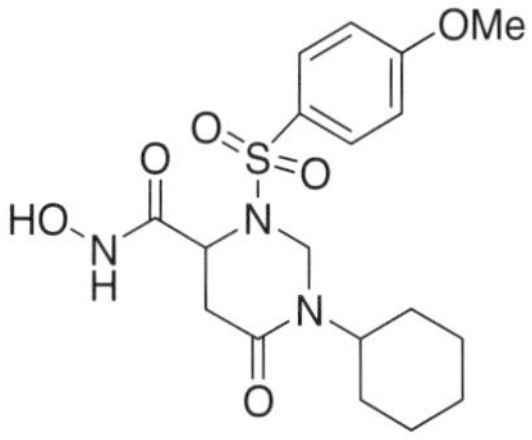

18
IC_{50} = 1.3 nM

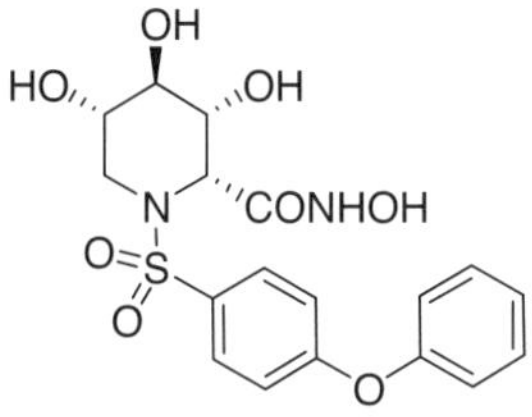

19
K_i = 0.06 nM

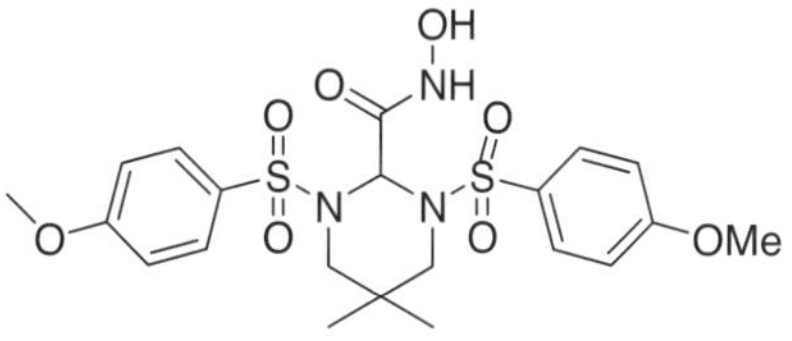

20
IC_{50} = 2.7 nM

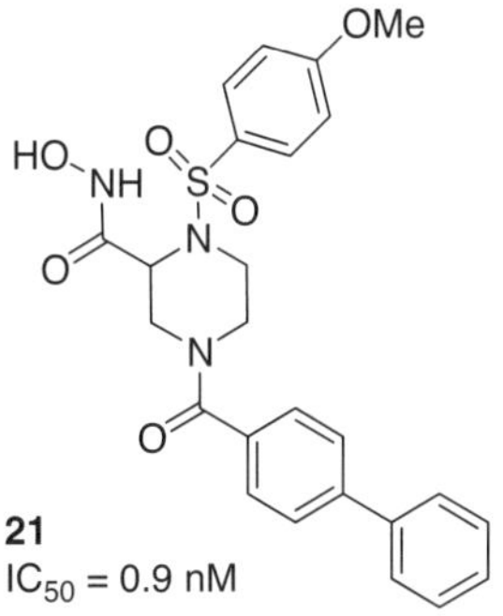

21
IC_{50} = 0.9 nM

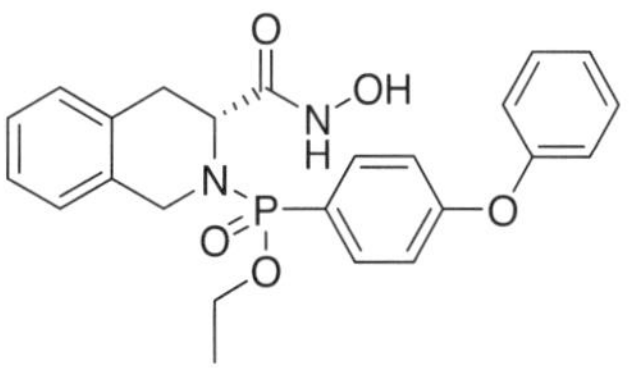

22
IC_{50} = 0.08 nM

MMP inhibitors constructed around nitrogen containing seven-membered ring scaffold such as diazepine have also been prepared as exemplified by compounds 23 and 24 [597,598]. Incorporation of the sulfonamide moiety into the ring (25 and 26) also led to the preparation of potent MMP inhibitors [599,600].

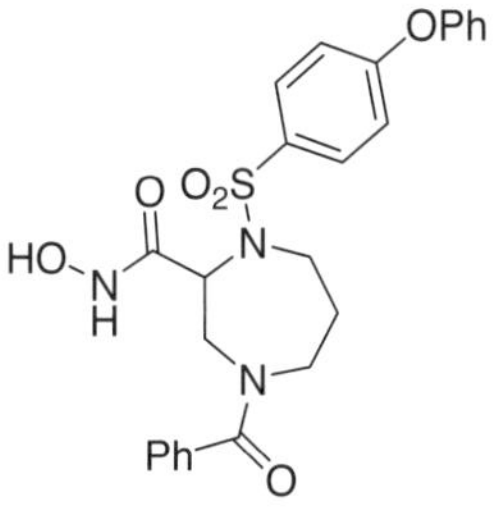

23
IC_{50} = 1.2 nM

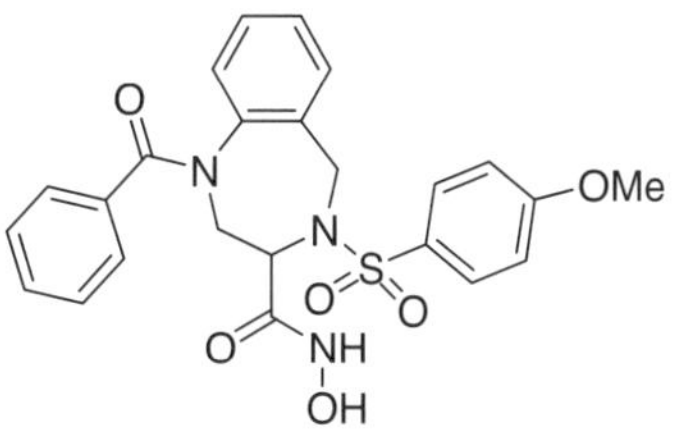

24
IC_{50} = 0.6 nM

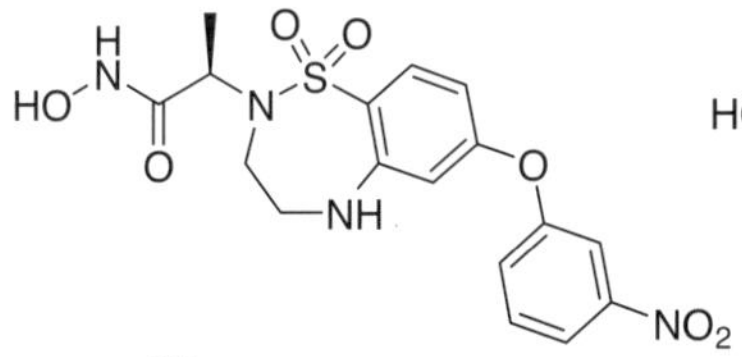

25
IC_{50} = 0.5 nM

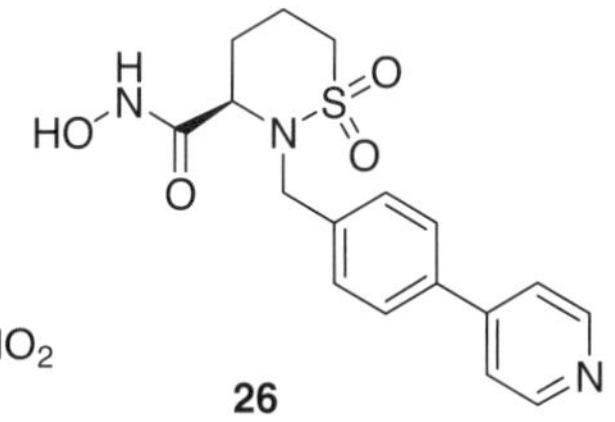

26
K_i = 10 nM

Potent MMP inhibitors have been derived from anthranilic acid. The prototype **27**, modeled after CGS-27023, has demonstrated some efficacy in cartilage degradation animal models [601]. Removal of the 3-pyridyl group and extension of the aryl sulfonamide in S1′ via a phenolic substituent yielded inhibitors such as **28** that display selectivity for MMP-9 over MMP-1 and MMP-13. Such compounds have also demonstrated some efficacy in cartilage degradation animal models [602]. Further optimization led to piperazine analog **29** that was shown to be more efficacious than CGS-27023 in cartilage degradation animal models, upon p.o. dosing, even at lower doses [603].

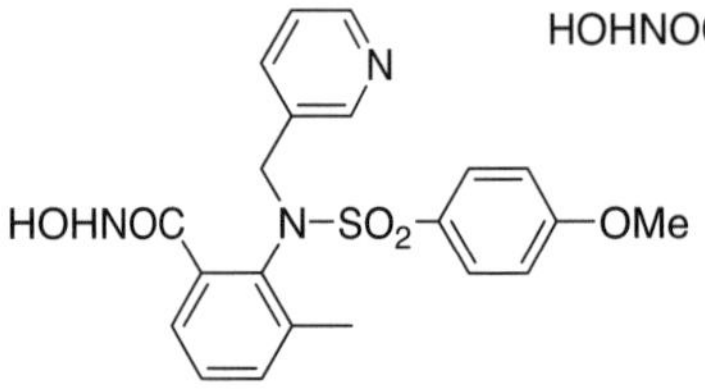

27
$IC_{50} = 5$ nM

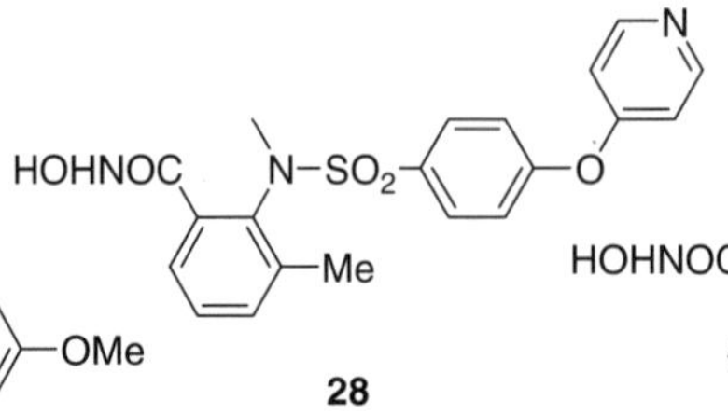

28
$IC_{50} = 2$ nM

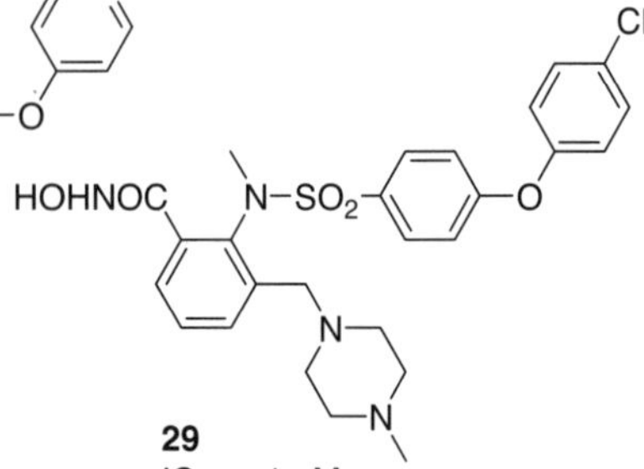

29
$IC_{50} = 1$ nM

8.1.4. Gem-Disubstituted and Spirocyclic Scaffold-Based Inhibitors Inhibitor **30** represents an early example of gem disubstitution along the succinic scaffold of Marimastat [604]. The concept was applied to sulfone templates derived from the original CGS-27023 scaffold, and then further refined to spirocyclic inhibitors. In the case of **31** [605] and **32** [606] the sulfonamide nitrogen atom has simply been deleted while in the case of **33** (RS-130,830), and **34** [607] a methylene spacer has been incorporated either in the place of the nitrogen or on the hydroxamic moiety side.

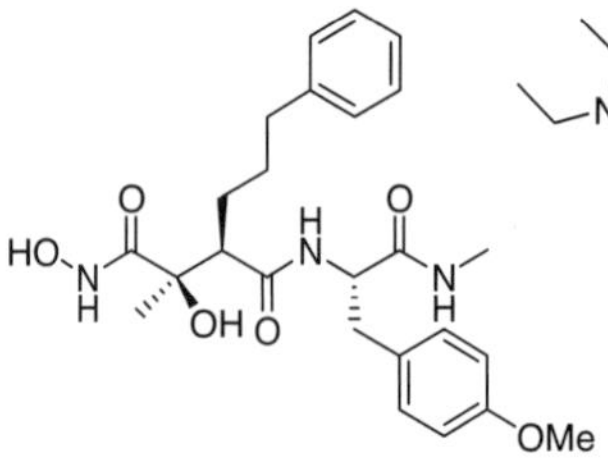

30
$K_i < 1$ nM

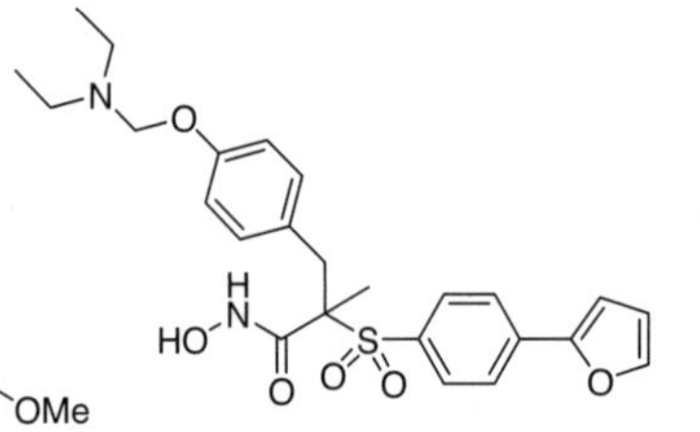

31
$IC_{50} = 4$ nM

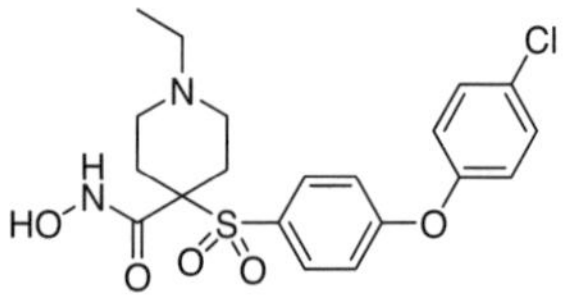

32
$IC_{50} = 1$ nM

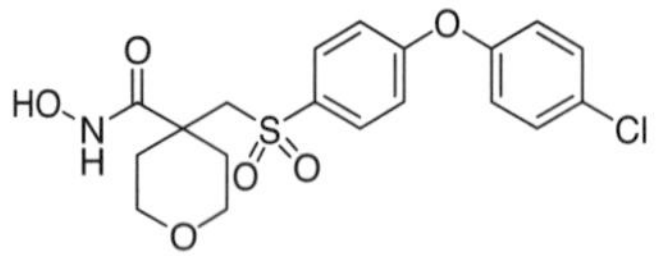

33: RS-130,830

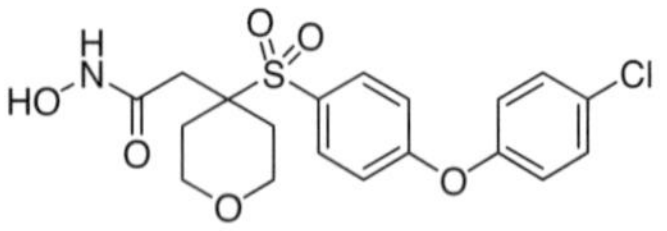

34
$IC_{50} = 0.065$ nM

8.1.5. Macrocyclic Inhibitors A number of potent macrocyclic MMP inhibitors have been disclosed, a few of which with activity against MMP-9 are illustrated here (**25–37**) [608–610].

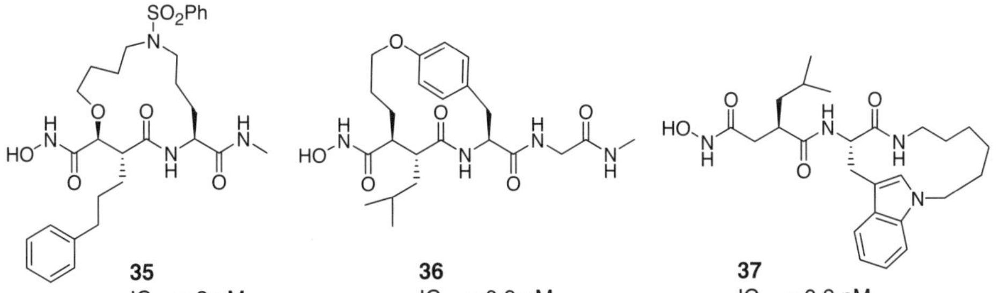

35
IC$_{50}$ = 2 nM

36
IC$_{50}$ = 0.9 nM

37
IC$_{50}$ = 0.2 nM

8.1.6. Oxalic Acid Hydroxamates Hydroxamates derived from oxalic acid have also been evaluated as MMP inhibitors (**38** and **39**), with limited success [611]. Inhibitor **39** also represents an early example of bis hydroxamate.

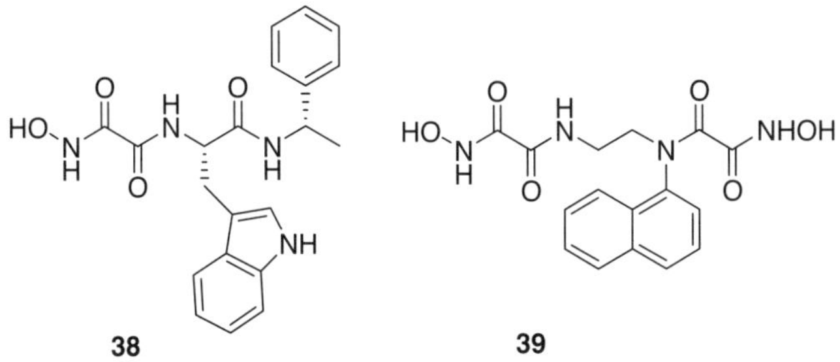

38
K_i = 100 nM

39
K_i = 100 nM

8.1.7. Bis-Hydroxamates Bis-hydroxamate **40** was shown to be a potent MMP-9 inhibitor and claimed to have reduced cellular cytotoxicity [612].

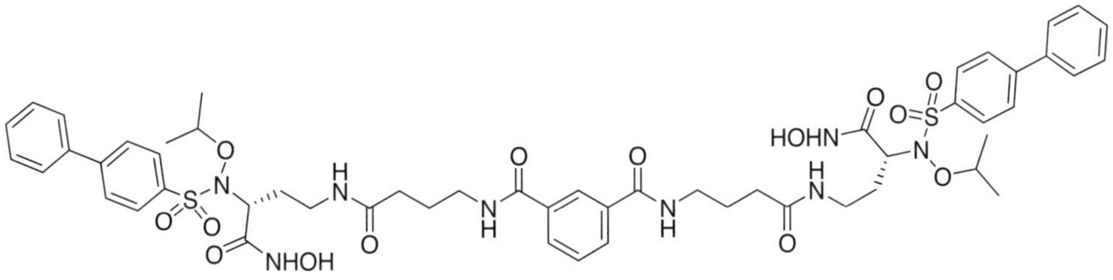

40
IC$_{50}$ = 8 nM

8.1.8. Reverse Hydroxamic Acids Reverse hydroxamic acids or formylated hydroxylamine derivatives have received much less attention as MMP-9 inhibitors than their hydroxamate counterparts, following are two such examples (**41** and **42**) as mimic of the succinic acid and sulfone templates [613,614].

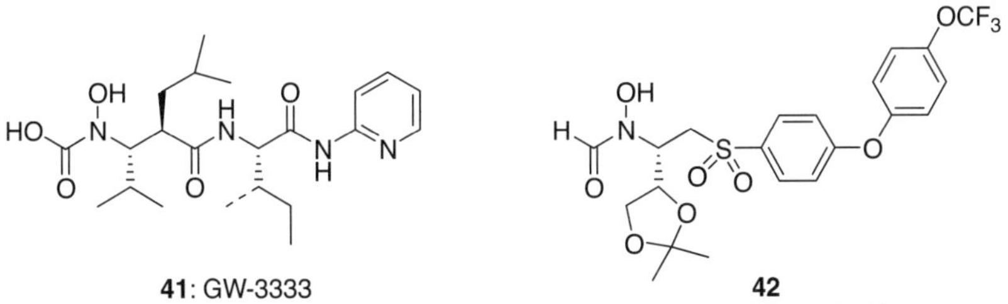

41: GW-3333
IC$_{50}$ = 16 nM

42
IC$_{50}$ = 1 nM

8.2. Carboxylic Acid-Derived Inhibitors

A number of investigators have evaluated carboxylic acids as replacements for hydroxamates. Because of the monodentate nature of the carboxylice acid zinc ligation, as opposed to bidentate for hydroxamates, a significant amount of potency is generally lost and has to be recovered by the expansion of the hydrophobic P1′ groups, deeper in the S1′ subpocket. Tanomastat (BAY 12-9566, **43**) has been one of the few carboxylic acid MMP inhibitor to be advanced to cancer and arthritis clinical trials that were eventually suspended [615]. Following are a few selected examples of carboxylic acid MMP-9 inhibitors. Compound **44** is derived from a peptidic scaffold [616], while **45** and **46** are based on the aminoacid sulfonamide template [617,619]. A sulfone-derived example (**47**) and a macrocyclic carboxylic acid have also been reported as MMP-9 inhibitors [619,620].

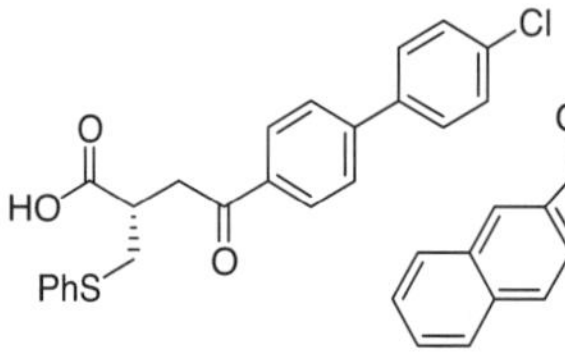

43: Tanomastat
BAY 12-9566
$IC_{50} = 301$ nM

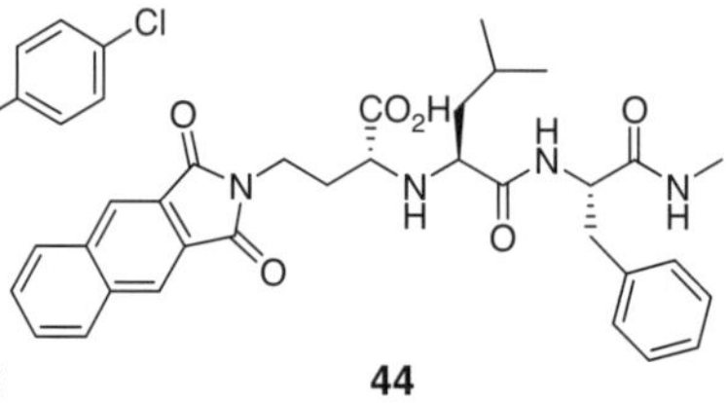

44
$IC_{50} = 91$ nM

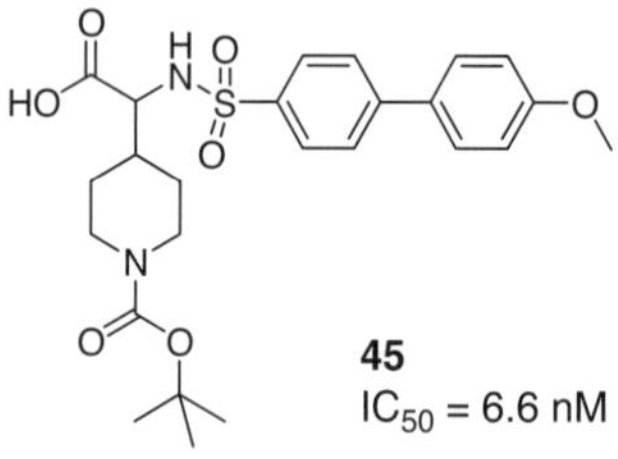

45
$IC_{50} = 6.6$ nM

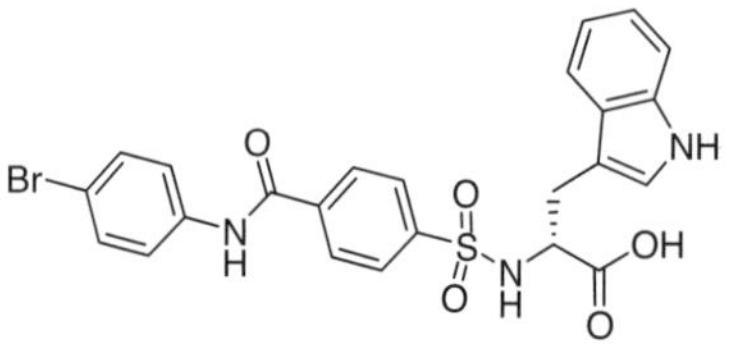

46
$IC_{50} = 2.7$ nM

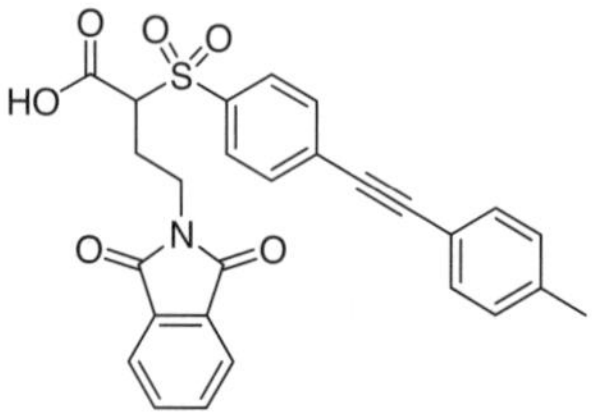

47
$IC_{50} = 0.5$ nM

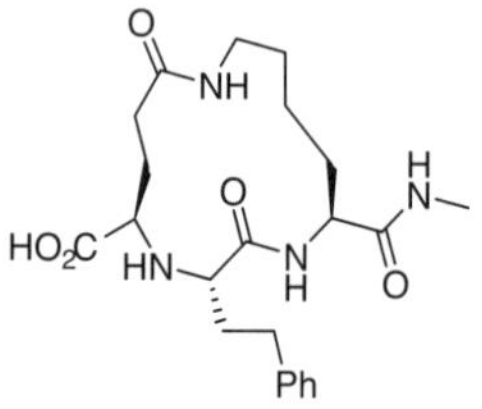

48
$K_i = 6600$ nM

8.3. Hydroxypyridinones

Alternate bidentate zinc binding ligands include hydroxypyridines such as **49** that is orally bioavailable and has demonstrated efficacy in a transient midcerebral artery occlusion mouse model of cerebral ischemia [621]. An hydroxypyridone variation (**50**) has also been reported as well [622].

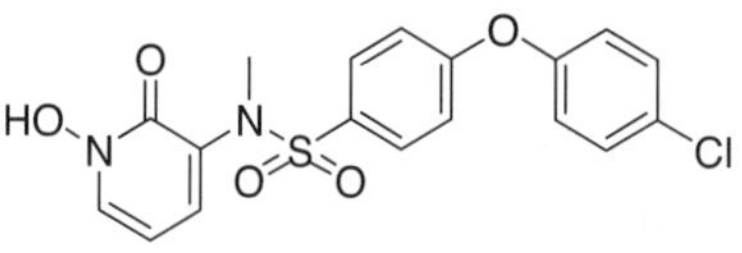

49
$K_i = 2.4$ nM

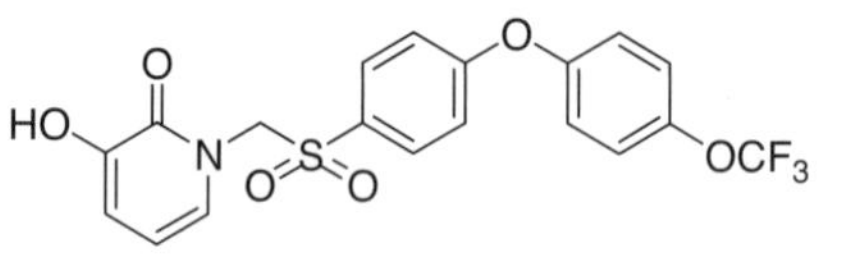

50
$K_i = 4$ nM

8.4. Pyrimidine-2,4,6-triones

The 5,5-disubstituted pyrimidine-2,4,6-trione zinc binding motif was identified during an HTS campaign and led to the preparation of potent MMP-9 inhibitors such as **51** and **52** [623–625].

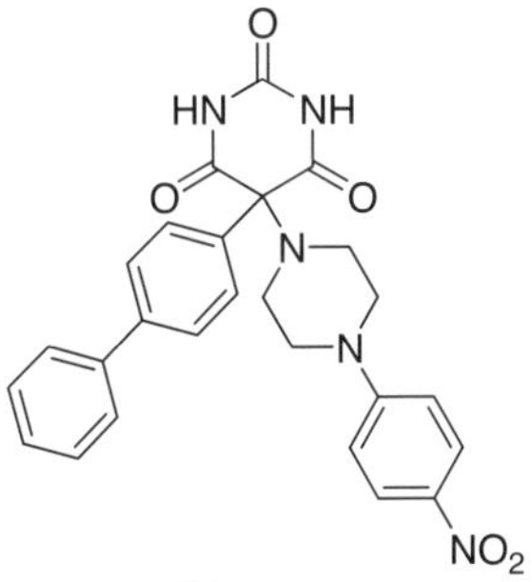

51
$K_i = 12$ nM

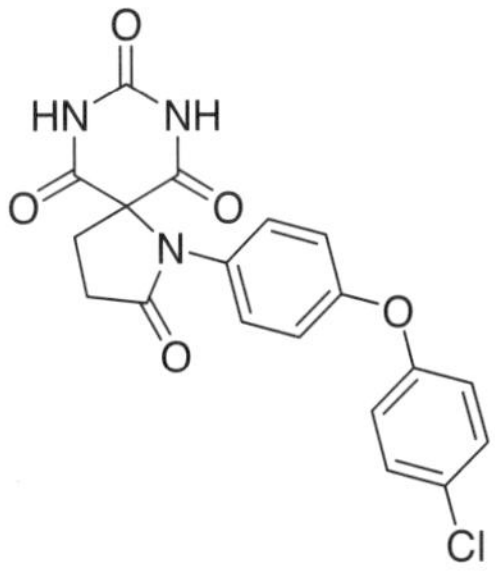

52
$K_i = 0.8$ nM

8.5. Thiols

The thiol group is a well-known zinc binding ligand and has been utilized in the preparation of numerous MMP inhibitors. D-2163 (**53**) originated from the Celltech group and was licensed by BMS to be advanced to cancer clinical trials [626]. Additional examples of potent thiol-derived MMP-9 inhibitors are represented by compounds **54–56** [627–629]. Nonpeptidic thiol-based inhibitor **57** was identified from combinatorial library work [630]. Inhibitor **58** displays a thio-thiodiazole moiety as the zinc ligand and maintains marginal potency [631].

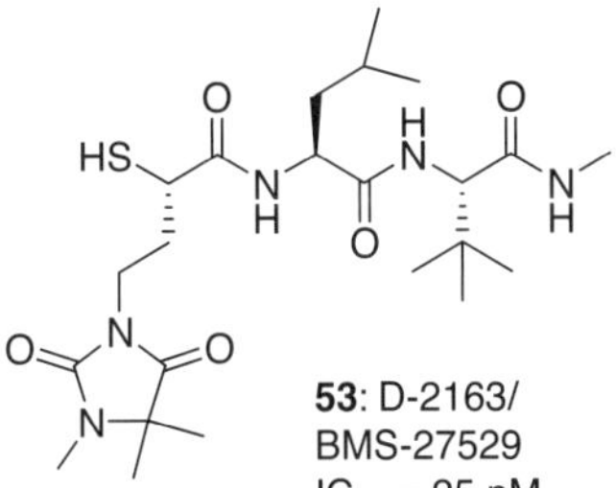

53: D-2163/
BMS-27529
$IC_{50} = 25$ nM

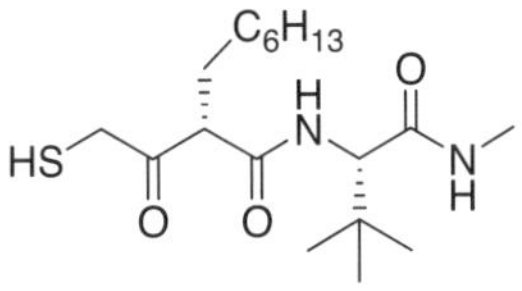

54
$IC_{50} = 0.3$ nM

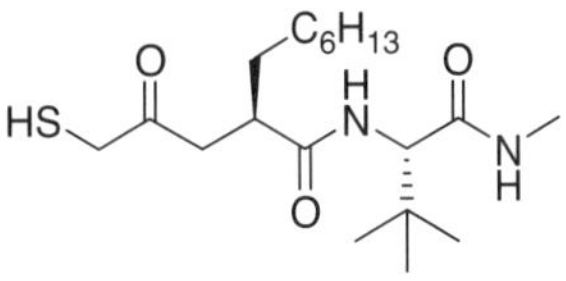

55
$IC_{50} = 0.1$ nM

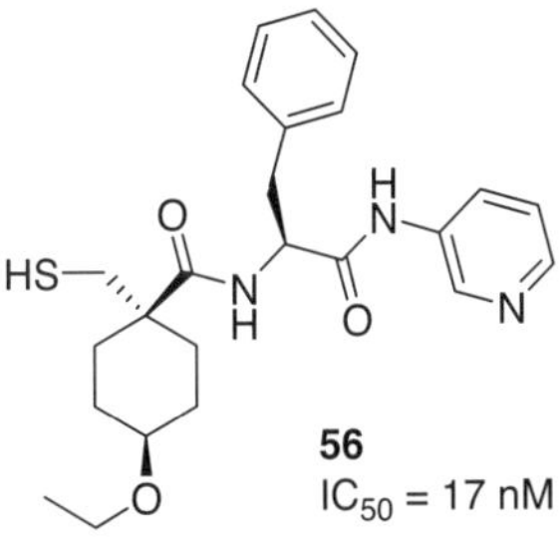

56
$IC_{50} = 17$ nM

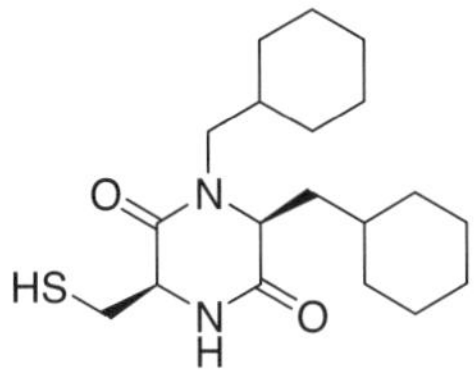

57
$IC_{50} = 113$ nM

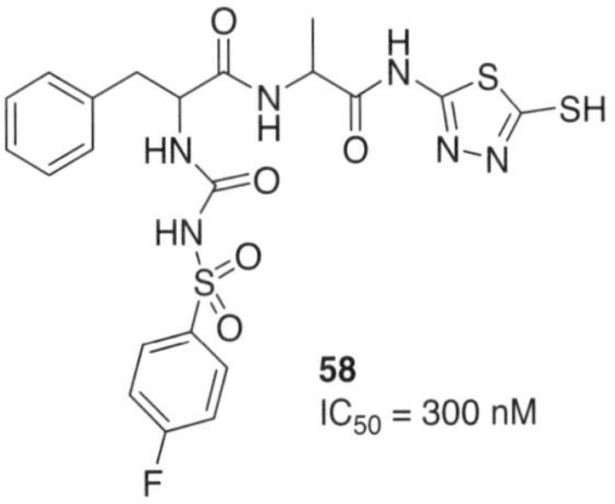

58
$IC_{50} = 300$ nM

8.6. Phosphorus-Derived Inhibitors

Phosphorus-derived zinc binding ligand containing MMP inhibitors have received much less attention than hydroxamates, for example. Compounds such as **59** utilizing a phosphinic aminoacid analog can span the prime and unprimed regions of the substrate binding site [632]. As observed before, extension of the P1′ group deep into S1′ allows for potent yet small and nonpeptidic inhibitors such as **60** [633].

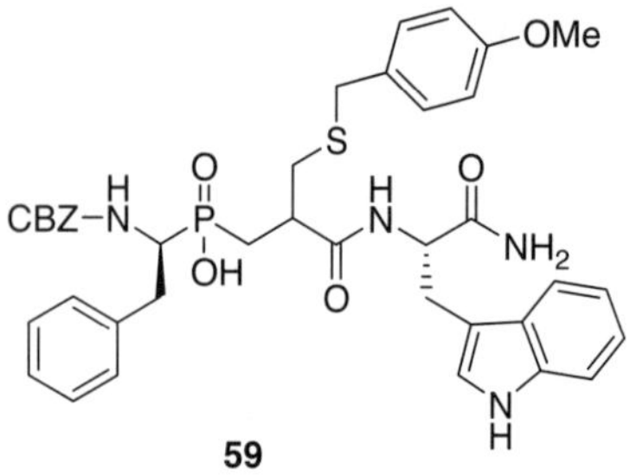

59
IC$_{50}$ = 3 nM

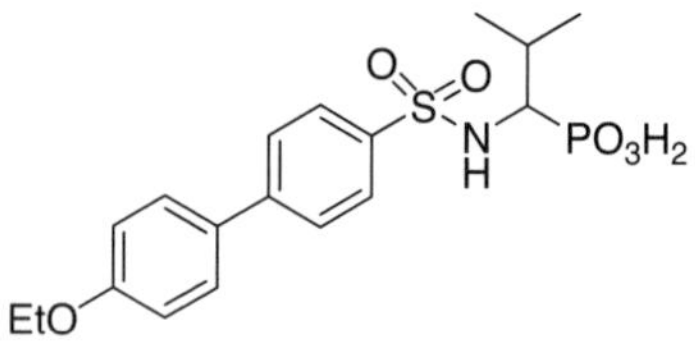

60
IC$_{50}$ = 0.6 nM

8.7. Miscellanous MMP-9 Inhibitors

Since acylsulfonamides are known to ligate zinc, it would seem likely that inhibitors such as **61** and **62** built around a similar concept would ligate zinc via the NH moiety [634,635]. The suicide substrate type inhibitor **63** displays good potency at MMP-9. Nucleophilic opening of the thiirane ring by Glu 404, following zinc ligation by the thiirane sulfur has been proposed as the mechanism of action [636–638].

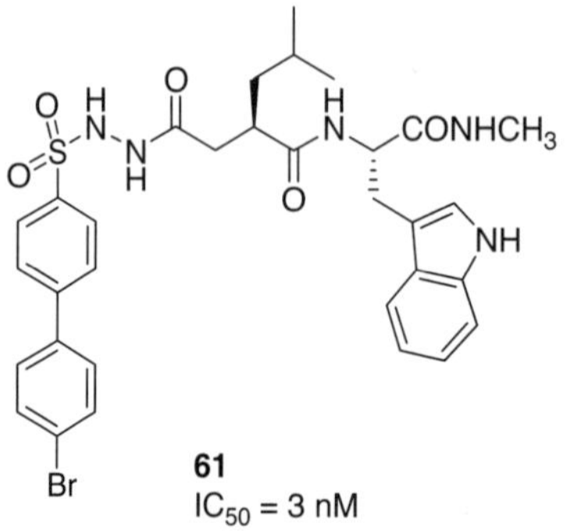

61
IC$_{50}$ = 3 nM

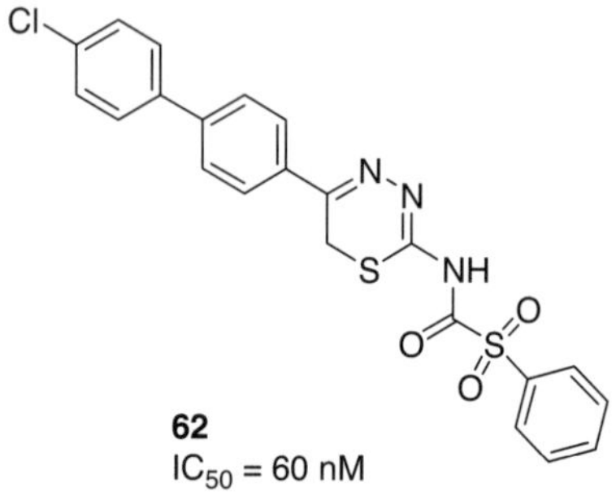

62
IC$_{50}$ = 60 nM

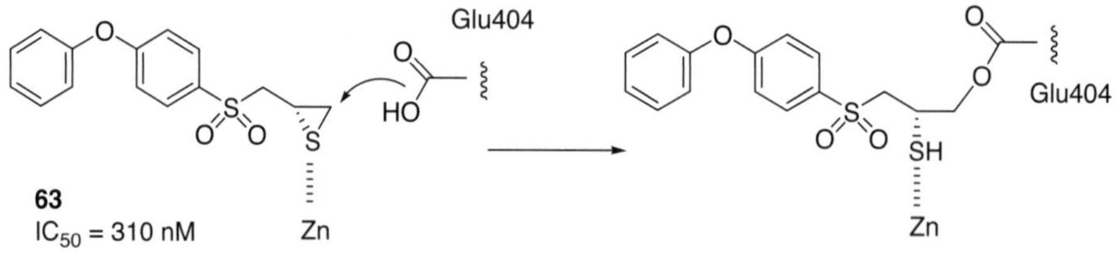

63
IC$_{50}$ = 310 nM

9. SRC KINASE INHIBITORS

Cerebral ischemia/reperfusion activates several protein kinase pathways. The Src family of kinases is hypothesized to be an intracellular hub for NMDA receptor regulation. Other experimental evidence that makes Src kinase worthy of investigation for stroke protection include the thrombin activation of Src; [639] Src phosphorylates and regulates matrix metalloproteinases [640]; and Src can effect vascular endothelial growth factor (VEGF) expression and activity [641]. Src kinase targets also include blood–brain barrier proteins that may be important in producing brain edema after reperfusion [642].

Clinical trials using Src kinase inhibitors for stroke have not occurred and the preclinical animal models that have been done, while showing promising results, have generally used nonspecific inhibitor molecules [643]. As summarized below, medicinal chemistry around Src kinase inhibitors has been active in other therapeutic areas. Therefore, one can hope that the future will bring the testing of these molecules in a stroke paradigm.

Inhibitors of Src kinase and their use in the field of stroke, osteoporosis, inflammation, oncology and immunological disorders have been reviewed recently [644–647]. Most of the findings related to Src kinase inhibition described in these reviews have been summarized here, along with recently published work, focusing mostly on structural diversity and/or significant clinical results.

9.1. Natural Derivatives

A number of natural products have been described as Src or Src family kinase (SFK) inhibitors. Examples of macrocyclic dieneones include herbimycin A (**1**), geldanamycin (**2**), and radicicol (**3**). Recently, homoisoflavanoids such as **4** have been isolated from a Src HTS compaign [648]. Src kinase inhibitors illustrated by **6** have recently been derived from the scaffold of the naturally occurring kinase inhibitor genistein (**5**) [649].

R_1 = OMe, R_2= H: **1**: Herbimycin A
R_1 = H, R_2= OMe: **2**: Geldanamycin

3: Radicicol

4

R_1 = H, R_2 = 4-Ph-OH: **5**: Genistein
R_1 = 3,4-Ph-$(OH)_2$, R_2 = OCOaa: **6**

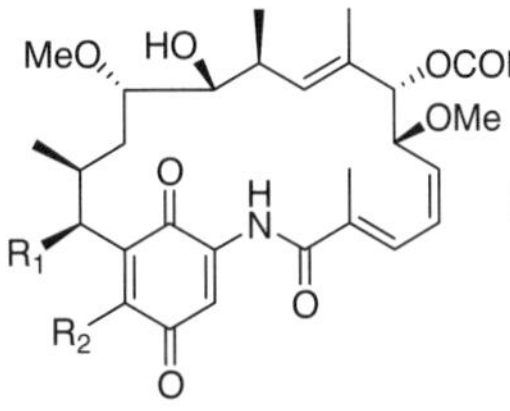

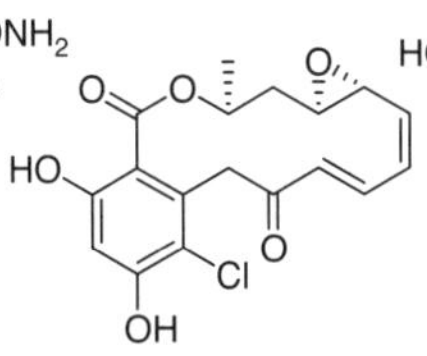

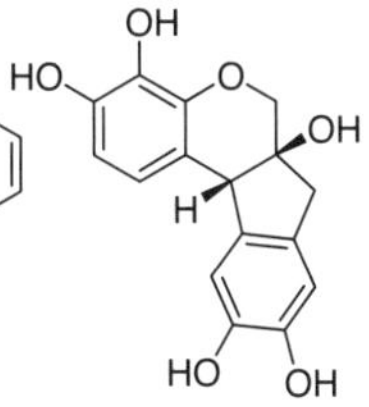

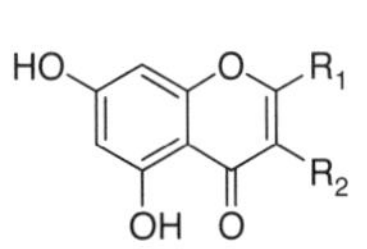

9.2. Purine-Based Derivatives

Purine-based inhibitors of Src have been described. NVP-AAK980 (**7**) has displayed anti-resorptive activitity in animal models [650].

Arethenyl derivatives such as **8** and **9** have also recently been described as dual Src/Abl tyrosine kinase inhibitors [651,652].

7: NVP-AAK_{980}

8: AP23464

9

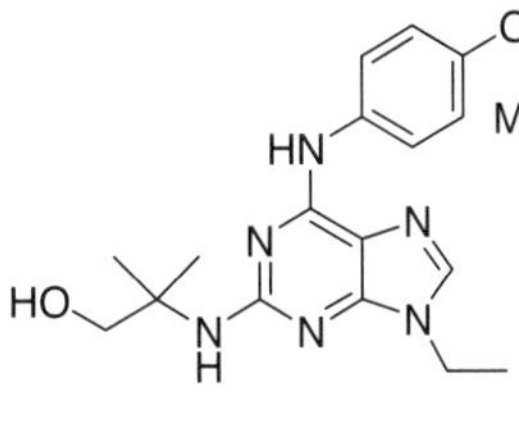

9.3. Pyrazole-Pyrimidine Inhibitors

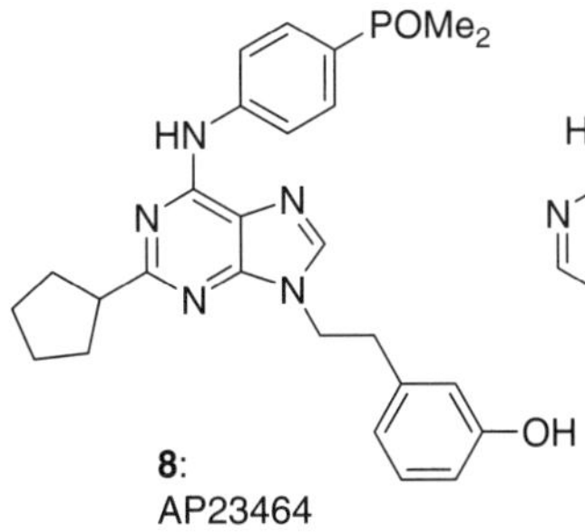

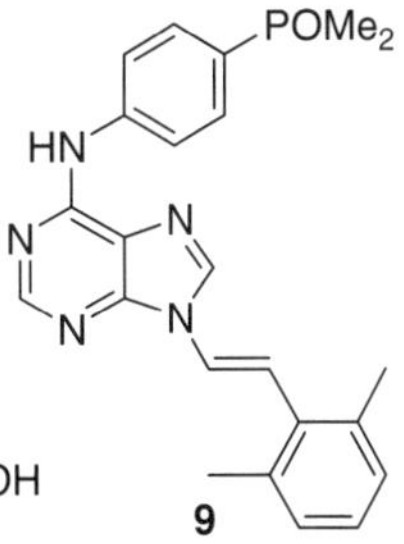

Further derived from ATP are the pyrazolo-pyrimidines-based inhibitors such as PP1 (**10**) and PP2 (**11**), both of which have displayed efficacy in animal models of stroke [653–657].

Peptide conjugates of general structure **12** have been shown to provide a synergistic inhibition effect as a possible result of conjugate and kinase domain interactions [658]. Further substitution of the pyrazolopyrimidine template has led to the identification of A-420983

(13) and A-770041 (14) with promising immunosuppressive properties [659,660]. Additional modifications to the template provided inhibitors 15 and 16 displaying significant activity in animal models of cancer [661–664].

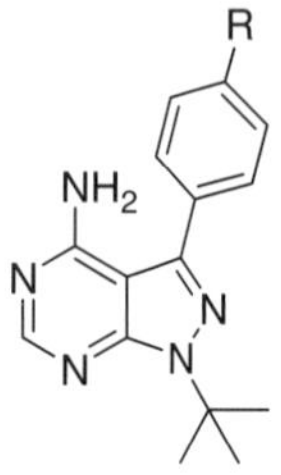

R = Me: 10: PP1
R = Cl 11: PP2

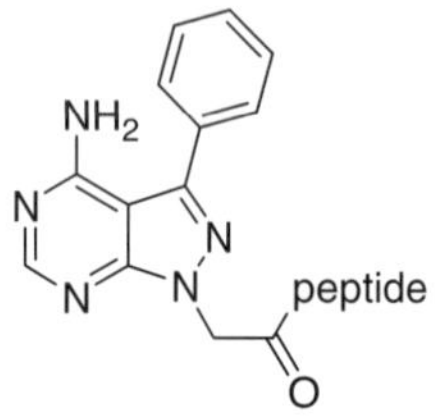

12

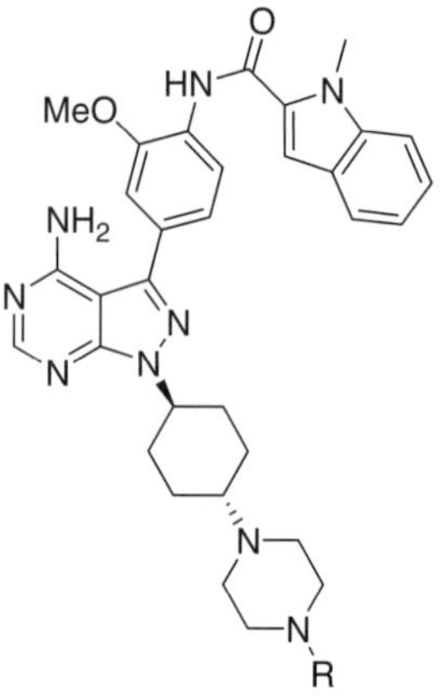

R = Me: 13: A-420983
R = Ac: 14: A-770041

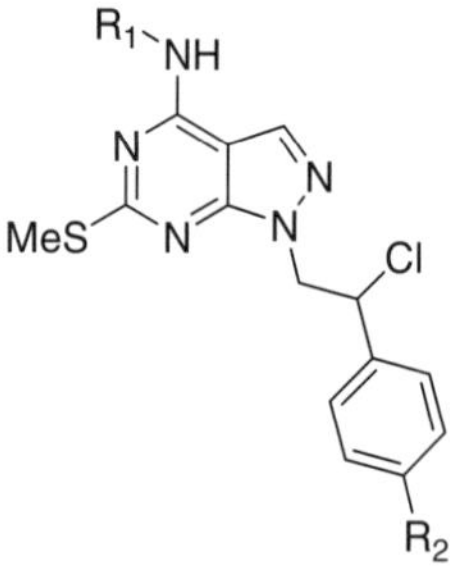

R₁ = 3-F-Ph, R₂ = F: 15
R₁ = CH₂Ph, R₂ = H: 16

9.4. Pyrrole-Pyrimidine and Other Fused Five-Membered Heterocycle Pyrimidine Inhibitors

Pyrrolopyrimidines such as CGP-76775 (17) represent selective Src kinase inhibitors with activity in animal models of osteoporosis [665]. Additional modifications of the original purine template have led to the discovery of different pyrrazoloprymidines as exemplified by 18 that displays efficacy in animal model of stroke [666]. Even more drastic modifications by the same group of researchers provided imidazopyrazine derivatives such as 19, also active in neuroprotection models [667]. Imidazopyrazine 19 was itself derived from BMS-297700 (20), an imidazoquinaxoline Lck inhibitor with anti-inflammatory activity [668].

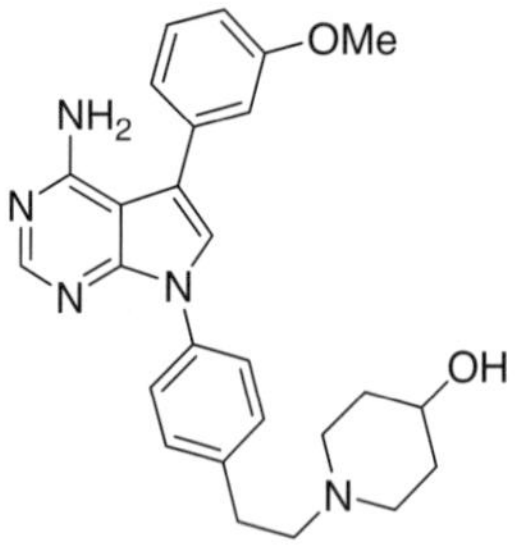

17: CGP-76775

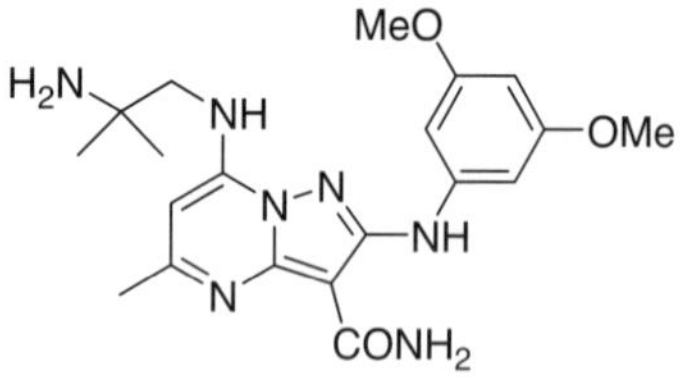

18

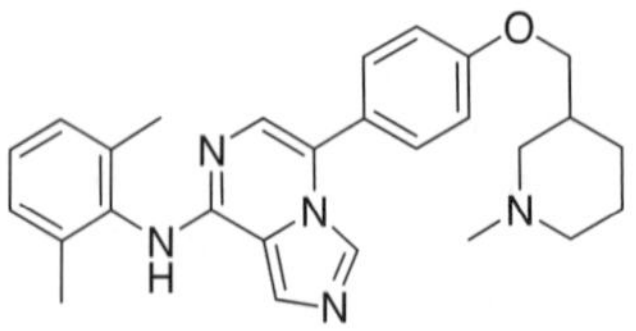

19

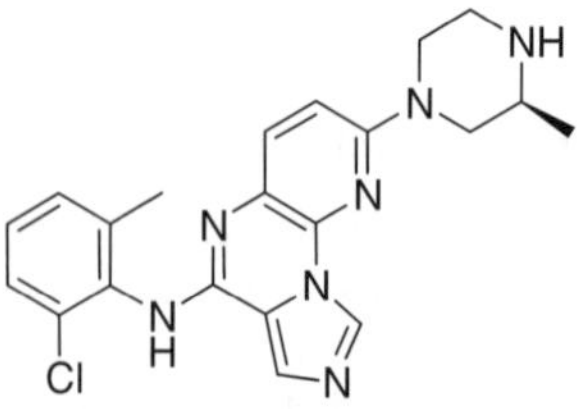

20: BMS-297700

9.5. Quinazoline-Derived Inhibitors

PD-153035 (**21**), an early EGFR kinase inhibitor, was derived from a quinazoline template [669]. This template eventually yielded successful cancer agents such as Iressa (**22**).

Elaboration of the original template toward selective Src kinase inhibition led to the identification of AZD-0530 (**23**) that also entered human clinical trials for cancer [670].

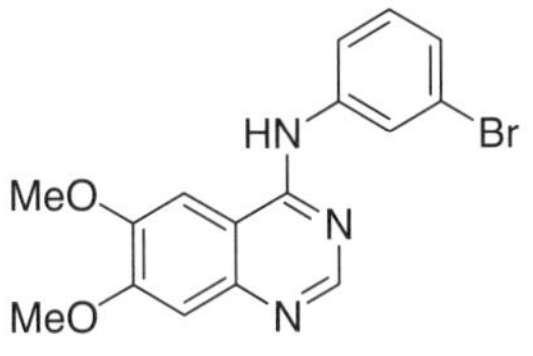

21: PD-153035

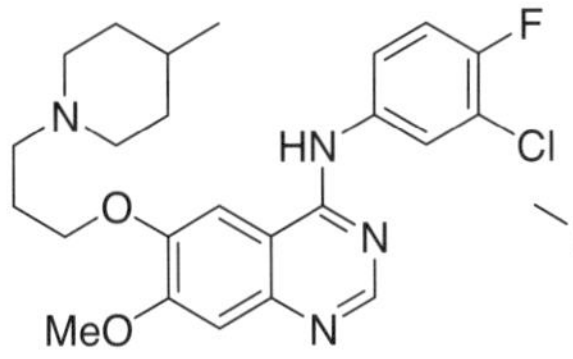

22: Iressa

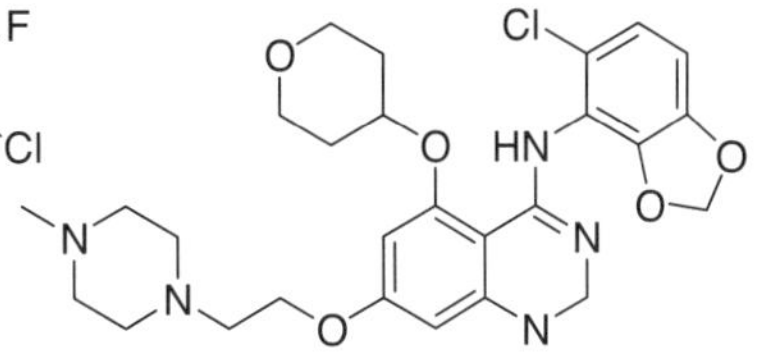

23: AZD-0530

9.6. Quinoline-Derived Inhibitors

Derived from the original quinazoline template, the 3-quinolinecarbonitrile core has yielded numerous Src kinase inhibitors [671]. From those efforts, SKI-606 (bosutinib, **24**) and SKS-927 (**25**) were identified and advanced to oncology clinical trials [672,673]. Derivative **27** was designed from analog **26** to perform as a dual inhibitor of Src kinase and iNOS [674,675].

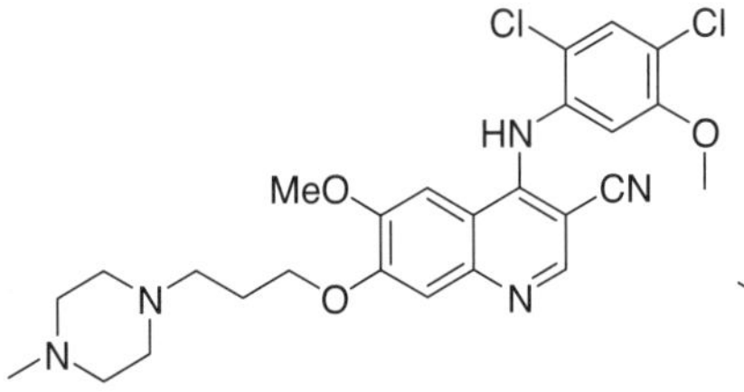

24: SKI-606 Bosutinib

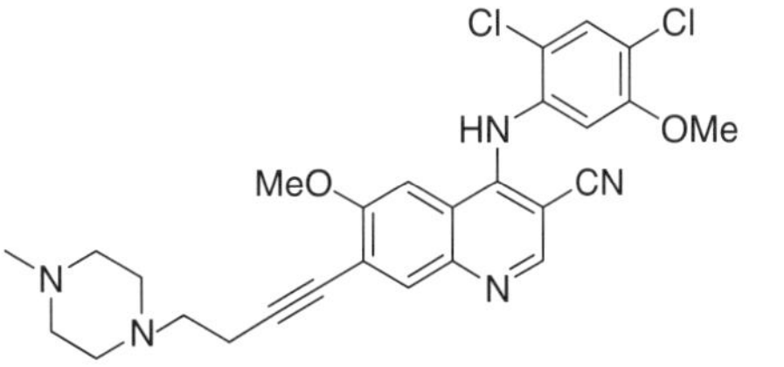

25: SKS-927

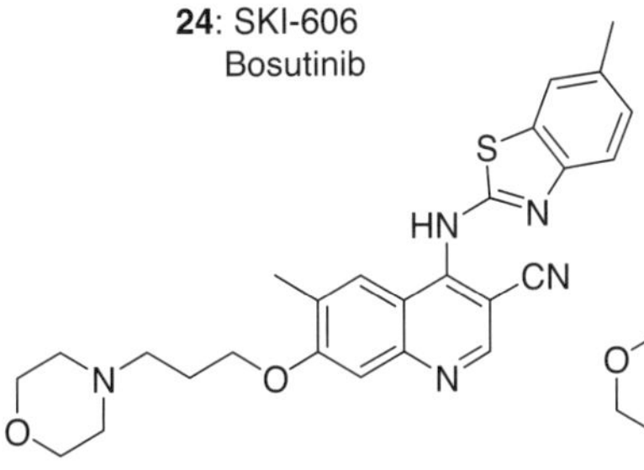

26: CPU-Y20

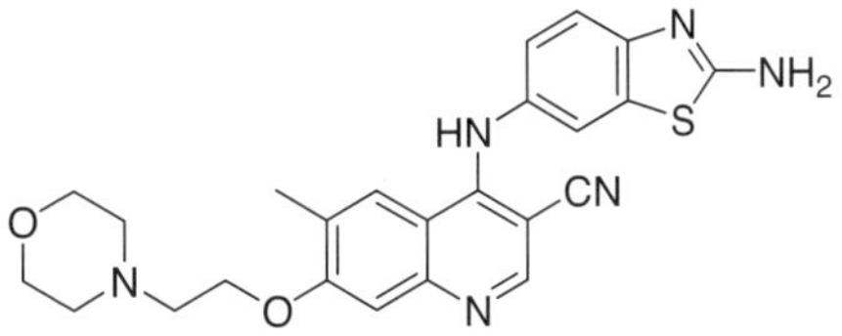

27

9.7. Pyrido-Pyrimidine and Benzotriazine-Derived Inhibitors

PD-180970 (**28**), based on a pyrido-pyrimidine scaffold has been reported as a potent Src kinase inhibitor and was shown to decrease human colon tumor growth [676]. Benzotriazines-derived inhibitors such as **29** and **30** have also demonstrated activity in human tumor cell lines and in animal models of tumor growth [677,678].

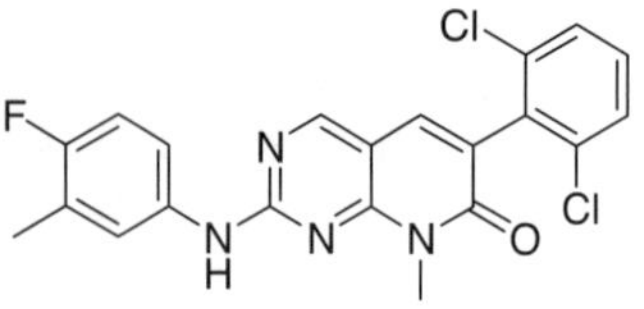

28: PD-180970

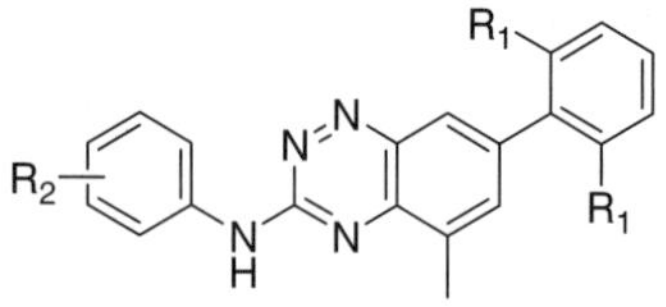

R_1 = Me, R_2 = 3-(2-(1-pyrrolidinyl)ethoxy): **29**
R_1 = Cl, R_2 = 4-(2-(1-pyrrolidinyl)ethoxy): **30**

9.8. 2-Aminothiazole Template

The aminothiazole scaffold for Src family kinase inhibition was discovered via HTS screening of the BMS sample collection. SAR optimization led to the identification of dasatinib (**31**) as a potent pan-Src kinase inhibitor that is now approved for the treatment of chronic myeloid leukemia [679,680].

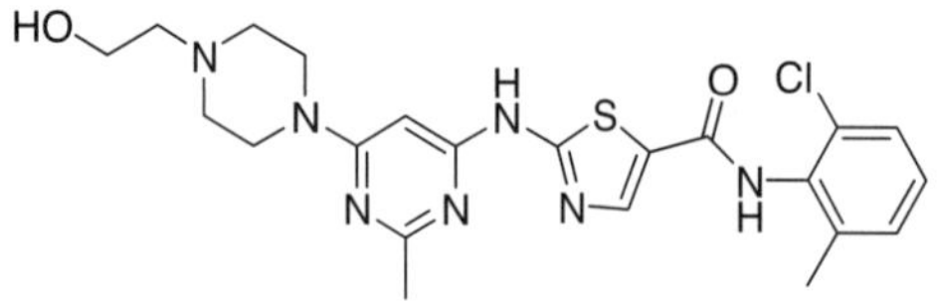

31: BMS-354825
Dasatinib

9.9. Indoline-2-One Derivatives

Researchers at Sugen have reported on an indolinone template from which potent Src kinase inhibitors such as SU-6656 could be derived [681,682].

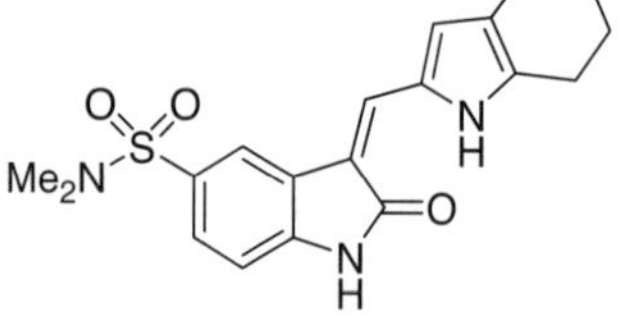

32: SU-6656

9.10. Substrate Binding Site Inhibitors

Because the substrate binding site of a kinase is much more unique than the ATP binding site, numerous efforts have been directed toward the identification of substrate binding site inhibitors in order to alleviate toxicities that might result from the cross-reactivity of classical ATP site inhibitors. Four types of substrate binding site inhibitors have been studied: peptides displaying Tyr residues to be phosphorylated, pseudo-substrate peptides, small molecules, and bisubstrates analogs [683]. Peptide substrates that interact with the area surrounding the phosphorylation site are typically weak inhibitors with limited selectivity and no proof of exclusive substrate site binding has been advanced. Efforts at replacing the Tyr residue or constraining such peptides by macrocyclization have been reported on. Following are a few examples of small-molecule substrate binding site inhibitor. β-Hydroxyisovalerylshikonin (**33**) was shown to inhibit Src autophosphorylation and to have a synergistic effect with imatinib on kinase inhibition [684–686]. Iminochromene derivative **34** was demonstrated to be a competitive substrate site inhibitor and as such displayed some selectivity against other Src family kinase members [687]. KX2-91 (**35**), a highly selective non-ATP competitive Src kinase inhibitor, has shown efficacy in cancer animal models and was the substrate site inhibitor to be advanced to human clinical trials.

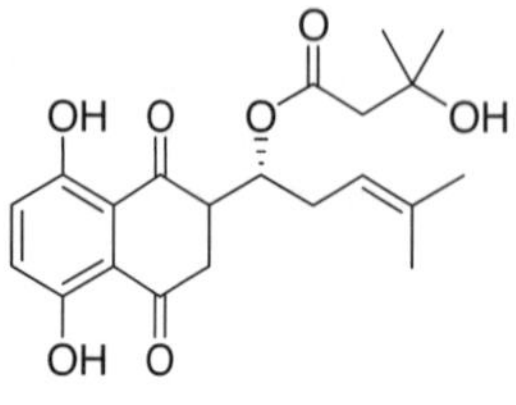

33 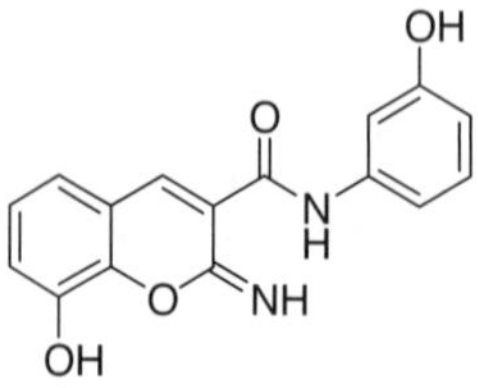   **34** 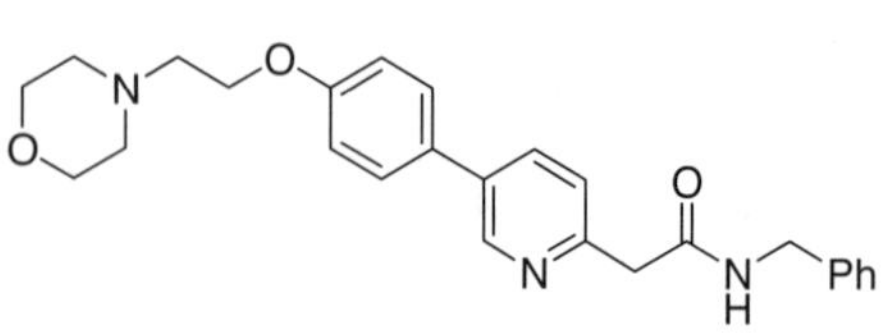   **35**: KX2-391

10. CONCLUSIONS AND FUTURE APPROACHES

While the focus of our chapter has been on the medicinal chemistry efforts surrounding stroke targets, some mention in this conclusion should be made of nonmedicinal chemistry approaches being pursued.

There is evidence that changes in intracellular and extracellular free brain magnesium concentrations following ischemia correlate with poor outcomes [688]. Consequently, the restoration of magnesium homeostasis in the brain, along with magnesium's known antiexcitotoxic actions and vascular effects, has been the rationale for the administration of magnesium as a neuroprotective treatment following cerebral ischemia [689]. Despite 23 positive preclinical (rodent) studies in which intravenous magnesium afforded a reduction in infarct volume [690], the IMAGES and FAST-MAG acute stroke clinical trials found magnesium to be largely ineffective [691,692].

Hypothermia as a mechanism of protecting brain function has received much attention in recent years. Why hypothermia is neuroprotective is not completely understood. It may be the result of multiple mechanisms occurring simultaneously that limit neuronal injury at several steps in the ischemic cascade. Proposed effects of hypothermia include reduced excitotoxicity, stabilized blood–brain barrier, anti-inflammatory effects, antioxidant effects, and reductions in cerebral metabolism [693]. The Copenhagen Stroke Study prospectively included 17 patients with acute ischemic stroke in a case–control study of early mild hypothermia by surface cooling [694]. Patients admitted within 12 h of symptom onset were cooled to 35.5°C. Compared to 56 historical controls, cases had no significant difference in 6-month Scandinavian Stroke Scale score or mortality. The first Cooling for Acute Ischemic Brain Damage (COOL AID) study was an open-label pilot study that enrolled 10 patients who were admitted with major ischemic stroke (NIHSS > 15) and treated with mild hypothermia (32°C) within 6 h of symptom onset for a mean duration of 47 h [695]. This study showed that induced hypothermia is feasible, can be initiated early, and is safe even after thrombolytic therapy.

Although clinical trials are underway that may address some unanswered questions regarding hypothermia, many areas of uncertainty may remain. These include the optimal target temperature for induced hypothermia, the therapeutic time window of initiation, the rate of cooling, the necessary duration of therapy, and the rate of rewarming. It is hoped that data from ongoing and future clinical trials will help to shed some light on these critical unknowns.

Limited, spontaneous functional recovery has been observed in a number of stroke patients and also in rodent models of cerebral ischemia. Inhibitory molecules contained in myelin (myelin-associated glycoprotein (MAG), Nogo, and oligodendrocyte myelin glycoprotein (OMgp)) are thought to contribute to the inability of the brain to recover from injury by preventing neurite outgrowth [696]. It is therefore reasonable to assume that alleviating these endogenous inhibitory factors will afford the brain's functional recovery efforts greater freedoms and opportunities to improve the long-term outcome of stroke patients. This thought has led to the development of (at least) two protein biologic approaches to stroke therapy. Antibodies against MAG [697] and Nogo [698] have shown promise in rodent models and are advancing toward the clinic and controlled, stroke trials.

Unlike treatments aimed at acute stroke which are designed to prevent the penumbral region from suffering cell loss, functional recovery targets aim to restore lost function either by enhancing the bodies own progenitor cell response or by encouraging synaptic plasticity in undamaged areas. Too often, researchers have not looked past the acute 3 h window. Rather than fully considering functional recovery, efforts are instead put on how quickly poststroke the patient can be accessed and how quickly can we deliver medicine. The benefits of functional recovery targets include the patient would be in a hospital, medically stabilized, the nature and location of the stroke, and the acute behavioral and cognitive affects are known. This allows a clearer picture of patient baseline and also allows for the enrollment of a better characterized more homogeneous patient population.

Pharmacological intervention theorized to help drive activity-induced learning and neural adaptations should be coupled with neurorehabilitation to develop training paradigms based on knowledge derived from preclinical research, clinical neuromedicine, and outcomes research [699]. Examples of past attempts include amphetamine [700], antidepressants [701], and Nogo-A [702]. While, like acute intervention strategies, functional recovery approaches have not resulted in FDA approved standards of care, it is easy to imagine how future, effective, stroke medical care, will involve distinct components—acute and chronic treatments.

"The results are promising and warrant further investigation." This mantra can be found in nearly all published reports for preclinical acute stroke treatments. This is a dangerous attitude that has led to the continued pursuit of targets and approaches that have consistently failed. Now, instead of rethinking our animal models or the nature of our medicinal approach to stroke, many prominent clinical reviews are calling for new endpoints and even new statistical methods to analyze the data. Increasingly, there are reports of smaller and smaller post-hoc patient sub-group analyses which tease out results that are "not significant but numerically superior." This invariably leads to the concluding claim that the target/molecule, "warrants further investigation."

In this review, we have summarized the state of medicinal chemistry for many targets where theorized biological relevance justified the expenditure of manpower and resources. None of these has yielded clinical therapeutic efficacy, but we have highlighted them here so that they may serve as useful jumping-off points for future forays into these areas. Given that greater than 100 stroke clinical trials have failed to show expected efficacy, despite strong results in animal models, future researchers should adopt a cautious attitude when initiating a stroke drug development program. Researchers need to invest time and effort early in the process to ensure that they understand the biology of their target; to ensure that biology is present in the animal model they choose; and most importantly to ensure that the clinical trial endpoints are reflective of the predictive consequence of therapeutic intervention.

REFERENCES

1. The National Institute of Neurological Disorders and Stroke rt-PA Stroke Study Group NEJM 1995;333:1581–1587.
2. Dirnagle et al. TINS 1997;22(9): 391–397.
3. Donnan GA, Baron JC, Ma H, Davis SM. Lancet Neurol 2009;8(3): 261–269.
4. Multicentre Acute Stroke Trial—Italy (MAST-I) Group. Lancet 1995;346(8989): 1509–1514.
5. Furlan A, Higashida R, Weschler L, Gent M, Rowley H, Kase C, et al. JAMA 1999;282: 2003–2011.
6. Hacke W, Furlan AJ, Al-Rawi Y, Davalos A, Fiebach JB, Gruber F, et al. Lancet Neurol 2008;18.
7. Hoyte L, Kaur J, Buchan AM. Exp Neurol 2004;188:200–204.
8. Chen SY, Winstein CJ. J Neurol Phys Ther 2009;33(1): 2–13.
9. Simon RP, et al. Science 1984;226:850.
10. Grotta J, et al. Stroke 1995;26:602.
11. Dyker AG, et al. Stroke 1999;30:2038.
12. Albers GW, et al. Stroke 1995;26:254.
13. Monaghan DT, Jane DE. Biology of the NMDA Receptor. Vol. 12. CRC Press, Boca raton, FL CODEN: 69LIRV ISBN: 978-1-4200-4414-0 edited by: Antonius M. VanDongen 2008; p 257–281.
14. Muir KW. Curr Opin Pharmacol 2006;6:5360.
15. Albensi BC. Curr Pharm Design 2007;13:3 185–3194.
16. Gerber AM, Vallano ML. Mini Rev Med Chem 2006;6:805–815.
17. Childers WE, Baudy RB. J Med Chem 2007;50:2557–2562.
18. McDonald JW, et al. Neurosci Lett 1990;109: 234–238.
19. Ionita CC, et al. Curr Med Chem Cent Nerv Syst Agents 2004;4:215–222.
20. Muir KW, et al. Lancet 2004;363:439–445.
21. Saver JL, et al. Stroke 2004;35:e106–e108.
22. Van Den berg WM, et al. Stroke 2005;36: 1011–1015.
23. Park CK. Ann Neurol 1988;24:543–551.
24. Albensi BC, et al. J Neurosci Res 2000;62: 177–185.

25. Lipton SA, Chen H-SV. Expert Opin Ther Targets 2005;9:427–429.
26. Chen H-SV, Lipton SA. J Neurochem 2006;97: 1611–1626.
27. Lapchak PA. Brain Res 1088;2006:141–147.
28. Chen H-SV, et al. Neuroscience 1998;86: 1121–1132.
29. Stieg PE. Eur J Pharmacol 1999;375:115–120.
30. Rao VL. Brain Res 2001;911:96–100.
31. Sobrado M. Neurosci Lett 2004;365:132–136.
32. Lee ST. J Cereb Blood Flow Metab 2006;26:536–544.
33. Lipton SA. NeuroRx 2004;1:101.
34. Albers GW, et al. J Am Med Assoc 2001;286: 2673.
35. Walker FO, et al. Clin Neuropharmacol 1989; 12:322–330.
36. Diener HC, et al. J Neurol 2002;249:561–568.
37. Davis SM, et al. Stroke 2000;31:347–354.
38. Morris GF, et al. J Neurosurg 1999;91: 737–743.
39. Kinney WA, et al. J Med Chem 1998;41:236.
40. Sun L, et al. J Pharmacol Exp Ther 2004;310: 563.
41. childers WE, et al. Drugs Fut 2002;27: 633–638.
42. Brandt MR, et al. J Pharmacol Exp Ther 2005; 313:1379–1386.
43. Sivaprakasam et al. ChemMedChem 2009;4: 110–117.
44. Conti P, et al. J Med Chem 2004;47:6740–6748.
45. Layer RT, et al. Curr Med Chem 2004;11: 3073–3084.
46. Catarzi D, et al. Curr Top Med Chem 2006;6: 809–821.
47. Jansen M, et al. Eur J Med Chem 2003;38: 661–670.
48. Madden K. Curr Med Res Opin 2002;18(Suppl 2): s27–s31.
49. Leeson PD, Iverson LL. J Med Chem 1994;37: 4053–4067.
50. Sacco RL, et al. J Am Med Assoc 2001;285: 1719–1728.
51. Lees KR, et al. Lancet 2000;355:1949–1954.
52. Haley EC, et al. Stroke 2005;36:1006–1010.
53. Baron BM, et al. J Med Chem 1005;48: 995–1018.
54. Nagata R, et al. Curr Top Med Chem 2006;6: 733–745.
55. Cai SX. Curr Top Med Chem 2006;6:651–662.
56. Nagata R, et al. Curr Top Med Chem 2006;6: 733–745.
57. Bare TM, et al. J Med Chem 2007;50: 3113–3131.
58. Nguyen KT, et al. ChemMedChem 2008;3: 1520–1524.Nguyen KT, et al. Bioorg Med Chem Lett 2009;19:502–507.
59. Berger ML, et al. Bioorg Med Chem Lett 2006; 16:2837–2841.
60. Layton ME, et al. Curr Top Med Chem 2006;6:697–709.
61. McCauley JA. Expert Opin Ther Pat 2005;15:389–407.
62. Chazot PL. Curr Med Chem 2004;11:389–396.
63. Doyle KM, et al. Behav Pharmacol 1998;9: 671–681.
64. Guscott MR, et al. Eur J Pharmacol 2003;476: 193–199.
65. Higgins GA, et al. Psychopharmacology 2005; 179:85–98.
66. Kundotriene J, et al. J Neurotrauma 2004;21: 83–93.
67. Wang CX, et al. Curr Drug Targets CNS Neurol Disord 2005;4:143–151.
68. Gotti B, et al. J Pharmacol Exp Ther 1988; 247:1211.
69. Toulmond S, et al. Brain Res 1993;620:32.
70. fisher G, et al. Abstr Soc Neurosci 1996;22: 1760.
71. Okiyama K, et al. J Neurotrauma 1997;14:211.
72. Tsuchida E, et al. J Neurotrauma 1997;14:409.
73. Di X, et al. Stroke 1997;28:2244.
74. bullock MR, et al. Ann NY Acad Sci 1999; 890:51.
75. Kemp JA, et al. Nat Neurosci 2002;5(Suppl): 1039.
76. Merchant RE, et al. Ann NY Acad Sci 1999;890: 42–50.
77. bullock MR, et al. Ann NY Acad Sci 1999;890:51.
78. Sang CN, et al. Abstr Soc Neurosci 2003; Abstract 814.9.
79. Tahirovic YA, et al. J Med Chem 2008;51: 5506–5521.
80. Hofner G, et al. Bioorg Med Chem Lett 2005;15:2231–2234.
81. Kawai M, et al. Bioorg Med Chem Lett 2007;17: 5558–5562.
82. Kawai M, et al. Bioorg Med Chem Lett 2007;17: 5333–5336.
83. McCauley JA, et al. J Med Chem 2004;47: 2089–2096.
84. Barta-Szalai J, et al. Bioorg Med Chem Lett 2004;14:3953–3956.

85. Wright JL, et al. J Med Chem 2000;43: 3408–3419.
86. Borza I, et al. Bioorg Med Chem Lett 2004;13: 3859–3861.
87. Borza I, et al. J Med Chem 2007;50:901–914.
88. liverton NJ, et al. J Med Chem 2007;50:807–819.
89. Curtis NR, et al. Bioorg Med Chem Lett 2003;13:693–696.
90. Claiborne CF, et al. Bioorg Med Chem Lett 2003;13:697–700.
91. Borza I, et al. Bioorg Med Chem Lett 2005;15:5439–5441.
92. Nguyen KT, et al. Bioorg Med Chem Lett 2007;17:3997–4000.
93. Alanine A, et al. Bioorg Med Chem Lett 2003;13:3155–3159.
94. Pinard E, et al. Bioorg Med Chem Lett 2002;12:2615–2619.
95. Buttelmann B, et al. Bioorg Med Chem Lett 2003;13:829–832.
96. Buttelmann B, et al. Bioorg Med Chem Lett 2003;13:1759–1762.
97. Borza I, et al. Bioorg Med Chem Lett 2007;17: 406–409.
98. Marinelli L, et al. ChemMedChem 2007;2: 1498–1510.
99. Gitto R, et al. ChemMedChem 2008;3: 1539–1548.
100. Rosini M, et al. J Med Chem 2008;51: 4381–4384.
101. Hunter AJ. Int Rev Neurobiol 1997;40:95–108.
102. Horn J, Limburg H. Stroke 2001;32:570–576.
103. Winquist Expert Opin Investig Drugs 2005;14: 579–592.
104. Takahara Curr Top Med Chem 2009;9: 377–395.
105. Niimi Recent Pat CNS Drug Discov 2009;4:96–111.
106. Nishizawa Life Sci 2001;69:369–381.
107. Bulaj G. Curr Pharm Des 2008;14:2462–2479.
108. Wells L. Nursing 2008;38:19–24.
109. Weber M. Presented at 217th National Meeting of American Chemical Society, March 21–25, Anaheim, CA 1999.
110. Szoke BG J. Med Chem 1999;42:4239–4249.
111. Vartanian MG. Bioorg Med Chem Lett 1999;9:907–912.
112. Toda M. Bioorg Med Chem Lett 2001;11: 2067–2070. Toda M. Bioorg Med Chem Lett 2002;12:915–918.
113. Ukai Y. Life Sci 2000;67:2331–2343.
114. Kimura K. Jpn J Pharmacol 1997;73:193–195.
115. Ukai Y. Brain Res 2001;890:162–169.
116. Ukai Y. Pharma Jpn 2003;18:1855.
117. Bostwick JR. Abstr Soc Neurosci 1999;25(Part 2): 896.
118. Yonaga M. Bioorg Med Chem Lett 2003;13: 919–922.
119. Snutch TP. WO01045709. 2001.
120. Snutch TP. WO06105670. 2006.
121. Chakravarty PK. WO07084394. 2007. Duffy JL. WO08066803. 2008.
122. Yao J. WO08008398. 2008.
123. Shoji M. Bioorg Med Chem 2006;14: 5333–5339.
124. Hunter AJ. Eur J Pharmacol 2000;401: 419–428.
125. O'Neill MJ. Eur J Pharmacol 2000;408: 241–248.
126. Tatsuta M. WO2007/125398.
127. Okada H. WO08143263. 2008.
128. James IF. Bioorg Med Chem Lett 2007;17: 662–667.
129. Shoji M. Bioorg Med Chem 2008;18: 4813–4816.
130. Lataste X. Drugs 1990;40:52–57.
131. Reivch M. Stroke 1989;20:1531–1537.
132. Donnan GA. Stroke 1990;30:1417–1423.
133. Wahlgren NG. Stroke 2000;31:1250–1255.
134. Limburg M. Stroke 2001;32:461–465.
135. The European Ad Hoc Consensus Group Cerbrovasc Dis 1998;8:59–72.
136. Horn J. Stroke 2001;32:2433–2438.
137. Gelmers HJ. N Engl J Med 1988;318:203–207.
138. Trust Study Group Lancet 1990;2,1205–1209.
139. The American Nimodipine Study Group Stroke 1992;23,3–8.
140. Horn J. Stroke 2001;32:461–465.
141. Saido TC, Kawashima S, Tani E, Yokota M. Neurosci Lett 1997;227(2): 75–78.
142. Rami A. Neurobiol Dis 2003;13(2): 75–88.
143. Saido TC, Sorimachi H, Suzuki K. FASEB J 1994;8(11): 814–822.
144. Rami A, Volkmann T, Agarwal R, Schoninger S, Nürnberger F, Saido TC, Winckler J. Neurosci Res 2003;47(4): 373–382.
145. Bartus RT, Baker KL, Heiser AD, Sawyer SD, Dean RL, Elliott PJ, Straub JA. J Cereb Blood Flow Metab 1994;14(4): 537–544.
146. Rami A, Agarwal R, Botez G, Winckler J. Brain Res 2000;866:299–312.

147. Markgraf CG, Velayo NL, Johnson MP, McCarty DR, Medhi S, Koehl JR, Chmielewski PA, Linnik MD. Stroke 1998;29(1): 152–158.
148. Wells GJ, Bihovsky GJ. Expert Opin Ther Pat 1998;8:1707–1727.
149. Ray SK. Curr Med Chem 2006;13:3425–3440.
150. Leung-Toung R, et al. Curr Med Chem 2006;13:547–581.
151. Aoyagi T, et al. J Antibiot 1969;22:283–286.
152. Lee KS, et al. Proc Natl Acad Sci USA 1991;88:7233.
153. Rami A, Krieglstein J. Brain Res 1993;609:67.
154. Tsujinaka T, et al. Biochem Biophys Res Commun 1988;153:1201.
155. Medhi S, et al. Biochem Biophys Res Commun 1988;157:1117.
156. Das A, et al. J Neurosci Res 2005;81:551.
157. Das A, et al. Brain Res 1084;2006:146.
158. Ray SK. Neuroscience 2006;139:577.
159. Hong SC, et al. Stroke 1994;25:663.
160. Li PA, et al. Neurosci Lett 1998;247:17.
161. Makgraph CG, et al. Stroke 1998;29:152.
162. Kawamura M, et al. Brain Res 1027;2005:59.
163. Sakamoto YR, et al. Curr Eye Res 2000;21: 571.
164. Tamada Y, et al. Curr Eye Res 2001;22:280.
165. Kupina NC, et al. J Neurotrauma 2001;18: 1229.
166. Inoue J, et al. J Med Chem 2003;46:868.
167. Chatterjee S, et al. J Med Chem 1998;41: 2663–2666.
168. Wells GJ, Tao M, Josef KA, Bihovsky R. J Med Chem 2001;44:3488–3503.
169. Abell AD, et al. J Med Chem 2007;50: 2916–2920.
170. Sanders ML, Donkor IO. Bioorg Med Chem Lett 2006;16:1965–1968.
171. Donkor IO, Korukonda R. Bioorg Med Chem Lett 2008;18:4806–4808.
172. Kondor I, et al. J Med Chem 2006;49: 5282–5290.
173. WO9621655. 1996.
174. Auvin S, et al. Bioorg Med Chem Lett 2004;14: 3825–3828; Bioorg Med Chem Lett2006;16: 1586–1589.
175. Pignol B, et al. J Neurochem 2006;98: 1217–1228.
176. Shirasaki Y, et al. J Med Chem 2006;49: 3926–3932.
177. US patent 5,639,732. 1997.
178. Morinaka AR, et al. J Am Chem Soc 1993;115: 1175.
179. Bartus RT, et al. J Cereb Blood Flow Metab 1994;14:537.
180. Herbeson S, et al. J Med Chem 1994;37:2918.
181. Bartus RT, et al. Stroke 1994;25:2265.
182. Li Z, et al. J Med Chem 1996;39:4089.
183. Shirazaki Y, et al. Bioorg Med Chem 2005;13:4473–4484.
184. Shirazaki Y, et al. Bioorg Med Chem 2006;14:5691–5698.
185. Shirazaki Y, et al. J Ocul Pharmacol Ther 2006;22:417.
186. Oka T, et al. Neuroscience 2006;141:2139.
187. Koumura A, et al. Neuroscience 2008;157: 309–318.
188. Lee KS, et al. Bioorg Med Chem 2005;15: 2857–2860.
189. Nam DH, et al. Bioorg Med Chem 2008;18: 205–209.
190. Zhang Y, et al. Bioorg Med Chem 2008; 19:502–507.
191. Lascop C, et al. Bioorg Med Chem 2005;15: 5176–5181.
192. Hanadar K, Chicharro et al. ChemMedChem 2006;1:710–714.
193. Hanada K, et al. Agric Biol Chem 1978;42: 523–528.
194. Inuzuka T, et al. Brain Res 1990;526:177–179.
195. Chatterjee S, et al. J Med Chem 1997;40: 3820–3828.
196. US patent 5,658,906. 1997.
197. Dolle RE, et al. J Med Chem 1995;38:220–222.
198. WO9837084. 1998.
199. Lin GD, et al. Nat Struct Biol 1997;4:539–547.
200. Farkas B, et al. Brain Res 1024;2004: 150–158.
201. Alnemri ES, Livingston DJ, Nicholson DW, Salvesen G, Thornberry NA, Wong WW, Yuan J. Cell 1996;87(2): 171.
202. Linton SD. Curr Top Med Chem 2005;5: 1697–1717.
203. Leung-Toung R, et al. Curr Med Chem 2006;13:547–581.
204. Weber IT, et al. Mini Rev Med Chem 2008;8:1154–1162.
205. Feng Y, et al. Neurosci Lett 2003;344:201.
206. Shahripour AB, et al. Bioorg Med Chem 2002;10:31–40.
207. Shahripour AB, et al. Bioorg Med Chem Lett 2001;11:2779–2782.

208. Vazquez J, et al. ChemMedChem 2008;3: 979–985.
209. Choong IC, et al. J Med Chem 2002;45: 5005–5022.
210. Erlanson DA, et al. Nat Biotechnol 2003;21: 308–314.
211. Allen DA, et al. Bioorg Med Chem Lett 2003;13:3651–3655.
212. Linton SD, et al. Bioorg Med Chem Lett 2002;12:2969–2971 and 2973–2975.
213. Siegmund B, Zeitz M. Drugs 2003;6: 154–158.
214. Ekici ZD, et al. J Med Chem 2004;47: 1889–1892.Ekici ZD, et al. J Med Chem 2006;49:5728–5749.
215. Li H, et al. Stroke 2000;31:176.
216. Rabuffetti M, et al. J Neurosci 2000;20:4398.
217. Fink K, et al. J Cereb Blood Flow Metab 1998;18:1071.
218. Gillardon F, et al. Neuroscience 1999;93:1219.
219. Knoblach SM, et al. J Cereb Blood Flow Metab 2004;24:1119–1132.
220. Yang W, et al. Br J Pharmacol 2003;140: 402–412.
221. Cai SX, et al. Bioorg Med Chem Lett 2004;14: 5295–5300.
222. Wang Y, et al. Bioorg Med Chem Lett 2005;15:1379–1383.
223. Wang Y, et al. Bioorg Med Chem Lett 2005;17: 6178–6182.
224. Holgen NC, et al. J Pharm Exp Ther 2001;297: 811–818.
225. Deckwerth TL, et al. Drug Dev Res 2001;52: 579–586.
226. McKerrow JH. Int J Parasitol 1999;29: 833–837.
227. Engle JC, et al. J Exp Med 1998;188:725–734.
228. Eichhold TH, et al. J Pharm Biomed Anal 1997;16:459–467.
229. Choong IC, et al. J Med Chem 2002;45: 5005–5022.
230. Erlanson DA, et al. Nat Biotechnol 2003;21: 308–314.
231. Allen DA, et al. Bioorg Med Chem Lett 2003; 13:3651–3655.
232. Linton SD, et al. Bioorg Med Chem Lett 2004;14:2685–2691.
233. Linton SD, et al. J Med Chem 2005;48: 6779–6782.
234. Mignon A, et al. Am J Respir Crit Care Med 1999;159:1308–1315.
235. Canbay A, et al. J Pharm Exp Ther 2003;308: 1191–1196.
236. Natori S, et al. Transplantation 1999;68: 89–96.
237. Natori S, et al. Liver Transpl 2003;9:278.
238. Holgen NC, et al. J Pharm Exp Ther 2004;309: 634–640.
239. Valentino KL, et al. Int J Clin Pharm Ther 2003;41:441–449.
240. Schiff ER, et al. Gastroenterelogy 2004;126 (Suppl 2):
241. Poordad FF. Curr Opin Investig Drugs 2004;5:1198–1204.
242. Baskin-Bey S, et al. Am J Transplant 2007;7:218–225.
243. Ullman BR, et al. Bioorg Med Chem Lett 2005;15:3632–3636.
244. Ueno H, et al. Bioorg Med Chem Lett 2009;19: 199–202.
245. Becker JW, et al. J Med Chem 2005;15: 1173–1180.
246. Grimm EL, et al. Bioorg Med Chem 2004;12: 845–851.
247. Isabel E, et al. Bioorg Med Chem Lett 2003;13: 2137–2140.
248. Han Y, et al. Bioorg Med Chem Lett 2005;15: 1173–1180.
249. Mellon C, et al. Bioorg Med Chem Lett 2004;47:2466–2474.
250. Isabel E, et al. Bioorg Med Chem Lett 2007;17: 1671–1674.
251. Hanet BH, et al. J Biol Chem 2002;277: 30128–30136.
252. Blomgrent K, et al. J Biol Chem 2001;276: 1091.
253. Chan SL, et al. J Neurosci Res 1999;58:167.
254. Rami A, et al. Brain Res 2000;866:299.
255. Ray SK, et al. Neuroscience 2006;139:577.
256. Higuchi M, et al. J Biol Chem 2005;280:15229.
257. Colucci J, et al. Br J Pharmacol 2004;141: 689–697.
258. Robertson GS, et al. Brain Pathol 2000;10: 283–292.
259. Schultz JB. Curr Opin Cent Peripher Nerv Syst Invest Drugs 2000;2:417–422.
260. Rideout H, et al. Histol Histopathol 2001;16:895–908.
261. Lee D, et al. J Biol Chem 2002;275: 16007–16014.
262. Lee D, et al. J Med Chem 2001;44:2015.
263. Chapman JG, et al. J Pharmacol 2002;456:59.

264. Chu W, et al. J Med Chem 2007;50:3751–3755.
265. Kravchenko DV, et al. Farmaco 2005;60:804.
266. Kravchenko DV, et al. J Med Chem 2005;48:3680.
267. Kravchenko DV, et al. Eur J Med Chem 2005;40:1377.
268. Kravchenko DV, et al. Bioorg Med Chem Lett 2005;15:1841.
269. Chen YH, et al. J Med Chem 2006;49:1613.
270. Ma XQ, et al. J Biochem Cell Biol 2007;85:56.
271. Zhang YH, et al. FEBS J 2006;273:4842.
272. Fattorusso R, et al. J Med Chem 2005;48:1649–1656.
273. Thanh Le G, et al. J Am Chem Soc 2006;128:12396–12397.
274. Brunk UT, Dalen H, Roberg K, Hellquist HB. Free Radic Biol Med 1997;23:616–626.
275. Yamashima T, Kohda Y, Tsuchiya K, Ueno T, Yamashita J, Yoshioka T, Kominami E. Eur J Neurosci 1998;10:1723–1733.
276. Yamashima T. Prog Neurobiol 2000;62:273–295.
277. Yoshida M, Yamashima T, Zhao L, Tsuchiya K, Kohda Y, Tonchev AB, Matsuda M, Kominami E. Acta Neuropathol 2002;104:267–272.
278. Frlan R. Curr Med Chem 2006;13:2309–2327.
279. Leung-Toung R, et al. Curr Med Chem 2006;13:547–581.
280. McConnell et al. J Med Chem 1993;36:1084.
281. Scheidt KA, et al. Bioorg Med Chem 1998;6:2477.
282. Yamashima T, et al. Nutr Neurosci 2001;4:389–397.
283. Katunuma N, et al. Biochem Biophys Res Commun 2000;267:850–854.
284. Abato P, et al. J Med Chem 1999;42:4001–4009.
285. Ando R, et al. Bioorg Med Chem 1999;7:571.
286. Potts KT, et al. Chem Rev 1974;74:189.
287. Bincoletto C, et al. Bioorg Med Chem 2005;13:3047.
288. Greenspan PD, et al. J Med Chem 2001;44:4524.
289. Greenspan PD, et al. Bioorg Med Chem Lett 2003;13:4121.
290. Yadav MR, et al. J Enzyme Inhib Med Chem 2008;23:190.
291. Baird IR, et al. Inorganica Chim Acta 2006;359:2736–2750.
292. Mosi R, et al. J Med Chem 2006;49:5262–5272.
293. Hanada K, et al. Agric Biol Chem 1978;42:523–528.
294. Tamai M, et al. J Pharmacobiodyn 1986;9:672.
295. Sumiya M, et al. Chem Pharm Bull 1992;40:299.
296. Yamashima T, et al. Eur J Neurosci 1998;10:1723–1733.
297. Tsuchiya K, et al. Exp Neurol 1999;155:187–194.
298. Yoshida M, et al. Acta Neuropathol 2002;104:267–272.
299. Tsubokawa T, et al. J Neursci Res 2006;84:832–840.
300. Tsubokawa T, et al. Stroke 2006;37:1888–1894.
301. Ray SK, et al. Brain Res 2001;916:115–126.
302. Miyahara T, et al. Jpn J Clin Pharmacol Ther 1985;16:537–546.
303. Fukushima K, et al. Toxicol Appl Pharmacol 1990;105:1–12.
304. Tachikura T, et al. Acta Pediatr Jpn 1990;3:495–501.
305. Toyo-oka T, et al. Arzneimittelforschung 1986;36:671–675.
306. Eilon GF, et al. Cancer Chemother Pharmacol 2000;25:1215.
307. Schirmeister T, et al. Bioorg Med Chem 2000;8:1281.
308. Martichonok V, et al. J Med Chem 1995;38:3078.
309. Schirmeister T, et al. J Med Chem 1999;42:560.
310. Vicick R, et al. ChemMedChem 2006;1:1126–1141.
311. Leung-Toung R, et al. Bioorg Med Chem 2003;11:5529–5537.
312. Honn K, et al. Science 1982;217:540.
313. Krantz A, et al. Biochemistry 1991;30:4678.
314. Dai I, et al. Biochemistry 2000;34:6498.
315. Gray J, et al. J Biol Chem 2001;276:32750–32755.
316. Seyfried DM, et al. Brain Res 2001;901:94–101.
317. Anagli J, et al. Biochem Biophys Res Commun 2008;366:86–91.
318. Zhou NE, et al. Bioorg Med Chem Lett 2003;13:139.
319. Zhou NE, et al. Bioorg Med Chem Lett 2002;12:3413.
320. Zhou NE, et al. Bioorg Med Chem Lett 2002;12:3417.
321. Epple R, et al. Bioorg Med Chem Lett 2007;17:1254–1259.

322. Cuhna RL, et al. Bioorg Med Chem Lett 2005;15:755.
323. Casini A, et al. J Med Chem 2008;51:6773–6781.
324. Lim IT, et al. J Am Chem Soc 2004;126:10271.
325. Darkins P, et al. Chem Biol Drug Des 2007;69:170–179.
326. Ettari R, et al. J Med Chem 2008;51:988–996.
327. Iley J, et al. Bioorg Med Chem Lett 2006;16:2738–2741.
328. Yamamoto A, et al. Biochim Biophys Acta 2002;139:244–251.
329. Wieczerzak E, et al. J Med Chem 2002;45:4202–4211.
330. Wieczerzak E, et al. J Pept Sci 2007;13:536–543.
331. Myers MC, et al. Bioorg Med Chem Lett 2008;18:210–214 and 3646–3651.
332. Schenker P, et al. Protein Sci 2008;17:2145–2155.
333. Murayama S, et al. Tetrahedron Lett 2008;49:4186–4188.
334. Hoang V, et al. Bioorg Med Chem Lett 2008;18:2083–2088.
335. Li Y, et al. Bioorg Med Chem Lett 2008;18:6130–6134.
336. Marquis R, et al. J Med Chem 2005;48:6870–6878.
337. Myers M, et al. Bioorg Med Chem Lett 2007;17:4761–4766.
338. Love S, Brain Pathol 1999;9:119–131.
339. Coyle JT, Puttfarcken P, Science 1993;262:689–695.
340. Dirnagl U, Iadecola C, Moskowitz MA, Trends Neurosci 1999;22:391–397.
341. Taylor JM, Crack PJ, Clin Exp Pharmacol Physiol 2004;31:397–406.
342. Sugawara T, Chan PH, Antioxid Redox Signal 2003;5:597–607.
343. Esterbauer H, Schaur RJ, Zollner H, Free Radic Biol Med 1991;11:81–128.
344. Cheung Y. Neurochem Res 1994;19:1557–1564.
345. Kuroda S. J Cereb Blood Flow Metab 1999;19:778–787.
346. Marshall JWB. Stroke 2003;34:2228–2233.
347. Lees KR. New Engl J Med 2006;354:588–600.
348. Kavanaugh RJ. Br J Anaesth 2001;86:110–119.
349. Lythgoe DJ. Br J Pharmacol 1990;100:454.
350. Mori E. Cerebrovasc Dis 1995;5:342–349.
351. Xue D. Stroke 1992;23:894–899.
352. Young W. Stroke 1988;19:1013–1019.
353. Hall ED. NeuroRx 2004;1:80–100.
354. Sena E. Stroke 2007;38:388–394.
355. The RANTTAS Investigators Stroke 1996;27:1453–1458.
356. van der Worp HB. Neurology 2002;58,133–135.
357. Parnham M. Exp Opin Invest Drugs 2000;9:607–619.
358. Noguchi N. Biochem Pharmacol 1992;44:39–44.
359. Herin GA. J Neurochem 2001;78:1307–1314.
360. Yamagata K. Neuroscience 2008;153:428–435.
361. Imai H. Free Radic Biol Med 2003;34:56–63.
362. Kalayci M. Neurochem Res 2005;30:403–410.
363. Lapchak PA. Stroke 2003;34:2013–2018.
364. Satoh T. Neurosci Lett 2004;371:1–5.
365. Salom JB. Eur J Pharmacol 2004;496:55–62.
366. Yamaguchi T. Stroke 1998;29:12–17.
367. Ogawa A. Cerebrovasc Dis 1999;9:112–118.
368. Kawai H. J Pharmacol Exp Ther 1997;281:921–927.
369. Watanabe T. J Pharmacol Exp 1994;268:1597–1604.
370. The Edaravone Acute Brain Infarction Study Group Cerebrovasc Dis 2003;15,222–229.
371. Toyoda K. J Neurol Sci 2004;221:11–17.
372. Lapchak PA. Expert Opin Emerg Drugs 2007;12:389–406.
373. Feng S;The Cochrane Collaboration. 2008; 3.
374. Hishida A. Clin Exp Nephrol 2007;11:292–296.
375. Lapchak PA. Expert Opin Emerg Drugs 2007;12:389–406.
376. Watanabe T. Cardiovasc Ther 2008;26:101–104.
377. Nick H. Curr Opin Chem Biol 2007;11:419–423.
378. Birch N. Expert Opin Ther Pat 2006;16:1533–1556.
379. Faa G. Coord Chem Rev 1999;184:291–310.
380. Selim MH. Ageing Res Rev 2004;3:345–353.
381. Winterbourn C. Toxicol Lett 1995;82:969–974.
382. Castellanos M. Brain Res 2002;952:1–6.
383. Millan M. Stroke 2006;38:90–95.
384. Freret T. Eur J Neurosci 2006;23:1757–1765.
385. Soloniuk DS. Surg Neurol 1992;38:110–113.
386. Kristian T. Stroke 1998;29:705–718.
387. Nicholson C. Proc Natl Acad Sci 1977;74:1287–1290.
388. Gribkoff VK. Nat Med 2001;7:471–477.
389. Gribkoff VK. Expert Opin Investig Drugs 2005;14:579–592.

390. Mackay KB. Curr Opin Investig Drugs 2001;2:820–823.
391. Romine JL. J Med Chem 2007;50:528–542.
392. Hewawasam P. Bioorg Med Chem Lett 2003; 1695–1698.
393. Green AR. Neuropharmacology 2000;39: 1483–1494.
394. Green AR. Pharmacol Ther 1998;80:123–147.
395. Riveros N. Neuroscience 1999;17:541–546.
396. Cross AJ. Br J Pharmacol 1995;114: 1625–1630.
397. Wahlgren NG. Stroke 1999;30:21–28.
398. Lyden P. Stroke 2002;33:122–129.
399. de Paulis T. Curr Opin Investig Drugs 2003;4:863–867.
400. Tateishi N. J Cereb Blood Flow Metabol 2002;22:723–734.
401. Matsui T. J Cereb Blood Flow Metab 2002;22: 711–722.
402. Pettigrew LC. J Neurol Sci 2006;251:50–56.
403. Pettigrew LC. J Neurol Sci 2006;251:57–61.
404. Chen JKL, et al. J Neurochem 1997;69: 232–245.
405. Elliasson MJ, et al. Nat Med 1997;3: 1089–1095.
406. Elliasson MJ, et al. Nat Med 1997;3: 1089–1095.
407. lo E, et al. Stroke 1998;29:830–836.
408. Woon ECY. Curr Med Chem 2005; 2373–2392.
409. Szabó C. Curr Med Chem 2003;10:321.
410. Costantino G. J Med Chem 2001;44:3786.
411. Banasik M. J Biol Chem 1992;267:1569.
412. Watson CY. Bioorg Med Chem 1998;6:721.
413. Yoshida S. J Antibiot 1991;44:111.
414. Chiarugi A. JPET 2003;305:943.
415. Ferraris D. Bioorg Med Chem 2003;11:3695.
416. Iwashita A. JPET 2004;310:1114.
417. Kinoshita T. FEBS Lett 2004;556:43.
418. White AW. J Med Chem 2000;43:4084.
419. Iwashita A. FEBS Lett 2005;579:1389.
420. Banasik M, J Biol Chem 1992;267:1569.
421. Ishida J. Bioorg Med Chem 2006;14:1378.
422. Pellicciari R. ChemMedChem 2008; 914.
423. Abdelkarim GE. Int J Mol Med 2001; 255.
424. Iwashita A. JPET 2004;310:425.
425. Menear KA. J Med Chem 2008; 6581.
426. Penning TD. J Med Chem 2009; 514.
427. Sherman B. WO2009064738. 2009.
428. Morrow DA. J Thromb Thrombolysis 2009; 359.
429. Shibata A. Oncogene 2005;24:1328.
430. Ryu H, et al. PNAS 2003;100:4281–4286.
431. Endres M, et al. J Neurosci 2000;20: 3175–3181.
432. Ren M, et al. J Neurochem 2004;89:1358–1367.
433. Fattorri J Med Chem 2008;51:1505–1529.
434. Miyata Curr Pharm Des 2008;42:529–544.
435. Ottow Annu Rep Med Chem 2007;42: 337–348.
436. Blumcke Expert Opin Investig Drugs 2008;17:169–184.
437. Monneret Eur J Med Chem 2005;40:1–13.
438. Glaser Biochem Pharmacol 2007;74:659–671.
439. Chuang DM. JPET 2007;321:892–901.
440. Chiarugi A. Mol Pharmacol 2006;70: 1876–1884.
441. Miller TA. Bio Med Chem Lett 2007;17: 3969–3971.
442. Michaelides MR. Bioorg Med Chem Lett 2003;13:3817–3820.
443. WO2006075888. 2006.
444. WO2005007091. 2004.
445. Joel SP. J Med Chem 2006;49:800–805.
446. Bhalla K. Clin Cancer Res 2006;12:4628–4635.
447. Kalvinish I. Helv Chim Acta 2005;88: 1630–1657.
448. Debono J. Clin Cancer Res 2008;14:804–810.
449. Ethirajulu K.20th EORTCNCI-AACR Symposium, Geneva, Switzerland, October 21–24 2008.
450. Kimoji K. WO2004063169. 2004.
451. Xiao XY. WO2004082638. 2004.
452. Bang YJ. WO200387066. 2003.
453. Dalrymple SA. Mol Cancer Ther 2006;5: 1309–1317.
454. Miyachi H. Bioorg Med Chem 2005;7: 4427–4431.
455. Miyachi H. Bioorg Med Chem 2006;8: 7625–7651.
456. Rambaldi A. Leukemia 2007; 1892–1900.
457. Meineke PT. WO2006017214. 2006.
458. Tibes U. WO2005121134. 2005.
459. Price S. WO2004013130. 2004.
460. Yin Z. WO2003070691. 2003.
461. Miyata N. Bioorg Med Chem 2004;14: 3313–3317.
462. Gill MIA. Bioorg Med Chem 2006;14: 3320–3329.
463. Hagenbeck A. Cancer Res 1993;53:3008–3014.

464. Papot S. Bioorg Med Chem 2008;16:8109–8116.
465. Kell J. Curr Opin Investig Drugs 2007;8:485–492.
466. Njar VCO. Bioorg Med Chem 2008;16:3352–3360.
467. Guidotti A. PNAS 2006;13:1587–1592.
468. Miller TA. Bioorg Med Chem Lett 2007;17:4619–4624.
469. Davidsen SK. Bioorg Med Chem Lett 2002;12:3443–3447.
470. Michaelides MR. Bioorg Med Chem Lett 2003;13:3901–3913.
471. Michaelides MR. Bioorg Med Chem Lett 2003;13:3331–3335.
472. Miyata N. J Med Chem 2009;52:2909–2922.
473. Celera Genomics. WO2006069096. 2006.
474. Chuang D. JPET 2007;321:892–901.
475. Jaye M. JPET 2003;307:720–728.
476. Matsunga S. BMCL 2004;14:2617–2620.
477. Yasuda YJ. J Antibiot 1990;43,1524–32.
478. Beppu T. J Biol Chem 1993;268:22429–22435.
479. Vincent SR, Progr Neurobiol 1994;42:129–160.
480. reviewed in JP, Bolanos A, Almeida V, Stewart S, Peuchen JM, Land JB, Clark SJR, Heales J Neurochem 1997;68:2227–2240.
481. Beckman JS, Beckman TW, Chen J, Marshall PA, Freeman BA, Proc Natl Acad Sci USA 1990;87:1620–1624.
482. Wilmot MR, Bath PMW. Expert Opin Investig Drugs 2003;12:455–470.
483. Tinkler AC, Wallace AW. Curr Top Med Chem 2006;6:77–92.
484. Erdal EP, et al. Curr Top Med Chem 2005;5:603–624.
485. Matter H, Kotsonis P. Curr Med Res Rev 2004;24:662–684.
486. Haynes WG, et al. J Hypertens 1993;11:1375–1380.
487. Albert J, et al. Acta Anaesthesiol Scand 1997;41:1104–1113.
488. White RP, et al. Stroke 1998;29:467–472.
489. McCall TB, et al. Br J Pharmacol 1991;102:234–238.
490. Moore WM, et al. J Med Chem 1994;37:3886–3888.
491. Shearer BG, et al. Bioorg Med Chem Lett 1997;7:1763–1768.
492. Lee Y, et al. Bioorg Med Chem 1999;7:1097–1104.
493. Garvey EP, et al. J Biol Chem 1997;272:4959–4963.
494. Parmentier S, et al. Br J Pharmacol 1999;127:546–552.
495. Maccallini C, et al. J Med Chem 2009;52:1481–1485.
496. Young RJ, et al. Bioorg Med Chem Lett 2000;10:597–600.
497. Alderton WK, et al. Biochem J 2001;357(Pt3):593–615.
498. Silverman RB. Acc Chem Res 2009;42:439–451.
499. Moore WM, et al. Bioorg Med Chem 1996;4:1559–1564.
500. Iadecola C, et al. Am J Physiol 1995;268:R286–R292.
501. Cockroft KM, et al. Stroke 1996;27:1393–1398.
502. Nagayama M, et al. J Cereb Blood Flow Metab 1998;18:1107–1113.
503. Bucala R, et al. Am J Kidney Dis 1995;26:875–888.
504. Martin NI, et al. J Med Chem 2008;51:924–931.
505. Garvey EP, et al. J Biol Chem 1994;269:26669–26676.
506. Ueda S, et al. Bioorg Med Chem 2004;12:4101–4116.
507. Ueda S, et al. Bioorg Med Chem Lett 2004;14:313–316.
508. Ueda S, et al. Bioorg Med Chem Lett 2005;15:1361–1366.
509. Southan GJ, et al. Biochem Pharmacol 1997;21:409–417.
510. Beaton HG, et al. Bioorg Med Chem Lett 2001;11:1023–1026.
511. Tinker AC, et al. J Med Chem 2003;56:913–916.
512. Boughton-Smith NK, et al. Arthritis Rheum 2002;46:S134.
513. salerno L, et al. Curr Pharm Des 2002;8:177–200.
514. O'Neill MJ, et al. Brain Res 2000;871:234–244.
515. Patman J, et al. Bioorg Med Chem Lett 2007;17:2540–2544.
516. Chabrier PE, et al. Proc Natl Acad Sci USA 1999;96:10824–10829.
517. Auvin S, et al. Bioorg Med Chem Lett 2003;13:209–212.
518. Faraci WS, et al. Br J Pharmacol 1996;119:1101–1108.
519. Southan GJ, et al. Pharmacol Commun 1996;7:275–286.

520. Hagmann WK, et al. Bioorg Med Chem Lett 2000;10:1975–1978.
521. Zhou D, et al. J Med Chem 2009;52:2443–2453.
522. Ijuin R, et al. Bioorg Med Chem 2006;14:3563–3570.
523. Lowe JA, et al. Bioorg Med Chem Lett 1999;9:2569–2572.
524. Lowe JA, et al. J Med Chem 2004;47:1575–1586.
525. Nason DM, et al. Bioorg Med Chem Lett 2004;14:4511–4514.
526. Connolly S, et al. J Med Chem 2004;47:3320–3323.
527. Lawton GR, et al. Bioorg Med Chem 2009;17:2371–2380.
528. Ji H, et al. J Med Chem 2009;52:779–797.
529. Ji H, et al. J Am Chem Soc 2008;130:3900–3914.
530. Jackson SA, et al. Bioorg Med Chem 2005;13:2723–2739.
531. Lopex Cara LC, et al. Eur J Med Chem 2009;44:2655–2666.
532. Takagi K, et al. Neurol Res 2001;23:662–668.
533. Sugimura M, et al. Eur J Pharmacol 1997;336:99–106.
534. Fukunaga K, et al. Biochem Pharmacol 2000;60:693–699.
535. Ashigushi A, et al. Neuroscience 2003;121:379–386.
536. Sato T, et al. J Pharmacol Exp Ther 2003;304:1042–1047.
537. Hashigushi A, et al. J Pharmacol Sci 2004;96:65–72.
538. Matter H, et al. Med Res Rev 2004;24:662–684.
539. Escott KJ, et al. J Cereb Blood Flow Metab 1998;18:281–287.
540. Kami H, et al. J Cereb Blood Flow Metab 1996;16:1153–1157.
541. Goyagi T, et al. Stroke 2001;32:1613–1620.
542. Yoshida T, et al. J Cereb Blood Flow Metab 1994;14:924–929.
543. Kelly PA, et al. J Cereb Blood Flow Metab 1995;15:766–773.
544. Boulouard M, et al. Bioorg Med Chem Lett 2007;17:3177–3180.
545. Handy RLC, et al. Br J Pharmacol 1995;116:2349–2350.
546. Handy RLC, et al. Life Sci 1997;60:389–394.
547. Haga KK, et al. Brain Res 2003;993:42–53.
548. Sennequier N, et al. J Biol Chem 1999;274:930–938.
549. Chida N, et al. Eur J Pharmacol 2005;509:71–76.
550. Ouyang J, et al. Transplantation 2005;79:1386–1392.
551. Blasko E, et al. J Biol Chem 2002;277:295–302.
552. Kolodziejski PJ, et al. Proc Natl Acad Sci USA 2004;101:18141–18146.
553. Szabolcs MJ, et al. Circulation 2002;106:2392–2396.
554. Ichinose F, et al. Am J Physiol Heart Circ Physiol 2003;285:H2524–H2530.
555. Enkhbaatar P, et al. Am J Respir Crit Care Med 2003;167:1021–1026.
556. Davey DD, et al. J Med Chem 2007;50:1146–1157.
557. Whitlow M, et al. Bioorg Med Chem Lett 2007;17:2505–2508.
558. Bonnefous C, et al. J Med Chem 2009;52:3047–3062.
559. Botta M, et al. ChemMedChem 2008;3:1580–1588.
560. Chakraborti S, Mandal M, Das S, Mandal A, Chakraborti T. Mol Cell Biochem 2003;253:269–285.
561. Nagase H, Meng Q, Malinovskii V, Huang W, Chung L, Bode W, Maskos K, Brew K. Ann NY Acad Sci 1999;878:1–11.
562. Mandal M, Mandal A, Das S, Chakraborti T, Sajal C. Mol Cell Biochem 2003;252:305–329.
563. Zhao BQ, Tejima E, Lo EH. Stroke 2007;38 (Suppl 2): 748–752.
564. Cunningham LA, Wetzel M, Rosenberg GA. Glia. 2005;50:329–339.
565. Wang X, Tsuji K, Lee SR, Ning M, Furie KL, Buchan AM, Lo EH. Stroke. 2004;35(11 Suppl 1): 2726–2730.
566. Lee SR, Tsuji K, Lee SR, Lo EH. J Neurosci 2004;24:671–678.
567. Gu Z, Cui J, Brown S, Fridman R, Mobashery S, Strongin AY, Lipton SA. J Neurosci 2005;25:6401–6408.
568. Montaner J, Rovira A, Molina CA, Arenillas JF, Ribo M, Chacon P, Monasterio J, Alvarez-Sabin J. J Cereb Blood Flow Metab 2003;23:1403–1407.
569. Rosell A, Ortega-Aznar A, Alvarez-Sabin J, Fernandez-Cadenas I, Ribo M, Molina CA, Lo EH, Montaner J. Stroke 2006;37:1399–1406.
570. Fingleton Curr Pharm Design 2007;13:333–346.
571. Rowan Expert Opin Ther Targets Dev 2008;12:1–18.

572. Skiles JW, et al. Curr Med Chem 2004;11:2911 Curr Med Chem2001;8:425; Ann Rep Med Chem2000;35:167–176.
573. Whittaker Chem Rev 1999;99:2735.
574. Matter Curr Opin Drug Discov Dev 2004;7: 513.
575. Tu G, et al. Curr Med Chem 2008;15:1388–1395.
576. Ngo J. Drugs Fut 1996;21:1215.
577. Beckett RP. Drug Discov Today 1996;1:16.
578. Broadhurst MJ, et al. Bioorg Med Chem Lett 1997;7:2299.
579. Hilbert H, et al. Tetrahedron 2001;57:7675.
580. Close DR. Annals Rheumatic Dis 2001;60 (Suppl 3): 62.
581. MacPherson LJ, et al. J Med Chem 1997;40: 2525.
582. Sorbera LA, et al. Drugs Fut 2000;25:150.
583. Nakatani S, et al. Bioorg Med Chem 2006;14:5402–5422.
584. Chollet A. Bioorg Med Chem 2002;10: 531–544.
585. Cheng X, et al. Curr Med Chem 2008;15: 368–373.
586. Hanessian S, et al. J Med Chem 2001;44:3066.
587. Rossello A, et al. Bioorg Med Chem Lett 2005;15:1321–1326.
588. O'Brien PM, et al. J Med Chem 2000;43:156.
589. Pikul S, et al. J Med Chem 1999;42:87.
590. Yang SM, et al. Bioorg Med Chem Lett 2008;18: 1135–1139 and 1140–1145.
591. Cheng X. Curr Med Chem 2008;15:374–385.
592. Pikul S, et al. Bioorg Med Chem Lett 2001;11:1009.
593. Moriyama H, et al. Bioorg Med Chem Lett 2003;13:2737–2740.Moriyama H, et al. J Med Chem 2004;47:1930.
594. Pikul S, et al. J Med Chem 1998;41:3568.
595. Cheng Y, et al. J Med Chem 2000;43:369.
596. Sawa M, et al. J Med Chem 2002;45:919.
597. Levin JI, et al. Bioorg Med Chem Lett 1998;8:2657–2662.
598. Nelson C, et al. Bioorg Med Chem Lett 2002;12:2867.
599. Cherney RJ, et al. J Med Chem 2003;46:1811.
600. Cherney RJ, et al. J Med Chem 2004;47: 2981–2983.
601. Levin JI, et al. Bioorg Med Chem Lett 2001;11:235–238 and 239–242.
602. Levin JI, et al. Bioorg Med Chem Lett 2001;11:2189–2192.
603. Levin JI, et al. Bioorg Med Chem Lett 2001;11:2975–2978.
604. Jacobson IC, et al. Bioorg Med Chem Lett 1998;8:837.
605. Aranapakam V, et al. J Med Chem 2003;46: 2361.
606. Aranapakam V, et al. J Med Chem 2003;46:2376.
607. Lovejoy B, et al. Nat Struct Biol 1999;6: 217.
608. Duan JJ-W, et al. Bioorg Med Chem Lett 1999;9:1453.
609. Xue CB, et al. J Med Chem 2001;44:2636.
610. Castelhano AL, et al. Bioorg Med Chem Lett 1995;5:1415.
611. Krumme D, et al. Bioorg Med Chem Lett 2002;12:933.
612. Rossello A, et al. Bioorg Med Chem Lett 2005;15:2311.
613. Carty J, et al. Inflamm Res 1999;48:229.
614. Wada CK, et al. J Med Chem 2002;45:219.
615. Moore M, et al. Proc Am Soc Clin Oncol 2000;19:240a.
616. Brown FK, et al. J Med Chem 1994;37:674.
617. Pikul S, et al. J Med Chem 2001;44:2499.
618. Tamura Y, et al. J Med Chem 1998;41:640.
619. Zhang Y-M, et al. Bioorg Med Chem Lett 2006;16:3096.
620. Cherney RJ, et al. J Med Chem 1998;41:1749.
621. Zhang Y-M, et al. Bioorg Med Chem Lett 2008;18:409.
622. Zhang Y-M. Bioorg Med Chem Lett 2008;18: 405.
623. Brandstetter et al. J Biol Chem 2001;276:17405.
624. Grams F, et al. Biol Chem 2001;382:1277.
625. Kim S-H, et al. Bioorg Med Chem Lett 2005;15:1101.
626. Douillard J, et al. Eur J Cancer 2001;37(Suppl 6): S19.
627. Campbell DA, et al. Bioorg Med Chem Lett 1998;8:1157.
628. Levin JI, et al. Bioorg Med Chem Lett 1998;8:1163.
629. Fink CA, et al. Bioorg Med Chem Lett 1999;9:195.
630. Szardenings AK, et al. J Med Chem 1998;41:2194.
631. Scozzafava A, et al. Bioorg Med Chem Lett 2002;12:2667.
632. Matziari M, et al. Org Lett 2001;3:659.
633. Biasone A, et al. Bioorg Med Chem 2007;15:791.

634. LeDour G, et al. Bioorg Med Chem 2008;16:8745.
635. Schroder J, et al. J Med Chem 2001;44:3231.
636. Lee M, et al. Org Lett 2005;7:4463.
637. Brown S, et al. J Am Chem Soc 2000;122:6799.
638. Kleifeld O, et al. J Biol Chem 2001;276:17125.
639. Thomas SM, Brugge JS. Annu Rev Cell Dev Biol 1997;13:513–609.
640. Shin EY, Ma EK, Kim CK, Kwak SJ, Kim EG. J Cancer Res Clin Oncol 2002;128:596–602.
641. Paul R, Zhang ZG, Eliceiri BP, Jiang Q, Boccia AD, Zhang RL, Chopp M, Cheresh DA, Schwartzberg PL, Hood JD, Leng J. Nat Med 2001;7:222–227.
642. Lee SW, Kim WJ, Choi YK, Song HS, Son MJ, Gelman IH, Kim YJ, Kim KW, Achen MG, Clauss M, Schnurch H, Risau W. Nat Med 2003;9:900–906.
643. Ardizzone TD, Zhan X, Ander BP, Sharp FR. Stroke 2007;May 38(5): 1621–1625.
644. Haydar SN, Hirst WD. Annu Rep Med Chem 2006;41:39–57.
645. Benati D, Baldari CT. Curr Med Chem 2008; 15:1154–1165.
646. Cao X, et al. Mini Rev Med Chem 2008;8: 1053–1063.
647. Parang K, Sun G. Curr Expert Opin Ther Pat 2005;15:1183–1207.
648. Lin L-G, et al. J Med Chem 2008;51: 4419–4429.
649. Huang H, et al. Eur J Med Chem 2009;44: 1982–1988.
650. Games R, et al. J Bone Miner Res 1999;14: S487.
651. Azam M, et al. Proc Natl Acad Sci 2006;103: 9244.
652. Wang Y, et al. Bioorg Med Chem Lett 2008;18:4907–4912.
653. Ardizzone TD, et al. Stroke 2007;38: 1621–1625.
654. Hou XY, et al. Neurosci Lett 2007;420: 235–239.
655. Lenmyr F, et al. Acta Neurol Scand 2004;110:175–179.
656. Karni R, et al. FEBS Lett 2003;537:47.
657. Warmuth M, et al. Blood 2003;101,664.
658. Kumar A, et al. ChemMedChem 2007;2: 1346–1360.
659. Borhani DW, et al. Bioorg Med Chem Lett 2004;14:2613–2616.
660. Burchat A, et al. Bioorg Med Chem Lett 2006;16:118–122.
661. Carraro F, et al. J Med Chem 2004;47:1595.
662. Carraro F, et al. J Med Chem 2006;49:1549.
663. Schenone S, et al. Eur J Med Chem 2008;43:2665–2676.
664. Santucci MA, et al. ChemMedChem 2009;4:118–126.
665. Missbach M, et al. Bone 1999;24:437–449.
666. Mukaiama H, et al. Chem Pharm Bull 2007;55:881–889.
667. Mukaiama H, et al. Bioorg Med Chem 2007;15:868–885.
668. Chen P, et al. J Med Chem 2004;47:4517–4529.
669. fry DW, et al. Science 1994;265:1093.
670. Hennequin L, et al. J Med Chem 2006;49: 6465–6488.
671. Boschelli DH. Curr Top Med Chem 2008;8:922–934.
672. Boschelli DH, et al. Drug Fut 2007;32: 481–490.
673. Boschelli DH, et al. Bioorg Med Chem Lett 2007;17:1358–1361.
674. Cao X, et al. Bioorg Med Chem Lett 2008;18:6206–6209.
675. Cao X, et al. Bioorg Med Chem 2008;16:5890.
676. Kraker AJ. Biochem Pharmacol 2000;60: 885–898.
677. Noronha G, et al. Bioorg Med Chem Lett 2006;16:5546–5550.
678. Noronha G, et al. Bioorg Med Chem Lett 2007;17:602–608.
679. Das J, et al. J Med Chem 2006;49:6819–6832.
680. Lombardo L, et al. J Med Chem 2004;47: 6658–6661.
681. Blake RA, et al. Mol Cell Biol 2000;20: 9018–9027.
682. Guan H, et al. Bioorg Med Chem Lett 2004;14: 187–190.
683. Ye G. Curr Opin Invest Drugs 2008;9:605–613.
684. Nakaya K, et al. Anticancer Drugs 2003;14: 683–693.
685. Hashimoto S, et al. Jpn J Cancer Res 2002;93:944–951.
686. Masuda Y, et al. Oncogene 2003;22: 1012–1023.
687. Wang CK, et al. Bioorg Med Chem Lett 1995;5:2423–2428.
688. Vande Linde AM, Chopp M. Metab Brain Dis 1991;6:199–206.
689. Muir KW. CNS Drugs 2001;15:921–930.
690. for review see, Meloni BP, Zhu H, Knuckey NW. Magnes Res 2006;Jun 19(2): 123–137.

691. Intravenous Magnesium Efficacy in Stroke (IMAGES) Study Investigators. (Intravenous Magnesium Efficacy in Stroke Trial.) Lancet 2004;363:439–445.
692. Saver JL, Kidwell C, Eckstein M, Starkman S, FAST-MAG Pilot Trial Investigators. Stroke 2004;35(5): e106–e108.
693. for review, Lazzaro MA, Prabhakaran S. Expert Opin Investig Drugs 2008;17(8): 1161–1174.
694. Copenhagen Stroke Study. Stroke 2000;31(9): 2251–2256.
695. Krieger DW, De Georgia MA, Abou-Chebl A, et al. Stroke 2001;32(8): 1847–1854.
696. Woolf, Bloechlinger. Science 2002;297(5584): 1132–1134.
697. Irving EA, Vinson M, Rosin C, Roberts JC, Chapman DM, Facci L, Virley DJ, Skaper SD, Burbidge SA, Walsh FS, Hunter AJ, Parsons AA. J Cereb Blood Flow Metab 2005;25(1): 98–107.
698. Walmsley AR, Mir AK. Curr Pharm Des 2007;13(24): 2470–2474.
699. Cramer SC. Ann Neurol 2008;63:549–560.
700. Walker-Batson D, Curtis S, et al. Stroke 2001;32:2093–2098.
701. Miyai I, Reding R. J Neurol Rehabil 1998;12: 5–13.
702. Papadopoulos CM, Tsai SY, et al. Cereb Cortex 2006;16:529–536.

THERAPEUTIC AND DIAGNOSTIC AGENTS FOR PARKINSON'S DISEASE

RAYMOND G. BOOTH[1]
JOHN L. NEUMEYER[2]
ROSS J. BALDESSARINI[2]
[1] University of Florida, Gainesville, FL
[2] Harvard Medical School, Boston, MA

1. INTRODUCTION

In 1817, British physician James Parkinson published "An Essay on the Shaking Palsy" [1] that first described clinical features of what is now the second most common neurodegenerative disorder. Parkinson's disease (PD) is characterized by resting tremor, disturbances of posture, and paucity or slowing of volitional movement. The primary etiology of PD is unknown, but its neuropathology is marked by progressive degeneration of pigmented neurons of midbrain and brainstem, mainly those that produce dopamine (DA) as a neurotransmitter in the midbrain substantia nigra and project to the forebrain extrapyramidal motor control center of the basal ganglia (caudate–putamen, and other components of the corpus striatum). There is also variable loss of other pigmented monoaminergic neurons in the brainstem, particularly those producing norepinephrine. Since the early 1960s, pharmacotherapy for PD has been based rationally on replacing the DA lost due to selective and idiopathic degeneration of DA neurons by giving large doses of its immediate metabolic precursor amino acid, L-dihydroxyphenylalanine (L-dopa). Later, synthetic DA receptor agonists and agents that inhibit the metabolic breakdown of DA (or L-dopa) were employed as well. Pharmacotherapy for PD, however, remains palliative and symptomatic, and does not address the still-elusive pathophysiological mechanisms underlying neuronal degeneration. Proposed mechanisms are based on several possibly convergent hypotheses encompassing genetic factors, oxidative metabolism associated with advancing age, and environmental factors. Current pharmacotherapy is limited in both effectiveness and tolerability, especially late in the progression of the disease. Improved symptomatic and anticipated curative pharmacotherapy depends on better understanding of the fundamental pathogenesis of PD. In this light, this chapter reviews medicinal–chemical aspects of available antiparkinsonism drugs, emerging treatments, and neuroradiological agents for diagnosis and monitoring the progression of PD.

2. PARKINSON'S DISEASE

Typically, PD presents in mid- or late life, most often at age between 55 and 65. It affects approximately 1–2% of the population elder than age 65; in the population elder than age 84, the incidence increases to 3–5% per year [2]. Prevalence of PD is expected to increase substantially over the next several decades, as the population ages [3]. PD presents as a classic tetrad of signs: (1) resting tremor that improves with voluntary activity, (2) bradykinesia or slow initiation and paucity of voluntary movements, (3) rigidity of muscle and joint motility, and (4) postural disturbances including falls. These signs vary in their early intensity, combinations, and progression among affected individuals. Some cases also show dyscontrol of autonomic functions mediated by the potentially affected central noradrenergic sympathetic nervous system, with losses of norepinephrine neurons of the locus coeruleus [4]. Dementia is about six times more common among elderly patients with PD, and, there can be other spontaneous or treatment-associated neuropsychiatric disturbances, including hallucinations and depression [5]. In fact, such neuropsychiatric symptoms lead to more nursing home placements than the motor dysfunctions of PD [6]. Although L-dopa pharmacotherapy has decreased morbidity and mortality, mortality among PD patients is still approximately 60% greater than in age-matched controls [7].

Parkinsonism, as a bradykinesia syndrome, represents the clinical outcome of etiologically diverse conditions. These include idiopathic degenerative disorders (including idiopathic Lewy body dementia and multiple system atrophy with dysautonomia [Shy-Drager syndrome]), infections (including postencephalitic parkinsonism, such as from von Economo's encephalitis lethargica, that arose with influenza epidemics of the early twentieth century), effects of neurotoxins (including certain heavy metals such as manganese, pyridiniums such as 1-methyl-4-phenylpyridinium (MPP$^+$) (Fig. 1), and the marine cyanobacteria product β-N-methylamino-L-alanine (BMAA) [8]. Bradykinesia and variable tremor also are commonly associated with the use of first-generation neuroleptics as well as some modern antipsychotic agents and other DA receptor

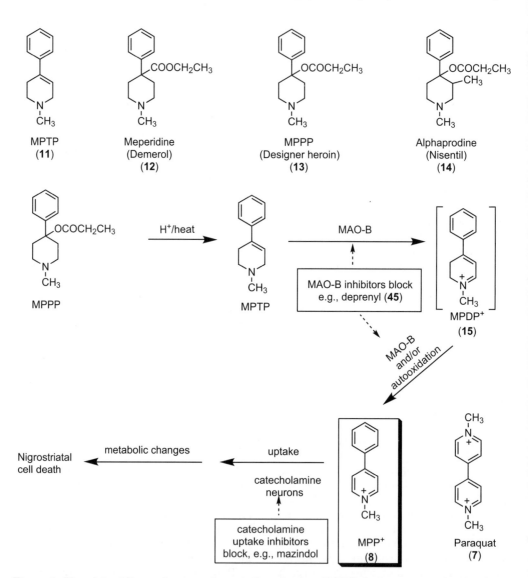

Figure 1. Phenylpiperidine analgesics and metabolic activation of MPTP. In the presence of acid and heat, MPPP forms MPTP that undergoes an MAO-B catalyzed two-electron oxidation to MPDP$^+$ MDPP$^+$ undergoes a two-electron auto-oxidation (also can be catalyzed by MAO-B) to MPP$^+$ that is accumulated into DA neurons by the DA transporter and subsequently into mitochondria where it disrupts cellular respiration, producing neuronal cell death. (This figure is available in full color at http://mrw.interscience.wiley.com/emrw/9780471266945/home.)

antagonists or DA-depleting agents [9,10]. The designation PD refers to the idiopathic disorder, to distinguish it from other parkinsonian syndromes ("parkinsonism").

Neuropathologically, PD is a slowly progressive neurodegenerative disorder of unknown cause that selectively affects the dopaminergic, extrapyramidal nigrostriatal pathway. The disease is characterized by gradual destruction of DA-containing neurons in the pars compacta component of the pigmented midbrain substantia nigra, leading to a deficiency of the neurotransmitter in DA nerve terminals of the corpus striatum [5]. Degenerative changes in the pigmented nuclei of the noradrenergic locus coeruleus region of the midbrain-pons also can occur, and remaining catecholamine cells typically acquire intraneuronal inclusions (Lewy bodies), whose development and significance remain unclear [4,5]. The discovery of DA deficiency in postmortem brain tissue of PD patients a half-century ago, with then-emerging knowledge of DA biosynthesis and metabolism, led to the rational prediction that L-dopa (1) (Fig. 2), the immediate metabolic precursor of DA (2) (Fig. 2), would be an effective palliative agent in PD [11].

2.1. Pathophysiology

The "basal ganglia" of the brain consist of five interconnected subcortical nuclei that span the telencephalon (forebrain), diencephalon, and mesencephalon (midbrain). These nuclei include the corpus neostriatum (caudate and putamen), globus pallidus, thalamus, subthalamic nucleus, and midbrain substantia nigra (pars compacta and pars reticulata). Medium-sized spiny neurons that produce the major inhibitory amino acid transmitter γ-aminobutyric acid (GABA) are principal neurons in the caudate–putamen. They receive dense input from descending corticostriatal projections mediated by the principal excitatory amino acid neurotransmitter L-glutamic acid, as well as a prominent dopaminergic input from the midbrain substantia nigra. The GABA-producing inhibitory neurons, as well as intrinsic acetylcholine-producing interneurons, of the caudate–putamen respond to DA input through several DA receptors (types D_1–D_4).

DA receptors are grouped into excitatory D_1-types (D_1 and much less prevalent D_5 receptors), and inhibitory D_2-types (D_2 with splice variants, and less abundant D_3 and D_4) subfamilies of membrane proteins. These peptides, composed of 387–477 amino acids in man, are typical of the superfamily of GTP-binding protein (G-protein)-associated membrane proteins that include most monoaminergic receptors and other physiologically important membrane proteins. Their structures are characterized by seven relatively hydrophobic, putative transmembrane regions linked by four extracellular and four intracellular loop segments, starting from an extracellular amino terminus, and extending to an intracytoplasmic carboxy end of the receptor polypeptide chain [12]. The third intracellular loop and carboxy-terminus segment vary most among DA receptor subtypes, and probably interact critically with excitatory or inhibitory G-proteins and intraneuronal molecular components of effector mechanisms, including adenylyl cyclase (stimulated through D_1-like and inhibited by D_2-like receptors) and phospholipase C (stimulated by D_2 and less abundant D_4, but inhibited by D_3 receptors) [12].

The inhibitory spiny GABAergic neurons send projections by two major pathways that appear to exert balanced regulatory influences on the ascending thalamocortical circuits mediating control of voluntary movement through descending corticospinal projections to the spinal ventral horn motor neurons that innervate skeletal muscles. A widely accepted, but tentative and possibly over-simplified model of the basic anatomical connections of this complex is summarized schematically [5,12–17] (Fig. 3).

Output pathways from neostriatum usually are considered to include a "direct" striatonigral pathway consisting of GABAergic neurons that project directly to the pars reticulata of the substantia nigra, as well as a prominent projection to the internal (medial) portion of the globus pallidus in human brain. Another "indirect" or striatopallidal pathway involves efferent GABAergic projections from striatum that communicate through the external (lateral) globus pallidus to the subthalamic nucleus. This inhibitory influence on the subthalamic nucleus is balanced against an excita-

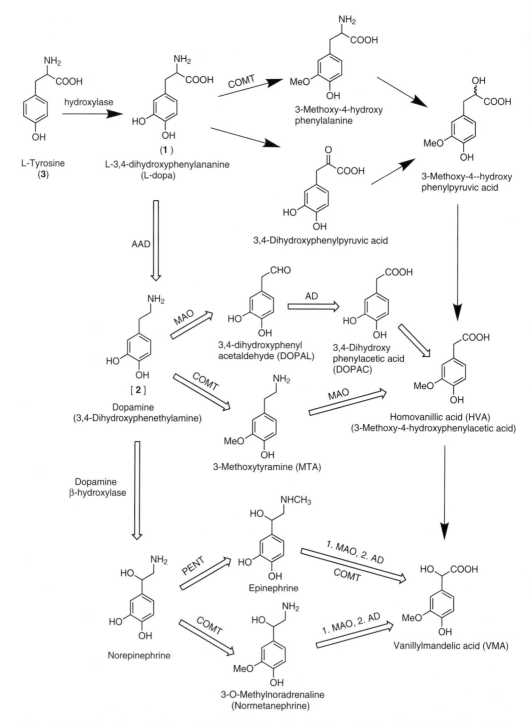

Figure 2. Pathways in metabolism of L-dopa and its major decarboxylated product DA. Heavy and light arrows indicate major or minor reactions. AD, aldehyde dehydrogenase; AAD, aromatic L-amino acid decarboxylase; COMT, catechol-*O*-methyltransferase; DH, DA β-hydroxylase; MAO, monoamine oxidase; PENT, phenylethanolamine-*N*-methyltransferase.

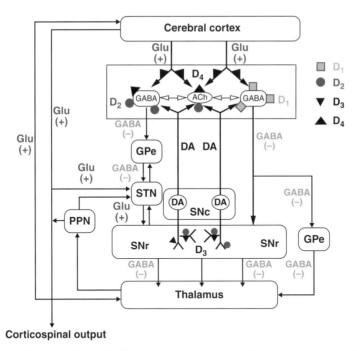

Figure 3. Circuitry of the basal ganglia. Shown are the major relationships to the neostriatum (caudate nucleus–putamen complex) with its prominent dopaminergic innervation (particularly of putamen in man) from the midbrain substantia nigra pars compacta (SNc), as well as descending control by corticostriatal glutamatergic projections. Dopamine exerts excitatory effects through D_1 receptors on efferent medium spiny neurons that are GABAergic and inhibitory to the substantia pars reticulata (SNr) and the internal portion of the globus pallidus (GPi), and limit a secondary inhibitory influence of nigrothalamic and pallidothalamic GABA neurons (that also express substance P [SP] and dynorphin [DYN] peptides) to facilitate the ascending excitatory glutamatergic (Glu) thalamocortical circuits. This short outflow loop from the neostriatum is paralleled by a long loop that involves D_2 DA receptors inhibitory to generally excitatory acetylcholinergic (ACh) interneurons (which are inhibited by GABA neurons) as well as to GABA efferents to the external portion of globus pallidus (GPi) and coexpress enkephalins (ENK). The globus projects GABAergic neurons that tonically inhibit excitatory glutamatergic neurons of the subthalamic nucleus (STN) that stimulate an inhibitory influence on the substantia nigra, internal globus pallidus, and PPN on the thalamus (including its ventral lateral nucleus) to yield a net reduction of thalamocortical activation. In Parkinson's disease, the nigrostriatal DA projections degenerate and dopaminergic influences in neostriatum are initially compromised and eventually lost. Loss of DA in PD reduces functions of both direct and indirect pathways to yield a net decrease in thalamocortical stimulation, with clinical bradykinesia. Excessive dopaminergic stimulation encountered in the treatment of PD increases thalamocortical activation, with clinical dyskinesias. This traditional model remains tentative and incomplete, and does not include simultaneous influences of the several D_1-like and D_2-like receptors at some GABAergic neurons in the neostriatum.

tory glutamatergic input from cerebral cortex. In turn, the subthalamic nucleus exerts an excitatory glutamatergic influence on the substantia nigra reticulata and internal (medial) globus pallidus.

Modified by these direct and indirect descending influences, the internal (medial) globus pallidus and pars reticulata send inhibitory GABAergic projections that modulate activity of ascending thalamic neurons, particularly in the ventral (mainly anterior and lateral) thalamic nuclei. The thalamic nuclei project ascending glutamatergic excitatory efferents to motor cerebral cortex, thus exerting a major regulatory influence over the descending corticospinal motor output pathway that controls the cholinergic spinal ventral motor horn cells innervating skeletal muscle [18].

DA modulates the activity of local and efferent inhibitory GABAergic neurons as well as acetylcholinergic interneurons of caudate–putamen [12,18]. Excitatory DA D_1-type receptors, together with the neuropeptides substance P and dynorphin, are expressed mainly by the striatonigral GABAergic neurons in the direct output pathway. In contrast, inhibitory DA D_2-type DA receptors, along with the neuropeptide enkephalin, are predominantly localized to striatopallidal GABAergic efferents of the indirect output pathway. These relationships lead to a complex role for DA in the basal ganglia. By stimulating excitatory D_1 receptors, DA appears to have a net facilitatory effect on the direct GABAergic pathway to internal (medial) globus pallidus and midbrain, which can diminish their inhibitory connections to thalamus, to *increase* ascending excitatory thalamocortical activity. In contrast, activation of D_2 receptors in the indirect pathway inhibits GABAergic neurons projecting to external (lateral) globus pallidus, thereby increasing pallidal inhibitory influence on the subthalamic nucleus. These effects result in a net inhibition of an excitatory glutamatergic link from the subthalamic nucleus to GPi/SNr that reduces pallidal GABAergic inhibition of thalamus. The outcome is to increase thalamocortical stimulation, opposite to the effect initiated by DA through the direct pathway (Fig. 3).

There are also important connections between the basal ganglia and the structures in the midbrain. The pedunculopontine nucleus (PPN) is in proximity to the mesencephalic locomotor area and consists primarily of cholinergic neurons (pars compacta) [19]. The pronounced cholinergic neuronal loss found in the PPN correlates with gait dysfunction and relative immobility (akinesia) in PD. In addition, hyperactivity of the subthalamic nucleus has been demonstrated in PD and is attributed to excitatory input primarily from thalamus glutamatergic neurons [20].

In sum, the overall effect of DA is to facilitate cortical excitation by thalamocortical glutamatergic projections through the direct pathway, but to decrease thalamocortical stimulation through the indirect pathway. Accordingly, in PD striatal DA deficiency alters the modulation of excitatory outflow from ventral thalamus to motor cortex [5,10]. Presumably, a balance of D_1- and D_2-mediated dopaminergic function would be optimal in restoring the functional losses that follow degeneration of the DA-producing neurons. Neurochemically, striatal DA deficiency seems to account for the major motor symptoms of PD, particularly bradykinesia. Cholinergic loss may result in dysfunctional modulation of dopaminergic circuits, as well. The mainstay of pharmacological treatment [5], continues to be replacement therapy with the α-amino acid, L-3,4-dihydroxyphenylalanine (L-dopa) (**2**) (Fig. 2), the immediate biochemical precursor of DA, discussed below, in an effort to produce nearly physiological agonism of both D_1 and D_2 DA receptors.

2.2. Etiology

2.2.1. Genetic Factors Although the neuropathology of PD is well defined, the primary cause of the neuronal degenerative changes involved remains unknown, confounding rational development of additional rational and novel symptomatic, prophylactic, or potentially curative pharmacotherapy. Several possibly convergent hypotheses are proposed regarding causes of PD, encompassing genetic factors, oxidative metabolism associated with advancing age, and environmental factors.

Given that several neurodegenerative disorders (including the hyperkinetic neuropsychiatric disorder Huntington's disease, marked by choreoathetosis, early psychiatric symptoms and progressive dementia) are genetically determined, researchers have investigated possible genetic influences in PD. Epidemiological studies have found that apart from age, a family history of PD is the strongest predictor of increased risk of the disorder [21,22]; however, shared environmental exposures in families must also be considered.

One familial form of PD is characterized by mutations in the α-synuclein gene, originally reported in a single large Italian family, three smaller Greek families, and a German pedigree [23,24]. The protein α-synuclein is a highly conserved, abundant, 140-amino-acid polypeptide expressed mainly in cerebral nerve terminals. Aggregation of α-synuclein molecules leads to pathological inclusions (synucleinopathy) that characterize many neurodegenerative disorders, including PD [25]. The function of α-synuclein is not well characterized, but it appears to play a role in regulating DA homeostasis, including, modulation of DA synthesis, release, and reuptake at nerve terminals. *In vitro* studies using neuronal cells [26] indicate that downregulation of α-synuclein enhances DA biosynthesis, suggesting that dysfunction of α-synuclein may lead to increased intraneuronal levels of DA that become neurotoxic (see below). Nevertheless, several studies failed to detect mutations in the α-synuclein gene in large family samples [27,28], or in studies of identical and heterozygous twins [29–31].

Mutations in four other genes—DJ-1, PINK1, parkin, and leucine-rich-repeat kinase 2 (LRRK-2)—are unequivocally associated with development of familial PD [32], and several other mutations also are implicated. Such mutations may lead to excessive production of damaged proteins or dysfunction of protein clearance mechanisms in the brain [33,34]. Under physiological conditions, damaged proteins usually are degraded and cleared by the ubiquitin–proteasome system (UPS). Increased production of damaged proteins or decreased UPS-mediated degradation and clearance may lead to protein aggregation and proteolytic stress, negatively affecting cellular structures and processes [35,36]. For example, the UPS protein parkin mediates engulfment of dysfunctional mitochondria by autophagosomes [37], and dysfunctional parkin may fail to remove dysfunctional mitochondria. PINK1 appears to function prior to parkin in the same pathway to maintain mitochondrial integrity and functioning in both muscles and dopaminergic neurons [38,39]. In cases of familial PD associated with mutations of LRRK-2 (a kinase encoding the protein dardarin), protein accumulation as well as Lewy body formation have been identified postmortem [40].

In summary, there is compelling evidence for apparently rare cases of genetically linked PD, primarily involving early onset of the disorder, however, most cases are not associated with known genetic mutations and are considered sporadic [41]. Notably, however, synucleinopathy has been detected in sporadic cases of PD [42], suggesting that abnormalities of protein aggregation may be relevant to the etiology of the more common "sporadic" forms of the disease. Moreover, although PD associated with genetic mutation involves fewer than 10% of cases, their study has facilitated understanding of molecular pathways that may lead to neurodegeneration, especially involving DA neurons.

2.2.2. Dopamine and Mitochondrial Oxidative Metabolism

Oxidative metabolism involving the synthesis and catabolism of DA has been implicated in the PD disease process through production of chemically reactive products, including, epoxides [43], free radicals [44], and quinones [45,46].

Synthesis of DA, as well as the catecholamine neurotransmitters norepineprine and epinephrine, proceed by initial 3-oxidation of the precursor amino acid L-tyrosine (3) to L-dopa in a rate-limiting step catalyzed by tyrosine hydroxylase (L-tyrosine-3-mono-oxygenase) (Fig. 2). The mechanism of this hydroxylation may involve direct insertion of an oxygen atom into L-tyrosine to form L-dopa, or through an intermediate arene epoxide (4) [43] (Fig. 4), as is case for metabolic conversion of benzene to phenol [47], and of estradiol to catechol-estrogens [48,49]. An analogous epoxide intermediate occurring during the conversion of tyrosine to L-dopa could alkylate nucleophilic functionalities on proteins, RNA, or DNA to produce neurotoxicity [49].

In addition, monoamine oxidase (MAO)-catalyzed oxidation of DA and other monoamines generates hydrogen peroxide (H_2O_2; see Equation 1 in Fig. 4), which can undergo a redox reaction with superoxide anion radical ($O_2^{\cdot-}$) in the Haber–Weiss reaction [50] to

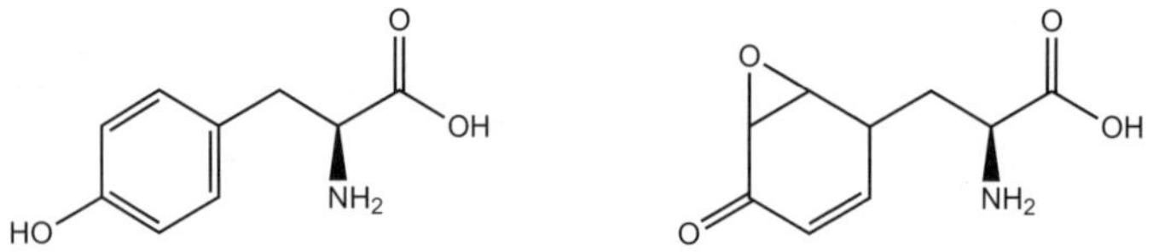

L-Tyrosine
(**3**)

L-Tyrosine epoxide
(**4**)

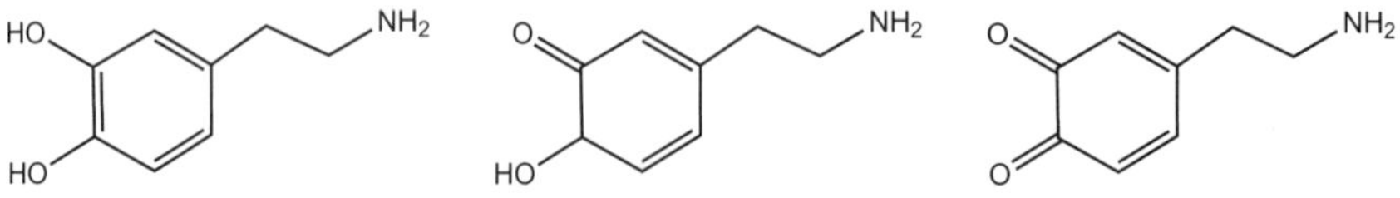

DA
(**2**)

DA semiquinone
(**5**)

DA quinone
(**6**)

Equation **1**:

$$RCH_2CH_2NH_2 + O_2 + H_2O \xrightarrow{MAO} RCH_2CHO + NH_3 + H_2O_2$$

Equation **2**:

$$H_2O_2 + O_2^{\bullet -} \longrightarrow O_2 + OH^- + OH^{\bullet}$$

Figure 4. Putative and known compounds involved in DA synthesis and catabolism, some of which may cause DA neuronal cell death.

form the extremely cytotoxic hydroxy ($HO^{\bullet}$) radical (see Equation 2 in Fig. 4) capable of causing lipid eroxidation.

MAO is expressed as two isoforms, A and B. Histochemical and immunohistochemical studies reveal that in human brain MAO-A is found primarily in noradrenergic neurons whereas MAO-B is found primarily in serotoninergic and histaminergic neurons as well as in glial cells (there is no consensus on which form may predominate in brain DA neurons; [51]). A compelling link between PD etiology and MAO is that levels of MAO-B are increased in the brains of PD patients as a consequence of gliosis [52]. Involvement of MAO-B in age-related neurological disorders that include PD also is suggested by observations that human and rodent MAO-B levels increase with aging [53,54]. Increased MAO-B levels would be expected not only to diminish DA levels but also to increase levels of potentially neurotoxic oxidation products (Fig. 4). Another link between MAO and PD is that 3,4-dihydroxyphenylacetaldehyde (DOPAL), the aldehyde that is the MAO-derived 2-electron oxidation product of DA (Equation 1 in Fig. 4), has been implicated in the aggregation of α-synuclein [55].

In addition to generating potentially neurotoxic products from MAO-catalyzed oxidation of DA, auto-oxidation of DA can yield electrophilic semiquinone (**5**) (Fig. 4) and quinone species (**6**) (Fig. 4). These reactive products of oxidation are of interest because they are cytotoxic and can alkylate protein sulfhydryl groups, including glutathione [46,56]. Manganese ion also can catalyze oxidation of DA to yield quinones implicated in manganese-associated parkinsonism [45]. Analysis of postmortem brain tissue from PD patients has found decreased levels of glutathione [57], increased lipid peroxidation [58,59], and increased oxidation of DNA [60] and proteins [61].

Auto-oxidation of catecholamines also leads to formation of the polymeric pigment neuromelanin that increases with age and is responsible for the dark coloration of

DA-producing cells in the substantia nigra and the norepinephrine neurons of the locus coeruleus [46]. The physiologic role of neuromelanin is poorly understood. The pigment is increasingly deposited in catecholaminergic neurons with advancing age, however, and it has been suggested that its accumulation in nigral neuronal cells eventually causes cell death [4].

Mitochondrial dysfunction also is hypothesized to play a role in PD neuronal cell death [62,63]. Postmortem samples of brain and peripheral tissue from PD patients indicate selective mitochondrial complex I deficiency, including in the substantia nigra and other sites [64,65] that might contribute to neuronal cell death in PD through decreased ATP synthesis and altered homeostasis of reactive species such as hydrogen peroxide, superoxide, and free hydroxyl radicals [62,63]. Mitochondrial dysfunction and altered oxidative metabolism of DA may converge, in that decreased energy production by mitochondria may impede vesicular storage of DA, leading to increased concentrations of free DA in cytoplasm that can auto-oxidize to quinones and semiquinones [66].

In summary, toxic chemical products associated with metabolism of DA or altered mitrochondrial function may contribute to the progressive loss of DA neurons that occurs normally with maturation and aging at a rate of approximately 13% per decade after age 45 [67–69]. Clinical symptoms of PD emerge as losses of DA neurons exceed 65% [70]. Conceivably, PD may result from neurodegenerative changes involving oxidative metabolism attributed to normal aging superimposed on neurotoxic insults that may occur at various ages. Such a two-factor pathophysiology might explain why PD is usually a progressive disorder of late onset [71]. Hypotheses concerning the initial pathological event have centered on environmental neurotoxicants, discussed below.

2.2.3. Environmental Factors

There is evidence to suggest that environmental toxicants may cause some parkinsonism syndromes [72,73]. This hypothesis seems consistent with the fact that PD is now the second most common degenerative neuropsychiatric disorder (after the dementias), in contrast to its evident rarity in previous centuries [74]. Even in disorders with strong familial association, a role of shared environmental exposures and other indirect factors may occur, in addition to any specific genetic contribution. Studies of some familial but idiopathic forms of PD-like neurodegenerative disorders in the Western Pacific (such as the prion disease Kuru and amyotrophic-sclerosis-PD-like syndromes associated with cycad plant toxins) do not support genetic or infectious etiologies, but leave open the possibility of neurotoxic factors [75,76]. A striking example of a probably neurotoxic disorder is the PD-like syndrome characterized by tremor and bradykinesia as well as dystonia and both cognitive and behavioral disturbances found among manganese miners in the Andes, and others exposed to the metal [77].

There is also evidence linking herbicides and pesticides to PD. For example, there is a remarkably high correlation between incidence of PD and use of pesticides in an agricultural region of Québec [78]. Of relevance here, N,N'-dimethyl-4,4'-bipyridinium dichloride (7) (Fig. 1), better known as Paraquat, is one of the most widely used herbicides in the world [79] and has a close structural resemblance to the selective DA neurotoxicant MPP^+ (8) (Fig. 1), that also was once marketed as the herbicide Cyperquat [73,80]. Like MPP^+ and the pesticide rotenone (see below), Paraquat inhibits mitochondrial complex I and cellular respiration, though only at concentrations (~ 10 mM) considerably higher than are required for MPP^+ ($\sim 10\,\mu$M) or rotenone (~ 10 nM) [80]. Some investigators have reported Paraquat-induced postmortem losses of nigral DA neurons and degeneration of striatal DA-terminals following decreased locomotor activity [81]. However, others found no evidence of neurotoxic changes after exposure to Paraquat [82], and Paraquat neurotoxicity, unlike MPP^+-associated parkinsonism, probably is not highly selective for DA neurons [80] (see below). Finally, the ability of Paraquat itself to gain entry to the brain remains uncertain [73].

Rotenone
(10)

Another potentially relevant neurotoxicant is the lipophilic pesticide–piscicide rotenone (**10**). Rotenone is an isoflavone, many of which are found in roots and stems of several plants. Of note, this association with plant products allows rotenone to be labeled as "natural" and to be used in organic food farming [79]. Rotenone easily crosses the blood–brain barrier and neuronal cell membranes. It appears to be capable of inhibiting mitochondrial complex I to cause highly selective degeneration of nigrostriatal DA neurons [73,80], producing hypokinesia and rigidity as well as accumulation of fibrillar cytoplasmic inclusions that contain ubiquitin and α-synuclein [83].

One of the best-characterized epidemiological findings in PD is its lower incidence in cigarette smokers than in nonsmokers [84]. Several mechanisms might account for this inverse correlation, which might enhance understanding of neuroprotection in neurodegenerative disorders and even lead to effective prophylactic or even curative pharmacotherapy. For example, activation of central nicotinic acetylcholine receptors by nicotine modulates DA neurotransmission and also reduces neural inflammation [85]. Moreover, other constituents in tobacco smoke are MAO inhibitors, including, farnesylacetone, a selective inhibitor of MAO-B [86], thought to be the predominant MAO isoform responsible for metabolism of DA in brain (glial cells). Of potential relevance to the pathophysiology of PD, prolonged nicotine exposure can increase activity of the UPS system [87] discussed in Section 2.2.1.

There also is an inverse correlation between risk of developing PD and coffee and caffeine consumption [88]. Caffeine is a nonselective adenosine A_1/A_2 receptor antagonist that reduces parkinsonism-like tremor in rodent models of PD [89]. The adenosine A_{2A} receptor subtype is expressed abundantly on medium spiny neurons, and activates intracellular kinases. Aberrant interneuronal signaling associated with these phosphate-adding enzymes has been linked to parkinsonism-like signs in rodent and primate models of PD, and with dyskinetic motor response produced by dopaminomimetic therapy in these models [90]. Adenosine A_{2A} receptor antagonists are proposed for pharmacotherapy of PD [91] and are undergoing clinical trials (see Section 3.2.2). Antagonism of adenosine A_1 receptors also may reduce parkinsonism-like tremor in rodents [89], but no clinical candidate anti-A_1 drugs have emerged.

In summary, there are several potential links between the PD and the environment, although their relevance to the clinical disorder remains uncertain. Some environmental factors, such as cigarette smoking and caffeine consumption, may protect against development of PD and might help to delineate biochemical mechanisms leading to effective pharmacotherapy for PD. Regarding the etiology and pathophysiology of PD, intriguing findings have arisen from epidemiologic studies of various apparent environmental or toxic risk factors for PD, including exposure to pesticides, herbicides, well water, wood pulp mills, and rural living [92,93]. Pathological mechanisms involving oxidative metabolism and environmental neurotoxicants may converge. A particularly well-studied example of such convergence involves the discovery that N-methyl-4-phenyl-1,2,3,6-tetrahydropyridine (MPTP) (**11**) (Fig. 1) produces a severe parkinsonism syndrome in humans and some laboratory animals that is remarkably similar to the idiopathic disease. As such, MPTP has been an extremely useful tool to study the etiology and pharmacotherapy of PD.

The cyclic tertiary amine MPTP (**11**) induces a form of parkinsonism in humans and lower primates similar in neuropathology and motor abnormalities to idiopathic PD [94–96]. The role of MPTP in parkinsonian disorders was revealed by a serendipitous series of events. In 1977, a young college student suddenly developed parkinsonian symptoms with severe rigidity, bradykinesia, and mutism. The early and abrupt onset of such symptoms was so atypical that the patient initially was thought to have catatonic schizophrenia. The subsequent diagnosis of parkinsonism was substantiated by a therapeutic response to L-dopa. He admitted having synthesized and used several illicit drugs. Chemical analysis of glassware that the patient used for chemical syntheses at his home revealed several pyridines, including MPTP (11), apparently formed as by-products during synthesis of the reverse ester of the narcotic analgesic meperidine (12) (Fig. 1) known as MPPP (N-methyl-4-propionoxy-4-phenylpiperidine) (13) (Fig. 1) or "designer heroin." MPPP is an analog of prodine, a non-opioid, synthetic analgesic, alphaprodine (14) (Fig. 1), itself an analog of the analgesic meperidine. Initially, it was unclear whether MPTP or other constituents of the injected mixture accounted for the neurotoxicity that produced parkinsonism.

After the patient returned home, he continued to abuse drugs and died of an overdose; autopsy revealed degeneration of the substantia nigra, the hallmark neuropathological feature of PD. Subsequently, other patients were identified with virtually identical parkinsonian symptoms who had also been taking intravenous injections of preparations of MPPP containing varying amounts of MPTP. In several patients, MPTP was the principal or sole constituent injected, implicating MPTP as a parkinsonism-producing neurotoxicant. The clinical and neuropathological features of MPTP-induced parkinsonism resemble idiopathic PD more closely than any other reported animal or human condition elicited by toxins, metals, viruses, or other means. Understanding the molecular pathophysiology of MPTP neurotoxicity has shed light on neurodegenerative mechanisms that may be present in idiopathic PD.

The chemical structure of MPTP suggests that the compound would be relatively chemically inert since it lacks a highly reactive functional group. However, MPTP can undergo metabolic activation to more reactive species. The oxidative enzyme, MAO-B, in brain tissue catalyzes the two-electron oxidation of MPTP at the allylic α-carbon to give the unstable intermediate product, 1-methyl-4-phenyl-2,3-dihydropyridinium (MPDP$^+$) (15) (Fig. 1). This intermediate undergoes a two-electron oxidation to the stable 1-methyl-4-phenylpyridinium species (8) (Fig. 1) via auto-oxidation, disproportionation, and MAO-B catalysis [97–99] (Fig. 1). MAO-B inhibitors can prevent MPTP-induced parkinsonism in primates [100]. Although a role for the unstable dihydropyridinium species MPDP$^+$ has not been ruled out, MPP$^+$ is currently considered the major metabolite of MPTP responsible for DA neuronal cell death.

The α-carbon oxidation of tertiary amines, such as MPTP, by MAO (type A or type B) is surprising in view of its assumed natural preference for monoamine substrates, notably including the neurotransmitters DA, norepinephrine, and serotonin. However, MAO might be of additional physiological importance in regulating the oxidation state of pyridine systems, such as those involving nucleic acids and NADH [101], that may be involved in the neurotoxicity of MPTP (see below).

Extensive metabolic, biochemical, and toxicologic investigations have established that the nigrostriatal neurodegenerative properties of MPTP are mediated by the MAO-B derived metabolite, MPP$^+$ This bioactivation reaction, however, probably proceeds outside of the target nigrostriatal DA neurons. The isoform of MAO, if any, inside DA neurons is not established [102,103]. Instead, MAO-B rich glial cells near striatal nerve terminals and nigral cell bodies probably oxidize MPTP to MPDP$^+$ The conjugate base MPDP presumably diffuses out of glial cells and subsequently oxidizes to MPP$^+$ that is sequestered into striatal dopaminergic nerve terminals via the DA neurotransporter, which accepts MPP$^+$ as a substrate (Fig. 1) [104]. Within DA neurons, MPP$^+$ is concentrated in the mitochondria, where it selectively inhibits complex I of the electron transport chain, inhibiting NADH oxidation and eventually depleting the nigrostriatal neuronal cell of ATP [105,106]. Thus, the currently accepted mechanism of nigrostriatal cell death induced by MPTP (via MPP$^+$) is inhibition of mitochondrial respiration [107,108].

Several sequential factors may account for selective damage of nigrostriatal DA neurons by MPTP (Fig. 1). First, MPTP binds selectively to MAO-B, which is highly concentrated in glial cells in human substantia nigra and corpus striatum. Then, MPP$^+$ produced from MPTP is selectively accumulated by DA neurotransporters into DA cell bodies in the SN and DA nerve terminals in striatum. Finally, within nigral cell bodies, MPP$^+$ binds to neuromelanin, and may be gradually released in a depot-like fashion, maintaining a toxic intracellular concentration of MPP$^+$ that inhibits mitochondrial respiration.

The serendipitous discovery and subsequent scientific investigation of the mechanism of parkinsonism produced by MPTP refocused study of the etiology and pathogenesis of idiopathic PD. For example, discovery of the selective ability of MPTP to induce nigral cell death has stimulated broad interest in identifying potential environmental or endogenous toxicants that may be causative agents in PD, as discussed above. Likewise, the mechanism of MPP$^+$ to cause DA cell death via inhibition of mitochondrial respiration provides support for theories involving mitochondrial dysfunction and oxidative metabolism in general (oxidative stress) in the etiology of idiopathic PD (see above). Delineation of the neurobiochemical mechanism of MPTP-induced parkinsonism also has led to new pharmacotherapeutic approaches aimed at slowing neurodegeneration in PD, focusing, notably, on MAO-B and oxidative stress. Clinical studies to evaluate the effectiveness of coadministration of an MAO-B inhibitor plus the antioxidant vitamin E to slow progression of neurodegeneration in PD, however, have not yielded encouraging results [109–111]. In fact, as indicated in the next section, despite progress toward understanding the etiology and pathophysiology of PD summarized above, no treatment has conclusively been proved to stop or even reliably slow the progressive neurodegeneration in the disease [112].

3. PARKINSON'S DISEASE PHARMACOTHERAPY

3.1. Dopaminergic Pharmacotherapy

Due to the failure of various neuroprotection strategies to provide unequivocal disease-modifying benefit for PD patients [112], the mainstay of PD pharmacotherapy continues to be palliative or symptomatic, involving replacement of the DA deficiency in striatum. This is accomplished currently by one or more of the following means: (1) augmentation of the synthesis of brain DA, (2) stimulation of dopamine release from presynaptic sites, (3) direct stimulation of dopamine receptors, (4) decreasing re-uptake of dopamine at presynaptic sites, or (5) decreasing metabolism of DA or its precursor L-dopa. Some of these treatments, notably, DA receptor agonists and MAO-B inhibitors (decrease DA metabolism), might also provide neuroprotective benefit, but such effects are not well established [112].

3.1.1. L-Dopa Therapy More than 40 years after its introduction, levodopa or L-dopa (**1**) (Fig. 2) remains the most effective symptomatic pharmacotherapy for PD [112]. Despite controversy regarding long-term efficacy, adverse effects, and even potential neurotoxicity of this amino acid precursor of DA, most PD patients derive a substantial benefit from L-dopa throughout their illness. Moreover, L-dopa increases life expectancy among PD patients, particularly if instituted early in the illness course [113].

In 1960, Ehringer and Hornykiewicz assayed DA in the brains of patients dying with PD and found that tissue concentrations of DA in the corpora striata of many of these patients averaged only 20% of normal [114]. Signs of PD in patients resembled behavioral changes in rats treated with reserpine or other amine-depleting agents. These findings led Birkmeyer and Hornykiewicz to administer high oral doses of racemic dopa to PD patients in Vienna in 1960 [115]. Subsequent clinical trials led by Barbeau in Montréal in the early 1960s [116] and by Cotzias in New York in the late 1960s [117] confirmed this effect of racemic dopa. Barbeau and Cotzias, also demonstrated the greater potency and safety of the physiological enantiomer L-dopa [118,119].

Development of L-dopa (**1**) (Fig. 2) as a therapeutic agent in PD is a rare example of a rationally predicted and logically pursued clinical treatment in a neuropsychiatric disorder, based on neurochemical pathology and basic pharmacological theory [11]. The effectiveness of L-dopa treatment requires its penetration into the central nervous system

(CNS) and local decarboxylation to DA. DA does not cross the blood–brain diffusion barrier because its amino moiety is protonated under physiological conditions ($pK_a = 10.6$ [NH_2]), making it excessively hydrophilic [120]. However, its precursor amino acid L-dopa is less basic ($pK_a = 8.72$ [NH_2]) and polar at physiological pH, and so more able to penetrate the CNS, in part facilitated by transport into brain with other aromatic and neutral aliphatic amino acids [120–122].

L-Dopa is normally a trace intermediary metabolite in the biosynthesis of catecholamines, formed from the essential amino acid L-tyrosine in a rate-limiting hydroxylation step by tyrosine hydroxylase (tyrosine-3-mono-oxygenase), a phosphorylation-activated cytoplasmic mono-oxygenase. L-Dopa is readily decarboxyated by the cytoplasmic enzyme L-aromatic amino acid decarboxylase ("dopa decarboxylase") to form DA (2). The effects observed after systemic administration of L-dopa have been attributed to its peripheral and cerebral metabolites, mainly DA, with much less conversion to norepinephrine by β-hydroxylation, or epinephrine formed by N-methylation of norepinephrine by phenylethanolamine-N-methyltransferase [120,122] (Fig. 2). A small amount of L-dopa is O-methylated by catechol-O-methyltranferase (COMT) to L-3-O-methyldopa (L-3-methoxytyrosine), which accumulates in the CNS because of its long half-life. However, most exogenous L-dopa is rapidly decarboxyated to DA in peripheral tissues, including liver, heart, lung, and kidney. Because only about 1% of an administered dose reaches the brain, L-dopa, by itself, has very limited dose effectiveness [123]. In humans, appreciable quantities of L-dopa enter the brain only when administered alone in doses (3–6 g daily) high enough to compensate for losses caused by peripheral metabolism.

L-Dopa peripheral decarboxylaton can be competitively inhibited [124] by coadministration of carbidopa (2S-3-[3,4-dihydroxyphenyl]-2-hydrazino-2-methylpropanoic acid (16) combined with levodopa in Sinemet and other products) or benserazide (2-amino-3-hydroxy-N'-[2,3,4-trihydroxybenzyl]-propanehydrazide (17) combined with L-dopa in Prolopa and others, in countries other than the United States). Such polar decarboxylase inhibitors do not appreciably penetrate the brain to inhibit cerebral decarboxylase, thus, markedly increasing the proportion of L-dopa that reaches the brain for conversion to DA, allowing for 2.5–30-fold lower dose (0.2–1.2 g/day) of L-dopa [17]. Patients with PD are typically started on a combination product, either alone or with other adjunctive agents discussed below. Products containing extended-release preparations of decarboxylase inhibitors should provide more sustained benefits with less "wearing off" of benefit after several hours, but the bioavailability and performance of these products is variable. In general, tissue uptake of L-dopa is highly dependent on competition with other aromatic and neutral aliphatic amino acids, and can be decreased substantially by a protein meal.

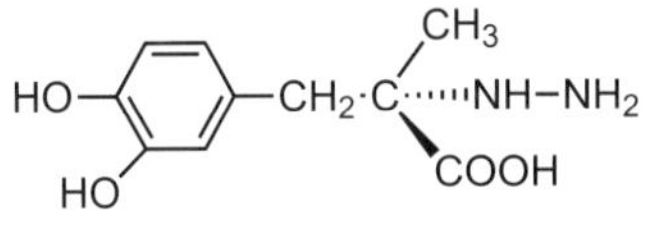

S-(-)-Carbidopa
(16)

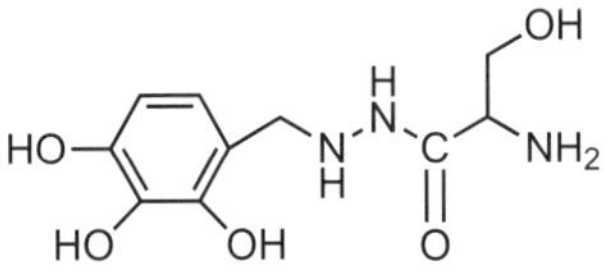

Benserazide
(17)

Pyridoxine (vitamin B_6) is the cofactor for aromatic amino acid decarboxylase. In high doses, B_6 can reverse the therapeutic effects of L-dopa by increasing peripheral decarboxylase activity. However, competitive blockade of peripheral decarboxylation with carbidopa or benserazide minimizes this potential effect of pyridoxine.

DA is relatively rapidly metabolized to its principal inactive by-products by MAO (particularly MAO-A in mitochondria of aminergic nerve terminals) and by extraneuronal COMT. Tissue concentrations of the methyl-donor cofactor of methylpherases, S-adenosyl-L-methionine (SAMe), can be depleted with large doses of L-dopa [125]. The main by-products of DA are 3,4-dihydroxyphenylacetic acid (DOPAC) and homovanillic acid (HVA) (3-methoxy-4-hydroxyphenylacetic acid) (Fig. 2).

Common adverse effects of L-dopa therapy are nausea and vomiting, possibly because of gastrointestinal irritation as well as stimulation by DA (and perhaps L-dopa) of the chemoreceptor trigger zone (CTZ) in the area postrema of the brainstem, an emesis-inducing center that is largely unprotected by the blood–brain diffusion barrier. An important advantage of combining L-dopa with a peripheral decarboxylase inhibitor, in association with the marked reduction of required doses of L-dopa, is less risk of emesis or other adverse effects, associated with peripheral formation of excess DA. These can include activation of peripheral adrenergic and DA receptors, in part by releasing endogenous adrenergic catecholamines [121], with a variety of cardiovascular effects. Theoretically, vasoconstriction and hypertension might occur by stimulation of peripheral α-adrenoceptors, tachycardia by stimulation of cardiac β-adrenoceptors, and direct renal and mesenteric vasodilatation by DA, although DA agonists and L-dopa usually induce hypotension. However, such effects are rarely encountered clinically with the use of a peripheral decarboxylase inhibitor with L-dopa [17].

After about 5 years of continuous treatment with L-dopa, at least half of PD patients develop fluctuating motor responses, and nearly three quarters do so by 15 years [126]. These fluctuations (so-called "on–off" effects) include "off" periods of immobility, and "on" periods with abnormal involuntary movements or dyskinesias (Fig. 3). These phenomena may reflect progression of the disease with more severe striatal nerve terminal degeneration and further loss of DA, along with increased sensitivity of its receptors.

Psychiatric disturbances such as hypersexuality, mania, visual hallucinations, and paranoid psychosis also are quite common, and sometimes severe adverse responses to treatment with L-dopa or direct DA agonists. Such behavioral disturbances probably reflect excessive stimulation of DA receptors in mesolimbic or mesocortical DA systems. They can greatly complicate clinical management of PD patients, including those with clinical depression commonly associated with PD, or dementia that sometimes arises in late stages of the disease. Modern antidepressants usually are well tolerated, with inconsistent and probably minor risk of worsening bradykinesia with serotonin-enhancing antidepressants [126]. Use of antipsychotic drugs, however, is limited to those with minimal risk of worsening bradykinesia and other aspects of extrapyramidal motor dysfunction [127,128]. Clozapine, though potentially toxic and relatively expensive, is best tolerated and may have particular efficacy for visual hallucinations in PD patients. Moderate doses of quetiapine and possibly of ziprasidone are sometimes tolerated, but their efficacy is not established, and other second-generation antipsychotic agents including olanzapine, risperidone, paliperidone, and even the DA partial agonist aripiprazole usually are poorly tolerated by PD patients owing to increased bradykinesia [127–131].

3.1.2. Drugs Targeting Dopamine Receptors

Dopamine Receptor Structure and Function
Recombinant DNA techniques have led to cloning and characterization of five different human DA receptors (number of amino acids in human): D_1 (446), $D_{2\ short}$ (414), $D_{2\ long}$ (443), D_3 (400), D_4 (387), and D_5 (477) [12]. DA receptors are members of the superfamily of G protein-coupled receptor (GPCR) cell-surface proteins. GPCRs have inherent structural flexibility and numerous thermodynamic conformations [132]. In addition to GPCR conformations that are constitutively active, agonist ligands can bind and stabilize or induce receptor conformations that lead to activation of the associated heterotrimeric (α, β, γ) guanine nucleotide binding (G) proteins [132]. Upon activation, the α-subunit of the G-protein releases GDP in exchange for binding of GTP, and, dissociation of the β/γ-dimer from the α-monomer occurs. Various intracellular signaling molecules then can be activated by the α- as well as β/γ-subunits to result in multiple physiological and/or psychological effects (and side effects, where drug therapy is concerned).

The same GPCR can couple to different Gα proteins to result in "multifunctional" signaling [133], which has been associated with "GPCR permissiveness" resulting from a high degree of flexibility in interactions among ligands, receptors, and G proteins [134]. Based on a "stimulus trafficking" hypothesis, GPCR multifunctional signaling requires heterogeneity of active receptor conformations, and some specificity of agonist ligands to induce, stabilize, or select among receptor conformations [135]. Of particular relevance for medicinal chemistry, structural parameters of agonist-ligands are very important determinants of GPCR conformation that influence the type of Gα protein and the signaling pathway activated. Thus, ligand stereochemistry or other more subtle structural parameters can result in differing, ligand-specific functional outcomes [136,137] that is referred to as "functional selectivity" [138–140].

A clinically relevant example of functional selectivity concerns the atypical antipsychotic drug aripiprazole that interacts with the dopamine D_2 receptor to produce antagonist, inverse agonist, or agonist functional effects, depending upon the D_2 receptor cellular milieu (e.g., G protein complement and concentration) and particular location (e.g., presynaptic versus postsynaptic, and, extrapyramidal versus limbic brain regions) [141]. Functionally selective agonists targeting D_2-type and/or perhaps D_1 DA receptors hold the promise of correcting motor deficits without necessarily causing psychiatric and other side effects. As mentioned above, however, aripiprazole, in fact, lacks efficacy for PD-related psychoses and exacerbates PD motor symptoms [128,130].

The thermodynamic flexibility of GPCRs manifests as thermal instability when attempts are made to extract the protein from lipid membranes with a detergent, making crystal generation difficult. Currently, the only human GPCR X-ray crystal structures solved are for the β_2-adrenergic [142] and adenosine A_{2A} [143] receptors. As there is currently no validated three-dimensional orientation of the amino acid residues at the ligand binding site for D1- or D2-type DA receptors, rational design of selective DA receptor agonists (and antagonists) is guided primarily by quantitative structure–activity relationship (QSAR) studies based on probe molecules, and ligand docking molecular modeling based on homology of DA receptors to the crystal structure of bovine rhodopsin, a particularly well-characterized GPCR [144–146]. DA molecular interactions with the β_2-adrenergic receptor have been extensively studied [147,148], and it is likely DA receptor homology models will be refined based on information gleaned from the β_2-receptor crystal structure.

Little information can be gained concerning the conformational requirements for DA receptor activation using DA itself because its ethylamine side chain has unlimited flexibility, and there is unrestricted rotation about the β-carbon–phenyl bond (Fig. 5). Accordingly, compounds in which the catechol ring and the amino-ethyl moiety of dopamine are held in rigid conformation have been synthesized to probe molecular determinants for receptor binding and activation. Such rigid analogs include aporphines, notably starting with the morphine acid-rearrangement product, R-(–)-apomorphine (**18**) (Fig. 5). The alkaloid apomorphine has been employed in experimental neuropharmacology since the late nineteenth century [149,150]. The *trans*-α-rotamer conformation of DA (Fig. 5) most closely aligns with apomorphine, as confirmed by the X-ray crystal structure of apomorphine [151]. In contrast, R-(–)-isoapomorphine (**19**) (Fig. 5) mimics the structure of DA in the *trans*-β-rotameric conformation, and has less DA-agonist activity than apomorphine. The analog R-(–)-1,2-dihydroxyaporphine (**20**) (Fig. 5), mimics the *cis*-α-rotamer conformation of DA and is inactive [152]. In addition to the aporphines, the semirigid aminotetralin, 2-amino-5,6-dihydroxy-1,2,3,4-tetrahydronaphthalene (A-5,6-DTN) (**21**) (Fig. 5), that has a *trans*-α-rotamer conformation between the benzene ring and the amino side chain, is a more potent DA agonist than its 6,7-dihydroxy congener (A-6,7-DTN) (**22**) (Fig. 5) with a *trans*-β-rotamer conformation [153]). Studies of these rigid dopaminomimetic compounds suggest that the preferred conformation of dopamine is the extended *trans* conformation (α- or β-rotamer). Results from mutagenesis, molecular modeling, and computa-

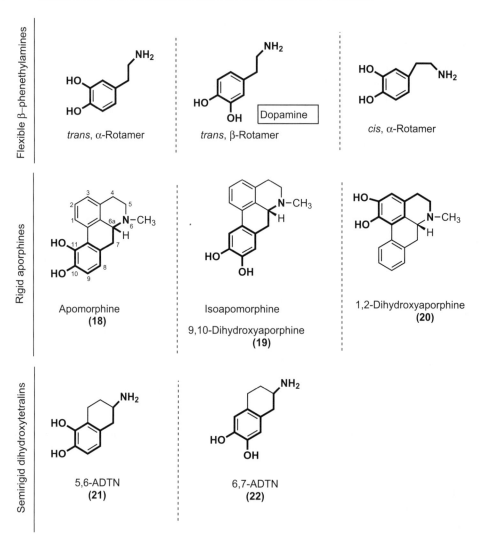

Figure 5. Conformations of dopamine in the *trans*-α-rotameric, *trans*-β-rotameric, and *cis*-rotameric forms, and their structural relationships to the rigid DA analogs apomorphine, isoapomorphine, and 1,2-dihydroxyaporphine. Also shown are the corresponding semirigid dihycroxyaminotetralin analogs of DA, A-5,6-DTN and A-6,7-DTN.

tional chemistry indicate that ligand activity at D_1 versus D_2 receptors is critically dependent on the position of protonated nitrogen moieties of candidate ligands (as in the *trans*-conformation of DA) that can support high-affinity ionic bonding at a proposed anionic aspartic acid residue in the third transmembrane α-helix (D3.32), as a preferred docking site of DA receptors [144,154,155].

Aporphine-Type Dopamine Receptor Agonists
R-(−)-Apomorphine hydrochloride was resurrected as a useful adjunct in the therapy of PD a decade ago [156], following years of neglect after promising early observations [117,157]. Lack of oral bioavailability, short duration of action, and potent central emetic action discouraged its clinical use. Nevertheless, in 1993, R-(−)-apomorphine received regulatory approval in the United Kingdom for control of refractory motor dysfunction and wide fluctuations in responses ("on–off" syndrome) to L-dopa or DA agonists [158–160]. Improved

motility in response to an acute challenge dose of apomorphine also can predict responsiveness to L-dopa treatment [161,162]. R-(–)-Apomorphine is an agonist for both D_1- and D_2-type DA receptors. With a pK_a of about 9, it is mostly protonated at physiological pH, but sufficiently lipophilic to cross the blood–brain barrier readily. Apomorphine can be administered subcutaneously by intermittent self-administration with a small self-injector (Penject), or continuous infusion with a portable minipump. Catechol-diester and methylenedioxy prodrugs of apomorphine limit first-pass metabolic inactivation, while retaining much of the activity of catecholaporphines or yielding the free catechols by metabolism of the prodrugs [163]. The 11-monohydroxy congener of the potent DA agonist R-(–)-N-n-propylnorapomorphine (**23**) (Fig. 6), R-(–)-N-n-propyl-11-hydroxynoraporphine (**24**) (Fig. 6), retains the critical free hydroxyl group analogous to the *meta*-3-hydroxy substituent of DA; the monohydroxyaporphine has potent dopaminergic activity and is orally bioavailable [163–166].

Several other agents with direct DA receptor agonist activity currently are used or are under clinical evaluation for treatment of PD (Fig. 6). These agents for the most part do not exploit known differences in molecular determinants for selective binding and function at D_1- versus D_2-type DA receptors [144,167,168], but most of those shown (Fig. 6) are at least partially D_2-selective.

Ergot-Type Dopamine Receptor Agonists Bromocriptine (**25**) (Fig. 6) is an ergot alkaloid-peptide that is a partial agonist at D_2 and D_3 DA receptors [169] that lacks appreciable activity at D_1- or D_4-receptors [170]. It was the first direct DA agonist to be employed in the treatment of PD, after its development for use at lower doses as a prolactin inhibitor [171]. Bromocriptine inhibits prolactin release from anterior pituitary mammotrophic cells that selectively express D_2 DA receptors. These receptors respond to DA produced in the arcuate nucleus of the hypothalamus and released at the median eminence into the hypophysioportal blood vessels, and carried to the pituitary to act as a prolactin-inhibitory hormone. Bromocriptine is an effective prolactin inhibitor at daily doses of 1–5 mg, for which it is used to treat hyperprolactinemia associated with pituitary adenomas, or to suppress prolactin output in prolactin-sensitive metastatic carcinoma of the breast. The D_2 partial agonist acts as an agonist at pituitary D_2 receptors that are normally in a high sensitivity state. At daily doses of 10–20 mg, bromocriptine and other D_2 partial agonist ergolines act as D_2 agonists with antiparkinson and perhaps mood-elevating effects. This agonism evidently reflects the supersensitized status of denervated DA receptors in PD [172,173]. Bromocriptine is absorbed after oral administration, but approximately 90% undergoes extensive first-pass hepatic metabolism; the remainder is hydrolyzed in the liver to inactive metabolites eliminated mainly in bile, and the overall elimination half-life is approximately 3 h.

Another ergoline, cabergoline (**26**) (Fig. 6), is a full agonist at D_2 receptors and a partial agonist at D_3- and D_4-receptors, without appreciable activity at D_1-type receptors [169,170], and a relatively long half-life of approximately 48 h [5]. Cabergoline was superior to placebo for treating motor signs of PD, but its comparative efficacy versus L-dopa is poorly documented [174]. Lisuride (**27**) (Fig. 6) is an ergoline partial agonist at D_2-, D_3-, and D_4-receptors with little activity at D_1-receptors [169,170]. This relatively short-acting agent is being evaluated in patients with advanced PD, using either patch or infusion delivery methods [5,175].

The peptide component of bromocriptine evidently is unnecessary for dopaminergic activity. Pergolide (**28**) (formerly, Permax®) was the first nonpeptide ergoline used successfully to treat PD, as well to inhibit release of prolactin [171,176]. Pergolide shows greater agonist effects at both D_2- and D_1-type DA receptors than bromocriptine does show, but it was withdrawn from clinical use due to association with valvular heart disease [177,178]. This adverse cardiac effect may be due to activation of serotonin 5-HT_{2B} receptors by pergolide, as has been hypothesized for other drugs with potent 5-HT_{2B} agonist activity, including, carbergoline (**26**) (Fig. 6) [179,180]. Moderate cardiac tricuspid valve regurgitation was more frequent in patients taking cabergoline repeatedly in relatively high doses for PD than in newly

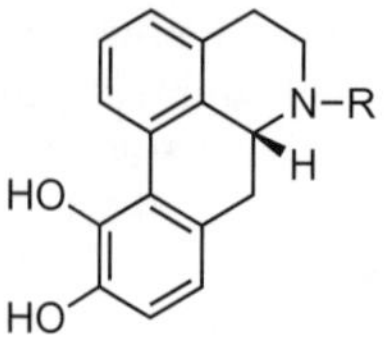

(18) R-(-)-Apomorphine, R = CH₃

(23) R-(-)-N-n-propylnorapomorphine (NPA), R = n-C₃H₇

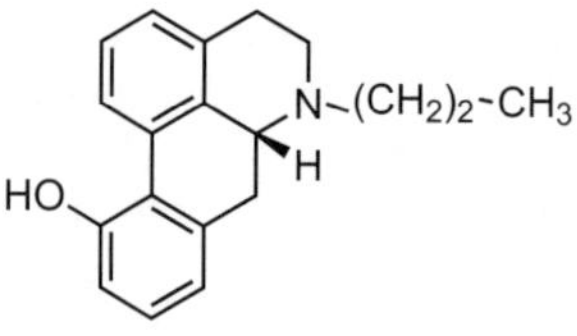

R-(-) N-n-propyl-11-hydroxynoraporphine
(24)

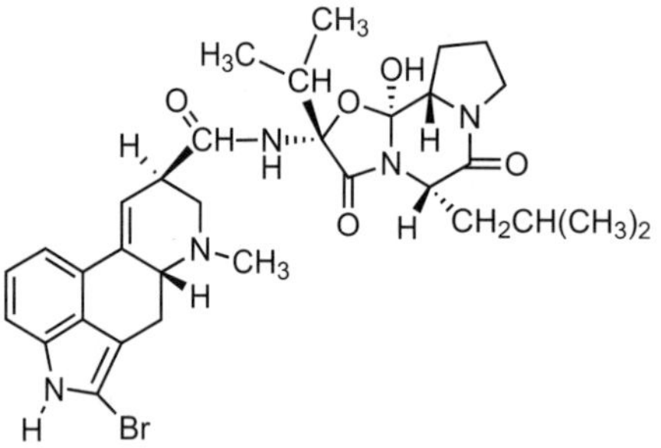

Bromocriptine (Parlodel)
(25)

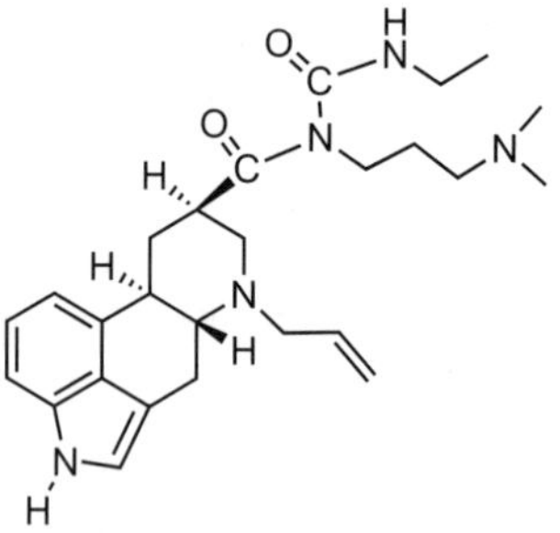

Cabergoline (Dostinex, Casbar)
(26)

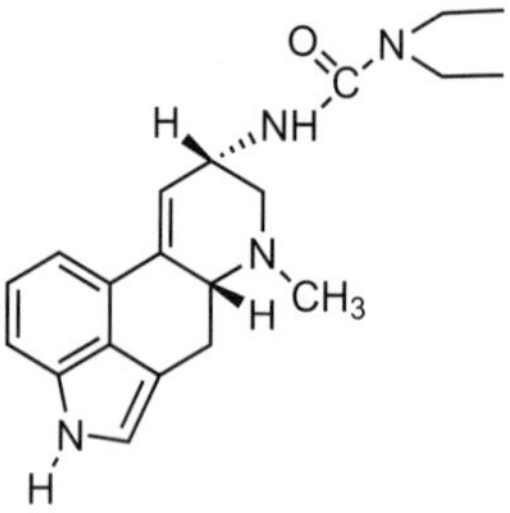

Lisuride
(27)

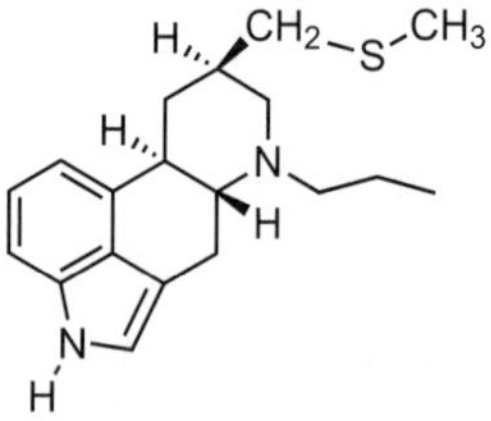

Pergolide (formerly, Permax)
(28)

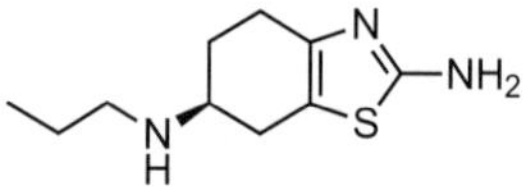

Pramipexole (Mirapex)
(29)

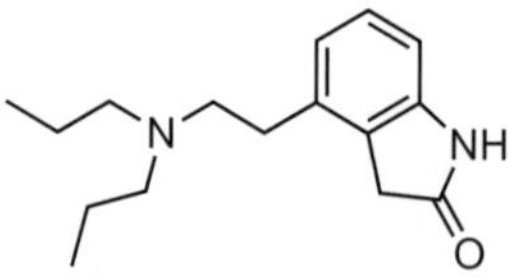

Ropinirole (Requip)
(30)

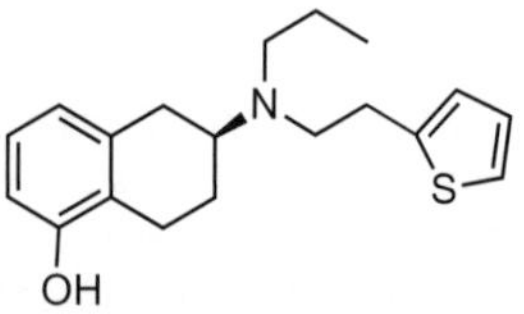

Rotigotine (Neupro)
(31)

Figure 6. Clinically used dopamine receptor agonists.

exposed or untreated controls [181]. However, doses of cabergoline used to treat hyperprolactinaemia are much lower [182] and may avoid valvular heart disease [182,183]. Ergolines such as bromocriptine (**25**) (Fig. 6) and lisuride (**27**) (Fig. 6) that lack 5-HT$_{2B}$ agonist activity do not seem to induce cardiac valvular damage [178,184,185].

Other Small-Molecule Dopamine Receptor Agonists Currently, pramipexole (**29**) (Fig. 6) and ropinirole (**30**) (Fig. 6) are among the most commonly prescribed direct DA agonists for PD in the United States [17]. They were introduced primarily for advanced stages of PD to limit fluctuations in response to L-dopa therapy, and as a "rescue" therapy when L-dopa became insufficiently effective. These direct D_2 DA-agonists are relatively well tolerated. Moreover, there has been concern that L-dopa might add further toxic damage to DA neurons through formation of reactive oxidized by-products. These characteristics have encouraged the use of these agents as first-line treatments, sometimes before L-dopa is added. An additional advantage of these agents is that they have relatively prolonged dopaminergic actions (long half-life), to provide more sustained clinical benefit with less risk of fluctuation of neurological status than with L-dopa, even as modified by cotreatment with inhibitors of its peripheral metabolism by decarboxylase and COMT. This impression is supported by controlled comparisons of both direct agonists with L-dopa [176]. A recent study in patients initially treated with pramipexole demonstrated a reduction in loss of radiolabeled striatal nerve terminals labeled by the DA-transporter-selective radioligand $[^{123}I]$-2-β-carbomethoxy-3-β-(4-iodophenyl)-tropane (β-CIT) (**31**) (see Fig. 7), a marker of DA neuron degeneration [186].

Rotigotine (**32**) (Fig. 6) is a relatively new, nonergoline, synthetic DA agonist that is administered by transdermal patch, and used to treat both early and advanced PD [187]. Like apomorphine, rotigotine is an agonist at D_1 as well as D_2 and D_3 receptors [188]. It is highly lipophilic and undergoes extensive first-pass hepatic clearance when given orally, encouraging use of the long-acting transdermal patch formulation [188]. Rotigotine provides benefits in PD both alone and combined with L-dopa and is better tolerated than other DA agonists [189–191].

The direct DA agonists generally produce similar adverse effects, including initial nausea and vomiting, postural hypotension, and fatigue. These effects are especially likely with the ergolines, which are started in low doses and increased slowly, as tolerated; ropinirole and pramipexole usually can be dosed more rapidly to clinically effective levels. Additional risks include psychotic reactions when DA agonists are given alone or with L-dopa. These reactions include hallucinations, delusions, and confusion, suggesting delirium, and are most likely in elderly PD patients with symptoms of dementia. Treatment is similar to psychotic reactions encountered during L-dopa therapy, and usually includes use of low doses of such antipsychotic agents as clozapine or quetiapine [126,173,192]. Adverse peripheral and central dopaminergic effects, including nausea, hypotension, and agitation also occur with ropinirole and pramipexole. They also can produce paradoxical somnolence as well as edema, and have been associated with uncommon narcolepsy-like, daytime sleep attacks, with potential risk during driving [193].

Experimental Dopamine D_1 Receptor Agonists
The opposing actions of the direct and indirect pathways in the basal ganglia (Fig. 3) suggest that coordinated neurotransmission requires activation of the direct pathway and attenuation of the indirect pathway involving DA neurotransmission in the basal ganglia. Such neuromodulation may require a balance of stimulatory actions at D_1-type receptors and inhibitory actions at D_2-type receptors. Consistent with a role for D_1 receptors in the direct output pathway, their stimulation represents a plausible pharmacotherapeutic approach for PD. Clinical trials in PD with early, selective D_1 partial agonists such as the benzergoline CY-208-243 (**33**) (Fig. 8) and the R-(+)-phenylbenzazepine SKF-38393 (**34**) (Fig. 8) found these drugs to be either short-acting or ineffective, suggesting that full D_1 agonist activity may be required [194,195]. Several analogs of SKF-38393 are full-efficacy D_1 agonists, including R-(+)-SKF-81297 (**35**) (Fig. 8) and its 6-halo derivatives, 6-Br-APB (**36**) (Fig. 8) and 6-Cl-APB (SKF-82958) (**37**) (Fig. 8) [196,197]. In MPTP-lesioned monkeys, 6-Cl-APB (**37**) produced antiparkinson effects [198], but its duration of action was less than 1 hour [198] and it produced severe dyskinetic effects [200]. Moreover, R-(+)-SKF-81297 (**35**), as well as, the benzophenanthridine dihydrexidine (**38**) (Fig. 8) [199], showed beneficial results only in monkeys with severe parkinsonism. It has

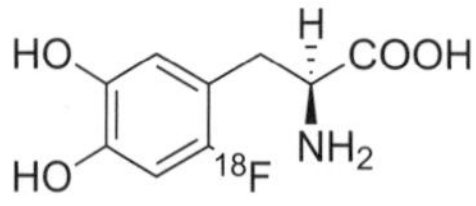

[¹⁸F] 6-fluoro-L-dopa
(62)

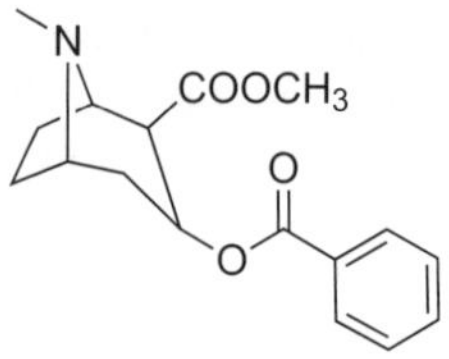

Cocaine
(63)

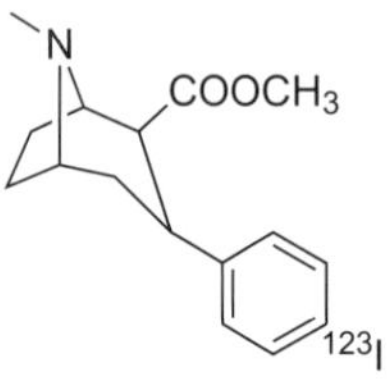

[¹²³I]-β-CIT
(DOPAScan)
(31)

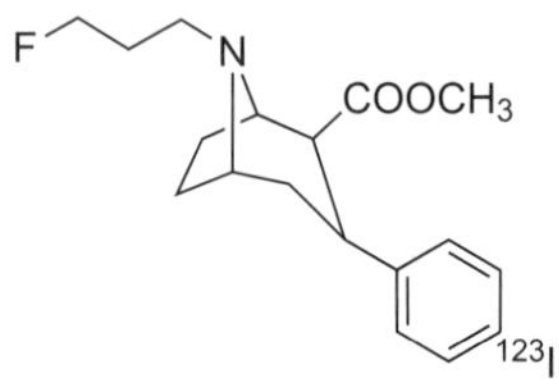

[¹²³I]-FP-CIT (Ioflupane)
(DATScan)
(64)

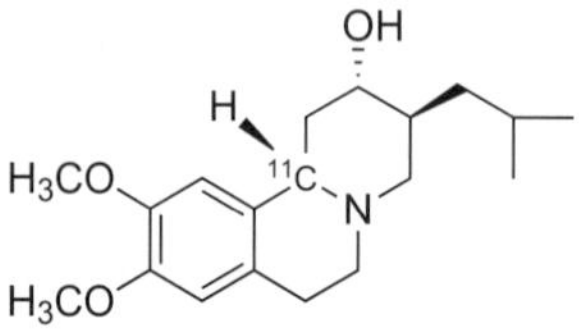

(+)-[¹¹C]-Dihydrotetrabenazine
(65)

Figure 7. Radioligands for dopamine neurons. These include [¹⁸F]-dopa and nonhydrolyzable, long-acting phenyltropane analogs of cocaine, which bind selectively to DA-transporter proteins and are highly selective markers of DA neurons. These agents are useful for imaging with [¹⁸F] for PET and [¹²³I] for SPECT.

been suggested that long-acting D_1 agonists may be most useful in late stages of PD [201].

Dihydrexidine (**38**) (Fig. 8) was the first full-efficacy D_1 agonist to be developed, although it also has some D_2-type activity [202,203]. In MPTP-lesioned monkeys, dihydrexidine essentially eliminated all parkinsonian signs, and this effect was fully blocked by a selective D_1 antagonist but not by a selective D_2 antagonist [204]. Continuous administration of dihydrexidine to rats for 2 weeks produced very little change in D_1 receptor density or D_1 receptor-mediated DA-stimulated adenylyl cyclase activity, suggesting that tolerance might not develop to its antiparkinson effects. In PD patients, however, dihydrexidine has a narrow therapeutic index, with dose-limiting adverse effects that include flushing, hypotension, and tachycardia after single intravenous doses [205].

A D_1 receptor pharmacophore model developed for dihydrexidine was used to design other novel molecular structures as full-efficacy D_1 agonists [206]. These D1 agonists include dinapsoline (**39**) (Fig. 8) [207] and several other analogs modeled on dihydrexidine (**38**) or dinapsoline (**39**). They include the dihydrexidine isostere A-86929 (**40**) (Fig. 8) and its diacetyl prodrug ABT-431 (**41**) (Fig. 8); both are full D_1 agonists with sustained antiparkinson effects in MPTP-lesioned monkeys [208]. In PD patients, ABT-431 was highly effective against bradykinesia, but produced dyskinesias [209].

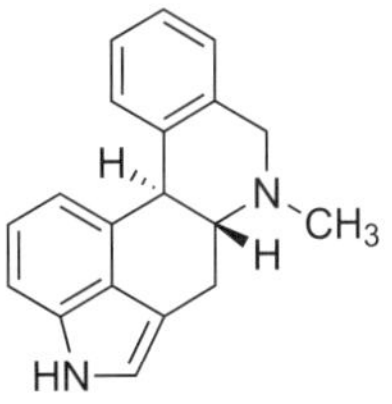

CY-208-243
(33)

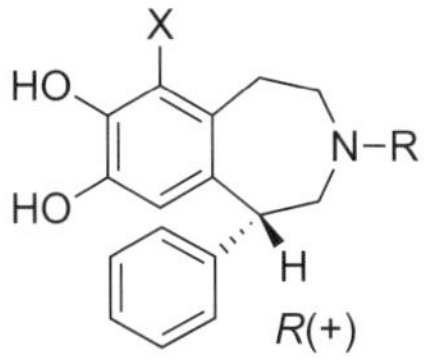

(34) X = R = H: SKF-38392
(35) X = Cl, R = H: SKF-81297
(36) X = Br, R = CH$_2$CH=CH$_2$: Br-APB
(37) X = Cl, R = CH$_2$CH=CH$_2$: SKF-82958, Cl-APB

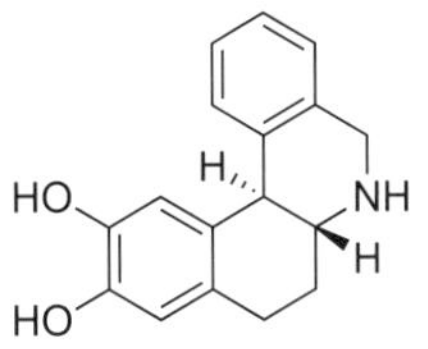

Dihydrexidine
(38)

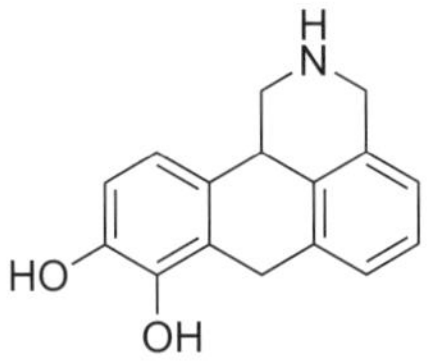

Dinapsoline
(39)

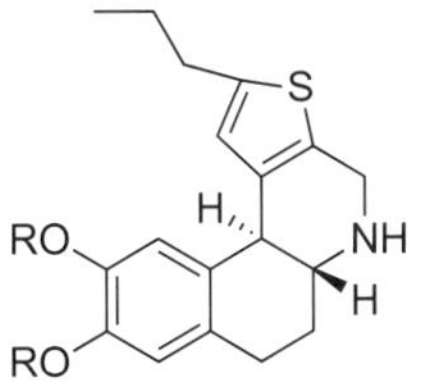

(40) R = H: A-86929
(41) R = CH$_3$CO: ABT-431

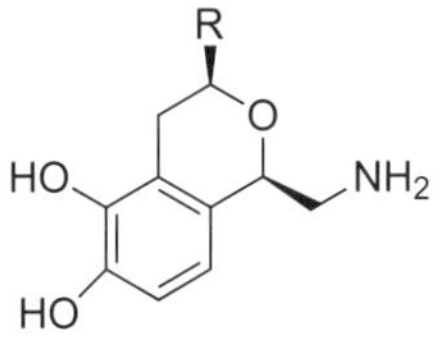

(42) R = C$_6$H$_5$: A-68930
(43) R = [adamantyl]: A-77636

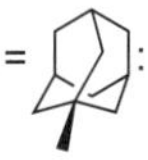

Figure 8. Representative experimental D$_1$ agonists. These agents have unproved or untested clinical utility in the treatment of PD.

Several isochromans also are full D$_1$ agonists. In primates pretreated with MPTP, an early compound of this type, A-68930 (42) (Fig. 8), produced seizures, but an analog, A-77636 (43) (Fig. 8), showed antiparkinson effects without inducing seizures [210]. However, A-77636 showed rapid desensitization to its beneficial effects [211], possibly related to its prolonged action (>20 h; [200]). Other conformationally restricted analogs containing the β-phenyldopamine pharmacophore continue to be reported, including, some with impressive D$_1$-type over D$_2$-type receptor agonist activity [212], however, *in vivo* studies with D$_1$-agonists have given disappointing results.

3.1.3. Monoamine Oxidase Inhibitors Given the role of MAO in catabolism of DA (Fig. 2), possibly, leading to neurotoxic oxidation products of both endogenous (Fig. 4) and exogenous (Fig. 1) substrates, MAO inhibition has the potential to boost levels of brain DA and prevent formation of damaging MAO-derived oxidation products. Furthermore, by reducing the oxidation of DA, MAO inhibitors can ex-

MAO inhibitors

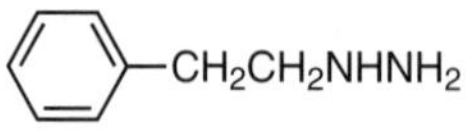

Phenelzine (Nardil)
(44)

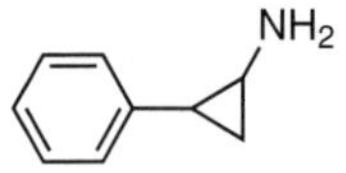

Tranylcypromine (Parnate)
(45)

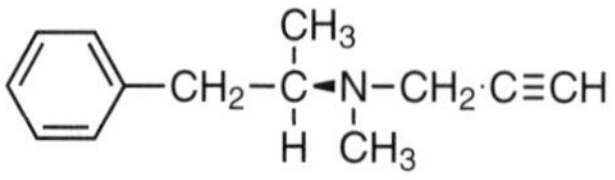

Selegiline (R-[-]-Deprenyl, Eldepryl)
(46)

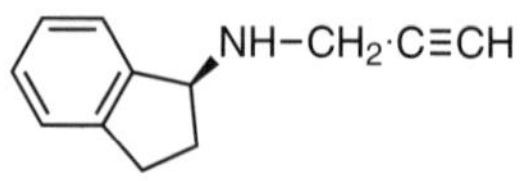

R-(+)-Rasagiline (Azilect)
(47)

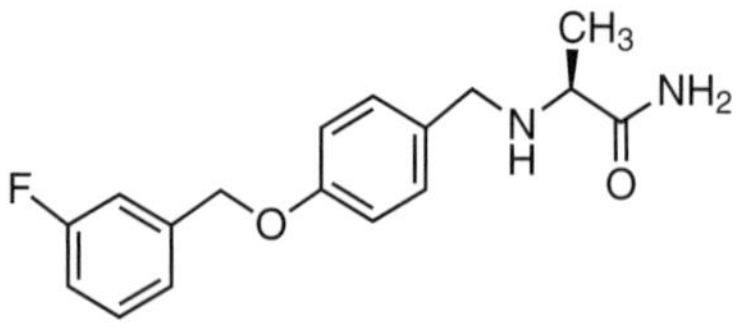

Safinamide
(48)

COMT inhibitors

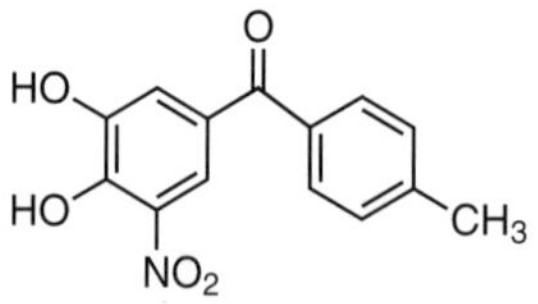

Tolcapone (Tasmar)
(49)

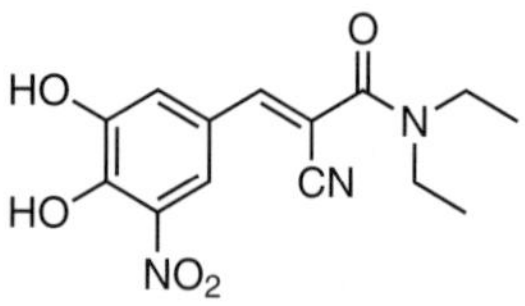

Entacapone (Comtan)
(50)

Figure 9. Monoamine oxidase (MAO) and catechol-O-methyltransferase (COMT) inhibitors.

tend the duration of response to L-dopa, and allow use of lower doses [109]. Long-acting, irreversible, nonselective MAO-A/B inhibitors, such as phenelzine (44) (Fig. 9) and tranylcypromine (45) (Fig. 9) [208,213] are contraindicated in combination with L-dopa because of the risk of inducing hypertensive crises and delirium [17]. Human brain MAO-A is found primarily in noradrenergic neurons and MAO-B is found primarily in serotoninergic or histaminergic neurons and in glial cells, and the MAO isozyme present in DA nerve terminal remains uncertain [51]. There is uncertainty as to whether to target MAO-A or -B for the treatment of PD, and less is known about the structural requirements for highly specific reversible MAO-A inhibitors [214] compared to selective MAO-B inhibitors, several of which produce clinical benefits in PD.

Selegiline (L-deprenyl; R-[-]-N,2-dimethyl-N-2-propynyl-phenethylamine) (46) (Fig. 9) and rasagiline (47) (Fig. 9) are propargylamine-type selective irreversible inhibitors of MAO-B. In addition to potentiating DA actions and allowing for reduction of L-dopa dose, it has been proposed that MAO-B inhibitors may prevent formation of neurotoxic oxidation products of DA and slow neurodegeneration in Parkinson's disease; however, data from recent clinical studies do not support this attractive "neuroprotective" hypothesis [109–112]. Nevertheless, MAO-B inhibitors have a beneficial effect on motor fluctuations because of their levodopa-sparing effect [109]. Selegiline and rasagiline undergo extensive hepatic metabolism. Selegiline is N-dealkylated via CYP2B6 and CYP2C19 to (–)-methamphetamine and, subsequently, to (–)-amphetamine, which has vasoactive activity similar to (+)-amphetamine [215]. The amphetamine metabolites of selegiline may contribute to its other pharmacological property of DA and norepinephrine reuptake inhibition, thus potentiating the pharmacological effects of L-dopa [107]. The amphetamine metabolites of selegiline have been associated with cardiovascular (orthostatic hypotension) and psychiatric (hallucinations) side effects. Rasagiline is N-dealkylated primarily by CYP1A2 to (R)-1-aminoindan, which does not have vasoactive activity [215]. Recently, rasagaline underwent clinical trials to carefully assess (controlled for confounding variables, both symptomatic and pharmacologic) its ability to provide neuroprotection [216]. Rasagiline at a dose of 1 mg/day met endpoints consistent with a disease modifying (neuroprotective) effect, but, rasagiline at 2 mg/day failed—the authors of the study hypothesize that the stronger symptomatic effect of the higher dose masked an underlying disease modifying effect.

Safinamide (**48**) (Fig. 9) is an α-aminoamide derivative that is a reversible selective MAO-B inhibitor shown to provide benefits in early PD [217,218]. In addition to its MAO-B inhibitor properties, safinamide also blocks voltage-dependent sodium and calcium channels, and, inhibits glutamate release [217]. It is currently in Phase 3 trials in the United States and in Europe [5].

3.1.4. Catechol-O-Methyltransferase Inhibitors

The peripheral metabolism of L-dopa given alone leads to very limited access of the amino acid to the CNS. It is rapidly decarboxylated by L-aromatic amino acid decarboxylase and 3-O-methylated by COMT. In addition to potentiating L-dopa with peripheral decarboxylase inhibitors including carbidopa and benserazide, there are also inhibitors of COMT. This methyl-pherase, with its methyl-donor cofactor SAMe, converts L-dopa and catecholamines preferentially to their m-methoxy derivatives (Fig. 2). These include 3-O-methyl-dopa and 3-O-methyl-DA (3-methoxytyramine, 3-methoxy-4-hydroxyphenethylamine), as well as the 3-O-methylated, deaminated compound HVA, the major final metabolite of DA in humans. Treatment with L-dopa can reduce tissue concentrations of SAMe [122], with uncertain consequences that should be limited by cotreatment with a COMT-inhibitor. COMT acts in both peripheral and cerebral tissues, although the effect of COMT inhibitors to potentiate L-dopa probably occurs mainly in peripheral tissues [17,198,219,220].

Tolcapone (**49**) (Fig. 9) and entacapone (**50**) (Fig. 9) are reversible COMT inhibitors. Examination of their chemical structures reveals obvious similarities, and the molecular mechanisms by which these drugs interact with human COMT are proposed to be similar [221]. Although the mechanisms of action and pharmacotherapeutic effects are similar for tolcapone and entacapone, they differ with respect to pharmacokinetic properties and adverse effects. Tolcapone has a relatively longer duration of action (8–12 h) and acts both in the brain and the periphery, whereas entacapone has a shorter duration of action (2 h) and acts mostly in the periphery. Some common adverse effects of these agents are predictable and attributable to increased brain DA (e.g., nausea, vivid dreams, confusion, and hallucinations). A potentially fatal adverse effect, however, occurs only with tolcapone—after marketing, three fatal cases of fulminant hepatic failure were observed, leading to its market withdrawal in some countries restriction in the United States to only those patients who have not responded to other therapies and who have appropriate monitoring for hepatic toxicity. The unforeseen hepatotoxicity associated with tolcapone has left entacapone as the only COMT inhibitor in wide clinical use [222], albeit, since the labeling restrictions in 1998, there have been no additional reports of hepatic fatality [223]. The mechanism by which liver damage is induced by tolcapone is believed to involve uncoupling of mitochondria oxidative phosphorylation, significantly reducing cellular generation of adenosine triphosphate [224,225]. Additionally, it was shown that tolcapone (but not entacapone) induces cytotoxic pro-oxidant radical formation in hepatocytes [226]. Finally, both COMT inhibitors may cause severe diarrhea and produce increased dyskinesias that may require a reduction in the dose of L-dopa [227].

3.2. Agents Acting on Nondopaminergic Systems

3.2.1. Anticholinergic Agents

Cholinergic interneurons in the striatum exert mainly excitatory effects on GABAergic output from the striatum (Fig. 3). Drugs that increase cholinergic neurotransmission (e.g., the cholinesterase inhibitor physostigmine and the direct agonist carbachol) have long been known to aggravate parkinsonism in humans, whereas centrally active muscarinic antagonists (such as the belladonna alkaloids, including atropine), have moderately beneficial effects [17,228]. Accordingly, before the discovery of L-dopa, drug therapy for parkinsonism depended primarily on the limited efficacy of the natural belladonna alkaloids and newer synthetic antimuscarinic alkaloids, as well as antihistamines that also exert central antimuscarinic actions (Fig. 10). Synthetic central anticholinergic agents include benztropine mesylate (**51**), biperiden (**52**), the antihistamine diphenhydramine (**53**), the phenothiazine ethopropazine (**54**), orphenadrine (**55**), procyclidine (**56**), and trihexylphenidyl (**57**). Such drugs continue to be used to control parkinsonism and other adverse extrapyramidal neurological effects of potent D_2-receptor antagonist antipsychotic agents, for which they are quite effective [17,126].

However, central antimuscarinic agents have limited therapeutic benefit in PD. They also exert a range of undesirable adverse effects because of their blockade of peripheral parasympathetic function and adverse CNS actions. These include dry mouth, impaired visual accommodation, constipation, urinary retention, and tachycardia. Adverse CNS effects include delirium, marked by confusion, memory impairment, and psychotic symptoms. Despite their relatively unfavorable benefit/risk ratio, these agents are still sometimes employed in the treatment of PD in combination with L-dopa, particularly to help control tremor [228].

3.2.2. Adenosine Receptor Antagonists

Adenosine is a nucleoside signaling molecule that acts at four GPCR subtypes, A_1, A_{2A}, A_{2B}, and A_3. The mRNA for A_{2A} receptor protein is highly concentrated in the striatum, nucleus accumbens, and olfactory tubercle, and colocalizes with D_2 receptor mRNA in these brain regions [229]. Activation of A_{2A} receptors inhibits GABA release in striatum and reduces GABA-mediated inhibition of striatal medium spiny output neurons [230]. Thus, antagonism of A_{2A} receptors is expected to increase GABA-mediated inhibition of the medium spiny output neurons to help compensate for the loss of DA D_1 receptor-stimulated GABA release and D_2 receptor-mediated inhibition of these neurons in PD [231] (see Fig. 3). Adenosine A_{2A} receptors also oppose the actions of D_1 and D_2 receptors on gene expression [232] and second-messenger systems and reduce the binding affinity of DA for D_2 receptors [233]. An A_{2A} antagonist presumably would block these A_{2A} receptor-mediated inhibitory effects on DA neurotransmission and perhaps provide benefit in PD. Activation of A_{2A} receptors also stimulates release of acetylcholine in striatum [234]. Because muscarinic acetylcholine receptor antagonists can ameliorate some signs of PD, A_{2A} receptor antagonists may exert additional benefits in PD by reducing striatal cholinergic neurotransmission. As indicated in Section 2.2.3, there is an inverse relationship between the risk of developing PD and the consumption of caffeine, a nonselective antagonist at adenosine A_1 and A_2 GPCRs [88,89].

Istradefylline (**58**) (formerly, KW 6002) (Fig. 11) is a potent and selective antagonist at adenosine A_{2A} receptors that was shown improving motor disability in primate models of PD [235]. A recent double-blind, placebo-controlled clinical trial [236] showed that in patients stabilized on L-dopa and other PD drug regimens, istradefylline-treated subjects had significant reductions in motor fluctuations and the drug was, in general, well tolerated [236]. It appears istradefylline will be a

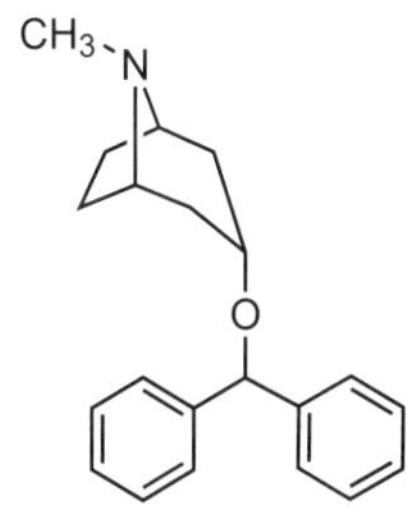

Benztropine mesylate (Cogentin)
(**51**)

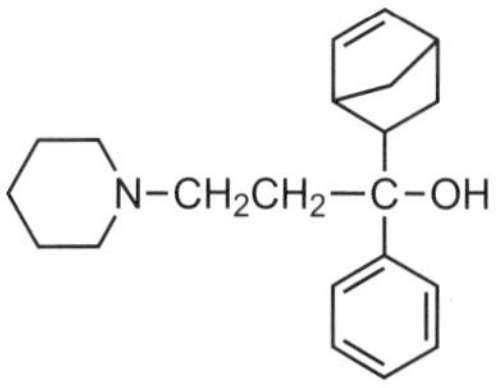

Biperiden (Akineton)
(**52**)

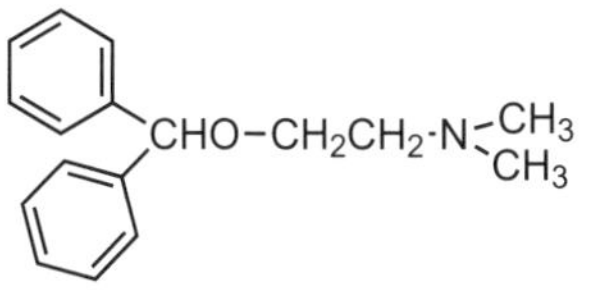

Diphenhydramine (Benadryl)
(**53**)

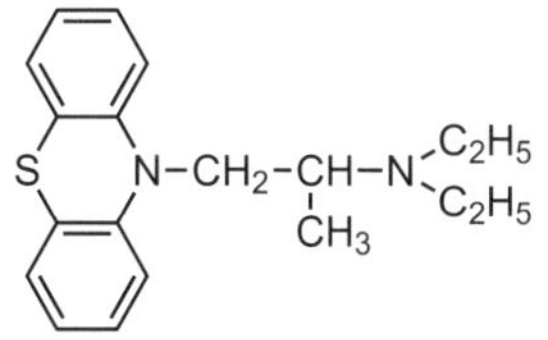

Ethopropazine (Parsidol)
(**54**)

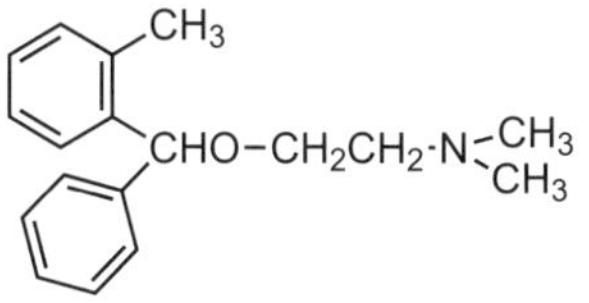

Orphenadrine (Disipal)
(**55**)

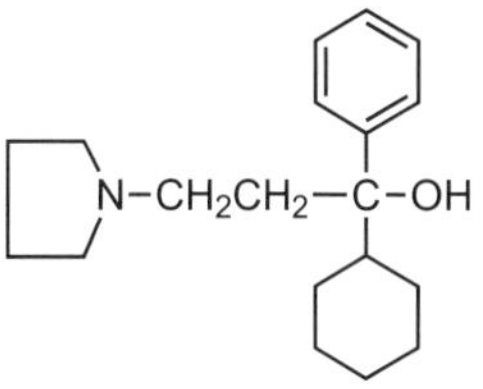

Procyclidine (Kemadrin)
(**56**)

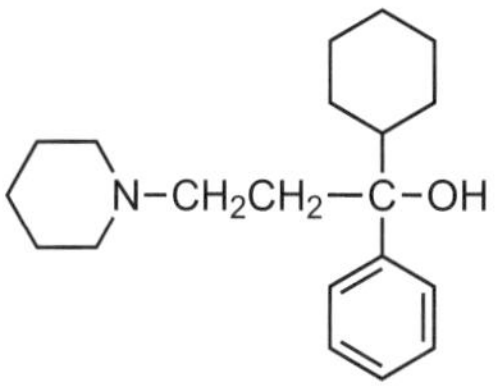

Trihexyphenidyl (Artane)
(**57**)

Figure 10. Agents with central antimuscarinic activity sometimes used to treat idiopathic or neuroleptic drug-induced parkinsonism.

useful medication in the treatment of PD. Studies with A_1 receptor antagonists lag behind, however, there is evidence they reduce parkinsonian-type tremor in rodents [89].

3.2.3. Serotonin 5-HT$_{1A}$ Agonists Dysfunction of neurotransmission mediated by 5-hydroxytryptamine (5-HT; serotonin) occurs in the basal ganglia of patients with PD, and excessive serotonergic transmission may contribute to dyskinesias associated with dopaminergic treatments [237]. The central and peripheral psychological and physiological effects of 5-HT are mediated by 14 serotonin

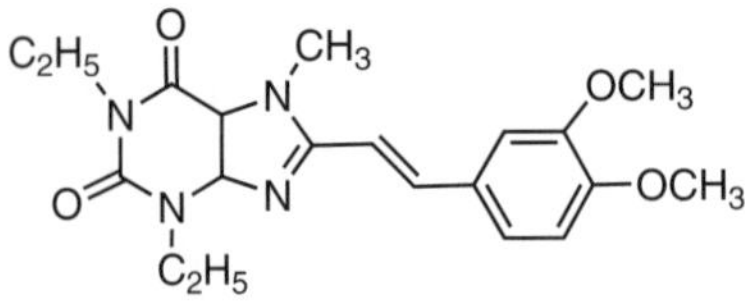

Istradefylline (formerly, KW 6002)
(58)

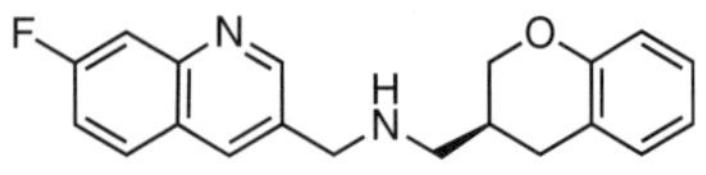

Sarizotan (formerly, EMD 128130)
(59)

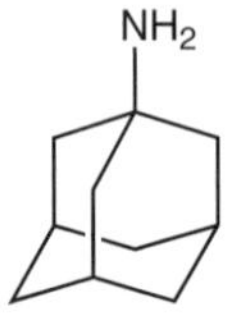

Amantadine (Symmetrel)
(60)

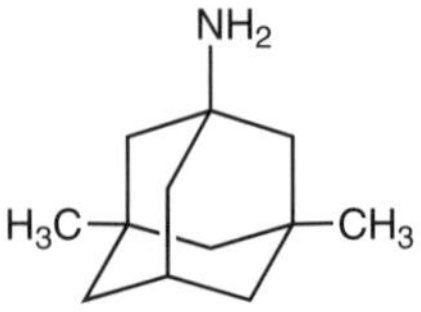

Memantine (Namenda)
(61)

Figure 11. Other PD drugs acting on nondopaminergic systems.

receptor subtypes grouped into the 5-HT_1–5-HT_7 families. The 5-HT_1 family consists of five GPCRs subtypes, 5-HT_{1A}, 5-HT_{1B}, 5-HT_{1D}, 5-HT_{1E}, and 5-HT_{1F} [238]. 5-HT_{1A} receptors are expressed presynaptically on 5-HT terminals, where they limit serotonin release as autoreceptors [239]. Their activation should decrease 5-HT release and perhaps alleviate dopaminergic dyskinesias in PD [240]. Because 5-HT_{1A} receptor stimulation can reverse parkinsonism-like catalepsy induced by haloperidol [241], 5-HT_{1A} receptor activation might also counteract losses of nigrostriatal DA neurotransmission in PD. Moreover, in patients with advanced PD, intact striatal 5-HT terminals are an important site of decarboxylation of exogenous L-dopa to DA. A 5-HT_{1A} agonist might act at striatal serotonergic terminals to limit release of DA produced by L-dopa treatment and released from 5-HT terminals as a "false transmitter."

Sarizotan (**59**) (formerly, EMD-128130) (Fig. 11) is an aminomethylchroman derivative that is fairly well characterized regarding its neurobiochemical activity [242]. It has very high affinity ($K_i \sim 0.05\,\text{nM}$) at human 5-HT_{1A} receptors at which it is a full-efficacy agonist relative to 5-HT and it is about 10 times more potent. Sarizotan also has appreciable affinity at human DA D_2-, D_3-, and D_4-receptors ($K_i \sim 2{,}5\,\text{nM}$), though 50 times less than at 5-HT_{1A} receptors. It is a partial agonist at D_2 but demonstrates no apparent activation of D_3- and D_4-receptors. Given by itself to MPTP-lesioned monkeys, sarizotan had no effect on the severity of motor deficits or on beneficial responses to L-dopa, but it reduced L-dopa-induced choreiform dyskinesias by more than 90% [239]. The lack of interaction of sarizotan with L-dopa in MPTP-lesioned monkeys is unsupportive of concern that it might limit release of DA from 5-HT terminals (via 5-HT_{1A} activation) or from DA nerve terminals via D_2 activation. The beneficial effects of sarizotan in parkinsonian monkeys seem specific to 5-HT_{1A} agonism, in that they were reversed by a selective 5-HT_{1A} antagonist. In PD patients with dyskinesias resulting from L-dopa therapy, adding sarizotan pharmacotherapy produced significant increases in periods of time without dyskinesia and significant reduction in periods of time with troublesome dyskinesias [243]. A subsequent clinical study that assessed improvement of dyskinesias gave mixed results, depending on the dyskinesia rating scale that was used, however, the drug proved to be safe and relatively well tolerated [244]. A recent study using the MPTP monkey PD model confirmed sarizotan produces a sustained antidyskinetic effect while maintaining L-dopa antiparkinsonian effects [245] suggesting the 5-HT_{1A}

agonist field may prove fruitful in PD drug discovery.

3.2.4. Glutamate Antagonists

The dyskinesias associated with L-dopa therapy may involve overactivity of thalamocortical excitatory glutamatergic input to the motor cerebral cortex (Fig. 3). In addition, excessive release of glutamate due to synaptic overactivity is hypothesized to lead to "excitotoxicity," resulting from excess neuronal Ca^{++} influx due to opening of N-methyl-D-aspartate (NMDA) ion channel receptors [246], at which glutamate is a coagonist (along with glycine).

The adamantane derivative, amantadine (**60**) (Fig. 11), and its dimethylated congener, memantine (**61**) (Fig. 11), are NMDA glutamate receptor antagonists that might provide neuroprotective effects [247]. Both have moderately beneficial effects early in PD, can enhance the effects of L-dopa, and perhaps limit the severity of dyskinesias induced by L-dopa therapy [248]. Also, memantine has been used as a spasmolytic agent in the treatment of both PD and dementia [249]. Amantadine was originally developed as an antiviral agent and its use in PD patients with influenza revealed unexpected improvement in PD symptoms [250]. The pK_a of this primary amine is 10.8, thus, it exists mainly in the protonated form at physiological pH. Adamantane and memantine, however, apparently can enter the brain because of their lipophilic cage-like structure that may provide resistance to metabolism by oxidative enzymes—these drugs are excreted mostly unchanged in the urine [251,252]. In addition, amantadine has some ability to release DA and norepinephrine from intraneuronal storage sites and to block reuptake of DA, and was initially considered a DA-potentiating agent for use in mild PD [253].

4. DIAGNOSTIC AGENTS FOR PARKINSON'S DISEASE

Even for an experienced clinical neurologist, diagnosis of PD can be difficult to confirm, especially in the early stages of this disease. Signs of PD vary markedly among patients and in the same person over time; disability can fluctuate dramatically, and progression of the disorder is unpredictable. In addition, a number of conditions mimic PD, and vary in their responses to antiparkinson drugs. Given these difficulties, brain imaging techniques are increasingly applied to diagnostic and neuropharmacological studies of brain function in PD patients. Positron emission tomography (PET) and single-photon emission computed tomography (SPECT) are sensitive methods employed in such studies.

Spatial resolution is greater with PET, but SPECT technology is less expensive and more widely accessible in many clinical settings. In addition, positron-emitting nuclides used in PET imaging have very short half-lives (^{11}C, 20 min; ^{18}F, 109 min) and usually require an on-site cyclotron for their production. SPECT nuclides have longer half-lives (^{123}I, 13 h; ^{99m}Tc, 6 h), often can be supplied commercially, and [^{99m}Tc]-labeled radioligands can be prepared locally as needed. Specifically, quantitative assessment of nigrostriatal presynaptic DA nerve terminal function by PET using [^{18}F]-labeled 6-fluoro-L-dopa [^{18}F] dopa) (**62**) (Fig. 7) has proved useful for the early diagnosis of PD [254].

Additional radioligands recently have been developed for probing DA transporter (DAT) proteins that are highly characteristic gene products of DA neurons and nerve terminals. Cocaine (**63**) (Fig. 7) binds to the DAT and other monoamine transporters, but limited DAT-specificity of radiolabeled cocaine and rapid hydrolysis at its benzoyl ester function makes it an impractical candidate for use in imaging [255]. However, linking the phenyl ring of cocaine directly to the tropane system yields nonhydrolyzable, long-acting phenyltropanes. Such compounds have proved to be potent psychostimulants; some have high affinity and varying selectivity for cerebral DAT, and have been developed as clinically useful radiopharmaceuticals (Fig. 7).

The first such agent was p-[^{123}I] phenyl-labeled 2-β-carbomethoxy-3-β-(4-iodophenyl)-tropane (**31**) ([^{123}I]-β-CIT or RTI-55), although this agent requires about 8 h for peak uptake before imaging, greatly limiting its practicability [186,256]. However, the radioiodinated N-3-fluoropropyl analog of β-CIT (N-3-fluoropropyl-2-β-carbomethoxy-3-β-(4-iodophenyl)-tropane) (**64**), also known as [^{123}I]FP-CIT or [^{123}I]ioflupane, has more favorable kinetics for clinical imaging in 1–2 h after injection of the radioligand [257]. A radioligand suitable for PET imaging is obtained by replacing the fluorine atom with [^{18}F] in FP-CIT [258].

Use of the protein vesicular monoamine transporter 2 (VMAT2) radioligand (+)-[^{11}C]-dihydrotetrabenazine (**65**) (Fig. 7) [259] as an investigational positron emission tomography radiotracer for DA neurons also has been reported [260]. This agent binds to VMAT2, the transporter that moves nerve terminal cytoplasmic DA into synaptic vesicles for storage and subsequent exocytotic release, thus, serving as a biomarker of DA neuronal loss in PD [261,262].

5. FUTURE DIRECTIONS

Research related to PD has been directed toward developing more effective and better-tolerated treatments, largely guided by the central role of DA in the pathophysiology of the disorder. Nevertheless, recent research has increasingly included efforts to clarify the pathophysiology of PD more broadly [15]. Regarding elucidation of the still-obscure primary cause of PD, emphasis is on understanding the mechanisms of DA neuronal cell death, including apparently genetically programmed death (apoptosis) that certainly involves mitochondria and may converge with hypotheses involving dysfunctional mitochondrial oxidation, neurotoxicants, and genetic mutations [263].

Several gene mutations have been linked to PD but most occur in uncommon familial cases. Thus, genetic testing is not likely to be useful for screening in the general population, where most cases are thought to occur sporadically. Guided by improved understanding of neuroanatomical and neurophysiological abnormalities in PD, there have been important advances in developing novel diagnostic agents for PD. Although neuroimaging is not required to make a diagnosis of PD [5], it may help in early identification of suspected cases. Increasing use of clinical neuroimaging with PET and SPECT techniques using brain-imaging agents to detect losses of DA neurons in PD *in vivo*, and perhaps application of functional magnetic resonance imaging (fMRI) methods are aiding early diagnosis and should enable monitoring of the progression of the disease. Ability to monitor the progression of PD should also encourage development of novel treatments aimed at slowing the progression of the disease.

Innovative neurosurgical PD treatment methods include application of deep brain stimulation [264]. Transplantation of fetal DA neurons into the striatum of PD patients can alleviate motor deficits, however, dyskinesias frequently result and pathological problems develop in the grafted cells such as reduced DA transporter and tyrosine hydroxylase expression [265]. Future DA neuronal cell replacement therapy almost certainly will involve use of embryonic stem cells to obtain large quantities of correctly differentiated midbrain DA neurons for transplantation [266].

Finally, the search for improved medicinal agents for the treatment of PD continues, greatly stimulated by the broadening range of leads targeting nondopaminerigc neurotransmission systems discussed above. With regard to DA, adenosine, and serotonin GPCR drug targets in PD, ligands targeting allosteric sites rather than the endogenous ligand binding (orthosteric) site offer the possibility of enhanced target selectivity (due to less conserved amino acid sequence), novel pharmacology, and/or, enhanced tolerance and safety (no intrinsic activity of allosteric ligand, but, it can modulate endogenous agonist activity) [267]. These and other novel medicinal chemistry approaches, in light of continued advances in delineating PD primary pathoetiology and recent renewed emphasis on collaboration between the basic and the clinical biomedical research scientists, likely will provide significantly improved symptomatic PD therapy, and perhaps halt or even reverse the associated neuronal degeneration.

ACKNOWLEDGMENTS

For reviewing this manuscript, helpful suggestions, and material, we thank Drs. Ludy Shih, Frank Tarazi, and Daniel Tarsy of Harvard Medical School, Boston, MA. We thank the Branfman Family Foundation and acknowledge support by NIH grants MH-068655, MH-081193, DA-023928, Bruce J. Anderson Foundation, and the McLean Private Donors Neuropharmacology Research Fund.

REFERENCES

1. Parkinson J. An Essay on the Shaking Palsy. London: Whittingham and Rowland for Sherwood, Neely and Jones; 1817.
2. Alves G, Forsaa EB, Pedersen KF, Dreetz Gjerstad M. Larsen JP. Epidemiology of Parkinson's disease. J Neurol 2008;255 (Suppl 5): 18–32.
3. Lilienfeld DE, Perl DP. Projected neurodegenerative disease mortality in the United States, 1990–2040. Neuroepidemiology 1993;12:219–228.
4. Forno LS. Neuropathology of Parkinson's disease. J Neuropathol Exp Neurol 1996;55: 259–72.
5. Olanow CW, Stern MB, Sethi K. The scientific and clinical basis for the treatment of Parkinson disease. Neurology 2009;72(21 Suppl 4): S1–S136.
6. Goetz CG, Stebbins GT. Risk factors for nursing home placement in advanced Parkinson's disease. Neurology. 1993;43:2227–2229.
7. Chen H, Zhang SM, Schwarzschild MA, Hernan MA, Ascherio A. Survival of Parkinson's disease patients in a large prospective cohort of male health professionals. Mov Disord 2006;21:1002–1007.
8. Banack SA, Cox PA. Biomagnification of cycad neurotoxins in flying foxes: implications for ALS-PDC in Guam. Neurology 2003;61: 387–389.
9. Dickman MS. von Economo encephalitis. Arch Neurol 2001;58:1696–1698.
10. Wenning GK, Quinn NP. Parkinsonism. Multiple system atrophy. Baillieres Clin Neurol 1997;6:187–204.
11. Carlsson A. Treatment of Parkinson's disease with L-dopa. The early discovery phase, and a comment on current problems. J Neural Transm 2002;109:777–787.
12. Baldessarini RJ, Tarazi FI. Brain dopamine receptors: a primer on their current status, basic and clinical. Harv Rev Psychiatry 1996;3:301–325.
13. Carlsson A, Waters N, Carlsson ML. Neurotransmitter interactions in schizophrenia: therapeutic implications. Biol Psychiatry 1999;46:1388–1395.
14. DeLong MR, Crutcher MD, Georgopoulos AP. Primate globus pallidus and subthalamic nucleus: functional organization. J Neurophysiol 1985;53:530–543.
15. DeLong MR, Wichmann T. Circuits and circuit disorders of the basal ganglia. Arch Neurol. 2007;64:20–24.
16. Parent A, Hazrati LN. Functional anatomy of the basal ganglia: the cortico-basal ganglia–thalamo–cortical loop. Brain Res Rev 1995;20:91–127.
17. Standaert DG, Young AB. Treatment of central nervous system degenerative disorders. In: Brunton LL, Lazo JS, Parker KL,editors. Goodman and Gilman's The Pharmacological Basis of Therapeutics. New York: McGraw-Hill; 2006. p 527–546.
18. Graybiel AM. Neurotransmitters and neuromodulators in the basal ganglia. Trends Neurosci 1990;13:244–254.
19. Pahapill PA, Lozano AM. The pedunculopontine nucleus and Parkinson's disease. Brain 2000;123(Pt 9): 1767–1783.
20. Orieux G, Francois C, Féger J, Yelnik J, Vila M, Ruberg M, Agid Y, Hirsch EC. Metabolic activity of excitatory parafascicular and pedunculopontine inputs to the subthalamic nucleus in a rat model of Parkinson's disease. Neuroscience 2000;97:79–88.
21. Gwinn-Hardy K. Genetics of parkinsonism. Mov Disord 2002;17:645–656.
22. Semchuk KM, Love EJ, Lee RG. Parkinson's disease: a test of the multifactorial etiologic hypothesis. Neurology 1993;43:1173–1180.
23. Kruger R, Kuhn W, Muller T, Woitalla D, Graeber M, Kösel S. Ala30Pro mutation in the gene encoding α-synuclein in Parkinson's disease. Nat Genet 1998;18:106–108.
24. Polymeropoulos MH, Lavedan C, Leroy E, Ide SE, Dehejia A, Dutra A. Mutation in the α-synuclein gene identified in families with Parkinson's disease. Science 1997;276:2045–2047.
25. Norris EH, Giasson BI, Lee VM. α-Synuclein: normal function and role in neurodegenerative diseases. Curr Top Dev Biol 2004;60:17–54.
26. Liu D, Jin L, Wang H, Zhao H, Zhao C, Duan C, Lu L, Wu B, Yu S, Chan P, Li Y, Yang H. Silencing alpha-synuclein gene expression enhances tyrosine hydroxylase activity in MN9D cells. Neurochem Res 2008;33:1401–1409.
27. Gasser T, Muller-Myhsok B, Wszolek ZK, Dürr A, Vaughan JR, Bonifati V. Genetic complexity

and Parkinson's disease. Science 1997;277: 388–389.
28. Scott WK, Staijich JM, Yamaoka LH, Speer MC, Vance JM, Roses AD, Pericak-Vance MA. Genetic complexity and Parkinson's disease. Deane Laboratory Parkinson Disease Research Group. Science 1997;277:387–388.
29. Tanner CM, Ottman R, Goldman SM, Ellenberg J, Chan P, Mayeux R, Langston JW. Parkinson disease in twins: an etiologic study. JAMA 1999;281:341–346.
30. Ward CD, Duvoisin RC, Ince SE, Nutt JD, Eldridge R, Calne DB. Parkinson's disease in 65 pairs of twins and in a set of quadruplets. Neurology 1983;33:815–824.
31. Wirdefeldt K, Gatz M, Schalling M, Pedersen NL. No evidence for heritability of Parkinson disease in Swedish twins. Neurology 2004; 63:305–311.
32. Jain S, Wood NW, Healy DG. Molecular genetic pathways in Parkinson's disease: a review. Clin Sci 2005;109:355–364.
33. Dodson MW, Guo M. Pink1, Parkin, DJ-1 and mitochondrial dysfunction in Parkinson's disease. Curr Opin Neurobiol 2007;17:331–337.
34. McNaught KS, Olanow CW. Protein aggregation in the pathogenesis of familial and sporadic Parkinson's disease. Neurobiol Aging 2006;27:530–545.
35. Ciechanover A, Brundin P. The ubiquitin–proteasome system in neurodegenerative diseases. Sometimes the chicken, sometimes the egg. Neuron 2003;40:427–446.
36. Goldberg AL. Protein degradation and protection against misfolded or damaged proteins. Nature 2003;426:895–899.
37. Narendra D, Tanaka A, Suen DF, Youle RJ. Parkin is recruited selectively to impaired mitochondria and promotes their autophagy. J Cell Biol 2008;183:795–803.
38. Clark IE, Dodson MW, Jiang C, et al. Drosophila pink1 is required for mitochondrial function and interacts genetically with parkin. Nature 2006;441:1162–1166.
39. Park J, Lee SB, Lee S, Kim Y, Song S, Kim S. Mitochondrial dysfunction in Drosophila PINK1 mutants is complemented by parkin. Nature 2006;441:1157–1161.
40. Biskup S, Mueller JC, Sharma M, Lichtner P, Zimprich A, Berg D, Wüllner U, Illig T, Meitinger T, Gasser T. Common variants of LRRK2 are not associated with sporadic Parkinson's disease. Ann Neurol 2005;58: 905–908.
41. Gasser T. Genetics of Parkinson's disease. Curr Opin Neurol 2005;18:363–369.
42. Baba M, Nakajo S, Tu PH, et al. Aggregation of α-synuclein in Lewy bodies of sporadic Parkinson's disease and dementia with Lewy bodies. Am J Pathol 1998;152:879–884.
43. Soloway AH. Med Hypotheses Res 2009;5: 19–26.
44. Chinta SJ, Andersen JK. Redox imbalance in Parkinson's disease. Biochim Biophys Acta 2008;1780:1362–1367.
45. Graham DG. Catecholamine toxicity: a proposal for the molecular pathogenesis of manganese neurotoxicity and Parkinson's disease. Neurotoxicology 1984;5:83–95.
46. Graham DG, Tiffany SM, Bell WR Jr, Gutknecht WF. Autoxidation versus covalent binding of quinones as the mechanism of toxicity of dopamine, 6-hydroxydopamine, and related compounds toward C1300 neuroblastoma cells in vitro. Mol Pharmacol 1978;14: 644–653.
47. Boyd DR, Hamilton JTG, Sharma ND, Harrison JS, McRoberts WC, Harper DB. Isolation of a stable benzene oxide from a fungal biotransformation and evidence for an "NIH shift" of the carbomethoxy group during hydroxylation of methyl benzoates. Chem Commun 2000; 1481–1482.
48. Menberu D, Nguyen PV, Onan KD, LeQuesne PW. Convenient syntheses of stereoisomeric 1,2-epoxyestr-4-en-3-ones, putative intermediates in estradiol metabolism. J Org Chem 1992;57:2065–2072.
49. Soloway AH. Potential endogenous epoxides of steroid hormones: initiators of breast and other malignancies? Med Hypoth 2007;69: 1225–1229.
50. Haber F, Weiss J. The catalysis of hydrogen peroxide. Naturwissenschaften 1932;20:948–950.
51. Saura J, Bleuel Z, Ulrich J, Mendelowitsch A, Chen K, Shih JC, Malherbe P, Da Prada M, Richards JG. Molecular neuroanatomy of human monoamine oxidases A and B revealed by quantitative enzyme radioautography and in situ hybridization histochemistry. Neuroscience 1996;70:755–774.
52. Youdim MBH, Edmondson D, Tipton KF. The therapeutic potential of monoamine oxidase inhibitors. Nat Rev 2006;7:295–309.
53. Fowler JS, Logan J, Volkow ND, Wang GJ, MacGregor RR, Ding YS. Monoamine oxidase: radiotracer development and human studies. Methods 2002;27:263–277.

54. Mallajosyula JK, Kaur D, Chinta SJ, Rajagopalan S, Rane A, Nicholls DG, Di Monte DA, Macarthur H, Andersen JK. MAO B elevation in mouse brain astrocytes results in Parkinson's pathology. PloS One 2008;3:e1616.

55. Burke WJ, Kumar VB, Pandey N, Panneton WM, Gan Q, Franko MW, O'Dell M, Li SW, Pan Y, Chung HD, Galvin JE. Aggregation of R-synuclein by DOPAL, the monoamine oxidase metabolite of dopamine. Acta Neuropathol 2008;115:193–203.

56. Jenner P. Oxidative stress in Parkinson's disease. Ann Neurol. 2003; 53(Suppl 3):S26–36.

57. Sian J, Dexter DT, Lees AJ, Daniel S, Agid Y, Javoy-Agid F, Jenner P, Marsden CD. Alterations in glutathione levels in Parkinson's disease and other neurodegenerative disorders affecting basal ganglia. Ann Neurol 1994; 36:348–355.

58. Dexter DT, Carter CJ, Wells FR, Javoy-Agid F, Agid Y, Lees A, Jenner P, Marsden CD. Basal lipid peroxidation in substantia nigra is increased in Parkinson's disease. J Neurochem 1989;52:381–389.

59. Dexter DT, Holley AE, Flitter WD, Slater TF, Wells FR, Daniel SE, Lees AJ, Jenner P, Marsden CD. Increased levels of lipid hydroperoxides in the parkinsonian substantia nigra: an HPLC and ESR study. Mov Disord. 1994;9:92–97.

60. Alam ZI, Daniel SE, Lees AJ, Marsden DC, Jenner P, Halliwell B. A generalised increase in protein carbonyls in the brain in Parkinson's but not incidental Lewy body disease. J Neurochem 1997;69:1326–1329.

61. Alam ZI, Jenner A, Daniel SE, Lees AJ, Cairns N, Marsden CD, Jenner P, Halliwell B. Oxidative DNA damage in the parkinsonian brain: an apparent selective increase in 8-hydroxyguanine levels in substantia nigra. J Neurochem 1997;69:1196–1203.

62. Beal MF. Mitochondria take center stage in aging and neurodegeneration. Ann Neurol 2005;58:495–505.

63. Shankar JC, Andersen JK. Redox imbalance in Parkinson's disease. Biochimica et Biophysica Acta 2008;1780:1362–1367.

64. Mann VM, Cooper JM, Krige D, Daniel SE, Schapira AH, Marsden CD. Brain, skeletal muscle and platelet homogenate mitochondrial function in Parkinson's disease. Brain 1992;115(Pt 2): 333–42.

65. Mizuno Y, Ohta S, Tanaka M, Takamiya S, Suzuki K, Sato T, Oya H, Ozawa T, Kagawa Y. Deficiencies in complex I subunits of the respiratory chain in Parkinson's disease. Biochem Biophys Res Commun 1989;163: 1450–1455.

66. Dauer W, Przedborski S. Parkinson's disease: mechanisms and models. Neuron 2003;39: 889–909.

67. Knoll J. Deprenyl (selegiline): the history of its development and pharmacological action. Acta Neurol Scand Suppl 1983;95:57–80.

68. Luo Y, Roth GS. The roles of dopamine oxidative stress and dopamine receptor signaling in aging and age-related neurodegeneration. Antioxid Redox Signal 2000;2:449–60.

69. Seeman P, Bzowej NH, Guan HC, Bergeron C, Becker LE, Reynolds GP, Bird ED, Riederer P, Jellinger K, Watanabe S, et al. Human brain dopamine receptors in children and aging adults. Synapse 1987;1: 399–404.

70. Riederer P, Woketich S. Time course of nigrostriatal degeneration in Parkinson's disease. A detailed study of influential factors in human brain amine analysis. J Neural Transm 1976;38:277–301.

71. Calne DB. Parkinson's disease is not one disease. Parkinsonism Rel Disord 2000;7: 3–7.

72. Drechsel DA, Patel M. Role of reactive oxygen species in the neurotoxicity of environmental agents implicated in Parkinson's disease. Free Radic Biol Med 2008;44:1873–1886.

73. Hatcher JM, Pennell KD, Miller GW. Parkinson's disease and pesticides: a toxicological perspective. Trends Pharmacol Sci 2008; 29:322–329.

74. Lockwood AH. Pesticides and parkinsonism: is there an etiological link? Curr Opin Neurol 2000;13:687–690.

75. Gajdusek DC, Salazar AM. Amyotrophic lateral sclerosis and parkinsonian syndromes in high incidence among the Auyu and Jakai people of West New Guinea. Neurology 1982;32:107–126.

76. Kuhn W, Müller T, Nastos I, Poehlau D. The neuroimmune hypothesis in Parkinson's disease. Rev Neurosci 1997;8:29–34.

77. Cotzias GC. Manganese in health and disease. Physiol Rev 1958;38:503–532.

78. Barbeau A, Roy M, Bernier G, Campanella G, Paris S. Ecogenetics of Parkinson's disease: prevalence and environmental aspects in rural areas. Can J Neurol Sci 1987;14:36–41.

79. Cicchetti F, Drouin-Ouellet J, Gross RE. Environmental toxins and Parkinson's disease: what have we learned from pesticide-induced animal models? Trends Pharmacol Sci 2009;30:475–483.

80. Miller GW. Paraquat: the red herring of Parkinson's disease research. Toxicol Sci 2007;100:1–2.
81. Brooks AI, Chadwick CA, Gelbard HA, Cory-Slechta DA, Federoff HJ. Paraquat elicited neurobehavioral syndrome caused by dopaminergic neuron loss. Brain Res 1999;823: 1–10.
82. Naylor JL, Widdowson PS, Simpson MG, Farnworth M, Ellis MK, Lock EA. Further evidence that the blood/brain barrier impedes paraquat entry into the brain. Hum Exp Toxicol 1995;14:587–954.
83. Betarbet R, Sherer TB, MacKenzie G, Garcia-Osuna M, Panov AV, Greenamyre JT. Chronic systemic pesticide exposure reproduces features of Parkinson's disease. Nat Neurosci 2000;3:1301–1306.
84. Tanner CM, Goldman SM, Aston DA, Ottman R, Ellenberg J, Mayeux R, Langston JW. Smoking and Parkinson's disease in twins. Neurology. 2002;58:581–588.
85. Ward RL, Lallemand F, de Witte P, Dexter DT. Neurochemical pathways involved in the protective effects of nicotine and ethanol in preventing the development of Parkinson's disease: potential targets for the development of new therapeutic agents. Prog Neurobiol 2008;85:135–147.
86. Khalil AA, Davies B, Castagnoli N Jr. Isolation and characterisation of a monoamine oxidase B selective inhibitor from tobacco smoke. Bioorg Med Chem 2006;14:3392–3398.
87. Chapman MA. Does smoking reduce the risk of Parkinson's disease through stimulation of the ubiquitin–proteasome system? Med Hypotheses 2009. (Epub ahead of print).
88. Hernan MA, Takkouche B, Caamano-Isorna F, Gestal- Otero JJ. A meta-analysis of coffee drinking, cigarette smoking, and the risk of Parkinson's disease. Ann Neurol 2002;52:276–284.
89. Trevitt J, Kawa K, Jalali A, Larsen C. Differential effects of adenosine antagonists in two models of parkinsonian tremor. Pharmacol Biochem Behav 2009;94:24–29.
90. Chase TN, Bibbiani F, Bara-Jimenez W, Dimitrova T, Oh-Lee JD. Translating A_{2A} antagonist KW6002 from animal models to parkinsonian patients. Neurology 2003;61(11 Suppl 6): S107–111.
91. Petzer JP, Castagnoli N Jr, Schwarzschild MA, Chen JF, Van der Schyf CJ. Dual-target-directed drugs that block monoamine oxidase B and adenosine A(2A) receptors for Parkinson's disease. Neurotherapeutics 2009;6:141–151.
92. Lai BC, Marion SA, Teschke K, Tsui JK. Occupational and environmental risk factors for Parkinson's disease. Parkinsonism Relat Disord 2002;8:297–309.
93. Tanner CM, Langston JW. Do environmental toxins cause Parkinson's disease? A critical review. Neurology 1990;40(Suppl 3): 17–30.
94. Burns RS, Chiueh CC, Markey SP, Ebert MH, Jacobowitz DM, Kopin IJ. A primate model of parkinsonism: selective destruction of dopaminergic neurons in the pars compacta of the substantia nigra by N-methyl-4-phenyl-1,2,3,6-tetrahydropyridine. Proc Natl Acad Sci USA 1983;80:4546–4550.
95. Davis GC, Williams AC, Markey SP, Ebert MH, Caine ED, Reichert CM, Kopin IJ. Chronic parkinsonism secondary to intravenous injection of meperidine analogues. Psychiatric Res 1979;1:249–254.
96. Langston JW, Ballard P, Tetrud JW, Irwin I. Chronic parkinsonism in humans due to a product of meperidine-analog synthesis. Science 1983;219:979–980.
97. Chiba K, Trevor A, Castagnoli N Jr. Metabolism of the neurotoxic tertiary amine, MPTP, by brain monoamine oxidase. Biochem Biophys Res Comm 1984;120:574–578.
98. Peterson LA, Caldera PS, Trevor A, et al. Studies on the 1-methyl-4-phenyl-2,3-dihydropyridinium species 2,3-MPDP$^+$, the monoamine oxidase catalyzed oxidation product of the nigrostriatal toxin 1-methyl-4-phenyl-1,2,3,6-tetrahydropyridine (MPTP). J Med Chem 1985;28:1432–1436.
99. Salach JI, Singer TP, Castagnoli N Jr, et al. Oxidation of the neurotoxic amine 1-methyl-4-phenyl-1,2,3,6-tetrahydropyridine (MPTP) by monoamine oxidases A and B and suicide inactivation of the enzymes by MPTP. Biochem Biophys Res Comm 1984;125: 831–835.
100. Langston JW, Irwin I, Langston EB, Forno LS. Pargyline prevents MPTP-induced parkinsonism in primates. Science 1984;225:1480–1482.
101. Snyder SH, D'Amato RJ. MPTP: a neurotoxin relevant to the pathophysiology of Parkinson's disease. The 1985 George C. Cotzias lecture. Neurology 1986;36:250–258.
102. Berry MD, Juorio AV, Paterson IA. The functional role of monoamine oxidases A and B in the mammalian central nervous system. Prog Neurobiol 1994;42:375–391.

103. Ikemoto K, Kitahamaa K, Seif I, Maeda T, DeMaeyer E, Valatxa J-L. Monoamine oxidase B (MAOB)-containing structures in MAOA-deficient transgenic mice. Brain Res 1997;771:121–132.

104. Javitch JA, D'Amato RJ, Strittmatter SM, Snyder SH. Parkinsonism-inducing neurotoxin, N-methyl-4-phenyl-1,2,3,6-tetrahydropyridine: uptake of the metabolite N-methyl-4-phenylpyridine by dopamine neurons explains selective toxicity. Proc Natl Acad Sci USA 1985;82:2173–2177.

105. Ramsay RR, McKeown KA, Johnson EA, Booth RG, Singer TP. Inhibition of NADH oxidation by pyridine derivatives. Biochem Biophys Res Comm 1987;146:53–60.

106. Vyas I, Heikkila RE, Nicklas WJ. Studies on the neurotoxicity of 1-methyl-4-phenyl-1,2,3,6-tetrahydropyridine: inhibition of NAD-linked substrate oxidation by its metabolite, 1-methyl-4-phenylpyridinium. J Neurochem 1986;46:1501–1507.

107. Castagnoli N Jr, Rimoldi JM, Bloomquist J, Castagnoli KP. Potential metabolic bioactivation pathways involving cyclic tertiary amines and aza-arenes. Chem Res Toxicol 1997;10:924–940.

108. Watanabe H, Muramatsu Y, Kurosaki R, et al. Protective effects of neuronal nitric oxide synthase inhibitor in mouse brain against MPTP neurotoxicity: an immunohistological study. Eur Neuropsychopharmacol 2004;14:93–104.

109. Macleod AD, Counsell CE, Ives N, Stowe R. Monoamine oxidase B inhibitors for early Parkinson's disease. Cochrane Database Syst Rev 2005;3:CD004898.

110. The Parkinson's Disease Study Group. Impact of tocopherol and deprenyl in DATATOP subjects not requiring levodopa. Ann Neurol 1996;39:29–36.

111. The Parkinson's disease study group. Impact of tocopherol and deprenyl in DATATOP subjects requiring levodopa. Ann Neurol 1996;39:37–45.

112. Olanow CW. Can we achieve neuroprotection with currently available anti-parkinsonian interventions? Neurology 2009;72(Suppl 7): S59–64.

113. Rajput AH, Uitti RJ, Offord KO. Timely levodopa administration prolongs survival in Parkinson's disease. Parkinsonism Relat Disord 1997;3:159–165.

114. Ehringer H, Hornykiewicz O. Distribution of noradrenaline and dopamine (3-hydroxytyramine) in the human brain and their behavior in diseases of the extrapyramidal system. Klin Wochenschr 1960;38:1236–1239.

115. Birkmayer W, Hornykiewicz O. The L-3,4-dioxyphenylalanine (DOPA)-effect in Parkinson-akinesia. Wien Klin Wochenschr 1961;73: 787–788.

116. Barbeau A. Biochemistry of Parkinson's disease. Proc 7th Int Cong Neurol 1961;2:925.

117. Cotzias GC, Van Woert MH, Schiffer LM. Aromatic amino acids and modification of parkinsonism. N Eng J Med 1967;276:374–379.

118. Cotzias GC, Papavasiliou PS, Gellene R. Modification of parkinsonism: chronic treatment with L-dopa. N Engl J Med 1969;280:337–45.

119. Barbeau A. L-dopa therapy in Parkinson's disease: a critical review of nine years' experience. Can Med Assoc J 1969;101:59–68.

120. Nagatsu T. Biochemistry of Catecholamines: The Biochemical Method. Baltimore: University Park Press; 1973. p 289–299.

121. Baldessarini RJ, Fischer JE. Substitute and alternative neurotransmitters in neuropsychiatric illness. Arch Gen Psychiatry 1977;34: 958–964.

122. Chalmers JP, Baldessarini RJ, Wurtman RJ. Effects of L-dopa on norepinephrine metabolism in the brain. Proc Natl Acad Sci USA 1971;68:662–666.

123. Vogel WH. Determination and physiological disposition of p-methoxyphenylethylamine in the rat. Biochem Pharmacolo 1970;19:2663–2665.

124. Burkard WP, Gey KF, Pletscher A. Inhibition of decarboxylase of aromatic amino acids by 2,3,4-trihydroxybenzylhydrazine and its seryl derivative. Arch Biochem Biophys 1964; 107:187–196.

125. Matthysse S, Baldessarini RJ. S-Adenosylmethionine and catechol-O-methyl-transferase in schizophrenia. Am J Psychiatry 1972;128: 1310–1312.

126. Miyawaki E, Lyons K, Pahwa R. Motor complications of chronic levodopa therapy in Parkinson's disease. Clin Neuropharmacol 1997;20:523–530.

127. Diederich NJ, Fénelon G, Stebbins G, Goetz CG. Hallucinations in Parkinson disease. Nat Rev Neurol 2009;5:331–342.

128. Tarsy D, Baldessarini RJ. Epidemiology of tardive dyskinesia: is risk declining with modern antipsychotics? Mov Disord 2006;21:589–598.

129. Zahodne LB, Fernandez HH. Pathophysiology and treatment of psychosis in Parkinson's disease: a review. Drugs Aging 2008;25:665–682.

130. Friedman JH, Berman RM, Goetz CG, Factor SA, Ondo WG, Wojcieszek J, Carson WH, Marcus RN. Open-label flexible-dose pilot study to evaluate the safety and tolerability of aripiprazole in patients with psychosis associated with Parkinson's disease. Mov Disord 2006;21:2078–2081.

131. Baldessarini RJ, Tarazi FI, Pharmacotherapy of psychosis and mania. In: Brunton LL, Lazo JS, Parker KL,editors. Goodman and Gilman's The Pharmacological Basis of Therapeutics. New York: McGraw-Hill; 2006. p 461–500.

132. Rosenbaum DM, Rasmussen SG, Kobilka BK. The structure and function of G-protein-coupled receptors. Nature. 2009;459:356–363.

133. Milligan G. Mechanisms of multifunctional signaling by G protein-linked receptors. Trends Pharmacol Sci 1993;14:239–244.

134. Raymond JR. Multiple mechanisms of receptor-G protein signaling specificity. Am J Physiol 1995;269:F141–158.

135. Kenakin T. Inverse, protean, and ligand-selective agonism: matters of receptor conformation. FASEB J 2001;15:598–611.

136. Gay EA, Urban JD. Nichols DE Oxford GS, Mailman RB. Functional selectivity of D_2 receptor ligands in a Chinese hamster ovary hD_{2L} cell line: evidence for induction of ligand-specific receptor states. Mol Pharmacol 2004;66:97–105.

137. Moniri NH, Covington-Strachan D, Booth RG. Ligand-directed functional heterogeneity of histamine H_1 receptors: novel dual-function ligands selectively activate and block H_1-mediated phospholipase C and adenylyl cyclase signaling. J Pharmacol Exp Ther 2004;311:274–281.

138. Kilts JD, Connery HS, Arrington EG, Lewis MM, Lawler CP, Oxford GS. Functional selectivity of dopamine receptor agonists. II. Actions of dihydrexidine in D_{2L} receptor-transfected MN9D cells and pituitary lactotrophs. J Pharmacol Exp Ther 2002;301:1179–1189.

139. Lawler CP, Prioleau C, Lewis MM, Mak C, Jiang D, Schetz JA. Interactions of the novel antipsychotic aripiprazole (OPC-14597) with dopamine and serotonin receptor subtypes. Neuropsychopharmacology 1999;20:612–627.

140. Mottola DM, Kilts JD, Lewis MM, Connery HS, Walker QD, Jones SR, Booth RG. Functional selectivity of dopamine receptor agonists. I. Selective activation of postsynaptic dopamine D_2 receptors linked to adenylate cyclase. J Pharmacol Exp Ther 2002;301:1166–7811.

141. Urban JD, Vargas GA, von Zastrow M, Mailman RB. Aripiprazole has functionally selective actions at dopamine D_2 receptor-mediated signaling pathways. Neuropsychopharmacology 2007;32:67–77.

142. Rasmussen SG, Choi HJ, Rosenbaum DM, Kobilka TS, Thian FS, Edwards PC, Burghammer M, Ratnala VR, Sanishvili R, Fischetti RF, Schertler GF, Weis WI, Kobilka BK. Crystal structure of the human beta2 adrenergic G-protein-coupled receptor. Nature 2007;450:383–387.

143. Jaakola VP, Griffith MT, Hanson MA, Cherezov V, Chien EY, Lane JR, Ijzerman AP, Stevens RC. The 2.6 angstrom crystal structure of a human A_{2A} adenosine receptor bound to an antagonist. Science 2008;322:1211–1217.

144. Fu W, Shen J, Luo X, Zhu W, Cheng J, Yu K, Briggs JM, Jin G, Chen K, Jiang H. Dopamine D1 receptor agonist and D_2 receptor antagonist effects of the natural product (-)-stepholidine: molecular modeling and dynamics simulations. Biophys J 2007;93:1431–1441.

145. Krebs A, Edwards PC, Villa C, Li J, Schertler GF. The three dimensional structure of bovine rhodopsin determined by electron cryomicroscopy. J Biol Chem 2003;278:50217–0225.

146. Okada T, Le Trong I, Fox BA, Behnke CA, Stenkamp RE, Palczewski K. X-ray defraction analysis of three-dimensional crystals of bovine rhodopsin obtained from mixed micelles. J Struct Biol 2000;130:73–80.

147. Kobilka BK, Deupi X. Conformational complexity of G-protein-coupled receptors. Trends Pharmacol Sci 2007;28:397–406.

148. Seifert R, Wenzel-Seifert K, Gether U, Kobilka BK. Functional differences between full and partial agonists: evidence for ligand-specific receptor conformations. J Pharmacol Exp Ther 2001;297:1218–1226.

149. Baldessarini RJ, Arana GW, Kula NS, Campbell A, Harding M.In: Corsini GU, Gian LG, editors. Apomorphine and Other Dopaminomimetics, Basic Pharmacology. Vol. I.New York: Raven Press; 1981. p 219–228.

150. Neumeyer JL, Baldessarini RJ. Apomorphine. new uses for and old drug. Pharmaceutical News 1997;4:12–16.

151. Giesecke J. The absolute configuration of apomorphine. Acta Crystol 1977;B33:302–303.

152. Neumeyer JL, McCarthy M, Battista S, et al. Aporphines 9. The synthesis and pharmacolo-

gical evaluations of (±)-9,10-dihyroxyaporphine, ([±]-isoapomorphine), (±)-, (−)-, and (+)-1,2-dihydroxyaporphine, and (+)-1,2,9,10-tetrahydroxyaprophine. J Med Chem 1973;16:1228.
153. Westerink BHC, Dijkstra D, Feenstra MGP, Grol CJ, Horn AS, Rollema H, Wirix E. Dopaminergic prodrugs: brain concentrations and neurochemical effects of 5,6- and 6,7-ADTN after administration as dibenzoyl esters. Eur J Pharmacol 1980;61:7–15.
154. Mansour A, Meng F, Meador-Woodruff JH, Taylor LP, Civelli O, Akil H. Site-directed mutagenesis of the human dopamine D_2 receptor. Eur J Pharmacol 1992;227:205–214.
155. Wilcox RE, Tseng T, Brusniak MY, Ginsburg B, Pearlman RS, Teeter M, DuRand C, Starr S, Neve KA. CoMFA-based prediction of agonist affinities at recombinant D1 vs D_2 dopamine receptors. J Med Chem 1998;41:4385–4399.
156. Stibe CM, Lees AJ, Kempster PA, Stern GM. Subcutaneous apomorphine in parkinsonian on–off oscillations. Lancet 1988;1:403–406.
157. Schwab RS, Amador LV, Lettvin JY. Apomorphine in Parkinson's disease. Trans Am Neurol Assoc 1951;56:251–253.
158. Colosimo C, Merello M, Albanese A. Clinical usefulness of apomorphine in movement disorders. Clin Neuropharmacol 1994;17: 243–259.
159. Mouradian MM, Chase TN.In: Marsden CD, Fahn S,editors. Movement Disorders. Oxford, UK: Butterworth; 1994. p 181–186.
160. Stocchi F. Use of apomorphine in Parkinson's disease. Neurol Sci 2008;29(Suppl 5): S383–386.
161. Frankel JP, Lees AJ, Kempster PA, Stern GM. Subcutaneous apomorphine in the treatment of Parkinson's disease. J Neurol Neurosurg Psychiatry. 1990;53:96–101.
162. Hughes AJ, Bishop S, Kleedorfer B, Turjanski N, Fernandez W, Lees AJ, Stern GM. Subcutaneous apomorphine in Parkinson's disease: response to chronic administration for up to five years. Mov Disord 1993;8:165–170.
163. Campbell A, Baldessarini RJ, Ram VJ, Neumeyer JL. Behavioral effects of (-)10,11-methylenedioxy-N-n-propylnoraporphine, an orally effective long-acting agent active at central dopamine receptors, and analogous aporphines. Neuropharmacology 1982;21:953–961.
164. Campbell A, Kula NS, Jeppsson B, Baldessarini RJ. Oral bioavailability of apomorphine in the rat with a portacaval venous anastomosis. Eur J Pharmacol 1980;67:139–142.
165. Campbell A, Baldessarini RJ, Gao Y, Zong R, Neumeyer JL. R(-) and S(+) stereoisomers of 11-hydroxy- and 11-methoxy-N-n-propylnoraporphine: central dopaminergic behavioral activity in the rat. Neuropharmacology 1990;29: 527–536.
166. Menon MK, Clark WG, Neumeyer JL. Comparison of the dopaminergic effects of apomorphine and (-)-N-n-propylnoraporphine. Eur J Pharmacol 1978;52:1–9.
167. Goddard WA 3rd, Abrol R. 3-Dimensional structures of G protein-coupled receptors and binding sites of agonists and antagonists. J Nutr 2007;137(6 Suppl 1): 1528S–1538S.
168. Si YG, Gardner MP, Tarazi FI, Baldessarini RJ, Neumeyer JL. Synthesis and dopamine receptor affinities of N-alkyl-11-hydroxy-2-methoxynoraporphines: N-alkyl substituents determine D_1 versus D_2 receptor selectivity. J Med Chem 2008;51:983–987.
169. Newman-Tancredi A, Cussac D, Audinot V, Nicolas JP, De Ceuninck F, Boutin JA, Millan MJ. Differential actions of antiparkinson agents at multiple classes of monoaminergic receptor. II. Agonist and antagonist properties at subtypes of dopamine D_2-like receptor and alpha(1)/alpha(2)-adrenoceptor. J Pharmacol Exp Ther 2002;303:805–814.
170. Millan MJ, Maiofiss L, Cussac D, Audinot V, Boutin JA, Newman-Tancredi A. Differential actions of antiparkinson agents at multiple classes of monoaminergic receptor. I. A multivariate analysis of the binding profiles of 14 drugs at 21 native and cloned human receptor subtypes. J Pharmacol Exp Ther 2002;303: 791–804.
171. Blanchet PJ. Rationale for use of dopamine agonists in Parkinson's disease: review of ergot derivatives. Can J Neurol Sci 1999;26 (suppl 2): S21–26.
172. Baldessarini RJ, Tarsy D. Dopamine and the pathophysiology of dyskinesias induced by antipsychotic drugs. Annu Rev Neurosci 1980;3:23–41.
173. Brandstädter D, Oertel WH. Treatment of drug-induced psychosis with quetiapine and clozapine in Parkinson's disease. Neurology 2002;58:160–161.
174. Hutton JT, Morris JL, Brewer MA. Controlled study of the antiparkinsonianian activity and tolerability of cabergoline. Neurology 1993;43: 613–616.
175. Stocchi F, Ruggieri S, Vacca L, Olanow CW. Prospective randomized trial of lisuride infusion versus oral levodopa in patients with

176. Rascol O, Brooks DJ, Korczyn AD, De Deyn PP, Clarke CE, Lang AE. A five-year study of the incidence of dyskinesia in patients with early Parkinson's disease who were treated with ropinirole or levodopa. 056 Study Group. N Engl J Med 2000;342:1484–1491.

177. Horvath J, Fross RD, Kleiner-Fisman G, Lerch R, Stalder H, Liaudat S, Raskoff WJ, Flachsbart KD, Rakowski H, Pache JC, Burkhard PR, Lang AE. Severe multivalvular heart disease: a new complication of the ergot derivative dopamine agonists. Mov Disord 2004; 19:656–662.

178. Roth BL. Drugs and valvular heart disease. N Engl J Med 2007;356:6–9.

179. Rothman RB, Baumann MH, Savage JE, Rauser L, McBride A, Hufeisen SJ, Roth BL. Evidence for possible involvement of $5-HT_{2B}$ receptors in the cardiac valvulopathy associated with fenfluramine and other serotonergic medications. Circulation 2000;102:2836–2841.

180. Setola V, Hufeisen SJ, Grande-Allen KJ, Vesely I, Glennon RA, Blough B, Rothman RB, Roth BL. 3,4-Methylenedioxymethamphetamine (MDMA, "Ecstasy") induces fenfluramine-like proliferative actions on human cardiac valvular interstitial cells in vitro. Mol Pharmacol 2003;63:1223–1229.

181. Colao A, Galderisi M, Di Sarno A, Pardo M, Gaccione M, D'Andrea M, Guerra E, Pivonello R, Lerro G, Lombardi G. Increased prevalence of tricuspid regurgitation in patients with prolactinomas chronically treated with cabergoline. J Clin Endocrinol Metab 2008;93: 3777–3784.

182. Wakil A, Rigby AS, Clark AL, Kallvikbacka-Bennett A, Atkin SL. Low dose cabergoline for hyperprolactinaemia is not associated with clinically significant valvular heart disease. Eur J Endocrinol 2008;159:R11–R14.

183. Vallette S, Serri K, Rivera J, Santagata P, Delorme S, Garfield N, Kahtani N, Beauregard H, Aris-Jilwan N, Houde G, Serri O. Long-term cabergoline therapy is not associated with valvular heart disease in patients with prolactinomas. Pituitary 2009;12:153–157.

184. Berger M, Gray JA, Roth BL. The expanded biology of serotonin. Annu Rev Med 2009; 60:355–66.

185. Hofmann C, Penner U, Dorow R, Pertz HH, Jähnichen S, Horowski R, Latté KP, Palla D, Schurad B. Lisuride, a dopamine receptor agonist with $5-HT_{2B}$ receptor antagonist properties: absence of cardiac valvulopathy adverse drug reaction reports supports the concept of a crucial role for $5-HT_{2B}$ receptor agonism in cardiac valvular fibrosis. Clin Neuropharmacol 2006;29:80–86.

186. Parkinson Study Group. Dopamine transporter brain imaging to assess the effects of pramipexole vs levodopa on Parkinson disease progression. JAMA. 2002;287:1653–1661.

187. Hutton JT, Metman LV, Chase TN, Juncos JL, Koller WC, Pahwa R. Transdermal dopaminergic D_2 receptor agonist therapy in Parkinson's disease with N-0923 TDS: a double-blind, placebocontrolled study. Mov Disord 2001;16:459–463.

188. Morgan JC, Sethi KD. Rotigotine for the treatment of Parkinson's disease. Expert Rev Neurother 2006;6:1275–1282.

189. LeWitt PA, Lyons KE, Pahwa R. SP 650 Study Group. Advanced Parkinson disease treated with rotigotine transdermal system: PREFER study. Neurology 2007;68: 1262–1267.

190. Poewe WH, Rascol O, Quinn N, Tolosa E, Oertel WH, Martignoni E. Efficacy of pramipexole and transdermal rotigotine in advanced Parkinson's disease: a double-blind, double-dummy, randomized controlled trial. Lancet Neurol 2007;6:513–520.

191. Watts RL, Jankovic J, Waters C, Rajput A, Boroojerdi B, Rao J. Randomized, blind, controlled trial of transdermal rotigotine in early Parkinson disease. Neurology 2007;68: 272–276.

192. Tarsy D, Baldessarini RJ, Tarazi FI. Effects of newer antipsychotics on extrapyramidal function. CNS Drugs 2002;16:23–45.

193. Frucht SJ, Greene PE, Fahn S. Sleep episodes in Parkinson's disease: a wake-up call. Mov Disord 2000;15:601–603.

194. Braun A, Fabbrini G, Mouradian MM, Serrati C, Barone P, Chase TN. Selective D-1 dopamine receptor agonist treatment of Parkinson's disease. J Neural Transm 1987; 68:41–50.

195. Emre M, Rinne UK, Rascol A, Lees A, Agid Y, Lataste X. Effects of a selective partial D1 agonist, CY 208-243, in de novo patients with Parkinson disease. Mov Disord 1992; 7:239–243.

196. Neumeyer JL, Baindur N, Niznik HB, Guan HC, Seeman P. (+/−)-3-Allyl-6-bromo-7,8-dihydroxy-1-phenyl-2,3,4,5-tetrahydro-1H-3-benzazepin, a new high-affinity D_1 dopamine receptor ligand: synthesis and structure-ac-

197. Neumeyer JL, Kula NS, Baldessarini RJ, Baindur N. Stereoisomeric probes for the D1 dopamine receptor: synthesis and characterization of R-(+) and S-(−) enantiomers of 3-allyl-7,8-dihydroxy-1-phenyl-2,3,4,5-tetrahydro-1H-3-benzazepine and its 6-bromo analogue. J Med Chem 1992;35:1466–1471.
198. Kuno S. Differential therapeutic effects of dopamine D_1 and D_2 agonists in MPTP-induced parkinsonian monkeys: clinical implications. Eur Neurol 1997;38Suppl1: 18–22.
199. Grondin R, Bédard PJ, Britton DR, Shiosaki K. Potential therapeutic use of the selective dopamine D_1 receptor agonist, A-86929: an acute study in parkinsonian levodopa-primed monkeys. Neurology 1997;49:421–426.
200. Andringa G, Lubbers L, Drukarch B, Stoof JC, Cools AR. The predictive validity of the drug-naive bilaterally MPTP-treated monkey as a model of Parkinson's disease: effects of L-dopa and the D1 agonist SKF 82958. Behav Pharmacol 1999;10:175–182.
201. Goulet M, Madras BK. D_1 dopamine receptor agonists are more effective in alleviating advanced than mild parkinsonism in 1-methyl-4-phenyl-1,2,3, 6-tetrahydropyridine-treated monkeys. J Pharmacol Exp Ther 2000; 292:714–724.
202. Brewster WK, Nichols DE, Riggs RM, Mottola DM, Lovenberg TW, Lewis MH, Mailman RB. trans-10,11-Dihydroxy-5,6,6a,7,8,12b-hexahydrobenzo[a]phenanthridine: a highly potent selective dopamine D_1 full agonist. J Med Chem 1990;33:1756–1764.
203. Lovenberg TW, Brewster WK, Mottola DM, Lee RC, Riggs RM, Nichols DE, Lewis MH, Mailman RB. Dihydrexidine, a novel selective high potency full dopamine D-1 receptor agonist. Eur J Pharmacol 1989;166:111–113.
204. Taylor JR, Lawrence MS, Redmond DE Jr, Elsworth JD, Roth RH, Nichols DE, Mailman RB. Dihydrexidine, a full dopamine D_1 agonist, reduces MPTP-induced parkinsonism in monkeys. Eur J Pharmacol 1991;199: 389–391.
205. Blanchet PJ, Fang J, Gillespie M, Sabounjian L, Locke KW, Gammans R, Mouradian MM, Chase TN. Effects of the full dopamine D_1 receptor agonist dihydrexidine in Parkinson's disease. Clin Neuropharmacol 1998;21339–343.
206. Mottola DM, Laiter S, Watts VJ, Tropsha A, Wyrick SD, Nichols DE, Mailman RB. Conformational analysis of D1 dopamine receptor agonists: pharmacophore assessment and receptor mapping. J Med Chem 1996;39: 285–296.
207. Ghosh D, Snyder SE, Watts VJ, Mailman RB, Nichols DE. 9-Dihydroxy-2,3,7,11b-tetrahydro-1H-naph[1,2,3-de]isoquinoline: a potent full dopamine D_1 agonist containing a rigid-beta-phenyldopamine pharmacophore. J Med Chem 1996;39:549–555.
208. Asin KE, Domino EF, Nikkel A, Shiosaki K. The selective dopamine D_1 receptor agonist A-86929 maintains efficacy with repeated treatment in rodent and primate models of Parkinson's disease. J Pharmacol Exp Ther 1997;281:454–459.
209. Rascol O, Nutt JG, Blin O, Goetz CG, Trugman JM, Soubrouillard C, Carter JH, Currie LJ, Fabre N, Thalamas C, Giardina WW, Wright S. Induction by dopamine D_1 receptor agonist ABT-431 of dyskinesia similar to levodopa in patients with Parkinson disease. Arch Neurol 2001;58:249–254.
210. Kebabian JW, Britton DR, DeNinno MP, Perner R, Smith L, Jenner P, Schoenleber R, Williams M. A-77636: a potent and selective dopamine D_1 receptor agonist with antiparkinsonian activity in marmosets. Eur J Pharmacol 1992;229:203–209.
211. Lin CW, Bianchi BR, Miller TR, Stashko MA, Wang SS, Curzon P, Bednarz L, Asin KE, Britton DR. Persistent activation of the dopamine D_1 receptor contributes to prolonged receptor desensitization: studies with A-77636. J Pharmacol Exp Ther 1996;276: 1022–1029.
212. Cueva JP, Giorgioni G, Grubbs RA, Chemel BR, Watts VJ. Nichols DE. trans-2,3-Dihydroxy-6a,7,8,12b-tetrahydro-6H-chromeno[3,4-c]isoquinoline: synthesis, resolution, and preliminary pharmacological characterization of a new dopamine D_1 receptor full agonist. J Med Chem 2006;49:6848–6857.
213. Baldessarini RJ, Drug therapy of depression and anxiety disorders. Pharmacotherapy of psychosis and mania. In: Brunton LL, Lazo JS, Parker KL,editors. Goodman and Gilman's The Pharmacological Basis of Therapeutics. New York: McGraw-Hill; 2006; p. 429–460.
214. Edmondson DE, Binda C, Wang J, Upadhyay AK,Mattevi A. Molecular and mechanistic properties of the membrane-bound mitochondrial monoamine oxidizes. Biochemistry 2009;48: 4220–4230.
215. Glezer S, Finberg JP. Pharmacological comparison between the actions of methamphetamine and 1-aminoindan stereoisomers on sym-

pathetic nervous function in rat vas deferens. Eur J Pharmacol 2003;472:173–177.
216. Olanow CW, Hauser RA, Jankovic J, Langston W, Lang A, Poewe W, Tolosa E, Stocchi F, Melamed E, Eyal E, Rascol O. A randomized, double-blind, placebo-controlled, delayed start study to assess rasagiline as a disease modifying therapy in Parkinson's disease (the ADAGIO study): rationale, design, and baseline characteristics. Mov Disord. 2008;23:2194–201.
217. Caccia C, Maj R, Calabresi M, Maestroni S, Faravelli L, Curatolo L, Salvati P, Fariello RG. Safinamide: from molecular targets to a new anti-Parkinson drug. Neurology 2006;67 (7 Suppl 2):S18–23.
218. Stocchi F, Arnold G, Onofrj M, Kwiecinski H, Szczudlik A, Thomas A, Bonuccelli U, Van Dijk A, Cattaneo C, Sala P, Fariello RG. Improvement of motor function in early Parkinson disease by safinamide. Neurology 2004;63:746–748.
219. Factor SA, Molho ES, Feustel PJ, Brown DL, Evans SM. Long-term comparative experience with tolcapone and entacapone in advanced Parkinson's disease. Clin Neuropharmacol 2001;24:295–299.
220. Teräväinen H, Rinne U, Gordin A. Catechol-O-methyltransferase inhibitors in Parkinson's disease. Adv Neurol 2001;86:311–25.
221. Lautala P, Ulmanen I, Taskinen J. Molecular mechanisms controlling the rate and specificity of catechol O-methylation by human soluble catechol Omethyltransferase. Mol Pharmacol 2001;59:393–402.
222. Gordin A, Kaakkola S, Teravainen H. Clinical advantages of COMT inhibition with entacapone—a review. J Neural Transm 2004;111: 1343–1363.
223. Olanow CW, Watkins PB. Tolcapone: an efficacy and safety review (2007). Clin Neuropharmacol 2007;3:287–94.
224. Haasio K, Nissinen E, Sopanen L. Different toxicological profile of two COMT inhibitors *in vivo*: the role of uncoupling effects. J Neural Transm 2002;109:1391–1401.
225. Korlipara LV, Cooper JM, Schapira AH. Differences in toxicity of the catechol-*O*-methyl transferase inhibitors, tolcapone and entacapone, to cultured human neuroblastoma cells. Neuropharmacology 2004;46:562–569.
226. Tafazoli S, Spehar DD, O'Brien PJ. Oxidative stress mediated idiosyncratic drug toxicity. Drug Metab Rev 2005;37:311–332.
227. Kurth MC, Adler CH, St. Hilaire MS, et al. Tolcapone improves motor function and reduces levodopa requirement in patients with Parkinson's disease experiencing motor fluctuations: a multicenter, double-blind, randomized, placebo-controlled trial. Neurology 1997; 48:81–87.
228. Felder CC, Bymaster FP, Ward J, DeLapp N. Therapeutic opportunities for muscarinic receptors in the central nervous system. J Med Chem 2000;43:4333–4353.
229. Fink JS, Weaver DR, Rivkees SA, Peterfreund RA, Pollack AE, Adler EM, Reppert SM. Molecular cloning of the rat A_2 adenosine receptor: selective co-expression with D_2 dopamine receptors in rat striatum. Brain Res Mol Brain Res 1992;14:186–195.
230. Mori A, Shindou T, Ichimura M, Nonaka H, Kase H. The role of adenosine A_{2a} receptors in regulating GABAergic synaptic transmission in striatal medium spiny neurons. J Neurosci 1996;16:605–611.
231. Richardson PJ, Kase H, Jenner PG. Adenosine A_{2A} receptor antagonists as new agents for the treatment of Parkinson's disease. Trends Pharmacol Sci 1997;18:338–344.
232. Pinna A, di Chiara G, Wardas J, Morelli M. Blockade of A_{2a} adenosine receptors positively modulates turning behaviour and c-Fos expression induced by D_1 agonists in dopamine-denervated rats. Eur J Neurosci 1996;8: 1176–1181.
233. Ferré S, O'Connor WT, Fuxe K, Ungerstedt U. The striopallidal neuron: a main locus for adenosine-dopamine interactions in the brain. J Neurosci 1993;13:5402–5406.
234. Kurokawa M, Koga K, Kase H, Nakamura J, Kuwana Y. Adenosine A_{2a} receptor-mediated modulation of striatal acetylcholine release *in vivo*. J Neurochem 1996;66:1882–1888.
235. Kanda T, Jackson MJ, Smith LA, Pearce RK, Nakamura J, Kase H, Kuwana Y, Jenner P. Adenosine A_{2A} antagonist: a novel anti-parkinson's agent that does not provoke dyskinesia in parkinsonian monkeys. Ann Neurol 1998;43:507–513.
236. Hauser RA, Shulman LM, Trugman JM, Roberts JW, Mori A, Ballerini R, et al. Study of istradefylline in patients with Parkinson's disease on levodopa with motor fluctuations. Mov Disord 2008;23:2177–85.
237. Melamed E, Zoldan J, Friedberg G, Ziv I, Weizmann A. Involvement of serotonin in clinical features of Parkinson's disease and complications of L-dopa therapy. Adv Neurol 1996;69:545–550.

238. Sanders-Bush E, Mayer SE. 5-Hydroxytryptamine (serotonin): receptor agonists and antagonists. In: Brunton LL, Lazo JS, Parker KL,editors. Goodman and Gilman's The Pharmacological Basis of Therapeutics. New York: McGraw-Hill; 2006. p 297–315.

239. Blier P, Piñeyro G, el Mansari M, Bergeron R, de Montigny C. Role of somatodendritic 5-HT autoreceptors in modulating 5-HT neurotransmission. Ann NY Acad Sci 1998; 861:204–216.

240. Bibbiani F, Oh JD, Chase TN. Serotonin 5-HT$_{1A}$ agonist improves motor complications in rodent and primate parkinsonian models. Neurology 2001;57:1829–1834.

241. Christoffersen CL, Meltzer LT. Reversal of haloperidol-induced extrapyramidal side effects in cebus monkeys by 8-hydroxy-2-(di-n-propylamino)tetralin and its enantiomers. Neuropsychopharmacology 1998;18: 399–402.

242. Bartoszyk GD, van Amsterdam C, Greiner HE, Rautenberg W, Russ H, Seyfried CA. Sarizotan, a serotonin 5-HT$_{1A}$ receptor agonist and dopamine receptor ligand: 1. Neurochemical profile. J Neural Transm 2004;111:113–126.

243. Olanow CW, Damier P, Goetz CG, Mueller T, Nutt J, Rascol O. Multicenter, open-label trial of sarizotan in Parkinson's disease patients with levodopainduced dyskinesias (SPLENDID study). Clin Neuropharmacol 2004;27:58–62.

244. Goetz CG, Damier P, Hicking C, Laska E, Müller T, Olanow CW, Rascol O, Russ H. Sarizotan as a treatment for dyskinesias in Parkinson's disease: a double-blind placebo-controlled trial. Mov Disord 2007;22: 179–186.

245. Grégoire L, Samadi P, Graham J, Bédard PJ, Bartoszyk GD, Di Paolo T. Low doses of sarizotan reduce dyskinesias and maintain antiparkinsonian efficacy of L-dopa in parkinsonian monkeys. Parkinsonism Relat Disord 2009;15:445–152.

246. Gardoni F, Di Luca M. New targets for pharmacological intervention in the glutamatergic synapse. Eur J Pharmacol 2006;545:2–10.

247. Merello M, Nouzeilles MI, Cammarota A, Leiguarda R. Effect of memantine (NMDA antagonist) on Parkinson's disease: a double-blind crossover randomized study. Clin Neuropharmacol 1999;22:273–276.

248. Paci C, Thomas A, Onofrj M. Amantadine for dyskinesia in patients affected by severe Parkinson's disease. Neurol Sci 2001;22:75–76.

249. Moryl E, Danysz W, Quack G. Potential antidepressive properties of amantadine, memantine and bifemelane. Pharmacol Toxicol 1993;72:394–397.

250. Schwab RS, Poskanzer DC, England AC Jr, Young RR. Amantadine in the treatment of Parkinson's disease. Review of more than two years' experience. JAMA 1972;222: 792–795.

251. Geldenhuys WJ, Malan SF, Bloomquist JR, Marchand AP, Van der Schyf CJ. Pharmacology and structure–activity relationships of bioactive polycyclic cage compounds: a focus on pentacycloundecane derivatives. Med Res Rev 2005;25:21–48.

252. Geldenhuys WJ, Malan SF, Bloomquist JR, Van der Schyf CJ. Structure–activity relationships of pentacycloundecylamines at the N-methyl-d-aspartate receptor. Bioorg Med Chem 2007;15:1525–1532.

253. Le DA, Lipton SA. Potential and current use of N-Methyl-D-aspartate (NMDA) receptor antagonists in diseases of aging. Drugs Aging 2001;18:717–724.

254. Garnett ES, Firnau G, Chan PK, Sood S, Belbeck LW. [18F]Fluoro-dopa, an analogue of dopa, and its use in direct external measurements of storage, degradation, and turnover of intracerebral dopamine. Proc Natl Acad Sci USA 1978;75:464–477.

255. Volkow ND, Fowler JS, Wang GJ, Logan J, Schlyer D, MacGregor R, Hitzemann R, Wolf AP. Decreased dopamine transporters with age in health human subjects. Ann Neurol 1994;36:237–239.

256. Neumeyer JL, Wang SY, Milius RA, Baldwin RM, Zea-Ponce Y, Hoffer PB, Sybirska E, al-Tikriti M, Charney DS, Malison RT. [123I]-2 beta-carbomethoxy-3 beta-(4-iodophenyl)tropane: high-affinity SPECT radiotracer of monoamine reuptake sites in brain. J Med Chem 1991;34:3144–3146.

257. Neumeyer JL, Wang S, Gao Y, Milius RA, Kula NS, Campbell A, Baldessarini RJ, Zea-Ponce Y, Baldwin RM, Innis RB. N-Omega-fluoroalkyl analogs of (1R)-2 beta-carbomethoxy-3 beta-(4-iodophenyl)-tropane (beta-CIT): radiotracers for positron emission tomography and single photon emission computed tomography imaging of dopamine transporters. J Med Chem 1994;37:1558–1561.

258. Chaly T, Dhawan V, Kazumata K, Antonini A, Margouleff C, Dahl JR, Belakhlef A, Margouleff D, Yee A, Wang S, Tamagnan G, Neumeyer JL, Eidelberg D. Radiosynthesis of [18F] N-3-fluoropropyl-2-beta-carbomethoxy-3-beta-(4-iodophenyl) nortropane and the first human

study with positron emission tomography. Nucl Med Biol 1996;23(8): 999– 1004.

259. Kilbourn M, Lee L, Vander Borght T, Jewett D, Frey K. Binding of alpha-dihydrotetrabenazine to the vesicular monoamine transporter is stereospecific. Eur J Pharmacol 1995;278: 249–252.

260. Bohnen NI, Albin RL, Koeppe RA, et al. Positron emission tomography of monoaminergic vesicular binding in aging and Parkinson disease. J Cereb Blood Flow Metab 2006;26: 1198–1212.

261. Martin WR, Wieler M, Stoessl AJ, Schulzer M. Dihydrotetrabenazine positron emission tomography imaging in early, untreated Parkinson's disease. Ann Neurol 2008;63:388–394.

262. Ravina B, Eidelberg D, Ahlskog JE, Albin RL, Brooks DJ, Carbon M. The role of radiotracer imaging in Parkinson disease. Neurology 2005;64:208–215.

263. Naoi M, Maruyama W, Yi H, Inaba K, Akao Y, Shamoto-Nagai M. Mitochondria in neurodegenerative disorders: regulation of the redox state and death signaling leading to neuronal death and survival. J Neural Transm 2009. (Epub ahead of print.) PMID: 19763773.

264. Benabid AL, Chabardes S, Torres N, Piallat B, Krack P, Fraix V, Pollak P. Functional neurosurgery for movement disorders: a historical perspective. Prog. Brain Res 2009;175:379–391.

265. Hedlund E, Perlmann T. Neuronal cell replacement in Parkinson's disease. J Intern Med 2009;266:358–371.

266. Friling S, Andersson E, Thompson LH, Jönsson ME, Hebsgaard JB, Nanou E, Alekseenko Z, Marklund U, Kjellander S, Volakakis N, Hovatta O, El Manira A, Björklund A, Perlmann T, Ericson J. Efficient production of mesencephalic dopamine neurons by Lmx1a expression in embryonic stem cells. Proc Natl Acad Sci USA 2009;106:7613–7618.

267. Conn PJ, Christopoulos A, Lindsley CW. Allosteric modulators of GPCRs: a novel approach for the treatment of CNS disorders. Nat Rev Drug Discov 2009;8:41–54.

OPIOID RECEPTOR LIGANDS

CHRISTOPHER R. MCCURDY[1]
THOMAS E. PRISINZANO[2]

[1] Department of Medicinal Chemistry, School of Pharmacy, The University of Mississippi, University, MS

[2] Department of Medicinal Chemistry, School of Pharmacy, University of Kansas, Lawrence, KS

1. INTRODUCTION

Since before recorded history, opium derived from the juice of the *Papaver somniferum* seedpod has been utilized to manage pain and has been a significant source of controversy and abuse. Still today, natural products isolated from opium and their derivatives are the mainstay of moderate to severe pain therapy comprising the most widely prescribed class of analgesics and the most widely abused. Licit and illicit usage regimens with these compounds, exemplified by morphine (1) (Fig. 1), are associated with severe and debilitating side effects. Of the array of side effects, constipation, addiction, and respiratory depression often limit the clinical usefulness of these powerful medicines. There has been an extensive effort for many decades to find novel compounds that retain the desired analgesic effects but are devoid of liabilities. Due to this massive effort, many compounds have been introduced into the scientific literature with opioid activity and several important agents are utilized clinically. These compounds have provided valuable information with regard to understanding the endogenous opioid system but still, there is no ideal drug.

Although the effects of opium have been realized from the beginning of recorded history, the true scientific understanding at the molecular level has only been under study for the past six decades. The idea that the effects of opium or opiates (now referred to collectively as opioids) were receptor mediated was first proposed by Beckett and Casy in 1954 [1]. This early description laid the foundation for work in the early 1970s that binding of these compounds to receptors in mammalian brain tissue was stereospecific [2–4]. Soon after this discovery, two seminal studies appeared that defined the endogenous opioid system. Martin and coworkers identified and classified three distinct types of opioid receptors [5,6] shaping the current understanding of opioid pharmacology. This discovery ignited a flurry of research and hypotheses that the liabilities of opioids could be eliminated through careful drug design and pharmacology. At about the same time, the first endogenous ligands were discovered to be pentapepties. The discovery of these peptides, leucine and methionine enkephalin (3 and 4) [7] were soon followed by the longer peptides dynorphin and β-endorphin.

Leucine enkephalin (3)   Tyr-Gly-Gly-Phe-Leu

Methionine enkephalin (4)   Try-Gly-Gly-Phe-Met

At this point, there were two distinct chemical classes of ligands and the term "opioid" was introduced to describe all compounds, regardless of structural class, that demonstrated opiate-like activity. With distinct chemical classes known, the next discoveries into the understanding at the molecular level came in the early 1990s with the cloning of the opioid receptors. This major breakthrough provided the molecular targets, or as Philip S. Portoghese described it as "the other piece of the puzzle," to begin the detailed understanding at the molecular level. Needless to say, the cloning of the delta, mu and kappa receptors only scratched the tip of the iceberg into the current understanding of the opioid system. Signaling cascades, receptor dimerization and oligomerization, and protein–protein interactions have followed to further complicate the roles of these receptors and how one might overcome the liabilities through ligand design.

The focus of this chapter will be on recent developments in the field of opioid research with summaries of the key features of the structure–activity relationships (SARs) of older compounds. Much of the early opioid SAR is discussed in detail in two comprehensive books on opioids published in 1986 [12,13].

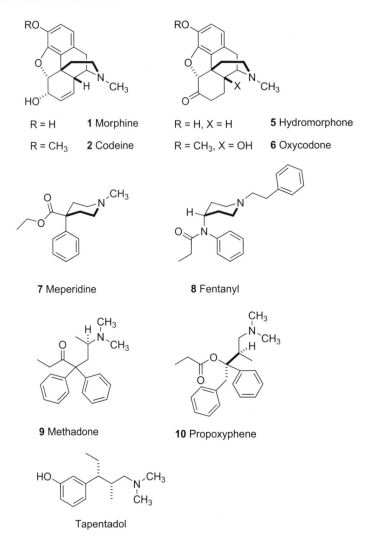

Figure 1. Structures of most commonly used clinical agents.

2. CLINICAL USE OF AGENTS

Opioid analgesics are the mainstay of prescription drug therapy for severe pain, cancer pain, and chronic pain. The widespread use of opioids in chronic, nonmalignant pain is still highly controversial due to the lack of substantial evidence from long-term controlled studies demonstrating effectiveness in this setting [17]. The clinical use of opioids in different types of pain has been reviewed previously [18]. Many opioids are combined with acetaminophen (APAP) or a nonsteroidal anti-inflammatory agents (such as aspirin (ASA) or ibuprofen (IBU)) for mild to moderate pain. These are generally utilized as a "first-line" therapy but if pain continues usually another narcotic analgesic is added to the regimen. This tiered approach for the treatment of cancer pain was introduced by the World Health Organization (WHO) and also serves as a meddle for the management of acute and chronic pain [19]. The next tier of treatment is reached when the pain escalates from moderate to severe. At this point, an opioid may be used as a single agent since they do not have a ceiling analgesic effect [20]. Adjuvant drugs, such as tricyclic antidepressants or anticonvulsants, may be added as a means to enhance the efficacy of opioids [17,21].

For continuous pain, analgesic agents are generally prescribed for use on a regular, around-the-clock basis using a long-acting analog or formulation. For acute pain or pain following surgery, often an immediate release, short-acting opioid is utilized. In addition, short-acting opioids with rapid-onset are utilized for "rescue" doses when breakthrough pain is problematic [21,22].

For those with addiction to opioids, a long-acting opioid analgesic is utilized such as buprenorphine or methadone. These agents are still not optimal as they are sort of step-down therapies. However, at this time, they continue to be the only treatments available along with intensive psychotherapy.

2.1. Current Drugs on the Market

The structures for the most commonly used clinical agents are shown in Fig. 1. Some of the opioid agonists used clinically (Table 1) such as morphine may be used as the sole agent for analgesia. Because of its rapid onset and short duration of action fentanyl and other 4-anilidopiperidines have been used extensively as adjuncts to anesthesia, while buprenorphine, methadone and its analog levomethadyl acetate (LAAM) are used as maintenance agents for individuals who are addicted to narcotics. Other agents such as loperamide or diphenoxylate are used primarily for their constipating side effect to treat diarrhea. Some drugs are used extensively in combination products (Table 2) for treatment of pain. Most of the clinically used agents are agonists at μ-opioid receptors (MOP). In contrast, mixed agonists/antagonists generally interact with two distinct opioid receptors to provide analgesic activity while exhibiting decreased potential for serious side effects such as respiratory depression and addiction (see Section 2.2). Pentazocine is a prototype for the mixed agonist/antagonist class and acts as an agonist at κ-opioid receptors generating analgesia (see Section 3.2); its antagonist activity at μ-opioid receptors significantly decreases or eliminates the potential for respiratory depression and addiction liability generated via μ receptor activation. The mixed agonists/antagonists shown in Table 3 find limited clinical utility, despite the analgesia resulting from activating κ-opioid receptors. The analgesia effect produced reaches a maximum despite increased drug dose (analgesic ceiling). Further, these drugs exhibit a different side-effect profile including dysphoria and hallucinations that appears to also be a result of κ-agonist activity [1].

Most recently introduced to the market is a mixed action opioid analgesic, tapentadol. This compound is approved for the treatment of moderate to severe acute pain and acts through the μ-opioid receptors as well as inhibiting the reuptake of norepinephrine [2]. It has also been noted to have a more tolerable side-effect profile [3,4]. Since this compound has a dual mechanism of action there is potential for use in chronic pain, although it is not approved for such. Physicians sometimes prescribe serotonin and/or norepinephrine reuptake inhibitors to help manage chronic pain.

Opioid agonists also have an antitussive effect due to the depression of the cough reflex. Thus some opioids, typically codeine or one of its derivatives, are used for their antitussive activity, predominantly in combination products. The antitussive effect is in part due to the interaction with opioid receptors at the cough center in the brain [1]. The dose required for antitussive activity, however, is lower than that required for analgesia; the opioid receptors involved in blocking the cough reflex are less sensitive to naloxone than those responsible for analgesia [1].

Opioid antagonists (Table 4), predominantly naloxone, are utilized clinically to reverse the effects of opiates in overdose or reversal of postoperative sedation. Naltrexone, which has oral bioavailability, is used for the treatment of narcotic addiction and alcohol dependence. As discussed in section "Constipation," peripherally selective antagonists are being evaluated for treatment of constipation and other gastrointestinal side effects associated with opioid agonist use.

Other opioid antagonists have recently been introduced to the market that are peripherally restricted in their actions. Alvimopan was the first introduced for the purpose of accelerating the time to upper and lower gastrointestinal recovery following partial

Table 1. Opioid Analgesics Used Clinically[a]

Chemical Class	Generic	Trade Name (Manufacturer)	Route of Administration	Equal Analgesic Dose (mg)[b]	
				i.m.	p.o.
4,5β-Epoxymorphinans	Morphine	Available as generic	p.o., i.v., i.m., s.c., rectal	10	60
	Codeine	Available as generic	p.o., i.v., i.m., s.c.	120	200
	Hydromorphone[c]	Dilaudid (Knoll), also available as generic	p.o., i.v., i.m., s.c., rectal	1.5	7.5
	Hydrocodone[c]				
	Oxymorphone	Numorphan (Endo Laboratories)	i.v., i.m., s.c., rectal	1	10[d]
	Oxycodone	OxyContin (Purdue Pharma LP), also available as generic	p.o.		30
Morphinans	Levorphanol	Levo-Dromoran (ICN), also available as generic	p.o.	2	4
Phenylpiperidines	Meperidine	Demerol (Sanofi-Synthelabo), also available as generic	p.o., i.v., i.m., s.c.	75	300
	Diphenoxylate[e]	Lomotil (Searle), also available as generic	p.o.		
	Difenoxin[f]	Motofen (Carnrick)	p.o.		
	Loperamide[g]	Imodrum A–D (McNeil-CPC), also available as generic	p.o.		
4-Anilidopiperidines	Fentanyl	Sublimaze (Taylor)	i.v.	0.1[h]	
		Fentanyl Oralet (Abbott)	Lozenges		
		Actiq (Abbott)	Lozenges on a stick		
		Duragesic (Janssen)	Transdermal patches		
	Alfentanil	Alfenta (Taylor)	i.v. with individualized dosing		
	Remifentanil	Ultive (Abbott)	i.v. with individualized dosing		
	Sufentanil	Sufentanil citrate (ESI Lederle), Sufenta (Taylor)	i.v. with individualized dosing	0.02[h]	
Acyclic analgesics	Methadone	Available as generic	i.m., s.c., p.o., with individualized dosing	10	20
	Levomethadyl actate	Orlaam (Roxane)	p.o. with individualized dosing		
	Propoxyphene	Darvon-N (napsylate) (Eli Lilly), available as generic (HCl)	p.o.		130–200[i]
	Tapentadol	Nucynta (Ortho-McNeil Johnson)	p.o.		50–100

[a] Agents currently marketed in the United States. Unless otherwise noted, these drugs are Class II controlled substances.
[b] Based on short-term use for acute pain.
[c] Used only as an antitussive product.
[d] Rectal administration.
[e] Class V narcotiv available only by prescription for treatment of diarrhea.
[f] Class IV narcotiv available only by prescription for treatment of diarrhea.
[g] A noncontrolled substance available both by prescription and over the counter for treatment of diarrhea.
[h] i.v. dose.
[i] Dose of 130 mg for the HCl salt and 200 mg for the napsylate salt.

Table 2. Examples of Narcotiv Agonists Currently Marketed as Oral Combination Products for Pain[a,b]

Trade of Common Name (Manufacturer)	Narcotiv Component	Nonnarcotiv Component
APAP with codeine (generic)	Codeine	APAP
ASA with codeine (generic)	Codeine	ASA
DHC Plus (Purdue Frederick)	Dihydrocodeine	APAP, caffeine[c]
Synalgos-DC (Wyeth-Ayerst)	Dihydrocodeine	ASA, caffeine
Vicodin (Knoll)	Hydrocodone	APAP
APAP with hydrocodone (generic)	Hydrocodone	APAP
Alor 5/500 (Atley)	Hydrocodone	ASA
Vicoprofen (Knoll)	Hydrocodone	Ibuprofen
Percocet (DuPont)	Oxycodone	APAP
APAP with oxycodone (generic)	Oxycodone	ASAP
Percodan (DuPont)	Oxycodone	ASA
ASA with oxycodone (generic)	Oxycodone	ASA
Mepergan Fortis (Wyeth-Ayerst)	Meperidine	Promethazine[d]
Darvocet-N (Eli Lilly)	Propoxyphene sapsylate	APAP
APAP with propoxyphene (generic)	Propoxyphene sapsylate	APAP
Wygesic (Wyeth-Ayerst)	Propoxyphene sapsylate	APAP
APAP with propoxyphene (generic)	Propoxyphene HCl	APAP
Darvon	Propoxyphene HCl	ASA, caffeine[c]

[a] Products currently marketed in the United states.
[b] Codeine, hydrocodone, and hydromorphone are use in combination products as antitussives.
[c] May be bebeficial in vascular headaches.
[d] Used for sedative effect.

large or small bowel resection surgery with primary anastomosis. Its use has been limited to the hospital setting and those hospitals must be registered to dispense it. Methylnaltrexone, administered by subcutaneous injection, was also introduced for the treatment of opioid-induced constipation in patients with advanced illness that do not respond to conventional laxative treatments.

2.2. Side Effects, Adverse Effects, Drug Interactions/Contraindications

In addition to analgesia, clinically used opioids display a plethora of biological effects.

Table 3. Narcotic Agonist–Anagonist[a]

Chemical Class	Generic Name	Trade Name (Manufacturer)	Controlled Substance Class	Route of Administration	Equivalent Dose (mg)[b]
6,14-endo-Etheno opiates	Buprenorphine	Buprenex (Reckitt & Colman)	Class V	i.v., i.m.	0.3
4,5α-Epoxy-morphinans	Nalbuphine	Nubain (DuPont)	NA	i.v., i.m., s.c.	10
Morphinans	Butorphanol	Stadol (Mead Johnson)	Class IV	i.v., i.m., nasal spray	2.5
Benzomorphans	Pentazocine[c]	Talwin (Sanofi Winthrop)	Class IV	i.v., i.m., s.c.	30
Aminotetralin	Dezocine	Dalgan (Astra)	NA	i.v., i.m., s.c.	10

[a] Agents currently marketed in the United States.
[b] Parenteral dose equivalent to 10 mg morphine.
[c] Also available as oral combination products with ASA (Talwin Compound), APAP (Talacen), or naloxone (Talwin NX) (Sanofi Winthrop)
NA: not applicable.

Table 4. Narcotiv Antagonists[a]

Generic Name	Trade Name (Manufacturer)	Route of Administration
Naloxone	Narcan (DuPont)	i.v., i.m., or s.c.
Naltrexone	ReVia (DuPont) Depade (Mallinckrodt)	p.o.
Nalmefene	Revex (Ohmeda)	i.v., i.m., or s.c.

[a] Agents currently marketed in the United States.

The most common side/adverse effects involve alterations of the nervous, respiratory, gastrointestinal, and integument systems (see Refs [1,5] and references cited therein for more detailed discussions). The most serious side effects associated with the majority of opioid analgesics are respiratory depression, addiction liability, and constipation that are associated with their agonist activity at μ-opioid receptors.

2.2.1. Central Side Effects

Respiratory Depression μ-Opioids used for analgesia result in slow breathing, constituting one of the most serious adverse effects [6]. Respiratory depression is caused at least in part by interaction of opioids with the respiratory center in the brain stem, causing a decreased response to carbon dioxide and thus depression of breathing rate [1]. Respiratory depression can occur at doses far lower that those that affect consciousness and increases progressively with increasing drug dose [1]. Mortality from opioid overdose is almost always a result of respiratory depression. Of important note is that the respiratory effects of sleep, which often accompanies pain relief, are additive with the depressant effects of the opioid analgesic on respiration [1]. The most profound respiratory depression occurs within 5–10 min post i.v. administration of morphine, and this effect occurs more rapidly as the lipophilicity of the narcotic analgesic increases [1].

Opioid-naïve patients with severe pain who require high doses of opioids are at highest risk for respiratory depression, while patients receiving chronic opioid therapy rarely experience this problem [7]. Fortunately, the occurrence of respiratory depression can often be circumvented with appropriate titration of opioid dose [6] unless there is underlying pulmonary dysfunction such as emphysema or severe obesity [1].

Tolerance, Dependence, and Addiction Liability Patients treated with long-term opioid therapy often develop tolerance and usually become physically dependent on narcotic analgesics as well. *Tolerance* results when exposure to a drug results in its decreased effectiveness with time and larger doses are required to achieve the same response [8]. *Physical dependence* is also an adaptive state that is characterized by specific withdrawal symptoms that occur upon abrupt cessation or significant reduction in the dose of the opioid or administration of an opioid antagonist [8]. *Addiction*, however, is distinct from physical dependence, and "the term addiction should never be used when physical dependence is meant" [6]. The American Academy of Pain Medicine, the American Pain Society, and the American Society of Addiction Medicine have written a consensus document that clearly outlines the recommended definitions for addiction, physical dependence, and tolerance related to the use of opioids for the treatment of pain [8]. According to the consensus document definitions, *addiction* is "a primary, chronic, neurobiologic disease, with genetic, psychosocial, and environmental factors influencing its development and manifestations. It is characterized by behaviors that include one or more or the following: impaired control over drug use, compulsive use, continued use despite harm, and craving." *Physical dependence* is defined as "a state of adaptation that is manifested by a drug class specific withdrawal syndrome that can be produced by abrupt cessation, rapid dose reduction, decreasing blood level of the drug, and/or administration of an antagonist" [8]. There is, however, a very low addiction potential for opioids used for pain management [9], on the order of only 3 cases per 10,000 patients (see Ref. [6] and references cited therein).

The pharmacological mechanisms responsible for the euphoria and rewarding behavior associated with μ-opioid analgesics and addiction liability remain uncertain [1]. These effects are distinct from analgesia [10]. Considerable evidence suggests that the rewarding effects result from interaction of opioid with dopaminergic pathways, particularly in the nucleus accumbens [1]. Activation of μ- or δ-opioid receptors results in the release of dopamine that results in the motivational effect [11]. In contrast, agonists interacting with κ-opioid receptors, naloxone and μ-selective antagonists inhibit dopamine release [12], producing aversion rather than motivation [13].

Sedation and Cognitive Impairment The initiation or dose escalation of narcotic analgesics may cause drowsiness and impair cognitive function. Tolerance usually develops fairly quickly to these side effects, but other medications that induce somnolence will produce an additive effect when taken concomitantly. If sedation remains problematic in order to achieve adequate analgesia, a psychostimulant such as caffeine, dexamphetamine, or methylphenidate may be added to counteract the side effect [6].

Nausea and Vomiting The most bothersome and unpleasant side effects for patients receiving opioids for pain are often the nausea and vomiting that has been associated with all clinically used μ-agonists. Emesis predominantly results from direct stimulation of the chemoreceptor trigger zone, yet the degree of effect depends on the individual [1]. Nausea and vomiting related to narcotic analgesia is occurs in 10–40% of patients [14]. Tolerance often develops to these side effects, however, and they often vanish with long-term use [6]. Nausea and vomiting can be treated using a variety of drugs such as transdermal scopolamine, hydroxyzine or a phenothiazine for movement-induced nausea [7], or metoclopramide or cisapride for patients with nausea and vomiting stemming from delayed gastric emptying. If the nausea and vomiting persists, steroid therapy with dexamethasone may be initiated or a serotonin $5-HT_3$ antagonist such as ondansetron utilized [6]. Another alternative is to try a different opioid, since there is significant individual variability in this side effect [7].

2.2.2. Other Side Effects

Constipation The most common side effect of long-term narcotic analgesia is constipation plus other gastrointestinal (GI) effects collectively referred to as opioid bowel dysfunction. The frequency of these side effects is very high (40–50% or more in patients receiving opioids [15–17]) and can become the limiting factor in opioid use. These effects are mediated predominantly by μ-receptors in the bowel [1]. The effects begin with delayed food digestion in the small intestine and decrease in peristaltic waves in the large intestine resulting in the retention of bowel contents. This is compounded by the enhanced tone of the anal sphincter and the reduction of the reflex relaxation in response to rectal distension. Tolerance does not usually develop to this side effect, and the patient on long-term opioid therapy remains chronically constipated.

Patients are generally started prophylactically on a regimen including a laxative such as bisacodyl or senna that increases bowel motility plus a stool softener such as docusate [6,7]. In patients refractory to laxatives, oral naloxone [6] has been successfully used as a therapeutic alternative for constipation without loss of analgesia [18]. Because of its central activity, however, higher doses of naloxone can decrease the analgesic effectiveness of the opiate and precipitate opioid withdrawal in some patients [18]. Peripherally selective antagonists offer the advantage of reversing the gastrointestinal and other peripheral side effects of narcotic analgesics without the potential for decreasing their central analgesic activity. Two peripherally selective antagonists, the quaternary derivative of naltrexone N-methylnaltrexone bromide (methylnaltrexone) (see Section 5.3.1) and the phenylpiperidine alvimopan (ADL 8-2698, LY246736) (see Section 5.7.1), are undergoing clinical trials for opioid-induced constipation [15–17,19]. Following both intravenous and oral administrations, methylnaltrexone reverses the opioid-induced delay in GI transit [15–17] and is effective in individuals receiving chronic opioid treatment (methadone users) as well as in healthy volunteers [15,17]. In clinical trials, oral alvimopan reverses the delay in GI transit following the administration of exogenous opioids to both opioid-naïve individuals and patients receiving chronic opioid treatment (both pain patients and individuals taking methadone for opioid addiction) [19,20]; in addition, it has been shown to speed the recovery of bowel function following abdominal surgery [21]. These agents are both currently available and approved by the U.S. Food and Drug Administration (FDA).

The constipating effect of orally administered opiates can be utilized for the treatment of diarrhea, as with camphorated tincture of opium (Paregoric or Parepectolin, which is a paragoric plus kaolin as an adsorbent and pectin as a demulcent) [22]. Two phenyl piperidine derivatives are used solely as antidiarrheal agents. Diphenoxylate, which is a congener of meperidine, is only available in combination with atropine, which has antispasmodic activity in the intestine. At therapeutic doses, diphenoxylate does not show any CNS effects, but at high doses it displays the typical opioid profile including euphoria. The carboxylic acid metabolite, difenoxylic acid (Motofen®) (Table 1) has activity similar to the parent [1]. Unlike diphenoxylate, the second opioid used for diarrhea, loperamide, does not exhibit pleasurable CNS effects even at large doses [1]; loperamide is a substrate for P-glycoprotein in the blood–brain barrier that excludes this drug from the CNS [23,24].

Pruritis Following administration of opioids, there may be urticaria at the injection site or generalized itching due to degranulation of mast cells resulting in histamine release. The itching is a common side effect and often one that results in severe patient distress [1]. The histamine release may also be partially responsible for the pruritus and sweating following drug administration as well as flushing due to blood vessels dilation of the skin [1]. Antihistamines may be utilized to combat the discomfort [7] or patients can be switched to either fentanyl or oxymorphone [25] that do not tend to cause histamine release. Opioid antagonists such as naloxone are effective in controlling the pruritus and can be used at low doses without loss of pain control [15,26]; the peripherally selective antagonist methylnaltrexone exhibits antipruritic efficacy without the potential to reverse morphine analgesia [27].

2.2.3. Contraindications Contraindications include hypersensitivity to opioids, head trauma or increased intracranial pressure, severe respiratory depression or compromised respiratory function, and potentially liver or renal insufficiency [28]. Whether or not morphine or other opioids are used depends on the severity of the contraindication, and the potential benefits must be weighed relative to the risk. Anaphylactoid reactions have been reported after morphine or codeine administered IV, although the reactions are rare [1]. Morphine may induce or exacerbate asthmatic attacks hence fentanyl may be a better choice is asthmatic patients [1]. Other relative contraindications to the use of narcotics also exist with respect to the potential for drug abuse [29]. While a history of substance abuse does not definitely preclude the use of opioids, it does necessitate careful vigilance. If the episode of abuse is active or recent, then another pain management strategy may be prudent. Consideration of the social network also requires consideration, especially if the patient lives with a substance abuser or has a home life conducive to enabling abuse.

2.2.4. Drug Interactions The pharmacological activity of opioids can be affected by a number of other drugs, including amphetamines, antihistamines, antidepressants, and antipsychotics (see Ref. [1]). Small doses of amphetamine significantly enhance the analgesic activity and euphoric effects of morphine and may counteract sedation. Diphenhydramine and hydroxyzine are antihistamines that exhibit modest analgesic activity themselves, and hydroxyzine has been shown to enhance the analgesic effects of low doses of narcotic analgesics [30]. The depressant effects of some opioids may be exaggerated and prolonged by monoamine oxidase inhibitors, tricyclic antidepressants and phenothiazines; the exact mechanism of action is not fully understood, but may involve metabolic or neurotransmitter alteration [1]. The antidepressants desimiprimine [31] and nefazodone [32] appear to enhance morphine-induced analgesia. The phenothiazine antipsychotics cannot only potentiate the analgesic effect of opioids but also increase respiratory depression and sedation [1].

Interactions with Cytochrome P450 Enzymes Drug interactions with opioid analgesics can

Table 5. Cytochrome P450 Isozymes and Opioid Substrates and Inhibitors

P450 Isozyme	Substrate	Inhibitor
1A2	Methadone	
2D6	Codeine	Codeine
	Hydrocodone	Methadone
	Meperidine	Propoxyphene
	Methadone	
	Morphine	
	Propoxyphene	
3A4	Alfentanil	Dextropropoxyphene
	Fentanyl	Propoxyphene
	Sufentanil	

also result from their interaction with cyctochrome P450 (CYP) isozymes, specifically 3A4 and 2D6 (Table 5) [33].

The CYP3A4 isozyme is responsible for the metabolism of a large number of endogenous compounds as well as a wide range of drugs [33]. Fentanyl, alfentanil, and sufentanil are substrates for CYP3A4, and therefore drugs that inhibit this enzyme, such as erythromycin, HIV protease inhibitors, or cimetidine, may result in oversedation or increased respiratory depression as well as prolonged duration of action of the opioid [34]. Conversely, more rapid metabolism of alfentanil or fentanyl may result when these agents are used in combination with rifampin that is an inducer of 3A4 [34].

Genetic polymorphism in CYP2D6 results in varied drug metabolism [35], with the lack of this isoenzyme affecting between 5–10% of Caucasians and 1–3% of African-Americans and Asians [35]. Individuals lacking 2D6 may display a larger response to some drugs and be at greater risk for toxicity due to their inability to metabolize certain drugs. Several opioids such as morphine, meperidine, and methadone are metabolized by 2D6 (Table 5), and interactions with drugs that induce 2D6 result in loss of opioid activity. Thus, rifampin, phenytoin, phenobarbital, primadone, and carbamazepine can all result in a significant reduction in opioid concentrations and concomitant use may require increasing the opioid dose. This genetic defect is also problematic when metabolism to an active metabolite occurs. Since CYP2D6 can convert codeine to morphine (see Section 2.3.2), 2D6 inhibitors especially quinidine can significant diminish the analgesic effects of codeine [34]. Ritonavir and cimetadine also inhibit metabolism by 2D6 and therefore can significantly increase the concentration of opioids metabolized by 2D6 (Table 5), including meperidine, methadone, and propoxyphene, which may result in toxicity [34].

2.3. Absorption, Distribution, Metabolism, and Elimination

2.3.1. Absorption and Distribution Opioid analgesics are available for administration by a variety of routes, including by subcutaneous and intramuscular injections and rectal suppositories as well as by oral and intravenous administration (see Tables 1, 3, and 4). The least invasive and safest route that provides adequate analgesia should be chosen [6]. Opioids are absorbed from the gastrointestinal tract or the rectal mucosa and are also readily absorbed into the bloodstream following subcutaneous or intramuscular injection. Some narcotics with increased lipophilicity may be absorbed through the nasal or buccal mucosa (butorphanol nasal spray and fentanyl lozenges, respectively). Fentayl is sufficiently lipophilic to be absorbed through the skin (fentanyl transdermal patch) [1,22].

Intravenous administration of opioids results in a rapid onset of action. The more lipophilic drugs show more rapid onset of action after subcutaneous administration due to differing rates of absorption and penetration into the CNS across the blood–brain barrier [1]. Intraspinal administration produces long-lasting analgesia, but the hydrophilicity of morphine can result in rostral spread of the drug in the spinal fluid. This may be problematic, as respiratory depression can occur up to 24 h after the last administered dose as a result of the drug reaching the respiratory control centers in the brain [1]. Fentanyl and hydromorphone are highly lipophilic and are rapidly absorbed by spinal tissue, and thus the rostral spread is significantly reduced resulting in a localized analgesic affect [1].

The binding of opioids to serum proteins is of considerable importance as it influences the distribution of the drug as well as metabolism and excretion [36]. The protein binding of morphine is low (35%), moderate with meperidine (70%), and high for methadone (90%) [37]. The high human plasma protein binding of methadone is well known [38]; the highest binding is to β-globulin III [39] and albumin [36,39]. The extensive protein binding of methadone is an important factor, as it significantly affects the amount of drug available in the plasma for penetration across the blood–brain barrier. The extensive protein binding may provide a reservoir for methadone to replenish the drug to the blood [40]. It has been suggested that the high protein binding of methadone accounts for its mild but extended withdrawal symptoms [1]. It is also fairly common for drug abusers (50% for heroin addicts) to have serum protein abnormalities, including elevated globulin and albumin levels, which may further complicate addiction treatment [40]. In addition, there may also be liver dysfunction or disease due to concomitant alcohol abuse that may account, in part, for the abnormal serum protein levels [40].

2.3.2. Metabolism and Elimination of Morphine and Derivatives
Most opioids are subjected to significant first-pass hepatic metabolism (see Ref. [1]).Thus, oral morphine has only approximately one-quarter bioavailabililty compared to a parenteral dose. The major route of metabolism is via glucuronic acid conjugation. Two major metabolites are formed with the conjugation of the glucuronic acid to either position 3 (M3G, **11**) (Fig. 2) (50%) or position 6 (M6G, **12**) (5–15%) of morphine, while only small amounts of the diglucuronide are formed [1,5]. M6G has pharmacological actions indistinguishable from those of morphine, yet it is twice as potent and is present in higher concentrations in plasma than morphine [41]. It has been suggested that M6G accounts for a significant portion of morphine's analgesic activity, especially with chronic use [42]. Dose adjustment is not required in mild hepatic disease, but excessive sedation can occur in cirrhotic patients due to the accumulation of M6G [42]. M3G has very low affinity for opioid receptors and does not contribute to analgesia, but it may contribute to neuroexcitatory side effects such as allodynia [43]. M3G has been reported to antagonize morphine [44], but this effect has not been consistently observed [45]. Metabolism by N-demethylation to normorphine is a minor metabolic pathway for morphine, while N-dealkylation is important to the metabolism of some morphine congeners (see Ref. [46] and references cited therein for an excellent overview of morphine metabolism and elimination). Codeine, levorphanol, oxycodone, and methadone have a high oral to parenteral potency ratio due to decreased first-pass hepatic metabolism [1].

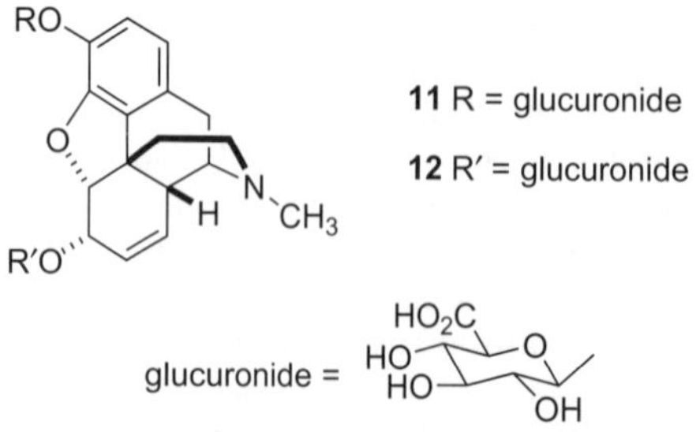

A small but significant amount of codeine ($\sim$10%) is metabolized to morphine via O-demethylation by the 2D6 isozyme of hepatic cytchrome P450 [47]. The resulting morphine is thought to be responsible the analgesic activity since codeine has very low affinity for opioid receptors [1]. Patients lacking CYP2D6, due to a genetic polymorphism, are unable to metabolize codeine to the morphine [34]. As noted above, there appears also to be variation in metabolic efficiency depending on ethnicity [48]. The predominant metabolite of codeine is the 6-glucruonide (C6G) that is renally excreted. One controversial report suggests that codeine analgesia is a result of C6G and not the morphine metabolite [49]. Hydrocodone is also metabolized to hydromorphone and oxymorphone is a minor active metabolite of oxycodone via hepatic P450 2D6 metabolism.

Naloxone undergoes extensive hepatic first-pass metabolism through glucuronidation. Naltrexone is metabolized to the active metabolite 6-naltrexol, which is less potent but has a prolonged half-life compared to the parent drug.

11 R = glucuronide
12 R' = glucuronide

glucuronide =

Figure 2. Major metabolites of morphine.

The opioids are predominantly excreted renally [1]. Morphine is eliminated by glomerular filtration from the kidney, predominantly as M3G or M6G [1]. The majority of excretion (>90%) takes place on the first day, but a small amount of enterohepatic circulation accounts for the presence of small amounts of morphine and metabolites for several days after the last drug dose of drug is given [1]. Both the free and the conjugated forms of codeine are excreted in the urine [1].

2.3.3. Metabolism and Elimination of Other Opioid Agents Meperidine is also metabolized in the liver. It is hydrolyzed directly either to meperidinic acid or to N-demethylated to normeperidine and then hydrolyzed to normeperidinic acid. The acid forms are conjugated then excreted [1].

Fentanyl is also hepatically metabolized and renally excreted. However, the congener remifentanil is metabolically distinct when compared to other members in its chemical or pharmacological class. Remifentanil is metabolized by plasma esterases to remifentanil acid, which is approximately 3000-fold less potent than the parent opioid [50].

Methadone undergoes significant hepatic metabolism by N-demethylation and cyclization to form pyrrolidines and pyrroline [1]. Propoxyphene is also hepatically metabolized predominantly by N-demethylation and renally eliminated. The metabolite norpropoxyphene is cardiotoxic and produces arrhthymias and pulmonary edema that have led to reports of cardiac arrest and death [42]. This is especially problematic due the long half-life of norpropoxyphene that accumulates with repeated doses of the parent drug. Methadone is excreted not only in the urine but also in the bile [1].

3. PHYSIOLOGY AND PHARMACOLOGY

3.1. Opioid Effects in the Central Nervous System and the Periphery

Opioid receptors are found both in the central nervous system and in the periphery. In the CNS, different types of opioid receptors (μ-, κ-, and δ-receptors—see below) exhibit distinct anatomical distributions (see Refs [51–53] for reviews), and there is considerable species variation in both relative receptor density and receptor distribution. Peripheral receptors mediate some effects of opioids, such as inhibition of gut motility, and for a number of years receptors from tissues such as the guinea pig ileum (GPI) formed the basis of standard bioassays used to assess compounds for opioid activity (see section on "*In Vitro* Assays for Efficacy"). Peripheral receptors have also been implicated in analgesia, particularly in cases of inflammation (see Refs [54–57] for reviews). Readers are referred to a comprehensive two volume series on opioids [58,59] plus more recent reviews [60–62] for detailed reviews of opioid pharmacology and physiology.

3.2. Multiple Opioid Receptor Types

3.2.1. Discovery of Multiple Opioid Receptor Types and Current Nomenclature Our understanding of opioid receptors has expanded considerably from the early assumption of a single opioid binding site to the characterization of multiple types of opioid receptors (see Refs [63–65] for reviews). The initial proposal of opioid receptors by Beckett and Casy in 1954 [66] assumed a single opioid binding site. Multiple opiate receptors were postulated as early as the 1960s by both Portoghese [67] and Martin [68], but it was behavioral studies in the chronic spinal dog by Martin and coworkers in the mid-1970s [69,70] that led to the classification of multiple opioid receptors. Based on the pharmacological profile of a variety of opioids, Martin proposed three types of opioid receptors, μ-, κ-, and σ-receptors, with morphine (**1**) (Fig. 1), ketocyclazocine (**13**), and SKF-10,047 (N-allylnorcyclazocine) (**14**), respectively, as the prototypical ligands (Fig. 3). The discovery of the enkephalins led to the proposal of a distinct opioid receptor type, the δ-receptor, for these opioid peptides [71]. The existence of three distinct opioid receptor types, the μ-, κ-, and δ-receptors, has now been clearly established and these receptors have been cloned (see below). σ-Receptors, however, are not considered opioid receptors since effects associated with this receptor are not reversed by opioid antagonists such as naloxone (see Ref [72]). Other opioid receptor types have also been proposed

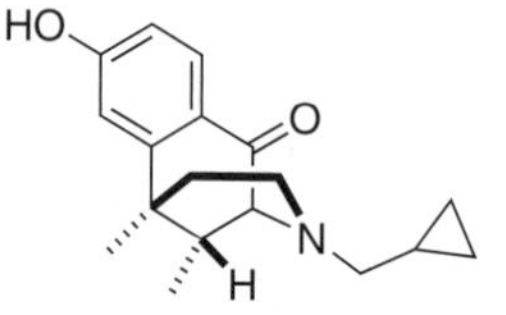

13 Ketocyclazocine

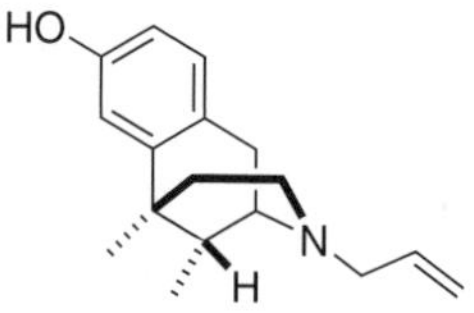

14 SKF 10,047
(*N*-Allylnormetazocine)

Figure 3. Agonists used by Martin and coworkers [69,70] to define κ- and σ-receptors. Morphine (**1**) was the prototypical ligand used to characterize μ-receptors.

(see Ref. [73]), but these receptor types are not universally accepted.

During attempts to clone the opioid receptors (see Section 3.2.4), a related receptor with high sequence homology was identified by several research groups (see Refs [74–77] for recent reviews). This receptor, referred to by Mollereau et al. as opioid-receptor-like 1 (ORL1) receptor [78], does not display affinity for classical opioid ligands, including naloxone. While distinctly different from opioid receptors, the ORL1 receptor interacts with opioid receptor system in the regulation of analgesia and other physiological effects. Details concerning the pharmacology of the ORL1 receptor and its endogenous ligand orphanin FQ/nociceptin (OFQ/N) [79,80] are discussed in Section 7.

In 1996, the International Union of Pharmacology (IUPHAR) recommended that OP1, OP2, and OP3 be used as the accepted names for δ-, κ-, and μ-receptors, respectively, [81] to replace the DOR, KOR, and MOR nomenclature typically used in the literature; OP4 was the proposed name for the related ORL1 receptor. This nomenclature has not gained widespread acceptance, however, and in 2000 the International Narcotic Research Conference [82] recommended a modified nomenclature DOP, KOP, MOP, and NOP for δ-, κ-, μ-, and ORL1 receptors, respectively, that is consistent with the nomenclature requirements of IUPHAR.

3.2.2. Signal Transduction Mechanisms

There is considerable evidence that opioid receptors are coupled to G-proteins and produce their effects through these proteins (see Refs [83,84] for reviews of opioid receptors and G-proteins). The structure of cloned opioid receptors is consistent with their belonging to this receptor superfamily (see below). G-proteins are heterotrimers, consisting of α-, β-, and γ-subunits, which bind guanine nucleotides to their α-subunit and catalyze the hydrolysis of GTP to GDP. G-proteins mediate the interaction of opioid and other receptors with a variety of effector systems, including adenylyl cyclase and ion channels. Numerous forms of G proteins have been identified, including G_i and G_o that inhibit adenylyl cyclase and G_s that stimulates adenylyl cyclase. Pertussis toxin inhibits G_i and G_o by ADP-ribosylation of the α-subunit, while cholera toxin persistently activates G_s. Thus, sensitivity to pertussis toxin is an indication of the involvement of G_i (or G_o) in the transduction mechanism, while sensitivity to cholera toxin is an indication of involvement of G_s.

Of the effector systems that have been implicated in the transduction mechanisms for opioid receptors, the best studied is opioid inhibition of adenylyl cyclase (see Refs [52,84–86] for reviews). Thus, binding of an agonist to opioid receptors inhibits the activity of adenylyl cyclase and decreases intracellular cAMP in a number of different tissues. Pertussis toxin sensitivity of opioid inhibition of adenylyl cyclase has been demonstrated in many systems, indicating the involvement of either G_i or G_o in the transduction mechanism. Agonist activation of all three types of cloned opioid receptors to the inhibition of adenylcyclase has been demonstrated (see Ref. [87] and references cited therein). There is also some evidence that μ- and δ-opioid receptors can stimulate adenylyl cyclase in certain tissues (see Refs [52,84] for reviews). There are conflicting reports on whether κ-opioid receptors stimulate or inhibit phosphatidylinositol turnover in some tissues (see Ref. [87]); δ- and μ-receptors, however, do not appear to be coupled to phosphatidylinositol turnover in neuroblastoma cell lines NG108-15 and SK-N-SH [88].

Opioid receptors can also be coupled to ion channels via G-proteins (see Refs [52,85,86,89] for reviews). All three receptor types can decrease voltage-dependent Ca^{2+} current. The coupling of opioid receptors to calcium channels involves a G-protein, and the actions of opioids on Ca^{2+} current are blocked by pertussis toxin, indicating involvement of G_i or G_o. Activation of μ- and δ-receptors can also increase K^+ conductance. Similar to the results found for calcium channels, potassium channel coupling to opioid receptors appears to involve a G-protein and is sensitive to pertussis toxin. Considerable evidence suggests that the effects on ion currents are due to direct coupling of opioid receptors to ion channels via G-proteins and are not related to inhibition of adenylyl cyclase (see Ref. [89]).

Agonist binding to opioid receptor also appears to activate the extracellular signal regulated kinase (ERK) cascade, which consists of three intracellular kinases: a mitogen-activated protein kinase (MAPK) kinase, a MAPK kinase, and a MAPK homolog (see Ref. [87]). This activation appears to be through G_i or G_o (see Ref. [87] and references cited therein).

3.2.3. Characterization of Opioid Receptors

Early evaluation of compounds for opioid activity relied on testing for antinociceptive activity *in vivo*. These pharmacological assays are often predictive of analgesic activity in humans, but the activity of compounds observed in these assays is affected by a variety of factors, including the route of administration of the compound, the ability of the compound to cross the blood–brain barrier into the CNS, the susceptibility of the compound to metabolism and pharmacokinetics, the choice of noxious stimulus, and the animal species and strain used for the assay (see Ref. [90]). The results of *in vitro* assays are not influenced by many of these factors that complicate *in vivo* assays. The pharmacological activity of opioids *in vitro* can still be complex, however, because more than one opioid receptor type is present in many tissues. Opioid receptors are present in a variety of peripheral tissues, and isolated tissue preparations, particularly the GPI and mouse vas deferens (MVD), have been routinely used to assess opioid activity. Radioligand binding assays for each of the opioid receptor types have been instrumental in the identification of selective opioids. With the cloning of the opioid receptors, assays for both opioid receptor affinity and efficacy can now be routinely performed using these cells that express only a single receptor type, greatly simplifying the interpretation of the results of the assays. Each of these types of assays and their utilization in characterizing compounds for opioid activity are discussed below.

Ligands Used to Characterize Opioid Receptors

Since the identification of multiple opioid receptor types, considerable effort has focused on developing more selective ligands for each of the receptor types (see Figs. 4–6 and Sections 5.3, 5.9, 5.10 and 6). Ligands commonly used to study the different receptor types include both nonpeptides and peptides (see Table 6). Morphinans such as morphine and the antagonists naloxone (**16**), naltrexone (**17**), and cyprodime (**18**) [91] are used to study μ-receptors (Fig. 4); naloxone and naltrexone retain significant affinity for δ- and κ-receptors and therefore at higher concentrations these compounds will antagonize all three receptor types. Several peptides, including the enkephalin analog DAMGO ([D-Ala^2, MeNPhe^4,glyol]enkephalin) (**15**), and CTOP and CTAP (**19** and **20**), somatostatin analogs that antagonize μ-receptors [92], are also used to characterize μ-receptors. Early studies of δ-opioid receptors utilized the enkephalin analog DADLE ([D-Ala^2,D-Leu^5]enkephalin) (**21**), but several more δ-selective enkephalin analogs, including DSLET and DTLET ([D-Ser^2,Leu^5,Thr^6] (**22**) and [D-Thr^2,Leu^5, Thr^6]enkephalin (**23**)) and particularly the cyclic analog DPDPE (*cyclo*[D-Pen^2,D-Pen^5]enkephalin) (**24**) (Fig. 5), are now used routinely to characterize δ-receptors; the naturally occurring deltorphins (e.g., **25**), peptides isolated from frog skin [93], exhibit marked δ-receptor selectivity. The nonpeptide agonists BW373U86 (**26**), SNC80 (**27**) [94], and TAN67 (**28**) [95,96] also exhibit very high

Agonist

Tyr-D-Ala-Gly-MeNPhe-NHCH$_2$OH

15 DAMGO

Antagonists

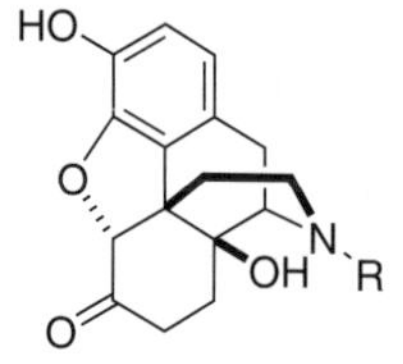

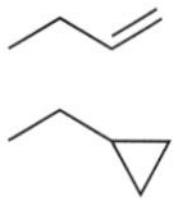

R= /\/ **16 Naloxone**

R= /\<| **17 Naltrexone**

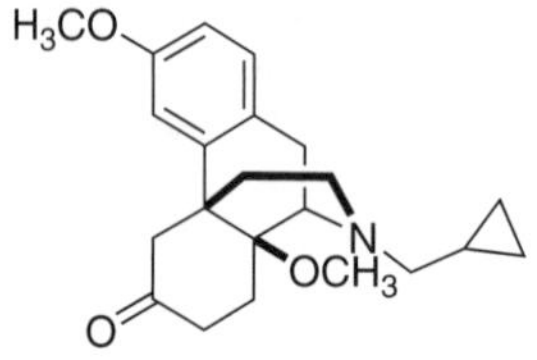

18 Cyprodime

D-Phe-Cys-Tyr-D-Trp-X-Thr-Pen-ThrNH$_2$

(Pen=penicillamine)

X=Orn **19 CTOP**

X=Arg **20 CTAP**

Figure 4. Ligands used to characterize μ-opioid receptors and subtypes.

selective for δ-receptors and are used frequently to study these receptors. δ-Receptor antagonists include the peptides ICI 174,864 (**29**) [97], TIPP (Tyr-Tic-Phe-PheOH, **30**) [98] and TIPP[Ψ] (Tyr-TicΨ[CH$_2$NH]Phe-PheOH) (**31**) [99], and the nonpeptide naltrindole (**32**) [100]. Recently, however, TIPP has been reported to exhibit agonist activity in adenylyl cyclase assays [101]. Early studies of κ-receptors used benzomorphans such as ethylketocyclazocine (EKC) (**33**), a close analog of k-tocyclazocine, and bremazocine (**34**) (Fig. 6), but the selectivity of these ligands for κ-receptors is very low (see Table 8). κ-Selective agonists such as U50,488 (**35**) [102], U69,593 (**36**) [103], and CI-977 (**37**) [104,105] and the κ-selective antagonist norBNI (**39**) [106,107] are now available. Recently, Portoghese and coworkers described GNTI (**41**) [108–110], which is a more κ-selective antagonist than norBNI, and which should be a useful pharmacological tool to study κ-receptors. The structure–activity relationships of these compounds are discussed in more detail in the sections below.

Radioligand Binding Assays The demonstration of stereospecific binding of tritiated ligands to opioid receptors in the early 1970s [111–113] paved the way for the subsequent development of radioligand binding assays for each of the opioid receptor types, which have been instrumental in the identification of selective opioids. Early studies often used nonselective ligands such as [^{3}H]etorphine ([^{3}H]-**42**) or [^{3}H]naloxone that labeled all types of opioid receptors, but today tritiated ligands selective for each receptor type are available (see Table 7 and Ref. [52,81]). [^{3}H]DAMGO ([^{3}H]-**15**) is most often used as the radioligand binding assays for μ-opioid receptors. [^{3}H]DADLE ([^{3}H]-**21**) was often used in early binding studies examining δ-opioid receptors, but studies now commonly use the more δ-selective ligand [^{3}H]DPDPE ([^{3}H]-**24**). Early binding studies of κ-opioid receptors were hampered by the low κ-selectivity of available tritiated ligands and the low levels of κ-receptors in rat brain (see below). One solution was to examine the binding of ligands such as [^{3}H]EKC ([^{3}H]-**33**) or [^{3}H]bremazocine ([^{3}H]-**34**) in the presence of unlabeled μ- and δ-selective ligands such as DAMGO and DPDPE in order to block μ- and δ-receptors in the tissue preparation. Since the highly κ-selective ligand U69,593 ([^{3}H]-**36**) is now available in tritiated form [103], it is routinely used in κ receptor binding assays; the κ-selective ligand CI-977 (**37**) [104,105] is also available in tritiated form. The affinity of a variety of opioids in radioligand binding assays for μ-, δ-, and κ-receptors are given in Table 8.

Prior to cloning of the opioid receptors affinity for these receptors was most commonly

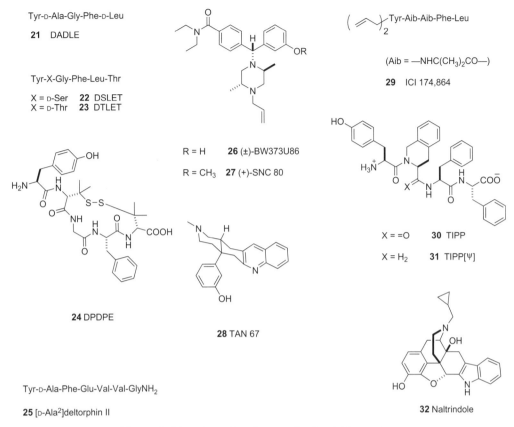

Figure 5. Ligands used to characterize δ-opioid receptors.

determined using homogenates or membrane fractions from rat, guinea pig, or mouse brain, which contain all three types of opioid receptors. The relative amounts of different opioid receptor types vary between species, however, particularly for κ-receptors. In rat brain κ-opioid receptors comprise only about 10–15% of the total number of opioid receptor sites [114], while in species such as guinea pig they represent approximately 30% of the total opioid receptor population [115]. Over 80% of the opioid receptors in the guinea pig cerebellum are κ-receptors [116], so this tissue was frequently used in κ-receptor binding assays. In contrast, the rabbit cerebellum contains predominantly (>70%) μ-opioid receptors [117]. Species differences also exist for δ-receptors, with mouse exhibiting substantially higher δ-receptor density than rat [118]. NG108-15 cells contain only δ-receptors and therefore have been used in radioligand binding assays for these receptors [119]. With the cloning of opioid receptors (Section 3.2.4), binding assays are now typically performed using these cells that express only a single receptor type.

In Vitro **Assays for Efficacy** Until the cloning of the opioid receptors, isolated tissue preparations, particularly smooth muscle preparations were used extensively to characterize opioids (see Refs [121,122] for reviews). The electrically stimulated GPI myenteric plexus-longitudinal muscle and MVD preparations have been the tissues used most extensively. The predominant effect of opioids is to inhibit smooth muscle contraction, generally by inhibiting the release of neurotransmitters; in the GPI and MVD acetylcholine and norepinephrine, respectively, are the neurotransmitters affected (see Ref. [122]).

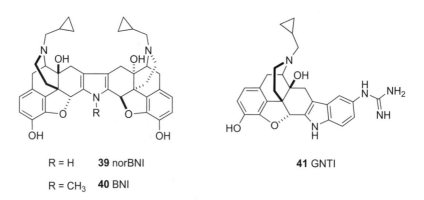

Figure 6. Ligands used to characterize κ-opioid receptors and subtypes.

In contrast to radioligand binding assays, which measure only opioid affinity, these bioassays provide information on the intrinsic activity of the compounds tested. The activity of compounds in these assays can be complex, however, because both tissues contain more than one opioid receptor type. The GPI contains both μ- and κ-receptors with little if any functional δ-receptors, while the MVD contains all three opioid receptor types. The activity of a variety of opioids in these tissues is given in Table 8. GPI and MVD preparations enriched in a single receptor population have been prepared using the affinity labels β-chlornaltrexamine (β-CNA) and β-funaltrexamine (β-FNA) [123,124] (see Section 5.11). Vas deferens from other species has been used to characterize opioids and appear

Table 6. Ligands Commonly Used to Study Different Opioid Receptor Types

Receptor Type	Agonists	Antagonists
μ	Morphine (**1**)	Naloxone (**16**)
	DAMGO (**15**)	Naltrexone (**17**)
		Cyprodime (**18**)
		CTOP (**19**),
		CTAP (**20**)
δ	DPDPE (**24**)	TIPP (**30**)
	DSLET, DTLET (**22, 23**)	TIPP[Ψ] (**31**)
	Deltorphins (**25**)	ICI 174864 (**29**)
	DADLE (**21**)	Naltrindole (**32**)
	BW373U86 (**26**)	
	SNC80 (**27**)	
	TAN67 (**28**)	
κ	U50,488 (**35**), U69,593 (**36**)	norBNI (**39**)
	CI-977 (**37**)	GNTI (**41**)
	EKC (**33**)	
	Bremazocine (**34**)	
	Dynorphin A (**38**) and derivatives	

Compounds with limited selectivity for a given receptor type are listed in parentheses (see Table 8 for affinities and opioid activities of these agents). See Figs. 1 and 4–6 for structures of these agents.

to contain predominantly a single receptor population. The rabbit vas deferens appears to contain only κ-receptors [125], while the hamster vas deferens contains only δ-receptors [126]. In the rat vas deferens β-endorphin is a potent inhibitor, which led to the proposal that this tissue contains an additional type of opioid receptor, the ε-receptor [127–129].

With the cloning of the opioid receptors *in vitro* functional assays using these receptors have been developed. These assays have utilized measurement of effects on signal transduction systems to determine the efficacy and potency of compounds being studied. Because opioid receptors are G-protein-coupled receptors, measurement of the stimulation of binding of the radiolabeled nonhydrolyzable analog of GTP [^{35}S]GTPγS [130,131] following interaction of a compound with the receptor can be used to determine the efficacy and potency. Inhibition of adenylyl cyclase has also been utilized as a functional assay to evaluate the efficacy and potency of opioid ligands at opioid receptors [132,133].

In Vivo **Evaluation of Opioids** A variety of antinociceptive assays utilizing different noxious stimuli and different animal species have been used to examine the activity of potential analgesics. Animal models for pain include models of acute pain (e.g., hot plate, tail flick, paw pressure, and writhing assays), inflammatory pain (e.g., the formalin test), chronic pain (e.g., adjuvant-induced arthritis), and neuropathic pain (e.g., nerve ligation) (see Table 2.1 of Ref. [134]). The different types of noxious stimuli commonly used in these antinociceptive assays include heat (e.g., hot plate and tail flick assays), pressure (e.g., tail pinch and paw pressure assays), chemical (writhing or abdominal constriction assay and the formalin test), and electrical (tail shock vocalization) stimuli. Among the most commonly used assays are the hot plate, tail flick, and writhing assays. In the hot plate test, the latency to various behavioral responses (e.g., forepaw

Table 7. Commercially Available Tritiated Opioid Receptor Ligands

Receptor Type			Nonselective
μ	δ	κ	(or Low Selectivity)
Morphine (**1**)	DPDPE (**24**)	U69,593 (**36**)	EKC (**33**)
DAMGO (**15**)	[D-Ala^2]deltorphin II (**25**)	CI-977 (**37**)	Diprenorphine (**43**)[a]
	[D-Ala^2,Ile^{5,6}]deltorphin II	Bremazocine (**34**)	Naloxone (**16**)[a]
	Naltrindole (**32**)[a]		
	DADLE (**21**)		

Compounds with limited selectivity for a given receptor type are listed in parentheses. Ligands commercially available from PerkinElmer Life Sciences, Amersham Biosciences or Tocris. Iodinated derivatives of β-endorphin and Met(O) enkephalin are also available.
[a] Antagonists.

Table 8. Opioid Receptor Affinities and Opioid Activity in the GPI and MVD of a Variety of Opioid Ligands Commonly Used to Study Opioid Receptors

Agonist	K_i (nM)			K_i Ratio	IC_{50} (nM)	
	μ	δ	κ	μ/δ/κ	GPI	MVD
Nonselective agonists:						
Etorphine (**42**)	1.0	0.56	0.23	4.3/2.4/1	0.08	0.40
(−)-EKC (**33**)	1.0	5.5	0.52	1.9/10.6/1	0.18^a	4.4^a
Agonists preferentially interacting with μ-receptors:						
Morphine (**1**)	1.8	90	317	1/50/175	28	478
Meperidine (**7**)	385	4,350	5,140	1/11/13	1,109	16,000
Fentanyl (**8**)	7.0	150	470	1/21/67	0.92	26
Methadone (**9**)	4.5	15	1,630	1/3.3/360	22	523
DAMGO (**15**)	1.9	345	6,090	1/180/3,200	4.5	33
Agonists preferentially interacting with δ-receptors:						
BW 373U86 (**26**)^b	46^c	0.92^c	—	50/1	143	0.2
SNC80 (**27**)^b	$2,470^c$	1.0^c	—	2,300/1	5,460	2.7
TAN67 (**28**)^d	2,320	1.1	1,790	2,070/1/1,600	26,000	6.6
DPDPE (**24**)	710	2.7	>15,000	260/1/>5,500	2,350	2.8
DSLET (**22**)	39	1.8	6,040	22/1/3,350	110	0.59
DADLE (**21**)	14	2.1	16,000	6.7/1/7,60	8.9	0.73
[D-Ala²]deltorphin II (**25**)	2,450	0.71	>10,000	3,450/1/>14,000	>3,000	0.32
Agonists preferentially interacting with κ-receptors:						
(−)-Bremazocine (**34**)	0.62	0.78	0.075	8.3/10.4/1	0.13^a	1.98^a
U50,488 (**35**)	435	9,200	0.69	630/13,300/1	16	11
U69,593 (**36**)	2,350	19,700	1.4	1,680/14,000/1	2.0	ND
CI 977 (**37**)	100	1,040	0.11	910/9,450/1	0.09	ND
[D-Pro¹⁰]Dyn A-(1–11)	0.56	2.3	0.029	19.3/79/1	3.3	ND

Antagonist	K_i (nM)			K_i Ratio	K_e (nM)		
	μ	δ	κ	μ/δ/κ	μ (GPI)	κ (GPI)	δ (MVD)
Naloxone (**16**)	1.8	23	4.8	1/13/2.7	1.9	18	49
Naltrexone (**17**)	1.1	6.6	8.5	1/6.0/7.7	0.36	4.4	12
Diprenorphine (**43**)	0.24	1.0	0.14	1.7/7.1/1	0.31	$0.5^{e,f}$	3.6
Antagonists referentially interacting with μ-receptors:							
Cyprodime (**18**)	9.4	356	176	1/38/19	31	1,160	6,110
CTOP (**19**)	1.7	>1,000	>1,000	1/>590/>590	16	444	NA (1 μM)
Antagonists referentially interacting with δ-receptors:							
Naltrindole (**32**)	11	0.12	18	92/1/150	22	100	0.27
ICI 174864 (**29**)	29,600	190	65,400	155/1/345	$>5,000^e$	$>5,000^e$	17
TIPP (**30**)^g	1,720	1.2	ND	1,430/1	ND	ND	3.0–5.9
TIPP[Ψ](**31**)^h	3,230	0.31	ND	10,500/1	ND	ND	2.1–2.9
Antagonists referentially interacting with κ-receptors:							
NorBNI (**39**)	14	10	0.34	41/29/1	25	0.05	16
GNTI (**41**)^i	22	46	0.18	125/257/1	30	0.20	NA (100 nM)

Data from Ref. [120] except where otherwise indicated.
^{a} Antagonist at μ- and δ-receptors in the rat and hamster vas deferens, respectively.
^{b} From Ref. [94].
^{c} IC_{50}
^{d} From Ref. [96].
^{e} Determined in the MVD.
^{f} Agonist in the GPI (IC_{50} = 1.4 nM).
^{g} From Ref. [98].
^{h} From Ref. [99].
^{i} From Ref. [109].
ND: not determined; NA: not active (at indicated dose).

or hindpaw licking) is measured when the animal is placed on a hot plate, typically set at 55°C, while the tail flick measures the time for the animal to flick its tail away from radiant heat focused on the tail. In the writhing or abdominal constriction assay the animal is injected intraperitoneally (i.p.) with an irritant (typically phenylquinone or acetic acid) and the dose of a test compound required to abolish the writhing syndrome determined. Based on comparison of the potencies of standard opioids in several pain models, it was concluded that the mouse abdominal constriction assay (using 0.4% acetic acid) was the most sensitive to opioids, while the mouse hot plate test (at 55°C) was the least sensitive; the rat tail flick test was intermediate (see Ref. [134]). For a detailed discussion of these antinociceptive assays, readers are referred to excellent recent reviews [134,135] that describe these procedures and discuss methodological issues with their execution and recent developments.

Activation of all three types, μ, δ, and κ, of opioid receptors can produce antinociception, but there are significant differences in the effects of activating different receptor types depending upon the noxious stimulus used and the animal species (see Refs [90,136] for excellent reviews; see particularly Table 1 of Ref. [90] for a summary of supraspinal opioid receptor involvement in antinociceptive assays). While μ-agonists are active against all types of noxious stimuli, the activity of κ-agonists depends on the type of stimulus. κ-Agonists are active against chemical and pressure stimuli, but they are inactive against electrical stimulation and their efficacy against thermal stimuli is dependent on the intensity [137]. δ-Agonists may be active against all four types of stimuli in mouse, depending on the route of administration (see Ref. [90]), but there are species differences. It has been reported that while δ-agonists are effective in both the tail flick and the hot plate assays in mice, in rats they are active in the hot plate, but not tail flick assays [138].

3.2.4. Opioid Receptor Structure and Molecular Biology
The first successful cloning of opioid receptors utilized a cDNA library from NG108-15 cells to clone δ-receptors [139,140]. Following expression of the cDNA library in COS cells, the cells were screened for [^{3}H] δ-ligand binding and the clone isolated. Yasuda et al. subsequently reported cloning of both mouse brain κ- and δ-receptors while trying to isolate cDNAs encoding somatostatin receptors [141], and Chen et al. reported cloning of rat μ-receptor [142]. Since then, the cloning of all three opioid receptor types from several different species, including human, have been reported (see Refs [63,64,81,143–146] for reviews). The human μ-, δ-, and κ-receptors exhibit 91–95% sequence homology to the corresponding type of receptor from rat and mouse [63]. Studies with antisense oligodeoxynucleotides to each type of cloned opioid receptor and in knockout mice lacking individual opioid receptors have confirmed the involvement of the cloned receptors in analgesia mediated by μ-, δ-, and κ-receptors (see Refs [143,147,148] for reviews).

The opioid receptors belong to the family of G-protein-coupled receptors. Based on the model for this family of receptors, the receptors contain extracellular regions including the N-terminus and extracellular loops, seven putative transmembrane (TM) regions, and intracellular regions including the C-terminus and intracellular loops (Fig. 7). Comparison of the sequences [149] indicates the highest sequence homology between the κ-, the μ-, and the δ-receptors in TM2, TM3, and TM7 (Fig. 8). TM2 and TM3 each contain a conserved Asp residue; the conserved Asp in TM3 is thought to interact with protonated amine groups on opioid ligands (see below). There are also high similarities in the intracellular loops; the third intracellular loop is thought to be involved in interactions with G-proteins. The second and third extracellular loop, TM1, and TMs 4–6 are less conserved. The largest structural diversity occurs in the extracellular N-terminus. Potential sites for possible posttranslational modification have been identified on the receptors. Two potential glycosylation sites are located in the N-terminal sequence. Two possible sites for protein kinase C phosphorylation are located in the C-terminus plus a third site is found in the third intracellular loop; a possible palmitoylation site is also located in the C-terminus. Conserved Cys residues in the first and second

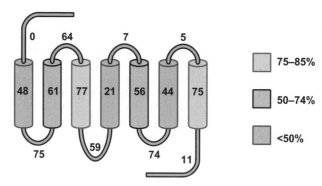

Figure 7. Schematic diagram of the protein structure of the three opioid receptors. Rectangles indicate transmembrane helices and numbers indicate the percent identical residues among the three opioid receptors in that segment. From Ref. [87]. (This figure is available in full color at http://mrw.interscience.wiley.com/emrw/9780471266945/home.)

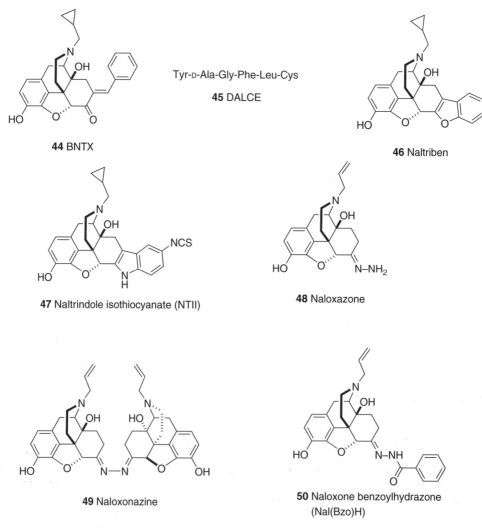

Figure 8. Ligands used to characterize opioid receptor subtypes.

extracellular loops may be involved in a disulfide linkage.

Mutagenesis Studies of Opioid Receptors A variety of chimeric receptors between different opioid receptor types have been prepared in attempts to identify regions of the receptors that are involved in ligand recognition and receptor type selectivity (see Refs [150,151] for recent reviews). What has emerged from these studies is a complex picture of the possible roles of different regions of the receptors. It has been suggested that the extracellular loops may play an exclusionary role rather than a direct role in ligand binding, acting as filters to prevent ligands selective for other receptor types from binding [152]. However, which loop serves this role can be different for different receptors excluding the same ligand. Thus, the study of μ/δ-chimeric receptors suggested that the first extracellular loop (EL1) causes the loss of high-affinity binding of the μ-opioid agonist DAMGO ([D-Ala^2,N-MePhe, glycol]enkephalin) to δ-receptors [153,154]; subsequent site-directed mutagenesis studies suggested that Lys108 in EL1 of the δ-receptor was the residue responsible [155]. In contrast, results for μ/κ-chimeric receptors suggest that EL3 may be responsible for the low binding affinity of DAMGO to κ-receptors [156,157]; site-directed mutagenesis identified Glu297, Ser310, Tyr312, and Tyr313 as the residues in the κ-receptor responsible for discriminating against DAMGO [158]. The involvement of different loops also appears to be ligand dependent (see Ref. [151]). The major determinant for binding of the μ-selective alkaloids appears to involve the region of TM5 through TM7 [154], and the binding of the δ-selective peptide DPDPE (**22a**) to δ-receptors also appears to involve this region near the C-terminus of the receptor [154]. EL3 also appears to be important for the selectivity of several other δ-selective compounds, both peptide and nonpeptide ligands for δ- over μ-receptors [159,160]; point mutations indicated that Trp284, Val296, and Val297 were the crucial residues in the EL3 of the δ-receptor for δ-selectivity [160]. Because of the number of acidic residues in EL2, it has been suggested that this region contributes directly to the affinity of κ-receptors for the opioid peptide dynorphin A through ionic interactions with basic residues in the C-terminus of this peptide. Examination of both κ/μ [161,162]- and κ/δ [163]-chimeric receptors supported the importance of EL2 for the high affinity of dynorphin A for κ-receptors. However, in a recent study in which three or four of the seven acidic residues in EL2 of κ-receptors were mutated to the corresponding amides, the mutant receptors still bound dynorphin A with high affinity, raising questions about the role of these acidic residues in Dyn A affinity for κ-receptors [164]. The study of chimeric receptors also suggests that different domains of a receptor may be involved in interactions with different ligands. Thus, in contrast to dynorphin A, the κ-selective nonpeptide agonists U50,488 and U69,593 appear to require the whole κ-receptor except for EL2 [161,162]. U69593 also appears to bind to κ-receptors differently than does naloxone, suggesting that agonists and antagonists may bind differently to this receptor [165].

Site-directed mutagenesis has been used to examine the roles of individual residues (see Ref. [150,151]). Thus, the conserved Asp in both TM2 and TM3 has been substituted with neutral residues [166–168] to explore their roles in opioid ligand binding. Results for the mutation of the Asp in TM2 suggested that agonists and antagonists may bind differently to their receptors. Substitution of the Asp in TM3 dramatically reduced the affinity of some, but not all, opioids [168], raising the possibility that this is not a universal counterion for opioid ligands. The His in TM6 has also been implicated in the binding of ligands to the μ-receptor [167], and this residue is conserved across the three receptor types. Mutation of Glu297 at the top of TM6 in κ-receptors was used to demonstrate the importance of this residue for the κ-receptor affinity and selectivity of the nonpeptide κ-selective antagonist norbinaltorphimine (nor-BNI) [169]; conversely the single mutation of the corresponding residue Lys303 in the μ-receptor to Glu imparts high affinity for norBNI [108]. Indeed, the close similarity between the opioid and the ORL1 receptors (see Section 7.2) has been clearly demonstrated by the ability to convert the ORL1 receptor to an opioid receptor with the mutation of only four residues in TM6 and TM7 [170]. Site-directed mutagenesis has also been used to examine possible points of attachment of the affinity label β-funaltrexamine (β-FNA) [171] (see Section 5.11.1).

Thus chimeric receptors and site-directed mutagenesis have provided tremendous insight into receptor structure and receptor–ligand interactions. However, the data from these studies must be interpreted with caution, particularly in cases involving loss of function [151]. Changes in the primary sequence of a receptor could have significant effects on the protein secondary or tertiary structure that in turn could affect the affinities of various ligands. Thus, whether observed changes in ligand affinity are due to the direct effect of the mutations or indirect effects due to altering the tertiary structure of the protein is not known.

Computational Models of Opioid Receptors
Because G-protein-coupled receptors are transmembrane proteins, until quite recently [172] structural information on these receptors was limited to low resolution (6–9 Å) structures from electron microscopy studies of rhodopsins. With the deduced amino acid sequences of the opioid receptors available, computational models of the three-dimensional structures of opioid receptors were developed based initially on these low resolution structures [160,173–182]. The rigid structures of many nonpeptide opioid ligands decreases the possible degrees of conformational freedom, decreasing the complexity of docking these compounds to opioid receptor models, and therefore several of the reports have described possible binding modes on these compounds. These include tifluadom-like benzodiazepine ligands at κ-opioid receptors [173], naltrexone-derived antagonists at κ- and δ-receptors [174], morphinans including morphine to the μ-receptor [176], arylacetamides to the κ-receptor [176,177,180], fentanyl analogs at the μ-receptor [176,181], and piperazine and piperidine derivatives at δ-receptors [176,182]. Modeling of the flexible opioid and opioid-like peptides bound to their receptors is more challenging, and therefore reports describing their receptor binding modes has been more limited. Models have been developed for the binding of the conformationally constrained δ-selective peptides JOM-13 and DPDPE to δ-opioid receptors [176], and the extracellular loops have been incorporated into computational models of κ-receptors [175] and ORL1 receptors [183] so that possible binding modes of Dyn A-(1–10) and OFQ-(1–13) to these receptors could be examined. Reports of computational models of opioid receptors derived from the X-ray structure of rhodopsin (2.8 Å resolution) continue to appear in the literature (see Refs [150,184]). Furthermore, with the recent report of the X-ray crystal structures of the β-adrenergic receptor, more reports are expected to appear.

Receptor Subtypes, Splice Variants, and Receptor Dimerization Prior to cloning of the opioid receptors subtypes of each of the three opioid receptors were proposed based on evidence from pharmacological assays (see Refs [52,90] for excellent reviews), but opioid receptors identified by cloning have consistently represented only a single subtype of each receptor type. Currently, there is still considerable debate on the existence and nature of some of these receptor subtypes.

There is considerable evidence from both functional assays and binding studies supporting the existence of two types of δ-opioid receptors (see Refs [52,185,186]). A key factor in the characterization of these δ-receptor subtypes was the availability of ligands selective for each of the proposed subtypes (Fig. 8). The $δ_1$-subtype was characterized by its preferential stimulation by the enkephalin analog DPDPE (**24**) and DADLE (**21**) and selective antagonism by the naltrindole analog BNTX (7-benzilidenenaltrexone) (**44**) [187] and the affinity label (see below) DALCE (**45**) [188]). The $δ_2$-subtype was distinguished by its stimulation by [D-Ala^2]deltorphin II (**25**) and DSLET (**22**) and antagonism by the benzofuran naltrindole analog naltriben (**46**) [189] and the irreversible antagonist naltrindole isothiocyanate (NTII) (**47**) [190]). δ-Receptor subtypes were also proposed by Rothman and coworkers that were differentiated by whether they were associated with a μ–δ-receptor complex ($δ_{cx}$) or are not associated with such a complex ($δ_{ncx}$) (see Ref. [191] for a review). These researchers used the irreversible ligands (see Section 5.11) FIT [192] or (+)-transSUPERFIT [193] to selectively acylate $δ_{ncx}$-receptors [194] and used the μ-receptor affinity label BIT [192] to deplete the membranes of $δ_{cx}$-receptors [195]. Based on binding and pharmacological studies Rothman and coworkers proposed that the δ_{ncx}- and the δ_{cx}-receptors corresponded to the δ_1- and δ_2-receptor, respectively [196,197]. Pharmacological characterization of the cloned δ-opioid receptor is consistent with classification as the δ_2-subtype [198], and antisense oligonucleotides differentially effect the two subtypes [199]. In δ-receptor knockout mice, binding of both ^{3}H-DPDPE and ^{3}H-[D-Ala^2] deltorphin II is absent [200], indicating that the proposed subtypes are not due to different gene products; selective δ-agonists still retain analgesic potency, suggesting the existence of a second δ-like analgesic system [200].

Multiple types of μ-opioid receptors have been proposed by Pasternak and coworkers. Two subtypes of μ-receptors, the μ_1-site, which was suggested to be a common high-affinity site for both nonpeptide opioids and opioid peptides, and the μ_2-site, which was proposed to correspond to the "traditional" μ-binding site [201], were initially postulated by these researchers (see Refs [52,202]). The hydrazone derivative of naloxone, naloxazone (**48**) [203], and its dimeric azine derivative naloxonazine (**49**) [204] have been used to study the putative μ_1-receptors (see Ref. [52]), which is thought to represent ~20% of the specific binding to rat brain membranes [202]. Following the cloning of the opioid receptors, Pasternak and coworkers have characterized multiple splice variants of the μ-receptor (see Refs [205,206] for reviews).

Three or more subtypes of κ-receptors have been postulated. In the early 1980s, several groups reported differences between the binding of [^{3}H]ethylketocyclazocine ([^{3}H]EKC) and [^{3}H]diprenorphine or [^{3}H]etorphine (after blockade of μ- and δ-receptors), suggesting that there might be κ-receptor subtypes; there was considerable debate, however, concerning the nature of these different binding sites (see Refs [52,207]). Subsequently, the κ-selective arylbenzacetamides U50,488 and U69,593 helped to clarify the definition of different κ-receptor subtypes, and two populations of binding sites, κ_1 versus κ_2, were differentiated based on their sensitivity and insensitivity, respectively, to these arylbenzacetamides [208]. The cloned κ-receptors appear to be the κ_1-subtype based on pharmacological characterization [198] and the analgesic activity of the U50,488 is abolished in κ-receptor knockout mice [209] (other effects of U50,488, hypolocomotion and dysphoric are decreased in these animals). Mice deficient in each of the opioid receptors and animals lacking all three opioid receptors have been examined for residual κ_2-receptor binding sites [210]; all of the residual non-κ_1-receptor labeling could be accounted for by μ- and δ-receptors, and the triple receptor deficient exhibited no residual binding of [^{3}H]bremazocine, indicating that no other gene product was involved in binding this opioid ligand. A third κ-receptor subtype, κ_3, has also been proposed based on studies using naloxone benzoylhydrazone, Nal(Bzo)H (**50**) [211].

Recently, an alternative explanation for opioid receptor subtypes, receptor dimerization, has appeared in the literature that could explain why different gene products have not been identified for the proposed receptor subtypes in spite of the evidence from pharmacological assays for their existence.

3.3. Physiology of Non-μ-Opioid Receptors

Because of the side effects associated with analgesics that interact with μ-opioid receptors, there has been considerable interest in developing opioid ligands that interact with other opioid receptors. In addition to analgesic activity, there are a number of other potential therapeutic applications of ligands, both agonist and antagonists, for non-μ-opioid receptors.

3.3.1. δ-Receptors
There has been considerable interest in δ-opioid agonists because they exhibit antinociceptive effects without the side effects associated with μ-opioid receptor agonists. Antinociceptive activity was first demonstrated with δ-selective opioid peptides (see Ref. [212] for a review), and more recently with nonpeptidic δ-selective agonists (see Refs [213–216] for reviews). Of particular interest is the activity of δ-agonists in inflammatory and neuropathic pain [214]. δ-Opioid receptors also modulate μ-opioid receptors, and as discussed above, one classification of δ-opioid receptor subtypes was based on their association with μ-opioid receptors. There is

now considerable evidence that interaction between the two receptor types can alter the activity of μ-opioid agonists. δ-Agonists can potentiate the analgesic activity of μ-agonists [217,218], and δ-antagonists can decrease the development of tolerance and dependence to morphine [219,220]. While the δ-selective antagonist naltrindole can block the development of tolerance to the analgesic effects of morphine [219], it does not block the beneficial development of tolerance to the respiratory depressant effects of morphine [221]. δ-Ligands can also reverse the respiratory depression caused by μ-agonists, and interestingly δ-agonists as well antagonists were reported to have similar effects on respiration [222]. Thus, δ-receptor ligands, particularly δ-antagonists, could have important therapeutic applications to minimize the deleterious effects of morphine. These findings have prompted the search for compounds that exhibit both μ-agonist and δ-antagonist activities, and peptidic ligands with this mixed activity have been reported by Schiller to produce potent analgesic effects without physical dependence and less tolerance than morphine [223]. Nonpeptide ligands with mixed activity have also recently been reported [224].

δ-Receptors may also modulate responses to substances of abuse other than opioids, including cocaine, amphetamines, and alcohol [213,214,216]. Several δ-agonists have been reported to be at least partially discriminate for cocaine (see Ref. [216]). There are numerous reports that δ-receptor antagonists can attenuate a number of the effects of cocaine (see Refs [213,214,216]), and an antisense oligodeoxynucleotide to δ-opioid receptors blocks cocaine-induced place preference in mice [225]. Thus, δ-antagonists could potentially be used in the treatment of cocaine abuse. One study, however, found that the δ-receptor antagonist naltrindole can potentiate the lethal effects of cocaine; this was only found after intracisternal administration and not after i.v. administration, so it is not clear whether this would limit the application of δ-receptor antagonists in treatment of cocaine abuse [226]. Similar results have been found for δ-agonists and antagonists in methamphetamine-induced place preference

(see Ref. [216]); one important finding is that the δ-agonist DADLE can protect against the neuronal damage caused by methamphetamine (see Refs [227,228] and references cited therein). There is considerable evidence that endogenous opioids and opioid receptors play important roles in the abuse of alcohol [229,230], and the opioid receptor antagonist naltrexone has been used to treat alcoholism. The involvement of δ-opioid receptors in alcohol abuse, however, remains unclear. Selective δ-receptor antagonists have been examined for their ability to decrease alcohol consumption in animal models, but the results have been mixed, with some studies reporting decreased alcohol consumption while others observed no effects (see Refs [216,230] for reviews). One possible explanation for this discrepancy could be differences in the genetic backgrounds of the animals employed.

δ-Receptors are also involved in a number of other biological effects, including both centrally and peripherally mediated effects (see Refs [213,214,216] for reviews). δ-Agonists are effective in learned helplessness animal models, suggesting a potential therapeutic application in depression [231]. Convulsant effects have been observed with nonpeptide δ-agonists, particularly BW373U86 (see Section 5.10.1), however, which could prevent their clinical use unless these adverse effects can be separated from their desired action (e.g., antinociceptive activity, see Ref. [232]). Recently several reports have appeared, which indicate that convulsions are not characteristics of all δ-agonists. Opioids, including δ-receptor ligands, have immunomodulatory effects [233]. δ-Agonists have been reported to stimulate immune responses, while δ-antagonists can cause immunosuppression (see Refs [213,214]). The immunomodulatory effects are still present in δ-receptor knockout mice [234], however, indicating that these effects are not mediated by these receptors.

3.3.2. κ-Receptors Like μ- and δ-receptors, activation of κ-opioid receptors can produce antinociceptive effects. However, the effectiveness of κ-opioid agonists as antinociceptive agents varies in different types of pain (see Ref. [136]), and κ-agonists are less effective in thermal antinocicpetive assays involving more intense stimuli.

κ-Agonists can also produce antinociceptive effects in inflammatory pain through interaction with peripheral κ-opioid receptors. κ-Agonists have been shown to have antiarthritic activity [235,236], and to produce antinociception in capsaicin-induced thermal allodynia through interaction with peripheral κ-receptors [237,238]. Since many painful conditions are associated with inflammation, there is significant potential for the application of κ-agonists as peripheral analgesics. Peripheral κ-opioid receptors also appear to be involved in visceral pain and the activity of κ-agonists is enhanced in the presence of chronic inflammation of the viscera [239,240], suggesting potential therapeutic applications of peripherally selective agents without the side effects (dysphoria) associated with many centrally acting κ-agonists.

In many cases, activation of central κ-opioid receptors opposes the activity of μ-opioid receptors (see Ref. [241] for a recent review). Several studies have reported that κ-agonists at doses that do not affect baseline pain threshold can antagonize the analgesic activity of μ-opioid agonists (see Ref. [241]). The κ-opioid peptide dynorphin has also been reported to improve the memory impairment induced by the μ-agonist DAMGO [242]. κ-Agonists also reduce tolerance to morphine in a variety of antinociceptive tests, and dynorphin can inhibit morphine withdrawal symptoms in morphine-dependent animals (see Ref. [241] and references cited therein). κ-Agonists generally lack reinforcing effects, and can abolish the reinforcing effects of morphine (see Ref. [241]). This effect of κ-agonists on morphine reward may be due to the opposing modulation of the mesolimbic dopamine system by μ- and κ-opioid receptors, where μ-agonists increase dopamine levels while κ-agonists decrease dopamine levels. κ-Agonists also generally have opposite subjective effects from μ-opioid agonists (dysphoria for κ-agonists versus euphoria for μ agonists). Chronic administration of morphine and other drugs with positive reinforcing effects increase brain levels of the endogenous κ-opioid peptide dynorphin (see Ref. [243] and references cited therein). It has been proposed [243] that this produces an imbalance in abstinent μ-opioid-dependent individuals and dysphoric mood states, which can result in the desire to take μ-opioid agonists to normalize mood. Thus, based on this "κ-overdrive" model, a κ-receptor antagonist could be useful to normalize this imbalance and treat opioid addiction [243].

κ-Opioid agonist effects on the dopamine system also have potential utility in the treatment of cocaine abuse. Cocaine blocks the reuptake of dopamine, and considerable evidence suggest that cocaine's reinforcing effects are mediated by these increases in extracellular dopamine (see Refs [244,245] and references cited therein). Since κ-agonists can decrease dopamine levels, they can act as functional antagonists of cocaine [244,245]. Several κ-agonists have been shown to decrease cocaine self-administration [244–248] (but see Ref. [249]) and κ-agonists can also attenuate many of the behavioral effects of cocaine (see Ref. [244]).

Central κ-opioid receptors also mediate other effects that could be therapeutically beneficial. κ-Agonists exhibit neuroprotective effects in stroke [250]; these effects are observed in experimental models of stroke even several hours after occlusion (see Ref. [251] and references cited therein), which is particularly promising for the potential therapeutic use of κ-agonists in stroke. Dynorphin improves memory dysfunction in an animal model of amnesia (see Ref. [241]). κ-Opioid agonists have also been shown to cause down-regulation of HIV expression in human microglial cells [252] and CD4+ lymphocytes [253], which raises the intriguing possibility of using κ-ligands as adjunctive therapy in HIV infected individuals; whether or not this will have clinical applicability remains to be determined.

3.4. Endogenous Opioid Peptides

During the mid-1970s the search for endogenous ligands for opioid receptors led to the discovery of peptides with opiate-like activity. The first opioid peptides reported were the pentapeptides leucine and methionine enkephalin (**3** and **4**) [254], followed shortly thereafter by dynorphin A (**38**) [255,256] and β-endorphin (**51**) [257]. Since these peptides are structurally distinct from the alkaloid opiates, the term "opioid" was introduced to describe all compounds, both nonpeptide and peptide, with opiate-like activity. These mammalian opioid peptides share a common N-terminal tetrapeptide sequence, but differ in their C-terminal residues (Fig. 9). They also differ in the receptor types with which they preferentially interact. While the enkephalins exhibit some preference for interacting with δ-receptors (Table 9), the dynorphins preferentially interact with κ-receptors; β-endorphin possesses high affinity for both μ- and δ-receptors. This led Goldstein to apply the "message–address" concept (see also Section 6.3) to the opioid peptides [258]. In Goldstein's proposal the common N-terminal tetrapeptide sequence constituted the "message" sequence responsible for activating opioid receptors,

Proenkephalin peptides

Leu-Enkephalin (**4**)   Tyr-Gly-Gly-Phe-Leu

Met-Enkephalin (**5**)   Tyr-Gly-Gly-Phe-Met

Met-Enkephalin-Arg^6-Phe^7 Tyr-Gly-Gly-Phe-Met-Arg-Phe

Met-Enkephalin-Arg^6-Phe^7-Leu^8 Tyr-Gly-Gly-Phe-Met-Arg-Phe-Leu

Prodynorphin peptides

Dynorphin A (**38**)   Tyr-Gly-Gly-Phe-Leu-Arg-Arg-Ile-Arg-Pro-Lys-Leu-Lys-Trp-Asp-Asn-Gln

Dynorphin B   Tyr-Gly-Gly-Phe-Leu-Arg-Arg-Gln-Phe-Lys-Val-Val-Thr

α-Neoendorphin   Tyr-Gly-Gly-Phe-Leu-Arg-Lys-Tyr-Pro-Lys

β-Neoendorphin   Tyr-Gly-Gly-Phe-Leu-Arg-Lys-Tyr-Arg-Pro

Pro-opiomelanocortin peptides

β-Endorphin (**51**)   Tyr-Gly-Gly-Phe-Met-Thr-Ser-Glu-Lys-Ser-Gln-Thr-Pro-Leu-Val-Thr-Leu-
Phe-Lys-Asn-Ala-Ile-Ile-Lys-Asn-Ala-Tyr-Lys-Lys-Gly-Glu

Endomorphins

Endomorphin 1 (**52**)   Tyr-Pro-Trp-PheNH$_2$

Endomorphin 2 (**53**)   Tyr-Pro-Phe-PheNH$_2$

Orphanin FQ/nociceptin and related peptides

Orphanin FQ/
Nociceptin (**54**)   Phe-Gly-Gly-Phe-Thr-Gly-Ala-Arg-Lys-Ser-Ala-Arg-Lys-Leu-Ala-Asn-Gln

Orphanin FQ 2   Phe-Ser-Glu-Phe-Met-Arg-Gln-Tyr-Leu-Val-Leu-Ser-Met-Gln-Ser-Ser-Gln

Nocistatin
(human)   Met-Pro-Arg-Val-Arg-Ser-Leu-Phe-Gln-Glu-Gln-Glu-Glu-Pro-Glu-Pro-Gly-
Met-Glu-Glu-Ala-Gly-Glu-Met-Glu-Gln-Lys-Gln-Leu-Gln

Figure 9. Mammalian opioid peptides and orphanin FQ/nociceptin.

Table 9. Opioid Receptor Affinities and Opioid Activity in the GPI and MVD of Endogenous Opioid Peptides

	K_i (nM)			K_i Ratio	IC_{50} (nM)	
	μ	δ	κ	μ/δ/κ	GPI	MVD
Peptide						
Proenkephalin peptides:						
Leu-enkephalin (3)	19	1.2	8,210	16/1/6,840	36	1.7
Met-enkephalin (4)	9.5	0.91	4,440	10/1/4,880	6.7	1.5
Met-enkephalin-Arg^6-Phe^7	3.7	9.4	93	1/2.5/25	10	5.3
Met-enkephalin-Arg^6-Gly^7-Leu^8	6.6	4.8	79	1.4/1/16	35	2.9
Prodynorphin peptides:						
Dynorphin A (38)	0.73	2.4	0.12	6.1/20/1	0.29	0.91
Dynorphin A-(1–8)	3.8	5.0	1.3	2.9/17/1	4.9	9.2
Dynorphin B	0.68	2.9	0.12	5.7/24/1	0.25	2.1
β-Neoendorphin	6.9	2.1	1.2	5.7/1.8/1	3.3	9.9
α-Neoendorphin	1.2	0.57	0.20	6.0/2.8/1	3.0	7.7
POMC:						
β$_h$-Endorphin (51)	2.1	2.4	96	1/1.1/46	62	40
Endomorphins^b:						
Endomorphin-1 (52)	0.36	1,506	5,428	1/4,183/15,077	3.6	—
Endomorphin-2 (53)	0.69	9,233	5,240	1/13,381/7,594	4	—

Data from Ref. [120] except where otherwise indicated. See Section 7.2 for data on orphanin FQ/nociceptin.
^{b} Data from Ref. [259].

while the unique C-terminal sequences functioned as "address" components to direct the peptides to particular opioid receptors.

Recently a new class of opioid peptides, the endomorphins (52 and 53) (Fig. 9), were discovered [259], which does not share the classical "message" sequence with other mammalian opioid peptides. In contrast to other mammalian opioid peptides, the endomorphins show high selectivity for their preferential receptor, the μ-receptor (Table 9). Since their discovery the pharmacology of these new mammalian opioid peptides have been studied extensively (see Ref. [260] for a detailed review).

Following the identification of the ORL1 receptor, two groups isolated a 17-residue peptide (Fig. 9) as the endogenous ligand for this receptor [79,80]. This peptide was referred to as orphanin FQ by one group, since it was the ligand for the orphan receptor (FQ are the N- and C-terminal residues, respectively, of the peptide) [79] and named nociceptin by the other group, since in the initial studies this peptide was reported to produce hyperalgesia [80]. While the N-terminal tetrapeptide sequence of orphanin FQ/nociceptin (OFQ/N) is similar to the classical opioid peptides, the presence of Phe rather than Tyr in position 1 is an important factor in the peptide's high selectivity for ORL1 receptors over opioid receptors (see Section 7.2).

The classical mammalian opioid peptides are derived from three distinct precursor proteins (Fig. 10) (see Refs [58,261] for reviews). Processing of the precursor proteins occurs at pairs of basic residues. β-Endorphin is derived from proopiomelanocortin (POMC), along with ACTH, α-MSH, and β-lipotropin (see Ref. [262] for a review). The enkephalins are derived from proenkephalin A (see Ref. [263] for a review). This protein contains four copies of Met-enkephalin flanked by pairs of basic residues along with one copy of Leu-enkephalin. In addition, extended Met-enkephalin derivatives Met-enkephalin-Arg^6-Phe^7 and Met-enkephalin-Arg^6-Gly^7-Leu^8 and longer peptides (e.g., peptides E and F and BAM-20)

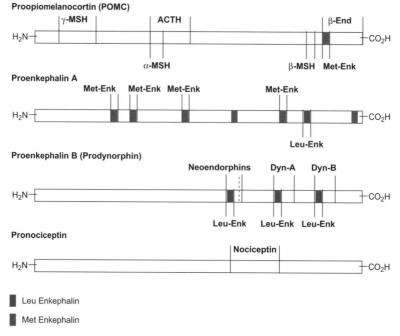

Figure 10. Peptide precursors and processing products. Taken from Ref. [87]. (This figure is available in full color at http://mrw.interscience.wiley.com/emrw/9780471266945/home.)

are obtained from proenkephalin A. The dynorphins, which contain a number of basic residues in the C-terminus, are derived from prodynorphin (also called proenkephalin B, see Ref. [264] for a review). In addition to dynorphins A and B, longer dynorphins (e.g., dynorphin 32, which contains both dynorphin A and B) and α- and β-neoendorphins are obtained from prodynorphin B. The precursor protein of the endomorphins has yet to be identified. The OFQ/N precursor protein prepronociceptin [80] has been characterized and also contains additional biologically active peptides related to OFQ/N, orphanin/nociceptin II [265] and nocistatin [266] (Fig. 10).

4. HISTORY: IDENTIFICATION OF MORPHINE AND EARLY ANALOGS

The effects of opium have been known for thousands of years, with written references dating as far back as the third century BC. Morphine was first isolated from opium in 1805 by Sertürner, who named it after Morpheus, the Greek god of sleep and dreams. The 3-methyl ether codeine (2) (Fig. 1) was subsequently isolated from the poppy plant *P. somniferum* in 1832. By the mid-1800s, the purified alkaloids morphine and codeine began to be used in place of crude opium preparations for medicinal purposes [1]. Because of its complex structure, it took more than a century following its isolation before the correct structure of morphine was proposed by Gulland and Robinson in 1925 [267]. It was not until the early 1950s that the proposed structure was confirmed by total synthesis by Gates and Tschudi [268,269]. The relative stereochemistry of the five chiral centers was determined by chemical synthesis [270] and X-ray crystallography in the early to mid-1950s [271]. The absolute configuration was subsequently proved by a combination of techniques [272–274].

The earliest model for how morphine and other analgesics interact with opioid receptors was proposed by Beckett and Casy (Fig. 11) [66]. Based upon the SAR of analgesics available, this model consisted of three sites: an anionic binding site that interacted with the protonated amine on the ligand, a

STRUCTURE–ACTIVITY RELATIONSHIPS OF NONPEPTIDE OPIOID ANALGESICS

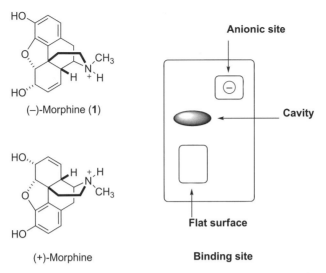

Figure 11. Beckett and Casy model for opioid receptor binding. Adapted from Ref. [66].

flat surface that interacted with the aromatic ring of the analgesic, and a cavity to accommodate the piperidine ring of the rigid opioid alkaloids. This model explained the stereospecificity of opioid receptors; while the natural (−)-isomer of morphine fits all three sites, in the unnatural (+)-isomer the projecting ring would not fit properly into the cavity (Fig. 11).

Synthetic efforts to modify the structure of morphine began with the synthesis of 3,6-diacetylmorphine (heroin) (55) (Fig. 12) in the late 1800s [275]. The first narcotic antagonist N-allylnorcodeine (57) was prepared in 1915 [276]. The subsequent demonstration of the antagonist activity of N-allylnormorphine (nalorphine) (56) in the 1940s stimulated the examination of other N-substituted morphine analogs for antagonist activity. The exploration of the SAR of morphine and its antagonist derivatives was actively pursued during the 1950s and 1960s, along with related synthetic morphinan and benzomorphan analogs. It was the differences observed in the patterns of pharmacological activity for different compounds that resulted in the proposal of multiple opioid receptors in the 1960s and their subsequent classification by Martin and coworkers in the mid-1970s [69,70] (see Section 3.2.1).

5. STRUCTURE–ACTIVITY RELATIONSHIPS OF NONPEPTIDE OPIOID ANALGESICS

5.1. Introduction

In a single chapter such as this, it is impossible to discuss the SAR of all of the compounds with

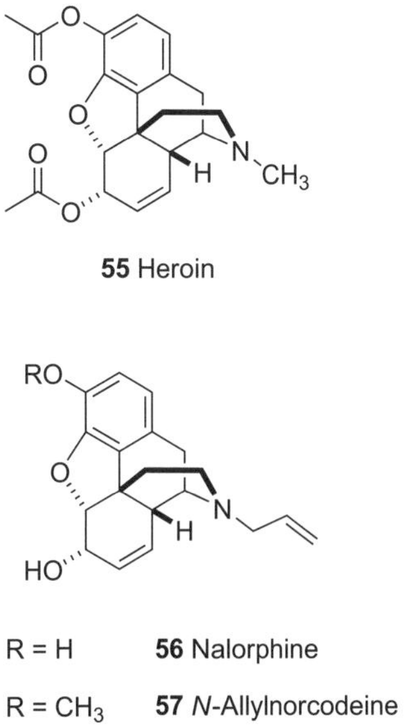

Figure 12. Early synthetic analogs of morphine.

opioid activity in detail. Two comprehensive books on opiates published in 1986 [277,278] cover much of the early literature, and there have been more recent reviews of general opioid SAR (see Refs [279–282] including the previous edition of this chapter [283]). Detailed reviews on ligands targeting specific receptors, namely, reviews on δ-receptor ligands [213,214], κ-receptor ligands [284,285], and on selective antagonists for each receptor type [286], have also recently been published. This chapter will attempt to highlight key structure–activity relationships for clinical agents or pharmacological tools in the field as well as provide a discussion of recent developments. Readers are referred to the reviews mentioned above for a more detailed discussion of opioid SAR and for additional references to the journal and patent literature.

Compounds with opioid activity are structurally diverse, ranging from rigid multicyclic compounds such as morphine to flexible acyclic analgesics such as methadone and the opioid peptides. The majority of the classical nonpeptide opioids falls into one of the five chemical classes: the 4,5α-epoxymorphinans, the morphinans, the benzomorphans, phenylpiperidines, and acyclic opioids. Conceptually, these different chemical classes can be related by a systematic dismantling of morphine nucleus (Fig. 13). As rings present in morphine are eliminated, conformational flexibility increases and changes in SAR occur. The effects on conformation and SAR will be discussed for each of these chemical classes in turn below, followed by descriptions of agonists selective for κ- and δ-receptors and of affinity labels, which interact with opioid receptors in a nonequilibrium manner.

5.2. 4,5α-Epoxymorphinans: Morphine and Derivatives

5.2.1. Morphine and Alkaloids from Opium

Morphine, **1**, is the principal alkaloid in opium, which is derived from the seed capsules of the poppy plant *P. somniferum*. Opium contains more than 50 alkaloids that fall into one of two chemical classes, the phenanthrene class including morphine and related derivatives and the benzylisoquinoline alkaloids such as papaverine (**58**) (Fig. 14) (see Ref. [288] for a complete listing of alkaloids present in opium). In addition to morphine, which on average accounts for 10–20% of opium, other related alkaloids to morphine found in opium

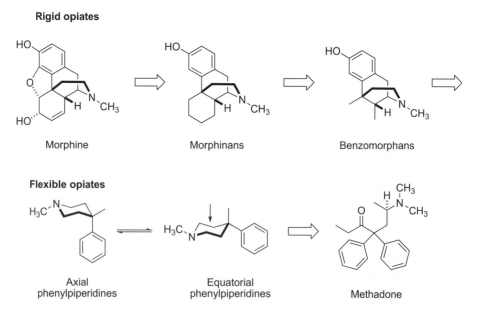

Figure 13. Systematic dismantling of morphine. Adapted from Ref. [287].

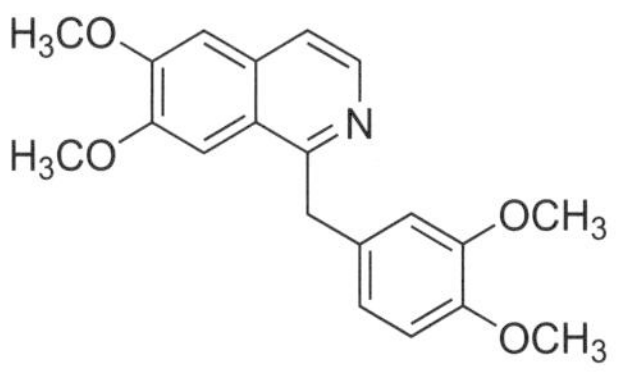

58 Papaverine

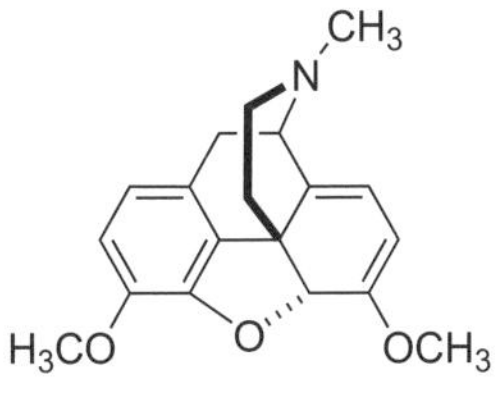

59 Thebaine

Figure 14. Alkaloids found in opium. The structures of morphine (**1**) and codeine (**2**) are shown in Fig. 1.

include codeine (**2**) (0.5%) and thebaine (**59**) (0.3%). While the latter is inactive as an analgesic, it is a key synthetic intermediate for the preparation of several potent analgesics (see Section 5.4); although present in low amounts in *P. somniferum,* it is the principal alkaloid found in another species of poppy, *Papaver bracteatum* [289].

One interesting finding has been the detection of morphine and related opiate alkaloids in vertebrates, including in a variety of mammalian tissues (see Refs [41,290–292] for reviews and see also Ref. [293] for a review of isolation techniques). The conversion of codeine and thebaine, which are known intermediates in the biosynthetic pathway in *P. somniferum*, to morphine in several tissues from rat supports the conclusion that the morphine found in animals is of endogenous origin. Endogenous morphine has been postulated to be involved in neural and immune regulation (see Ref. [292]), but the levels detected in mammalian tissues are low (low pmol/g [292]) in comparison to those of the endogenous opioid peptide levels [294], so what physiological roles such endogenous opiate alkaloids play are still unclear.

The conformation of morphine was determined from X-ray studies (see Ref. [295] for a review). The overall shape of the molecule is

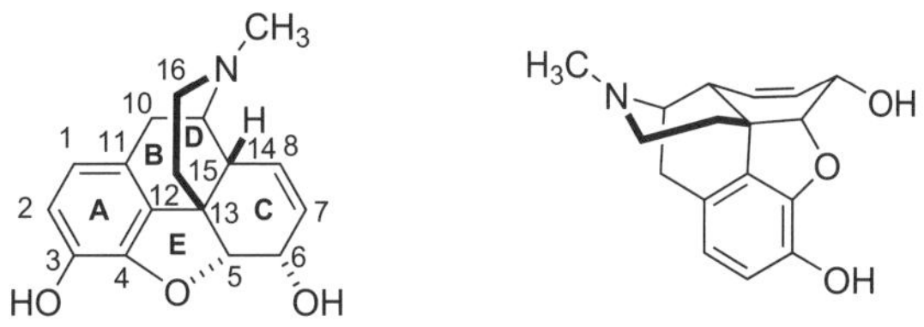

Figure 15. Structure of morphine.

a three-dimensional "T," with the A, B, and E rings forming the vertical portion of the "T" and the C and D rings forming the top of the "T" (Fig. 15). The piperdine (D) ring is in the energetically favored chair conformation, but the C-ring of morphine is in a boat conformation, which places the 6α-hydroxyl in the equatorial position (see Ref. [295]). (+)-Morphine, the enantiomer of the naturally occurring (−)-morphine, has been synthesized (see Refs [280,296]) and is devoid of analgesic activity.

5.2.2. Morphine Derivatives The synthesis of 3,6-diacetylmorphine (heroin) (**55**) in the late 1800s [275] marked the beginning of synthetic efforts to modify the structure of morphine, and since then numerous structural modifications have been made to morphine (see Refs [296,297] for detailed reviews). While modifications have been made to all portions of the molecule, the focus has been in three regions: the phenol at position 3, the C ring, and the basic nitrogen. The phenol, while not critical for activity, enhances opioid receptor affinity [298]. Masking of the phenolic hydroxyl group generally decreases opioid activity. Methylation of the 3-phenol of morphine gives codeine (**2**) (Fig. 1), which has approximately 10–20% the potency of morphine. As noted in Section 2.3.2, a small but significant percentage (~10%) of codeine is O-demethylated to morphine, accounting for its analgesic activity. Heroin, on the other hand, is approximately twice as potent as morphine. Heroin penetrates the blood–brain barrier very rapidly (68% uptake of heroin versus unmeasurable uptake of morphine 15 s after carotid injection into rats [299]), and this may partially account for its greater potency. Heroin is rapidly deacetylated to give 6-acetylmorphine, which has similar potency to morphine.

As discussed in Section 2.3.2, the 6-glucuronide metabolite of morphine **12** is about twice as potent as morphine [41] and may account for a significant portion of morphine's analgesic activity, especially with chronic use [42].

Much of the early synthetic efforts focused on modifications of the C ring. Changing the B–C ring juncture in morphine or codeine from *cis* to *trans* decreases potency 2–10-fold [300]. This is understandable since such a change causes severe distortion in the C ring of *trans*-morphine. Reduction of the C_7–C_8 double bond of morphine and codeine yield dihydromorphine (**60**) (Fig. 16) and dihydrocodeine (**61**), respectively, which have similar to slightly increased (15–20%) potencies compared to their parent compounds. Oxidation of the C_6 hydroxyl in morphine decreases analgesic potency, but when this same modification is made to dihydromorphine or dihydrocodeine to give hydromorphone (**5**) and hydrocodone (**62**), respectively, opioid potency is enhanced. This modification also alters the conformation of the C ring of the opiate (see Fig. 16). While the C ring of morphine and other derivatives containing a 6α-alcohol (**60** and **61**) is in a boat conformation, in oxymorphone **63** and the antagonist derivative naloxone **16** (see below), the C ring is in the chair conformation (see Fig. 16 and Ref. [295]). Numerous additional modifications have been made to the 6-position of 4,5α-epoxymorphinans (see Refs [296,297]). These include analogs that can bind irreversibly to opioid receptors and selective antagonists in which additional rings are attached at the 6- and 7-positions; these types of opiates are discussed in Sections 5.11 and 5.3, respectively, later in the chapter.

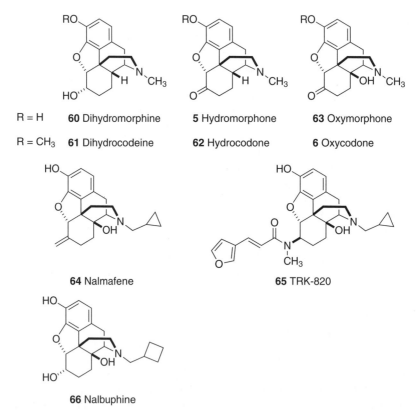

Figure 16. C ring derivatives of morphine. Structures of hydromorphone (**5**) and oxycodone (**6**) from Fig. 1 are included for comparison. Nalmefene (**64**) is an antagonist, TRK-820 (**65**) is a κ-selective agonist, and nalbuphine (**66**) is a mixed agonist/antagonist.

Zwitterionic groups have also been attached at the 6-position to produce peripherally selective derivatives with decreased ability to penetrate the CNS [301]. A 14β-hydroxyl group can be introduced into the opiate skeleton by oxidation of thebaine (see Refs. [296,297]) and this modification can enhance potency. The 14β-hydroxylated derivatives oxymorphone (**63**) and oxycodone (**6**) are potent analgesics, with oxymorphone exhibiting approximately 5–10 times the potency of morphine. The analgesic activity of these morphine derivatives along with a variety of other opioids, are compared in Table 1 in Section 2.1.

5.3. N-Substituted 4,5α-Epoxymorphinans: Opioid Antagonists

5.3.1. Introduction A critical determinant of the type of activity observed for morphine derivatives is the identity of the nitrogen substituent, and variation in this group yields compounds ranging in activity from pure agonists to pure antagonists. Removal of the N-methyl of morphine to give normorphine decreases potency (the relative potency of normorphine, however, is dependent on both the route of administration and the species examined—see Ref. [302]). While some nitrogen substituents such as phenethyl enhance agonist acitvity, other nitrogen substituents, notably propyl, allyl, cyclopropylmethyl (CPM) and cyclobutylmethyl (CBM), impart antagonist activity to the compounds. The replacement of the N-methyl group of morphine and codeine by an allyl group yields N-allylnormorphine (nalorphine) (**56**), and N-allylnorcodeine (**57**) (Fig. 12), respectively. Although the first narcotic antagonist N-allylnorcodeine was prepared in 1915 [276], it was the demonstration of the antagonist activity of nalorphine in the 1940s that stimulated the examination of other N-substituted morphine analogs for antagonist activity (see Ref. [302]). Nalorphine is inactive as an analgesic in animals, but it is an effective analgesic in humans [303]. Although it was thought that mixed agonist/antagonists such as nalorphine were agonists at κ-receptors and antagonists at μ-receptors, these drugs may be more accurately described as partial agonists at both κ- and μ-receptors (see Ref. [304]). Nalorphine produces intense psychotomimetic effects and dysphoria, which precluded its clinical use as an analgesic. However, it was used as an antidote for opioid overdose until the introduction of naloxone in the late 1960s.

Modification of the N-substituent of the potent morphine analog oxymorphone has led to several compounds with antagonist activity. The N-allyl derivative of oxymorphone, naloxone (**16**) (Fig. 4) was the first example of a pure opioid antagonist essentially devoid of agonist activity, and is 7–10 times as potent an antagonist as nalorphine (see Ref. [302]). Naloxone has a short duration of action and is used for the treatment of narcotic overdoses (see Section 2.1). Although naloxone exhibits some preference for μ-receptors, it also antagonizes δ- and κ-receptors as well (see Table 8), and sensitivity to antagonism by naloxone is routinely used as a criterion for opioid receptor involvement in an observed response. The N-CPM derivative naltrexone (**17**) is a pure antagonist that is more potent than naloxone, has a longer duration of action and is orally active, making it more useful for the treatment of former opiate addicts. The two antagonists have different metabolic routes, which account for the differences in duration of action and doses required (see Section 2.3.2 and Ref. [302]). The 6-methylene analog of naltrexone nalmefene (**64**) is also used clinically as an antagonist (see Table 4) to reverse the effect opioids and has longer duration of action than naltrexone [22]. Interestingly, at cloned μ-opioid receptors nalmefene is an inverse agonist, while naloxone and naltrexone are neutral antagonists (the latter two compounds display inverse agonist activity if the cells were pretreated with morphine) [305]. The 6β-fluoro analog of naltrexone cyclofoxy has been prepared in tritiated form [306]. Recently, the 6β-substituted naltrexamine derivative TRK-820 (**65**) was reported to be a κ-selective agonist ($K_i = 3.5$ nM, K_i (κ/μ) ratio 15) (see Table 10) that acts as a full agonist at κ-receptors, partial agonist at μ-receptors and a weak antagonist at ORL1 receptors (see Section 7.2) [307,308]. The N-CBM analog is a mixed agonist/antagonist, and the 6α-alcohol derivative, nalbuphine (**66**), is used clinically (see Table 3). Examination of diastereomeric pairs of nalorphine and naloxone

Table 10. Opioid Receptor Affinities and Opioid Activity in the GPI and MVD of Benzomorphan Derivatives and TRK-820

Agonist	K_i (nM)			K_i Ratio	IC_{50} (nM)	
	κ	μ	δ	κ/μ/δ	GPI	MVD
(−)-EKC (**33**)	0.52	1.0	5.5	1/1.9/10.6	0.18^a	4.4^a
(−)-Bremazocine (**34**)	0.075	0.62	0.78	1/8.3/10.4	0.13^a	1.98^a
Mr2034 (**97**)	0.45	0.66	5.8	1/1.4/13	0.77^a	20^a
TRK-820 (**65**)^b	3.5	53	1200	1/15/340	0.0048	0.036

Antagonist	K_i (nM)			K_i ratio	K_e (nM)		
	κ	μ	δ	κ/μ/δ	μ (GPI)	κ (GPI)	δ (MVD)
(−)-Win 44,441 (**93**)	0.69	0.67	6.4	1/1/9.6	0.67	2.8	15
Mr2266 (**98**)	0.28	1.3	2.7	1/4.6/9.6	1.5	1.3	14

Data for EKC and bremazocine from Table 8 are included for comparison. Data from Ref. [120] except where otherwise indicated.
^{a}Antagonist at μ- and δ-receptors in the rat and hamster vas deferens, respectively.
^{b}From Refs [307,308].

quaternary ammonium salts (which have chiral nitrogens) found that the diastereomers with the N-allyl group equatorial were the more potent antagonists [309,310]. The quaternary derivatives of several narcotic antagonists have been used to explore the central versus peripheral actions of opioids. These compounds generally have much lower affinity for opioid receptors and antagonist potency (1–12%) relative to the corresponding tertiary antagonists [311], and in some cases (e.g., N-methylnaloxone) the quaternary antagonist or an active metabolite appears to enter the CNS after peripheral administration (see Ref. [311]). The peripheral selectivity of the quaternary antagonist methylnaltrexone [312] appears to be greater than that of N-methylnaloxone or N-methylnalorphine [313], most likely due to the higher resistance of methylnaltrexone to N-demethylation [314]. The extent of this metabolic pathway varies among species; in humans no appreciable N-demethylation has been observed, which probably has a significant impact on the peripheral selectivity of this drug. Methylnaltrexone has recently been approved for the treatment of opioid-induced constipation (see section "Constipation,").

While the above agents have been useful pharmacological tools as well as therapeutic agents, these compounds preferentially antagonize μ-opioid receptors. Antagonists selective for all three opioid receptor types are valuable tools in understanding both the pharmacological effects of opioid agonists and the physiological effects of the endogenous opioid peptides. In the past decade, Portoghese and coworkers have synthesized numerous naltrexone derivatives with additional groups and ring systems attached to the C ring. Depending upon the modifications made analogs selective for both κ- and δ-receptors have been prepared (see Refs [286,315–318] for reviews).

5.3.2. κ-Receptor Selective Antagonists Portoghese and coworkers used the bivalent ligand approach to design κ-selective ligands. In this design strategy, it was envisioned that two pharmacophores could bridge two neighboring opioid recognition sites if they were connected by an appropriate spacer. This should then lead to substantial increases in potency because of the proximity of the second pharmacophore to the neighboring site and hence a large increase in the local concentration of the pharmacophore (see Ref. [316]). The first κ-selective antagonist TENA (**68**) (Fig. 17) was prepared by connecting two naltrexamine pharmacophores with a spacer obtained from triethylene glycol [319,320]. Both the monovalent analog with a single pharmacophore and an analog with a longer spacer were much

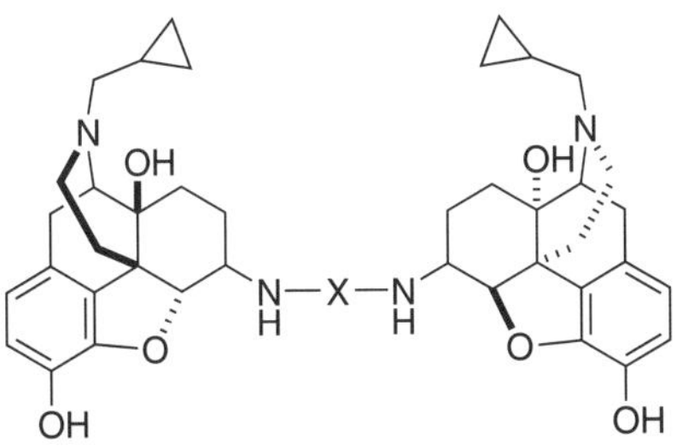

67

68 X = –(CH₂CH₂O)₂CH₂CH₂– (TENA)

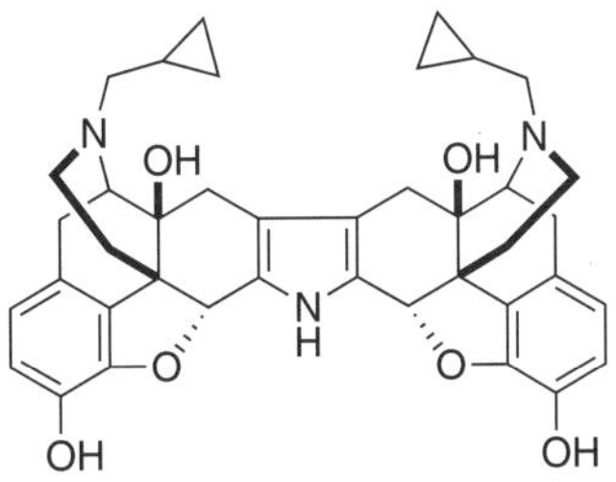

69

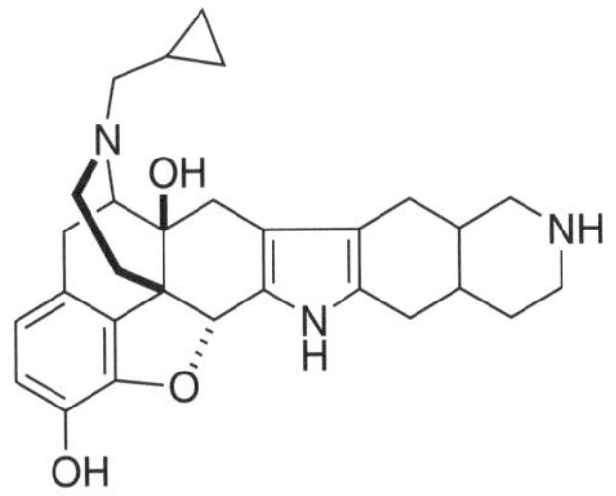

70

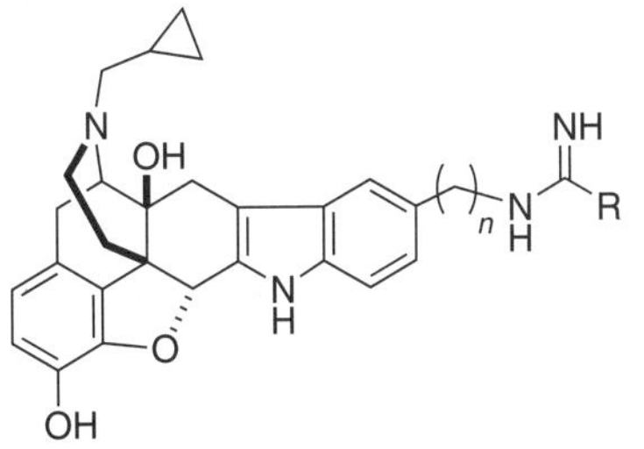

71 n = 1 or 2

R = CH₃ or (CH₂)₃CH₃

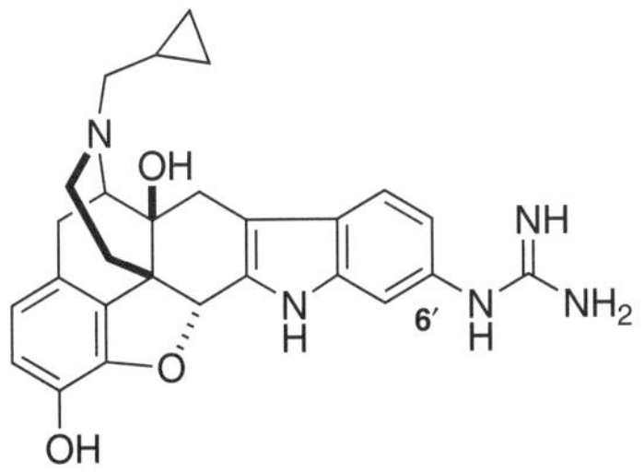

72

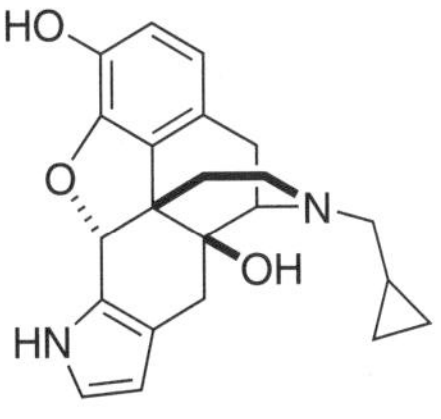

73

Figure 17. κ-Receptor selective antagonists and related compounds. The structures of norBNI (**39**), binaltorphimine (**40**), and GNTI (**41**) are shown in Fig. 6.

less potent κ-antagonists, suggesting that TENA contained the appropriate spacer length for κ-receptor affinity and selectivity. Subsequently, glycyl units were used in the spacer, making it possible to easily vary the chain length of the spacer; a central succinyl (X = -Gly$_n$–COCH₂CH₂CO–Gly$_n$- in **67**) [321] or fumaryl group (-Gly$_n$–COCH=CHCO–Gly$_n$- in **67**) [322] were used in the spacer. While the optimum length of the spacer for interaction with μ-receptors was similar in the two series (two glycyl units, i.e., $n = 2$, in each half of the spacer), the optimum length of the spacer for interaction with κ-receptors was different in the two series [322]. In the series with the succinyl spacer, the shortest spacer (with no glycyl units, i.e., $n = 0$) yielded the most potent κ-antagonist, while in the series with the more rigid fumaryl spacer a much longer spacer (two glycyl units, $n = 2$, in each half) was required for maximal κ-antagonist activity.

Incorporation of a rigid pyrrole spacer led to the synthesis of the κ-selective antagonists norbinaltorphimine (norBNI) (**39**) (Fig. 6) and binaltorphimine (**40**) [106,107]. These compounds were much more selective for antagonism of κ-receptors than TENA [107], and norBNI has been routinely used to determine whether κ-opioid receptors are involved in an observed activity. A more detailed understanding of how these compounds interact with opioid receptors has come from examining their structure–activity relationships and molecular modeling. The monovalent ligand with only the pyrrole ring attached (**73**) is not selective for κ-receptors, suggesting that the pyrrole ring functions as a spacer and does not contribute to κ-selectivity [323]. This conclusion is supported by comparison of different spacers; while the thiophene analog of norBNI in which a sulfur replaces the indole nitrogen, which has a very similar structure, retains κ-receptor selectivity, a pyran derivative with a very different spacer looses κ-selectivity [324].

While the κ-antagonists described above were designed using the bivalent ligand approach, subsequent studies revealed that these ligands do not appear to bridge two opioid binding sites, but that a basic amino group in the second pharmacophore interacts with a subsite on the κ-opioid receptor. Thus, the meso isomer of norBNI (**69**) was also found to be a potent κ-antagonist, indicating that only one pharmacophore is required for interaction with κ-receptors [325]. This suggested that the second half of the bivalent ligand might be mimicking the basic C-terminal "address" sequence of dynorphin, which imparts κ-receptor affinity to the peptide, and that simpler derivatives could be used in place of the second pharmacophore. A simplified analog of BNI that does not contain the second aromatic ring (**70**) is twice as selective as norBNI for κ-receptors [326]. The basicity of the second nitrogen is important for κ-receptor selectivity in both norBNI analogs [327] and the simplified analog **70** [326]. Examination of the binding of norBNI to mutated κ-receptors [169] suggests that this nitrogen interacts with Glu297 at the top of TM6 on the receptor; a significant (70-fold) increase in affinity for the μ-receptor mutated at the corresponding position ([K303E]) [328] is consistent with this conclusion. A computational model of norBNI bound to κ-receptors has been developed [174], and in this model the second nitrogen forms an ion pair with Glu297. With the recent proposal that opioid receptors may exist as dimers (see Section 7.1), however, Portoghese is revisiting these bivalent ligands and how they may interact with dimeric receptors [318].

These results led to the preparation of analogs of the δ-selective antagonist naltrindole (see below) with basic alkylamidino groups (**71**) [109,329,330] and guanidinium groups (GNTI) (**41**) [108–110] attached to the indole ring as κ-selective antagonists. Like norBNI, the decrease in binding affinity of GNTI to [E297K] mutated κ-receptors and increase in affinity for the corresponding [K303E] mutated μ-receptors and the [W284E] mutated δ-receptors suggests that the guanidinium group of GNTI interacts with Glu297 on κ-opioid receptors [108,331]. Interestingly, shifting the position of the guanidinium group from the 5′-position in GNTI to the 6′-position (**72**), results in a potent κ-agonist [332].

5.3.3. δ-Receptor Selective Antagonists and Related Compounds
Portoghese and coworkers used the "message–address" concept to design naltrexone derivatives selective for δ-receptors (see Ref. [333]). The "message–address" concept, which was originally described by Schwyzer for ACTH [334], was applied to the opioid peptide dynorphin by Chavkin and Goldstein [258]. In this model, the "message" consists of the amino acids in the peptide, which are responsible for activating the receptor and producing a biological response, while the "address" component is the portion of the molecule, which enhances affinity for a given receptor type. Portoghese postulated that in the endogenous enkephalins the Tyr[1] residue functions as the "message" and the phenyl ring of Phe[4] functions as both part of the "message" and the "address"; the intervening Gly[2]-Gly[3] then functions as a spacer. In naltrexone analogs, the naltrexamine moiety of these antagonists functioned as the "message" portion and a phenyl ring was attached to the pyrrole spacer as the address." This led to the synthesis of naltrindole (NTI) (**32**) (Fig. 5) [100,335], the first nonpeptide δ-selective antagonist. In addition to being δ-selective, this naltrexone analog is about 500 times as potent as the δ-selective peptide antagonist ICI 174,864 (**29**) (Fig. 5) [335]. In a computational

model of NTI bound to δ-opioid receptors [174], the indole moiety is directed toward a hydrophobic pocket formed by residues from TM6 and TM7, which includes two residues (Trp284 in TM6 and Leu299 in TM7 of the mouse δ-receptor) unique to the δ-opioid receptor. Site-directed mutagenesis studies of these positions in δ-, μ-, and κ-receptors were consistent with these positions contributing to the δ-receptor affinity and selectivity of NTI [331]. Substitution of Ala in the position corresponding to Leu299 in μ- and κ-receptors increased the affinity for NTI by 21- and 96-fold, respectively [331]. Substitution of Trp284 in the δ-receptor had a smaller effect, with mutation to Glu [331] and Lys [160] (the residues found in the corresponding position in κ- and μ-receptors, respectively) decreasing the affinity of NTI 9- and 15-fold (mutation to Ala decreased NTI affinity 5-fold [160]).

Fluorescent derivatives of naltrindole have been prepared. The fluorescent analog in which fluorescein was attached via a tetraglycine spacer to 7'-amino-NTI has been prepared [336]; its fluorescence is blocked by NTI, indicating specific binding to δ-receptors. Recently, a fluorogenic "reporter affinity label" derivative of naltrindole, PNTI, has been prepared from 7'-amino-NTI, which in contrast to its reversible counterpart, is an agonist in the MVD [337] (see Section 5.11).

Naltrindole analogs containing other heterocycle spacers have also been examined. The benzofuran analog naltriben (NTB) (**46**) (Fig. 8) is also a δ-selective antagonist [189]; as discussed in section "Receptor Subtypes, Splice Variants, and Receptor Dimerization," subsequent studies indicated that this analog is a selective $δ_2$-antagonist [338,339].

The aromatic ring in the "address" portion of naltrindole and its analogs is important for antagonist activity at δ-receptors. The tetrahydroindole derivative is much less potent, as are several 6-aryl derivatives [189]. The 7(*E*)-benzylidine analog of naltrexone, 7-benzylidenenaltrexone (BNTX) (**44**) (Fig. 8), however, is a potent δ-antagonist [187,339]. It has 100-fold greater affinity for [^{3}H]DPDPE sites ($δ_1$) than DSLET ($δ_2$) sites, and therefore is a selective $δ_1$-antagonist [187]; it is also a selective antagonist for $δ_1$-receptors in the mouse spinal cord in antinociceptive assays [339].

Substitution on the phenyl ring of BNTX with either *o*-methoxy or *o*-chloro resulted in analogs with increased antagonist potency and δ-receptor selectivity in smooth muscle assays, but *in vivo* in the tail flick assay these analogs exhibited similar selectivity but lower potency than BNTX [340]. Both the *E*- and the corresponding *Z*-isomers of a series of aryl analogs of BNTX have been prepared [341], and the *Z*-isomers found to have higher δ-receptor selectivity; the (*Z*)-1-naphthyl derivative exhibited the highest δ-receptor affinity ($K_i = 0.7$ nM versus 6.2 nM for BNTX) and selectivity for δ- over μ-receptors.

The relative position of the "address" aromatic ring in these analogs is important. Those analogs with the phenyl ring in the same region of space as the phenyl ring of the indole in NTI are active as δ-antagonists [342]; a large decrease in activity was observed for indole regioisomers of NTI that have the aromatic ring in a different relative position [343]. Attachment of the phenyl ring directly to the 7-position of the morphinan system resulted in analogs with decreased δ-receptor potency and selectivity [344]. Peripherally selective naltrindole analogs have been prepared by conjugating amino acids to 7-carboxynaltrindole (**74**) (Fig. 18); these derivatives are δ-selective antagonists in smooth muscle assays and $δ_1$-selective antagonists *in vivo* [345]. Benzylation of the indole nitrogen of NTI, by contrast, results in a selective and long-lasting $δ_2$-antagonist **75** [346]. Attachment of fluoresceineamine via a thiourea linkage to the *para*-position of the benzyl ring of *N*-benzyl-naltrindole resulted in an analog with potent antagonist activity *in vitro*. The inability to block the fluorescence by several δ-selective ligands, including NTI, however, suggested that the fluorescent analog exhibits high non-specific binding, possibly due to the high lipophilicity of the fluorophore [347].

A number of other naltrindole analogs have been synthesized, including numerous ones by Rice and coworkers. While most of the modifications examined decreased δ-opioid receptor affinity, in several cases, for example, the *N*-2-methylallyl naltrindole analog [348], the 3-methyl ether of naltrindole [349] and the corresponding ring-opened 4-hydroxy-3-methoxyindolemorphinan [350],

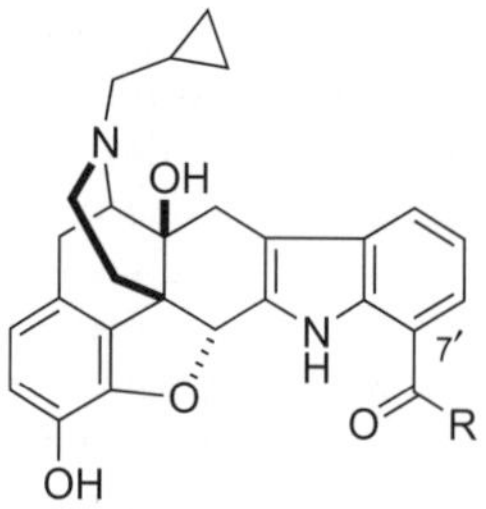

74 R = Aminoacid

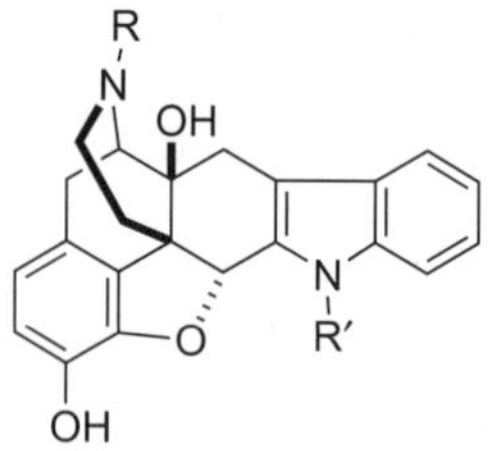

R = CPM, R' = CH2Ph   **75** N-Benzylnaltrindole
R = —CH₂CH—cyclohexyl, R' = H   **76**
R = CH₃, R' = H   **77** Oxymorphindole

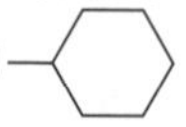

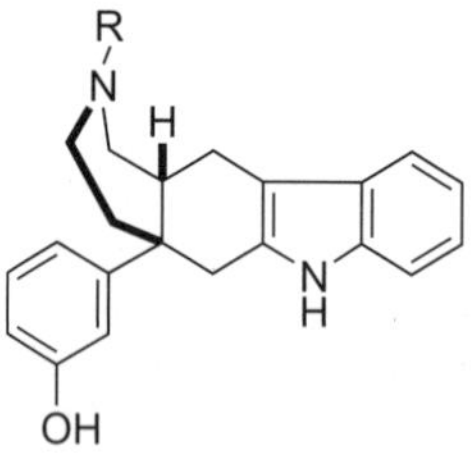

R = CH₂CH₃   **78** SB 205588
R = CPM   **79** SB 206848

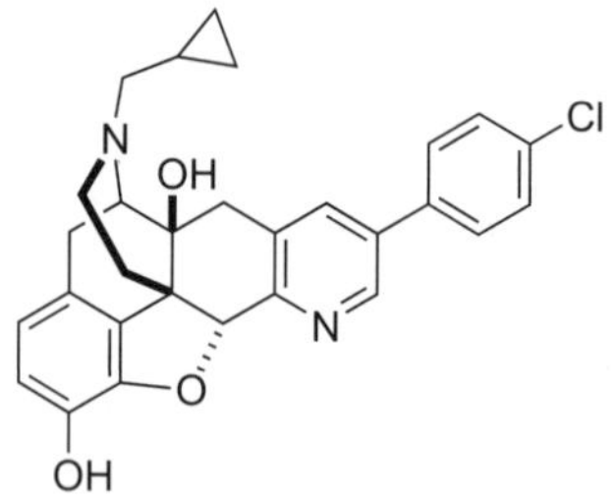

80 SoRI9409

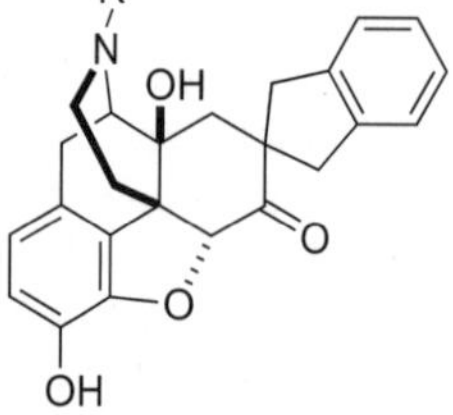

R = CH3   **81** SIOM
R = CPM   **82** SINTX

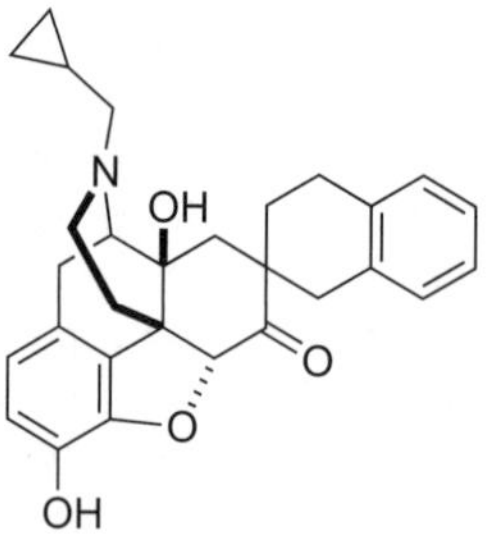

83

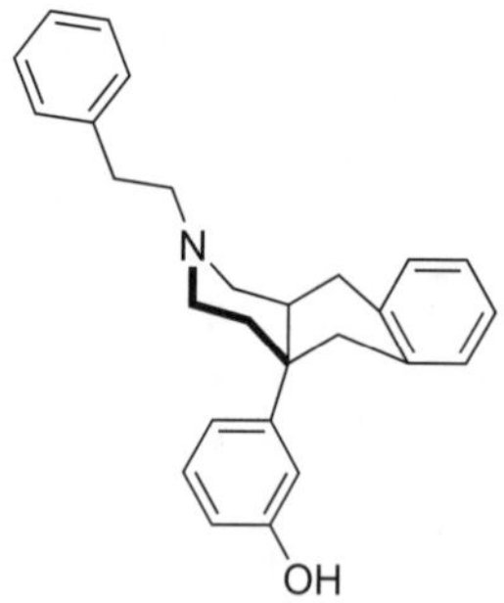

84

Figure 18. δ-Receptor antagonists and agonists related to naltrindole (**32**). Other δ-selective ligands, including naltrindole, are shown in Figures 5 and 8.

the selectivity for δ-receptors increased. Replacement of the CPM group by the 2-methylallyl in NTB and SoRI 9409 (see below), but not BNTX or SIOM (see below), resulted in compounds that retained reasonable δ-receptor affinity and selectivity [351]. The 14-alkoxy derivatives, with or without a 5-methyl group, also exhibited high δ-receptor selectivity in the MVD assay [352,353]. Interestingly the N-cyclohexylethyl derivative (**76**) derivative is a μ-selective agonist [354]. Additional analogs of naltrindole have been reported in the patent literature (see Refs. [213,214] for reviews).

Fragmentation of the indolomorphinan structure of NTI by removal of the 4,5α-epoxy and 10-methylene groups resulted in a series of octahydroisoquinolines (see Refs [213,355]). The compounds that are the analogs of NTI with a five-membered ring spacer (SB 205588B (**78**) and SB 206848 (**79**)) (Fig. 18) are δ-receptor antagonists, while compounds with a six-membered ring as the spacer are δ-agonists (see Section 5.10.2). Interestingly, the related compound **84** without the pyrrole ring spacer and the *cis* configuration of the ring juncture is an antagonist at μ- and κ-receptors [356].

As discussed in Section 3.3.1, δ-receptor antagonists can decrease the development of tolerance and dependence to morphine, and therefore there is considerable interest in the development of compounds exhibiting both δ-antagonist activity and μ-agonist activity. Anathan and coworkers identified naltrindole analogs with μ-agonist activity as well as δ-receptor antagonist activity [224,357]. The 7'-phenoxy naltrindole derivative exhibits weak ($IC_{50} = 450$ nM) agonist activity in the GPI, while retaining potent δ-receptor antagonist activity ($K_e = 0.25$ nM) in the MVD [357]. The naltrindole analog SoRI 9409 (**80**) bearing a phenyl ring attached to a pyridine rather than an indole ring [224] is a δ-receptor antagonist, but unexpectedly also exhibits μ-opioid agonist activity in the GPI assay [224]; interestingly, the corresponding derivative without the chlorine is a μ-receptor antagonist. Further examination of SoRI 9409 in nociceptive assays indicated that the compound is a weak partial agonist in the high intensity (55C) tail flick test and weak full agonist in the acetic acid writhing assay [224,358]; studies *in vivo* were also consistent with activity as a mixed partial μ-agonist/δ-antagonist. In contrast to morphine, repeated doses of SoRI 9409 did not produce tolerance [224,358], and SoRI 94094 suppressed withdrawal signs when coadministered with naloxone to morphine-dependent animals. In [^{35}S]GTPγS assays, however, SoRI 9409 exhibits no μ-agonist activity and instead acts as an antagonist at μ- and κ- as well as a δ-receptors; at this time, the reason for this difference between the *in vitro* and *in vivo* results is not clear.

Attempts have also been made to prepare agonist analogs of NTI by modification of the basic nitrogen. The *N*-methyl analog oxymorphindole (OMI) (**77**) (Fig. 18) is a partial agonist ($IC_{50} = 100$ nM) [335], while the unsubstituted and the *N*-phenethyl derivatives are full agonists ($IC_{50} = 85-180$ nM) in the MVD [359]. These derivatives either did not exhibit any δ-antagonist activity or were only weak δ-antagonists in this smooth muscle preparation. *In vivo* these analogs, along with naltrindole, exhibit antinociceptive activity [359,360]. Based on antagonism by selective antagonists, the antinociceptive activity of all of the compounds except the unsubstituted derivative appears to be mediated by κ- rather than δ-receptors (the unsubstituted derivative was not antagonized by any of the selective antagonists). At lower doses these compounds also exhibit antagonism at δ-receptors *in vivo*. Modification of the indole structure of OMI to change the orientation of the phenyl ring yields the 7-spiroindane derivative 7-spiroindanyloxymorphone (SIOM) (**81**), which is a δ_1-agonist [361]. Portoghese et al. suggested that the phenyl ring of the indane ring system adopts a conformation similar to that of the phenyl ring of Phe^4 of the δ-selective enkephalin analog DPDPE (**24**). (Linking the C-terminal sequence Phe-LeuOMe of leucine enkephalin to oxymorphanone (but not to naltrexone) also significantly increases affinity for δ-receptors [362].) Like NTI, the results of site-directed mutagenesis of positions corresponding to Trp^{284} (in TM6) and Leu^{299} (in TM7) of the δ-receptor in δ-, μ-, and κ-receptors were consistent with these positions being involved in the δ-receptor affinity and selectivity of SIOM [331]. Replacing the *N*-methyl group of SIOM with a CPM group gives SINTX (**82**) that is a potent δ-antagonist [342]. The benzospiroindanyl derivative of SINTX exhibits improved selectivity for δ- over μ-receptors, although lower potency *in vitro* compared to SINTX and is a δ_1-antagonist *in vivo* [363]. The 14-hydroxyl group contributes to both δ-agonist and antagonist activities, and its removal from SIOM and NTI decreases agonist and antagonist potency, respectively [364]. The 7-spirobenzocyclohexyl derivatives of SIOM are also potent δ-agonists, but the corresponding analogs of SINTX (**83**) (both α- and β-isomers) are μ-selective antagonists [365].

5.4. Diels–Alder Adducts

Thebaine (**59**), which is present in large amounts from another species of poppy *P. bracteatum* [289], serves as the precursor for the 6,14-*endo*etheno opiates. While the natural levorotatory isomer of thebaine is inactive as an analgesic, recently it was reported that the (+) isomer exhibits significant antinociceptive activity [366]; both isomers exhibit some affinity for opioid receptors (the (+) isomer for μ-receptors and the (−) isomer for δ-receptors), but the affinities were very low (micromolar).

Diels–Alder reaction of thebaine with various electrophiles yields compounds (Fig. 19) that have extremely high potencies, over a 1000-fold higher than morphine in some cases (see Refs [367,368]). X-ray [369] and NMR [370] analysis of 19-propylthevinol, the 3-methyl ether of etorphine (**42**), indicates that the 6,14-etheno bridge is held "inside" (*endo*) the tetrahydrothebaine ring system and below the plane (α) of the C_7–C_8 bond (see Fig. 19 and Ref. [280]); the C ring is held in a boat conformation by the 6,14-*endo*etheno bridge.

The C_7-substituent in these compounds is in the α-configuration and in many cases contains a chiral center. The stereochemistry at C_{19} can have significant effect on potency of the derivatives (see Ref. [280]); generally, the diastereomer with the *R* configuration at C_{19} is the more potent isomer with the differences in potencies of the diastereomers sometimes exceeding 100-fold.

A variety of C_{19} alcohol derivatives have been prepared (see Refs [367,368]), and three of these compounds are frequently used in opioid research. Etorphine (**42**) is a potent analgesic, over 1000-fold more potent than morphine, which has been widely used to immobilize animals, including large game animals. It exhibits high affinity for all three opioid receptor types (see Table 8), and

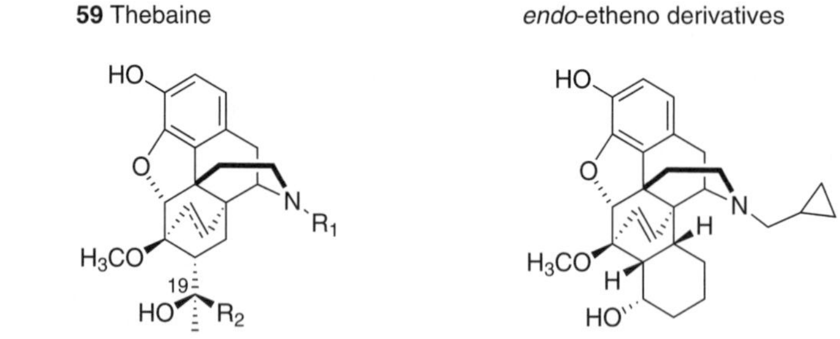

42: Etorphine
$R_1 = CH_3$; $R_2 = (CH_2)_2CH_3$; bond = double

43: Diprenorphine
$R_1 = CPM$; $R_2 = CH_3$; bond = single

85: Buprenorphine
$R_1 = CPM$; $R_2 = C(CH_3)_3$; bond = single

86 BU48

Figure 19. Diels–Alder reaction to give 6,14-*endo*etheno derivatives of thebaine (**59**) and the structures of buprenorphine (**85**) and BU48 (**86**). The structure of diprenorphine (**43**) is included for comparison.

therefore [³H]etorphine has been used as a universal tritiated ligand for opioid receptors (see section "Radioligand Binding Assays"). The N-cyclopropylmethyl 6,14-ethano derivatives diprenorphine (**43**) and buprenorphine (**85**) (Fig. 19), which differ only in the identity of one of the alkyl groups attached to C_{19} (see Refs [371,372] for structural studies of these compounds), exhibit antagonist and partial agonist activities, respectively. Diprenorphine also has high affinity for all three opioid receptor types and its tritiated form has been used as a universal tritiated ligand for opioid receptors (see section "Radioligand Binding Assays"); it antagonizes μ-, δ-, and κ-ligands in the MVD (see Table 8). Buprenorphine, which is used clinically, is a potent partial agonist at μ-receptors with antagonist (or partial agonist) activity at κ-receptors [373]. This very lipophilic agent dissociates slowly from opioid receptors, and its complex pharmacology is not completely understood (see Ref. [304]). Its unique pharmacological profile offers several clinical advantages; it causes less severe respiratory depression than full agonists and has less abuse potential. It can suppress withdrawal symptoms in addicted individuals undergoing withdrawal from opiates and thus has been used in the maintenance of these patients. Because of its partial agonist activity, however, it can also precipitate withdrawal symptoms in those addicted to opiates. The 18,19-dehydro derivative of buprenorphine HS-599 exhibits higher affinity, selectivity, and potency at μ-receptors than the parent compound [374]. A series of buprenorphine analogs were prepared in which C_{20} was constrained in a five-membered ring in order to assess the influence of the orientation of the C_{20} hydroxyl on activity; while the configuration of this hydroxyl did not affect the binding affinity, it did influence κ-receptor efficacy and potency [375]. One of these novel ring constrained analogs BU48 (**86**) exhibits an unusual pattern of pharmacological activity, producing δ-opioid receptor mediated convulsions but not δ-receptor mediated antinociception [232].

5.5. Morphinans

Morphinans differ in structure from morphine and other 4,5α-epoxymorphinans in that they lack the 4,5α-ether bridge. While the 4,5α-epoxymorphinans are typically prepared from naturally occurring alkaloids, the morphinans are generally made from racemic materials and therefore must be resolved to obtain individual isomers. Racemorphan [376] was one of the first morphinans studied. Following resolution the analgesic activity was found exclusively in the *levo* isomer, levorphanol (**87**) (Fig. 20) [377], which has the identical configuration as morphine and is about four to five times as potent, while the *dextro* isomer dextrorphan (**89**) has negligible opioid activity. Levorphanol and dextrorphan were used in one of the first attempts by Goldstein to demonstrate stereospecific binding to opioid receptors [378]. Dextrorphan, and particularly its 3-methyl ether dextromethorphan (**90**), have significant antitussive activity. In derivatives with an unsubstituted 6-position such as dextromethorphan, the C ring exists in the chair conformation [379]. In contrast to the morphine derivatives, the morphinans with a *trans* B–C ring juncture are potent analgesics (see Ref. [280]).

Substitution on the nitrogen of morphinans has similar effects on activity as found in derivatives of morphine (see Ref. [380]). Thus, replacement of the N-methyl with groups such as an allyl or CPM group to yield levallorphan and cyclorphan (**88**), respectively, imparts antagonist activity at μ-receptors. As indicated above for nalorphine, these compounds were thought to be mixed agonists/antagonists with agonist activity at κ-receptors, but may be more accurately described as partial agonists at both κ- and μ-receptors (see Ref. [304]). Neumeyer and coworkers have recently described the receptor binding affinities of a series of N-substituted morphinans as potential therapeutics for cocaine abuse [381,382]. A comparison of cyclorphan with its cyclobutylmethyl analog *in vivo* illustrates the effect small changes in the N-substituent can have on the activity in this class of compounds. Interestingly, cyclorphan (**88**) produced antinociception through δ- as well as κ-receptors and was a μ-receptor antagonist; while the N-cyclobutylmethyl derivative was also a κ-agonist in antinociceptive assays, it was an agonist at μ-receptors but was without effect, either agonist or antagonist, at δ-receptors [381].

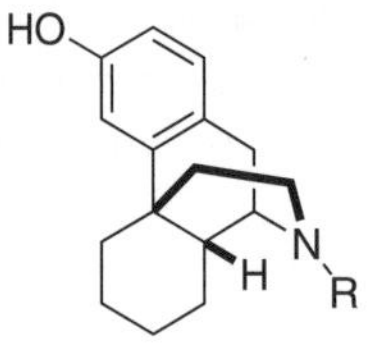

| R = CH₃ | **87** Levorphanol |
| R = CPM | **88** Cyclorphan |

| R = H | **89** Dextrophan |
| R = CH₃ | **90** Dextromethorphan |

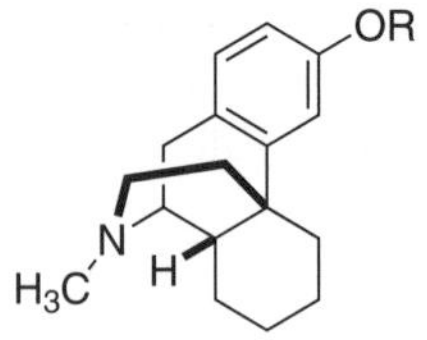

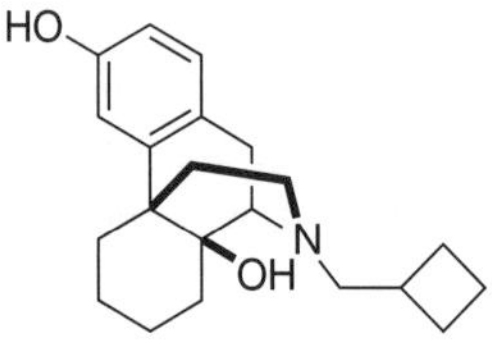

91 Butorphanol

Figure 20. Morphinan derivatives.

As is the case for morphine derivatives, 14-hydroxylation yields potent analogs. The N-cyclobutylmethyl derivative with a 14β-hydroxyl group, butorphanol (**91**) is one of the mixed agonists/antagonists used clinically in the United States (Table 3); in addition to a parental formulation, a nasal spray formulation of this drug was introduced in 1992. Since 4,14-dimethoxy-N-methylmorphinan-6-one proved to be a very potent agonist, Schmidhammer et al. prepared the corresponding N-CPM derivative cyprodime (**18**) (Fig. 4) [91]. Cyprodime is a pure μ-receptor antagonist without any agonist activity [91] that has enhanced μ-receptor selectivity compared to naloxone and naltrexone (see Table 8).

5.6. Benzomorphans

Further structural simplification by elimination of the C ring of the morphinan structure yields the benzomorphans (see Fig. 21), which are also known as benzazocines.

Although the benzomorphan ring system was first synthesized in 1947 by Barltrop [383], it was the synthesis of 2,5-dimethylbenzomorphan (**92**) (R = R′ = CH₃) by May and Murphy [384] that began the investigation

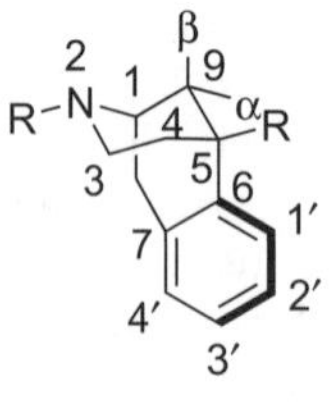

92 6,7-Benzomorphan numbering

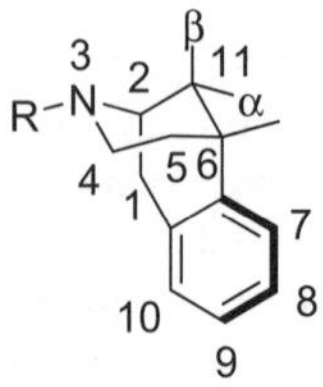

Benzazocine numbering
(*Chemical Abstracts*)

Figure 21. Numbering systems for benzomorphans based on benzomorphan and benzazocine nomenclature (used in *Chemical Abstracts*). The more common benzomorphan numbering will be used in this chapter.

into the synthesis and pharmacology of this structural family. As is the case for the morphinans, the benzomorphans are prepared synthetically and therefore are obtained as racemic mixtures. A number of these racemates have been resolved and the activities of the individual isomers examined (see Refs. [280,385] for reviews, also Ref. [386]). The active isomers are the *levo* isomers, which have the same absolute configuration at the bridgehead carbons as morphine (i.e., $1R,5R$) (see Refs [280,385]), although in some cases the *dextro* isomers retain weak activity.

The majority of the benzomorphan derivatives was prepared prior to the classification of multiple opioid receptors and were characterized using *in vivo* assays. In the classification of multiple opioid receptors, Martin and coworkers [69,70] used the benzomorphan ketocyclazocine (13) (Fig. 3) as the prototypical ligand to define κ-receptors (see Section 3.2.1), and subsequent characterization indicated that a variety of benzomorphans had high affinity for κ-receptors. Several of these benzomorphans, both agonists and antagonists, were important ligands in early studies of κ-receptors. The selectivity of these benzomorphans for κ-receptors is generally low, however (see Table 10), and much more selective agonists and antagonists, that is, the arylacetamide agonists (Section 5.9) and antagonists norBNI and GNTI (Section 5.3.2) are now available for studying κ-receptors.

A variety of benzomorphans have been prepared with various combinations of alkyl substituents (methyl, ethyl, propyl) at the 5- and 9-positions (see Refs [280,385]). The alkyl group at the 9-position can be oriented either α, in which the substituents in the 5- and 9-positions are *cis*, or β, in which these substituents are *trans* (see 92) (Fig. 21). The synthetically minor β-isomers, which have the opposite stereochemistry from that of the corresponding position in morphine (C_8), are more potent than the α-isomers as antinociceptive agents (see Refs [280,385]). Attachment of a 3-alkanone side chain at the 9β-position of metazocine yielded a series of potent compounds that range from pure agonists to pure antagonists, depending upon the length of the side chain [387]. One of these derivatives, WIN 44,441 (93) (Fig. 22) is a potent κ-receptor antagonist that has been used to characterize κ-receptors; the active isomer is the *levo* isomer, WIN 44,441-3 [388]. It is also a potent antagonist at μ-receptors, however, and does not exhibit selectivity for κ- over μ-receptors (see Table 10). In the benzomorphan derivatives introduction of a 9β-hydroxyl in the 9α-methylbenzomorphans, which corresponds to the 14-hydroxyl in oxymorphone and naloxone, does not enhance the agonist activity in the *N*-methyl derivatives (see Ref. [280]), but in benzomorphans with other substituents on the nitrogen, for example, cyclopropylmethyl or dimethylallyl (see below), a 9β-hydroxyl enhances antagonist potency 3–10-fold [389].

As in the morphinan series, a variety of substituents on the nitrogen have been examined (see Refs [280,385,386]). Metazocine (94), with an *N*-methyl substituent analogous to morphine, exhibits agonist activity, and the *N*-phenethyl analog phenazocine shows increased analgesic potency. Replacing the *N*-methyl with groups such as allyl or CPM led to compounds with antagonist or mixed agonist/antagonist activity (generally, agonist activity at κ-receptors and antagonist activity at other opioid receptors, similar to nalorphine) (see Ref. [120]). The orientation of the alkyl group at position 9 influences the relative agonist versus antagonist activities of these compounds, with the orientation of this group affecting antagonist potency less than agonist potency (see Ref. [280]). A number of these derivatives exhibit dysphoric and psychotomimetic effects, limiting their clinical usefulness. Thus, the *N*-allyl derivative *N*-allylnormetazocine (SKF 10,047) (14) (Fig. 3) was the prototypical ligand used by Martin and coworkers [69,70] to characterize σ-receptors and exhibits psychotomimetic and dysophoric effects in humans (see Ref. [390]); based on animal studies, these adverse effects appear to reside in the *dextro* isomer [391]. The *N*-dimethylallyl derivative pentazocine (95), however, produces considerably less dysphoria than cyclazocine [304] and is the only benzomorphan used clinically (see Table 3). The N-CPM analog cyclazocine (96) is a potent mixed opioid agonist/antagonist, but its considerable psychotomimetic effects prevent its clinical use (see Ref. [390]). As indicated above, the 8-keto derivative ketocyclazocine

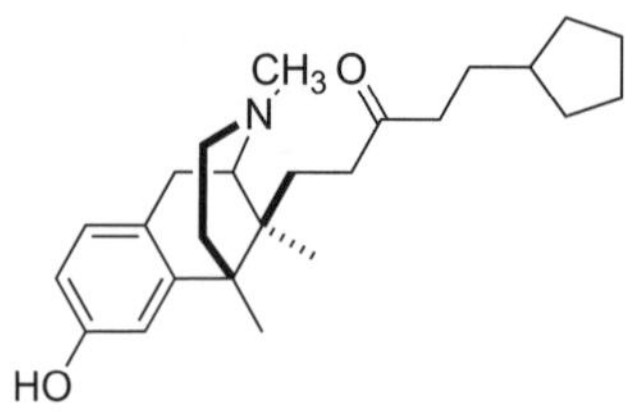

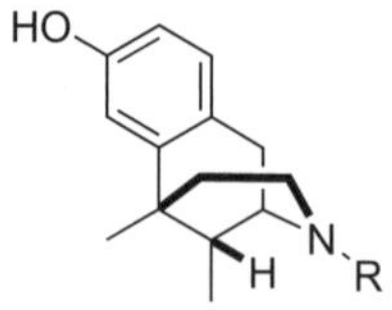

R = CH$_3$	**94** Metazocine
R = CH$_2$CH=CH$_2$	**14** SKF 10,047
R = CH$_2$CH=CH$_2$(CH$_3$)$_2$	**95** Pentazocine
R = CPM	**96** Cyclazocine

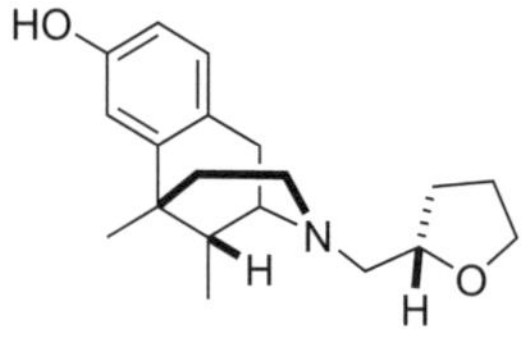

93 WIN 44,441

97 Mr2034

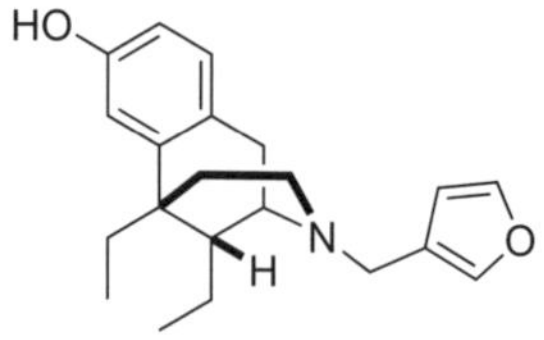

98 Mr2266

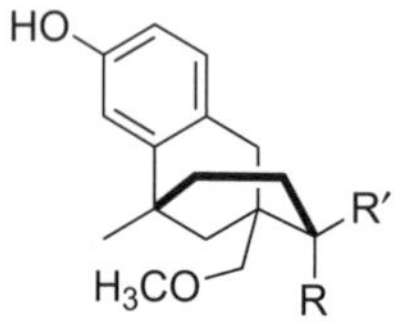

99 R = NH$_2$, R' = H
100 R = H, R' = NH$_2$

Figure 22. Benzomorphan derivatives.

was used in the initial characterization of κ-receptors by Martin and coworkers; the 5-ethyl, 8-keto derivative EKC (**33**) (Fig. 6) has also been used in a variety of studies of κ-receptors. The closely related benzomorphan (−)-bremazocine (**34**) (Fig. 6) [392] exhibits some preference for binding to κ-receptors (μ/κ ratio = 8.2) (see Table 10) and thus has been used in its tritiated form in radioligand binding assays for κ-receptors (see section "Radioligand Binding Assays"); in smooth muscle assays, these compounds exhibit mixed agonist/activity with agonist activity at κ-receptors and antagonist activity at μ- and δ-receptors (see Ref. [120]).

A series of analogs with tetrahydrofurfuryl [393,394] and furfuryl [395] groups on the nitrogen have been utilized as κ-receptor agonists and antagonists. The tetrahydrofurfuryl derivative Mr2034 (**97**) (absolute configuration 1R,5R,9R,2"S [394]) exhibits high affinity for κ-receptors and has been used as a κ-agonist, but it binds equally well to μ-opioid receptors (see Table 10); like other benzomorphan mixed agonist/antagonists, it exhibits agonist activity at κ-receptors, but antagonist activity at μ- and δ-receptors in smooth muscle assays (see Ref. [120]). Although this *levo* isomer is inactive at σ-PCP receptors, it also produces dysphoric and psychotomimetic effects in humans that are antagonized by naloxone, suggesting that in this case these adverse effects are mediated by κ-receptors [390]. In the case of the furyl-substituted analogs the chain length affects the type of activity observed; thus the N-2-furfuryl-derivative exhibits antagonist activity while the longer N-[2-(2-furyl)ethyl] derivative is a potent analgesic [395]. Mr2266 (**98**), the *levo* isomer of the 5,9-diethyl N-2-furfuryl analog, has been used as a κ-receptor antagonist in a variety of studies, particularly in early studies when more κ-selective antagonists were not available. Similar to other benzomorphans, its selectivity for κ-receptors is low (μ/κ ratio 4.6) (see Table 10) and in smooth muscle assays it antagonizes both μ- and κ-receptors at similar concentrations (see Table 10).

The requirement for a phenol for opioid receptor interaction was recently revisited in a series of cyclazocine analogs in which the hydroxyl was replaced by a primary, secondary, or tertiary amino group [396] and several of these analogs retained high affinity for opioid receptors; the same modifications to morphine decreased μ-receptor affinity by at least 35-fold [397].

A series of analogs of the benzomorphans have been prepared in which the amino group is exocyclic (**99** and **100**) [398]. In contrast to the benzomorphans and other rigid opiates, the receptor binding of **99** and its analogs exhibit almost no stereoselectivity (K_i (μ) = 2.0 and 2.2 nM for the (+) and (−) isomers); *in vivo* both isomers of **99** were inactive. Racemic **100** exhibits ∼10-fold increase in κ-receptor affinity (K_i = 6.6 nM) compared to the isomers of **99** while retaining nanomolar affinity for μ-receptors (K_i = 2.0 nM); this compound is a full κ-agonist *in vivo*.

5.7. Piperidine Derivatives

Further structural simplification by elimination of the B ring of the benzomorphans yields the phenylpiperidines (Fig. 13). This disrupts the fused 3-ring system found in the morphinans and benzomorphans, resulting in much more flexible compounds. Thus, while the B ring in the rigid opioid alkaloids fixes the phenyl ring in an axial orientation relative to the piperidine (D) ring, without the B ring the phenyl ring can be either axial or equatorial, depending upon the substitution pattern on the piperidine ring (see Fig. 13). Thus, the structure–activity relationships for the phenylpiperidines can be complex, with differences in SAR observed for different groups within this structural family; these differences can partly be explained by conformational differences and the orientation of the phenyl ring.

The prototype of this class meperidine (**7**) (Fig. 1) was discovered serendipitously in 1939 during examination of compounds for antispasmodic activity [399]. Subsequent examination of meperidine verified that it was an analgesic, with 10–20% the potency of morphine. Later, it was found that meperidine's affinity for opioid receptors is very low (0.2% that of morphine); however meperidine penetrates into the brain more readily and reaches much higher (600-fold) concentrations in the brain than does morphine [400].

The piperidine analgesics can be classified into groups based on the substitution on C_4 (Fig. 23). In the case of the 4-arylpiperidines, the second substituent can be attached to the piperidine ring via either a carbon (e.g., meperidine) (**7**) or an oxygen atom (e.g., α- and β-prodine) (**101** and **102**). Another group of piperidine analgesics are the extremely potent 4-anilidopiperidines (e.g., fentanyl) (**8**) (Fig. 1). Meperidine, α-prodine, and fentanyl are all μ-opioid agonists [401]. All portions of the piperidine analgesics—the N-substituent, the phenyl ring, the piperidine ring, and the C_4 substituent—have been modified. Modifications made to the piperidine ring include substitutions, particularly 3-alkyl substitutions, that affect the preferred conformation of these compounds, introduction of conformational constraints to fix the conformation of this ring and the orientation of the 4-aryl ring, and expansion or contraction of the piperidine ring. It is not possible to cover all of the details of the structure–activity relationships of the piperidine analgesics in this chapter, and thus selected modifications will be discussed here. More detailed discussions of the SAR are given in the comprehensive books on opiates [277,402] and extensive reviews by Casy [280,281,403]; the reviews by Casy discuss conformational studies in considerable detail.

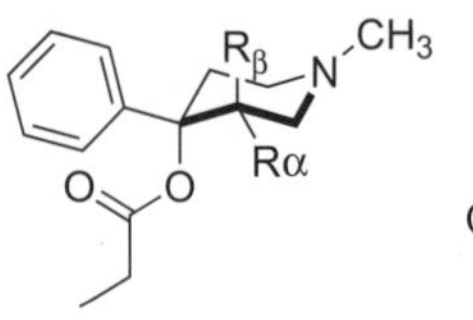

614   OPIOID RECEPTOR LIGANDS

	Rα	Rβ	
	CH₃	H	**101** α-Prodine
	H	CH₃	**102** β-Prodine
	CH₂CH=CH₂	H	**103** α-Allylprodine
	H	CH₂CH=CH₂	**104** β-Allylprodine

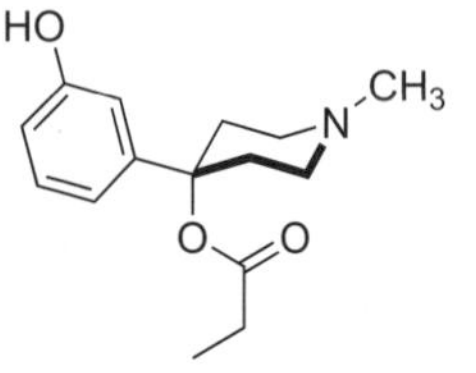

105 Ketobemidone

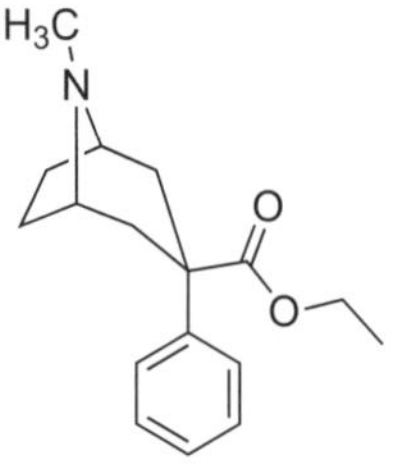

106

Figure 23. Examples of piperidine analgesics with different substituents on C_4. The structures of meperidine (**7**) and fentanyl (**8**) are shown in Fig. 1.

5.7.1. 4-Arylpiperidines with a Carbon Substituent at C_4

The carbon substituents at C_4 of the 4-arylpiperidines can be a carbalkoxy (e.g., meperidine) (**7**) (Fig. 1), a ketoxy (ketobemidone) (**105**) (Fig. 23) or alkyl substituent. In the case of meperidine (also known as pethidine), examination by X-ray crystallographic [404,405] and NMR [406] techniques indicates that the phenyl ring is equatorial in the major conformer, although in computational studies the calculated energy differences between the equatorial and the axial phenyl conformations of meperidine is small (0.6–0.7 kcal/mol) [407].

A large number of analogs of meperidine have been prepared (see Ref. [402]). The 4-substituent in meperidine is optimal for analgesic activity; lengthening or shortening of the ester chain decreases activity, and except for ketobemidone, substitution of the ester by an amide or a ketone or hydrolysis to the acid reduces activity. Introduction of a m-hydroxyl on the phenyl ring of meperidine to give bemidone enhances activity. This substitution, however, can have complex effects on biological activity, and depending upon the nitrogen substitution, both potent agonists and antagonists have been produced in this series (see Ref. [402]). A wide variety of N-substituents have been examined (see Ref. [402]). Several phenylalkyl groups, for example, phenethyl (in pheneridine) and p-aminophenethyl (in anileridine), increase potency; N-allyl or cyclopropylmethyl groups, however, do not generate antagonists in the meperidine series. Introduction of a 3-methyl group into the piperidine ring yields two isomers, with the 3β-methyl isomer (with the methyl and phenyl cis) 8–10 times more potent than the α isomer [408]; the 3β-methyl group should increase the population of the axial 4-phenyl conformer [280]. In conformationally constrained tropane derivatives of meperidine [409,410], the small difference in potency between the α- and the β-phenyl isomers (the α-phenyl derivative **106** is only three to four times the potency of the β-phenyl isomer) suggested that the analgesic activity of meperidine analogs is not very sensitive to the conformation of the phenyl ring. In the α-phenyl analog (**106**), in which the phenyl ring adopts a pseudo-axial orientation [410], introduction of a m-hydroxyl group decreases potency [411], suggesting that even though the phenyl ring in these meperidine derivatives adopts a similar conformation to that found in morphine, the interactions of the phenyl ring with the receptor are different for these two types of compounds.

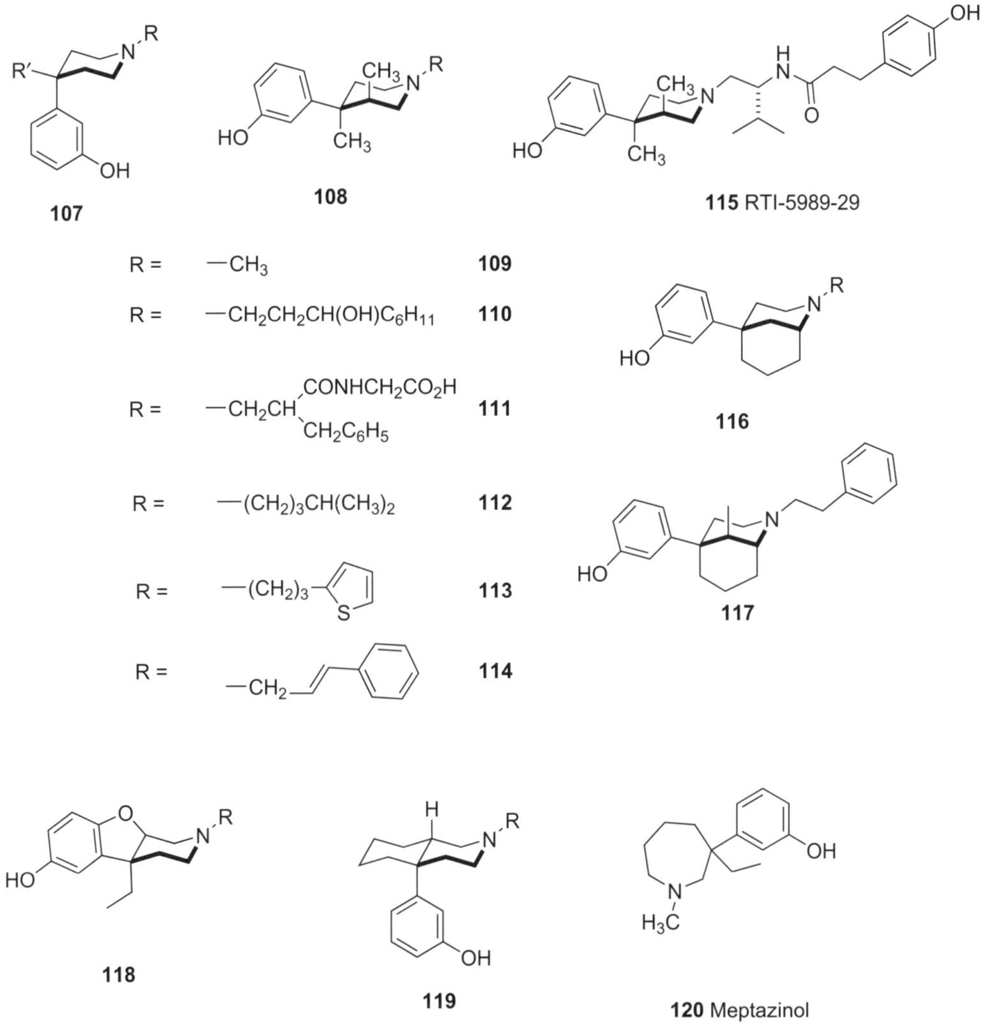

STRUCTURE–ACTIVITY RELATIONSHIPS OF NONPEPTIDE OPIOID ANALGESICS 615

107

108

115 RTI-5989-29

R =	—CH$_3$	**109**
R =	—CH$_2$CH$_2$CH(OH)C$_6$H$_{11}$	**110**
R =	—CH$_2$CH(CONHCH$_2$CO$_2$H)(CH$_2$C$_6$H$_5$)	**111**
R =	—(CH$_2$)$_3$CH(CH$_3$)$_2$	**112**
R =	—(CH$_2$)$_3$-(thienyl)	**113**
R =	—CH$_2$-CH=CH-C$_6$H$_5$	**114**

116

117

118

119

120 Meptazinol

Figure 24. 4-Alkyl-4-arylpiperidines.

Constrained meperidine analog (**106**)

A variety of 4-arylpiperidines without an oxygen functionality on the second substituent at C$_4$ (**107**) (Fig. 24) have been prepared. All of the active derivatives contain a *m*-hydroxyl group on the phenyl ring [403]. The 4-alkyl derivatives favor an axial aryl conformation (**107**) [406,412,413]. Introduction of a 3-methyl group, however, can alter the preferred conformation, with the 3β-methyl isomer favoring the equatorial aryl conformation (**108**) [412,413]. The structure–activity relationships found in the more rigid opiates, particularly with regards to the nitrogen substituent and its effects on agonist versus antagonist activity, do not hold in the 4-alkyl-4-arylpiperidine series. In the case of 4-alkyl-piperidines where the axial aryl conformation is preferred, for example, the *N*-allyl derivative of 4-propyl-4-(*m*-hydroxyphenyl)piperidine (**107**) (R = allyl) is an agonist with only weak antagonist activity [412,413]. In contrast the 1,3β,4-trimethyl derivative (**109**) is an antagonist rather than an agonist [414].

In this series, the 3-methyl group is critical for antagonist activity and an N-allyl or cyclopropylmethyl group did not increase antagonist potency [414]. Further variation of the nitrogen substituent of the *trans*-3,4-dimethylphenylpiperidines (108) has resulted in the synthesis of several potent analogs with antagonist activity at both μ- and κ-receptors [415–418], including compounds such as LY255582 (110), which has potent anorectant activity, and the peripherally selective antagonist LY246736 (111) (now designated AD 8-2698 or alvimopan) [417], which is undergoing clinical trials for the treatment of gastrointestinal motility disorders (see Ref. [19] and section "Constipation"). Potent antagonists that exhibit some selectivity for κ-receptors ((LY227053) (112), LY253886 (113) [418], and 115 [419]), along with μ-selective antagonists (114) [420], have also been identified.

Conformationally constrained derivatives of the 4-alkyl-4-arylpiperidines in which the aryl ring is constrained into both axial and equatorial conformations have been prepared. Arylmorphans (116) are constrained analogs in which the aryl ring is equatorial (see Ref. [280,403]). The N-methyl derivative is equipotent with morphine following subcutaneous administration to mice [421]; the configuration of the more potent *dextro* enantiomer of this analgesic (the 1S,5R isomer [422]) is related stereochemically to the more potent enantiomer of 4-arylpiperidines such as α- and β-prodine (101 and 102) rather than the rigid benzomorphan (−)-metazocine [280]. The receptor affinities of both isomers for different receptor subtypes have also been examined [423]. Consistent with the stereochemical relationships, the N-allyl and N-CPM derivatives are agonists with little if any antagonist activity [424]. Introduction of a 9β-methyl group, which is comparable to the 3-methyl substituent of the *trans*-3,4-dimethylphenylpiperidines, resulted in compounds with antagonist activity; the N-phenethyl derivative (117) was much more potent than the N-methyl analog [425]. These derivatives still retain rotational freedom around the phenyl-piperidine bond. Further constraint of the 5-arylmorphans with an ether bridge between the aryl ring and the morphan ring systems yielded compounds with low affinity for opioid receptors (see Ref. [280]), but the constrained benzofuro [2,3-c]pyridin-6-ols (118) in which the phenyl ring is constrained to a dihedral angle of 92° [423] retain μ-opioid receptor affinity and are potent analgesics [426]. In the *trans*-4a-aryldecahydroisoquinolines (119), in which the aryl ring is constrained to the axial conformation, the N-methyl derivative is twice the potency of morphine; interestingly while the N-methyl derivative is selective for μ-receptors, the N-CPM derivative exhibits a slight preference for κ-receptors [427].

Active analogs have also been prepared by shifting the alkyl and aryl substituents to the 3-position. All active derivatives of the 3-methyl-3-arylpiperines contain the *m*-hydroxyl on the phenyl ring [280]. Although the N-methyl derivatives are weak analgesics, derivatives with an N-arylalkyl substituent are significantly more potent [277]. Some derivatives with N-allyl or N-CPM groups behave as antagonists, similar to the morphinan series. Ring contracted and expanded analogs have also been prepared, including the mixed agonist/antagonist meptazinol (120). The pharmacological effects of this compound are somewhat unusual (see Ref. [428]) and may involve both opiate and cholinergic mechanisms [429]. It has been proposed that the opioid actions of meptazinol are mediated by μ_1-receptors [428].

5.7.2. 4-Arylpiperidines with an Oxygen Substituent at C_4
Reversal of the ester in meperidine gives MPPP (N-methyl-4-phenyl-4-propionoxypiperidine) (121) and increases analgesic activity 5–10-fold (see Ref. [402]). First described in 1943 [430], this compound took on new significance in the late 1970s when the drug began appearing as a "designer drug" (a compound with a minor structural change that was prepared in attempts to evade laws regulating controlled substances) and sold on the streets as a heroin substitute. During the synthesis of MPPP in clandestine drug laboratories, however, dehydration occurred during the acylation of the 4-phenyl-4-piperidol, resulting in the formation of MPTP (N-methyl-4-phenyl-1,2,3,6-tetrahydropyridine) (122) (see Fig. 25). People who took the contaminated drug exhibited symptoms of parkinsonism,

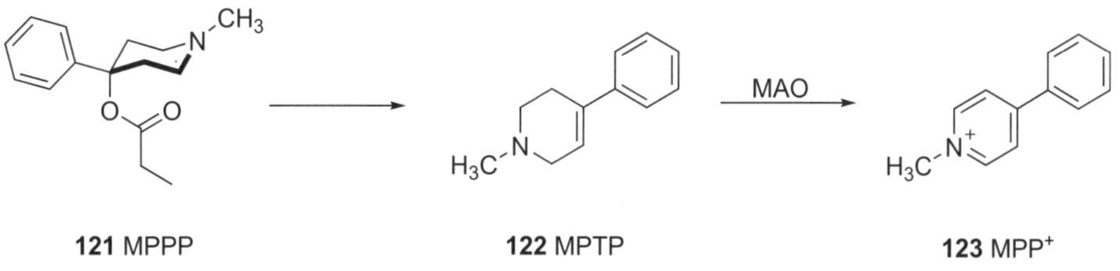

Figure 25. Formation of MPTP and MPP+ from MPPP.

even in subjects in their twenties [431]. The neurotoxic effects of MPTP are thought to be due to its metabolism to MPP+ (**123**) (N-methyl-4-phenylpyridinium) by monoamine oxidase and destruction of dopaminergic neurons, resulting in the parkinsonism-like symptoms (see Ref. [432]).

A wide variety of modifications have been made to MPPP, including modification of the ester functionality, variation in the nitrogen substituent, substitution on the piperidine ring, etc. (see Ref. [402]). The propionyl chain is the optimum length, and hydrolysis of the ester to the corresponding alcohols usually results in compounds with little if any activity. Several N-phenylalkyl substitutions such as phenethyl increase potency; in some cases (i.e., the anilinoethyl derivative, $(CH_2)_2NHC_6H_5$ [433]), the resulting compound can be more than 1000-fold more potent than meperidine.

A variety of C_3 alkylated derivatives of the reversed esters of meperidine have been examined (see Ref. [402]). Incorporation of a methyl group at C_3 yields α- and β-prodine (**101** and **102**) (Fig. 23) [434]. The β-isomer (**102**) is five times the potency of the α-isomer. X-ray cyrstallography and NMR indicate that the preferred conformation of the prodines is the chair form with the phenyl ring equatorial (see Refs [280,435]). Computational studies also indicate a preference for this conformer; in the case of the reversed esters, the energy differences between the equatorial and the axial conformers (1.9–3.4 kcal/mol) are much greater than the differences between the two conformers of meperidine (**7**) and ketobemidone (**105**) [407].

Larger alkyl groups (ethyl, propyl, and allyl) are also tolerated at the 3-position, but for these derivatives the α-isomer is more potent than the β-isomer. For the α-isomers the more potent enantiomer has the 3R,4S stereochemistry [436]. This led to the suggestion that the prodine derivatives present the pro-4R face of the molecule to the receptor (see Fig. 26) (see Ref. [280]); thus, substitution on the 4R side interferes with drug–receptor interactions while substituents on the 4S-side are not involved in interaction with the receptor and are well tolerated. A variety of other methyl- and dimethyl-substituted derivatives of the reversed esters have been prepared, resulting in compounds with a wide range of potencies and with different orientations of the phenyl ring depending upon the substitution pattern (see Refs. [280,402,437]); positions that tolerate methyl substitution are

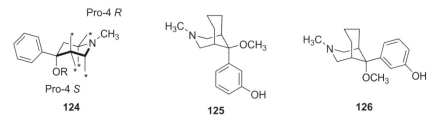

Figure 26. Pro-4R and pro-4S face of prodine derivatives. Positions that tolerate methyl substitution are indicated by an asterisk.

indicated in structure **124** (Fig. 26) by an asterisk. Very potent analogs have been prepared by bridging the 3- and 5-positions with a trimethylene chain (**125** and **126**) [438]. Both the α-isomer (**125**, in which the aryl ring is axial) and the β-isomer (**126**, in which the aryl ring is equatorial) have similar potencies; introduction of an *m*-hydroxyl group on the phenyl ring results in a dramatic (400–1000-fold) increase in the potency of both isomers. The corresponding *m*-hydroxyphenyl-3,7-diazabicyclanes also exhibited significant antinociceptive activity [439].

In contrast to meperidine and other phenylpiperidines with a C_4 carbon substituent discussed above, introduction of a *m*-hydroxyl group on the phenyl ring of the allylprodines (**103** and **104**) (Fig. 23) results in inactive compounds [440]. This led Portoghese to propose an alternative mode of interaction to explain how α-allylprodine (**103**) and other phenylpiperidines in which the phenyl ring is equatorial can interact with the same receptors as morphine and other rigid opiates where the phenyl ring is axial. In the initial bimodal binding model proposed by Portoghese [67], the amine of different opiates was postulated to interact with a common anionic site and the rest of the molecule then would pivot around the nitrogen to bind in one of the two possible orientations. The bimodal binding model was subsequently modified to include a second lipophilic site [440] where the equatorial phenyl ring of phenylpiperidines such as α-allylprodine (**103**) was proposed to interact.

A substantial reduction in analgesic potency was also observed for the *m*-hydroxyl derivative of the 2α-methyl reversed ester in which the preferred conformation of the phenyl ring is axial [441], suggesting that there can also be differences in receptor interaction for the axial aryl moiety in this derivative and in morphine. This in turn may be due to the differences in the relative orientation of the aryl rings in the rigid morphine versus the phenylpiperidine derivatives in which the aryl ring is free to rotate.

5.7.3. 4-Anilidopiperidines 4-Anilidopiperidines, in which an amido nitrogen functionality is positioned between C_4 of the piperidine ring and the phenyl ring, are extremely potent analgesics. Fentanyl (**8**) (Fig. 1), the prototype of this class, is almost 500 times the potency of meperidine and 200–500 times the potency of morphine (see Ref. [442]). Like meperidine, fentanyl is a μ-agonist [401]. Because of its rapid onset and short duration of action, fentanyl and other 4-anilidopiperidines have been used extensively in anesthesia. In 1991, fentanyl became available as a transdermal patch for the treatment of chronic pain, and recently lozenge formulations have become available (see Table 1).

The preferred conformation of fentanyl and other 4-anilidopiperidines has been examined by X-ray crystallography, NMR, and computational methods (see Refs [181,281,443]). In the crystal structure, the anilido moiety adopts an equatorial conformation (see **8**) (Fig. 1) [444]. Several conformationally constrained analogs of fentanyl have been prepared (see the previous edition of this chapter [283] for a detailed review). Generally, conformationally constrained derivatives in which the phenyl is constrained in the β-orientation are inactive (see Ref. [445]), but in conformationally constrained tropane analogs of fentanyl the β-analog (**128**) (Fig. 27), in which the anilide group is equatorial, is more potent than the α-isomer (**127**), in which the propananilido moiety is pseudo-equatorial [446]. The high potency of the conformationally constrained spirane derivative **129** provides evidence that the phenyl ring may be oriented α when fentanyl binds to opioid receptors. Additional conformationally constrained 1,5-benzoxazepine derivatives constrained via cyclization between the ortho position of the phenyl ring and the 3-position (**130**) are also potent analgesics [447].

A large number of modifications have been made to fentanyl (see Refs. [281,403,442,445]). Unlike the phenylpiperidines, a second substituent at C_4 is not required, but the addition of a polar carbon substituent (CO_2CH_3 or CH_2OCH_3) at this position to give, for example, carfentanil (**131**) enhances potency 27-fold over fentanyl and results in compounds 7800 times the potency of morphine [448]. In the case of the anilido side chain, the propionamide is the optimum length [445]; the methoxyacetamides are also potent analgesics

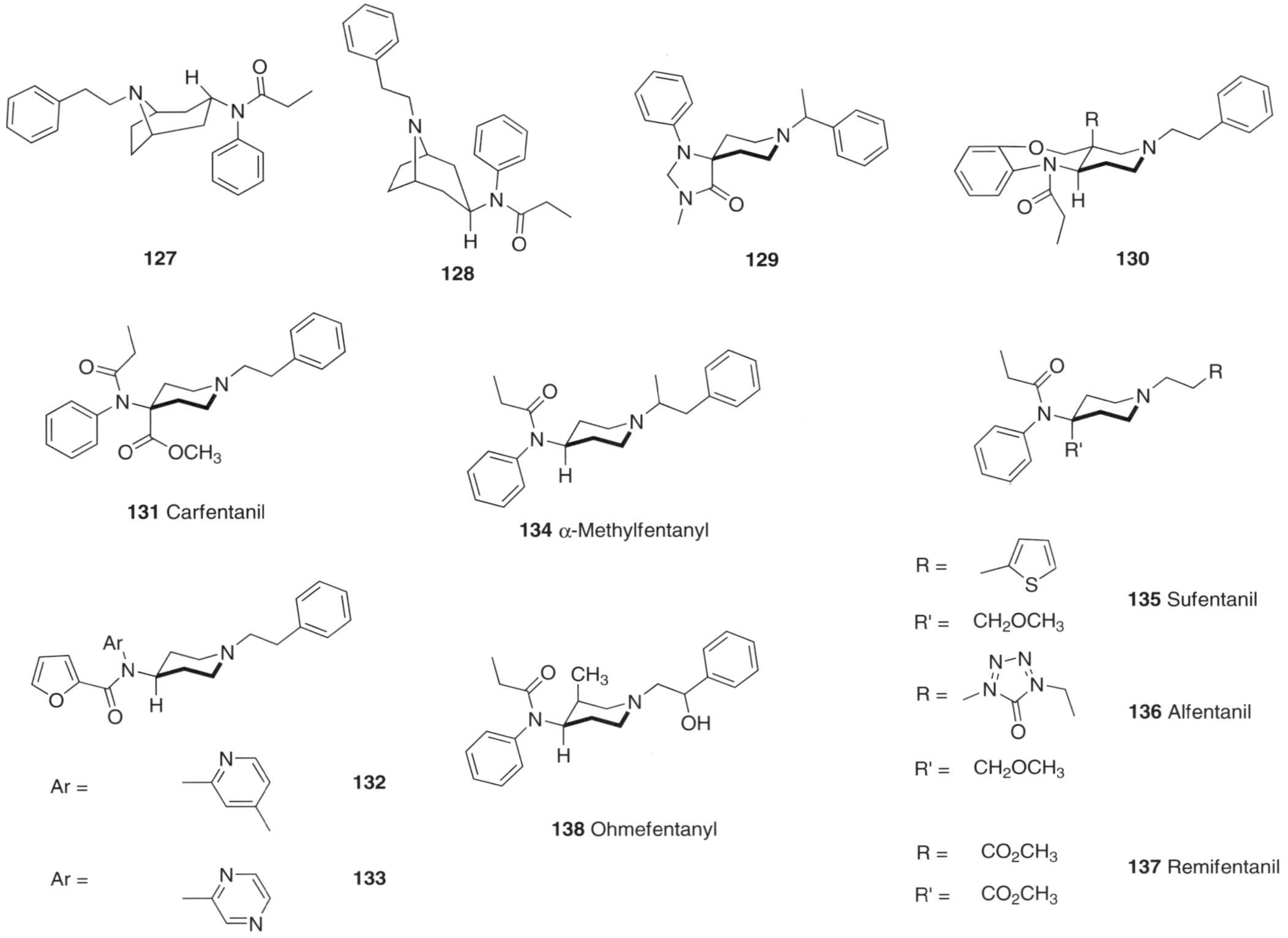

Figure 27. Fentanyl analogs.

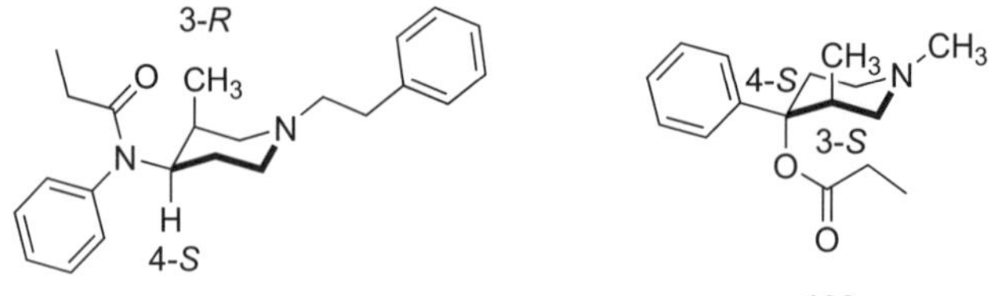

Figure 28. Comparison of the 3R,4S isomer of cis-3-methylfentanyl (**139**) and the 3S,4S isomer of β-prodine (**102**).

(see Ref. [403]). Replacement of the propionamide with 2- and 3-fumaramide resulted in compounds with antagonist activity against both morphine-induced analgesia and respiratory depression (**132**) or against respiratory depression alone (**133**) [449]; the flat aromatic furan ring appears to be important for antagonist activity. Previous attempts to produce antagonist derivatives of fentanyl by replacement of the piperidino phenethyl group with allyl or CPM were unsuccessful [446,450], and thus it appears that changes should be made around the amide nitrogen rather than to the basic piperidino substituent in order to impart antagonist activity to fentanyl analogs [447].

A variety of changes to the nitrogen substituent on the piperidine ring have been made, which produce potent analgesics. The branched chain derivative α-methylfentanyl (**134**) has achieved notoriety as the street drug "China White" [451]. The phenethyl substituent is the optimum length, but potency can be enhanced by replacement of the phenyl ring with a heterocycle such as the 2-thienyl ring [448]. Combination of the heterocycle substitution on the piperidino chain and a polar carbon substituent attached to position 4 has yielded the clinically used agents sufentanil (**135**) and alfentanil (**136**) (Table 1), which like fentanyl are used as adjuncts in anesthesia. Although less potent than fentanyl and sufentanil, alfentanil has a faster onset and shorter duration of action than these other 4-anilidopiperidines. The rapid onset of action of alfentanil may be due to its physiochemical properties; alfentanil is a much weaker base ($pK_a = 6.5$) than fentanyl ($pK_a = 8.4$) and sufentanil ($pK_a = 8.0$), so that a higher proportion of the unprotonated amine would be available to penetrate the blood–brain barrier [403]. Other heterocyclic substitutions have been made on the piperidino chain [447], resulting in some compounds with less respiratory depression. Incorporation of an ester functionality into the piperidino substituent to give remifentanil (**137**) (Table 1) results in a compound that is 30-fold more potent than alfentanil, and which has very rapid onset (1.6 min) and offset (5.4 min) that is independent of the duration of administration [452,453]. This is due to the rapid hydrolysis by esterases to the acid derivative, which has very low analgesic activity and is rapidly excreted.

Introduction of a methyl substituent in the 3-position of the piperidine ring results in chiral compounds. The racemic cis derivative of fentanyl is more potent than the trans derivative [454]. For the cis racemate, the dextro isomer is 120 times the potency of the (−)-isomer; the absolute stereochemistry of the (+)-cis isomer is 3R,4S [455]. The analog of cis-3-methylfentanyl with a hydroxyl group on the N-phenethyl chain ohmefentanyl (**138**) is extremely potent (7000 times morphine) in antinociceptive assays and exhibits remarkable selectivity for μ-receptors (27,000-fold selectivity for μ- versus δ-receptors [456]). Comparisons of the 3R,4S isomer of cis-3-methylfentanyl (**139**) and the 3S,4S isomer of β-prodine (**102**) (Fig. 28)[1] suggest that although these compounds are both μ-agonists, they represent different classes of ligands that have different modes of interaction with opioid receptors [281]. Fentanyl derivatives containing propyl and allyl substituents

---

[1] There is an error in stereochemistry at position 3 of β-prodine in Ref. [281], p 504.

at C_3 exhibit significant differences in SAR from the corresponding prodine analogs (see Ref. [281]), supporting this conclusion. A series of cis/trans pairs of 3-methylfentanyl analogs have been reported [457], and in some cases in which the anilido ring had an ortho substituent, the trans isomers were more potent than the cis isomers. This may be due to steric hindrance between the ortho substituent and the 3-methyl group in the cis isomers interfering with the phenyl ring adopting an α-orientation (see above) [281]. As discussed above, the conformationally constrained 1,5-benzoxazepine derivatives linked via the ortho position on the phenyl ring and the 3-methyl substituent (130) are active in antinociceptive assays [447].

Recently, two models of cis-(+)-3-methylfentanyl (139) and other 4-anilidopiperidines docked to μ-opioid receptors have been described [176,181]. The binding mode for cis-3-methylfentanyl in the two models is different, which has been attributed to different conformations of this compound used for docking to the receptor [181]. Comparison of cis-3-methylfentanyl to N-phenethylnormorphine suggested that there was considerable overlap in the region of the receptor occupied by the N-phenethyl groups, but that there was no overlap in the region in the receptor binding pocket occupied by the N-phenyl ring of cis-3-methylfentanyl and the phenol ring of the morphine analog (see Fig. 5 in Ref. [181]).

Affinity label derivatives of fentanyl and (+)-cis-3-methylfentanyl FIT [192,458] and super-FIT [455], which have an isothiocyanate group on the phenyl ring of the piperidino substituent, have been prepared. Unlike the reversible parent compounds that are μ-agonists, these affinity labels irreversibly bind to δ-receptors. These compounds are discussed further in Section 5.11 on affinity labels.

5.8. Acyclic Analgesics

Methadone (9) (Fig. 1) and its analogs can be viewed as ring-opened derivatives of the phenylpiperidines. As was the case for meperidine, methadone was not discovered by this systematic structural approach, but was instead identified in Germany during World War II while looking for compounds with antispasmodic activity [459]. These acyclic analgesics are highly flexible molecules capable of adopting a multitude of conformations (see Ref. [281]). Methadone is a potent analgesic, and the racemate is approximately twice the potency of morphine and 5–10 times the potency of meperidine (see Ref. [460]). A principal use of methadone, however, has been in the maintenance of individuals addicted to narcotics. It has relatively good oral activity and long duration of action, permitting once-daily dosing, and is reported to produce less euphoria than morphine. The major metabolic pathway for methadone involves N-demethylation, but these derivatives are not stable and undergo cyclization to inactive metabolites via intramolecular Schiff base formation (see Ref. [460]). Methadone is also metabolized to an active metabolite methadol (see below).

Methadone and its isomer isomethadone (140) (Fig. 29), which was also obtained from early syntheses of methadone (see Ref. [460]), each contain an asymmetric center. The more active isomers for both compounds are the (−) isomers (see Refs [281,460]). The (−) enantiomer of methadone, which has the R-configuration, has 7–50 times higher potency in antinociceptive assays and greater than 10-fold higher affinity for opioid receptors than the (+) isomer. In the case of isomethadone, which is slightly less potent than methadone, the more active (−) enantiomer has the S-configuration and is 40 times the potency of the (+) isomer (see Refs [281,460]). NMR, circular dichroism [461], and molecular modeling studies [462] suggest that isomethadone may be less flexible than methadone due to the proximity of the methyl group to the phenyl rings in isomethadone.

These flexible analgesics exhibit their own distinct SAR (see Refs. [281,460,463]). A variety of nitrogen substituents have been examined, and while larger acyclic groups markedly decrease or abolish activity, compounds containing a cyclic substituent, such as pyrrolidine, piperidine, or morpholine, on the nitrogen retain activity. The two carbon chain length between the quaternary carbon and the nitrogen is the optimal length, and lengthening the chain abolishes activity. Removal of the methyl group from the alkyl chain to give the achiral compound normethadone

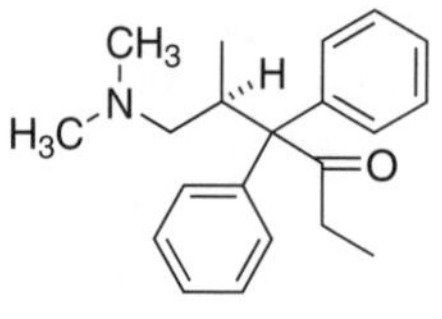

140 Isomethadone

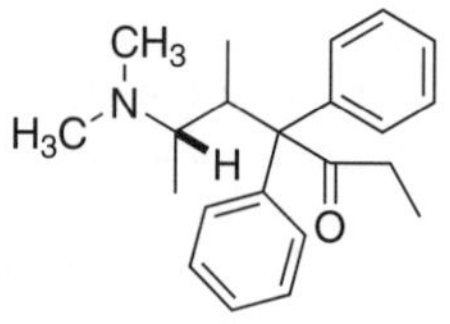

141 *erythro*-5-Methylmethadone

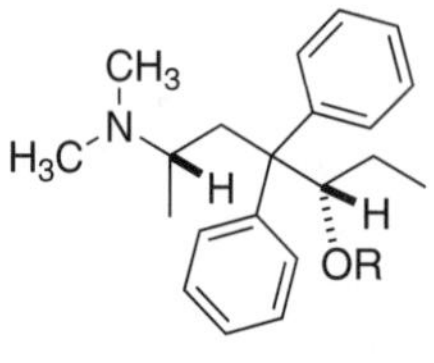

R = H **142** α-Methadol

R = Ac **143** *levo*-α-Acetylmethadol (LAAM)

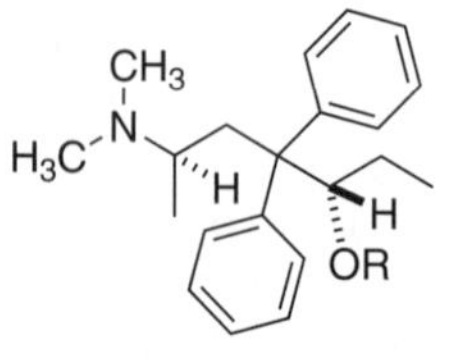

144 β-Methadol

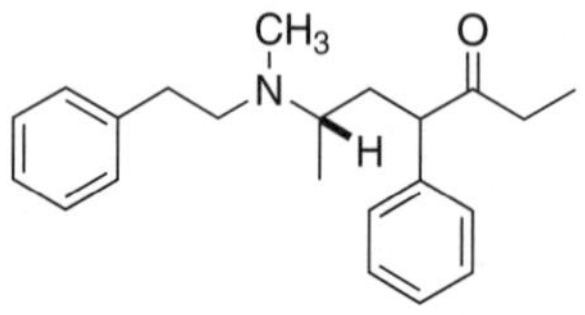

145 Diampromide

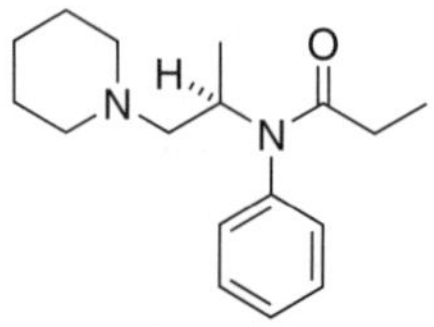

146 Phenampromide

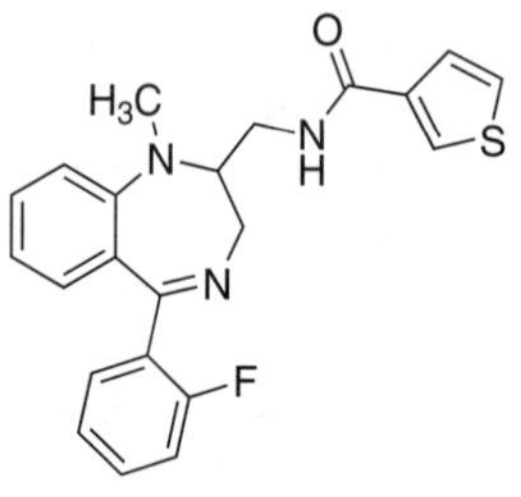

147 Tifluadom

Figure 29. Methadone analogs.

decreases potency 6–10-fold relative to methadone and isomethadone. Introduction of a second methyl group at position 5 yields *erythro*- and *threo* -5-methylmethadone [464]. The *erythro* form (**141**) is five times the potency of methadone, while the *threo* form is inactive. The more active *levo* isomer of *erythro*-5-methylmethadone has the 5S,6S configuration [465]. Interestingly, one of the isomers of the inactive *threo* racemate, with the 5S,6R configuration, combines the configurations found in the more active enantiomers of methadone and isomethadone [464]. NMR analysis suggests that the *erythro* form exhibits greater conformational flexibility than the *threo* form. The authors suggested that the marked difference in analgesic activity may be due to different conformational preferences of the *threo* and *erythro* forms [464], and that the conformation observed for *erythro*-5-methylmethadone in the solid state [466] is the active conformation of *erythro*-5-methylmethadone, as well as for (−)-methadone and (−)-isomethadone.

The ketone side chain is also important for activity (see Ref. [460]) and changing the length of this chain decreases activity. Reduction of the ketone to the two possible alcohols, α- and β-methadol (**142** and **144**) decreases activity, but activity can be restored by acetylation [467]; the resulting acetates are more potent than the parent ketones. The more active methadol isomers, (−)-α- and (−)-β-methadol (with absolute configurations 3S,6S and 3S,6R, respectively [468]), both have the same 3S configuration, suggesting that the stereochemistry around the alcohol is more important in these derivatives. The more active 3S,6R isomers of β-methadol and β-acetylmethadol are derived from the more active R-(−) enantiomer of methadone. An interesting reversal in enantioselectivity occurs in the α-series; while the more potent 3S,6S isomer of α-methadol is derived from the less active S-(+) isomer of methadone, acetylation reverses the enantioselectivity so that the more potent isomer is the 3R,6R-(+) isomer [468]. *levo*-α-Acetylmethadol (LAAM) (**143**) has a longer duration than methadone, requiring dosing only once every 3 days, and is being used in the United States for maintenance of individuals addicted to narcotics. Like methadone the major route of metabolism of this compound is *N*-demethylation. In the case of the methadols and acetylmethadols, these secondary amines derivatives are active, with potencies similar to the parent tertiary amines, and probably contribute to the activity and longer duration of action of LAAM (see Ref. [460]). The ketone of methadone has also been replaced with a variety of other functional groups, including esters, ethers and amides (see Ref. [460]). In the acyloxy series, propionyloxyisomethadone shows significant activity [469]; further modification of this compound yields propoxyphene (see below).

Most modifications of the phenyls in the diphenyl fragment of methadone result in substantial loss of analgesic activity (see Ref. [460]). In normethadone analogs, replacement of one of the phenyl rings by a benzyl group abolishes activity [470], but in the isomethadone analog with a propionoxy group in place of the ketone this modification results in propoxyphene (**10**) (Fig. 1) that has modest analgesic activity (approximately one-tenth the potency of methadone) [469] (see Table 1). The replacement of one of the phenyl rings by a benzyl group introduces a second chiral center into this molecule. The active (+) isomer has the 2S,3R stereochemistry, which is the same absolute configuration at C_3 as in the active (−)-isomethadone (see Refs [281,460]). *N*-Arylpropionamide analgesics such as the methadone analog diampromide (**145**) and the isomethadone analog phenampromide (**146**) contain a single aromatic ring attached to a nitrogen, analogous to the 4-anilidopiperidines, and are exceptions to the requirement for a second phenyl ring. Diampromide is somewhat more potent than phenampromide, with a potency between meperidine and morphine. The more active isomer of phenampromide is the R-(−) isomer (which is the same configuration as the more active (−)-isomethadone), while the more active isomer of diampromide has the S configuration, which is opposite that of (−)-methadone (see Refs [281,460]). This led Portoghese to suggest that methadone and diampromide probably differ in their modes of interaction with opioid receptors [67].

5.9. κ-Selective Agonists

There has been considerable interest in developing κ-agonists as analgesics that would not have the side effects characteristic of morphine and other μ-opiates (e.g., respiratory depression and addiction), and therefore numerous κ-selective compounds have been reported over the last two decades (see Refs [284,285,471] for reviews). As noted above (Section 3.3.2), while κ-receptors can mediate analgesia, there are differences between the effects mediated by μ- and κ-receptors (see Ref. [136] for a review). There has also been interest in κ-selective compounds as potential neuroprotective and anticonvulsants agents (see Refs [250,472] for reviews).

While some compounds with κ-agonist activity are used clinically as analgesics (e.g., pentazocine, nalbuphine, and butorphanol) (Table 3), the use of many κ-agonists has been severely limited by centrally mediated side effects, mainly dysphoria and sedation, associated with most of these compounds [136]. The ability of κ-agonists to produce analgesia in inflammatory pain through interaction with peripheral κ-opioid receptors has stirred interest in the development of peripherally selective κ-agonists (see Section 5.9.2).

5.9.1. Centrally Acting Agonists Benzomorphan ligands such as EKC (**33**) and bremazocine (**34**) (Fig. 6) have been used to study κ-opioid receptors, but these compounds exhibit low selectivity for these receptors (K_i ratio (μ/κ) = 1.9 for EKC and 8.3 for bremazocine) (see Table 10). As discussed above (Section 5.3.1), the 4,5α-epoxymorphinan TRK-820 (**65**) (Fig. 16) is also a potent agonist at κ-opioid receptors that exhibits modest selectivity for these receptors (K_i ratio (μ/κ) = 15) (see Table 10). The benzodiazepine tifluadom (**147**) reported by Römer [473] is also a κ-opioid receptor agonist, but it too exhibits low selectivity for κ-receptors (K_i ratio (μ/κ) = 5.4) (see Table 11). Racemic tifluadom exhibits only very weak affinity for benzodiazepine receptors (IC_{50} = 4.1 μM) [474]; the (−)-isomer has greater affinity for opioid receptors and is more selective for opioid over benzodiazepine receptors than the (+)-isomer [474]. Tifluadom also exhibits high affinity for peripheral CCK receptors (IC_{50} = 29 nM for the (−)-isomer) and is an antagonist at these receptors [475]. Additional analogs of tifluadom with comparable affinity and somewhat greater selectivity have been reported that are devoid of affinity for CCK receptors [173].

The first κ-selective nonpeptide derivative identified was the benzacetamide U50,488 (**35**) (Fig. 6) that was found while examining cycloalkane-1,2-diamines as antidepressants [102]. This compound exhibits high κ-selectivity in binding assays (IC_{50} ratio (μ/κ) = 256 [103]) (see Table 11) and produces analgesia via κ-receptors *in vivo* [476,477].

U50,488, which was initially characterized as the racemic mixture, was resolved [478], and the *levo* isomer found to have greater affinity for κ-receptors [479] (see Table 11). The absolute stereochemistry of the *levo* isomer was determined by X-ray crystallography of an intermediate to be 1S,2S [478]. The protective effects of U50,488 against the temporary bilateral carotid occlusion model of cerebral ischemia also reside predominantly in the *levo* isomer [480], consistent with the hypothesis that κ-receptors may be involved in these protective effects.

Introduction of a spiro ether group on the cyclohexane ring was one of the earliest modifications to U50,488 reported, giving (−) U69,593 (**36**) (Fig. 6) [103] and spiradoline (U62,066) (**148**) (Fig. 30) [481]. U69,593 exhibits improved κ-receptor selectivity over U50,488 (IC_{50} ratio (μ/κ) = 484 [103]) (see Table 11). The tritiated form of U69,593 was prepared by catalytic tritiation of the aromatic chlorines of U62,066 [103], and this highly selective tritiated κ-ligand has been used extensively in radioligand binding assays (see section "Radioligand Binding Assays"). The X-ray structure of U69,593 has been reported [482].[2] As is the case with U50,488, the (−) isomer of spiradoline is much more potent (>30-fold) than the (+) isomer in analgesic assays [481] and possesses much greater affinity for κ-receptors (see Table 11) [483]; the (+) isomer displays μ-receptor selectivity (see Table 11) [483]. Like U50,488 spiradoline exhibits neuroprotective effects and is even more effective than U50,488 in reducing postischemic necrosis of the vulnerable hippocampal CA_1 neurons [484]. Further examination of C_4 and/or C_5 methyl ether and spiro tetrahydrofuran substituents indicated that optimal κ-receptor selectivity was obtained when the oxygen was in the equatorial (β) orientation at C_4 of the ring [104].

---

[2] However, as noted by Rees [471], there is a discrepancy between the X-ray structure as drawn in this paper (5R,7S,8S) and that indicated in the title (5S,7S,8S), which has led to some confusion concerning the absolute stereochemistry of U69,593.

Table 11. Opioid Receptor Affinities, κ-Selectivity, and Analgesic Activity of κ-Opioid Ligands

Compound	Receptor Affinity K_i (nM)		K_i Ratio	Analgesic Activity MPE$_{50}$ (mg/kg)^c		References
	κ^a	μ^b		i.v.	p.o.	
Tifluadom (147)	4.1	22	5.4			[120]
U50,488 (35)	0.69	435	630	1.96	9.6	
(−) isomer	0.89					[479]
(+) isomer	299					[479]
U69,593 (36)	0.67	2460	3670	0.67	18.4	
Spiradoline						
U62,066 (148)	0.35	43.7	125	0.38	48.5	
(−) isomer	0.31	84	271			[483]
(+) isomer	1360	9.8	0.0072			[483]
PD 117302 (149)	0.50	399	798	1.41	25.3	
(−) isomer	0.39	414	1060			[483]
(−) CI-977 (37)	0.11	99.6	905	0.02	1.8	
DuP 747 (150)	6^d	304^d	50.6	0.46 s.c.^{e} 6.2^e		[491]
S,S isomer	7^d	750^d	107	0.15 s.c.^{e} 5.2^e		[492]
Niravoline						
RU 51599 (151)	0.41	699	1700	—	—	[530]
BRL 52537A (155)	0.24	1,560	6,500	0.05 s.c.^{m} —		[500]
BRL 52656A (156)	0.57	2,340	4,100	0.11 s.c.^{m} —		[500]
GR 89,696 (158)	1.15^g	0.65^g	0.56	0.0005 s.c.^{h}		[504,508]
R-84760 (159)	0.44^f	297^f	681	0.0013 s.c.^{e} 0.013p.o.^{e}		[503]
Apadoline						
RP 60180 (160)	0.55^i	57.1	104^j	—	—	[511]
ICI 197067 (161)	6.3^k	11,800^k	1,870	0.05 s.c.^{l} —		[512]
ICI 199441 (162)	6.9^k	4,500^k	652	0.004 s.c.^{l} —		[512]
HZ2 (163)	15^n	>1,000^n	>65	—	—	[518]

Data from Ref. [105] except where otherwise indicated.
^{a} Inhibition of [^{3}H]U69,593 binding in guinea pig forebrain, except where otherwise noted.
^{b} Inhibition of [^{3}H]DAMGO binding in guinea pig forebrain, except where otherwise noted.
^{c} Rat paw pressure test, MPE$_{50}$ is the dose required to produce 50% of the maximum possible analgesic effect, except where otherwise indicated.
^{d} U50,488 K_i = 6 and 825 nM in κ- and μ-assays, respectively [491].
^{e} Mouse phenylquinone writhing assay ED$_{50}$ s.c. and p.o., respectively. ED$_{50}$ for U50,488 = 1.2 mg/kg s.c. and 13 mg/kg p.o. [491]; 0.47 mg/kg s.c. and 27 mg/kg p.o. [503].
^{f} IC$_{50}$ values for U50,488 = 7.59 nM and 571 nM for κ- and μ-binding, respectively.
^{g} Binding determined in monkey cortical membranes.
^{h} Mouse acetylcholine writhing assays ED$_{50}$ s.c. ED$_{50}$ for U50,488 = 0.41 mg/kg s.c.
^{i} Inhibition of [^{3}H]EKC binding in guinea pig cerebellum.
^{j} K_i (δ) = 1.6 nM ([^{3}H]EKC binding in NG-108-15 cells), K_i ratio (δ/κ) = 2.9.
^{k} IC$_{50}$ values for inhibition of [^{3}H]bremazocine binding in guinea pig brain minus cerebellum and [^{3}H]naloxone binding in rat brain for κ- and μ-receptors, respectively. U50,488 IC$_{50}$ = 95.5 nM and 14,200 nM, respectively.
^{l} ED$_{50}$ in mouse acetic acid abdominal constriction assay. U50,488 ED$_{50}$ = 1.1 mg/kg s.c.
^{m} ED$_{50}$ mouse tail flick assay. U50,488 ED$_{50}$ = 1.9 mg/kg s.c.
^{n} Inhibition of [^{3}H]CI-977 and [^{3}H]naloxone binding in rat brain membranes for κ- and μ-receptors, respectively.

From the initial compounds, it was clear that the spacing between the amide and the aromatic ring system in the N-methylamide side chain is critical for κ-receptor activity. While phenylacetamide derivatives (i.e., −NCH$_3$COCH$_2$Ar) such as U50,488 and its analogs exhibit κ-activity, benzamide derivatives (i.e., −NHCOAr and −NCH$_3$COAr) exhibit morphine-like effects [102,485]. In both series, N-methyl substitution increases potency over the unsubstituted secondary amides [102,486].

Numerous variations have been examined in all portions of the U50,488 and spiradoline

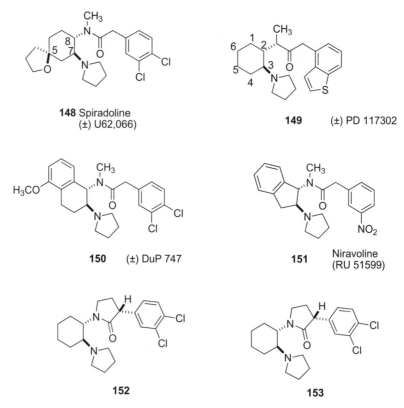

Figure 30. κ-Receptor selective agonists. The structures of U50,488 (**35**), U69,593 (**36**), and CI-977 (**37**) are given in Fig. 6.

structures (see Refs [284,285,471] and the previous edition of this chapter [283] for detailed reviews). Figure 31 shows the general structure (**154**) common to most of the analogs and a summary of the SAR of U50,488 [284]. In addition to varying the spacing between the amide and the aromatic ring, other modifications in the N-methylamide side chain of

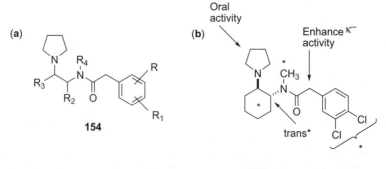

Figure 31. (a) General structure (**154**) common to most of the analogs of U50,488. (b) Summary of SAR for analgesic activity of U50,488 (**35**). From Ref. [284].

U50,488 and spiradoline examined include replacement of the amide linkage, substitution on the methylene between the amide and the aromatic ring, and varying the identity of the aromatic ring system. The amide linkage appears to be important because substitution of this linkage with a variety of replacements (reversed amide, reduced N-methyl amide, or ester) all caused significant decreases in κ-receptor affinity [486]; only the N-methylthioamide derivative retained significant affinity, and even this minor substitution of sulfur for oxygen resulted in approximately a 15-fold decrease in affinity. While substitution of a methyl group α to the carbonyl results in a large (200-fold) decrease in κ-receptor affinity, conformational constraint involving linking the α-carbon to either the aromatic group (**152**) [487] or the amide nitrogen (**153**) [488] resulted in analogs with high affinity for κ-receptors.

Variations in the aromatic ring system of the N-methylamide side chain have led to two well-characterized agents. Replacement of the dichlorophenyl ring with either 4-benzo[b]thiophene to give (±)PD 117302 (**149**) (Fig. 30) or 4-benzofuran [489] yields potent compounds with high κ-receptor selectivity (see Table 11). Again the (−)-S,S enantiomer exhibits much greater affinity for κ-receptors than the R,R isomer (see Table 11) and is the one exhibiting analgesic activity. Similar replacement of the phenyl ring of U69,593 with the 4-benzofuran ring system yields (−)-CI-977 (enadoline) (**37**) (Fig. 6) [104,105], which is also one of the most potent and highly selective κ-ligand (see Table 8) [104,105]. Tritiated CI-977 is commercially available (see Table 7) and frequently used in radioligand binding assays.

Numerous variations and substitutions on the cyclohexane ring have also been reported. The configuration of the 1,2-diamino amide on the cyclohexane ring is critical for κ-receptor affinity and activity. κ-Ligands such as U50,488 and its analogs all have the *trans* configuration; isomers with the *cis* configuration have weak affinity for κ-receptors and instead have much higher affinity for the (+) -3-PPP [(+)-3-(3-hydroxyphenyl)-N-(1-propyl)piperidine] or σ-binding site [479]. For the amine side chain, the pyrrolidine ring is the optimal substituent; changing the ring size decreased κ-receptor affinity and selectivity and "opening" the ring to give the N,N-diethyl analog resulted in almost a 500-fold decrease in κ receptor affinity [489]. Attachment of aromatic rings to the cyclohexane ring has been examined by several groups [490–492]. By combining the structures of U50,488 and 2-aminotetralin, researchers at DuPont Merck developed tetrahydronaphthyl analogs as κ-agonists [491,492]; a 5-methyl ether substituent gave optimal activity and led to the identification of DuP 747 (**150**) as a new κ-analgesic. Analysis of the two enantiomers of racemic DuP 747 indicated that again the (+)-S,S isomer was the one with high κ-receptor affinity and analgesic activity [492]. The related compound niravoline (RU 51599) (**151**), in which a cyclopentane ring replaces the cyclohexane ring [493,494], has been studied for its diuretic effects in rats with cirrhosis [495–497] and in brain edema in animal models of ischemia and stroke, traumatic brain injury, and brain tumors (see Refs [498,499]).

Alternative ring systems to the 1,2-diamine-substituted cyclohexane skeleton have also been examined (Fig. 32). Researchers from Glaxo and SmithKline have examined both piperidine and piperazine analogs as κ-agonists (see Refs [283–285] for detailed reviews). Researchers at SmithKline in Italy identified a common pharmacophore (−N−C−C−N−COCH$_2$Ar) in κ-selective ligands, with a torsion angle of 60° around the C−C bond [500], which led them to develop piperidine derivatives, such as BRL 52537A (**155**) and BRL 52656A (**156**). The torsion angle found for the N$_2$-C$_1$-C$_9$-N$_{10}$ pharmacophore by X-ray and NMR analysis of these compounds was approximately 60°, similar to that in the proposed pharmacophore [500]. In both the piperidine and the related tetrahydroisoquinoline series (e.g., **157**), the (−)-S isomer was the active enantiomer [500–502]. A compound with the related thiazine ring system (R-84760) (**159**) [503] has also been reported to have κ-opioid receptor activity. R-84760 is an extremely potent κ-agonist, with a potency of 2.5–20 times that of CI-977 in several different assays; the stereochemistry of R-84760, the most active isomer,

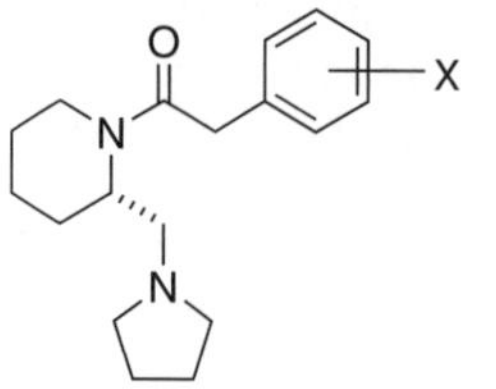

X = 3,4-diCl **155** (S) BRL 52537A

X = 4-CF$_3$ **156** (S) BRL 52656A

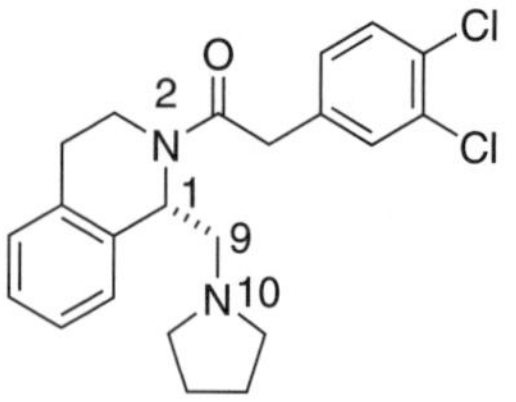

157

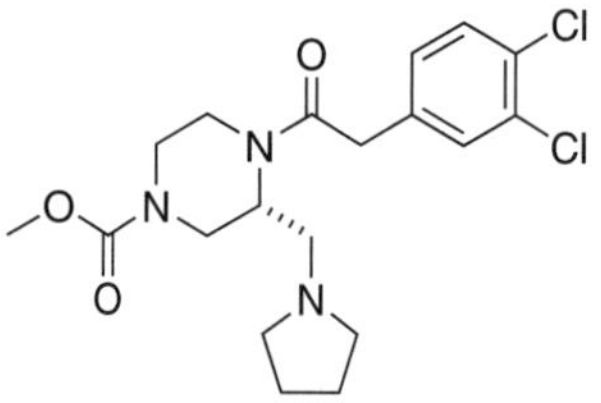

158 (R,S) GR 89696
(R) GR 103545

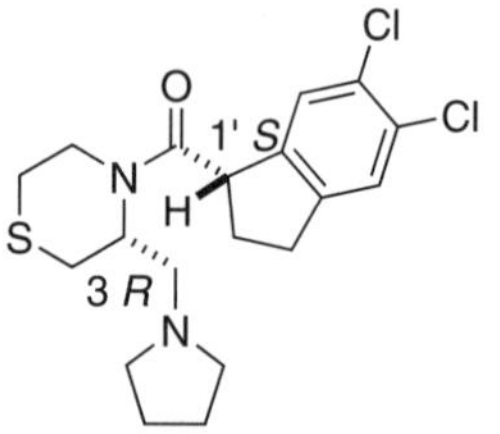

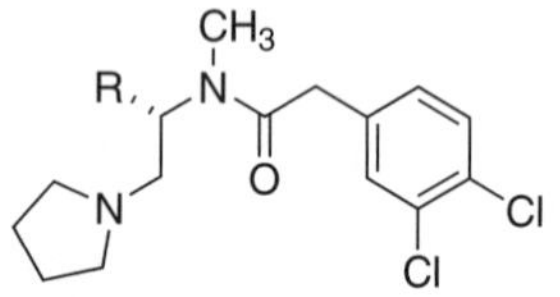

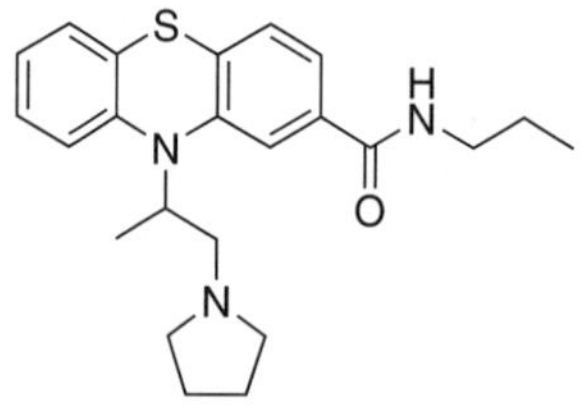

R = CH(CH$_3$)$_2$ **161** (S) ICI 197067

R = —⟨phenyl⟩   **162** (S) ICI 199441

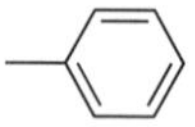

160 Apadoline
(RP 60180)

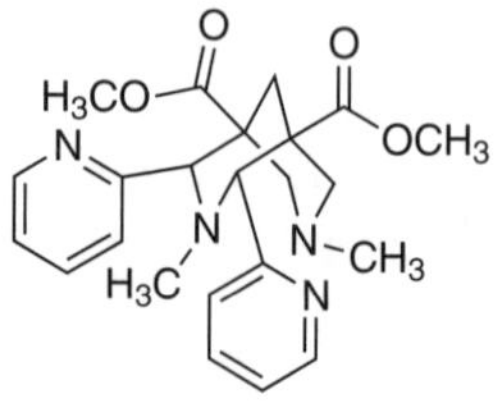

163 HZ2

Figure 32. κ-Receptor selective agonists with alternative ring systems and open chain analogs of the 1,2-diamine-substituted cyclohexane ring.

is 3R,1'S. Because of the "relative inacessibility" of the 2,4-substituted piperidine ring system, the Glaxo group evaluated piperazine analogs where N$_4$ substiutents could be more easily introduced [504]. This led to the development of racemic GR 89696 and its R-isomer GR 103545 (**158**) [504,505]. Neuroprotective effects have been demonstrated for GR

89696 [506]. GR 89696 has been examined in the guinea pig hippocampus and reported to be an agonist at the κ_2-receptor subtype but an antagonist at κ_1-receptors [507]. Recent studies in rhesus monkeys [508] found that GR 89,696 exhibited considerable affinity for μ-receptors (Table 11) in binding assays, but low efficacy and potency (100-fold lower than at κ-receptors) in [^{35}S]GTPγS assays. *In vivo* the effects of GR 89,696 were less sensitive than U50,488 or U69,593 to antagonism by naltrexone and insensitive to norBNI, consistent with action through the postulated κ_2-receptors. Interestingly, substitution of a methyl group α to the pyrrolidine nitrogen had a marked effect on the affinity of the derivatives for μ-opioid receptors, with the (*S,S*) isomer exhibiting subnamolar affinity for μ- as well as κ-receptors while the (*S,R*) and (*R,S*) exhibited very low (>1 μM) affinity for μ-receptors [509]. Addition of a 3-hydroxyl to the pyrrolidine ring led to compounds with significant peripheral selectivity (see below) [510]. The phenothioazine derivative apadoline (RP 60180) (**160**) [511] exhibits 100-fold selectivity for κ receptors over μ-receptors, but it possesses only a small (3-fold) selectivity for κ- over δ-receptors.

Compounds with κ-receptor activity that do not contain the cyclohexane ring have also been identified (Fig. 32). While replacement of the cyclohexane ring with an unsubstituted ethyl chain results in compounds with weak κ-receptor affinity, a substituent α to the amide functionality yields compounds with high affinity for κ-receptors [512,513]. Examination of both isomers indicated that a methyl group at the 1-position with the *S*-configuration was active and had reasonable affinity for κ-receptors, while the isomer with the *R*-configuration at this position was inactive [513]. Conformational analysis suggested that only those compounds capable of adopting a conformation similar to U50,488 were κ-agonists. The most potent analogs were the isopropyl (ICI 197067) (**161**) and phenyl derivatives (ICI 199441) (**162**) (Fig. 32). Fluorescent derivatives of ICI 199441 have been prepared by attachment of fluorescein isothiocyanate via a spacer to the *meta*- or *para*-position of the central phenyl ring [514]. Attachment of an acidic functionality to the *meta*-position of the central phenyl ring of ICI 199441 gave ICI 204448 that has limited ability to penetrate the CNS [515] (see below).

U50,488 and its analogs are structurally distinct from the benzomorphan κ-opiates and are considerably more flexible than are the rigid alkaloid alkaloids. This raises questions concerning their bioactive conformation and how they bind to opioid receptors compared to the more rigid alkaloid opiates. Early studies [516–518] attempted to identify possible bioactive conformations of U50,488 and its congeners by conformational analysis of the ligands and superimposition of the arylacetamides with benzomorphans. With the cloning of opioid receptors and development of computational models of opioid receptors (see section "Computational Models of Opioid Receptors"), several groups [173,176,177,180] have proposed possible binding modes for the arylacetamides docked to the κ-opioid receptor. While all of the models of the arylacetamides docked to κ-opioid receptors assumed an interaction of the protonated amine of the ligand with Asp 138, there are significant differences in other proposed receptor–ligand interactions in these models (see Ref. [180] for a detailed comparison of the models). These results illustrate the complexity of modeling these more flexible ligands and determining how they interact with their receptors.

HZ2 (**163**), which is structurally unrelated to the arylacetamide κ-selective agonists, exhibits reasonable κ-receptor affinity ($K_i = 15$ nM) with low affinity for μ-receptors ($K_i > 1000$ nM) [518]. NMR and molecular modeling studies were performed to compare this novel bicyclononanone to the arylacetmides and ketocyclazocine [518]. The compound is reported to be have potent antinociceptive activity, and like other κ-agonists be active against inflammatory pain [519]. The quaternary methiodide derivative retains high κ-receptor affinity, and thus may be useful as a peripheral κ-agonist [520].

Several of the κ-selective ligands have undergone testing in humans [521–528] (see Ref. [472] for a review). Side effects associated with κ-receptors include sedation and dysphoric effects [136]. The dysphoric effects are of particular concern, and have severely limited the usefulness of the majority of centrally acting κ-agents. Many of the older nonselective compounds possessed high affinity for σ and PCP sites, raising the possibility that these sites might contribute to the dysphoric effects. The benzomorphan Mr2034 (**97**), which is inactive at these sites, [390] and the κ-selective agonist spiradoline (**148**) also produce naloxone-sensitive dysphoric effects [529], however, indicating that κ-receptors mediate pschychotomimetic effects. Clinical trials of spiradoline for the treatment of pain were discontinued [472]. Initial evaluation in humans of enadoline (CI-977) (**37**) (Fig. 6) for its diuretic effects found that the dysphoric effects attributable to the drug were minimal and not considered clinically significant [522], but in a subsequent study of its use in postsurgical pain adverse neuropyschiatric events led to early termination of the study [524]. Enadoline has, however, been granted orphan drug status for the treatment of severe head injuries [472]. Apadoline (RP 60180) (**160**) has been evaluated in an experimental human pain model and reported to cause less side effects than the clinically used agent pentazocine [526]. Niravoline (**151**) was examined for its aquaretic effects in patients with cirrhosis; the highest doses examined induced personality disorders, but moderate doses produced the desired aquaretic effects and were well tolerated [528]. However, clinical development of this agent has been discontinued (see Ref. [472]). Clinical investigation of several centrally acting κ-selective agents (apadoline, DuP 747, enadoline, and the 4,5-epoxymorphinan TRK-820) has continued [285].

5.9.2. Peripherally Acting Agonists Concern over centrally mediated side effects has prompted attempts to develop peripherally selective κ-opioid agonists. Peripheral opioid receptors can mediate analgesia, particularly in cases of inflammation (see Refs [55–57] for reviews). In order to limit penetration of the blood–brain barrier (BBB), generally polar and/or charged funtionalities have been introduced into the compounds (see Ref. [531] for a review). Several modifications to the κ-selective agonist ICI199441 (**162**) (Fig. 32) have been reported to restrict the compounds access to the CNS. In ICI 204448 (**164**) (Fig. 33), an acid functionality was introduced on the central phenyl ring [515], and Portoghese and coworkers prepared aspartic acid conjugates of ICI 199441 (**165**) [532] to decrease penetration into the CNS. Investigators at the Adolor Corporation have investigated various substituents on the phenylacetamide side-chain phenyl ring and found that the 4-trifluoromethyl group resulting in an analog with high κ-receptor affinity and less central activity compared to the parent ICI 199441 [533]; incorporation of additional functionalities into the central phenyl ring (e.g., structure **166**) were examined to further restrict the compounds to the periphery [534]. Researchers at Merck prepared EMD 60400 (**167**) [535] by introducing an amino substituent on the phenyl ring of the phenylacetamide side chain and a 3-hydroxyl group on the pyrrolidine ring; researchers at Glaxo have prepared analogs of GR 103545 with reduced penetration of the BBB using a similar approach [510]. In the case of BRL 52974 (**169**) [536] an imidazole ring was attached to the piperidine ring in order to increase hydrophilicity.

In asimadoline (EMD 61753) (**168**) [537] an additional phenyl ring was attached α to the amide on the phenylacetamide side chain, which increases lipophilicity. This compound is also a peripherally selective κ-agonist [538]; studies in knockout mice lacking P-gylcoprotein indicate that asmidaloline is transported by P-glycoprotein, and that transport by this protein limits the compound's penetration of the BBB [539]. As noted by Barber et al. [538], amphiphilic structures such asimadoline generally have greater oral activity than hydrophilic compounds; consistent with this they found that asimadoline exhibits much greater potency by the oral route than did ICI 204448 (ID$_{50}$ in pressure nociception in inflammatory hyperalgesia: ID$_{50}$ for asimadoline = 0.2 mg/kg s.c. and 3.1 mg/kg p.o., versus ID$_{50}$ for ICI 204448 = 0.8 mg/kg s.c. and 30 mg/kg p.o.). Oral absorption of asimadoline was not significantly altered in the P-glycoprotein knockout mice, indicating that the intestinal P-glycoprotein did not impede absorption after oral administration [539]. Unexpectedly, increases in pain were reported in clinical trials of this compound in patients after knee surgery [540]. Subsequent investigation of this compound in inflammation in the rat

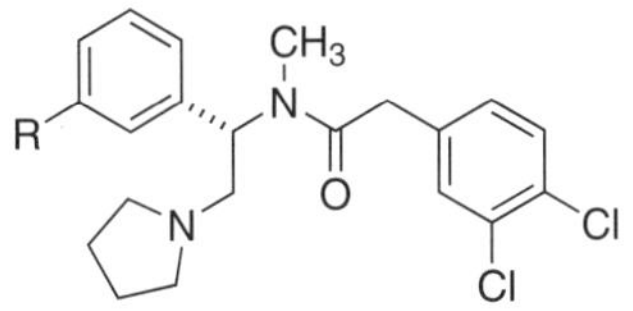

R = OCH$_2$CO$_2$H	**164**	(R,S) ICI 204448
R = NH-L/D-Asp	**165**	

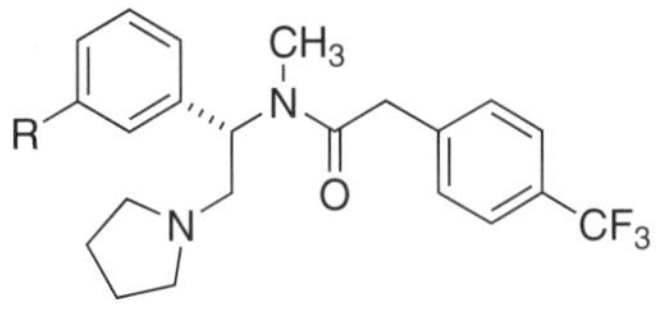

166 R = OCH$_2$CO$_2$H

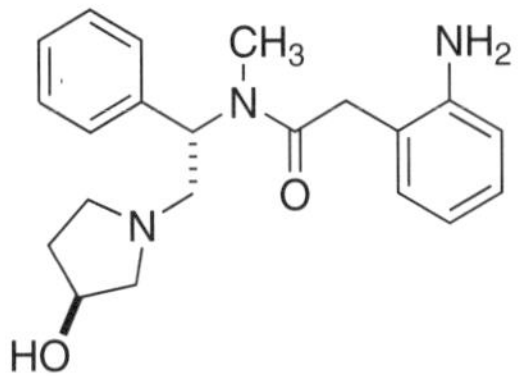

167 EMD 60400

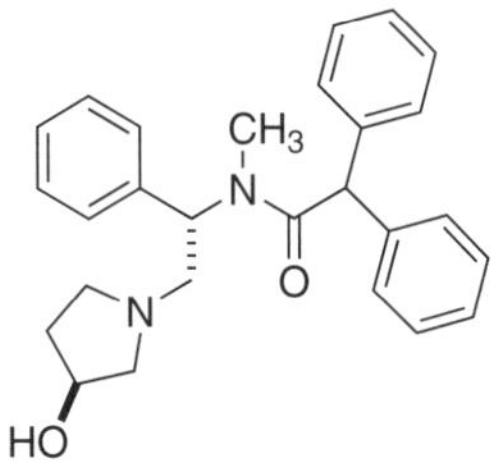

168 Asimadoline (EMD 61753)

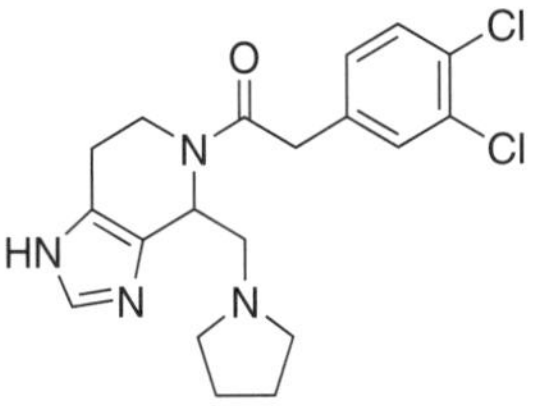

169 BRL 52974

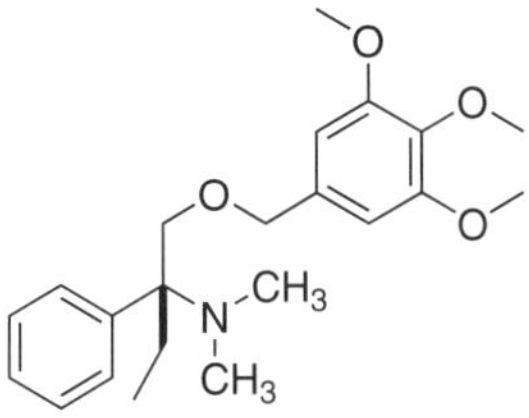

170 Fedotozine

Figure 33. Peripheral κ-receptor selective agonists.

found that while the analgesic effects of asimadoline were κ-opioid receptor modiated, the adverse hyperalgesic and proinflammatory effects observed were not mediated by opioid receptors [540].

Fedotozine (Jo-1196) (**170**), which is structurally related to the acyclic κ-agonists, has *in vivo* antinociceptive effects on duodenal pain that appear to be mediated by peripheral κ-opioid receptors, but the compound is inactive after central administration [541]. In binding assays (in dog myeteric plexus), however, this compound exhibits similar affinity ($K_i = 0.3$–$0.8\,\mu$M) for all three types of opioid receptors [542]. Unlike other κ-agonists, fedotozine does not induce diuresis following either s.c. or i.c.v. administration [543]. Fedotozine also fails to substitute for either U50,488 or morphine in animals trained to discriminate these drugs [544]. The main effects demonstrated for fedotozine have been in the gastrointestinal tract (see Ref. [545] for a detailed review of the pharmacology of fedotozine), and therefore this compound has been investigated clinically for the treatment of digestive disorders characterized by abdominal pain, namely, dyspepsia and irritable bowel syndrome (see Ref. [545] for a review). In Phase II/III studies for both disease states [546–548], significant improvement of symptoms in the patients treated with fedotozine compared to placebo was reported. Fedotozine also relieves visceral hypersensitivity to gastric and colonic distention, which are often observed in dyspepsia and irritable bowel syndrome, respectively [549,550]. Although clinical trials of fedotozine have been discontinued (see Ref. [472]), peripherally selective κ-agonists remain a potentially important therapeutic target for treatment of these digestive disorders.

5.10. δ-Selective Agonists

5.10.1. BW373U86, SNC80, and Analogs

Initially, all of the δ-selective agonists were peptide derivatives. The first nonpeptide agonist selective for δ-receptors, BW373U86 (**26**) (Fig. 5), was discovered by screening [551]. The pharmacology of this compound has been examined in considerable detail (see Refs [213,215] for reviews). This compound has only modest selectivity for δ-receptors in binding assays (see Table 12). In antinociceptive assays in mice, it appears to function as a partial agonist at both δ- and μ-receptors. The activity of BW373U86 is highly dependent on the route of administration, with effects at the spinal level mediated by δ-receptors and supraspinal effects involving interactions with μ-receptors [552]. In monkeys, BW373U86 did not produce antinociceptive effects in the warm-water tail-withdrawal assay following subcutaneous administration [553]. This compound also produces convulsant effects in both mice and monkeys [553–555], which appeared to be mediated by δ-receptors.

BW373U86 is a racemic mixture and this could complicate its pharmacological profile. Therefore, Rice and coworkers undertook the synthesis and characterization of isomers of BW373U86 [94]. The isomers with the R-configuration at the benzylic carbon exhibited greater affinity for δ-receptors than did the isomers with the S-configuration. One compound, SNC80 (**27**) (Fig. 5), the methyl ether of (+)-BW373U86, exhibited marked selectivity for δ- over μ-receptors in binding assays and smooth muscle assays (see Table 12) [94,556], making it the most δ-selective nonpeptide agonist reported. SNC80 is a systemically active δ-receptor selective agonist, with its antinociceptive actions produced via interaction with both $δ_1$- and $δ_2$-, but not μ-opioid, receptors [556]. SNC80 is more effective in nociceptive assays than BW373U86, and consistent with its higher selectivity for δ-receptors, the antinociceptive effects of SNC80 appear to be mediated only by δ-receptors (see Ref. [215]). Brief, nonlethal seizures were observed in mice only at very high doses of SNC80 (100 mg/kg) [556]; this may be due to metabolism of

Table 12. Opioid receptor affinities, δ selectivity and opioid activity in the MVD of δ opioid agonists. Data for BW 373U86, SNC 80 and TAN 67 from Table 8 are included for comparison

Agonist	IC_{50}^{c} or K_i (nM)			IC_{50} ratio	IC_{50} (nM)	
	δ	μ	κ	(μ/δ)	MVD	Reference
BW373U86 (**26**)	0.92^c	46^c	–	50	0.2^a	[94]
SNC 80 (**27**)	1.0^c	2500^c	–	2300	2.7^a	[94]
	2.9	2500	–	860	2.7	[558]
172	0.87^c	3800^c	7500	4370	–	[565]
(-) SL-3111 (**174**)	4.1^c	7700^c	–	1900	360	[566]
(-) **175**	5.6	2620	1450	470	–	[569]
176	0.4	5000	–	14,000	–	[572]
177	1.2	1200	–	980	–	[573]
(±) TAN 67 (**28**)	1.1	2300	1800	2070	6.6^b	[96]
(SB 205607)	0.67	110	450	170	160	[577]
(−) TAN 67 (SB 213698)	0.47	70	270	150	62	[577]
(+) TAN 67 (SB 213697)	11	815	> 1000	74	> 1000	[577]
(−) **178**	0.9	129	1300	140	26	[574]

[a] Data for BW373U86, SNC80, and TAN67 from Table 8 are included for comparison.
[b] IC_{50} in the GPI are 143 and 5500 nM for BW373U86 and SNC 80, respectively.
[c] IC_{50} in the GPI is 26,500 nM.

SNC80 to a BW373U86-like compound [557]. In monkeys, SNC80 does not cause convulsions at doses up to 32 mg/kg, suggesting it may be safer than BW373U86 [215].

Several groups have explored the SAR of SNC80. Rice and coworkers have examined substitutions for the methoxy group on the phenyl ring [558], modifications of the amide functionality [559], modifications to the piperazine ring [560], and different substitutions on N^4 of the piperazine ring [561]. The methoxy group on the one phenyl ring of SNC80 could be substituted with other groups with retention of δ-receptor affinity and selectivity; the derivative without a substituent at this position has higher δ-affinity and selectivity ($IC_{50} = 0.94$ nM, IC_{50} (μ/δ) ratio >2660) than SNC80 [558]. The amide functionality is particularly sensitive to modification, with the N,N-dialkylbenzamide derivatives having higher δ-receptor affinity than the monosubstituted or unsubstituted derivatives [559]. This suggests that the amide group is an important structural feature for interaction with δ-receptors. Modifications can be made to the piperazine ring, including removal of the methyl groups and replacement of N^1 by carbon, with retention of reasonable δ-receptor affinity and selectivity [560,562,563]. A series of simplified piperazine derivatives were prepared, including **171** (Fig. 34), with improved δ-receptor selectivity (IC_{50} ratio = 1240 versus 245 for SNC80) and increased metabolic stability over SNC80 [564]. A piperidinylidene derivative of SNC80 with a double bond to the benzylic carbon exhibits higher δ-receptor affinity than the corresponding piperidine derivative [560]. Researchers at AstraZeneca have examined an extensive series of these piperidinylidene analogs [565] and identified derivatives, for example, **172**, which exhibited extremely high selectivity for δ-receptors (see Table 12) and were considerably more stable to degradation by rat liver microsomes than SNC80 [565]. Interestingly, the analog DPI 2505 (**173**) in which the 2-methyl group is shifted to the 3-position has been reported to be an antagonist [222]. The basic N^4 nitrogen is critical for δ-receptor binding, and opening the piperizine ring also results in large decreases (>60-fold) in δ-receptor affinity [560]. A variety of alkyl substituents on N^4 of the piperazine ring are tolerated by the δ-receptor [561,563], but SNC80 analogs containing a saturated alkyl group exhibit decreased efficacy compared to SNC80 in the [^{35}S]GTPγS assay; the N-cyclopropylmethyl analog is also a partial agonist [561]. Thus, the SAR of the nitrogen substituent of SNC80 for δ-receptor affinity is distinctly different from the SAR of this group in the morphinans for interaction with μ-receptors.

Hruby and coworkers designed a series of piperazine derivatives as peptidomimetic analogs of the δ-selective peptide [(2S,2R)-TMT^1]DPDPE (TMT = 3,2′,6′-trimethyltyrosine) [566,567], with SL-3111 (**174**) (see Table 12) as the lead compound. Although SL-3111 is structurally similar to SNC80, the large decrease in δ-receptor affinity upon methylation of the phenol and the high affinity of SL-3111 for the [W284L] mutated human δ-receptor is in contrast to the results for SNC80 and more closely parallel those for the peptide p-Cl-DPDPE, suggesting that the binding profile of SL-3111 is more similar to that of the peptide than to SNC80 [566].

A series of 4-aminopiperidine derivatives was designed by researchers at the Research Triangle Institute by transposition of the N^4 of the piperazine ring and the benzylic carbon [568–570], with the cis (3S,4R) isomer (**175**) exhibiting the highest affinity. This transposition decreased δ-receptor affinity while increasing μ-receptor affinity, such that the piperidines exhibited somewhat lower δ-selectivity than the piperazine derivatives (see Table 12). A large series of these 4-aminopiperidine analogs was also prepared by researchers at R.W. Johnson Pharmaceutical Institute and subjected to CoMFA analysis [182]. Recently, a series of constrained 4-diarylaminotropane derivatives (**176**) were reported by two groups [571,572], with the unsubstituted derivative exhibiting high δ-receptor affinity and exceptional δ-selectivity (see Table 12) [572]; the N-allyl and related derivatives exhibit decreased efficacy and antagonist activity in the [^{35}S]GTPγS assay [571]. In an alternative approach chosen to yield achiral δ-agonists without the complicated stereochemistry of SNC80, researchers

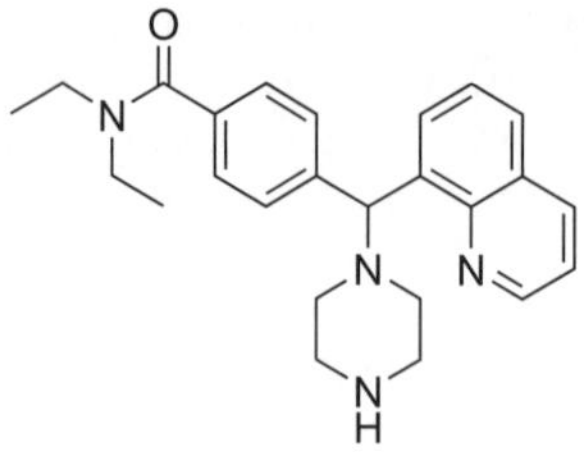

171

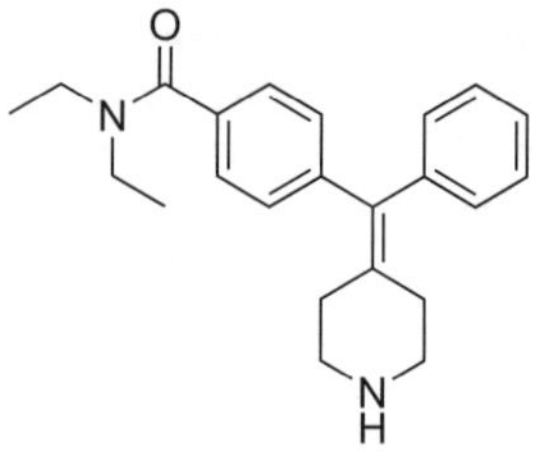

172

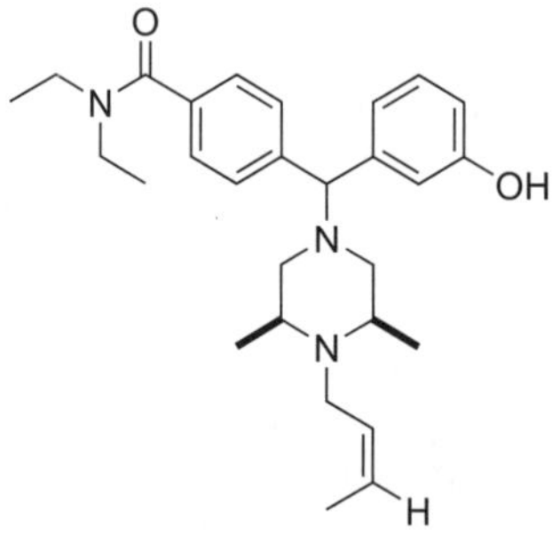

173 DPI 2505

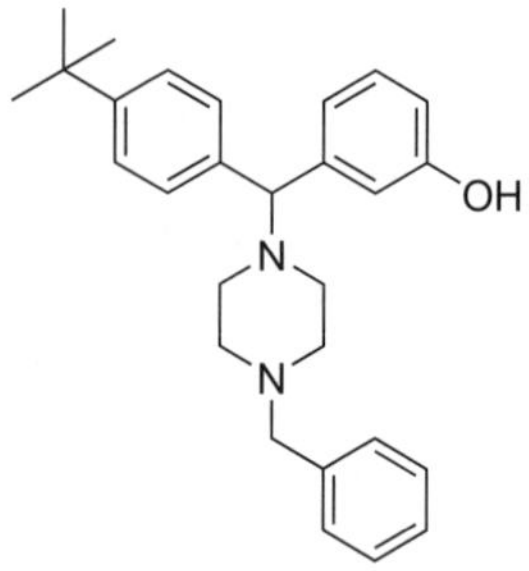

174 SL-3111

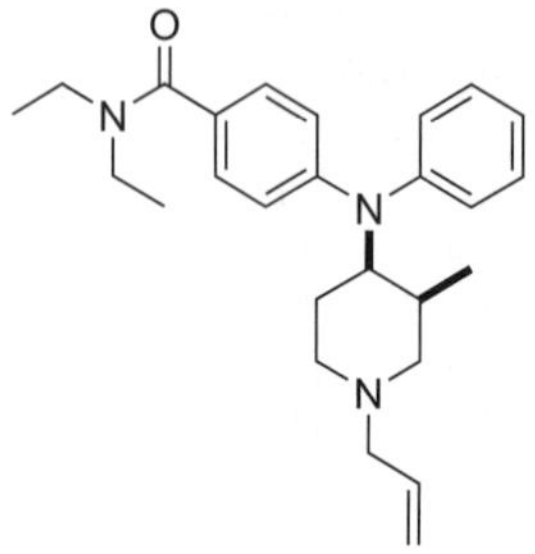

175

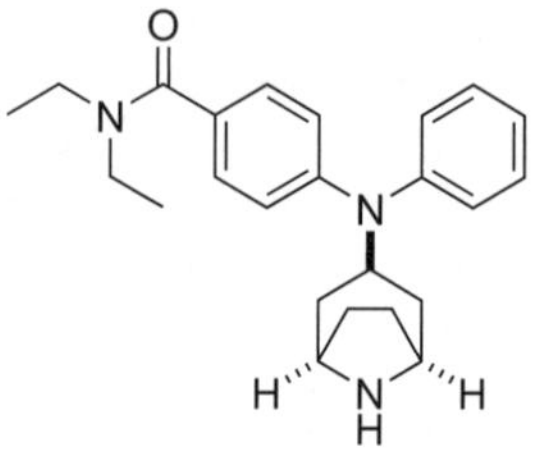

176

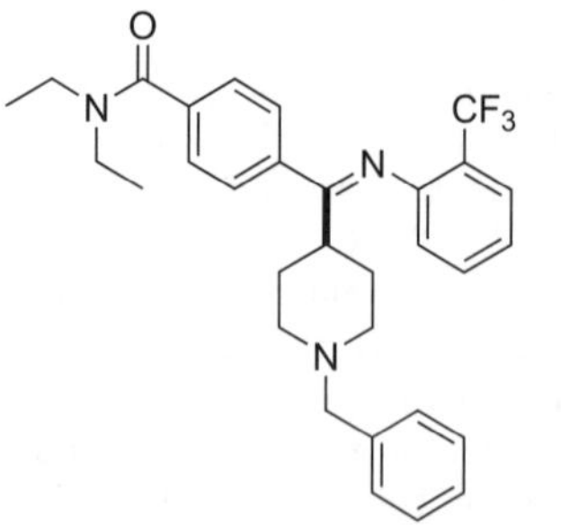

177

Figure 34. SNC80 analogs.

at R.W. Johnson also prepared a series of piperazinyl benzamidines [573]; the highest affinity ligand (**177**) exhibited affinity and selectivity similar to SNC80 (see Table 12). A number of other SNC80 analogs described in the patent literature have been reviewed by Dondio [213,214].

One question is how SNC80 interacts with δ-opioid receptors compared to the more classical morphinan ligands. Dondio et al. [574] compared SNC80 to the δ-antagonist SB 205588 (**78**) (Section 5.3.3). In this model, the basic N^4 nitrogen and the oxygenated phenyl ring of SNC80 were superimposed on the corresponding groups in SB 205588, with the centroid of the second phenyl ring of SNC80 overlapped with the pyrrole ring of SB 205588. Based on this model, it was hypothesized that the amide group of SNC80 might function as a nonaromatic δ "address" and thus be responsible for the δ-selectivity of SNC80. Loew and coworkers in their pharmacophoric model for a wide range of δ-receptor ligands also overlaid the basic N^4 nitrogen and oxygenated phenyl ring of SNC80 with the corresponding groups in the epoxymorphinans, with the amide occupying the third site in the 3-point model for δ-selective opioid recognition [575]. Coop and Jacobson, however, developed a 4-point pharmacophoric model based on a series of 4,5α-epoxymorphinans with high affinity for δ-receptors and found that SNC80 did not fit the model, suggesting that SNC80 does not bind to the δ-receptor in the same orientation as oxymorphindole [576]. Mutational analysis of δ-opioid receptors has found that three residues, Trp284 at the top of TMVI and Val296 and Val297 in the third extracellular loop, are crucial for the δ-receptor affinity of SNC80 as well as other δ-receptor agonists [160]. Computational models of BW373U86 [176] and 4-aminopiperidine analogs of SNC80 [182] docked to the δ-receptor have been described. In the models the basic nitrogen and oxygenated phenyl ring of these compounds occupy similar regions in the receptors as the corresponding groups in the epoxymorphinans, although there are some differences in the exact location of the two types of compounds and orientation of their phenyl rings (see Ref. [176]). In both models, the benzamide ring occupies a region at the TM/extracellular interface, and interacts with one or more of the residues identified as critical from mutagenesis studies. Comparison of the docking of SIOM with the 4-aminopiperidine compounds suggested that the benzamide ring occupies a similar region to the spiroindane of SIOM, consistent with the pharmacophoric models, and thus functions as an "address" to target the compounds to the δ-receptor [182].

5.10.2. Other δ-Receptor Agonists Attempts to identify δ-selective agonists related to the epoxymorphinans have concentrated in two areas. Modification of the nitrogen substituent of NTI resulted in the identification of SIOM as a $δ_1$-agonist (see Section 5.3.3). Other δ-selective agonists identified have been octahydroisoquinolines. As discussed above (Section 5.3.3), octahydroisoquinolines are either antagonists or agonists, depending upon the ring size of the spacer. While compounds with five-membered ring spacers are δ-receptor antagonists, introduction of a six-membered ring spacer, explored by both Japanese and Italian researchers, resulted in a new class of δ-receptors agonists [95,355,577] (see Refs [96,213] for reviews), with TAN67 (**28**) (Fig. 5) [95,96] being the most extensively studied. The rationale for the design of TAN67 by the Japanese group involved making the phenol ring freely rotatable by removing the 4,5-epoxy and 10-methylene functionalities of the epoxymorphinans and utilizing a heteroatom capable of forming a hydrogen bond with the receptor as an additional pharmacophoric group [96]. Racemic TAN67 shows high affinity and selectivity for δ-receptors (Table 12) [96], and is a potent agonist in cloned δ-human cells [578], and is a potent δ-agonist in cloned human cells [578] and in the MVD [96]. *In vivo* racemic TAN67 exhibits antinociceptive activity in the acetic acid writhing assay, but not the tail flick test, in normal mice (TAN67 was active in the tail flick assay in diabetic mice [579]); antagonism by BNTX but not naltriben suggested that TAN67 produced its antinociceptive effects through $δ_1$-receptors [580]. The pharmacological effects of the two enantiomers of TAN67 are distinctly different (see Ref. [581] for a review). (−) TAN67 (also named SB 213698 by the Italian group), which has the same absolute configuration as morphine, shows high affinity and selectivity for δ-receptors (Table 12), and is a full agonist in the MVD [577], while the (+) isomer is inactive *in vitro*. *In vivo* (–)TAN67 exhibits antinociceptive activity in the tail flick assay following both i.c.v. [582] and i.t. [583] administration, which appears to be mediated by δ_1-receptors. While (+)TAN67 was inactive following i.c.v. administration [582], after i.t. administration (+)TAN67 produced nociceptive behavior [583]; interestingly, the effects of (+)TAN67 were blocked by both naltrindole and (–)TAN67 [581]. (–)TAN67 exhibits decreased affinity for the [W284L] mutated δ-receptor compared to the wild-type receptor, but the magnitude of the decrease is less than for SNC80 [584]. In contrast, this mutation increases the intrinsic activity of SNC80 (indicated by an increase in the maximum [^{35}S]GTPγS binding), but it decreases the intrinsic activity of (–)TAN67, suggesting that these compounds may interact with different active receptor conformations [584]. In a recent report of a pharamacophoric model for the δ-opioid receptor, low-energy conformations of TAN67 were identified in which the phenol, the basic amine and the second aromatic amine could be superimposed on the corresponding groups in OMI and SIOM [585].

Based on their comparison of the antagonist SB 205588 and SNC80 and the resulting hypothesis that the amide of SNC80 might function as a nonaromatic address, Dondio and coworkers prepared pyrrolooctohyroisoquinolines **178** (Fig. 35) lacking the second aromatic ring that exhibit high δ-receptor affinity and selectivity (see Table 12) [574]. The analog in which the pyrrole ring was in the opposite orientation (**179**) also retained high δ-receptor affinity ($K_i = 3$ nM), suggesting that this ring functions as a spacer [586].

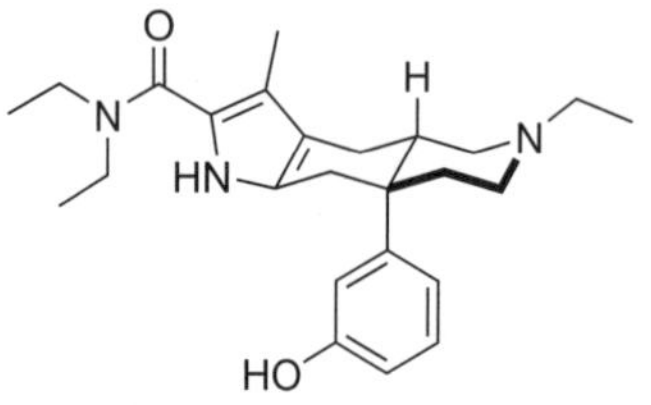

178

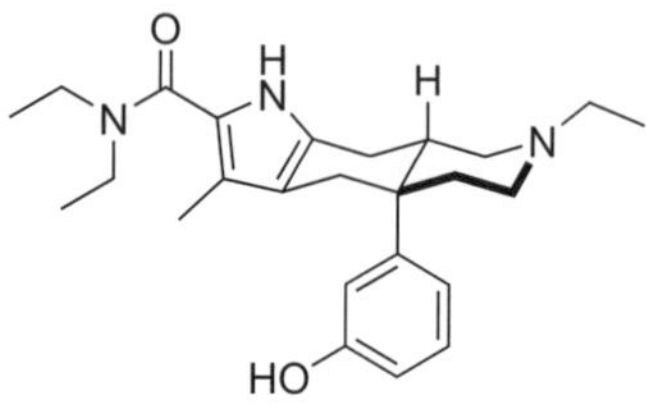

179

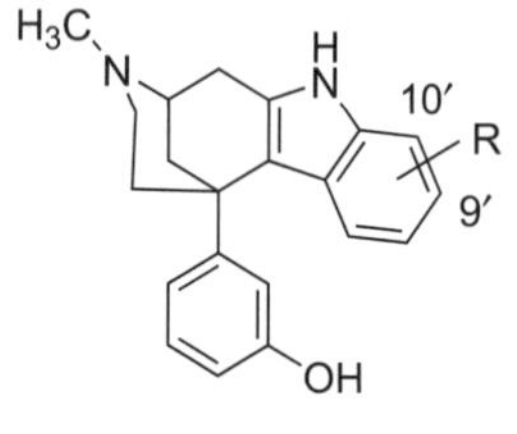

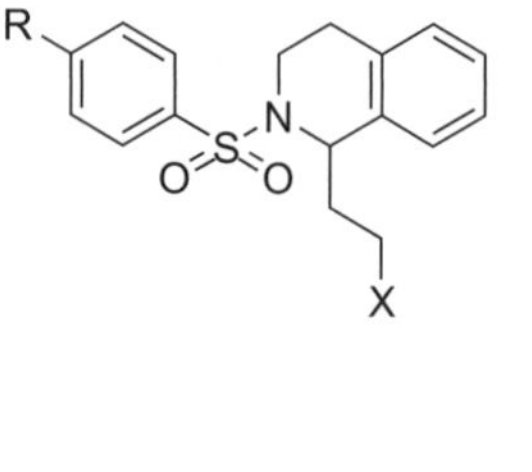

180 R = H

182 $R = CH_3$ X=

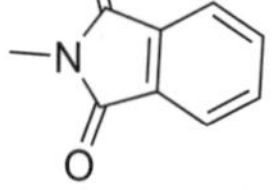

181 $R = OCH_3$

183 $R= (CH_2)_3CH_3$ X=

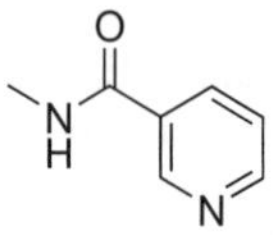

Figure 35. Other nonpeptide δ-receptor selective agonists.

An attempt has also been made to convert 5-(3-hydroxyphenyl)-2-methylmorphan, which has negligible δ-receptor affinity, to δ-selective agonists using the "message–address" concept [587]. Addition of an indole ring system to the morphan skeleton to give **180** increased δ-receptor affinity more than 140-fold (IC$_{50}$ = 7 nM); this structural changes had little effect on μ-receptor affinity, however, so that the selectivity of this compound for δ-receptors was low (IC$_{50}$ ratio (μ/δ) = 4.2). Based on differences between the SAR of SNC80 and the indole phenylmorphan derivatives, Rice and coworkers postulated that the indole phenyl ring, not the phenol, of the indole phenylmorphans might be structurally analogous to the methoxyphenyl group on SNC80, and therefore prepared a series of methoxy-substituted derivatives of **180**; the C9'- and C10'-substituted derivatives (**181**) (Fig. 35) exhibited somewhat higher δ-receptor affinity and potency, but the selectivity for δ-receptors was still low (IC$_{50}$ ratio (μ/δ) = 18 and 7, respectively) [588].

An additional type of nonpeptidic δ-agonist, **182**, which lacks a basic nitrogen, was identified by high-throughput screening [589]. Further structural modification of this lead compound to improve its water solubility led to a series of amide derivatives (e.g., **183**) with modest δ-receptor affinity (IC$_{50}$ = 37–256 nM) and in vivo potency comparable to TAN67 [590].

5.11. Nonpeptide Affinity Labels Used to Study Opioid Receptors

Affinity labels, ligands that interact with receptors in a nonequilibrium manner, are useful pharmacological tools to study opioid receptors and receptor–ligand interactions. Tritiated affinity labels have been useful in receptor isolation and for determination of the molecular weights of solubilized receptors (see Refs [73,86] for reviews). These compounds can be used to irreversibly block one or more receptor type in tissues containing multiple receptor types so that the remaining receptors can be studied in isolation. The covalent binding of affinity labels can be used to study receptor–ligand interactions. Thus, Liu-Chen and coworkers have characterized the binding of the affinity labels β-funaltrexamine (β-FNA) and SUPERFIT to μ- and δ-receptors, respectively, using a combination of molecular biology approaches and protein isolation (see below).

There are two types of affinity labels based on the type of reactive functionality. Electrophilic affinity labels contain an electrophilic group that can react with nucleophiles on the receptor. Photoaffinity labels are converted to highly reactive intermediates, most often a nitrene or carbene, upon exposure to light of the appropriate wavelength and these reactive species then can react covalently with the receptors. Reaction of affinity labels with receptors involves a two-step mechanism (Fig. 36) [316,591]. Initially, the ligand binds to the receptor reversibly, followed in the second step by covalent bond formation between the reactive functionality on the ligand and a group on the receptor. Depending upon the nature of the reactive functionality on the affinity label, the second step can enhance receptor selectivity. While an electrophilic affinity label may bind reversibly to more than one receptor type, covalent bond formation requires the proper juxtaposition of an appropriate nucleophile on the receptor with the electrophilic group on the ligand (see Fig. 36), so that covalent binding can occur to only one type of receptor; examples of electrophilic affinity labels that exhibit such enhanced selectivity are β-FNA (β-funaltrexamine) (**186**) and naltrindole isothiocyanate (NTII) (**47**) (Fig. 8) (see below). In the case of photoaffinity labels, however, the reactivity of the photolyzed intermediate is so high that they can react with almost any residues on the receptor, and therefore the selectivity is determined only by the first reversible step in the mechanism.

A variety of affinity labels, mostly nonpeptide ligands, have been prepared for opioid receptors. Detailed reviews of the structure–activity relationships for affinity labels have been published (see the previous edition of this chapter [283] and Refs [316,591–593]). Therefore, the discussion below will focus on those ligands that have been most useful in the characterization of opioid receptors, and on recent reports of new affinity label derivatives.

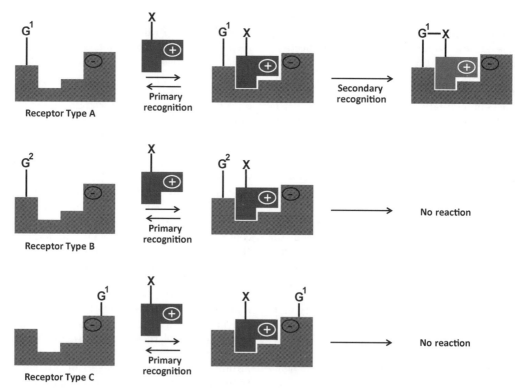

Figure 36. Two-step mechanism for covalent binding of an affinity label containing a selective electrophilic group X with receptor type A. Although receptor types A–C have similar topographic features that lead to reversible binding (1° recognition), they differ with respect to the reactivity of the receptor-based nucleophiles (G^1 and G^2) and their locations. Only with receptor type A is the nucleophile G^1 reactive with respect to X and within covalent binding distance (2° recognition). From Ref. [316]. (This figure is available in full color at http://mrw.interscience.wiley.com/emrw/9780471266945/home.)

5.11.1. Morphine and Naltrexone Derivatives

A variety of morphine and naltrexone derivatives have been prepared that incorporate a reactive functionality, often at the 6-position (Fig. 37). Incorporation of a nitrogen mustard at the 6β-position of naltrexamine by Portoghese and coworkers yielded β-chlornaltrexamine (β-CNA) (**184**) [594,595], the first successful opioid antagonist affinity label. This compound is a potent affinity label that because of the reactivity of the nitrogen mustard blocks all opioid receptor types (see Ref. [591]). It has been a useful tool in studying opioid receptors and has been utilized to single out a specific receptor type in tissues containing multiple receptor types. The desired receptor type for study can be protected by incubation with a ligand selective for that receptor and the tissue subsequently treated with β-CNA to irreversibly block the remaining opioid receptors [123,124,596]. Only one of the chloroethyl groups of β-CNA is required for irreversible antagonist activity [597]. The nitrogen mustard analog of oxymorphone, β-chloroxymorphamine (β-COA) (**185**) [598,599], has also been prepared. It is a potent irreversible agonist in the GPI and appears to bind irreversibly to opioid receptors *in vivo*; *in vivo* it initially produces analgesia followed by a long-lasting antagonism of morphine analgesia [599,600].

In order to obtain affinity labels selective for a single receptor type, Portoghese incorporated less reactive electrophiles at the 6β-position leading to the preparation of β-funaltrexamine (β-FNA) (**186**) [601]. This compound illustrates how the second step in the mechanism of affinity labels can enhance selectivity; while β-FNA interacts reversibly

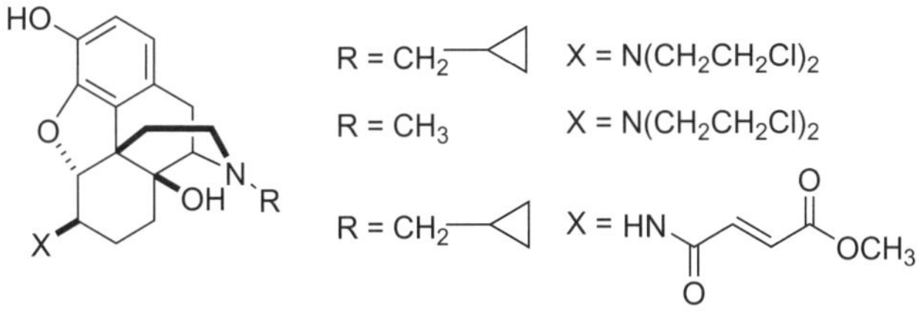

184 β-Chlornaltrexamine (β-CNA) R = CH₂–cyclopropyl, X = N(CH₂CH₂Cl)₂
185 β-Chloroxymorphamine (β-COA) R = CH₃, X = N(CH₂CH₂Cl)₂
186 β-Funaltrexamine (β-FNA) R = CH₂–cyclopropyl, X = HN–CH=CH–C(O)OCH₃ (fumaramide)

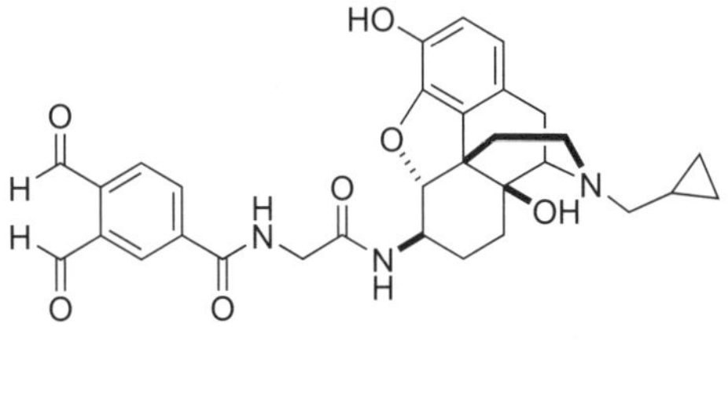

187

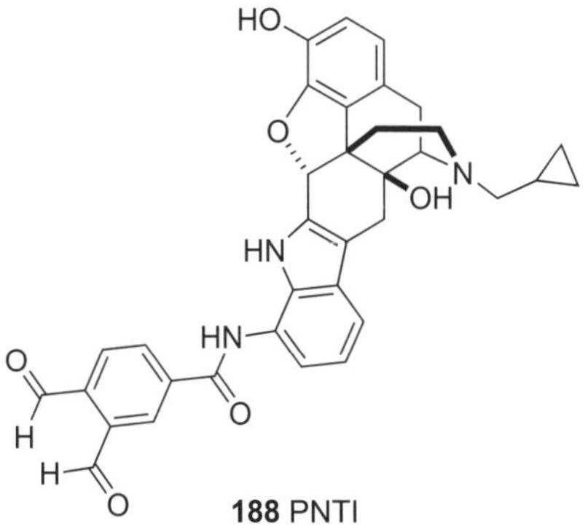

188 PNTI

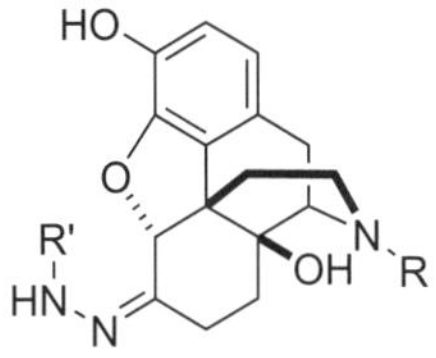

189 R = allyl, CPM, or CH₃

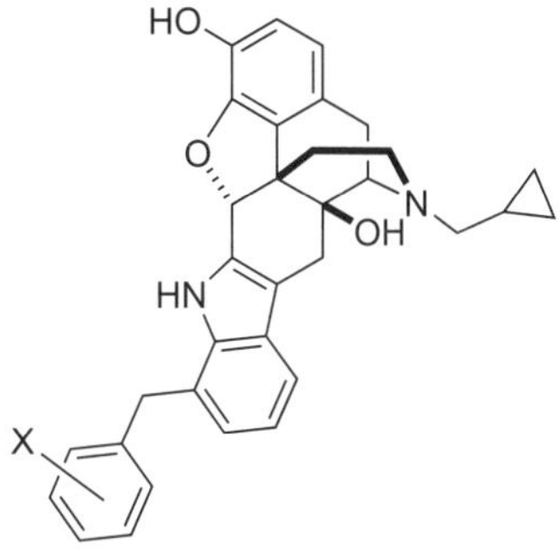

190 X = NHCOCH₂Br or NCS

Figure 37. Affinity label derivatives of naloxone, morphine and naltrindole.

with κ-receptors, where it is an agonist, it is an irreversible antagonist at μ-receptors. (There are conflicting reports on whether the effects of β-FNA on δ-receptors are irreversible [602,603].) The orientation and configuration of the fumaramide functionality are important for irreversible binding to μ-opioid receptors; neither the 6β-maleimide with a *cis* double bond nor the α-FNA with the electrophile in the 6α-position are irreversible μ-antagonists [602]. Examination of the conformations of α- and β-FNA by X-ray diffraction [604] offers insight into the differences in reactivity of these two isomers. The conformations of the ring system in the two compounds are almost identical except for the C ring, which is in a twist-boat conformation for the α-isomer and a chair conformation for β-FNA (see **186**). The fumarate group is then equatorial in both compounds, which when the morphinan skeletons are superimposed places the fumarate double bond in α-FNA more than 2 Å away from the double bond in β-FNA and in the wrong orientation for reaction with a nucleophile on the receptor.

The binding of β-FNA to μ-opioid receptors has been examined by Liu-Chen and coworkers using a combination of molecular biology approaches and protein isolation. These studies illustrate the utility of using affinity labels to study receptor–ligand interactions. The binding of [^{3}H]β-FNA to μ/κ-receptor chimeras suggested the region of the receptor from the middle of the third extracellular loop (EL3) to the C-terminus was necessary for irreversible binding to μ-receptors [605]. Subsequent isolation and partial purification of the labeled receptor and digestion with CNBr, however, located the point of attachment of β-FNA to the EL2-TM5 region of the receptor; subsequent site-directed mutagenesis of residues in this region indicated that the point of attachment was Lys233 at the EL2-TM 5 interface, which is a conserved residue among the opioid receptors [171]. Thus, the selectivity of β-FNA irreversible binding for μ-opioid receptors appears to be due to differences in the tertiary structure of the receptor, not the primary sequence. This illustrates the subtleties of receptor–ligand interactions and the importance of examining receptor–ligand interactions directly.

Recently, Portoghese and coworkers reported the phthalaldehyde derivatives of 6β-naltrexamine (**187**) and naltrindole (**188**) as "reporter affinity labels" [337,606]. Reaction of the phthalaldehyde group with an amine and thiol (from Lys and Cys side chains, respectively, in the receptor) results in a fluorescent isoindole; detection of fluorescence indicates that covalent reaction has occurred. In contrast to naltrindole, the phtalaldehyde derivative PNTI (**188**) is an agonist in the mouse vas deferens, leading to the proposal that the covalent binding of PNTI to δ-opioid receptors results in a conformational change in the receptor and agonist activity [337].

Pasternak and coworkers have prepared a series of hydrazone derivatives of the 6-ketone of naloxone, naltrexone, and oxymorphone (**189**) (Fig. 37) [203,204,607]. Prolonged actions in vivo and nonequilibrium binding in vitro have been reported for several of these compounds. The hydrazone naloxazone (**48**) (Fig. 8) [203] and the corresponding azine naloxonazine (**49**) [204] have been used to characterize the postulated μ$_1$-receptor subtype (see Ref. [202]); the azines are 20–40-fold more potent than the corresponding hydrazones [204]. Studies with [^{3}H]naloxazone suggested that a portion of the binding may involve covalent interaction with μ$_1$-receptors [608]; other researchers, however, have not found evidence for irreversible binding to opioid receptors [609,610]. The acylhydrazone naloxone benzoylhydrazone (NalBzoH) (**50**) (Fig. 8) exhibits extremely slow dissociation from μ-receptors ("pseudo-irreversible" binding) that may be related to interactions with a G-protein [611]; it also binds reversibly to κ-receptors and the tritiated form has been used to characterize the proposed κ$_3$-receptor subtype (see section "Receptor Subtypes, Splice Variants, and Receptor Dimerization") [211,612].

Portoghese and coworkers have prepared derivatives of the δ-selective antagonist naltrindole containing reactive functionalities as affinity labels for δ-receptors. An isothiocyanate group was incorporated at the 5′-position of naltrindole to give naltrindole 5′-isothiocyanate (NTII) (**47**) (Fig. 8) [190], which is a potent and selective nonequilibrium δ-receptor antagonist. NTII antagonizes the antinociceptive activities of [D-Ala^2]deltorphin II and DSLET, but not that of DPDPE, and therefore was proposed to be a selective δ$_2$-receptor antagonist [613]. Electrophilic moieties, either an isothiocyanate or a haloacetamide have also been incorporated into the indole N-benzyl aromatic ring of the δ$_2$-selective antagonist N-benzylnaltrindole (BNTI) to give **190** [614]. Interestingly, the *meta*-substituted isothiocyanate derivative was an irreversible δ-agonist in the MVD; the ortho- and para-substituted isothiocyanates and the haloacetamides were δ-receptor antagonists that exhibited time-dependent increases in antagonism consistent with covalent interaction with δ-receptors. In vivo, these compounds were less selective for δ$_2$- over δ$_1$-receptors than BNTI.

A number of 14β-amino-substituted derivatives containing electrophilic groups (Fig. 38) were prepared by Archer and coworkers (see the previous edition of this chapter [283] for a detailed review). Reactive functionalities that have been attached to the 14β-amino group include bromoacetamide, thioglycolamide, and cinnamoyl groups. The naltrxamine derivative clocinnamox or C-CAM (**191**) [615] has been the most extensively studied and is a potent long-lasting μ-antagonist [616–618]. Binding studies indi-

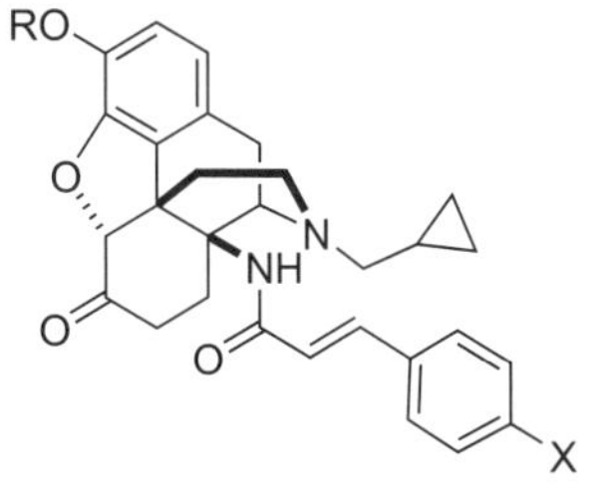

R = H, X = Cl	**191**	Clocinnamox (C-CAM)
R = CH₃, X = Cl	**192**	MC-CAM
R = H, X = CH₃	**193**	M-CAM

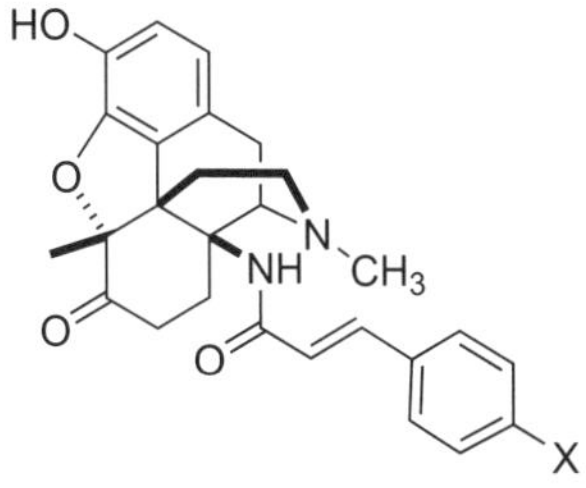

X = NO₂	**194**	MET-CAMO
X = Cl	**195**	MET- Cl-CAMO

Figure 38. Affinity label derivatives of naloxone and morphine containing a reactive functionality at the 14β-position.

cated that clocinnamox selectively decreases the density of μ-receptors [618], but not of δ- or κ-receptors [619], without affecting receptor affinity, as would be expected for an irreversible ligand. A subsequent examination of [³H]clocinnamox binding to mouse brain membranes, however, [620] found that the binding was fully reversible, although half of the binding dissociated very slowly ($t_{1/2} = 11$ h). The p-nitro-substituted derivative with a 5β-methyl group MET-CAMO (**194**) was the first N-methyl derivative reported with long-lasting μ-receptor selective antagonist activity with no agonist activity [621]; it appears to bind irreversibly to μ-receptors [622,623]. Like MET-CAMO the corresponding p-chloro substituted, 5β-methyl derivative MET-Cl-CAMO (**195**) is also a long-lasting μ-receptor antagonist with no agonist activity that appears to bind irreversibly to μ-receptors [624]. The p-methyl derivative of clocinnamox methocinnamox or M-CAM (**193**) has been reported to be a more μ-receptor selective, long-lasting antagonist in vivo than clocinnamox or β-FNA [625], although in standard binding assays, such as clocinnamox, its selectivity for μ-receptors is very low (K_i ratio 3–8). Recently, the relative importance of the 3-hydroxyl group to the opioid receptor affinity of clocinnamox has been examined by preparing both a series of 3-alkyl ether derivatives [626] and the 3-deoxy analog [627]. Interesting in the 3-alkyl ether derivatives the identity of the alkyl group affects efficacy. In vivo the O-methyl derivative MC-CAM (**192**) is a μ-partial agonist, the propargyl ether is a potent agonist, and the cyclopropyl-methyl is a long-lasting antagonist with little agonist activity; both the methyl and the propargyl ethers exhibit delayed long-lasting antagonist activity. The 3-deoxy analog exhibits high μ-opioid receptor affinity comparable to clocinnamox, indicating that the C₃-hydroxyl does not play a significant role in the binding of these 14β-cinnamoyl epoxymorphinans to opioid receptors; the deoxy derivative exhibits greater selectivity for μ- over δ-receptors than clocinnamox.

5.11.2. Other Nonpeptide Affinity Labels

Rice and coworkers have prepared a variety of affinity labels based on the structures of etonitazine, fentanyl, and oripavine (Fig. 39) [192,458]. The etonitazine derivative

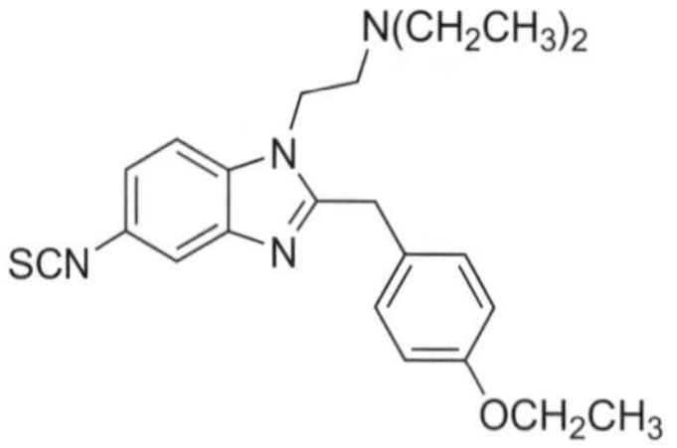

196 BIT

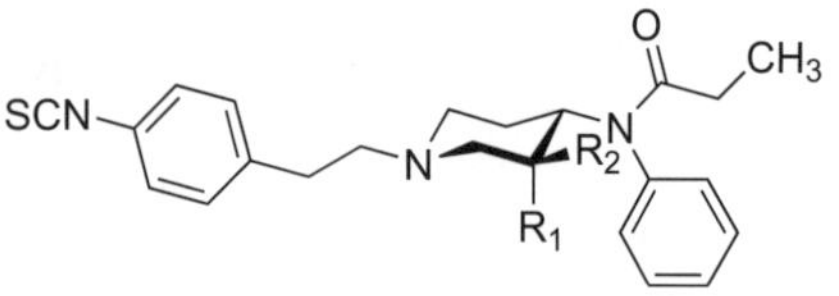

R₁ = R₂ = H   **197** FIT
R₁ = CH₃, R₂ = H   **198** SUPERFIT
R₁ = H, R₂ = CH₃   **199** *trans*SUPERFIT

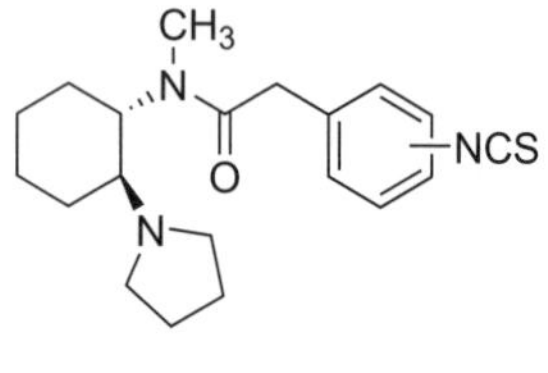

201

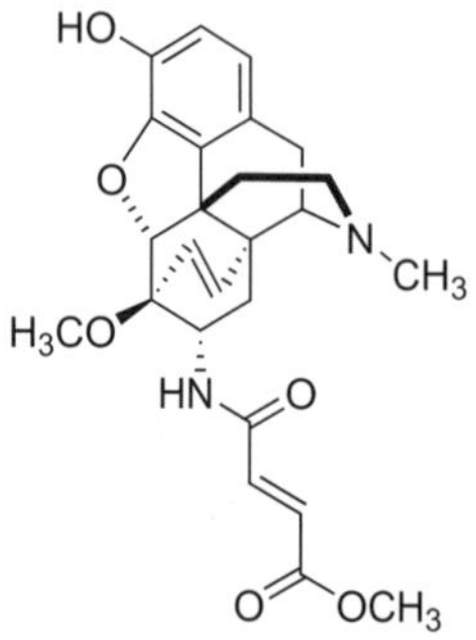

200 FAO

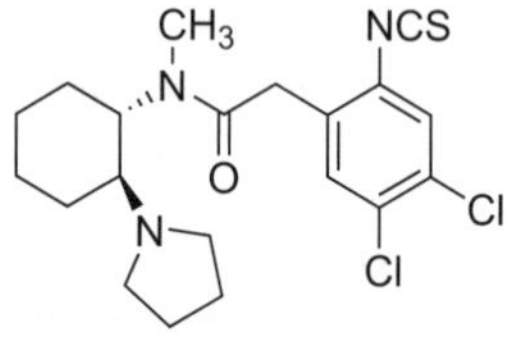

202 UPHIT

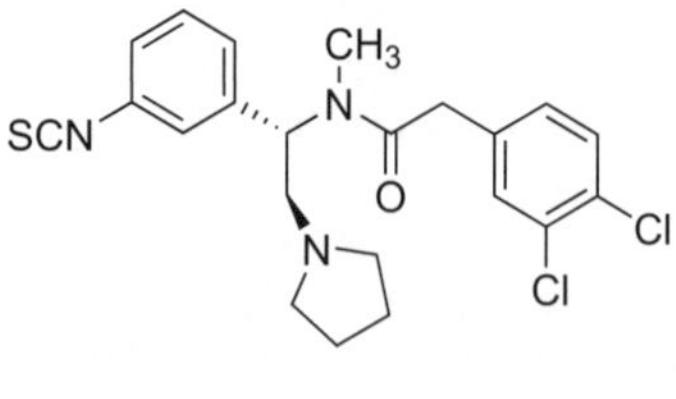

203 DIPPA (S)

Figure 39. Other nonpeptide affinity labels.

BIT (**196**) selectively inactivates μ-receptors, while the fentanyl derivative FIT (fentanyl isothiocyanate) (**197**) and the oripavine derivative FAO (fumaramido orpavine) (**200**) selectively inactivate δ-receptors. The selective alkylation of δ-receptors by FIT, which is a derivative of the μ-selective ligand fentanyl, illustrates how much the alkylation step can influence the receptor selectivity of affinity labels. Both BIT and FIT have been prepared in tritiated form [628] and [^{3}H]FIT used to specifically label a 58 kDa protein from NG108-15 cells [629]. The enantiomeric pair of the *cis*-3-methyl derivatives of FIT were synthesized and one of these isomers, the (+)-3R,4S enantiomer, SUPERFIT (**198**) found to be a very potent (5–10 times the potency of FIT) and selective irreversible ligand for δ-receptors [455]. SUPERFIT was used to purify δ-receptors from NG108-15 cells to apparent homogeneity [630]. Liu-Chen and coworkers have used similar approaches to those described above for β-FNA to study of the binding of SUPERFIT to δ-opioid receptors. The results for wash-resistant inhibition of binding to μ/δ-chimeric receptors suggested that the segment from the start of the first intracellular loop to the middle of TM3 of δ-receptors is important for the selective irreversible binding of SUPERFIT [631]. The enantiomeric pair of the trans-3-methylfentanyl isothiocyanates has also been prepared [193]. The (+)-(3S,4S) isomer was a δ-selective acylating agent in vitro with similar potency to SUPERFIT, and has been used to selectively deplete $δ_{ncx}$-binding sites (see section "Receptor Subtypes, Splice Variants, and Receptor Dimerization") [194].

Several derivatives of κ-selective compounds containing an isothiocyanate have been prepared as potential irreversible ligands for κ-receptors (Fig. 39). de Costa et al. prepared analogs of U-50,488 (**201** and **202**) [632–634]. In the series of enantiomerically pure analogs of **201**, the (−)-1S,2S isomers generally exhibited wash-resistant inhibition of binding of [^{3}H]U69,593 to guinea pig brain membranes, but none of the compounds irreversibly inhibited the binding of [^{3}H]bremazocine to either guinea pig or rat brain membranes [632,634], supporting the conclusion of heterogeneity of κ-receptors. The (−) o-isothiocyanate isomer of **201** exhibited selective wash-resistant inhibition of [^{3}H]U69593 binding, and was the most potent in vitro, but it failed to produce any irreversible inhibition of κ-receptors following i.c.v. injection into guinea pig brain. This led de Costa and coworkers to prepare UPHIT (**202**), the chlorine-containing analog of **201**, in order to improve affinity and selectivity [633]. In contrast to **201**, this compound inhibited binding to κ-receptors following i.c.v. administration [633]. In vivo in mice UPHIT antagonizes antinociception produced by U69593, but not by bremazocine, providing additional supporting evidence for κ-receptor subtypes [635].

Isothiocyanate derivatives of the κ-selective agonist ICI 199,441 have also been described. Chang et al. prepared the m-isothiocyanate derivative DIPPA (**203**) [636,637], which exhibits wash-resistant inhibition of [^{3}H]U69,593 binding and long-lasting κ-receptor antagonism in vivo. Liu-Chen and coworkers examined the binding of the corresponding p-isothiocyanate derivative to μ/κ-chimeric receptors and determined that the region from TM3 to the C-terminus of the κ-receptor is important for the binding of this compound [638]. Nelson and coworkers incorporated the isothiocyanate in the phenylacetamide phenyl ring of ICI 199,441 [639]. These isothiocyanate analogs all exhibited wash-resistant inhibition of binding, while the parent compound with an unsubstituted phenylacetamide phenyl ring was completely removed by washing; the lead compound ICI-199,441, with chlorines on this ring but without the isothiocyanate group, however, was not completely removed by the washing procedure.

A number of photoaffinity label derivatives of opiates have also been prepared (see the previous edition of this chapter [283] and Ref. [591] for more detailed reviews). Azide derivatives of a number of different opiates, including etonitazene, carfentanil, and 6α- and 6β-substituted naltrexamine derivatives, have been prepared (see Ref. [283]). A significant problem with using opioid photoaffinity labels is the sensitivity of opioid receptors to destruction by short wavelength UV-irradiation [640]. Incorporation of a nitro group into the aromatic ring bearing the azide functionality shifts the absorption maximum so that photolysis can be conducted at longer wavelengths where little if any photodestruction of opioid receptors occurs.

5.12. Miscellaneous Nonpeptide Opioids

A variety of compounds in other chemical classes have also been identified, which have analgesic activity (see Ref. [463]). In addition to the κ-receptor selective arylacetamides, such as U50,488 and related compounds, the benzodiazepine tifluadom (**147**) is a κ-agonist (see Section 5.9). The benzimidazole

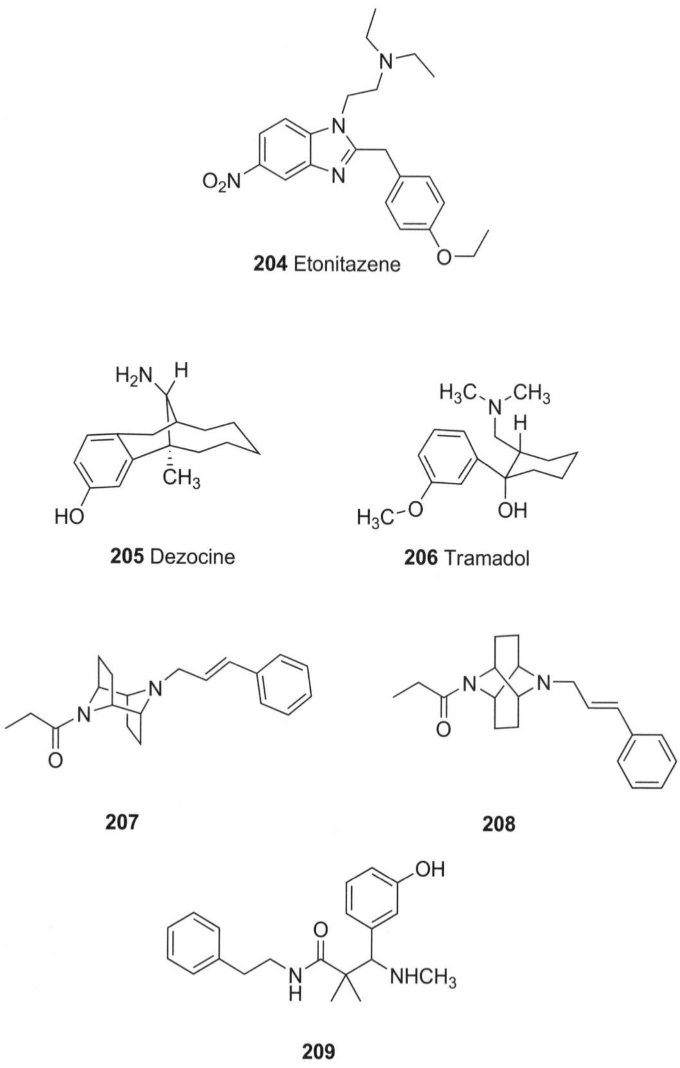

Figure 40. Miscellaneous nonpeptide opiates.

etonitazene (**204**) (Fig. 40) is a potent analgesic, approximately 1500 times the potency of morphine in mice [641], and is a μ-receptor agonist. The aminotetralin dezocine (**205**) bears some resemblance to the benzomorphans and contains a phenol and basic amino group; in contrast to other opiates, however, the amine group is a primary amine. Dezocine is the *levo* isomer of the β-epimer; the (+) isomer is inactive. This clinically used analgesic is a mixed agonist/antagonist [642] (see Table 3) with partial μ-agonist activity; it may also have activity at κ-receptors (see Ref. [304]). The cyclohexane derivative tramadol (Ultram®, **206**) is an atypical analgesic (see Refs [304,643,644] for reviews) that appears to produce analgesia through both opioid and nonopioid mechanisms [645]. Tramadol antinociception appears to be mediated through both activation of μ-opioid receptors, where it exhibits low affinity ($K_i = 2\,\mu M$), and by inhibition of monoamine uptake.

Examination of the isomers of tramadol [646] found that while the (+) enantiomer had higher affinity for μ-receptors ($K_i = 1.3\,\mu M$), the activities of the isomers were complementary and synergistic. Other recently reported compounds with opioid receptor affinity and activity include pyrrolidinylnaphthalenes, which are structurally related to heterosteroids [647]. Highly constrained tricyclic piperazine derivatives (**207** and **208**) have been prepared, which are structurally related to the 4-anilidopiperidines and that exhibit reasonable affinity ($K_i = 10$ and 7 nM, respectively) and selectivity (K_i ratio (μ/δ/κ) = 1/230/300 and 1/71/110) for μ-receptors; *in vivo* **208** is sixfold more potent analgesic than morphine [648]. 3-Amino-3-phenylpropionamide derivatives have been prepared as small molecule mimics of the peptide antagonists CTOP and CTAP and analogs identified with high affinity for μ- and κ-opioid receptors (e.g., **209**) [649].

6. OPIOID PEPTIDE ANALOGS

6.1. Introduction

The identification of the opioid peptides in the mid-1970s opened up a whole new area for the development of opioid receptor ligands. The endogenous opioid peptides, both mammalian and amphibian (see below), have served as lead compounds that have been extensively modified to enhance potency, receptor type selectivity, stability, and/or decrease conformational flexibility. Peptide ligands selective for both μ- and δ-opioid receptors are found in more than one type of peptide sequence. Thus, some enkephalin analogs, as well as the recently discovered endomorphins, analogs of the peptide β-casomorphin (derived from casein) and the amphibian peptide dermorphin, preferentially interact with μ-opioid receptors (see Sections 6.2.1, 6.4.1 and 6.5.1). Other enkephalin analogs, as well as analogs of the amphibian peptides the deltorphins, preferentially interact with δ-opioid receptors (see Sections 6.2.2 and 6.5.2). For κ-receptors, the endogenous opioid peptides identified to date been limited to one class of peptides, the dynorphins (see Section 6.3). Peptides with affinity for opioid receptors that have sequences completely different from those of the endogenous opioid peptides have also been identified. Analogs of somatostatin, which are μ-opioid receptor antagonists, have been prepared and novel peptides with opioid receptor affinity have been identified using combinatorial approaches (see Section 6.6). Affinity label derivatives of opioid peptide, which bind irreversibly to opioid receptors, have also been identified (see Section 6.7). In addition to preparing analogs of opioid peptides, inhibitors of opioid peptide metabolism have been developed as an indirect approach to utilizing opioid peptides as potential therapeutic agents (see Section 6.8).

Much of the early SAR of the enkephalins has been discussed [650,651]. Subsequent reviews of opioid peptides include those by Hruby [652,653] and Schiller [654,655], which contain extensive tabular data. An issue of *Biopolymers (Peptide Science)* (Vol. 51, issue number 6, 1999) is devoted solely to reviews of peptide and peptidomimetic ligands for opioid receptors. The reader is referred to these reviews for additional references to the literature.

In addition to the four classes of opioid peptides discussed above, other peptides of mammalian origin with opioid activity have also been identified. β-Casomorphin (**210**), obtained by enzymatic digestion of the milk protein casein [656,657], exhibits some selectivity for μ-receptors. Human β-casomorphin (**211**) differs from the bovine sequence in two positions; the human β-casomorphin pentapeptide and tetrapeptide fragments are less potent than the corresponding bovine peptides [658]. Other peptides with affinity for opioid receptors include peptides derived from hemoglobin (see Ref. [659]).

Bovine β-casomorphin (**210**)   Tyr-Pro-Phe-Pro-Gly-Pro-Ile
Human β-casomorphin (**211**)   Tyr-Pro-Phe-Val-Glu-Pro-Ile

Dermorphin (**212**) Tyr-D-Ala-Phe-Gly-Tyr-Pro-SerNH$_2$

Deltorphin (dermenkephalin, deltorphin A, **213**) Tyr-D-Met-Phe-His-Leu-Met-AspNH$_2$

[D-Ala^2]deltorphin I (deltorphin C, **214**) Tyr-D-Ala-Phe-Asp-Val-Val-GlyNH$_2$

[D-Ala^2]deltorphin II (deltorphin, **25**) Tyr-D-Ala-Phe-Glu-Val-Val-GlyNH$_2$

Figure 41. Opioid peptides from frog skin.

6.1.1. Opioid Peptides from Amphibian Skin

Based on their finding, amphibian skin peptides that were counterparts to other mammalian bioactive peptides, Erspamer and coworkers examined amphibian skin for opioid peptides (see Ref. [660] for a review). This led first to the isolation and characterization of dermorphin (**212**) (Fig. 41), which is a μ-selective peptide (see Table 13), from the skin of South American Phyllomedusinae hylid frogs in the early 1980s [661]. Inspection of the sequence of one of the cloned cDNAs for the precursor of dermorphin suggested the existence of another heptapeptide with a similar N-terminal sequence [662]. This then led to the isolation of deltorphin (also called dermenkephalin or deltorphin A) (**213**) (Fig. 41), the first δ-selective amphibian opioid peptide, from these frogs [663,664]. Synthesis confirmed that the amino acid in position 2 of deltorphin was D-methionine rather than L-methionine [663,665,666]. Two additional peptides [D-Ala^2]deltorphin I (also referred to as deltorphin C) (**214**) (Fig. 41) and [D-Ala^2]deltorphin II (also referred to as deltorphin B) (**25**) (Fig. 5) were subsequently discovered [93], which exhibited even greater δ-receptor affinity and exceptional selectivity (see Table 13), making them the most selective of the naturally occurring opioid peptides. [D-Ala^2]Deltorphin II has been used to study the proposed δ$_2$-receptor subtype (see section "Receptor Subtypes, Splice Variants, and Receptor Dimerization"). The pharmacology of these amphibian opioid peptides has been discussed in a recent review [667].

The unique feature of these amphibian skin opioid peptides is the sequence between the important aromatic residues. In contrast to the enkephalins and other mammalian opioid peptides that contain the Gly-Gly dipeptide sequence between Tyr and Phe, the amphibian opioid peptides contain a single D-amino acid (see Fig. 41), which apparently arises from a posttranslational conversion of the L-amino acid to its D-isomer [662]. The identification of these unusual opioid peptides expanded our understanding of the structural requirements for interaction with opioid receptors and provided new lead compounds for further modification (see Sections 6.5.1 and 6.5.2).

6.2. Enkephalin Analogs

The enkephalins have been the most extensively modified of the opioid peptides, and thousands of analogs of these pentapeptides

Table 13. Opioid Receptor Affinities and Opioid Activity in the GPI and MVD of Peptides from Amphibian Skin

	K_i (nM)^b		K_i Ratio	IC$_{50}$ (nM)	
Peptide	μ	δ	μ/δ/κ	GPI	MVD
Dermorphin (**212**)	0.70	62	1/89/>14,000	1.4	2.4
Deltorphin (**213**)	1,630	2.4	6,800/1/>4,100	5,000	1.4
[D-Ala^2]Deltorphin I (**214**)	3,150	0.15	21,000/1/>66,000	>1,500	0.21
[D-Ala^2]Deltorphin II (**25**)	2,450	0.71	3,400/1/>14,000	>3,000	0.32

Data from Ref. [93].
$^b K_i > 10,000$ nM for κ-receptors for all of the amphibian skin peptides.

have been prepared (see Refs [650–655] for reviews). The naturally occurring enkephalins exhibit some selectivity for δ-receptors (see Table 9), but these peptides are rapidly degraded by a variety of peptidases (see Section 6.8). Therefore, one major goal of structural modification of these small peptides has been to increase metabolic stability. Depending upon the nature of the modifications made, both μ- and δ-selective enkephalin derivatives have been prepared (enkephalin derivatives generally have very low affinity for κ-opioid receptors). These derivatives have included both linear peptides and conformationally constrained derivatives. Conformational constraints have included cyclizations between residues in the peptide chain and the local constraints by incorporation of an amino acid whose side-chain conformation is restricted.

Early SAR studies (see Ref. [650] for a review) identified important structural features of the enkephalins and which positions could be readily modified. The importance of Tyr^1 for opioid activity was apparent from the large decreases in potency when this residue was substantially modified. A D-amino acid in position 2 was one of the early modifications examined in order to decrease the cleavage of the Tyr-Gly bond by aminopeptidases [668]. Incorporation of a D-amino acid at this position increases potency at both μ- and δ-receptors and is found in the vast majority of enkephalin derivatives; an L-amino acid at position 2 significantly decreases potency at both receptors. In the enkephalins, the 3-position is very intolerant of substitution, and therefore enkephalin derivatives generally retain a glycine at this position. There are significant differences between μ- and δ-receptors, however, in the structural requirements for residues in positions 4 and 5, and frequently modifications in these positions have been used to impart selectivity for one of these opioid receptor types (see below).

6.2.1. μ-Selective Enkephalin Analogs

Linear Analogs Enkephalin analogs with substantial structural changes from the endogenous peptides, particularly in the C-terminus, retain affinity for μ-opioid receptors and can exhibit improved μ-receptor selectivity. Thus, the aromatic moiety of Phe^4 is not an absolute requirement for interaction with μ-receptors, and a cyclohexane ring [669] or leucine side chain [670,671] is tolerated in this position. Significant variation is also tolerated in residue 5 and the C-terminus by μ-receptors, and modifications in this region of the peptides have been very useful in differentiating μ- versus δ-receptors. Thus, the C-terminus can be amidated, reduced or eliminated with retention of μ receptor affinity and often with a substantial increase in μ-receptor selectivity.

One of the most commonly used μ-selective ligands DAMGO (15) (Fig. 4) [672] contains a reduced C-terminus and is available in tritiated form. Substitution of the glyol functionality of DAMGO with Met(O)ol (to give FK 33824) increases δ-receptor affinity 10-fold, decreasing μ-receptor selectivity [120]. Other related tetrapeptide analogs with a modified C-terminus syndyphalin-25 (Tyr-D-Met-Gly-NMePheol) [673] and LY 164929 (NMeTyr-D-Ala-Gly-NEtPhe[Ψ(CH$_2$NMe$_2$)]) (215) (Fig. 42) with a reduced C-terminal amide [674] exhibit significantly higher selectivity for μ receptors than DAMGO (Table 14). Shorter tripeptide amide analogs with a branched alkyl chain in place of the second phenyl ring also exhibit μ-receptor selectivity, although their potency in the GPI is significantly lower (IC$_{50}$ = 240–320 nM) [675].

Conformationally Constrained Analogs Local restriction of conformational freedom can be accomplished by incorporating conformationally constrained amino acids (see Refs [676,677] for reviews). In the case of the enkephalins, incorporation of 2-amino-6-hydroxy-2-tetralincarboxylic acid (Hat) (Fig. 43) in place of Tyr^1 in [Leu^5]enkephalin methyl ester results in a μ-selective compound (IC$_{50}$ ratio (MVD)/(GPI) = 5.1), whereas the analog containing 2-hydroxy-5-hydroxy-2-indanecarboxylic acid (Hai) is virtually inactive [678]. Incorporation of 2′,6′-dimethyltyrosine (Dmt) (Fig. 43) in position 1 of [Leu^5]enkephalin results in large increases in both μ- and δ-affinity and an analog with some selectivity for μ-receptors [679], while incorporation of 2′,6′-dimethylphenylalanine (Dmp) (Fig. 43) in position 4 results in increased μ-receptor selectivity as a result of decreased δ-receptor affinity [680]. Interestingly, combining the

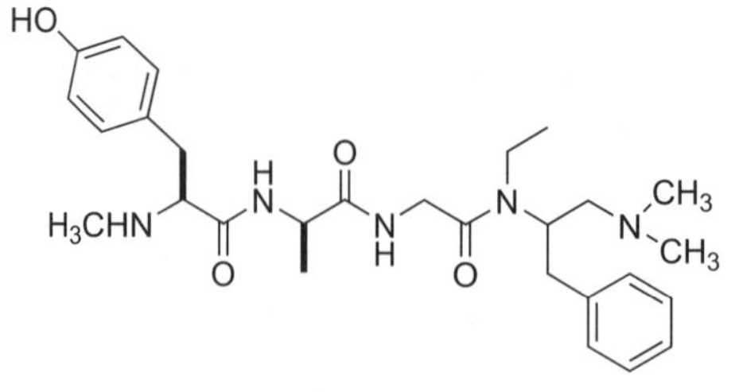

215 LY 164929

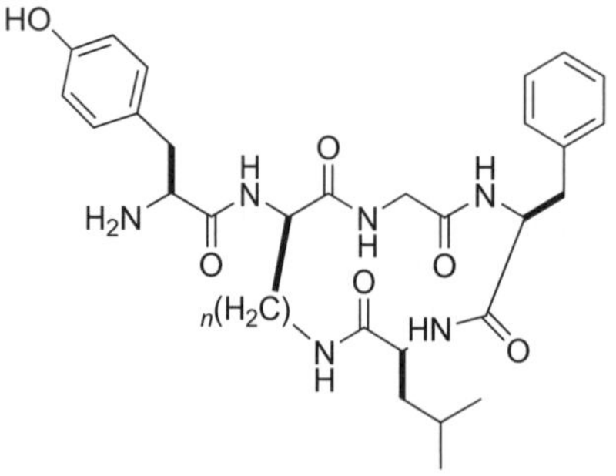

216 n = 2
217 n = 4

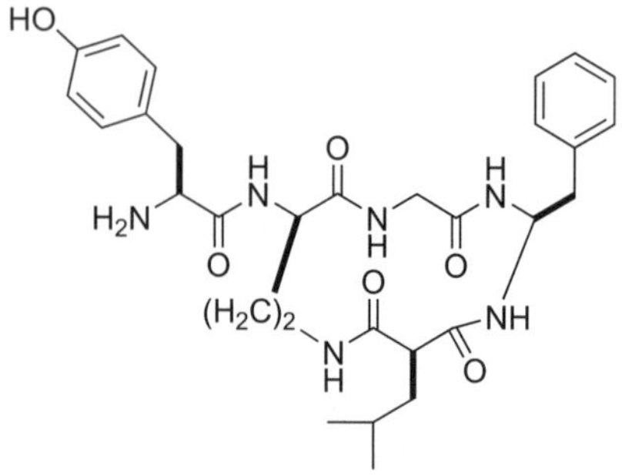

218

Figure 42. μ-Receptor selective enkephalin analogs. The structure of DAMGO (**15**) is shown in Fig. 4.

two modifications results in a weak antagonist (pA$_2$ = 6.90) with some selectivity (20-fold) for μ-receptors [680].

Enkephalin derivatives selective for μ-receptors have also been prepared by cyclization between a D-amino acid and the C-terminus. Tyr-cyclo[N$^\gamma$-Dab-Gly-Phe-Leu] (**216**) (Fig. 42) (Dab = α,γ-diaminobutyric acid) [681] exhibits both high potency and μ-receptor selectivity (Table 14), while the corresponding linear

Table 14. Opioid Receptor Affinities and Opioid Activity in the GPI and MVD of μ-Selective Enkephalin Analogs

Peptide	K_i (nM)		K_i Ratio	IC$_{50}$ (nM)		Reference
	μ	δ	(δ/μ)	GPI	MVD	
DAMGO (**15**)	1.9	345	180	4.5	33	[120]
Syndyphalin-25						
(Tyr-D-Met-Gly-MePheol)	0.29^a	1250^a	4300	0.0025	—	[691]
LY 164929 (**215**)	0.6^a	900^a	1500	—	—	[674]
Tyr-cyclo[D-X-Gly-Phe-Leu]						
X = Dab (**216**)	13.8	115	8.3	14.1	81.4	[683]
X = Lys (**217**)	12.4	14.6	1.2	4.8	141	[683]
Tyr-cyclo[D-Lys-GlyΨ[CSNH]Phe-Leu]						
	4.55	654	44	24	186	[689]
Tyr-cyclo[D-Glu-Gly-gPhe-D-Leu]						
(**218**)	11.0	389	35	19.4	313	[690]

Data for DAMGO from Table 8 is included for comparison.
^{a}IC$_{50}$ values.

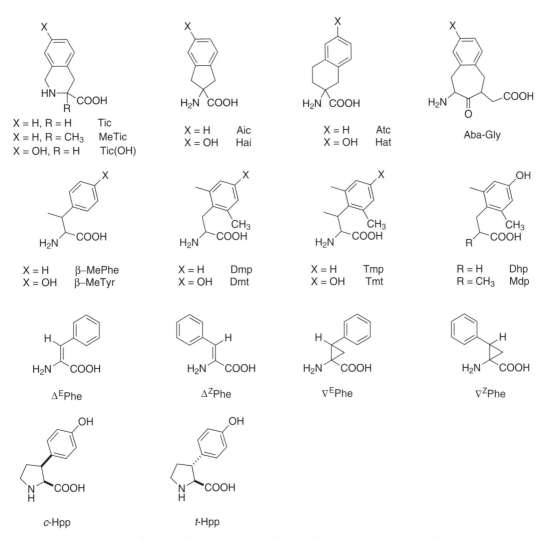

Figure 43. Conformationally constrained phenylalanine and tyrosine analogs.

peptide [D-Dab², Leu⁵]enkephalinamide is not μ-receptor selective [682]. This was the first demonstration that receptor selectivity could be imparted by conformational restriction. Related analogs with D-Orn or D-Lys in position 2 also exhibit high potency for μ-receptors but decreased selectivity in radioligand binding assays (see Table 14) [683]. The conformations of these peptides have been examined by both computational methods [684–687] and NMR spectroscopy [686,688]. As expected, the ring structure reduces conformational flexibility, but some flexibility in the ring remains, particularly for the larger ring sizes, and different intramolecular hydrogen bond patterns have been proposed for the cyclic structures. Modifications to the peptide bonds in these cyclic peptides, for example, thioamide replacement of the Gly³-Phe⁴ peptide bond [689] or partial retro-inverso analogs of **216** Tyr-*cyclo*[D-Glu-Gly-gPhe-D-Leu] (for example, **218**) (Fig. 42) [690], can further enhance μ-receptor selectivity (see Table 14).

6.2.2. δ-Selective Enkephalin Analogs

Linear Peptides Differences in the SAR of enkephalin analogs for interaction with δ- versus μ-opioid receptors have been exploited to develop δ-selective derivatives. Thus, more hydrophilic D-amino acids, such as D-Ser or K-Thr, can be incorporated into position 2 to impart δ-receptor selectivity, while μ-opioid receptors prefer a more hydrophobic residue in this position [671]. Whereas μ-receptors can accommodate an aliphatic residue in position 4 of enkephalins, δ-receptors generally require an aromatic moiety in this position. At the C-terminus δ-receptors prefer a free carboxylic acid, and lengthening of the peptide with a residue such as Thr can result in increased δ-selectivity.

Several linear enkephalin analogs have enhanced selectivity for δ-opioid receptors (Tables 8 and 15). DADLE (**21**) (see Fig. 5 and Table 8) [692] was an early analog prepared that shows a slight preference for δ-opioid receptors. Rogues and coworkers prepared several [Leu^5,Thr^6]enkephalin analogs containing D-Ser, D-Thr, or a derivative in position 2 that have greater selectivity for these receptors. Thus, incorporation of D-Ser in position 2 that yields DSLET (**22**) [693] and substitution of D-Thr in this position to give DTLET (**23**) [694] enhances δ-receptor selectivity. X-ray structures for both DADLE [695] and DTLET [696] have been reported; in both cases, a single bend conformation is found in the crystals (see Ref. [697] for a review). The steric bulk of the residues in positions 2 and/or 6 was increased by incorporating the *t*-butyl ether derivatives of D-Ser and/or Thr. The resulting derivatives DSTBULET ([D-Ser(*Ot*Bu)^2,Leu^5,Thr^6]enkephalin) and BUBU ([D-Ser(*Ot*Bu)^2,Leu^5,Thr(*Ot*Bu)^6]enkephalin) [693] exhibited decreased affinity for μ-receptors and thus enhanced selectivity for δ-receptors. Replacement of the ether in the side chain at position 2 by a sulfur to give BUBUC ([D-Cys(*St*Bu)^2,Leu^5,Thr(*Ot*Bu)^6]enkephalin) [698] further decreases affinity for μ-receptors, resulting in an analog with a 1000-fold selectivity for δ- over μ-receptors (Table 15).

The penetration of the BBB by enkephalin analogs is relatively low and therefore their analgesic activity following systemic administration is relatively weak [699]. Therefore, modifications have been made to enkephalin analogs in an attempt to enhance BBB penetration. Attachment of D-glucose via an *O*-linkage to Ser^6 in the DTLET analog Tyr-D-Thr-Gly-Phe-Leu-SerNH$_2$ resulted in a peptide, which retains similar opioid receptor affinity and potency in smooth muscle assays as the parent peptide (see Table 15) [699]. Glycosylation led to significant increases in both enzymatic stability and BBB permeability, resulting in significantly improved analgesia following i.v. administration [700].

Conformationally Constrained Analogs Local restriction of conformation in the linear enkephalins has included incorporation of dehydroamino acids, for example, dehydrophenylalanine (ΔPhe) (Fig. 43) and cyclopropylmethylphenylalanine (∇Phe), into the peptides (see Ref. [701] for a review). In the case of the dehydrophenylalanine (ΔPhe^4) derivatives of [D-Ala^2,Leu^5]enkephalin, the *Z*-isomer exhibits 150–260 higher affinity for δ- and μ-receptors than the *E*-isomer [702]. For the peptides containing cyclopropylphenylalanine (∇Phe), one of the *Z*-isomers exhibits high affinity for [^{3}H]naloxone binding sites in rat brain [703], while the (2*R*,3*S*) isomer of ∇^{E}Phe exhibits high affinity ($K_i =$ 13 nM) and selectivity (K_i ratio = 250) for δ-opioid receptors [704]; the latter compound, however, was essentially inactive in smooth muscle assays (IC$_{50}$ = 2000–4000 nM).

δ-Selective enkephalin analogs have been prepared by cyclization via disulfide bond formation. Cyclic enkephalin analogs containing D- or L-Cys residues in positions 2 and 5 were first synthesized by Sarantakis [705] and Schiller [706]. These initial cyclic enkephalin analogs exhibit only a slight preference for δ- over μ-receptors, but introduction of methyl groups on the β-carbons of the cysteine residues by incorporation of penicillamine (Pen) in positions 2 and/or 5 markedly enhances δ-receptor selectivity [707–710]. The most δ-selective compounds in the series were the bis-penicillamines derivatives DPDPE (*cyclo*[D-Pen^2,D-Pen^5]enkephalin) (**24**) (Fig. 5) and DPLPE (*cyclo*[D-Pen^2,L-Pen^5]enkephalin) [710]. The individual contributions of the β-methyl groups in residue 2 to δ-receptor

Table 15. Opioid Receptor Affinities and Opioid Activity in the GPI and MVD of δ-Selective Enkephalin Analogs

Peptide	K_i (nM) δ	K_i (nM) μ	K_i Ratio (μ/δ)	IC$_{50}$ (nM) MVD	IC$_{50}$ (nM) GPI	References
DSLET (22)	1.8	39	22	0.59	110	
DTLET (23)	2.7	34	13	0.41	68	
DSTBULET	2.8	370	130	1.1	1,800	
BUBU	1.7	480	280	0.6	2,790	
BUBUC	2.9	2,980	1,030	0.05	13,300	[698]
[D-Thr^2,Leu^5]enkephalin-Ser(OR)^6						
R = H	2.4	4.0	1.7	2.7	25	[699]
R = D-Glucose	3.4	8.2	2.4	1.6	34	[699]
DPDPE (24)	2.7	710	260	2.8	2,350	
DPLPE	2.8	660	235	4.1	3,000	
[Phe(p-Cl)^4]DPDPE	1.6	780	490	0.89	8,300	
cyclo[D-Pen^2,Cys^5]enkephalin-Phe^6	1.4^a	280^a	200	0.016	82.7	[717]
cyclo[D-Pen^2,Phe(p-X)^4,Pen^5]enkephalin-Phe^6						
X = F	0.43^a	1650^a	3800	0.016	740	[718]
X = Br	0.20^a	4200^a	21,000	0.18	3,400	[718]
[2′MeTyr^1]DPDPE	0.89^a	1,170^a	1320	4.5	430	[723]
[(2S,3R)-Tmt^1]DPDPE	5.0	4,300	850	1.8	Antagonist^b	[724,725]
[(2S,3S)-Tmt^1]DPDPE	210	720	3	170	290	[724,725]
[Hat^1]DPDPE	2.36^a	1,440^a	610	22	3,500	[723]
[(2S,3S)-β-MePhe^4]DPDPE	10	14,000	1,400	39	57,400	[729]
ICI 174,864 (29)	190	29,600	155	17^c	>5,000^c	
N,N-Dibenzyl Leu-enkephalin	78	1,600	20	180^c	NA	[734,735]
[(2S)-Mdp^1,D-Ala^2,Leu^5]enkephalin amide 12	190	16		28^c	154^c	[736]

Data for DSLET, DTLET, DPDPE, and ICI 174,864 from Table 8 are included for comparison. Data from Ref. [120] except where otherwise indicated.
^{a} IC$_{50}$ values.
^{b} IC$_{50}$ = 5 μM.
^{c} Antagonist, K_e values (nM).
NA: not active.

affinity and selectivity were examined by preparing cyclo[(3S)Me-D-Cys^2,L-Pen^5]- and cyclo[(3R)Me-D-Cys^2,L-Pen^5]enkephalin [711]. The β-methyl groups in the 2-position had only a minor effect on δ-opioid receptor affinity, and the similar affinity for μ-receptors of cyclo[(3S)Me-D-Cys^2,D-Pen^5]- and cyclo[D-Cys^2,D-Pen^5]enkephalin suggested that adverse steric interactions with the pro-R methyl group are responsible for the low μ-receptor affinity, and therefore δ-receptor selectivity, of DPDPE.

Additional modifications to DPDPE can further enhance δ-receptor affinity and/or selectivity (see Ref. [653] for a review, including tables of analogs). Halogen substitution on the para-position of the phenylalanine ring in position 4 enhances both δ-receptor potency and selectivity [712], and [Phe(p-Cl)^4]DPDPE has been used in tritiated form in radioligand binding assays. Halogenation can also increase blood–brain barrier permeability [713,714]. [L-Ala^3]DPDPE has been prepared and exhibits an unusual spectrum of activity [715]. In the mouse vas deferens [L-Ala^3]DPDPE is a potent (IC$_{50}$ = 12 nM) δ-selective agonist, while at central δ-receptors it is a moderately potent δ$_1$-receptor antagonist and weak agonist with no apparent affects on central δ_2-receptors. The conformation of [L-Ala^3] DPDPE has been compared to that of DPDPE [716], and the differences in conformation around residues 2 and 3 were proposed to explain the differences in efficacy observed for the two analogs. C-Terminal extension of cyclo[D-Pen^2,L-Cys^5]enkephalin) (DPLCE) with phenylalanine resulted in an analog with extremely high potency in the MVD (IC$_{50}$ = 0.016 nM) [717]. A similar extension of cyclo [D-Pen^2,L-Pen^5]enkephalin (DPLPE) had only a modest effect on δ-receptor affinity or potency in the MVD, but halogenation of the resulting DPLPE-Phe resulted in increased δ-receptor affinity and exceptionally high δ-receptor selectivity and potency (see Table 15) [718]; the p-chloro and p-bromo derivatives also exhibited enhanced penetration of the BBB [714].

Glycopeptide derivatives of cyclo[D-Cys^2, Cys^5]enkephalin and DPDPE have been prepared by attachment of Ser(β-D-glucose)-GlyNH$_2$ to the C-terminus [719]. After peripheral administration, the cyclo[D-Cys^2,Cys^5]enkephalin derivative exhibits significant analgesic activity that appears to be centrally mediated, indicating that the glycopeptide penetrates the blood–brain barrier. The penetration of the BBB was initially postulated to be through interaction with the glucose transporter GLUT-1, but subsequent studies have proved this to be incorrect (see Ref. [720] and references cited therein). In a series of cyclo [D-Cys^2,D-Cys^5,Ser^6,Gly^7]enkephalin amide analogs in which the sugar attached to Ser^6 was varied, the nature of the sugar affected analgesic potency following i.v. administration; the α-glucose derivative showed the highest analgesic potency and δ-receptor affinity of the glycosylated analogs examined [720]. Recently, a prodrug derivative of DPDPE in which PEG (polyethylene glycol) is attached to the N-terminus has been reported to also exhibit enhanced analgesia compared to DPDPE following i.v. administration; this prodrug has very weak affinity for δ-opioid receptors, but is converted to DPDPE in vivo [721].

The conformation of DPDPE has been examined by NMR and computational methods (see Ref. [722] and references cited therein) and an X-ray structure has been obtained [722]. There is generally good agreement between these studies on the conformation of the 14-membered ring, but there is still considerable conformational flexibility around Tyr^1 in DPDPE. This is evident in the crystal where three distinct structures, with essentially identical conformations for the 14-membered ring but with different conformations for Tyr^1, were found [722]. The activity of the Hat^1, β-MeTyr^1, and Tmt derivatives (see below) suggested that the preferred conformation of the side chain of residue 1 in DPDPE is trans [723–725]. Based on the activity of the β-MePhe^4 derivatives (see below), the proposed side-chain conformation for Phe^4 is gauche (−) ($\chi_1 = -60°$) [726].

Therefore, other modifications to DPDPE have been made in order to incorporate additional conformational constraint into the aromatic residues in the peptide. Constrained Tyr analogs, including 2'-methyltyrosine, Dmt, β-methyltyrosine, 2',6',β-trimethyltyrosine (Tmt) (Fig. 43), and Hat [723–725,727], have been incorporated at position 1. Substitution of Dmt enhanced affinity at both δ- and μ-receptors and increased in vitro and in vivo potencies [727]. Peptides containing 2'-MeTyr and Hat were also potent analogs with high δ-receptor selectivity [723] (see Table 15). Of the four peptides containing β-MeTyr, the (2S,3R)-β-MeTyr^1 derivative has the highest affinity for δ-receptors and the greatest potency in the MVD [723]. In the Tmt derivatives, which combine methyl groups on the phenyl ring with the β-methyl substitution, only the peptide containing the (2S,3R) isomer of Tmt retaining high δ-receptor potency and selectivity [724,725], while the peptide containing the diastereomer (2S,3S) exhibits much lower δ-receptor affinity and selectivity (Table 15). Interestingly, in vivo [(2S,3S)-Tmt^1]DPDPE is the more potent analog in the tail flick test following i.c.v. administration [728]. The antinociceptive activity of both peptides is antagonized by β-FNA as well as by DALCE [728], suggesting that in contrast to DPDPE the antinocieptive activity of both these Tmt analogs is partially mediated by μ-receptors as well as δ_1-receptors; these results are surprising for [(2S,3R)-Tmt¹]DPDPE, given its high selectivity for δ-receptors in binding assays.

Further conformational constraint in DPDPE has also been examined by incorporation of constrained residues in position 4. All four isomers of β-methylphenylalanine have been incorporated into DPDPE [729]. [(2S,3S)-β-MePhe⁴]DPDPE (see Table 15) exhibits the highest δ-receptor affinity of the four peptides, although it is 6–10-fold less potent than DPDPE; it also exhibits the highest selectivity for δ- over μ-receptors. Incorporation of a *p*-nitro group into (2S,3S)-β-MePhe⁴ increases δ-receptor affinity and potency 6–10-fold, so that [(2S,3S)-β-MePhe (*p*-NO₂)⁴]DPDPE exhibits similar potency at δ-receptors as DPDPE. Further constraint of Phe⁴ by incorporation of 2′,6′,β-trimethylphenylalanine (Tmp) (Fig. 43), however, results in decreased δ-receptor binding and selectivity; in the case of the (2S,3S) isomer affinity for the μ-receptor increases, resulting in a peptide with 10-fold selectivity for μ-receptors [730].

Enkephalin Analogs with Antagonist Activity at δ Receptors The first antagonists for δ-opioid receptors were prepared by *N*,*N*-dialkylation of the N-terminus of enkephalins. The first δ-selective antagonists reported were *N*,*N*-diallyl Leu-enkephalin methyl ester and the derivative ICI 154,129 containing a thioether in place of the peptide bond between residues 3 and 4 [731], but the potency of these peptides was weak (K_e against Leu-enkephalin in the MVD = 254 nM for ICI 154,129). Substitution of Aib (α-aminoisobutyric acid) in positions 2 and 3 of *N*,*N*-diallyl Leu-enkephalin to give ICI 174,864 (**29**) (Fig. 5) (Table 15) enhanced potency approximately 10-fold and δ-receptor selectivity at least 5-fold compared to ICI 154,129 [97], and so ICI 174,864 has been frequently used as a δ-selective antagonist in a number of pharmacological studies. The structure–activity relationships for the *N*,*N*-diallyl derivatives as δ-antagonists are distinctly different from those of the unsubstituted agonist series [732]. Thus, the *N*,*N*-diallyl derivatives of a variety of potent enkephalin agonists such as DADLE are weak nonselective antagonists. Also while ICI 154,129 is a selective δ-antagonist, the corresponding [Gly³Ψ(CH₂S)Phe⁴,Leu⁵]enkephalin is virtually inactive as an agonist. Other modifications have been made to the *N*,*N*-diallyl enkephalins, including replacement of Gly²-Gly³ with a rigid 4-aminobenzoic acid spacer [733], which yields a δ-antagonist with similar potency to ICI 154,129. Substitution of benzyl groups for the allyl groups in *N*,*N*-diallyl Leu-enkephalin enhances δ-antagonist potency, while *N*,*N*-dialkylation with other alkyl groups (e.g., phenethyl) result in derivatives with agonist activity [734]. Interestingly, while ICI-174,864 is more potent than *N*,*N*-dibenzyl Leu-enkephalin in the MVD [734], the reverse was true in binding studies using cloned δ-receptors [735], suggesting possible differences in δ-receptors in the two preparations. *N*,*N*-Dialkyl Leu-enkephalin analogs bearing a reactive functionality recently have been described [735] (see Section 6.7).

Recently, the Leu-enkephalin analog containing a novel analog of 2′,6′-dimethyltyrosine lacking a basic amine (2S)-2-methyl-3-(2,6-dimethyl-4-hydroxyphenyl)propanoic acid, (2S)-Mdp) (Fig. 43) was reported; this peptide is an antagonist in both the MVD and GPI smooth muscle preparations, with five-fold higher potency in the MVD [736].

6.2.3. Dimeric Enkephalin Analogs A number of dimeric enkephalin analogs have been prepared (Fig. 44). Their receptor selectivity depends on the peptide sequences and the length of the spacer between the two peptides. The tripeptide dimer (DTRE)₂ (Tyr-D-Ala-Gly-NH)₂(CH₂)₂) (**219**) (Table 16) is μ-selective, while the corresponding monovalent tripeptide Tyr-D-Ala-Gly-NH₂ (TRE) is much less potent and only moderately μ-selective [737]. Replacement of the Tyr residue in one-half of DTRE₂ with Phe or D-Tyr significantly decreased potency, which suggested that μ-opioid receptors contain two similar sites that are in proximity to one another [738].

In the tetrapeptide series biphalin ((Tyr-D-Ala-Gly-Phe-NH)₂) (**220**), with only a hydrazine spacer, is not selective for μ- or δ-receptors (Table 16) [739]. This peptide exhibits potent analgesic activity *in vivo* [739,740], even though its potency in smooth muscle assays is modest (Table 16). Biphalin is equipotent

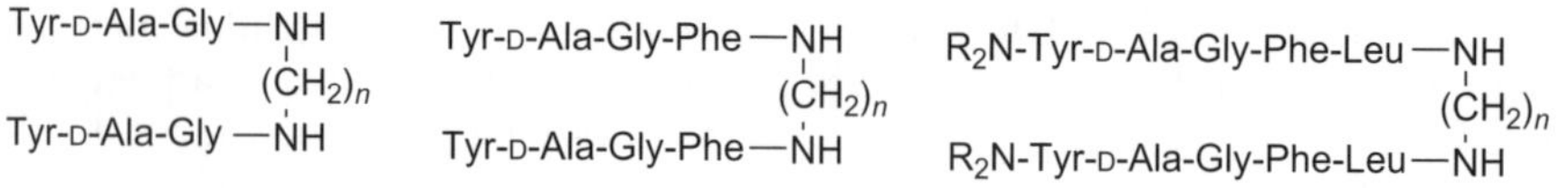

Tyr-D-Ala-Gly—NH
       |
       $(CH_2)_n$
Tyr-D-Ala-Gly—NH

219 $n = 2$ (DTRE)$_2$

Tyr-D-Ala-Gly-Phe—NH
            |
            $(CH_2)_n$
Tyr-D-Ala-Gly-Phe—NH

220 $n = 0$ Biphalin
221 $n = 12$ (DTE)$_{12}$

R_2N-Tyr-D-Ala-Gly-Phe-Leu—NH
                      |
                      $(CH_2)_n$
R_2N-Tyr-D-Ala-Gly-Phe-Leu—NH

222 R = H, $n = 2$ (DPE)2
223 R = $-CH_2CH=CH_2$, $n = 2$

Figure 44. Dimeric enkephalin analogs.

with morphine, but produces little if any physical dependence, following systemic (i.p.) administration. The truncated dimer Tyr-D-Ala-Gly-Phe-NHNH(Phe) retains high binding affinity and potency in smooth muscle assays, indicating that the entire sequence of biphalin is not required for activity [741]. Modification of the residue in the 4-position can alter opioid receptor affinities and impart some selectivity for either μ- or δ-receptors. Thus, introduction of L-β-MePhe or naphthylalanine in this position results in analogs that preferentially interact with μ-receptors (see Table 16) [742], while p-halogenation or nitration of the phenyl rings enhances δ-receptor affinity more than μ-receptor affinity [743]. *Para* chlorination also increases penetration across the BBB [744,745]. *In situ* brain perfusion studies of [^{125}Tyr]biphalin suggest that part of the transport of this compound may utilize the large neutral amino acid carrier [746].

Other dimeric enkephalin derivatives have been prepared, which exhibit δ-receptor selectivity. The dimer (DTE)$_{12}$ (**221**) with a long spacer ($n = 12$) between the two tetrapeptides is a δ-selective ligand (the monomer Tyr-D-Ala-Gly-Phe-NH$_2$ is μ-selective) [747]. In the pentapeptide series (DPE)$_n$, the dimer with $n = 2$ (**222**) exhibits the greatest δ-selectivity in this series [748], but the selectivity of this peptide is low (IC$_{50}$ ratio (μ/δ) = 6.5). Dimeric analogs with antagonist activity have also been examined. The dimeric derivative of the antagonist N,N-diallyl Leu-enkephalin **223** is approximately ninefold more potent than the monomeric N,N-diallyl-Tyr-Gly-GlyPhe-LeuNHEt [749], but the antagonist activity of the truncated dimer N,N-diallyl-Tyr-Gly-Gly-Phe-Leu-NHCH$_2$CH$_2$NH(AcPhe-Leu) suggests that the enhanced activity of the full dimer is not due to bridging two δ-receptor binding sites.

6.3. Dynorphin Analogs

Dynorphin has not been nearly as well studied as other smaller opioid peptides. Although several peptides with high affinity for κ-receptors are obtained from prodynorphin (see Section 3.4), SAR studies have focused on derivatives of dynorphin A (see Refs [653,750] for reviews). Dynorphin A is a heptadecapeptide, but dynorphin A-(1–13) accounts for essentially all of the activity of the larger peptide [751]. Further truncation of dynorphin A-(1–13) from the C-terminus identified the basic residues Arg^7 and Lys^{11} as important for κ-receptor potency and selectivity [258]. Therefore, typically dynorphin A-(1–13) or A-(1–11) have been used as the parent peptide for further modification.

Table 16. Opioid Receptor Affinities and Opioid Activity in the GPI and MVD of Dimeric Enkephalin Analogs

Peptide	K_i (nM)		K_i Ratio	IC$_{50}$ (nM)		
	δ	μ	(δ/μ)	MVD	GPI	References
(DTRE)$_2$ (**219**)	14,000	34	410	2,300	410	[737,738]
Biphalin (**220**)	5.2	2.8	1.8	8.8	27	[742]
[(2S,3S)-β-MePhe$^{4,4'}$]Biphalin	110	1.3	85	180	21	[742]
DTE$_{12}$ (**221**)	10	38	0.26	38	1,000	[747]

The possible conformations of dynorphin A have been studied by a variety of spectral techniques (see Ref. [750] for a review). Like other linear peptides, a variety of conformations have been observed for dynorphin A that depend upon the experimental conditions. Schwyzer [752] proposed a "membrane-assisted" model for dynorphin's interaction with κ-receptors in which the N-terminal "message" sequence adopts an α-helical structure from residues 1–9 when it binds to the receptor. NMR studies of dynorphin A bound to dodecylphosphocholine micelles [753,754] observed a helical structure in the N-terminal portion of the peptide, supporting this proposal. This helical structure of dynorphin A has been docked to a computational model of κ-opioid receptors [175].

Studies of dynorphin A have been complicated by its metabolic lability [755]. In addition to inactivation by peptidase cleavage in the N-terminus, cleavage in the C-terminus yields shorter active peptides, which may have different receptor selectivity profiles from the parent peptide. Dynorphin A-[1–8], which is the predominant form of dynorphin A present in rat brain [756,757], is less selective for κ-receptors than the longer peptides (see Ref. [758]). C-Terminal amidation enhances the metabolic stability of dynorphin A-(1–13) [755], and therefore this modification is typically incorporated into dynorphin A analogs. The peptide bonds in dynorphin A-(1–11)NH$_2$ have been replaced by reduced amide bonds to increase metabolic stability [759,760]; this modification was well tolerated by κ-receptors in the C-terminal "address" sequence, but led to decreased opioid receptor affinity when incorporated in the "message" sequence. An analog of dynorphin A-(1–8) E2078 ([N-MeTyr^1,N-MeArg^7, D-Leu^8]dynorphin A-(1–8)NHEt) [761,762] containing modifications to stabilize the peptide to metabolism has been studied extensively in vivo. It exhibits a slight preference for κ-receptors in binding assays (see Table 17) and produces analgesia following both intravenous and subcutaneous administration [761–763], apparently via spinal κ-receptors [764].

MeTyr-Gly-Gly-Phe-Leu-Arg-NMeArg-D-LeuNHEt
224 E2078

6.3.1. Linear Analogs Early structural modifications were made in the C-terminal "address" [258] sequence of dynorphin A and focused on the nonbasic residues (see Refs [653,750]. D-Pro has been incorporated in place of Pro^{10} in both dynorphin A-(1–11) [765] and A-(1–13) [766] and reported to enhance κ-receptor selectivity (see Table 17). Ala and D-Ala have been substituted for Ile^8 with retention of κ-receptor affinity and selectivity [766,767]. Replacement of the basic residues individually by N^ε-acetylated lysine [768] indicated that substitution of Arg^9 by a nonbasic residue is well tolerated, and that the basicity of Arg^6 is not required for interaction with κ-receptors, but contributes to κ-receptor selectivity by decreasing μ-receptor affinity. Analogs in which Arg^6 or Arg^7 were replaced by norleucine also retained high κ-receptor affinity, but exhibited decreased κ-selectivity [769]. A C-terminal extended peptide dynorphin A-(1–13)-Tyr-Leu-Phe-Asn-Gly-Pro (dynorphin Ia) (**225**), based on the structure of a dynorphin-related peptide purified from bovine adrenal medulla, was synthesized [770] and found to be more selective for κ-receptors than dynorphin A-(1–13); additional structural modifications (D-Leu^8 and/or N-methylation of Tyr^1), were made in order to reduce the motor effects observed with dynorphin Ia [771]. Labeled derivatives of dynorphin A have been prepared by attaching functionalities to the C-terminus (DAKLI, [Arg^{11,13}]dynorphin A-(1–13)NH(CH$_2$)$_5$NH-R, where R is fluorescein, ^{125}I-labeled Bolton Hunter reagent or biotin) [772]) or to the side chain of Lys^{13} (biotin) in dynorphin A-(1–13) amide [773].

Tyr-Gly-Gly-Phe-Leu-Arg-Arg-Ile-Arg-Pro-Lys-Leu-Lys-Tyr-Leu-Phe-Asn-Gly-Pro
225 Dynorphin Ia

Several studies have focused on modifications in the N-terminal sequence of dynorphin A. Interestingly, the most κ-selective derivatives of dynorphin A have been prepared by

Table 17. Opioid Receptor Affinities and Opioid Activity in the GPI and MVD of Analogs of Dynorphin A (Dyn A)

	K_i (nM)			K_i Ratio	IC_{50} (nM)		
	κ	μ	δ	κ/μ/δ	GPI	MVD	References
Peptide:							
Dyn A-(1–13)	0.15	1.3	4.1	1/8.6/27	1.7	78	[769]
Dyn Ia (**225**)	0.25	6.7	71	1/26/280	0.5	236	[770]
[D-Pro^{10}]Dyn A-(1–11)	0.032	2.0	7.5	1/62/230	—	—	[765]
E2078 (**224**)	1.9	4.5	27.2	1/320/14	0.3	7.4	[761]
[N-allyl,D-Pro^{10}]Dyn A-(1–11)	0.049	11	450	1/220/9,160	18	—	[774,798]
[N-CPM, D-Pro^{10}]Dyn A-(1–11)	0.020	9.6	560	1/480/28,000	2.2	—	[774,798]
[N-benzyl,D-Pro^{10}]Dyn A-(1–11)	0.029	31	175	1/1,070/6,080	990	—	[774,798]
[Ala^3]Dyn A-(1–11)NH$_2$	1.1	210	730	1/190/660	1.7	—	[776]
[D-Ala^3]Dyn A-(1–11)NH$_2$	0.76	260	1,000	1/350/1,300	8.1	—	[776]
[(R)-Atc^4,D-Ala^8]Dyn A-(1–11)NH$_2$	0.89	33	>10,000	1/37/>6,000	—	—	[779]
[(S)-Atc^4,D-Ala^8]Dyn A-(1–11)NH$_2$	9.5	88	>10,000	1/9/>1,000	—	—	[779]
cyclo[Cys^5,Cys^{11}]Dyn A-(1–11)NH$_2$	0.28	0.27	1.6	1/1/6	1,080	420	[782]
cyclo[Cys^5,D-Pen^{11}]Dyn A-(1–11)NH$_2$	1.1	31	240	1/28/220	690	—	[784]
cyclo[D-Asp^2,Dap^5]Dyn A-(1–13)NH$_2$	0.22	0.49	10	1/2/46	0.16	—	[788]
cyclo[D-Asp^5,Dap^8]Dyn A-(1–13)NH$_2$	8.0	75	3,300	1/9/400	>5,000	—	[789,790]
cyclo[D-Asp^6,Dap^9]Dyn A-(1–13)NH$_2$	2.6	4.4	48	1/2/19	46	—	[789,790]
cyclo[D-Asp^3,Lys^7]Dyn A-(1–11)NH$_2$	4.9	310	130	1/64/27	600	—	[791]
Antagonist analogs:							
[N,N-diallyl,D-Pro^{10}]Dyn A-(1–11)	21	135	350	1/6.5/17	190^b	—	[796]
[N,N-diCPM,D-Pro^{10}]Dyn A-(1–11)	0.19	3.9	166	1/21/880^c	—	—	[798]
[Pro^3]Dyn A-(1–11)NH$_2$	2.7	5,700	8,800	1/2,100/3,260	244,494^a	—	[778]
Arodyn	10	1,700	5,800	1/170/580^c	—	—	[803]

$^a K_e$ values against U50,488 and a dynorphin analog, respectively.
$^b K_e$ values against a dynorphin analog.
^{c} Reverses the agonist activity of Dyn A-(1–13)NH$_2$ in an adenylyl cyclase assay using cloned κ-receptors.

modifications in this "message" sequence rather than in the C-terminal "address" sequence. N-terminal monoalkylation of [D-Pro^{10}]dynorphin A-(1–11) with an allyl, cyclopropylmethyl, or benzyl group results in marked enhancement in κ-receptor affinity by decreasing μ-receptor affinity (Table 17) [774]; N,N-dialkylation results in analogs with antagonist activity (see below). An alanine scan [767] verified the importance of Tyr^1 and Phe^4 in dynorphin A-(1–13) for potency in smooth muscle assays and receptor affinity. While incorporation of D-amino acids in position 2 can enhance metabolic stability and yields analogs with high affinity for κ-receptors, this modification significantly increases μ-receptor affinity and thus yields μ-selective analogs [258,775]. In contrast, substitution of either Ala or D-Ala in position 3 of dynorphin A-(1–11)NH$_2$ markedly enhances κ-receptor selectivity (Table 17) [776]; incorporation of other amino acids in this position, however, are generally less well tolerated [777]. Recently, [Pro^3]Dyn A-(1–11)NH$_2$ was reported to be a highly selective κ-receptor antagonist (see below) [778].

6.3.2. Conformationally Constrained Analogs

Local constraints have been incorporated into dynorphin A. The conformationally constrained phenylalanine derivative Atc (2-aminotetralin-2-carboxylic acid, Fig. 43) has been incorporated in [D-Ala^8]dynorphin A-(1–11)NH$_2$ [779]. Interestingly, the peptide containing (R)-Atc, which corresponds to a conformationally constrained D-Phe analog, possesses higher affinity for κ- and μ-receptors than the peptide containing the S-isomer (Table 17), even though the D-Phe^4 analog exhibits relatively low affinity for these receptors (K_i = 8.9 and 146 nM, respectively). Both peptides exhibit negligible affinity for δ-receptors. A novel conformational constraint, 4-aminocyclohexanecarboxylic acid, has been incorporated into dynorphin A-(1–13) amide in place of Gly^2-Gly^3 [780]. While the affinities of the two peptides for κ-receptors is significantly reduced compared to the parent peptide, it is interesting to note that the peptides containing the cis and trans isomers of 4-aminocyclohexanecarboxylic acid exhibit similar affinity for κ-receptors.

Cyclic derivatives of dynorphin A have been prepared by formation of both disulfide and lactam linkages (see Ref. [653] for a review including extensive tables). The first cyclic dynorphin analog reported was cyclo[D-Cys^2, Cys^5]dynorphin A-(1–13) [781], which is a more potent agonist in smooth muscle assays than dynorphin A-(1–13). Subsequently, a variety of cyclic dynorphin analogs containing a disulfide linkage in the C-terminus were described [782–784] (see Ref. [653] for a tabular summary of these analogs). Except for the analogs cyclized through a D-amino acid in position 5, these peptides with a constraint in the C-terminus retain high affinity for κ-receptors in radioligand binding assays. Three analogs cyclized between positions 5 and 11 (cyclo[Cys^5,D-Pen^{11}]-, cyclo[$Pen^{5,11}$]-, cyclo[Pen^5,D-Pen^{11}]dynorphin A-(1–11)NH_2) show increased κ-receptor selectivity in the radioligand binding assays compared to the linear parent peptide (Table 17). Interestingly, a number of these cyclic disulfide-containing peptides, particularly those in which position 5 is involved in the disulfide linkage, show much lower potency in the GPI than expected from the binding assays; these discrepancies between the two assays were postulated to be due to different receptor subtypes present in the brain versus peripheral tissues [782–784]. A conformational search of cyclo[Cys^5,Cys^{11}]- and cyclo[Cys^5,D-Ala^8,Cys^{11}]dynorphin A-(1–11)NH_2 found a low-energy α-helical conformation for the disulfide-bridged ring in these peptides, suggesting that this conformation may be important for the potency of these peptides at central κ-receptors [785]. When cyclo[Cys^5,Cys^{11}]dynorphin A-(1–11)NH_2 bound to micelles was examined by NMR and molecular dynamics, conformations resulting from cis–trans isomerism around the Arg^9-Pro^{10} amide bond were observed [786], with the trans isoform exhibiting a β-turn from residues Cys^5 to Ile^8, while the cis isoform contained a type III β-turn from residues Arg^7 to Pro^{10}.

Conformational constraint via lactam formation has been examined in both the N-terminus and the C-terminus of dynorphin A [787–791]. The initially reported derivatives exhibited low K_e values for antagonism by naloxone (K_e = 1.5–4.5 nM) in the GPI, suggesting that these peptides preferentially interacted with μ-receptors (affinities for κ-receptors were not reported) [787]. More recently, two series of dynorphin A-(1–13) amide analogs constrained though a lactam linkage between residues in either i and $i + 3$ positions [788–790] or i and $i + 4$ positions [791] have been reported (Table 17). These constraints were chosen to be compatible with the helical conformation proposed by Schwyzer as the bioactive conformation for the N-terminal sequence of dynorphin A [752]. For derivatives constrained between a D-Asp in position i and Dap (α,β-diaminopropionic acid) in position $i + 3$, the constraint is well tolerated by κ-receptors between positions 2 and 5 [788] and also between positions 6 and 9 [789,790], while moving the constraint to residues 3 and 6 markedly decreases affinity for opioid receptors [789,790]. Consistent with the results for the cyclic disulfide-containing analogs cyclized through position 5, cyclo[D-Asp^5,Dap^8]dynorphin A-(1–13)NH_2 is a weak agonist in the GPI while retaining reasonable affinity for κ-receptors in the radioligand binding assays [789,790]. For the i to $i + 4$ derivatives [791], cyclo[D-Asp^3,Lys^7]dynorphin A-(1–11)NH_2 exhibited the highest κ-receptor affinity and selectivity (Table 17); the cyclic peptide cyclo[Lys^5,D-Asp^9]dynorphin A-(1–11)NH_2 also exhibited nanomolar affinity for both κ- and μ-receptors. The synthesis of novel N-terminal to side-chain cyclic dynorphin A analogs has recently been reported [792,793].

6.3.3. Dynorphin A Analogs with Antagonist Activity

Early attempts to prepare κ-selective antagonists by modification of dynorphin A met with limited success, and the analogs reported generally exhibited weak antagonist activity, residual agonist activity, and/or low selectivity for κ-receptors [794,795]. In recent years, antagonist analogs with improved pharmacological profiles have been identified. Modifications of Tyr^1 have resulted in dynorphin A analogs with antagonist activity. Incorporation of Phe in place of Tyr^1 in [D-Ala^8]dynorphin A-(1–11)NH$_2$ resulted in a peptide, which antagonized dynorphin A (K_e = 30–65 nM) in the GPI, but still exhibited weak agonist activity [769]. N,N-Terminal dialkylation of dynorphin A fragments with either allyl or CPM groups resulted in analogs with antagonist activity [132,796–798]; the N,N-di(cyclopropylmethyl) derivative exhibits higher κ-receptor selectivity than the N,N-diallyl peptide [132,798]. Recently, novel analogs of 2′,6′-dimethyltyrosine that lack a basic amine (Dhp and Mdp) (Fig. 43) have been incorporated in place of Tyr^1 in dynorphin A-(1–11)NH$_2$, resulting in [(2S)-Mdp^1]dynorphin A-(1–11)NH$_2$ (dynantin) (Fig. 45), which is a highly κ-selective peptide that is a potent antagonist at κ-opioid receptors (see Table 17) [799]. [Pro^3]Dyn A-(1–11)NH$_2$ was also reported to be a highly selective, but relatively weak, κ-receptor antagonist (see Table 17) [778].

Chimeric dynorphin A analogs, in which the N-terminal "message" sequence of dynorphin A is replaced by small peptides with opioid antagonist activity (Fig. 45), have also been prepared as potential antagonists. Addition of the C-terminal "address" sequence in these peptides significantly enhances κ-receptor affinity, but the receptor preference of the N-terminal sequence still often predominates. Thus, incorporation of the δ-receptor antagonist TIPP (Tyr-Tic-Phe-Phe) in place of the N-terminal sequence of [D-Pro^{10}]dynorphin A-(1–11)NH$_2$ or the corresponding acid resulted in peptides with affinity and antagonist activity at κ-receptors, but which still preferentially bind to δ-receptors [800,801]. Incorporation of only Tic^2 in position 2 produced similar results for the dynorphin A-(1–11)NH$_2$ derivative [802] or a slight (twofold) preference for δ- over κ-receptors for the [D-Pro^{10}]dynorphin A-(1–11) analog [801]; incorporation of N-MePhe in position 2 of [D-Pro^{10}]dynorphin A-(1–11) resulted in a greater (eightfold) preference for κ- over δ-receptors but little selectivity for κ- over μ-receptors [801]. Attachment of the C-terminal "address" sequence from [D-Ala^8]dynorphin A-(1–11)NH$_2$ to the acetylated hexapeptide [Arg^6]acetalin, which is a μ-receptor antagonist, to give the peptide extacet increases κ-receptor affinity 65-fold resulting in nanomolar affinity for κ-receptors (K_i = 6.6 nM), but it still preferentially binds to μ-receptors (K_i = 1.1 nM) [801]. Examination of analogs of this lead peptide using a combinatorial library led to the identification of arodyn, a novel dynorphin A analog with higher affinity, much higher selectivity and antagonist activity at cloned κ-receptors (see Table 17) [803]. A Boc-protected tetrapeptide derived from a sequence found in venom of the Philippine cobra (*Naja philippinensis*) has been reported to have weak antagonist activity at κ-receptors [804]. Attachment of the C-terminal "address" sequence of [D-Ala^8]dynorphin A-(1–11)NH$_2$ to this sequence resulted in a novel acetylated dynorphin A analog, JVA-901 (now referred to as venorphin), which exhibits antagonist activity and greatly enhanced affinity for cloned κ-receptors [805]. In spite of the structural similarities to dynorphin A, the SAR of venorphin is completely different from dynorphin A, suggesting that these two peptides interact with κ-receptors in different ways [806,807]. In venorphin, only Trp^3, and not Tyr^1 or a basic functionality, is required in the N-terminal "message" sequence for high affinity for κ-receptors [806], and Arg^9, which is not important for the κ-receptor affinity of dynorphin A, is the most critical residue in the C-terminus [807].

Dynantin	(2S)-Mdp-Gly-Gly-Phe-Leu-Arg-Arg-Ile-Arg-Pro-LysNH$_2$
TIPP-Dyn A	Tyr-Tic-Phe-Phe-Leu-Arg-Arg-Ile-Arg-Pro-LysNH$_2$
Extacet	AcArg-Phe-Met-Trp-Met-Arg-Arg-D-Ala-Arg-Pro-LysNH$_2$
Arodyn	AcPhe-Phe-Phe-Arg-Leu-Arg-Arg-D-Ala-Arg-Pro-LysNH$_2$
JVA-901/venorphin	AcTyr-Lys-Trp-Trp-Leu-Arg-Arg-D-Ala-Arg-Pro-LysNH$_2$

Figure 45. Novel analogs of dynorphin A.

Table 18. Opioid Receptor Affinities and Opioid Activity in the GPI and MVD of μ-Selective Opioid Peptide Analogs

Peptide	K_i (nM) μ	K_i (nM) δ	K_i Ratio (δ/μ)	IC$_{50}$ (nM) GPI	IC$_{50}$ (nM) MVD	Reference
β-Casomorphin analogs:						
Morphiceptin (**226**)	63	30,000	475	318	4,800	[810]
PL017 (**227**)	11	7,250	660	34	240	[120]
Tyr-*cyclo*[D-Orn-Phe-D-Pro-Gly] (**232**)	0.88	13.2	15	2.1	4.9	[822]
Tyr-*cyclo*[D-Orn-2-Nal-D-Pro-Gly] (**233**)	5.9	17.2	2.9	384	$K_e = 200$–270^a	[822]
Tyr-*cyclo*[D-Orn-2-Nal-D-Pro-D-Ala]	0.76	72	95	600	$K_e = 5.4$–6.0^a	[825]
Dmt-*cyclo*[D-Orn-2-Nal-D-Pro-Gly] (**234**)	0.46	0.46	1	7.9	$K_e = 2.1$–3.3^a	[827]
CHO-Dmt-*cyclo*[D-Orn-2-Nal-D-Pro-Gly]		218	33	0.15	$K_e = 216^a$ $K_e = 16^a$	[826]
Dermorphin analogs:						
TAPP (Tyr-D-Ala-Phe-PheNH$_2$)	1.5	625	409	255	780	[828]
DALDA (**242**)	1.7	19,200	11,400	3.23	800	[828]
[Dmt^1]DALDA (**243**)^c	0.14	2,100	14,700	1.4	23	[829]
Tyr-*cyclo*[D-Orn-Phe-Asp]NH$_2$ (**244**)	10.4	2,220	213	36.2	3,880	[830]
Tyr-*cyclo*[D-Orn-Phe-Glu]NH$_2$	0.98	3.21	3.3	1.2	1.1	[831]
Tyr-*cyclo*[D-Orn-Aic-Glu]NH$_2$	4.21	209	50	7.21	36.5	[831]
JOM-6 (**246**)	0.29	24.8	86	—	—	[832]
JH-54 ([Phe^1]JOM-6, **247**)	1.36	1020	750	9.1	—	[832]
[D-Hat^1]JOM-6 (**248**)	0.39	58	150	b	b	[833]

^{a} Antagonist.
^{b} EC$_{50}$ = 1.4 and 1500 nM in GTPγS assays using cloned μ- and δ-receptors, respectively.
^{c} K_i (κ) = 22 nM.

6.4. Opioid Peptides with the Tyr-Pro-Phe Sequence

6.4.1. β-Casomorphin Analogs and the Endomorphins

β-Casomorphin was identified over 20 years ago, so a number of analogs of this peptide have been reported. The heptapeptide β-casomorphin (**210**), although exhibiting some selectivity for μ-receptors, is a weak opioid agonist (IC$_{50}$ in the GPI = 57 μM [808]). Shortening the peptide to the pentapeptide and tetrapeptide increases potency [808,809], so generally analogs have been prepared of one of these smaller fragments. Conversion of the tetrapeptide to the C-terminal amide to give morphiceptin (**226**) (see Table 18) [809] substantially increases both potency and μ-receptor selectivity. Further modification of morphiceptin by incorporation of D-Pro at position 4 and N-methylation at position 3 yields PL017 (**227**) [810], which is significantly more potent than morphiceptin and more μ-selective (see Table 18). D-Pro or D-Pip (pipecolic acid) in position 4 of β-casomorphin (1–5) also enhances potency in the GPI [811].

Morphiceptin (**226**) Tyr-Pro-Phe-ProNH$_2$
PL017 (**227**) Tyr-Pro-NMePhe-D-ProNH$_2$

cis/trans-Isomerization occurs around amide bonds involving the nitrogen of proline and other *N*-alkyl amino acid residues. NMR studies of both morphiceptin and PL017 found that while the major isomers were the all-*trans* isomers, the second most common isomer (25%) had a *cis* amide bond between Tyr^1 and Pro^2 [812]. Therefore, Goodman and coworkers incorporated 2-aminocyclopentane carboxylic acid (2-Ac^{5}c) (**228**) (Fig. 46) into position 2 of morphiceptin analogs to eliminate

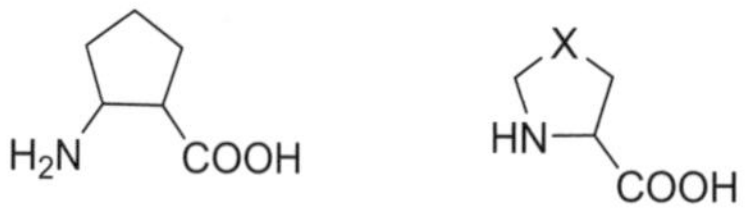

228 2-Ac⁵c

229 X = S or O

Figure 46. Proline analogs.

possible *cis/trans* isomerization [813,814]. Of the four possible stereoisomers only morphiceptin analogs containing *cis*-(1S,2R)-2-Ac⁵c, which exhibits a structure similar to the *cis* conformation of the Tyr-Pro amide bond, are potent opioids, while the analogs with *cis*-(1R,2S)-2-Ac⁵c, which is similar to morphiceptin with the Tyr-Pro bond in a *trans* configuration, were inactive. This led Goodman and coworkers to propose that the *cis* conformation around the Tyr-Pro amide bond is required for the opioid activity of morphiceptin and its analogs [814]. A more detailed model for the bioactive conformation of morphiceptin was developed based on comparison of a series of active and inactive analogs [815].

Like morphiceptin and PL017, cis–trans isomerization occurs around the Tyr-Pro amide bond in endomorphin-1, with similar populations (75% *trans*, 25% *cis*) found for the two conformations [816]. Based on structural comparison of the *cis* and *trans* conformations of endomorphin-1 to other μ- and δ-selective opioid peptides, Podlogar et al. proposed that the *trans* conformation was the bioactive form [816]. Recently, Schiller and coworkers reported endomorphin-2 and morphiceptin analogs containing a pseudo-proline derivative in place of Pro² [817]; analogs of both peptides containing the dimethylated pseudo-prolines Xaa[Ψ^{Me,Me} Pro)] (Xaa = Cys or Ser) (**229**) (Fig. 46), which exist almost exclusively in the *cis* conformation, retain μ-opioid receptor affinity and agonist activity, indicating that the *cis* conformation around the Tyr-Pro amide bond is the bioactive form. Pro has been proposed to function as a stereochemical spacer, and inversion of its stereochemistry in endomorphin-1 essentially abolishes agonist activity in the GPI [818]. [D-Pro²]endomorphin-2, however, was a much more potent analgesic than endomorphin-2 in the tail flick assay following i.c.v. administration; the analgesia produced by both peptides was reversed by naloxone [819].

Potent cyclic derivatives of β-casomorphin (1–5) have been prepared by cyclization between a D-amino acid in position 2 and the C-terminus. Tyr-*cyclo*[D-Orn-Phe-Pro-Gly] (**230**) (Fig. 47) and Tyr-*cyclo*[D-Orn-Phe-D-Pro-Gly] (**232**) exhibit high affinity in μ-receptor

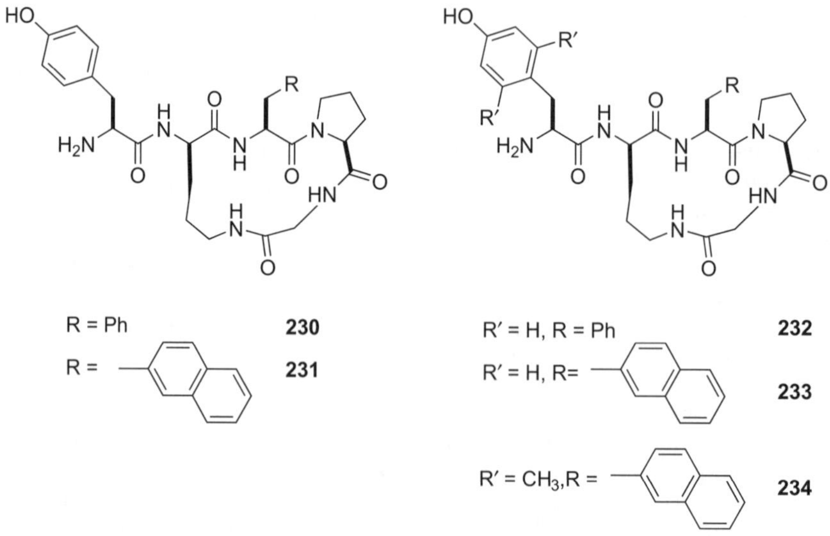

R = Ph **230**
R = (naphthyl) **231**

R' = H, R = Ph **232**
R' = H, R = (naphthyl) **233**

R' = CH₃, R = (naphthyl) **234**

Figure 47. Cyclic β-casomorphin analogs.

binding assays and are potent agonists in smooth muscle assays (Table 18) [820]. Peptides Tyr-cyclo[D-X-Phe-D-Pro] (X = D-Orn or D-Lys) in which Gly^5 has been removed, resulting in a smaller ring size, exhibited similar potency in the GPI as the corresponding peptides with a larger ring, but decreased potency in the MVD [821]. Substitution of Phe^3 with 2-naphthylalanine (2-Nal) in the D-Pro-containing peptides gave Tyr-cyclo[D-X-2-Nal-D-Pro-Gly] (X = D-Orn (**233**) (Table 18) or D-Lys) that are agonists in the GPI but antagonists in the MVD [822], and thus are mixed μ-agonists/δ-antagonists. Examination of Tyr-cyclo[D-Orn-X-D-Pro-Gly] (X = Phe (**232**) or 2-Nal (**233**)) and Tyr-cyclo[D-Orn-2-Nal-Pro-Gly] (**231**) by NMR indicated that the overall backbone conformations and preferred side-chain conformations were roughly similar for these peptides [823]. Interestingly, substitution of 1-naphthylalanine in position 3 of **232** yields a compound with full agonist activity in the MVD [822]. Substitution with other bulky aromatic amino acids was generally well tolerated by δ-receptors, but drastically decreased μ-receptor affinity [824]. Replacement of Gly^5 in **233** with sarcosine or D-Ala significantly enhances δ-receptor affinity and antagonist potency (see Table 18) [825]. Recently, derivatives of Dmt-cyclo[D-Orn-2-Nal-D-Pro-Gly] (**234**) lacking a basic N-terminus were reported (see Table 18); these peptides exhibit antagonist activity at both δ- and μ-receptors [826].

6.4.2. TIPP and Related Peptides Exploration of a series of tetrapeptides consisting solely of aromatic residues led Schiller and coworkers to identify TIPP (Tyr-Tic-Phe-Phe) (**30**) (Fig. 5) [98]. TIPP is a potent δ-antagonist in the MVD with very high selectivity for δ-receptors (Table 19). (Recently, however, TIPP and its analog TIPP[Ψ] (**31**) (Fig. 5) (see below) were reported to exhibit agonist activity in adenylyl cyclase assays using cells containing both endogenous and transfected δ-opioid receptors [101].) The tripeptide derivatives TIP and TIPNH$_2$ are also δ-receptor antagonists in the MVD. In contrast, the tetrapeptide Tyr-D-Tic-Phe-PheNH$_2$ with D-Tic in position 2 is a potent agonist that is selective for μ-receptors. The amide derivative of TIPP, in addition to δ-antagonist activity in the MVD, is a full agonist in the GPI (see Table 19), and was the first compound reported to be a mixed μ-agonist/δ-antagonist [98].

The conformations of TIPP have been examined by molecular mechanics and X-ray crystallography (see Ref. [834] for a review). Examination of the δ-antagonist TIP and the μ-agonist Tyr-D-Tic-PheNH$_2$ by molecular mechanics found compact structures for both peptides, but with different patterns of aromatic ring stacking [835]. Superimposition of low-energy conformations of these two peptides found that the Phe^3 residues were on opposite sides of the plane defined by the Tic residue, providing a possible explanation for the differences in activity observed for the two peptides. Comparison of TIP to the nonpeptide δ-antagonist naltrindole found good spatial overlap of the N-terminal amine and aromatic rings in the peptide with the corresponding groups in the alkaloid. In this model, the Tyr-Tic peptide bond is *trans*. In an alternative model based on the weak dipeptide δ-antagonist Tyr-TicNH$_2$ this bond is *cis* [836]. Further examination of the possible conformations of a series of TIPP analogs, including derivatives of TIPP[Ψ] (**31**), by molecular mechanics found low-energy conformations consistent with the model containing a *trans* Tyr-Tic bond, but the model with a *cis* amide bond still remained plausible [837]. Other modeling studies involving TIPP found conformations consistent with both the *trans* [838,839] and the *cis* conformations [839]. Therefore, Schiller and coworkers prepared TyrΨ[CH$_2$NH]Tic-Phe-PheOH and TyrΨ[CH$_2$NH]MeTic-Phe-PheOH (MeTic = 3-methyl-1,2,3,4-tetrahydroisoquinoline-3-carboxylic acid) (Fig. 43) in which a *cis* amide bond between the first two residues is sterically forbidden [840]. The δ-receptor affinity and antagonist activity of these peptides, although 20–40-fold lower than TIPP, is consistent with a *trans* conformation as the bioactive conformation for TIPP. Interestingly, these modifications substantially increased μ-receptor affinity, so that TyrΨ[CH$_2$NH]MeTic-Phe-PheOH was a nonselective antagonist at both μ- and δ-receptors (less than a twofold difference in K_e values in the GPI and MVD).

Table 19. Opioid receptor affinities and opioid activity in the GPI and MVD of TIPP and selected analogues.[a]

Peptide	K_i (nM) δ	K_i (nM) μ	K_i ratio (μ/δ)	K_e (nM) MVD	K_e (nM) GPI	Reference
TIPP (30)	1,720	1.2	1,430	3.0–5.9		[98]
TIPPNH$_2$	3.0	79	26	14–18	IC$_{50}$=1700[b]	[98]
Tyr-D-Tic-Phe-PheNH$_2$	7.3	520	71	IC$_{50}$=454[b]	IC$_{50}$=37[b]	[98]
TIPP[Ψ] (31)	0.31	3230	10,500	2.1–2.9		[99]
DIPP (236)	0.25	141	570	0.15	IC$_{50}$=770[b]	[848]
DIPPNH$_2$(237)	0.12	1.2	10	0.20	IC$_{50}$=18[b]	[863]
DIPPNH$_2$[Ψ] (239)	0.45	0.94	2	0.54	IC$_{50}$=7.7[b]	[863]
TICP (Tyr-Tic-Cha-Phe)	0.61	3600	5,900	0.44		[844]
TICP[Ψ]	0.26	1050	4,000	0.22		[844]
	1.6[d]	890[d]	560			[854]
(2S,3R)Tmt-Tic-OH	9.3	35,000	3,800	1.8	[e]	[858]
Dmt-D-Phe-NH$_2$	15	3.6	4.4		60–310	[861]
240[e]	9.35	0.22	0.023		[f]	[862]
NMeTyr-Tic-NHCH$_2$CHPh$_2$	0.98	29	29	IC$_{50}$=3.8[b]	IC$_{50}$=3600[b]	[843,863]
Tyr-Tic-NHCH$_2$CH(Ph)CO$_2$Et	0.57	890	1560	IC$_{50}$=1.3[b]		[843]
N,N-Me$_2$-Dmt-Tic-NH-1-adamantane	0.16	1.1	9	0.87	IC$_{50}$=16[b]	[865]
241[g]	0.17	62	360	[g]	[g]	[855]

[a] Data for TIPP and TIPP[Ψ] from Table 8 is included for comparison. Data for a number of other analogues are given in Ref. [834].
[b] Agonist
[c] Not active
[d] IC$_{50}$
[e] 30% inhibition at 30 μM
[f] K_i (κ) = 68 nM; EC$_{50}$ = 0.18 nM for GTPγS binding at cloned μ receptors
[g] K_i (κ) = 1.3 nM; EC$_{50}$ = 0.65 and 6.9 for GTPγS binding at cloned δ and κ receptors, respectively. Inactive at μ receptors.

During examination of TIPP by NMR in deuterated DMSO spontaneous degradation via diketopiperazine formation occurred [841]. This led Schiller and coworkers to prepare TIPP[Ψ] (31) (Fig. 5) and the tripeptide TIP[Ψ] [99] containing a reduced peptide bond between Tic^2 and Phe^3. These analogs exhibit increased δ-receptor antagonist potency in the MVD and higher δ-receptor affinity compared to the parent peptides and exceptional δ-receptor selectivity (see Table 19). This modification also enhances the metabolic stability of the peptides. While the diketopiperazine of Tyr-Tic is inactive, the Dmt analog cyclo[Dmt-Tic] (235) (Fig. 48) exhibits δ-receptor affinity (K_i = 9.6 nM) and weak antagonist activity (K_e = 3.98 μM) in the MVD [842].

A variety of structural modifications have been made to all positions of TIPP (see Refs [834,843] for recent reviews), and these can have profound effect on the activity profile of these peptides. Numerous substitutions have been made in position 3, and the results indicate that an aromatic residue is not required in this position for δ-antagonist activity. Thus, the cyclohexylalanine (Cha) analogs of both TIPP and TIPP[Ψ] (TICP and TICP[Ψ], respectively) are approximately 10-fold more potent as δ-receptor antagonists than the parent peptides [844] (see Table 19). Interestingly, incorporation of D-amino acids containing a β-methyl group can have a profound effect on efficacy at δ-receptors [845,846]. Thus, the (2R,3R)β-MePhe^3 derivative of TIPPNH$_2$ is a δ-antagonist while the (2R,3S)β-MePhe^3 isomer is a δ-agonist, and incorporation of β-Me-D-Tic^2 derivatives in Tyr-D-Tic-Phe-PheNH$_2$ converts the peptide from an agonist to an antagonist. Iodination of Tyr^1 in TIPP converts the peptide from an

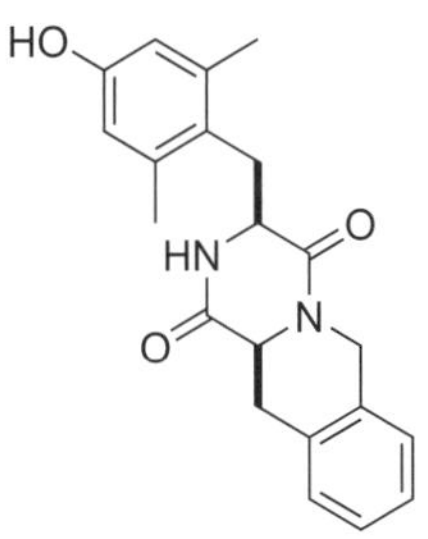

235

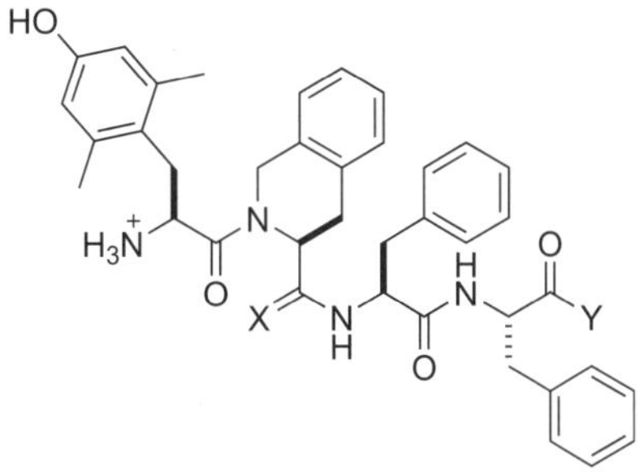

X = =O, Y = OH 236 DIPP
X = =O, Y = NH_2 237 $DIPPNH_2$
X = H_2, Y = OH 238 DIPP[ψ]
X = H_2, Y = NH_2 239 DIPP[ψ]NH_2

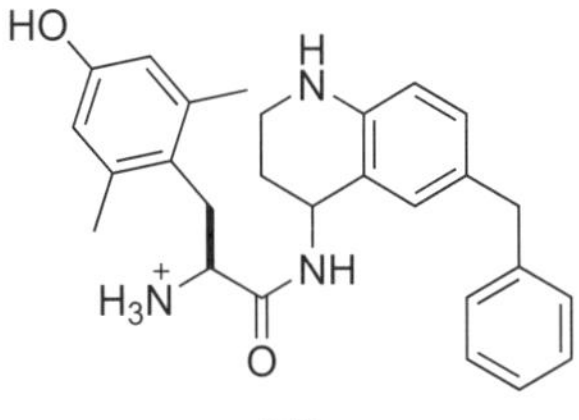

240

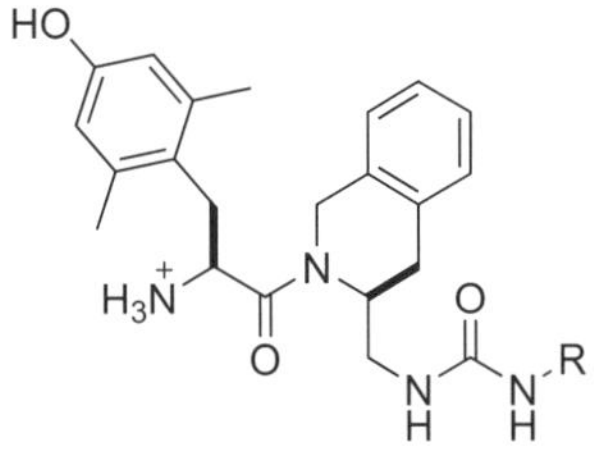

241 R = *t*-Bu

Figure 48. Analogs of Dmt-Tic and TIPP.

antagonist to a δ-selective agonist [847]; the Tyr(3'-I)[1] analogs of TIPP[Ψ], TICP, and TIP, however, are δ-antagonists, which illustrate the subtleties of the effects of modifications on efficacy at δ-receptors [834,843].

Substitution of Dmt in position 1 of TIPP and analogs yielded DIPP (Dmt-Tic-Phe-Phe) (**236**) (Fig. 48) and related peptides. DIPP is an extremely potent δ-antagonist (Table 19) and like $TIPPNH_2$ is also a full agonist in the GPI [848,849]. As noted earlier, the nonpeptide naltrindole can prevent the development of morphine tolerance and dependence in mice [219], and therefore compounds with mixed μ- agonist/δ-antagonist activity could have therapeutic potential. Therefore, Schiller and coworkers also incorporated Dmt into position 1 of $TIPPNH_2$ and TIPP[Ψ]NH_2 to enhance μ-agonist activity; the resulting $DIPPNH_2$[Ψ] (**239**) was the first compound with balanced μ-agonist/δ-antagonist properties [223,850]. $DIPPNH_2$[Ψ] is a potent analgesic following i.c.v. administration (three times more potent than morphine) and produces less acute tolerance than morphine and no physical dependence with chronic administration [223]. The tripeptide analogs Dmt-Tic-Ala-X (where X = OH or NH_2) also exhibit high affinity for δ-receptors and are potent antagonists in the MVD; the amide derivative retains appreciable μ-receptor affinity and is a weak agonist in the GPI [851]. Replacement of Dmt[1] with *N*,2′,6′-trimethyltyrosine (NMeDmt) in $DIPPNH_2$ decreases efficacy at μ-receptors, so that NMeDmt-Tic-Phe-PheNH_2 is a partial agonist and NMeDmt-TicΨ[CH_2NH]Phe-PheNH_2 is an antagonist in the GPI [223].

Based on the hypothesis that the "message" domain in these δ-antagonist peptides consisted of only the Tyr-Tic dipeptide rather than the tripeptide Tyr-Tic-Phe, Temussi and coworkers synthesized Tyr-L/D-Tic-NH_2 and Tyr-L/D-Tic-AlaNH_2 [836]. Although they exhibited much lower potency than TIPP and TIP, the shorter peptides containing Tyr-L-Tic were δ-selective antagonists; peptides containing the Tyr-D-Tic sequence were nonselective agonists. Lazarus and coworkers have explored the structure–activity relationships of Tyr-Tic in considerable detail (see Refs [834,852,853] for reviews, including extensive tables of analogs). The dipeptide Dmt-Tic-OH was initially reported to have exceptionally high δ-receptor affinity and selectivity [851], although other researchers subsequently reported lower δ-receptor affinity and selectivity [854,855] (see Table 19). Modifications examined have included N-terminal alkyaltion, which enhances δ-receptor antagonism [856]. Thus, N,N-Me$_2$-Dmt-Tic-OH and N,N-Me$_2$-Dmt-Tic-Ala-OH exhibit enhanced δ-antagonist potency in the MVD ($K_e = 0.2$–0.3 nM) [856]; also these derivatives cannot undergo diketopiperazine formation. In the GTPγS assay, N,N-Me$_2$-Dmt-Tic-NH$_2$ is a full inverse agonist, while Dmt-Tic-OH is a partial inverse agonist and Dmt-Tic-NH$_2$ is a neutral antagonist [857]. (2S,3R)Tmt-Tic-OH, with a β-methyl group added to Dmt [858], is also a full inverse agonist in this assay [859]. Derivatives of Dmt-Tic with substitutions on Tic^2, with or without a C-terminal acid functionality, have also recently been reported [854,860]. Interestingly, substitution of Tic by D-Phe to give Dmt-D-PheNH$_2$ resulted in a μ-receptor antagonist (although its selectivity was relatively low) [861], while substitution of Tic by 6-benzyl-1,2,3,4-tetrahydroquinoline-4-amine to give **240** resulted in a μ-selective agonist (Table 19) [862].

C-Terminal extension of Tyr-Tic and Dmt-Tic can alter efficacy at δ-receptors, depending on the substitution. Thus, the efficacy of the dipeptide amides Tyr-Tic-NH(CH$_2$)$_n$Ph depend on the length of the amide chain (antagonist for $n = 1$ or 3, agonist for $n = 2$) [843]. In Tyr-Tic-NH(CH$_2$)$_2$Ph substitution on the phenyl ring by p-fluorine or chlorine or replacement of this phenyl ring by a cylohexyl ring converts the peptide from an agonist back to an antagonist [843]. In contrast, ortho substitution on the phenyl ring with chlorine enhances δ-agonist potency 10-fold [843,863]. Introduction of a second phenyl group and N-terminal methylation also enhance agonist potency, so that NMeTyr-Tic-NHCH$_2$CHPh$_2$ is an extremely potent δ-agonist (Table 19). Substitution of the β-carbon of the phenethyl group in Tyr-Tic-NH(CH$_2$)$_2$Ph with CO$_2$Et results in the most selective δ-agonist within this class (Table 19) [843]. Removal of the C-terminal acid from Tyr-Tic-AtcOH functionality converts the peptide from an antagonist to an agonist [843].

C-terminal extension of Tyr-Tic and Dmt-Tic can also enhance μ-receptor affinity. The dipeptide derivative Dmt-Tic-NH(CH$_2$)$_3$Ph is a δ-receptor antagonist and μ-receptor agonist [864]. Interestingly, Dmt-Tic-NH-1-adamantane is a weak agonist at δ-receptors and potent agonist at μ-receptors, although its affinity for the two receptors is similar [865]. N,N-Me$_2$-Dmt-Tic-NH-1-adamantane is both a potent δ-receptor antagonist and a potent μ-receptor agonist (Table 19) [865]. Interestingly, modification of the C-terminus of Dmt-Tic to yield amine, urea and thiourea derivatives (for example) (**241**) results in compounds with nanomolar affinity for κ- as well as μ- and δ-receptors, and agonist activity at all three receptor types [855].

6.5. Opioid Peptides with the Tyr-D-aa-Phe Sequence

6.5.1. Dermorphin Analogs and Related μ-Selective Peptides

Linear Analogs As indicated above (Section 6.1.1), the amphibian heptapeptide dermorphin is a potent μ-selective peptide, and thus it has served as the lead compound for structural modification (see Ref. [660] for a review). Dermorphin contains a C-terminal amide, and deamidation of dermorphin or its shorter fragments to the corresponding acids decreases affinity for μ-receptors and potency in the GPI. Examination of shorter fragments indicated that the N-terminal tetrapeptide Tyr-D-Ala-Phe-GlyNH$_2$ maintains significant opioid activity, and therefore a variety of tetrapeptide amide derivatives have been prepared (see below). The further truncated tripeptide amide Tyr-D-Ala-PheNH$_2$ also retains significant opioid activity (25% of the potency of Met-enkephalin) [866], and could be further simplified by removing the C-terminal amide to give dipeptide aryl amides with maintenance of opioid activity. Tyr-D-Ala phenylpropylamide (DAPPA) was further modified by incorporation of Dmt in position 1 to give SC-39566 [867], which is orally active [868].

Extension of the dermorphin sequence with residues from the precursor sequence also yields potent μ-selective peptides. While introduction of the additional residues through the Glu^9 or Ala^{10} residues decreases μ-receptor affinity, apparently due to introduction of the acidic Glu^9 residue, further extension of the sequence increases μ-receptor affinity as basic residues are incorporated [869]; the resulting pentadecapeptide (Tyr-D-Ala-Phe-Gly-Tyr-Pro-Ser-Gly-Glu-Ala-Lys-Lys-Ile-Lys-ArgNH$_2$) has exceptional affinity for μ-receptors ($K_i = 2.0$ pM), but this C-terminal extension does not enhance potency in the GPI. Dimeric derivatives of dermorphin fragments have been prepared by bridging two monomers with hydrazine or diamines of various lengths [870]. Di-dermorphin$_0$ ((dermorphinNH-)$_2$), in which two dermorphin molecules are linked via hydrazine, exhibits fivefold higher affinity for μ-receptors and similar μ-receptor selectivity as dermorphin. Di-tetra-dermorphin$_2$ (Tyr-D-Ala-Phe-GlyNH-CH$_2$)$_2$) and other dimeric derivatives exhibit greater affinity for δ-receptors, and hence decreased μ-receptor selectivity compared to the monomer.

Glycosylated derivatives of dermorphin have been prepared by attaching β-D-glucose to the hydroxyl of a Ser or Thr or galactose via a C-α linkage to Ala in position 7 [871–873]. While glycosylation decreased μ-receptor affinity twofold, the penetration of the blood–brain barrier was significantly higher for the glycosylated derivatives and they exhibited twice the antinociceptive activity of dermorphin [872,873]. The enhanced BBB penetration by the C-α-galactoside analog suggested that the glucose transporter was not involved in the transport [872].

Amino acid substitutions have been examined in every position of dermorphin (see Ref. [660] for a review). An alanine scan of the peptide indicated that substitutions in positions 4, 6, and 7 are well tolerated, while substitution particularly in positions 1 and 2, but also in positions 3 or 5, results in large decreases in potency in the GPI [874]. A D-amino acid in position 2 is important for activity, and the L-Ala^2 peptide is virtually inactive (<0.1% the potency of dermorphin). Tetrapeptide analogs containing D-methionine sulfoxide in position 2 are also potent μ-selective agonists [875].

Substitutions for Gly^4 are well tolerated, particularly in the tetrapeptide derivatives. Sarcosine (NMeGly, Sar) at position 4 in tetrapeptide derivatives enhances opioid activity in antinociceptive assays [876]. Substitution of Phe in position 4 of the tetrapeptide amide yields the dermorphin/enkephalin hybrid TAPP (Tyr-D-Ala-Phe-PheNH$_2$) [828], which is a potent μ-selective agonist (see Table 18). This peptide can also be considered an analog of endomorphin-2, although it was synthesized several years before the discovery of the endomorphins [259]. TAPP analogs in which either Phe^3 or Phe^4 were nitrated were prepared to examine whether the Phe^3 aromatic ring of dermorphin interacts with a different subsite on opioid receptors than the Phe^4 aromatic ring of the enkephalins. Examination of Phe(p-NO$_2$)-containing analogs of dermorphin, morphiceptin, and various enkephalin analogs had found that p-nitro substitution in Phe^3 of dermorphin and morphiceptin causes large decreases in potency in the GPI, while this substitution in Phe^4 of μ- and δ-selective enkephalin analogs enhances potency in smooth muscle assays [877]. Incorporation of Phe(p-NO$_2$) in position 3 of TAPP decreases receptor affinity while incorporation in position 4 increases affinity, consistent with the concept of two distinct receptor subsites for these aromatic residues on opioid receptors [878]. Bulky aromatic amino acids such as tryptophan or naphthylalanine are well tolerated in positions 3 and 4 of TAPP and yield even more lipophilic peptides [878].

Incorporation of D-Arg in position 2 of dermorphin and tetrapeptide analogs yields analogs that are potent opioids in antinociceptive assays in mice [876,879]. The tetrapeptide derivative Tyr-D-Arg-Phe-Sar (TAPS) is a potent opioid in antinociceptive assays and causes respiratory stimulation rather than respiratory depression that is antagonized by naloxonazine [880]; TAPS also antagonizes the respiratory depression caused by dermorphin. Based on these results, TAPS has been postulated to be a μ$_1$-agonist and a μ$_2$-antagonist *in vivo* [880]. In contrast, incorporation of L-Tic in position 2 of dermorphin converts the peptide to a δ-receptor antagonist [881].

Schiller and coworkers postulated that positively charged ligands should display μ-receptor selectivity [828] based on Schwyzer's proposal that μ-receptors are located in an anionic membrane compartment [882]. Consistent with this concept, they found that incorporation of a positively charged residue in position 4 of tetrapeptide dermorphin derivatives enhances μ-receptor selectivity by decreasing affinity for δ-receptors [828]. Combination of a D-Arg in position 2 with a second basic residue in position 4 yielded the polar peptide Tyr-D-Arg-Phe-LysNH$_2$ (DALDA) (**242**) that carries a net charge of +3 [828] and exhibits exceptional μ-receptor selectivity in binding assays (see Table 18). Quaternization of the side-chain amine of Lys^4 in DALDA and related analogs is well tolerated, and the resulting analogs retain potent *in vivo* antinociceptive activity in the mouse writhing assay following subcutaneous administration [883]. The antinoceptive effects of DALDA and the quaternary derivatives are substantially reduced by the quaternized antagonist N-methyllevallorphan, suggesting that these peptide analogs have a high degree of peripheral antinociceptive activity in this assay. The distribution of DALDA to the CNS is limited [884], and the antinociceptive activity of DALDA in the hot plate test after subcutaneous administration is low [885]. The *in vivo* distribution of DALDA has been examined in pregnant sheep [886], and DALDA was not detected in any of the fetal plasma samples. This is in contrast to meperidine and morphine that undergo rapid transfer across the placenta to the fetus, and suggested that DALDA could be a promising opioid for obstetrical use. Replacement of Tyr^1 by Dmt results in a peptide [Dmt^1]DALDA (**243**) [829] with 10-fold higher affinities for both μ and δ-receptors and 200-fold higher affinity for κ-receptors [887]. [Dmt^1]DALDA is 220 and 3000 times more potent than DALDA and morphine in the rat tail flick test following i.t. administration. [Dmt^1]DALDA also inhibits norepinephrine uptake in spinal cord synaptosomes, and this dual action may contribute to its antinociceptive potency Both [Dmt^1] DALDA and DALDA exhibit longer durations of antinociceptive action than morphine following i.t. administration to rats (7 and 13 h, respectively, versus 3 h) [887]. In sheep, these peptides have longer elimination half-lives (1.5 and 1.8 h, respectively) than either morphine (20–30 min) or the much more hydrophobic μ-selective peptide DAMGO (15 min) [888]. In contrast to morphine and DALDA, [Dmt^1]DALDA did not cause respiratory depression at the doses examined [829], suggesting that it could be a drug candidate for intrathecal analgesia.

X-D-Arg-Phe-LysNH$_2$
X = Tyr DALDA (**242**)
X = Dmt Dmt^1]DALDA (**243**)

Conformationally Constrained Analogs Dermorphin analogs containing a local constraint have also been prepared by incorporation of a dipeptide mimetic in place of Phe^3-Gly^4. The Aba^3-Gly^4 (see Fig. 43) analog of dermorphin retains high affinity at μ-receptors and potency in the GPI, while this structural modification increases δ-receptor affinity and potency in the MVD 17–25-fold [889]. A number of heterocycles are also tolerated as bond replacements for the Phe^3-Gly^4 peptide bond [890].

Cyclic tetrapeptide analogs of dermorphin with the structure Tyr-*cyclo*[D-X-Phe-Y]NH$_2$ have been prepared by Schiller and coworkers. The 13-membered ring cyclic peptide Tyr-*cyclo*[D-Orn-Phe-Asp]NH$_2$**244** (Fig. 49) exhibits high selectivity for μ-receptors (see Table 18), while the more flexible peptide Tyr-*cyclo*[D-Lys-Phe-Glu]NH$_2$ with a 15-membered ring is nonselective [830]. Peptides Tyr-*cyclo*[D-Asp-Phe-Orn]NH$_2$ [891] and Tyr-*cyclo*[D-Asp-Phe-Dab]NH$_2$ (Dab, α,γ-diaminobutyric acid) [892] in which the lactam linkage is reversed also exhibit high affinity and μ-selectivity. The antiparallel cyclic dimers (Tyr-D-Orn-Phe-AspNH$_2$)$_2$ and (Tyr-D-Asp-Phe-OrnNH$_2$)$_2$, obtained as byproducts during the synthesis of the cyclic monomers, have similar affinities for both μ- and δ-receptors [891].

Various modifications to Phe^3 in these cyclic peptides have also been examined. Many of the modifications (e.g., *p*-nitro or *N*-methyl substitution, or shortening the side chain) decrease μ-receptor affinity by 25-fold or more [892]. While the cyclic structure restricts

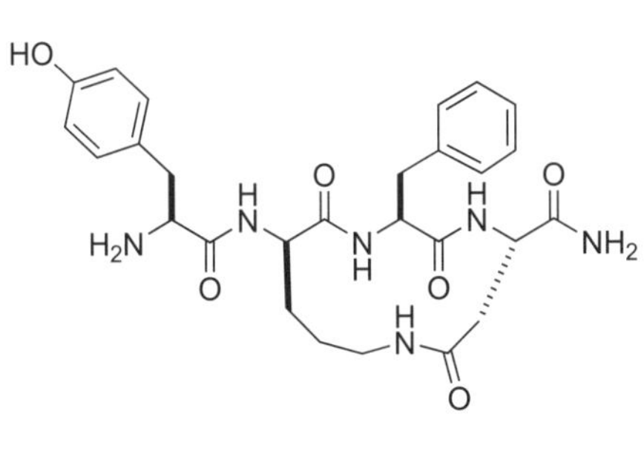

244

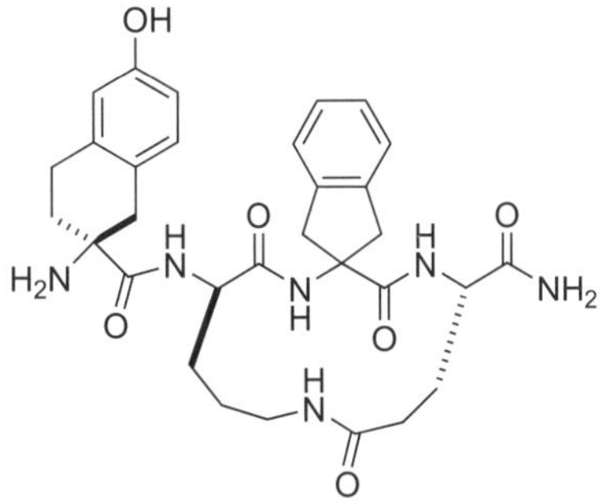

245

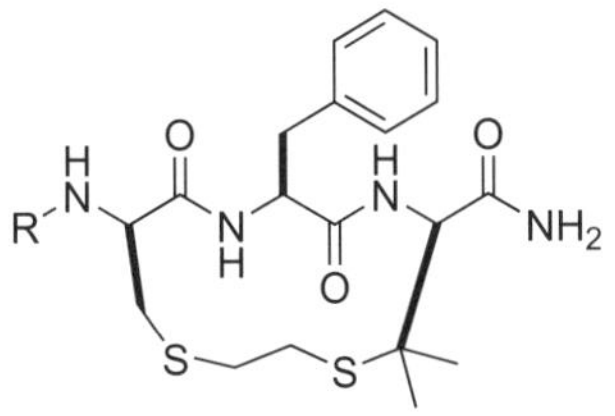

R = Tyr	**246** JOM-6
R = Phe	**247** JH-54
R = D-Hat	**248**

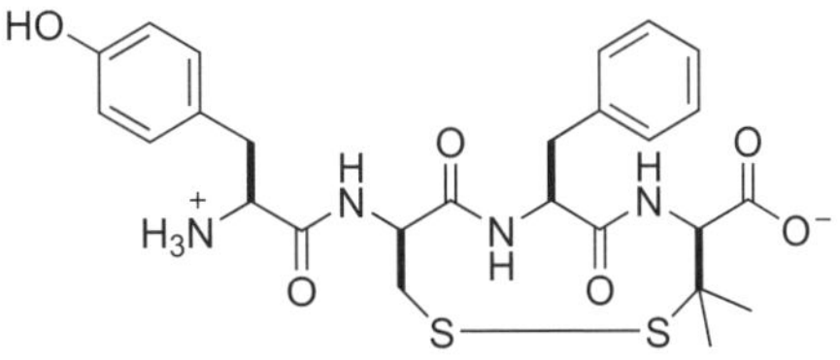

249 JOM-13

Figure 49. Conformationally constrained dermorphin analogs.

the conformation of the peptide backbone, there is still considerable conformational flexibility around the Tyr^1 and Phe^3 side chains [893]. Conformational constraint of Phe^3 by incorporation of 2-aminoindan-2-carboxylic (Aic) (Fig. 43) into the relatively nonselective peptide (K_i ratio (δ/μ) = 3) Tyr-*cyclo*[D-Orn-Phe-Glu]NH$_2$ markedly enhances µ-receptor selectivity (see Table 18) [831]. This is a direct consequence of conformational restriction because incorporation of the acyclic derivatives α-methylphenylalanine or 2′-methylphenylalanine in position 3 does not significantly change µ-receptor selectivity. Interestingly, the peptides containing L- and D-2-aminotetralin-2-carboxylic (Atc) (Fig. 43) in position 3 had similar µ-receptor affinity and high µ-receptor selectivity [831]. The restricted conformation of the side chain of residue 3 in the Aic analogs was apparent from comparisons of molecular dynamics simulations for the Phe^3 versus Aic^3-containing peptides [894]. Further conformational constraint

of Tyr-*cyclo*[D-Orn-Aic-Glu]NH$_2$ by replacement of Tyr^1 with L- or D-Hat yields peptides with only two freely rotatable bonds (**245**) (Fig. 49) [894]; interestingly, both diastereomers exhibit reasonable µ-receptor affinity and selectivity (K_i (µ) = 20–30 nM, K_i ratio (δ/μ) = 12–19).

Cyclic pentapeptide derivatives of Tyr-*cyclo*[D-Dab-Phe-Phe-(D or L)Leu] containing retro-inverso modifications have been prepared by Goodman and coworkers [895]. The parent peptides Tyr-*cyclo*[D-Dab-Phe-Phe-D/L-Leu] are potent agonists in the GPI with relatively low µ-receptor selectivity (IC$_{50}$ ratio (MVD/GPI) = 4.4 and 12 for the L-Leu and D-Leu analogs, respectively). The analog Tyr-*cyclo*[D-Dab-Phe-gPhe-*S*-mLeu] containing a reversed amide bond between Phe^4 and Leu^5 is also a µ-receptor selective agonist. Interestingly, reversal of a second amide bond in the side-chain lactam linkage to give Tyr-*cyclo*[D-Glu-Phe-gPhe-rLeu] results in a δ-selective derivative (IC$_{50}$ ratio (GPI/MVD) = 11) that exhibits a considerably different conformation from that exhibited by the other retro-inverso analogs that preferentially interact with μ-receptors [896].

In the case of peptides, cyclized through a disulfide or dithioether linkage, the receptor selectivity depends on the linkage. The cyclic disulfide analog Tyr-*cyclo*[D-Cys-Phe-Cys]NH$_2$ exhibits 10-fold greater affinity for δ-receptors than the corresponding lactam Tyr-*cyclo*[D-Asp-Phe-Orn]NH$_2$ and is therefore less μ-receptor selective (IC$_{50}$ ratio (δ/μ) 34) [892]. The cyclic tetrapeptide JOM-13 Tyr-*cyclo*[D-Cys-Phe-D-Pen]OH is a δ-selective agonist [897] (see below), but incorporation of an ethyl group between the two sulfurs and amidation of the C-termination results in the μ-selective peptide JOM-6 (**246**) [897] (Table 18). Comparison of JOM-13 and its μ-selective amide derivative JH-42 that contains *E*-dehydrophenalanine in position 3 docked to computational models of δ- and μ-receptors, respectively, found key differences [176]. The Phe^3 side chain of the peptide adopts the *gauche* (−) (approximately −60°) conformation and interacts with Leu300 in TM7 in the δ-receptor, while the presence of Trp in the corresponding position in the μ-receptor causes a shift of the entire peptide within the binding pocket and reorientation of the Phe^3 side chain from gauche to the *trans* conformation. The C-terminal acid in JOM-13 forms an ion pair with Lys214 in TM5 of the δ-receptor in the model, but because of the shift in the peptide in the binding site its C-terminus is in close contact with Glu229 in the μ-receptor and therefore amidation of the C-terminus removes unfavorable electrostatic repulsion. The shift of the peptide lengthened the distance between the phenol of Tyr^1 of the peptide and His297 in the μ receptor binding site compared to the corresponding distance in the δ binding pocket, suggesting that the hydrogen bond between these two groups may be less important for binding to δ-receptors. Mosberg and coworkers therefore prepared JH-54 (**247**), the Phe^1 analog of JOM-6, and found that it retained high affinity for μ-receptors but greatly reduced affinity for δ-receptors (Table 18), consistent with the expected results from the modeling [832]; this peptide is a potent full agonist in the GPI. Even the aromaticity of the residue in position 1 was not critical, and the Cha^1 (cyclohexylalanine) analog of JOM-6 retains moderate μ-receptor affinity ($K_i = 32$ nM) and agonist potency (EC$_{50} = 59$ nM in the GTPγS assay) [898]. Incorporation of conformationally restricted tyrosine analogs in position 1 resulted in potent μ-agonists (see [D-Hat^1]JOM-6) (**248**) (Table 18), but the potency changes in the GTPγS assay did not always correlate with affinity, suggesting that the conformation required for receptor activation versus binding were different [833].

6.5.2. Deltorphin Analogs and Related Peptides

Linear Analogs Among the naturally occurring deltorphins, [D-Ala^2]deltorphin I and II are more potent and more δ-selective than deltorphin (see Table 13) and have a distinctly different C-terminal sequence [93]. Like dynorphin, the deltorphins have been divided into two domains, the *N*-terminal "message" region and C-terminal "address" sequence. There are differences in the structure–activity relationships for deltorphin and [D-Ala^2]deltorphin I and II, particularly in the C-terminal "address" domain, and therefore the SAR in this region of the peptides will be discussed separately below. [D-Ala^2]deltorphin I and II are also metabolically more stable than deltorphin, apparently due to the branched amino acid present in position 5 of the former peptides [899].

A number of studies have examined possible conformations of the deltorphins (see Refs [900–902] for reviews). A β-turn in the N-terminus has been observed for both the μ-selective peptide dermorphin and the δ-selective deltorphins, while differences in the conformations of the C-terminal sequences have been described (see Refs [900–902] and references cited therein). These studies are complicated, however, by the inherent conformational flexibility of linear peptides, and very different conformations can be observed for a given peptide (see, for example, Ref. [903]).

Amino acid substitution has provided information on the contributions of different residues in the deltorphin sequence to δ-receptor affinity and selectivity. A variety of

Table 20. Opioid Receptor Affinities and Opioid Activity in the GPI and MVD of Deltorphin Analogs

Peptide	K_i (nM) δ	K_i (nM) μ	K_i Ratio μ/δ	IC_{50} (nM) GPI	IC_{50} (nM) MVD	Reference
[D-Ala²]Deltorphin I (**214**)	0.15	3,150	21,000	>1,500	0.21	
[D-Ala²]Deltorphin II (**25**)	0.71	2,450	3,400	>3,000	0.32	
[Dmt¹,D-Ala²]deltorphin II	0.13	1.0	7.7	300	—	[921]
[(2S,3S)Tmt¹,D-Ala²]deltorphin I	3.0	17,000	5,740	3,840	0.66	[725]
[Tic²]deltorphin I	6.49	9,230	1,420	>65,000	22.8^a	[904]
[D-Ala²,Nle³]deltorphin I	10	1,020	100	—	—	[922]
[D-Ala²,D/L-Atc³]deltorphin I	5.36	670	125	1,820	0.115	[904]
[D-Ala²,D/L-Atc³]deltorphin I	6.52	1,410	215	1,380	0.178	[904]
[D-Ala²,Phe⁴]deltorphin I	1.5	3.9	2.6	—	—	[916]
[N-nBuGly⁶]deltorphin	0.04	820	18,000	580	0.56	[919]
JOM-13	0.74	54	70	460	4.2	[923]
[t-Hpp¹]JOM-13	0.66	110	170	770	1.6	[923]
[c-Hpp¹]JOM-13	2.4	720	300	12,000	75	[923]
[Δ^Z-Phe³]JOM-13	2.4	780	330	—	—	[924]
cyclo[D-Cys²,Cys⁵]deltorphin I	0.87	5.5	5.9	2.98	0.23	[925]
cyclo[D-Cys²,Pen⁵]deltorphin I	2.2	3,760	1,700	1,100	0.25	[925]
cyclo[D-Pen²,Pen⁵]deltorphin I	3.7	26,000	7,000	68,000	6.30	[925]

Data for [D-Ala²]deltorphin I and II from Table 13 are included for comparison.
^{a} Partial agonist.

substitutions have been made in each position of the deltorphins, resulting in hundreds of analogs reported in the literature. Only selected analogs and structural modifications will be discussed here to illustrate key SAR; for more detailed discussion, readers are referred to recent detailed reviews [901,902] that include extensive tables of analogs.

One of the first structural modifications that was examined in the deltorphins was the chirality of residue 2. Along with the initial descriptions of the deltorphin sequence, the peptides with both D-Met and L-Met in position 2 were synthesized to determine which was the naturally occurring sequence [663,665,666]. While the peptide containing D-Met exhibited high δ-receptor affinity and selectivity, the analog with L-Met² had very low affinity and potency in the MVD, indicating the importance of a D-amino acid in this position. As was found for deltorphin, the D-Ala in position 2 of [D-Ala²]deltorphin I is also important for δ-receptor affinity and selectivity, and the L-Ala² analog has extremely weak activity in the MVD (IC_{50} > 1 μM) [93]. The exception to the requirement for a D-amino acid in position 2 is [Tic²]deltorphin I, which retains high affinity and exhibits increased selectivity for δ-receptors (Table 20) [904]; this substitution decreases efficacy, however, and in the MVD this analog is a partial agonist. In deltorphin substitution of D-Ala, as is found in [D-Ala²]deltorphin I and II, for D-Met is well tolerated at δ-receptors, although it increases μ-receptor affinity and potency in the GPI [905,906]. In contrast, substitution of D-Met in position 2 of [D-Ala²] deltorphin I or II causes a large decrease in δ-receptor affinity and selectivity [905,906].

Modifications to the aromatic residues in the N-terminal "message" sequence have been examined, mostly in [D-Ala²]deltorphin I or II. Replacing either Tyr¹ or Phe³ by a D-amino causes a large decrease in δ-receptor affinity [906]. As with most opioid peptides, other modifications of Tyr¹ are also generally not tolerated, except for substitutions by conformationally restricted derivatives of Tyr (see section "Conformationally Constrained Analogs"). A variety of structural modifications have been examined at position 3 of [D-Ala²]deltorphin I, including incorporation of aromatic, heterocyclic, and nonaromatic amino acids (see Ref. [901]) and conformationally constrained derivatives (see section

"Conformationally Constrained Analogs"). The effects of substitution on the phenyl ring of Phe^3 vary with the substituent. *para*-Substitution with a halogen, but not an amine, nitro, hydroxyl or methyl group, in [D-Ala^2]deltorphin I is well tolerated by δ-receptors, and *p*-bromo substitution enhances δ-receptor selectivity in both binding and smooth muscle assays [907–909]. The heterocyclic phenylalanine analog 3-(2-thienyl)alanine is well tolerated in position 3 by δ-receptors, but other heterocyclic aromatic amino acids, such as the pyridylalanine derivatives or His, significantly decrease δ-receptor affinity [909]. An aromatic residue is not required in position 3 of [D-Ala^2]deltorphin I, and the analogs with cyclohexylalanine (Cha) and even Leu or norleucine (Nle) exhibit δ-receptor affinity only six- to sevenfold lower than that of the parent peptide (Table 20). QSAR analysis of the effect of substitution in position 3 suggested that the binding site around this side chain is very similar for δ- and μ-receptors [910]. Substitution of either the α-carbon or the nitrogen of Phe^3 with a methyl group, which can alter the conformation of the peptide backbone, is not tolerated [908].

C-Terminal truncation of the deltorphins can affect both δ-receptor affinity and selectivity. Deamidation of both [D-Ala^2]deltorphin I and II causes significant decreases in δ-receptor affinity (25–50-fold) and δ-selectivity [905,911,912]. Although there is one report of a large (90-fold) decrease in δ-receptor affinity upon conversion of deltorphin to the C-terminal acid [911], there are other reports that deamidation of this heptapeptide causes only a small decrease [913], or even a slight increase [912], in δ-receptor affinity. Shortening of either [D-Ala^2]deltorphin I or deltorphin from the C-terminus causes progressive decreases in δ-receptor affinity and selectivity and potency in the MVD [905,912,913]. The hexapeptide and pentapeptide derivatives of [D-Ala^2]deltorphin I retain δ-receptor selectivity [905]; the corresponding deltorphin fragments, while retaining high affinity for δ-receptors, are nonselective [913]. The N-terminal tetrapeptide amide derivative of deltorphin is selective for μ-receptors [914,915] (the tetrapeptide acid exhibits low affinity for both μ- and δ-receptors [915]), indicating that the C-terminal "address" sequence is capable of changing receptor type selectivity. Indeed, swapping the C-terminal tripeptide sequence of dermorphin and deltorphin reverses the receptor selectivity profile (i.e., [Leu^5,Met^6,Asp^7]-dermorphin is δ-selective [913]). The N-terminal tetrapeptide amide fragment of [D-Ala^2]deltorphin I (Tyr-k-Ala-Phe-AspNH$_2$) shows a slight preference for μ-receptors [905,912], but its affinity (K_i = 100–195 nM) and selectivity for μ-receptors is much lower compared to the N-terminal tetrapeptide amide fragment of deltorphin (Tyr-D-met-Phe-HisNH$_2$, K_i (μ) = 8.0 nM).

In the C-terminal "address" sequence of [D-Ala^2]deltorphin I and II considerable attention has focused on the role of the acidic residue in position 4. Replacement of Asp and Glu in this position with Asn and Gln, respectively, results in compounds that maintain δ-receptor affinity, but have enhanced μ-receptor affinity and thus significantly decreased δ-receptor selectivity [905,911,912]. A variety of other amino acid substitutions in this position are well-tolerated by δ-receptors but significantly enhance μ-receptor affinity [905,916]. Thus, the negative charge at position 4 is not necessary for δ-receptor interaction, but contributes significantly to δ-receptor selectivity by interfering with interaction with μ-receptors. Consistent with this, substitution of Asp/Glu by His, the residue found in this position of deltorphin, does not affect δ-receptor affinity but increases μ-receptor affinity [905,916]. Replacement of Asp/Glu by hydrophobic residues (e.g., Phe (see Table 20), α-aminobutyric (Abu) or α-aminoisobutyric acid (Aib), and the conformationally restricted 1-aminocycloalkane derivatives) substantially increases μ-receptor affinity, resulted in analogs with nanomolar affinity for both δ- and μ-receptors, which has been referred to as "opioid infidelity" [900]. QSAR analysis of substitutions in this position has also been performed [910], and suggests that the receptor binding site for this residue is quite different for δ- versus μ-receptors. Analysis of binding to δ-receptors suggested at least a partial positive charge in the binding pocket, with both electrostatic and van der Waals forces, but not hydrogen bonding, contributing to receptor binding.

In positions 5 and 6 of [D-Ala²]deltorphin II, replacement of Val⁵ and Val⁶ individually by Ala suggested that Val⁵ is more important than Val⁶ for δ-receptor selectivity [917]. Replacement of valine in both positions by norleucine, Ile or γ-methylleucine enhances both δ-receptor affinity and selectivity four and eightfold [917]. Replacing Gly⁷ in [D-Ala²]deltorphin I with Asp, which is found in this position in deltorphin, decreases affinity for both δ- and μ-receptors approximately 10-fold [916].

Amino acid substitutions in the C-terminus of deltorphin have also focused mainly on the charged residues His⁴ and Asp⁷. As was found for Asp⁴/Glu⁴ in [D-Ala²]deltorphin I and II, His⁴ appears to play an important role in the δ-receptor selectivity, but not affinity, of deltorphin. Thus, a variety of amino acid substitutions for His⁴, including Gly and aromatic amino acids, are well tolerated by δ-receptors, but they enhance μ-receptor affinity and thus decrease δ-receptor selectivity [912]. Interestingly, substitution by Lys markedly decreases δ-receptor affinity [912], and substitution by Asp, as is found in [D-Ala²]deltorphin I, decreases both δ- and μ-receptor affinity [905,911,912]. Replacement of His⁴ by D-His only decreases δ-receptor affinity and selectivity four- to sixfold, while des-His4 analogs have markedly lower δ-receptor affinities and selectivities [906], suggesting that His⁴ may play a role in maintaining the necessary spatial orientation of the C-terminal sequence of deltorphin relative to the N-terminus. Further modifications of His⁴ have been examined [918]. N-Methylation on either nitrogen of the imidazole enhances δ-receptor selectivity, while N^α-methylation of His⁴ decreases both δ-receptor affinity and selectivity. Substitution of the conformationally constrained residue Tic for His⁴ is well tolerated by δ-receptors, but decrease δ-receptor selectivity [918]. The charge on Asp⁷ also does not appear to be critical for interaction with δ-receptors, but contributes to δ-receptor selectivity by adversely affecting μ-receptor affinity. Thus, replacement of Asp⁷ by a nonacidic amino acid decreases δ-receptor affinity by less than two- to threefold, but enhances μ-receptor affinity so that δ-receptor selectivity decreases [911,912,916].

A number of residues have been incorporated in position 5 of deltorphin, and generally the δ-receptor affinities of these analogs correlate with the hydrophobicity of the residue in this position (see Ref. [901]). Branched hydrophobic amino acids Val, Ile, and γ-methyl-leucine have been incorporated into this position of deltorphin to enhance metabolic stability; these analogs have similar δ-receptor affinities and selectivities as the parent peptide [899]. N-Alkyl amino acids were also introduced at position 6 in order decrease metabolism of the Leu⁵-Met⁶ bond; these analogs possess the desired stability to degradation by both rat brain and plasma, and the N-nBuGly⁶ analog exhibits exceptional selectivity for δ-receptors (Table 20) [919]. A variety of other substitutions for Met⁶ in deltorphin are well tolerated (see Ref. [901]), suggesting that residue 5 is more important for receptor interaction. However, incorporation of an additional acidic residue in position 6 decreases δ-receptor affinity [911,920].

Conformationally Constrained Analogs The conformationally restricted tyrosine analogs Dmt and Tmt (Fig. 43) have been incorporated into position 1 of [D-Ala²]deltorphin I or II. Substitution of Dmt for Tyr in [D-Ala²]deltorphin II markedly enhances μ-receptor affinity, resulting in a compound with nanomolar affinity for both receptors (Table 20) [921]. In contrast, [D-Ala²]deltorphin I analogs containing either of the 2S-isomers of Tmt, which has the additional β-methyl group, exhibited high δ-receptor affinity and selectivity; the (2S,3S) isomer exhibited greater δ-receptor selectivity than the parent peptide (IC$_{50}$ ratio 5740 versus 3500) [725].

Linear peptides containing conformationally constrained Phe analogs in position 3 have also been prepared. The conformationally restricted phenylalanine analogs Aic and both isomers of Atc (Fig. 43) are well tolerated at this position by δ-receptors, although the Aic³ analog is less selective for δ-receptors than are the Atc³ derivatives or [D-Ala²]deltorphin I [904,908]. The similar potencies of both diastereomers of [D-Ala²,Atc³]deltorphin I is in sharp contrast to the large (13–50-fold) decrease in potency of the D-Phe³ analog compared to [D-Ala²]deltorphin I, suggesting that the receptor binding mode of the D-Atc³ analog may be different from that of the D-Phe^3 derivative [904]. The configuration of Atc in the diastereomeric peptides was not determined in these initial studies; but was in subsequent derivatives of [D-Ala^2,Ile^{5,6}]deltorphin I and II [926]. The four isomers of β-MePhe have been incorporated into both [D-Ala^2]deltorphin I and deltorphin [927]. The peptides containing the 2S-isomers exhibit higher δ-receptor affinity and agonist activity in the MVD than the analogs with the 2R-isomers for both [D-Ala^2]deltorphin I and deltorphin, consistent with the preference for L- over D-Phe in this position. The stereochemistry of the methyl group at the β-position, however, had different effects on the δ-affinity and agonist activity of the two peptides, with the 3S-isomer preferred in [D-Ala^2]deltorphin I and the 3R-isomer preferred in deltorphin. Substitution with the dipeptide mimetic Aba^3-Gly^4 is also reasonably well tolerated in [D-Ala^2]deltorphin II (IC$_{50}$ = 5.0 nM) [889]. Other constrained residues in position 3, that is, Tic or N-MePhe, result in drastic decreases in δ-receptor affinity [904,908].

As in the case of cyclic enkephalin analogs, conformationally constrained δ-selective peptides with the sequence Tyr-D-X-Phe have also been prepared by cyclization via a disulfide bond. Cyclization between D-Cys in position 2 and D-Pen in position 4 yielded the tetrapeptide derivative Tyr-cyclo[D-Cys-Phe-D-Pen]OH (JOM-13) (**249**) that contains a fairly rigid 11-membered cyclic structure and that exhibits high affinity and selectivity for δ-receptors (Table 20) [897]. Conformational analysis by NMR, X-ray crystallography, and computational analysis [928] indicated a single preferred conformation for the cyclic tripeptide portion of JOM-13, but the key pharmacophoric elements, the exocyclic Tyr^1 and the Phe^3 side chains, still exhibit conformational flexibility.

A variety of substitutions have been made for Phe^3 in JOM-13 [929–932]. Comparison of Phe^3 substitutions in JOM-13 to the same substitutions for Phe^4 in DPDPE found that modifications that would be expected to affect the conformation of the peptide backbone, for example, incorporation of N-MePhe or D-Phe, had deleterious effects on δ-receptor affinity and potency in the MVD in both series [929];

the effects of these substitutions was generally greater in the pentapeptide series, however, which may be due to the greater rigidity, and therefore reduced susceptibility to conformational perturbation, of the tetrapeptide. Other substitutions that affect only the side chain of residue 3, particularly homophenylalanine (Hfe), had significantly different effects on δ-receptor affinity in the two series. Hfe substitution does not adversely affect δ-receptor affinity or potency in the MVD when it is incorporated into JOM-13 (although it does increase μ-receptor affinity 20-fold so that δ-receptor selectivity decreases), but it decreases δ-receptor affinity and potency in the MVD 100- and 25-fold when incorporated into DPDPE [929]. *para*-Substitution of Phe^3 with flourine, chlorine or a nitro group in JOM-13 is well tolerated by δ-receptors, similar to the results found for DPDPE analogs [930]. An aromatic residue at position 3 is not essential for δ-receptor interaction, and incorporation of cyclohexylalanine (Cha) at position 3 of JOM-13 causes less than a threefold decrease in δ-receptor affinity [931]; not surprisingly, this modification also enhances μ-receptor affinity so that [Cha^3]JOM-13 exhibits only a fourfold lower K_i for δ-receptors than μ-receptors. The conformation of residue 3 has been restricted by incorporation of β-MePhe [932] and dehydrophenylalanine (ΔPhe) [924] in this position. Both the (3S) and the (3R) diastereomers of L-β-MePhe were compatible with interaction with δ-receptors, with the peptide containing the (2S,3S) isomer exhibiting approximately eightfold higher affinity than the peptide containing the (2S,3R) isomer. Surprisingly, the peptide containing the (2R,3R) isomer of D-β-MePhe also exhibits high affinity and selectivity for δ-receptors. Examination of this latter peptide by molecular modeling suggested a side-chain conformation of Phe^3 in JOM-13 with χ_1 approximately −60° when the peptide interacts with δ-receptors [932]. Based on these results, it was predicted that the JOM-13 analog containing Δ^{Z}Phe (Fig. 43), but not ΔPhe^E, could be superimposed on the resulting proposed bioactive conformation of JOM-13; as predicted [Δ^{Z}Phe^3]JOM-13 exhibited nanomolar affinity and higher δ-receptor selectivity than JOM-13, while incorporation of Δ^{E}Phe in position 4 resulted in a 60-fold decrease in δ-receptor affinity [924].

Mosberg and coworkers have also examined conformationally constrained tyrosine derivatives in position 1 of JOM-13 [923,933]. Incorporation of *trans*- and *cis*-3-(4′-hydroxyphenyl)proline (*t*-Hpp and *c*-Hpp, respectively) (Fig. 43) yielded the analogs with the highest δ-receptor affinity and selectivity (Table 20), with the peptides containing Hat and Hai (2-hydroxy-2-aminoindan-2-carboxylic acid) exhibiting modest δ-receptor affinities and selectivity. Molecular modeling of the active Tyr^1-substituted analogs suggested that in the receptor-bound conformation the side chain of residue 1 is in the *trans* ($\chi_1 =$ 180°) conformation and that the exocyclic peptide group is in an extended conformation (Ψ_1 and $\Phi_2 = \sim 160°$) [923]; a second low-energy conformation differing in Φ_2 ($\sim 70°$) also still remained a possibility.

These results led to a pharmacophoric model of the bioactive conformation of JOM-13 [923,932]. In order to distinguish between the two remaining possible bioactive conformations for JOM-13, these conformations were then compared and superimposed with DPDPE and other δ-receptor ligands (TIPP, and the alkaloids SIOM and OMI) [838]. Similar arrangements in space of the key pharmacophoric elements (the tyramine moiety and the second aromatic ring) were found with the other ligands for JOM-13 in the proposed bioactive conformation (with $\Phi_2 \sim 160°$), but not the alternate (with $\Phi_2 \sim 70°$) conformation. Different orientations were observed for the second aromatic ring in agonists versus antagonists, but the same orientation of this aromatic ring was found for both peptide and alkaloid ligands with the same type of activity (i.e., agonists JOM-13, DPDPE, and SIOM). Subsequently, JOM-13 was docked to a computational model of the δ-receptor [934]; in contrast to the modeling based on superimposition of different ligands only the alternate conformation ($\Phi_2 \sim 70°$), and not the proposed conformation with $\Phi_2 \sim 160°$, was compatible with the receptor binding pocket. These results are another indication that the receptor interactions may differ for different ligands, and caution against a simplistic view of a single common binding mode for different ligands [934].

Cyclic deltorphin and [D-Ala^2]deltorphin I and II analogs cyclized by a disulfide linkage between residues 2 and 5 have also been reported. *cyclo*[D-Cys^2,Cys^5]deltorphin I exhibits affinity for δ-receptors similar to that of [D-Ala^2]deltorphin I, but the cyclic peptide exhibits 400-fold higher μ-receptor affinity and thus very low δ-receptor selectivity (Table 20) [935]. Substitution of Pen in position 5 or preparation of the [D-Pen^2,Pen^5] derivative in the deltorphin I series decreases μ-receptor affinity 730- and 5000-fold, respectively, while only decreasing δ-receptor affinity by 3.5–6-fold [925], thus restoring high δ-receptor selectivity (Table 20). A series of [D-Cys^2,L/D-Pen^5] derivatives of [D-Ala^2]deltorphin II and deltorphin have also been prepared [936]; these analogs, including some with D-Pen in position 5, exhibit nanomolar affinity for δ-receptors, but their selectivity for δ- over μ-receptors is much lower (generally <60-fold) than the [D-Pen^2,Pen^5] peptides.

Schiller and coworkers have prepared derivatives of the μ-selective lactam Tyr-*cyclo*[D-Orn-Phe-Asp] with the C-terminal Val-Val-GlyNH$_2$ sequence from [D-Ala^2]deltorphin [904]. This C-terminal extension reduces μ-receptor affinity and increases δ-receptor affinity 10-fold, resulting in similar affinities for the two receptors. The peptide Tyr-*cyclo*[D-Lys-Phe-Asp]-Val-Val-GlyNH$_2$ with a slightly larger ring size exhibits eightfold higher δ-receptor affinity and a slight preference for δ- over μ-receptors.

6.6. Other Peptides with High Affinity for Opioid Receptors

6.6.1. μ-Receptor Antagonists Derived from Somatostatin
Potent μ-opioid antagonists have been identified that are derivatives of somatostatin rather than of an opioid peptide (see Ref. [653] for a review). Somatostatin exhibits low affinity for opioid receptors, and the potent somatostatin analog SMS-201,995 (D-Phe-*cyclo*[Cys-Phe-D-Trp-Lys-Thr-Cys]-Thr-ol) was found to be an antagonist at μ-opioid receptors [937]. Further structural modification yielded a series of peptides with the general structure D-Phe-*cyclo*[Cys-Tyr-D-Trp-X-

Table 21. Opioid Receptor Affinities and Opioid Activity in the GPI and MVD of Peptide Antagonists

Peptide	Ki (nM) μ	Ki (nM) δ	Ki Ratio δ/μ	pA2 GPI (μ)
CTP	3.7	1,150	310	7.1
CTOP (19)	4.3	5,600	1,300	6.4
CTAP (20)	2.1	5,310	2,530	7.1
TCTP	1.2	9,320	7,770	8.1
TCTOP	1.4	20,400	11,400	7.4
TCTAP	1.2	1,270	1,000	8.7

Data from Ref. [941].

Thr-Pen]-ThrNH$_2$, where X = Lys, Orn, or Arg in CTP, CTOP (19) (Fig. 4), and CTAP (20), respectively [92,938], which exhibit greatly reduced affinity for somatostatin receptors and enhanced affinity and selectivity for μ-receptors. Analysis of CTP by NMR suggested that the central Tyr-D-Trp-Lys-Thr sequence adopts a type II′ β-turn [939]. The conformation of these peptides was further restricted and μ-receptor selectivity further enhanced by incorporation of the constrained amino acid D-Tic in position 1; the resulting peptides TCTP, TCTOP, and TCTAP are potent μ-antagonists with exceptional selectivity for μ-receptors (Table 21) [940,941]. Incorporation of either isomer of β-Me-D-Phe in position 1 decreases μ-receptor affinity and selectivity 10- and 40-fold, respectively [942].

Other μ-receptor antagonists have been identified from combinatorial libraries (see below).

6.6.2. Peptides and Peptidomimetics from Combinatorial Libraries Mixture-based combinatorial peptide libraries have been extensively explored by Houghten and coworkers and have led to the identification of a variety of peptides with affinity for opioid receptors (see Ref. [943] for a review). Early hexapeptide libraries, deconvoluted by binding to μ-opioid receptors and either iterative deconvolution [944] or positional scanning [945] (see Ref. [943] for a description of these deconvolution techniques), identified sequences related to the enkephalins. Other nonacetylated peptide libraries have identified more varied peptides, some of which resemble opioid peptides and some of which do not. Iterative deconvolution of a hexapeptide library resulted in identification of both Tyr-Pro-Phe-Gly-Phe-XNH$_2$ (X = one of 20 natural amino acids), reminiscent of the sequence of β-casomorphin, and the unrelated peptides Trp-Trp-Pro-Lys-His-XNH$_2$ [946] with affinity for μ-receptors. Hexapeptide libraries have also been examined for affinity for δ-receptors, resulting in the identification by positional scanning of several peptides that share some similarity to the enkephalins (Tyr-X-Y-Z-Leu-ValNH$_2$, where X-Y-Z = Gly-Met-His, His-Gly-Trp [947], and Gly-Phe-His [943]).

A variety of novel acetylated peptides with sequences unrelated to known opioid peptides have been identified for all three opioid receptors from combinatorial libraries (see Ref. [943]. Initially, Houghton and coworkers used μ-receptor binding as the screening assay to identify peptides from an acetylated hexapeptide amide library with affinity for μ-receptors. These peptides, termed acetalins (250) (Fig. 50), are potent μ-receptor antagonists in the GPI [948]. Further exploration of the acetylated hexapeptide library resulted in the additional identification of AcPhe-Arg-Trp-Trp-Tyr-XNH$_2$ and AcArg-Trp-Ile-Gly-Trp-XNH$_2$; AcPhe-Arg-Trp-Trp-Tyr-MetNH$_2$ is an agonist while AcArg-Trp-Ile-Gly-Trp-ArgNH$_2$ is an antagonist at μ-receptors [946]. Screening of an acetylated hexapeptide library of all D-amino acids led to the identification of Ac-arg-phe-trp-ile-asn-lysNH$_2$ (D-amino acids indicated by use of all small letters for the amino acid) as a ligand for μ-opioid receptors; interestingly, this peptide is a potent agonist that produces analgesia after peripheral administration [949]. A nonacetylated all D-amino acid hexapeptide library also yielded peptides (ile-phe-trp-tyr-argNH$_2$ and ile-met-ser-trp-trp-glyNH$_2$) with affinity for μ-opioid receptors [943]. An acetylated decapeptide library was used to identify AcTyr-Arg-Thr-Arg-Tyr-Arg-Tyr-Arg-Arg-ArgNH$_2$ and AcArg-Gly-Trp-Phe-His-Tyr-Lys-Pro-Lys-ArgNH$_2$ as ligands for κ-opioid receptors [943]; like dynorphin A, these peptides have a number of basic residues in their C-terminal sequences.

A single tetrapeptide library containing both L- and D-amino acids, including a number

AcArg-Phe-Met-Trp-Met-XNH$_2$
250 X = Arg, Lys, or Trp Acetalins

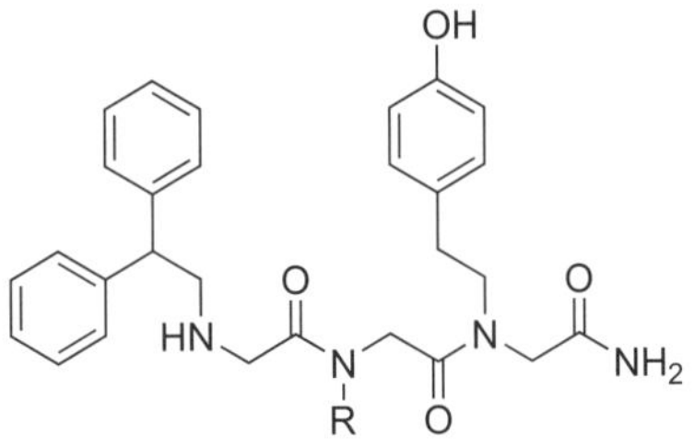

Figure 50. Selected peptides and peptidomimetics derived from combinatorial libraries.

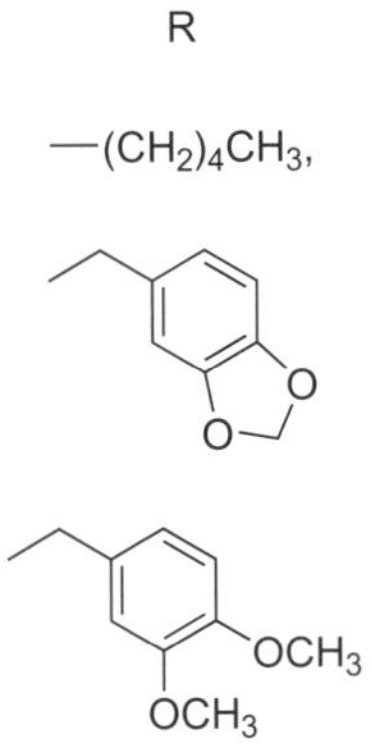

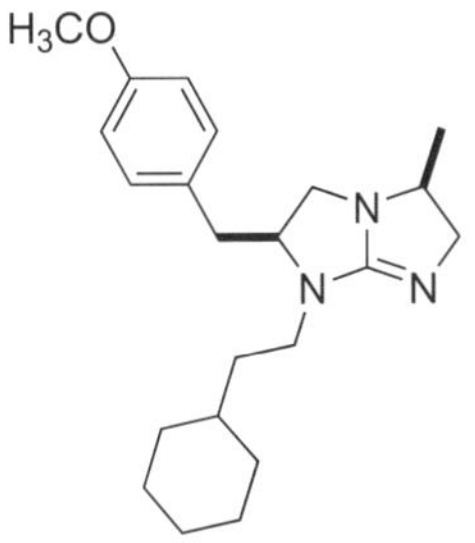

of unnatural amino acids (50 amino acids total), has been screened by positional scanning for affinity for all three types of opioid receptors [133]. This led to the identification of peptides with nanomolar affinity for each receptor type. Based on screening of the mixtures, an aromatic L-amino acid (Tyr for μ-receptors, and Tyr or Trp for δ-receptors) was incorporated in position 1 in the individual peptides prepared for these two receptor types. In the 4 position, an aromatic L-amino acid was also incorporated in peptides with affinity for μ-receptors, while either Arg or Trp were found in this position in peptides with δ-receptor affinity. Interestingly, some of the peptides prepared as ligands for the δ-receptor exhibited higher affinity for μ-receptors. The most unusual results were those obtained for the κ-receptor. In this case, at each position the mixtures exhibiting the highest affinity contained a D-amino acid in the defined position, so that the resulting peptides contained all D-amino acids. The peptides with the highest affinity for μ- and κ-receptors were tested in adenylyl cyclase assays and found to be full agonists.

Peptidomimetic ligands for opioid receptors have also been identified from combinatorial libraries. Screening of an N-(substituted)glycine peptoid library for affinity for μ-opioid receptors yielded novel structures (**251**) with high affinity for these receptors ($K_i = 6$–46 nM) [950]. A library of dipeptide amides with alkyl substituents on both the interior and the C-terminal amides have been prepared, and high-affinity agonists for all three opioid receptors identified from the library (see Ref. [943]). Peptide libraries can also be further modified ("libraries from libraries" [951]) to yield new potential ligands for receptors. Thus, an acetylated hexapeptide library was subjected to exhaustive reduction, and the resulting polyamine library screened by positional scanning for affinity for μ-opioid receptors [952]. The polyamine with the highest affinity was the fully reduced heptamine of AcTyr-Tyr-Phe-Pro-Thr-MetNH$_2$ (IC$_{50}$ = 4 nM). The fifth and sixth residues could be truncated to yield the pentamine without loss of affinity; this derivative is an antagonist in both the GPI and the MVD ($K_e = 13.6$ and 163 nM, respectively). Further modifications have been made to the pentamine to enhance μ-receptor affinity (see Ref. [943]943). A bicyclic guanidine library has also been prepared and following screening for binding at κ-receptors, a ligand with modest affinity (**252**, IC$_{50}$ = 37 nM) was identified [953].

6.7. Peptide Affinity Label Derivatives

Peptide-based affinity labels have been principally photoaffinity labels, including the azide derivatives of several enkephalin analogs and CTP (see Refs [283,591] for detailed reviews). As noted above (Section 5.11.2), the use of photoaffinity labels, however, has been limited because opioid receptors are susceptible to inactivation by UV-irradiation [640].

The preparation of opioid peptide derivatives containing electrophilic affinity labels has been limited to a few compounds. The chloromethyl ketone DALECK (Tyr-D-Ala-Gly-Phe-LeuCH$_2$Cl) [954,955] is one of the best studied electrophilic peptide derivatives. It selectively alkylates μ-opioid receptors [956,957], and [^{3}H]DALECK has been used to label and characterize these receptors [956,958,959]. DAMK (Tyr-D-Ala-Gly-MePheCH$_2$Cl), the chloromethyl ketone derivative of DAMGO, has also been prepared and binds selectively to μ-opioid receptors [960]. The chloromethyl ketone of Dyn A-(1–10) [961] inhibits binding to frog brain membranes in a wash-resistant manner, although the affinity of this peptide for κ-receptors is relatively weak (apparent IC$_{50}$ for irreversible blockade ~10 μM). The C-terminal maleamide derivative of DSLET has recently been reported to exhibit wash-resistant inhibition of binding to δ-receptors at micromolar concentrations [962]. Enkephalin analogs containing melphalan (Mel), the nitrogen mustard derivative of p-aminophenylalanine, have also been prepared (see Refs [283,963]). Recently, the first isothiocyanate derivatives of opioid peptides were described [735,964]; the Phe(p-N=C=S)^4 derivative of N,N-dibenzyl enkephalin and both the Phe(p-N=C=S)^3 and the Phe(p-N=C=S)^4 derivatives of TIPP exhibited wash-resistant inhibition of binding to cloned δ-opioid receptors. The (Phe(p-bromoacetamide)^4 derivative of TIPP is even more potent than the p-isothiocyanate derivative at inhibiting binding to cloned δ-opioid receptors in a wash-resistant manner [965].

Thiol-containing derivatives of opioid peptides have been prepared, which potentially can form disulfide linkages with cysteine residues on opioid receptors. DALCE ([D-Ala^2, Leu^5, Cys^6]enkephalin) (45) (Fig. 8) [188] binds with high affinity to δ-receptors and has been used to characterize δ-receptor subtypes (see section "Receptor Subtypes, Splice Variants, and Receptor Dimerization"). 3-Nitro-2-pyridinesulphenyl (Npys)-containing derivatives of cysteine have been incorporated into opioid peptides to yield potential affinity label derivatives [966–969]. S-Activated enkephalin analogs containing the Npys group attached to the C-terminus label μ-opioid receptors in a dose-dependent manner [966], while the Npys-protected derivative of DALCE reacts with δ-receptors [968]. Incorporation of Cys(Npys) into position 8 of dynorphin A-(1–9) yields a peptide that decreases the B_{max} value, but does not affect the K_d value, for binding to κ-receptors [967]; recovery of binding following treatment with dithiothreotol suggests that the dynorphin analog binds to κ-receptors though a disulfide bond. [D-Ala^2,Cys(Npys)^8]-dynorphin A-(1–9) amide is reported to label all three opioid receptors, while [D-Ala^2,Cys(Npys)^{12}] dynorphin A-(1–13) amide apparently labels μ- and δ-, but not κ-, receptors [969].

7. ORL1 RECEPTOR AND ITS ENDOGENOUS LIGAND ORPHANIN FQ/NOCICEPTIN

7.1. Introduction

During attempts to clone different opioid receptor types several laboratories isolated a cDNA for a receptor with high homology to opioid receptors (see Refs [74,76,77] for recent reviews). Since this receptor, referred to by Mollereau et al. as opioid-receptor-like 1 receptor [78], did not display affinity for classical opioid ligands, it was classified as an "orphan receptor." Subsequently, two groups isolated a 17-residue peptide as the endogenous ligand for this receptor (see Fig. 9) [79,80]. This peptide was referred to as orphanin FQ by one group, since it was the ligand for the orphan receptor (F and Q are the N- and C-terminal residues, respectively, of the peptide) [79] and named nociceptin by the other group, since in the initial studies this peptide was reported to produce hyperalgesia [80]. The OFQ/N precursor protein prepronociceptin [80] contains additional biologically active peptides related to OFQ/N. Nocistatin (Fig. 9) [266] blocks a number of the effects of OFQ/N (see Ref. [970] for a review), but it does not interact with ORL1 receptors. A second 17-residue peptide, referred to as orphanin/nociceptin II, is also found in prepronociceptin [265], along with a longer peptide OFQ/N$_{160-187}$ [971]; these peptides, however, have not been as well characterized as OFQ/N (see Ref. [74] for a recent review).

Like opioid receptors, the ORL1 receptor is a G-protein-coupled receptor, consisting of seven TM regions plus extracellular and intracellular domains. The ORL1 receptor exhibits high sequence homology to opioid receptors, with 60–62% identity for the whole transmembrane domain (residues 52–342 in the human hORL1 receptor) [77] and higher homology in TMs 2, 3, and 7. There is also high sequence homology with opioid receptors in the intracellular loops, consistent with ORL1 receptors coupling to the same G proteins as opioid receptors. Splice variants of the ORL1 gene have been reported (see Refs [74,77] for reviews), and the results of some pharmacological studies have suggested receptor heterogeneity (see Ref. [74]), but the existence of subtypes of the ORL1 has not been firmly established.

The ORL1 receptor exhibits high selectivity for its endogenous ligand, and has very low affinity for most opioid ligands. Site-directed mutagenesis and chimeric receptor studies have examined possible structural reasons for this selectivity of the ORL1 receptor. Point mutations of only four residues in TM6 (VQV$^{276-278}$-IHI) and TM7 (T^{302}-I) of the ORL1 receptor were sufficient to impart binding affinity for Dyn A fragments without affecting the affinity or potency of OFQ/N [170]. The additional mutation of a residue in TM5 (A^{213}-K) enhanced affinity for several opioid alkaloids, particularly antagonists [972], but this mutant no longer bound OFQ/N. Alanine mutation of several TM residues that are conserved with opioid receptors yielded mutant receptors with reduced affinity for OFQ/N [973], suggesting that the binding pocket in the ORL1 receptor may be similar to that found in opioid receptors. Alanine replacement of Gln286 at the C-terminus of TM6 in hORL1 results in a mutant to which OFQ/N still binds with high affinity, but which cannot mediate inhibition of forskolin-stimulated cAMP formation [973], implicating this residue in ORL1 receptor signal transduction.

OFQ/N and the ORL1 receptor exhibit the greatest similarity to dynorphin A and κ-opioid receptors, respectively, but OFQ/N exhibits very low affinity for κ-receptors and conversely Dyn A exhibits low affinity for ORL1 receptors. In order to study the structural reasons for this selectivity, chimeric receptors have been constructed between ORL1 and κ-receptors [974,975]. Replacement of the N-terminal region of the κ-receptor up through the top of TM3 with the corresponding sequence of the ORL1 receptor imparted affinity for OFQ/N, but low potency in an adenylyl cyclase assay, without affecting the binding or potency of Dyn A. Further incorporation of extracellular loop 2 (EL2) from the ORL1 receptor restored efficacy for OFQ/N, again without affecting the ability of Dyn A to bind and activate the receptor. Thus, EL2 appears to be required for activation of ORL1 receptors, but not κ-opioid receptors.

Based on the experimental data, a computational model of OFQ/N bound to ORL1 receptors has been proposed [183]. In this model, the N-terminal sequence containing the two Phe residues binds in a highly conserved pocket formed by TMs 3, 5, 6, and 7, which is similar to that proposed for opioid receptors. Residues 5-7 (T-G-A) of OFQ/N are then positioned at the TM-EL2 interface in a largely nonconserved region; unfavorable side-chain interactions in this region of the receptor are then used to explain the selectivity of ORL1 for OFQ/N over Dyn A. The positively charged C-terminus of the peptide is proposed to make multiple contacts with the highly acidic EL2.

7.2. Physiological and Pharmacological Effects

Since their discovery interest in this receptor and its endogenous ligand has increased exponentially. There have been a number of excellent reviews covering the complex pharmacology of this system (see Refs [74–76] for recent ones), including a special issue of the journal *Peptides* (Vol. 21, issue number 7, 2000) devoted solely to the ORL1 receptor and OFQ/N.

Consistent with the sequence similarities between the ORL1 and the opioid receptors, activation of the ORL1 receptor triggers the same signal transduction mechanisms as utilized by the opioid receptors. Thus, activation of ORL1 receptors inhibits both forskolin-stimulated adenylyl cyclase and Ca^{++} currents and activates several other effectors, including inward rectifying K^+ channels, protein kinase C, mitogen activated protein kinase (MAP kinase) and phospholipase C (see Ref. [76] for a review).

There has been considerable controversy over the roles of ORL1 receptors and orphanin FQ/nociceptin (OFQ/N) in response to painful stimuli (see Refs [74,76,976] for recent reviews). When administered by intracerebroventricular (i.c.v.) injection, OFQ/N was initially reported to produce hyperalgesia [79,80], hence the name nociceptin for the endogenous peptide ligand. These effects, however, were subsequently classified as antianalgesic effects [76,976] based on reevaluation of the controls and the effects of stress-induced analgesia accompanying the experimental procedures. Effects reported for OFQ/N administered i.c.v. have ranged from hyperalgesia, analgesia, antianalgesic activity or a combination of these effects [74,76,976]. A similar range of activities has been reported following spinal (intrathecal) administration [74,76,977]. A number of factors appear to influence these often contradictory findings, including the noxious stimuli studied, the species and strain of animal used, the dose of OFQ/N, stress, and the physiological state of the animal (see Ref. [74] for a detailed discussion). The most robust and consistently observed effects of OFQ/N are the antianalgesic effects following supraspinal administration. OFQ/N acts as a functional antagonist of opioid receptors and blocks analgesia produced by a wide variety of opioids (see Refs [74,76]). At the spinal level, the predominant effect of OFQ/N appears to be inhibitory, resulting in analgesia and/or antihyperalgesia/antiallodynia (see Refs [74,76,977]). Several studies have reported antihyperalgesic or antiallodynic activity for OFQ/N in rat models of inflammation and nerve injury [76,977]. Since morphine appears to have reduced effectiveness in treating neuropathic pain following nerve injury, the activity of OFQ/N in this type of pain could have important therapeutic implications. OFQ/N has also been implicated in the development of tolerance to morphine (see Refs [74,76]).

OFQ/N and the ORL1 receptor are also involved in a number of other physiological effects (see Refs [74,76]). One of the most significant effects is the anxiolytic activity of OFQ/N [978], which has been postulated to be one of OFQ/N's most fundamental actions, and may help explain the effects of OFQ/N on other phenomena, for example, locomotion, reward, and feeding [74]. A small molecule ORL1 agonist has also demonstrated anxiolytic activity [979], demonstrating an important potential therapeutic application of these compounds. Like opioids, OFQ/N inhibits electrically induced contractions in the GPI and MVD smooth muscle preparations; these effects are not naloxone-reversible, indicating that they are not mediated by opioid receptors (see Ref. [76] for a review).

7.3. Structure–Activity Relationships of OFQ/N and Other Peptidic Ligands for the ORL1 Receptor

Shortly after the identification of the endogenous ligand, several research groups began exploring the SAR of this peptide. Several recent reviews [75,980,981] have discussed the details of the SAR of OFQ/N, so the discussion here will focus on some of the key findings.

Truncation studies have been performed to identify the minimum sequence required for ORL1 affinity and activation. These studies revealed that, like opioid peptides, the N-terminal aromatic residue was important for biological activity [982,983], although one study [982] has reported that the further N-terminal truncated fragments OFQ/N-(6-17) and OFQ/N-(12–17) retain affinity and activity for ORL1 receptors. In contrast to dynorphin A, where shorter fragments retain opioid activity [258], 13 of the 17 residues of OFQ/N appear to be required for ORL1 receptor affinity and activation [982,983]. The amide derivative of OFQ-(1–13) is a considerably more potent agonist in the mouse vas deferens assay than the acid derivative, apparently due to decreased metabolism [981]; therefore, this fragment is typically used as the parent structure in further SAR studies (see below). OFQ/N-(1–11), however, has been reported to be active both *in vitro* [984] and *in vivo* [985], despite its low affinity for cloned ORL1 receptors in binding assays, resulting in the proposal of receptor heterogeneity for OFQ/N [984]. Results from binding studies have also suggested receptor heterogeneity (see Ref. [74]).

A number of analogs of OFQ/N have been examined for their pharmacological activity (see Refs [75,980,981] for recent reviews). Shortly after identification of the endogenous ligand, the results of an alanine scan of OFQ/N were reported, identifying Phe^1, Phe^4, and Arg^8 as critical residues in the sequence [982,986]. Phe^1 can be replaced by tyrosine [983,986], resulting in an analog that retains affinity and potency at ORL1 receptors, but also exhibits increased affinity and activity at opioid receptors, particularly κ- and μ-receptors (see Refs [980,981]). Unlike in the opioid peptides, an aromatic amino acid is not required in position 1, and Phe^1 can be replaced by the aliphatic residues cyclohexylalanine and leucine [987]; in contrast an aromatic residue is required in position 4, and replacement of Phe^4 with an aliphatic residue results in loss of activity [987]. An Arg in position 8 appears to be critical, and replacement with Lys results in large decreases (>100-fold) in affinity and potency [981]. Incorporation of Tyr in place of Leu^{14} was used to obtain an analog that could be radioiodinated or tritiated for use in radioligand binding assays [79]; both labeled derivatives are commercially available.

A series of chimeric peptides between OFQ/N and Dyn A were prepared in order to explore the structural reasons for the selectivity of OFQ/N for ORL1 over κ-opioid receptors [988]. The results from this study suggested that residues Thr^5 and Gly^6, in addition to Phe^1, are responsible for the activity and selectivity of OFQ/N for ORL1 receptors; the chimera Dyn A-(1–5)/OFQ/N (6–17) **253** was able to bind and activate both ORL1 and κ-opioid receptors.

Tyr-Gly-Gly-Phe-Leu – Gly-Ala-Arg-Lys-Ser-Ala-Arg-Lys-Leu-Ala-Asn-Gln

253 DynA-(1–5)/OFQ/N (6–17)

Early studies of the pharmacology of the ORL1 system were hindered by the lack of antagonists for this receptor. Therefore, there was considerable excitement in the field when the first report of an antagonist appeared in the literature [989]. The reduced amide derivative of OFQ/N [$Phe^1(\psi CH_2NH)Gly^2$] OFQ/N-(1–13)NH_2 **254** (Fig. 51, referred to as [F/G]NC(1–13)NH_2 by Calo and coworkers), which was synthesized to protect the peptide from metabolism by aminopeptidases, was initially reported to be an antagonist of OFQ/N (1–13)NH_2 in smooth muscle preparations [989]. Subsequent examination of this compound in a variety of assays, however, indicated that the activity observed depended on the assay, and that while **254** was an antagonist in some assays, it was a partial or full agonist in a number of other assays, including forskolin-stimulated cAMP accumulation in CHO cells expressing cloned ORL1 (see Refs [75,980] for detailed reviews). Subsequently, the N-substituted glycine analog [$Nphe^1$]OFQ/N (1–13)NH_2**255** was reported to be an ORL1 receptor antagonist [990]. Although the potency of the compound is weak ($pA_2 > 6$ in most assays), it is an antagonist in all of the assays examined to date (see Refs [75,980] for reviews).

The utilization of combinatorial libraries has resulted in the identification of peptidic and peptidomimetic ligands for the ORL1 receptor, which are not structurally related to OFQ/N. Houghten and coworkers identified acetylated hexapeptides with high affinity for the ORL1 receptor, with Ac-RYYRXK-NH_2**256** (X = W or I) having the highest affinity [991]; in most assays these peptides exhibit partial agonist activity (see Ref. [980]). A combinatorial library of conformationally constrained peptides resulted in the identification of III-BTD (**257**) that exhibits moderate affinity ($K_i = 24$ nM) but only modest selectivity (K_i ratio (ORL1/μ/κ/δ) = 1/4.6/6.1/22) for ORL1 receptors; this compound acts as an antagonist at ORL1 receptors while it is exhibits partial agonist activity at opioid receptors [992]. The related conformationally

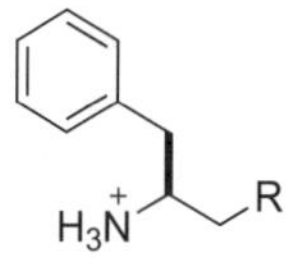

254 [F/G]NC(1–13)NH₂

R = Gly-Gly-Phe-Thr-Gly-Ala-Arg-Lys-Ser-Ala-Arg-LysNH₂

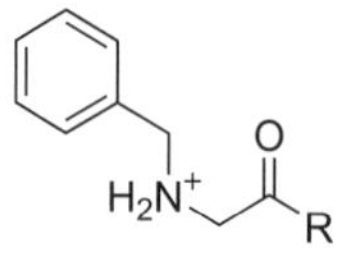

255 [Nphe¹]OFQ/N (1–13)NH₂

Ac-Arg-Tyr-Tyr-Arg-X-LysNH₂

256 X = Trp or Ile

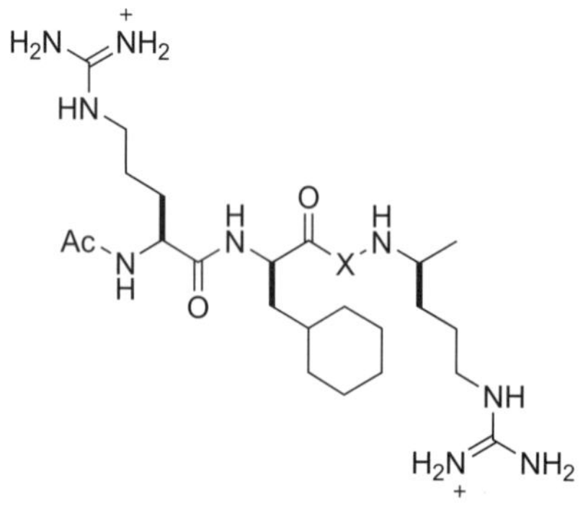

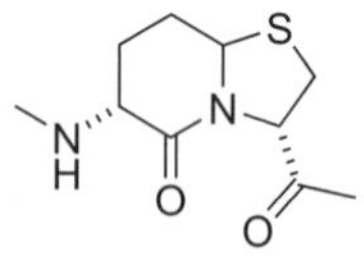

257 III-BTD

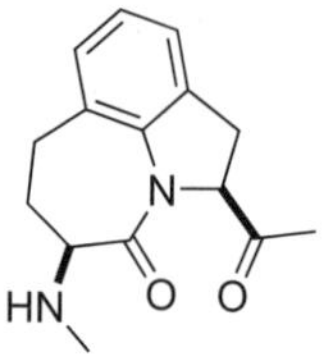

258 III-Haic

Figure 51. Peptidic ligands for ORL1 receptors.

constrained peptide III-Haic (**258**) was nonselective, exhibiting modest affinity for ORL1 and the three opioid receptors (K_i = 50–125 nM) [992].

7.4. Nonpeptide Ligands for the ORL1 Receptor

Because of the potential therapeutic applications of ORL1 receptor ligands, there has been considerable interest in identifying nonpeptidic compounds that interact with this receptor. Several opiates have been reported to exhibit some affinity for ORL1 receptors (see Refs [980,981]) with some opiates, most notably the fentanyl analog lofentanyl [983], naloxone benzoylhydrazone (NalBzOH) (see Refs [980,981]), the naltrexamine derivative TRK-820 (**65**) (Fig. 16) [308], and buprenorphine [993], exhibiting reasonable affinity and/or potency. Reports by groups from the pharmaceutical industry of both nonpeptidic selective agonists and antagonists have begun to appear in the scientific and patent literature, with leads identified from screening assays (see Refs [980,981,994] for recent reviews). Starting from a lead **259** (Fig. 52), identified by high-throughput screening as a nonselective ORL1 ligand, a group from Hoffmann-La Roche explored a number of 1-phenyl-1,3,8-triaza-spiro[4.5]decan-4-ones to identify more selective and potent ORL1 agonists [995–998]. The pharmacology of one of these compounds, Ro 64-6198 (**260**), has been examined in considerable detail [979,998–1001], including determination of

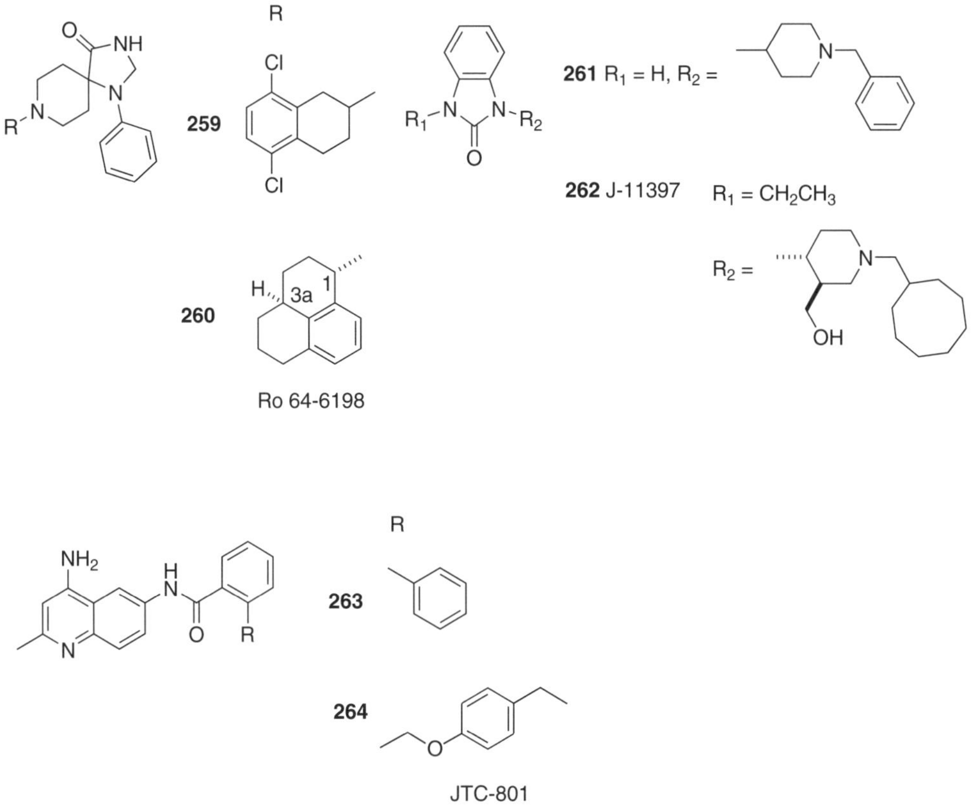

Figure 52. Nonpeptidic ligands for ORL1 receptors.

the affinities of the different stereoisomers for opioid receptors as well as ORL1 receptors [998]. The stereochemistry at positions 1 and 3a had significant effects on ORL1 receptor affinity, while the affinities for the opioid receptors were comparable for the different isomers; the *1S,3aS* isomer exhibits the highest affinity for ORL1 receptors ($pK_i = 9.41$) and therefore the highest selectivity for ORL1 over opioid receptors [998]. This compound is a full agonist at ORL1 receptors and produces dose-dependent anxiolytic effects in several rat models of anxiety, with an efficacy and potency following systemic administration comparable to those of benzodiazepine anxiolytics such as diazepam [979].

Researchers from the Banyu Tsukuba Research Institute reported the first nonpeptide antagonist for ORL1 receptors [1002]. Starting from a lead 261 from their chemical library, which again exhibited reasonable affinity but poor selectivity for ORL1 receptors, these researchers identified J-113397 [262] that exhibits high affinity ($K_i = 1.8$ nM) and high selectivity for ORL1 receptors over opioid receptors [1002–1004]. This compound is an antagonist of OFQ/N in vitro and in vivo following subcutaneous administration [1003,1004]. A series of 4-aminoquinolines [264] has also been examined for antagonist activity, based on the weak affinity of the lead compound 271 in a random screen [1005]. JTC-801 [264] was selected because it showed the best bioavailability. Following oral administration, this compound was shown to antagonize the effects of OFQ/N and to produce an analgesic effect that was not antagonized by naloxone; this compound is reported to be undergoing clinical trials [1005].

8. RECENT DEVELOPMENTS

8.1. Salvia Divinorum and Neoclerodane Diterpenes

Salvinorin A (**265**) (Fig. 53), a nonnitrogenous naturally occurring neoclerodane diterpene from the "magic mint" *Salvia divinorum*, has been identified as a highly selective κ-opioid agonist [1006–1008]. This discovery was one that changed the paradigm of opioid structure–activity requirements as the compound is completely neutral and has no nitrogen. Various studies have been carried out on this template to create a better understanding of the SAR around this unique template [1009]. Determinants of binding modes have been carried out by two groups with similar results indicating a novel binding mode compared to classical nitrogen containing opioids.

Neoclerodane **265** produces short but intense hallucinations in humans with a potency similar to lysergic acid diethyl amide (LSD) [1010,1011]. This activity was interesting given its lack of structural similarity to other psychotomimetic substances but remarkable given that **265** was found to be a potent and selective agonist for KOP receptors [1008]. Additional pharmacological studies have provided further evidence that **265** acts as a KOP agonist. Salvinorin A has been found to (1) produce a discriminative effect similar to other KOP agonists in both nonhuman primates [1012] and rats [1013], (2) elicit antinociception in mice that is blocked

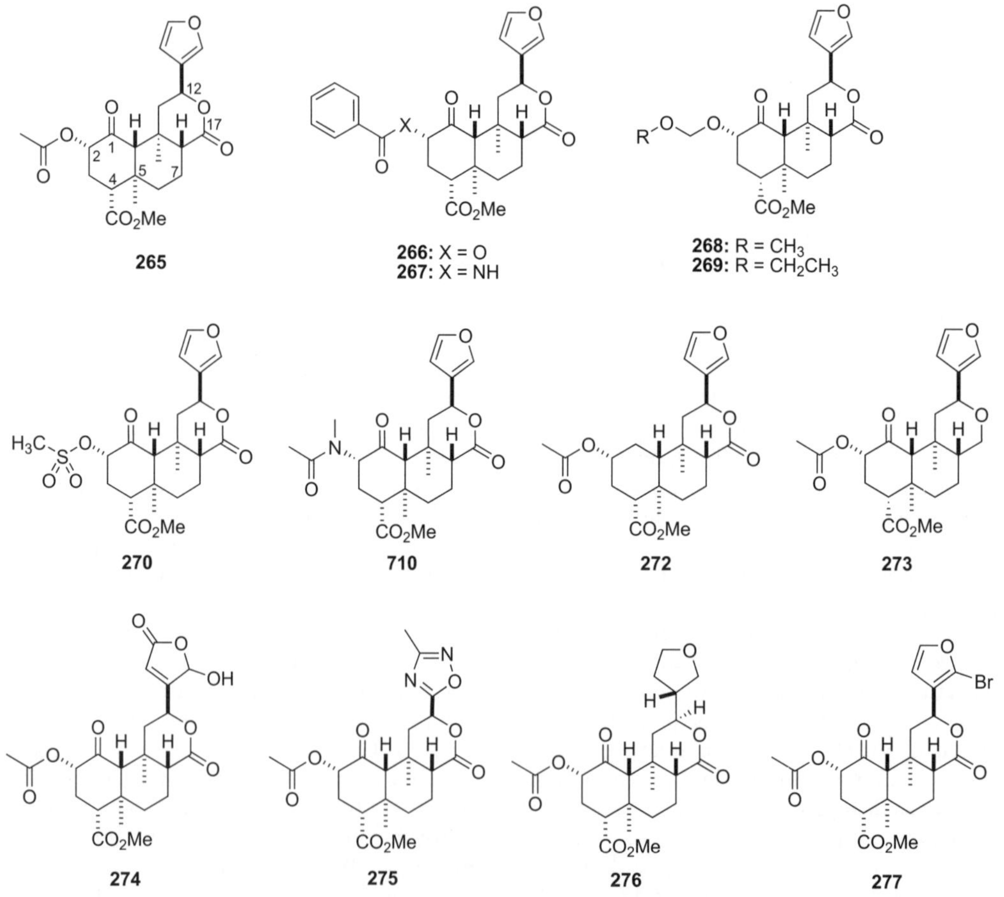

Figure 53. Selected neoclerodane diterpenes.

norBNI [1014,1015], (3) produce an aversive response in the conditioned place preference assay [1016], (4) decreases dopamine levels in the caudate putamen of mice [1016], (5) dose dependently increases immobility in the forced swim test [1017], (6) disrupts climbing behavior on an inverted screen task [1018], (7) blocks the locomotor-stimulant effects of cocaine [1019], and (8) not exert DOM-like effects in nonhuman primates [1020]. Furthermore, **265** has been found to decreases mesostriatal neurotransmission of dopamine [1021]. In contrast to other KOP agonists, **265** does not cause diuresis in rats, likely due to its short duration of action [1022]. However, differences have been seen in the interaction of **265** and other KOP agonists with respect to the behavioral responses to cocaine [1021].

Additional work in nonhuman primates found that **265** acts as a high-efficacy KOP agonist in a translationally viable neuroendocrine biomarker assay and produces facial relaxation and ptosis that can be detected within 1–2 min of injection [1023,1024]. Pharmacokinetic studies indicate that the half-life of **265** in nonhuman primates is approximately 30 min [1025,1026] and other studies in baboons using carbon-11 labeled **265** [1027] were highly consistent with this observation [1027].

The growing body of evidence that **265** is a potent and selective KOP agonist [1008] has challenged the belief that a basic nitrogen is required for binding and activity at opioid receptors. Therefore, **265** is a unique tool to study opioid receptors. In light of these findings, efforts have begun to explore the SAR of this compound at opioid receptors [1009].

Several agonists, partial agonists, and antagonists of the different opioid receptor subtypes have been identified by systematically modifying the structure of **265**. Much of the focus has been on semisynthetic modification at the C-2 position. Replacement of the C-2 acetate of this compound with a benzoyl group led to the MOP agonist herkinorin (**266**) [1028]. Herkinorin (**266**) does not promote β-arrestin-2 recruitment to the MOP receptors, unlike most MOP agonists [1029,1030]. In addition, **266** does not lead to the internalization of MOP receptors [1029,1030]. Introduction of a benzamide (**267**) group at this position led to an increase in affinity and activity compared to **266** [1029].

The replacement of the C-2 acetate with a methoxymethyl (MOM) or ethoxymethyl (EOM) ether provided analogs with increased efficacy and potency compared to **265** [1031,1032]. MOM-salvinorin B (**268**) was found to dose-dependently elicit antinociception in the hot plate test in rats [1032]. The analgesic response of **268** was attenuated by norBNI [1032]. This compound was reported to be more potent as a KOP agonist than **265** [1031]. EOM-salvinorin B (**269**) was also reported to be more potent as a KOP agonist than **265** [1031]. Substitution of a mesylate (**270**) at this position led to a slight increase in potency compared to **265** [1028]. The C-2 *N*-methylacetamide analog (**271**) of **265** was reported to be more metabolically stable both *in vitro* and *in vivo*, and to be orally active [1033]. Removal of the ketone (**272**) at C-1 led to a reduction in affinity and potency at KOP receptors [1034]. The C-4 position has also been investigated [1035–1038]. The only well-tolerated changes at the C-4 position were replacement of the carbomethoxy group with amides of amino acids [1036]. Removal of the C-17 lactone carbonyl (**273**) had no significant effect on binding compared to **265** but led to a decrease in activity at KOP receptors [1038].

Efforts have also made to modify the furan ring at C-12 [1038–1040]. Salvidivin A (**274**) was isolated from *S. divinorum* [1041], and was identified as a KOP antagonist and is the first naturally occurring diterpene to display this activity [1040]. Replacement of the furan ring with a 4-methyl-1,3,5-oxadiazole (**275**) resulted in antagonist activity at MOP and KOP receptors (1040). Reduction of the furan ring to tetrahydrosalvinorin A (**276**) resulted in a partial agonist at KOP receptors (1040). Bromination of the furan ring resulted in an analog (**277**) that was equipotent with **265** (1040).

In tandem with structural modifications to the **265** skeleton, *in vitro* and *in silico* studies have provided insights into the unique binding mode of neoclerodane diterpenes. Initial studies identified tyrosine residues in helices 2 and 7 as stabilizers of **265** binding, and noted that these groups are unique stabilizing moieties [1042]. Later studies, supported by *in silico* models, indicated that KOP receptor selectivity is conferred by valine residues in helix 2 [1043]. More recently, opioid receptor chimera studies led to the development of a model to describe the selectivity of **266** for MOP over KOP [1044]. In combination with SAR derived from additional chemical modifications to **265**, these studies will aid the design of future opioid receptor probes.

9. THINGS TO COME

Several advancements in opioid pharmacology during recent years could have a significant impact on the types of compounds used as narcotic analgesics in the future. The involvement of peripheral opioid receptors in inflammatory pain (see Section 3.3) has resulted in the search for peripherally selective analgesics (see Section 5.9.2) that would be free from serious centrally mediated side effects. The challenge in this area may be to obtain compounds that do not cross the blood–brain barrier, but which still are orally bioavailable.

The ability of δ-receptor antagonists to decrease the development of tolerance and dependence to morphine (see Section 3.3.1) has considerable therapeutic potential and has prompted the search for compounds with both μ-agonist and δ-antagonist activities. While peptide derivatives with both activities have been studied in some detail (see Section 6.4.2), nonpeptide opioids with this activity profile have only been reported recently. This approach of using δ-antagonism together with μ-agonism, either combined in a single drug or by coadministering two agents, to minimize the side effects of μ-agonists is very exciting. This approach is still in its infancy, however, so its therapeutic application in clinical trials still remains to be demonstrated.

There has also been considerable advancement in the identification of agents that produce analgesic effects through different mechanisms than interaction with opioid receptors (see Refs [1045–1048] for reviews). Thus, NMDA (*N*-methyl-D-aspartate) antagonists, GABA (γ-aminobutyric acid) agonists, nicotinic acetylcholine receptor agonists (e.g., epibatadine), antagonists of substance P at NK-1 receptors, and a number of other compounds targeting different receptors exhibit antinociceptive activity in animal models. These compounds do not cause the side effects associated with the clinically used opiates, but they have their own distinct side-effect profiles, which in some cases, for example, NMDA antagonists (which cause psychotomimetic effects), have resulted in termination of clinical studies (see Ref [1046]). Also the promising antinociceptive activity observed in animal models has not always translated into clinical efficacy in humans. Thus, a number of NK-1 receptor antagonists are highly effective in animal models of pain and exhibit excellent pharmacokinetic profiles in humans, but have failed in phase II clinical trials for treatment of pain and migraine (see Ref. [1046]). Thus, which of these novel targets may ultimately result in clinically used agents for pain remains to be determined.

A major unmet therapeutic need is effective treatments for substance abuse, including abuse of opioids, cocaine, and amphetamines. Results for ligands interacting with either δ- or κ-receptors hold some promise in this area (see Section 3.3), but whether such agents will be clinically useful remains to be demonstrated. Initial studies in humans have been promising (e.g., an improved positive response of opioid-dependent individuals to a "functional" κ-antagonist (buprenorphine in the presence of naltrexone to block μ-agonist activity) as compared to naltrexone alone [243]). A central problem in treating drug addiction is the vulnerability to relapse during abstinence [1049]. Behavioral studies have shown that presentation of drug-associated cues, drug priming, and acute footshock stress each increased drug self-administration [1050–1052]. It is believed that release of dynorphins, the endogenous agonists for KOP receptors, may mediate a component of stress-induced drug craving in reinstatement models (models of drug relapse) [1049]. Studies have shown that interference of the KOP system by pretreatment with antagonists or gene disruption of KOP receptor attenuates the reinstatement of extinguished drug-taking behavior [1049,1053,1054].

The biological basis of mood is not understood. Most research on mood and affective states has focused on the roles of brain systems containing monoamines (e.g., dopamine, norepinephrine, and serotonin). However, it is becoming clear that endogenous opioid systems in the brain may also be involved in the regulation of mood [1055]. Interference of the KOP system has also been shown to produce effects in animal models often used to study psychiatric illness. KOP receptor antagonists significantly decrease immobility and increase swimming time in the forced swim stress test similar to the antidepressant desipramine in rats [1053]. Furthermore, KOP antagonists have anxiolytic-like effects in models of unlearned and learned fear in rats [1056]. This suggests that KOP antagonists may have utility in treating depression and anxiety, as well as drug relapse. Other research suggests that KOP agonists might be useful for mania, and KOP partial agonists might be useful for mood stabilization [1057].

Stress is a complex human experience having both positive and negative motivational properties. When chronic and uncontrollable, the adverse effects of stress on human health are considerable and yet poorly understood. The dysphoric properties of chronic stress are encoded by the endogenous opioid peptide dynorphin acting on specific stress-related neuronal circuits [1058,1059]. Using different forms of stress presumed to evoke dysphoria in mice, it was found that repeated forced swim and inescapable footshock both produced aversive behaviors that were blocked by a KOP receptor antagonist and absent in mice lacking dynorphin [1058]. Injection of corticotropin-releasing factor (CRF) or urocortin III, key mediators of the stress response, produced place aversion that was also blocked by dynorphin gene deletion or KOP antagonism. Recent results indicate that KOP receptors are poised to presynaptically inhibit diverse afferent signaling to the locus ceruleus [1060]. This is a novel and potentially powerful means of regulating the locus ceruleus-norepinephrine system that can impact on forebrain processing of stimuli and the organization of behavioral strategies in response to environmental stimuli. This suggests that KOP receptors are a novel target for alleviating symptoms of opiate withdrawal, stress-related disorders, or disorders characterized by abnormal sensory responses, such as autism.

While norBNI has been extensively used to study κ-opioid receptors, its pharmacological properties are not optimal, and it exhibits a much longer than expected half-life *in vivo* [1061]. As noted above, further study of its structure–activity relationships identified GNTI as a κ-antagonist. While GNTI has increased potency *in vivo* compared to norBNI, it also has a slow onset of action and a long half-life *in vivo* [1062]. Recently, Thomas et al. identified several novel κ-opioid receptor antagonists from several classes of opioids [1063–1066]. Among the compounds identified was JDTic, ((3R)-7-hydroxy-N-(1S)-1-[(3R,4R)-4-(3-hydroxyphenyl)-3,4-dimethyl-1-piperidinyl]methyl-2-methylpropyl-1,2,3,4-tetrahydro-3-isoquinolinecarboxamide, was identified as being a more potent κ-antagonist than norBNI [1064]. Biological studies of JDTic have shown that (1) it blocks κ-agonist-induced antinociception in mice and squirrel monkey [1067], (2) antagonizes κ-agonist-induced diuresis in rats [1067], and (3) significantly reduced footshock-induced reinstatement of cocaine responding in rats and decreased immobility and increased swimming time in the forced swim stress test similar to the antidepressant desipramine [1053]. However, like other κ-antagonists mentioned above, JDTic has a slow onset and extremely long duration of action [1068]. Thus, there is a pressing need for a truly short-acting KOP antagonist as a pharmaceutical and research tool. Recently, the cyclic peptide zyklophin has been identified as short-acting KOP antagonist *in vivo* [1069]. This suggests that novel structural motifs may also deliver additional short-acting antagonists.

The goal of identifying potent analgesics free of the side effects of morphine and other narcotics has remained elusive. As more information continues to become available about opioid receptor structure, opioid pharmacology, and related systems, this will provide new challenges to medicinal chemists to prepare compounds with unique pharmacological profiles. With the diversity of structures exhibiting affinity for opioid receptors, there is still ample opportunity for the development of new therapeutic agents that hopefully will bring us closer to the goal of identifying optimal opioid agonists and antagonists.

ACKNOWLEDGMENTS

The authors would like to thank the authors of the Narcotic Analgesics chapter of previous editions for the groundwork they provided to this edition. Without their valuable contributions, this chapter would not have been possible. Furthermore, the authors thank Christopher Cunningham, Kimberly Lovell, Anthony Lozama, Katherine Smith, and Tamara Vasilijecik for their technical assistance.

REFERENCES

1. Gutstein H, Akil H. Opioid analgesics. In: Hardman J, Limbird L, Gilman A, editors. Goodman and Gilman's The Pharmacological Basis of Therapeutics. 10th ed. New York: McGraw-Hill Medical Publishing Division; 2001. p 569–619.
2. Tzschentke TM, Christoph T, Kogel B, Schiene K, Hennies HH, Englberger W, Haurand M, Jahnel U, Cremers TI, Friderichs E, De Vry J. (−)-(1R,2R)-3-(3-dimethylamino-1-ethyl-2-methyl-propyl)-phenol hydrochloride (tapentadol HCl): a novel mu-opioid receptor agonist/norepinephrine reuptake inhibitor with broad-spectrum analgesic properties. J Pharmacol Exp Ther 2007;323:265–276.
3. Tapentadol (Nucynta): a new analgesic. Med Lett Drugs Ther 2009;51:61–62.
4. Tzschentke TM, Jahnel U, Kogel B, Christoph T, Englberger W, De Vry J, Schiene K, Okamoto A, Upmalis D, Weber H, Lange C, Stegmann JU, Kleinert R. Tapentadol hydrochloride: a next-generation, centrally acting analgesic with two mechanisms of action in a single molecule. Drugs Today 2009;45: 483–496.
5. Mather LE, Smith MT. Clinical Pharmacology and Adverse Effects. In: Stein C, editor. Opioids in Pain Control: Basic and Clinical Aspects. Cambridge: Cambridge University Press; 1999. p 188–211.
6. Cherny N. Opioid analgesics: comparative features and prescribing guidelines. Drugs 1996;51:713–737.
7. Society AP. Principles of Analgesic Use in the Treatment of Acute Pain and Cancer Pain. 4th ed. Glenview, IL: American Pain Society; 1999. p 64.
8. Savage S, Covington EC, Heit HA, Hunt J, Joranson D, Schnoll SH. Definitions Related to the Use of Opioids for the Treatment of Pain. Glenview, IL: American Academy of Pain Medicine, American Pain Society, American Society of Addiction Medicine; 2001.
9. Aronoff GM. Opioids in chronic pain management: is there a significant risk of addiction?. Curr Rev Pain 2000;4:112–121.
10. Koob GF, Bloom FE. Cellular and molecular mechanisms of drug dependence. Science 1988;242:175–723.
11. Devine DP, Leone P, Pocock D, Wise RA. Differential involvement of ventral tegmental mu, delta, and kappa opioid receptors in modulation of basal mesolimbic dopamine release: in vivo microdialysis studies. J Pharmacol Exp Ther 1993;266:1236–1246.
12. Mulder A, Schoffelmeer A. Multiple opioid receptors and presynaptic modulation of neurotransmitter release in the brain. In: Herz A, editor. Opioids I. Volume 104/I, Berlin: Springer-Verlag; 1993. p 125–144.
13. Cooper S. Interactions between endogenous opioids and dopamine: implications for reward and aversion. In: Willner P, Scheel-Kruger J, editors. The Mesolimbic Dopamine System: From Motivation to Action. Chichester, UK: Wiley; 1991. p 331–366.
14. Campora E, Merlini L, Pace M, Bruzzone M, Luzzani M, Gottlieb A, Rosso R. The incidence of narcotic induced emesis. J Pain Symptom Manage 1991;6:428–430.
15. Friedman JD, Dello Buono FA. Opioid Antagonists in the treatment of opioid-induced constipation and pruritus. Ann Pharmacother 2001;35:85–91.
16. Pappagallo M. Incidence, prevalence, and management of opioid bowel dysfunction. Am J Surg 2001;182(Suppl 5A): 11S–18S.
17. Foss JF. A review of the potential role of methylnaltrexone in opioid bowel dysfunction. Am J Surg 2001;182(Suppl 5A): S19–S26.
18. Sykes NP. An investigation of the ability of oral naloxone to correct opioid-related constipation in patients with advanced cancer. Palliat Med 1996;10:135–144.
19. Schmidt WK. Alvimpoan* (ADL 8-2698) is a novel peripheral opioid antagonist. Am J Surg 2001;182(Suppl 5A): 27S–38S.
20. Liu SS, Hodgson PS, Carpenter RL, Fricke JR. ADL 8-2698, a trans-3,4-dimethyl-4-(3-

hydroxyphenyl) piperidine, prevents gastrointestinal effects of intravenous morphine without affecting analgesia. Clin Pharmacol Ther 2000;68:66–71.
21. Taguchi A, Sharma N, Saleem RM, Sessler DI, Carpenter RL, Seyedsadr M, Kurz A. Selective postoperative inhibition of gastrointestinal opioid receptors. N Engl J Med 2001;345:935–940.
22. Drug Facts and Comparisons. St. Louis, MO: Wolters Kluwer; 2001.
23. Schinkel AH, Wagenaar E, Mol CAAM, van Deemter L. P-glycoprotein in the blood–brain barrier of mice influences the brain penetration and pharmacological activity of many drugs. J Clin Invest 1996;97:2517–2524.
24. Sadeque AJM, Wandel C, He H, Shah S, Wood AJJ. Increased drug delivery to the brain by P-glycoprotein inhibition. Clin Pharmacol Ther 2000;68:231–237.
25. Rogers AG. Considering histamine release in prescribing opioid analgesics. J Pain Symptom Manage 1991;6:44–45.
26. Kjellberg F, Tramer MR. Pharmacological control of opioid-induced pruritus: a quantitative systematic review of randomized trials. Eur J Anasethesiol 2001;18:346–357.
27. Yuan C-S, Foss JF, O'Conner M, Osinski J, Roizen MF, Moss J. Efficacy of orally administered methylnaltrexone in decreasing subjective effects after intravenous morphine. Drug Alcohol Depend 1998;52:161–165.
28. Lacy CF, Armstrong LL, Goldman MP, Lance LL. Drug Information Handbook. Cleveland: Lexi-Comp, Inc.; 2000.
29. Marcus D. Treatment of nonmalignant chronic pain. Am Fam Physician 2000;61: 1331–1338.
30. Rumore MM, Schlichting DA. Clinical efficacy of antihistamines as analgesics. Pain 1986;25:7–22.
31. Levine JD, Gordon NC, Smith R, McBryde R. Desipramine enhances opiate postoperative analgesia. Pain 1986;27:45–49.
32. Pick GG, Paul D, Eison MS, Pasternak GW. Potentiation of opioid analgesia by the antidepressant nefazodone. Eur J Pharmacol 1992;211:375–381.
33. Michalets E. Update: clinically significant cytochrome P-450 drug interactions. Pharmacotherapy 1998;18:84–112.
34. Maurer PM, Bartowski RR. Drug Interactions of clinical significance with opioid analgesics. Drug Saf 1993;8:30–48.
35. Slaughter RL, Edwards DJ. Recent advances: the cytochrome P450 enzymes. Ann Pharmacother 1995;29:619–624.
36. Olsen GD. Methadone binding to human plasma albumin. Science 1972;176:525–526.
37. Bernards CM. Clinical implications of physiochemical properties of opioids. In: Stein C, editor. Opioids in Pain Control: Basic and Clinical Aspects. Cambridge: Cambridge University Press; 1999. p 166–187.
38. Olsen GD. Methadone binding to human plasma proteins. Clin Pharmacol Ther 1973;14: 338–343.
39. Judis J. Binding of codeine, morphine, and methadone to human serum proteins. J Pharm Sci 1977;66:802–806.
40. Kreek MJ. Methadone in treatment: physiological and pharmacological issues. In: Dupont RL, Goldstein A, O'Donnell J, editors. Handbook on Drug Abuse. Washington, DC: US Goverment Printing Office; 1979. p 57–86.
41. Benyhe S. Minireview—Morphine: new aspects in the study of an ancient compound. Life Sci 1994;55:969–979.
42. Barkin R, Barkin D. Pharmacologic management of acute and chronic pain: focus on drug interactions and patient-specific pharmacotherapeutic selection. South Med J 2001; 94:756–812.
43. Smith MT. Neuroexcitatory effects of morphine and hydromorphone: evidence implicating the 3-glucuronide metabolites. Clin Exp Pharmacol Physiol 2000;27:524–528.
44. Smith MT, Watt JA, Cramond T. Morphine-3-glucuronide: a potent antagonist of morphine analgesia. Life Sci 1990;47:579–585.
45. Ulens C, Baker L, Ratka A, Waumans D, Tytgat J. Morphine-6beta-glucuronide and morphine-3-glucuronide, opioid receptor agonists with different potencies. Biochem Pharmacol 2001;62:1273–1282.
46. McQuay HJ, Moore RA. Opioid problems, and morphine metabolism and excretion. In: Dickenson A, Besson J-M, editors. The Pharmacology of Pain. Volume 130, New York: Springer; 1997. p 335–360.
47. Eichelbaum M, Evert B. Influence of pharmacogenetics on drug disposition and response. Clin Exp Pharmacol Physiol 1996;1996: 983–985.
48. Caraco Y, Sheller J, Wood AJ. Impact of ethnic origin and quinidine coadministration on codeine's disposition and pharmacokinetic

effects. J Pharmacol Exp Ther 1999;290: 413–422.

49. Vree TB, van Dongen RT, Koopman-Kimenai PM. Codeine analgesia is due to codeine-6-glucuronide, not morphine. Int J Clin Pract 2000;45:395–398.

50. Burkle H, Dunbar S, Van Aken H. Remifentanil: a novel, short-acting, μ-opioid. Anesth Analg 1996;83:646–651.

51. Mansour A, Fox CA, Akil H, Watson SJ. Opioid-receptor mRNA expression in the rat CNS: anatomical and functional implications. Trends Neurosci 1995;18:22–29.

52. Fowler CJ, Fraser GL. Invited review: mu-, delta-, kappa-opioid receptors and their subtypes. A critical review with emphasis on radioligand binding experiments. Neurochem Int 1994;24:401–426.

53. Mansour A, Watson SJ. Anatomical distribution of opioid receptors in mammalians: an overview. In: Herz A, Akil H, Simon EJ, editors. Opioids I. Volume 104/I, Berlin: Springer-Verlag; 1993. p 79–105.

54. Stein C, Cabot PJ, Schafer M. Peripheral analgesia: mechanisms and clinical implications. In: Stein C, editor. Opioids in Pain Control: Basic and Clinical Aspects. Cambridge: Cambridge University Press; 1999. p 96–108.

55. Stein C. The control of pain in peripheral tissue by opioids. N Engl J Med 1995;332: 1685–1690.

56. Stein C. Peripheral mechanisms of opioid analgesia. In: Herz A, Akil H, Simon EJ, editors. Opioids II. Volume 104/II, Berlin: Springer-Verlag; 1993. p 91–103.

57. Barber A, Gottschlich R. Opioid angonists and antagonists: an evaluation of their peripheral actions in inflammation. Med Res Rev 1992;12:525–562.

58. Herz A, Akil H, Simon EJ. Opioids I. Volume 104/I, Berlin: Springer-Verlag; 1993. p 815.

59. Herz A, Akil H, Simon EJ. Opioids II. Volume 104/II, Berlin: Springer-Verlag; 1993. p 825.

60. Stein C. Opioids in pain control: basic and clinical aspects. Cambridge: Cambridge University Press; 1999. p 359.

61. Ossipov M, Lai J, Malan TP Jr, Porreca F. Recent advances in the pharmacology of opioids. In: Sawynok J, Cowan A, editors. Novel Aspects of Pain Management: Opioids and Beyond. New York: Wiley-Liss; 1999. p 49–71.

62. Ossipov M, Malan TP Jr, Lai J, Porreca F. Opioid pharmacology of acute and chronic pain. In: Dickenson A, Besson J-M, editors. The Pharmacology of Pain, Volume 130, Berlin: Springer; 1997. p 305–333.

63. Knapp RJ, Malatynska E, Collins N, Fang L, Wang JY, Hruby VJ, Roeske WR, Yamamura HI. Molecular biology and pharmacology of cloned opioid receptors. FASEB J 1995;9: 516–525.

64. Satoh M, Minami M. Molecular pharmacology of the opioid receptors. Pharmacol Ther 1995;68:343–364.

65. Simon EJ, Hiller JM. Opioid peptides and opioid receptors. In: Siegel GJ, Agranoff BW, Albers RW, Molinoff PB, editors. Basic Neurochemistry: Molecular, Cellular, and Medical Aspects. 5th ed. New York: Raven Press; 1994. p 321–339.

66. Beckett AH, Casy AF. Synthetic analgesics: stereochemical considerations. J Pharm Pharmacol 1954;6:986–1001.

67. Portoghese PS. A new concept on the mode of interaction of narcotic analgesics with receptors. J Med Chem 1965;8:609–616.

68. Martin WR. Opioid antagonists. Pharmacol Rev 1967;19:463–521.

69. Martin WR, Eades CG, Thompson JA, Huppler RE, Gilbert PE. The effects of morphine- and nalorphine-like drugs in the nondependent and morphine-dependent chronic spinal dog. J Pharmacol Exp Ther 1976;197: 517–532.

70. Gilbert PE, Martin WR. The effects of morphine- and nalorphine-like drugs in the nondependent, morphine-dependent and cyclazocine-dependent chronic spinal dog. J Pharmacol Exp Ther 1976;198:66–82.

71. Lord JAH, Waterfield AA, Hughes J, Kosterlitz HW. Endogenous opioid peptides: multiple agonists and receptors. Nature 1977;267: 495–499.

72. Quirion R, Chicheportiche R, Contreras PC, Johnson KM, Lodge D, Tam SW, Woods JH, Zukin SR. Classification and nomenclature of phencyclidine and sigma receptor sites. Trends Neurosci 1987;10:444–446.

73. Simon EJ, Gioannini TL. Opioid receptor multiplicity: isolation, purification, and chemical characterization of binding sites. In: Herz A, Akil H, Simon EJ, editors. Opioids I. Volume 104/I, Berlin: Springer-Verlag; 1993. p 3–26.

74. Mogil JS, Pasternak GW. The molecular and behavioral pharmacology of the orphanin

FQ/nociceptin peptide and receptor family. Pharmacol Rev 2001;53:381–415.
75. Calò G, Guerrini R, Rizzi A, Salvadori S, Regoli D. Review—Pharmacology of Nociceptin and its receptor: a novel therapeutic target. Br J Pharmacol 2000;129: 1261–1283.
76. Harrison LM, Grandy DK. Opiate modulating properties of nociceptin/orphanin FQ. Peptides 2000;21:151–172.
77. Meunier J, Mouledous L, Topham CM. The nociceptin (ORL1) receptor: molecular cloning and functional architecture. Peptides 2000;21:893–900.
78. Mollereau C, Parmentier M, Mailleux P, Butour J-L, Moisand C, Chalon P, Caput D, Vassart G, Meunier J-C. ORL1, a novel member of the opioid receptor family: cloning, functional expression and localization. FEBS Lett 1994;341:33–38.
79. Reinscheid RK, Nothacker H-P, Bourson A, Ardati A, Henningsen RA, Bunzow JR, Grandy DK, Langen H, Monsma FJ, Civelli O. Orphanin FQ: a neuropeptide that activates an opioidlike G-protein-coupled receptor. Science 1995;270:792–794.
80. Meunier J-C, Mollereau C, Toll L, Suaudeau C, Moisand C, Alvinerle P, Butour J-L, Guillemot J-C, Ferrara P, Monsarrat B, Mazarguil H, Vassart G, Parmentier M, Costentin J. Isolation and structure of the endogenous agonist of opioid receptor-like ORL1 receptor. Nature 1995;377:532–535.
81. Dhawan BN, Cesselin F, Raghubir R, Reisine T, Bradley PB, Portoghese PS, Hamon M. International union of pharmacology. XII. Classification of opioid receptors. Pharmacological Reviews 1996;48:567–592.
82. Opioid nomenclature proposal; revised IUPHAR opioid receptor nomenclature subcommittee proposal.International Narcotics Research Conference, Seattle, WA, 2000.
83. Cox BM. Opioid receptor-G protein interactions: acute and chronic effects of opioids. In: Herz A, Akil H, Simon EJ, editors. Opioids I. Volume 104/I, Berlin: Springer-Verlag; 1993. p 145–188.
84. Childers SR. Opioid receptor-coupled second messenger systems. In: Herz A, Akil H, Simon EJ, editors. Opioids I. Volume 104/I, Berlin: Springer-Verlag; 1993. p 189–216.
85. Barnard EA, Simon J. Opioid receptors. In: Hucho F, editor. Neurotransmitter Receptors. Volume 24, Amsterdam: Elsevier; 1993. p 297–323.
86. Loh HH, Smith AP. Molecular characterization of opioid receptors. Annu Rev Pharmacol Toxicol 1990;30:123–147.
87. Akil H, Owens C, Gutstein H, Taylor L, Curran E, Watson S. Endogenous opioids: overview and current issues. Drug Alcohol Depend 1998;51:127–140.
88. Yu VC, Sadee W. Phosphotidylinositol turnover in neuroblastoma cells: regulation by bradykinin, acetylcholine, but not μ- and δ-opioid receptors. Neuroscience Lett 1986;71: 219–223.
89. North RA. Opioid actions on membrane ion channels. In: Herz A, Akil H, Simon EJ, editors. Opioids I. Volume 104/I, Berlin: Springer-Verlag; 1993. p 773–797.
90. Porreca F, Burks TF. Supraspinal opioid receptors in antinociception. In: Herz A, Akil H, Simon EJ, editors. Opioids II. Volume 104/II, Berlin: Springer-Verlag; 1993. p 21–51.
91. Schmidhammer H, Burkard WP, Eggstein-Aeppli L, Smith CFC. Synthesis and biological evaluation of 14-alkoxymorphinans. 2. (−)-N-(Cyclopropylmethyl)-4,14-dimethoxymorphinan-6-one, a selective μ opioid receptor antagonist. J Med Chem 1989;32: 418–421.
92. Pelton JT, Kazmierski W, Gulya K, Yamamura HI, Hruby VJ. Design and synthesis of conformationally constrained somatostatin analogues with high potency and specificity for μ opioid receptors. J Med Chem 1986;29: 2370–2375.
93. Erspamer V, Melchiorri P, Falconieri-Erspamer G, Negri L, Corsi R, Severini C, Barra D, Simmaco M, Kreil G. Deltorphins: a family of naturally occurring peptides with high affinity and selectivity for δ opioid binding sites. Proc Natl Acad Sci USA 1989;86:5188–5192.
94. Calderon SN, Rothman RB, Porreca F, Flippen-Anderson JL, McNutt RW, Xu H, Smith LE, Bilsky EJ, Davis P, Rice KC. Probes for narcotic receptor mediated phenomena. 19. Synthesis of (+)-4-[(α R)-α-((2S,5R)-4-Allyl-2,5-dimethyl-1- piperazinyl)-3-methoxybenzyl]-N,N-diethylbenzamide (SNC 80): a highly selective, nonpeptide δ opioid receptor agonist. J Med Chem 1994;37:2125–2128.
95. Nagase HH, Wakita H, Kawai K, Endoh T, Matsura H, Tanaka C, Takezawa Y. Synthesis of non-peptidic delta opioid agonists and their structure–activity relationships. Jpn J Pharmacol 1994;64(Suppl I): 35.
96. Nagase H, Kawai K, Hayakawa J, Wakita H, Mizusuna A, Matsuura H, Tajima C,

Takezawa Y, Endoh T. Rational drug design and synthesis of a highly selective nonpeptide delta-opioid agonist, (4aS^*,12aR^*)-4a-(3-hydroxyphenyl)-2-methyl- 1,2,3,4,4a,5,12,12a-octahydropyrido[3,4-b]acridine (TAN-67). Chem Pharm Bull 1998;46:1695–1702.

97. Cotton R, Giles MG, Miller L, Shaw JS, Timms D. ICI 174864: a highly selective antagonist for the opioid δ-receptor. Eur J Pharmacol 1984;97:331–332.

98. Schiller PW, Nguyen TM-D, Weltrowska G, Wilkes BC, Marsden BJ, Lemieux C, Chung NN. Differential stereochemical requirements of μ vs δ opioid receptors for ligand binding and signal transduction: development of a class of potent and highly δ-selective peptide antagonists. Proc Natl Acad Sci USA 1992;89:11871–11875.

99. Schiller PW, Weltrowska G, Nguyen TM-D, Wilkes BC, Chung NN, Lemieux C. TIPP[Ψ]: a highly potent and stable pseudopeptide δ opioid receptor antagonist with extraordinary δ selectivity. J Med Chem 1993;36: 3182–3187.

100. Portoghese PS, Sultana M, Nagase H, Takemori AE. Application of the message-address concept in the design of highly potent and selective non-peptide δ opioid receptor antagonists. J Med Chem 1988;31:281–282.

101. Martin NA, Terruso MT, Prather PL. Agonist activity of the δ-antagonists TIPP and TIPP-Ψ in cellular models expressing endogenous or transfected δ-opioid receptors. J Pharmacol Exp Ther 2001;298:240–248.

102. Szmuszkovicz J, Von Voigtlander PF. Benzeneacetamide amines: structurally novel non-mu opioids. J Med Chem 1982;25:1125–1126.

103. Lahti RA, Mickelson MM, McCall JM, von Voigtlander PF. [^{3}H]U-69593, a highly selective ligand for the opioid κ receptor. Eur J Pharmacol 1985;109:281–284.

104. Halfpenny PR, Horwell DC, Hughes J, Hunter JC, Rees DC. Highly selective κ-opioid analgesics. 3. Synthesis and structure–activity relationships of novel N-[2-(1-pyrrolidinyl)-4- or -5-substituted-cyclohexyl]arylacetamide derivatives. J Med Chem 1990;33: 286–291.

105. Hunter JC, Leighton GE, Meecham KG, Boyle SJ, Horwell DC, Rees DC, Hughes J. CI-977, a novel and selective agonist for the κ-opioid receptor. Br J Pharmacol 1990;101: 183–189.

106. Portoghese PS, Lipkowski AW, Takemori AE. Bimorphinans as highly selective, potent κ opioid receptor antagonists. J Med Chem 1987;30:238–239.

107. Portoghese PS, Lipkowski AW, Takemori AE. Binaltorphimine and nor-binaltorphimine, potent and selective κ-opioid receptor antagonists. Life Sci 1987;40:1287–1292.

108. Jones RM, Hjorth SA, Schwartz TW, Portoghese PS. Mutational evidence for a common κ antagonist binding pocket in the wild-type κ and mutant μ[K303E] opioid receptors. J Med Chem 1998;41:4911–4914.

109. Stevens WC, Jones RM, Subramanian G, Metzger TG, Ferguson DM, Portoghese PS. Potent and selective indolomorphinan antagonists of the kappa-opioid receptors. J Med Chemi 2000;43:2759–2769.

110. Jones RM, Portoghese PS. 5′-Guanidinonaltrindole, a highly selective and potent kappa-opioid receptor antagonist. Eur J Pharmacol 2000;396:49–52.

111. Pert CB, Snyder SH. Opiate receptor: demonstration in nervous tissue. Sci 1973;179: 1011–1014.

112. Simon EJ, Hiller JM, Edelman I. Stereospecific binding of the potent narcotic analgesic [^{3}H]etorphine to rat-brain homogenate. Proc Natl Acad Sci USA 1973;70:1947–1949.

113. Terenius L. Stereospecific interaction between narcotic analgesics and a synaptic plasma membrane fraction of rat cerebral cortex. Acta Pharmacol Toxicol 1973;32: 317–320.

114. Gillan MGC, Kosterlitz HW. Spectrum of the μ-, δ - and κ-binding sites in homogenates of rat brain. Br J Pharmacol 1982;77:461–469.

115. Kosterlitz HW, Paterson SJ, Robson LE. Characterization of the k-subtype of the opiate receptor in the guinea pig brain. Br J Pharmacol 1981;73:939–949.

116. Robson LE, Foote RW, Maurer R, Kosterlitz HW. Opioid binding sites of the κ-type in guinea pig cerebellum. Neuroscience 1984; 12:621–627.

117. Meunier J-C, Kouakou Y, Puget A, Moisand C. Multiple opiate binding sites in the central nervous system of the rabbit. Mol Pharmacol 1983;24:23–29.

118. Yoburn BC, Lutfy K, Candido J. Species-differences in mu-opioid and delta-opioid receptors. Eur J Pharmacol 1991;193:105–108.

119. Blume AJ, Shorr J, Finberg JPM, Spector S. Proc Natl Acad Sci USA 1977;74:4927–4931.

120. Corbett AD, Paterson SJ, Kosterlitz HW. Selectivity of ligands for opioid receptors.

In: Herz A, Akil H, Simon EJ, editors. Opioids I. Volume 104/I, Berlin: Springer-Verlag; 1993. p 645–679.
121. Smith JAM, Leslie FM. Use of organ systems for opioid bioassay. In: Herz A, Akil H, Simon EJ, editors. Opioids I. Volume 104/I, Berlin: Springer-Verlag; 1993. p 53–78.
122. Leslie FM. Methods used for study of opioid receptors. Pharmacol Rev 1987;39:197–249.
123. Chavkin C, Goldstein A. Demonstration of a specific dynorphin receptor in guinea pig myenteric plexus. Nature 1981;291:591–593.
124. Ward SJ, Portoghese PS, Takemori AE. Improved assays for the assessment of κ and δ-properties of opioid ligands. Eur J Pharmacol 1982;85:163–170.
125. Oka T, Negishi K, Suda M, Matsumiya T, Inazu T, Ueki M. Rabbit vas deferens: a specific bioassay for opioid κ-receptor agonists. Eur J Pharmacol 1980;73:235–236.
126. McKnight AT, Corbett AD, Marcoli M, Kosterlitz HW. The opioid receptors in the hamster vas deferens are of the δ-type. Neuropharmacology 1985;24:1011–1017.
127. Lemaire S, Magnan J, Regoli D. Rat vas deferens: a specific bioassay for endogenous opioid peptides. Br Jo Pharmacol 1978;64:327–329.
128. Wüster M, Schulz R, Herz A. Specificity of opioids towards the μ, δ and ε-opiate receptors. Neurosci Lett 1978;15:193–198.
129. Schulz R, Faase E, Wüster M, Herz A. Selective receptors for β-endorphin on the rat vas deferens. Life Sci 1979;24:843–849.
130. Traynor JR, Nahorski SR. Modulation by μ-opioid agonists of guanosine-5′-O-(3-[^{35}S]thio)triphosphate binding to membranes from human neuroblastoma SH-SY5Y Cells. Mol Pharmacol 1995;47:848–854.
131. Selley DE, Sim LJ, Xizo R, Liu Q, Childers SR. mu-opioid receptor-stimulated guanosine-5′-O-(γ-thio)-triphosphate binding in rat thalamus and cultured cell lines: signal transduction mechanisms underlying agonist efficacy. Mol Pharmacol 1997;51:87–96.
132. Soderstrom K, Choi H, Aldrich JV, Murray TF. N-alkylated derivatives of [D-Pro^{10}]dynorphin A-(1-11) are high affinity partial agonists at the cloned rat κ-opioid receptor. Eur J Pharmacol 1997;338:191–197.
133. Dooley CT, Ny P, Bidlack JM, Houghten RA. Selective ligands for the μ, δ, and κ opioid receptors identified from a single mixture based tetrapeptide positional scanning combinatorial library. J Biol Chem 1998;273:18848–18856.
134. Cowen A. Animal models of pain. In: Sawynok J, Cowen A, editors. Novel Aspects of Pain Management: Opioids and Beyond. New York: Wiley-Liss; 1999. p 21–47.
135. Tjølsen A, Hole K. Animal models of analgesia. In: Dickenson A, Besson J-M, editors. The Pharmacology of Pain. Volume 130, Berlin: Springer; 1997. p 1–20.
136. Millan MJ. Kappa-opioid receptors and analgesia. Trends Pharmacol Sci 1990;11:70–76.
137. Millan MJ. Kappa-opioid receptor-mediated antinociception in the Rat. 1. Comparative actions of mu-opioids and kappa-opioids against noxious thermal, pressure and electrical stimuli. J Pharmacol Exp Ther 1989;251:334–341.
138. Heyman JS, Vaught JL, Raffa RB, Porreca F. Can supraspinal δ receptors mediate antinociception?. Trends Pharmacol Sci 1988;9:134–138.
139. Evans CJ, Keith DE Jr, Morrison H, Magendzo K, Edwards RH. Cloning of a delta opioid receptor by functional expression. Science 1992;258:1952–1955.
140. Kieffer BL, Befort K, Gaveriaux-Ruff C, Hirth CG. The δ-opioid receptor: isolation of a cDNA by expression cloning and pharmacological characterization. Proc Natl Acad Sci USA 1992;89:12048–12052.
141. Yasuda K, Raynor K, Kong H, Breder CD, Takeda J, Reisine T, Bell GI. Cloning and Functional Comparison of κ and δ opioid receptors from mouse brain. Proc Natl Acad Sci USA 1993;90:6736–6740.
142. Chen Y, Mestek A, Liu J, Hurley JA, Yu L. Molecular cloning and functional expression of a μ-opioid receptor from rat brain. Mol Pharmacol 1993;44:8–12.
143. Gaveriaux-Ruff C, Kieffer BL. Opioid receptors: gene structure and function. In: Stein C, editor. Opioids in Pain Control: Basic and Clinical Aspects. Cambridge: Cambridge University Press; 1999. p 1–20.
144. Kieffer BL. Molecular aspects of opioid receptors. In: Dickenson A, Besson J-M, editors. The Pharmacology of Pain. Volume 130, Berlin: Springer; 1997. p 281–303.
145. Uhl GR, Childers S, Pasternak G. An opiate-receptor gene family reunion. Trends Neurosci 1994;17:89–93.
146. Reisine T, Bell GI. Molecular biology of opioid receptors. Trends Neurosci 1993;16:506–510.

147. Kieffer BL. Opioids: first lessons from knockout mice. Trends Pharmacol Sci 1999;20:19–26.
148. Pasternak GW, Standifer KM. Mapping of opioid receptors using antisense oligodeoxynucleotides: correlating their molecular biology and pharmacology. Trends Pharmacol Sci 1995;16:344–350.
149. Chen Y, Mestek A, Liu J, Yu L. Molecular cloning of a rat κ opioid receptor reveals sequence similarities to the μ and δ opioid receptors. Biochem J 1993;295:625–628.
150. Chaturvedi K, Christoffers KH, Singh K, Howells RD. Structure and regulation of opioid receptors. Biopolymers 2000;55:334–346.
151. Law PY, Wong YH, Loh HH. Mutational analysis of the structure and function of opioid receptors. Biopolymers 1999;51:440–455.
152. Metzger TG, Ferguson DM. On the role of extracellular loops of opioid receptors in conferring ligand selectivity. FEBS Lett 1995;375:1–4.
153. Onogi T, Minami M, Katao Y, Nakagawa T, Aoki Y, Toya T, Katsumata S, Satoh M. DAMGO, a mu-opioid receptor selective agonist, distinguishes between mu- and delta-opioid receptors around their first extracellular loops. FEBS Lett 1995;357:93–97.
154. Fukuda K, Kato S, Mori K. Location of regions of the opioid receptor involved in selective agonist binding. J Biol Chem 1995;270:6702–6709.
155. Minami M, Nakagawa T, Seki T, Onogi T, Aoki Y, Katao Y, Katsumata S, Satoh M. A Single residue, Lys108, of the δ-opioid receptor prevents the μ-opioid-selective ligand [D-Ala^2,N-MePhe^4, Gly-ol^5]enkephalin from binding to the δ-opioid receptor. Mol Pharmacol 1996;50:1413–1422.
156. Minami M, Onogi T, Nakagawa T, Katao Y, Aoki Y, Katsumata S, Satoh M. DAMGO, a mu-opioid receptor selective ligand, distinguishes between mu- and kappa-opioid receptors at a different region from that for the distinction between mu- and delta-opioid receptors. FEBS Lett 1995;364:23–27.
157. Xue J-C, Chen C, Zhu J, Kunapuli SP, de Riel KJ, Yu L, Liu-Chen L-Y. The third extracellular loop of the μ opioid receptor is important for agonist selectivity. J Biol Chem 1995;270:12977–12979.
158. Seki T, Minami M, Nakagawa T, Ienaga Y, Morisada A, Satoh M. DAMGO recognized four residues in the third extracellular loop to discriminate between μ- and κ-opioid receptors. Eur J Pharmacol 1998;350:301–310.
159. Varga E, Li X, Stropova D, Zalewska T, Landsman RS, Knapp RJ, Malatynska E, Kawai K, Mizusura A, Nagase H, Calderon SN, Rice R, Hruby VJ, Roeske WR, Yamamura HI. The third extracellular loop of the human δ-opioid receptor determines selectivity of δ-opioid agonists. Mol Pharmacol 1996;50:1619–1624.
160. Valiquette M, Vu HK, Yue SY, Wahlestedt C, Walker P. Involvement of Tyr-284, Val-296, and Val-297 of the human δ-opioid receptor in binding of δ-selective ligands. J Biol Chem 1996;271:18789–18796.
161. Wang JB, Johnson PS, Wu JM, Wang WF, Uhl GR. Human κ opiate receptor second extracellular loop elevates dynorphin's affinity for human μ/κ chimeras. J Biol Chem 1994;269:25966–25969.
162. Xue JC, Chen CG, Zhu JM, Kunapuli S, Deriel JK, Yu L, Liu-Chen L-Y. Differential binding domains of peptide and non-peptide ligands in the cloned rat κ opioid receptor. J Biol Chem 1994;269:30195–30199.
163. Meng F, Hoversten MT, Thompson RC, Taylor L, Watson SJ, Akil H. A chimeric study of the molecular basis of affinity and selectivity of the kappa and the delta opioid receptors: potential role of extracellular domains. J Biol Chem 1995;270:12730–12736.
164. Ferguson DM, Kramer S, Metzger TG, Law PY, Portoghese PS. Isosteric replacement of acidic with neutral residues in extracellular loop-2 of the κ-opioid receptor does not affect dynorphin A(1-13) affinity and function. J Med Chem 2000;43:1251–1252.
165. Kong HY, Raynor K, Yano H, Takeda J, Bell GI, Reisine T. Agonists and antagonists bind to different domains of the cloned κ opioid receptor. Proc Natl Acad Sci USA 1994;91:8042–8046.
166. Kong H, Raynor K, Yasuda K, Moe ST, Portoghese PS, Bell GI, Reisine T. A single residue, aspartic acid 95, in the δ opioid receptor specifies selective high affinity agonist binding. J Biol Chem 1993;268:23055–23058.
167. Surratt CK, Johnson PS, Moriwaki A, Seidleck BK, Blaschak CJ, Wang JB, Uhl GR. μ-Opiate receptor: charged transmembrane domain amino acids are critical for agonist recognition and intrinsic activity. J Biol Chem 1994;269:20548–20553.
168. Befort K, Tabbara L, Bausch S, Chavkin C, Evans CE, Kieffer BL. The conserved aspar-

tate residue in the third putative transmembrane domain of the δ-opioid receptor is not the anionic counterpart for cationic opiate binding but is a constituent of the receptor binding site. Mol Pharmacol 1996;49: 216–233.

169. Hjorth SA, Thirstrup K, Grandy DK, Schwartz TW. Analysis of selective binding epitopes for the kappa-opioid receptor antagonist nor-binaltorphimine. Mol Pharmacol 1995;47:1089–1094.

170. Meng F, Taylor LP, Hoversten MT, Ueda Y, Ardati A, Reinscheid RK, Monsma FJ, Watson SJ, Civelli O, Akil H. Moving from the orphanin FQ receptor to an opioid receptor using four point mutations. J Biol Chem 1996;71:32016–32020.

171. Chen C, Yin J, de Riel JK, DesJarlais RL, Raveglia LF, Zhu J, Liu-Chen L-Y. Determination of the amino acid residue involved in [^{3}H]β-funaltrexamine covalent binding in the cloned rat μ opioid receptor. J Biol Chem 1996;271:21422–21429.

172. Palczewski K, Kumasaka T, Hori T, Behnke CA, Motoshima H, Fox BA, Le Trong I, Teller DC, Okada T, Stenkamp RE, Yamamoto M, Miyano M. Crystal structure of rhodopsin: a G-protein-coupled receptor. Science 2000; 289:739–745.

173. Cappelli A, Anzini M, Vomero S, Menziani MC, De Benedetti PG, Sbacchi M, Clarke GD, Mennuni L. Synthesis, biological evaluation, and quantitative receptor docking simulations of 2-[(acylamino)ethyl]-1,4-benzodiazepines as novel tifluadom-like ligands with high affinity and selectivity for κ-opioid receptors. J Med Chem 1996;39:860–872.

174. Metzger TG, Paterlini MG, Portoghese PS, Ferguson DM. Application of the message–address concept to the docking of naltrexone and selective naltrexone-derived opioid antagonists into opioid receptors models. Neurochem Res 1996;21:1287–1294.

175. Paterlini MG, Portoghese PS, Ferguson DM. Molecular simulation of dynorphin A-(1-10) binding to extracellular loop 2 of the κ-opioid receptor. A model for receptor activation. J Med Chem 1997;40:3254–3262.

176. Pogozheva ID, Lomize AL, Mosberg HI. Opioid receptor three-dimensional structures from distance geometry calculations with hydrogen bonding constraints. Biophys J 1998; 75:612–634.

177. Subramanian G, Paterlini MG, Larson DL, Portoghese PS, Ferguson DM. Conformational analysis and automated receptor docking of selective arylacetamide-based κ-opioid agonists. J Med Chem 1998;41:4777–4789.

178. Filizola M, Carteni-Farina M, Perez JJ. Molecular modeling study of the differential ligand–receptor interaction at the mu, delta and kappa opioid receptors. J Comput Aided Mol Des 1999;13:397–407.

179. Filizola M, Laakkonen L, Loew GH. 3D modeling, ligand binding and activation studies of the cloned mouse delta, mu, and kappa opioid receptors. Protein Eng 1999;12:927–942.

180. Lavecchia A, Greco G, Novellino E, Vittorio F, Ronsisvalle G. Modeling of k-opioid receptor/agonist interactions using pharmacophore-based and docking simulations. J Med Chem 2000;43:2124–2134.

181. Subramanian G, Paterlini MG, Portoghese PS, Ferguson DM. Molecular docking reveals a novel binding site model for fentanyl at the μ-opioid receptor. J Med Chem 2000;43: 381–391.

182. Podlogar BL, Poda GI, Demeter DA, Zhang SP, Carson JR, Neilson LA, Reitz AB, Ferguson DM. Synthesis and evaluation of 4-(N,N-diarylamino)piperidines with high selectivity to the delta-opioid receptor: a combined 3D-QSAR and ligand docking study. Drug Des Discov 2000;17:34–50.

183. Topham CM, Mouledous L, Poda G, Maigret B, Meunier JC. Molecular modeling of the ORL1 receptor and its complex with nociceptin. Protein Eng 1998;11:1163–1179.

184. Huang P, Li J, Chen C, Visiers I, Weinstein H, Liu-Chen LY. Functional role of a conserved motif in TM6 of the rat μ opioid receptor: constitutively active and inactive receptors result from substitutions of Thr6.34(279) with Lys and Asp. Biochemistry 2001;40: 13501–13509.

185. Zaki PA, Bilsky EJ, Vanderah TW, Lai J, Evans CJ, Porreca F. Opioid receptor types and subtypes: the delta receptor as a model. Annu Rev Pharmacol Toxicol 1996;36: 379–401.

186. Traynor JR, Elliott J. δ-Opioid receptor subtypes and cross-talk with μ-receptors. Trends Pharmacol Sci 1993;14:84–86.

187. Portoghese PS, Sultana M, Nagase H, Takemori AE. A highly selective δ$_1$-opioid receptor antagonist: 7-benzylidenenaltrexone. Eur J Pharmacol 1992;218:195–196.

188. Bowen WD, Hellewell SB, Kelemen M, Huey R, Stewart D. Affinity Labeling of δ-opiate receptors using [D-Ala^2, Leu^5, Cys^6]enkepha-

lin. Covalent attachment via thiol-disulfide exchange. J Biol Chem 1987;262: 13434–13439.
189. Portoghese PS, Nagase H, Maloneyhuss KE, Lin CE, Takemori AE. Role of spacer and address components in peptidomimetic δ opioid receptor antagonists related to naltrindole. J Med Chem 1991;34:1715–1720.
190. Portoghese PS, Sultana M, Takemori AE. Naltrindole 5′-isothiocyanate: a nonequilibrium, highly selective delta-opioid receptor antagonist. J Med Chem 1990;33:1547–1548.
191. Rothman RB, Holaday JW, Porreca F. Allosteric coupling among opioid receptors: evidence for an opioid receptor complex. In: Herz A, Akil H, Simon EJ, editors. Opioids I. Volume 104/I, Berlin: Springer-Verlag; 1993. p 217–237.
192. Rice KC, Jacobson AE, Burke TRJ, Bajwa BS, Streaty RA, Klee WA. Irreversible ligands with high selectivity toward δ or μ opiate receptors. Science 1983;220:314–316.
193. Kim C-H, Rothman RB, Jacobson AE, Mattson MV, Bykov V, Streaty RA, Klee WA, George C, Long JB, Rice KC. Probes for narcotic receptor mediated phenomena. 15. (3S,4S)-(+)-trans-3-Methylfentanyl isothiocyanate, a potent site-directed acylating agent for the δ opioid receptors in vitro. J Med Chem 1989;32:1392–1398.
194. Rothman RB, Mahboubi A, Bykov V, Kim CH, Jacobson AE, Rice KC. Probing the opioid receptor complex with (+)-trans-superfit. 1. Evidence that (D-Pen2,D-Pen5)enkephalin interacts with high-affinity at the delta-Cx binding-site. Peptides 1991;12:359–364.
195. Rothman RB, Long JB, Bykov V, Jacobson AE, Rice KC, Holaday JW. β-FNA binds irreversibly to the opiate receptor complex: in vivo and in vitro evidence. J Pharmacol Exp Ther 1988;247:405–416.
196. Xu H, Partilla JS, de Costa BR, Rice KC, Rothman RB. Differential binding of opioid peptides and other drugs to two subtypes of opioid δ_{ncx} binding sites in mouse brain: further evidence for δ receptor heterogeneity. Peptides 1993;14:893–907.
197. Cha XY, Xu H, Rice KC, Porreca F, Lai J, Ananthan S, Rothman RB. Opioid peptide receptor studies. 1. Identification of a novel δ-opioid receptor binding site in rat brain membranes. Peptides 1995;16:191–198.
198. Raynor K, Kong HY, Chen Y, Yasuda K, Yu L, Bell GI, Reisine T. Pharmacological characterization of the cloned κ-, δ-, and μ-opioid receptors. Mol Pharmacol 1994;45:330–334.
199. Negri L, Lattanzi R, Borsodi A, Toth G, Salvadori S. Differential knockdown of delta-opioid receptor subtypes in the rat brain by antisense oligodeoxynucleotides targeting mRNA. Antisense Nucleic Acid Drug Dev 1999;9:203–211.
200. Zhu Y, King MA, Schuller AGP, Nitsche JF, Reidl M, Elde RP, Unterwald E, Pasternak GW, Pintar JE. Retention of supraspinal delta-like analgesia and loss of morphine tolerance in δ opioid receptor knockout mice. Neuron 1999;24:243–252.
201. Wolozin BL, Pasternak GW. Classification of Multiple morphine and enkephalin binding sites in the central nervous system. Proc Natl Acad Sci USA 1981;78:6181–6185.
202. Pasternak GW, Wood PJ. Minireview. Multiple mu opiate receptors. Life Sci 1986;38: 1889–1898.
203. Pasternak GW, Childers SR, Snyder SH. Naloxazone, a long-acting opiate antagonist: effects on analgesia in intact animals and on opiate receptor binding in vitro. J Pharmacol Exp Ther 1980;214:455–462.
204. Hahn EF, Carroll-Buatti M, Pasternak GW. Irreversible opiate aonists and antagonists: the 14-hydroxydihydromorphinone azines. J Neurosci 1982;2:572–576.
205. Pasternak GW. Insights into mu opioid pharmacology. The role of mu opioid receptor subtypes. Life Sci 2001;68:2213–2219.
206. Pasternak GW. Incomplete cross tolerance and multiple mu opioid peptide receptors. Trends Pharmacol Sci 2001;22:67–60.
207. Akil H, Watson SJ. Cloning of kappa opioid receptors: functional significance and future directions. In: Bloom FE, editor. Neuroscience: from the Molecular to the Cognitive. Volume 100, Amsterdam: Elsevier Science Publ B.V. 1994. p 81–86.
208. Zukin RS, Eghbali M, Olive D, Unterwald EM, Tempel A. Characterization and Visualization of rat and guinea pig brain κ opioid receptors: evidence for κ_1 and κ_2 receptors. Proc Natl Acad Sci USA 1988;85:4061–4065.
209. Simonin F, Valverde O, Smadja C, Slowe S, Kitchen I, Dierich A, Le Meur M, Roques BP, Maldonado R, Kieffer BL. Disruption of the kappa-opioid receptor gene in mice enhances sensitivity to chemical visceral pain, impairs pharmacological actions of the selective kappa-agonist U-50,488H and attenuates morphine withdrawal. EMBO J 1998;17:886–897.

210. Simonin F, Slowe S, Becker JAJ, Matthes HWD, Filliol D, Chluba J, Kitchen I, Keiffer BL. Analysis of [³H]bremazocine binding in single and combinatorial opioid receptor knockout mice. Eur J Pharmacol 2001;414: 189–195.

211. Clark JA, Liu L, Price M, Hersh B, Edelson M, Pasternak GW. Kappa-opiate receptor multiplicity: evidence for 2 U50,488-sensitive kappa-1 subtypes and a novel kappa-3 subtype. J Pharmacol Exp Ther 1989;251:461–468.

212. Quock R, Burkey T, Varga E, Hosohata Y, Hosohata K, Cowell S, Slate C, Ehlert F, Roeske W, Yamamura H. The delta-opioid receptor: molecular pharmacology, signal transduction, and the determination of drug efficacy. Pharmacol Rev 1999;51:503–532.

213. Dondio G, Ronzani S, Petrillo P. Non-peptide δ opioid agonists and antagonists. Expert Opin Ther Pat 1997;7:1075–1098.

214. Dondio G, Ronzani S, Petrillo P. Non-peptide δ opioid agonists and antagonists (Part II). Exp Opin Ther Patents 1999;9:353–374.

215. Negus SS, Gatch MB, Mello NK, Zhang X, Rice K. Behavioral effects of the delta-selective opioid agonist SNC80 and related compounds in rhesus monkeys. J Pharmacol Exp Ther 1998;286:362–375.

216. Coop A, Rice KC. Role of δ-opioid receptors in biological processes. Drug News Perspect 2000;13:481–487.

217. Porreca F, Takemori AE, Sultana M, Portoghese PS, Bowen WD, Mosberg HI. Modulation of mu-mediated antinociception in the mouse involves opioid delta-2 receptors. J Pharmacol Exp Ther 1992;263:147–52.

218. He L, Lee NM. Delta opioid receptor enhancement of mu opioid receptor-induced antinociception in spinal cord. J Pharmacol Exp Ther 1998;285:1181–1186.

219. Abdelhamid EE, Sultana M, Portoghese PS, Takemori AE. Selective blockage of delta-opioid receptors prevents the development of morphine-tolerance and dependence in mice. J Pharmacol Exp Ther 1991;258: 299–303.

220. Fundytus ME, Schiller PW, Shapiro M, Weltrowska G, Coderre TJ. Attenuation of morphine tolerance and dependence with the highly selective delta-opioid receptor antagonist TIPP[psi]. Eur J Pharmacol 1995;286: 105–8.

221. Hepburn MJ, Little PJ, Gingras J, Kuhn CM. Differential effects of naltrindole on morphine-induced tolerance and physical dependence in rats. J Pharmacol Exp Ther 1997;281:1350–1356.

222. Su YF, McNutt RW, Chang KJ. Delta-opioid ligands reverse alfentanil-induced respiratory depression but not antinociception. J Pharmacol Exp Ther 1998;287:815–823.

223. Schiller PW, Fundytus ME, Merovitz L, Weltrowska G, Nguyen TM, Lemieux C, Chung NN, Coderre TJ. The opioid μ agonist/δ antagonist DIPP-NH$_2$[Ψ] produces a potent analgesic effect, no physical dependence, and less tolerance than morphine in rats. J Med Chem 1999;42:3520–3526.

224. Ananthan S, Kezar HS 3rd, Carter RL, Saini SK, Rice KC, Wells JL, Davis P, Xu H, Dersch CM, Bilsky EJ, Porreca F, Rothman RB. Synthesis, opioid receptor binding, and biological activities of naltrexone-derived pyrido- and pyrimidomorphinans. J Med Chem 1999;42:3527–3538.

225. Suzuki T, Tsuji M, Ikeda H, Misawa M, Narita M, Tseng LF. Antisense oligodeoxynucleotide to delta opioid receptors blocks cocaine-induced place preference in mice. Life Sci 1997;60:283–288.

226. Patterson AB, Gordon FJ, Holtzman SG, Ananthan S, Kezar HS 3rd, Carter RL, Saini SK, Rice KC, Wells JL, Davis P, Xu H, Dersch CM, Bilsky EJ, Porreca F, Rothman RB. Naltrindole, a selective delta-opioid receptor antagonist, potentiates the lethal effects of cocaine by a central mechanism of action. Eur J Pharmacol 1997;333:47–54.

227. Hayashi T, Hirata H, Asanuma M, Ladenheim B, Tsao LI, Cadet JL, Su TP. Delta opioid peptide [D-Ala2, D-Leu5]enkephalin causes a near complete blockade of the neuronal damage caused by a single high dose of methamphetamine: examining the role of p53. Synapse 2001;39:305–312.

228. Tsao LI, Hayashi T, Su TP. Blockade of dopamine transporter and tyrosine hydroxylase activity loss by [D-Ala(2), D-Leu(5)]enkephalin in methamphetamine-treated CD-1 mice. Eur J Pharmacol 2000;404:89–93.

229. Herz A. Endogenous opioid systems and alcohol addiction. Psychopharmacology 1997; 129:99–111.

230. Town T, Schinka J, Tan J, Mullan M, Herz A. The opioid receptor system and alcoholism: a genetic perspective. Eur J Pharmacol 2000;410:243–248.

231. Broom DC, Jutkiewicz EM, Folk JE, Traynor JR, Rice KE, Woods JH. Non-peptidic δ-opioid receptor agonists reduce immobility in the

forced swim assay in rats. Neuropsychopharmacology, 2002;26:744–755.
232. Broom DC, Guo L, Coop A, Husbands SM, Lewis JW, Woods JH, Traynor JR. BU48: a novel buprenorphine analog that exhibits δ-opioid-mediated convulsions but not δ-opioid-mediated antinociception in mice. J Pharmacol Exp Ther 2000;294:1195–1200.
233. McCarthy L, Wetzel M, Sliker JK, Eisenstein TK, Rogers TJ, Su YF, McNutt RW, Chang KJ. Opioids, opioid receptors, and the immune response. Drug Alcohol Depend 2001; 62:111–123.
234. Gaveriaux-Ruff C, Filliol D, Simonin F, Matthes HW, Kieffer BL. Immunosuppression by delta-opioid antagonist naltrindole: delta- and triple mu/delta/kappa-opioid receptor knockout mice reveal a nonopioid activity. J Pharmacol Exp Ther 2001;298: 1193–8.
235. Wilson JL, Nayanar V, Walker JS. The site of anti-arthritic action of the kappa-opioid, U50488H, in adjuvant arthritis: importance of local administration. Br J Pharmacol 1996;118:1754–1760.
236. Binder W, Walker JS. Effect of the peripherally selective kappa-opioid agonist, asimadoline, on adjuvant arthritis. Br J Pharmacol 1998;124:647–654.
237. Ko M-C, Butelman ER, Woods JH. Activation of peripheral κ opioid receptors inhibits capsaicin-induced thermal nociception in rhesus monkeys. J Pharmacol Exp Ther 1999;289: 378–385.
238. Ko MCH, Willmont KJ, Burritt A, Hruby VJ, Woods JH. Local inhibitory effects of dynorphin A-(1-17) on capsaicin-induced thermal allodynia in rhesus monkeys. Eur J Pharmacol 2000;402:69–76.
239. Gebhart GF, Segupta JN, Su X. Opioids in visceral pain. In: Stein C, editor. Opioids in pain control: basic and clinical aspects. Cambridge: Cambridge University Press; 1999. p 325–334.
240. Gebhart GF, Su X, Joshi S, Ozaki N, Sengupta JN. Peripheral opioid modulation of visceral pain. Ann NY Acad Sci 2000;909:41–50.
241. Pan ZZ. μ-Opposing actions of the κ-opioid receptors. Trends Pharmacol Sci 1998;19: 94–98.
242. Itoh J, Ukai M, Kameyama T. Dynorphin A-(1-13) potently improves the impairment of spontaneous alternation performance induced by the mu-selective opioid receptor agonist DAMGO in mice. J Pharmacol Exp Ther 1994;269:15–21.
243. Rothman RB, Gorelick DA, Heishman SJ, Eichmiller PR, Hill BH, Norbeck J, Liberto JG. An open-label study of a functional opioid K antagonist in the treatment of opioid dependence. J Subst Abuse Treat 2000;18: 277–281.
244. Mello N, Negus SS. Interactions between kappa opioid agonists and cocaine. Ann NY Acad Sci 2000;909:104–132.
245. Schenk S, Partridge B, Shippenberg TS. U69,593, a kappa-opioid agonist, decreases cocaine self-administration and decreases cocaine-produced drug-seeking. Psychopharmacology 1999;144:339–346.
246. Mello N, Negus SS. Effects of kappa opioid agonists on cocaine- and food-maintained responding by rhesus monkeys. J Pharmacol Exp Ther 1998;286:812–814.
247. Mello N, Negus SS. Effects of kappa opioid agonists on cocaine self-administration by rhesus monkeys. J Pharmacol Exp Ther 1997;282:44–55.
248. Glick SD, Maisonneuve IM, Raucci J, Archer S. Kappa-opioid inhibition of morphine and cocaine self-administration. Brain Res 1995;681:147–152.
249. Walsh SL, Geter-Douglas B, Strain EC, Bigelow GE. Enadoline and butorphanol: evaluation of κ-agonists on cocaine pharmacodynamics and cocaine self-adminstration in humans. J Pharmacol Exp Ther 2001;299: 147–158.
250. Tortella FC, Decoster MA. Kappa opioids: therapeutic considerations in epilepsy and CNS injury. Clin Neuropharmacol 1994;17: 403–416.
251. Baskin DS, Widmayer MA, Browning JL, Heizer ML, Schmidt WK. Evaluation of delayed treatment of focal cerebral ischemia with three selective κ-opioid agonists in cats. Stroke 1994;25:2047–2053.
252. Chao CC, Gekker G, Hu S, Sheng WS, Shark KB, Bu D-F, Archer S, Bidlack JM, Peterson PK. κ-Opioid receptors in human microglia downregulate human immunodeficiency virus 1 expression. Proc Natl Acad Sci USA 1996;93:8051–8056.
253. Peterson PK, Gekker G, Lokensgard JR, Bidlack JM, Chang AC, Fang X, Portoghese PS. Kappa-opioid receptor agonist suppression of HIV-1 expression in CD4+ lymphocytes. Biochem Pharmacol 2001;61:1145–1151.
254. Hughes J, Smith TW, Kosterlitz HW, Fothergill LA, Morgan BA, Morris HR. Identification of two related pentapeptides from the

brain with potent opiate agonist activity. Nature 1975;258:577–579.
255. Teschemacher H, Opheim KE, Cox BM, Goldstein A. A peptide-like substance from pituitary that acts like morphine. 1. Isolation. Life Sci 1975;16:1771–1776.
256. Cox BM, Opheim KE, Teschemacher H, Goldstein A. A Peptide-like substance from pituitary that acts like morphine. 2. Purification and properties. Life Sci 1975;16:1777–1782.
257. Li CH, Chung D. Isolation and structure of an untriakontapeptides with opiate activity from camel pituitary galnds. Proc Natl Acad Sci USA 1976;73:1145–1148.
258. Chavkin C, Goldstein A. A specific receptor for the opioid peptide dynorphin: structure–activity relationships. Proc Natl Acad Sci USA 1981;78:6543–6547.
259. Zadina JE, Hackler L, Ge L-J, Kastin AJ. A potent and selective endogenous agonist for the µ-opiate receptor. Nature 1997;386:499–502.
260. Horvath G. Endomorphin-1 and endomorphin-2: pharmacology of the selective endogenous µ-opioid receptor agonists. Pharmacol Ther 2000;88:437–463.
261. Höllt V. Opioid peptide processing and receptor selectivity. Ann Rev Pharmacol Toxicol 1986;26:59–77.
262. Young E, Bronstein D, Akil H. Proopiomelanocortin biosynthesis, processing and secretion: functional implications. In: Herz A, Akil H, Simon EJ, editors. Opioids I. Volume 104/I, Berlin: Springer-Verlag; 1993. p 393–421.
263. Rossier J. Biosynthesis of enkephalins and proenkephalin-derived peptides. In: Herz A, Akil H, Simon EJ, editors. Opioids I. Volume 104/I, Berlin: Springer-Verlag; 1993. p 423–447.
264. Day R, Trujillo KA, Akil H. Prodynorphin biosynthesis and posttranslational processing. In: Herz A, Akil H, Simon EJ, editors. Opioids I. Volume 104/I, Berlin: Springer-Verlag; 1993. p 449–470.
265. Florin S, Suaudeau C, Meunier JC, Costentin J. Orphan neuropeptide NocII, a putative pronociceptin maturation product, stimulates locomotion in mice. Neuroreport 1997;8:705–7.
266. Okuda-Ashitaka E, Minami T, Tachibana S, Yoshihara Y, Nishiuchi Y, Kimura T, Ito S. Nocistatin, a peptide that blocks nociceptin action in pain transmission. Nature 1998;392:286–289.

267. Gulland JM, Robinson R. The constitution of codeine and thebaine. Mem Proc Manchester Lit Phil Soc 1925;69:79.
268. Gates M, Tschudi G. The synthesis of morphine. J Am Chem Soc 1952;74:1109–1110.
269. Gates M, Tschudi G. The synthesis of morphine. J Am Chem Soc 1956;78:1380–1393.
270. Rapoport H, Lavigne JB. Stereochemical studies in the morphine series. The relative configuration at carbons thirteen and fourteen. J Am Chem Soc 1953;75:5329–5334.
271. Mackay M, Hodgkin DC. A crystallographic examination of the structure of morphine. J Chem Soc 1955; 3261–3267.
272. Bentley KW, Cardwell HME. The morphine-thebaine group of alkaloids. Part V. The Absolute stereochemistry of the morphine, benzylisoquinoline, aporphine, and tetrahydroberberine alkaloids. J Chem Soc 1955; 3252–3260.
273. Kalvoda J, Buchschacher P, Jeger O. Über die absolute Konfiguration des Morphins und verwandter Alkaloide. Helv Chim Acta 1955;38:1847–1856.
274. Kartha G, Ahmed FR, Barnes WH. Refinement of the crystal structure of codeine hydrobromide dihydrate, and establishment of the absolute configuration of the codeine molecule. Acta Crystallogr 1962;15:326–333.
275. Wright CRA. On the action of organic acids and their anhydrides on the natural alkaloids. J Chem Soc 1874;27:1031.
276. Pohl J. Uber das N-allylnorcodein, einen antagonisten des morphins. Z Exp Pathol Ther 1915;17:370.
277. Casy AF, Parfitt RT. Opioid analgesics: chemistry and receptors. New York: Plenum Press; 1986. p 518.
278. Lenz GR, Evans SM, Walters DE, Hopfinger AJ. Opiates. Orlando FL: Academic Press, Inc. 1986. p 560.
279. Archer S. Chemistry of nonpeptide opioids. In: Herz A, Akil H,, Simon EJ, editors. Opioids I. Volume 104/I, Berlin: Springer-Verlag; 1993. p 241–277.
280. Casy AF, Dewar GH. Opioid ligands. Part 1. In: The Steric Factor in Medicinal Chemistry. Dissymmetric Probes of Pharmacological Receptors. New York: Plenum Press; 1993. p 429–501.
281. Casy AF, Dewar GH. Opioid ligands. Part 2. In: The Steric Factor in Medicinal Chemistry. Dissymmetric Probes of Pharmacological

Receptors. New York: Plenum Press; 1993. p 503–548.

282. Reitz AB, Jetter MC, Wild KD, Raffa R. Centrally acting analgesics. Ann Rep Med Chem 1995;30:11–20.

283. Aldrich JV. Analgesics. In: Wolff ME, editor. Burger's Medicinal Chemistry and Drug Discovery. Therapeutic Agents. 5th ed. Volume 3, New York: John Wiley & Sons, Inc. 1996. p 321–441.

284. Szmuszkovicz J. U-50,488 and the κ receptor: a personalized account covering the period 1973 to 1990. Prog Drug Res. 1999;52:167–195.

285. Szmuszkovicz J. U-50,488 and the κ receptor. Part II. 1991 to 1998. Prog Drug Res 2000;53: 1–51.

286. Schmidhammer H. Opioid receptor antagonists. Prog Med Chem 1998;35:83–132.

287. Thorpe DH. Opiate structure and activity: a guide to understanding the receptor. Anesth Anal 1984;63:143–151.

288. Lenz GR, Evans SM, Walters DE, Hopfinger AJ. Morphine and its analogs. In: Opiates. Orlando FL: Academic Press, Inc. 1986. p 1–28.

289. Fairbairn JW, Helliwell K. *Papaver bracteum* Lindley: thebaine content in relation to plant development. J Pharm Pharmacol 1977;29: 65–69.

290. Stefano GB, Scharrer B. Endogenous morphine and related opiates, a new class of chemical messengers. Adv Neuroimmunol 1994;4:57–67.

291. Spector S, Donnerer J. Presence of endogenous opiate alkaloids in mammalian tissues. In: Herz A, Akil H, Simon EJ, editors. Opioids I. Volume 104/I, Berlin: Springer-Verlag; 1993. p 295–304.

292. Stefano GB, Gouman Y, Casares F, Cadet P, Fricchione GI, Rialas C, Peter D, Sonetti D, Guarna M, Welters ID, Bianchi C. Endogenous morphine. Trends Neurosci 2000;23: 436–442.

293. Leung MK. Biochemical isolation and detection of morphine. Adv Neuroimmunol 1994;4:93–103.

294. Kosterlitz HW. Biosynthesis of morphine in the animal kingdom. Nature 1987;330:606.

295. Lenz GR, Evans SM, Walters DE, Hopfinger AJ. Physical chemistry, molecular modeling, and QSAR analysis of the morphine, morphinans, and benzomorphan analgesics. In: Opiates. Orlando FL: Academic Press, Inc. 1986. p 166–187.

296. Lenz GR, Evans SM, Walters DE, Hopfinger AJ. Synthesis and structure–activity relationships of morphine, codeine, and related alkaloids. In: Opiates. Orlando FL: Academic Press, Inc. 1986. p 45–165.

297. Casy AF, Parfitt RT. 4,5-Epoxymorphinans. In: Opioid Analgesics: Chemistry and Receptors. Plenum Press: New York; 1986. p 9–104.

298. Reden J, Reich MF, Rice KC, Jacobson AE, Brossi A, Streaty RA, Klee WA. Deoxymorphines: role of the phenolic hydroxyl in antinociception and opiate receptor interactions. J Med Chem 1979;22:256–259.

299. Olendorf WH, Hyman S, Braun L, Olendorf SZ. Blood–brain barrier: penetration of morphine, codeine, heroin, and methadone after carotid injection. Science 1972;178:984–986.

300. Kugita H, Takeda M, Inoue H. Synthesis of B/C *trans*-fused morphine structures. V. Pharmacological summary of trans-morphine derivatives and an improved synthesis of trans-codeine. J Med Chem 1970;13:973–975.

301. Botros S, Lipkowski AW, Larson DL, Stark PA, Takemori AE, Portoghese PS. Opioid agonist and antagonist activities of peripherally selective derivatives of naltrexamine and oxymorphamine. J Med Chem 1989;32: 2068–2071.

302. Lenz GR, Evans SM, Walters DE, Hopfinger AJ. Synthesis and structure–activity relationships of morphine, codeine, and related alkaloids. In: Opiates. Orlando FL: Academic Press, Inc. 1986. p 72–79.

303. Lasagna L, Beecher HK. J Pharmacol Exp Ther 1954;112:356.

304. Bowdle TA, Nelson WL. Clinical pharmacology of partial agonist, mixed agonist-antagonist, and antagonist opioids. In: Bowdle TA, Horita A, Kharasch ED, editors. The Pharmacological Basis of Anesthesiology: Basic Science and Practical Applications. New York: Churchill Livingstone; 1994. p 121–148.

305. Wang D, Raehal KM, Bilsky EJ, Sadée W. Inverse agonists and neutral antagonists at μ opioid receptor (MOR): possible role of basal receptor signaling in narcotic dependence. J Neurochem 2001;77:1590–1600.

306. Ostrowski NL, Burke TR Jr, Rice KC, Pert A, Pert CB. The Pattern of [^{3}H]cyclofoxy retention in rat brain after *in vivo* injection corresponds to the *in vitro* opiate receptor distribution. Brain Res 1987;402:257–286.

307. Nagase H, Hayakawa J, Kawamura K, Kawai K, Takezawa Y, Matsura H, Tajima C, Endo T. Discovery of a Structurally novel opioid

κ-agonist derived from 4,5-epoxymorphinan. Chem Pharm Bull 1998;46:366–369.
308. Seki T, Awamura S, Kimura C, Ide S, Sakano K, Minami M, Nagase H, Satoh M. Pharmacological properties of TRK-820 on cloned μ-, δ- and κ-opioid receptors and nociceptin receptor. Eur J Pharmacol 1999;376:159–167.
309. Kobylecki RJ, Lane AC, Smith CFC, Wakelin LPG, Cruse WBT, Egert E, Kennard O. N-Methylnalorphine: definition of N-allyl conformation for antagonism at the opiate receptor. J Med Chem 1982;25:1278–1280.
310. Bianchetti A, Nisato D, Sacilotto R, Dragonetti M, Picerno N, Tarantino A, Manara L, Angel LM, Simon EJ. Quaternary derivatives of narcotic antagonists: stereochemical requirements at the chiral nitrogen for in $vito$ and in $vivo$ activity. Life Sci 1983;33(Suppl 1): 415–418.
311. Brown DR, Goldberg LI. Review. The use of quaternary narcotic antagonists in opiate research. Neuropharmacology 1986;24: 181–191.
312. Valentino RJ, Herling S, Woods JH, Medzihradsky F, Merz H. Quaternary naltrexone: evidence for the central mediation of discriminative stimulus effects of narcotic agonists and antagonists. J Pharmacol Exp Ther 1981;217:652–659.
313. Valentino RJ, Katz JL, Medzihradsky F, Woods JH. Receptor binding, antagonist and withdrawal precipitating properties of opiate antagonists. Life Sci 1983;32:2887–2896.
314. Kotake AN, Kuwahara SK, Burton E, McCoy CE, Goldberg LI. Variations in demethylation of N-methylnaltrexone in mice, rats, dogs, and humans. Xenobiotica 1989;19:1247–1254.
315. Takemori AE, Portoghese PS. Selective natrexone-derived opioid receptor antagonists. Ann Rev Pharmacol Toxicol 1992;32:239–269.
316. Portoghese PS. The role of concepts in structure–activity relationship studies of opioid ligands. J Med Chem 1992;35:1927–1937.
317. Portoghese PS. Selective nonpeptide opioid antagonists. In: Herz A, Akil H, Simon EJ, editors. Opioids I. Volume 104/I, Berlin: Springer-Verlag; 1993. p 279–293.
318. Portoghese PS. From models to molecules: opioid receptor dimers, bivalent ligands, and selective opioid receptor probes. J Med Chem 2001;44:2259–2269.
319. Erez M, Takemori AE, Portoghese PS. Narcotic antagonistic potency of bivalent ligands which contain β-naltrexamine. Evidence for bridging between proximal recognition sites. J Med Chem 1982;25:847–849.
320. Portoghese PS, Takemori AE. TENA, a selective kappa opioid receptor antagonist. Life Sci 1985;36:801–805.
321. Portoghese PS, Larson DL, Sayre LM, Yim CB, Ronsisvalle G, Tam SW, Takemori AE. Opioid agonist and antagonist bivalent ligands. The relationship between spacer length and selectivity at multiple opioid receptors. J Med Chem 1986;29:1855–1861.
322. Portoghese PS, Ronsisvalle G, Larson DL, Takemori AE. Synthesis and opioid antagonist potencies of naltrexamine bivalent ligands with conformationally restricted spacers. J Med Chem 1986;29:1650–1653.
323. Portoghese PS, Nagase H, Lipkowski AW, Larson DL, Takemori AE. Binaltorphimine-related bivalent ligands and their κ opioid receptor antagonist selectivity. J Med Chem 1988;31:836–841.
324. Portoghese PS, Garzon-Aburbeh A, Nagase H, Lin C-E, Takemori AE. Role of the spacer in conferring κ opioid receptor selectivity to bivalent ligands related to norbinaltorphimine. J Med Chem 1991;34:1292–1296.
325. Portoghese PS, Nagase H, Takemori AE. Only one pharmacophore is required for the κ opioid antagonist selectivity of norbinaltorphimine. J Med Chem 1988;31:1344–1347.
326. Lin C-E, Takemori AE, Portoghese PS. Synthesis and κ-opioid antagonist selectivity of a norbinaltorphimine congener. Identification of the address moiety required for κ-antagonist activity. J Med Chem 1993;36: 2412–2415.
327. Portoghese PS, Lin C-E, Farouz-Grant F, Takemori AE. Structure–activity relationship of N17′-substituted norbinaltorphimine congeners. Role of the N17′ basic group in the interaction with a putative address subsite on the κ opioid receptor. J Med Chem 1994;37:1495–1500.
328. Larson DL, Jones RM, Hjorth SA, Schwartz TW, Portoghese PS. Binding of norbinaltorphimine (norBNI) congeners to wild-type and mutant mu and kappa opioid receptors: molecular recognition loci for the pharmacophore and address components of kappa antagonists. J Med Chem 2000;43:1573–1576.
329. Olmsted SL, Takemori AE, Portoghese PS. A Remarkable change of opioid receptor selectivity on the attachment of a peptidomimetic κ address element to the δ antagonist, naltrindole: 5′-[(N²-alkylamidino)methyl]naltrindole

derivatives as a novel class of κ opioid receptor antagonists. J Med Chem 1993;36:179–180.
330. Jales AR, Husbands SM, Lewis JW. Selective κ-opioid antagonists related to naltrindole. effect of side-chain spacer in the 5'-amidinoalkyl series. Bioorg Med Chem Lett 2000;10:2259–2261.
331. Metzger TG, Paterlini MG, Ferguson DM, Portoghese PS. Investigation of the selectivity of oxymorphone- and naltrexone-derived ligands via site-directed mutagenesis of opioid receptors: exploring the "address" recognition locus. J Med Chem 2001;44: 857–862.
332. Sharma SK, Jones RM, Metzger TG, Ferguson DM, Portoghese PS. Transformation of a kappa-opioid receptor antagonist to a kappa-agonist by transfer of a guanidinium group from the 5'- to 6'-position of naltrindole. J Med Chem 2001;44:2073–9.
333. Portoghese PS. An Approach to the design of receptor-type-selective non-peptide antagonists of peptidergic receptors: δ opioid antagonists. J Med Chem 1991;34:1757–1762.
334. Schwyzer R. ACTH: a short introductory review. Ann NY Acad Sci 1977;297:3–26.
335. Portoghese PS, Sultana M, Takemori AE. Design of peptidomimetic δ-opioid receptor antagonists using the message address concept. J Med Chem 1990;33:1714–1720.
336. Kshirsagar T, Nakano AH, Law PY, Elde R, Portoghese PS. NTI4F: a non-peptide fluorescent probe selective for functional delta opioid receptors. Neurosci Lett 1998;249: 83–6.
337. Le Bourdonnec B, El Kouhen R, Poda G, Law PY, Loh HH, Ferguson DM, Portoghese PS. Covalently Induced activation of the δ opioid receptor by a fluorogenic affinity label, 7'-(phthalaldehydecarboxamido)naltrindole (PNTI). J Med Chem 2001;44: 1017–1020.
338. Sofuoglu M, Portoghese PS, Takemori AE. Differential antagonism of delta-opioid agonists by naltrindole and its benzofuran analog (NTB) in mice: evidence for delta-opioid receptor subtypes. J Pharmacol Exp Ther 1991;257:676–680.
339. Sofuoglu M, Portoghese PS, Takemori AE. 7-Benzylidenenaltrexone (BNTX): a selective $δ_1$ opioid receptor antagonist in the mouse spinal cord. Life Sci 1993;52:769–775.
340. Ohkawa S, Portoghese PS. 7-Arylidenenaltrexones as selective delta1 opioid receptor antagonists. J Med Chem 1998;41: 4177–4180.
341. Palmer RB, Upthagrove AL, Nelson WL. (E)- and (Z)-7-Arylidenenaltrexones: synthesis and opioid receptor radioligand displacement assays. J Med Chem 1997;40:749–753.
342. Portoghese PS, Sultana M, Moe ST, Takemori AE. Synthesis of naltrexone-derived δ-opioid antagonists. Role of conformation of the δ address moiety. J Med Chem 1994;37: 579–585.
343. Portoghese PS, Ohkawa S, Moe ST, Takemori AE. Synthesis and δ-opioid receptor antagonist activity of a naltrindole analogue with a regioisomeric indole moiety. J Med Chem 1994;37:1886–1888.
344. Gao P, Larson DL, Portoghese PS. Synthesis of 7-arylmorphinans. Probing the "address" requirements for selectivity at opioid delta receptors. J Med Chem 1998;41:3091–3098.
345. Portoghese PS, Farouz-Grant F, Sultana M, Takemori AE. 7'-Substituted amino acid conjugates of naltrindole. Hydrophilic groups as determinants of selective antagonism of $δ_1$ opioid receptor-mediated antinociception in mice. J Med Chem 1995;38:402–407.
346. Korlipara VL, Takemori AE, Portoghese PS. N-Benzylnaltrindoles as long-acting δ-opioid receptor antagonists. J Med Chem 1994;37: 1882–1885.
347. Korlipara V, Ells J, Wang J, Tam S, Elde R, Portoghese PS. Fluorescent N-benzylnaltrindole analogues as potential delta opioid receptor selective probes. Eur J Med Chem 1997;32:171–174.
348. McLamore S, Ullrich T, Rothman RB, Xu H, Dersch C, Coop A, Davis P, Porreca F, Jacobson AE, Rice KC. Effect of N-alkyl and N-alkenyl substituents in noroxymorphindole, 17-substituted-6,7-dehydro-4,5alpha-epoxy-3,14-dihydroxy-6,7: 2',3'-indolom orphinans, on opioid receptor affinity, selectivity, and efficacy. J Med Chem 2001;44:1471–4.
349. Coop A, Pinto J, Wang L, McCullough K, Rothman RB, Dersch C, Jacobson AE, Rice KC. Delta opioid binding selectivity of 3-ether analogs of naltrindole. Bioorg Med Chem Lett 1999;9:3435–3438.
350. Coop A, Rothman RB, Dersch C, Partilla J, Porreca F, Davis P, Jacobson AE, Rice KC. Delta opioid affinity and selectivity of 4-hydroxy-3-methoxyindolomorphinan analogues related to naltrindole. J Med Chem 1999;42: 1673–1679.
351. Ullrich T, Dersch CM, Rothman RB, Jacobson AE, Rice KC. Derivatives of 17-(2-methylallyl)-substituted noroxymorphone: variation of the delta address and its effects on affinity and selectivity for the delta opioid receptor. Bioorg Med Chem Lett 2001;11:2883–2885.

352. Schmidhammer H, Daurer D, Wieser M, Monory K, Borsodi A, Elliott J, Traynor JR. Synthesis and biological evaluation of 14-alkoxymorphinans. 14. 14-Ethoxy-5-methyl substituted indolomorphinans with δ opioid receptor selectivity. Bioorg Med Chem Lett 1997;7:151–156.

353. Schmidhammer H, Krassnig R, Greiner E, Schütz J, White A, Berzetei-Gurske IP. Synthesis and biological evaluation of 14-alkoxymorphinans. Part 15. Novel δ opioid receptor antagonists with high affinity and selectivity in the 14-alkoxy-substituted indolomorphinan series. Helvetica Chimica Acta 1998;81:1064–1069.

354. Coop A, Jacobson AE, Aceto MD, Harris LS, Traynor JR, Woods JH, Rice KC. N-Cyclohexylethyl-N-noroxymorphindole: a mu-opioid preferring analogue of naltrindole. Bioorg Med Chem Lett 2000;10:2449–2451.

355. Dondio G, Clarke GD, Giardina G, Petrillo P, Rapalli L, Ronzoni S, Vecchietti V. Potent and selective non-peptidic delta opioid ligands based on the novel heterocycle-condensed octahydroisoquinoline structure. Regul Pept 1994; 53 (Suppl1) S43–S44.

356. Thomas JB, Mascarella SW, Burgess JP, Xu H, McCullough KB, Rothman RB, Flippen-Anderson JL, George CF, Cantrell BE, Zimmerman DM, Carroll FI. N-Substituted octahydro-4a-(3-hydroxyphenyl)-10a-methyl-benzo[g]isoquinolines are opioid receptor pure antagonists. Bioorg Med Chem Lett 1998;8:3149–3152.

357. Ananthan S, Johnson CA, Carter RL, Clayton SD, Rice KC, Xu H, Davis P, Porreca F, Rothman RB. Synthesis, opioid receptor binding, and bioassay of naltrindole analogues substituted in the indolic benzene moiety. J Med Chem 1998;41:2872–2881.

358. Wells JL, Bartlett JL, Ananthan S, Bilsky EJ. In vivo pharmacological characterization of SoRI 9409, a nonpeptidic opioid μ-agonist/δ-antagonist that produces limited antinociceptive tolerance and attenuates morphine physical dependence. J Pharmacol Exp Ther 2001;297:597–605.

359. Portoghese PS, Larson DL, Sultana M, Takemori AE. Opioid agonist and antagonist activities of morphindoles related to naltrindole. J Med Chem 1992;35:4325–4329.

360. Takemori AE, Sultana M, Nagase H, Portoghese PS. Agonist and antagonist selectivities of ligands derived from naltrexone and oxymorphone. Life Sci 1992;50:1491–1495.

361. Portoghese PS, Moe ST, Takemori AE. A selective δ_1 opioid receptor agonist derived from oxymorphone. Evidence for separate recognition sites for δ_1 opioid receptor agonists and antagonists. J Med Chem 1993;36:2572–2574.

362. Lipkowski AW, Tam SW, Portoghese PS. Peptides as receptor selectivity modulators of opiate pharmacophores. J Med Chem 1986;29:1222–1225.

363. Ohkawa S, DiGiacomo B, Larson DL, Takemori AE, Portoghese PS. 7-Spiroindanyl derivatives of naltrexone and oxymorphone as selective ligands for delta opioid receptors. J Med Chem 1997;40:1720–1725.

364. Kshirsagar TA, Fang X, Portoghese PS. 14-Desoxy analogues of naltrindole and 7-spiroindanyloxymorphone: the role of the 14-hydroxy group at delta opioid receptors. J Med Chem 1998;41:2657–2660.

365. Fang X, Larson DL, Portoghese PS. 7-Spirobenzocyclohexyl derivatives of naltrexone, oxymorphone, and hydromorphone as selective opioid receptor ligands. J Med Chem 1997;40:3064–3070.

366. Aceto MD, Harris LS, Abood ME, Rice KC. Stereoselective mu- and delta-opioid receptor-related antinociception and binding with (+)-thebaine. Eur J Pharmacol 1999;365:143–147.

367. Lenz GR, Evans SM, Walters DE, Hopfinger AJ. Synthesis and structure–activity relationships of morphine, codeine, and related alkaloids. In: Opiates. Orlando FL: Academic Press, Inc. 1986. p 101–154.

368. Casy AF, Parfitt RT. 4,5-Epoxymorphinans. In: Opioid Analgesics: Chemistry and Receptors. New York: Plenum Press; 1986. p 69–83.

369. van den Hende JH, Nelson NR. The crystal and molecular structure of 7α-(1-(R)-Hydroxy-1-methylbutyl)-6,14-endo-ethenotetrahydrothebaine hydrobromide (19-propylthevinol hydrobromide). J Am Chem Soc 1967;89:2901–2905.

370. Fulmor W, Lancaster JE, Morton GO, Brown JJ, Howell CF, Nora CT, Hardy RA. Nuclear magnetic resonance studies in the 6,14-endo-ethenotetrhydrothebaine series. J Am Chem Soc 1967;89:3322–3330.

371. Mazza SM, Erickson RH, Blake PR, Lever JR. Two-dimensional homonuclear and heteronuclear correlation NMR studies of diprenorphine: a prototypic 6α,14-endo-ethanotetrahydrothebaine. Magn Reson Chem 1990;28:675–681.

372. Amato ME, Bandoli G, Grassi A, Nicolini M, Pappalardo GC. Molecular determinants for drug–receptor interactions. Part 14. X-ray structure and theoretical conformational (MNDO, MM2) studies of the narcotic dualist buprenorphine hydrochloride. J Mol Struct 1991;236:411–425.

373. Richards ML, Sadee W. Buprenorphine is an antagonist at the κ opioid receptor. Pharm Res 1985;4:178.

374. Lattanzi R, Negri L, Giannini E, Schmidhammer H, Schutz J, Improta G. HS-599: a novel long acting opioid analgesic does not induce place-preference in rats. Br J Pharmacol 2001;134:441–447.

375. Traynor JR, Guo L, Coop A, Lewis JW, Woods JH. Ring constrained orvinols as analogs of buprenorphine. J Pharmacol Exp Ther 1999;191:1093–1099.

376. Schnider O, Grüssner A. Synthese von Oxymorphinanen. Helv Chim Acta 1949;32:821–828.

377. Benson WM, Stefko PL, Randall LO. Comparative pharmacology of levorphan, racemorphan and dextrorphan and related methyl ethers. J Pharmacol Exp Ther 1953;109:189–200.

378. Goldstein A, Lowney LI, Pal BK. Stereospecific and nonspecific interactions of the morphine congener levorphanol in subcellular fractions of mouse brain. Proc Natl Acad Sci USA 1971;68:1742–1747.

379. Gylbert L. Carlstrom D. The structure and absolute configuration of (+)-3-methoxy-N-methylmorphinan (dextromethorphan) hydrobromide monohydrate. Acta Crystallogr 1977;B33:2833–2837.

380. Lenz GR, Evans SM, Walters DE, Hopfinger AJ. The morphinans. In: Opiates. Orlando FL: Academic Press, Inc. 1986. p 188–249.

381. Neumeyer JL, Bidlack JM, Zong R, Bakthavachalam V, Gao P, Cohen DJ, Negus SS, Mello NK. Synthesis and opioid receptor affinity of morphinan and benzomorphan derivatives: mixed κ agonists and μ agonists/antagonists as potential pharmacotherapeutics for cocaine dependence. J Med Chem 2000;43:114–122.

382. Neumeyer JL, Gu X-H, van Vliet A, DeNunzio NJ, Rusovici DE, Cohen DJ, Negus SS, Mello NK, Bidlack JM. Mixed κ agonists and μ agonists/antagonists as potential pharmacotherapeutics for cocaine abuse: synthesis and opioid receptor binding affinity of N-substituted derivatives of morphinan. Bioorg Med Chem Lett 2001;11:2735–2740.

383. Barltrop JA. Syntheses in the morphine series. Part I. Derivatives of bicyclo[3:3:1]-2-azanonane. J Chem Soc 1947; 399–401.

384. May EL, Murphy JG. Structures related to morphine. III. Synthesis of an analog of N-methylmorphinan. J Org Chem 1955;20:257–263.

385. Lenz GR, Evans SM, Walters DE, Hopfinger AJ. The benzomorphans. In: Opiates. Orlando FL: Academic Press, Inc. 1986. p 250–317.

386. May EL, Jacobson AE, Mattson MV, Traynor JR, Woods JH, Harris LS, Bowman ER, Aceto MD. Synthesis and in vitro and in vivo activity of (−)-(1R,5R,9R)- and (+)-(1S,5S,9S)-N-alkenyl-, -N-alkynyl-, and -N-cyanoalkyl-5,9-dimethyl-2′-hydroxy-6,7-benzomorphan homologues. J Med Chem 2000;43:5030–5036.

387. Michne WF, Lewis TR, Michalec SJ, Pierson AK, Gillan MGC, Paterson SJ, Robson LE, Kosterlitz HW. Novel developments of N-methylbenzomorphan narcotic antagonists. In: Van Ree JM, Terenius L, editors. Characteristics and Function of Opioids. Volume 4, Amsterdam: Elsevier; 1978. p 197–206.

388. Wood PL, Piplapil C, Thakur M, Richard JW. WIN 44,441: a stereospecific and long-acting narcotic antagonist. Pharm Res 1984;1:46–48.

389. Albertson NF. Effect of 9-hydroxylation on benzomorphan antagonist activity. J Med Chem 1975;18:619–620.

390. Pfeiffer A, Brantl V, Herz A, Emrich HM. Psychotomimesis mediated by κ opiate receptors. Science 1986;233:774–776.

391. Brady KT, Balster RL, May EL. Stereoisomers of N-allylnormetazocine: phencyclidine-like behavioral effects in squirrel monkeys and rats. Science 1982;215:178–180.

392. Römer D, Büscher H, Hill RC, Maurer R, Petcher TJ, Welle HBA, Bakel HCCK, Akkerman AM. Bremazocine: a potent, long-acting opiate kappa-agonist. Life Sci 1980;27: 971–978.

393. Merz H, Stockhaus K, Wick H. Stereoisomeric 5,9-dimethyl-2′-hydroxy-2-tetrahydro-furfuryl- 6,7-benzomorphans, strong analgesics with non-morphine-like action profiles. J Med Chem 1975;18:996–1000.

394. Merz H, Stockhaus K. N-[(Tetrahydrofuryl)alkyl] and N-(alkoxyalkyl) derivatives of (−)-normetazocine, compounds with differentiated opioid action profiles. J Med Chem 1979;22:1475–1483.

395. Merz H, Langbein A, Stockhaus K, Walther G, Wick H. Structure–Activity relationships in narcotic antagonists with N-furylmethyl substituents. In: Braude MC, Harris LS, May EL, Smith JP, Villarreal JE, editors. Narcotic Antagonists. Volume 8, New York: Raven Press; 1974. p 91–107.

396. Wentland MP, Xu G, Cioffi CL, Ye Y, Duan W, Cohen DJ, Colasurdo AM, Bidlack JM. 8-Aminocyclazocine analogues: synthesis and structure–activity relationships. Bioorg Med Chem Lett 2000;10:183–187.

397. Wentland MP, Duan W, Cohen DJ, Bidlack JM. Selective protection and functionalization of morphine: synthesis and opioid receptor properties of 3-amino-3-desoxymorphine derivatives. J Med Chem 2000;43:3558–3565.

398. Schultz AG, Wang A, Alva C, Sebastian A, Glick SD, Deecher DC, Bidlack JM. Asymmetric syntheses, opioid receptor affinities, and antinociceptive effects of 8-amino-5,9-methanobenzocyclooctenes, a new class of structural analogues of the morphine alkaloids. J Med Chem 1996;39:1956–1966.

399. Eisleb O, Schaumann O. Dolantin, ein neuartiges Spasmolytikum und Analgetikum. Deut Med Wochenschr 1939;65:967–968.

400. Larson DL, Portoghese PS. Relationship between analgetic ED_{50} dose and time-couse brain levels of N-alkylnormeperidine homologues. J Med Chem 1976;19:16–19.

401. Magnan J, Paterson SJ, Tavani A, Kosterlitz HW. The binding spectrum of narcotic analgesic drugs with different agonist and antagonist properties. Naunyn Schmiedebergs Arch Pharmacol 1982;319:197–205.

402. Lenz GR, Evans SM, Walters DE, Hopfinger AJ. Piperidine analgesics. In: Opiates. Orlando FL: Academic Press, Inc. 1986. p 318–376.

403. Casy AF. Opioid receptors and their ligands: recent developments. Adv Drug Res 1989;18:178–289.

404. van Koningsveld H. The crystal and molecular structure of pethidine hydrobromide. Recl Trav Chim Pays Bas 1970;89:375–378.

405. Tillack JV, Seccombe RC, Kennard CHL, Oh PWT. Analgetics. Part I. The crystal structure of pethidine hydrochloride, 4-carbethoxy-1-methyl-4-phenylpiperidine hydrochloride, $C_{15}H_{21}NO_2 \cdot HCl$. Recl Trav Chim Pays Bas 1974;93:164–165.

406. Casy AF, Dewar GH, Al-Deeb OAA. Conformational equilibra of hydrochloride salts of pethidine, ketobemidone, and related central analgesics of the 4-arylpiperidine class. J Chem Soc Perkin Trans II 1989; 1243–1247.

407. Froimowitz M. Conformation-activity study of 4-phenylpiperidine analgesics. J Med Chem 1982;25:1127–1133.

408. Casy AF, Chatten LG, Khullar KK. Synthesis and stereochemistry of 3-methyl analogues of pethidine. J Chem Soc C 1969; 2491–2495.

409. Bell MR, Archer S. Ethyl 3α-phenyltropane-3β-carboxylate and related compounds. J Am Chem Soc 1960;82:4638–4641.

410. Daum SJ, Martini CM, Kullnig RK, Clarke RL. Analgesic activity of the epimeric tropane analogs of meperidine. A physical and pharmacological study. J Med Chem 1975;18:496–501.

411. Casy AF, Dewar GH, Pascoe RA. Opioid properties of some derivatives of pethidine based on tropane. J Pharm Pharmacol 1992;44:787–790.

412. Loew GH, Lawson JA, Uyeno ET, Toll L, Frenking G, Polgar W, Ma LYY, Camerman N, Camerman A. Strucure–activity studies of morphine fragments. I. 4-Alkyl-4-(m-hydroxy-phenyl)-piperidines. Mol Pharmacol 1988;34:363–376.

413. Casy AF, Dewar GH, Al-Deeb OAA. Stereochemical studies of the 4-alkyl-4-arylpiperidine class of opioid ligand. Magn Reson Chem 1989;27:964–972.

414. Zimmerman DM, Nickander R, Horng JS, Wong DT. New structural concepts for narcotic antagonists defined in a 4-phenylpiperidine series. Nature 1978;275:332–334.

415. Zimmerman DM, Leander JD, Cantrell BE, Reel JK, Snoddy J, Mendelsohn LG, Johnson BG, Mitch CH. Structure activity relationships of trans-3,4-dimethyl-4-(3-hydroxyphenyl)-piperidine antagonists for mu and kappa opioid receptors. J Med Chem 1993;36:2833–2841.

416. Mitch CH, Leander JD, Mendelsohn LG, Shaw WN, Wong DT, Cantrell BE, Johnson BG, Reel JK, Snoddy JD, Takemori AE, Zimmerman DM. 3,4-Dimethyl-4-(3-hydroxyphenyl)piperidines: opioid antagonists with potent anorectant activity. J Med Chem 1993;36:2842–2850.

417. Zimmerman DM, Gidda JS, Cantrell BE, Schoepp DD, Johnson BG, Leander JD. Discovery of a potent, peripherally selective trans-3,4-dimethyl-4-(3-hydroxyphenyl)piperidine opioid antagonist for the treatment of gastrointestinal motility disorders. J Med Chem 1994;37:2262–2265.

418. Cohen ML, Mendelsohn LG, Mitch CH, Zimmerman DM. Use of the mouse vas deferens to determine μ, δ, and κ receptor affinities of opioid antagonists. Receptor 1994;4:43–53.

419. Thomas JB, Fall MJ, Cooper JB, Rothman RB, Mascarella SW, Xu H, Partilla JS, Dersch CM, McCullough KB, Cantrell BE, Zimmerman DM, Carroll FI. Identification of an opioid kappa receptor subtype-selective N-substituent for (+)-(3R,4R)-dimethyl-4-(3-hydroxyphenyl)piperidine. J Med Chem 1998;41:5188–97.

420. Thomas JB, Mascarella SW, Rothman RB, Partilla JS, Xu H, McCullough KB, Dersch CM, Cantrell BE, Zimmerman DM, Carroll FI. Investigation of the N-substituent conformation governing potency and mu receptor subtype-selectivity in (+)-(3R,4R)-dimethyl-4-(3-hydroxyphenyl)piperidine opioid antagonists. J Med Chem 1998;41:1980–90.

421. May EL, Murphy JG. Structures related to morphine. IV. m-Substituted phenylcyclohexane derivatives. J Org Chem 1955;20:1197–1201.

422. Cochran TG. Stereochemistry and Absolute configuration of the analgesic agonist-antagonist (−)-5-m-hydroxyphenyl-2-methylmorphan. J Med Chem 1974;17:987–989.

423. Froimowitz M, Pick CG, Pasternak GW. Phenylmorphans and analogs: opioid receptor subtype selectivity and effect of conformation on activity. J Med Chem 1992;35:1521–1525.

424. Ong HH, Oh-ishi T, May EL. Phenylmorphan agonists-antagonists. J Med Chem 1974;17:133–134.

425. Thomas JB, Zheng X, Mascarella SW, Rothman RB, Dersch CM, Partilla JS, Flippen-Anderson JL, George CF, Cantrell BE, Zimmerman DM, Carroll FI. N-Substituted 9beta-methyl-5-(3-hydroxyphenyl)morphans are opioid receptor pure antagonists. J Med Chem 1998;41:4143–4149.

426. Hutchison AJ, de Jesus R, Williams M, Simke JP, Neale RF, Jackson RH, Ambrose F, Barbaz BJ, Sills MA. Benzofuro[2,3-c]pyridin-6-ols: synthesis, affinity for opioid–receptor subtypes, and antinociceptive activity. J Med Chem 1989;32:2221–2226.

427. Zimmerman DM, Cantrell BE, Swartzendruber JK, Jones ND, Mendelsohn LG, Leander JD, Nickander RC. Synthesis and analgesic properties of N-substituted trans-4a-aryldecahydroisoquinolines. J Med Chem 1988;31:555–560.

428. Spiegel K, Pasternak GW. Meptazinol: a novel mu-1 selective opioid analgesic. J Pharmacol Exp Ther 1984;228:414–419.

429. Bill DJ, Hartley JE, Stephens RJ, Thomson AM. The antinociceptive activity of meptazinol depends on both opiate and cholinergic mechanisms. Br J Pharmacol 1983;79:191–199.

430. Jensen KA, Lindquist F, Rekling E, Wolfbrandt CG. A new type of piperidine derivative with analgesic and spasmolytic activity. Dan Tidsskr Farm 1943;17:173–182.

431. Langston JW, Ballard P, Tetrud JW, Irwin I. Chronic parkinsonism in humans due to a product of meperidine-analog synthesis. Science 1983;219:979–980.

432. Neumeyer JL, Booth RG. Drugs used to treat neuromuscular disorders: antiparkinsonism agents and skeletal muscle relaxants. In: Foye WO, Lemke TL, Williams DA, editors. Principles of Medicinal Chemistry. 4th ed. Baltimore: Williams & Wilkins; 1995. p 232–246.

433. Carabateas PM, Grumbach L. Strong analgesics. Some 1-substituted 4-phenyl-4-propionoxypiperidines. J Med Pharm Chem 1962;5:913–919.

434. Ziering A, Lee J. Piperidine derivatives. V. 1,3-Dialkyl-4-aryl-4-acyloxypiperidines. J Org Chem 1947;12:911–914.

435. Lenz GR, Evans SM, Walters DE, Hopfinger AJ. Physical chemistry, molecular modeling, and QSAR analysis of the arylpiperidine analgesics. In: Opiates. Orlando FL: Academic Press, Inc. 1986. p 377–399.

436. Larson DL, Portoghese PS. Stereochemical studies on medicinal agents. 12. The distinction of enantiotopic groups in the interaction of 1-methyl-4-phenyl-4-propionoxypiperidine with analgetic receptors. J Med Chem 1973;16:195–198.

437. Casy AF, Ogungbamila FO. The solute conformation of opioid ligands of the 4-arylpiperidine class: α-promedol and related compounds. Magn Reson Chem 1992;30:969–976.

438. Froimowitz M, Salva P, Hite GJ, Gianutsos G, Suzdak P, Heyman R. Conformational properties of α- and β-azabicyclane opiates. The effect of conformation on pharmacological activity. J Comput Chem 1984;5:291–298.

439. Salva PS, Hite GJ, Heyman RA, Gainutsos G. 3,7-Diazabicyclane: a new narcotic analgesic. J Med Chem 1986;29:2111–2113.

440. Portoghese PS, Alreja BD, Larson DL. Allylprodine analogues as receptor probes. Evi-

dence that phenolic and nonphenolic ligands interact with different subsites on identical opioid receptors. J Med Chem 1981;24: 782–787.
441. Casy AF, Dewar GH, Pascoe RA. Phenolic analogues of diastereoisomeric 2-methyl reversed esters of pethidine. J Pharm Pharmacol 1989;41:209–211.
442. Lenz GR, Evans SM, Walters DE, Hopfinger AJ. Piperidine analgesics. In: Opiates. Orlando FL: Academic Press, Inc. 1986. p 362–366.
443. Cometta-Morini C, Macguire PA, Loew GH. Molecular determinants of μ receptor recognition for the fentanyl class of compounds. Mol Pharmacol 1992;41:185–196.
444. Peeters OM, Blaton NM, de Ranter J, van Herk AM, Goubitz K. Crystal and molecular structure of N-[1-(2-phenylethyl)-4-piperidinylium]-N-phenylpropanamide (fentanyl) citrate-toluene solvate. J Crystallogr Mol Struct 1979;9:153–161.
445. Casy AF, Parfitt RT. Fentanyl and the 4-anilinopiperidine group of analgesics. In: Opioid Analgesics: Chemistry and Receptors. New York: Plenum Press; 1986. p 287–302.
446. Riley TN, Bagley JR. 4-Anilidopiperidine analgesics. 2. A study of the conformational aspects of the analgesic activity of the 4-anilidopiperidines utilizing isomeric N-substituted 3-(propananilido)nortropane analogues. J Med Chem 1979;22:1167–1171.
447. Bagley JR, Kudzma LV, Lalinda NL, Colapret JA, Huang B-S, Lin B-S, Jerussi TP, Benvenga MJ, Doorley BM, Ossipov MH, Spaulding TC, Spencer HK, Rudo FG, Wynn RL. Evolution of the 4-anilidopiperidine class of opioid analgesics. Med Res Rev 1991;11: 403–436.
448. Van Daele PGH, De Bruyn MFL, Boey JM, Sanczuk S, Agten JTM, Janssen PAJ. Synthetic analgesics: N-(1-[2-Arylethyl]-4-substituted 4-piperidinyl) N-arylalkanamides. Arzneimittelforschung 1976;26: 1521–1531.
449. Bagley JR, Wynn RL, Rudo FG, Doorley BM, Spencer HK, Spaulding T. New 4-(heteroanilido)piperidines, structurally related to the pure opioid agonist fentanyl, with agonist and/or antagonist properties. J Med Chem 1989;32:663–671.
450. Finney ZG, Riley TN. 4-Anilidopiperidine analgesics. 3. 1-Substituted 4-(Propananilido) perhydroazepines as ring-expanded analgoues. J Med Chem 1980;23:895–899.
451. Cheng MT, Kruppa GH, McLafferty FW, Cooper DA. Structural information from tandem mass spectrometry for China White and related fentanyl derivatives. Anal Chem 1982;54:2204–2207.
452. Remifentanil hydrochloride. Drugs Fut 1994;19:1088–1092.
453. Patel SS, Spencer CM. Remifentanil. Drugs 1996;52:417–427.
454. Van Bever WFM, Niemegeers CJE, Janssen PAJ. Synthetic analgesics. Synthesis and pharmacology of the diastereoisomers of N-[3-methyl-1-(2-phenylethyl)-4-piperidyl]-N-phenylpropanamide and N-[3-methyl-1-(1-methyl-2-phenylethyl)-4-piperidyl]-N-phenylpropanamide. J Med Chem 1974;17: 1047–1051.
455. Burke TR, Jacobson AE, Rice KC, Silverton JV, Simonds WF, Streaty RA, Klee WA. Probes for narcotic receptor mediated phenomena. 12. cis-(+)-3-Methylfentanyl isothiocyanate, a potent site-directed acylating agent for δ opioid receptors. Synthesis, absolute configuration, and receptor enantioselectivity. J Med Chem 1986;29: 1087–1093.
456. Rothman RB, Xu H, Seggel M, Jacobson AE, Rice KC, Brine GA, Carroll FI. RTI-4614-4: an analog of (+)-cis-3-methylfentanyl with a 27,000-fold binding selectivity for mu versus delta opioid binding-sites. Life Sci 1991;48: PL111–PL116.
457. Lalinde N, Moliterni J, Wright D, Spencer HK, Ossipov MH, Spaulding TC, Rudo FG. Synthesis and pharmacological evaluation of a series of new 3-methyl-1,4-disubstituted-piperidine analgesics. J Med Chem 1990;33:2876–2882.
458. Burke TR, Bajwa BS, Jacobson AE, Rice KC, Streaty RA, Klee WA. Probes for narcotic receptor mediated phenomena. 7. Synthesis and pharmacological properties of irreversible ligands for μ and δ opiate receptors. J Med Chem 1984;27:1570–1574.
459. Kleiderer EC. Rice JB, Conquest V, Williams JH. Pharmaceutical activities at the I. G. Farbenindustrie Plant, Höchst-am-Main, Germany; 981; Washington, DC: U. S. Department of Commerce, Office of the Publication Board; 1945.
460. Lenz GR, Evans SM, Walters DE, Hopfinger AJ. Open-chain analgesics. In: Opiates. Orlando FL: Academic Press, Inc. 1986. p 400–447.
461. Henkel JG, Bell KH, Portoghese PS. Stereochemical studies on medicinal agents. 16. Conformational studies of methadone and isomethadone utilizing circular dischroism

and proton magnetic resonance. J Med Chem 1974;17:124–129.
462. Froimowitz M. Conformation-activity study of methadone and related compounds. J Med Chem 1982;25:689–696.
463. Casy AF, Parfitt RT. Methadone and related 3,3-diphenylpropylamines. In: Opioid Analgesics: Chemistry and Receptors. New York: Plenum Press; 1986. p 303–332.
464. Henkel JG, Berg EP, Portoghese PS. Stereochemical studies on medicinal agents. 21. Investigation of the role of conformational factors in the action of diphenylpropylamines. Synthesis and analgetic potency of 5-methylmethdone diastereomers. J Med Chem 1976;19:1308–1314.
465. Portoghese PS, Poupaert JH, Larson DL, Groutas WC, Meitzner GD, Swenson DC, Smith GD, Duax WL. Synthesis, X-ray crystallographic determination, and opioid activity of *erythro*-5-methylmethadone enantiomers. Evidence which suggests that μ and δ opioid receptors possess different stereochemical requirements. J Med Chem 1982;25: 684–688.
466. Duax WL, Smith GD, Griffin JF, Portoghese PS. Methadone conformation and opioid activity. Science 1983;220:417–418.
467. Eddy NB, May EL, Mosettig E. Chemistry and pharmacology of the methadols and acetylmethadols. J Org Chem 1952;17:321–326.
468. Casy AF, Hassan MMA. Configurational influences in methadol and normethadol analgetics. J Med Chem 1968;11:601–603.
469. Pohland A, Sullivan HR. Analgesics: esters of 4-dialkylamino-1,2,-diphenyl-2-butanols. J Am Chem Soc 1953;75:4458–4461.
470. Bockmühl M, Ehrhart G. Über eine neue Klasse von spasmolytisch und analgetisch wirkenden Verbindungen I. Liebigs Ann Chem 1948;561:52–85.
471. Rees DC. Chemical structures and biological activities of non-peptide selective kappa opioid ligands. Prog Med Chem 1992;29: 109–139.
472. Barber A, Gottschlich R. Novel developments with selective, non-peptidic kappa-opioid receptor agonists. Expert Opin Investig Drugs 1997;6:1351–1368.
473. Römer D, Büscher HH, Hill RC, Maurer R, Petcher TJ, Zeugner H, Benson W, Finner E, Milkowski W, Thies PW. An opioid benzodiazepine. Nature 1982;298:759–760.
474. Kley H, Scheidemantel U, Bering B, Müller WE. Reverse stereoselectivity of opiate and benzodiazepine receptors for the opioid benzodiazepine tifluadom. Eur J Pharmacol 1983;87:503–504.
475. Chang RSL, Lotti VJ, Chen TB, Keegan ME. Tifluadom, a κ-opiate agonist, acts as a peripheral cholecystokinin receptor antagonist. Neurosci Lett 1986;72:211–214.
476. Lahti RA, VonVoigtlander PF, Barsuhn C. Properties of a selective kappa agonist, U-50,488. Life Sci 1982;31:2257–2260.
477. Von Voigtlander PF, Lahti RA, Ludens JH. U-50,488: a selective and structurally novel non-mu (kappa) opioid agonist. J Pharmacol Exp Ther 1983;224:7–12.
478. de Costa B, George C, Rothman RB, Jacobson AE, Rice KC. Synthesis and absolute configuration of optically pure enantiomers of a κ-opioid receptor selective agonist. FEBS Lett 1987;223:335–339.
479. de Costa BR, Bowen WD, Hellewell SB, George C, Rothman RB, Reid AA, Walder JM, Jacobson AE, Rice KC. Alterations in the stereochemistry of the κ-selective opioid agonist U50,488 result in high-affinity σ ligands. J Med Chem 1989;32:1996–2002.
480. Tang AH. Protection from cerebral ischemia by U-50,488E, a specific kappa opioid analgesic agent. Life Sci 1985;37:1475–1482.
481. Von Voigtlander PF, Lewis RA. Analgesic and mechanicstic evaluation of spiradoline, a potent kappa opioid. J Pharmacol Exp Ther 1988;246:259–262.
482. Doi M, Ishida T, Inoue M. Conformational characteristics of opioid kappa-receptor agonist: crystal structure of (5S,7S,8S)-(−)-*N*-methyl-*N*-[7-(1-pyrrolidinyl)-1-oxaspiro[4.5]dec-8-yl]benzeneacetamide (U69,593), and conformational comparison with some κ-agonists. Chem Pharm Bull 1990;38:1815–1818.
483. Meecham KG, Boyle SJ, Hunter JC, Hughes J. An *in vitro* profile of activity for the (+) and (−) enantiomers of spiradoline and PD117302. Eur J Pharmacol 1989;173: 151–157.
484. Hall ED, Pazara KE. Quantiative analysis of effects of κ-opioid agonists on postischemic hippocampal CA$_1$ Neuronal Necrosis in gerbils. Stroke 1988;19:1008–1012.
485. Cheney BV, Szmuskovicz J, Lahti RA, Zichi DA. Factors affecting binding of *trans*-*N*-[2-(methylamino)cyclohexyl]benzamides at the primary morphine receptor. J Med Chem 1985;28:1853–1864.
486. Halfpenny PR, Hill RG, Horwell DC, Hughes J, Hunter JC, Johnson S, Rees DC. Highly selective κ-opioid analgesics. 2. Synthesis

and structure–activity relationships of novel N-[(2-aminocyclohexyl)aryl]actamide derivatives. J Med Chem 1989;32:1620–1626.

487. Halfpenny PR, Horwell DC, Hughes J, Humblet C, Hunter JC, Neuhaus D, Rees DC. Highly selective κ opioid analgesics. 4. Synthesis of some conformationally restricted naphthalene derivatives with high receptor affinity and selectivity. J Med Chem 1991;34: 190–194.

488. Cheng C-Y, Lu H-Y, Lee F-M, Tam SW. Synthesis of (1′,2′-trans)-3-phenyl-1-[2′-(N-pyrrolidinyl)cyclohexyl]-pyrrolid-2-ones as κ-selective opiates. J Pharm Sci 1990;79: 758–762.

489. Clark CR, Halfpenny PR, Hill RG, Horwell DC, Hughes J, Jarvis TC, Rees DC, Schofield D. Highly selective κ opioid analgesics. Synthesis and structure–activity relationships of novel N-[(2-aminocyclohexyl)aryl]acetamide and N-[(2-aminocyclohexyl)aryloxy]acetamide derivatives. J Med Chem 1988;31: 831–836.

490. Freeman JP, Michalson ET, D'Andrea SV, Baczynskyj L, VonVoigtlander PF, Lahti RA, Smith MW, Lawson CF, Scahill TA, Mizsak SA, Szmuszkovicz J. Naphtho and benzo analogues of the κ opioid agonist trans- (±)-3,4-dichloro-N-methyl-N-[2-(1-pyrrolidinyl)cyclohexyl]benzeneacetamide. J Med Chem 1991;34:1891–1896.

491. Rajagopalan P, Scribner RM, Pennev P, Schmidt WK, Tam SW, Steinfels GF, Cook L. DuP 747: a new potent, kappa opioid analgesic. Synthesis and pharmacology. Bioorg Med Chem Lett 1992;2:715–720.

492. Rajagopalan P, Scribner RM, Pennev P, Mattei PL, Kezar HS, Cheng CY, Cheeseman RS, Ganti VR, Johnson AL, Wuonola MA, Schmidt WK, Tam SW, Steinfels GF, Cook L. DuP 747: SAR study. Bioorg Med Chem Lett 1992;2:721–726.

493. Hamon G, Fortin M, Le Martret O, Jouquey S, Vincent JC, Bichet DG. Pharmacological profile of viravoline, a new aquaretic compound. Am J Soc Nephrol 1994;5:272.

494. Hamon G, Clemence F, Fortin M. Niravoline hydrochloride, a novel selective kappa agonist with marked aquaretic properties but low analgesic potency. Br J Pharmacol 1995;114:310P.

495. Bosch-Marce M, Jimenez W, Angeli P, Leivas A, Claria J, Graziotto A, Arroyo V, Rivera F, Rodes J. Aquaretic effect of the kappa-opioid agonist RU 51599 in cirrhotic rats with ascites and water retention. Gastroenterology 1995;109:217–223.

496. Moreau R, Cailmail S, Hamon G, Lebrec D. Renal and haemodynamic responses to a novel kappa opioid receptor agonist, niravoline (RU 51, 599), in rats with cirrhosis. J Gastroenterol Hepatol 1996;11:857–863.

497. Bosch-Marce M, Poo JL, Jimenez W, Bordas N, Leivas A, Morales-Ruiz M, Munoz RM, Perez M, Arroyo V, Rivera F, Rodes J. Comparison of two aquaretic drugs (niravoline and OPC-31260) in cirrhotic rats with ascites and water retention. J Pharmacol Exp Ther 1999;289:194–201.

498. Gueniau C, Oberlander C. The kappa opioid agonist niravoline decreases brain edema in the mouse middle cerebral artery occlusion model of stroke. J Pharmacol Exp Ther 1997;282:1–6.

499. Nagao S, Bemana I, Kuratani H, Takahashi E, Nakamura T. Niravoline, a selective kappa-opioid receptor agonist effectively reduces elevated intracranial pressure. Exp Brain Res 2000;130:338–344.

500. Vecchietti V, Giordani A, Giardina G, Colle R, Clarke GD. (2S)-1-(Arylacetyl)-2-(aminomethyl)piperidine derivatives: novel, highly selective κ opioid analgesics. J Med Chem 1991;34:397–403.

501. Vecchietti V, Clarke GD, Colle R, Giardina G, Petrone G, Sbacchi M. (1S)-1-(Aminomethyl)-2-(arylacetyl)-1,2,3,4-tetrahydroisoquinoline and heterocycle-condensed tetrahydropyridine derivatives: members of a novel class of very potent κ opioid analgesics. J Med Chem 1991;34:2624–2633.

502. Colle R, Clarke GD, Dondio G, Giardina G, Petrone G, Sbacchi M, Vecchietti V. Enantiospecificity of kappa-receptors: comparison of racemic compounds and active enantiomers in 2 novel series of kappa-agonist analgesics. Chirality 1992;4:8–15.

503. Fujibayashi K, Sakamoto K, Watanabe M, Iizuka Y. Pharmacological properties of R-84760, a novel kappa-opioid receptor agonist. Eur J Pharmacol 1994;261:133–140.

504. Naylor A, Judd DB, Lloyd JE, Scopes DIC, Hayes AG, Birch PJ. A potent new class of κ-receptor agonist: 4-substituted 1-(arylacetyl)-2-[(dialkylamino)methyl]piperazines. J Med Chem 1993;36:2075–2083.

505. Hayes AG, Birch PJ, Hayward NJ, Sheehan MJ, Rogers H, Tyers MB, Judd DB, Scopes DIC, Naylor A. A series of novel, highly potent and selective agonists for the κ-opioid receptor. Br J Pharmacol 1990;101:944–948.

506. Birch PJ, Rogers H, Hayes AG, Hayward NJ, Tyers MB, Scopes DIC, Naylor A, Judd DB. Neuroprotective actions of Gr89696, a highly potent and selective κ-opioid receptor agonist. Br J Pharmacol 1991;103:1819–1823.

507. Caudle RM, Mannes AJ, Iadarola MJ. GR89,696 is a kappa-2 opioid receptor agonist and a kappa-1 opioid receptor antagonist in the guinea pig hippocampus. J Pharmacol Exp Ther 1997;283:1342–1349.

508. Butelman ER, Ko MC, Traynor JR, Vivian JA, Kreek MJ, Woods JH. GR89,696: a potent kappa-opioid agonist with subtype selectivity in rhesus monkeys. J Pharmacol Exp Ther 2001;298:1049–1059.

509. Soukara S, Maier CA, Predoiu U, Ehret A, Jackisch R, Wunsch B. Methylated analogues of methyl (R)-4-(3,4-dichlorophenylacetyl)-3-(pyrrolidin-1-ylmethyl)piperazine-1-carboxylate (GR-89,696) as highly potent kappa-receptor agonists: stereoselective synthesis, opioid-receptor affinity, receptor selectivity, and functional studies. J Med Chem 2001;44:2814–2826.

510. Birch PJ, Hayes AG, Johnson MR, Lea TA, Murray PJ, Rogers H, Scopes DIC. Preparation and evaluation of some hydrophilic phenylacetylpiperazines as peripherally selective κ-opioid receptor agonists. Bioorg Med Chem Lett 1992;2:1275–1278.

511. Fardin V, Jolly A, Flamand O, Curruette A, Laduron P-M, Garret C. Rp 60180, a new phenothiazine with high affinity for opiate kappa bindings sites in animal and human brain. Eur J Pharmacol 1990;183:2332.

512. Costello GF, Main BG, Barlow JJ, Carroll JA, Shaw JS. A novel series of potent and selective agonists at the opioid κ-receptor. Eur J Pharmacol 1988;151:475–478.

513. Costello GF, James R, Shaw JS, Slater AM, Stutchbury NCJ. 2-(3,4-Dichlorophenyl)-N-methyl-N-[2-1-pyrrolidinyl)-1-substituted-ethyl]-acetamides: the use of conformational analysis in the development of a novel series of potent opioid κ agonists. J Med Chem 1991;34:181–189.

514. Chang AC, Chao CC, Takemori AE, Gekker G, Hu S, Peterson PK, Portoghese PS. Arylacetamide-derived fluorescent probes: synthesis, biological evaluation, and direct fluorescent labeling of kappa opioid receptors in mouse microglial cells. J Med Chem 1996;39:1729–1735.

515. Shaw JS, Carroll JA, Alcock P, Main BG. ICI 204448: a κ-opioid agonist with limited access to the CNS. Br J Pharmacol 1989;96:986–992.

516. Froimowitz M, DiMeglio CM, Makriyannis A. Conformational preferences of the κ-selective opioid agonist U50488. A combined molecular mechanics and nuclear magnetic resonance study. J Med Chem 1992;35:3085–3094.

517. Higginbottom M, Nolan W, O'Toole J, Ratcliffe GS, Rees DC, Roberts E. The design and synthesis of kappa opioid ligands based on a binding model for kappa agonists. Bioorg Med Chem Lett 1993;3:841–846.

518. Brandt W, Drosihn S, Haurand M, Holzgrabe U, Nachtsheim C. Search for the pharmacophore in kappa-agonistic diazabicyclo[3.3.1]nonan-9-one-1,5-diesters and arylacetamides. Arch Pharm 1996;329:311–323.

519. Kögel B, Christoph T, Friderichs E, Hennies H-H, Matthiesen T, Schneider J, Holzgrabe U. HZ2 a selective κ-opioid agonists. CNS Drug Rev 1998;4:54–70.

520. Siener T, Cambareri A, Kuhl U, Englberger W, Haurand M, Kogel B, Holzgrabe U. Synthesis and opioid receptor affinity of a series of 2,4-diaryl-substituted 3,7-diazabicylononanones. J Med Chem 2000;43:3746–3751.

521. Peters GR, Ward NJ, Antal EG, Lai PY, deMaar EW. Diuretic actions in man of a selective kappa opioid agonist: U-62,066E. J Pharmacol Exp Ther 1987;240:128–131.

522. Reece PA, Sedman AJ, Rose S, Wright DS, Dawkins R, Rajagopalan R. Diuretic effects, pharmacokinetics, and safety of a new centrally acting kappa-opioid agonist (CI-977) in humans. J Clin Pharmacol 1994;34:1126–1132.

523. Pande AC, Pyke RE, Greiner M, Cooper SA, Benjamin R, Pierce MW. Analgesic efficacy of the kappa-receptor agonist, enadoline, in dental surgery pain. Clin Neuropharmacol 1996;19;92–97.

524. Pande AC, Pyke RE, Greiner M, Wideman GL, Benjamin R, Pierce MW. Analgesic efficacy of enadoline versus placebo or morphine in postsurgical pain. Clin Neuropharmacol 1996;19:451–456.

525. Bellissant E, Denolle T, Sinnassamy P, Bichet DG, Giudicelli JF, Lecoz F, Gandon JM, Allain H. Systemic and regional hemodynamic and biological effects of a new kappa-opioid agonist, niravoline, in healthy volunteers. J Pharmacol Exp Ther 1996;278: 232–242.

526. Lotsch J, Ditterich W, Hummel T, Kobal G. Antinociceptive effects of the kappa-opioid receptor agonist RP 60180 compared with

pentazocine in an experimental human pain model. Clin Neuropharmacol 1997;20: 224–233.
527. Ur E, Wright DM, Bouloux PM, Grossman A. The effects of spiradoline (U-62066E), a kappa-opioid receptor agonist, on neuroendocrine function in man. Br J Pharmacol 1997;120:781–784.
528. Gadano A, Moreau R, Pessione F, Trombino C, Giuily N, Sinnassamy P, Valla D, Lebrec D. Aquaretic effects of niravoline, a kappa-opioid agonist, in patients with cirrhosis. J Hepatol 2000;32:38–42.
529. Peters G, Gaylor SG. Human central nervous system (CNS) effects of a selective kappa opioid agonist. Am Soc Clin Pharmacol Ther 1989;45:130.
530. Brooks DP, Velente M, Petrone G, Depalma PD, Sbacchi M, Clarke GD. Comparison of the Water diuretic activity of kappa receptor agonists and a vasopressin receptor antagonist in dogs. J Pharmacol Exp Ther 1997;280: 1176–1183.
531. Giardina G, Clarke GD, Grugni M, Sbacchi M, Vecchietti V. Central and peripheral analgesic agents: chemical strategies for limiting brain penetration in kappa-opioid agonists belonging to different chemical classes. Farmaco 1995;50:405–418.
532. Chang AC, Cowan A, Takemori AE, Portoghese PS. Aspartic acid conjugates of 2-(3,4-dichlorophenyl)-N-methyl-N-[(1S)-1(3-aminophenyl)-2-(1-pyrrolidi nyl) ethyl]acetamide: kappa opioid receptor agonists with limited access to the central nervous system. J Med Chem 1996;39:4478–4482.
533. Kumar V, Marella MA, Cortes-Burgos L, Chang AC, Cassel JA, Daubert JD, DeHaven RN, DeHaven-Hudkins DL, Gottshall SL, Mansson E, Maycock AL. Arylacetamides as peripherally restricted kappa opioid receptor agonists. Bioorg Med Chem Lett 2000;10: 2567–2570.
534. Guo D, Kumar V, Maycock A, DeHaven RN, Daubert JD, Cassel JA, Gauntner EK, DeHaven-Hudkins DL, Gottshall SL, Greiner S, Koblish M, Little PJ. Synthesis and evaluation of novel peripheral κ opioid receptor agonists. Abstr Am Chem Soc 2000;219: MEDI 263.
535. Barber A, Bartoszyk GD, Greiner HE, Mauler F, Murray RD, Seyfried CA, Simon M, Gottschlich R, Harting J, Lues I. Central and peripheral actions of the novel κ-opioid receptor agonist, EMD 60400. Br J Pharmacol 1994;111:843–851.

536. Brooks DP, Giardina G, Gellai M, Dondio G, Edwards RM, Petrone G, DePalma PD, Sbacchi M, Jugus M, Misiano P, Wang Y-X, Clarke GD. Opiate receptors within the blood–brain barrier mediate kappa agonist-induced water diuresis. J Pharmacol Exp Ther 1993;266: 164–171.
537. Gottschlich R, Ackermann KA, Barber A, Bartoszyk GD, Greiner HE. EMD-61753 as a favourable representative of structurally novel arylacetamido-type κ opiate receptor agonists. Bioorg Med Chem Lett 1994;4: 677–682.
538. Barber A, Bartoszyk GD, Bender HM, Gottschlich R, Greiner HE, Harting J, Mauler F, Minck KO, Murray RD, Simon M, Seyfried CA. A Pharmacological Profile of the novel, peripherally-selective κ-opioid receptor agonist, EMD 61753. Br J Pharmacol 1994;113: 1317–1327.
539. Jonker JW, Wagenaar E, van Deemter L, Gottschlich R, Bender HM, Dasenbrock J, Schinkel AH. Role of blood–brain barrier P-glycoprotein in limiting brain accumulation and sedative side-effects of asimadoline, a peripherally acting analgaesic drug. Br J Pharmacol 1999;127:43–50.
540. Machelska H, Pfluger M, Weber W, Piranvisseh-Volk M, Daubert JD, Dehaven R, Stein C. Peripheral effects of the kappa-opioid agonist EMD 61753 on pain and inflammation in rats and humans. J Pharmacol Exp Ther 1999; 290:354–361.
541. Diop L, Riviere PJ, Pascaud X, Junien JL. Peripheral kappa-opioid receptors mediate the antinociceptive effect of fedotozine (correction of fetodozine) on the duodenal pain reflex inrat. Eur J Pharmacol 1994;271: 65–71.
542. Allescher HD, Ahmad S, Classen M, Daniel EE. Interaction of trimebutine and Jo-1196 (fedotozine) with opioid receptors in the canine ileum. J Pharmacol Exp Ther 1991;257: 836–842.
543. Soulard CD, Guerif S, Payne A, Dahl SG. Differential effects of fedotozine compared to other kappa agonists on diuresis in rats. J Pharmacol Exp Ther 1996;279:1379–1385.
544. Broqua P, Wettstein JG, Rocher MN, Riviere PJ, Dahl SG. The discriminative stimulus properties of U50,488 and morphine are not shared by fedotozine. Eur Neuropsychopharmacol 1998;8:261–266.
545. Delvaux M. Pharmacology and clinical experience with fedotozine. Expert Opin Investig Drugs 2001;10:97–110.

546. Fraitag B, Homerin M, Hecketsweiler P. Double-blind dose-response multicenter comparison of fedotozine and placebo in treatment of nonulcer dyspepsia. Dig Dis Sci 1994;39:1072–1077.

547. Dapoigny M, Abitbol JL, Fraitag B. Efficacy of peripheral kappa agonist fedotozine versus placebo in treatment of irritable bowel syndrome. A multicenter dose-response study. Dig Dis Sci 1995;40:2244–2249.

548. Read NW, Abitbol JL, Bardhan KD, Whorwell PJ, Fraitag B. Efficacy and safety of the peripheral kappa agonist fedotozine versus placebo in the treatment of functional dyspepsia. Gut 1997;41:664–668.

549. Coffin B, Bouhassira D, Chollet R, Fraitag B, De Meynard C, Geneve J, Lemann M, Willer JC, Jian R. Effect of the kappa agonist fedotozine on perception of gastric distension in healthy humans. Aliment Pharmacol Ther 1996;10:919–925.

550. Delvaux M, Louvel D, Lagier E, Scherrer B, Abitbol JL, Frexinos J. The kappa agonist fedotozine relieves hypersensitivity to colonic distention in patients with irritable bowel syndrome. Gastroenterology 1999;116: 38–45.

551. Chang K-J, Rigdon GC, Howard JL, McNutt RW. A novel, potent and selective nonpeptidic delta opioid receptor agonist BW373U86. J Pharmacol Exp Ther 1993;267:852–857.

552. Wild KD, McCormick J, Bilsky EJ, Vanderah T, McNutt RW, Chang K-J, Porreca F. Antinociceptive actions of BW373U86 in the mouse. J Pharmacol Exp Ther 1993;267: 858–865.

553. Negus SS, Butelman ER, Chang K-J, DeCosta B, Winger G, Woods JH. Behavioral effects of the systemically active delta opioid agonist BW373U86 in rhesus monkeys. J Pharmacol Exp Ther 1994;270:1025–1034.

554. Comer SD, Hoenicke EM, Sable AI, McNutt RW, Chang K-J, de Costa BR, Mosberg HI, Woods JH. Convulsive effects of systemic administration of the delta opioid agonist BW373U86 in mice. J Pharmacol Exp Ther 1993;267:888–895.

555. Dykstra LA, Schoenbaum GM, Yarbrough J, McNutt R, Chang K-J. A novel delta opioid agonist, BW373U86, in squirrel monkeys responding under a schedule of shock titration. J Pharmacol Exp Ther 1993;267: 875–882.

556. Bilsky EJ, Calderon SN, Wang T, Bernstein RN, Davis P, Hruby VJ, McNutt RW, Rothman RB, Rice KC, Porreca F. SNC 80, A selective, nonpeptidic and systemically active opioid delta agonist. J Pharmacol Exp Ther 1995;273:359–366.

557. Schetz JA, Calderon SN, Bertha CM, Rice K, Porreca F, Yu H, Prisinzano T, Dersch CM, Marcus J, Rothman RB, Jacobson AE, Rice KC. Rapid in vivo metabolism of a methylether derivative of (+/−)-BW373U86: the metabolic fate of [^{3}H]SNC121 in rats. J Pharmacol Exp Ther 1996;279:1069–1076.

558. Calderon SN, Rice KC, Rothman RB, Porreca F, Flippen-Anderson JL, Kayakiri H, Xu H, Becketts K, Smith LE, Bilsky EJ, Davis P, Horvath R. Probes for narcotic receptor mediated phenomena. 23. Synthesis, opioid receptor binding, and bioassay of the highly selective delta agonist (+)-4-[(alpha$R$)-alpha-((2S,5R)-4-allyl-2,5-dimethyl-1-piperazinyl)-3-methoxybenzyl]-N,N-diethylbenzamide (SNC 80) and related novel nonpeptide delta opioid receptor ligands. J Med Chem 1997;40:695–704.

559. Katsura Y, Zhang X, Homma K, Rice KC, Calderon SN, Rothman RB, Yamamura HI, Davis P, Flippen-Anderson JL, Xu H, Becketts K, Foltz EJ, Porreca F. Probes for narcotic receptor-mediated phenomena. 25. Synthesis and evaluation of N-alkyl-substituted (alpha-piperazinylbenzyl)benzamides as novel, highly selective delta opioid receptor agonists. J Med Chem 1997;40:2936–2947.

560. Zhang X, Rice KC, Calderon SN, Kayakiri H, Smith L, Coop A, Jacobson AE, Rothman RB, Davis P, Dersch CM, Porreca F. Probes for narcotic receptor mediated phenomena. 26. Synthesis and biological evaluation of diarylmethylpiperazines and diarylmethylpiperidines as novel, nonpeptidic delta opioid receptor ligands. J Med Chem 1999;42: 5455–5463.

561. Furness MS, Zhang X, Coop A, Jacobson AE, Rothman RB, Dersch CM, Xu H, Porreca F, Rice KC. Probes for narcotic receptor-mediated phenomena. 27. Synthesis and pharmacological evaluation of selective delta-opioid receptor agonists from 4-[(alpha$R$)-alpha-(2S,5R)-4-substituted-2,5-dimethyl-1-piperazinyl-3-methoxybenzyl]-N,N-diethylbenzamides and their enantiomers. J Med Chem 2000;43:3193–3196.

562. Cottney J, Rankovic Z, Morphy JR. Synthesis of novel analogues of the delta opioid ligand SNC-80 using REM resin. Bioorg Med Chem Lett 1999;9:1323–1328.

563. Barn DR, Bom A, Cottney J, Caulfield WL, Morphy JR. Synthesis of novel analogues of

the delta opioid ligand SNC-80 using AlCl3-promoted aminolysis. Bioorg Med Chem Lett 1999;9:1329–1334.

564. Plobeck N, Delorme D, Wei ZY, Yang H, Zhou F, Schwarz P, Gawell L, Gagnon H, Pelcman B, Schmidt R, Yue SY, Walpole C, Brown W, Zhou E, Labarre M, Payza K, St-Onge S, Kamassah A, Morin PE, Projean D, Ducharme J, Roberts E. New diarylmethylpiperazines as potent and selective nonpeptidic delta opioid receptor agonists with increased in vitro metabolic stability. J Med Chem 2000;43:3878–3894.

565. Wei ZY, Brown W, Takasaki B, Plobeck N, Delorme D, Zhou F, Yang H, Jones P, Gawell L, Gagnon H, Schmidt R, Yue SY, Walpole C, Payza K, St-Onge S, Labarre M, Godbout C, Jakob A, Butterworth J, Kamassah A, Morin PE, Projean D, Ducharme J, Roberts E. N,N-Diethyl-4-(phenylpiperidin-4-ylidenemethyl) benzamide: a novel, exceptionally selective, potent delta opioid receptor agonist with oral bioavailability and its analogues. J Med Chem 2000;43:3895–3905.

566. Liao S, Alfaro-Lopez J, Shenderovich MD, Hosohata K, Lin J, Li X, Stropova D, Davis P, Jernigan KA, Porreca F, Yamamura HI, Hruby VJ. De novo design, synthesis, and biological activities of high-affinity and selective non-peptide agonists of the delta-opioid receptor. J Med Chem 1998;41:4767–4776.

567. Alfaro-Lopez J, Okayama T, Hosohata K, Davis P, Porreca F, Yamamura HI, Hruby VJ. Exploring the structure–activity relationships of 1-(4-tert-butyl-3'-hydroxy)benzhydryl-4-benzylpiperazine (SL-3111), a high-affinity and selective delta-opioid receptor nonpeptide agonist ligand. J Med Chem 1999;42:5359–5368.

568. Thomas JB, Herault XM, Rothman RB, Burgess JP, Mascarella SW, Xu H, Horel RB, Dersch CM, Carroll FI. (+/−)-4-[(N-allyl-cis-3-methyl-4-piperidinyl)phenylamino]-N,N-diethylbenza mide displays selective binding for the delta opioid receptor. Bioorg Med Chem Lett 1999;9:3053–3056.

569. Thomas JB, Atkinson RN, Herault XM, Rothman RB, Mascarella SW, Dersch CM, Xu H, Horel RB, Carroll FI. Optically pure (−)-4-[(N-allyl-3-methyl-4-piperidinyl)phenyl-amino]-N,N-diethylbenzami de displays selective binding and full agonist activity for the delta opioid receptor. Bioorg Med Chem Lett 1999;9:3347–3350.

570. Thomas JB, Herault XM, Rothman RB, Atkinson RN, Burgess JP, Mascarella SW, Dersch CM, Xu H, Flippen-Anderson JL, George CF, Carroll FI. Factors influencing agonist potency and selectivity for the opioid delta receptor are revealed in structure–activity relationship studies of the 4-[(N-substituted-4-piperidinyl)arylamino]-N,N-diethylbenzamides. J Med Chem 2001;44:972–987.

571. Thomas JB, Atkinson RN, Rothman RB, Burgess JP, Mascarella SW, Dersch CM, Xu H, Carroll FI. 4-[(8-Alkyl-8-azabicyclo[3.2.1]octyl-3-yl)-3-arylanilino]-N,N-diethylbenza mides: high affinity, selective ligands for the delta opioid receptor illustrate factors important to antagonist activity. Bioorg Med Chem Lett 2000;10:1281–1284.

572. Boyd RE, Carson JR, Codd EE, Gauthier AD, Neilson LA, Zhang SP. Synthesis and binding affinities of 4-diarylaminotropanes, a new class of delta opioid agonists. Bioorg Med Chem Lett 2000;10:1109–1111.

573. Nortey SO, Baxter EW, Codd EE, Zhang SP, Reitz AB. Piperazinyl benzamidines: synthesis and affinity for the delta opioid receptor. Bioorg Med Chem Lett 2001;11:1741–1743.

574. Dondio G, Ronzoni S, Eggleston DS, Artico M, Petrillo P, Petrone G, Visentin L, Farina C, Vecchietti V, Clarke GD. Discovery of a novel class of substituted pyrrolooctahydroisoquinolines as potent and selective delta opioid agonists, based on an extension of the message-address concept. J Med Chem 1997;40:3192–3198.

575. Huang P, Kim S, Loew G. Development of a common 3D pharmacophore for δ-opioid recognition from peptides and non-peptides using a novel computer program. J Comput Aided Mol Des 1997;11:21–28.

576. Coop A, Jacobson AE. The LMC delta opioid recognition pharmacophore: comparison of SNC80 and oxymorphindole. Bioorg Med Chem Lett 1999;9:357–362.

577. Dondio G, Clarke GD, Giardina G, Petrillo P, Petrone G, Ronzoni S, Visentin L, Vecchietti V. The role of the "spacer" in the octahydroisoquinoline series: discovery of SB 213698, a non-peptidic, potent and selective delta opioid agonists. Analgesia 1995;1:394–399.

578. Knapp RJ, Landsman R, Waite S, Malatynska E, Varga E, Haq W, Hruby VJ, Roeske WR, Nagase H, Yamamura HI. Properties of TAN-67, a nonpeptidic delta-opioid receptor agonist, at cloned human delta- and mu-

opioid receptors. Eur J Pharmacol 1995; 291:129–134.

579. Kamei J, Saitoh A, Ohsawa M, Suzuki T, Misawa M, Nagase H, Kasuya Y. Antinociceptive effects of the selective non-peptidic δ-opioid receptor agonist TAN-67 in diabetic mice. Eur J Pharmacol 1995;276:131–135.

580. Suzuki T, Tsuji M, Mori T, Misawa M, Endoh T, Nagase H. Effects of a highly selective nonpeptide δ opioid receptor agonist, TAN-67, on morphine-induced antinociception in mice. Life Sci 1995;57:155–168.

581. Nagase H, Yajima Y, Fujii H, Kawamura K, Narita M, Kamei J, Suzuki T. The pharmacological profile of delta opioid receptor ligands, (+) and (–) TAN-67 on pain modulation. Life Sci 2001;68:2227–2231.

582. Kamei J, Kawai K, Mizusuna A, Saitoh A, Morita K, Narita M, Tseng JL, Nagase H. Supraspinal δ_1-opioid receptor-mediated antinociceptive properties of (–)-TAN-67 in diabetic mice. Eur J Pharmacol 1997;322: 27–30.

583. Tseng JL, Narita M, Mizoguchi H, Kawai K, Mizusuna A, Kamei J, Suzuki T, Nagase H. Delta-1 opioid receptor-mediated antinociceptive properties of a nonpeptidic delta opioid receptor agonist, (–)TAN-67, in the mouse spinal cord. J Pharmacol Exp Ther 1997;280:600–605.

584. Hosohata Y, Varga EV, Stropova D, Li X, Knapp RJ, Hruby VJ, Rice KC, Nagase H, Roeske WR, Yamamura HI. Mutation W284L of the human delta opioid receptor reveals agonist specific receptor conformations for G protein activation. Life Sci 2001;68: 2233–2242.

585. Shenderovich MD, Liao S, Qian X, Hruby VJ. A three-dimensional model of the delta-opioid pharmacophore: comparative molecular modeling of peptide and nonpeptide ligands. Biopolymers 2000;53:565–580.

586. Dondio G, Ronzoni S, Petrillo P, DesJarlais RL, Raveglia LF. Pyrrolooctahydroisoquinolines as potent and selective δ opioid receptor ligands: SAR analysis and docking studies. Bioorg Med Chem Lett 1997;7:2967–2972.

587. Bertha CM, Flippen-Anderson JL, Rothman RB, Porreca F, Davis P, Xu H, Beckketts K, Cha X-Y, Rice KC. Probes for narcotic receptor-mediated phenomena. 20. Alteration of opioid receptor subtype selectivity of the 5-(3-hydroxyphenyl)morphans by application of the message-address concept: preparation of δ-opioid receptor ligands. J Med Chem 1995;38:1523–1537.

588. Bertha CM, Ellis M, Flippen-Anderson JL, Porreca F, Rothman RB, Davis P, Xu H, Becketts K, Rice KC. Probes for narcotic receptor-mediated phenomena. 21. Novel derivatives of 3-(1,2,3,4,5,11-hexahydro-3-methyl-2,6-methano-6H-azocino[4,5-b]indol-6-yl)-phenols with improved delta opioid receptor selectivity. J Med Chem 1996;39: 2081–2086.

589. Barn DR, Morphy JR. Solid-phase synthesis of cyclic imides. J Comb Chem 1999;1: 151–156.

590. Barn DR, Caulfield WL, Cottney J, McGurk K, Morphy JR, Rankovic Z, Roberts B. Parallel synthesis and biological activity of a new class of high affinity and selective delta-opioid ligand. Bioorg Med Chem 2001;9: 2609–2624.

591. Takemori AE, Portoghese PS. Affinity labels for opioid receptors. Annu Rev Pharmacol Toxicol 1985;25:193–223.

592. Newman AH. Irreversible ligands for drug receptor characterization. Annu Rep Med Chem 1989;25:271–280.

593. Newman AH. Irreversible ligands as probes for drug receptors. NIDA Res Monogr 1991;112:256–283.

594. Portoghese PS, Larson DL, Jiang JB, Takemori AE, Caruso TP. 6b-[N,N-Bis(2-chloroethyl)amino]-17-(cyclopropylmethyl)-4,5α-epoxy-3,14-dihydroxymorphinan (chlornaltrexamine), a potent opioid receptor alkylating agent with ultralong narcotic antagonist activity. J Med Chem 1978;21:598–599.

595. Portoghese PS, Larson DL, Jiang JB, Caruso TP, Takemori AE. Synthesis and Pharmacologic characterization of an alkylating analogue (chlornaltrexamine) of naltrexone with ultralong-lasting narcotic antagonist properties. J Med Chem 1979;22:168–173.

596. James IF, Chavkin C, Goldstein A. Preparation of brain membranes containing a single type of opioid receptor highly selective for dynorphin. Proc Natl Acad Sci USA 1982; 79:7570–7574.

597. Schoenecker JW, Takemori AE, Portoghese PS. Opioid agonist and antagonist activities of monofunctional nitrogen mustard analogues of β-chlornaltrexamine. J Med Chem 1987;30:933–935.

598. Caruso TP, Takemori AE, Larson DL, Portoghese PS. Chloroxymorphamine, an opioid receptor site-directed alkylating agent having narcotic agonist activity. Science 1979; 204:316–318.

599. Caruso TP, Larson DL, Portoghese PS, Takemori AE. Pharmacological studies with an

alkylating narcotic agonist chloroxymorphamine and antagonist, chlornaltrexamine. J Pharmacol Exp Ther 1980;213:539–544.
600. Larson AA, Armstrong MJ. Morphine analgesia after intrathecal administration of a narcotic agonist, chloroxymorphamine and antagonist, chlornaltrexamine. Eur J Pharmacol 1980;68:25–31.
601. Portoghese PS, Larson DL, Sayre LM, Fries DS, Takemori AE. A Novel opioid receptor site directed alkylating agent with irreversible narcotic antagonist and reversible agonistic activities. J Med Chem 1980;23:233–234.
602. Sayre LM, Larson DL, Fries DS, Takemori AE, Portoghese PS. Importance of C-6 chirality in conferring irreversible opioid antagonism to naltrexone-derived affinity labels. J Med Chem 1983;26:1229–1235.
603. Hayes AG, Sheehan MJ, Tyers MB. Determination of the receptor selectivity of opioid agonists in the guinea-pig ileum and mouse vas deferens by use of β-Funaltrexamine. Br J Pharmacol 1985;86:899–904.
604. Griffin JF, Larson DL, Portoghese PS. Crystal structure of α- and β-funaltrexamine: conformational requirement of the fumarate moiety in the irreversible blockage of μ opioid receptors. J Med Chem 1986;29: 778–783.
605. Chen C, Xue J-C, Zhu J, Chen Y-W, Kunapuli S, de Riel JK, Yu L, Liu-Chen L-Y. Characterization of irreversible binding of β-funaltrexamine to the cloned rat μ opioid receptor. J Biol Chem 1995;270:17866–17870.
606. Le Bourdonnec B, El Kouhen R, Lunzer MM, Law PY, Loh HH, Portoghese PS. Reporter affinity labels: an o-phthalaldehyde derivative of β-naltrexamine as a fluorogenic ligand for opioid receptors. J Med Chem 2000;43:2489–2492.
607. Pasternak GW, Hahn EF. Long-Acing opiate agonists and antagonists: 14-hydroxydihydromorphinone hydrazones. J Med Chem 1980;23:674–676.
608. Johnson N, Pasternak GW. Binding of [^{3}H]Naloxonazine to rat brain membranes. Mol Pharmacol 1984;26:477–483.
609. Koman A, Kolb VM, Terenius L. Prolonged receptor blockade by opioid receptor probes. Pharma Res 1986;3:56–60.
610. Cruciani RA, Lutz RA, Munson PJ, Rodbard D. Naloxonazine Effects on the interaction of enkephalin analogs with mu-1, mu and delta opioid binding sites in rat brain membranes. J Pharmacol Exp Ther 1987;242:15–20.
611. Price M, Gistrak MA, Itzhak Y, Hahn EF, Pasternak GW. Receptor binding of [^{3}H]naloxone benzoylhydrazone: a reversible κ and slowly dissociable μ opiate. Mol Pharmacol 1989;35:67–74.
612. Cheng J, Roques BP, Gacel GA, Huang E, Pasternak GW. Kappa-3 opiate receptor-binding in the mouse and rat. Eur J Pharmacol 1992;226:15–20.
613. Jiang Q, Takemori AE, Sultana M, Portoghese PS, Bowen WD, Mosberg HI, Porreca F. Differential antagonism of opioid delta antinociception by [D-Ala^2, Leu^5, Cys^6]enkephalin and naltrindole 5′-isothiocyanate: evidence for delta receptor subtypes. J Pharmacol Exp Ther 1991;257:1069–1075.
614. Korlipara VL, Takemori AE, Portoghese PS. Electrophilic N-benzylnaltrindoles as δ-opioid receptor antagonists. J Med Chem 1995;38:1337–1343.
615. Lewis JW, Smith C, McCarthy P, Walter D, Kobylecki R, Myers M, Haynes A, Lewis C, Waltham K. New 14-aminomorphinones and codeinones. NIDA Res Monogr 1988;90:136–143.
616. Aceto MD, Bowman ER, May EL, Harris LS, Woods JH, Smith CB, Medzihradsky F, Jacobson AE. Very Long-acting narcotic antagonists: The 14β-p-substituted cinnamoylaminomorphinones and their partial μ agonist codeinone relatives. Arzneim Forsch Drug Res 1989;39:570–575.
617. Comer SD, Burke TF, Lewis JW, Woods JH. Clocinnamox: a novel systemically-active, irreversible opioid antagonist. J Pharmacol Exp Ther 1992;262:1051–1056.
618. Burke TF, Woods JH, Lewis JW, Medzihradsky F. Irreversible opioid antagonist effects of clocinnamox on opioid analgesia and mu receptor binding in mice. J Pharmacol Exp Ther 1994;271:715–721.
619. Zernig G, Burke T, Lewis JW, Woods JH. Clocinnamox blocks only mu receptors irreversibly: binding evidence. Regul Pept 1994;54:343–344.
620. Zernig G, van Bemmerl BC, Lewis JW, Woods JH. Characterization of [^{3}H]-clocinnamox binding in mouse brain membranes. Analgesia 1995;1:874–877.
621. Jiang Q, Sebastian A, Archer S, Bidlack JM. 5β-Methyl-14β-(p-nitrocinnamoylamino)-7,8-dihydromorphinone: a long-lasting μ-opioid receptor antagonist devoid of agonist properties. Eur J Pharmacol 1993;230:129–130.

622. Sebastian A, Bidlack JM, Jiang Q, Deecher D, Teitler M, Glick SD, Archer S. 14β-[p-Nitrocinnamoyl)amino]morphinones, 14β-[p-nitrocinnamoyl)amino]-7,8-dihydromorphinones, and their codeinone analogues: Synthesis and receptor activity. J Med Chem 1993;36: 3154–3160.

623. Jiang Q, Sebastian A, Archer S, Bidlack JM. 5β-Methyl-14β-(p-nitrocinnamoylamino)-7,8-dihydromorphinone and its corresponding N-cyclopropylmethyl analog, N-cyclopropylmethylnor-5β-methyl-14β-(p-nitrocinnamoylamino)-7,8-dihydromorphinone: mu-selective irreversible opioid antagonists. J Pharmacol Exp Ther 1994;268:1107–1113.

624. McLaughlin JP, Sebastian A, Archer S, Bidlack JM. 14β-Chlorocinnamoylamino derivatives of metopon: long-term μ-opioid receptor antagonists. Eur J Pharmacol 1997;320:121–129.

625. Broadbear JH, Sumpter TL, Burke TF, Husbands SM, Lewis JW, Woods JH, Traynor JR. Methocinnamox is a potent, long-lasting, and selective antagonist of morphine-mediated antinociception in the mouse: comparison with clocinnamox, β-funaltrexamine, and β-chlornaltrexamine. J Pharmacol Exp Ther 2000;294:933–940.

626. Husbands SM, Sadd J, Broadbear JH, Woods JH, Martin J, Traynor JR, Aceto MD, Bowman ER, Harris LS, Lewis JW. 3-Alkyl ethers of clocinnamox: delayed long-term μ-atagonists with variable efficacy. J Med Chem 1998;41:3493–3498.

627. Derrick I, Neilan CL, Andes J, Husbands SM, Woods JH. 3-Deoxyclocinnamox: the first high-affinity, nonpeptide μ-opioid antagonist lacking a phenolic hydroxyl group. J Med Chem 2000;43:3348–3350.

628. Burke TR, Rice KC, Jacobson AE, Simonds WF, Klee WA. Probes for narcotic receptor mediated phenomena. 8. Tritiation of irreversible mu and delta specific opioid receptor affinity ligands to high specific activity. J Labelled Comp Radiopharm 1984;21: 693–702.

629. Klee WA, Simonds WF, Sweat FW, Burke TRJ, Jacobson AE, Rice KC. Identification of a Mr58000 glycoprotein subunit of the opiate receptor. FEBS Lett 1982;150:125–128.

630. Simonds WF, Burke TRJ, Rice KC, Jacobson AE, Klee WA. Purification of the opiate receptor of NG108-15 neuroblastoma-glioma hybrid cells. Proc Nat Acad Sci USA 1985;82:4974–4978.

631. Zhu J, Yin J, Law P-Y, Claude PA, Rice KC, Evans CJ, Chen C, Yu L, Liu-Chen L-Y. Irreversible binding of cis-(+)-3-methylfentanyl isothiocyanate to the δ opioid receptor and determination of its binding domain. J Biol Chem 1996;271:1430–1434.

632. de Costa BR, Rothman RB, Bykov V, Jacobson AE, Rice KC. Selective and enantiospecific acylation of κ opioid receptors by (1S,2S)-trans-2-isothiocyanato-N-methyl-N-[2-(1-pyrrolidinyl)cyclohexyl]benzeneacetamide. Demonstration of κ receptor heterogeneity. J Med Chem 1989;32:281–283.

633. de Costa BR, Band L, Rothman RB, Jacobson AE, Bykov V, Pert A, Rice KC. Synthesis of an affinity ligand ('UPHIT') for in vivo acylation of the κ-opioid receptor. FEBS Lett 1989;249: 178–182.

634. de Costa BR, Rothman RB, Bykov V, Band L, Pert A, Jacobson AE, Rice KC. Probes for narcotic receptor mediated phenomena. 17. Synthesis and evaluation of a series of trans-3,4-dichloro-N-methyl-N-[2-(1-pyrrolidinyl)cyclohexyl]benzeneacetamide (U50,488) related isothiocyanate derivatives as opioid receptor affinity ligands. J Med Chem 1990;33: 1171–1176.

635. Horan P, de Costa BR, Rice KC, Porecca F. Differential antagonism of U69,593- and Bremazocine-induced antinociception by (−)-UPHIT: evidence of kappa opioid receptor multiplicity in mice. J Pharmacol Exp Ther 1991;257:1154–1161.

636. Chang A-C, Takemori AE, Portoghese PS. 2-(3,4-Dichlorophenyl)-N-methyl-N-[(1S)1-(3- isothiocyanatophenyl)-2-(1-pyrrolidinyl) ethyl]acetamide: an opioid receptor affinity label that produces selective and long-lasting κ antagonism in mice. J Med Chem 1994;37: 1547–1549.

637. Chang A-C, Takemori AE, Ojala WH, Gleason WB, Portoghese PS. κ-Opioid receptor selective affinity labels: elctrophilic benzeneacetamides as κ-selective opioid antagonists. J Med Chem 1994;37:4490–4498.

638. Chen C, Yin J, Li JG, Xue JC, Weerawarna SA, Nelson WL, Liu-Chen LY. Irreversible binding of N-methyl-N-[(1S)-1-(4-isothiocyanatophenyl)-2-(1-pyrrolidinyl)ethyl-3,4-dichlorophenylacetamide to the cloned rat kappa opioid receptor. Life Sci 1997;61: 787–794.

639. Weerawarna SA, Davis RD, Nelson WL. Isothiocyanate-substituted κ-selective opioid receptor ligands derived from N-methyl-N-

[(1S)-1-phenyl-2-(1-pyrrolidinyl)ethyl]-phenylacetamide. J Med Chem 1994;37: 2856–2864.

640. Glasel JA, Venn RF. The sensitivity of opiate receptors and ligands to short wavelength ultraviolet light. Life Sci 1981;29: 221–228.

641. Eddy NB. Chemical structure and action of morphine like analgesics and related substances. Sixth Lister Memorial lecture. Chem Ind (London) 1959:1462–1469.

642. O'Brien JJ, Benfield P. Dazocine: a preliminary review of its pharmacodynamic and pharmacokinetic properties, and therapeutic efficacy. Drugs 1989;38:226.

643. Lee CR, McTavish D, Sorkin EM. Tramadol: a preliminary review of its pharmacodynamic and pharmacokinetic properties, and therapeutic potential in acute and chronic pain states. Drugs 1993;46:313–340.

644. Scott LJ, Perry CM. Tramadol: a review of its use in perioperative pain. Drugs 2000;60: 139–176.

645. Raffa RB, Friderichs E, Reimann W, Shank RP, Codd EE, Vaught JL. Opioid and nonopioid components independently contribute to the mechanism of action of tramadol, an 'atypical' opioid analgesic. J Pharmacol Exp Ther 1992;260:275–285.

646. Raffa RB, Friderichs E, Reimann W, Shank RP, Codd EE, Vaught JL, Jacoby HI, Selve N. Complementary and Synergistic antinociceptive interaction between the enantiomers of tramadol. J Pharmacol Exp Ther 1993;267: 331–340.

647. Collina S, Azzolina O, Vercesi D, Sbacchi M, Scheideler MA, Barbieri A, Lanza E, Ghislandi V. Synthesis and antinociceptive activity of pyrrolidinylnaphthalenes. Bioorg Med Chem 2000;8:1925–1930.

648. Vianello P, Albinati A, Pinna GA, Lavecchia A, Marinelli L, Borea PA, Gessi S, Fadda P, Tronci S, Cignarella G. Synthesis, molecular modeling, and opioid receptor affinity of 9,10-diazatricyclo[4/2/1/1^{2,5}]decanes and 2,7-diazatricyclo[4.4.0.0^{3,8}]decanes structurally related to 3,8-diazabicyclo[3.2.1]octanes. J Med Chem 2000;43:2115–2123.

649. Allen MP, Blake JF, Bryce DK, Haggan ME, Liras S, McLean S, Segelstein BE. Design, synthesis and biolgical evaluation of 3-amino-3-phenylpripionamide derivatives as novel opioid receptor ligands. Bioorg Med Chem Lett 2000;10:523–526.

650. Morley JS. Structure–activity relationships of enkephalin-like peptides. Ann Rev Pharmacol Toxicol 1980;20:81–110.

651. Udenfried SM, Meienhofer J. Opioid Peptides: Biology, Chemistry, and Genetics. Volume 6, Orlando: Academic Press; 1984.

652. Hruby VJ, Gehrig CA. Recent developments in the design of receptor specific opioid peptides. Med Res Rev 1989;9:343–401.

653. Hruby VJ, Agnes RS. Conformation-activity relationships of opioid peptides with selective activities at opioid receptors. Biopolymers 1999;51:391–410.

654. Schiller PW. Development of receptor-specific opioid peptide analogues. Prog Med Chem 1991;28:301–340.

655. Schiller PW. Development of receptor-selective opioid peptide analogs as pharmacologic tools and as potential drugs. In: Herz A, Akil H, Simon EJ, editors. Opioids I. Volume 104/I, Berlin: Springer-Verlag; 1993. p 681–710.

656. Brantl V, Teschemacher H, Henschen A, Lottspeich F. Novel opioid peptides derived from casein (β-casomorphins). I. Isolation from bovine casein peptone. Hoppe Seylers Z Physiol Chem 1979;360:1211–1216.

657. Henschen A, Lottspeich F, Brantl V, Teschemacher H. Novel opioid peptides derived from casein (β-casomorphins). II. Structure of Active Components from bovine casein peptone. Hoppe Seylers Z Physiol Chem 1979;360: 1217–1224.

658. Brantl V. Novel opioid peptides derived from human β-casein: human β-casomorphins. Eur J Pharmacol 1985;106:213–214.

659. Teschemacher H. Atypical opioid peptides. In: Herz A, Akil H, Simon EJ, editors. Opioids I. Volume 104/I, Berlin: Springer-Verlag; 1993. p 499–528.

660. Erspamer V. The opioid peptides of the amphibian skin. Int J Dev Neurosci 1992;10: 3–30.

661. Montecucchi PC, de Castiglione R, Piani S, Gozzini L, Erspamer V. Amino acid composition and sequence of dermorphin, a novel opiate-like peptide from the skin of *Phyllomdusa sauvagei*. Int J Pept Protein Res 1981;17:275–283.

662. Richter K, Egger R, Kreil G. D-Alanine in the frog skin peptide dermorphin is derived from L-alanine in the precursor. Science 1987;238: 200–202.

663. Kreil G, Barra D, Simmaco M, Erspamer V, Erspamer GF, Negri L, Severini C, Corsi R,

Melchiorri P. Deltorphin, a novel amphibian skin peptide with high selectivity and affinity for δ opioid receptors. Eur J Pharmacol 1989;162:123–128.

664. Mor A, Delfour A, Sagan S, Amiche M, Pradelles P, Rossier J, Nicolas P. Isolation of dermenkephalin from amphibian skin, a high-affinity delta-selective opioid heptapeptide containing a D-amino acid residue. FEBS Lett 1989;255:269–274.

665. Lazarus LH, Wilson WE, de Castiglione R, Guglietta A. Dermorphin gene sequence peptide with high affinity and selectivity for d-opioid receptors. J Biol Chem 1989;264:3047–3050.

666. Amiche M, Sagan S, Mor A, Delfour A, Nicolas P. Dermenkephalin (Tyr-D-Met-Phe-His-Leu-Met-Asp-NH_2): a potent and fully specific agonist for the δ opioid receptor. Mol Pharmacol 1989;35:774–779.

667. Negri L, Melchiorri P, Lattanzi R. Pharmacology of amphibian opiate peptides. Peptides 2000;21:1639–1647.

668. Hambrook JM, Morgan BA, Rance MJ, Smith CFC. Mode of deactivation of the enkephalins by rat and human plasma and rat brain homogenates. Nature 1976;262:782–783.

669. Audigier Y, Mazarguil H, Gout R, Cros J. Structure–activity relationships of enkephalin analogs at opiate and enkephalin receptors: correlation with analgesia. Eur J Pharmacol 1980;63:35–46.

670. Roques BP, Gacel C, Fournie-Zaluski M-C, Senault B, LeComte J-M. Demonstration of the crucial role of the phenylalanine moiety in enkephalin analogues for differential recognition of the μ- and δ-receptors. Eur J Pharmacol 1979;60:109–110.

671. Fournie-Zaluski M-C, Gacel G, Maigret B, Premilat S, Roques BP. Structural requirements for specific recognition of μ and δ opiate receptors. Mol Pharmacol 1981;20:484–491.

672. Handa BK, Lane AC, Lord JAH, Morgan BA, Rance MJ, Smith CFC. Analogues of β-LPH_{61-64} possessing selective agonist activity at μ-opiate receptors. Eur J Pharmacol 1981;70:531–540.

673. Kiso Y, Yamaguchi M, Akita T, Moritoki H, Takei M, Nakamura H. Simple tripeptide hydroxyalkylamides exhibit surprisingly high and long-lasting opioid activities. Naturwissenschaften 1981;68:210–212.

674. Shuman RT, Hynes MD, Woods JH, Gesellchen P. A highly selective in vitro μ-opioid agonist with atypical in vivo pharmacology. In: Rivier JE, Marshall GR, editors. Peptides: Chemistry, Structure, Biology. Leiden: ESCOM; 1990. p 326–328.

675. Gacel G, Zajac JM, Delay-Goyet P, Daugé V, Roques BP. Investigation of the structural parameters involved in the μ and δ opioid receptor discrimination of linear enkephalin-related peptides. J Med Chem 1988;31:374–383.

676. Hruby VJ, Li G, Haskell-Luevano C, Shenderovich M. Design of peptides, proteins and peptidomimetics in chi space. Biopolymers 1997;43:219–266.

677. Hruby VJ. Design in topographical space of peptide and peptidomimetic ligands that affect behavior. A chemist's glimpse at the mind–body problem. Acc Chem Res 2001;34:389–397.

678. Deeks T, Crooks PA, Waigh RD. Synthesis and analgesic properties of two leucine-enkephalin analogues containing a conformationally restrained N-terminal tyrosine Residue. J Med Chem 1983;26:762–765.

679. Sasaki Y, Suto T, Ambo A, Ouchi H, Yamamoto Y. Biological properties of opioid peptides rplacing Tyr at position 1 by 2,6-dimethyl-Tyr. Chem Pharm Bull 1999;47:1506–1509.

680. Sasaki Y, Hirabuki M, Ambo A, Ouchi H, Yamamoto Y. Enkephalin analogues with 2′,6′-dimethylphenylalanine replacing phenylalanine in position 4. Bioorg Med Chem Lett 2001;11:327–329.

681. DiMaio J, Schiller PW. A cyclic enkephalin analog with high in vitro opiate activity. Proc Natl Acad Sci USA 1980;77:7162–7166.

682. Schiller PW, DiMaio J. Opiate receptor subclasses differ in their conformational requirements. Nature 1982;297:74–76.

683. DiMaio J, Nguyen TM-D, Lemieux C, Schiller PW. Synthesis and pharmacological characterization in vitro of cyclic enkephalin analogues: effect of conformational constraints on opiate receptor selectivity. J Med Chem 1982;25:1432–1438.

684. Hall D, Pavitt N. Conformation of a cyclic tetrapeptide related to an analog of enkephalin. Biopolymers 1984;23:1441–1445.

685. Hall D, Pavitt N. Conformation of cyclic analogs of enkephalin. III. Effect of varying ring size. Biopolymers 1985;24:935–945.

686. Mammi NJ, Hassan M, Goodman M. Conformational analysis of a cyclic enkephalin analogue by ^{1}HNMR and computer simulations. J Am Chem Soc 1985;107:4008–4013.

687. Maigret B, Fournié-Zaluski M-C, Roques BP, Premilat S. Proposals for the μ-active conformation of the enkephalin analog Tyr-cyclo(-N^γ-D-A_2-bu-Gly-Phe-Leu-). Mol Pharmacol 1986;29:314–320.

688. Kessler H, Hölzemann G, Zechel C. Peptide conformations. 33. Conformational analysis of cyclic enkephalin analogs of the type Tyr-cyclo(N^ω-Xxx-Gly-Phe-Leu-). Int J Pept Protein Res 1985;25:267–279.

689. Sherman DB, Spatola AF, Wire WS, Burks TF, Nguyen TM-D, Schiller PW. Biological activities of cyclic enkephalin pseudopeptides containing thioamides as amide bond replacements. Biochem Biophys Res Commun 1989;162:1126–1132.

690. Berman JM, Goodman M, Nguyen TM-D, Schiller PW. Cyclic and acyclic partial retro-inverso enkephalinamides: mu receptor selective enzyme resistant analogs. Biochem Biophys Res Commun 1983;115:864–870.

691. Quirion R. Syndyphalin SD-25: a highly selective ligand for μ opiate receptors. FEBS Lett 1982;141:203–206.

692. Bedell CR, Clark RB, Hardy GW, Lowe LA, Ubatuba FB, Vane JR, Wilkinson S, Chang K-J, Cuatrecasas P, Miller RJ. Structural requirements for opioid activity of analogues of the enkephalins. Proc R Soc Lond 1977;198:249–265.

693. Gacel G, Daugé V, Breuze P, Delay-Goyet P, Roques BP. Development of conformationally constrained linear peptides exhibiting a high affinity and pronounced selectivity for δ opioid receptors. J Med Chem 1988;31:1891–1897.

694. Zajac JM, Gacel G, Petit F, Dodey P, Rossignol P, Roques BP. Deltakephalin, Tyr-D-Thr-Gly-Phe-Leu-Thr: a new highly potent and fully specific agonist for opiate δ-receptors. Biochem Biophys Res Commun 1983;111:390–397.

695. Flippen-Anderson JL, George C, Deschamps JR. In: Hodges RS, Smith JA, editors. Peptides: Structure and Biology. Proceedings of the 13th American Peptide Symposium, Leiden, the Netherlands: ESCOM; 1994. p 490–492.

696. Flippen-Anderson JL, Deschamps JR, Ward KB, George C, Houghten R. Crystal structure of deltakephalin: a delta-selective opioid peptide with a novel beta-bend-like conformation. Int J Pept Protein Res 1994;44:97–104.

697. Deschamps JR, George C, Flippen-Anderson JL. Structural studies of opioid peptides: a review of recent progress in X-ray diffraction studies. Biopolymers 1996;40:121–139.

698. Gacel G, Fellion E, Baamonde A, Daugé V, Roques BP. Synthesis, biochemical and pharmacological properties of BUBUC, a Highly selective and systemically active agonist for in vivo studies of δ opioid receptors. Peptides 1990;11:983–988.

699. Bilsky EJ, Egleton RD, Mitchell SA, Palian MM, Davis P, Huber JD, Jones H, Yamamura HI, Janders J, Davis TP, Porreca F, Hruby VJ, Polt R. Enkephalin glycopeptide analogues produce analgesia with reduced dependence liability. J Med Chem 2000;43: 2586–2590.

700. Egleton RD, Mitchell SA, Huber JD, Palian MM, Polt R, Davis TP. Improved blood–brain barrier penetration and enhanced analgesia of an opioid peptide by glycosylation. J Pharmacol Exp Ther 2001;299:967–972.

701. Hansen PE, Morgan BA. Structure–Activity relationships in enkephalin peptides. In: Udenfriend S, Meienhofer J, editors. The Peptides: Analysis, Synthesis, Biology. Volume 6, Opioid Peptides: Biology, Chemistry, and Genetics. Orlando: Academic Press Inc. 1984. p 269–321.

702. Nitz TJ, Shimohigashi Y, Costa T, Chen H-C, Stammer CH. Synthesis and receptor binding affinity of both E- and Z-dehydrophenylalanine^4 enkephalins. Int J Pept Protein Res 1986;27:522–529.

703. Mapelli C, Kimura H, Stammer CH. Synthesis of four diastereomeric enkephalins incorporating cyclopropyl phenylalanine. Int J Pept Protein Res 1986;28:347–359.

704. Shimohigashi Y, Costa T, Pfeiffer A, Herz A, Kimura H, Stammer CH. $\triangledown^E$Phe^4-Enkephalin analogs. Delta receptors in rat brain are different from those in mouse vas deferens. FEBS Lett 1987;222:71–74.

705. Sarantakis D. Analgesic polypeptide. US patent 4,148,786. 1979.

706. Schiller PW, Eggimann B, DiMaio J, Lemieux C, Nguyen TM-D. Cyclic enkephalin analogs containing a cystine bridge. Biochem Biophys Res Commun 1981;101:337–343.

707. Mosberg HI, Hurst R, Hruby VJ, Galligan JJ, Burks TF, Gee K, Yamamura HI. [D-Pen^2, L-Cys^5]enkephalinamide and [D-Pen^2,D-Cys^5] enkephalinamide, conformationally constrained cyclic enkephalinamide analogs with δ receptor specificity. Biochem Biophy Res Commun 1982;106:506–512.

708. Mosberg HI, Hurst R, Hruby VJ, Galligan JJ, Burks TF, Gee K, Yamamura HI. Conforma-

tionally constrained cyclic enkephalin analogs with pronounced delta opioid receptor agonist selectivity. Life Sci 1983;32: 2565–2569.

709. Mosberg HI, Hurst R, Hruby VJ, Gee K, Akiyama K, Yamamura HI, Galligan JJ, Burks TF. Cyclic penicillamine containing enkephalin analogs display profound delta receptor selectivies. Life Sci 1983;33(Suppl 1): 447–450.

710. Mosberg HI, Hurst R, Hruby VJ, Gee K, Yamamura HI, Galligan JJ, Burks TF. Bis-penicillamine enkephalins possess highly improved selectivity toward δ opioid receptors. Proc Natl Acad Sci USA 1983;80: 5871–5874.

711. Mosberg HI, Haaseth RC, Ramalingham K, Mansour A, Akil H, Woodard RW. Role of steric interactions in the delta opioid receptor selectivity of [D-Pen^2,D-Pen^5]enkephalin. Int J Pept Protein Res 1988;32:1–8.

712. Toth G, Kramer TH, Knapp R, Lui G, Davis P, Burks TF, Yamamura HI, Hruby VJ. [D-Pen2, D-Pen5]enkephalin analogues with increased affinity and selectivity for δ opioid receptors. J Med Chem 1990;33:249–253.

713. Weber SJ, Greene DL, Sharma SD, Yamamura HI, Kramer TH, Burks TF, Hruby VJ, Hersh LB, Davis TP. Distribution and analgesia of [^{3}H][D-Pen^2,D-Pen^5]enkephalin and two halogenated analogs after intravenous administration. J Pharmacol Exp Ther 1991; 259:1109–1117.

714. Gentry CL, Egleton RD, Gillespie T, Abbruscato TJ, Bechowski HB, Hruby VJ, Davis TP. The effect of halogenation on blood–brain barrier permeability of a novel peptide drug. Peptides 1999;20:1229–1238.

715. Haaseth RC, Horan PJ, Bilsky EJ, Davis P, Zalewska T, Slaninova J, Yamamura HI, Weber SJ, Davis TP, Porreca F, Hruby VJ. [L-Ala^3]DPDPE: a new enkephalin analog with a unique opioid receptor activity profile. Further evidence of δ-opioid receptor multiplicity. J Med Chem 1994;37:1572–1577.

716. Collins, N, Flippen-Anderson, JL, Haaseth, RC, Deschamps, JR, George, C, Kövér, K, Hruby, VJ. Conformational determinants of agonist versus antagonist properties of [D-Pen^2,D-Pen^5]enkephlain (DPDPE) analogs at opioid receptors. Comparison of X-ray crystallographic structure, solution ^{1}H NMR data, and molecular dynamic simulations of [L-Ala^3]DPDPE and [D-Ala^3]DPDPE. J Am Chem Soc, 1996;118:2143–2152.

717. Bartosz-Bechowski H, Davis P, Zalewska T, Slaninova J, Porreca F, Yamamura HI, Hruby VJ. Cyclic enkephalin analogs with exceptional potency at peripheral delta opioid receptors. J Med Chem 1994;37:146–150.

718. Hruby VJ, Bartosz-Bechowski H, Davis P, Slaninova J, Zalewska T, Stropova D, Porreca F, Yamamura HI. Cyclic enkephalin analogues with exceptional potency and selectivity for delta-opioid receptors. J Med Chem 1997;40:3957–3962.

719. Polt R, Porreca F, Szabo LZ, Bilsky EJ, Davis P, Abbruscato TJ, Davis TP, Horvath R, Yamamura HI, Hruby VJ. Glycopeptide enkephalin analogues produce analgesia in mice: evidence for penetration of the blood–brain barrier. Proc Natl Acad Sci USA 1994;91:7114–7118.

720. Egleton RD, Mitchell SA, Huber JD, Janders J, Stropova D, Polt R, Yamamura HI, Hruby VJ, Davis TP. Improved bioavailability to the brain of glycosylated Met-enkephalin analogs. Brain Res 2000;881:37–46.

721. Witt KA, Huber JD, Egleton RD, Roberts MJ, Bentley MD, Guo L, Wei H, Yamamura HI, Davis TP. Pharmacodynamic and pharmacokinetic characterization of poly(ethylene glycol) conjugation to met-enkephalin analog [D-Pen2, D-Pen5]-enkephalin (DPDPE). J Pharmacol Exp Ther 2001;298:848–856.

722. Flippen-Anderson JL, Hruby VJ, Collins N, George C, Cudney B. X-ray Structure of [D-Pen^2,D-Pen^5]enkephalin, a Highly potent, δ opioid receptor-selective compound: comparisons with proposed solution conformations. J Am Chem Soc 1994;116: 7523–7531.

723. Toth G, Russell KC, Landis G, Kramer TH, Fang L, Knapp R, Davis P, Burks TF, Yamamura HI, Hruby VJ. Ring substituted and other conformationally constrained tyrosine analogues of [D-Pen^2,D-Pen^5]enkephalin with δ opioid receptor selectivity. J Med Chem 1992;35:2384–2391.

724. Qian XH, Kover KE, Shenderovich MD, Lou BS, Misicka A, Zalewska T, Horvath R, Davis P, Bilsky EJ, Porreca F, Yamamura HI, Hruby VJ. Newly discovered stereochemical requirements in the side-chain conformation of delta opioid agonists for recognizing opioid delta receptors. J Med Chem 1994;37: 1746–1757.

725. Qian X, Shenderovich MD, Kövér KE, Davis P, Horváth R, Zalewska T, Yamamura HI, Porreca F, Hruby VJ. Probing the stereochemical requirements for receptor recognition of δ opioid agonists through topographic modifications in position 1. J Am Chem Soc 1996; 118:7280–7290.

726. Nikiforovich GV, Hruby VJ, Prakaskh O, Gehrig CA. Topographical requirements for δ-selective opioid peptides. Biopolymers 1991;31:941–955.
727. Hansen DW, Stapelfeld A, Savage MA, Reichman M, Hammond DL, Haaseth RC, Mosberg HI. Systemic analgesic activity and δ-opioid selectivity in [2,6-dimethyl-Tyr^1,D-Pen^2,D-Pen^5]enkephalin. J Med Chem 1992;35: 684–687.
728. Bilsky EJ, Qian X, Hruby VJ, Porreca F. Antinociceptive activity of [beta-methyl-2',6'-dimethyltyrosine(1)]-substituted cyclic [D-Pen(2), D-Pen(5)]enkephalin and [D-Ala(2), Asp(4)]deltorphin analogs. J Pharmacol Exp Ther 2000;293:151–158.
729. Hruby VJ, Toth G, Gehrig CA, Kao L-F, Knapp R, Lui GK, Yamamura HI, Kramer TH, Davis P, Burks TF. Topographically designed analogues of [D-Pen^2,D-Pen^5]enkephalin. J Med Chem 1991;34:1823–1830.
730. Witt KA, Slate CA, Egleton RD, Huber JD, Yamamura HI, Hruby VJ, Davis TP. Assessment of stereoselectivity of trimethylphenylalanine analogues of delta-opioid [D-Pen(2), D-Pen(5)]-enkephalin. J Neurochem 2000;75: 424–435.
731. Shaw JS, Miller L, Turnbull MJ, Gormley JJ, Morley JS. Selective antagonists at the opiate delta-receptor. Life Sci 1982;31:1259–1262.
732. Belton P, Cotton R, Giles MB, Gormley JJ, Miller L, Shaw JS, Timms D, Wilkinson A. Divergent structure–activity relationships in series of enkephalin agonists and cognate antagonists. Life Sciences 1983;33(Suppl I): 443–446.
733. Thornber CW, Shaw JS, Miller L, Hayward CF, Morley JS, Timms D, Wilkinson A. New δ-receptor antagonists. NIDA Res Monogr 1986;75:177–180.
734. Aldrich Lovett J, Portoghese PS. N, N-Dialkylated leucine enkephalins as potential δ opioid receptor antagonists. J Med Chem 1987;30:1144–1149.
735. Maeda DY, Ishmael JE, Murray TF, Aldrich JV. Synthesis and evaluation of n, n-dialkyl enkephalin-based affinity labels for δ opioid receptors. J Med Chem 2000;43:3941–3948.
736. Lu Y, Weltrowska G, Lemieux C, Chung NN, Schiller PW. Stereospecific synthesis of (2S)-2-methyl-3-(2',6'-dimethyl-4'-hydroxyphenyl)-propionic Acid (Mdp) and its incorporation into an opioid peptide. Bioorg Med Chem Lett 2001;2001:323–325.
737. Lutz RA, Cruciani RA, Shimohigashi Y, Costa T, Kassis S, Munson PJ, Rodbard D. Increased affinity and selectivity of enkephalin tripeptide (Tyr-D-Ala-Gly) Dimers. Eur J Pharmacol 1985;111:257–261.
738. Shimohigashi Y, Ogasawara T, Koshizaka T, Waki M, Kato T, Izumiya N, Kurono M, Yagi K. Interaction of dimers of inactive enkephalin fragments with μ opiate receptors. Biochem Biophys Res Commun 1987;146: 1109–1115.
739. Lipkowski AW, Konecka AM, Sroczynska I. Double enkephalins: synthesis, activity on guinea-pig ileium, and analgesic effect. Peptides 1982;3:697–670.
740. Horan PJ, Mattia A, Bilsky EJ, Weber S, Davis TP, Yamamura HI, Malatynska E, Appleyard SM, Slaninova J, Misicka A, Lipkowski AW, Hruby VJ, Porreca F. Antinociceptive profile of biphalin, a dimeric enkephalin analog. J Pharmacol Exp Ther 1993;265:1446–1454.
741. Lipkowski AW, Misicka A, Davis P, Stropova D, Janders J, Lachwa M, Porreca F, Yamamura HI, Hruby VJ. Biological activity of fragments and analogues of the potent dimeric opioid peptide, biphalin. Bioorg Med Chem Lett 1999;9:2763–2766.
742. Li G, Haq W, Xiang L, Lou BS, Hughes R, De Leon IA, Davis P, Gillespie TJ, Romanowski M, Zhu X, Misicka A, Lipkowski AW, Porreca F, Davis TP, Yamamura HI, O'Brien DF, Hruby VJ. Modifications of the 4,4'-residues and SAR studies of biphalin, a highly potent opioid receptor active peptide. Bioorg Med Chem Lett 1998;8:555–560.
743. Misicka A, Lipkowski AW, Horvath R, Davis P, Porreca F, Yamamura HI, Hruby VJ. Structure–activity relationship of biphalin. The synthesis and biological activities of new analogues with modifications in positions 3 and 4. Life Sci 1997;60:1263–1269.
744. Weber SJ, Abbruscato TJ, Brownson EA, Lipkowski AW, Polt R, Misicka M, Haaseth RC, Bartosz H, Hruby VJ, Davis TP. Assessment of an in vitro blood–brain barrier model using several [Met^5]enkephalin opioid analogs. J Pharmacol Exp Ther 1993;266: 1649–1655.
745. Abbruscato TJ, Williams SA, Misicka A, Lipkowski AW, Hruby VJ, Davis TP. Blood-to-central nervous system entry and stability of biphalin, a unique double-enkephalin analog, and its halogenated derivatives. J Pharmacol Exp Ther 1996;276:1049–1057.

746. Abbruscato TJ, Thomas SA, Hruby VJ, Davis TP. Brain and spinal cord distribution of biphalin: correlation with opioid receptor desnisty and mechanism of CNS entry. J Neurochem 1997;69:1236–1245.

747. Shimohigashi Y, Costa T, Chen H-C, Rodbard D. Dimeric tetrapeptide enkephalins display extraordinary selectivity for the δ opiate receptor. Nature 1982;297:333–335.

748. Shimohigashi Y, Costa T, Matsura S, Chen H-C, Rodbard D. Dimeric enkephalins display enhanced affinity and selectivity for the delta opiate receptor. Mol Pharmacol 1982;21:558–563.

749. Thornber CW, Shaw JS, Miller L, Hayward CF. Dimeric opioid antagonists. NIDA Res Monogr 1986;75:181.

750. Naqvi T, Haq W, Mathur KB. Structure–activity relationship studies of dynorphin A and related peptides. Peptides 1998;19:1277–1292.

751. Goldstein A, Fischli W, Lowney LI, Hunkapiller M, Hood L. Porcine pituitary dynorphin: complete amino acid sequence of the biologically active heptadecapeptide. Proc Natl Acad Sci USA 1981;78:7219–7223.

752. Schwyzer R. Estimated Conformation, Orientation, and Accumulation of Dynorphin A-(1-13)-tridecapeptide on the Surface of Neutral Lipid Membranes. Biochemistry 1986;25:4281–4286.

753. Kallick D. Conformation of dynorphin A(1-17) bound to dodecylphosphocholine micelles. J Am Chem Soc 1993;115:9317–9318.

754. Tessmer MR, Kallick DA. NMR and structural model of dynorphin A-(1-17) Bound to dodecylphosphocholine micelles. Biochemistry 1997;36:1971–1981.

755. Leslie FM, Goldstein A. Degradation of dynorphin-(1-13) by membrane-bound rat brain enzymes. Neuropeptides 1982;2:185–196.

756. Weber E, Evans CJ, Barchas JD. Predominance of the amino-terminal octapeptide fragment of dynorphin in rat brain regions. Nature 1982;299:77–79.

757. Cone RI, Weber E, Barchas JD, Goldstein A. Regional distribution of dynorphin and neo-endorphin peptides in rat brain, spinal cord, and pituitary. J Neurosci 1983;3:2146–2152.

758. Goldstein A. Biology and chemistry of the dynorphin peptides. In: Udenfried S, Meienhofer J, editors. The Peptides: Analysis, Synthesis, Biology. Opioid Peptides: Biology, Chemistry, and Genetics. Volume 6, Orlando: Academic Press; 1984. p 95–145.

759. Meyer JP, Davis P, Lee KB, Porreca F, Yamamura HI, Hruby VJ. Synthesis using a Fmoc-based strategy and biological activities of some reduced peptide bond pseudopeptide analogues of dynorphin A. J Med Chem 1995;38:3462–3468.

760. Meyer JP, Gillespie TJ, Hom S, Hruby VJ, Davis TP. In vitro stability of some reduced peptide bond pseudopeptide analogues of dynorphin A. Peptides 1995;16:1215–1219.

761. Yoshino H, Nakazawa T, Arakawa Y, Kaneko T, Tsuchiya Y, Matsunaga M, Araki S, Ikeda M, Yamatsu K, Tachibana S. Synthesis and structure–activity relationships of dynorphin A-(1-8) amide analogues. J Med Chem 1990;33:206–212.

762. Yoshino H, Kaneko T, Arakawa Y, Nakazawa T, Yamatsu K, Tachibana S. Synthesis and structure–activity relationships of [MeTyr^1, MeArg^7]-dynorphin A(1-8)-OH analogues with substitution at position 8. Chem Pharm Bull 1990;38:404–406.

763. Nakazawa T, Furuya Y, Kaneko T, Yamatsu K, Yoshino H, Tachibana S. Analgesia produced by E-2078, a Systemically active dynorphin analog, in mice. J Pharmacol Exp Ther 1990;252:1247–1254.

764. Nakazawa T, Furuya Y, Kaneko T, Yamatsu K. Spinal kappa-receptor-mediated analgesia of E-2078, a systemically active dynorphin analog, in mice. J Pharmacol Exp Ther 1991;256:76–81.

765. Gairin JE, Gouarderes C, Mazarguil H, Alvinerie P, Cros J. [D-Pro^{10}]Dynorphin-(1-11) is a highly potent and selective ligand for κ opioid receptors. Eur J Pharmacol 1985;106:457–458.

766. Lemaire S, Lafrance L, Dumont M. Synthesis and biological activity of dynorphin-(1-13) and analogs substituted in positions 8 and 10. Int J Pept Protein Res 1986;27:300–305.

767. Turcotte A, Lalonde J-M, St Pierre S, Lemaire S. Dynorphin-(1-13). I. Structure–function relationships of Ala-containing analogs. Int J Pept Protein Res 1984;23:361–367.

768. Snyder KR, Story SC, Heidt ME, Murray TF, DeLander GE, Aldrich JV. Effect of modification of the basic residues of dynorphin A-(1-13) amide on κ opioid receptor selectivity and opioid activity. J Med Chem 1992;35:4330–4333.

769. Kawasaki AM, Knapp RJ, Walton A, Wire WS, Zalewska T, Yamamura HI, Porreca F,

Burks TF, Hruby VJ. Synthesis, opioid binding affinities, and potencies of dynorphin A Analogues substituted in positions 1, 6, 7, 8 and 10. Int J Pept Protein Res 1993;42: 411–419.

770. Martinka GP, Jhamandas K, Sabourin L, Lapierre C, Lemaire S. Dynorphin-A-(1-13)-Tyr^{14}-Leu^{15}-Phe^{16}-Asn^{17}-Gly^{18}-Pro^{19}: a potent and selective κ-opioid peptide. Eur J Pharmacol 1991;196:161–167.

771. Shukla VK, Kemaire S, Ibrahim IH, Cyr TD, Chen Y, Michelot R. Design of potent and selective dynorphin A related peptides devoid of supraspinal motor effects in mice. Can J Physiol Pharmacol 1993;71:211–216.

772. Goldstein A, Nestor JJ, Naidu A, Newman SR. "DAKLI". A multipurpose ligand with high affinity and selectivity for dynorphin (κ opioid) binding sites. Proc Natl Acad Sci USA 1988;85:7375–7379.

773. Hochhaus G, Patthy A, Schwietert R, Santi DV, Sadee W. [Biocytin^{13}]dynorphin A 1-13 amide: a potential probe for the κ opioid receptor. Pharm Res 1988;5:790–794.

774. Choi H, Murray TF, DeLander GE, Caldwell V, Aldrich JV. N-terminal alkylated derivatives of [D-Pro^{10}]dynorphin A-(1-11) are highly selective for κ-opioid receptors. J Med Chem 1992;35:4638–4639.

775. Story SC, Murray TF, DeLander GE, Aldrich JV. Synthesis and opioid activity of 2-substituted dynorphin A-(1-13) amide analogues. Int J Pept Protein Res 1992;40:89–96.

776. Lung F-DT, Meyer J-P, Li G, Lou B-S, Stropova D, Davis P, Yamamura HI, Porreca F, Hruby VJ. Highly κ receptor-selective dynorphin A analogues with modifications in position 3 of dynorphin A(1-11)-NH$_2$. J Med Chem 1995;38:585–586.

777. Lung F-DT Meyer J-P, Lou B-S, Xiang L, Li G, Davis P, De Leon IA, Yamamura HI, Porreca F, Hruby VJ. Effects of modifications of residues in position 3 of dynorphin A(1-11)-NH$_2$ on κ receptor selectivity and potency. J Med Chem 1996;39:2456–2460.

778. Schlechtingen G, Zhang L, Maycock A, DeHaven RN, Daubert JD, Cassel J, Chung NN, Schiller PW, Goodman M. [Pro^3]Dyn A(1-11) NH$_2$: a dynorphin analogue with high selectivity for the κ opioid receptor. J Med Chem 2000;43:2698–2702.

779. Aldrich JV, Zheng Q, Murray TF. Dynorphin A analogues containing a conformationally constrained phenylalanine analogue in position 4: reversal of preferred stereochemistry for opioid receptor affinity and discrimination of κ vs. δ receptors. Chirality 2001;13: 125–129.

780. Snyder KR, Murray TF, DeLander GE, Aldrich JV. Synthesis and opioid activity of dynorphin A-(1-13)NH$_2$ analogues containing cis- and trans-4-aminocyclohexanecarboxylic acid. J Med Chem 1993;36:1100–1103.

781. Schiller PW, Eggimann B, Nguyen TM-D. Comparative structure–function studies with analogs of dynorphin-(1-13) and [Leu^5]enkephalin. Life Sci 1982;31:1777–1780.

782. Kawasaki AM, Knapp RJ, Kramer TH, Wire WS, Vasquez OS, Yamamura HI, Burks TF, Hruby VJ. Design and synthesis of highly potent and selective cyclic dynorphin-A analogues. J Med Chem 1990;33:1874–1879.

783. Kawasaki AM, Knapp RJ, Kramer TH, Walton A, Wire WS, Hashimoto S, Yamamura HI, Porreca F, Burks TF, Hruby VJ. Design and synthesis of highly potent and selective cyclic dynorphin A analogs. 2. New analogs. J Med Chem 1993;36:750–757.

784. Meyer JP, Collins N, Lung FD, Davis P, Zalewska T, Porreca F, Yamamura HI, Hruby VJ. Design, synthesis, and biological properties of highly potent cyclic dynorphin A analogues. Analogues cyclized between positions 5 and 11. J Med Chem 1994;37: 3910–3917.

785. Collins N, Hruby VJ. Prediction of the conformational requirements for binding to the κ-opioid receptor and its subtypes. 1. Novel α-helical cyclic peptides and their role in receptor selectivity. Biopolymers 1994;34: 1231–1241.

786. Tessmer MR, Meyer J-P, Hruby VJ, Kallick DA. Structural model of a cyclic dynorphin A analog bound to dodecylphospholine micelles by NMR and restrained molecular dynamics. J Med Chem 1997;40:2148–2155.

787. Schiller PW, Nguyen TM-D, Lemieux C. Synthesis and opioid activity profiles of cyclic dynorphin analogs. Tetrahedron 1988;44: 733–743.

788. Arttamangkul S, Murray TF, DeLander GE, Aldrich JV. Synthesis and opioid activity of conformationally constrained dynorphin a analogues. 1. Conformational constraint in the "message" sequence. J Med Chem 1995;38:2410–2417.

789. Arttamangkul S, Murray TF, DeLander GE, Aldrich JV. Synthesis and opioid activity of conformationally constrained dynorphin A analogues. Regul Pept 1994;54:13–14.

790. Arttamangkul S, Ishmael JE, Murray TF, Grandy DK, DeLander GE, Kieffer BL, Aldrich JV. Synthesis and opioid activity of conformationally constrained dynorphin A analogues. 2. Conformational constraint in the "address" sequence. J Med Chem 1997;40:1211–1218.

791. Lung F-DT Collins N, Stropova D, Davis P, Yamamura HI, Porreca F, Hruby VJ. Design, synthesis, and biological activities of cyclic lactam peptide analogues of dynorphin A(1-11)-NH$_2$. J Med Chem 1996;39:1136–1141.

792. Vig B, Murray TF, Aldrich JV. Synthesis of novel N-terminal cyclic dynorphin A analogues: strategies and side reactions. In: Peptides: Chemistry and Biology. 2002; in press.

793. Vig B. Synthesis and pharmacological evaluation of dynorphin A analogs constrained in the "message" sequence. Ph.D. Thesis, University of Maryland Baltimore, Baltimore, 2001.

794. Gairin JE, Mazarguil H, Alvinerie P, St Pierre S, Meunier J-C, Cros J. Synthesis and biological activities of dynorphin A analogues with opioid antagonist properties. J Med Chem 1986;29:1913–1917.

795. Lemaire S, Turcotte A. Synthesis and biological activity of analogs of dynorphin-A(1-13) substituted in positions 2 and 4: design of [Ala^2, Trp^4]-Dyn-A(1-13) as a putative selective opioid antagonist. Can J Physiol Pharmacol 1986;64:673–678.

796. Gairin JE, Mazarguil H, Alvinerie P, Botanch C, Cros J, Meunier J-C. N,N-Diallyl-tyrosyl Substitution confers antagonist properties on the κ-selective opioid peptide [D-Pro^{10}]dynorphin A-(1-11). Br J Pharmacol 1988;95: 1023–1030.

797. Lemaire, S, Parent, P, Lapierre, C, Michelot, R. N,N-Diallylated analogs of dynorphin A-(1–13) as potent antagonists for the endogenous peptide and selective κ opioid analgesics. In: Quirion R, Jhamandas K, Gianoulakis C, editors. International Narcotics Research Conference (INRC) '89. Volume 328, New York: Alan R. Liss, Inc. 1990. p 77–80.

798. Choi H, Murray TF, DeLander GE, Schmidt WK, Aldrich JV. Synthesis and opioid activity of [D-Pro^{10}]dynorphin A-(1-11) analogues with N-terminal alkyl substitution. J Med Chem 1997;40:2733–2739.

799. Lu Y, Nguyen TM-D, Weltrowska G, Berezowska I, Lemieux C, Chung NN, Schiller PW. [2',6'-Dimethyltyrosine]dynorphin A(1-11) NH$_2$ analogues lacking an N-terminal amino group: potent and selective κ opioid antagonists. J Med Chem 2001;44:3048–3053.

800. Schmidt R, Chung NN, Lemieux C, Schiller PW. Tic(2)-substitution in dermorphin, deltorphin I and dynorphin A analogs: effect on opioid receptor binding and opioid activity in vitro. Regul Pept 1994;54:259–260.

801. Kulkarni, SN, Choi, H, Murray, TF, DeLander, GE, Aldrich, JV. The use of the message–address concept in the design of potential antagonists based on dynorphin A. In: Kaumaya TP, Hodges RS, editors. Peptides: Chemistry, Structure and Biology. West Midlands, England: Mayflower Scientific Ltd. 1996. p 655–656.

802. Guerrini R, Capasso A, Marastoni M, Bryant SD, Cooper PS, Lazarus LH, Temussi PA, Salvadori S. Rational design of dynorphin A analogues with delta-receptor selectivity and antagonism for delta- and kappa-receptors. Bioorg Med Chem 1998;6:57–62.

803. Bennett MA, Murray TF, Aldrich JV. Structure–activity relationship studies of arodyn, a novel dynorphin A-(1–11)) analog. In: Peptides: The wave of the future, American peptide society, San Diego, CA, 2001; pp 894–895.

804. Orosz G, Ronai AZ, Bajusz S, Medzihradszky K. N-terminally protected penta- and tetrapeptide opioid antagonists based on a pentapeptide sequence found in the venom of Philippine cobra. Biochem Biophys Res Commun 1994;202:1285–1290.

805. Wan Q, Murray TF, Aldrich JV. A novel acetylated analogue of dynorphin A-(1-11) amide as a κ opioid receptor antagonist. J Med Chem 1999;42:3011–3013.

806. Aldrich, JV, Wan, Q, Murray, TF. Novel opioid peptides as kappa opioid receptor antagonists. In: Barany G, Fields G, editors. Peptides: Chemistry and Biology. Leiden: ESCOM; 2000. p 616–618.

807. Sasiela C, Bennett MA, Murray TF, Aldrich JV. C-terminal structure–activity relationships for the novel opioid peptide JVA-901 (Venorphin). In Peptides: The wave of the future, American peptide society, San Diego, CA, 2001; pp 689–690.

808. Brantl V, Teschemacher H, Bläsig J, Henschen A, Lottspeich F. Opioid activities of β-casomorphins. Life Sci 1981;28: 1903–1909.

809. Chang K-J, Killian A, Hazum E, Cuatrecasas P, Chang J-K. Morphiceptin (NH$_4$-Tyr-Pro-Phe-Pro-CONH$_2$): A potent and specific agonist for morphine (μ) receptors. Science 1981;212:75–77.

810. Chang K-J, Wei ET, Killian A, Chang K-J. Potent morphiceptin analogs: structure–activity relationships and morphine-like activities. J Pharmacol Exp Ther 1983;227: 403–408.

811. Matthies H, Stark H, Hartrodt B, Ruethrich H-L, Spieler H-T, Barth A, Neubert K. Derivatives of β-casomorphins with high analgesic potency. Peptides 1984;5:463–470.

812. Goodman M, Mierke DF. Configurations of morphiceptins by ^{1}H and ^{13}C NMR spectroscopy. J Am Chem Soc 1989;111:3489–3496.

813. Mierke DF, Nossner G, Schiller PW, Goodman M. Morphiceptin analogs containing 2-aminocyclopentane carboxylic acid as a peptidomimetic for proline. Int J Pep Protein Res 1990;35:35–45.

814. Yamazaki T, Pröbstl A, Schiller PW, Goodman M. Biological and conformational studies of [Val^4]morphiceptin and [D-Val^4] morphiceptin analog incorporating cis-2-aminocyclopentane carboxylic acid as a peptidomimetic for proline. Int J Pept Protein Res 1991;37:364–381.

815. Yamazaki T, Ro S, Goodman M, Chung NN, Schiller PW. A Topochemical approach to explain morphiceptin bioactivity. J Med Chem 1993;36:708–719.

816. Podlogar BP, Paterlini MG, Ferguson DM, Leo GC, Demeter DA, Brown FK, Reitz AB. Conformational analysis of the endogenous mu-opioid agonist endomorphin-1 using NMR spectroscopy and molecular modeling. FEBS Lett 1998;439:13–20.

817. Keller M, Boissard C, Patiny L, Chung NN, Lemieux C, Mutter M, Schiller PW. Pseudoproline-containing analogues of morphiceptin and endomorphin-2: evidence for a cis Tyr-Pro amide bond in the bioactive conformation. J Med Chem 2001;44:3896–903.

818. Paterlini MG, Avitabile F, Ostrowski BG, Ferguson DM, Portoghese PS. Stereochemical requirements for receptor recognition of the mu-opioid peptide endomorphin-1. Biophys J 2000;78:590–599.

819. Shane R, Wilk S, Bodnar RJ. Modulation of endormorphin-2-induced analgesia by dipeptidyl peptidase IV. Brain Res 1999;815: 278–286.

820. Schmidt R, Neubert K, Barth A, Liebmann C, Schnittler M, Chung NN, Schiller PW. Structure–activity relationships of cyclic β-casomorphin-5 analogues. Peptides 1991;12:1175–1180.

821. Vogel D, Schmidt R, Hartung K, Demuth HU, Chung NN, Schiller PW. Cyclic morphiceptin analogs: cyclization studies and opioid activities in vitro. Int J Pept Protein Res 1996;48:495–502.

822. Schmidt R, Vogel D, Mrestani-Klaus C, Brandt W, Neubert K, Chung NN, Lemieux C, Schiller PW. Cyclic β-casomorphin analogues with mixed μ agonist δ antagonist properties: synthesis, pharmacological characterization, and conformational aspects. J Med Chem 1994;37:1136–1144.

823. Mrestani-Klaus C, Brandt W, Schmidt R, Neubert K, Schiller PW. Proton NMR conformational analysis of cyclic beta-casomorphin analogues of the type Tyr-cyclo[N omega-D-Orn-Xaa-Yaa-Gly-]. Arch Pharm 1996;329: 133–142.

824. Schmidt R, Wilkes BC, Chung NN, Lemieux C, Schiller PW. Effect of aromatic amino acid substitutions in the 3-position of cyclic beta-casomorphin analogues on mu-opioid agonist/ delta-opioid antagonist properties. Int J Pept Protein Res 1996;48:411–419.

825. Carpenter KA, Schiller PW, Schmidt R, Wilkes BC. Distinct conformational preferences of three cyclic beta-casomorphin-5 analogs determined using NMR spectroscopy and theoretical analysis. Int J Pept Protein Res 1996;48:102–111.

826. Schiller PW, Berezowska I, Nguyen TM, Schmidt R, Lemieux C, Chung NN, Falcone-Hindley ML, Yao W, Liu J, Iwama S, Smith AB 3rd, Hirschmann R. Novel ligands lacking a positive charge for the δ- and μ-opioid receptors. J Med Chem 2000;43: 551–559.

827. Schmidt, R, Chung, NN, Lemieux, C, Schiller, PW. Development of cycylic casomorphin analogs with potent δ antagonist and balanced mixed μ agonist/δ antagonist properties. In: Kaumaya TP, Hodges RS, editors. Peptides: Chemistry, Structure, Biology. West Midlands, England: Mayflower Scientific Ltd. 1996. p 645–646.

828. Schiller PW, Nguyen TM-D, Chung NN, Lemieux C. Dermorphin analogues carrying an increased positive net charge in their "message" domain display extremely high μ opioid receptor selectivity. J Med Chem 1989;32:698–703.

829. Schiller PW, Nguyen TM, Berezowska I, Dupuis S, Weltrowska G, Chung NN, Lemieux C. Synthesis and in vitro opioid activity profiles of DALDA analogues. Eur J Med Chem 2000;35:895–901.

830. Schiller PW, Nguyen TM-D, Maziak L, Lemieux C. A novel cyclic opioid peptide analog

showing high preference for μ-receptors. Biochem Biophys Res Commun 1985;127: 558–564.

831. Schiller PW, Weltrowska G, Nguyen TM-D, Lemieux C, Chung NN, Marsden BJ, Wilkes BC. Conformational restriction of the phenylalanine residue in a cyclic opioid peptide analogue: effects on receptor selectivity and stereospecificity. J Med Chem 1991;34: 3125–3132.

832. Mosberg HI, Ho JC, Sobczyk-Kojiro K. A high affinity, mu-opioid receptor-selective enkephalin analogue lacking an N-terminal tyrosine. Bioorg Med Chem Lett 1998;8: 2681–2684.

833. McFadyen IJ, Ho JC, Mosberg HI, Traynor JR. Modifications of the cyclic mu receptor selective tetrapeptide Tyr-c[D-Cys-Phe-D-Pen]NH$_2$ (Et): effects on opioid receptor binding and activation. J Pept Res 2000;55: 255–261.

834. Schiller PW, Weltrowska G, Berezowska I, Nguyen TM, Wilkes BC, Lemieux C, Chung NN. The TIPP opioid peptide family: development of δ antagonists, δ agonists, and mixed μ agonist/δ antagonists. Biopolymers 1999;51:411–425.

835. Wilkes BC, Schiller PW. Theoretical conformational analysis of the opioid δ antagonist H-Tyr-Tic-Phe-OH and the μ agonist H-Tyr-D-Tic-Phe-NH$_2$. Biopolymers 1994;34: 1213–1219.

836. Temussi PA, Salvadori S, Amodeo P, Bianchi C, Guerrini R, Tomatis R, Lazarus LH, Picone D, Tancredi T. Selective Opioid Dipeptides. Biochemical and Biophysical Research Communications 1994;198: 933–939.

837. Wilkes BC, Schiller PW. Comparative analysis of various proposed models of the receptor-bound conformation of H-Tyr-Tic-Phe-OH related delta-opioid antagonists. Biopolymers 1995;37:391–400.

838. Lomize AL, Pogozheve ID, Mosberg HI. Development of a model for the δ-opioid receptor pharmacophore. 3. Comparison of the cyclic tetrapeptide, tyr-c[D-Cys-Phe-D-Pen]OH with other conformationally constrained δ-receptor selective ligands. Biopolymers 1996;38: 221–234.

839. Chao T-M, Perez JJ, Loew GH. Characterization of the bioactive form of linear peptide antagonists at the δ-opioid receptor. Biopolymers 1996;38:759–768.

840. Wilkes BC, Nguyen TM, Weltrowska G, Carpenter KA, Lemieux C, Chung NN, Schiller PW. The receptor-bound conformation of H-Tyr-Tic-(Phe-Phe)-OH-related delta-opioid antagonists contains all *trans* peptide bonds. J Pept Res 1998;51:386–94.

841. Marsden BJ, Nguyen TM-D, Schiller PW. Spontaneous degradation via diketopiperazine formation of peptides containing a tetrahydoisoquinoline-3-carboxylic acid residue in the 2-position of the peptide sequence. Int J Pept Protein Res 1993;41:313–316.

842. Balboni G, Guerrini R, Salvadori S, Tomatis R, Bryant SD, Bianchi C, Attila M, Lazarus LH. Opioid diketopiperazines: synthesis and activity of a prototypic class of opioid antagonists. Biol Chem 1997;378:19–29.

843. Schiller PW, Weltrowska G, Schmidt R, Berezowska I, Nguyen TM, Lemieux C, Chung NN, Carpenter KA, Wilkes BC. Subtleties of structure-agonist versus antagonist relationships of opioid peptides and peptidomimetics. J Recept Signal Transduct Res 1999;19: 573–588.

844. Schiller PW, Weltrowska G, Nguyen TM-D, Lemieux C, Chung NN, Zelent B, Wilkes BC, Carpenter KA. Structure-agonist/antagonist activity relationships of TIPP analogs. In: Kaumaya TP, Hodges RS, editors. Peptides: Chemistry, Structure, Biology. West Midlands, England: Mayflower Scientific Ltd. 1996. p 609–611.

845. Mannekens E, Tourwé D, Venderstichele S, Diem TNT, Tóth G, Péter A, Chung NN, Schiller PW. Synthesis of the diastereomers of β-Me-Tyr and β-Me-Phe and their effect on the biological properties of the delta opioid receptor antagonist TIPP. Lett Pept Sci 1995;2:190–192.

846. Tourwé D, Mannekens E, Diem TNT, Verheyden P, Jaspers H, Tóth G, Péter A, Kertész I, Török G, Chung NN, Schiller PW. Side chain methyl substitution in the δ-opioid receptor antagonist TIPP has an important effect on the activity profile. J Med Chem 1998;41: 5167–5176.

847. Lee PHK, Nguyen TMD, Chung NN, Schiller PW, Chang KJ. Tyrosine-iodination converts the delta-opioid peptide antagonist TIPP to an agonist. Eur J Pharmacol 1995;280: 211–214.

848. Schiller PW, Weltrowska G, Nguyen TMD, Chung N, Lemieux C, Wilkes BC. TIPP analogs – highly selective δ opioid antagonists with subnanomolar potency and first known compounds with mixed μ agonist/δ antagonist properties. Regul Pept 1994; 53(suppl1) S63–S64.

849. Schiller PW, Nguyen TM-D, Weltrowska G, Wilkes BC, Marsden BJ, Schmidt R, Lemieux

C, Chung NN. TIPP opioid peptides: development of extraordinarily potent and selective δ antagonists and observation of astonishing structure–instrinsic activity relationships. In: Hodges RS, Smith JA, editors. Peptides: Chemistry, Structure and Biology. Leiden: ESCOM; 1994. p 514–516.
850. Schiller PW, Weltrowska G, Nguyen TM, Lemieux C, Chung NN, Wilkes BC. A highly potent TIPP-NH$_2$ analog with balanced mixed mu agonist/delta antagonist properties. Regul Pept 1994;54:257–258.
851. Salvadori S, Attila M, Balboni G, Bianchi C, Bryant SD, Crescenzi O, Guerrini R, Picone D, Tancredi T, Temussi PA, Lazarus LH. δ-Opioidmimetic antagonists: prototypes for designing a new generation of ultraselective opioid peptides. Mol Med 1995;1:678–689.
852. Bryant SD, Salvadori S, Cooper PS, Lazarus LH. New delta-opioid antagonists as pharmacological probes. Trends Pharmacol Sci 1998;19:42–46.
853. Lazarus LH, Bryant SD, Cooper PS, Guerrini R, Balboni G, Salvadori S. Design of δ-opioid peptide antagonists for emerging drug applications. Drug Discov Today 1998;3:284–294.
854. Pagé D, McClory A, Mischki T, Schmidt R, Butterworth J, St-Onge S, Labarre M, Payza K, Brown W. Novel Dmt-Tic Dipeptide analogues as selective delta-opioid receptor antagonists. Bioorg Med Chem Lett 2000;10: 167–170.
855. Pagé D, Naismith A, Schmidt R, Coupal M, Labarre M, Gosselin M, Bellemare D, Payza K, Brown W. Novel C-terminus modifications of the Dmt-Tic motif: a new class of dipeptide analogues showing altered pharmacological profiles toward the opioid receptors. J Med Chem 2001;44:2387–2390.
856. Salvadori S, Balboni G, Guerrini R, Tomatis R, Bianchi C, Bryant SD, Cooper PS, Lazarus LH. Evolution of the Dmt-Tic pharmacophore: N-terminal methylated derivatives with extraordinary delta opioid antagonist activity. J Med Chem 1997;40:3100–3108.
857. Labarre M, Butterworth J, St-Onge S, Payza K, Schmidhammer H, Salvadori S, Balboni G, Guerrini R, Bryant SD, Lazarus LH. Inverse agonism by Dmt-Tic analogues and HS 378, a naltrindole analogue. Eur J Pharmacol 2000;406:R1–R3.
858. Liao S, Lin J, Shenderovich MD, Han Y, Hasohata K, Davis P, Qiu W, Porreca F, Yamamura HI, Hruby VJ. The stereochemical requirements of the novel δ-opioid selective dipeptide antagonist TMT-Tic. Bioorg Med Chem Lett 1997;7:3049–3052.
859. Hosohata K, Burkey TH, Alfaro-Lopez J, Hruby VJ, Roeske WR, Yamamura HI. (2S,3R) TMT-L-Tic-OH is a potent inverse agonist at the human delta-opioid receptor. Eur J Pharmacol 1999;380:R9–R10.
860. Santagada V, Balboni G, Caliendo G, Guerrini R, Salvadori S, Bianchi C, Bryant SD, Lazarus LH. Assessment of substitution in the second pharmacophore of Dmt-Tic analogues. Bioorg Med Chem Lett 2000;10: 2745–2748.
861. Capasso A, Amodeo P, Balboni G, Guerrini R, Lazarus LH, Temussi PA, Salvadori S. Design of mu selective opioid dipeptide antagonists. FEBS Lett 1997;417:141–144.
862. Wang C, McFadyen IJ, Traynor JR, Mosberg HI. Design of a high affinity peptidomimetic opioid agonist from peptide pharmacophore models. Bioorg Med Chem Lett 1998;8: 2685–2688.
863. Schiller PW, Schmidt R, Weltrowska G, Berezowska I, Nguyen TM-D, Dupuis S, Chung NN, Lemieux C, Wilkes BC, Carpenter KA. Conformationally constrained opioid peptide analogs with novel activity profiles. Letters in Pept Sci 1998;5:209–214.
864. Schiller PW, Weltrowska G, Schmidt R, Nguyen TM-D, Berezowska I, Lemieux C, Chung NN, Carpenter KA, Wilkes BC. Four different types of opioid peptides with mixed μ agonist/δ antagonist properties. Analgesia 1995;1:703–706.
865. Salvadori S, Guerrini R, Balboni G, Bianchi C, Bryant SD, Cooper PS, Lazarus LH. Further studies on the Dmt-Tic pharmacophore: hydrophobic substituents at the C-terminus endow delta antagonists to manifest mu agonism or mu antagonism. J Med Chem 1999;42:5010–5019.
866. Vavrek RJ, Hsi L-H, York EJ, Hall ME, Stewart JM. Minimum structure opioids: dipeptide and tripeptide analogs of the enkephalins. Peptides 1981;2:303–308.
867. Hammond DL, Mazur RH, Hansen DW Jr, Pilipauskas DR, Bloss J, Drower E. Analgesic activity of SC-39566. Pain 1987; 30(suppl 1): 253S.
868. Hammond DL, Stapelfeld A, Drower EJ, Savage MA, Tam L, Mazur RH. Antinociception produced by oral, subcutaneous or intrathecal administration of SC-39566, an opioid dipeptide arylalkylamide, in the rodent. J Pharmacol Exp Ther 1994;268:607–615.

869. Ambo A, Sasaki Y, Suzuki K. Synthesis of carboxyl-terminal extension analogs of dermorphin and evaluation of their opioid receptor-binding and opioid activities. Chem Pharma Bull 1994;42:888–891.

870. Lazarus LH, Guglietta A, Wilson WE, Irons BJ, de Castiglione R. Dimeric dermorphin analogues as µ-receptor probes on rat brain membranes. Correlation between central µ-receptor potency and suppression of gastric acid secretion. J Biol Chem 1989;264: 354–362.

871. Tomatis R, Marastoni M, Balboni G, Guerrini R, Capasso A, Sorrentino L, Santagada V, Caliendo G, Lazarus LH, Salvadori S. Synthesis and pharmacological activity of deltorphin and dermorphin-related glycopeptides. J Med Chem 1997;40:2948–2952.

872. Negri L, Lattanzi R, Tabacco F, Scolaro B, Rocchi R. Glycodermorphins: opioid peptides with potent and prolonged analgesic activity and enhanced blood–brain barrier penetration. Br J Pharmacol 1998;124:1516–1522.

873. Negri L, Lattanzi R, Tabacco F, Orru L, Severini C, Scolaro B, Rocchi R. Dermorphin and Deltorphin glycosylated analogues: synthesis and antinociceptive activity after systemic administration. J Med Chem 1999;2:400–402.

874. Darlak K, Grzonka Z, Krzáscik P, Janicki P, Gumulka SW. Structure–activity studies of dermorphin. The role of side chains of amino acid residues on the biological activity of dermorphin. Peptides 1984;5:687–689.

875. Salvadori S, Marastoni M, Balboni G, Sarto GP, Tomatis R. Synthesis and opioid activity of dermorphin tetrapeptides bearing D-methionine oxide at position 2. J Med Chem 1986;29:889–894.

876. Sasaki Y, Matsui M, Fujita H, Hosono M, Taguchi M, Suzuki K, Sakurada S, Sato T, Sakurada T, Kisara K. The analgesic effect of D-Arg²-dermorphin and its N-terminal tetrapeptide analogs after subcutaneous administration in mice. Neuropeptides 1985;5: 391–394.

877. Schiller PW, Nguyen TM-D, DiMaio J, Lemieux C. Comparison of µ-, δ- and κ-receptor binding sites through phamacologic evaluation of p-nitrophenylalanine analogs of opioid peptides. Life Sci 1983;33(Suppl I): 319–322.

878. Schiller PW, Nguyen TM-D, Lemieux C. New types of opioid peptide analogs showing high µ-receptor selectivity and preference for either central or peripheral sites. In: Jung G, Bayer E, editors. Peptides 1988. Berlin: DeGruyter; 1989. p 613.

879. Kisara K, Sakurada S, Sakurada T, Sasaki Y, Sato T, Suzuki K, Watanabe H. Dermorphin analogues contianing D-kyotorphin: structure–antinociceptive relationships in mice. Br J Pharmacol 1986;87;183–189.

880. Paakkari P, Paakkari I, Vonhof S, Feuerstein G, Sirén A-L. Dermorphin analog Tyr-D-Arg²-Phe-sarcosine-induced opioid analgesia and respiratory stimulation: the role of mu_1-receptors?. J Pharmacol Exp Ther 1993;266: 544–550.

881. Tancredi T, Salvadori S, Amodeo P, Picone D, Lazarus LH, Bryant SD, Guerrini R, Marzola G, Temussi PA. Conversion of enkephalin and dermorphin into delta-selective opioid antagonists by single-residue substitution. Eur J Biochem 1994;224:241–247.

882. Schwyzer R. Molecular mechanism of opioid receptor selection. Biochemistry 1986;25: 6335–6342.

883. Sasaki Y, Watanabe Y, Ambo A, Suzuki K. Synthesis and biological properties of quaternized N-methylation analogs of D-Arg-2-dermorphin tetrapeptide. Bioorg Med Chem Lett 1994;4:2049–2054.

884. Samii A, Bickel U, Stroth U, Pardridge WM. Blood–brain barrier transport of neuropeptides: analysis with a metabolically stable dermorphin analogue. Am J Physiol 1994; 267:E124–E131.

885. Schiller PW, Nguyen TM, Chung NN, Dionne G, Martel R. Peripheral antinociceptive effect of an extremely mu-selective polar dermorphin analog (DALDA). Prog Clin Biol Res 1990;328;53–56.

886. Szeto HH, Clapp JF, Desiderio DM, Schiller PW, Grigoriants OO, Soong Y, Wu D, Olariu N, Tseng JL, Becklin R. *In vivo* disposition of dermorphin analog (DALDA) in nonpregnant and pregnant sheep. J Pharmacol Exp Ther 1998;284:61–65.

887. Shimoyama M, Shimoyama N, Zhao GM, Schiller PW, Szeto HH. Antinociceptive and respiratory effects of intrathecal H-Tyr-D-Arg-Phe-Lys-NH₂ (DALDA) and [Dmt1] DALDA. J Pharmacol Exp Ther 2001;297:364–371.

888. Szeto HH, Lovelace JL, Fridland G, Soong Y, Fasolo J, Wu D, Desiderio DM, Schiller PW. *In vivo* pharmacokinetics of selective mu-opioid peptide agonists. J Pharmacol Exp Ther 2001;298:57–61.

889. Tourwe D, Verschueren K, Frycia A, Davis P, Porreca F, Hruby VJ, Toth G, Jaspers H, Verheyden P, Van Binst G. Conformational restriction of Tyr and Phe side chains in

opioid peptides: information about preferred and bioactive side-chain topology. Biopolymers 1996;38:1–12.
890. Borg S, Vollinga RC, Labarre M, Payza K, Terenius L, Luthman K. Design, synthesis, and evaluation of Phe-Gly mimetics: heterocyclic building blocks for pseudopeptides. J Med Chem 1999;42:4331–4342.
891. Schiller PW, Nguyen TM-D, Lemieux C, Maziak LA. Synthesis and activity profiles of novel cyclic opioid peptide monomers and dimers. J Med Chem 1985;28:1766–1771.
892. Schiller PW, Nguyen TM-D, Maziak LA, Wilkes BC, Lemieux C. Structure–activity relationships of cyclic opioid peptide analogues containing a phenylalanine residue in the 3-position. J Med Chem 1987;30: 2094–2099.
893. Wilkes BC, Schiller PW. Theoretical Conformational analysis of a μ-selective cyclic opioid peptide analog. Biopolymers 1987;26: 1431–1444.
894. Schiller PW, Weltrowska G, Nguyen TM-D, Lemieux C, Chung NN, Wilkes BC. The use of conformational restriction in the development of opioid peptidomimetics. In: Giralt E, Andreu D, editors. Peptides 1990. Leiden: ESCOM; 1991. p 621.
895. Said-Nejad OE, Felder ER, Mierke DF, Yamazaki T, Schiller PW, Goodman M. 14-Membered cyclic opioids related to dermorphin and their partially retro-inverso modified analogs. 1. Synthesis and biological-activity. Int J Pept and Protein Res 1992;39; 145–160.
896. Yamazaki T, Mierke DF, Said-Nejad OE, Felder ER, Goodman M. 14-Membered cyclic opioids related to dermorphin and their partially retro-inverso modified analogs. 2. Preferred conformations in solution as studied by H-1-NMR spectroscopy. Int J Pept Protein Res 1992;39:161–181.
897. Mosberg HI, Omnaas JR, Medzihradsky F, Smith GB. Cyclic disulfide- and dithioether-containing opioid tetrapeptides: development of a ligand with high delta opioid receptor selectivity and affinity. Life Sci 1988;43: 1013–1020.
898. McFadyen IJ, Sobczyk-Kojiro K, Schaefer MJ, Ho JC, Omnaas JR, Mosberg HI, Traynor JR. Tetrapeptide derivatives of [D-Pen(2), D-Pen(5)]-enkephalin (DPDPE) lacking an N-terminal tyrosine residue are agonists at the μ-opioid receptor. J Pharmacol Exp Ther 2000;295:960–966.
899. Sasaki Y, Chiba T, Ambo A, Suzuki K. Degradation of deltorphins and their analogs by rat brain synaptosomal peptidases. Chem Pharm Bull 1994;42:592–594.
900. Lazarus LH, Bryant SD, Salvadori S, Attila M, Sargent Jones L. Opioid infidelity: novel opioid peptides with dual high affinity for delta- and mu-receptors. Trends Neurosci 1996;19:31–35.
901. Heyl DL, Schullery SE. Developments in the structure–activity relationships for the δ-selective opioid peptides of amphibian skin. Curr Med Chem 1997;4:117–150.
902. Lazarus LH, Bryant SD, Cooper PS, Salvadori S. What peptides these deltorphins be. Prog Neurobiol 1999;57:377–420.
903. Duchesne D, Naim M, Nicolas P, Baron D. Folding trends in a flexible peptide: two-dimensional NMR Study of deltorphin-I, a δ-selective opioid heptapeptide. Biochem Biophys Res Commun 1993;195:630–636.
904. Schiller PW, Weltrowska G, Nguyen TM-D, Wilkes BC, Chung NN, Lemieux C. Conformationally restricted deltorphin analogues. J Med Chem 1992;35;3956–3961.
905. Melchiorri P, Negri L, Falconieri-Erspamer G, Severini C, Corsi R, Soaje M, Erspamer V, Barra D. Structure–activity relationships of the δ-opioid-selective agonists, deltorphins. Eur J Pharmacol 1991;195:201–207.
906. Lazarus LH, Salvadori S, Balboni G, Tomatis R, Wilson WE. Stereospecificity of amino-acid side-chains in deltorphin defines binding to opioid receptors. J Med Chem 1992;35: 1222–1227.
907. Salvadori S, Bianchi C, Lazarus LH, Scaranari V, Attila M, Tomatis R. para-Substituted Phe^3 deltorphin analogues: enhanced selectivity of halogenated derivatives for δ opioid receptors. J Med Chem 1992;35; 4651–4657.
908. Salvadori S, Bryant SD, Bianchi C, Balboni G, Scaranari V, Attila M, Lazarus LH. Phe^3-substituted analogues of deltorphin C. Spatial Conformation and topography of the aromatic ring in peptide recognition by δ opioid receptors. J Med Chem 1993;36:3748–3756.
909. Heyl DL, Dandabathula M, Kurtz KR, Mousigian C. Opioid receptor binding requirements for the delta-selective peptide deltorphin I: Phe^3 replacement with ring-substituted and heterocyclic amino acids. J Med Chem 1995;38:1242–1246.
910. Schullery SE, Mohammedshah T, Makhlouf H, Marks EL, Wilenkin BS, Escobar S, Mousigian C, Heyl DL. Binding to delta and mu opioid receptors by deltorphin I/II analogues

910. (cont.) modified at the Phe3 and Asp4/Glu4 side chains: a report of 32 new analogues and a QSAR study. Bioorg Med Chem 1997;5: 2221–2234.

911. Lazarus LH, Salvadori S, Santagada V, Tomatis R, Wilson WE. Function of negative charge in the "address domain" of deltorphins. J Med Chem 1991;34:1350–1355.

912. Salvadori S, Marastoni M, Balboni G, Borea PA, Morari M, Tomatis R. Synthesis and structure–activity relationships of deltorphin analogues. J Med Chem 1991;34: 1656–1661.

913. Sagan S, Amiche M, Delfour A, Mor A, Camus A, Nicolas P. Molecular determinants of receptor affinity and selectivity of the natural Δ-opioid agonist, dermenkephalin. J Biol Chem 1989;264:17100–17106.

914. Sagan S, Amiche M, Delfour A, Camus A, Mor A, Nicolas P. Differential contribution of C-Terminal region of dermorphin and dermenkephalin to opioid-sites selection and binding potency. Biochem Biophys Res Commun 1989;163:726–732.

915. Lazarus LH, Salvadori S, Tomatis R, Wilson WE. Opioid receptor selectivity reversal in deltorphin tetrapeptide analogues. Biochem Biophys Res Commun 1991;178:110–115.

916. Sagan S, Charpentier S, Delfour A, Amiche M, Nicolas N. The aspartic acid in deltorphin I and dermenkephalin promotes targeting to δ-opioid receptor independently of receptor binding. Biochem Biophys Res Commun 1992;187:1203–1210.

917. Sasaki Y, Ambo A, Suzuki K. [D-Ala^2]deltorphin II analogs with high affinity and selectivity for delta-opioid receptor. Biochem Biophys Res Commun 1991;180:822–827.

918. Salvadori S, Guerrini R, Forlani V, Bryant SD, Attila M, Lazarus LH. Prerequisite for His^4 in deltorphin A for high delta opioid receptor selectivity. Amino Acids 1994;7: 291–304.

919. Sasaki Y, Chiba T. Novel Deltorphin heptapeptide analogs with potent δ agonist, δ antagonist, or mixed μ antagonist/δ agonist porperties. J Med Chem 1995;38:3995–3999.

920. Charpentier S, Sagan S, Naim M, Delfour A, Nicolas P. Mechanism of delta-opioid receptor selection by the address domain of dermenkephalin. Eur J Pharmacol 1994;266:175–180.

921. Guerrini R, Capasso A, Sorrentino L, Anacardio R, Bryant SD, Lazarus LH, Attila M, Salvadori S. Opioid receptor selectivity alteration by single residue replacement: synthesis and activity profile of [Dmt1]deltorphin B. Eur J Pharmacol 1996;302:37–42.

922. Heyl DL, Bouzit H, Mousigian C. Structural requirements for binding to the δ opioid receptor: alkyl replacements at the third residue of deltorphin I. Lett Pept Sci 1995;2: 277–284.

923. Mosberg HI, Lomize AL, Wang C, Kroona H, Heyl DL, Sobczyk-Kojiro K, Ma W, Mousigian C, Porreca F. Development of a model for the δ opioid receptor pharmacophore. 1. Conformationally restricted Tyr1 replacements in the cyclic δ receptor selective tetrapeptide Tyr-c[D-Cys-Phe-D-Pen]OH (JOM-13). J Med Chem 1994;37:4371–4383.

924. Mosberg HI, Dua RK, Pogozheva ID, Lomize AL. Development of a model for the delta-opioid receptor pharmacophore. 4. Residue 3 dehydrophenylalanine analogues of Tyr-c[D-Cys-Phe-D-Pen]OH (JOM-13) confirm required gauche orientation of aromatic side chain. Biopolymers 1996;39:287–296.

925. Misicka A, Lipkowski AW, Horvath R, Davis P, Yamamura HI, Porreca F, Hruby VJ. Design of cyclic deltorphins and dermenkephalins with a disulfide bridge leads to analogues with high selectivity for delta-opioid receptors. J Med Chem 1994;37:141–145.

926. Tóth G, Darula Z, Péter A, Fülöp F, Tourwé D, Jaspers H, Verheyden P, Böcskey Z, Tóth A, Borsodi A. Conformationally constrained deltorphin analogs with 2-aminotetralin-2-carboxylic acid in position 3. J Med Chem 1997;40:990–995.

927. Misicka A, Cavagnero S, Horvath R, Davis P, Porreca F, Yamamura HI, Hruby VJ. Synthesis and biological properties of β-MePhe^3 analogues of deltorphin I and dermenkephalin: influence of biased χ^1 of Phe^3 residues on peptide recognition for δ-opioid receptors. J Pept Res 1997;50:48–54.

928. Lomize AL, Flippenanderson JL, George C, Mosberg HI. Conformational analysis of the delta-receptor-selective, cyclic opioid peptide, Tyr-cyclo[D-Cys-Phe-D-Pen]OH (JOM-13): comparison of X-ray crystallographic structures, molecular mechanics simulations, and H-1 NMR data. J Am Chem Soc 1994;116: 429–436.

929. Mosberg HI, Heyl DL, Haaseth RC, Omnaas JR, Medzihradsky F, Smith CB. Cyclic dermorphin-like tetrapeptides with delta-opioid receptor selectivity. 3. Effect of residue 3 modification on in vitro opioid activity. Mol Pharmacol 1990;38:924–928.

930. Heyl DL, Mosberg HI. Substitution on the Phe³ Aromatic ring in cyclic δ opioid receptor-selective dermorphin/deltorphin tetrapeptide analogues: electronic and lipophilic requirements for receptor affinity. J Med Chem 1992;35:1535–1541.

931. Heyl DL, Mosberg HI. Modification of the Phe³ Aromatic moiety in delta receptor-selective dermorphin/deltorphin-related tetrapeptides: effects on opioid receptor binding. Int J Pept Protein Res 1992;39: 450–457.

932. Mosberg HI, Omnaas JR, Lomize AL, Heyl DL, Nordan I, Mousigian C, Davis P, Porreca F. Development of a Model for the δ opioid receptor pharmacophore. 2. Conformationally restricted Phe³ Replacements in the cyclic δ receptor selective tetrapeptide Tyr-c[D-Cys-Phe-D-Pen]OH (JOM-13). J Med Chem 1994;37:4384–4391.

933. Mosberg HI, Kroona HB. Incorporation of a novel conformationally restricted tyrosine analog into a cyclic, δ opioid receptor selective tetrapeptide (JOM-13) enhances δ receptor binding affinity and selectivity. J Med Chem 1992;35:4498–5000.

934. Mosberg HI. Complementarity of delta opioid ligand pharmacophore and receptor models. Biopolymers 1999;51:426–439.

935. Misicka A, Nikiforovich G, Lipkowski AW, Horvath R, Davis P, Kramer TH, Yamamura HI, Hruby VJ. Topographical requirements for delta opioid ligands: the synthesis and biological properties of a cyclic analogue of deltorphin I. Bioorg Med Chem Lett 1992;2:547–552.

936. Mosberg, HI, Kroona, HB, Omnaas, JR, Sobczyk-Kojiro, K, Bush, P, Mousigian, C. Cyclic deltorphin analogues with high δ opioid receptor affinity and selectivity. In: Hodges RS, Smith JA, editors. Peptides: Chemistry, Structure and Biology. Leiden: ESCOM; 1994. p 514–516.

937. Maurer R, Gaehwiler BH, Buescher HH, Hill RC, Roemer D. Opiate Antagonistic properties of an octapeptide somatostatin analog. Proc Natl Acad Sci USA 1982;79: 4815–4817.

938. Pelton JT, Gulya K, Hruby VJ, Duckles SP, Yamamura HI. Conformationally restricted analogs of somatostatin with high μ-opiate receptor specificity. Proc Natl Acad Sci USA 1985;82:236–239.

939. Pelton JT, Whalon M, Cody WL, Hruby VJ. Conformation of D-Phe-Cys-Tyr-D-Trp-Lys-Thr-Pen-Thr-NH₂ (CTP-NH₂) a highly selective mu-opioid antagonist peptide, by ¹H and ¹³C NMR. Int J Pept Protein Res 1988;31: 109–115.

940. Kazmierski W, Hruby VJ. A new approach to receptor ligand design: synthesis and conformation of a new class of potent and highly selective μ opioid antagonists utilizing tetrahydroisoquinoline carboxylic acid. Tetrahedron 1988;44:697–710.

941. Kazmierski W, Wire WS, Lui GK, Knapp RJ, Shook JE, Burks TF, Yamamura HI, Hruby VJ. Design and synthesis of somatostatin analogues with topographical properties that lead to highly potent and specific μ opioid receptor antagonists with greatly reduced binding at somatostatin receptors. J Med Chem 1988;31:2170–2177.

942. Bonner GG, Davis P, Stropova D, Ferguson R, Yamamura HI, Porreca F, Hruby VJ. Opioid peptides: simultaneous delta agonism and mu antagonism in somatostatin analogues. Peptides 1997;18:93–100.

943. Dooley CT, Houghten RA. New opioid peptides, peptidomimetics, and heterocyclic compounds from combinatorial libraries. Biopolymers 1999;51:379–390.

944. Houghten RA, Dooley CT. The use of synthetic peptide combinatorial libraries for the determination of peptide ligands in radioreceptor assays: opioid peptides. Bioorg Med Chem Lett 1993;3:405–412.

945. Dooley CT, Houghton RA. The use of positional scanning synthetic peptide combinatorial libraries for the rapid determination of opioid receptor ligands. Life Sciences 1993; 52:1509–1517.

946. Dooley CT, Kaplan RA, Chung NN, Schiller PW, Bidlack JM, Houghten RA. Six highly active mu-selective opioid peptides identified from two synthetic combinatorial libraries. Pept Res 1995;8:124–137.

947. Pinilla C, Appel JR, Blondelle SE, Dooley CT, Eichler J, Ostresh JM, Houghten RA. Versatility of positional scanning synthetic combinatorial libraries for the identification of individual compounds. Drug Dev Res 1994; 33:133–145.

948. Dooley CT, Chung NN, Schiller PW, Houghten RA. Acetalins: opioid receptor antagonists determined through the use of synthetic peptide combinatorial libraries. Proc Natl Acad Sci USA 1993;90:10811–10815.

949. Dooley CT, Chung NN, Wilkes BC, Schiller PW, Bidlack JM, Pasternak GW, Houghten RA. An all D-amino acid opioid peptide with

central analgesic activity from a combinatorial library. Science 1994;266:2019–2022.
950. Zuckermann RN, Martin EJ, Spellmeyer DC, Stauber GB, Shoemaker KR, Kerr JM, Figliozzi GM, Goff DA, Siani MA, Simon RJ, Banville SC, Brown EG, Wang L, Richter LS, Moos WH. Discovery of nanomolar ligands for 7-transmembrane G-protein-coupled receptors from a diverse N-(substituted)glycine peptoid library. J Med Chem 1994;37: 2678–2685.
951. Ostresh JM, Husar GM, Blondelle SE, Dörner B, Weber PA, Houghten RA. Proc Natl Acad Sci USA 1994;91:11138–11142.
952. Dooley CT, Houghten RA. Identification of mu-selective polyamine antagonists from a synthetic combinatorial library. Analgesia 1995;1:400–404.
953. Houghten RA, Pinilla C, Appel JR, Blondelle SE, Dooley CT, Eichler J, Nefzi A, Ostresh JM. Mixture-based synthetic combinatorial libraries. J Med Chem 1999;42:3743–3778.
954. Pelton JT, Johnston RB, Balk J, Schmidt CJ, Roche EB. Synthesis and biological activity of chloromethyl ketones of leucine enkephalin. Biochem Biophys Res Commun 1980;97: 1391–1398.
955. Venn RF, Barnard EA. A potent peptide affinity reagent for the opiate receptor. J Biol Chem 1981;256:1529–1532.
956. Newman EL, Borsodi A, Toth G, Hepp F, Barnard EA. Mu-Receptor Specificity of the opioid peptide irreversible reagent, [^{3}H]DALECK. Neuropeptides 1986;8: 305–315.
957. Benyhe S, Hepp J, Szucs M, Simon J, Borsodi A, Medzihradszky K, Wollemann M. Irreversible labelling of rat brain opioid receptors by enkephalin chloromethyl ketones. Neuropeptides 1986;8:173–181.
958. Newman EL, Barnard EA. Identification of an opioid receptor subunit carrying the μ binding site. Biochemistry 1984;23: 5385–5389.
959. Szücs M, Belcheva M, Simon J, Benye S, Tóth G, Hepp J, Wollemann M, Medzihradszky K. Covalent labeling of opioid receptors with ^{3}H-D-Ala^2-Leu^5-enkephalin chloromethyl ketone. I. Binding characteristics in rat brain membranes. Life Sci 1987;41:177–184.
960. Benyhe S, Hepp J, Simon J, Borsodi A, Medzihradszky K, Wollemann M. Tyr-D-Ala-Gly-(Me)Phe-chloromethyl ketone: a mu specific affinity label for the opioid receptors. Neuropeptides 1987;9:225–235.
961. Benyhe S, Ketevan A, Simon J, Hepp J, Medzihradszky K, Borsodi A. Affinity labeling of frog brain opioid receptors by dynorphin$_{(1-10)}$ chloromethyl ketone. Neuropeptides 1997;31: 52–59.
962. Szatmari I, Orosz Z, Medzihradszky K, Borsodi A. Affinity labeling of delta opioid receptors by an enkephalin-derivative alkylating agent, DSLET-Mal. Biochem Biophys Res Commun 1999;265:513–519.
963. Sartani N, Szatmári I, Orosz G, Rónai AZ, Medzihradszky K, Borsodi A, Benyhe S. Irreversible labelling of the opioid receptors by a melphalan-substituted [Met^5]enkephalin-Arg-Phe derivative. Eur J Pharmacol 1999;373:241–249.
964. Maeda DY, Berman F, Murray TF, Aldrich JV. Synthesis and evaluation of isothiocyanate derivatives of the δ-opioid receptor antagonist Tyr-Tic-Phe-Phe (TIPP) as potential affinity labels for δ-opioid receptors. J Med Chem 2000;43:5044–5049.
965. Kumar V, Murray TF, Aldrich JV. Solid phase synthesis and evaluation of Tyr-Tic-Phe-Phe(p-NHCOCH$_2$Br)([Phe(p-NHCOCH$_2$Br)^4] TIPP, a potent affinity label for the δ opioid receptors. J Med Chem 2002;45:3820–3823.
966. Shimohigashi Y, Takada K, Sakamoto H, Matsumoto H, Yasunaga T, Kondo M, Ohno M. Discriminative affinity labeling of opioid receptors by enkephalin and morphiceptin analogs containing 3-nitro-2-pyridinesulphenyl-activated thiol residues. J Chromatogr 1992;597:425–428.
967. Koike K, Takayanagi I, Matsueda R, Shimohigashi Y. Interaction of dynorphin derivative containing 3-nitro-2-pyridinesulfenyl (Npys) group with opioid receptors. J Pharmacobio Dyn 1991;14:S103–S103.
968. Yasunaga T, Motoyama S, Nose T, Kodama H, Kondo M, Shimohigashi Y. Reversible affinity labeling of opioid receptors via disulfide bonding: discriminative labeling of μ and δ subtypes by chemically activated thiol-containing enkephalin analogs. J Biochem 1996;120:459–465.
969. Shirasu N, Kuromizu T, Nakao H, Chuman Y, Nose T, Costa T, Shimohigashi Y. Exploration of universal cysteines in the binding sites of three opioid receptor subtypes by disulfide-bonding affinity labeling with chemically activated thiol-containing dynorphin A analogs. J Biochem 1999;126:254–259.
970. Okuda-Ashitaka E, Ito S. Nocistatin: a novel neuropeptide encoded by the gene for the nociceptin/orphanin FQ precursor. Peptides 2000;21:101–109.
971. Mathis JP, Rossi GC, Pellegrino MJ, Jimenez C, Pasternak GW, Allen RG. Carboxyl term-

inal peptides derived from prepro-orphanin FQ/nociceptin (ppOFQ/N) are produced in the hypothalamus and possess analgesic bioactivities. Brain Res 2001;895:89–94.

972. Meng F, Ueda Y, Hoversten MT, Taylor LP, Reinscheid RK, Monsma FJ, Watson SJ, Civelli O, Akil H. Creating a functional opioid alkaloid binding site in the orphanin FQ receptor through site-directed mutagenesis. Mol Pharmacol 1998;53:772–777.

973. Mouledous L, Topham CM, Moisand C, Mollereau C, Meunier J-C. Functional inactivation of the nociceptin receptor by alanine substitution of glutamine 286 at the C terminus of transmembrane VI: evidence from a site-directed mutagenesis study of the ORL1 receptor transmembrane-binding domain. Mol Pharmacol 2000;57:495–502.

974. Lapalu S, Moisand C, Butour J-L, Mollereau C, Meunier J-C. Different domains of the ORL1 and κ-opioid receptors are involved in recognition of nociceptin and dynorphin A. FEBS Lett 1998;427:296–300.

975. Mollereau C, Mouledous L, Lapalu S, Cambois G, Moisand C, Butour J-L, Meunier J-C. Distinct mechanisms for activation of the opioid receptor-like 1 and κ-opioid receptors by nociceptin and dynorphin A. Mol Pharmacol 1999;55:324–331.

976. Grisel JE, Mogil JS. Effects of supraspinal orphanin FQ/nociceptin. Peptides 2000;21: 1037–1045.

977. Xu X, Grass S, Hao J, Xu IS, Wiesenfeld-Hallin Z. Nociceptin/orphanin FQ in spinal nociceptive mechanisms under normal and pathological conditions. Peptides 2000;21: 1031–1036.

978. Jenck F, Moreau JL, Martin JR, Kilpatrick GJ, Reinscheid RK, Monsma FJ Jr, Nothacker HP, Civelli O. Orphanin FQ acts as an anxiolytic to attenuate behavioral responses to stress. Proc Natl Acad Sci USA 1997;94: 14854–14858.

979. Jenck F, Wichmann J, Dautzenberg FM, Moreau J-L, Ouagazzal AM, Martin JR, Lundstrom K, Cesura AM, Poli SM, Roever S, Kolczewski S, Adam G, Kilpatrick GA. Synthetic agonist at the orphanin FQ/nociceptin receptor ORL1: anxiolytic profile in the rat. Proc Natl Acad Sci USA 2000;97: 4938–4943.

980. Calò G, Bigoni R, Rizzi A, Guerrini R, Salvadori S, Regoli D. Nociceptin/orphanin FQ receptor ligands. Peptides 2000;21:935–947.

981. Guerrini R, Calo G, Rizzi A, Bigoni R, Rizzi D, Regoli D, Salvadori S. Structure–activity relationships of nociceptin and related peptides: comparison with dynorphin A. Peptides 2000;21:923–933.

982. Dooley CT, Houghten RA. Orphanin FQ: receptor binding and analog structure–activity relationships in rat brain. Life Sci 1996;59: 23–29.

983. Butour J-L, Moisand C, Mazarguil H, Mollereau C, Meunier J-C. Recognition and activation of the opioid receptor-like ORL1 receptor by nociceptin, nociceptin analogs and opioids. Eur J Pharmacol 1997;321:97–103.

984. Mathis JP, Ryan-Moro J, Chang A, Hom JS, Scheinberg DA, Pasternak GW. Biochemical evidence for orphanin FQ/nociceptin receptor heterogeneity in mouse brain. Biochem Biophys Res Commun 1997;230:462–465.

985. Rossi GC, Leventhal L, Bolan E, Pasternak GW. Pharmacological characterization of orphanin FQ/nociceptin and its fragments. J Pharmacol Exp Ther 1997;282: 858–865.

986. Reinscheid RK, Ardati A, Monsma FJ, Civelli O. Structure–activity relationship studies of the novel neuropeptide orphanin FQ. J Biol Chem 1996;271:14163–14168.

987. Guerrini R, Calò G, Rizzi A, Bianchi C, Lazaraus L, Salvadori S, Temusi PA, Regoli D. Address and message sequences for the nociceptin receptor: a structure–activity study of nociceptin-(1-13)-peptide amide. J Med Chem 1997;40:1789–1793.

988. Lapalu S, Moisand C, Mazarguil H, Cambois G, Mollereau C, Meunier J-C. Comparison of the structure–activity relationships of nociceptin and dynorphin A using chimeric peptides. FEBS Lett 1997;417:333–336.

989. Guerrini R, Calo G, Rizzi A, Bigoni R, Bianchi C, Salvadori S, Regoli D. A new selective antagonist of the nociceptin receptor. Br J Pharmacol 1998;123:163–165.

990. Calò G, Guerrini R, Bigoni R, Rizzi A, Marzola G, Okawa H, Bianchi C, Lambert DG, Salvadori S, Regoli D. Characterization of [Nphe^1]NC(1-13)NH$_2$, a new selective nociceptin receptor antagonist. Br J Pharmacol 2000;129: 1183–1193.

991. Dooley CT, Spaeth CG, Berzetei-Gurske IP, Craymer K, Adapa ID, Brandt SR, Houghten RA, Toll L. Binding and in vitro activities of peptides with high affinity for the nociceptin/orphanin FQ receptor. ORL1. J Pharmacol Exp Ther 1997;283:735–741.

992. Becker JAJ, Wallace A, Garzon A, Ingallinella P, Bianchi E, Cortese R, Simonin F, Kieffer BL, Pessi A. Ligands for κ-opioid

and ORL1 receptors identified from a conformationally constrained peptide combinatorial library. J Biol Chem 1999;274: 27513–27522.
993. Wnendt S, Kruger T, Janocha E, Hildebrandt D, Englberger W. Agonistic effect of buprenorphine in a nociceptin/OFQ receptor-triggered reporter gene assay. Mol Pharmacol 1999;56:334–338.
994. Barlocco D, Cignarella G, Giardina G, Toma L. The opioid-receptor-like 1 (ORL-1) as a potential target for new analgesics. Eur J Med Chem 2000;35:275–282.
995. Wichmann J, Adam G, Röver S, Cesura AM, Dautzenberg FM, Jenck F. 8-Acenaphthen-1-yl-1-phenyl-1,3,8-triaza-spiro[4,5]decan-4-one derivatives as orphanin FQ receptor agonists. Bioorg Med Chem Lett 1999;9: 2343–2348.
996. Röver S, Adam G, Cesura AM, Galley G, Jenck F, Monsma FJJ, Wichmann J, Dautzenberg FM. High-affinity, non-peptide agonists for the ORL1 (orphanin FQ/nociceptin) receptor. J Med Chem 2000;43:1329–1338.
997. Röver S, Wichmann J, Jenck F, Adam G, Cesura AM. ORL1 receptor ligands: structure–activity relationships of 8-cycloalkyl-1-phenyl-1,3,8-triaza-spiro[4,5]decan-4-ones. Bioorg Med Chem Lett 2000;10: 831–834.
998. Wichmann J, Adam G, Rover S, Hennig M, Scalone M, Cesura AM, Dautzenberg FM, Jenck F, Gomes I, Jordan BA, Gupta A, Rios C, Trapaidze N, Devi LA. Synthesis of (1S,3aS)-8-(2,3,3a,4,5,6-hexahydro-1H-phenalen-1-yl)-1-phenyl-1,3,8-triaza-spiro[4.5]decan-4-one, a potent and selective orphanin FQ (OFQ) receptor agonist with anxiolytic-like properties. Eur J Med Chem 2000;35: 839–851.
999. Dautzenberg FM, Wichmann J, Higelin J, Py-Lang G, Kratzeisen C, Malherbe P, Kilpatrick GJ, Jenck F. Pharmacological characterization of the novel nonpeptide orphanin FQ/nociceptin receptor agonist Ro 64-6198: rapid and reversible desensitization of the ORL1 receptor in vitro and lack of tolerance in vivo. J Pharmacol Exp Ther 2001;298: 812–819.
1000. Higgins GA, Grottick AJ, Ballard TM, Richards JG, Messer J, Takeshima H, Pauly-Evers M, Jenck F, Adam G, Wichmann J. Influence of the selective ORL1 receptor agonist, Ro64-6198, on rodent neurological function. Neuropharmacology 2001;41:97–107.
1001. Rizzi D, Bigoni R, Rizzi A, Jenck F, Wichmann J, Guerrini R, Regoli D, Calo G. Effects of Ro 64-6198 in nociceptin/orphanin FQ-sensitive isolated tissues. Naunyn Schmiedebergs Arch Pharmacol 2001;363:551–555.
1002. Kawamato H, Ozaki S, Itoh Y, Miyaji M, Arai S, Nakashima H, Kato T, Ohta H, Iwasawa Y. Discovery of the first potent and selective small molecule opioid receptor-like (ORL1) antagonist: 1-[(3R,4R)-1-cyclooctylmethyl-3-hydroxymethyl-4-piperidyl]-3-ethyl-1,3-dihydro-2H-benzimidazol-2-one (J-113397). J Med Chem 1999;42:5061–5063.
1003. Ozaki S, Kawamato H, Itoh Y, Miyaji M, Iwasawa Y, Ohta H. A Potent and highly selective nonpeptidyl nociceptin/orphanin FQ receptor (ORL1) antagonist: J-113397. Eur J Pharmacol 2000;387:R17–R18.
1004. Ozaki S, Kawamoto H, Itoh Y, Miyaji M, Azuma T, Ichikawa D, Nambu H, Iguchi T, Iwasawa Y, Ohta H. In vitro and in vivo pharmacological characterization of J-113397, a potent and selective non-peptidyl ORL1 receptor antagonist. Eur J Pharmacol 2000;402:45–53.
1005. Shinkai H, Ito T, Iida T, Kitao Y, Yamada H, Uchida I. 4-Aminoquinolines: novel nociceptin antagonists with analgesic activity. J Med Chem 2000;43:4667–4677.
1006. Ortega A, Blount JF, Manchand PS. Salvinorin, a new trans-neoclerodane diterpene from Salvia divinorum (labiatae). J Chem Soc Perkin Trans 1 1982; 2505–2508.
1007. Valdes LJ III, Butler WM, Hatfield GM, Paul AG, Koreeda M. Divinorin A, a psychotropic terpenoid, and divinorin B from the hallucinogenic Mexican mint Salvia divinorum. J Org Chem 1984;49:4716–4720.
1008. Roth BL, Baner K, Westkaemper R, Siebert D, Rice KC, Steinberg S, Ernsberger P, Rothman RB. Salvinorin A: a potent naturally occurring nonnitrogenous kappa opioid selective agonist. Proc Natl Acad Sci USA 2002;99:11934–11939.
1009. Prisinzano TE, Rothman RB. Salvinorin A analogs as probes in opioid pharmacology. Chem Rev 2008;108:1732–1743.
1010. Siebert DJ. Salvia Divinorum and salvinorin A: new pharmacological findings. J Ethnopharmacol 1994;43:53–56.
1011. Sheffler DJ, Roth BL. Salvinorin A: the 'magic mint' hallucinogen finds a molecular target in the kappa opioid receptor. Trends Pharmacol Sci 2003;24:107–109.
1012. Butelman ER, Harris TJ, Kreek MJ. The plant-derived hallucinogen, salvinorin A,

produces kappa-opioid agonist-like discriminative effects in rhesus monkeys. Psychopharmacology 2004;172:220–224.

1013. Willmore-Fordham CB, Krall DM, McCurdy CR, Kinder DH. The hallucinogen derived from *Salvia divinorum*, salvinorin A, has [kappa]-opioid agonist discriminative stimulus effects in rats. Neuropharmacology 2007;53:481–486.

1014. John TF, French LG, Erlichman JS. The antinociceptive effect of salvinorin A in mice. Eur J Pharmacol 2006;545:129–133.

1015. McCurdy CR, Sufka KJ, Smith GH, Warnick JE, Nieto MJ. Antinociceptive profile of salvinorin A, a structurally unique kappa opioid receptor agonist. Pharmacol Biochem Behav 2006;83:109–113.

1016. Zhang Y, Butelman ER, Schlussman SD, Ho A, Kreek MJ. Effects of the plant-derived hallucinogen salvinorin A on basal dopamine levels in the caudate putamen and in a conditioned place aversion assay in mice: agonist actions at kappa opioid receptors. Psychopharmacology 2005;179:551–558.

1017. Carlezon WA Jr, Beguin C, Dinieri JA, Baumann MH, Richards MR, Todtenkopf MS, Rothman RB, Ma Z, Lee DY, Cohen BM. Depressive-like effects of the κ-opioid receptor agonist salvinorin A on behavior and neurochemistry in rats. J Pharmacol Exp Ther 2006;316:440–447.

1018. Fantegrossi WE, Kugle KM, Valdes LJ 3rd, Koreeda M, Woods JH. Kappa-opioid receptor-mediated effects of the plant-derived hallucinogen, salvinorin A, on inverted screen performance in the mouse. Behav Pharmacol 2005;16:627–633.

1019. Chartoff EH, Potter D, Damez-Werno D, Cohen BM, Carlezon WA Jr. Exposure to the selective [kappa]-opioid receptor agonist salvinorin A modulates the behavioral and molecular effects of cocaine in rats. Neuropsychopharmacology 2008;33:2676–2687.

1020. Li J-X, Rice KC, France CP. Discriminative stimulus effects of 1-(2,5-dimethoxy-4-methylphenyl)-2-aminopropane in rhesus monkeys. J Pharmacol Exp Ther 2008;324: 827–833.

1021. Gehrke B, Chefer V, Shippenberg T. Effects of acute and repeated administration of salvinorin A on dopamine function in the rat dorsal striatum. Psychopharmacology 2008;197:509–517.

1022. Inan, S, Lee, DY, Liu-Chen, LY, Cowan, A. Comparison of the diuretic effects of chemically diverse kappa opioid agonists in rats: nalfurafine, U50,488H, and salvinorin A. Naunyn Schmiedebergs Arch Pharmacol 2009;379:263–270.

1023. Butelman ER, Mandau M, Tidgewell K, Prisinzano TE, Yuferov V, Kreek MJ. Effects of salvinorin A, a κ-opioid hallucinogen, on a neuroendocrine biomarker assay in nonhuman primates with high κ-receptor homology to humans. J Pharmacol Exp Ther 2007;320: 300–306.

1024. Butelman ER, Prisinzano TE, Deng H, Rus S, Kreek MJ. Unconditioned behavioral effects of the powerful κ-opioid hallucinogen, salvinorin A: fast onset and entry into cerebrospinal fluid. J Pharmacol Exp Ther 2009;328:588–597.

1025. Schmidt MD, Schmidt MS, Butelman ER, Harding WW, Tidgewell K, Murry DJ, Kreek MJ, Prisinzano TE. Pharmacokinetics of the plant-derived κ-opioid hallucinogen salvinorin A in nonhuman primates. Synapse 2005;58:208–210.

1026. Schmidt MS, Prisinzano TE, Tidgewell K, Harding WW, Butelman ER, Kreek MJ, Murry DJ. Determination of salvinorin A in body fluids by high proformance liquid chromatography: atmospheric pressure chemical ionization. J Chromatogr B 2005;818: 221–225.

1027. Hooker JM, Xu Y, Schiffer W, Shea C, Carter P, Fowler JS. Pharmacokinetics of the potent hallucinogen, salvinorin A in primates parallels the rapid onset and short duration of effects in humans. NeuroImage 2008;41:1044–1050.

1028. Harding WW, Tidgewell K, Byrd N, Cobb H, Dersch CM, Butelman ER, Rothman RB, Prisinzano TE. Neoclerodane diterpenes as a novel scaffold for mu opioid receptor ligands. J Med Chem 2005;48:4765–4771.

1029. Tidgewell K, Groer CE, Harding WW, Lozama A, Schmidt M, Marquam A, Hiemstra J, Partilla JS, Dersch CM, Rothman RB, Bohn LM, Prisinzano TE. Herkinorin analogues with differential beta-arrestin-2 interactions. J Med Chem 2008;51:2421–2431.

1030. Groer CE, Tidgewell K, Moyer RA, Harding WW, Rothman RB, Prisinzano TE, Bohn LM. An opioid agonist that does not induce mu-opioid receptor–arrestin interactions or receptor internalization. Mol Pharmacol 2007;71:549–557.

1031. Munro TA, Duncan KK, Xu W, Wang Y, Liu-Chen LY, Carlezon WA Jr, Cohen BM, Beguin

C. Standard protecting groups create potent and selective kappa opioids: salvinorin B alkoxymethyl ethers. Bioorg Med Chem 2008;16:1279–1286.

1032. Wang Y, Chen Y, Xu W, Lee DY, Ma Z, Rawls SM, Cowan A, Liu-Chen LY. 2-Methoxymethyl-salvinorin B is a potent kappa opioid receptor agonist with longer lasting action in vivo than salvinorin A. J Pharmacol Exp Ther 2008;324:1073–1083.

1033. Beguin C, Potter DN, Dinieri JA, Munro TA, Richards MR, Paine TA, Berry L, Zhao Z, Roth BL, Xu W, Liu-Chen LY, Carlezon WA Jr, Cohen BM. N-Methylacetamide analog of salvinorin A: a highly potent and selective kappa-opioid receptor agonist with oral efficacy. J Pharmacol Exp Ther 2008;324: 188–195.

1034. Holden KG, Tidgewell K, Marquam A, Rothman RB, Navarro H, Prisinzano TE. Synthetic studies of neoclerodane diterpenes from Salvia divinorum: exploration of the 1-position. Bioorg Med Chem Lett 2007;17: 6111–6115.

1035. Beguin C, Richards MR, Li JG, Wang Y, Xu W, Liu-Chen LY, Carlezon WA Jr, Cohen BM. Synthesis and in vitro evaluation of salvinorin A analogues: effect of configuration at C(2) and substitution at C(18). Bioorg Med Chem Lett 2006;16:4679–4685.

1036. Lee DY, He M, Kondaveti L, Liu-Chen LY, Ma Z, Wang Y, Chen Y, Li JG, Beguin C, Carlezon WA Jr, Cohen B. Synthesis and in vitro pharmacological studies of C(4) modified salvinorin A analogues. Bioorg Med Chem Lett 2005;15:4169–4173.

1037. Lee DY, He M, Liu-Chen LY, Wang Y, Li JG, Xu W, Ma Z, Carlezon WA Jr, Cohen B. Synthesis and in vitro pharmacological studies of new C(4)-modified salvinorin A analogues. Bioorg Med Chem Lett 2006;16: 5498–5502.

1038. Munro TA, Rizzacasa MA, Roth BL, Toth BA, Yan F. Studies toward the pharmacophore of salvinorin A, a potent kappa opioid receptor agonist. J Med Chem 2005;48:345–348.

1039. Harding WW, Schmidt M, Tidgewell K, Kannan P, Holden KG, Dersch CM, Rothman RB, Prisinzano TE. Synthetic studies of neoclerodane diterpenes from Salvia divinorum: selective modification of the furan ring. Bioorg Med Chem Lett 2006;16: 3170–3174.

1040. Simpson DS, Katavic PL, Lozama A, Harding WW, Parrish D, Deschamps JR, Dersch CM, Partilla JS, Rothman RB, Navarro H, Prisinzano TE. Synthetic studies of neoclerodane diterpenes from Salvia divinorum: preparation and opioid receptor activity of salvinicin analogues. J Med Chem 2007;50: 3596–3603.

1041. Shirota O, Nagamatsu K, Sekita S. Neo-clerodane diterpenes from the hallucinogenic sage Salvia divinorum. J Nat Prod 2006;69: 1782–1786.

1042. Yan F, Mosier PD, Westkaemper RB, Stewart J, Zjawiony JK, Vortherms TA, Sheffler DJ, Roth BL. Identification of the molecular mechanisms by which the diterpenoid salvinorin A binds to κ-opioid receptors. Biochemistry 2005;44:8643–8651.

1043. Vortherms TA, Mosier PD, Westkaemper RB, Roth BL. Differential helical orientations among related G protein-coupled receptors provide a novel mechanism for selectivity. Studies with salvinorin A and the kappa-opioid receptor. J Biol Chem 2007;282: 3146–3156.

1044. Kane BE, McCurdy CR, Ferguson DM. Toward a structure-based model of salvinorin A recognition of the kappa-opioid receptor. J Med Chem 2008;51:1824–1830.

1045. Sawynok J, Cowan A. Novel Aspects of Pain management: Opioids and Beyond. New York: Wiley-Liss; 1999. p 373.

1046. Williams M, Kowaluk EA, Arneric SP. Emerging molecular approaches to pain therapy. J Med Chem 1999;42:1481–1500.

1047. Eglen RM, Hunter JC, Dray A. Ions in the fire: recent ion-channel research and approaches to pain therapy. Trends Pharmacol Sci 1999;20:337–342.

1048. Kowaluk EA, Arneric SP. Novel molecular approaches to analgesia. Annu Rep Med Chem 1998;33:11–20.

1049. Redila V, Chavkin C. Stress-induced reinstatement of cocaine seeking is mediated by the kappa opioid system. Psychopharmacology 2008;200:59–70.

1050. Shaham Y, Shalev U, Lu L, de Wit H, Stewart J. The reinstatement model of drug relapse: history, methodology and major findings. Psychopharmacology 2003;168:3–20.

1051. Bossert JM, Ghitza UE, Lu L, Epstein DH, Shaham Y. Neurobiology of relapse to heroin and cocaine seeking: an update and clinical implications. Eur J Pharmacol 2005;526: 36–50.

1052. Epstein D, Preston K, Stewart J, Shaham Y. Toward a model of drug relapse: an assessment of the validity of the reinstatement procedure. Psychopharmacology 2006;189: 1–16.

1053. Beardsley PM, Howard JL, Shelton KL, Carroll FI. Differential effects of the novel kappa opioid receptor antagonist, JDTic, on reinstatement of cocaine-seeking induced by footshock stressors vs cocaine primes and its antidepressant-like effects in rats. Psychopharmacol 2005;183:118–126.

1054. Carey AN, Borozny K, Aldrich JV, McLaughlin JP. Reinstatement of cocaine place-conditioning prevented by the peptide kappa-opioid receptor antagonist arodyn. Eur J Pharmacol 2007;569:84–89.

1055. Carlezon WA Jr, Beguin C, Knoll AT, Cohen BM. Kappa-opioid ligands in the study and treatment of mood disorders. Pharmacol Ther 2009;123:334–343.

1056. Knoll AT, Meloni EG, Thomas JB, Carroll FI, Carlezon WA Jr. Anxiolytic-like effects of kappa-opioid receptor antagonists in models of unlearned and learned fear in rats. J Pharmacol Exp Ther 2007;323:838–845.

1057. Ma J, Ye N, Lange N, Cohen BM. Dynorphinergic GABA neurons are a target of both typical and atypical antipsychotic drugs in the nucleus accumbens shell, central amygdaloid nucleus and thalamic central medial nucleus. Neuroscience 2003;121:991–998.

1058. Land BB, Bruchas MR, Lemos JC, Xu M, Melief EJ, Chavkin C. The dysphoric component of stress is encoded by activation of the dynorphin kappa-opioid system. J Neurosci 2008;28:407–414.

1059. Carey AN, Lyons AM, Shay CF, Dunton O, McLaughlin JP. Endogenous {kappa} opioid activation mediates stress-induced deficits in learning and memory. J Neurosci 2009;29: 4293–4300.

1060. Kreibich A, Reyes BAS, Curtis AL, Ecke L, Chavkin C, Bockstaele EJV, Valentino RJ. Presynaptic inhibition of diverse afferents to the locus ceruleus by {kappa}-opiate receptors: a novel mechanism for regulating the central norepinephrine system. J Neurosci 2008;28:6516–6525.

1061. Ko MCH, Johnson MD, Butelman ER, Willmont KJ, Mosberg HI, Woods JH. Intracisternal nor-binaltorphimine distinguishes central and peripheral kappa opioid antinociception in rhesus monkeys. J Pharmacol Exp Ther 1999;291:1113–1120.

1062. Negus SS, Mello NK, Linsenmayer DC, Jones RM, Portoghese PS. Kappa opioid antagonist effects of the novel kappa antagonist 5′-guanidinylnaltrindole (GNTI) in an assay of schedule-controlled behavior in rhesus monkeys. Psychopharmacol 2002;163: 412–419.

1063. Thomas JB, Fall MJ, Cooper JB, Rothman RB, Mascarella SW, Xu H, Partilla JS, Dersch CM, McCullough KB, Cantrell BE, Zimmerman DM, Carroll FI. Identification of an opioid kappa receptor subtype-selective N-substituent for (+)-(3R,4R)-dimethyl-4-(3-hydroxyphenyl)piperidine. J Med Chem 1998;41:5188–5197.

1064. Thomas JB, Atkinson RN, Rothman RB, Fix SE, Mascarella SW, Vinson NA, Xu H, Dersch CM, Lu Y, Cantrell BE, Zimmerman DM, Carroll FI. Identification of the First *trans*-(3R,4R)-dimethyl-4-(3-hydroxyphenyl)piperidine derivative to possess highly potent and selective opioid kappa receptor antagonist activity. J Med Chem 2001;44:2687–2690.

1065. Thomas JB, Atkinson RN, Namdev N, Rothman RB, Gigstad KM, Fix SE, Mascarella SW, Burgess JP, Vinson NA, Xu H, Dersch CM, Cantrell BE, Zimmerman DM, Carroll FI. Discovery of an opioid kappa receptor selective pure antagonist from a library of N-substituted 4β-methyl-5-(3-hydroxyphenyl)morphans. J Med Chem 2002;45: 3524–3530.

1066. Thomas JB, Atkinson RN, Vinson NA, Catanzaro JL, Perretta CL, Fix SE, Mascarella SW, Rothman RB, Xu H, Dersch CM, Cantrell BE, Zimmerman DM, Carroll FI. Identification of (3R)-7-hydroxy-N-((1S)-1-[[(3R,4R)-4-(3-hydroxyphenyl)-3,4-dimethyl-1-piperidinyl]methyl]-2-methylpropyl)-1,2,3,4-tetrahydro-3-isoquinolinecarboxamide as a novel potent and selective opioid kappa receptor antagonist. J Med Chem 2003;46:3127–3137.

1067. Carroll I, Thomas JB, Dykstra LA, Granger AL, Allen RM, Howard JL, Pollard GT, Aceto MD, Harris LS. Pharmacological properties of JDTic: a novel kappa-opioid receptor antagonist. Eur J Pharmacol 2004;501: 111–119.

1068. Metcalf MD, Coop A. Kappa opioid antagonists: past successes and future prospects. AAPS J 2005;7:E704–E722.

1069. Aldrich JV, Patkar KA, McLaughlin JP. Zyklophin, a systemically active selective kappa opioid receptor peptide antagonist with short duration of action. Proc Natl Acad Sci USA 2009;106:18396–18401.

PET AND SPECT IN DRUG DEVELOPMENT

Andy Welch[1]
Sally Pimlott[2]

[1] School of Medical Sciences, University of Aberdeen, Aberdeen, UK
[2] West of Scotland Radionuclide Dispensary, NHS Greater Glasgow and Clyde, Glasgow, UK

1. INTRODUCTION

The radionuclide imaging techniques of positron emission tomography (PET) and single photon emission computed tomography (SPECT) are the most sensitive techniques for imaging function *in vivo*. They are also truly translational techniques in that the same imaging biomarkers, acquisition protocols, and image processing techniques can be used in animal models and in man. The extremely high sensitivity (of the order of picomolar) comes from the relationship between the activity (defined as disintegrations per second or becquerels (Bq)) and mass of a radioactive isotope. The specific activity is defined as

$$SA = \frac{4.2 \times 10^{23}}{A T_{1/2}} \text{Bq/g}$$

where A is the atomic mass number for the isotope and $T_{1/2}$ is the half-life.

For example, 400 MBq (10 mCi) of ^{18}F $\sim 1.1 \times 10^{-10}$ g. In practice, activities of a few MBq can be detected by a modern scanner. Therefore, PET and SPECT are described as tracer imaging techniques in which "trace" (subpharmacological) amounts of a radiolabeled imaging biomarker are administered.

While there are some physical differences between PET and SPECT, in practice, the two techniques are mainly differentiated by the isotopes (and hence the chemistry) that can be used. In the following sections, we will outline the physical differences between the two techniques and the isotopes that can be used. We will then discuss the potential roles in drug development and give examples of where the techniques have been successfully applied.

2. THE PHYSICS OF PET AND SPECT

Both PET and SPECT rely on the detection of gamma rays by a scanner. In the case of SPECT, the gamma rays are emitted directly by the isotope used. In the case of PET, the radioactive isotopes emit positrons. Positrons are positively charged (antimatter equivalents of) electrons. These positrons travel a short distance (of the order of a few mm) in tissue before combining with an electron. This combination, referred to as an annihilation event, results in all the mass of the positron and electron being converted into energy. In order to conserve energy and momentum, this energy is produced in the form of two 511 keV gamma rays emitted at approximately 180° to each other. Three photons can also be produced, but this is rare.

Both PET and SPECT scanners effectively measure the total activity along lines through the object. The process of converting this "projection data" into a three-dimensional image is referred to as image reconstruction. A full discussion on image reconstruction is beyond the scope of this book. However, it is known that an image can be reconstructed from a set of parallel projections acquired at different angles covering at least 180 degrees (see Fig. 1) [1]. In SPECT, the direction information is obtained by placing a collimator in front of the gamma camera. This collimator is simply an absorber (lead or tungsten) with lots of holes. In PET, the direction information is obtained by exploiting the fact that two gamma rays are emitted at 180° to each other. The subject is surrounded by detectors and an event is registered if two detectors detect a gamma ray within a small interval of time, known as the timing window (see Fig. 1). This technique is referred to as electronic collimation (as opposed to the physical collimation used in SPECT).

There are various parameters that can be used to describe the performance of a PET or SPECT system, but two of the most important parameters are the sensitivity and resolution of the system. For SPECT, these two parameters are primarily defined by the collimator. Making the holes in the collimator larger will improve the sensitivity of the system, but reduce the resolution and vice versa. For PET,

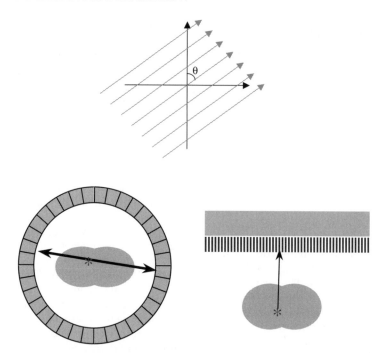

Figure 1. Top: The definition of a parallel projection, which is the line integral of the object obtained along a set of parallel lines at some angle θ. It is known that an image can be reconstructed from sets of parallel projections obtained over an angular range of at least 180°. Bottom: The mechanism by which PET (left) and SPECT (right) cameras obtain projection data. (This figure is available in full color at http://mrw.interscience.wiley.com/emrw/9780471266945/home.)

the sensitivity is primarily a function of the number of detectors used and their performance characteristics. The resolution is determined by the size of the detector elements and by two physical factors, positron range and noncolinearity. Positron range refers to the small distance that the positron moves before combining with an electron. This range is typically of the order of a few millimeters and is a function of the isotope used. Noncolinearity refers to the fact that the two gamma rays are rarely emitted at exactly 180° to each other as the positron and electron are usually not quite at rest when they combine. This noncolinearity produces an uncertainty in the location of the annihilation, which is larger for larger scanners (see Fig. 2).

In practice, sensitivity is usually the most important parameter for a PET or SPECT system as there is an inherent link between sensitivity and resolution. Both PET and SPECT systems rely on counting gamma rays and radioactive decay obeys Poisson statistics, which means that the standard deviation in the number of gamma rays detected is the square root of the number. So, the signal-to-noise ratio increases as the number of events detected increases. In practice, both PET and SPECT systems are count limited and, without extra smoothing (which reduces resolution), produce unacceptably noisy images. Increasing sensitivity reduces the amount of extra smoothing that is required and hence improves resolution. Since electronic collimation is inherently more sensitive than physical collimation, PET systems are generally around an order of magnitude more sensitive than SPECT systems. For clinical systems, this translates into better resolution of ~5 mm for PET compared with ~10 mm for SPECT. For preclinical systems, where the positron range becomes more significant, the resolution of PET and SPECT systems is similar (~1 mm). SPECT systems can theoretically achieve higher resolutions than PET, although the extra sensitivity of PET means that this rarely happens in practice.

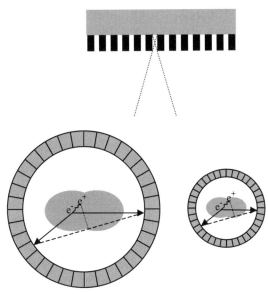

Figure 2. Schematic showing the main factors affecting resolution of SPECT (top) and PET (bottom) scanners. For SPECT, the main factor is the collimator and resolution depends on the size of the holes and distance from the collimator. For PET, there are two factors: the distance traveled by the positron before annihilation (which is a function of the isotope) and the fact that the gamma rays are not emitted at exactly 180°. The effect of this noncolinearity depends on the size of the scanner. (This figure is available in full color at http://mrw.interscience.wiley.com/emrw/9780471266945/home.)

3. ISOTOPES USED IN PET AND SPECT

While there are some differences in the physics of PET and SPECT systems (described above) the primary difference stems from the isotopes used, which determines the radiopharmaceuticals that can be developed. There is a large range of isotopes that can be used. In fact, either gamma or positron-emitting forms of almost any element can be produced. However, the physical characteristics of most isotopes make them difficult to use in practice. The primary characteristics that need to be considered are the half-life of the isotope and what is emitted during radioactive decay. Ideally, the physical half-life of the isotope should be comparable to the biological half-life of the process being imaged. For a SPECT isotope, we also desire that the energy of the gamma rays produced is in the range 100–200 keV as conventional gamma cameras are designed to operate in this range. Higher energy gamma rays require thicker detectors, which have poorer resolution. Lower energy gamma rays experience more attenuation and are less likely to escape from the subject, although this is less of an issue for small animals where lower energy gamma rays can be used. For PET, we desire that the primary mode of decay be positron emission and that there are not too many other gamma rays emitted, or at least that these are not emitted at energies close to the 511 keV annihilation gamma rays as that would reduce the quality of the image due to detection of erroneous "random" coincidences (where two independent gamma rays are detected close together and falsely recorded as an annihilation event).

Some details on PET and SPECT isotopes are shown in Table 1. The two most commonly used positron-emitting isotopes are ^{11}C and ^{18}F. Carbon is present naturally in biologically occurring compounds and fluorine atoms are small thus reducing the problem of inserting a large alien marker into a lead compound that may alter the basic pharmacology of the compound. However, the short half-life can restrict imaging and lead to more complicated modeling. Furthermore, the short half-life of the positron-emitting isotopes limits the availability of PET imaging, requiring an

Table 1. Common PET and SPECT Isotopes

	PET		SPECT	
Common Isotopes	^{11}C	^{18}F	^{123}I	^{99m}Tc
Relative sizes	Biologically natural	Small	Large	Very large—requiring chelating moiety
Half-life	20 min	110 min	13.2 h	6 h
Relative availability	Requires on-site cyclotron	Availability limited	Availability limited	Readily from a ^{99m}Tc generator

on-site cyclotron, and makes radiolabeling protocols challenging.

SPECT uses lower energy gamma-emitting radioisotopes, such as ^{123}I and ^{99m}Tc. In contrast to PET, these SPECT isotopes have a considerably longer half-life that makes SPECT tracers more available for imaging. The major advantage of ^{99m}Tc is that it can be produced from the decay of its longer lived parent nuclide, ^{99}Mo, on commercially available generator systems, making it widely available to any nuclear medicine department worldwide. The longer half-life of ^{123}I allows transportation of ^{123}I radiotracers over further distances making them more commercially viable. However, addition of the larger SPECT radioisotopes into a biological compound can be problematic and restrict the pharmacology and biodistribution kinetics of the labeled compound compared to the parent compound.

4. CURRENT AND EMERGING APPLICATIONS

PET and SPECT imaging can be used across the field of medicine to provide information about disease mechanisms, diagnosis of disease, identification of drug targets, and monitoring of therapeutic response. There are numerous radiotracers that have been developed to enable us to visualize various targets.

4.1. PET Imaging

PET has become an established diagnostic tool in most parts of the Western world over the past 10 years. While PET offers the potential to probe a wide range of physiological processes through targeted biomarkers; in practice, the vast majority of PET scans involve the use of a single tracer, 2-[^{18}F]*fluoro*-2-deoxy-D-glucose (FDG), to image cancer. In fact, it could be argued that the growth of PET as a diagnostic tool is almost entirely due to the value of FDG as a biomarker of cancer. FDG-PET (usually combined with CT in a hybrid PET-CT scanner) has been shown to be a useful tool in the management of a wide variety of cancers [2]. Those where it is deemed cost effective vary from country to country but in general it is used for staging and restaging of solid tumors (e.g., lung, lymphoma, colorectal, etc.). The strength of FDG as a biomarker of cancer is its sensitivity, which stems from the wide range of processes that have been shown to affect its uptake. These processes include hypoxia, cell proliferation rate, microvascularization, glycolysis, membrane transport, growth pattern, inflammatory response, cell density, tumor volume, tumor grade, tissue heterogeneity, necrosis, fibrosis, and blood glucose concentration [3]. The high sensitivity of FDG as a biomarker of cancer also makes it suitable for monitoring response to therapy as virtually all therapies will affect at least one of the processed biomarkers that govern FDG uptake. However, care is needed here as uptake is affected by inflammatory responses, which can mask the response of the tumor to the therapy. The main weaknesses of FDG are its low specificity (due to the wide range of processes that affect its uptake) and high uptake in some normal tissues (e.g., brain and bladder). The latter limits its usefulness in certain tumor types (e.g., gliomas and prostate cancers). These weaknesses have prompted research into more specific biomarkers of cancer. While [C-11]methionine, a marker of protein synthesis, has been used successfully for imaging brain tumors, most effort has been focused on the development of biomarkers for cell proliferation and apoptosis [4,5]. This research is driven by the fact that a majority of new anticancer drugs target molecular pathways to inhibit cell proliferation rather than stimulate cell death, resulting in tumor stasis rather than tumor shrinkage.

The most widely used PET proliferation marker is ^{18}F-fluorothymidine (FLT). The metabolism of FLT is directly linked to nuclear DNA synthesis as it is a good substrate for thymidine kinase 1 (but not for thymidine kinase 2). It is worth noting that FLT is not a true proliferative marker as it is not incorporated into DNA. However, FLT uptake has been shown to correlate well with other proliferation measurements (such as K_i-67 score) in lung, brain, and breast cancer [6], suggesting that it is a good surrogate marker of proliferation. A number of "true" PET proliferative markers, that is, ones that are incorporated into DNA, have been developed (generally based on labeled nucleosides). The most promising of these is 1-(2′-deoxy-2′-fluoro-β-D-arabinofuranosyl)thymine (FMAU), which has been labeled with both ^{11}C and ^{18}F. While ^{11}C-FMAU tended to produce poor quality images [7], ^{18}F-FMAU generally produced clear images of tumors [8]. FMAU has also been shown to have higher uptake in tumors than FLT, but with a higher background signal [9]. Interestingly, FMAU and FLT have a different biodistribution in normal tissues making them complimentary in terms of the tumors that can be visualized. While FMAU has a higher liver, kidney, and myocardial uptake than FLT, it has significantly lower marrow uptake and a slower rate excretion through the kidneys. Therefore, for example, FMAU may be more suitable for imaging bone metastasis or for imaging the pelvic region.

Attempts to develop an effective PET marker of apoptosis have so far met with limited success. Most PET markers developed so far have been based on annexin V, caspase 3, or Aposense. While promising results have been obtained in vitro with compounds based on caspase 3 and Aposense [10,11], annexin V based compounds have been most widely studied. These have been labeled with both ^{18}F and ^{124}I for PET and with other isotopes of iodine for SPECT. The most promising PET candidate appears to be annexin V-128, which is a self-chelating mutant [12].

Outside of cancer, the main applications of PET are in the fields of cardiovascular imaging and neuroimaging. While FDG-PET is still considered by some people to be the gold standard for assessing cardiac viability, it is being largely supplanted by MRI in the clinic [13]. Recent PET efforts in the field of cardiovascular imaging have focused on the development of a biomarker for unstable plaque [14]. However, despite significant activity, no robust methods have been reported to date.

One of the most promising areas for PET is neuroimaging. Radiotracer techniques (i.e., PET and SPECT) are inherently suited to receptor studies due to the tracer principle, which means that good quality images can be obtained by administering subpharmacological doses of the radiopharmaceutical. Research in this area has been stimulated by the development of high-resolution scanners designed for imaging small animal models of disease and a number of clinical applications (most notably in Alzheimer's disease (AD) and Parkinson's disease) have started to emerge [15].

4.2. SPECT Imaging

For years, nuclear medicine imaging, using tracers radiolabeled with ^{99m}Tc and other gamma-emitting photons has provided an essential routine diagnostic tool, in the fields of oncology, cardiology, neurology, and so on. Technetium radiotracers include [^{99m}Tc]-methylene diphosphonate ([^{99m}Tc]-MDP) for bone imaging, [^{99m}Tc]-HMPAO for imaging regional cerebral blood flow, useful in the diagnosis of Alzheimer's disease, [^{99m}Tc]-tetrofosmin for myocardial perfusion imaging in cardiac disorders, and many more with numerous applications across the medical field.

Fewer iodinated SPECT ligands have been translated into routine clinical use, possibly due to its limited availability compared to ^{99m}Tc; however, ^{123}I does offer advantages when radiolabeling small drug molecules. One example of an iodinated radiotracer is the noradrenaline transporter ligand, ^{123}I labeled meta-iodobenzylguanidine ([^{123}I]-mIBG, shown in Fig. 3) that was originally developed for imaging the adrenal medulla [16], rapidly followed by the use of this tracer as a marker and therapeutic agent for neuroendocrine-derived tumors such as neuroblastoma. More recently, [^{123}I]-mIBG has also been used as a marker of myocardial sympathetic neuronal integrity in various cardiac diseases [17]. Another successful iodinated radiopharmaceutical is the dopamine transporter ligand, 2β-carbomethoxy-3β3-(4-iodophenyl)-N-(3-fluoropropyl)nortropane ([^{123}I]-FP-CIT or DaTScan, Fig. 3) [18], a marker of functional dopaminergic terminals. This tracer has proven to be useful in patients with clinically uncertain parkinsonian syndromes, aiding the differential diagnosis of essential tremor from other parkinsonian syndromes related to idiopathic Parkinson's disease, multiple system entropy, and progressive supranuclear palsy [19]. Imaging the dopamine transport

742 PET AND SPECT IN DRUG DEVELOPMENT

[^{123}I]-mIBG

[^{123}I]-FP-CIT

Figure 3. Examples of radioiodinated SPECT radiotracers.

with DaTScan can also be used in the differential diagnosis of dementia with Lewy bodies from Alzheimer's disease [20,21].

As with PET, SPECT scanners have recently been coupled with CT scanners to produce hybrid machines. Besides providing an anatomical reference for the functional image, the CT images can also be used for attenuation and scatter correction, improving quantitative accuracy (Fig. 4). These scanners have been shown to have improved performance compared with planar scintigraphy or conventional SPECT in bone imaging, somatostatin-receptor imaging, parathyroid imaging, adrenal gland imaging, and sentinel node imaging, suggesting a possible resurgence of SPECT in these clinical areas [22].

4.3. Development of Small Animal Imaging

The development of new radiopharmaceuticals is essential for expanding the usefulness of PET and SPECT imaging. Both techniques offer the potential of imaging a wide range of functional processes through the development of new highly selective biomarkers. However, translating these new radiopharmaceuticals from bench to bedside can be as challenging as for novel drugs. As with drug development, evaluation of novel candidates in living animals is crucial to this translational process. The development of small animal imaging cameras allows the determination of *in vivo* characteristics such as tracer biodistribution and pharmacokinetics more efficiently, improving both the drug and radiopharmaceutical development process. Established imaging biomarkers can also be used in preclinical models to test the efficacy of novel therapeutics. Such biomarkers could be as straightforward as using FDG to demonstrate the efficacy of an anticancer drug at the preclinical stage (e.g., see Ref. [23]) or it may involve more specific ligands, as described below.

5. DRUG DEVELOPMENT APPLICATIONS

PET and SPECT imaging agents can be used as biomarkers of drug action, providing essential information early in the drug development process. As a result, PET and SPECT imaging has contributed to drug development in terms of rational drug dosing, biodistribution of drug, therapeutic rationale for drug use, mechanism of drug action, and treatment response.

The information gained from PET and SPECT imaging depends on the radiolabeled agent used. The drug of interest itself, a selective ligand binding to the target or a biomarker of the disease process can be radiolabeled for use in drug development imaging studies. These strategies and examples are discussed below.

5.1. A Radiolabeled Drug

The potential to radiolabel the drug of interest itself with ^{11}C for PET imaging, without alter-

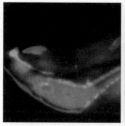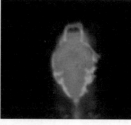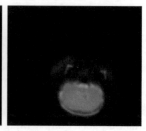

Figure 4. Hybrid PET/CT images of FDG uptake in a mouse brain, acquired on a Suinsa Argus preclinical PET/CT scanner. (See color insert.)

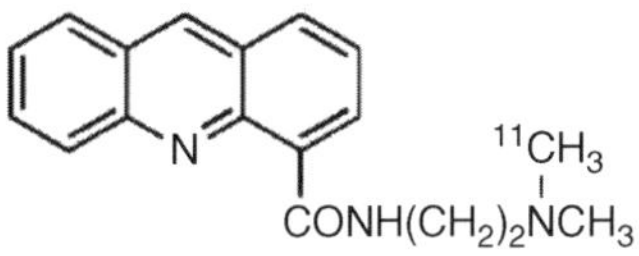

Figure 5. ^{11}C-Radiolabeled anticancer agent N-[2-(dimethylamino)ethyl]acridine-4-carboxamide.

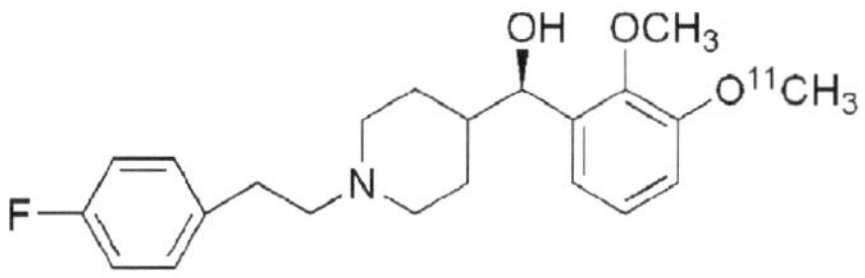

Figure 6. ^{11}C-Radiolabeled M100907, a selective 5-HT$_{2a}$ antagonist.

ing its chemical structure, is particularly useful in assessing the ability of a drug to reach target tissues in the living human body. These studies not only determine the biodistribution of the drug throughout the body but also provide information on the pharmacokinetics of the drug.

An example is a PET study with the ^{11}C-radiolabeled anticancer agent N-[2-(dimethylamino)ethyl]acridine-4-carboxamide (DACA) (Fig. 5). In this study, investigators used [^{11}C]-DACA imaging to provide biodistribution and pharmacokinetic information of the drug in cancer patients [24]. Results showed radiotracer uptake in several tumor types, lower than expected brain uptake compared with other tissues studied, low vertebral body uptake, and high myocardial uptake. The studies also showed that the compound was extensively metabolized in humans and therefore more stable analogs should be investigated. This was one of the first studies to investigate the tumor and normal tissue pharmacokinetics of an anticancer agent before conventional Phase I trials and highlights the ability to use only microdoses of labeled drug that were unlikely to be toxic, to provide important information early in the drug development process.

A further example is a Phase I clinical PET study of the selective 5-HT$_{2a}$ antagonist M100907, using ^{11}C- radiolabeled M100907 (Fig. 6). This study demonstrated that greater than 90% occupancy of the 5-HT$_{2a}$ receptor is reached at a dose of 20 mg per day [25] therefore ensuring adequate receptor occupancy at a given dose.

5.2. A Radiolabeled Ligand that Binds to Drug Target

Radiolabeling drugs isotopically (without changing the drugs chemical structure) can be challenging. Carbon-11 is the most commonly used isotope for radiolabeling drugs isotopically, due to its abundance in all drug candidates. However, the short half-life (20.4 min) of ^{11}C limits the chemical routes available for its incorporation, making it costly and time consuming for an individual drug candidate. Alternatively, the development or the use of an existing radiolabeled ligand that binds to the target and hence can be displaced by the drug under investigation has a number of following advantages:

(1) The radioisotope used is not restricted to the elements present in compounds allowing longer lived PET and SPECT isotopes to be investigated.
(2) A radioligand can be designed to have the appropriate characteristics for the imaging modality of choice, such as metabolism, pharmacokinetics, and so on.
(3) Such a radioligand can be used to investigate many drugs for the same target.

A radiolabeled ligand that binds to the target of interest is particularly useful for drugs designed to treat psychiatric disorders, where clinical endpoints are limited, enabling the determination of brain uptake and specific receptor occupancy. The monoamine system is the most widely studied system in the brain using PET and SPECT imaging, due to the abundance of tracers for monoamine targets. One example is a PET study with pindolol (Fig. 7) and the serotonin type 1A receptor tracer, [^{11}C]-WAY100635 (Fig. 7). This study showed the dose of pindolol used in previous clinical trials was too low to produce an adequate clinical response, possibly accounting

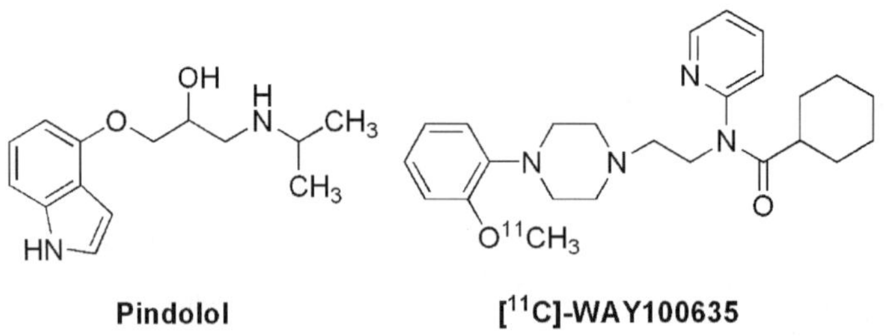

Figure 7. Pindolol and [^{11}C]-WAY100635, a serotonin type 1A receptor tracer.

for their inconsistent results [26]. Thus, this study provided essential information informing future clinical trials on the dose of pindolol required for adequate serotonin type 1A receptor occupancy.

A more recent study using the dopamine D2 receptor radiotracers [^{11}C]-raclopride (Fig. 8) and [^{11}C]-FLB457 (Fig. 8) provided information on striatal and extrastriatal dopamine D2 receptor dose occupancy in schizophrenia patients treated with paliperidone ER [27]. Results suggest that paliperidone ER at a dose of 6–9 mg provides an estimated level of dopamine D2 receptor occupancy between 70% and 80%.

Other neuroreceptor systems have also been investigated. The radiotracer neurokinin-1 receptor tracer, [^{18}F]-SPA-RQ (Fig. 9), was used to investigate aprepitant as an antidepressant in the treatment of major depressive disorder [28]. The study found dosing regimens provided continuously high levels of neurokinin-1 receptor blockade over 8 weeks, but no efficacy, providing evidence against the use of neurokinin-receptor antagonism as an antidepressant mechanism.

SPECT imaging has also been used effectively to aid drug development. One example is a SPECT imaging study where the dopamine D2 receptor SPECT tracer [^{123}I]-IBZM (Fig. 10) was used to show action of atypical antipsychotic (olanzapine) at D2 receptor [29].

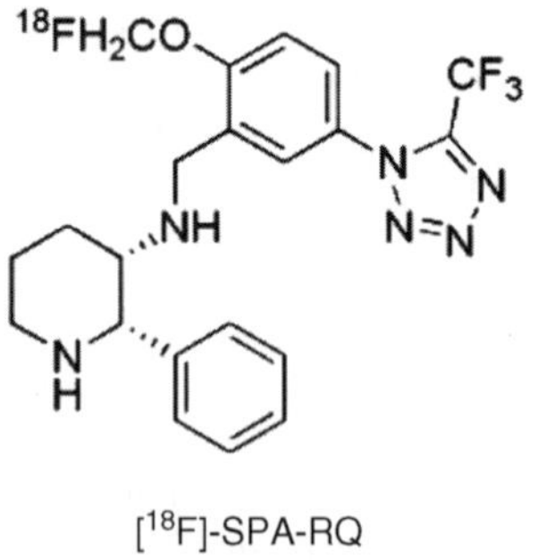

Figure 9. [^{18}F]-SPA-RQ, a radiotracer neurokinin-1 receptor tracer.

5.3. A Radiolabeled Biomarker of Disease

A biomarker of the disease itself, such as increased glucose metabolism and cell proliferation in cancer or the presence of inflammation, can be used to assess drug efficacy in the treatment of disease. PET imaging of glucose metabolism with [^{18}F]-FDG has proven to be a useful method in the field of cancer treatment development. For example, [^{18}F]-FDG imaging

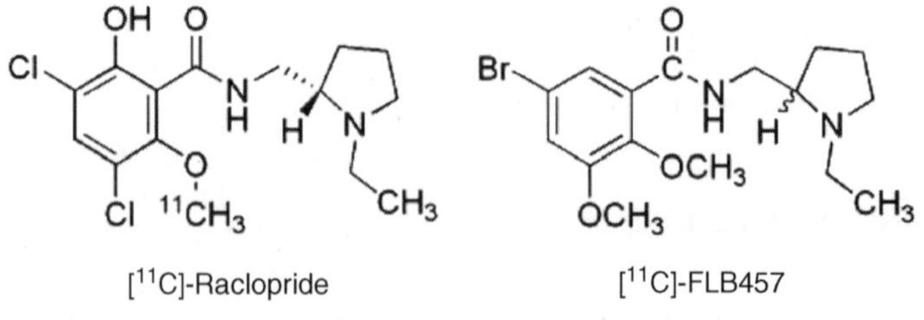

Figure 8. [^{11}C]-Raclopride and [^{11}C]-FLB457, dopamine D2 receptor radiotracers.

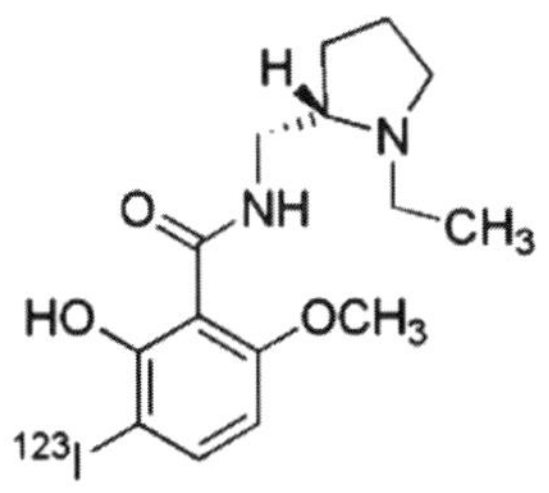

[¹²³I]-IBZM

Figure 10. [¹²³I]-IBZM, a dopamine D2 receptor SPECT tracer.

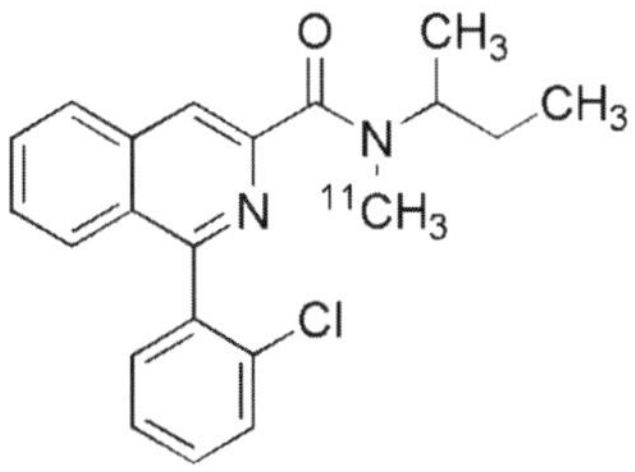

[¹¹C]-PK11195

Figure 11. [¹¹C]-PK11195, a biomarker of neuroinflammation.

has been used in patients with gastrointestinal stromal tumors, identifying early those that are responding well to treatment with imatinib mesylate [30]. The early detection of the efficacy of treatments allows decisions on the course of treatment to be taken earlier.

The ability to image apoptosis after cancer therapy can also be used to assess drug efficacy. A Phase I clinical trial investigated the use of SPECT imaging with ^{99m}Tc-recombinant human annexin V as a measure of apoptosis to predict response to chemotherapy [31]. Results showed a significant increase in uptake posttreatment with chemotherapy, predicting at least a partial response.

The PET radiotracer [¹¹C]-PK11195 (Fig. 11) that binds to the peripheral benzodiazepine receptor (PBR) can be used as a biomarker of neuroinflammation due to increased binding following microglial activation [32]. Neuroinflammation is suspected to play a role in numerous neurological disorders, such as multiple sclerosis and Alzheimer's disease [33]. [¹¹C]-PK11195 imaging of neuroinflammation can therefore be used to monitor disease severity/progression and has the potential to aid drug development and early clinical trials. However, [¹¹C]-PK11195 has several limitations due to its high nonspecific binding and poor signal to noise that complicates its quantification. More recently, there have been numerous studies investigating improved PBR ligands, labeled with ¹¹C, ¹⁸F, and ¹²³I (reviewed in Ref. [34]). Hence, the full usefulness of imaging neuroinflammation in drug development is still to be determined.

In Alzheimer's disease, amyloid imaging radiotracers, such as ¹¹C-radiolabeled Pittsburgh Compound B ([¹¹C]-PIB) and 2-(1-{6-[(2-[¹⁸F]fluoroethyl)(methyl)amino]-2-naphthyl} ethylidene)malononitrile ([¹⁸F]-FDDNP) (Fig. 12), may in the future prove to be useful as biomarkers of therapeutic outcome in clinical trials of novel dementia medications. Elevated amyloid content can be detected *in vivo* using these tracers in early AD, preceding any functional impairment [35]. Therefore, these tools may be useful not only in monitoring new drug therapies but also in identifying the populations of patients that may benefit from early intervention with these therapies, thereby having a significant impact on disease management (reviewed in Ref. [36]). To date, however, clinical studies have only shown significant differences in well-characterized

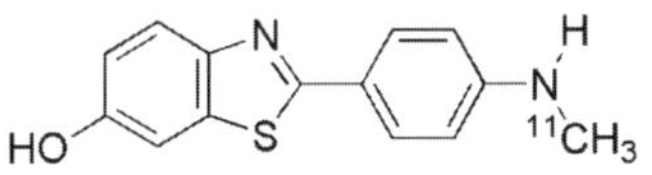

[¹¹C]PIB

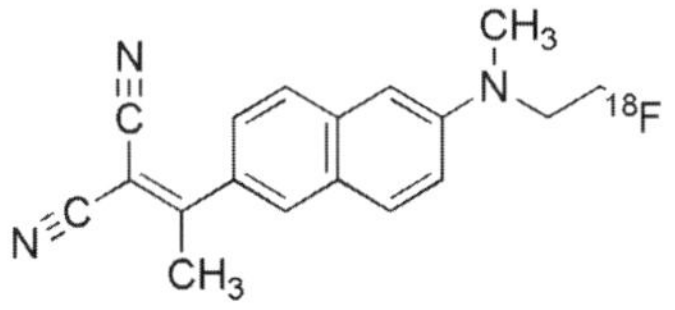

[18F]FDDNP

Figure 12. [¹¹C]-PIB and [¹⁸F]-FDDNP, amyloid imaging radiotracers.

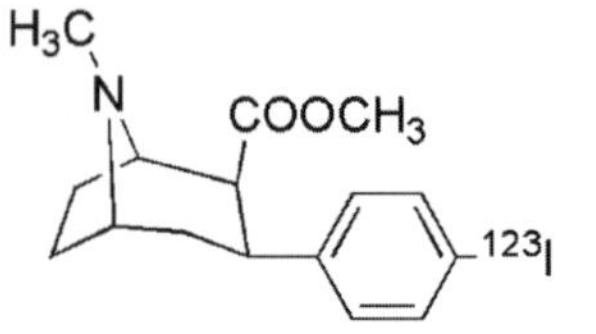

Figure 13. [^{123}I]-β-CIT, a dopamine and serotonin transporter radiotracer.

AD and healthy controls, with no evidence that these biomarkers can correlate with disease progression and severity (for some examples, see Refs [37–44]–). Further investigation is therefore required before the usefulness of these agents can be fully determined.

Imaging of the dopaminergic system in Parkinson's disease has highlighted that results of biomarker imaging studies are not always as expected. A study by Fahn et al. found treatment with L-dopa improved clinical outcome, compared to placebo. However, in contrast, reduction in the binding of the dopamine transporter radiotracer, [^{123}I]β-carbomethoxy-3-β-(4 iodophenyl)tropane ([^{123}I]-β-CIT) (Fig. 13) in the striatum was greater in patients with the higher doses of L-dopa, suggesting either that levodopa accelerates the loss of nigrostriatal dopamine nerve terminals or that its pharmacological effects modify the dopamine transporter [45]. Therefore, highlighting that the mechanism and therefore interpretation of biomarker imaging studies needs to be carefully investigated before a biomarker can be used in the drug development process.

6. CONCLUSION

Currently, PET and SPECT imaging techniques in drug development are hampered by the limited number of radiotracers available. As new biological targets are discovered, efforts are often focused on designing compounds as drug therapies and not as radiotracers. It is often difficult to find compounds with the appropriate radiotracer characteristics that can be imaged and analyzed using the technology available. For imaging to be useful in drug discovery, radiotracers need to be developed in parallel with drug candidates. The pooling of resources from both pharmaceutical companies and academia can only be beneficial to this process.

REFERENCES

1. Herman GT. Image reconstruction from projections. Real Time Imaging 1995;1(1): 3–18.
2. Papathanassiou D, Bruna-Muraille C, Liehn JC, Nguyen TD, Cure H. Positron emission tomography in oncology: present and future of PET and PET/CT. Crit Rev Oncol Hematol 2009;72(3): 239–254.
3. Bos R, Van der Hoeven JJM, Van der Wall E, van der Groep P, van Diest PJ, Comans EFI, Joshi U, Semenza GL, Hoekstra OS, Lammertsma AA, Molthoff CFM. Biologic correlates of ^{18}Fluorodeoxyglucose uptake in human breast cancer measured by positron emission tomography. J Clin Oncol 2002;20:379–387.
4. Barwick T, Bencherif B, Mountz JM, Avril N. Molecular PET and PET/CT imaging of tumour cell proliferation using F-18 fluoro-L-thymidine: a comprehensive evaluation. Nucl Med Commun 2009;30(12): 908–917.
5. Faust A, Hermann S, Wagner S, Haufe G, Schober O, Schafers M, Kopka K. Molecular imaging of apoptosis in vivo with scintigraphic and optical biomarkers: a status report. Anticancer Agents Med Chem 2009;9(9): 968–985.
6. Salskov A, Tammisetti VS, Grierson J, Vesselle H. FLT: measuring tumor cell proliferation in vivo with positron emission tomography and 3′-deoxy-3′-[^{18}F]fluorothymidine. Semin Nucl Med 2007;(37)(6): 429–439.
7. Lu L, Samuelsson L, Bergstrom M, Sato K, Fasth K-J, Langstrom B. Rat studies comparing ^{11}C-FMAU, ^{18}F-FLT, and ^{76}Br-BFU as proliferation markers. J Nucl Med 2002;43(12): 1688–1698.
8. Sun H, Sloan A, Mangner TJ, Vaishampayan U, Muzik O, Collins JM, Douglas K, Shields AF. Imaging DNA synthesis with [^{18}F]FMAU and positron emission tomography in patients with cancer. Eur J Nucl Med Mol Imaging 2005;32(1): 15–22.
9. Nishii R, Volgin AY, Mawlawi O, Mukhopadhyay U, Pal A, Bornmann W, Gelovani JG, Alauddin MM. Evaluation of 2′-deoxy-2′-[^{18}F]fluoro-5-methyl-1- β-l-arabinofuranosyluracil ([^{18}F]-l-FMAU) as a PET imaging agent for cellular proliferation: comparison with

[^{18}F]-d-FMAU and [^{18}F]FLT. Eur J Nucl Med Mol Imaging 2008;35(5): 990–998.

10. Zhou D, Chu W, Rothfuss J, Zeng C, Xu J, Jones L, Welch MJ, Mach RH. Synthesis, radiolabeling, and in vivo evaluation of an ^{18}F-labeled isatin analog for imaging caspase-3 activation in apoptosis. Bioorg Med Chem Lett 2006;16 (19): 5041–5046.

11. Aloya R, Shirvan A, Grimberg H, Reshef A, Levin G, Kidron D, Cohen A, Ziv I. Molecular imaging of cell death in vivo by a novel small molecule probe. Apoptosis 2006;11(12): 2089–2101.

12. Tait JF, Smith C, Levashova Z, Patel B, Blankenberg FG, Vanderheyden J-L. Improved detection of cell death in vivo with annexin V radiolabeled by site-specific methods. J Nucl Med 2006;47(9): 1546–1583.

13. Sawada SG. Positron emission tomography for assessment of viability. Curr Opin Cardiol 2006;21:464–468.

14. van der Vaart MG, Meerwaldt R, Slart RHJA, van Dam GM, Ti RA, Zeebregts CJ. Application of PET/SPECT imaging in vascular disease. Eur J Vasc Endovasc Surg 2008;35:507–513.

15. Hammoud DA, Hoffman JM, Pomper MG. Molecular neuroimaging: from conventional to emerging techniques. Radiology 2007;245: 21–42.

16. Wieland DM, Wu J, Brown LE, Mangner TJ, Swanson DP, Beierwaltes WH. Radiolabeled adrenergic neuron-blocking agents: adrenomedullary imaging with [^{131}I]iodobenzylguanidine. J Nucl Med 1980;21:349–353.

17. Rufini V, Shulkin B. The evolution in the use of MIBG in more than 25 years of experimental and clinical applications. Q J Nucl Med Mol Imaging 2008;52:341–350.

18. Neumeyer JL, Campbell A, Wang S, Gao Y, Milius RA, Kula NS, et al. N-ω-Fluoroalkyl analogs of (1R)-2β-carbomethoxy-3beta-(4-iodophenyl)tropane (β-CIT): radiotracers for positron emission tomography and single photon emission computed tomography imaging of dopamine transporters. J Med Chem 1994;37:1558–1561.

19. Catafau AM, Tolosa E. Impact of dopamine transporter SPECT using ^{123}I-ioflupane on diagnosis and management of patients with clinically uncertain parkinsonian syndromes. Mov Disord 2004;19:1175–1182.

20. McKeith I, O'Brien J, Walker Z, Tatsch K, Booij J, Darcourt J, et al. Sensitivity and specificity of dopamine transporter imaging with ^{123}I-FP-CIT SPECT in dementia with Lewy bodies: a phase III, multicentre study. Lancet Neurol 2007;6:305–313.

21. Walker Z, Jaros E, Walker RW, Lee L, Costa DC, Livingston G, et al. Dementia with Lewy bodies: a comparison of clinical diagnosis, FP-CIT single photon emission computed tomography imaging and autopsy. J Neurol Neurosurg Psychiatry 2007;78:1176–1181.

22. Buck AK, Nekolla S, Ziegler S, Beer A, Krause BJ, Herrmann K, Scheidhauer K, Wester H-J, Rummeny EJ, Schwaiger M, Drzezga A. SPECT/CT. J Nucl Med 2008;49:1305–1319.

23. Tseng JR, Kang KW, Dandekar M, Yaghoub S, Lee JH, Christensen JG, Muir S, Vincent PW, Michaud NR, Gambhir SS. Preclinical efficacy of the c-Met inhibitor CE-355621 in a U87 MG mouse xenograft model evaluated by ^{18}F-FDG small-animal PET. J Nucl Med 2008;49(1): 129–134.

24. Saleem A, Harte RJ, Matthews JC, Osman S, Brady F, Luthra SK, et al. Pharmacokinetic evaluation of N-[2-(dimethylamino)ethyl]acridine-4-carboxamide in patients by positron emission tomography. J Clin Oncol 2001;19: 1421–1429.

25. Talvik-Lotfi M, Nyberg S, Nordstrom AL, Ito H, Halldin C, Brunner F, et al. High 5HT$_{2A}$ receptor occupancy in M100907-treated schizophrenic patients. Psychopharmacology 2000;148: 400–403.

26. Rabiner EA, Bhagwagar Z, Gunn RN, Sargent PA, Bench CJ, Cowen PJ, et al. Pindolol augmentation of selective serotonin reuptake inhibitors: PET evidence that the dose used in clinical trials is too low. Am J Psychiatry 2001;158:2080–2082.

27. Arakawa R, Ito H, Takano A, Takahashi H, Morimoto T, Sassa T, et al. Dose-finding study of paliperidone ER based on striatal and extrastriatal dopamine D2 receptor occupancy in patients with schizophrenia. Psychopharmacology 2008;197:229–235.

28. Keller M, Montgomery S, Ball W, Morrison M, Snavely D, Liu G, et al. Lack of efficacy of the substance p (neurokinin1 receptor) antagonist aprepitant in the treatment of major depressive disorder. Biol Psychiatry 2006;59:216–223.

29. Pilowsky LS, Busatto GF, Taylor M, Costa DC, Sharma T, Sigmundsson T, et al. Dopamine D2 receptor occupancy in vivo by the novel atypical antipsychotic olanzapine: a ^{123}I IBZM single photon emission tomography (SPET) study. Psychopharmacology 1996;124:148–153.

30. Van den Abbeele AD, Badawi RD. Use of positron emission tomography in oncology and

its potential role to assess response to imatinib mesylate therapy in gastrointestinal stromal tumors (GISTs). Eur J Cancer 2002;38(Suppl 5): S60–S65.

31. Belhocine T, Steinmetz N, Hustinx R, Bartsch P, Jerusalem G, Seidel L, et al. Increased uptake of the apoptosis-imaging agent (99m)Tc recombinant human annexin V in human tumors after one course of chemotherapy as a predictor of tumor response and patient prognosis. Clin Cancer Res 2002;8:2766–2774.

32. Galiegue S, Tinel N, Casellas P. The peripheral benzodiazepine receptor: a promising therapeutic drug target. Curr Med Chem 2003;10: 1563–1572.

33. Banati RB. Visualising microglial activation *in vivo*. Glia 2002;40:206–217.

34. Chauveau F, Boutin H, Van Camp N, Dolle F, Tavitian B. Nuclear imaging of neuroinflammation: a comprehensive review of [^{11}C]PK11195 challengers. Eur J Nucl Med Mol Imaging 2008;35:2304–2319.

35. Pike KE, Savage G, Villemagne VL, Ng S, Moss SA, Maruff P, et al. Beta-amyloid imaging and memory in non-demented individuals: evidence for preclinical Alzheimer's disease. Brain 2007;130:2837–2844.

36. Xiong KL, Yang QW, Gong SG, Zhang WG. The role of positron emission tomography imaging of [beta]-amyloid in patients with Alzheimer's disease. Nucl Med Commun 2010;31:4–11.

37. Klunk WE, Engler H, Nordberg A, Wang Y, Blomqvist G, Holt DP, et al. Imaging brain amyloid in Alzheimer's disease with Pittsburgh Compound-B. Annu Neurol 2004;55: 306–319.

38. Buchner RL, Snyder AZ, Shannon BJ, LaRossa G, Sachs R, Fotenos AF, et al. Molecular, structural, and functional characterisation of Alzheimer's disease: evidence for a relationship between default activity, amyloid, and memory. J Neurosci 2005;25:7709–7717.

39. Kemppainen NM, Aalto S, Wilson IA, Nagren K, Helin S, Bruck A, et al. Voxel-based analysis of PET amyloid ligand [^{11}C]PIB uptake in Alzheimer's disease. Neurology 2006;67: 1575–1580.

40. Small GW, Kepe V, Ercoli LM, Siddarth P, Bookheimer SY, Miller KJ, et al. Pet of brain amyloid and tau in mild cognitive impairment. N Eng J Med 2006;355:2652–2663.

41. Rowe CC, Ng S, Ackermann U, Gong SJ, Pike K, Savage G, et al. Imaging beta-amyloid burden in aging and dementia. Neurology 2007;68: 1718–1725.

42. Jack CR, Lowe VJ, Senjem ML, Weigand SD, Kemp BJ, Shieung MM, et al. ^{11}C PIB and structural MRI provide complementary information in imaging of Alzheimer's disease and amnestic mild cognitive impairment. Brain 2008;131:665–680.

43. Linazasoro G. Imaging beta-amyloid burden in aging and dementia. Neurology 2008;70: 1649–1650.

44. Jagust WJ, Landau SM, Shaw LM, Trojanowski JQ, Koeppe RA, Reiman EM, et al. Relationships between biomarkers in aging and dementia. Neurology 2009;73:1193–1199.

45. Fahn S, Oakes D, Shoulson I, Kieburtz K, Rudolph A, Lang A, et al. Levodopa and the progression of Parkinson's disease. N Engl J Med 2004;351:2498–2508.

INDEX

A-63930, cognitive enhancement, 41–42
A-412997, cognitive enhancement, 41–42
A-582941, cognitive enhancement, 35–36
Absorption, opioid receptors, 577–578
ABT compounds, nicotinic agonists:
  ABT-089, 35–36
  ABT-418, 27
ACAT inhibitors, Alzheimer's disease, 406–407, 411
Acetazolamide, anticonvulsants, 133–134, 151–152
Acetaminophen, migraine therapy, 268
Acetylcholine (ACh):
  analogs, 64
  cholinergic nervous system, 61–62
  cognitive dysfunction and, 20–21
Acetylcholinesterase inhibitors:
  cognitive enhancement, 30–32
  schizophrenia therapy, 32
Acquired immunodeficiency syndrome (AIDS), cognitive impairment, epidemiology, 15, 21
Acyclic hydroxamate inhibitors, stroke therapy, 503–504
Acyclic opioid analgesics, structure-activity relationship, 621–623
Acylated piperidines, stroke therapy, 454–455

Acylguanidines, BACE (β-site cleavage enzyme) inhibitors, 351–353
Acyloxymethyl ketones, stroke therapy, cathepsin B inhibitors, 475
Acyl sulfonamides, stroke therapy, MMP inhibitors, 510
AD Assessment Scale (ADAS-COG), cognitive dysfunction assessment, 21–22
Adatanserin, serotonin receptor targeting, 244–247
Addiction liability, opioid receptors, 574–575
Adenosine receptors:
  A_{2A} receptor, Parkinson's disease therapy, 538–540
  cognitive enhancement, 46–47
  Parkinson's disease therapy:
    and A_{2A} receptor, 538–540
    antagonists, 552–553
ADMET (absorption, distribution, metabolism, excretion, and toxicity):
  opioid receptors, 577–579
  psychostimulants, 97–99
α-Adrenergic receptor, antidepressant pharmacology, 247
β_3-Adrenergic receptor, antidepressant targeting, 247
β-Adrenergic receptor, antidepressant

efficacy assessment, 226
Adrenergic receptors, antidepressant targeting, 247
Adrenocorticotropic hormone (ACTH), cognitive enhancement, 49–50
Adriamycin, Alzheimer's disease, tau-targeted therapy, 426–427
Adrogolide, cognitive enhancement, 41–42
Adverse drug reactions (ADRs):
  antidepressants, 229–234
  bupropion, 234
  monoamine oxidase inhibitors, 230
  nefazodone, 234–235
  reboxetine, 235
  serotonin/norepinephrine reuptake inhibitors, 234
  serotonin-selective reuptake inhibitors, 232–234
  trazodone, 234
  tricyclic antidepressants, 231–232
  antipsychotics, 165–169
  animal models, 175–176
  anticholinergic effects, 167–168
  antipsychotic discontinuation, 169
  cardiovascular effects, 167
  extrapyramidal symptoms, 165–166
  hematologic effects, 168–169
  hepatic effects, 169

Adverse drug reactions (ADRs) (*Continued*)
  metabolic effects, 168
  neuroleptic malignant syndrome, 167
  sedation, 168
  seizures, 168
  sexual dysfunction, 169
  psychostimulant therapy, 92, 95–97
ADX-63365, cognitive enhancement, 42–43
AF267B muscarinic receptor, cognitive enhancement, 32–33
Affinity labeling, opioid receptors:
  nonpeptide analgesics, 637–643
  peptide derivatives, 675–676
Aggregation, Alzheimer's disease, tau-targeted therapies, 423–429
Agmatine, bupropion, 255
Agonist compounds:
  cholinergic agonists, 63–72
    acetylcholine analogs, 64
    muscarinic agonists, structure and classification, 64–68
    nicotinic agonists, 68–72
  cognitive enhancement, 34–36
  neoclerodane diterpene, 682–683
  opioid receptors:
    chemical structure, 571, 573
    δ-opioid selective agonists, 632–637
    κ-opioid receptors:
      physiology, 592–593
      selective agonists, 623–632
    multiple receptor definition, 579–580
    μ-opioid receptors, 593
  Parkinson's disease therapy:
    aporphine-type dopamine receptor agonists, 544–545
    D_1 receptor agonists, 547–549
    ergot-type receptor agonists, 545–546
    serotonin 5-HT_{1A} receptor agonists, 553–555
    small-molecule dopamine receptor agonists, 547
  serotonin receptor targeting, 242–244
  migraine therapy, 268–272

stroke therapy, $GABA_A$ agonist, 483
Agranulocytosis, clozapine side effect, 178
Akathisia, antipsychotic side effect, 165–166
Albendazole, Alzheimer's disease, 431–432
Aldehydes, stroke therapy:
  calpain inhibitors, 461–464
  caspase 3/9 inhibitors, 466–467
  cathepsin B inhibitors, 472–473
Alkaloids:
  ergot alkaloids, migraine therapy, 267–268
  morphine and derivatives, 599–601
  opium, 598–599
Allosteric proteins:
  central nervous system drug discovery, 5–6
  cognitive enhancement, 32–33
  muscarinic receptors and, 68
Aloisine, Alzheimer's disease, tau-targeted therapies, 420–422
Alpha-alkyl substituent, amphetamines, 108
Alpha-carbon atom stereochemistry, amphetamines, 108
Alpha oxo/thio/amino-methyl ketones, stroke therapy, 469–470
Alvameline muscarinic receptor, cognitive enhancement, 32–33
Alzheimer's disease:
  acetylcholinesterase inhibitor therapy, 31–32
  amyloid therapies, 334–366
    advanced glycosylation end products inhibitors, 365
    beta-amyloid reduction, 365–366
    fibrillization inhibitors, 361–365
    HMG-CoA reductase inhibitors, 359–361
    research background, 329–330
    α secretase modulators, 357–358
    β secretase inhibitors:
      clinical, 356–357

nonpeptidomimetic compounds, 347–356
      preclinical, 345–347
    γ secretase inhibitors, 358–359
      clinical, 341–345
      preclinical, 334–341
    secretases, amyloid precursor protein processing, 330–334
  cognitive dysfunction in, 20
  cognitive enhancement drugs, 15–16
  cognitive impairment, epidemiology, 15
  immunotherapy approaches, 366–380
    advantages/disadvantages, 369–370
    anti-Aβ antibody mechanisms, 367–368
    bapineuzumab, 370, 378
    Aβ immunization, 366, 368–369
    current clinical trials, 370–380
    gammaguard liquid, 379
    MABT5102A, 379–380
    PF-4360365, 378–379
    preclinical/clinical trials, 369
    R1450, 379
    solaneuzmab, 378
    TgAPP behavioral and learning deficits, 366–367
  muscarinic receptor activity, 66–68
  non-amyloid therapies:
    metabolic syndrome targeting, 405–417
      cholesterol-related approaches, 405–414
      insulin-related approaches, 414–417
    research background, 405
    tau pathology targeting, 417–432
      aggregation, 423–429
      glycosylation, 422–423
      microtubule stabiliization, 430
      phosphorylation, 418–422
      protein levels, 430–432
Amantidine:
  Parkinson's disease therapy, 554–555
  stroke therapy, 450

INDEX

AMG-517 antagonist, migraine therapy, 307–308
Amidines, stroke therapy:
　iNOS/nNOS analogs, 497–498
　NR2B antagonist compounds, 455
Amino acids:
　migraine therapy, CGRP antagonists, 288–290
　psychostimulant physiology and pharmacology, 105–106
3-Aminobenzamide PARP, stroke therapy, 484–487
Aminohydantoins, BACE (β-site cleavage enzyme) inhibitors, 353–356
α-Amino-3-hydroxy-5-methyl-4-isoxazole propionic acid (AMPA):
　anticonvulsants, ligand-gated ion channels, 129
　antipsychotic glutamatergics, 205–206
　cognitive enhancement:
　　glutamate, 41–42
　　potentiators, 43–44
　glutmate modulation, migraine therapy, 291–295
　long-term potentiation, 24–25
2-Aminophenylamides, HDAC inhibitors, stroke therapy, 492
Aminopyrazoles, Alzheimer's disease, tau-targeted therapies, 420–422
Aminopyridines, stroke therapy, 498–499
Aminosteroid neuromuscular blockers, nicotinic receptor antagonists, 78–80
2-Aminothiazoles, stroke therapy, 514
Aminothienopyridine family, Alzheimer's disease, tau-targeted therapy, 429
Amisulpride, pharmacokinetics, 199–200
Amitriptyline:
　chemical structure, 222
　historical background, 248–249
　migraine prevention, 318–319
　monoamine transporter inhibition, 238–241, 240–241

　side effects and adverse reactions, 223, 231–232
Amlodipine, stroke therapy, 459–460
Amoxepine:
　chemical structure, 222
　structure-activity relationship, 182–183, 250–251
Amphetamine psychosis, 95
　antipsychotic therapy, animal models, 170–171
　schizophrenia pathology, 162–163
Amphetamines:
　ADMET properties, 97–99
　alpha-alkyl substituent, 108
　alpha-carbon stereochemistry, 108
　aromatic ring substitution, 108–109
　cognitive enhancement, 51–52
　historical development, 90–91
　nitrogen substituents, 107
　physiology and pharmacology, 100–106
　receptor-based drug design, central nervous system effects, 2–3
　side chain modification, 107–108
　side effects, adverse reactions and drug interactions, 92, 95–97
　structure-activity relationship, 106–109
Amphibian skin, opioid peptides, 646
β-Amyloid peptide (Aβ):
　Alzheimer's cognitive deficits, 16
　Alzheimer's disease epidemiology, 329–330
　anti-Aβ antibody mechanisms, 367–368
　fibrillization inhibitors, 361–365
　immunization against, 366
　insulin related therapy, Alzheimer's disease, 415–417
　secretase processing, 330–334
　small-molecule inhibitors, 362–365
Amyloid cascade hypothesis, Alzheimer's disease

　epidemiology, 329–330
Amyloid precursor protein (APP):
　Alzheimer's disease epidemiology, 329–330
　secretase processing, 330–334
AN1792 Aβ peptide, Alzheimer's immunization therapy, 368–369
Analog design:
　acetylcholine, 64
　anticonvulsants, 149
　morphine, identification and early history, 596–597
　muscarinic antagonists, 72–77
　opioid peptide analogs, 645–676
　　affinity-labeled derivatives, 675–676
　　amphibian skin peptides, 646
　　antagonist dynorphin A analogs, 657–658
　　combinatorial library peptides and peptidomimetics, 674–675
　　dynorphin analogs, 654–658
　　conformational constraints, 656–657
　　linear analogs, 655–656
　　enkephalin analogs, 646–654
　　　δ-selective analogs, 649–653
　　　dimeric analogs, 653–654
　　　μ-selective analogs, 646–649
　　somatostatin μ-receptor antagonists, 673–674
　　Tyr-D-aa-Phe sequence, 664–673
　　　deltorphin analogs, 668–673
　　　dermorphin analogs and μ-selective peptides, 664–668
　　Tyr-Pro-Phe sequence, 659–664
　　　β-casomorphin analogs and endomorphins, 659–661
　　　TIPP and related peptides, 661–664
　opioid receptors, history, 596–597
γ-secretase inhibitors, 334–335
stroke therapy, iNOS/nNOS analogs, 495–502

INDEX

Anatoxin-A, nicotinic agonists, 70
Angiotensin IV, cognitive enhancement, 50
Angiotensin-converting enzyme (ACE) inhibitors, migraine prevention, 319–320
4-Anilidopiperidines, opioid analgesics, 618–621
Animal studies:
  Alzheimer's disease immunotherapy, 366–367
  antidepressants, monoamine transporter inhibition, 240–241
  antipsychotics, 170–176
    behavioral models, 172–173
    dopamine model, 170–171
    electroencephalogram, 174–175
    genetic models, 173–174
    NMDA model, 171–172
    side effects models, 175–176
  BACE (β-site cleavage enzyme), 331–333
  PET/SPECT systems in, 742
Aniracetam, cognitive enhancement, 34
Anorexigenics, therapeutic applications, 92–94
Anpirtoline, serotonin receptor targeting, 245–247
Antagonist compounds:
  antidepressants, serotonin antagonists, 225–229
  antipsychotic agents:
    histamine H3 receptor antagonists, 206–207
    neurokinin receptor antagonists, 208
    serotonin-dopamine antagonists, 177–179
  cholinergic antagonists:
    muscarinic antagonists, 72–77
    nicotinic antagonists, 77–80
    ganglionic antagonists, 78
    neuromuscular antagonists, 78–80
    structure and classification, 72–80
  competitive receptor antagonists, stroke therapy, 451
  CRF antagonists, 257
  ionotropic glutamate receptors:

anticonvulsants, 129
  migraine therapy, 291–297
migraine therapy:
  calcitonin gene-related peptide antagonists, 272–290
  azepinone binding affinity, 281–286
  benzodiazepine binding affinity, 277–281
  BIBN4096 BS, 273, 275–277
  BMS-694153, 288–290
  MK-0974, 277, 279–288
  neuropeptide-induced blood flow, 286–288
  receptor binding affinities, 273–275
  unnatural amino acids, 289–290
  vanilloid receptor antagonists, 307–312
NMDA receptors:
  antidepressants, 256
  stroke therapy, 450–455
  competitive receptor antagonists, 451–453
  ion channel blockers, 450–451
  NR2B-specific noncompetitive receptor antagonists, 453–455
opioid analgesics:
  chemical structure, 571–574
  δ-opioid receptors, 604–608
  enkephalin analogs, 653
  dynorphin A analogs, 657–658
  κ-opioid receptors, 602–604
  μ-opioids:
    selective antagonists, 607
    somatostatin derivation, 673–674
Parkinson's disease:
  adenosine receptor antagonists, 552–553
  glutamate antagonists, 555
stroke therapy:
  NMDA receptors, 450–455
  competitive receptor antagonists, 451–453
  ion channel blockers, 450–451
  NR2B-specific noncompetitive

receptor antagonists, 453–455
voltage-dependent calcium channels, 455–460
substance P-NK1 receptor antagonist, 256–257
vanilloid receptor antagonists, migraine therapy, 307–312
voltage-gated calcium channels, 455–460
Anthraquinones, Alzheimer's disease, tau-targeted therapy, 427
Anti-Aβ antibodies, mechanisms of action, 367–368
Antibody-based therapeutics, anti-Aβ antibodies, 367–368
Anticholinergic agents:
  antipsychotics, 167–168
  Parkinson's disease therapy, 552
Anticonvulsants. *See also* Epilepsy
  barbiturates, 139–141
  benzodiazepines, 144–147
  carabersat, 155–156
  carisbamate, 155
  classification, 134–139
  felbamate, 152–154
  GABA analogs, 149
  ganaxolone, 156
  hydantoins, 141
  iminostilbenes, 142–144
  lacosamide, 154
  lamotrigine, 148–149
  migraine prevention, 315–316
  molecular targeting, 121–134
    connexins, 134
    enzymes, 132–134
    G-protein-coupled receptors, 134
    ligand-gated ion channels, 126–130
    neurotransmitter transporters and synaptic proteins, 130–132
    voltage-gated ion channels, 122–126
  pyrrolidin-2-ones, 149–151
  research background, 121
  retigabine, 154
  rufinamide, 154–155
  stiripenol, 155
  succinimides, 142

sulfonamides, 151–152
tiagabine, 152
trimethadione, 142
valproic acid, 147–148
Antidepressants:
  classification and application, 219–221
  clinical efficacy, 221–229
  dosing regimens, 219–221
  fluoxetine, historical background, 248–249
  historical background, 248–249
  mechanistic classification, 221
  migraine prevention, 318–319
  nondepression applications, 229
  pharmacokinetics, 235–237
    serotonin-selective reuptake inhibitors, 236
    tricyclic antidepressants, 235–236
  physiology and pharmacology, 237–248
    α-adrenergic receptors, 247
    monoamine oxidase, 247–248
    monoamine transporters, 237–241
    serotonergic agents, 241–247
  recent and future developments, 255–258
    monoaminergic drugs, 255–256
    neuropeptide receptor agonists/antagonists, 256–258
    NMDA receptor antagonists, 256
  research background, 219
  side effects and adverse reactions, 229–234
    bupropion, 234
    monoamine oxidase inhibitors, 230
    nefazodone, 234–235
    reboxetine, 235
    serotonin/norepinephrine reuptake inhibitors, 234
    serotonin-selective reuptake inhibitors, 232–234
    trazodone, 234
    tricyclic antidepressants, 231–232
  structure-activity relationship and metabolism, 249–255

bupropion, 254–255
monoamine oxidase inhibitors, 251–253
serotinergic agents, 253–254
tricyclic antidepressants, 249–251
Antiemetics, migraine therapy, 297–298
Anti-inflammatory compounds. *See also* Nonsteroidal anti-inflammatory drugs (NSAIDs)
  migraine therapy:
    NSAIDs/acetaminophen, 268
    triptans, 269–272
Antioxidants, stroke therapy, 479–482
  deferoxamine, 482
  ebselen, 481
  edavarone, 481
  NXY-059, 479–480
  tirlazad, 480
Antipsychotics:
  animal models, 170–176
    behavioral models, 172–173
    dopamine model, 170–171
    electroencephalogram, 174–175
    genetic models, 173–174
    NMDA model, 171–172
    side effects models, 175–176
  drug discovery and development, 200–210
    AMPA receptors, 205–206
    cannabinoids, 209–210
    dopamine receptor subtypes, 200–201
    glutamatergic agents, 202
    histamine H_3 receptor, 206–207
    metabotropic glutamate receptors, 203–204
    neurokinin 3 receptor antagonists, 208
    neuronal nicotinic acetylcholine receptor, 207
    NMDA receptors, 204
    phosphodiesterase inhibitors, 208–209
    serotonin receptor subtypes, 201–202
  FDA-approved compounds, 176–180
    atypical compounds, 177–179
    conventional drugs, 176–177
    indication sfor, 170

migraine therapy, 297–298
pharmacokinetics, biodistribution, and drug-drug interactions, 185–200
  benzisoxazoles, 195–196
  benzoisothiazoles, 196–198
  butyrophenones, 189–190
  diarylbutylamines, 190
  dibenzodiazepines, 191–192
  dibenzothiepine, 194
  dibenzoxazepines, 189
  dibenzthiazepines, 193–194
  dihydroindolone, 190–191
  phenothiazines, 186–188
  phenylindole, 199
  quinolinone, 198
  substituted benzamide, 199–200
  thiobenzodiazepine, 192–193
  thioxanthenes, 188–189
receptor-based drug design, 3–4
research background, 161
"rich pharmacology" principle, 4–5
schizophrenia:
  clinical applications, 169–170
  epidemiology, 161–162
  pathology, 162–165
  side effects, 165–169
    animal models, 175–176
    anticholinergic effects, 167–168
    antipsychotic discontinuation, 169
    cardiovascular effects, 167
    extrapyramidal symptoms, 165–166
    hematologic effects, 168–169
    hepatic effects, 169
    metabolic effects, 168
    neuroleptic malignant syndrome, 167
    sedation, 168
    seizures, 168
    sexual dysfunction, 169
  structure-activity relationships, 180–185
    atypical tricyclic neuroleptics, 180–183
    benzisoxazole/benzithazole neuroleptics, 183–185
Anxiolytic agents, serotonin receptor targeting, 242–244

Apadoline, κ-opioid receptor
    agonists, 630
Apaxifylline, cognitive
    enhancement, 48–49
Apnea of prematurity,
    psychostimulant
    therapy, 92
Apolipoprotein A-1, Alzheimer's
    disease therapy,
    413–414
Apolipoprotein E (ApoE),
    Alzheimer's disease
    therapy:
  cholesterol targeting, 413–414
  HMG-CoA reductase inhibitors,
    359–361
Apomorphine, antipsychotic
    therapy, animal
    models, 170–171
Apoptosis, Parkinson's disease,
    mitochondrial
    dysfunction, 537
Aporphine-type dopamine
    receptor agonists,
    Parkinson's disease
    therapy, 544–545
Appetite suppressants,
    therapeutic
    applications, 92–94
Aptiganel, stroke therapy, 451
Aquifex aeolicus bacterial leucine
    transporter,
    psychostimulant
    pharmacology and,
    114
Arecoline, muscarinic activity,
    66–67
L-Arginine, nitric oxide
    enhancement, stroke
    therapy, 495–496
Aripiprazole:
  dopamine model, 170–171
  FDA approval, 178–179
  functional selectivity, 8–9
  pharmacokinetics, 198
  side effects, 168
AR-R15896AR, stroke therapy,
    451
AR-R18565 scaffold, stroke
    therapy, 457
Aryl-linked hydroxymate
    inhibitor, stroke
    therapy, 490–491
4-Arylpiperidines, opioid
    derivatives, 615–621
  carbon substituent, 614–616
  oxygen substitutent, 616–618

Asenapine antipsychotic, 201
Asimadoline κ-opioid receptor
    agonists, 630–632
Aspirin, migraine therapy, 268
Assay techniques, cognitive
    function assessment,
    26–29
Atomexetine, cognitive function
    assessment, 27
Atorvastatin, Alzheimer's disease
    therapy, preclinical/
    clinical testing,
    359–361
Atracurium, nicotinic
    antagonists, 79–80
Atropine:
  cholinergic nervous system,
    61–62
  muscarinic antagonist, 72
Attention deficit-hyperactivity
    disorder (ADHD),
    psychostimulant
    therapy, 91–92
  side effects, adverse reactions
    and drug interactions,
    95
Atypical antipsychotics:
  FDA-approved compounds,
    176–180
  pharmacokinetics, 191–200
    benzisoxazoles, 195–196
    benzoisothiazoles, 196–198
    dibenzodiazepines,
      191–192
    dibenzothiepine, 194
    dibenzthiazepines, 193–194
    phenylindole, 199
    quinolinone, 198
    substituted benzamide,
      199–200
    thienobenzodiazepine,
      192–193
  structure-activity
    relationships,
    180–185
Axial phenylpiperidines, 598
Azepines:
  nicotinic agonists, 68–69
  γ-secretase inhibitors, 336–337
Azepinone, migraine therapy,
    CGRP antagonists,
    280–288
Azetidines, nicotinic antagonist,
    68–69
Aziridine, stroke therapy,
    cathepsin B inhibitors,
    474

Azure A and B compounds,
    Alzheimer's disease,
    tau-targeted therapy,
    424–425

BACE (β-site cleavage enzyme):
  acylguanidines, 351–353
  amyloid precursor protein
    processing, 330–334
  chemical structure, 331–333
  clinical testing, 356–357
  guanidines, aminohydantoins,
    353–356
  insulin related therapy,
    Alzheimer's disease,
    416–417
  nonpeptidomimetic inhibitors,
    347–356
  preclinical testing, 345–347
  transition-state isosteres and
    statines, 347–351
Bapineuzamab, Alzheimer's
    immunization
    therapy, 370, 378
Barbiturates, anticonvulsant
    agents, 139–141
Basal ganglia, Parkinson's
    disease,
    pathophysiology,
    531–534
Batimastat hydroxymate
    inhibitors, stroke
    therapy, 503
Beckett-Casey opioid receptor
    binding model,
    597
Befloxatone, structure-activity
    relationship and
    metabolism, 252–253
Behavioral assessment:
  Alzheimer's disease
    immunotherapy,
    366–367
  antipsychotic therapy, animal
    models, 172–173
  cognitive function, 25–29
Belinostat, hydroxamate
    inhibitors, stroke
    therapy, 489–490
Benserazide, Parkinson's
    disease, dopaminergic
    pharmacotherapy,
    541–542
Benzacetamide κ-opioid receptor
    agonist, 624–632
Benzamide substitution,
    antipsychotic

pharmacokinetics, 199–200
AMPA receptor glutmatergics, 205–206
Benzisoxazole:
  pharmacokinetics, 195–196
  structure-activity relationship, 183–185
Benzithiazole, structure-activity relationship, 183–185
Benzodiazepines:
  anticonvulsant applications, 135–139, 144–147
  migraine therapy, telcagepant calcitonin gene-related peptide antagonist, 277–288
  zotepine interaction, 194
Benzoisothiazoles, pharmacokinetics, 196–198
Benzomorphans:
  numbering systems, 610–611
  opioid receptor affinity, 601–602
  structure-activity relationship, 598–599, 610–613
Benzothiadiazines, AMPA receptor glutmatergics, 206
Benzothiazoles, NOS inhibitors, migraine therapy, 302–306
Benzothiophene, migraine therapy, 288–290
Benzotriazine inhibitors, stroke therapy, 513–514
Benzotriazole, calpain inhibitor, stroke therapy, 465
Benzphetamine, structure-activity relationship, 107
Benztropine mesylate, Parkinson's disease therapy, 552–553
N-Benzyl-4-aminopiperidines, NOS inhibitors, migraine therapy, 303–306
Benzyl carbamates, stroke therapy, 454–455
Besonprodil, stroke therapy, 454
β-blockers:
  antidepressants, efficacy assessment, 225–229
  migraine prevention, 313–315

Bethanechol, structure and function, 64
BF2.649 histamine, cognitive enhancement, 36–37
BIBN4096 calcitonin gene-related peptide antagonist, migraine therapy, 273, 275–277
Bicifadine, stroke therapy, 450
Binaltorphimine, κ-opioid selective antagonist, 602–604
Biodistribution, antipsychotics, 185–200
  benzisoxazoles, 195–196
  benzoisothiazoles, 196–198
  butyrophenones, 189–190
  diarylbutylamines, 190
  dibenzodiazepines, 191–192
  dibenzothiepine, 194
  dibenzoxazepines, 189
  dibenzthiazepines, 193–194
  dihydroindolone, 190–191
  phenothiazines, 186–188
  phenylindole, 199
  quinolinone, 198
  substituted benzamide, 199–200
  thiobenzodiazepine, 192–193
  thioxanthenes, 188–189
Biosynthetic pathways, morphine derivatives, 599
Biperiden:
  muscarinic antagonists, 72–74
  Parkinson's disease therapy, 552–553
Bis-hydroxymate inhibitors, stroke therapy, 507
Black-box model of cognition, 17–19
BMS-191011 potassium channel, stroke therapy, 482–483
BMS 204352/MaxiPost potassium channel, stroke therapy, 482
BMS-694153 calcitonin gene-related peptide antagonist, migraine therapy, 288–290
Boronates, HDAC inhibitors, stroke therapy, 494
Bradykinesia syndrome, pathology, 529–530
Brain-derived neurotrophic factor (BDNF), antidepressant

efficacy assessment, 226–229
Brain imaging studies:
  depression, 277–279
  migraine pathology, 267
Brasofensine, monoamine transporter inhibition, 240–241
Brivaracetam, anticonvulsant applications, 136–139, 150–151
Brofaromine, structure-activity relationship and metabolism, 252–253
Bromocriptine, Parkinson's disease dopaminergic therapy, 545–546
Bupropion:
  chemical structure, 224
  monoamine transporter inhibition, 239–241
  side effects and adverse reactions, 234
  structure-activity relationship, 251, 254–255
Buspirone, serotonin receptor targeting, 242–244
Butyric acid, stroke therapy, 494
Butyrophenones, pharmacokinetics, 189–190
Butyrylcholinesterase (BuChE), cognitive enhancement, 30–32
BW373U86 δ-opioid selective agonist, 632–635
BW 1370U87 RIMA, structure-activity relationship and metabolism, 252–253

Cabergoline, Parkinson's disease dopaminergic therapy, 545–546
Caffeine:
  cognitive enhancement, 47–48, 51–52
  Parkinson's disease and, 538–540
  physiology and pharmacology, 105–106
  side effects, adverse reactions and drug interactions, 96–97
  structure and properties, 90

Calcitonin gene-related peptide (CGRP) antagonists, migraine therapy, 272–290
  azepinone binding affinity, 281–286
  benzodiazepine binding affinity, 277–281
  BIBN4096 BS, 273, 275–277
  BMS-694153, 288–290
  future research issues, 320–321
  MK-0974, 277, 279–288
  neuropeptide-induced blood flow, 286–288
  receptor binding affinities, 273–275
  unnatural amino acids, 289–290
Calcium/calmodulin-dependent kinase (CAMKII), long-term potentiation, 24–25
Calcium channel blockers, migraine prevention, 316–318
Calcium ion channels:
  anticonvulsants, 124
  opioid receptor coupling, 581
  stroke therapy, 455–460
    N-type voltage-dependent calcium channel, 456–459
Calmodulin, stroke therapy, nNOS inhibition, 499–500
Calpain inhibitors, stroke therapy, 460–465
  irreversible inhibitors, 464–465
  reversible inhibitors, 461–464
  unconventional inhibitors, 465
Calpeptin, stroke therapy, 461–462
Cambridge Neurophysiological Test Automated Battery (CANTAB), cognitive dysfunction assessment, 22
Cannabinoids:
  antipsychotic agents, 209–210
  cognitive enhancement, 47–50
Capsaicin, migraine therapy, vanilloid receptor antagonists, 307–313
Capsazepine, migraine therapy, 307–308

Carabersat, anticonvulsant applications, 136–139, 155–156
Carbamazepine:
  anticonvulsant applications, 135–139, 142–144
  aripiprazole interactions, 198
  clozapine pharmacokinetics, 192
  risperidone interactions, 195–196
  sertindole interaction, 199
Carbidopa, Parkinson's disease, dopaminergic pharmacotherapy, 541–542
Carbonic anhydrase inhibitors, anticonvulsants, 132–134
Carboxylic acids, stroke therapy:
  HDAC inhibitors, 494
  MMP inhibitors, 508
Cardiovascular system:
  antidepressant side effects, tricyclic antidepressants, 231–232
  antipsychotic side effects, 167, 179–180
Carisbamate, anticonvulsant applications, 136–139, 155
β-Casomorphin analogs and endomorphins, opioid peptide analogs, Tyr-Pro-Phe sequence, 659–661
Caspase 3/9 inhibitors, stroke therapy, 465–471
  aldehydes, 466–467
  alpha oxo/thio/amino-methyl ketones, 469–470
  halomethyl ketones, 467–469
  irreversible inhibitors, 467–470
  isatin-derived inhibitors, 470–471
  masked aldehydes, 467
  quinoline-derived inhibitors, 471
  reversible inhibitors, 466–467
Catalepsy, antipsychotic side effects, 175
Catecholamines, Parkinson's disease, auto-oxidation, 536–537

Catechol-O-methyltransferase: Parkinson's disease, dopaminergic therapy, 550–552
  schizophrenia genetics, 174
Cathepsin B inhibitors, stroke therapy, 471–478
  irreversible inibitors, 473–477
  reversible inibitors, 472–473
  non-cysteine design, 477–478
Cathepsin L inhibitors, stroke therapy, 478
(−)Cathinone, ephedrine stimulant, 90
Central nervous system (CNS):
  cognitive dysfunction indications and diagnostic criteria, 19–21
  drug development and discovery:
    allosteric receptor sites, 5–6
    cotransmission, 5
    functional selectivity and receptor targeting, 7–9
    future research issues, 9–11
    receptor-selective drugs, 1–3
    research background, 1
    rich pharmacology concept, 4–5
  neurotransmitters, overview, 61–62
  opioid effects, 579
  stimulants:
    ADMET properties, 97–99
    alpha-alkyl substituent, 108
    alpha-carbon stereochemistry, 108
    amphetamines, 106–109
    aromatic ring substitution, 108–109, 112–113
    C(2) substituent, 111–112
    C(3) ester linkage, 112
    caffeine, 90
    cocaine, 109–113
    ephedra and khat, 89–90
    historical background, 90–91
    methylphenidate, 109
    nitrogen substituents, 107, 110–111
    N-substituents, 110
    physiology and pharmacology, 100–106
    recent and future development, 113–115

receptor classification and function, 102–106
research background, 89
side chain length, 107
side effects, adverse effects, drug interactions, 92–97
structure-activity relationships, 106–113
therapeutic applications, 91–92
tropane ring preservation, 113
Cephalon calpain inhibitor, stroke therapy, 462
Cevimeline muscarinic receptor:
analog development, 65–66
cognitive enhancement, 32–33
Chemotypes:
antipsychotics, phenothiazines, 186–189
migraine therapy, calcitonin gene-related peptide antagonists, 287–288
Chloride ion channel, anticonvulsants, 126
Chloromethyl ketones, stroke therapy, cathepsin B inhibitors, 475
Chlorpromazine:
ADMET properties, 186–189
central nervous system effects, 3–5
FDA approval, 176–177
historical background, 248
nonpsychotic applications, 170
pharmacokinetics, biodistribution, and drug-drug interactions, 185–186
side effects, 165–166, 168
Cholesterol, Alzheimer's disease, metabolic syndrome targeting, 405, 411–414
apolipoproteins E and AI, 413–414
esterification, 406, 411
synthesis, 406
turnover, 411–413
Choline acetyltransferase (ChAT), cognitive dysfunction and, 30–31
Cholinergic agonists, 63–72
acetylcholine analogs, 64

muscarinic agonists, structure and classification, 64–68
nicotinic agonists, 68–72
Cholinergic antagonists:
muscarinic antagonists, 72–77
nicotinic antagonists, 77–80
ganglionic antagonists, 78
neuromuscular antagonists, 78–80
structure and classification, 72–80
Cholinergic neurons:
cognitive dysfunction and, 30–31
defined, 61–62
Cholinesterase inhibitors:
cognitive enhancement, 30–32
memory assessment and, 28–29
Cinamide piperidine derivatives, stroke therapy, 454–455
Cinamidines, stroke therapy, NR2B antagonist compounds, 455
Cinnamoyl linkers, hydroxamate inhibitors, stroke therapy, 488–490
Ciproxifan, cognitive function assessment, 27
Cisatracurium, nicotinic antagonists, 79–80
Citalopram:
chemical structure, 223
historical background, 249
pharmacokinetics, 236
serotonin receptor targeting, 244–247
side effects and adverse reactions, 233–234
Civamide, migraine therapy, 307–308
Clevidipine, stroke therapy, 459–460
Clinical trials:
Alzheimer's immunization therapy, 369
antidepressant efficacy assessment, 221–229
BACE (β-site cleavage enzyme) inhibitors, 356–357
fibrillization inhibitors, 361–365
γ-secretase inhibitors, 341–345
Clinidipine, stroke therapy, 458–459

Clioquinol β-amyloid peptide inhibitor, 363
Clobazam, anticonvulsant applications, 144–147
Clomipramine:
chemical structure, 222
historical background, 249
migraine prevention, 318–319
side effects and adverse reactions, 231–232
Clonazepam, anticonvulsant applications, 144–147
Clonidine, cognitive enhancement, 41–42
Clorazepate, anticonvulsant applications, 144–147
Clothiapine, structure-activity relationship, 182–183
Clozapine:
behavioral models, 172–173
central nervous system effects, 4–5
dopamine model, 170–171
FDA approval, 178
pharmacokinetics, 191–192
side effects, 168–169, 178
structure-activity relationship, 180–185
Cocaine:
ADMET properties, 98–99
κ-opioid receptor effects, 593
monoamine transporter inhibition, 240–241
Parkinson's disease dopaminergic therapy, 547–549
physiology and pharmacology, 100–106
side effects, adverse reactions and drug interactions, 96
structure-activity relationship, 109–113
structure and properties, 90
Codeine:
identification and early history, 596–597
morphine metabolism and elimination, 578
opium source, 599
synthesis of, 599–600
Cognition:
defined, 15
disease-related dysfunction, 15
ADAS-COG assessment scale, 21–22
Alzheimer's disease, 20

Cognition (*Continued*)
  CANTAB assessment scale, 22
  disease models, 25–26
    HIV-associated dementia, 21
    indications and diagnostic criteria, 19–21
    lifestyle factors, 21
    MATRICS assessment scale, 22
    Parkinson's disease, 21
    schizophrenia, 20–21, 162
    stroke/vascular dementia, 21
  enhancement, 15–17
    adenosine modulator, 46–47
    cannabinoids, 47–48
    cholinesterase inhibitors, 30–32
    dopamine, 39–40
    GABA$_A$ receptor, 45–46
    glutamate, 41–45
    histamine, 35–37
    muscarinics, 32–33
    neuronal nicotinic agonists, 34–35
    neuropeptides, 48–50
    neurotropic agents, 47
    NMDA modulators, 33–34
    norepinephrine, 41
    phosphodiesterase inhibitors, 50–51
    serotonin, 37–39
    small-molecule approaches, 30–51
    smart drugs, 51–52
    summary and future research issues, 52–53
  functional schematic, 17–19
  memory and, 17–19
  opioid-related impairment, 575
  preclinical behavioral assessment, 25–29
    5-choice serial reaction time task, 26–27
    Morris Water Maze, 28–29
    set shifting, 29–30
    social recognition and novel object recognition, 28
  substrates of, 22–25
    molecular substrates, 24–25
    neuroanatomical substrates, 22–24
    synaptic plasticity, 25
Combinatorial libraries, opioid peptides and peptidomimetics, 674–675
Computational techniques, opioid receptors, 590
Conanotokins, stroke therapy, 451
Conditioned avoidance:
  antipsychotic therapy, 173
  cognitive assessment, 28
Conformational constraints:
  deltorphin analogs, opioid peptides, 671–673
  dermorphin analogs, μ-opioids, 666–668
  dynorphin analogs, 656–657
  enkephalin analogs:
    δ-opioid analogs, 650–653
    μ-opioid analogs, 647–649
Connexins, anticonvulsants, 134
Constipation, opioid-related effects, 575–576
Construct validity, cognitive function assessment, 25–29
Contextual freezing task, memory assessment, 28
Contextual memory, conditioned fear, 27–28
Continuous performance tasks (CPTs), cognitive function assessment, 26–27
Contraindications, opioids, 576
Cortical spreading depression (CSD), migraine pathology, 266–267
Corticotropin-releasing factor (CRF), antagonists, 257
Cotransmission, central nervous system drug discovery, 5
Coumarin, stroke therapy, 499
CP-122,288 triptan compound, migraine therapy, 270–272
CX-546, cognitive function assessment, 27–28
Cyclic AMP response element binding protein (CREB), antidepressant efficacy assessment, 226–229
Cyclic carbamates, stroke therapy, cathepsin B inhibitors, 475–476
Cyclic hydroxyamate inhibitors, stroke therapy, 504–506
Cyclin-dependent kinase inhibitors, kinase 5 inhibitors, Alzheimer's disease, tau-targeted therapies, 418–420
Cyclohexane isosteres, κ-opioid receptor agonists, 629–632
Cyclometallated complexes, stroke therapy, cathepsin B inhibitors, 473
Cyclooxygenase (COX) inhibitors, 268
Cyclopropenones, stroke therapy, cathepsin B inhibitors, 472–473
Cycloserine, cognitive enhancement, 45
Cyclothiazide, cognitive enhancement, 43–44
CYP1A2 enzyme, antipsychotics, phenothiazines, 186–189
CYP2D6 enzyme:
  antipsychotics, 185–186
  phenothiazines, 186–189
  opioid interaction with, 577
CYP3A4 enzyme:
  antipsychotics, 185–186
  phenothiazines, 186–189
  migraine therapy, CGRP antagonists, 288–290
  opioid interaction with, 577
Cys-loop family, nicotinic acetylcholine receptors, 63
Cysteine proteases, stroke therapy, cathepsin B inhibitors, absence of, 477–478
Cytisine, nicotinic agonists, 70
Cytochrome P450 enzymes:
  antidepressants and:
    pharmacokinetics, 236
    serotonin-selective reuptake inhibitors, 234, 236
  antipsychotic pharmacokinetics, 185–200
  opioid interactions with, 576–577

Dale's principle, central nervous system drug discovery, 5
DAMGO agonist, opioid receptor mutagenesis, 589–590
Darifenacin, muscarinic antagonists, 74–77
DATScan, Parkinson's disease dopaminergic therapy, 547–549
Daunorubicin, Alzheimer's disease, tau-targeted therapy, 426–427
Decahydroisoquinoline-3-carboxylic acids, glutmate modulators, migraine therapy, 292–297
Decamethonium, nicotinic receptor blockade, 77–78
Deferoxamine, stroke therapy, 482
Deltorphin analogs, opioid peptides, 668–673
Dementia-related psychosis, antipsychotic medications, 170
Depression:
  brain imaging studies, 227–229
  diagnostic criteria and agent selection, 219–221
  epidemiology, 219
  long-term depression, 25
  psychostimulant therapy, 92
Dermorphin analogs, opioid peptides, 664–668
Desipramine:
  chemical structure, 222
  historical background, 248–249
  monoamine transporter inhibition, 240–241
  pharmacokinetics, 235–236
  serotonin receptor targeting, 241–242
Desmodus rotundus salivary plasminogen activator alpha 1, stroke therapy, 448–449
Desvelafaxine:
  chemical structure, 223
  side effects and adverse reactions, 234
Dextromethorphan, stroke therapy, 451
Dezocline nonpeptide opiate, 643–645

Diabetes mellitus, Alzheimer's disease, insulin-related therapy, 414–417
Diagnostic techniques:
  cognitive dysfunction, 21–22
  Parkinson's disease, 555–556
Dialdehydes, Alzheimer's disease, tau-targeted therapy, 427–428
Diarylbutylamines, pharmacokinetics, 190
Diazabicyclo[3.0.3] compound, nicotinic agonists, 71–72
Diazepam:
  anticonvulsant applications, 144–147
  cognitive enhancement, 45–46
  zotepine interaction, 194
Diazepines, γ-secretase inhibitors, 336–337
Diazomethyl ketones, stroke therapy, cathepsin B inhibitors, 475
Dibenzoazepines, pharmacokinetics, 189
Dibenzodiazepines, pharmacokinetics, 191–192
Dibenzothiepine, pharmacokinetics, 194
Dibenzthiazepines, pharmacokinetics, 193–194
Diclofenac, migraine therapy, 268
Diels-Alder adducts, opioid analgesics, 608–609
Diet, antidepressants and, monoamine oxidase inhibitors, 230
Diethylpropion:
  ADMET properties, 99
  structure-activity relationship, 106–109
Difluorophenacylalanine (DAPT), γ-secretase inhibitors, 335
Dihydrexidine:
  cognitive enhancement, 41–42
  functional selectivity, 8–9
  Parkinson's disease dopaminergic therapy, 548–549
Dihydroergotamine (DHE), migraine therapy, 267–268

Dihydroindolone, pharmacokinetics, 190–191
Dihydropyridine derivatives, stroke therapy, 459–460
(+)-[^{11}C]-Dihydrotetrabenazine, Parkinson's disease dopaminergic therapy, 547–549
Diketopiperazine, irreversible inhibitors, stroke therapy, 465
Diltiazem, migraine prevention, 318
Dimeric structures:
  opioid receptors, 590–591
  enkephalin analogs, 653–654
  stroke therapy, NOS inhibitors, 501
Dinapsoline, Parkinson's disease dopaminergic therapy, 548–549
Dioxolane, muscarinic activity, 64–66
Diphenhydramine, Parkinson's disease therapy, 552–553
5,5-Diphenyl-barbituric acid, anticonvulsant applications, 140–141
DISC1 genes, schizophrenia assessment, 173–174
Discodermolide, Alzheimer's disease, tau-targeted therapy, 430
Disease characteristics:
  cognitive function, 25–29
  radiolabled biomarkers, 744–746
Dissociation constant, atypical tricyclic neuroleptics, 180–183
Distribution, opioid receptors, 577–578
Dizocilpine, stroke therapy, 450
DMCM receptor, cognitive enhancement, 45–46
DMXB-A, cognitive enhancement, 35–36
Donepezil, cognitive enhancement, 15–16, 31
Donor-acceptor motifs, stroke therapy, NOS inhibitors, 502

Dopamine (DA):
  antipsychotic effects:
    animal models, 170–171
    receptor subtype techniques, 200–201
    structure-activity relationship, 180–185
  cocaine structure-activity relationship, 111–113
  cognitive enhancement, 39–41
  conformations of, 543–544
  D_1 receptor agonists, Parkinson's disease dopaminergic therapy, 547–549
  D_2 receptor antagonists:
    antipsychotic medications, 170
    central nervous system effects, 3–5
    radiolabeled ligands for, 744
  Parkinson's disease:
    dopaminergic pharmacotherapy, 540–552
    catechol-O-methyltransferase inhibitors, 551–552
    L-dopa therapy, 540–542
    monoamine oxidase inhibitors, 549–551
    receptor-targeted compounds, 542–549
    aporphine-type receptor agonists, 544–545
    dopamine receptor structure and function, 542–544
    ergot-type receptor agonists, 545–546
    experimental D_1 receptor agonists, 547–549
    small-molecule agonists, 547
  mitochondrial metabolism and, 535–537
  pathophysiology, 531–534
  psychostimulant pharmacology, 100–106
  recent developments, 114–115
  structure-activity relationship, 106–109
  receptor structure and function, 542–544

schizophrenia pathology, 162–163
latent inhibition, 172–173
Dopamine- and cAMP-regulated phosphoprotein of MW 32,000 (DARPP-32), serotonin receptor targeting, 245–247
Dopamine transporter (DAT):
  antidepressant pharmacology, 238–241
  Parkinson's disease diagnosis, 555–556
DOPAScan, Parkinson's disease dopaminergic therapy, 547–549
Doramapimod kinase inhibitor, Alzheimer's disease, tau-targeted therapies, 420–422
Dothiepin:
  chemical structure, 222
  side effects and adverse reactions, 231–232
Doxacurium, nicotinic antagonists, 79–80
Doxepine:
  chemical structure, 222
  side effects and adverse reactions, 231–232
  structure-activity relationship, 250–251
Doxorubicin, Alzheimer's disease, tau-targeted therapy, 426–427
Drug development and discovery:
  antipsychotics, 200–210
    AMPA receptors, 205–206
    cannabinoids, 209–210
    dopamine receptor subtypes, 200–201
    glutamatergic agents, 202
    histamine H_3 receptor, 206–207
    metabotropic glutamate receptors, 203–204
    neurokinin 3 receptor antagonists, 208
    neuronal nicotinic acetylcholine receptor, 207
    NMDA receptors, 204
    phosphodiesterase inhibitors, 208–209
    serotonin receptor subtypes, 201–202

central nervous system:
  allosteric receptor sites, 5–6
  cotransmission, 5
  functional selectivity and receptor targeting, 7–9
  future research issues, 9–11
  receptor-selective drugs, 1–3
  research background, 1
  rich pharmacology concept, 4–5
stroke therapy:
  current research, 447–449
  future research and development, 515–516
Drug discrimination model, antipsychotic drugs, 173
Drug-drug interactions:
  antidepressants:
    efficacy assessment, 225–229
    monoamine oxidase inhibitors, 230
    serotonin-selective reuptake inhibitors, 232–234
    tricyclic antidepressants, 231–232
  antipsychotics, 185–200
    benzisoxazoles, 195–196
    benzoisothiazoles, 196–198
    butyrophenones, 189–190
    diarylbutylamines, 190
    dibenzodiazepines, 191–192
    dibenzothiepine, 194
    dibenzoxazepines, 189
    dibenzthiazepines, 193–194
    dihydroindolone, 190–191
    phenothiazines, 186–188
    phenylindole, 199
    quinolinone, 198
    substituted benzamide, 199–200
    thiobenzodiazepine, 192–193
    thioxanthenes, 188–189
  bupropion, 255
  nefazodone, 234–235
  opioids, 576–577
  psychostimulant therapy, 92, 95–97
  trazodone, 234
Duloxetine:
  chemical structure, 223
  migraine prevention, 318–319
  monoamine transporter inhibition, 238–241
  side effects and adverse reactions, 234

DuP747 κ-opioid receptor agonist, 627–632
Dye-drug interactions, β-amyloid peptide inhibitors, 364–365
Dynorphins:
  analog structures, 654–659
  antagonists, dynorphin A analogs, 657–658
  conformational constraints, 656–657
  linear analogs, 655–656
  κ-opioid receptor interaction, 594–596
  "message-address" concept, 604–608
  opioid receptors, precursors and processing products, 595–596
Dysbindin, schizophrenia assessment, 173

Ebselen, stroke therapy, 481
Edaravone, stroke therapy, 481
Efficacy assessment:
  antidepressants, 221–229
  antipsychotics, animal models, 170–172
  opioid receptors:
    in vitro assays, 583–585
    in vivo assays, 585–587
EGb-761 β-amyloid peptide inhibitor, 363–365
Electroencephalography (EEG), schizophrenia assessment, 174–175
Electrophiles:
  opioid analgesics, nonpeptide affinity labeling, 638–643
  stroke therapy, 465
Elimination, opioid receptors, 578–579
Eliprodil, stroke therapy, 453
ELND005 β-amyloid peptide inhibitor, 364–365
EMD 60400 κ-opioid receptor agonists, 630–632
EMDT receptor agonist, serotonin receptor targeting, 246–247
Emodin, Alzheimer's disease, tau-targeted therapy, 426–427
Enadoline κ-opioid receptor agonist, 627–632

Endomorphins:
  evolution of, 595–596
  opioid peptide analogs, Tyr-Pro-Phe sequence, 659–661
β-endorphins, opioid receptor affinity, 594–596
Endothelial nitric oxide synthase (eNOS), migraine therapy, 298–302
Enkephalins:
  δ-opioids:
    analogs, 650–653
    antagonist activity, 653
    receptor interaction, 594–596
  μ-opioid analogs, 647–649
  opioid peptide analogs, 646–647
Entacapone, Parkinson's disease, dopaminergic therapy, 550–552
Environmental factors, Parkinson's disease, 537–540
Enzymes, anticonvulsants, 132–134
Epalrestat, Alzheimer's disease, tau-targeted therapy, 427
Ephedra, stimulant effect, 89–90
Ephedrine, stimulant effect, 89–90
Epibatidine, nicotinic agonists, 69–70
Epilepsy. See also Anticonvulsants
  classification, 122
  epidemiology, 121
4,5α-Epoxymorphinans, structure-activity relationship, 598–601
N-substituted 4,5α-Epoxymorphinans, opioid antagonists, 601–608
Epoxysuccinates, stroke therapy:
  calpain inhibitors, 464–465
  cathepsin B inhibitors, 473–474
Ergoline, Parkinson's disease dopaminergic therapy, 545–546
Ergot alkaloids:
  migraine therapy, 267–268
  Parkinson's disease dopaminergic therapy, 545–546
Escitalopram, chemical structure, 223

Eslicarbazepine acetate, anticonvulsant applications, 136–139, 143–144
Esterification, cholesterol, in Alzheimer's disease, 406–407, 411
Eterobarb, anticonvulsant applications, 140–141
Ethopropazine, Parkinson's disease therapy, 552–553
Ethosuximide, anticonvulsant applications, 135–139
Etonitazine, nonpeptide affinity labeling, 641–644
Etorphine, Diels-Alder adducts, 608–609
Evoked related potentials (ERPs), schizophrenic patients, 174–175
Excitatory amino acid receptors, stroke therapy, 450–478
  calpain inhibitors, 460–465
    irreversible inhibitors, 464–465
    reversible inhibitors, 461–464
    unconventional inhibitors, 465
  caspase 3/9 inhibitors, 465–471
    aldehydes, 466–467
    alpha oxo/thio/amino-methyl ketones, 469–470
    halomethyl ketones, 467–469
    irreversible inhibitors, 467–470
    isatin-derived inhibitors, 470–471
    masked aldehydes, 467
    quinoline-derived inhibitors, 471
    reversible inhibitors, 466–467
  cathepsin B inhibitors, 471–478
    irreversible inibitors, 473–477
    reversible inibitors, 472–473
    non-cysteine design, 477–478
  NMDA antagonists, 450–455
    competitive receptor antagonists, 451–453
    ion channel blockers, 450–451

Excitatory amino acid receptors, stroke therapy (*Continued*)
  NR2B-specific noncompetitive receptor antagonists, 453–455
  voltage-dependent calcium channels, 455–460
Excitatory amino acid transporters (EEATs), anticonvulsants, 131–132
Executive function:
  neuranatomical substrates, 23–24
  set shifting assessment of, 28–30
Exifone, Alzheimer's disease, tau-targeted therapy, 425–426
Extracellular signal regulated kinase (ERK), opioid receptor signaling, 581
Extrapyramidal symptoms (EPSs):
  antipsychotic side effects, 165–166
  dopamine-based models, 171

Familial hemiplegic migraine, epidemiology, 265–266
Faramptator, cognitive enhancement, 44
Fedotozine κ-opioid receptor agonists, 631–632
Felbamate, anticonvulsant applications, 135–139, 152–154
Fenofibrate, γ-secretase modulators, 358–359
Fentanyl:
  4-anilidopiperidine analogs, 618–621
  chemical structure, 570
  metabolism and elimination, 579
  nonpeptide affinity labeling, 641–643
Fentanyl isothiocyanate (FIT), nonpeptide affinity labeling, 642–643
Fibrilllization inhibitors, Alzheimer's disease therapy, preclinical/clinical testing, 361–365
  peptide derivation, 362
  small-molecule inhibitors, 362–365
First-generation compounds, anticonvulsants, 135–139
5-Choice Serial Reaction Time Task (5-CSRTT), cognitive function assessment, 26–27
Flavin monooxygenase, antipsychotics, phenothiazines, 186–189
Flesinoxan, serotonin receptor targeting, 242–244
Flexible opiates, 598–599
FLIPR antagonist activity, migraine therapy, 308–312
Flumazenil, cognitive enhancement, 45–46
Flunarizine, migraine prevention, 318
Fluorofelbamate, anticonvulsant applications, 153–154
6-Fluoro-L-dopa, Parkinson's disease dopaminergic therapy, 547–549
Fluoromethyl ketones, stroke therapy, caspase inhibitors, 468–469
Fluoxetine:
  chemical structure, 223
  clozapine pharmacokinetics, 192
  historical background, 248–249
  migraine prevention, 318–319
  monoamine transporter inhibition, 238–241
  pharmacokinetics, 236
  pindolol interaction, 225–229
  serotonin receptor targeting, 242
  side effects and adverse reactions, 233–234
  structure-activity relationship, 251
Fluperlapine, structure-activity relationship, 182–183
Fluphenazine, pharmacokinetics, 187–189
Fluvoxamine:
  chemical structure, 223
  clozapine pharmacokinetics, 192
  historical background, 249
  pharmacokinetics, 236
  side effects and adverse reactions, 233–234
Food and Drug Administration (FDA), antipsychotics, approved compounds, 176–180
  atypical compounds, 177–179
  conventional drugs, 176–177
Food effects:
  antidepressants, monoamine oxidase inhibitors, 230
  migraine pathology and, 266–267
FR247304 PARP inhibitor, stroke therapy, 486–487
β-Funaltrexamine (β-FNA), nonpeptide affinity labeling, opioid analgesics, 638–643
Functional selectivity, receptor-based drug design, central nervous system drugs, 7–9

Gabapentin:
  anticonvulsant applications, 135–139, 149
  migraine prevention, 318
Galanin receptor ligands, antidepressant effects, 257–258
Galantamine, cognitive enhancement, 15–16, 31
GAL receptors, cognitive enhancement, 49
γ-aminobutyric acid (GABA) receptors:
  anticonvulsants:
    analogs, 149
    barbiturates, 139–141
    benzodiazepines, 136–139, 144–147
    classification, 135–139
    enzymes, 132–134
    ligand-gated ion channels, $GABA_A$ receptors, 126–129
    synaptic membrane transporters, 130–132
    voltage-gated ion channels, 123–126

cognitive enhancement, GABA$_A$, 45–46
Parkinson's disease, pathophysiology, 531–534
schizophrenia pathology, 163–164
stroke therapy, GABA$_A$ agonist, 483
Gammaguard Liquid, Alzheimer's immunization therapy, 378
Ganaxolone, anticonvulsant applications, 136–139, 156
Ganglionic nicotinic antagonists, 78
Gantacurium, nicotinic antagonists, 80
Gavestinel, stroke therapy, 452–453
Geldanamycin, Alzheimer's disease, 431–432
Gem-disubstituted scaffold-based inhibitors, stroke therapy, 506
Genetics:
 central nervous system disorders, 3–4
 Parkinson's disease etiology and, 534–535
 schizophrenia pathology, 165
 antipsychotic drug assessment, 173–174
Genomics, central nervous system disorders, 3–4
Gepirone, serotonin receptor targeting, 242–244
Glaucoma, pilocarpine therapy, 67–68
Glucuronidation, morphine metabolism and elimination, 578–579
Glutamate:
 anticonvulsants, ionotropic glutamate receptors, 129
 cognitive enhancement, 41–42
 modulators, migraine therapy, 290–297
 ionotropic receptor antagonists, 291–297
 receptor properties, 291
 Parkinson's disease therapy, antagonist agents, 555

schizophrenia pathology, 163–164
site antagonists, stroke therapy, 451
Glutamatergic agents, antipsychotics, 202–206
Glycemic control, antipsychotic side effects and, 168
Glycine:
 anticonvulsants, ligand-gated ion channels, 130
 cognitive enhancement, 44–45
 site antagonists, stroke therapy, 451–453
Glycogen synthase kinase 3 (GSK3), Alzheimer's disease:
 insulin-related therapy, 415
 tau-targeted therapies, 418–420
Glycopyrrolate, muscarinic antagonists, 72–77
Glycosaminoglycan mimetics, β-amyloid peptide inhibitors, 362–365
Glycosylation:
 Alzheimer's disease, tau-targeted therapies, 422–423
 end products inhibitors, β-amyloid peptide, 365
GNTI, κ-opioid selective antagonist, 602–604
Gossypentin, Alzheimer's disease, tau-targeted therapy, 425–426
G-protein coupled receptor (GPCR):
 anticonvulsants, 134
 antidepressant efficacy assessment, 226–229
 antipsychotics, structure-activity relationships, 180–183
 central nervous system drug discovery, 5–6
 dopamine receptor structure and function, 542–544
G proteins, opioid receptor signal transduction, 580–581
Growth factors, antidepressant efficacy and, 228–229
GSK 189254, cognitive function assessment, 27–28
Guanfacine, cognitive enhancement, 41–42

Guanidines:
 BACE (β-site cleavage enzyme), 353–356
 stroke therapy, 497
GW273629 iNOS inhibitor, migraine therapies, 300–302
GW274150 iNOS inhibitor, migraine therapies, 300–302

Half-life estimates, antidepressants, 235–237
 structure-activity relationship and metabolism, 252–253
Hallucinogens, receptor-based drug design, central nervous system effects, 2–3
Halomethyl ketones and electrophiles, stroke therapy:
 calpain inhibitors, 465
 caspase inhibitors, 467–469
Haloperidol:
 dopamine model, 170–171
 FDA approval, 176–177
 nonpsychotic applications, 170
 pharmacokinetics, 189–190
Hamilton rating scale for depression (HAM-D), antidepressant efficacy assessment, 221–229
Hematologic effects, antipsychotics, 168–169
Heme-derived materials, stroke therapy, NOS inhibition, 501
Hepatotoxicity, psychostimulants, 95–96
Heroin, synthetic analog, 597–600
Heteroaryl-linked hydroxymate inhibitor, stroke therapy, 490–491
Heterocyclic derivatives, stroke therapy, 458–459
High-affinity ion channel blockers, stroke therapy, 450

High-throughput screening (HTS), γ-secretase inhibitors, 335
Histamine, cognitive enhancement, 35–37
Histamine H3 receptor antagonists:
  antipsychotic agents, 206–207
  cognitive enhancement, 36–37
Histone deacetylase (HDAC) inhibitors, stroke therapy, 487–495
  2-aminophenylamides, 492–493
  boronates and silanediols, 494
  carboxylic acids, 494
  hydroxamate inhibitors, 488–491
  ketones, 493
  thiol inhibitors, 492
Homatropine, muscarinic antagonist, 72
Human immunodeficiency virus (HIV) infection, dementia, 21
Huperzine A, cognitive enhancement, 31–32
Hydantoins, anticonvulsant applications, 141
Hydrophobicity, cocaine structure-activity relationship, 111–113
Hydroxamic acid inhibitors:
  histone deacetylase inhibitors, stroke therapy, 488–491
    aryl and heteroaryl linker, 490–491
    cinnamoyl linker, 488–490
    linear side-chain linker, 488
  stroke therapy:
    bis-hydroxymates, 507
    histone deacetylase inhibitors, 488–491
      aryl and heteroaryl linker, 490–491
      cinnamoyl linker, 488–490
      linear side-chain linker, 488
    matrix metalloproteinase inhibitors, 502–507
    oxalic acid hydroxymates, 507
    reverse hydroxyamic acids, 507

Hydroxybupropion, structure-activity relationship, 251, 254–255
7-Hydroxychloropromazine, pharmacokinetics, biodistribution, and drug-drug interactions, 185–186
Hydroxyethyl ureas (HEUs), γ-secretase inhibitors, 334–335
Hydroxymethylglutaryl coenzyme A (HMG-CoA), cholesterol biosynthesis, Alzheimer's disease, 406
Hydroxymethylglutaryl coenzyme A (HMG-CoA) reductase inhibitors, Alzheimer's disease therapy, preclinical/clinical testing, 359–361
Hydroxypyridinones, stroke therapy, 508
9-Hydroxy risperidone, pharmacokinetics, 195–196
Hymenialdisine, Alzheimer's disease, tau-targeted therapies, 420–422
Hyoscyamine, muscarinic antagonist, 72
Hypericin, Alzheimer's disease, tau-targeted therapy, 425–426
Hyperpolarization-activated cyclic nucleotide cation (HCN) channels, anticonvulsants, 125–126

Ibuprofen, migraine therapy, 268
Idebenone, cognitive enhancement, 15–16
Ifenprodil, stroke therapy, 453
Imidazoles, γ-secretase modulators, 359
Imidazolidine-2,4-diones, anticonvulsant applications, 141
Imidazoline derivatives:
  α-adrenergic receptor targeting, 247

  bupropion interaction, 255
Iminostilbenes, anticonvulsant applications, 142–144
Imipramine:
  chemical structure, 222
  historical background, 248
  migraine prevention, 318–319
  monoamine transporter inhibition, 238–241
  pharmacokinetics, 235–236
  serotonin receptor targeting, 241–242
  side effects and adverse reactions, 231–232
Immunotherapy approaches, Alzheimer's disease, 366–380
  advantages/disadvantages, 369–370
  anti-Aβ antibody mechanisms, 367–368
  bapineuzamab, 370–378
  Aβ immunization, 366, 368–369
  current clinical trials, 370–380
  gammaguard liquid, 379
  MABT5102A, 379–380
  PF-4360365, 378–379
  preclinical/clinical trials, 369
  R1450, 379
  solaneuzmab, 378
  TgAPP behavioral and learning deficits, 366–367
Indanylspiroazaoxindole-based calcitonin gene-related peptide antagonist, migraine therapy, 287–288
Indirubin, Alzheimer's disease, tau-targeted therapies, 420–422
Indole-2-carboxylic derivatives, stroke therapy, 452–453
Indoline-2-one derivatives, stroke therapy, 514
Inducible nitric oxide synthase (iNOS), migraine therapy, 298–302
INOS inhibitors:
  migraine therapy, 300–302
  stroke therapy, 495–502
In silico screening, BACE (β-site cleavage enzyme) inhibitors, 352–353
Insulin-related therapy, Alzheimer's disease, 414–417

Intrinsic efficacy, receptor-based drug design, central nervous system, 7
In vitro studies, opioid efficacy assessment, 583–585
In vivo studies, opioid efficacy assessment, 585–587
Ion channels:
  noncompetitive blockers, stroke therapy, 450–451
  opioid receptors, signal transduction, 580–581
  voltage-gated ion channels, anticonvulsants, 122–126
    calcium channels, 124
    chloride channels, 126
    hyperpolarization-activated cyclic nucleotide cation channels, 125–126
    potassium channels, 124–125
    sodium channels, 123–124
Ionotropic glutamate receptors:
  anticonvulsants, ligand-gated ion channels, 129
  migraine therapy, antagonists, 291–297
Ioxistatin, stroke therapy, cathepsin B inhibitors, 473–474
Ipratropium, muscarinic antagonist, 72
Iproniazid, as antidepressant, 248
Ipsaspirone, serotonin receptor targeting, 242–244
Irreversible inhibitors, stroke therapy:
  calpains, 464–465
  caspase, 467–469
  cathepsin B inhibitors, 473–477
Isatin-derived caspase inhibitors, stroke therapy, 470–471
Ischemic cascade, stroke epidemiology and pathology, 447–449
Isoclothiapine, structure-activity relationship, 182–183
Isoclozapine, structure-activity relationship, 182–183
Isomethodone, structure-activity relationship, 621–623
Isophthalamide ligands, BACE (β-site cleavage enzyme) inhibitors, 347–351
Isoquinolones, calpain inhibitors, stroke therapy, 464
Isothiourea analogs, stroke therapy, 497
Isotope labeling, PET/SPECT systems, 739–740
Isovaleramide, migraine prevention, 316
Ispronicline antipsychotic, 207
Isradipine, stroke therapy, 459–460
Istradefylline:
  cognitive enhancement, 48–49
  Parkinson's disease therapy, 552–554
Ixabepilone, Alzheimer's disease, tau-targeted therapy, 430

JL-13 antipsychotic drug, structure-activity relationship, 182–183
JN403 agonist, cognitive enhancement, 35–36
JOM-13 tetrapeptide derivative, deltorphin analogs, opioid peptides, 672–673

Ketamines:
  antipsychotic effects, NMDA model, 171–172
  stroke therapy, 450
Ketoamides, calpain, stroke therapy, 463–464
Ketoconazole, sertindole interaction, 199
Ketocyclazocine:
  morphinan derivatives, 611–612
  multiple opioid receptors, 579–580
Ketones:
  methadone analogs, 623
  stroke therapy:
    alpha oxo/thio/amino-methyl ketones, 469–470
    calpain inhibitors, 463–464
    caspase inhibitors, 467–469
    cathepsin B inhibitors, 472–473, 475–478
    halomethyl ketones, 465, 467–469
    HDAC inhibitors, 493
Ketorolac, migraine therapy, 268

Khat, stimulant effect, 89–90
Kinase inhibitors, antidepressant efficacy and, 228–229
Kynurenic acid:
  schizophrenia pathology, 163–164
  stroke therapy, 452–453

L-685458 γ-secretase inhibitor, 334–335
Lacidipine, stroke therapy, 459–460
Lacosamide, anticonvulsant applications, 136–139, 154
Lactams, antidepressants, structure-activity relationship and metabolism, 252–253
β-Lactams, stroke therapy:
  calpain irreversible inhibitors, 465
  cathepsin B inhibitors, 475–476
Lamotrigine, anticonvulsant applications, 135–139, 148–149
Latent inhibition, antipsychotic therapy, 172
Lead compounds:
  BACE (β-site cleavage enzyme) inhibitors, 347
  migraine therapy, calcitonin gene-related peptide antagonists, 287–288
Learning deficits, Alzheimer's disease immunotherapy, 366–367
Lecozotan, cognitive enhancement, 37–40
Leucine-rich-repeat kinase 2 (LRKK-2), Parkinson's disease genetics, 535
Leupeptin, stroke therapy, 461–462
  cathepsin B inhibitors, 472
Levetiracetam:
  anticonvulsant applications, 135–139, 149–151
  cognitive enhancement, 34
Levodopa (L-dopa), Parkinson's disease:
  dopaminergic pharmacotherapy, 540–542
  pathophysiology, 531–534

Lewy body dementia (LBD), 21
Licostinel, stroke therapy, 452–453
Lifestyle, cognitive dysfunction and, 21
Ligand-binding domain (LBD), central nervous system drug discovery, 5–6
Ligand-gated ion channels (LGICs), anticonvulsants, 126–130
  $GABA_A$ receptors, 126–129
  ionotropic glutamate receptors, 129
Ligand-receptor interactions:
  opioid-receptor-like 1/orphanin FQ/nociceptin peptide, 680–681
  opioid receptors, 581–588
Linear peptides:
  deltorphin analogs, opioid peptides, 668–671
  dermorphin analogs, μ-opioids, 664–666
  dynorphin analogs, 655–656
  enkephalin analogs:
    δ-opioid analogs, 650
    μ-opioid analogs, 647
Linker compounds:
  glutmate modulators, migraine therapy, 292–297
  hydroxamate HDAC inhibitors, stroke therapy, 488–491
Lisuride, Parkinson's disease dopaminergic therapy, 545–546
Liver:
  antidepressant effects on, monoamine oxidase inhibitors, 230
  antipsychotic side effects on, 169
Liver X receptors (LXRs), Alzheimer's disease, cholesterol turnover, 411–413
Locomotor effects, antipsychotic therapy, animal models, 170–171
Lofepramine:
  chemical structure, 222
  monoamine transporter inhibition, 240–241
  pharmacokinetics, 235–236

Long-chain arylpiperazines (LCAPs),
  antidepressant actions, 242–244
  structure-activity relationships and metabolism, 253–254
Long-term depression (LTD), 25
Long-term memory, 19
Long-term potentiation (LTP), 24–25
"Loose" neuroleptics, structure-activity relationship, 185
Lorazepam, anticonvulsant applications, 144–147
Lovastatin, Alzheimer's disease therapy, preclinical/clinical testing, 359–361
Loxapine:
  migraine therapy, 297–298
  pharmacokinetics, 189
  structure-activity relationship, 180–183, 250–251
L-type calcium channel antagonist, stroke therapy, 459–460
LY293558 glutamate modulator, migraine therapy, 291–295
LY382884 glutamate modulator, migraine therapy, 295–297
LY-426965 antagonist, efficacy assessment, 225–229
LY466195 glutamate modulator, migraine therapy, 295–297
Lysergic acid diethylamide (LSD), receptor-based drug design, central nervous system effects, 2–3

MABT5102A monoclonal antibody, Alzheimer's immunization therapy, 378–379
Macrocyclic inhibitors, stroke therapy, 506–507
Maleimide analogs, Alzheimer's disease, tau-targeted therapies, 420–422
Maprotiline:
  chemical structure, 222
  historical background, 248–249

  serotonin receptor targeting, 241–242
  side effects and adverse reactions, 231–232
  structure-activity relationship, 250–251
Marimastat hydroxymate inhibitors, stroke therapy, 503
Masked aldehydes, stroke therapy, 463
  caspase 3/9 inhibitors, 467
Matrix metalloproteinase (MMP) inhibitors, stroke therapy, 502–510
  carboxylic acid-derived inhibitors, 508
  hydroxamic acid-derived inhibitors, 502–507
  hydroxypyridinones, 508
  phosphorus-derived inhibitors, 509–510
  pyramidine-2,4,6-triones, 508–509
  thiols, 509
Maxi-K channel, stroke therapy, 482–483
m-Chlorophenylpiperazine (m-CPP):
  antidepressant pharmacokinetics, 236–237
  serotonin receptor targeting, 242–244
MDL-28170 calpain inhibitor, stroke therapy, 461–462
Measurement and Treatment Research to Improve Cognition in Schizophrenia (MATRICS), cognitive dysfunction assessment, 20–22, 162
Mecamylamine nicotinic antagonist, 78
Memantine:
  cognitive enhancement, 15–16, 33–34
  Parkinson's disease therapy, 554–555
  stroke therapy, 450
Memory:
  assay techniques, 26–29
  cognition and, 17–19

molecular substrates, 24–25
neuranatomical substrates, 23
Meperidine:
  chemical structure, 570
  metabolism and elimination, 579
  piperidine derivatives, 614–621
"Message-address" concept:
  δ-opioid selective agonists, 637
  opioid receptor selective antagonists, 604–608
Metabolic syndrome, Alzheimer's disease, 405, 411–417
  cholesterol-related approaches, 405, 411–414
  insulin-related approaches, 414–417
Metabolism:
  antipsychotic effects on, 168
  opioid receptors, 578–579
Metabotropic glutamate receptors:
  anticonvulsants, 134
  antipsychotic glutamatergics, 203
  cognitive enhancement, 42–43
  psychostimulant physiology and pharmacology, 101–106
  schizophrenia pathology, 164
Metaclopramide, migraine therapy, 297–298
Metal-protein attenuating compounds (MPACs), β-amyloid peptide inhibitors, 363–365
Metazocine, morphinan derivatives, 611
Metformin, Alzheimer's disease, insulin related therapy, 415–417
Methacholine, structure and function, 64
Methadol, structure-activity relationship, 621–623
Methadone:
  acyclic analogs, 621–623
  chemical structure, 570
  metabolism and elimination, 579
  structure-activity relationship, 598
Methamphetamine:
  ADMET properties, 97–99
  historical development, 90–91
  nitrogen substituents, 107
  physiology and pharmacology, 100–106
  side chain modification, 107
  side effects, adverse reactions and drug interactions, 92, 95–97
  structure-activity relationship, 106–109
Methantheline, muscarinic antagonists, 72–74
Methoctramine, muscarinic antagonists, 76–77
Methylene blue β-amyloid peptide inhibitors, 364–365
  tau aggregation, 423–429
Methylphenidate:
  ADMET properties, 98–99
  cognitive function assessment, 26–27
  physiology and pharmacology, 103–106
  side effects, adverse reactions and drug interactions, 95
  structure-activity relationship, 109
1-Methyl-4-phenylpyridinium (MPP+), Parkinson's disease and, 530, 539–540
N-Methyl-4-phenyl-1,2,3,6-tetrahydropyridine (MPTP), Parkinson's disease and, 538–540
N-Methyl-4-propionoxy-4-phenylpiperidine (MPPP):
  opioid derivatives, piperidine reversal, 616–618
  Parkinson's disease and, 539–540
Methylthioninium chloride (MTC), Alzheimer's disease, tau-targeted therapy, 424–425
Metoprolol, migraine prevention, 314–315
Mianserin:
  α-adrenergic receptor targeting, 247
  cognitive enhancement, 37–40
  serotonin receptor targeting, 241–242
Michael acceptors, stroke therapy:
  calpain inhibitors, 464
  cathepsin B inhibitors, 477
Microtubule-affinity-regulating kinase (MARK), Alzheimer's cognitive deficits, 16
Microtubules, Alzheimer's disease, tau-targeted therapy, 430
Midazolam, anticonvulsant applications, 144–147
Migraine:
  epidemiology, 265–266
  pathology, 266–267
  therapeutic agents:
    antiemetics/antipsychotics, 297–298
    calcitonin gene-related peptide antagonists, 272–290
      azepinone binding affinity, 281–286
      benzodiazepine binding affinity, 277–281
      BIBN4096 BS, 273, 275–277
      BMS-694153, 288–290
      MK-0974, 277, 279–288
      neuropeptide-induced blood flow, 286–288
      receptor binding affinities, 273–275
      unnatural amino acids, 289–290
    ergot alkaloids, 267–268
    future research issues, 320–321
    glutamate modulators, 290–297
      ionotropic receptor antagonists, 291–297
      receptor properties, 291
    nitric oxide synthase inhibitors, 298–306
      iNOS inhibitors, 300–302
      isoforms, 298–299
      nNOS inhibitors, 302–306
      tool compounds, 299–300
    nonsteroidal anti-inflammatory drugs and paracetamol, 268
    preventive therapies, 312–320
      angiotensin converting enzyme inhibitors, 319–320
      anticonvulsants, 315–316
      antidepressants, 318–319
      β-blockers, 313–315

Migraine (*Continued*)
  calcium channel blockers, 316–318
  serotonin agonists, 268–272
   anti-inflammatory/vasoconstriction mechanisms, 270–272
   second-generation triptans, 270
   sumatriptan, 268–270
  vanilloid receptor antagonists, 307–312
   capsazepine, civamide, and AMG-517, 307–308
   SB-705498, 308–312
Mild cognitive impairment (MCI), epidemiology, 15
Milnacipran, structure-activity relationship, 251
Mirtazepine:
  α-adrenergic receptor targeting, 247
  chemical structure, 224
  pharmacokinetics, 237
Mitochondrial metabolism, Parkinson's disease, dopamine and, 535–537
Mitogen-activated protein kinase (MAPK) inhibitors:
  Alzheimer's disease, tau-targeted therapies, 420–422
  opioid receptor signaling, 581
Mivacurium, nicotinic antagonists, 79–80
Mixed-function ligands, serotonin receptor targeting, 243–244
MK-0974 calcitonin gene-related peptide antagonist, migraine therapy, 277, 279–288
MK-3207 calcitonin gene-related peptide antagonist, migraine therapy, 287–288
Moclobemide:
  chemical structure, 223
  physiology and pharmacology, 248
  side effects and adverse reactions, 230
  structure-activity relationship and metabolism, 252–253

Modafinil:
  memory assessment, 28–30
  psychostimulant physiology and pharmacology, 114
Molecular biology, opioid receptors, 587–591
Molecular memory substrates, 24–25
Molecular targeting, anticonvulsants, 121–134
  connexins, 134
  enzymes, 132–134
  G-protein-coupled receptors, 134
  ligand-gated ion channels, 126–130
  neurotransmitter transporters and synaptic proteins, 130–132
  voltage-gated ion channels, 122–126
Molindone, pharmacokinetics, 190–191
Monoamine oxidase (MOA) inhibitors:
  chemical structure, 223
  historical background, 248
  Parkinson's disease:
   dopamine-mitochondrial metabolism and, 535–537
   dopaminergic therapy, 549–551
   etiology, 539–540
  pharmacokinetics, 236
  physiology and pharmacology, 247–248
  recent and future research issues, 255–256
  side effects and adverse reactions, 230
  structure-activity relationship and metabolism, 251–253
Monoamine transporter, antidepressant pharmacology, 237–241
Monoamine uptake inhibitors, psychostimulant physiology and pharmacology, 100–106
Monoclonal antibodies (mAB), Alzheimer's immunization therapy, 370–381

Morin β-amyloid peptide inhibitors, 364–365
Morphinans, structure-activity relationship, 598, 609–610
Morphine and derivatives:
  ADMET properties, 578–579
  C-ring derivatives, 600–601
  dismantling of, 598
  identification and early history, 596–597
  nonpeptide affinity labeling, 638–641
  structure-activity relationship, 599–601
  synthetic analogs, 597–598
Morris Water Maze test:
  Alzheimer's disease assessment, cholesterol levels, 405–406
  spatial memory working memory assessment, 27–28
MPPP, Parkinson's disease and, 539–540
Multiple opioid receptors, discovery and development, 579–580
Muscarinic acetylcholine receptor (mAChR) antagonists, structure-activity relationships, 72–77
Muscarinic agonists, 64–68
Muscarinic receptors:
  antipsychotic effects on, 167–168
  structure-activity relationship, 185
  cholinergic nervous system, 61–62
  cognitive enhancement, 32–33
  serotonin-selective reuptake inhibitors and, 233–234
  structure and classification, 62
Muscarone agonist, 64–65
Mutagenicity, opioid receptors, 589–590
Myricetin, Alzheimer's disease, tau-targeted therapy, 425–426

N-acetyl-aspartyl glutamate (NAAG),

INDEX 769

schizophrenia pathology, 163–164
Nalorphine:
  structure-activity relationship, 601
  synthetic analog, 597–598
Naloxone:
  antagonist activity, 601
  chemical structure, 571
  morphine metabolism and elimination, 578
  nonpeptide affinity labeling, 640–643
Naltrexamine derivative:
  antagonist activity, 601
  nonpeptide affinity labeling, 640–643
  TENA κ-opioid selective antagonist, 602–604
Naltrexone:
  antagonist activity, 601
  δ-opioid receptor antagonist, 604–608
  nonpeptide affinity labeling, 638–641
Naltrindole:
  δ-opioid receptor antagonist, 604–608
  μ-opioid receptor antagonist, 607
  nonpeptide affinity labeling, 638–643
NAN-190 antagonist, efficacy assessment, 225–229
Naproxen, migraine therapy, 268
Narcolepsy, psychostimulant therapy, 91–92
Natural products, β-amyloid peptide inhibitors, 363–365
NC-531 glycosaminoglycan mimetic, 362
Nebracetam, cognitive enhancement, 34
Nefazodone:
  chemical structure, 224
  pharmacokinetics, 236–237
  serotonin receptor targeting, 241–242
  side effects and adverse reactions, 234–235
Neoclerodane diterpene, current developments, 682–683
Neramexane, stroke therapy, 450

NeurAxon nNOS inhibitor, migraine therapies, 302–306
Neuregulin 1, schizophrenia genetics, 174
Neuroanatomy, cognition and, 22–23
Neurofibrillary tangle (NFT), Alzheimer's disease, tau-targeted therapies, 417–432
Neurokinins:
  cognitive enhancement, 49
  radiolabeled ligands for, 744
  receptor antagonists, antipsychotic agents, 208
Neuroleptic antipsychotics, structure-activity relationships, 180–185
Neuroleptic malignant syndrome (NMS), antipsychotic side effects, 167
Neuromuscular nicotinic antagonists, 78–80
Neuronal nicotinic acetylcholine receptors (nAChRs), cognitive enhancement, 34–35
Neuronal nitric oxide synthase (nNOS), migraine therapies, 299, 302–306
Neuropeptides:
  antidepressant effects, 256–258
  calcitonin gene-related peptide antagonists, migraine therapy, 286–288
  cognitive enhancement, 48–49
  schizophrenia pathology, 165
Neurotransmitters:
  anticonvulsants, transporters and synaptic proteins, 130–132
  antidepressants:
    effect on, 220–221
    side effects and adverse reactions, 231–234
  schizophrenia pathology, 164–165
Neurotransmitter sodium symporter (NSS) family, psychostimulant physiology and pharmacology, 100–106
Neurotrophic factors:
  Alzheimer's disease, tau-targeted therapies, 422
  antidepressant efficacy assessment, 226–229
Neurotropic agents, cognitive enhancement, 47
Neutropenia, antipsychotic side effects, 168–169
Nicergoline, cognitive enhancement, 41–42
Nicotine:
  cholinergic nervous system, 61–62
  cognitive enhancement, 34–36
Nicotinic acetylcholine receptor (nAChR):
  anticonvulsants, ligand-gated ion channels, 129–130
  antipsychotic agents, 207
  nicotinic agonists, 68–72
  nicotinic antagonists, 77–80
    ganglionic antagonists, 78
    neuromuscular antagonists, 78–80
  schizophrenia pathology, 165
  structure and classification, 63
Nicotinic acid receptor agonists, cognitive enhancement, 34–35
Nicotinic agonists, structure and function, 68–72
Nifedipine, migraine prevention, 318
Nimetazepam, anticonvulsant applications, 144–147
Nimodipine:
  migraine prevention, 318
  stroke therapy, 459–460
Niravoline κ-opioid receptor agonist, 627–632
Nisoxetine, structure-activity relationship, 251
Nitrazepam, anticonvulsant applications, 144–147
Nitric oxide (NO), stroke therapy, INOS/NNOS inhibitors, 495–502
  amidines, 497–498
  aminopyridines, 498–499
  L-arginine analogs, 495–497
  calmodulin site, 499–500
  coumarin derivatives, 499
  dimerization inhibitors, 501

Nitric oxide (Continued)
  guanidine and isothiourea analogs, 497
  heme site, 501
  non-arginine sites, 499
  NOS inhibitor/NO donor compounds, 502
  tetrahydrobiopeterin site, 500
Nitric oxide synthases (NOS):
  migraine therapy, nitric oxide synthase inhibitors, 298–306
    iNOS inhibitors, 300–302
    isoforms, 298–299
    nNOS inhibitors, 302–306
    tool compounds, 299–300
  stroke therapy, INOS/NNOS inhibitors, 495–502
Nitriles, stroke therapy, cathepsin B inhibitors, 473
Nitrogen, psychostimulant structure-activity relationship, 107
  cocaine, 109–110
NK1 receptor, substance P-NK1 receptor antagonist, 256–257
NMED-160 scaffold, stroke therapy, 457–458
N-methyl-D-aspartate (NMDA) receptors:
  anticonvulsants, ligand-gated ion channels, 129
  antidepressants, 256
  antipsychotic drugs:
    animal models, 171–172
    glutmatergics, 204–205
    latent inhibition, 172
  cognitive enhancement:
    glutamate, 41–42
    glycine, 44–45
  glutmate modulation, migraine therapy, 291–295
  long-term potentiation, 24–25
  modulators, 33–34
  schizophrenia pathology, 163–164
  stroke therapy, 450–455
    competitive receptor antagonists, 451–453
    ion channel blockers, 450–451
    NR2B-specific noncompetitive receptor antagonists, 453–455

NNOS inhibitors:
  migraine therapy, 302–306
  stroke therapy, 495–502
Non-amyloid therapies, Alzheimer's disease:
  metabolic syndrome targeting, 405–417
    cholesterol-related approaches, 405–414
    insulin-related approaches, 414–417
    research background, 405
    summary, 408–410
  tau pathology targeting, 417–432
    aggregation, 423–429
    glycosylation, 422–423
    microtubule stabiliization, 430
    phosphorylation, 418–422
    protein levels, 430–432
Nonpeptide opioid analgesics:
  opioid-receptor-like 1/orphanin FQ/nociceptin peptide, 680–681
  structure-activity relationships, 597–645
    acyclic analgesics, 621–623
    affinity labeling, 637–643
    morphine-naltrexone derivatives, 638–641
  agonists:
    δ-selective agonists, 632–637
    κ-selective agonists, 623–632
    centrally-acting agonists, 624–630
    peripherally-acting agonists, 630–632
  antagonists, 601–608
    δ-opioid receptors, 604–608
    κ-opioid receptors, 602–604
  benzomorphans, 610–613
  Diels-Alder adducts, 608–609
  miscellaneous opioids, 643–645
  morphinans, 609–610
  morphine derivatives, 599–601
  opium-derived morphine and alkaloids, 598–599
  ORL1 receptor, endogenous ligand orphanin FQ/nociceptin, 681–682

  piperidine derivatives, 613–621
    4-anilidopiperidines, 618–621
    4-arylpiperidines:
      carbon substituent, 614–616
      oxygen substituent, 616–618
Nonpeptidic chemotypes:
  BACE (β-site cleavage enzyme) inhibitors, 347–356
  γ-secretase inhibitors, 337–341
Nonsteroidal anti-inflammatory drugs (NSAIDs):
  migraine therapy, 268
  γ-secretase modulators, 358–359
Nootropics, cognitive enhancement, 34
NorBNI, κ-opioid selective antagonist, 602–604
Norclozapine, cognitive enhancement, 32–33
Nordiazepam, anticonvulsants, 144–147
Norepinephrine:
  antidepressants, neurotransmission enhancement, 220–221
  cognitive enhancement, 41
  psychostimulant pharmacology, 100–106
  structure-activity relationship, 106–109
Norepinephrine-selective reuptake inhibitors (NSRIs):
  chemical structure, 223
  historical background, 248–249
  pharmacokinetics, 237
  side effects and adverse reactions, 223–224, 234–235
  structure-activity relationship, 251
Norepinephrine transporter (NET), antidepressant pharmacology, 237–241
Norfluoxetine:
  serotonin receptor targeting, 242–244
  structure-activity relationship, 251

Nortriptyline:
  chemical structure, 222
  migraine prevention, 318–319
  serotonin receptor targeting, 241–242
Novel object recognition, cognitive function assessment, 27
NR2B NMDA subunit, noncompetitive receptor antagonist, stroke therapy, 453–455
NS-7 scaffold, stroke therapy, 457
NS1738, cognitive function assessment, 27–28
N-type voltage-dependent calcium channel, stroke therapy, 456–459
NXY-059 antioxidant, stroke therapy, 479–480

Obesity, antipsychotic effects, 168
Octahydroisoquinolines, δ-opioid selective agonists, 635–637
Off label drug use, antipsychotic medications, 170
O-GlcNAcase inhibitors, Alzheimer's disease, tau-targeted therapies, 422–423
Olanzapine:
  FDA approval, 178
  pharmacokinetics, 192–193
  side effects, 168–169
  structure-activity relationship, 180–183
Olcegepant, migraine therapy, 273, 275–277
(−)-Oleocanthal, Alzheimer's disease, tau-targeted therapy, 429
ONO2506 GABA$_A$ agonist, stroke therapy, 483–484
Opioid-receptor-like 1 (ORL1) receptor:
  discovery and nomenclature, 580
  endogenous peptides, 594–596
  nonpeptide ligands, 680–681
  physiology and pharmacology, 677–678
  research background, 676–677
  structure-activity relationship, 678–680

Opioids. See also specific subtypes, e.g. κ-opioid receptors
  ADMET properties, 577–579
  morphine and derivatives, 578–579
  clinical applications, 570–579
  current drugs, 571–573
  cytochrome P450 enzyme interactions, 576–577
  future research issues, 684–686
  morphine and analogs, historical background, 596–597
  nonpeptide analgesics, structure-activity relationships, 597–645
    acyclic analgesics, 621–623
    affinity labeling, 637–643
    morphine-naltrexone derivatives, 638–641
  agonists:
    δ-selective agonists, 632–637
    κ-selective agonists, 623–632
    centrally-acting agonists, 624–630
    peripherally-acting agonists, 630–632
  antagonists, 601–608
    δ-opioid receptors, 604–608
    κ-opioid receptors, 602–604
  benzomorphans, 610–613
  Diels-Alder adducts, 608–609
  miscellaneous opioids, 643–645
  morphinans, 609–610
  morphine derivatives, 599–601
  opium-derived morphine and alkaloids, 598–599
  ORL1 receptor, endogenous ligand orphanin FQ/nociceptin, 681–682
  piperidine derivatives, 613–621
    4-anilidopiperidines, 618–621
    4-arylpiperidines:
      carbon substituent, 614–616
      oxygen substituent, 616–618

ORL1 receptor, endogenous ligand orphanin FQ/nociceptin, 676–681
  nonpeptide ligands, 681–682
  physiology and pharmacology, 677–678
  research background, 676–677
  structure-activity relationship, 678–680
peptide analogs, 645–676
  affinity-labeled derivatives, 675–676
  amphibian skin peptides, 646
  antagonist dynorphin A analogs, 657–658
  combinatorial library peptides and peptidomimetics, 674–675
  dynorphin analogs, 654–658
    conformational constraints, 656–657
    linear analogs, 655–656
  enkephalin analogs, 646–654
    δ-selective analogs, 649–653
    dimeric analogs, 653–654
    μ-selective analogs, 646–649
  somatostatin μ-receptor antagonists, 673–674
  Tyr-D-aa-Phe sequence, 664–673
    deltorphin analogs, 668–673
    dermorphin analogs and μ-selective peptides, 664–668
  Tyr-Pro-Phe sequence, 659–664
    β-casomorphin analogs and endomorphins, 659–661
    TIPP and related peptides, 661–664
physiology and pharmacology, 579–597
  central nervous system effects, 579
  computational models, 590
  δ-opioid receptors, 591–592
  endogenous peptides, 593–596
  κ-opioid receptors, 592–593

Opioids (*Continued*)
  ligand characterization, 581–583
  multiple opioid receptors, 579–580
  mutagenesis studies, 589–590
  non-μ-opioid receptors, 591–593
  signal transduction mechanisms, 580–581
  structure and molecular biology, 587–589
  subtypes, splice variants, and dimerization, 590–591
  in vitro efficacy assays, 583–585
  in vivo evaluation, 585–587
  recent developments, 682–684
  receptor-based drug design, central nervous system, 2–3
  research background, 569
  salvia divinorum and neoclerodane diterpenes, 682–684
  side effects, 573–576
    constipation, 575–576
    contraindications, 576
    drug-drug interactions, 576–577
    nausea and vomiting, 575
    pruritis, 576
    respiratory depression, 574
    sedation and cognitive impairment, 575
    tolerance, dependence, and addition liabilities, 574–575
  subtype classification, 586
δ-Opioids:
  chemical structure, 571
  discovery and nomenclature, 579–580
  β-endorphin affinity, 594–596
  enkephalin analogs, 650–653
  ligand characterization, 582–583
  mutagenesis studies, 589–590
  nonpeptide affinity labeling, 640–643
  physiology, 591–592
  selective agonists, 632–637
  selective antagonists, 604–608
  signal transduction, 580–581
  subtypes, splice variants, and dimerization, 590–591
  Tyr-Pro-Phe sequence, 661–664
κ-Opioids:
  agonist compounds, 623–632
  benzomorphan derivatives, 611–612
  chemical structure, 571
  discovery annd nomenclature, 579–580
  ligand characterization, 581–583
  nonpeptide affinity labeling, 642–643
  physiology, 592–593
  selective antagonists, 602–604
  signal transduction, 580–581
  subtypes, splice variants, and dimerization, 591
μ-Opioids:
  chemical structure, 571
  dermorphin analogs, peptide selectivity, 664–668
  discovery and nomenclature, 579–580
  endomorphin affinity, 595–596
  β-endorphin affinity, 594–596
  enkephalin analogs, 647–649
  ligand characterization, 581–583
  mutagenesis studies, 589–590
  selective antagonists, 607
  somatostatin derivation, 673–674
  signal transduction, 580–581
  subtypes, splice variants, and dimerization, 591
Opium, morphine derivatives, 599–60
Organoruthenium, stroke therapy, cathepsin B inhibitors, 476–477
Organotellurium, stroke therapy, cathepsin B inhibitors, 476–477
Oripavine, nonpeptide affinity labeling, 641–643
Orphanin FQ/nociceptin peptide:
  nonpeptide ligands, 680–681
  opioid receptor affinity, 595–596
  physiology and pharmacology, 677–678
  research background, 676–677
  structure-activity relationship, 678–680
Orphenadrine, Parkinson's disease therapy, 552–553
Orthostatic hypotension:
  antidepressant side effect, 231–232
  antipsychotic side effects, 167
Osanetant, cognitive enhancement, 48–49
Overdose:
  antidepressants and, tricyclic antidepressants, 231–232
  serotonin-selective reuptake inhibitors, 232–234
Oxadiazole, muscarinic activity, 66–67
Oxalic acid hydroxymates, stroke therapy, 507
Oxamide piperidine derivatives, stroke therapy, 454–455
Oxathiolane analog, muscarinic activity, 65
Oxazepam, anticonvulsants, 144–147
Oxazolidine, nicotinic agonists, 71
Oxcarbazepine, anticonvulsant applications, 135–139, 142–144
Oxigon β-amyloid peptide inhibitor, 363–365
Oxiracetam, cognitive enhancement, 34
Oxorhenium complexes, stroke therapy, cathepsin B inhibitors, 473
Oxotremorine, muscarinic activity, 68
Oxybutynin, muscarinic antagonists, 74–77
Oxymorphone:
  antagonist activity, 601
  nonpeptide affinity labeling, 640–643
Oxyphenonium, muscarinic antagonists, 72–77

Paclitaxel, Alzheimer's disease, tau-targeted therapy, microtubule stabilization, 430
Pain management, opioid receptors, 570–571
Paliperidone:
  FDA approval, 178
  pharmacokinetics, 195–196

Panobinostat, hydroxamate inhibitors, stroke therapy, 489–490
Papaverine:
  antipsychotic agents, 208–209
  memory assessment, 28–30
  structure-activity relationship, 599
Paracetamol. See Acetaminophen
Parkinsonism:
  drug-induced, antipsychotic side effects, 165–166
  MPTP and, 538–540
  pathology, 529–530
Parkinson's disease:
  cognitive dysfunction in, 21
  diagnostic agents, 555–556
  dopaminergic pharmacotherapy, 540–552
    catechol-O-methyltransferase inhibitors, 551–552
    L-dopa therapy, 540–542
    monoamine oxidase inhibitors, 549–551
    receptor-targeted compounds, 542–549
    aporphine-type receptor agonists, 544–545
    dopamine receptor structure and function, 542–544
    ergot-type receptor agonists, 545–546
    experimental D_1 receptor agonists, 547–549
    small-molecule agonists, 547
  epidemiology, 529
  etiology, 534–540
    dopamine/mitochondrial oxidative mechanism, 535–537
    environmental factors, 537–540
    genetic factors, 534–535
  future research issues, 556
  nondopaminergic pharmacotherapy, 552–555
    adenosine receptor antagonists, 552–553
    anticholinergic agents, 552
    glutamate antagonists, 555
    serotonin 5-HT_{1A} agonists, 553–555
  pathology, 529–534

Paroxetine:
  chemical structure, 223
  clozapine interaction, 192
  pindolol interaction, 225–229
  side effects and adverse reactions, 233–234
Partial agonists, opioids, 601–608
Passive avoidance task, memory assessment, 28
Paullones, Alzheimer's disease, tau-targeted therapies, 420–422
Paw test, antipsychotic side effects, 175
PD-150606 calpain inhibitor, stroke therapy, 465
PD151307 peptide ligand, N-type voltage-dependent calcium channels, stroke therapy, 456–459
PD-168077, cognitive enhancement, 41–42
Pedunculopontine nucleus (PPN), Parkinson's disease, pathophysiology, 534
Pemoline, side effects, adverse reactions and drug interactions, 95
Pentazocine, chemical structure, 571
Penthienate, muscarinic antagonists, 72–77
Pentolinium nicotinic antagonist, 78
Peptide isosteres:
  fibrillization inhibitors, 362
  γ-secretase inhibitors, 334–337
Peptides:
  cognitive enhancement, 50
  N-type voltage-dependent calcium channels, stroke therapy, 456–457
  opioid analogs, 645–676
    affinity-labeled derivatives, 675–676
    amphibian skin peptides, 646
    antagonist dynorphin A analogs, 657–658
    combinatorial library peptides and peptidomimetics, 674–675
    dynorphin analogs, 654–658
      conformational constraints, 656–657

linear analogs, 655–656
enkephalin analogs, 646–654
  δ-selective analogs, 649–653
  dimeric analogs, 653–654
  μ-selective analogs, 646–649
  somatostatin μ-receptor antagonists, 673–674
  Tyr-D-aa-Phe sequence, 664–673
    deltorphin analogs, 668–673
    dermorphin analogs and μ-selective peptides, 664–668
  Tyr-Pro-Phe sequence, 659–664
    β-casomorphin analogs and endomorphins, 659–661
    TIPP and related peptides, 661–664
opioid-receptor-like 1/orphanin FQ/nociceptin peptide, 678–680
opioid receptors:
  endogenous peptides, 593–596
  precursors and processing products, 595–596
Peptidomimetics:
  BACE (β-site cleavage enzyme) inhibitors, 345–347
  opioid peptides, from combinatorial libraries, 674–675
  stroke therapy, 462–463
  nNOS inhibitors, 496
Perampanel glutamate modulator, migraine therapy, 296–298
Perception, cognition and, 17–19
Pergolide, Parkinson's disease dopaminergic therapy, 545–546
Perlapine, structure-activity relationship, 180–183
Perospirone, pharmacokinetics, 197–198
Peroxisome proliferator-activated receptors (PPARs), g-agonists, Alzheimer's disease, insulin related therapy, 415–417

Perphenazine, pharmacokinetics, 187–189
Perzinfotel, stroke therapy, 451
Pesticides, Parkinson's disease and, 537–540
Pexacerfont antidepressant, CRF antagonist, 257
Pharmacokinetics:
  antidepressants, 235–237
    serotonin-selective reuptake inhibitors, 236
    tricyclic antidepressants, 235–236
  antipsychotics, 185–200
    benzisoxazoles, 195–196
    benzoisothiazoles, 196–198
    butyrophenones, 189–190
    diarylbutylamines, 190
    dibenzodiazepines, 191–192
    dibenzothiepine, 194
    dibenzoxazepines, 189
    dibenzthiazepines, 193–194
    dihydroindolone, 190–191
    phenothiazines, 186–188
    phenylindole, 199
    quinolinone, 198
    substituted benzamide, 199–200
    thiobenzodiazepine, 192–193
    thioxanthenes, 188–189
  migraine therapy:
    ergot alkaloids, 267–268
    triptans, 270–272
Pharmacology studies:
  antidepressants, 237–248
    α-adrenergic receptors, 247
    monoamine oxidase, 247–248
    monoamine transporters, 237–241
    serotonergic agents, 241–247
  antipsychotics, animal models, 170–172
  opioid-receptor-like 1/orphanin FQ/nociceptin peptide, 677–678
  opioid receptors, 579–597
    central nervous system effects, 579
    computational models, 590
    δ-opioid receptors, 591–592
    endogenous peptides, 593–596
    κ-opioid receptors, 592–593
    ligand characterization, 581–583
    multiple opioid receptors, 579–580

mutagenesis studies, 589–590
non-μ-opioid receptors, 591–593
signal transduction mechanisms, 580–581
structure and molecular biology, 587–589
subtypes, splice variants, and dimerization, 590–591
in vitro efficacy assays, 583–585
in vivo evaluation, 585–587
psychostimulants, 100–106
Pharmacophores, stroke therapy, PARP inhibitors, 485–487
Phase 3 clinical trials, stroke therapy, 447–449
Phencyclidine (PCP):
  cognitive function assessment, 26–27
  schizophrenia pathology, 163–164
  NMDA model, 171–172
Phendimetrazine, structure-activity relationship, 106–109
Phenelzine:
  chemical structure, 223
  Parkinson's disease, dopaminergic therapy, 549–551
  physiology and pharmacology, 248
  side effects and adverse reactions, 230
  structure-activity relationship and metabolism, 252–253
Phenmetrazine, structure-activity relationship, 106–109
Phenobarbital, anticonvulsant applications, 135–141
Phenothiazines:
  Alzheimer's disease, tau-targeted therapy, 424–425
  pharmacokinetics, biodistribution, and drug-drug interactions, 186–189
Phenserine, cognitive enhancement, 32

Phentermine, amphetamine alpha-alkyl substituent, 108
N-Phenylamines, Alzheimer's disease, tau-targeted therapy, 427
Phenyl esters, cocaine structure-activity relationship, 111–113
Phenylindole, pharmacokinetics, 199
Phenylpiperidine analgesics:
  flexible opiates, 598
  MPTP activation, 530
Phenylthiazolyl-hydrazines, Alzheimer's disease, tau-targeted therapy, 427–428
Phenytoin:
  anticonvulsant applications, 135–139, 141
  clozapine pharmacokinetics, 192
  sertindole interaction, 199
Phobic conditions, antidepressant applications, 229
Phosphinate, calpain inhibitor, stroke therapy, 465
Phosphodiesterase inhibitors:
  antidepressant efficacy assessment, 226–229
  antipsychotic agents, 208–209
  cognitive enhancement, 50–51
Phosphorus, stroke therapy, MMP inhibitors, 509–510
Phosphorylation, Alzheimer's disease, tau-targeted therapies, 418–422
Phthalocyanine, Alzheimer's disease, tau-targeted therapy, 425–426
Physostigmine, memory assessment and, 28–30
Pilocarpine, muscarinic activity, 67–68
Pimagedine, stroke therapy, 497
Pimozide, pharmacokinetics, 190
Pindolol:
  antidepressant interaction, efficacy assessment, 225–229
  radiolabeled ligands for, 743–744
PINK1 gene, Parkinson's disease genetics, 535

Pinoxepin, structure-activity relationship, 250–251
Pioglitazone, Alzheimer's disease, insulin related therapy, 415–417
Piperazines:
  muscarinic antagonists, 76–77
  opioid derivatives, κ-opioid receptor agonist, 627–632
  stroke therapy:
    cathepsin B inhibitors, 473–474
    piperidine/piperazine scaffolds, 457–458
Piperidines:
  nicotinic agonists, 68–69
  opioid derivatives, 613–621
    4-anilidopiperidines, 618–621
    4-arylpiperidines:
      carbon substituent, 614–616
      oxygen substituent, 616–618
  κ-opioid receptor agonist, 627–632
  stroke therapy, piperidine/piperazine scaffolds, 457–458
Piperidolate, muscarinic antagonists, 72–74
Piracetam:
  anticonvulsant applications, 149–151
  cognitive enhancement, 15–16, 34
Placebo response:
  antidepressant efficacy assessment, 221–229
  migraine therapy, calcitonin gene-related peptide antagonists vs., 286–288
Plasma drug profile, migraine therapies, 298–299
Platelets, serotonin-selective reuptake inhibitor effect on, 232–234
Poly(ADP-ribose) polymerase (PARP) inhibition, stroke therapy, 484–487
Polyamines, muscarinic antagonists, 76–77

Polymorphism associations, opioid drug interactions, 577
Polyphenols, Alzheimer's disease, tau-targeted therapy, 425–427
Porphyrins, Alzheimer's disease, tau-targeted therapy, 425–427
Positron-emission tomography (PET):
  drug discovery and development applications:
    disease markers, radiolabeling of, 744–746
    imaging studies, 740–741
    radiolabeled drugs, 742–743
    research background, 737
    small animal imaging, 742
    targeted drugs, radiolabeled ligands, 743–744
  isotopes in, 739–740
  Parkinson's disease diagnosis, 555–556
  physics of, 737–739
Post-synaptic density proteins (PSDs), long-term potentiation, 24–25
Potassium channels:
  anticonvulsants, 124–125
  stroke therapy, maxi-K channel, 482–483
Pozanicline, cognitive enhancement, 35–36
Pramipexole, Parkinson's disease dopaminergic therapy, 546–547
Pravastatin, Alzheimer's disease therapy, preclinical/clinical testing, 359–361
Preattentive processing, cognition and, 17–19
Preclinical testing:
  Alzheimer's immunization therapy, 369
  BACE (β-site cleavage enzyme), 345–351
  cognitive function, 25–29
  fibrillization inhibitors, 361–365
  HMG-CoA reductase inhibitors, 359–361
  α-secretase inhibitors, 357–358

γ-secretase inhibitors, 334, 340–341
Predictive validity, cognitive function assessment, 25–29
Prefrontal cortex:
  antipsychotics, dopamin receptor subtypes, 200–201
  memory and, 23–24
Pregabalin, anticonvulsant applications, 135–139, 149
Prepulse inhibition, antipsychotic therapy, 172
Preventive therapy, migraines, 312–320
  angiotensin converting enzyme inhibitors, 319–320
  anticonvulsants, 315–316
  antidepressants, 318–319
  β-blockers, 313–315
  calcium channel blockers, 316–318
Primidone, anticonvulsant applications, 135–141
Prinomastat inhibitor, stroke therapy, 503–505
Probenecid, olanzepine interaction, 193
Processing, cognition and, 19
Prochlorperazine:
  FDA approval, 176–177
  migraine therapy, 297–298
Procyclidine:
  muscarinic antagonists, 72–77
  Parkinson's disease therapy, 552–553
Prodine derivatives, opioid analgesics, 617–618
Prolactin response, antipsychotic side effects, 175–176
Proline analogs, opioid peptide analogs, Tyr-Pro-Phe sequence, 660–661
Proline dehydrogenase, schizophrenia genetics, 174
Propoxyphene:
  chemical structure, 570
  metabolism and elimination, 579
Propranolol, migraine prevention, 314–315
N-(n)-Propyldihydrexidine, functional selectivity, 8–9

INDEX

Protein kinase A (PKA):
Alzheimer's cognitive deficits, 16
antidepressant efficacy assessment, 227–229
Protriptyline, chemical structure, 222
Prucalopride, serotonin receptor targeting, 246–247
Psilocybin, cognitive enhancement, 37–40
Psychogenic response, stimulants, 92, 95–97
Psychostimulants:
ADMET properties, 97–99
alpha-alkyl substituent, 108
alpha-carbon stereochemistry, 108
amphetamines, 106–109
aromatic ring substitution, 108–109, 112–113
C(2) substituent, 111–112
C(3) ester linkage, 112
caffeine, 90
cocaine, 109–113
ephedra and khat, 89–90
historical background, 90–91
methylphenidate, 109
nitrogen substituents, 107, 110–111
N-substituents, 110
physiology and pharmacology, 100–106
recent and future development, 113–115
receptor classification and function, 102–106
research background, 89
side chain length, 107
side effects, adverse effects, drug interactions, 92–97
structure-activity relationships, 106–113
therapeutic applications, 91–92
tropane ring preservation, 113
Purine derivatives, stroke therapy, Src kinase inhibitors, 511
Pyrazole derivatives, stroke therapy, pyrazole-pyrimidine inhibitors, 511–512
Pyrazole-pyrimidine inhibitors, stroke therapy, 511–512

Pyrido[3,4-d]azepines, nicotinic agonist, 68–69
Pyridone-based calcitonin gene-related peptide antagonists, migraine therapy, 287–288
Pyrido-pyrimidine inhibitors, stroke therapy, 513–514
Pyridoxine, Parkinson's disease, dopaminergic pharmacotherapy, 541–542
Pyrimidine-2,4,6-triones, stroke therapy, 508
Pyrimidines, stroke therapy:
N-type voltage-dependent calcium channel scaffolds, 457
pyrazole-pyrimidine inhibitors, 511–512
pyrido-pyrimidine inhibitors, 513–514
pyrrole-pyrimidine inhibitors, 512
Pyrrole-pyrimidine inhibitors, stroke therapy, 512
Pyrrolidine-2,5-diones, anticonvulsant applications, 142, 149–151

QT prolongation, antipsychotic side effects, 167
Quercitin β-amyloid peptide inhibitors, 364–365
Quetiapine:
FDA approval, 178–179
pharmacokinetics, 193–194
side effects, 169
structure-activity relationship, 180–183
Quinaxoline-2,3-dione, stroke therapy, 452–453
Quinazolines, stroke therapy, Src kinase inhibitors, 513
Quinazolone, cognitive enhancement, 37
Quinicrine mustards, Alzheimer's disease, tau-targeted therapy, 424–425
Quinolines, stroke therapy:
caspase inhibitors, 471
Src kinase inhibitors, 513
Quinolinone, pharmacokinetics, 198

Quinolone, irreversible inhibitors, stroke therapy, 465
Quinuclidine:
muscarinic antagonists, 75–77
nicotinic agonists, 71

R1450 monoclonal antibody, Alzheimer's immunization therapy, 378
R-84760 κ-opioid receptor agonist, 627–632
Racemic antidepressants, current and future research, 255–256
Radiolabeled compounds:
disease biomarkers, 744–746
Parkinson's disease diagnosis, 555–556
in PET/SPECT systems, 742–743
targeted drug development, 743–744
Radioligand binding assays:
κ-opioid receptor agonists, 624–632
opioid receptors, 582–583
Rasagiline, Parkinson's disease, dopaminergic therapy, 549–551
Reboxetine:
chemical structure, 223
cognitive enhancement, 41–42
monoamine transporter inhibition, 239–241
pharmacokinetics, 237
side effects and adverse reactions, 235
Receptor-based drug design:
antidepressants, serotonergic agents, 241–247
central nervous system, 1–3
functional selectivity, 7–9
Parkinson's disease, dopaminergic pharmacotherapy, 542–549
aporphine-type receptor agonists, 544–545
dopamine receptor structure and function, 542–544
ergot-type receptor agonists, 545–546
experimental D_1 receptor agonists, 547–549
small-molecule agonists, 547

Receptors:
  calcitonin gene-related peptide antagonist, migraine therapy, 273
  psychostimulant physiology and pharmacology, 102–106
Recombinant prourokinase (rpro-UK), stroke therapy, 448
Regulator of G-protein signaling 4 (RGS-4), schizophrenia genetics, 174
Respiration, depression, opioid receptor effects on, 574
Retigabine, anticonvulsant applications, 136–139, 154
Reversible inhibitors, stroke therapy:
  calpains, 461–463
  caspase 3/9 inhibitors, 466–467
  cathepsin B inhibitors, 472–473
  non-cystein design, 477–478
  hydroxamic acids, 507
Reversible inhibitors of monoamine oxidase-A (RIMAS), structure-activity relationship and metabolism, 252–253
Rhodanines, Alzheimer's disease, tau-targeted therapy, 428–429
"Rich pharmacology," central nervous system drug design, 4–5
Rigid opiates, structure-activity relationship, 598
Rimonabant:
  antipsychotic agents, 209–210
  cognitive enhancement, 48–49
Ring systems:
  amphetamines, aromatic ring substitution, 108–109
  cocaine structure-activity relationship, 112–113
  morphine derivatives, 599–601
    benzomorphans, 610–611
  opioid receptors:
    δ-opioid receptor antagonist, 605–608
    κ-opioid receptor agonists, 624–632
  tricyclic antidepressants, 249–251

Risperidone:
  behavioral models, 172–173
  cognitive enhancement, 37–40
  FDA approval, 178
  pharmacokinetics, 195–196
  side effects, 169
  structure-activity relationship, 182–185
Ritanserin, serotonin receptor targeting, 241–242
Rivastigmine, cognitive enhancement, 15–16, 31
Ro-25-6981, stroke therapy, 453
Robalzotan, efficacy assessment, 225–229
Rolipram:
  antidepressant efficacy assessment, 226–229
  antipsychotic agents, 208–209
  memory assessment, 28–30
Roprinole, Parkinson's disease dopaminergic therapy, 546–547
Rosiglitazone, Alzheimer's disease, insulin related therapy, 415–417
Rotigotine, Parkinson's disease dopaminergic therapy, 545–546
Rufinamide, anticonvulsant applications, 136–139, 154–155
RVX-208 apolipoprotein target, Alzheimer's disease therapy, 413–414

S32006 serotonin receptor antagonist, 242–244
Safinamide, Parkinson's disease, dopaminergic therapy, 549–551
Salvia divinorum, current developments, 682–683
SAR102279 receptor antagonist, 257
Sarcosine, cognitive enhancement, 45
Saredutant substance P-NK1 receptor antagonist, 257
Sarizotan, Parkinson's disease therapy, 554–555

SB-271046:
  cognitive function assessment, 27–28
  serotonin receptor targeting, 246–247
SB-705498 vanilloid receptor antagonist, migraine therapy, 308–312
Scaffold structures, stroke therapy:
  gem-disubstituted and spirocyclic inhibitors, 506
  piperidine/piperazine scaffolds, 457–458
  pyrimidine-based, 457
Schaffer collateral-CA1 synapse, long-term potentiation, 24–25
Schizophrenia:
  acetylcholinesterase inhibitor therapy, 32
  antipsychotic therapy, 169–170
  animal models, 170–176
  behavioral models, 172–173
  drug discrimination model, 173
  genetic-based models, 173–174
  cognitive deficits, 162
  cognitive dysfunction in, 20–21
  dopamine therapy, cognitive enhancement, 39–41
  dopamin receptor subtypes, 200–201
  electroencephalography assessment, 174–175
  epidemiology, 161–162
  negative symptoms, 161–162
  nicotinic agonists, 71
  pathology, 162–165
  positive symptoms, 161
  receptor-based drug design, 3
Scopolamine, muscarinic antagonist, 72
Second-generation compounds:
  anticonvulsants, 135–139
  migraine therapy, triptans, 270
α-Secretase:
  amyloid precursor protein processing, 330–334
  chemical structure, 334
  modulators, preclinical/clinical testing, 357–358
β-Secretase. *See* BACE (β-site cleavage enzyme)

γ-Secretase:
   amyloid precursor protein processing, 330–334
   chemical structure, 333–334
   inhibitors (GSIs):
      clinical research, 341–345
      preclinical research, 334, 340–341
   modulators, preclinical/clinical testing, 358–359
   nonpeptidomimetic inhibitors, 337–340
   peptide isostere compounds, 334–337
      azepine/diazepine-derived inhibitors, 336–337
      transition-state analogs, 334–335
Secretases. See also BACE (β-site cleavage enzyme)
   amyloid precursor protein processing, 330–334
Sedation:
   antipsychotic side effects, 168
   opioid receptors, 575
Seizures:
   antipsychotic side effects, 168
   bupropion and, 234
Selegiline:
   Parkinson's disease, dopaminergic therapy, 549–551
   structure-activity relationship and metabolism, 251–253
Seletracetam, anticonvulsant applications, 150–151
Selfotel, stroke therapy, 451
Semagestat γ-secretase inhibitor, clinical testing, 341–345
Semantic memory, 19
Sensitization mechanisms, psychostimulant physiology and pharmacology, 105–106
Sercloremine, structure-activity relationship and metabolism, 252–253
Serdaxin antidepressant, 255–256
D-Serine, cognitive enhancement, 45
Serotonergic agents:
   current and future research, 255–256
   receptor-based drug design, 241–247
   structure-activity relationship and metabolism, 253–254
Serotonin-dopamine antagonists, FDA approved compounds, 177–179
Serotonin/norepinephrine-selective reuptake inhibitors (SNRIs):
   chemical structure, 223
   historical background, 249
   migraine prevention, 319
   monoamine transporter inhibition, 238–241
   side effects and adverse reactions, 223, 234
Serotonin receptors:
   5-HT_{1A} receptors:
      antidepressant targeting, 242–244
      cognitive enhancement, 37–38
      Parkinson's disease therapy, 553–555
   5-HT_{1B} receptors, 245–247, 253–254
   5-$HT_{1D/1B}$ receptors, 244–247, 253–254
   migraine therapy, 269–272
   5-HT_{2A} receptors, cognitive enhancement, 38–40
   5-HT_{2C} receptors, 243–244, 253–254
   5-HT_2 receptors, 241–242, 253–254
   5-HT_3 receptors, 246–247, 253–254
   5-HT_4 receptors, cognitive enhancement, 38–40
   5-HT_6 receptors, cognitive enhancement, 39–40
   5-HT_7 receptors, 247
      cognitive enhancement, 39–40
   agonists, migraine therapy, 268–272
      anti-inflammatory/vasoconstriction mechanisms, 270–272
      second-generation triptans, 270
      sumatriptan, 268–270
   antagonists, antidepressant design, 225–229
antidepressants:
   current and future research, 255–256
   efficacy assessment, 221–229
   neurotransmission enhancement, 220–221
   populations and subtypes, 241–247
antipsychotic structure-activity relationship, 180–185
   receptor subtypes, 201–202
cognitive enhancement, 37–40
migraine therapy:
   agonists, 268–272
      anti-inflammatory/vasoconstriction mechanisms, 270–272
      second-generation triptans, 270
      sumatriptan, 268–270
   NOS inhibitor binding affinity, 305–306
Parkinson's disease therapy, 553–555
psychostimulant physiology and pharmacology, 103–106
schizophrenia pathology, 164–165
Serotonin-selective reuptake inhibitors (SSRIs):
   chemical structure, 223
   dosing regimens, 219–221
   historical background, 249
   migraine prevention, 319
   monoamine transporter inhibition, 238–241
   pharmacokinetics, 236
   serotonin receptor targeting, 244–247
   side effects and adverse reactions, 223–224, 232–234
   structure-activity relationship and metabolism, 249–251
Serotonin transporter (SERT)
   gene, antidepressant applications, 229
   pharmacology, 238–241
Sertindole:
   cognitive enhancement, 37–40
   FDA approval, 179–180
   pharmacokinetics, 199
   structure-activity relationship, 182–185

Sertraline:
  chemical structure, 223
  historical background, 249
  pharmacokinetics, 236
  structure-activity relationship, 251
Set shifting, executive function assessment, 28–30
Sexual dysfunction:
  antidepressants, 233–234
  antipsychotic side effects, 169
Short-term memory, 18–19
Sibutramine, monoamine transporter inhibition, 239–241
Side chain modification:
  amphetamines, 108
  hydroxamate HDAC inhibitors, stroke therapy, 488
  psychostimulant structure-activity relationship, 107
Side effects:
  antidepressants, 229–234
    bupropion, 234
    monoamine oxidase inhibitors, 230
    nefazodone, 234–235
    reboxetine, 235
    serotonin/norepinephrine reuptake inhibitors, 234
    serotonin-selective reuptake inhibitors, 232–234
    trazodone, 234
    tricyclic antidepressants, 231–232
  antipsychotics, 165–169
    animal models, 175–176
    anticholinergic effects, 167–168
    antipsychotic discontinuation, 169
    cardiovascular effects, 167
    extrapyramidal symptoms, 165–166
    hematologic effects, 168–169
    hepatic effects, 169
    metabolic effects, 168
    neuroleptic malignant syndrome, 167
    sedation, 168
    seizures, 168
    sexual dysfunction, 169
  L-dopa therapy, Parkinson's disease, 542
  opioid receptors, 573–576
  psychostimulant therapy, 92, 95–97
Signal transduction, opioid receptors, 580–581
Silanediols, HDAC inhibitors, stroke therapy, 494
Simvastatin, Alzheimer's disease therapy, preclinical/clinical testing, 359–361
Single-photon emission computed tomography (SPECT):
  drug discovery and development applications:
    disease markers, radiolabeling of, 744–746
    imaging studies, 741–742
    radiolabeled drugs, 742–743
    research background, 737
    small animal imaging, 742
    targeted drugs, radiolabeled ligands, 743–744
    isotopes in, 739–740
  Parkinson's disease diagnosis, 555–556
Site-directed mutagenesis, opioid receptors, 589–590
SKF-81297, cognitive enhancement, 42
Sleep deprivation, cognitive enhancement, 51–52
SM-18400 tricyclic derivative, stroke therapy, 452–453
Small-molecule compounds:
  β-amyloid peptide inhibitors, 362–365
  cognitive enhancement, 30–51
    adenosine modulator, 46–47
    cannabinoids, 47–48
    cholinesterase inhibitors, 30–32
    dopamine, 39–40
    GABA$_A$ receptor, 45–46
    glutamate, 41–45
    histamine, 35–37
    historical background, 30–31
    muscarinics, 32–33
    neuronal nicotinic agonists, 34–35
    neuropeptides, 48–50
    neurotropic agents, 47
    NMDA modulators, 33–34
    norepinephrine, 41
    phosphodiesterase inhibitors, 50–51
    serotonin, 37–39
    smart drugs, 51–52
    summary and future research issues, 52–53
  Parkinson's disease dopamine receptor agonists, 547
  stroke therapy, N-type voltage-dependent calcium channels, 456–459
Smart drugs, cognitive enhancement, 51–52
Smoking, Parkinson's disease and, 538–540
SNC80 δ-opioid selective agonist, 632–635
Social Recognition and Novel Object Recognition, working memory assessment, 27
Sodium valproate, migraine prevention, 315–316
Sodium voltage-gated ion channel, anticonvulsants, 123–124
Sofinicline, cognitive enhancement, 35–36
Solifenacin, muscarinic antagonists, 74–77
Somatostatin, μ-opioid receptor antagonist derivation, 673–674
Spatial memory, Morris Water Maze assessment, 27–28
Spiradoline κ-opioid receptor agonist, 624–632
Spirocyclic scaffold-based inhibitors, stroke therapy, 506
Spirodioxolane analog, muscarinic activity, 66
7-Spiroindanyloxymorphone (SIOM), opioid antagonists, 607–608
Spirotramine, muscarinic antagonists, 76–77
Splice variants, opioid receptors, 590–591
Src kinase inhibitors, stroke therapy, 510–514
  2-aminothiazole template, 514
  indoline-2-one derivatives, 514
  natural derivatives, 511
  purine-based derivatives, 511

Src kinase inhibitors, stroke therapy (*Continued*)
  pyrazole-pyrimidine inhibitors, 511–512
  pyrido-pyrimidine and benzotriazine-derived inhibitors, 513
  pyrrole-pyrimidine/fused fivemembered heterocycle pyrimidine inhibitors, 512
  quinazoline-derived inhibitors, 513
  quinoline-derived inhibitors, 513
  substrate binding site inhibitors, 514
SSR-180711, cognitive enhancement, 35–36
Statin drugs, Alzheimer's disease, cholesterol synthesis, 406
Statines, BACE (β-site cleavage enzyme) inhibitors, 347–351
Stereochemistry, amphetamines, 108
Steroidal analogs, nicotinic antagonists, 79–80
Stimulants, central nervous system:
  ADMET properties, 97–99
  alpha-alkyl substituent, 108
  alpha-carbon stereochemistry, 108
  amphetamines, 106–109
  aromatic ring substitution, 108–109, 112–113
  C(2) substituent, 111–112
  C(3) ester linkage, 112
  caffeine, 90
  cocaine, 109–113
  ephedra and khat, 89–90
  historical background, 90–91
  methylphenidate, 109
  nitrogen substituents, 107, 110–111
  N-substituents, 110
  physiology and pharmacology, 100–106
  recent and future development, 113–115
  receptor classification and function, 102–106
  research background, 89
  side chain length, 107

  side effects, adverse effects, drug interactions, 92–97
  structure-activity relationships, 106–113
  therapeutic applications, 91–92
  tropane ring preservation, 113
Stiripentol, anticonvulsant applications, 136–139, 155
Streptokinase (SK), stroke therapy, 447–448
Stress, depression and, 229
Stroke:
  drug therapy:
    antioxidants, 479–482
    deferoxamine, 482
    ebselen, 481
    edavarone, 481
    NXY-059, 479–480
    tirlazad, 480
    current products, 447–449
    excitatory amino acid receptors, 450–478
    calpain inhibitors, 460–465
    caspase 3/9 inhibitors, 465–471
    cathepsin B inhibitors, 471–478
    NMDA antagonists, 450–455
    voltage-dependent calcium channels, 455–460
  future research and development, 515–516
  GABA$_A$ agonist, 483
  HDAC inhibitors, 487–495
    2-aminophenylamides, 492–493
    boronates and silanediols, 494
    carboxylic acids, 494
    hydroxamate inhibitors, 488–491
    ketones, 493
    thiol inhibitors, 492
  INOS/NNOS inhibitors, 495–502
    amidines, 497–498
    aminopyridines, 498–499
    L-arginine analogs, 495–497
    calmodulin site, 499–500
    coumarin derivatives, 499
    dimerization inhibitors, 501

    guanidine and isothiourea analogs, 497
    heme site, 501
    non-arginine sites, 499
    NOS inhibitor/NO donor compounds, 502
    tetrahydrobiopeterin site, 500
  maxi K-channel mechanism, 482–483
  MMP-9 inhibitors, 502–510
    carboxylic acid-derived inhibitors, 508
    hydroxamic acid-derived inhibitors, 502–507
    hydroxypyridinones, 508
    phosphorus-derived inhibitors, 509–510
    pyramidine-2,4,6-triones, 508–509
    thiols, 509
  ONO-2506, 483–484
  PARP inhibitors, 484–487
  SRC kinase inhibitors, 510–514
    2-aminothiazole template, 514
    indoline-2-one derivatives, 514
    natural derivatives, 511
    purine-based derivatives, 511
    pyrazole-pyrimidine inhibitors, 511–512
    pyrido-pyrimidine and benzotriazine-derived inhibitors, 513
    pyrrole-pyrimidine/fused fivemembered heterocycle pyrimidine inhibitors, 512
    quinazoline-derived inhibitors, 513
    quinoline-derived inhibitors, 513
    substrate binding site inhibitors, 514
  ischemic damage and intervention points, 447–449
  vascular dementia and, 21
Structural biology studies, opioid receptors, 587–591
Structure-activity relationship (SAR):
  antidepressants, 249–255

bupropion, 254–255
monoamine oxidase
  inhibitors, 251–253
serotinergic agents, 253–254
tricyclic antidepressants,
  249–251
antipsychotics, 180–185
atypical tricyclic
  neuroleptics, 180–183
benzisoxazole/benzithazole
  neuroleptics, 183–185
muscarinic antagonists, 72–77
nonpeptide opioid analgesics,
  597–645
acyclic analgesics, 621–623
affinity labeling, 637–643
  morphine-naltrexone
    derivatives, 638–641
agonists:
  δ-selective agonists,
    632–637
  κ-selective agonists,
    623–632
    centrally-acting
      agonists, 624–630
    peripherally-acting
      agonists, 630–632
antagonists, 601–608
  δ-opioid receptors, 604–608
  κ-opioid receptors,
    602–604
benzomorphans, 610–613
Diels-Alder adducts, 608–609
miscellaneous opioids,
  643–645
morphinans, 609–610
morphine derivatives,
  599–601
opium-derived morphine and
  alkaloids, 598–599
ORL1 receptor, endogenous
  ligand orphanin FQ/
  nociceptin, 681–682
piperidine derivatives,
  613–621
  4-anilidopiperidines,
    618–621
  4-arylpiperidines:
    carbon substituent,
      614–616
    oxygen substituent,
      616–618
opioid-receptor-like 1/orphanin
  FQ/nociceptin peptide,
  678–680
psychostimulants, 106–113
  amphetamine, 106–109

cocaine, 109–113
methylphenidate, 109
Strychnine, anticonvulsants,
  glycine receptors, 130
Suberoylanilide hydroxamic acid
  (SAHA), stroke
  therapy, 488
Substance abuse, opioid
  receptors, 571
  dependence-addiction liability,
    574–575
Substance P-NK1 receptor
  antagonist,
  antidepressant effects,
  256–257
Substrates:
  inhibitors, stroke therapy,
    binding site inhibitors,
    514
  psychostimulant physiology
    and pharmacology,
    101–106
Succinimides, anticonvulsant
  applications, 142
Succinylcholine, nicotinic
  antagonists, 78–80
Suicide, antidepressants and:
  serotonin-selective reuptake
    inhibitors, 234
  tricyclic antidepressants,
    231–232
Sulfonamides:
  anticonvulsant applications,
    151–152
  γ-secretase inhibitors, 337–340,
    343–345
Sulfoxide derivatives, muscarinic
  activity, 65
Sumatriptan, migraine therapy,
  269–272
SUPERFIT fentanyl enatiomer,
  nonpeptide affinity
  labeling, 642–643
Synaptic monoamines,
  psychostimulant
  physiology and
  pharmacology,
  100–106
Synaptic plasticity, cognitive
  function and, 25
Synaptic proteins,
  anticonvulsants,
  130–132
Synthetic drug development,
  opioids, 597
α-Synuclein protein, Parkinson's
  disease etiology, 535

T2000 barbiturate,
  anticonvulsant
  applications, 140–141
Tacrine, cognitive enhancement,
  15–16, 30
Talampanel, anticonvulsant
  applications, 136–139,
  144–147
Talnetant, cognitive
  enhancement, 48–49
Talsaclidine, muscarinic activity,
  67–68
TAN67 δ-opioid selective
  agonists, 635–637
Tapentadol, chemical structure,
  570–571
Tardive dyskinesia (TD),
  antipsychotic side
  effect, 165–166
  vacuous chewing movements
  model, 175
Tarenflurbil, γ-secretase
  modulators, 358–359
Targeted drug development:
  Alzheimer's disease:
    metabolic syndrome
      targeting, 405–417
    cholesterol-related
      approaches, 405–414
    insulin-related
      approaches,
      414–417
    research background, 405
    summary, 408–410
    tau pathology targeting,
      417–432
      aggregation, 423–429
      glycosylation, 422–423
      microtubule stabiliization,
        430
      phosphorylation, 418–422
      protein levels, 430–432
  anticonvulsants, molecular
    targeting, 121–134
    connexins, 134
    enzymes, 132–134
    G-protein-coupled receptors,
      134
    ligand-gated ion channels,
      126–130
    neurotransmitter
      transporters and
      synaptic proteins,
      130–132
    voltage-gated ion channels,
      122–126

Targeted drug development (*Continued*)
radiolabeled ligands for, 743–744
stroke therapy:
current research, 447–449
future research, 515–516
HDAC inhibitors:
hydroxamate inhibitors, 488–491
ketones, 493
thiol inhibitors, 492
INOS/NNOS inhibitors, 495–502
amidines, 497–498
aminopyridines, 498–499
L-arginine analogs, 495–497
calmodulin site, 499–500
coumarin derivatives, 499
dimerization inhibitors, 501
guanidine and isothiourea analogs, 497
heme site, 501
non-arginine sites, 499
NOS inhibitor/NO donor compounds, 502
tetrahydrobiopterin site, 500
maxi K-channel mechanism, 482–483
MMP-9 inhibitors, 502–510
carboxylic acid-derived inhibitors, 508
hydroxamic acid-derived inhibitors, 502–507
hydroxypyridinones, 508
phosphorus-derived inhibitors, 509–510
pyramidine-2,4,6-triones, 508–509
thiols, 509
ONO-2506, 483–484
PARP inhibitors, 484–487
SRC kinase inhibitors, 510–514
2-aminothiazole template, 514
indoline-2-one derivatives, 514
natural derivatives, 511
purine-based derivatives, 511
pyrazole-pyrimidine inhibitors, 511–512
pyrido-pyrimidine and benzotriazine-derived inhibitors, 513
pyrrole-pyrimidine/fused fivemembered heterocycle pyrimidine inhibitors, 512
quinazoline-derived inhibitors, 513
quinoline-derived inhibitors, 513
substrate binding site inhibitors, 514
Tau protein:
Alzheimer's disease:
cognitive deficits, 16
pathology, 417–432
aggregation, 423–429
glycosylation, 422–423
microtubule stabiliization, 430
phosphorylation, 418–422
protein levels, 430–432
targeted therapy for, 430–432
TC-5619, cognitive enhancement, 35–36
Tebanicline, nicotinic agonists, 69
Telcagepant calcitonin gene-related peptide antagonist, migraine therapy, 277, 279–288
Temazepam, anticonvulsant applications, 144–147
TENA κ-opioid selective antagonist, 602–604
Tetrahydrobiopterin, stroke therapy, NOS inhibition, 500
Tezampanel, migraine therapy, 295
TgAPP mice, Alzheimer's disease immunotherapy, 366–367
Thebaine:
Diels-Alder adducts, 608–609
opium source, 599
Theophylline, cognitive enhancement, 47–48
Thiadiazole, stroke therapy, cathepsin B inhibitors, 475
Thiadiazolidine-dione, Alzheimer's disease, tau-targeted therapies, 420–422
Thiamet G, Alzheimer's disease, tau-targeted therapy, 423
Thienobenzodiazepine, pharmacokinetics, 192–193
Thiocarbocyanines, Alzheimer's disease, tau-targeted therapy, 427
Thiols, stroke therapy:
HDAC inhibitors, 491
MMP inhibitors, 509
Thionine, Alzheimer's disease, tau-targeted therapy, 424–425
Thiophenoamidine-benzothiazoles, NOS inhibitors, migraine therapy, 303–306
Thiophenoamidine-indoles, migraine therapy, NOS inhibitor binding affinity, 305–306
Thioridazine, pharmacokinetics, 187–189
Thiothixene, pharmacokinetics, 188–189
Thioxanthenes, pharmacokinetics, 188–189
Third-generation compounds, anticonvulsants, 136–139
Tiagabine, anticonvulsant applications, 135–139, 152
Tianeptine, structure-activity relationship, 255
Tiotropium, muscarinic antagonist, 72
TIPP peptide, opioid peptide analogs, Tyr-Pro-Phe sequence, 661–664
Tirlazad, stroke therapy, 480
Tissue plasminogen activator (t-PA), stroke therapy, 447–448
Tolcapone, Parkinson's disease, dopaminergic therapy, 550–552
Tolerability assessment, opioid receptors, 574–575
Tolterodine, muscarinic antagonists, 74–77
Toluidine blue, Alzheimer's disease, tau-targeted therapy, 424–425

Tonabersat, migraine prevention, 316
Topiramate:
  anticonvulsants, 133–139, 151–152
  migraine prevention, 315–316
Torsades des pointes, antipsychotic side effects, 167, 179–180
Tramadol nonpeptide opiate, 643–645
Tramiprosate glycosaminoglycan mimetic, 362
Trance amine receptor, schizophrenia genetics, 174
Transcription factors, antidepressant efficacy assessment, 226–229
Transition-state isosteres:
  BACE (β-site cleavage enzyme) inhibitors, 347–351
  γ-secretase inhibitors, 334–335
Transmembrane AMPA receptor regulatory proteins (TARPs), long-term potentiation, 24–25
Transmembrane domains (TMDs), opioid receptors, 587–591
Transport proteins:
  anticonvulsants, 130–132
  antidepressant pharmacology, 237–248
  cocaine structure-activity relationship, 109, 111–113
  psychostimulant physiology and pharmacology, 100–106
  recent developmenta, 113–114
Tranylcypromine:
  chemical structure, 223
  Parkinson's disease, dopaminergic therapy, 549–551
  physiology and pharmacology, 248
  side effects and adverse reactions, 230
  structure-activity relationship and metabolism, 252–253
"Trapping blockers," stroke therapy, 450

Traxoprodil, stroke therapy, 453
Trazodone:
  chemical structure, 224
  pharmacokinetics, 236–237
  serotonin receptor targeting, 241–242
  side effects and adverse reactions, 234
Triazoles, γ-secretase modulators, 359
Tricyclic antidepressants (TCAs):
  chemical structure, 222
  dosing regimens, 219–221
  monoamine transporter inhibition, 238–241
  pharmacokinetics, 235–236
  side effects and adverse reactions, 231–233
  stroke therapy, 452–453
  structure-activity relationship and metabolism, 249–251
Tricyclic neuroleptics, structure-activity relationships, 180–183
Tricyclic pharmacophores, stroke therapy, PARP inhibitors, 485–487
Trifluoperazine, pharmacokinetics, 187–189
Trigeminovascular system (TGVS):
  migraine pathology, 266–267
  calcitonin gene-related peptide antagonists, 272–290
  vanilloid receptor antagonists, 307–312
  migraine therapy, nonsteroidal anti-inflammatory and acetaminophen compounds, 268
Trihexyphenidyl, Parkinson's disease therapy, 552–553
Trimethadione, anticonvulsant applications, 135–139, 142–143
Trimipramine:
  chemical structure, 222
  side effects and adverse reactions, 231–232
Tripitramine, muscarinic antagonists, 76–77

Triptans, migraine therapy, 269–273
Trocade hydroxymate inhibitors, stroke therapy, 503
Tropane ring, cocaine structure-activity relationship, 112–113
Trospium, muscarinic antagonists, 74–77
D-Tubocurarine, nicotinic receptor blockade, 77–80
Tyr-D-aa-Phe sequence, opioid peptide analogs, 664–673
  deltorphin analogs, 668–673
  dermorphin analogs and μ-selective peptides, 664–668
Tyrosine ureas, muscarinic antagonists, 76–77
Tyr-Pro-Phe sequence, opioid peptide analogs, 659–664
  β-casomorphin analogs and endomorphins, 659–661
  TIPP and related peptides, 661–664

UB-165 nicotinic agonist, 70
Ubiquitin-protasome system, Parkinson's disease genetics, 535
Uptake transporters, psychostimulant pharmacology, 114
Urea-based peptidomimetic inhibitors, stroke therapy, 462–463

Vacuous chewing movements, antipsychotic side effects, 175
Valproic acid:
  anticonvulsant applications, 135–139, 147–148
  migraine prevention, 315–316
  stroke therapy, 494
Valrocemide, anticonvulsant applications, 136–139
Vanilloid receptor antagonists, migraine therapy, 307–312
  capsazepine, civamide, and AMG-517, 307–308
  SB-705498, 308–312

Varenicline:
    cognitive enhancement, 35–36
    nicotinic agonists, 70
Vascular dementia, epidemiology, 21
Vascular endothelial growth factor (VEGF), stroke therapy, Src kinase inhibitors, 510–514
Vascular-only hypothesis, migraine pathology, 266–267
Vasoconstriction, migraine therapy, triptans, 270–272
Venlafaxine:
    chemical structure, 223
    migraine prevention, 318–319
    monoamine transporter inhibition, 238–241
    side effects and adverse reactions, 234
Verapamil:
    migraine prevention, 318
    stroke therapy, 459–460
Vesicular glutamate transporters, anticonvulsants, 132
Vesicular monoamine transporter (VMAT):
    antidepressant pharmacology, 238–241
    Parkinson's disease diagnosis, 556

Vigabatrin, anticonvulsant applications, 135–139, 149
Vilazodone SSRI, 256
Volinanserin, cognitive enhancement, 37–40
Voltage-gated ion channels:
    anticonvulsants, 122–126
    calcium channels, 124
    chloride channels, 126
    hyperpolarization-activated cyclic nucleotide cation channels, 125–126
    potassium channels, 124–125
    sodium channels, 123–124
    migraine pathology and, 266–267
    stroke therapy:
        calcium channels, 455–460
        potassium channels, 482–483

WAY-129 prodrug, stroke therapy, 451
WAY-100635 serotonin antagonist:
    efficacy assessment, 225–229
    memory assessment and, 28–30
WAY-252623 LXR agonist, Alzheimer's disease, cholesterol turnover, 411–413
Wisconsin card sorting test, executive function assessment, 28–30

Withdrawal effects, antipsychotics, 169
Working memory, 18–19
    Morris Water Maze assessment, 27–28
    Social Recognition and Novel Object Recognition assessment, 27

Xanomeline muscarinic receptor:
    activity of, 67–68
    cognitive enhancement, 32–33

ZD-9379 scaffold, stroke therapy, 453
Zimelidine, historical background, 249
Ziprasidone:
    FDA approval, 178–179
    pharmacokinetics, 196–198
    structure-activity relationship, 182–185
Zolmitriptan, migraine therapy, calcitonin gene-related peptide antagonists vs, 287–288
Zonisamide:
    anticonvulsant applications, 133–139, 151–152
    migraine prevention, 316
Zotepine, pharmacokinetics, 194
Zuclopenthixol antipsychotic, 201

CUMULATIVE INDEX

A4F separation technique, protein therapeutics, **3:**324
A-5021 ganciclovir congener, **7:**229–230
A-63930 small molecule compound, **8:**41–42
A-412997 small molecule compound, **8:**41–42
A-582941 small molecule compound, **8:**35–36
Abacavir:
  HIV reverse transcriptase inhibition, **7:**141–145
  research background, **7:**139–140
Abbreviated new drug application (ANDA):
  FDA guidelines:
    approval, **3:**92
    delays and exclusivities, **3:**86
    generic drug review and approval, **3:**87–91
    historical background, **3:**67
    labeling, **3:**91
    preapproval inspection, **3:**91–92
    withdrawal of approval, **3:**92
  patent infringement and, **3:**151–152
  submission requirements, **3:**85, 88
ABCB1, single nucleotide polymorphisms, **1:**189–190
Abciximab, antiplatelet activity, **4:**445–446

AbeAdo nucleotide transporter trypanocide, **7:**577–580
Ab initio calculations:
  oral drug melting point, **3:**30–31
  quantitative structure-activity relationship, **1:**23
Absolute novelty principle:
  patent protection, **3:**110–111
  U.S.C.§102 provision, **3:**134
Absorption:
  ADMET structural properties, **2:**52–57
    chemical/gastrointestinal stability, **2:**57
    permeability, **2:**54–57
    solubility, **2:**52–54
  androgens, **5:**158–159
    anabolic agents, **5:**186–188
  antiandrogen compounds, **5:**193
  apical sodium-dependent bile acid transporter enhancement, **2:**472–473
  chirality and, **1:**138–140
  glucocorticoid anti-inflammatory compounds, **5:**41, 43, 45–53
  corneal penetration, **5:**53
  enzymatic metabolism, **5:**53–54
  intestinal absorption, **5:**41, 43, 45
  intra-articular administration, **5:**51–53

  percutaneous absorption, **5:**46–51
  water-soluble esters, **5:**53
  membrane proteins, drug delivery targeting, **2:**443–444
  opioid receptors, **8:**577–578
  oral drug delivery:
    Biopharmaceutics Classification System, **3:**353–355
    blood absorption, **2:**748
    class III drugs, **3:**50–52
    class V drugs, barriers, **3:**52–54
    prodrug development, **3:**223–226
    CK-2130 prodrug case study, **3:**269–276
    quantitative structure-activity relationship, human intestinal absorption, **1:**65
    rate constant, **2:**751
    type 2 diabetes therapies, dipeptidyl peptidase 4 inhibitors, **4:**45–46
ABT compounds:
  ABT-737, brain tumor therapy, **6:**258
  nicotinic agonists:
    ABT-089, **8:**35–36
    ABT-418, **8:**27
  peptidomimetics, **1:**230–231
Abuse of discretion standard, infringement of patent proceedings, **3:**164

ABYFG C-glycoside derivative, cocrystal engineering, 3:211–212
Acarbose, type 2 diabetes therapy, 4:5–8
ACAT inhibitors, Alzheimer's disease, 8:406–407, 411
ACC-9358 analogs, soft drug design, 2:114–115
ACE2 carboxypeptidase, angiotensin-converting enzyme inhibitors, 4:292–293
Acetazolamide:
  anticonvulsants, 8:133–134, 151–152
  glaucoma therapy, 5:598–600
Acetic acid:
  side chains, thyroid hormone receptor β-selective agonists, structure-based design, 2:687–688
  topical antimicrobial compounds, 7:496
Acetominophen:
  drug metabolism, oxidation/reduction reactions, 1:419
  migraine therapy, 8:268
  toxicophore reactive metabolites, aniline metabolism, 2:306–307
Acetophenones, chemokine receptor-5 (CCR5) antagonists, 7:111–116
Acetylation:
  drug metabolism, 1:430–432
  structural genomics, aromatic amines and hydrazines, 1:585–586
Acetylcholine (ACh):
  analogs, 8:64
  cholinergic nervous system, 8:61–62
  chronic obstructive pulmonary disease, muscarinic acetylcholinergic receptor antagonists, 5:766–767
  cognitive dysfunction and, 8:20–21

Acetylcholinesterase inhibitors:
  cognitive enhancement, 8:30–32
  multitarget drug development: AChE-plus, Alzheimer's disease, 1:261, 263
  optimization strategies, 1:268–273
  schizophrenia therapy, 8:32
Acetyl coenzyme A:
  biotin metabolism, 5:688
  drug metabolism, 1:430–431
N-Acetylcysteine, glutathione conjugation, 1:434–435
Acid-base equilibrium, salt compounds, 3:384–388
  pH solubility profile, 3:385–388
  weak acids, 3:384–385
  weak bases, 3:384–385
  zwitterions, 3:385, 387
Acids:
  salt compounds, acid-base equilibrium, 3:384–385
  topical antimicrobial compounds, 7:496
Acitretin:
  adverse events, 5:400–401
  chemical structure, 5:395–396
  disease therapies, 5:396–400
    cancer, 5:400
    psoriasis, 5:396–399
    skin diseases, 5:399–400
  pharmacology/metabolism, 5:401
  retinoic nuclear receptors, 5:370–372
Acne vulgaris:
  adapalene therapy, 5:410–411
  all-trans-retinoic acid therapy, 5:382–383
  13-cis-retinoic acid therapy, 5:388
  vitamin A pharmacology, 5:649–651
Acquired immunodeficiency syndrome (AIDS):
  chemokine receptor-5 (CCR5) antagonists, HIV-1 entry blockade:
    anilide core, 7:124–131
    binding mode and mechanism, 7:133–134

piperazine/piperidine-based antagonists, 7:110–116
research background, 7:107–110
spiro diketopiperazine-based antagonists, 7:131–133
summary and future research issues, 7:134–135
tri-substituted pyrrolidine antagonists, 7:121–124
tropane-based antagonists, 7:116–121
cognitive impairment, epidemiology, 8:15, 21
epidemiology, 7:75–76
FDA reforms and, 3:68–69
HIV protease inhibitor therapies:
  amprenavir (agenerase), 7:22, 26–30
  atazanavir sulfate (Reyataz), 7:36–42
  darunavir (Prezista), 7:54–60
  design criteria, 7:2–4
  evolution of, 7:1
  first-generation inhibitors, 7:3–30
  fosamprenavir calcium (Lexiva/Telzir), 7:42–45
  indinavir (crixivan), 7:12–20
  lopinavir/ritonavir (Kaletra), 7:31–36
  nelfinavir mesylate (viracept), 7:20–22
  next-generation inhibitors, 7:60–64
  ritonavir (norvir), 7:10–12
  saquinavir mesylate (invirase), 7:3, 5–10
  second-generation inhibitors, 7:31–60
  tipranavir (Aptivus), 7:45–54
HIV reverse transcriptase inhibitors:
  nonnucleoside reverse transcriptase inhibitors:
    clinical potential, 7:155–156
    current compounds, 7:148–152

research background,
  7:139–140
resistance limitations,
  7:152–155
nucleoside reverse
  transcriptase
  inhibitors:
  clinical potential,
    7:155–156
  current compounds,
    7:141–145
  research background,
    7:139–140
nucleotide reverse
  transcriptase
  inhibitors:
  clinical potential,
    7:155–156
  current compounds,
    7:145–148
  research background,
    7:139–140
  research background,
    7:139–140
  target identification,
    7:140–141
9-*cis*-retinoic acid therapy,
  5:421–424
Acrenocorticotropic hormone
  (ACTH), adrenal
  steroidogenesis,
  5:20–22
Acridine, antitrypanosomal
  polyamine
  metabolism, 7:590
Acridinyl hydrazide, plasmepsin
  antimalarial agents,
  7:642–644
Acridone antimalarial agents,
  cellular respiration
  and electron
  transport, 7:658
Acridone-based inosine
  monophosphate
  dehydrogenase
  inhibitors,
  5:1046–1047
ACT-078573 orexin receptor
  antagonist, insomnia
  therapy, 5:720–721
Actin-myosin interactions,
  congestive heart
  failure therapy,
  mechanism-targeted
  studies, 4:493–494
Actinomycetes, genome mining,
  2:209–210

Actinomycins:
  anti-cancer activity, 2:235–239
  DNA-targeted compounds,
    dactinomycin, 6:3–8
Actin-protein complexes,
  structural
  comparisons,
  2:338–339
Activated soft drugs:
  defined, 2:80
  pharmacological activity,
    2:128–129
Active metabolites, soft drug
  design, 2:80, 127–128
Active pharmaceutical
  ingredients (APIs):
  age-related macular
    degeneration therapy,
    5:616
  commercial-scale operations,
    continuous
    microfluidic reactors,
    3:13–14
  crystal engineering, research
    background,
    3:187–190
  large-scale synthesis, 3:1
  polymorph crystallization,
    3:190–192
  case studies, 3:196–198
  salt compounds:
    basic properties, 3:381–382
    screening process, 3:388–396
    scale-up operations, 3:1–14
Active site-directed
  anticoagulants,
  4:394–398
Activity-based protein profiling
  (ABPP),
  chemogenomics, 2:581
"Activity cliffs," quantitative
  structure-activity
  relationship modeling,
  1:511
Activity outliers, quantitative
  structure-activity
  relationship, 1:43–46
Activity ranking, quantitative
  structure-activity
  relationship
  validation, 1:50–51
Acute coronary syndrome (ACS):
  antiplatelet agents, aspirin,
    4:420
  epidemiology, 4:409–410
  platelet function in,
    4:410–411

risk stratification and
  therapeutic
  guidelines, 4:411
Acute myelogenous leukemia
  (AML):
  bexarotene therapy, 5:443
  Flt3 inhibitor trerapy, 6:301
Acute promyelocytic leukemia:
  all-*trans*-retinoic acid therapy,
    5:376–377, 380–382
  Am80 therapy, 5:414–417
  NRX195183 analog therapy,
    5:420
  9-*cis*-retinoic acid therapy,
    5:424–425
Acute respiratory stress
  syndrome (ARDS),
  inhaled nitric oxide
  therapy, 5:276–278
Acyclic hydroxamate inhibitors,
  stroke therapy,
  8:503–504
Acyclic nucleoside analogs:
  human herpesvirus type 6
    (HHV-6) therapy,
    7:244
  third-generation compounds,
    7:228–229
Acyclic nucleoside phosphates
  (ANPs):
  anti-DNA virus agents,
    7:231–239
  adenine derivatives,
    7:235–236
  8-azapurine derivatives,
    7:237
  2,6-diaminopurine
    derivatives, 7:236
  guanine derivatives, 7:236
  new generation prodrugs,
    7:238–239
  purine N6-substituted amino
    derivatives, 7:237–238
  tissue-targeted prodrugs,
    7:239
  human herpesvirus type 6
    (HHV-6) therapy,
    7:244
Acyclic opioid analgesics,
  structure-activity
  relationship,
  8:621–623
Acyclovir:
  anti-DNA virus:
    antimetabolites, 7:228–229
    herpes simplex virus
      therapy, 7:242–243

Acyclovir (Continued)
varicella-zoster virus therapy, 7:243
selective toxicity, 2:554–557
2-Acylaminothiazole derivative, cyclin-dependent kinase inhibitors, 6:313–314
Acylated piperidines, stroke therapy, 8:454–455
Acylation:
cephalosporins, 7:305–306
drug metabolism, 1:430–432
Acylguanidines, BACE (β-site cleavage enzyme) inhibitors, 8:351–353
Acyl halides, combinatorial chemistry, 1:296
Acyloxymethyl ketones, stroke therapy, cathepsin B inhibitors, 8:475
Acyl sulfonamides:
EP$_3$ receptor antagonists, 5:855–858
EP$_4$ receptor antagonists, 5:880–884
stroke therapy, MMP inhibitors, 8:510
ADAM algorithm, in silico screening, protein flexibility, side-chain conformation, 2:867–868
Adamantyl derivatives, dipeptidyl peptidase 4 (DPP-4) inhibition, 4:51–54
Adapalene:
adverse effects, 5:412–413
chemical structure, 5:408–409
pharmacology/metabolism, 5:414
retinoic nuclear receptors, 5:370–372
therapeutic applications, 5:410–412
vitamin A pharmacology, 5:649
Adaprolol, soft drug design, 2:83–85
AD Assessment Scale (ADAS-COG), cognitive dysfunction assessment, 8:21–22
Adatanserin, serotonin receptor targeting, 8:244–247
Addiction liability, opioid receptors, 8:574–575

Addition order, bench-scale experiments, 3:6–7
Additives, high-throughput screening, assay optimization, 3:412–413
Additivity, quantitative structure-activity relationship, biological parameters, 1:31–32
Adefovir:
current development, 7:145–148
dipivoxil derivative:
acyclic nucleoside phosphate compound, 7:235–236
hepatitis B virus therapy, 7:246–247
research background, 7:139–140
Adenine derivatives:
acyclic nucleoside phosphates, 7:235–236
ligand-based virtual screening, pharmacophore formation, inhibitor compounds, 2:4–5
Adenoid cystic carcinoma, bexarotene therapy, 5:443
Adenosine diphosphate (ADP):
irreversible antagonists, 4:426–435
clopidogrel, 4:427–433
prasugrel, 4:433–435
ticlopidine, 4:426–427
reversible antagonists, 4:435–437
Adenosine monophosphate activated protein kinase (AMPK):
fructose-1,6-biphosphatase inhibitors, 4:18–20
metformin mechanism of action and, 4:4
Adenosine receptors:
A$_{2A}$ receptor:
chronic obstructive pulmonary disease therapy, 5:793–794
Parkinson's disease therapy, 8:538–540
allosteric protein model, 1:394
cognitive enhancement, 8:46–47
Parkinson's disease therapy:

and A$_{2A}$ receptor, 8:538–540
antagonists, 8:552–553
Adenosine triphosphate (ATP):
antitubercular agents, cofactor biosynthesis targeting, 7:728–729
cholesterol biosynthesis, 5:15–16
kinase inhibitors and, 6:295–296
type 2 diabetes therapy, sulfonylureas, 4:10
Adhesins, receptor-mediated endocytosis, 2:451–452
Adipogenesis, peroxisome proliferator-activated receptors:
γ-agonists, 4:100–102
selective PPARγ modulators, PPARγ partial agonists, 4:107–115
ADMET (absorption, distribution, metabolism, excretion, and toxicity):
absorption properties, 2:52–57, 65–66
chemical/gastrointestinal stability, 2:57
lipophilicity assays, 2:65
permeability, 2:55–57, 66
solubility, 2:52–55, 65–66
aminoglycosides, 7:419–423
antitubercular agents:
capreomycin, 7:764
diarylquinoline (TMC207), 7:785
ethambutol, 7:755
ethionamide, 7:772
fluoroquinolones, 7:775–776
isoniazid, 7:742–743
linezolid, 7:786
nitroimidazoles, 7:781–782
para-Aminosalicylic acid, 7:767
pyrazinamide, 7:759
rifabutin, 7:749–750
rifamycin antibacterials, 7:748
rifapentine, 7:750–751
SQ109, 7:783
thiacetazone, 7:773–774
assessment rules, 2:63–69
bioavailability rules (Veber's rules), 2:64
BioPrint® database, 1:487–489, 502

chirality, **1:**137–143
CP-690,550
    immunosuppressive compound, **5:**1057–1059
distribution properties, **2:**59–61
    organ barriers, **2:**61
    plasma protein binding, **2:**59–61
    tissue binding, **2:**61
excretion properties, **2:**61
FDA preclinical testing guidelines, **3:**71–72
fingolimod compound, **5:**1054–1055
G-protein coupled receptor, homology modeling, **2:**292–293
inosine monophosphate dehydrogenase inhibitors:
    mizoribine, **5:**1049
    mycophenolate mofetil, **5:**1047–1049
    mycophenolate sodium, **5:**1049
lead discovery applications:
    optimization phase, **2:**64
    research background, **2:**47
    in vitro data, **2:**47–51
macrolides, **7:**438–441
metabolic properties, **2:**57–59, 66–68
    enzymatic stability, **2:**67
    metabolite identification, **2:**67–68
    plasma stability, **2:**59
    stability, **2:**57–59, 66–67
multiobjective optimization, drug discovery:
    blood brain barrier, solubility, and ADMET CYP2D6, **2:**269–273
    research background, **2:**259–261
opioid receptors, **8:**577–579
permeability:
    basic principles, **3:**367
    classification, **3:**370–371
    distribution, **3:**375–376
    excretion, **3:**373–374
    fraction absorption, **3:**369–370
    gastrointestinal absorption, solubility *vs.* permeability, **3:**374–375
    metabolism, **3:**371–373
    organ accumulation and toxicity, **3:**377
    transport interactions and polymorphism, **3:**367–369
physicochemical property assays, **2:**65–66
pK_a assays, **2:**65
PK parameters, **2:**50
prodrug properties, **3:**220–226
    CK-2130 prodrug case study, **3:**269–276
    classical drug discovery paradime, **3:**226–227
    current drug discovery paradigm, **3:**226, 228–229
    design principles, **3:**230
    metabolic issues, **3:**231–233
profile properties and activity, **2:**50
psychostimulants, **8:**97–99
quantitative structure-activity relationship, **1:**62–68
    blood-brain barrier, **1:**65
    excretion, **1:**67
    human intestinal absorption, **1:**65
    metabolism, **1:**66–67
    oral bioavailability, **1:**66
    permeability, **1:**64–65
    plasma/serum protein binding, **1:**66
    solubility, **1:**62–64
    toxicity, **1:**67–68
    volume of distribution, **1:**66
receptor-based drug design, **2:**524–525
rule of five (Lipinski's rules), **2:**63–64
in silico assessment tools, **2:**64
sirolimus compounds, **5:**1034–1035
structural properties, **2:**51–63
    hydrogen bonding, **2:**51–52
    ionizability, **2:**52
    lipophilicity, **2:**51
tacrolimus compounds, **5:**1022–1023
tetracyclines, **7:**409–410
toxicity properties, **2:**61–63, 68–69
    arrhythmia assays, **2:**69
    cytotoxicity, **2:**62, 68
    drug-drug interactions, **2:**63, 68–69
    genotoxicity/mutagenicity, **2:**62, 68
    reactive metabolites, **2:**68
    teratogenicity, **2:**62, 68
type 2 diabetes therapies:
    α-glycoside inhibitors (AGIs), **4:**6–8
    dipeptidyl peptidase 4 inhibitors, **4:**45–47
    glinides, **4:**12–13
    metformin, **4:**3–5
    sulfonylureas, **4:**10
in vitro property assays, **2:**64–65
in vivo studies, **2:**50–51
Administration protocols, prodrug development, **3:**221–226
Adornments:
    combinatorial libraries, **1:**297
    small molecule libraries, **1:**299–300
ADP-ribose (ADR), poly(ADP-ribosyl)ation biochemistry, **6:**151–157
Adrenaline, chronic obstructive pulmonary disease bronchodilators, **5:**768–769
Adrenal steroids, steroidogenesis and biosynthesis, **5:**19–22
α-Adrenergic receptor:
    agonists, glaucoma therapy, **5:**597–600
    antidepressant pharmacology, **8:**247
β-Adrenergic receptor. *See also* β-blockers
    agonists, enantiomer activity, **1:**131
    antagonists, glaucoma therapy, **5:**596–600
    antidepressant efficacy assessment, **8:**226
$β_2$-Adrenergic receptor agonists:
    bronchodilators, chronic obstructive pulmonary disease, **5:**769–772, 777–779
    soft drug design, **2:**103–104
$β_3$-Adrenergic receptor, antidepressant targeting, **8:**247

β-Adrenergic receptor/signaling pathway, congestive heart failure therapy, 4:507
Adrenergic receptors:
 antidepressant targeting, 8:247
 multitarget drug optimization, 1:268–273
 single nucleotide polymorphisms:
  B_1-adrenergic receptor, 1:192
  B_2-adrenergic receptor, 1:192–193
Adrenocorticoids:
 activity enhancement factors, 5:69–70, 73
 steroidogenesis, 5:19–22
Adrenocorticotropic hormone (ACTH), cognitive enhancement, 8:49–50
Adrenomedullin, biological actions and receptor pharmacology, 4:540
Adriamycin, Alzheimer's disease, tau-targeted therapy, 8:426–427
Adrogolide, cognitive enhancement, 8:41–42
Adverse drug reactions (ADRs). *See also* Side effects
 acitretin, 5:400–401
 adapalene, 5:412–413
 AHPC retinoid-related compound, 5:498–499
 AHPN retinoid-related compound, 5:498–499
 all *trans*-retinoic acid, 5:384–386
 Am80 therapy, 5:419–420
 angiotensin-converting enzyme inhibitors, 4:281–282
 antidepressants, 8:229–234
  bupropion, 8:234
  monoamine oxidase inhibitors, 8:230
  nefazodone, 8:234–235
  reboxetine, 8:235
  serotonin/norepinephrine reuptake inhibitors, 8:234
  serotonin-selective reuptake inhibitors, 8:232–234
  trazodone, 8:234
  tricyclic antidepressants, 8:231–232

anti-inflammatory corticosteroids, 5:41, 43–44
antipsychotics, 8:165–169
 animal models, 8:175–176
 anticholinergic effects, 8:167–168
 antipsychotic discontinuation, 8:169
 cardiovascular effects, 8:167
 extrapyramidal symptoms, 8:165–166
 hematologic effects, 8:168–169
 hepatic effects, 8:169
 metabolic effects, 8:168
 neuroleptic malignant syndrome, 8:167
 sedation, 8:168
 seizures, 8:168
 sexual dysfunction, 8:169
antitubercular agents:
 capreomycin, 7:764
 D-cycloserine, 7:771
 diarylquinoline (TMC207), 7:785
 ethambutol, 7:755
 ethionamide, 7:772–773
 fluoroquinolones, 7:776
 isoniazid, 7:743
 linezolid, 7:786
 nitroimidazoles, 7:782
 para-Aminosalicylic acid, 7:767
 pyrazinamide, 7:759
 rifabutin, 7:750
 rifamycin antibacterials, 7:748–749
 streptomycin, 7:761
 thiacetazone, 7:774
bexarotene therapy, 5:450–453
BioPrint® database, 1:488489
 safety, selectivity, and promiscuity profiles, 1:499–501
3-Cl-AHPC retinoid, 5:499
etretinate, 5:401–402
female sex hormones, 5:219–221
N-(4-hydroxyphenyl) retinamide (4-HPR), 5:476–478
linezolid, 7:544
9cUAB30 RXR-selective retinoid therapy, 5:457

NRX 194204 RXR selective analog, 5:463–464
NRX195183 analogs, 5:421
psychostimulant therapy, 8:92, 95–97
quinolones/fluoroquinolones, 7:539–540
9-*cis*-retinoic acid therapy, 5:426–427
13-*cis*-retinoic acid therapy, 5:393–395
SHetA2 retinoid-related compound, 5:505–506
sirolimus, 5:1036–1037
sulfonamides and sulfones, 7:521–522
summary of, 1:441–443
tazarotene, 5:406–408
Advicor, LDL cholesterol lowering mechanisms, 4:317–318
ADX-63365, cognitive enhancement, 8:42–43
AEB071. *See* Sotrastaurin
AF267B muscarinic receptor, cognitive enhancement, 8:32–33
Affinity capillary electrophoresis-mass spectrometry, drug screening, 1:113–114
Affinity chromatography:
 chemogenomics, 2:581
 protein therapeutics:
  analytic development, 3:320
  downstream processing, 3:304
  protein A, 3:308–309
Affinity chromatography-mass spectrometry, screening applications, 1:111–112
Affinity labeling:
 opioid receptors:
  nonpeptide analgesics, 8:637–643
  peptide derivatives, 8:675–676
 triarylethylenes, 5:252–253
Agenerase. *See* Amprenavir
Age-related macular degeneration (AMD), 5:604–623
 clinical trials, agents in, 5:614–623

current therapeutics,
    5:613–614
disease pathology, 5:604–610
epidemiology, 5:604–610
future research issues,
    5:622–623
in vivo preclinical screening
    models, 5:610–613
wet *vs*. dry classification,
    5:604–605
in vivo screening models,
    5:610–613
Aggregation:
  Alzheimer's disease, tau-
    targeted therapies,
    8:423–429
  protein therapeutics
    formulation and
    delivery, 3:313–314
Aggrenox, antiplatelet activity,
    4:450
Agitation functions, pilot plant
    development, 3:8–9
Agmatine, bupropion, 8:255
AGN193109 13-*cis*-retinoic acid
    therapy, chemical
    structure, 5:388
Agonist compounds. *See also*
    specific Agonists
  cholinergic agonists, 8:63–72
    acetylcholine analogs, 8:64
    muscarinic agonists,
      structure and
      classification, 8:64–68
    nicotinic agonists, 8:68–72
  chronic obstructive pulmonary
    disease therapy:
    adenosine A_{2a} receptor
      agonists, 5:793–794
    β_2-adrenergic receptor
      agonists, 5:769–772
    cognitive enhancement,
      8:34–36
  estrogen receptors, potency and
    efficacy, 5:237
  farnesoid X receptor:
    medicinal chemistry, 4:161
    tool molecules, 4:157–158
  glaucoma therapy:
    α-adrenergic receptor
      agonists, 5:597–600
    cholinergic muscarinic
      agonists, 5:599–600
  insomnia therapy:
    melatonin receptor agonists,
      5:719
    serotonin receptors, 5:718

liver X receptors:
  anti-inflammatory
    properties, 4:147–148
  clinical trials, 4:154–156
  first-generation agonists,
    4:147
  glucose homeostasis, 4:148
  lipid metabolism, 4:148–150
  medicinal chemistry,
    4:151–154
  reverse cholesterol transport,
    4:147
neoclerodane diterpene,
    8:682–683
nuclear hormone receptors,
    structural biology,
    4:82–83
opioid receptors:
  chemical structure, 8:571,
    573
  δ-opioid selective agonists,
    8:632–637
  κ-opioid receptors:
    physiology, 8:592–593
    selective agonists,
      8:623–632
  multiple receptor definition,
    8:579–580
  μ-opioid receptors, 8:593
Parkinson's disease therapy:
  aporphine-type dopamine
    receptor agonists,
    8:544–545
  D_1 receptor agonists,
    8:547–549
  ergot-type receptor agonists,
    8:545–546
  serotonin 5-HT$_{1A}$ receptor
    agonists, 8:553–555
  small-molecule dopamine
    receptor agonists,
    8:547
partial agonists:
  nuclear hormone receptors,
    structural biology,
    4:82–83
  occupancy theory,
    2:496–497
  opioids, 8:601–608
  peroxisome proliferator-
    activated receptors:
    selective PPARα
      modulators/PPARα
      partial agonists, 4:97
    selective PPARδ
      modulators/PPARδ
      partial agonists, 4:122

selective PPARγ
  modulators PPARγ
  partial agonists,
  4:106–115
peroxisome proliferator-
  activated receptors:
  α-agonists, 4:88–99
    clinical data, 4:96–99
    medicinal chemistry,
      4:90–96
    preclinical biology, 4:88–90
  α/δ-dual agonists, 4:136–137
  α/γ-dual agonists, 4:124–136
  α/γ?δ-pan agonists,
    4:140–145
  δ-agonists, 4:117–124
    biology, 4:117–118
    clinical data, 4:121–122
    medicinal chemistry,
      4:118–121
  γ-agonists, 4:100–106
    biology, 4:100–102
    clinical data, 4:104–106
    medicinal chemistry,
      4:102–104
  γ/δ-dual agonists, 4:137–140
platelet activation, 4:413–414
prostaglandin/thromboxane
  receptors:
  CRTH2 receptor, 5:907–909
  DP$_1$ receptor, 5:900–901
  EP$_1$ receptor, 5:812–813
  EP$_2$ receptor, 5:846–850
    mixed EP$_2$/E$_4$ agonists,
      5:850–852
  EP$_3$ receptor, 5:854–855
  EP$_4$ receptor, 5:872–883
    lactam stereochemistry,
      5:873–878
    thienopyrimidine-based
      compounds, 5:878–880
    ω-chain derivatives,
      5:872–876
  FP receptor, 5:897–899
  IP receptor, 5:922–939
    cyclohexene isosteres,
      5:933, 935
    FR181157 compound,
      5:932, 934–935
    FR193262 compound,
      5:932, 936
    heterocyclic isosteres,
      5:926–930
    nonprostanoid agonists,
      pyrazole derivatives,
      5:922–924
    oxazole moiety, 5:924–927

Agonist compounds (*Continued*)
  oxime derivatives, **5:**932, 937
  propyl analog, **5:**930, 932
  pyrazole derivative, **5:**930, 932
  pyridazinone derivative, **5:**932–933
  tetrahydronaphthyl derivative, **5:**930–931
  thromboxane synthase, **5:**935, 938–939
  TRA-418, **5:**939
  thromboxane A$_2$ receptor, **5:**886–887
protein-protein interactions, **2:**337–340
RAR/RXR panagonists, **5:**421–429
  adverse effects, **5:**426–427
  pharmacology/metabolism, **5:**428–429
  therapeutic applications, **5:**421–426
receptor-based drug design, inverse agonist, **2:**501–503
renin-angiotensin system, **4:**520
selective toxicity, **2:**541–542
serotonin receptor targeting, **8:**242–244
  migraine therapy, **8:**268–272
sphingosine 1-phosphate (S1P), multiple sclerosis therapy, **5:**574–578
stroke therapy, GABA$_A$ agonist, **8:**483
thrombopoietin receptors:
  eltrombopag, **4:**584–586
  romiplostim, **4:**583–584
thyroid hormone receptor agonists:
  structural properties, **4:**201
  therapeutic potential, **4:**197–198
Agranulocytosis, clozapine side effect, **8:**178
AH-13205 EP$_2$ receptor agonist, **5:**846–848
AHPC retinoid-related compound:
  adverse effects, **5:**498–499
  anticancer activities, **5:**497–498
  apoptosis, **5:**489–497
  chemical structure, **5:**483

development and discovery, **5:**483–497
discovery and development:
  adverse effects, **5:**499
  analog design, **5:**499–500
  anticancer activities, **5:**497–498
  DNA damage, **5:**492–493
  lysosomal membrane permeability, **5:**492
  retinoic nuclear receptors, **5:**372
AHPN retinoid-related compound:
  adverse effects, **5:**498–499
  anticancer activities, **5:**497–498
  development and discovery, **5:**483–497
  drug development:
    adverse effects, **5:**498–499
    analog design, **5:**499–500
    anticancer activities, **5:**497–498
  targeted drug development, **5:**483–497
    apoptosis, **5:**489–497
    gene expression, protein expression, and protein activity, **5:**494–497
    small heterodimer partner, **5:**487–489
  target identification, **5:**486–497
    apoptosis, AHPN/AHPC comparisons, **5:**489–497
    endoplasmic reticulum stress-associated apoptosis, **5:**492
    gene expression, protein expression, and protein activity modulation, **5:**494–497
    small heterodimer partner, **5:**487–489
Akathisia, antipsychotic side effect, **8:**165–166
Aklyl sulfonamide derivatives, prostaglandin/thromboxane EP$_1$ antagonists, **5:**819–820
Akt signaling pathway:
  brain tumor therapy, **6:**249–250

survival signaling inhibitors, **6:**323–324
Alanine-scanning mutagenesis, recombinant DNA technology, epitope mapping, **1:**544
Ala-Val-Pro-Ile, peptidomimetics, **1:**235–237
Albendazole, Alzheimer's disease, **8:**431–432
Alcohol dehydrogenase (ADH), drug metabolism, functionalization reactions, **1:**412
Alcoholism:
  phosphodiesterase inhibitor therapy, **5:**716
  thiamine deficiency, **5:**670
Alcohols, antimicrobial agents, **7:**492–493
Alcohol series (ROH), quantitative structure-activity relationships, multilinear regression analysis, **1:**33–37
Aldehyde dehydrogenases (ALDHs), drug metabolism, functionalization reactions, **1:**412
Aldehydes:
  combinatorial chemistry, **1:**296
  hemoglobin oxygen delivery, allosteric effectors, **4:**617–619
  stroke therapy:
    calpain inhibitors, **8:**461–464
    caspase 3/9 inhibitors, **8:**466–467
    cathepsin B inhibitors, **8:**472–473
  topical antimicrobial compounds, **7:**495
Aldosterone, steroidogenesis, **5:**20–22
Aleglitazar, α/γ-dual agonists, **4:**135–136
Alemtuzumab:
  multiple sclerosis therapy, **5:**572
  resistance to, **6:**369
Alendronate, osteoporosis antiresorptive therapy, **5:**733–734

Aliphatic oxidation, toxicophore reactive metabolites, aniline nitrogen oxidation, metabolism via, **2:**307–311
Aliskiren:
congestive heart failure therapy, ongoing Phases I, II, and III trials, **4:**496–499
renin inhibitor:
animal studies, **4:**250–251
clinical trials, **4:**251–252
congestive heart failure, **4:**252–253
discovery, **4:**250
drug-drug interactions, **4:**253
end organ protection, **4:**252
pharmacokinetics, **4:**253
research background, **4:**239
Alkaloids:
analog drug design, **1:**168–170
cinchona alkaloids, quinine antimalarials, **7:**605–607
dual topoisomerase I/II inhibitors, **6:**103–105
ergot alkaloids, migraine therapy, **8:**267–268
lead drug discoveries, natural product sources, **2:**527–528
morphine and derivatives, **8:**599–601
opium, **8:**598–599
vinca alkaloids, **6:**35–40
chemical structure, **6:**35–37
vinblastine, **6:**37–39
vincristine, **6:**39
vinorelbine, **6:**39–40
Alkanes, quantitative structure-activity relationship, hydrophobic interactions, **1:**10–19
Alkenyldiarylmethanes (ADAMs) reverse transcriptase inhibitors, **7:**152–155
Alkyl amines, renin inhibitors, **4:**254–255
Alkylamino derivatives, streptomycin, RNA targets, **5:**971
4-Alkylaminophenols, *N*-(4-hydroxyphenyl) retinamide (4-HPR) analogs, **5:**482–483
Alkylating agents:

anabolic analogs, **5:**181–182
DNA-targeted chemotherapeutic compounds, **6:**83–95
cyclopropylindoles, **6:**91–93
mustards, **6:**83–89
nitrosoureas, **6:**93–94
platinum complexes, **6:**89–91
triazenes, **6:**94–95
soft drug design, **2:**129
Alkylation reactions, acyclic nucleoside phosphates, **7:**233
Alkyloxime derivatives, eye-targeted chemical delivery systems, **2:**158–160
Alkylphenols, antimicrobial agents, **7:**493–494
N-Alkyl-tetrahydropyridines, toxicophore reactive metabolites, **2:**322–323
Alkynes, toxicophore reactive metabolites, **2:**320–321
Allergic rhinitis, soft drug design:
flucortin butyl, **2:**97
loteprednol etabonate and analogs, **2:**96
Allergies, multitarget drug development, histamine H_1-plus, **1:**259–262
Allosteric proteins:
central nervous system drug discovery, **8:**5–6
cognitive enhancement, **8:**32–33
computational approaches, **1:**390
drug development and, **1:**371
future trends, **1:**394–395
target identification, **1:**390–395
effector therapeutic potential, **1:**385–386
glycogen phosphorylase, **1:**394
G-protein coupled receptors, **1:**392
hemoglobin, **1:**377–379
transport proteins, **1:**391–392
hemoglobin oxygen delivery:
basic principles, **4:**610–612
right shifter therapeutics, **4:**615–619

HIV proteins, **1:**393–394
insomnia therapeutic agents, **5:**717–718
intrinsically disordered proteins and conformational changes, **1:**386–387
ion channels and neuroreceptors, **1:**393
kinases, **1:**392
Koshland-Nemethy-Filmer/ Dalziel-Engle model, **1:**375–377
ligand features, **1:**379–380
molecular dynamics, **1:**390
molecular mechanisms, **1:**381–385
Monod-Wyman Changeaux model, **1:**375–376
muscarinic receptors and, **8:**68
phosphatases, **1:**392
proteases, **1:**393
protein-protein interactions, **1:**392–393
receptor-based drug design:
constitutive receptor activity, **2:**498–499
modulators, **2:**499–501
ROSETTALIGAND algorithm, **1:**390
site discovery protocols, **1:**387–390
high-throughput screening, **1:**387–388
multiple solvent crystal structure, **1:**388–390
NMR localization, **1:**390
theoretical background and modeling, **1:**371–387
transition structure, **1:**380–381
transport proteins, hemoglobin, **1:**391–392
All-*trans*-retinoic acid (ATRA):
adverse effects, **5:**384–386
basic properties, **5:**376–377, 380–384
breast cancer, retinol level decrease, **5:**476–477
chemistry, **5:**644
disease therapies, **5:**376–377, 380–384
cutaneous diseases, **5:**382–384
hematologic malignancies, **5:**376–377, 380–382
pharmacology/metabolism, **5:**386–387

All-*trans*-retinoic acid (ATRA)
(*Continued*)
retinoic nuclear receptors,
5:370–373
systemic treatments,
5:384–385
topical treatments, **5**:385
N-(All-*trans*-retinoyl) L-proline,
N-(4-hydroxyphenyl)
retinamide (4-HPR)
analogs, **5**:483
Allyl amino carbamate, tri-
substituted
pyrrolidine chemokine
receptor-5 (CCR5)
antagonist, **7**:122–124
Almorexant, insomnia therapy,
5:720–721
Alogliptin, type 2 diabetes
therapy:
ADMET properties, **4**:45–47
physiology and pharmacology,
4:43–45
Aloisine, Alzheimer's disease,
tau-targeted
therapies, **8**:420–422
Alosentron, soft analog
development, **2**:118
Alovudine:
anti-DNA virus
antimetabolites,
7:226–227
current development, **7**:142–145
Alpha-alkyl substituent,
amphetamines, **8**:108
Alpha-carbon atom
stereochemistry,
amphetamines, **8**:108
α-glycoside inhibitors (AGIs),
type 2 diabetes, **4**:5–8
Alpha oxo/thio/amino-methyl
ketones, stroke
therapy, **8**:469–470
Alprostadil soft drug design,
virtual soft analog
library, **2**:134–135
Altered peptide ligands (APLs),
multiple sclerosis,
antigen-specific
therapy, **5**:579–580
Alternative dispute resolution,
infringement of patent
proceedings,
3:164–165
Alvameline muscarinic receptor,
cognitive
enhancement, **8**:32–33

Alvocidib cyclin-dependent
kinase inhibitor,
medicinal chemistry
and classification,
6:310–315
Alzheimer's disease:
acetylcholinesterase inhibitor
therapy, **8**:31–32
acitretin therapy, **5**:399–400
amyloid therapies, **8**:334–366
advanced glycosylation end
products inhibitors,
8:365
beta-amyloid reduction,
8:365–366
fibrillization inhibitors,
8:361–365
HMG-CoA reductase
inhibitors, **8**:359–361
research background,
8:329–330
α secretase modulators,
8:357–358
β secretase inhibitors:
clinical, **8**:356–357
nonpeptidomimetic
compounds, **8**:347–356
preclinical, **8**:345–347
γ secretase inhibitors,
8:358–359
clinical, **8**:341–345
preclinical, **8**:334–341
secretases, amyloid
precursor protein
processing, **8**:330–334
beta-secretase inhibitors,
1:327–328
cognitive dysfunction in, **8**:20
cognitive enhancement drugs,
8:15–16
cognitive impairment,
epidemiology, **8**:15
immunotherapy approaches,
8:366–380
advantages/disadvantages,
8:369–370
anti-Aβ antibody
mechanisms,
8:367–368
bapineuzamab, **8**:370, 378
Aβ immunization, **8**:366,
368–369
current clinical trials,
8:370–380
gammaguard liquid, **8**:379
MABT5102A, **8**:379–380
PF-4360365, **8**:378–379

preclinical/clinical trials,
8:369
R1450, **8**:379
solaneuzmab, **8**:378
TgAPP behavioral and
learning deficits,
8:366–367
multitarget drug development,
AChE-plus, **1**:261, 263
muscarinic receptor activity,
8:66–68
non-amyloid therapies:
metabolic syndrome
targeting, **8**:405–417
cholesterol-related
approaches,
8:405–414
insulin-related
approaches, **8**:414–417
research background, **8**:405
tau pathology targeting,
8:417–432
aggregation, **8**:423–429
glycosylation, **8**:422–423
microtubule stabiliization,
8:430
phosphorylation,
8:418–422
protein levels, **8**:430–432
single gene pharmacogenetics,
1:193–195
Am80 retinoid:
chemical structure, **5**:414–415
HL-60 myeloid leukemia cell
differentiation, **5**:416
pharmacology/metabolism,
5:419–420
retinoic nuclear receptors,
5:370–372
therapeutic applications,
5:414–420
Amantadine:
anti-HCV effect, **7**:174
Parkinson's disease therapy,
8:554–555
selective toxicity, **2**:556–557
stroke therapy, **8**:450
Amber codons, recombinant DNA
technology, **1**:546
AMBER scoring function,
semiempirical
techniques, **2**:623
Ambler β-lactamase classification
system, **7**:275–279
AMD3100 bicyclam, structure
and properties,
4:586–587

Amdoxovir, current development, 7:143–145
American trypanosomiasis:
  current therapies, 7:573
  epidemiology, 7:565–566
  pathology, 7:571–573
AMG 517 antagonist:
  cocrystal engineering, 3:210–211
  migraine therapy, 8:307–308
Amides:
  chemokine receptor-5 (CCR5) antagonist development, 7:112–116
  geldanamycin Hsp90 inhibitor derivatives, 6:389–391
  novobiocin Hsp90 inhibitor, side chain optimization, 6:434–436
  plasmepsin antimalarial agents, 7:641–644
  poly(ADP-ribose)polymerase inhibitors, 6:163–167
  prostaglandin/thromboxane EP_1 antagonists, 5:818–820, 836–837
  tropane-based chemokine receptor-5 (CCR5) antagonist, 7:117–121
Amidines:
  antitrypanosomal agents, 7:578–580
  stroke therapy:
    iNOS/nNOS analogs, 8:497–498
    NR2B antagonist compounds, 8:455
  topical antimicrobial agents, 7:498
Amidoerythromycins, recent developments, 7:446–456
Amidothiazoles, cyclin dependent kinase/p25 inhibitor, 1:329–330
Amikacin:
  antitubercular agents, 7:761–762
  chemical structure, 7:416–419
  history and biosynthesis, 7:423–424
  pharmacology and ADMET properties, 7:422–423
Amine-based platinum complexes, chirality and toxicity, 1:144–145
Amines:
  drug metabolism, acetylation/acylation, 1:431–432
  methylation reactions, drug metabolism, 1:423–425
Amino acids:
  antimalarial drugs, heme metabolism and salvage, 7:634–644
  antitubercular agents, targeted biosynthesis, 7:726–727
  cyclosporin A structure, 5:1007–1008
  drug metabolism, conjugation reactions, 1:432
  HIV protease processing sites, 7:2
  migraine therapy, CGRP antagonists, 8:288–290
  peptide conformation, local constraints, 1:212–214
  prodrug moieties, 3:251–252
  psychostimulant physiology and pharmacology, 8:105–106
  solid-phase organic synthesis, peptide arrays, 1:280–291
  tri-substituted pyrrolidine chemokine receptor-5 (CCR5) antagonist, 7:123–124
α-Amino acid derivatives, dipeptidyl peptidase 4 (DPP-4) inhibitors, 4:48–56
  hybrid α/β-amino acid analogs, 4:59
  product-like reversible inhibitors, 4:54–56
  substrate-like irreversible inhibitors, 4:48–49
  substrate-like reversible inhibitors, 4:49–50
  transition state-like covalent, reversible inhibitors, 4:50–54
β-Amino acid derivatives:
  dipeptidyl peptidase 4 (DPP-4) inhibitors, 4:57–59
  hybrid α/β-amino acid analogs, 4:59
Aminoacyl-tRNA synthetases, antitubercular agents, 7:725–726
3-Aminobenzamide PARP, stroke therapy, 8:484–487
1-Amino-4-benzylphthalazines, brain tumor therapy, hedgehog signaling pathway, 6:260–261
4-Aminobiphenyl adduct formation, toxicophore reactive metabolites, aniline nitrogen oxidation, 2:307–311
Aminocandin, development, 2:231–232
7-Aminocephalosporanic acid derivatives, cephalsporin development, 7:299–300
2-Amino-6-chloropurine, anti-DNA virus antimetabolites, 7:228–229
Aminoglutethimide:
  androgen antagonists, 5:197
  aromatase inhibition, 5:231–232
Aminoglycosides:
  antibiotic resistance, 7:424–427
  antitubercular agents, kanamycin and amikacin, 7:761–762
  biosynthesis, 7:423–424
  current compounds and clinical applications, 7:415–419
  historical development, 2:222, 7:423–424
  Leishmaniasis therapy, paramomycin sulfate, 7:576–577
  natural product sources, 7:405–406
  pharmacology and mechanism of action, 7:420–423
  recent drug development, 7:427–432
  RNA targets:
    neomycin- and gentamicin-type aminoglycosides, 5:971–979
    toxicity, 5:979–981
  selective toxicity, 2:551–553
  side effects, toxicity, and contraindications, 7:419–420

Aminohydantoins, BACE (β-site cleavage enzyme) inhibitors, **8**:353–356
α-Amino-3-hydroxy-5-methyl-4-isoxazole propionic acid (AMPA):
 anticonvulsants, ligand-gated ion channels, **8**:129
 antipsychotic glutmaterics, **8**:205–206
 cognitive enhancement:
  glutamate, **8**:41–42
  potentiators, **8**:43–44
 glutmate modulation, migraine therapy, **8**:291–295
 long-term potentiation, **8**:24–25
Amino inositol moieties, aminoglycoside antibiotics, **7**:415–416, 423–424
5-Amino-isoquinolinone (AIQ), poly(ADP-ribose) polymerase inhibitors, **6**:164–167
Aminomethyl heterocycles, dipeptidyl peptidase 4 (DPP-4) inhibition, **4**:60–61
3-Amino nocardicinic acid derivatives, nocardicin synthesis, **7**:324–326
6-Amino-penicillanic acid (6-APA):
 derivatives, structure-activity relationship, **7**:289–292
 historical development of, **7**:288–289
2-Aminophenylamides, HDAC inhibitors, stroke therapy, **8**:492
Aminopiperidine heterocycles, dipeptidyl peptidase 4 (DPP-4) inhibition, **4**:61–64
Aminopterin, cancer therapy, **6**:106–109
2-Aminopurine dioxolane (APD), anti-DNA virus antimetabolite, **7**:227
Aminopyrazoles, Alzheimer's disease, tau-targeted therapies, **8**:420–422
Aminopyridines, stroke therapy, **8**:498–499

Aminopyrrolidine inhibitors, dipeptidyl peptidase 4 (DPP-4) inhibition, **4**:63–64
Aminoquinoline prodrugs: glutathione inhibitors, antimalarial drugs, **7**:662–663
γ and δ lactams, antimalarial cell cycle regulation, **7**:635–637
Aminosteroid neuromuscular blockers, nicotinic receptor antagonists, **8**:78–80
Amino-tetrahydroquinones, CRTH2 receptor antagonist, **5**:918–920
2-Aminothiazoles:
 stroke therapy, **8**:514
 toxicophore reactive metabolites, **2**:319–320
Aminothiazolylphenyl derivatives, herpes simplex virus therapy, **7**:242–243
Aminothienopyridizine family, Alzheimer's disease, tau-targeted therapy, **8**:429
Amiodarone, soft drug analogs, **2**:115–117
Amisulpride, pharmacokinetics, **8**:199–200
Amitriptyline:
 chemical structure, **8**:222
 historical background, **8**:248–249
 migraine prevention, **8**:318–319
 monoamine transporter inhibition, **8**:238–241, 240–241
 side effects and adverse reactions, **8**:223, 231–232
Amlodipine:
 LDL cholesterol lowering mechanisms, atorvastatin combined with, **4**:318
 stroke therapy, **8**:459–460
Amodiaquine:
 antimalarial therapy, **7**:611–613

 combined artesunate experimental agents, **7**:627
 toxicophore reactive metabolites, aniline metabolism, aromatic ring oxidation, **2**:306
Amorphous solids, oral dosage systems, **3**:29
Amoxepine:
 chemical structure, **8**:222
 structure-activity relationship, **8**:182–183, 250–251
Amoxycillin:
 β-lactamase inhibitors, **7**:361–363
 penicillin resistance and development of, **7**:291–292
Amphetamine psychosis, **8**:95
 antipsychotic therapy, animal models, **8**:170–171
 schizophrenia pathology, **8**:162–163
Amphetamines:
 ADMET properties, **8**:97–99
 alpha-alkyl substituent, **8**:108
 alpha-carbon stereochemistry, **8**:108
 aromatic ring substitution, **8**:108–109
 cognitive enhancement, **8**:51–52
 historical development, **8**:90–91
 nitrogen substituents, **8**:107
 physiology and pharmacology, **8**:100–106
 receptor-based drug design, central nervous system effects, **8**:2–3
 side chain modification, **8**:107–108
 side effects, adverse reactions and drug interactions, **8**:92, 95–97
 structure-activity relationship, **8**:106–109
Amphibian skin, opioid peptides, **8**:646
Amphiphilic molecules, nanoscale drug delivery:
 polymers, **3**:473–480
 self-assembled low molecular weight structures, **3**:469–473

Amphiphilic pyridinium ion A$_2$E, age-related macular degeneration, **5:**609–610
Amphotericin B:
  development of, **2:**231
  Leishmaniasis therapy, **7:**576–577
  selective toxicity, **2:**559–560
Amphoteric surfactants, topical antimicrobial agents, **7:**501
Ampicillin, penicillin resistance and development of, **7:**291–292
Amplification, hemostasis, **4:**367
Amprenavir (APV):
  binding interactions, **7:**26, 29
  chemical structure, **7:**22, 26, 28
  efficacy and tolerability, **7:**28–29
  fosamprenavir prodrug, **7:**42–43
  HIV-1 protease inhibitors, **1:**310–311
  optimization, **7:**26, 28
  pharmacokinetics, **7:**28, 30
  resistance profile, **7:**29
Amrubicin, topoisomerase II inhibitors, **6:**100–102
Amsacrine, topoisomerase II inhibitors, **6:**100–102
β-Amyloid peptide (Aβ):
  Alzheimer's cognitive deficits, **8:**16
  Alzheimer's disease epidemiology, **8:**329–330
  anti-Aβ antibody mechanisms, **8:**367–368
  fibrillization inhibitors, **8:**361–365
  immunization against, **8:**366
  insulin related therapy, Alzheimer's disease, **8:**415–417
  secretase processing, **8:**330–334
  small-molecule inhibitors, **8:**362–365
Amyloid cascade hypothesis, Alzheimer's disease epidemiology, **8:**329–330

Amyloid precursor protein (APP):
  Alzheimer's disease epidemiology, **8:**329–330
  beta-secretase inhibitors (BACE1), **1:**327–328
  secretase processing, **8:**330–334
AN1792 Aβ peptide, Alzheimer's immunization therapy, **8:**368–369
Anabolic agents:
  androgens:
    absorption, distribution, and metabolism, **5:**186–188
    abuse of, **5:**188–189
    current marketed compounds, **5:**176–177
    structure-activity relationships, **5:**179–186
    alkylated analogs, **5:**181–182
    dehydro derivatives, **5:**180
    deoxy/heterocyclic-fused analogs, **5:**183–185
    esters and ethers, **5:**185–186
    hydroxy and mercapto derivatives, **5:**182–183
    nor derivatives, **5:**179
    oxo, thia, and azo derivatives, **5:**183
  osteoporosis therapy:
    basic principles, **5:**753
    parathyroid hormones, **5:**753–755
    selective androgen receptor modulator, **5:**760
    statins, leptin, and neuropeptide Y, **5:**755
    vitamin D, **5:**759–760
    Wnt signaling pathway, **5:**755–759
    bone remodeling, **5:**756–757
    canonical pathway, **5:**756
    genetic models and bone target cells, **5:**757–758
    therapeutics development, **5:**758–759
Anacetrapib, atherosclerosis, HDL elevation, **4:**341, 343–344

Anaerobic infection:
  chemotherapeutic agents, **7:**502–503
  combined therapeutic regimens, **7:**505
  linezolid antibacterial activity, **7:**542
Analog design:
  acetylcholine, **8:**64
  aminoglycoside antibiotics, **7:**429–432
  androgens:
    alkylated analogs, **5:**181–182
    deoxy/heterocyclic-fused analogs, **5:**183–185
  anticonvulsants, **8:**149
  anti-DNA virus agents:
    bicyclic pyrimidine nucleoside analogs, **7:**223–225
    inorganic diphosphate analogs, **7:**240–241
  antimalarial drugs:
    antifolates, **7:**647–648
    apicoplast targeting, **7:**663–665
    primaquine, **7:**614
    quinine, **7:**605–607
  antiparasitic/antiprotozoal agents, **7:**579–580
    parasitic glucose metabolism and glycosomal transport inhibition, **7:**585–586
    polyamine metabolism, **7:**586–590
  atherosclerosis, HDL elevation:
    anthranilic acid analogs, **4:**348–349
    niacin analogs, **4:**346–348
    xanthine and babituric acid analogs, **4:**349–350
  basic principles, **1:**167
  bioisosterism, **1:**171
  categories, **1:**167–168
  chronic obstructive pulmonary disease:
    CXCR2 receptor antagonistst, **5:**784–785
    nitric oxide synthases, **5:**785–786
    diaryl urea-based inhibitors, **5:**1043–1045
  direct analogs, **1:**170–171

Analog design (*Continued*)
  dual topoisomerase I/II inhibitors, bis analogs, **6:**105–106
  estradiol:
    17α-ethynyl analogs, **5:**222–224
    steroidal analogs, **5:**244–245
  existing drugs, **1:**170
  functional analogs, **1:**174–177
  future trends, **1:**178–179
  geldanamycin Hsp90 inhibitors, **6:**392–393
  HIV protease inhibitors, indinavir sulfate development, **7:**14, 17
  inosine monophosphate dehydrogenase inhibitors, C2-MAD analog, **5:**1044–1047
  macrolides, **7:**441–444, 446–456
  marine microbial compounds, bryostatins and dolastatins, **2:**243–245
  metabolites, **1:**170
  morphine, identification and early history, **8:**596–597
  muscarinic antagonists, **8:**72–77
  mycophenolic acid, **5:**1039–1042
  natural compounds, **1:**168–170
  nuclear magnetic resonance:
    FK506 analog binding to FKBP, **2:**410–412
    ligand structure limitations, **2:**390
  opioid peptide analogs, **8:**645–676
    affinity-labeled derivatives, **8:**675–676
    amphibian skin peptides, **8:**646
    antagonist dynorphin A analogs, **8:**657–658
    combinatorial library peptides and peptidomimetics, **8:**674–675
    dynorphin analogs, **8:**654–658
      conformational constraints, **8:**656–657
      linear analogs, **8:**655–656
    enkephalin analogs, **8:**646–654
      δ-selective analogs, **8:**649–653
      dimeric analogs, **8:**653–654
      μ-selective analogs, **8:**646–649
    somatostatin μ-receptor antagonists, **8:**673–674
    Tyr-D-aa-Phe sequence, **8:**664–673
      deltorphin analogs, **8:**668–673
      dermorphin analogs and μ-selective peptides, **8:**664–668
    Tyr-Pro-Phe sequence, **8:**659–664
      β-casomorphin analogs and endomorphins, **8:**659–661
      TIPP and related peptides, **8:**661–664
  opioid receptors, history, **8:**596–597
  oxazolidinones, **7:**544–548
  penicillin modification, **7:**295
  pharmacophore formation:
    bioactive conformation, **1:**469–471
    constrained synthesis, **1:**468–469
  prodrug developmenet, esmolol case study, **3:**259–268
  prostaglandin analogs, glaucoma intraocular pressure reduction, **5:**592–600
  prostaglandin/thromboxane EP$_1$ antagonists, **5:**820–824
  radicicol Hsp90 inhibitor, **6:**399–405
  retinoids:
    N-(4-hydroxyphenyl) retinamide (4-HPR), **5:**480–483
    9cUAB30 RXR-selective retinoids, **5:**458
    19-norpregnane analogs, **5:**254–255
    receptor-selective retinoids, NRX195183 analog, **5:**420
    SHetA2 retinoid-related compound, **5:**506–507
  RNA targets:
    erythromycin, **5:**984–990
    linezolid, **5:**990–993
    streptomycin, **5:**965–971
    tetracycline, **5:**981–984
  γ-secretase inhibitors, **8:**334–335
  sirolimus, triene region modification, **5:**1030–1033
  soft drug development, **2:**80, 110–111
  steroidal analogs:
    antiandrogens, **5:**190–192
    structure-activity relationships, **5:**170–174
  stroke therapy, iNOS/nNOS analogs, **8:**495–502
  structural analogs, **1:**171–174
  tacrolimus compounds, peptidyl-prolyl isomerase, **5:**1016–1021
  tetracyclines, **7:**414–415
  triarylethylene fused-ring analogs, **5:**249–253
  virtual screening and scaffold hopping, **1:**177–178
  virtual soft analog library, alprostadil soft drug design example, **2:**134–135
  vitamin D analogs, **5:**654–660
Analytical development:
  defined, **3:**289–290
  protein therapeutics, **3:**317–327
    bioactivity assays, **3:**321–322
    biophysical characterization, **3:**323–325
    electrophoresis, **3:**320
    future trends, **3:**326–327
    immunoassays, **3:**321
    in-process/product release testing, **3:**318–320
    mass spectrometry, **3:**322–323
    nucleic acid testing, **3:**320–321
Anaplastic astrocytomas, histologic classification, **6:**230–231
Anastrozole:
  aromatase inhibition, **5:**232

breast cancer resistance, 6:369–370
Anatoxin-A, nicotinic agonists, 8:70
Androgen binding protein (ABP), androgen absorption and distribution, 5:159
Androgen-responsive elements (AREs), biological mechanisms, 5:165–168
Androgens. *See also* specific hormones, e.g., Testosterone
  absorption and distribution, 5:158–159
  anabolic agents:
    absorption, distribution, and metabolism, 5:186–188
    abuse of, 5:188–189
    current marketed compounds, 5:176–177
    structure-activity relationships, 5:179–186
      alkylated analogs, 5:181–182
      dehydro derivatives, 5:180
      deoxy/heterocyclic-fused analogs, 5:183–185
      esters and ethers, 5:185–186
      hydroxy and mercapto derivatives, 5:182–183
      nor derivatives, 5:179
      oxo, thia, and azo derivatives, 5:183
    therapeutic applications and bioassays, 5:178–179
    toxicities, 5:188
  antagonists:
    antiandrogens, 5:190–193
    current marketed compounds, 5:189–190
    enzyme inhibitors, 5:193–200
  basic properties, 5:153
  biosynthesis, 5:22–25, 155–158
  distribution, 5:159–164
  future research issues, 5:200–201
  mechanism of action, 5:164–168
  metabolism, 5:26–27, 159–164
    oxidative metabolism, 5:162–164
    reductive metabolism, 5:159–162

  nonsteroidal androgens, 5:174–177
    absorption, metabolism, and distribution, 5:175–177
    toxicities, 5:176
  occurrence and physiological roles, 5:154–155
  receptor structure, 5:164–168
  steroidal progesterone receptor ligands, 5:254–255
  research background, 5:153–154
  steroidal androgens:
    halo derivatives, 5:171–172
    methylated derivatives, 5:171
    structure-activity relationships, 5:170–174
    synthetic compounds, 5:168–176
    current marketed compounds, 5:168–169
    therapeutic applications and bioassays, 5:169–170
Androstadiene 3-carboxylic acids, androgen antagonists, 5:195
Androstanes:
  antiandrogens, 5:191–192
  antiglucocorticoids, 5:114–120
Androstenedione:
  aromatase inhibition, 5:231–232
  biosynthesis, 5:22–25
  female sex hormone biosynthesis, 5:230–233
  metabolism, 5:27
Androsterone, basic structure, 5:153–154
Anemias:
  erythropoietin therapy, 4:571–572
  folic acid deficiency, 5:696
  vitamin B_6 deficiency, 5:681, 683
  vitamin B_{12} (cobalamin) deficiency, 5:700
Anesthetics:
  enzymatic hydrolysis, 2:79–80
  soft drug design, 2:125–126
Ang-(1-7) receptor, angiotensin pharmacology, 4:518

Angiogenesis. *See also* Antiangiogenic inhibitors
  Am80 inhibition, 5:417
  brain tumors, 6:242–243
  SHetA2 retinoid-related compound inhibition, 5:505
Angiotensin-converting enzyme (ACE) inhibitors:
  ACE2 carboxypeptidase, 4:292–293
  ACE gene, 4:286–287
  binding domains, 4:287–288
  captopril, 4:269–271
  clinical applications, 4:280–285
    adverse effects, 4:281–282
    cardiovascular disease risk reduction, 4:284–285
    congestive heart failure, 4:282–284
    hypertension, 4:282
    marketed inhibitor compounds, 4:280–281
    post-acute myocardial infarction, 4:284
    vasopeptidase inhibition, 4:285
  current research, renin-angiotensin system, 4:285–286
  domain selectivity, 4:288–289
  future research issues, 4:293
  insertion/deletion polymorphism, 1:191
  medicinal chemistry, 4:271–279
    carboxyl dipeptides, 4:273–274
    hydroxamic acid, 4:274–275
    mercaptans, 4:272–273
    natural products, 4:276
    peptide analogs, 4:275–276
    phosphinic/phosphonic acids, 4:274
    vasopeptidase inhibition, 4:276–279
  migraine prevention, 8:319–320
  multitarget drug development, ACE-plus for hypertension, 1:257, 259–260
  pharmacology:
    angiotensin II blockade, 4:279

Angiogenesis (*Continued*)
bradykinin potentiation, 4:279–280
peptide potentiation, 4:280
pharmacophore-based ligand libraries, lead drug discoveries, 2:528–529
prodrug development, 3:242
pseudopeptide analogs, 1:303
renin, angiotensin, and angiotensin-converting enzymes, 4:267–268
research background, 4:267–271
selective toxicity, antagonist structures, 2:542–543
soft drug design, 2:118–119
somatic enzyme localization, 4:287
structure and function, 4:520
three-dimensional structure, 4:289–292
venom peptides, bradykinin, and teprotide, 4:268–269
Angiotensinogen, renin inhibitor development, 4:243
Angiotensins:
AT$_1$ receptors, 4:518
agonists/antagonists, 4:520
angiotensin-converting enzyme inhibitors, 4:267–268
antagonists:
allosteric proteins, 1:388
multitarget drug development:
dual AT$_1$/ET$_A$ antagonist, 1:269–273
hypertension, 1:257, 259–260
pharmacology, 4:518
AT$_2$ receptor, 4:518
angiotensin-converting enzyme inhibitors:
blockade pharmacology, 4:279
captopril development, 4:270–271
antagonists:
EP$_3$ receptor antagonists, 5:856–858
selective PPARγ modulators, PPARγ partial agonists, 4:115

AT$_4$ receptor, cognitive enhancement, 8:50
biological actions and receptor pharmacology, 4:516–518
synthesis and metabolism, 4:516–518
Anhydrolide compounds, macrolide development, 7:452–456
Anhydrotetracyclines, history and biosynthesis, 7:411–413
Anidulafungin:
development, 2:231–232
selective toxicity, 2:559, 562
Anilides:
chemokine receptor-5 (CCR5) antagonist core structure, 7:124–131
histone deacetylase inhibitors, 6:66–70
chidamide, 6:68, 70
entinostat (MS-275), 6:67
mocetinostat (MGCD-0103), 6:67–70
tacedinaline (CI-994), 6:66–67
4-Anilidopiperidines, opioid analgesics, 8:618–621
Anilines, toxicophore reactive metabolites, 2:303–311
aniline nitrogen oxidation, metabolism via, 2:306–311
aromatic ring oxidation, metabolism via, 2:303–306
9-Anilinoacridines, antimalarial topoisomerase inhibitors, 7:650–652
Animal drugs, regulatory guidelines, 3:97
Animal studies. *See also* Knockout mouse models; Transgenic mouse models
AHPN compound, anticancer activities, 5:497–498
aliskiren renin inhibitor, 4:250–251
all *trans*-retinoic acid, adverse effects, 5:385–386

Alzheimer's disease immunotherapy, 8:366–367
Am80 therapy, 5:419
antidepressants, monoamine transporter inhibition, 8:240–241
anti-HCV agents, 7:175
antipsychotics, 8:170–176
behavioral models, 8:172–173
dopamine model, 8:170–171
electroencephalogram, 8:174–175
genetic models, 8:173–174
NMDA model, 8:171–172
side effects models, 8:175–176
antitubercular drug development and discovery, 7:735–737
BACE (β-site cleavage enzyme), 8:331–333
bexarotene therapy, 5:441–447, 449–450, 453
brain cancer modeling, 6:272–273
congestive heart failure therapy, clinical congruency, 4:489–491
glaucoma screening models, 5:590–591
N-(4-hydroxyphenyl)retinamide (4-HPR) therapy, 5:468–469
multiple sclerosis pathology, 5:567
9cUAB30 RXR-selective retinoid, 5:455–456
NRX194204 RXR selective analog, 5:460–464
NRX195183 analogs, 5:420
PET/SPECT systems in, 8:742
protease-activated receptor-1 antagonists, 4:440–441
psoriasis, 5:397
9-*cis*-retinoic acid therapy, 5:428–429
13-*cis*-retinoic acid therapy, 5:392
SHetA2 retinoid-related compound, 5:502–507
tazarotene therapy, 5:407
vitamin dose-response data, 5:643

Anionic surfactants, topical antimicrobial agents, **7**:500–501
Aniracetam, cognitive enhancement, **8**:34
Annexin I, glucocorticoid anti-inflammatory signaling mechanisms, **5**:63–65
Anorexigenics, therapeutic applications, **8**:92–94
Anpirtoline, serotonin receptor targeting, **8**:245–247
Antagonist compounds. *See also* specific antagonists
 β-adrenergic receptor antagonists, glaucoma therapy, **5**:596–600
 androgens:
  antiandrogens, **5**:190–193
  current marketed compounds, **5**:189–190
  enzyme inhibitors, **5**:193–200
 antidepressants, serotonin antagonists, **8**:225–229
 antiglucocorticoids, **5**:112–120
 antipsychotic agents:
  histamine H3 receptor antagonists, **8**:206–207
  neurokinin receptor antagonists, **8**:208
  serotonin-dopamine antagonists, **8**:177–179
 chemokine receptor-5 (CCR5) antagonists, HIV-1 entry blockade:
  anilide core, **7**:124–131
  binding mode and mechanism, **7**:133–134
  piperazine/piperidine-based antagonists, **7**:110–116
  research background, **7**:107–110
  spiro diketopiperazine-based antagonists, **7**:131–133
  summary and future research issues, **7**:134–135
  tri-substituted pyrrolidine antagonists, **7**:121–124
  tropane-based antagonists, **7**:116–121
 chemokine receptor CXCR4 antagonist-plerixafor, **4**:586–587
 cholinergic antagonists:
  muscarinic antagonists, **8**:72–77
  nicotinic antagonists, **8**:77–80
  ganglionic antagonists, **8**:78
  neuromuscular antagonists, **8**:78–80
  structure and classification, **8**:72–80
 chronic obstructive pulmonary disease:
  CXCR2 antagonists, **5**:784–785
  muscarinic acetylcholinergic receptor antagonists, **5**:766–769
 competitive receptor antagonists, stroke therapy, **8**:451
 CRF antagonists, **8**:257
 EP$_3$ receptor antagonists, **4**:450–451
 estrogen receptors:
  diarylethylenes, **5**:248–249
  potency and efficacy, **5**:237
 farnesoid X receptor, **4**:161
 glycoprotein IIb/IIIa receptors, **4**:444–449
  abciximab, **4**:445–446
  eptifibatide, **4**:446–447
  intravenous antagonists, **4**:445–447
  oral antagonists, **4**:448–449
  tirofiban, **4**:447
 insomnia therapy:
  orexin receptor antagonists, **5**:720–723
  serotonin receptors, **5**:718
 ionotropic glutamate receptors:
  anticonvulsants, **8**:129
  migraine therapy, **8**:291–297
 migraine therapy:
  calcitonin gene-related peptide antagonists, **8**:272–290
  azepinone binding affinity, **8**:281–286
  benzodiazepine binding affinity, **8**:277–281
  BIBN4096 BS, **8**:273, 275–277
  BMS-694153, **8**:288–290
  MK-0974, **8**:277, 279–288
  neuropeptide-induced blood flow, **8**:286–288
  receptor binding affinities, **8**:273–275
  unnatural amino acids, **8**:289–290
  vanilloid receptor antagonists, **8**:307–312
 multiple sclerosis therapy:
  small-molecule VLA-4 antagonists, **5**:578–579
  sphingosine 1-phosphate (S1P), **5**:577–578
 NMDA receptors:
  antidepressants, **8**:256
  stroke therapy, **8**:450–455
  competitive receptor antagonists, **8**:451–453
  ion channel blockers, **8**:450–451
  NR2B-specific noncompetitive receptor antagonists, **8**:453–455
 nuclear hormone receptors, structural biology, **4**:82–83
 opioid analgesics:
  chemical structure, **8**:571–574
  δ-opioid receptors, **8**:604–608
  enkephalin analogs, **8**:653
  dynorphin A analogs, **8**:657–658
  κ-opioid receptors, **8**:602–604
  μ-opioid receptors:
   selective antagonists, **8**:607
   somatostatin derivation, **8**:673–674
 osteoporosis antiresorptive therapy, calcitonin and integrin antagonists, **5**:730
 Parkinson's disease:
  adenosine receptor antagonists, **8**:552–553
  glutamate antagonists, **8**:555

Antagonist compounds
(*Continued*)
partial antagonists, occupancy theory, **2:**496–497
peroxisome proliferator-activator receptors:
PPARα antagonist, **4:**99–100
PPARδ antagonists, **4:**122–124
PPARγ antagonists, **4:**115–117
phosphodiesterase-III inhibitors, **4:**449–450
aggrenox, **4:**450
cilostazol, **4:**449–450
dipyridamole, **4:**450
milrinone, **4:**450
progesterones, **5:**239
prostaglandin/thromboxane receptors:
CRTH2 receptor, **5:**909–920
DP_1, dual pharmacology, **5:**920
indole acetic acids, **5:**910–914
miscellaneous compounds, **5:**918
monoaryl acetic acids, **5:**914
nonacidic antagonists, **5:**918–920
phenoxyacetic acid, **5:**914–918
Ramatroban analogs, **5:**909–910
DP_1 receptor, **5:**899–907
acidic antagonists, **5:**905
clinical trials, **5:**907
CRTH2 receptor dual pharmacology, **5:**920
fused-tricyclic antagonists, **5:**903–905
hemi-prostanoid antagonists, **5:**901–902
indole acetic acid, **5:**902–903
nonacid-based antagonists, **5:**905–907
EP_1 receptor, **5:**813–846
amide replacements, **5:**818–820, 836–837
AstraZeneca derivatives, **5:**816–817
benzene derivatives, **5:**829, 831, 834

benzoic acid derivatives, **5:**823, 827–832
carbamates, **5:**836–838
cyclopentene derivatives, **5:**829, 831, 834
diacyl hydrazide replacement, **5:**815–817
glycine sulfonamide derivatives, **5:**837, 840–845
Ono derivatives, **5:**817–824
pyrazole derivatives, **5:**831, 835–838
pyrrole derivatives, **5:**826, 828–832
Searle derivatives, **5:**813–816
sulfonylureas, **5:**820–821, 825
thiophene derivatives, **5:**821, 823, 826, 829
EP_2 receptors, **5:**852–854
EP_3 receptors, **5:**855–872
acylsulfonamide derivatives, **5:**858–858
angiotensin II derivatives, **5:**856–858
hyperalgesia/edema modeling, **5:**858, 861
indole analogs, **5:**861, 863–869
thiophene derivatives, **5:**858, 860–861
EP_4 receptor, **5:**880–883
FP receptors, **5:**899
IP receptor, **5:**939–943
aryl guanidine derivatives, **5:**949–941
compound 500, **5:**940–942
pyrimidines, **5:**942–943
synthesis, **5:**811–812
thromboxane A_2 receptor, **5:**886–896
4-oxazolylcarboxamide analogs, **5:**886–889
phenol derivatives, **5:**888–889, 891
racemic 1,3-dioxolane derivatives, **5:**887–889, 891
semicarbazone compounds, **5:**886–889
sulfonamide analogs, **5:**887–890
tetrahydronaphthalene template, **5:**891, 893

thromboxane synthase inhibitors, **5:**894–896
trimetoquinol, **5:**891, 893–894
UK-147,535 derivative, **5:**894
protein-protein interactions, **2:**337–340
purinergic receptor antagonists, antiplatelet agents, **4:**423–437
cangrelor, **4:**435–436
clopidogrel, **4:**427–433
elinogrel, **4:**437
irreversible ADP antagonists, **4:**426–435
$P2X_1$ receptor, **4:**423–424
$P2Y_1$ receptor, **4:**424–425
$P2Y_{12}$ antagonists, **4:**425–426
$P2Y_{12}$ receptor, **4:**425–437
prasugrel, **4:**433–435
reversible ADP antagonists, **4:**435–437
ticagrelor, **4:**436–437
ticlopidine, **4:**426–427
receptor-based drug design, **2:**496–497, 502–503
efficacy parameters, **2:**502–503
renin-angiotensin system, **4:**520
selective toxicity, **2:**542–544
stroke therapy:
NMDA receptors, **8:**450–455
competitive receptor antagonists, **8:**451–453
ion channel blockers, **8:**450–451
NR2B-specific noncompetitive receptor antagonists, **8:**453–455
voltage-dependent calcium channels, **8:**455–460
substance P-NK1 receptor antagonist, **8:**256–257
thrombin receptor (protease-activated receptor-1) antagonists, **4:**437–445
thromboxane antagonists, **4:**419

dual synthase inhibitor and
  antagonist, 4:423
thromboxane A$_2$ receptor
  antagonists, 4:422
thyroid hormone receptors:
  structural properties,
    4:201–202
  therapeutic potential,
    4:198–199
  thyromimetic structure-
    activity relationship,
    4:219–222
vanilloid receptor antagonists,
  migraine therapy,
  8:307–312
vitamin K antagonist:
  anticoagulant therapy, 4:365
  inhibition mechanisms,
    4:369–370
  molecular mechanisms of
    action, 4:380–385
voltage-gated calcium
  channels, 8:455–460
Antagonist-induced dissociation
  assay, protein-protein
  interactions,
  2:345–359
Antedrugs. *See also* Prodrugs
  defined, 2:76
  glucocorticoid anti-
    inflammatory
    compounds, 5:45–46
    C-20 carbonyl alterations,
      5:102–103
  steroid soft drug design,
    2:99
Anthracyclines:
  anti-cancer activity, 2:235–239,
    6:17–22
    daunorubicin, 6:19–20
    doxorubicin, 6:20–22
    epirubicine, 6:22
    valrubicin, 6:22–23
  serum albumin binding,
    3:452–454
Anthralin, antimicrobial activity,
  7:495
Anthranilic acid, atherosclerosis,
  HDL elevation,
  4:348–349
Anthrapyrazoles, topoisomerase
  II inhibitors, 6:102
Anthraquinones, Alzheimer's
  disease, tau-targeted
  therapy, 8:427
Anti-Aβ antibodies, mechanisms
  of action, 8:367–368

Antiandrogens:
  ADMET properties, 5:193
  marketed compounds, 5:189
  structure-activity
    relationships,
    5:190–193
  therapeutic applications,
    5:190–193, 221–222
Antiangiogenic inhibitors:
  central nervous system tumors,
    6:272
  resistance to, 6:368
  vascular endothelial growth
    factor, 6:307–310
Antiarrhythmic drugs:
  chirality and distribution,
    1:140–141
  soft drug design, 2:114–117
    ACC-9358 analogs,
      2:114–115
    amiodarone analogs,
      2:115–117
Antibiotic/antibacterial drugs.
  See also
    Aminoglycosides;
    Antimalarials;
    Antitubercular
    agents; Antiviral
    compounds; β-
    Lactams; Macrolides;
    Tetracyclines
  adapalene therapy combined
    with, 5:411
  antimalarial agents,
    interaction,
    7:621–624
  microbial compounds,
    2:221–231
    antifungals, 2:230–231
    glypeptides, 2:225–227
    β-lactams, 2:222–223
    lipopeptides, 2:227–228
    macrolides, 2:228–230
    tetracyclines, 2:223–225
  natural products, 7:405–406
  prodrug development,
    3:242–246
  RNA targets:
    aminoglycoside toxicity,
      5:979–981
    chloramphenicol, 7:462–463
    erythromycin and analogs,
      5:984–990
    fusidic acid, 7:464
    future research issues,
      5:993–996
    lincosamides, 7:460–462

linezolid and analogs,
  5:990–993
neomycin- and gentamicin-
  type aminoglycosides,
  5:971–979
oxazolidinones, 7:464
research background, 5:963,
  7:405–406
ribosome structure and
  function, 5:964
streptogramins, 7:457–460
streptomycin and analogs,
  5:965–971
tetracycline and analogs,
  5:981–984
selective toxicity and
  comparative
  biochemistry,
  2:548–553
soft drugs design, 2:111–114
  L-carnitine esters, 2:113–114
  cetylpyridinium analogs,
    2:111–112
  long-chain esters, betaine/
    choline, 2:114
  quaternary analogs,
    2:111–113
steroid transformation, 5:7–13
synthetic agents:
  future research issues,
    7:552–553
  hybrid agents, 7:550–552
  research background,
    7:483–484
  systemic antibacterials,
    7:501–553
    anaerobic infections,
      7:502–503
    chemoprophylaxis, 7:505
    combination therapies,
      7:504–505
    dihydrofolate reductase
      inhibitors, 7:531–534
    drug resistance, 7:502, 504
    folate inhibitors,
      7:506–534
    genetic factors, 7:503
    host reactions, 7:503
    methenamine, 7:550
    nitrofurans, 7:548–550
    oxazolidinones, 7:540–548
    pharmacology, 7:501–504
    quinolones and
      fluoroquinolones,
      7:535–540
    research background,
      7:501

Antibiotic/antibacterial drugs
(*Continued*)
  sulfonamides and sulfones,
    7:506–531
  tissue factors, **7**:503–504
  treatment guidelines,
    7:504–506
  topical agents, **7**:484–501
    acids, **7**:496
    alcohols, **7**:492–493
    amphoteric surfactants,
      7:501
    anionic surfactants,
      7:500–501
    cationic surfactants,
      7:499–500
    cellular targets, **7**:488
    diarylureas, amidines, and
      biguanides,
      7:498–499
    dyes, **7**:498
    epoxides and aldehydes,
      7:495–496
    evaluation, **7**:489–490
    halogens and halophores,
      7:490–492
    heavy metals, **7**:497–498
    historical development,
      7:490
    kinetics, **7**:488–489
    oxidizing agents,
      7:496–497
    phenols, **7**:493–495
    resistance mechanisms,
      7:488
    selective toxicity, **7**:484,
      488
    terminology, **7**:484
Antibiotic-producing organisms,
  recombinant DNA
  technology, **1**:539
Antibody-based therapeutics:
  anti-Aβ antibodies, **8**:367–368
  recombinant DNA technology,
    1:543
Antibody-directed enzyme
  prodrug therapy
  (ADEPT):
  DNA-targeted therapeutics,
    6:120–123
  prodrug design, **3**:229
Anticancer agents. *See also*
  Cancer therapy
  microbial compounds,
    2:235–243
    actinomycins,
      anthracyclines,

bleomycins,
  enediynes, **2**:235–239
erbstatin and lavendustin-
  related molecules,
  2:239–240
rapamycins and epothilones,
  2:240–243
tyrphostins, **2**:239–240
prodrug development,
  3:246–248
Anticholesterolemics, microbial
  compounds, **2**:231–235
  combination agents, **2**:234–235
  HMG-CoA reductase inhibitors,
    2:234
Anticholinergic agents:
  antipsychotics, **8**:167–168
  Parkinson's disease therapy,
    8:552
  soft drug design, **2**:107–109
  quaternary analogs,
    2:110–111
Anticoagulants:
  basic principles, **4**:365
  hemostasis, **4**:365–368
    amplification, **4**:367
    clotting initiation, **4**:366–367
    coagulation factors,
      4:365–366
    intrinsic regulators, **4**:368
    propagation, **4**:368
  new molecule development,
    4:390–400
    dabigatran and rivaroxaban,
      4:398–400
    de novo nonsaccharide
      heparin mimics,
      4:392–394
    indirect anticoagulants:
      antithrombin
        conformational
        activation, **4**:390–392
      bridging mechanism, **4**:392
      enzymatic engineering,
        4:392
      heparin-based, **4**:392
    site-directed anticoagulants,
      4:394–398
  therapeutic applications,
    4:368–390
    clinical approval, **4**:368–370
    direct anticoagulants, **4**:374
    direct thrombin inhibitors,
      4:385–390
    heparin molecular
      mechanisms,
      4:374–380

indirect anticoagulants,
  4:371–374
inhibition mechanisms,
  4:370–371
vitamin K antagonists,
  molecular
  mechanisms,
  4:380–385
Anticonvulsants. *See also*
  Epilepsy
  barbiturates, **8**:139–141
  benzodiazepines, **8**:144–147
  carabersat, **8**:155–156
  carisbamate, **8**:155
  classification, **8**:134–139
  felbamate, **8**:152–154
  GABA analogs, **8**:149
  ganaxolone, **8**:156
  hydantoins, **8**:141
  iminostilbenes, **8**:142–144
  lacosamide, **8**:154
  lamotrigine, **8**:148–149
  migraine prevention,
    8:315–316
  molecular targeting, **8**:121–134
    connexins, **8**:134
    enzymes, **8**:132–134
    G-protein-coupled receptors,
      8:134
    ligand-gated ion channels,
      8:126–130
    neurotransmitter
      transporters and
      synaptic proteins,
      8:130–132
    voltage-gated ion channels,
      8:122–126
  pyrrolidin-2-ones, **8**:149–151
  research background, **8**:121
  retigabine, **8**:154
  rufinamide, **8**:154–155
  stiripenol, **8**:155
  succinimides, **8**:142
  sulfonamides, **8**:151–152
  tiagabine, **8**:152
  trimethadione, **8**:142
  valproic acid, **8**:147–148
Antidepressants:
  classification and application,
    8:219–221
  clinical efficacy, **8**:221–229
  dosing regimens, **8**:219–221
  fluoxetine, historical
    background,
    8:248–249
  historical background,
    8:248–249

mechanistic classification, **8**:221
migraine prevention, **8**:318–319
nondepression applications, **8**:229
pharmacokinetics, **8**:235–237
  serotonin-selective reuptake inhibitors, **8**:236
  tricyclic antidepressants, **8**:235–236
physiology and pharmacology, **8**:237–248
  α-adrenergic receptors, **8**:247
  monoamine oxidase, **8**:247–248
  monoamine transporters, **8**:237–241
  serotonergic agents, **8**:241–247
recent and future developments, **8**:255–258
  monoaminergic drugs, **8**:255–256
  neuropeptide receptor agonists/antagonists, **8**:256–258
  NMDA receptor antagonists, **8**:256
research background, **8**:219
side effects and adverse reactions, **8**:229–234
  bupropion, **8**:234
  monoamine oxidase inhibitors, **8**:230
  nefazodone, **8**:234–235
  reboxetine, **8**:235
  serotonin/norepinephrine reuptake inhibitors, **8**:234
  serotonin-selective reuptake inhibitors, **8**:232–234
  trazodone, **8**:234
  tricyclic antidepressants, **8**:231–232
structure-activity relationship and metabolism, **8**:249–255
  bupropion, **8**:254–255
  monoamine oxidase inhibitors, **8**:251–253
  serotinergic agents, **8**:253–254
  tricyclic antidepressants, **8**:249–251

Anti-DNA virus agents, **7**:221–247
  acyclic nucleoside phosphates, **7**:231–239
    adenine derivatives, **7**:235–236
    8-azapurine derivatives, **7**:237
    2,6-diaminopurine derivatives, **7**:236
    guanine derivatives, **7**:236
    new generation prodrugs, **7**:238–239
    purine N6-substituted amino derivatives, **7**:237–238
    tissue-targeted prodrugs, **7**:239
  antimetabolites, **7**:221–230
    bicyclic pyrimidine nucleoside analogs, **7**:223–224
    1,3-dioxolane derivatives, **7**:227–228
    first-generation structure-activity relationship, **7**:222–225
    human heptatitis virus type B, **7**:226–227
    nucleoside sugar moiety, **7**:224–225
    1,3-oxathiolane derivatives, **7**:227–228
    research background, **7**:221–222
    second-generation compounds, **7**:225–226
    third-generation nucleosides, **7**:228–230
  defined, **7**:221
  inorganic diphosphate analogs, **7**:240–241
  nucleoside prodrugs, **7**:230–231
  nucleotide prodrugs, **7**:231
  posttranscriptional methylation inhibitors, **7**:241
  research background, **7**:221
  viral DNA-polymerase nonnucleotide inhibitors, **7**:239–240
  virus assembly inhibitors, **7**:241
Antiemetics, migraine therapy, **8**:297–298

Antiestrogens, clinical applications:
  adverse effects and precautions, **5**:221
  current drugs, **5**:219–221
Antifolates:
  antimalarial drugs, **7**:617–618
  DNA targeted compounds, **7**:644–648
  antitubercular agents, cofactor biosynthesis targeting, **7**:727–728
  DNA-targeted chemotherapeutic compounds, **6**:106–109
  synthetic antibacterial agents, **7**:506–534
    dihydrofolate reductase inhibitors, **7**:531–534
    sulfonamides/sulfones, **7**:506–531
Antifungal compounds:
  antibiotics, **2**:230–231
  phenotypic screening, **2**:208
  selective toxicity, **2**:558–559, 562
Antigen-presenting cells (APCs):
  central nervous system tumors, vaccine development, **6**:269–270
  immune function, **5**:1002–1006
Antigen-specific therapies, multiple sclerosis, **5**:579–580
Antiglucocorticoids:
  nonsteroidal anti-inflammatory compounds, **5**:121–134
  structure and function, **5**:112–120
  nonsteroidal, **5**:121–125
Anti-HCV agents:
  future research issues, **7**:204
  nonnucleoside inhibitors:
    benzothiadiazine, **7**:202–203
    HCV-796, **7**:203–204
    NS5B inhibitors, **7**:201–204
    NX5B inhibitors, **7**:201–204
      benzothiadiazine, **7**:202–203
      PF-00868554, **7**:201–202
    PF-00868554, **7**:201–202
  NS3/4A protease inhibitors, **7**:176–195
    boceprevir, **7**:186–191
    ciluprevir, **7**:176–181
    ITMN-191, **7**:191–192

Anti-HCV agents (*Continued*)
  MK7009, **7:**193–195
  telaprevir, **7:**181–186
  TMC435350, **7:**193
  nucleoside inhibitors,
    7:195–201
  MK0608, **7:**199–201
  NM283 (valopicitabine),
    7:195–196
  PSI-6130 (Pharmasset),
    7:196–197
  R1479/1626, **7:**197–199
Antihistamines, selective toxicity,
    2:564–566
Anti-HIV compounds:
  chirality, **1:**139–140
  scaffold properties,
    1:152–155
  quantitative structure-activity
    relationship, **1:**60–62
Antihypertensives, renin
    inhibitors:
  aliskiren:
    animal studies, **4:**250–251
    clinical trials, **4:**251–252
    congestive heart failure,
      4:252–253
    discovery, **4:**250
    drug-drug interactions, **4:**253
    end organ protection, **4:**252
    pharmacokinetics, **4:**253
  anticipated hypertension and
    end organ benefit,
    4:242–243
  drug discovery process,
    4:243–250
  hydroxyethylene renin
    inhibitors, **4:**245–247
  hydroxyl transition-state
    isostere renin
    inhibitors, **4:**243–244
  norstatine renin inhibitors,
    4:247
  peptide/reduced peptide
    inhibitors, **4:**243
  statin-based renin inhibitors,
    4:244–245
  vicinal diol renin inhibitors,
    4:247–250
  future research issues,
    4:255–256
  hypertension epidemiology,
    4:240
  nonpeptidic chemotypes,
    4:253–255
    alkylamines, **4:**254–255
    diaminopyrimidines, **4:**254

piperidines/piperazines,
    4:253–254
preclinical models, **4:**242
pro(renin) receptor, **4:**240–241
renin-angiotensin-aldosterone
    physiology, **4:**240–241
renin biochemistry/
    biosynthesis, **4:**241
research background,
    4:239–240
Anti-infective libraries:
  fluoroquinolone, **1:**303–304
  oxazolidinone, **1:**303–305
Anti-inflammatory compounds.
    See also Nonsteroidal
    anti-inflammatory
    drugs (NSAIDs)
  glucocorticoids:
    absorption effects, **5:**41, 43,
      45–53
    corneal penetration, **5:**53
    enzymatic metabolism,
      5:53–54
    intestinal absorption, **5:**41,
      43, 45
    intra-articular
      administration,
      5:51–53
    percutaneous absorption,
      5:46–51
    water-soluble esters, **5:**53
  adverse effects, **5:**41, 43–44
  antiglucocorticoids,
    5:112–120
  nonsteroidal, **5:**121–125
  biosynthesis and
    metabolism, **5:**56–59
  clinical applications, **5:**37–40
  currently available
    corticosteroids,
    5:39–40, 42
  drug distribution effects,
    5:53–56
  drug receptor affinity,
    5:65–70
  future research issues,
    5:133–134
  mechanism of action, **5:**62–63
  nomenclature, **5:**36–37
  nonsteroidal dissociated
    glucocorticoids,
    5:125–133
  nonsteroidal glucocorticoids,
    5:121
  nuclear structure-activity
    relationships,
    5:82–112

C-1 alterations, **5:**83
C-2 alterations, **5:**83–86
C-3 alterations, **5:**86–87
C-4 alterations, **5:**87–88
C-5 alterations, **5:**88
C-6 alterations, **5:**88–89
C-7 alterations, **5:**89–90
C-8 alterations, **5:**91
C-11 alterations, **5:**91–92
C-12 alterations, **5:**92–95
C-15 alterations, **5:**95
C-16 alterations, **5:**95–99
C-17 alterations, **5:**99–102
C-20 alterations,
    5:102–103
C-21 alterations,
    5:104–112
6,7-disubstituted
    compounds, **5:**91
pharmacological testing,
    5:70–71
quantitative structure-
    activity relationships,
    5:71–82
  conformational changes,
    5:76–77
  de novo constants, **5:**71–72
  electronic characteristics,
    5:77
  Hansch type analyses,
    5:72–74
  neural network
    predictions, **5:**74–75
  receptor binding affinity,
    LinBiExp model,
    5:75–76
  three-dimensional QSARs,
    5:77–82
receptor modulators,
    5:120–121
receptor structure, **5:**58,
    60–61
relative potencies, **5:**84–85
research background,
    5:35–36
signaling mechanisms,
    5:63–65
steroidal dissociated
    glucocorticoids,
    5:120–121
high-density lipoproteins,
    4:334
migraine therapy:
  NSAIDs/acetominophen,
    8:268
  triptans, **8:**269–272
nitric oxides, **5:**272–278

Antimalarials:
  antibiotics, 7:621–624
  antifolates, 7:617–618
  artemisinins, 7:614–617
  biguanides and pyrimethamine, 7:617
  drug discovery and development, 7:627–672
    apicoplast targets, 7:663–668
    cell cycle regulation, 7:627–634
    cellular and oxidative stress management, 7:658–663
    cellular respiration and electron transport chain, 7:655–658
    DNA synthesis, repair, and regulation, 7:644–655
    heme metabolism and amino acid salvage, 7:634–644
    targeted drug systems, 7:668–672
  experimental agents, 7:625–627
  halofantrine, 7:619
  hydroxynaphtoquinones, 7:619
  malaria epidemiology, 7:603–605
  pyronaridine, 7:619–621
  quinolines, 7:605–614
    amodiaquine, 7:611–613
    chloroquine, 7:607–610
    mefloquine, 7:610–611
    primaquine, 7:613–614
    quinine, 7:605–607
  sulfonamides and sulfones, 7:617
Antimetabolites:
  DNA-targeted chemotherapeutic compounds, 6:106–113
    antifolates, 6:106–109
    purine analogs, 6:112–113
    pyrimidine analogs, 6:109–112
  DNA-targeted therapeutic compounds, 7:221–230
    bicyclic pyrimidine nucleoside analogs, 7:223–224
    1,3-dioxolane derivatives, 7:227–228
    first-generation structure-activity relationship, 7:222–225
    human heptatitis virus type B, 7:226–227
    nucleoside sugar moiety, 7:224–225
    1,3-oxathiolane derivatives, 7:227–228
    research background, 7:221–222
    second-generation compounds, 7:225–226
    third-generation nucleosides, 7:228–230
  sulfonamides and sulfones, 7:506–507
Antimicrobials. See Microbial compounds
Antimycin, peptidomimetics, 1:233–234
Antimycobacterial drugs, selective toxicity, 2:557–558
Antioxidants:
  high-density lipoproteins, 4:334
  nitric oxide effects, 5:273–278
  nitric oxide enhancement, 5:291–294
  stroke therapy, 8:479–482
    deferoxamine, 8:482
    ebselen, 8:481
    edavarone, 8:481
    NXY-059, 8:479–480
    tirlazad, 8:480
  supplement controversy, 5:643–644
Antiplatelet agents:
  aspirin:
    primary prevention, 4:421
    resistance, 4:421
    secondary prevention, 4:421
    thromboxane A_2 biosynthesis, 4:419–420
  dual thromboxane A_2 synthase inhibitor/thromboxane A_2 receptor antagonist, 4:423
  EP_3 receptor antagonists, 4:450–451
  epinephrine receptors, 4:444
  future research issues, 4:451–453
  glycoprotein IIb/IIIa antagonists, 4:444–449
    abciximab, 4:445–446
    eptifibatide, 4:446–447
    intravenous antagonists, 4:445–447
    oral antagonists, 4:448–449
    tirofiban, 4:447
  marketed agents, 4:418–419
  phosphodiesterase-III inhibitors, 4:449–450
    aggrenox, 4:450
    cilostazol, 4:449–450
    dipyridamole, 4:450
    milrinone, 4:450
  purinergic receptor antagonists, 4:423–437
    cangrelor, 4:435–437
    clopidogrel, 4:427–433
    elinogrel, 4:437
    irreversible ADP antagonists, 4:426–435
    $P2X_1$ receptor, 4:423–424
    $P2Y_1$ receptor, 4:424–425
    $P2Y_{12}$ antagonists, 4:425–426
    $P2Y_{12}$ receptor, 4:425–437
    prasugrel, 4:433–435
    reversible ADP antagonists, 4:435–437
    ticagrelor, 4:436–437
    ticlopidine, 4:426–427
  structure and function, 4:409
  thrombin receptor (protease-activated receptor-1) antagonists, 4:437–444
    activation mechanism, 4:438–439
    E555 antagonist, 4:443–444
    intracellular signaling, 4:439
    knockout animal studies, 4:440–441
    peptidomimetic antagonists, 4:441–442
    preclinical pharmacology, 4:439–440
    SCH 530348, 4:443
    small-molecule antagonists, 4:442–443
  thromboxane A_2 receptor antagonists, 4:422
  thromboxane A_2 synthase inhibitor, 4:422–423
  thromboxane antagonists, 4:419
Antiprogestins:
  adverse effects and precautions, 5:221

Antiprogestins (*Continued*)
current drugs, **5:**219–221
physiology and pharmacology, **5:**238–239
steroidal progesterone receptor ligands, **5:**254–255
Antiproliferative studies, acitretin, **5:**399
Antiproteolide, microbial metabolite production, **2:**250
Antiprotozoal/antiparasitic agents. *See also* Protozoan infections
kinetoplastic protozoan infections, **7:**566–577
American trypanosomiasis (Chagas' disease), **7:**571–573
biochemistry, **7:**567
glucose metabolism and glycosomal transport inhibitors, **7:**582–585
human African trypanosomiasis (sleeping sickness), **7:**567–571
leishmaniasis, **7:**574–577, 590–593
lipid metabolism inhibitors, **7:**580–581
nucleotide transporters, **7:**577–580
polyamine pathway targeting, **7:**585–590
protease inhibitors, **7:**581–582
parasitic epidemiology, **7:**565–566
Antipsychotics:
animal models, **8:**170–176
behavioral models, **8:**172–173
dopamine model, **8:**170–171
electroencephalogram, **8:**174–175
genetic models, **8:**173–174
NMDA model, **8:**171–172
side effects models, **8:**175–176
drug discovery and development, **8:**200–210
AMPA receptors, **8:**205–206
cannabinoids, **8:**209–210
dopamine receptor subtypes, **8:**200–201

glutamatergic agents, **8:**202
histamine H_3 receptor, **8:**206–207
metabotropic glutamate receptors, **8:**203–204
neurokinin 3 receptor antagonists, **8:**208
neuronal nicotinic acetylcholine receptor, **8:**207
NMDA receptors, **8:**204
phosphodiesterase inhibitors, **8:**208–209
serotonin receptor subtypes, **8:**201–202
FDA-approved compounds, **8:**176–180
atypical compounds, **8:**177–179
conventional drugs, **8:**176–177
indication sfor, **8:**170
migraine therapy, **8:**297–298
pharmacokinetics, biodistribution, and drug-drug interactions, **8:**185–200
benzisoxazoles, **8:**195–196
benzoisothiazoles, **8:**196–198
butyrophenones, **8:**189–190
diarylbutylamines, **8:**190
dibenzodiazepines, **8:**191–192
dibenzothiepine, **8:**194
dibenzoxazepines, **8:**189
dibenzthiazepines, **8:**193–194
dihydroindolone, **8:**190–191
phenothiazines, **8:**186–188
phenylindole, **8:**199
quinolinone, **8:**198
substituted benzamide, **8:**199–200
thiobenzodiazepine, **8:**192–193
thioxanthenes, **8:**188–189
receptor-based drug design, **8:**3–4
research background, **8:**161
"rich pharmacology" principle, **8:**4–5
schizophrenia:
clinical applications, **8:**169–170
epidemiology, **8:**161–162
pathology, **8:**162–165

selective toxicity, **2:**569
side effects, **8:**165–169
animal models, **8:**175–176
anticholinergic effects, **8:**167–168
antipsychotic discontinuation, **8:**169
cardiovascular effects, **8:**167
extrapyramidal symptoms, **8:**165–166
hematologic effects, **8:**168–169
hepatic effects, **8:**169
metabolic effects, **8:**168
neuroleptic malignant syndrome, **8:**167
sedation, **8:**168
seizures, **8:**168
sexual dysfunction, **8:**169
structure-activity relationships, **8:**180–185
atypical tricyclic neuroleptics, **8:**180–183
benzisoxazole/benzithazole neuroleptics, **8:**183–185
Antisense oligonucleotides:
LDL cholesterol lowering agents, **4:**319
natural products lead generation, phenotypic screening, **2:**208
receptor-based drug design, **2:**531–533
Antiseptics, testing methods for, **7:**489–490
Antithrombin binding domain (ABD), indirect anticoagulants, **4:**392
Antithrombins:
antiplatelet agents, **4:**411–412
conformational activation, indirect anticoagulants, **4:**390–392
heparin inhibition mechanisms, **4:**370–371
molecular mechanism of action, **4:**374–380
nonsaccharide activators, **4:**393–395
serine proteases, combinatorial chemistry, **1:**332–333

Antitubercular agents:
  chemotherapeutic agents, 7:737–786
  capreomycin, 7:762–764
  D-cycloserine, 7:767–771
  ethambutol, 7:751–755
  ethionamide, 7:771–773
  first-line agents, 7:738–759
  fluoroquinolines, 7:774–776
  isoniazid, 7:738–744
  kanamycin and amikacin, 7:761–762
  para-aminosalicylic acid, 7:764–767
  pyrazinamide, 7:755–759
  rifamycin antibacterials, 7:744–751
  streptomycin, 7:759–761
  thiacetazone, 7:773–774
  drug discovery and development, 7:735–737
  linezolid, 7:785–786
  nitroimidazoles, 7:776–782
  preclinical and clinical research, 7:776–786
  SQ109, 7:782–783
  TMC207, 7:783–785
  drug resistance mechanisms and, 7:733–735
  epidemiology, 7:713–716
  second-line agents, 7:760–776
  structural genomics, 1:587
  targeted drug development for, 7:719–733
  amino acid biosynthesis, 7:726–727
  cell wall biosynthesis targeting, 7:719–724
  cofactor biosynthesis, 7:727–729
  cytochrome P450 monooxygenase targeting, 7:729
  DNA synthesis and repair targeting, 7:724–725
  intermediary metabolism and respiration, 7:729–730
  lipid metabolism, 7:730
  mycobactin biosynthesis and iron acquisition, 7:730–731
  mycothiol biosynthesis, 7:731
  protein synthesis and degradation, 7:725–726

  signal transduction targeting, 7:732–733
  terpenoid biosynthesis, 7:731–732
Antiviral compounds. *See also* Antibiotic/antibacterial drugs; Anti-DNA virus agents
  N-(4-hydroxyphenyl) retinamide (4-HPR) therapy, diabetes, 5:474–475
  selective toxicity, 2:553–557
Anxiolytic agents, serotonin receptor targeting, 8:242–244
Apadoline, κ-opioid receptor agonists, 8:630
Apaxifylline, cognitive enhancement, 8:48–49
Apical sodium-dependent bile acid transporter (ASBT), drug delivery, 2:467–475
Apicidin, antimalarial HDAC inhibitor, 7:653–655
Apicoplast, antimalarial agents, 7:621–624
  fatty acid biosynthesis, 7:663–665
  isoprenoid biosynthesis, 7:665–668
Aplaviroc chemokine receptor-5 (CCR5) antagonist, 7:132–133
Apnea of prematurity, psychostimulant therapy, 8:92
Apolipoprotein A-1:
  Alzheimer's disease therapy, 8:413–414
  ApoA-I$_{Milano}$, atherosclerosis, HDL elevation, 4:335–336
  atherosclerosis, HDL elevation, 4:335–337
  ApoA-I$_{Milano}$, 4:335–336
  reconstituted HDL, 4:336
Apolipoprotein B-100 (apoB-100):
  high-density lipoproteins, 4:332–333
  low-density lipoprotein cholesterol lowering agents, antisense oligonucleotides, 4:319

Apolipoprotein E (ApoE), Alzheimer's disease therapy:
  cholesterol targeting, 8:413–414
  HMG-CoA reductase inhibitors, 8:359–361
Apomorphine, antipsychotic therapy, animal models, 8:170–171
Apoptosis:
  AHPN targeted drug development:
    analog design and cell effects, 5:489–497
    extrinsic mediation, 5:491
    therapeutic applications, 5:489–497
    TR3 expression, nuclear export, and mitochondrial association, 5:492
  Parkinson's disease, mitochondrial dysfunction, 8:537
  peptidomimetics, Bcl-2 inhibitors, 1:231–233
  protein targeting, multiple drug resistance, 6:370–371
Aporphine-type dopamine receptor agonists, Parkinson's disease therapy, 8:544–545
Apparent volume of distribution, drug delivery system bioavailability, 2:751
Appeals process, infringement of patent proceedings, 3:163–164
Appetite suppressants, therapeutic applications, 8:92–94
Applicability domain, quantitative structure-activity relationship models, 1:521–522
  ensemble modeling and consensus prediction, 1:524–525
  ligand-based virtual screening, 2:13–14
Apramycin, chemical structure, 7:419

Apricitabine:
  anti-DNA virus antimetabolite, 7:227
  current development, 7:142–145
Aptamers, recombinant DNA technology, drug development, 1:549
Aptiganel, stroke therapy, 8:451
Aptivus. See Tipranavir
Aqueous solubility:
  gastrointestinal absorption, permeability vs., 3:374–375
  oral drug physicochemistry, 3:32–33
  salt compounds, 3:381–382
  sulfonamides, 7:518–519
Aquifex aeolicus bacterial leucine transporter, psychostimulant pharmacology and, 8:114
AR-102 prostaglandin analog, glaucoma therapy, 5:600–602
Ara-A vidarabine®, anti-DNA virus agents, antimetabolites, 7:222–225
Arabinogalactan, antitubercular agents, biosynthesis, 7:720–722
Arachidonic acid:
  prostaglandin/thromboxane synthesis, 5:809–812
  recombinant DNA technology, drug efficacy and personalized medicine, 1:542–543
Arbekacin, chemical structure, 7:419
Area under the curve (AUC) measurements:
  chemical delivery systems, site-targeting index, 2:140–141
  intravenous drug delivery:
    continuous intravenous infusion, 2:762
    repeated intravenous bolus injections, 2:758–759
    repeated short infusions, constant flow rate over finite period, 2:765–766

oral drug delivery, sustained-release dosage forms, in vitro-in vivo correlations, 2:787–791
permeability, ADMET interactions, transport interactions, 3:368–369
protein therapeutics, 3:324
Arecoline, muscarinic activity, 8:66–67
L-Arginine:
  chronic obstructive pulmonary disease, nitric oxide synthases, 5:786–787
  nitric oxide enhancement, 5:291–294
  stroke therapy, 8:495–496
Arginine vasopressin (AVP), structure and function, 4:520–523
Arilloxazine inhibitors, antimalarial drugs, 7:662–663
Aripiprazole:
  dopamine model, 8:170–171
  FDA approval, 8:178–179
  functional selectivity, 8:8–9
  pharmacokinetics, 8:198
  side effects, 8:168
Aristolochic acid, toxicophore risk assessment, 2:325–326
Aromatase inhibitors:
  androgens:
    antagonists, 5:197–199
    oxidative metabolism, 5:162–164
  estrogens:
    adverse effects and precautions, 5:221
    current drugs, 5:219–221
    female sex hormone biosynthesis, 5:231–232
Aromatic amines, structural genomics, 1:585–586
Aromatic amino acid residues, protein-protein interactions, 2:339–340
Aromatic oxidation, toxicophore reactive metabolites, aniline nitrogen oxidation, metabolism via, 2:307–311

O-Aroyl hydroxylamines, dipeptidyl peptidase 4 (DPP-4) inhibition, 4:49
AR-R15896AR, stroke therapy, 8:451
AR-R18565 scaffold, stroke therapy, 8:457
Arrhythmias:
  ADMET toxicity properties, 2:63, 69
  thyroid hormone receptor antagonists, 4:199
ARRY-520 kinesin spindle protein inhibitor, 6:209–210
Arsenic, drug metabolism, oxidation/reduction reactions, 1:420–421
Arsenic trioxide (ATO), acute promyelocytic leukemia therapy, 5:376–377
Artemisinins, antimalarial therapy, 7:614–617
  OZ277 artemisinin analog, 7:660
Artemisone experimental antimalarial compound, 7:625–627
Artesunate, antimalarial drugs:
  combined mefloquine/amodiaquine experimental agents, 7:625–627
  experimental agents, 7:625–627
  quinine combined with, 7:607
Artificial erythrocytes, oxygen delivery, 4:640–642
Artificial neural networks (ANNs):
  ligand-based virtual screening, quantitative structure-activity relationships, 2:11–13
  quantitative structure-activity relationship, 1:38–39
  glucocorticoid anti-inflammatory compounds, 5:74–75
Artificial red blood cells (ARBCs), oxygen delivery, 4:641–642
Aryl compounds, geldanamycin Hsp90 inhibitor derivatives, 6:389–391

Aryl guanidine derivatives, prostaglandin/thromboxane receptors, IP receptor antagonist, **5:**940–941
Arylhydroxylamines, acetylation/acylation, **1:**431–432
Aryl-linked hydroxymate inhibitors:
stroke therapy, **8:**490–491
toxicophore reactive metabolites, **2:**313
4-Arylpiperidines, opioid derivatives, **8:**615–621
carbon substituent, **8:**614–616
oxygen substitutent, **8:**616–618
Arylpropionic acids, drug metabolism, conjugation reactions, **1:**433–434
Aryl sulfonamide derivatives, quantitative structure-activity relationship, **1:**60–62
Aryl sulfonamidoindane analogs, chirality and toxicity, **1:**144
Arylsulfonamidothiazoles, type 2 diabetes therapy, 11β-hydroxysteroid dehydrogenase-1 inhibitors, **4:**23–25
Arzoxifene:
diarylethylene antagonists, **5:**248–249
future research issues and information retrieval, **5:**256
Ascomycin, nuclear magnetic resonance, FK506 analog binding to FKBP, **2:**410–412
Ascorbic acid. *See* Vitamin C (ascorbic acid)
Asenapine antipsychotic, **8:**201
Asimadoline κ-opioid receptor agonists, **8:**630–632
Asparaenomycin A, occurrence, structural variations and chemistry, **7:**333–335
Aspartate transcarbamoylase, allosteric proteins, **1:**382–386

Aspirin:
antiplatelet activity:
acute coronary syndrome, **4:**420
primary prevention, **4:**421
secondary prevention, **4:**421
thromboxane A_2 biosynthesis, **4:**419–420
migraine therapy, **8:**268
polymorphism, **3:**192–193
prodrug development, metabolic issues, **3:**232–233
resistance, mechanisms of, **4:**421
Assay techniques:
androgens:
anabolic agents, **5:**178–179
synthetic androgens, **5:**169–170
carboxylesterases, preclinical properties, **3:**239
cognitive function assessment, **8:**26–29
congestive heart failure therapy:
ex vivo cell-based biomarker assays, **4:**501
PET/SPECT ligands, predictive biomarkers, **4:**501
predictive hemodynamic/biochemical biomarker assays, **4:**500–501
target-engagement assays, **4:**501–502
drug development, mass spectrometry applications, **1:**119–121
functional assays, receptor-based drug design, **2:**523
HCV replication assays, **7:**174–175
high-throughput screening, enzymatic assays:
adaptation mechanisms, **3:**422–423
additives, stabilizers, and other cofactors, **3:**412–413
basic principles, **3:**401, 406
buffers and pH, **3:**410–412
configuration, **3:**414–423

substrate/enzyme concentration, **3:**414–418
substrate selection, **3:**420
unconventional setup, **3:**418–420
continuous/discontinuous assays, **3:**406–407
direct, indirect, and coupled assays, **3:**407–409
drug targets, **3:**401–402
expression and purification, **3:**402–406
hydrolases, **3:**425–426
isomerases, **3:**427–428
ligases, **3:**428
lyases, **3:**426–427
monovalent/divalent salts, **3:**412
optimization conditions, **3:**409–414
oxidoreductases, **3:**423–424
temperature, **3:**413–414
transferases, **3:**424–425
liver X receptors, **4:**149–150
nuclear hormone receptors, **4:**80–82
cotransfection assay, **4:**80
full-length receptor cotransfection assay, **4:**80–81
GAL4-LBD cotransfection assay, **4:**82
ligand displacement assay, **4:**80
receptor binding assays, **2:**521–523
retinoid nuclear receptor selectivity, **5:**373–378
Assessment rules, ADMET properties, **2:**63–69
Astellas patents, prostaglandin/thromboxane EP_1 antagonists, **5:**840, 843
Asthma therapy:
carbon monoxide inhalation therapy, **5:**303
soft drug design, $β_2$-agonists, **2:**103–104
AstraZeneca ZD prostaglandin/thromboxane EP_1 antagonists, structure-activity relationship, **5:**816–817

Astrocytic tumors (astrocytomas), histologic classification, **6:**230–234
Asymmetric dimethylarginine (ADMA), nitric oxide effects, **5:**294
Asymptotic q^2 rule, quantitative structure-activity relationship validation, **1:**48
AT7519 cyclin-dependent kinase inhibitor, **6:**314–315
Atazanavir sulfate:
  binding interactions, **7:**37, 39–40
  chemical structure, **7:**36
  efficacy and tolerability, **7:**40–42
  optimization, **7:**36–38
  pharmacokinetics, **7:**40
  resistance profile, **7:**42
ATG-003/mecamylamine hydrochloride, age-related macular degeneration therapy, **5:**622
Atherosclerosis. *See also* Cardiovascular disease
  bexarotene therapy, **5:**448
  epidemiology, **4:**331–332
  farnesoid X receptors, **4:**159–160
  HDL elevation:
    antioxidant/anti-inflammatory properties, **4:**334
    apolipoprotein A-1 and mimetics, **4:**335–337
    ApoA-I$_{Milano}$, **4:**335–336
    reconstituted HDL, **4:**336
    basic principles, **4:**331
    cholesteryl ester transfer protein inhibitors, **4:**339–345
      anacetrapib, **4:**343–344
      clinical trials, **4:**341–344
      current and future development, **4:**344–345
      dalcetrapib, **4:**342–343
      functional analysis, **4:**344
      torcetrapib, **4:**341–342
    innate immunity, **4:**334–335
    lipase inhibitors, **4:**337–340
      endothelial lipase, **4:**338–339
      hepatic lipase, **4:**337–338
    nicotinic acid receptor agonists, **4:**345–350
      anthranilic acid and related analogs, **4:**348–349
      niacin analogs, **4:**346–348
      xanthine and barbituric acid analogs, **4:**349–350
    plasma lipoproteins, **4:**332–333
    reverse cholesterol transport, **4:**333–334
  LDL cholesterol:
    current therapeutic agents, **4:**305–318
      bile acid sequestrants, **4:**315–316
      cholesterol absorption inhibitors, **4:**310–315
      combination products, **4:**316–318
      HMG-CoA reductase inhibitors, **4:**306–310
    emerging therapeutic agents, **4:**318–320
      antisense oligonucleotide, apoB100, **4:**319
      microsomal triglyceride transfer protein inhibitors, **4:**319–320
      PCSK9 inhibition, **4:**318–319
    research background, **4:**303–305
  liver X receptors, **4:**147–148
  peroxisome proliferator-activated receptors:
    α-agonists, **4:**89–99
    δ-agonists, **4:**118
    γ-agonists, **4:**102
  platelet function, **4:**409–410
  risk factors, **4:**331–332
Atmospheric pressure chemical ionization (APCI), liquid chromatography-mass spectrometry, **1:**101–105
Atomexetine, cognitive function assessment, **8:**27
Atomic-level molecular dynamics, docking and scoring techniques, flexibility computation, **2:**639
Atorvastatin:
  Alzheimer's disease therapy, preclinical/clinical testing, **8:**359–361
  anticholesterolemics, **2:**233–234
  ezetimibe and, **2:**234
  cholesterol biosynthesis, **5:**15–16
  LDL cholesterol lowering mechanisms, **4:**307–308
  amlodipine besylate combination, **4:**318
Atovaquone antimalarial agent, **7:**619–620
  cellular respiration and electron transport, **7:**655–658
ATP binding cassette (ABC) transporters:
  ezetimibe mechanism of action, **4:**315
  high-density lipoproteins, reverse cholesterol transport, **4:**333–334
  liver X receptors, **4:**147
  resistance targeting and mediation, **6:**371–376
    breast cancer resistance protein, **6:**372–373
    membrane structure, **6:**374–376
    multidrug resistance protein 1, **6:**372
    P-glycoprotein, **6:**371–372
    structure-activity and quantitative structure-activity relationship studies, **6:**374
  structure and classification, **2:**456–458
ATP-competitive kinesin spindle protein inhibitor, **6:**211–214
  GlaxoSmithKline/Cytokinetics compounds, **6:**211–213
  Merck compounds, **6:**213–214
Atracurium, nicotinic antagonists, **8:**79–80
Atrasentan, brain tumor therapy, **6:**268

Atrial natriuretic peptide, structure and function, **4:**533–537
Atropine:
cholinergic nervous system, **8:**61–62
chronic obstructive pulmonary disease bronchodilators, **5:**766–767
muscarinic antagonist, **8:**72
soft drug design, anticholinergics, **2:**109
Atropisomer:
defined, **1:**127
structural properties, **1:**133–137
Attention deficit-hyperactivity disorder (ADHD):
psychostimulant therapy, **8:**91–92
side effects, adverse reactions and drug interactions, **8:**95
soft drug design, methylphenidate, **2:**104–106
Attrited compounds:
BioPrint analysis, **1:**498–499
chemotype evaluation, **1:**496–498
congestive heart failure therapy, **4:**480–483
Atypical antipsychotics:
FDA-approved compounds, **8:**176–180
pharmacokinetics, **8:**191–200
benzisoxazoles, **8:**195–196
benzoisothiazoles, **8:**196–198
dibenzodiazepines, **8:**191–192
dibenzothiepine, **8:**194
dibenzthiazepines, **8:**193–194
phenylindole, **8:**199
quinolinone, **8:**198
substituted benzamide, **8:**199–200
thienobenzodiazepine, **8:**192–193
structure-activity relationships, **8:**180–185
Auristatin PE, marine sources, **2:**244–245
Aurora kinase inhibitors:
drug resistance, **6:**366

medicinal chemistry and clinical trials, **6:**316–321
Australia, patent requirements, **3:**122
Autocorrelation vectors, three-dimensional pharmacophores, **1:**471
AutoDock algorithm, docking and scoring techniques, **2:**613–614
Autoimmune diseases:
immunosuppressive drugs, research background, **5:**1001–1002
multiple sclerosis research, **5:**566–567
Autolysins, β-lactam mechanism of action, cell wall structural organization, **7:**266–269
Automated assisted screening, salt compound selection, **3:**395–396
Automation:
combinatorial chemistry, **1:**348
high-throughput screening, enzymatic assays, **3:**406
nuclear magnetic resonance screening, **2:**434
AWD-12-281 PDE inhibitor, chronic obstructive pulmonary disease therapy, **5:**783–784
Axial phenylpiperidines, **8:**598
Axitinib inhibitor, antiangiogenic properties, **6:**309–310
Axtirome, structure-activity relationship, **4:**206–207
2-Azabenzimidazoles, toxicophore reactive metabolites, aniline nitrogen oxidation, **2:**309–311
Azabicyclo[3.3.1] template, smoking cessation therapy, **5:**723–724
8-Azapurine derivatives, acyclic nucleoside phosphate, **7:**237
Azathioprine:
chemical structure, **5:**1050–1051

mechanism of action, **5:**1050–1051
multiple sclerosis therapy, **5:**569
research background, **5:**1050
therapeutic applications, **5:**1051–1053
Azauracil thyromimetic, structure-activity relationship, **4:**215–216
Azauricil side chains, thyroid hormone receptor β-selective agonists, structure-based design, **2:**689
6-Azauridine prodrug, lipophilicity increase, **2:**442–443
AZD4877 kinesin spindle protein inhibitor, ispenesib and, **6:**203–204
Azelaic acid, topical antimicrobial compounds, **7:**496
Azepines:
nicotinic agonists, **8:**68–69
plasmepsin antimalarial agents, **7:**644
γ-secretase inhibitors, **8:**336–337
Azepinone, migraine therapy, CGRP antagonists, **8:**280–288
Azetidines:
nicotinic antagonist, **8:**68–69
prostaglandin EP$_2$ receptor antagonists, **5:**852–854
Azidothymidine (AZT). *See also* Zidovudine
chemical delivery systems, **2:**141–142
Aziridine:
antimalarial cell cycle regulation, falcipain scaffold structures, **7:**635–637
estrogen receptor interaction, **5:**252–253
stroke therapy, cathepsin B inhibitors, **8:**474
Azithromycin:
analog design, **7:**444
antimalarial agents, **7:**621–624
experimental agents, **7:**625–627
quinine combined with, **7:**607

Azithromycin (*Continued*)
  pharmacology and ADMET
    properties, **7:**439–441
  RNA targets, **5:**984–990
Azithromycin, selective toxicity,
    2:552–553
Azo derivatives, anabolic agents,
    5:183
Aztreonam, biological activity,
    7:327–329
Azure A and B compounds,
    Alzheimer's disease,
    tau-targeted therapy,
    8:424–425

Baby hamster kidney (BHK) cells,
    protein therapeutics,
    3:293–294
Bacampicillin, penicillin
    resistance and
    development of,
    7:291–292
BACE (β-site cleavage enzyme):
  acylguanidines, **8:**351–353
  amyloid precursor protein
    processing, **8:**330–334
  chemical structure, **8:**331–333
  clinical testing, **8:**356–357
  guanidines, aminohydantoins,
    8:353–356
  insulin related therapy,
    Alzheimer's disease,
    8:416–417
  nonpeptidomimetic inhibitors,
    8:347–356
  preclinical testing, **8:**345–347
  transition-state isosteres and
    statines, **8:**347–351
Backbone structure:
  HIV protease inhibitor design,
    7:3
  mobility, *in silico* screening,
    protein flexibility,
    protein-ligand
    interactions,
    2:868–870
  peptide conformation, local
    constraints,
    1:210–214
  steroids, **5:**3–4
Back propagation algorithm
    (BBP), quantitative
    structure-activity
    relationship, **1:**39
Back propagation neural
    networks (BNNs),
    quantitative
    structure-activity
    relationship, **1:**39
Bacteria. *See also* Microbial
    compounds
  β-lactam mechanism of action:
    cell wall components,
    7:261–262
  cell wall structural
    organization,
    7:266–269
Bacterial genomes, natural
    products lead
    generation, **2:**209–210
BAL30072 monobactam, **7:**329
BAL30076 monobactam,
    7:329–331
Balaglitazone, selective PPARγ
    modulators, PPARγ
    partial agonists,
    4:108–115
Banoxantrone, hypoxia-activated
    bioreductive prodrugs,
    6:117–120
Bapineuzamab, Alzheimer's
    immunization
    therapy, **8:**370, 378
Barbiturates, anticonvulsant
    agents, **8:**139–141
Barbituric acid analogs,
    atherosclerosis, HDL
    elevation, **4:**349–350
Basal ganglia, Parkinson's
    disease,
    pathophysiology,
    8:531–534
Base-excision repair (BER)
    pathway:
  poly(ADP-ribose)polymerase-1
    (PARP-1) single-
    strand break repair,
    6:160–162
  poly(ADP)-ribosylation
    biochemistry,
    6:151–157
Bases, salt compounds, acid-base
    equilibrium, **3:**384–385
Basiliximab, transplantation
    rejection prevention,
    5:1001–1002
Batch processing:
  commercial-scale operations,
    continuous
    microfluidic reactors,
    3:13–14
  protein therapeutics, upstream
    process development,
    3:298–300
Batimastat hydroxymate
    inhibitors, stroke
    therapy, **8:**503
Batten disease,
    N-(4-hydroxyphenyl)
    retinamide (4-HPR)
    therapy, diabetes,
    5:474
BAY 57-1293 anti-DNA virus
    agent, herpes simplex
    virus therapy,
    7:242–243
Bayesian models, BioPrint®
    database,
    1:501–502
BBR 3464 platinum complex,
    cross-linked
    interstrand DNA
    targeting, **6:**91
β-cell function, type 2 diabetes,
    4:1–2
B cells, immune function,
    5:1002–1006
Bcl-2 inhibitors:
  hit seeking combinatorial
    libraries, **1:**312–315
  peptidomimetics,
    1:231–233
Bcr-Abl inhibitors:
  chronic myelogenous leukemia,
    6:298–301
  drug-resistant mutation,
    6:354–356
  small-molecule imatinib
    mesylate, resistance
    to, **6:**364–366
  tyrosine kinase inhibitors,
    selective toxicity,
    2:563
BE18257B antagonist:
  ligand-based design, nuclear
    magnetic resonance,
    2:382–383
  peptidomimetic design,
    1:224–230
Becatecarin, topoisomerase II
    inhibitors, **6:**102
Beckett-Casey opioid receptor
    binding model,
    8:597
Beclomethasone dipropionate/
    monopropionate
    (BDP/BMP), chronic
    obstructive pulmonary
    disease inhalation
    therapy, **5:**774–775,
    777–779

Befloxatone, structure-activity relationship and metabolism, **8:**252–253
Behavioral assessment:
  Alzheimer's disease immunotherapy, **8:**366–367
  antipsychotic therapy, animal models, **8:**172–173
  cognitive function, **8:**25–29
Behet's disease, 13-*cis*-retinoic acid therapy, **5:**395
Belatacept, transplantation rejection prevention, **5:**1001–1002
Belinostat:
  hydroxamate inhibitors, stroke therapy, **8:**489–490
  structure and development, **6:**58–59
Belotecan, biological activity and side effects, **6:**97–98
Bench-scale experiments, scale-up operations, **3:**4–7
  pilot plant scale-up, **3:**7–8
  reaction solvent selection, **3:**5–6
  reaction temperature, **3:**6
  reaction times, **3:**6
  solid-state requirements, **3:**7
  stoichiometry and addition order, **3:**6–7
"Bench-to-bedside" predictive models, congestive heart failure therapy, **4:**482–483
Bendamustine, cancer therapy, **6:**83–84
Benign prostatic hyperplasia (BPH):
  finasteride chemical reactions, **5:**11–12
  phosphodiesterase inhibitors, **5:**716
Benserazide, Parkinson's disease, dopaminergic pharmacotherapy, **8:**541–542
Benzacetamide κ-opioid receptor agonist, **8:**624–632
Benzaldehydes, antimalarial falcipain inhibitors, **7:**637
Benzalkonium chloride, cationic surfactants, **7:**500

Benzamides:
  poly(ADP-ribose)polymerase inhibitors, **6:**163–167
  substitution, antipsychotic pharmacokinetics, **8:**199–200
  AMPA receptor glutmatergics, **8:**205–206
1-Benzazepines, chemokine receptor-5 (CCR5) antagonist, anilide core, **7:**127–131
Benzene derivatives, prostaglandin/thromboxane EP$_1$ antagonists, **5:**829, 831, 834
Benzimidazole:
  DP$_1$ receptor antagonists, **5:**905–906
  poly(ADP-ribose)polymerase inhibitors, **6:**164–167
Benzisoxazole:
  Hsp90 inhibitors, **6:**431
  pharmacokinetics, **8:**195–196
  structure-activity relationship, **8:**183–185
Benzithiazole, structure-activity relationship, **8:**183–185
Benznidazole, American trypanosomiasis therapy, **7:**573
Benzodiazepines:
  anticonvulsant applications, **8:**135–139, 144–147
  antimalarial drugs:
    falcipain inhibitors, **7:**635–637
    farnesylation, **7:**634
  combinatorial library, **1:**300–302
  HDM2-p53 interaction antagonists, **1:**326–327
  insomnia therapy, **5:**711–712, 716–718
  migraine therapy, telcagepant calcitonin gene-related peptide antagonist, **8:**277–288
  pharmacological activity, **1:**439–441
  privileged pharmacophores, lead drug discoveries, **2:**529–530

protein-protein interactions, p53/MDM2 compounds, **2:**346–352
soft drug analogs, **2:**124–125
zotepine interaction, **8:**194
Benzodioxolanes, toxicophore reactive metabolites, **2:**314–315
Benzoic acids (BAs):
  prostaglandin/thromboxane EP$_1$ antagonists, **5:**823, 827–829, 832
  quantitative structure-activity relationship, electronic parameters, **1:**2–8
Benzoisothiazoles, pharmacokinetics, **8:**196–198
Benzomorphans:
  numbering systems, **8:**610–611
  opioid receptor affinity, **8:**601–602
  structure-activity relationship, **8:**598–599, 610–613
Benzopyranoquinolines, nonsteroidal glucocorticoid receptor modulators, **5:**128–134
Benzoquinone ansamycins, geldanamycin Hsp90 inhibitor, **6:**397–399
Benzothiadiazines:
  AMPA receptor glutmatergics, **8:**206
  NS5B inhibitors, anti-HCV agent, **7:**202–203
Benzothiazole ketone, dipeptidyl peptidase 4 (DPP-4) inhibition, **4:**49–50
Benzothiazoles, NOS inhibitors, migraine therapy, **8:**302–306
Benzothiazones, antitubercular agents, arabinogalactan biosynthesis, **7:**722
Benzothiophene, migraine therapy, **8:**288–290
Benzotriazine inhibitors, stroke therapy, **8:**513–514
Benzotriazole, calpain inhibitor, stroke therapy, **8:**465

Benzoxazoles:
poly(ADP-ribose)polymerase inhibitors, **6:**164–167
smoking cessation therapy, **5:**724–725
structure-based design, **2:**698–700
Benzoyl peroxide:
adapalene combination therapy, **5:**410–411
topical antimicrobial compounds, **7:**496–497
Benzphetamine, structure-activity relationship, **8:**107
Benztropine mesylate, Parkinson's disease therapy, **8:**552–553
Benzyl alcohol, antimicrobial agents, **7:**492–493
N-Benzyl-4-aminopiperidines, NOS inhibitors, migraine therapy, **8:**303–306
Benzyl carbamates, stroke therapy, **8:**454–455
Benzylpenicillin, chemical delivery system, **2:**142–143
Benzyl sulfides, DP$_1$ receptor antagonists, **5:**905–906
Berenil antitrypanosomal agent, **7:**578–580
Beriberi, thiamine deficiency, **5:**668–670
Besonprodil, stroke therapy, **8:**454
Best mode principle, patent specifications, **3:**126, 130–131
β-blockers. *See also* β-Adrenergic receptor antagonists
antidepressants, efficacy assessment, **8:**225–229
eye-targeted chemical delivery systems, oxime/methoxime analogs, **2:**158–160
glaucoma therapy, **5:**596–600
migraine prevention, **8:**313–315
quantitative structure-activity relationship, **1:**59–60
soft drug development, **2:**81–88

adaprolol, **2:**83–85
esmolol, **2:**85
landiolol, **2:**85–86
vasomolol, flestolol, et al, **2:**86
β-turn motifs, chemogenomics, protein secondary structure mimetics, **2:**577–578
Betaine, soft drug design, **2:**113
Betamethasone:
C-21 carbonyl alterations, **5:**106–112
chemical reactions, **5:**11–14
potency estimations, **5:**93–95
Betamipron, biological activity, **7:**351–353
Beta-secretase inhibitors (BACE1), hit seeking combinatorial libraries, **1:**327–328
Beta-turns, first-generation peptidomimetics, **1:**215–216
Bethanechol, structure and function, **8:**64
Bevacizumab:
age-related macular degeneration therapy, **5:**615–616
glioblastoma therapies, **6:**232–233
resistance to, **6:**368–369
VEGFR2 inhibitors, antiangiogenic properties, **6:**307–310
Bexarotene:
adverse effects, **5:**450–453
chemical structure, **5:**429–435
pharmacology/metabolism, **5:**453
retinoic nuclear receptors, **5:**369–370, 372
therapeutic applications, **5:**435–450
atherosclerosis, **5:**448
cancer, **5:**435–448
cardiovascular disease, **5:**448–449
dermatitis, **5:**448–449
diabetes, **5:**449
psoriasis, **5:**449
schizophrenia, **5:**449–450
vitamin A pharmacology, **5:**649
Beyond our Borders Initiative (FDA), **3:**69

Bezafibrate:
hemoglobin oxygen delivery, allosteric effectors, **4:**615–619
peroxisome proliferator-activator receptors, α-agonists, **4:**96–99
BF2.649 histamine, cognitive enhancement, **8:**36–37
Biapenem, development of, **7:**354–356
BIBN4096 calcitonin gene-related peptide antagonist, migraine therapy, **8:**273, 275–277
BIBW-2992 erbB family inhibitor, **6:**302–305
Bicifadine, stroke therapy, **8:**450
Bicyclic β-lactamase inhibitors:
lactam inhibitors, **7:**371
penems, **7:**364–368
Bicyclic pyrimidine nucleoside analogs (BCNA):
anti-DNA virus antimetabolite, **7:**223–224
varicella-zoster virus therapy, **7:**243
Bicyclic scaffolds, peptidomimetics, **1:**235–239
Bicyclo[3.1.0]hexane, chirality and toxicity, **1:**144
Bicycloproline inhibitors, teleprevir anti-HCV agent, **7:**185–186
Bidimensional structures, thyroid hormone receptor β-selective agonists, structure-based design, **2:**685–691
Biguanides:
antimalarial drugs, **7:**617–618
antitrypanosomal polyamine metabolism, **7:**588–590
topical antimicrobial agents, **7:**498–499
type 2 diabetes therapy, **4:**2–3
BILA 2157 BS renin inhibitor, **4:**246, 249
Bile acid response element (BARE), drug delivery, **2:**468–475
Bile acids:
farnesoid X receptor, **4:**156–157
metabolism, **4:**157

sequestrants, LDL cholesterol lowering mechanisms, **4:**315–316
Biliary excretion, permeability, ADMET interactions, **3:**373
Bilirubin, serum albumin binding, **3:**442–451
BILN-2061 protease inhibitor. *See* Ciluprevir anti-HCV protease inhibitor
Bimatoprost, glaucoma intraocular pressure reduction, **5:**592–600
Binaltorphimine, κ-opioid selective antagonist, **8:**602–604
Binary complex residence time, receptor-based drug design, **2:**504
Binary quantitative structure-activity relationships, ligand-based virtual screening, **2:**11–13
Binding activity:
  allosteric proteins, **1:**387–392
  angiotensin-converting enzyme, **4:**286–288
  chemokine receptor-5 (CCR5) antagonist, **7:**133–134
  G-protein coupled receptor, homology modeling, lead generation, **2:**289–292
  heparin molecular mechanism of action, **4:**377–380
  HIV protease inhibitors, **7:**2–3
    amprenavir, **7:**26, 29
    atazanavir sulfate, **7:**37, 39–40
    darunavir, **7:**55–57
    indinavir sulfate, **7:**14, 19
    lopinavir/ritonavir, **7:**32, 34
    nelfinavir mesylate, **7:**20, 22, 25
    saquinavir mesylate (SQV), **7:**7–8
    tipranavir, **7:**46
  Hsp90 inhibitors, **6:**383–386, 425–429
    C-terminal nucleotide binding site:
      cisplatin, **6:**440–441
      elucidation, **6:**430–431
      epigallocatechin-3-gallate, **6:**441

GHLK protein family, **6:**429–430
novobiocin inhibitor, **6:**431–440
derrubone, **6:**425–426
gedunin and celastrol, **6:**428–429
high-throughput screening and modeling, **6:**429
naphthoquinones, **6:**426–427
small-molecule Hsp90/HOP interactions, **6:**427
taxol, **6:**427–428
ligand-macromolecular interactions, structure-based thermodynamic analysis, **2:**703–705
macromolecule-ligand interactions:
  FK506 analog, **2:**410–412
  nuclear magnetic resonance studies, **2:**407–409
serum albumin structural survey, **3:**440–442
  anthracyclines, **3:**452–454
  bilirubin binding, **3:**442–451
  podophyllotoxin derivatives, **3:**454–456
structural genomics research, **1:**583–584
Binding affinity evaluation:
  quantum mechanics/molecular mechanics:
    advantages, **2:**719–723
    linear interaction energy method, **2:**729–730
    PBSA/GBSA approaches, **2:**730–733
    pertubation, **2:**723–729
      free-energy perturbation, **2:**723–728
      thermodynamic integration, **2:**728–729
    research background, **2:**717–719
  in silico screening, protein flexibility, **2:**870–871
Binding efficiency index:
  empirical scoring, **2:**625–628
  nuclear magnetic resonance, ligand-based drug design, **2:**433
Binding pockets:
  dihydropteroate synthase, **7:**528–531

Hsp90 inhibitors, N-terminal ATP-binding pocket, **6:**386–425
  chimeric geldanamycin and radicicol, **6:**405–410
  geldanamycin, **6:**386–399
  high-throughput small-molecule inhibitors, **6:**417–421
  purine-based inhibitors, **6:**410–416
  radicicol analogs, **6:**399–405
  shepherdin, **6:**416
  structure-based design, **6:**421–425
kinase inhibitors, multiconformational catalytic domain, **6:**346–349
kinesin spindle protein inhibitors:
  EMD-534085, **6:**210–211
  monastrol, **6:**193–197
Bioactive conformation:
  chemokine receptor CXCR4 antagonist-plerixafor, **4:**586–587
  drug metabolism, **1:**439
  erythropoietin, **4:**571
  granulocyte colony stimulating factor, **4:**576–577
  granulocyte-macrophage colony stimulating factor, **4:**573–574
  interleukin-11, **4:**578–580
  ligand-based drug design, nuclear magnetic resonance, **2:**377–383
  peptidomimetics, **1:**215
  pharmacophore generation, **1:**469–471
  stem cell factor, **4:**581–582
  thrombopoietins, **4:**583–586
Bioactivity assays, protein therapeutics, analytic development, **3:**321–322
Bioadhesives, oral drug systems, class III bioadhesives, **3:**52
Bioavailability:
  ADMET properties assessment, **2:**64
  Biopharmaceutics Classification System, regulatory practices, **3:**355–356

Bioavailability (*Continued*)
  defined, 2:749
  drug delivery systems, mathematical modeling:
    definitions, 2:746–751
    diffusion-controlled dosage, plasma drug profile, 2:801–816
    erosion-controlled dosage, plasma drug profile, 2:816–828
    future research issues, 2:851–852
    intravenous drug administration, 2:751–771
    oral dosage:
      immediate-release, 2:771–787
      sustained release, 2:787–791
    research background, 2:745–746
    sustained release *in vitro* testing, 2:791–801
    tissue-based drug transfer, 2:838–851
    transdermal systems, 2:828–838
  heparin molecular mechanism of action, 4:378–380
  HIV protease inhibitors, design criteria for, 7:3
  metropolol experiment, transdermal drug delivery, 2:832–833
  permeability, ADMET interactions, oral bioavailability, 3:372
  pharmaceutical cocrystals, 3:198–213
  polymorph cocrystals, 3:196–198
Bioconversion mechanisms, prodrug design, 3:252–254
Biodistribution, antipsychotics, 8:185–200
  benzisoxazoles, 8:195–196
  benzoisothiazoles, 8:196–198
  butyrophenones, 8:189–190
  diarylbutylamines, 8:190
  dibenzodiazepines, 8:191–192
  dibenzothiepine, 8:194
  dibenzoxazepines, 8:189
  dibenzthiazepines, 8:193–194
  dihydroindolone, 8:190–191
  phenothiazines, 8:186–188
  phenylindole, 8:199
  quinolinone, 8:198
  substituted benzamide, 8:199–200
  thiobenzodiazepine, 8:192–193
  thioxanthenes, 8:188–189
Biodiversity, natural products lead generation, 2:191–192
  harvesting techniques, 2:199–200
Bioequivalence (BE):
  Biopharmaceutics Classification System, regulatory practices, 3:355–356
  generic drug guidelines, 3:85, 91
Biofenthrin, quantitative structure-activity relationships, 1:73–74
Biofilms, chemotherapeutic agents, 7:503
Bioisosteric design:
  direct analog, 1:170–172
  peptide libraries, 1:290–291
  plasmepsin antimalarial agents, 7:639–644
  thyromimetic structure-activity relationship, 4:218–219
Biological assays, ADMET applications, 2:48–49
Biological fingerprints:
  basic properties, 1:483–484
  CEREP Bioprint database, 1:487–489
  drug analyses and targeting, 1:496–498
  drug discovery applications, 1:499–501
  future trends and applications, 1:502–503
  hit/lead compound selection, 1:489–491
  ligand-based virtual screening, 2:3
  polypharmacology, drug development, 1:493–496
  in silico approaches, 1:501–502
  structure-based pharmacophore modeling, 2:682–683
  tool compounds and target validation, 1:491–493
Biological parameters:
  drug metabolism, 1:447–449
  quantitative structure-activity relationship, 1:31–32
  in vivo interactions, 1:59–60
Biological spectra technique, safety, selectivity, and promiscuity profiles, 1:499–501
Biologic Control Act, 3:93
Biologics, transplantation rejection prevention, 5:1001–1002
Biologics license applications (BLAs), FDA guidelines, 3:94–95
Biology-oriented synthesis (BIOS), chemogenomics, 2:580
Bioluminescence imaging, brain tumor modeling, 6:273–274
Biomolecular interactions, docking/scoring paradigms, 2:594–596
Biopharmaceuticals, drug delivery system bioavailability goals, 2:746–747
Biopharmaceutics Classification System (BCS):
  crystal engineering, 3:187–190
  drug ADMET/PK interactions, Permeability-Based Classification System, 3:370–371
  drug disposition classification system, 3:361–362
  future trends and issues, 3:362–363
  gastrointestinal absorption, permeability *vs.* solubility, 3:374–375
  oral drug delivery, 3:41–55
    absorption, 3:353–355
    class I drugs, 3:42, 354
    class II drugs, 3:42–50, 354
    class III drugs, 3:50–52, 355
    class IV drugs, 3:52, 355
    class V drugs, 3:52–55
    global market, 3:353–363
    provisional classification process, 3:356–362
      biowaiver monographs, 3:360–361

literature data, **3:**356–359
in silico calculations, **3:**359–360
regulatory practices, **3:**355–356
Biopharmaceutics Drug Disposition Classification System (BDDCS):
drug ADMET/PK interactions, **3:**371
evolution of, **3:**361–362
Biophysical techniques, protein therapeutics, **3:**323–325
BioPrint® database:
attrited compound analysis, **1:**498–499
biological space *vs.* chemical structure, **1:**485–487
chemotype development, **1:**484–487
compounds *vs.* assays, **1:**483–484
data flowchart, **1:**487–489
development of, **1:**483
hit/lead compound selection, **1:**489–491
polypharmacology of drugs, **1:**493–496
safety, selectivity, and promiscuity profiles, **1:**499–501
in silico methods, **1:**501–502
tool compounds and target validation, **1:**491–493
Bioseparation, protein therapeutics, downstream processing, **3:**301–311
Biospectra analysis, chemogenomics, ligand-based data analysis, **2:**585
Biosynthetic pathways:
aminoglycoside resistance, **7:**425–427
anthracyclines, **6:**22–24
antifolates, **6:**106–109
antitubercular agents, cell wall biosynthesis, **7:**719–724
DNA-targeted compounds:
bleomycin, **6:**11
dactinomycin, **6:**6–8
plicamycin, **6:**17
docetaxel/taxotere, **6:**34–36

female sex hormones, **5:**230–233
glucorticoid anti-inflammatory compounds, **5:**56–59
macrolide drug development, **7:**446–456
morphine derivatives, **8:**599
natural products lead generation, **2:**210–212
thyroid hormones, **4:**192–193
Biosynthetic penicillin, development of, **7:**288–289
Biotechnology:
drug development, FDA guidelines for, **3:**93–96
intellectual property:
future research issues, **3:**186
miscellaneous protections, **3:**185–186
patents:
enforcement, **3:**148–165
alternative dispute resolution, **3:**162–163
claim construction proceedings, **3:**162
defenses to infringement, **3:**154–157
discovery phase, **3:**160–161
geographic scope, **3:**148
infringement determination, **3:**152–154
parties to suit, **3:**150–152
proceedings commencement, **3:**159–160
remedies for infringement, **3:**157–159
summary judgment, **3:**161
trial and appeal processes, **3:**148–150, 162–164
global patent protection, **3:**107–108, 165–171
international agreements, **3:**165–168
international laws and regulations, **3:**169–171

PCT patent practice, **3:**168–169
protection and strategy, **3:**105–118
absolute novelty principle, **3:**110–111
application publication, **3:**114–115
first to invent *vs.* first to file, **3:**108–110
global strategy, **3:**107–108, 165–171
patent term, **3:**111–114
PTO prosecution strategy, **3:**115–118
trade secrets and, **3:**182–183
requirements, **3:**118–148
corrections, **3:**146–148
interference, **3:**142–146
new and nonobvious inventions, **3:**133–138
patentable subject matter, international requirements, **3:**121–123
provisional applications, **3:**127–133
PTO procedures, **3:**138–142
specifications, **3:**123–127
U.S. requirements, **3:**118–121
research background, **3:**101–105
trademarks, **3:**171–179
global rights, **3:**177–178
Lanham Act protections, **3:**178–179
as marketing tools, **3:**172
oppositions and cancellations, **3:**176–177
preservation of rights through proper use, **3:**177
registration process, **3:**174–176
selection criteria, **3:**172–174
trade secrets, **3:**179–184
defined, **3:**180
enforcement, **3:**181–182
Freedom of Information Acts, **3:**183–184
global protection, **3:**184

Biotechnology (*Continued*)
   patent protection and, 3:182–183
   protection requirements, 3:180–181
Biotin, 5:685–691
   antitubercular agents, cofactor biosynthesis targeting, 7:728
   chemistry, 5:686
   deficiency, 5:688, 691
   hypervitaminosis biotin, 5:691
   metabolism, 5:686–690
Biotransformation:
   drug metabolism, global expert systems, 1:447
   xenobiotic metabolism, 1:407–408
Biowaiver monographs, Biopharmaceutics Classification System, 3:360–361
Biperiden:
   muscarinic antagonists, 8:72–74
   Parkinson's disease therapy, 8:552–553
Biphenylsulfonacetic acid derivatives, human papillomavirus therapy, 7:242
Biricodar kinase inhibitor, drug resistance, 6:372
Bis analogs, dual topoisomerase I/II inhibitors, 6:105–106
Bisaryloxime ethers, structure-based design, 2:700–701
Bischalcones, antimalarial topoisomerase inhibitors, 7:650–652
Bis-hydroxymate inhibitors, stroke therapy, 8:507
Bisphenols, antimicrobial agents, 7:494–495
Bisphosphonates:
   nitrogen-containing, structural genomics, 1:584–585
   osteoporosis antiresorptive therapy, 5:733–734
   selective toxicity, 2:564
Bivalent linear/cyclic Smac mimetics, peptidomimetic design, 1:237–240

Black-box model of cognition, 8:17–19
Bladder cancer, N-(4-hydroxyphenyl) retinamide (4-HPR) therapy, 5:465, 468–469
BLEEP1/BLEEP2 mean-field scoring functions, empirical scoring, 2:630–631
Bleomycins:
   anti-cancer activity, 2:235–239
   biosynthesis, 6:11
   chemical structure, 6:8
   contraindications and side effects, 6:8
   current and future research issues, 6:12
   molecular mechanisms, 6:12
   pharmacokinetics, 6:8–11
   therapeutic applications, 6:8
Blister fluid drug transport, 2:844–848
Block copolymers, nanoscale drug delivery, 3:475–479
Blood. *See also* Hematopoietic agents
   drug absorption:
     oral administration, 2:748
     repeated intravenous bolus injection, 2:754–755
     transdermal drug delivery, 2:831–832
   thyroid hormone transport, 4:193
Blood-brain barrier (BBB):
   brain-targeting chemical delivery systems, 2:136–140
   central nervous system tumors, 6:270–271
   estradiol chemical delivery system, 2:144–149
   multiobjective drug discovery optimization, solubility, 2:268–273
   multiple sclerosis pathology, 5:566–567
   permeability, ADMET interactions:
     distribution, 3:375–376
     transport interactions, 3:368–369
   quantitative structure-activity relationship, ADMET parameters, 1:65

Blood substitutes:
   hemoglobin-based oxygen carriers, 4:625–643
   artificial erythrocytes, 4:640–642
   conjugated hemoblogins, 4:635–637
   cross-linked hemoglobin, 4:628–631
   nitric oxide reactions, 4:612–614
   polymerized hemoglobins, 4:631–635
   recombinant hemoglobins, 4:637–640
   redox/radical reactions, 4:614–615
   inhaled nitric oxide therapy and, 5:277–278
BMS-182874, peptidomimetic design, 1:229–230
BMS-191011 potassium channel, stroke therapy, 8:482–483
BMS 204352/MaxiPost potassium channel, stroke therapy, 8:482
BMS-387032 cyclin-dependent kinase inhibitor, medicinal chemistry and classification, 6:311–312, 314
BMS-644950 compound, anticholesterolemic mechanisms, 2:234
BMS-687453 compound, peroxisome proliferator-active receptors, α-agonist derivation, 4:93–96
BMS-694153 calcitonin gene-related peptide antagonist, migraine therapy, 8:288–290
Boceprevir anti-HCV agent, 7:186–191
Boltzmann jump strategy, pharmacophore formation, bioactive conformation, 1:470
Bombesin, structure and function, 4:549
Bondi atomic volumes, quantitative structure-activity relationships, 1:26

Bone:
  calcification changes, 13-*cis*-retinoic acid therapy, **5:**394
  osteoporosis:
    anabolic therapy, Wnt pathway and bone remodeling, **5:**756–757
    antiresorptives, mechanistic studies, **5:**729–742
  thyroid hormone effects, **4:**196
Bone mineral density (BMD), osteoporosis antiresorptive therapy:
  bisphosphonates, **5:**733–734
  cathepsin K inhibitors, **5:**731–732
  mechanistic studies, **5:**729–730
  selective estrogen receptor modulators, **5:**739–742
Bone morphogenetic proteins (BMPs):
  brain tumor therapy, **6:**267
  osteoporosis anabolic therapy, Wnt pathway and bone target cells, **5:**757–758
Bootstrapping, quantitative structure-activity relationship validation, **1:**48
Boric acid, topical antimicrobial compounds, **7:**496
Boronates, HDAC inhibitors, stroke therapy, **8:**494
Boronic acid derivatives:
  dipeptidyl peptidase 4 (DPP-4) inhibition, **4:**50–54
  proteosome inhibitors, bortezomib clinical trials, **6:**177–180
Bortezomib, discovery and development, clinical trials, **6:**177–180
Bosentan, peptidomimetic design, **1:**228–229
Bosutinib:
  multitarget drug development, cancer therapy, **1:**263–267
  second-generation development, drug resistance and, **6:**364–365

tumor metastases inhibition, **6:**325–327
Boundary conditions, diffusion-based drug release, **2:**792–793
Bound ligands, three-dimensional pharmacophores, **1:**462–465
BP 897, dopamine D3 receptor ligands, **1:**315–317
BQ-123, peptidomimetic design, **1:**225–226
BQ-788, peptidomimetic design, **1:**225–226
Bradykinesia syndrome, pathology, **8:**529–530
Bradykinin:
  angiotensin-converting enzyme inhibitors:
    potentiating peptides, **4:**268–269
    potentiation pharmacology, **4:**279–280
  B2 antagonists, peptidomimetic design, **1:**223–224
  biological actions and receptor pharmacology, **4:**545–547
  dipeptidyl peptidase 4 inhibitors, **4:**41–42
  structure and function, **4:**544–547
B-Raf inhibitors, structure and properties, **6:**305–306
Braf kinase, chemogenomics, **2:**587–588
Brain cancer. *See also* specific tumors, e.g. Glioma
  angiogenesis, **6:**242–243
  astrocytic tumors, **6:**230
  ependymoma, **6:**233
  epidemiology and prevalence, **6:**223–224
  N-(4-hydroxyphenyl) retinamide (4-HPR) therapy, **5:**465, 469–470, 478–480
  imaging studies, **6:**273–274
  invasiveness, **6:**241–242
  meningiomas, **6:**237
  metastatic, **6:**240
  modeling approaches, **6:**272–273
  neuroepithelial tumors, **6:**229–230

oligodendroglial tumors, **6:**233–234
primary brain tumors, **6:**227–234
therapies, **6:**244–270
  blood-brain barrier, **6:**270–271
  bone morphogenetic proteins, **6:**267
  chemokine receptors, **6:**264–265
  cytotoxic agents, **6:**244
  epidermal growth factor receptors, **6:**246–247
  estrogen receptors, **6:**265
  growth factors, receptors, and signaling pathways, **6:**245
  GSK3β inhibitors, **6:**263–264
  heat shock protein inhibitors, **6:**267–268
  hedgehog signaling pathway, **6:**259–263
  HGF/MET sytems, **6:**247–248
  hypoxia inducible factor pathway, **6:**253–255
  integrins, **6:**255–256
  MDM2/MDMX pathway, **6:**257–258
  mTOR signaling, **6:**250–252
  nuclear hormone receptors, **6:**265
  obstacles to, **6:**270
  p53 signaling pathway, **6:**256–257
  peroxisome proliferator-activated receptors, **6:**265–267
  phosphatase and tensin homolog (PTEN), **6:**258–259
  phosphodiesterase 4, **6:**265
  phosphoinositide 3-kinase inhibitors, **6:**249–250
  PI3K/AKT signaling pathway, **6:**249
  platelet-derived growth factor and receptor, **6:**249
  protein kinase C, **6:**245–246
  RAS/RAF/MEK/ERK signaling, **6:**252
  retinoid X receptor, **6:**267
  Rho/ROCK and MAPK signaling pathways, **6:**252–253

Brain cancer (*Continued*)
  small-molecule compounds, **6:**268
  targeted delivery, **6:**271–272
  transforming growth factor-β, **6:**267
  vaccines for, **6:**268–270
  vascular endothelial growth factor inhibitors, **6:**248
  vascular endothelial growth factor kinase inhibitors, **6:**248–249
  Wnt signaling pathway, **6:**263
 tumor classification, **6:**228–229
  astrocytic tumors, **6:**230
  ependymoma, **6:**233
  neuroepithelial tumors, **6:**229–230
  oligodendroglial tumors, **6:**233–234
  WHO grading system (I-IV), **6:**230–233
 tumor initiation, **6:**224–227
 WHO grading system (I-IV), **6:**230–233
Brain-derived neurotrophic factor (BDNF), antidepressant efficacy assessment, **8:**226–229
Brain imaging studies:
 depression, **8:**277–279
 migraine pathology, **8:**267
Brain natriuretic peptide (BNP):
 biological actions and pharmacology, **4:**536
 congestive heart failure therapy:
  disease biomarkers, **4:**504–505
  specific target/mechanism perturbation, **4:**494–495
  structure and function, **4:**533–537
  therapeutic potential, **4:**537
Brain-targeted chemical delivery systems, **2:**136–140
 molecular packaging, **2:**150–155
  kyotorphin analogs, **2:**153–154
  leu-enkephalin analogs, **2:**151–153
  redox analogs, **2:**154–155
  TRH analogs, **2:**153

Brasofensine, monoamine transporter inhibition, **8:**240–241
BRCA1/BRCA2 genes, poly(ADP-ribose)polymerase (PARP) inhibitors, cancer therapy, **6:**168–170
Breast cancer:
 Am80 therapy, **5:**417
 bexarotene therapy, **5:**443–447
 endocrine resistance in, **6:**369–370
 N-(4-hydroxyphenyl) retinamide (4-HPR) therapy, **5:**465–466, 468–470, 478–480
 peptidomimetic analogs, **5:**480–483
 kinesin spindle protein inhibitors, ispenesib, **6:**198–204
 9cUAB30 RXR-selective retinoid, **5:**455–456
 NRX 194204 RXR selective analog, **5:**460–461
 NRX 195183 analog therapy, **5:**420
 9-*cis*-retinoic acid therapy, **5:**425
 13-*cis*-retinoic acid therapy, **5:**389–390
 selective estrogen receptor modulators, **5:**741–742
Breast cancer resistance protein (MXR), drug resistance mechanisms, **6:**372–373
Brefeldin A (BFA) metabolite, transcytosis, **2:**448
Bridging group "X," thyromimetic structure-activity relationship, **4:**213–214
Bridging mechanism:
 heparin molecular mechanisms, **4:**374–380
 indirect anticoagulants, **4:**392
Brimonidine, glaucoma therapy, **5:**598–600
Brinzolamide, glaucoma therapy, **5:**598–600
 combined protocols, **5:**602

Brivanib, antiangiogenic properties, **6:**308–310
Brivaracetam, anticonvulsant applications, **8:**136–139, 150–151
Brivudin (BVDU):
 anti-DNA virus antimetabolite, **7:**223–224
 isodideoxynucleoside congener, **7:**227
 varicella-zoster virus therapy, **7:**243
Brofaromine, structure-activity relationship and metabolism, **8:**252–253
Bromocriptine, Parkinson's disease dopaminergic therapy, **8:**545–546
Bronchial mucus, drug delivery through, **2:**839–844
Bronchodilators:
 chronic obstructive pulmonary disease, **5:**766–773
  β_2-adrenergic receptor agonists, **5:**769–772
  combined therapies, **5:**772
  methylxanthines, **5:**772–773
  muscarinic M$_3$ receptor antagonists, **5:**766–769
 soft drug design, anticholinergics, **2:**108–109
Brönsted catalysis law, quantitative structure-activity relationship, **1:**2–8
Bryostatins, marine sources, **2:**243–245
BSI-201 chemotherapeutic agent, poly(ADP-ribose) polymerase (PARP) inhibitor coadjuvants, **6:**169–170
Budapest Treaty, drug patent requirements, **3:**126–127
Budesonide:
 chronic obstructive pulmonary disease inhalation therapy, **5:**774–775, 777–779
 soft drug design, etiprednol dicloacetate and analogs, **2:**96–97

Buffers, high-throughput screening, assay optimization, 3:410–412
Buformin, structure-activity relationships, 4:3
Bufuralol analogs, soft drug design, 2:86–88
  active metabolites, 2:80, 127–128
Bupropion:
  chemical structure, 8:224
  monoamine transporter inhibition, 8:239–241
  side effects and adverse reactions, 8:234
  structure-activity relationship, 8:251, 254–255
Buriedness, hydrogen bonding, and binding energy (BHB) scoring method, structure-based virtual screening, 2:19–20
Buspirone, serotonin receptor targeting, 8:242–244
Butane 1,4-diamine scaffolds, tri-substituted pyrrolidine chemokine receptor-5 (CCR5) antagonist, 7:121–124
Butaprost, EP_2 receptor agonists, 5:846–848
Butirosin:
  chemical structure, 7:419
  history and biosynthesis, 7:423–424
Butixocort 21-propionate, soft drug design, 2:100
Butyric acid, stroke therapy, 8:494
Butyrophenones, pharmacokinetics, 8:189–190
Butyrylcholinesterase (BuChE), cognitive enhancement, 8:30–32
BW373U86 δ-opioid selective agonist, 8:632–635
BW 1370U87 RIMA, structure-activity relationship and metabolism, 8:252–253

C-1 carbonyl, glucocorticoid anti-inflammatory compound alterations, 5:83
C-2 carbonyl, glucocorticoid anti-inflammatory compound alterations, 5:83–86
C2-MAD analog, nucleoside inhibition, 5:1044–1047
C-3 carbonyl:
  glucocorticoid anti-inflammatory compound alterations, 5:86–87
  glucocorticoid biosynthesis and metabolism, 5:58
C-4 carbonyl, glucocorticoid anti-inflammatory compound alterations, 5:87–88
C-6 carbonyl, glucocorticoid anti-inflammatory compound alterations, 5:88–89
C-6 hydroxylation, glucocorticoid biosynthesis and metabolism, 5:58
C-6-substituted penicillins, penicillin resistance and development of, 7:292–295
C-7 carbonyl, glucocorticoid anti-inflammatory compound alterations, 5:89–90
C-8 carbonyl, glucocorticoid anti-inflammatory compound alteration, 5:91
C-11 carbonyl, glucocorticoid anti-inflammatory compound alteration, 5:91–92
C-12 carbonyl, glucocorticoid anti-inflammatory compound alteration, 5:92–95
C-15 carbonyl, glucocorticoid anti-inflammatory compound alteration, 5:95
C-16 carbonyl, glucocorticoid anti-inflammatory compound alteration, 5:95–99

C-17 carbonyl:
  glucocorticoid anti-inflammatory compound alteration, 5:99–102
  glucocorticoid biosynthesis and metabolism, 5:58
C-20 carbonyl:
  glucocorticoid anti-inflammatory compound alterations, 5:102–103
  glucocorticoid biosynthesis and metabolism, 5:58
C-20 oxyprednisolonate 21-esters, soft drug design, 2:100
C-21 carbonyl, glucocorticoid anti-inflammatory compound alterations, 5:104–112
Cabacephems, structure-activity relationships, 7:260–261
Cabergoline, Parkinson's disease dopaminergic therapy, 8:545–546
Caco-2 assay:
  ADMET permeability, 2:66
  membrane proteins, transcytosis, 2:448
Cade testing procedure, antiseptic testing, 7:490
CADEX process, serum albumin structural survey, 3:439–440
Caffeine:
  cognitive enhancement, 8:47–48, 51–52
  Parkinson's disease and, 8:538–540
  physiology and pharmacology, 8:105–106
  side effects, adverse reactions and drug interactions, 8:96–97
  structure and properties, 8:90
Calcineurin inhibitors:
  soft drug design, 2:119–121
    cyclosporine analogs, 2:120–121
    tacrolimus analogs, 2:121
  transplantation rejection prevention:
    cyclosporin A binding, 5:1008–1009

Calcineurininhibitors(*Continued*)
cyclosporin-cyclophilin-calcineurin structure, **5**:1009
tacrolimus, **5**:1014–1016
Calcitonin antagonists, osteoporosis antiresorptive therapy, **5**:730
Calcitonin gene-related peptide (CGRP) antagonists:
biological actions and receptor pharmacology, **4**:538–540
migraine therapy, **8**:272–290
azepinone binding affinity, **8**:281–286
benzodiazepine binding affinity, **8**:277–281
BIBN4096 BS, **8**:273, 275–277
BMS-694153, **8**:288–290
future research issues, **8**:320–321
MK-0974, **8**:277, 279–288
neuropeptide-induced blood flow, **8**:286–288
receptor binding affinities, **8**:273–275
unnatural amino acids, **8**:289–290
structure and function, **4**:537–540
Calcitriol analogs, soft drug design, **2**:101–102
Calcium:
thromboxane A_2 receptor mobilization, **5**:896–897
vitamin D regulation, **5**:653, 655
Calcium-activated potassium channels, carbon monoxide therapeutic effects, **5**:296–299
Calcium/calmodulin-dependent kinase (CAMKII), long-term potentiation, **8**:24–25
Calcium channel blockers:
ATP binding cassette (ABC) transporters, **6**:371–372
migraine prevention, **8**:316–318
selective toxicity and stereochemistry, **2**:546–547

soft drug design, **2**:126–127
Calcium homeostasis, congestive heart failure therapy, mechanism-targeted studies, **4**:493–494
Calcium ion channels:
anticonvulsants, **8**:124
opioid receptor coupling, **8**:581
stroke therapy, **8**:455–460
N-type voltage-dependent calcium channel, **8**:456–459
Calcium phosphate, protein therapeutics, transfection, **3**:295
Calcium sensing receptor (CaSR), osteoporosis therapy, parathyroid hormone secretion/calcilytics, **5**:754–755
Calibration validation, commercial-scale operations, **3**:12
Calicheamicin, anti-cancer activity, **2**:238–239
Calmodulin, stroke therapy, nNOS inhibition, **8**:499–500
Calpain inhibitors, stroke therapy, **8**:460–465
irreversible inhibitors, **8**:464–465
reversible inhibitors, **8**:461–464
unconventional inhibitors, **8**:465
Calpeptin, stroke therapy, **8**:461–462
Cambridge Neurophysiological Test Automated Battery (CANTAB), cognitive dysfunction assessment, **8**:22
Cambridge Structural Database (CSD):
cocrystal design, **3**:194–196
crystal engineering, **3**:189–190
Camptothecins:
enzyme inhibition, **6**:23–29
irinotecan, **6**:26
metabolism, **6**:25–26
topotecan, **6**:26–29
serum albumin binding, **3**:451–453
in vitro studies, **3**:458–462
topoisomerase inhibitors, **6**:96–106

antimalarial drugs, **7**:650–652
Topo I inhibitors, **6**:96–98
Canada, patent requirements, **3**:122
Cancer therapy:
brain tumors:
angiogenesis, **6**:242–243
astrocytic tumors, **6**:230
ependymoma, **6**:233
imaging studies, **6**:273–274
invasiveness, **6**:241–242
metastatic, **6**:240
modeling approaches, **6**:272–273
neuroepithelial tumors, **6**:229–230
oligodendroglial tumors, **6**:233–234
WHO grading system (I-IV), **6**:230–233
central nervous system cancers, **6**:243–270
blood-brain barrier and compound design, **6**:270–271
cranial/spinal nerve tumors, **6**:237–239
embryonal tumors, **6**:234–235
epidemiology and prevalence, **6**:223–224
germ cell tumors, **6**:235
histological classification, **6**:228–229
meningioma, **6**:237
microenvironment, **6**:227
obstacles to, **6**:270
pituitary tumors, **6**:235–237
primary lymphoma, **6**:239–240
recurrent tumors, **6**:240–241
signaling pathways, **6**:227–228
targeted delivery, **6**:271–272
DNA-targeted therapeutics:
alkylating agents, **6**:83–95
cyclopropylindoles, **6**:91–93
mustards, **6**:83–89
nitrosoureas, **6**:93–94
platinum complexes, **6**:89–91
triazenes, **6**:94–95
antimetabolites, **6**:106–113
antifolates, **6**:106–109
purine analogs, **6**:112–113

pyrimidine analogs,
6:109–112
cytotoxic agents, 6:3–17
bleomycin, 6:8–12
dactinomycin, 6:3–8
enzyme inhibitors, 6:17–31
anthracyclines, 6:17–23
camptothecins, 6:23–29
isopodophyllotoxins,
6:29–31
future research issues,
6:40–41
mitomycin (mutamycin),
6:12–16
plicamycin, 6:16–17
research background,
6:1–3
tubulin polymerization/
depolymerization
inhibition, 6:31–40
dimeric vinca alkaloids,
6:35–40
taxus diterpenes,
6:32–35
research background, 6:83
topoisomerase inhibitors,
6:95–106
dual topo I/II inhibitors,
6:103–106
Topo II inhibitors,
6:98–102
Topo I inhibitors, 6:96–98
tumor-activated prodrugs,
6:113–124
ADEPT prodrugs,
6:120–123
hypoxia-activated
bioreductive prodrugs,
6:114–120
drug resistance mechanisms:
endocrine resistance, breast
cancer, 6:369–370
multiple drug resistance,
6:370–376
ABC transporter
mediation, 6:371–376
apoptosis protein
targeting, 6:370–371
stem cells, 6:376–377
new strategies for, 6:377–378
research background,
6:361–363
small-molecule tyrosine
kinase inhibitor
resistance, 6:364–369
stem cells and multiple drug
resistance, 6:376–377

targeted drug development,
6:363–370
tyrosine kinase inhibitor
resistance, 6:364
Hsp90 inhibitors, 6:441–442
kinesin spindle protein
inhibitors:
ATP-competitive inhibitors,
6:211–214
dihydropyrazoles, 6:204–210
future research issues,
6:214–215
"induced fit" pocket
inhibitors, 6:210–211
ispinesib, 6:198–204
mitosis targeting, 6:191–192
monastrol, 6:193–197
quinazolinone-based
inhibitors, 6:197–202
multitarget drug development,
multikinase
inhibitors, 1:261,
263–267
peroxisome proliferator-
activated receptors,
6:160–162
chemotherapy/radiotherapy
coadjuvants,
6:168–170
future research issues,
6:170–171
inhibitors, 6:168
monotherapy, 6:170
poly(ADP-ribose)
glycohydrolase and
inhibitors, 6:170
poly(ADP-ribose)
glycohydrolase, 6:170
retinoids:
acitretin, 5:400
adapalene, 5:412
AHPC compound, 5:497–498
AHPN compound, 5:497–498
Am80 compound, 5:417–419
bexarotene, 5:435–448,
450–453
3-Cl-AHPC, 5:499
etretinate, 5:401–402
N-(4-hydroxyphenyl)
retinamide,
5:464–473, 478–480
9cUAB30 RXR-selective
retinoid, 5:455–456
NRX 194204 RXR selective
analog, 5:460–461
NRX195183 analog,
5:420

9-*cis*-retinoic acid therapy,
5:424–425
13-*cis*-retinoic acids,
5:388–392
vitamin A pharmacology,
5:649
serum albumin:
chemical properties,
3:437–438
therapeutic applications,
3:456–464
clofibrate pharmacokinetic
modulation,
3:457–464
pharmacokinetic studies,
3:462–464
in vitro studies, 3:458–462
X-ray structural analysis,
3:438–456
anthracyclines, 3:452–454
bilirubin binding,
3:442–451
CADEX categories,
3:439–440
camptothecins, 3:451–453
drug-binding sites,
3:440–442
podophyllotoxin
derivatives, 3:454–456
SHetA2 retinoid-related
compound, 5:502–505
temsirolimus, everolimus, and
deforolimus, 5:1037
ubiquitin-proteasome pathway,
6:176–177
vitamin D analogs, 5:655–658
Cand5/bevasiranib sodium, age-
related macular
degeneration therapy,
5:619–622
Candidate ranking, soft drugs,
computer-aided
techniques, 2:131–132
Cangrelor, antiplatelet activity,
4:435–436
Cannabinoids:
antipsychotic agents,
8:209–210
cognitive enhancement,
8:47–50
multiple sclerosis therapy,
5:580
receptor agonists, G-protein
coupled receptor,
homology modeling,
2:293
soft drug design, 2:123–124

Capecitabine:
  ispenesib combined therapy, **6:**200–204
  pyrimidine antimetabolite analogs, **6:**110–112
Capillary wall drug diffusion, **2:**847–848
Capreomycin, antitubercular agents, **7:**762–764
Capsaicin, migraine therapy, vanilloid receptor antagonists, **8:**307–313
Capsazepine, migraine therapy, **8:**307–308
Captopril:
  angiotensin-converting enzyme inhibitors, drug discovery and development, **4:**269–271
  methylation reactions, **1:**423–425
  pseudopeptide analogs, **1:**303
  soft drug design, **2:**118–119
Carabersat, anticonvulsant applications, **8:**136–139, 155–156
Carba-analogy principle, acyclic nucleoside analogs, **7:**229
Carbacephems, structure-activity relationship, **7:**310–311
Carbacyclin, prostaglandin IP receptor, **5:**922–923
Carbamates:
  antibody-directed enzyme prodrug therapy, **6:**121–123
  prostaglandin/thromboxane EP$_1$ antagonists, **5:**836–838
Carbamazepine:
  anticonvulsant applications, **8:**135–139, 142–144
  aripiprazole interactions, **8:**198
  clozapine pharmacokinetics, **8:**192
  cocrystal engineering, **3:**198, 208
  risperidone interactions, **8:**195–196
  sertindole interaction, **8:**199
Carbamide peroxide, topical antimicrobial compounds, **7:**496

Carbapenams/carbapenems:
  biosynthesis, **7:**335–338
  discovery of, **7:**333–345
  exploratory structure-activity relation ship, **7:**353–354
  β-lactamase inhibitors, **7:**366–368
  1β-methylcarbapenems, **7:**340–345
  natural products, occurrence, structural variations and chemistry, **7:**333–335
  pharmacokinetics, **7:**282
  structure-activity relationships, **7:**260–261
  trinems and polycyclic carbapenems, **7:**345–353
Carbapenemases, classification, **7:**277–279
Carbenicillin, penicillin resistance and development of, **7:**291–292
Carbenoxolone, type 2 diabetes therapy, **4:**22–25
Carbidopa, Parkinson's disease, dopaminergic pharmacotherapy, **8:**541–542
Carbohydrates:
  chirality and malabsorption, **1:**146–147
  combinatorial libraries, **1:**291–292
  thyroid hormone metabolism, **4:**195
Carbon:
  chirality at, **1:**127
  drug metabolism:
    oxidation and reduction, **1:**413–416
    two-carbon chain elongation, **1:**434
Carbon-carbon double bonds:
  glucocorticoid biosynthesis and metabolism, **5:**58
  riboflavin metabolism, **5:**673
  steroid structure, **5:**3–4
Carbon framework analysis, ligand-based drug design, nuclear magnetic resonance, **2:**376–377

Carbonic anhydrase inhibitors:
  anticonvulsants, **8:**132–134
  glaucoma therapy, **5:**598–600
  hit seeking combinatorial libraries, **1:**311–312
Carbon monoxide (CO), **5:**295–308
  alternative delivery systems, **5:**307–308
  inhalation therapy, **5:**299–305
  prodrugs and releasers, **5:**305–307
  production and biological effects, **5:**266, 296–299
  therapeutic approaches, **5:**267–269
Carbon monoxide releasing molecules (CORMs):
  prodrugs and releasers, **5:**305–307
  transfusion and tissue delivery, **5:**307–308
^{13}C NMR spin-lattice relaxation, NS3/4A protease inhibitors, ciluprevir, **7:**179–181
Carboplatin, ispenesib combined therapy, **6:**200–204
Carborane pharmacophores, structure-based design, **2:**701
Carboxyalkyl dipeptides, ACE inhibitor design, **4:**273–274
Carboxyethylpyrrole (CEP) adducts, age-related macular degeneration, **5:**609–610
γ-Carboxylase, vitamin K antagonists, **4:**380–385
Carboxylation reactions, biotin, **5:**686–690
Carboxylesterases:
  bioconversion mechanisms, **3:**252–254
  prodrug development, **3:**233–241
    classification, **3:**233–234
    function and distribution, **3:**234
    molecular biology, **3:**234, 237–238
    preclinical assays, **3:**239, 241
    substrates and inhibitors, **3:**235–240

structure-metabolism
relationships,
3:254–256
Carboxylic acid acyl glucuronides,
toxicophore reactive
metabolites,
2:323–324
Carboxylic acids:
HIV integrase, discovery and
elaboration, two-metal
chelation elaboration,
7:82–97
prodrug moieties, **3:**250–252
stroke therapy:
HDAC inhibitors, **8:**494
MMP inhibitors, **8:**508
Carboxylic dimers, cocrystal
design, **3:**195–196
Carboxylic ester hydrolases, soft
drug development,
2:77–80
Carcinogenicity:
chirality and, **1:**145–146
estrogen receptor ligands,
5:226–228
Cardiac contractility efficiency:
carbon monoxide inhalation
therapy, **5:**304
congestive heart failure
therapy, mechanism-
targeted studies,
4:492–494
Cardiac troponins, congestive
heart failure therapy,
disease biomarkers,
4:505
Cardiotoxicity:
antimalarial drugs,
chloroquine, **7:**610
biomarkers:
drug-induced congestive
heart failure, **4:**503
topoisomerase II inhibitors,
6:101–102
Cardiovascular disease. *See also*
Atherosclerosis;
Congestive heart
failure (CHF)
age-related macular
degeneration risk and,
5:604–605
Am80 inhibition, **5:**417–418
angiotensin-converting enzyme
inhibitors, risk
reduction, **4:**284–285
aspirin primary prevention,
4:421

bexarotene therapy, **5:**449
congestive heart failure,
thyromimetic therapy,
4:198
hypothyroidism, thyromimetic
therapy, **4:**197
peroxisome proliferator-
activator receptors:
α-agonists, **4:**97–99
α?δ-dual agonists, **4:**136–137
α/γ-dual agonists, **4:**125–135
γ-agonists, **4:**104–106
phosphodiesterase inhibitors,
5:716
type 2 diabetes, **4:**2
Cardiovascular system:
antidepressant side effects,
tricyclic
antidepressants,
8:231–232
antipsychotic side effects,
8:167, 179–180
thyroid hormone effects,
4:195–196
Carfilzomib proteasome inhibitor,
clinical trials, **6:**180
Carisbamate, anticonvulsant
applications,
8:136–139, 155
Carmoterol, chronic obstructive
pulmonary disease
bronchodilators,
5:771–772
L-Carnitine esters, soft drug
design, **2:**113–114
Carotene:
hypercarotenosis, **5:**648
retinol (vitamin A) uptake
and metabolism,
5:646
Carpetimycin A, occurrence,
structural variations
and chemistry,
7:333–335
Carrier testing, topical
antimicrobial agent
evaluation, **7:**489
Carumonam, biological activity,
7:328–329
Carvedilol, glucuronidation/
glucosidation,
1:428–430
β-Casomorphin analogs and
endomorphins, opioid
peptide analogs,
Tyr-Pro-Phe sequence,
8:659–661

Caspase 3/9 inhibitors, stroke
therapy, **8:**465–471
aldehydes, **8:**466–467
alpha oxo/thio/amino-methyl
ketones, **8:**469–470
halomethyl ketones, **8:**467–469
irreversible inhibitors,
8:467–470
isatin-derived inhibitors,
8:470–471
masked aldehydes, **8:**467
quinoline-derived inhibitors,
8:471
reversible inhibitors,
8:466–467
Caspase 3 inhibitors, hit seeking
combinatorial
libraries, **1:**315–317
Caspofungin, development, **2:**231
Cassette, defined, **1:**280
Catalepsy, antipsychotic side
effects, **8:**175
Catalytic cleft,
multiconformational
kinase catalytic
domain, **6:**346–349
Catechins, tetracycline
development, **7:**415
Catecholamines, Parkinson's
disease, auto-
oxidation, **8:**536–537
Catechol-*O*-methyltransferase:
chronic obstructive pulmonary
disease
bronchodilators, β$_2$-
adrenergic receptor
agonists, **5:**769–772
Parkinson's disease,
dopaminergic therapy,
8:550–552
schizophrenia genetics, **8:**174
Catechols:
cephalosporins, **7:**305–306
methylation reactions,
1:423–425
Cathepsin B inhibitors, stroke
therapy, **8:**471–478
irreversible inhibitors,
8:473–477
reversible inhibitors,
8:472–473
non-cysteine design,
8:477–478
Cathepsin K inhibitors,
osteoporosis
antiresorptive
therapy, **5:**730–732

Cathepsin L inhibitors, stroke therapy, **8**:478
Cathepsin S, combinatorial chemistry, **1**:330–332
(–)Cathinone, ephedrine stimulant, **8**:90
Cationic surfactants, topical antimicrobial agents, **7**:499–500
CAVEBASE technique, chemogenomics, **2**:587–588
Caveolin-independent endocytosis, lipid rafts, membrane proteins, **2**:449–450
CB-181963 cephalosporin, **7**:311–312
CBP-38560 renin inhibitor, **4**:245–247
"C-clamp" in protease inhibitors, boceprevir NS3/4A anti-HCV agent, **7**:188–191
CD4 molecules, protein therapeutics, recombinant DNA technology, **1**:540–541
CD26 molecules, dipeptidyl peptidase 4 inhibitors, **4**:40–41
cDNA, protein therapeutics, expression vectors, **3**:294–295
CDOCKER algorithm, docking and scoring techniques, molecular dynamics, **2**:615
CEBS Microarray Database, quantitative structure-activity relationship modeling, **1**:514
Cediranib:
  antiangiogenic properties, **6**:308–310
  brain tumor angiogenesis, **6**:242–243
  glioblastoma therapies, **6**:232–233
Cefaclor, second-generation development, **7**:302–306
Cefadroxil, development of, **7**:301–306

Cefamandole, second-generation development, **7**:302–306
Cefazidime, third-generation development, **7**:303–306
Cefazolin, development of, **7**:301–306
Cefdinir, third-generation development, **7**:302–306
Cefepime, third-generation development, **7**:303–306
Cefipirome, third-generation development, **7**:303–306
Cefixime, third-generation development, **7**:303–306
Cefmetazole, development of, **7**:307
Cefoperzone, third-generation development, **7**:303–306
Cefotaxim, third-generation development, **7**:302–306
Cefotetan, second-generation development, **7**:302–306
Cefoxitin:
  β-lactamase inhibitors, **7**:360
  methylthiolation, **7**:307
  second-generation development, **7**:302–306
Cefpodoxime proxetil, third-generation development, **7**:302–306
Cefradine, first-generation development, **7**:301–306
Cefsulodin, third-generation development, **7**:303–306
Ceftaroline fosamil, development of, **7**:315–316
Ceftazidime, third-generation development, **7**:302–306
Ceftibuten, third-generation development, **7**:303–306
Ceftobiprole, development of, **7**:316–317

Ceftriaxome, third-generation development, **7**:303–306
Celastrol, Hsp90 inhibitor binding, **6**:428–429
Celesticetin, ribosome targeting, **7**:462
Cell-based biomarker assays, congestive heart failure therapy, **4**:501
Cell culture:
  AHPN compound, anticancer activities, **5**:497–498
  all *trans*-retinoic acid, adverse effects, **5**:385–386
  Am80 therapy, **5**:419
  bexarotene therapy, **5**:439, 442–443, 447–448
  brain cancer modeling, **6**:272–273
  HCV replication assays, **7**:174–175
  N-(4-hydroxyphenyl) retinamide (4-HPR), **5**:469–473
  9cUAB30 RXR-selective retinoid, **5**:456–457
  NRX194204 RXR selective analog, **5**:460–464
  NRX195183 analogs, **5**:420
  protein therapeutics, **3**:293–294
  attachment-dependent culture, **3**:296–297
  banking systems, **3**:296
  suspension culture, **3**:298
  psoriasis, **5**:397–398
  receptor-based drug design, **2**:523–524
  9-*cis*-retinoic acid therapy, **5**:428–429
  13-*cis*-retinoic acid therapy, **5**:392
  SHetA2 retinoid-related compound, **5**:502–505
  tazarotene therapy, **5**:406
Cell cycle kinase inhibitors, **6**:310–321
  antimalarial drug development, **7**:627–634
  farnesylation, **7**:630–634
  protein kinases, **7**:627–630
  aurora kinase inhibitors, **6**:316–321
  checkpoint kinase inhibitors, **6**:315–316

cyclin-dependent kinase
inhibitors, **6:**310–315
polo-like kinase inhibitors,
6:316
Cell differentiation and
proliferation, Am80
effects, **5:**418–419
Cell division, vitamin D
regulation, **5:**655–658
Cell-specific targeting,
recombinant DNA
technology, **1:**551
Cell-surface receptors:
monoclonal antibodies,
selective toxicity,
2:560, 562
receptor-mediated endocytosis,
membrane proteins,
2:447–448
Cellular adhesion molecules:
brain cancer therapy, hypoxia-
inducible factor
pathway, **6:**254–255
recombinant DNA technology,
1:560–561
Cellular FLICE inhibitory protein
(c-FLIP), SHetA2
retinoid-related
compound, **5:**502
Cellular information processing,
receptor-based drug
design, **2:**492–493
Cellular interactions,
quantitative
structure-activity
relationship, **1:**57–59
Cellular respiration, antimalarial
drugs, **7:**655–658
Cellular stress, antimalarial
drugs, **7:**658–663
glutathione biosynthesis,
7:660–663
PfaATP6, **7:**658–660
Cellular targeting, topical
antimicrobial agents,
7:488
Cell wall biosynthesis:
antitubercular agents,
7:719–724
arabinogalactan, **7:**720–722
mycolic acid, **7:**722–723
peptidoglycans, **7:**720
polyketide-derived lipids,
7:723–724
porins, **7:**724
β-lactam mechanisms of action,
7:266–269

CEM-101, macrolidic antibiotics,
2:228–229
Centchroman analog,
triarylethylene, **5:**251
Central nervous system (CNS):
cancers
brain tumors:
angiogenesis, **6:**242–243
astrocytic tumors, **6:**230
ependymoma, **6:**233
imaging studies, **6:**273–274
invasiveness, **6:**241–242
metastatic, **6:**240
modeling approaches,
6:272–273
neuroepithelial tumors,
6:229–230
oligodendroglial tumors,
6:233–234
WHO grading system
(I-IV), **6:**230–233
cranial/spinal nerve tumors,
6:237–239
embryonal tumors,
6:234–235
epidemiology and prevalence,
6:223–224
germ cell tumors, **6:**235
histological classification,
6:228–229
meningioma, **6:**237
microenvironment, **6:**227
pituitary tumors, **6:**235–237
primary lymphoma,
6:239–240
recurrent tumors, **6:**240–241
signaling pathways,
6:227–228
therapies, **6:**243–270
blood-brain barrier and
compound design,
6:270–271
obstacles to, **6:**270
targeted delivery,
6:271–272
cognitive dysfunction
indications and
diagnostic criteria,
8:19–21
drug development and
discovery:
allosteric receptor sites,
8:5–6
cotransmission, **8:**5
functional selectivity and
receptor targeting,
8:7–9

future research issues,
8:9–11
receptor-selective drugs,
8:1–3
research background, **8:**1
rich pharmacology concept,
8:4–5
neurotransmitters, overview,
8:61–62
opioid effects, **8:**579
stimulants:
ADMET properties, **8:**97–99
alpha-alkyl substituent,
8:108
alpha-carbon
stereochemistry, **8:**108
amphetamines, **8:**106–109
aromatic ring substitution,
8:108–109, 112–113
C(2) substituent, **8:**111–112
C(3) ester linkage, **8:**112
caffeine, **8:**90
cocaine, **8:**109–113
ephedra and khat, **8:**89–90
historical background,
8:90–91
methylphenidate, **8:**109
nitrogen substituents, **8:**107,
110–111
N-substituents, **8:**110
physiology and
pharmacology,
8:100–106
recent and future
development,
8:113–115
receptor classification and
function, **8:**102–106
research background,
8:89
side chain length, **8:**107
side effects, adverse effects,
drug interactions,
8:92–97
structure-activity
relationships,
8:106–113
therapeutic applications,
8:91–92
tropane ring preservation,
8:113
thyroid hormone effects,
4:196
Centrifugation, mammalian
protein purification,
harvest operations,
3:306–307

Centroid:
combinatorial libraries, **1:**297
defined, **1:**280
small molecule libraries, **1:**299–300
CEP-9722 chemotherapeutic agent, poly(ADP-ribose)polymerase (PARP) inhibitor coadjuvants, **6:**169–170
Cephabacins M, development of, **7:**306–307
Cephalexin:
antibiotics, **2:**222–223
cephalosporin development, **7:**298–299, 301–306
intestinal peptide transporter (PepT1) building blocks, **2:**463–467
Cephalon calpain inhibitor, stroke therapy, **8:**462
Cephalosporins:
7-aminocephalosporanic acid derivatives, **7:**299–300
antibiotics, **2:**222–223
vancomycin and, **2:**227
carbacephalosporins, **7:**310–311
combinatorial libraries, **1:**305
D-Ala-D-Ala mimicry, Tipper-Strominger hypothesis, **7:**265–266
discovery of, **7:**295–296
7α-formamido derivatives, **7:**307–308
7α-methoxy cephalosporins, **7:**307
new compounds, **7:**311–317
CB-181963, **7:**311–312
ceftaroline fosamil, **7:**315–316
ceftobiprole, **7:**316–317
CXA-101, **7:**311–312
LB-11058, **7:**311–312
RWJ-442831, **7:**312–313
S-3578, **7:**313–314
TD-1792, **7:**314–315
oxacephalosporins, **7:**308–310
penicillin sulfoxide-cephalosporin conversion, **7:**298–299
pharmacokinetics, **7:**281–280
selective toxicity, **2:**550–553
structure-activity relationship, **7:**300–306

7α-substituted cephalosporins, **7:**306–307
total synthesis, **7:**296–298
Cephalothin, development of, **7:**301–306
Cephamycins, development of, **7:**306–307
Cephapirin, first-generation development, **7:**301–306
Cephem sulfones, β-lactamase inhibitors, **7:**368–370
Ceramic hydroxyapatite chromatography, protein therapeutics, downstream processing, **3:**304
Ceramide pathway, *N*-(4-hydroxyphenyl) retinamide (4-HPR), **5:**475–476
Cerivastatins, LDL cholesterol lowering mechanisms, **4:**309
Cerubidine. *See* Daunorubicin
Cervical cancer:
N-(4-hydroxyphenyl) retinamide (4-HPR) therapy, **5:**466, 470–471
9-*cis*-retinoic acid therapy, **5:**425
Cethromycin, macrolidic antibiotics, **2:**228–229
Cetirizine, selective toxicity, **2:**565
Cetuximab:
growth factor signaling inhibitor, **6:**302–305
resistance to, **6:**369
Cetylpyridinium analogs:
cationic surfactants, **7:**500
soft drug design, **2:**111–113
Cevimeline muscarinic receptor:
analog development, **8:**65–66
cognitive enhancement, **8:**32–33
CGS-21680 adenosine A_{2a} receptor agonist, chronic obstructive pulmonary disease therapy, **5:**794
CGS-23425 thyromimetic, axtirome structure-activity relationship, **4:**208

Chagas' disease. *See* American trypanosomiasis
Chalcones:
antimalarial topoisomerase inhibitor, **7:**651–652
p53/MDM2 antagonists, **2:**348–349
"Chameleon" steric parameter, quantitative structure-activity relationship, **1:**20–21
Change control program, commercial-scale operations, **3:**12–13
Charge state, ligand-based drug design, nuclear magnetic resonance, **2:**384–385
Checkpoint kinase inhibitors, medicinal chemistry and clinical trials, **6:**315–316
Chelation region interatom distance model, HIV integrase, two-metal chelation elaboration, **7:**82–97
Chelators:
HIV integrase, metal chelator inhibitors:
active sites, **7:**80–81
pharmacophore elaboration, **7:**81–97
tetracyclines, **7:**408–409
Chelerythrine, peptidomimetics, **1:**233, 235
Chelocardin, history and biosynthesis, **7:**412–413
Chemical degradation, protein therapeutics formulation and delivery, **3:**311–314
Chemical delivery systems (CDS):
defined, **2:**75
prodrugs *vs.*, **2:**76
retrometabolic drug design:
basic principles, **2:**134–136
benzylpenicillin, **2:**142–143
brain-targeting compounds, **2:**136–140
redox analogs, **2:**154–155
cyclodextrin complexes, **2:**149–150
estradiol, **2:**144–149
eye-targeting site-specific compounds, **2:**155–160

oxime/methoxime
β-blocker analogs,
2:158–160
ganciclovir, **2**:142–143
molecular packaging,
2:150–155
brain-targeting redox
analogs, **2**:154–155
kyotorphin analogs,
2:153–154
leu-enkephalin analogs,
2:151–153
TRH analogs, **2**:153
prodrugs *vs.*, **2**:76
receptor-based transient-
anchor compounds,
2:160–161
site-targeting index and
targeting
enhancement factors,
2:140–141
zidovudine, **2**:141–142
Chemical ionization (CI), mass
spectrometry
development, **1**:98
Chemically advanced template
search (CATS),
functional analog
design, **1**:175–177
Chemical reactivity, toxicophores,
2:301–302
Chemical shift, nuclear magnetic
resonance, ligand-
based designs,
2:370–371
bioactive peptides, **2**:376–383
charge state, **2**:384–385
drug screening, chemical-shift
perturbation, **2**:419,
421
line-shape and relaxation data,
2:387–389
macromolecule-ligand
interactions,
2:395–402
macromolecule-ligand
mapping, **2**:402–403
Chemical stability, ADMET
structural properties,
2:57
Chemistry-driven drug discovery,
recombinant DNA
technology, **1**:537–538
Chemogenomics:
affinity chromatography/
activity-based protein
profiling, **2**:581

basic principles, **2**:573–574
biological principles, **2**:580–582
chemical space theory,
2:574–580
cofactor-based discovery, **2**:578
diversity-oriented/biology-
oriented syntheses,
2:578–580
molecular informatics,
2:582–588
information systems,
2:582–583
ligand-based data analysis
and predictive
modeling, **2**:583–585
structure-based data
analysis and
predictive modeling,
2:585–588
privileged scaffolds,
2:575–577
protein family targeted
libraries, **2**:574–575
protein secondary structure
mimetics, **2**:577–578
quantitative structure-activity
relationship modeling,
1:514
receptor-based drug design,
2:516–519
structural genomics, **1**:589–595
epigenetics, **1**:594–595
kinase inhibitor protein
structures, **1**:590–591
kinase program, **1**:591–594
yeast three-hybrid screens,
2:581–582
Chemokine receptor-3 (CCR3),
quantitative
structure-activity
relationship, five-
dimensional models,
1:42–43
Chemokine receptor-5 (CCR5)
antagonists, HIV-1
entry blockade:
anilide core, **7**:124–131
binding mode and mechanism,
7:133–134
piperazine/piperidine-based
antagonists,
7:110–116
research background,
7:107–110
spiro diketopiperazine-based
antagonists,
7:131–133

summary and future research
issues, **7**:134–135
tri-substituted pyrrolidine
antagonists,
7:121–124
tropane-based antagonists,
7:116–121
Chemokine receptors:
brain tumor therapy, **6**:264–265
dipeptidyl peptidase 4
inhibitors, **4**:42
Chemoprophylaxis:
antitubercular agents,
isoniazid, **7**:740–743
chemotherapeutic agents,
7:505
Chemotherapy:
antimalarial drugs:
antibiotics combinations,
7:621–624
artemisinins, **7**:614–617, 617
experimental agents,
7:625–627
farnesylation, **7**:634
antitubercular agents,
7:737–786
capreomycin, **7**:762–764
D-cycloserine, **7**:767–771
ethambutol, **7**:751–755
ethionamide, **7**:771–773
first-line agents, **7**:738–759
fluoroquinolines, **7**:774–776
isoniazid, **7**:738–744
kanamycin and amikacin,
7:761–762
mycolic biosynthesis,
7:722–723
para-aminosalicylic acid,
7:764–767
pyrazinamide, **7**:755–759
rifamycin antibacterials,
7:744–751
streptomycin, **7**:759–761
thiacetazone, **7**:773–774
tuberculosis pathogenesis,
7:719
brain cancer, ependymoma,
6:233
combination therapies,
7:504–505
antimalarials, **7**:617,
621–627
HIV reverse transcriptase
inhibitors,
7:155–156
neutrophil restoring agents,
4:588–589

Chemotherapy (*Continued*)
parasitic infections:
historical background, **7:**565–566
Leishmaniasis, **7:**575–577
poly(ADP-ribose)polymerase (PARP) inhibitor coadjuvants, **6:**169–170
synthetic agents:
future research issues, **7:**552–553
hybrid agents, **7:**550–552
systemic synthetic antibacterials, **7:**501–553
anaerobic infections, **7:**502–503
chemoprophylaxis, **7:**505
combination therapies, **7:**504–505
dihydrofolate reductase inhibitors, **7:**531–534
drug resistance, **7:**502, 504
folate inhibitors, **7:**506–534
genetic factors, **7:**503
host reactions, **7:**503
methenamine, **7:**550
nitrofurans, **7:**548–550
oxazolidinones, **7:**540–548
pharmacology, **7:**501–504
quinolones and fluoroquinolones, **7:**535–540
research background, **7:**501
sulfonamides and sulfones, **7:**506–531
tissue factors, **7:**503–504
treatment guidelines, **7:**504–506
Chemotypes:
antipsychotics, phenothiazines, **8:**186–189
attrited compounds:
analysis, **1:**498–499
targeting, **1:**496–498
basic properties, **1:**484–487
drug analyses and targeting, **1:**496–498
drug discovery applications, **1:**499–501
future trends and applications, **1:**502–503
hit/lead compound selection, **1:**489–491

insomnia therapy, orexin receptor antagonists, **5:**721
ligand-based virtual screening, **2:**3
migraine therapy, calcitonin gene-related peptide antagonists, **8:**287–288
peroxisome proliferator-activator receptors, γ-agonists, **4:**104
polypharmacology, drug development, **1:**493–496
in silico approaches, **1:**501–502
tool compounds and target validation, **1:**491–493
ChemScore function:
empirical scoring, **2:**625–628
structure-based virtual screening, **2:**18
Chenodeoxycholic acid (CDCA), drug delivery, **2:**468–475
Chidamide, histone deacetylase inhibitors, **6:**68, 70
Chimeric structures, Hsp90 inhibitors, geldanamycin and radicicol chimeras, **6:**405–410
Chinese hamster ovary (CHO) cells, protein therapeutics, **3:**293–294
Chirality:
ADME properties, **1:**137–143
absorption/permeability, **1:**138–140
distribution, **1:**140–141
excretion, **1:**142–143
metabolism, **1:**141–142
defined, **1:**127
drug design, **1:**149–155
drug metabolism, **1:**443–444
drug-receptor interactions, **1:**127–129
enantiomer activity, **1:**130–131
future trends and applications, **1:**159–160
GPCR ligands, **1:**132
ion channel ligands, **1:**132–133
lipid combinatorial library, **1:**292–293
nitrogen, **1:**148
noncarbon atoms, **1:**147–149

Pfeiffer's rule, **1:**129–120
phosphorus, **1:**148–149
regulatory issues, **1:**155–156
scale-up operations, **3:**4
sulfur, **1:**147–148
switching mechanism, **1:**156–159
toxicity effects, **1:**143–147
amine-based platinum complexes, **1:**144–145
carcinogenicity, **1:**145–146
clinical chemistry, **1:**146–147
environmental toxicity, **1:**147
hERG activity, **1:**143–144
Chiral sulfoxides, chemokine receptor-5 (CCR5) antagonist, anilide core, **7:**127–131
Chitosans, membrane protein permeability enhancement, **2:**441
Chlorambucil, cancer therapy, **6:**83–84
Chloramines, soft drug design, **2:**128–129
Chloramphenicol:
ribosome targeting, **7:**462–463
RNA targets, **5:**993–996
toxicophore reactive metabolites, **2:**312–313
Chlorhexidine, topical antimicrobial agents, **7:**499
3-Cl-AHPC retinoid:
adverse effects, **5:**499
anticancer activities, **5:**498
targeted drug development, **5:**499–500
therapeutic applications, **5:**489–497
Chloride ion channel, anticonvulsants, **8:**126
Chlorine, antimicrobial agents, **7:**490–492
Chlorobenzilate, soft drug design, **2:**106–107
2-[4-(4-Chloro-2-fluorphenoxy) phenyl]pyrimidine-4-carboxamide (CFPPC), cocrystal engineering, **3:**210
Chlorofusin, p53/MDM2 derivatives, **2:**352–359
Chloromethyl ketones, stroke therapy, cathepsin B inhibitors, **8:**475

Chlorophores, antimicrobial agents, **7**:490–492
Chloroquine:
  antimalarial therapy, **7**:607–610
  experimental agents, **7**:625–627
Chlorotrianesene:
  oxidative metabolism, **5**:225–226
  structure-activity relationships, **5**:249–253
Chloroxine, antimicrobial activity, **7**:494
Chlorpheniramine, selective toxicity, **2**:565
Chlorpromazine:
  ADMET properties, **8**:186–189
  antitrypanosomal polyamine metabolism, **7**:589–590
  central nervous system effects, **8**:3–5
  FDA approval, **8**:176–177
  historical background, **8**:248
  nonpsychotic applications, **8**:170
  pharmacokinetics, biodistribution, and drug-drug interactions, **8**:185–186
  side effects, **8**:165–166, 168
Chlortetracycline:
  chemical structure, **7**:407
  history and biosynthesis, **7**:410–413
Cholecalciferol:
  analogs, **5**:656–657
  chemistry, **5**:651, 653
  photochemistry, **5**:655–656
Cholecystokinin-2 receptor (CCK-2R):
  chirality properties, **1**:151–155
  enantiomer activity, **1**:132–133
Cholelithiasis, vitamin K deficiency, **5**:668
Cholesterol:
  Alzheimer's disease, metabolic syndrome targeting, **8**:405, 411–414
  apolipoproteins E and AI, **8**:413–414
  esterification, **8**:406, 411
  synthesis, **8**:406
  turnover, **8**:411–413

androgen biosynthesis, **5**:23–24, 156–158
antifungals, selective toxicity, **2**:558–559
atherosclerosis, LDL-lowering drugs:
  current therapeutic agents, **4**:305–318
  bile acid sequestrants, **4**:315–316
  cholesterol absorption inhibitors, **4**:310–315
  combination products, **4**:316–318
  HMG-CoA reductase inhibitors, **4**:306–310
  emerging therapeutic agents, **4**:318–320
  antisense oligonucleotide, apoB100, **4**:319
  microsomal triglyceride transfer protein inhibitors, **4**:319–320
  PCSK9 inhibition, **4**:318–319
  research background, **4**:303–305
biosynthesis, **5**:13, 15–16
delivery systems, apical sodium-dependent bile acid transporter enhancement, **2**:474–475
estradiol biosynthesis, **5**:24–25
farnesoid X receptors, **4**:158–160
glucocorticoid biosynthesis and metabolism, **5**:56–59
liver X receptors:
  lipid metabolism, **4**:148–150
  reverse cholesterol transport, **4**:147
peroxisome proliferator-activated receptors, δ-agonists, **4**:117–118
pregnenolone conversion, **5**:16–17, 19
thyroid hormone metabolism, **4**:195
axitirome structure-activity relationship, **4**:207
receptor agonist effects, **4**:197–198
Cholesteryl ester transfer protein inhibitors (CETP),

  atherosclerosis, HDL elevation, **4**:340–345
  anacetrapib, **4**:343–344
  clinical trials, **4**:341–344
  current and future development, **4**:344–345
  dalcetrapib, **4**:342–343
  functional analysis, **4**:344
  torcetrapib, **4**:341–342
Cholestyramine 16, LDL cholesterol lowering, **4**:316
Choline, soft drug design, **2**:113
Choline acetyltransferase (ChAT), cognitive dysfunction and, **8**:30–31
Cholinergic agonists, **8**:63–72
  acetylcholine analogs, **8**:64
  muscarinic agonists:
    glaucoma therapy, **5**:599–600
    clinical trials, **5**:602
    structure and classification, **8**:64–68
  nicotinic agonists, **8**:68–72
Cholinergic antagonists:
  muscarinic antagonists, **8**:72–77
  nicotinic antagonists, **8**:77–80
    ganglionic antagonists, **8**:78
    neuromuscular antagonists, **8**:78–80
  structure and classification, **8**:72–80
Cholinergic neurons:
  cognitive dysfunction and, **8**:30–31
  defined, **8**:61–62
Cholinesterase inhibitors:
  cognitive enhancement, **8**:30–32
  memory assessment and, **8**:28–29
Choroidal neovascularization (CNV), age-related macular degeneration, **5**:604–605
in vivo screening model, **5**:612
Chromatography, protein therapeutics, in-process and product release testing, **3**:318–320

Chromatography hydrophobicity index (CHI), quantitative structure-activity relationship, partition coefficients, hydrophobic interactions, **1:**13–14

Chromene, structure-activity relationship, **5:**251–252

Chromenoquinoline, tissue selectivity, **5:**242

Chromenoquinolone, nonsteroidal progesterone receptor ligands, **5:**255–256

Chromenotriazolopyridines, protein-protein interactions, **2:**353–359

Chromofiltration, combinatorial chemistry, **1:**345–347

Chronic fatigue syndrome, human herpesviruses and, **7:**244

Chronic lymphocytic leukemia (CLL), cyclin-dependent kinase inhibitor therapies, **6:**311–315

Chronic myelogenous leukemia (CML):
 Bcr-Abl inhibitor therapy, **6:**298–301
 multitarget drug development, multikinase inhibitors, **1:**261, 263
 small-molecule imatinib mesylate, resistance to, **6:**364–366

Chronic obstructive pulmonary disease (COPD):
 adenosine A_{2a} receptor agonists, **5:**793–794
 bronchodilators, **5:**766–773
  β_2-adrenergic receptor agonists, **5:**769–772
  combined therapies, **5:**772
  methylxanthines, **5:**772–773
  muscarinic M_3 receptor antagonists, **5:**766–769
 CXCR2 receptor antagonists, **5:**784–785

epidemiology and pathophysiology, **5:**765–766
future therapeutic development, **5:**794–795
IKK-2 inhibitors, **5:**787–789
inhalation therapy:
 carbon monoxide, **5:**304–305
 corticosteroids, **5:**773–780
 matrix metalloproteinase inhibitors, **5:**791–792
 neutrophil elastase inhibitors, **5:**790–791
 nitric oxide synthase inhibitors, **5:**785–786
 p38 mitogen-activated protein kinase inhibitors, **5:**789–790
 phosphodiesterase inhibitors, **5:**780–784

Chronic renal failure, vitamin D analogs, **5:**655

Chylomicrons:
 structure and function, **4:**332
 vitamin E uptake and metabolism, **5:**660–662

Chymotrypsins, quantitative structure-activity relationship, isolated receptor interactions, **1:**54–56

CI-992 renin inhibitor, **4:**246, 249

CI-994. *See* Tacedinaline

Ciclesonide (CIC):
 chronic obstructive pulmonary disease inhalation therapy, **5:**775–776
 soft drug design, **2:**101

Cidofovir:
 acyclic nucleoside phosphates, **7:**233–234
 cytomegalovirus therapy, **7:**243–244
 human papillomavirus therapy, **7:**241–242
 poxvirus therapy, **7:**245

Cilastatin, biological activity, **7:**351–353

Ciliary neurotrophic factor (CNTF), age-related macular degeneration therapy, **5:**613–614

Cilomilast:
 chronic obstructive pulmonary disease therapy, **5:**781–784
 selective toxicity, **2:**567

Cilostazol, antiplatelet activity, **4:**449–450

Ciluprevir anti-HCV protease inhibitor, **7:**176–181, 193–195

Cimetidine:
 chemistry-driven drug discovery, **1:**537–538
 recombinant DNA technology, receptor-based targeting mechanism, **1:**556–560
 selective toxicity, **2:**566

Cinamide piperidine derivatives, stroke therapy, **8:**454–455

Cinamidines, stroke therapy, NR2B antagonist compounds, **8:**455

Cinnamoyl linkers, hydroxamate inhibitors, stroke therapy, **8:**488–490

Ciprofloxacin:
 antibacterial agents, **7:**535–540
 antimalarial drugs:
  combination therapies, **7:**621, 623–624
  topoisomerase inhibitors, **7:**650–652
 diffusion-based drug release, plasma drug profile, **2:**803–808
 intravenous drug delivery, single dose (bolus injection), **2:**752–754
 tissue-based drug delivery, lung and bronchial mucus, **2:**839–844

Ciprokiren renin inhibitor, **4:**246, 249

Ciproxifan, cognitive function assessment, **8:**27

Circular dichroism (CD), protein therapeutics, biophysical assessment, **3:**324–325

Cisatracurium, nicotinic antagonists, **8:**79–80

Cisplatin:
  DNA-targeted chemotherapeutic compounds, **6:**89–91
  Hsp90 inhibitor, C-terminal binding sight, **6:**440–441
Citalopram:
  chemical structure, **8:**223
  historical background, **8:**249
  pharmacokinetics, **8:**236
  serotonin receptor targeting, **8:**244–247
  side effects and adverse reactions, **8:**233–234
Citizens petition, generic drug approval, FDA guidelines, **3:**86–87
L-Citrulline, nitric oxide enhancement, **5:**291–294
Civamide, migraine therapy, **8:**307–308
CK-2130 prodrug case study, cardiotonicity, **3:**268–276
CK929866SB-743921 kinesin spindle protein inhibitor, ispenesib and, **6:**202–203
cKIT inhibitors, chemogenomics, **1:**590–951
Cladribine:
  antimetabolite analogs, **6:**112–113
  multiple sclerosis therapy, **5:**570
Claims requirements:
  infringement of patents:
    alternative dispute resolution, **3:**164–165
    appeal process, **3:**163–164
    claim construction, **3:**162
    commencement of proceedings, **3:**159–160
    defenses to, **3:**154–157
    determination of, **3:**152–154
    discovery process, **3:**160–161
    licensee rights, **3:**150–152
    remedies for, **3:**157–159
    summary judgment, **3:**161
    trial, **3:**162–163
    U.S. trial and appellate courts, **3:**148–150
  patent specifications, **3:**126–127

composition of matter claims, obviousness standards and, **3:**137–138
Clarithromycin:
  hemiketalization, **7:**447–456
  pharmacology and ADMET properties, **7:**439–441
  RNA targets, **5:**984–990
  selective toxicity, **2:**552–553
Class I drugs, oral drug physicochemistry, high solubility/high permeability, **3:**41, 354
Class II drugs, oral drug physicochemistry:
  cocrystallization, **3:**47–48
  complexation, **3:**46–47
  lipid technologies, **3:**48–50
  low solubility/high permeability, **3:**42–50, 354
  metastable forms, **3:**45–46
  particle size reduction, **3:**44–45
  precipitation inhibition, **3:**45
  salt compounds, **3:**42, 44
  solid dispersion, **3:**46
Class III drugs, oral drug physicochemistry, **3:**50–52
Class IV drugs, oral drug physicochemistry, **3:**52, 355
Class V drugs, oral drug physicochemistry, **3:**52–55
Clathrin-independent endocytosis, lipid rafts, membrane proteins, **2:**449–450
Clavulanate:
  antibiotics, **2:**222–223
  β-lactamase inhibitors, **7:**361–363
Clavulanic acid:
  β-lactamase inhibitors, discovery and inactivation, **7:**360–363
  monobactam activity, **7:**329–331
Cleaning validation, commercial-scale operations, **3:**12
Clearance, drug delivery and bioavailability, **2:**748–749
  systemic clearance, **2:**751

Clevidipine:
  soft drug design, **2:**126–127
  stroke therapy, **8:**459–460
Clevudine:
  anti-DNA virus antimetabolites, **7:**226–227
  hepatitis B virus therapy, **7:**246–247
Clindamycin, antimalarial drugs, **7:**621–624
  quinine combined with, **7:**607
Clinical chemistry, chirality and, **1:**146–147
Clinical efficacy, single nucleotide polymorphisms, drug target genes, **1:**190–193
Clinical trials:
  age-related macular degeneration therapy, **5:**616–622
  aliskiren renin inhibitor, **4:**251–252
  Alzheimer's immunization therapy, **8:**369
  antidepressant efficacy assessment, **8:**221–229
  antiglaucoma therapeutics, **5:**600–602
  antitubercular agents, SQ109, **7:**783
  aurora kinase inhibitors, **6:**316–321
  BACE (β-site cleavage enzyme) inhibitors, **8:**356–357
  cangrelor, **4:**435–436
  checkpoint kinase inhibitors, **6:**315–316
  cholesteryl ester transfer protein inhibitors, **4:**341–344
  chronic obstructive pulmonary disease therapy, phosphodiesterase inhibitors, **5:**781–784
  congestive heart failure therapy:
    ongoing Phases I, II, and III trials, **4:**496–500
    recent disappoints and emerging therapies, **4:**478–480
  farnesoid X receptor, **4:**161–163
  FDA guidelines, **3:**76–78

Clinical trials: (*Continued*)
   research subjects guidelines, **3:**80
   site guidelines, **3:**80, 82
fibrillization inhibitors, **8:**361–365
kinesin spindle protein inhibitors, ispenesib, **6:**198–204
liver X receptor agonists, **4:**154–156
NS3/4A protease inhibitors:
   ciluprevir anti-HCV agent, **7:**178–181
   teleprevir anti-HCV agent, **7:**186
perfluorocarbons, oxygen delivery, **4:**622–623
peroxisome proliferator-activator receptors:
   α-agonists, **4:**96–99
   α/γ-dual agonists, **4:**135–136
   δ-agonists, **4:**121–122
   γ-agonists, **4:**104–106
pharmacogenomics, **1:**197–198
polo-like kinase inhibitors, **6:**316–317
proteosome inhibitors, **6:**177–180
retinoids:
   acitretin, **5:**398–399
   adapalene therapy, **5:**411–414
   all-*trans*-retinoic acid therapy, **5:**385–386
   Am80 therapy, **5:**414–420
   bexarotene, **5:**435–445
   N-(4-hydroxyphenyl) retinamide (4-HPR), **5:**464–473, 476–478, 478–480
   NRX 194204 RXR selective analog, **5:**460–464
   NRX195183 analog therapy, **5:**420
   recent trials, **5:**507–516
   9-*cis*-retinoic acid therapy, **5:**422
   13-*cis*-retinoic acid therapy, **5:**389–392
   tazarotene, **5:**402–408
γ-secretase inhibitors, **8:**341–345
selective estrogen receptor modulators, **5:**740–741

structural genomics, drug targeting research, **1:**579–580
ticagrelor, **4:**436–437
Clinidipine, stroke therapy, **8:**458–459
Clioquinol β-amyloid peptide inhibitor, **8:**363
Clk kinases, structural genomics, **1:**592
Clobazam, anticonvulsant applications, **8:**144–147
Clobetasol propionate, soft drug design, **2:**99–100
Clofibrate:
   hemoglobin oxygen delivery, allosteric effectors, **4:**615–619
   peroxisome proliferator-activator receptors, α-agonists, **4:**90–96
   serum albumin modulation, **3:**457–464
CLOGP values:
   quantitative structure-activity relationship, partition coefficient calculations, **1:**14–19
   receptor-based drug design, ADMET properties assessment, **2:**525
Clomiphene:
   adverse effects and precautions, **5:**221
   current compounds and applications, **5:**220–221
   ovulation modulation, **5:**243
   oxidative metabolism, **5:**226
   structure-activity relationships, **5:**249–253
Clomipramine:
   antitrypanosomal polyamine metabolism, **7:**589–590
   chemical structure, **8:**222
   historical background, **8:**249
   migraine prevention, **8:**318–319
   side effects and adverse reactions, **8:**231–232
Clonazepam, anticonvulsant applications, **8:**144–147

Clonidine, cognitive enhancement, **8:**41–42
Cloning techniques:
   angiotensin-converting enzyme, **4:**285–286
   protein therapeutics, **3:**296
   recombinant DNA technology, receptor-based targeting mechanism, **1:**559–560
Clopidogrel:
   antiplatelet activity, **4:**427–433
   clinical trials, **4:**428–432
   future research issues, **4:**451–452
   primary prevention trial, **4:**430–432
   resistance to, **4:**432–433
   safety and tolerability, **4:**432
Clorazepate, anticonvulsant applications, **8:**144–147
Cloretazine, biological activity and side effects, **6:**94
Closed system conditions, receptor-based drug design, residence times, **2:**504
Closure systems, protein therapeutics formulation, material interactions, **3:**315
Clothiapine, structure-activity relationship, **8:**182–183
Clotiapine, analog design, **1:**170–174
Clotting factors:
   hemostasis and initiation of, **4:**366–367
   vitamin K carboxylation, **5:**665–668
Clozapine:
   behavioral models, **8:**172–173
   biological fingerprint, **1:**495–496
   central nervous system effects, **8:**4–5
   dopamine model, **8:**170–171
   FDA approval, **8:**178
   multitarget drug development, dopamine D_2-plus for schizophrenia, **1:**257–258
   pharmacokinetics, **8:**191–192

side effects, **8:**168–169, 178
structure-activity relationship, **8:**180–185
Cluster analysis:
hit/lead compounds, BioPrint® database, **1:**490–491
natural products lead generation, bacterial genomes, **2:**210
quantitative structure-activity relationship:
compound selection, **1:**33
K-means clustering, **1:**49
in silico screening, protein flexibility, rigid cluster normal mode analysis, **2:**864–865
c-Met kinase inhibitors:
brain tumor therapy, **6:**247–248
structure-based design, hydrogen bonding, **6:**353–354
tumor metastases, **6:**326, 328–330
C-nitroso compounds, nitric oxide donors and prodrugs, **5:**286–287
Coactivator dependent receptor ligand assay (CARLA), nuclear hormone receptors, **4:**84–86
Coactivators:
nuclear hormone receptors, **4:**83–84
fluorescence resonance energy transfer recruitment, **4:**84–86
thyroid hormone receptors, **4:**191
Coagulation. *See also* Anticoagulants
enzymes:
hemostasis, **4:**365–366
intrinsic regulators, **4:**368
molecular mechanisms, **4:**385–390
platelet function, **4:**409–410
vitamin K, **5:**665–668
Coarse-grained molecular dynamics, docking and scoring techniques, flexibility computation, **2:**639
Coartem® antimalarial agent, **7:**625–627

Cobalamin. *See* Vitamin B$_{12}$ (cobalamin)
Cobalamin-intrinsic factor complex, receptor-mediated endocytosis, **2:**454
Cobalt protoporphyrin IX (CoPRIX), heme oxygenase upregulation, **5:**308
Cocaine:
ADMET properties, **8:**98–99
monoamine transporter inhibition, **8:**240–241
κ-opioid receptor effects, **8:**593
Parkinson's disease dopaminergic therapy, **8:**547–549
physiology and pharmacology, **8:**100–106
side effects, adverse reactions and drug interactions, **8:**96
structure-activity relationship, **8:**109–113
structure and properties, **8:**90
Cocrystallization:
evolution of, **3:**192–196
glucokinase activators, type 2 diabetes therapy, **4:**16–17
Hsp90 inhibitors:
chimeric radicicol and geldanamycin structures, **6:**405–410
pyrazole-scaffold Hsp90 inhibitors, **6:**418–419
radicicol Hsp90 inhibitor, **6:**400–405
NS3/4A protease inhibitors, ciluprevir, **7:**178–181
oral drug physicochemistry, class II drugs, **3:**47–48
peroxisome proliferator-activated receptors, α?γ-dual agonists, **4:**129–135
pharmaceutical case studies, **3:**198–213
AMG 517 antagonist, **3:**210–211
carbamazepine, **3:**198, 208
C-glycoside derivative, **3:**211–212
fluoxetine hydrochloride, **3:**208–209
itraconazole, **3:**209–210

melamine and cyanuric acid, **3:**213
monophospate salt I, **3:**212–213
sildenafil, **3:**211
sodium channel blockers, **3:**210
polymorphs, solvates, and hydrates, **3:**196
structural properties, **3:**188–190
Codeine:
identification and early history, **8:**596–597
morphine metabolism and elimination, **8:**578
opium source, **8:**599
synthesis of, **8:**599–600
Coding single nucleotide polymorphisms, classification, **1:**182–183
Coenzyme A:
biotin metabolism, **5:**686–690
drug metabolism, conjugation reactions, **1:**432
vitamin B$_{12}$ (cobalamin), **5:**697, 699
Cofactor-based drug discovery:
antitubercular agents, targeted biosynthesis, **7:**727–729
chemogenomics, **2:**578
high-throughput screening, assay optimization, **3:**412–413
Cognition:
defined, **8:**15
disease-related dysfunction, **8:**15
ADAS-COG assessment scale, **8:**21–22
Alzheimer's disease, **8:**20
CANTAB assessment scale, **8:**22
disease models, **8:**25–26
HIV-associated dementia, **8:**21
indications and diagnostic criteria, **8:**19–21
lifestyle factors, **8:**21
MATRICS assessment scale, **8:**22
Parkinson's disease, **8:**21
schizophrenia, **8:**20–21, 162
stroke/vascular dementia, **8:**21

Cognition: (*Continued*)
  enhancement, 8:15–17
    adenosine modulator, 8:46–47
    cannabinoids, 8:47–48
    cholinesterase inhibitors, 8:30–32
    dopamine, 8:39–40
    $GABA_A$ receptor, 8:45–46
    glutamate, 8:41–45
    histamine, 8:35–37
    muscarinics, 8:32–33
    neuronal nicotinic agonists, 8:34–35
    neuropeptides, 8:48–50
    neurotropic agents, 8:47
    NMDA modulators, 8:33–34
    norepinephrine, 8:41
    phosphodiesterase inhibitors, 5:716, 8:50–51
    serotonin, 8:37–39
    small-molecule approaches, 8:30–51
    smart drugs, 8:51–52
    summary and future research issues, 8:52–53
  functional schematic, 8:17–19
  memory and, 8:17–19
  opioid-related impairment, 8:575
  preclinical behavioral assessment, 8:25–29
    5-choice serial reaction time task, 8:26–27
    Morris Water Maze, 8:28–29
    set shifting, 8:29–30
    social recognition and novel object recognition, 8:28
  substrates of, 8:22–25
    molecular substrates, 8:24–25
    neuroanatomical substrates, 8:22–24
    synaptic plasticity, 8:25
Colesevelam, LDL cholesterol lowering mechanisms, 4:316–318
Collagenase, nuclear magnetic resonance studies, 2:412–414
Collander equation, quantitative structure-activity relationship,
  hydrophobic parameters, 1:9–12
Collision-induced dissociation (CID):
  mass spectrometry development, 1:97–98
  tandem mass spectrometry, drug development applications, 1:105–106
Colon cancer, bexarotene therapy, 5:444, 447
Colon-specific drug delivery, glucocorticoid anti-inflammatory compounds, 5:45
Colony stimulating factors (CSFs), research background, 4:569–570
Column chromatography, combinatorial chemistry, 1:346–347
Combinatorial chemistry:
  analytical problems, 1:349
  antimalarial drugs:
    chloroquine, 7:609–610
    quinine combined therapies, 7:607
  automation, 1:348
  definitions, 1:280
  future trends, 1:350–352
  historical background, 1:277–280
  informatics and data handling, 1:349–350
  large arrays *vs*. pulsed-iterative compound libraries, 1:296–297
  linear, convergent, and multicomponent reactions, 1:345
  lipopeptide antibiotics, 2:227–228
  microwave acceleration, 1:348–349
  molecular diversity *vs*. druggability, 1:295
  natural products, 1:297–298
  patents, 1:350
  prodrug discovery, 3:226, 228–229
  purification, 1:345–347
    "pure" compounds, 1:347
    testing mixtures, 1:347
  recombinant DNA technology, biosynthesis, 1:547–549
  resins and solid supports, 1:343–344
  retrometabolic drug development, 2:73–74
  reverse chemical genetics and, 1:298–299
  sample handling and storage, 1:349
  scaffolds, centroids, and adornments, 1:297
  solid-phase organic synthesis, 1:280–294
    libraries, 1:294–343
    lipid arrays, 1:293–294
    mixed drug synthesis, 1:345
    nucleoside arrays, 1:292–293
    oligosaccharide arrays, 1:291–292
    peptide arrays, 1:280–291
  solution-phase synthesis:
    lead optimization, 1:344–345
    libraries, 1:294–343
    mixed drug synthesis, 1:345
    synthetic success, 1:347–348
  tri-substituted pyrrolidine chemokine receptor-5 (CCR5) antagonist, 7:123–124
Combinatorial libraries:
  docking and scoring computations, 2:601–618
    point complementarity, 2:603–604
  encoding and identification, 1:110–111
  liquid chromatography-mass spectrometry purification, 1:107–108
  opioid peptides and peptidomimetics, 8:674–675
  peptide arrays, 1:280–291
  pulsed ultrafiltration-mass spectrometry, 1:116–119
  solid- and solution-phase libraries, 1:294–343
  structure and purity confirmation, 1:108–110
  virtual screening, 2:29–33

privileged structures,
2:30–31
protein family targeting,
2:31–33
Combinatorial quantitative
structure-activity
relationship:
integrated predictive modeling
workflow modeling,
1:517–520
model acceptability thresholds,
1:522–525
Combrestatin phosphate,
antibody-directed
enzyme prodrug
therapy, DNA-
targeted therapeutics,
6:120–123
Commercial-scale operations:
nevirapine case study, 3:21–23
scale-up of, 3:10–13
chemical safety, 3:13
environmental controls, 3:13
in-process controls, 3:11
processing equipment
requirements, 3:11
validation, 3:12–13
trade secrets limitations and,
3:180–181
Common technical document
(CTD), generic drug
approvals, 3:90–91
Communicable disease,
chemotherapeutic
agents and, 7:505–506
Compactin, anticholesterolemics,
2:232
Comparability techniques,
protein therapeutics
process development,
3:325–326
Comparative biochemistry:
antiprotozoan/antiparasitic
agents, kinetoplastid
chemical structure,
7:566–567
selective toxicity:
antibiotics, 2:548–553
antifungal drugs, 2:558–559
antihistamines, 2:564–566
antimycobacterial drugs,
2:557–558
antiviral compounds,
2:553–557
basic principles, 2:545
monoamine oxidase
inhibitors, 2:563–564

selective estrogen receptor
modulators,
2:566–567
selective serotonin reuptake
inhibitors, 2:569
tyrosine kinase inhibitors,
2:562–563
Comparative cytology, selective
toxicity:
antibiotics, 2:550–553
antimycobacterial drugs,
2:557–558
basic principles, 2:545–546
monoclonal antibodies,
2:559–562
Comparative distribution,
selective toxicity:
antihistamines, 2:564–566
basic principles, 2:545
bisphosphonates, 2:564
levobupivacaine, 2:564
nonbenzodiazepine sedatives,
2:568
phosphodiesterase inhibitors,
2:567–568
poton pump inhibitors, 2:566
selective estrogen receptor
modulators,
2:566–567
selective serotonin reuptake
inhibitors, 2:569
Comparative molecular field
analysis (CoMFA):
combinatorial quantitative
structure-activity
relationship modeling,
1:519–520
quantitative structure-activity
relationship, 1:39–43
glucocorticoid anti-
inflammatory
compounds, 5:77–80
integrated predictive
modeling workflow,
1:516–525
structure-based drug design,
computer-aided
techniques, 2:681–682
Comparative molecular similarity
index analysis
(CoMSIA):
quantitative structure-activity
relationship,
1:40–43
structure-based drug design,
computer-aided
techniques, 2:681–682

Comparative molecular surface
analysis (COMSA),
quantitative
structure-activity
relationships,
glucocorticoid anti-
inflammatory
compounds, 5:75
Comparative quantitative
structure-activity
relationship (C-
QSAR), 1:68–74
database development,
1:68–72
data mining for models, 1:69,
72–73
Comparative stereochemistry,
selective toxicity:
basic principles, 2:546–547
levobupivacaine, 2:564
Compartment of uncoupling
receptor and ligand
(CURL), receptor-
mediated endocytosis,
membrane proteins,
2:447
Complement factor H (CFH)
protein, age-related
macular degeneration,
5:605–606
Complement inhibitors, age-
related macular
degeneration therapy,
5:621–622
Complexation, oral drug
physicochemistry,
class II drugs, 3:46–47
Composition of matter claims,
patent obviousness
standards and,
3:137–138
Compound 500 IP receptor
antagonist,
prostaglandin/
thromboxane
receptors, 5:940–942
Compound 502 IP receptor
antagonist,
prostaglandin/
thromboxane
receptors, 5:938–942
Compound databases, receptor-
based drug design,
2:525
Compound selection, quantitative
structure-activity
relationship, 1:32–33

Compound similarity, ligand-based virtual screening, 2:2–3
"Comprising" terminology, infringement of patents, 3:152–154
Computational techniques:
allosteric protein sites, 1:390–392
docking/scoring approaches: homology modeling, 2:596–597
molecular dynamics, flexibility, 2:638–639
protein-protein interactions, 2:343–344, 354–359
opioid receptors, 8:590
pharmacophore generation, 1:469–470
Computer-aided design:
quantum mechanics/molecular mechanics, binding affinity evaluation, 2:720–723
receptor-based drug design, structure-activity relationships, 2:520
soft drugs, 2:130–133
candidate ranking, 2:131–132
hydrolytic liability, 2:132–133
structure generation, 2:131
structure-based drug design, 2:679–683
molecular docking, 2:679–680
pharmacophore modeling, 2:682–683
quantitative structure-activity relationships, 2:681–682
virtual screening, 2:680–681
Computer-assisted molecular design (CAMD), quantitative structure-activity relationship, 1:1
multidimensional models, 1:39–43
Conanotokins, stroke therapy, 8:451
Concentration profiles:
high-throughput screening, assay configuration, 3:414–418
oral drug delivery:

immediate-release dosage, 2:772–776
repeated dosages, 2:780–784
plasma drug profile:
diffusion-based drug release, sustained-release drug delivery, 2:815
erosion-controlled drug release, 2:827–828
Concerted models:
allosteric proteins, 1:375–376
receptor-based drug design, allosteric transition, 2:499–501
CONCOORD method, in silico screening, protein flexibility, 2:863–864
Conditioned avoidance:
antipsychotic therapy, 8:173
cognitive assessment, 8:28
Confidence index, quantitative structure-activity relationship models, applicability domains, 1:522
Configurational entropy, in silico screening, protein flexibility, 2:871
Conformational activation:
antithrombin, 4:390–392
heparin molecular mechanisms, 4:376–380
Conformational analysis:
cyclosporin A, 5:1007–1008
glucocorticoid anti-inflammatory compounds, 5:76–77
ligand-based design, nuclear magnetic resonance, 2:383–384
pharmacophores, 1:459
bioactive conformation, 1:469–471
ligand preparation, 1:466–468
in silico screening, protein flexibility, side-chain flexibility, 2:867–868
space differences, chemotype development, 1:486–487
triarylethylenes, 5:252
Conformational constraints:
deltorphin analogs, opioid peptides, 8:671–673

dermorphin analogs, μ-opioids, 8:666–668
dynorphin analogs, 8:656–657
enkephalin analogs:
δ-opioid analogs, 8:650–653
μ-opioid analogs, 8:647–649
Conformational polymorphism:
crystal engineering, 3:191–192
protein therapeutics, biophysical assessment, 3:324–325
Congestive heart failure (CHF):
aliskiren renin inhibitors, 4:252–253
angiotensin-converting enzyme inhibitors, 4:282–284
CK-2130 prodrug case study, 3:268–276
direct efficacy readouts, 4:502
drug development:
bench-to-bedside candidates and attrition development, 4:482–485
drug attrition, 4:480–482
emerging therapies, 4:479–480
critical evaluation, 4:483
epidemiology and prevalence, 4:479–480
etiology, 4:477
future research issues, 4:507
iontophoresis methodologies, 4:506
management approaches, 4:478–479
noninvasive imaging, 4:505–506
patient selection and stratification, 4:506–507
pharmacodynamic activity, 4:502–503
cardiotoxicity biomarkers, 4:503
dose selection, 4:502–503
polymorphisms and phenotypes, 4:507
predictive biomarkers, disease modification, 4:503–506
response-to-treatment studies, 4:507
target-compound interaction: predictive biomarkers:

ex vivo cell based biomarker assays, **4:**501
hemodynamic/biochemical biomarker assays, **4:**500–501
PET/SPECT ligands, **4:**501
target-engagement assays, **4:**501–502
target validation, **4:**483–500
clinical congruency, **4:**489–490
clinical importance, **4:**483, 485–487
current clinical trials, **4:**497–498
knockout/transgenic mouse models, **4:**488–490
mechanism perturbation, **4:**494–495
mechanism-related risk and reliability, **4:**500
mechanistic approaches, cardiac contractility/myocardial oxygen demand, **4:**492–494
normal patient mechanisms *vs.*, **4:**495–496
novel target *vs.* existing therapy, **4:**496, 499–500
pharmacodynamics/pharmacokinetics, **4:**491–492
preclinical models, **4:**486–491
tissue studies, species and disease states, **4:**488
thyromimetic therapy, **4:**198
Congreve rule, small druggable molecules, **1:**296
Conjugated derivatives, geldanamycin Hsp90 inhibitor, **6:**393–397
Conjugated hemoglobins, structure and properties, **4:**635–637
Conjugate gradient (CG) algorithm, quantitative structure-activity relationship, **1:**39
Conjugation reactions:
defined, **1:**406
drug metabolism, **1:**421–439
acetylation/acylation, **1:**429–432

amino acid formation, **1:**432
arylpropionic acids, **1:**433–434
coenzyme A, **1:**432–434
glucuronidation/glucosidation, **1:**426–429
glutathione, **1:**434–438
hybrid lipid/sterol ester formation, **1:**432–433
methylation, **1:**423–425
β-oxidation and two-carbon chain reactions, **1:**434
phosphate esters and hydrazones, **1:**438–439
sulfonation, **1:**425–426
equine estrogen metabolism, **5:**223–224, 232–233
β-lactam antibacterial drug resistance, **7:**269–281
conjugative transposons, **7:**270–271
plasmids, **7:**270
Connexins, anticonvulsants, **8:**134
Consensus interactions:
docking and scoring techniques:
hybrid methods, **2:**617–618
scoring functions, **2:**631–632
pharmacophore sources, **1:**459–461
quantitative structure-activity relationship model ensembles, **1:**522–525
structure-based virtual screening, **2:**23
transport proteins, *N*-glycosylation, **2:**460–461
"Consisting essentially of" terminology, infringement of patents, **3:**152–154
"Consisting of" terminology, infringement of patents, **3:**152–154
Constipation, opioid-related effects, **8:**575–576
Constitutive receptor activity, receptor-based drug design, **2:**498–499
Constrained geometric simulation (CGS), *in silico* screening, protein flexibility, **2:**863–864

Constructionist approach, quantitative structure-activity relationship, partition coefficient calculations, **1:**14–19
Construct validity, cognitive function assessment, **8:**25–29
Container material interactions, protein therapeutics formulation, **3:**315
Contaminants, protein therapeutics, downstream processing, **3:**301–303
Contextual freezing task, memory assessment, **8:**28
Contextual memory, conditioned fear, **8:**27–28
Continuation-in-part (CIP) application, drug discovery and biotechnology patentsd:
current practices, **3:**104–105
written description requirements, **3:**123–124
Continuous assays, high-throughput screening, **3:**407–408
Continuous intravenous infusion, drug delivery and bioavailability, **2:**760–762
Continuous microfluidic reactors, commercial-scale operations, **3:**13–14
Continuous performance tasks (CPTs), cognitive function assessment, **8:**26–27
Contraindications:
aminoclycosides, **7:**419–420
macrolides, **7:**438
opioids, **8:**576
tetracycline antibiotics, **7:**407–409
Convection-enhanced delivery, central nervous system tumor therapies, **6:**271–272
Convergent reactions, lead optimization, **1:**345

Convex hull method, outlier detection, quantitative structure-activity relationship, **1:**44
Cooling systems, pilot plant development, **3:**8
Cooperativity models, allosteric proteins, **1:**375–377, 379–380
Copaxone®, multiple sclerosis therapy, **5:**570–571
Copolymers, nanoscale drug delivery, **3:**475–479
Copyright protections, drug discovery and, **3:**185
Corepressors:
nuclear hormone receptors, **4:**83–84
thyroid hormone receptors, **4:**191
Corneal penetration, glucocorticoid anti-inflammatory compounds, **5:**53
Coronary artery bypass graft (CABG), acute coronary syndrome risk, **4:**411
Coronary artery disease (CAD), platelet function, **4:**409–410
Correct classification rate (CCR), quantitative structure-activity relationship modeling, data curation applications, **1:**528
Corresponding two-dimensional homonuclear spectroscopy (COSY), basic principles, **2:**371
Corrin ring structure, vitamin B_{12} (cobalamin), **5:**697, 699–701
Cortical spreading depression (CSD), migraine pathology, **8:**266–267
Corticosteroid binding globulin (CBG):
androgen absorption and distribution, **5:**158–159
glucocorticoid anti-inflammatory distribution, **5:**53–55

Corticosteroids:
chronic obstructive pulmonary disease inhalation therapy, **5:**773–780
combined β_2-receptor agonists, **5:**777–779
monotherapy, **5:**776–777
resistance, **5:**779–780
enzymatic metabolism, **5:**53–56
metabolism, **5:**25–26
multiple sclerosis therapy, **5:**568
pharmacological profiles, **5:**101–102
receptor affinity, **5:**66–70
soft drug design, **2:**90–101
antedrug steroids, **2:**99
etiprednol dicloacetate and analogs, **2:**94, 96–97
flucortin butyl, **2:**97
17-furoate androstadienes, **2:**99
glucocorticoid γ-lactones, **2:**98–99
itrocinonide, **2:**97–98
loteprednol etabonate and analogs, **2:**90, 92–96
steroidogenesis, **5:**20–22
topical compounds, current availability, **5:**39–40
Corticotropin-releasing factor (CRF):
adrenal steroidogenesis, **5:**20–22
antagonists, **8:**257
Cortienic acid-based steroids, soft drug design:
etiprednol dicloacetate and analogs, **2:**96–97
flucortin butyl, **2:**97
loteprednol etabonate and analogs, **2:**90, 92–96
Cortisol. *See* Hydrocortisone
cortisol-cortisone conversion, **5:**56–59
enzymatic metabolism, **5:**53–55
fluoro substitutions, **5:**83–85
glucocorticoid anti-inflammatory compound alterations, C-21 carbonyl alterations, **5:**104–112
percutaneous absorption, **5:**46–51
skeletal changes in, **5:**82–83
type 2 diabetes therapy, 11β-hydroxysteroid dehydrogenase-1 inhibitors, **4:**22–25

Cortisone:
cortisol-cortisone conversion, **5:**56–59
enzymatic metabolism, **5:**53
synthesis, **5:**7–9
type 2 diabetes therapy, 11β-hydroxysteroid dehydrogenase-1 inhibitors, **4:**22–25
Cosmegen. *See* Dactinomycin
Cost-effectiveness analysis (CEA):
drug development:
basic principles, **3:**345
decision-analytic approaches, **3:**346–349
pharmacogenomics and personalization, **3:**349
postlaunch analysis, **3:**349–350
regulatory approval, **3:**345–346
pharmacogenomics, **1:**199
Cotransfection assays, nuclear hormone receptors, **4:**80–82
Cotransmission, central nervous system drug discovery, **8:**5
Coumarin:
ring system, novobiocin Hsp90 inhibitor optimization, **6:**436–440
stroke therapy, **8:**499
Counterions, salt compound selection, **3:**389–391
Coupled assays, high-throughput screening, **3:**408–409
Coupling constants, ligand-based design, nuclear magnetic resonance, **2:**384
Court of Federal Claims (U.S.), patent infringement suits, **3:**149–150
Covalent binding, MDM2/MDM4 antagonists, **2:**357–359
Covalent bond interactions, quantitative structure-activity relationship, **1:**29–31

CP-122,288 triptan compound, migraine therapy, 8:270–272
CP320626 compound, allosteric protein model, 1:394
CP-690,550 immunosuppressive compound, 5:1055–1059
  chemical structure, 5:1055
  mechanism of action, 5:1055–1057
  research background, 5:1055
CP-778,875 compound, peroxisome proliferator-activator receptors, α-agonists, 4:96–97
cPrPMEDAP anti-DNA virus agent, 7:242
CPU86017 compound, chirality at, 1:148
Cranial nerve tumors, classification and therapy, 6:237–239
C-reactive protein, age-related macular degeneration, 5:604–606
Creatine kinase biomarkers, peroxisome proliferator-activator receptors, α-agonists, 4:96–97
Cre/loxP system, protein therapeutics, 3:296
Cresols, antimicrobial agents, 7:493–494
Critical micelle concentration (CMC):
  nanoscale drug delivery, self-assembled low molecular weight amphiphiles, 3:472–473
  oral drug physicochemistry, class II drugs, micelle solubilization, 3:48–49
Critical packing parameter, nanoscale drug delivery, self-assembled low molecular weight amphiphiles, 3:469–473
Crixivan. *See* Indinavir
Cross-linked DNA-targeted compounds, platinum complexes, 6:91

Cross-linked hemoglobin:
  HemAssist®, 4:628–630
  oxygen carrier functions, 4:628–631
  polynitroxylated hemoglobin, 4:630
Cross-linked polymeric nanoparticles, nanoscale drug delivery, 3:481
Cross-reactivity, PPARδ-agonists, 4:121
Cross-relaxation-enhanced polarization transfer (CRINEPT), receptor-based drug design, macromolecular compounds, 2:393–394
Cross-relaxation-induced polarization transfer (CRIPT), receptor-based drug design, macromolecular compounds, 2:393–394
Cross-validation test, quantitative structure-activity relationship, 1:46
  internal validation, 1:47–48
  model validation, 1:520–521
  outlier detection, 1:45
Cruzain, parasitic protease inhibitors, 7:581–582
Cryptophycins, marine sources, 2:245
Crystal engineering:
  cocrystals, 3:189–190, 192–196, 198–213
  future research issues, 3:213–214
  glucokinase activators, type 2 diabetes therapy, 4:16–17
  hydrates and solvates, 3:189
  NS3/4A protease inhibitors:
    boceprevir anti-HCV agent, 7:188–191
    ciluprevir, 7:178–181
  peroxisome proliferator-activator receptors, γ-agonists, 4:101–102
  polymorphs, 3:188, 190–192, 196–198
  protein therapeutics, 3:316–317
  research background, 3:187–190

  salts, 3:188–189
Crystalline solids:
  estrogen receptor topography, 5:234–236
  oral drug delivery, 3:25–26
  salt compounds, 3:381–383
CS-917 phosphonic diamide, type-2 diabetes therapy, 4:19–20
C-terminal nucleotide binding site, Hsp90 inhibitors:
  cisplatin, 6:440–441
  elucidation, 6:430–431
  epigallocatechin-3-gallate, 6:441
  GHLK protein family, 6:429–430
  novobiocin inhibitor, 6:431–440
C-type natriuretic peptide (CNP), structure and function, 4:533–537
Cubic side calculations, oral drug delivery, sustained release dosages, 2:796, 799–800
Curacin A, marine sources, 2:245
Curation protocols, quantitative structure-activity relationship modeling, 1:514–516
Current Good Laboratory Practices (cGLP), FDA guidelines, 3:81
Current Good Manufacturing Practices (cGMP), FDA guidelines, 3:81–82
Cutaneous diseases:
  adapalene, 5:412
  all-*trans*-retinoic acid therapy, 5:382–384
  9-*cis*-retinoic acid therapy, 5:425–426
Cutaneous T-cell lymphoma (CTCL):
  all-*trans*-retinoic acid therapy, 5:382
  bexarotene therapy, 5:435–439, 443–444, 450–451
CX-546, cognitive function assessment, 8:27–28
CXA-101 cephalosporin, 7:311–312
CXCR2 receptor antagonists, chronic obstructive pulmonary disease therapy, 5:784–785

CXCR4 receptor antagonist:
  brain tumor therapy, **6:**264–265
  chemokine receptor-5 (CCR5)
    antagonists, HIV-1
    entry blockade and,
    7:107–110
  plerixafor, structure and
    properties, **4:**586–587
Cyanobacteria, natural products
  lead generation,
  terrestrial and marine
  habitats, **2:**197–199
Cyanuric acid, cocrystal
  engineering, **3:**213
Cyclic amides, poly(ADP-ribose)
  polymerase inhibitors,
  6:164–167
Cyclic AMP response element
  binding protein
  (CREB):
  antidepressant efficacy
    assessment,
    8:226–229
  KIX conformation, **1:**376–377
Cyclic carbamates, stroke
  therapy, cathepsin B
  inhibitors, **8:**475–476
Cyclic guanosine monophosphate
  (cGMP):
  chronic obstructive pulmonary
    disease,
    phosphodiesterase
    inhibitors, **5:**781–784
  nitric oxide effects, **5:**271–278
  therapeutic enhancement,
    5:291–294
  phosphodiesterase inhibitors,
    erectile dysfunction
    therapy, **5:**712
Cyclic hydroxyamate inhibitors,
  stroke therapy,
  8:504–506
Cyclic peptides, global
  conformational
  constraints,
  1:214–215
Cyclic urea inhibitors, nuclear
  magnetic resonance
  studies, **2:**416–419
Cyclin-dependent kinase
  inhibitors:
  antimalarial cell cycle
    regulation,
    7:628–630
  chemogenomics, protein family
    targeted libraries,
    2:575

kinase 5 inhibitors:
  Alzheimer's disease, tau-
    targeted therapies,
    8:418–420
  hit seeking combinatorial
    libraries, **1:**329–330
  ligand-based virtual screening,
    pharmacophore
    formation, **2:**4–5
  medicinal chemistry and
    classification,
    6:310–315
  multidrug resistance and,
    6:377–378
  in silico screening, protein
    flexibility, backbone
    mobility, **2:**869–870
Cyclobutenedione derivatives,
  thyromimetic
  structure-activity
  relationship,
  heterocyclic rings,
  4:211
Cyclodextrin complexes:
  chemical delivery systems,
    2:149–150
  oral drug physicochemistry,
    class II drugs, **3:**47
Cycloguanil:
  antimalarial antifolates,
    7:644–648
  antimalarial therapy,
    7:617–618
Cyclohexane isosteres, κ-opioid
  receptor agonists,
  8:629–632
Cyclohexene isosteres, IP
  receptor agonist,
  5:933, 935
Cyclohexenylamines, dipeptidyl
  peptidase 4 (DPP-4)
  inhibition, **4:**64–65
Cyclohexenylguanine, anti-DNA
  virus antimetabolites,
  7:225–226
Cyclohexyl (pyran) analogs:
  sirolimus compound,
    5:1033–1034
  tacrolimus compounds,
    5:1018–1019
Cyclohexyl inhibitors, dipeptidyl
  peptidase 4 (DPP-4)
  inhibition, **4:**64–65
Cyclometallated complexes,
  stroke therapy,
  cathepsin B inhibitors,
  8:473

Cyclooxygenase (COX) inhibitors,
  8:268
  aspirin inactivation:
    aspirin resistance and,
      4:421
    thromboxane A_2 receptor
      biosynthesis,
      4:419–420
  multitarget drug development,
    COX-plus,
    inflammatory pain,
    1:267
  recombinant DNA technology,
    drug efficacy and
    personalized
    medicine,
    1:541–543
  selective toxicity, **2:**544
Cyclopamine, brain tumor
  therapy, **6:**260
Cyclopentene derivatives,
  prostaglandin/
  thromboxane
  receptors:
  EP_1 antagonists, **5:**829, 831,
    833
  FP receptor agonists,
    5:897–899
Cyclopenylglycine series,
  dipeptidyl peptidase 4
  (DPP-4) inhibition,
  4:51–54
Cyclophilin, cyclosporin-
  cyclophilin-
  calcineurin structure,
  5:1009–1010
Cyclophosphamide:
  cancer therapy, DNA-targeted
    mechanisms, **6:**83–84,
    86
  multiple sclerosis therapy,
    5:569–570
Cycloproparadicicol, analog
  design, **6:**402–403
Cyclopropavir:
  Epstein-Barr virus therapy,
    7:244
  human herpesvirus type 6
    (HHV-6) therapy,
    7:244
Cyclopropenones, stroke therapy,
  cathepsin B inhibitors,
  8:472–473
Cyclopropylindoles, DNA-
  targeted
  chemotherapeutics,
  6:91–93

Cycloserine:
  antitubercular agents, peptidylglycans biosynthesis, **7:**720
  cognitive enhancement, **8:**45
Cyclosporin A:
  amino acid composition and physical properties, **5:**1007
  conformational analysis, **5:**1007–1008
  discovery, **5:**1006–1007
  transplantation rejection prevention, **5:**1006–1013
  mechanism of action, **5:**1008–1010
  sirolimus combined with, **5:**1036
  structure-activity relationships, **5:**1010–1011
Cyclosporins:
  multiple sclerosis therapy, **5:**568
  recombinant DNA technology, targeting mechanisms, **1:**555–556
  soft drug analogs, **2:**120–121
Cyclothiazide, cognitive enhancement, **8:**43–44
Cylindrical radius/height calculations, oral drug delivery, sustained release dosages, **2:**796, 799
CYP1A1 enzyme, amodiaquine antimalarial therapy, **7:**614
CYP1A2 enzyme:
  antipsychotics, phenothiazines, **8:**186–189
  clopidogrel resistance, **4:**432–433
CYP2C8 enzyme, amodiaquine antimalarial therapy, **7:**614
CYP2C9 enzyme:
  all *trans*-retinoic acid interaction, **5:**386
  clopidogrel resistance, **4:**432–433
  pharmacogenomics, **1:**186
  toxicophore reactive metabolites, aniline metabolism, aromatic ring oxidation, **2:**303–306
  type 2 diabetes therapies, glinides, **4:**12–13
CYP2C19 enzyme:
  antitubercular agents, isoniazid interactions, **7:**743–744
  clopidogrel resistance, **4:**432–433
CYP2D6 enzyme:
  ADMET properties, metabolic stability, **2:**58–59
  antipsychotics, **8:**185–186
  phenothiazines, **8:**186–189
  BioPrint profiles, **1:**502
  HIV protease inhibitors: indinavir sulfate pharmacokinetics, **7:**17
  ritonavir, **7:**12
  multiobjective drug discovery optimization, blood-brain barrier and, **2:**268–273
  opioid interaction with, **8:**577
  tropane-based chemokine receptor-5 (CCR5) antagonist, **7:**116–121
CYP3A enzyme:
  ADMET properties, metabolic stability, **2:**58–59
  antitubercular agents, isoniazid interactions, **7:**743–744
  HIV protease inhibitors, ritonavir, **7:**12
  P-glycoprotein metabolism, **2:**461–463
CYP3A4 enzyme:
  all *trans*-retinoic acid interaction, **5:**386
  antipsychotics, **8:**185–186
  phenothiazines, **8:**186–189
  clopidogrel resistance, **4:**432–433
  drug resistance inhibition, **6:**372
  HIV protease inhibitors, indinavir sulfate pharmacokinetics, **7:**17
  migraine therapy, CGRP antagonists, **8:**288–290
  opioid interaction with, **8:**577
  type 2 diabetes therapies, glinides, **4:**12–13
CYP enzyme inhibition, ADMET toxicity properties, drug-drug interactions, **2:**63, 68–69
Cypionate, current compounds and applications, **5:**220–221
Cyproheptadine, glucuronidation/glucosidation, **1:**429–430
Cyproteron acetate antiandrogen, **5:**190–192
Cys-loop family, nicotinic acetylcholine receptors, **8:**63
Cys-scanning mutagenesis, transport proteins, **2:**460
Cystathionine-beta-synthase (CBS), hydrogen sulfide production, **5:**309–312
Cystathionine-gamma-lyase (CSE), hydrogen sulfide production, **5:**309–312
L-cysteine, hydrogen sulfide production, **5:**266, 309–312
Cysteine proteases. *See also* specific Cathepsin inhibitors
  antimalarial cell cycle regulation:
    falcipain inhibitors, **7:**635–637
    farnesylation, **7:**630–634
    hit seeking combinatorial libraries, **1:**330–332
    parasitic protease inhibitors, **7:**581–582
  stroke therapy, cathepsin B inhibitors, absence of, **8:**477–478
Cysteinylation, protein therapeutics formulation and delivery, **3:**313–314
Cystic fibrosis, *N*-(4-hydroxyphenyl) retinamide (4-HPR) therapy, **5:**473–474

Cytisine:
  nicotinic agonists, **8:**70
  smoking cessation therapy, **5:**723–724
Cytochrome P450 enzymes:
  ADMET properties, metabolism, **2:**57–59
  adrenal steroidogenesis, **5:**22
  all *trans*-retinoic acid interactions, **5:**386
  antidepressants and:
    pharmacokinetics, **8:**236
    serotonin-selective reuptake inhibitors, **8:**234, 236
  antimalarial drugs, artemisinins, **7:**617
  antipsychotic pharmacokinetics, **8:**185–200
  antitubercular agents, monooxygenase targeting, **7:**729
  drug metabolism:
    functionalization reactions, **1:**409–413
    sp^3-carbon atom oxidation and reduction, **1:**413–415
  estradiol biosynthesis, **5:**25
  immunosuppressive drugs, cyclosporin A, **5:**1012
  opioid interactions with, **8:**576–577
  oral dosage forms, class V drugs, **3:**54
  P-glycoprotein metabolism, **2:**461–463
  pharmacogenomics, **1:**183–187
    CYP2C9 substrate, **1:**186
    CYP2C19 substrate, **1:**186–187
    CYP2D6 substrate, **1:**184–186
    CYP3A4/5/7, **1:**187
  prodrug development, CK-2130 prodrug case study, **3:**274–276
  quantitative structure-activity relationship, metabolism parameters, **1:**66–67
  receptor-based drug design, receptor classification, **2:**506
  selective PPARγ modulators, PPARγ partial agonists, **4:**113–115
  sirolimus mediation, **5:**1034–1035
  soft drug development, **2:**77
  serotonin receptor agonists, **2:**117–118
  steroidogenesis, cholesterol-pregnenolone conversion, **5:**16–17, 19
  tetracycline selective toxicity, **2:**552–553
  toxicophore reactive metabolites:
    alkynes, **2:**320–321
    aniline metabolism, aromatic ring oxidation, **2:**303–306
  toxicophores, tight binding mechanisms, **2:**324–325
Cytokines. *See also* specific cytokines, e.g., Stem cell factor
  future research issues, **4:**588–590
  immune function, **5:**1003–1006
  investigational new agents, **4:**587–588
  receptor-based drug design, non-GCPR-like cytokine receptors, **2:**512–513
  research background, **4:**569–570
  soft drug design, inhibitors, **2:**121–122
  tuberculosis pathogenesis, **7:**718–719
Cytokinetics, ATP-competitive kinesin spindle protein inhibitors, **6:**211–213
Cytomegalovirus (CMV):
  anti-DNA virus agents, **7:**243–244
  chemical delivery systems therapy, ganciclovir, **2:**142
Cytosine arabinose, antimetabolite analogs, **6:**109–112
Cytosine deaminase, gene-directed enzyme prodrug therapy, **6:**123–124
Cytotoxic agents:
  brain tumors:
    stem cell initiation, **6:**225–227
    temozolomide, **6:**244
  DNA targeting compounds, **6:**3–17
    bleomycin, **6:**8–12
    dactinomycin, **6:**3–8
    mitomycin (mutamycin), **6:**12–16
    plicamycin, **6:**16–17
  enzyme inhibitors, **6:**17–31
    anthracyclines, **6:**17–23
    camptothecins, **6:**23–29
    isopodophyllotoxins, **6:**29–31
  future research issues, **6:**40–41
  research background, **6:**1–3
  tubulin polymerization/depolymerization inhibition, **6:**31–40
    dimeric vinca alkaloids, **6:**35–40
    taxus diterpenes, **6:**32–35
Cytotoxicity assays, ADMET toxicity properties, **2:**62, 68

Dabigatran, active site-directed inhibitors, **4:**398–400
Dacarbazine, DNA-targeted compounds, **6:**94–95
Daclizumab:
  multiple sclerosis therapy, **5:**572
  transplantation rejection prevention, **5:**1001–1002
Dactinomycin:
  biosynthesis, **6:**6–8
  chemical structure, **6:**3
  contraindications and side effects, **6:**4
  medicinal chemistry, **6:**4–5
  molecular mechanisms, **6:**5–7
  pharmacokinetics, **6:**4–7
  therapeutic applications, **6:**3–4
Dactylocyclines, history and biosynthesis, **7:**412–413
Daily values (DVs), vitamins, **5:**658, 661
D-Ala-D-Ala mimicry, Tipper-Stromiger hypothesis, penicillin proteins and, **7:**264–266

Dalbavancin, antibacterial
    mechanism,
    2:226–227
Dalcetrapib, atherosclerosis,
    HDL elevation,
    4:341–343
Dale's principle, central nervous
    system drug discovery,
    8:5
DAMGO agonist, opioid receptor
    mutagenesis,
    8:589–590
Dapagliflozin, type 2 diabetes
    therapies, **4:**14
Dapivirine HIV reverse
    transcriptase
    inhibitor, **7:**150–152
Dapsone:
    current applications,
      7:521
    folate inhibition, **7:**527
    historical background, **7:**514
Daptomycin, antibiotic
    mechanism,
    2:227–228
Darifenacin, muscarinic
    antagonists, **8:**74–77
Dark-adaptation, 13-*cis*-retinoic
    acid side effects, **5:**394
Darunavir (DRV):
    binding interactions,
      7:55–57
    chemical structure, **7:**52
    design criteria, **7:**3
    efficacy and tolerability, **7:**57,
      59
    optimization, **7:**52–55
    pharmacokinetics, **7:**57–58
    resistance profile, **7:**59–60
DARWIN algorithm, docking and
    scoring techniques,
    2:613–614
Dasatinib:
    chronic myelogenous leukemia
      therapy, Bcr-Abl
      inhibitors, **6:**299–301
    second-generation
      development, drug
      resistance and,
      6:364–365
    selective toxicity, **2:**563
    structure-based design:
      multiconformational
        catalytic domain,
        6:346–349
      protein targeting,
        6:352–354

Data analysis:
    chemogenomics, molecular
      information systems,
      2:582–583
    combinatorial chemistry,
      1:349–350
    quantitative structure-activity
      relationship modeling,
      1:514–516
    data set complexity,
      1:512–516
    structural genomics research,
      1:581
    virtual screening, pre- and
      postprocessing,
      2:25–29
Database development:
    chemogenomics, molecular
      information systems,
      2:582–583
    comparative quantitative
      structure-activity
      relationship,
      1:68–72
    organic compounds, ligand-
      based virtual
      screening,
      pharmacophore
      formation, **2:**4–5
    receptor-based drug design:
      compound databases, **2:**525
      nomenclature databases,
        2:504–505
Data curation, quantitative
    structure-activity
    relationship modeling,
    1:514–516
    experimental applications,
      1:527–528
Data mining, comparative
    quantitative
    structure-activity
    relationship, **1:**69,
    72–73
DATA R-106168 HIV reverse
    transcriptase
    inhibitor, **7:**151–152
DATScan, Parkinson's disease
    dopaminergic therapy,
    8:547–549
Daunomycin. *See* Daunorubicin
Daunorubicin:
    Alzheimer's disease, tau-
      targeted therapy,
      8:426–427
    anti-cancer activity,
      2:236–239

chemical structure and
    therapeutic
    applications, **6:**19–21
D-Carba-T HIV reverse
    transcriptase
    inhibitor, **7:**145
D-Cycloserine (DCS),
    7:767–771
    ADMET properties,
      7:770–771
    adverse effects, **7:**771
    historical background,
      7:767–768
    resistance mechanisms,
      7:768–770
    structure-activity relationship,
      7:768
DD carboxylate residue triad,
    HIV integrase and,
    7:78–79
Deacetoxycephalosporin C,
    biosynthesis, **7:**295
Deamidation, protein
    therapeutics
    formulation and
    delivery, **3:**312–314
Debrisoquine/sparteine
    hydroxylase,
    pharmacogenomics,
    1:184–186
DEBS combinatorial library,
    recombinant DNA
    technology,
    1:547–549
Decahydroisoquinoline-3-
    carboxylic acids,
    glutmate modulators,
    migraine therapy,
    8:292–297
Decamethonium, nicotinic
    receptor blockade,
    8:77–78
Decaprenylphosphoryl arabinose
    (DPA), antitubercular
    agents,
    arabinogalactan
    biosynthesis, **7:**722
Decision analysis, cost-
    effectiveness analysis
    of drug development,
    3:346–349
Decision Forest method,
    quantitative
    structure-activity
    relationship models,
    applicability domains,
    1:522

Decision trees, cost-effectiveness analysis of drug development, 3:347–349
Deferoxamine, stroke therapy, 8:482
DEFGH pentasaccharide sequence:
 antithrombin conformational activation, 4:390–392
 heparin molecular mechanism of action, 4:375–380
Deforolimus:
 ADMET properties, 5:1035
 cancer therapy, 5:1037
 cyclohexyl analog design, 5:1033–1034
Degradation pathways, protein therapeutics formulation and delivery, 3:311–314
Dehydro derivatives, anabolic agents, 5:180
Dehydroepiandrosterone (DHEA):
 androgen biosynthesis, 5:24–25
 basic structure, 5:153–154
Dehydropeptidase (DHP-1) enzyme:
 carbapenem pharmacokinetics, 7:282, 350–353
 doripenem development, 7:356
Delavirdine HIV reverse transcriptase inhibitor, 7:148–152
Deletion mutagenesis, receptor-based drug design, 2:514
Deltorphin analogs, opioid peptides, 8:668–673
Dementia-related psychosis, antipsychotic medications, 8:170
Demethylchlortetracycline:
 chemical structure, 7:407
 history and biosynthesis, 7:410–413
Denaturation, protein therapeutics formulation and delivery, 3:313–314
Dendrimers, nanoscale drug delivery, 3:482–483
Dendritic cells, central nervous system tumors, vaccine development, 6:269–270

De novo drug design:
 docking and scoring computations, 2:599–632
 glucocorticoid anti-inflammatory compounds, quantitative structure-activity relationships, 5:71–72
 G-protein coupled receptor, homology modeling, 2:283–287
 heparin nonsaccharide mimics, 4:392–394
 Pareto ligand designer, 2:266–273
 peroxisome proliferator-activated receptors, α/γ-dual agonists, 4:126–135
 recombinant DNA technology, 1:553–556
Dense tubular system, platelet activation, 4:416–418
Density functional theory (DFT):
 free-energy perturbation, binding affinity evaluation, 2:727–728
 quantum mechanics/molecular mechanics, 2:737–738
Deoxy analogs, anabolic agents, 5:183–185
1-Deoxy-D-xylulose 5-phosphate (DOXP) pathway, apricoplast targeting, antimalarial agents, 7:665–668
Deoxyerythronolide B synthase (DEBS), macrolides, 7:441–444
Deposition process, infringement of patent proceedings, 3:161
Depression:
 brain imaging studies, 8:227–229
 diagnostic criteria and agent selection, 8:219–221
 epidemiology, 8:219
 long-term depression, 8:25
 psychostimulant therapy, 8:92
 13-*cis*-retinoic acid side effects, 5:394
 SERT-plus multitarget drug development, 1:254–257

Depth filtration, mammalian protein purification, harvest operations, 3:307
Derived acridine-4-carboxamide analog (DACA), dual topoisomerase I/II inhibitors, 6:104–105
Dermorphin analogs, opioid peptides, 8:664–668
Derrubone Hsp90 inhibitor, 6:425–426
Designed multiple ligands (DMLs):
 historical background, 1:254
 multitarget drug development, 1:251–252
 ACE-plus and angiotensin-1 receptor (AT_1) antagonist, 1:259–260
 AChE-plus, Alzheimer's disease, 1:261, 263
 dopamine D_2-plus for schizophrenia, 1:257–258
 histamine H_1-plus for allergies, 1:259–262
 knowledge-based approaches, 1:253–254
 screening procedures, 1:252–253
 SERT-plus multitarget drug development, 1:254–257
 optimization strategies, 1:267–273
 physicochemical/pharmacokinetic properties, 1:269–273
Desipramine:
 chemical structure, 8:222
 historical background, 8:248–249
 monoamine transporter inhibition, 8:240–241
 pharmacokinetics, 8:235–236
 serotonin receptor targeting, 8:241–242
Desirability-based multiobjective optimization (MOOP-DESIRE), drug discovery, 2:264–266
Desirability optimization methodology (DOM), drug discovery, 2:262–263

Desmodus rotundus salivary plasminogen activator alpha 1, stroke therapy, 8:448–449
Desnoviase benzamide derivative, novobiocin Hsp90 inhibitor optimization, 6:439–440
Desogestrel, biotransformation, 5:230
Desorption ionization techniques, mass spectrometry, drug discovery applications, 1:99–100
Desvelafaxine:
  chemical structure, 8:223
  side effects and adverse reactions, 8:234
Detachment strategies, peptide libraries, 1:288–291
Developmental Therapeutics Program (DTP), quantitative structure-activity relationship modeling, 1:514
Dexamethasone:
  chemogenomics, yeast three-hybrid screening, 2:581–582
  colon-specific drug delivery, 5:45
Dexelvucitabine:
  anti-DNA virus antimetabolites, 7:226–227
  current development, 7:142–145
Dextromethorphan:
  selective toxicity and stereochemistry, 2:546
  stroke therapy, 8:451
Dextrothyroxine, structure-activity relationship, 4:203
Dezocline nonpeptide opiate, 8:643–645
"DFG-Asp-in/alphaC-Glu-out" conformation, structural genomics, 1:593–594
DFG motif, multiconformational kinase catalytic domain, 6:346–349
DFIRE potential, empirical scoring, 2:630–631

DG-041 EP_3 receptor antagonist, antiplatelet activity, 4:450–451
Diabetes mellitus. *See also* Maturity onset diabetes of the young type-2 (MODY-2); Type 2 diabetes
  Alzheimer's disease, insulin-related therapy, 8:414–417
  Am80 therapy, 5:418
  current therapies, 4:2–13
    α-glucosidase inhibitors, 4:5–8
    biguanides, 4:2–5
    glinides, 4:10–13
    metformin, 4:3–5
    sulfonylureas, 4:8–10
  dipeptidyl peptidase 4 inhibitors, 4:39–65
  emerging therapies, 4:13–27
    fructose-1,6-biphosphatase inhibitors, 4:18–20
    glucokinase activators, 4:15–17
    GPR119 agonists, 4:20–21
    11β-hydroxysteroid dehydrogenase-1 inhibitors, 4:22–25
    SIRT1 activators, 4:25–27
    sodium-dependent glucose cotransporter inhibitors, 4:13–14
  epidemiology, 4:1–2
  farnesoid X receptors, 4:160
  hepatocyte nuclear factor4 and, 4:163–164
  N-(4-hydroxyphenyl) retinamide (4-HPR) therapy, 5:474
  NRX 194204 RXR selective analog, 5:461–462
  peroxisome proliferator-activator receptors:
    α-agonists, 4:97–99
    γ-agonists, 4:100–102
  13-*cis*-retinoic acid side effects, 5:394–395
  thyromimetic therapy, 4:198
Diacyl hydrazide, prostaglandin/thromboxane EP_1 antagonists, structure-activity relationships, 5:813–817
Diagnostic techniques:

  cognitive dysfunction, 8:21–22
  Parkinson's disease, 8:555–556
  recombinant DNA technology, aptamer-based techniques, 1:549–550
Dialdehydes, Alzheimer's disease, tau-targeted therapy, 8:427–428
Diallyldisulfide, hydrogen sulfide releasing molecules, 5:314–315
Diallyltrisulfide, hydrogen sulfide releasing molecules, 5:314–315
Diaminopimelic acid (DAP), antitubercular agents, peptidylglycans biosynthesis, 7:720
2,6-Diaminopurine (DAP) derivative:
  acyclic nucleoside phosphate, 7:236
  hepatitis B virus therapy, 7:2456
Diaminopurine dioxolane (DAPD), anti-DNA virus antimetabolite, 7:227
Diaminopyrimidine renin inhibitors, 4:254–255
Diarylbutylamines, pharmacokinetics, 8:190
Diarylethylenes:
  estrogen antagonists, 5:248–249
  estrogen receptor affinity, 5:247–248
Diarylpyridines, reverse transcriptase inhibitors, 7:154–155
Diarylquinoline (TMC207) antitubercular agent, 7:783–785
Diarylureas:
  inhibitors, 5:1042–1045
  topical antimicrobial agents, 7:498
Diastereoselective synthesis, trinem formation, 7:345–350
Diastolic dysfunction models, preclinical target validation, congestive heart failure therapy, 4:487

Diazabicyclo[3.0.3] compound,
nicotinic agonists,
8:71–72
Diazeniumdiolates, nitric oxide
donors and prodrugs,
5:281–283
Diazepam:
anticonvulsant applications,
8:144–147
cognitive enhancement,
8:45–46
zotepine interaction, **8:**194
Diazepines, γ-secretase
inhibitors, **8:**336–337
Diazomethyl ketones, stroke
therapy, cathepsin B
inhibitors, **8:**475
Dibekacin, chemical structure,
7:419
Dibenzoazepines,
pharmacokinetics,
8:189
Dibenzodiazepines,
pharmacokinetics,
8:191–192
Dibenzothiepine,
pharmacokinetics,
8:194
Dibenzthiazepines,
pharmacokinetics,
8:193–194
DIBRT thyromimetic, structure-
activity relationship,
4:219
Dichlorodiphenyltrichloroethane
(DDT), soft drug
design,
chlorobenzilate
analog, **2:**106–107
6,8-Dichloropurine heterocyclic
scaffold, ERK-1/
RasGap dual
inhibitors, **1:**320–321
Diclofenac:
hydrogen sulfide releasing
molecules, **5:**314–315
methylation reactions,
1:423–425
migraine therapy, **8:**268
SNO-diclofenac, combined
nitric oxide donors and
prodrugs, **5:**288–290
structure-based design,
2:696–698
toxicophore reactive
metabolites, aniline
metabolism, aromatic
ring oxidation,
2:303–306
Didanosine, **7:**141–145
Diels-Alder adducts:
opioid analgesics, **8:**608–609
sirolimus, triene region
modification,
5:1032–1033
(Z,Z)-Dienestrol, **5:**247–248
Diet:
antidepressants and,
monoamine oxidase
inhibitors, **8:**230
N-(4-hydroxyphenyl)
retinamide (4-HPR)
therapy and, **5:**479
vitamins in, **5:**638
Dietary reference intakes (DRIs),
vitamins, **5:**638–644
biotin, **5:**691
folic acid, **5:**697
niacin, **5:**679
pantothenic acid, **5:**685
riboflavin (vitamin B_2), **5:**673
thiamine (vitamin B_1), **5:**673
vitamin A, **5:**648
vitamin B_6 family, **5:**684–685
vitamin B_{12} (cobalamin), **5:**701
vitamin C (ascorbic acid), **5:**705
vitamin D family, **5:**658–659
vitamin E, **5:**665
vitamin K, **5:**668
Diethylaminoethanol (DEAE),
protein therapeutics,
attachment-
dependent cell culture,
0
Diethylpropion:
ADMET properties, **8:**99
structure-activity relationship,
8:106–109
Diethylstilbestrol:
biotransformation, **5:**226
electrostatic interactions, **5:**236
structure-activity
relationships, vicinal
diarylethylenes,
5:247–248
Differential line broadening
(DLB), NS3/4A
protease inhibitors,
ciluprevir, **7:**177–181
Differential scanning calorimetry
(DSC):
oral drug melting point,
3:30–31
protein therapeutics, **3:**325
Differential scanning calorimetry
(DSC), ligand-
macromolecular
interactions,
structure-based
thermodynamic
analysis, **2:**706–712
Diffusion-based drug release:
plasma drug profile, **2:**801–816
calculation methods,
2:802–803
controlled-release dosage,
2:812–816
erosion-controlled drug
release vs., **2:**821–822
nomenclature, **2:**801
repeated doses, **2:**808–812
single-dose results,
2:803–808
sustained-release dosage
forms, *in vitro* testing,
2:792–798
erosion-controlled release
and, **2:**801
tissue-based drug delivery:
capillary wall drug diffusion,
2:847–848
endocarditis, **2:**849–851
lung and bronchial mucus,
2:839–844
Diffusion coefficients, nuclear
magnetic resonance
studies, drug
screening
applications,
2:426–427
Diflomotecan, biological activity
and side effects, **6:**98
Diflunisal, sulfonation reactions,
1:426
Difluorophenacylalanine (DAPT),
γ-secretase inhibitors,
8:335
Dihydrexidine:
cognitive enhancement,
8:41–42
functional selectivity, **8:**8–9
Parkinson's disease
dopaminergic therapy,
8:548–549
Dihydroergotamine (DHE),
migraine therapy,
8:267–268
Dihydrofolate reductase (DHFR),
protein therapeutics:
dhfr- development, **3:**293–294
transfection, **3:**295–296

Dihydrofolate reductase
inhibitors (DHFRIs):
antibiotic selective toxicity and comparative biochemistry, **2**:548–553
discovery, **7**:483
folate inhibition, **7**:531–534
antimalarial antifolates, **7**:617, 644–648
inhibitors, hit seeking combinatorial libraries, **1**:317, 319
Leishmaniasis therapy, **7**:592–593
nuclear magnetic resonance studies, **2**:414–415
quantitative structure-activity relationship:
isolated receptor interactions, **1**:51–57
receptor interactions, **1**:28–31
recombinant DNA technology, drug targeting applications, **1**:553–556
soft drug design, **2**:119
sulfonamide/sulfone synergism, **7**:527–528
Dihydrogeldanamycin derivatives, geldanamycin Hsp90 inhibitor, **6**:388–391
Dihydroindolone, pharmacokinetics, **8**:190–191
Dihydroorotate dehydrogenase (DHOD) inhibitors, antimalarial drugs, **7**:648–650
cellular respiration and electron transport, **7**:655–658
Dihydropteroate synthase (DHPS), folate inhibition, **7**:528–531
1,4-Dihydropyridines (DHP), privileged pharmacophores, lead drug discoveries, **2**:531
Dihydropyrazoles, kinesin spindle protein inhibitors, **6**:204–210
ARRY-520, **6**:209–210
LY2523355 compound, **6**:210

MK-0731, **6**:204–209
Dihydropyridine analogs:
combinatorial libraries, **1**:302–303
stroke therapy, **8**:459–460
Dihydropyrimidine dehydrogenase (DPD), single nucleotide polymorphisms, **1**:187–188
Dihydropyrroles, kinesin spindle protein inhibitors, **6**:204–209
5α-Dehydrotestosterone:
androgen biosynthesis, **5**:24–25
basic structure, **5**:153
biological mechanisms, **5**:164–168
metabolism, **5**:26–27
(+)-[^{11}C]-Dihydrotetrabenazine, Parkinson's disease dopaminergic therapy, **8**:547–549
Dihydroxyaromatics, toxicophore reactive metabolites, **2**:313–314
6α,7α-Dihydroxycortisone 21-acetate, **5**:91
1,2-Dihydroxyethylene group, plasmepsin antimalarial agents, **7**:641–644
Dihydroxyphenyl amide Hsp inhibitors, fragment-based design, **6**:424–425
Dihydroxypyrimidine, raltegravir development, **7**:89–90, 92–97
1α,25-Dihydroxyvitamin D_3, soft drug design, **2**:101–102
Diiodotyrosine (DIT), thyroid hormone biosynthesis, **4**:193
Diketo acid derivatives, HIV integrase, discovery and elaboration, **7**:81–97
Diketopiperazine, irreversible inhibitors, stroke therapy, **8**:465
Diketopiperazine-based chemokine receptor-5 (CCR5) antagonist, **7**:131–133

Diltiazem, migraine prevention, **8**:318
Dimensionality, combinatorial quantitative structure-activity relationship, integrated predictive modeling workflow modeling, **1**:519–520
Dimensionless numbers:
oral drug delivery:
immediate-release dosage, **2**:772
master curves, **2**:781–784
sustained-release dosage forms, **2**:795
repeated intravenous bolus injections, **2**:756
Dimeric structures:
aminoglycoside antibiotics, **7**:429–432
geldanamycin Hsp90 inhibitor, **6**:393–397
glucocorticoid receptors, **5**:62
novobiocin Hsp90 inhibitor, **6**:434
opioid receptors, **8**:590–591
enkephalin analogs, **8**:653–654
stroke therapy, NOS inhibitors, **8**:501
vinca alkaloids, **6**:35–40
X-linked inhibitor of apoptosis, peptidomimetics, **1**:237–240
DIMIT compound, thyroid hormone receptor β-selective agonists, structure-based design, **2**:685–691
Dinapsoline, Parkinson's disease dopaminergic therapy, **8**:548–549
Dioxolane, muscarinic activity, **8**:64–66
1,3-Dioxolane derivatives:
anti-DNA virus antimetabolite, **7**:227–228
hepatitis B virus therapy, **7**:2456
1-(β-D-Dioxolane)thymine (DOT) HIV reverse transcriptase inhibitor, **7**:143–145

Dipeptide analogs:
  EP$_4$ receptor antagonists,
      5:881–883
  first-generation
      peptidomimetics,
      1:215–216
  peptide conformation, local
      constraints, **1:**213–214
  peptide libraries, **1:**288–291
Dipeptidyl peptidase 4 (DPP-4)
      inhibitors:
  ADMET properties, **4:**45–47
  alternate functions, **4:**40–42
  future directions, **4:**65
  inhibitor binding, **4:**47
  physiology and pharmacology,
      4:43–45
  related enzymes, **4:**42
  research background, **4:**39–40
  selectivity parameters, **4:**47–48
  side effects, **4:**45
  structure-activity
      relationships,
      4:48–65
    α-amino acid derivatives,
        4:48–56
    β-amino acid derivatives,
        4:57–59
    cyclohexyl- and
        cyclohexenylamines,
        4:64–65
    fluoroolefin inhibitors,
        4:56–57
    heterocyclic derivatives,
        4:59–64
    hybrid α/β-amino acid
        derivatives, **4:**59
Diphenhydramine:
  Parkinson's disease therapy,
      8:552–553
  selective toxicity, **2:**565
5,5-Diphenyl-barbituric acid,
      anticonvulsant
      applications,
      8:140–141
Diphenyloxazole derivatives, EP$_4$
      receptor antagonists,
      5:880–883
Diphenylsulfide derivatives,
      antitrypanosomal
      polyamine
      metabolism, **7:**590
Dipyridamole, antiplatelet
      activity, **4:**450
Direct analogs:
  defined, **1:**167–168
  production of, **1:**170–171

Direct assays, high-throughput
      screening, **3:**408–409
Directly observed therapy, short-
      course (DOTS)
      regimen, tuberculosis
      management,
      7:713–716
  rifamycin antibacterials,
      7:747
Direct thrombin inhibitors (DTIs):
  active site-directed inhibitors,
      4:395–398
  heparin-based indirect
      anticoagulants, **4:**392
  limitations of, **4:**374
  molecular mechanisms of
      action, **4:**385–390
  structural properties,
      4:369–371
Dirithromycin, chemical
      structure, **7:**438
DISC1 genes, schizophrenia
      assessment,
      8:173–174
Discodermolide, Alzheimer's
      disease, tau-targeted
      therapy, **8:**430
Discontinuous assays, high-
      throughput screening,
      3:407–408
DISCO program, pharmacophore
      formation, bioactive
      conformation, **1:**470
DISCOtech program,
      pharmacophore
      formation, bioactive
      conformation, **1:**470
Discovery process, infringement
      of patents, **3:**160–161
Discretes, defined, **1:**280
Disease characteristics:
  cognitive function, **8:**25–29
  congestive heart failure
      therapy, biomarkers
      and disease
      modulation,
      4:503–506
  diffusion-based drug release,
      plasma drug profile,
      repeated dose
      delivery, **2:**811–812
  radiolabled biomarkers,
      8:744–746
Disinfectant testing:
  historical background, **7:**490
  topical antimicrobial agents,
      7:489

Dispersion methods, oral drug
      physicochemistry,
      class II drugs, solid
      dispersion, **3:**46
2,3-Disphosphoglycerate (DPG),
      hemoglobin oxygen
      delivery, **4:**615–619
Dissociated glucocorticoids:
  nonsteroidal, **5:**125–134
  steroidal, **5:**120–125
Dissociation constant:
  acid-base equilibrium, salt
      compounds, **3:**384–388
  ADMET structural properties,
      2:52
    solubility, **2:**53–55
  atypical tricyclic neuroleptics,
      8:180–183
  receptor-based drug design,
      residence times, **2:**504
  receptor binding assays,
      2:521–523
Dissolution:
  oral drug delivery:
    absorption, **3:**354–355
    physicochemistry evaluation,
        3:33
    sustained-release dosage
        forms:
      in vitro-in vivo
        correlations,
        2:787–791
      in vitro tests, **2:**793
  permeability, ADMET
      interactions,
      3:374–375
Distance between r^2 and q^2,
      quantitative
      structure-activity
      relationship:
  outlier detection, **1:**45
  validation, **1:**46–47
Distance cutoff value,
      quantitative
      structure-activity
      relationship models,
      applicability domains,
      1:522
Distance geometry, docking and
      scoring computations,
      2:604–605
Distribution:
  ADMET properties, **2:**59–61
    organ barriers, **2:**61
    plasma protein binding,
        2:59–61
    tissue binding, **2:**61

androgens, **5**:158–159
  anabolic agents, **5**:186–188
  antiandrogen compounds, **5**:193
  chirality and, **1**:140–141
  opioid receptors, **8**:577–578
  permeability, ADMET interactions and, **3**:375–376
  prodrug development, **3**:223–226
  quantitative structure-activity relationship, **1**:62–68
  type 2 diabetes therapies, dipeptidyl peptidase 4 inhibitors, **4**:45–46
DITBA thyromimetic, structure-activity relationship, **4**:205
Ditekiren renin inhibitor, **4**:244–247
Diuretics, steroid hydroxylation, **5**:9–10
Divalent salts, high-throughput screening, assay optimization, **3**:412
DIVALI algorithm, docking and scoring techniques, **2**:613–614
Diversity-based ligand libraries, lead drug discoveries, **2**:531
Diversity-oriented synthesis (DOS), chemogenomics, **2**:578–580
Dizocilpine, stroke therapy, **8**:450
DMCM receptor, cognitive enhancement, **8**:45–46
DMP-323 compound, nuclear magnetic resonance studies, **2**:416–419
DMXB-A, cognitive enhancement, **8**:35–36
DNA binding domain (DBD):
  glucocorticoid receptor structure, **5**:62
  nuclear hormone receptors, **4**:78–80
  nuclear magnetic resonance studies, macromolecule-ligand interactions, **2**:405–410
    stoichiometry and kinetics, **2**:406–407

poly(ADP-ribose)polymerase-1 (PARP-1), **6**:159–160
retinoid nuclear receptors, **5**:369–373
steroid hormone receptors, **5**:30–31
thyroid hormone receptors, **4**:189–190
thyroid hormone receptors, β-selective agonists, structure-based design, **2**:684–685
toxicophores, **2**:325
DNA damage, AHPN apoptosis, **5**:492–493
DNA methylation, pyrimidine antimetabolite analogs, **6**:111–112
DNA polymerase IIIC-fluoroquinolone hybrid antibacterial, **7**:551–552
DNA-targeted therapeutics:
  anti-DNA virus agents, **7**:221–247
    acyclic nucleoside phosphates, **7**:231–239
      adenine derivatives, **7**:235–236
      8-azapurine derivatives, **7**:237
      2,6-diaminopurine derivatives, **7**:236
      guanine derivatives, **7**:236
      new generation prodrugs, **7**:238–239
      purine N6-substituted amino derivatives, **7**:237–238
      tissue-targeted prodrugs, **7**:239
    antimetabolites, **7**:221–230
    inorganic diphosphate analogs, **7**:240–241
    nucleoside prodrugs, **7**:230–231
    nucleotide prodrugs, **7**:231
    posttranscriptional methylation inhibitors, **7**:241
    research background, **7**:221
    viral DNA-polymerase nonnucleotide inhibitors, **7**:239–240
    virus assembly inhibitors, **7**:241

antimalarial agents, **7**:644–655
  antifolates, **7**:644–648
  dihydroorotate dehydrogenase, **7**:648–650
  histone deacetylase, **7**:652–655
  topoisomerase, **7**:650–652
antitubercular agents, **7**:724–725
chiral compounds, **1**:154–155
cytotoxic agents, **6**:3–17
  bleomycin, **6**:8–12
  dactinomycin, **6**:3–8
  mitomycin (mutamycin), **6**:12–16
  plicamycin, **6**:16–17
Epstein-Barr virus, **7**:244
hepadnaviruses, **7**:246–247
herpes simplex virus, **7**:242–243
human cytomegalovirus, **7**:243–244
human herpesvirus type 6, **7**:244
human papillomavirus, **7**:241–242
polyomaviruses, **7**:241
poxviruses, **7**:244–245
synthetic chemotherapeutic compounds:
  alkylating agents, **6**:83–95
    cyclopropylindoles, **6**:91–93
    mustards, **6**:83–89
    nitrosoureas, **6**:93–94
    platinum complexes, **6**:89–91
    triazenes, **6**:94–95
  antimetabolites, **6**:106–113
    antifolates, **6**:106–109
    purine analogs, **6**:112–113
    pyrimidine analogs, **6**:109–112
  research background, **6**:83
  topoisomerase inhibitors, **6**:95–106
    dual topo I/II inhibitors, **6**:103–106
    Topo II inhibitors, **6**:98–102
    Topo I inhibitors, **6**:96–98
  tumor-activated prodrugs, **6**:113–124
    ADEPT prodrugs, **6**:120–123

DNA-targeted therapeutics:
(*Continued*)
  hypoxia-activated bioreductive prodrugs, **6:**114–120
  varicella-zoster virus, **7:**243
Docetaxel:
  ispenesib combined therapy, **6:**200–204
  structure and therapeutic applications, **6:**34–35
DOCK algorithm, docking and scoring computations, **2:**606–608
  force field scoring, **2:**620–622
Docked conformer-based alignment (DCBA), quantitative structure-activity relationship, multidimensional models, **1:**41–43
Docking techniques:
  antimalarial compounds:
    antifolates, **7:**644–648
    apicoplast targeting, **7:**665
    plasmepsin antimalarial agents, **7:**642–644
  drug discovery applications, **2:**599–632
    algorithm development, **2:**599–600
    basic principles and concepts, **2:**593–596
    benefits, **2:**643
    current research issues, **2:**632–641
    entropy effects, **2:**633
    flexibility, **2:**637–640
    hydrophobic effects, **2:**634–636
    isomerization and tautomerization, **2:**636–637
    limitations, **2:**642–643
    model scoring functions, **2:**618–632
    multiple solutions, **2:**641
    structural searching functions, **2:**601–618
    target structure requirements, **2:**596–598
    user guides, **2:**641–642
  G-protein coupled receptor, homology modeling, ligand docking, **2:**289–292
  macromolecule-ligand interactions, nuclear Overhauser effect spectroscopy, **2:**404–405
  multiple solutions, **2:**641
  protein-protein interactions:
    computational docking, **2:**343–344
    scaffold structures, **2:**350–359
  quantum mechanics/molecular mechanics, binding affinity evaluation, **2:**723, 733–735
  in silico screening, protein flexibility, protein-ligand interactions, **2:**865–870
  structure-based drug design, computer-aided molecular docking, **2:**679–680
Docosahexaenoic acid (DHA):
  age-related macular degeneration, **5:**609–610
  chemical structure, **5:**605–606
Doctrine of equivalents, infringement of patents, **3:**153–155
Dolasetron, chirality, **1:**443–444
Dolastatins, marine sources, **2:**243–245
Domain-selective inhibitors, angiotensin-converting enzyme, **4:**286–289
Domain threshold calculations, quantitative structure-activity relationship modeling, ligand-based virtual screening, **2:**14
Donepezil, cognitive enhancement, **8:**15–16, 31
Donor-acceptor motifs:
  kinase inhibitors, **6:**295–296
  stroke therapy, NOS inhibitors, **8:**502
Dopamine (DA):
  antipsychotic effects:
    animal models, **8:**170–171
    receptor subtype techniques, **8:**200–201
    structure-activity relationship, **8:**180–185
  cocaine structure-activity relationship, **8:**111–113
  cognitive enhancement, **8:**39–41
  conformations of, **8:**543–544
  D_1 receptor agonists, Parkinson's disease dopaminergic therapy, **8:**547–549
  D_2-plus, multitarget drug development, schizophrenia, **1:**257–258
  D_2 receptor antagonists:
    antipsychotic medications, **8:**170
    central nervous system effects, **8:**3–5
    radiolabeled ligands for, **8:**744
  D_3 receptor ligands, hit seeking combinatorial libraries, **1:**315, 318
  Parkinson's disease:
    dopaminergic pharmacotherapy, **8:**540–552
    catechol-*O*-methyltransferase inhibitors, **8:**551–552
    L-dopa therapy, **8:**540–542
    monoamine oxidase inhibitors, **8:**549–551
    receptor-targeted compounds, **8:**542–549
    aporphine-type receptor agonists, **8:**544–545
    dopamine receptor structure and function, **8:**542–544
    ergot-type receptor agonists, **8:**545–546
    experimental D_1 receptor agonists, **8:**547–549
    small-molecule agonists, **8:**547
  mitochondrial metabolism and, **8:**535–537
  pathophysiology, **8:**531–534

prodrug development,
    3:223–226
  metabolic issues,
    3:231–233
  psychostimulant
    pharmacology,
    8:100–106
  recent developments,
    8:114–115
  structure-activity
    relationship,
    8:106–109
  receptor structure and function,
    8:542–544
  schizophrenia pathology,
    8:162–163
  latent inhibition,
    8:172–173
Dopamine- and cAMP-regulated
    phosphoprotein of MW
    32,000 (DARPP-32),
    serotonin receptor
    targeting, 8:245–247
Dopamine transporter (DAT):
  antidepressant pharmacology,
    8:238–241
  ligand-based virtual screening,
    2:7–8
  Parkinson's disease diagnosis,
    8:555–556
DOPAScan, Parkinson's disease
    dopaminergic therapy,
    8:547–549
Doppler imaging, congestive
    heart failure, disease
    biomarkers,
    4:505–506
D-optimal onion design (DOODs),
    quantitative
    structure-activity
    relationship
    validation, 1:50
Doramapimod kinase inhibitor:
  Alzheimer's disease, tau-
    targeted therapies,
    8:420–422
  structure-based design, protein
    targeting, 6:351–354
Doripenem:
  development of, 7:356
  occurrence, structural
    variations and
    chemistry, 7:333–335
Dorzolamide, glaucoma therapy,
    5:598–600
Dosage forms:
  oral drug delivery:

biopharmaceutics
    classification system,
    3:41–43
  class I drugs, 3:42
  class II drugs, 3:42–50
  class III drugs, 3:50–52
  class IV drugs, 3:52, 355
  class V drugs, 3:52–55
  vitamin A, 5:647–648
  vitamin D, 5:658
  vitamin E, 5:664
Dosage forms, WHO definition,
    2:746
Dose number, gastrointestinal
    absorption,
    permeability vs.
    solubility, 3:374–375
Dose-response data:
  biological fingerprints,
    1:483–484
  quantitative structure-activity
    relationship,
    biological parameters,
    1:31–32
  vitamins, 5:640, 643
Dose-response data, receptor-
    based drug design,
    occupancy theory,
    2:494–497
Dose selection, congestive heart
    failure therapy,
    pharmacodynamic-
    kinetic parameters,
    4:502–503
Dothiepin:
  chemical structure, 8:222
  side effects and adverse
    reactions, 8:231–232
"Double prodrug," anti-DNA
    virus agents,
    antimetabolites,
    7:223–225
Double-stranded breaks:
  Et743 compounds, 2:247
  poly(ADP-ribose)polymerase-1
    (PARP-1),
    6:161–162
Downstream processing:
  defined, 3:289
  protein therapeutics,
    3:301–311
  classifications, 3:303–304
  design criteria, 3:303
  E. coli protein purification,
    3:304–306
  future trends,
    3:309–311

impurities and
    contaminants,
    3:301–303
  mammalian protein
    purification,
    3:306–309
Doxacurium, nicotinic
    antagonists, 8:79–80
Doxepine:
  chemical structure,
    8:222
  side effects and adverse
    reactions, 8:231–232
  structure-activity relationship,
    8:250–251
Doxorubicin:
  Alzheimer's disease, tau-
    targeted therapy,
    8:426–427
  antibody-directed enzyme
    prodrug therapy,
    peptidase enzymes,
    6:120–121
  anti-cancer activity, 2:236–239
  cardiomyopathy, carbon
    monoxide inhalation
    therapy, 5:303
  nanoscale drug delivery,
    liposomes,
    3:472–473
  serum albumin binding,
    3:452–454
  structure and therapeutic
    effects, 6:20–22
  topoisomerase II inhibitors,
    6:98–102
Doxycycline:
  ADMET properties, 7:409–410
  antimalarial drugs,
    7:621–624
  quinine combined with,
    7:607
  chemical structure, 7:407
  history and biosynthesis,
    7:410–413
  research background,
    7:406–407
  stability, side effects, and
    contraindications,
    7:408–409
DPP-IV activity and/or structure
    homologues (DASH)
    proteins, dipeptidyl
    peptidase 4 inhibitors,
    4:42
Dramamine, selective toxicity,
    2:565–566

Drug delivery systems. *See also* Chemical delivery systems (CDS); Targeted drug development
  age-related macular degeneration therapy, **5**:613–623
  bioavailability, mathematical modeling:
    definitions, **2**:746–751
    diffusion-controlled dosage, plasma drug profile, **2**:801–816
    erosion-controlled dosage, plasma drug profile, **2**:816–828
    future research issues, **2**:851–852
    immediate-release oral dosage, **2**:771–787
    intravenous drug administration, **2**:751–771
    research background, **2**:745–746
    sustained release *in vitro* testing, **2**:791–801
    sustained release oral dosage, **2**:787–791
    tissue-based drug transfer, **2**:838–851
    transdermal systems, **2**:828–838
  crystal engineering:
    cocrystals, **3**:189–190, 192–196, 198–213
    future research issues, **3**:213–214
    hydrates and solvates, **3**:189
    polymorphs, **3**:188, 190–192, 196–198
    research background, **3**:187–190
    salts, **3**:188–189
  glucocorticoid anti-inflammatory compounds:
    corneal penetration, **5**:53
    enzymatic metabolism, **5**:53–54
    intestinal absorption, **5**:41, 43, 45
    intra-articular administration, **5**:51–53
    percutaneous absorption, **5**:46–51
    water-soluble esters, **5**:53
  membrane proteins:
    apical sodium-dependent bile acid transporter, **2**:467–475
    ATP-binding cassette and solute carrier genetic superfamilies, **2**:456–461
    classification and structures, **2**:454–455
    Cys-scanning mutagenesis, **2**:460
    excimer fluorescence, **2**:461
    future research issues, **2**:476
    insertion scanning, **2**:460
    integral structure analysis, **2**:459
    intestinal peptide transporter example, **2**:463–467
    N-glycosylation and epitope scanning mutagenesis, **2**:460–461
    passive diffusion, **2**:444–446
      kinetics, **2**:445–446
    permeability enhancement, **2**:439–444
      absorption enhancement, transporter targeting, **2**:443–444
      lipid-based oral delivery, **2**:441
      penetration enhancers, **2**:439–441
      prodrugs, **2**:441–443
    P-glycoprotein example, **2**:461–463
    receptor-mediated endocytosis, **2**:446–454
      bacterial adhesins and invasins, **2**:451–452
      bacterial and plant toxins, **2**:452
      cell surface receptor structure, **2**:447–448
      endocytosis, **2**:446–447
      immunoglobulin transport, **2**:450–451
      lectins, **2**:452–453
      lipid rafts, **2**:449–450
      oral absorption systems, **2**:450–453
      potocytosis, **2**:448–449
      transcytosis, **2**:448
      viral hemagglutinins, **2**:452
      vitamins and metal ions, **2**:453–454
    research background, **2**:439
    site-directed chemical cleavage, **2**:461
    structural models and substrate design, **2**:459
    transporter structures, therapeutic applications, **2**:457–459
  nanotechnology:
    amphiphilic polymers, **3**:473–480
    basic principles, **3**:469
    carbon nanotubes, **3**:483–484
    dendrimers, **3**:482–483
    polymer-drug conjugates, **3**:480–481
    self-assembled low molecular weight amphiphiles, **3**:469–473
    water-insoluble and crosslinked polymeric nanoparticles, **3**:481
  protein therapeutics, **3**:311–317
    alternative modes, **3**:315–316
    container materials interactions, **3**:315
    degradation pathways, **3**:311–314
    future trends, **3**:316–317
    stability parameters, **3**:314–315
  structural genomics research, **1**:581–584
Drug development and discovery. *See also* Targeted drug development
  ADMET applications, **2**:47–48
  AHPC compound, **5**:483–497
  AHPN compound, **5**:483–497
  allosteric targets, **1**:390–392
    future trends, **1**:394–395
  anti-HCV agents, antiviral resistance and, **7**:175–176
  antimalarials, **7**:627–672
    apicoplast targets, **7**:663–668
    cell cycle regulation, **7**:627–634

cellular and oxidative stress
  management,
  7:658–663
cellular respiration and
  electron transport
  chain, **7:**655–658
DNA synthesis, repair, and
  regulation, **7:**644–655
heme metabolism and amino
  acid salvage,
  7:634–644
targeted drug systems,
  7:668–672
antiparasitic/antiprotozoal
  agents, **7:**577–593
  glucose metabolism/
    glycosomal transport
    inhibitors, **7:**582–585
  leishmaniasis therapies,
    7:590–593
  lipid metabolism inhibitors,
    7:580–581
  nucleotide transporters,
    7:577–580
  polyamine pathway
    targeting, **7:**585–590
  protease inhibitors,
    7:581–582
antiplatelet agents, **4:**418–419
antipsychotics, **8:**200–210
  AMPA receptors, **8:**205–206
  cannabinoids, **8:**209–210
  dopamine receptor subtypes,
    8:200–201
  glutamatergic agents, **8:**202
  histamine H$_3$ receptor,
    8:206–207
  metabotropic glutamate
    receptors, **8:**203–204
  neurokinin 3 receptor
    antagonists, **8:**208
  neuronal nicotinic
    acetylcholine receptor,
    8:207
  NMDA receptors, **8:**204
  phosphodiesterase
    inhibitors, **8:**208–209
  serotonin receptor subtypes,
    8:201–202
antitubercular agents,
  7:735–737
  linezolid, **7:**785–786
  nitroimidazoles, **7:**776–782
  preclinical and clinical
    research, **7:**776–786
  SQ109, **7:**782–783
  TMC207, **7:**783–785

bexarotene, **5:**429–435
central nervous system:
  allosteric receptor sites,
    8:5–6
  cotransmission, **8:**5
  functional selectivity and
    receptor targeting,
    8:7–9
  future research issues,
    8:9–11
  receptor-selective drugs,
    8:1–3
  research background,
    8:1
  rich pharmacology concept,
    8:4–5
chemogenomics, **1:**589–595
chirality in, **1:**149–155
cost-effectiveness analysis:
  basic principles, **3:**345
  decision-analytic
    approaches,
    3:346–349
  pharmacogenomics and
    personalization, **3:**349
  postlaunch analysis,
    3:349–350
  regulatory approval,
    3:345–346
diabetes therapy, type 2
  diabetes:
  current and emerging
    therapies,
    4:1–27
  dipeptidyl peptidase 4
    inhibitors, **4:**39–65
docking techniques:
  basic principles and concepts,
    2:593–596
  benefits, **2:**643
  current research issues,
    2:632–641
  entropy effects, **2:**633
  flexibility, **2:**637–640
  hydrophobic effects,
    2:634–636
  isomerization and
    tautomerization,
    2:636–637
  limitations, **2:**642–643
  multiple solutions, **2:**641
  scoring techniques,
    2:599–632
    model scoring functions,
      2:618–632
    structural searching
      functions, **2:**601–618

target structure
  requirements,
  2:596–598
user guides, **2:**641–642
erectile dysfunction,
  phosphodiesterase
  inhibitors, **5:**712–715
intellectual property law:
  future research issues, **3:**186
  miscellaneous protections,
    3:185–186
  patents:
    enforcement, **3:**148–165
      alternative dispute
        resolution, **3:**162–163
      claim construction
        proceedings, **3:**162
      defenses to
        infringement,
        3:154–157
      discovery phase,
        3:160–161
      geographic scope, **3:**148
      infringement
        determination,
        3:152–154
      parties to suit,
        3:150–152
      proceedings
        commencement,
        3:159–160
      remedies for
        infringement,
        3:157–159
      summary judgment,
        3:161
      trial and appeal
        processes, **3:**148–150,
        162–164
    global patent protection,
      3:107–108, 165–171
      international
        agreements,
        3:165–168
      international laws and
        regulations,
        3:169–171
      PCT patent practice,
        3:168–169
    protection and strategy,
      3:105–118
      absolute novelty
        principle, **3:**110–111
      application publication,
        3:114–115
      first to invent *vs.* first to
        file, **3:**108–110

Drug development and discovery. (*Continued*)
  global strategy, **3:**107–108, 165–171
  patent term, **3:**111–114
  PTO prosecution strategy, **3:**115–118
  trade secrets and, **3:**182–183
  requirements, **3:**118–148
    corrections, **3:**146–148
    interference, **3:**142–146
    new and nonbovious inventions, **3:**133–138
    patentable subject matter, U.S. and international requirements, **3:**118–123
    provisional applications, **3:**127–133
    PTO procedures, **3:**138–142
    specifications, **3:**123–127
  research background, **3:**101–105
  trademarks, **3:**171–179
    global rights, **3:**177–178
    Lanham Act protections, **3:**178–179
    as marketing tools, **3:**172
    oppositions and cancellations, **3:**176–177
    preservation of rights through proper use, **3:**177
    registration process, **3:**174–176
    selection criteria, **3:**172–174
  trade secrets, **3:**179–184
    defined, **3:**180
    enforcement, **3:**181–182
    Freedom of Information Acts, **3:**183–184
    global protection, **3:**184
    patent protection and, **3:**182–183
    protection requirements, **3:**180–181
mass spectrometry:
  affinity capillary electrophoresis, **1:**113–114
  affinity chromatography-mass spectrometry screening, **1:**111–112
  assays, **1:**119–121
  combinatorial compound structure and purity, **1:**108–110
  combinatorial libraries, LC-MS applications, **1:**107–108
  compound encoding and identification, **1:**110–111
  current trends and recent developments, **1:**106–121
  desorption ionization techniques, **1:**99–100
  electron impact, **1:**97–98
  frontal affinity chromatography, **1:**114–115
  future trends, **1:**121–122
  gas chromatography-mass spectrometry, **1:**99
  gel permeation chromatography-mass spectrometry, **1:**113
  liquid chromatography-mass spectrometry, **1:**100–105
  mass analyzers, **1:**98
  pulsed ultrafiltration-mass spectrometry, **1:**116–119
  research background, **1:**97
  screening applications, **1:**111–121
  solid-phase screening, **1:**115–116
  tandem mass spectrometry, **1:**105–106
  web site addresses and information sources, **1:**122–123
multiobjective optimization:
  ADMET CYP 2D6, **2:**268–269
  blood-brain-barrier and solubility, **2:**268–269
  clinical applications, **2:**263–266
  heptatoxicity, **2:**269
  overview of techniques, **2:**260–263
  Pareto ligand designer, **2:**266–273
  research background, **2:**259–260
non-life-threatening disorders, **5:**711–726
nuclear hormone receptors, future research issues, **4:**164–167
nuclear magnetic resonance and:
  current applications, **2:**375–376
  ligand-based design, **2:**375–390
    analog limitations, **2:**390
    charge state, **2:**384–385
    conformational analysis, **2:**383–384
    line-shape and relaxation data, **2:**387–389
    pharmacophore modeling, **2:**389–390
    structure elucidation, **2:**376–383
      bioactive peptides, **2:**377–383
      natural products, **2:**376–377
    tautomeric equilibria, **2:**385–387
  receptor-based design, **2:**390–419
    chemical-shift mapping, **2:**402–403
    dihydrofolate reductase inhibitors, **2:**414–419
    DNA-binding drugs, **2:**405–410
    isotope editing/filtering, **2:**403–404
    macromolecular structure, **2:**391–394
    matrix metalloproteinases, **2:**412–414
    NOE docking, **2:**404–405
    protein-ligand interactions, **2:**394–419
    titrations, **2:**403
  research background, **2:**367–369
pharmacophores, **1:**458–459
protein-protein interactions:
  computational docking, **2:**343–344
  fragment-based approach, **2:**340–342

future research issues,
  2:354–359
high-throughput screening,
  2:337–340
hotspot technique, 2:340
p53/MDM2/MDM4 functions,
  2:344–359
research background,
  2:335–337
structure-based approach,
  2:342–343
receptor-based design:
  cellular information
    processing,
    2:492–493
  chemical space, 2:519
  classes, 2:505–514
  complexes and allosteric
    modulators,
    2:499–501
  compound properties,
    2:519–525
    ADMET parameters,
      2:524–525
    databases, 2:525
    receptor-ligand
      interaction, 2:520–523
    source materials,
      2:523–524
    structure-activity
      relationships,
      2:519–520
  drug receptors, 2:514
  dynamics, 2:503
  efficacy parameters,
    2:501–503
  functional genomics,
    2:515–519
  future research issues,
    2:533–534
  G-protein-coupled receptors,
    2:504–509
    genetic variation,
      2:514–515
    signaling modulators,
      2:508–509
  intracellular receptors, 2:512
  lead compound discovery,
    2:525–533
    biologicals and antisense
      compounds, 2:531–533
    diversity-based ligand
      libraries, 2:531
    high-throughput
      screening, 2:526–527
    natural product sources,
      2:527–528

pharmacophore-based
  ligand libraries,
  2:528–529
privileged
  pharmacophores,
  2:529–530
ligand-gated ion channels,
  2:509–511
molecular biology, 2:514
neurotransmitters:
  binding proteins, 2:514
  transporters, 2:513–514
nomenclature, 2:504–505
non-GPCR-linked cytokine
  receptors, 2:512–513
nuclear magnetic resonance,
  2:390–419
  chemical-shift mapping,
    2:402–403
  dihydrofolate reductase
    inhibitors, 2:414–419
  DNA-binding drugs,
    2:405–410
  isotope editing/filtering,
    2:403–404
  macromolecular structure,
    2:391–394
  matrix metalloproteinases,
    2:412–414
  NOE docking, 2:404–405
  protein-ligand
    interactions,
    2:394–419
  titrations, 2:403
orphan receptors, 2:513
receptor residence time,
  2:503–504
research background,
  2:491–492
RNA targets, 2:512
steroid receptor superfamily,
  2:512
theoretical background,
  2:493–498
  constitutive receptor
    activity, 2:498–499
  occupancy theory,
    2:494–497
  rate theory, 2:497
  ternary complex model,
    2:497–498
9-*cis*-Retinoic acid, 5:421–422
stroke therapy:
  current research, 8:447–449
  future research and
    development,
    8:515–516

structural genomics research,
  1:581–595
antituberculosis drugs, 1:587
aromatic amine and
  hydrazine drugs
  acetylation, 1:585–586
chemogenomics, 1:589–595
  epigenetics, 1:594–595
  kinase inhibitor protein
    structures, 1:590–591
  kinase program, 1:591–594
data dissemination, methods,
  and reagents, 1:581
delivered structures impact,
  1:581–584
fragment cocrystallography
  techniques, sleeping
  sickness target,
  1:587–588
future trends, 1:595
nitrogen-containing
  bisphosphonates,
  1:584–585
poly(ADP) ribosylation
  inhibition, 1:586–587
Drug discrimination model,
  antipsychotic drugs,
  8:173
Drug-drug interactions:
acitretin, 5:401
ADMET toxicity properties,
  2:63, 68–69
aliskiren renin inhibitors, 4:253
all *trans*-retinoic acid,
  5:385–386
antidepressants:
  efficacy assessment,
    8:225–229
  monoamine oxidase
    inhibitors, 8:230
  serotonin-selective reuptake
    inhibitors, 8:232–234
  tricyclic antidepressants,
    8:231–232
antipsychotics, 8:185–200
  benzisoxazoles, 8:195–196
  benzoisothiazoles, 8:196–198
  butyrophenones, 8:189–190
  diarylbutylamines, 8:190
  dibenzodiazepines,
    8:191–192
  dibenzothiepine, 8:194
  dibenzoxazepines, 8:189
  dibenzthiazepines,
    8:193–194
  dihydroindolone, 8:190–191
  phenothiazines, 8:186–188

Drug-drug interactions:
(*Continued*)
phenylindole, **8:**199
quinolinone, **8:**198
substituted benzamide,
8:199–200
thiobenzodiazepine,
8:192–193
thioxanthenes, **8:**188–189
antitubercular agents:
isoniazid, **7:**743–744
rifamycin antibacterials,
7:749
bupropion, **8:**255
β-lactam antibacterial drugs,
7:282–284
macrolides, **7:**438
nefazodone, **8:**234–235
opioids, **8:**576–577
psychostimulant therapy, **8:**92,
95–97
9-*cis*-retinoic acid therapy,
5:428
13-*cis*-retinoic acids, **5:**395
trazodone, **8:**234
vitamin B$_6$ and, **5:**681–684
vitamin D, **5:**660
Drug-eluting coronary stents,
sirolimus,
zotarolimus, and
everolimus, **5:**1037
Druggable molecule libraries:
chemogenomics, **2:**574–575
receptor-based drug design:
chemical and biological
space, **2:**519
functional genomics,
2:516–519
small-molecule compounds:
benzodiazepines,
1:300–302
solid- and solution-phase
synthesis, **1:**294–343
Drug Guru system, multiobjective
drug discovery
optimization,
2:266–273
Drug-macromolecule
interactions,
structure-based
thermodynamic
analysis, **2:**702–712
binding activity, **2:**703–705
binding measurement,
2:706–707
calorimetry results,
2:711–712

enthalpy-entropy
compensation,
2:705–706
fragment-based design,
2:709–711
isothermal titration
calorimetry,
2:707–709
solvation thermodynamics,
ligand-receptor
interactions, **2:**711
Drug receptors, receptor-based
drug design,
classification and
examples, **2:**514
Drug resistance:
aminoglycoside antibiotics,
7:416–419, 424–427
antibiotic selective toxicity and,
2:553
anti-HCV agents, **7:**175–176
antimalarial drugs:
chloroquine, **7:**610
experimental agents,
7:625–627
mefloquine, **7:**611
quinine, **7:**607
antitubercular agents:
capreomycin, **7:**764
D-cycloserine, **7:**768–770
diarylquinoline (TMC207),
7:784–785
ethambutol, **7:**754–755
ethionamide, **7:**771–772
fluoroquinolones, **7:**775
isoniazid, **7:**740–742
kanamycin and amikacin,
7:762
linezolid, **7:**786
mechanisms of, **7:**733–735
nitroimidazoles, **7:**781
para-Aminosalicylic acid,
7:765–767
pyrazinamide, **7:**758–759
rifamycin antibacterials,
7:748
SQ109, **7:**783
streptomycin, **7:**760–761
thiacetazone, **7:**773–774
aspirin, **4:**421
ATP binding cassette (ABC)
transporters,
resistance targeting
and mediation,
6:371–376
breast cancer resistance
protein, **6:**372–373

membrane structure,
6:374–376
multidrug resistance protein
1, **6:**372
P-glycoprotein, **6:**371–372
structure-activity and
quantitative
structure-activity
relationship studies,
6:374
cancer therapy:
endocrine resistance, breast
cancer, **6:**369–370
mechanisms, **6:**361–363
multiple drug resistance,
6:370–376
ABC transporter
mediation, **6:**371–376
apoptosis protein
targeting, **6:**370–371
stem cells, **6:**376–377
new strategies for, **6:**377–378
small-molecule tyrosine
kinase inhibitor
resistance, **6:**364–369
targeted drug development,
6:363–370
tyrosine kinase inhibitor
resistance, **6:**364
chemotherapeutic agents,
7:502, 504
combination therapies,
7:504–505
future research issues,
7:552–553
chloramphenicol, **7:**463
clopidogrel, **4:**432–433
corticosteroids, chronic
obstructive pulmonary
disease inhalation
therapy, **5:**779–780
dihidrofolate reductase
inhibitors, **7:**534
dihydropteroate synthase,
7:530–531
fusidic acid, **7:**464
glycopeptide antibiotics,
2:226–227
HIV protease inhibitors:
amprenavir, **7:**29
atazanavir sulfate, **7:**42
darunavir, **7:**59–60
design criteria for, **7:**3
fosamprenavir calcium, **7:**45
indinavir sulfate, **7:**18, 20
nelfinavir mesylate, **7:**22
ritonavir, **7:**12

saquinavir mesylate (SQV), **7:**8, 10
tipranavir, **7:**51–52
β-lactam antibacterial drugs, **7:**269–281
acquisition and spread of resistance, **7:**269–271
conjugative transposons, **7:**270–271
E. faecium PBP-5, **7:**273
enterobacteria porin loss, **7:**278
methicillin-resistant staphylococci mechanism and genetics, **7:**272–273
mosaic gene formation, resistant PBPs, **7:**271–272
N. gonorrhoeae PBPs, **7:**272
outer membrane permeability and active efflux decrease, **7:**279–281
P. aeruginosa resistance, **7:**280–281
penicillin binding protein modification, **7:**271–274
plasmids, **7:**272
S. pneumoniae PBP-2x, **7:**273–274
linezolid RNA targets, **5:**990–993
linezolids, **7:**543–544
lopinavir/ritonavir, **7:**34, 36
macrolide antibiotics, **7:**444–446
multidrug-resistant tuberculosis, epidemiology, **7:**713–716
nonnucleoside reverse transcriptase inhibitors, **7:**152–155
quinolones/fluoroquinolones, **7:**539
second-line kinase inhibitors, conformational targeting, **6:**354–356
streptogramins, **7:**459–460
sulfonamides, **7:**509–514
tetracycline antibiotics, **2:**223–225, **7:**406–407, 413–414
topical antimicrobial agents, **7:**488

Drug response, single nucleotide polymorphisms, **1:**181–182
Drugs:
FDA definition of, **3:**53
WHO definition of, **2:**746
DrugScore algorithm, empirical scoring, **2:**630–631
Drug transporter genes, single nucleotide polymorphisms, **1:**189–190
Drusen deposits, age-related macular degeneration, **5:**605
Drying operations, pilot plant development, **3:**9–10
DSSTox database, quantitative structure-activity relationship modeling, **1:**513–514
"Dual quats," cationic surfactants, **7:**500
Dual-substituent parameter (DSP) equation, quantitative structure-activity relationship, electronic parameters, **1:**8
Duloxetine:
chemical structure, **8:**223
migraine prevention, **8:**318–319
monoamine transporter inhibition, **8:**238–241
side effects and adverse reactions, **8:**234
structural analysis, **1:**489–491
DuP747 κ-opioid receptor agonist, **8:**627–632
Dutasteride, androgen antagonists, **5:**195
Dutogliptin, type 2 diabetes therapy, ADMET properties, **4:**45–47
DXOP reductoisomerase (DXR), apricoplast targeting, antimalarial agents, **7:**665–668
DXOP synthase (DXS), apricoplast targeting, antimalarial agents, **7:**665–668
Dye-drug interactions:

β-amyloid peptide inhibitors, **8:**364–365
parasitic glucose metabolism and glycosomal transport inhibition, **7:**585–586
sulfonamide development, **7:**507–508
synthetic antibacterials, **7:**483–484
topical antimicrobial agents, **7:**498
Dynamic light scattering (DLS), protein therapeutics, **3:**324
Dynamic moisture sorption, oral drug physicochemistry, **3:**38–39
Dynorphins:
analog structures, **8:**654–659
antagonists, dynorphin A analogs, **8:**657–658
conformational constraints, **8:**656–657
linear analogs, **8:**655–656
κ-opioid receptor interaction, **8:**594–596
"message-address" concept, **8:**604–608
opioid receptors, precursors and processing products, **8:**595–596
Dysbindin, schizophrenia assessment, **8:**173
Dyspigmentation, all-*trans*-retinoic acid therapy, **5:**383–384

E3 ubiquitin ligases, proteasome inhibitors, **6:**177
E555 thrombin receptor antagonist, **4:**443–444
side effects, **4:**452–453
Eadie-Hofstee plot, allosteric proteins, **1:**379–380
eBay doctrine, infringement of patent claims, **3:**158–159
Ebselen, stroke therapy, **8:**481
Echinocandins, selective toxicity, **2:**559, 560, 562
ECL2 conformation, G-protein coupled receptor, homology modeling, **2:**285–287

ECO-02301 compound, natural products lead generation, bacterial genomes, **2:**210–211

Econazole, selective toxicity, **2:**558

Ecteinascidin, natural products lead generation, terrestrial and marine habitats, **2:**197–199

Eczema, 9-*cis*-retinoic acid therapy, **5:**425–426

Ed4T HIV reverse transcriptase inhibitor, **7:**142–145

Edaravone, stroke therapy, **8:**481

Edatrexate, antifolate DNA-targeted chemotherapeutics, **6:**106–109

Edema modeling, prostaglandin/thromboxane EP$_3$ receptor antagonists, **5:**858, 861

Edman sequencing, bioactive peptides:
 libraries, **1:**288–291
 nuclear magnetic resonance, **2:**377–383

EDP-420, macrolidic antibiotics, **2:**228–229

Efaproxiral, hemoglobin oxygen delivery, allosteric effectors, **4:**619

Efavirenz HIV reverse transcriptase inhibitor, **7:**148–152

EFdA HIV reverse transcriptase inhibitor, **7:**142–145

Effective hydrogen charge (EHC), quantitative structure-activity relationship, **1:**27–28

Effective Prediction Domain, quantitative structure-activity relationship models, **1:**521–522

Effector molecules, allosteric targets, **1:**390–392

Efficacy assessment:
 antidepressants, **8:**221–229
 antipsychotics, animal models, **8:**170–172
 estrogen receptors, **5:**236–237
 HIV protease inhibitors:
  amprenavir, **7:**28–29
  atazanavir sulfate, **7:**40–42
  darunavir, **7:**57, 59
  fosamprenavir calcium, **7:**44–45
  indinavir sulfate, **7:**17–18
  lopinavir/ritonavir, **7:**32, 34, 36
  nelfinavir mesylate, **7:**22
  ritonavir, **7:**12
  saquinavir mesylate (SQV), **7:**8
  tipranavir, **7:**48–49, 51
 nuclear hormone receptors, ligand-receptor affinity, **4:**84–86
 opioid receptors:
  in vitro assays, **8:**583–585
  in vivo assays, **8:**585–587
 phosphodiesterase 5 inhibitors, **5:**715–716
 receptor-based drug design:
  measurement techniques, **2:**501–503
  occupancy theory, **2:**496–497
  recombinant DNA technology, **1:**540–543

Efflux mechanism:
 ADMET permeability, **2:**66
 transporters, **2:**55–56
 aminoglycoside resistance, **7:**424–427
 antitubercular resistance, **7:**734–735
 macrolide resistance, **7:**445–446
 streptogramin resistance, **7:**459–460
 tetracycline microbial resistance, **7:**413–415

Eflornithine, human African trypanosomiasis therapy, **7:**571

EGb-761 β-amyloid peptide inhibitor, **8:**363–365

eHITS docking program, empirical scoring, **2:**627–628

Eicosanoids, peroxisome proliferator-activated receptors, α-agonists, **4:**88–99

Einstein-Sutherland equation, quantitative structure-activity relationship, **1:**20–21

Elastic network models (ENMs), *in silico* screening, protein flexibility, **2:**864–865

Electroencephalography (EEG), schizophrenia assessment, **8:**174–175

Electronic parameters:
 drug metabolism, **1:**445
 glucocorticoid anti-inflammatory compounds, **5:**77
 electronic eigenvalue molecular descriptor, **5:**80
 quantitative structure-activity relationship, **1:**2–8

Electron impact (EI), mass spectrometry development, **1:**97–98

Electron-rich heterocycles, toxicophore reactive metabolites, **2:**314–317

Electron transport, antimalarial drugs, **7:**655–658

Electrophiles:
 estrogen receptor ligands, **5:**226–228
 opioid analgesics, nonpeptide affinity labeling, **8:**638–643
 stroke therapy, **8:**465

Electrophoretic techniques, protein therapeutics, analytic development, **3:**320

Electroporation, protein therapeutics, transfection, **3:**295

Electroretinography (ERG), age-related macular degeneration, *in vivo* screening models, **5:**610–613

Electrospray technology:
 dihydrofolic acid reductase inhibitor combinatorial chemistry, **1:**317, 319
 liquid chromatography-mass spectrometry, **1:**101–105

Electrostatic interactions:
 estrogen receptor topography, **5:**234–236
 progesterone receptors, **5:**237–238

quantitative structure-activity relationship, 1:29–31
glucocorticoid anti-inflammatory compounds, 5:78–80
Electrostatic potential, CK-2130 prodrug case study, 3:270–276
Elimination:
drug bioavailability, 2:747–748
intravenous administration, 2:754
oral drug delivery, immediate-release dosage, 2:775–776
repeated short infusions, constant flow rate over finite period, 2:763–765, 768–770
intravenous drug delivery, single dose (bolus injection), 2:754
opioid receptors, 8:578–579
oral dosage forms, class V drugs, 3:53
pharmacokinetic parameters, 2:747
rate constant determination, 2:750
type 2 diabetes therapies, dipeptidyl peptidase 4 inhibitors, 4:46–47
Elinafide, dual topoisomerase I/II inhibitors, bis analogs, 6:105–106
Elinogrel, antiplatelet activity, 4:437
Eliprodil, stroke therapy, 8:453
Ellipticine derivatives, antimalarial topoisomerase inhibitors, 7:650–652
ELND005 β-amyloid peptide inhibitor, 8:364–365
Eltrombopag olamine, structure and properties, 4:584–586
Elvigtegravir, HIV integrase, two-metal chelation model, 7:95–97
Elvucitabine:
anti-DNA virus antimetabolites, 7:226–227
current development, 7:142–145

Embelin, peptidomimetics, 1:237–239
EMD 60400 κ-opioid receptor agonists, 8:630–632
EMD-534085 kinesin spindle protein inhibitor, "induced fit" binding pockets, 6:210–211
EMD compounds, peptidomimetic design, 1:227–229
EMDT receptor agonist, serotonin receptor targeting, 8:246–247
Emericellamide A, plant metabolites, 2:251
Em family methyltransferases, combinatorial library, 1:307–308
Emodin, Alzheimer's disease, tau-targeted therapy, 8:426–427
Emopamil binding protein (EBP), functional analog design, 1:176–177
Empirical data, quantitative structure-activity relationship modeling, 1:509–510
Empirical scoring:
drug discovery, 2:623–628
functions table, 2:646–649
structure-based virtual screening, 2:18
Emtricitabine:
anti-DNA virus antimetabolite, tenofovir double combination drug regimen, 7:226
HIV reverse transcriptase inhibition, 7:141–145
research background, 7:139–140
Emulsions, oral drug physicochemistry, class II drugs, 3:48–49
Enablement requirement, patent specifications, 3:124–126
Enadoline κ-opioid receptor agonist, 8:627–632
Enalapril, soft drug design, 2:118–119
Enalkiren renin inhibitor, 4:246–248
Enantiomers, chiral activity, 1:130–131

Enantioselective multidimensional capillary gas chromatography-mass spectrometry (Enantio-MDGC-MS), chirality and, 1:147
Enantioselective synthesis, chemokine receptor-5 (CCR5) antagonists, 7:111–116
Encoded representations, pharmacophores, three-dimensional structures, 1:471
Endocarditis, tissue-based drug delivery, 2:849–851
Endocrine resistance, breast cancer, 6:369–370
Endogenous substances:
analog drug design, 1:169–170
drug metabolism, conjugation reactions, 1:421–423
enzymes, prodrug design, 3:229
serum albumin drug transport system:
basic principles, 3:437–438
structural survey, 3:438–456
therapeutic applications, 3:456–464
Endomorphins:
evolution of, 8:595–596
opioid peptide analogs, Tyr-Pro-Phe sequence, 8:659–661
Endoplasmic reticulum, stress-associated apoptosis, AHPN activation, 5:492
End organ protection:
inhaled nitric oxide therapy, 5:277–278
renin inhibitors, 4:242–243
aliskiren, 4:252
renin targeting, 4:240
β-Endorphins, opioid receptor affinity, 8:594–596
Endothelial lipase, atherosclerosis, HDL elevation, 4:338–340
Endothelial nitric oxide synthase (eNOS), migraine therapy, 8:298–302

Endothelin-converting enzyme (ECE):
angiotensin-converting enzyme inhibitors, 4:278–279
metalloprotease inhibitors, 1:325–327

Endothelins (ETs):
antagonists (EAs):
biological actions and receptor pharmacology, 4:525–527
multitarget drug development:
ACE-plus and angiotensin-1 receptor (AT$_1$) antagonist, 1:259–260
dual AT$_1$/ET$_A$ antagonist, 1:269–273
peptidomimetic design, 1:224–230
biological actions and receptor pharmacology, 4:524–527
endothelin-1 (ET-1):
glaucoma screening, 5:591
metalloprotease inhibitors, 1:325–327
structure and function, 4:523–526
target validation, congestive heart failure therapy, pharmacodynamics/pharmacokinetics, 4:491–492
lead compounds, ligand-based design:
nuclear magnetic resonance, 2:382–383
tautomerization, 2:385–387
structure and function, 4:523–526

Endothelium-dependent relaxation factor, nitric oxide effects, 5:270–274

Enediynes, anti-cancer activity, 2:235–239

Energetic binding affinity, *in silico* screening, protein flexibility, 2:870

Enkephalins:
δ-opioids:
analogs, 8:650–653
antagonist activity, 8:653

receptor interaction, 8:594–596
μ-opioid analogs, 8:647–649
opioid peptide analogs, 8:646–647
peptidomimetics, 1:207–208

eNOS enhancers, 5:294

Enoxacin, antibacterial agents, 7:535–540

Enoyl-(acyl-carrier-protein) reductase (InhA), antituberculosis drugs, structural genomics, 1:587

Enrichment factors, structure-based virtual screening, 2:20–22

Ensemble distance geometry, pharmacophore formation, bioactive conformation, 1:470

Ensemble modeling:
docking technique, flexibility modeling, 2:640
quantitative structure-activity relationship models, consensus prediction, 1:522–525

Entacapone, Parkinson's disease, dopaminergic therapy, 8:550–552

Entecavir:
anti-DNA virus antimetabolites, 7:225–226
hepatitis B virus therapy, 7:246–247

Enteric-coated mycophenolate sodium, transplant surgery immunosuppression, 5:1050

Enterobacteria, β-lactams, porin loss and, 7:280

Enterococcus faecium, β-lactam antibacterial drug resistance, PBP-5 modification, 7:273

Enterocytes, receptor-mediated endocytosis, 2:450

Enterohepatic recycling, estrogen glucuronide conjugation, 5:228–229

Enthalpy, ligand-macromolecular interactions, structure-based thermodynamic analysis, 2:703–712

Entinostat, histone deacetylase inhibitors, 6:67

Entropy:
docking techniques, 2:631–633
ligand-macromolecular interactions, structure-based thermodynamic analysis, 2:703–712
quantum mechanics/molecular mechanics, binding affinity evaluation, 2:718–719
in silico screening, protein flexibility, 2:871

Environmental factors:
chirality and toxicity and, 1:147
commercial-scale operations and control, 3:13
Parkinson's disease, 8:537–540

Enzastaurin, brain tumor therapy, 6:245–246

Enzymatic assays, high-throughput screening:
adaptation mechanisms, 3:422–423
additives, stabilizers, and other cofactors, 3:412–413
basic principles, 3:401, 406
buffers and pH, 3:410–412
configuration, 3:414–423
substrate/enzyme concentration, 3:414–418
substrate selection, 3:420
unconventional setup, 3:418–420
continuous/discontinuous assays, 3:406–407
direct, indirect, and coupled assays, 3:407–409
drug targets, 3:401–402
expression and purification, 3:402–406
hydrolases, 3:425–426
isomerases, 3:427–428
ligases, 3:428
lyases, 3:426–427
monovalent/divalent salts, 3:412
optimization conditions, 3:409–414
oxidoreductases, 3:423–424
temperature, 3:413–414
transferases, 3:424–425

Enzymatic hydrolysis, soft drug development, **2**:77–80
Enzyme catalysis, xenobiotic reactions, **1**:405–406
Enzyme inhibitors:
 androgen antagonists, **5**:193–200
 cytotoxic agents, **6**:17–31
  anthracyclines, **6**:17–23
  camptothecins, **6**:23–29
  isopodophyllotoxins, **6**:29–31
 histone deacetylases:
  anilides, **6**:66–70
  crystalline structure, **6**:51–53
  fatty acids, **6**:73
  future research issues, **6**:73–75
  hydroxamic acid-based compounds, **6**:53–66
  natural products, **6**:70–72
  research background, **6**:49–50
 β-lactamases, **7**:358–372
  bicyclic lactam inhibitor, **7**:371
  carbapenems, **7**:366–368
  clavulanic acid disovery and inactivation, **7**:360–363
  6-heteroarylmethylene penems, **7**:364–366
  modified penams, penam sulfones, and cephem sulfones, **7**:368–370
  monolactams, **7**:371
  nonlactam inhibitors, **7**:372
  oxapenems, structural modification, **7**:370–371
  penam sulfones, **7**:363–364, 367–369
 peptidomimetics, **1**:217–222
  HIV-1 protease, **1**:218–220
  Ras farnesyl transferase, **1**:220–222
 recombinant DNA technology, drug targeting applications, **1**:553–556
 topoisomerase inhibitors, **6**:95–106
  dual topo I/II inhibitors, **6**:103–106
  Topo II inhibitors, **6**:98–102
  Topo I inhibitors, **6**:96–98

Enzymes:
 ADMET metabolic stability assays, **2**:67
 allosteric protein theory, **1**:371–372
  functional domains, **1**:382–383
 aminoglycoside resistance, **7**:424–427
 anticonvulsants, **8**:132–134
 drug metabolism:
  functionalization reactions, **1**:409–413
  cytochromes P450, **1**:410–411
  hydrolases, **1**:412–413
  oxidoreductases, **1**:411–413
 glucocorticoid anti-inflammatory compounds, **5**:53
 recombinant DNA technology, drug targeting applications, **1**:552–556
 streptogramin resistance, **7**:459–460
Enzyme-substrate (ES) complex, receptor-based drug design, **2**:494–498
EP$_3$ receptor antagonist, **4**:450–451
Epalrestat, Alzheimer's disease, tau-targeted therapy, **8**:427
Ependymoma, histologic classification, **6**:233
Ephedra, stimulant effect, **8**:89–90
Ephedrine, stimulant effect, **8**:89–90
Epibatidine, nicotinic agonists, **8**:69–70
Epidermal growth factor receptor (EGFR):
 brain tumor therapy, **6**:246–247
 erlotinib, **6**:251
 vaccine development, **6**:269–270
 growth factor signaling inhibitors, **6**:302–305
 kinase inhibitors, structure-based design:
  protein targeting, **6**:350–354
  resistance, **6**:366–368

 multitarget drug development, cancer therapy, **1**:265–267
Epidermal growth factor receptor-2 (EGFR-2), recombinant DNA technology, drug efficacy and personalized medicine, **1**:542–543
Epigallocatechin-3-gallate (EGCG):
 Hsp90 inhibitor, C-terminal binding sight, **6**:441
 tetracycline development, **7**:415
Epigenetics, structural genomics, **1**:594–595
Epilepsy. *See also* Anticonvulsants
 classification, **8**:122
 epidemiology, **8**:121
Epimerization, tetracyclines, **7**:408–409
Epinephrine, glaucoma therapy, **5**:597–600
Epinephrine receptors, antiplatelet agents, **4**:444
Epirubicin, structure and therapeutic effects, **6**:22–23
Epitope mapping, recombinant DNA technology, **1**:543–544
Epitope scanning mutagenesis, transport proteins, **2**:460–461
Eplivanserin, insomnia therapy, **5**:718
Epothilones, anti-cancer activity, **2**:240–243
Epotirome thyromimetic, structure-activity relationship, **4**:209–210
Epoxide hydrolase inhibitors, hit seeking combinatorial libraries, **1**:321–322
Epoxides:
 topical antimicrobial compounds, **7**:495
 trinem derivatives, **7**:348–349
4,5α-Epoxymorphinans, structure-activity relationship, **8**:598–601

N-substituted 4,5α-
Epoxymorphinans,
opioid antagonists,
8:601–608
Epoxysuccinates, stroke therapy:
calpain inhibitors, **8**:464–465
cathepsin B inhibitors,
8:473–474
Epristeride, androgen
antagonists, **5**:195
Epstein-Barr virus, anti-DNA
virus agentts, **7**:244
Eptifibatide, antiplatelet activity,
4:446–447
Equilibrium partitioning,
ADMET lipophilicity
assays, **2**:65
Equilin:
equine estrogen genesis,
5:232–233
structure-activity relationship,
5:246
Equine estrogens:
conjugated metabolic fate,
5:223–224
genesis of, **5**:232–233
Equipment/systems qualification,
commercial-scale
operations, **3**:12
erbB-family kinases, growth
factor signaling
inhibitors, **6**:302–305
Erbstatin, anti-cancer activity,
2:239–240
Erectile dysfunction:
drug development and
discovery, overview,
5:711–712
phosphodiesterase inhibitors:
pharmacokinetic and efficacy
comparisons,
5:715–716
sildenafil, **5**:712–713
tadalafil, **5**:714
vardenafil, **5**:713
Ergocalciferol:
chemistry, **5**:651, 653
chronic renal failure, analogs
for, **5**:655–656
Ergoline, Parkinson's disease
dopaminergic therapy,
8:545–546
Ergosterol, selective toxicity,
2:558–559
Ergot alkaloids:
migraine therapy,
8:267–268

Parkinson's disease
dopaminergic therapy,
8:545–546
ERK-1/RasGap dual inhibitors,
hit seeking
combinatorial
libraries, **1**:320–321
Erlotinib:
brain tumor therapy, **6**:247–248
EGFR-driven gliomas, **6**:251
growth factor signaling
inhibitors, **6**:302–305
structure-based design, protein
targeting, **6**:350–354
Erosion-controlled drug release:
oral drug delivery, sustained-
release dosage forms,
2:798–801
plasma drug profile, **2**:816–828
calculation methods,
2:818–819
diffusion-controlled release
vs., **2**:821–822
dosage form predictions,
2:826–828
nomenclature, **2**:816–818
repeated doses, **2**:822–825
single dose calculations,
2:819–822
Ertapenem:
biological activity, **7**:351–353
development of, **7**:354
occurrence, structural
variations and
chemistry, **7**:333–335
Erythro-9-(2-hyroxy-3-nonyl)
adenine (EHNA):
pulsed ultrafiltration-mass
spectrometry
identification,
1:117–119
structural analogs, **1**:168
Erythrocytes, artificial
erythrocytes,
4:640–642
Erythromycin:
chemical structure, **7**:433–438
macrolidic antibiotics,
2:228–229
pharmacology and ADMET
properties, **7**:438–441
RNA targets, **5**:984–990
mechanism of action, **5**:985
structure-activity
relationship,
5:985–990
selective toxicity, **2**:552–553

side effects, toxicity, and
contraindications,
7:438
structure-activity relationship,
7:441–444
Erythromyclamine, chemical
structure, **7**:438
Erythropoietin (EPO):
bioactivity, **4**:571
pharmacokinetics, **4**:572
physical properties, **4**:570
preparations, **4**:570–571
receptor-based drug design,
2:531–533
side effects, **4**:572–573
therapeutic indications,
4:571–572
Escherichia coli, protein
purification,
3:304–306
Escitalopram:
chemical structure, **8**:223
selective toxicity, **2**:569
Eslicarbazepine acetate,
anticonvulsant
applications,
8:136–139, 143–144
Esmolol:
enantiomers, chirality,
1:138–139
prodrug development,
3:258–268
soft drug design, **2**:85
homo-metoprolol analog,
2:133–134
Esomeprazole, selective toxicity,
2:566
E-state index, quantitative
structure-activity
relationships, **1**:24
Esterases, enzymatic hydrolysis,
2:78–80
Esterification, cholesterol, in
Alzheimer's disease,
8:406–407, 411
Esters:
anabolic agents, **5**:185–186
androgen derivatives,
5:171
glucocorticoid anti-
inflammatory
compounds:
C-21 carbonyl alterations,
5:106–112
percutaneous absorption,
5:49–52
water-soluble esters, **5**:53

prodrug development,
    3:242–248
esmolol case study,
    3:260–268
soft drug design, **2:**113–114
cytokine inhibitors,
    2:121–122
Estradiol:
  adverse effects and
      precautions, **5:**221
  biosynthesis, **5:**24–25
  chemical delivery system
      design, **2:**144–149
  current compounds and
      applications,
      5:219–221
  17α-ethynyl analogs,
      5:222–224
  metabolism, *in vivo* reversible
      monoesters, **5:**224
  oxidation and reduction,
      5:221–222
  soft drug design, **2:**102–103
  steroidal analogs, **5:**244–245
Estrane progestins:
  biotransformation, **5:**229–230
  steroidal progesterone receptor
      ligands, **5:**253–255
Estriol, soft drug design,
    2:102–103
Estrogen receptor-like family. *See also* Selective estrogen receptor modulators
  receptor-based drug design,
      2:512
Estrogen replacement therapy (ERT):
  adverse effects and
      precautions, **5:**221
  current drugs, **5:**219–221
  equine estrogens, **5:**232–233
  tissue specificity, estrogens and
      progestins, **5:**242
Estrogen response elements (EREs):
  estrogen receptor molecular
      endocrinology, **5:**236
  tissue specificity, estrogens and
      progestins, **5:**241–242
Estrogens:
  aromatization, **5:**230–231
  biosynthesis, **5:**24–25
  chemical delivery system
      design, **2:**144–149
  clinical applications:
    adverse effects and
        precautions, **5:**221

current drugs, **5:**219–221
environmental/dietary sources,
    5:239–240
metabolism, **5:**28–29
  conjugated equine estrogens,
      5:223–224
  glucuronide conjugation,
      5:228–229
  ligand metabolites,
      electrophilicity/
      carcinogenicity,
      5:226–228
  oxidative metabolism,
      5:225–226
multiple sclerosis therapy,
    5:573–574
physiology and pharmacology,
    5:233–243
potency-efficacy assessment,
    5:236–237
receptors (*See also* Selective
    estrogen receptor
    modulators)
  affinity limitations, **5:**246
  brain tumor therapy, **6:**265
  breast cancer resistance,
      6:369–370
  isoform distribution, **5:**236
  molecular endocrinology,
      5:236
  nonsteroidal ligands,
      5:246–247
  structure-activity
      relationships,
      5:244–247
  tissue specificity, **5:**240–242
  topography, **5:**233–236
  triarylethylenes, **5:**249–253
  vicinal diarylethylenes,
      5:247–248
soft drug design, **2:**102–103
Estrone, soft drug design,
    2:102–103
Estrone, structure-activity
    relationship, **5:**246
Eszopiclone:
  insomnia therapy, **5:**716–718
  selective toxicity, **2:**568
Et743 compound, marine sources,
    2:246–247
Eterobarb, anticonvulsant
    applications,
    8:140–141
Ethambutol:
  antitubercular agents:
    ADMET properties, **7:**755
    adverse reactions, **7:**755

arabinogalactan
    biosynthesis, **7:**722
historical background, **7:**751
mechanism of action,
    7:753–754
resistance mechanisms,
    7:754–755
structure-activity
    relationship,
    7:751–753
selective toxicity, **2:**557
Ethanols, antimicrobial agents,
    7:492–493
Ethers, anabolic agents,
    5:185–186
Ethical issues,
    pharmacogenomics,
    1:199–200
Ethionamide, antitubercular
    agents:
  ADMET properties, **7:**772
  adverse effects, **7:**772–773
  antibacterial activity, **7:**771
  historical background, **7:**771
  mycolic biosynthesis, **7:**723
  resistance mechanisms,
      7:771–772
Ethopropazine, Parkinson's
    disease therapy,
    8:552–553
Ethosuximide, anticonvulsant
    applications,
    8:135–139
Ethylenediamines, antimalarial
    drugs, farnesylation,
    7:632–634
Ethynodiol acetate, steroidal
    progesterone receptor
    ligands, **5:**254
Ethynyl estradiol:
  adverse effects and
      precautions, **5:**221
  current compounds and
      applications,
      5:220–221
  tissue specificity, **5:**240–242
17α-Ethynyl estradiol analogs:
  metabolic fate, **5:**222–224
  steroidal analogs, **5:**245
Etiprednol dicloacetate and
    analogs, soft drug
    design, **2:**92–94,
    96–97
Etomidate, enzymatic hydrolysis,
    2:79–80
Etonitazine, nonpeptide affinity
    labeling, **8:**641–644

Etoposide:
  antibody-directed enzyme prodrug therapy, **6:**120–123
  structure and therapeutic applications, **6:**30–31
Etorphine, Diels-Alder adducts, **8:**608–609
Etravirine:
  HIV reverse transcriptase inhibition, **7:**148–152
  selective toxicity, **2:**554–557
Etretinate, structure and clinical applications, **5:**402
Eudismic ratio (ER), chiral compounds, **1:**129–130
EUDOC system, docking and scoring computations, **2:**605
  force field scoring, **2:**621–622
Eugenol, antimicrobial agents, **7:**493–494
Euglycemia, type 2 diabetes, **4:**1–2
Eukaryotic expression:
  high-throughput screening, enzymatic assays, **3:**402–406
  protein therapeutics, upstream processing, **3:**291–294
European Patent Convention (EPC), **3:**166–168
Evaluation techniques, pharmacophores, **1:**471
Event reporting, FDA guidelines, **3:**81
Everolimus:
  ADMET properties, **5:**1034–1035
  brain cancer therapy, **6:**251–252
  cancer therapy, **5:**1037
  cyclohexyl analog design, **5:**1033–1034
  drug-eluting coronary stents, **5:**1037
  organ transplantation applications, **5:**1035–1037
Evoked related potentials (ERPs), schizophrenic patients, **8:**174–175
Evolutionary divergence, β-lactam mechanisms of action, penicillin binding proteins (PBPs), **7:**266
Evolutionary programming (EP), docking and scoring techniques, **2:**614
Evolutionary trace method (ETM), receptor-based drug design, GPCR signal modulation, **2:**509
Ewing's sarcoma, N-(4-hydroxyphenyl) retinamide (4-HPR) therapy, **5:**471
Exatecan, biological activity and side effects, **6:**98
Excessive daytime sleepiness, chirality at sulfur and, **1:**148
Excimer fluorescence, transport proteins, **2:**461
Excipients:
  oral dosage forms, class V drugs, **3:**54–55
  protein therapeutics, **3:**326
  sustained-release dosage, **2:**746
Excitatory amino acid receptors:
  anticonvulsants, **8:**131–132
  stroke therapy, **8:**450–478
    calpain inhibitors, **8:**460–465
      irreversible inhibitors, **8:**464–465
      reversible inhibitors, **8:**461–464
      unconventional inhibitors, **8:**465
    caspase 3/9 inhibitors, **8:**465–471
      aldehydes, **8:**466–467
      alpha oxo/thio/amino-methyl ketones, **8:**469–470
      halomethyl ketones, **8:**467–469
      irreversible inhibitors, **8:**467–470
      isatin-derived inhibitors, **8:**470–471
      masked aldehydes, **8:**467
      quinoline-derived inhibitors, **8:**471
      reversible inhibitors, **8:**466–467
    cathepsin B inhibitors, **8:**471–478
      irreversible inibitors, **8:**473–477
      reversible inibitors, **8:**472–473
      non-cysteine design, **8:**477–478
    NMDA antagonists, **8:**450–455
      competitive receptor antagonists, **8:**451–453
      ion channel blockers, **8:**450–451
      NR2B-specific noncompetitive receptor antagonists, **8:**453–455
      voltage-dependent calcium channels, **8:**455–460
Exclusivity, generic drug approval, FDA guidelines, **3:**86–88
Excretion:
  ADMET properties, **2:**61
  chirality and, **1:**142–143
  drug delivery and bioavailability, **2:**748–749
  permeability, ADMET interactions, **3:**373–374
  prodrug development, **3:**224–226
  quantitative structure-activity relationship, **1:**67
  testosterone metabolites, **5:**161–162
Executive function:
  neuranatomical substrates, **8:**23–24
  set shifting assessment of, **8:**28–30
Exemestane:
  androgen antagonists, **5:**198–199
  aromatase inhibition, **5:**232
  breast cancer resistance, **6:**369–370
Exhaustive/systematic approaches, docking and scoring computations, **2:**605
Exifone, Alzheimer's disease, tau-targeted therapy, **8:**425–426
Ex parte patent procedures:
  infringement of patent claims, defense based on, **3:**155–156

U.S. PTO requirements, **3:**138–142
Experimental autoimmune encephalomyelitis (EAE):
  Am80 effects, **5:**418
  multiple sclerosis pathology, **5:**567, 579–580
"Experimental use" defense, infringement of patent claims, **3:**156
Expert systems, biotransformation prediction, **1:**447
Expert testimony, infringement of patent proceedings, **3:**162–163
Expressed sequence tags (ESTs), patent requirements, **3:**120–123
Expression augmenting sequence element (EASE) sequence, protein therapeutics, expression vectors, **3:**294–295
Expression systems:
  high-throughput screening, enzymatic assays, **3:**402–406
  protein therapeutics, upstream processing, **3:**291–294
Expression vectors, protein therapeutics, upstream processing, **3:**294–295
Extended-spectrum β-lactamases (ESBLs):
  classification, **7:**277–279
  monobactam activity, **7:**329–331
Extensively drug-resistant tuberculosis (XDR-TB):
  antitubercular agents, peptidylglycans biosynthesis, **7:**720
  epidemiology, **7:**714–716
Extracellular signal regulated kinase (ERK), opioid receptor signaling, **8:**581
Extracorporeal membrane oxygenation (ECMO), inhaled nitric oxide and, **5:**276–278

Extrapyramidal symptoms (EPSs):
  antipsychotic side effects, **8:**165–166
  dopamine-based models, **8:**171
Extrinsically-mediated apoptosis, AHPN targeted drug development, **5:**491
Ex vivo studies, congestive heart failure therapy:
  cell-based biomarker assays, **4:**501
  predictive hemodynamic/biochemical biomarker assays, **4:**501
Eye, basic anatomy, **5:**587–588
Eye-targeted drug design. *See also* Ophthalmic agents
  chemical delivery systems, site-specific compounds, **2:**155–160
  oxime/methoxime β-blocker analogs, **2:**158–160
  glucocorticoid anti-inflammatory compounds, corneal penetration, **5:**53
  N-(4-hydroxyphenyl) retinamide (4-HPR) therapy, diabetes, **5:**474
  soft drugs:
    adaprolol, **2:**83–85
    cortienec acid-based steroids, **2:**90, 92–96
    L-653,328 compound, **2:**86
Ezetimibe:
  anticholesterolemic mechanisms, **2:**234
  LDL cholesterol lowering mechanisms, **4:**311–315

Factor analysis (FA), quantitative structure-activity relationship, **1:**38
Factor VII, serine proteases, **1:**332–333
Factor Xa:
  active site-directed inhibitors, **4:**395–398
  direct thrombin inhibitors mechanism of action, **4:**385–390

Fadrozole, aromatase inhibition, **5:**231–232
Failure analysis:
  quantitative structure-activity relationship modeling, **1:**510–511
  13-*cis*-retinoic acid therapy, **5:**393
Falcipains, antimalarial drug development, **7:**634–637
Famciclovir:
  anti-DNA virus agents, **7:**230–231
  varicella-zoster virus therapy, **7:**243
Familial hemiplegic migraine, epidemiology, **8:**265–266
Faramptator, cognitive enhancement, **8:**44
Farglitazar, peroxisome proliferator-activator receptors, γ-agonists, **4:**104
Farnesoid X receptor (FXR):
  agonist molecules, **4:**157–158, 161
  antagonists, **4:**161
  atherosclerosis, **4:**159–160
  basic properties, **4:**156–157
  bile acid metabolism, **4:**157
  clinical studies, **4:**161–163
  drug delivery, **2:**468–475
  future research issues, **4:**167
  glucose homeostasis, **4:**160
  HDL cholesterol, **4:**159
  LDL cholesterol, **4:**159
  lipid regulation, **4:**158–159
  medicinal chemistry, **4:**161
  structural biology, **4:**160–161
  triglycerides, **4:**158–159
  vascular tissues and inflammation, **4:**160
Farnesylation, antimalarial cell cycle regulation, **7:**630–634
Farnesyl pyrophosphate synthase (FPPS), nitrogen-containing bisphosphonates, structural genomics, **1:**584–585
Farnesyltransferase inhibitors, hit seeking combinatorial libraries, **1:**322–323

Faropenem, development of, 7:321–323
Faslodex, chemical reactions, 5:11, 13–14
Fast atom bombardment (FAB):
  liquid chromatography-mass spectrometry, 1:100–105
  mass spectrometry, 1:99–100
Fast exchange conditions, nuclear magnetic resonance, macromolecule-ligand interactions, 2:398–399
"Fast-followers," analog drug design, 1:170
Fatty acids:
  antimalarial agents, biosynthesis, apicoplast targeting, 7:663–665
  histone deacetylase inhibition, 6:73
  peroxisome proliferator-activated receptors:
    α-agonists, 4:88–99
    δ-agonists, 4:117–118
    γ-agonists, 4:101–102
Fatty acid synthases (FAS-1/FAS-II), antitubercular agents:
  isoniazid resistance, 7:740–743
  mycolic biosynthesis, 7:723
FCPF6 circular fingerprint, BioPrint® database, 1:501–502
Fed-batch systems, protein therapeutics:
  comparability analysis, 3:325–326
  upstream process development, 3:298–300
Federal Courts Improvement Act (FCIA), patent infringement suits, 3:150
Federal Declaratory Judgments Act, patent infringement and, 3:152
Fedotozine κ-opioid receptor agonists, 8:631–632
Felbamate:
  anticonvulsant applications, 8:135–139, 152–154
  toxicophore filtering, 2:327

Female sex hormones. *See* Sex hormones
Female sexual dysfunction (FSD), estradiol chemical delivery system, 2:144–149
Fenfluramine, drug metabolism, oxidation/reduction reactions, 1:420–421
Fenofibrate, γ-secretase modulators, 8:358–359
Fenofibrate intervention and event lowering in diabetes (FIELD) data, peroxisome proliferator-activator receptors, α-agonists, 4:97–99
Fenofibric acid derivatives, peroxisome proliferator-activator receptors, α-agonists, 4:90–96
Fentanyl:
  4-anilidopiperidine analogs, 8:618–621
  chemical structure, 8:570
  metabolism and elimination, 8:579
  nonpeptide affinity labeling, 8:641–643
Fentanyl isothiocyanate (FIT), nonpeptide affinity labeling, 8:642–643
Ferrodoxin, steroidogenesis, cholesterol-pregnenolone conversion, 5:16–17, 19
Ferrodoxin reductase, steroidogenesis, cholesterol-pregnenolone conversion, 5:16–17, 19
Festo I and *II* decisions, infringement of patents, 3:154
Fexofenadine, selective toxicity, 2:565
Fibrate compounds:
  anticholesterolemics, 2:235
  peroxisome proliferator-activator receptors, α-agonists, 4:89–99

Fibril formation inhibitors, structure-based design, 2:693–702
Fibrillization inhibitors, Alzheimer's disease therapy, preclinical/clinical testing, 8:361–365
  peptide derivation, 8:362
  small-molecule inhibitors, 8:362–365
Fick's laws of passive diffusion:
  ADMET structural properties, solubility, 2:54–55
  diffusion-based drug release, 2:792–793
  membrane proteins, 2:445–446
  oral drug delivery:
    absorption, 3:353–355
    permeablity, 3:34–35
Finasteride:
  androgen antagonists, 5:194–195
  chemical reactions, 5:11, 13
  toxicophore chemical reactivity, 2:301–302
Fingerprints for ligands and proteins (FLAP) approach, structure-based pharmacophore modeling, 2:683
Fingolimod (FTY720), transplant rejection inhibition, 5:1053–1055
FIRST analysis methods, *in silico* screening, protein flexibility, 2:864–865
First-generation compounds:
  anticonvulsants, 8:135–139
  cephalosporins, 7:301–306
  HIV integrase, discovery and elaboration, 7:83–97
  HIV protease inhibitors, 7:3–30
    amprenavir (agenerase), 7:22, 26–30
    indinavir (crixivan), 7:12–20
    nelfinavir mesylate (viracept), 7:20–22
    ritonavir (norvir), 7:10–12
    saquinavir mesylate (invirase), 7:3, 5–10
    peptidomimetics, 1:215–216
    teleprevir anti-HCV agent, 7:181–186
First-line agents, antitubercular chemotherapeutics, 7:737–759

First-pass metabolism, oral
 dosage forms, class V
 drugs, **3:**53
First to invent *vs.* first to file
 principle, patent
 protection, **3:**108–110
Fischer statistics *(F)*, quantitative
 structure-activity
 relationship
 validation, **1:**47
FITTED algorithm, docking and
 scoring techniques,
 2:613–614
5-Choice Serial Reaction Time
 Task (5-CSRTT),
 cognitive function
 assessment, **8:**26–27
502(b)(2) applications, FDA
 guidelines, **3:**92
Five-dimensional molecules,
 quantitative
 structure-activity
 relationship, **1:**42–43
Fixed charge models, quantum
 mechanics/molecular
 mechanics, binding
 affinity evaluation,
 2:720–723
FK64 compound, selective PPARγ
 modulators, PPARγ
 partial agonists, **4:**108,
 111–115
FK-228. *See* Romidepsin
FK506. *See* Tacrolimus (FK506)
FK binding protein (FKBP):
 recombinant DNA technology,
 targeting
 mechanisms,
 1:555–556
 sirolimus-FKBP 12 complex:
 mechanism of action,
 5:1025–1029
 ternary structure,
 5:1029–1030
 tacrolimus-FKBP-calcineurin
 structure,
 5:1014–1016,
 1018–1021
Flavin monooxygenase:
 antipsychotics, phenothiazines,
 8:186–189
 drug metabolism,
 functionalization
 reactions, **1:**411–412
Flavonoids:
 physiology and pharmacology,
 5:240

structure-based drug design,
 2:701–702
Flesinoxan, serotonin receptor
 targeting, **8:**242–244
Flestolol, soft drug design, **2:**86
Flexibility analysis:
 docking and scoring functions,
 2:637–640
 ensemble docking, **2:**640
 induced fit, **2:**640
 molecular dynamics,
 2:638–639
 NMR spectroscopy, **2:**638
 side-chain simulations, **2:**640
 x-ray crystallography, **2:**638
 in silico screening, protein
 flexibility, **2:**863–864
Flexible opiates, **8:**598–599
FlexX system, docking and
 scoring computations,
 2:607–608
 empirical scoring, **2:**625–628
 water conservation, **2:**635–636
FLIPR antagonist activity,
 migraine therapy,
 8:308–312
Flobufen, quantitative structure-
 activity relationships,
 1:73–74
FLOG algorithm, docking/scoring
 computations:
 force field scoring, **2:**620–622
 point complementarity
 methods, **2:**602–604
Flomoxef, development of,
 7:309–310
Flow properties, salt compounds,
 3:382
Floxacrine antimalarial agent,
 cellular respiration
 and electron
 transport, **7:**655–658
Floxamine antimalarial drug,
 cellular respiration
 and electron
 transport, **7:**657–658
Flp/FRT system, protein
 therapeutics, **3:**296
Flt3 inhibitors, acute
 myelogenous
 leukemia, **6:**301–302
Flucortin butyl, soft drug design,
 2:97
Fludarabine, antimetabolite
 analogs, **6:**112–113
Flufenamic acid, structure-based
 design, **2:**694–696

Fluidized bed reactors, protein
 therapeutics,
 3:297–298
Flumazenil, cognitive
 enhancement,
 8:45–46
Flunarizine, migraine
 prevention, **8:**318
Flunisolide, chronic obstructive
 pulmonary disease
 inhalation therapy,
 5:774–775
Flunoprost, platelet aggregation,
 thromboxane A_2
 receptor agonists,
 5:886–887
Fluocinolone:
 age-related macular
 degeneration therapy,
 5:622
 receptor affinity, **5:**69–70
Fluorenone, antitrypanosomal
 polyamine
 metabolism, **7:**590
Fluorescence resonance energy
 transfer (FRET):
 HIV reverse transcriptase
 inhibitor targeting,
 7:140–141
 nuclear hormone receptors,
 coactivator
 recruitment, **4:**84–86
Fluorescence spectroscopy,
 protein therapeutics,
 3:325
Fluorination, macrolide
 development,
 7:453–456
^{19}F nuclear magnetic resonance,
 drug screening, **2:**430
Fluorine atoms for biochemical
 screening (FABS),
 drug screening
 applications,
 2:430–431
Fluoro analogs, dipeptidyl
 peptidase 4 (DPP-4)
 inhibition, **4:**51–54
Fluorofelbamate, anticonvulsant
 applications,
 8:153–154
6-Fluoro-L-dopa, Parkinson's
 disease dopaminergic
 therapy, **8:**547–549
Fluorometholone, C-21 carbonyl
 alterations,
 5:104–112

Fluoromethyl ketones, stroke therapy, caspase inhibitors, **8:**468–469
Fluoroolefin inhibitors, dipeptidyl peptidase 4 inhibition, **4:**56–57
Fluoroquinolones:
  adverse reactions, **7:**540
  antibacterial drugs, **7:**535–540
  antimalarial drugs:
    combination therapies, **7:**621, 623–624
    topoisomerase inhibitors, **7:**652
  antitubercular agents:
    ADMET properties, **7:**775–776
    adverse effects, **7:**776
    DNA-targeted synthesis and repair, **7:**724–725
    history, **7:**774–775
    mechanism of action, **7:**775
    resistance mechanisms, **7:**775
  classification, **7:**536–537
  combinatorial libary, **1:**303–304
  discovery, **7:**483–484
  DNA polymerase IIIC-fluoroquinolone hybrid, **7:**551–552
  drug resistance, **7:**539
  historical background, **7:**483–484
  hybrid fluoroquinolone-oxazolidinone, **7:**551
  mechanism of action, **7:**539–539
  pharmacokinetics, **7:**539–540
  selective toxicity, **2:**552–553
  structure-activity relationship, **7:**537–538
5-Fluorouracil, antimetabolite analogs, **6:**109–112
Fluorous phase, defined, **1:**280
Fluosol, oxygen delivery mechanisms, **4:**623
Fluoxetine:
  chemical structure, **8:**223
  chiral switch in, **1:**158–159
  clozapine pharmacokinetics, **8:**192
  historical background, **8:**248–249
  migraine prevention, **8:**318–319
  monoamine transporter inhibition, **8:**238–241

pharmacokinetics, **8:**236
pindolol interaction, **8:**225–229
serotonin receptor targeting, **8:**242
side effects and adverse reactions, **8:**233–234
structure-activity relationship, **8:**251
Fluoxetine hydrochloride, cocrystal engineering, **3:**208–209
Fluperlapine, structure-activity relationship, **8:**182–183
Fluphenazine, pharmacokinetics, **8:**187–189
Flurazepam, insomnia therapy, **5:**711, 716–718
Flurithromycin, chemical structure, **7:**438
Flutamide, antiandrogen compounds, **5:**192–193
Fluticasone propionate:
  chronic obstructive pulmonary disease inhalation therapy, **5:**774–775, 777–779
  soft drug design, **2:**100–101
Fluvoxamine:
  chemical structure, **8:**223
  clozapine pharmacokinetics, **8:**192
  historical background, **8:**249
  pharmacokinetics, **8:**236
  side effects and adverse reactions, **8:**233–234
  soft analog development, **2:**118
Focal adhesion kinase (FAK) inhibitors, tumor metastases, **6:**326, 328
Fodor lithographic technique, peptide libraries, **1:**284–291
Folate inhibitors:
  antimalarial drugs, **7:**617–618
  DNA targeted compounds, **7:**644–648
  antitubercular agents, cofactor biosynthesis targeting, **7:**727–728
  DNA-targeted chemotherapeutic compounds, **6:**106–109
  synthetic antibacterial agents, **7:**506–534
  dihydrofolate reductase inhibitors, **7:**531–534

sulfonamides/sulfones, **7:**506–531
Folate receptors:
  antibiotic selective toxicity and comparative biochemistry, **2:**548–549
  receptor-mediated endocytosis, **2:**453
Folic acid, **5:**691–697
  antitubercular agents, cofactor biosynthesis targeting, **7:**727–728
  chemistry, **5:**691–692
  deficiency, **5:**696
  hypervitaminosis folic acid, **5:**696–697
  uptake and metabolism, **5:**691–698
Follicle-stimulating hormone (FSH):
  agonists, hit seeking combinatorial libraries, **1:**328–329
  androgen biosynthesis, **5:**22–24
  estradiol chemical delivery system, **2:**147–149
  ovulation modulation mechanism, **5:**242–243
  testosterone biosynthesis, **5:**155–158
Fondaparinux, antithrombin conformational activation, **4:**390–392
Food and Drug Administration (FDA):
  Amendments Act (FDAAA), **3:**69
  antipsychotics, approved compounds, **8:**176–180
    atypical compounds, **8:**177–179
    conventional drugs, **8:**176–177
  chemistry, manufacturing and control information, **3:**74–75
  clinical site guidelines, **3:**80
  event reportin guidelines, **3:**81
  IND applications, **3:**72–74
  inspections guide lines, **3:**81
  Institutional Review Board, **3:**80–81
  meetings guidelines, **3:**80
  Modernization Act (FDAMA), **3:**67–68

new drug application
  guidelines, **3**:78
new drug definitions, **3**:82
  sponsor rights, **3**:80
pharmacology and toxicology
  information, **3**:75
problems and challenges of,
  3:80
prodrug approvals, **3**:242
protocols, **3**:74
regulatory role
  biotechnology-derived drugs,
    3:93–96
  generic drug review and
    approval, **3**:84–92
  historical background,
    3:63–69
  new drug approval process,
    3:71–82
  Orange Book, **3**:92–93
  organizational structure,
    3:70
  over the counter drug
    approval, **3**:82–84
  postapproval process, **3**:96
  preclinical testing, **3**:71–72
  review process, **3**:70–71
  statutes, **3**:69–70
  United States
    Pharmacopoeia and
    National Formulary,
    3:96–97
retinoid drugs, approved
  compounds, **5**:508–509
reviewing process, **3**:78–80
trade secrets limitations and,
  3:179–180
Food effects:
  antidepressants, monoamine
    oxidase inhibitors,
    8:230
  migraine pathology and,
    8:266–267
Force field scoring:
  drug discovery, **2**:619–622
  functions table, **2**:644
  quantum mechanics/molecular
    mechanics, binding
    affinity evaluation,
    2:720–723
  structure-based virtual
    screening, **2**:17–18
Formadicins, structure and
  properties, **7**:326–327
Formaldehyde, topical
  antimicrobial
  compounds, **7**:495

7α-Formamido cephalosporin
  derivatives,
  7:307–308
Formoterol, chronic obstructive
  pulmonary disease
  bronchodilators,
  5:770–771, 777–779
Formulation development:
  defined, **3**:289
  N-(4-hydroxyphenyl)
    retinamide (4-HPR),
    5:478
  immunosuppressive drugs,
    cyclosporin A,
    5:1011–1012
  prodrugs, **3**:221–226
  protein therapeutics,
    3:311–317
    alternative modes, **3**:315–316
    container materials
      interactions, **3**:315
    degradation pathways,
      3:311–314
    future trends, **3**:316–317
    stability parameters,
      3:314–315
Fortimicin A, chemical structure,
  7:419
Fosamprenavir calcium:
  bioavailability, **7**:26
  chemical structure, **7**:42–43
  efficacy and tolerability,
    7:44–45
  optimization, **7**:43–44
  pharmacokinetics, **7**:44
  resistance profile, **7**:45
Foscarnet, anti-DNA virus
  agents, **7**:240–241
Fosmidomycin, apricoplast
  targeting,
  antimalarial agents,
  7:666–668
Four-dimensional molecules:
  combinatorial quantitative
    structure-activity
    relationship modeling,
    1:519–520
  quantitative structure-activity
    relationship,
    1:41–43
Fourier transform analysis,
  nuclear magnetic
  resonance, **2**:367–369
Fourier transform ion cyclotron
  resonance (FTICR),
  mass spectrometry
  development, **1**:98

Four-point attachment model,
  chiral compounds,
  1:128–129
Fourth-generation compounds,
  cephalosporins,
  7:303–306
FP receptor-based theory,
  glaucoma intraocular
  pressure reduction,
  5:594–600
FR181157 compound, IP receptor
  agonist, **5**:932,
  934–935
FR193262 compound, IP receptor
  agonist, **5**:932, 936
FR247304 PARP inhibitor, stroke
  therapy, **8**:486–487
Fraction absorbed (f_a),
  permeability, ADMET
  interactions,
  3:369–370
Fraction of variance, quantitative
  structure-activity
  relationship:
  outlier detection, **1**:45
  validation, **1**:46–47
Fragment-based approach (FBA):
  designed multiple ligands,
    1:273
  docking and scoring
    computations,
    2:605–608
  Hsp90 inhibitors, **6**:423–425
    dihydroxyphenyl amide
      inhibitors, **6**:424–425
    tetrahydro-4H-carbazol-4-
      one inhibitors,
      6:423–424
  ligand-macromolecular
    interactions,
    structure-based
    thermodynamic
    analysis, isothermal
    titration calorimetry,
    2:709–712
  nuclear magnetic resonance
    screening, ligand
    properties, **2**:432–433
  protein-protein interactions,
    2:340–342
  structure-activity relationships
    by nuclear magnetic
    resonance, drug
    screening
    applications,
    2:421–424
  virtual screening, **2**:25

Fragment cocrystallography,
structural genomics,
sleeping sickness
targeting, **1:**587–588
Framework combination:
designed multiple ligands
optimization,
1:269–273
multitarget drug development,
knowledge-based
approaches,
1:253–254
France, patent requirements,
3:122
Fraud defense, infringement of
patent claims,
3:155–156
FRB (FKBP12-rapamycin
binding) domain,
sirolimus-FKBP 12
complex:
mechanism of action,
5:1025–1029
ternary structure, **5:**1029–1030
FRED approach, docking and
scoring computations:
empirical scoring, **2:**627–628
point complementarity
methods, **2:**603–604
Freedom of Information Act
(FOIA), trade secrets,
3:183–184
Free-energy perturbation (FEP),
quantum mechanics/
molecular mechanics,
binding affinity
evaluation, **2:**723–728
Free fatty acid receptor 1 ligands:
G-protein coupled receptor,
homology modeling,
2:286–287
virtual screening, **2:**33–34
Free induction decay (FID),
nuclear magnetic
resonance, **2:**370–371
Free-Wilson method, quantitative
structure-activity
relationship,
multilinear regression
analysis, **1:**36–37
Fried-Borman enhancement
factors, glucocorticoid
anti-inflammatory
compounds, **5:**69–70,
72–73
Frontal affinity chromatography-
mass spectrometry,
drug screening,
1:114–115
Frozen density functional theory
(FDFT), free-energy
perturbation, binding
affinity evaluation,
2:727–728
Fructose-1,6-biphosphatase
inhibitors, type 2
diabetes therapy,
4:18–20
FTDOCK program, docking/
scoring computations,
point
complementarity
methods, **2:**602–604
FtsZ bacterial proteins, antibiotic
selective toxicity,
2:553
FTY720 fungal metabolite,
multiple sclerosis
therapy, **5:**575–578
Fujita-Nishioka treatment,
quantitative
structure-activity
relationship,
electronic parameters,
1:7–8
Fukuyama-Mitsunobu sequence,
motilin receptor
antagonist,
1:307–310
Full-length receptor
cotransfection assay,
nuclear hormone
receptors, **4:**80–81
Fulvestrant, breast cancer
resistance,
6:369–370
β-Funaltrexamine (β-FNA),
nonpeptide affinity
labeling, opioid
analgesics, **8:**638–643
Functional analogs:
defined, **1:**168
production, **1:**174–177
Functional antagonism, receptor-
based drug design,
efficacy parameters,
2:502–503
Functional domains. *See also*
specific domains, e.g.,
Ligand-binding
domaink
allosteric proteins, **1:**382–383
retinoid nuclear receptors,
5:369–373
Functional groups, CK-2130
prodrug case study,
3:270–276
Functionalization reaction:
defined, **1:**405–406
drug metabolism, **1:**408–421
carbon oxidation and
reduction, **1:**413–416
enzymes, **1:**409–413
cytochromes P450,
1:410–411
hydrolases, **1:**412–413
oxidoreductases,
1:411–412
hydration and hydrolysis,
1:421
nitrogen oxidation and
reduction, **1:**416–418
oxidative cleavage,
1:419–421
sulfur oxidation and
reduction, **1:**418–419
Functional selectivity, receptor-
based drug design:
central nervous system drugs,
8:7–9
efficacy parameters, **2:**503
Fungi:
drug development from,
2:250–251
steroid hydroxylation, **5:**10
Fungicides, quantitative
structure-activity
relationships,
1:73–74
Furan, toxicophore reactive
metabolites,
2:314–317
Furoate, chronic obstructive
pulmonary disease
inhalation therapy,
5:778–779
17-Furoate androstadienes, soft
drug design, **2:**99
Fused lactam system, toxicophore
reactive metabolites,
aniline nitrogen
oxidation, **2:**310–311
Fused-ring analogs:
DP$_1$ receptor antagonists,
5:903–905
triarylethylenes, **5:**249–253
Fusidic acid, ribosome targeting,
7:464
FV-100 anti-DNA virus agent,
varicella-zoster virus
therapy, **7:**243

G444137 prodrug, hydrogen
  sulfide releasing
  molecules, **5**:314–315
GABA$_A$/benzodiazepine(BZ)
  receptor, receptor-
  based drug design,
  ligand-gated ion
  channels, **2**:510–511
Gabapentin:
  anticonvulsant applications,
    8:135–139, 149
  migraine prevention, **8**:318
Gabazine, functional analog
  design, **1**:174–177
Gain-of-function mutations:
  glucokinase, type 2 diabetes
    therapies,
    4:15–17
  receptor-based drug design,
    G-protein coupled
    receptor, **2**:515
GAL4-LBD cotransfection assay,
  nuclear hormone
  receptors, **4**:82
GALAHAD program,
  pharmacophore
  structural properties,
  bioactive
  conformation,
  1:470–471
Galanin receptor ligands,
  antidepressant effects,
  8:257–258
Galantamine, cognitive
  enhancement,
  8:15–16, 31
Gallbladder, drug delivery, apical
  sodium-dependent
  bile acid transporter
  enhancement, **2**:474
GAL receptors, cognitive
  enhancement, **8**:49
γ-aminobutyric acid (GABA)
  receptors:
  anticonvulsants:
    analogs, **8**:149
    barbiturates, **8**:139–141
    benzodiazepines, **8**:136–139,
      144–147
    classification, **8**:135–139
    enzymes, **8**:132–134
    ligand-gated ion channels,
      GABA$_A$ receptors,
      8:126–129
    synaptic membrane
      transporters,
      8:130–132

voltage-gated ion channels,
  8:123–126
cognitive enhancement,
  GABA$_A$, **8**:45–46
Parkinson's disease,
  pathophysiology,
  8:531–534
receptor-based drug design,
  allosteric modulators,
  2:500–501
schizophrenia pathology,
  8:163–164
soft drug analogs,
  benzodiazepine,
  2:124–125
stroke therapy, GABA$_A$ agonist,
  8:483
Gammaguard Liquid,
  Alzheimer's
  immunization
  therapy, **8**:378
GAMMA program,
  pharmacophore
  structural properties,
  bioactive
  conformation,
  1:470–471
Ganaxolone, anticonvulsant
  applications,
  8:136–139, 156
Ganciclovir:
  anti-DNA virus agents:
    antimetabolites, **7**:229
    cytomegalovirus therapy,
      7:243–244
    herpes simplex virus
      therapy, **7**:242–243
  chemical delivery systems,
    2:142–143
Ganglionic nicotinic antagonists,
  8:78
Gantacurium, nicotinic
  antagonists, **8**:80
GAR-936 compound,
  development of,
  2:224–225
Gas chromatography-mass
  spectrometry (GC-
  MS), drug discovery
  applications,
  1:99
Gasotransmitters. *See also*
  specific compounds, e.
  g., Nitric oxide
  basic properties, **5**:265
  carbon monoxide,
    5:295–308

alternative delivery systems,
  5:307–308
inhalation therapy,
  5:299–305
prodrugs and releasers,
  5:305–307
production and biological
  effects, **5**:266, 296–299
therapeutic approaches,
  5:267–269
comparison of compounds,
  5:266
hydrogen sulfide, **5**:309–316
  alternative delivery systems,
    5:315–316
  inhalation therapy,
    5:312–313
  injectable formulations,
    5:313–314
  prodrugs and combined
    releasing molecules,
    5:314–315
  production and biological
    effects, **5**:266, 309–312
  therapeutic approaches,
    5:267–269
nitric oxide, **5**:265–294
  donors and prodrugs,
    5:278–290, 294
    C-nitroso class, **5**:286–287
    combined donors,
      5:287–290
    diazeniumdiolates,
      5:281–283
    S-nitrosothiols, **5**:283–285
    organic nitrates,
      5:278–280
    sodium nitroprusside,
      5:280–281
    sydnoimines, **5**:285–286
  inhalation therapy,
    5:274–278
  production and biological
    effects, **5**:265–266,
    270–274
  therapeutic applications,
    5:267–269
  indirect enhancement
    approaches,
    5:291–294
  therapeutic applications,
    5:267–269
GASP program, pharmacophore
  structural properties,
  1:459–460
  bioactive conformation,
  1:470–471

Gastrin-releasing peptide (GRP), structure and function, **4:**549
Gastrointestinal stability:
  ADMET structural properties, **2:**57
  soft drug design, serotonin receptor agonists, **2:**117–118
Gatifloxacin, antibacterial agents, **7:**535–540
Gaussian network models (GNM), *in silico* screening, protein flexibility, **2:**864–865
Gavestinel, stroke therapy, **8:**452–453
GC-14 thyromimetic, structure-activity relationship, **4:**219–220
GC-24 thyromimetic, structure-activity relationship, **4:**215
GDC-0449, brain tumor therapy, hedgehog signaling pathway, **6:**262
GDC-0941 PI3K inhibitor, survival signaling pathways, **6:**322–323
Gedunin, Hsp90 inhibitor binding, **6:**428–429
Geftinib:
  brain tumor therapy, **6:**247
  growth factor signaling inhibitors, **6:**302–305
Gelatinase, nuclear magnetic resonance studies, **2:**412–414
Geldanamycin:
  Alzheimer's disease, **8:**431–432
  antimalarial drug development, **7:**671–672
  Hsp90 inhibitor, **6:**386–399
  analog design, **6:**392–395
  benzoquinone ansamycins, **6:**397–399
  chemical structure, **6:**386–388
  chimeric structures, **6:**405–410
  dihydrogeldanamycin derivatives, **6:**388–391
  semisynthetic derivatives, **6:**391–392
Gel permeation chromatography (GPC), mass spectrometry, screening applications, **1:**113
Gemcitabine, antitubercular agents, DNA-targeted synthesis and repair, **7:**725
gem-Dialkylbenzothiadiazines, anti-HCV agents, **7:**202–203
Gem-disubstituted scaffold-based inhibitors, stroke therapy, **8:**506
GEMDOCK approach, empirical scoring, **2:**627–628
Gemfibrozil, peroxisome proliferator-activator receptors, α-agonists, **4:**90–97
Gene amplification, protein therapeutics, **3:**295–296
Gene-directed enzyme prodrug therapy (GDEPT):
  DNA-targeted compounds, **6:**123–124
  Leishmaniasis therapy, **7:**593
  prodrug design, **3:**229
Gene expression:
  AHPN modulation, **5:**494–497
  Am80 modulation, **5:**419
  N-(4-hydroxyphenyl) retinamide (4-HPR) regulation, **5:**476
  peroxisome proliferator-active receptors:
    α-agonist derivation, **4:**94–96
    γ-agonists, **4:**101–102
  tazarotene therapy, **5:**402–405
Gene induction, chiral compounds, **1:**155
Generalized-Born model and solvent accessible surface area (GBSA), quantum mechanics/molecular mechanics binding affinity, **2:**731–733
Generic drugs:
  biologics, FDA guidelines, **3:**96
  FDA approval process, **3:**84–92
  3:505(b)(2) provision, **3:**92
  ANDA approvals and exclusivities, **3:**86
  ANDA review process, **3:**87–91
  bioequivalence, **3:**85
  citizens petition, **3:**86–87
  Hatch-Waxman Amendment provisions, **3:**67, 85–86
  historical background, **3:**85
  labeling, **3:**91
  patient term extensions, **3:**87
  pharmaceutically equivalent compounds, **3:**84–85
  preapproval inspection, **3:**91–92
  withdrawal of approval, **3:**92
Gene therapy:
  nanoscale dendrimer delivery systems, **3:**482–483
  nitric oxide enhancement, **5:**294
  plasmid DNA:
    angiogenesis, **6:**463
    autoimmune diseases, **6:**463–464
    chemical delivery, **6:**467–480
    electroporation gene transfer, **6:**464–465
    emulsion delivery, **6:**480–481
    functional applications, **6:**461
    gene gun mechanisms, **6:**465
    gene transfer and long-term expression, **6:**458–460
    hydrodynamic injection, **6:**466
    laser irradiation, **6:**466
    myocardial delivery, **6:**461–463
    naked direct delivery, **6:**458–464
    nonviral vector toxicity, **6:**481–482
    physical delivery, **6:**464–466
    research background, **6:**457–458
    therapeutic applications, **6:**460–461
    ultrasound gene transfer, **6:**465–466
Genetic algorithms, docking and scoring techniques, **2:**611–614
Genetically-engineered drug discovery:
  osteoporosis anabolic therapy, Wnt pathway and bone target cells, **5:**757–758
  recombinant DNA technology, **1:**546–561
  aptamer-based diagnostics, **1:**549–550
  aptamer-based drugs, **1:**549

cell-specific targeting, 1:551
cellular adhesion proteins, 1:560–561
combinatorial biosynthesis and microbe reengineering, 1:547–549
enzyme-based drug targets, 1:552–556
phage display, 1:550–552
phage library preparation, 1:550–551
purified proteins vs. phage display, 1:551
receptor-based drug targets, 1:556–560
screening reagents, 1:546–547
small interfering RNA, 1:550
structural biology reagents, 1:552
vector-mediated delivery, siRNA, 1:550
in vitro evolution (SELEX), 1:549
in vivo phage display, 1:551–552
Genetics:
central nervous system disorders, 8:3–4
chemotherapeutic agents and, 7:503–504
Parkinson's disease etiology and, 8:534–535
schizophrenia pathology, 8:165
antipsychotic drug assessment, 8:173–174
single nucleotide polymorphisms, 1:181–182
Genome mining, natural products lead generation, 2:209–210
Genome replication, hepatitis C virus, 7:173–174
Genome scanning, natural products lead generation, 2:212
Genome sequencing:
microbial metabolites, 2:248–250
Mycobacterium tuberculosis, 7:719–720
natural products lead generation,

biosynthetic pathways, 2:210–212
patent requirements, 3:120–123
single nucleotide polymorphisms, 1:181–182
structural genomics focus on, 1:572–573
Genomics:
central nervous system disorders, 8:3–4
defined, 1:181–182
receptor-based drug design, 2:515–519
Genotoxicity assays, ADMET toxicity properties, 2:62, 68
Gentamicin:
chemical structure, 7:416–419
history and biosynthesis, 7:424
resistance to, 7:425–427
RNA targets, 5:971–979
mechanism of action, 5:972–973
structure-activity relationship:
subclass rings I and II, 5:973–978
subclass rings III and IV, 5:979
Geographic jurisdiction, patent enforcement, 3:148
Geometry-based simulation, *in silico* screening, protein flexibility, 2:863–864
Gepirone, serotonin receptor targeting, 8:242–244
Germany, patent requirements, 3:122
Germ cell nuclear factor-like family, receptor-based drug design, 2:512
Germ cell tumors, classification and therapy, 6:235
Germinomas, central nervous system, 6:235
Gestodene, biotransformation, 5:230
Geysen pins and wells, peptide libraries, 1:283–291
GFT505 selective PPARα modulator (SPPARαM),

peroxisome proliferator-activator receptors, 4:97
GHKL protein family, Hsp90 inhibitors, C-terminal nucleotide binding site, 6:429–430
Gibbs free energy:
docking/scoring paradigms, 2:596
ligand-macromolecular interactions, structure-based thermodynamic analysis, 2:705–712
Gimatecan, biological activity and side effects, 6:98
GKA50 glucokinase activator, type 2 diabetes therapy, 4:17
Glatiramer acetate, multiple sclerosis therapy, 5:570–571
Glaucoma:
disease pathology and etiology, 5:588–590
ophthalmic agents, 5:588–603
clinical trials, antiglaucoma therapeutics, 5:600–602
current clinical treatments, 5:592–600
future research issues, 5:602–603
intraocular pressure reduction, 5:591–592
in vivo preclinical screening models, 5:590–591
pilocarpine therapy, 8:67–68
Glaucoma filtration surgery (GFS), glaucoma intraocular pressure reduction, 5:592
GlaxoSmithKline ATP-competitive kinesin spindle protein inhibitors, 6:211–213
Gleevic®:
antimalarial cell cycle regulation, 7:627–630
cranial and spinal nerve tumors, 6:238–239
Glinides, type 2 diabetes therapy, 4:10–13

Glioblastoma:
  histologic classification, **6**:231–233
  hypoxia inducible factor (HIF) pathway, **6**:253–255
  mTOR therapies, **6**:251–252
  PI3K/AKT signaling pathway therapies, **6**:249–250
  stem cell initiation, **6**:225–227
Glioblastoma multiforme (GBM):
  histologic classification, **6**:231–233
  hypoxia-inducible factor pathway, **6**:255
  temozolomide therapy, **6**:244
  Wnt signaling pathway, **6**:263
Glioma:
  classification, **6**:228–230
  hypoxia inducible factor (HIF) pathway, **6**:253–255
  invasiveness, **6**:241–242
  stem cell initiation, **6**:225–227
Glitazones, peroxisome proliferator-activator receptors, γ-agonists, **4**:102–104
Gli transcription factor, brain tumor therapy, hedgehog signaling pathway, **6**:261–263
Globalization of drug discovery:
  intellectual property law:
    international agreements, **3**:165–168
    patententable subject matter requirements, **3**:121–123
    patent strategies, **3**:107–108, 165–171
    U.S. patent applications fron non-U.S. applicants, **3**:133
    trademark rights, international regulations, **3**:177–178
    trade secrets protection, **3**:184
Glomerulonephritis:
  NRX 194204 RXR selective analog, **5**:462
  NRX195183 analog therapy, **5**:420
Glove juice test, antiseptic testing, **7**:489–490
Glucagon-like peptide-1 (GLP-1):
  dipeptidyl peptidase 4 inhibitors, **4**:39–40

type-2 diabetes therapy, **4**:20–21
Glucagon-like peptide-2 (GLP-2), dipeptidyl peptidase 4 inhibitors, **4**:41–42
Glucan inhibitors, antifungal agents, **2**:231
Glucocorticoid γ-lactones, soft drug design, **2**:98–99
Glucocorticoid response elements (GREs):
  corticosteroid inhalation therapy, chronic obstructive pulmonary disease, **5**:773–774
  soft drug design, etiprednol dicloacetate and analogs, **2**:97
Glucocorticoids:
  anti-inflammatory compounds:
    absorption effects, **5**:41, 43, 45–53
    corneal penetration, **5**:53
    enzymatic metabolism, **5**:53–54
    intestinal absorption, **5**:41, 43, 45
    intra-articular administration, **5**:51–53
    percutaneous absorption, **5**:46–51
    water-soluble esters, **5**:53
  adverse effects, **5**:41, 43–44
  antiglucocorticoids, **5**:112–120
    nonsteroidal, **5**:121–125
  biosynthesis and metabolism, **5**:56–59
  clinical applications, **5**:37–40
  currently available corticosteroids, **5**:39–40, 42
  drug distribution effects, **5**:53–56
  drug receptor affinity, **5**:65–70
  future research issues, **5**:133–134
  mechanism of action, **5**:62–63
  nomenclature, **5**:36–37
  nonsteroidal dissociated glucocorticoids, **5**:125–133
  nonsteroidal glucocorticoids, **5**:121

nuclear structure-activity relationships, **5**:82–112
  C-1 alterations, **5**:83
  C-2 alterations, **5**:83–86
  C-3 alterations, **5**:86–87
  C-4 alterations, **5**:87–88
  C-5 alterations, **5**:88
  C-6 alterations, **5**:88–89
  C-7 alterations, **5**:89–90
  C-8 alterations, **5**:91
  C-11 alterations, **5**:91–92
  C-12 alterations, **5**:92–95
  C-15 alterations, **5**:95
  C-16 alterations, **5**:95–99
  C-17 alterations, **5**:99–102
  C-20 alterations, **5**:102–103
  C-21 alterations, **5**:104–112
  6,7-disubstituted compounds, **5**:91
pharmacological testing, **5**:70–71
quantitative structure-activity relationships, **5**:71–82
  conformational changes, **5**:76–77
  de novo constants, **5**:71–72
  electronic characteristics, **5**:77
  Hansch type analyses, **5**:72–74
  neural network predictions, **5**:74–75
  receptor binding affinity, LinBiExp model, **5**:75–76
  three-dimensional QSARs, **5**:77–82
receptor modulators, **5**:120–121
relative potencies, **5**:84–85
research background, **5**:35–36
signaling mechanisms, **5**:63–65
steroidal dissociated glucocorticoids, **5**:120–121
glaucoma risk and, **5**:589–590
receptor structure, anti-inflammatory compounds, **5**:58, 60–61
steroidogenesis, **5**:20–22

Glucokinase (GK):
  chirality properties, 1:152–155
  type 2 diabetes therapies, activators, 4:15–17
β-D-glucopyranosyl, colon-specific absorption, 5:45–46
Glucose-dependent insulinotropic polypeptide (GIP), dipeptidyl peptidase 4 inhibitors, 4:39–40
Glucose metabolism:
  farnesoid X receptors, 4:160
  liver X receptors, 4:148
  parasitic inhibitors, 7:582–585
  peroxisome proliferator-activated receptors, α-agonists, 4:89–99
Glucose sensitive insulin release (GSIR), type 2 diabetes therapies, glucokinase activation, 4:15–17
Glucosidation, drug metabolism, 1:426–430
Glucuronidases, antibody-directed enzyme prodrug therapy, 6:121–122
Glucuronidation:
  drug metabolism, 1:426–430
  estrogen receptor conjugation, 5:228–229
  glucocorticoid biosynthesis and metabolism, 5:58
  morphine metabolism and elimination, 8:578–579
GLUT5 transporter, protein therapeutics, upstream process development, 3:300–301
Glutamate:
  anticonvulsants, ionotropic glutamate receptors, 8:129
  cognitive enhancement, 8:41–42
  modulators, migraine therapy, 8:290–297
  ionotropic receptor antagonists, 8:291–297
  receptor properties, 8:291
  Parkinson's disease therapy, antagonist agents, 8:555

schizophrenia pathology, 8:163–164
  site antagonists, stroke therapy, 8:451
Glutamate dehydrogenase, allosteric proteins, 1:382–385
  high-throughput screening, 1:387–392
Glutamatergic agents, antipsychotics, 8:202–206
Glutamic acid, vitamin K carboxylation, 5:665–668
Glutamine synthase, protein therapeutics, cell line expression, 3:294
Glutaraldehyde, topical antimicrobial compounds, 7:495
Glutathione:
  drug metabolism, conjugation reactions, 1:434–438
  reductase inhibitors, antimalarial drugs, 7:660–663
Glutathione-S-transferase (GST), antimalarial drugs, 7:662–663
Glycan synthesis, oligosaccharide compound libraries, 1:292
Glycation, protein therapeutics formulation and delivery, 3:313–314
Glycemic control:
  antipsychotic side effects and, 8:168
  type 2 diabetes, 4:2
Glycine:
  anticonvulsants, ligand-gated ion channels, 8:130
  cognitive enhancement, 8:44–45
  site antagonists, stroke therapy, 8:451–453
  sulfonamide derivatives, prostaglandin/thromboxane EP_1 antagonists, 5:837, 840–845
Glycogen phosphorylase (GP), allosteric protein model, 1:394

Glycogen synthase kinase 3 (GSK3), Alzheimer's disease:
  insulin-related therapy, 8:415
  tau-targeted therapies, 8:418–420
Glycols, antimicrobial agents, 7:492–493
Glycopeptides, antibiotics, 2:225–227
Glycoprotein IIb/IIIa receptors:
  antagonists, 4:444–449
  abciximab, 4:445–446
  eptifibatide, 4:446–447
  intravenous antagonists, 4:445–447
  oral antagonists, 4:448–449
  tirofiban, 4:447
  future research issues, 4:451–452
  platelet aggregation, 4:414–417
  recombinant DNA technology, 1:560–561
Glycoproteins. See also P-glycoprotein; Variable surface glycoprotein (VSG)
  platelet adhesion, 4:412–413
Glycopyrrolate:
  chronic obstructive pulmonary disease bronchodilators, 5:768–769
  muscarinic antagonists, 8:72–77
Glycosaminoglycan mimetics, β-amyloid peptide inhibitors, 8:362–365
C-Glycoside derivative, cocrystal engineering, 3:211–212
α-Glycosidic bond, poly(ADP-ribosyl)ation biochemistry, 6:153–157
Glycosomal transport, parasitic inhibitors, 7:582–585
Glycosylated hemoglobin (HbA1c), type 2 diabetes:
  α-glycoside inhibitors, 4:8
  epidemiology, 4:2
  metformin clinical pharmacology, 4:4–5
  sulfonylurea therapy, 4:10

Glycosylation:
  Alzheimer's disease, tau-targeted therapies, 8:422–423
  end products inhibitors, β-amyloid peptide, 8:365
  streptomycin, RNA targets, 5:971
Glycosyl phosphatidylinositols, parasite lipid metabolism inhibitors, 7:580–581
Glycylcycline, history and biosynthesis, 7:412–413
GNTI, κ-opioid selective antagonist, 8:602–604
GOLD algorithm:
  chronic obstructive pulmonary disease bronchodilators, short-acting β_2-adrenergic receptor agonists, 5:770–771
  docking and scoring techniques, 2:612–614
  semiempirical techniques, 2:622–623
  in silico screening, protein flexibility, side-chain conformation, 2:867–868
"Goldilocks" compounds, prodrug design, esmolol case study, 3:264–268
Gonadotropin releasing hormone (GnRH), ovulation modulation, 5:242–243
Good clinical practices (GCPs), FDA Twenty-first Century Initiative, 3:68–69
Gossypentin, Alzheimer's disease, tau-targeted therapy, 8:425–426
Gossypol, androgen antagonists, 5:200
GPI-1046 tacrolimus compounds, structure-activity relationships, 5:1020–1021
GPR119 agonists, type-2 diabetes therapy, 4:20–21
G-protein coupled receptor (GPCR):
  allosteric protein model, 1:392

anticonvulsants, 8:134
antidepressant efficacy assessment, 8:226–229
antipsychotics, structure-activity relationships, 8:180–183
brain tumor therapy, Wnt signaling pathway, 6:263
central nervous system drug discovery, 8:5–6
chemogenomics:
  ligand-based data analysis, 2:584–585
  privileged scaffolds, 2:575–577
  protein family targeted libraries, 2:574–575
chemokine receptor-5 (CCR5) antagonists, HIV-1 entry blockade and, 7:108–110
designed multiple ligands, 1:268–273
dopamine receptor structure and function, 8:542–544
enantiomer activity, 1:132
GPR-119 agonists, type-2 diabetes therapy, 4:21
growth hormone secretagogue receptor antagonists, 1:333–334
homology model:
  applications, 2:288
  construction, 2:284–287
  evolution, 2:283–284
  future trends, 2:287–288
  lead generation, 2:289–292
  lead optimization, 2:292–293
  receptor structure/function, 2:288–289
  research background, 2:279–283
  strengths and limitations, 2:293–294
ligand-based virtual screening:
  pharmacophore formation, 2:4
  quantitative structure-activity relationships, 2:11–13
niacin analogs, atherosclerosis, HDL elevation, 4:346–348

peptidomimetics, ligand structures, 1:222–230
  bradykinin B2 antagonists, 1:223–224
  endothelin antagonists, 1:224–230
platelet activation, 4:415–416
prostaglandin/thromboxane synthesis, 5:809–812
receptor-based drug design:
  allosteric modulators, 2:500–501
  classification, 2:505–506
  family schematic, 2:506–508
  genetic variation, 2:514–515
  nomenclature, 2:504–505
  receptor binding assays, 2:522–523
  receptor sources, 2:523–524
  signaling modulators, 2:508–509
receptor-based transient-anchor chemical delivery systems, 2:160–161
recombinant DNA technology, receptor-based targeting mechanism, 1:557–560
single nucleotide polymorphisms, B_1-adrenergic receptor, 1:192
G proteins, opioid receptor signal transduction, 8:580–581
Graham v. John Deere Co. test, patent obviousness principle, 3:134–135
Gram-negative/gram-positive bacteria:
  aminoglycoside antibiotics, 7:415–419
  cephalosporin efficacy and, 7:302–306
  β-lactam mechanism of action:
    cell wall structural organization, 7:266–269
    outer membrane permeability and active efflux, resistance based on, 7:279–280
  streptogramins, 7:457–460
  synthetic antibacterial agents, 7:483–484

cationic surfactants,
7:499–500
nitrofurans, 7:549–550
phenols, 7:494–495
tetracycline antibiotics,
7:406–407
Granulocyte colony stimulating
actor (G-CSF),
structure and
properties, 4:575–578
Granulocyte-macrophage colony
stimulating factor
(GM-CSF), structure
and properties,
4:573–575
Graph sets, crystal engineering,
evolution of,
3:189–190
GRID probes:
chemotype development,
1:484–487
structure-based
pharmacophore
modeling, 2:683
Griseofulvin, development of,
2:230–231
Growth factors. See also specific
growth factors, e.g.,
Vascular endothelial
growth factor (VEGF)
antidepressant efficacy and,
8:228–229
brain tumor therapy, 6:245
RAS/RAF/MEK/ERK
signaling, 6:252
breast cancer resistance,
6:369–370
monoclonal antibodies,
selective toxicity,
2:560, 562
signaling inhibitors, 6:301–307
B-Raf inhibitors, 6:305–306
erbB-family kinases,
6:302–305
MEK inhibitors, 6:306–307
Growth hormone releasing
hormone (GHRH),
dipeptidyl peptidase
4 inhibitors,
4:41–42
Growth hormone secretagogue
receptor (GHS-R)
antagonists, hit
seeking combinatorial
libraries, 1:333–334
GS-9148 reverse transcriptase
inhibitor, 7:148

GS-9219 acyclic nucleoside
phopshate compound,
anti-DNA virus
agents, 7:238–239
GSK3β inhibitors, brain tumor
therapy, 6:263–264
GSK0060 PPARδ antagonists,
4:122–124
GSK-159797, chronic obstructive
pulmonary disease
inhalation therapy,
5:779
GSK 189254, cognitive function
assessment,
8:27–28
GSK690693 protein kinase B
inhibitor, 6:323–324
Guanfacine, cognitive
enhancement,
8:41–42
Guanidines:
antitrypanosomal agents,
7:578–580
polyamine metabolism,
7:588–590
BACE (β-site cleavage enzyme),
8:353–356
biguanide therapies, 4:2–3
stroke therapy, 8:497
Guanine derivative, acyclic
nucleoside phosphate,
7:236
Guanylyl cyclase:
carbon monoxide therapeutic
effects, 5:297–299
nitric oxide enhancement,
5:294
Gut-wall metabolism,
permeability, ADMET
interactions, 3:372
GW-6471 peroxisome
proliferator-activator
receptorα antagonist,
4:99–100
GW-9135 compound, α/γ/δ-pan
agonists, 4:141–145
GW-9662 compound, peroxisome
proliferator-activator
receptorγ antagonists,
4:116–117
GW27410 nitric oxide synthase
inhibitor, chronic
obstructive pulmonary
disease, 5:787
GW273629 iNOS inhibitor,
migraine therapies,
8:300–302

GW274150 iNOS inhibitor,
migraine therapies,
8:300–302
GW-501516, PPARδ-agonist,
4:117–122,
137–140
GW-590735, peroxisome
proliferator-active
receptors, α-agonist
derivation, 4:93–96
GW-677954 compound, α/γ/δ-pan
agonists, 4:141–145

H_2-antagonists, chemistry-driven
drug discovery,
1:537–538
H-142 renin inhibitor, 4:243
HADDOCK algorithm, docking
and scoring
techniques, Monte
Carlo simulation,
2:610–611
Hairpin polyamide concept, minor
groove targeting
compounds, 6:88
Half-life estimates:
antidepressants, 8:235–237
structure-activity
relationship and
metabolism,
8:252–253
drug delivery and
bioavailability, 2:748
intravenous drug delivery,
single dose (bolus
injection), 2:752–754
plasma drug profile:
diffusion-based drug release,
sustained-release
drug delivery,
2:812–815
erosion-controlled drug
release, 2:826–827
prodrug design, esmolol case
study, 3:264–268
sulfonamides, 7:509–514
Hallucinogens, receptor-based
drug design, central
nervous system
effects, 8:2–3
α-Halo ketones, combinatorial
chemistry, 1:296
Haloalkenes, drug metabolism,
glutathione
conjugation, 1:438
Halo derivatives, androgens,
5:171–174

Halofantrine antimalarial agent, **7:**619
Halogens:
 antimicrobial agents, **7:**490–492
 natural products lead generation, **2:**212–213
Halomethyl ketones and electrophiles, stroke therapy:
 calpain inhibitors, **8:**465
 caspase inhibitors, **8:**467–469
Haloperidol:
 dopamine model, **8:**170–171
 FDA approval, **8:**176–177
 nonpsychotic applications, **8:**170
 pharmacokinetics, **8:**189–190
Halophores, antimicrobial agents, **7:**490–492
Hamiltion rating scale for depression (HAM-D), antidepressant efficacy assessment, **8:**221–229
Hammett equation:
 quantitative structure-activity relationship:
  electronic parameters, **1:**2–8
  hydrophobic parameters, **1:**9–19
 quantitative structure-activity relationships, multilinear regression analysis, **1:**33–34
Hansch analysis, quantitative structure-activity relationship modeling, **1:**36–37, 509–510
 glucocorticoid anti-inflammatory compounds, **5:**72–74
Hantzsch methodology, dihydropyridine analogs, **1:**302–303
Haplotypes, single nucleotide polymorphisms, B$_2$-adrenergic receptor, **1:**193
Hard drugs:
 defined, **2:**75
 soft drugs *vs.*, **2:**76–77
Hardware requirements, nuclear magnetic resonance screening, **2:**434

Harmonic analysis, *in silico* screening, protein flexibility, **2:**864–865
Harvest operations, mammalian protein purification, **3:**306–307
Hatch-Waxman Act, generic drug approval, **3:**85–88
HCV-796 protease inhibitor, anti-HCV agents, **7:**203–204
HCV pseudotype particles (HCVpp), anti-HCV agents, **7:**175
HDM2-p53 interaction antagonists, hit seeking combinatorial libraries, **1:**326–327
Head-and-neck cancer:
 N-(4-hydroxyphenyl) retinamide (4-HPR) therapy, **5:**471–472
 13-*cis*-retinoic acid therapy, **5:**390
Healthcare policies, pharmacogenomics, **1:**199–200
Heart, thyroid hormone effects, **4:**195–196
Heating systems, pilot plant development, **3:**8
Heat shock proteins (HSPs):
 antimalarial targeted drugs, **7:**671–672
 brain tumor therapy, inhibitors, **6:**267–268
 drug resistance, **6:**366
 glucocorticoid receptor structure, **5:**62
 heme-oxygenases expression, carbon monoxide therapeutic effects, **5:**296–299
Hsp90 inhibitors:
 applications, **6:**441–444
  cancer therapy, **6:**441–442
  neurodegenerative diseases, **6:**442–444
 cancer markers, **6:**384
 C-terminal nucleotide binding site:
  cisplatin, **6:**440–441
  elucidation, **6:**430–431
  epigallocatechin-3-gallate, **6:**441
  GHLK protein family, **6:**429–430

 novobiocin inhibitor, **6:**431–440
N-terminal ATP-binding pocket, **6:**386–425
 benzisoxazole inhibitors, **6:**421
 chimeric geldanamycin and radicicol, **6:**405–410
 fragment-based screening, **6:**423–425
 geldanamycin, **6:**386–399
 high-throughput small-molecule inhibitors, **6:**417–421
 isoxazole-scaffold inhibitors, **6:**419–421
 purine-based inhibitors, **6:**410–416
 pyrazole-scaffold inhibitors, **6:**417–419
 radicicol analogs, **6:**399–405
 shepherdin, **6:**416
 structure-based design, **6:**421–425
 protein folding process, **6:**384–385
 protein structure and function, **6:**385–386
 research background, **6:**383
 tumor cell selectivity, **6:**385
 unidentified binding site inhibitors, **6:**425–429
 derrubone, **6:**425–426
 gedunin and celastrol, **6:**428–429
 high-throughput screening and modeling, **6:**429
 naphthoquinones, **6:**426–427
 small-molecule Hsp90/HOP interactions, **6:**427
 taxol, **6:**427–428
 steroid hormone actions, **5:**27, 29–30
Heat-transfer correlation, agitation functions, pilot plant development, **3:**8–9
Heavy metals, topical antimicrobial compounds, **7:**497–498
Hedgehog signaling pathway:
 brain tumor therapy, **6:**259–263

central nervous system tumors, **6**:228
Helical amphipathic peptides, reconstituted HDL, atherosclerosis, **4**:337
Hemagglutinins, receptor-mediated endocytosis, **2**:452
HemAssist®, structure and properties, **4**:628–630
Hematologic effects, antipsychotics, **8**:168–169
Hematologic malignancy, all-*trans*-retinoic acid therapy, **5**:376–377, 380–382
Hematopoietic agents:
  basic properties, **4**:567–570
  chemokine receptor CXCR4 antagonist-plerixafor, **4**:586–587
  erythropoietin, **4**:570–572
  future research issues, **4**:588–590
  granulocyte colony stimulating factor, **4**:575–578
  granulocyte-macrophage colony stimulating factor, **4**:572–575
  interleukin-11, **4**:578–580
  investigational agents, **4**:587–588
  stem cell factor, **4**:580–582
  thrombopoietin, **4**:582–586
Heme-derived materials:
  antimalarial drugs:
    amino acid salvage, **7**:634–644
    quinines, **7**:605–607
  antitubercular agent targeting, **7**:732
  oxygenases, carbon monoxide therapeutic effects, **5**:295–299
  oxygen carriers, **4**:642–643
  carbon monoxide therapeutic effects, **5**:307–308
  stroke therapy, NOS inhibition, **8**:501
Hemi-prostanoid DP$_1$ antagonists, **5**:901–903
Hemodynamic function biomarkers,

congestive heart failure therapy, **4**:506
Hemoglobin:
  allosteric protein model, **1**:377–379
  tertiary two-state model, **4**:610–612
  transport proteins, **1**:391–392
  antimalarial heme metabolism and amino acid salvage, **7**:635–644
    falcipain inhibitors, **7**:635–637
    plasmepsins, **7**:637–644
  lipid vesicles, artificial erythrocytes, **4**:640–642
  oxygen delivery, **4**:625–643
    artificial erythrocytes, **4**:640–642
    conjugated hemoblogins, **4**:635–637
    cross-linked hemoglobin, **4**:628–631
    nitric oxide reactions, **4**:612–614
    polymerized hemoglobins, **4**:631–635
    recombinant hemoglobins, **4**:637–640
    redox/radical reactions, **4**:614–615
    structure and function, **4**:610–612
Hemoglobin Polytaur, recombinant design, **4**:639–640
Hemolink polymerized hemoglobin, structure and properties, **4**:632–633
Hemopure®, structure and properties, **4**:633–634
Hemorrhagic disease of the newborn, vitamin K deficiency, **5**:668
Hemospan® conjugated hemoglobin, **4**:636–637
Hemostasis, anticoagulants, **4**:365–368
  amplification, **4**:367
  clotting initiation, **4**:366–367
  coagulation factors, **4**:365–366
  intrinsic regulators, **4**:368
  propagation, **4**:368

Henderson-Hasselbach equation, ADMET structural properties, **2**:52
  solubility, **2**:53–55
Hepadnaviruses, anti-DNA virus agents, **7**:235, 246–247
Heparin-induced thrombocytopenia (HIT), indirect anticoagulants and, **4**:372–374
Heparins:
  indirect anticoagulants, **4**:371–374, 392
  inhibition mechanisms, **4**:370–371
  limitations of, **4**:371–374
  molecular mechanism of action, **4**:374–380
  nonsaccharide mimics, *de novo* design, **4**:392–394
Hepatic lipase, atherosclerosis, HDL elevation, **4**:337–338
Hepatitis B virus (HBV), anti-DNA virus agents:
  antimetabolites, **7**:226–227
  hepadnaviruses, **7**:246–247
  lamivudine cytosine derivatives, **7**:226
Hepatitis C infection:
  anti-HCV agents:
    future research issues, **7**:204
    nonnucleoside inhibitors:
      benzothiadiazine, **7**:202–203
      HCV-796, **7**:203–204
      NX5B inhibitors, **7**:201–204
      PF-00868554, **7**:201–202
    NS3/4A protease inhibitors, **7**:176–195
      boceprevir, **7**:186–191
      ciluprevir, **7**:176–181
      ITMN-191, **7**:191–192
      MK7009, **7**:193–195
      telaprevir, **7**:181–186
      TMC435350, **7**:193
    nucleoside inhibitors, **7**:195–201
      MK0608, **7**:199–201
      NM283 (valopicitabine), **7**:195–196
      PSI-6130 (Pharmasset), **7**:196–197
      R1479/1626, **7**:197–199
  epidemiology, **7**:169–170

Hepatitis C virus (HCV):
    animal models, 7:175
    antiviral resistance, 7:175–176
    classification, 7:170–171
    genome replication, 7:173–174
    life cycle and intervention targets, 7:171–172
    protein expression, 7:172
    replication assays, cell culture for, 7:174–175
Hepatocyte growth factor/mesenchymal-epithelial transition factor (HGF/MET), brain tumor therapy, 6:247–248
Hepatocyte nuclear factor 4, basic principles, 4:163–164
Hepatoprotection, NRX 194204 RXR selective analog, 5:462–463
Hepatotoxicity:
    macrolides, 7:447–456
    multiobjective optimization, drug discovery, blood brain barrier, solubility, and ADMET CYP2D6, 2:269–273
    permeability, ADMET interactions, drug transport mechanisms, 3:368–369
    psychostimulants, 8:95–96
HepDirect technology, tissue-targeted acyclic nucleoside phosphate prodrug development, 7:239
HEPT derivatives:
    current development, 7:148–152
    research background, 7:140
HERCEPTIN. See Trastuzumab
hERG channel activity:
    chirality and toxicity, 1:143–144
    potassium channel blocking, ADMET toxicity properties, 2:69
    receptor-based drug design, efficacy parameters, 2:503
    tropane-based chemokine receptor-5 (CCR5) antagonist, 7:118–121

Heroin, synthetic analog, 8:597–600
Herpes simplex virus (HSV) anti-DNA virus agents:
    antimetabolites, 7:222–225
    compounds and prodrugs, 7:242–243
HETA nucleotide transporter trypanocide, 7:577–580
Heteroaryl-linked hydroxymate inhibitor, stroke therapy, 8:490–491
6-Heteroarylmethylene penems, β-lactamase inhibitors, 7:364–366
Heteroaryl templates:
    chemokine receptor-5 (CCR5) antagonist development, 7:115–116
    five-member ring replacement, 7:128–131
    chronic obstructive pulmonary disease, CXCR2 receptor antagonistst, 5:784–785
Heterocyclic derivatives:
    anabolic agents, 5:183–185
    dipeptidyl peptidase 4 (DPP-4) inhibition, 4:59–64
    aminomethyl heterocycles, 4:60–61
    aminopiperidine/piperazine heterocycles, 4:61–62
    aminopyrrolidine/aminopiperidine inhibitors, 4:63–64
    estrogen receptors, triarylheterocycles, 5:253
HIV integrase, naphthyridine integrase inhibitors, 7:86–97
    prostaglandin/thromboxane receptors:
    EP_1 antagonists, 5:819–820
    IP receptor agonist, 5:926–930
    stroke therapy, 8:458–459
    thyromimetic structure-activity relationship, 4:210–211
    azauracil compounds, 4:215–216

Heterologous interactions, allosteric proteins, 1:383–386
Heteronuclear multiple-bond correlation (HMBC), ligand-based drug design, nuclear magnetic resonance, 2:376–377
Heteronuclear single quantum coherence (HSQC) spectroscopy:
    basic principles, 2:371
    ligand-based drug design, nuclear magnetic resonance, 2:376–377
Heterotropic effects, allosteric proteins, 1:385–389
Hexachlorophene, synthetic antibacterial agents, 7:494–495
Hexhydropyranoquinoline (HHPQ) scaffold, EMD-534085 kinesin spindle protein inhibitor, "induced fit" binding pockets, 6:210–211
High-affinity ion channel blockers, stroke therapy, 8:450
High-affinity rolipram binding site (HARBS), chronic obstructive pulmonary disease therapy, 5:783–784
High-density lipoprotein (HDL):
    atherosclerosis, elevation:
        antioxidant/anti-inflammatory properties, 4:334
        apolipoprotein A-1 and mimetics, 4:335–337
        ApoA-I$_{Milano}$, 4:335–336
        reconstituted HDL, 4:336
    basic principles, 4:331
    cholesteryl ester transfer protein inhibitors, 4:340–345
        anacetrapib, 4:343–344
        clinical trials, 4:341–344
        current and future development, 4:344–345
        dalcetrapib, 4:342–343
        functional analysis, 4:344
        torcetrapib, 4:341–342

innate immunity, **4**:334–335
lipase inhibitors, **4**:337–340
  endothelial lipase, **4**:338–339
  hepatic lipase, **4**:337–338
nicotinic acid receptor agonists, **4**:345–350
  anthranilic acid and related analogs, **4**:348–349
  niacin analogs, **4**:346–348
  xanthine and barbituric acid analogs, **4**:349–350
plasma lipoproteins, **4**:332–333
reverse cholesterol transport, **4**:333–334
farnesoid X receptors, **4**:159
peroxisome proliferator-activated receptors:
  α-agonists, **4**:88–99
  α/δ-dual agonists, **4**:136–137
  α/γ-dual agonists, **4**:125–135
  δ-agonists, **4**:117–118
thyroid hormone metabolism, **4**:195
  receptor agonist effects, **4**:197–198
vitamin E uptake and metabolism, **5**:662
Highest occupied molecular orbital (HOMO), quantitative structure-activity relationships, **1**:24–25
High-performance liquid chromatography (HPLC):
  combinatorial library purification, **1**:107–110
  drug development:
    current trends, **1**:106–107
    liquid chromatography-mass spectrometry, **1**:100–105
    natural products libraries characterization, **2**:202–203
    nuclear magnetic resonance and, **2**:374
High-resolution magic angle spinning (HR-MAS), nuclear magnetic resonance studies, drug screening applications, **2**:425–426
High-temperature requirement A-1 (HTRA01) serine proteases, age-related macular degeneration, **5**:607–610
High-throughput docking (HTD), G-protein coupled receptor, homology modeling, **2**:290–292
High-throughput formulation (HTF), protein therapeutics, **3**:316, 327
High-throughput screening (HTS):
  allosteric proteins, **1**:387–392
  antimalarial drugs, dihydroorotate dehydrogenase inhibitors, **7**:648–650
  chemogenomics, **2**:580–582
  enzymatic assays:
    adaptation mechanisms, **3**:422–423
    additives, stabilizers, and other cofactors, **3**:412–413
    basic principles, **3**:401, 406
    buffers and pH, **3**:410–412
    configuration, **3**:414–423
      substrate/enzyme concentration, **3**:414–418
      substrate selection, **3**:420
      unconventional setup, **3**:418–420
    continuous/discontinuous assays, **3**:406–407
    direct, indirect, and coupled assays, **3**:407–409
    drug targets, **3**:401–402
    expression and purification, **3**:402–406
    hydrolases, **3**:425–426
    isomerases, **3**:427–428
    ligases, **3**:428
    lyases, **3**:426–427
    monovalent/divalent salts, **3**:412
    optimization conditions, **3**:409–414
    oxidoreductases, **3**:423–424
    temperature, **3**:413–414
    transferases, **3**:424–425
  Hsp90 inhibitors:
    modeling and identification, **6**:429
    small-molecule compounds, **6**:417
  ligand-based virtual screening, kinase inhibitor enrichment, **2**:14
  multiobjective optimization, drug discovery, **2**:259–261
  natural products lead generation, **2**:205–207
  nuclear magnetic resonance studies, ligand properties, **2**:431–434
  plasmepsin antimalarial agents, **7**:642–644
  prodrug discovery, **3**:226, 228–229
  protein-protein interactions, antagonist/agonist discovery, **2**:337–340
  quantitative structure-activity relationship models, **1**:525
  receptor-based drug design, lead drug discoveries, **2**:526–527
  retrometabolic drug development, **2**:73–74
  γ-secretase inhibitors, **8**:335
  triaging, virtual screening data, **2**:27
Hill plots:
  allosteric proteins, **1**:379–380
  allosteric protein theory, **1**:373–375
HINT algorithm:
  allosteric proteins, **1**:383–388
  docking techniques, water conservation, **2**:635–636
  empirical scoring, **2**:628
HipHop algorithm, pharmacophore formation:
  bioactive conformation, **1**:470
  catalysts, **1**:459–461
  ligand-based virtual screening, **2**:4
Hirudin:
  limitations of, **4**:374
  structural properties, **4**:369–370
Histamine, cognitive enhancement, **8**:35–37

Histamine H$_1$-plus, multitarget drug development, allergies, **1**:259–262
Histamine H3 receptor antagonists:
  antipsychotic agents, **8**:206–207
  cognitive enhancement, **8**:36–37
  G-protein coupled receptor, homology modeling, **2**:293
Histone deacetylase (HDAC) inhibitors:
  anilides, **6**:66–70
    chidamide, **6**:68, 70
    entinostat (MS-275), **6**:67
    mocetinostat (MGCD-0103), **6**:67–70
    tacedinaline (CI-994), **6**:66–67
  antimalarial drugs, **7**:653–655
  brain tumor therapy, hedgehog signaling pathway, **6**:262–263
  chemogenomics, **1**:589–595
  classification, **6**:50–51
  corticosteroid inhalation therapy, chronic obstructive pulmonary disease, **5**:774
  crystalline structure, **6**:51–53
  drug resistance, **6**:366
  fatty acids, **6**:73
  future research issues, **6**:73–75
  hydroxamic acid compounds, **6**:53–66
    current clinical trials and drug development, **6**:53–54
    ITF-2357, **6**:60–61
    JNJ-26481585 second-generation HDACi, **6**:64–66
    NVP-LAQ-824 and LBH-589, **6**:55, 57–58
    PCI-024781, **6**:60–62
    PXD-101, **6**:58–59
    R306465 first-generation HDACi, **6**:62–64
    SB939, **6**:59–60
    suberoyl hydroxamic acid analogs, **6**:53, 55–56
  multitarget drug development, cancer therapy, **1**:266–267
  natural products, **6**:70–72

largazole, **6**:72
romidepsin (FK-228), **6**:70–72
research background, **6**:49–50
stroke therapy, **8**:487–495
  2-aminophenylamides, **8**:492–493
  boronates and silanediols, **8**:494
  carboxylic acids, **8**:494
  hydroxamate inhibitors, **8**:488–491
  ketones, **8**:493
  thiol inhibitors, **8**:492
Histone methyltransferases, structural genomics, human proteome coverage, **1**:578–579
Hit/lead compound selection:
  ADMET properties, **2**:49–50
  BioPrint® database, **1**:489–491
  receptor-based drug design, structure-activity relationships, **2**:519–520
  spiro diketopiperazine-based chemokine receptor-5 (CCR5) antagonist, **7**:131–133
Hit seeking combinatorial libraries, **1**:305–343
  B-cell lymphocyte/leukemia-2 inhibitors, **1**:312–315
  beta-secretase inhibitors, **1**:327–328
  carbonic anhydrase II inhibitors, **1**:311–312
  caspase 3 inhibitors, **1**:315–317
  cyclin dependent kinase 5/p25 inhibitors, **1**:329–330
  cysteine proteases, **1**:330–332
  dihydrofolic acid reductase inhibitors, **1**:317, 319
  dopamine D3 receptor ligands, **1**:315, 318
  epoxide hydrolase inhibitors, **1**:321–322
  farnesyltransferase inhibitors, **1**:322–323
  follicle-stimulating hormone agonists, **1**:328–329
  growth hormone secretagogue receptor antagonists, **1**:333–334
  HDM2-p53 interaction antagonists, **1**:326–327

HIV-1 protease inhibitors, **1**:310–311
human papillomavirus 6 E1 helicase inhibitors, **1**:322–324
5-hydroxytryptamine 2A receptor antagonists, **1**:311–312
integrins, selectins, and protein-protein interactions, **1**:341–343
ion channels, **1**:336–337
kinase inhibitors, **1**:337–340
melanin-concentrating hormone 1 receptor antagonists, **1**:324–325
metabotropic glutamate receptor 5, **1**:318–320
metalloprotease inhibitors, **1**:325–327
methionine aminopeptidase-2 inhibitors, **1**:315–317, 319
motilin receptor antagonists, **1**:307–310
nuclear hormone receptors, **1**:334–336
P2X$_7$ nucleotide receptor antagonists, **1**:324–326
protein tyrosine phosphatases, **1**:339–342
RasGap and ERK-1 dual inhibitors, **1**:320–321
serine proteases, **1**:332–333
Hit triaging, virtual screening data, **2**:26
HIV integrase:
  active site biochemistry, **7**:80–81
  diketo acid discovery and two-metal chelation pharmacophore elaboration, **7**:81–97
  future research issues, **7**:97
  integration process, **7**:78–79
  research background, **7**:75–76
HIV protease inhibitors (HIV-1 protease inhibitors):
  Hit seeking combinatorial libraries, **1**:310–311
HIV/AIDS therapy:
  amprenavir (agenerase), **7**:22, 26–30

atazanavir sulfate (Reyataz), 7:36–42
darunavir (Prezista), 7:54–60
design criteria, 7:2–4
evolution of, 7:1
first-generation inhibitors, 7:3–30
fosamprenavir calcium (Lexiva/Telzir), 7:42–45
indinavir (crixivan), 7:12–20
lopinavir/ritonavir (Kaletra), 7:31–36
nelfinavir mesylate (viracept), 7:20–22
next-generation inhibitors, 7:60–64
ritonavir (norvir), 7:10–12
saquinavir mesylate (invirase), 7:3, 5–10
second-generation inhibitors, 7:31–60
tipranavir (Aptivus), 7:45–54
ligand-macromolecular interactions, structure-based thermodynamic analysis, 2:705–712
nuclear magnetic resonance studies, 2:415–419
peptidomimetics, 1:218–220
plasmepsin antimalarial agents, 7:640–644
receptor-based design, nuclear magnetic resonance, 2:391–394
renin inhibitors and, 4:239
structural biology, 7:1–2
transition-state isosteres, 7:3–4
HIV reverse transcriptase inhibitors:
nonnucleoside reverse transcriptase inhibitors:
clinical potential, 7:155–156
current compounds, 7:148–152
research background, 7:139–140
resistance limitations, 7:152–155
nucleoside reverse transcriptase inhibitors:
clinical potential, 7:155–156
current compounds, 7:141–145
research background, 7:139–140
nucleotide reverse transcriptase inhibitors:
clinical potential, 7:155–156
current compounds, 7:145–148
research background, 7:139–140
research background, 7:139–140
target identification, 7:140–141
HOE14 (icatibant), peptidomimetic design, 1:223–224
Hoechst ligand, DNA binding drugs, nuclear magnetic resonance studies, 2:405–410
Hollow fiber reactors, protein therapeutics, 3:297
Hologram quantitative structure-activity relationship (HQSAR), oral bioavailability, 1:66
Homatropine, muscarinic antagonist, 8:72
Homing receptors, recombinant DNA technology, cellular adhesion proteins, 1:561
Homogeneity, high-throughput screening, enzymatic assays, 3:406
Homologous recombination, poly (ADP-ribose) polymerase (PARP) inhibitors:
cancer therapy, 6:168
PARP-1 structures, 6:161–162
Homolog-scanning mutagenesis, recombinant DNA technology, epitope mapping, 1:543–544
Homology modeling:
antimalarial antifolates, 7:644–648
ATP-binding cassette transporters, 6:375–376
G-protein coupled receptor (GPCR):
applications, 2:288
construction, 2:284–287
evolution, 2:283–284
future trends, 2:287–288
lead generation, 2:289–292
lead optimization, 2:292–293
receptor structure/function, 2:288–289
research background, 2:279–283
strengths and limitations, 2:293–294
structure-based virtual screening, protein-ligand interactions, 2:14–16
Homo-metoprolol analog, esmolol soft drug design, 2:133–134
Hormone binding domain (HBD), glucocorticoid receptor structure, 5:60, 62
Hormone-responsive elements (HREs), steroid hormone actions, 5:29–30
Hormone therapy. See also Estrogen replacement therapy (ERT)
chemical delivery system design, 2:144–149
osteoporosis antiresorptive therapy, bisphosphonates, 5:733–734
selective toxicity, 2:541
soft drug design, estrogens, 2:102–103
Host interactions, chemotherapeutic agents, 7:503–504
Hotspot techniques, protein-protein interactions:
drug discovery, 2:340
scaffold structures, 2:354–359
Houghton "tea bag" technique, peptide libraries, 1:284–291
HPA-23 reverse transcriptase inhibitor, research background, 7:139–140
HPMPA HIV reverse transcriptase inhibitor, 7:146–148
Hsp90 organizing protein (HOP), Hsp90 inhibitors interruption, 6:427

Human African trypanosomiasis (HAT):
 biochemistry and pathology, **7:**568–569
 current drug treatments, **7:**569–571
 epidemiology, **7:**565–566
 new drug development for, **7:**577–590
Human cytomegalovirus (HCMV). *See* Cytomegalovirus (CMV)
Human embryonic kidney (HEK-293) cells, protein therapeutics, **3:**293–294
Human growth hormone (hGH), recombinant DNA technology, epitope mapping, **1:**543–544
Human herpesvirus type 6 (HHV-6), anti-DNA virus agents, **7:**244
Human immunodeficiency virus (HIV) infection:
 allosteric protein model, **1:**393–394
 antiviral compounds, selective toxicity, **2:**554–557
 chemical delivery systems therapy, zidovudine, **2:**141–142
 chemokine receptor-5 (CCR5) antagonists, HIV-1 entry blockade:
  anilide core, **7:**124–131
  binding mode and mechanism, **7:**133–134
  piperazine/piperidine-based antagonists, **7:**110–116
  research background, **7:**107–110
  spiro diketopiperazine-based antagonists, **7:**131–133
  summary and future research issues, **7:**134–135
  tri-substituted pyrrolidine antagonists, **7:**121–124
  tropane-based antagonists, **7:**116–121
 dementia, **8:**21

epidemiology, **7:**75–76
HIV protease inhibitor therapies:
 amprenavir (agenerase), **7:**22, 26–30
 atazanavir sulfate (Reyataz), **7:**36–42
 darunavir (Prezista), **7:**54–60
 design criteria, **7:**2–4
 evolution of, **7:**1
 first-generation inhibitors, **7:**3–30
 fosamprenavir calcium (Lexiva/Telzir), **7:**42–45
 indinavir (crixivan), **7:**12–20
 lopinavir/ritonavir (Kaletra), **7:**31–36
 nelfinavir mesylate (viracept), **7:**20–22
 next-generation inhibitors, **7:**60–64
 ritonavir (norvir), **7:**10–12
 saquinavir mesylate (invirase), **7:**3, 5–10
 second-generation inhibitors, **7:**31–60
 tipranavir (Aptivus), **7:**45–54
HIV reverse transcriptase inhibitors:
 nonnucleoside reverse transcriptase inhibitors:
  clinical potential, **7:**155–156
  current compounds, **7:**148–152
  research background, **7:**139–140
  resistance limitations, **7:**152–155
 nucleoside reverse transcriptase inhibitors:
  clinical potential, **7:**155–156
  current compounds, **7:**141–145
  research background, **7:**139–140
 nucleotide reverse transcriptase inhibitors:
  clinical potential, **7:**155–156
  current compounds, **7:**145–148

  research background, **7:**139–140
 research background, **7:**139–140
 target identification, **7:**140–141
 quantitative structure-activity relationship, anti-HIV activity, **1:**60–62
 recombinant DNA technology: *de novo* drug development, **1:**553–556
 Tat inhibitors, **1:**546–547
 retroviral replication cycle, **7:**76–78
 tuberculosis epidemiology and, **7:**713–716
Human intestinal absorption, quantitative structure-activity relationship, ADMET parameters, **1:**65
Human monocyte binding, chirality properties, **1:**151–155
Human papillomavirus (HPV):
 anti-DNA viral agents, **7:**241–242
 vaccine development and outcomes, cost-effectiveness analysis of, **3:**347–349
Human papillomavirus 6 (HPV6) E1 helicase inhibitors, hit seeking combinatorial libraries, **1:**322–324
Human prolactin (hPRL), recombinant DNA technology, epitope mapping, **1:**544
Human recombinant estrogen receptors (hERs), topography, **5:**233–236
Human T-cell lymphotropic virus type I (HTLV-1), Am80 therapy, **5:**416–417
Humidity, oral drug physicochemistry, moisture uptake, **3:**37–39
Huperzine A, cognitive enhancement, **8:**31–32

Hybrid/alternative scoring methods:
  docking and scoring techniques, consensus docking, **2:**617–618
  structure-based virtual screening, **2:**19–20
Hybrid antibacterial agents, **7:**550–552
Hybrid chemical biological descriptors, quantitative structure-activity relationship models, **1:**525
Hydantoins:
  anticonvulsant applications, **8:**141
  toxicophore reactive metabolites, **2:**322
Hydgromycin, chemical structure, **7:**419
Hydralazine, toxicophore reactive metabolites, alkynes, **2:**320–321
Hydrates:
  cocrystal design, **3:**196
  crystal engineering, defined, **3:**189
  oral dosage systems, **3:**27–28
  salt compounds, **3:**382
Hydration, drug metabolism, **1:**421
Hydrazines:
  drug metabolism, conjugation reactions, **1:**439
  structural genomics, **1:**585–586
  toxicophore reactive metabolites, **2:**320–321
Hydrocortisone:
  metabolic pathways, soft drug design, **2:**90–91, 94–96
  prosoft drug design, **2:**129–130
  steroidogenesis, **5:**20–22
Hydrogenation reactions, agitation functions, pilot plant development, **3:**8–9
Hydrogen bonds:
  ADMET structural properties, **2:**51–52
  estrogen receptor topography, **5:**234–236
  glucocorticoid anti-inflammatory compound alteration, C-15 alterations, **5:**95
  kinase inhibitors, structure-based design, protein targeting, **6:**353–354
  quantitative structure-activity relationship:
    indicator variables, **1:**22–23
    multidimensional models, **1:**40–43
    receptor interactions, **1:**29–31
  quantum mechanics/molecular mechanics, binding affinity evaluation, **2:**718–719
Hydrogen charge (HC) descriptors, quantitative structure-activity relationship, **1:**27–28
Hydrogen peroxide, topical antimicrobial compounds, **7:**496–497
Hydrogen sulfide, **5:**309–316
  alternative delivery systems, **5:**315–316
  inhalation therapy, **5:**312–313
  injectable formulations, **5:**313–314
  prodrugs and combined releasing molecules, **5:**314–315
  production and biological effects, **5:**266, 309–312
  therapeutic approaches, **5:**267–269
Hydrolases:
  drug metabolism, **1:**412–413
  high-throughput screening, enzymatic assays, **3:**425–426
Hydrolysis, drug metabolism, **1:**421
Hydrolytic degradation, soft corticosteroids, **2:**90
Hydrolytic lability, soft drugs, computer-aided techniques, **2:**132–133
Hydrophobic interaction chromatography (HIC), protein therapeutics:
  analytic development, **3:**320
  downstream processing, **3:**304

Hydrophobicity:
  cocaine structure-activity relationship, **8:**111–113
  docking techniques, **2:**634–636
  entropy and, **2:**631
  oral drug physicochemistry, **3:**37–39
  quantitative structure-activity relationship, **1:**8–19
  nonlinear models, **1:**34–36
  partition coefficient:
    calculation methods, **1:**14–19
    measurement methods, **1:**12–14
  quantum mechanics/molecular mechanics, binding affinity evaluation, **2:**717–719
Hydroquinone derivatives, geldanamycin Hsp-90 inhibitor, **6:**390–391
11β-Hydroxysteroid dehydrogenase-1 inhibitors, type 2 diabetes therapy, **4:**22–25
Hydroxamic acid inhibitors:
  ACE inhibitor design, **4:**274–275
  histone deacetylase inhibitors, **6:**53–66
    current clinical trials and drug development, **6:**53–54
    ITF-2357, **6:**60–61
    JNJ-26481585 second-generation HDACi, **6:**64–66
    NVP-LAQ-824 and LBH-589, **6:**55, 57–58
    PCI-024781, **6:**60–62
    PXD-101, **6:**58–59
    R306465 first-generation HDACi, **6:**62–64
    SB939, **6:**59–60
  stroke therapy, **8:**488–491
    aryl and heteroaryl linker, **8:**490–491
    cinnamoyl linker, **8:**488–490
    linear side-chain linker, **8:**488
    suberoyl hydroxamic acid analogs, **6:**53, 55–56

Hydroxamic acid inhibitors
    (*Continued*)
  stroke therapy:
    bis-hydroxymates, **8:**507
    histone deacetylase
        inhibitors, **8:**488–491
      aryl and heteroaryl linker,
          8:490–491
      cinnamoyl linker,
          8:488–490
      linear side-chain linker,
          8:488
    matrix metalloproteinase
        inhibitors, **8:**502–507
    oxalic acid hydroxymates,
        8:507
    reverse hydroxyamic acids,
        8:507
Hydroxyanilines, toxicophore
    reactive metabolites,
    2:303–311
Hydroxyapatite chromatography,
    protein therapeutics,
    downstream
    processing, **3:**304
Hydroxybupropion, structure-
    activity relationship,
    8:251, 254–255
7-Hydroxychloropromazine,
    pharmacokinetics,
    biodistribution, and
    drug-drug
    interactions,
    8:185–186
Hydroxy derivatives, anabolic
    agents, **5:**182–183
Hydroxyethylene renin
    inhibitors,
    4:245–247
Hydroxyethyl ureas (HEUs), γ-
    secretase inhibitors,
    8:334–335
Hydroxylamines/
    hydroxylamides, drug
    metabolism:
  glucuronidation/glucosidation,
    1:428–430
  sulfonation reactions,
    1:426
Hydroxylated polychlorinated
    biphenyls (OH-PCBs),
    structure-based
    design, **2:**700
Hydroxylation:
  antiglucocorticoids, **5:**113
  steroid biotransformation, **5:**7,
    9–10

Hydroxyl transition-state isostere
    renin inhibitors,
    4:243–244
Hydroxymate-containing ligands,
    nuclear magnetic
    resonance, matrix
    metalloproteinases,
    2:412–414
Hydroxymethyl derivatives,
    trinem formation,
    7:349–350
Hydroxymethylglutaryl
    coenzyme A (HMG-
    CoA):
  cholesterol biosynthesis, **5:**13,
    15–16
  Alzheimer's disease, **8:**406
  reductase inhibitors (*See also*
    Statin drugs)
    Alzheimer's disease therapy,
        preclinical/clinical
        testing, **8:**359–361
    anticholesterolemic
        mechanisms,
        2:231–234
    combined compounds,
        2:234–235
    chemistry-driven drug
        discovery, **1:**537–538
    chemotype evaluation,
        1:496–497
    chirality in, **1:**150–155
    LDL cholesterol lowering
        mechanisms,
        4:306–310
Hydroxynaphtoquinone
    antimalarial agent,
    7:619–620
N-(4-Hydroxyphenyl)4-Oxo-
    retinamide analog,
    5:483
N-(4-Hydroxyphenyl)retinamide
    (4-HPR)
  adverse effects, **5:**476–478,
    484–485
  analogs, **5:**480–483
  chemical structure, **5:**464
  mechanism of action, **5:**474–476
  pharmacology/metabolism,
    5:478–480
  therapeutic applications,
    5:464–475
    cancer therapy, **5:**464–473,
        478–480
    cystic fibrosis, **5:**473–474
    diabetes mellitus, **5:**474
    eye diseases, **5:**474

    pediatric neuronal ceroid
        lipofuscinosis,
        5:474–475
3=Hydroxy-2-phosphonyl-
    methoxypropyl
    (HPMP) derivatives,
    acyclic nucleoside
    phosphates,
    7:232–239
17α-Hydroxyprogesterone
    caproate,
    biotransformation,
    5:229–230
4-Hydroxyproline, teleprevir anti-
    HCV agent
    modifications,
    7:181–186
Hydroxypropylamines,
    plasmepsin
    antimalarial agents,
    7:640–644
Hydroxypyridinones, stroke
    therapy, **8:**508
9-Hydroxy risperidone,
    pharmacokinetics,
    8:195–196
5-Hydroxytryptamine 2A
    receptor antagonists,
    hit seeking
    combinatorial
    libraries, **1:**311–312
Hymenialdisine, Alzheimer's
    disease, tau-targeted
    therapies, **8:**420–422
Hyoscyamine, muscarinic
    antagonist, **8:**72
Hyperalgesia modeling,
    prostaglandin/
    thromboxane EP_3
    receptor antagonists,
    5:858, 861
Hypercalcemia, vitamin D
    overdose, **5:**656
Hypercarotenosis, **5:**648
Hypercholesterolemia, single
    gene
    pharmacogenetics,
    1:195–196
Hyperglycemia, type 2 diabetes,
    4:1–2
Hypericin, Alzheimer's disease,
    tau-targeted therapy,
    8:425–426
Hyperlipidemia, 13-*cis*-retinoic
    acid therapy, **5:**393
Hyperpolarization-activated
    cyclic nucleotide

cation (HCN) channels, anticonvulsants, 8:125–126
Hypertension. See also Antihypertensives
angiotensin-converting enzyme inhibitors, 4:282
multitarget drug development, ACE-plus and angiotensin-1 receptor (AT$_1$) antagonist, 1:257, 259–260
renin inhibitors:
aliskiren clinical trials, 4:251–252
anticipated hypertension and end-organ benefit, 4:242–243
renin targeting, 4:240
Hyperthyroidism:
obesity effects, 4:198
thyroid hormone receptor antagonists, 4:199
Hypertriglyceridemia:
bexarotene therapy, 5:451
NRX 194204 RXR selective analog, 5:464
13-cis-retinoic acid therapy, 5:393
Hypervitaminoses:
biotin, 5:691
folic acid, 5:696–697
niacin, 5:677–678
pantothenic acid, 5:685
pyridoxine, 5:684
riboflavin, 5:673
thiamine, 5:670
vitamin A, 5:647–648
vitamin B$_{12}$ (cobalamin), 5:700–701
vitamin C (ascorbic acid), 5:703–705
vitamin D, 5:658
vitamin E, 5:662, 664
vitamin K, 5:668
Hypothyroidism:
bexarotene therapy, 5:450–453
epidemiology, 4:197
N-(4-hydroxyphenyl) retinamide, 5:464
Hypoxia-activated bioreductive prodrugs, DNA-targeted therapeutics, tumor-activation, 6:114–120

Hypoxia inducible factor (HIF) pathway, brain tumor therapy, 6:253–255

Ibuprofen:
drug metabolism, conjugation reactions, 1:433–434
migraine therapy, 8:268
IC$_{50}$ values, receptor binding assays, 2:521–523
ICI 118551 compound, receptor-based drug design, efficacy parameters, 2:502–503
ICI-164384 analog, thyromimetic structure-activity relationship, 4:213–214
Idarubicin, serum albumin binding, 3:452–454
Idebenone, cognitive enhancement, 8:15–16
Idoxuridine, antimetabolites, anti-DNA virus agents, 7:222–225
Ifenprodil, stroke therapy, 8:453
IFREDA algorithm, in silico screening, protein flexibility, backbone mobility, 2:868–870
Ignition prevention, commercial-scale operations, 3:11
IKK inhibitors, chronic obstructive pulmonary disease, 5:787–789
Ileal bile acid binding protein (iBABP), drug delivery, 2:468–475
ILS-920 nitrosobenzene adduct, sirolimus compound, 5:1032–1033
Imatinib:
chronice myelogenous leukemia therapy, small-molecule imatinib mesylate, resistance to, 6:364–366
chronic myelogenous leukemia therapy, 6:298–301
multitarget drug development, cancer therapy, 1:261, 263–267
selective toxicity, 2:562–563

structure-based design:
drug-resistant mutation, 6:354–356
multiconformational catalytic domain, 6:346–349
protein targeting, 6:351–354
Imidazoacridinones, topoisomerase II inhibitors, 6:102
Imidazoles:
co-crystalline structures, 2:354–356
γ-secretase modulators, 8:359
Imidazolidine-2,4-diones:
anticonvulsant applications, 8:141
thyromimetic structure-activity relationship, heterocyclic rings, 4:210–211
Imidazoline derivatives:
α-adrenergic receptor targeting, 8:247
bupropion interaction, 8:255
Leishmaniasis therapy, 7:591–593
P2X$_7$ nucleotide receptor antagonists, 1:324–326
ion channels, 1:337
Iminoquinone, toxicophore reactive metabolites, aniline metabolism, aromatic ring oxidation, 2:304–306
Iminostilbenes, anticonvulsant applications, 8:142–144
Imipenem, occurrence, structural variations and chemistry, 7:333–335
Imipramine:
chemical structure, 8:222
historical background, 8:248
migraine prevention, 8:318–319
monoamine transporter inhibition, 8:238–241
pharmacokinetics, 8:235–236
serotonin receptor targeting, 8:241–242
side effects and adverse reactions, 8:231–232
Imiquimod, Leishmaniasis therapy, 7:591–593

Immediate-release drug delivery, oral administration and bioavailability, 2:771–787
   area under the curve measurements, 2:785–786
   concentration profile calculation, 2:773–774
   dimensionless numbers, 2:772
   drug compound characteristics, 2:776
   intravenous delivery vs., 2:786–787
   master curves with dimensionless numbers, 2:781–784
   multiple dose administration, 2:776–784
   nomenclature, 2:771–772
   pharmacokinetics, plasma drug profile, 2:774–776
   repeated dose principles and parameters, 2:777–781
   transport principle, 2:772–773
Immobilized artificial membrane (IAM) method, quantitative structure-activity relationship, partition coefficients, hydrophobic interactions, 1:13–14
Immune function:
   Am80 effects, 5:418
   chemotherapeutic agent interface with, 7:503
   dipeptidyl peptidase 4 inhibitors, 4:40–41
   multiple sclerosis therapy, 5:567–568
   NRX 194204 RXR selective analog, 5:463
   overview, 5:1002–1006
Immunoadhesins, recombinant DNA technology, 1:540–541
Immunoassays, protein therapeutics, analytic development, 3:321
Immunoglobulins:
   immune function, 5:1002–1006
   membrane proteins, receptor-mediated endocytosis, 2:450–451

Immunomodulators, multiple sclerosis therapy, opportunistic approaches, 5:574
Immunophilins, nuclear magnetic resonance studies, 2:410–412
Immunoproteasome-specific inhibitors:
   basic properties, 6:175
   mechanism of action, 6:184–185
Immunosuppressive drugs:
   multiple sclerosis therapy:
      current therapies, 5:568–572
      opportunistic approaches, 5:574
   pipeline projects, 5:1053–1060
   recombinant DNA technology, targeting mechanisms, 1:555–556
   soft drug design:
      calcineurin inhibitors, 2:119–121
      loteprednol etabonat and analogs, 2:96
   transplantation rejection prevention:
      azathioprine, 5:1050–1053
      cycolsporin A, 5:1006–1013
      deforolimus, 5:1037
      diaryl urea inhibitors, 5:1042–1044
      everolimus, 5:1035–1037
      immune system overview, 5:1002–1006
      inosine monophosphate dehydrogenase inhibitors, 5:1038–1050
      mizoribine, 5:1044–1045, 1049–1050
      mycophenolate mofetil, 5:1047–1049
      mycophenolate sodium, 5:1049–1050
      mycophenolic acid analogs, 5:1039–1042
      pimecrolimus, 5:1023–1024
      research background, 5:1001–1002
      ribavarin, 5:1044–1046
      sirolimus, 5:1024–1037
      tacrolimus, 5:1013–1024
      temsirolimus, 5:1037
      zotarolimus, 5:1037

Immunotherapy approaches, Alzheimer's disease, 8:366–380
   advantages/disadvantages, 8:369–370
   anti-Aβ antibody mechanisms, 8:367–368
   bapineuzamab, 8:370–378
   Aβ immunization, 8:366, 368–369
   current clinical trials, 8:370–380
   gammaguard liquid, 8:379
   MABT5102A, 8:379–380
   PF-4360365, 8:378–379
   preclinical/clinical trials, 8:369
   R1450, 8:379
   solaneuzmab, 8:378
   TgAPP behavioral and learning deficits, 8:366–367
Improper use principle, trademark preservationand, 3:177
Impurities, protein therapeutics, downstream processing, 3:301–303
Inactive metabolites, glucocorticoid anti-inflammatory compounds, C-16 alterations, 5:96–99
Inactive metabolites, soft drug development, 2:80
   anticholinergics, 2:107–108
INCB13739 inhibitor, type 2 diabetes therapy, 4:24–25
Inclusion bodies (IBs), protein purification, E. coli systems, 3:304–306
Inclusion complexes, oral drug physicochemistry, class II drugs, 3:47
Incremental construction approach, docking and scoring computations, 2:605–608
Incretin, type-2 diabetes therapy:
   dipeptidyl peptidase 4 inhibitors, 4:40–42
   GPR119 agonists, 4:20–21
Indacaterol, chronic obstructive pulmonary disease bronchodilators, 5:771–772

Indanylspiroazaoxindole-based calcitonin gene-related peptide antagonist, migraine therapy, **8:**287–288
Indazoles, liver X receptor agonists, **4:**152–154
Indeglitazar, α/γ/δ-pan agonists, **4:**141–145
Indicator variables, quantitative structure-activity relationship, **1:**21–23
Indinavir, **7:**12–20
Indinavir sulfate:
  binding interactions, **7:**14, 19
  chemical structure, **7:**12, 14
  efficacy and tolerability, **7:**17–18
  optimization, **7:**12, 14, 16–18
  pharmacokinetics, **7:**14, 17–18, 20–21
  resistance profile, **7:**18, 20
Indirect anticoagulants, **4:**371–374
  antithrombin conformational activation, **4:**390–392
  enzymatic engineering, **4:**392–393
  limitations, **4:**371–374
  molecular mechanism of action, **4:**374–380
  structural properties, **4:**368–370
Indirect assays, high-throughput screening, **3:**408–409
Indirubin, Alzheimer's disease, tau-targeted therapies, **8:**420–422
Indole-2-carboxylic derivatives, stroke therapy, **8:**452–453
Indole acetic acid:
  CRTH2 receptor antagonist, **5:**910–914, 918
  DP$_1$ receptor antagonists, **5:**902–903
Indole derivatives:
  chronic obstructive pulmonary disease:
    CXCR2 receptor antagonistst, **5:**784–785
    IKK inhibitors, **5:**788–789
  prostaglandin/thromboxane EP$_3$ receptor antagonists, **5:**861, 863–869

side chains, thyroid hormone receptor β-selective agonists, structure-based design, **2:**689
thyromimetic structure-activity relationship, **4:**212–213, 218
Indoline-2-one derivatives, stroke therapy, **8:**514
Indolo-imidazoles, MDM2/MDM4 antagonists, **2:**357–359
Indomethacin, CRTH2 receptor agonist, **5:**909–910
Induced fit model:
  binding pockets, kinesin spindle protein inhibitors:
    EMD-534085, **6:**210–211
    monastrol, **6:**193–197
  docking and scoring techniques, **2:**640
  G-protein coupled receptor, homology modeling, **2:**286–287
  receptor-based drug design, allosteric transition, **2:**499–501
Inducible nitric oxide synthase (iNOS), migraine therapy, **8:**298–302
Inequitable conduct defense, infringement of patent claims, **3:**155–156
Inflammation:
  carbon monoxide inhalation therapy, **5:**303
  farnesoid X receptors, **4:**160
  multiple sclerosis therapy, **5:**566–568
  multitarget drug development, COX-plus, **1:**267
Inflammatory bowel disease (IBD), 13-*cis*-retinoic acid side effects, **5:**395
Informatics, combinatorial chemistry, **1:**349–350
Informational macromolecules, solid-phase organic synthesis, **1:**280–294
  lipid arrays, **1:**293–294
  nucleoside arrays, **1:**292–293
  oligosaccharide arrays, **1:**291–292
  peptide arrays, **1:**280–291
Information Disclosure Statement (IDS), U.S.

PTO procedures, **3:**140–142
Information protection, trade secrets limitations and, **3:**180–181
Infringement claims:
  patents:
    alternative dispute resolution, **3:**164–165
    appeal process, **3:**163–164
    claim construction, **3:**162
    commencement of proceedings, **3:**159–160
    defenses to, **3:**154–157
    determination of, **3:**152–154
    discovery process, **3:**160–161
    licensee rights, **3:**150–152
    remedies for, **3:**157–159
    summary judgment, **3:**161
    trial, **3:**162–163
    U.S. trial and appellate courts, **3:**148–150
  trademarks, **3:**173–177
Infusion rate changes, repeated short infusions, constant flow rate over finite period, **2:**770–771
Inhalation therapy:
  carbon monoxide, **5:**299–305
  chronic obstructive pulmonary disease, corticosteroids, **5:**773–780
  hydrogen sulfide, **5:**312–313
  nitric oxide, **5:**271, 274–278
Inhibition Detection Limit (IDL), high-throughput screening, assay configuration, **3:**420
Inhibitor binding (IB). *See also* Binding sites
  HIV protease inhibitors, **7:**2–3
  serum albumin structural survey, **3:**440–451
    anthracyclines, **3:**452–454
    bilirubin binding, **3:**442–451
    podophyllotoxin derivatives, **3:**454–456
    sulfonamides and sulfones, **7:**524–527
  type 2 diabetes therapy, dipeptidyl peptidase 4 inhibitors, **4:**47

Inhibitor of apoptosis proteins
(IAPs),
peptidomimetics,
1:230–240
Inhibitor-resistant β-lactamases,
classification,
7:277–279
Inhibitors (generally). *See also*
HIV protease
inhibitors (HIV-1
protease inhibitors);
specific inhibitors, e.g.,
Protease inhibitors
antiparasitic/antiprotozoal
lipid metabolism
inhibitors,
7:580–581
carboxylesterases, **3:**235–240
INH-NAD(P) adduct, isoniazid
resistance, **7:**740–743
Innate immunity, high-density
lipoproteins,
4:334–335
Inorganic diphosphate analogs,
anti-DNA virus
agents, **7:**240–241
Inosine monophosphate
dehydrogenase
(IMDPH) inhibitors:
acridone-based inhibitors,
5:1046–1047
ADMET properties:
mizoribine, **5:**1049
mycophenolic acid analogs,
5:1047–1049
biosynthetic pathways,
5:1039–1042
chemical structure, **5:**1038
diaryl urea-based inhibitors,
5:1042–1044
enteric-coated compounds,
5:1050
IMPDH-MPA structure, **5:**1039
mechanism of action,
5:1038–1039
mycophenolic acid analogs,
5:1039–1042,
1047–1050
nucleoside inhibitors,
5:1044–1046
mizoribine, **5:**1044,
1049–1050
ribavirin and C2-MAD,
5:1044–1046
research background, **5:**1038
therapeutic applications,
5:1049–1050

INOS inhibitors:
migraine therapy, **8:**300–302
stroke therapy, **8:**495–502
Inositol 1,4,5-triphosphate
receptor-associated
cGMP kinase
substrate (IRAG),
nitric oxide effects,
5:271–278
INPHARMA method, drug
screening, nuclear
Overhauser effect
analysis, **2:**429
In-process controls (IPCs):
bench-scale experiments, **3:**5
commercial-scale operations,
3:11
protein therapeutics,
3:317–320
Insect cells, protein therapeutics,
3:292–294
Insecticides, soft drug design,
2:106–108
In silico screening. *See also*
Virtual screening
ADMET property assessment,
2:64
antitrypanosomal polyamine
metabolism,
7:589–590
BACE (β-site cleavage enzyme)
inhibitors, **8:**352–353
Biopharmaceutics
Classification System,
provisional
classification,
3:359–361
BioPrint profiling, **1:**501–502
chemogenomics, ligand-based
data analysis,
2:584–585
chimeric radicicol and
geldanamycin Hsp90
inhibitors, **6:**406–410
drug metabolism,
biotransformation
prediction, **1:**445–447
prodrug design, **3:**230
protein flexibiligy:
energetic contributions,
binding affinity, **2:**870
entropic contributions,
binding affinity, **2:**871
future research issues,
2:871–872
mobility determination,
2:861–865

flexibility and geometry-
based analysis,
2:863–864
harmonic analysis,
2:864–865
molecular dynamics
simulations,
2:862–863
protein-ligand docking,
2:865–870
backbone mobility,
2:868–870
side-chain conformational
analysis, **2:**867–868
receptor plasticity,
2:860–870
research background,
2:859–860
structural genomics research,
1:582
toxicophore filtering, **2:**327
Insomnia, drug development and
discovery, **5:**716–721
orexin receptor antagonists,
5:720–723
overview, **5:**711
serotonergic antagonists and
inverse agonists,
5:718
Inspections, FDA guidelines, **3:**81
Institutional Review Board (IRB)
(FDA), guidelines,
3:80–81
Insulin-like growth factor-1
receptor kinase
inhibitors, survival
signaling pathways,
6:321–322
Insulin-related therapy,
Alzheimer's disease,
8:414–417
Insulin secretagogues, type 2
diabetes therapy,
4:8–10
INT-131/AMG-131/T-131
compound, selective
PPARγ modulators,
PPARγ partial
agonists, **4:**108,
111–115
Integral membrane proteins,
analysis of, **2:**459
Integrase. *See* HIV integrase
Integrated predictive modeling
workflow,
quantitative
structure-activity

relationship modeling, **1**:516–525
applicability domains, **1**:521–522
combinatorial modeling, **1**:518–520
hybrid chemical biological descriptors, **1**:525
model ensembles and consensus prediction, **1**:522–525
model validation, **1**:520–521
Integration chemistry, HIV integrase and, **7**:78–79
Integrin antagonists, osteoporosis antiresorptive therapy, **5**:730
Integrin-linked kinase (ILK), brain cancer therapy, **6**:250
Integrins:
brain tumor invasiveness, **6**:241–242
hit seeking combinatorial libraries, **1**:341–343
platelet adhesion, **4**:412–413
recombinant DNA technology, cellular adhesion proteins, **1**:560–561
Intellectual property law. *See also* Patents
drug discovery and biotechnology:
future research issues, **3**:186
miscellaneous protections, **3**:185–186
patents:
enforcement, **3**:148–165
alternative dispute resolution, **3**:162–163
claim construction proceedings, **3**:162
defenses to infringement, **3**:154–157
discovery phase, **3**:160–161
geographic scope, **3**:148
infringement determination, **3**:152–154
parties to suit, **3**:150–152
proceedings commencement, **3**:159–160
remedies for infringement, **3**:157–159
summary judgment, **3**:161
trial and appeal processes, **3**:148–150, 162–164
global patent protection, **3**:107–108, 165–171
international agreements, **3**:165–168
international laws and regulations, **3**:169–171
PCT patent practice, **3**:168–169
protection and strategy, **3**:105–118
absolute novelty principle, **3**:110–111
application publication, **3**:114–115
first to invent *vs.* first to file, **3**:108–110
global strategy, **3**:107–108, 165–171
patent term, **3**:111–114
PTO prosecution strategy, **3**:115–118
trade secrets and, **3**:182–183
requirements, **3**:118–148
corrections, **3**:146–148
interference, **3**:142–146
new and nonobvious inventions, **3**:133–138
patentable subject matter, international requirements, **3**:121–123
patentable subject matter, U.S. and international requirements, **3**:118–123
provisional applications, **3**:127–133
PTO procedures, **3**:138–142
specifications, **3**:123–127
research background, **3**:101–105
trademarks, **3**:171–179
global rights, **3**:177–178
Lanham Act protections, **3**:178–179
as marketing tools, **3**:172
oppositions and cancellations, **3**:176–177
preservation of rights through proper use, **3**:177
registration process, **3**:174–176
selection criteria, **3**:172–174
trade secrets, **3**:179–184
defined, **3**:180
enforcement, **3**:181–182
Freedom of Information Acts, **3**:183–184
global protection, **3**:184
patent protection and, **3**:182–183
protection requirements, **3**:180–181
Intent requirements, infringement of patent claims, **3**:155–156
Intent-to-use provisions, trademark registration, **3**:175–176
Intercellular adhesion molecule (ICAM), recombinant DNA technology, cellular adhesion proteins, **1**:560–561
Interference in patent claims, **3**:142–146
Interferon-beta, multiple sclerosis therapy, **5**:571
Interleukin-2 signal transduction pathway, CP-690,550 immunosuppressive compound, **5**:1055–1057
Interleukin-5 inhibitor, soft drug design, **2**:121–122
Interleukin-8, chronic obstructive pulmonary disease therapy, **5**:784–785
Interleukin-11 (IL-11), structure and properties, **4**:578–580

Inter-ligand nuclear Overhauser effect (ILOE) spectroscopy, screening procedures, **2**:429
Intermediate exchange conditions, nuclear magnetic resonance, macromolecule-ligand interactions, **2**:399–402
Intermolecular interactions: quantitative structure-activity relationship, **1**:29–31
three-dimensional pharmacophores, bound ligands, **1**:464–465
Internal ribosomal entry site (IRES), protein therapeutics, expression vectors, **3**:294–295
Internal ribosome entry site (IRES), hepatitis C viral expression, **7**:172
Internal validation, quantitative structure-activity relationship, **1**:47–48
International Conference on Harmonization (ICH), drug development guidelines, **3**:97
International patent agreements, **3**:165–168
International Trade Commission (ITC) (U.S.), patent infringement suits, **3**:149–150
Intestinal absorption:
ADMET structural properties, solubility, **2**:54–55
apical sodium-dependent bile acid transporter enhancement, **2**:472–473
glucocorticoid anti-inflammatory compounds, **5**:41, 43, 45–46
oral dosage forms, class V drugs, **3**:53
permeability vs. solubility/dissolution, **3**:374–375
Intestinal excretion, permeability, ADMET interactions, **3**:373–374
Intestinal peptide transporter (PepT1), drug delivery, **2**:463–467
Intoplicine, dual topoisomerase I/II inhibitors, **6**:103–105
Intra-articular administration, glucocorticoid anti-inflammatory compounds, **5**:51–53
Intracellular signaling:
carbon monoxide therapeutic efffects, **5**:297–299
hydrogen sulfide, **5**:310–312
platelet activation, **4**:415–417
endogenous regulatory mechanisms, **4**:418
receptor-based drug design, **2**:512
thrombin receptor activation, **4**:439
Intramolecular dynamics: macromolecule-ligand interactions, nuclear magnetic resonance studies, **2**:409
neomycin/gentamicin aminoglycosides, RNA targets, **5**:973–978
Intraocular pressure (IOP):
eye-targeted chemical delivery systems, oxime/methoxime analogs, **2**:158–160
glaucoma, disease pathology and etiology, **5**:588–591
soft drug targeting:
β-blockers, **2**:83–86
cortienic acid-based steroid, **2**:95–96
therapeutic reduction, **5**:591–592
in vivo preclinical screening models, **5**:590–591
Intravenous drug delivery. *See also* Parenteral drug delivery
bioavailability:
continuous intravenous infusion, **2**:760–762
oral drug delivery vs., **2**:786–787
pharmacokinetic stage, **2**:747
repeated bolus injections, **2**:754–760
repeated short infusions, constant flow rate over finite period, **2**:763–771
area under curve measurements, **2**:765–766
dosage and time period changes, **2**:768–770
infusion-elimination stages, **2**:763–765
infusion rate changes, **2**:770–771
patient adaptation and intervariability, **2**:766–768
single dose (bolus injection), **2**:751–754
glycoprotein IIb/IIIa receptor antagonists, **4**:445–447
protein therapeutics, **3**:316
Intrinsically disordered proteins, conformational changes, **1**:386–387
Intrinsic efficacy, receptor-based drug design, central nervous system, **8**:7
Intrinsic solubility, ADMET structural properties, **2**:52–53
Invasins, receptor-mediated endocytosis, **2**:451–452
Inverase. *See* Saquinavir mesylate (SQV) protease inhibitor
Inverse agonist, receptor-based drug design, efficacy parameters, **2**:502–503
Investigational new drugs (INDs):
FDA guidelines, **3**:72–74
additional and relevant information guidelines, **3**:76
previous human experience, **3**:75–76
hematopoietic agents, **4**:587–588
In vitro studies:
ADMET applications, **2**:47–48
property assessment, **2**:64–65
carbapenems, **7**:350–353

monobactam activity, 7:329–331
opioid efficacy assessment, 8:583–585
oral drug delivery, sustained-release dosage forms, 2:787–791
  diffusion parameters, 2:792–798
    boundary conditions, 2:792–793
    cubic side calculations, 2:796
    cylindrical radius/height calculations, 2:796
    dimensionless numbers, 2:795
    dissolution tests, 2:793
    erosion-controlled release vs., 2:801
    parallelepiped side calculations, 2:795–796
    shape-volume effects, 2:796
    spherical release kinetics, 2:795–798
    thin film kinetics, 2:794–795
  erosion-controlled release, 2:798–801
    cubic side calculations, 2:799–800
    cylindrical radius/height calculations, 2:799
    diffusion-release vs., 2:801
    parallelepiped side calculations, 2:799
    sheet thickness kinetics, 2:799
    spherical radius calculations, 2:799
  nomenclature, 2:791–792
prodrug design, ADMET issues, 3:230
recombinant DNA technology, selection (SELEX), 1:549–550
serum albumin drug transport, 3:458–462
thyromimetic structure-activity relationship, 4:204
transdermal drug delivery, metropolol experiment, 2:832–833

xenobiotic metabolism, 1:407–408
In vivo studies:
  ADMET applications, compound selection and planning, 2:50–51
  age-related macular degeneration screening models, 5:610–613
  glaucoma screening models, 5:590–591
  linezolid antibacterial activity, 7:542
  opioid efficacy assessment, 8:585–587
  oral drug delivery, sustained-release dosage forms, 2:787–791
  quantitative structure-activity relationship, 1:59–60
  recombinant DNA technology, phage display, 1:551–552
  renin inhibitors, 4:242
  side effect analysis, BioPrint® database, 1:492–493
  thyromimetic structure-activity relationship, 4:204
  tigecycline analogs, 7:414–415
  transdermal drug delivery, metropolol experiment, 2:833
Iodine:
  antimicrobial agents, 7:492
  thyroid hormone biosynthesis, 4:192–193
Ion channels:
  hit seeking combinatorial libraries, 1:336–337
  ligands:
    allosteric protein model, 1:393
    combinatorial chemistry, 1:336–337
    enantiomer activity, 1:132–133
    noncompetitive blockers, stroke therapy, 8:450–451
  opioid receptors, signal transduction, 8:580–581
  thyroid hormone effects, 4:195–196

voltage-gated ion channels, anticonvulsants, 8:122–126
  calcium channels, 8:124
  chloride channels, 8:126
  hyperpolarization-activated cyclic nucleotide cation channels, 8:125–126
  potassium channels, 8:124–125
  sodium channels, 8:123–124
Ion exchange chromatography (IEC), protein therapeutics:
  analytic development, 3:319
  downstream processing, 3:304
Ionizability:
  ADMET structural properties, 2:52
  docking and scoring computations, 2:636–637
Ionization constants:
  oral drug physicochemistry, 3:35–37
  quantitative structure-activity relationship, 1:2–8
Ion mobility mass spectrometry (IMMS), drug development and, 1:122
Ion mobility spectrometry (IMS), drug development and, 1:122
Ionotropic glutamate receptors:
  anticonvulsants, ligand-gated ion channels, 8:129
  migraine therapy, antagonists, 8:291–297
Ion pairing, oral drug delivery, class III drugs, 3:51–52
Iontophoresis, congestive heart failure therapy, hemodynamic function biomarkers, 4:506
Ion trapping time-of-flight (TOF) instrumentation, mass spectrometry development, 1:98
Ioxistatin, stroke therapy, cathepsin B inhibitors, 8:473–474

IPI-504 Hsp90 inhibitor, geldanamycin derivatives, **6:**390–391
IPI-926 compound, brain tumor therapy, hedgehog signaling pathway, **6:**260
Ipratropium, muscarinic antagonist, **8:**72
Ipratropium bromide, chronic obstructive pulmonary disease bronchodilators, **5:**767
Iproniazid:
  as antidepressant, **8:**248
  antitubercular agents, history, **7:**738
  toxicophore reactive metabolites, hydrazine metabolism, **2:**320–321
Ipsaspirone, serotonin receptor targeting, **8:**242–244
IQU0528 HIV reverse transcriptase inhibitor, **7:**150–152
Irinotecan (CPT-11):
  meningioma therapy, **6:**237
  structure and applications, **6:**26
Iron acquisition, antitubercular agents:
  mycobactin targeting and biosynthesis, **7:**730–731
  terpenoid targeting and biosynthesis, **7:**731–732
Iron-dependent repressor (IdeR), antitubercular agents, mycobactin targeting and biosynthesis, **7:**731
Irreversible inhibitors, stroke therapy:
  calpains, **8:**464–465
  caspase, **8:**467–469
  cathepsin B inhibitors, **8:**473–477
Isatin-chloroquine conjugate, antimalarial cell cycle regulation, falcipain inhibitors, **7:**635–637
Isatin-derived caspase inhibitors, stroke therapy, **8:**470–471
Ischemia-reperfusion injury:

carbon monoxide inhalation therapy, **5:**301
phosphodiesterase inhibitors, **5:**716
Ischemic cascade, stroke epidemiology and pathology, **8:**447–449
Isobenzofuranone, mycophenolic acid analog modification, **5:**1039–1040
Isoclothiapine, structure-activity relationship, **8:**182–183
Isoclozapine, structure-activity relationship, **8:**182–183
Isocyanates, drug metabolism, glutathione conjugation, **1:**438
Isodideoxynucleosides, anti-DNA virus antimetabolites, **7:**226–227
Isoellipicinium derivatives, antimalarial topoisomerase inhibitors, **7:**650–652
Isoindolone derivatives, protein-protein interactions, **2:**349–352
Isolated receptor interactions, quantitative structure-activity relationship applications, **1:**51–57
Isoleucine derivatives, dipeptidyl peptidase 4 (DPP-4) inhibition, **4:**50–54
Isoleucyl thiazolidide, type 2 diabetes therapy, **4:**42
Isomerases, high-throughput screening, enzymatic assays, **3:**427–428
Isomers and isomerization:
  protein therapeutics formulation and delivery, **3:**312–314
  xenobiotic metabolism, **1:**406–407
Isomethodone, structure-activity relationship, **8:**621–623
Isoniazid:
  antitubercular agents:
    ADMET properties, **7:**742–743
    adverse effects, **7:**743

    antibacterial activity, **7:**739
    drug interactions, **7:**743–744
    history, **7:**738
    mechanism of action, **7:**739–740
    mycolic biosynthesis, **7:**723
    resistance to, **7:**740–742
    structure-activity relationship, **7:**738–739
  selective toxicity, **2:**557
  vitamin B_6 interaction, **5:**681–684
Isopenicillin N, biosynthesis, **7:**295
Isopentenyl pyrophosphate, cholesterol biosynthesis, **5:**15–16
Isophthalamide ligands, BACE (β-site cleavage enzyme) inhibitors, **8:**347–351
Isopodophyllotoxins, enzyme inhibition, **6:**29–31
Isoprenoids, apricoplast targeting, antimalarial agents, **7:**665–668
Isopropanol, antimicrobial agents, **7:**492–493
Isoquine antimalarial agent, **7:**625–627
Isoquinolines, antimalarial cell cycle regulation, falcipain inhibitors, **7:**635–637
Isoquinolinones, poly(ADP-ribose)polymerase inhibitors, **6:**164–167
Isoquinolones:
  calpain inhibitors, stroke therapy, **8:**464
  protein-protein interactions, **2:**351–352
Isosorbide mononitrate, nitric oxide donors and prodrugs, **5:**278–280
Isothermal titration calorimetry (ITC), ligand-macromolecular interactions, structure-based thermodynamic analysis, **2:**704–712
Isothiourea analogs, stroke therapy, **8:**497

Isotope labeling:
nuclear magnetic resonance:
macromolecule-ligand interactions:
slow exchange conditions, 2:397–398
titrations, 2:403–404
matrix metalloproteinases, 2:412–414
receptor-based drug design, 2:392–394
PET/SPECT systems, 8:739–740
Isotretinoin, vitamin A pharmacology, 5:649
Isovaleramide, migraine prevention, 8:316
Isoxazole-scaffold inhibitors, Hsp90 inhibitors, 6:419–421
Ispemicin, chemical structure, 7:419
Ispenesib, kinesin spindle protein inhibitors, 6:198–204
CK929866SB-743921 and, 6:202–203
Ispronicline antipsychotic, 8:207
Isradipine, stroke therapy, 8:459–460
Istradefylline:
cognitive enhancement, 8:48–49
Parkinson's disease therapy, 8:552–554
Italy, patent requirements, 3:122
Iteration number, multiobjective optimization, drug discovery, blood brain barrier, solubility, and ADMET CYP2D6, 2:269–273
ITF-2357 histone deacetylase inhibitor, structure and development, 6:60–61
ITMN-191 anti-HCV agent, 0
Itraconazole, cocrystal engineering, 3:209–210
Itrocinonide, soft drug design, 2:97–98
ITScore function, empirical scoring, 2:630–631
ITU R-100943 HIV reverse transcriptase inhibitor, 7:151–152

Ixabepilone, Alzheimer's disease, tau-targeted therapy, 8:430

Janus kinase 3 (JAK3):
combinatorial chemistry, 1:338–340
CP-690,550 immunosuppressive compound, 5:1055–1057
Japan, patent requirements, 3:122
Jaundice, vitamin K deficiency, 5:668
JL-13 antipsychotic drug, structure-activity relationship, 8:182–183
JN403 agonist, cognitive enhancement, 8:35–36
JNJ-26481585 second-generation histone deacetylase inhibitor, 6:64–66
JNJ-26854165 compound, protein-protein interactions, 2:352
JOM-13 tetrapeptide derivative, deltorphin analogs, opioid peptides, 8:672–673
JSM6427, brain cancer therapy, hypoxia-inducible factor pathway, 6:254–255
JTP-426467 compound, PPARγ antagonist, 4:117
"Junk" peptides:
aminoglycoside resistance, 7:424–427
macrolide pharmacology, 7:440–441

Kabiramide C complex, protein-protein interactions, 2:338–339
Kahalide F, marine microbial sources, 2:247
Kahn-Ingold-Prelog rules, anti-DNA virus agents, acyclic nucleoside phosphates, 7:232–239
Kanamycin:
analog design, 7:429–432
antitubercular agents, 7:761–762

history and biosynthesis, 7:423–424
pharmacology and ADMET properties, 7:422–423
resistance, 7:425–427
Kan Reactors, defined, 1:280
Kaposi's sarcoma, 9-*cis*-retinoic acid therapy, 5:422–424
Karenitecin, biological activity and side effects, 6:98
Karplus relationship, ligand-based drug design, 2:384
KatG inactivity, antitubercular resistance mechanisms, 7:734–735
isoniazid, 7:739–743
KB-141 thyromimetic, structure-activity relationship, 4:208–209, 216–218
Kefauver-Harris Amendments, over the counter drug approval, 3:83–84
KEGG database, quantitative structure-activity relationship modeling, 1:514
Kennard-Stone (KS) selection, quantitative structure-activity relationship validation, 1:50
Ketal formation, macrolides, 7:439–441
16α,17α-Ketals:
glucocorticoid biosynthesis and metabolism, 5:58–59
percutaneous absorption, 5:46–49
Ketamines:
antipsychotic effects, NMDA model, 8:171–172
stroke therapy, 8:450
Ketoamides:
calpain, stroke therapy, 8:463–464
teleprevir anti-HCV agent, 7:183–186
Ketoconazole:
androgen antagonists, 5:197
CK-2130 prodrug case study, 3:274–276
selective toxicity, 2:558
sertindole interaction, 8:199

Ketocyclazocine:
  morphinan derivatives, 8:611–612
  multiple opioid receptors, 8:579–580
Ketolides, recent developments, 7:447–456
Ketones:
  methadone analogs, 8:623
  stroke therapy:
    alpha oxo/thio/amino-methyl ketones, 8:469–470
    calpain inhibitors, 8:463–464
    caspase inhibitors, 8:467–469
    cathepsin B inhibitors, 8:472–473, 475–478
    halomethyl ketones, 8:465, 467–469
    HDAC inhibitors, 8:493
Ketorolac, migraine therapy, 8:268
Khat, stimulant effect, 8:89–90
Kidney, ADMET excretion properties, 2:61
Kinase catalytic domain, small-molecule inhibitor design, 6:346–349
Kinase inhibitors. *See also* Protease inhibitors; Tyrosine kinase inhibitors
  antidepressant efficacy and, 8:228–229
  antimalarial cell cycle regulation, 7:627–634
    farnesylation, 7:630–634
    protein kinases, 7:627–630
  Bcr-Abl inhibitors, chronic myelogenous leukemia therapy, 6:298–301
  brain tumor therapy, 6:245
    vascular endothelial growth factor, 6:248–249
  cell cycle inhibitors, 6:310–321
    aurora kinase inhibitors, 6:316–321
    checkpoint kinase inhibitors, 6:315–316
    cyclin-dependent kinase inhibitors, 6:310–315
    polo-like kinase inhibitors, 6:316
  chemogenomics, 1:590–951
  classification and properties, 6:295–298

clinical candidates *vs.* marketed drugs, 6:297–298
donor-acceptor motifs, 6:295–296
FDA-approved compounds, 6:296–297
Flt3 inhibitors, acute myelogenous leukemia, 6:301
future research issues, 6:330–331
glaucoma therapy, clinical trials, 5:601–602
growth factor signaling inhibitors, 6:301–307, 302–307
  B-Raf inhibitors, 6:305–306
  erbB-family kinases, 6:302–305
  MEK inhibitors, 6:306–307
hit seeking combinatorial libraries, 1:337–340
cyclin dependent kinase/p25 inhibitor, 1:329–330
ERK-1/RasGap dual inhibitors, 1:320–321
ligand-based virtual screening, high-throughput screening enrichment, 2:14
polypharmacology, BioPrint® database, 1:494–496
structure-based design:
  future research issues, 6:356–357
  multiconformational catalytic domain:
    binding cleft, 6:346–348
    binding pockets, 6:348–349
    protein targeting, 6:350–354
    second-line inhibitors, drug-resistant mutation, 6:354–356
    research background, 6:345–346
survival signaling inhibitors, 6:321–325
  insulin-like growth factor-1 receptor inhibitors, 6:321
  mTOR inhibitors, 6:324–325
  phosphoinoside-3 kinase inhibitors, 6:321–323
  protein kinase B inhibitors, 6:323–324

tumor invasion/metastasis inhibitors, 6:325–330
  c-Met kinase inhibitors, 6:326–330
  focal adhesion kinase inhibitors, 6:326
  Src kinase inhibitors, 6:325–326
  VEGFR2 inhibitors, antiangiogenic agents, 6:307–310
Kinase insert domain receptor (KDR), quantitative structure-activity relationship, multidimensional models, 1:41–43
Kinases:
  allosteric protein model, 1:392
  gene-directed enzyme prodrug therapy, 6:123–124
  structural genomics, human proteome coverage, 1:577–579
Kinesin spindle protein (KSP):
  basic properties, 6:191
  inhibitors:
    ATP-competitive inhibitors, 6:211–214
    dihydropyrazoles, 6:204–210
    future research issues, 6:214–215
    "induced fit" pocket inhibitors, 6:210–211
    ispinesib, 6:198–204
    monastrol, 6:193–197
    quinazolinone-based inhibitors, 6:197–202
  mitosis-targeted therapy, 6:191–192
  structure and mechanism, 6:192–193
Kinetic parameters:
  membrane protein passive diffusion, 2:444–446
  nuclear magnetic resonance, macromolecule-ligand interactions, 2:394–402
    DNA binding, 2:406–407
  oral drug delivery, sustained-release dosage forms, 2:794–798
  topical antimicrobial agents, 7:488–489
  transdermal drug delivery, 2:836–838

Kinetic solubility assays, ADMET structural properties, **2:**65–66
Kinetoplastic protozoan infections, **7:**566–577
American trypanosomiasis (Chagas' disease), **7:**571–573
biochemistry, **7:**567
epidemiology, **7:**565–566
glucose metabolism and glycosomal transport inhibitors, **7:**582–585
human African trypanosomiasis (sleeping sickness), **7:**567–571
leishmaniasis, **7:**574–577, 590–593
lipid metabolism inhibitors, **7:**580–581
nucleotide transporters, **7:**577–580
polyamine pathway targeting, **7:**585–590
protease inhibitors, **7:**581–582
Kit tyrosine kinase, structure-based design:
drug-resistant mutation, **6:**354–356, 367–368
protein targeting, **6:**351–354
Klotz plot, allosteric proteins, **1:**379–381
K-means clustering, quantitative structure-activity relationship validation, **1:**49
k-nearest neighbor (k-NN). *See also* Lazy learning
kNN-like method
docking techniques, hydrophobicity, **2:**634–636
quantitative structure-activity relationship:
data curation applications, **1:**528
glucocorticoid anti-inflammatory compounds, **5:**74
ligand-based virtual screening, **2:**11–13
modeling principles, **1:**525
KNI-272 inhibitor, nuclear magnetic resonance studies, **2:**418–419

Knockout mouse models, target validation, congestive heart failure therapy, **4:**488
Knowledge-based approaches
algorithm functions, **2:**628–631
functions table, **2:**650–651
multitarget drug development, **1:**253–254
structure-based virtual screening, **2:**18–19
Kohonen's self-organizing neural network (KohNN), quantitative structure-activity relationship validation, **1:**49
Koshland-Nemethy-Filmer/Dalziel-Engel model, allosteric protein theory, **1:**375–379
KR-62980 compound, selective PPARγ modulators, PPARγ partial agonists, **4:**114–115
Krebs cycle:
antitubercular agents, intermediary metabolism and respiration, **7:**729–730
biotin metabolism, **5:**686–690
parasitic glucose metabolism and glycosomal transport inhibition, **7:**582–585
Kynurenic acid:
schizophrenia pathology, **8:**163–164
stroke therapy, **8:**452–453
Kyotorphin analogs, chemical delivery systems, **2:**153–154

L-685458 γ-secretase inhibitor, **8:**334–335
L-689502 lead compound, indinavir sulfate development, **7:**12, 14–15
Labeling, FDA requirements, **3:**91
Lacidipine, stroke therapy, **8:**459–460
Lacosamide, anticonvulsant applications, **8:**136–139, 154

Lactacystin, proteasome inhibitors, **6:**183–184
β-Lactamases:
antibiotic inhibitors, **2:**222–223, **7:**358–372
bicyclic lactam inhibitor, **7:**371
carbapenems, **7:**366–368
clavulanic acid disovery and inactivation, **7:**360–363
6-heteroarylmethylene penems, **7:**364–366
modified penams, penam sulfones, and cephem sulfones, **7:**368–370
monolactams, **7:**371
nonlactam inhibitors, **7:**372
oxapenems, structural modification, **7:**370–371
penam sulfones, **7:**363–364, 367–369
antibody-directed enzyme prodrug therapy, **6:**121–123
classification, **7:**274–279
clinical relevance, **7:**274
penicillin binding protein evolution and, **7:**274
recombinant DNA technology, **1:**545–546
toxicophore chemical reactivity, **2:**301–302
Lactams:
antidepressants, structure-activity relationship and metabolism, **8:**252–253
poly(ADP-ribose)polymerase (PARP) inhibitor coadjuvants, **6:**169–170
prostaglandin/thromboxane receptors, EP$_4$ agonists, **5:**873–878
tricyclic quinoline lactam integrase inhibitors, **7:**89–91
β-Lactams. *See also* β-Lactamases; Carbapenems; Cephalosporins; specific compounds, e. g., Penicillins
antibacterial drugs, **2:**222–223
carbapenems, **7:**282

β-Lactams. (*Continued*)
cephalosporins, **7:**265–266, 281–282, 295–310
clinical applications, **7:**281–287
future research and development, **7:**372–373
indications for use, **7:**284–287
mechanism of action:
bacterial cell wall components, **7:**261–262
cell wall structural organization, **7:**266–269
penicillin binding proteins, **7:**262–264
Tipper-Strominger hypothesis, D-Ala-D-Ala mimicry, **7:**264–266
penicillins, **7:**265–266, 281, 287–295
pharmacokinetics, **7:**281–282
research background, **7:**259
side effects and drug interactions, **7:**282–284
structural classification, **7:**259–260
chemistry, **7:**287–372
drug delivery, intestinal peptide transporter (PepT1), **2:**463–467
monocyclic nonantibacterial pharmacophore, **7:**332–333
N-thiol monocyclic β-lactams, **7:**331–332
prodrug approvals, **3:**242–246
resistance mechanisms, **7:**269–281
acquisition and spread of resistance, **7:**269–271
conjugative transposons, **7:**270–271
E. faecium PBP-5, **7:**273
enterobacteria porin loss, **7:**280
methicillin-resistant staphylococci mechanism and genetics, **7:**272–273
mosaic gene formation, resistant PBPs, **7:**271–272
N. gonorrhoeae PBPs, **7:**272
outer membrane permeability and active efflux decrease, **7:**279–281
P. aeruginosa resistance, **7:**280–281
penicillin binding protein modification, **7:**271–274
plasmids, **7:**270
S. pneumoniae PBP-2x, **7:**273–274
selective toxicity, **2:**550–553
stroke therapy:
calpain irreversible inhibitors, **8:**465
cathepsin B inhibitors, **8:**475–476
γ and δ Lactams, antimalarial cell cycle regulation, **7:**635–637
Lactic acidosis, type 2 diabetes, metformin therapy, **4:**5
Lamivudine:
anti-DNA virus antimetabolites, cytosine derivatives, **7:**226
hepatitis B virus therapy, **7:**246–247
HIV reverse transcriptase inhibition, **7:**141–145
research background, **7:**139–140
Lamotrigine, anticonvulsant applications, **8:**135–139, 148–149
Landiolol, soft drug design, **2:**85–86
Lanham Act (U.S.):
false designation of trademarks and, **3:**178–179
trademark selection guidelines, **3:**172–174
Lansoprazole, selective toxicity, **2:**566
Lapatinib:
cranial and spinal nerve tumors, **6:**238–239
growth factor signaling inhibitors, **6:**302–305
multitarget drug development, cancer therapy, **1:**265–267
structure-based design, multiconformational catalytic domain, **6:**346–349
Lapinone antimalarial agent, **7:**619–620
Large-scale synthesis:
nevirapine case study, **3:**14–23
commercial production and process optimization, **3:**20–23
medicinal chemistry synthetic route, **3:**15–16
pilot-plant scale-up:
chemical development, **3:**16–17
process development, **3:**17–20
research background, **3:**14–15
research background, **3:**1
scale-up operations, **3:**1–14
basic principles, **3:**1
bench-scale experiments, **3:**4–7
reaction solvent selection, **3:**5–6
reaction temperature, **3:**6
reaction times, **3:**6
solid-state requirements, **3:**7
stoichiometry and addition order, **3:**6–7
chiral requirements, **3:**4
commercial-scale operations, **3:**10–13
chemical safety, **3:**13
environmental controls, **3:**13
in-process controls, **3:**11
processing equipment requirements, **3:**11
validation, **3:**12–13
continuous microfluidic reactors, **3:**13–14
pilot plant scale-up, **3:**7–10
agitation, **3:**8–9
drying and solid handling, **3:**9–10
heating and cooling, **3:**8
liquid-solid separations, **3:**9
safety issues, **3:**10

process analytical
 technology, 3:14
route selection, 3:3–4
synthetic strategies, 3:1–4
Largezole, histone deacetylase
 inhibition, 6:72
Laropiprant, anticholesterolemic
 mechanisms, 2:234
Laryngeal dysplasia, 13-*cis*-
 retinoic acid therapy,
 5:390
Laser trabeculoplasty (LTP),
 glaucoma intraocular
 pressure reduction,
 5:591–592
Lasofoxifene:
 future research issues and
 information retrieval,
 5:256
 structure-activity relationship,
 5:252
Latanoprost, glaucoma
 intraocular pressure
 reduction, 5:592–600
Latent inhibition, antipsychotic
 therapy, 8:172
Laurylpyridinium chlorides,
 cationic surfactants,
 7:500
Lavendustin-related molecules,
 anti-cancer activity,
 2:239–240
Lawesson's reagent, hydrogen
 sulfide releasing
 molecules, 5:314–315
Law of mass action (LMA),
 receptor-based drug
 design, 2:494
Lazy learning *k*NN-like method,
 quantitative
 structure-activity
 relationship models,
 applicability domains,
 1:522
LB-11058 cephalosporin,
 7:311–312
LBH-589 histone deacetylase
 inhibitor, structure
 and development,
 6:55, 57–58
LDE-225 small molecule, brain
 tumor therapy,
 hedgehog signaling
 pathway, 6:262
Lead drug discoveries:
 ADMET applications, 2:47–48
 assessment rules, 2:64

hit/lead compound selection,
 2:49–50
structure modification and
 optimization, 2:50
BACE (β-site cleavage enzyme)
 inhibitors, 8:347
G-protein coupled receptor,
 homology modeling,
 2:289–292
optimization techniques,
 2:292–293
migraine therapy, calcitonin
 gene-related peptide
 antagonists,
 8:287–288
natural products:
 biosynthetic pathways,
 2:210–213
 future research issues,
 2:208–213
 genome mining,
 2:209–210
 high-throughput screening,
 2:205–207
 hits analysis, 2:207
 libraries, 2:200–204
 attributes, 2:201–202
 complexity, 2:200–201
 mass spectrometry
 analysis, 2:202–203
 nuclear magnetic
 resonance
 characterization,
 2:203–204
 screening platforms,
 2:205–209
 phenotypic screening,
 2:207–208
 sources, 2:191–200
 biodiversity harvesting,
 2:199
 metagenomics, 2:199–200
 microbial sources,
 2:193–199
 terrestrial and marine
 habitats, 2:193–199
 virtual screening, 2:213
receptor-based drug design,
 2:525–533
 biologicals and antisense
 compounds, 2:531–533
 diversity-based ligand
 libraries, 2:531
 high-throughput screening,
 2:526–527
 natural product sources,
 2:527–528

pharmacophore-based ligand
 libraries, 2:528–529
privileged pharmacophores,
 2:529–530
Learning deficits, Alzheimer's
 disease
 immunotherapy,
 8:366–367
Leave-one-out (LOO) cross-
 validation study,
 quantitative
 structure-activity
 relationship:
 integrated predictive modeling
 workflow, 1:516–525
 multidimensional models,
 1:40–43
 recent trends, 1:510–511
Lecozotan, cognitive
 enhancement,
 8:37–40
Lectins, receptor-mediated
 endocytosis,
 2:452–453
Left-side ether substituent,
 chemokine receptor-5
 (CCR5) antagonist,
 anilide core,
 7:126–131
Leishmania major:
 genome sequencing,
 7:565–566
 life cycle, 7:575
 lipid metabolism inhibitors,
 7:580–581
Leishmaniasis:
 current therapies, 7:575–577
 drug discovery and
 development,
 7:590–593
 epidemiology, 7:565–566
 pathology, 7:574–575
Lenapicillin, penicillin resistance
 and development of,
 7:291–292
Leptins, osteoporosis therapy,
 5:755
Lestaurtinib, acute myelogenous
 leukemia therapy,
 6:301–302
Lethality dose (LD_{50}), FDA
 preclinical testing
 guidelines, 3:71–72
Letrozole:
 aromatase inhibition, 5:232
 breast cancer resistance,
 6:369–370

Leucine-rich-repeat kinase 2 (LRKK-2), Parkinson's disease genetics, 8:535
Leu-enkephalin analogs, chemical delivery systems, 2:151–153
Leukemias. *See also* specific leukemias
  bexarotene therapy, 5:443, 447
  N-(4-hydroxyphenyl) retinamide (4-HPR) therapy, 5:472
  9cUAB30 RXR-selective retinoid therapy, 5:456–457
LEUKINE®, structure and properties, 4:573–575, 588–590
Leukotrienes, selective toxicity, antagonist structures, 2:543–544
Leupeptin, stroke therapy, 8:461–462
  cathepsin B inhibitors, 8:472
Levenberg-Marquardt (L-M) algorithm, quantitative structure-activity relationship, 1:39
Leverage outliers, quantitative structure-activity relationship, 1:43–46
Levetiracetam:
  anticonvulsant applications, 8:135–139, 149–151
  cognitive enhancement, 8:34
Levobupivacaine, selective toxicity, 2:564
Levodopa (L-dopa):
  Parkinson's disease:
    dopaminergic pharmacotherapy, 8:540–542
    pathophysiology, 8:531–534
    prodrug development, 3:223–226
Levofloxacin, antibacterial agents, 7:535–540
Levomethorphan, selective toxicity and stereochemistry, 2:546
Levonorgestrel, current compounds and applications, 5:220–221
Lewy body dementia (LBD), 8:21

Lexiva. *See* Fosamprenavir calcium
LFA701 statin, multiple sclerosis therapy, 5:573
LFER, quantitative structure-activity relationship, 1:2–8
LG100641 compound, selective PPARγ modulators, PPARγ partial agonists, 4:107–115
Liability issues, congestive heart failure therapy, risk/liability evaluation, 4:500
Liarozole, androgen antagonists, 5:197
Liberation of drug, drug delivery system bioavailability, 2:747
Licensing agreements, international patent laws, 3:169–171
Licochalcone A antimalarial, cellular respiration and electron transport, 7:655–658
Licostinel, stroke therapy, 8:452–453
Lifestyle, cognitive dysfunction and, 8:21
"Lifestyle" drugs, overview, 5:711–712
Ligand-based design:
  catalysts, pharmacophores, 1:458–463, 472
  chemogenomics:
    affinity chromatography/activity-based protein profiling, 2:581
  basic principles, 2:573–574
  biological principles, 2:580–582
  chemical space theory, 2:574–580
  cofactor-based discovery, 2:578
  diversity-oriented/biology-oriented syntheses, 2:578–580
  molecular informatics, 2:582–588
    information systems, 2:582–583
    ligand-based data analysis and predictive modeling, 2:583–585

structure-based data analysis and predictive modeling, 2:585–588
privileged scaffolds, 2:575–577
protein family targeted libraries, 2:574–575
protein secondary structure mimetics, 2:577–578
receptor-based drug design, 2:516–519
yeast three-hybrid screens, 2:581–582
G-protein coupled receptor, homology modeling, docking mechanism, 2:289–292
nuclear magnetic resonance, 2:375–402
  analog limitations, 2:390
  charge state, 2:384–385
  conformational analysis, 2:383–384
  efficiency properties, 2:433
  fragment-based properties, 2:432–433
  high-throughput screening, 2:431–432
  library design, 2:431–434
  line-shape and relaxation data, 2:387–389
  pharmacophore modeling, 2:389–390
  structure elucidation, 2:376–383
    bioactive peptides, 2:377–383
    natural products, 2:376–377
  tautomeric equilibria, 2:385–387
Pareto designer, 2:266–273
structure-activity relationships:
  pharmacophore formation, 1:465–466
  pharmacophores, 1:465–477
    computational techniques, 1:469–471
    constrained analog synthesis, 1:468–469
    limitations, 1:471
    preparation, 1:466–468
virtual screening techniques, 2:2–14
  compound similarity, 2:2–3

efficiency assessment,
2:26–27
kinase inhibitor high-
throughput screening,
2:14
machine learning
applications,
2:10–14
algorithms for QSAR
models, 2:10–13
applicability domain
concept, 2:13–14
QSAR model generation
and validation, 2:10
pharmacophores, 2:3–10
dopamine transporter
inhibitors, 2:7–8
novel PPAR ligands, 2:9
organic compound
databases, 2:4–5
research background, 2:4
three-dimensional ligand-
based generation, 2:6
three-dimensional
receptor-based
generation, 2:8–9
three-dimensional
screening, 2:9–10
two-dimensional
searching, 2:6
Ligand binding domain (LBD):
estrogen receptor, 5:234–236
molecular endocrinology,
5:236
tissue specificity, 5:241–242
nuclear hormone receptors,
4:77–79
assay techniques, 4:80–82
peroxisome proliferator-
activated receptors:
α-agonists, 4:88–99
α/γ-dual agonists, 4:129–135
γ-agonists, 4:101–102
medicinal chemistry, 4:87–88
progesterone receptors,
5:237–238
progestins, tissue specificity,
5:241–242
retinoid nuclear receptors,
5:370–373
selective estrogen receptor
modulators,
osteoporosis
antiresorptive
therapy, 5:735–742
steroid hormone receptors,
5:30–31

thyroid hormone receptors,
4:189–189, 200–202
β-selective agonists,
structure-based
design, 2:684–685
Ligand-binding domain (LBD),
central nervous
system drug discovery,
8:5–6
Ligand binding pockets (LBPs),
retinoic nuclear
receptors, 5:372–373
Ligand displacement assay,
nuclear hormone
receptors, 4:80
Ligand efficiency indices,
multiobjective
optimization, drug
discovery, 2:260–261
Ligand-gated ion channels
(LGICs):
anticonvulsants, 8:126–130
$GABA_A$ receptors,
8:126–129
ionotropic glutamate
receptors, 8:129
receptor-based drug design:
allosteric modulators,
2:500–501
classification, 2:505–506,
509–511
functional assays, 2:523
genetic variation, 2:515
Ligand-macromolecular
interactions:
nuclear magnetic resonance,
2:394–419
chemical-shift mapping,
2:402–403
dihydrofolate reductase,
2:414–415
DNA binding experiments,
2:405–410
docking experiments,
2:404–405
HIV protease, 2:415–419
immunophilins, 2:410–412
kinetics and timescale
parameters,
2:394–402
matrix metalloproteinases,
2:412–414
titration, 2:403–404
structure-based
thermodynamic
analysis, 2:702–712
binding activity, 2:703–705

binding measurement,
2:706–707
calorimetry results,
2:711–712
enthalpy-entropy
compensation,
2:705–706
fragment-based design,
2:709–711
isothermal titration
calorimetry,
2:707–709
solvation thermodynamics,
ligand-receptor
interactions, 2:711
Ligand-receptor interactions:
allosteric protein theory,
1:379–380
nuclear hormone receptors,
4:84–86
opioid-receptor-like 1/orphanin
FQ/nociceptin peptide,
8:680–681
opioid receptors, 8:581–588
quantitative structure-activity
relationship, 1:28–31
receptor-based drug design,
2:494–498
binding dynamics, 2:503
constitutive receptor activity,
2:498–499
definition, 2:520–523
macromolecule-ligand
interactions,
2:394–419
sphingosine 1-phosphate (S1P),
multiple sclerosis
therapy, 5:575–578
LigandScout tools, ligand-based
virtual screening,
pharmacophore
formation, 2:4
Ligand-targed SAR matrices,
chemogenomics:
data analysis and predictive
modeling, 2:583–585
research background,
2:573–574
Ligases, high-throughput
screening, enzymatic
assays, 3:428
LIGIN function, docking/scoring
computations, point
complementarity
methods, 2:602–604
Ligscore1 function, empirical
scoring, 2:627–628

LinBiExp method, relative receptor binding affinity predictions, glucocorticoid anti-inflammatory compounds, **5:**75–76
Lincomycin, ribosome targeting, **7:**460–462
Lincosamides, ribosome targeting, **7:**460–462
Linear combinatorial synthesis: lead optimization, **1:**345
small molecule libraries, **1:**299–300
Linear interaction energy (LIE) method, quantum mechanics/molecular mechanics binding affinity, **2:**729–731
Linear modeling, quantitative structure-activity relationships: glucocorticoid anti-inflammatory compounds, **5:**74
multilinear regression analysis, **1:**33–34
Linear peptides:
deltorphin analogs, opioid peptides, **8:**668–671
dermorphin analogs, μ-opioids, **8:**664–666
dynorphin analogs, **8:**655–656
enkephalin analogs:
δ-opioid analogs, **8:**650
μ-opioid analogs, **8:**647
Linear pharmacokinetics, defined, **2:**751
Linear scaling, quantum mechanics/molecular mechanics, binding affinity evaluation, **2:**721–723
Linear solvation energy relationships (LSERs), quantitative structure-activity relationship, **1:**22–23
Line-shape analysis, ligand-based design, **2:**387–389
LineWeaver Burk plot, allosteric proteins, **1:**379–381
Linezolid:
antitubercular agents, **7:**785–786
microbial resistance, **7:**543–544
pharmacokinetics and metabolism, **7:**543
ribosome targeting, **7:**464
RNA targeting, **7:**542–543
RNA targets, **5:**990–993
synthetic antibacterial agents, **7:**541–544
toxicity and adverse effects, **7:**544
virulence factor effects, **7:**543
Linifanib, brain cancer therapy, hypoxia inducible factor (HIF) pathway, **6:**254–255
Linker compounds:
aminoglycosides, **7:**432
benzodiazepines, **1:**301–302
chemokine receptor-5 (CCR5) antagonist development, **7:**113–116
anilide core structure, **7:**125–131
defined, **1:**280
glutmate modulators, migraine therapy, **8:**292–297
histone deacetylase inhibitiors, siberoyl hydroxamic acid structures, **6:**53–56
hydroxamate HDAC inhibitors, stroke therapy, **8:**488–491
peroxisome proliferator-activated receptors:
α-agonist derivation, **4:**90–96
selective PPARγ modulators, PPARγ partial agonists, **4:**111–115
thyroid hormone receptor β-selective agonists, structure-based design, **2:**687
Linsidomine, nitric oxide donors and prodrugs, **5:**285–286
Lipase inhibitors, atherosclerosis, HDL elevation, **4:**337–340
endothelial lipase, **4:**338–339
hepatic lipase, **4:**337–338
Lipid arrays:
drug metabolism, conjugation reactions, **1:**432–433
solid-phase organic synthesis, **1:**293–294
Lipid II formation, penicillin binding proteins, **7:**263–264
Lipid metabolism:
antiparasitic/antiprotozoal inhibitors, **7:**580–581
antitubercular agents: polyketide-derived lipids, **7:**723–724
targeting of, **7:**730
Lipid rafts, membrane proteins, receptor-mediated endocytosis, **2:**449–450
Lipid solubility:
brain-targeting chemical delivery systems, **2:**136–140
oral drug delivery, **2:**441
oral drug physicochemistry, class II drugs, **3:**48–50
sulfonamides, **7:**518–519
Lipinski's rules. *See* Rule of five (Lipinski)
Lipofection, protein therapeutics, transfection, **3:**295
Lipopeptides, antibiotics, **2:**227–228
Lipophilicity:
ADMET structural properties, **2:**51, 65
amide antiviral activity, tropane-based chemokine receptor-5 (CCR5) antagonist, **7:**117–121
antifolate DNA-targeted chemotherapeutics, **6:**109
drug metabolism and, **1:**444–445
ligand-macromolecular interactions, structure-based thermodynamic analysis, **2:**705
membrane proteins, drug delivery strategies, prodrug development, **2:**441–443
oral drug physicochemistry:
partition coefficient, **3:**31–32
permeability, **3:**34–35
tetracyclines, **7:**409–410
Lipoproteins. *See also* Apolipoproteins; High-density lipoprotein;

Low-density lipoprotein; Very low-density lipoprotein
liver X receptor agonists, 4:148–150
nitric oxide effects, 5:272–278
peroxisome proliferator-activated receptors:
α-agonists, 4:88–99
δ-agonists, 4:117–118
plasma lipoproteins, atherosclerosis and, 4:332–333
structure and function, 4:303–305
thyroid hormone metabolism, 4:195
receptor agonists, therapeutic applications, 4:197–198
Liposomes:
nanoscale drug delivery, 3:469–473
oral drug physicochemistry, class II drugs, 3:49
5-Lipoxygenase polymorphisms, 1:191–192
multitarget drug development, histamine H_1-plus for allergies, 1:260–262
15-Lipoxygenase inhibitors, ligand-based virtual screening, 2:3
Liquid chromatography-mass spectrometry (LC-MS):
combinatorial chemistry, 1:346–347
combinatorial libraries, purification, 1:107–108
drug development, 1:100–105
log P determination, 1:119–120
screening applications, affinity chromatography, 1:111–112
Liquid chromatography-nuclear magnetic resonance (LC-NMR), development and instrumentation, 2:374
Liquid formulation, protein therapeutics stability, 3:314–315

Liquid secondary ion mass spectrometry (LSIMS), drug development, 1:99–100
Liquid-solid separation, pilot plant development, 3:9
Lisuride, Parkinson's disease dopaminergic therapy, 8:545–546
Literature searches:
Biopharmaceutics Classification System, 3:356–359
quantitative structure-activity relationships, 1:507–509
single nucleotide polymorphisms, 1:198
virtual screening, 2:1–2
Liver:
ADMET excretion properties, 2:61
antidepressant effects on, monoamine oxidase inhibitors, 8:230
antipsychotic side effects on, 8:169
drug delivery, apical sodium-dependent bile acid transporter enhancement, 2:474
NRX 194204 RXR selective analog, hepatoprotection, 5:462–463
permeability, ADMET interactions:
distribution, 3:376
drug transport mechanisms, 3:368–369
metabolism, 3:368–369, 371–372
thyroid hormone effects, 4:196–197
Liver cancer:
Am80 therapy, 5:417
N-(4-hydroxyphenyl) retinamide (4-HPR) therapy, 5:468, 472
Liver X receptors (LXRs):
adverse lipid effects, 4:148–150
Alzheimer's disease, cholesterol turnover, 8:411–413
atherosclerosis, 4:147–148
basic properties, 4:145–146

clinical data, 4:154–156
first-generation agonists, 4:147
future research issues, 4:167
glucose homeostasis, 4:148
lipid metabolism, 4:148
medicinal chemistry, 4:151–154
modulator identification, structure-based virtual screening, 2:23–25
peroxisome proliferator-active receptors, α-agonist derivation, 4:93–96
structural biology, 4:150
Lividomycin, chemical structure, 7:419
LKB1 tumor suppressor kinase, metformin mechanism of action, 4:4
Loc387715/Arms2 gene locus, age-related macular degeneration, 5:607–608
Lock-in mechanism:
cyclodextrin chemical delivery systems, 2:149–150
estradiol chemical delivery system, 2:144–149
Locomotor effects, antipsychotic therapy, animal models, 8:170–171
Lofepramine:
chemical structure, 8:222
monoamine transporter inhibition, 8:240–241
pharmacokinetics, 8:235–236
Log P determination, liquid chromatography-mass spectrometry, 1:119–120
Lomerizine, quantitative structure-activity relationships, 1:73–74
Long-acting β_2-adrenergic receptor agonists (LABAs), chronic obstructive pulmonary disease bronchodilators, 5:770–771, 777–779
Long-acting muscarinic receptor antagonists (LAMAs), chronic obstructive pulmonary disease bronchodilators, 5:768–769

Long-chain arylpiperazines (LCAPs),
  antidepressant actions, 8:242–244
  structure-activity relationships and metabolism, 8:253–254
Long-term depression (LTD), 8:25
Long terminal repeat (LTR):
  HIV integrase and, 7:78–79
  HIV regulation, recombinant DNA technology, 1:546–547
Long-term memory, 8:19
Long-term potentiation (LTP), 8:24–25
"Loose" neuroleptics, structure-activity relationship, 8:185
Lopinavir/ritonavir protease inhibitor:
  binding interactions, 7:32, 34
  chemical structure, 7:31
  efficacy and tolerability, 7:32, 34, 36
  optimization, 7:31–33
  pharmacokinetics, 7:32, 34–35
  resistance profile, 7:34, 36
Loratidine, selective toxicity, 2:565
Lorazepam, anticonvulsant applications, 8:144–147
Lorentz-Lorenz equation:
  quantitative structure-activity relationship, steric parameters, 1:20–21
  quantitative structure-activity relationships, polarizability indices, 1:24–25
Losartan, bioisosterism, 1:171–174
Lossen rearrangement, toxicophore reactive metabolites, arylhydroxyamic acids, 2:313
Loss-of-function mutations:
  glucokinase, type 2 diabetes therapies, 4:15–17
  receptor-based drug design, GPCR genetic variation, 2:515
Lost profits remedy provisions, infringement of patent claims, 3:157–159

Loteprednol etabonate and analogs, soft drug design, 2:90, 92–96
Lovastatin:
  Alzheimer's disease therapy, preclinical/clinical testing, 8:359–361
  anticholesterolemics, 2:233
  niacin combination, 2:234
  cholesterol biosynthesis, 5:15–16
  LDL cholesterol lowering mechanisms, 4:307
  multiple sclerosis therapy, 5:573
Loviride HIV reverse transcriptase inhibitor, 7:151–152
Low-affinity rolipram binding site (LARBS), chronic obstructive pulmonary disease therapy, 5:783–784
Low-density lipoproteins (LDLs):
  atherosclerosis, cholesterol-lowering compounds:
    current therapeutic agents, 4:305–318
    bile acid sequestrants, 4:315–316
    cholesterol absorption inhibitors, 4:310–315
    combination products, 4:316–318
    HMG-CoA reductase inhibitors, 4:306–310
    emerging therapeutic agents, 4:318–320
    antisense oligonucleotide, apoB100, 4:319
    microsomal triglyceride transfer protein inhibitors, 4:319–320
    PCSK9 inhibition, 4:318–319
    research background, 4:303–305
  farnesoid X receptors, 4:159
  liver X receptor agonists, 4:148–149
  peroxisome proliferator-activator receptors:
    α-agonists, 4:96–99
    α/δ-dual agonists, 4:136–137
  thyroid hormone metabolism, 4:195

  receptor agonist effects, 4:197–198
  vitamin E uptake and metabolism, 5:662
Lowest unoccupied molecular orbital (LUMO), quantitative structure-activity relationships, 1:24–25
Low exposure, toxicophore risk assessment, 2:325–326
Low-molecular-weight heparin (LMWH):
  limitations of, 4:373–374
  molecular mechanism of action, 4:375–380
Loxapine:
  migraine therapy, 8:297–298
  pharmacokinetics, 8:189
  structure-activity relationship, 8:180–183, 250–251
L-type calcium channel antagonist, stroke therapy, 8:459–460
Luminal degradation, oral dosage forms, class V drugs, 3:53
Lung cancer:
  bexarotene therapy, 5:439–443, 450–453
  N-(4-hydroxyphenyl) retinamide (4-HPR) therapy, 5:466, 472, 479
  NRX 194204 RXR selective analog, 5:460–461
  13-cis-retinoic acid therapy, 5:390
  SHetA2 retinoid-related compound, 5:502
Lung tissue, drug delivery through, 2:839–844
Luteinizing hormone (LH):
  androgen biosynthesis, 5:22–24
  estradiol chemical delivery system, 2:144–149
  ovulation modulation mechanism, 5:242–243
  testosterone biosynthesis, 5:155–158
Luteinizing hormone-releasing hormone (LHRH):
  androgen biosynthesis, 5:22–24
  estradiol chemical delivery system, 2:144–149

receptor-mediated endocytosis, vitamin B$_{12}$, **2**:454
testosterone biosynthesis, **5**:155–158
LY293558 glutamate modulator, **8**:291–295
LY382884 glutamate modulator, **8**:295–297
LY-426965 antagonist, **8**:225–229
LY466195 glutamate modulator, migraine therapy, **8**:295–297
LY518674 compound, peroxisome proliferator-activator receptors, α-agonists, **4**:97–99
LY2523355 kinesin spindle protein inhibitor, **6**:210
17,20-Lyase inhibitors, androgen antagonists, **5**:197
Lyases, high-throughput screening, enzymatic assays, **3**:426–427
Lymphocytes, recombinant DNA technology, cellular adhesion proteins, **1**:561
Lymphomas:
bexarotene therapy, **5**:443–444
N-(4-hydroxyphenyl) retinamide (4-HPR), **5**:472
primary central nervous system tumors, **6**:239–240
Lymphomatoid papulosis, bexarotene therapy, **5**:444
Lyophilization, protein therapeutics formulation, **3**:315
Lysergic acid diethylamide (LSD), receptor-based drug design, central nervous system effects, **8**:2–3
Lysosomal membrane permeability, AHPN apoptosis, **5**:492

MABT5102A monoclonal antibody, Alzheimer's immunization therapy, **8**:378–379
Macebecin Hsp90 inhibitors, **6**:397–399

Machine learning applications, ligand-based virtual screening, **2**:10–14
algorithms for QSAR models, **2**:10–13
applicability domain concept, **2**:13–14
QSAR model generation and validation, **2**:10
Macrocyclic inhibitors:
proteasome inhibitors, **6**:183–184
stroke therapy, **8**:506–507
tacrolimus, structure-activity relationship, **5**:1021–1022
Macrolides:
current compounds and clinical applications, **7**:433–438
history, biosynthesis and structure-activity relationships, **7**:441–444
marine sources, **2**:246
natural product sources, **7**:405–406
pharmacology and mechanism of action, **7**:438–441
radicicol Hsp90 inhibitor, **6**:402–405
recent drug development, **7**:446–456
resistance to, **7**:444–446
RNA targets, erythromycin, **5**:984–990
selective toxicity, **2**:552–553
side effects, toxicity, and contraindications, **7**:438
structure and function, **2**:228–229
Macromolecular compounds, nuclear magnetic resonance, **2**:368
ligand-macromolecular interactions, **2**:394–419
chemical-shift mapping, **2**:402–403
dihydrofolate reductase, **2**:414–415
DNA binding experiments, **2**:405–410
docking experiments, **2**:404–405
HIV protease, **2**:415–419

immunophilins, **2**:410–412
kinetics and timescale parameters, **2**:394–402
matrix metalloproteinases, **2**:412–414
titration, **2**:403–404
receptor-based design, **2**:391–394
Macugen (pegaptanib sodium injection), aptamer-based recombinant DNA technology, **1**:549
Madrid Protocol, international trademark rights, **3**:178
Magnesium ions, HIV integrase integration, **7**:79–81
Magnetic resonance imaging, brain tumor modeling, **6**:273–274
Magnetization transfer, nuclear magnetic resonance studies, drug screening applications, **2**:424–426
Mahalanobis distance, quantitative structure-activity relationship, outlier detection, **1**:44
Major facilitator (MFS) superfamily, structure and classification, **2**:456–458
Major histocompatibility complex (MHC):
immune function, **5**:1002–1006
tuberculosis pathogenesis, **7**:718–719
Malaria. *See also* Antimalarials; *Plasmodium falciparum*
clinical features and pathogenicity, **7**:605
epidemiology, **7**:565, 603–605
tetracycline efficacy, **7**:415
Malate dehydrogenase, allosteric proteins, **1**:381–383
Malathion, soft drug design, **2**:107–108
Maleimide analogs:
Alzheimer's disease, tau-targeted therapies, **8**:420–422

Maleimide analogs: (*Continued*)
conjugated hemoglobins, 4:635–637
sotrastaurin immunosuppressive compound, 5:1059–1060
Male sex hormones. *See* Sex hormones
Malignant peripheral nerve sheath tumors (MPNSTs), classification and therapy, 6:238–239
Mammalian cell lines, protein therapeutics:
purification, 3:306–309
upstream processing, 3:293–294
MAPK phosphatase, glucocorticoid anti-inflammatory signaling mechanisms, 5:64–65
Maprotiline:
chemical structure, 8:222
historical background, 8:248–249
serotonin receptor targeting, 8:241–242
side effects and adverse reactions, 8:231–232
structure-activity relationship, 8:250–251
Maraviroc compound, tropane-based chemokine receptor-5 (CCR5) antagonist, 7:120–121
Maribavir:
anti-DNA viral agents, 7:240–241
cytomegalovirus therapy, 7:244
Marimastat hydroxymate inhibitors, stroke therapy, 8:503
Marine habitats:
microbial compounds, 2:243–248
bryostatins, bolastatins, and analogues, 2:243–245
cryptophycins, 2:245
curacin A, 2:245
Et743, 2:246–247
peloruside A, 2:246
secondary metabolites/derivatives, 2:247–248

natural products lead generation, microbial compounds, 2:193–199
Marketed drugs:
androgens:
anabolic agents, 5:176–177
antiandrogens, 5:189–190
synthetic androgens, 5:168–169
angiotensin-converting enzyme inhibitors, 4:280–282
antiplatelet agents, 4:418–419
LDL cholesterol-lowering agents, 4:303–305
prodrugs, 3:241–248
examples, 3:242, 244–248
FDA approvals, 3:242–243
prevalence, 3:241–242
trademarks and, 3:172
Markman claim, infringement of patent proceedings, 3:159–162
Masked aldehydes, stroke therapy, 8:463
caspase 3/9 inhibitors, 8:467
Mass analyzers, mass spectrometry development, 1:98
Mass spectrometry (MS):
drug development:
affinity capillary electrophoresis, 1:113–114
affinity chromatography-mass spectrometry screening, 1:111–112
assays, 1:119–121
combinatorial compound structure and purity, 1:108–110
combinatorial libraries, LC-MS applications, 1:107–108
compound encoding and identification, 1:110–111
current trends and recent developments, 1:106–121
desorption ionization techniques, 1:99–100
electron impact, 1:97–98
frontal affinity chromatography, 1:114–115
future trends, 1:121–122

gas chromatography-mass spectrometry, 1:99
gel permeation chromatography-mass spectrometry, 1:113
liquid chromatography-mass spectrometry, 1:100–105
mass analyzers, 1:98
pulsed ultrafiltration-mass spectrometry, 1:116–119
research background, 1:97
screening applications, 1:111–121
solid-phase screening, 1:115–116
tandem mass spectrometry, 1:105–106
web site addresses and information sources, 1:122–123
natural products libraries characterization, 2:202–203
peptide libraries, 1:288–291
protein therapeutics, 3:322–323
Master cell bank (MCB), protein therapeutics, 3:296
Master curve calculations:
oral drug delivery, immediate-release dosages, 2:781–784
repeated intravenous bolus injections, 2:756–759
Material interactions, protein therapeutics formulation, container closure materials, 3:315
Materiality requirement, infringement of patent claims, 3:155–156
Maternal transport mechanisms, immunoglobulins, 2:450–451
Mathematical modeling:
cost-effectiveness analysis of drug development, 3:346–349
drug delivery system bioavailability:
definitions, 2:746–751
diffusion-controlled dosage, plasma drug profile, 2:801–816

erosion-controlled dosage, plasma drug profile, 2:816–828
future research issues, 2:851–852
immediate-release oral dosage, 2:771–787
intravenous drug administration, 2:751–771
research background, 2:745–746
sustained release *in vitro* testing, 2:791–801
sustained release oral dosage, 2:787–791
tissue-based drug transfer, 2:838–851
transdermal systems, 2:828–838
Matrix-assisted laser desorption ionization (MALDI):
combinatorial library/natural product encoding and identification, 1:110–111
mass spectrometry, drug development, 1:99–100
solid-phase mass spectrometry, 1:115–116
Matrix metalloproteinase (MMP) inhibitors:
brain cancer therapy, hypoxia-inducible factor pathway, 6:254–255
chronic obstructive pulmonary disease therapy, 5:791–792
glaucoma intraocular pressure reduction, 5:592
prostaglandin analogs, 5:596–600
multiple sclerosis therapy, 5:580
nuclear magnetic resonance studies, 2:412–414
quantum mechanics/molecular mechanics binding affinity, linear interaction energy approaches, 2:730–731
soft drug design, 2:122–123
stroke therapy, 8:502–510
carboxylic acid-derived inhibitors, 8:508

hydroxamic acid-derived inhibitors, 8:502–507
hydroxypyridinones, 8:508
phosphorus-derived inhibitors, 8:509–510
pyramidine-2,4,6-triones, 8:508–509
thiols, 8:509
tazarotene therapy, 5:406
Maturity onset diabetes of the young type-2 (MODY-2):
glucokinase activation, 4:15–17
hepatocyte nuclear factor4 and, 4:163–164
Maxacalcitol, soft drug design, 2:101–102
Maxi-K channel, stroke therapy, 8:482–483
Maximum-tolerated dose, all *trans*-retinoic acid, 5:384–386
MAX/MIN values, high-throughput screening, assay configuration, 3:422–423
MB07803 prodrug, type-2 diabetes therapy, 4:19–20
MB07811 prodrug, thyromimetic structure-activity relationship, 4:211–212
MCDOCK algorithm, docking and scoring techniques, Monte Carlo simulation, 2:609–611
M-cells, receptor-mediated endocytosis, 2:450
m-Chlorophenylpiperazine (*m*-CPP):
antidepressant pharmacokinetics, 8:236–237
serotonin receptor targeting, 8:242–244
MC minimization (MCM) technique, docking and scoring techniques, Monte Carlo simulation, 2:610–611
MDL27695 antiparasitic/antiprotozoal agent, polyamine metabolism, 7:586–590

MDL-28170 calpain inhibitor, stroke therapy, 8:461–462
MDM2/MDM4:
brain cancer therapy, 6:256–258
protein-protein interactions, 2:344–352
MDMX protein, brain cancer therapy, 6:257–258
ME-1036 carbapenem, 7:357–358
Measurement and Treatment Research to Improve Cognition in Schizophrenia (MATRICS), cognitive dysfunction assessment, 8:20–22, 162
MeBmt analogs, cyclosporin A, structure-activity relationships (SARs), 5:1011
Mecamylamine nicotinic antagonist, 8:78
Mechanism-targeted studies:
β-lactamase inhibitors, clavulanic acid, 7:361–363
chemistry-driven drug discovery, 1:537–538
congestive heart failure therapy:
contractility efficiency and myocardial oxygen demand, 4:492–494
normal subjects *vs.* CHF patients, 4:495–496
pharmacodynamics/pharmacokinetics, 4:491–492
risk/liability evaluation, 4:500
specific target/mechanism perturbation, 4:494–495
osteoporosis antiresorptives:
bisphosphonates, 5:733–734
calcitonin and integrin antagonists, 5:730
cathepsin K inhibitors, 5:730–732
clinical trials, 5:740–741
epidemiology, 5:729–730
future research and applications, 5:742

Mechanism-targeted studies: (*Continued*)
   OPG/RANKL/RANK inhibitors, **5:**732–733
   preclinical studies, **5:**739–740
   selective estrogen receptor modulators, **5:**734–742
   quantitative structure-activity relationship modeling, **1:**526–529
   13-*cis*-retinoic acid therapy, **5:**388
Medical Structural Genomics of Pathogenic Protozoa (MSGPP), genome-centered research, **1:**572–573
Medicinal chemistry:
   angiotensin-converting enzyme inhibitors, **4:**271–279
     carboxyl dipeptides, **4:**273–274
     hydroxamic acid, **4:**274–275
     mercaptans, **4:**272–273
     natural products, **4:**276
     peptide analogs, **4:**275–276
     phosphinic/phosphonic acids, **4:**274
     vasopeptidase inhibition, **4:**276–279
   anti-HCV agents, **7:**176–204
   NS3/4A protease inhibitors, **7:**176–195
     boceprevir, **7:**186–191
     ciluprevir, **7:**176–181
     ITMN-191, **7:**191–192
     MK7009, **7:**193–195
     telaprevir, **7:**181–186
     TMC435350, **7:**193
   camptothecins, **6:**28–29
   CP-690,550 immunosuppressive compound, **5:**1057
   DNA-targeted compounds:
     bleomycin, **6:**9–10
     dactinomycin, **6:**4
     mitomycin, **6:**13–14, 15–16
     plicamycin, **6:**17
   farnesoid X receptors, **4:**161
   fingolimod development, **5:**1054–1055
   liver X receptors, **4:**151–154
   nevirapine synthesis, **3:**15–16
   nuclear hormone receptors, drug discovery applications, **4:**164–167
   peroxisome proliferator-activated receptors, **4:**87–88
     α-agonists, **4:**90–96
     α/γ-dual agonists, **4:**124–135
     α/γ/δ-pan agonists, **4:**140–145
     δ-agonists, **4:**118–121
     γ-agonists, **4:**102–104
   proteasome inhibitors, **6:**181–184
   structural genomics, PDB development of, **1:**582–584
   vinblastine, **6:**39
Medium design and composition, protein therapeutics, upstream process development, **3:**299–300
Medium-throughput screening (MTS), chemogenomics, **2:**580–582
Medroxyprogesterone acetate:
   biotransformation, **5:**229–230
   current compounds and applications, **5:**220–221
Medrysone, C-21 carbonyl alterations, **5:**104–112
Medulloblastoma:
   classification and therapy, **6:**234–235
   hedgehog signaling pathway, **6:**259–260
Mefloquine, antimalarial therapy, **7:**610–611
   combined artesunate experimental agents, **7:**627
Megalomicins, analog design, **7:**444
Megesterol acetate, steroidal progesterone receptor ligands, **5:**253
Meglumine antimonoiate, Leishmaniasis therapy, **7:**575–577
MEK inhibitors:
   antimalarial cell cycle regulation, **7:**630
   structure and properties, **6:**306–307
Melagatran, active site-directed inhibitors, **4:**397–398
Melamine, cocrystal engineering, **3:**213
Melanin-concentrating hormone 1 receptor (MCHR1) antagonists, hit seeking combinatorial libraries, **1:**324–325
Melanoma:
   bexarotene therapy, **5:**447
   N-(4-hydroxyphenyl) retinamide (4-HPR), **5:**472
Melarsen oxide, human African trypanosomiasis therapy, **7:**570–571
Melarsopol:
   human African trypanosomiasis therapy, **7:**570–571
   nucleotide transporter trypanocide development, **7:**577–580
Melatonin receptors, insomnia therapy, **5:**717–719
Melphalan, cancer therapy, **6:**83–84
Melting point, oral drug delivery, **3:**30–31
Memantine:
   cognitive enhancement, **8:**15–16, 33–34
   glaucoma therapy, clinical trials, **5:**602
   Parkinson's disease therapy, **8:**554–555
   stroke therapy, **8:**450
Membrane insertion scanning, transport proteins, **2:**460
Membrane proteins:
   drug transport:
     future research issues, **2:**476
     passive diffusion, **2:**444–446
     kinetics, **2:**445–446
     permeability enhancement, **2:**439–444
     absorption enhancement, transporter targeting, **2:**443–444
     lipid-based oral delivery, **2:**441
     penetration enhancers, **2:**439–441
     prodrugs, **2:**441–443

receptor-mediated endocytosis, **2**:446–454
  bacterial adhesins and invasins, **2**:451–452
  bacterial and plant toxins, **2**:452
  cell surface receptor structure, **2**:447–448
  endocytosis, **2**:446–447
  immunoglobulin transport, **2**:450–451
  lectins, **2**:452–453
  lipid rafts, **2**:449–450
  oral absorption systems, **2**:450–453
  potocytosis, **2**:448–449
  transcytosis, **2**:448
  viral hemagglutinins, **2**:452
  vitamins and metal ions, **2**:453–454
research background, **2**:439
transport proteins:
  apical sodium-dependent bile acid transporter, **2**:467–475
  ATP-binding cassette and solute carrier genetic superfamilies, **2**:456
  classification and structures, **2**:454–455
  Cys-scanning mutagenesis, **2**:460
  excimer fluorescence, **2**:461
  insertion scanning, **2**:460
  integral structure, **2**:459
  intestinal peptide transporter example, **2**:463–467
  models and substrate design, **2**:459
  N-glycosylation and epitope scanning mutagenesis, **2**:460–461
  P-glycoprotein example, **2**:461–463
  site-directed chemical cleavage, **2**:461
  therapeutic implications, **2**:457–459
receptor-based drug design, macromolecular compounds, **2**:393–394

structural genomics research, **1**:580–581
"Membranous web" assembly, hepatitis C viral genome, **7**:173–174
Memory:
  assay techniques, **8**:26–29
  cognition and, **8**:17–19
  molecular substrates, **8**:24–25
  neuranatomical substrates, **8**:23
  phosphodiesterase inhibitors, **5**:716
Menaquinone:
  antibiotic selective toxicity, **2**:553
  antitubercular agents, cofactor biosynthesis targeting, **7**:728
Meningiomas, classification and therapy, **6**:237
Menoctone antimalarial agent, **7**:619–620
  cellular respiration and electron transport, **7**:655–658
Mepacrine, antimalarial therapy, **7**:607–610
Meperidine:
  chemical structure, **8**:570
  metabolism and elimination, **8**:579
  piperidine derivatives, **8**:614–621
Mephenytoin hydroxylase, pharmacogenomics, **1**:186–187
Mercaptans, ACE inhibitor design, **4**:272–273
Mercapto derivatives, anabolic agents, **5**:182–183
Mercaptohexanoic acid analogs, peroxisome proliferator-active receptors, α-agonist derivation, **4**:93–96
6-Mercaptopurine, antimetabolite analogs, **6**:112–113
Merck ATP-competitive kinesin spindle protein inhibitor, **6**:213–214
Mercury compounds, topical antimicrobial agents, **7**:497

Merimepodib, structure and properties, **5**:1042–1043
Mer kinases, structural genomics, **1**:592–593
Meromycolates, antitubercular agents, mycolic biosynthesis, **7**:722–723
Meropenem:
  occurrence, structural variations and chemistry, **7**:333–335
  synthesis and development, **7**:343–345
Mesalamine, hydrogen sulfide releasing molecules, **5**:314–315
"Message-address" concept:
  δ-opioid selective agonists, **8**:637
  opioid receptor selective antagonists, **8**:604–608
Mestranol, current compounds and applications, **5**:220–221
Metabolic disease, multitarget drug development, PPAR-plus, **1**:261, 264
Metabolic syndrome:
  Alzheimer's disease, **8**:405, 411–417
  cholesterol-related approaches, **8**:405, 411–414
  insulin-related approaches, **8**:414–417
  peroxisome proliferator-activator receptors, PPARγ/δ antagonists, **4**:137–140
Metabolism. *See also* Retrometabolic drug design
  acitretin, **5**:401
  adapalene, **5**:414
  ADMET properties, **2**:57–59
  assays, **2**:66–68
  all *trans*-retinoic acid, **5**:385–386
  Am80 therapy, **5**:419–420
  androgens, **5**:159–164
    anabolic agents, **5**:186–188
    oxidative metabolism, **5**:162–164

Metabolism. (*Continued*)
   reductive metabolism, **5**:159–162
   antiandrogen compounds, **5**:193
   antipsychotic effects on, **8**:168
   antitubercular agents, intermediary metabolism and respiration targeting, **7**:729–730
   bexarotene therapy, **5**:453
   bioactivity, **1**:439–443
   biotin, **5**:686–690
   chirality and, **1**:141–142
   conjugation reactions, **1**:421–439
     acetylation/acylation, **1**:429–432
     amino acid formation, **1**:432
     arylpropionic acids, **1**:433–434
     coenzyme A, **1**:432–434
     glucuronidation/glucosidation, **1**:426–429
     glutathione, **1**:434–438
     hybrid lipid/sterol ester formation, **1**:432–433
     methylation, **1**:423–425
     β-oxidation and two-carbon chain reactions, **1**:434
     phosphate esters and hydrazones, **1**:438–439
     sulfonation, **1**:425–426
   definitions and concepts, **1**:405
   drug delivery and bioavailability, **2**:748
   etretinate, **5**:402
   folic acid, **5**:691–698
   functionalization reactions, **1**:408–421
     carbon oxidation and reduction, **1**:413–416
     enzymes, **1**:409–413
       cytochromes P450, **1**:410–411
       hydrolases, **1**:412–413
       oxidoreductases, **1**:411–412
     hydration and hydrolysis, **1**:421
     nitrogen oxidation and reduction, **1**:416–418
     oxidative cleavage, **1**:419–421
     sulfur oxidation and reduction, **1**:418–419
   glucorticoid anti-inflammatory compounds, **5**:56–59
   N-(4-hydroxyphenyl) retinamide (4-HPR), **5**:478–480
   linezolid, **7**:543
   mass spectrometry assays, **1**:120–121
   9cUAB30 RXR-selective retinoid therapy, **5**:457–458
   nuclear magnetic resonance analysis, **2**:368
   opioid receptors, **8**:578–579
   oral dosage forms, class V drugs, **3**:53
   pantothenic acid, **5**:685–687
   permeability, ADMET interactions:
     gut-wall metabolism, **3**:372
     hepatic metabolism, **3**:371–372
     oral bioavailability, **3**:372
     polymorphism and, **3**:367–369
     renal metabolism, **3**:372–373
   pharmacogenomics, **1**:183–189
     cytochrome p450 system, **1**:183–187
     dihydropyrimidine dehydrogenase, **1**:187–188
     N-acetyltransferase-2, **1**:188
     thiopurine methyltransferase, **1**:188–189
   pharmacological activity, **1**:439–441
   prodrug development, **3**:224–226
     design issues, **3**:230–233
     rule-of-one metabolism, **3**:248–250
   protein therapeutics, upstream process development, **3**:300
   pulsed ultrafiltration-mass spectrometry screening, **1**:118–119
   quantitative structure-activity relationship, **1**:66–67
   research background, **1**:405
   9-*cis*-retinoic acid therapy, **5**:428–429
   13-*cis*-retinoic acid therapy, **5**:395
   retinol (vitamin A), **5**:644–646
   SHetA2 retinoid-related compound, **5**:506
   steroids, **5**:25–27
     androgens, **5**:26–27
     corticosteroids, **5**:25–26
     estrogen, **5**:27, 29
     progesterone, **5**:26–27
   structure-metabolism relationships, **1**:443–449
     biological factors, **1**:447–449
     biotransformation, global expert systems, **1**:447
     chirality, **1**:443–444
     lipophilicity, **1**:444–445
     three-dimensional quantitative structure-metabolism relationships, **1**:445–447
   sulfonamides, **7**:522–523
   sulfones, **7**:523
   tazarotene, **5**:408
   thiamine (vitamin B_1), **5**:670–672
   thyroid hormones:
     carbohydrates, **4**:195
     lipids, **4**:195
     metabolic rate, **4**:194
     proteins, **4**:195
     receptor agonists, modulation effects, **4**:197–198
   toxicity, **1**:441–443
   toxicophores, alternative pathways, **2**:326
   type 2 diabetes therapies, dipeptidyl peptidase 4 inhibitors, **4**:46–47
   vitamin B_{12} (cobalamin), **5**:697, 699
   vitamin C (ascorbic acid), **5**:703
   vitamins:
     human studies, **5**:643
     vitamin D, **5**:651
   xenobiotic reactions, **1**:405–408
     enzyme/metabolic reactions, **1**:407–408
     pharmacodynamics, **1**:407
     specificity and selectivity, **1**:406–407
     in vitro studies, **1**:407–408

Metabolites:
ADMET metabolic stability, identification assays, **2:**67–68
analog drug design, **1:**170
estrogen receptor ligands, **5:**226–228
genomic influences, **2:**248–250
glucocorticoid anti-inflammatory compounds, **5:**96–99
N-(4-hydroxyphenyl) retinamide (4-HPR), **5:**479–480
natural products lead generation, terrestrial and marine habitats, **2:**195–199
pharmacological activity, **1:**439–441
plant metabolites, **2:**250–251
progesterone, **5:**229–230
soft drug development:
active metabolite-based compounds, **2:**80, 127–128
inactive metabolite-based compounds, **2:**80–81
testosterone excretion, **5:**161–162
Metabotropic glutamate receptors:
anticonvulsants, **8:**134
antipsychotic glutamatergics, **8:**203
cognitive enhancement, **8:**42–43
psychostimulant physiology and pharmacology, **8:**101–106
receptor 5 (mGluR5), hit seeking combinatorial libraries, **1:**318–320
schizophrenia pathology, **8:**164
Metaclopramide, migraine therapy, **8:**297–298
Metagenomics, natural products lead generation, **2:**199–200
Metaglidasen/halofenate, selective PPARγ modulators, PPARγ partial agonists, **4:**108–113
Metal chelator inhibitors, HIV integrase active sites, **7:**80–81

pharmacophore elaboration, **7:**81–97
Metal-heteroatom chelation model, HIV integrase, two-metal chelation elaboration, **7:**82–97
Metal ions, receptor-mediated endocytosis, **2:**453–454
Metallo-β-lactamases, thiol inhibition, **7:**372
Metalloprotease inhibitors, hit seeking combinatorial libraries, **1:**325–327
Metal-protein attenuating compounds (MPACs), β-amyloid peptide inhibitors, **8:**363–365
Metastable drug forms, oral drug physicochemistry, **3:**45–46
Metastases:
brain tumors, **6:**240
animal models, **6:**273
kinase inhibitors, **6:**325–330
c-Met kinase inhibitors, **6:**326, 328–330
focal adhesion kinase inhibitors, **6:**326, 328
Src kinase inhibitors, **6:**325–328
Metazocine, morphinan derivatives, **8:**611
Metdetomidine enantiomers, chirality, **1:**133–137
Metformin:
ADMET properties, **4:**3–5
Alzheimer's disease, insulin related therapy, **8:**415–417
type 2 diabetes, **4:**3
Methacholine, structure and function, **8:**64
Methadol, structure-activity relationship, **8:**621–623
Methadone:
acyclic analogs, **8:**621–623
chemical structure, **8:**570
chirality and metabolism, **1:**141–142
metabolism and elimination, **8:**579
structure-activity relationship, **8:**598
Methamphetamine:
ADMET properties, **8:**97–99

historical development, **8:**90–91
nitrogen substituents, **8:**107
physiology and pharmacology, **8:**100–106
side chain modification, **8:**107
side effects, adverse reactions and drug interactions, **8:**92, 95–97
structure-activity relationship, **8:**106–109
Methanoprolinenitriles, dipeptidyl peptidase 4 (DPP-4) inhibition, **4:**51–54
Methantheline, muscarinic antagonists, **8:**72–74
Methazolamide, glaucoma therapy, **5:**598–600
Methchloramine, cancer therapy, **6:**83–85
Methemoglobin formation, nitric oxide effects, **5:**273–278
Methenamine, structure and properties, **7:**550
Methicillin-resistant *Staphylococcus aureus* (MRSA):
ceftobiprole activity against, **7:**316–317
chemogenomics, diversity-oriented synthesis, **2:**579–580
dihydrofolate reductase inhibitors, **7:**534
glycopeptide antibiotics, **2:**225–227
β-lactam antibacterial drug resistance, PBP modification, mechanism and genetics, **7:**272–273
linezolid antibacterial activity, **7:**541–544
lipopeptide antibiotics, **2:**227–228
meropenem activity, **7:**344–345
RNA targets, linezolid, **5:**990–993
tetracycline antibiotics and, **2:**225
Methicillin-resistant *Staphylococcus epidermis* (MRSE), linezolid RNA targets, **5:**990–993

Methimazole (MMI), hyperthyroidism, **4:**199
Methionine aminopeptidase-2 (MetAP-2) inhibitors, hit seeking combinatorial libraries, **1:**315–317, 319
Methoctramine, muscarinic antagonists, **8:**76–77
Methods validation, commercial-scale operations, **3:**12
Methotrexate:
  chemogenomics, yeast three-hybrid screening, **2:**581–582
  folic acid uptake and metabolism, **5:**691–698
    antifolate cancer therapy, **6:**106–109
    multiple sclerosis therapy, **5:**568–569
  nuclear magnetic resonance studies, **2:**414–415
  protein therapeutics:
    cell line development, **3:**293–294
    transfection and gene amplification, **3:**295–296
  selective toxicity and comparative biochemistry, **2:**549
Methoxime analogs, eye-targeted chemical delivery systems, **2:**158–160
7α-Methoxy cephalosporins, development of, **7:**307
Methoxychlor, physiology and pharmacology, **5:**239–240
1-Methyl-4-phenyl-pyridinium (MPP+), selective toxicity, **2:**563–564
Methylated derivatives, androgens, **5:**171
Methylation:
  chemokine receptor-5 (CCR5) antagonists, **7:**111–116
  drug metabolism, **1:**423–425
  posttranscriptional methylation inhibitors, **7:**241

Methylbenzethonium chloride, cationic surfactants, **7:**500
1β-Methylcarbapenems, synthesis, **7:**340–345
β-Methylcrotonyl CoA carboxylase, biotin metabolism, **5:**688–690
Methylene blue β-amyloid peptide inhibitors, **8:**364–365
  tau aggregation, **8:**423–429
Methylene chloride, carbon monoxide prodrugs, **5:**305–307
Methylguanine methyltransferase (MGMT), glioblastoma therapy, **6:**231–232
  temozolomide, **6:**244
3-Methylindoles, toxicophore reactive metabolites, **2:**317–318
[4-(methylnitrosamino)-1-(3-pyridyl)-1-butaneone] (NNK), chirality and carcinogenicity, **1:**145–146
Methylphenidate:
  ADMET properties, **8:**98–99
  cognitive function assessment, **8:**26–27
  physiology and pharmacology, **8:**103–106
  side effects, adverse reactions and drug interactions, **8:**95
  structure-activity relationship, **8:**109
Methylphenidate, soft drug design, **2:**104–106
Methylphenols, antimicrobial agents, **7:**493–494
1-Methyl-4-phenylpyridinium (MPP+), Parkinson's disease and, **8:**530, 539–540
N-Methyl-4-phenyl-1,2,3,6-tetrahydropyridine (MPTP), Parkinson's disease and, **8:**538–540
N-Methyl-4-propionoxy-4-phenylpiperidine (MPPP):
  opioid derivatives, piperidine reversal, **8:**616–618

Parkinson's disease and, **8:**539–540
Methylscopalamine, soft drug design, anticholinergics, **2:**109
Methylsulfonamide heptanoic acid analogs, EP_2 receptor agonist, **5:**849–850
1-Methyl-1,2,3,6-tetrahydropyridine (MPTP), selective toxicity, **2:**563–564
Methylthioninium chloride (MTC), Alzheimer's disease, tau-targeted therapy, **8:**424–425
Methyl trap hypothesis, vitamin B_{12} (cobalamin), **5:**697, 700
Methyltriazolopthalazine, conjugation reactions, **1:**439
Methylxanthines, chronic obstructive pulmonary disease bronchodilators, **5:**772, 778–779
Methymycin, chemical structure, **7:**436–438
"Me-too drugs":
  direct analogs, **1:**167–168
  existing drug analogs, **1:**170
Metoprolol:
  chirality, **1:**139–140
  migraine prevention, **8:**314–315
  soft drug design, **2:**81–83
Metropolis algorithm, docking and scoring techniques, Monte Carlo simulation, **2:**609–611
Metropolol compound, transdermal drug delivery, **2:**832–835
Mevalonate, cholesterol biosynthesis, **5:**15–16
Mevastatin, anticholesterolemics, **2:**232
MGCD-0103. *See* Mocetinostat
MI-63 compound, brain tumor therapy, **6:**257–258
MI-219 compound, protein-protein interactions, **2:**352–353

Mianserin:
  α-adrenergic receptor targeting, 8:247
  chirality, 1:443–444
  cognitive enhancement, 8:37–40
  serotonin receptor targeting, 8:241–242
Mibefradil, functional analog design, 1:176–177
Micafungin:
  development, 2:231–232
  selective toxicity, 2:559, 562
Micelle preparation:
  nanoscale drug delivery, 3:469–473
  oral drug physicochemistry, class II drugs, 3:48–49
Michael acceptors:
  combinatorial chemistry, 1:296
  stroke therapy:
    calpain inhibitors, 8:464
    cathepsin B inhibitors, 8:477
Michaelis-Menten equation:
  chirality and drug metabolism, 1:443–444
  high-throughput screening, assay configuration, 3:420
  receptor-based drug design:
    allosteric modulators, 2:499–501
    occupancy theory, 2:494–497
Microbe reengineering, recombinant DNA technology, 1:547–549
Microbial compounds:
  antibacterial drug development, 2:221–231
    antifungals, 2:230–231
    glypeptides, 2:225–227
    β-lactams, 2:222–223
    lipopeptides, 2:227–228
    macrolides, 2:228–230
    tetracyclines, 2:223–225
  anticancer agents, 2:235–243
    actinomycins, anthracyclines, bleomycins, enediynes, 2:235–239
    erbstatin and lavendustin-related molecules, 2:239–240
    rapamycins and epothilones, 2:240–243
    tyrphostins, 2:239–240

  anticholesterolemics, 2:231–235
    combination agents, 2:234–235
    HMG-CoA reductase inhibitors, 2:234
  chemotherapy, basic principles, 7:501–502
  marine sources, 2:243–248
    bryostatins, bolastatins, and analogues, 2:243–245
    cryptophycins, 2:245
    curacin A, 2:245
    Et743, 2:246–247
    peloruside A, 2:246
    secondary metabolites/derivatives, 2:247–248
  metabolite genomics, 2:248–250
  natural products lead generation, 2:193–199
  plant metabolites, 2:250–251
  receptor-mediated endocytosis:
    adhesins and invasins, 2:451–452
    plant toxins, 2:452
  sources, 2:221
  sulfonamides and sulfones, 7:523–524
  topical antimicrobial agents:
    resistance mechanisms, 7:488
    selective toxicity, 7:484, 488
Microcalorimetry, ligand-macromolecular interactions, structure-based thermodynamic analysis, 2:702–712
Microcarriers, protein therapeutics, attachment-dependent cell culture, 3:296–297
Microemulsions, oral drug physicochemistry, class II drugs, 3:49
Microenvironmental factors, central nervous system tumors, 6:227
Microfiltration, mammalian protein purification, harvest operations, 3:307
Microorganism depositories, drug patent requirements, 3:126–127

Microsomal triglyceride transfer protein (MTP), atherosclerosis therapy, 4:319–320
Microsomes, ADMET metabolic stability assays, 2:67
Microtubule-affinity-regulating kinase (MARK), Alzheimer's cognitive deficits, 8:16
Microtubules:
  Alzheimer's disease, tau-targeted therapy, 8:430
  kinesin spindle protein structure:
    crystallographic analysis, 6:192–193
    mitosis-targeted therapies, 6:191–192
Microwave-accelerated techniques, combinatorial chemistry, 1:348–349
  cephalosporins, 1:305
  nucleoside libraries, 1:292–293
  P2X$_7$ nucleotide receptor antagonists, 1:325–326
Midazolam, anticonvulsant applications, 8:144–147
Midesteine, chronic obstructive pulmonary disease therapy, 5:791
Midostaurin, acute myelogenous leukemia therapy, 6:301–302
Mifeprestone:
  antiglucocorticoid mechanisms, 5:113–114
  physiology and pharmacology, 5:239
  recombinant DNA technology, receptor-based targeting mechanism, 1:560
Miglitol, type 2 diabetes therapy, 4:5–8
Migraine:
  epidemiology, 8:265–266
  pathology, 8:266–267
  therapeutic agents:
    antiemetics/antipsychotics, 8:297–298

Migraine: (Continued)
  calcitonin gene-related peptide antagonists, 8:272–290
    azepinone binding affinity, 8:281–286
    benzodiazepine binding affinity, 8:277–281
    BIBN4096 BS, 8:273, 275–277
    BMS-694153, 8:288–290
    MK-0974, 8:277, 279–288
    neuropeptide-induced blood flow, 8:286–288
    receptor binding affinities, 8:273–275
    unnatural amino acids, 8:289–290
  ergot alkaloids, 8:267–268
  future research issues, 8:320–321
  glutamate modulators, 8:290–297
    ionotropic receptor antagonists, 8:291–297
    receptor properties, 8:291
  nitric oxide synthase inhibitors, 8:298–306
    iNOS inhibitors, 8:300–302
    isoforms, 8:298–299
    nNOS inhibitors, 8:302–306
    tool compounds, 8:299–300
  nonsteroidal anti-inflammatory drugs and paracetamol, 8:268
  preventive therapies, 8:312–320
    angiotensin converting enzyme inhibitors, 8:319–320
    anticonvulsants, 8:315–316
    antidepressants, 8:318–319
    β-blockers, 8:313–315
    calcium channel blockers, 8:316–318
  serotonin agonists, 8:268–272
    anti-inflammatory/vasoconstriction mechanisms, 8:270–272

second-generation triptans, 8:270
sumatriptan, 8:268–270
vanilloid receptor antagonists, 8:307–312
capsazepine, civamide, and AMG-517, 8:307–308
SB-705498, 8:308–312
Mild cognitive impairment (MCI), epidemiology, 8:15
Milnacipran, structure-activity relationship, 8:251
Milrinone, antiplatelet activity, 4:450
Miltefosine, Leishmaniasis therapy, 7:576–577, 590–593
Milveterol, chronic obstructive pulmonary disease bronchodilators, 5:771–772
Mineralocorticoids, steroidogenesis, 5:20–22
Miniaturization, high-throughput screening, enzymatic assays, 3:406
Minimization, docking and scoring techniques, Monte Carlo simulation, 2:608–611
Minimum daily requirements (MDRs), vitamins, 5:638–640
Minocycline:
  ADMET properties, 7:409–410
  chemical structure, 7:407
  history and biosynthesis, 7:410–413
  research background, 7:407
  stability, side effects, and contraindications, 7:408–409
Minor groove targeting, DNA-targeted chemotherapeutic compounds:
  cyclopropylindoles, 6:91–93
  mustards, 6:87–88
Minoxidil, sulfonation reactions, 1:426
Mirincamycin antimalarial agent, 7:625–627

Mirtazepine:
  α-adrenergic receptor targeting, 8:247
  chemical structure, 8:224
  pharmacokinetics, 8:237
Mitemcinal macrolide compound, 7:456
Mitglinide, type 2 diabetes therapy, 4:10–13
Mithracin. See Plicamycin
Mithramycin. See Plicamycin
Mitochondrial ATO-sensitive potassium channel, chirality, enantiomer activity, 1:130–131
Mitochondrial metabolism:
  AHPN-induced apoptosis, 5:490–497
  Parkinson's disease, dopamine and, 8:535–537
  toxicity, anti-DNA virus antimetabolites, 7:226–227
Mitogen-activated protein kinase-activated protein kinase 2 (MAP-KAPK2), quantitative structure-activity relationship, multidimensional models, 1:40–43
Mitogen-activated protein kinase (MAPK) inhibitors:
  Alzheimer's disease, tau-targeted therapies, 8:420–422
  antimalarial cell cycle regulation, 7:630
  brain tumor therapy, signaling pathways, 6:252–253
  carbon monoxide therapeutic effects, 5:298–299
  chronic obstructive pulmonary disease therapy, 5:789–790
  drug resistance, 6:366
  growth factor signaling pathways, 6:301–307
  opioid receptor signaling, 8:581
Mitomycin:
  chemical structure, 6:12–13
  contraindications and side effects, 6:13
  hypoxia-activated bioreductive prodrugs, 6:116–120
  medicinal chemistry, 6:13–16

molecular mechanisms,
  6:14–15
pharmacokinetics, **6:**13
therapeutic applications, **6:**13
Mitosis, therapeutic targeting,
  6:191–192
Mitoxantrone:
  multiple sclerosis therapy,
    5:570
  topoisomerase II inhibitors,
    6:99–102
Mitozolomide, DNA-targeted
  compounds, **6:**94–95
Mitsunobu conditions, acyclic
  nucleoside
  phosphates, **7:**233
MIV-170 HIV reverse
  transcriptase
  inhibitor, **7:**150–152
MIV-210 HIV reverse
  transcriptase
  inhibitor, **7:**142–145
Mivacurium, nicotinic
  antagonists, **8:**79–80
Mixed-function ligands, serotonin
  receptor targeting,
  8:243–244
Mixture design, nuclear magnetic
  resonance, **2:**433–434
Mizoribine:
  ADMET properties, **5:**1049
  nucleoside inhibitors,
    5:1044–1046
  research background, **5:**1038
  therapeutic applications,
    5:1050
MK-0533 compound, selective
  PPARγ modulators,
  PPARγ partial
  agonists, **4:**114–115
MK0608 anti-HCV agent,
  7:199–201
MK-0731 kinesin spindle protein
  inhibitor, **6:**204–209
MK886 peroxisome proliferator-
  activator receptorα
  antagonist, **4:**99–100
MK-0974 calcitonin gene-related
  peptide antagonist,
  migraine therapy,
  8:277, 279–288
MK-3207 calcitonin gene-related
  peptide antagonist,
  migraine therapy,
  8:287–288
MK-4965 reverse transcriptase
  inhibitor, **7:**155

MK7009 protease inhibitor, anti-
  HCV agents,
  7:193–195
MKC-442 HIV reverse
  transcriptase
  inhibitor, **7:**148–152
MNDO, quantitative structure-
  activity relationships,
  1:23
Mobility determination, *in silico*
  screening, protein
  flexibility,
  2:861–865
Mocetinostat, histone deacetylase
  inhibitors, **6:**67–70
Moclobemide:
  chemical structure, **8:**223
  physiology and pharmacology,
    8:248
  side effects and adverse
    reactions, **8:**230
  structure-activity relationship
    and metabolism,
    8:252–253
Modafinil:
  memory assessment, **8:**28–30
  psychostimulant physiology
    and pharmacology,
    8:114
Modeling framework,
  quantitative
  structure-activity
  relationship modeling,
  1:509–510
Model scoring functions, drug
  discovery, **2:**618–632
  consensus scoring,
    2:631–632
  empirical methods, **2:**623–628
  force field-based methods,
    2:619–622
  knowledge-based methods,
    2:628–631
  semiempirical methods,
    2:622–623
Modified dispersion model
  (MDM), permeability,
  ADMET interactions,
  fraction absorbed (f_a)
  vs., **3:**369–370
MOGA program, pharmacophore
  structural properties,
  bioactive
  conformation,
  1:470–471
Moieties, prodrug design,
  3:250–252

Moisture uptake, oral drug
  physicochemistry,
  3:37–39
Molecular biology:
  camptothecins, **6:**27–28
  carboxylesterases, **3:**234,
    237–238
  DNA-targeted
    chemotherapeutic
    compounds, mustards,
    6:87
  DNA-targeted compounds:
    bleomycin, **6:**11
    dactinomycin, **6:**5–7
    mitomycin, **6:**14–15
    plicamycin, **6:**17
  HIV integrase active sites,
    7:80–81
  opioid receptors, **8:**587–591
  peroxisome proliferator-
    activated receptors,
    6:157–159
  receptor-based drug design,
    2:514
  biologicals and antisense
    compounds, **2:**531–533
  recombinant DNA technology:
    future trends, **1:**561–562
  receptor-based targeting
    mechanism,
    1:556–560
  vinblastine, **6:**39
Molecular descriptors,
  quantitative
  structure-activity
  relationships, **1:**23–28
  database development, **1:**68–72
  DCW (SMILES), **1:**26–27
  effective hydrogen charge,
    1:27–28
  polarizability, **1:**24–25
  polar surface area, **1:**25–26
  topological descriptors, **1:**23–24
  van der Waals volumes, **1:**26
Molecular diffusion, nuclear
  magnetic resonance
  studies, drug
  screening
  applications,
  2:426–427
Molecular diversity, druggability
  vs., **1:**295
Molecular dynamics (MD):
  docking and scoring techniques,
    2:614–616
  flexibility computation,
    2:638–639**

Molecular dynamic (*Continued*)
structure-based drug design, **2**:679–680
G-protein coupled receptor, homology modeling, **2**:286–287
quantitative structure-activity relationship:
hydrophobic interactions, **1**:11–19
multidimensional models, **1**:41–43
in silico screening, protein flexibility, **2**:862–863
backbone mobility, **2**:869–870
Molecular informatics, chemogenomics, **2**:582–588
information systems, **2**:582–583
ligand-based data analysis and predictive modeling, **2**:583–585
structure-based data analysis and predictive modeling, **2**:585–588
Molecular interaction fields (MIFs):
chemogenomics, **2**:587–588
chemotype development, **1**:484–487
drug metabolism, **1**:447
RNA targets:
linezolid, **5**:990–993
neomycin/gentamicin aminoglycosides, RNA targets, **5**:973–978
streptomycin, **5**:968–971
structure-based pharmacophore modeling, **2**:682–683
Molecular Libraries Roadmap Initiative, quantitative structure-activity relationship modeling, **1**:513–514
Molecular mechanics (MM) force fields. *See also* Quantum mechanics/ molecular mechanics (QM/MM)
drug discovery, **2**:619–622
ionization and tautomerization, **2**:637
quantum mechanics/molecular mechanics, binding affinity evaluation, **2**:717–723
Molecular memory substrates, **8**:24–25
Molecular modeling, drug metabolism, **1**:445–447
Molecular orbital (MO) calculations, quantum mechanics/molecular mechanics, **2**:736–737
Molecular packaging, chemical delivery systems, **2**:150–155
brain-targeting redox analogs, **2**:154–155
kyotorphin analogs, **2**:153–154
leu-enkephalin analogs, **2**:151–153
TRH analogs, **2**:153
Molecular targeting:
anticonvulsants, **8**:121–134
connexins, **8**:134
enzymes, **8**:132–134
G-protein-coupled receptors, **8**:134
ligand-gated ion channels, **8**:126–130
neurotransmitter transporters and synaptic proteins, **8**:130–132
voltage-gated ion channels, **8**:122–126
chirality:
drug-receptor interactions, **1**:127–129
enantiomer activity, **1**:130–131
Molecular weight (MW):
protein-protein interactions, **2**:335–337
quantitative structure-activity relationship, steric parameters, **1**:20–21
Molindone, pharmacokinetics, **8**:190–191
Molsidomine, nitric oxide donors and prodrugs, **5**:285–286
Mometasone furoate, chronic obstructive pulmonary disease inhalation therapy, **5**:775–776
Monastrol kinesin spindle protein inhibitor, "induced fit" binding pockets, **6**:193–197
Monoamine oxidase (MOA) inhibitors:
chemical structure, **8**:223
historical background, **8**:248
Parkinson's disease:
dopamine-mitochondrial metabolism and, **8**:535–537
dopaminergic therapy, **8**:549–551
etiology, **8**:539–540
pharmacokinetics, **8**:236
physiology and pharmacology, **8**:247–248
prodrug development, **3**:224–226
recent and future research issues, **8**:255–256
selective toxicity, **2**:563–564
side effects and adverse reactions, **8**:230
structure-activity relationship and metabolism, **8**:251–253
Monoamine transporter, antidepressant pharmacology, **8**:237–241
Monoamine uptake inhibitors, psychostimulant physiology and pharmacology, **8**:100–106
Monoaryl acetic acid, CRTH2 receptor antagonist, **5**:914
Monobactams:
β-lactamase inhibitors, **7**:371
development of, **7**:324
natural products and synthetic derivatives, **7**:327–329
new compounds, **7**:329–333
structure-activity relationships, **7**:260–261
structure determination and synthesis, **7**:327
Monoclonal antibodies (mAb):
Alzheimer's immunization therapy, **8**:370–381
anti-cancer activity, **2**:238–239
mammalian protein purification, **3**:306–309

multiple sclerosis therapy, 5:572–573
receptor-based drug design, 2:533
recombinant DNA technology, antibody-based therapeutics, 1:543
selective toxicity, 2:559–562
targeted chemotherapy, resistance and, 6:363–370
tyrosine kinase targeting, 6:368–369
Monocyclic nonantibacterial pharmacophore, 7:332–333
Monocyte chemotactic proteins (MCP), soft drug design, cytokine inhibitors, 2:122
Monod-Wyman-Changeaux hemoglobin model:
  allosteric protein theory, 1:375–379
  binding site location, 1:388–392
  oxygen delivery, 4:610–612
Monoiodothyronamine (T_1AM), 3,5,3′,5′-tetraiodo-L-tyronine ($L-T_4$) uptake and metabolism, 4:194
Monoiodotyrosine (MIT), thyroid hormone biosynthesis, 4:193
Monooxygenases:
  antitubercular agents, cytochrome P450 targeting, 7:729
  drug metabolism, functionalization reactions, 1:411–412
Monophenols, antimicrobial agents, 7:493–494
Monophosphate salt I, cocrystal engineering, 3:212–213
Monovalent salts, high-throughput screening, assay optimization, 3:412
Monte Carlo simulation:
  allosteric protein sites, 1:390
  docking and scoring techniques, 2:608–611
  quantum mechanics/molecular mechanics, 2:736

Morin β-amyloid peptide inhibitors, 8:364–365
Morphinans, structure-activity relationship, 8:598, 609–610
Morphine and derivatives:
  ADMET properties, 8:578–579
  C-ring derivatives, 8:600–601
  dismantling of, 8:598
  identification and early history, 8:596–597
  nonpeptide affinity labeling, 8:638–641
  structure-activity relationship, 8:599–601
  synthetic analogs, 8:597–598
Morris Water Maze test:
  Alzheimer's disease assessment, cholesterol levels, 8:405–406
  spatial memory working memory assessment, 8:27–28
Mosaic gene formation, β-lactam antibacterial drug resistance, PBP modification, 7:271–272
Motesanib, antiangiogenic properties, 6:309–310
Motilin receptor antagonists, hit seeking combinatorial libraries, 1:307–310
Mouse myeloma (NS0) cells, protein therapeutics, 3:293–294
Moxalactam, development of, 7:309–310
Moxifloxacin, antibacterial agents, 7:535–540
Mozibil, structure and properties, 4:586–587
MRE-0094 adenosine A_{2a} receptor agonist, chronic obstructive pulmonary disease therapy, 5:793–794
mRNA:
  protein therapeutics:
    AHPN compound expression and activation, 5:494–497
    expression vectors, 3:294–295
    steroid hormone actions, 5:29–30

MS-275. See Entinostat
MsbA lipid transporter, homology modeling, 6:374–376
M-Score function, empirical scoring, 2:630–631
MT2 melatonin receptor, G-protein coupled receptor, homology modeling, 2:288–289
mTOR signaling pathway:
  brain cancer therapy, 6:250–252
  cranial and spinal nerve tumors, 6:238–239
  inhibitors, 6:324–325
  sirolimus-FKBP 12 complex, 5:1026–1029
Mucocutaneous diseases, adapalene, 5:412
Mucosal drug delivery, protein therapeutics, 3:315–316
Multiangle laser light scattering (MALLS), protein therapeutics, 3:323–324
Multicomponent reactions (MCRs):
  lead optimization, 1:345
  protein-protein interactions:
    high-throughput screening, 2:338–340
    p53/MDM2 compounds, 2:346–359
Multidimensional models, quantitative structure-activity relationship, 1:39–43
Multidimensional nuclear magnetic resonance, basic principles, 2:370–371
Multidrug resistance protein 1 (MRP1; ABCC1), drug resistance mechanisms, 6:372
Multidrug-resistant tuberculosis (MDR-TB), epidemiology, 7:713–716
Multikinase inhibitors (MKIs), multitarget drug development, cancer therapy, 1:261, 263–267

Multilinear regression analysis
(MRA), quantitative
structure-activity
relationships, 1:33–37
  Free-Wilson and other
    approaches, 1:36–37
  linear models, 1:33–34
  nonlinear models, 1:34–36
Multiobjective evolutionary
    graph algorithm
    (MEGA), drug
    discovery, 2:264–266
Multiobjective optimization, drug
    discovery:
  ADMET CYP 2D6, 2:268–269
  blood-brain-barrier and
    solubility, 2:268–269
  clinical applications, 2:263–266
  heptatoxicity, 2:269
  overview of techniques,
    2:261–263
  Pareto ligand designer,
    2:266–273
  research background,
    2:259–261
Multiple active site correction
    (MASC), structure-
    based virtual
    screening, 2:23
Multiple dose administration,
    oral drug delivery,
    immediate-release
    dosage, 2:776–784
Multiple drug resistance:
  apoptosis-based protein
    targeting, 6:370–371
  cancer therapies, 6:370–376
    ATP binding cassette (ABC)
      transporters,
      6:371–376
    stem cells, 6:376–377
Multiple myeloma, osteoporosis
    anabolic therapy, Wnt
    pathway and bone
    remodeling,
    5:756–757
Multiple opioid receptors,
    discovery and
    development,
    8:579–580
Multiple parallel synthesis,
    combinatorial
    chemistry:
  analytical problems, 1:349
  automation, 1:348
  definitions, 1:280
  future trends, 1:350–352

historical background,
    1:277–280
informatics and data handling,
    1:349–350
libraries:
  encoding and identification,
    1:110–111
  liquid chromatography-mass
    spectrometry
    purification,
    1:107–108
  pulsed ultrafiltration-mass
    spectrometry,
    1:116–119
  solid- and solution-phase
    libraries, 1:294–343
  structure and purity
    confirmation,
    1:108–110
  linear, convergent, and
    multicomponent
    reactions, 1:345
microwave acceleration,
    1:348–349
patents, 1:350
peptide arrays, 1:282–291
purification, 1:345–347
  "pure" compounds, 1:347
  testing mixtures, 1:347
  resins and solid supports,
    1:343–344
  sample handling and storage,
    1:349
solid-phase organic synthesis,
    1:280–294
  libraries, 1:294–343
  lipid arrays, 1:293–294
  mixed drug synthesis, 1:345
  nucleoside arrays, 1:292–293
  oligosaccharide arrays,
    1:291–292
  peptide arrays, 1:280–291
solution-phase synthesis:
  lead optimization,
    1:344–345
  libraries, 1:294–343
  mixed drug synthesis, 1:345
synthetic success, 1:347–348
Multiple sclerosis:
  animal models, 5:567
  current therapies, 5:568–572
  azathioprine, 5:569
  cladribine, 5:570
  corticosteroids, 5:568
  cyclosphosphamide,
    5:569–570
  cyclosporin, 5:568

glatiramer acetate,
    5:570–571
interferon-beta, 5:571
methotrexate, 5:568–569
mitoxantrone, 5:570
natalizumab, 5:571–72
epidemiology, 5:565
future research issues, 5:580
novel emerging therapies,
    5:574–580
  antigen-specific therapy,
    5:579–580
  cannabinoid system, 5:580
  matrix metalloproteinases,
    5:580
  small-molecule VLA-4
    antagonists,
    5:578–579
  sphingosine 1-phosphate and
    receptors, 5:574–578
opportunistic emerging
    therapies, 5:572–573
  estrogens, 5:573–574
  immunomodulators/
    immunosuppressants,
    5:574
  monoclonal antibodies,
    5:572–573
  neuroprotectants, 5:574
  PPAR agonists, 5:574
  small molecules, 5:573–574
  statins, 5:573
pharmacology of, 5:565–567
therapeutic strategies,
    5:567–568
Multiple solvent crystal structure
    method, allosteric
    proteins, binding site
    location, 1:388–392
Multitarget drug development
    (MTDD):
  ACE-plus and AT_1-plus for
    hypertension,
    1:257–259
  AChE-plus for Alzheimer's
    disease, 1:261–263
  COX-plus for inflammatory
    pain, 1:267
  dopamine D_2-plus for
    schizophrenia, 1:257
  histamine H_1-plus for allergies,
    1:259–261
  lead compound criteria,
    1:252–254
  knowledge-based
    approaches,
    1:253–254

screening methods,
1:252–253
multikinase inhibitors for
cancer, **1:**261–267
optimization strategies,
1:267–273
desired activity ratio,
1:267–268
undesired activity removal,
1:268–269
physicochemical/
pharmacokinetic
properties, **1:**269–273
PPAR-plus for metabolic
diseases, **1:**261, 264
research background,
1:251–252
SERT-plus for depression,
1:254–257
Muraglitazar, single nucleotide
polymorphisms, **4:**166
Muscarinic acetylcholine receptor
(mAChR) antagonists,
structure-activity
relationships, **8:**72–77
Muscarinic acetylcholinergic
receptor (mAChR)
antagonists:
chronic obstructive pulmonary
disease, **5:**766–769
combined therapies, **5:**772
recombinant DNA technology,
receptor-based
targeting mechanism,
1:556–560
soft drug design,
anticholinergics,
2:107–109
subtype selectivity/
stereoselectivity,
1:135–137
Muscarinic agonists, **8:**64–68
Muscarinic receptor antagonist/
β_2-receptor agonist
(MABA), chronic
obstructive pulmonary
disease
bronchodilators, **5:**772
Muscarinic receptors:
antipsychotic effects on,
8:167–168
structure-activity
relationship, **8:**185
cholinergic nervous system,
8:61–62
cognitive enhancement,
8:32–33

M1/M2 receptors, chemokine
receptor-5 (CCR5)
antagonist
development,
7:113–116
serotonin-selective reuptake
inhibitors and,
8:233–234
structure and classification,
8:62
Muscarone agonist, **8:**64–65
Muscimol, analog drug design,
1:168–170
Mustards:
DNA-targeted
chemotherapeutic
compounds, **6:**83–89
antibody-directed enzyme
prodrug therapy,
peptidase enzymes,
6:120–121
biology and side effects, **6:**87
mechanism and structure-
activity relationship,
6:85–87
minor-groove targeting,
6:87–88
hypoxia-activated bioreductive
prodrugs, **6:**117–120
Mutagenicity:
ADMET toxicity properties,
2:62, 68
NRX 194204 RXR selective
analog, **5:**464
opioid receptors, **8:**589–590
Mutamycin. See Mitomycin
MVIIA synthetic compound
(Prialt®), structural
determination,
2:377–383
Mycobacterium tuberculosis:
drug resistance mechanisms in,
7:733–734
genome sequencing, **7:**719–720
life cycle, **7:**717–719
structure and classification,
7:716–717, 719
targeted drug development for,
7:719–733
Mycobactin, antitubercular agent
targeting and
biosynthesis,
7:730–731
para-Aminosalicylic acid, **7:**767
Mycolates, antitubercular agents,
mycolic biosynthesis,
7:722–723

Mycolic acid, antitubercular
agents:
biosynthesis, **7:**722–723
isoniazid resistance, **7:**740–743
Mycophenolate mofetil (MMF):
ADMET properties,
5:1047–1049
chemical structure,
5:1038–1039
research background, **5:**1038
therapeutic applications,
5:1049
Mycophenolate sodium (MPS):
ADMET properties, **5:**1049
chemical structure,
5:1038–1039
enteric-coated formulation,
5:1050
research background, **5:**1038
Mycophenolic acid (MPA):
ADMET properties,
5:1047–1049
analog design, **5:**1039–1042
chemical structure,
5:1038–1039, 1041
mechanism of action,
5:1038–1039
mizoribine combined with,
nucleoside inhibition,
5:1045
research background, **5:**1038
Mycosis fungoides, tazarotene
therapy, **5:**407
Mycothiols, antitubercular
agents, targeting and
biosynthesis, **7:**731
Myelin basic protein (MBP),
multiple sclerosis:
animal models, **5:**567
antigen-specific therapy,
5:579–580
current therapeutic targeting,
5:571–572
pathology, **5:**566
Myeloperoxidase, drug
metabolism,
functionalization
reactions, **1:**412
Myocardial infarction:
angiotensin-converting enzyme
inhibitors, **4:**284
aspirin secondary prevention,
4:421
esmolol prodrug design,
3:260–268
Myocardial oxygen demand,
congestive heart

failure therapy, mechanism-targeted studies, **4**:492–494
Myricetin, Alzheimer's disease, tau-targeted therapy, **8**:425–426
Myriocin, fingolimod development, **5**:1053–1055
Myristoylated and palmitoylated serine/threonine kinase 1 (MPSK1), structural genomics, **1**:591–594

$N1$-activating groups, monobactam activity, **7**:331
N-acetyl-aspartyl glutamate (NAAG), schizophrenia pathology, **8**:163–164
N-acetyltransferase-2 (NAT2), single nucleotide polymorphisms, **1**:188
N-acylated aminopenicillins, penicillin resistance and development of, **7**:291–292
Nafoxidine, structure-activity relationship, **5**:252
NAG-NAM glycan pair, penicillin binding proteins, **7**:262–264
Naïve Bayesian (NB) modeling, chemogenomics, ligand-based data analysis, **2**:583–585
Nalidixic acid analogs, antibacterial agents, **7**:535–540
Nalorphine:
 structure-activity relationship, **8**:601
 synthetic analog, **8**:597–598
Naloxone:
 antagonist activity, **8**:601
 chemical structure, **8**:571
 morphine metabolism and elimination, **8**:578
 nonpeptide affinity labeling, **8**:640–643
Naltrexamine derivative:
 antagonist activity, **8**:601
 nonpeptide affinity labeling, **8**:640–643

TENA κ-opioid selective antagonist, **8**:602–604
Naltrexone:
 antagonist activity, **8**:601
 δ-opioid receptor antagonist, **8**:604–608
 nonpeptide affinity labeling, **8**:638–641
Naltrindole:
 δ-opioid receptor antagonist, **8**:604–608
 μ-opioid receptor antagonist, **8**:607
 nonpeptide affinity labeling, **8**:638–643
L-NAME, chronic obstructive pulmonary disease, **5**:786–787
Namenamicin, natural products lead generation, terrestrial and marine habitats, **2**:197–199
NAN-190 antagonist, efficacy assessment, **8**:225–229
Nanotechnology, drug delivery systems:
 amphiphilic polymers, **3**:473–480
 basic principles, **3**:469
 carbon nanotubes, **3**:483–484
 dendrimers, **3**:482–483
 polymer-drug conjugates, **3**:480–481
 self-assembled low molecular weight amphiphiles, **3**:469–473
 water-insoluble and crosslinked polymeric nanoparticles, **3**:481
Naphthalamide, poly(ADP-ribose)polymerase inhibitors, **6**:167
1,4-Naphthoquinone scaffold, glutathione inhibitors, antimalarial drugs, **7**:662–663
Naphthoquinones, Hsp 90 inhibitors, **6**:426–427
Naphthyridine heterocyclic scaffold, HIV integrase, discovery and elaboration, **7**:85–97
Naproxen, migraine therapy, **8**:268

Narcolepsy:
 chirality at sulfur and, **1**:148
 psychostimulant therapy, **8**:91–92
Natalizumab, multiple sclerosis therapy, **5**:571–572, 578–579
Nateglinide, type 2 diabetes therapy, **4**:10–13
Natriuretic peptides. *See also* specific NPs, e.g., Brain natriuretic peptide
 biological actions and receptor pharmacology, **4**:533–537
 recombinant DNA technology, receptor-based targeting mechanism, **1**:558–560
 structural genomics, **1**:585–586
 structure and function, **4**:533–537
 therapeutic potential, **4**:537
Natural products. *See also* Microbial compounds
 β-amyloid peptide inhibitors, **8**:363–365
 analog drug design, **1**:168–170
 angiotensin-converting enzyme inhibitors, **4**:276
 antibiotics, research background, **7**:405–406
 carbapenems:
  chemical synthesis, **7**:336
  occurrence, structural variations and chemistry, **7**:333–335
 combinatorial chemistry, **1**:297–298
 testing mixtures, **1**:347
 cytotoxic agents:
  DNA targeting compounds, **6**:3–17
   bleomycin, **6**:8–12
   dactinomycin, **6**:3–8
   mitomycin (mutamycin), **6**:12–16
   plicamycin, **6**:16–17
  enzyme inhibitors, **6**:17–31
   anthracyclines, **6**:17–23
   camptothecins, **6**:23–29
   isopodophyllotoxins, **6**:29–31
  future research issues, **6**:40–41

research background, **6:**1–3
tubulin polymerization/
depolymerization
inhibition, **6:**31–40
dimeric vinca alkaloids,
6:35–40
taxus diterpenes, **6:**32–35
histone deacetylase inhibitors,
6:70–72
largazole, **6:**72
romidepsin (FK-228),
6:70–72
lead drug discoveries:
biosynthetic pathways,
2:210–213
future research issues,
2:208–213
genome mining, **2:**209–210
high-throughput screening,
2:205–207
hits analysis, **2:**207
libraries, **2:**200–204
attributes, **2:**201–202
complexity, **2:**200–201
mass spectrometry
analysis, **2:**202–203
nuclear magnetic
resonance
characterization,
2:203–204
screening platforms,
2:205–209
phenotypic screening,
2:207–208
sources, **2:**191–200
biodiversity harvesting,
2:199
metagenomics, **2:**199–200
microbial sources,
2:193–199
terrestrial and marine
habitats,
2:193–199
virtual screening, **2:**213
ligand-based drug design,
nuclear magnetic
resonance, **2:**376–377
mass spectrometry encoding
and identification,
1:110–111
monobactams, **7:**325–327
p53/MDM2 derivatives,
2:350–352
proteasome inhibitors:
clinical trials, **6:**179–180
medicinal chemistry,
6:183–184

proteins, recombinant DNA
technology, **1:**538–539
receptor-based drug design:
biologicals and antisense
compounds, **2:**531–533
lead drug discoveries,
2:527–528
receptor sources, **2:**523–524
soft drug development,
2:129–130
statins, LDL cholesterol
lowering mechanisms,
4:306–310
structure-based drug design:
flavonoids, **2:**701–702
stilbenes, **2:**701
vitamins, **5:**655
Navelbine. *See* Vinrelbine
NC-531 glycosaminoglycan
mimetic, **8:**362
Ncotinamide adenine
dinucleotide (NAD):
allosteric proteins, **1:**383–387
niacin chemistry, **5:**677–679
Near-infrared spectroscopy,
quantitative
structure-activity
relationship,
hydrophobic
interactions, **1:**11–19
Nebracetam, cognitive
enhancement, **8:**34
Nefazodone:
chemical structure, **8:**224
pharmacokinetics, **8:**236–237
serotonin receptor targeting,
8:241–242
side effects and adverse
reactions, **8:**234–235
Negative allosteric modulator
(NAM), receptor-based
drug design, **2:**500
Negative antagonist, receptor-
based drug design,
efficacy parameters,
2:502–503
Neisseria gonorrhoeae, β-lactam
antibacterial drug
resistance, PBP
modification, **7:**272
Nek2 kinases, structural
genomics, **1:**592–583
Nelfinavir mesylate:
binding interaction, **7:**20, 22, 25
chemical structure, **7:**20, 22
efficacy and tolerability, **7:**22
optimization, **7:**20, 23–24

pharmacokinetics, **7:**22, 26–27
resistance profile, **7:**22
Neoclerodane diterpene, current
developments,
8:682–683
Neomycin:
aminoglycosides, RNA targets,
5:971–979
mechanism of action,
5:972–973
structure-activity
relationship:
subclass rings I and II,
5:973–978
subclass rings III and IV,
5:979
chemical structure, **7:**416–419
history and biosynthesis,
7:423–424
Neonatal transport mechanisms,
immunoglobulins,
2:450–451
Nepafenac, age-related macular
degeneration therapy,
5:613–614
Nephelometric assay, ADMET
solubility, **2:**65–66
Nephrotoxicity, aminoglycosides,
7:419–420
Neramexane, stroke therapy,
8:450
Neratinib erbB family inhibitor,
6:302–305
Nerve growth factor IB-like
family, receptor-based
drug design, **2:**512
Nesiritide, congestive heart
failure therapy,
mechanism-targeted
studies, **4:**494
Netilmicin:
chemical structure, **7:**416–419
history and biosynthesis,
7:423–424
Netoglitazone compound,
selective PPARγ
modulators, PPARγ
partial agonists,
4:107–115
NEULASTA®, structure and
properties, **4:**575–578,
588–590
NEUMEGA, structure and
properties, **4:**578–580
NEUPOGEN®, structure and
properties, **4:**575–578,
588–590

Neural stem cells, brain tumor initiation, **6:**225–226
NeurAxon nNOS inhibitor, migraine therapies, **8:**302–306
Neuregulin 1, schizophrenia genetics, **8:**174
Neuroanatomy, cognition and, **8:**22–23
Neuroblastoma:
  N-(4-hydroxyphenyl) retinamide (4-HPR) therapy, **5:**466, 469, 472–473
  13-*cis*-retinoic acid therapy, **5:**390–391
Neurodegenerative diseases, Hsp90 inhibitor therapy, **6:**442–444
Neuroectodermal tumors:
  classification and therapy, **6:**234–235
  N-(4-hydroxyphenyl) retinamide (4-HPR) therapy, **5:**473
Neuroepithelial tumors, classification, **6:**230–231, 235
Neurofibrillary tangle (NFT), Alzheimer's disease, tau-targeted therapies, **8:**417–432
Neurofibroma, classification and therapy, **6:**237–239
Neurofibromatosis, central nervous system tumors, **6:**238–239
Neurokinin-1 (NK-1), antiplatelet activity, **4:**453
Neurokinin A receptor, tachykinin pharmacology, **4:**542–544
Neurokinin B receptor, tachykinin pharmacology, **4:**542–544
Neurokinins:
  cognitive enhancement, **8:**49
  radiolabeled ligands for, **8:**744
  receptor antagonists, antipsychotic agents, **8:**208
Neuroleptic antipsychotics, structure-activity relationships, **8:**180–185

Neuroleptic malignant syndrome (NMS), antipsychotic side effects, **8:**167
Neuromedin B, structure and function, **4:**549
Neuromuscular nicotinic antagonists, **8:**78–80
Neuronal nicotinic acetylcholine receptors (nAChRs), cognitive enhancement, **8:**34–35
Neuronal nitric oxide synthase (nNOS), migraine therapies, **8:**299, 302–306
Neuropathy, type 2 diabetes, **4:**2
Neuropeptides:
  antidepressant effects, **8:**256–258
  calcitonin gene-related peptide antagonists, migraine therapy, **8:**286–288
  cognitive enhancement, **8:**48–49
  schizophrenia pathology, **8:**165
Neuropeptide Y (NPY):
  biological actions and receptor pharmacology, **4:**527–531
  osteoporosis therapy, **5:**755
  structure and function, **4:**526–531
Neuroprotectants:
  glaucoma therapy, clinical trials, **5:**602
  Hsp90 inhibitor therapy, **6:**442–444
  multiple sclerosis therapy, **5:**574
Neuroprotectin 1 metabolite, age-related macular degeneration, **5:**609–610
Neuroprotection models, glaucoma screening, **5:**591
Neuroreceptors, allosteric protein model, **1:**393
Neurotensin, structure and function, **4:**549–550
Neurotransmitters:
  anticonvulsants, transporters and synaptic proteins, **8:**130–132
  antidepressants:
    effect on, **8:**220–221

    side effects and adverse reactions, **8:**231–234
  receptor-based drug design:
    binding proteins, **2:**514
    transporters, **2:**513–514
  schizophrenia pathology, **8:**164–165
Neurotransmitter sodium symporter (NSS) family, psychostimulant physiology and pharmacology, **8:**100–106
Neurotrophic factors:
  Alzheimer's disease, tau-targeted therapies, **8:**422
  antidepressant efficacy assessment, **8:**226–229
Neurotropic agents, cognitive enhancement, **8:**47
Neutral antagonists, receptor-based drug design, efficacy parameters, **2:**501–503
Neutral cooperativity, receptor-based drug design, allosteric modulators, **2:**500–501
Neutral endopeptidase (NEP), multitarget drug development, ACE-plus and angiotensin-1 receptor (AT_1) antagonist, **1:**257, 259–260
Neutropenia, antipsychotic side effects, **8:**168–169
Neutrophil elastase inhibitors, chronic obstructive pulmonary disease therapy, **5:**790–791
Neutrophils, recombinant DNA technology, cellular adhesion proteins, **1:**561
"Never in mitosis" (NIMA) protein family, structural genomics, human proteome coverage, **1:**578–579
Nevirapine:
  HIV reverse transcriptase inhibition, **7:**148–152

large-scale synthesis case study, **3**:14–23
  commercial production and process optimization, **3**:20–23
  medicinal chemistry synthetic route, **3**:15–16
  pilot-plant scale-up: chemical development, **3**:16–17
  process development, **3**:17–20
  research background, **3**:14–15
New and nonobvious invention standard:
  patent requirements and, **3**:133–138
  U.S.C.§ 103 provision, **3**:134–138
New Animal Drug Application (NADA), regulatory guidelines, **3**:97
New drug applications (NDAs):
  FDA guidelines, **3**:78
  new drug definitions, **3**:82
  over the counter drug approval, **3**:83–84
  patent infringement and, **3**:151–152
  submission requirements, **3**:85, 88
New molecular entities (NMEs), prodrug development, **3**:276–279
Nexium, chirality at sulfur and, **1**:127, 147–148
  switching mechanism, **1**:156–159
Next-generation compounds:
  HIV integrase inhibitors, **7**:95–97
  HIV protease inhibitors, **7**:60–64
NFκB:
  chronic obstructive pulmonary disease therapy, IKK inhibitors, **5**:787–789
  liver X receptor agonists, **4**:147–148
  proteasome inhibitors, **6**:177
N-glucuronidation, drug metabolism, **1**:429–430
N-glycosylation, transport proteins, **2**:460–461

NH-3 thyromimetic, structure-activity relationship, **4**:220–221
Niacin, **5**:677–679
  atherosclerosis, HDL elevation:
    analogs, **4**:346–348
    basic properties, **4**:345–346
  chemistry, **5**:677–679
  deficiency, **5**:677
  hypervitaminosis niacin, **5**:677–678
  prostaglandin/thromboxane EP_1 antagonists, **5**:823, 827–828
Niacinamide, **5**:679–681
Niacin equivalents (NEs), defined, **5**:637
NIC002 antinicotine vaccine, smoking cessation, **5**:725–726
Nicergoline, cognitive enhancement, **8**:41–42
Nicorandil, combined nitric oxide donors and prodrugs, **5**:287–288
Nicotinamide:
  antitubercular agents, isoniazid, **7**:738–740
  chemistry, **5**:679–681
  chemokine receptor-5 (CCR5) antagonist development, **7**:112–116
  deficiency, **5**:659
  methylation reactions, **1**:423–425
  poly(ADP-ribose)polymerase-1 (PARP-1) domain organization, **6**:160
  poly(ADP-ribosyl)ation biochemistry, **6**:153–157
Nicotinamide adenine dinucleotide (NAD):
  antitubercular agents, cofactor biosynthesis targeting, **7**:728
  poly(ADP-ribosyl)ation biochemistry, **6**:151–157
  binding sites, **6**:162–167
Nicotine:
  cholinergic nervous system, **8**:61–62
  cognitive enhancement, **8**:34–36

Nicotinic acetylcholine receptor (nAChR):
  anticonvulsants, ligand-gated ion channels, **8**:129–130
  antipsychotic agents, **8**:207
  ligand-gated ion channels, **2**:510–511
  nicotinic agonists, **8**:68–72
  nicotinic antagonists, **8**:77–80
    ganglionic antagonists, **8**:78
    neuromuscular antagonists, **8**:78–80
  schizophrenia pathology, **8**:165
  smoking cessation therapy, varenicline, **5**:723–725
  structure and classification, **8**:63
Nicotinic acid. *See* Niacin
Nicotinic acid receptor agonists:
  atherosclerosis, HDL elevation, **4**:345–350
  anthranilic acid and related analogs, **4**:348–349
  niacin analogs, **4**:346–348
  xanthine and barbituric acid analogs, **4**:349–350
  cognitive enhancement, **8**:34–35
Nicotinic agonists, structure and function, **8**:68–72
NicVax antinicotine vaccine, smoking cessation, **5**:725–726
Niemann-Pick C1 like 1 (NPC1L1) gene, ezetimibe mechanism of action, **4**:314–315
Nifedipine, migraine prevention, **8**:318
Nifurtimox, American trypanosomiasis therapy, **7**:573
Night blindness, vitamin A chemistry, **5**:647
NIH Roadmap, combinatorial chemistry, **1**:298–299
Nilotinib:
  chronic myelogenous leukemia therapy, Bcr-Abl inhibitors, **6**:298–301
  second-generation development, drug resistance and, **6**:364–365
  selective toxicity, **2**:563

Nilotinib (*Continued*)
structure-based design, drug-resistant mutation, **6:**355–356
Nilutamide, antiandrogen compounds, **5:**192–193
Nimetazepam, anticonvulsant applications, **8:**144–147
Nimodipine:
migraine prevention, **8:**318
stroke therapy, **8:**459–460
9cUAB30 RXR-selective retinoid, **5:**453–458
chemical structure, **5:**453–455
therapeutic applications, **5:**455–457
Nipradilol, combined nitric oxide donors and prodrugs, **5:**287–288
Niravoline κ-opioid receptor agonist, **8:**627–632
Nisoxetine, structure-activity relationship, **8:**251
Nitacrine-*N*-oxide, hypoxia-activated bioreductive prodrugs, **6:**115–120
Nitidine, dual topoisomerase I/II inhibitors, **6:**103–105
Nitrazepam, anticonvulsant applications, **8:**144–147
Nitric oxide (NO), **5:**265–294
antitubercular agents:
nitroimidazoles, **7:**780–781
tuberculosis pathogenesis, **7:**718–719
donors and prodrugs, **5:**278–290, 294
C-nitroso class, **5:**286–287
combined donors, **5:**287–290
diazeniumdiolates, **5:**281–283
S-nitrosothiols, **5:**283–285
organic nitrates, **5:**278–280
sodium nitroprusside, **5:**280–281
sydnoimines, **5:**285–286
hemoglobin reactions, **4:**612–614
cross-linked hemoglobin, **4:**630–631
recombinant hemoglobin, **4:**638–639
hydrogen sulfide biological effects, **5:**311–312
inhalation therapy, **5:**274–278

multitarget drug development, COX-plus for inflammatory pain, **1:**267
production and biological effects, **5:**265–274
stroke therapy, INOS/NNOS inhibitors, **8:**495–502
amidines, **8:**497–498
aminopyridines, **8:**498–499
L-arginine analogs, **8:**495–497
calmodulin site, **8:**499–500
coumarin derivatives, **8:**499
dimerization inhibitors, **8:**501
guanidine and isothiourea analogs, **8:**497
heme site, **8:**501
non-arginine sites, **8:**499
NOS inhibitor/NO donor compounds, **8:**502
tetrahydrobiopeterin site, **8:**500
therapeutic applications, **5:**267–269
indirect enhancement approaches, **5:**291–294
Leishmaniasis therapy, **7:**591–593
Nitric oxide synthases (NOS):
chronic obstructive pulmonary disease inhibitors, **5:**785–786
migraine therapy, nitric oxide synthase inhibitors, **8:**298–306
iNOS inhibitors, **8:**300–302
isoforms, **8:**298–299
nNOS inhibitors, **8:**302–306
tool compounds, **8:**299–300
production and biological effects, **5:**265, 270–274
stroke therapy, INOS/NNOS inhibitors, **8:**495–502
Nitriles:
cathepsin S combinatorial chemistry, **1:**330–332
dipeptidyl peptidase 4 (DPP-4) inhibition, **4:**50–54
stroke therapy, cathepsin B inhibitors, **8:**473
Nitrite, nitric oxide enhancement, **5:**294

Nitro-alkyls, toxicophore reactive metabolites, **2:**312–313
Nitroaromatics:
hypoxia-activated bioreductive prodrugs, **6:**115–120
toxicophore reactive metabolites, **2:**312–313
Nitro-aspirin, combined nitric oxide donors and prodrugs, **5:**288–290
Nitrofenac, combined nitric oxide donors and prodrugs, **5:**288–290
Nitrofurans, synthetic antibacterials, **7:**548–550
Nitrogen:
bisphosphonates, structural genomics, **1:**584–585
chirality at, **1:**148
drug metabolism, oxidation and reduction, **1:**416–418
psychostimulant structure-activity relationship, **8:**107
cocaine, **8:**109–110
toxicophore reactive metabolites, aniline metabolism, **2:**306–311
Nitroglycerine, nitric oxide donors and prodrugs, **5:**278–280
Nitro-hydrocortisone, combined nitric oxide donors and prodrugs, **5:**288–290
Nitroimidazoles, antitubercular agents, **7:**776–782
ADMET properties, **7:**781–782
adverse effects, **7:**782
history, **7:**776
mechanism of action, **7:**780–781
resistance mechanisms, **7:**781
structure-activity relationships, **7:**776–780
Nitro-moxysylyte, combined nitric oxide donors and prodrugs, **5:**289–290
Nitro-pravastatin, combined nitric oxide donors and prodrugs, **5:**289–290
Nitro-prednisolone, combined nitric oxide donors and prodrugs, **5:**288–290

S-Nitrosoalbumin, nitric oxide donors and prodrugs, **5:**285
S-Nitrosocysteine, nitric oxide donors and prodrugs, **5:**284–285
S-Nitrosoglutathione (GSNO), nitric oxide donors and prodrugs, **5:**283–285
S-Nitroso-penicillamine (SNAP), nitric oxide donors and prodrugs, **5:**284–285
S-Nitrosothiols:
nitric oxide donors and prodrugs, **5:**283–285
nitric oxide effects, **5:**273–278
Nitrosoureas, DNA-targeted compounds, **6:**93–94
S-Nitrosylation, nitric oxide effects, **5:**273–278
Nitro-yohimbine, combined nitric oxide donors and prodrugs, **5:**289–290
NK1 receptor:
substance P-NK1 receptor antagonist, **8:**256–257
targeted development, **2:**516–519
NM283 anti-HCV agent, **7:**195–196
NMED-160 scaffold, stroke therapy, **8:**457–458
N-methanocarbathymidine, poxvirus therapy, **7:**245
N-methylation reactions, drug metabolism, **1:**423–425
N-methyl-D-aspartate (NMDA) receptors:
anticonvulsants, ligand-gated ion channels, **8:**129
antidepressants, **8:**256
antipsychotic drugs:
animal models, **8:**171–172
glutmatergics, **8:**204–205
latent inhibition, **8:**172
cognitive enhancement:
glutamate, **8:**41–42
glycine, **8:**44–45
glutmate modulation, migraine therapy, **8:**291–295
ligand-gated ion channels, receptor-based drug design, **2:**510–511
long-term potentiation, **8:**24–25

metabotropic glutamate receptor 5 combinatorial chemistry, **1:**318–320
modulators, **8:**33–34
schizophrenia pathology, **8:**163–164
stroke therapy, **8:**450–455
competitive receptor antagonists, **8:**451–453
ion channel blockers, **8:**450–451
NR2B-specific noncompetitive receptor antagonists, **8:**453–455
N-(methylenecyclopentyl) thymine, anti-DNA virus antimetabolites, **7:**225–226
NNC 61-5920 compound, PPARδ-agonist, **4:**119–121
NNOS inhibitors:
migraine therapy, **8:**302–306
stroke therapy, **8:**495–502
Nocardicins:
biological activity and structure-activity relationship, **7:**326
chemical structure, **7:**324–325
formadicins, **7:**326–327
Noisomes, nanoscale drug delivery, **3:**469–473
Nolatrexed, lipophilic antifolate DNA-targeted chemotherapeutics, **6:**109
Nomegestrol acetate, ovulation modulation, **5:**242–243
Nomenclature databases, receptor-based drug design, **2:**504–505
Non-amyloid therapies, Alzheimer's disease:
metabolic syndrome targeting, **8:**405–417
cholesterol-related approaches, **8:**405–414
insulin-related approaches, **8:**414–417
research background, **8:**405
summary, **8:**408–410
tau pathology targeting, **8:**417–432

aggregation, **8:**423–429
glycosylation, **8:**422–423
microtubule stabiliization, **8:**430
phosphorylation, **8:**418–422
protein levels, **8:**430–432
Nonbenzodiazepine sedatives, selective toxicity, **2:**568
Non-β-lactam inhibitor, β-lactamases, **7:**371–372
Noncoding single nucleotide polymorphisms, classification, **1:**183
Noncompliance in patients:
glaucoma intraocular pressure reduction, **5:**591–592
plasma drug profile:
diffusion-based drug release, repeated dose delivery, **2:**810–812
erosion-controlled drug release, **2:**824–826
Nonenzymatic reactions, drug metabolism, **1:**405–406
Non-GCPR-linked cytokine receptors, receptor-based drug design, **2:**512–513
Nonglycosamino pentasaccharides, antithrombin conformational activation, **4:**390–392
Nonhemoglobin, heme-based oxygen delivery, **4:**642–643
Non-Hodgkin's lymphoma:
N-(4-hydroxyphenyl) retinamide (4-HPR), **5:**472
primary central nervous system tumors, **6:**239–240
9-cis-retinoic acid therapy, **5:**425
Nonhomologous end-joining (NHEJ) pathway, poly (ADP-ribose) polymerase-1 (PARP-1), **6:**161–162
Nonhygroscopic properties, salt compounds, **3:**381

Noninvasive imaging, congestive heart failure therapy: disease biomarkers, **4:**505–506
hemodynamic biomarkers, iontophoresis methodologies, **4:**506
Non-life-threatening disorders, drug development and discovery, **5:**711–726
Nonlinear models, quantitative structure-activity relationships, multilinear regression analysis, **1:**34–36
Nonlinear pharmacokinetics, defined, **2:**751
Nonnucleoside inhibitors, anti-HCV agents:
benzothiadiazine, **7:**202–203
HCV-796, **7:**203–204
NX5B inhibitors, **7:**201–204
PF-00868554, **7:**201–202
Nonnucleoside reverse transcriptase inhibitors (NNRTIs):
allosteric protein model, **1:**393–394
research background, **7:**139–140
Nonnucleotide inhibitors, viral DNA-polymerase, **7:**239–240
NONOates, nitric oxide donors and prodrugs, **5:**281–283
Nonpeptide opioid analgesics:
opioid-receptor-like 1/orphanin FQ/nociceptin peptide, **8:**680–681
structure-activity relationships, **8:**597–645
acyclic analgesics, **8:**621–623
affinity labeling, **8:**637–643
morphine-naltrexone derivatives, **8:**638–641
agonists:
δ-selective agonists, **8:**632–637
κ-selective agonists, **8:**623–632
centrally-acting agonists, **8:**624–630
peripherally-acting agonists, **8:**630–632
antagonists, **8:**601–608

δ-opioid receptors, **8:**604–608
κ-opioid receptors, **8:**602–604
benzomorphans, **8:**610–613
Diels-Alder adducts, **8:**608–609
miscellaneous opioids, **8:**643–645
morphinans, **8:**609–610
morphine derivatives, **8:**599–601
opium-derived morphine and alkaloids, **8:**598–599
ORL1 receptor, endogenous ligand orphanin FQ/nociceptin, **8:**681–682
piperidine derivatives, **8:**613–621
4-anilidopiperidines, **8:**618–621
4-arylpiperidines:
carbon substituent, **8:**614–616
oxygen substituent, **8:**616–618
Nonpeptidic chemotypes:
BACE (β-site cleavage enzyme) inhibitors, **8:**347–356
plasmepsin antimalarial agents, **7:**642–644
renin inhibitors, **4:**253–255
alkylamines, **4:**254–255
diaminopyrimidines, **4:**254
piperidines/piperazines, **4:**253–254
γ-secretase inhibitors, **8:**337–341
Nonprostanoid agonists, IP receptors, **5:**922–924
Nonribosomal peptides, natural products lead generation, **2:**193, 210
"Nonsense proteins," aminoglycoside pharmacology, **7:**420–423
Nonsolid drug forms, oral dosage systems, **3:**29
Nonsteroidal androgens:
antiandrogens, **5:**192–193
structure-activity relationships, **5:**174–176
Nonsteroidal anti-inflammatory drugs (NSAIDs):
antiglucocortucoids, **5:**121–134

fibril formation inhibitors, structure-based design, **2:**694–702
glucocorticoids, **5:**121–134
migraine therapy, **8:**268
multitarget drug development, **1:**267
recombinant DNA technology, drug efficacy and personalized medicine, **1:**541–543
γ-secretase modulators, **8:**358–359
Nonsteroidal estrogen receptor ligands, structure-activity relationships, **5:**246–247
Nonsteroidal liver receptor agonists, **4:**152–154
Nonsteroidal progesterone receptor ligands, structure-activity relationships, **5:**255–256
Nontuberculous mycobacteria (NTM), structure and classification, **7:**716–717
Nootropics, cognitive enhancement, **8:**34
Noradrenaline, chronic obstructive pulmonary disease bronchodilators, **5:**768–769
NorBNI, κ-opioid selective antagonist, **8:**602–604
Norbornane derivatives, protein-protein interactions, **2:**349–352
Norclozapine, cognitive enhancement, **8:**32–33
Nor derivatives, anabolic agents, **5:**179
Nordiazepam, anticonvulsants, **8:**144–147
Norepinephrine:
antidepressants, neurotransmission enhancement, **8:**220–221
cognitive enhancement, **8:**41
psychostimulant pharmacology, **8:**100–106

structure-activity
relationship,
8:106–109
Norepinephrine-selective
reuptake inhibitors
(NSRIs):
chemical structure, **8:**223
historical background,
8:248–249
pharmacokinetics, **8:**237
side effects and adverse
reactions, **8:**223–224,
234–235
structure-activity relationship,
8:251
Norepinephrine transporter
(NET), antidepressant
pharmacology,
8:237–241
Norethindrone:
biotransformation, **5:**229–230
chemical reactions, **5:**11–12
ovulation modulation,
5:242–243
physiology and pharmacology,
5:238–239
steroidal progesterone receptor
ligands, **5:**254
Norethynodrel, steroidal
progesterone receptor
ligands, **5:**254
Norflenfluramine, drug
metabolism,
oxidation/reduction
reactions, **1:**420–421
Norfloxacin:
antibacterial agents, **7:**535–540
antimalarial drugs:
combination therapy, **7:**621,
623–624
topoisomerase inhibitors,
7:650–652
quantitative structure-activity
relationships, **1:**73–74
Norfluoxetine:
serotonin receptor targeting,
8:242–244
structure-activity relationship,
8:251
Norgestrel:
biotransformation, **5:**229–230
steroidal progesterone receptor
ligands, **5:**254
Normal mode analyses (NMA), *in
silico* screening,
protein flexibility,
2:864–865

Normal tension glaucoma (NTG),
disease pathology and
etiology, **5:**589–590
19-Norpregnane analogs,
steroidal progesterone
receptor ligands,
5:254–255
Norstatine renin inhibitors,
4:245–247
North American Free Trade
Agreement (NAFTA),
patent interference
limitations, **3:**144–146
Nortriptyline:
chemical structure, **8:**222
migraine prevention,
8:318–319
serotonin receptor targeting,
8:241–242
Norvir. *See* Ritonavir
Novel object recognition, cognitive
function assessment,
8:27
Novobiocin:
chemical structure, **6:**430–431
Hsp90 C-terminal inhibition,
6:431–440
parallel library approach,
6:432–433
Nplate, structure and properties,
4:583–584
npmA gene, aminoglycosides,
RNA targets,
5:973–978
NR2B NMDA subunit,
noncompetitive
receptor antagonist,
stroke therapy,
8:453–455
NRX 194204 RXR selective
analog, **5:**458–464
adverse effects, **5:**463–464
chemical structure, **5:**458–459
pharmacology and metabolism,
5:464
therapeutic applications,
5:460–464
NRX 195183 retinoic receptor-
selective analog,
5:370–372, 420
NS3/4A protease inhibitors, anti-
HCV agents,
7:176–195
boceprevir, **7:**186–191
ciluprevir, **7:**176–181
ITMN-191, **7:**191–192
MK7009, **7:**193–195

telaprevir, **7:**181–186
TMC435350, **7:**193
NS5A phosphoprotein:
antiviral resistance and,
7:175–176
hepatitis C viral genome,
7:173–174
NS5B phosphoprotein,
nonnucleoside
inhibitors, anti-HCV
agents, **7:**201–204
NS-7 scaffold, stroke therapy,
8:457
NS1738, cognitive function
assessment, **8:**27–28
NSC 333003 compound, brain
tumor therapy,
6:258
N-terminal ATP-binding pocket,
Hsp90 inhibitors,
6:386–425
chimeric geldanamycin and
radicicol, **6:**405–410
geldanamycin, **6:**386–399
high-throughput small-
molecule inhibitors,
6:417–421
isoxazole-scaffold inhibitors,
6:419–421
purine-based inhibitors,
6:410–416
pyrazole-scaffold inhibitors,
6:417–419
radicicol analogs, **6:**399–405
shepherdin, **6:**416
structure-based design,
6:421–425
N-terminal domains (NTDs):
G-protein coupled receptor,
homology modeling,
2:280–284
N-terminal transactivation
domain (AF-1), thyroid
hormone receptors,
4:189–190
N-terminal nucleophilic (NTN)
hydrolases,
proteasome inhibitors,
6:175
N-thiol monocyclic β-lactams,
7:331–332
N-type voltage-dependent
calcium channel,
stroke therapy,
8:456–459
Nuclear export, AHPN-induced
apoptosis, **5:**492

Nuclear hormone receptors (NHRs):
  allosteric proteins, binding site location, **1:**390
  assays, **4:**80–82
  brain tumor therapy, **6:**265–267
  classification, **4:**77–78
  coactivators/corepressors, **4:**83–84
    fluorescence resonance energy transfer recruitment, **4:**84–86
  drug discovery applications, **4:**164–166
  farnesoid X receptor:
    agonist molecules, **4:**157–158, 161
    antagonists, **4:**161
    atherosclerosis, **4:**159–160
    basic properties, **4:**156–157
    bile acid metabolism, **4:**157
    clinical studies, **4:**161–163
    glucose homeostasis, **4:**160
    HDL cholesterol, **4:**159
    LDL cholesterol, **4:**159
    lipid regulation, **4:**158–159
    medicinal chemistry, **4:**161
    structural biology, **4:**160–161
    triglycerides, **4:**158–159
    vascular tissues and inflammation, **4:**160
  hepatocyte nuclear factor α, **4:**163–164
  hit seeking combinatorial libraries, **1:**334–336
  ligand affinity and efficacy, **4:**84
  ligand biology, **4:**82–83
  liver X receptors:
    adverse lipid effects, **4:**148–150
    atherosclerosis, **4:**147–148
    basic properties, **4:**145–146
    clinical data, **4:**154–156
    first-generation agonists, **4:**147
    glucose homeostasis, **4:**148
    lipid metabolism, **4:**148
    medicinal chemistry, **4:**151–154
    structural biology, **4:**150
  nomenclature, **4:**78
  peroxisome proliferator-activated receptors:
    α-agonists, **4:**88–99
      clinical data, **4:**96–99
      medicinal chemistry, **4:**90–96
      preclinical biology, **4:**88–90
    α-antagonists, **4:**99–100
    α/δ-dual agonists, **4:**136–137
    α/γ-dual agonists, **4:**124–136
    α/γ/δ-pan agonists, **4:**140–145
    biology, **4:**86
    current development, **4:**166–167
    δ-agonists, **4:**117–124
      biology, **4:**117–118
      clinical data, **4:**121–122
      medicinal chemistry, **4:**118–121
    δ-antagonists, **4:**122–124
    γ-agonists, **4:**100–106
      biology, **4:**100–102
      clinical data, **4:**104–106
      medicinal chemistry, **4:**102–104
    γ/δ-dual agonists, **4:**137–140
    γ-antagonists, **4:**115–117
    medicinal chemistry, **4:**87–88
    modulators/PPARα partial agonists, **4:**97
    modulators PPARγ partial agonists, **4:**106–115
    safety issues, **4:**145
  retinoid nuclear receptors, **5:**369–378
    selectivity, **5:**373–376
  retinoid X receptor:
    liver X receptors and, **4:**146
    peroxisome proliferator-activator receptors, γ-agonists, **4:**101–102
    thyroid hormone receptors, heterodimerization, **4:**190–191
  REV-ERBs receptors, **4:**164
  structural biology, **4:**78–80
    DNA binding and response elements, **4:**79–80
Nuclear localization signal (NLS), poly(ADP-ribose) polymerase-1 (PARP-1) domain organization, **6:**159–160
Nuclear magnetic resonance (NMR):
  basic principles, **2:**367–369
  docking and scoring functions, flexibility, **2:**638
  drug discovery:
    ligand-based design, **2:**375–402
      analog limitations, **2:**390
      charge state, **2:**384–385
      conformational analysis, **2:**383–384
      line-shape and relaxation data, **2:**387–389
      pharmacophore modeling, **2:**389–390
      structure elucidation, **2:**376–383
        bioactive peptides, **2:**377–383
        natural products, **2:**376–377
      tautomeric equilibria, **2:**385–387
    receptor-based design, **2:**390
      chemical-shift mapping, **2:**402–403
      dihydrofolate reductase inhibitors, **2:**414–419
      DNA-binding drugs, **2:**405–410
      isotope editing/filtering, **2:**403–404
      macromolecular structure, **2:**391–394
      matrix metalloproteinases, **2:**412–414
      NOE docking, **2:**404–405
      protein-ligand interactions, **2:**394–402
      titrations, **2:**403
    research background, **2:**367–369, 374–375
  drug screening:
    chemical-shift perturbation, **2:**419, 421–422
    ^{19}F-NMR, **2:**430
    fragment-based approach, **2:**432–433
    hardware and automation, **2:**434
    library design, **2:**431–434
    ligand efficiency, **2:**433
    magnetization transfer, **2:**424–426
    mixture design, **2:**433–434
    molecular diffusion, **2:**426–427
    NOE experiments, **2:**427–428
    overview, **2:**419–420
    relaxation, **2:**427–429
    screening methods evaluation, **2:**430–431
    spin labels, **2:**429–430

structure-activity
  relationships,
  2:421–424
instrumentation, **2:**371–374
intrinsically disordered
  proteins, **1:**386–387
natural products libraries
  characterization,
  2:203–204
peptide libraries, **1:**288–291
quantitative structure-activity
  relationship:
  hydrophobic interactions,
    1:11–19
  receptor interactions,
    1:29–31
  recombinant DNA technology,
    reagent development,
    1:552
Nuclear Overhauser effect (NOE)
  spectroscopy:
  basic principles, **2:**370–371
  ligand-based design:
    bound ligand conformation,
      2:390
    conformational analysis,
      2:384
    Prialt® (MVIIA) pain
      medication, **2:**377–382
  macromolecule-ligand
    interactions:
    DNA binding, **2:**405–410
    docking experiments,
      2:404–405
    matrix metalloproteinases,
      2:412–414
    screening procedures,
      2:427–429
Nucleic acids:
  protein therapeutics, analytic
    development,
    3:320–321
  receptor-based drug design,
    nuclear magnetic
    resonance, **2:**392–394
Nucleoside 2-deoxyribosyl-
  transferase (NDRT),
  structural genomics,
  fragment
  cocrystallography for
  drug development,
  1:587–588
Nucleoside arrays, solid-phase
  organic synthesis,
  1:292–293
Nucleoside inhibitors:
  anti-HCV agents, **7:**195–201

MK0608, **7:**199–201
NM283 (valopicitabine),
  7:195–196
PSI-6130 (Pharmasset),
  7:196–197
R1479/1626, **7:**197–199
human herpesvirus type 6
  (HHV-6) therapy,
  7:244
inosine monophosphate
  dehydrogenase,
  5:1044–1046
mizoribine, **5:**1044,
  1049–1050
ribavirin and C2-MAD,
  5:1044–1046
nucleoside reverse
  transcriptase
  inhibitors (NRTIs):
  clinical potential, **7:**155–156
  current compounds,
    7:141–145
  research background,
    7:139–140
Nucleoside linkage mechanisms,
  anti-DNA virus
  antimetabolites,
  7:225–226
Nucleoside prodrugs, anti-DNA
  virus agents,
  7:230–231
Nucleoside sugar moiety
  modification, anti-
  DNA virus
  antimetabolites,
  7:224–225
Nucleotide binding domains
  (NBDs), ATP binding
  cassette (ABC)
  transporters,
  resistance targeting,
  6:371
Nucleotide prodrugs:
  anti-DNA virus agents, **7:**231
  antitubercular agents, DNA-
    targeted synthesis and
    repair, **7:**725
  human herpesvirus type 6
    (HHV-6) therapy,
    7:244
Nucleotide reverse transcriptase
  inhibitors (NtRTIs):
  clinical potential, **7:**155–156
  current compounds,
    7:145–148
  research background,
    7:139–140

Nucleotide transporters,
  trypanocide drug
  development and
  discovery, **7:**577–580
Numb-associated kinases (NAK),
  structural genomics,
  1:591–594
Number of valence electrons
  (NVE), quantitative
  structure-activity
  relationships,
  polarizability indices,
  1:25
Nutilin-3 compound:
  brain cancer therapy,
    6:257–258
  protein-protein interactions,
    2:346–352
Nutrient intake, vitamin
  deficiency and, **5:**643
NVP-BEZ235 inhibitor:
  brain cancer therapy,
    6:250–251
  survival signaling pathways,
    6:322–323
NVP-LAQ-824 histone
  deacetylase inhibitior,
  structure and
  development, **6:**55,
  57–58
NXL104 non-β-lactam inhibitor,
  7:371
NXN compounds, protein-protein
  interactions, p53/
  MDM2 antagonists,
  2:351–352
NXY-059 antioxidant, stroke
  therapy, **8:**479–480
Nystatin, development of,
  2:231

Obatoclax, peptidomimetics,
  1:233, 236
Obesity:
  antipsychotic effects, **8:**168
  peroxisome proliferator-
    activated receptors:
    α/γ/δ-pan agonists,
      4:141–145
    δ-agonists, **4:**118
    γ-agonists, **4:**100–102
  thyromimetic therapy, **4:**198
  type 2 diabetes, incidence and
    epidemiology,
    4:1–2
Obviousness standards, patent
  law and, **3:**134–138

Occupancy theory, receptor-based drug design, **2**:494–497
congestive heart failure therapy, mechanism-targeted studies, **4**:492
Octahydroisoquinolines, δ-opioid selective agonists, **8**:635–637
Octanol-water partition coefficients:
oral drug delivery, physicochemistry evaluation, **3**:34–35
quantitative structure-activity relationship, hydrophobic interactions, **1**:10–19
Octoxynol, antimicrobial agents, **7**:493
Ocular hypertensives (OHT), glaucoma disease pathology and etiology, **5**:589–590
ODC (outlier detection by distance toward training set compounds) method, outlier detection, quantitative structure-activity relationship, **1**:45
Off label drug use, antipsychotic medications, **8**:170
Off-target activity:
hit/lead compounds, BioPrint profiling, **1**:491
tool compounds, BioPrint® database, **1**:492–493
Ofloxacin, antibacterial agents, **7**:535–540
O-GlcNAcase inhibitors, Alzheimer's disease, tau-targeted therapies, **8**:422–423
O-glucuronidation, drug metabolism, **1**:428–430
Oil/water partition coefficient:
membrane proteins, drug delivery strategies, prodrug development, **2**:441–443
oral drug physicochemistry, **3**:31–32
Olanzapine:
FDA approval, **8**:178

pharmacokinetics, **8**:192–193
side effects, **8**:168–169
structure-activity relationship, **8**:180–183
Olaparib, poly(ADP-ribose) polymerase (PARP) inhibitor coadjuvants, **6**:169–170
Olcegepant, migraine therapy, **8**:273, 275–277
Olefinic bonds, **1**:416
(-)-Oleocanthal, Alzheimer's disease, tau-targeted therapy, **8**:429
Oligodendrogliomas:
classification and therapy, **6**:233–234
stem cell initiation, **6**:225–227
Oligosaccharide arrays, solid-phase organic synthesis, **1**:291–292
Olivanic acids, carbapenem synthesis, **7**:333–339
Omeparazole, chiral switch in, **1**:156–159
Omeprazole, selective toxicity, **2**:566
O-methylation reactions, drug metabolism, **1**:423–425
Onapristone, physiology and pharmacology, **5**:239
Oncovin. See Vincristine
One-at-a-time synthesis, small molecule libraries, **1**:299–300
One-dimensional nuclear magnetic resonance, limitations of, **2**:370–371
ONO2506 GABA$_A$ agonist, stroke therapy, **8**:483–484
Ono prostaglandin/thromboxane EP$_1$ antagonists, structure-activity relationship, **5**:817–824
OPC-67683 antitubercular agent, **7**:776–782
ADMET properties, **7**:781–782
adverse effects, **7**:782
history, **7**:776
mechanism of action, **7**:780–781
resistance mechanisms, **7**:781
structure-activity relationships, **7**:776–780

"Open-ring" purine derivatives, acyclic nucleoside phosphates, **7**:237–238
Ophthalmic agents:
age-related macular degeneration, **5**:604–623
clinical trials, agents in, **5**:614–623
current therapeutics, **5**:613–614
disease pathology, **5**:604–610
future research issues, **5**:622–623
in vivo preclinical screening models, **5**:610–613
antimetabolites, anti-DNA virus agents, **7**:222–225
cidofovir therapy, **7**:243
future research issues, **5**:622–623
glaucoma, **5**:588–603
clinical trials, antiglaucoma therapeutics, **5**:600–602
current clinical treatments, **5**:592–600
disease pathology, **5**:588–590
future research issues, **5**:602–603
intraocular pressure reduction, **5**:591–592
in vivo preclinical screening models, **5**:590–591
glucocorticoid anti-inflammatory compounds, corneal penetration, **5**:53
N-(4-hydroxyphenyl) retinamide (4-HPR) therapy, **5**:474
research and development, **5**:587–588
Opioid-receptor-like 1 (ORL1) receptor:
discovery and nomenclature, **8**:580
endogenous peptides, **8**:594–596
nonpeptide ligands, **8**:680–681
physiology and pharmacology, **8**:677–678
research background, **8**:676–677

structure-activity relationship, 8:678–680
Opioids. *See also* specific subtypes, e.g. κ-opioid receptors
ADMET properties, 8:577–579
morphine and derivatives, 8:578–579
analgetics, soft drug design, 2:88–90
clinical applications, 8:570–579
current drugs, 8:571–573
cytochrome P450 enzyme interactions, 8:576–577
future research issues, 8:684–686
morphine and analogs, historical background, 8:596–597
nonpeptide analgesics, structure-activity relationships, 8:597–645
acyclic analgesics, 8:621–623
affinity labeling, 8:637–643
morphine-naltrexone derivatives, 8:638–641
agonists:
δ-selective agonists, 8:632–637
κ-selective agonists, 8:623–632
centrally-acting agonists, 8:624–630
peripherally-acting agonists, 8:630–632
antagonists, 8:601–608
δ-opioid receptors, 8:604–608
κ-opioid receptors, 8:602–604
benzomorphans, 8:610–613
Diels-Alder adducts, 8:608–609
miscellaneous opioids, 8:643–645
morphinans, 8:609–610
morphine derivatives, 8:599–601
opium-derived morphine and alkaloids, 8:598–599
ORL1 receptor, endogenous ligand orphanin FQ/nociceptin, 8:681–682
piperidine derivatives, 8:613–621

4-anilidopiperidines, 8:618–621
4-arylpiperidines:
carbon substituent, 8:614–616
oxygen substituent, 8:616–618
ORL1 receptor, endogenous ligand orphanin FQ/nociceptin, 8:676–681
nonpeptide ligands, 8:681–682
physiology and pharmacology, 8:677–678
research background, 8:676–677
structure-activity relationship, 8:678–680
peptide analogs, 8:645–676
affinity-labeled derivatives, 8:675–676
amphibian skin peptides, 8:646
antagonist dynorphin A analogs, 8:657–658
combinatorial library peptides and peptidomimetics, 8:674–675
dynorphin analogs, 8:654–658
conformational constraints, 8:656–657
linear analogs, 8:655–656
enkephalin analogs, 8:646–654
δ-selective analogs, 8:649–653
dimeric analogs, 8:653–654
μ-selective analogs, 8:646–649
somatostatin μ-receptor antagonists, 8:673–674
Tyr-D-aa-Phe sequence, 8:664–673
deltorphin analogs, 8:668–673
dermorphin analogs and μ-selective peptides, 8:664–668
Tyr-Pro-Phe sequence, 8:659–664

β-casomorphin analogs and endomorphins, 8:659–661
TIPP and related peptides, 8:661–664
physiology and pharmacology, 8:579–597
central nervous system effects, 8:579
computational models, 8:590
δ-opioid receptors, 8:591–592
endogenous peptides, 8:593–596
κ-opioid receptors, 8:592–593
ligand characterization, 8:581–583
multiple opioid receptors, 8:579–580
mutagenesis studies, 8:589–590
non-μ-opioid receptors, 8:591–593
signal transduction mechanisms, 8:580–581
structure and molecular biology, 8:587–589
subtypes, splice variants, and dimerization, 8:590–591
in vitro efficacy assays, 8:583–585
in vivo evaluation, 8:585–587
recent developments, 8:682–684
receptor-based drug design:
central nervous system, 8:2–3
GPCR signal modulation, 2:508–509
research background, 8:569
salvia divinorum and neoclerodane diterpenes, 8:682–684
side effects, 8:573–576
constipation, 8:575–576
contraindications, 8:576
drug-drug interactions, 8:576–577
nausea and vomiting, 8:575
pruritis, 8:576
respiratory depression, 8:574
sedation and cognitive impairment, 8:575
tolerance, dependence, and addition liabilities, 8:574–575
subtype classification, 8:586

δ-Opioids:
  chemical structure, **8:**571
  discovery and nomenclature, **8:**579–580
  β-endorphin affinity, **8:**594–596
  enkephalin analogs, **8:**650–653
  ligand characterization, **8:**582–583
  mutagenesis studies, **8:**589–590
  nonpeptide affinity labeling, **8:**640–643
  physiology, **8:**591–592
  selective agonists, **8:**632–637
  selective antagonists, **8:**604–608
  signal transduction, **8:**580–581
  subtypes, splice variants, and dimerization, **8:**590–591
  Tyr-Pro-Phe sequence, **8:**661–664
κ-Opioids:
  agonist compounds, **8:**623–632
  benzomorphan derivatives, **8:**611–612
  chemical structure, **8:**571
  discovery annd nomenclature, **8:**579–580
  ligand characterization, **8:**581–583
  nonpeptide affinity labeling, **8:**642–643
  physiology, **8:**592–593
  selective antagonists, **8:**602–604
  signal transduction, **8:**580–581
  subtypes, splice variants, and dimerization, **8:**591
μ-Opioids:
  chemical structure, **8:**571
  dermorphin analogs, peptide selectivity, **8:**664–668
  discovery and nomenclature, **8:**579–580
  endomorphin affinity, **8:**595–596
  β-endorphin affinity, **8:**594–596
  enkephalin analogs, **8:**647–649
  ligand characterization, **8:**581–583
  mutagenesis studies, **8:**589–590
  selective antagonists, **8:**607
  somatostatin derivation, **8:**673–674
  signal transduction, **8:**580–581

subtypes, splice variants, and dimerization, **8:**591
Opium, morphine derivatives, **8:**599–60
OPOT-80 macrolide compound, **7:**455–456
Optimization in drug development:
  anti-HCV agents, benzothiadiazine NS5B inhibitors, **7:**202–203
  high-throughput screening, assay optimization, **3:**409–414
  HIV protease inhibitors:
    atazanavir sulfate, **7:**36–38
    darunavir, **7:**52–55
    fosamprenavir calcium, **7:**43–44
    indinavir sulfate, **7:**12, 14, 16–18
    lopinavir/ritonavir, **7:**31–33
    nelfinavir mesylate, **7:**20, 24–25
    ritonavir, **7:**10
    saquinavir mesylate, **7:**5–7
    tipranavir, **7:**46–49
  lead development, ADMET rules for, **2:**64
  nevirapine case study, **3:**21–23
  spiro diketopiperazine-based chemokine receptor-5 (CCR5) antagonist, **7:**131–133
Optro® recombinant hemoglobin (rHb1.1), structure and properties, **4:**638
Oral cancer, N-(4-hydroxyphenyl) retinamide (4-HPR) therapy, **5:**469
Oral contraceptives:
  estrogen/progestin combinations, ovulation modulation mechanism, **5:**242–243
  steroid chemical reactions, norethindrone, **5:**11
Oral drug delivery:
  bioavailability:
    blood absorption, **2:**748
    immediate release forms, **2:**771–787
    area under the curve measurements, **2:**785–786

concentration profile calculation, **2:**773–774
dimensionless numbers, **2:**772
drug compound characteristics, **2:**776
intravenous delivery *vs.*, **2:**786–787
master curves with dimensionless numbers, **2:**781–784
multiple dose administration, **2:**776–784
nomenclature, **2:**771–772
pharmacokinetics, plasma drug profile, **2:**774–776
repeated dose principles and parameters, **2:**777–781
transport principle, **2:**772–773
pharmacokinetic stage, **2:**747
quantitative structure-activity relationship, ADMET parameters, **1:**66
sustained-release dosage forms:
  in vitro-in vivo correlations, **2:**787–791
  in vitro tests, **2:**791–801
Biopharmaceutics Classification System, **3:**41–43
  biowaiver monographs, **3:**360–361
  global market, **3:**353–363
  class I drugs, **3:**42
  class II drugs, **3:**42–50
  class III drugs, **3:**50–52
  class IV drugs, **3:**52, 355
  class V drugs, **3:**52–55
  dosage form development strategies, **3:**40–55
glycoprotein IIb/IIIa receptor antagonists, **4:**448–449
membrane proteins:
  lipid-based mechanisms, **2:**441
  receptor-mediated endocytosis, **2:**450
nonsolid forms, **3:**29

permeability, ADMET
interactions, oral
bioavailability, **3**:372
pharmacokinetic stage, **2**:747
physicochemical property
evaluation, **3**:29–40
aqueous solubility, **3**:32–33
dissolution rate, **3**:33
hygroscopicity, **3**:37–39
ionization constant, **3**:35–37
melting point, **3**:30–31
partition coefficient, **3**:31–32
permeability, **3**:33–35
rule of five, **3**:34–35
stability, **3**:39–40
prodrug properties, **3**:220–226
protein therapeutics,
3:315–316
solid form selection criteria,
3:25–29
amorphous solids, **3**:29
polymorphs, **3**:26–27
pseudopolymorphs, **3**:27–29
salts, **3**:26
solvates, **3**:28–29
Oral leukoplakia:
N-(4-hydroxyphenyl)
retinamide (4-HPR),
5:466, 478–479
13-cis-retinoic acid therapy,
5:391
Orange Book (FDA), drug
approval data, **3**:92–93
Orexin receptor antagonists,
insomnia therapy,
5:720–723
Organ barriers, ADMET
distribution
properties, **2**:61
permeability and toxicity, **3**:377
Organ failure, carbon monoxide
inhalation therapy,
5:300–302
Organic anion transporters
(OATPs), single
nucleotide
polymorphisms, **1**:189
Organic cation transporter-1
(OCT1), metformin
metabolism, **4**:4
Organic chemistry, vitamins,
5:638
Organic nitrates, nitric oxide
donors and prodrugs,
5:278–280
Organization for Economic Cooperation and

Development (OECD),
quantitative
structure-activity
relationship modeling
principles, **1**:508,
528–529
Organoruthenium, stroke
therapy, cathepsin B
inhibitors, **8**:476–477
Organotellurium, stroke therapy,
cathepsin B inhibitors,
8:476–477
Organ storage fluids, carbon
monoxide transfusion
and tissue delivery,
5:307–308
Oripavine, nonpeptide affinity
labeling, **8**:641–643
Oritavancin, antibacterial
mechanism,
2:226–227
OrM3 compound, chronic
obstructive pulmonary
disease
bronchodilators,
5:768–769
Ornithine decarboxylase (ODC)
inhibitors, human
African
trypanosomiasis
therapy, **7**:571
OROS® dosage form, oral drug
delivery, sustained-
release dosage forms,
in vitro-in vivo
correlations,
2:789–791
Orphan Drug Act, **3**:67–68
Orphanin FQ/nociceptin peptide:
nonpeptide ligands, **8**:680–681
opioid receptor affinity,
8:595–596
physiology and pharmacology,
8:677–678
research background,
8:676–677
structure-activity relationship,
8:678–680
Orphan receptors, receptor-based
drug design:
classification, **2**:513
nomenclature, **2**:504–505
Orphenadrine, Parkinson's
disease therapy,
8:552–553
Ortho-biphenyl carboxamides,
brain tumor therapy,

hedgehog signaling
pathway, **6**:260–261
Ortho-dihydroxyaromatics,
toxicophore reactive
metabolites,
2:313–314
Orthogonality:
defined, **1**:280
small molecule libraries,
1:299–300
Orthostatic hypotension:
antidepressant side effect,
8:231–232
antipsychotic side effects, **8**:167
Orthosteric site, receptor-based
drug design, **2**:500
Osanetant, cognitive
enhancement, **8**:48–49
Oseltamivir, selective toxicity,
2:556–557
Osteoporosis:
anabolic therapy:
basic principles, **5**:753
parathyroid hormones,
5:753–755
selective androgen receptor
modulator, **5**:760
statins, leptin, and
neuropeptide Y, **5**:755
vitamin D, **5**:759–760
Wnt signaling pathway,
5:755–759
bone remodeling,
5:756–757
canonical pathway, **5**:756
genetic models and bone
target cells, **5**:757–758
therapeutics development,
5:758–759
mechanistic antiresorptive
therapy:
bisphosphonates, **5**:733–734
calcitonin and integrin
antagonists, **5**:730
cathepsin K inhibitors,
5:730–732
clinical trials, **5**:740–741
epidemiology, **5**:729–730
future research and
applications, **5**:742
OPG/RANKL/RANK
inhibitors, **5**:732–733
preclinical studies,
5:739–740
selective estrogen receptor
modulators,
5:734–742

Osteoporosis-pseudoglioma syndrome (OPPG), osteoporosis anabolic therapy, Wnt pathway and bone remodeling, **5:**756–757
Osteoprotegerin (OPG) inhibitors, osteoporosis antiresorptive therapy, **5:**732–733
Ostwald's step rule, polymorph crystallization, **3:**190–192
OT-551 topical agent, age-related macular degeneration therapy, **5:**622
Ototoxicity:
aminoglycosides, **7:**419–420
quinine antimalarials, **7:**607
Outlier detection, quantitative structure-activity relationship, **1:**43–46
convex hull method, **1:**44
distance between r^2 and q^2, **1:**45
Mahalanobis distance, **1:**44
OCD method, **1:**45
potential functions, **1:**44
R-NN curves, **1:**44–45
smallest half-volume method, **1:**44
standard deviation test, **1:**45–46
uncertainty method, **1:**44
X-residual, **1:**44
Ovalicin, transplantation rejection prevention, **5:**1006–1007
Ovarian cancer:
N-(4-hydroxyphenyl) retinamide (4-HPR) therapy, **5:**466, 468–469, 473, 479–480
SHetA2 retinoid-related compound, **5:**502–503
Overdose:
antidepressants and, tricyclic antidepressants, **8:**231–232
serotonin-selective reuptake inhibitors, **8:**232–234
Over the counter (OTC) drugs, FDA approval, **3:**82–84
Ovulation modulation, estrogen/progestin combinations, **5:**242–243
OWFEG approach, force field scoring, **2:**621–622
Oxacarbenium ions, poly(ADP-ribosyl)ation biochemistry, **6:**153–157
Oxacephalosporins, development of, **7:**308–310
Oxadiazole carboxamides, BioPrint profile, **1:**491–493
Oxadiazole ketone, dipeptidyl peptidase 4 (DPP-4) inhibition, **4:**49–50
Oxadiazoles:
muscarinic activity, **8:**66–67
prostaglandin/thromboxane EP_1 antagonists, **5:**840, 842
Oxalic acid hydroxymates, stroke therapy, **8:**507
Oxamide piperidine derivatives, stroke therapy, **8:**454–455
Oxandrolone, anabolic agents, **5:**183
Oxapenems:
β-lactamase inhibitors, structural modification, **7:**370–371
structure-activity relationships, **7:**260–261
Oxaphems, structure-activity relationships, **7:**260–261
Oxathiazol-2-one, antitubercular agents, protasome synthesis and degradation, **7:**726
1,3-Oxathiolane derivatives, anti-DNA virus antimetabolite, **7:**227–228
Oxathiolane analog, muscarinic activity, **8:**65
Oxazepam, anticonvulsants, **8:**144–147
Oxazoles:
prostaglandin/thromboxane receptors, IP receptor agonists, **5:**924–927
structure-based design, **2:**698–700
Oxazolidine, nicotinic agonists, **8:**71
Oxazolidinones:
anti-infective libraries, **1:**303–305
historical background, **7:**484
hybrid fluoroquinolone-oxazolidinone, **7:**551
macrolide development, **7:**451–456
ribosome targeting, **7:**464
synthetic antibacterial agents, **7:**540–548
future research and development, **7:**547–548
linezolid, **7:**541–544
structure-activity relationship, **7:**544–547
4-Oxazolylcarboxamide analogs, thromboxane A_2 receptor antagonist, **5:**886–889
Oxcarbazepine, anticonvulsant applications, **8:**135–139, 142–144
Oxene, drug metabolism, functionalization reactions, **1:**409–413
Oxidation:
protein therapeutics formulation and delivery, **3:**312–314
tetracyclines, **7:**409
topical antimicrobial compounds, **7:**496
β-Oxidation, drug metabolism, conjugation reactions, **1:**434
Oxidative cleavage reactions:
drug metabolism, **1:**419–421
hemoglobin oxygen delivery, **4:**614–615
Oxidative enzymes, gene-directed enzyme prodrug therapy, **6:**124
Oxidative metabolism:
androgens, **5:**162–164
antimalarial drugs, **7:**658–663
glutathione biosynthesis, **7:**660–663
PfaATP6, **7:**658–660
triarylethylenes, **5:**225–226

Oxidoreductases:
  drug metabolism, functionalization reactions, 1:411–412
  high-throughput screening, enzymatic assays, 3:423–424
  Vitamin K oxidoreductase, 4:381–385
Oxigon β-amyloid peptide inhibitor, 8:363–365
Oxime derivatives:
  prostaglandin/thromboxane receptors, IP receptor agonist, 5:932, 937
  radicicol Hsp90 inhibitor, 6:400–405
Oxime derivatives, eye-targeted chemical delivery systems, 2:158–160
  β-blocker analogs, 2:158–160
Oximonam, monobactam activity, 7:331–332
Oxiprenaline, chronic obstructive pulmonary disease bronchodilators, 5:769–770
Oxiracetam, cognitive enhancement, 8:34
Oxitropium, chronic obstructive pulmonary disease bronchodilators, 5:767
Oxorhenium complexes, stroke therapy, cathepsin B inhibitors, 8:473
Oxotremorine, muscarinic activity, 8:68
Oxyacetic acid side chains, thyroid hormone receptor β-selective agonists, structure-based design, 2:687–688
Oxybutynin:
  chiral switch in, 1:158–159
  muscarinic antagonists, 8:74–77
Oxycyte®, oxygen delivery mechanisms, 4:625
Oxygen delivery:
  allosteric effectors, 4:615–619
    efaproxiral, 4:619
    right shifters, 4:615–619
  future research issues, 4:643–644
  hemoglobin:
    basic properties, 4:610–612
    nitric oxide reactions, 4:612–614
  oxygen carriers, 4:625–643
    artificial erythrocytes, 4:640–642
    conjugated hemoblogins, 4:635–637
    cross-linked hemoglobin, 4:628–631
    polymerized hemoglobins, 4:631–635
    recombinant hemoglobins, 4:637–640
    redox/radical reactions, 4:614–615
  nonhemoglobin, heme-based oxygen carriers, 4:642–643
  perfluorocarbons, 4:619–625
    basic properties, 4:620–621
    fluosol, 4:623
    Oxycyte®, 4:625
    Oxyfluor®, 4:624
    Oxygent, 4:624–625
    Perftoran, 4:623–624
    pharmacokinetics, pharmacodynamics, and toxicity, 4:621–622
    PHER-O2®, 4:624
  research background, 4:610–612
Oxygen-induced retinopathy (OIR) model, wet age-related macular degeneration, *in vivo* screening models, 5:612
Oxygent, oxygen delivery mechanisms, 4:624–625
Oxyhemoglobin oxidation, nitric oxide effects, 5:273–278
Oxymorphone:
  antagonist activity, 8:601
  nonpeptide affinity labeling, 8:640–643
Oxyphenonium, muscarinic antagonists, 8:72–77
Oxytetracycline group:
  history and biosynthesis, 7:412–413
  research background, 7:407
Oxytocin, agonist/antagonist, 4:522–523

OZ277 artemisinin analog, antimalarial compounds, 7:660
OZ439 antimalarial agent, 7:625–627

P2X$_1$ receptor, antiplatelet agents, 4:423–424
P2X$_7$ nucleotide receptor antagonists:
  hit seeking combinatorial libraries, 1:324–326
  ion channels, 1:336–337
P2X receptor family, receptor-based drug design, ligand-gated ion channels, 2:510–511
P2Y$_1$ antagonist, antiplatelet agents, 4:424–425
P2Y$_1$ receptor, antiplatelet agents, 4:424
P2Y$_{12}$ antagonist, antiplatelet agents, 4:425–426
P2Y$_{12}$ receptor, antiplatelet agents, 4:423
p7 protein, hepatitis C viral genome replication, 7:173–174
p25 protein, cyclin dependent kinase 5/p25 inhibitors, hit seeking combinatorial libraries, 1:329–330
p38 mitogen-activated protein kinase inhibitors, chronic obstructive pulmonary disease inhibitors, 5:789–790
p53 protein:
  AHPN compound expression and activation, 5:494–497
  allosteric protein model, 1:394
  HDM2-p53 interaction antagonists, 1:326–327
  protein-protein interactions, drug discovery, 2:344–359
  signaling pathway, brain cancer therapy, 6:256–257
PA-824 antitubercular agent, 7:776–782
  ADMET properties, 7:781–782
  adverse effects, 7:782
  history, 7:776

p53 protein (*Continued*)
  mechanism of action, **7:**780–781
  resistance mechanisms, **7:**781
  structure-activity relationships, **7:**776–780
Packing polymorphism, crystal engineering, **3:**191–192
Paclitaxel:
  Alzheimer's disease, tau-targeted therapy, microtubule stabilization, **8:**430
  structure and therapeutic applications, **6:**32–33
Pain management:
  opioid receptors, **8:**570–571
  phosphodiesterase inhibitors, **5:**716
Paliperidone:
  FDA approval, **8:**178
  pharmacokinetics, **8:**195–196
Palosuran, congestive heart failure therapy, pharmacodynamic-kinetic dose selection, **4:**502–503
Pan-AKT inhibitor, multitarget drug development, cancer therapy, **1:**266–267
Panipenem:
  biological activity, **7:**351–353
  synthesis and development, **7:**340–341
Panobinostat:
  histone deacetylase inhibitor, structure and development, **6:**58
  hydroxamate inhibitors, stroke therapy, **8:**489–490
Pantothenate, antitubercular agents, cofactor biosynthesis targeting, **7:**727
Pantothenic acid, **5:**685–687
Papaverine:
  antipsychotic agents, **8:**208–209
  memory assessment, **8:**28–30
  structure-activity relationship, **8:**599
Para-aminobenzoic acid (PABA), prodrug development, prontosil case study, **3:**258

Para-Aminosalicylic acid (PAS), antitubercular agents, **7:**764–767
Paracellular permeability, ADMET structural properties, **2:**56–57
Paracetamol. *See* Acetaminophen
Para-dihydroxyaromatics, toxicophore reactive metabolites, **2:**313–314
Parallel artificial membrane permeability assay (PAMPA), ADMET permeability, **2:**66
Parallelepiped side calculations, oral drug delivery, sustained-release dosage forms, **2:**795–796, 799
Paramomycin sulfate, Leishmaniasis therapy, **7:**576–577, 590–593
Paraoxonase 1 (PON1), high-density lipoproteins, **4:**334
Parasitic infections. *See also* Antiprotozoal/antiparasitic agents
  epidemiology, **7:**565–566
Parathyroid hormone (PTH), osteoporosis therapy, **5:**753–755
  secretion/calcilytics, **5:**754–755
Parent data set, quantitative structure-activity relationship validation, **1:**49
Parenteral drug delivery, hydrogen sulfide, **5:**313–314
Pareto optimization, drug discovery:
  clinical applications, **2:**263–266
  ligand designer, **2:**266–273
  research background, **2:**262–263
Paris Convention for the Protection of Industrial Property, **3:**165–166
Parkinsonism:
  drug-induced, antipsychotic side effects, **8:**165–166
  MPTP and, **8:**538–540
  pathology, **8:**529–530

Parkinson's disease:
  cognitive dysfunction in, **8:**21
  diagnostic agents, **8:**555–556
  dopaminergic pharmacotherapy, **8:**540–552
    catechol-*O*-methyltransferase inhibitors, **8:**551–552
    L-dopa therapy, **8:**540–542
    monoamine oxidase inhibitors, **8:**549–551
    receptor-targeted compounds, **8:**542–549
      aporphine-type receptor agonists, **8:**544–545
      dopamine receptor structure and function, **8:**542–544
      ergot-type receptor agonists, **8:**545–546
      experimental D_1 receptor agonists, **8:**547–549
      small-molecule agonists, **8:**547
  epidemiology, **8:**529
  etiology, **8:**534–540
    dopamine/mitochondrial oxidative mechanism, **8:**535–537
    environmental factors, **8:**537–540
    genetic factors, **8:**534–535
  future research issues, **8:**556
  nondopaminergic pharmacotherapy, **8:**552–555
    adenosine receptor antagonists, **8:**552–553
    anticholinergic agents, **8:**552
    glutamate antagonists, **8:**555
    serotonin 5-HT_{1A} agonists, **8:**553–555
  pathology, **8:**529–534
Paromyomycin:
  current applications, **7:**416
  pharmacology and ADMET properties, **7:**421–423
Paroxetine:
  chemical structure, **8:**223
  clozapine interaction, **8:**192
  pindolol interaction, **8:**225–229
  selective toxicity, **2:**569
  side effects and adverse reactions, **8:**233–234

Partial agonists:
  nuclear hormone receptors, structural biology, 4:82–83
  occupancy theory, 2:496–497
  opioids, 8:601–608
  peroxisome proliferator-activated receptors:
    selective PPARα modulators/PPARα partial agonists, 4:97
    selective PPARδ modulators/PPARδ partial agonists, 4:122
    selective PPARγ modulators PPARγ partial agonists, 4:106–115
Partial antagonists, occupancy theory, 2:496–497
Partial least squares regression (PLSR), quantitative structure-activity relationship, 1:37
Partial least squares (PLS) technique, ligand-based virtual screening, quantitative structure-activity relationships, 2:11–13
Particle size parameters, oral dosage systems, class II drugs, low solubility/high permeability, 3:44–45, 354
Particulate removal, protein therapeutics, downstream processing, 3:303–304
Partition coefficient:
  oil/water partition coefficient, membrane proteins, drug delivery strategies, 2:441–443
  oral drug physicochemistry, 3:31–32
  quantitative structure-activity relationship:
    calculation methods, 1:14–19
    hydrophobic parameters, 1:8–19
    potentiometric measurements, 1:12–13
    reversed-phase chromatography measurements, 1:13–14
    shake-flask measurement, 1:12
    measurement methods, 1:12–14
Parvaquone antimalarial agent, 7:619–620
Passerini reaction, small molecule libraries, 1:300
Passive avoidance task, memory assessment, 8:28
Passive diffusion:
  ADMET structural properties, permeability, 2:55
  membrane proteins, 2:444–446
  oral drug delivery, permeablity, 3:34–35
Patch characteristics and preparation, transdermal drug delivery, metropolol experiment, 2:832
Patent and Trademark Office (PTO) (U.S.):
  copending applications, 3:109–110
  corrections of patents, 3:146–148
  intellectual property law, 3:103–105
  interference procedures, 3:142–146
  patent delay classifications, 3:113–114
  patent procedures at, 3:138–142
  patent prosecution strategy, 3:115–118
  patent protection, 3:106–107
  provisional applications to, 3:127–133
  trademark registration, 3:174–175
  trademark selection guidelines, 3:173–174
Patent Cooperation Treaty (PCT), 3:167–169
Patents:
  design patents:
    drug discovery and, 3:185
    protection and strategy, 3:105–118
  drug discovery and (See also Intellectual property law; specific types of patents, e.g., Utility patents)
  applicants from non-U.S. countires, 3:133
  combinatorial chemistry, 1:350
  delays classifications, 3:113–114
  enforcement, 3:148–165
    alternative dispute resolution, 3:162–163
    claim construction proceedings, 3:162
    defenses to infringement, 3:154–157
    discovery phase, 3:160–161
    geographic scope, 3:148
    infringement determination, 3:152–154
    parties to suit, 3:150–152
    proceedings commencement, 3:159–160
    remedies for infringement, 3:157–159
    summary judgment, 3:161
    trial and appeal processes, 3:148–150, 162–164
  first to invent vs. first to file distinction, 3:103–104, 108–110
  generic drug approval, FDA term extension guidelines, 3:87
  global patent protection, 3:107–108, 165–171
    international agreements, 3:165–168
    international laws and regulations, 3:169–171
    PCT patent practice, 3:168–169
  historical background, 3:101–105
  protection and strategy, 3:105–118
    absolute novelty principle, 3:110–111, 134
    application publication, 3:114–115
    first to invent vs. first to file, 3:108–110
    global strategy, 3:107–108, 165–171
    patent term, 3:111–114

Patents (*Continued*)
  PTO prosecution strategy, **3:**115–118
  trade secrets and, **3:**182–183
  reexamination process, **3:**107
  requirements, **3:**118–148
    best mode requirement, **3:**126, 130–131
    claims, **3:**126–127
    corrections, **3:**146–148
    enablement requirement, **3:**124–126
    interference, **3:**142–146
    new and nonbovious inventions, **3:**133–138
    patentable subject matter, international requirements, **3:**121–123
    patentable subject matter, U.S. and international requirements, **3:**118–123
    provisional applications, **3:**127–133
    PTO procedures, **3:**138–142
    specifications, **3:**123–127
    statutory requirements, **3:**129–130
    written description, **3:**123–124
  time limits for, **3:**106–107, 111–114
  trade secrets *vs.*, **3:**182–183
Patient adaptation:
  repeated intravenous bolus injections, **2:**759–760
  repeated short infusions, constant flow rate over finite period, **2:**766–771
Patient intervariability:
  congestive heart failure therapy, selection and stratification criteria, **4:**506–507
  drug bioavailability, repeated intravenous bolus injection, **2:**756
  plasma drug profile:
    diffusion-based drug release: repeated dose delivery, **2:**809–812
    single dose delivery, **2:**806–808

erosion-controlled drug release, **2:**819–825
repeated short infusions, constant flow rate over finite period, **2:**766–771
Paullones, Alzheimer's disease, tau-targeted therapies, **8:**420–422
Paw test, antipsychotic side effects, **8:**175
Pazopanib inhibitor, antiangiogenic properties, **6:**309–310
PCI-024781 histone deacetylase inhibitor, structure and development, **6:**60–62
PCSK9 inhibition, LDL cholesterol lowering agents, **4:**318–319
PD012527, peptidomimetic design, **1:**226–227
PD 145065, peptidomimetic design, **1:**225–226
PD-150606 calpain inhibitor, stroke therapy, **8:**465
PD151307 peptide ligand, N-type voltage-dependent calcium channels, stroke therapy, **8:**456–459
PD166326 kinase inhibitor, structure-based design, protein targeting, **6:**352–354
PD-168077, cognitive enhancement, **8:**41–42
PD 0332991 cyclin-dependent kinase inhibitor, medicinal chemistry and classification, **6:**313–314
PDE4 inhibitors, brain tumor therapy, **6:**265
PDSPK_i Database, quantitative structure-activity relationship modeling, **1:**514
Pediatric neuronal ceroid lipofuscinosis, *N*-(4-hydroxyphenyl) retinamide (4-HPR) therapy, diabetes, **5:**474
Pediatric Rule (FDA), **3:**68

Pedunculopontine nucleus (PPN), Parkinson's disease, pathophysiology, **8:**534
Pefloxacin, antimalarial drugs combined with, **7:**621, 623–624
Pegaptanib, age-related macular degeneration therapy, **5:**615
Pegylated interferon-α (PEG-IFN-α):
  hepatitis C therapy, **7:**170
  R1479/1626 anti-HCV agent combined with, **7:**197–199
Pellagra, niacin deficiency, **5:**659
Peloruside A, marine sources, **2:**246
Pemetrexed, antifolate DNA-targeted chemotherapeutics, **6:**107–109
Pemoline, side effects, adverse reactions and drug interactions, **8:**95
Penam sulfones, β-lactamase inhibitors:
  discovery and properties, **7:**363–364, 367
  structural modification, **7:**368–370
Penciclovir, anti-DNA virus agents:
  antimetabolites, **7:**229
  herpes simplex virus therapy, **7:**242–243
Pendant amphiphilic polymers, nanoscale drug delivery, **3:**474–475, 479
Penems. *See also* Carbapenems; Oxapenems
  biological properties, **7:**321–322
  β-lactamase inhibitors: 6-heteroarylmethylene penems, **7:**364–366
  structural modification, **7:**368–370
  development of, **7:**317–324
  early candidates, **7:**321
  faropenem, **7:**321–323
  structure-activity relationship, **7:**260–261
  sulopenem, **7:**323–324
  synthesis of, **7:**317–321

Penetration enhancers, membrane protein permeability, **2**:439–441
Penicillinases, penicillin resistance and development of, **7**:289–292
Penicillin binding proteins (PBPs):
  antitubercular agents, peptidylglycans biosynthesis, **7**:720
  β-lactamases, biological evolution, **7**:274
  β-lactams:
    antibacterial drug resistance, native PBP modification, **7**:271–274
    E. faecium PBP-5, **7**:273
    methicillin staphylococci mechanism and genetics, **7**:272–273
    mosaic gene formation, **7**:271–272
    N. gonorrhoeae PBPs, **7**:272
    S. pneumoniae PBP-2x, **7**:273–274
  mechanisms of action, **7**:262–264
  cell wall structural organization, **7**:266–269
  Tipper-Strominger hypothesis, D-Ala-D-Ala mimicry, **7**:264–266
  monobactams, **7**:327–329
Penicillin G:
  historical development, **2**:221–222, **7**:287–288
  penicillinase development and resistance to, **7**:289–290
Penicillins:
  antibiotic compounds, **2**:222–223
  D-Ala-D-Ala mimicry, Tipper-Strominger hypothesis, **7**:265–266
  medicinal chemistry, **7**:287–295
    6-APA derivatives, structure-activity relationship, **7**:289–292
  biosynthetic compounds, **7**:288–289
  C-6-substituted penicillins, **7**:292–295
  isopenicillin N and deacetoxy-cephalosporin C, **7**:295
  modifications, **7**:295
  pharmacokinetics, **7**:281
  selective toxicity and comparative biology, **2**:549–553
Penicillin-sulfoxide-cephalosporin conversion, **7**:298–299
Penicillin V, penicillin resistance and development of, **7**:290–292
Pentaerythritol tetranitrate, nitric oxide donors and prodrugs, **5**:278–280
Pentamidine:
  human African trypanosomiasis therapy, **7**:569–571
  nucleotide transporter trypanocide development, **7**:577–580
Pentazocine, chemical structure, **8**:571
Penthienate, muscarinic antagonists, **8**:72–77
Pentolinium nicotinic antagonist, **8**:78
Pentose phosphate pathways, parasitic glucose metabolism and glycosomal transport inhibition, **7**:584–585
Pentostatin, antimetabolite analogs, **6**:112–113
Pepstatin A protease inhibitor, antimalarial agents, **7**:639–644
Peptidases, antibody-directed enzyme prodrug therapy, **6**:120–121
Peptide aldehydes, proteasome inhibitors, medicinal chemistry, **6**:181–184
Peptide epoxyketone proteasome inhibitor:
  clinical trials, **6**:180
  medicinal chemistry, **6**:183–184
Peptides:
  ACE inhibitor analogs, **4**:275–276
  pharmacological actions, **4**:280
  bioactive conformation, **1**:215
    nuclear magnetic resonance, **2**:377–383
  cognitive enhancement, **8**:50
  conformational constraints:
    global constraints, **1**:214–215
    local constraints, **1**:210–214
  HIV protease cleavage mechanisms, **7**:2–3
  isosteres:
    fibrillization inhibitors, **8**:362
    γ-secretase inhibitors, **8**:334–337
    tacrolimus structure-activity relationships, **5**:1017–1021
  N-type voltage-dependent calcium channels, stroke therapy, **8**:456–457
  opioid analgesics, **8**:645–676
    affinity-labeled derivatives, **8**:675–676
    amphibian skin peptides, **8**:646
    antagonist dynorphin A analogs, **8**:657–658
    combinatorial library peptides and peptidomimetics, **8**:674–675
    dynorphin analogs, **8**:654–658
      conformational constraints, **8**:656–657
      linear analogs, **8**:655–656
    endogenous peptides, **8**:593–596
    enkephalin analogs, **8**:646–654
      δ-selective analogs, **8**:649–653
      dimeric analogs, **8**:653–654
      μ-selective analogs, **8**:646–649
    precursors and processing products, **8**:595–596
    somatostatin μ-receptor antagonists, **8**:673–674
    Tyr-D-aa-Phe sequence, **8**:664–673

Peptides (*Continued*)
  deltorphin analogs, 8:668–673
  dermorphin analogs and μ-selective peptides, 8:664–668
    Tyr-Pro-Phe sequence, 8:659–664
    β-casomorphin analogs and endomorphins, 8:659–661
    TIPP and related peptides, 8:661–664
  opioid-receptor-like 1/orphanin FQ/nociceptin peptide, 8:678–680
  recombinant DNA technology, phage display vs., 1:551–552
  renin inhibitor development, 4:243
  solid-phase organic synthesis, informational peptide arrays, 1:280–291
  structural properties, 1:208–209
Peptide vinylsulfone proteasome inhibitors, medicinal chemistry, 6:181–184
Peptidoglycans, bacterial:
  antitubercular agents, cell wall biosynthesis, 7:720
  β-lactam mechanism of action, cell wall structural organization, 7:266–269
Peptidomimetics:
  angiotensin-converting enzyme inhibitors, 4:275–277
  BACE (β-site cleavage enzyme) inhibitors, 8:345–347
  defined, 1:207–208
  design case studies, 1:217
  design process, 1:209–210
  enzyme inhibitors, 1:217–222
    HIV-1 protease, 1:218–220
    Ras farnesyl transferase, 1:220–222
  first-generation design, 1:215–216
  G-protein coupled receptor ligands, 1:222–230
    bradykinin B2 antagonists, 1:223–224
    endothelin antagonists, 1:224–230

N-(4-hydroxyphenyl) retinamide (4-HPR) analogs, 5:480–483
opioid peptides, from combinatorial libraries, 8:674–675
protein-protein inhibitors, 1:230–240
  Bcl-2 inhibitors, 1:231–233
  XIAP inhibitors, 1:233–240
  research background, 1:207–208
  second-generation design, 1:217
stroke therapy, 8:462–463
nNOS inhibitors, 8:496
thrombin receptor antagonists, 4:441–442
Peptidyl-prolyl isomerase (PPIase), tacrolimus binding mechanism, 5:1014–1016
  structure-activity relationship, 5:1016–1021
Peptidyl vinyl sulfones, falcipain inhibitors, 7:635–637
Peptstatin, renin inhibitors, 4:244–245
Peracetic acid, topical antimicrobial compounds, 7:497
Perampanel glutamate modulator, migraine therapy, 8:296–298
Perception, cognition and, 8:17–19
Percutaneous absorption, glucocorticoid anti-inflammatory compounds, 5:46–51
  C-16 alterations, 5:96–99
Percutaneous coronary intervention (PCI), acute coronary syndrome risk, 4:411
Perfluorocarbons (PFCs), oxygen delivery, 4:619–625
  basic properties, 4:620–621
  fluosol, 4:623
  Oxycyte®, 4:625
  Oxyfluor®, 4:624
  Oxygent, 4:624–625
  Perftoran, 4:623–624
  pharmacokinetics, pharmacodynamics, and toxicity, 4:621–622

PHER-02®, 4:624
Perftoran, oxygen delivery mechanisms, 4:623–624
Pergolide, Parkinson's disease dopaminergic therapy, 8:545–546
Perhexiline analogs, soft drug design, 2:126
Perlapine, structure-activity relationship, 8:180–183
Permanent neonatal diabetes mellitus (PNDM), glucokinase activation, 4:15–17
Permeability:
  ADMET-PK drug interactions:
    basic principles, 3:367
    classification, 3:370–371
    distribution, 3:375–376
    excretion, 3:373–374
    fraction absorption, 3:369–370
    gastrointestinal absorption, solubility vs. permeability, 3:374–375
    metabolism, 3:371–373
    organ accumulation and toxicity, 3:377
    transport interactions and polymorphism, 3:367–369
  ADMET properties:
    assays, 2:66
    cell-based bioassays, 2:48–49
    structural properties, 2:55–57
  Biopharmaceutics Drug Disposition Classification System, 3:361–362
  brain-targeting chemical delivery systems, 2:136–140
  membrane protein enhancement, 2:439–444
    absorption enhancement, transporter targeting, 2:443–444
    lipid-based oral delivery, 2:441
    penetration enhancers, 2:439–441
    prodrugs, 2:441–443

oral drug delivery:
  Biopharmaceutics
    Classification System,
    3:41–42, 357–362
    class I drugs, **3:**42, 354
    class II drugs, **3:**42–50, 354
    class III drugs, **3:**50–52, 355
    class IV drugs, **3:**52, 355
  physicochemistry evaluation,
    3:33–35
  sulfonamides, **7:**517–518
Permeability-Based
  Classification System
  (PCS), drug ADMET/
  PK interactions,
  3:370–371
Permeability coefficient:
  chirality and, **1:**138–140
  quantitative structure-activity
    relationship:
    ADMET parameters, **1:**64–65
    outlier detection, **1:**46
Permeation enhancers, oral drug
  delivery, class III
  drugs, **3:**51
Pernicious anemia, vitamin B_{12}
  (cobalamin)
  deficiency, **5:**700
Perospirone, pharmacokinetics,
  8:197–198
Peroxisome proliferator-activated
  receptors (PPARs):
  α-agonists, **4:**88–99
    clinical data, **4:**96–99
    medicinal chemistry, **4:**90–96
    preclinical biology, **4:**88–90
  α-antagonists, **4:**99–100
  α/δ-dual agonists, **4:**136–137
  α/γ-dual agonists, **4:**124–136
    α-agonist derivation, **4:**91–95
  α/γ/δ-pan agonists, **4:**140–145
  biology, **4:**86
  brain tumor therapy, **6:**265–267
  carbon monoxide therapeutic
    effects, **5:**298–299
  combinatorial chemistry,
    1:335–336
  current development,
    4:166–167
  δ-agonists, **4:**117–124
    α-agonist derivation, **4:**90–96
    biology, **4:**117–118
    clinical data, **4:**121–122
    medicinal chemistry,
      4:118–121
  δ-antagonists, **4:**122–124
  γ-agonists, **4:**100–106

biology, **4:**100–102
clinical data, **4:**104–106
medicinal chemistry,
  4:102–104
g-agonists, Alzheimer's
  disease, insulin
  related therapy,
  8:415–417
γ/δ-dual agonists, **4:**137–140
γ-antagonists, **4:**115–117
ligand-based virtual screening,
  2:9
medicinal chemistry,
  4:87–88
drug discovery and,
  4:164–167
multiple sclerosis therapy,
  5:574
multitarget drug development:
  fragment-based drug
    discovery, **1:**273
  PPAR-plus for metabolic
    diseases, **1:**261, 264
partial agonists:
  selective PPARα modulators/
    PPARα partial
    agonists, **4:**97
  selective PPARγ modulators
    PPARγ partial
    agonists, **4:**106–115
prostaglandin/thromboxane
  synthesis,
  5:811–812
quantitative structure-activity
  relationship, partial
  least squares
  regression, **1:**37
safety issues, **4:**145
selective modulators, **4:**83
structural biology, **4:**77–78
Peroxisome proliferator response
  elements (PPREs),
  structural biology,
  4:86
Peroxynitrite, nitric oxide and
  formation of,
  5:273–278
Perphenazine, pharmacokinetics,
  8:187–189
Persistent hyperinsulinemic
  hypoglycemia of
  infancy (PHHI),
  glucokinase
  activation, **4:**15–17
Personalized drug design:
  cost-effectiveness analysis,
    3:349

recombinant DNA technology,
  1:540–543
Personnel training, commercial-
  scale operations,
  3:12
Perturbation approaches,
  quantum mechanics/
  molecular mechanics,
  binding affinity
  evaluation, **2:**722–723
  free-energy perturbation,
    2:723–728
  thermodynamic integration,
    2:728–729
Perzinfotel, stroke therapy, **8:**451
Pestalone, harvesting techniques,
  2:199–200
Pesticides:
  Parkinson's disease and,
    8:537–540
  soft drug design, **2:**106–108
Pexacerfont antidepressant, CRF
  antagonist, **8:**257
PF-00868554 nonnucleoside
  inhibitor, anti-HCV
  agent, **7:**201–202
PF-03187207 FP receptor agonist,
  glaucoma therapy,
  5:600–602
PfATP6 enzyme inhibitor,
  antimalarial drugs,
  7:658–660
Pfeiffer's rule, chiral compounds,
  1:129–130
PfENR inhibitor, apicoplast
  targeting, fatty acid
  biosynthesis,
  7:663–665
PfGR enzyme, antimalarial
  therapy, **7:**660–663
PfHSP70 antimalarial compound,
  7:672
PfHSP90, antimalarial drug
  development,
  7:671–672
Pfmkr homolog, antimalarial cell
  cycle regulation,
  7:628–630
Pfnek-1 kinase, antimalarial cell
  cycle regulation,
  7:628–630
PfPK5 kinase, antimalarial cell
  cycle regulation,
  7:630
PfPK7 kinase, antimalarial cell
  cycle regulation,
  7:630

P-glycoprotein:
ATP binding cassette (ABC) transporters, resistance mechanisms, **6:**371–374
efflux mechanism, oral dosage forms, class V drugs, **3:**54
homology modeling, **6:**374–376
permeability, ADMET interactions, transport interactions, **3:**368–369
regulators, substrates, and inhibitors, **2:**461–463
PHA-739358 aurora kinase inhibitor, **6:**320–321
Phage display, recombinant DNA technology, **1:**550–552
library preparation, **1:**550–552
Pharmaceutical equivalence, generic drug guidelines, **3:**84–85
Pharmaceutical industry:
cocrystal engineering applications, **3:**193–196
recombinant DNA technology applications, **1:**538–539
Pharmacodynamics:
CK-2130 prodrug case study, **3:**272–273
congestive heart failure therapy:
predictive biomarkers, **4:**502–503
target validation, **4:**491–492
corticosteroids, **5:**55–56
derfined, **1:**405
drug bioavailability, **2:**748
immunosuppressive drugs, cyclosporin A, **5:**1011–1012
perfluorocarbons, oxygen delivery, **4:**621–622
peroxisome proliferator-activated receptors, α-agonists, **4:**89–99
xenobiotic metabolism, **1:**407
Pharmacogenetics, defined, **1:**182
Pharmacogenomics:
cost-effectiveness analysis of drug development and, **3:**349
defined, **1:**182

recombinant DNA technology, drug efficacy and personalized medicine, **1:**541–543
single nucleotide polymorphisms, **1:**183–196
care quality improvement, **1:**199–200
clinical trials, **1:**197–198
complexity/cost issues, **1:**199
drug metabolizing enzymes, **1:**183–189
cytochrome p450 system, **1:**183–187
dihydropyrimidine dehydrogenase, **1:**187–188
N-acetyltransferase-2, **1:**188
thiopurine methyltransferase, **1:**188–189
drug transporter genes, **1:**189–190
ethical issues, **1:**199–200
literature sources, **1:**198
single gene studies, disease progression, **1:**193–196
target genes and clinical efficacy, **1:**190–193
Pharmacokinetics:
ADMET *in vivo* dosing optimization, **2:**51
aliskiren renin inhibitors, **4:**253
antidepressants, **8:**235–237
serotonin-selective reuptake inhibitors, **8:**236
tricyclic antidepressants, **8:**235–236
antipsychotics, **8:**185–200
benzisoxazoles, **8:**195–196
benzoisothiazoles, **8:**196–198
butyrophenones, **8:**189–190
diarylbutylamines, **8:**190
dibenzodiazepines, **8:**191–192
dibenzothiepine, **8:**194
dibenzoxazepines, **8:**189
dibenzthiazepines, **8:**193–194
dihydroindolone, **8:**190–191
phenothiazines, **8:**186–188
phenylindole, **8:**199
quinolinone, **8:**198

substituted benzamide, **8:**199–200
thiobenzodiazepine, **8:**192–193
thioxanthenes, **8:**188–189
antitubercular agents, rifamycin antibacterials, **7:**746–747
BioPrint® database, safety, selectivity, and promiscuity profiles, **1:**500–501
congestive heart failure therapy, target validation, **4:**491–492
corticosteroids, **5:**55–56
CP-690,550 immunosuppressive compound, **5:**1057–1059
derfined, **1:**405
DNA-targeted compounds:
bleomycin, **6:**8–9
dactinomycin, **6:**4–5
mitomycin, **6:**13
plicamycin, **6:**16–17
docetaxel/taxotere, **6:**34–35
drug delivery system bioavailability, **2:**747
oral drug delivery, immediate-release dosage, **2:**774–776
parameter determination, **2:**749–750
enzyme inhibitors:
daunorubicin, **6:**19–21
doxorubicin, **6:**22
erythropoietin, **4:**572
fingolimod, **5:**1054–1055
granulocyte colony stimulating factor, **4:**577–578
granulocyte-macrophage colony stimulating factor, **4:**574–575
HIV protease inhibitors:
amprenavir, **7:**28, 30
atazanavir sulfate, **7:**40
darunavir, **7:**57–58
fosamprenavir calcium, **7:**44
indinavir sulfate, **7:**14, 17–18, 20–21
lopinavir/ritonavir, **7:**32, 34–35
nelfinavir mesylate, **7:**22, 26–27
ritonavir, **7:**10, 12–13

saquinavir mesylate (SQV), **7:**7–9
tipranavir, **7:**46, 48, 50
immunosuppressive drugs, cyclosporin A, **5:**1011–1012
interleukin-11, **4:**580
intravenous drug delivery, single dose (bolus injection), **2:**752–754
β-lactam antibacterial drugs, **7:**281–282
linear/nonlinear, **2:**751
linezolid, **7:**543
migraine therapy:
ergot alkaloids, **8:**267–268
triptans, **8:**270–272
nitrofurans, **7:**549
paclitaxel/taxol, **6:**33
perfluorocarbons, oxygen delivery, **4:**621–622
permeability, ADMET interactions:
basic principles, **3:**367
classification, **3:**370–371
distribution, **3:**375–376
excretion, **3:**373–374
fraction absorption, **3:**369–370
gastrointestinal absorption, solubility vs. permeability, **3:**374–375
metabolism, **3:**371–373
organ accumulation and toxicity, **3:**377
transport interactions and polymorphism, **3:**367–369
phosphodiesterase 5 inhibitors, **5:**715–716
quinine antimalarials, **7:**607
quinolones/fluoroquinolones, **7:**539–540
serum albumin modulation, **3:**457–464
clofibrates, **3:**462–464
stem cell factor, **4:**582
sulfonamides, **7:**509–514, 522–523
sulfones, **7:**523
thrombopoietins, **4:**584–586
topotecan, **6:**26–27
type 2 diabetes therapies, sulfonylureas, **4:**10
vinblastine, **6:**39

Pharmacology studies:
acitretin, **5:**401
adapalene, **5:**414
ADMET in vivo dosing optimization, **2:**51
all trans-retinoic acid, **5:**386–387
Am80 therapy, **5:**419–420
aminoglycosides, **7:**420–423
antidepressants, **8:**237–248
α-adrenergic receptors, **8:**247
monoamine oxidase, **8:**247–248
monoamine transporters, **8:**237–241
serotonergic agents, **8:**241–247
antipsychotics, animal models, **8:**170–172
bexarotene therapy, **5:**453
etretinate, **5:**402
FDA guidelines, **3:**75
N-(4-hydroxyphenyl) retinamide (4-HPR), **5:**478–480
macrolides, **7:**438–441
multiple sclerosis therapies, **5:**565–567
9cUAB30 RXR-selective retinoid therapy, **5:**457–458
opioid-receptor-like 1/orphanin FQ/nociceptin peptide, **8:**677–678
opioid receptors, **8:**579–597
central nervous system effects, **8:**579
computational models, **8:**590
δ-opioid receptors, **8:**591–592
endogenous peptides, **8:**593–596
κ-opioid receptors, **8:**592–593
ligand characterization, **8:**581–583
multiple opioid receptors, **8:**579–580
mutagenesis studies, **8:**589–590
non-μ-opioid receptors, **8:**591–593
signal transduction mechanisms, **8:**580–581
structure and molecular biology, **8:**587–589

subtypes, splice variants, and dimerization, **8:**590–591
in vitro efficacy assays, **8:**583–585
in vivo evaluation, **8:**585–587
prodrug design, esmolol case study, **3:**265–268
psychostimulants, **8:**100–106
9-cis-retinoic acid therapy, **5:**428–429
13-cis-retinoic acid therapy, **5:**395
SHetA2 retinoid-related compound, **5:**506
tazarotene, **5:**408
tetracyclines, **7:**409–410
Pharmacophores:
aminoglycoside antibiotics, **7:**429–432
applications, **1:**458–459
defined, **1:**457
evaluation methods, **1:**471
historical background, **1:**457–458
HIV integrase, two-metal chelation elaboration, **7:**81–97
identification techniques, **1:**461–471
ligand-based virtual screening, **2:**3–10
dopamine transporter inhibitors, **2:**7–8
lead drug discoveries, **2:**527–531
novel PPAR ligands, **2:**9
organic compound databases, **2:**4–5
research background, **2:**4
three-dimensional generation, **2:**9–10
ligand-based, **2:**6
receptor-based, **2:**8–9
two-dimensional searching, **2:**6
ligand structure-activity relationships, **1:**465–477
computational techniques, **1:**469–471
constrained analog synthesis, **1:**468–469
limitations, **1:**471
preparation, **1:**466–468

Pharmacophores (*Continued*)
  monocyclic nonantibacterial pharmacophore, **7**:332–333
  nuclear magnetic resonance, ligand-based designs, modeling and conformational analysis, **2**:389–390
  peroxisome proliferator-activated receptors:
    α?γ-dual agonists, **4**:125–135
    α/γ/δ-pan agonists, **4**:141–145
    medicinal chemistry, **4**:87–88
  poly(ADP-ribose)polymerase inhibitors, **6**:162–167
  proteasome inhibitors:
    bortezomib clinical trials, **6**:178–180
    medicinal chemistry, **6**:181–184
  protein-protein interactions, three-finger p53/MDM2 antagonist model, **2**:356–359
  quinolones/fluoroquinolones, **7**:537–538
  receptor-based drug design, lead drug discoveries, **2**:527–531
  soft drug development:
    cortienec acid-based steroids, **2**:92–96
    inactive metabolites, **2**:81
    sources, **1**:459–461
  stroke therapy, PARP inhibitors, **8**:485–487
  structural comparisons, **1**:477
  structure-based modeling, computer-aided techniques, **2**:682–683
  sulfonylureas, **4**:8–10
  tetracycline antibiotics, **2**:223–225
Pharmacophoric hypothesis, peptidomimetics, **1**:210
PharmID program, pharmacophore formation, bioactive conformation, **1**:470
Phase 1 clinical trials:
  antimalarial agents, **7**:625–627
  FDA guidelines, **3**:77
  prasugrel, **4**:433–434
Phase 2 clinical trials:
  antimalarial drugs, **7**:625–627
  FDA guidelines, **3**:77
  prasugrel, **4**:434
Phase 3 clinical trials:
  antimalarial agents, **7**:625–627
  FDA guidelines, **3**:77
  prasugrel, **4**:434–435
  stroke therapy, **8**:447–449
Phase 4 clinical trials, FDA guidelines, **3**:77–78
Phase trafficking, defined, **1**:280
Phenantridinone (PND), poly (ADP-ribose) polymerase inhibitors, **6**:166–167
Phencyclidine (PCP):
  cognitive function assessment, **8**:26–27
  schizophrenia pathology, **8**:163–164
  NMDA model, **8**:171–172
Phendimetrazine, structure-activity relationship, **8**:106–109
Phenelzine:
  chemical structure, **8**:223
  Parkinson's disease, dopaminergic therapy, **8**:549–551
  physiology and pharmacology, **8**:248
  side effects and adverse reactions, **8**:230
  structure-activity relationship and metabolism, **8**:252–253
Phenethicillin, penicillin resistance and development of, **7**:290–292
Phenformin, structure-activity relationships, **4**:3
Phenmetrazine, structure-activity relationship, **8**:106–109
Phenobarbital, anticonvulsant applications, **8**:135–141
Phenols:
  antimicrobial agents, **7**:493–495
  coefficients, topical antimicrobial agent testing, **7**:489
  hydroxyl groups, prodrug moieties, **3**:250–252
  ring analogs, *N*-(4-hydroxyphenyl) retinamide (4-HPR), **5**:480–482
  thromboxane A_2 receptor antagonists, **5**:888–889, 891
Phenothiazines:
  Alzheimer's disease, tau-targeted therapy, **8**:424–425
  pharmacokinetics, biodistribution, and drug-drug interactions, **8**:186–189
Phenotype/polymorphism associations, congestive heart failure therapy, **4**:507
Phenotypic screening, natural products lead generation, **2**:207–208
Phenoxyacetic acid:
  CRTH2 receptor antagonists, **5**:914–918
  DP_1 receptor antagonists, **5**:905
Phenserine, cognitive enhancement, **8**:32
Phentermine, amphetamine alpha-alkyl substituent, **8**:108
Phenylacetamides, type 2 diabetes therapy, **4**:15–17
Phenylacetic acids (PAs), quantitative structure-activity relationship, electronic parameters, **1**:2–8
N-Phenylamines, Alzheimer's disease, tau-targeted therapy, **8**:427
Phenyl esters, cocaine structure-activity relationship, **8**:111–113
Phenylindole, pharmacokinetics, **8**:199
Phenylnapthyl linker, thyromimetic structure-activity relationship, **4**:214–215
Phenylpiperidine analgesics:
  flexible opiates, **8**:598
  MPTP activation, **8**:530

Phenyl sulfonamide, prostaglandin/thromboxane EP$_1$ antagonists, 5:819–821
Phenyl tertiary amine ligands, liver X receptors, 4:151–154
Phenylthiazolyl-bearing hydroxymates, antimalarial HDAC inhibitor, 7:654–655
Phenylthiazolyl-hydrazines, Alzheimer's disease, tau-targeted therapy, 8:427–428
Phenylurenyl calchones, falcipain inhibitors, 7:635–637
Phenytoin:
 anticonvulsant applications, 8:135–139, 141
 clozapine pharmacokinetics, 8:192
 glucuronidation/glucosidation, 1:428–430
 sertindole interaction, 8:199
PHER-02®, oxygen delivery mechanisms, 4:624
Philadelphia chromosome, chronic myelogenous leukemia therapy, small-molecule tyrosine kinase inhibitors, resistance mechanism, 6:264–266
pH levels:
 high-throughput screening, assay optimization, 3:410–412
 salt compounds, acid-base equilibrium, 3:385–388
Phlorizin, type 2 diabetes therapies, 4:13–14
Phobic conditions, antidepressant applications, 8:229
Phomallenic acids, phenotypic screening, 2:208–209
Phosphatase and tensin homolog (PTEN), brain tumor therapy, 6:258–259
 signaling pathways, 6:245, 251
 Wnt signaling pathway, 6:263
Phosphatases:
 allosteric protein model, 1:392

antibody-directed enzyme prodrug therapy, DNA-targeted therapeutics, 6:120–123
Phosphinate, calpain inhibitor, stroke therapy, 8:465
Phosphinic acids, ACE inhibitor design, 4:274
3'-Phosphoadenosine 5'-phosphosulfate (PAPS), drug metabolism, sulfonation reactions, 1:425–426
Phosphodiesterase inhibitors:
 antidepressant efficacy assessment, 8:226–229
 antipsychotic agents, 8:208–209
 benign prostatic hypertrophy, 5:716
 BioPrint profile, 1:496, 498
 chronic obstructive pulmonary disease, 5:780–784
 CK-2130 prodrug case study, 3:268–276
 cognitive enhancement, 8:50–51
 erectile dysfunction:
  drug development and discovery, 5:711–712
  pharmacokinetic and efficacy comparisons, 5:715–716
  sildenafil, 5:712–713
  tadalafil, 5:714
  vardenafil, 5:713
 nitric oxide effects, 5:272–278, 294
 PDE-III antiplatelet agents, 4:449–450
  aggrenox, 4:450
  cilostazol, 4:449–450
  dipyridamole, 4:450
  milrinone, 4:450
 pulmonary hypertension, 5:715–716
 selective toxicity, 2:567–568
 therapeutic applications, 5:715–716
 virtual screening data, scaffold enumeration, 2:27–29
3-Phosphoglycerate dehydrogenase

(PGDH), allosteric proteins, 1:382–388
Phosphoinositide 3-kinase inhibitors:
 brain cancer therapy, 6:249–250
 resistance mechanisms, 6:367–368
 survival signaling pathways, 6:321–323
Phosphoinositide pathway:
 FP receptor antagonists, 5:899
 platelet activation, 4:415–418
Phospholipase pathway, platelet activation, 4:415–418
Phosphonic acids, ACE inhibitor design, 4:274
Phosphonoacetic acid (PAA), anti-DNA viral agents, 7:240
Phosphonoformic acid (PFA), anti-DNA virus agents, 7:240–241
Phosphonylmethoxyethyl (PME) derivatives, anti-DNA virus agents, 7:232–239
Phosphorane synthesis, penems, 7:317–321
Phosphorus:
 chirality at, 1:127, 148–149
 drug metabolism, oxidation/reduction reactions, 1:420–421
 stroke therapy, MMP inhibitors, 8:509–510
Phosphorylation:
 Alzheimer's disease, tau-targeted therapies, 8:418–422
 drug metabolism, conjugation reactions, 1:438–439
Photocoagulation therapy, age-related macular degeneration, 5:614–615
Photodamage:
 adapalene therapy, 5:411
 all-*trans*-retinoic acid therapy, 5:383–384
 9-*cis*-retinoic acid therapy, 5:425–426
 tazarotene therapy, 5:406
Photodegradation, protein therapeutics formulation and delivery, 3:313–314

Photodynamic therapy (PDT), age-related macular degeneration, **5**:614–615
Phototoxicity, tetracyclines, **7**:409
Phthalocyanine, Alzheimer's disease, tau-targeted therapy, **8**:425–426
Physicochemical property assays: ADMET property assessment, **2**:65–66
salt compounds, **3**:382
Physostigmine:
analog drug design, **1**:168–170
memory assessment and, **8**:28–30
Phytochemicals, proteasome inhibitors, **6**:184
Phytoestrogens, physiology and pharmacology, **5**:240
Phytohemagglutinins, receptor-mediated endocytosis, **2**:452–453
Pigment epithelium-derived factor (PEDF), age-related macular degeneration therapy, **5**:622
Pilocarpine:
glaucoma therapy, **5**:599–600
muscarinic activity, **8**:67–68
Pilocytic astrocytomas, histologic classification, **6**:230
Pilot plant scale-up, **3**:7–10
agitation, **3**:8–9
drying and solid handling, **3**:9–10
heating and cooling, **3**:8
liquid-solid separations, **3**:9
nevirapine case study, **3**:16–20
safety issues, **3**:10
Pilot screening, high-throughput screening, assay configuration, **3**:423
Pimagedine, stroke therapy, **8**:497
Pimavanserin, insomnia therapy, **5**:718
Pimecrolimus:
ADMET properties, **5**:1023
therapeutic applications, **5**:1023–1024
Pimozide, pharmacokinetics, **8**:190
PIM Ser/Thr kinase, structural genomics, **1**:593–594

Pindolol:
antidepressant interaction, efficacy assessment, **8**:225–229
radiolabeled ligands for, **8**:743–744
PINK1 gene, Parkinson's disease genetics, **8**:535
Pinoxepin, structure-activity relationship, **8**:250–251
Pioglitazone:
Alzheimer's disease, insulin related therapy, **8**:415–417
peroxisome proliferator-activator receptors: future drug discoveries, **4**:165–167
γ-agonists, **4**:102–106
Pipecolyl ester analogs, tacrolimus compounds, **5**:1018–1019
Pipeline Pilot protocol, multiobjective drug discovery optimization, **2**:264–273
Piperacillin, penicillin resistance and development of, **7**:291–292
Piperazines:
carbapenem synthesis, **7**:344–345
chemokine receptor-5 (CCR5) antagonists, **7**:110–116
diketopiperazine-based chemokine receptor-5 (CCR5) antagonist, **7**:131–133
dipeptidyl peptidase 4 (DPP-4) inhibition, **4**:61–62
muscarinic antagonists, **8**:76–77
nonpeptidic renin inhibitors, **4**:253–254
opioid derivatives, κ-opioid receptor agonist, **8**:627–632
stroke therapy:
cathepsin B inhibitors, **8**:473–474
piperidine/piperazine scaffolds, **8**:457–458

Piperidines:
chemokine receptor-5 (CCR5) antagonists, **7**:110–116
tropane-based chemokine receptor-5 (CCR5) antagonist, **7**:118–121
nicotinic agonists, **8**:68–69
nonpeptidic renin inhibitors, **4**:253–254
opioid derivatives, **8**:613–621
4-anilidopiperidines, **8**:618–621
4-arylpiperidines:
carbon substituent, **8**:614–616
oxygen substituent, **8**:616–618
κ-opioid receptor agonist, **8**:627–632
stroke therapy, piperidine/piperazine scaffolds, **8**:457–458
Piperidolate, muscarinic antagonists, **8**:72–74
π-values, quantitative structure-activity relationship, partition coefficient calculations, **1**:14–19
Piracetam:
anticonvulsant applications, **8**:149–151
cognitive enhancement, **8**:15–16, 34
polymorphism, **3**:192
Piragliatin, type 2 diabetes therapy, **4**:15–17
Piriqualone, atropisomerism, **1**:133–137
Piritrexim, lipophilic antifolate DNA-targeted chemotherapeutics, **6**:109
Pituitary adenylate cyclase-activating peptide (PACAP), structure and function, **4**:547–548
Pituitary tumors, classification and therapy, **6**:235–237
Pivaloate, α-adrenergic receptor agonists, **5**:597–600
Pivaloyl ester prodrug, lipophilicity increase, **2**:442–443

Pivampicillin, penicillin
resistance and
development of,
7:291–292
Pixantrone, topoisomerase II
inhibitors, **6:**101–102
PK parameters, ADMET
liabilities, **2:**50
Place and join techniques, docking
and scoring
computations,
2:606–608
Placebo response:
antidepressant efficacy
assessment,
8:221–229
migraine therapy, calcitonin
gene-related peptide
antagonists *vs.*,
8:286–288
Placental bone morphogenetic
protein (PLAB)
N-(4-hydroxyphenyl)
retinamide (4-HPR),
5:475
Plant metabolites, drug
development,
2:250–251
Plant patents, protection and
strategy, **3:**105–118
Plant toxins, receptor-mediated
endocytosis, **2:**452
Plasma drug profile:
diffusion-controlled dosage
forms, **2:**801–816
calculation methods,
2:802–803
controlled-release dosage,
2:812–816
nomenclature, **2:**801
repeated doses, **2:**808–812
single-dose results,
2:803–808
erosion-controlled dosage
forms, **2:**816–828
calculation methods,
2:818–819
dosage form predictions,
2:826–828
nomenclature, **2:**816–818
repeated doses, **2:**822–825
single dose calculations,
2:819–822
N-(4-hydroxyphenyl)
retinamide (4-HPR),
5:478–479
migraine therapies, **8:**298–299

oral drug delivery:
immediate-release dosage,
2:774–776
sustained-release dosage
forms, numerical
calculations,
2:790–791
repeated intravenous bolus
injection, master curve
and dimensionless
numbers, **2:**756
tissue-based drug delivery,
lung and bronchial
mucus, **2:**840–844
transdermal drug delivery,
2:830–832
metropolol experiment,
2:833–834
Plasma protein binding:
ADMET distribution
properties, **2:**59–61, 66
quantitative structure-activity
relationship, **1:**66
Plasma stability, ADMET
properties, **2:**59
Plasmepsins, antimalarial heme
metabolism and amino
acid salvage,
7:637–644
Plasmids:
antitubercular resistance
mechanisms,
7:733–735
β-lactam antibacterial drug
resistance, **7:**270
Plasminogen activator inhibitors
(PAIs):
metformin mechanism of
action, **4:**4
recombinant DNA technology,
1:540
Plasmodial surface anion channel
(PSAC), antimalarial
targeted drugs,
7:668–672
Plasmodium falciparum:
epidemiology, **7:**565
life cycle, **7:**603–605
PfATP6 expression, **7:**658–660
protein kinase inhibition,
7:627–630
tetracycline efficacy, **7:**415
Plasmodium protein
farnesyltransferase
(P*f*PFT), antimalarial
cell cycle regulation,
7:631–634

Plasticity analysis, *in silico*
screening, protein
flexibility, protein-
ligand interactions,
2:865–867
Platelet activating factor (PAF),
mechanisms of, **4:**414
Platelet-derived growth factor
(PDGF) receptor:
brain tumor therapy, **6:**249
multitarget drug development,
multikinase
inhibitors, cancer
therapy, **1:**264–267
Platelets:
activation mechanisms,
4:410–418
adhesion mechanisms,
4:412–413
aggregation mechanisms,
4:412–418
endogenous regulation,
activation control,
4:418
functional anatomy, **4:**412
intracellular activation,
4:416–418
nitric oxide and aggregation of,
5:272–278
secretion mechanisms, **4:**415
serotonin-selective reuptake
inhibitor effect on,
8:232–234
structure and function, **4:**409
Platencin, phenotypic screening,
2:208–209
Platensimycin, phenotypic
screening, **2:**208–209
Platinum complexes:
chirality and toxicity,
1:144–145
DNA-targeted
chemotherapeutic
compounds, **6:**89–91
Plerixafor compounds, structure
and properties,
4:586–587
Pleuromutilin antibiotics,
2:229–230
Plicamycin, structure and
properties, **6:**16–17
Pluracidomycin A, occurrence,
structural variations
and chemistry,
7:333–335
Pluropitin, combinatorial
chemistry, **1:**320–321

PMDTA HIV reverse transcriptase inhibitor, **7:**147–148
PMDTT HIV reverse transcriptase inhibitor, **7:**147–148
PMEO-5-Me-DAPy HIV reverse transcriptase inhibitor, **7:**146–148
PMF scoring function, structure-based virtual screening, **2:**18–19
Podophyllotoxin derivatives, serum albumin binding, **3:**454–456
Point complementarity, docking and scoring computations, **2:**602–604
Point mutations, tyrosine kinase inhibitors, selective toxicity, **2:**563
Poisson-Boltzmann electrostatics approach:
 docking and scoring computations, ionization and tautomerization, **2:**636–637
 empirical scoring, **2:**626–628
Poisson-Boltzmann solvent accessible surface area (PBSA), quantum mechanics/molecular mechanics binding affinity, **2:**731–733
Polarity, oral drug physicochemistry, **3:**34–35
Polarizability indices, quantitative structure-activity relationships, **1:**24–25
Polar surface area (PSA):
 quantitative structure-activity relationships, molecular descriptors, **1:**25–26
 tropane-based chemokine receptor-5 (CCR5) antagonist, **7:**118–121
Polishing steps, mammalian protein purification, **3:**307–309

Polo-like kinase 1 inhibitors:
 medicinal chemistry and clinical trials, **6:**316–317
 multidrug resistance and, **6:**377–378
Poly(ADP-ribose)glycohydrolase (PARG):
 cancer therapy, **6:**170
 poly(ADP) ribosylation biochemistry, **6:**151–157
 research background, **6:**151
Poly(ADP-ribose) polymerase (PARP) inhibition:
 cancer therapy, **6:**160–162, 168
 chemotherapy/radiotherapy coadjuvants, **6:**168–170
 future research issues, **6:**170–171
 inhibitors, **6:**168
 monotherapy, **6:**170
 poly(ADP-ribose) glycohydrolase and inhibitors, **6:**170
 chemotherapy/radiotherapy coadjuvants, **6:**168–170
 molecular biology, **6:**157–159
 monotherapy, **6:**170
 PARP-1:
  biochemistry, **6:**151–157
  cancer therapy, **6:**160–162
  domain organization and catalytic site structure, **6:**159–160
  domain organization and structure, **6:**159–160
  molecular biology, **6:**157–159
  research background, **6:**151
  selectivity, **6:**167–168
 PARP-2, molecular biology, **6:**158–159
 PARP-3, molecular biology, **6:**158–159
 poly(ADP) ribosylation:
  biochemistry, **6:**151–157
  structural genomics, **1:**586–587
 research background, **6:**151
 selective inhibitors, **6:**167–168
 stroke therapy, **8:**484–487
 structure and classification, **6:**162–167

Polyamine metabolism:
 antiparasitic/antiprotozoal agents, **7:**585–590
 kinetoplastid chemical structure, **7:**566
 Leishmaniasis therapy, **7:**591–593
Polyamines, muscarinic antagonists, **8:**76–77
Polycyclic carbapenems, chemical structure, **7:**345–350
Polyethylene glycol (PEG):
 conjugated hemoglobins, **4:**635–637
 membrane protein passive diffusion, **2:**444–446
Polyglutamates, folic acid uptake and metabolism, **5:**691–696, 698
PolyHeme, structure and properties, **4:**631–632
Polyketides (PKS). *See also* specific polyketides, e. g., Tetracyclines
 antibiotics, **7:**405–406
 antitubercular agents, lipid derivation, **7:**723–724
 natural products lead compounds, **2:**193, 199, 210–212
Polyketide synthases, antitubercular agents:
 lipid derivation, **7:**724
 mycobactin targeting and biosynthesis, **7:**730–731
 mycolic biosynthesis, **7:**723
Polymerase chain reaction (PCR), recombinant DNA technology, **1:**547
Polymeric immunoglobulin transport, receptor-mediated endocytosis, **2:**451
Polymeric transitions, allosteric proteins, **1:**381–383
Polymerized hemoglobins:
 Hemolink, **4:**632–633
 Hemopure®, **4:**633–634
 PolyHeme, **4:**631–632
 structure and function, **4:**631–635
 zero-link polymer, **4:**634
Polymer matrices, plasma drug profile:
 diffusion-based drug release, **2:**808–809

erosion-controlled drug release, 2:823–824
Polymers:
  central nervous system tumors, targeted drug delivery, 6:271–272
  nanoscale drug delivery:
    amphiphilic molecules, 3:473–480
    polymer-drug conjugates, 3:480–481
    water-insoluble polymers and cross-linked polymeric nanoparticles, 3:481
  poly(ADP-ribosyl)ation biochemistry, branching reaction, 6:155–157
  tubulin polymerization/depolymerization inhibition, 6:31–40
    dimeric vinca alkaloids, 6:35–40
    taxus diterpenes, 6:32–35
Polymersomes, nanoscale drug delivery, 3:476–477
Polymicrobial infection, prevention, 7:505
Polymorphic crystals:
  case studies, 3:196–198
  cocrystal design, 3:196
  crystal engineering:
    defined, 3:188
    evolution of, 3:190–192
  oral dosage systems, 3:26–27
Polymorphism associations:
  congestive heart failure therapy, 4:507
  opioid drug interactions, 8:577
  psoriasis, 5:397–398
Polynitroxylated hemoglobin, structure and properties, 4:630
Polyomaviruses, anti-DNA viral agents, 7:241
Polypharmacology studies:
  BioPrint® database, 1:493–496
  future research issues, 1:502–503
Polyphenols, Alzheimer's disease, tau-targeted therapy, 8:425–427
Polyunsaturated fatty acid (PUFA), vitamin E uptake and metabolism, 5:662

Pool and split synthesis, peptide libraries, 1:285–291
Porins, antitubercular agents, 7:724
Porphyrins, Alzheimer's disease, tau-targeted therapy, 8:425–427
Positive allosteric modulator (PAM), receptor-based drug design, 2:500
Positron-emission tomography (PET):
  congestive heart failure therapy:
    disease biomarkers, 4:505–506
    predictive ligand biomarkers, 4:501
  drug discovery and development applications:
    disease markers, radiolabeling of, 8:744–746
    imaging studies, 8:740–741
    radiolabeled drugs, 8:742–743
    research background, 8:737
    small animal imaging, 8:742
    targeted drugs, radiolabeled ligands, 8:743–744
  isotopes in, 8:739–740
  Parkinson's disease diagnosis, 8:555–556
  physics of, 8:737–739
Postapproval process:
  cost-effectiveness analysis, 3:349–350
  FDA guidelines, 3:96
Postenzymatic reactions, drug metabolism, 1:405–406
Postoperative ileus, carbon monoxide inhalation therapy, 5:302–303
Post-synaptic density proteins (PSDs), long-term potentiation, 8:24–25
Posttranscriptional methylation inhibitors, anti-DNA viral agents, 7:241
Posttransplantation lymphoproliferative disorder (PTLD), bexarotene therapy, 5:444
Potassium channels:

anticonvulsants, 8:124–125
carbon monoxide therapeutic effects, 5:296–299
stroke therapy, maxi-K channel, 8:482–483
type 2 diabetes therapy, sulfonylureas, 4:10
Potassium permanganate, topical antimicrobial compounds, 7:497
Potency assays:
  estrogen receptors, 5:236–237
  glucocorticoid anti-inflammatory compounds, 5:47–48, 56
  C-20 carbonyl alterations, 5:102–103
  structural changes, 5:84
  topical agents, 5:85
Potency optimization:
  G-protein coupled receptor, homology modeling, 2:292–293
  HIV protease inhibitors, saquinavir, 7:5–7
  protein therapeutics, analytic development, 3:321–322
  tigecycline analogs, 7:414–415
Potential functions, outlier detection, quantitative structure-activity relationship, 1:44
Potential of mean force (PMF) algorithm, empirical scoring, 2:629–631
Potentiometry, quantitative structure-activity relationship, partition coefficients, hydrophobic interactions, 1:12–13
Potocytosis, membrane proteins, 2:448–449
Power x-ray diffraction (PXRD), salt compound selection:
  automated assisted screening, 3:396–397
  crystallinity, 3:381
  manual screening, 3:390–395
  selection process, 3:396
Poxviruses, anti-DNA virus agents, 7:244–245

Pozanicline, cognitive
  enhancement, 8:35–36
Pramipexole, Parkinson's disease
  dopaminergic therapy,
  8:546–547
Prasugrel, antiplatelet activity,
  4:433–435
Pravastatin:
  Alzheimer's disease therapy,
    preclinical/clinical
    testing, 8:359–361
  anticholesterolemics, 2:233
  fibrate compounds, 2:235
  cholesterol biosynthesis,
    5:15–16
  LDL cholesterol lowering
    mechanisms, 4:307
Preapproval inspections, FDA
  requirements, 3:91
Preattentive processing,
  cognition and, 8:17–19
Precipitation, oral drug
  physicochemistry,
  class II drugs,
  inhibition, 3:45
Preclinical testing:
  Alzheimer's immunization
    therapy, 8:369
  BACE (β-site cleavage enzyme),
    8:345–351
  cognitive function, 8:25–29
  FDA regulations, 3:71–72
  fibrillization inhibitors,
    8:361–365
  HMG-CoA reductase inhibitors,
    8:359–361
  α-secretase inhibitors,
    8:357–358
  γ-secretase inhibitors, 8:334,
    340–341
  selective estrogen receptor
    modulators,
    5:739–740
  target validation, congestive
    heart failure therapy,
    4:486–491
Predictive biomarkers:
  congestive heart failure:
    direct efficacy readouts,
      4:502
    disease modification,
      4:503–506
    drug development:
      bench-to-bedside
        candidates and
        attrition development,
        4:482–485

drug attrition, 4:480–482
emerging therapies,
  4:479–480
critical evaluation, 4:483
epidemiology and prevalence,
  4:477–478
etiology, 4:477
future research issues, 4:507
iontophoresis methodologies,
  4:506
management approaches,
  4:478–479
noninvasive imaging,
  4:505–506
patient selection and
  stratification,
  4:506–507
pharmacodynamic activity,
  4:502–503
cardiotoxicity biomarkers,
  4:503
dose selection, 4:502–503
polymorphisms and
  phenotypes, 4:507
response-to-treatment
  studies, 4:507
target-compound interaction:
  ex vivo cell based
    biomarker assays,
    4:501
  hemodynamic/biochemical
    biomarker assays,
    4:500–501
  PET/SPECT ligands, 4:501
  target-engagement assays,
    4:501–502
target validation, 4:483–500
  clinical congruency,
    4:489–490
  clinical importance, 4:483,
    485–487
  current clinical trials,
    4:497–498
  knockout/transgenic
    mouse models,
    4:488–490
  mechanism perturbation,
    4:494–495
  mechanism-related risk
    and reliability, 4:500
  mechanistic approaches,
    cardiac contractility/
    myocardial oxygen
    demand, 4:492–494
  normal patient
    mechanisms *vs.*,
    4:495–496

  novel target *vs.* existing
    therapy, 4:496,
    499–500
  pharmacodynamics/
    pharmacokinetics,
    4:491–492
  preclinical models,
    4:486–491
  tissue studies, species and
    disease states, 4:488
  utilitarian classification,
    4:483–485
Predictive modeling,
  chemogenomics:
  ligand-based data analysis,
    2:583–585
  structure-based data analysis,
    2:585–588
Predictive power, quantitative
  structure-activity
  relationship models,
  1:51
  virtual screening applications,
    1:526–527
Predictive validity, cognitive
  function assessment,
  8:25–29
Prednisolone, colon-specific
  delivery, 5:45
Prefractionated libraries, natural
  products lead
  generation, 2:201
Prefrontal cortex:
  antipsychotics, dopamin
    receptor subtypes,
    8:200–201
  memory and, 8:23–24
Pregabalin, anticonvulsant
  applications,
  8:135–139, 149
Pregnane compounds:
  antiandrogens, 5:190–192
  pregnandiol,
    biotransformation,
    5:229–230
  steroidal progesterone receptor
    ligands, 5:253–255
Pregnane X receptor antagonists,
  multiobjective
  optimization,
  2:260–261
Pregnenolone:
  adrenocorticoid formation,
    5:19–22
  androgen biosynthesis,
    5:23–24, 156–158
  steroidogenesis, 5:16–17, 19

Pre-IND phase, FDA regulations, 3:71
Premafloxacin, polymorph design, 3:196–198
Prepulse inhibition, antipsychotic therapy, 8:172
Prescription Drug User Fee Act (PDUFA), 3:67–68
President's Emergency Plan for AIDS, 3:69
Pressure perturbation calorimetry (PPC), ligand-macromolecular interactions, structure-based thermodynamic analysis, 2:706–712
Pressure requirements, commercial-scale operations, 3:11
Presystemic elimination, oral dosage forms, class V drugs, 3:53
Prevalence, prodrug marketing, 3:241–242
Preventive maintenance, commercial-scale operations, 3:12
Preventive therapy, migraines, 8:312–320
  angiotensin converting enzyme inhibitors, 8:319–320
  anticonvulsants, 8:315–316
  antidepressants, 8:318–319
  β-blockers, 8:313–315
  calcium channel blockers, 8:316–318
Prezista. See Darunavir (DRV)
Prialt® (MVIIA) pain medication, nuclear magnetic resonance structure:
  bioactivity analysis, 2:377–382
  charge state, 2:384–385
Prilosec, chiral switch in, 1:156–159
Prima facie obviousness, patent specification and, 3:135–138
Primaquine, antimalarial therapy, 7:613–614
Primary central nervous system lymphoma (PCNSL), classification and therapy, 6:239–240

Primidone, anticonvulsant applications, 8:135–141
Principal component analysis (PCA):
  attrited compounds, BioPrint profile vs., 1:498–499
  quantitative structure-activity relationship validation, 1:49–50
Principal component regression analysis (PCRA), quantitative structure-activity relationship, 1:37–38
Prinomastat inhibitor, stroke therapy, 8:503–505
Privileged structures:
  chemogenomics, scaffold structures, 2:575–577
  pharmacophores, lead drug discoveries, 2:529–531
  prodrug designesmolol case study, 3:267–268
  prostaglandin/thromboxane EP$_1$ antagonists, 5:831, 835–838
  small druggable molecules, 1:296
  virtual screening, 2:30–31
Probenecid, olanzepine interaction, 8:193
Procaspase-9, kinase inhibitors, 1:342–343
Procaterol, soft drug design, 2:103–104
Process analytical technology (PAT), commercial-scale operations, 3:14
Process chromatography, protein therapeutics, downstream processing, 3:303–304
Process claims, patent obviousness standards and, 3:138
Process flow diagramas (PFDs), commercial-scale operations, 3:11
Processing, cognition and, 8:19
Processing systems and equipment:
  commercial-scale operations, 3:11
  nevirapine case study, 3:17–20

oral drug systems, class V drug design, 3:54–55
protein therapeutics, 3:289–291
  comparability, 3:325–326
  reagent clearance and excipient testing, 3:326
Process intensification, protein therapeutics, current and future applications, 3:309–310
Process validation, commercial-scale operations, 3:12
Prochlorperazine:
  FDA approval, 8:176–177
  migraine therapy, 8:297–298
PROCOGNATE database, structural genomics research, 1:582
Procyclidine:
  muscarinic antagonists, 8:72–77
  Parkinson's disease therapy, 8:552–553
Prodine derivatives, opioid analgesics, 8:617–618
Prodipine 7, dipeptidyl peptidase 4 (DPP-4) inhibition, 4:49
PRODOCK, docking and scoring techniques, Monte Carlo simulation, 2:611
Prodrugs:
  anti-DNA virus agents:
    acyclic nucleoside phophates, next-generation compounds, 7:238–239
  "double prodrug" antimetabolites, 7:223–225
  nucleosides, 7:230–231
  nucleotides, 7:231
  antitubercular agents:
    isoniazid, 7:739–740
    resistance mechanisms, 7:734–735
  carbon monoxide, 5:305–307
  carboxylesterases, 3:233–241
    classification, 3:233–234
    function and distribution, 3:234
    molecular biology, 3:234, 237–238

Prodrugs (*Continued*)
  preclinical assays, **3**:239, 241
  substrates and inhibitors, **3**:235–240
  case studies, **3**:257–276
    CK-2130, **3**:268–276
    esmolol, **3**:258–268
    prontosil, **3**:257–258
  chemical delivery systems *vs.*, **2**:76
  CYP2D6 substrate reactions, **1**:185–186
  defined, **2**:442, **3**:219–220
  design principles, **3**:229–233
    ADMET issues, **3**:230
    bioconversion mechanisms, **3**:252–254
    drug targeting, **3**:229
    metabolic issues, **3**:230–233
    moieties, **3**:250–252
    "rule-of-one" metabolism, **3**:248–250
    structure-metabolism relationships, **3**:265–256
  DNA-targeted therapeutics, tumor-activation, **6**:113–124
    ADEPT prodrugs, **6**:120–123
    hypoxia-activated bioreductive prodrugs, **6**:114–120
  drug discovery paradigms, **3**:226–229
  early examples and applications, **3**:220–226
  future trends, **3**:276
  hydrogen sulfide, **5**:314–315
  marketed prodrugs, **3**:241–248
    examples, **3**:242, 244–248
    FDA approvals, **3**:242–243
    prevalence, **3**:241–242
  membrane proteins, drug delivery strategies, **2**:441–443
  nitric oxides, **5**:278–290, 294
    C-nitroso class, **5**:286–287
    combined donors, **5**:287–290
    diazeniumdiolates, **5**:281–283
    S-nitrosothiols, **5**:283–285
    organic nitrates, **5**:278–280
    sodium nitroprusside, **5**:280–281
    sydnoimines, **5**:285–286
  oral drug delivery, class III drugs, **3**:50–51
  penicillin resistance and development of, **7**:291–292
  pharmacologically active products, **3**:276–279
  soft drugs:
    comparisons, **2**:75–76, 80
    natural compounds, **2**:129–130
Product-like reversible inhibitors, dipeptidyl peptidase 4 inhibition, **4**:54–56
Product release testing, protein therapeutics, **3**:318–320
Progesterone:
  biosynthesis, **5**:22–23
  biotransformation, metabolites, **5**:229–230
  hydroxylation, **5**:7, 9–11
  metabolism, **5**:26–27
  receptors:
    affinity, **5**:68–70
    antiglucocorticoids, **5**:114–120
    isoform structure and function, **5**:237–238
    ligand activity, **5**:238
    nonsteroidal progesterone receptor ligands, **5**:255–256
    steroidal ligands, **5**:253–255
Progestins:
  adverse effects and precautions, **5**:221–222
  antiandrogens, **5**:190–192
  clinical applications:
    adverse effects and precautions, **5**:221
    current drugs, **5**:219–221
    environmental/dietary sources, **5**:239–240
    nonsteroidal progesterone receptor ligands, **5**:255–256
    physiology and pharmacology, **5**:238–239
    steroidal progesterone receptor ligands, **5**:254
    tissue specificity, **5**:240–242
Proguanil, antimalarial therapy, **7**:617–618

Prokaryotic expression:
  high-throughput screening, enzymatic assays, **3**:402–406
  protein therapeutics, upstream processing, **3**:291–294
Prolactin response, antipsychotic side effects, **8**:175–176
Prolactin secreting adenomas, classification and therapy, **6**:236–237
Proline analogs, opioid peptide analogs, Tyr-Pro-Phe sequence, **8**:660–661
Proline dehydrogenase, schizophrenia genetics, **8**:174
cis-trans Prolyl isomerase, tacrolimus binding mechanism, **5**:1015–1016
PROMACTA®, **4**:584–586
Promegestone, steroidal progesterone receptor ligands, **5**:253–254
Promiscuity of compounds:
  broad profiling data, **1**:499–501
  polypharmacology, BioPrint® database, **1**:493–496
  thyromimetic selectivity, **4**:205–206
Promyelocytic leukemia zinc finger (PLZF)-RARα chimera, acute promyelocytic leukemia therapy, **5**:380–382
Prontosil, prodrug development, **3**:257–258
Prontosil rubrum, discovery, **7**:483, 506–509
"Proof-of-concept" clinical studies:
  chemokine receptor-5 (CCR5) antagonist development, **7**:113–116
  osteoporosis therapy, parathyroid hormone secretion/calcilytics, **5**:754–755
  target validation, congestive heart failure therapy, normal subjects *vs.* CHF patients, **4**:496
Propagation, hemostasis, **4**:368

Property data sets, quantitative structure-activity relationship modeling, 1:513–514
Propicillin, penicillin resistance and development of, 7:290–292
Propionyl coenzyma A carboxylase, biotin metabolism, 5:688
Propoxyphene:
chemical structure, 8:570
metabolism and elimination, 8:579
Propranolol, migraine prevention, 8:314–315
Propyl analog, prostaglandin/thromboxane receptors, IP receptor agonist, 5:930, 932
N-(n)-Propyldihydrexidine, functional selectivity, 8:8–9
Propylthiouracil (PTU):
hyperthyroidism, 4:199
3,5,3′,5′-tetraiodo-L-tyronine (L-T$_4$), uptake and metabolism, 4:193–194
Prorenin, structure and function, 4:518–519
Prorenin receptor (PRR), structure and physiology, 4:241–242
"Prosecution history estoppel," infringement of patents, 3:154
Prostaglandin endoperoxide (PGH$_2$), thromboxane antagonists, 4:419–422
Prostaglandins:
analogs, glaucoma intraocular pressure reduction:
clinical trials, emerging compounds, 5:600–602
current therapies, 5:592–600
CRTH2 receptor, 5:907–911
agonists, 5:907–909
antagonists, 5:909–920
DP$_1$, dual pharmacology, 5:920
indole acetic acids, 5:910–914
miscellaneous compounds, 5:918

monoaryl acetic acids, 5:914
nonacidic antagonists, 5:918–920
phenoxyacetic acid, 5:914–918
ramatroban analogs, 5:909–910
DP$_1$ receptor, 5:899–907
agonists, 5:900–901
antagonists, 5:899–907
acidic antagonists, 5:905
clinical trials, 5:907
CRTH2 receptor dual pharmacology, 5:920
fused-tricyclic antagonists, 5:903–905
hemi-prostanoid antagonists, 5:901–902
indole acetic acid, 5:902–903
nonacid-based antagonists, 5:905–906
PGD$_2$ synthesis, 5:899–900
EP$_1$ receptor
agonists, 5:812–813
antagonists, 5:813–846
amide replacements, 5:818–820, 836–837
AstraZeneca derivatives, 5:816–817
benzene derivatives, 5:829, 831, 834
benzoic acid derivatives, 5:823, 827–832
carbamates, 5:836–838
cyclopentene derivatives, 5:829, 831, 833
diacyl hydrazide replacement, 5:815–817
glycine sulfonamide derivatives, 5:837, 840–845
Ono derivatives, 5:817–824
pyrazole derivatives, 5:831, 835–838
pyrrole derivatives, 5:826, 828–832
Searle derivatives, 5:813–816
sulfonylureas, 5:820–821, 825
thiophene derivatives, 5:821, 823, 826, 829

EP$_2$ receptor:
agonists, 5:846–850
antagonists, 5:852–854
mixed EP$_2$/E$_4$ agonists, 5:850–852
EP$_3$ receptor, 5:854–872
agonists, 5:854–855
antagonists, 5:855–872
acylsulfonamide derivatives, 5:855–858
angiotensin II derivatives, 5:856–858
hyperalgesia/edema modeling, 5:858, 861
indole analogs, 5:861, 863–869
thiophene derivatives, 5:858, 860–861
EP$_4$ receptor, 5:872–883
agonists, 5:872–880
lactam stereochemistry, 5:873–878
thienopyrimidine-based compounds, 5:878–880
ω-chain derivatives, 5:872–876
antagonists, 5:880–883
FP receptor, 5:897–899
IP receptor, 5:921–943
agonists, 5:922–939
cyclohexene isosteres, 5:933, 935
FR181157 compound, 5:932, 934–935
FR193262 compound, 5:932, 936
heterocyclic isosteres, 5:926–930
nonprostanoid agonists, pyrazole derivatives, 5:922–924
oxazole moiety, 5:924–927
oxime derivatives, 5:932, 937
propyl analog, 5:930, 932
pyrazole derivative, 5:930, 932
pyridazinone derivative, 5:932–933
tetrahydronaphthyl derivative, 5:930–931
thromboxane synthase, 5:935, 938–939
TRA-418, 5:939
antagonists, 5:939–943
aryl guanidine derivatives, 5:940–941

Prostaglandins (Continued)
  compound 500, **5:**940–942
  compound 502, **5:**942
  pyrimidines, **5:**942–943
  receptors, basic properties, **5:**809–812
  recombinant DNA technology, drug efficacy and personalized medicine, **1:**542–543
Prostamide receptor theory, glaucoma intraocular pressure reduction, **5:**594–600
Prostate cancer:
  bexarotene therapy, **5:**448
  N-(4-hydroxyphenyl) retinamide (4-HPR) therapy, **5:**468, 473, 479
Protease-activated receptor-1 (PAR-1) antagonists:
  antiplatelet agents, **4:**437–444
  side effects, **4:**452
Protease inhibitors. See also HIV protease inhibitors (HIV-1 protease inhibitors)
  allosteric protein model, **1:**393
  anti-HCV agents, **7:**176–204
    nonnucleoside inhibitors, **7:**201–202
    benzothiadiazine, **7:**202–203
    HCV-796, **7:**203–204
    NX5B inhibitors, **7:**201–204
    PF-00868554, **7:**201–202
  NS3/4A protease inhibitors, **7:**176–195
    boceprevir, **7:**186–191
    ciluprevir, **7:**176–181
    ITMN-191, **7:**191–192
    MK7009, **7:**193–195
    telaprevir, **7:**181–186
    TMC435350, **7:**193
  nucleoside inhibitors, **7:**195–201
    MK0608, **7:**199–201
    NM283 (valopicitabine), **7:**195–196
    PSI-6130 (Pharmasset), **7:**196–197
    R1479/1626, **7:**197–199
  cysteine proteases, **1:**330–332
  HIV-1 combinatorial library, **1:**310–311
  parasitic proteases, **7:**581–582
  plasmepsins, aspartyl protease inhibitors, **7:**637–644
  recombinant DNA technology, de novo drug development, **1:**553–556
  serine proteases, **1:**332–333
Proteasome:
  antitubercular agents, synthesis and degradation, **7:**726
  basic structure, **6:**175
Proteasome inhibitors:
  clinical trials, **6:**177–180
  future research issues, **6:**184–185
  immunoproteasome-specific inhibitors, **6:**184
  medicinal chemistry, **6:**181–184
Protein A affinity chromatography, protein therapeutics, **3:**308–309
  mimetics research, **3:**310
Protein Data Bank (PDB), structural genomics:
  clinical drug targets, **1:**579–580
  human proteome coverage, **1:**577–579
  medicinal chemistry universal rules, **1:**582–584
  membrane protein research, **1:**580–581
  research strategies, **1:**571–572
  structural space coverage, **1:**577
Protein engineering, recombinant DNA technology, **1:**539–546
  antibody-based therapeutics, **1:**543
  drug efficacy and personalized medicine, **1:**540–543
  epitope mapping, **1:**543–544
  future trends, **1:**545–546
  second-generation protein therapeutics, **1:**540–543
Protein ensemble theory, constitutive receptor activity, **2:**498–499
Protein family targeted libraries, chemogenomics, **2:**574–575

Protein farnesyl transferase (PFT), antimalarial cell cycle regulation, **7:**630–634
Protein folding:
  Hsp90 inhibitors, **6:**384–385
  protein purification, E. coli systems, **3:**305–306
Protein kinase A (PKA):
  Alzheimer's cognitive deficits, **8:**16
  antidepressant efficacy assessment, **8:**227–229
Protein kinase B (PKB):
  brain tumor therapy, **6:**249–250
  survival signaling inhibitors, **6:**323–324
Protein kinase C (PKC):
  brain tumor therapy, **6:**245–246
  chronic myelogenous leukemia therapy, Bcr-Abl inhibitors, **6:**298–301
  metformin mechanism of action, **4:**4
Protein kinases, antimalarial cell cycle regulation, **7:**627–630
Protein-ligand interactions:
  nuclear magnetic resonance, **2:**394–419
    kinetics and timescale parameters, **2:**394–402
  quantum mechanics/molecular mechanics, binding affinity evaluation, **2:**717–719
  in silico screening, protein flexibility, docking techniques, **2:**865–870
  structural genomics, **1:**573–577
    Protein Data Bank rules, **1:**583–584
  structure-based virtual screening, **2:**14–16
Protein-protein interactions (PPIs):
  allosteric protein model, **1:**392–393
  chemogenomics, secondary structure mimetics, **2:**577–578
  drug discovery:
    computational docking, **2:**343–344

fragment-based approach,
  2:340–342
future research issues,
  2:354–359
high-throughput screening,
  2:337–340
hotspot technique, 2:340
p53/MDM2/MDM4 functions,
  2:344–354
research background,
  2:335–337
structure-based approach,
  2:342–343
hit seeking combinatorial
  libraries, 1:341–343
peptidomimetics, 1:230–240
  Bcl-2 inhibitors, 1:231–233
  XIAP inhibitors, 1:233–240
Protein(s). *See also* specific
    proteins, e.g., Heat-
    shock proteins
  antitubercular agents,
    synthesis and
    degradation,
    7:725–726
  chemogenomics, kinase
    inhibitor structure,
    1:590–951
  family targeting, virtual
    screening, 2:31–32
  hepatitis C viral expression,
    7:172
  kinase inhibitor targeting,
    structure-based
    design, 6:350–354
  recombinant DNA technology:
    cellular adhesion proteins,
      1:560–561
    purification, phage display
      vs., 1:551
    second-generation
      compounds, 1:540
  in silico screening flexibility:
    energetic contributions,
      binding affinity,
      2:870
    entropic contributions,
      binding affinity, 2:871
    future research issues,
      2:871–872
    mobility determination,
      2:861–865
    flexibility and geometry-
      based analysis,
      2:863–864
    harmonic analysis,
      2:864–865

molecular dynamics
  simulations,
  2:862–863
protein-ligand docking,
  2:865–870
backbone mobility,
  2:868–870
side-chain conformational
  analysis, 2:867–868
receptor plasticity,
  2:860–870
research background,
  2:859–860
structural genomics, family-
  based genome
  coverage, 1:573–577
structure-based virtual
  screening, structure
  and flexibility,
  2:14–16
sulfonamide binding,
  7:519–520
thyroid hormone metabolism,
  4:195
Protein Structure Initiative (PSI),
    structural genomics,
    research strategies,
    1:571–572
Protein structure similarity
    clustering (PSSC),
    chemogenomics,
    biology-oriented
    synthesis, 2:580
Protein therapeutics:
  AHPN compounds,
    5:494–497
  analytical methods, 3:317–327
  bioactivity assays,
    3:321–322
  biophysical characterization,
    3:323–325
  electrophoresis, 3:320
  future trends, 3:326–327
  immunoassays, 3:321
  in-process/product release
    testing, 3:318–320
  mass spectrometry,
    3:322–323
  nucleic acid testing,
    3:320–321
  downstream processing,
    3:301–311
    classifications, 3:303–304
    design criteria, 3:303
    E. coli protein purification,
      3:304–306
    future trends, 3:309–311

impurities and
  contaminants,
  3:301–303
mammalian protein
  purification,
  3:306–309
formulation and delivery,
  3:311–317
  alternative modes, 3:315–316
  container materials
    interactions, 3:315
  degradation pathways,
    3:311–314
  future trends, 3:316–317
  stability parameters,
    3:314–315
process development,
  3:289–291
comparability, 3:325–326
reagent clearance and
  excipient testing,
  3:326
research background, 3:289
upstream processing,
  3:291–301
  cell culture, 3:298–300
  expression plurality,
    3:291–294
  formats, 3:296–298
  future trends, 3:300–301
  recombinant cell lines,
    3:294–296
Protein tyrosine kinases (PTKs),
    combinatorial
    chemistry, 1:339–342
Protein tyrosine phosphatases
    (PTPs):
  antituberculosis drugs,
    structural genomics,
    1:587
  hit seeking combinatorial
    libraries, 1:339–342
Proteomics, structural genomics
    research, human
    proteome coverage,
    1:577–579
Proteomimetics, p53/MDM2
    derivatives,
    2:349–352
Proton position modeling, docking
    and scoring
    computations,
    ionization and
    tautomerization,
    2:636–637
Proton pump inhibitors, selective
    toxicity, 2:566

Protozoan infections. *See also* Antiprotozoal/antiparasitic agents
  epidemiology, 7:565–566
  kinetoplastic protozoan infections, 7:566–577
    American trypanosomiasis (Chagas' disease), 7:571–573
    biochemistry, 7:567
    glucose metabolism and glycosomal transport inhibitors, 7:582–585
    human African trypanosomiasis (sleeping sickness), 7:567–571
    leishmaniasis, 7:574–577, 590–593
    lipid metabolism inhibitors, 7:580–581
    nucleotide transporters, 7:577–580
    polyamine pathway targeting, 7:585–590
    protease inhibitors, 7:581–582
Protriptyline, chemical structure, 8:222
Provisional patent applications, requirements for, 3:127–133
Prucalopride, serotonin receptor targeting, 8:246–247
Pruvanserin, insomnia therapy, 5:718
Pseudomonnas aeruginosa:
  cephalosporin activity against, 7:305–306
  β-lactam antibacterial drug resistance, 7:280–281
  penicillin resistance and, 7:291–292
Pseudopeptides:
  captopril-related analogs, 1:303
  peptidomimetic design, PD 245065, 1:226–230
Pseudopolymorphs, oral dosage systems, 3:27–29
Pseudo-receptor model, structure-based virtual screening, protein-ligand interactions, 2:14–16
Pseudo-symmetry properties, HIV integrase, two-metal binding inhibitors, 7:86–87
Pseudotumor cerebri, 13-*cis*-retinoic acid therapy, 5:393
PSI-6130 anti-HCV agent, 7:196–197
Psilocybin, cognitive enhancement, 8:37–40
PSN-GK1 compound, type 2 diabetes therapy, 4:15–17
Psoriasis:
  acitretin therapy, 5:396–399
  animal studies, 5:397
  arthritis, 5:397–399
  bexarotene therapy, 5:449
  comorbid conditions, 5:396–397
  cutaneous psoriasis, 5:396–397
  incidence and prevalence, 5:396
  soft drug design, β_2-agonists, 2:103–104
  tazarotene therapy, 5:402–408
  vitamin A pharmacology, 5:649, 651
  vitamin D analogs, 5:655–656
Psychogenic response, stimulants, 8:92, 95–97
Psychostimulants:
  ADMET properties, 8:97–99
  alpha-alkyl substituent, 8:108
  alpha-carbon stereochemistry, 8:108
  amphetamines, 8:106–109
  aromatic ring substitution, 8:108–109, 112–113
  C(2) substituent, 8:111–112
  C(3) ester linkage, 8:112
  caffeine, 8:90
  cocaine, 8:109–113
  ephedra and khat, 8:89–90
  historical background, 8:90–91
  methylphenidate, 8:109
  nitrogen substituents, 8:107, 110–111
  N-substituents, 8:110
  physiology and pharmacology, 8:100–106
  recent and future development, 8:113–115
  receptor classification and function, 8:102–106
  research background, 8:89
  side chain length, 8:107
  side effects, adverse effects, drug interactions, 8:92–97
  soft drug design, 2:104–106
  structure-activity relationships, 8:106–113
  therapeutic applications, 8:91–92
  tropane ring preservation, 8:113
Pterin binding, dihydropteroate synthase, 7:528–531
PubChem Substance database, quantitative structure-activity relationship modeling, 1:513–514
Pulmonary hypertension:
  inhalation therapy:
    carbon monoxide, 5:299–305
    nitric oxide, 5:274–278
  sildenafil therapy, 5:715–716
Pulsed-iterative compound libraries, combinatorial chemistry, 1:296–297
Pulsed ultrafiltration-mass spectrometry, drug screening, 1:116–119
Purification process:
  combinatorial chemistry, 1:345–347
  high-throughput screening, enzymatic assays, 3:402–406
  protein therapeutics:
    downstream processing, 3:301–311
    E. coli systems, 3:304–306
    mammalian cell lines, 3:306–309
Purified natural product libraries, natural products lead generation, 2:201
Purine derivatives:
  acyclic nucleoside phosphates:
    N6-substituted amino derivatives, 7:237
    "open-ring" derivatives, 7:237–238
  Hsp90 inhibitors, N-terminal ATP-binding pocket, 6:410–416
  stroke therapy, Src kinase inhibitors, 8:511

Purine nucleoside phosphorylase (PNP), antimetabolite analogs, **6:**113
Purinergic receptor antagonists, antiplatelet agents, **4:**423–437
  cangrelor, **4:**435–437
  clopidogrel, **4:**427–433
  elinogrel, **4:**437
  irreversible ADP antagonists, **4:**426–435
  P2X$_1$ receptor, **4:**423–424
  P2Y$_1$ receptor, **4:**424–425
  P2Y$_{12}$ antagonists, **4:**425–426
  P2Y$_{12}$ receptor, **4:**425–437
  prasugrel, **4:**433–435
  reversible ADP antagonists, **4:**435–437
  ticagrelor, **4:**436–437
  ticlopidine, **4:**426–427
Purines, antimetabolite analogs, **6:**112–113
PX-866 PI3K inhibitor, survival signaling pathways, **6:**323
PXD-101 histone deacetylase inhibitor, structure and development, **6:**58–59
Pyramax® antimalarial agent, **7:**625–627
Pyrazinamide:
  ADMET properties, **7:**759
  adverse reactions, **7:**759
  antibacterial activity, **7:**756
  drug resistance, **7:**758–759
  historical background, **7:**755–756
  mechanism of action, **7:**756, 758
  selective toxicity, **2:**557
  structure-activity relationship, **7:**756–758
Pyrazole derivatives:
  HIV integrase, two-metal chelation model, **7:**95–97
  Hsp90 inhibitor scaffolds, **6:**417–419
  prostaglandin/thromboxane receptors:
    EP$_1$ antagonists, **5:**831, 835–837
    IP receptor agonist, **5:**930, 932
    IP receptor agonsts, **5:**922–924

stroke therapy, pyrazole-pyrimidine inhibitors, **8:**511–512
Pyrazole-pyrimidine inhibitors, stroke therapy, **8:**511–512
Pyrazolidinone, mixed EP$_2$/E$_4$ agonists, **5:**850–852
Pyrazoloacridine, dual topoisomerase I/II inhibitors, **6:**103–105
Pyrazolols, poly(ADP-ribose) polymerase inhibitors, **6:**166–167
Pyridazinone derivative, prostaglandin/thromboxane receptors, IP receptor agonist, **5:**932–933
Pyrido[3,4-*d*]azepines, nicotinic agonist, **8:**68–69
Pyridone-based calcitonin gene-related peptide antagonists, migraine therapy, **8:**287–288
Pyrido-pyrimidine inhibitors, stroke therapy, **8:**513–514
Pyridoxalated hemoglobin polyoxyethylene, conjugated hemoglobins, **4:**635–637
Pyridoxal phosphate (PLP), chemistry, **5:**679, 681–682
Pyridoxine:
  biochemistry, **5:**679–682
  hypervitaminosis, **5:**684
  Parkinson's disease, dopaminergic pharmacotherapy, **8:**541–542
Pyridylvinyl thyromimetic, structure-activity relationship, **4:**220–221
Pyrimethamine, antimalarial drugs, **7:**617–618
  antifolates, **7:**647–648
Pyrimidine-2,4,6-triones, stroke therapy, **8:**508
Pyrimidines:
  antimetabolite analogs, **6:**109–112
  prostaglandin/thromboxane receptors:

DP$_1$ receptor antagonists, **5:**905–906
IP receptor antagonist, **5:**942–943
stroke therapy:
  N-type voltage-dependent calcium channel scaffolds, **8:**457
  pyrazole-pyrimidine inhibitors, **8:**511–512
  pyrido-pyrimidine inhibitors, **8:**513–514
  pyrrole-pyrimidine inhibitors, **8:**512
Pyrimidinone, raltegravir development, **7:**89–90, 92–97
Pyroglutamate formation, protein therapeutics formulation and delivery, **3:**313–314
Pyronaridine antimalarial agent, **7:**619–621
Pyrrole derivatives:
  prostaglandin/thromboxane EP$_1$ antagonists, **5:**826, 828–832
  toxicophore reactive metabolites, **2:**314–317
Pyrrole-pyrimidine inhibitors, stroke therapy, **8:**512
Pyrrolidine-2,5-diones, anticonvulsant applications, **8:**142, 149–151
Pyrrolidine ring compounds:
  dipeptidyl peptidase 4 inhibition, **4:**50–56
  tri-substituted pyrrolidine chemokine receptor-5 (CCR5) antagonist, **7:**121–124
Pyrrolobenz(pyrido) oxazepinones, reverse transcriptase inhibitors, **7:**153–155
Pyruvate carboxylase, biotin metabolism, **5:**686–690
PZ-601 carbapenem, **7:**357

q^2 values, quantitative structure-activity relationship, model validation, **1:**520–521

QseC membrane embedded
    sensor histidine
    kinase, antibiotic
    selective toxicity and,
    2:553
QT prolongation:
  antipsychotic side effects, **8**:167
  chirality and toxicity,
    1:143–144
QT prolongation, soft drug
    development,
    serotonin receptor
    agonists, **2**:117–118
Quadrupole time-of-flight (qTOF)
    instrumentation,
    mass spectrometry
    development, **1**:98
Quality-adjusted life years
    (QALY), cost-
    effectiveness analysis
    of drug development,
    3:345
Quality by design (QbD), protein
    therapeutics, **3**:327
Quality control issues,
    pharmacogenomics,
    1:199
Quality factor *(Q)*, quantitative
    structure-activity
    relationship
    validation, **1**:47
Quantitative high-throughput
    screening (qHTS):
  natural products lead
    generation, **2**:206–207
  receptor-based drug design,
    lead drug discoveries,
    2:526–527
Quantitative polymerase chain
    reaction (QPCR),
    protein therapeutics,
    analytic development,
    3:321
Quantitative structure-activity
    relationship (QSAR).
    See also Structure-
    activity relationship
    (SAR)
  applications, **1**:51–68
    ADMET properties, **1**:62–68
      blood-brain barrier, **1**:65
      excretion, **1**:67
      human intestinal
        absorption, **1**:65
      metabolism, **1**:66–67
      oral bioavailability, **1**:66
      permeability, **1**:64–65

plasma/serum protein
    binding, **1**:66
solubility, **1**:62–64
toxicity, **1**:67–68
volume of distribution, **1**:66
anti-HIV activity, **1**:60–62
cellular interactions, **1**:57–59
isolated receptor
    interactions, **1**:51–57
in vivo interactions, **1**:59–60
artificial neural networks,
    1:38–39
ATP binding casette
    transporter inhibitors,
    6:374
camptothecins, **6**:29
chiral compounds, Pfeiffer's
    rule, **1**:129–130
comparative QSAR, **1**:68–74
  database development,
    1:68–72
  data mining for models, **1**:69,
    72–73
  current trends in, **1**:73–74
data set complexity and data
    curation
    requirements,
    1:512–516
  modern data sets, model
    construction,
    1:512–514
  quality control, **1**:514–516
dependent variables, **1**:28–33
  biological parameters,
    1:31–32
  compound selection,
    1:32–33
  receptors, **1**:28–31
drug metabolism, lipophilicity,
    1:445
glucocorticoid anti-
    inflammatory
    compounds, **5**:71–82
  conformational changes,
    5:76–77
  de novo constants, **5**:71–72
  electronic characteristics,
    5:77
  Hansch type analyses,
    5:72–74
  neural network predictions,
    5:74–75
  receptor binding affinity,
    LinBiExp model,
    5:75–76
  three-dimensional QSARs,
    5:77–82

G-protein coupled receptor,
    homology modeling,
    2:292–293
historical development, **1**:1–2
independent variables, **1**:2–28
  electronic parameters, **1**:2–8
  hydrophobic parameters,
    1:8–19
  partition coefficient:
    calculation methods,
      1:14–19
    measurement methods,
      1:12–14
  molecular descriptors,
    1:23–28
    DCW (SMILES), **1**:26–27
    effective hydrogen charge,
      1:27–28
    polarizability, **1**:24–25
    polar surface area, **1**:25–26
    topological descriptors,
      1:23–24
    van der Waals volumes,
      1:26
  steric parameters, **1**:20–21
integrated predictive modeling
    workflow, **1**:516–525
  applicability domains,
    1:521–522
  combinatorial modeling,
    1:518–520
  hybrid chemical biological
    descriptors, **1**:525
  model ensembles and
    consensus prediction,
    1:522–525
  model validation, **1**:520–521
investigative protocols,
    1:508–509
ligand-based virtual screening:
  machine learning
    algorithms, **2**:10–13
  model generation and
    validation, **2**:10
model applications, **1**:526–529
  experimental data curation,
    1:527–528
  future research issues,
    1:529–530
  OECD principles and
    regulatory acceptance,
    1:528–529
  validated/predictive models,
    virtual screening
    applications,
    1:526–527
model failure, **1**:510–511

modeling framework,
1:509–510
multidimensional models,
1:39–43
multilinear regression
analysis, 1:33–37
Free-Wilson and other
approaches, 1:36–37
linear models, 1:33–34
nonlinear models, 1:34–36
outlier detection, 1:43–46
convex hull method, 1:44
distance between r^2 and q^2,
1:45
Mahalanobis distance, 1:44
OCD method, 1:45
potential functions, 1:44
R-NN curves, 1:44–45
smallest half-volume
method, 1:44
standard deviation test,
1:45–46
uncertainty method, 1:44
X-residual, 1:44
partial least squares
regression, 1:37
pharmacophores, 1:457–463
bioactive conformations,
1:471
principal component
regression, 1:37–38
quantum mechanics/molecular
mechanics, binding
affinity evaluation,
2:722–723
recent research on, 1:507–510
receptor-based drug design,
2:520
soft drug design, quaternary
anticholinergics,
2:110–111
structure-based drug design,
computer-aided
techniques, 2:681–682
success of, 1:74
sulfonamides, 7:515–518
validation, 1:46–51
Quantitative structure-
metabolism
relationship (QSMR):
drug metabolism, 1:445–447
soft drugs, computer-aided
techniques,
2:132–133
Quantitative structure-property
relationships (QSPR).
See also Quantitative

structure-activity
relationship (QSAR)
historical development, 1:1–2
polarizability indices, 1:24–25
receptor-based drug design,
2:520
validation, 1:520–521
Quantum mechanics/molecular
mechanics (QM/MM):
binding affinity evaluation:
advantages, 2:719–723
linear interaction energy
method, 2:729–730
PBSA/GBSA approaches,
2:730–733
pertubation, 2:723–729
free-energy perturbation,
2:723–728
thermodynamic
integration, 2:728–729
research background,
2:717–719
combined methodologies,
2:737–738
docking and scoring
computations:
binding affinity evaluation,
2:733–735
ionization and
tautomerization,
2:637
methodologies, 2:736–738
sampling methods, 2:736
Quantum-refined force field
molecular dynamics
(QMFF-MD), scoring
functions, 2:734–735
Quaternary derivatives, soft drug
design:
anticholinergics, 2:108–112
antimicrobials, 2:111–114
Quaternary salts:
cationic surfactants, 7:500
chemokine receptor-5 (CCR5)
antagonists, anilide
core structure,
7:124–131
dual topoisomerase I/II
inhibitors, 6:103–105
Quercitin β-amyloid peptide
inhibitors, 8:364–365
Quetiapine:
FDA approval, 8:178–179
pharmacokinetics, 8:193–194
side effects, 8:169
structure-activity relationship,
8:180–183

QUIK rule, quantitative
structure-activity
relationship
validation, 1:47
Quinaxoline-2,3-dione, stroke
therapy, 8:452–453
Quinazolines:
kinesin spindle protein
inhibitors, 6:197–200
quantitative structure-activity
relationship, isolated
receptor interactions,
1:54–56
stroke therapy, Src kinase
inhibitors, 8:513
Quinazolone, cognitive
enhancement, 8:37
Quinestrol, current compounds
and applications,
5:220–221
Quinicrine mustards,
Alzheimer's disease,
tau-targeted therapy,
8:424–425
Quinine, antimalarial activity,
7:605–607
Quinolines:
antimalarials, 7:605–614
amodiaquine, 7:611–613
chloroquine, 7:607–610
experimental agents,
7:625–627
mefloquine, 7:610–611
primaquine, 7:613–614
quinine, 7:605–607
stroke therapy:
caspase inhibitors, 8:471
Src kinase inhibitors, 8:513
Quinolinone, pharmacokinetics,
8:198
Quinolone carboxylic acid, HIV
integrase, two-metal
chelation model,
7:94–97
Quinolones:
antibacterial agents,
7:535–540
irreversible inhibitors, stroke
therapy, 8:465
selective toxicity, 2:552–553
Quinones, hypoxia-activated
bioreductive prodrugs,
6:115–120
Quinuclidine:
muscarinic antagonists,
8:75–77
nicotinic agonists, 8:71

QXP algorithm, force field scoring, **2:**621–622

R^2 values, quantitative structure-activity relationship, model validation, **1:**520–521
R547 cyclin-dependent kinase inhibitor, **6:**313–315
R1450 monoclonal antibody, Alzheimer's immunization therapy, **8:**378
R1479/1626 anti-HCV agent, **7:**197–199
R7227 protease inhibitor. *See* ITMN-191 anti-HCV agent
R-84760 κ-opioid receptor agonist, **8:**627–632
R306465 first-generation histone deacetylase inhibitor, **6:**62–64
Racemic 1,3-dioxolane derivatives, thromboxane A_2 receptor antagonists, **5:**887–889, 891
Racemic antidepressants, current and future research, **8:**255–256
Racivir®:
  anti-DNA virus antimetabolite, **7:**227
  current development, **7:**142–145
Radamide analogs, chimeric radicicol and geldanamycin Hsp90 inhibitors, **6:**408–410
Radanamycin analogs, chimeric radicicol and geldanamycin Hsp90 inhibitors, **6:**408–410
Radical reactions, hemoglobin oxygen delivery, **4:**614–615
Radicicol Hsp90 inhibitor:
  analog design, **6:**399–405
  brain tumor therapy, **6:**268
  chimeric structures, **6:**405–410
Radiolabeled compounds:
  disease biomarkers, **8:**744–746
  Parkinson's disease diagnosis, **8:**555–556
  in PET/SPECT systems, **8:**742–743

prostaglandin/thromboxane receptors:
  DP_1 antagonists, **5:**899–907
  EP_1 antagonists, **5:**843, 846
  targeted drug development, **8:**743–744
Radioligand binding assays:
  κ-opioid receptor agonists, **8:**624–632
  opioid receptors, **8:**582–583
Radiotherapy, poly(ADP-ribose) polymerase (PARP) inhibitor coadjuvants, **6:**169–170
Raf kinases, signaling inhibitors, **6:**305–306
Raloxifene:
  adverse effects and precautions, **5:**221
  current compounds and applications, **5:**219–221
  diarylethylene antagonists, **5:**248–249
  discovery, **5:**243–244
  electrostatic interactions, **5:**236
  osteoporosis antiresorptive therapy, **5:**734–742
  selective toxicity, **2:**567
  tissue specificity, **5:**240–242
Raltegravir:
  development of, **7:**89–90, 92–97
  selective toxicity, **2:**554–557
Raltitrexed, antifolate DNA-targeted chemotherapeutics, **6:**107–109
Raman spectroscopy, salt compound selection, automated assisted screening, **3:**396
Ramatroban analogs:
  CRTH2 receptor antagonist, **5:**909–911
  thromboxane A_2 receptor antagonists, **5:**888–889, 891–892
Ramelteon, insomnia therapy, **5:**717–719
Ramoplanin, antibiotic mechanism, **2:**227–228
Random selection, quantitative structure-activity relationship validation, **1:**49

Random walk process, quantitative structure-activity relationships, nonlinear models, **1:**34–36
"Random walk" profile:
  prodrug development, rule-of-one metabolism, **3:**248–250
  prodrug properties, **3:**220–226
Ranibizumab, age-related macular degeneration therapy, **5:**615–616
Ranitidine:
  analog drug design, **1:**170
  polymorph crystallization, **3:**190–192
Rank by rank strategy, consensus scoring, **2:**631–632
Rank by vote strategy, consensus scoring, **2:**631–632
RANTES binding assay, chemokine receptor-5 (CCR5) antagonist development, **7:**114–116
  anilide core structures, **7:**129–131
Rapamycins. *See also* Sirolimus
  age-related macular degeneration therapy, **5:**622
  anti-cancer activity, **2:**240–243
  cranial and spinal nerve tumors, **6:**238–239
  mTOR inhibitors, **6:**324–325
RARα-X fusion protein, acute promyelocytic leukemia therapy, **5:**380–382
RAR/RXR panagonists, **5:**421–429
  adverse effects, **5:**426–427
  pharmacology/metabolism, **5:**428–429
  therapeutic applications, **5:**421–426
Rasagiline, Parkinson's disease, dopaminergic therapy, **8:**549–551
Ras farnesyl transferase, peptidomimetics, **1:**220–222
RasGap and ERK-1 dual inhibitors, hit seeking

combinatorial libraries, **1:**320–321
kinase inhibitors, **1:**340–342
RAS/RAF/MEK/ERK signaling pathway:
  brain cancer therapies, **6:**252
  growth factor signaling inhibitors, **6:**301–307
Ras signaling pathway, brain cancer therapy, **6:**252
Rate constants:
  elimination, drug delivery system bioavailability, **2:**750–751
  macromolecule-ligand interactions, **2:**394–402
  oral drug delivery, immediate-release dosage, plasma drug profile absorption, **2:**774–776
Rate theory, receptor-based drug design, **2:**497
RBx-10558, anticholesterolemic mechanisms, **2:**234
Reaction solvents, bench-scale experiments, **3:**5–6
Reaction temperature, bench-scale experiments, **3:**6
Reaction time, bench-scale experiments, **3:**6
Reactive metabolites:
  ADMET toxicity properties, **2:**62, 68
  toxicophores, **2:**303–324
    N-alkyl-tetrahydropyridines, **2:**322–323
    alkynes, **2:**320
    2-aminohiazoles, **2:**319–320
    anilines and hydroxyanilines, **2:**303–311
    arylhydroxamic acids, **2:**313
    benzodioxolanes, **2:**314
    carboxylic acid acyl glucuronides, **2:**323–324
    electron-rich heterocycles, **2:**314–317
    hydantoins, **2:**322
    hydrazines, **2:**320–321
    3-methylindoles, **2:**317–318
    nitro-aromatics/nitro-alkyls, **2:**312–313
    ortho/para-dihydroxyaromatics, **2:**313–314
  screening for, **2:**326–327
  thiazolidinediones, **2:**322
  thioamides, thioureas, and thiocarbamates, **2:**318–319
Reactive oxygen species (ROS), *N*-(4-hydroxyphenyl) retinamide (4-HPR):
  cancer therapy, **5:**465–473
  generation mechanism, **5:**474
Reactive oxygen species (ROSs), adverse drug reactions, **1:**441–443
Reagent development:
  protein therapeutics, clearance process, **3:**326
  recombinant DNA technology:
    screening applications, **1:**546–547
    structural biology applications, **1:**552
    structural genomics research, **1:**581
Reboxetine:
  chemical structure, **8:**223
  cognitive enhancement, **8:**41–42
  monoamine transporter inhibition, **8:**239–241
  pharmacokinetics, **8:**237
  side effects and adverse reactions, **8:**235
Receptor activator of nuclear factor-κB ligand (RANKL), osteoporosis antiresorptive therapy, **5:**732–733
Receptor-based drug design:
  antidepressants, serotonergic agents, **8:**241–247
  cellular information processing, **2:**492–493
  central nervous system, **8:**1–3
  functional selectivity, **8:**7–9
  chemical space, **2:**519
  classes, **2:**505–514
  complexes and allosteric modulators, **2:**499–501
  compound properties, **2:**519–525
    ADMET parameters, **2:**524–525
    databases, **2:**525
    receptor-ligand interaction, **2:**520–523
    source materials, **2:**523–524
    structure-activity relationships, **2:**519–520
  congestive heart failure therapy, mechanism-targeted studies, **4:**492
  constitutive receptor activity, **2:**498–499
  drug receptors, **2:**514
  dynamics, **2:**503
  efficacy parameters, **2:**501–503
  functional genomics, **2:**515–519
  future research issues, **2:**533–534
  G-protein-coupled receptors, **2:**504–509
    genetic variation, **2:**514–515
    signaling modulators, **2:**508–509
  intracellular receptors, **2:**512
  lead compound discovery, **2:**525–533
    biologicals and antisense compounds, **2:**531–533
    diversity-based ligand libraries, **2:**531
    high-throughput screening, **2:**526–527
    natural product sources, **2:**527–528
    pharmacophore-based ligand libraries, **2:**528–529
    privileged pharmacophores, **2:**529–530
  ligand-gated ion channels, **2:**509–511
  molecular biology, **2:**514
  neurotransmitter binding proteins, **2:**514
  neurotransmitter transporters, **2:**513–514
  nomenclature, **2:**504–505
  non-GPCR-linked cytokine receptors, **2:**512–513
  nuclear magnetic resonance, **2:**390–419
    chemical-shift mapping, **2:**402–403
    dihydrofolate reductase inhibitors, **2:**414–419
    DNA-binding drugs, **2:**405–410
    isotope editing/filtering, **2:**403–404
    macromolecular structure, **2:**391–394

Receptor-based drug design (*Continued*)
  matrix metalloproteinases, **2:**412–414
  NOE docking, **2:**404–405
  protein-ligand interactions, **2:**394–419
  titrations, **2:**403
  orphan receptors, **2:**513
  Parkinson's disease, dopaminergic pharmacotherapy, **8:**542–549
  aporphine-type receptor agonists, **8:**544–545
  dopamine receptor structure and function, **8:**542–544
  ergot-type receptor agonists, **8:**545–546
  experimental D_1 receptor agonists, **8:**547–549
  small-molecule agonists, **8:**547
  receptor residence time, **2:**503–504
  research background, **2:**491–492
  RNA targets, **2:**512
  steroid receptor superfamily, **2:**512
  theoretical background, **2:**493–498
    occupancy theory, **2:**494–497
    rate theory, **2:**497
    ternary complex model, **2:**497–498
  three-dimensional pharmacophores, ligand-based virtual screening, **2:**8–9
  transient-anchor chemical delivery systems, **2:**160–161
Receptor-ligand interactions. *See* Ligand-receptor interactions
Receptor-mediated endocytosis (RME):
  human African trypanosomiasis therapy, **7:**569–571
  membrane proteins, **2:**446–454
  absorption enhancement, **2:**444
  bacterial adhesins and invasins, **2:**451–452

bacterial and plant toxins, **2:**452
cell surface receptor structure, **2:**447–448
endocytosis, **2:**446–447
immunoglobulin transport, **2:**450–451
lectins, **2:**452–453
lipid rafts, **2:**449–450
oral absorption systems, **2:**450–453
potocytosis, **2:**448–449
transcytosis, **2:**448
transport mechanisms, **2:**446–447
viral hemagglutinins, **2:**452
vitamins and metal ions, **2:**453–454
transport mechanisms, **2:**446–447
Receptors. *See also* specific receptors, e.g., Nuclear hormone receptors (NHRs)
calcitonin gene-related peptide antagonist, migraine therapy, **8:**273
chiral compounds, drug-receptor interactions, **1:**127–129
G-protein coupled receptor, homology modeling, structure/function, **2:**288–289
plasticity, *in silico* screening, protein flexibility:
  mobility determination, **2:**861–865
  protein-ligand docking, **2:**865–870
  research background, **2:**859–860
psychostimulant physiology and pharmacology, **8:**102–106
quantitative structure-activity relationship, **1:**28–31
isolated receptor interactions, **1:**51–57
recombinant DNA technology, drug targeting mechanisms, **1:**556–560
RECEPTOR second-generation program, pharmacophore

formation, bioactive conformation, **1:**470
Receptor subtype occupancy theory, congestive heart failure therapy, mechanism-targeted studies, **4:**492
Receptor tyrosine kinases (RTKs):
  brain tumor therapy, **6:**245
  PDGF receptor, **6:**249
  RAS/RAF/MEK/ERK signaling, **6:**252
  EGFR inhibitor, resistance mechanisms, **6:**367–368
Recombinant DNA technology:
  chemistry-driven drug discovery, **1:**537–538
  evolution, **1:**538
  future trends in, **1:**561–562
  genetically-engineered drug discovery, **1:**546–561
    aptamer-based diagnostics, **1:**549–550
    aptamer-based drugs, **1:**549
    cell-specific targeting, **1:**551
    cellular adhesion proteins, **1:**560–561
    combinatorial biosynthesis and microbe reengineering, **1:**547–549
    enzyme-based drug targets, **1:**552–556
    phage display, **1:**550–552
    phage library preparation, **1:**550–551
    purified proteins *vs.* phage display, **1:**551
    receptor-based drug targets, **1:**556–560
    screening reagents, **1:**546–547
    small interfering RNA, **1:**550
    structural biology reagents, **1:**552
    vector-mediated delivery, siRNA, **1:**550
    in vitro evolution (SELEX), **1:**549
    in vivo phage display, **1:**551–552
  peptide libraries, **1:**285–291
  protein engineering and site-directed mutagenesis, **1:**539–546

antibody-based therapeutics, **1:**543
drug efficacy and personalized medicine, **1:**540–543
epitope mapping, **1:**543–544
future trends, **1:**545–546
second-generation protein therapeutics, **1:**540–543
protein therapeutics, upstream processing, **3:**294–295
receptor-based drug design, **2:**514
therapeutic milestones, **1:**538–539
Recombinant hemoglobins (rHB), structure and properties, **4:**637–640
Recombinant prourokinase (rpro-UK), stroke therapy, **8:**448
Recommended daily allowances (RDAs), vitamins, dietary reference intakes vs., **5:**640–644
Reconstituted HDL, atherosclerosis, HDL elevation, **4:**336
Recovery, structure-based virtual screening, enrichment factors, **2:**22
Recurrent cancer, central nervous system tumors, **6:**240–241
Red blood cells:
artificial red blood cells, **4:**641–642
permeability, ADMET interactions, **3:**376
Redox analogs, brain-targeted chemical delivery systems, **2:**154–155
Redox reactions:
drug metabolism, glutathione conjugation, **1:**434–438
hemoglobin oxygen delivery, **4:**614–615
kinetoplastid chemical structure, **7:**566–567
5α-Reductase:
androgen antagonists, **5:**194–196
quantitative structure-activity relationship, isolated

receptor interactions, **1:**56–57
Reductase enzymes, gene-directed enzyme prodrug therapy, **6:**124
Reductionist approach, quantitative structure-activity relationship, partition coefficient calculations, **1:**14–19
Reduction to practice principle, patent protection, **3:**109–110
Reductive metabolism, androgens, **5:**159–162
Redundancy issues, BioPrint® database, **1:**489
Refold concentrations, protein therapeutics, **3:**310
Regioisomers, acyclic nucleoside phosphates, 8-azapurine derivatives, **7:**237
Regioselectivity, drug metabolism, sp^3-carbon atom oxidation and reduction, **1:**414–415
Regulator of G-protein signaling 4 (RGS-4), schizophrenia genetics, **8:**174
Regulatory issues. See also specific drug legislation, e.g., Orphan Drug Act
animal dugs, **3:**97
Biopharmaceutics Classification System, **3:**355–356
chiral compounds, **1:**155–156
cost-effectiveness analysis of drug development, **3:**345–346
FDA role in biotechnology-derived drugs, **3:**93–96
generic drug review and approval, **3:**84–92
historical background, **3:**63–69
new drug approval process, **3:**71–82
Orange Book, **3:**92–93

organizational structure, **3:**70
over the counter drug approval, **3:**82–84
postapproval process, **3:**96
review process, **3:**70–71
statutes, **3:**69–70
United States Pharmacopoeia and National Formulary, **3:**96–97
International Conference on Harmonization, **3:**97
quantitative structure-activity relationship modeling principles, **1:**508, 528–529
Relative receptor binding affinity (rRBA):
antiglucocorticoids, **5:**112–120
glucocorticoid anti-inflammatory compounds:
LinBiExp method, **5:**75–76
nonsteroidal glucocorticoids/antiglucocorticoids, **5:**123–134
Relaxation data:
ligand-based design, **2:**387–389
nuclear magnetic resonance studies, drug screening applications, **2:**427–429
Relaxin, structure and function, **4:**549–550
Remedies in infringement of patent claims, **3:**157–159
Remifentamil, soft drug design, **2:**90
Remikiren renin inhibitor, **4:**246, 248–249
structure and function, **4:**520–522
Remogliflozin etabonate, type 2 diabetes therapies, **4:**14
Renal cancer:
9-*cis*-retinoic acid therapy, **5:**425
SHetA2 retinoid-related compound, **5:**504–505

Renal system. *See also* Chronic renal failure
permeability, ADMET interactions:
excretion, **3**:373
metabolism, **3**:372–373
Renal transplants, sirolimus immunosuppressor and, **5**:1036–1037
Renin-angiotensin-aldosterone system (RAAS):
aliskiren clinical trials, **4**:251–252
angiotensin-converting enzyme inhibitors, **4**:267–268, 285–286
congestive heart failure therapy, phenotype/polymorphism associations, **4**:507
physiology, **4**:240–241
renin inhibitors:
anticipated hypertension and end organ benefit, **4**:242–243
preclinical models, **4**:242
vasoactive peptides, **4**:516–520
angiotensins, **4**:516–518
inhibitors, antagonists, and agonists, **4**:520
prorenin and renin, **4**:518–519
Renin inhibitors:
aliskiren:
animal studies, **4**:250–251
clinical trials, **4**:251–252
congestive heart failure, **4**:252–253
discovery, **4**:250
drug-drug interactions, **4**:253
end organ protection, **4**:252
pharmacokinetics, **4**:253
anticipated hypertension and end organ benefit, **4**:242–243
drug discovery process, **4**:243–250
hydroxyethylene renin inhibitors, **4**:245–247
hydroxyl transition-state isostere renin inhibitors, **4**:243–244
norstatine renin inhibitors, **4**:247
peptide/reduced peptide inhibitors, **4**:243

statin-based renin inhibitors, **4**:244–245
vicinal diol renin inhibitors, **4**:247–250
future research issues, **4**:255–256
hypertension epidemiology, **4**:240
nonpeptidic chemotypes, **4**:253–255
alkylamines, **4**:254–255
diaminopyrimidines, **4**:254
piperidines/piperazines, **4**:253–254
plasmepsin antimalarial agents, **7**:642–644
preclinical models, **4**:242
pro(renin) receptor, **4**:241–242
renin-angiotensin-aldosterone physiology, **4**:240–241
renin biochemistry/biosynthesis, **4**:241
research background, **4**:239–240
structure and function, **4**:520
Renin structure and function, renin-angiotensin system, **4**:518–519
Renova®, properties and applications, **5**:376
Repaglinide, type 2 diabetes therapy, **4**:10–13
Repeated bolus injections, intravenous drug delivery and bioavailability, **2**:754–760
area under curve calculations, **2**:758–759
concentration peaks and troughs, calculation, **2**:755–756
dose time intervals, master curve, **2**:756–757
master curve calculations, **2**:756–758
patient adaptation, **2**:759–760
patient intervariability, **2**:756
pharmacokinetic parameters, **2**:756
plasma drug profiles, master curve and dimensionless numbers, **2**:756
Repeated dose plasma drug profiles:

diffusion-based drug release, **2**:808–812
erosion-controlled drug release, **2**:822–825
Repeated oral drug delivery:
area under the curve measurements, **2**:785–786
dosage principles and parameters, **2**:777–784
Repeated short infusions, constant flow rate over finite period, **2**:763–771
area under curve measurements, **2**:765–766
dosage and time period changes, **2**:768–770
drug concentration measurements, **2**:763–765
infusion-elimination stages, **2**:763–765
infusion rate changes, **2**:770–771
patient adaptation and intervariability, **2**:766–768
second infusion-elimination stage, **2**:763
Replacement therapy protocols, selective toxicity, **2**:541
Replication, hepatitis C viral genome, **7**:173–174
Request for continued examination (RCE), drug patent delays, **3**:114
Research trends and quality:
quantitative structure-activity relationships, **1**:507–509
structural genomics, **1**:571–572
Reservatrol, type 2 diabetes therapy, **4**:26–27
Residence times:
ligand-macromolecular interactions, structure-based thermodynamic analysis, isothermal titration calorimetry, **2**:707–712

oral dosage forms, class III bioadhesives, **3:**52
receptor-based drug design, **2:**503–504
Residual factor (R factor), docking/scoring techniques, **2:**597–598
Resin synthesis:
  combinatorial chemistry, **1:**343–344
  peptide libraries, **1:**286–291
Resistance to thyroid hormone (RTH), TRIAC thyromimetic and, **4:**204
Resolution, docking/scoring techniques, **2:**597–598
Resorcinolic macrolides, radicicol Hsp90 inhibitor and, **6:**402–405
Resorcinols, antimicrobial agents, **7:**494
Respiration:
  antitubercular agent targeting of, **7:**729–730
  depression, opioid receptor effects on, **8:**574
Response elements (REs):
  nuclear hormone receptors, **4:**79–80
  peroxisome proliferator response elements, **4:**86
  retinoid nuclear receptors, **5:**369, 371–373
  thyroid hormones, **4:**190
Response-to-treatment studies, congestive heart failure therapy, **4:**507
Restasis, soft drug design, **2:**120–121
Retigabine, anticonvulsant applications, **8:**136–139, 154
Retin-A®, properties and applications, **5:**376
Retinaldehyde, olefin isomerization, **5:**587–588
9-cis-Retinoic acid:
  chemical structure, **5:**421–422
  retinoic nuclear receptors, **5:**372
13-cis-retinoic acid:
  adverse effects, **5:**393–395
  basic properties, **5:**387–388
  disease therapies, **5:**388–392
  pharmacology and metabolism, **5:**395
  vitamin A pharmacology, **5:**649
Retinoic acid metabolism-blocking agents (RAMBAs), development of, **5:**386–387
Retinoic acid receptor (RAR):
  N-(4-hydroxyphenyl) retinamide (4-HPR), independence mechanism, **5:**474–476
  RAR/RXR panagonists, **5:**421–429
    adverse effects, **5:**426–427
    chemical structure, **5:**421–422
    pharmacology/metabolism, **5:**428–429
    therapeutic application, AIDS-related Kaposi's sarcoma, **5:**422–424
    therapeutic applications, **5:**421–426
  receptor-selective compounds:
    acitretin, **5:**395–400
    adapalene, **5:**408–414
    Am80, **5:**414–420
    etretinate, **5:**401–402
    NRX 195183 analog, **5:**420
    13-cis-retinoic acid, **5:**387–395
  receptor-selective retinoids, **5:**376–421
    all-trans-retinoic acid, **5:**376–387
  selectivity, **5:**373–376
  subtypes, **5:**369–378
    binding affinities, **5:**372–373, 377–379
    transcriptional activation activity values, **5:**372–379
  vitamin A chemistry, **5:**644–647
Retinoic acid receptor responder (RARRES) gene, tazarotene discovery, **5:**402–405
Retinoic X receptor (RXR):
  brain tumor therapy, **6:**267
  discovery, **5:**369
  liver X receptors and, **4:**146
  peroxisome proliferator-activator receptors, γ-agonists, **4:**101–102
  RAR/RXR panagonists, **5:**421–429
    adverse effects, **5:**426–427
    pharmacology/metabolism, **5:**428–429
    therapeutic application, AIDS-related Kaposi's sarcoma, **5:**422–424
    therapeutic applications, **5:**421–426
  selectivity, **5:**373–376
  bexarotene, **5:**429–453
  subtypes, **5:**369, 371–373, 376–378
    binding affinities and transcriptional activation, **5:**376–378
  thyroid hormone receptors, heterodimerization, **4:**190–191
  vitamin A chemistry, **5:**644
Retinoid-inducible gene (RIG), tazarotene discovery, **5:**402–405
Retinoid-related molecules (RRMs), **5:**372
Retinoids:
  chemical structure., **5:**369–370
  in clinical trials, **5:**507–516
  FDA currently-approved drugs, **5:**508–509
  future research issues, **5:**507, 509, 516
  mechanisms of action and clinical status, **5:**516
  nuclear receptors, **5:**369–378
  selectivity, **5:**373–376
  RAR/RXR panagonists, **5:**421–429
  9-cis retinoic acid, **5:**421–429
  retinoic acid receptor-selective compounds, **5:**376–377, 380–421
    acitretin, **5:**395–400
    adapalene, **5:**408–414
    all-trans-retinoic acid, **5:**376–377, 380–384
    Am80, **5:**414–420
    etretinate, **5:**401–402
    NRX 195183 analog, **5:**420
    13-cis-retinoic acid, **5:**387–395
  retinoid-related compounds, **5:**464–507
    N-(4-hydroxyphenyl) retinamide, **5:**464–483

Retinoids (*Continued*)
　SHetA2 compound, 5:500–507
　ST1926 (AHPC) compound, 5:483–500
　RXR-selective retinoids, 5:429–464
　　9cUAB30, 5:453–458
　　bexarotene, 5:429–453
　　NRX 194204, 5:458–464
　translational status, 5:507, 509, 516
　vitamin A pharmacology, 5:648–649
Retinoid X receptor-like family, receptor-based drug design, 2:512
Retinol (vitamin A), 5:644–649
　biochemical functions and deficiency, 5:646–647
　chemistry, 5:644
　dosage forms, 5:647
　N-(4-hydroxyphenyl) retinamide (4-HPR) reduction, 5:476–477
　hypercarotenosis, 5:648
　hypervitaminosis A, 5:647–648
　pharmacologically active retinoids, 5:648–649
　uptake and metabolism, 5:644–646
Retinol-binding protein (RBP):
　N-(4-hydroxyphenyl) retinamide (4-HPR) therapy, diabetes, 5:474
　retinol (vitamin A) uptake and metabolism, 5:646
Retrometabolic drug design (RMDD):
　basic principles, 2:75
　chemical delivery systems:
　　basic principles, 2:134–136
　　benzylpenicillin, 2:142–143
　　brain-targeting compounds, 2:136–140
　　　redox analogs, 2:154–155
　　cyclodextrin complexes, 2:149–150
　　estradiol, 2:144–149
　　eye-targeting site-specific compounds, 2:155–160
　　oxime/methoxime β-blocker analogs, 2:158–160
　　ganciclovir, 2:142–143

　　molecular packaging, 2:150–155
　　brain-targeting redox analogs, 2:154–155
　　kyotorphin analogs, 2:153–154
　　leu-enkephalin analogs, 2:151–153
　　TRH analogs, 2:153
　prodrugs *vs.*, 2:76
　receptor-based transient-anchor compounds, 2:160–161
　site-targeting index and targeting enhancement factors, 2:140–141
　zidovudine, 2:141–142
research background, 2:73–74
soft drugs:
　activated compounds, 2:128–129
　active metabolite-based compounds, 2:127–128
　analog compounds, 2:80, 110–111
　anesthetics, 2:125–126
　angiotensin converting enzyme inhibitors, 2:118–119
　antiarrhythmic agents, 2:114–117
　　ACC-9358 analogs, 2:114–115
　　amiodarone analogs, 2:115–117
　anticholinergics, 2:107–109
　antimicrobials, 2:111–114
　　L-carnitine esters, 2:113–114
　　cetylpyridinium analogs, 2:111–112
　　long-chain esters, betaine/choline, 2:114
　　quaternary analogs, 2:111–113
　benzodiazepine analogs, 2:124–125
　β₂-agonists, 2:103–104
　β-blockers, 2:80–88
　　adaprolol, 2:83–85
　　esmolol, 2:85
　　landiolol, 2:85–86
　　vasomolol, flestolol, et al, 2:86
　bufuralol analogs, 2:86–88

　calcineurin inhibitors, 2:119–121
　　cyclosporine analogs, 2:120–121
　　tacrolimus analogs, 2:121
　calcitriol analogs, 2:101–102
　calcium channel blockers, 2:126–127
　cannabinoids, 2:123–124
　computer-aided design, 2:130–133
　　candidate ranking, 2:131–132
　　hydrolytic liability, 2:132–133
　　structure generation, 2:131
　corticosteroids, 2:90–101
　　antedrug steroids, 2:99
　　etiprednol dicloacetate and analogs, 2:96–97
　　flucortin butyl, 2:97
　　17-furoate androstadienes, 2:99
　　glucocorticoid γ-lactones, 2:98–99
　　itrocinonide, 2:97–98
　　loteprednol etabonate and analogs, 2:90–96
　cytokine inhibitors, 2:121–122
　dihydrofolate reductase inhibitors, 2:119
　enzymatic hydrolysis, 2:77–80
　esmolol homo-metoprolol analog, 2:133–134
　estrogens, 2:102–103
　hard drugs *vs.*, 2:75
　inactive metabolite-based design, 2:80–81
　insecticides/pesticides, 2:106–107
　　chlorobenzilate, 2:106–107
　　malathion, 2:107
　matrix metalloproteinase inhibitors, 2:122–123
　opioid analgesics, 2:88–90
　pharmacological activity, 2:76–77
　prodrugs *vs.*, 2:75–76
　prosoft drugs, 2:129–130
　psychostimulants, 2:104–106
　serotonin receptor agonists, 2:117–118
　virtual library, 2:134–135
terminology, 2:75–76

Retrosynthetic analysis, retrometabolic drug design, **2:**75
Retroviruses:
  replication cycle, HIV/AIDS and, **7:**76–78
  reverse transcriptase, HIV reverse tranascriptase inhibitor targeting, **7:**140–141
Rev-erb compounds, basic principles, **4:**164
Reverse chemical genetics, combinatorial chemistry, **1:**298–299
Reverse cholesterol transport:
  high-density lipoproteins, **4:**333–334
  liver X receptors, **4:**147
Reversed-phase chromatography, quantitative structure-activity relationship, partition coefficients, hydrophobic interactions, **1:**13–14
Reverse phase high-performance liquid chromatography (RP-HPLC):
  combinatorial chemistry, **1:**346–347
  protein therapeutics, analytic development, **3:**319–320
Reverse small molecule design, recombinant DNA technology, **1:**545–546
Reverse transcriptase. *See* HIV reverse transcriptase inhibitors
Reversible hydrolysis, camptothecin lactone, **3:**452–453
Reversible inhibitors, stroke therapy:
  calpains, **8:**461–463
  caspase 3/9 inhibitors, **8:**466–467
  cathepsin B inhibitors, **8:**472–473
  non-cystein design, **8:**477–478
  hydroxamic acids, **8:**507
Reversible inhibitors of monoamine oxidase-A (RIMAS), structure-activity relationship and metabolism, **8:**252–253
Reyataz. *See* Atazanavir sulfate
Rhabdoid tumors, *N*-(4-hydroxyphenyl) retinamide (4-HPR) therapy, **5:**473
Rho-associated coiled-coil kinase inhibitors, glaucoma therapy, clinical trials, **5:**601–602
Rhodanines:
  Alzheimer's disease, tau-targeted therapy, **8:**428–429
  analogs, apicoplast targeting, **7:**665
Rhodopsin cycle, vitamin A chemistry, **5:**646–647
Rho/ROCK signaling pathway, brain tumor therapy, **6:**252–253
Ribavirin:
  combined pegylated interferon-α/R1479/1626 anti-HCV agent, **7:**198–199
  nucleoside inhibition, **5:**1044–1046
Riboflavin (vitamin B$_2$), **5:**673–676
  antitubercular agents, cofactor biosynthesis targeting, **7:**727
  chemistry, **5:**673
  deficiency, **5:**673
  hypervitaminosis riboflavin, **5:**673
  metabolism, **5:**673–677
  receptor-mediated endocytosis, **2:**453
  uptake and biochemistry, **5:**673–674
Ribosomal ambiguity (*Ram* state), streptomycin, RNA targets, **5:**966
Ribosomes:
  aminoglycoside pharmacology, **7:**420–423
  antibiotic targeting, **7:**457–464
  chloramphenicol, **7:**462–463
  fusidic acid, **7:**464
  lincosamides, **7:**460–462
  oxazolidinones, **7:**464
  research background, **7:**405–406
  streptogramins, **7:**457–460
  antitubercular agents, synthesis and degradation, **7:**725–726
  linezolid targeting, **7:**542–543
  macrolide pharmacology, **7:**440–441
  structure and function, **5:**964–965
  X-ray analysis of, **7:**405–406
Ribostamycin, chemical structure, **7:**419
Richmond-Sykes β-lactamase classification system, **7:**275–279
"Rich pharmacology," central nervous system drug design, **8:**4–5
Rickets, vitamin D deficiency, **5:**649–651, 655
Rifabutin:
  ADMET properties, **7:**749–750
  antitubercular agents, **7:**744–751
Rifampicin, antimalarial drugs combined with, **7:**621–624
Rifampin:
  antitubercular agents, **7:**744–751
  selective toxicity, **2:**557–558
Rifamycin antibacterials, antitubercular agents, **7:**744–751
  ADMET properties, **7:**748
  adverse effects, **7:**748–749
  drug interactions, **7:**749
  history, **7:**744–745
  mechanism of action, **7:**746–748
  resistance mechanisms, **7:**748
  structure-activity relationship, **7:**745–746
Rifapentine, antitubercular agents, **7:**744–751
  ADMET properties, **7:**750–751
Right shifter therapeutics, hemoglobin oxygen delivery, allosteric effectors, **4:**615–619
Rigid cluster normal mode analysis (RCNMA), *in silico* screening, protein flexibility, **2:**864–865
Rigid opiates, structure-activity relationship, **8:**598

RIKEN project, structural
genomics, 1:572
human proteome coverage,
1:577–579
protein family-based genome
coverage, 1:573–577
Rilpivirine HIV reverse
transcriptase
inhibitor, 7:151–152
Rimantadine, selective toxicity,
2:556–557
Rimexolone, glucocorticoid anti-
inflammatory
compound, C-17
alterations, 5:99–102
Rimonabant:
antipsychotic agents,
8:209–210
cognitive enhancement,
8:48–49
Ring-opening metathesis
polymerization
(ROMP) resins,
defined, 1:280
Ring systems:
aminoglycoside antibiotics,
7:429–432
amphetamines, aromatic ring
substitution,
8:108–109
atropisomers, 1:127–128
cephalosporin compounds,
7:306–310
chemokine receptor-5 (CCR5)
antagonists, anilide
core structure,
7:124–131
cocaine structure-activity
relationship,
8:112–113
epidermal growth factor
receptor signaling
inhibitors, 6:304–305
fusidic acid, 7:464
glucocorticoid biosynthesis and
metabolism, 5:58
antiglucocorticoids, 5:113
hypoxia-activated bioreductive
prodrugs, 6:119–120
β-lactams, structure-activity
relationship,
7:259–261
macrolides, 7:441–444,
448–456
morphine derivatives,
8:599–601
benzomorphans, 8:610–611

novobiocin Hsp90 inhibitor
optimization,
6:436–440
NS3/4A protease inhibitors,
ciluprevir, 7:179–181
opioid receptors:
δ-opioid receptor antagonist,
8:605–608
κ-opioid receptor agonists,
8:624–632
oxazolidinones, 7:544–547
RNA targets:
neomycin/gentamicin
aminoglycosides,
5:973–978
tetracyclines, 5:982–983
steroid structure, 5:2–4
tetracyclines, 7:407, 411–413
thyromimetic structure-
activity relationship:
heterocyclic rings, 4:210–212
prime ring modification,
4:221–222
ring-closure mechanisms,
4:212–213
tricyclic antidepressants,
8:249–251
RIP renin inhibitor, 4:243
Risk analysis:
acute coronary syndrome, 4:411
congestive heart failure
therapy, risk/liability
evaluation, 4:500
toxicophores, 2:325
Risperidone:
behavioral models, 8:172–173
cognitive enhancement,
8:37–40
FDA approval, 8:178
pharmacokinetics, 8:195–196
side effects, 8:169
structure-activity relationship,
8:182–185
RITA (bisthiphenefurane),
protein-protein
interactions,
2:349–352
Ritanserin, serotonin receptor
targeting, 8:241–242
Ritonavir (RTV):
chemical structure, 7:10
design criteria for, 7:3
lopinavir/ritonavir second-
generation compound,
7:31–36
optimization, 7:10–11
pharmacokinetics, 7:10, 12–13

protease inhibitor, 7:10–12
Rituximab, resistance
mechanisms, 6:369
Rivaroxaban, active site-directed
inhibitors, 4:398–400
Rivastigmine, cognitive
enhancement,
8:15–16, 31
RNA-dependent RNA polymerase
(RdRp), nucleoside
inhibitors, anti-HCV
agents, 7:195–201
RNA-induced silencing complex
(RISC), age-related
macular degeneration
therapy, 5:619–622
RNA polymerase, antitubercular
agents, rifamycin
targeting, 7:747–748
RNA targets:
antibiotic/antibacterial drugs:
chloramphenicol, 7:462–463
fusidic acid, 7:464
lincosamides, 7:460–462
oxazolidinones, 7:464
research background,
7:405–406
streptogramins, 7:457–460
antitubercular agents:
ribosome synthesis and
degradation,
7:725–726
rifamycins, 7:747–748
receptor-based drug design,
2:512
therapeutic agents:
aminoglycoside toxicity,
5:979–981
erythromycin and analogs,
5:984–990
future research issues,
5:993–996
linezolid and analogs,
5:990–993,
7:542–543
neomycin- and gentamicin-
type aminoglycosides,
5:971–979
research background, 5:963
ribosome structure and
function, 5:964–965
streptomycin and analogs,
5:965–971
tetracycline and analogs,
5:981–984
R-NN curves, outlier detection,
quantitative

structure-activity relationship, **1:**44–45
Ro5-3335 inhibitor, recombinant DNA technology, screening applications, **1:**546–547
Ro 24-7429 inhibitor, recombinant DNA technology, screening applications, **1:**546–547
Ro-25-6981, stroke therapy, **8:**453
Ro 46-2005, peptidomimetic design, **1:**227–229
RO0281675 compound, type 2 diabetes therapy, **4:**15–17
Robalzotan, efficacy assessment, **8:**225–229
Robotic systems, structural genomics research, **1:**581
Robustness, quantitative structure-activity relationship modeling, recent improvements, **1:**511
Roflumilast:
chronic obstructive pulmonary disease therapy, **5:**782–784
selective toxicity, **2:**567–568
Rolipram:
antidepressant efficacy assessment, **8:**226–229
antipsychotic agents, **8:**208–209
brain tumor therapy, **6:**265
chronic obstructive pulmonary disease therapy, **5:**783
memory assessment, **8:**28–30
Roller bottle cell cultures, protein therapeutics, **3:**297
Romidepsin, histone deacetylase inhibitors, **6:**70–72
Romiplostim, structure and properties, **4:**583–584
Rondomycin:
chemical structure, **7:**407
history and biosynthesis, **7:**410–413
stability, side effects, and contraindications, **7:**408–409

Roprinole, Parkinson's disease dopaminergic therapy, **8:**546–547
ROSETTALIGAND algorithm, allosteric protein sites, **1:**390
Rosiglitazone:
Alzheimer's disease, insulin related therapy, **8:**415–417
peroxisome proliferator-activator receptors: future drug discoveries, **4:**165–167
γ-agonists, **4:**102–104
selective PPARγ modulators, PPARγ partial agonists, **4:**107–115
Rosuvastatin/choline fenofibrate, anticholesterolemics, **2:**235
Rotamer libraries, *in silico* screening, protein flexibility, side-chain conformation, **2:**867–868
Rotational isomeric model:
chemokine receptor-5 (CCR5) antagonist development, **7:**115–116
peptides, **1:**209
Rotigotine, Parkinson's disease dopaminergic therapy, **8:**545–546
Route selection criteria, scale-up operations, **3:**3–4
Roxithromycin, chemical structure, **7:**438
ROY pharmaceutical intermediate, polymorphism, **3:**191–192
RPE cells, age-related macular degeneration, **5:**604–610
RU-486. *See* Mifeprestone
RU-44790, monobactam activity, **7:**331–332
Rubidomycin. *See* Daunorubicin
Rubitecan, biological activity and side effects, **6:**97–98
Rufinamide, anticonvulsant applications, **8:**136–139, 154–155
Rule of five (Lipinski):

ADMET properties assessment, **2:**63–64
receptor-based drug design, **2:**524–525
natural products lead generation, **2:**202
oral drug physicochemistry, permeability, **3:**34–35
RNA targeted antibiotics, **5:**993–996
small druggable molecules, **1:**295–296
Rule-of-one metabolism, prodrug design, **3:**248–250
Rule of three compliance, nuclear magnetic resonance screening procedures, ligand properties, **2:**432–433
RVX-208 apolipoprotein target, Alzheimer's disease therapy, **8:**413–414
RWJ-442831 cephalosporin, **7:**312–314

S-3578 cephalosporin, **7:**313–314
S32006 serotonin receptor antagonist, **8:**242–244
Sabarubicin, topoisomerase II inhibitors, **6:**100–102
S-adenosyl-L-methionine (SAM), drug metabolism, **1:**423–425
Safe harbor provisions, infringement of patent claims, **3:**156–157
Safety catch linkers, defined, **1:**280
Safety issues:
α-glycoside inhibitors, **4:**7–8
biological fingerprinting, **1:**499–501
clopidogrel, **4:**432
commercial-scale operations, **3:**13
metformin therapy, **4:**5
peroxisome proliferator-activated receptors, **4:**145
pilot plant development, **3:**10
Safinamide, Parkinson's disease, dopaminergic therapy, **8:**549–551
Safratoxin peptides, amino acid sequencing, **4:**523–526

Salbutamol, chronic obstructive pulmonary disease bronchodilators, **5:**769–770
Salicyluric acid, drug metabolism, conjugation reactions, **1:**433–434
Salinosporamide A:
 marine microbial sources, **2:**248
 proteasome inhibitor, clinical trials, **6:**179–180
Salmeterol, chronic obstructive pulmonary disease bronchodilators, **5:**770–771, 777
Salt compounds:
 crystal engineering:
  defined, **3:**188–189
  monophosphate salt I cocrystals, **3:**212–213
 high-throughput screening, assay optimization, monovalent/divalent salts, **3:**412
 oral dosage systems, **3:**26, 28
  class II drugs, low solubility/high permeability, **3:**42, 44, 354
 screening and selection:
  acid-base equilibrium, **3:**384–388
  active pharmaceutical ingredients, **3:**388–396
  automated assisted screening, **3:**395–396
  basic properties, **3:**381–382
  counterion selection, **3:**389–390
  crystallization, **3:**383–384
  limitations, **3:**382–383
  manual process, **3:**391–395
  physicochemical properties, **3:**383
  selection technology, **3:**396–397
  solvent selection, **3:**389–391
Salvia divinorum, current developments, **8:**682–683
Sample handling and storage, combinatorial chemistry, **1:**349
Sancycline, history and biosynthesis, **7:**411–413

Sandimmune, soft drug design, **2:**120–121
SANDOCK technique, docking/scoring computations, point complementarity methods, **2:**603–604
Sanfetrinem, chemical structure, **7:**346–350
Sanofi-Aventis salt selection, DP$_1$ receptor antagonists, **5:**906–907
SANT-2 small molecule, brain tumor therapy, hedgehog signaling pathway, **6:**260–261
Saquinavir mesylate (SQV) protease inhibitor:
 binding site mechanisms, **7:**3, 7
 chemical structure, **7:**4–5
 efficacy and tolerability, **7:**8
 HIV/AIDS therapy, **7:**3, 5–10
 optimization strategies, **7:**5–7
 pharmacokinetics, **7:**7–9
 resistance profile, **7:**8, 10
 transition-state mimetic screening, **7:**5
SAR102279 receptor antagonist, **8:**257
Saracatinib, tumor metastases inhibition, **6:**325–327
Saralasin antagonist, structure and function, **4:**520
Sarasar®, farnesyltransferase inhibitors, **1:**322–323
Sarcoplasmic/endoplasmic reticulum calcium ATPase (SERCA) family, antimalarial drugs, **7:**658–660
Sarcosine, cognitive enhancement, **8:**45
Saredutant substance P-NK1 receptor antagonist, **8:**257
Sarizotan, Parkinson's disease therapy, **8:**554–555
Saturation curves, allosteric protein theory, **1:**371–373
Saturation transfer difference (STD), nuclear magnetic resonance studies, drug screening applications, **2:**424–426

Saxagliptin, type 2 diabetes therapy:
 physiology and pharmacology, **4:**43–45
 transition state-like covalent, reversible inhibitors, **4:**51–54
SB939 histone deacetylase inhibitor, structure and development, **6:**59–60
SB225002 CXCR2 receptor antagonist, chronic obstructive pulmonary disease, **5:**784–785
SB-269970 tool compound, BioPrint® database, **1:**491–493
SB-271046:
 cognitive function assessment, **8:**27–28
 serotonin receptor targeting, **8:**246–247
SB272844 CXCR2 receptor antagonist, chronic obstructive pulmonary disease, **5:**784–785
SB-705498 vanilloid receptor antagonist, migraine therapy, **8:**308–312
Scaffold hopping:
 functional analog design, **1:**175–177
 growth hormone secretagogue receptor antagonists, **1:**334
 prostaglandin/thromboxane EP$_1$ antagonists, **5:**837, 839
 virtual screening data, **2:**27–29
Scaffold structures:
 antimalarial cell cycle regulation, **7:**628–634
 falcipain inhibitors, **7:**635–637
 chemogenomics, privileged scaffolds, **2:**575–577
 combinatorial libraries, **1:**297
 defined, **1:**280
 Hsp90 inhibitors:
  novobiocin C-terminal inhibition, parallel library approach, **6:**432–433
  purine-based Hsp90 inhibitors, **6:**410–416

pyrazole-scaffold inhibitors, **6:**417–419
Janus Kinase 3 (JAK3) combinatorial chemistry, **1:**338–340
peptidomimetics, local constraints, **1:**214
peroxisome proliferator-activated receptors, selective PPARγ modulators, PPARγ partial agonists, **4:**111–115
poly(ADP-ribose)polymerase inhibitors, **6:**162–167
prodrug design, esmolol case study, **3:**263–268
protein-protein interactions, **2:**351–359
quantitative structure-activity relationship models, virtual screening applications, **1:**526–527
stroke therapy:
  gem-disubstituted and spirocyclic inhibitors, **8:**506
  piperidine/piperazine scaffolds, **8:**457–458
  pyrimidine-based, **8:**457
Scale-up operations, large-scale synthesis, **3:**1–14
  basic principles, **3:**1
  bench-scale experiments, **3:**4–7
  reaction solvent selection, **3:**5–6
  reaction temperature, **3:**6
  reaction times, **3:**6
  solid-state requirements, **3:**7
  stoichiometry and addition order, **3:**6–7
  chiral requirements, **3:**4
  commercial-scale operations, **3:**10–13
  chemical safety, **3:**13
  environmental controls, **3:**13
  in-process controls, **3:**11
  processing equipment requirements, **3:**11
  validation, **3:**12–13
  continuous microfluidic reactors, **3:**13–14
  pilot plant scale-up, **3:**7–10
  agitation, **3:**8–9
  drying and solid handling, **3:**9–10

heating and cooling, **3:**8
liquid-solid separations, **3:**9
safety issues, **3:**10
process analytical technology, **3:**14
process trends and technologies, **3:**13–14
route selection, **3:**3–4
synthetic strategies, **3:**1–4
Scanning-mutagenesis mapping, recombinant DNA technology, epitope mapping, **1:**544
Scatchard binding constants, chiral compounds, **1:**154–155
Scatchard plot, allosteric proteins, **1:**379–380
SCFscore algorithm, empirical scoring, **2:**628
SCH-351125 chemokine receptor-5 (CCR5) antagonists, development and discovery, **7:**110–116
SCH 503034 protease inhibitor. *See* Boceprevir anti-HCV agent
Sch527123 CXCR2 receptor antagonists, chronic obstructive pulmonary disease, **5:**786
SCH 530348 thrombin receptor antagonist, **4:**441
  side effects, **4:**452
Schaffer collateral-CA1 synapse, long-term potentiation, **8:**24–25
Schild plot regression, receptor-based drug design, occupancy theory, **2:**495–497
Schizophrenia:
  acetylcholinesterase inhibitor therapy, **8:**32
  antipsychotic therapy, **8:**169–170
  animal models, **8:**170–176
  behavioral models, **8:**172–173
  drug discrimination model, **8:**173
  genetic-based models, **8:**173–174
  cognitive deficits, **8:**162
  cognitive dysfunction in, **8:**20–21

dopamine therapy, cognitive enhancement, **8:**39–41
dopamin receptor subtypes, **8:**200–201
electroencephalography assessment, **8:**174–175
epidemiology, **8:**161–162
metabotropic glutamate receptor 5 combinatorial chemistry, **1:**318–320
multitarget drug development, dopamine D₂-plus, **1:**257–258
negative symptoms, **8:**161–162
nicotinic agonists, **8:**71
pathology, **8:**162–165
positive symptoms, **8:**161
receptor-based drug design, **8:**3
targeted drug design, functional genomics, 518-519
Schizophrenia, targeted drug design, functional genomics, 518-519
Schwannoma, classification and therapy, **6:**237–239
Schwannomatosis, classification and therapy, **6:**238–239
Scopolamine, muscarinic antagonist, **8:**72
SCORE1 function, empirical scoring, **2:**624–628
Scoring functions:
  drug discovery applications:
    basic principles and concepts, **2:**593–596
    benefits, **2:**643
    current research issues, **2:**632–641
    entropy effects, **2:**633
    flexibility, **2:**637–640
    hydrophobic effects, **2:**634–636
    isomerization and tautomerization, **2:**636–637
    limitations, **2:**642–643
    multiple solutions, **2:**641
    scoring techniques, **2:**599–632
      model scoring functions, **2:**618–632
      structural searching functions, **2:**601–618

Scoring functions (*Continued*)
    target structure requirements, **2**:596–598
    user guides, **2**:641–642
    quantum mechanics/molecular mechanics binding affinity, **2**:733–735
structure-based virtual screening, **2**:16–25
    comparison, consolidation and consensus, **2**:22–23
    empirical scoring, **2**:18
    enrichment factors, **2**:20–22
    force-field scoring, **2**:17–18
    hybrid/alternative scoring, **2**:19–20
    knowledge-based scoring, **2**:18–19
    liver X receptor modulator example, **2**:23–25
Screening libraries:
    natural products lead generation, **2**:200–204
    high-throughput screening, **2**:205–207
    phenotypic screening, **2**:207–208
    nuclear magnetic resonance studies, **2**:431–434
Screening procedures:
    functional analog design, **1**:175–177
    mass spectrometry, **1**:111–121
    multitarget drug development, **1**:252–253
    nuclear magnetic resonance:
        chemical-shift perturbation, **2**:419, 421–422
        ^{19}F-NMR, **2**:430
        fragment-based approach, **2**:432–433
        hardware and automation, **2**:434
        library design, **2**:431–434
        ligand efficiency, **2**:433
        magnetization transfer, **2**:424–426
        mixture design, **2**:433–434
        molecular diffusion, **2**:426–427
        NOE experiments, **2**:427–429
        overview, **2**:368, 419, 421
        relaxation, **2**:427–429
        screening methods evaluation, **2**:430–431
        spin labels, **2**:429–430
        structure-activity relationships, **2**:421–424
    reagents, recombinant DNA technology, **1**:546–547
    toxicophores, reactive metabolite formation, **2**:326–327
ScreenScore function, consensus scoring, **2**:632
Seagate doctrine, infringement of patent claims, **3**:158–159
Searle prostaglandin EP$_1$ receptor antagonists, **5**:813–815
Secondary meaning principle, trademark selection guidelines, **3**:173–174
Secondary metabolites:
    cytotoxic agents, research background, **6**:1–3
    genomic influences, **2**:248–250
    marine microbial sources, **2**:247–248
Secondary protein structure, protein therapeutics, biophysical assessment, **3**:324–325
Second-generation compounds:
    anticonvulsants, **8**:135–139
    anti-DNA virus antimetabolites, **7**:225–226
    cephalosporins, **7**:301–306
    HIV protease inhibitors, **7**:31–60
        atazanavir sulfate (Reyataz), **7**:36–42
        darunavir (Prezista), **7**:54–60
        fosamprenavir calcium (Lexiva/Telzir), **7**:42–45
        lopinavir/ritonavir (Kaletra), **7**:31–36
        tipranavir (Aptivus), **7**:45–54
    kinase inhibitors:
        drug resistance:
            conformational targeting, **6**:354–356
            P-glycoprotein inhibition, **6**:372
            small-molecule tyrosine kinase inhibitors, **6**:364–366
        purine-based Hsp90 inhibitors, **6**:413–416
        pyrazole-scaffold Hsp90 inhibitors, **6**:419
    migraine therapy, triptans, **8**:270
    peptidomimetics, **1**:217
    protein therapeutics, recombinant DNA technology, **1**:540
Second-line agents, antitubercular chemotherapeutics, **7**:760–776
Second mitochondrial-derived activator of caspases (Smac), peptidomimetics, XIAP inhibitors, **1**:233–240
α-Secretase:
    amyloid precursor protein processing, **8**:330–334
    chemical structure, **8**:334
    modulators, preclinical/clinical testing, **8**:357–358
β-Secretase. *See* BACE (β-site cleavage enzyme)
γ-Secretase:
    amyloid precursor protein processing, **8**:330–334
    chemical structure, **8**:333–334
    inhibitors (GSIs):
        clinical research, **8**:341–345
        preclinical research, **8**:334, 340–341
    modulators, preclinical/clinical testing, **8**:358–359
    nonpeptidomimetic inhibitors, **8**:337–340
    peptide isostere compounds, **8**:334–337
    azepine/diazepine-derived inhibitors, **8**:336–337
    transition-state analogs, **8**:334–335
Secretases. *See also* BACE (β-site cleavage enzyme)
    amyloid precursor protein processing, **8**:330–334
Sedation:
    antipsychotic side effects, **8**:168
    opioid receptors, **8**:575
Segmental labeling, receptor-based drug design, macromolecular compounds, **2**:393–394

Seizures:
  antipsychotic side effects, **8:**168
  bupropion and, **8:**234
Seleciclib cyclin-dependent kinase inhibitor, medicinal chemistry and classification, **6:**310–315
Selectins:
  hit seeking combinatorial libraries, **1:**341–343
  recombinant DNA technology, cellular adhesion proteins, **1:**561
Selective androgen receptor modulators (SARMs):
  osteoporosis anabolic therapy, **5:**760
  structure and properties, **5:**174–177
Selective estrogen receptor modulators (SERMs):
  breast cancer resistance, **6:**369–370
  current drugs, **5:**219–221
  discovery, **5:**243–244
  future research issues and information retrieval, **5:**256
  glucuronide conjugation, **5:**228–229
  nuclear hormone receptors, **4:**83
  osteoporosis antiresorptive therapy, **5:**734–742
  breast cancer, **5:**741–742
  clinical trials, **5:**740–741
  future research issues, **5:**742
  preclinical trials, **5:**739–740
  selective toxicity, **2:**566–567
  tissue specificity, **5:**240–242
Selective PPARα modulators (SPPARαMs):
  drug discovery and, **4:**164–167
  peroxisome proliferator-activator receptors:
  α-agonists, **4:**96–97
  PPARα partial agonists, **4:**97
Selective PPARγ modulators (SPPARγMs):
  drug discovery and, **4:**164–167
  PPARγ partial agonists, **4:**106–115
Selective serotonin reuptake inhibitors (SSRIs):
  multitarget drug development: optimization to SNRI, **1:**267–268
  SERT-plus for depression, **1:**254–257
selective toxicity, **2:**569
Selective somatostatin receptor subtype 5 antagonists, chemogenomics, ligand-based data analysis, **2:**584–585
Selective toxicity:
  antibiotics, **2:**548–553
  antifungal drugs, **2:**558–559
  antihistamines, **2:**564–566
  antimycobacterial drugs, **2:**557–558
  antivirals, **2:**553–557
  bisphosphonates, **2:**564
  comparative biochemistry, **2:**545
  comparative cytology, **2:**545–546
  comparative distribution, **2:**545
  comparative stereochemistry, **2:**546–547
  future research issues, **2:**569
  levobpuivacaine, **2:**564
  monoamine oxidase inhibitors, **2:**563–564
  monoclonal antibodies, **2:**559–562
  nonbenzodiazepine sedatives, **2:**568
  phosphodiesterase inhibitors, **2:**567–568
  proton pump inhibitors, **2:**566
  research background, **2:**541
  selective estrogen receptor modulators, **2:**566–567
  selective serotonin reuptake inhibitors, **2:**569
  therapeutic agent categorization, **2:**541–544
  topical antimicrobial agents, **7:**484, 488
  tyrosine kinase inhibitors, **2:**562–563
Selectivity. *See also* Functional selectivity
  broad profiling data, **1:**499–501
  dipeptidyl peptidase 4 inhibitors, structural biology, **4:**47–48
  G-protein coupled receptor, homology modeling, **2:**292–293
  Hsp90 inhibitors, tumor cell selectivity, **6:**385
  poly(ADP-ribose)polymerase (PARP) inhibitors, **6:**167–168
  receptor-based drug design, structure-activity relationships, **2:**520
  retinoid nuclear receptors, **5:**373–376
  selective toxicity principles, **2:**545–547
  sulfonamides and sulfones, **7:**527
  thyromimetics, **4:**205–206
  xenobiotic metabolism, **1:**406–407
Selegiline:
  Parkinson's disease, dopaminergic therapy, **8:**549–551
  structure-activity relationship and metabolism, **8:**251–253
Seletracetam, anticonvulsant applications, **8:**150–151
SELEX (*in vitro* selection), recombinant DNA technology, **1:**549–550
Self-assembled low molecular weight amphiphiles, drug delivery systems, **3:**469–473
  polymerization, **3:**477
Self-assembled monolayers for matrix-assisted laser desorption ionization (SAMDI), drug screening, **1:**116
Self-emulsifying drug delivery systems (SEDDS), oral drug physicochemistry, class II drugs, **3:**49–50
Self-methylation:
  aminoglycoside resistance, **7:**424–427
  macrolide resistance, **7:**445–446
  streptogramin resistance, **7:**460

Self-micro-emulsifying drug delivery system (SMEDDS), oral drug physicochemistry, class II drugs, **3:**49–50
Self-organizing map (SOM), ligand-based virtual screening, quantitative structure-activity relationships, **2:**11–13
Selfotel, stroke therapy, **8:**451
Semagestat γ-secretase inhibitor, clinical testing, **8:**341–345
Semantic memory, **8:**19
Semicarbazone compounds, thromboxane A_2 receptor antagonists, **5:**886–889
Semiempirical techniques:
 model scoring, **2:**622–623
 quantum chemical methods, quantitative structure-activity relationships, **1:**23
 scoring functions, **2:**645
Sensitivity, high-throughput screening, enzymatic assays, **3:**406
Sensitization mechanisms, psychostimulant physiology and pharmacology, **8:**105–106
Separation techniques, combinatorial chemistry, **1:**345–347
Septic shock, carbon monoxide inhalation therapy, **5:**301–302
Sequence-based phylogeny (SBP), chemogenomics, ligand-based data analysis, **2:**585
Sequence-independent phylogeny (SIP), chemogenomics, ligand-based data analysis, **2:**585
Sequential metabolism, brain-targeting chemical delivery systems, **2:**136–140
Sequential modeling, allosteric proteins, **1:**376–377
Sequential multicolumn chromatography (SMCC), protein therapeutics, **3:**310–311
Sercloremine, structure-activity relationship and metabolism, **8:**252–253
Serdaxin antidepressant, **8:**255–256
Sergliflozin etabonate, type 2 diabetes therapies, **4:**14
Serine β-lactamases:
 inhibitors, classification, **7:**359–360
 mechanisms of, **7:**274–277
D-Serine, cognitive enhancement, **8:**45
Serine proteases:
 age-related macular degeneration, **5:**607–610
 hepatitis C viral expression, **7:**172
 hit seeking combinatorial libraries, **1:**332–333
Serine/threonine protein kinases (STPK), antitubercular agent targeting, **7:**732–733
Serine trap technology, teleprevir anti-HCV agent, **7:**181–186
Serotonergic agents:
 current and future research, **8:**255–256
 receptor-based drug design, **8:**241–247
 structure-activity relationship and metabolism, **8:**253–254
Serotonin-dopamine antagonists, FDA approved compounds, **8:**177–179
Serotonin/norepinephrine-selective reuptake inhibitors (SNRIs):
 chemical structure, **8:**223
 historical background, **8:**249
 migraine prevention, **8:**319
 monoamine transporter inhibition, **8:**238–241
 side effects and adverse reactions, **8:**223, 234
Serotonin receptors:

5-HT_{1A} receptors:
 antidepressant targeting, **8:**242–244
 cognitive enhancement, **8:**37–38
 Parkinson's disease therapy, **8:**553–555
5-HT_{1B} receptors, **8:**245–247, 253–254
5-$HT_{1D/1B}$ receptors, **8:**244–247, 253–254
 migraine therapy, **8:**269–272
5-HT_{2A}, receptor combinatorial library, **1:**311–312
5-HT_{2A} receptors, cognitive enhancement, **8:**38–40
5-HT_{2C} receptors, **8:**243–244, 253–254
5-HT_2 receptors, **8:**241–242, 253–254
5-HT_3 receptors, **8:**246–247, 253–254
5-HT_4 receptors, cognitive enhancement, **8:**38–40
5-HT_6 receptors, cognitive enhancement, **8:**39–40
5-HT_7 receptors, **8:**247
 cognitive enhancement, **8:**39–40
 agonists, migraine therapy, **8:**268–272
 anti-inflammatory/vasoconstriction mechanisms, **8:**270–272
 second-generation triptans, **8:**270
 sumatriptan, **8:**268–270
agonists, soft drug design, **2:**117–118
antagonists, antidepressant design, **8:**225–229
antidepressants:
 current and future research, **8:**255–256
 efficacy assessment, **8:**221–229
 neurotransmission enhancement, **8:**220–221
 populations and subtypes, **8:**241–247
antipsychotic structure-activity relationship, **8:**180–185
 receptor subtypes, **8:**201–202
BioPrint profile, **1:**491–493

chemotype evaluation, 1:496–497
cognitive enhancement, 8:37–40
insomnia therapy, antagonists and inverse agonists, 5:718
migraine therapy:
  agonists, 8:268–272
  anti-inflammatory/vasoconstriction mechanisms, 8:270–272
  second-generation triptans, 8:270
  sumatriptan, 8:268–270
NOS inhibitor binding affinity, 8:305–306
Parkinson's disease therapy, 8:553–555
platelet activation, 4:414
psychostimulant physiology and pharmacology, 8:103–106
schizophrenia pathology, 8:164–165
SERT-plus multitarget drug development, 1:254–257
Serotonin-selective reuptake inhibitors (SSRIs):
  chemical structure, 8:223
  dosing regimens, 8:219–221
  historical background, 8:249
  migraine prevention, 8:319
  monoamine transporter inhibition, 8:238–241
  pharmacokinetics, 8:236
  serotonin receptor targeting, 8:244–247
  side effects and adverse reactions, 8:223–224, 232–234
  structure-activity relationship and metabolism, 8:249–251
Serotonin transporter (SERT) gene, antidepressant applications, 8:229
  pharmacology, 8:238–241
Sertindole:
  cognitive enhancement, 8:37–40
  FDA approval, 8:179–180
  pharmacokinetics, 8:199
  structure-activity relationship, 8:182–185

Sertraline:
  chemical structure, 8:223
  historical background, 8:249
  pharmacokinetics, 8:236
  selective toxicity, 2:569
  structure-activity relationship, 8:251
Serum albumin:
  androgen absorption and distribution, 5:158–159
  chemical properties, 3:437–438
  heme-based oxygen delivery, 4:642–643
  therapeutic applications, 3:456–464
    clofibrate pharmacokinetic modulation, 3:457–464
    pharmacokinetic studies, 3:462–464
    $in\ vitro$ studies, 3:458–462
  X-ray structural analysis, 3:438–456
    anthracyclines, 3:452–454
    bilirubin binding, 3:442–451
    CADEX categories, 3:439–440
    camptothecins, 3:451–453
    drug-binding sites, 3:440–442
    podophyllotoxin derivatives, 3:454–456
Serum profiles:
  N-(4-hydroxyphenyl) retinamide (4-HPR), 5:478–479
  vitamin intake, 5:643
Serum protein binding (SPB), quantitative structure-activity relationship, 1:66
Sesquiterpenes, Leishmaniasis therapy, 7:592–593
Set shifting, executive function assessment, 8:28–30
Sex hormone binding globulin (SHBG), androgen absorption and distribution, 5:158–159
Sex hormones. See also Estrogens; specific hormones, e.g. Androgens
  female sex hormones:
    basic properties, 5:219
    biosynthesis, 5:230–233

aromatase inhibitors, 5:231–232
equine estrogen genesis, 5:232–233
steroid aromatization, 5:230–231
clinical applications:
  adverse effects and precautions, 5:221
  current drugs, 5:219–221
drug metabolism:
  conjugated equine estrogens, 5:223–224
  diethylstilbestrol biotransformation, 5:226
  electrophilic (carcinogenic) estrogen receptor ligand metabolites, 5:226–228
  estradiol 17α-ethynyl analogs, 5:222–224
  glucuronide conjugation, 5:228–229
  oxidation and reduction, 5:221–222
  triarylethylene oxidative metabolism, 5:225–226
  $in\ vivo$ reversible monoesters, 5:224
future research issues, 5:256
information retrieval concerning, 5:256
physiology and pharmacology, 5:233–243
  environmental/dietary estrogens and progestins, 5:239–240
  estrogen activity assessment, 5:236–237
  estrogen receptors:
    endocrinology, 5:236
    topography, 5:233–236
    ubiquitous isoform distribution, 5:236
  ovulation modulation, 5:242–243
  progesterone receptors, 5:237–238
    ligand activity, 5:238
  progestin/antiprogestin effects, 5:238–239
  tissue-selective estrogens/progestins, 5:240–242

Sex hormones (*Continued*)
  progesterone
    biotransformation and analogs, **5:**229–230
  selective estrogen receptor modulators, **5:**243–244
  structure-activity relationships:
    diarylethylene estrogen antagonists, **5:**248–249
    estrogen receptor affinity, **5:**246
    nonsteroidal estrogen receptor ligands, **5:**246–247
    progesterone nonsteroidal PR ligands, **5:**255–256
    progesterone steroidal PR ligands, **5:**253–255
    steroidal analogs, **5:**244–245
      7α-substitution, **5:**245
      11β-substitution, **5:**245
      estrone-equilin comparisons, **5:**246
      nonbenzenoid seteroid (-like) estrogen receptor ligands, **5:**246
    triarylethylenes, **5:**249–253
    triarylheterocycles, **5:**253
    vicinal diarylethylenes, **5:**247–248
male sex hormones:
  absorption and distribution, **5:**158–159
  anabolic agents:
    absorption, distribution, and metabolism, **5:**186–188
    abuse of, **5:**188–189
    current marketed compounds, **5:**176–177
    structure-activity relationships, **5:**179–186
      alkylated analogs, **5:**181–182
      dehydro derivatives, **5:**180
      deoxy/heterocyclic-fused analogs, **5:**183–185
      esters and ethers, **5:**185–186
      hydroxy and mercapto derivatives, **5:**182–183
      nor derivatives, **5:**179
      oxo, thia, and azo derivatives, **5:**183
    therapeutic applications and bioassays, **5:**178–179
    toxicities, **5:**188
  antagonists:
    antiandrogens, **5:**190–193
    current marketed compounds, **5:**189–190
    enzyme inhibitors, **5:**193–200
  basic properties, **5:**153
  biosynthesis, **5:**22–25, 155–158
  distribution, **5:**159–164
  future research issues, **5:**200–201
  mechanism of action, **5:**164–168
  metabolism, **5:**26–27, 159–164
    oxidative metabolism, **5:**162–164
    reductive metabolism, **5:**159–162
  nonsteroidal androgens, **5:**174–177
    absorption, metabolism, and distribution, **5:**175–177
    toxicities, **5:**176
  occurrence and physiological roles, **5:**154–155
  receptor structure, **5:**164–168
  research background, **5:**153–154
  steroidal androgens:
    halo derivatives, **5:**171–172
    methylated derivatives, **5:**171
    structure-activity relationships, **5:**170–174
    synthetic compounds, **5:**168–176
      current marketed compounds, **5:**168–169
      therapeutic applications and bioassays, **5:**169–170
Sexual dysfunction:
  antidepressants, **8:**233–234
  antipsychotic side effects, **8:**169
Shake-flask measurements, quantitative structure-activity relationship, hydrophobic interactions, partition coefficients, **1:**12
Shape analysis:
  oral drug delivery, sustained release dosage forms, **2:**796
  plasma drug profile:
    diffusion-based drug release, **2:**803–805
    erosion-controlled drug release, **2:**819–822
Shepherdin Hsp90 inhibitor, **6:**416
SHetA2 retinoid-related compound:
  adverse effects, **5:**505–506
  chemical structure, **5:**500–501
  pharmacology and metabolism, **5:**506
  therapeutic applications, **5:**502–505
Short bowel syndrome (SBS), chirality and, **1:**146–147
Short-term memory, **8:**18–19
Sibutramine, monoamine transporter inhibition, **8:**239–241
Side chain modification:
  amphetamines, **8:**108
  antimalarial drugs, chloroquine, **7:**609–610
  docking and scoring techniques, **2:**640
  hydroxamate HDAC inhibitors, stroke therapy, **8:**488
  novobiocin Hsp90 inhibitor, **6:**434–436
  oxazolidinones, **7:**547–548
  peptide conformation, local constraints, **1:**211–214
  psychostimulant structure-activity relationship, **8:**107
  in silico screening, protein-ligand interactions, **2:**867–868

steroid structure, **5**:3–4
  cleavage mechanisms,
    5:10–11
  thyroid hormone receptor β-
    selective agonists,
    structure-based
    design, **2**:687–688
Side effects. *See also* ADMET
  (absorption,
  distribution,
  metabolism, excretion,
  and toxicity); Adverse
  drug reactions
  aminoclycosides, **7**:419–420
  antidepressants, **8**:229–234
    bupropion, **8**:234
    monoamine oxidase
      inhibitors, **8**:230
    nefazodone, **8**:234–235
    reboxetine, **8**:235
    serotonin/norepinephrine
      reuptake inhibitors,
      8:234
    serotonin-selective reuptake
      inhibitors, **8**:232–234
    trazodone, **8**:234
    tricyclic antidepressants,
      8:231–232
  antiplatelet agents, **4**:451–453
  antipsychotics, **8**:165–169
    animal models, **8**:175–176
    anticholinergic effects,
      8:167–168
    antipsychotic
      discontinuation, **8**:169
    cardiovascular effects, **8**:167
    extrapyramidal symptoms,
      8:165–166
    hematologic effects,
      8:168–169
    hepatic effects, **8**:169
    metabolic effects, **8**:168
    neuroleptic malignant
      syndrome, **8**:167
    sedation, **8**:168
    seizures, **8**:168
    sexual dysfunction, **8**:169
  chemokine receptor CXCR4
    antagonist-plerixafor,
    4:587
  dipeptidyl peptidase 4
    inhibitors, **4**:45
  DNA-targeted compounds:
    antibody-directed enzyme
      prodrug therapy,
      6:123
    antifolates, **6**:108–109
    bleomycin, **6**:8
    cyclopropylindoles, **6**:92–93
    dactinomycin, **6**:4
    mitomycin, **6**:13
    mustards, **6**:87
    nitrosoureas, **6**:94
    platinum complexes, **6**:91
    plicamycin, **6**:16
    pyrimidine antimetabolite
      analogs, **6**:110–112
    triazines, **6**:95
  docetaxel/taxotere, **6**:34–35
  dual topoisomerase I/II
    inhibitors, **6**:104–105
  enzyme inhibitors:
    daunorubicin, **6**:19–20
    doxorubicin, **6**:21–22
  erythropoietin, **4**:572–573
  gene-directed enzyme prodrug
    therapy, **6**:124
  granulocyte colony stimulating
    factor, **4**:578
  granulocyte-macrophage
    colony stimulating
    factor, **4**:575
  hypoxia-activated bioreductive
    prodrugs, **6**:118–120
  immunosuppressive drugs,
    cyclosporin A,
    5:1012–1013
  β-lactam antibacterial drugs,
    7:282–284
  L-dopa therapy, Parkinson's
    disease, **8**:542
  macrolides, **7**:438
  opioid receptors, **8**:573–576
  paclitaxel/taxol, **6**:33
  peroxisome proliferator-
    activator receptors, γ-
    agonists, **4**:104–106
  psychostimulant therapy, **8**:92,
    95–97
  stem cell factor, **4**:582
  tetracycline antibiotics,
    7:407–409
  thrombopoietins, **4**:584–586
  topoisomerase inhibitors, Topo
    I inhibitors, **6**:97–98
  topotecan, **6**:26–27
  vinblastine, **6**:38–39
  vinorelbine, **6**:40
Side population, cancer stem cells,
  multidrug resistance
  and, **6**:376–377
Sigma factors, antitubercular
  agent targeting, **7**:732
Signaling molecules:
  glucocorticoid anti-
    inflammatory
    compounds, **5**:63–65
  peptide libraries, **1**:288–291
  receptor-based drug design, G-
    protein coupled
    receptors, **2**:508–509
Signaling pathways:
  central nervous system tumors,
    6:227–228
  growth factors:
    brain tumor therapy, **6**:245
    kinase inhibitors, **6**:301–307
  hedgehog signaling pathway,
    brain tumor therapy,
    6:259–263
  hypoxia-induced pathway,
    6:253–255
  kinase inhibitors:
    growth factors, **6**:301–307
    survival signaling inhibitors,
      6:321–325
      insulin-like growth factor-1
        receptor inhibitors,
        6:321
      mTOR inhibitors,
        6:324–325
      phosphoinoside-3 kinase
        inhibitors, **6**:321–323
      protein kinase B inhibitors,
        6:323–324
  monoclonal antibodies,
    resistance
    mechanisms, **6**:369
  mTOR signaling pathway:
    brain cancer therapy,
      6:250–252
    cranial and spinal nerve
      tumors, **6**:238–239
  p53 protein, brain cancer
    therapy, **6**:256–257
  PI3K/AKT pathway, brain
    tumor therapy,
    6:249–250
  RAS/RAF/MEK/ERK signaling,
    brain tumor therapy,
    6:252
  Rho/ROCK and MAPK
    signaling, brain tumor
    therapy, **6**:252–253
  Wnt signaling pathway,
    osteoporosis anabolic
    therapy, **5**:755–759
    bone remodeling, **5**:756–757
    canonical pathway, **5**:756
    genetic models and bone
      target cells, **5**:757–758

Signaling pathways (*Continued*)
  therapeutics development,
    5:758–759
Signal peptide peptidases (SPP)
    family, antimalarial
    agents, **7:**668–672
Signal transduction:
  antitubercular agent targeting,
    7:732–733
  CP-690,550
    immunosuppressive
    compound,
    5:1055–1057
  opioid receptors, **8:**580–581
  recombinant DNA technology,
    immunosuupressive
    drug targeting,
    1:555–556
Silanediols, HDAC inhibitors,
    stroke therapy, **8:**494
Sildenafil:
  chirality in, **1:**149–150
  cocrystal engineering, **3:**211
  erectile dysfunction therapy:
    chemical structure, **5:**711
    structure-activity
      relationships,
      5:712–713
  hydrogen sulfide releasing
    molecules,
    5:314–315
  pulmonary hypertension
    therapy,
    5:715–716
  selective toxicity,
    2:567–568
  structural analog, **1:**167–168
Silicon, drug metabolism,
    oxidation/reduction
    reactions, **1:**420–421
Silver salts, topical antimicrobial
    agents, **7:**497–498
Similarity descriptors:
  chemogenomics, ligand-based
    data analysis,
    2:584–586
  virtual screening, scaffold
    hopping, **2:**29
Simulated annealing, docking
    and scoring
    techniques, Monte
    Carlo simulation,
    2:608–611
Simulated moving bed (SMB)
    techniques, protein
    therapeutics,
    3:310–311

Simvastatin:
  Alzheimer's disease therapy,
    preclinical/clinical
    testing, **8:**359–361
  anticholesterolemics, **2:**233
    niacin combination, **2:**234
  cholesterol biosynthesis,
    5:15–16
  LDL cholesterol lowering
    mechanisms, **4:**307
SIN-1, nitric oxide donors and
    prodrugs, **5:**285–286
Single dose drug delivery:
  intravenous administration,
    2:751–754
  oral administration,
    immediate-release
    dosage, **2:**772–776
  plasma drug profile:
    diffusion-based drug release,
      2:803–808
    erosion-controlled drug
      release, **2:**819–822
Single gene pharmacogenetics,
    disease progression,
    1:193–196
Single-molecule conformers,
    chirality, **1:**136–137
Single nucleotide polymorphisms
    (SNPs):
  age-related macular
    degeneration,
    5:605–607
  classification, **1:**182–183
  clinical relevance, **1:**196
  defined, **1:**182
  historical background,
    1:181–182
  nuclear hormone receptors,
    drug discovery,
    4:165–167
  patent requirements,
    3:120–123
  pharmacogenomics, **1:**183–196
    care quality improvement,
      1:199–200
    clinical trials, **1:**197–198
    complexity/cost issues, **1:**199
    drug metabolizing enzymes,
      1:183–189
    cytochrome p450 system,
      1:183–187
    dihydropyrimidine
      dehydrogenase,
      1:187–188
    N-acetyltransferase-2,
      1:188

    thiopurine
      methyltransferase,
      1:188–189
    drug transporter genes,
      1:189–190
    ethical issues, **1:**199–200
    literature sources, **1:**198
    single gene studies, disease
      progression,
      1:193–196
    target genes and clinical
      efficacy, **1:**190–193
Single-objective approach, drug
    discovery, **2:**262–263
Single-photon emission computed
    tomography (SPECT):
  congestive heart failure
    therapy:
    disease biomarkers,
      4:505–506
    predictive ligand biomarkers,
      4:501
  drug discovery and
    development
    applications:
    disease markers,
      radiolabeling of,
      8:744–746
    imaging studies, **8:**741–742
    radiolabeled drugs,
      8:742–743
    research background,
      8:737
    small animal imaging, **8:**742
    targeted drugs, radiolabeled
      ligands, **8:**743–744
    isotopes in, **8:**739–740
  Parkinson's disease diagosis,
    8:555–556
Single-strand break (SSB) repair,
    poly(ADP-ribose)
    polymerase-1 (PARP-
    1), **6:**160–162
Sirolimus, **5:**1025–1037
  ADMET properties,
    5:1034–1035
  chemical structure,
    5:1024–1025
  drug-eluting coronary stents,
    5:1037
  FKBP12-sirolimus-FRB
    ternary structure,
    5:1029–1030
  mechanism of action,
    5:1025–1029
  non-CYP450 metabolites,
    5:1034

organ transplantation applications, 5:1035–1037
structure-activity relationships, 5:1030–1034
cyclohexyl region modifications, 5:1033–1034
triene region modifications, 5:1030–1033
SIRT1 activators, type 2 diabetes therapy, 4:25–27
Sisomicin, chemical structure, 7:419
Sitafloxacin, antibacterial agents, 7:535–540
Sitagliptin, type 2 diabetes therapy:
ADMET properties, 4:45–47
physiology and pharmacology, 4:43–45
Site-directed chemical cleavage, transport proteins, 2:461
Site-directed mutagenesis:
G-protein coupled receptor, homology modeling, 2:288–289
opioid receptors, 8:589–590
receptor-based drug design, 2:514
recombinant DNA technology, 1:539–546
antibody-based therapeutics, 1:543
drug efficacy and personalized medicine, 1:540–543
epitope mapping, 1:543–544
future trends, 1:545–546
second-generation protein therapeutics, 1:540–543
Site-exposure enhancement factor (SEF), chemical delivery systems, 2:140–141
SITESORTER database, chemogenomics, structure-based data analysis predictive modeling, 2:585–588
Site-specific enzyme-targeted chemical delivery systems, eye-targeted drug design, 2:155–160
Site-targeting index, chemical delivery systems, 2:140–141
Six-dimensional molecules, quantitative structure-activity relationship, 1:43
Size-based nanometer filtration, mammalian protein purification, 3:309
Size exclusion chromatography (SEC), protein therapeutics, 3:319
SJ749 antagonist, brain cancer therapy, hypoxia-inducible factor pathway, 6:254–255
Skeletal muscle, thyroid hormone effects, 4:196
SKF-81297, cognitive enhancement, 8:42
SKF-94901 thyromimetic, structure-activity relationship, 4:203
Skin:
percutaneous absorption, glucocorticoid anti-inflammatory compounds, 5:46–51
thyroid hormone effects, 4:197
transdermal drug delivery, 2:830
Skin disease:
cancer, 13-*cis*-retinoic acid therapy, 5:392
noncancerous:
acitretin therapy, 5:399–400
bexarotene therapy, 5:448–449
etretinate therapy, 5:401–402
13-*cis*-retinoic acid therapy, 5:393
tazarotene therapy, 5:405–406
SLCO1B1, single nucleotide polymorphisms, 1:189
Sleep deprivation, cognitive enhancement, 8:51–52
Sleeping sickness. See Human African trypanosomiasis
structural genomics, fragment cocrystallography for drug development, 1:587–588
SLIDE algorithm, docking and scoring computations, 2:608
empirical scoring, 2:626–628
water conservation, 2:635–636
Slow exchange conditions, nuclear magnetic resonance, macromolecule-ligand interactions, 2:396–398
SM-18400 tricyclic derivative, stroke therapy, 8:452–453
Smallest half-volume (SHV) method, outlier detection, quantitative structure-activity relationship, 1:44
Small heterodimer partner (SHP), AHPN targeted drug development, 5:487–489
Small interfering RNA (siRNA):
age-related macular degeneration therapy, 5:620–622
recombinant DNA technology, 1:550
Small-molecule compounds:
age-related macular degeneration therapy, 5:617–622
β-amyloid peptide inhibitors, 8:362–365
antibiotic selective toxicity and, 2:553
brain tumor therapy:
atrasentan, 6:268
hedgehog signaling pathway, 6:260–263
HGF/MET compounds, 6:247–248
hypoxia-inducible factor pathway, 6:254–255
MDM2/MDMX, 6:257–258
PI3K/AKT pathway, 6:249–250
cancer therapy:
targeted chemotherapy, resistance and, 6:363–370

Small-molecule compounds (*Continued*)
  tumor metastases, c-Met kinase inhibitors, 6:328–330
  tyrosine kinase inhibitors, resistance mechanisms, 6:364–366
chemogenomics:
  chemical principles, 2:574
  yeast three-hybrid screening, 2:581–582
chemokine receptor-5 (CCR5) antagonists, HIV-1 entry blockade:
  anilide core, 7:124–131
  binding mode and mechanism, 7:133–134
  piperazine/piperidine-based antagonists, 7:110–116
  research background, 7:107–110
  spiro diketopiperazine-based antagonists, 7:131–133
  summary and future research issues, 7:134–135
  tri-substituted pyrrolidine antagonists, 7:121–124
  tropane-based antagonists, 7:116–121
chronic obstructive pulmonary disease therapy, ne773utrophil elastase inhibitors
cognitive enhancement, 8:30–51
  adenosine modulator, 8:46–47
  cannabinoids, 8:47–48
  cholinesterase inhibitors, 8:30–32
  dopamine, 8:39–40
  GABA$_A$ receptor, 8:45–46
  glutamate, 8:41–45
  histamine, 8:35–37
  historical background, 8:30–31
  muscarinics, 8:32–33
  neuronal nicotinic agonists, 8:34–35
  neuropeptides, 8:48–50

neurotropic agents, 8:47
NMDA modulators, 8:33–34
norepinephrine, 8:41
phosphodiesterase inhibitors, 8:50–51
serotonin, 8:37–39
smart drugs, 8:51–52
summary and future research issues, 8:52–53
druggable molecule libraries:
  benzodiazepines, 1:300–302
  solid- and solution-phase synthesis, 1:294–343
glaucoma therapy, clinical trials, 5:600–602
Hsp90 inhibitors, 6:427–429
  high-throughput screening identification, 6:417
kinesin spindle protein inhibitors, crystallographic analysis, 6:192–193
MDM2/MDM4 antagonists, 2:356–359
multiple sclerosis therapy:
  opportunistic approaches, 5:573–574
  small-molecule VLA-4 antagonists, 5:578–579
nuclear magnetic resonance and, 2:368
Parkinson's disease dopamine receptor agonists, 8:547
protein kinase drugs, structure-based design, 6:345–346
protein therapeutics and, 3:290–291
quantitative structure-activity relationship modeling, 1:513–514
recombinant DNA technology, reverse small molecule design, 1:545–546
renin inhibitor development, 4:243
  peptide/reduced-peptide inhibitors, 4:243
stroke therapy, N-type voltage-dependent calcium channels, 8:456–459
structural genomics, family-based genome coverage, 1:573–577

thrombin receptor antagonists, 4:442–444
type 2 diabetes therapy:
  glucokinase activators, 4:15–17
  GPR-119 agonists, 4:21
  11β-hydroxysteroid dehydrogenase-1 inhibitors, 4:23–25
Small-molecule growth (SMoG96) algorithm, knowledge-based scoring, 2:629–631
Smart drugs, cognitive enhancement, 8:51–52
SMILES notation, quantitative structure-activity relationship:
  DCW descriptor, 1:26–27
  partition coefficient calculations, 1:14–19
Smoking, Parkinson's disease and, 8:538–540
Smoking cessation, drug development and discovery:
  antinicotine vaccines, 5:725–726
  overview, 5:711, 721, 723
  varenicline, 5:723–725
SNC80 δ-opioid selective agonist, 8:632–635
S-nitrosohemoglobin (SNO-Hb):
  conjugated hemoglobins, 4:636–637
  formation of, 4:612–614
SNO-captopril, combined nitric oxide donors and prodrugs, 5:289–290
SNO-diclofenac, combined nitric oxide donors and prodrugs, 5:288–290
SNRI, multitarget drug development, 1:267–268
Sobetirome thyromimetic:
  GC-14 analog, 4:219–220
  ICI-164384 analog, 4:213–214
  NH-3 analog, 4:220
  structure-activity relationship, 4:207–208
Social Recognition and Novel Object Recognition, working memory assessment, 8:27

Sodeglitazar, α/γ/δ-pan agonists, 4:141–145
Sodium channel blockers, cocrystal engineering, 3:210
Sodium nitroprusside, nitric oxide donors and prodrugs, 5:280–281
Sodium stibogluconate, Leishmaniasis, 7:575–577
Sodium valproate, migraine prevention, 8:315–316
Sodium voltage-gated ion channel, anticonvulsants, 8:123–124
Sodum-dependent glucose cotransporter inhibitors (SGLT1/SGLT2), type 2 diabetes therapies, 4:13–14
Sofinicline, cognitive enhancement, 8:35–36
Soft drugs (SDs):
  defined, 2:75
  glucocorticoid anti-inflammatory compounds, C-21 carbonyl alterations, 5:106–112
  hard drugs vs., 2:75, 77
  prodrug developmenet, esmolol case study, 3:258–268
  retrometabolic drug design:
    activated compounds, 2:128–129
    active metabolite-based compounds, 2:127–128
    analog compounds, 2:80, 110–111
    anesthetics, 2:125–126
    angiotensin converting enzyme inhibitors, 2:118–119
    antiarrhythmic agents, 2:114–117
      ACC-9358 analogs, 2:114–115
      amiodarone analogs, 2:115–117
    anticholinergics, 2:107–109
    antimicrobials, 2:111–114
      L-carnitine esters, 2:113–114
      cetylpyridinium analogs, 2:111–112
      long-chain esters, betaine/cholne, 2:114
      quaternary analogs, 2:111–113
    benzodiazepine analogs, 2:124–125
    β_2-agonists, 2:103–104
    β-blockers, 2:80–88
      adaprolol, 2:83–85
      esmolol, 2:85
      landiolol, 2:85–86
      vasomolol, flestolol, et al, 2:86
    bufuralol analogs, 2:86–88
    calcineurin inhibitors, 2:119–121
      cyclosporine analogs, 2:120–121
      tacrolimus analogs, 2:121
    calcitriol analogs, 2:101–102
    calcium channel blockers, 2:126–127
    cannabinoids, 2:123–124
    computer-aided design, 2:130–133
      candidate ranking, 2:131–132
      hydrolytic liability, 2:132–133
      structure generation, 2:131
    corticosteroids, 2:90–101
      antedrug steroids, 2:99
      etiprednol dicloacetate and analogs, 2:96–97
      flucortin butyl, 2:97
      17-furoate androstadienes, 2:99
      glucocorticoid γ-lactones, 2:98–99
      itrocinonide, 2:97–98
      loteprednol etabonate and analogs, 2:90, 92–96
    cytokine inhibitors, 2:121–122
    dihydrofolate reductase inhibitors, 2:119
    enzymatic hydrolysis, 2:77–80
    esmolol homo-metoprolol analog, 2:133–134
    estrogens, 2:102–103
    hard drugs vs., 2:75
    inactive metabolite-based design, 2:80–81
    insecticides/pesticides, 2:106–107
    chlorobenzilate, 2:106–107
    malathion, 2:107
    matrix metalloproteinase inhibitors, 2:122–123
    opioid analgetics, 2:88–90
    pharmacological activity, 2:76–77
    prodrugs vs., 2:75–76
    prosoft drugs, 2:129–130
    psychostimulants, 2:104–106
    serotonin receptor agonists, 2:117–118
    virtual library, 2:134–135
Soft tissue calcification, 13-cis-retinoic acid therapy, 5:394
Solid drug forms:
  handling systems, pilot plant development, 3:9–10
  oral dosage systems, 3:25–29
    amorphous solids, 3:29
    dispersion mechanisms, class II drugs, 3:46
    polymorphs, 3:26–27
    pseudopolymorphs, 3:27–29
    salts, 3:26
    solvates, 3:28–29
Solid phase extraction, liquid chromatography-nuclear magnetic resonance and, 2:374
Solid-phase mass spectrometry, drug screening, 1:115–116
Solid-phase organic synthesis (SPOS):
  defined, 1:280
  follicle-stimulating hormone agonists, 1:328–329
  informational macromolecules, 1:280–294
    lipid arrays, 1:293–294
    nucleoside arrays, 1:292–293
    oligosaccharide arrays, 1:291–292
    peptide arrays, 1:280–291
  libraries, 1:294–343
  mixed drug synthesis, 1:345
  resin properties, 1:343–344
  small, druggable molecule libraries, 1:294–343
Solid-state requirements, bench-scale experiments, 3:7

Solid state stability, oral drug physicochemistry, **3:**40
Solid tumors:
  N-(4-hydroxyphenyl) retinamide (4-HPR) therapy, **5:**468, 479
  13-*cis*-retinoic acid therapy, **5:**392
Solifenacin, muscarinic antagonists, **8:**74–77
Solubility:
  ADMET structural properties, **2:**52–55, 65–66
  Biopharmaceutics Classification System, oral drug delivery, **3:**41–52, 355, 357–362
  multiobjective drug discovery optimization, blood-brain barrier and, **2:**268–273
  oral drug delivery:
    aqueous solubility, **3:**32–33
    class I drugs, **3:**42, 354
    class II drugs, **3:**42–50, 354
    class III drugs, **3:**50–52, 355
    class IV drugs, **3:**52, 355
  permeability, ADMET interactions, **3:**374–375
  quantitative structure-activity relationship, ADMET properties, **1:**62–64
  salt compounds, acid-base equilibrium, pH solubility, **3:**384–388
  sulfonamides, **7:**518–519
Soluble epoxide hydrolase (sEH), combinatorial chemistry, urea derivatives, **1:**321–322
Soluble guanylyl cyclase (sGC), nitric oxide, **5:**271–278
Solute carrier organic anion transporter (SLCO), single nucleotide polymorphisms, **1:**189
Solute carrier (SLC) superfamily, structure and classification, **2:**456–458
Solution-phase synthesis:
  lead optimization, **1:**344–345
  mixed drug development, **1:**345
  nucleoside arrays, **1:**293
  resin properties, **1:**344

small, druggable molecule libraries, **1:**294–343
Solution recrystallization, salt compound selection, **3:**391–395
Solutions, oral drug physicochemistry:
  class II drugs, lipid solutions, **3:**48
  stability, **3:**40
Solvates:
  cocrystal design, **3:**196
  crystal engineering, defined, **3:**189
  oral dosage systems, **3:**28–29
Solvation effects, ligand-macromolecular interactions, structure-based thermodynamic analysis, **2:**711
Solvent accessible surface area (SASA), quantum mechanics/molecular mechanics binding affinity, **2:**731–733
Solvent selection:
  bench-scale experiments, **3:**5–6
  high-throughput screening, assay configuration, **3:**421–422
  salt compound screening, **3:**389–393
Somatostatin:
  µ-opioid receptor antagonist derivation, **8:**673–674
  pituitary adenoma therapy, analog compounds, **6:**236–237
  structure and function, **4:**548–549
Sontochin, antimalarial therapy, **7:**609
Sorafenib signaling inhibitor:
  antiangiogenic properties, **6:**307–310
  resistance to, **6:**368
  structure and properties, **6:**305–306
Sorivudine, anti-DNA virus antimetabolite, **7:**223–225
Sotrastaurin, transplant rejection immunosuppression, **5:**1059–1060
sp^2-carbon atoms, drug metabolism:

glutathione conjugation, **1:**437–438
oxidation and reduction, **1:**415–416
sp^3-carbon atoms, drug metabolism:
  glutathione conjugation, **1:**436–438
  oxidation and reduction, **1:**413–415
Spatial memory, Morris Water Maze assessment, **8:**27–28
Spatial separate/spatially addressed techniques, peptide libraries, **1:**283–291
Species-related differences, drug metabolism studies, **1:**447–449
Specificity, xenobiotic metabolism, **1:**406–407
SPECITOPE algorithm:
  docking and scoring computations, **2:**608
  in silico screening, protein flexibility, side-chain conformation, **2:**867–868
Spectinomycin:
  chemical structure, **7:**416–419
  pharmacology and ADMET properties, **7:**423
Sphere exclusion:
  quantitative structure-activity relationship, integrated predictive modeling workflow, **1:**517–525
  quantitative structure-activity relationship validation, **1:**50
Spherical kinetics, oral drug delivery, sustained-release dosage forms, **2:**795–799
Sphingosine 1-phosphate (S1P) and receptors:
  brain cancer therapy, hypoxia-inducible factor pathway, **6:**254–255
  fingolimod development, **5:**1053–1055
  multiple sclerosis therapy, **5:**574–578

Spinal nerve tumors, classification and therapy, 6:237–239
Spin-labeling techniques, drug screening, 2:429–430
Spin labels attached to protein side chains to identify interacting compounds (SLAPSTIC), drug screening, 2:429–430
Spiradoline κ-opioid receptor agonist, 8:624–632
Spirocyclic compounds, protein-protein interactions, 2:352–354
Spirocyclic scaffold-based inhibitors, stroke therapy, 8:506
Spiro diketopiperazine-based chemokine receptor-5 (CCR5) antagonist, 7:131–133
Spirodioxolane analog, muscarinic activity, 8:66
7-Spiroindanyloxymorphone (SIOM), opioid antagonists, 8:607–608
Spiro-indolone compounds, CRTH2 receptor antagonist, 5:913–914
Spirothiazolidines, prosoft drug design, 2:130
Spirotramine, muscarinic antagonists, 8:76–77
Splice variants, opioid receptors, 8:590–591
Split and mix synthesis, peptide libraries, 1:285–291
S. pneumoniae, β-lactam antibacterial drug resistance, PBP-2x modification, 7:273–274
Sponsorship, FDA definition, 3:71
SQ109 antitubercular, 7:782–783
SQ-83360 monobactam, 7:331–332
Squalene, cholesterol biosynthesis, 5:16–18
Squamous cell carcinoma (SCC), acitretin, 5:400
Squared leave-one-out-cross-validation correlation coefficient, quantitative structure-activity relationship modeling, 1:523–525

Src kinase inhibitors:
stroke therapy, 8:510–514
2-aminothiazole template, 8:514
indoline-2-one derivatives, 8:514
natural derivatives, 8:511
purine-based derivatives, 8:511
pyrazole-pyrimidine inhibitors, 8:511–512
pyrido-pyrimidine and benzotriazine-derived inhibitors, 8:513
pyrrole-pyrimidine/fused fivemembered heterocycle pyrimidine inhibitors, 8:512
quinazoline-derived inhibitors, 8:513
quinoline-derived inhibitors, 8:513
substrate binding site inhibitors, 8:514
tumor metastases, 6:325–328
SSR-180711, cognitive enhancement, 8:35–36
ST-246 poxvirus inhibitor, 7:245
ST1926 compound. See AHPC compound
Stability:
ADMET properties:
chemical/gastrointestinal stability, 2:57
metabolic stability, 2:58–59, 66–67
oral drug physicochemistry, 3:39–40
protein therapeutics formulation and delivery, 3:314–315
tetracycline antibiotics, 7:407–409
Stabilizers, high-throughput screening, assay optimization, 3:412–413
Standard deviation, quantitative structure-activity relationship:
outlier detection, 1:45
validation, 1:46

STAT-C drugs, current research, 7:170
Static light scattering (SLS), protein therapeutics, 3:323–324
Statin drugs. See also Hydroxymethylglutaryl coenzyme A (HMG-CoA) reductase inhibitors
Alzheimer's disease, cholesterol synthesis, 8:406
cholesterol biosynthesis, 5:15–16
LDL cholesterol lowering mechanisms, 4:306–310
ezetimibe combination therapy, 4:311–312
natural products, 4:306–310
synthetic products, 4:307–309
tissue specificity, 4:309–310
multiple sclerosis therapy, 5:573
osteoporosis therapy, 5:755
peptide libraries, pseudo-peptides, 1:289–291
peroxisome proliferator-activator receptors, α-agonists, 4:96–97
plasmepsins, aspartyl protease inhibitors, 7:637–644
Statines:
BACE (β-site cleavage enzyme) inhibitors, 8:347–351
renin inhibitors, 4:244–245
Statistical diagnostics, quantitative structure-activity relationship validation, 1:46–47
Statistical molecular design (SMD), quantitative structure-activity relationship validation, 1:49–50
Statutory invention registration (SIR), drug discovery and, 3:185–186
Stavudine:
HIV reverse transcriptase inhibition, 7:141–145
research background, 7:139–140

Stay provisions, generic drug approval, FDA guidelines, **3**:86
Stem cell factor, structure and properties, **4**:582–584
Stem cells. *See also* Hematopoietic agents
  brain tumor initiation, **6**:224–227
  cancer therapy, multiple drug resistance, **6**:376–377
STEMGEN®, **4**:581–582
Stepwise synthesis, acyclic nucleoside phosphates, **7**:233
Stereocenter removal, chiral compounds, **1**:153–155
Stereochemistry:
  amphetamines, **8**:108
  chirality and drug metabolism, **1**:443–444
  combinatorial libraries, **1**:297
  selective toxicity, basic principles, **2**:546–547
Stereoisomers:
  atropisomers, defined, **1**:127–128
  selective toxicity:
    basic principles, **2**:546–547
    levobupivacaine, **2**:564
    xenobiotic metabolism, **1**:406–407
Steric hindrance:
  prodrug design, **3**:254–255
    esmolol case study, **3**:265–268
  toxicophore reactive metabolites, electron-rich heterocycles, **2**:316–317
  tropane-based chemokine receptor-5 (CCR5) antagonist, **7**:118–121
Steric parameters:
  glucocorticoid anti-inflammatory compounds, **5**:76–80
  quantitative structure-activity relationship, **1**:20–21
STERIMOL parameters, quantitative structure-activity relationship, **1**:20–21
Steroidal analogs:
  androgens:
    antiandrogens, **5**:190–192

structure-activity relationships, **5**:170–174
dissociated glucocorticoids, **5**:120–121
estradiols, **5**:244–245
nicotinic antagonists, **8**:79–80
Steroidogenesis:
  steroid biochemistry, **5**:13–25
  adrenal steroids, **5**:19–22
  androgen biosynthesis, **5**:22–24
  cholesterol biosynthesis, **5**:13, 15–16
  estradiol biosynthesis, **5**:24–25
  pregnenolone formation, **5**:16–19
  progesterone biosynthesis, **5**:22
  testosterone biosynthesis, **5**:156–158
Steroidogenic factor-like family, receptor-based drug design, **2**:512
Steroid receptor superfamily, receptor-based drug design, **2**:512
Steroids. *See also* specific steroids, e.g., Norethindrone
  aromatization, female sex hormone biosynthesis, **5**:230–231
  biochemistry:
    adrenal steroids, **5**:19–22
    androgen biosynthesis, **5**:22–24
    cholesterol biosynthesis, **5**:13, 15–16
    estradiol biosynthesis, **5**:24–25
    hormone activation, **5**:27, 29–31
    metabolism, **5**:25–27
      androgens, **5**:26–28
      corticosteroids, **5**:25–26
      estrogen, **5**:27, 29
    pregnenolone formation, **5**:16–19
    progesterone biosynthesis, **5**:22
    steroidogenesis, **5**:13–25
  chemistry:
    betamethasone, **5**:11
    cortisone synthesis, **5**:7–9
    faslodex, **5**:11, 13–14
    finasteride, **5**:11, 13

microbial transformation, **5**:7–13
  hydroxylations, **5**:7, 9–10
  side-chain cleavage, **5**:10–11
  nomenclature, **5**:4–6
  norethindrone, **5**:11–12
  research background, **5**:1–2
  structure and physical properties, **5**:2–4
  lipid combinatorial library, **1**:293
  receptor proteins, **5**:30–31
  soft drug design, "antedrug" compounds, **2**:99
  structural analog production, **1**:172–174
Steroid xenobiotic receptor (SXR), bexarotene reactions, **5**:452–453
Sterol esters, drug metabolism, conjugation reactions, **1**:432–433
Stilbenes, structure-based drug design, **2**:701
Stimulants, central nervous system. *See also* Psychostimulants
  ADMET properties, **8**:97–99
  alpha-alkyl substituent, **8**:108
  alpha-carbon stereochemistry, **8**:108
  amphetamines, **8**:106–109
  aromatic ring substitution, **8**:108–109, 112–113
  C(2) substituent, **8**:111–112
  C(3) ester linkage, **8**:112
  caffeine, **8**:90
  cocaine, **8**:109–113
  ephedra and khat, **8**:89–90
  historical background, **8**:90–91
  methylphenidate, **8**:109
  nitrogen substituents, **8**:107, 110–111
  N-substituents, **8**:110
  physiology and pharmacology, **8**:100–106
  recent and future development, **8**:113–115
  receptor classification and function, **8**:102–106
  research background, **8**:89
  side chain length, **8**:107
  side effects, adverse effects, drug interactions, **8**:92–97

structure-activity
 relationships,
 8:106–113
therapeutic applications,
 8:91–92
tropane ring preservation,
 8:113
Stiripentol, anticonvulsant
 applications,
 8:136–139, 155
Stirred-tank reactor (STRs),
 protein therapeutics,
 suspension cell
 cultures, 3:298
Stoichiometry:
 bench-scale experiments, 3:6–7
 DNA binding, nuclear magnetic
 resonance studies,
 2:406–407
Streptogramins, ribosome
 targeting, 7:457–460
Streptokinase (SK), stroke
 therapy, 8:447–448
Streptomycin:
 antitubercular agents,
 7:759–761
 development of, 7:416–419
 history and biosynthesis,
 7:423–424
 pharmacology and ADMET
 properties, 7:422–423
 RNA targets, 5:965–971
 mechanism of action,
 5:965–966
 structure-activity
 relationship,
 5:967–971
Streptozotocin, biological activity
 and side effects, 6:94
Stress, depression and, 8:229
Stroke:
 drug therapy:
 antioxidants, 8:479–482
 deferoxamine, 8:482
 ebselen, 8:481
 edavarone, 8:481
 NXY-059, 8:479–480
 tirlazad, 8:480
 current products, 8:447–449
 excitatory amino acid
 receptors, 8:450–478
 calpain inhibitors,
 8:460–465
 caspase 3/9 inhibitors,
 8:465–471
 cathepsin B inhibitors,
 8:471–478

NMDA antagonists,
 8:450–455
voltage-dependent calcium
 channels, 8:455–460
future research and
 development,
 8:515–516
GABA$_A$ agonist, 8:483
HDAC inhibitors, 8:487–495
 2-aminophenylamides,
 8:492–493
 boronates and silanediols,
 8:494
 carboxylic acids, 8:494
 hydroxamate inhibitors,
 8:488–491
 ketones, 8:493
 thiol inhibitors, 8:492
INOS/NNOS inhibitors,
 8:495–502
 amidines, 8:497–498
 aminopyridines,
 8:498–499
 L-arginine analogs,
 8:495–497
 calmodulin site, 8:499–500
 coumarin derivatives,
 8:499
 dimerization inhibitors,
 8:501
 guanidine and isothiourea
 analogs, 8:497
 heme site, 8:501
 non-arginine sites, 8:499
 NOS inhibitor/NO donor
 compounds, 8:502
 tetrahydrobiopeterin site,
 8:500
maxi K-channel mechanism,
 8:482–483
MMP-9 inhibitors, 8:502–510
 carboxylic acid-derived
 inhibitors, 8:508
 hydroxamic acid-derived
 inhibitors, 8:502–507
 hydroxypyridinones, 8:508
 phosphorus-derived
 inhibitors, 8:509–510
 pyramidine-2,4,6-triones,
 8:508–509
 thiols, 8:509
ONO-2506, 8:483–484
PARP inhibitors, 8:484–487
SRC kinase inhibitors,
 8:510–514
 2-aminothiazole template,
 8:514

indoline-2-one derivatives,
 8:514
natural derivatives, 8:511
purine-based derivatives,
 8:511
pyrazole-pyrimidine
 inhibitors, 8:511–512
pyrido-pyrimidine and
 benzotriazine-derived
 inhibitors, 8:513
pyrrole-pyrimidine/fused
 fivemembered
 heterocycle
 pyrimidine inhibitors,
 8:512
quinazoline-derived
 inhibitors, 8:513
quinoline-derived
 inhibitors, 8:513
substrate binding site
 inhibitors, 8:514
ischemic damage and
 intervention points,
 8:447–449
vascular dementia and, 8:21
Stromal cell-derived factor 1α,
 dipeptidyl peptidase 4
 inhibitors, 4:42
Stromelysin, nuclear magnetic
 resonance studies,
 2:412–414
Structural alerts. See
 Toxicophores
Structural analogs:
 direct analogs, 1:167–168
 production, 1:171–174
Structural biology studies:
 dipeptidyl peptidase 4
 inhibitors, 4:47–48
 farnesoid X receptors,
 4:160–161
 hepatitis C virus, 7:170–176
 HIV protease inhibitors, 7:1–2
 liver X receptors, 4:150
 nuclear hormone receptors,
 4:77–79
 agonists, antagonists, and
 partial agonists,
 4:82–83
 assay techniques, 4:80–82
 peroxisome proliferator-
 activated receptors,
 4:86
 selective modulators, 4:83
opioid receptors, 8:587–591
peroxisome proliferator-
 activated receptors:

Structural biology studies
(*Continued*)
α-agonists, **4:**88–90
δ-agonists, **4:**117–118
γ-agonists, **4:**100–102
recombinant DNA technology,
reagent development,
1:552
Structural classification of
natural products
(SCONP),
chemogenomics,
biology-oriented
synthesis, **2:**580
Structural genomics:
basic principles, **1:**571
current technology, **1:**571
drug discovery, current and
future trends,
1:581–595
antituberculosis drugs, **1:**587
aromatic amine and
hydrazine drugs
acetylation, **1:**585–586
chemogenomics, **1:**589–595
epigenetics, **1:**594–595
kinase inhibitor protein
structures, **1:**590–591
kinase program, **1:**591–594
data dissemination, methods,
and reagents, **1:**581
delivered structures impact,
1:581–584
fragment cocrystallography
techniques, sleeping
sickness target,
1:587–588
future trends, **1:**595
nitrogen-containing
bisphosphonates,
1:584–585
poly(ADP) ribosylation
inhibition,
1:586–587
drug-like inhibitors, recent
developments,
1:574–576
genome-centric approach,
1:572–573
ligand chemistry, **1:**576–577
Protein Data Base
contributions,
1:577–581
clinical drug targets,
1:579–580
human proteome coverage,
1:577–579

membrane proteins,
1:580–581
structure space coverage,
1:577
protein family-based human
genome coverage,
1:573–576
research strategies, **1:**571–572
Structural Genomics Consortium
(SGC):
genome-centered research,
1:572–573
kinase program, **1:**591–594
nitrogen-containing
bisphosphonates,
1:585
protein family-based genome
coverage, **1:**573–577
research strategies, **1:**571–572
Structure-activity relationship
(SAR). *See also*
Quantitative
structure-activity
relationship (QSAR)
aminoglycoside antibiotics,
7:428–432
antidepressants, **8:**249–255
bupropion, **8:**254–255
monoamine oxidase
inhibitors, **8:**251–253
serotinergic agents,
8:253–254
tricyclic antidepressants,
8:249–251
antimalarial drugs:
amodiaquine, **7:**611–613
artemisinins, **7:**616–617
chloroquine, **7:**609–610
farnesylation, **7:**631–634
quinine, **7:**605–607
antimetabolites, anti-DNA
virus agents, first
generation
compounds, **7:**222–225
antipsychotics, **8:**180–185
atypical tricyclic
neuroleptics,
8:180–183
benzisoxazole/benzithazole
neuroleptics,
8:183–185
antitubercular agents:
capreomycin, **7:**762–764
diarylquinoline (TMC207),
7:784
ethambutol, **7:**751–753
ethionamide, **7:**771

isoniazid, **7:**738–739
linezolid, **7:**786
nitroimidazoles, **7:**776–780
para-Aminosalicylic acid,
7:765
pyrazinamide, **7:**756–758
rifamycins, **7:**745–746
SQ109, **7:**782–783
apical sodium-dependent bile
acid transporter,
2:470–471
carbapenems, **7:**353–354
cationic surfactants, **7:**500
chemistry-driven drug
discovery, **1:**537–538
chemogenomics:
molecular information
systems, **2:**582–583
research background,
2:573–574
dihydrofolate reductase
inhibitors, **7:**532–534
G-protein coupled receptor,
homology modeling,
2:292–293
HIV integrase:
dihydroxypyrimidine
derivatives, **7:**89–90,
92–97
N-methyl pyrimidinone
derivatives, **7:**89–90,
94
two-metal chelation
elaboration, **7:**82–97
β-Lactams, **7:**259–261
cephalosporins, **7:**300–306
macrolides, **7:**441–444
monobactams, **7:**327
muscarinic antagonists,
8:72–77
nitrofurans, **7:**549
nocardicins, **7:**326
nonpeptide opioid analgesics,
8:597–645
acyclic analgesics, **8:**621–623
affinity labeling, **8:**637–643
morphine-naltrexone
derivatives, **8:**638–641
agonists:
δ-selective agonists,
8:632–637
κ-selective agonists,
8:623–632
centrally-acting
agonists, **8:**624–630
peripherally-acting
agonists, **8:**630–632

antagonists, 8:601–608
  δ-opioid receptors,
    8:604–608
  κ-opioid receptors,
    8:602–604
  benzomorphans, 8:610–613
  Diels-Alder adducts,
    8:608–609
  miscellaneous opioids,
    8:643–645
  morphinans, 8:609–610
  morphine derivatives,
    8:599–601
  opium-derived morphine and
    alkaloids, 8:598–599
  ORL1 receptor, endogenous
    ligand orphanin FQ/
    nociceptin, 8:681–682
  piperidine derivatives,
    8:613–621
  4-anilidopiperidines,
    8:618–621
  4-arylpiperidines:
    carbon substituent,
      8:614–616
    oxygen substituent,
      8:616–618
nuclear magnetic resonance
  and:
  drug screening applications,
    2:421–424
  evolution of, 2:367–369
opioid-receptor-like 1/orphanin
  FQ/nociceptin peptide,
  8:678–680
oxazolidinones, 7:544–548
penicillins, 6-amino-
  penicillanic acid
  derivatives, 7:289–292
peptides, 1:209–210
prodrug moieties, 3:251–252
  esmolol case study,
    3:262–268
psychostimulants, 8:106–113
  amphetamine, 8:106–109
  cocaine, 8:109–113
  methylphenidate, 8:109
quinolones/fluoroquinolones,
  7:537–538
receptor-based drug design,
  2:493–498
  compound properties,
    2:519–520
  residence times, 2:504
serum albumin, 3:438–456
sulfonamides and sulfones,
  7:515

teleprevir anti-HCV agent,
  7:183–186
Structure-activity relationships
  (SARs):
  androgens:
    anabolic agents, 5:179–186
    alkylated analogs,
      5:181–182
    dehydro derivatives, 5:180
    deoxy/heterocyclic-fused
      analogs, 5:183–185
    esters and ethers,
      5:185–186
    hydroxy and mercapto
      derivatives, 5:182–183
    nor derivatives, 5:179
    oxo, thia, and azo
      derivatives, 5:183
    synthetic androgens,
      5:170–174
  antibody-directed enzyme
    prodrug therapy,
    DNA-targeted
    therapeutics,
    6:120–121
  antifolate DNA-targeted
    chemotherapeutics,
    6:108
  ATP binding casette
    transporter inhibitors,
    6:374
  biguanides, 4:2–3
  dipeptidyl peptidase 4 (DPP-4)
    inhibitors, 4:48–65
    α-amino acid derivatives,
      4:48–56
    β-amino acid derivatives,
      4:57–59
    cyclohexyl- and
      cyclohexenylamines,
      4:64–65
    fluoroolefin inhibitors,
      4:56–57
    heterocyclic derivatives,
      4:59–64
    hybrid α/β-amino acid
      derivatives,
      4:59
  DNA-targeted
    chemotherapeutic
    compounds:
    cyclopropylindoles,
      6:92–93
    mustards, 6:85–87
    nitrosoureas, 6:94
    platinum complexes,
      6:90–91

female sex hormones:
  diarylethylene estrogen
    antagonists,
    5:248–249
  estrogen receptor affinity,
    5:246
  nonsteroidal estrogen
    receptor ligands,
    5:246–247
  progesterone nonsteroidal
    PR ligands, 5:255–256
  progesterone steroidal PR
    ligands, 5:253–255
  steroidal analogs, 5:244–245
    7α-substitution, 5:245
    11β-substitution, 5:245
  estrone-equilin
    comparisons, 5:246
  nonbenzenoid seteroid(-
    like) estrogen receptor
    ligands, 5:246
  triarylethylenes, 5:249–253
  triarylheterocycles, 5:253
  vicinal diarylethylenes,
    5:247–248
fingolimod (FTY720),
  5:1054–1055
fructose-1,6-biphosphatase
  inhibitors, 4:18–20
gene-directed enzyme prodrug
  therapy, 6:123–124
Hsp90 inhibitors:
  geldanamycin Hsp90
    inhibitor, 6:389–391
  novobiocin Hsp90 inhibitor
    optimization,
    6:436–440
  purine-based Hsp90
    inhibitors, 6:415–416
hypoxia-activated bioreductive
  prodrugs, 6:116–120
immunosuppressive drugs:
  cyclosporin A, 5:1010–1011
  tacrolimus, 5:1016–1022
peroxisome proliferator-
  activated receptors:
  α/δ-dual agonists, 4:136–137
  α/γ-dual agonists, 4:126–135
  α/γ/δ-pan agonists,
    4:140–145, 141–145
  PPARδ-agonist, 4:119–121
  selective PPARγ modulators,
    PPARγ partial
    agonists, 4:111–115
prostaglandin/thromboxane
  receptors:
  EP$_1$ antagonists:

Structure-activity relationships (*Continued*)
amide replacements, **5**:818–820, 836–837
AstraZeneca derivatives, **5**:816–817
benzene derivatives, **5**:829, 831, 832
benzoic acid derivatives, **5**:823, 827–832
carbamates, **5**:836–838
cyclopentene derivatives, **5**:829, 831, 833
diacyl hydrazide replacement, **5**:815–817
glycine sulfonamide derivatives, **5**:837, 840–845
Ono derivatives, **5**:817–824
pyrazole derivatives, **5**:831, 835–838
pyrrole derivatives, **5**:826, 828–832
Searle derivatives, **5**:813–816
sulfonylureas, **5**:820–821, 825
thiophene derivatives, **5**:821, 823, 826, 829
EP_2 receptor:
agonists, **5**:846–850
antagonists, **5**:852–854
EP_3 receptor, antagonists:
acylsulfonamide derivatives, **5**:855–858
angiotensin II derivatives, **5**:856–858
hyperalgesia/edema modeling, **5**:858, 861
indole analogs, **5**:861, 863–869
thiophene derivatives, **5**:858, 860–861
EP_4 receptor:
agonists:
lactam stereochemistry, **5**:873–878
thienopyrimidine-based compounds, **5**:878–880
ω-chain derivatives, **5**:872–876
antagonists, **5**:880–883
mixed EP_2/E_4 agonists, **5**:850–852
purine antimetabolite analogs, **6**:112–113
pyrimidine antimetabolite analogs, **6**:109–112
radicicol Hsp90 inhibitor, **6**:405
RNA targets:
erythromycin, **5**:985–990
linezolid, **5**:992–993
neomycin/gentamicin aminoglycosides:
subclass rings I and II, **5**:973–978
subclass rings III and IV, **5**:979
tetracyclines, **5**:982–983
sildenafil, **5**:712–713
sirolimus compounds, **5**:1030–1034
triene region modification, **5**:1030–1033
streptomycin, RNA targets, **5**:967–971
tacrolimus, **5**:1016–1022
PPIase analogs, **5**:1016–1021
tadalafil, **5**:714
thyroid hormones and thyromimetics, **4**:202–222
axitirome, **4**:206–207
azauracil compounds, **4**:215–216
bioisosteric replacement, obligatory R'_4-hydroxy group, **4**:218–219
bridging group X, **4**:213
cytochrome P450-activated MB07811 prodrug, **4**:211–212
DITBA, **4**:205
early thyromimetics, **4**:205–222
epotirome, **4**:209–210
future research issues, **4**:222
heterocyclic rings, **4**:210–211
ICI164384, **4**:213–214
indole derivatives, **4**:212–213
KB-141, **4**:208–209, 216–218
L-T_3 and L-T_4, **4**:202–203
pharmacological selectivity strategies, **4**:205–206
phenylnaphthyl linker, **4**:214–215
prime ring modification, **4**:221–222
recent thyromimetics, **4**:205–222
receptor isoform position variation, **4**:206–212, 215
ring-closure, receptor positions, **4**:212–213, 218
ring substitution, R'_3 and R'_4 positions, **4**:219–221
SKF-94901, **4**:203
sobetirome, **4**:207–208
TRIAC, **4**:204
topoisomerase II inhibitors, **6**:99–102
vardenafil, **5**:713–714
Structure-based design (SBD):
ADMET properties, **2**:50
chemogenomics, data analysis and predictive modeling, **2**:585–588
chiral compounds, **1**:153–155
computer-aided techniques, **2**:679–683
molecular docking, **2**:679–680
pharmacophore modeling, **2**:682–683
quantitative structure-activity relationship, **2**:681–682
virtual screening, **2**:680–681
current trends, **2**:677–679
docking/scoring computations:
basic principles and concepts, **2**:593–599
point complementarity methods, **2**:602–604
drug-macromolecule interactions, thermodynamic analysis-, **2**:702–712
binding activity, **2**:703–705
binding measurement, **2**:706–707
calorimetry results, **2**:711–712
enthalpy-entropy compensation, **2**:705–706
fragment-based design, **2**:709–711
isothermal titration calorimetry, **2**:707–709
solvation thermodynamics, ligand-receptor interactions, **2**:711
HIV protease inhibitors, **7**:3
Hsp90 inhibitors, **6**:421–423

kinase inhibitors:
  future research issues,
    6:356–357
  multiconformational
    catalytic domain:
      binding cleft, **6:**346–348
      binding pockets, **6:**348–349
      protein targeting,
        6:350–354
      second-line inhibitors,
        drug-resistant
        mutation, **6:**354–356
    research background,
      6:345–346
liver X receptors, **4:**151–154
multitarget drug development,
  AChE-plus for
  Alzheimer's disease,
  1:261, 263
nuclear magnetic resonance:
  ligand-based design,
    2:375–402
    analog limitations, **2:**390
    charge state, **2:**384–385
    conformational analysis,
      2:383–384
    line-shape and relaxation
      data, **2:**387–389
    pharmacophore modeling,
      2:389–390
    structure elucidation,
      2:376–383
      bioactive peptides,
        2:377–383
      natural products,
        2:376–377
    tautomeric equilibria,
      2:385–387
  receptor-based design,
    2:390–419
    chemical-shift mapping,
      2:402–403
    dihydrofolate reductase
      inhibitors, **2:**414–419
    DNA-binding drugs,
      2:405–410
    isotope editing/filtering,
      2:403–404
    macromolecular structure,
      2:391–394
    matrix metalloproteinases,
      2:412–414
    NOE docking, **2:**404–405
    protein-ligand
      interactions,
      2:394–402
    titrations, **2:**403

optimization iterative process,
  2:677–679
protein-protein interactions,
  2:342–343
RNA targets:
  aminoglycoside toxicity,
    5:979–981
  erythromycin and analogs,
    5:984–990
  future research issues,
    5:993–996
  linezolid and analogs,
    5:990–993
  neomycin- and gentamicin-
    type aminoglycosides,
    5:971–979
  research background, **5:**963
  ribosome structure and
    function, **5:**964–965
  streptomycin and analogs,
    5:965–971
  tetracycline and analogs,
    5:981–984
in silico screening, protein
  flexibility:
  receptor plasticity,
    2:860–870
  mobility determination,
    2:861–865
  protein-ligand docking,
    2:865–870
  research background,
    2:859–860
soft drugs, computer-aided
  techniques, **2:**131
thyroid hormone receptor β-
  selective agonists,
  2:684–692
  hormone binding site,
    2:684–685
  TR:T3 crystal complexes,
    2:685–691
transthyretin amyloidosis
  inhibitors, **2:**692–702
  natural products and
    derivatives, **2:**701–702
  NSAIDs, **2:**694–702
  TTR fibril formation
    inhibitors, **2:**693–702
  TTR T4 binding site,
    2:692–693
virtual screening, **2:**14–25
  fragment-based screening,
    2:25
  protein structures and
    flexibility, **2:**14–16
  scoring functions, **2:**16–25

comparison, consolidation
  and consensus,
  2:22–23
empirical scoring, **2:**18
enrichment factors,
  2:20–22
force-field scoring, **2:**17–18
hybrid/alternative scoring,
  2:19–20
knowledge-based scoring,
  2:18–19
liver X receptor modulator
  example, **2:**23–25
Structure-metabolism
  relationships (SMRs):
  drug metabolism, **1:**443
  prodrug design, **3:**254–256
Strychnine, anticonvulsants,
  glycine receptors,
  8:130
ST segment elevation myocardial
  infarction (STEMI),
  acute coronary
  syndrome risk, **4:**411
Subcellular localization,
  glucocorticoid receptor
  structure, **5:**62
Suberoylanilide hydroxamic acid
  (SAHA):
  histone deacetylase inhibition,
    6:53, 55–56
  stroke therapy, **8:**488
Substance abuse:
  anabolic agents, **5:**188
  opioid receptors, **8:**571
  dependence-addiction
    liability, **8:**574–575
Substance P:
  dipeptidyl peptidase 4
    inhibitors, **4:**41–42
  receptor, tachykinin
    pharmacology,
    4:542–544
Substance P-NK1 receptor
  antagonist,
  antidepressant effects,
  8:256–257
Substantial evidence standard,
  infringement of patent
  proceedings, **3:**164
Substituent constants,
  quantitative
  structure-activity
  relationship:
  electronic parameters, **1:**7–8
  partition coefficient
    calculations, **1:**14–19

Substitution/omission method, peptide libraries, 1:286–291
Substrate-like irreversible inhibitors, dipeptidyl peptidase 4 (DPP-4) inhibition, 4:48–49
Substrate-like reversible inhibitors, dipeptidyl peptidase 4 (DPP-4) inhibition, 4:49–50
Substrates:
  carboxylesterases, 3:235–239
  high-throughput screening, assay configuration, 3:414–420
  HIV protease inhibitor design, 7:2–3
  inhibitors:
    allosteric proteins, 1:371–373
    dipeptidyl peptidase 4 inhibitors, 4:41–42
    stroke therapy, binding site inhibitors, 8:514
    xenobiotic metabolism, 1:406–407
  psychostimulant physiology and pharmacology, 8:101–106
  transport protein models and design of, 2:459
"Substrate switching," congestive heart failure, disease biomarkers, 4:505–506
Succinimides, anticonvulsant applications, 8:142
Succinylcholine:
  enzymatic hydrolysis, 2:79–80
  nicotinic antagonists, 8:78–80
  pharmacological activity, 1:440–441
Sugar attachments:
  aminoglycoside antibiotics, 7:429–432
  macrolide development, 7:447–456
Suicide, antidepressants and:
  serotonin-selective reuptake inhibitors, 8:234
  tricyclic antidepressants, 8:231–232
Suicide hypothesis, poly(ADP-ribosyl)ation biochemistry, 6:155–157

Sulbactam, β-lactamase inhibitors, 7:360, 363–364
Sulbenicillin, penicillin resistance and development of, 7:291–292
Sulfacetamide, discovery of, 7:509
Sulfadiazine, discovery of, 7:509
Sulfadimethoxine, glucuronidation/glucosidation, 1:428–430
Sulfamerazine, discovery of, 7:509
Sulfamethoxazole, discovery of, 7:514
Sulfamethoxypyridazine, discovery of, 7:509–514
Sulfanilamide:
  discovery, 7:483
  early development, 7:508–514
  Leishmaniasis therapy, 7:592–593
Sulfapyridine, discovery, 7:508–509
Sulfathiazole, discovery, 7:509–514
Sulfinpyrazone, glucuronidation/glucosidation, 1:429–430
Sulfonamides:
  adverse reactions, 7:521–522
  analog design:
    EP$_2$ receptor agonist, 5:849–850
    tacrolimus compounds, 5:1019–1020
    thromboxane A$_2$ receptor antagonists, 5:887–890
  anticonvulsant applications, 8:151–152
  antimalarial agents, 7:617
  biological activity and structure-activity relationship, 7:515
  clinical applications, 7:520–521
  current applications, 7:520–521
  dihydrofolate reductase inhibitor synergism, 7:527–528
  discovery, 7:483
  historical background, 7:507, 514
  p53/MDM2 derivatives, 2:350–352

  permeability, 7:517–518
  pharmacokinetics, 7:509–514, 522–523
  protein binding, 7:519–520
  quantitative structure-activity relationship, 7:515–518
  γ-secretase inhibitors, 8:337–340, 343–345
  selective toxicity and comparative biochemistry, 2:548–549
  selectivity, 7:527
  solubility, 7:518–519
  synthetic folate inhibitors, 7:506–531
Sulfonation, drug metabolism, 1:425–426
Sulfones:
  adverse reactions, 7:521–522
  antimalarial agents, 7:617
  falcipain inhibitors, 7:635–637
  clinical applications, 7:520–521
  current applications, 7:520–521
  dihydrofolate reductase inhibitor synergism, 7:527–528
  historical background, 7:507–514
  permeability, 7:517–518
  pharmacokinetics and metabolism, 7:523
  protein binding, 7:519–520
  selectivity, 7:527
  solubility, 7:518–519
  structure-activity relationship, 7:515
  synthetic folate inhibitors, 7:506–531
Sulfonilamide, prodrug development, prontosil case study, 3:258
Sulfonyl halides, combinatorial chemistry, 1:296
Sulfonylureas:
  prostaglandin/thromboxane receptors:
    EP$_1$ antagonists, 5:820–821, 825
    EP$_4$ receptor antagonists, 5:881–883, 885
  structural analogs, 1:167–168
  type 2 diabetes therapy, 4:8–10

Sulfoxide derivatives, muscarinic activity, 8:65
Sulfur:
  chirality at, 1:127, 147–148
  drug metabolism, oxidation and reduction reactions, 1:418–419
Sulfur-containing compounds, soft drug design, 2:100
Sulopenem, development of, 7:323–324
Sumatriptan, migraine therapy, 8:269–272
Summary judgment, infringement of patents, 3:161
Sunitinib:
  antiangiogenic properties, 6:307–310
  brain tumor angiogenesis, 6:242–243
  combinatorial chemistry, 1:277–279
  multitarget drug development, cancer therapy, 1:264–267
  resistance to, 6:368
  structure-based design, protein targeting, 6:352–354
Super agonists, occupancy theory, 2:496–497
Supercritical fluid chromatography, combinatorial chemistry, 1:346–347
SUPERFIT fentanyl enatiomer, nonpeptide affinity labeling, 8:642–643
Superoxide radicals:
  American trypanosomiasis therapy, 7:573
  nitric oxide enhancement, 5:291–294
Supersaturation, class II drugs, precipitation inhibition, 3:45
"Superstatins," LDL cholesterol lowering mechanisms, 4:308–309
Support vector network (SVN), ligand-based virtual screening, quantitative structure-activity relationships, 2:11–13
Supramolecular heterosynthons: cocrystal design, 3:194–196

evolution of, 3:189–190
Supramolecular homosynthons:
  cocrystal design, 3:195–196
  evolution of, 3:189–190
Supramolecular synthons:
  cocrystal design, 3:194–197
  crystal engineering, evolution of, 3:189–190
  polymorphism, 3:191–192
Suramin:
  human African trypanosomiasis therapy, 7:569–571
  nucleotide transporter trypanocide development, 7:577–580
  research background, 7:139–140
Surfactants:
  nanoscale drug delivery, noisomes, 3:473
  oral drug physicochemistry, class II drugs, micelle solubilization, 3:48–50
  perfluorocarbons, oxygen delivery, 4:621–622
  topical antimicrobial agents:
    amphoteric surfactants, 7:501
    anionic surfactants, 7:500–501
    cationic surfactants, 7:499–500
Surrogate peptides, combinatorial libraries, 1:289–291
Survival signaling inhibitors, 6:321–325
  insulin-like growth factor-1 receptor inhibitors, 6:321
  mTOR inhibitors, 6:324–325
  phosphoinoside-3 kinase inhibitors, 6:321–323
  protein kinase B inhibitors, 6:323–324
Suspensions:
  oral drug physicochemistry, class II drugs, 3:48
  protein therapeutics, cell cultures, 3:298
Sustained-release drug delivery:
  diffusion-based drug release, plasma drug profile, 2:812–816
  half-time expression, 2:803

excipients, 2:746
oral administration:
  in vitro-in vivo correlations, 2:787–791
  in vitro testing, 2:791–801
    diffusion control, 2:792–798
    erodible drug release, 2:798–801
    nomenclature, 2:791–792
Sustained viral response (SVR), anti-HCV agents, 7:170
Swain Lupton, field-inductive constant, quantitative structure-activity relationship, electronic parameters, 1:7–8
Sweden, patent requirements, 3:122
Switching mechanisms, chiral compounds, 1:156–159
Sydnonimines, nitric oxide donors and prodrugs, 5:285–286
Synaptic monoamines, psychostimulant physiology and pharmacology, 8:100–106
Synaptic plasticity, cognitive function and, 8:25
Synaptic proteins, anticonvulsants, 8:130–132
Synercide, ribosome targeting mechanism, 7:457–460
Synthalin antitrypanosomal agent, 7:578–580
Synthetic drug development:
  androgens, 5:168–176
  absorption, metabolism, and distribution, 5:175–176
  current marketed compounds, 5:168–169
  ester derivatives, 5:171
  halo derivatives, 5:171–174
  methylated derivatives, 5:171
  nonsteroidal structure-activity relationships, 5:174–175
  therapeutic applications and bioassays, 5:169–170

Synthetic drug development
    (*Continued*)
  toxicities, **5:**176
  antibacterial agents:
    systemic antibacterials,
        7:501–553
      anaerobic infections,
          7:502–503
      chemoprophylaxis, **7:**505
      combination therapies,
          7:504–505
      dihydrofolate reductase
          inhibitors, **7:**531–534
      drug resistance, **7:**502, 504
      folate inhibitors,
          7:506–534
      genetic factors, **7:**503
      host reactions, **7:**503
      methenamine, **7:**550
      nitrofurans, **7:**548–550
      oxazolidinones, **7:**540–548
      pharmacology, **7:**501–504
      quinolones and
          fluoroquinolones,
          7:535–540
      research background,
          7:501
      sulfonamides and sulfones,
          7:506–531
      tissue factors, **7:**503–504
      treatment guidelines,
          7:504–506
    topical agents, **7:**484–501
      acids, **7:**496
      alcohols, **7:**492–493
      amphoteric surfactants,
          7:501
      anionic surfactants,
          7:500–501
      cationic surfactants,
          7:499–500
      cellular targets, **7:**488
      diarylureas, amidines, and
          biguanides, **7:**498–499
      dyes, **7:**498
      epoxides and aldehydes,
          7:495–496
      evaluation, **7:**489–490
      halogens and halophores,
          7:490–492
      heavy metals, **7:**497–498
      historical development,
          7:490
      kinetics, **7:**488–489
      oxidizing agents,
          7:496–497
      phenols, **7:**493–495

  resistance mechanisms,
      7:488
  selective toxicity, **7:**484,
      488
  terminology, **7:**484
combinatorial chemistry,
    1:347–348
Geldanamycin Hsp90
    inhibitors, **6:**391–392
LDL cholesterol lowering
    statins, **4:**307–309
monobactams, **7:**327–329
opioids, **8:**597
proteasome inhibitors,
    6:182–183
steroid research, **5:**6–7
Synthons, acyclic nucleoside
    phosphates, **7:**233
α-Synuclein protein, Parkinson's
    disease etiology, **8:**535
Systemic drug delivery:
  apical sodium-dependent bile
      acid transporter
      enhancement, **2:**474
  clearance, **2:**751
Systolic dysfunction models,
    preclinical target
    validation, congestive
    heart failure therapy,
    4:487

T2000 barbiturate,
    anticonvulsant
    applications,
    8:140–141
T2384 compound, selective
    PPARγ modulators,
    PPARγ partial
    agonists, **4:**108,
    111–115
Tabu search strategy, docking and
    scoring techniques,
    2:616–617
Tacedinaline, histone deacetylase
    inhibitors,
    6:66–67
Tachykinins:
  biological actions and receptor
      pharmacology,
      4:542–544
  structure and function,
      4:540–544
Tacrine, cognitive enhancement,
    8:15–16, 30
Tacrolimus:
  ADMET properties,
      5:1022–1023

  calcineurin binding,
      5:1014–1015
  chemical structure,
      5:1013–1014
  cis-trans prolyl isomerase
      activity, **5:**1015–1016
  discovery and development,
      5:1013
  drug resistance, biricodar
      kinase inhibitor, **6:**372
  FKBP-calcineurin structure,
      5:1016
  mechanism of action,
      5:1014–1015
  nuclear magnetic resonance
      studies, **2:**410–412
  structure-activity
      relationships by
      nuclear magnetic
      resonance, **2:**421–424
  recombinant DNA technology,
      targeting
      mechanisms,
      1:555–556
  soft drug design, **2:**121
  structure-activity
      relationships,
      5:1016–1022
    PPIase analogs, **5:**1016–1021
  therapeutic applications,
      5:1023–1024
Tadalafil:
  erectile dysfunction therapy,
      chemical structure,
      5:711
  selective toxicity, **2:**567–568
  structure-activity
      relationships,
      5:714
Tafenoquine antimalarial agent,
    7:625–627
Tafluposide, dual topoisomerase
    I/II inhibitors,
    6:103–105
Tafluprost, glaucoma therapy,
    5:600–602
Taft's steric constant,
    quantitative
    structure-activity
    relationship:
  electronic parameters, **1:**7–8
  hydrophobic parameters,
      1:9–19
  steric parameters, **1:**20–21
TAK-220 chemokine receptor-5
    (CCR5) antagonist,
    7:130–131

TAK-652 chemokine receptor-5 (CCR5) antagonist, 7:129–131
TAK-779 chemokine receptor-5 (CCR5) antagonist, 7:124–131
Talabostat, dipeptidyl peptidase 4 (DPP-4) inhibition, 4:50–54
Talampanel, anticonvulsant applications, 8:136–139, 144–147
Talampicillin, penicillin resistance and development of, 7:291–292
Tallimustine, minor groove targeting compound, 6:88
Talnetant, cognitive enhancement, 8:48–49
Talsaclidine, muscarinic activity, 8:67–68
Tamoxifen:
  adverse effects and precautions, 5:221
  breast cancer resistance, 6:369–370
  current compounds and applications, 5:219–221
  CYP2D6 substrate reactions, 1:186
  discovery, 5:243–244
  nuclear hormone receptors, 4:83
  osteoporosis antiresorptive therapy, 5:734–742
  oxidative metabolism, 5:225–226
  pharmacological activity, 1:440–441
  receptor ligand electrophilicity/carcinogenicity, 5:227–228
  selective toxicity, 2:567
  structure-activity relationships, 5:249–253
  tissue specificity, 5:240–242
TAN67 δ-opioid selective agonists, 8:635–637
Tandem mass spectrometry (MS-MS):
  drug development applications, 1:105–106
  electron impact technology and, 1:97–98
Tandutinib, acute myelogenous leukemia therapy, 6:301–302
Tangential flow microfiltration, mammalian protein purification, harvest operations, 3:307
TA-NIC antinicotine vaccine, smokng cessation, 5:725–726
Tankyrases, molecular biology, 6:158–159
Tapentadol, chemical structure, 8:570–571
Tardive dyskinesia (TD), antipsychotic side effect, 8:165–166
  vacuous chewing movements model, 8:175
Tarenflurbil, γ-secretase modulators, 8:358–359
Targeted drug development. See also DNA-targeted therapeutics; Mechanism-targeted studies; Multitarget drug development (MTDD); Retrometabolic drug design
  AHPN compounds, 5:483–497
    apoptosis, 5:489–497
    cell effects, 5:486–487
    gene expression, protein expression, and protein activity, 5:494–497
    small heterodimer partner, 5:487–489
  allosteric targets, 1:390–392
  future trends, 1:394–395
  Alzheimer's disease:
    metabolic syndrome targeting, 8:405–417
    cholesterol-related approaches, 8:405–414
    insulin-related approaches, 8:414–417
    research background, 8:405
    summary, 8:408–410
    tau pathology targeting, 8:417–432
  aggregation, 8:423–429
  glycosylation, 8:422–423
  microtubule stabiliization, 8:430
  phosphorylation, 8:418–422
  protein levels, 8:430–432
anticonvulsants, molecular targeting, 8:121–134
  connexins, 8:134
  enzymes, 8:132–134
  G-protein-coupled receptors, 8:134
  ligand-gated ion channels, 8:126–130
  neurotransmitter transporters and synaptic proteins, 8:130–132
  voltage-gated ion channels, 8:122–126
antimalarial agents:
  apicoplast targets, 7:663–668
  cell cycle regulation, 7:627–634
  cellular and oxidative stress management, 7:658–663
  cellular respiration and electron transport chain, 7:655–658
  DNA synthesis, repair, and regulation, 7:644–655
  heme metabolism and amino acid salvage, 7:634–644
  targeted drug systems, 7:668–672
antitubercular agents, 7:719–733
  amino acid biosynthesis, 7:726–727
  cell wall biosynthesis targeting, 7:719–724
  cofactor biosynthesis, 7:727–729
  cytochrome P450 monooxygenase targeting, 7:729
  DNA synthesis and repair targeting, 7:724–725
  intermediary metabolism and respiration, 7:729–730
  lipid metabolism, 7:730

Targeted drug development (*Continued*)
  mycobactin biosynthesis and iron acquisition, 7:730–731
  mycothiol biosynthesis, 7:731
  protein synthesis and degradation, 7:725–726
  signal transduction targeting, 7:732–733
  terpenoid biosynthesis, 7:731–732
cancer chemotherapy, resistance and, 6:363–370
  ATP binding cassette (ABC) transporters, 6:371–376
central nervous system tumors, 6:271–272
congestive heart failure:
  target-compound interaction:
    ex vivo cell based biomarker assays, 4:501
    hemodynamic/biochemical biomarker assays, 4:500–501
    PET/SPECT ligands, 4:501
    target-engagement assays, 4:501–502
  target validation, 4:483–500
    clinical congruency, 4:489–490
    clinical importance, 4:483, 485–487
    current clinical trials, 4:497–498
    knockout/transgenic mouse models, 4:488–490
    mechanism perturbation, 4:494–495
    mechanism-related risk and reliability, 4:500
    mechanistic approaches, cardiac contractility/myocardial oxygen demand, 4:492–494
    normal patient mechanisms *vs.*, 4:495–496
    novel target *vs.* existing therapy, 4:496, 499–500

pharmacodynamics/pharmacokinetics, 4:491–492
preclinical models, 4:486–491
tissue studies, species and disease states, 4:488
DNA-targeted compounds, cytotoxic agents, 6:3–17
  bleomycin, 6:8–12
  dactinomycin, 6:3–8
  mitomycin (mutamycin), 6:12–16
  plicamycin, 6:16–17
high-throughput screening, enzymatic assays, 3:401–402
kinase inhibitors, structure-based design, 6:350–354
membrane transporters, absorption enhancement, 2:443–444
mitosis targeting, 6:191–192
molecular targeting chirality:
  drug-receptor interactions, 1:127–129
  enantiomer activity, 1:130–131
multiple drug resistance, apoptosis-based protein targeting, 6:370–371
multiple sclerosis therapy, 5:572–573
  antigen-specific therapy, 5:579–580
prodrug design, 3:229
  esmolol case study, 3:260–268
radiolabeled ligands for, 8:743–744
receptor-based drug design, functional genomics, 2:515–519
RNA targets:
  aminoglycoside toxicity, 5:979–981
  antibiotic ribosome targeting, 7:457–464
  chloramphenicol, 7:462–463
  fusidic acid, 7:464
  lincosamides, 7:460–462
  oxazolidinones, 7:464

research background, 7:405–406
streptogramins, 7:457–460
erythromycin and analogs, 5:984–990
future research issues, 5:993–996
linezolid and analogs, 5:990–993
neomycin- and gentamicin-type aminoglycosides, 5:971–979
research background, 5:963–964
ribosome structure and function, 5:964–965
streptomycin and analogs, 5:965–971
tetracycline and analogs, 5:981–984
stroke therapy:
  current research, 8:447–449
  future research, 8:515–516
  HDAC inhibitors:
    hydroxamate inhibitors, 8:488–491
    ketones, 8:493
    thiol inhibitors, 8:492
  INOS/NNOS inhibitors, 8:495–502
    amidines, 8:497–498
    aminopyridines, 8:498–499
    L-arginine analogs, 8:495–497
    calmodulin site, 8:499–500
    coumarin derivatives, 8:499
    dimerization inhibitors, 8:501
    guanidine and isothiourea analogs, 8:497
    heme site, 8:501
    non-arginine sites, 8:499
    NOS inhibitor/NO donor compounds, 8:502
    tetrahydrobiopeterin site, 8:500
  maxi K-channel mechanism, 8:482–483
  MMP-9 inhibitors, 8:502–510
    carboxylic acid-derived inhibitors, 8:508
    hydroxamic acid-derived inhibitors, 8:502–507
    hydroxypyridinones, 8:508

phosphorus-derived
inhibitors, 8:509–510
pyramidine-2,4,6-triones,
8:508–509
thiols, 8:509
ONO-2506, 8:483–484
PARP inhibitors, 8:484–487
SRC kinase inhibitors,
8:510–514
2-aminothiazole template,
8:514
indoline-2-one derivatives,
8:514
natural derivatives,
8:511
purine-based derivatives,
8:511
pyrazole-pyrimidine
inhibitors, 8:511–512
pyrido-pyrimidine and
benzotriazine-derived
inhibitors, 8:513
pyrrole-pyrimidine/fused
fivemembered
heterocycle
pyrimidine inhibitors,
8:512
quinazoline-derived
inhibitors, 8:513
quinoline-derived
inhibitors, 8:513
substrate binding site
inhibitors, 8:514
tissue-targeted acyclic
nucleoside phosphate
prodrugs, 7:239
topical antimicrobial agents,
7:488
transplantation rejection
prevention,
5:1001–1002
validation, BioPrint® database
tools, 1:491–493
Target-engagement assays,
congestive heart
failure therapy,
4:501–502
Target hopping, chemogenomics,
2:587–588
Targeting enhancement factors,
chemical delivery
systems, 2:140–141
Target-related affinity profiling
(TRAP),
chemogenomics,
ligand-based data
analysis, 2:584–585

TAS 103 dual topoisomerase I/II
inhibitors, 6:103–105
Tasimelteon, insomnia therapy,
5:719
TASSER procedure, G-protein
coupled receptor,
homology modeling,
2:286–287
Tat protein, recombinant DNA
technology, HIV
inhibitors, 1:546–547
Tau protein:
Alzheimer's disease:
cognitive deficits, 8:16
pathology, 8:417–432
aggregation, 8:423–429
glycosylation, 8:422–423
microtubule stabiliization,
8:430
phosphorylation,
8:418–422
protein levels, 8:430–432
targeted therapy for, 8:430–432
Tautomerization:
docking and scoring
computations,
2:636–637
ligand-based drug design,
nuclear magnetic
resonance, 2:385–387
Taxol:
Hsp90 inhibitor binding,
6:427–428
ligand-based design:
Karplus relationship, 2:384
nuclear magnetic resonance,
2:376–377
structure and therapeutic
applications, 6:32–33,
36
Taxotere, structure and
therapeutic
applications,
6:34–35
Taxus diterpenes:
docetaxel/taxotere, 6:34–35
paclitaxel/taxol, 6:32–33
Tazarotene:
chemical structure, 5:402
disease therapy, 5:402–408
gene expression, 5:402–405
retinoic nuclear receptors,
5:370–372
Tazarotene-induced gene (TIG),
5:402–405
Tazobactam, β-lactamase
inhibitors, 7:360, 364

TBC-11251 compound,
peptidomimetic
design, 1:229–230
TC-5619, cognitive enhancement,
8:35–36
T-cell receptor (TCR), immune
function, 5:1002–1006
T-cells:
dipeptidyl peptidase 4
inhibitors, 4:40–41
tuberculosis pathogenesis,
7:718–719
TD-1792, antibacterial
mechanism, 2:227
TD-1792 cephalosporin,
7:314–315
"Teaching, suggestion, or
motivation" (TSM)
test, patent
obviousness principle
and, 3:135–138
Tebanicline, nicotinic agonists,
8:69
Tebipenem pivoxil, 7:358
Telavancin, antibacterial
mechanism,
2:226–227
Telbivudine:
anti-DNA virus
antimetabolites,
7:226–227
hepatitis B virus therapy,
7:246–247
Telcagepant calcitonin gene-
related peptide
antagonist, migraine
therapy, 8:277,
279–288
Teleprevir anti-HCV agent,
7:181–186
Telithromycin:
macrolidic antibiotics,
2:228–229
pharmacology, 7:441
recent developments,
7:447–456
RNA targets, 5:990–993
Telzir. See Fosamprenavir
calcium
TEM-1 β-lactamase, recombinant
DNA technology,
1:545
Tematropium, soft drug design,
anticholinergics, 2:109
Temazepam, anticonvulsant
applications,
8:144–147

Temisirolimus:
  ADMET properties, **5:**1035
  cancer therapy, **5:**1019
  cyclohexyl analog design, **5:**1033–1034
Temozolomide:
  brain tumors, **6:**244
  DNA-targeted compounds, **6:**94–95
  glioblastoma therapy, **6:**231–233
  poly(ADP-ribose)polymerase (PARP) inhibitor coadjuvants, **6:**169–170
Temperature factor, docking/scoring techniques, **2:**597–598
Temperature requirements:
  commercial-scale operations, **3:**11
  high-throughput screening, assay optimization, **3:**413–414
Temsirolimus:
  brain cancer therapy, **6:**250–251
  mTOR inhibitors, **6:**324–325
TENA κ-opioid selective antagonist, **8:**602–604
Teniposide, structure and therapeutic applications, **6:**31
Tenofovir:
  anti-DNA virus antimetabolite, emtricitabine double combination drug regimen, **7:**226
  clinical potential, **7:**155–156
  current development, **7:**145–148
  disoproxil fumarate derivative, **7:**235–236
  hepatitis B virus therapy, **7:**247
  research background, **7:**139–140
  selective toxicity, **2:**554–557
Tentoxin, chirality, **1:**136–138
Teprotide, angiotensin-converting enzyme inhibitors, **4:**268–269
Teratogenicity:
  acitretin, **5:**400–401
  adapalene, **5:**413
  ADMET toxicity properties, **2:**62, 68
  bexarotene, **5:**453
  9-*cis*-retinoic acid therapy, **5:**427
  13-*cis*-retinoic acid therapy, **5:**394
Terbutaline, soft drug design, **2:**103–104
Terfenadine, chirality and metabolism, **1:**142
Terlakiren renin inhibitor, **4:**245, 247
Ternary complex model (TCM), receptor-based drug design, **2:**497–498
  constitutive receptor activity, **2:**498–499
Terpenoids, antitubercular agents, targeting and biosynthesis, **7:**731–732
Terphenyl derivative, protein-protein interactions, **2:**349–352
Terrestrial habitats, natural products lead generation, microbial compounds, **2:**193–199
Tertiary protein structure:
  chemokine receptor-5 (CCR5) antagonist, anilide core, **7:**126–131
  protein therapeutics, biophysical assessment, **3:**324–325
Testosterone:
  absorption and distribution, **5:**158–159
  anabolic agents, **5:**179–186
  basic structure, **5:**153
  biological mechanisms, **5:**164–168
  biosynthesis, **5:**155–158
  ester derivatives, **5:**171
  halo derivatives, **5:**171–174
  metabolism, **5:**27–28
  oxidative metabolism, **5:**162–164
  reductive metabolism, **5:**159–162
  methylated derivatives, **5:**171
  occurrence and physiology, **5:**154–155
  steroidogenesis and biosynthesis, **5:**22–25
  structure-activity relationships, **5:**170–174
  urinary metabolites, **5:**161
Testosterone-estradiol binding globulin (TEBG), androgen absorption and distribution, **5:**158–159
Tethering mechanism:
  motilin receptor antagonist, **1:**307–310
  small molecule libraries, **1:**299–300
Tetomilast, chronic obstructive pulmonary disease therapy, **5:**782–784
Tetracyclic compounds:
  poly(ADP-ribose)polymerase inhibitors, **6:**167
  topoisomerase II inhibitors, **6:**102
Tetracyclines:
  antimalarial drugs, **7:**621–624
  chemical stability, side effects, toxicity and contraindications, **7:**407–409
  current compounds and clinical applications, **7:**406–407
  history and biosynthesis, **7:**410–413
  microbial resistance to, **7:**413–414
  natural products, **7:**405–406
  pharmacology and mechanism of action, **7:**409–410
  recent developments in, **7:**414–415
  RNA targets, **5:**981–984
    mechanism of action, **5:**982
    structure-activity relationship, **5:**982–983
  selective toxicity, **2:**552–553
  structure and function, **2:**223–225
Tetrahydro-4H-carbazol-4-one Hsp90 inhibitors, fragment-based design, **6:**423–424
Tetrahydrobiopeterin, stroke therapy, NOS inhibition, **8:**500

Tetrahydrofolate (THF):
  dihydrofolate reductase
    inhibitors,
    7:532–534
  folic acid uptake and
    metabolism,
    5:691–696, 698
Tetrahydroisoquinoline (THIQ),
    teleprevir anti-HCV
    agent, **7**:183–186
Tetrahydronaphthalene
    template,
    thromboxane A_2
    receptor antagonists,
    5:891, 893
Tetrahydronaphthyl derivative,
    prostaglandin/
    thromboxane
    receptors, IP receptor
    agonist, **5**:930–931
Tetrahydroquinolines:
  antimalarial drugs,
    farnesylation, **7**:634
  reverse transcriptase
    inhibitors, **7**:154–155
3,5,3′,5′-Tetraiodo-L-tyronine
    (L-T_4):
  biosynthesis, **4**:192–193
  blood transport, **4**:193
  hyperthyroidism and, **4**:199
  regulation, **4**:192
  structure-activity relationship,
    4:202–203
  thyromimetics, research
    background, **4**:189
  uptake and metabolism,
    4:193–194
Tetramic acids, protein-protein
    interactions,
    2:349–359
Tetrapeptides, peptidomimetics,
    1:235–240
TetR transcriptional regulators,
    tetracycline microbial
    resistance, **7**:414
Tezampanel, migraine therapy,
    8:295
TG-100801, age-related macular
    degeneration therapy,
    5:613–614, 623
TgAPP mice, Alzheimer's disease
    immunotherapy,
    8:366–367
Thebaine:
  Diels-Alder adducts,
    8:608–609
  opium source, **8**:599

Theiler's murine
    encephalomyelitis
    virus (TMEV),
    multiple sclerosis
    pathology, **5**:567
Theophylline:
  chronic obstructive pulmonary
    disease
    bronchodilators,
    5:772–773, 779–780
  cognitive enhancement,
    8:47–48
  methylation reactions,
    1:423–425
Therapeutic index (TI):
  plant metabolites, **2**:749
  retrometabolic drug
    development, **2**:74
  topical antimicrobial agents,
    selective toxicity,
    7:488
Thermal behavior, salt
    compounds, **3**:381
Thermodynamic integration (TI),
    quantum mechanics/
    molecular mechanics
    binding affinity,
    2:728–729
Thermodynamics, drug-
    macromolecule
    interactions,
    2:702–712
Thermospray, liquid
    chromatography-mass
    spectrometry,
    1:101–105
Thiacetazone, antitubercular
    agents:
  history, **7**:738
  second-line applications,
    7:773–774
Thia derivatives, anabolic agents,
    5:183
Thiadiazole, stroke therapy,
    cathepsin B inhibitors,
    8:475
Thiadiazolidine-dione,
    Alzheimer's disease,
    tau-targeted
    therapies, **8**:420–422
Thiamet G, Alzheimer's disease,
    tau-targeted therapy,
    8:423
Thiamine (vitamin B_1),
    5:668–673
  chemistry, **5**:669–670
  deficiencies, **5**:670

hypervitaminosis thiamine,
    5:670
uptake and metabolism,
    5:670–672
Thiamine chloride hydrochloride,
    chemistry, **5**:670
Thiamine pyrophosphate (TPP),
    formation and
    metabolism,
    5:670–672
Thiazole acids, DP_1 receptor
    antagonists, **5**:905
Thiazole sulfonamide,
    prostaglandin/
    thromboxane EP_1
    antagonists,
    5:820–824
Thiazolidinediones:
  peroxisome proliferator-
    activator receptors, γ-
    agonists, **4**:100–106
  survival signaling pathways,
    6:323
  thyromimetic structure-
    activity relationship,
    heterocyclic rings,
    4:210–211
  toxicophore reactive
    metabolites, **2**:322
Thienamycin:
  β-lactamase inhibitors,
    7:360
  occurrence, structural
    variations and
    chemistry, **7**:333–339
  trinem activity, **7**:350–353
Thienobenzodiazepine,
    pharmacokinetics,
    8:192–193
Thienopyrimidines,
    prostaglandin/
    thromboxane
    receptors, EP_4
    agonists, **5**:878–880
Thin film kinetics, oral drug
    delivery, sustained-
    release dosage forms,
    2:794–795, 799
Thioamides, toxicophore reactive
    metabolites,
    2:318–319
Thiocarbamates, toxicophore
    reactive metabolites,
    2:318–319
Thiocarbocyanines, Alzheimer's
    disease, tau-targeted
    therapy, **8**:427

Thiocarboxanilide UC-781 HIV reverse transcriptase inhibitor, **7:**150–152
6-Thioguanine, antimetabolite analogs, **6:**112–113
Thiolactomycin:
 antimalarial drugs, apicoplast targeting, **7:**665
 parasite lipid metabolism inhibitor, **7:**580–581
Thiols:
 drug metabolism:
  methylation reactions, **1:**423–425
  oxidation and reduction reactions, **1:**418–419
 stroke therapy:
  HDAC inhibitors, **8:**491
  MMP inhibitors, **8:**509
Thionine, Alzheimer's disease, tau-targeted therapy, **8:**424–425
Thiophene derivatives:
 prostaglandin/thromboxane EP$_1$ antagonists, **5:**821, 823, 826, 829
 prostaglandin/thromboxane EP$_3$ receptor antagonists, **5:**858, 860–861
 toxicophore reactive metabolites, **2:**314–317
Thiophenoamidine-benzothiazoles, NOS inhibitors, migraine therapy, **8:**303–306
Thiophenoamidine-indoles, migraine therapy, NOS inhibitor binding affinity, **8:**305–306
Thiopurine methyltransferase (TPMT):
 deficiency, azathioprine therapy, **5:**1051–1053
 single nucleotide polymorphisms, **1:**188–189
Thioridazine, pharmacokinetics, **8:**187–189
Thiosemicarbazones, antimalarial falcipain inhibitors, **7:**637
Thiothixene, pharmacokinetics, **8:**188–189
Thioureas:
 quantitative structure-activity relationship, **1:**61–62
 toxicophore reactive metabolites, **2:**318–319
Thiovir®, anti-DNA virus agent, **7:**240
Thioxanthenes, pharmacokinetics, **8:**188–189
Third generation compounds:
 anticonvulsants, **8:**136–139
 anti-DNA virus antimetabolites, **7:**228–229
 cephalosporin development, **7:**302–306
 kinase inhibitors, multidrug resistance, **6:**372
Three-dimensional quantitative structure-metabolism relationships (3D-QSMRs), drug metabolism, **1:**445–447
Three-dimensional structures:
 angiotensin-converting enzyme, **4:**289–292
 combinatorial quantitative structure-activity relationship modeling, **1:**519–520
 pharmacophores, **1:**457–464
  biological targets, **1:**474–476
  bound ligands, **1:**462–465
  encoded representations, **1:**471
  historical background, **1:**457–458
  ligand-based virtual screening, **2:**5–10
  quantitative structure-activity relationship, **1:**39–43
  limitations, **1:**74
  quantitative structure-activity relationships, glucocorticoid anti-inflammatory compounds, **5:**77–80
Three-finger p53/MDM2 antagonist model, protein-protein interactions, **2:**354–355
Three-point attachment model, chiral compounds, **1:**127–129

Thrombin:
 active site-directed inhibitors, **4:**395–398
 direct thrombin inhibitors mechanism of action, **4:**385–390
 heparin molecular mechanisms, **4:**374–380
 platelet activation, **4:**414–415
 receptor (protease-activated receptor-1) antagonists, **4:**437–438
  activation mechanisms, **4:**438–439
  animal studies, **4:**440–441
  future research issues, **4:**452–453
  peptidomimetics, **4:**441–442
  pharmacology, **4:**439–440
  small-molecule antagonists, **4:**442–444
 thrombosis role of, **4:**411
Thrombin binding domain (TBD), indirect anticoagulants, **4:**392
Thrombin receptor activating peptide (TRAP), thrombin receptor antagonists, **4:**442–444
Thrombopoietin (TPO):
 receptor agonist:
  eltrombopag, **4:**584–586
  romiplostim, **4:**583–584
 structure and properties, **4:**582–586
Thrombosis:
 acute coronary syndrome, **4:**409–410
 antiplatelet agents, antithrombotic action, **4:**411–412
 platelet function, **4:**410–411
 risk stratification and therapeutic guidelines, **4:**411
Thromboxane antagonists, antiplatelet agents, **4:**419
Thromboxane receptors:
 A$_2$ receptor:
  antagonists, **4:**422
  aspirin biosynthesis, **4:**419–420

dual synthase inhibitor and
antagonist, 4:423
multitarget drug
development,
histamine H_1-plus for
allergies, 1:259–262
synthase inhibitor,
4:422–423
basic properties, 5:809–812
DP_1 receptor:
agonists, 5:900–901
antagonists, 5:899–907
acidic antagonists, 5:905
clinical trials, 5:907
fused-tricyclic antagonists,
5:903–905
hemi-prostanoid
antagonists,
5:901–902
indole acetic acid,
5:902–903
nonacid-based
antagonists,
5:905–906
EP_1 receptor:
agonists, 5:812–813
antagonists, 5:813–846
amide replacements,
5:818–820, 836–837
AstraZeneca derivatives,
5:816–817
benzene derivatives, 5:829,
831, 834
benzoic acid derivatives,
5:823, 827–832
carbamates, 5:836–838
cyclopentene derivatives,
5:829, 831, 833
diacyl hydrazide
replacement,
5:815–817
glycine sulfonamide
derivatives, 5:837,
840–845
Ono derivatives, 5:817–824
pyrazole derivatives,
5:831, 835–838
pyrrole derivatives, 5:826,
828–832
SC-19220 series,
5:813–814
Searle derivatives,
5:813–816
sulfonylureas, 5:820–821,
825
thiophene derivatives,
5:821, 823, 826, 829

EP_2 receptor:
agonists, 5:846–850
antagonists, 5:852–854
mixed EP_2/E_4 agonists,
5:850–852
EP_3 receptor, 5:854–872
agonists, 5:854–855
antagonists, 5:855–872
acylsulfonamide
derivatives, 5:855–858
angiotensin II derivatives,
5:856–858
hyperalgesia/edema
modeling, 5:858, 861
indole analogs, 5:861,
863–869
thiophene derivatives,
5:858, 860–861
EP_4 receptor, 5:872–883
agonists, 5:872–883
lactam stereochemistry,
5:873–878
thienopyrimidine-based
compounds, 5:878–880
ω-chain derivatives,
5:876–876
antagonists, 5:880–883
FP receptor, 5:897–899
thromboxane A_2 receptor,
5:883–897
agonists, 5:886–887
antagonists, 5:886–896
4-oxazolylcarboxamide
analogs, 5:886–889
phenol derivatives,
5:888–889, 891
racemic 1,3-dioxolane
derivatives,
5:887–889, 891
Ramatroban analogs,
5:888–889, 891–892
semicarbazone
compounds, 5:886–889
sulfonamide analogs,
5:887–890
tetrahydronaphthalene
template, 5:891, 893
thromboxane synthase
inhibitors, 5:894–896
trimetoquinol, 5:891,
893–894
UK-147,535 derivative,
5:894
calcium mobilization,
5:896–897
Thromboxane synthase
inhibitors:

prostaglandin/thromboxane
receptors, IP receptor
agonist, 5:935,
938–939
thromboxane A_2 receptor
antagonists,
5:894–896
Thymidine analog mutations:
antimetabolites, anti-DNA
virus agents, first
generation
compounds, 7:222–225
HIV reverse transcriptase
inhibitors, 7:144–145
Thymidylate synthase (TS):
antifolate DNA-targeted
chemotherapeutics,
6:106–109
antimalarial antifolates,
7:644–648
pyrimidine antimetabolite
analogs, 6:109–110
Thyroglobulin (Tg), thyroid
hormone biosynthesis,
4:192–193
Thyroid cancer:
bexarotene therapy, 5:444, 447
13-cis-retinoic acid therapy,
5:392
Thyroid hormone receptor-like
family, receptor-based
drug design, 2:512
Thyroid hormones:
bone and skeletal muscle, 4:196
carbohydrate metabolism,
4:195
central nervous system, 4:196
classification and basic
properties, 4:189
heart and cardiovascular
system, 4:195–196
lipid metabolism, 4:195
liver, 4:196–197
metabolic rate, 4:194–195
nuclear magnetic resonance,
ligand-based designs,
line-shape and
relaxation data,
2:387–389
physiology, 4:192–194
biosynthesis, 4:192–193
blood transport, 4:193
regulation, 4:192
uptake and metabolism,
4:193–194
protein metabolism, 4:195
receptors, 4:200–202

Thyroid hormones (*Continued*)
agonists, **4:**197–198, 200
antagonists, **4:**198–199, 201–202
β-selective agonists, structure-based design, **2:**684–692
hormone binding site, **2:**684–685
TR:T3 crystal complexes, **2:**685–691
domain structure, **4:**189–190
intervention, **4:**206
isoforms, **4:**191–192, 206
ligand binding domain, **4:**200–201
molecular mechanism, **4:**189–190
domain structure, **4:**189–190
receptor isoforms, **4:**191–192
receptors, **4:**189
transcriptional regulation, **4:**190–191
R_1-position variations, **4:**206–207
retinoic nuclear receptors, **5:**371–372
skin, **4:**197
therapeutic potential:
endogenous hormones, **4:**197
receptor agonists, **4:**197–198
receptor antagonists, **4:**198–199
thyromimetics structure-activity relationship, **4:**202–222
axitirome, **4:**206–207
azauracil compounds, **4:**215–216
bioisosteric replacement, obligatory R'_4-hydroxy group, **4:**218–219
bridging group X, **4:**213
cytochrome P450-activated MB07811 prodrug, **4:**211–212
DITBA, **4:**205
early thyromimetics, **4:**205–222
epotirome, **4:**209–210
future research issues, **4:**222
heterocyclic rings, **4:**210–211
ICI164384, **4:**213–214
indole derivatives, **4:**212–213
KB-141, **4:**208–209, 216–218

L-T_3 and L-T_4, **4:**202–203
pharmacological selectivity strategies, **4:**205–206
phenylnaphthyl linker, **4:**214–215
prime ring modification, **4:**221–222
recent thyromimetics, **4:**205–222
receptor isoform position variation, **4:**206–212, 215
ring-closure, receptor positions, **4:**212–213, 218
ring substitution, R'_3 and R'_4 positions, **4:**219–221
SKF-94901, **4:**203
sobetirome, **4:**207–208
TRIAC, **4:**204
Thyroid releasing hormone (TRH), regulation and physiology, **4:**192
Thyroid response elements (TREs), transcriptional regulation, **4:**190
Thyroid stimulating hormone (TSH):
hypothyroidism, **4:**197
regulation and physiology, **4:**192
Thyroid stimulating hormone receptor (TSHR), loss-of-function mutation, **2:**515
Thyromimetics:
discovery tools for, **4:**205
research background, **4:**189
structure-activity relationship, **4:**202–222
axitirome, **4:**206–207
azauracil compounds, **4:**215–216
bioisosteric replacement, obligatory R'_4-hydroxy group, **4:**218–219
bridging group X, **4:**213
cytochrome P450-activated MB07811 prodrug, **4:**211–212
DITBA, **4:**205
early thyromimetics, **4:**205–222
epotirome, **4:**209–210
future research issues, **4:**222

heterocyclic rings, **4:**210–211
ICI164384, **4:**213–214
indole derivatives, **4:**212–213
KB-141, **4:**208–209, 216–218
L-T_3 and L-T_4, **4:**202–203
pharmacological selectivity strategies, **4:**205–206
phenylnaphthyl linker, **4:**214–215
prime ring modification, **4:**221–222
recent thyromimetics, **4:**205–222
receptor isoform position variation, **4:**206–212, 215
ring-closure, receptor positions, **4:**212–213, 218
ring substitution, R'_3 and R'_4 positions, **4:**219–221
SKF-94901, **4:**203
sobetirome, **4:**207–208
TRIAC, **4:**204
therapeutic applications, **4:**197–199
Thyronamine (T$_0$AM), 3,5,3',5'-tetraiodo-L-tyronine (L-T_4) uptake and metabolism, **4:**194
Thyroperoxidase (TPO), thyroid hormone regulation and biosynthesis, **4:**192–193
Thyrotoxicosis, thyroid hormone receptor antagonists, **4:**199
Thyroxine, nuclear magnetic resonance, ligand-based designs, line-shape and relaxation data, **2:**387–389
Thyroxine binding globulin (TBG), thyroid hormone blood transport, **4:**193
Thyroxine binding prealbumin (TBPA), thyroid hormone blood transport, **4:**193
Thyroxine response elements (TREs), estrogen tissue specificity, **5:**241–242
Tiagabine, anticonvulsant applications, **8:**135–139, 152

Tianeptine, structure-activity relationship, **8:**255
TIBO derivatives:
 current development, **7:**148–152
 research background, **7:**140
Tibolone, physiology and pharmacology, **5:**239
Tibotec/Medivir protease inhibitor. *See* TMC435350 anti-HCV agent
Ticagrelor:
 antiplatelet activity, **4:**436–437
 future research issues, **4:**452
Ticarcillin, penicillin resistance and development of, **7:**291–292
Ticlopidine, antiplatelet activity, **4:**426–427
Tigecycline:
 ADMET properties, **7:**410
 clinical applications and development, **7:**406–407
 history and biosynthesis, **7:**411–413
 recent developments in, **7:**414–415
 stability, side effects, and contraindications, **7:**408–409
Tigemonam, monobactam activity, **7:**329–331
Time intervals between doses, repeated intravenous bolus injections, **2:**756–757
Timentin, β-lactamase inhibitors, **7:**361–363
Time-of-flight (TOF) instrumentation, mass spectrometry development, **1:**98
 combinatorial library structure and purity confirmation, **1:**109–110
Timescale regimes, nuclear magnetic resonance, macromolecule-ligand interactions, **2:**394–402
Timolol, glaucoma intraocular pressure reduction, **5:**596–600

Tiotropium:
 chronic obstructive pulmonary disease bronchodilators, **5:**767
 muscarinic antagonist, **8:**72
Tipifarnib, parasite lipid metabolism inhibitor, **7:**580–581
TIPP-204 PPARδ antagonists, **4:**122–124
Tipper-Strominger hypothesis, D-Ala-D-Ala mimicry, penicillin proteins and, **7:**264–266
TIPP peptide, opioid peptide analogs, Tyr-Pro-Phe sequence, **8:**661–664
Tipranavir:
 binding interactions, **7:**46, 50
 chemical structure, **7:**45–46
 efficacy and tolerability, **7:**48–49, 51
 optimization, **7:**46–49
 pharmacokinetics, **7:**46, 48, 50
 resistance profile, **7:**51–52
Tipredane:
 pharmacological profiles, **5:**101–102
 soft drug design, **2:**100
Tirapazamine, hypoxia-activated bioreductive prodrugs, **6:**115–120
Tirlazad, stroke therapy, **8:**480
Tirofiban, antiplatelet activity, **4:**447
Tissue associated antigens (TIAs), central nervous system tumors, vaccine development, **6:**269–270
Tissue factor (TF):
 clotting initiation, **4:**366–367
 serine proteases, **1:**332–333
Tissue plasminogen activator (t-PA):
 recombinant DNA technology, **1:**540
 stroke therapy, **8:**447–448
Tissue specificity:
 ADMET distribution properties, **2:**61, 66
 blister fluid drug transport, **2:**844–848
 endocarditis drug transfer, **2:**849–851

 estrogens and progestins, **5:**240–242
 LDL cholesterol lowering statins, **4:**309–310
 lungs and bronchial mucus, **2:**839–844
 mathematical/numerical calculations, **2:**840–842
 pharmacokinetic parameters, **2:**841–844
 nomenclature, **2:**838–839
 recombinant DNA technology, **1:**547
 thyromimetic selectivity, **4:**205–206
 tissue-targeted acyclic nucleoside phosphate prodrugs, **7:**239
 vitamins, **5:**638
Titration properties, nuclear magnetic resonance, macromolecule-ligand interactions, **2:**403–404
Tivirapine HIV reverse transcriptase inhibitor, **7:**148–152
TLR3 protein apoptosis, age-related macular degeneration therapy, **5:**621–622
TMC435350 anti-HCV agent, **7:**193
Tobramycin:
 chemical structure, **7:**416–419
 current applications, **7:**416
Toceranib, brain tumor therapy, **6:**249
Tocopherols:
 antioxidant supplement controversy, **5:**643–644
 biochemical function, **5:**662–664
 chemistry, **5:**660
 deficiencies, **5:**662
 dosage forms, **5:**664–665
 hypervitaminosis E, **5:**662, 664
 uptake and metabolism, **5:**660–662
Tocotrienols:
 antioxidant supplement controversy, **5:**643–644

Tocotrienols (*Continued*)
  biochemical function, **5**:662–664
  chemistry, **5**:660
  deficiencies, **5**:662
  dosage forms, **5**:664–665
  hypervitaminosis E, **5**:662, 664
  uptake and metabolism, **5**:660–662
Tofimilast, chronic obstructive pulmonary disease therapy, **5**:783–784
Tolcapone:
  Parkinson's disease, dopaminergic therapy, **8**:550–552
  toxicophore reactive metabolites, **2**:312–313
Tolerability assessment:
  clopidogrel, **4**:432
  HIV protease inhibitors:
    amprenavir, **7**:28–29
    atazanavir sulfate, **7**:40–42
    darunavir, **7**:59
    fosamprenavir calcium, **7**:44–45
    indinavir sulfate, **7**:17–18
    lopinavir/ritonavir, **7**:32, 34, 36
    nelfinavir mesylate, **7**:22
    ritonavir, **7**:12
    saquinavir mesylate (SQV), **7**:8
    tipranavir, **7**:48–49, 51
  opioid receptors, **8**:574–575
Toll-like receptor (TLR):
  adapalene therapy, **5**:411
  carbon monoxide therapeutic effects, **5**:298–299
  immune function, **5**:1002–1006
Tolterodine, muscarinic antagonists, **8**:74–77
Toluidine blue, Alzheimer's disease, tau-targeted therapy, **8**:424–425
Tomopenem, development of, **7**:354–357
Tonabersat, migraine prevention, **8**:316
Tool compounds, BioPrint® database, **1**:491–493
Topical agents:
  all-*trans*-retinoic acid, **5**:385
  corticosteroids, **5**:49–50
  glucocorticoid anti-inflammatory compounds, **5**:85
OT-551, age-related macular degeneration, **5**:622
synthetic antibacterials, **7**:484–501
  acids, **7**:496
  alcohols, **7**:492–493
  amphoteric surfactants, **7**:501
  anionic surfactants, **7**:500–501
  cationic surfactants, **7**:499–500
  cellular targets, **7**:488
  diarylureas, amidines, and biguanides, **7**:498–499
  dyes, **7**:498
  epoxides and aldehydes, **7**:495–496
  evaluation, **7**:489–490
  halogens and halophores, **7**:490–492
  heavy metals, **7**:497–498
  historical development, **7**:490
  kinetics, **7**:488–489
  oxidizing agents, **7**:496–497
  phenols, **7**:493–495
  resistance mechanisms, **7**:488
  selective toxicity, **7**:484, 488
  terminology, **7**:484
Topiramate:
  anticonvulsants, **8**:133–139, 151–152
  migraine prevention, **8**:315–316
Topoisomerase inhibitors:
  antimalarial drugs, **7**:650–652
  synthetic chemotherapeutic compounds, **6**:95–106
    dual topo I/II inhibitors, **6**:103–106
    Topo II inhibitors, **6**:98–102
    Topo I inhibitors, **6**:96–98
Topological descriptors, quantitative structure-activity relationships, **1**:23–24
  polar surface area (TPSA), **1**:25–26
Topotecan:
  brain cancer therapy, **6**:254–255
  poly(ADP-ribose)polymerase (PARP) inhibitor coadjuvants, **6**:169–170
  structure and applications, **6**:26–29
Torcetrapib, atherosclerosis, HDL elevation, **4**:341–342
Toremifene:
  current compounds and applications, **5**:219–221
  oxidative metabolism, **5**:226
  structure-activity relationships, **5**:249–253
TOR proteins, sirolimus-FKBP 12 complex, mechanism of action, **5**:1026–1029
Torsades des pointes, antipsychotic side effects, **8**:167, 179–180
Tosyl compounds, novobiocin Hsp90 inhibitor optimization, **6**:439–440
Total correlation spectroscopy (TOCSY), bioactive peptides, **2**:377–383
ToxCast project, quantitative structure-activity relationship modeling, **1**:513–514
Toxicity:
  ADMET properties, **2**:61–63, 68–69
  arrhythmias, **2**:63
  cytotoxicity, **2**:62
  drug-drug interactions, **2**:63
  genotoxicity/mutagenicity, **2**:62
  permeability, ADMET/PK interactions, **3**:377
  teratogenicity, **2**:62
  allosteric protein inhibition, **1**:407
  aminoglycosides, **7**:419–420
  RNA targets, **5**:979–981
  androgens:
    anabolic agents, **5**:188
    synthetic androgens, **5**:176
  antiandrogen compounds, **5**:193
  anti-DNA virus antimetabolites, **7**:226–227
  antimalarial drugs:
    artemisinins, **7**:617

chloroquine, 7:610
mefloquine, 7:611
primaquine, 7:614
quinine, 7:607
chirality, 1:143–147
  amine-based platinum complexes, 1:144–145
  carcinogenicity, 1:145–146
  clinical chemistry, 1:146–147
  environmental toxicity, 1:147
  hERG activity, 1:143–144
chloramphenicol, 7:462–463
CK-2130 prodrug case study, 3:273–276
dipeptidyl peptidase 4 inhibitors, 4:42
drug metabolism, 1:441–443
FDA information guidelines, 3:75
linezolid, 7:544
macrolides, 7:438
perfluorocarbons, oxygen delivery, 4:621–622
prodrug development, 3:225–226
pulsed ultrafiltration-mass spectrometry screening, 1:118–119
quantitative structure-activity relationship:
  ADMET properties, 1:67–68
  effective hydrogen charge, 1:27–28
  ensemble modeling and consensus prediction, 1:523–525
retrometabolic drug development, 2:74
selective toxicity:
  antibiotics, 2:548–553
  antifungal drugs, 2:558–559
  antihistamines, 2:564–566
  antimycobacterial drugs, 2:557–558
  antivirals, 2:553–557
  bisphosphonates, 2:564
  comparative biochemistry, 2:545
  comparative cytology, 2:545–546
  comparative distribution, 2:545
  comparative stereochemistry, 2:546–547
  future research issues, 2:569
  levobpuivacaine, 2:564

monoamine oxidase inhibitors, 2:563–564
monoclonal antibodies, 2:559–562
nonbenzodiazepine sedatives, 2:568
phosphodiesterase inhibitors, 2:567–568
proton pump inhibitors, 2:566
research background, 2:541
selective estrogen receptor modulators, 2:566–567
selective serotonin reuptake inhibitors, 2:569
therapeutic agent categorization, 2:541–544
topical antimicrobial agents, 7:484, 488
tyrosine kinase inhibitors, 2:562–563
single nucleotide polymorphisms, 1:181–182
structural alerts:
  alternative metabolic pathways, 2:326
  chemical reactivity, 2:301–302
  CYP450 enzyme binding, 2:324–325
  DNA binding, 2:325
  low exposure benefits, 2:325–326
  reactive metabolites, 2:303–324
    N-alkyl-tetrahydropyridines, 2:322–323
    alkynes, 2:320
    2-aminohiazoles, 2:319–320
    anilines and hydroxyanilines, 2:303–311
    arylhydroxamic acids, 2:313
    benzodioxolanes, 2:314
    carboxylic acid acyl glucuronides, 2:323–324
    electron-rich heterocycles, 2:314–317
    hydantoins, 2:322
    hydrazines, 2:320–321

    3-methylindoles, 2:317–318
    nitro-aromatics/nitro-alkyls, 2:312–313
    ortho/para-dihydroxyaromatics, 2:313–314
    screening for, 2:326–327
    thiazolidinediones, 2:322
    thioamides, thioureas, and thiocarbamates, 2:318–319
  research background, 2:301
  risk analysis, 2:325–327
Toxicity data sets, quantitative structure-activity relationship modeling, 1:513–514
Toxicophores:
  alternative metabolic pathways, 2:326
  chemical reactivity, 2:301–302
  combinatorial chemistry, 1:296
  CYP450 enzyme binding, 2:324–325
  DNA binding, 2:325
  drug metabolism, 1:441–443
  low exposure benefits, 2:325–326
  reactive metabolites, 2:303–324
    N-alkyl-tetrahydropyridines, 2:322–323
    alkynes, 2:320
    2-aminohiazoles, 2:319–320
    anilines and hydroxyanilines, 2:303–311
    arylhydroxamic acids, 2:313
    benzodioxolanes, 2:314
    carboxylic acid acyl glucuronides, 2:323–324
    electron-rich heterocycles, 2:314–317
    hydantoins, 2:322
    hydrazines, 2:320–321
    3-methylindoles, 2:317–318
    nitro-aromatics/nitro-alkyls, 2:312–313
    ortho/para-dihydroxyaromatics, 2:313–314
    screening for, 2:326–327
    thiazolidinediones, 2:322
    thioamides, thioureas, and thiocarbamates, 2:318–319

Toxicophores (*Continued*)
risk analysis, **2:**325–327
in silico filtering, **2:**327
TR3 expression, AHPN-induced apoptosis, **5:**492
TRA-418 IP receptor agonist, prostaglandin/thromboxane receptors, **5:**939
Trabectedin:
marine sources, **2:**246–247
microbial sources, **2:**235
Traceless linker technology, benzodiazepine library, **1:**301–302
Trademarks, drug discovery and biotechnology, **3:**171–179
global rights, **3:**177–178
Lanham Act protections, **3:**178–179
as marketing tools, **3:**172
oppositions and cancellations, **3:**176–177
preservation of rights through proper use, **3:**177
registration process, **3:**174–176
selection criteria, **3:**172–174
Trade secrets, drug discovery and biotechnology, **3:**179–184
defined, **3:**180
enforcement, **3:**181–182
Freedom of Information Acts, **3:**183–184
global protection, **3:**184
patent protection and, **3:**182–183
protection requirements, **3:**180–181
Traffic lights, virtual screening data, **2:**26
Tramadol nonpeptide opiate, **8:**643–645
Tramiprosate glycosaminoglycan mimetic, **8:**362
Trance amine receptor, schizophrenia genetics, **8:**174
Transactivation, glucocorticoid receptor structure, **5:**62–63
Transcription activation factors (TAFs), estrogen receptor topography, **5:**234–236

Transcriptional regulation, thyroid hormones, **4:**190–191
Transcription factors:
antidepressant efficacy assessment, **8:**226–229
retinoid nuclear receptors, **5:**369
Transcytosis, receptor-mediated endocytosis, membrane proteins, **2:**448
Transdermal drug delivery:
bioavailability, pharmacokinetic stage, **2:**747
future research issues, **2:**835–836
system characteristics, **2:**836–838
liposomes, nanoscale technology, **3:**472–473
mechanisms, **2:**829–830
metropolol experiment, **2:**832–835
nomenclature, **2:**828–829
pharmacokinetic stage, **2:**747
plasma drug profile, **2:**830–832
protein therapeutics, **3:**316
skin properties, **2:**830
Transfected cell lines, receptor-based drug design, **2:**524
Transfection, protein therapeutics, **3:**295–296
Transferases:
drug metabolism, **1:**421–423
high-throughput screening, enzymatic assays, **3:**424–425
Transferred nuclear Overhauser effect (TrNOE) technique:
ligand-based design, **2:**390
screening procedures, **2:**428–429
Transferrin:
receptor-mediated endocytosis, **2:**454
transcytosis, **2:**448
Transfer RNA (tRNA), recombinant DNA technology, **1:**546

Transforming growth factor-β, brain tumor therapy, **6:**267
Transfusion therapy, carbon monoxide, **5:**307–308
Transgenic mouse models:
protein therapeutics, upstream processing, **3:**291–294
target validation, congestive heart failure therapy, **4:**488
Transition-state isosteres:
BACE (β-site cleavage enzyme) inhibitors, **8:**347–351
HIV protease inhibitors:
design criteria, **7:**3–4
saquinavir mimetic screening, **7:**5
γ-secretase inhibitors, **8:**334–335
Transition state-like covalent, reversible inhibitors, dipeptidyl peptidase 4 (DPP-4) inhibition, **4:**50–54
Transition structure, allosteric proteins, **1:**380–381
Translational medicine:
congestive heart failure:
direct efficacy readouts, **4:**502
drug development:
bench-to-bedside candidates and attrition development, **4:**482–485
drug attrition, **4:**480–482
emerging therapies, **4:**479–480
critical evaluation, **4:**483
epidemiology and prevalence, **4:**477–478
etiology, **4:**477
future research issues, **4:**507
iontophoresis methodologies, **4:**506
management approaches, **4:**478–479
noninvasive imaging, **4:**505–506
patient selection and stratification, **4:**506–507
pharmacodynamic activity, **4:**502–503
cardiotoxicity biomarkers, **4:**503

dose selection, **4:**502–503
polymorphisms and
  phenotypes, **4:**507
predictive biomarkers,
  disease modification,
  4:503–506
response-to-treatment
  studies, **4:**507
target-compound interaction:
  predictive biomarkers:
    ex vivo cell based
      biomarker assays,
      4:500
    hemodynamic/
      biochemical
      biomarker assays,
      4:500–501
    PET/SPECT ligands,
      4:501
    target-engagement assays,
      4:501–502
target validation, **4:**483–500
  clinical congruency,
    4:489–490
  clinical importance, **4:**483,
    485–487
  current clinical trials,
    4:497–498
  knockout/transgenic
    mouse models,
    4:488–490
  mechanism perturbation,
    4:494–495
  mechanism-related risk
    and reliability, **4:**500
  mechanistic approaches,
    cardiac contractility/
    myocardial oxygen
    demand, **4:**492–494
  normal patient
    mechanisms *vs.*,
    4:495–496
  novel target *vs.* existing
    therapy, **4:**496,
    499–500
  pharmacodynamics/
    pharmacokinetics,
    4:491–492
  preclinical models,
    4:486–491
  tissue studies, species and
    disease states,
    4:488
predictive biomarkers,
  utilitarian
  classification,
  4:483–485

prodrug development, prontosil
  case study, **3:**258
Translocation pathway, P-
  glycoprotein, **6:**376
7-Transmembrane receptors
  (7TMRs):
  chemogenomics, privileged
    scaffolds, **2:**576–577
  G-protein coupled receptor,
    homology modeling,
    2:279–283
  receptor-based drug design,
    2:505–506
  GPCR topology, **2:**506–508
Transmembrane AMPA receptor
  regulatory proteins
  (TARPs), long-term
  potentiation, **8:**24–25
Transmembrane domains
  (TMDs):
  ATP binding cassette (ABC)
    transporters, **6:**371
  opioid receptors, **8:**587–591
  receptor-mediated endocytosis,
    membrane proteins,
    2:447–448
Transplantation surgery:
  carbon monoxide inhalation
    therapy, **5:**300–301
  immunosuppressive drugs,
    rejection prevention:
    azathioprine, **5:**1050–1053
    cycolsporin A, **5:**1006–1013
    deforolimus, **5:**1037
    diaryl urea inhibitors,
      5:1042–1044
    everolimus, **5:**1035–1037
    immune system overview,
      5:1002–1006
    inosine monophosphage
      dehydrogenase
      inhibitors,
      5:1038–1050
    mizoribine, **5:**1044–1045,
      1049–1050
    mycophenolate mofetil,
      5:1047–1049
    mycophenolate sodium,
      5:1049–1050
    mycophenolic acid analogs,
      5:1039–1042
    pimecrolimus, **5:**1023–1024
    research background,
      5:1001–1002
    ribavarin, **5:**1044–1046
    sirolimus, **5:**1024–1037
    tacrolimus, **5:**1013–1024

    temsirolimus, **5:**1037
    zotarolimus, **5:**1037
Transport proteins:
  anticonvulsants, **8:**130–132
  antidepressant pharmacology,
    8:237–248
  cocaine structure-activity
    relationship, **8:**109,
    111–113
  drug development, **1:**391–392
    apical sodium-dependent bile
      acid transporter,
      2:467–475
    ATP-binding cassette and
      solute carrier genetic
      superfamilies, **2:**456
    classification and structures,
      2:454–455
    Cys-scanning mutagenesis,
      2:460
    excimer fluorescence, **2:**461
    insertion scanning, **2:**460
    integral structure, **2:**459
    intestinal peptide
      transporter example,
      2:463–467
    models and substrate design,
      2:459
    N-glycosylation and epitope
      scanning
      mutagenesis,
      2:460–461
    P-glycoprotein example,
      2:461–463
    site-directed chemical
      cleavage, **2:**461
    therapeutic implications,
      2:457–459
  permeability, ADMET
    interactions,
    2:55–56,
    3:367–369
  psychostimulant physiology
    and pharmacology,
    8:100–106
  recent developmenta,
    8:113–114
Transposons, β-lactam
  antibacterial drug
  resistance,
  conjugative
  transposons,
  7:270–271
Transthyretin (TTR):
  amyloidosis inhibitors,
    structure-based
    design, **2:**692–702

Transthyretin (*Continued*)
  natural products and
    derivatives, **2:**701–702
  NSAIDs, **2:**694–702
  TTR fibril formation
    inhibitors, **2:**693–702
  TTR T4 binding site,
    2:692–693
  thyroid hormone blood
    transport, **4:**193
Transverse relaxation-optimized
    spectroscopy
    (TROSY):
  development and
    instrumentation,
    2:373–374
  receptor-based drug design,
    macromolecular
    compounds, **2:**393–394
Tranylcypromine:
  chemical structure, **8:**223
  Parkinson's disease,
    dopaminergic therapy,
    8:549–551
  physiology and pharmacology,
    8:248
  side effects and adverse
    reactions, **8:**230
  structure-activity relationship
    and metabolism,
    8:252–253
"Trapping blockers," stroke
    therapy, **8:**450
Trastuzumab:
  geldanamycin Hsp90 inhibitor,
    6:396–397
  growth factor signaling
    inhibitors, **6:**302–305
  recombinant DNA technology,
    drug efficacy and
    personalized
    medicine, **1:**543
  resistance to, **6:**369
Travoprost, glaucoma intraocular
    pressure reduction,
    5:592–600
Traxoprodil, stroke therapy,
    8:453
Trazodone:
  chemical structure, **8:**224
  pharmacokinetics, **8:**236–237
  serotonin receptor targeting,
    8:241–242
  side effects and adverse
    reactions, **8:**234
Tretinoin, vitamin A
    pharmacology, **5:**649

TRH analogs, chemical delivery
    systems, **2:**153
TRIAC thyromimetic, structure-
    activity relationship,
    4:204
Triaging of hit/lead compounds,
    virtual screening data,
    2:26–27
Triamcinolone:
  hydroxylation and discovery of,
    5:9
  intra-articular administration,
    5:52–53
  percutaneous absorption,
    5:46–51
Triarylethylenes:
  adverse effects and
    precautions, **5:**221
  oxidative metabolism,
    5:225–226
  structure-activity
    relationships,
    5:249–253
Triarylheterocycles, estrogen
    receptor interaction,
    5:253
Triazines:
  DNA-targeted compounds,
    6:94–95
  quantitative structure-activity
    relationship, isolated
    receptor interactions,
    1:51–57
Triazolam, insomnia therapy,
    5:711, 716–718
Triazoles:
  aromatase inhibition, **5:**232
  γ-secretase modulators, **8:**359
  tropane-based chemokine
    receptor-5 (CCR5)
    antagonist, **7:**119–121
Z-Triazolylmethylenepenem, β-
    lactamase inhibitors,
    7:365–366
Triazolyl-phenyl compounds,
    antimalarial HDAC
    inhibitor, **7:**654–655
Tribactams. *See* Trinems
Trichostatin A, histone
    deacetylase inhibitor
    binding, **6:**51–53
Triclosan:
  antitubercular agents, mycolic
    biosynthesis, **7:**723
  apicoplast targeting, fatty acid
    biosynthesis,
    7:663–665

Tricyclic antidepressants (TCAs):
  chemical structure, **8:**222
  dosing regimens, **8:**219–221
  monoamine transporter
    inhibition, **8:**238–241
  pharmacokinetics, **8:**235–236
  side effects and adverse
    reactions, **8:**231–233
  stroke therapy, **8:**452–453
  structure-activity relationship
    and metabolism,
    8:249–251
Tricyclic neuroleptics, structure-
    activity relationships,
    8:180–183
Tricyclic pharmacophores, stroke
    therapy, PARP
    inhibitors, **8:**485–487
Tricyclic quinoline lactam
    integrase inhibitors,
    development of,
    7:89–91
Trifluoperazine,
    pharmacokinetics,
    8:187–189
Trifluridine, antimetabolites,
    anti-DNA virus
    agents, **7:**222–225
Trigeminovascular system
    (TGVS):
  migraine pathology,
    8:266–267
  calcitonin gene-related
    peptide antagonists,
    8:272–290
  vanilloid receptor
    antagonists,
    8:307–312
  migraine therapy, nonsteroidal
    anti-inflammatory
    and acetaminophen
    compounds, **8:**268
Triglycerides:
  farnesoid X receptor, **4:**158–159
  liver X receptors, **4:**148–150
  microsomal transfer protein
    inhibitors, **4:**319–320
  peroxisome proliferator-
    activated receptors:
    α-agonists, **4:**88–99
    α/δ-dual agonists, **4:**136–137
    α/γ-dual agonists, **4:**125–135
  type 2 diabetes, **4:**1–2
Trigonelline targetor system,
    brain-targeting
    chemical delivery
    systems, **2:**136–140

Trihexyphenidyl, Parkinson's
  disease therapy,
  8:552–553
3,5,3'-Triiodo-L-thyronine (L-T$_3$):
  biosynthesis, **4**:192–193
  blood transport, **4**:193
  hyperthyroidism and,
    4:198–199
  regulation, **4**:192
  structure-activity relationship,
    4:202–203
    DITBA thyromimetic, **4**:205
    TRIAC thyromimetic, **4**:204
  thyromimetics, research
    background, **4**:189
Trimethadione, anticonvulsant
  applications,
  8:135–139, 142–143
Trimethoprim:
  antibacterial activity, **7**:534
  antimalarial antifolates,
    7:647–648
  dihydrofolic acid reductase
    inhibitor
    combinatorial
    chemistry, **1**:317, 319
  nuclear magnetic resonance
    studies, **2**:414–415
  selective toxicity and
    comparative
    biochemistry,
    2:548–549
Trimethoprim-sulfamethoxazole
  combination,
  antibacterial activity,
  7:534
Trimethoxybenzoylyohimbine
  (TMBY), P-
  glycoprotein
  metabolism,
  2:462–463
Trimethylammonium
  methylketone 3,
  dipeptidyl peptidase 4
  (DPP-4) inhibition,
  4:48–49
Trimetoquinol, thromboxane A$_2$
  receptor antagonists,
  5:891, 893–894
Trimetrexate, lipophilic
  antifolate DNA-
  targeted
  chemotherapeutics,
  6:109
Trimetrexate, nuclear magnetic
  resonance studies,
  2:414–415

Trimipramine:
  chemical structure, **8**:222
  side effects and adverse
    reactions, **8**:231–232
Trinems:
  biological properties, **7**:350–353
  β-lactamase inhibitors, **7**:368
  chemical structure, **7**:345–350
Trioxifene, discovery, **5**:243–244
Tripitramine, muscarinic
  antagonists, **8**:76–77
Triptans:
  direct analog design, **1**:170–171
  migraine therapy, **8**:269–273
Tri-substituted pyrrolidine
  chemokine receptor-5
  (CCR5) antagonist,
  7:121–124
Trocade hydroxymate inhibitors,
  stroke therapy, **8**:503
Troglitazone, peroxisome
  proliferator-activator
  receptors, γ-agonists,
  4:102–104
Tropane-based chemokine
  receptor-5 (CCR5)
  antagonist, **7**:116–121
Tropane ring, cocaine structure-
  activity relationship,
  8:112–113
Trospium, muscarinic
  antagonists,
  8:74–77
Trypanocide drug development
  and discovery
  dinucleotide transporters,
    7:577–580
  lipid metabolism inhibitors,
    7:580–581
  polyamine metabolism,
    7:586–590
Trypanosoma brucei/
  Trypanosoma cruzi:
  genome sequencing,
    7:565–566
  kinetoplastic protozoan
    infections,
    7:566–577
  life cycle, **7**:569, 571–572
  lipid metabolism inhibitors,
    7:580–581
  potential new agents against,
    7:577–590
Trypanosome alternative oxidase
  (TAO), kinetoplastid
  chemical structure,
  7:566–567

Trypanosomiasis. *See* American
  trypanosomiasis;
  Human African
  trypanosomiasis
Trypanothione, kinetoplastid
  chemical structure,
  7:566–567
Trypsin-like coagulation
  enzymes, active site-
  directed
  anticoagulants,
  4:394–398
Tryptophan, niacin synthesis,
  5:637
Tryrphostins, anti-cancer
  activity, **2**:239–240
TTNPB retinoid:
  bexarotene selectivity,
    5:429–435
  NRX 194204 RXR-selective
    retinoid, **5**:458–461
  SHetA2 retinoid-related
    compound, **5**:500–501
Tuberculosis:
  chemotherapeutic agents for,
    7:737–786
  capreomycin, **7**:762–764
  D-cycloserine, **7**:767–771
  ethambutol, **7**:751–755
  ethionamide, **7**:771–773
  first-line agents, **7**:738–759
  fluoroquinolines, **7**:774–776
  isoniazid, **7**:738–744
  kanamycin and amikacin,
    7:761–762
  para-aminosalicylic acid,
    7:764–767
  pyrazinamide, **7**:755–759
  rifamycin antibacterials,
    7:744–751
  streptomycin, **7**:759–761
  thiacetazone, **7**:773–774
  drug discovery and
    development,
    7:735–737
  linezolid, **7**:785–786
  nitroimidazoles, **7**:776–782
  preclinical and clinical
    research, **7**:776–786
  SQ109, **7**:782–783
  TMC207, **7**:783–785
  drug resistance mechanisms
    and, **7**:733–735
  epidemiology, **7**:713–716
  pathogenesis, **7**:717–719
  targeted drug development for,
    7:719–733

Tuberculosis
    amino acid biosynthesis, 7:726–727
    cell wall biosynthesis targeting, 7:719–724
    cofactor biosynthesis, 7:727–729
    cytochrome P450 monooxygenase targeting, 7:729
    DNA synthesis and repair targeting, 7:724–725
    intermediary metabolism and respiration, 7:729–730
    lipid metabolism, 7:730
    mycobactin biosynthesis and iron acquisition, 7:730–731
    mycothiol biosynthesis, 7:731
    protein synthesis and degradation, 7:725–726
    signal transduction targeting, 7:732–733
    terpenoid biosynthesis, 7:731–732
D-Tubocurarine, nicotinic receptor blockade, 8:77–80
Tubulin polymerization/depolymerization inhibition, cytotoxic agents, 6:31–40
    dimeric vinca alkaloids, 6:35–40
    taxus diterpenes, 6:32–35
Tumor-activated prodrugs, DNA-targeted therapeutics, 6:113–124
    ADEPT prodrugs, 6:120–123
    hypoxia-activated bioreductive prodrugs, 6:114–120
Tumorigenesis:
    brain tumor stem cells, 6:224–227
    poly(ADP)-ribosylation, 6:161–162
Two-component drug systems, antitubercular agent targeting, 7:732
Two-dimensional echocardiography, congestive heart failure, disease biomarkers, 4:505–506

Two-dimensional molecules: chemotype development, 1:484–487
    combinatorial quantitative structure-activity relationship modeling, 1:519–520
    pharmacophores, 1:466–467
    ligand-based virtual screening, 2:6
    quantitative structure-activity relationship, limitations, 1:74
Two-dimensional nuclear magnetic resonance spectroscopy:
    basic principles, 2:371
    Prialt® (MVIIA) pain medication, 2:377–382
Two-metal chelation pharmacophore, HIV integrase:
    active site biochemistry, 7:80–81
    diketo acid discovery and elaboration, 7:81–97
Tylosin, chemical structure, 7:436–438
Type 2 diabetes. *See also* Maturity onset diabetes of the young type-2 (MODY-2)
    current therapies, 4:2–13
    α-glycoside inhibitors, 4:5–8
    biguanides, 4:2–3
    glinides, 4:10–13
    metformin, 4:3–5
    sulfonylureas, 4:8–10
    dipeptidyl peptidase 4 inhibitors:
        ADMET properties, 4:45–47
        alternate functions, 4:40–41
        future directions, 4:65
        inhibitor binding, 4:47
        physiology and pharmacology, 4:43–45
        related enzymes, 4:42
        research background, 4:39–40
        selectivity parameters, 4:47–48
        side effects, 4:45
        structure-activity relationships, 4:48–65
        α-amino acid derivatives, 4:48–56

β-amino acid derivatives, 4:57–59
cyclohexyl- and cyclohexenylamines, 4:64–65
fluoroolefin inhibitors, 4:56–57
heterocyclic derivatives, 4:59–64
hybrid α/β-amino acid derivatives, 4:59
emerging therapies, 4:13–27
    fructose-1,6-biphosphatase inhibitors, 4:18–20
    glucokinase activators, 4:15–17
    GPR119 agonists, 4:20–21
    11β-hydroxysteroid dehydrogenase-1 inhibitors, 4:22–25
    SIRT1 activators, 4:25–27
    sodium-dependent glucose cotransporter inhibitors, 4:13–14
epidemiology, 4:1–2
farnesoid X receptors, 4:160
pathogenesis, 4:39
peroxisome proliferator-activator receptors:
    α-agonists, 4:97–99
    α/γ-dual agonists, 4:124–125
    α/γ/δ-pan agonists, 4:141–145
    γ-agonists, 4:100–106
    PPARγ antagonists, 4:115–117
    PPARγ/δ antagonists, 4:137–140
    selective PPARγ modulators, PPARγ partial agonists, 4:106–115
13-*cis*-retinoic acid side effects, 5:394–395
Tyr-D-aa-Phe sequence, opioid peptide analogs, 8:664–673
    deltorphin analogs, 8:668–673
    dermorphin analogs and μ-selective peptides, 8:664–668
Tyrosine kinase inhibitors:
    angiogenic properties, 6:307–310
    cancer therapy, resistance to, 6:364–369
    glioblastoma therapy, 6:232–233

growth factor signaling pathways, **6:**301–307
selective toxicity, **2:**562–563
Tyrosine ureas, muscarinic antagonists, **8:**76–77
Tyr-Pro-Phe sequence, opioid peptide analogs, **8:**659–664
β-casomorphin analogs and endomorphins, **8:**659–661
TIPP and related peptides, **8:**661–664

UB-165 nicotinic agonist, **8:**70
Ubiquitin-conjugating enzymes, proteasome inhibitors, **6:**175
Ubiquitin-proteasome pathway (UPP):
 basic structure, **6:**175
 Parkinson's disease genetics, **8:**535
 targeted therapy, cancer and other disease, **6:**176–177
UDP-galactopyranose mutase (UGM), antitubercular agents, **7:**722
Ugi multicomponent reaction:
 small molecule libraries, **1:**300
 spiro diketopiperazine-based chemokine receptor-5 (CCR5) antagonist, **7:**131–133
UGT1/UGT2 gene family, drug metabolism, glucuronidation/glucosidation, **1:**427–430
UK-147,535 derivative, thromboxane A_2 receptor antagonists, **5:**894
Ulcerative colitis, soft drug design, **2:**100
Ultrahigh pressure liquid chromatography (UHPLC), combinatorial library purification, **1:**109–110
Ultrarapid metabolizers (UM), CYP2D6 substrate, **1:**185–186

Ultrashourt-action (USA), soft drug design, adaprolol, **2:**83–85
Uncertainty method, outlier detection, quantitative structure-activity relationship, **1:**44
Uncialamycin, anti-cancer activity, **2:**239
Undecanediamidine antitrypanosomal agent, **7:**578–580
Unfractionated heparin (UFH):
 molecular mechanism of action, **4:**375–380
 structural properties, **4:**371–374
UniProt database, structural genomics research, human proteome coverage, **1:**577–579
United Kingdom, patent requirements, **3:**122
United States Pharmacopoeia and National Formulary, **3:**96–97
Unit-to-unit relationships, oral drug delivery, sustained-release dosage forms, *in vitro* studies, **2:**787–791
Universal screening compound library, theoretical background, **1:**294–295
Unoprostone isopropyl ester, glaucoma intraocular pressure reduction, **5:**592–600
Unstable angina/non-ST-segment elevation myocardial infarction (UA/NSTEMI), acute coronary syndrome risk, **4:**411
Upstream processing:
 defined, **3:**289
 protein therapeutics, **3:**291–301
 cell culture, **3:**298–300
 expression plurality, **3:**291–294
 formats, **3:**296–298
 future trends, **3:**300–301

reagent clearance and excipient testing, **3:**326
recombinant cell lines, **3:**294–296
Uptake transporters:
 ADMET structural properties, permeability, **2:**56
 psychostimulant pharmacology, **8:**114
Urea-based peptidomimetic inhibitors, stroke therapy, **8:**462–463
Urea derivatives, geldanamycin Hsp90 inhibitor, **6:**389–391
Urea structures, diaryl urea-based inhibitors, **5:**1042–1044
Ureido penicillin, monobactam biological activity and, **7:**328–329
Urotensin-II (undecapeptide ligand) (U-II):
 biological actions and receptor pharmacology, **4:**531–533
 structure and function, **4:**531–533
 target validation, congestive heart failure therapy:
  congestive heart failure therapy, **4:**485–487
  ex vivo cell-based biomarker assays, **4:**501
  normal subjects *vs.* CHF patients, **4:**495–496
  pharmacodynamic-kinetic dose selection, **4:**502–503
  pharmacodynamics/pharmacokinetics, **4:**491–492
  species and disease state impacts, **4:**488
  specific target/mechanism perturbation, **4:**494–495
U.S. Courts, patent trials in, **3:**148–150
Ustiloxins A-F, chirality, **1:**134–137
Utility patents:
 protection and strategy, **3:**105–118
 requirements, **3:**120–121

Vaccine development:
  antinicotine vaccines, smoking cessation, **5**:725–726
  central nervous system tumors, **6**:268–270
  multiple sclerosis therapy, antigen-specific therapy, **5**:579–580
Vacuous chewing movements, antipsychotic side effects, **8**:175
Vaginal microsides, HIV reverse transcriptase inhibitors, **7**:156
Valaciclovir:
  herpes simplex virus therapy, **7**:242–243
  varicella-zoster virus therapy, **7**:243
Valganciclovir:
  cytomegalovirus therapy, **7**:243–244
  human herpesvirus 6 therapy, **7**:244
VALIDATE function, empirical scoring, **2**:624–628
Validation:
  commercial-scale operations, **3**:12–13
  infringement of patent claims, defense based on, **3**:155–156
  quantitative structure-activity relationship, **1**:46–51
  integrated predictive modeling workflow, **1**:516, 520–521
  virtual screening models, **1**:526–527
  structure-based virtual screening, **2**:23
Valopicitabine anti-HCV agent, **7**:195–196
Valproic acid:
  anticonvulsant applications, **8**:135–139, 147–148
  histone deacetylase inhibition, **6**:73
  migraine prevention, **8**:315–316
  mitochondrial β-oxidation, **1**:434–435
  stroke therapy, **8**:494
Valrocemide, anticonvulsant applications, **8**:136–139

Valrubicin, structure and therapeutic effects, **6**:22–23
Valspodar, drug resistance inhibition, **6**:372
Vancomycin:
  atropisomerism, **1**:133–137
  glycopeptide antibiotics, **2**:225–227
Vancomycin-resistant enterocci (VRE), linezolid RNA targets, **5**:990–993
van der Waals volume:
  allosteric protein sites, **1**:390
  quantitative structure-activity relationships, **1**:26
Vandetanib:
  antiangiogenic properties, **6**:308–310
  brain tumor angiogenesis, **6**:242–243
  multitarget drug development, cancer therapy, **1**:266–267
Vanilloid receptor antagonists, migraine therapy, **8**:307–312
  capsazepine, civamide, and AMG-517, **8**:307–308
  SB-705498, **8**:308–312
Vardenafil:
  erectile dysfunction therapy, chemical structure, **5**:711
  structure-activity relationships, **5**:713–714
Vardenafil, selective toxicity, **2**:567–568
Varenicline:
  chemical structure, **5**:711
  cognitive enhancement, **8**:35–36
  nicotinic agonists, **8**:70
  smoking cessation therapy, **5**:723–725
Variable surface glycoprotein (VSG), human African trypanosomiasis, **7**:568–569
Varicella zoster virus (VZV), anti-DNA virus agents:
  bicyclic pyrimidine nucleoside analogs, **7**:223–225
  compounds and prodrugs, **7**:243
Vascular dementia, epidemiology, **8**:21

Vascular endothelial growth factor (VEGF):
  antiangiogenic inhibitors, **6**:307–310
  resistance to, **6**:368
  glioblastoma therapies, **6**:232–233
  stroke therapy, Src kinase inhibitors, **8**:510–514
Vascular endothelial growth factor (VEGF) receptor:
  age-related macular degeneration therapy, **5**:615–623
  brain tumor therapy:
    inhibitors, **6**:248
    kinase inhibitors, **6**:248–249
    multitarget drug development, multikinase inhibitors, cancer therapy, **1**:264–267
Vascular-only hypothesis, migraine pathology, **8**:266–267
Vascular tissue:
  farnesoid X receptors, **4**:160
  injury, carbon monoxide inhalation therapy, **5**:302
Vasoactive intestinal constrictor (VIC), structure and function, **4**:523–526
Vasoactive intestinal peptide (VIP), structure and function, **4**:547–548
Vasoactive peptides:
  adrenomedullin, **4**:540
  bombesin, gastrin-releasing peptide, neuromedin B, **4**:549
  bradykinin, **4**:544–547
  calcitonin gene-related peptides, **4**:537–540
  endothelin, **4**:523–526
  natriuretic peptide family, **4**:533–537
  neuropeptide Y, **4**:526–531
  neurotensin, **4**:549–550
  oxytocin, **4**:522–523
  relaxin, **4**:550
  renin-angiotensin system, **4**:516–520
    angiotensins, **4**:516–518
    inhibitors, antagonists, and agonists, **4**:520

prorenin and renin,
4:518–519
somatostatin, 4:548–549
structure and function, 4:513
tachykinins, 4:540–544
urotensin-II, 4:531–533
vasoactive intestinal peptide,
related peptides,
4:547–548
vasopressin, 4:520–523
Vasoconstriction, migraine
therapy, triptans,
8:270–272
Vasomolol, soft drug design, 2:86
Vasopeptidase inhibition,
angiotensin-
converting enzyme
inhibitors, 4:276–279
clinical trials, 4:285
Vasopressin type-2 receptor,
allosteric protein
model, 1:394
Vatalanib:
antiangiogenic properties,
6:309–310
brain tumor angiogenesis,
6:242–243
Vault poly(ADP-ribose)
polymerase (V-PARP),
molecular biology,
6:158–159
Veber rule, small druggable
molecules, 1:295–296
Veber's rules, ADMET properties
assessment, 2:64
Vector-mediated small
interfering RNA
(siRNA), recombinant
DNA technology,
1:550
Venlafaxine:
chemical structure, 8:223
migraine prevention,
8:318–319
monoamine transporter
inhibition,
8:238–241
optimization of, 1:267–273
side effects and adverse
reactions, 8:234
Venom peptides, angiotensin-
converting enzyme
inhibitors,
4:268–269
Vent treatment requirements,
commercial-scale
operations, 3:11

Verapamil:
ATP binding cassette (ABC)
transporters,
6:371–372
migraine prevention, 8:318
stroke therapy, 8:459–460
Verapamil, selective toxicity and
stereochemistry,
2:546–547
Verkhivker's algorithm,
knowledge-based
scoring, 2:629–631
Verteporfin, age-related macular
degeneration therapy,
5:614–615
Vertex. See Teleprevir anti-HCV
agent
Very-low density lipoproteins
(VLDL):
atherosclerosis and,
4:331–333
liver X receptors, 4:148–150
structure and function,
4:303–305
vitamin E uptake and
metabolism,
5:662
Vesanoid, acute promyelocytic
leukemia therapy,
5:376–377
Vesicle polymerization, nanoscale
drug delivery, 3:477,
479–480
Vesicular glutamate
transporters,
anticonvulsants, 8:132
Vesicular monoamine transporter
(VMAT):
antidepressant pharmacology,
8:238–241
Parkinson's disease diagnosis,
8:556
Vibrational modes, in silico
screening, protein
flexibility, backbone
mobility, 2:868–870
Vicinal diarylethylenes,
structure-activity
relationships,
5:247–248
Vicinal diol renin inhibitors,
4:246–250
Vicriviroc compound, chemokine
receptor-5 (CCR5)
antagonist
development,
7:115–116

Vigabatrin, anticonvulsant
applications,
8:135–139, 149
Vilazodone SSRI, 8:256
Vildagliptin, type 2 diabetes
therapy:
ADMET properties, 4:45–47
physiology and pharmacology,
4:43–45
Vinblastine, structure and
therapeutic
applications, 6:37–39
Vinca alkaloids:
chemical structure, 6:35–37
vinblastine, 6:37–39
vincristine, 6:39
vinorelbine, 6:39–40
Vincasar PFS. See Vincristine
Vincristine, structure and
therapeutic
applications, 6:38–39
Vindisine, analog design, 6:40
Vinflunine, analog design, 6:40
Vinorelbine, structure and
therapeutic
applications, 6:39–40
Vinylogous cyclic carbonates,
prodrug development,
3:246–247
Viracept. See Nelfinavir mesylate
Viral DNA-polymerase,
nonnucleotide
inhibitors, 7:239–240
Viral Enzymes Involved in
Replication (VIZIER)
project, genome-
centered research,
1:573
Viral hemagglutinins, receptor-
mediated endocytosis,
2:452
Virginiamycin, ribosome
targeting mechanism,
7:457–460
Virtual screening. See also In
silico methods
chemogenomics, molecular
informatics,
2:587–588
combinatorial libraries,
2:29–33
privileged structures,
2:30–31
protein family targeting,
2:31–33
data pre- and postprocessing,
2:25–27

Virtual screening (*Continued*)
  high throughput screening
    triaging, **2:**27
  hit triaging, **2:**25
  ligand efficiency, **2:**25–26
  traffic lights, **2:**25
defined, **2:**1–2
free fatty acid receptor 1
    ligands, **2:**33–34
G-protein coupled receptor,
    homology modeling,
    lead generation,
    2:289–292
ligand-based techniques,
    2:2–14
  compound similarity, **2:**2–3
  kinase inhibitor high-
    throughput screening,
    2:14
  machine learning
    applications, **2:**10–14
  algorithms for QSAR
    models, **2:**10–13
  applicability domain
    concept, **2:**13–14
  QSAR model generation
    and validation, **2:**10
  pharmacophores, **2:**3–10
  dopamine transporter
    inhibitors, **2:**7–8
  novel PPAR ligands, **2:**9
  organic compound
    databases, **2:**4–5
  research background, **2:**4
  three-dimensional ligand-
    based generation, **2:**6
  three-dimensional
    receptor-based
    generation, **2:**8–9
  three-dimensional
    screening, **2:**9–10
  two-dimensional
    searching, **2:**6
natural products lead
    generation, **2:**213
protein-protein interactions,
    p53/MDM2
    antagonists,
    2:352–359
quantitative structure-activity
    relationship modeling,
    1:510–511
  data set complexity,
    1:512–516
  validated and predictive
    models, **1:**526–527
scaffold hopping, **2:**27–29

PDE5 inhibitor enumeration
    example, **2:**27–29
similarity descriptors, **2:**29
structure-based techniques,
    2:14–25
  computer-aided drug design,
    2:680–681
  fragment-based screening,
    2:25
  protein structures and
    flexibility, **2:**14–16
  scoring functions, **2:**16–25
    comparison, consolidation
    and consensus,
    2:22–23
  empirical scoring, **2:**18
  enrichment factors,
    2:20–22
  force-field scoring, **2:**17–18
  hybrid/alternative scoring,
    2:19–20
  knowledge-based scoring,
    2:18–19
  liver X receptor modulator
    example, **2:**23–25
  synergies, **2:**33–34
Virtual soft analog library,
    alprostadil soft drug
    design, **2:**134–135
Virulence factors, linezolid
    inhibition, **7:**543
Virus assembly inhibitors, anti-
    DNA virus agents,
    7:241
Virus-directed enzyme prodrug
    therapy (VDEPT),
    prodrug design, **3:**229
Viruses. *See also* Retroviruses;
    specific viruses, e.g.,
    Hepatitis C virus
  mammalian protein
    purification,
    3:308–309
Visual pigmentation, vitamin A
    chemistry, **5:**646–647
Vitamin A (retinol):
  all *trans*-retinoic acid and,
    5:376
  biochemical functions and
    deficiency, **5:**646–647
  chemistry, **5:**644
  dosage forms, **5:**647
  N-(4-hydroxyphenyl)
    retinamide (4-HPR)
    reduction, **5:**476–477
  hypercarotenosis, **5:**648
  hypervitaminosis A, **5:**647–648

pharmacologically active
    retinoids, **5:**648–649
13-*cis*-retinoic acid
    interactions, **5:**395
uptake and metabolism,
    5:644–646
Vitamin B$_1$. *See* Thiamine
    (vitamin B$_1$)
Vitamin B$_2$. *See* Riboflavin
    (vitamin B$_2$)
Vitamin B$_6$, **5:**679–665
  deficiency, **5:**681, 683
  drug interactions, **5:**681–682,
    684
  hypervitaminosis pyridoxine,
    5:684
  pyridoxal phosphate
    biochemistry, **5:**679,
    681–682
  uptake and metabolism,
    5:679–680
Vitamin B$_{12}$ (cobalamin), **5:**697,
    699–701
  biochemistry, **5:**699–701
  chemistry, **5:**697, 699
  deficiency, **5:**700
  hypervitaminosis B$_{12}$, **5:**700
  uptake and metabolism, **5:**697,
    699
Vitamin C (ascorbic acid),
    5:701–705
  antioxidant supplement
    controversy,
    5:643–644
  ascorbic acid deficiency,
    5:703
  chemistry, **5:**701–702, 704
  hypervitaminosis C, **5:**703, 705
  uptake and metabolism, **5:**703
Vitamin D, **5:**649–660. *See also*
    Cholecalciferol;
    Ergocalciferol
  analogs, **5:**655
  biochemical function,
    5:652–657, 660
  calcium regulation,
    5:653, 655
  chemistry, **5:**649–651
  dosage forms, **5:**658
  drug interactions, **5:**660
  hypervitaminosis D, **5:**658
  osteoporosis anabolic therapy,
    5:759–760
  uptake and metabolism,
    5:651–654
  vitamin D$_3$ receptor:
    biochemical function, **5:**652

cholecalciferol
    photochemistry,
        5:655–656
    retinoic nuclear receptors,
        5:371–372
Vitamin E, **5:**660–665. *See also*
        Tocopherols;
        Tocotrienols
    antioxidant supplement
        controversy,
        5:643–644
    biochemical function,
        5:662–664
    chemistry, **5:**660
    deficiencies, **5:**662
    dosage forms, **5:**664–665
    hypervitaminosis E, **5:**662, 664
    uptake and metabolism,
        5:660–662
Vitamin K:
    antagonist:
        anticoagulant therapy, **4:**365
        inhibition mechanisms,
            4:369–370
        molecular mechanisms of
            action, **4:**380–385
    biochemistry and deficiency,
        5:665–666, 668
    chemistry, **5:**665
    hypervitaminosis K, **5:**668
    uptake and metabolism, **5:**665
Vitamin K oxidoreductase
        (VKOR), mechanism of
        action, **4:**381–385
Vitamins:
    age-related macular
        degeneration therapy,
        5:613–614
    animal studies, **5:**643
    antioxidant supplements,
        5:643–644
    biotin, **5:**685–691
        chemistry, **5:**686
        deficiency, **5:**688, 691
        hypervitaminosis biotin,
            5:691
        metabolism, **5:**686–690
    daily requirements, **5:**638
    deficiencies, **5:**638–640
        nutrient intake, **5:**643
    defined, **5:**637
    dietary reference intakes and
        daily values,
        5:638–644
        biotin, **5:**691
        folic acid, **5:**697
        niacin, **5:**679

pantothenic acid, **5:**685
riboflavin (vitamin B_2), **5:**673
thiamine (vitamin B_1), **5:**673
vitamin A, **5:**648
vitamin B_6 family, **5:**684–685
vitamin B_{12} (cobalamin),
    5:701
vitamin C (ascorbic acid),
    5:705
vitamin D family, **5:**658–659
vitamin E, **5:**665
vitamin K, **5:**668
dietary sources, **5:**638
dose determinations, **5:**640, 643
folic acid, **5:**691–697
    chemistry, **5:**691–692
    deficiency, **5:**696
    hypervitaminosis folic acid,
        5:696–697
    uptake and metabolism,
        5:691–696, 698
human biochemical pathways,
    5:637–638, 643
metabolic balance studies,
    5:643
natural sources, **5:**637
niacin-nicotinamide, **5:**677–679
    chemistry, **5:**677–679
    deficiency, **5:**677
    hypervitaminosis niacin,
        5:677–678
organic chemistry, **5:**638
pantothenic acid, **5:**685–687
receptor-mediated endocytosis,
    2:453–454
    vitamin B_{12}, **2:**454
retinol family, **5:**644–649
    biochemical functions and
        deficiency, **5:**646–647
    chemistry, **5:**644
    dosage forms, **5:**647
    hypercarotenosis, **5:**648
    hypervitaminosis A,
        5:647–648
    pharmacologically active
        retinoids, **5:**648–649
    uptake and metabolism,
        5:644–646
riboflavin (vitamin B_2),
    5:673–677
    chemistry, **5:**673
    deficiency, **5:**673
    hypervitaminosis riboflavin,
        5:673
    metabolism, **5:**673–677
    uptake and biochemistry,
        5:673–674

serum levels, **5:**643
therapeutic applications,
    5:637–638, 643
thiamine (vitamin B_1),
    5:668–673
    chemistry, **5:**669–670
    deficiencies, **5:**670
    hypervitaminosis thiamine,
        5:670
    uptake and metabolism,
        5:670–672
vitamin B_6 family, **5:**679–685
    deficiency, **5:**681, 683
    drug interactions, **5:**681–682,
        684
    hypervitaminosis pyridoxine,
        5:684
    pyridoxal phosphate
        biochemistry, **5:**679,
        681–682
    uptake and metabolism,
        5:679–680
vitamin B_{12} (cobalamin), **5:**697,
    699–701
    biochemistry, **5:**699–701
    chemistry, **5:**697, 699
    deficiency, **5:**700
    hypervitaminosis B_{12}, **5:**700
    uptake and metabolism,
        5:697, 699
vitamin C (ascorbic acid),
    5:701–705
    ascorbic acid deficiency,
        5:703
    chemistry, **5:**701–702, 704
    hypervitaminosis C, **5:**703,
        705
    uptake and metabolism,
        5:703
vitamin D family,
    5:649–660
    biochemical function,
        5:652–657, 660
    chemistry, **5:**649–651
    dosage forms, **5:**658
    drug interactions, **5:**660
    hypervitaminosis D, **5:**658
    uptake and metabolism,
        5:651–652
vitamin E family, **5:**660–665
    biochemical function,
        5:662–664
    chemistry, **5:**660
    deficiencies, **5:**662
    dosage forms, **5:**664–665
    hypervitaminosis E, **5:**662,
        664

Vitamins (Continued)
uptake and metabolism, 5:660–662
vitamin K family, 5:665–668
biochemistry and deficiency, 5:665–666, 668
chemistry, 5:665
hypervitaminosis K, 5:668
uptake and metabolism, 5:665
VLA-4 antagonists, multiple sclerosis therapy, 5:578–579
Voglibose, type 2 diabetes therapy, 4:5–8
Volinanserin:
cognitive enhancement, 8:37–40
insomnia therapy, 5:718
Voltage-gated ion channels:
anticonvulsants, 8:122–126
calcium channels, 8:124
chloride channels, 8:126
hyperpolarization-activated cyclic nucleotide cation channels, 8:125–126
potassium channels, 8:124–125
sodium channels, 8:123–124
migraine pathology and, 8:266–267
stroke therapy:
calcium channels, 8:455–460
potassium channels, 8:482–483
Volume of distribution, quantitative structure-activity relationship, 1:66
Volume similarity, dosage shapes, oral drug delivery, sustained release dosage forms, 2:796
Vorinostat:
histone deacetylase inhibition, 6:53, 55–56
multitarget drug development, cancer therapy, 1:266–267
Vorozole, aromatase inhibition, 5:232
VX-950 protease inhibitor. See Teleprevir anti-HCV agent

Vytorin, LDL cholesterol lowering mechanisms, 4:317–318

W1807, allosteric protein model, 1:394
Warfarin:
CYP2C9 substrate, pharmacogenomics, 1:186
selective toxicity and stereochemistry, 2:547
soft analog development, 2:118
Water-insoluble polymers, nanoscale drug delivery, 3:481
Water-LOGSY, nuclear magnetic resonance studies, drug screening applications, 2:426
Water solubility. See Aqueous solubility
Water solvents, docking techniques, 2:634–636
WAY-129 prodrug, stroke therapy, 8:451
WAY-100635 serotonin antagonist:
efficacy assessment, 8:225–229
memory assessment and, 8:28–30
Way-124466 sirolimus compound, triene region modification, 5:1032–1033
WAY-252623 LXR agonist, Alzheimer's disease, cholesterol turnover, 8:411–413
ω-chain derivatives, prostaglandin/thromboxane receptors:
EP$_4$ agonists, 5:876–876
FP receptor agonists, 5:897–898
Western blotting, protein therapeutics, AHPN compound expression and activation, 5:494–497
WHO histological classification system, central nervous system tumors, 6:230–234

"Willful infringement" damages, infringement of patent claims, 3:158–159
Wisconsin card sorting test, executive function assessment, 8:28–30
Withdrawal effects, antipsychotics, 8:169
Withdrawal of approval, FDA requirements, 3:92
Wittig-type cyclization:
faropenem, 7:322–323
penems, 7:319–320
Wnt signaling pathway:
brain tumor therapy, 6:263
central nervous system tumors, 6:228
osteoporosis anabolic therapy, 5:755–759
bone remodeling, 5:756–757
canonical pathway, 5:756
genetic models and bone target cells, 5:757–758
therapeutics development, 5:758–759
"Wobble" base phenomenon, aminoglycoside pharmacology, 7:420–423
WOMBAT/WOMBAT-PK data sets, quantitative structure-activity relationship modeling, 1:513–514
Woodward's synthesis:
cephalosporin C, 7:296–298
penems, 7:317–321
Working agreements, international patent laws, 3:169–171
Working cell bank (WCB), protein therapeutics, 3:296
Working memory, 8:18–19
Morris Water Maze assessment, 8:27–28
Social Recognition and Novel Object Recognition assessment, 8:27
World Trade Organization (WTO), patent interference limitations, 3:144–146

Xanomeline muscarinic receptor: activity of, 8:67–68

cognitive enhancement, 8:32–33
Xanthine analogs, atherosclerosis, HDL elevation, 4:349–350
X-CScore method, consensus scoring, 2:632
Xenobiotic reactions:
  biological factors, 1:447–449
  drug metabolism, 1:405–408
  enzyme/metabolic reactions, 1:407–408
  pharmacodynamics, 1:407
  specificity and selectivity, 1:406–407
  in vitro studies, 1:407–408
Ximelagatran, active site-directed inhibitors, 4:398
X-linked inhibitor of apoptosis (XIAP), peptidomimetics, 1:233–240
X-RARα fusion protein, acute promyelocytic leukemia therapy, 5:377, 380–382
X-ray crystallography:
  docking and scoring functions, flexibility, 2:638
  G-protein coupled receptor, homology modeling, 2:281–284
  kinesin spindle protein structure, 6:192–193
  dihydropyrazole inhibitors, 6:204–210
  "induced fit" binding pockets, 6:193–197
X-ray diffraction (XRD) analysis, quantitative structure-activity relationship, hydrophobic interactions, 1:11–19
X-residual, outlier detection, quantitative structure-activity relationship, 1:44
X-SCORE approach, empirical scoring, 2:626–628

Yalkowsky-Valvani equation, oral drug physicochemistry, aqueous solubility, 3:33
Yeast systems:
  chemogenomics, three-hybrid screening, 2:581–582
  natural products lead generation, phenotypic screening, 2:208
  protein therapeutics, 3:292–294
Ynolides, cyclopropardicicol analogs, 6:402–403
Yondelis®, marine sources, 2:246–247
Y-randomization test, quantitative structure-activity relationship validation, 1:48–49
Yukawa-Tsuno approach, quantitative structure-activity relationship, electronic parameters, 1:7–8

Zalcitabine, 7:141–145
Zaleplon:
  insomnia therapy, 5:716–718
  selective toxicity, 2:568
Zanamivir, selective toxicity, 2:556–557
Zankiren renin inhibitor, 4:246, 248
Zaprinast, erectile dysfunction therapy, 5:712–713
ZD-9379 scaffold, stroke therapy, 8:453
Zebularine, "double prodrug," anti-DNA virus antimetabolites, 7:223–225
Zero-linked polymers, polymerized hemoglobin, 4:634
Zidovudine:
  chemical delivery systems, 2:141–142

conjugation reactions, 1:439
discovery of, 7:75–76
hepatitis C therapy, 7:169–170
HIV reverse transcriptase inhibition, 7:139–140, 141–145
Thiovir® interaction with, 7:240
Zilpaterol, chirality, enantiomer activity, 1:131
Zimelidine, historical background, 8:249
Zinc binding group (ZBG), histone deacetylase inhibitors, 6:53–54
  future drug development, 6:74–75
Zinc finger proteins, poly(ADP-ribose)polymerase-1 (PARP-1) domain organization, 6:159–160
Ziprasidone:
  FDA approval, 8:178–179
  pharmacokinetics, 8:196–198
  structure-activity relationship, 8:182–185
ZK216348 nonsteroidal glucocorticoid receptor agonist, 5:128–134
Zolmitriptan, migraine therapy, calcitonin gene-related peptide antagonists vs, 8:287–288
Zolpidem:
  functional analog design, 1:176–177
  insomnia therapy, 5:716–718
  selective toxicity, 2:568
Zonisamide:
  anticonvulsant applications, 8:133–139, 151–152
  migraine prevention, 8:316
Zonula occludens toxin (ZOT), membrane protein permeability enhancement, 2:440–441

Zopiclone:
   functional analog design, **1:**176–177
   insomnia therapy, **5:**716–718
Zotarolimus compound:
   cyclohexyl analog design, **5:**1033–1034
   drug-eluting coronary stents, **5:**1037
Zotepine, pharmacokinetics, **8:**194
Zuckerman synthesis, benzodiazepine library, **1:**302–303
Zuclopenthixol antipsychotic, **8:**201

Zwanzig equation, free-energy perturbation, binding affinity evaluation, **2:**724–728
Zwitterions, salt compounds, acid-base equilibrium, **3:**385, 387